THE GENETIC CODE (mRNA)

Second position

First position	U	C	A	G	Third position
U	UUU Phe	UCU Ser	UAU Tyr	UGU Cys	U
U	UUC Phe	UCC Ser	UAC Tyr	UGC Cys	C
U	UUA Leu	UCA Ser	UAA Stop	UGA Stop	A
U	UUG Leu	UCG Ser	UAG Stop	UGG Trp	G
C	CUU Leu	CCU Pro	CAU His	CGU Arg	U
C	CUC Leu	CCC Pro	CAC His	CGC Arg	C
C	CUA Leu	CCA Pro	CAA Gln	CGA Arg	A
C	CUG Leu	CCG Pro	CAG Gln	CGG Arg	G
A	AUU Ile	ACU Thr	AAU Asn	AGU Ser	U
A	AUC Ile	ACC Thr	AAC Asn	AGC Ser	C
A	AUA Ile	ACA Thr	AAA Lys	AGA Arg	A
A	AUG Met/start	ACG Thr	AAG Lys	AGG Arg	G
G	GUU Val	GCU Ala	GAU Asp	GGU Gly	U
G	GUC Val	GCC Ala	GAC Asp	GGC Gly	C
G	GUA Val	GCA Ala	GAA Glu	GGA Gly	A
G	GUG Val	GCG Ala	GAG Glu	GGG Gly	G

SOME PREFIXES USED IN THE INTERNATIONAL SYSTEM OF UNITS

10^9	giga	G
10^6	mega	M
10^3	kilo	k
10^{-1}	deci	d
10^{-2}	centi	c
10^{-3}	milli	m
10^{-6}	micro	μ
10^{-9}	nano	n
10^{-12}	pico	p
10^{-15}	femto	f

CONVERSION FACTORS

Energy: 1 joule = 10^7 ergs = 0.239 cal
1 cal = 4.184 joule

Length: 1 nm = 10 Å = 1×10^{-7} cm = 1×10^{-9} m

Mass: 1 kg = 1000 g = 2.2 lb
1 lb = 453.6 g

Pressure: 1 atm = 760 torr = 14.696 psi
1 torr = 1 mm Hg

Temperature: °K = °C + 273
°C = (5/9)(°F − 32)

Volume: 1 L = 1×10^{-3} m^3 = 1000 cm^3

PHYSICAL CONSTANTS

Name	Symbol	SI Units	cgs Units
Avogadro's number	N	6.022137×10^{23}/mol	6.022137×10^{23}/mol
Boltzmann constant	k	1.38066×10^{-23}J/°K	1.38066×10^{-16} erg/°K
Curie	Ci	3.7×10^{10} d/s	3.7×10^{10} d/s
Electron charge	e	1.602177×10^{-19} coulomb[b]	4.80321×10^{-10} esu
Faraday constant	F	96485 J/V · mol	9.6485×10^{11} erg/V · m
Gas constant[a]	R	8.31451 J/°K · mol	8.31451×10^{7} erg/°K · m
Light speed (vacuum)	c	2.99792×10^{8} m/s	2.99792×10^{10} cm/s
Planck's constant	h	6.626075×10^{-34} J · s	6.626075×10^{-27} erg · s

[a]Other values of R: 1.9872 cal/K · mol = 0.082 L · atm/°K · mol
[b]1 coulomb = 1 J/V

USEFUL EQUATIONS

Henderson–Hasselbalch equation	$\mathrm{pH} = \mathrm{p}K_a + \log([\mathrm{A}^-]/[\mathrm{HA}])$
Michaelis–Menten equation	$V = V_{max}[\mathrm{S}]/(K_M + [\mathrm{S}])$
Free energy change under non-standard-state conditions	$\Delta G = \Delta G^\circ + RT \ln([\mathrm{C}][\mathrm{D}]/[\mathrm{A}][\mathrm{B}])$
Free energy change and standard reduction potential	$\Delta G^{\circ\prime} = -nF\Delta E_0'$
Reduction potentials in a redox reaction	$\Delta E_0' = E_0'(\text{acceptor}) - E_0'(\text{donor})$
Proton motive force	$\Delta p = \Delta\Psi - 2.3RT\,\Delta\mathrm{pH}/F$
Passive diffusion of a charged species	$\Delta G = G_2 - G_1 = RT \ln(C_2/C_1) + ZF\Delta\Psi$

Biochemistry

Of Related Interest
From the Benjamin/Cummings Series
In the Life Sciences and Chemistry

Biochemistry and Cell Biology

W. M. Becker, L. J. Kleinsmith, and J. Hardin
The World of the Cell, Fourth Edition (2000)

R. Boyer
Modern Experimental Biochemistry,
Third Edition (2000)

D. J. Holme and H. Peck
Analytical Biochemistry, Third Edition
(1998)

Ecology and Evolution

M. G. Barbour, J. H. Burk, W. D. Pitts,
F. S. Gilliam, and M. W. Schwartz
Terrestrial Plant Ecology, Third Edition
(1999)

C. J. Krebs
Ecological Methodology, Second Edition
(1999)

C. J. Krebs
Ecology: The Experimental Analysis of Distribution and Abundance, Fourth Edition
(1994)

J. W. Nybakken
Marine Biology: An Ecological Approach,
Fourth Edition (1997)

E. R. Pianka
Evolutionary Ecology, Sixth Edition (2000)

Molecular Biology and Genetics

R. J. B. King
Cancer Biology (1996)

M. V. Bloom, G. A. Freyer, and D. A. Micklos
Laboratory DNA Science (1996)

J. D. Watson, N. H. Hopkins, J. W. Roberts,
J. A. Steitz, and A. M. Weiner
Molecular Biology of the Gene, Fourth Edition
(1987)

R. J. Brooker
Genetics: Analysis and Principles (1999)

J. P. Chinnici and D. J. Matthes
Genetics: Practice Problems and Solutions
(1999)

R. P. Nickerson
Genetics: A Guide to Basic Concepts and Problem Solving (1990)

P. J. Russell
Fundamentals of Genetics, Second Edition
(2000)

P. J. Russell
Genetics, Fifth Edition (1998)

Biochemistry

THIRD EDITION

Christopher K. Mathews
OREGON STATE UNIVERSITY

K. E. van Holde
OREGON STATE UNIVERSITY

Kevin G. Ahern
OREGON STATE UNIVERSITY

An Imprint of Addison Wesley Longman, Inc.

San Francisco • Reading, Massachusetts • New York • Harlow, England
Don Mills, Ontario • Sydney • Madrid • Amsterdam

Publisher: Robin Heyden
Sponsoring Editor: Ben Roberts
Production Editor, Art Coordinator: Lisa Weber
Managing Editor: Joan Marsh
Marketing Manager: Jennifer Schmidt
Director of Marketing: Stacy Treco
Prepress and Manufacturing Supervisor: Lillian Hom
Publishing Assistant: Anthony J. Asaro
Publishing Associate: Claudia Herman
Developmental Editor: John Murdzek
Copyeditor: Mary Prescott
Proofreader: Anita Wagner
Indexer: Shane-Armstrong Information Systems
Text Designers: Bruce Kortebein, Kathleen Cunningham
Developmental Artist (Second Edition): Darwen Hennings
Artists: Graham Johnson, Precision Graphics, Karl Miyajima, Thompson Type
Permissions Editors: Marcy Lunetta, Charles Morris
Photo Editor: Roberta Spieckerman Associates
Compositor: Thompson Type
Cover Designer: Yvo Riezebos Design
Cover Artist: Graham Johnson

Library of Congress Cataloging-in-Publication Data

Mathews, Christopher K., 1937–
Biochemistry / Christopher K. Mathews, K. E. van Holde, Kevin G. Ahern.—3rd ed.
p. cm.
Includes bibliographical references and index.
ISBN 0-8053-3066-6
1. Biochemistry. I. Title. II. Van Holde, K. E. (Kensal Edward), 1928–
III. Ahern, Kevin G.

QD415 .M34 1999
572 21—dc21 99-043683

1 2 3 4 5 6 7 8 9 10—VH—03 02 01 00 99

Benjamin/Cummings, an imprint of Addison Wesley Longman
1301 Sansome Street
San Francisco, CA 94111

Credits and Acknowledgments

Figures 1.6, 2.5, 2.8, 2.10, 2.11, 2.13, 2.14, 4.15, 4.16, 4.20, 5.12, 6.2, 6.5, 6.13, 7.5, 7.7, 7.10, 7.20, 7.33, 9.19, 9.20, 9.21, 10.22, 10.23, and 11.11 were rendered by Graham Johnson.

Part Opening Images Part I: © K. R. Porter/Photo Researchers, Inc. Part II: Rendered by Graham Johnson. Part III: Courtesy of John Peters of Utah State University; adapted from J. W. Peters et al., *Science* (1998) 282: 1853–1858, with permission from *Science.* Part IV: Courtesy of S. Yoshikawa from T. Tsukihara et al., *Science* (1996) 272:1136, with permission from *Science.* Part V: Courtesy of T. J. Richmond; reprinted by permission from K. Luger, A. W. Mader, R. K. Richmond, D. F. Sargent, and T. J. Richmond, *Nature* (1997) 389:251–260. © 1997 Macmillan Magazines, Ltd.

Figure 15.19 is republished with permission of the Proceedings of the National Academy of Sciences, USA, 2101 Constitution Ave., NW, Washington, D.C., 20418, and is reproduced by permission of the publisher via Copyright Clearance Center.

In credit lines throughout this book, The American Association for the Advancement of Science has been abbreviated to "AAAS."

Biacore is a registered trademark of Biacore International AB.

MICROMAX is a trademark of NEN® Life Science Products.

Roundup and Roundup Ready cottonseed are registered trademarks of Monsanto Company.

To our students,
who continue to teach us biochemistry
by showing us that the best way
to learn a subject is to teach it.

About the Authors

CHRISTOPHER K. MATHEWS is Distinguished Professor and Chairman of the Department of Biochemistry and Biophysics at Oregon State University. He holds a B.A. from Reed College (1958) and a Ph.D. from The University of Washington (1962). He served on the faculties of Yale University and The University of Arizona before assuming his present position in 1978. His major research interest is the coordination between DNA precursor biosynthesis and DNA replication. Dr. Mathews was an Eleanor Roosevelt International Cancer Fellow at the Karolinska Institute in Stockholm in 1984–85 and Tage Erlander Guest Professor at Stockholm University in 1994–95. Dr. Mathews has published over 150 scientific papers dealing with molecular virology, metabolic regulation, nucleotide enzymology, and biochemical genetics. He is the author of *Bacteriophage Biochemistry* (1971) and coeditor of *Bacteriophage T4* (1983) and *Structural and Organizational Aspects of Metabolic Regulation* (1990). His teaching experience includes undergraduate, graduate, and medical school biochemistry courses.

K. E. VAN HOLDE is Distinguished Professor Emeritus of Biophysics and Biochemistry at Oregon State University. He earned his B.A. (1949) and Ph.D. (1952) from the University of Wisconsin. Over many years, Dr. van Holde's major research interest has been the structure of chromatin; his work resulted in the award of an American Cancer Society Research Professorship in 1977. He has been at Oregon State University since 1967, and was named Distinguished Professor in 1988. He is a member of the National Academy of Sciences and the American Academy of Arts and Sciences, and has received Guggenheim, NSF, and EMBO fellowships. He is the author of over 200 scientific papers and three books in addition to this volume: *Physical Biochemistry* (1971, 1985), *Chromatin* (1988), and *Principles of Physical Biochemistry* (1998). He was also coeditor of *The Origins of Life and Evolution* (1981). His teaching experience includes undergraduate and graduate chemistry, biochemistry and biophysics, and also the Physiology and Molecular Biology Course at The Marine Biological Laboratory at Woods Hole.

KEVIN G. AHERN is a Senior Instructor in the Department of Biochemistry and Biophysics at Oregon State University. He received his B.S (1976) and M.S. (1981) degrees from Oklahoma State University and his Ph.D. (1986) from Oregon State University. His interests include the use of computers in scientific research and teaching. Since 1994, Dr. Ahern has been on the faculty at Oregon State University, where he pioneered the development of Web-based courses for teaching biochemistry. He is a contributing editor to *Science* magazine, a columnist for *Genetic*

Engineering News, and a freelance writer for numerous online and printed publications. Dr. Ahern served as editor of *Biotechnology Software and Internet Journal* from 1987 to 1998 and is the founder of DaVinci Press Ink (www.davincipress.com), a scientific software consulting firm. He has edited two books, *Biotechnology Software Reports—Computer Applications for Molecular Biologists* and *The Biotechnology Software Directory, A Buyer's Guide.* Dr. Ahern's teaching experience includes undergraduate and graduate biochemistry courses.

Preface

PLANNING THE THIRD EDITION OF A WELL-RECEIVED BIOCHEMISTRY TEXTBOOK would seem a simple matter. Certainly, improvements are needed: those suggested by users of the text, as well as essential changes identified by the authors. Despite the greatest care, the inevitable pesky errors have come to light and must be corrected. But surely, if the text has worked well, the chief task should be updating. So we thought—until we began to contemplate the explosive expansion of biochemical knowledge in the few years since the second edition of *Biochemistry* appeared. How can this new information be accommodated without producing a textbook so densely packed with information as to be unusable by students, most of whom are taking a first course in biochemistry?

It became apparent that a qualitatively different approach was required. Without doubt, this approach must harness the power of the computer to access and organize the vast information on genomics, proteomics, gene sequences, and protein structures present in ever-expanding databases. It is crucial that the text form a seamless whole with the auxiliary information resources and that these resources be closely keyed to the text. Furthermore, these resources must be designed and produced in collaboration with one who has not only the technical skills but also experience in teaching biochemistry, ideally from previous editions of our textbook. In other words, to realize our goals, it was essential to recruit a third author.

The two of us involved in the first two editions (i.e., van Holde and Mathews) were extraordinarily fortunate to enlist Dr. Kevin Ahern as our third author. Dr. Ahern has a broad and deep knowledge of bioinformatics, honed by experience as the Editor of the journal *Biotechnology Software & Internet Journal* and as a Contributing Editor for *Science* with particular responsibility for computer applications in biological sciences. Of equal importance, Dr. Ahern teaches biochemistry courses in our department using this textbook.

Dr. Ahern's participation makes the third edition of *Biochemistry* quite a new enterprise, even though the book itself resembles the earlier editions in essential ways. We think of the text as serving two functions. As in the past, we have tried to create a readable and usable text, to guide students through an introductory biochemistry course that is usually, but not always, of one year's duration. But, in combination with the Electronic Study Guide designed by Dr. Ahern, the text has become a doorway to the vast and continuously evolving world of biochemistry. Our goal has been to create a media resource that will help and guide students long after they have completed their first course in biochemistry.

MEDIA RESOURCES

The Electronic Study Guide must serve two primary purposes. (1) It must enhance the ability of students to understand biochemistry, and (2) it must provide access to a body of knowledge far greater than could be covered in the printed pages of the textbook. To maximize accessibility and ease of use, we chose to provide all materials in a common hypertext format (HTML) and programming language (Java) that could be accessed easily via any of the common browsers, such as Netscape Navigator or Microsoft Internet Explorer, on virtually any personal computer. Our Electronic Study Guide has two parts: one stored on fixed media, namely the CD-ROM accompanying this text, and the other on a dynamic, continually updated Web site. Thus, items that we expect will require little or no updating over the life of the textbook are provided on the CD-ROM, but items that require more frequent updating will be kept on the Addison Wesley Longman Web site at www.awlonline.com/mathews.

The Electronic Study Guide CD-ROM that is packaged with each new copy of this text contains Outline, Concepts, Terminology and Quizzing sections. The Outline section follows the chapter organization of the text and contains hyperlinks to lead students at the click of a button to other points of relevant information, such as chemical structures and enzymatic reactions. The Concepts sections provide short, hyperlinked summaries of the important ideas of each chapter. The Terminology section contains summaries and definitions of all of the important terms in each chapter. The Quizzing section, organized by chapter, employs a database of questions and answers to enable students to test their knowledge of biochemical structures, names, enzymes, reactions, terms, and concepts.

The Benjamin/Cummings Science Digital Library for Biochemistry is a CD-ROM for faculty, designed to enhance teaching from the textbook. The figures on this CD can be used to create transparencies and slides and to build PowerPoint presentations and personal Web pages. The Science Digital Library CD is available to qualified adopters of this book. For more information, please contact your Benjamin/Cummings sales representative.

To keep pace with the development of biochemical knowledge, we have created a dynamic Web site that will be continually updated throughout the lifetime of the textbook. In this way we can keep users of our text abreast of exciting developments as they happen. Our Web site incorporates all of the essential features of both the student and faculty CD-ROMs, as well as Current News in biochemistry and biotechnology, a search engine for locating pages via user entry of keywords, and a Links section containing a vast collection of hyperlinks.

EVOLUTION OF THE TEXT

Within the text, our overall organization remains the same—that is, we begin with structure and mechanism, followed by intermediary metabolism and then biological information processing. Nevertheless, a few significant changes have been made. Chapter 3 (Bioenergetics) was shortened by transferring one section (ATP as Energy Currency) to Chapter 12 (Introduction to Metabolism). This change removed some redundancy and placed this material in appropriate juxtaposition with metabolism.

Many sections were either created or rewritten to amplify new concepts and discoveries. Examples include expanded discussions of protein folding and chaperone proteins (Chapter 6), molecular motors (Chapter 8), structures of mitochondrial respiratory complexes (Chapter 15), reactive oxygen species and human disease (Chapter 15), biochemical insights into obesity (Chapter 18), folic acid and the heart (Chapter 20), neurotransmitters and psychopharmacology (Chapter 21), new signal transduction pathways (Chapter 23), structures and mechanisms of

DNA polymerases (Chapter 24), genetic recombination mechanisms (Chapter 25), and eukaryotic gene expression (Chapter 28). This new material has been folded into the format that has worked well in previous editions, namely an early introduction of nucleic acid structure (to clarify presentations of protein structure and function), an emphasis on the experimental roots of biochemistry, a continual emphasis on energy relationships in biochemistry, and a stepwise approach to complex metabolic pathways (introduction, followed by overview and then details, and ending with reiteration of overview and discussion of regulation).

TOOLS OF BIOCHEMISTRY

Because the methods used by biochemists are continually evolving, in each edition we add new techniques in this section and delete those that are no longer of major use. In the Third Edition we have added one new Tools section, "Methods for Detecting and Analyzing Protein–Protein Interactions." We have deleted the section on Maxam–Gilbert DNA sequencing (although this historically important technique is outlined where we present transcript mapping techniques in Chapter 26), and we have also deleted a section that describes identification of N- and C-termini of polypeptides. The section on protein sequence analysis (Chapter 5) still indicates how N-termini are identified as part of automated sequencing protocols.

We have expanded other Tools sections to include new variants of older techniques, such as laser scanning confocal microscopy (Chapter 1) and the synthesis of combinatorial arrays of peptides (Chapter 5). The section on mass spectrometry (Chapter 6) has also been expanded to reflect the growing importance and versatility of this technique.

END-OF-CHAPTER PROBLEMS

Quantitative problems and discussion-type questions are among the most valuable learning resources in a text, and users consistently ask for more problems and complete answers. Most chapters in the Third Edition have two to four new problems each, with brief answers to all problems given at the end of the book.

ADDITIONAL SUPPLEMENTS

Along with the CD-ROM and Web site, another important learning aid is the *Complete Solutions Manual* (0-8053-3074-7), written by Joshua Hicks and Christopher Stoner, two graduate students in our department at Oregon State University. This printed manual contains fully worked solutions to all end-of-chapter problems in the text. Also, a set of 107 full-color *Transparency Acetates* (0-8053-3068-2) provides representative samples of illustrations from the text and can be used as lecture aids. The *Complete Solutions Manual* and the *Transparency Acetates* are available to qualified adopters of this book. For more information about these supplements or any of the media resources, please contact your Benjamin/Cummings sales representative or our customer service department at (800) 282-0693.

ACKNOWLEDGMENTS

Each of the three editions of this book has found us working with a completely different editorial and production team at Addison Wesley Longman. We have particularly enjoyed working with this team. Their professionalism and commitment to the project made it a pleasure to work with them and allowed us to write this text, create the media resources, and see the project to completion in nearly one year less than the time originally envisaged. Heading the team was Ben Roberts, Chemistry Editor, who oversaw both editorial and production phases of the work. Ben's consistent enthusiasm for the project was a powerful motivating force. Dr. John Murdzek, Developmental Editor, was a perceptive and persistent critic. John holds a Ph.D. in chemistry, which allowed him to interact with us as a fellow scientist and helped him to detect errors and ambiguities that would certainly have been missed by an editor less well trained in contemporary science.

Claudia Herman, Publishing Associate, helped in many ways to keep the project moving forward, including lining up all of the reviewers and making sure that we received the reviews in timely fashion. Once the text went into production, we had the pleasure of working with Lisa Weber, Production Editor. In her ability to keep abreast of myriad details and to coordinate all phases of the production process, Lisa reminded us of a chess player who can play twenty games simultaneously and win all of them. Lisa was assisted by Photo Researcher Roberta Spieckerman, who saved us an enormous amount of work by contacting authors and copyright holders of the numerous figures that we reproduced or adapted from published sources. Anita Wagner, our proofreader, made us aware of many substantive issues, as well as typographical errors, during her perceptive reading of the page proofs. We are grateful to many colleagues, acknowledged in the appropriate figure legends, who interacted with the authors directly in providing molecular graphics and artwork. We are grateful also to Copy Editor Mary Prescott, the only team member to have worked also on the First Edition of this text. Thanks also to Lynn Armstrong and Charlotte Shane for preparing the index.

Special thanks go to artist Graham Johnson, who designed a strikingly beautiful and original cover illustration and who brilliantly turned many of our sketched ideas into attractive and informative figures. Thanks also to many biochemists, acknowledged individually on pages xii–xiv, who were engaged by the publisher to review early drafts of our manuscripts. All of their suggestions were appreciated, even those that we were unable to follow.

Special thanks are due to two reviewers, Drs. Kimberley Waldron (Regis College) and Robert Ludwig (University of California, Santa Cruz), both of whom read the entire text at the page proof stage, catching errors and ambiguities that had been unaccountably missed by all the authors and other reviewers.

Kevin Ahern's efforts in developing the Electronic Study Guide depended heavily upon Oregon State University student Jody Franke, who assisted with linking and HTML issues; Heather Wycoff, who helped with graphics; and Hamid Ghanadan, for consultation regarding HTML issues. Thanks to these three from the three of us.

As always, our most persistent critics and supporters were our wives—Kate Mathews, Barbara van Holde, and Indira Rajagopal. The love and support that they consistently offered were the most important elements in seeing this project to a timely and satisfying conclusion.

CHRISTOPHER K. MATHEWS K. E. VAN HOLDE KEVIN G. AHERN

Reviewers

REVIEWERS (THIRD EDITION)

Roger Acey, California State University, Long Beach
Hugh Akers, Lamar University
Charles Allen, University of Florida
James Allen, Arizona State University
Dean Appling, University of Texas, Austin
Roy Baker, University of Toronto
Steven Blanke, University of Houston
Linda Bloom, Arizona State University
Albert Bobst, University of Cincinnati
Rodney Boyer, Hope College
Alexander Brownie, SUNY Buffalo
Tom Buckley, University of Victoria, Canada
Jeffrey Cohlberg, California State University, Long Beach
Richard Drake, University of Arkansas for Medical Sciences
Patricia Draves, University of Central Arkansas
Kristin Eckert, Pennsylvania State University
Jeremy Evans, Washington State University
David Fahrney, Colorado State University
James Franzen, University of Pittsburgh
Jeffrey Hayes, University of Rochester
Colleen Jonsson, New Mexico State University
Thomas Jue, University of California, Davis
Jason Kahn, University of Maryland, College Park
Harvey Knull, University of North Dakota
Robert Kuchta, University of Colorado, Boulder
Robley Light, Florida State University
Dennis Lohr, Arizona State University
Robert Ludwig, University of California, Santa Cruz
Theo Macrides, RMIT University, Australia
Susan Martinis, University of Houston
Douglas McAbee, California State University, Long Beach
Graham Parslow, University of Melbourne, Australia
Mulchand Patel, SUNY Buffalo
Kevin Plaxco, University of California, Santa Barbara
Leigh Plesniak, University of San Diego
Linda Roberts, California State University, Sacramento
Andrew Robertson, University of Iowa
John Sadleir, University of Western Australia, Australia
Wilma Saffran, CUNY Queens College
Charles Samuel, University of California, Santa Barbara
Pearl Tsang, University of Cincinnati
Elizabeth Vierling, University of Arizona
Kimberley Waldron, Regis University
Leigh Ward, University of Queensland, Australia
Randy Weselake, University of Lethbridge, Canada
E. J. Wood, University of Leeds, England

STUDENT REVIEWERS (SECOND EDITION)

Dennis Ahern, State College, PA
Jasmeet Bajaj, University Park, PA
Ken Bradley, Santa Cruz, CA
Ann Brolly, Santa Cruz, CA
Jay Chaplin, Santa Cruz, CA
Kirk Egge, Stillwater, MN
Felicia Goodvum, Blacksburg, VA
Jeanette Harlow, Santa Cruz, CA
Mara Jeffress, Santa Cruz, CA
Tabinda Khan, Santa Cruz, CA
Scott Kiss, Spruce Grove, Alberta, Canada
Emile Le Blanc, Edmonton, Alberta, Canada
Jonathan Lo Cicero, Sherwood Park, Alberta, Canada
Mihir Mooi, University Park, PA
Sam Pejham, Los Gatos, CA
Keri Pomella, Eaton Park, FL
Arthur Roberson, Memphis, TN
Terrance Strobaugh, Jr., Altoona, PA
Tim Sturgeon, University Park, PA
Trevor Swartz, Santa Cruz, CA
Ruth Thatcher, Blacksburg, VA
Krista Watson, Edmonton, Alberta, Canada

GRADUATE STUDENT PROBLEM CHECKERS (SECOND EDITION)

Wei Chen, Tucson, AZ
Brian Orendel, Tucson, AZ
Eric Peterson, Denver, CO
Ziaoquing You, Tucson, AZ

REVIEWERS (SECOND EDITION)

Genia S. Albrecht, Cornell University
Lisa T. Alty, Washington and Lee University
Lars Backman, University of Umeå, Sweden
Bruce R. Banks, University of North Carolina, Greensboro
Frank O. Brady, University of South Dakota, School of Medicine, Vermillion

Barbara Brennessel, Wheaton College
William A. Bridger, University of Alberta, Edmonton, Canada
Ronald Brosemer, Washington State University
Ray Brown, Wayne State University
Michael F. Bruist, Vassar College
Andrew Buchman, Pennsylvania State University
George E. Bunce, Virginia Polytechnic Institute and State University, Blacksburg
Jim Chambers, University of Texas, San Antonio
Scott Champney, East Tennessee State University
Derek A. Chignell, Wheaton College
William Currier, University of Vermont
James Davenport, University of Tennessee, Memphis
Daniel Davison, University of Houston
Michael Dennis, Eastern Montana College
Ruth L. Dusenbery, Wayne State University
Lehman L. Ellis, University of New Orleans
Mary Lou Ernst-Fonberg, East Tennessee State University
Nancy Federspiel, University of Idaho
Gerald W. Feigenson, Cornell University
Blaise Frost, West Chester State University
Thomas E. Goyne, Valparaiso University
Ralph J. Henderson, Louisiana State University, School of Medicine, Shreveport
E. Clifford Herrmann, Loma Linda University
George Hoch, Emeritus, University of Rochester
Ching-hsien Huang, University of Virginia, School of Medicine
Richard Ikeda, Georgia Institute of Technology
Ralph A. Jacobson, California Polytechnic State University
Peter C. Kahn, Rutgers University
Teh-Hui Kao, Pennsylvania State University, University Park
Thomas A. Keevil, Southern Oregon State University
Marvin L. Kientz, Sonoma State University
George B. Kitto, University of Texas, Austin
James A. Knopp, North Carolina State University
Torsten Kristensen, University of Aarhus, Langelandsgate, Denmark
Rick Krueger, University of South Carolina, Spartanburg
Robert D. Kuchta, University of Colorado, Boulder
LeRoy R. Kuehl, University of Utah, Medical School
Franklin R. Leach, Oklahoma State University
Harold G. Martinson, University of California, Los Angeles
Celia Marshak, San Diego State University
Ronald W. McCune, Idaho State University
Kelly Meckling-Gill, University of Guelph, Ontario, Canada
Armando John Merola, Ohio State University, College of Medicine, Columbus
Julie T. Millard, Colby College
Richard Morgan, University of Alberta, Edmonton, Canada
Kim Kusk Mortensen, University of Aarhus, Langelandsgate, Denmark
Bradley B. Olwin, Purdue University
David H. Peyton, Portland State University
Robert L. Potter, University of South Florida
Roger Rice, Wye College, University of London, England
Joe M. Ross, Central State University
Alan H. Rowe, Norfolk State University
Dennis P. Ryan, Hofstra University
Robert Ryan, University of Alberta, Edmonton, Canada
Marvin Salin, Mississippi State University
Dorothy Elaine Schumm, Ohio State University
Thomas W. Sneider, Colorado State University
Hans Sperling-Petersen, University of Aarhus, Langelandsgate, Denmark
Thomas Squier, University of Kansas
Pamela C. Stacks, San Jose State University
Paul Stein, College of St. Scholastica
Rune Stjernholm, Tulane University, Medical Center
Bik-Kwoon Tye, Cornell University
Theodorus van Es, Rutgers University
Harry van Keulen, Cleveland State University
Charles J. Waechter, University of Kentucky, School of Medicine
Gary A. Weisman, University of Missouri, Columbia
Peter Wejksnora, University of Wisconsin, Milwaukee
Michael Wells, University of Arizona
William R. Widger, University of Houston
Ann Wood, University of Puget Sound
Les Wynston, California State University, Long Beach
Lee A. Young, Lincoln, MA
Daniel M. Ziegler, University of Texas, Austin

REVIEWERS (FIRST EDITION)

Hugh Akers, Lamar University
David F. Albertini, Tufts University
Mark Alper, University of California, Berkeley
Dean R. Appling, University of Texas at Austin
Thomas O. Baldwin, Texas A&M University
Clinton E. Ballou, University of California, Berkeley
Wayne M. Becker, University of Wisconsin, Madison
Helen M. Berman, Fox Chase Cancer Center
Loran L. Bieber, Michigan State University
Robert Blankenship, Arizona State University
John W. Bodnar, Northeastern University
Rodney F. Boyer, Hope College
Robert B. Buchanan, University of California, Berkeley
Neil A. Campbell, San Bernardino Valley College
W. Scott Champney, East Tennessee State University
John Coffin, Tufts University
Jeffrey A. Cohlberg, California State University, Long Beach
Anne Dell, Imperial College, London
John Elam, Florida State University
Donald M. Engleman, Yale University

David E. Fahrney, Colorado State University
Richard E. Fine, Boston University School of Medicine
William H. Fuchsman, Oberlin College
Reginald H. Garrett, University of Virginia
Arthur M. Geller, University of Tennessee, Memphis
John H. Golbeck, Portland State University
Lowell P. Hager, University of Illinois, Urbana-Champaign
Gerald W. Hart, Johns Hopkins University
Standish C. Hartman, Boston University
Glenn A. Herrick, University of Utah
John W. B. Hershey, University of California, Davis
C. H. W. Hirs, University of Colorado Health Sciences Center
Laura L. M. Hoopes, Occidental College
Joyce E. Jentoft, Case Western Reserve University
Howard M. Jernigan, Jr., University of Tennessee, Memphis
Kenneth A. Johnson, Pennsylvania State University
G. Barrie Kitto, University of Texas
Gunter B. Kohlhaw, Purdue University
Sydney R. Kushner, University of Georgia
Tomas M. Laue, University of New Hampshire
Timothy M. Lohman, Texas A&M University
Kenneth J. Longmuir, University of California, Irvine
Ponzy Lu, University of Pennsylvania
Joan Lusk, Brown University
Judith K. Marquis, Boston University School of Medicine
Rowena G. Matthews, University of Michigan, Ann Arbor
William R. McClure, Carnegie Mellon University
David B. McKay, University of Colorado, Boulder
David Mount, University of Arizona
Burton L. Nesset, Pacific Lutheran University
Merle S. Olson, University of Texas Health Science Center, San Antonio
Stanley Parsons, University of California, Santa Barbara
Mulchand S. Patel, Case Western Reserve University
David M. Prescott, University of Colorado, Boulder
David G. Priest, Medical University of South Carolina
Norbert O. Reich, University of California, Santa Barbara
Thomas Schleich, University of California, Santa Cruz
Earl Shrago, University of Wisconsin, Madison
Elizabeth R. Simons, Boston University School of Medicine
Gerald R. Smith, Fred Hutchinson Cancer Research Institute
Thomas W. Sneider, Colorado State University
Lewis Stevens, Stirling University
Phyllis R. Strauss, Northeastern University
Charles C. Sweeley, Michigan State University, East Lansing
Robert L. Switzer, University of Illinois at Urbana-Champaign
Buddy Ullman, Oregon Health Sciences University
Dennis E. Vance, University of Alberta
Andrew H.-J. Wang, Massachusetts Institute of Technology
Peter J. Wejksnora, University of Wisconsin, Milwaukee
Beulah M. Woodfin, University of New Mexico

Brief Contents

Part IV Dynamics of Life: Energy, Biosynthesis, and Utilization of Precursors 445

Part V Information 875

Detailed Contents

PART I

The Realm of Biochemistry 1

CHAPTER 6

The Three-Dimensional Structure of Proteins 161

CHAPTER 7

Protein Function and Evolution 212

CHAPTER 8

Proteins in Motion: Contractile Systems and Molecular Motors *256*

CHAPTER 9

Carbohydrates *278*

CHAPTER 10

Lipids, Membranes, and Cellular Transport *315*

PART III

Dynamics of Life: Catalysis and Control of Biochemical Reactions *359*

CHAPTER 11

Enzymes: Biological Catalysts *360*

CHAPTER 16

Carbohydrate Metabolism II: Biosynthesis 560

CHAPTER 17

Photosynthesis 594

CHAPTER 18

Lipid Metabolism I: Fatty Acids, Triacylglycerols, and Lipoproteins 627

CHAPTER 19

Lipid Metabolism II: Membrane Lipids, Steroids, Isoprenoids, and Eicosanoids 667

CHAPTER 20

Metabolism of Nitrogenous Compounds: Principles of Biosynthesis, Utilization, and Turnover 707

CHAPTER 21

Metabolism of Nitrogenous Compounds: Amino Acids, Porphyrins, and Neurotransmitters 746

PART V

Information 875

CHAPTER 24

Information Copying: Replication 876

CHAPTER 25

Information Restructuring: Restriction, Repair, Recombination, Rearrangement, and Amplification 926

CHAPTER 26

Information Readout: Transcription 980

CHAPTER 27

Information Decoding: Translation 1027

CHAPTER 28

Eukaryotic Genes and Their Expression 1068

Tools of Biochemistry

PART 1

The Realm of Biochemistry

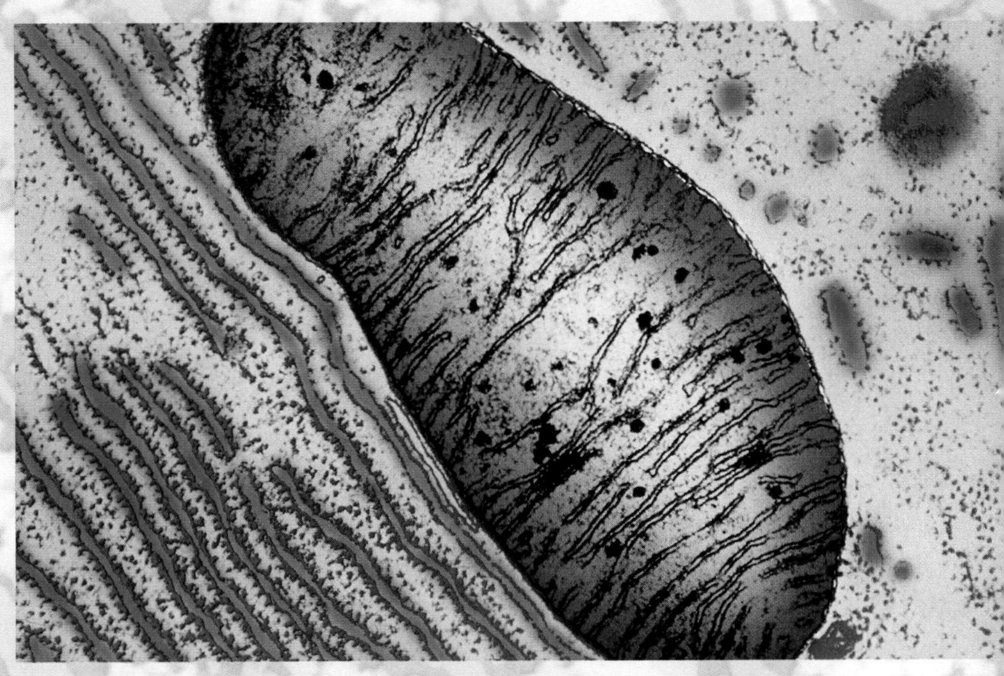

CHAPTER 1

The Scope of Biochemistry

THE BIOLOGICAL SCIENCES HAVE BEEN UNDERGOING A REVOLUTION, AND BIOchemistry has been at the heart of that revolution. Nowhere is this better illustrated than in the remarkable number of Nobel Prizes in Chemistry and in Medicine or Physiology that have been won by biochemists in recent years. A typical example is the award of the 1988 Nobel Prize for Medicine or Physiology, to Gertrude Elion and George Hitchings of the United States and Sir James Black of Great Britain for their leadership in inventing new drugs. Elion and Hitchings had developed chemical analogs of nucleic acids and vitamins, now used to treat leukemia, bacterial infections, malaria, gout, herpes viral infections, and AIDS; Black developed beta blockers that are used to reduce the risk of heart attack and to treat diseases such as asthma. Three of these drugs are described in Figure 1.1. **6-Mercaptopurine,** an analog of a nucleic acid precursor, inhibits the uncontrolled DNA replication associated with proliferation of white blood cells in leukemia. **3′-Azido-2′,3′-dideoxythymidine** (AZT), an analog of the nucleotide thymidine, was originally synthesized as a potential anticancer drug but has been used mostly in treating human immunodeficiency virus (HIV), the infectious agent that causes AIDS. **Isoproterenol,** an analog of the hormone **epinephrine** (also called **adrenaline**), mimics the actions of epinephrine, which transmits signals to the heart, lungs, and other vital organs.

These drugs were not discovered through random organic chemical synthesis. Instead, they were *designed,* as a consequence of several decades of accumulated knowledge in central areas of biochemistry—protein structure and function, nucleic acid synthesis, enzyme mechanisms, receptors and metabolic control, vitamins and coenzymes, and comparative biochemistry (the study of chemical differences among organisms). The impact of biochemistry is felt in all of our lives through discoveries such as these and in the way these discoveries fuel the growth of all of the life sciences.

Two more recent prizes give further evidence for the breadth of the impact of biochemistry on all of the life sciences. In 1997, the Chemistry Prize was shared by three—the American Paul Boyer and the British J. Walker, for their discovery of the "rotary engine" that generates the energy-carrying compound ATP, and the Danish J. Skou, for his studies of the "pump" that drives sodium and potassium

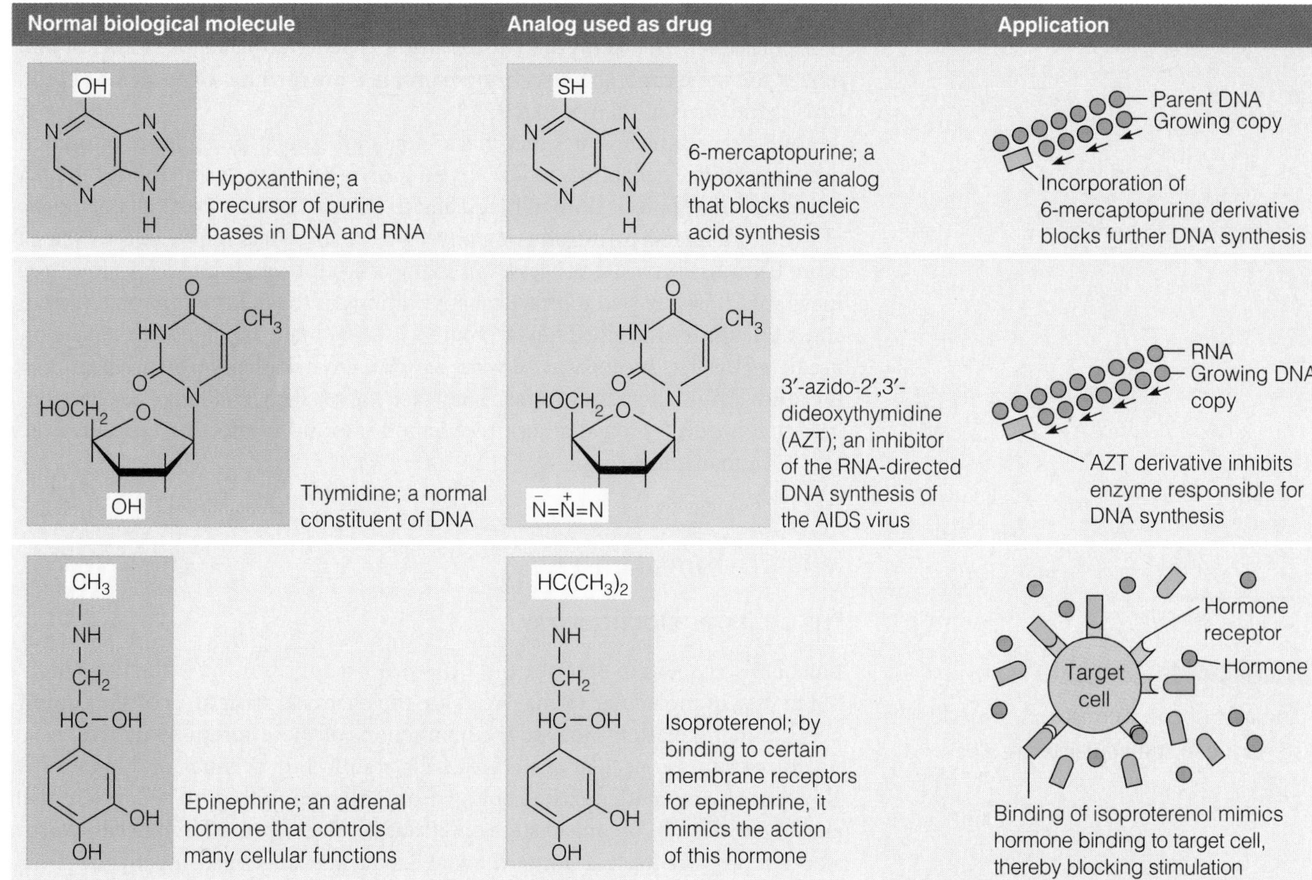

FIGURE **1.1**

Medical applications of biochemistry. Three examples of metabolic analogs designed by biochemists and used as important drugs.

across membranes. In the same year, the Prize in Medicine or Physiology went to Stanley Prusiner, for his studies on the agent responsible for "mad cow" disease.

Two factors contribute both to the excitement in today's biochemistry and to its impact on other life sciences. First, it is now well established that living matter obeys the same fundamental physical laws that govern all matter. Therefore, the full power of modern chemical and physical theory can now be brought to bear on biological problems. Second, incredibly powerful new research techniques are permitting scientists to ask questions about the basic processes of life that could not have been imagined even a few years ago.

Living matter, exemplified by the cell depicted in Figure 1.2, is composed of chemical substances, and any biological function can be described in terms of the structures and reactions of those substances. To understand any life process, therefore, demands understanding of the chemistry of that process. This is true even for very complex biological processes—such as evolution, differentiation, and

FIGURE **1.2**

The complexity of living matter. This colored micrograph of a one-celled organism, the photosynthetic protozoan *Euglena,* reveals the structural complexity that can exist within a single cell. The nucleus, which contains most of the cell's DNA, is the brown area at right center. Above and to the left of the nucleus are the green chloroplasts, the sites of photosynthetic energy generation. At the top left is the whiplike flagellum that propels the organism. The functioning of a cell, no matter how complex, can be described in terms of physical and chemical reactions.

behavior—that until recently were not thought to be amenable to analysis at the molecular level. The fact is that biochemistry is penetrating all disciplines of biology. At the same time, chemists from many areas are turning to biochemical problems as the focus of their research.

The research techniques that make these new studies possible are of tremendous power. They enable us to identify an individual by analyzing the genetic material in a single hair or drop of blood and to change the most basic characteristics of living organisms by introducing novel genes. These techniques allow us to explore the most intricate processes in living cells and watch molecules in action. Inevitably, they have led to practical applications in the fields of pharmacy, medicine, agriculture, and allied fields. Progress is being made much faster than anyone predicted even as recently as 10 years ago. We envy students who are beginning the study of biochemistry today. They face the exciting challenge of entering a field that is developing at an unprecedented rate and is affecting virtually every area of human endeavor.

What Is Biochemistry?

THE GOALS OF BIOCHEMISTRY

The aim of biochemistry is to understand life in molecular terms.

Biochemistry seeks to describe the structure, organization, and functions of living matter in molecular terms. What are the chemical structures of the components of living matter? How do the interactions of these components give rise to organized supramolecular structures, cells, multicellular tissues, and organisms? How does living matter extract energy from its surroundings in order to remain alive? How does an organism store and transmit the information it needs to grow and to reproduce itself accurately? What chemical changes accompany the reproduction, aging, and death of cells and organisms? How are chemical reactions controlled inside living cells? These are the kinds of questions being asked by biochemists; the search for the answers is the study of the chemistry of life.

Biochemistry can be divided into three principal areas: (1) the **structural chemistry** of the components of living matter and the relationship of biological function to chemical structure; (2) **metabolism,** the totality of chemical reactions that occur in living matter; and (3) the chemistry of processes and substances that store and transmit biological information. The third area is also the province of **molecular genetics,** a field that seeks to understand heredity and the expression of genetic information in molecular terms.

THE ROOTS OF BIOCHEMISTRY

Early biochemists had to overcome the view that living matter and nonliving matter were fundamentally different.

$$\overset{+}{N}H_4NCO^- \longrightarrow H_2N{-}\overset{\overset{\displaystyle O}{\|}}{C}{-}NH_2$$

Ammonium cyanate **Urea**

Biochemistry had its origins as a distinct field of study in the early nineteenth century, with the pioneering work of Friedrich Wöhler (Figure 1.3). Prior to Wöhler's time it was believed that the substances in living matter were somehow qualitatively different from those in nonliving matter and did not behave according to the known laws of physics and chemistry. In 1828 Wöhler showed that urea, a substance of biological origin, could be synthesized in the laboratory from the inorganic compound ammonium cyanate. As Wöhler phrased it in a letter to a colleague, "I must tell you that I can prepare urea without requiring a kidney or an animal, either man or dog." This was a shocking statement in its time, for it breached the presumed barrier between the living and the nonliving.

Even after Wöhler's demonstration, a persuasive viewpoint called **vitalism** held that, if not the compounds, at least the reactions of living matter could occur only in living cells. According to this view, biological reactions took place through the action of a mysterious "life force" rather than physical and chemical processes. The vitalist dogma was shattered in 1897, when two German brothers, Eduard and Hans Buchner, found that extracts from broken—and thoroughly dead—yeast

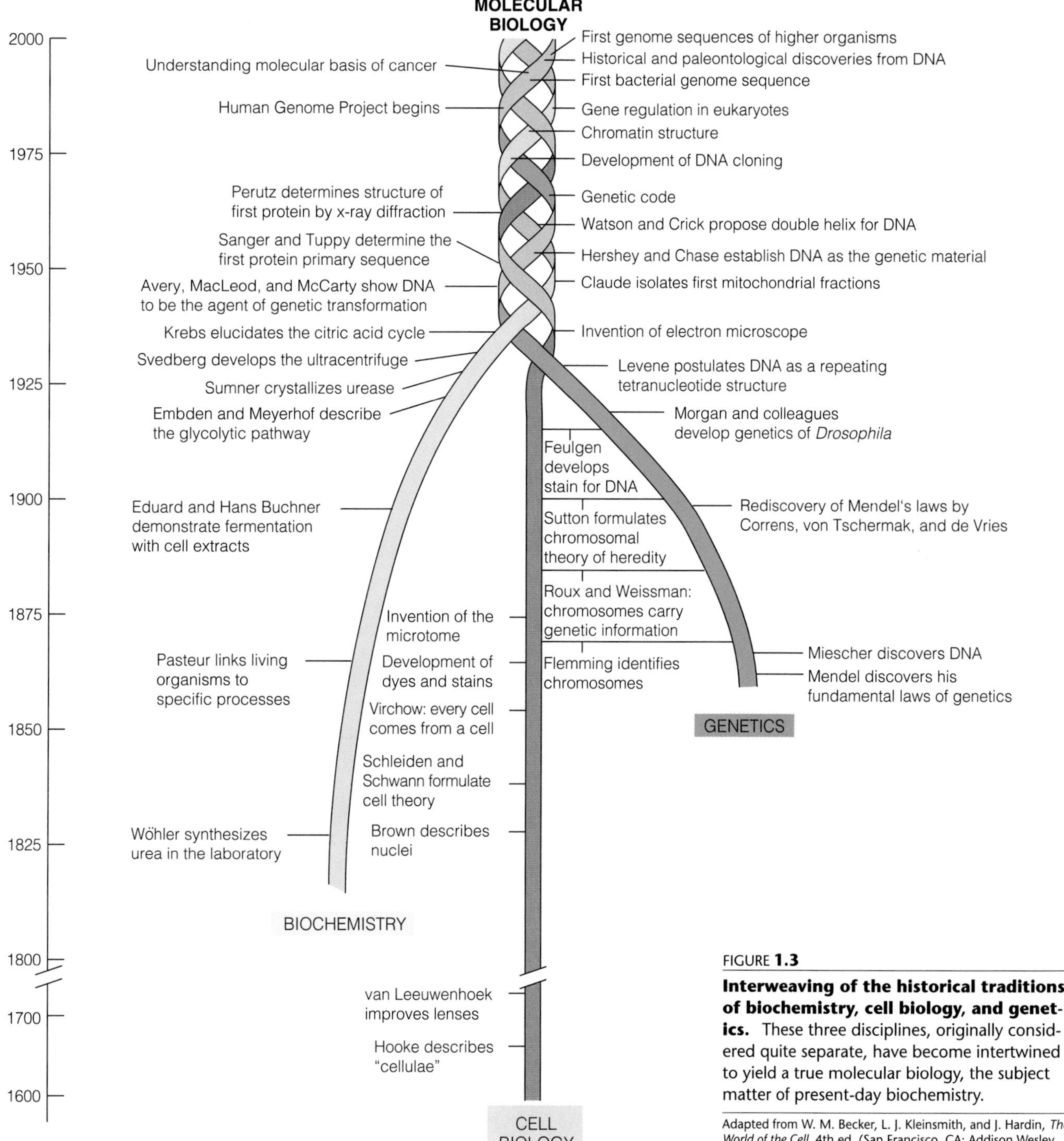

FIGURE **1.3**

Interweaving of the historical traditions of biochemistry, cell biology, and genetics. These three disciplines, originally considered quite separate, have become intertwined to yield a true molecular biology, the subject matter of present-day biochemistry.

Adapted from W. M. Becker, L. J. Kleinsmith, and J. Hardin, *The World of the Cell*, 4th ed. (San Francisco, CA: Addison Wesley Longman, 2000). © Addison Wesley Longman, Inc.

cells could carry out the entire process of fermentation of sugar into ethanol. This discovery opened the door to analysis of biochemical reactions and processes **in vitro** (Latin, "in glass"), meaning in a test tube rather than **in vivo**, in living matter. In succeeding decades many other metabolic reactions and reaction pathways were reproduced in vitro, allowing identification of reactants and products and of the **enzymes**, or biological catalysts, that promoted each biochemical reaction.

The nature of biological catalysis remained the last refuge of the vitalists, who held that the structures of enzymes (or "ferments") were far too complex to be described in chemical terms. But in 1926 J. B. Sumner showed that the protein urease, an enzyme from jack beans, could be crystallized like any organic compound. Although proteins have large and complex structures, they are just organic compounds, and their structures can be determined by the methods of chemistry. This discovery marked the final fall of vitalism.

In parallel with developments in biochemistry, **cell biologists** had been continually refining knowledge of cellular structure. Beginning with Robert Hooke's first observation of cells in the seventeenth century, steady improvements in microscopic techniques led to the understanding that the cell was a complex, compartmental structure (see Figure 1.2). *Chromosomes* were discovered in 1875 by Walter Flemming and identified as genetic elements by 1902. The development of the electron microscope, between about 1930 and 1950, provided a whole new level of insight into cellular structure. Now subcellular organelles like *mitochondria* and *chloroplasts* could be studied, and it was realized that specific biochemical processes were localized in these subcellular particles.

Although developments in the first half of the twentieth century revealed in broad outline the chemical structures of biological materials, identified the reactions in many metabolic pathways, and localized these reactions within the cell, biochemistry remained an incomplete science. We knew that the uniqueness of an organism is determined by the totality of its chemical reactions. However, we had little understanding of how those reactions are controlled in living tissue or of how the information that regulates those reactions is stored, transmitted when cells divide, and processed when cells differentiate.

The idea of the **gene,** a unit of hereditary information, was first proposed in the mid-nineteenth century by Gregor Mendel. By about 1900, cell biologists realized that genes must be found in chromosomes, which are composed of proteins and nucleic acids. During the next few decades, the new science of **genetics** provided increasingly detailed knowledge of the patterns of inheritance and development. However, until the mid-twentieth century no one had isolated a gene or determined its chemical composition. Nucleic acids had been isolated in 1869 by Friedrich Miescher, but their chemical structures were poorly understood, and in the early 1900s they were thought to be simple substances, fit only for structural roles in the cell. Most biochemists believed that only the proteins were structurally complex enough to carry genetic information.

That belief was dead wrong. Experiments in the 1940s and early 1950s proved conclusively that **deoxyribonucleic acid (DNA)** is the bearer of genetic information. One of the most important advances in the history of science occurred in 1953, when James Watson and Francis Crick described the double-helical structure of DNA. This concept immediately suggested ways in which information could be encoded in the structure of molecules and transmitted intact from one generation to the next.

Molecular biology is a fusion of biochemistry, cell biology, and genetics.

At about this point the strands of scientific development shown in Figure 1.3—biochemistry, cell biology, and genetics—became inextricably interwoven, and the new science of **molecular biology** emerged. The distinction between molecular biology and biochemistry is not always clear, because both disciplines take as their ultimate aim the complete definition of life in molecular terms. The term *molecular biology* is often used in a narrower sense, to denote the study of nucleic acid structure and function and the genetic aspects of biochemistry—an area we might more properly call *molecular genetics.* Molecular biology and biochemistry are perhaps distinguished more readily by the orientations of their practitioners than by the research problems being addressed. It can be said that biochemists think like chemists and molecular biologists think like biologists. Even this distinction is somewhat artificial, since successful scientists in either field must use the approaches of all relevant disciplines, including chemistry, biology, and physics. In fact, three of the most powerful research techniques used by biochemists were developed by physicists: **electron microscopy,** which has revealed remarkable details of cellular structure (see Tools of Biochemistry 1A), and **x-ray diffraction** and

nuclear magnetic resonance, which have revealed the precise three-dimensional structures of huge biological molecules (see Tools of Biochemistry 4A and 6A).

BIOCHEMISTRY AS A DISCIPLINE AND AN INTERDISCIPLINARY SCIENCE

Biochemistry draws its major themes from many disciplines—from organic chemistry, which describes the properties of biomolecules; from biophysics, which applies the techniques of physics to study the structures of biomolecules; from medical research, which increasingly seeks to understand disease states in molecular terms; from nutrition, which has illuminated metabolism by describing the dietary requirements for maintenance of health; from microbiology, which has shown that single-celled organisms and viruses are ideally suited for the elucidation of many metabolic pathways and regulatory mechanisms; from physiology, which investigates life processes at the tissue and organism levels; from cell biology, which describes the biochemical division of labor within a cell; and from genetics, which describes mechanisms that give a particular cell or organism its biochemical identity. Biochemistry draws strength from all of these disciplines, and it nourishes them in return; it is truly an interdisciplinary science.

Biochemistry is also a distinct discipline, with its own identity. It is distinctive in its emphasis on the structures and reactions of biomolecules, particularly on enzymes and biological catalysis; on the elucidation of metabolic pathways and their control; and on the principle that life processes can be understood through the laws of chemistry. As you read this book, keep in mind both the uniqueness of biochemistry as a separate discipline and the absolute interdependence of biochemistry and other physical and life sciences.

Biochemistry as a Chemical Science

Though we often describe biochemistry as a life science and relate its developments to the history of biology, it remains first and foremost a chemical science. In order to understand the impact of biochemistry on biology, you must understand the chemical elements of living matter and the complete structures of many biological compounds—amino acids, sugars, lipids, nucleotides, vitamins, and hormones—and their behavior during metabolic reactions. You will need to know the stoichiometry and mechanisms of a large number of reactions. In addition, an understanding of the basic thermodynamic principles is essential for learning how plants derive energy from sunlight and how animals derive energy from food.

To understand biochemistry, one must first study basic chemistry.

All forms of life, from the smallest bacterial cell to a human being, are constructed from the same chemical elements, which in turn make up the same types of molecules; the chemistry of living matter is similar throughout the biological world. Undoubtedly, this continuity in biochemical processes reflects the common evolutionary ancestry of all cells and organisms. Let us begin a preliminary examination of the composition of living matter, starting with the chemical elements.

THE CHEMICAL ELEMENTS OF LIVING MATTER

Life is a phenomenon of the second generation of stars. This rather strange-sounding statement is based on the fact that life, as we conceive it, can come into being only when certain elements—carbon, hydrogen, oxygen, nitrogen, and phosphorus (C, H, O, N, and P)—are abundant. The very early universe was made almost entirely of hydrogen and helium, for only these simplest elements were produced in the condensation of matter following the primeval explosion, or "big bang." The first generation of stars contained no heavier elements from which to form planets. As these early stars matured over billions of years, they burned their hydrogen and helium in thermonuclear reactions. These reactions produced

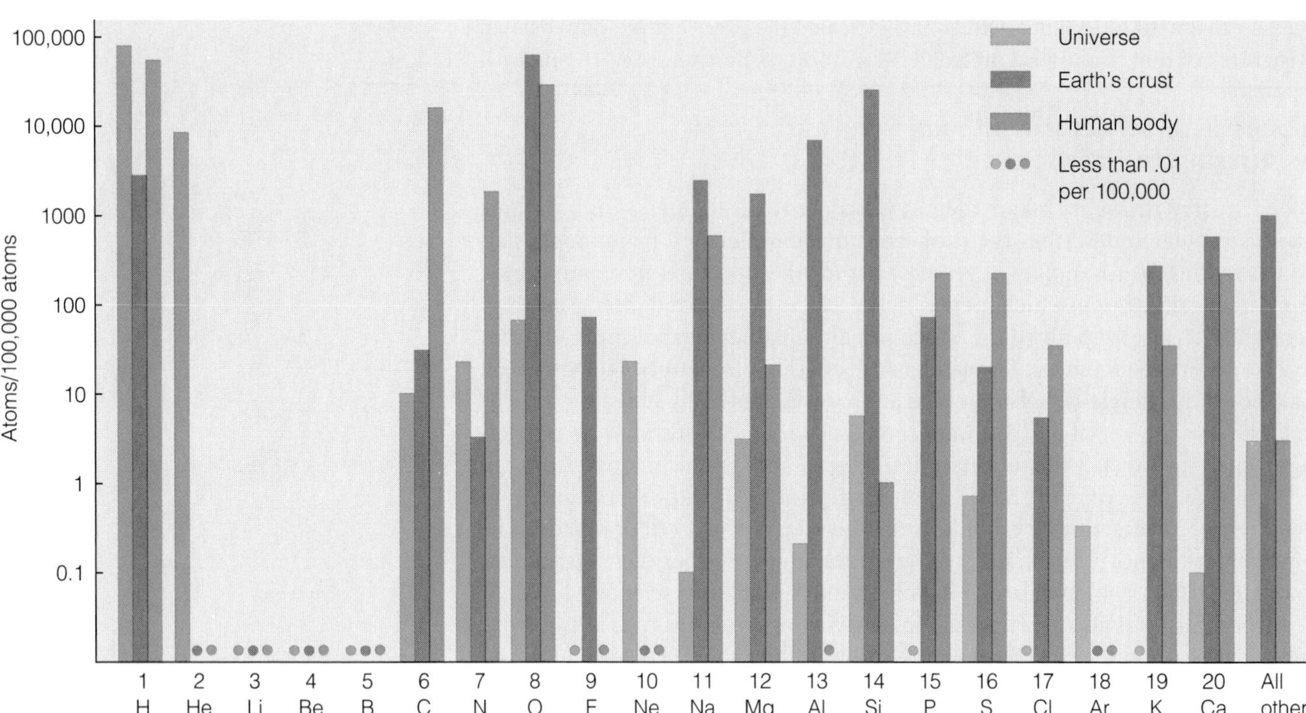

FIGURE **1.4**

Composition of the universe, the earth's crust, and the human body. Amounts are expressed as number of atoms of each element per 100,000 atoms. Note that the scale is logarithmic. On a linear scale, H and He would greatly dominate in the universe, O and Si in the earth's crust, and H, C, N, and O in the body. A circle indicates less than 0.01 atom of the element per 100,000 atoms.

heavier elements—first carbon, nitrogen, and oxygen and later all the other members of the periodic table. As large stars matured, they became unstable and exploded as novas and supernovas, spreading the heavier elements through the cosmic surroundings. This matter condensed again to form second-generation stars, complete with planetary systems rich in the heavier elements. Our universe, which is now rich in second-generation stars, has the approximate elemental composition shown in Figure 1.4. In such a universe there is the possibility of life, and we know that on at least one planet, of one solar system, life does exist.

Why are elements heavier than H and He essential for life? The answer is that life, as we can imagine it, requires large and complex molecular structures. These can be formed only from certain elements and can be stable only under restricted environmental conditions. A universe of hydrogen and helium has no chemistry. Nor is chemistry possible in the heat of stars, where all compounds are broken into their elements. In environments as cold as the moon or space, a slow, simple chemistry may occur, but we cannot envision the formation of molecules as complex as proteins or nucleic acids. Only in the temperate environment of an appropriate planet, enriched with elements capable of forming complicated compounds, can life arise.

Life depends primarily on a few elements (C, H, O, N), although many others are used to a lesser extent.

Living creatures on the earth are composed mainly of a very few elements, principally carbon, hydrogen, oxygen, and nitrogen (C, H, O, N), as shown in Figure 1.4. Notice that these are also, with the addition of helium and neon, the most abundant elements in the universe. Helium and neon, inert gases, are not equipped for a role in life processes; they do not form stable compounds, and they are readily lost from planetary atmospheres.

The abundance of oxygen and hydrogen in organisms is explained partly by the major role of water in life on earth. We live in a highly aqueous world, and, as we shall see in Chapter 2, the solvent properties of water are indispensable in biochemical processes. The human body, in fact, is about 70% water. The elements C, H, O, and N are important to life because of their strong tendencies to form covalent bonds. In particular, the stability of carbon–carbon bonds and the possibility of forming single, double, or triple bonds give carbon the versatility to be part of an enormous diversity of chemical compounds.

But life is not built on these four elements alone. Many other elements are necessary for terrestrial organisms, as you can see in Table 1.1. A "second tier" of essential elements includes sulfur and phosphorus, which form covalent bonds, and the ions Na^+, K^+, Mg^{2+}, Ca^{2+}, and Cl^-. Sulfur is an important constituent of proteins, and phosphorus plays essential roles in energy metabolism and the structure of nucleic acids. Beyond the first two tiers of elements (which correspond roughly to the most abundant elements of the first two rows of the periodic table), we come to those that play quantitatively minor, but often indispensable, roles. As Table 1.1 shows, most of these third- and fourth-tier elements are metals, and some serve as aids to catalysis of biochemical reactions. In succeeding chapters we shall encounter many examples of the importance of these trace elements to life.

TABLE **1.1** **Elements found in organisms**

Element	Comment
First Tier	
Carbon (C)	Most abundant in *all organisms*
Hydrogen (H)	
Nitrogen (N)	
Oxygen (O)	
Second Tier	
Calcium (Ca)	Much less abundant but found in *all organisms*
Chlorine (Cl)	
Magnesium (Mg)	
Phosphorus (P)	
Potassium (K)	
Sodium (Na)	
Sulfur (S)	
Third Tier	
Cobalt (Co)	Metals present in small amounts in *all organisms* and essential to life
Copper (Cu)	
Iron (Fe)	
Manganese (Mn)	
Zinc (Zn)	
Fourth Tier	
Aluminum (Al)	Found in or required by *some organisms* in trace amounts
Arsenic (As)	
Boron (B)	
Bromine (Br)	
Chromium (Cr)	
Fluorine (F)	
Gallium (Ga)	
Iodine (I)	
Molybdenum (Mo)	
Nickel (Ni)	
Selenium (Se)	
Silicon (Si)	
Tungsten (W)	
Vanadium (V)	

BIOLOGICAL MOLECULES

The complexity of life processes requires that many of the molecules that participate in these processes be enormous. The most extreme example is DNA. Consider, for instance, the DNA molecules released from one human chromosome, as shown in Figure 1.5. The long, looped thread you see corresponds to just two enormous molecules, each with a molecular weight of about 20 billion daltons (a dalton [Da] is 1/12 the mass of a carbon-12 atom, 1.66×10^{-24} g). Even a simple organism such as the single-celled bacterium *Escherichia coli* contains a DNA molecule with a molecular weight of about 2 billion Da. Protein molecules are generally much smaller, but they are still large, with a typical protein having a mass of 50,000 Da. To give an idea of the complexity of such a molecule, Figure 1.6 shows the three-dimensional structure of one protein molecule as revealed by x-ray crystallography.

FIGURE **1.5**

The DNA from a single human chromosome. Most of the chromosomal proteins have been removed, leaving only a protein "skeleton" from which enormous loops of DNA emerge. There are just two DNA molecules in this chromosome. The enlargement to the right shows the long DNA fiber in more detail. The fiber is about 2nm thick.

Courtesy of J. Paulson and U. K. Laemmli, *Cell* 12 (1977):817: © 1997 Cell Press.

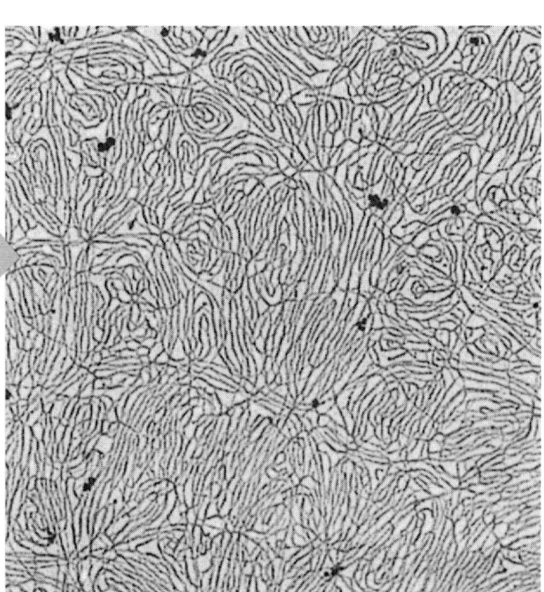

FIGURE **1.6**

A protein molecule. Shown here is a model of an immunoglobulin molecule—a molecule that serves as an antibody in the immune reaction. It has a molecular weight of approximately 150,000 Da. The molecule is made of two identical halves, or subunits.

Adapted from Eduardo A. Padlan, *Mol. Imm.* (1994) 31:169–217. © 1994 with permission from Elsevier Science.

Many of the important molecules in cells are enormous.

These giant molecules, or **macromolecules,** constitute a large fraction of the mass of any cell. As we shall see in detail in later chapters, there are good reasons for some biological materials to be so large. DNA molecules, for example, can be thought of as "tapes" from which genetic information is read out in a linear fashion. Since the amount of information needed to specify the structure of a multicellular organism is very great, these tapes must be extremely long. In fact, the DNA molecules in a single human cell, if stretched end to end, would reach a length of about 2 meters.

The synthesis of such large molecules poses an interesting challenge to the cell. If the cell functioned like an organic chemist carrying out a complex laboratory synthesis bit by bit, millions of different types of reactions would be involved, and thousands of intermediates would accumulate. Instead, cells use a modular approach for constructing large molecules. All such structures are **polymers,** made by joining prefabricated units, or **monomers.** The monomers of a given type of macromolecule are of limited diversity and are linked together, or **polymerized,** by identical mechanisms. A simple example is the carbohydrate **cellulose** (Figure 1.7a), a major constituent of the cell walls of plants. Cellulose is a polymer made by joining thousands of molecules of glucose, a simple sugar; in this polymer all of the chemical linkages between the monomers are identical. Covalent links between glucose units are formed by removing a water molecule between two adjoining glucose molecules; the portion of a glucose molecule remaining in the chain is called a glucose **residue.**

The major biopolymers include the nucleic acids, the proteins, and the polysaccharides. All are polymers, made up of one or more kinds of monomer units.

Because cellulose is a polymer of a simple sugar, or **saccharide,** it is called a **polysaccharide.** This particular polymer is constructed from identical monomeric units, so it is called a **homopolymer.** In contrast, many polysaccharides and all nucleic acids and proteins are **heteropolymers,** polymers constructed from a number of different kinds of monomer units. Nucleic acids (Figure 1.7b) are polymers of four **nucleotides,** so nucleic acids are also called **polynucleotides.** Similarly, proteins (Figure 1.7c) are assembled from combinations of 20 different **amino acids.** Protein chains are called **polypeptides,** a term derived from the **peptide bond** that joins two amino acids together.

Polymers form much of the structural and functional machinery of the cell. Polysaccharides serve both as structural components, such as cellulose, and as reserves of biological energy, such as **starch,** another type of glucose polymer found in plants. The nucleic acids, DNA and RNA, participate in information storage, transmission, and expression. DNA, or deoxyribonucleic acid, serves principally as a storehouse of genetic information, while the chemically similar **RNA,** or **ribonucleic acid,** is involved in the readout of information stored in DNA. Proteins,

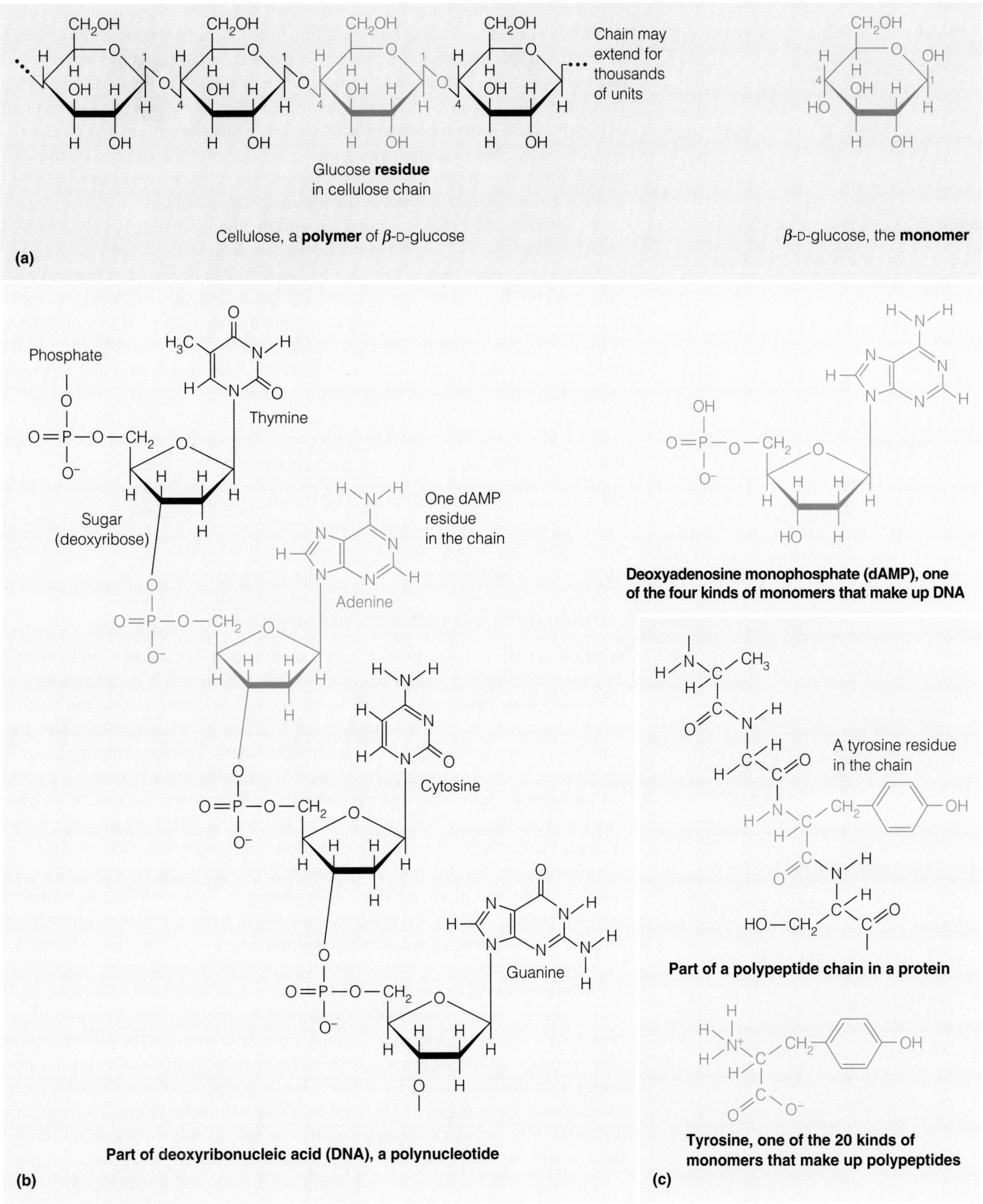

FIGURE **1.7**

Examples of polymers. **(a)** A carbohydrate. The carbohydrate cellulose is a polymer of β-D-glucose monomers. **(b)** A nucleic acid. The nucleic acids, DNA and RNA, are polymers of nucleotides. Part of a DNA molecule is shown, along with one of its monomers, dAMP. **(c)** A polypeptide. Protein chains, or polypeptides, are polymers of amino acids. Part of a polypeptide is shown, along with one of its monomers, tyrosine.

which have far more structural diversity than polysaccharides or nucleic acids, perform a more diverse set of biological functions. Some play structural roles, such as keratin in hair and skin and collagen in connective tissue. Others act as transport substances, an outstanding example being hemoglobin, the oxygen-carrying protein of blood. Proteins may transmit information between distant parts of an organism, as do protein **hormones** and the cell surface **receptors** that receive the hormone signals, or they may defend an organism against infection, as do the **antibodies.** Most important of all, proteins function as enzymes, catalyzing the thousands of chemical reactions that occur within an individual cell.

Lipids form the major constituents of biological membranes.

In addition to these macromolecules and the many small molecules involved in metabolism and as monomers in macromolecular synthesis, there is one other extremely important class of cellular constituents. The **lipids** are a chemically diverse group of compounds that are classified together because of their hydrocarbon-rich structures, which give them very low solubility in the aqueous environment of the cell. This low solubility equips lipids for one of their most important functions—to serve as the major structural element of the **membranes** that surround cells and partition them into various compartments.

Biochemistry as a Biological Science

We must never lose sight of the fact that it is the chemistry of *life* that concerns us here. The complex chemical substances and reactions that we have introduced have their significance as parts of living matter and life processes. To see biochemistry from this perspective, we should begin by asking, What is life?

DISTINGUISHING CHARACTERISTICS OF LIVING MATTER

What distinguishes living matter from nonliving matter? The first attribute that may come to mind is the sheer *complexity* of even the simplest living creature. But complexity is not enough; many things in the universe are complex but nonliving. A dead mouse is, for a little while at least, almost as complicated as a live one. Nor can we distinguish living from nonliving matter simply on the basis of *motility,* for many completely nonmotile things (like mushrooms) are very much alive.

Life depends on creating and duplicating order in a chaotic environment. This ordering uses energy.

The major quality that distinguishes life is the constant *renewal* of a highly ordered structure, often accompanied by an increase in the complexity of that structure. Organisms create an elegant molecular order within themselves and pass a pattern of that order on to descendent organisms. This creation and duplication of order, out of often chaotic surroundings, is unique to life, and it seems to fly in the face of one of the fundamental laws of the universe, the **second law of thermodynamics.** One way of stating the second law is this: The overall disorder of the universe continually increases. But the second law does not require that disorder increase everywhere, at all times; living creatures are like little whirlpools of order in the stream of the universe. The laws of thermodynamics do require, however, that this local creation of order and complexity in matter be paid for by the continual expenditure of energy. This is why living organisms must forever take energy from their surroundings—either from sunlight, as plants do, or from foodstuffs derived from plants or other animals, as animals do. To obtain energy, any organism must interact with its surroundings, and in many cases it must act on them, as we humans do. No organism can survive if isolated from its environment.

Finally, and in a sense most important of all, life is *self-replicating.* It has been suggested that the most primitive "organisms," at the dawn of life, were nothing more than giant molecules that could direct their own replication. The reproductive process has become exceedingly complex through evolution, but its basis remains the same: Information describing the structure of an organism is passed from one generation to the next. Thus, although every individual creature must ultimately lose its private battle with chaos, life itself continues.

How did this remarkable process we call life arise? We do not know, but we know that it is truly ancient, almost as old as the earth itself. The earth condensed from cosmic dust about 4.5 billion years ago, but recognizable traces of living microbes have been dated to 3.8 billion years, only 700 million years later. It is possible that some of these earliest organisms (or proto-organisms) utilized preformed chemical building blocks. For example, traces of amino acids have been found in meteorites, proof that such substances can be generated abiotically.

Whatever their origin, we know that the earliest organisms must have lived an anaerobic existence, for the earth was devoid of free oxygen. Indeed, it is believed that all of the oxygen now present in the earth's atmosphere is the product of photosynthesis by algae and plants. It probably required 1–2 billion years for the present oxygen level to accumulate. Life has not only occupied this planet, it has remade it.

THE UNIT OF BIOLOGICAL ORGANIZATION: THE CELL

One of the first major discoveries in biology was Robert Hooke's observation (1665) that plant tissues (in this case, cork) were divided into tiny compartments, which he called *cellulae,* or **cells.** By 1840, improved observations on many tissues led Theodor Schwann to propose that all organisms exist as either single cells or aggregates of cells. More than a century of careful study by biologists has confirmed this hypothesis.

All living creatures are composed of cells. All cells are roughly the same size.

Furthermore, cells, from whatever organism, are of quite similar size. Most bacterial cells are about 1–2 μm in diameter, and most cells of higher organisms are only about 10–20 times larger, plant cells being somewhat larger than animal cells. There are, to be sure, exceptions: There are very small bacteria (0.2 μm), and there are unusual cells like those in the nervous systems of vertebrates, some of which may be over 1 m long. But compared with the overall range of sizes of natural organisms (Figure 1.8), all cell sizes are much alike.

In both plants and animals, the size of the cells bears no relationship to the size of the organism. An elephant and a flea have cells of about the same size; the elephant just has more of them. Why is such uniformity in cell size maintained? A clue can be found in the fact that the surface/volume ratio for an object of a given shape depends on its size (Figure 1.9). The complex chemical processes in a cell and the large molecules that participate in them require a significant volume. Yet the cell must also exchange substances with its surroundings to support the active metabolism within. Too large a cell will not have enough surface for this exchange to occur, unless it is highly elongated like a vertebrate nerve cell, increasing the surface/volume ratio. Bacterial cells are smaller than the cells of higher organisms because bacterial metabolism is simpler. Viruses, which are even smaller than bacteria, do not have a metabolism of their own but exist as parasites in the cells they invade.

The two great classes of organisms have different cell types. Prokaryotic cells are uncompartmentalized; eukaryotic cells have membrane-bound organelles.

Because cells are the universal units of life, let us examine them more closely. Major differences between cell structures define the two great classes of organisms—prokaryotic and eukaryotic. The **prokaryotes,** which are always unicellular, include the true **bacteria** (eubacteria) and an ancient class called **archaebacteria.** A typical prokaryotic organism is shown schematically in Figure 1.10. Prokaryotic cells are surrounded by a plasma membrane and usually by a rigid cell wall as well. Within the membrane is the **cytoplasm,** which contains the **cytosol**—a semi-liquid concentrated solution or suspension—and the structures suspended within it. In prokaryotes the cytoplasm is not divided into compartments, and the genetic information is in the form of one or more DNA molecules that exist free in the cytosol. Also suspended in the cytosol are the **ribosomes,** which constitute the molecular machinery for protein synthesis. The surface of a prokaryotic cell may carry **pili,** which aid in attaching the organism to other cells or surfaces, and **flagella,** which enable it to swim.

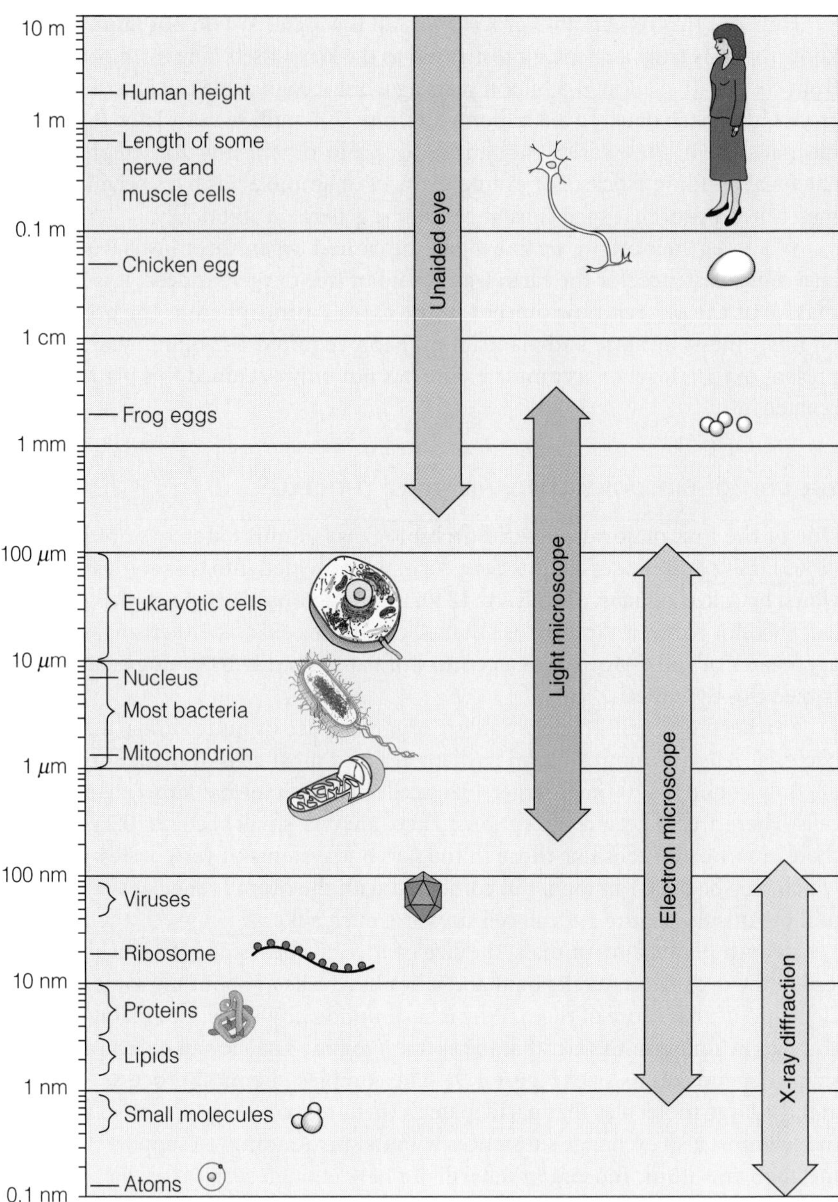

FIGURE **1.8**

Range of sizes of objects studied by biochemists and biologists. Note that the scale is logarithmic. The approximate ranges of different methods for observing structure are indicated by the arrows.

FIGURE **1.10** (*facing page*)

Prokaryotic cells. (a) Schematic view of a representative bacterial cell. The DNA molecule that constitutes most of the genetic material is coiled up in a region called the nucleoid, which shares the fluid interior of the cell (the cytoplasm) with ribosomes (which synthesize proteins), other particles, and a large variety of dissolved molecules. The cell is bounded by a plasma membrane, outside which is usually a fairly rigid cell wall. Many bacteria also have a gelatinous outer capsule. Projecting from the surface may be pili, which attach the cell to other cells or surfaces, and one or more flagella, which enable the cell to swim through a liquid environment. **(b)** Electron micrograph of a bacterial cell. This micrograph was made from a thin section of a dividing cell of the bacterium *Bacillus coagulans*. The light areas represent the two nucleoids, and the dark granules are ribosomes.

(b) Courtesy of S. C. Holt, University of Texas Health Center/BPS.

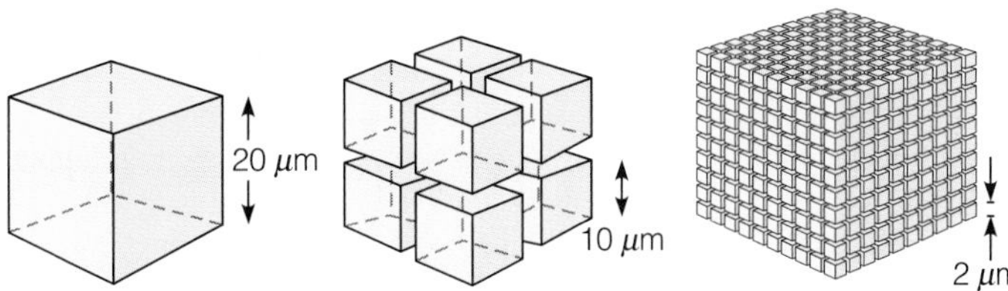

Length of one side	20 μm	10 μm	2 μm
Total surface area (height × width × number of sides × number of cubes)	2400 μm^2	4800 μm^2	24,000 μm^2
Total volume (length × width × height × number of cubes)	8000 μm^3	8000 μm^3	8000 μm^3
Surface area to volume ratio (surface area ÷ volume)	0.3	0.6	3.0

FIGURE **1.9**

How the surface/volume ratio depends on size. If we divide a given volume into smaller and smaller elements, the ratio of surface to volume increases dramatically. Thus, small objects have a greater surface/volume ratio than larger ones and can communicate with their environment better.

From W. M. Becker, L. J. Kleinsmith, and J. Hardin, *The World of the Cell*, 4th ed. (San Francisco, CA: Addison Wesley Longman, 2000) © Addison Wesley Longman, Inc.

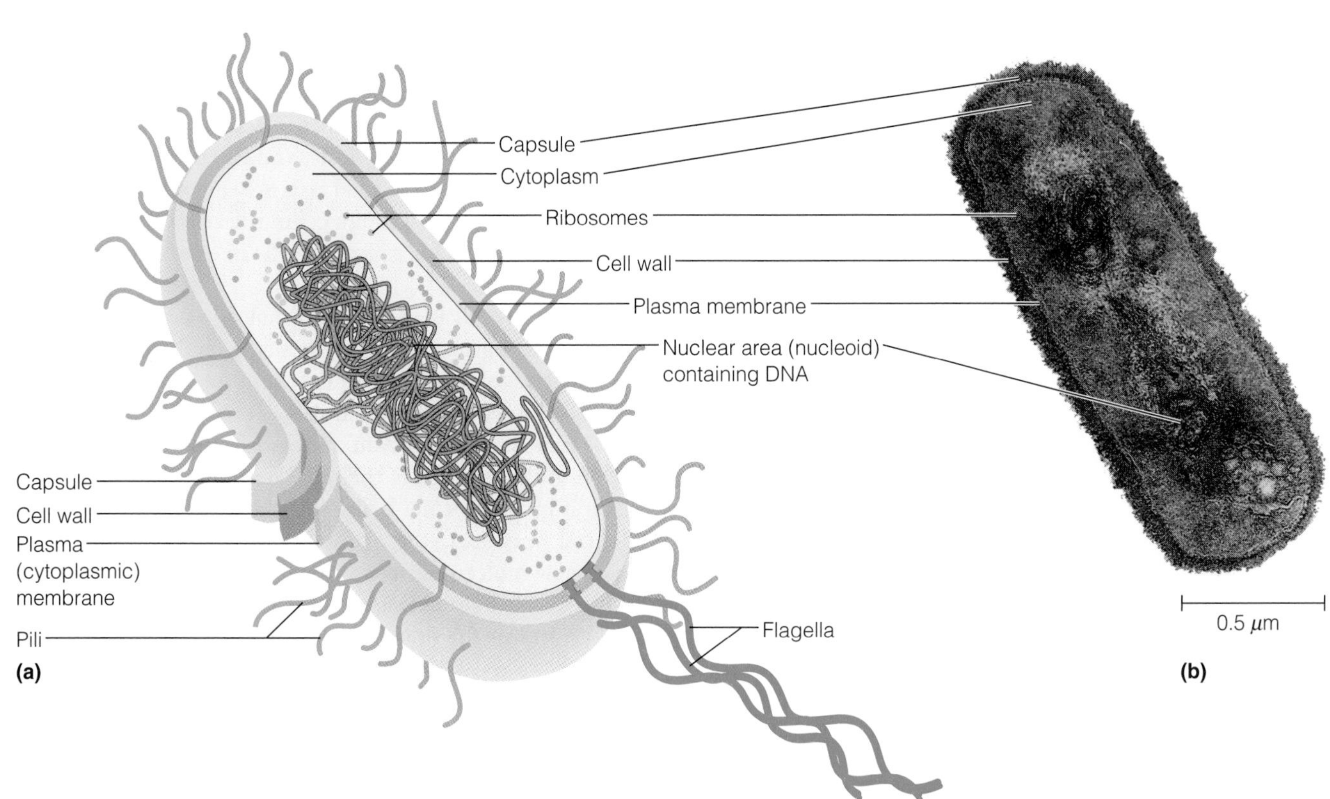

TABLE **1.2 Comparison of some properties of prokaryotic and eukaryotic cells**

	Prokaryotic Cells	Eukaryotic Cells
Size	0.2–5 μm in diameter	Most are 10–50 μm in diameter
Internal compartmentalization	No	Yes, with several different kinds of organelles
Containment of DNA	Free in cytoplasm as nucleoid	In nucleus, condensed with proteins into multiple chromosomes
Ploidy[a]	Usually haploid	Almost always diploid or polyploid
Mechanism of cell replication	Simple division following DNA replication	Mitosis in somatic cells, meiosis in gametes[b]

[a]The term *ploidy* refers to the number of copies of the genetic information carried by each cell. Haploid cells have one copy, diploid cells two, and polyploid cells more than two.

[b]In mitosis the diploid state is retained by chromosome duplication. This occurs in most *somatic,* or "body," cells of organisms. In the cells that produce gametes (sperm or ova) there is a somewhat different process called meiosis, which leads to a haploid state.

All other organisms are called **eukaryotes.** These include the multicellular plants and animals, as well as the unicellular and simple multicellular organisms called protozoans, fungi, and algae. A few of the many differences between eukaryotes and prokaryotes are listed in Table 1.2. Most eukaryotic cells are larger (by 10–20 fold) than prokaryotic cells, but they compensate for their large size by being *compartmentalized.* Their specialized functions are carried out in **organelles**—membrane-surrounded structures lying within the surrounding cytoplasm.

Schematic views of idealized animal and plant cells are shown in Figure 1.11. Major organelles common to most eukaryotic cells are the **mitochondria,** which specialize in oxidative metabolism; the **endoplasmic reticulum,** a folded membrane structure rich in ribosomes and where much protein synthesis occurs; the **Golgi complex,** membrane-bound chambers that function in secretion and the intracellular shuttling of new proteins; and the **nucleus.** The nucleus of a eukaryotic cell contains the cell's genetic information, encoded in DNA that is packaged into **chromosomes.** A portion of this DNA is subpackaged into a dense region within the nucleus called the **nucleolus.** Surrounding the nucleus is a **nuclear envelope,** pierced by pores through which the nucleus and cytoplasm communicate.

There are also organelles specific to plant or animal cells. For example, animal cells contain digestive bodies called **lysosomes,** which are lacking in plants. Plant cells have **chloroplasts,** the sites of photosynthesis; and usually a large, water-filled **vacuole.** Furthermore, whereas most animal cells are surrounded only by a **plasma membrane,** plant cells often have a tough cellulosic **cell wall** outside the membrane. **Basal bodies** act as anchors for cilia or flagella in cells that have those appendages.

It is often useful to think of the cell as a factory, an analogy we shall frequently use in later chapters. Membranes enclose the whole structure and separate different organelles, which can be thought of as departments with specialized functions. The nucleus, for example, is the central administration. It contains in its DNA a library of information for cellular structures and processes, and it issues instructions for proper regulation of the business of the cell. The chloroplasts and mitochondria are power generators (the former being solar, the latter fuel-burning). The cytoplasm is the general work area, where protein machinery (enzymes) carries out the formation of new molecules from imported raw materials. Special molecular channels in the membranes between compartments and between the cell and its surroundings monitor the flow of molecules in appropriate directions. Like factories, cells tend to specialize in function; for example, many of the cells of higher organisms are largely devoted to the production and export of one or a few molecular products.

The eukaryotic cells depicted in Figure 1.11 are idealized representatives of their classes. Figure 1.12, which presents microscopic views of several real cell

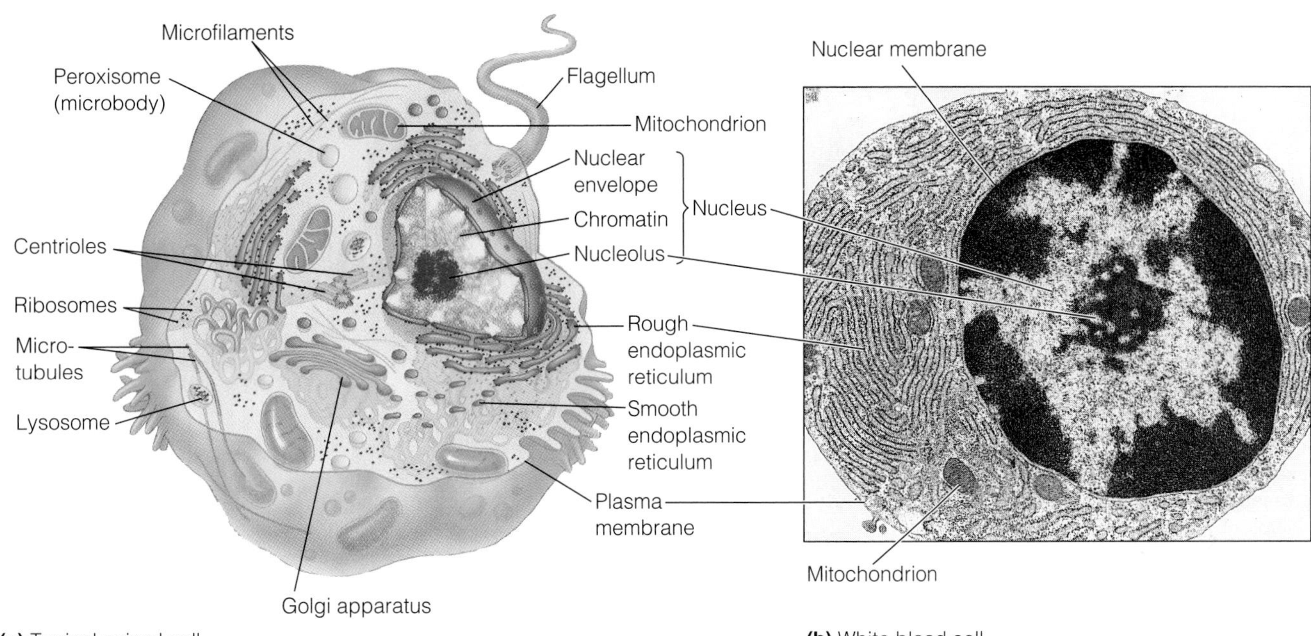

(a) Typical animal cell

(b) White blood cell

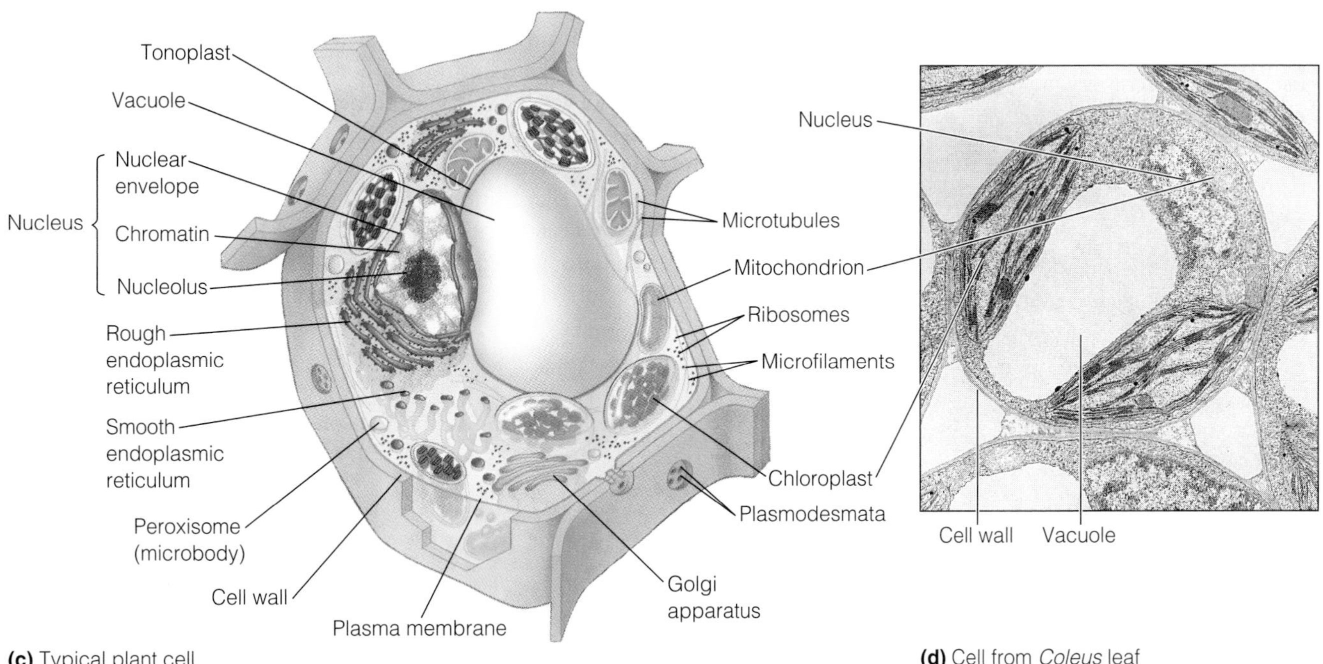

(c) Typical plant cell

(d) Cell from *Coleus* leaf

FIGURE **1.11**

Eukaryotic cells. (a) Schematic view of a representative animal cell. **(b)** Electron micrograph of an animal cell. The micrograph shows a thin section of a white blood cell. **(c)** Schematic view of a typical plant cell. Note the chloroplasts, large vacuole, and rigid cell wall, which distinguish this cell from an animal cell. **(d)** Electron micrograph of a plant cell. The micrograph shows a thin section of a *Coleus* leaf cell.

(a, c) From Neil Campbell, Jane Reece, and Larry Mitchell, *Biology,* 5th ed. (Menlo Park, CA: Addison Wesley Longman, 1999) © Addison Wesley Longman, Inc.; **(b)** Courtesy of R. Rodewald, University of Virginia/BPS; **(d)** W. P. Wergin; courtesy of E. H. Newcomb, University of Wisconsin–Madison.

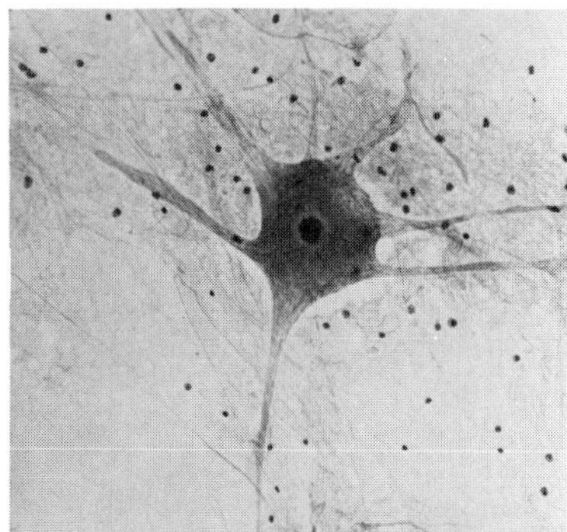
(a) A nerve cell from the human spinal cord

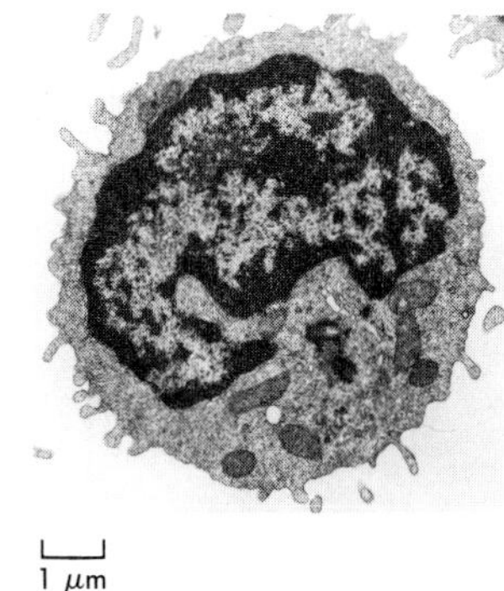

(b) An immature B lymphocyte

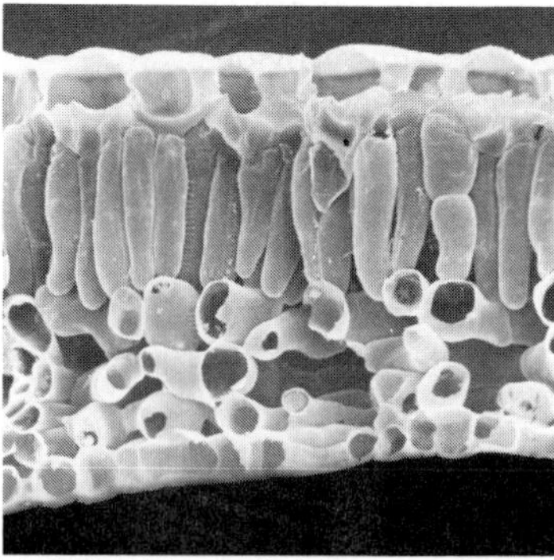
(c) Cells in the leaf of the water fern *Salvinia*

FIGURE **1.12**

Examples of cells. Panel **(a)** is by light microscopy; panel **(b)** is by transmission electron microscopy; panel **(c)** is by scanning electron microscopy.

(a) © Ed Reschke; **(b)** Courtesy of Dr. Dorothea Zucker-Franklin, New York University School of Medicine; **(c)** Courtesy of Judith Croxdale, University of Wisconsin, Madison.

types, gives a better idea of the diversity of forms that have evolved to suit various organismal needs. Diverse as these cells may be, they are all composed of the same kinds of molecules and macromolecules we discussed in the preceding section. Figure 1.13 gives an idea of how these types of molecules are distributed in a cell.

WINDOWS ON CELLULAR FUNCTION: THE VIRUSES

Viruses are intracellular parasites that use cellular machinery and energy sources to replicate.

In analyzing metabolism and the processing of genetic information, biochemists have been aided immeasurably by the **viruses.** Viruses are not cellular and are therefore described as "biological entities" rather than organisms; they are intracellular parasites that can grow only by invading cells. Viruses usually consist of one molecule of nucleic acid (either DNA or RNA) wrapped in an envelope made largely or completely of protein. The envelope is specialized to allow the virus particle to enter particular plant, animal, or bacterial cells. Figure 1.14 shows the structures of several representative viruses.

Because viruses contain no metabolic machinery of their own, they must use that of the host cell to reproduce. By studying the conditions necessary for virus

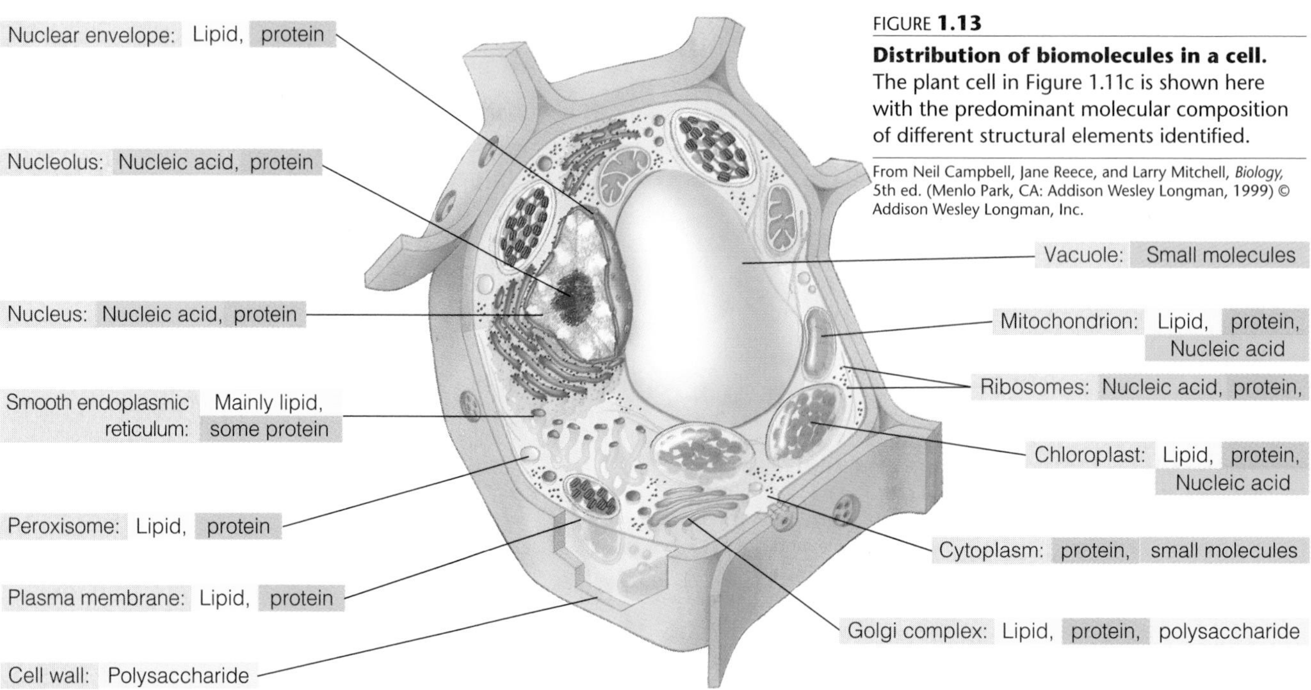

FIGURE **1.13**

Distribution of biomolecules in a cell. The plant cell in Figure 1.11c is shown here with the predominant molecular composition of different structural elements identified.

From Neil Campbell, Jane Reece, and Larry Mitchell, *Biology*, 5th ed. (Menlo Park, CA: Addison Wesley Longman, 1999) © Addison Wesley Longman, Inc.

replication, we can obtain simple tests as to how this cellular machinery operates. Viruses thus provide useful windows onto the cellular functions that they co-opt during infection. For example, the smallest DNA-containing viruses replicate their DNA using only host-cell enzymes. Because of their small size, these viral DNA molecules can be isolated and characterized much more easily than the giant DNA molecules in cellular chromosomes. Larger viruses stimulate the formation of new enzymes after infection, with viral genes specifying the structures of these enzymes. Studies of these enzymes have illuminated genetic regulatory mechanisms, and they have identified targets for treating viral diseases.

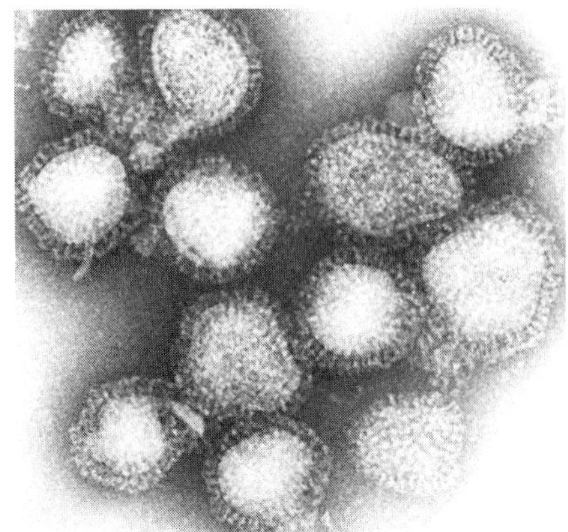

(a) Influenza virus

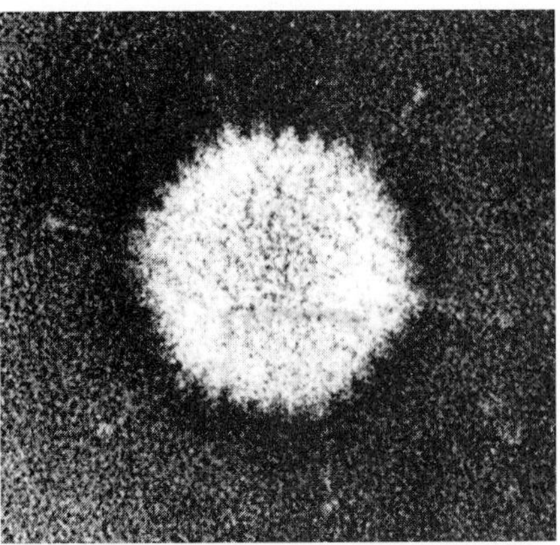

(b) Adenovirus

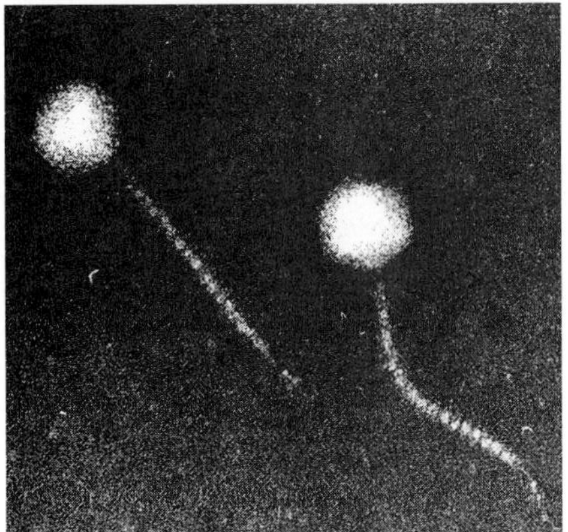

(c) Bacteriophage lambda

FIGURE **1.14**

Examples of viruses.

(a, b) Courtesy of Frederick A. Murphy, Centers for Disease Control and Prevention; **(c)** Courtesy of Dr. R. C. Williams, University of California, Berkeley.

New Tools in the Biological Revolution

From its very beginnings, biochemistry has been much more an experimental science than a theoretical one. Knowledge has developed more from attempts to explain laboratory observations than from experiments designed to test theories. The Buchner brothers, for example, did not set out to prove that fermentation of sugar to ethanol could occur outside the cell. Rather, their chance observation that this occurred spurred the developments that led to our understanding of carbohydrate metabolism.

The experimental nature of biochemistry has two important corollaries. First, in order to understand biochemistry, you must understand the critical experiments and experimental techniques on which our common understanding rests. That is why this book includes separate descriptions of experimental methodologies—the Tools of Biochemistry sections following many chapters. Second, the spectacular growth of biochemical knowledge over the past four decades is due directly to the development of many powerful research methodologies—the armament of the biological revolution. Some of this development is chronicled in Figure 1.15.

An example which will be familiar to you is gene cloning, which dates from about 1973. Our ability to isolate any desired gene, to read its information, and to manipulate it so that any gene product can be formed in essentially unlimited amounts has transformed all of biology. However, gene cloning is but one of a long series of experimental advances that have transformed biochemistry and biology in equally fundamental ways. For example, the introduction of radioisotopes was essential to the elucidation of most metabolic pathways; x-ray crystallography permitted the structural localization of every atom within a protein structure; and two-dimensional gel electrophoresis permitted the visualization and quantitation of virtually all of the thousands of protein molecules present in a cell. The amount of information being generated by the new methodologies is so vast that it would be almost unusable were it not for parallel advances in computer compilation and processing of data.

The Uses of Biochemistry

Biochemistry is a research discipline, but the results of biochemical research are used extensively in the world outside the laboratory—in agriculture, medical sciences, nutrition, and many other fields. In clinical chemistry, biochemical measurements on people help diagnose illnesses and monitor responses to treatment. To cite just one example, liver disease is now routinely diagnosed and monitored by measurements of blood levels of enzymes called **transaminases** and of a hemoglobin breakdown product called **bilirubin**.

Pharmacology and toxicology are concerned with the effects of external chemical substances on metabolism. Drugs and poisons usually act by interfering with

2000
- Gene analysis on microchips
- Single-molecule dynamics

1995
- Targeted gene disruption
- Atomic force microscopy

1990
- Scanning tunneling microscopy
- Amplification of DNA: polymerase chain reaction

1985
- Pulsed field electrophoresis
- Transgenic animals
- Automated oligonucleotide synthesis
- Site-directed mutagenesis of cloned genes

1980
- Automated micro-scale protein sequencing
- Rapid DNA sequence determination
- Monoclonal antibodies

1975
- Southern blotting
- Two-dimensional gel electrophoresis
- Gene cloning
- Restriction cleavage mapping of DNA molecules

1970
- Rapid methods for enzyme kinetics

1965
- High-performance liquid chromatography
- Polyacrylamide gel electrophoresis
- Solution hybridization of nucleic acids
- X-ray crystallographic protein structure determination

1960
- Zone sedimentation velocity centrifugation
- Equilibrium gradient centrifugation
- Liquid scintillation counting

1955
- First determination of the amino acid sequence of a protein
- X-ray diffraction of DNA fibers

1950
- Radioisotopic tracers used to elucidate reactions

1945

FIGURE **1.15**

Recent developments in experimental techniques in biochemistry.

specific metabolic pathways. A good example is the antibiotic penicillin, which kills bacteria by inhibiting an enzyme that synthesizes an essential polysaccharide of the bacterial cell wall. Because animal cells do not synthesize these polysaccharides, they are not harmed by this inhibitor, which can therefore be used therapeutically. A particularly exciting prospect in contemporary biochemistry is that of creating so-called designer drugs. If the target site for action of a drug is a protein enzyme or receptor, determining the detailed molecular structure of that target allows us to design inhibitors that bind to it with great selectivity. Early products of this *drug architecture* are now being tested. In the mid-1980s the detailed molecular structures of viruses began to be determined, opening the possibility of rational design of antiviral agents as well.

Herbicides and pesticides, in many instances, act in similar ways—by blocking enzymes or receptors in the target organism. The first generations of these toxic agents (such as DDT) were so nonspecific in their effects that organisms other than the target populations were often affected, resulting in unforeseen and often severe damage to the environment. Furthermore, the indiscriminate use of these agents gave rise to resistant target populations, so that an ever-increasing number of toxic substances needed to be used. Biochemistry is involved in understanding the actions of herbicides and pesticides, in increasing their selectivity, and in understanding and dealing with mechanisms by which target organisms become resistant to them. Thus, biochemistry has become an important component of environmental science.

Finally, through gene engineering and cloning, remarkable progress in agriculture is to be expected in the near future. Already, genetically engineered plants are in production, and whole-animal cloning has been successful with creatures as large as sheep.

SUMMARY

The aim of the science of biochemistry is to explain life in molecular terms, because we now realize that there is no fundamental difference between living matter and nonliving matter. Modern biochemistry draws on knowledge from chemistry, cell biology, and genetics and uses techniques borrowed from physics. Discoveries in all of these sciences have contributed to the development of a true molecular biology.

Although biochemistry deals with organisms, cells, and cellular components, it is fundamentally a chemical science. The basic chemistry involved is that of carbon, hydrogen, oxygen, and nitrogen, but organisms use many other elements in smaller quantities. Many of the important biological substances are giant molecules that are polymers of simpler monomer units. Such biopolymers include polysaccharides, proteins, and nucleic acids. Lipids form the fourth major group of biologically important substances.

The distinguishing feature of living matter is the use of energy to create and duplicate orderly structure. All living organisms are composed of one or more cells, and these cells are quite similar in size. However, the two great classes of organisms, prokaryotes and eukaryotes, have fundamentally different cellular structures: Prokaryotic cells are uncompartmentalized, lacking the membrane-bound organelles characteristic of eukaryotic cells. Viruses are intracellular parasites, essentially carriers of nucleic acid; to duplicate themselves, they use the reproductive machinery and energy sources of the host cell.

Biochemistry is an experimental science, and the remarkable recent advances in biochemistry are due in large part to the development of powerful new laboratory techniques. Biochemistry has had major impacts on medicine, agriculture, nutrition, ecology, and many other facets of life.

TOOLS OF BIOCHEMISTRY 1A

Microscopy at Many Levels

The Light Microscope and Its Limitations

All students of science are familiar with the light microscope (Figure 1A.1)—most have used one. This was the instrument that made cell biology possible. In fact, the field can be said to have begun with Hooke's pioneering microscopic studies. Generations of biologists followed Hooke with steadily improved instruments. But as biology has looked deeper into the details of life, the light microscope has reached its limits.

To understand why these limits exist, we must consider a quantity called the **resolution** of a microscope. The resolution (r) is quantitatively defined as the minimum distance between two objects that can just be distinguished as separate. It is given by the equation

$$r = \frac{0.61\lambda}{n \sin \alpha} \qquad (1A.1)$$

Here λ is the wavelength of the radiation used, n is the refractive index of the medium between the sample and the objective lens, and α is the **angular aperture** of the objective lens. The quantity $\sin \alpha$ is basically a measure of the radiation-gathering power of the lens system. Resolution depends primarily on wavelength, because the objects must be comparable in size to the wavelength in order to perturb the waves sufficiently to convey information.

The angular apertures of the best light microscopes are about 70°, so even if deep blue light of wavelength 450 nm is used and the medium between the sample and the objective lens is air ($n = 1$), we get

$$r = \frac{0.61 \times 450}{1.0 \times \sin 70°} \cong 300 \text{ nm} = 0.3\ \mu\text{m} \qquad (1A.2)$$

This value represents the practical limit of resolution for light microscopy. A bit more resolution can be gained by going into the near ultraviolet, but absorption of this light by cellular materials limits its usefulness. Photographic images can be enlarged, but there is no sense in magnifying an image beyond the point where its resolution is just what the eye can resolve. Because our eyes can resolve images about 0.3–0.6 mm apart, the best light microscopes have a useful maximum magnifying power of about 1000–2000 (magnifying 0.3 μm by 2000 gives 0.6 mm). Further magnification of the image does not help—the fuzziness just gets bigger. To make a major advance, it was necessary to use radiation of much shorter wavelength, radiation that we cannot see but that can produce a photographic image. Thus the **electron microscope** was born in the 1930s.

Transmission Electron Microscopy

There are several types of electron microscopes. The first type to be used was the **transmission electron microscope (TEM)**, so called because it detects electrons that have been transmitted through a sample. The transmission electron microscope is compared with the light microscope in Figure 1A.1. An electron beam is emitted from a tungsten filament and accelerated by an electric field. Magnetic lenses focus the beam, as glass lenses focus a beam of light in the conventional microscope. The key to the higher resolution is that electrons, like the photons of light, have both a particle-like and a wavelike nature. A photon or an electron moving with an energy E is characterized by a wavelength

$$\lambda = \frac{h}{E} \qquad (1A.3)$$

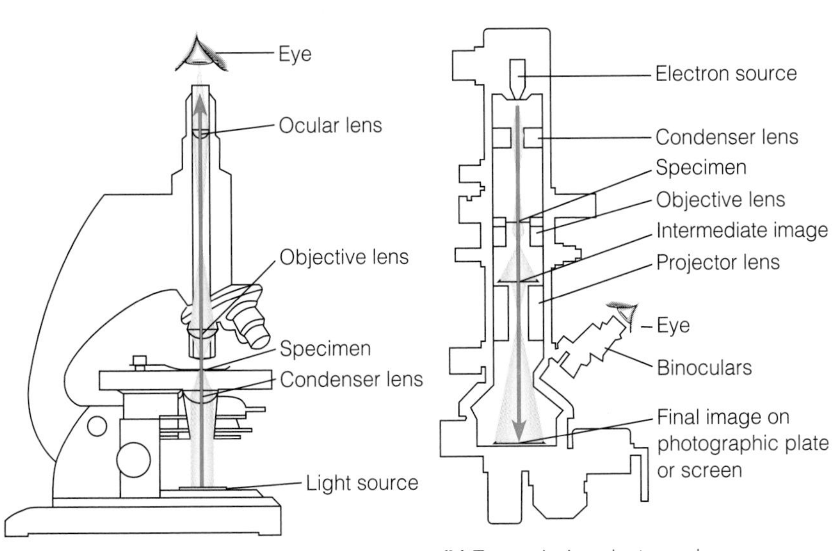

FIGURE **1A.1**

Structure of the optical microscope and the transmission electron microscope. The two images are not to the same scale; the electron microscope is much larger than a conventional light microscope.

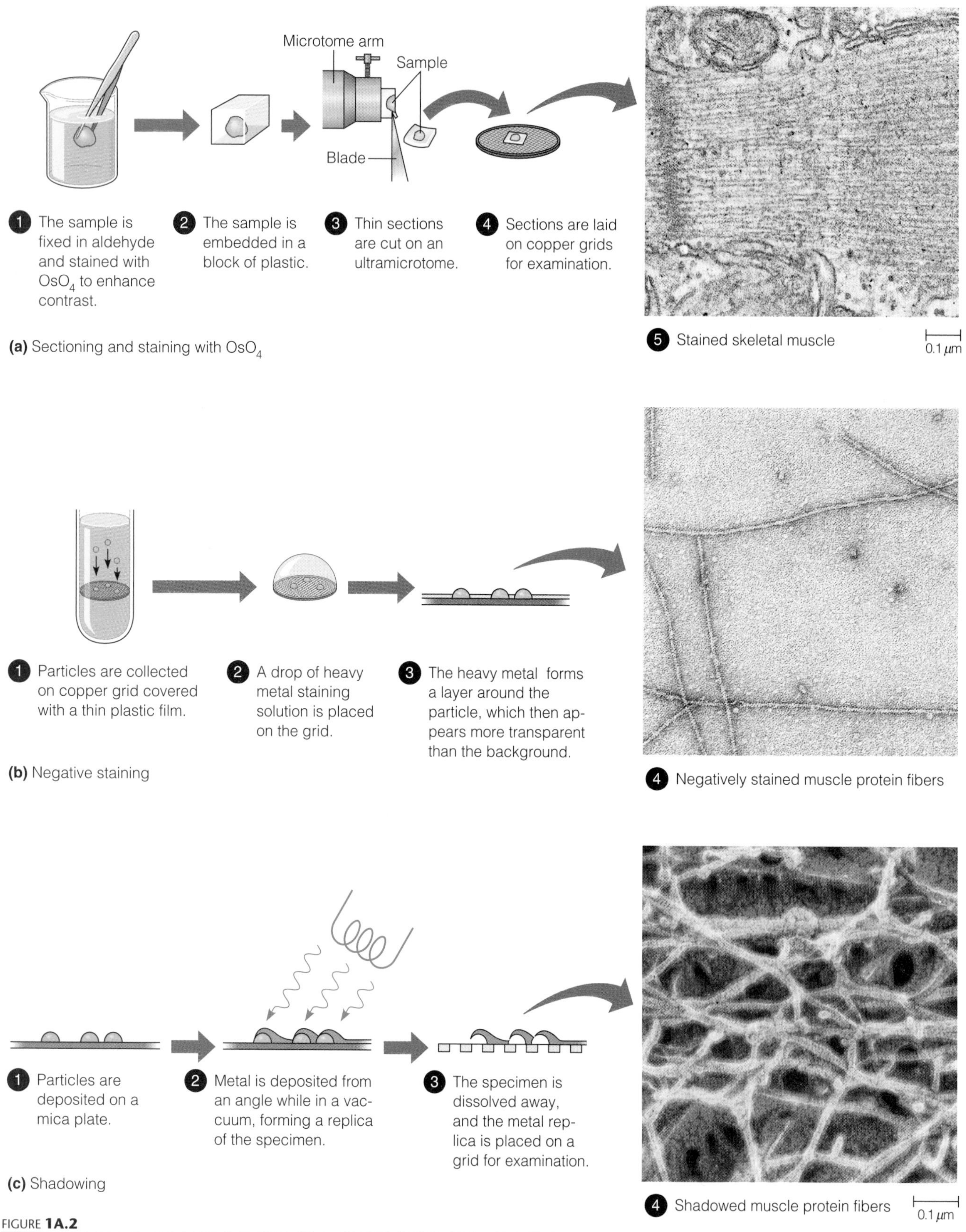

Microtome arm

Sample

Blade

1 The sample is fixed in aldehyde and stained with OsO_4 to enhance contrast.

2 The sample is embedded in a block of plastic.

3 Thin sections are cut on an ultramicrotome.

4 Sections are laid on copper grids for examination.

5 Stained skeletal muscle

0.1 μm

(a) Sectioning and staining with OsO_4

1 Particles are collected on copper grid covered with a thin plastic film.

2 A drop of heavy metal staining solution is placed on the grid.

3 The heavy metal forms a layer around the particle, which then appears more transparent than the background.

4 Negatively stained muscle protein fibers

(b) Negative staining

1 Particles are deposited on a mica plate.

2 Metal is deposited from an angle while in a vacuum, forming a replica of the specimen.

3 The specimen is dissolved away, and the metal replica is placed on a grid for examination.

4 Shadowed muscle protein fibers

0.1 μm

(c) Shadowing

FIGURE **1A.2**

Three methods of preparing samples for transmission electron microscopy.

Courtesy of T. Pollard and P. Maupin from *Electron Microscopy in Biology,* Vol. II, J. D. Griffith, ed.; © 1982 by John Wiley and Sons, Inc.

where h is Planck's constant (6.626×10^{-27} erg·s = 6.626×10^{-34} J·s). When electrons are accelerated by 50,000–100,000 volts between the cathode and the anode, their wavelengths are much shorter than that of visible light—in fact, less than 1 nm. This wavelength would predict a resolution of better than 1 nm for a transmission electron microscope. Practical considerations give an operational limit of about 2 nm for most instruments. Still, this resolution is about 100 times finer than even the best optical microscope can accomplish: A good transmission electron microscope can usefully magnify to over 100,000 times.

Clear as this advantage may be, transmission electron microscopy has some disadvantages. The electron beam requires that a high vacuum be maintained throughout the instrument, including the sample chamber. This, in turn, means that only completely dried samples can be examined. Although many methods for careful fixation and drying have been devised, there is always the possibility of inducing changes in samples. Living structures, of course, cannot be examined.

Some of the methods used to prepare samples for transmission electron microscopy are shown in Figure 1A.2. The electron energies in most transmission microscopes do not allow penetration of thick samples (>100 nm). Thus, cell samples must be fixed, stained, and sliced very thin, using an **ultramicrotome** (Figure 1A.2a). Particles like viruses and large molecules can be deposited directly on a thin film supported by a copper grid. But the contrast between such a particle and the background is not sufficient, so the sample is usually **negatively stained** (Figure 1A.2b) or **shadowed** (Figure 1A.2c). Other special techniques such as *freeze fracturing* and *freeze etching* are discussed in later chapters.

Scanning Electron Microscopy

A quite different kind of technique is called **scanning electron microscopy (SEM)**. A schematic diagram of a scanning electron microscope is shown in Figure 1A.3. Here, the electron beam is scanned back and forth across the sample, in a pattern generated by the scan generator and beam deflector, and secondary electrons emitted from the point at which the beam impinges on the sample surface are picked up by a detector. The image is then displayed on a video screen, whose surface is scanned in register with the scanning of the sample. The SEM method does not have the resolution of TEM, but it is excellent for obtaining extremely clear views of the surfaces of minute objects, as you can see in Figure 1A.4. Preparation for SEM studies does not require sectioning, but the specimen must be fixed and dried to be stable in the high vacuum and is usually coated with a thin layer of gold to aid in emission of secondary electrons.

Another technique should be mentioned, although it is as yet employed in only a limited number of laboratories: **scanning transmission electron microscopy (STEM)**. In this method the electron beam is scanned over the specimen, as in SEM, but it is detected in transmission. The method has the advantage that unstained, unfixed specimens can sometimes be used. Furthermore, the absorption of electrons of different energies provides information about the composition of different portions of the sample.

Laser Scanning Confocal Microscopy

Aside from the problems of resolution mentioned on page 21, there exists another fundamental limitation to the use of conventional

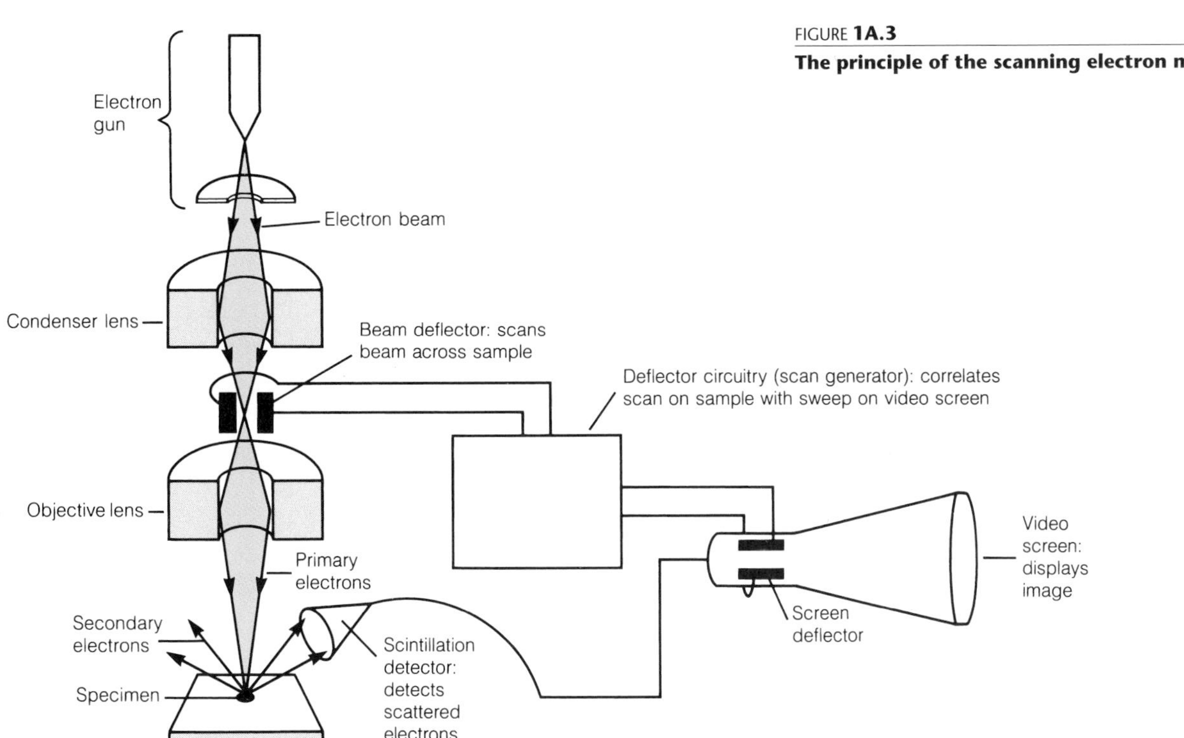

FIGURE **1A.3**
The principle of the scanning electron microscope.

light microscopy to study internal structure in cells and other biological samples. At high resolution (~0.3 μm) the depth of focus of a light microscope will be about 3 μm. Superposition of images of material in this thick slice will obscure detail. To get around this problem, the confocal microscope was developed. As shown in Figure 1A.5, a light beam (preferably from a laser) is focused into a very small volume at the desired level within the sample. Reflected or fluorescent light from this spot is brought back to a detector, through a pinhole that excludes light scattered from other regions. The position of the illuminated spot is scanned back and forth through the sample, always at the same level. The image that is electronically built up in this way represents a very thin, highly resolved "slice" through the sample. It can also be repeated at different levels to build up a three-dimensional image.

The method is most powerful when fluorescence detection is used, for then specifically labeled structures or substances can be precisely located within a cell. Because the method is relatively nondestructive, it can be used to follow dynamic processes in living cells. It has been employed, for example, to pinpoint the places within a cell nucleus where active replication of DNA is taking place. The rapid development of more versatile and discriminating fluorescent probes is making confocal microscopy a major technique in cellular biochemistry.

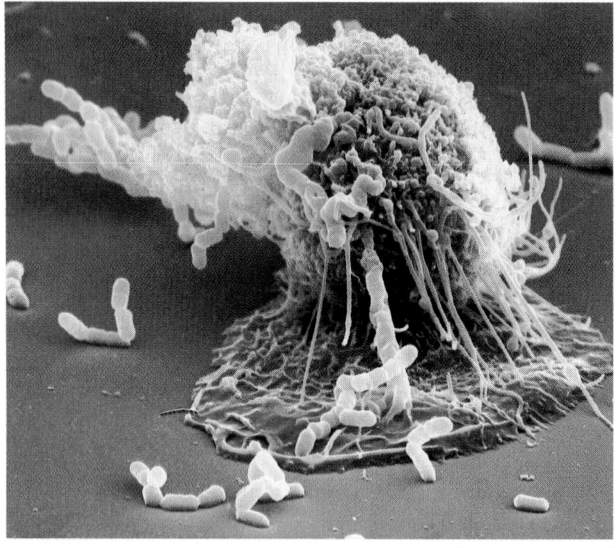

FIGURE **1A.4**

A scanning electron micrograph showing phagocytosis. A macrophage is engulfing several sausage-shaped *E. coli* in this image, which is magnified 4300×.

Scanning Tunneling and Atomic Force Microscopy

Very recently, a remarkable new kind of microscope has been developed. **Scanning tunneling microscopy** uses a very fine, electrically charged metal tip, which is scanned across the sample. As

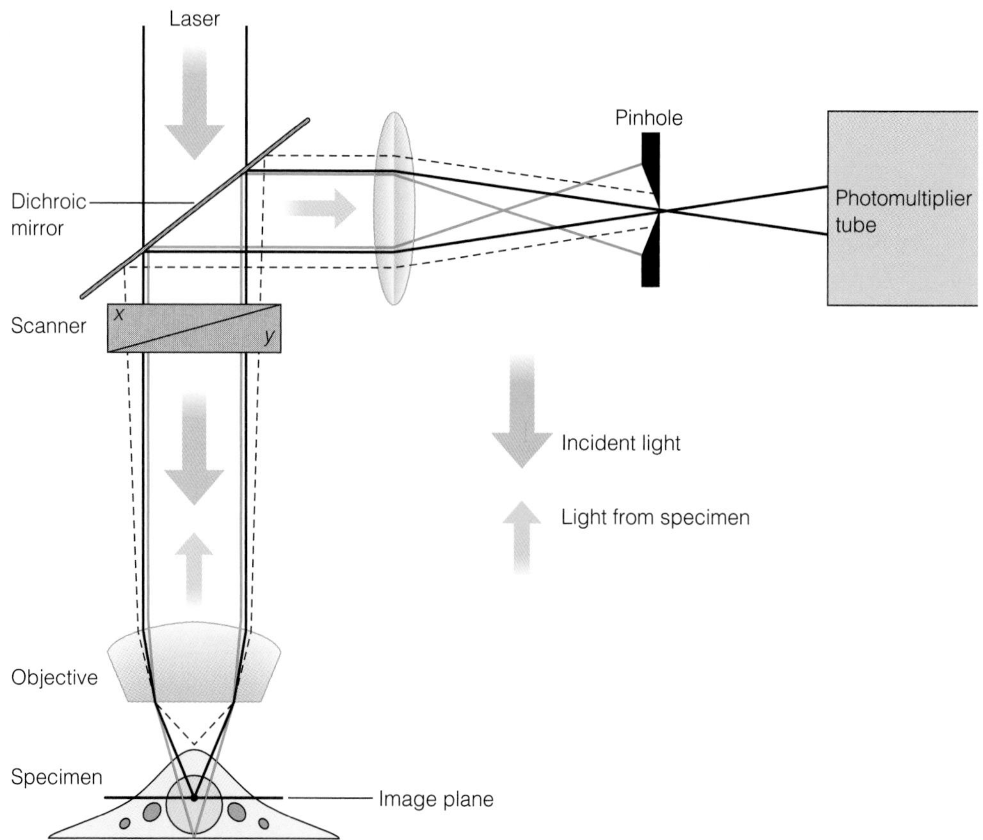

FIGURE **1A.5**

Diagram illustrating the principle of laser scanning confocal microscopy. A laser beam is passed through an *x*–*y* scanner, collimated to a small spot by the objective lens, and scanned across the specimen. Fluoresced light is collected by the objective and directed by a dichroic mirror (a mirror that reflects the fluorescent light but not the shorter-wavelength laser light) to a pinhole aperture placed in the conjugate image plane. Light originating from the specimen plane of focus passes through the pinhole to a photomultiplier detector. Light from above or below the specimen focal plane strikes the walls of the aperture and is not transmitted.

From J. Lemasters, et al., *Optical Microscopy: Emerging Methods and Applications*, edited by B. Herman and J. Lemasters, pp. 339–354. (San Diego: Academic Press, 1996). © 1993 Academic Press, Inc.

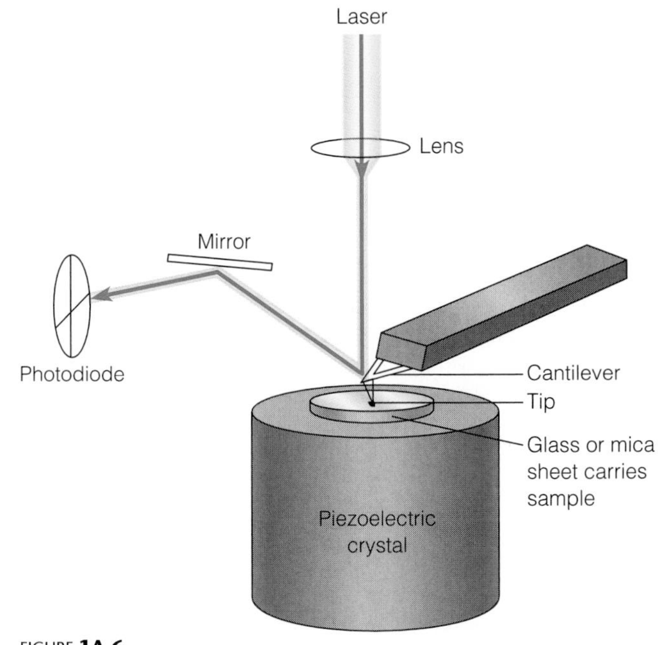

FIGURE **1A.6**

The principle of the atomic force microscope. Power to the piezoelectric crystal is regulated to move the sample up and down and keep the tip at constant height as the sample is scanned.

electrons leak (tunnel) between the tip and the surface supporting the sample, the resistance they encounter varies according to the height of microscopic objects lying on the surface. The resulting fluctuations in current produce a video display of the surface with a resolution comparable to that of an electron microscope. In **atomic force microscopy** (Figure 1A.6), an extremely sharp tip is either dragged or tapped back and forth across the sample, and its up-and-down motion is detected by the deflection of a laser beam reflected off the cantilever which holds the tip. This motion is greatly amplified to give a contour map of the object. Both of these techniques have the enormous advantage over electron microscopy that wet, even immersed, samples can be studied. Both techniques are still under development and have fascinating potential.

References

Corle, T. R., and G. S. Kino (eds.) (1998) *Confocal Scanning Optical Microscopy and Related Imaging Systems.* Academic Press, San Diego.

Engel, A. (1991) Biological application of scanning probe microscopy. *Annu. Rev. Biophys. & Biophys. Chem.* 20:79–108.

Herman, B., and J. J. Lemasters (eds.) (1996) *Optical Microscopy: Emerging Methods and Applications.* Academic Press, San Diego. A collection of short papers on a wide variety of new microscopic methods.

Watt, I. M. (1985) *The Principles and Practice of Electron Microscopy.* Cambridge University Press, Cambridge, England.

CHAPTER 2

The Matrix of Life: Weak Interactions in an Aqueous Environment

THE MACROMOLECULES THAT PARTICIPATE IN THE STRUCTURAL AND FUNCTIONAL matrix of life are immense structures held together by strong, covalent bonds. Yet covalent bonding alone cannot begin to describe the complexity of molecular structure in biology. Much weaker interactions are responsible for most of the elegant cellular architecture visible in the electron micrographs of Chapter 1. These are the **noncovalent interactions,** also called *noncovalent forces* or *noncovalent bonds,* between ions, molecules, and parts of molecules.

Consider the macromolecules we discussed in Chapter 1. The linear sequence of the nucleotide residues in a strand of DNA is maintained by covalent bonds. But DNA also has a highly specific three-dimensional structure, which is stabilized by noncovalent interactions between different parts of the molecule. Similarly, every kind of protein is made up of covalently linked amino acids but is also folded into a specific molecular conformation by noncovalent forces. Proteins interact with other protein molecules or with DNA to form still higher levels of organization, ultimately leading to cells, tissues, and whole organisms. All of this complexity is accounted for by a myriad of noncovalent interactions within and between macromolecules.

Noncovalent interactions are the weak interactions among ions, molecules, and parts of molecules.

What makes noncovalent interactions so important in biology and biochemistry? The key is seen in Figure 2.1, which compares noncovalent and covalent bond energies. The *covalent* bonds most important in biology (such as C—C and C—H) have bond energies in the range of 300–400 kJ/mol. Biologically important *noncovalent* bonds are 10 to 100 times weaker. It is their very weakness that makes noncovalent bonds so essential, for it allows them to be continually broken and re-formed in the dynamic molecular interplay that is life. This interplay depends on rapid exchanges of molecular partners, which could not occur if intermolecular forces were so strong as to lock the molecules in conformation and in place.

Thus, if we are to understand life, we must know something about noncovalent interactions. Furthermore, we must know how such interactions behave in an aqueous environment, for every cell in every organism on earth is bathed in and permeated by water. This is as true for creatures living in the most arid deserts as for those in the depths of the sea. Water is the major constituent of organisms—70% or more of the total weight, in most cases.

This chapter first describes noncovalent interactions, then shows that the properties of water have a profound effect on those interactions.

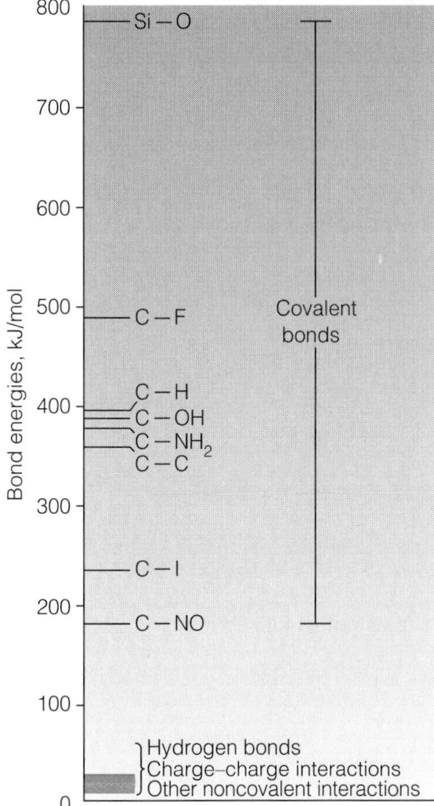

FIGURE **2.1**

Covalent and noncovalent bond energies. Energies typical of noncovalent bonds are about one to two orders of magnitude weaker than energies of the covalent bonds commonly found in biochemical compounds.

The Nature of Noncovalent Interactions

Molecules and ions can interact noncovalently in a number of different ways, as described in this section and summarized in Figure 2.2. All of these noncovalent interactions are fundamentally electrostatic in nature—meaning that they all depend on the forces that electrical charges exert on one another—although this may not be immediately obvious in some cases.

Noncovalent interactions always involve electrical charges.

CHARGE–CHARGE INTERACTIONS

The simplest noncovalent interaction is the electrostatic interaction between a pair of charged particles. Many of the molecules in cells, including macromolecules like DNA and proteins, carry a net electrical charge. In addition to these molecules, the cell contains an abundance of small ions, both cations like Na^+, K^+, and Mg^{2+} and anions like Cl^- and $HOPO_3{}^{2-}$. All of these charged entities exert forces on one another (see Figure 2.2a). The force between a pair of charges q_1 and q_2, separated *in a vacuum* by a distance r, is given by **Coulomb's law:**

$$F = k\frac{q_1 q_2}{r^2} \tag{2.1}$$

where k is a constant whose value depends on the units used.* If q_1 and q_2 have the same sign, F is positive, so a positive value corresponds to repulsion. If one charge is + and the other −, F is negative, signifying attraction. It is such charge–charge interactions that stabilize a crystal of a salt, like that shown in Figure 2.3.

The biological environment, of course, is not a vacuum. In a cell, charges are always separated by water or by other molecules or parts of molecules. The existence of this **dielectric medium** between charges has the effect of screening them from one another, so that the actual force is always less than that given by equation (2.1). This screening effect is expressed by inserting a dimensionless number, the **dielectric constant** (ε), in equation (2.1):

$$F = k\frac{q_1 q_2}{\varepsilon r^2} \tag{2.2}$$

Every substance that acts as a dielectric medium has a characteristic value of ε; the higher this value, the weaker the force between the separated charges. The dielectric constant of water is high, approximately 80, whereas organic substances usually have much lower values, in the range of 1 to 10. We shall see presently the reason for the high value of ε in water, but its major consequence is that charged particles interact rather weakly with one another in an aqueous environment unless they are very close together.

Coulomb's law is an expression of *force;* that is, it is a quantitative description of an interaction. However, every interaction involves a change in *energy,* and because we are concerned with the energy changes in biological processes, we are

*In the c.g.s. (centimeter–gram–second) system, with charges in electrostatic units, k is unity. In this book we use the SI, or international, system of units. Here q_1 and q_2 are in coulombs (C), r is in meters (m), and $k = 1/(4\pi\varepsilon_0)$. The quantity ε_0 is the *permittivity of a vacuum* and has the value $8.85 \times 10^{-12}\ J^{-1}C^2m^{-1}$, where J is the energy unit joules. F is in newtons (N).

Type of Interaction	Model	Example	Dependence of Energy on Distance
(a) Charge–charge Longest-range force; nondirectional	+ –	$-\overset{+}{N}H_3$ $\quad$ $^{-}O_2C-$	$1/r$
(b) Charge–dipole Depends on orientation of dipole	+ q^- q^+	$-\overset{+}{N}H_3$ $\quad$ q^- O q^+ (H, H)	$1/r^2$
(c) Dipole–dipole Depends on mutual orientation of dipoles	q^- q^+ q^- q^+	q^- O q^+ (H, H) $\quad$ q^- O q^+ (H, H)	$1/r^3$
(d) Charge–induced dipole Depends on polarizability of molecule in which dipole is induced	+ q^- q^+	$-\overset{+}{N}H_3$ $\quad$ q^- q^+ (benzene ring)	$1/r^4$
(e) Dipole–induced dipole Depends on polarizability of molecule in which dipole is induced	q^- q^+ q^- q^+	q^- O + (H, H) $\quad$ q^- q^+ (benzene ring)	$1/r^5$
(f) Dispersion Involves mutual synchronization of fluctuating charges	q^+ q^- / q^- q^+	q^+ q^- / q^- q^+ (stacked rings)	$1/r^6$
(g) van der Waals repulsion Occurs when outer electron orbitals overlap		– / –	$1/r^{12}$
(h) Hydrogen bond Charge attraction + partial covalent bond	— q^+ · · · q^- Donor Acceptor	N—H···O=C Hydrogen bond length	Length of bond fixed

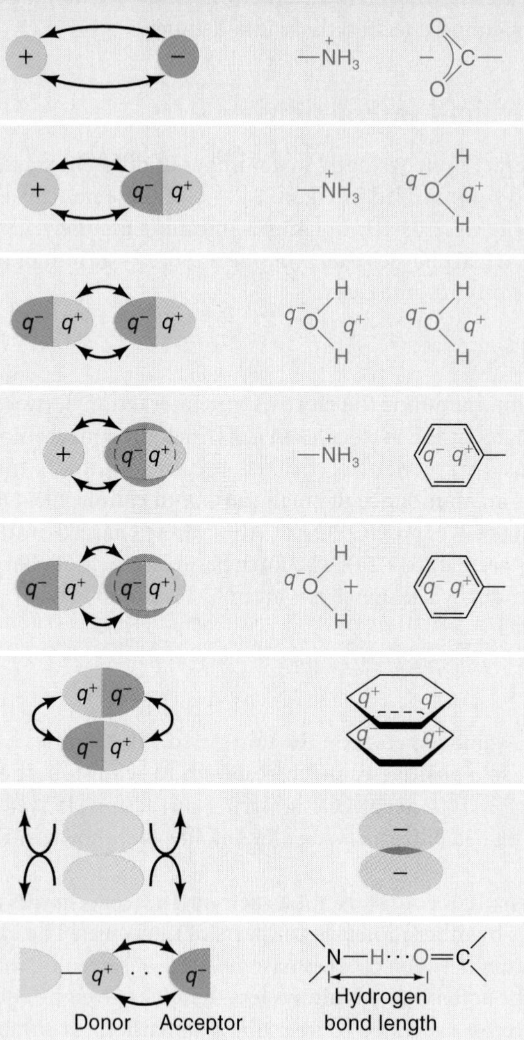

FIGURE **2.2**

Types of noncovalent interactions. The induced dipole **(d, e)** and the dispersion forces **(f)** depend on a distortion of the electron distribution in a nonpolar atom or molecule. The symbols q^-, q^+ denote a fraction of an electron or proton charge.

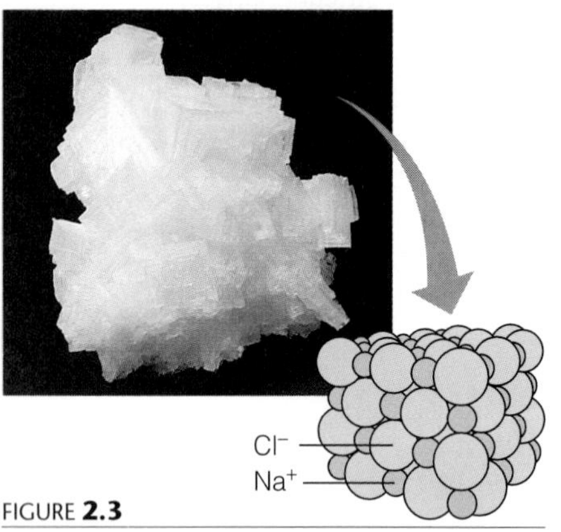

FIGURE **2.3**

Charge–charge interactions in an ionic crystal. Ionic crystals are held together by charge–charge interactions between positive and negative ions. In a sodium chloride crystal, each sodium ion is surrounded by six chloride ions, and each chloride ion is surrounded by six sodium ions.

Photo courtesy of Stephen Frisch.

particularly interested in the **energy of interaction** (U). This is the energy required to separate two charged particles from a distance r to an infinite distance—in other words, to pull them apart working against the electrostatic force. The energy of interaction is given by equation (2.3), which is similar to equation (2.2):

$$U = k\,\frac{q_1 q_2}{\varepsilon r} \qquad (2.3)$$

As with force, the energy of an oppositely charged pair q_1 and q_2 is always negative, signifying attraction, but U approaches zero as r becomes very large.

Two major characteristics of charge–charge interactions emerge from equations (2.2) and (2.3). First, the force between the charges is wholly *nondirectional*, depending only on the *distance* of separation. Second, the energy of interaction varies quite gradually with the distance—it is inversely proportional to the first power of r. These characteristics distinguish charge–charge interactions from some of the other noncovalent interactions we shall describe. Charge–charge interactions often occur within or between biopolymers—for example, in the attraction between amino and carboxyl groups, as shown in Figure 2.2a.

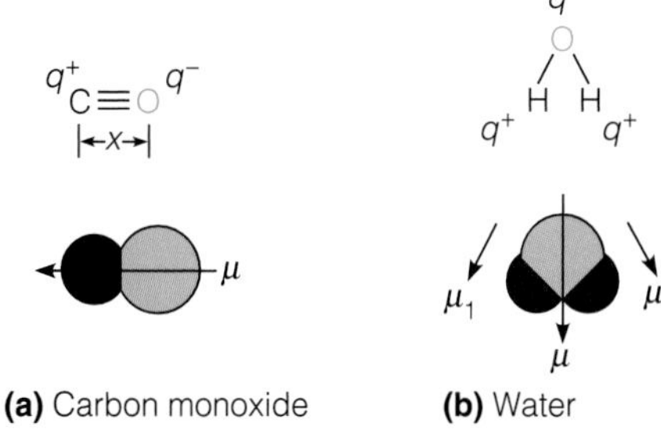

(a) Carbon monoxide **(b)** Water

FIGURE **2.4**

Dipolar molecules. (a) Carbon monoxide: the excess negative charge (q^-) on the oxygen together with corresponding positive charge (q^+) on the carbon produces a dipole moment directed along the O—C axis. **(b)** Water: the excess negative charge on O together with the excess positive charge on each H produces two moments, μ_1 and μ_2, directed along the O—H bonds. Their vector sum (μ) represents the dipole moment of the molecule.

PERMANENT AND INDUCED DIPOLE INTERACTIONS

Molecules that carry no *net* charge may nevertheless have an asymmetric internal distribution of charge. For example, the electron distribution of the uncharged carbon monoxide molecule is such that the oxygen end is slightly more negative than the carbon end (Figure 2.4a). Such a molecule is called *polar*, or a **permanent dipole,** and is said to have a **permanent dipole moment** (μ). The dipole moment expresses the magnitude of a molecule's polarity. If a linear molecule like CO has fractional charges q^+ and q^-, separated by a distance x, the dipole moment is a vector directed toward q^+, whose magnitude is

$$\mu = qx \tag{2.4}$$

In molecules with a more complex shape, like water, the dipole moment is a vector sum of moments along bonds (Figure 2.4b). Water has a significant μ, because electrons are drawn from the hydrogens toward the oxygen, as a consequence of electronegativity of the oxygen.

Some dipole moment values are given in Table 2.1. Notice the large values for glycine and glycylglycine. At neutral pH, the amino acid glycine exists as the ion $^+H_3NCH_2COO^-$, which has both a positive and a negative ionized group. Thus,

Some molecules are dipoles and interact because they possess dipole moments.

TABLE **2.1 Dipole moments of some molecules**

Molecule	Formula	Dipole Moment (D)[a]
Carbon monoxide	C≡O	0.12
Carbon dioxide	O=C=O	0
Water	H_2O (O with two H)	1.83
ortho-Dichlorobenzene	$C_6H_4Cl_2$ (Cl, Cl ortho)	2.59
para-Dichlorobenzene	Cl—C_6H_4—Cl	0
Glycine	$H_3\overset{+}{N}—CH_2—COO^-$	16.7
Glycylglycine	$H_3\overset{+}{N}—CH_2—C(=O)—N(H)—CH_2COO^-$	28.6

[a]The common units of dipole moment are *debyes;* 1 debye (D) equals 3.34×10^{-30} C m.

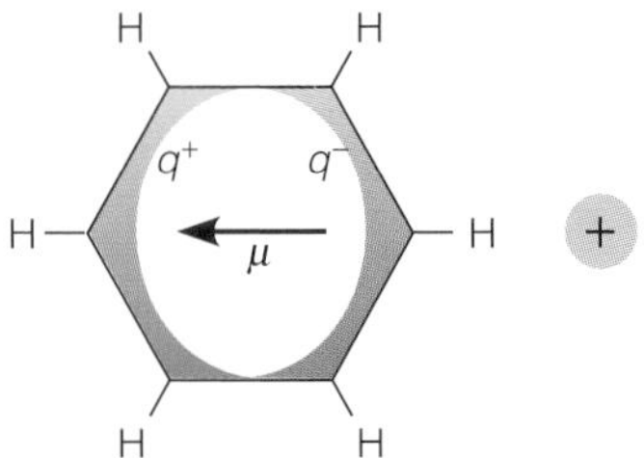

(a) Induction of a dipole in benzene by a positively charged ion

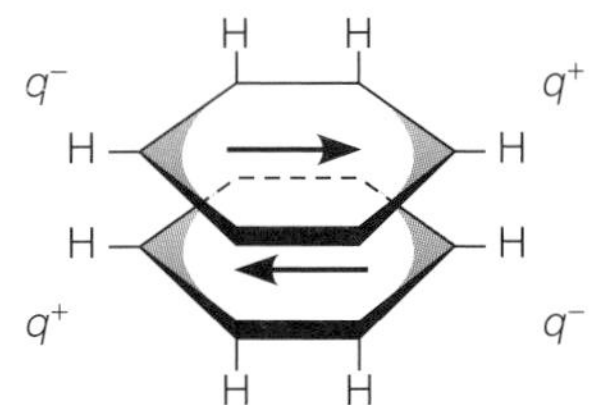

(b) Dispersion forces between two benzene molecules

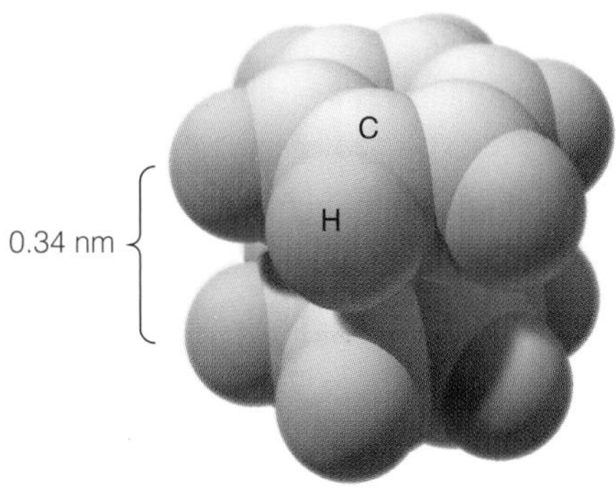

(c) Space-filling model of molecules in (b)

FIGURE **2.5**

Induced dipoles and dispersion forces. **(a)** Benzene has neither a net charge nor a permanent dipole moment, but a nearby charge can induce a redistribution of electrons within the benzene ring, producing an induced moment (μ). **(b)** Planar molecules like benzene have a strong tendency to stack, because fluctuations in the electron clouds of the stacked rings interact with one another, producing a dispersion force. **(c)** Although the molecules approach closely, they do not interpenetrate.

whole electron charges are separated by the length of the molecule, accounting for the large μ. In glycylglycine, which is made by covalently linking two glycine molecules, the dipole moment is nearly twice as big, because the charge separation is almost twice as large. Molecules with large dipole moments are said to be highly polar.

Note also in Table 2.1 that molecules must be asymmetric to have dipole moments: Compare C≡O with O=C=O, or *o*-dichlorobenzene with *p*-dichlorobenzene. In carbon dioxide and *p*-dichlorobenzene, the dipole vectors are equal in magnitude but oppositely directed, so that their effects cancel each other.

In the aqueous environment of a cell, a permanent dipole can be attracted by a nearby ion (a *charge–dipole interaction*) or by another permanent dipole (a *dipole–dipole interaction*). These **permanent dipole interactions** are described in Figure 2.2b and c. Unlike the simple charge–charge interactions described earlier, dipole interactions depend on the orientation of the dipoles. Furthermore, they are shorter-range interactions: The energy of a charge–dipole interaction is proportional to $1/r^2$ and that of a dipole–dipole interaction to $1/r^3$. Thus, a pair of permanent dipoles must be quite close together before the interaction becomes strong. The permanent dipole moment of water contributes to its interaction with other water molecules and explains why it readily dissolves ionic substances like NaCl.

Molecules that do not have permanent dipole moments can become dipolar in the presence of an electric field. The field may be externally imposed, as in a laboratory instrument, or it may be produced by a neighboring charged or dipolar particle. A molecule in which a dipole can be so induced is said to be **polarizable.** Aromatic rings, for example, are very polarizable, because the electrons can easily be displaced in the plane of the ring, as shown in Figure 2.5a. Interactions of polarizable molecules are called **induced dipole interactions.** An anion or a cation may induce a dipole in a polarizable molecule and thereby be attracted to it (a *charge–induced dipole interaction*, Figure 2.2d), or a permanent dipole may do the same (a *dipole–induced dipole interaction*, Figure 2.2e). These induced dipole interactions are even shorter range than permanent dipole interactions, with energies of interaction proportional to $1/r^4$ and $1/r^5$, respectively.

Even two molecules that have neither a net charge nor a permanent dipole moment can attract one another if they are close enough (Figure 2.2f). The distribution of electronic charge in a molecule is never static, but fluctuates. When two molecules approach very closely, they synchronize their charge fluctuations so as to give a net attractive force. Such intermolecular forces, which can be thought of as mutual dipole induction, are called **dispersion forces.** Their attractive energy varies as the inverse sixth power of the distance, so dispersion forces are significant only at very short range. They can become particularly strong when two planar molecules can stack on one another, as shown in Figure 2.5b and c. We shall encounter many examples of such interactions in the internal packing of molecules like proteins and nucleic acids.

The rather weak intermolecular and intramolecular forces involving dipoles and induced dipoles are often given the general designation of *van der Waals* forces.

MOLECULAR REPULSION AT EXTREMELY CLOSE APPROACH: THE VAN DER WAALS RADIUS

Molecules may attract one another by noncovalent forces but cannot interpenetrate: van der Waals radii determine molecular surfaces.

When molecules or atoms that do not have covalent bonds between them come so close together that their outer electron orbitals begin to overlap, there is a mutual repulsion (Figure 2.2g). This repulsion increases very rapidly as the distance between their centers (r) decreases; it can be approximated as proportional to r^{-12}. If we combine this repulsive energy with one or more of the kinds of attractive energy described previously, we see that the total energy of noncovalent interaction (U) of a pair of atoms, molecules, or ions will vary with distance of

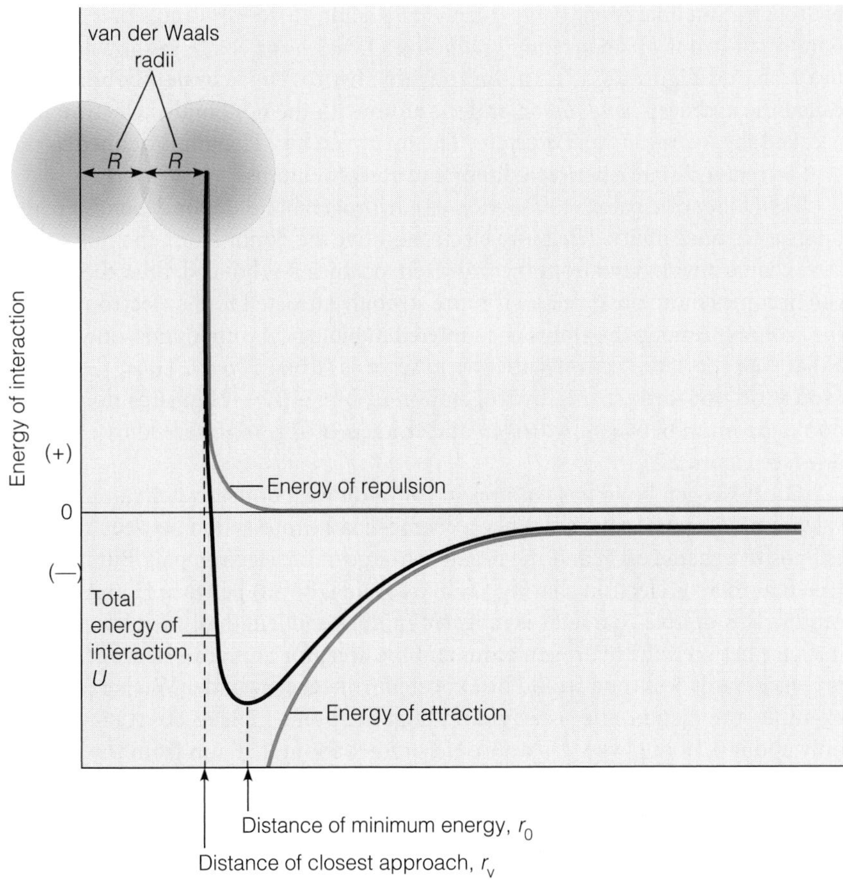

FIGURE **2.6**

Noncovalent interaction energy of two approaching particles. The interaction energy of two atoms, molecules, or ions is graphed versus the distance between their centers, *r*. The total interaction energy (*U*) at any distance is the sum of the energy of attraction and the energy of repulsion. As the distance between the particles decreases (reading *right to left* along the *x*-axis), both the attractive energy (−) and the repulsive energy (+) increase, but at different rates. At first the longer-range attraction dominates, but then the repulsive energy increases so rapidly that it acts as a barrier, defining the distance of closest approach (r_v) and the van der Waals radii (*R*). The position of minimum energy (r_0) is usually very close to r_v.

their separation (r) in the manner depicted in Figure 2.6. Two points on the graph should be noted. First, there is a minimum in the energy curve, at position r_0. This minimum corresponds to the most stable distance between the centers of the two particles. If we allow them to approach each other, this is how close they will come. Second, the repulsive potential rises so steeply at shorter distances that it acts as a "wall," effectively barring approach closer than the distance r_v. This distance defines the so-called **van der Waals radius, *R*,** the effective radius for closest molecular packing. For a pair of identical spherical molecules, $r_v = 2R$; for molecules with van der Waals radii R_1 and R_2, $r_v = R_1 + R_2$.

Real molecules, of course, are not spherical objects like those depicted in Figure 2.6. Because large biological molecules all have complicated shapes, it is useful to extend the concept of van der Waals radius to atoms or groups of atoms within a molecule. The values for van der Waals radii given in Table 2.2 represent the distances of closest approach for another atom or group. When we attempt to depict complex molecules in a "space filling" manner, we represent each atom by a sphere with its appropriate van der Waals radius (see Figure 2.5c). In this case, the van der Waals radii of the carbon atoms (0.17 nm) mean that the planes of the two stacked rings cannot be closer than 0.34 nm.

TABLE **2.2** **van der Waals radii of some atoms and groups of atoms**

	R (nm)
Atoms	
H	0.12
O	0.14
N	0.15
C	0.17
S	0.18
P	0.19
Groups	
—OH	0.14
$—NH_2$	0.15
$—CH_2—$	0.20
$—CH_3$	0.20
Half-thickness of aromatic ring	0.17

HYDROGEN BONDS

One specific kind of noncovalent interaction, the **hydrogen bond,** is of the greatest importance in biochemistry. The structure and properties of many biological molecules and of water, the universal biological solvent, are determined largely by this type of bond. A hydrogen bond is an interaction between a covalently bonded

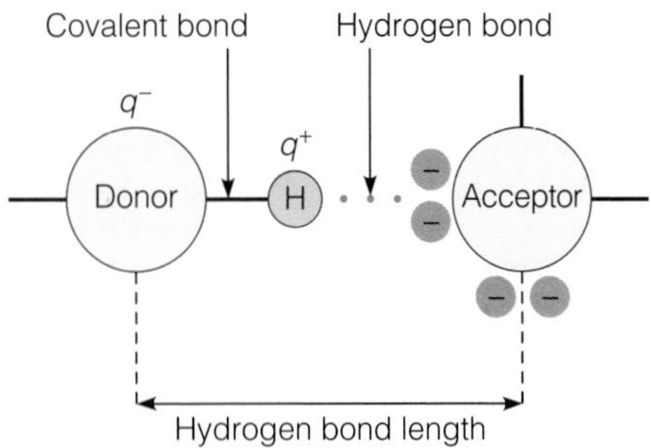

FIGURE **2.7**

The hydrogen bond. The figure shows an idealized H bond that might exist, for example, in —O—H···O=. Although the H bond is between H and the acceptor, the H-bond length is defined as the distance between donor and acceptor.

hydrogen atom on a *donor group* (like —O—H or >N—H) and a pair of nonbonded electrons on an *acceptor group* (like O=C— or N≦), as shown in Figure 2.2h and Figure 2.7. The atom to which hydrogen is covalently bonded is called the *hydrogen bond donor*, and the atom with the nonbonded electron pair is called the *hydrogen bond acceptor*. The hydrogen bond length is defined as the center–center distance between donor and acceptor atoms.

The ability of an atom to function as a hydrogen bond donor depends greatly on its electronegativity. The more electronegative the donor atom, the more negative charge it withdraws from the hydrogen to which it is bonded; thus, the hydrogen becomes more positive and is more strongly attracted to the electron pair of the acceptor. Among the atoms encountered in biological compounds, only O and N have appropriate electronegativities to serve as strong donors. Thus, ≥C—H groups do not form strong hydrogen bonds, but —O—H groups do. In this book, the bond between hydrogen and the acceptor is represented by a dotted line (see Figure 2.7).

The hydrogen bond has features in common with both covalent and noncovalent interactions. In part, it is like a charge–charge interaction between the partial positive charge on H and the negative charge of the electron pair. But it is also true that there is electron sharing (as in a covalent bond) between H and the acceptor. This double character is reflected in the bond length of the H bond. The distance between the hydrogen atom and the acceptor atom in a hydrogen bond is considerably less than would be expected from their van der Waals radii. For example, the distance between H and O in the bond >N—H···O=C< is only about 0.19 nm, whereas we would predict about 0.26 nm from the sum of the van der Waals radii given in Table 2.2. On the other hand, a *covalent* H—O bond has a length of only 0.10 nm. The entire hydrogen bond, including the covalent bond with nitrogen, is about 0.29 nm long. The bond lengths of some particularly strong hydrogen bonds are listed in Table 2.3. Note that the bond lengths are fixed, as they are for covalent (but not for other noncovalent) bonds.

Hydrogen bonds are the strongest, most specific noncovalent interactions.

The energy of H bonds is considerably higher than that of most other noncovalent bonds, in keeping with their partially covalent character (see Figure 2.1). Hydrogen bonds are also like covalent bonds in being highly directional: The donor H bond tends to point directly at the acceptor electron pair. The impor-

TABLE 2.3 Major types of hydrogen bonds found in biologically important molecules

Donor···Acceptor	Bond Length[a] (nm)	Comment
—O—H···O(—H)<	0.28 ± 0.01	H bond formed in water
—O—H···O=C<	0.28 ± 0.01	Bonding of water to other molecules often involves these
>N—H···O(—H)<	0.29 ± 0.01	Bonding of water to other molecules often involves these
>N—H···O=C<	0.29 ± 0.01	Very important in protein and nucleic acid structures
>N—H···N≦	0.31 ± 0.02	Very important in protein and nucleic acid structures
>N—H···S<	0.37	Relatively rare; weaker than above

[a]Defined as distance from center of donor atom to center of acceptor atom. For example, in the N—H···O=C< bond it is the N—O distance.

tance of this directionality is seen in the role that hydrogen bonds play in organizing a regular biochemical structure such as the α helix in proteins (Figure 2.8). This is but one example of many we shall encounter in which hydrogen bonds stabilize ordered structure in large molecules.

The several kinds of noncovalent interactions we have described are individually weak, but when many are present within a given macromolecule or between macromolecules, their energies can sum to an impressive total—often several hundreds of kilojoules. Thus, such interactions can account for the stability of macromolecular structures. At the same time, the ease with which individual noncovalent bonds can be broken and re-formed gives these structures a flexibility necessary to their function.

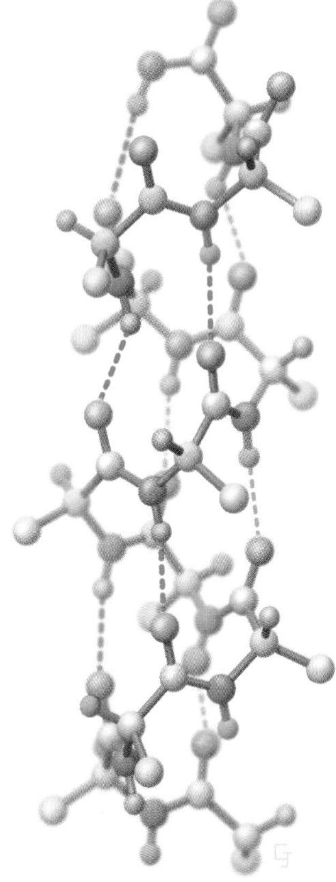

Key:
- Nitrogen
- Oxygen
- Carbon
- Side chain of amino acid
- Hydrogen
- ••• Hydrogen bond

FIGURE **2.8**

Hydrogen bonding in biological structure. This example is a portion of a protein in an α-helical conformation. The α-helix, a common structural element in proteins, is maintained by $>N-H \cdots O=C<$ hydrogen bonds between groups in the protein chain.

The Role of Water in Biological Processes

The chemical and physical processes of life require that molecules be able to move about, encounter one another, and change partners frequently in the complicated processes of metabolism and synthesis. A fluid environment allows molecular mobility, and water not only is the most abundant fluid on earth but also is admirably suited to this purpose. To see why, we must examine the properties of water in some detail.

THE STRUCTURE AND PROPERTIES OF WATER

Although we tend to take its properties for granted, water is really a most curious substance. Table 2.4, which contrasts H_2O (molecular weight 18) with other compounds of comparable molecular weight, reveals a remarkable fact. Most such low-molecular-weight compounds are gases at room temperature and have much lower boiling points than water does. Why is water unique? The answer lies mainly in the strong tendency of water molecules to form hydrogen bonds with other water molecules.

The electron arrangement of a single water molecule is shown in Figure 2.9a. Of the six electrons in the outer orbitals of the oxygen atom, two are involved in covalent bonds to the hydrogens. The other four electrons exist in nonbonded pairs, which are excellent hydrogen bond acceptors. The —OH groups in water are strong hydrogen bond donors. Thus each water molecule is *simultaneously* a hydrogen bond donor and a hydrogen bond acceptor, and a sample of water is a dynamic network of H-bonded molecules (Figure 2.9b). As a consequence, vaporization of water requires an unusual amount of energy for a molecule of its size. Both the heat of vaporization and the boiling point of water are therefore unusually high (Table 2.4), and water remains in the liquid state at temperatures characteristic of much of the earth's surface.

The hydrogen bonding between water molecules becomes most regular and clearly defined when water freezes to ice, creating a rigid tetrahedral molecular lat-

TABLE **2.4 Properties of water compared to some other low-molecular-weight compounds**

Compound	Molecular Weight	Melting Point (°C)	Boiling Point (°C)	Heat of Vaporization (kJ/mol)
CH_4	16.04	−182	−164	8.16
NH_3	17.03	−78	−33	23.26
H_2O	18.02	0	+100	40.71
H_2S	34.08	−86	−61	18.66

FIGURE **2.9**

Hydrogen bonding in water. (a) The electron structure of an individual water molecule. The nonbonded electron pairs of the two orbitals act as H acceptors. **(b)** Bonding between water molecules. Each molecule acts as both an H donor and an H acceptor, allowing clusters of water molecules to form.

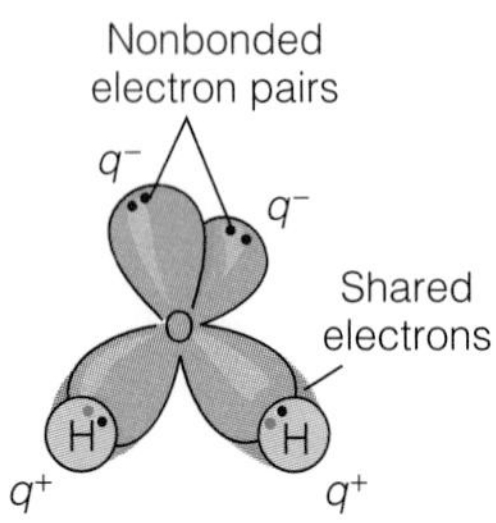

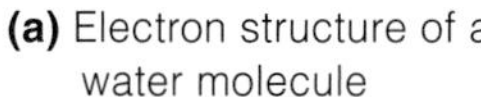

(a) Electron structure of a water molecule

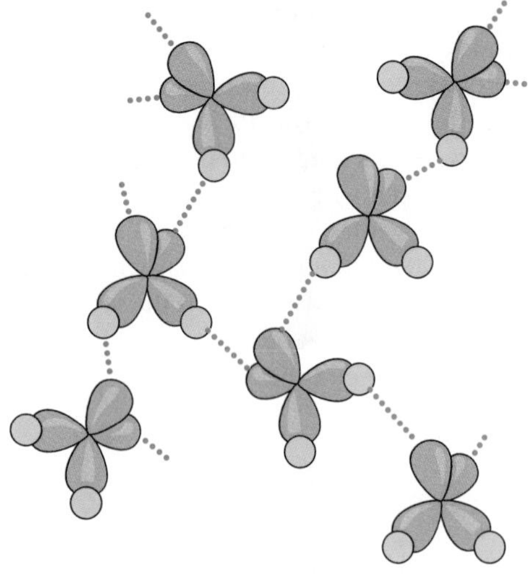

(b) Hydrogen bonding in water

Most of the unique properties of water are due to its hydrogen bonding potential and its polar nature.

tice in which each molecule is H-bonded to four others (Figure 2.10a). The lattice structure is only partially dismantled when ice melts, and some long-range order persists even at higher temperatures. The structure of liquid water has been described as "flickering clusters" of hydrogen bonds, with remnants of the ice lattice continually breaking and re-forming as the molecules move about (Figure 2.10b). The rather open structure of the ice lattice accounts for another of water's unusual properties—liquid water is denser than its solid form, because when the lattice breaks down, molecules can move closer together. This seemingly trivial fact is of the utmost importance for life on earth. If water behaved like most substances and became denser when frozen, the ice formed on lake and ocean surfaces each winter would sink to the bottom. There, insulated by the overlying layers, it would have accumulated over the ages, and most of the water on earth would by now have become locked up in ice. It seems doubtful that life could have evolved if liquid water at 0°C were 9% less dense than it is, for then ice would sink.

Other unusual properties of water, listed in Table 2.5, are also readily explained in terms of its molecular structure. Relative to most organic liquids, water has a high *viscosity*—a consequence of the interlocked, hydrogen-bonded structure. This cohesiveness also accounts for the high *surface tension* of water. The high dielectric constant of water, mentioned previously, results from its dipolar character. An electric field generated between two ions causes extensive orientation of intervening water dipoles and a significant amount of induced polarization. These oriented dipoles contribute to a counterfield, reducing the effective electrostatic force between the two ions.

TABLE **2.5 Important properties of liquid water compared with those of *n*-pentane, a nonpolar, non-hydrogen-bonding liquid**[a]

Property	Water	*n*-Pentane
Molecular weight (g/mol)	18.02	72.15
Density (g/cm^3)	0.997	0.626
Boiling point (°C)	100	36.1
Dielectric constant	78.54	1.84
Viscosity (g/cm·s)	0.890×10^{-2}	0.228×10^{-2}
Surface tension (dyne/cm)	71.97	17

[a]All data are for 25°C.

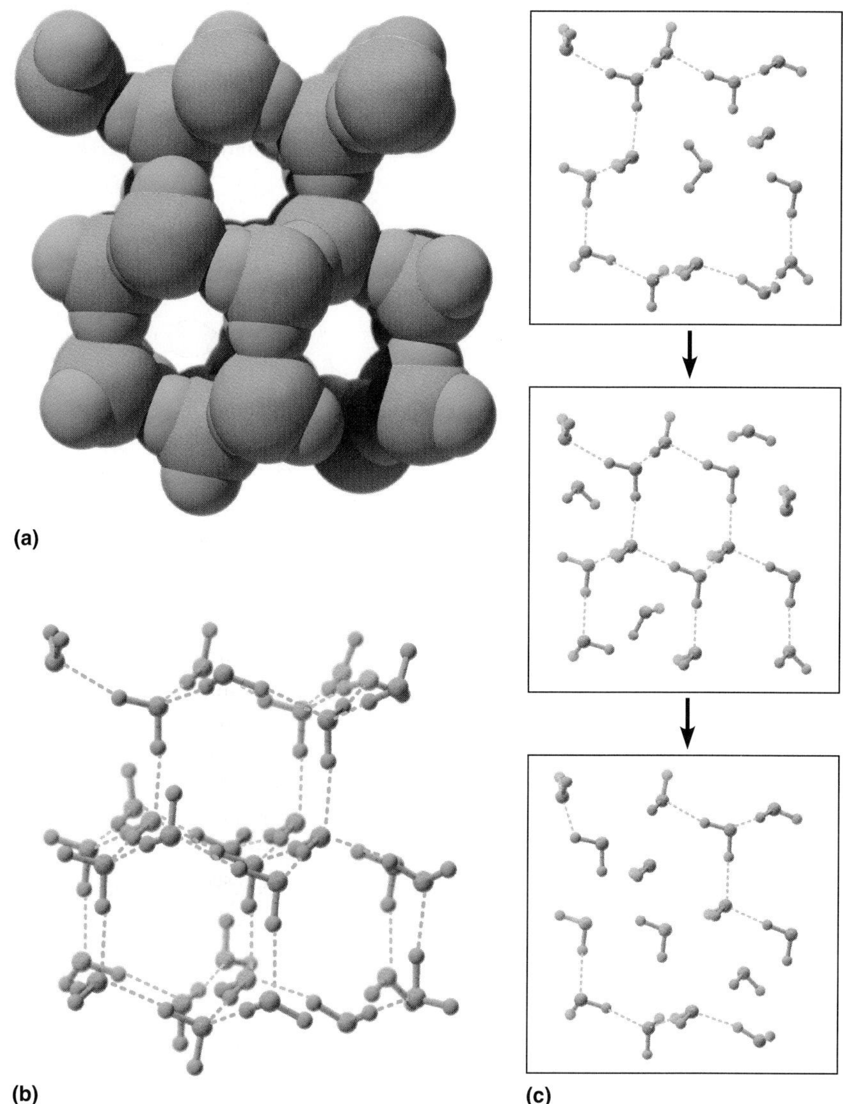

FIGURE **2.10**

Water as a molecular lattice. (a) A space-filling model of the structure of ice. Ice is a molecular lattice formed by indefinite repetition of a tetrahedral hydrogen-bonding pattern. Each molecule acts as a donor to two others and as an acceptor from two others. Because of the length of the hydrogen bonds, the structure is a relatively open one, which accounts for the low density of ice. **(b)** A skeletal model of the ice lattice. **(c)** The structure of liquid water. When ice melts, the regular tetrahedral lattice is broken, but substantial portions of it remain, especially at low temperatures. In liquid water, flickering clusters of molecules are held together by hydrogen bonds that continually break and re-form. In this schematic "motion picture," successive frames represent changes occurring in picoseconds (10^{-12} s).

(a–b) Adapted from L. Pauling and P. Pauling, *Chemistry* (New York, NY: W. H. Freeman, 1975), p. 289.

WATER AS A SOLVENT

Water is an excellent solvent because of its hydrogen bonding potential and its polar nature.

The processes of life require a wide variety of ions and molecules to move about in proximity, that is, to be soluble in a common medium. Water serves as the universal intracellular and extracellular medium, thanks primarily to the two properties of water we have been discussing: its tendency to form hydrogen bonds and its dipolar character. Substances that can take advantage of these properties so as to readily dissolve in water are called **hydrophilic,** or "water loving."

Hydrophilic Molecules in Aqueous Solution

Molecules with groups capable of forming hydrogen bonds tend to hydrogen-bond with water. Thus, water readily dissolves hydroxyl compounds, amines, sulfhydryl compounds, esters, ketones, and a wide variety of other organic compounds. When molecules that contain internal hydrogen bonds (such as the α helix shown in Figure 2.8) dissolve in water, some or all of their internal H bonds may be exchanged for H bonds to H_2O (Figure 2.11).

But it is not only hydrogen-bond acceptors or donors that dissolve well in water. In contrast to most organic liquids, water is an excellent solvent for ionic compounds. Substances like sodium chloride, which exist in the solid state as very stable lattices of ions, dissolve readily in water. The explanation lies in the dipolar nature of the water molecule. The interactions of the water dipoles with cations and anions in aqueous solution cause the ions to become **hydrated,** that is, surrounded by shells of water molecules called **hydration shells** (Figure 2.12). The propensity of many ionic compounds like NaCl to dissolve in water can be ac-

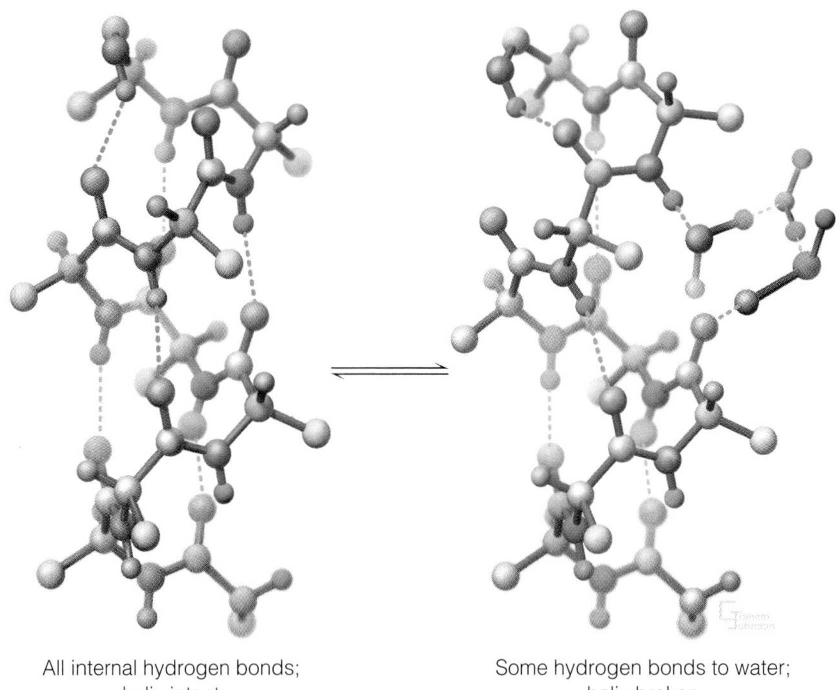

FIGURE **2.11**

Exchange of internal hydrogen bonds for water hydrogen bonds. A section of a protein molecule like that in Figure 2.8 is shown here replacing some of its intramolecular hydrogen bonds with hydrogen bonds to solvent water. Because of this exchange, molecules with internal hydrogen bonds dissolve readily in water.

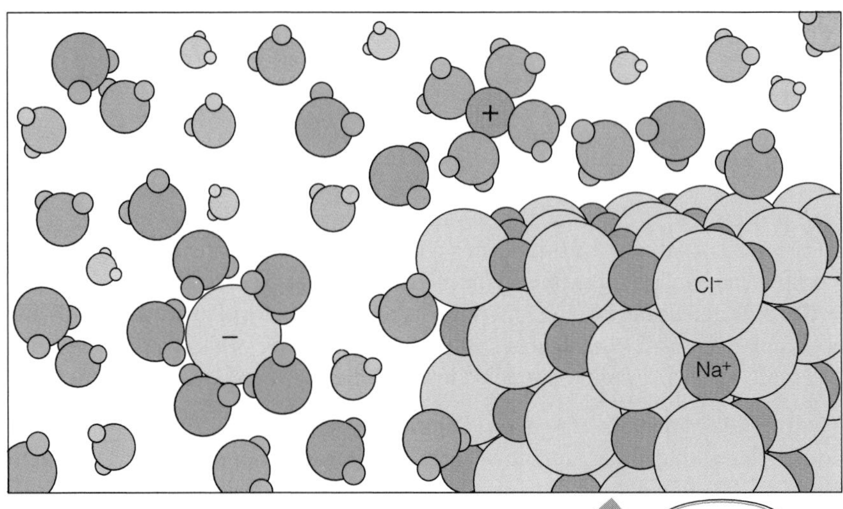

FIGURE **2.12**

Hydration of ions in solution. A salt crystal is shown dissolving in water. As sodium and chloride ions leave the crystal, the noncovalent interaction between these ions and the dipolar water molecules produces a hydration shell around each ion. The energy released in this interaction helps overcome the charge–charge interactions stabilizing the crystal.

counted for largely by two factors. First, the formation of hydration shells is energetically favorable. Second, the high dielectric constant of water screens and decreases the electrostatic force between oppositely charged ions that would otherwise pull them back together.

The dipolar nature of the water molecule also contributes to water's ability to dissolve such nonionic, but polar, organic molecules as phenols, esters, and amides. These molecules often have large dipole moments, and interaction with the water dipole favors their solubility in water.

Hydrophobic Molecules in Aqueous Solution

The solubility of hydrophilic substances depends on their energetically favorable interaction with water molecules. It is therefore not surprising that substances like hydrocarbons, which are nonpolar and nonionic and cannot form hydrogen bonds, show only limited solubility in water. Molecules that behave in this way are called **hydrophobic** or "water fearing." However, energy is not the only factor limiting their solubility. When hydrophobic molecules do dissolve, they do not form hydration shells as hydrophilic substances do. Instead, the regular water lattice forms icelike **clathrate** structures, or "cages," about nonpolar molecules (Figure 2.13). This ordering of water molecules, which extends well beyond the cage, corresponds to a decrease in the *entropy*, or randomness, of the mixture (see Chapter 3). The decrease in entropy contributes to the low solubility of hydrophobic substances in water. It also accounts for the well-known tendency of hydrophobic substances to form aggregates in water—we have all seen oil form droplets when we shake it with vinegar. Surrounding two hydrophobic molecules with two separate cages requires more ordering than surrounding both with a single cage, so the hydrophobic molecules tend to cluster.

Amphipathic Molecules in Aqueous Solution

A most interesting and important class of molecules exhibits both hydrophilic and hydrophobic properties simultaneously. Such **amphipathic** substances include fatty acids and detergents (Figure 2.14). Amphipathic molecules have a head group that is strongly hydrophilic, coupled to a hydrophobic tail—usually a hydrocarbon. When we attempt to dissolve them in water, amphipathic substances

Amphipathic substances are molecular schizophrenics—part of the molecule is hydrophilic and part is hydrophobic.

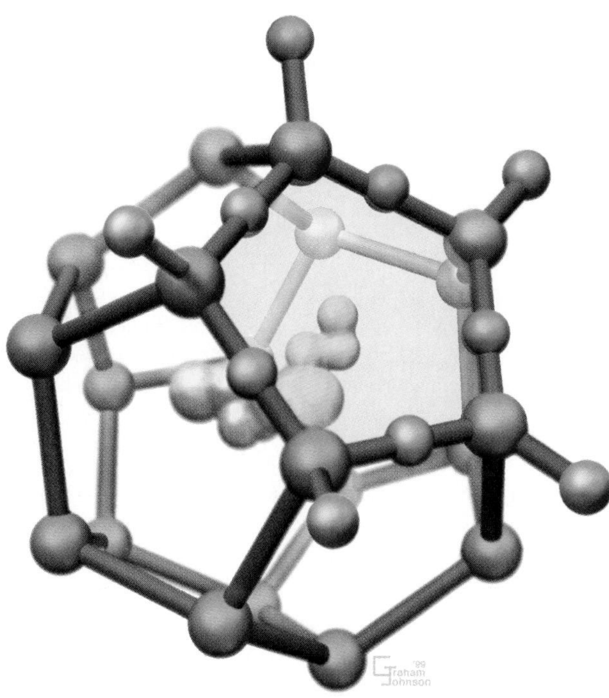

FIGURE **2.13**

One unit of clathrate structure surrounding a hydrophobic molecule (yellow). Oxygen atoms are shown in red. Hydrogens are shown for one pentagon of oxygens. The ordered structure may extend considerably further into the surrounding water.

Hydrophilic head groups

Hydrophobic tail groups

Dodecanoate ion (a fatty acid)

Dodecyl sulfate (a detergent)

Phosphatidylethanolamine (a phospholipid)

FIGURE **2.14**

Amphipathic molecules. These three examples illustrate the "schizophrenic" structure of amphipathic molecules, which have a hydrophilic head group attached to a hydrophobic tail.

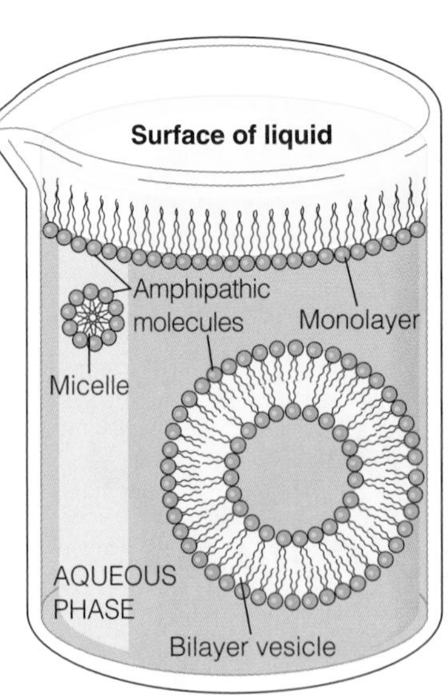

(a) Structures formed in water

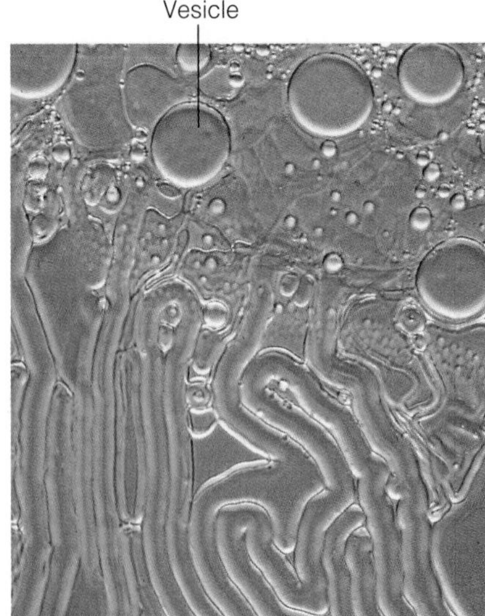

(b) Vesicle formation

FIGURE **2.15**

Interactions of amphipathic molecules with water. (a) Structures that can occur when amphipathic substances are mixed with water include a monolayer on the water surface, a micelle, and a bilayer vesicle, a hollow sphere with water both inside and out. In each case, the hydrophilic head groups are in contact with the aqueous phase, whereas the hydrophobic tails associate with one another. **(b)** When phospholipids are mixed with water, the amphipathic molecules aggregate to form films similar to biological membranes. Agitation causes the film to break up into vesicles.

(b) Courtesy of D. W. Deamer and P. B. Armstrong, University of California, Davis.

form one or more of the structures shown in Figure 2.15a. For example, they may form a **monolayer** on the water surface, with only the head groups immersed. Alternatively, if the mixture is vigorously stirred, **micelles** (spherical structures formed by a single layer of molecules) or **bilayer vesicles** may form. In such cases the hydrocarbon tails of the molecules tend to lie in roughly parallel arrays, which allows them to interact via van der Waals forces. The polar or ionic head groups are strongly hydrated by the water around them. Most important to biochemistry is the fact that amphipathic molecules are the basis of the biological **membrane bilayers** that surround cells and form the partitions between cellular compartments. These bilayers are made primarily from phospholipids, such as that shown in Figure 2.14. Figure 2.15b shows the formation of synthetic membranes from phospholipids. We have much more to say about phospholipids and membranes in Chapter 10.

Ionic Equilibria

Virtually all biochemical reactions occur in an aqueous environment; the exceptions are the few that occur within the hydrophobic interiors of membranes. The many substances dissolved in the aqueous cytosol and extracellular body fluids include free ions like K^+, Cl^-, and Mg^{2+}, as well as molecules and macromolecules carrying ionizable groups. The behavior of all these molecules in biochemical processes depends strongly on their state of ionization. Thus, it is important that we review briefly some aspects of ionic equilibrium, particularly acid–base equilibria and the ionization of water.

ACIDS AND BASES: PROTON DONORS AND ACCEPTORS

Recall from general elementary chemistry that acids are proton donors and bases are proton acceptors. A **strong acid** dissociates almost completely into protons and anions. For example, HCl dissociates almost completely in water to yield H^+ and Cl^-, so that the H^+ concentration in the solution is almost exactly equal to the molar concentration of HCl added. Similarly, NaOH is called a **strong base** because it ionizes entirely to give OH^- ions, which are powerful proton acceptors.

Most of the acidic and basic substances encountered in biochemistry are **weak acids** or **weak bases,** which dissociate only partially. In an aqueous solution of a weak acid there is a measurable equilibrium between the acid and its *conjugate base*, the substance that can accept a proton to re-form the acid. Examples of weak acids and their conjugate bases are given in Table 2.6. Note that these bases do not necessarily contain —OH groups, but they increase the OH^- concentration of a solution by extracting a proton from water.

Many biological molecules are weak acids or weak bases.

The weak acids listed in Table 2.6 vary greatly in strength, that is, in their tendency to donate protons. This variation in acid strength is indicated by the range of values of K_a and pK_a, which we will discuss shortly. The stronger the acid, the weaker its conjugate base; in other words, the more strongly an acid tends to donate a proton, the more weakly its conjugate base tends to accept a proton and re-form the acid.

IONIZATION OF WATER AND THE ION PRODUCT

Although water is essentially a neutral molecule, it does have a slight tendency to ionize; in fact, it can act as both a very weak acid and a very weak base. The most correct way to understand this ionization reaction is to note that one water molecule can transfer a proton to another to yield a hydronium ion (H_3O^+) and a hydroxyl ion (OH^-), so that water is both the proton donor and the proton acceptor:

$$H_2O + H_2O \rightleftharpoons H_3O^+ + OH^-$$

This is really an oversimplification, for the transferred proton may be associated with different clusters of water molecules to yield species like $H_5O_2^+$ and $H_7O_3^+$. Protons in aqueous solution are very mobile, with the charge hopping from one water molecule to another with a period of about 10^{-15} second.

TABLE **2.6 Some weak acids and their conjugate bases**

Acid (Proton Donor)		Conjugate Base (Proton Acceptor)		pK_a	K_a (M)
HCOOH Formic acid	$\rightleftharpoons$	$HCOO^-$ Formate ion	$+H^+$	3.75	1.78×10^{-4}
CH_3COOH Acetic acid	$\rightleftharpoons$	CH_3COO^- Acetate ion	$+H^+$	4.76	1.74×10^{-5}
$CH_3CH(OH)—COOH$ Lactic acid	$\rightleftharpoons$	$CH_3CH(OH)—COO^-$ Lactate ion	$+H^+$	3.86	1.38×10^{-4}
H_3PO_4 Phosphoric acid	$\rightleftharpoons$	$H_2PO_4^-$ Dihydrogen phosphate ion	$+H^+$	2.14	7.24×10^{-3}
$H_2PO_4^-$ Dihydrogen phosphate ion	$\rightleftharpoons$	HPO_4^{2-} Monohydrogen phosphate ion	$+H^+$	6.86	1.38×10^{-7}
HPO_4^{2-} Monohydrogen phosphate ion	$\rightleftharpoons$	PO_4^{3-} Phosphate ion	$+H^+$	12.4	3.98×10^{-13}
H_2CO_3 Carbonic acid	$\rightleftharpoons$	HCO_3^- Bicarbonate ion	$+H^+$	6.37	4.27×10^{-7}
HCO_3^- Bicarbonate ion	$\rightleftharpoons$	CO_3^{2-} Carbonate ion	$+H^+$	10.25	5.62×10^{-11}
C_6H_5OH Phenol	$\rightleftharpoons$	$C_6H_5O^-$ Phenolate ion	$+H^+$	9.89	1.29×10^{-10}
$\overset{+}{N}H_4$ Ammonium ion	$\rightleftharpoons$	NH_3 Ammonia	$+H^+$	9.25	5.62×10^{-10}

Phosphoric acid series   Carbonic acid series

For practical purposes, it almost always suffices to describe the ionization process in a much simpler way,

$$H_2O \rightleftharpoons H^+ + OH^-$$

as long as we remember that a proton *never* exists in aqueous solution as a free ion—it is always associated with one or more water molecules. Whenever we write a reaction involving H^+, we are really referring to a *hydrated* proton.

The equilibrium described by the previous equation can be expressed in terms of the **ion product** of water, which is 10^{-14} M^2 at 25°C:

$$[H^+][OH^-] = K_w = 1 \times 10^{-14}\ \text{M}^2 \tag{2.5}$$

The symbol M here and elsewhere designates moles per liter. Because the ion product is a constant, $[H^+]$ and $[OH^-]$ cannot vary independently. If we change either $[H^+]$ or $[OH^-]$ by adding acidic or basic substances to water, the other concentration must change accordingly. A solution with a high $[H^+]$ has a low $[OH^-]$, and vice versa.

If we take pure water to which no acidic or basic substances have been added, all the H^+ and OH^- ions must come from the dissociation of the water itself. Under these circumstances the concentrations of H^+ and OH^- must be equal, so at 25°C

$$[H^+] = [OH^-] = 1 \times 10^{-7}\ \text{M} \tag{2.6}$$

and the solution is said to be *neutral*, that is, neither acidic nor basic. However, the ion product depends on temperature, so a neutral solution does not always have $[H^+]$ and $[OH^-]$ of exactly 10^{-7}M. For example, at human body temperature (37°C) the concentrations of H^+ and OH^- ions in a neutral solution are each 1.6×10^{-7} M.

THE pH SCALE AND THE PHYSIOLOGICAL pH RANGE

To avoid working with negative powers of 10, we almost always express hydrogen ion concentration as pH, defined* as

$$\mathrm{pH} = -\log\,[\mathrm{H}^+] \qquad (2.7)$$

The higher the $[H^+]$ of a solution, the lower the pH, so a low pH describes an acid solution. On the other hand, a low $[H^+]$ must be accompanied by a high $[OH^-]$, as indicated by equation (2.5), so a high pH describes a basic solution.

A diagrammatic scale of pH values is shown in Figure 2.16, with the values for some well-known solutions indicated. Note that most body fluids have pH values in the range 6.5–8.0, which is often referred to as the **physiological pH range.** Most biochemistry occurs in this region of the scale.

Most biological reactions occur between pH 6.5 and pH 8.0.

Because of the sensitivity of biochemical processes to even small pH changes, monitoring pH is essential in most biochemical experiments. Today almost all pH measurements are made with glass electrode pH meters. The electrode generates an electrical potential, which depends on the H^+ concentration; this is converted by the instrument into a pH reading.

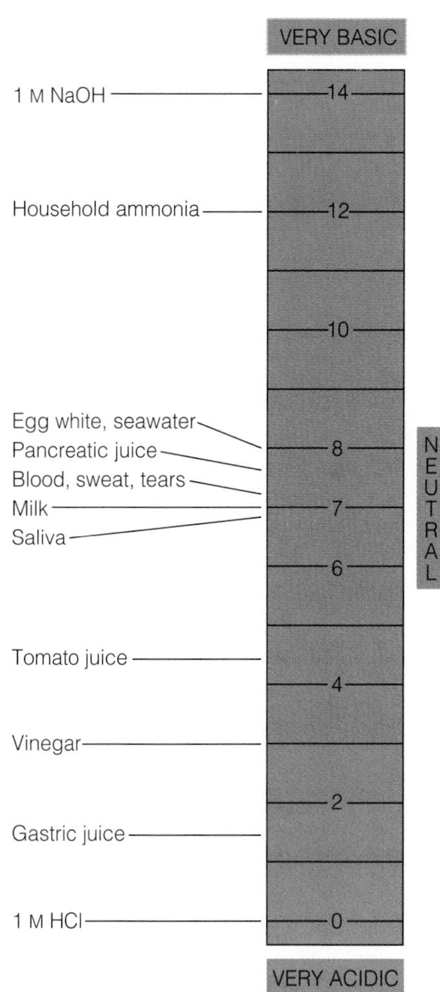

FIGURE **2.16**

The pH scale and the physiological pH range. The pH values of some common substances and body fluids are listed, with NaOH at the basic end of the range shown here and HCl at the acid end. The pH of most body fluids is in the neutral range between 6.5 and 8.0, where most physiological processes occur. There are a few exceptions, like gastric juice, which has a pH between 1 and 2.

WEAK ACID AND BASE EQUILIBRIA

Many biologically important compounds contain weakly acidic and basic groups. Very large protein molecules, for example, carry on their surfaces both acidic (e.g., carboxylate) and basic (e.g., amino) groups. The response of such groups to pH changes in or near the physiological range is often of considerable importance to their function. For example, the catalytic efficiency of many enzymes depends critically on the ionization state of certain groups, so these catalysts are effective only in defined pH ranges. We need to review the dissociation equilibria of weak acids and bases to understand such effects. To do so, consider the examples given in Table 2.6.

K_a and pK_a

Each of the reactions shown in Table 2.6 can be written as the dissociation of an acid. This dissociation may take several forms, depending on the substance involved:

$$\mathrm{HA^+} \rightleftharpoons \mathrm{A} + \mathrm{H^+}$$
$$\mathrm{HA} \rightleftharpoons \mathrm{A^-} + \mathrm{H^+}$$
$$\mathrm{HA^-} \rightleftharpoons \mathrm{A^{2-}} + \mathrm{H^+}$$

Note that in some cases the conjugate base has a negative charge and in other cases it does not, but in *all* cases it has one less positive charge than the acid. For convenience, we will always write such reactions as $HA \rightleftharpoons H^+ + A^-$. The equilibrium constant for the dissociation of a weak acid (often called the **dissociation constant**) is defined as

$$K_a = \frac{[\mathrm{H^+}][\mathrm{A^-}]}{[\mathrm{HA}]} \qquad (2.8)$$

*To be more exact in our definition of pH, we should use hydrogen ion *activity* rather than concentration, activity being a dimensionless number equal to the effective concentration. However, the distinction between molar concentration and activity is almost always neglected in biochemistry. This approach is appropriate in most biochemical experiments, which are usually conducted in dilute solutions, where activity and concentration become almost equal.

The larger K_a is, the greater is the tendency for the acid to dissociate. Thus, the larger K_a is, the stronger the acid.

The strength of acids is usually expressed in terms of the pK_a value:

$$pK_a = -\log K_a \tag{2.9}$$

A convenient way to express the strength of an acid is by its pK_a; the smaller the pK_a, the stronger the acid.

Because pK_a is the negative logarithm of K_a, a numerically small value of pK_a corresponds to a strong acid, and a numerically large value to a weak acid. Both K_a and pK_a values are given for the acids listed in Table 2.6.

Some acids, such as the phosphoric and carbonic acids in Table 2.6, are capable of losing more than one proton. These acids are called **polyprotic acids**. The successive dissociations involve separate steps, with separate pK_a values; thus polyprotic acids exist in several different states of ionization.

A CLOSER LOOK AT pK_a VALUES: FACTORS AFFECTING ACID DISSOCIATION

The tendency of a particular acid to dissociate results from a specific balance of the factors favoring and opposing dissociation. We can understand some of the factors affecting pK_a in light of our earlier discussion of the solvent properties of water.

The dissociation of an acid results in hydration of the proton and, in most cases, of the conjugate base as well. Because hydration is energetically favorable and decreases the attraction between the ions, it favors the dissociation of most acids. Exceptions are positively charged acids like $\overset{+}{N}H_4$, which dissociate to produce an uncharged conjugate base. In these cases it is the acid that is hydrated and thus stabilized; this is one reason $\overset{+}{N}H_4$ is such a weak acid.

Opposing the dissociation of an acid is the electrostatic attraction between the proton and a negatively charged conjugate base. This effect can be seen by comparing the successive pK_a values for phosphoric acid dissociation in Table 2.6. As the charge on the conjugate base rises in going from $H_2PO_4^-$ to PO_4^{3-}, the pK_a rises as well; HPO_4^{2-} is a very weak acid.

These examples are important, for they show how environmental effects can influence pK_a values. When we investigate proteins, we will find that the pK_a values for supposedly identical groups can vary widely, depending on the local molecular environment.

TITRATION OF WEAK ACIDS: THE HENDERSON–HASSELBALCH EQUATION

How does the pH of a solution change as we vary the base/acid ratio for a weak acid? This question is important to biochemists, for many of the small molecules and macromolecules of the cell ionize weakly. We can answer this question by taking the negative logarithm of both sides of equation (2.8) and rearranging:

$$-\log [H^+] = -\log K_a + \log \frac{[A^-]}{[HA]} \tag{2.10}$$

Substituting pH for $-\log [H^+]$ and pK_a for $-\log K_a$ gives us the *Henderson–Hasselbalch equation*, which relates the pH of a solution to the conjugate base/undissociated acid ratio:

$$pH = pK_a + \log \frac{[A^-]}{[HA]} \tag{2.11}$$

For example, if we have a solution of formic acid, equation (2.11) becomes

$$pH = 3.75 + \log \frac{[HCOO^-]}{[HCOOH]} \tag{2.12}$$

If we know the pH of the solution, we can calculate the conjugate base/undissociated acid ratio. Or we could rearrange the equation to use the ratio to calculate the pH of the solution.

The usefulness of the Henderson–Hasselbalch equation can be seen in its application to titrations: It shows exactly how pH changes as the base is added to an acidic solution or vice versa. Suppose we want to titrate a 1 M solution of formic acid with sodium hydroxide. First, we must ask: What is the pH of the solution made by dissolving 1 mol formic acid in 1 L of water? This can be calculated from equation (2.8), if we note that virtually all of the H^+ in such a solution comes from the formic acid, rather than from water, and that dissociation of one formic acid molecule gives one H^+ and one $HCOO^-$. If we denote their concentration by x, then equation (2.8) becomes

The Henderson–Hasselbalch equation describes the change in pH during titration of a weak acid or a weak base.

$$K_a = 1.78 \times 10^{-4} = \frac{[H^+][HCOO^-]}{[HCOOH]} = \frac{x^2}{1 - x} \quad (2.13)$$

To get an exact answer, we would have to solve this quadratic equation (see Problem 4). For weak acids, however, the quantity x is much less than the concentration of total acid added (in this case 1 M). As a result, in such cases we can neglect x in the denominator, giving as a good approximation

$$K_a \cong x^2 \quad (2.14)$$

In this example, this yields

$$x = [H^+] = [HCOO^-] = 1.33 \times 10^{-2}\ \text{M} \quad (2.15)$$

Note that only about 1% of the acid has dissociated, so our approximation is quite reasonable. It would not be if the acid were more dilute, for then a larger fraction would dissociate.

The preceding calculation tells us that the initial pH is about 1.9. Now, what happens when we begin to add a solution of NaOH? As NaOH is added, it dissociates completely into Na^+ and OH^-. However, the hydroxyl ions are in equilibrium with protons according to the relation $K_w = [H^+][OH^-]$, so addition of OH^- removes protons from the solution. As protons are removed, more formic acid must dissociate in an effort to replace them, to satisfy the equilibrium relationship given by equation (2.13). This means that the ratio $[HCOO^-]/[HCOOH]$ increases continuously. Applying the Henderson–Hasselbalch equation (2.11), we see that the pH must also increase continuously as the titration proceeds. At the midpoint of the titration, half of the original formic acid has been neutralized. That means that half is still present in the acid form and half is present as the conjugate base, so $[A^-]/[HA] = 1$. The Henderson–Hasselbalch equation then becomes

$$\text{pH} = \text{p}K_a + \log 1 = \text{p}K_a \quad (2.16)$$

Thus, the pH of a weak acid at the midpoint of its titration curve has the same value as its $\text{p}K_a$. This can be confirmed experimentally, as shown by the titration curves of two acids, formic acid and ammonium ion, in Figure 2.17. The titration curves in Figure 2.17 plot the measured pH against *moles of base added per mole of acid originally present.* Note that over much of the titration curve the pH lies within one pH unit below or above the $\text{p}K_a$. If we know the $\text{p}K_a$ of a weak acid, we know the midpoint of the titration; this defines the pH range in which most of the addition of base will be required to complete the titration.

It should be emphasized that titration curves are reversible. If we were to take the final solution, at high pH, and begin adding a strong acid like HCl, the same curve would be retraced in the opposite direction.

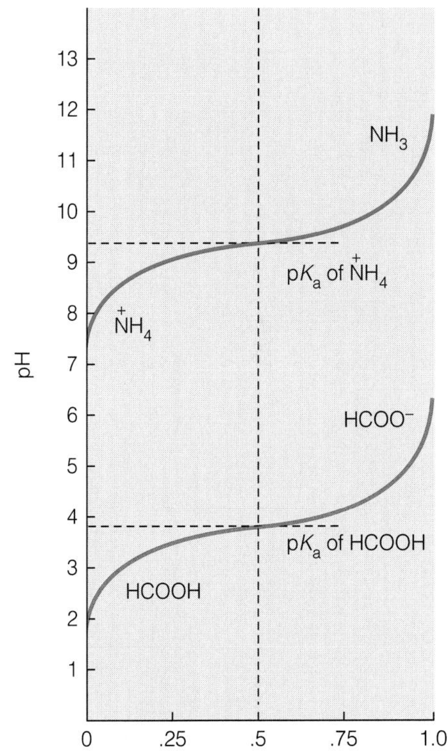

FIGURE **2.17**

Titration curves of weak acids. The curves for titration of formic acid (HCOOH) and ammonium ion ($\overset{+}{N}H_4$) show the change in pH as moles of base are added. Note that pH = pK_a at the half-titration point of each substance. Most of the titration occurs in a range of about one pH unit on either side of this value. In this range, the change in pH with moles of base added is minimal; thus, this is the best buffering range.

BUFFER SOLUTIONS

Buffer solutions function because the pH of a weak acid–base solution is least sensitive to added acid or base near the pK_a.

If we look at Figure 2.17 in a different way, another important point emerges. In the pH range near the pK_a, the pH changes only a little with each increment of base or acid added. In fact, the pH is least changed per increment of acid or base just at the pK_a. This is the principle behind **buffering** of solutions by the use of weak acid–base mixtures, a technique used in virtually every biochemical experiment.

Suppose a biochemist wishes to study a reaction at pH 4.00. The reaction may be one that generates or consumes protons. To prevent the pH from drifting too much during the reaction, the experimenter should use a buffer solution consisting of a specific mixture of a weak acid and its conjugate base. In this example a formic acid–formate buffer would be a good choice, because the pK_a of formic acid (3.75) is close to the pH value required. An acetic acid–acetate mixture would not be so satisfactory, because the pK_a of acetic acid (4.76) is nearly 1 pH unit away. The ratio of formate ion to formic acid required can be calculated from the Henderson–Hasselbalch equation:

$$4.00 = 3.75 + \log \frac{[HCOO^-]}{[HCOOH]} \tag{2.17}$$

In order to calculate the base/acid ratio, we simply subtract 3.75 from 4.00, and take the antilogarithm of both sides of equation (2.17):

$$\frac{[HCOO^-]}{[HCOOH]} = 10^{0.25} = 1.78 \tag{2.18}$$

Such a buffer could be made, for example, by mixing equal volumes of 0.1 M formic acid and 0.178 M sodium formate. Alternatively, a buffer solution could be prepared by titrating a solution of formic acid to pH 4.00 with sodium hydroxide.

Because it is desirable to study many biochemical reactions near physiological pH, there is a particular need for mixtures that buffer the pH in the pH range 6.5–8.0. Of the acid–base pairs listed in Table 2.6, only mixtures of dihydrogen phosphate ion and monohydrogen phosphate ion, or possibly carbonic acid and bicarbonate ion, would be satisfactory. Phosphate buffers are often used in experiments, but they cannot serve under all circumstances, because phosphate is consumed or produced in some biochemical reactions. Furthermore, both phosphate- and carbonate-containing solutions precipitate some ions (Ca^{2+}, for example) that may be needed in the reaction. Therefore, a number of other naturally occurring and synthetic compounds are employed as buffers in this range. Examples are given in Table 2.7.

Organisms must maintain the pH inside cells and in most bodily fluids within the narrow pH range of about 6.5 to 8.0. We have already mentioned two buffer sys-

TABLE **2.7 Some buffers commonly employed for biochemical studies**

Buffer Substance (Acid Form)	Common Name	pK_a
Cacodylic acid	—	6.2
2,2-Bis(hydroxymethyl)2,2′,2″ nitrilotriethanol	BISTRIS	6.5
Piperazine-*N*,*N*′-bis(2-ethanesulfonic acid)	PIPES	6.8
Imidazole	—	7.0
N′-2-Hydroxyethylpiperazine-*N*′,2-ethanesulfonic acid	HEPES	7.6
Tris(hydroxymethyl)aminomethane	Tris	8.3

tems of great importance for biological pH control. The dihydrogen phosphate–monohydrogen phosphate system, with a pK_a of 6.86, plays a major role in controlling intracellular pH, for phosphate is abundant in cells. In blood, which contains dissolved CO_2 as a waste product of metabolism, the carbonic acid–bicarbonate system, pK_a 6.37, provides considerable buffering capacity. The bicarbonate–carbonate system is actually more complex than indicated in Table 2.6, because of the reaction of dissolved CO_2:

Organisms use buffer systems to maintain the pH of cells and body fluids in the appropriate range.

$$CO_2 + H_2O \rightleftharpoons HCO_3^- + H^+$$

In addition to these and other small-molecule buffers, proteins play a major role in the control of pH in organisms. As shown in Chapter 5, proteins contain many weakly acidic or basic groups, and some of these have pK_a values near 7.0. Because proteins are abundant both in cells and in body fluids like blood and lymph, pH buffering is very strong.

MOLECULES WITH MULTIPLE IONIZING GROUPS: AMPHOLYTES, POLYAMPHOLYTES, AND POLYELECTROLYTES

So far, we have considered molecules containing only one or a few weakly acidic or basic groups. But many molecules contain multiple ionizing groups and display more complex behavior during titration.

A molecule that contains groups with both acidic and basic pK_a values is called an **ampholyte.** Consider, for example, the molecule glycine: $H_2N—CH_2—COOH$. Glycine is an α-amino acid, one of a group of important amino acids encountered in Chapter 5 as constituents of proteins. The pK_a values of the carboxylate and amino groups on glycine are 2.3 and 9.6, respectively. If we dissolved glycine in a very acidic solution (say, pH 1.0), both the amino group and the carboxylate group would be protonated; and the molecule would have a net charge of +1. If the pH were increased (for example, by adding NaOH), proton dissociation would occur in the following sequence:

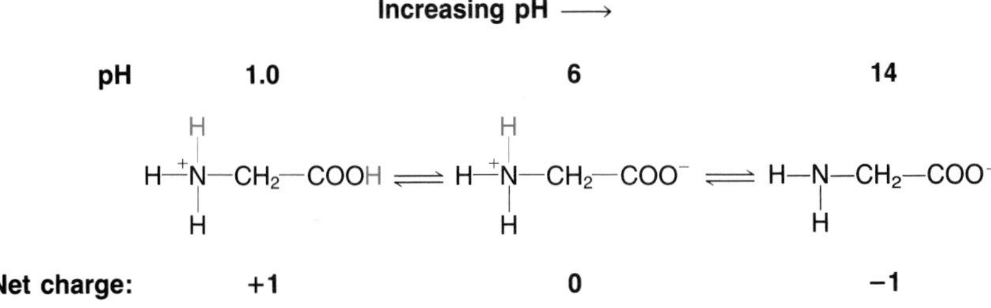

Thus, the titration of glycine occurs in two steps as the more acidic carboxylate and the less acidic amino groups successively lose their protons. Glycine can therefore serve as a good buffer in two quite different pH ranges, as shown in Figure 2.18. In each range we can describe the titration curve by applying the Henderson–Hasselbalch equation to the appropriate ionizing group. At low pH the principal form of glycine has a net charge of +1; at high pH the principal form has a net charge of −1. The relative concentrations of the three forms are shown in Figure 2.19.

The situation near neutral pH is an interesting one. In this region, most glycine is in the form $H_3\overset{+}{N}—CH_2—COO^-$, which has a net charge of zero. An ampholyte in this state, with equal numbers of positive and negative charges, is called a *zwitterion.* However, there is only one point within this pH region where the *average* charge on glycine is zero. At this pH, called the **isoelectric point (pI),**

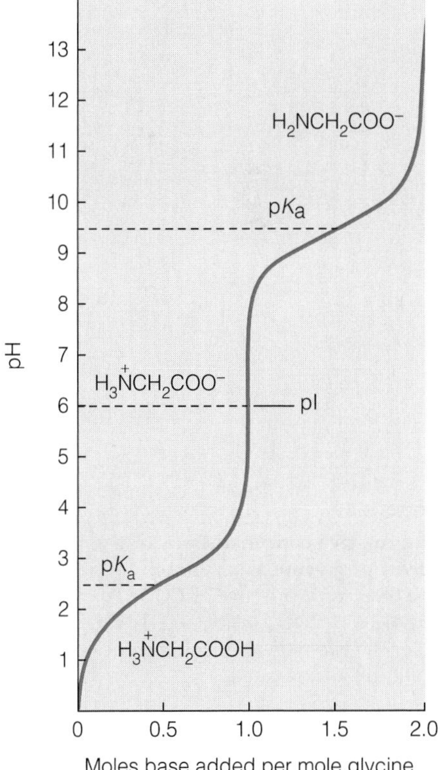

FIGURE **2.18**

Titration of the ampholyte glycine. Since two groups with quite different pK_a values can be titrated, this is a two-step titration curve. The calculated isoelectric point, pI, is shown.

most of the glycine molecules are in the zwitterion form, with very small but exactly equal amounts of $H_3\overset{+}{N}—CH_2—COOH$ and $H_2N—CH_2—COO^-$ molecules. We can calculate the isoelectric point by applying the Henderson–Hasselbalch equation to both of the ionizing groups. If we call the pH at the isoelectric point pI, we have

$$\text{pI} = \text{p}K_{\text{COOH}} + \log \frac{[\text{H}_3\overset{+}{\text{N}}\text{CH}_2\text{COO}^-]}{[\text{H}_3\overset{+}{\text{N}}\text{CH}_2\text{COOH}]} \qquad (2.19\text{a})$$

An ampholyte's isoelectric point is the pH at which average charge, for all forms of the molecule, is zero.

and

$$\text{pI} = \text{p}K_{\overset{+}{\text{N}}\text{H}_3} + \log \frac{[\text{H}_2\text{NCH}_2\text{COO}^-]}{[\text{H}_3\overset{+}{\text{N}}\text{CH}_2\text{COO}^-]} \qquad (2.19\text{b})$$

Adding these equations (remembering that the sum of the logarithms of two quantities is the logarithm of their product) gives

$$2\ \text{pI} = \text{p}K_{\text{COOH}} + \text{p}K_{\overset{+}{\text{N}}\text{H}_3} + \log \frac{[\text{H}_2\text{NCH}_2\text{COO}^-]}{[\text{H}_3\overset{+}{\text{N}}\text{CH}_2\text{COOH}]} \qquad (2.20)$$

But because $[H_2NCH_2COO^-] = [H_3\overset{+}{N}CH_2COOH]$ at pI, the term to the far right is equal to log 1, or zero, so

$$\text{pI} = \frac{\text{p}K_{\text{COOH}} + \text{p}K_{\overset{+}{\text{N}}\text{H}_3}}{2} \qquad (2.21)$$

The result is simple in this case: For a molecule with only two ionizable groups, the pI is simply the average of the two pK_a's. If we insert the values given on page 45, we obtain pI = 5.95 for glycine. Actually, as Figure 2.19 shows, glycine will be almost entirely in the zwitterion form, with zero charge, from about pH 4 to about pH 8.

Large molecules such as proteins can have *many* acidic and basic groups. Such molecules are called **polyampholytes.** With more than two charged groups present, the calculation of pI becomes more complicated. However, as long as the molecule has both positively and negatively charged groups, it always has an isoelectric point, at which the net average charge is zero. If acidic groups predominate, the pI will be low; if basic groups predominate, the pI will be high. In Chapter 5 we find that this is an important consideration in dealing with solutions of proteins.

We can determine the pI of an ampholyte or a polyampholyte experimentally,

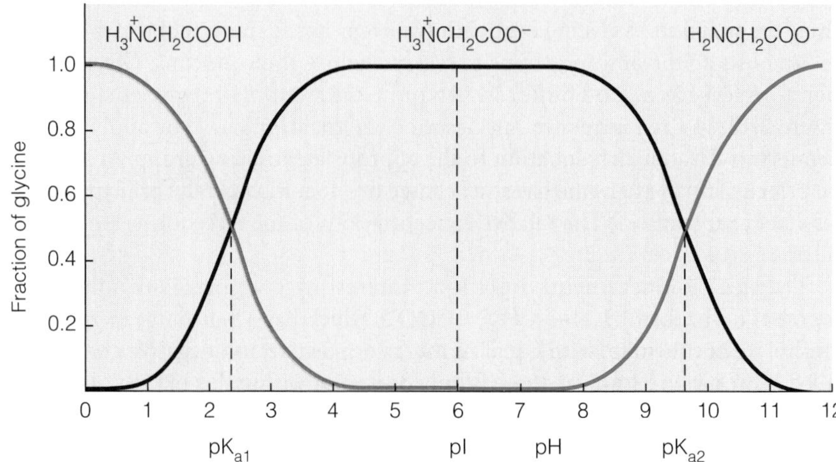

FIGURE **2.19**

The relative concentrations of the three forms of glycine as a function of pH. The three forms are $H_3\overset{+}{N}CH_2COOH$, shown in red; $H_3\overset{+}{N}CH_2COO^-$, black; and $H_2NCH_2COO^-$, blue. The two pK_a values and the isoelectric point (pI) are indicated.

using the simple method of *electrophoresis.* When an electric field is applied to a solution of such ions, positively charged molecules migrate toward the cathode and negatively charged ones toward the anode. At its isoelectric point, an ampholyte moves in *neither* direction, because it has zero net charge. In the method called **isoelectric focusing,** ampholytes move through a pH gradient, each coming to rest at its own isoelectric point. In this way, the ampholytes are separated and their isoelectric points determined (see Tools of Biochemistry 2A).

Some macromolecules, called **polyelectrolytes,** carry multiples of only positive or only negative charge. Strong polyelectrolytes, like the negatively charged nucleic acids (see Chapter 4), are ionized over a wide pH range. In addition, there are weak polyelectrolytes, like polylysine, a polymer of the amino acid lysine:

Polylysine

When a number of weakly ionizing groups are carried on the same molecule, the pK_a of each group is influenced by the state of ionization of the others. In a molecule like polylysine, the first protons are more easily removed than the last, because the strong positive charge on the fully protonated molecule helps drive protons away. Conversely, a molecule that develops a strong negative charge as protons are removed gives up the last ones only with difficulty, as their pK_a values become unusually high.

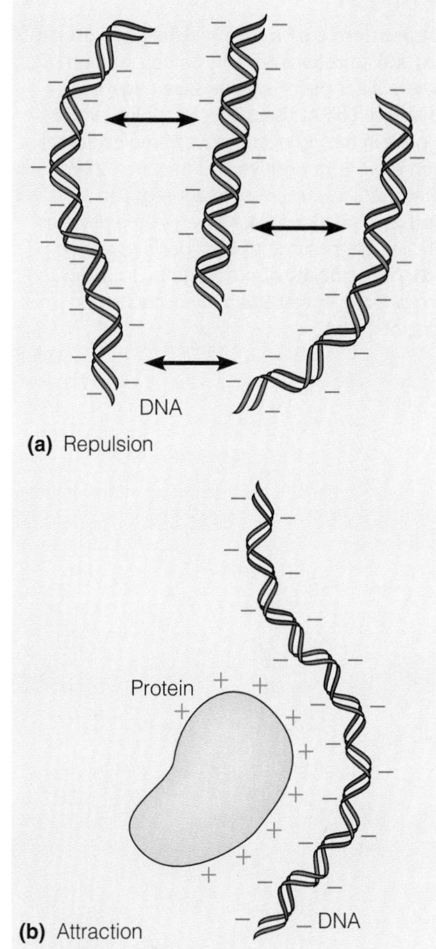

FIGURE **2.20**

Electrostatic interactions between macroions. (a) Repulsion. DNA molecules, with many negative charges, strongly repel one another in solution. **(b)** Attraction. If DNA is mixed with a positively charged protein, these molecules have a strong tendency to associate.

Interactions between Macroions in Solution

Large polyelectrolytes such as nucleic acids and polyampholytes such as proteins are classed together as **macroions.** Depending on the solution pH, they may carry a substantial net charge. The electrostatic forces of attraction or repulsion between such charged particles play a major role in determining their behavior in solution.

Interactions between macroions are greatly influenced by pH and the small ions in the solution.

SOLUBILITY OF MACROIONS AND pH

Because macroions of like net charge repel one another, nucleic acid molecules tend to remain separated in solution (Figure 2.20a). For the same reason, proteins tend to be soluble when they have a net charge, that is, at pH values above or below their isoelectric points. On the other hand, if positively and negatively charged macromolecules are mixed, electrostatic attraction makes them tend to associate with one another (Figure 2.20b). Many proteins interact strongly with DNA; most of these turn out to be positively charged (see Figure 2.20b). A striking example is found in the chromosomes of higher organisms, in which the negatively charged DNA is strongly associated with positively charged proteins called *histones* to form the complex called chromatin (discussed in Chapter 28).

A more subtle type of electrostatic interaction may cause the molecules of a particular protein to self-associate at the isoelectric pH (Figure 2.21). For example, the common milk protein β-lactoglobulin is a polyampholyte with an isoelectric point of about 5.3. Above or below this pH, the molecules all have either negative or positive charges and repel one another. This protein is therefore very soluble at either acidic or basic pH. At the isoelectric point the net charge is zero, but each molecule still carries surface patches of both positive and negative charge. The charge–charge interactions, together with other kinds of intermolecular interactions such as van der Waals forces, make the molecules tend to clump

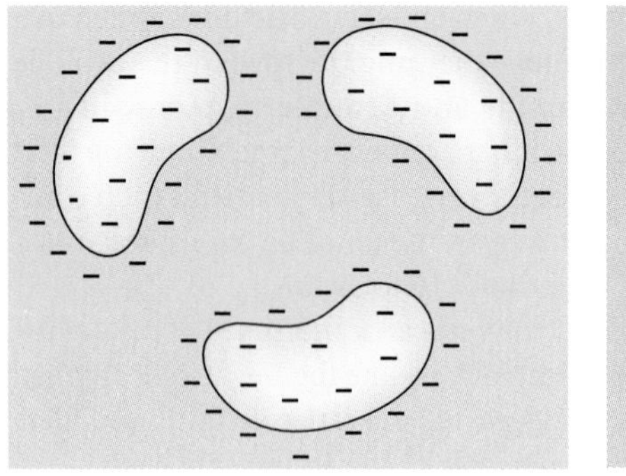

(a) High pH: protein soluble (deprotonated)

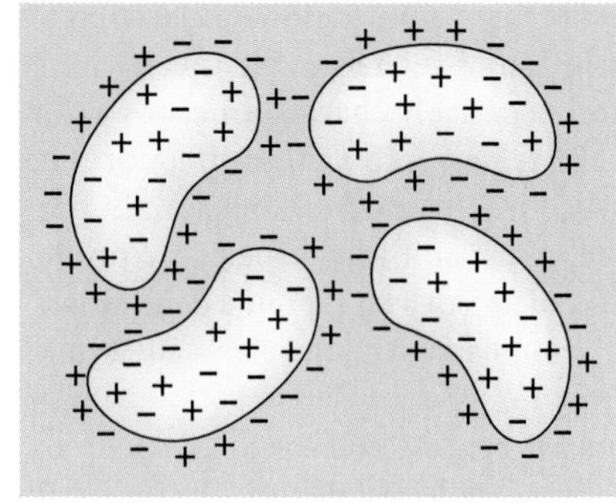

(b) Isoelectric point: protein aggregates

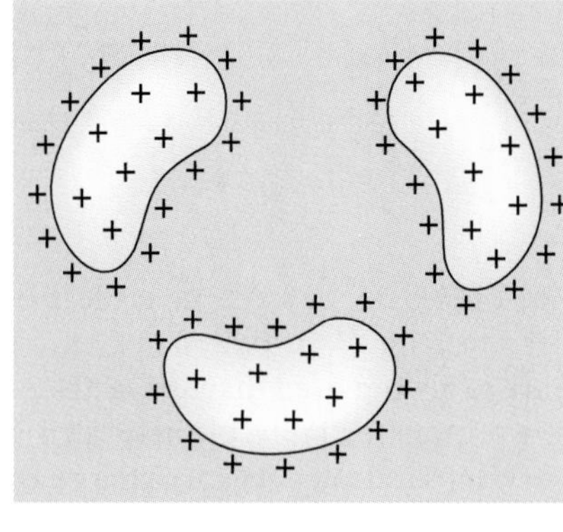

(c) Low pH: protein soluble (protonated)

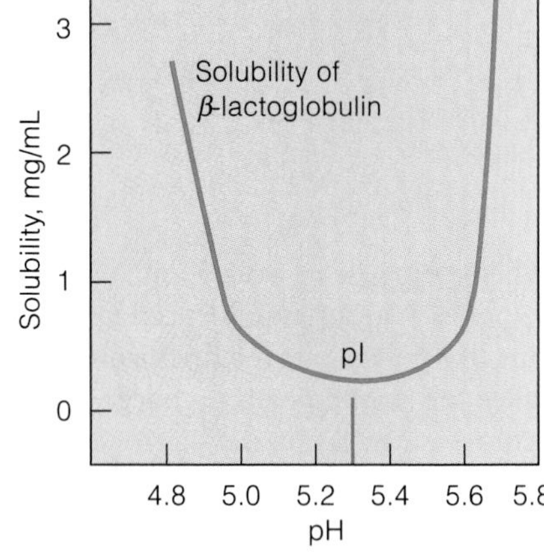

(d) Solubility of β-lactoglobulin

FIGURE **2.21**

Dependence of protein solubility on pH. **(a)** Most proteins are very soluble at high pH, where all of their molecules are negatively charged. **(b)** At the isoelectric point, where a protein has no *net* charge, its molecules retain regions of positive and negative charge on their surfaces, resulting in aggregation and precipitation. **(c)** At low pH the proteins are soluble because of their positive charge. **(d)** The solubility of β-lactoglobulin with varying pH; the lowest solubility occurs at the isoelectric point.

together and precipitate. Therefore, β-lactoglobulin, like many other proteins, has minimum solubility at its isoelectric point (Figure 2.21d).

THE INFLUENCE OF SMALL IONS: IONIC STRENGTH

The interactions of macroions are strongly modified by the presence of small ions, such as those from salts dissolved in the same solution. Each macroion collects about it a **counterion atmosphere** enriched in oppositely charged small ions, and this cloud of ions tends to screen the molecules from one another (Figure 2.22a). Obviously, the larger the concentration of small ions present, the more effective this electrostatic screening will be; however, the precise relationship of screening to concentration is rather complex. A quantitative expression of the screening effect for spherical macroions, proposed by P. Debye and E. Hückel, is stated in terms of an effective radius (r) of the counterion atmosphere. This radius may be taken as a measure of the distance at which two macroions "sense" one another's presence. According to the Debye–Hückel theory,

$$r = \frac{K}{I^{1/2}} \tag{2.22}$$

where K is a constant that depends on the dielectric constant of the medium and the temperature, and I is a function of the concentration called the **ionic strength.** The ionic strength is defined as follows:

$$I = \tfrac{1}{2} \sum_i \mathrm{M}_i Z_i^2 \tag{2.23}$$

where the sum is taken over all small ions in the solution. For each ion type, M_i

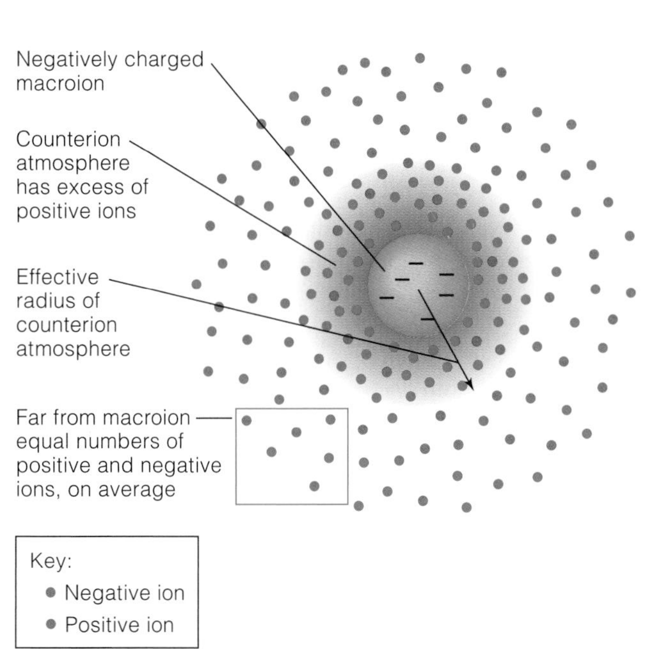

(a) The counterion atmosphere

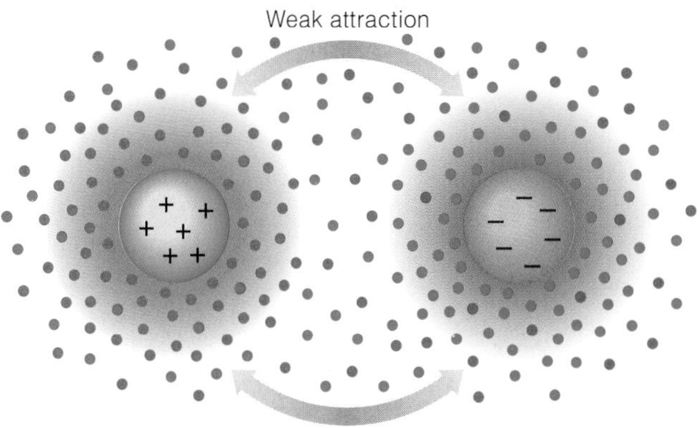

(b) The influence of ionic strength

FIGURE **2.22**

The influence of small ions on macroion interactions. **(a)** When a macroion (in this example, negatively charged) is placed in an aqueous salt solution, small ions of the opposite sign tend to cluster about it, forming a counterion atmosphere. There are more cations than anions near the macroanion shown here; far away from the macroion, the average concentrations of cations and ions are equal. **(b)** At low ionic strength, the counterion atmosphere is diffuse and interferes little with the interactions of the macroions. At high ionic strength, the counterion atmosphere is concentrated about the macroions and greatly reduces their interactions.

is its molarity and Z_i is its stoichiometric charge. For a 1:1 electrolyte like NaCl we have $Z_{Na^+} = +1$ and $Z_{Cl^-} = -1$, and since $M_{Na^+} = M_{Cl^-} = M_{NaCl}$, we find that $I_{NaCl} = M_{NaCl}$. Thus, ionic strength equals salt molarity for 1:1 electrolytes, but this is not true if multivalent ions (e.g., Mg^{2+} or SO_4^{2-}) are involved. Multivalent ions make greater individual contributions to the ion atmosphere than do monovalent ions, as reflected in the fact that the *square* of the ion charge is included in calculating the ionic strength. For these electrolytes, $I > M$.

The effects of the ionic strength of the medium on the interaction between charged macroions can be summarized as shown in Figure 2.22b. At very low ionic strength, the counterion atmosphere is highly expanded and diffuse, and screening is ineffective. In such a solution, macroions attract or repel one another strongly. If the ionic strength is increased, the counterion atmosphere shrinks and becomes concentrated about the macroion, and the attractive interactions between positive and negative groups are effectively screened.

The screening effect of the counterion atmosphere helps explain a general observation concerning protein solubility: Increasing ionic strength (up to a point) increases solubility even at the isoelectric point. This effect of putting proteins into solution by increasing the salt concentration is called **salting in.**

Raising the salt concentration to very high levels (several molar, for example) introduces another, opposite effect. In very concentrated salt solutions, much of the water that would normally solvate and help solubilize the protein molecule is bound up in the hydration shells of the numerous salt ions, preventing sufficient hydration of the protein. Thus, at extremely high salt concentration, the solubility of a protein again decreases, an effect called **salting out.** Because different proteins respond differently to these two effects, salting in and salting out are often used to purify proteins.

The effects of ionic interactions on the behavior of biological macromolecules mean that the biochemist must pay close attention to both ionic strength and pH. Experimenters usually use a neutral salt (like NaCl or KCl) to control the ionic strength of a solution, as well as a buffer to control the pH. In determining the amount of salt to add, they often try to mimic the ionic strengths of cell and body fluids. Although ionic strengths vary from one cell type or fluid to another, a value of 0.1 to 0.2 M is often appropriate in biochemical experiments.

SUMMARY

A variety of weak, noncovalent interactions occur among ions, molecules, and parts of molecules in the cell. These interactions, which are 10 to 100 times weaker than most covalent interactions, include charge–charge interactions and the interactions of permanent and induced dipoles. Molecules never interpenetrate, because attraction between them is countered by repulsion when their electron orbitals begin to overlap; a molecule's van der Waals radius is its distance of closest approach to another molecule. The hydrogen bond is the strongest noncovalent interaction, and it shares some features (directionality, specificity) with covalent bonds.

Water is the essential milieu of life. Most of the unique properties of water as a substance are accounted for by its polarity and hydrogen bonding, properties that also make it an excellent solvent. Polar, hydrogen-bonding, and ionic substances dissolve easily in water and are called hydrophilic, whereas other compounds dissolve in water to only a limited extent and are called hydrophobic. Amphipathic molecules, which have both polar and nonpolar parts, form distinctive structures such as monolayers, vesicles, and micelles when in contact with water. Such molecules form the membrane bilayers that surround cells and cellular compartments.

The ionization of weak acids and bases is of major importance in biochemistry. Most of the important processes occur in the pH range between 6.5 and 8.0, called the physiological range. The behavior of weak acids and their conjugate bases is described by the Henderson–Hasselbalch equation, which relates the conjugate base/undissociated acid ratio to pH and pK_a. Titration curves show that the pH change with added acid or base is most gradual in the range near the pK_a of the acid; this is the basis for preparation of buffer solutions.

An ampholyte has both acidic and basic ionizing groups; the molecules can have a net positive, zero, or negative charge, depending on solution pH. A polyampholyte has many acidic and basic groups. The isoelectric point of an ampholyte or a polyampholyte is the pH at which the average net charge of the molecules is zero. Polyelectrolytes have multiple ionizing groups with a single kind of charge. The behavior of macroions (polyampholytes and polyelectrolytes) in solution depends on pH and on the presence of small ions, which screen the macroions from each other's charges. The magnitude of screening depends on the ionic strength of the solution and is described quantitatively by the Debye–Hückel theory.

REFERENCES

Noncovalent Interactions

Burley, S. K., and G. A. Petsko (1988) Weakly polar interactions in proteins. *Adv. Protein Chem.* 39:125–189. Contains an excellent (though fairly sophisticated) treatment of weak interactions in general.

Eisenberg, D., and D. Crothers (1979) *Physical Chemistry with Applications to the Life Sciences.* Benjamin/Cummings, Redwood City, Calif. In addition to a thorough description of covalent bonding, Chapter 11 contains an excellent discussion of dipole moments, polarizability, and noncovalent interactions. Chapter 8 contains some useful material on electrolyte solutions, at a more advanced level than this book. This is, in our opinion, the best physical chemistry book for biochemists.

van Holde, K. E., W. C. Johnson, and P. S. Ho (1998) *Principles of Physical Biochemistry.* Prentice Hall, Upper Saddle River, N.J. Covers most of the topics in this chapter in considerably more depth.

Water

Eigen, M., and L. DeMaeyer (1959) Hydrogen bond structure, proton hydration, and proton transfer in aqueous solutions. In: *The Structure of Electrolyte Solutions,* edited by W. J. Hamer, pp. 64–85. Wiley, New York. Although not recent, this remains an excellent, interesting review.

Hagler, A. T., and J. Moult (1978) Computer simulation of solvent structure around biological macromolecules. *Nature* 272:222.

Kamb, B. (1968) Ice polymorphism and the structure of water. In: *Structural Chemistry and Molecular Biology,* edited by A. Rich and N. Davidson, pp. 507–542. Freeman, San Francisco. A review of theories of water structure.

Tanford, C. (1980) *The Hydrophobic Effect. Formation of Micelles and Biological Membranes.* Wiley, New York. A classic study of hydrophobicity.

Ionic Equilibria

Edsall, J. T., and J. Wyman (1958) *Biophysical Chemistry,* Vol. 1. Academic Press, New York. An excellent in-depth treatment. Here you can find extensive discussions of polyprotic dissociation, isoelectric points for polyampholytes, and so forth.

Williams, V. R., W. L. Mattice, and H. B. Williams (1978) *Basic Physical Chemistry for the Life Sciences.* Freeman, New York. Chapter 5 has a good discussion of pH measurement and buffers.

PROBLEMS

In this and following chapters, more difficult problems are indicated by an asterisk.

1. A chloride ion and a sodium ion are separated by a center–center distance of 0.5 nm. Calculate the interaction energy (the energy required to pull them infinitely far apart) if the medium between them is (a) water and (b) *n*-pentane (see Table 2.5). Express the energy in joules per mole of ion pairs. [Note: The charge on an electron is 1.602×10^{-19} C, Avogadro's number is 6.02×10^{23} molecules/mol.]

2. Rank the following in terms of expected dipole moment, and explain your choice:

H_2S, CCl_4, $H_3\overset{+}{N}—CH_2—COO^-$, $H_3\overset{+}{N}—CH_2—CH_2—CH_2—COO^-$

3. The accompanying graph depicts the interaction energy between two water molecules situated so that their dipole moments are parallel and pointing in the same direction. Sketch an approximate curve for the interaction between two water molecules oriented with *antiparallel* dipole moments.

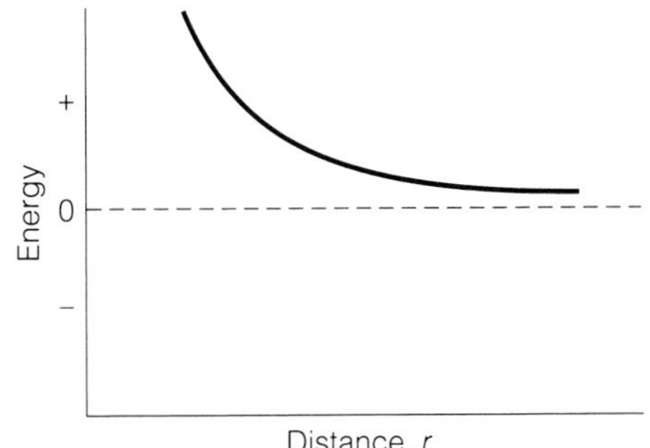

*4. (a) Solve the quadratic equation referred to in equation (2.13) for the more general case where the total concentration of acid ($A_0 = [HA] + [A^-]$) has any value A_0.
(b) One can avoid solution of the quadratic equation by using a method of successive approximations, starting with $x^2 \cong A_0K_a$. Explain how this would be done.

5. What is the pH of each of the following solutions?
(a) 0.35 M hydrochloric acid
(b) 0.35 M acetic acid
(c) 0.035 M acetic acid
[Note that the approximate method used in equation (2.14) will not give a good answer at low A_0.]

6. The weak acid HA is 2% ionized (dissociated) in a 0.20 M solution.
(a) What is K_a for this acid?
(b) What is the pH of this solution?

*7. Calculate the pH values and draw the titration curve for the titration of 500 mL of 0.010 M acetic acid (pK_a 4.76) with 0.010 M KOH.

8. What is the pH of the following buffer mixtures?
(a) 1 M acetic acid plus 0.5 M sodium acetate
(b) 0.3 M phosphoric acid plus 0.8 M KH_2PO_4

*9. (a) Suppose you wanted to make a buffer of exactly pH 7.00 using KH_2PO_4 and Na_2HPO_4. If you had a solution of 0.1 M KH_2PO_4, what concentration of Na_2HPO_4 would you need?
(b) Now assume you wish to make a buffer at the same pH, using the same substances, but want the total phosphate molarity ($[HPO_4^{2-}] + [H_2PO_4^-]$) to equal 0.3. What concentrations of the KH_2PO_4 and Na_2HPO_4 would you use?

10. A 500-mL sample of a 0.100 M formate buffer, pH 3.75, is treated with 5 mL of 1.00 M KOH. What is the pH following this addition?

11. You need to make a buffer whose pH is 7.0, and you can choose from the weak acids shown in Table 2.6 on page 40. Briefly explain your choice.

12. Describe the preparation of 2.00 L of 0.100 M glycine buffer, pH 9.0, from glycine (molecular weight in the zwitterionic form—75.07 g/mol) and 1.00 M NaOH. The appropriate pK_a of glycine is 9.6.

13. Carbon dioxide is dissolved in blood (pH 7.4) to form a mixture of carbonic acid and bicarbonate. Neglecting free CO_2, what fraction will be present as carbonic acid? Would you expect a significant amount of carbonate (CO_3^{2-})?

*14. The efficiency of a buffer in resisting changes in pH upon addition of base or acid is referred to as the *buffer capacity.* One way to define this quantity, which we shall call *B*, is $B = dx/d\text{pH}$, where x is the number of moles of base added to a weak acid. Note that the Henderson–Hasselbalch equation may be written as

$$\text{pH} = pK_a + \log \frac{x}{A_0 - x}$$

where A_0 is the total concentration of acid.
(a) Obtain an expression for *B*, in terms of A_0 and the fraction of acid titrated *f*, where $f = x/A_0$. [Hint: It is easiest to determine $d\text{pH}/dx$ and take the reciprocal.]
(b) At what value of *f* is buffer capacity maximal? (The easiest way to determine this is to make a graph of *B* versus *f*.)
(c) What is the maximal value of *B*?
(d) What is the effect of A_0 on *B*?

*15. The amino acid *arginine* ionizes according to the following scheme:

I $\underset{+H^+}{\overset{pK_a = 1.82,\ -H^+}{\rightleftharpoons}}$ II $\underset{+H^+}{\overset{pK_a = 8.99,\ -H^+}{\rightleftharpoons}}$ III $\underset{+H^+}{\overset{pK_a = 13.2,\ -H^+}{\rightleftharpoons}}$ IV

I: $H_3N^+-CH(COOH)-(CH_2)_3-NH-C(NH_2)=N^+H_2$

II: $H_3N^+-CH(COO^-)-(CH_2)_3-NH-C(NH_2)=N^+H_2$

III: $H_2N-CH(COO^-)-(CH_2)_3-NH-C(NH_2)=N^+H_2$

IV: $H_2N-CH(COO^-)-(CH_2)_3-NH-C(NH_2)=N-H$

Calculate the isoelectric point of arginine. You can neglect contributions from form I. Why?

16. It is possible to make a buffer that functions well near pH 7, using citric acid, which contains only carboxylate groups. Explain.

$$\begin{array}{c} CH_2-CO_2H \\ | \\ HO-C-CO_2H \\ | \\ CH_2-CO_2H \end{array}$$

Citric acid

17. A student is carrying out a biological preparation that requires 1 M NaCl to maintain an ionic strength of 1.0. The student chooses to use 1.0 M ammonium sulfate instead. Why is this a serious error?

*18. What is the optimum pH to separate a mixture of lysine, arginine, and cysteine using electrophoresis? Draw the structures of the three amino acids *in the protonation state that would predominate at the pH you have chosen.* See Table 5.1 on page 128. For each amino acid, indicate the net charge at the chosen pH as well as the direction of migration and relative mobility in the electric field.

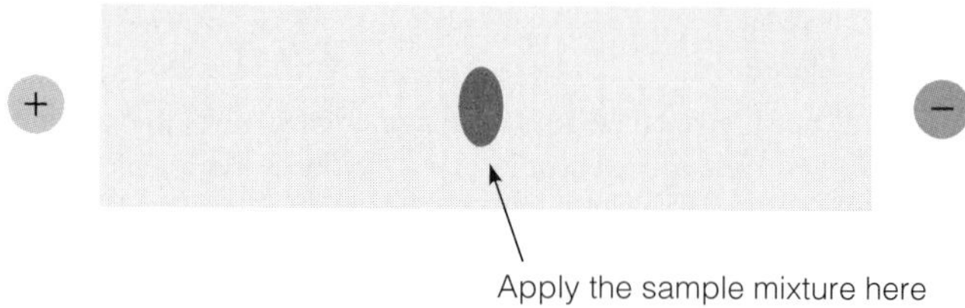

19. Suppose you have two genetic variants of a large protein that differ only in that one contains a histidine (side chain $pK_a = 6.4$) when the other has a valine (uncharged side chain).
(a) Which would be better for separation—gel electrophoresis or isoelectric focusing? Why?
(b) What pH would you choose for the separation?

TOOLS OF BIOCHEMISTRY 2A

Electrophoresis and Isoelectric Focusing

General Principles

When an electric field is applied to a solution, solute molecules with a net positive charge migrate toward the cathode, and molecules with a net negative charge move toward the anode. This migration is called **electrophoresis.** The velocity of the molecules depends on two factors. Driving the motion is the force $q\mathscr{E}$ exerted by the electric field on the particle, where q is the molecule's charge (in coulombs) and $\mathscr{E}$ is the electrical field strength (in volts per meter). Resisting the motion is the frictional force fv exerted on the particle by the medium, where v is the velocity of the particle and f is the **frictional coefficient,** which depends on the size and shape of the molecules. Large or asymmetric molecules encounter more frictional resistance than small or compact ones and consequently have larger frictional coefficients.

When the electric field is turned on, the molecule quickly accelerates to a velocity at which these forces balance and then moves steadily at this rate. The steady velocity is determined by the balance of forces:

$$fv = q\mathscr{E} \tag{2A.1}$$

We can rewrite this equation as $v/\mathscr{E} = q/f$ in order to express the rate of motion per unit of field strength, $v/\mathscr{E}$. This ratio is called the **electrophoretic mobility** (μ) of the molecule:

$$\mu = \frac{v}{\mathscr{E}} = \frac{q}{f} = \frac{Ze}{f} \tag{2A.2}$$

On the right-hand side of this equation, we have expressed the charge on the molecule as the product of the unit of electron (or proton) charge (e) times the number of unit charges, Z (a positive or negative integer). Because f depends on molecular size and shape, equation (2A.1) tells us that the mobility of a molecule depends on its charge and on the molecular dimensions.* Because ions and macroions differ in both respects, their behavior

*Equation (2A.2) is actually an approximation, for it neglects effects of the ion atmosphere. See van Holde, Johnson, and Ho (References) for more detail.

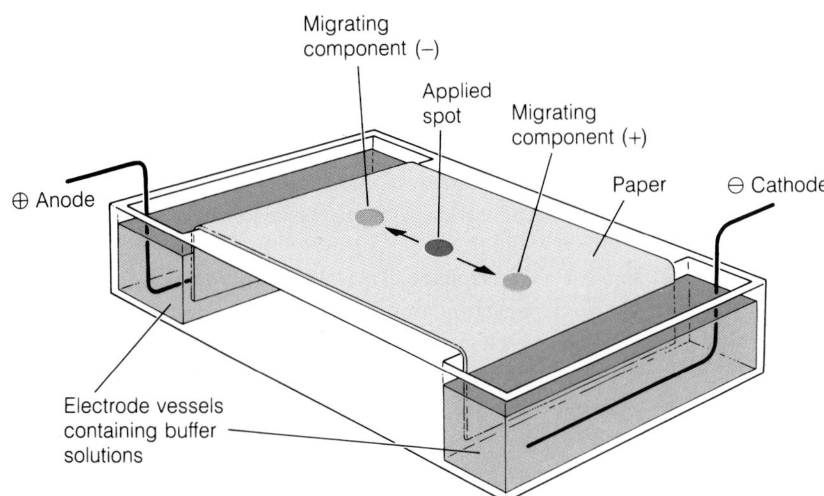

FIGURE **2A.1**
Paper electrophoresis.

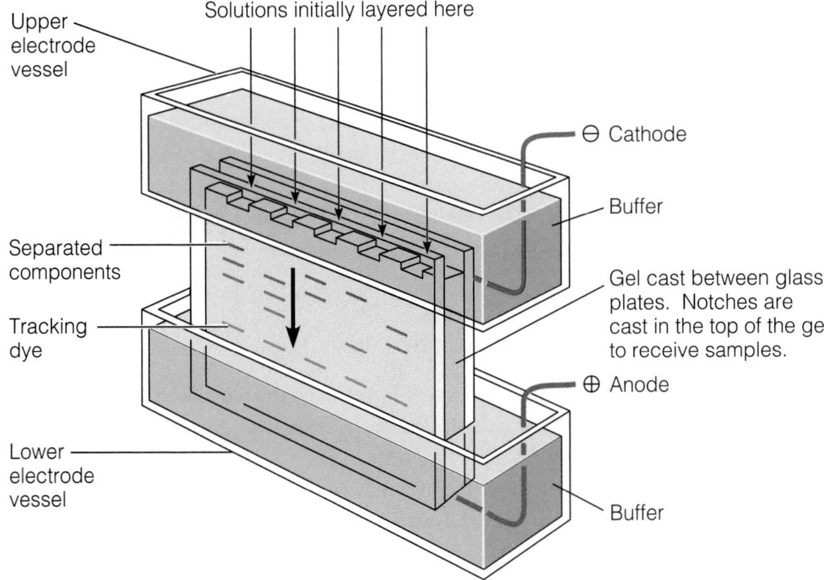

FIGURE **2A.2**
Gel electrophoresis.

in an electric field provides a powerful way of separating them. Electrophoretic separation is one of the most widely used methods in biochemistry.

Paper Electrophoresis and Gel Electrophoresis

Although electrophoresis can be carried out free in solution, it is more convenient to use some kind of *supporting medium.* The two most commonly used supporting media, paper and gel, are shown in Figures 2A.1 and 2A.2. **Paper electrophoresis** (Figure 2A.1) is often used for separating mixtures of small charged molecules. A piece of filter paper, moistened with a buffer solution to control the pH, is stretched between two electrode vessels. A drop of the mixture to be analyzed is placed on the paper, and the electric current is turned on. After the molecules have migrated for a sufficient time—usually several hours—the paper is removed, dried, and stained with a dye that colors the substances to be examined. Each kind of charged molecule in the mixture will have migrated a certain distance toward either the anode or the cathode, depending on its charge and dimensions, and will show up as a stained spot on the paper at the new position. Usually the spots can be identified by comparison with a set of standards run on the same paper. If the unknown substances are radioactive, the spots can be cut out and their radioactivity measured by scintillation counting (see Tools of Biochemistry 12A).

Gel electrophoresis (Figure 2A.2) is a technique much used with proteins and nucleic acids. In fact, it is probably the single most important method in molecular biology. A gel containing the appropriate buffer solution is cast as a thin slab between glass plates. Common gel-forming materials are polyacrylamide, a water-soluble, cross-linked polymer, and agarose, a polysaccharide. The slab is placed between electrode compartments, with the bottom selected as anode or cathode, depending on whether

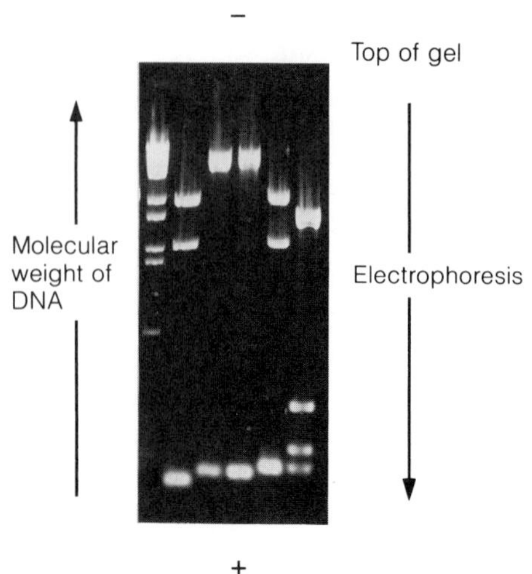

FIGURE **2A.3**
Gel showing separation of DNA fragments.
Courtesy of David Helfman.

anions or cations are being separated. A small amount of a solution of each sample is carefully pipetted into one of several precast notches on top of the gel. Usually glycerol and a water-soluble cationic or anionic "tracking" dye are added to the sample. The glycerol makes the sample solution dense, so that it does not mix into the buffer solution in the upper electrode chamber. The dye migrates faster than most macroions, so the experimenter is easily able to follow the progress of the experiment. The current is turned on until the tracking dye band is near the bottom of the slab. The gel is then removed from between the glass plates and is usually stained with a dye that binds to proteins or nucleic acids. At this point, a photograph is taken of the gel for a permanent record. Because the protein or nucleic acid mixture was applied as a narrow band at the top of the gel, components migrating with different mobilities appear as narrow bands on the gel, although the bands may be broadened somewhat by diffusion. Certain techniques (see Tools of Biochemistry References) make it possible to sharpen the bands even further, so that individual types of macroions appear as narrow lines on the gel; Figure 2A.3 shows an example of separation of DNA fragments by this method. The *relative mobility* of each component is calculated from the distance it has moved relative to the tracking dye.

Principles of Separation in Gel Electrophoresis

When electrophoresis is carried out in a gel or other supporting medium, the mobility is lower than would be expected from equation (2A.1) because the gel or other matrix exhibits a molecular sieving effect. This can be seen by graphing mobility as a function of the concentration of the gel (Figure 2A.4a). A graph of log μ versus percent gel is usually linear; this is called a **Ferguson plot.** The limiting mobility approached as percent gel approaches zero is called the **free mobility;** it should be given (approximately) by equation (2A.1). The steepness of the Ferguson plot depends on the size and shape of the macroion, for it reflects the difficulty the macroion experiences in passing through the molecular mesh of the gel.

As a result of these several factors, different kinds of molecules can exhibit widely different behaviors in gel electrophoresis (Figure 2A.4b). However, certain simple cases are of great importance. Polyelectrolytes like DNA or the polylysine molecule shown in the figure on page 47 have one unit charge on each residue, so each molecule has a charge (Ze) proportional to its molecular length. But the frictional coefficient (f) also increases with molecular length, so to a first approximation, a macroion whose charge is proportional to its length has a *free* mobility almost independent of its size. In a mixture of such molecules, the molecular sieving effect determines the relative mobilities at any given gel concentration (Figure 2A.4c), and the sieving effect is proportional to molecular length or molecular weight. This means that we can neatly separate molecules of this kind on the basis of *size alone* by gel electrophoresis, as shown in Figure 2A.3. For extended molecules like nucleic acids, the relative mobility is often approximately a linear function of the logarithm of the molecular weight (Figure 2A.4d). Usually, standards of known molecular weight are electrophoresed in one or more lanes on the gel. The molecular weight is then read from a graph like that in Figure 2A.4d prepared from the standards.

Isoelectric Focusing

Yet another gel electrophoresis technique allows separation of molecules purely on the basis of their charge characteristics. A polyampholyte will migrate in an electric field like other ions, if it has a net positive or negative charge. At its isoelectric point, however, its net charge is zero, and it is attracted to neither the anode nor the cathode. If we use a gel with a stable pH gradient covering a wide pH range, each polyampholyte molecule migrates to the position of its isoelectric point and accumulates there. We can establish such a gradient by using a mixture of low-molecular-weight ampholytes as the gel buffer. This method of separation, called isoelectric focusing, produces distinct bands of accumulated polyampholytes and can resolve molecules with very small differences in isoelectric point (Figure 2A.5). Since the pH of each portion of the gel is known, isoelectric focusing can also be used to determine the isoelectric point of a particular polyampholyte.

Other Methods

We discuss other electrophoretic techniques in later chapters and Tools sections. For example, in Tools of Biochemistry 5C, we describe the use of SDS gel electrophoresis for protein separation and determination of molecular weight. What we have presented here is only a brief overview of a widely applied technique. For further information, consult the references on page 56.

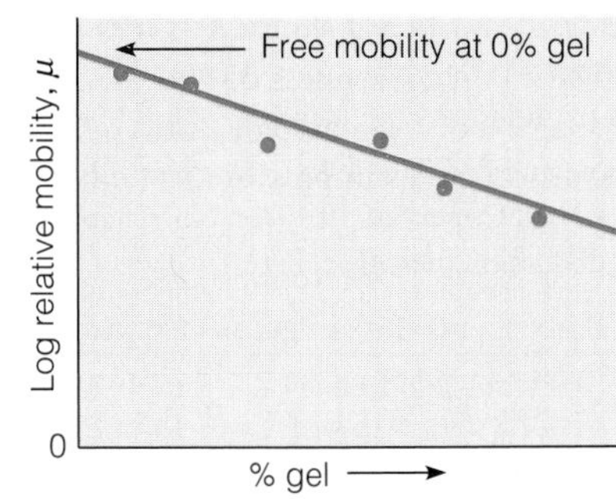

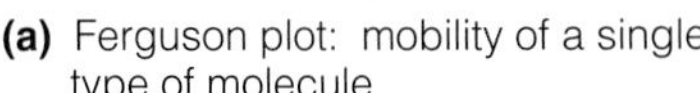
(a) Ferguson plot: mobility of a single type of molecule

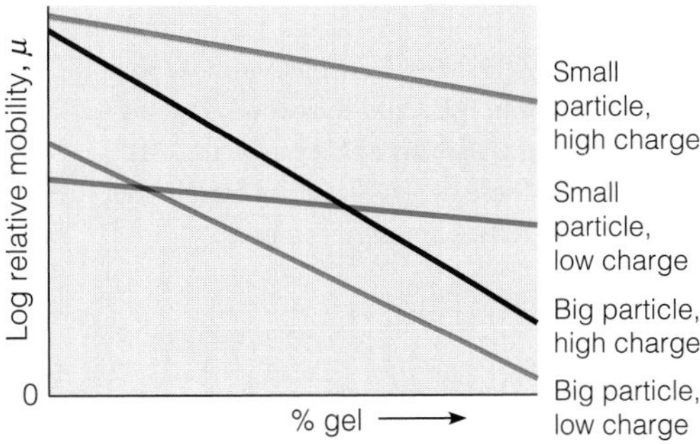

(b) Representative Ferguson plot for different kinds of molecules

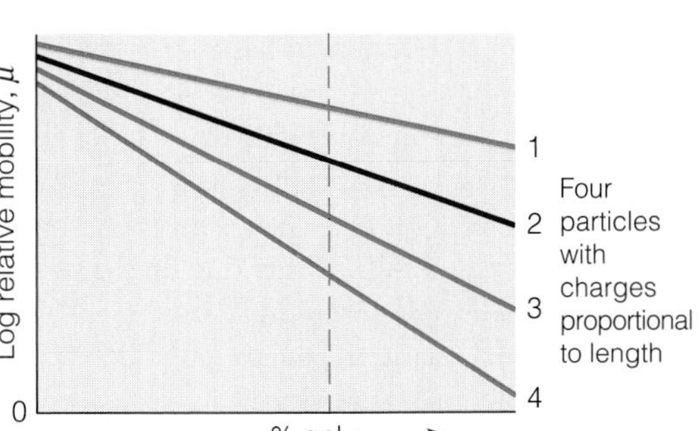

(c) Ferguson plot observed when charge is proportional to length

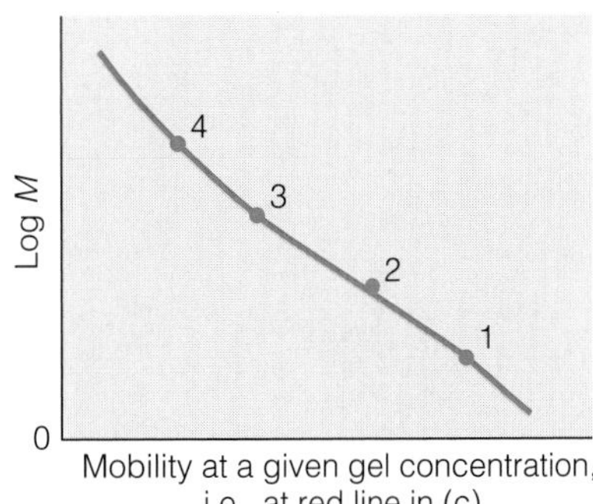

(d) Relationship between molecular weight (M) and mobility, at a given gel concentration for molecules like those shown in (c)

FIGURE **2A.4**

Mobility of the particles in gel electrophoresis. The mobility of the particles varies with the concentration of the gel. A Ferguson plot graphs the log of the relative mobility (μ) against the percent gel in the matrix. **(a)** Ferguson plot for a single type of molecule. Extending the plot to 0% gel gives the theoretical free mobility of the molecule. **(b)** Ferguson plot for four molecules of different size and charge. Note that free mobilities depend more on charge than on size but the slope depends mainly on size. **(c)** Ferguson plot for molecules with charges proportional to their lengths. The molecules are numbered in order of increasing length and charge. The free mobilities of such molecules are almost the same, but the longer the molecule, the more it is slowed by increasing gel concentration. **(d)** Plot of the relationship between molecular weight and mobility. The log of the molecular weight (M) of the four molecules shown in (c) is plotted against their mobilities at a single gel concentration. When this type of graph is prepared from standards, it can be used to determine the molecular weights of separated molecules.

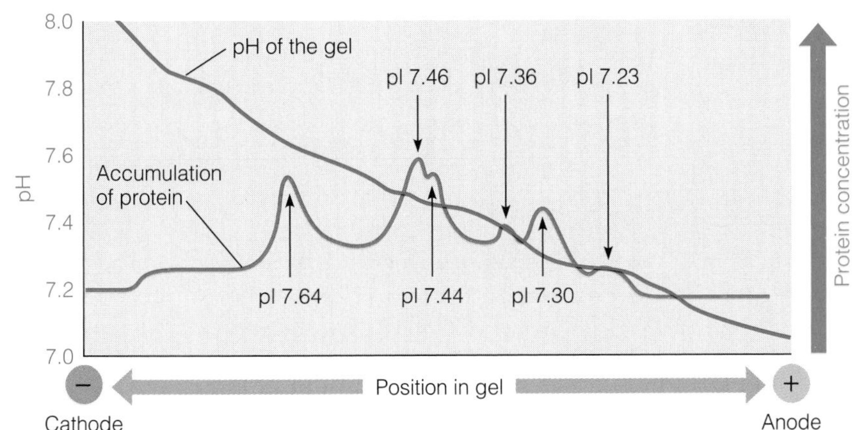

FIGURE **2A.5**

Isoelectric focusing of polyampholytes. A mixture of variants of the polyampholyte hemoglobin is placed on a gel with a pH gradient. When an electric field is applied, each variant protein migrates to its own isoelectric point.

References

Hames, B. D., and D. Rickwood, eds. (1981) *Gel Electrophoresis of Proteins.* IRL Press, Oxford, Washington, D.C., and Rickwood, D., and B. D. Hames, eds. (1982) *Gel Electrophoresis of Nucleic Acids.* IRL Press, Oxford, Washington, D.C. These two volumes are extremely useful laboratory manuals for gel electrophoresis techniques.

Osterman, L. A. (1984) *Methods of Protein and Nucleic Acids Research,* Vol. 1, Parts 1 and 2. Springer-Verlag, New York. A comprehensive summary of electrophoresis and isoelectric focusing.

van Holde, K. E., W. C. Johnson, and P.-S. Ho (1998) *Principles of Physical Biochemistry.* Prentice Hall, Upper Saddle River, N.J. Chapter 5 contains a more detailed discussion than given here.

CHAPTER 3

The Energetics of Life

A LIVING CELL IS A DYNAMIC STRUCTURE. IT GROWS, IT MOVES, IT SYNTHESIZES complex macromolecules, and it selectively shuttles substances in and out and between compartments. All of this activity requires energy, so every cell and every organism must obtain energy from its surroundings and expend it as efficiently as possible. Plants gather energy from sunlight; animals use the energy stored in plants or other animals that they consume. The processing of this energy, tapping as much of it as possible to do the things a cell or organism needs to do, is what much of biochemistry is about. Most of the elegant molecular machinery that exists in every cell is dedicated to this task.

Because of the central role of energy in life, it is appropriate that we begin a study of biochemistry with an introduction to **bioenergetics**—the quantitative analysis of how organisms gain and use energy. Bioenergetics may be regarded as a special part of the general science of energy transformations, which is called **thermodynamics.** In this chapter we shall review just a bit of that field, choosing only concepts that are important to the biochemist or biologist.

Bioenergetics describes how organisms gain, channel, and use energy.

Energy, Heat, and Work

A word we shall often use in our discussion is **system.** In this context, a system is any part of the universe that we choose for study. It can be a single bacterial cell, a Petri dish containing nutrient and millions of cells, the whole laboratory in which this dish rests, the earth, or the entire universe. A system must have defined boundaries, but otherwise there are few restrictions. The system may be *isolated,* and thus unable to exchange energy and matter with its surroundings; it may be *closed,* able to exchange energy but not matter; or it may be *open,* so that both energy and matter can pass in and out.

INTERNAL ENERGY AND THE STATE OF A SYSTEM

Any system contains a certain amount of **internal energy,** which we denote by *E.* It is important for our understanding to be specific as to what this internal energy includes. The system's atoms and molecules have kinetic energy of motion

The internal energy of a system includes all forms of energy that can be exchanged via simple (nonnuclear) physical processes or chemical reactions.

and energy of vibration and rotation. In addition, we include all of the energy stored in the chemical bonds between atoms and the energy of noncovalent interactions between molecules. We should include, in fact, any kind of energy that might be changed by chemical or nonnuclear physical processes. We need not include energy stored in the atomic nucleus, for this is unchanged in any chemical or biochemical reaction. The internal energy is a function of the **state** of a system. The thermodynamic state is defined by prescribing the amounts of all substances present and any two of the following three variables: the temperature (T), the pressure on the system (P), and the volume of the system (V). It is essentially a recipe for producing the system in a defined way. For example, a system composed of 1 mole of O_2 gas in 1 liter at 273 K has a defined state and therefore a definite internal energy value. This value is independent of any past history of the system.

Unless a system is isolated, it can exchange energy with its surroundings and thereby change its internal energy; we define this change as ΔE. For a closed system, this exchange can happen in only two ways. First, **heat** may be transferred to or from the system. Second, the system may do **work** on its surroundings or have work done on it. Work can take many forms. It may include expansion of the system against an external pressure such as expansion of the lungs, electrical work such as that done by a battery or required for the pumping of ions across a membrane, expansion of a surface against surface tension, flexing of a flagellum to propel a protozoan, or lifting of a weight by contraction of a muscle. In all of these examples, a force is exerted against a resistance to produce a displacement, so work is done.

Note that heat and work are not *properties* of the system. They may be thought of as "energy in transit" between the system and its surroundings. Certain conventions have been adopted to describe these ways of exchanging energy:

1. We denote heat by the symbol q. A positive value of q indicates that heat is absorbed by the system from its surroundings. A negative value means that heat flows from the system to its surroundings.
2. We denote work by the symbol w. A positive value of w indicates that work is done by the system on its surroundings. A negative value means that the surroundings do work on the system.

All of this may seem excessively abstract, yet it bears the most direct relationship to the everyday functioning of our bodies. When we ingest a foodstuff like glucose, we metabolize it, ultimately oxidizing it to CO_2 and water. A defined energy change (ΔE) is associated with oxidizing a gram of glucose, and some of the released energy is available for our use. We expend some of this available energy in heat generation (to maintain body temperature, for example) and some in various kinds of work. These latter kinds of work include not only the obvious ones, like walking and breathing, but other more subtle kinds—sending impulses along nerves, pumping ions across membranes, and so forth.

THE FIRST LAW OF THERMODYNAMICS

A system's internal energy can change only by the exchange of heat or work with the surroundings. This is the first law of thermodynamics.

Because the internal energy of a system can change only by heat or work exchanges with the surroundings, the *change in internal energy* must be given by

$$\Delta E = q - w \tag{3.1}$$

This equation, which holds for all processes, expresses the **first law of thermodynamics.** This first law is simply a bookkeeping rule, a statement of the conservation of energy. When a physical process or a chemical reaction has occurred, we can total up the incomes and expenditures of energy, and the books must balance. Energy can be gained and spent in different ways, but at least in chemical processes, it cannot be created or destroyed. Consider, for example, some process

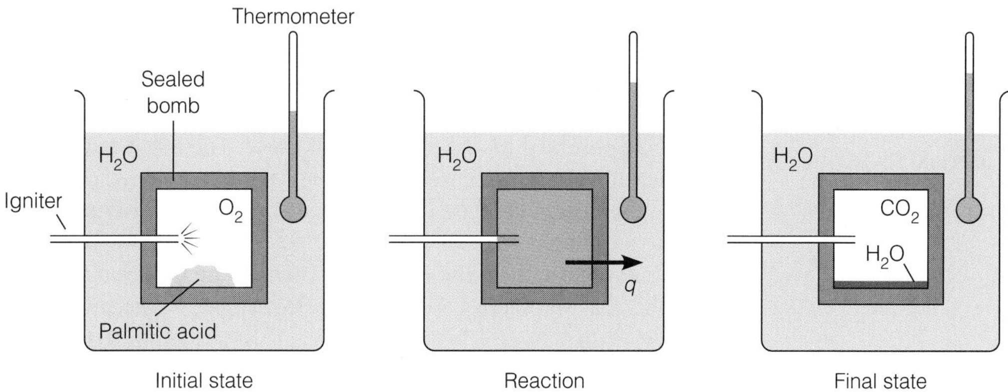

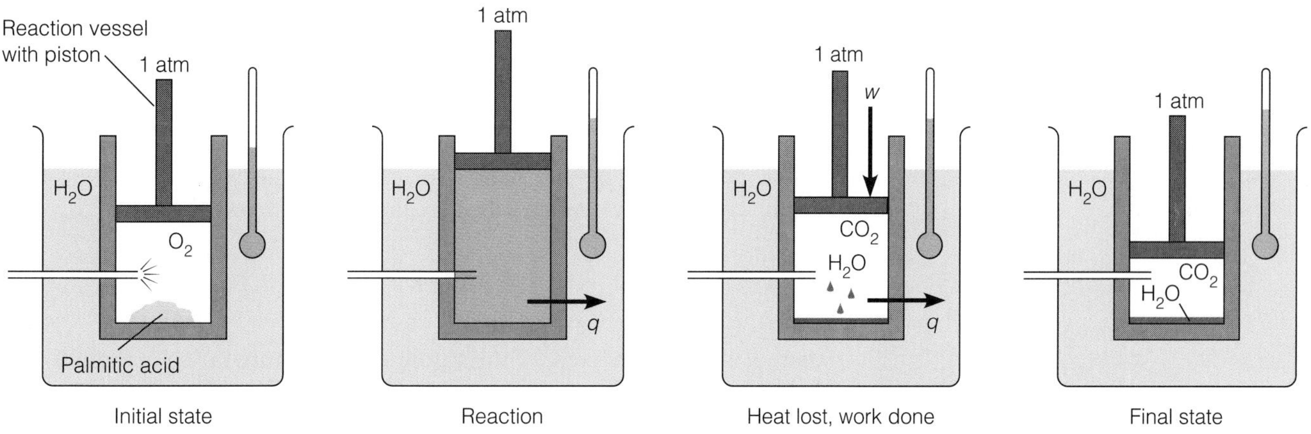

FIGURE **3.1**

Exchange of heat and work in constant-volume and constant-pressure reactions. A single reaction, oxidation of 1 mole of a fatty acid, is carried out under two sets of conditions. **(a)** The reaction occurs in a sealed vessel, or "bomb." Heat (q) is transferred to the surrounding water bath and is measured by the small increase in temperature of the water. No work is done, because the system is at constant volume. **(b)** The reaction vessel is fitted with a piston held at 1 atm pressure. During the reaction, heating of the gas in the vessel causes the piston to be pushed up. However, the reaction results in a decrease in the number of moles of gas, so after the vessel and gas have cooled to the water temperature, the volume of the gas is smaller than the initial volume. Thus, net work is done *on* the system, and the total amount of heat delivered to the bath is slightly more than in (a).

in which a certain amount of heat is absorbed by a system, while the system does an exactly equivalent amount of work on its surroundings. In this case, both q and w are positive, and $q = w$, so $\Delta E = 0$. This agrees with common sense: If so much energy went in as heat and an equal quantity came out as work, then the energy within the system must be unchanged.

Changes in internal energy, as for any function of state, depend only on the initial and final states of a system and are independent of the path. But the amounts of heat and work exchanged in any process depend very much on the conditions we impose. To make this idea concrete, let us consider a specific chemical reaction—the complete oxidation of 1 mole of a fatty acid, palmitic acid:

$$CH_3(CH_2)_{14}COOH\ (\text{solid}) + 23O_2\ (\text{gas}) \longrightarrow 16CO_2\ (\text{gas}) + 16H_2O\ (\text{liquid})$$

The oxidation of palmitic acid is in fact an important biochemical reaction that takes place, in a much more indirect way, in our bodies when we metabolize fats. We shall consider running this reaction in two different ways, as shown in Figure 3.1. In Figure 3.1a the reaction is carried out by igniting the mixture in a sealed vessel (a "bomb" calorimeter) immersed in a water bath. The reaction, under these conditions, is being carried out at constant volume. We can measure the heat passed from the reaction vessel (the system) to the water bath (the surroundings) by the temperature change in the bath, knowing the mass of water and the heat capacity (per gram) of water. Because the reaction vessel has a fixed volume, no work has been done against the surroundings or by the surroundings. Therefore, from equation (3.1),

$$\Delta E = q \tag{3.2}$$

The heat evolved in a reaction at constant volume is equal to ΔE.

The total heat that is transferred from the reaction vessel to the surroundings just equals the change in internal energy, and that energy change results mainly from the changes in chemical bonding that occurred during the reaction. The presently accepted unit for heat, work, and energy is the **joule (J).*** For the above reaction, the value observed for ΔE is -9941.4 kJ/mol. The negative sign corresponds to the fact that the reaction releases energy stored in chemical bonds. The energy within the system *decreased* as this bond energy was transferred as heat out into the surroundings.

Now suppose the same reaction is carried out at a constant pressure of 1 atmosphere, as shown in Figure 3.1b. In this case, the system is free to either expand or contract, and it finally contracts by an amount proportional to the decrease in the number of moles of gas, which went from 23 to 16 moles during the reaction. (We neglect the relatively tiny volume of solids and liquids.) The decrease in gas volume means that a certain amount of work has been done by the surroundings on the system. This can be calculated in the following way.

When volume (V) is changed against a constant pressure (P),

$$w = P\,\Delta V \tag{3.3}$$

To calculate w, we may make an approximation. We assume that the initial and final temperatures of the system are essentially the same (say 25°C, or 298 K) and that the gases are ideal. We may then use the simple gas law, $PV = nRT$. This gives

$$\Delta V = \Delta n\,\frac{RT}{P} \tag{3.4}$$

where R is the gas constant, T the absolute temperature in kelvins, and Δn the change in number of moles of gas per mole of palmitic acid oxidized. Then, inserting (3.4) in (3.3), we obtain

$$w = \Delta n\,RT \tag{3.5}$$

Since we would like w in joules per mole, we use $R = 8.314$ J/K·mol in equation (3.5), resulting in $w = -17.3$ kJ/mol palmitate.

The heat evolved in this constant-pressure combustion will then be

$$\begin{aligned} q &= \Delta E + w = \Delta E + P\,\Delta V = \Delta E + \Delta n\,RT \\ &= -9941.4\text{ kJ/mol} - 17.3\text{ kJ/mol} = -9958.7\text{ kJ/mol} \end{aligned} \tag{3.6}$$

A slightly greater amount of heat is released to the surroundings under these *constant-pressure* conditions than under the *constant-volume* conditions of Figure 3.1a. The reason for this should be clear: Under constant-pressure conditions the surroundings can do work on the system, and this work (called *PV* work) reappears as extra heat released from the system to the surroundings.

Although the heat and work exchanged with the surroundings depend on path, it is important to remember that ΔE does not—it depends only on the initial and final states.

ENTHALPY

Most chemical reactions in the laboratory and virtually all biochemical processes occur under conditions more nearly approximating constant pressure than con-

*In the past, biochemists tended to express energy, heat, and work in *calories* or *kilocalories*. However, the International System of Units (SI units) *joules* and *kilojoules* are now replacing these. Conversion is easy: 1 cal = 4.184 J. Similarly, 1 kcal (kilocalorie, or 10^3 calories) = 4.184 kJ (kilojoules). A complication arises from the fact that the "calorie" referred to in dietetics is really a kilocalorie.

stant volume. If we are interested in the heat obtainable by oxidizing palmitic acid in an animal, then the heat evolved at constant pressure is what we want to know. As we showed in equation (3.6), this heat is not exactly equal to ΔE, because of the PV work done.

The heat evolved in a reaction at constant pressure is equal to the enthalpy, ΔH.

Therefore, to express the heat change in a constant-pressure reaction, we need another function of state. We define a new quantity, the **enthalpy,** which we give the symbol ***H:***

$$H = E + PV \tag{3.7}$$

Because E and PV are functions of state, H is also a function of state. The change ΔH depends only on the initial and final states of the process for which it is calculated. For reactions at constant pressure, ΔH is defined as follows:

$$\Delta H = \Delta E + P\,\Delta V \tag{3.8}$$

The value of ΔH is the same as the amount of heat (q) calculated in equation (3.6). In other words, *when the heat of a reaction is measured at constant pressure, it is ΔH that is determined.*

The energy changes you will find tabulated throughout this book and other books on biochemistry will almost always be given as ΔH values. That is most appropriate, for in vivo these reactions occur under nearly constant-pressure conditions. If a nutritionist wishes to know the energy available from the oxidation of palmitic acid in the body, for example, ΔH is the appropriate quantity.

The enthalpy change in a reaction is the energy change of most interest to biochemists.

Measuring changes in energy like ΔE and ΔH is of practical use to biochemists and dieticians simply because these quantities depend only on the initial and final states of the system. The oxidation of a fatty substance like palmitic acid occurs very differently in the human body than it does in a calorimeter like that shown in Figure 3.1. Nevertheless, the values of ΔE and ΔH for the process are *exactly* the same in both pathways, because a quantity like ΔE or ΔH depends only on the final and initial states. Thus, the calorimeter provides an exact measurement of the energy available to a human from each gram of palmitic acid.

The average human requires the expenditure of about 6000 kJ (roughly 1500 kcal or 1500 of the "large calories" used in dietetics) just to sustain bodily functions. With moderate exercise, this need may easily double.

Although we have pointed out the distinction between ΔE and ΔH, we should emphasize that for most biochemical reactions the quantitative difference between them is of little consequence. Most of these reactions occur in solution and do not involve the consumption or formation of gases. The volume changes are thus exceedingly small, and $P\,\Delta V$ is a tiny quantity relative to ΔE or ΔH. Even for the example given, the oxidation of palmitic acid, the difference between ΔH and ΔE is only 0.2%. Thus, we are justified in most cases in thinking of ΔH as a direct measure of the energy change in a process, and we commonly refer to ΔH as the energy change.

Entropy and the Second Law of Thermodynamics

THE DIRECTION OF PROCESSES

However useful the first law may be for keeping track of energy changes in processes, it cannot give us one very important piece of information: What is the *favored direction* for a process? It cannot answer questions like these:

> We place an ice cube in a glass of water at room temperature. It melts. Why doesn't the rest of the water freeze instead?
>
> We place an ice cube in a jar of carefully supercooled water. The whole freezes. Why?

We touch a lit match to a piece of paper. The paper burns to carbon dioxide and water. Why can't we mix carbon dioxide and water to form paper?

Reversible processes occur always near a state of equilibrium; irreversible processes drive toward equilibrium.

One characteristic of such processes is their *irreversibility* under the given conditions. An ice cube in a glass of room-temperature water will continue to melt—there is no way to turn that process around without major changes in the conditions. But there *is* a *reversible* way to melt ice—to have it in contact with water at 0°C. Then, adding a bit of heat will melt a bit, taking away a little heat will cause a bit of the water to freeze. A **reversible** process like melting ice at 0°C is always almost at equilibrium. The **irreversible** processes we just described happen when systems are set up far from an equilibrium state. They then drive *toward* a state of equilibrium.

An irreversible process is also often called a spontaneous process, but we prefer the word *favorable.* The word *spontaneous* tends to imply, perhaps falsely, that the process is rapid. Thermodynamics has nothing to say about how fast processes will be, but it can indicate which direction is favored. The melting of ice, not freezing, is favored at 25°C and 1 atm pressure. Here, the result accords with intuition; you would not expect the ice cube to grow, or even remain unmelted, when placed in 25°C water. Either result would be most definitely an unfavorable process.

Knowing whether a process is reversible, favorable, or unfavorable is vital to bioenergetics. This information can be expressed most succinctly by the *second law of thermodynamics,* which tells us which processes are thermodynamically favorable. In order to present the second law, we must consider a new concept—entropy.

ENTROPY

Why do chemical and physical processes have thermodynamically favored directions? A first guess at an explanation might be that systems simply go toward a lowest-energy state. Water runs downhill, losing energy as it spontaneously falls in the earth's gravitational field; the oxidation of palmitic acid, like the burning of paper, releases energy as heat. Certainly, energy minimization is the key to the favored direction for *some* processes. But such an explanation cannot account for the melting of ice at 25°C; in fact, energy is *absorbed* in that process. Another, very different factor must be at work, and a simple experiment gives a clear indication of what this factor may be. If we carefully layer pure water on top of a sucrose solution, we will observe as time passes that the whole system becomes more and more uniform (Figure 3.2). Eventually the sucrose molecules will be evenly distributed throughout the container. Though there is practically no energy change, the process is clearly a favorable one. We know from experience that the opposite process (self-segregation of the sucrose molecules into a portion of the solution volume) never occurs. What is clearly important here is that *systems of molecules have a natural tendency to randomization.*

FIGURE **3.2**

Diffusion as an entropy-driven process. The gradual mixing of a dilute sucrose solution and pure water is the result of random movement of their molecules. We can visualize the increase in entropy if we imagine the volume of the two liquids to be made up of cells, each big enough to hold one sucrose molecule. **(a)** Initially, the sucrose solution is at equilibrium, because its N_A molecules are distributed randomly into its N_I cells. **(b)** When a layer of pure water is added without mixing, the system is no longer at equilibrium. It has become more ordered, with all the occupied cells located in one-half of the solution. **(c)** As sucrose and water molecules continue to move randomly, their arrangement becomes less ordered, because every cell has an equal chance of being occupied. Eventually, the solution reaches a new equilibrium, with sucrose molecules randomly distributed throughout. The drive toward equilibrium is a consequence of the tendency for entropy to increase. A system would never go spontaneously from state (c) to state (b).

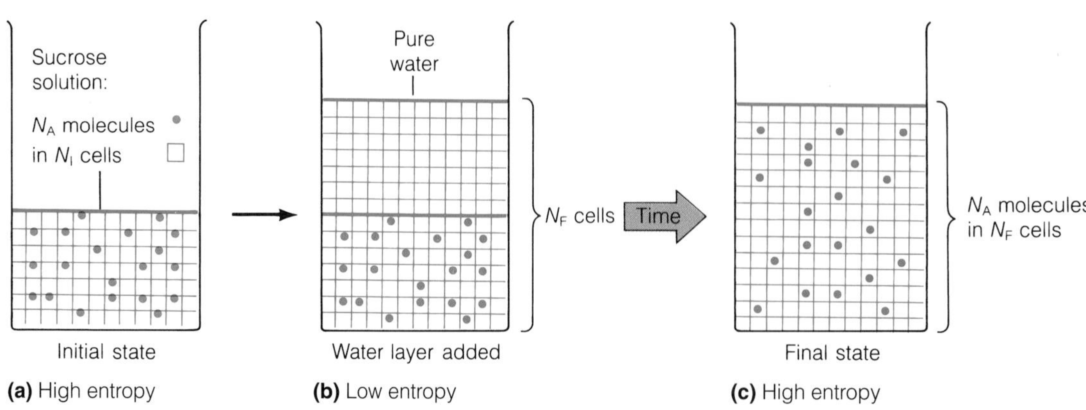

The degree of randomness or disorder of a system is measured by a function of state called the **entropy** (S). There are a number of ways of defining entropy, but the most useful for our applications depends on the fact that a given thermodynamic state may have many *substates* of equal energy. Those substates correspond, for example, to different ways in which molecules can be arranged or distributed within the system. If the thermodynamic state has a number (W) of substates of equal energy, the entropy (S) is defined as

$$S = k \ln W \tag{3.9}$$

where k is the **Boltzmann constant,** the gas constant R divided by Avogadro's number. A consequence of this definition is that entropy is seen to be a measure of disorderliness, for there will always be many more ways of putting a large number of molecules into a disorderly arrangement than into an orderly one. Therefore, the entropy of an ordered state is lower than that of a disordered state of the same system. In fact, the minimal value of entropy (zero) is found *only* for a perfect crystal at the absolute zero of temperature (0 K or −273°C). The process of diffusion evens out the concentrations in our sucrose solution simply because there are more ways to distribute molecules over a large volume than over a small one. To make the concept of entropy a bit more familiar, consider the examples given in Table 3.1.

Entropy is a measure of the randomness or disorder in a system.

TABLE **3.1 Examples of low-entropy and high-entropy states**

Low Entropy	High Entropy
Ice, at 0°C	Water, at 0°C
A diamond, at 0 K	Carbon vapor, at 1,000,000 K
A protein molecule in its regular, native structure	The same protein molecule in an unfolded, random coil state
A Shakespearean sonnet	A random string of letters
A bank manager's desk	A professor's desk

THE SECOND LAW OF THERMODYNAMICS

The preceding example shows that the driving force toward equilibrium for an isolated system is just the increase in entropy. This universal truth is stated as the **second law of thermodynamics:** *The entropy of an isolated system will tend to increase to a maximum value.* The entropy of such a system will not decrease—sucrose will never "de-diffuse" into a corner of the solution. This simply reflects our commonsense understanding that things, if left alone, will not get more orderly.

The second law of thermodynamics states that the entropy of an isolated system will tend to increase to a maximum value.

Free Energy: The Second Law in Open Systems

The form of the second law as stated in the previous section is not very useful to biologists or biochemists, because we never deal with isolated systems. Every biological system (cell, organism, or population, for example) is open to exchange energy and matter with its environment. Because living systems can exchange energy with their surroundings, both energy *and* entropy changes will take place in many reactions, and both must be of importance in determining the direction of thermodynamically favorable processes. For such systems we need *a function of state that includes both energy and entropy.* There are several such functions, but the one of importance in biochemistry is the **Gibbs free energy** (G) or, as we shall call it, the **free energy.** This function of state combines an enthalpy term, which measures the energy change at constant pressure, and an entropy term, which takes into account the importance of randomization. The Gibbs free energy is defined as

The free energy change for a process at constant temperature and pressure is $\Delta G = \Delta H - T\Delta S$.

$$G = H - TS \tag{3.10}$$

where T is the absolute temperature. For a free energy change ΔG in a system at constant temperature and pressure, we can write

$$\Delta G = \Delta H - T\Delta S \tag{3.11}$$

We can gain an insight into the meaning of free energy by considering the factors that experience shows to make processes favorable. We said that a decrease

A thermodynamically favored process tends in the direction that minimizes free energy (results in a negative ΔG); this is one way of stating the second law of thermodynamics.

in energy (ΔH is negative) and/or an increase in entropy (ΔS is positive) are typical of favorable processes. Either of these conditions will tend to make ΔG negative. In fact, another way to state the second law of thermodynamics—the most important way for our purposes—is this: *The criterion for a favorable process in a nonisolated system, at constant temperature and pressure, is that ΔG be negative.* Conversely, a positive ΔG means that a process is *not* favorable, but the *reverse* of that process *is* favorable. Processes accompanied by negative free energy changes are said to be **exergonic;** those for which ΔG is positive are **endergonic.** You will see these terms frequently throughout this book.

Now suppose that the ΔH and $T\Delta S$ terms in the free energy equation just balance one another. In this case $\Delta G = 0$, and the process is not favored to go either forward or backward. In fact, the system is in equilibrium. Under these conditions, the process is reversible; that is, it can be displaced in either direction by an infinitesimal push one way or the other. These simple but important rules about free energy changes are summarized in Table 3.2.

TABLE **3.2 Free energy rules**

If ΔG is ...	The process is ...
Negative	Thermodynamically favored
Zero	Reversible; at equilibrium
Positive	Thermodynamically unfavored; reverse process is favored

AN EXAMPLE OF THE INTERPLAY OF ENTHALPY AND ENTROPY: THE TRANSITION BETWEEN LIQUID WATER AND ICE

To make these ideas more concrete, let us consider in detail a process we mentioned before, the transition between liquid water and ice. This familiar example demonstrates the interplay of enthalpy and entropy in determining the state of a system. In an ice crystal there is a maximum number of hydrogen bonds between the water molecules (see page 34 in Chapter 2). When ice melts, some of these bonds must be broken. The enthalpy difference between ice and water corresponds almost entirely to the energy that must be put into the system to break these hydrogen bonds. As Figure 3.3 shows, the enthalpy change for the transition ice $\longrightarrow$ water is positive, as would be expected from the preceding argument.

The entropy change in melting arises primarily because liquid water is a more random structure than ice. In an ice crystal, each water molecule has a fixed place in the lattice and binds to its neighbor in the same way as every other water molecule. On the other hand, molecules in liquid water are continually moving, ex-

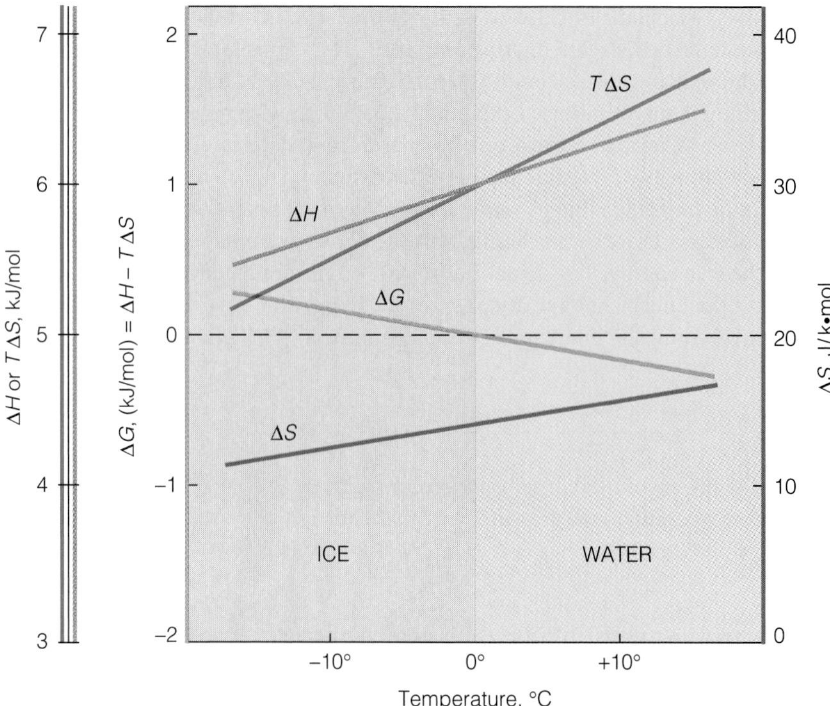

FIGURE **3.3**

Interplay of enthalpy and entropy in the ice-to-water transition. For the ice-to-water transition, ΔH and ΔS are both positive and approximately constant over a large temperature range. The increase in $T\Delta S$ with increasing temperature means that ΔG decreases from a positive to a negative value. At 0°C, the ΔH and $T\Delta S$ curves cross and ΔG is zero.

changing hydrogen-bond partners as they go (compare Figures 2.10a and 2.10b). Figure 3.3 shows that the entropy change for melting ice to water is a positive quantity, corresponding to the increase in randomness. If we calculate the free energy change ($\Delta G = \Delta H - T\Delta S$) for the ice $\longrightarrow$ water transition, we find the following: At low temperatures, ΔH dominates and ΔG is positive. For example, at $-10°C$ we find $\Delta G = +213$ J/mol. This means that the transition ice $\longrightarrow$ water is *not* favorable under these conditions. The opposite transition (water $\longrightarrow$ ice) *is* favorable and therefore irreversible, as we can see from the behavior of supercooled water. If we disturb supercooled water, or add a minute ice crystal to start the process of freezing, the entire sample will freeze. No infinitesimal change we can make will reverse this process.

At a temperature above 0°C, say +10°, an ice cube will irreversibly melt. Again, we would predict this outcome from the data, for if we calculate ΔG at +10°C, we obtain $\Delta G = -225$ J/mol. The sign of ΔG is now negative because the $-T\Delta S$ term dominates when T becomes large enough. At +10°C the process (ice $\longrightarrow$ water) is favorable and irreversible.

But note what happens at 0°C. At this temperature, the ΔH and $T\Delta S$ terms exactly balance and $\Delta G = 0$. A zero value for ΔG is the condition for equilibrium, and we know that ice and water are in equilibrium at 0°C. The change is now reversible; when ice and liquid water are together at 0°C, we can melt a bit more ice by adding an infinitesimal amount of heat. Alternatively, we can take a minute amount of heat away from the system and freeze a bit more water. The melting point of a substance is simply the temperature at which the curves for ΔH and $T\Delta S$ cross; at this temperature the energetically favored process of freezing is in balance with the entropically favored process of melting. Neither ΔH nor ΔS alone can tell us what will happen, but their combination, $\Delta H - T\Delta S$, prescribes exactly which form of water is stable at any temperature.

At the melting point of any substance, enthalpy and entropy contributions to ΔG balance, and $\Delta G = 0$.

THE INTERPLAY OF ENTHALPY AND ENTROPY: A SUMMARY

For all chemical and physical processes it is the competition of enthalpy and entropy terms that determines the favorable direction. As Figure 3.4 shows, in some processes the enthalpy change dominates; in others the entropy change is more important. Furthermore, because ΔS is multiplied by T in equation (3.11), the favorable direction will depend on the temperature. We have seen one example, the melting of ice, but quite different scenarios are possible depending on the signs of

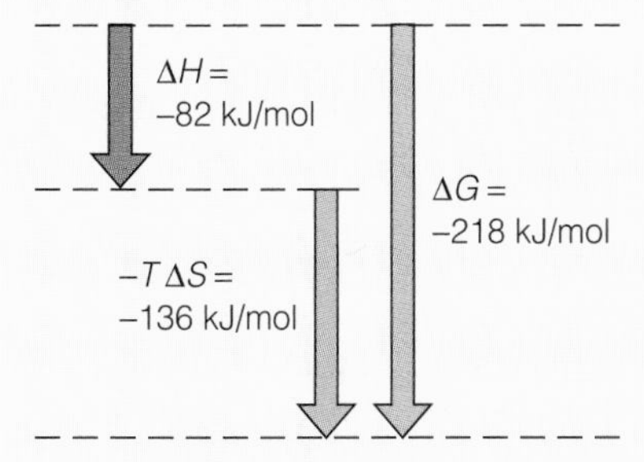

(a) Fermentation of glucose to ethanol

$C_6H_{12}O_6(s) \rightarrow 2C_2H_5OH(l) + 2CO_2(g)$

Both enthalpy and entropy changes favor the reaction.

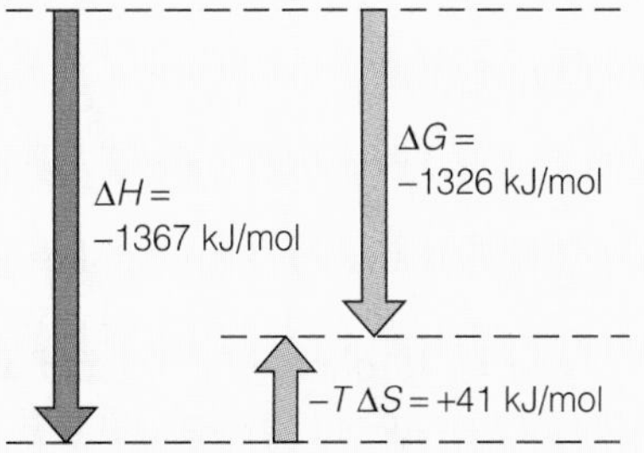

(b) Combustion of ethanol

$C_2H_5OH(l) + 3O_2(g) \rightarrow 2CO_2(g) + 3H_2O(l)$

Enthalpy favors this reaction, but entropy opposes it. We could call this an "enthalpy-driven" reaction. If water *vapor* were the product, an entropy increase would favor the reaction as well.

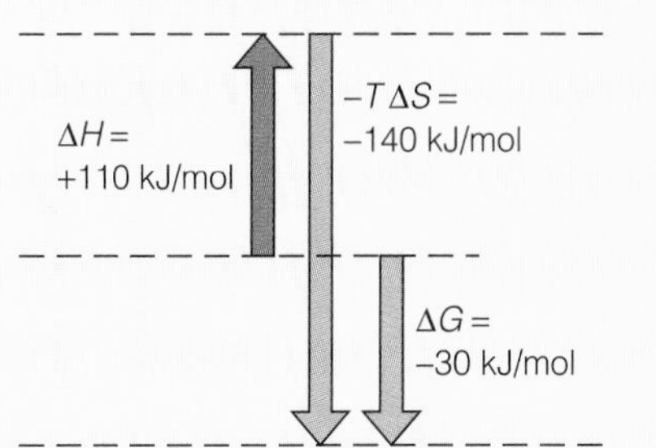

(c) Decomposition of nitrogen pentoxide

$N_2O_5(s) \longrightarrow 2NO_2(g) + 1/2O_2(g)$

This is a somewhat unusual chemical reaction in that it is "entropy-driven." The reaction actually absorbs heat but is favored by the large entropy increase resulting from the formation of gaseous products.

FIGURE **3.4**

Contribution of enthalpy and entropy to several processes. Each of these processes has a negative free energy change, but the change is accomplished in different ways. (Note that the arrows in the diagrams are not to scale.) **(a)** ΔH negative, $-T\Delta S$ negative. When glucose is fermented to ethanol, enthalpy decreases and entropy increases, so both the enthalpy and the entropy changes favor this reaction. **(b)** ΔH negative, $-T\Delta S$ positive. When ethanol burns, enthalpy and entropy both decrease. The negative ΔH favors this reaction, but the positive $-T\Delta S$ opposes it, so we could call this an enthalpy-driven reaction. If water *vapor* were the product, an entropy increase would favor the reaction because of the larger entropy of water vapor. **(c)** ΔH positive, $-T\Delta S$ negative. When nitrogen pentoxide decomposes, enthalpy and entropy both increase. This chemical reaction is somewhat unusual, in that it is entropy driven. Although the reaction absorbs heat, it is favored by the large entropy increase resulting from the formation of gaseous products.

TABLE **3.3 How the signs of ΔH and ΔS determine the effect of temperature on processes or reactions**

ΔH	ΔS	Low T	High T
+	+	ΔG positive; not favored	ΔG negative; favored
+	−	ΔG positive; not favored	ΔG positive; not favored
−	+	ΔG negative; favored	ΔG negative; favored
−	−	ΔG negative; favored	ΔG positive; not favored

ΔH and ΔS. Table 3.3 lists the possibilities. Note that when ΔH is negative and ΔS is positive, ΔG must always be negative, so the reaction is favored at all temperatures. The reverse is true when ΔH is positive and ΔS is negative; ΔG is always positive, and the reaction is not favored at any temperature.

Favorable processes are not necessarily rapid.

Two matters that frequently cause confusion should be cleared up at this point. First, we must emphasize a point we have mentioned already: The favorability of a process has nothing to do with its rate. Students frequently assume that favorable processes are rapid, but this is not necessarily so. A reaction may have a large negative free energy change but still proceed at a slow rate because of obstacles to the kinetic reaction process. A surprising example of this situation is the simple reaction C (diamond) $\longrightarrow$ C (graphite). The free energy change for this transformation, at room temperature, is −2.88 kJ/mol. Thus, diamond is unstable. Yet the reaction is imperceptibly slow, for it is very difficult for the rigid crystal lattice to change its form. Diamonds are *not* forever—just almost. A catalyst may increase the rate for some reactions, but the favored direction is always dictated by ΔG and is independent of whether the reaction is catalyzed or not. We will see later that the protein catalysts called enzymes select from among thermodynamically favorable reactions the particular ones that are to be catalyzed.

Life involves a temporary decrease in entropy, paid for by the expenditure of energy.

Second, the entropy of an *open* system can decrease. We have just seen that this happens whenever water freezes. More important to us in this course is that decreases in entropy happen all the time in living organisms. An organism takes in foodstuffs, often in the form of disorganized small molecules, and from them it builds enormous, complex, highly ordered macromolecules like proteins and nucleic acids. From these macromolecules, it constructs elegantly structured cells, tissues, and organs. All of this activity involves a tremendous entropy decrease. The implication of equation (3.11) is that entropy can decrease in a favored process, but only if this change is accompanied by a large enthalpy decrease. *Energy must be expended to pay the price of organization.* This exchange really is what life is all about. Living organisms spend energy to overcome entropy. For these life processes to proceed, the *overall* free energy changes in the organism must be negative. Life is an irreversible process. An organism that comes to equilibrium with its surroundings is dead.

There is an even deeper philosophical implication of bioenergetics. The universe as a whole is an isolated system. *The entropy of the whole universe must be increasing.* It follows that each of us, as a living organism that locally and temporarily decreases entropy, must produce somewhere in the world around us an increase in entropy. As we metabolize food, for example, we give off heat and increase random molecular motion around us. In a sense, we buy our lives through the entropic death of the universe.

FREE ENERGY AND USEFUL WORK

Why is the quantity G called *free* energy? The reason is that ΔG represents the portion of an energy change (ΔH) that is *available*, or free, to do useful work (that is, work beyond the $P\,\Delta V$ work of expansion). Knowing that ΔG measures the maximum amount of *useful* work that can be obtained from a chemical process is of great importance to biochemistry, because useful work includes the work involved in muscle contraction and cell motility, in transport of ions and molecules, in signal transmission, and in tissue growth, to give a few examples.

Recall that ΔH is the total energy change in a reaction, including the $P\,\Delta V$ work. The equation $\Delta G = \Delta H - T\Delta S$ indicates that a part of ΔH is always dissipated as heat, as expressed by the $T\Delta S$ term, and is therefore not available to do other things. Whatever the process may be, at least the amount of energy represented by $T\Delta S$ must be unavailable. The remainder, ΔG, is potentially available for other needs, although how much of it is actually used for work depends on the path of the process. The **efficiency** of a biochemical process is defined as the ratio of the work actually accomplished to the maximum work expected from the change in free energy.

The free energy change, ΔG, is a measure of the maximum useful work obtainable from any reaction.

We can use the concept of ΔG as a measure of useful work to show how the free energy change in an oxidation–reduction reaction is related to the voltage developed by the (real or hypothetical) electrochemical cell in which that reaction occurs. The work done in moving Q coulombs of charge against a negative potential ($-E$ volts) is just $w = -QE$. If the reaction moves n moles of electrons, the result will be

$$\Delta G = -nFE \tag{3.12}$$

where F is the number of coulombs per mole of electrons, 96,485 C/mol, or J/V·mol. If the reaction is carried out under standard state conditions (1 M for reactants and products; see below) the equation relates $\Delta G°$ for the standard state reduction potential, E_0:

$$\Delta G° = -nFE_0 \tag{3.13}$$

In later chapters, we shall frequently talk about oxidation–reduction reaction in terms of E_0 values, which are easily converted to $\Delta G°$.

Free Energy and Concentration

The sign of the free energy change in a process tells us whether that process or its reverse is thermodynamically favorable. The magnitude of ΔG is an indication of how far the process is from equilibrium, and how much useful work may be obtained from it. Clearly, ΔG is a quantity of fundamental importance in determining which processes will or will not occur in a cell, and what they may be used for. But to express these ideas quantitatively, in terms of changes in the concentrations of substances, we need to answer this question: How does the free energy of a system depend on the concentrations of various components in a mixture?

CHEMICAL POTENTIAL

The relationship between free energy and the concentrations of mixture components can be expressed very simply: If we have a mixture containing a moles of component A, b moles of component B, and so on, we may write

$$G = a\overline{G}_A + b\overline{G}_B + c\overline{G}_C + \cdots \tag{3.14}$$

The quantities $\overline{G}_A$, $\overline{G}_B$, and so forth are called the **partial molar free energies** or **chemical potentials** of the various components. Each represents the contribution,

The chemical potential of a substance measures that substance's contribution to the system's free energy.

per mole, of a particular component to the total free energy of the system. (In some texts, chemical potential is given the symbol μ.) We shall make the approximation, which is usually valid for dilute solutions, that each of the chemical potentials depends only on the concentration of the substance in question. For dilute solutions, $\overline{G}_A$, $\overline{G}_B$, etc. turn out to be simple logarithmic functions of the *activities* of the corresponding substance. The **activity (*a*)** is a dimensionless quantity that measures the effective concentration of a substance, describing its contribution to the free energy of the system. We find

$$\overline{G}_A = \overline{G}_A^\circ + RT \ln a_A \qquad (3.15a)$$
$$\overline{G}_B = \overline{G}_B^\circ + RT \ln a_B$$
$$\text{etc.}$$

For dilute solutions the activity of each solute component can be taken as a number approximately equal to the molar concentration of that component. At very low concentration they become numerically exactly the same. This approximation will be fine for almost all biochemical applications. We can then forget about activities and rewrite equation (3.15a) in terms of concentrations:

$$\overline{G}_A = \overline{G}_A^\circ + RT \ln [A] \qquad (3.15b)$$
$$\overline{G}_B = \overline{G}_B^\circ + RT \ln [B]$$
$$\text{etc.}$$

where $[A]$, $[B]$, etc. are equal to the molar concentrations of the components. Note what happens when the concentration equals 1 M. The logarithmic term then vanishes because ($\ln 1 = 0$) and, for example, $\overline{G}_A = \overline{G}_A^\circ$. This shows that $\overline{G}_A^\circ$, $\overline{G}_B^\circ$, etc. are reference, or standard state, values of the chemical potential. We always express the chemical potentials with respect to a **standard state.** In solutions, the standard state for each *solute* component is 1 M concentration; the standard state for the solvent is pure solvent. In each case, $a = 1$ in the standard state. In equations (3.15), T is the absolute temperature and R is the gas constant. In all calculations we will do, $R = 8.314$ J/K·mol.

The importance of equations (3.15) is that they allow us to apply general thermodynamic principles to practical problems. In particular, they allow us to predict the favored directions for real processes under defined conditions. One biochemical process to which these principles are particularly relevant is diffusion through membranes, a process we shall use as an example here, and discuss in much more detail in later chapters. We know from experience that if a substance can diffuse across a membrane, it will do so in such a direction as to make the concentrations on the two sides equal. Now we will see whether our thermodynamics can justify this.

AN EXAMPLE OF HOW CHEMICAL POTENTIAL IS USED: A CLOSE LOOK AT DIFFUSION THROUGH A MEMBRANE

Suppose we have two solutions of substance A separated by a membrane through which A can pass (Figure 3.5). Assume that in region 1 the concentration is initially $[A]_1$ and in region 2 the concentration is $[A]_2$. To determine in which direction diffusion of A will be favored, imagine transferring 1 mole of A from region 1 to region 2. We must assume that the volumes of regions 1 and 2 are so large that the concentrations are not changed appreciably by the transfer of 1 mole. Now we calculate the free energy changes in regions 1 and 2 by using equation (3.15b). Recall that we are taking 1 mole of A out of region 1, where its concen-

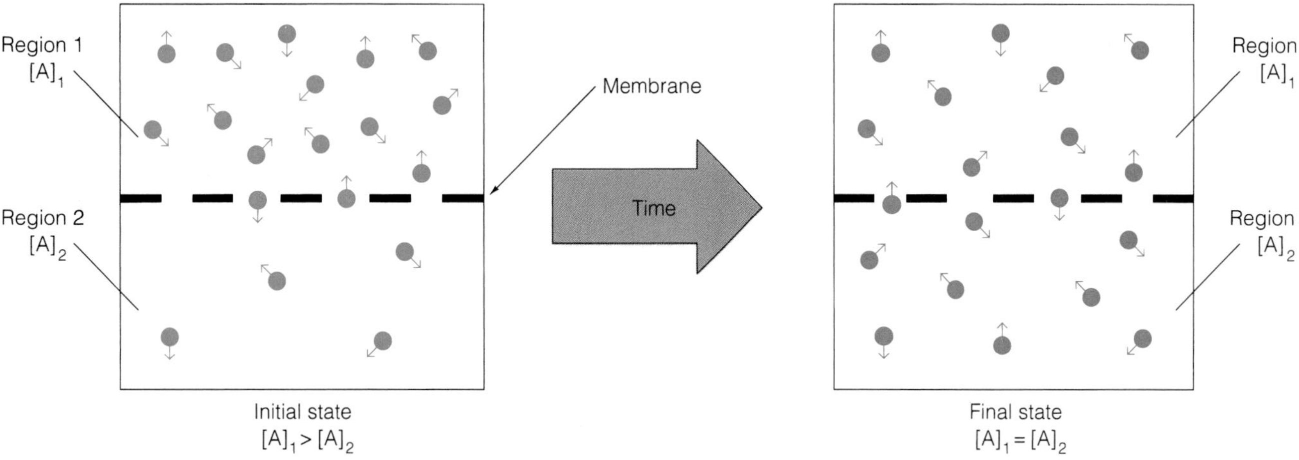

FIGURE **3.5**

Equilibration across a membrane. Two solutions of A, of concentrations $[A]_1$ and $[A]_2$, are separated by a membrane through which A can pass in either direction. If the initial concentration is higher in region 1, the chemical potential of A will also be higher in that region, and net transport will occur from region 1 to region 2 until equal concentrations (and chemical potentials) are obtained.

tration is $[A]_1$, and putting it into region 2, where its concentration is $[A]_2$. From region 1 we are removing material, therefore

$$\Delta G_1 = -G_A^\circ - RT \ln[A]_1 \tag{3.16a}$$

Since we are putting material into region 2, we have

$$\Delta G_2 = +G_A^\circ + RT \ln[A]_2 \tag{3.16b}$$

The *total* free energy change is

$$\Delta G = \Delta G_1 + \Delta G_2 = RT(\ln[A]_2 - \ln[A]_1) = RT \ln([A]_2/[A]_1) \tag{3.17}$$

Note that the ΔG° terms have canceled out.

We will draw on equation (3.17) throughout our later discussions of the transfer of substances across cell membranes. At the moment, it is important to note the rules that the equation sets down for such transfers:

1. If $[A]_2$ is less than $[A]_1$, ΔG is negative. Transfer from region 1 to region 2 is favorable.
2. If $[A]_2$ is greater than $[A]_1$, ΔG is positive. Transfer from region 1 to region 2 is not favorable, but transfer in the opposite direction is.

Thus, we have the following conclusion: If a substance can pass through a membrane, the direction of favorable transfer will always be from the region of high concentration to the region of low concentration. More generally, we can say that a substance will diffuse spontaneously from a region where its chemical potential is high to one where it is low. Thus, chemical potential plays much the same role for chemical substances that electrical potential does for electrons—it is the driving force. Furthermore, suppose $[A]_1 = [A]_2$. In this case, equation (3.17) states that $\Delta G = 0$. Therefore, the process is reversible; the system is at equilibrium. Thus, the equilibrium state is the state in which the concentrations of A on the two sides of the membrane have become equal. From whatever distribution of A we may start with, this equilibrium state of equal concentration will be approached.

We shall encounter situations in living cells that seem, at first glance, to violate this rule. There are cases in which substances pass readily from regions of low concentration to regions of high concentration. But we shall find that, in such circumstances, the necessary free energy price is paid by *coupling* the transport process to thermodynamically favorable chemical reactions. Cells may seem to avoid the laws of thermodynamics, but they never really do so.

Free Energy and Chemical Reactions: Chemical Equilibrium

THE FREE ENERGY CHANGE AND THE EQUILIBRIUM CONSTANT

Perhaps the most important use of the chemical potential is to describe quantitatively the free energy changes accompanying chemical reactions under various conditions. We are thereby able to predict the favorable directions for reactions. Suppose we have a reaction like

$$aA + bB \rightleftharpoons cC + dD$$

Even though the reaction can proceed in either direction, we have written it with C and D on the right, so we call them *products* and A and B *reactants*. We wish to calculate the change in free energy that occurs when *a* moles of A and *b* moles of B form *c* moles of C and *d* moles of D, each at some given concentration.* For the reverse reaction, we need only change the sign of the calculated free energy change.

The free energy change must be the free energy of the products minus that of the reactants.

$$\Delta G = G\,(\text{products}) - G\,(\text{reactants}) \qquad (3.18)$$

By the rule given in equation (3.14), we can write these free energies in terms of the chemical potentials of the substances, each multiplied by the number of moles involved. For our example,

$$\Delta G = c\overline{G}_C + d\overline{G}_D - a\overline{G}_A - b\overline{G}_B \qquad (3.19)$$

Here we are simply stating that the driving force for the reaction is the total free energy of the products minus that of the reactants.

Now, using (3.15b) we insert the appropriate expressions for $\overline{G}_C$ and so forth in terms of concentrations and obtain

$$\Delta G = cG_C^\circ + cRT\ln[C] + dG_D^\circ + dRT\ln[D] - aG_A^\circ - aRT\ln[A] - bG_B^\circ - bRT\ln[B] \qquad (3.20a)$$

$$\Delta G = cG_C^\circ + dG_D^\circ - aG_A^\circ - bG_B^\circ + RT\ln[C]^c + RT\ln[D]^d - RT\ln[A]^a - RT\ln[B]^b \qquad (3.20b)$$

$$\Delta G = \Delta G^\circ + RT(\ln[C]^c + \ln[D]^d - \ln[A]^a - \ln[B]^b) \qquad (3.20c)$$

The free energy change in a chemical reaction depends on the standard state change and on the concentrations of reactants and products.

In going from (3.20a) to (3.20c), we have done two things: grouped the G° terms into ΔG° and made use of rearrangements like: $aRT\ln[A] = RT\ln[A]^a$. The group of G° terms (ΔG°) has a simple meaning; because G° is the free energy per mole of a substance in the standard state (1 M), ΔG° represents the **standard state free energy change** in the reaction. It is the free energy change that would be observed if *a* moles of A and *b* moles of B, each at 1 M concentration, formed *c* moles of C and *d* moles of D, each at 1 M. The terms containing logarithms in equation (3.20c) can be combined, because each is multiplied by the same factor, *RT*. Therefore, we obtain the following for the total ΔG:

$$\Delta G = \Delta G^\circ + RT\ln\left(\frac{[C]^c[D]^d}{[A]^a[B]^b}\right) \qquad (3.21a)$$

*How do we carry out a finite amount of reaction while keeping concentrations of both reactants and products constant? Two ways can be imagined. First, the total amounts of reactants and products could be so enormous that a finite reaction would not appreciably change concentration. (We chose this kind of model in discussing diffusion across a membrane.) Alternatively, we could imagine hypothetical processes that would remove products and add reactants so as to keep concentrations unchanged. This often happens in living cells.

This can be remembered more easily as

$$\Delta G = \Delta G^\circ + RT \ln \frac{[\text{products}]}{[\text{reactants}]} \tag{3.21b}$$

if we keep in mind that each product and reactant concentration is raised to the power equal to its stoichiometry in the reaction. The quantity ΔG represents the free energy change when a moles of A (at concentration [A]) and b moles of B (at [B]) make c moles of C (at [C]) and d moles of D (at [D]). These concentrations may be anything we want them to be. When all are 1 M, equation (3.21a) reduces to $\Delta G = \Delta G^\circ$, as we would expect. The importance of equation (3.21a) is that it allows us to calculate ΔG under any conditions we wish.

Suppose, now, that the reaction has come to equilibrium. In that case, two things must be true. First, the concentrations in the factor in parentheses in equation (3.21a) must be equilibrium concentrations, and therefore the factor in parentheses is now the *equilibrium constant K* for the reaction:

$$K = \left(\frac{[C]^c[D]^d}{[A]^a[B]^b} \right)_{eq} \tag{3.22}$$

Second, if the system is at equilibrium, ΔG must equal zero. In this case, equation (3.21a) reduces to

$$0 = \Delta G^\circ + RT \ln \left(\frac{[C]^c[D]^d}{[A]^a[B]^b} \right)_{eq} \tag{3.23}$$

or

$$-\Delta G^\circ = RT \ln K \tag{3.24}$$

This can be rearranged as

$$K = e^{-\Delta G^\circ/RT} \tag{3.25}$$

Equations (3.24) and (3.25) express an important relationship between ΔG° and the equilibrium constant that we shall use many times. These equations make it possible, for example, to use data from tables of standard state free energy changes to predict equilibrium constants for reactions.

The equilibrium constant K can be calculated from the standard state free energy change.

Equation (3.21a) may best be thought of in the following way. ΔG° represents a reference value for the free energy change, whereby the intrinsic free energy changes in different reactions can be compared under equivalent circumstances (1 M concentrations). The magnitude of this term tells us the equilibrium constant. The second (concentration-dependent) term in equation (3.21a) represents the extra free energy change (+ or −) involved if we were to carry out the reaction at some other, arbitrary set of concentrations. Tabulations of ΔG° values for different reactions are common, but in applying such data to biochemical problems, we must always keep in mind that it is ΔG, as determined by the actual concentrations in the cell, rather than ΔG° that determines whether or not a reaction is favored in vivo.

FREE ENERGY CALCULATIONS: A BIOCHEMICAL EXAMPLE

To make the application of these somewhat abstract ideas a bit clearer, let us consider an example—a very simple but important biochemical reaction, the isomerization of glucose-6-phosphate into fructose-6-phosphate:

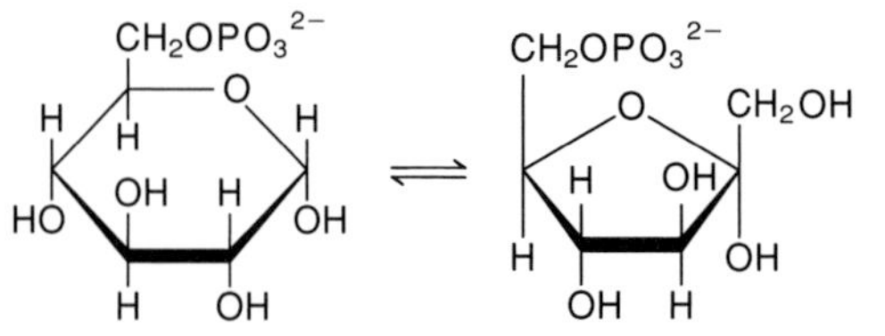

Glucose-6-phosphate Fructose-6-phosphate

which may be written more compactly as

$$\text{Glucose-6-phosphate} \rightleftharpoons \text{Fructose-6-phosphate} \qquad \Delta G° = +1.7 \text{ kJ/mol}$$

$$\text{G6P} \rightleftharpoons \text{F6P}$$

This is the second step in the glycolytic pathway, which is discussed in Chapter 13. The reaction is clearly endergonic under standard conditions. In other words, the system is not at equilibrium when G6P and F6P are both at 1 M; in fact, the reverse reaction is favored under these conditions because $\Delta G°$ is positive (+1.7 kJ/mol). Therefore, the equilibrium must lie to the left, with a higher concentration of G6P than F6P. We can express this quantitatively by calculating the equilibrium constant from equation (3.25). Using the above value for $\Delta G°$, and assuming a temperature of 25°C (298 K), we obtain:

$$K = e^{-\Delta G°/RT} = e^{-(1700 \text{ J/mol})/(8.314 \text{ J/K·mol})(298 \text{ K})} = 0.504 = \left(\frac{[\text{F6P}]}{[\text{G6P}]}\right)_{eq} \qquad (3.26)$$

where $([\text{F6P}]/[\text{G6P}])_{eq}$ is the equilibrium ratio of the concentration of fructose-6-phosphate to that of glucose-6-phosphate. The fact that $K < 1$ is another way of saying that the equilibrium lies to the left. But we can do better than this and show just what the composition of the equilibrium mixture is. The total substance in the reaction must be in the form of either F6P or G6P. We can write

$$K = \frac{[\text{F6P}]_{eq}}{[\text{G6P}]_{eq}} = \frac{[\text{F6P}]_{eq}/\text{total concentration}}{[\text{G6P}]_{eq}/\text{total concentration}} = \frac{(f_{\text{F6P}})_{eq}}{(f_{\text{G6P}})_{eq}} \qquad (3.27)$$

where f_{F6P} and f_{G6P} represent the *fractions* of the total material in each form. Since

$$f_{\text{G6P}} = 1 - f_{\text{F6P}} \qquad (3.28)$$

we have

$$K = \frac{(f_{\text{F6P}})_{eq}}{1 - (f_{\text{F6P}})_{eq}} \qquad (3.29)$$

Using 0.504 as the value of K, we can solve equation (3.29) to yield $(f_{\text{F6P}})_{eq}$ = 0.335. In other words, at equilibrium 33.5% of the sugar will be in the form of fructose-6-phosphate, and the remaining 66.5% will be in the form of glucose-6-phosphate.

A further insight into the power of such analysis is provided by the application of equation (3.21) to this problem:

$$\Delta G = \Delta G° + RT \ln \frac{[\text{F6P}]}{[\text{G6P}]} \qquad (3.30a)$$

$$\Delta G = \Delta G° + RT \ln \left(\frac{f_{\text{F6P}}}{1 - f_{\text{F6P}}}\right) \qquad (3.30b)$$

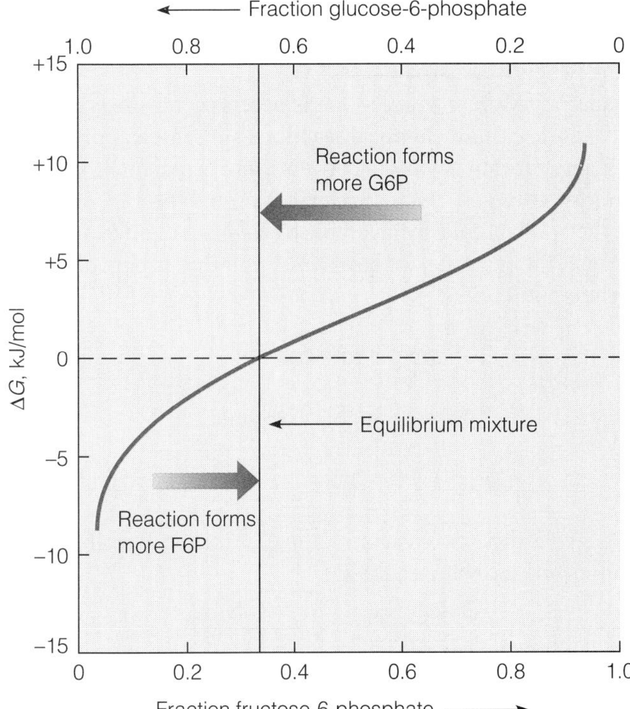

FIGURE **3.6**

Free energy change as a function of the reaction mixture composition. The red curve shows ΔG for the conversion of 1 mole glucose-6-phosphate to 1 mole fructose-6-phosphate as a function of the fraction of either component. If the fraction of fructose-6-phosphate is initially higher than its equilibrium value, the reaction will proceed to the left (forming more G6P); if the fraction of fructose-6-phosphate is initially lower than its equilibrium value, the reaction will proceed to the right (forming more F6P). In either case, the reaction proceeds toward equilibrium.

What we are doing now is calculating the ΔG value when f_{F6P} (and therefore f_{G6P}) has *any* value we wish. We can pick a value and ask what the free energy change for converting 1 mole of glucose-6-phosphate to 1 mole of fructose-6-phosphate would be if the fraction f_{G6P} were set at that value. The results are graphed in Figure 3.6; their significance can be seen from the following examples.

1. Suppose we made a mixture with $f_{\mathrm{F6P}} = 0.2$. Under these conditions, equation (3.30b) shows that ΔG is negative (-1.76 kJ/mol). The reaction to the right is thermodynamically favored. Fructose-6-phosphate will be formed at the expense of glucose-6-phosphate until f_{F6P} increases to the equilibrium value, 0.335. At this point, the free energy difference driving the reaction will have vanished, and the process will have come to equilibrium.
2. If the initial mixture were prepared instead with $f_{\mathrm{F6P}} = 0.8$, the initial value for ΔG would be $+5.11$ kJ/mol. The reverse reaction would now be favored, and fructose-6-phosphate would be converted into glucose-6-phosphate until the equilibrium state was attained.

This example shows exactly how thermodynamically favored (irreversible) processes are related to equilibrium. Whenever a system is displaced from equilibrium, it will spontaneously proceed in whichever direction is necessary to move toward the equilibrium state. If the system is not at equilibrium, whichever direction leads toward equilibrium will have a negative ΔG. We may consider the negative ΔG to be the driving force for the reaction.

High-Energy Phosphate Compounds: Free Energy Sources in Biological Systems

The central role of free energy changes in determining the favorable directions for chemical reactions is of the greatest importance in biochemistry, for every metabolic pathway must, overall, be a thermodynamically favored process. In a similar way, such physical processes as transport of ions and contraction of muscles also require a free energy source. These latter can be considered examples of the "useful work" that free energy can provide.

Thermodynamically unfavored reactions or processes may become favored if coupled to strongly exergonic reactions.

Very often, a particular reaction or process that is necessary for life is in itself endergonic. Examples include certain chemical reactions and physical processes (like transport through a membrane *against* a concentration gradient) that have large positive free energy changes. Such intrinsically unfavorable reaction processes can be made thermodynamically favorable by *coupling* them to strongly favored reactions. Suppose, for example, we have a reaction that is part of an essential pathway but is endergonic:

$$\mathrm{A} \rightleftharpoons \mathrm{B} \qquad \Delta G° = +10 \text{ kJ/mol}$$

At the same time, suppose another process is highly exergonic:

$$\mathrm{C} \rightleftharpoons \mathrm{D} \qquad \Delta G° = -30 \text{ kJ/mol}$$

If the cell can manage to *couple* these two reactions, the $\Delta G°$ for the overall process will be the algebraic sum of the individual values:

$$\begin{array}{rl} \mathrm{A} \rightleftharpoons \mathrm{B} & \Delta G° = +10 \text{ kJ/mol} \\ \mathrm{C} \rightleftharpoons \mathrm{D} & \Delta G° = -30 \text{ kJ/mol} \\ \hline \mathrm{A + C} \rightleftharpoons \mathrm{B + D} & \Delta G° = -20 \text{ kJ/mol} \end{array}$$

Equilibrium for the overall process now lies far to the right; the consequence is that B is efficiently produced from A.

Coupling of endergonic reactions or processes to exergonic reactions is one of the most important principles in biochemistry. Coupling of highly favored to unfavored processes is used not only to drive countless reactions but also to transport materials across membranes, transmit nerve impulses, contract muscles, and carry out other physical changes.

HIGH-ENERGY PHOSPHATE COMPOUNDS AS ENERGY SHUTTLES

High-energy phosphate compounds have very large negative free energies of hydrolysis.

Driving processes by coupling is so common that it must mean that the cell contains a number of compounds (like the previously described hypothetical C) that can undergo reactions with large negative free energy changes. In later chapters you will encounter a variety of such compounds; they include phosphate anhydrides, enol phosphates, some thioesters, and compounds containing N—P bonds. Such substances can be thought of as shuttles of free energy in the cell. The most important of these high-energy compounds are certain phosphates, which can undergo hydrolytic release of their phosphate groups in aqueous solution. A number of such compounds and their hydrolysis reactions are shown in Figure 3.7. You will encounter all of these important substances in later chapters on metabolism. Some of these substances, like **phosphoenolpyruvate (PEP), creatine phosphate (CP)**, and **adenosine triphosphate (ATP)**, have very negative standard state free energies of hydrolysis. ATP is perhaps the most important of these compounds, and the one you will encounter most often in this book. The structure and hydrolysis reactions of ATP are shown in Figure 3.8. Hydrolysis of ATP is highly exergonic, with $\Delta G°' = -31$ kJ/mol. (See page 78 for the precise definition of $\Delta G°'$.) This value corresponds to an equilibrium constant greater than 10^5. Such an equilibrium lies so far to the right that ATP hydrolysis can be considered essentially irreversible.

Figure 3.7 also shows that whereas some of these phosphate hydrolysis reactions are truly high-energy processes, others are not. In the past, biochemists

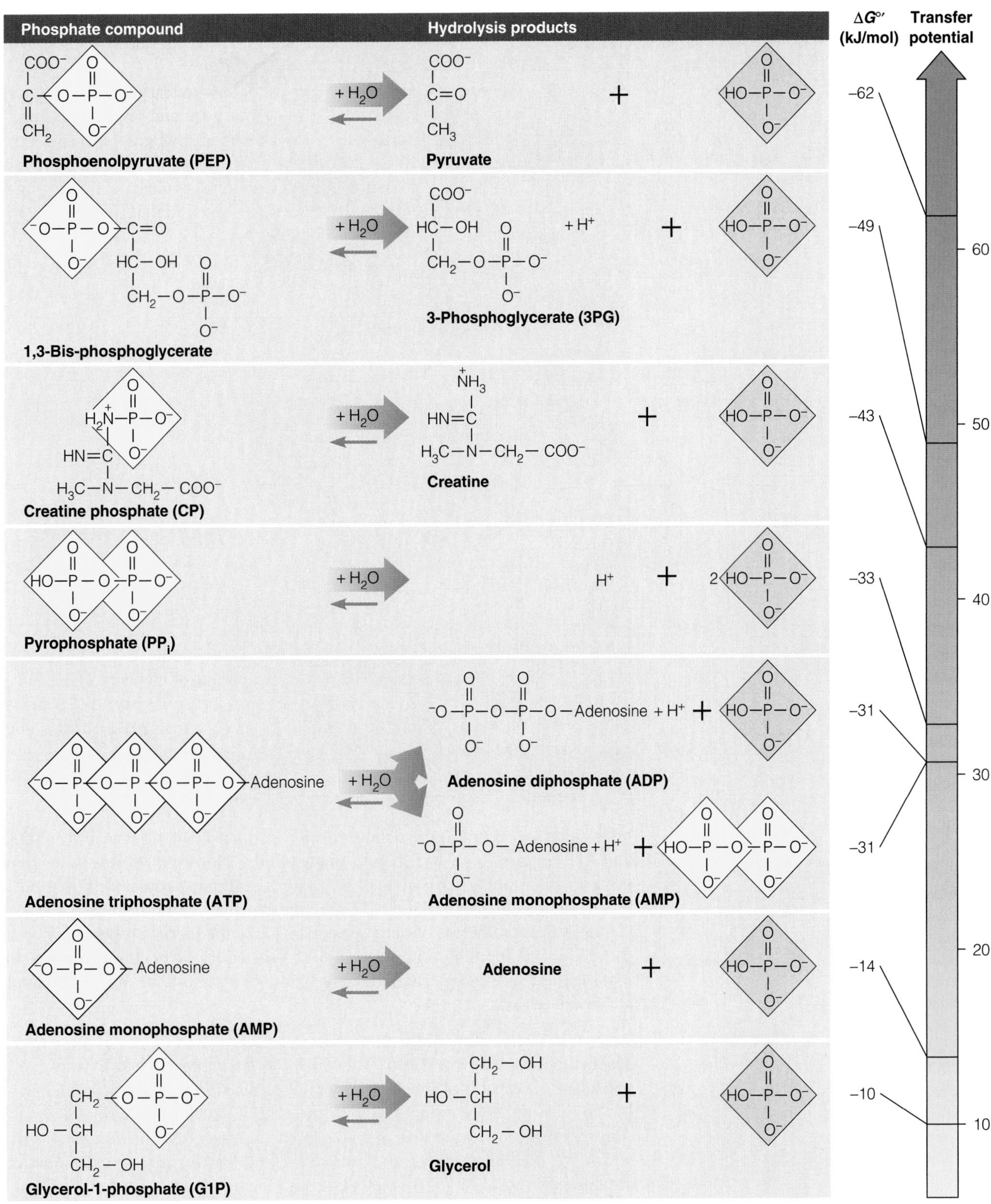

FIGURE **3.7**

Hydrolysis reactions for some biochemically important phosphate compounds. The labile phosphate group of each compound is shown in yellow. The more stable reaction product P_i is in gray. A scale of phosphate transfer potential is shown to the right.

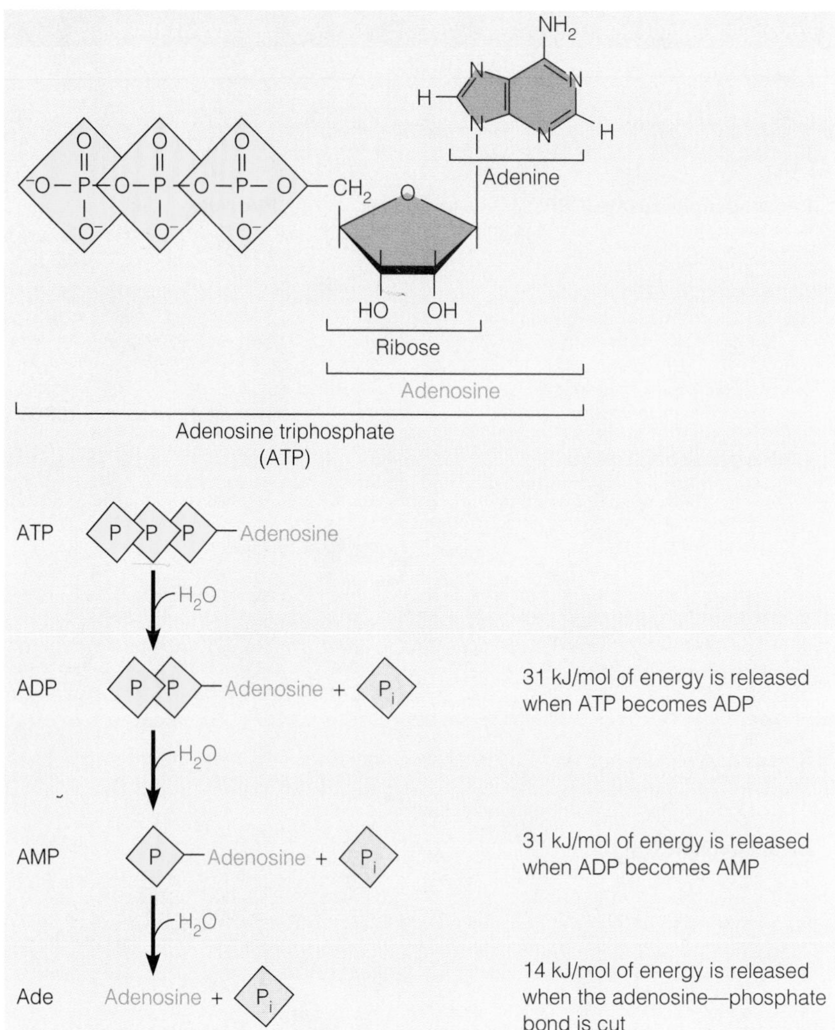

FIGURE **3.8**

The ATP molecule and its hydrolysis reactions. Throughout this book, Ⓟ represents the tetrahedral phosphate group.

often said that compounds such as phosphoenolpyruvate, creatine phosphate, and ATP contain "high-energy phosphate bonds." However, we now know that this concept is not strictly correct, for there is nothing special about the phosphate bonds in these compounds. Rather, the explanation for the high free energies of hydrolysis of some of these substances lies in the special properties of both the reactants and products in the reactions. A variety of factors can contribute to making these free energy changes large. Those that seem to be most important are described next.

Resonance Stabilization of the Phosphate Products

The **orthophosphate** ion (HPO_4^{2-}), which is often abbreviated as $\mathbf{P_i}$ (inorganic phosphate), is capable of a wide variety of resonance forms. Both the bound proton and the oxygen bonding should be thought of as delocalized, so a more appropriate way to write the structure is as shown in Figure 3.9. The multiple forms, which are of equal energy, contribute to the high entropy of such a resonance structure (see equation (3.9)). Not all of these forms are possible when the phosphate is bound in an ester. Consequently, release of the orthophosphate results in an entropy increase in the system and is therefore favored. Resonance stabilization applies in *all* of the phosphate hydrolysis reactions described in Figure 3.7.

Additional Hydration of the Hydrolysis Products

Release of the phosphate residue from its bonded state allows greater opportunities for hydration, especially when both products are charged. Recall from Chap-

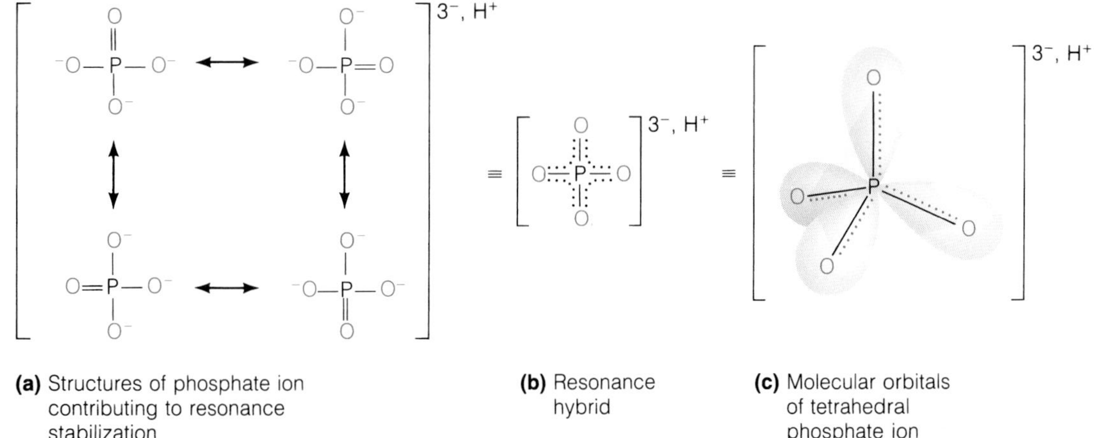

(a) Structures of phosphate ion contributing to resonance stabilization

(b) Resonance hybrid

(c) Molecular orbitals of tetrahedral phosphate ion

FIGURE **3.9**

Resonance stabilization of orthophosphate, HPO_4^{2-} (P_i). The resonance delocalization of the charge of the orthophosphate ion is depicted here in three ways. **(a)** In this somewhat artificial depiction, the four resonance forms of the phosphate ion are shown with the H^+ not assigned permanently to any one of the four oxygen atoms. **(b)** In this more exact depiction of resonance, the dotted lines represent partial bonds. **(c)** This depiction shows the physical significance of partial bonds. The phosphate ion is a tetrahedral structure with four equivalent bonds.

ter 2 that ions are highly hydrated in aqueous solution and that such hydration is an energetically favored state.

Electrostatic Repulsion Between Charged Products

In the hydrolysis of phosphoenolpyruvate, bisphosphoglycerate, adenosine triphosphate, and pyrophosphate, both products of hydrolysis carry a negative charge. The repulsion between these ionic products strongly favors the hydrolysis reaction.

Enhanced Resonance Stabilization or Tautomerization of Product Molecules

Hydrolysis is sometimes favored not only by resonance stabilization of the orthophosphate but also by the ability of the other product to adopt more molecular forms. This contributes to the hydrolysis of phosphoenolpyruvate, for pyruvate can tautomerize:

$$\begin{matrix} COO^- \\ | \\ C{=}O \\ | \\ CH_3 \end{matrix} \rightleftharpoons \begin{matrix} COO^- \\ | \\ C{-}OH \\ \| \\ CH_2 \end{matrix}$$

whereas PEP cannot undergo the same tautomerization.

Release of a Proton in Buffered Solutions

In some of the reactions listed in Figure 3.7 a proton is released. Therefore, the hydrogen ion concentration (i.e., the pH) will influence the equilibrium. Note that we have designated the standard state free energies of hydrolysis for these reactions as $\Delta G°'$. The superscript *prime* signifies that the standard state is taken at pH 7, near the pH value maintained for almost all physiological processes (see Chapter 2). Consider how we would write the true ΔG for a reaction like the hydrolysis of ATP to ADP, under circumstances in which both H_2O and H^+ concentrations could be varied:

$$ATP^{4-} + H_2O \rightleftharpoons ADP^{3-} + HPO_4^{2-} + H^+$$

Then, by equation (3.21), we should have

$$\Delta G = \Delta G° + RT \ln\left(\frac{[ADP^{3-}][HPO_4^{2-}][H^+]}{[ATP^{4-}][H_2O]}\right) \tag{3.31}$$

But since the hydrolysis reactions in vivo always occur in very dilute aqueous solution, near pH 7, it is inappropriate to retain the concentrations of H_2O and H^+ as variables. In dilute solution, the solvent (water) is close to its standard state, so its activity equals unity (see page 68). Because living cells are usually buffered at a pH very close to 7.0, we may consider the hydrogen ion concentration to be maintained at about 10^{-7} M. Using the properties of logarithms, equation (3.31) may be conveniently rewritten as

$$\Delta G = \Delta G^\circ + RT \ln\left(\frac{[\text{ADP}][\text{P}_i]}{[\text{ATP}]}\right) + RT \ln \frac{(10^{-7})}{(1)} \tag{3.32}$$

The third term on the right is essentially constant. We may insert it into a redefined ΔG°, which we shall call $\Delta G^{\circ\prime}$:

$$\Delta G = \Delta G^{\circ\prime} + RT \ln\left(\frac{[\text{ADP}][\text{P}_i]}{[\text{ATP}]}\right) \tag{3.33}$$

where

$$\Delta G^{\circ\prime} = \Delta G^\circ + RT \ln (10^{-7}) \tag{3.34}$$

Standard free energy changes for biochemical reactions are defined by $\Delta G^{\circ\prime}$, where the water concentration is assumed constant and pH = 7.0.

This definition is the true meaning of the $\Delta G^{\circ\prime}$ values given in Figure 3.7. The fact that the hydrolysis of ATP normally occurs in the presence of a vast excess of H_2O, and under conditions in which the hydrogen ion concentration is kept very low, contributes enormously to the favorability of the reaction. The former condition is true of hydrolysis reactions in general; the latter is important whenever protons are released.

PHOSPHATE TRANSFER POTENTIAL

The phosphate transfer potential shows which compounds can phosphorylate others at standard conditions.

There is another useful way in which we can think about the $\Delta G^{\circ\prime}$ values for various high-energy phosphate compounds. As Figure 3.7 shows, these values form a scale of **phosphate transfer potentials.** The potential is defined simply as $-\Delta G^{\circ\prime}$ of hydrolysis. Each compound is capable of driving the phosphorylation of compounds lower on the scale, provided that a suitable coupling mechanism is available. Consider, for example, the following reactions, which are written in the somewhat abbreviated form we will frequently use:

(1) Hydrolysis of phosphoenolpyruvate	PEP $\rightleftharpoons$ pyruvate + P_i	$\Delta G^{\circ\prime} = -62$ kJ/mol
(2) Phosphorylation of adenosine diphosphate	ADP + P_i $\rightleftharpoons$ ATP	$\Delta G^{\circ\prime} = +31$ kJ/mol
(1) + (2): Coupled phosphorylation of ADP by PEP	PEP + ADP $\rightleftharpoons$ pyruvate + ATP	$\Delta G^{\circ\prime} = -31$ kJ/mol

Thus, phosphoenolpyruvate, having the very high phosphate transfer potential of 62 kJ/mol, is capable of adding a phosphate group to ADP in a thermodynamically favored process. ATP can pass this phosphate on to glucose, because the phosphate transfer potential of glucose-6-phosphate lies still farther down the scale:

(1) Hydrolysis of ATP	ATP $\rightleftharpoons$ ADP + P_i	$\Delta G^{\circ\prime} = -31$ kJ/mol
(2) Phosphorylation of glucose	Glucose + P_i $\rightleftharpoons$ glucose-6-phosphate	$\Delta G^{\circ\prime} = +14$ kJ/mol
(1) + (2): Coupled phosphorylation of glucose by ATP	ATP + glucose $\rightleftharpoons$ ADP + glucose-6-phosphate	$\Delta G^{\circ\prime} = -17$ kJ/mol

These examples emphasize how ATP can act as a versatile phosphate transfer agent through coupled reactions. In each case, the coupling is accomplished by having the reactions take place on the surface of a large protein molecule, an enzyme. We shall study enzymes in detail in Chapter 11 and find that they can both facilitate such coupling and accelerate the reactions. In Chapter 12 we examine the metabolic role of ATP in much more depth.

SUMMARY

Bioenergetics is that branch of thermodynamics that deals with energy acquisition, exchange, and utilization in organisms. The internal energy (E) of a system includes all energy that can be exchanged by nonnuclear processes: the energy of motion of atoms and molecules and the energy of chemical bonds and noncovalent interactions. E is determined by the state of the system and can be changed only by exchange of heat or work with the surroundings ($\Delta E = q - w$). This is the first law of thermodynamics. Under constant-volume conditions, $q = \Delta E$. Under constant-pressure conditions, $q = \Delta E + P\Delta V = \Delta H$, where H denotes the enthalpy ($H = E + PV$). In biochemistry, ΔH is more important than ΔE.

Processes may be reversible (near equilibrium) or irreversible (far from equilibrium). The thermodynamically favored direction of a reaction (the direction that leads toward equilibrium) is determined by changes in both the enthalpy (H) and the entropy (S, a measure of randomness). The free energy, $G = H - TS$, takes both into account. The criterion for a favorable process is that the free energy change, $\Delta G = \Delta H - T\Delta S$, be negative; this is one statement of the second law of thermodynamics. The ice-to-water transition demonstrates the importance of temperature (T) in determining reaction direction. At the melting point, solid and liquid are in equilibrium ($\Delta G = 0$). The entropy of an open system can decrease, as in freezing of water, but only if the enthalpy decreases. Thus, organisms must constantly expend energy to maintain organization. In every energy transfer, some part of the energy (ΔH) is lost as heat ($T\Delta S$), so ΔG is a measure of the energy that is potentially available for useful work.

To apply thermodynamic relationships in chemical problems, we use chemical potential, which relates the concentration of each substance to its contribution to the total free energy of the system. From the chemical potential we obtain the equations governing free energy changes in chemical reactions:

$$\Delta G = \Delta G^\circ + RT \ln \frac{[\text{products}]}{[\text{reactants}]}$$

and $\Delta G^\circ = -RT \ln K$. Here ΔG° is the standard state free energy change and K is the equilibrium constant.

Reactions that are not thermodynamically favored may nevertheless be driven if coupled to reactions that have large negative ΔG values. In living systems, the hydrolysis of certain phosphate compounds is frequently used for this purpose. The phosphate transfer potential ranks these compounds according to their ability to phosphorylate other compounds under standard conditions. ATP, the most important of the compounds, is generated in the energy-producing metabolic pathways and is used to drive many reactions.

REFERENCES

This chapter has presented a very abbreviated treatment of thermodynamics. For the student who wishes more background in this field and more information about its applications to biochemistry, we recommend the following books.

Eisenberg, D., and D. Crothers (1979) *Physical Chemistry with Applications to the Life Sciences.* Benjamin/Cummings, Redwood City, Calif. A very fine physical chemistry text, written by two physical biochemists. Strongly recommended, for it contains many biochemical applications of physical-chemical principles not found in most physical chemistry texts.

Klotz, I. (1986) *Introduction to Biomolecular Energetics.* Academic Press, New York. A brief introduction to thermodynamics for biochemists. Some excellent examples and explanations.

Morowitz, H. J. (1970) *Entropy for Biologists.* Academic Press, New York. Broader than its title suggests, this is a good, concise thermodynamics text.

van Holde, K. E., W. C. Johnson, and P.-S. Ho (1998) *Principles of Physical Biochemistry.* Prentice Hall, Upper Saddle River, N.J. Chapters 2–4 extend the applications of thermodynamics to biochemistry.

For a sophisticated discussion of the effect of ionic conditions on the free energy changes in phosphate ester hydrolysis, see the following article.

Alberty, R. A. (1992) Equilibrium calculations on systems of biochemical reactions at specified pH and pMg. *Biophys. Chem.* 42:117–131.

PROBLEMS

1. The enthalpy change (heat of fusion, ΔH_f) for the transition ice $\longrightarrow$ water at 0°C and 1 atm pressure is +6.01 kJ/mol. The change in volume when 1 mole of ice is melted is -1.625 cm^3/mol = -1.625×10^{-6} m^3/mol. Calculate the difference between ΔH_f and ΔE_f for this process, and express it as a percentage of ΔE_f. [Note: 1 atm = 1.013×10^5 N/m^2 in SI units.]

*2. Given the following reactions and their enthalpies:

	ΔH (kJ/mol)
$H_2(g) \longrightarrow 2H(g)$	+436
$O_2(g) \longrightarrow 2O(g)$	+495
$H_2(g) + \frac{1}{2}O_2(g) \longrightarrow H_2O(g)$	−242

(a) Devise a way to calculate ΔH for the reaction

$$H_2O(g) \longrightarrow 2H(g) + O(g)$$

(b) From this, estimate the H—O bond energy.

3. The decomposition of crystalline N_2O_5

$$N_2O_5(s) \longrightarrow 2NO_2(g) + \tfrac{1}{2}O_2(g)$$

is an example of a reaction that is thermodynamically favored even though it absorbs heat. At 25°C we have the following values for the standard state enthalpy and free energy changes of the reaction:

$$\Delta H° = +109.6 \text{ kJ/mol}$$

$$\Delta G° = -30.5 \text{ kJ/mol}$$

(a) Calculate $\Delta S°$ at 25°C.
(b) Why is the entropy change so favorable for this reaction?
(c) Calculate $\Delta E°$ for this reaction at 25°C.
(d) Why is $\Delta H°$ greater than $\Delta E°$?

4. The combustion of glucose to CO_2 and water is a major source of energy in aerobic organisms. It is a reaction favored mainly by a large negative enthalpy change.

$$C_6H_{12}O_6(s) + 6O_2(g) \longrightarrow 6CO_2(g) + 6H_2O(l)$$

$$\Delta H° = -2816 \text{ kJ/mol} \qquad \Delta S° = +181 \text{ J/K·mol}$$

(a) At 37°C, what is the value for $\Delta G°$?
(b) In the overall reaction of aerobic metabolism of glucose, 38 moles of ATP are produced from ADP for every mole of glucose oxidized. Calculate the standard state free energy change for the *overall* reaction when glucose oxidation is coupled to the formation of ATP.
(c) What is the *efficiency* of the process in terms of the percentage of the available free energy change captured in ATP?

*5. The first reaction in glycolysis is the phosphorylation of glucose:

$$P_i + \text{glucose} \rightleftharpoons \text{glucose-6-phosphate}$$

This is a thermodynamically unfavorable process, with $\Delta G°' = +14$ kJ/mol.
(a) In a liver cell at 37°C the concentrations of both phosphate and glucose are normally maintained at about 5 mM each. What would the *equilibrium* concentration of glucose-6-phosphate be, according to the above data?
(b) This very low concentration of the desired product would be very unfavorable for glycolysis. In fact, the reaction is coupled to ATP hydrolysis to give the overall reaction

$$\text{ATP} + \text{glucose} \rightleftharpoons \text{glucose-6-phosphate} + \text{ADP}$$

What is $\Delta G°'$ for the reaction now?
(c) If, in addition to the constraints on glucose concentration listed previously, we have in the liver cell [ATP] = 3 mM and [ADP] = 1 mM, what is the equilibrium concentration of glucose-6-phosphate? The answer you will obtain is an absurdly high value for the cell and in fact is never approached in reality. Explain why.

6. In another key reaction in glycolysis, dihydroxyacetone phosphate (DHAP) is isomerized into glyceraldehyde-3-phosphate (G3P):

$$\underset{\textbf{DHAP}}{CH_2OH{-}C(=O){-}CH_2OPO_3^{2-}} \rightleftharpoons \underset{\textbf{G3P}}{H{-}C(=O){-}HC(OH){-}CH_2OPO_3^{2-}} \qquad \Delta G°' = +7.5 \text{ kJ/mol}$$

Because $\Delta G°'$ is positive, equilibrium lies to the left.
(a) Calculate the equilibrium constant, and the equilibrium fraction of G3P from the above, at 37°C.
(b) In the cell, depletion of G3P makes the reaction proceed. What will ΔG be if the concentration of G3P is always kept at $\frac{1}{100}$ of the concentration of DHAP?

7. A protein molecule, in its folded native state, has *one* favored conformation. But when it is denatured, it becomes a random coil, with many possible conformations.
(a) What must be the sign of ΔS for the change native $\longrightarrow$ denatured?
(b) Will the contribution of ΔS to the free energy change be + or −? What requirement does this impose on ΔH if proteins are to be stable structures?

8. When a hydrophobic substance like a hydrocarbon is dissolved in water, a clathrate cage of ordered water molecules is formed about it (see page 37 in Chapter 2). What do you expect the sign of ΔS to be for this process? Explain your answer.

*9. It is observed that most protein molecules go from their defined, folded state into their random-coil, denatured state as temperature is increased.
(a) Given what you know about ΔH and ΔS, explain why this is reasonable. [Hint: Consider Problem 7.]
(b) Sometimes, however, proteins denature as temperature is *decreased.* How might this be explained? [Hint: Consider Problem 8.]

*10. Suppose a reaction has $\Delta H°$ and $\Delta S°$ values independent of temperature. Show from this, and equations given in this chapter, that

$$\ln K = \frac{-\Delta H°}{RT} + \frac{\Delta S°}{R}$$

where K is the equilibrium constant. How could you use values of K determined at different temperatures to determine $\Delta H°$ for the reaction?

11. The following data give the ion product (K_w; see equation (2.5) on page 40) for water at various temperatures:

T(°C)	K_w (M^2)
0	1.14×10^{-15}
25	1.00×10^{-14}
30	1.47×10^{-14}
37	2.56×10^{-14}

(a) Using the results from Problem 10, calculate $\Delta H°$ for the ionization of water.

(b) Use these data, and the ion product at 25°C, to calculate $\Delta S°$ for water ionization.

12. The phosphate transfer potentials for glucose-1-phosphate and glucose-6-phosphate are 21 kJ/mol and 14 kJ/mol, respectively.

(a) What is the equilibrium constant for this reaction at 25°C?

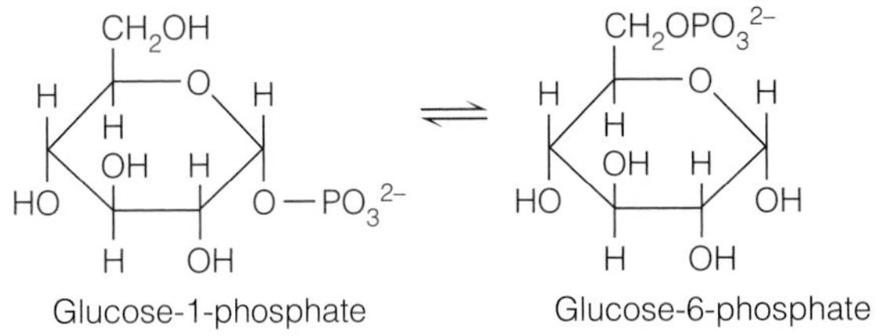

Glucose-1-phosphate $\rightleftharpoons$ Glucose-6-phosphate

(b) If a mixture were prepared containing 1 M glucose-6-phosphate and 1×10^{-3} M glucose-1-phosphate, what would be the thermodynamically favored direction for the reaction?

13. We give a value of the standard state free energy change for ATP hydrolysis (Figure 3.7) to be -31 kJ/mol. Would you expect the same value for the reaction $ATP^{4-} + H_2O \longrightarrow ADP^{2-} + P_i^{2-}$? Explain.

14. As you take each breath, you expel about 0.5 L against a pressure of 1 atm.

(a) If you breathe about 30 times per minute, how much work do you do in this way each day? (Neglect any work involved in inhaling.)

(b) Using the $\Delta G°$ value for ATP hydrolysis, what is the minimum number of moles of ATP you will utilize per day just in breathing?

15. Undergoing moderate activity, an average person will generate about 350 kJ of heat per hour. Using the heat of combustion of palmitic acid (equation (3.6)) as an approximate value for fatty substances, estimate how many grams of fat would be required per day to sustain this level, if all were burned for heat.

*16. The major difference between a protein molecule in its native state and in its denatured state lies in the number of conformations available. The native, folded state can be thought to have one conformation. The unfolded state can be estimated to have three possible orientations about each bond between residues.

(a) For a protein of 100 residues, estimate the entropy change per mole upon denaturation.

(b) What must be the enthalpy change accompanying denaturation to allow the protein to be half-denatured at 50°C?

(c) Will the fraction denatured increase or decrease with increasing temperature?

17. A cell exists in a glucose solution of 0.01 M. The concentration of glucose inside the cell is 0.1 M.

(a) What would be the free energy change involved in transporting 10^{-6} mole of glucose from the medium into the cell? Assume $T = 37°C$.

(b) What would the free energy change be if the intracellular and extracellular concentrations were 1 mM and 10 mM, respectively?

(c) If the process were coupled to ATP hydrolysis, how many moles of ATP would have to be hydrolyzed in order to make this process favorable?

PART II

Molecular Architecture of Living Matter

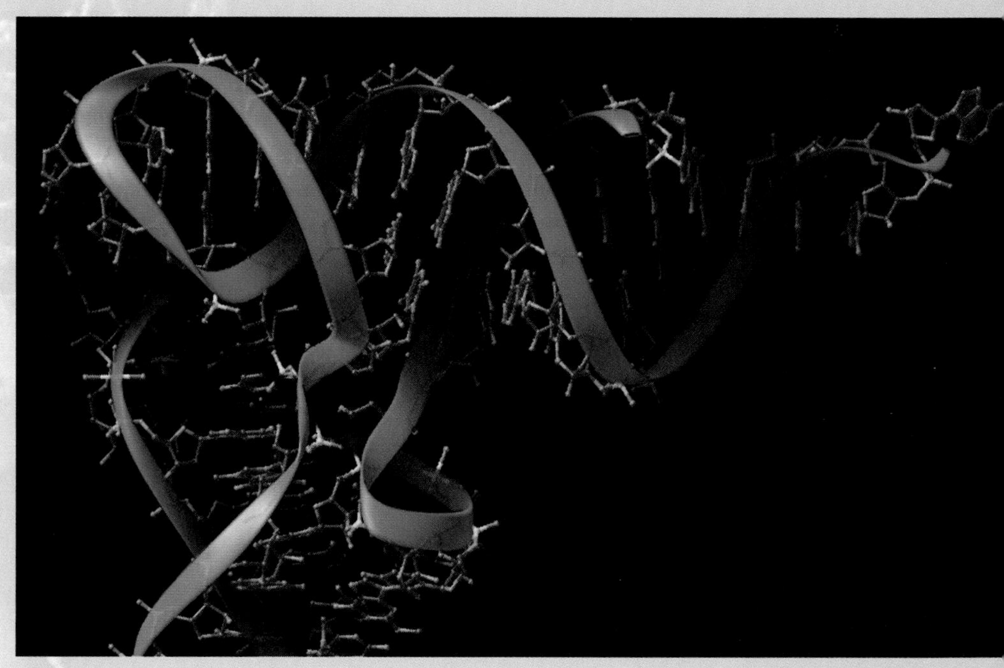

CHAPTER 4

Nucleic Acids

IN THE NEXT SEVERAL CHAPTERS WE INTRODUCE THREE MAJOR CLASSES OF BIO-molecules—nucleic acids, proteins, and polysaccharides. Together, they make up a large part of all living matter. They direct and carry out that vast array of chemical processes that constitute life. Although they are remarkably different in structure, nucleic acids, proteins, and polysaccharides have certain features in common. As we saw in Chapter 1, these substances exist as macromolecules, some of them truly giant. We shall find all of these macromolecules to be polymers; each type is made by the linking of a limited number of kinds of monomer units. In each case the sequence of units is specific.

The Nature of Nucleic Acids

It is appropriate to begin this section with the **nucleic acids,** for in a certain sense they are the most fundamental constituents of the living cell. It seems probable that life itself began its evolution with nucleic acids, for only they, of all biological substances, carry the remarkable potential for self-duplication. At the present, nucleic acids act as the repositories and transmitters of genetic information for every cell, tissue, and organism. The blueprint for an organism is encoded in its nucleic acid, in gigantic molecules like that shown in Figure 1.5 (page 9). Much of an organism's physical development throughout life is programmed in these molecules. The proteins that its cells will make and the functions that they will perform are all recorded on this molecular tape.

In this chapter, and in the several that follow, we first describe the nucleic acids and then provide a brief introduction to the ways in which they preserve and transmit genetic information. The details of these processes are covered in Part V of this book, but it is important that we consider, at the beginning, the role that nucleic acids play in the formation of proteins and cellular structure.

THE TWO TYPES OF NUCLEIC ACID: DNA AND RNA

There are two types of nucleic acid, **ribonucleic acid** (RNA) and **deoxyribonucleic acid** (DNA). As Figure 4.1 shows, each is a polymeric chain, in which the monomer units are connected by covalent bonds. Structures of the monomer units of RNA and DNA are shown here:

Phosphate, Ribose — **Repeating unit of ribonucleic acid (RNA)**

Phosphate, 2′-deoxyribose — **Repeating unit of deoxyribonucleic acid (DNA)**

In each case the monomer unit contains a five-carbon sugar, **ribose** in RNA and **2′-deoxyribose** in DNA (shown in blue in the structures). The carbon atoms are designated by primes (1′, 2′, etc.) to distinguish them from atoms in the bases. The difference between the two sugars lies solely in the 2′ hydroxyl group on ribose in RNA, which is replaced by hydrogen in DNA. The connection between

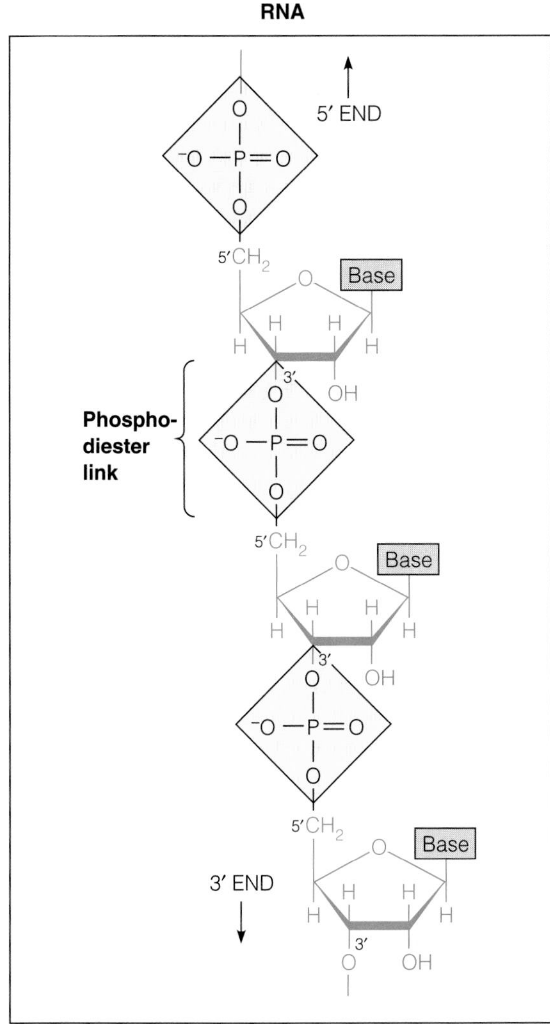

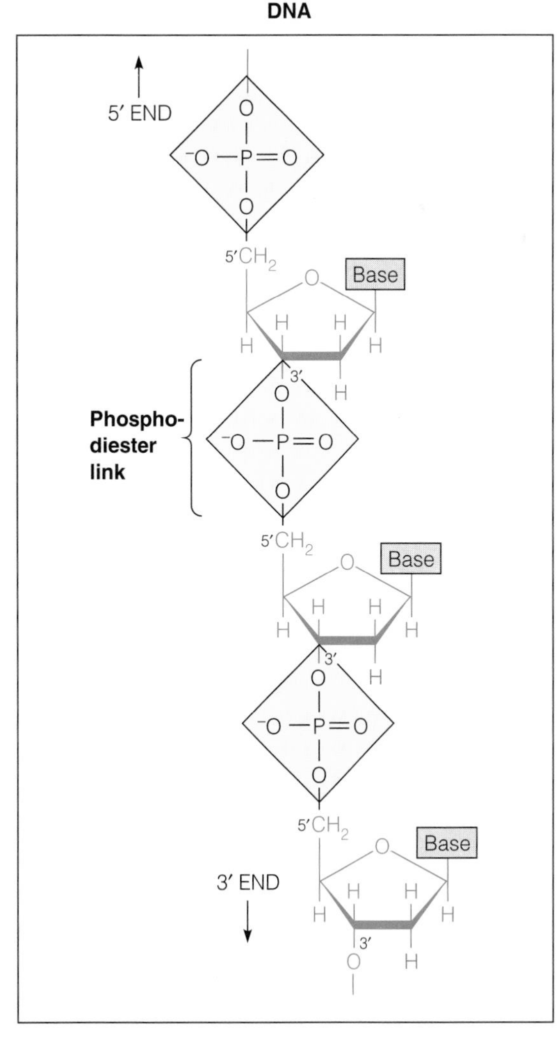

FIGURE **4.1**

Chemical structures of ribonucleic acid (RNA) and deoxyribonucleic acid (DNA). The ribose–phosphate or deoxyribose–phosphate backbone of each chain is shown in detail. The bases shown schematically here are detailed in Figure 4.2.

Both DNA and RNA are polynucleotides. RNA has the sugar ribose; DNA has deoxyribose.

The nucleic acid bases are of two kinds: the purines, adenine and guanine, and the pyrimidines, cytosine, thymine, and uracil. RNA and DNA use the same bases, except RNA uses uracil where DNA uses thymine.

successive monomer units in nucleic acids is through a phosphate residue attached to the hydroxyl on the 5′ carbon of one unit and the 3′ hydroxyl of the next one. This forms a **phosphodiester link** between two sugar residues (Figure 4.1). In this way, long nucleic acid chains, sometimes containing hundreds of millions of units, are built up. The phosphate group is a strong acid, with a pK_a of about 1; this is why DNA and RNA are called nucleic *acids.* Every residue in a DNA or RNA molecule carries a negative charge at physiological pH.

The phosphodiester-linked sugar residues form the backbone of the nucleic acid molecule. By itself, the backbone is a repetitious structure, incapable of encoding information. The importance of the nucleic acids in information storage and transmission derives from their being **heteropolymers.** Each monomer in the chain carries a heterocyclic base, always linked to the 1′ carbon of the sugar (see Figure 4.1). The structures of the major bases found in the nucleic acids are shown in Figure 4.2. There are two types of heterocyclic bases, called **purines** and **pyrimidines.** DNA has two purines, **adenine** (A) and **guanine** (G), and two pyrimidines, **cytosine** (C) and **thymine** (T). RNA has the same bases except that **uracil** (U) replaces thymine. RNA and, to a lesser extent, DNA also contain a small fraction of chemically modified bases. We will discuss these modified bases in later sections of this chapter.

DNA and RNA can each be regarded as a polymer made from four kinds of monomers. The monomers are phosphorylated ribose or deoxyribose molecules

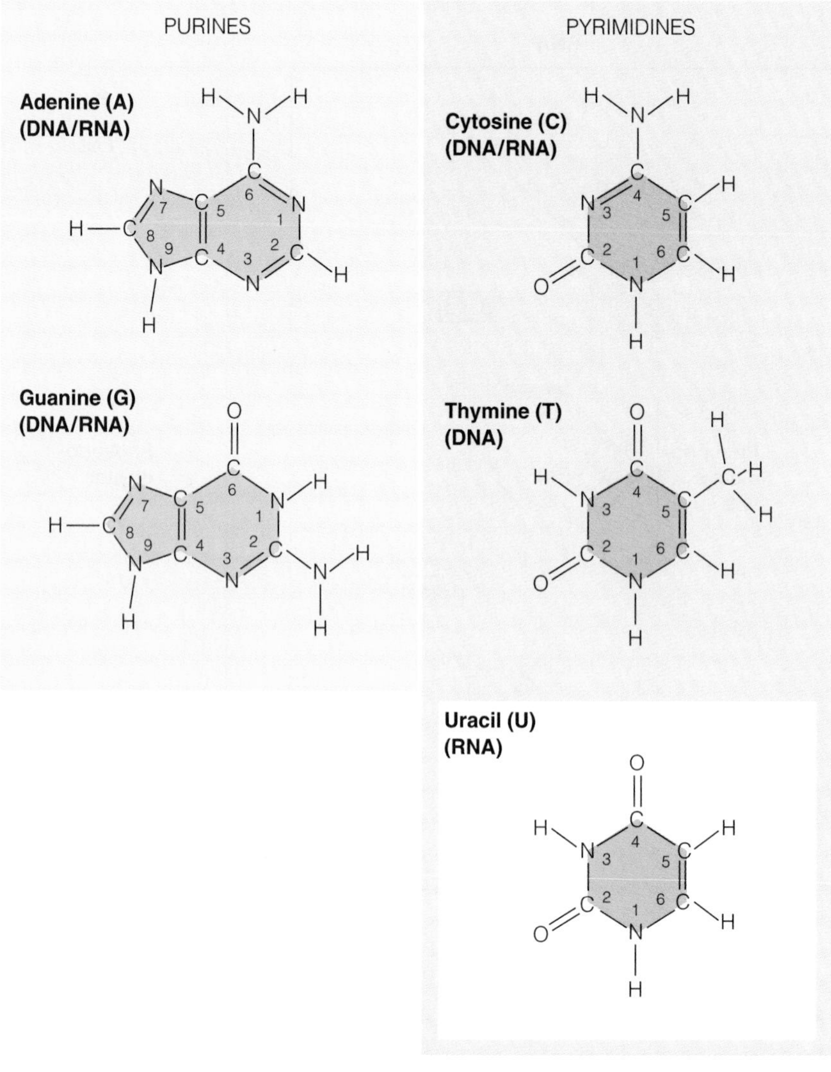

FIGURE **4.2**

Purine and pyrimidine bases found in DNA and RNA. DNA always contains the bases A, G, C, T, whereas RNA always has A, G, C, U. Thymine is simply 5-methyluracil.

with purine or pyrimidine bases attached to their 1′ carbons. In purines the attachment is at nitrogen 9, in pyrimidines at nitrogen 1. The bond between the 1′ carbon of the sugar and the base nitrogen is referred to as a **glycosidic bond.** These monomers are called **nucleotides.** Each nucleotide can be considered the 5′-monophosphorylated derivative of a sugar-base adduct called a **nucleoside** (Figure 4.3). Thus, the nucleotides could also be called *nucleoside 5′-monophosphates.* You have already encountered one of these molecules in Chapter 3: adenosine 5′ monophosphate, or AMP.

Since all of the nucleic acids may be regarded as polymers of nucleotides, they are often referred to by the generic name **polynucleotides.** Small polymers, containing only a few residues, are called **oligonucleotides.**

PROPERTIES OF THE NUCLEOTIDES

Nucleotides are quite strong acids; the primary ionization of the phosphate occurs with a pK_a of approximately 1.0. Both secondary ionization of the phosphate and protonation or deprotonation of some of the groups on the bases within the nucleotides can be observed at pH values quite close to neutrality (Table 4.1.)

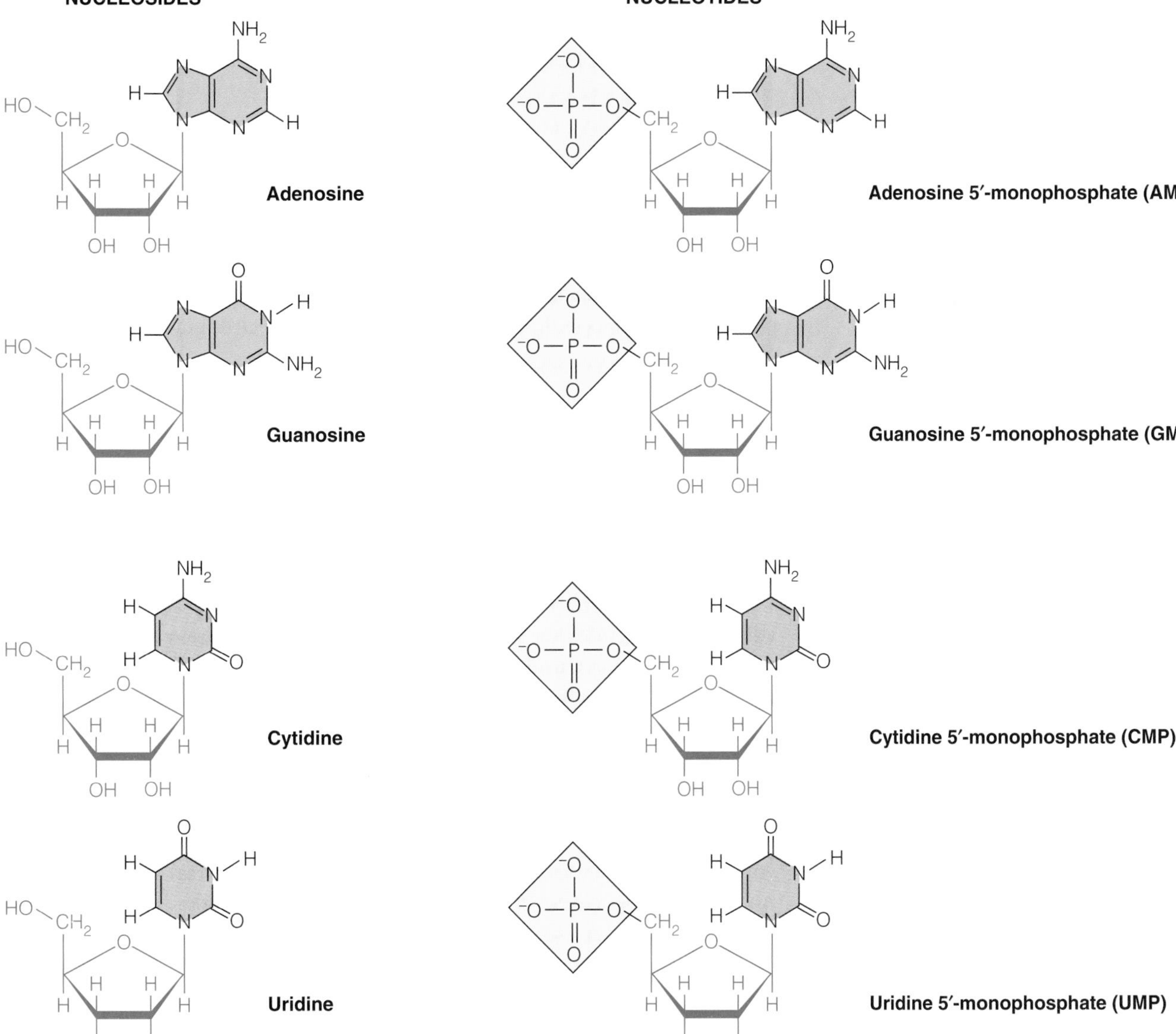

FIGURE **4.3**

Nucleosides and nucleotides. The ribonucleosides and ribonucleotides are shown here; the deoxyribonucleosides and deoxyribonucleotides are identical, except that they lack the 2′OH, and except that T in DNA substitutes for U in RNA. Each nucleoside is formed by coupling ribose or deoxyribose to a base. The nucleotides, which can be considered the monomer units of nucleic acids, are the 5′-monophosphates of the nucleosides. Nucleoside phosphates with phosphorylation at other hydroxyl groups exist, but they are not found in nucleic acids.

TABLE 4.1 Ionization constants of ribonucleotides expressed as pK_a values

	Phosphate		Base	
	Primary Ionization: $HO{-}P(=O)(OH){-}R \rightleftharpoons HO{-}P(=O)(O^-){-}R + H^+$	Secondary Ionization: $HO{-}P(=O)(O^-){-}R \rightleftharpoons {}^-O{-}P(=O)(O^-){-}R + H^+$		
	pK_{a1}	pK_{a2}	pK_a	Reaction (as Loss of Proton from)
5′ AMP	0.9	6.1	3.8	N-1
5′ GMP	0.7	6.1	2.4	N-7
			9.4	N-1
5′ UMP	1.0	6.4	9.5	N-3
5′ CMP	0.8	6.3	4.5	N-3

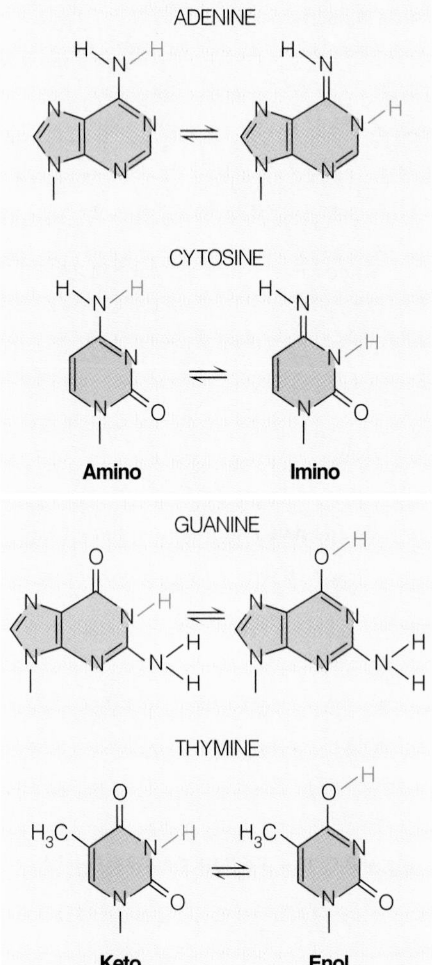

FIGURE **4.4**

Tautomerization of the bases. The most stable (and therefore common) forms are shown on the left. The less common imino and enol forms, shown on the right, are found in some special base interactions. Still other tautomers (not shown here) are possible.

The bases are also capable of conversion between **tautomeric** forms. Tautomeric forms, or **tautomers**, are structural isomers differing in the location of their hydrogen atoms and double bonds. The major forms are those shown in Figure 4.2, but G, T, and U can partially isomerize to enol forms, A and C to imino forms, as shown in Figure 4.4.

As a consequence of the conjugated double-bond systems in the purine and pyrimidine rings, the bases and all of their derivatives (nucleosides, nucleotides, and nucleic acids) absorb light strongly in the near-ultraviolet region of the spectrum. This absorption depends somewhat on pH, because of the ionization reactions in the bases; representative spectra at neutral pH for ribonucleotides are depicted in Figure 4.5. This strong absorbance is often used for quantitative determination of nucleic acids because it makes it possible to measure nucleic acid concentrations at the microgram/mL level by spectrophotometry (see Tools of Biochemistry 6A).

Ultraviolet light can also have chemically damaging effects on DNA, leading, for example, to skin cancer.

STABILITY AND FORMATION OF THE PHOSPHODIESTER LINKAGE

If we compare the structures of the nucleotides shown in Figure 4.3 with the polynucleotide chains depicted in Figure 4.1, we see that, in principle, a polynucleotide could be generated from its nucleotide monomers by elimination of a water molecule between each pair of monomers. That is, we might imagine adding another nucleotide residue to a polynucleotide chain by the dehydration reaction shown in Figure 4.6. However, the free energy change in this hypothetical reaction is quite positive, about +25 kJ/mol; therefore equilibrium lies far to the side of hydrolysis of the phosphodiester bond in the aqueous environment of the cell. Hydrolysis of polynucleotides to nucleotides is the thermodynamically favored process.

We encounter here the first of many examples of the **metastability** of biologically important polymers. Metastable compounds are thermodynamically favored to break down, but do so only very slowly unless the reaction is catalyzed. According to the free energy change involved, polynucleotides should hydrolyze under the conditions existing in living cells, but their hydrolysis is exceedingly slow unless catalyzed. This characteristic is of the greatest importance, for it ensures that the DNA in cells is sufficiently stable to serve as a useful repository of genetic information. In dehydrated conditions, DNA is so stable that it has even

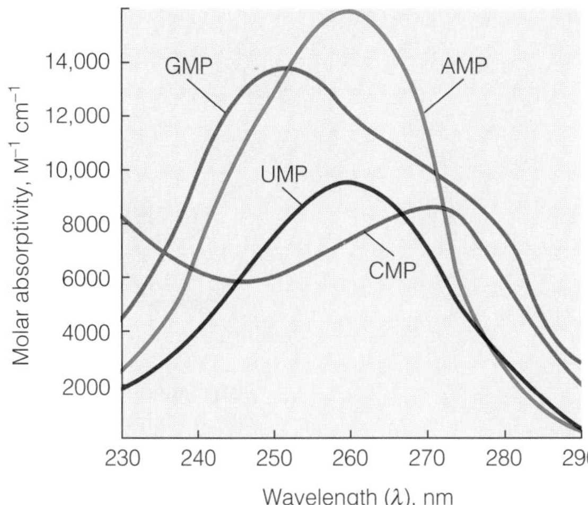

FIGURE **4.5**

Ultraviolet spectra of ribonucleotides. The dimensions of the absorption coefficients are $\text{M}^{-1}\text{cm}^{-1}$. Thus a 10^{-4} M solution of UMP would have an absorbance of 0.95 at 260 nm in a 1-cm-thick cuvette. (Absorbance = molar absorptivity × light path in cm × molar concentration; see Tools of Biochemistry 6A.)

Data from A. L. Lehninger, D. L. Nelson, and M. M. Cox, *Principles of Biochemistry,* 2nd ed. (New York: Worth, 1993). © 1993, 1982 by Worth Publishers, Inc. Used with permission.

been possible to recover fragments of DNA molecules from some ancient fossils. When catalysts *are* present, however, hydrolysis can be exceedingly rapid in aqueous solution. Acid catalysis leads to hydrolysis of the phosphodiester bonds in RNA, yielding a mixture of nucleotides. In both RNA and DNA, the glycosidic bond between the base and the sugar is also hydrolyzed; a mixture of bases, phosphoric acid, and ribose (or deoxyribose) is produced. RNA, but not DNA, is also labile in alkaline solution; treatment with 0.1 M alkali yields a mixture of 2′ and 3′ nucleoside phosphates. Finally, and most important biologically, the enzymes called **nucleases** catalyze the hydrolysis of phosphodiester bonds in both RNA and DNA. Your body is able to break down and utilize polynucleotides in the foodstuffs you consume because your digestive system contains nucleases. Examples of such enzymes are described in Chapter 11.

The unfavorable thermodynamics of the hypothetical dehydration reaction shown in Figure 4.6 leads us to ask: If polynucleotides cannot be synthesized in vivo by the direct elimination of water, how are they actually made? The answer is that their synthesis involves the high-energy nucleoside or deoxynucleoside *triphosphates.* Although the process as it occurs in cells is quite complex, the basic reaction is simple. Instead of the dehydration reaction of Figure 4.6, what happens in living cells is the reaction shown in Figure 4.7. The nucleoside monophosphate being added to the growing chain is presented as a *nucleoside triphosphate,* like ATP or deoxy ATP (dATP), and pyrophosphate is released in the reaction. We can calculate the free energy change for this reaction by noting that it can be considered the sum of two reactions—hydrolysis of a nucleoside triphosphate and formation of a phosphodiester link by elimination of water:

				$\Delta G^{\circ\prime}$
	Nucleoside triphosphate + H_2O	$\rightleftharpoons$	nucleoside monophosphate + pyrophosphate (PP_i)	−31 kJ/mol
	(Polynucleotide chain)$_N$ + nucleoside monophosphate	$\rightleftharpoons$	(polynucleotide chain)$_{N+1}$ + H_2O	+ 25 kJ/mol
Sum:	(Polynucleotide chain)$_N$ + nucleoside triphosphate	$\rightleftharpoons$	(polynucleotide chain)$_{N+1}$ + pyrophosphate (PP_i)	−6 kJ/mol

The coupled reaction is favorable because the *net* $\Delta G^{\circ\prime}$ is negative. The reaction is further favored because the hydrolysis of the pyrophosphate product (PP_i) to orthophosphate (P_i) has a $\Delta G^{\circ\prime} = -33$ kJ/mol. Thus, the pyrophosphate is readily removed, driving the synthesis reaction even further to the right and yielding an overall $\Delta G^{\circ\prime}$ of -39 kJ/mol. Polynucleotide synthesis is an example of a principle we emphasized in Chapter 3—the use of favorable reactions to drive thermodynamically unfavorable ones.

FIGURE **4.6** *(left)*

Formation of a polynucleotide by a hypothetical dehydration reaction. We might imagine that a polynucleotide could be formed directly from nucleoside monophosphates by removal of water, as shown here, but this dehydration reaction is thermodynamically unfavorable. The reverse reaction, hydrolysis, is favored. Note that in this and subsequent figures we adopt a somewhat more compact way of representing the sugar-phosphate backbone.

FIGURE **4.7** *(right)*

How polynucleotides are actually formed. In this reaction, each monomer is presented as a nucleoside triphosphate to be added to the chain. Cleavage of the nucleoside triphosphate provides the free energy that makes the reaction thermodynamically favorable. The enzymes catalyzing such reactions are called **polymerases.**

Polynucleotide with *N* residues

+

Deoxynucleoside monophosphate

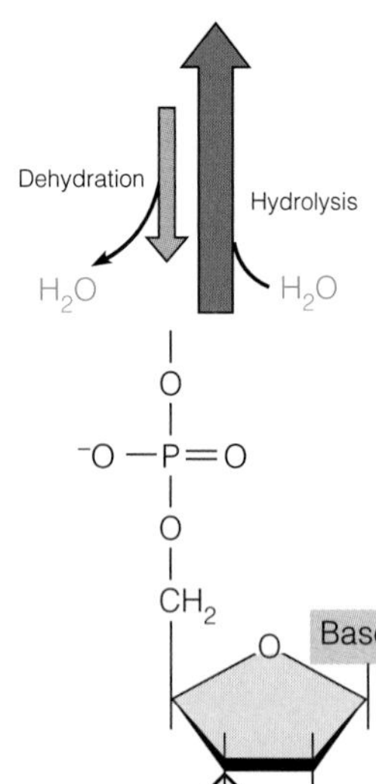

Polynucleotide with *N* + 1 residues

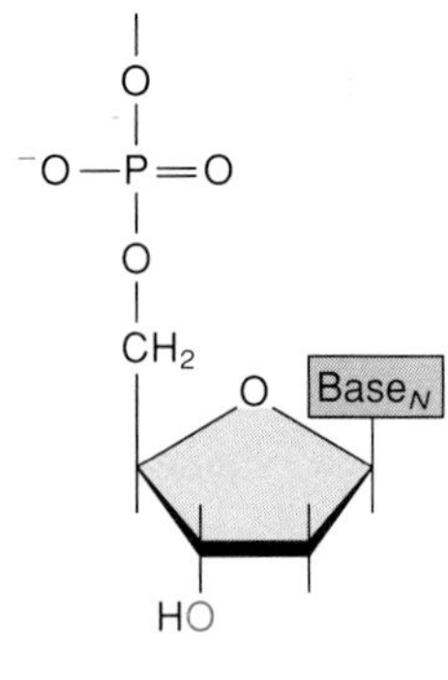

Polynucleotide with *N* residues

+

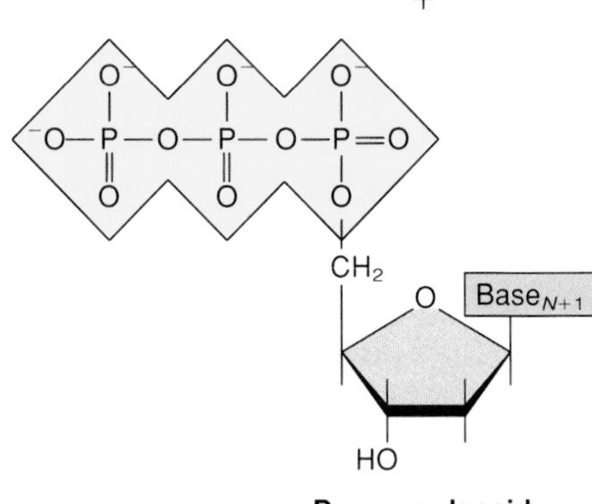

Deoxynucleoside triphosphate

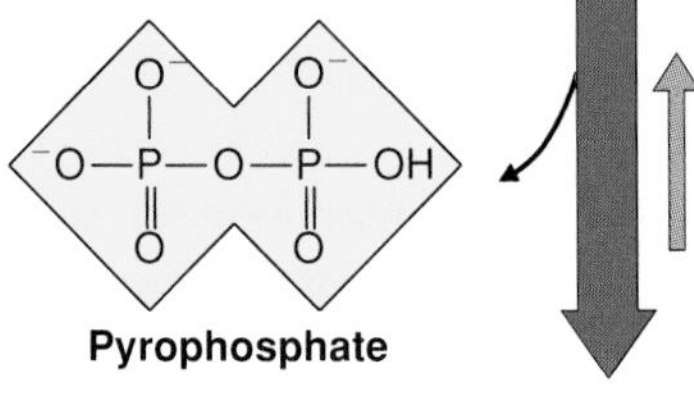

Pyrophosphate

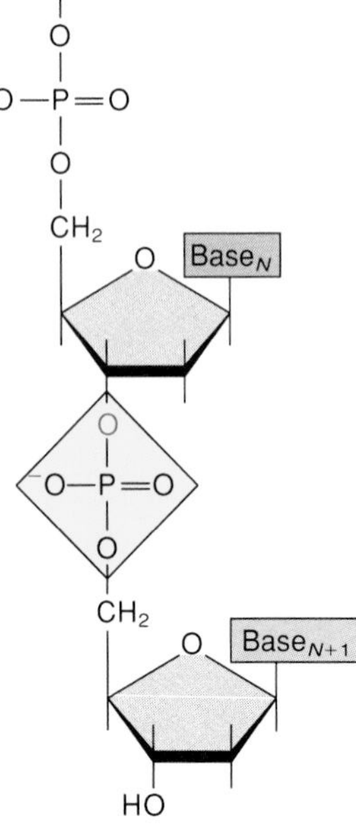

Polynucleotide with *N* + 1 residues

It is important to appreciate how the energetics of such processes fit into the overall scheme of life. An organism obtains energy—either from photosynthesis, if it is a plant, or from metabolism of foodstuffs—and stores part of this energy by generating ATP, GTP, dATP, dGTP, and similar high-energy compounds. It uses these compounds in turn as energy sources to drive the synthesis of macromolecules like DNA, RNA, and proteins. This use of triphosphates as the energy currency of the cell is a theme that you will see repeated again and again throughout this book.

Primary Structure of Nucleic Acids

THE NATURE AND SIGNIFICANCE OF PRIMARY STRUCTURE

A closer examination of Figure 4.1 reveals two important features of all polynucleotides:

1. A polynucleotide chain has a *sense* or *directionality.* The phosphodiester linkage between monomer units is between the 3′ carbon of one monomer and the 5′ carbon of the next. Thus, the two ends of a linear polynucleotide chain are distinguishable. One end normally carries an unreacted 5′ phosphate, the other end an unreacted 3′ hydroxyl group.
2. A polynucleotide chain has *individuality,* determined by the sequence of its bases—that is, the *nucleotide sequence.* This sequence is called the **primary structure** of that particular nucleic acid.

If we want to describe a particular polynucleotide sequence (either DNA or RNA), it is exceedingly awkward and wholly unnecessary to draw the molecule in its entirety as in Figure 4.1. Accordingly, some compact nomenclatures have been devised. If we state that we are describing a DNA molecule or an RNA molecule, then most of the structure is understood. We can then abbreviate a small DNA molecule as follows:

Every naturally occurring polynucleotide has a defined sequence, its primary structure.

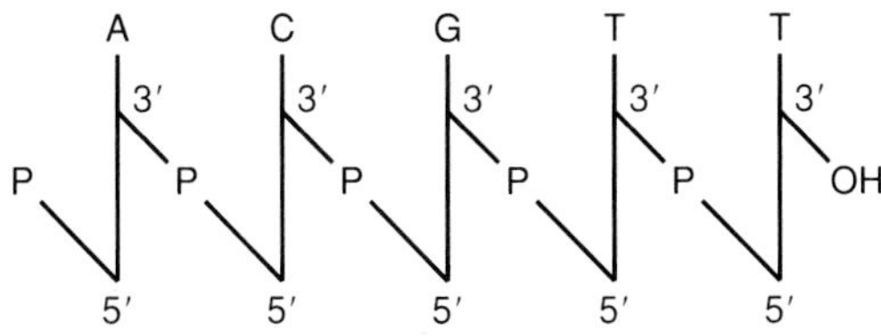

This notation shows (1) the sequence of nucleotides, by their letter abbreviations (A, C, G, T); (2) that all phosphodiester links are between 3′ hydroxyls and 5′ phosphates; and (3) that this particular molecule has a phosphate group at its 5′ end and an unreacted 3′ hydroxyl at its 3′ end. It also tells us it is a DNA sequence, not RNA, because it has T, not U.

If all of the phosphodiester links can be assumed to link a 3′ hydroxyl to a 5′ phosphate (as is usually the case), a more compact notation is possible for the same molecule:

pApCpGpTpT

The 3′ —OH group is understood to be present and unreacted. Were there a phosphate on the 3′ end and an unreacted hydroxyl on the 5′ end, we would write

ApCpGpTpTp

Finally, if we are concerned *only* with the sequence of bases in the molecule, as will often be the case, we can write it still more compactly as

ACGTT

Note that the sequence of a polynucleotide chain is usually written, by convention, with the 5′ end to the left and the 3′ end to the right.

The primary structure of DNA encodes genetic information.

The main importance of primary structure, or sequence, is that *genetic information is stored in the primary structure of DNA*. A *gene* is nothing more than a particular DNA sequence, encoding information in a four-letter language in which each "letter" is one of the bases.

DNA AS THE GENETIC SUBSTANCE: EARLY EVIDENCE

The search for the substance of which genes are made has a long history. In the late 1800s, shortly after the German biochemist Friedrich Miescher had first isolated DNA from salmon sperm, some scientists suspected that DNA might be the genetic material. But subsequent studies showing that DNA contained only four kinds of monomers seemed to deny that it could have such a complicated role. Early researchers thought it more likely that genes were made of proteins, for proteins were beginning to be recognized as extremely complex molecules. For most of the first half of the twentieth century, nucleic acids were considered to be merely some kind of structural material in the cell nucleus.

Between 1944 and 1952 a series of crucial experiments clearly pointed to DNA as the genetic material. In 1944 Oswald Avery, Colin MacLeod, and Maclyn McCarty found that the DNA from pathogenic strains of the bacterium *Pneumococcus* could be transferred into nonpathogenic strains, making them pathogenic (Figure 4.8a). The transformation was genetically stable; succeeding generations of bacteria retained the new characteristics. However, it was an elegant experiment by Alfred Hershey and Martha Chase that finally convinced many scientists. Hershey and Chase studied the infection of the bacterium *Escherichia coli* by a bacterial virus, the bacteriophage T2. Making use of the fact that the bacteriophage proteins contain sulfur but little phosphorus and that the bacteriophage DNA contains phosphorus but no sulfur, they labeled T2 bacteriophage with the radioisotopes ^{35}S and ^{32}P (Figure 4.8b). They then showed that when the bacteriophage attached to *E. coli*, it was mainly the ^{32}P (and hence the bacteriophage DNA) that was transferred to the bacteria. Even if the residual protein part of the bacteriophage was shaken off the bacteria, the inserted DNA alone was sufficient to direct the formation of new bacteriophage.

Through these and similar experiments, it was generally recognized by 1952 that DNA must be the genetic substance. But how could it carry the enormous amount of information that a cell needed, how could it transmit that information to the cell, and, above all, how could it be accurately replicated in cell division? The answers to these questions came only after one of the most momentous discoveries in the history of science. In 1953 James Watson and Francis Crick proposed a structure for DNA that opened a whole new world of molecular biology.

Secondary and Tertiary Structure of Nucleic Acids

THE DOUBLE HELIX

Watson and Crick sought the answers to the questions posed above in the three-dimensional structure of DNA. For some time, a number of laboratories had been investigating fibers drawn from concentrated DNA solutions, using the technique of **x-ray diffraction** (see Tools of Biochemistry 4A). Watson and Crick, working at Cambridge University in England, had access to DNA diffraction patterns photographed by Rosalind Franklin, a researcher in the laboratory of Maurice Wilkins at King's College, London. The critical photographs were some of the best patterns from wet DNA fibers that had yet been obtained. They clearly

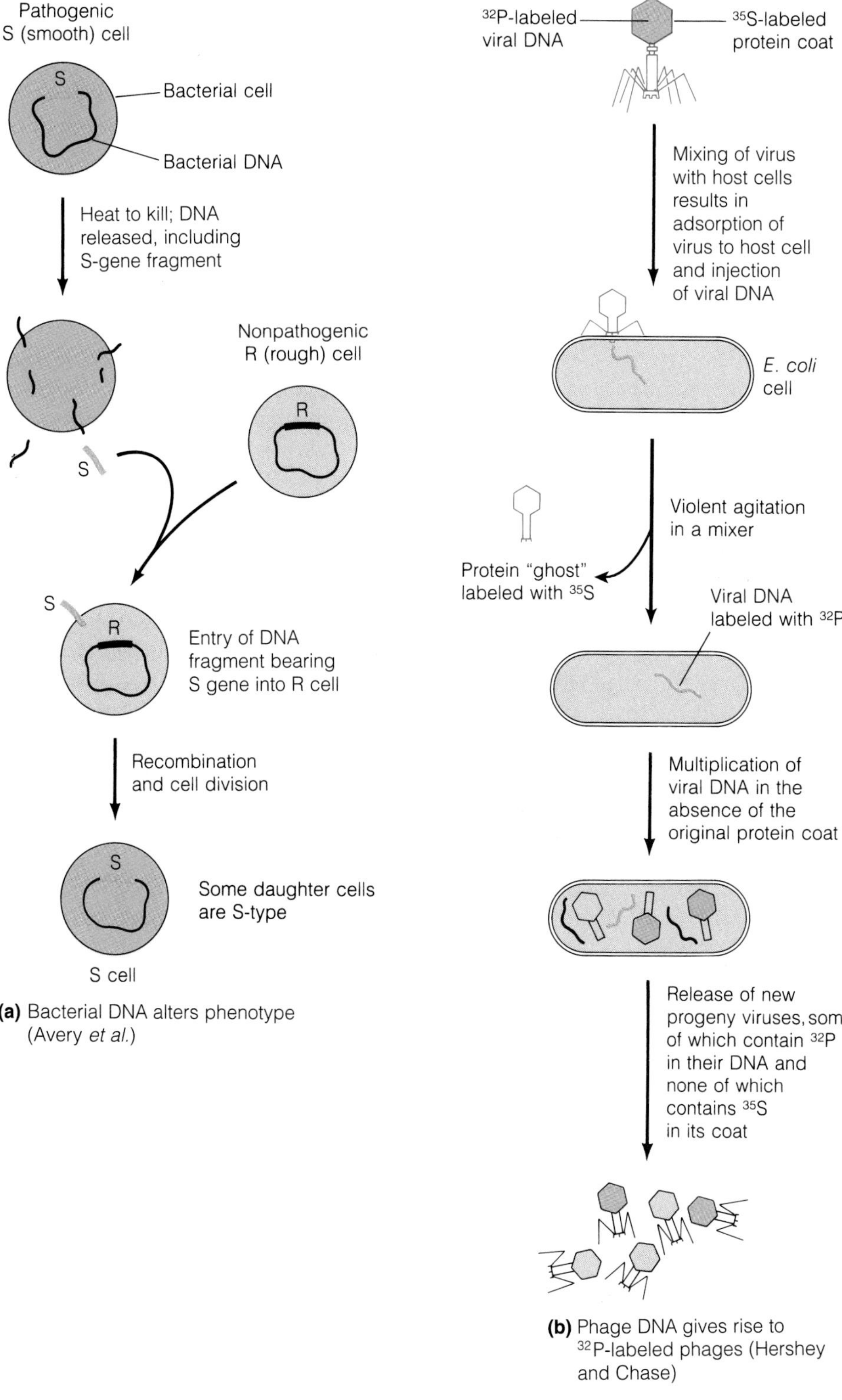

FIGURE **4.8**

Experiments that showed DNA to be the genetic substance. (a) Avery et al. showed that nonpathogenic pneumococci could be made pathogenic by transfer of DNA from a pathogenic strain. **(b)** Hershey and Chase showed that it is the transfer of just the viral DNA from a virus to a bacterium that gives rise to new viruses.

showed that the DNA in the fibers must have some kind of regular, repetitive three-dimensional structure. We refer to such regular folding in polymers as *secondary structure,* as distinguished from the primary structure, which is simply the sequence of residues.

Watson and Crick quickly recognized that the DNA fiber diffraction exhibited a cross pattern typical of a helical secondary structure (Figure 4.9). They noted that because the layer line spacing was one-tenth of the pattern repeat, there must

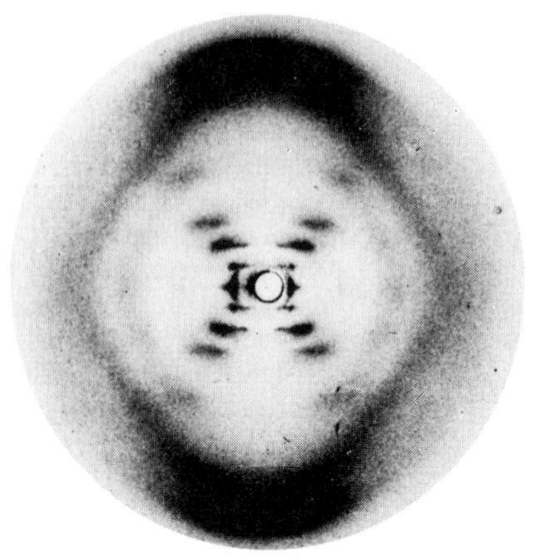

FIGURE **4.9**

Evidence for the structure of DNA. This photograph, taken by Rosalind Franklin, shows the x-ray diffraction pattern produced by wet DNA fibers. It played a key role in the elucidation of DNA structure. The cross pattern indicates a helical structure, and the strong spots at top and bottom correspond to a helical rise of 0.34 nm. The layer line spacing is one-tenth of the distance from the center to either of these spots, showing that there are 10 base pairs per repeat.

Reprinted by permission from R. E. Franklin and R. Gosling, *Nature* (1953) 171:740; © 1953 Macmillan Magazines, Ltd.

be 10 residues per turn (see Tools of Biochemistry 4A). Data on the density of the fibers suggested that there must be *two* DNA strands in each helical molecule. So far, only direct scientific deductions had been made from the data. The great leap of intuition that Watson and Crick made was the realization that a two-strand helix could be stabilized by hydrogen bonding between bases on opposite strands if the bases were paired in *one* particular way—the A-T and G-C pairs shown in Figure 4.10. With this pairing, strong hydrogen bonds are formed between the bases. Furthermore, distances between the 1′ carbons of the deoxyribose moieties of A-T and of G-C are the same—about 1.1 nm in each case (Figure 4.10a). This pairing arrangement meant that the double helix could be regular in diameter, an impossibility if purines paired with purines or pyrimidines with pyrimidines.

The Watson–Crick model for DNA was a two-strand, antiparallel double helix with 10 base pairs per turn. Pairing was A-T and G-C.

In the Watson–Crick model, the hydrophilic phosphate–deoxyribose backbones of the helix were on the outside, in contact with the aqueous environment, and the base pairs were stacked on one another with their planes perpendicular to the helix axis. Two views of such a structure are shown in Figure 4.10b and c. (The figure shows a recent, refined model, based on better data than Watson and Crick had available: The bases are not exactly perpendicular to the helix axis, and the sugar conformation is slightly different from that proposed by Watson and Crick.) Stacking of the bases, as shown in Figure 4.10b, allows strong van der Waals interaction between them; this is often referred to as "stacking interaction." Each base pair is rotated by 36°, that is, ⅒ of a 360° rotation, with respect to the next to accommodate 10 base pairs in each turn of the helix. The diffraction pattern showed the repeat distance to be about 3.4 nm, so the helix *rise*, that is, the distance between base pairs, had to be about 0.34 nm (Figure 4.10c). This distance is just twice the van der Waals thickness of a planar ring (see Table 2.2 on page 31), so the bases are closely packed within the helix, as shown in a space-filling model (Figure 4.11). The model also shows that although the bases are inside, they can be approached through two deep spiral grooves called the *major* and *minor* grooves. The major groove gives more direct access to the bases; the minor groove faces the sugar backbone. Building molecular models of two-strand DNA structures soon convinced Watson and Crick that the DNA strands must run in opposite directions. This arrangement can be seen clearly in Figure 4.10c. The model Watson and Crick presented was for a right-hand helix, although at that time evidence for the sense (direction of the turn) of the helix was weak. Their guess proved to be correct.

As is often the case with a good theory or model, the Watson–Crick structure also explained other data that had not been understood until then. The biochemist Erwin Chargaff, who had measured the relative amounts of A, T, G, and C in DNAs from many organisms, had noted the perplexing fact that A and T were almost always present in nearly equal quantities, as were G and C (Table 4.2). If most DNA in cells was double-stranded, with the Watson–Crick base pairing, then *Chargaff's rule* followed as a natural consequence.

The complementary, two-strand structure of DNA explains how the genetic material can be replicated.

The Watson–Crick model not only explained the structure of DNA and Chargaff's rule but also carried implications that went to the very heart of biology. Since A always pairs with T, and G always pairs with C, the two strands are *complementary*. If the strands could be separated and new DNA synthesized along each, following the same base-pairing rule, two double-strand DNA molecules would be obtained, each an *exact* copy of the original (Figure 4.12). This **self-replication** is precisely the property that the genetic material must have: When a cell divides, two complete copies of the genetic information carried in the original cell must be produced. In their 1953 paper announcing the model (see References), Watson and Crick expressed this idea in what may have been the most understated scientific prediction ever made: "It has not escaped our notice that the specific pairing we have postulated immediately suggests a possible copying mechanism for the genetic material."

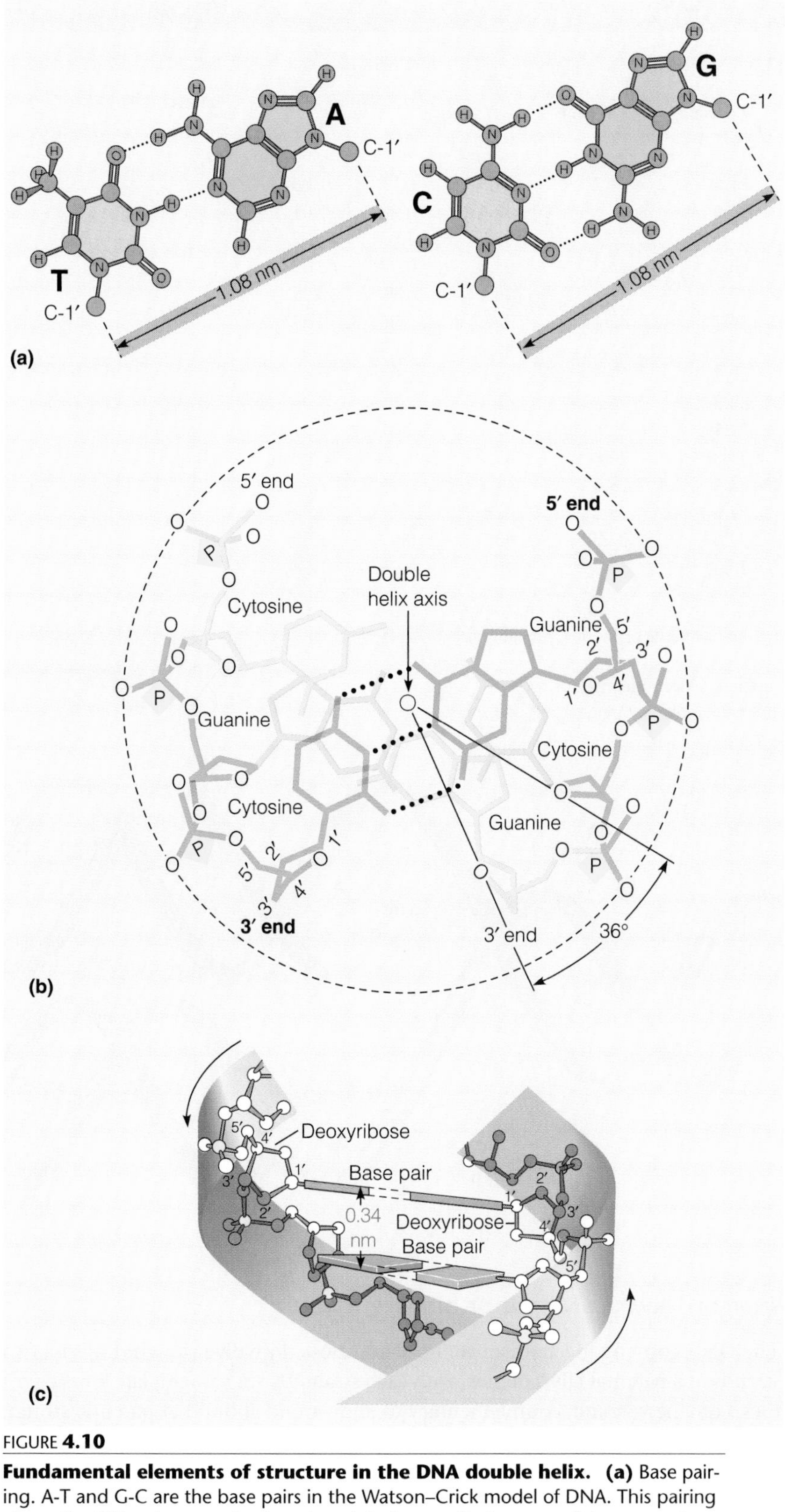

FIGURE **4.10**

Fundamental elements of structure in the DNA double helix. (a) Base pairing. A-T and G-C are the base pairs in the Watson–Crick model of DNA. This pairing allows the C1′ carbons on the two strands to be exactly the same distance apart for both pairs. **(b)** Stacking of the base pairs. This view *down* the helix axis shows how the base pairs stack on one another, with each pair rotated 36° with respect to the next. **(c)** Distance between the base pairs. A side view of the base pairs shows the 0.34-nm distance between them. This distance is called the rise of the helix.

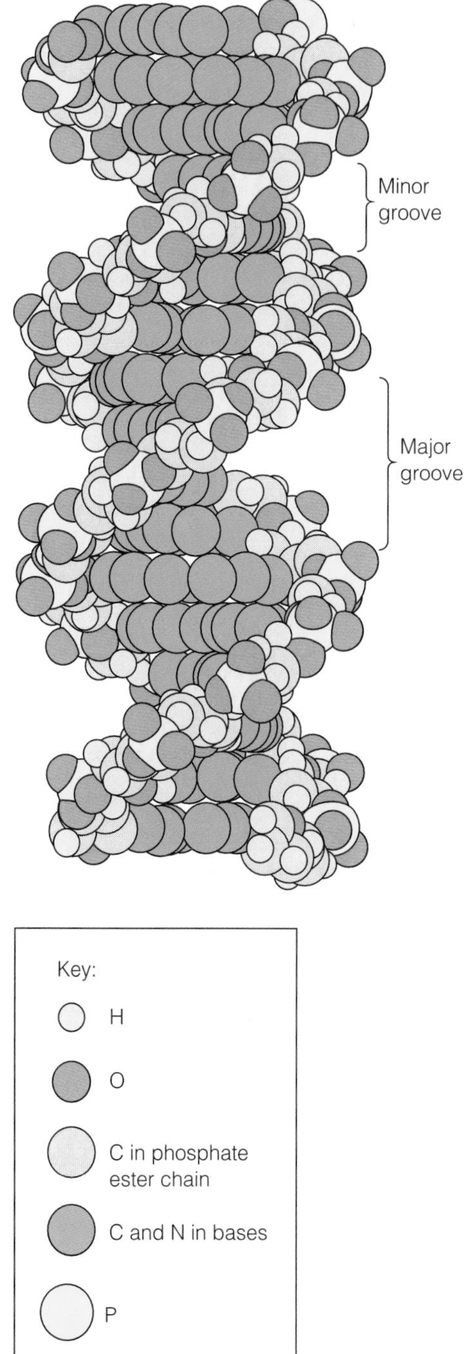

FIGURE **4.11**

A space-filling model of DNA. The DNA molecule as modeled by Watson and Crick is shown here with each atom given its van der Waals radius. This model shows more clearly than Figure 4.10 how closely the bases are packed within the helix. The major and minor grooves are indicated.

From J. D. Watson, N. H. Hopkins, J. W. Roberts, J. A. Steitz, and A. M. Weiner, *Molecular Biology of the Gene*, 4th ed. (Redwood City, CA: Benjamin/Cummings Publishing Co., 1987). © 1987 James D. Watson.

TABLE **4.2 Base compositions of DNAs from various organisms**

	Mol % of Bases					Ratios	
Source	Adenine (A)	Guanine (G)	Cytosine[a] (C)	Thymine (T)	(G + C)	A/T	G/C
Bacteriophage ϕX174	24.0	23.3	21.5	31.2	44.8	0.77[b]	1.08[b]
Bacteriophage T7	26.0	23.8	23.6	26.6	47.4	0.98	1.01
Escherichia coli B	23.8	26.8	26.3	23.1	53.2	1.03	1.02
Neurospora	23.0	27.1	26.6	23.3	53.8	0.99	1.02
Corn (maize)	26.8	22.8	23.2	27.2	46.1	0.99	0.98
Tetrahymena	35.4	14.5	14.7	35.4	29.2	1.00	0.99
Octopus	33.2	17.6	17.6	31.6	35.2	1.05	1.00
Drosophila	30.7	19.6	20.2	29.5	39.8	1.03	0.97
Starfish	29.8	20.7	20.7	28.8	41.3	1.03	1.00
Salmon	28.0	22.0	21.8	27.8	44.1	1.01	1.01
Frog	26.3	23.5	23.8	26.8	47.4	1.00	0.99
Chicken	28.0	22.0	21.6	28.4	43.7	0.99	1.02
Rat	28.6	21.4	21.6	28.4	42.9	1.01	1.00
Calf	27.3	22.5	22.5	27.7	45.0	0.99	1.00
Human	29.3	20.7	20.0	30.0	40.7	0.98	1.04

Source: Data taken from H. E. Sober (ed.), *Handbook of Biochemistry*, 2nd ed. (Chemical Rubber Publishing Co., 1970). Values for higher organisms vary slightly from one tissue to another, probably as a result of experimental error.

[a]Amount includes, for some organisms, a few percent of a modified base, 5-methylcytosine.

[b]This bacteriophage has a single-strand DNA, which need not follow Chargaff's rule.

FIGURE **4.12**

A model for DNA replication. Each strand acts as a template for a new, complementary strand. When copying is complete, there will be two double-strand daughter DNA molecules, each identical in sequence to the parent molecule. The actual process is rather complicated, but the basic principle is illustrated here.

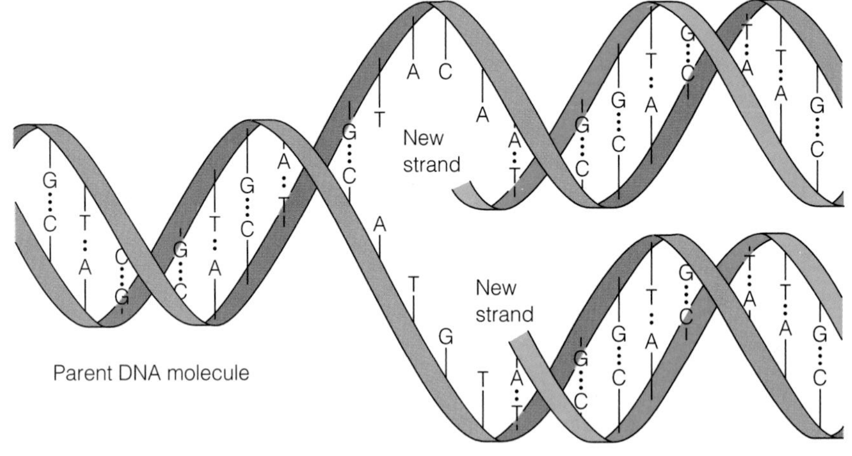

Meselson and Stahl proved that DNA replicates semiconservatively.

SEMICONSERVATIVE NATURE OF DNA REPLICATION

The DNA copying mechanism we have mentioned involves unwinding the two strands of a parental DNA duplex, with each strand serving as template for the synthesis of a new strand, complementary to and wound about that parental strand. Complete replication of a DNA molecule would yield two "daughter" duplexes, each consisting of one-half parental DNA (one strand of the original duplex) and one-half new material. This mode of replication is called **semiconservative,** because half of the original material is conserved in each of the two copies (Figure 4.13). It is distinguished from two other possible modes: **conservative,** in which one of the two daughter duplexes is the conserved parental duplex while the other is synthesized *de novo*, and **dispersive,** in which parental material is scattered through the structures of the daughter duplexes.

The first experimental test of this model came in 1958, when Matthew Meselson and Franklin Stahl realized that molecules that differ in density by very small amounts could be separated from each other by centrifugation in density gradients. In this method, a density gradient is created by centrifuging a concentrated solution of a heavy metal salt like cesium chloride (CsCl). Nucleic acid molecules of different densities suspended in such a gradient will each migrate to the point where the solution density equals their own. This technique allowed Meselson and Stahl to follow the fate of density-labeled DNA through several rounds of replication, with the results shown in Figure 4.14. The density label was applied by growing the bacterium *E. coli* in a medium containing the heavy isotope of nitrogen, ^{15}N, for many generations, so that DNA achieved a higher density through extensive substitution of ^{15}N for ^{14}N in its purine and pyrimidine bases. When isolated and centrifuged to equilibrium at pH 7.0, this DNA formed a single band in a region of the gradient corresponding to a density of 1.724 g/mL (Figure 4.14a, first graph). By contrast, when DNA from bacteria grown in light medium (containing ^{14}N) was similarly analyzed, it banded at a density of 1.710 g/mL (Figure 4.14a, second graph).

When density-labeled bacteria grown in heavy medium were transferred to the light medium, the DNA isolated after one generation of growth banded exclusively at an intermediate density, 1.717 g/mL (Figure 4.14a, third graph). This result is expected if the newly replicated DNA is a *hybrid* molecular species, consisting of one-half parental material and one-half new DNA (synthesized in light medium). If these bacteria were cultured for an additional generation in light medium, two equal-sized bands were seen, one light and one of hybrid density (Figure 4.14a, fourth graph), as expected if the hybrid-density DNA underwent a second round of semiconservative replication.

These results were consistent with the idea that each replicated chromosome contains one parental strand and one daughter strand, but the data did not preclude alternative forms of semiconservative replication, involving the breaking of DNA strands. These models were ruled out by centrifugal analysis of the density-labeled DNAs at pH 12, where DNA strands separate (Figure 4.14b). When centrifuged to equilibrium after bacterial growth in heavy medium followed by one generation in light medium, the DNA formed two bands: one light (equal to that seen on analysis of ^{14}N-grown bacteria) and one heavy (Figure 4.14b, third graph). DNA analyzed after a second round of replication in light medium showed three-fourths light and one-fourth heavy material (Figure 4.14b, fourth graph). The inescapable conclusion was that the replicative hybrid contains one complete strand of parental DNA and one complete strand of newly synthesized DNA.

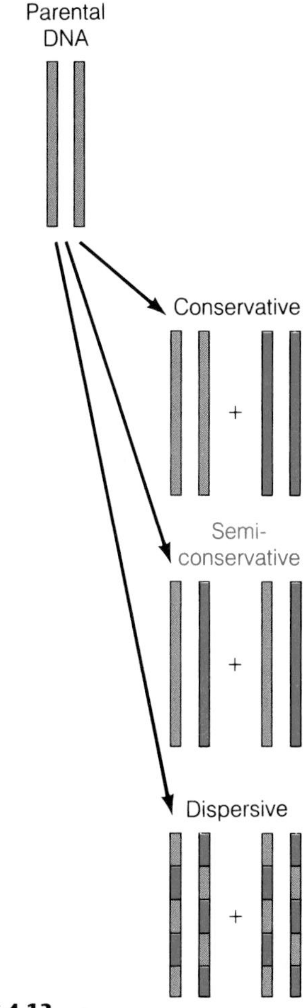

FIGURE **4.13**

Three models of DNA replication. Experimental evidence supports the semiconservative model. Brown = parental DNA, blue = new DNA.

ALTERNATIVE NUCLEIC ACID STRUCTURES: B AND A HELICES

At the time Watson and Crick proposed their model, two quite different x-ray diffraction patterns had already been obtained for DNA, indicating that the molecule can exist in more than one form. The **B form**, which is seen in DNA fibers prepared under conditions of high humidity, is shown in Figures 4.10, 4.11, 4.15a, and 4.15b. Watson and Crick chose to study the B form, because they correctly expected it to be the form found in the aqueous milieu of the cell. DNA fibers prepared under conditions of low humidity have a different structure, the so-called **A form** (Figure 4.15c and d). Although a B helix is indeed the form of DNA found in cells, the A helix is also important biologically. Double-strand RNA molecules always form the A structure, and so do **DNA–RNA hybrid molecules,** which are formed by the pairing of one DNA strand with one RNA strand. Thus, two major kinds of secondary structures exist in polynucleotides. As we shall see later in this chapter, a number of other kinds of secondary structures are possible under special circumstances, including the left-hand Z form mentioned in Table 4.3.

The two major forms of polynucleotide secondary structure are called A and B. Most DNA is in the B form; RNA–RNA and DNA–RNA helices are A form.

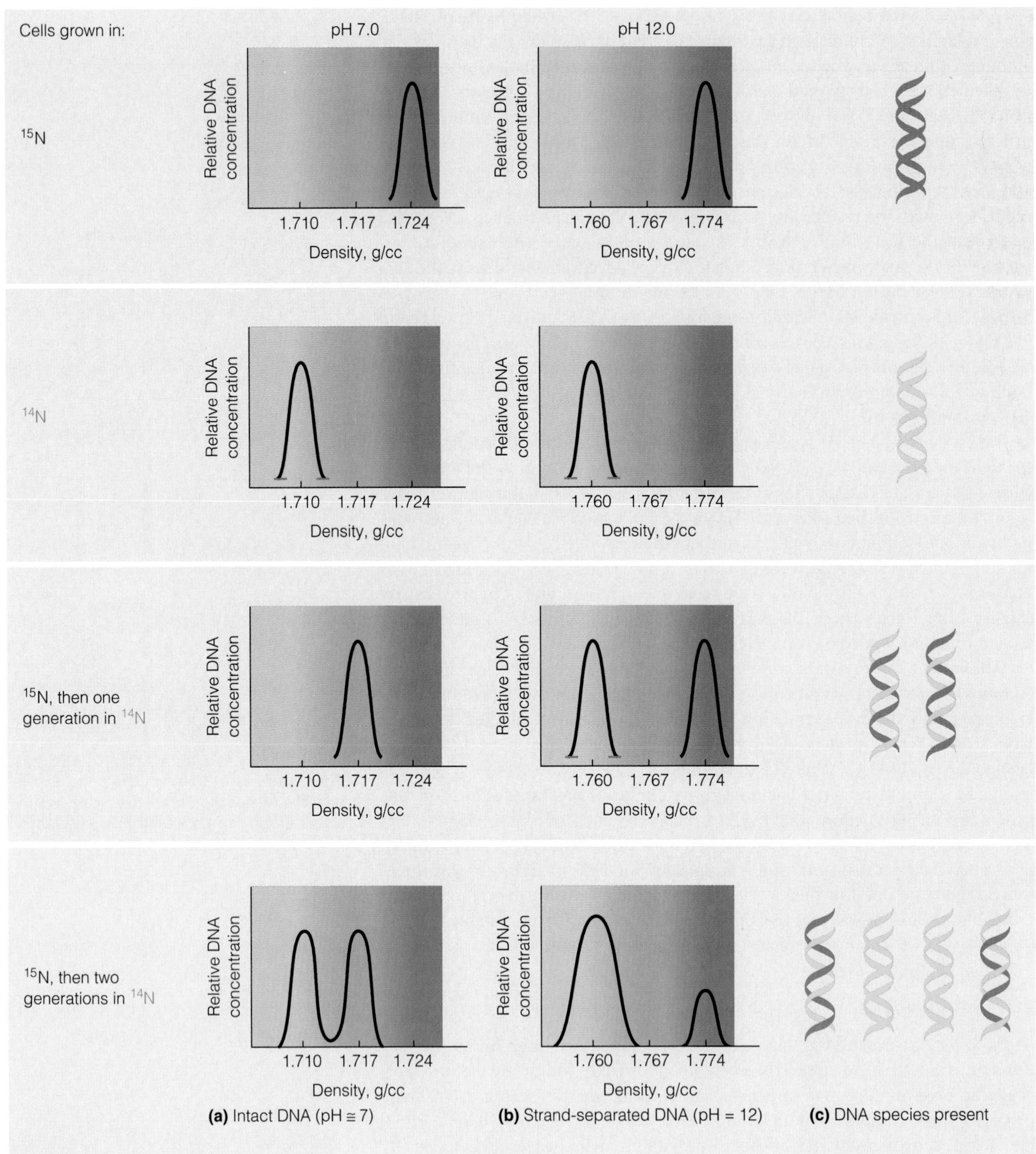

FIGURE **4.14**

The Meselson–Stahl experiment proves DNA replicates semiconservatively. The banding patterns shown were obtained in density gradients under two pH conditions. At pH 7, the DNA is double-stranded; at pH 12 the strands are separated.

As Figure 4.15 and Table 4.3 show, the A and B forms are very different, although both are right-hand helices. In the B helix the bases lie close to the helix axis, which passes between the hydrogen bonds (note the end-on views of the helices in Figure 4.15a and c). In the A helix the bases lie farther to the outside and are strongly tilted with respect to the helix axis. The surfaces of the helices are also different. In the B helix the major and minor grooves are quite distinguishable, whereas in the A helix the two grooves are more nearly equal in width.

TABLE **4.3 Parameters of polynucleotide helices**

	A Form	B Form	Z Form
Direction of helix rotation	Right	Right	Left
Number of residues per turn (n)	11	10	12 (6 dimers)
Rotation per residue ($= 360°/n$)	33°	36°	−60° per dimer; ~−30° per residue
Rise[a] in helix per residue (h)	0.255 nm	0.34 nm	0.37 nm
Pitch[a] of helix ($= nh$)	2.8 nm	3.4 nm	4.5 nm

[a]For definitions of *rise* and *pitch* of a helix, see Tools of Biochemistry 4A.

FIGURE **4.15**

Comparison of the two major forms of DNA. The structures of B-DNA and A-DNA, as deduced from recent fiber diffraction studies, are shown here in both end-on and side views.

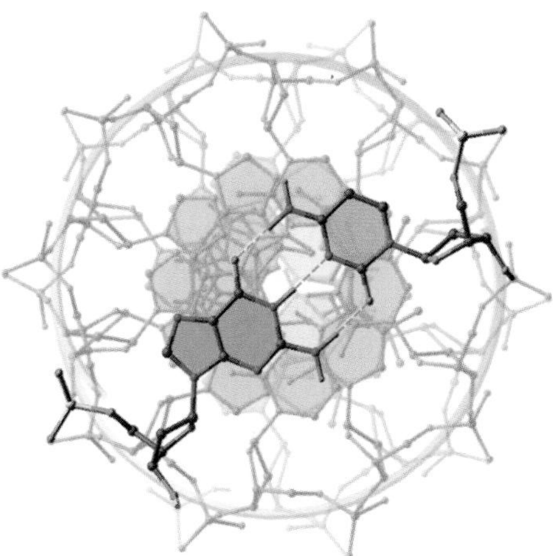

(a) B-DNA, end-on view

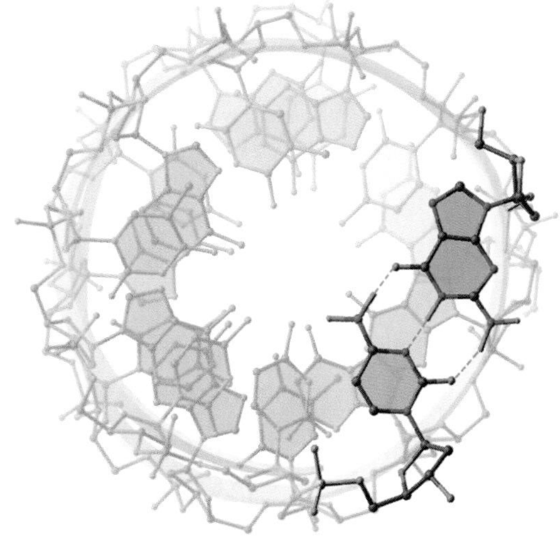

(c) A-DNA, end-on view

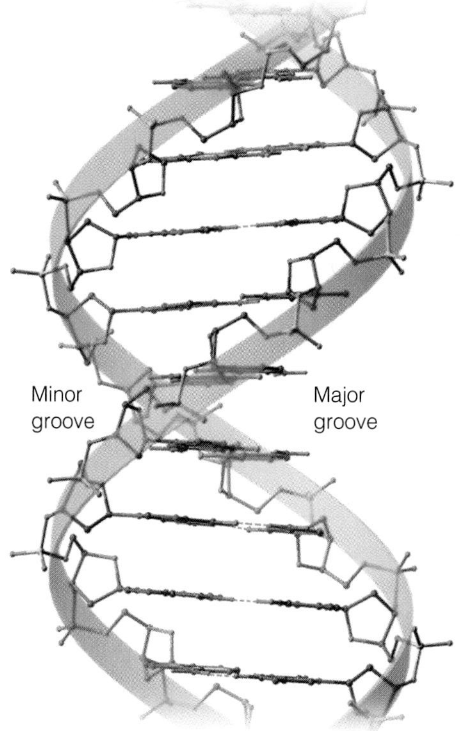

(b) B-DNA, side view

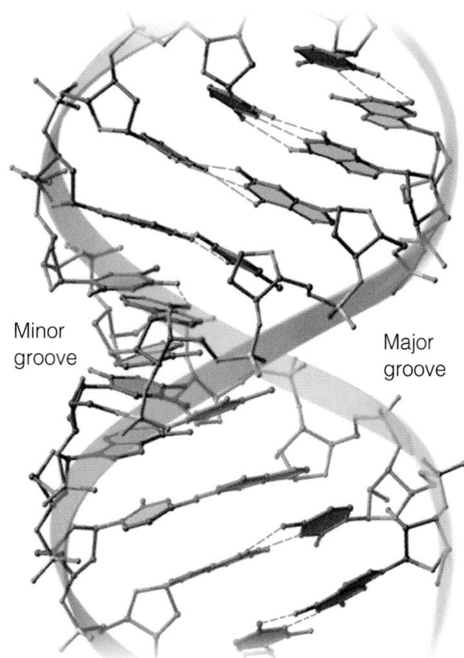

(d) A-DNA, side view

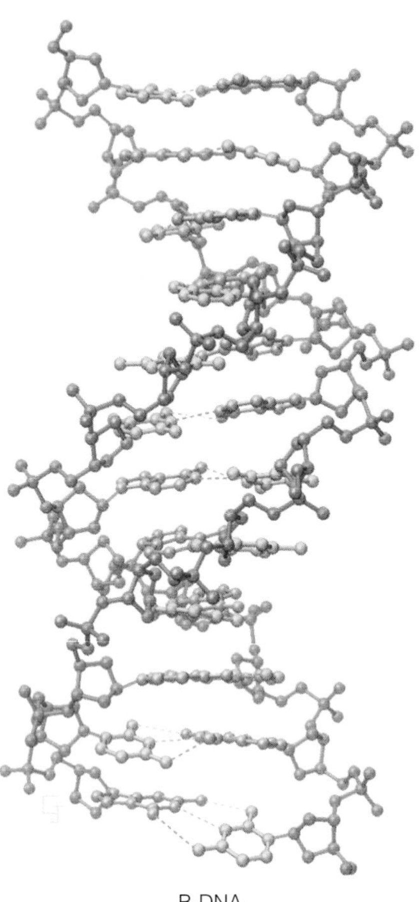

FIGURE **4.16**

The structure of B-DNA from studies of molecular crystals. Note the local distortions of the idealized structure shown in Figures 4.11 and 4.15a and b.

Adapted from R.E. Dickenson, *Sci. Am.* December 1983, pp. 100–104;

All fiber diffraction studies, including those that provided the information just described, suffer from a major limitation. In analyzing fiber patterns, researchers do not directly determine the details of a nucleic acid secondary structure. Instead, they propose models that best account for the positions and intensities of the spots on the diffraction pattern (see Tools of Biochemistry 4A). This approach is necessary because fibers are never perfect crystals, and there is always some ambiguity in the interpretation of their diffraction patterns. Therefore, a major advance was made when R. E. Dickerson and his colleagues succeeded in *crystallizing* a small double-strand DNA fragment, which had the sequence

$$
\begin{array}{l}
^{5'}\text{CGCGAATTCGCG}^{3'} \\
^{3'}\text{GCGCTTAAGCGC}^{5'}
\end{array}
$$

Molecular crystallography of this fragment and other small DNA fragments has given us much more detailed information concerning the secondary structure of polynucleotides. The results of such a study on B-DNA are shown in Figure 4.16. Here we can show DNA molecules with the position of every atom clearly specified.

A first major point that emerges from the molecular crystal studies is that the models drawn from the fiber patterns represent oversimplifications of the structures. The real structure of B-DNA involves local variations in the angle of rotation between base pairs, the sugar conformation, the tilt of the bases, and even the rise distance. If you examine Figure 4.16 carefully, you can see many distortions from the idealized structures. Nucleic acid secondary structure is not homogeneous. It varies in response to the local sequence and can be changed by interaction with other molecules. The parameters given for various forms of DNA in Table 4.3 should therefore be thought of as *average* values, from which considerable local deviation is possible.

If one examines structures like that shown in Figure 4.16 very closely, another departure from the original Watson–Crick model becomes apparent. Many DNA molecules are slightly *bent;* that is, the helix axis does not follow a straight line. The degree and directions of bending depend in a complicated way on DNA sequence. They can also be strongly influenced by the interaction of DNA with various protein molecules; we see examples of this in other chapters.

The molecular crystallographic studies also provide a possible explanation of why B-DNA is favored in an aqueous environment. The B form of DNA, but not A-DNA, can accommodate a spine of water molecules lying in the minor groove. The hydrogen bonding between these water molecules and the DNA may confer stability to the B form. According to this hypothesis, when this water is removed (as in fibers at low humidity), the B form becomes less stable than the A form.

Why, then, do double-strand RNA and DNA–RNA hybrids always adopt the A form? The answer probably lies in the extra hydroxyl group on the ribose in RNA. This hydroxyl interferes sterically in the B form by lying too close to the phosphate and carbon 8 on the adjacent base. Therefore, RNA *cannot* adopt the B form, even under conditions in which hydration might favor it. In DNA the hydroxyl is replaced by hydrogen, and such steric hindrance does not occur.

DNA AND RNA MOLECULES IN VIVO

Most DNA molecules in vivo are double-stranded; many are closed circles. Most double-strand circular DNA molecules are supercoiled.

We have described some of the major features of DNA and RNA. But in what forms do these molecules exist in the living cell? Most of the DNA in most organisms is double-stranded, with the two strands being complementary, although some DNA viruses carry single-strand DNA molecules (Table 4.4). The prevalence of B and A forms of polynucleotides in vivo is as you might expect from the conditions under which these conformations are stable. Since cells contain much water, we would expect most of the double-strand DNA to be in the B form or something very like it. There is evidence that B-DNA dissolved in solution is only

TABLE 4.4 Properties of some naturally occurring DNA molecules

Source	Single Strand (SS) or Double Strand (DS)	Circular or Linear	Number of Base Pairs (bp) or Bases (b)	Molecular Mass (Da)	Length[b]	% (G + C)
Simian virus 40 (genome)[a]	DS	Circular	5243 bp	3.293×10^6	1.78 μm	40.80
Bacteriophage φX174 (genome)	SS	Circular	5386 b	1.664×10^6	—[d]	44.76
Bacteriophage M13 (genome)	SS	Circular	6407 b	1.977×10^6	—[d]	40.75
Cauliflower mosaic virus (genome)	DS	Circular	8031 bp	4.962×10^6	2.73 μm	40.19
Adenovirus AD-2 (genome)	DS	Linear	35,937 bp	2.221×10^7	12.2 μm	55.20
Epstein–Barr virus (genome)	DS	Circular	172,282 bp	1.065×10^8	58.6 μm	59.94
Bacteriophage T4 (genome)	DS	Linear	168,899 bp	1.062×10^8	57.4 μm	35.30
Bacterium *E. coli* (genome)	DS	Circular	4,639,221 bp	2.869×10^9	1.57 mm	50.80
Fruit fly (*Drosophila melanogaster*) (one chromosome)[c]	DS	Linear	$\sim 6.5 \times 10^7$ bp	$\sim 4.3 \times 10^{10}$	~2 cm	~40

[a]The term *genome* designates the total DNA to specify the genetic information for an organism.

[b]Calculated for double-strand DNA of known sequence: 0.34 nm × the number of base pairs (assumes B form).

[c]This molecule has not been completely sequenced, so numbers of base pairs, molecular weights, and % (G + C) cannot be given exactly.

[d]The lengths of single-strand DNAs are not well defined; they depend very much on solvent conditions.

a little different in conformation from the B-form seen in fiber preparations, having about 10.5 base pairs per turn instead of the expected 10.0. Double-strand RNA, as mentioned earlier, always exists in the A form.

The DNA molecules found in organisms vary over an enormous range of sizes. Some bacteria carry circular *plasmid* DNA molecules (see Figure 4.18) that can be as small as a few thousand base pairs. On the other hand, some DNAs, like those in the chromosomes of eukaryotes, are immense molecules. The DNA from one *Drosophila* (fruit fly) chromosome has a molecular weight of about 4×10^{10} g/mol and would be 2 cm long if fully extended.

Circular DNA and Supercoiling

Another important feature of naturally occurring DNA molecules is illustrated in Figure 4.17a and b: Many of them are *circular.* This means that they do not have free 5′ or 3′ ends. The circles may be small, as in bacteriophage φX174 DNA (Figure 4.17a), or immense, as in *E. coli* DNA (Figure 4.17b), and they may involve either a single strand or two strands intertwined in a B-form double helix. Not all DNA molecules are circular, however. Figure 4.17c shows the linear DNA of a virus, bacteriophage T2. Human chromosomes also contain giant linear DNA molecules.

There is a special aspect to circular DNA molecules that you might not have expected: Most are **supercoiled.** What this means can be seen by considering the group of molecules shown in Figure 4.18. The figure shows both relaxed and supercoiled forms of plasmid DNA molecules. A relaxed circle can lie flat on a plane surface. A supercoiled molecule cannot. In addition to the twist of the DNA strands about one another, a supercoiled molecule has extra twists in the helix axis itself—the helix axis crosses over itself one or more times. Three-dimensional structure, such as supercoiling, that involves a higher-order folding of elements of regular secondary structure is called the **tertiary** structure of a polymer.

The high-order folding of a biopolymer's secondary structure is called its tertiary structure.

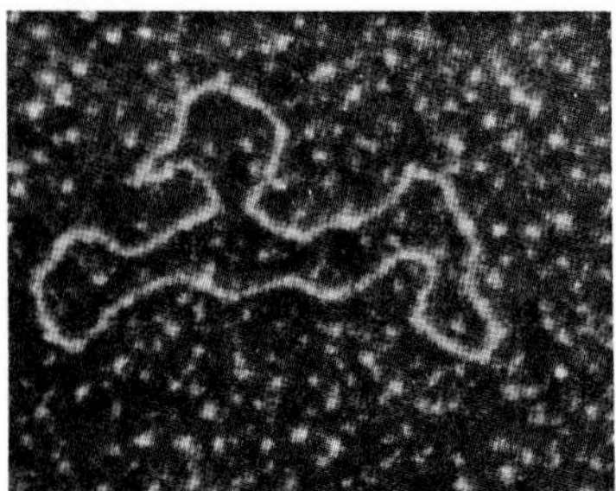

(a) Viral single-strand DNA (circular)

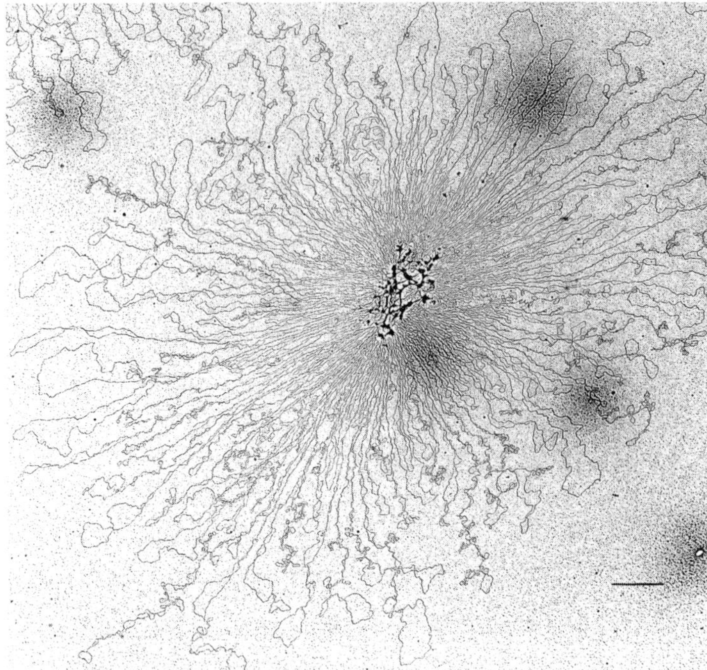

(b) Bacterial double-strand DNA (circular)

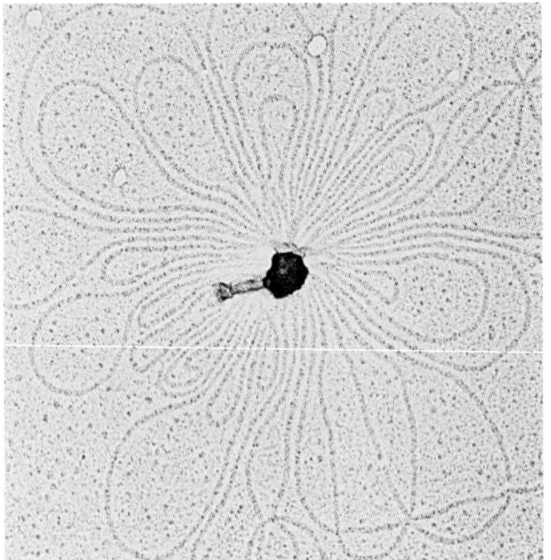

(c) Bacteriophage double-strand DNA (linear)

FIGURE **4.17**

Circular and linear nucleic acid molecules as seen by the electron microscope. (a) The circular single-strand DNA of the small bacteriophage ϕX174. **(b)** The large circular double-strand DNA of *E. coli*. This molecule exists as a number of supercoiled loops bound to a protein matrix. **(c)** The single linear double-strand DNA molecule of bacteriophage T2. The bacteriophage has been lysed, and its DNA has spilled out. One of the two ends is off screen to the lower right. The other is presumably buried in the phage head.

(**a**) Courtesy of Drs. D. Dressler and K. Koths; (**b**) Courtesy of R. Kavenoff, Designergenes Ltd./BPS; (**c**) © Biology Media/Photo Researchers.

Supercoiling is by no means rare. Rather, it is the usual state of closed circular DNA molecules. Most naturally occurring circular DNA molecules have left-hand superhelical twists, but it is possible to form DNA molecules with right-hand superhelicity. By convention, we call right supercoiling **positive** and left supercoiling **negative.**

Most DNA molecules found in vivo are left-hand supercoils.

The DNA molecules shown in Figure 4.18 differ only in their topology; therefore they are called **topoisomers.** Topoisomers can be interconverted *only* by cutting and resealing the DNA. Cells have enzymes capable of doing this. These enzymes are called **topoisomerases,** and they regulate the superhelicity of natural DNA molecules.

When superhelical turns (either left-hand or right-hand) are introduced into a previously relaxed DNA molecule, the molecule is put under strain. Therefore, energy must be expended to make a DNA molecule supercoiled. Prokaryotic cells like *E. coli* have a special topoisomerase called **DNA gyrase.** This enzyme introduces left-hand superhelical turns in a reaction driven by ATP hydrolysis. Some topoisomerases, which can only relax supercoiled DNA, do not require ATP. We discuss topoisomerases in more detail in later chapters. The energy stored in circular DNAs by twisting them into supercoils may have major effects on DNA conformation; we describe such effects later in this chapter after we present a quantitative theory of supercoiling.

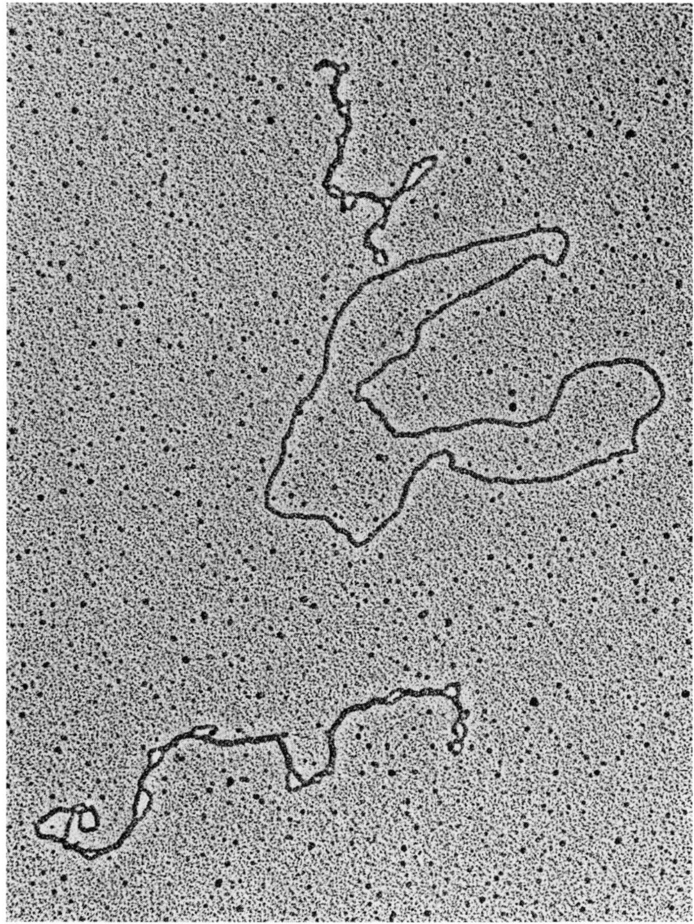

FIGURE **4.18**

Relaxed and supercoiled DNA molecules. Electron micrograph showing three mitochondrial DNA molecules. All three are of identical sequence and contain about 16,500 bp each. However, the molecule in the center is relaxed, whereas those at top and bottom are tightly supercoiled.

Courtesy of Dr. D. A. Clayton.

Structure of Single-Strand Polynucleotides

Single-strand polynucleotides can exhibit a variety of structures, depending on their sequences and the solution conditions. At high temperature or in the presence of denaturing substances, most will be largely in a random coil form, as depicted in Figure 4.19a. Such a structure is characterized by flexibility and freedom of rotation about backbone bonds, which leads to a floppy, constantly changing form. However, under conditions closer to those found in vivo, the stacking interactions will tend to form regions of single-chain, stacked-base helix (Figure 4.19b). Furthermore, most naturally occurring nucleic acid sequences contain regions of self-complementarity between which base pairing is possible. Here the molecule can loop back upon itself to form a double-strand structure, as diagrammed in Figure 4.19c. A more complex example is given in Figure 4.20, which portrays the structure of **transfer RNA (tRNA)**, a kind of RNA involved in protein synthesis (see next section). Here we find not only A-type secondary structure from the folding of the chain back on itself but also more complex folding of such helices together. Thus, the tRNA molecule possesses a defined *tertiary* structure, a higher-order folding that gives it a defined shape and internal arrangement necessary for its function.

RNA molecules are usually single-stranded, but most have self-complementary regions that form hairpin structures, and some have well-defined tertiary structures.

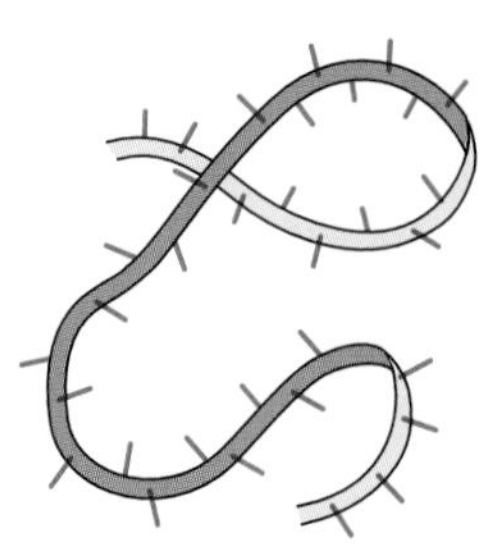
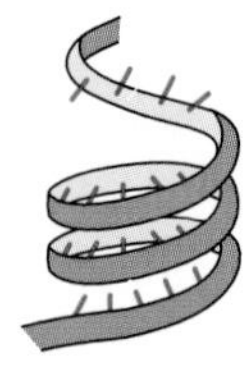
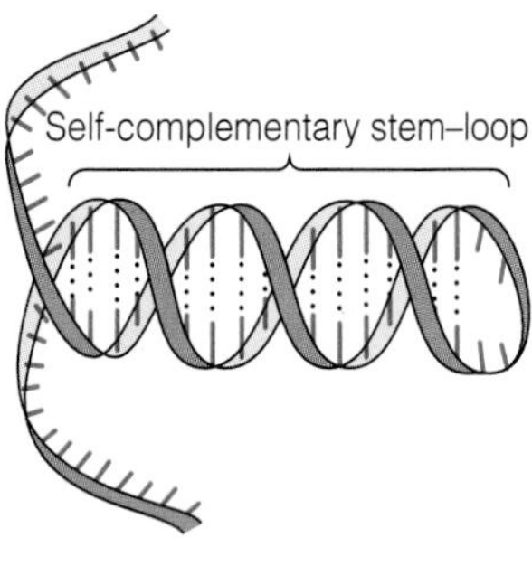

(a) Random coil

(b) Stacked-base structure (single-strand helix)

(c) Hairpin formation in self-complementary region (double helix)

FIGURE **4.19**

Conformations of single-strand nucleic acids. (a) The random coil structure of denatured single strands. There is flexibility of rotation of residues and no specific structure. **(b)** Stacked-base structure adapted by non–self-complementary single strands under "native" conditions. Bases stack to pull the chain into a helix, but there is no H-bonding. **(c)** Hairpin structures formed by self-complementary sequences; the chain folds back on itself to make a stem–loop structure.

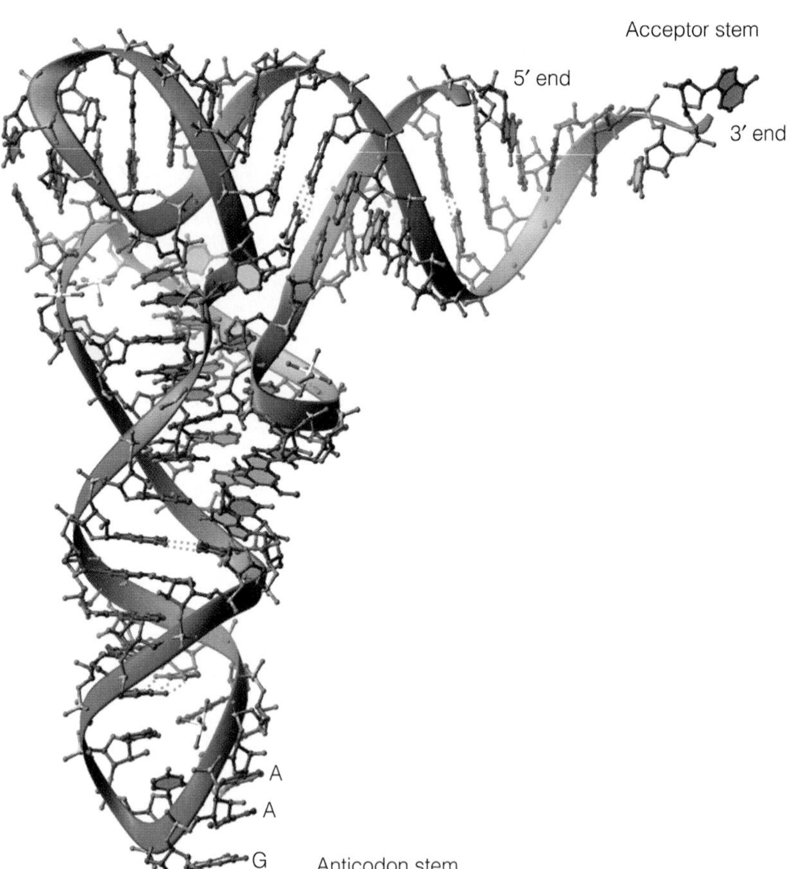

FIGURE **4.20**

The tertiary structure of a transfer RNA as determined by x-ray diffraction. This molecule is the tRNA that transfers the amino acid phenylalanine into proteins synthesized in yeast cells. The path of the phosphate-ribose backbone is traced by the purple ribbon. Note that there are some regions of triple-base bonding.

The Biological Functions of Nucleic Acids: A Preview of Molecular Biology

We have emphasized that the fundamental role of the nucleic acids is the storage and transmission of genetic information, and we will continue to develop this theme throughout the book. In Part V we describe in detail how nucleic acids are passed from parent cell to daughter cell (or from an organism to its descendents) and how they direct biochemical processes and the formation of other complex molecules, such as proteins. In this chapter and the next, we will present a preliminary overview of these nucleic acid functions. From this, you will gain some appreciation of the relationships between nucleic acid and protein structures, of evolution at the molecular level, and of our ability to modify microbes, plants, and animals through genetic engineering.

GENETIC INFORMATION STORAGE: THE GENOME

Every organism carries in each of its cells at least one copy of the total genetic information needed by that organism. This is referred to as the *genome.* Usually, the genomic information is coded into double-strand DNA, but some viruses use single-strand DNA or even RNA (see Table 4.4). Genomes vary enormously in size; the smallest viruses need only a few thousand bases (b) or base pairs (bp), whereas the human genome consists of about 2×10^9 bp of DNA, distributed in 23 chromosomes.

Recent years have seen remarkable advances in our ability to determine DNA or RNA sequences. In 1977, Maxam and Gilbert (see References) devised a chemical sequencing method involving selective cleavage at A, T, G, or C residues, followed by separation of the fragments by gel electrophoresis. This remarkable technique allowed researchers to begin the exploration of genomic information. The Maxam–Gilbert method has been largely supplanted by a technique that uses enzymes to generate oligonucleotide fragments started and terminated at specific base positions. This method, developed by Fred Sanger, is described in detail in Chapter 25. Today, this enormously fast, automated technique allows the sequencing of large genomes. As Table 4.4 shows, the entire sequence of the *E. coli* genome (4,693,221 bp) is now known. Even more remarkable is the recent sequencing of the yeast genome, which consists of 12,147,777 bp distributed between 16 chromosomes and the mitochondrion. The genomes of many other organisms, including humans, are in the process of being sequenced.

In every organism a substantial fraction of the genomic DNA is capable of being transcribed, or "read," to allow the expression of its information in directing synthesis of RNA and protein molecules. The segments that can be transcribed are called **genes.** The DNA in each cell of every organism contains at least one copy (and sometimes several) of the gene carrying the information to make each protein that the organism requires. In addition, there are genes (often reiterated manyfold) for the many specific functional RNA molecules, such as the transfer RNA (tRNA) shown in Figure 4.20. Like proteins, these RNAs play specific roles as part of the cell's machinery. (See Table 4.5.)

REPLICATION: DNA TO DNA

The process that is essential for the continued life of each organism and for the continuation of life on earth is DNA replication. Replication passes on the genetic information from cell to cell and from generation to generation. The essence of the process is depicted in Figure 4.12—a complementary copy of each strand of duplex DNA is made, usually resulting in two identical copies of the original duplex. The process is highly accurate in copying—making less than 1 error in 10^8 bases—but occasionally mistakes are made. These contribute to the mutations that have allowed the evolution of life to ever more complex forms.

The replication of DNA is accomplished by a complex of enzymes. These enzymes are described in detail in later chapters. Each enzyme complex, centered on a protein called *DNA polymerase,* has multiple functions. It opens the double-strand DNA and guides the pairing of each incoming deoxyribonucleoside triphosphate with its complementary partner on the strand to be copied. It then catalyzes the formation of the phosphodiester bond to link this residue to the new growing chain. In many cases, the enzyme complex also checks or "proofreads" the addition before proceeding to add the next residue, which contributes to the high overall accuracy of replication. DNA replication is discussed in much more detail in Chapters 24 and 28.

Replication is the copying of both strands of a duplex DNA to produce two identical DNA duplexes.

TABLE **4.5 Properties of some naturally occurring RNA molecules**

Source (Organism)	Designation	Function	Size (b or bp)
tRNA (transfer RNA)			
E. coli	$tRNA^{Leu}$	Transfers leucine in protein synthesis	87 b
Yeast	$tRNA^{Phe}$	Transfers phenylalanine in protein synthesis	76 b
Rat	$tRNA^{Ser}$	Transfers serine in protein synthesis	85 b
rRNA (ribosomal RNA)			
E. coli	5S RNA	Part of ribosome structure	120 b
	16S RNA	Part of ribosome structure	1542 b
	23S RNA	Part of ribosome structure	2904 b
mRNA (messenger RNA)			
Chicken	$mRNA_{LYS}$	Messenger RNA for protein lysozyme	584 b
Rat	$mRNA_{SA}$	Messenger RNA for protein serum albumin	~2030 b
vRNA (viral RNA)			
Polio virus	Polio RNA	Genome of the virus	7440 b
Cytoplasmic polyhedrosis virus of tussock moth	CPV RNA	Genome of the virus	Ten double-strand molecules, ~890 to ~5150 bp

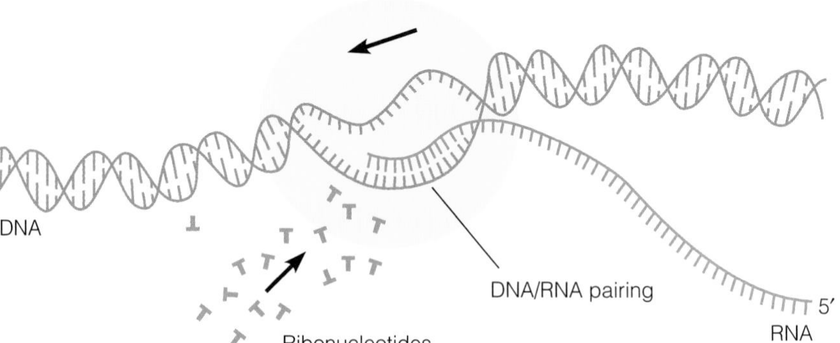

FIGURE **4.21**

The basic principle of transcription. An enzyme (RNA polymerase) travels along a DNA molecule, opening the double strand and making an RNA transcript by adding one ribonucleotide at a time. It copies the oligonucleotide sequence from only one of the two DNA strands. After the enzyme passes, the DNA rewinds.

TRANSCRIPTION: DNA TO RNA

Transcription is the copying of a DNA strand into a complementary RNA molecule.

Expression of the genetic information always involves as a first step the **transcription** of genes into complementary RNA molecules. This production of specific RNA molecules is easy to visualize. Just as a DNA strand can direct replication, it can equally well direct transcription, the formation of a complementary RNA strand (Figure 4.21). Of course, the monomers required in transcription are different from those required in replication. Instead of the deoxyribonucleoside triphosphates, the ribonucleoside triphosphates ATP, GTP, CTP, and UTP are needed to make RNA. (Note that U in the new RNA pairs with A in the DNA template.) DNA transcription, like DNA replication, requires a special set of enzyme catalysts known as *RNA polymerases.* We discuss these enzymes, and the details of the processes they catalyze, in Chapters 24, 25, 26, and 28.

TRANSLATION: RNA TO PROTEIN

Transcription alone is sufficient for the production of the many functional RNA molecules of the cell, such as the tRNAs or ribosomal RNAs listed in Table 4.5. However, the synthesis of specific proteins, under the direction of specific genes, is a more complex matter. The problem, as we see in Chapter 5, is that proteins are polymers made from 20 different kinds of amino acid monomers. Because there are only four different nucleotide monomer types in DNA, there cannot be a one-to-one relationship between the sequence of nucleotides in a DNA molecule and the sequence of amino acids in a protein. Rather, the linear sequence of bases which constitutes the protein-coding information is "read" by the cell in blocks of three nucleotide residues, or **codons,** each of which specifies a different amino acid. The set of rules that specifies which nucleic acid codons correspond to which amino acids is known as the **genetic code.** We describe this code in Chapter 5 after we have described the amino acids and the structure of proteins.

Although the information for all protein sequences is coded in DNA, the production of proteins does not proceed directly from DNA. For the information to be converted from the DNA sequences of the genes into amino acid sequences of proteins, special RNA molecules are needed as intermediates. Complementary copies of the genes to be expressed are transcribed from DNA in the form of **messenger RNA (mRNA)** molecules (see Table 4.5), so called because they carry information from DNA to the protein-synthesizing machinery of the cell. The protein-making machinery includes tRNA molecules, special enzymes, and **ribosomes,** which are RNA–protein complexes where the assembly of new proteins takes place. This **translation** of RNA information is outlined in Figure 4.22. (We present the main features of translation in Chapter 5 and describe it in detail in Chapters 27 and 28.) The flow of genetic information in the cell can be summarized by the simple schematic diagram shown in Figure 4.23.

In translation, an RNA sequence dictates a protein sequence.

As we show in the following chapters, proteins are the major structural and functional molecules in most cells. What a cell is like and what it can do depend largely on the proteins it contains. These, in turn, are dictated by the information stored in the cell's DNA, transcribed into mRNA, and expressed by the protein-synthesizing machinery. If we use the analogy of a cell as a factory, the proteins are the working machinery. The master plans for this machinery are stored in a central

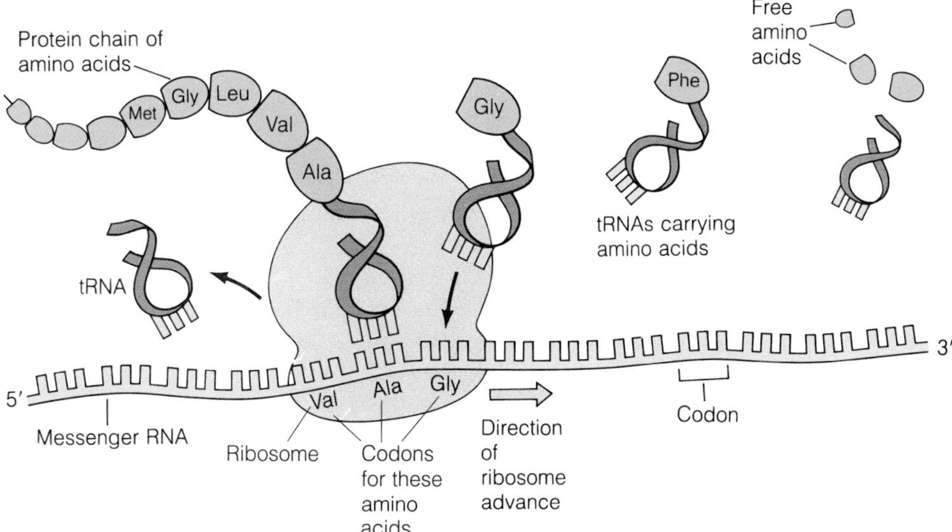

FIGURE **4.22**

The basic principle of translation. A messenger RNA molecule is bound to a ribosome, and transfer RNA molecules bring amino acids to the ribosome one at a time. Each tRNA identifies the appropriate codon on the mRNA and adds this amino acid to the growing protein chain. The ribosome travels along the mRNA, so that the genetic message can be read and translated into a protein.

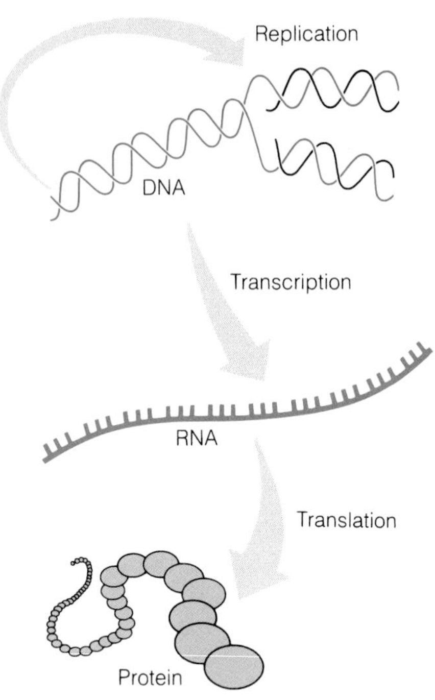

FIGURE **4.23**

The flow of genetic information in a typical cell. DNA can both replicate and be transcribed into RNA. Messenger RNAs are translated into protein sequences.

repository (the DNA of the cell nucleus). Copies of certain plans (mRNAs) are sent out from time to time as new protein machinery or replacements are needed. In addition, there is continual need for specialized RNAs like tRNAs.

MANIPULATING DNA

As observers of the cell's operations, scientists have learned to read the DNA plans and now understand the code by which it is translated. This knowledge has given us the ability to make genetic changes in living organisms, via **recombinant DNA** techniques. These techniques, which involve the insertion of novel DNA into the genome of an organism, have completely revolutionized biology and biochemistry. To discuss them in detail at this point would be premature, for they depend on complicated enzymatic procedures that we have yet to describe. You will find a detailed description of recombinant DNA technology in Chapter 25. It is, however, appropriate to list here a few of the kinds of things that can now be accomplished:

1. Individual genes or other desired DNA sequences can be retrieved from any part of the genome of a higher organism and **cloned** (reproduced identically) in bacteria such as *E. coli.* This technique allows the large-scale production of particular DNA regions, to be used, for example, in determination of nucleotide sequences in DNA. Almost all of our knowledge of DNA sequences depends on this technique.
2. DNA carrying specific genes of higher organisms can be made to direct the expression of proteins in bacteria, allowing the production of large amounts of some normally difficult to purify eukaryotic proteins. Medically important examples include insulin and blood-clotting factors, which can now be made more cheaply and safely.
3. The cloned DNAs can be modified in desired ways and their modified translation products studied, helping us to understand how proteins function.
4. In some cases, the modified DNAs can be reinserted into the organism from which they came, to allow study of the physiological and biochemical effects of the modification of the proteins they code for.
5. Genes from one kind of organism can be inserted into the genome of another kind, to produce new traits.

Recombinant DNA techniques allow us to modify the genetic content of organisms.

Such methods will clearly lead—and have already led—to enormous increases in our understanding of biochemical and biological processes. Even more significant in the long term may be their potential for human manipulation of the genetic information in bacteria, plants, and animals. Already, organisms as complex as mice and sheep have been cloned. Human cloning is probably possible. Like most great technical advances, genetic engineering may be used for good or ill. The ethical questions involved will be debated for a long time. However these debates are resolved, the knowledge now exists, and humanity—in fact, life itself—has passed a new threshold. For the first time in the long history of evolution, that process itself can now be consciously controlled in precise ways.

Plasticity of Secondary and Tertiary DNA Structure

In preceding sections we described the B and A forms of DNA, and we briefly mentioned supercoiling of circular DNA. It is now recognized that a number of special secondary and tertiary structures exist. These structures depend in many cases on the presence of special primary structures, and in some cases they can be stabilized by the presence of supercoiling. To begin, then, we must look more closely into the nature of supercoiled DNA.

CHANGES IN TERTIARY STRUCTURE: A CLOSER LOOK AT SUPERCOILING

In order to get a quantitative feeling for what supercoiling means, examine the "thought experiment" described in Figure 4.24. Consider the linear DNA molecule shown in Figure 4.24a. It contains enough base pairs to make exactly 10 helical turns; we say it has a **twist** (T) = 10.0. Now suppose we took this DNA molecule, laid it flat on a surface, and brought the strand ends together 5′ to 3′ (Figure 4.24a). Since the DNA molecule contains just enough base pairs to make 10 complete turns of the helix, the 5′ and 3′ ends of each strand would be in position to meet one another, and we could covalently join them. (Certain enzymes called *ligases* commonly found in cells can do this joining.) This connection would make a relaxed circle that could still lie flat on the surface (Figure 4.24b). The covalently closed molecule now has another property—the strands are interlinked 10 times. We say that it has a **linking number** (L) of 10.

Now suppose instead that before joining the ends of this right-hand double helix, we made one turn to the left, unwinding by one turn and thereby reducing the linking number to 9. This change would place a strain on the circular DNA, which could respond in two different ways:

1. The molecule could remain flat on the surface with its twist reduced to 9 ($T = 9$, as shown in Figure 4.24c). The normal 10.5 bp per turn of the helix would then be increased to 11.67. The strain on this DNA molecule is expressed in its "unnatural" twist; it is said to be *underwound.* (If we had added an extra turn to the right instead of left, the circle would also be strained, but it would be *overwound*, with 11 twists and 9.54 bp/turn.)
2. The molecule could restore its original twist ($T = 10$) by wrapping about itself in one negative (left-hand) superhelical turn, as shown in Figure 4.24d. The number of such superhelical turns is called the **writhe** (W); in

FIGURE **4.24**

Forming a DNA supercoil. We can imagine making a linear DNA circular in several ways: **(a)** A linear DNA is laid flat. It has 105 base pairs and 10.5 bp/turn, so there are 10 turns; the twist $T = 10$. **(b)** Because there is an integral number of turns, we can link the 5′ and 3′ ends together without twisting, to form a relaxed circle that lies flat. The linking number $L = 10$ and the writhe $W = 0$. **(c)** If we reduce the number of turns before joining the ends, by unwinding the DNA one turn, the closed circle will have $L = 9$. If we still require the DNA molecule to lie flat, the twist of the helix must change. We now have $L = 9$, $T = 9$, and $W = 0$. With $T = 9$ in 105 bp, the DNA is forced into a conformation with 105/9 = 11.67 bp/turn. **(d)** Rather than change its twist, the strained DNA molecule may writhe, or supercoil, so that $L = 9$, $T = 10$, and $W = -1$. Strained molecules usually supercoil rather than change their twist.

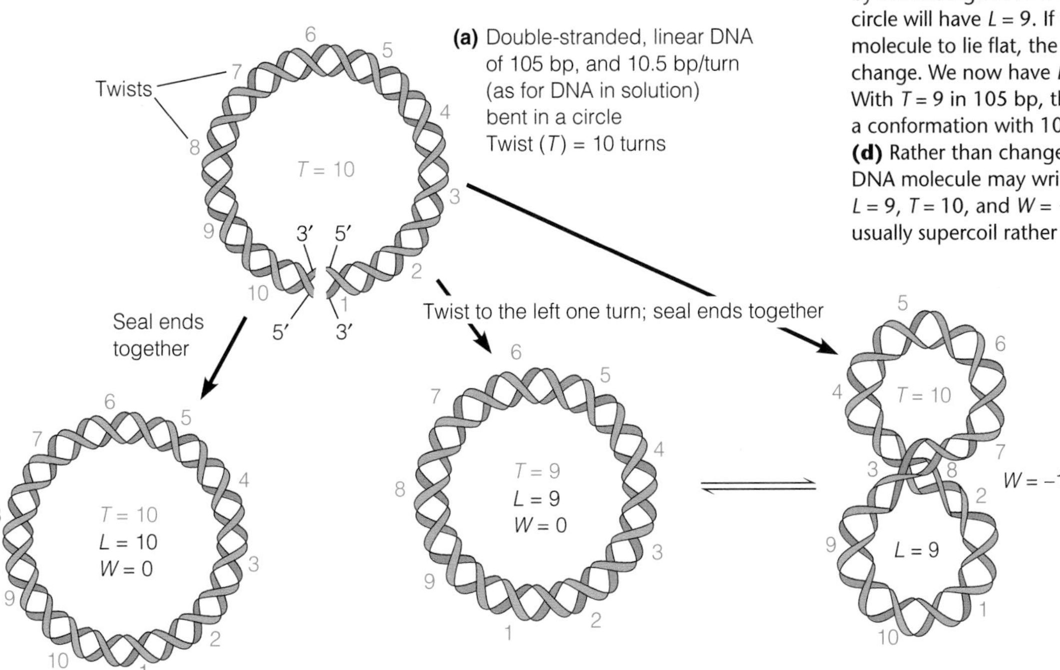

(b) Unstrained circle: Double-stranded circular DNA
Linking number (L) = 10
Twist (T) = 10 turns
bp/turn = 10.5
Writhe (W) = 0

(c) Strained circle: Double-stranded circular DNA
Linking number (L) = 9
Twist (T) = 9 turns
bp/turn = 11.67
Writhe (W) = 0

(d) Supercoil: Double-stranded DNA
Linking number (L) = 9
Twist (T) = 10 turns
bp/turn = 10.5
Writhe (W) = −1

this case $W = -1$. (If we had added one twist to the right, the molecule could form a positive, or right-hand, writhe with $W = +1$; in the first response, with no supercoiling, $W = 0$.)

When we untwisted one turn and then joined, we produced a molecule in which the two strands were interlinked only 9 times; $L = 9$. The quantity L will remain unchanged however we may distribute the strain of underwinding between twist and writhe. Thus, the two forms found in Figure 4.24c and 4.24d both have $L = 9$.

Note that in every case, the linking number is the algebraic sum of T and W:

$$L = T + W \tag{4.1}$$

The linking number is always the algebraic sum of twist and writhe.

The only way in which L can be changed is by cutting, twisting, and rejoining the circular molecule—by means of topoisomerases, for example. When we added the extra twist to the left at the beginning of our experiment, we changed the value of L by -1 ($\Delta L = -1$). The strain imposed by putting ΔL extra turns in the DNA helix distributes itself between a change in twist and a change in writhe:

$$\Delta L = \Delta T + \Delta W \tag{4.2}$$

where ΔL is positive or negative, depending on whether the extra twists (ΔT) and superhelical turns (ΔW) are right-handed (positive) or left-handed (negative).

The superhelix density, σ, is a quantitative measure of the intensity of supercoiling.

The superhelicity of DNA molecules is often expressed in terms of the **superhelix density,** $\sigma = \Delta L/L_0$, where L_0 is the linking number for the DNA in the relaxed state. Many naturally occurring DNA molecules have superhelix densities of about -0.06. To get an idea of what this means, consider a hypothetical DNA molecule of 10,500 bp, which is in the solution B form, with 10.5 bp/turn. Then L_0 is 10,500 bp/(10.5 bp/turn), or 1000 turns. Each DNA strand crosses the other 1000 times in the relaxed circle. If the topoisomerase gyrase (see page 102) twisted the molecule to a superhelical density of -0.06, then $\Delta L = -0.06\ L_0$, or $\Delta L = -60$. This change could be accommodated, for example, by the helix axis writhing about itself 60 times in a left-hand sense, which would correspond to $\Delta W = -60$, $\Delta T = 0$; the molecule would have 60 left-hand superhelical turns. Alternatively, the twist of the molecule could change so that it had 940 turns in 10,000 bp ($T = 940$) or 10,500/940 = 11.2 bp/turn. This would correspond to $\Delta W = 0$, $\Delta T = -60$. Although any combination of ΔT and ΔW that sums to -60 could occur, real molecules release strain mainly by writhing into superhelical turns, because it is easier to bend long DNA than to untwist it. This explains the contorted structures of the double-strand circular DNA molecules in Figures 4.17b and 4.18.

Differences in supercoiling can be detected by gel electrophoresis. As described in Tools of Biochemistry 2A, the rate at which a molecule can be electrophoresed through a gel matrix depends on its dimensions; hence the more compact superhelical forms will move faster than the relaxed form. Figure 4.25 shows electrophoretic patterns for supercoiled DNA molecules that are being progressively relaxed by the action of the enzyme topoisomerase. Thus, gel electrophoresis allows us to separate topoisomers of a given DNA.

UNUSUAL SECONDARY STRUCTURES OF DNA

Most of the DNA and RNA in cells can be described as having one of the three secondary structures—random coil (which is really a lack of secondary structure), B form, or A form. But these three do not nearly exhaust the conformational possibilities of these remarkable molecules. In the remainder of this chapter we shall consider some of the alternative structures that have been recognized in recent years, and we will examine the conditions under which conformational transitions between them take place. We will see that supercoiling often plays a dominant role in these transitions.

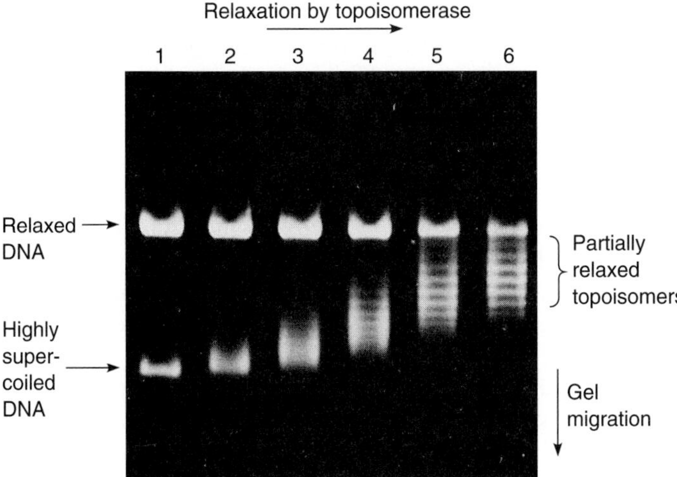

FIGURE **4.25**

Gel electrophoresis demonstrating DNA supercoiling. Lane 1: A mixture of relaxed and highly supercoiled DNA. Lanes 2 to 6: The progress of relaxation catalyzed by the enzyme topoisomerase. Samples have been taken at successive times after adding the enzyme. Individual topoisomers are resolved as individual bands on the gel. The highly supercoiled material, which forms a densely packed series of overlapping bands at the bottom, gradually disappears.

Courtesy of J. C. Wang.

Left-Hand DNA (Z-DNA)

Z-DNA is a left-hand helix with alternate purine/pyrimidine bases in alternate *syn*/*anti* conformation.

Since both the A and B forms of polynucleotide helices are right-handed (see Figure 4.15), the discovery of a left-hand form in 1979 caused considerable surprise. Alexander Rich and his colleagues carried out x-ray diffraction studies of crystals of the small deoxyoligonucleotide

$$^{5'}\text{CGCGCG}^{3'}$$
$$^{3'}\text{GCGCGC}^{5'}$$

and determined that it is a double-strand helix with G-C base pairing, as they had expected. However, the data were consistent only with a peculiar *left-hand* structure they called **Z-DNA.** A model for a long DNA molecule in the Z conformation is shown in Figure 4.26.

In addition to the reverse sense of the helix, Z-DNA exhibits other structural peculiarities. There are, in polynucleotides, two most stable orientations of the bases with respect to their deoxyribose rings. They are called *syn* and *anti:*

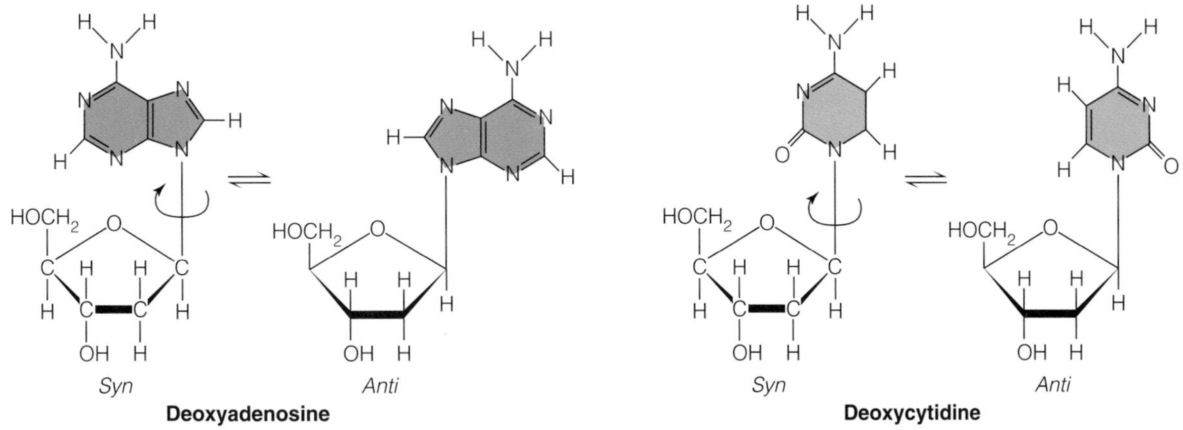

Syn ⇌ *Anti*
Deoxyadenosine

Syn ⇌ *Anti*
Deoxycytidine

In both A- and B-form polynucleotides, all bases are in the *anti* orientation. In Z-DNA, however, the *pyrimidines* are always *anti,* but the *purines* are always *syn.* Because Z-DNA is most often found in polynucleotides with alternating purines and pyrimidines in each strand (such as that shown earlier), the base orientations will alternate. Parameters for Z-DNA (see Table 4.3, page 99) reflect this characteristic, in that the repeating unit is not one base pair but two base pairs. Furthermore, this alternation gives a zigzag pattern to the phosphates; hence the name Z-DNA (see Figure 4.26).

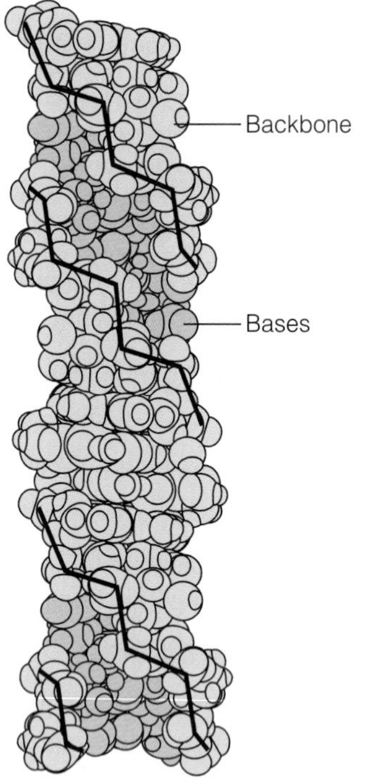

FIGURE **4.26**

Z-DNA. The structure of Z-DNA as determined by single-crystal x-ray diffraction studies. Compare this left-hand form of DNA with the similar, space-filling model of B-DNA in Figure 4.11. The single groove of Z-DNA is shown in green. The black line follows the zigzag phosphate backbone.

Courtesy of Dr. A. H.-J. Wang from *Cold Spring Harbor Symp. Quant. Biol.* (1982) 47:41.

There is now abundant evidence that Z-DNA exists in vivo. However, the exact role played by Z-DNA in vivo is still an open question. It is perhaps significant that methylation of cytosines on carbon 5, a fairly common modification in vivo, favors Z-DNA.

Hairpins and Cruciforms

We have already encountered examples of "hairpin" structures—first in Figure 4.19c and then in the transfer RNA shown in Figure 4.20. In each of these single-strand molecules, self-complementarity in the base sequence allows the chain to fold back on itself and form a base-paired, antiparallel helix. The schematic structure of a tRNA shown in Figure 4.27 depicts how a particular base sequence accomplishes this folding.

Double hairpins, often called **cruciform** (cross-like) structures, can be formed in some DNA sequences. A special kind of sequence, referred to as **palindromic,** is required. The word *palindrome* is of literary origin and usually refers to a statement that reads the same backward and forward, such as "Able was I ere I saw Elba." As used in descriptions of DNA, the word refers to segments of complementary strands that are the exact (or almost exact) reverse of one another. Such a DNA sequence is shown in Figure 4.28, along with the two conformations available to it. In most instances, formation of the cruciform structure leaves a few bases unpaired at the ends of the hairpins. This means that under normal circumstances the cruciform will be less stable than the extended structure. As we shall see later in this chapter, one effect of superhelical strain is to stabilize cruciforms.

Triple Helices and H-DNA

It has long been recognized that certain homopolymers can form *triple* helices. The first to be discovered was the ribonucleic acid structure

$$\text{poly(U)} \cdot \text{poly(A)} \cdot \text{poly(U)}$$

Later, it was observed that deoxy triplets like $T \cdot A \cdot T$ and $C^+ \cdot G \cdot C$ (where C^+ is a protonated cytosine) can also be formed. Such structures involve, in addition to the normal Watson–Crick base pairing, the *Hoogsteen-type* base pairing shown in Figure 4.29. It is now recognized that many polynucleotide strands, including

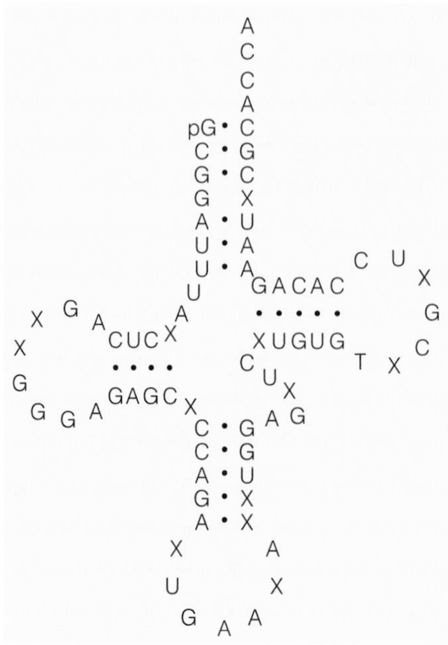

FIGURE **4.27**

How self-complementarity dictates the tertiary structure of tRNA. A schematic drawing of the tRNA shown in Figure 4.20. X's represent unusual or modified bases (see Chapter 27). The molecule has three hairpin arms resulting from self-complementarity. These arms fold into the tertiary structure of Figure 4.20.

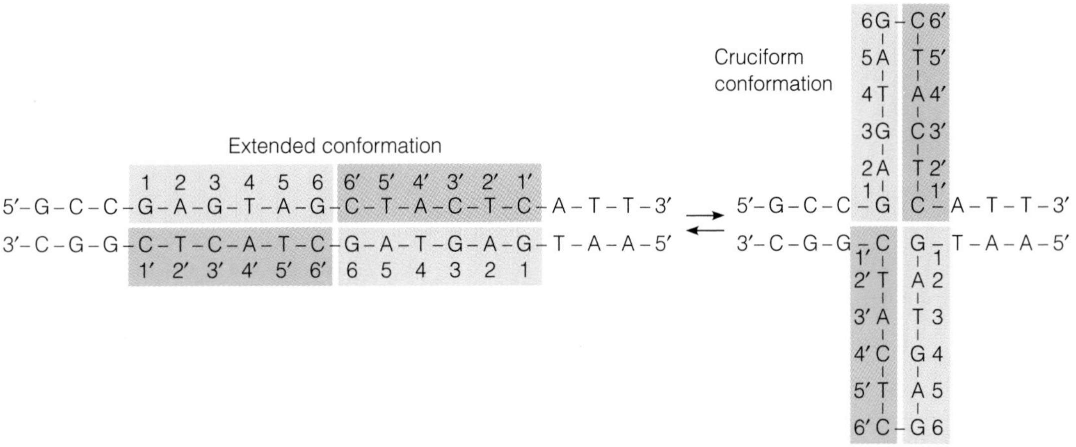

FIGURE **4.28**

A palindromic DNA sequence. A palindrome is strandwise symmetrical about a center of symmetry. Note that in the portion in brown and blue, both blue segments read the same 5′ ⟶ 3′, as do both brown segments. The sequence is shown in its extended and cruciform conformations. Two bases that will pair with each other in the cruciform conformation are given the same number.

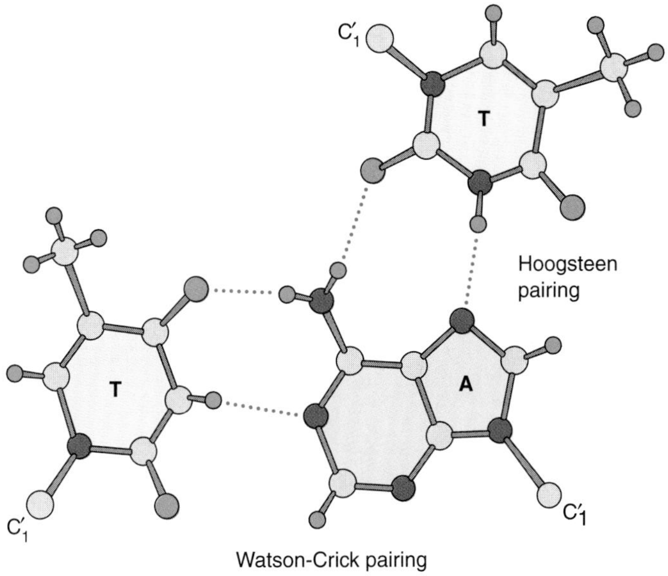

FIGURE **4.29**

Base-pairing in one type of DNA triple helix. Both normal Watson–Crick pairing and the unusual Hoogsteen pairing occur on the same A residue.

some of nonrepeating sequence, can enter into such triple helices. An unusual structure that can incorporate a triple helix is shown in Figure 4.30. It is called H-DNA and requires a molecule that has one all-pyrimidine strand (Pyr) and one all-purine strand (Pur). In the example shown, the structure

···TCTCTCTCTCTC···

···AGAGAGAGAGAG···

can exist either in the normal, double-strand form, or with one strand looped back to form a triple helix (with $C^+ \cdot G \cdot C$ and $G \cdot A \cdot G$ triplets), leaving the other (Pur) strand unpaired. Again, the loss of base pairing should make this structure unstable, except in situations where strain in the DNA molecule favors it.

Even such bizarre structures as H-DNA do not exhaust the repertoire of polynucleotides. There is evidence, for example, for the existence of four-strand structures involved in DNA recombination. However, we leave consideration of such structures to Part V.

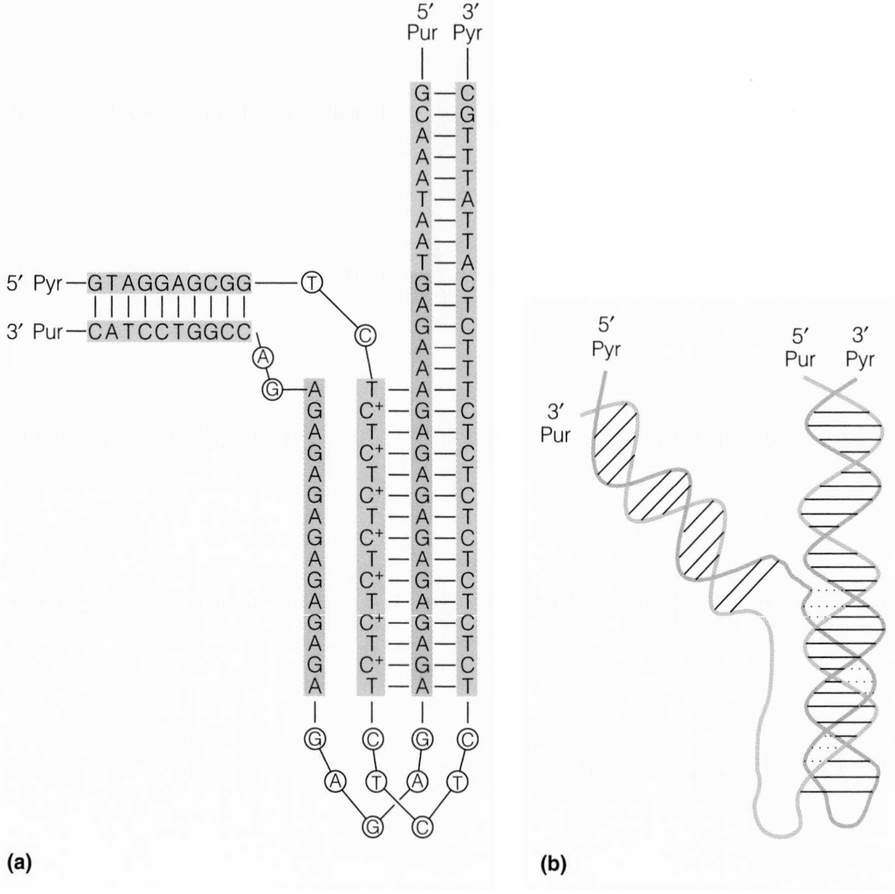

FIGURE **4.30**

H-DNA. An H-DNA region has one all-purine (blue) and one all-pyrimidine (brown) strand, allowing it to form a triple-stranded helix by doubling back. Some segments contain both purines and pyrimidines (green). **(a)** The nucleotide sequence shown here is one that could give rise to H-DNA. A segment of the purine strand is shown bonded to two different segments of the pyrimidine strand. **(b)** The bonding shown in (a) gives rise to a triple-stranded helix, shown schematically here.

Adapted from E. Paleček, *Crit. Rev. Biochem. Mol. Biol.*, G. D. Fasman, ed. (1991) 26, 2:151–226. (Boca Raton, FL: CRC Press, 1991), with permission.

Stability of Secondary and Tertiary Structure

THE HELIX-TO-RANDOM-COIL TRANSITION: NUCLEIC ACID DENATURATION

The major polynucleotide secondary structures (the A and B forms) are relatively stable for RNA and DNA, respectively, under physiological conditions. Yet they must not be *too* stable, because very important biochemical processes—DNA replication and transcription, for example—require that the double-helix structure be opened up. When it extends over large regions, this loss of secondary structure is called **denaturation** (Figure 4.31). Competing factors create a balance between structured and unstructured forms of nucleic acids.

Two major factors favor dissociation of double helices into randomly coiled single chains. The first is the electrostatic repulsion between the chains. At physiological pH, every residue on a DNA or RNA molecule carries a negative charge on the phosphate group. Even though this charge is partially neutralized by small counterions (like K^+, Na^+, and Mg^{2+}) present in the medium, a substantial net negative charge remains on each chain in the helix and tends to drive the two chains apart. Therefore, high ionic strength tends to stabilize the double helix.

A more subtle factor favoring denaturation is that the random coil structure has a higher entropy. This higher entropy is due to the greater randomness of the denatured form, with its many possible configurations. Consider equation (3.9) on page 63 ($S = k \ln W$): If a rigid double helix separates into two flexible random coils, the number of configurations accessible to the molecule greatly increases (Figure 4.31a); therefore the entropy increases. The free energy change in going

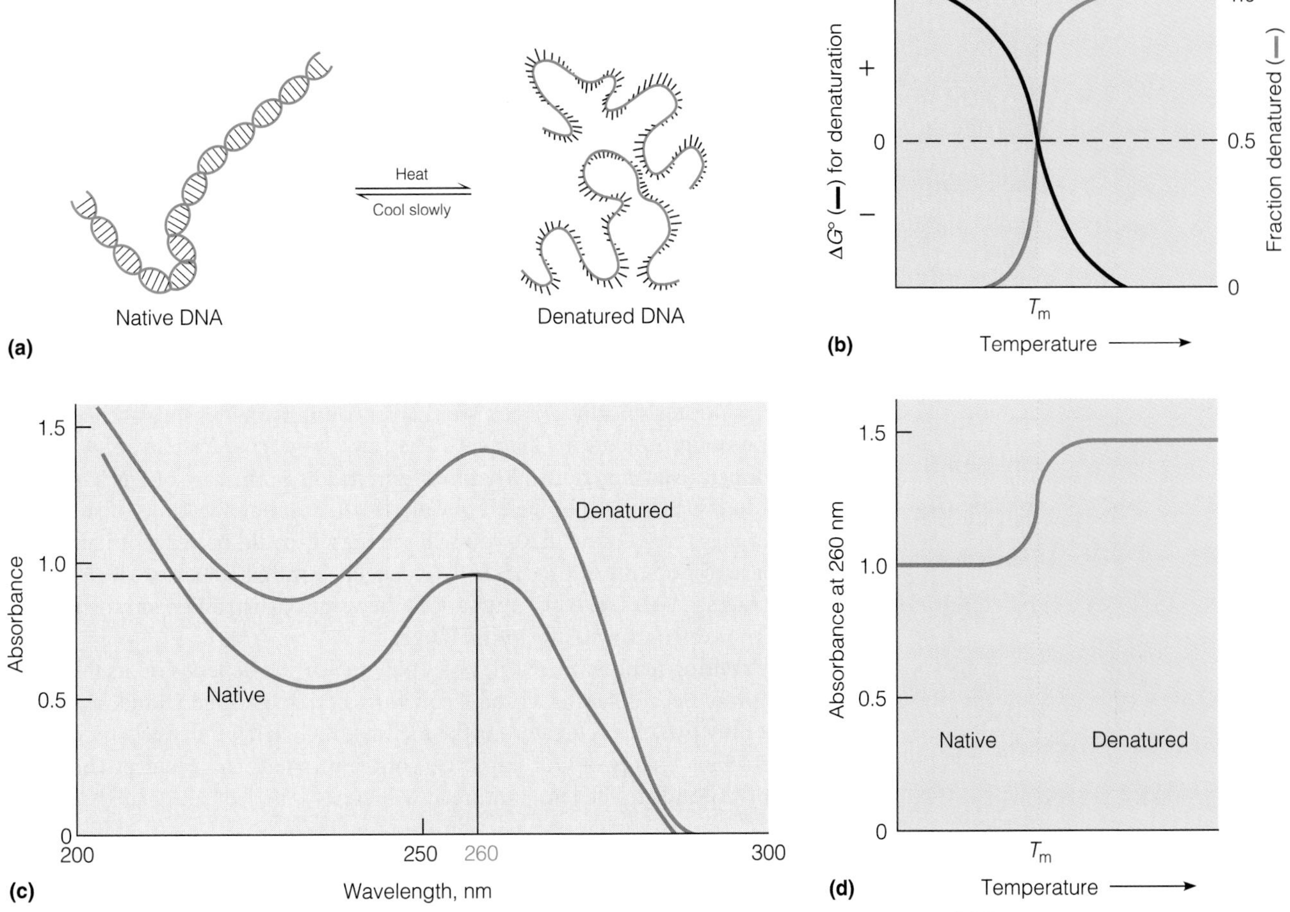

FIGURE **4.31**

Denaturation of DNA. **(a)** When native (double-strand) DNA is heated above its "melting" temperature, it is denatured (separates into single strands). The two random-coil strands have a higher entropy than the double helix. **(b)** At low T, ΔG is positive and denaturation of DNA is not favored. As T increases, $-T\,\Delta S$ overcomes ΔH, making ΔG negative and denaturation favorable. The midpoint of the curve marks the "melting" temperature, T_m, of DNA. **(c)** Absorbance spectra of native and denatured DNA show that native DNA absorbs less light than denatured DNA, with the maximum difference occurring at a wavelength of 260 nm. The hypochromicity of double-strand DNA can be used to distinguish between native and denatured forms. **(d)** The change in absorbance can be used to follow the denaturation of DNA as temperature increases. An abrupt increase in absorbance, corresponding to the sudden "melting" of DNA, is seen at T_m.

from a regular two-strand polynucleotide secondary structure (such as B-form DNA) to individual random-coil strands is given by the usual formula:

$$\Delta G = \Delta H - T\Delta S \quad \text{(helix} \rightleftharpoons \text{random coil)} \tag{4.3}$$

Because ΔS is positive, the term $(-T\,\Delta S)$ makes a negative contribution to the free energy change, favoring denaturation.

Thus, two factors favor the helix $\longrightarrow$ coil transition: the higher randomness of the random coil ($\Delta S > 0$) and the electrostatic repulsion between chains ($\Delta H_{el} < 0$). If the double-strand helical structure is to be stable under any conditions, ΔG for the unfolding reaction must be positive. Therefore we must look for a large positive contribution from ΔH to compensate for the factors just mentioned. The sources of such a positive ΔH are the hydrogen bonds between the base pairs and van der Waals interactions between stacked bases. In fact, the planar bases stack upon one another in van der Waals contact. Much energy must be expended to break these bonds and interactions, and hence the total ΔH is positive.

Because ΔH and ΔS in equation (4.3) are both positive, the sign of ΔG will change as T is increased. At low temperature, the term $T\,\Delta S$ will be less than ΔH; ΔG will be >0, and the helix will be stable. But as the temperature is increased, $T\,\Delta S$ will become greater than ΔH, and ΔG will become negative. Thus, at higher temperatures the double-strand structure becomes unstable and will fall apart (Figure 4.31b).

It is possible to follow this denaturation process by observing the absorbance of ultraviolet light of wavelength about 260 nm in a DNA solution. As mentioned on page 89, all nucleotides and nucleic acids absorb light strongly in this wavelength region. When the nucleotides are polymerized into a polynucleotide, and the bases packed into a helical structure, the absorption of light is reduced (Figure 4.31c). This phenomenon, called *hypochromism,* results from close interaction of the light-absorbing purine and pyrimidine rings. If the secondary structure is lost, the absorbance increases and becomes closer to that of a mixture of the free nucleotides. Therefore, raising the temperature of a DNA solution, with accompanying breakdown of the secondary structure, will result in an absorbance change like that shown in Figure 4.31d.

The remarkable thing about this helix-to-random-coil transition is that it is so sharp. It occurs over a very small temperature range, almost like the melting of ice into water, as described in Chapter 3. Therefore, nucleic acid denaturation is sometimes referred to as a *melting* of the polynucleotide double helix, even though the term is not technically correct. We shall encounter similar abrupt changes in configuration in proteins in Chapter 6. They are always characteristic of what are called **cooperative transitions.** What this term means in the case of DNA or RNA is that a double helix cannot melt bit by bit. If you examine the kinds of structure shown in Figures 4.11 and 4.20, you will see that it would be very difficult for a single base to pop out of the stacked, hydrogen-bonded structure. Rather, the whole structure holds together until it is at the verge of instability and then denatures over a very narrow temperature range.

The "melting temperature" (T_m) of a polynucleotide depends on its (G + C)/(A + T) ratio. Because each G-C base pair forms three hydrogen bonds and each A-T pair only two, ΔH is greater for the melting of GC-rich polynucleotides. The greater stacking energy of G-C pairs also contributes to the difference. The value of T_m corresponds to the temperature at which $\Delta G = 0$ (see Figure 4.31b and d). Thus,

AT-rich regions melt more easily than GC-rich regions.

$$0 = \Delta H - T_m \Delta S \tag{4.4}$$

or

$$T_m = \frac{\Delta H}{\Delta S} \tag{4.5}$$

On a per-base-pair basis, ΔS is about the same for all polynucleotides, but ΔH depends on base composition, as just described. This is why T_m increases with increasing G + C content. Figure 4.32 shows a graph of T_m versus percent (G + C) for a number of naturally occurring DNAs.

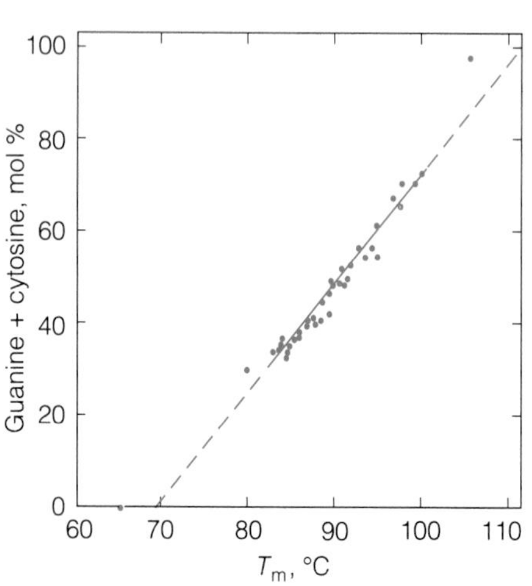

FIGURE **4.32**

Effect of base-pair composition on the denaturation temperature of DNA. The graph shows the rise in "melting" temperature of DNA as its percent (G + C) increases.

Data from J. Marmur and P. Doty, *J. Mol. Biol.* (1962) 5:120.

SUPERHELICAL ENERGY AND CHANGES OF DNA CONFORMATION

The storage of energy in DNA supercoiling is analogous to the energy required to wind up (supercoil) rubber bands.That is, the first few turns are easy, but the energy required, per turn, increases as the winding gets tighter. In fact, for DNA the amount of free energy stored in supercoiling (ΔG_{sc}) is proportional to the *square* of the superhelical density σ:

$$\Delta G_{sc} = K\sigma^2 \tag{4.6}$$

where K is a constant. Note that ΔG_{sc} is zero when the DNA is relaxed ($\sigma = 0$) and increases with either positive or negative supercoiling. Furthermore, since the square of σ is involved, the energy required to add an extra turn increases as each turn is added, and it becomes about equal to the energy that one ATP can provide when $\sigma \cong \pm 0.06$. This is at least one reason why superhelix density is limited to about this value in vivo. Highly supercoiled DNA has a lot of stored energy, which can be reduced by *any* process that decreases the superhelix density. For example, suppose we have a circular DNA molecule with negative superhelical turns. If one repeat of the DNA helix (10 base pairs) were to unwind and locally melt, this change would be the equivalent of a ΔT of -1. Then, to compensate, one negative superhelical turn could be removed ($\Delta W = +1$); that is, some of the writhing has been compensated by local unwinding. Converting a segment of B-DNA to Z-DNA would be even more efficient, requiring less energy. Each 10 bp of DNA that goes from B to Z goes from a twist of +1 (one right-hand twist) to approximately -1 (one left-hand twist). This single change amounts to $\Delta T = -2$, which would allow relaxation of 2 negative superhelical turns ($\Delta W = +2$).

Formation of cruciform structures also relaxes superhelical DNA, because every base pair put into a cruciform hairpin is essentially removed from superhelical strain. Likewise, H-DNA formation, which leaves part of one strand unpaired, has the same effect.

In other words, imposing high levels of superhelical torsion on a DNA molecule can promote any one of the following changes: local melting, Z-DNA formation, cruciform extension, formation of H-DNA regions, and quite likely formation of some other special conformations we have not yet discovered. Which of these will happen depends on what special sequences are present in the DNA circle under stress. For example:

Energy stored in supercoiled DNA can be used to drive structural transformation.

- The presence of AT-rich regions, which melt more easily than GC-rich regions, may favor local melting.
- Alternate purine/pyrimidine tracts (like ···CGCGCG)··· favor Z-DNA formation, especially if the C's are methylated on carbon 5.
- Palindromes allow cruciform extension.
- A segment that is primarily purines in one strand and pyrimidines in the other may permit H-DNA formation.

Superhelical stress is both widespread and controlled in DNA molecules found in cells. As we shall see in later chapters, the special structures described here play diverse roles in the regulation of gene expression. The idea that genes may be turned on and off by changes in supercoiling is an intriguing one.

SUMMARY

There are two kinds of nucleic acids, DNA and RNA. Each is a polynucleotide, a polymer of four kinds of nucleoside 5′-phosphates, connected by links between 3′ hydroxyls and 5′ phosphates. RNA has the sugar ribose; DNA has deoxyribose. The phosphodiester linkage is inherently unstable, but it hydrolyzes only very slowly in the absence of catalysts. Each naturally occurring nucleic acid has a defined sequence, or primary structure. Early evidence indicated that DNA might be the genetic material, but it was not until Watson and Crick elucidated its two-strand secondary structure in 1953 that it became obvious how DNA might direct its own replication. The structure they proposed involved specific pairing between A and T and between G and C. The helix is right-handed, with 10 base pairs (bp) per turn. Such a structure can replicate in a semiconservative manner, as demonstrated by Meselson and Stahl in 1958. Other forms of polynucleotide structures exist, of which the most important is the A form, found in RNA–RNA and DNA–RNA double helices. In vivo, most DNA is double-stranded; some molecules are circular. Most of the circular DNA molecules found in nature are supercoiled. Most RNA is single-stranded, but it may fold back to form hairpins and other well-defined tertiary structures.

The biological functions of nucleic acids may be briefly summarized as follows: DNA contains the genetic information, which is transcribed into RNA. Some of these RNA molecules act as messengers to direct protein synthesis. The messenger RNA is translated on a particle called a ribosome, using the genetic code, to produce proteins. Modern molecular biological techniques allow us to manipulate DNA to make new proteins and modify existing ones.

Supercoiling of DNA can be expressed in terms of twist (T) and writhe (W). These terms are related to the linking number (L) by $L = T + W$. To form superhelical coiling requires the expenditure of ATP energy, using an enzyme called gyrase. Gyrase is one of a class of topoisomerases; others relax supercoiled DNA.

Polynucleotides can form a number of unusual structures, including left-hand DNA (Z-DNA), cruciforms, and in some cases triple helices. The secondary structures of polynucleotides can be changed in various ways. The helix can "melt," which involves strand separation. This change is easiest for regions rich in A-T pairs. Energy stored in superhelical DNA may promote local DNA melting or changes to a variety of alternative structures, such as Z-DNA, cruciforms, or a particular triple-helical structure called H-DNA.

REFERENCES

General

Bates, A. D., and A. Maxwell (1993) *DNA Topology.* Oxford University Press, New York. A very clear, helpful little book.

Dickerson, R. E. (1983) The DNA helix and how it is read. *Sci. Am.* 249(6):94–111. A short, popularly written, but very lucid description of DNA structure.

Saenger, W. (1984) *Principles of Nucleic Acid Structure.* Springer-Verlag, New York. This reference provides much greater detail concerning nucleic acid structure than is given in this book.

van Holde, K. E., W. C. Johnson, and P.-S. Ho (1998) *Principles of Physical Biochemistry.* Prentice Hall, Upper Saddle River, N.J. Has much more on nucleic acid stability and structural transitions.

Watson, J. D., N. H. Hopkins, J. W. Roberts, J. A. Steitz, and A. M. Weiner (1987) *Molecular Biology of the Gene,* 4th ed. Benjamin/Cummings, Redwood City, Calif. A fine general reference.

Historical

Avery, O. T., C. M. MacLeod, and M. McCarty (1944) Studies on the chemical transformation of pneumococcal types. *J. Exp. Med.* 79:137–158. The pioneering study that lent credence to the idea that DNA is the genetic substance.

Hershey, A. D., and M. Chase (1952) Independent function of viral protein and nucleic acid on growth of bacteriophage. *J. Gen. Physiol.* 36:39–56. The convincing evidence that DNA is the genetic material.

Judson, H. (1979) *The Eighth Day of Creation.* Simon & Schuster, New York. A detailed, fascinating account of the development of modern ideas about nucleic acids.

Maxam, A. M., and W. Gilbert (1977) A new method for sequencing DNA. *Proc. Natl. Acad. Sci. USA* 74:560–564.

Meselson, M., and F. Stahl (1958) The replication of DNA in *Escherichia coli. Proc. Natl. Acad. Sci. USA* 44:671–682. An example of a beautifully designed and executed experiment.

Sayre, A. (1978) *Rosalind Franklin and DNA.* W. W. Norton, New York. An account of the contributions of this remarkable scientist.

Watson, J. D. (1968) *The Double Helix.* Atheneum, New York (trade and paperback editions); New American Library, New York (paperback). A delightful and illuminating recounting of the elucidation of the DNA structure.

Watson, J. D., and F. H. C. Crick (1953) Molecular structure of nucleic acids. A structure for deoxyribose nucleic acid. *Nature* 171:737–738. Two pages that shook the world.

Specialized Papers of Importance

Bauer, W. R., F. H. C. Crick, and J. H. White (1980) Supercoiled DNA. *Sci. Am.* 243(1):118–133. A readable description of supercoiling.

Lilley, P. M. J. (1984) DNA: Sequence structure and supercoiling. *Biochem. Soc. Trans.* 12:127–140. An analysis of the effects of supercoiling.

Maniatis, T. R., C. Hardison, E. Lacey, J. Lauer, C. O'Connell, D. Quon, G. K. Sim, and A. Efstratiadis (1978) The isolation of structural genes from libraries of eukaryotic DNA. *Cell* 15:687–701. A pioneering paper on gene manipulation.

Paleček, E. (1991) Local supercoil-stabilized DNA structures. *Crit. Rev. Biochem. Mol. Biol.* 26:161–226. A very comprehensive, yet readable review.

Saenger, W., W. N. Hunter, and O. Kennard (1986) DNA conformation is determined by economics in the hydration of phosphate groups. *Nature* 324:385–388.

Vologodskii, A. V., and N. R. Cozzarelli (1994) Conformational and thermodynamic properties of supercoiled DNA. *Annu. Rev. Biophys. Biomol. Struct.* 23:609–643.

Wells, R., D. A. Collier, J. C. Hanvey, M. Shimizu, and F. Wohlrab (1988) The chemistry and biology of unusual DNA structures adopted by oligopurine–oligopyrimidine sequences. *FASEB J.* 2:2939–2949. H-DNA and similar structures are discussed.

Wing, R. M., H. R. Drew, T. Takano, C. Brodka, S. Tanaka, K. Itakura, and R. E. Dickerson (1980) Crystal structure analysis of a complete turn of B-DNA. *Nature* 287:755–758. First crystallographic study of a B-DNA structure.

PROBLEMS

1. A viral DNA is analyzed and found to have the following base composition, in mole percent: A = 32, G = 16, T = 40, C = 12.
   (a) What can you immediately conclude about this DNA?
   (b) What kind of secondary structure do you think it would have?

2. Given the following sequence for one strand of a double-strand oligonucleotide:

$^{5'}$ACCGTAAGGCTTTAG$^{3'}$

(a) Write the sequence for the complementary DNA strand.
(b) Suppose you knew that the strand shown above had phosphate on both ends. Using an accepted nomenclature, write the sequence so as to show this.
(c) Write the sequence of the RNA complementary to the strand shown above.

*3. Some naturally occurring polynucleotide sequences are *palindromic;* that is, they are self-complementary about an axis of symmetry. Such a sequence is

···TCAAGTCCATGGACTTGG

AGTTCAGGTACCTGAACC

Show how this structure might form a double hairpin, or cruciform, conformation. Indicate the center of symmetry in the sequence and the bounds of the cruciform.

4. The *E. coli* genome has a superhelical density in vivo of about -0.06. Assuming the DNA has 10.5 bp/turn, what is the expected writhing number of the *E. coli* genome?

*5. Given the following sequence for an RNA molecule, find a secondary structure that will be maximally stable.

GUCCAGCCAUUGCGUUCGCAAUGGC

6. The largest of the double-strand RNA molecules of cytoplasmic polyhedrosis virus contains 5150 bp (see Table 4.5). How long do you expect this molecule to be if extended?

7. A circular, double-strand DNA contains 2100 base pairs. The solution conditions are such that DNA has 10.5 bp/turn.
(a) What is L_0 for this DNA?
(b) The DNA is found to have 12 left-hand superhelical turns. What is the superhelix density σ?

*8. In a supercoiled DNA, a stretch of about 20 base pairs changes from the B form to the Z form. What is the change in (a) T, (b) L, and (c) W?

9. Of the DNA molecules listed in Table 4.4, which would you expect to have the highest and lowest T_m?

10. A scientist isolates the DNA genome from a virus. She attempts to carry out a melting analysis but finds only 10% hypochromicity.
(a) Suggest an explanation for the low value.
(b) Why do you think she finds *this* much?

11. A particular double-strand DNA has, under the conditions used in Figure 4.32, a melting point of 94°C. Estimate the base composition (in mole percent) of this DNA.

*12. A variant of B-form DNA has been reported to exist in the presence of Li^+ ions. This form, called C-DNA, is found by x-ray diffraction to have 9⅓ base pairs per turn.
(a) How many base pairs are contained in one repeat of this structure? How many turns in one repeat?
(b) Is C-DNA twisted more or less tightly than B-DNA?
(c) Would high superhelix density favor or disfavor C-DNA over B-DNA?

*13. A closed circular supercoiled DNA is relaxed by treatment with topoisomerase. No matter how much enzyme is used, or how long the experiment is run, the experimenter always finds a gel electrophoresis pattern indicating some DNA with one, two, and three superhelical turns in addition to the relaxed (nicked) circle (see figure). Suggest an explanation for this observation.

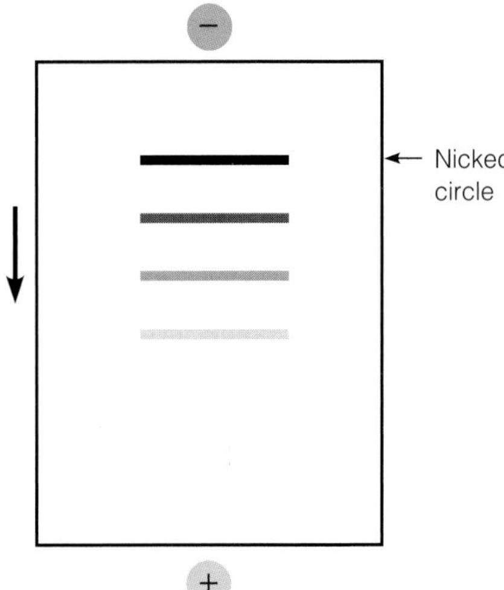

*14. The dye ethidium is a planar molecule that can *intercalate* into double-strand DNA. This means that it slips between adjacent stacked base pairs. In doing so, it unwinds the DNA helix by about 26° for every ethidium bound.
(a) If ethidium were added to relaxed, closed circular DNA, would negative or positive writhing be expected? Explain.
(b) What would the effect on writhing be if the DNA were nicked in one strand?
(c) It is observed that progressive addition of ethidium to negatively supercoiled DNA has a peculiar effect: First the electrophoretic mobility decreases, but with further addition of ethidium it again increases. Explain.

TOOLS OF BIOCHEMISTRY 4A

An Introduction to X-Ray Diffraction

Only a few decades ago, virtually nothing was known about the three-dimensional structures of nucleic acids, proteins, and polysaccharides. Today, largely as a result of the technique of x-ray diffraction, many of these molecules are understood at a level of detail that would have astounded the biochemists of 1950. The method is complicated and it is possible to give only a brief introduction here, describing what is measured and what can be obtained.

When radiation of any kind passes through a regular, repeating structure, *diffraction* is observed. This means that radiation scattered by the repeating elements in the structure shows reinforcement of the scattered waves in certain specific directions and weakening of the waves in other directions. A simple example is given in Figure 4A.1, which shows radiation being scattered from a row of equally spaced atoms. Only in certain directions will the scattered waves be in phase and therefore constructively interfere with (reinforce) one another. In all other directions they will be out of phase and destructively interfere with one another. Thus, a **diffraction pattern** is generated. For the diffraction pattern to be sharp, it is essential that the wavelength of the radiation used be somewhat shorter than the regular spacing between the elements of the structure. This is why x-rays are used in studying molecules, for x-rays typically have a wavelength of only a few tenths of a nanometer. If the regular spacing in the object being studied is large (as in a window screen), we can observe exactly the same phenomenon with visible light, which has a wavelength thousands of times longer than x-rays. We will find that a point source, seen through a window screen, gives a rectangular diffraction pattern of spots.

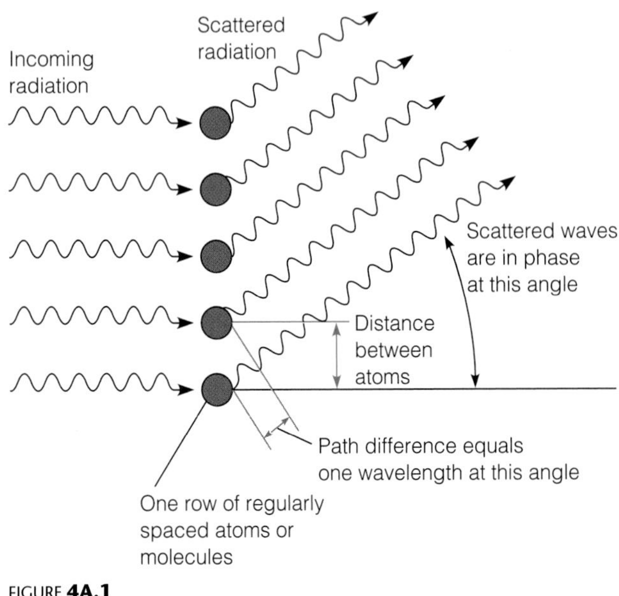

FIGURE **4A.1**
Diffraction from a very simple structure—a row of atoms or molecules.

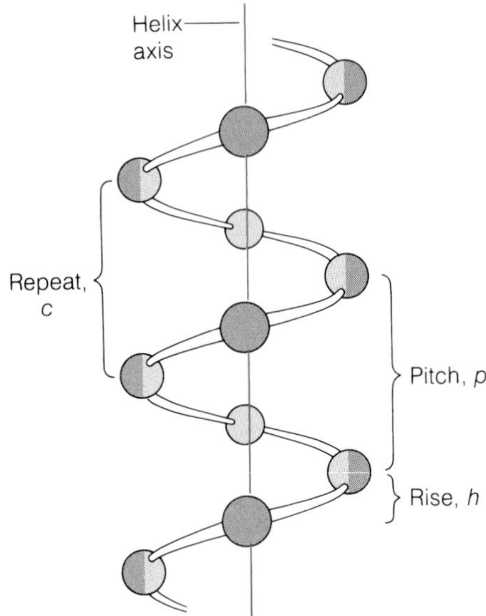

FIGURE **4A.2**
A simple helical molecule.

The rule relating the periodic spacings in object and diffraction pattern is simple: Short spacings in the periodic structure correspond to large spacings in the diffraction pattern, and vice versa. In addition, by determining the relative intensities of different spots, we can tell how matter is distributed within each repeat of the structure.

Fiber Diffraction

We consider first the diffraction from helical molecules, aligned approximately parallel to the axis of a stretched fiber. A helical molecule, like the one shown schematically in Figure 4A.2, is characterized by certain parameters:

The *repeat* (c) of the helix is the distance parallel to the axis in which the structure exactly repeats itself. The repeat contains some integral number (m) of polymer residues. In Figure 4A.2, $m = 4$.

The *pitch* (p) of the helix is the distance parallel to the helix axis in which the helix makes one turn. If there is an integral number of residues per turn (as here), the pitch and repeat are equal.

The *rise* (h) of the helix is the distance parallel to the axis from the level of one residue to the next, so $h = c/m$. If we think of a spiral staircase as an example of a helix, the rise is the height of each step and the pitch is the distance from where one is standing to the corresponding spot directly overhead.

Suppose we wish to investigate a polymer with the helical structure shown in Figure 4A.2. A fiber is pulled from a concentrated solution of the polymer. Stretching the fiber further will produce approximate alignment of the long helical molecules with the fiber axis. The fiber is then placed in an x-ray beam, and

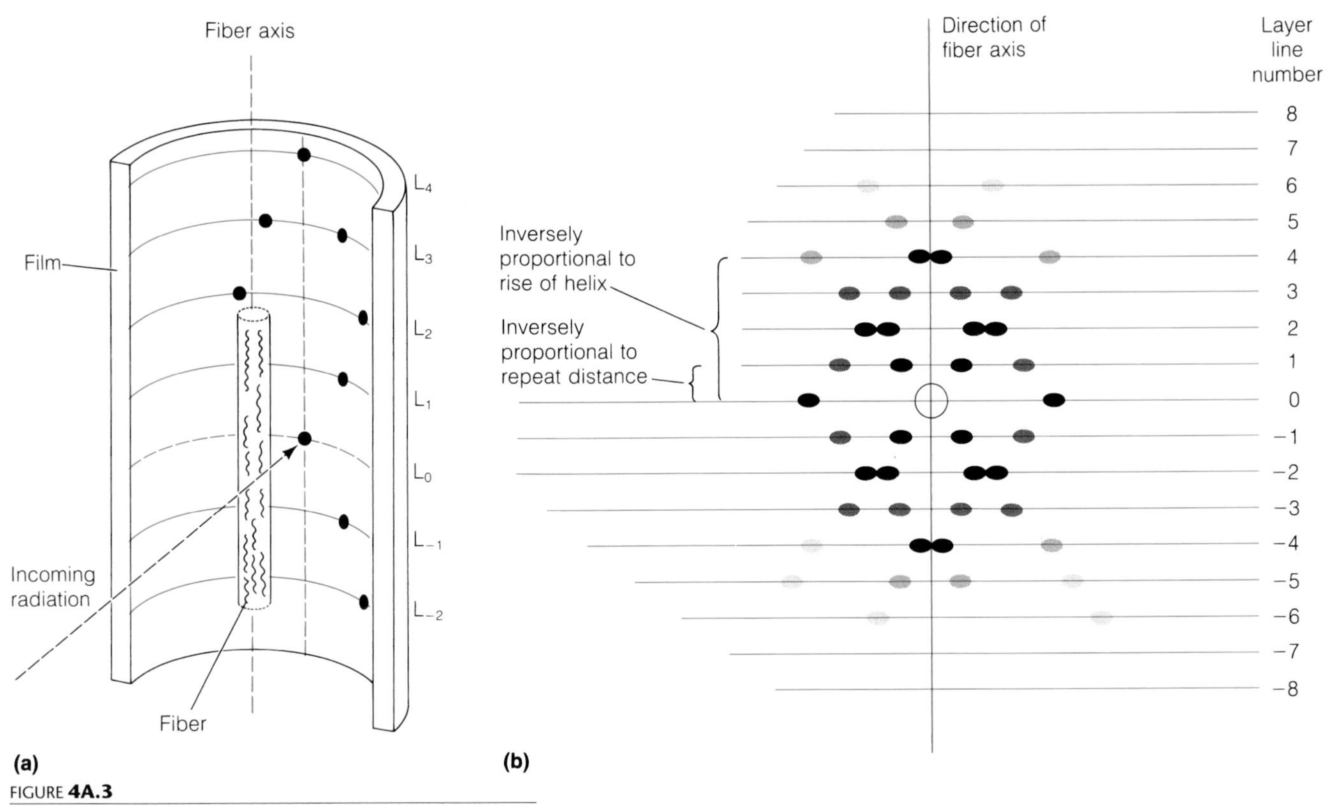

FIGURE **4A.3**

Diffraction from fibers. (a) A fiber in an x-ray beam. **(b)** The diffraction pattern.

a photographic film is positioned behind it, as shown in Figure 4A.3a. The diffraction pattern, which consists of spots or short arcs, will look like that in Figure 4A.3b. It can be read as follows: According to the mathematics of diffraction theory, a helix always gives rise to this kind of cross-shaped pattern. Therefore, we know we are dealing with a helical structure. The spots all lie on lines perpendicular to the fiber axis; these are called **layer lines.** The spacing between these lines is inversely proportional to the repeat of the helix, *c*, which in this case equals the pitch. Note that the cross pattern repeats itself on every fourth layer line. This repetition pattern tells us that there are exactly 4.0 residues per turn in the helix. Thus, the rise in the helix is $c/4$. This is the kind of evidence that told Watson and Crick that B-DNA was a helix with 10 residues per turn.

The information above is given directly by the pattern. To find out exactly how all of the atoms in each residue are arranged in each repeat, a more detailed analysis is necessary. Usually, a model is made using the correct repeat, pitch, and rise. Model making is simplified because we know approximate bond lengths and the angles between many chemical bonds. The model must also be inspected to see that no two atoms approach closer than their van der Waals radii. From such a model, the intensities of the various spots can be predicted. These predictions are compared with the observed intensities, and the model is readjusted until a best fit is obtained. The initial determination of the structure of DNA was done in just this way. As you can see from Figure 4.9, real fiber diffraction patterns are not as neat as the idealized example, mainly because of incomplete alignment of the molecules.

Crystal Diffraction

To study molecular crystals such as those formed by small oligonucleotides, molecules like tRNA, and globular proteins, the experimenter faces a rather different problem from the study of helical fibers and proceeds in a quite different way. A schematic drawing of such a crystal is shown in Figure 4A.4. The repeating unit is now the **unit cell,** which may contain one, two, or more molecules. The unit cell may be thought of as the basic building

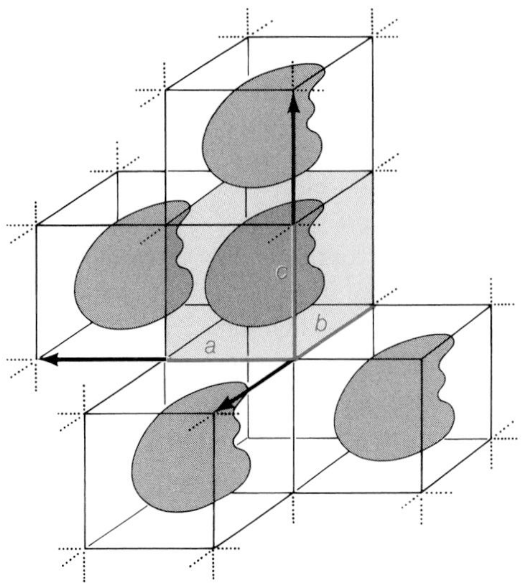

FIGURE **4A.4**

Schematic drawing of a molecular crystal.

block of the crystal. Repetition of the unit cell in three dimensions (marked by arrows on the figure) creates the whole crystal. A simple two-dimensional analog of the crystal unit cell is the repeating pattern in wallpaper. No matter how random a wallpaper pattern may seem, if you stare at it long enough, you can always find a unit that, by repetition, fills the entire wall.

Just as in fiber diffraction, passing an x-ray beam through a molecular crystal produces a diffraction pattern. The pattern shown in Figure 4A.5 was obtained from a crystal of a small DNA. Again, the spacing of the spots allows us to determine the repeating distances in the periodic structure—in this case the *x*, *y*, and *z* dimensions of the unit cell labeled *a*, *b*, and *c* in Figure 4A.4. But the important information in crystal diffraction studies is just how the atoms are arranged *within* each unit cell, for that arrangement describes the molecule. Again, this information is contained in the relative intensities of the diffraction spots in a pattern like that shown in Figure 4A.5. But in crystal diffraction, more exact information can be extracted than from a fiber diffraction pattern, because the corresponding molecules in each unit cell of the crystal are of the same shape and are oriented in the same way. In fiber diffraction, the helical molecules may all have their long axes pointed in the same direction, but they are rotated randomly about these axes. This difference in exactness of arrangement can be appreciated by comparing the sharpness of the crystal diffraction pattern shown in Figure 4A.5 with the fiber pattern depicted in Figure 4.9.

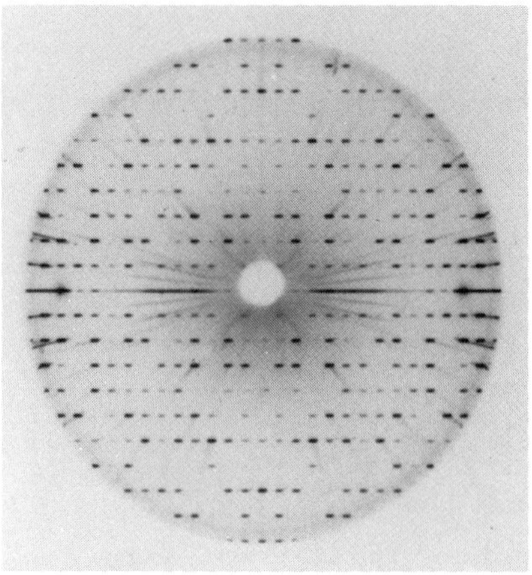

FIGURE **4A.5**
Diffraction pattern produced by a molecular crystal of a small DNA.
Courtesy of C. A. Frederick.

After obtaining the diffraction pattern from a molecular crystal, the experimenter measures the intensities of a large number of the spots. If the molecule being studied is a small one, it is possible to proceed in much the same manner as with fiber patterns. A structure is guessed at, and expected intensities are calculated and compared with the observed intensities. The structure is refined until the relative intensities of all spots are correctly predicted. However, such a procedure won't work with a molecule as complex as the tRNA shown in Figure 4.20—there is simply no way to guess such a structure.

Why not proceed directly from spot intensities to the structure? The difficulty is that some of the information contained in the spot intensities is hidden. To greatly simplify a complex problem, we may say that it is as if the quantities that the experimenter needed in order to deduce the structure (which are called **structure factors**) were the square roots of the intensities.* If the intensity has a value of, say, 25, the investigator knows that the number needed is +5 or −5. But which? This sort of quandary is the essence of the *phase problem*, which prevented progress in large-molecule crystallography for many years. One way of solving the problem was discovered in the early 1950s. Suppose a heavy metal atom can be introduced into some point in the molecule in such a way that the molecule and crystal are otherwise unchanged. This process is called an **isomorphous replacement.** Now suppose the heavy metal contributes a value of +2 to the structure factor for the spot we were just discussing. If the original value was +5, its new value is +7 and its square is 49. If the original value was −5, the value now becomes −3 and its square is 9. The investigator takes a diffraction photograph of the crystal with the heavy metal inserted. If the new crystal has an intensity of 9 for this spot, the original structure factor must have been −5, not +5. Although an oversimplification, this example gives the essence of the method. Usually multiple isomorphous replacements are necessary to determine the phases of the structure factors.

Given structure factors for all of the spots, the investigator can calculate the positions of all atoms in the unit cell. What is actually calculated is an **electron density** distribution (Figure 4A.6), but this amounts to the same thing, for regions of high electron density are where the atoms are. In the particular view shown in Figure 4A.6, we are looking at a two-dimensional "slice" through the three-dimensional electron density distribution.

It is now appropriate to review the steps that must be taken to determine the three-dimensional structure of a macromolecule from crystal diffraction studies:

1. Obtain satisfactory crystals. This step is often the hardest part of the procedure, for the crystals must be of good quality and at least a few tenths of a millimeter in minimum dimension. Crystals that are too small will not give sharp diffraction patterns. Getting macromolecules to crystallize well is still more of an art than a science.
2. Record the diffraction pattern from the crystal, and measure the intensities of many of the spots.
3. Find some way to make isomorphous replacements in the molecule. Usually two or more replacements are required.
4. Repeat steps 1 and 2 for each isomorphous derivative.
5. Calculate structure factors and, from them, the electron density distribution. These calculations are usually done on a large computer.

In most cases, the investigator will first carry through this analysis with a relatively small number of spots. This procedure will give a *low-resolution* structure. If all is going well, more spots will be measured and the calculations refined to give higher resolution. With the best crystals, it is now possible to obtain resolutions of about 0.10 nm. This resolution is sufficient to identify

*For the more mathematically sophisticated reader, we note that the structure factors are usually complex numbers and can thus be represented as vectors in the complex plane. What are determined from the intensities are their *amplitudes*, but what is not known are their *phases.*

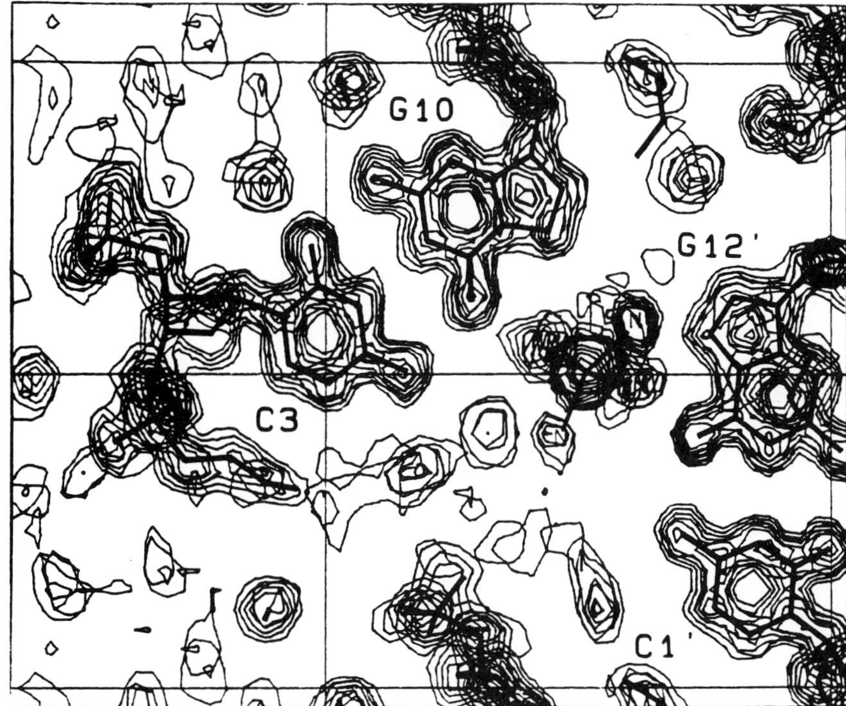

FIGURE **4A.6**

Part of an electron density map derived from the DNA crystal diffraction pattern in Figure 4A.5.

Reprinted by permission from F. L. Suddath, *Nature* (1974) 248:20–24, fig. 3, p. 22. © 1974 Macmillan Magazines Ltd.

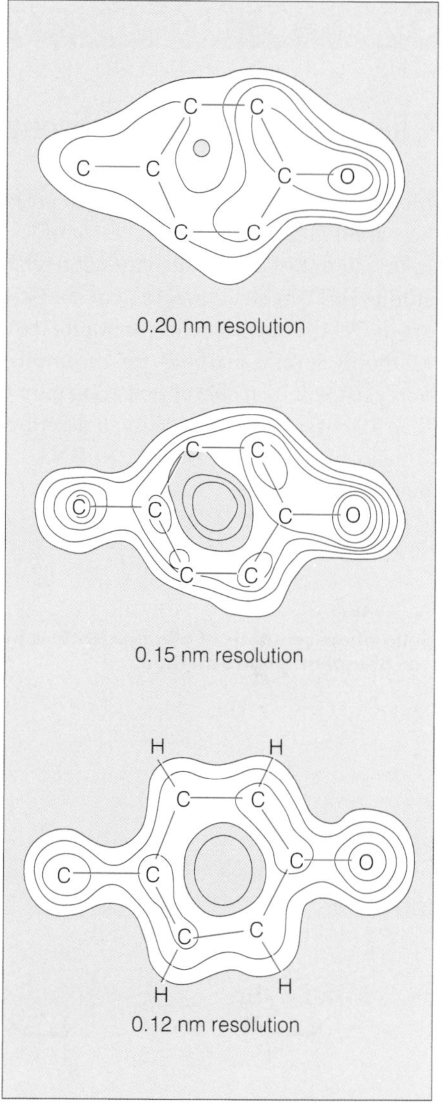

FIGURE **4A.7**

Effect of increased resolution on molecular detail observed by x-ray diffraction.

From K. D. Watenpaugh, L. K. Sieker, L. H. Jensen, *J. Mol. Biol.* (1980) 138:615–633. © 1980 Academic Press.

individual groups and even some atoms and to show how they interact with one another. The detail in the phenolic ring of a protein side chain revealed at different resolutions is shown in Figure 4A.7.

Most of the detailed three-dimensional structures of biological macromolecules shown in this book have been determined by x-ray diffraction studies of crystals. At present, hundreds of such structures are known. This knowledge represents an enormous amount of labor in many laboratories, but the results allow us to understand macromolecular function at a level that would have been unbelievable only a short time ago.

Reference

van Holde, K. E., W. C. Johnson, and P.-S. Ho (1998) *Principles of Physical Biochemistry,* Chapter 6. Prentice Hall, Upper Saddle River, N.J. A more detailed treatment of x-ray diffraction of biopolymers.

TOOLS OF BIOCHEMISTRY 4B

Chemical Synthesis of Oligonucleotides

Modern molecular biology has been greatly stimulated by the new ability to synthesize oligonucleotides of defined sequence in the laboratory. These oligonucleotides have been used in x-ray studies of DNA structure, in gene modification, and as primers in DNA sequence determination (see Chapter 25, Dideoxy Method). Several methods for accomplishing such syntheses now exist, some capable of preparing pure polynucleotides more than 100 bases in length. We shall describe one such technique, which can be used in automated **DNA synthesizers,** or "gene machines."

First, several requirements must be met to obtain a pure product in high yields:

1. The residues must be added one at a time, and the reactions must be very efficient. If even a few growing chains in the reaction mixture fail to add the expected residue at each step, an impure product will result.
2. Conditions must be found to make the reactions thermodynamically favorable. This can be accomplished by using nonaqueous solvents and activated, highly reactive forms of the nucleotides. The reactions can be further facilitated by the use of condensing agents to remove any water from the reaction.

FIGURE **4B.1**

Solid-phase synthesis of oligonucleotides by the phosphoramidite method.

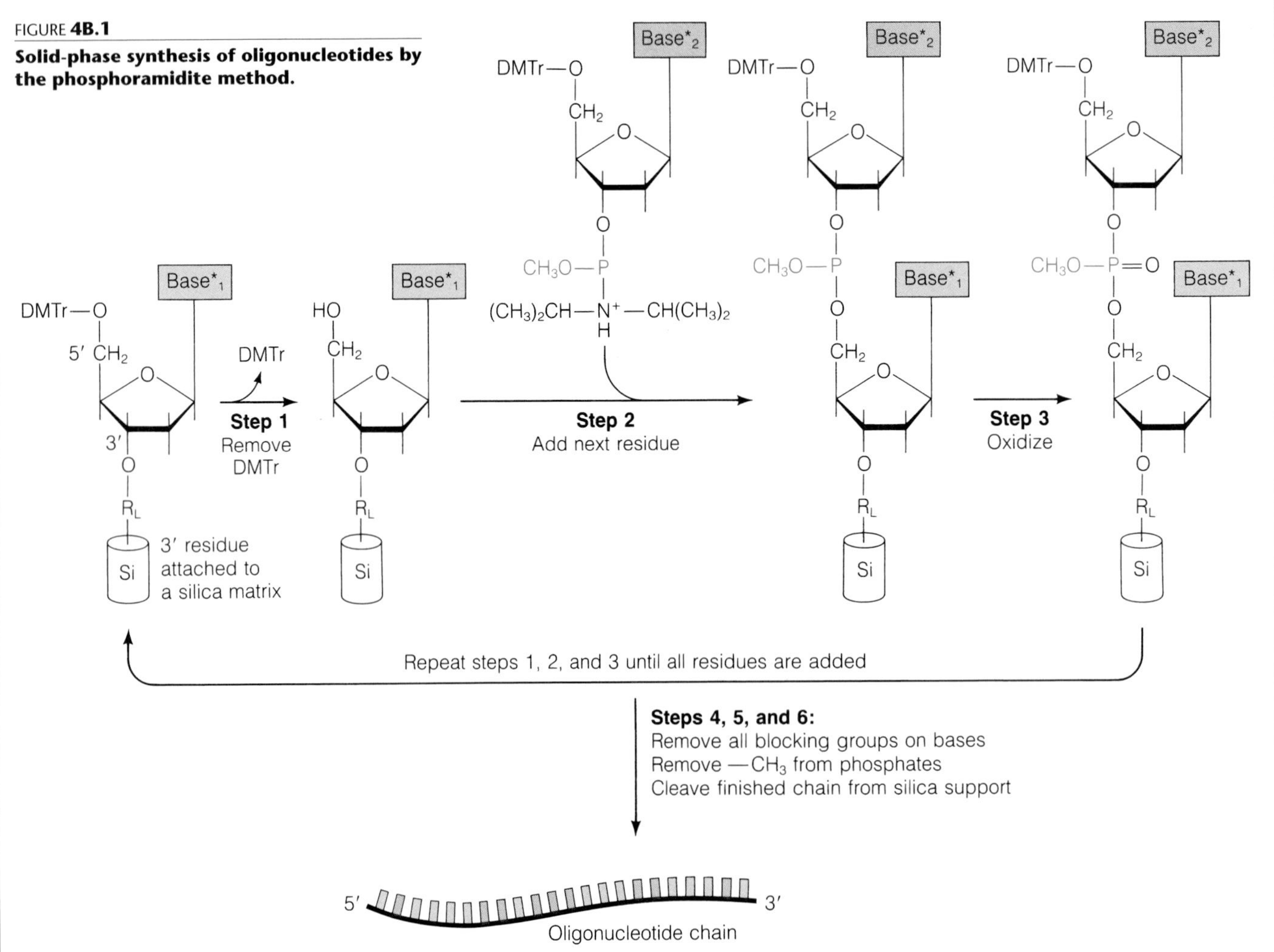

3. To avoid unwanted side reactions, all potentially reactive groups on the molecules, except for those where reaction is desired, must be chemically blocked.

Suppose we wish to synthesize a deoxyribonucleic acid with the following specific sequence:

ATTGACTTAGCC

Figure 4B.1 (page 124) illustrates one method. The monomers used are *phosphoramidites,* highly reactive molecules with *trivalent* phosphorus.

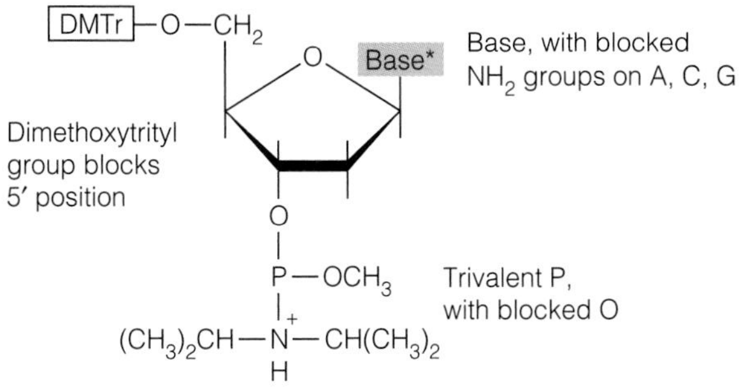

DMTr—O—CH_2 ... Base* — Base, with blocked NH_2 groups on A, C, G

Dimethoxytrityl group blocks 5′ position

O—P—OCH_3 — Trivalent P, with blocked O

$(CH_3)_2CH$—N^+(H)—$CH(CH_3)_2$

Note that potentially reactive groups are blocked so that side reactions will not occur. The structures of commonly used blocking groups are

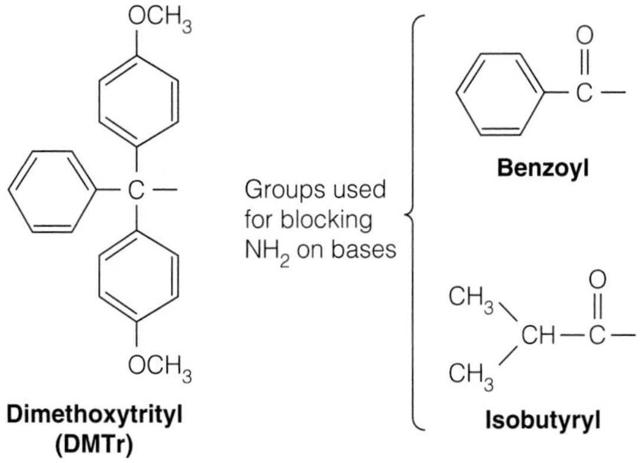

OCH_3 / C— / OCH_3

Dimethoxytrityl (DMTr)

Groups used for blocking NH_2 on bases

Benzoyl

CH_3, CH_3 CH—C(=O)—

Isobutyryl

As shown in Figure 4B.1, the nucleotide selected to be at the 3′ end of the chain is coupled through its 3′ hydroxyl to a solid silica support. In the first step of the synthesis, the DMTr group is cleaved off. Then the next monomer is added in a protonated form. To stabilize the product, the trivalent phosphorus is oxidized to the pentavalent state, forming a *phosphotriester.* This process is repeated, in a wholly automated fashion, until the entire sequence is completed. Finally, the blocking groups are removed and the chain is hydrolyzed from the silica support.

The necessity for high-reaction yields at each step can be seen from the following calculation. In the example above, 11 addition steps are required. If the yield were 90% at each step, the final yield of the desired product would be only 31%. In practice, stepwise yields of about 98% can be routinely accomplished, which would give an 80% final yield.

References

Beaucage, S. L., and M. H. Caruthers (1981) Deoxynucleoside phosphorimidates: A new class of key intermediates for deoxypolynucleotide synthesis. *Tetrahedron Lett.* 22:1859–1862.

Mattencci, M. D., and M. H. Caruthers (1981) Synthesis of deoxyoligonucleotides on a polymer support. *J. Am. Chem. Soc.* 103:3185–3191.

CHAPTER 5

Introduction to Proteins: The Primary Level of Protein Structure

WE HAVE SEEN THAT ONE CLASS OF BIOPOLYMERS, THE NUCLEIC ACIDS, STORES and transmits the genetic information of the cell. Much of that information is expressed in another class of biopolymers, the **proteins.** Proteins play an enormous variety of roles: Some carry out the transport and storage of small molecules; others make up a large part of the structural framework of cells and tissues. Muscle contraction, the immune response, and blood clotting are all mediated by proteins. Perhaps the most important of all proteins are the **enzymes**—the catalysts that promote the tremendous variety of reactions that channel metabolism into essential pathways. Each type of cell in every organism has several thousand kinds of proteins to serve these many functions.

In keeping with the multiplicity of their functions, proteins are extremely complex molecules. This you can see by a glance at Figure 5.1, which depicts the molecular structure of myoglobin, a protein used to store oxygen in animal tissues. In this and the following three chapters we analyze in detail the structure of a number of proteins, including myoglobin. We will see that each protein has a logical, functional structure of its own, as well as certain features in common with all other proteins. Structures that may appear at first glance to be hopelessly complex are readily comprehensible when we understand the way in which their forms are dictated by their functions. We begin with a description of the units of chemical structure common to all proteins, the amino acids.

Amino Acids

STRUCTURE OF THE α-AMINO ACIDS

Proteins are polymers of 20 different kinds of α-amino acids.

All proteins are polymers, and the monomers that combine to make them are **$\boldsymbol{\alpha}$-amino acids.** A representative α-amino acid, valine, is shown in Figure 5.2a. The amino group is attached to the α-carbon, the carbon next to the carboxyl group; hence the name α-amino acid. To the α-carbon of every amino acid are also attached a *hydrogen atom* and a *side chain* (R). Different α-amino acids are distinguished by their different side chains. We can write the general structure for an α-amino acid in the fashion shown to the left in Figure 5.2b. However, this

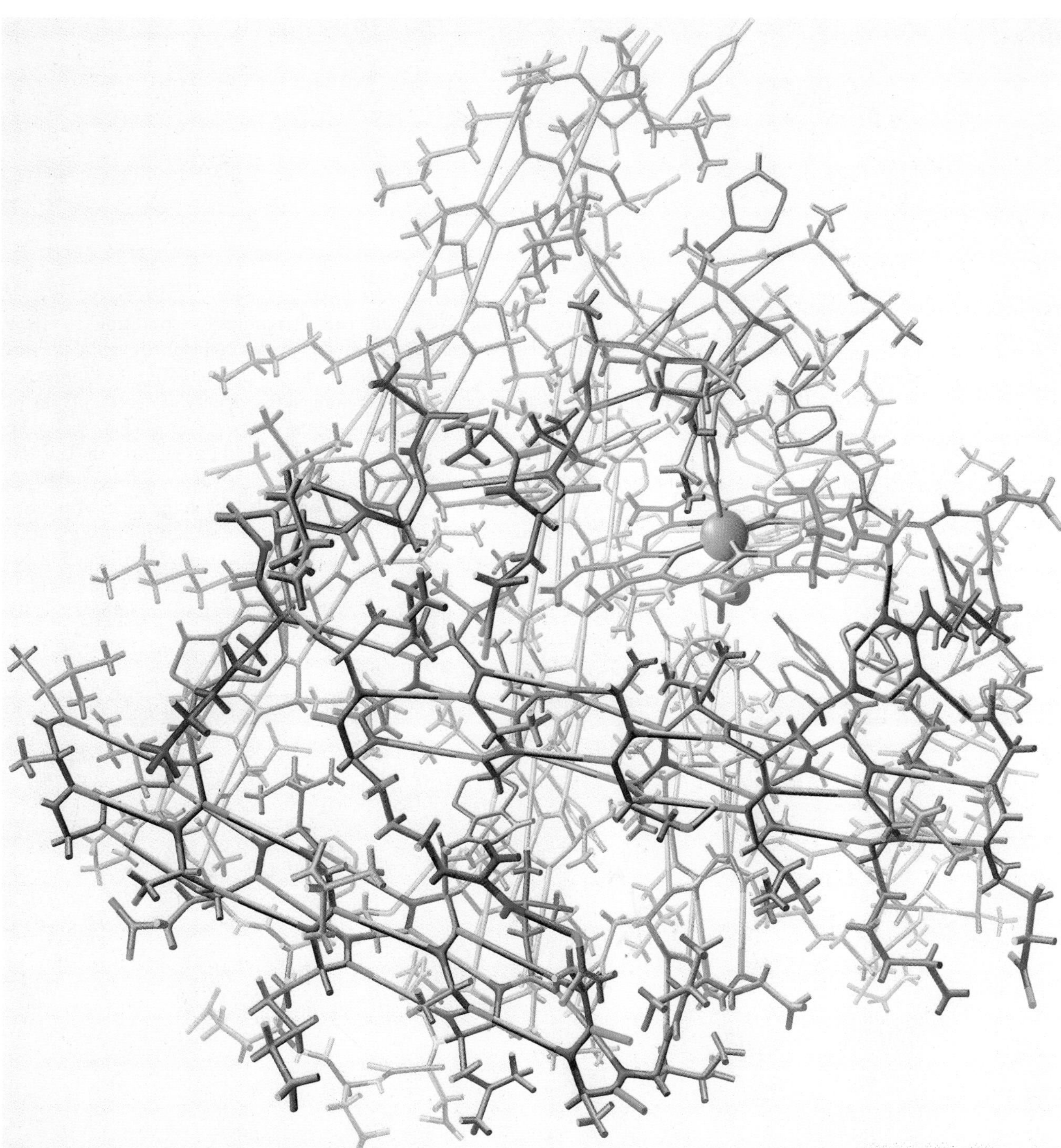

FIGURE **5.1**

The three-dimensional structure of myoglobin. This historic painting, by Irving Geis in collaboration with John Kendrew, portrays as a stick model the first protein whose structure was deduced by x-ray diffraction—sperm whale myoglobin. It depicts, therefore, our first indication of the complexity and specificity of the three-dimensional structure of proteins.

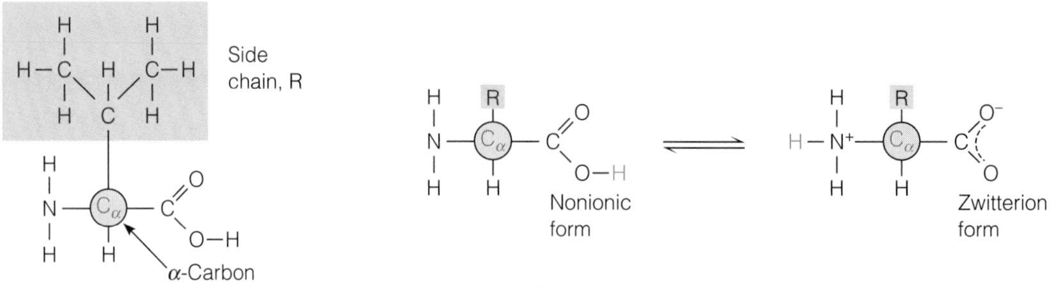

Side chain, R

α-Carbon

(a) Valine

Nonionic form

Zwitterion form

(b) Generalized amino acid, forming zwitterion at neutral pH

FIGURE **5.2**

The structure of α-amino acids. (a) Valine, a representative α-amino acid, has a carboxyl group, an amino group, and a hydrogen on the α-carbon, as well as a side chain (R) that gives it its unique properties. **(b)** This generalized α-amino acid is shown both in its nonionic form, and as the zwitterion at neutral pH. Under physiological conditions, amino acids exist as zwitterions in which the carboxyl group has lost a proton and the amino group has gained one. Note that the negative charge on the carboxyl is delocalized between the two oxygen atoms.

representation, although chemically correct, ignores the conditions in vivo. As pointed out in Chapter 2, most biochemistry occurs in the physiological pH range near neutrality. The pK_a's of the carboxyl and amino groups of the α-amino acids are about 2 and 10, respectively. Therefore, in the vicinity of neutral pH the carboxylate group will have lost a proton, and the amino group will have picked up a proton, to yield the *zwitterion* form shown on the right in Figure 5.2b. This is the form in which we will customarily write amino acid structures.

Twenty different kinds of amino acids are coded for in the genes of all organisms and incorporated into proteins. The complete structures of these amino acids are shown in Figure 5.3; other important data are given in Table 5.1. Note that proline has been set apart in Figure 5.3. Proline is a cyclic amino acid, because the side chain is bonded back to the nitrogen atom, forming a ring. It is sometimes incorrectly referred to as an imino acid.

TABLE **5.1 Properties of the amino acids found in proteins**

Name	Abbreviations	pK_a of α-COOH Group	pK_a of α-NH_3^+ Group	pK_a of Ionizing Side Chain[a]	Residue[b] Mass (daltons)	Occurrence[c] in Proteins (mol %)
Alanine	A, Ala	2.3	9.7	—	71.08	9.0
Arginine	R, Arg	2.2	9.0	12.5	156.20	4.7
Asparagine	N, Asn	2.0	8.8	—	114.11	4.4
Aspartic acid	D, Asp	2.1	9.8	3.9	115.09	5.5
Cysteine	C, Cys	1.8	10.8	8.3	103.14	2.8
Glutamine	Q, Gln	2.2	9.1	—	128.14	3.9
Glutamic acid	E, Glu	2.2	9.7	4.2	129.12	6.2
Glycine	G, Gly	2.3	9.6	—	57.06	7.5
Histidine	H, His	1.8	9.2	6.0	137.15	2.1
Isoleucine	I, Ile	2.4	9.7	—	113.17	4.6
Leucine	L, Leu	2.4	9.6	—	113.17	7.5
Lysine	K, Lys	2.2	9.0	10.0	128.18	7.0
Methionine	M, Met	2.3	9.2	—	131.21	1.7
Phenylalanine	F, Phe	1.8	9.1	—	147.18	3.5
Proline	P, Pro	2.0	10.6	—	97.12	4.6
Serine	S, Ser	2.2	9.2	—	87.08	7.1
Threonine	T, Thr	2.6	10.4	—	101.11	6.0
Tryptophan	W, Trp	2.4	9.4	—	186.21	1.1
Tyrosine	Y, Tyr	2.2	9.1	10.1	163.18	3.5
Valine	V, Val	2.3	9.6	—	99.14	6.9

[a]Approximate values found for side chains on the *free* amino acids.

[b]To obtain the mass of the amino acid itself, add the mass of a mole of water, 18.02 g. The values given are for neutral side chains; slightly different values will apply at pH values where protons have been gained or lost from the side chains.

[c]Average for a large number of proteins. Individual proteins can show large deviations from these values.

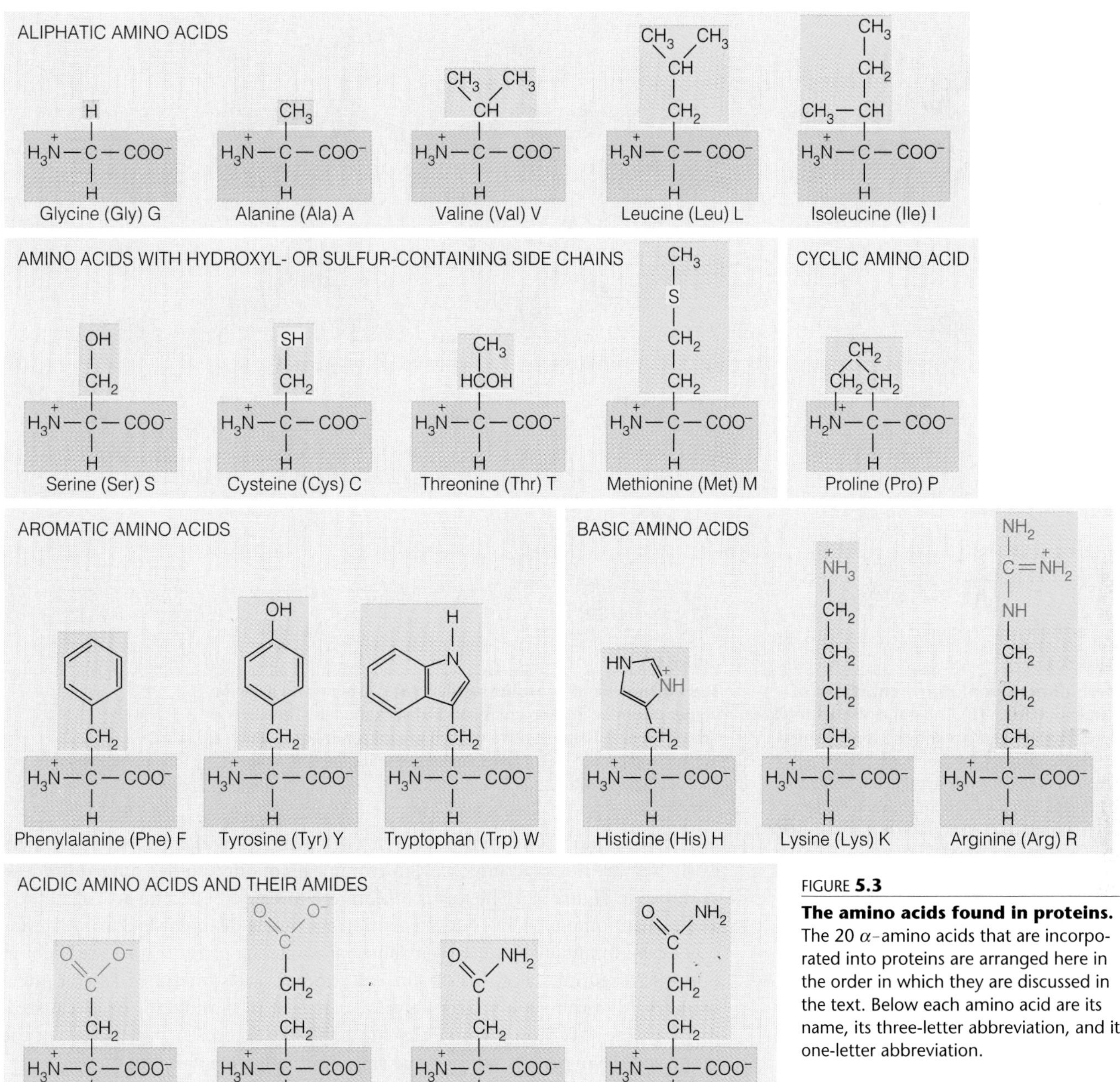

FIGURE **5.3**

The amino acids found in proteins. The 20 α-amino acids that are incorporated into proteins are arranged here in the order in which they are discussed in the text. Below each amino acid are its name, its three-letter abbreviation, and its one-letter abbreviation.

STEREOCHEMISTRY OF THE α-AMINO ACIDS

Planar formulas like those shown in Figure 5.3 fail to reveal some important features of amino acid structure. Bonding about the α-carbon is tetrahedral, as you would expect, so the three-dimensional representation of an amino acid shown in Figure 5.4a is more realistic. The same feature can be represented more compactly in the way shown in Figure 5.4b. Here the solid wedges represent bonds coming forward from the page and the dashed wedges represent bonds receding behind the page.

Whenever a carbon atom has four different substituents attached to it, forming an asymmetric molecule, the carbon is said to be **chiral** or a **center of chirality** or a **stereocenter.** It is also often simply called an **asymmetric carbon.** If a

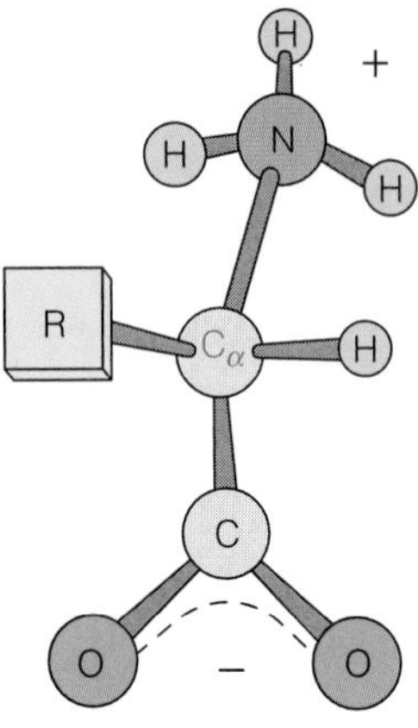

(a) α-Amino acid

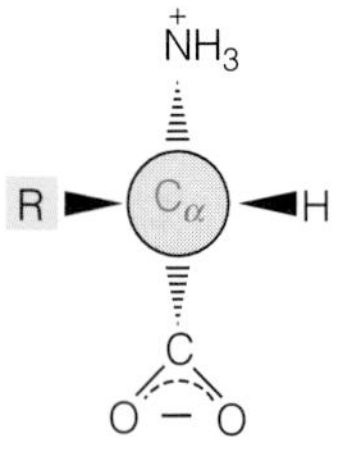

(b) Compact representation

FIGURE **5.4**

Three-dimensional representations of α-amino acids. (a) This ball-and-stick model shows the bond angles and the three-dimensional arrangement of the atoms. The α-carbon is chiral, with tetrahedral bonding. **(b)** In this compact rendition, solid wedges represent bonds coming forward, and shaded wedges represent receding bonds.

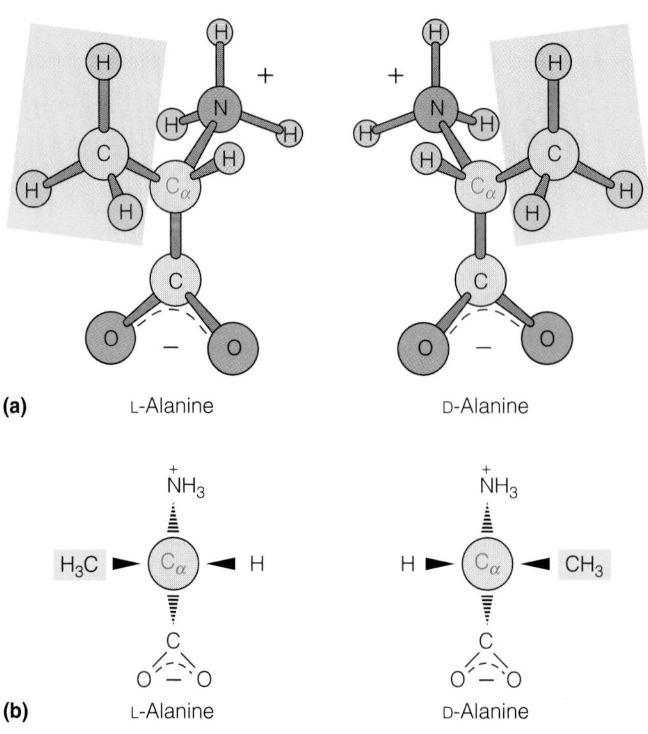

FIGURE **5.5**

Stereoisomers of α-amino acids. (a) L-Alanine and its enantiomer D-alanine are shown as ball-and-stick models. The alanine side chain is —CH_3. The two models are mirror images, which are not superimposable. **(b)** The same two enantiomers in a more compact representation.

molecule contains one asymmetric carbon, two distinguishable **stereoisomers** exist; these are nonsuperimposable mirror images of one another, or **enantiomers,** as shown in Figure 5.5. The forms of alanine shown in that figure are called the L and D enantiomers.* The L and D enantiomers can be distinguished from one another experimentally because their solutions rotate the plane of polarized light in opposite directions. For this reason, enantiomers are sometimes called **optical isomers.** All amino acids except glycine can exist in D and L forms, because in each case the α-carbon is chiral. Glycine is the sole exception, because two of the groups on the α-carbon are the same (H), eliminating the chirality.

All α-amino acids except glycine contain an asymmetric α-carbon and thus have L and D enantiomers. Only L-enantiomers are found in proteins.

The important fact for our purposes is that *all of the amino acids incorporated by organisms into proteins are of the* L *form.* Although it is evident that random mixtures of D and L amino acids could not form protein structures as precise as that shown in Figure 5.1, the absolute preference for the L isomer is puzzling. Indeed, we shall find that each of the three major classes of biological macromolecules exhibits a strong preference for one stereoisomer class or the other. Most naturally occurring polysaccharides involve D-sugars, as do DNA and RNA. It may be that mutual interaction between these substances has established these relationships very early in the evolution of life. But why was a particular set of enantiomers chosen at all? It is hard to see how L-amino acids have any inherent superiority over their D-isomers for biological function. Indeed, D-amino acids exist in nature, and some play important biochemical roles (some examples are given in Table 5.2), but they are never found in proteins.

*Those who are familiar with modern organic chemistry will know that there are two systems for distinguishing stereoisomers—the older D–L system and the newer, more comprehensive *R–S* system. Both are discussed in more detail in Chapter 9.

TABLE 5.2 Some biologically important amino acids not found in proteins

Name	Formula	Biochemical Source, Function
β-Alanine	$H_3\overset{+}{N}-CH_2-CH_2-COO^-$	Found in the vitamin pantothenic acid and in some important natural peptides
D-Alanine	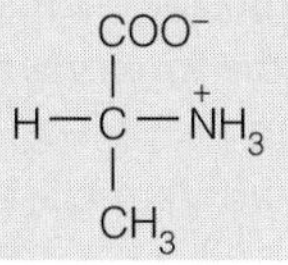 COO^-; $H-C-\overset{+}{N}H_3$; CH_3	In polypeptides in some bacterial cell walls
γ-Aminobutyric acid	$H_3\overset{+}{N}-CH_2-CH_2-CH_2-COO^-$	Brain, other animal tissues; functions as neurotransmitter
D-Glutamic acid	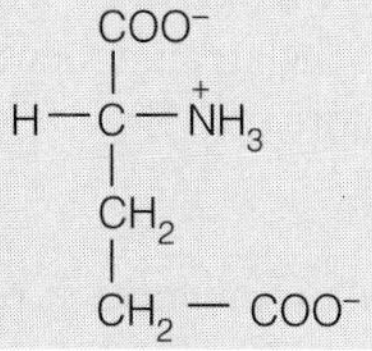 COO^-; $H-C-\overset{+}{N}H_3$; CH_2; CH_2-COO^-	In polypeptides in some bacterial cell walls
L-Homoserine	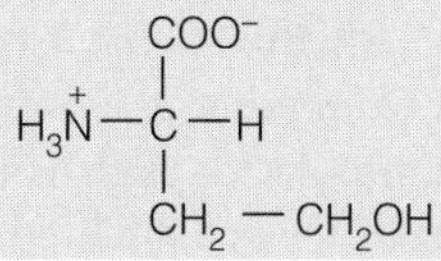 COO^-; $H_3\overset{+}{N}-C-H$; CH_2-CH_2OH	Many tissues; an intermediate in amino acid metabolism
L-Ornithine	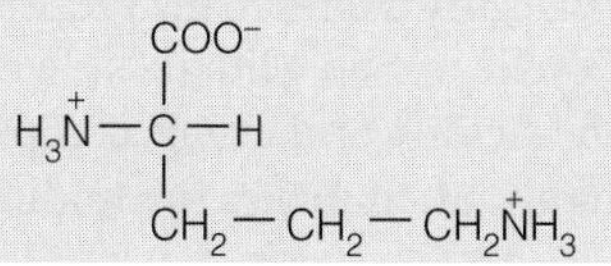 COO^-; $H_3\overset{+}{N}-C-H$; $CH_2-CH_2-CH_2\overset{+}{N}H_3$	Many tissues; an intermediate in arginine synthesis
Sarcosine	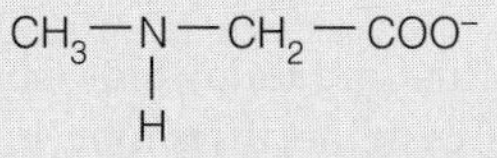 $CH_3-N(H)-CH_2-COO^-$	Many tissues; intermediate in amino acid synthesis
L-Thyroxine	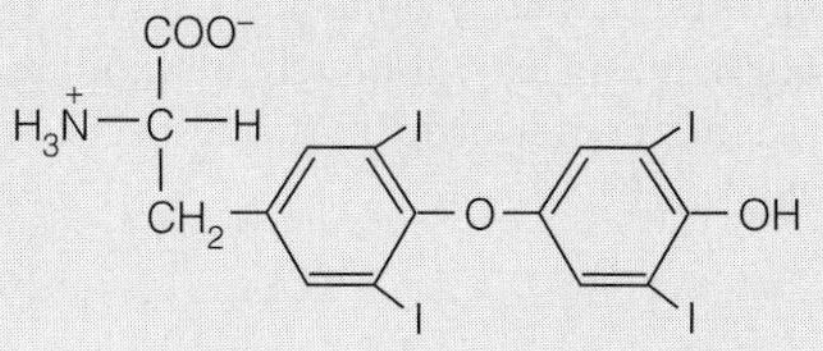 COO^-; $H_3\overset{+}{N}-C-H$; CH_2–(3,5-diiodophenyl)–O–(3,5-diiodophenyl)–OH	Thyroid gland; is thyroid hormone (I = iodine)

Many scientists have attempted to provide explanations for this "handedness preference" in biology. Most point to an intrinsic asymmetry in the behavior of subnuclear particles, a kind of asymmetry that gives electrons emitted in β decay a preferential left-hand spin. Such influences are very weak but might, in a competition between primitive organisms using L or D protein, give a slight advantage to one or the other. After billions of generations, even a small advantage can become overwhelming.

Recently, using modern synthetic methods as described in Tools of Biochemistry 5D, it has been possible to synthesize "proteins" using all D-amino acids. These structures are the mirror images of the corresponding natural proteins. One such D-protein synthesized is the mirror image of a protease (a protein-cleaving enzyme) from the human immunodeficiency virus, HIV (see References). Whereas its natural L-counterpart cleaves natural L-proteins, this synthetic enzyme will cleave only those containing D-amino acids. Thus, "D-life" would be possible.

For most of our considerations in this and following chapters, the chirality of amino acids is not a critical point. Therefore we shall generally not draw their structures so as to indicate chirality but rather, as in Figure 5.3, write them with amino group to left, carboxyl to right, and R group at the top.

PROPERTIES OF AMINO ACID SIDE CHAINS: CLASSES OF α-AMINO ACIDS

The 20 amino acids contain, in their 20 different side chains, a remarkable collection of chemical groups. It is this diversity of the monomers that allows proteins to exhibit such a great variety of structures and properties. If we examine Figure 5.3, it becomes evident that there are several different classes of side chains, distinguished by their dominant chemical features. These features include hydrophobic or hydrophilic character, polar or nonpolar nature, and presence or absence of ionizable groups. Many ways have been proposed to group the amino acids into classes, but none is wholly satisfactory. We shall discuss the amino acids in an order that makes them easy to remember: the arrangement shown in Figure 5.3, which proceeds from the simplest to the more complex.

The variety of side chains on amino acids allows proteins enormous versatility in structure.

Amino Acids with Aliphatic Side Chains

Glycine, alanine, valine, leucine, and **isoleucine** have aliphatic, or **alkane,** side chains. As we progress from left to right along the top row of Figure 5.3, the R group becomes more extended and more hydrophobic. Isoleucine, for example, has a much greater tendency to transfer from water to a hydrocarbon solvent than does alanine. The more hydrophobic amino acids such as isoleucine are usually found *within* a protein molecule, where they are shielded from water. **Proline,** which is difficult to fit into any category, shares many properties with the aliphatic amino acids. Although it is a cyclic amino acid, its side chain has a primarily aliphatic character. However, the rigidity of the ring, compared with the flexibility of most amino acid side chains, often makes the folding of proline residues into protein structures difficult.

Amino Acids with Hydroxyl- or Sulfur-Containing Side Chains

In this category we can place **serine, cysteine, threonine,** and **methionine.** These amino acids, because of their weakly polar side chains, are generally more hydrophilic than their aliphatic analogs, although methionine is fairly hydrophobic. Among the group, cysteine is noteworthy in two respects. First, the side chain can ionize at moderately high pH:

$$\mathrm{H_3\overset{+}{N}{-}\underset{H}{\overset{SH-CH_2}{C}}{-}COO^-} \;\xrightleftharpoons{pK_a\,=\,8.3}\; \mathrm{H_3\overset{+}{N}{-}\underset{H}{\overset{S^- -CH_2}{C}}{-}COO^-} + \mathrm{H^+}$$

Second, oxidation can occur between pairs of cysteine side chains to form a **disulfide bond:**

$$2\;\mathrm{H_3\overset{+}{N}{-}CH(CH_2{-}SH){-}COO^-} \;\underset{\text{reduction}}{\overset{\text{oxidation}}{\rightleftharpoons}}\; \mathrm{H_3\overset{+}{N}{-}CH(COO^-){-}CH_2{-}S{-}S{-}CH_2{-}CH(COO^-){-}\overset{+}{N}H_3} + 2H^+ + 2e^-$$

Cysteine **Cystine**

The product of this oxidation is given the name **cystine**. We do not list it among the 20 amino acids because cystine is always formed by oxidation of two cysteine side chains and is not coded for by DNA. The presence of such disulfide bonds between cysteine residues in proteins often plays an important structural role.

Aromatic Amino Acids

Three amino acids, **phenylalanine, tyrosine,** and **tryptophan,** carry aromatic side chains. Phenylalanine, together with the aliphatic amino acids valine, leucine, and isoleucine, is one of the most hydrophobic amino acids. Tyrosine and tryptophan have some hydrophobic character as well, but it is tempered by the polar groups in their side chains. In addition, tyrosine can ionize at high pH:

$$HO-C_6H_4-CH_2-CH(NH_2)-COO^- \xrightleftharpoons{pK_a\,=\,10.1} {}^-O-C_6H_4-CH_2-CH(NH_2)-COO^- + H^+$$

The aromatic amino acids, like most compounds carrying conjugated rings, strongly absorb light in the near-ultraviolet region of the spectrum (Figure 5.6). This characteristic is frequently used for the analytical detection of proteins, by measuring absorption at 280 nm.

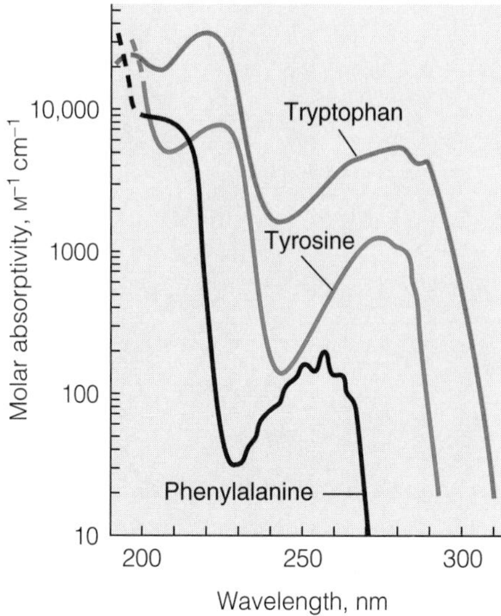

FIGURE **5.6**

Absorption spectra of the aromatic amino acids in the near-ultraviolet region. Tryptophan (red) and tyrosine (blue) account for most of the UV absorbance by proteins in the region around 280 nm. Note that the absorptivity scale is logarithmic; tryptophan absorbs hundreds of times more strongly than phenylalanine (black). However, compared with nucleic acids, amino acids absorb only weakly in the UV; see Figure 4.5 for comparison.

From D. Wetlaufer, *Adv. Protein Chem.* (1962) 17:303–390.

Basic Amino Acids

Histidine, lysine, and **arginine** carry basic groups in their side chains. They are represented in Figure 5.3 in the form that exists at pH values near neutrality. Histidine is the least basic of the three, and as its titration curve (Figure 5.7a) shows, the imidazole ring in the side chain of the free amino acid loses its proton at about pH 6. (pK_a values for the side chains of free amino acids are given in Table 5.1.) When histidine is incorporated into proteins, the pK_a is raised to about 7 (Table 5.3). Because the histidine side chain can exchange protons near physiological pH, it often plays a role in enzymatic catalysis involving proton transfer. Lysine and arginine are more basic amino acids, and as their pK_a values indicate (Tables 5.1 and 5.3), their side chains are always positively charged under physiological conditions.

The basic amino acids are strongly polar, and as a consequence they are usually found on the exterior surfaces of proteins, where they can be hydrated by the surrounding aqueous environment.

Acidic Amino Acids and Their Amides

Aspartic acid and **glutamic acid** are the only amino acids that carry negative charges at pH 7; they are depicted in the anionic forms in Figure 5.3. The titration curve of aspartic acid is shown in Figure 5.7b. The pK_a values of the acidic amino acids are so low (see Table 5.3) that even when the amino acids are incorporated into proteins, the negative charge on the side chain is retained under physiological conditions. Hence, these amino acid residues are often referred to as **aspartate** and **glutamate** (i.e., the conjugate bases rather than the acids).

Companions to aspartic and glutamic acids are their amides, **asparagine** and **glutamine.** Unlike their acidic analogs, asparagine and glutamine have uncharged side chains, although they are decidedly polar. Like the basic and acidic amino acids, they are definitely hydrophilic and tend to be on the surface of a protein molecule, in contact with the surrounding water.

TABLE **5.3** **Typical ranges observed for pK_a values of groups in proteins**

Group Type	Typical pK_a Range[a]
α-Carboxyl	3.5–4.0
Side chain carboxyls of aspartic and glutamic acids	4.0–4.8
Imidazole (histidine)	6.5–7.4
Cysteine (—SH)	8.5–9.0
Phenolic (tyrosine)	9.5–10.5
α-Amino	8.0–9.0
Side chain amino (lysine)	9.8–10.4
Guanidinyl (arginine)	~12

[a] Values outside these ranges are observed. For example, side chain carboxyls have been reported with pK_a values as high as 7.3.

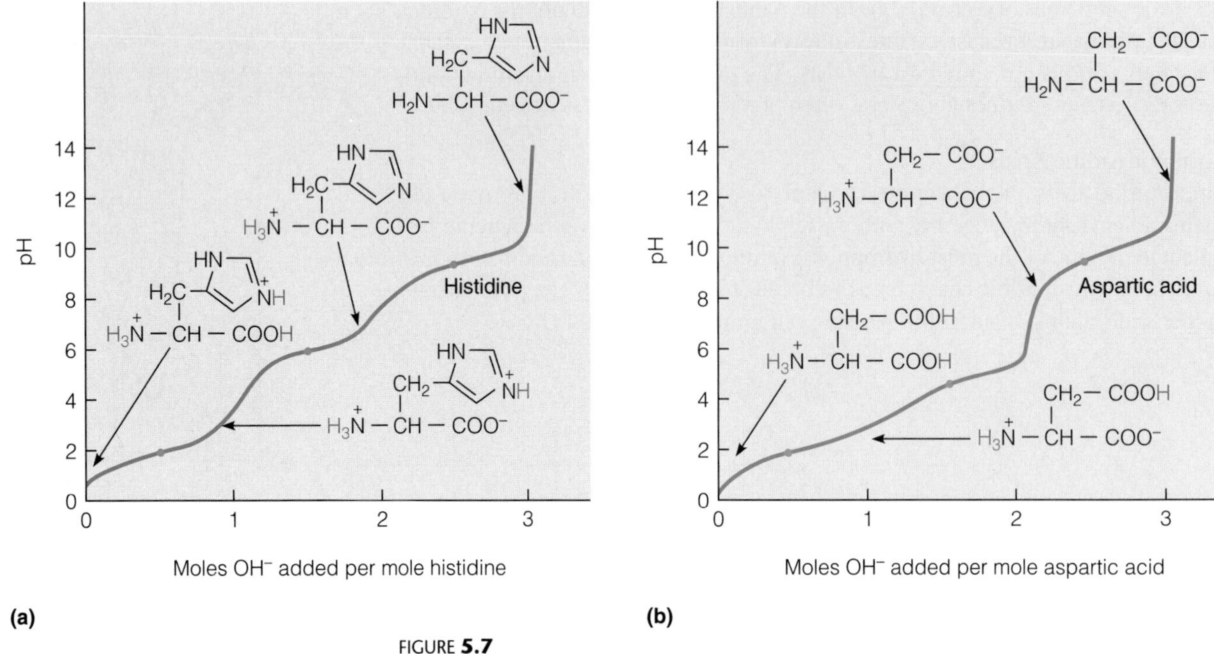

FIGURE **5.7**

Titration curves of amino acids with ionizing side chains. The dots correspond to pK_a values, and the forms predominating at different pH values are shown. Labile hydrogens are shown in red. **(a)** Histidine. **(b)** Aspartic acid. In each case it is presumed that the starting solution had been adjusted to pH < 2 by the addition of H^+ to the dissolved amino acid.

MODIFIED AMINO ACIDS

Sometimes amino acid side chains are modified after incorporation into a protein.

We have now considered all of the 20 amino acids that are coded for in DNA and are incorporated directly into proteins. The repertoire of side chain groups in proteins is embellished, however, by the ability of certain amino acids to occasionally become chemically modified after they are assembled into proteins. Depictions of several such *modified amino acids* follow, with the modifying group shown in red.

Phosphoserine **4-Hydroxyproline** **δ-Hydroxylysine** **γ-Carboxyglutamic acid**

We shall consider these again when we encounter specific proteins in which such modification has occurred.

The amino acids found in proteins are by no means the only ones to occur in living organisms. Many other amino acids play important roles in metabolism. A partial list is given in Table 5.2. Note that not all of them are α-amino acids, and not all of the α-amino acids are the L-enantiomers. We shall encounter all of the amino acids in this table again in later chapters.

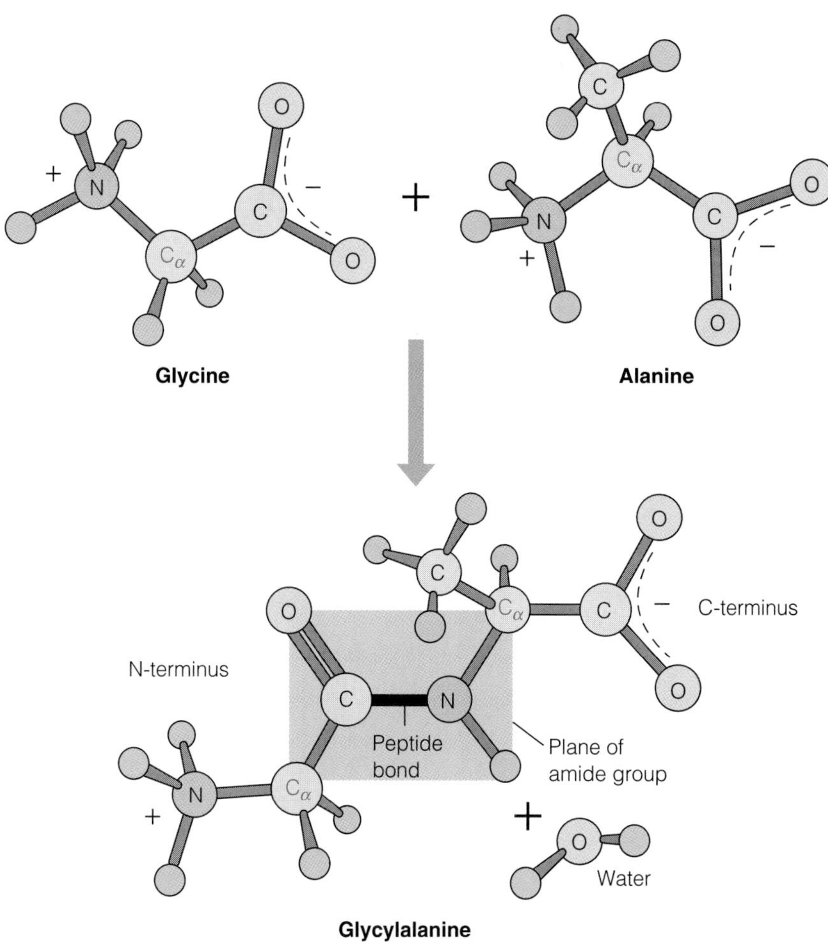

FIGURE **5.8**

Formation of a peptide. When two amino acids join, a peptide bond forms. Here the dipeptide glycylalanine (Gly–Ala) is depicted as being formed by removal of a water molecule when glycine is linked to alanine.

Peptides and the Peptide Bond

PEPTIDES

Amino acids can be covalently linked together by formation of an **amide bond** between the α-carboxyl group on one amino acid and the α-amino group on another. This bond is often referred to as a **peptide bond,** and the products formed by such a linkage are called **peptides.** The formation of a peptide bond between glycine and alanine is shown in Figure 5.8. The product in this case is called a **dipeptide,** because two amino acids have been combined. The reaction can be viewed as simple elimination of a water molecule between the carboxyl group of one amino acid and the amino group of the other. Note that the reaction in Figure 5.8 still leaves an $H_3\overset{+}{N}$— group available on one end of the dipeptide and an unreacted carboxyl group on the other. Thus, the reaction could in principle be continued by adding, for example, glutamic acid to one end and lysine to the other to yield the **tetrapeptide** shown in Figure 5.9. As each amino acid is added to the chain, another molecule of water must be eliminated. The portion of each amino acid remaining in the chain is called an **amino acid residue.** Thus, the alanyl residue in the tetrapeptide in Figure 5.9 is

$$-\underset{\displaystyle H}{\underset{|}{N}}-\underset{\displaystyle H}{\underset{|}{\overset{\displaystyle CH_3}{\overset{|}{C}}}}-\overset{\displaystyle O}{\overset{\|}{C}}-$$

Oligopeptides and polypeptides are formed by polymerization of amino acids. All proteins are polypeptides.

Chains containing only a few amino acid residues (like a tetrapeptide) are collectively referred to as **oligopeptides.** If the chain is very long, it is called a

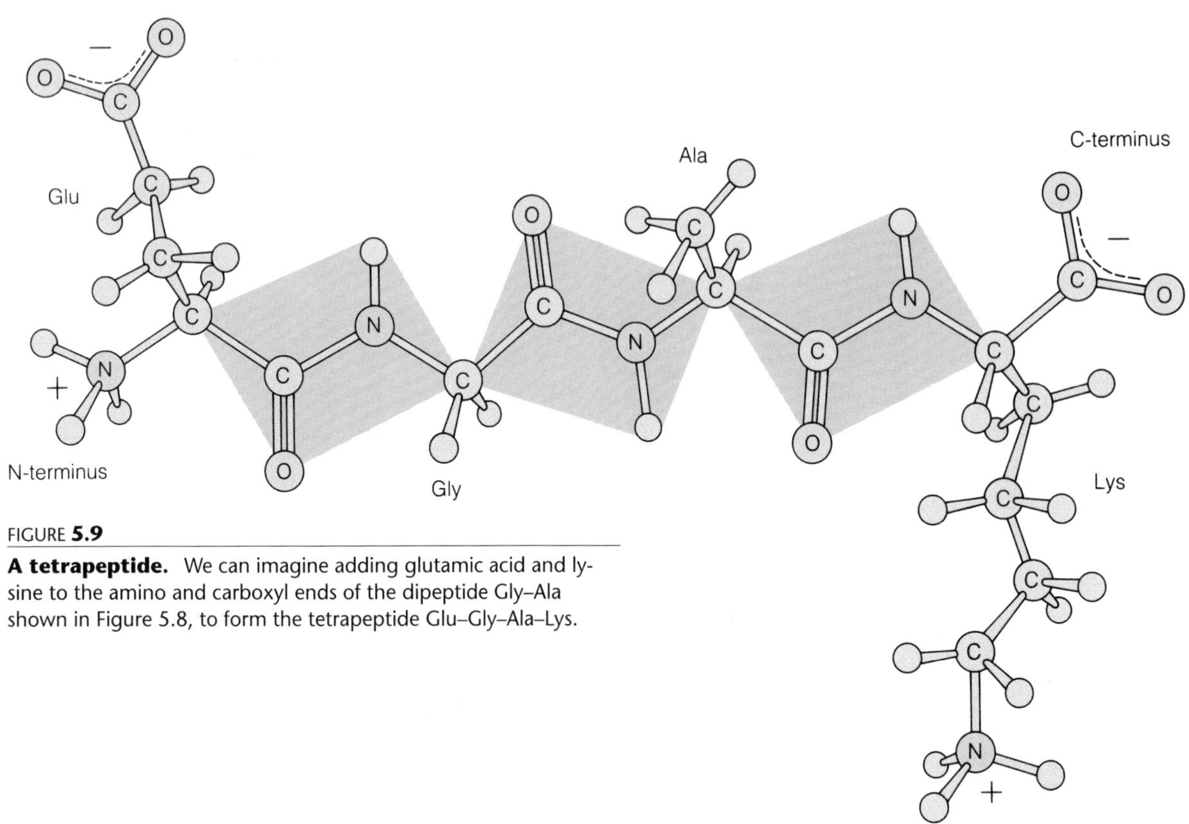

FIGURE **5.9**

A tetrapeptide. We can imagine adding glutamic acid and lysine to the amino and carboxyl ends of the dipeptide Gly–Ala shown in Figure 5.8, to form the tetrapeptide Glu–Gly–Ala–Lys.

N-Formyl group

N-Acetyl group

C-terminal amide

FIGURE **5.10**

Groups that may block N- or C-termini in proteins. Blocking of the N-terminus by a formyl or acetyl group is more common than modification of the C-terminus to an amide.

polypeptide. Most oligopeptides and polypeptides retain an unreacted amino group at one end (called the **amino terminus** or **N-terminus**) and an unreacted carboxyl at the other end (the **carboxyl terminus** or **C-terminus**). Exceptions are certain small **cyclic oligopeptides,** in which the N- and C-termini have been linked. In addition, many proteins have N-termini blocked by *N*-formyl or *N*-acetyl groups, and a few have C-terminal carboxyls that have been modified to amides (Figure 5.10).

In writing the sequence of an oligopeptide or polypeptide, it would be awkward to spell out all the amino acid residue names. Therefore, biochemists usually write such sequences in terms of either the three-letter abbreviations or the one-letter abbreviations given in Figure 5.3. For example, the oligopeptide shown in Figure 5.9 could be written as either

Glu–Gly–Ala–Lys

or

EGAK

Note that the convention is to *always* write the N-terminus to the left, and the C-terminus to the right.

All proteins are polypeptides. This is why understanding the nature of polypeptides and the peptide bond is so important a part of biochemistry.

POLYPEPTIDES AS POLYAMPHOLYTES

In addition to the free amino group at the N-terminus and the free carboxyl group at the C-terminus, polypeptides usually contain some amino acids that have ionizable groups on their side chains. These various groups have a wide range of pK_a

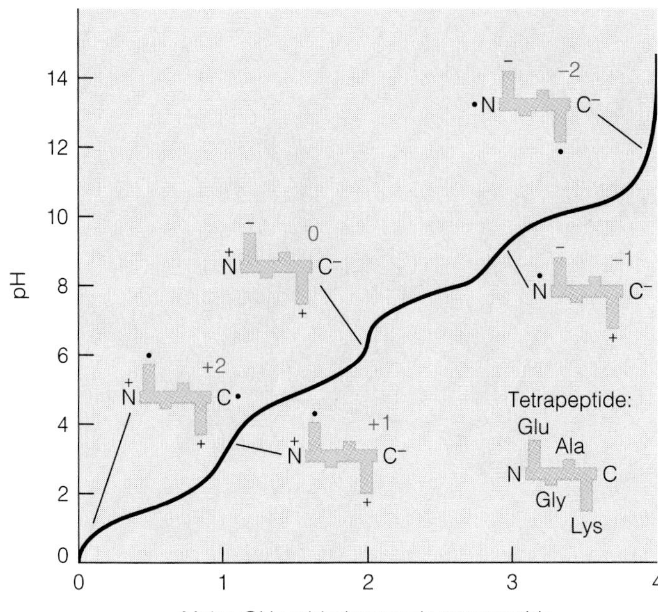

FIGURE **5.11**

Polyampholytic behavior of a tetrapeptide. This titration curve for the tetrapeptide Glu–Gly–Ala–Lys (Figure 5.9) shows the major forms present at several pH values. Groups that may be charged are indicated by +, − , or •, depending on whether they carry a positive, negative, or zero charge at a given pH. Net charges are shown in red.

values, as shown in Table 5.3, but are all weakly acidic or basic groups. Thus, polypeptides are excellent examples of the polyampholytes described in Chapter 2.

Amino acids, peptides, and proteins are ampholytes; each has an isoelectric point.

The kind of behavior that is seen during the titration of an oligopeptide or a polypeptide is exemplified by the tetrapeptide (Glu–Gly–Ala–Lys) in Figure 5.9. We can imagine starting with the tetrapeptide in a very acidic solution, say pH 0. At this pH, which is below the pK_a of any of the groups present, all of the ionizable residues will be in their protonated forms:

COOH
CH_2
CH_2 H O CH_3 H
CH N C CH N COOH
$H_3\overset{+}{N}$ C CH_2 N C CH
O H O $CH_2—CH_2—CH_2—CH_2—\overset{+}{N}H_3$
Glu **Gly** **Ala** **Lys**

All amino groups are positively charged, and each carboxyl has zero charge. Therefore, the whole molecule has a charge of +2 at this pH.

If we now imagine removing protons from the solution (by titrating with NaOH, for example), the various groups will lose protons at pH values in the vicinity of their pK_a values. The progress of this titration is shown in Figure 5.11. As protons are removed, raising the pH, more groups become deprotonated. The positive charge decreases and passes through zero at the isoelectric point (see page 45 in Chapter 2). As more base is added, the molecule becomes negatively charged, ultimately reaching a net charge of −2 at very high pH.

We will find these effects of changing pH to be of importance in biochemistry in general and protein chemistry in particular. Sometimes even a small shift in pH will significantly alter the constellation of charges with which a protein molecule faces its environment and will thereby significantly modify its behavior. Solubility of many proteins is minimal at the isoelectric point, because the molecules no longer repel one another when their net charge is zero (see Figure 2.21).

The fact that different proteins and oligopeptides have different net charges at a given pH is often used to advantage in their separation, either by electrophoresis

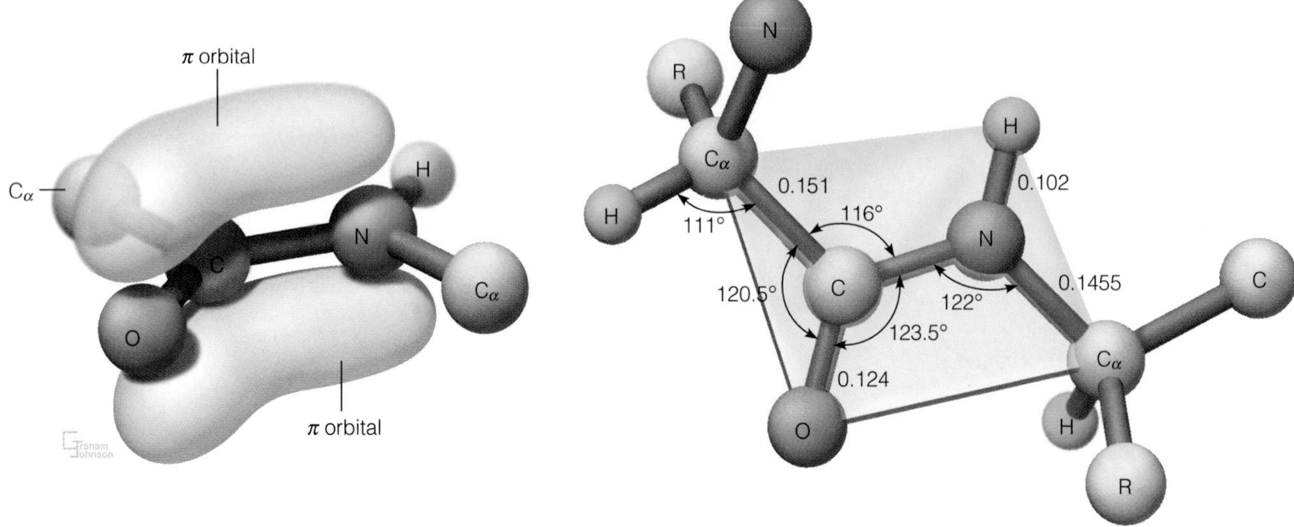

(a) Partial double-bond character of peptide bond

(b) Bond angles and lengths

FIGURE **5.12**

Structure of the peptide bond. (a) Delocalization of the π-electron orbitals over O—C—N accounts for the partial double-bond character of the C—N bond. **(b)** The presently accepted values for bond angles and lengths are shown here. Bond lengths are in nanometers (nm).

The peptide bond is nearly planar, and the *trans* form is favored.

(see Tools of Biochemistry 2A) or by ion-exchange chromatography (see Tools of Biochemistry 5A).

THE STRUCTURE OF THE PEPTIDE BOND

Now let us examine the nature of the bond that has been formed in linking amino acids together. In the dipeptide of Figure 5.8 (Gly–Ala), the blue-shaded portion contains the peptide bond. This amide bond, which is found between every pair of residues in a protein, has some properties very important to protein structure. For example, almost invariably the —C═O and —N—H bonds are nearly parallel; in fact the atoms O, C, N, and H are usually coplanar. There is little twisting possible around the C—N bond, because the peptide bond has a substantial fraction of double-bond character. The peptide bond can be considered a resonance hybrid of two forms:

$$\mathrm{O{=}C{-}N{-}H} \longleftrightarrow \mathrm{O^{-}{-}C{=}N^{+}{-}H}$$

or

$$\mathrm{O^{q-}\cdots C\cdots N^{q+}{-}H}$$

A schematic depiction of the electron density about the peptide bond is shown in Figure 5.12a, and bond lengths and angles are given in Figure 5.12b.

Even though they are coplanar, the group of atoms about the peptide bond could exist in two possible configurations, *trans* and *cis:*

trans *cis*

In fact, the *trans* form is usually favored, for in the *cis* configuration bulky R groups on adjacent α-carbons can sterically interfere. The major exception is the bond in the sequence X—Pro, where X is any other amino acid. In this bond the *cis* configuration is sometimes allowed, although the *trans* configuration is still favored about 4:1.

STABILITY AND FORMATION OF THE PEPTIDE BOND

The peptide bond is metastable. Proteins hydrolyze in aqueous solution when a catalyst is present.

Figure 5.8 and our discussion of it implied that a peptide bond could be formed by the elimination of a water molecule between two amino acids. In fact, in an aqueous environment this process is not thermodynamically favored. The free energy change for this reaction at room temperature in aqueous solution is about +10 kJ/mol. Therefore, the favored reaction under these conditions is the *hydrolysis* of the peptide bond:

$$H_3\overset{+}{N}-CH_2\text{(H)}-C(=O)-NH-CH(CH_3)-COO^- \xrightarrow[H_2O]{H_2O} H_3\overset{+}{N}-CH_2-COO^- + H_3\overset{+}{N}-CH(CH_3)-COO^-$$

with equilibrium lying well to the right. However, the uncatalyzed reaction is exceedingly slow at physiological pH and temperature. Like polynucleotides, polypeptides are metastable, hydrolyzing rapidly only under extreme conditions or when catalysts are present.

Peptide hydrolysis can be catalyzed in several ways. A general method, which cleaves all peptide bonds, is boiling in strong mineral acid (usually 6 M HCl). More specific catalysis is provided by **proteolytic enzymes** or **proteases.** Many such enzymes exhibit specificity as to which bonds they will cleave; a number of them are listed in Table 5.4. Some of these enzymes are secreted into the digestive tracts of animals, where they break down proteins for further digestion. Others, such as papain, are found in certain plant tissues. The existence of a battery of such enzymes, with specific cutting sites, is of great utility to the biochemist, for they allow the cleavage of polypeptides in well-defined ways. A nonenzymatic reaction that

TABLE **5.4 The sequence specificities of some proteolytic enzymes**

N-terminal ··· —NH—CH(R_1)—C(=O)—▲—NH—CH(R_2)—C(=O)— ··· C-terminal

Enzyme	Preferred Site[a]	Source
Trypsin	R_1 = Lys, Arg	From digestive systems of animals, many other sources
Chymotrypsin	R_1 = Tyr, Trp, Phe, Leu	Same as trypsin
Thrombin	R_1 = Arg	From blood; involved in coagulation
V-8 protease	R_1 = Asp, Glu	From *Staphylococcus aureus*
Prolyl endopeptidase	R_1 = Pro	Lamb kidney, other tissues
Subtilisin	Very little specificity	From various bacilli
Carboxypeptidase A	R_2 = C-terminal amino acid	From digestive systems of animals
Thermolysin	R_2 = Leu, Val, Ile, Met	From *Bacillus thermoproteolyticus*

[a]The residues indicated are those next to which cleavage is most likely. Note that in some cases preference is determined by the residue on the N-terminal side of the cleaved bond (R_1) and sometimes by the residue to the C-terminal side (R_2). Generally, proteases do not cleave where proline is on the other side of the bond. Even prolyl endopeptidase will not cleave if R_2 = Pro.

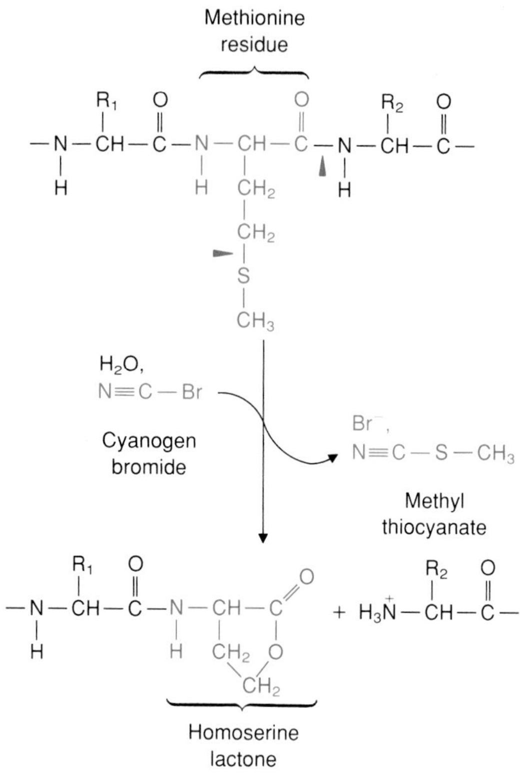

FIGURE **5.13**

The cyanogen bromide reaction. This reaction specifically cleaves the peptide bond to the carboxyl side of methionine in any polypeptide and converts Met to homoserine lactone, an internal ester involving homoserine carboxyl and hydroxyl groups. The cleavage sites are indicated by ▲.

cleaves a specific peptide bond uses the reagent cyanogen bromide (BrC≡N). Cyanogen bromide specifically cleaves the peptide bond at the carboxyl side of methionine residues (Figure 5.13). We will see later how these specific cleavage reactions can aid in determining the sequence of residues in a protein.

As with polynucleotides, the thermodynamic instability of polypeptides raises the question of how they can be synthesized in the aqueous medium of the cell. You may already have guessed the answer—that coupling of the synthetic reaction to the hydrolysis of high-energy phosphate compounds is required. In fact, every amino acid must be activated by an ATP-driven reaction before it can be incorporated into proteins. We shall give a brief outline of the process later in this chapter. First it is appropriate to describe the most important class of polypeptides—the proteins.

Proteins: Polypeptides of Defined Sequence

Every protein has a unique, defined amino acid sequence—its primary structure.

Proteins are not just polypeptides—they are polypeptides of defined sequence. Every protein has a defined order of amino acid residues. As with the nucleic acids, this sequence is referred to as the *primary structure* of the protein. In later chapters we shall see that it is this fundamental level of structure upon which higher levels of organization are based.

Figure 5.14 shows the amino acid sequence of sperm whale *myoglobin*, the protein whose three-dimensional structure we saw in Figure 5.1. Also listed is the sequence of human myoglobin, the protein that serves the same oxygen storage function in humans. Two points are immediately obvious from examination of these sequences. First, proteins are *long* polypeptides. Sperm whale myoglobin contains 153 amino acids; so does human myoglobin. Yet these are among the smaller proteins; some proteins have sequences extending for many hundreds or even thousands of amino acid residues. Second, although the two myoglobin sequences are similar, they are not identical. Their similarity is sufficient for each to serve the same biochemical purpose; therefore we call each a *myoglobin*. But they

FIGURE **5.14**

The amino acid sequences of sperm whale myoglobin and human myoglobin. Single-letter abbreviations are used here for the amino acids; numbering of the amino acids starts at the N-terminus. Of the 153 amino acid residues, 128 (84%) are identical in humans and whales. If we include the 16 conservative substitutions (e.g., isoleucine for leucine), the two proteins are 94% homologous.

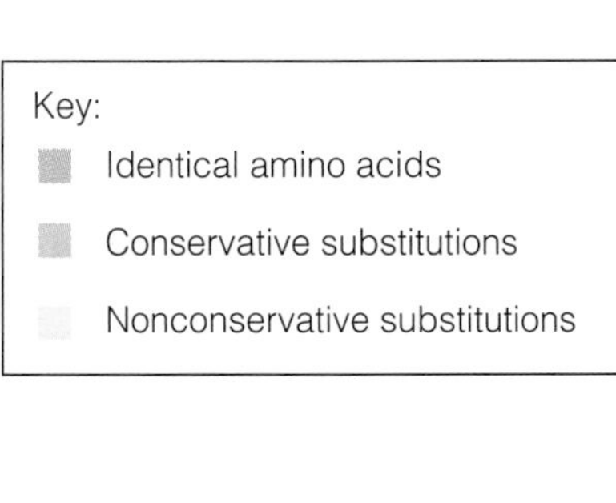

Number	1	2	3	4	5	6	7	8	9	10	11	12	13	14	15
Human	G	L	S	D	G	E	W	Q	L	V	L	N	V	W	G
Whale	V	L	S	E	G	E	W	Q	L	V	L	H	V	W	A

Number	16	17	18	19	20	21	22	23	24	25	26	27	28	29	30
Human	K	V	E	A	D	I	P	G	H	G	Q	E	V	L	I
Whale	K	V	E	A	D	V	A	G	H	G	Q	D	I	L	I

Number	31	32	33	34	35	36	37	38	39	40	41	42	43	44	45
Human	R	L	F	K	G	H	P	E	T	L	E	K	F	D	K
Whale	R	L	F	K	S	H	P	E	T	L	E	K	F	D	R

Number	46	47	48	49	50	51	52	53	54	55	56	57	58	59	60
Human	F	K	H	L	K	S	E	D	E	M	K	A	S	E	D
Whale	F	K	H	L	K	T	E	A	E	M	K	A	S	E	D

Number	61	62	63	64	65	66	67	68	69	70	71	72	73	74	75
Human	L	K	K	H	G	A	T	V	L	T	A	L	G	G	I
Whale	L	K	K	H	G	V	T	V	L	T	A	L	G	A	I

Number	76	77	78	79	80	81	82	83	84	85	86	87	88	89	90
Human	L	K	K	K	G	H	H	E	A	E	I	K	P	L	A
Whale	L	K	K	K	G	H	H	E	A	E	L	K	P	L	A

Number	91	92	93	94	95	96	97	98	99	100	101	102	103	104	105
Human	Q	S	H	A	T	K	H	K	I	P	V	K	Y	L	E
Whale	Q	S	H	A	T	K	H	K	I	P	I	K	Y	L	E

Number	106	107	108	109	110	111	112	113	114	115	116	117	118	119	120
Human	F	I	S	E	C	I	I	Q	V	L	Q	S	K	H	P
Whale	F	I	S	E	A	I	I	H	V	L	H	S	R	H	P

Number	121	122	123	124	125	126	127	128	129	130	131	132	133	134	135
Human	G	D	F	G	A	D	A	Q	G	A	M	N	K	A	L
Whale	G	N	F	G	A	D	A	Q	G	A	M	N	K	A	L

Number	136	137	138	139	140	141	142	143	144	145	146	147	148	149	150	151	152	153
Human	E	L	F	R	K	D	M	A	S	N	Y	K	E	L	G	F	Q	G
Whale	E	L	F	R	K	D	I	A	A	K	Y	K	E	L	G	Y	Q	G

are not quite the same, for many millions of years have passed since sperm whales and humans had a common ancestor. Proteins evolve, and they evolve by changes in their amino acid sequences. Some of these are called **conservative** changes; they conserve the nature of the side chain (Asp for Glu, for example). Other, **nonconservative** changes (Asp for Ala, for example) may have more serious consequences.

We must also emphasize the uniqueness of a given protein in a particular species of organism. Every sample of sperm whale myoglobin, taken from any sperm whale, has the same amino acid sequence (unless, by rare chance, a sample is taken from a whale that carries a mutated myoglobin gene).

Biochemists have come to understand such a complex structure as the myoglobin molecule bit by bit, by analyzing successively higher levels of complexity in the protein structure. To begin any such study of a protein, it is necessary to prepare the protein in a pure form, free of contamination by other proteins or other cellular substances. Methods for doing this are described in Tools of Biochemistry 5A. Traditionally, the next step after purifying a protein would be to determine its amino acid composition, that is, the relative amounts of the different amino acids

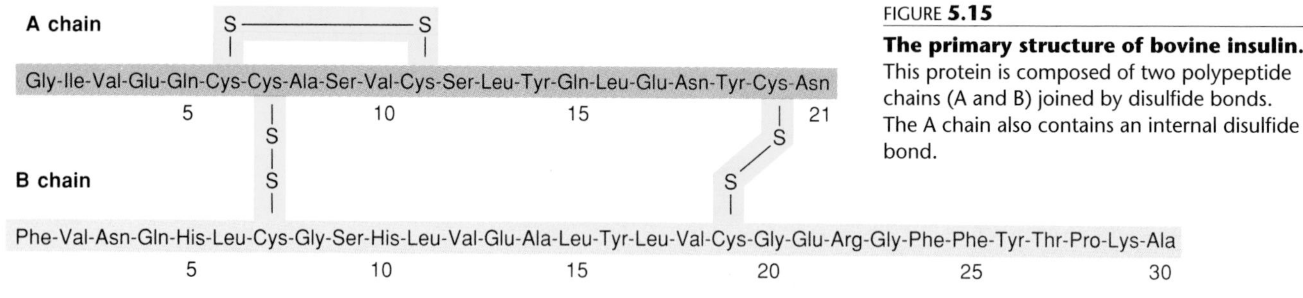

FIGURE **5.15**

The primary structure of bovine insulin. This protein is composed of two polypeptide chains (A and B) joined by disulfide bonds. The A chain also contains an internal disulfide bond.

in the protein. Table 5.1 gives average results for a large number of proteins, although individual proteins differ widely in composition. To determine a protein's amino acid composition, the polypeptide is hydrolyzed into its constituent amino acids by treatment with 6 M hydrochloric acid at 105°C. The mixture of amino acids can then be separated by ion-exchange chromatography and the amount of each measured. Just how this is done is described in Tools of Biochemistry 5B.

Composition determination gives limited information about a protein. What is much more important to the biochemist or biologist is the *sequence* of the amino acids. Methods have been developed that allow the sequences of even very small amounts (micrograms) of purified proteins to be determined quickly. The most important of these is based on the **Edman degradation** reaction and is described in detail in Tools of Biochemistry 5C. Briefly, it involves a sequential series of reactions that remove residues from the polypeptide chain one at a time, starting from the N-terminus. The reactions and isolation of amino acid products are well suited to automated operation. However, the method becomes unreliable for long chains, so large polypeptides have to be cut into specific shorter pieces. For this, reactions like those described in Table 5.4 can be employed.

Some proteins contain two or more polypeptide chains held together by noncovalent or covalent forces.

A complication arises if the protein being studied contains more than one polypeptide chain. These may be held together by noncovalent forces, as in the protein *hemoglobin*, which is made up of four myoglobin-like chains (see Chapter 7). Alternatively, chain connection can be by covalent bonds such as disulfide bonds. An example of such a protein is the hormone **insulin** (Figure 5.15). Tools of Biochemistry 6B describes methods for detecting such multichain proteins and how to separate them. If the bonding is covalent, chemical cleavage is necessary; see Tools of Biochemistry 5C for an example.

Although direct sequencing of proteins, as described here, has given us much of our enormous store of protein sequence information, biochemists are more and more turning to gene sequencing for such information. As we explained in Chapter 4, the primary structure of every protein is dictated by a particular gene. Because we now know the code that relates DNA sequence to protein sequence, determination of the nucleotide sequence of a gene (or more often, the sequence of the messenger RNA read from that gene) allows us to read the corresponding protein sequence. It should be kept in mind, however, that gene sequencing tells us only the sequence *as synthesized*. As we will see, there are often postsynthetic modifications of the polypeptide chain that are not revealed in this way and can be found only by direct sequencing.

The techniques for identifying protein-defining sequences in the genome and for retrieving, cloning, and sequencing the genes are discussed in Part V of this text.

From Gene to Protein

THE GENETIC CODE

In Chapter 4 we stated that the DNA sequences of genes are transcribed into messenger RNA molecules, which are in turn translated into proteins. But there are only 4 kinds of nucleotides in DNA, each of which transcribes to a particular nucleotide in RNA, and there are 20 kinds of amino acids. Obviously, a 1:1 correspondence between nucleotide and amino acid is impossible. In fact, triplets of nucleotides (**codons**) are used to code for each amino acid, allowing 4^3, or 64, different combinations. This number is more than enough to code for 20 amino acids, so most amino acids have multiple codons. The genetic code is nearly universal—that is, virtually all organisms use the same codons to translate their genomes into proteins. The few exceptions are scattered through the biological kingdoms. We discuss these and other details of the genetic code in Chapter 27.

The genetic code specifies RNA triplets that correspond to each amino acid residue.

Figure 5.16 depicts the genetic code in terms of the mRNA triplets that correspond to the different amino acid residues. Three triplets—UAA, UAG, and UGA—do not code for any amino acids but serve as "stop" signals to end translation at the C-terminus of the chain. The codon AUG, which normally codes for methionine, also serves as a "start" signal. When starting a polypeptide chain, AUG directs the placement of *N*-formylmethionine (in prokaryotes) or methionine (in eukaryotes) at the N-terminal position (Figure 5.17). The implication is that all prokaryotic proteins should start with *N*-formylmethionine, and eukaryotic proteins with methionine. Often this is true, though in many cases the N-terminal residue or even several residues are cleaved off in the cell by specific proteases

First position	Second position: U		C		A		G		Third position
U	UUU	Phe	UCU	Ser	UAU	Tyr	UGU	Cys	U
	UUC	Phe	UCC	Ser	UAC	Tyr	UGC	Cys	C
	UUA	Leu	UCA	Ser	UAA	Stop	UGA	Stop	A
	UUG	Leu	UCG	Ser	UAG	Stop	UGG	Trp	G
C	CUU	Leu	CCU	Pro	CAU	His	CGU	Arg	U
	CUC	Leu	CCC	Pro	CAC	His	CGC	Arg	C
	CUA	Leu	CCA	Pro	CAA	Gln	CGA	Arg	A
	CUG	Leu	CCG	Pro	CAG	Gln	CGG	Arg	G
A	AUU	Ile	ACU	Thr	AAU	Asn	AGU	Ser	U
	AUC	Ile	ACC	Thr	AAC	Asn	AGC	Ser	C
	AUA	Ile	ACA	Thr	AAA	Lys	AGA	Arg	A
	AUG	Met/start	ACG	Thr	AAG	Lys	AGG	Arg	G
G	GUU	Val	GCU	Ala	GAU	Asp	GGU	Gly	U
	GUC	Val	GCC	Ala	GAC	Asp	GGC	Gly	C
	GUA	Val	GCA	Ala	GAA	Glu	GGA	Gly	A
	GUG	Val	GCG	Ala	GAG	Glu	GGG	Gly	G

FIGURE **5.16**

The genetic code. The table is arranged so that users can quickly find any amino acid from the three bases of the mRNA codon (written in the 5′ ⟶ 3′ direction). For example, to find the amino acid corresponding to the codon 5′ AUC 3′, we look first in the A row, then in the U column, and then in the C space, and find the amino acid to be Ile.

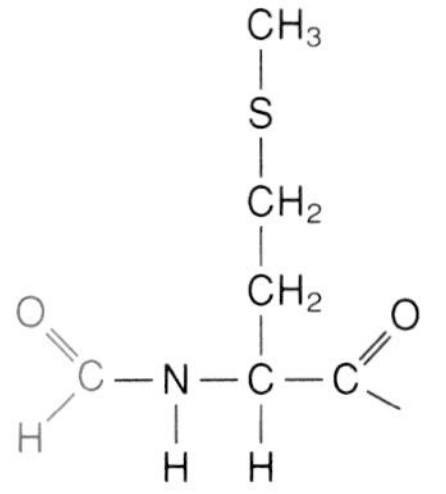

FIGURE **5.17**

***N*-Formylmethionine.** This amino acid residue initiates prokaryotic translation. It is coded by AUG when that triplet appears near the start of a message.

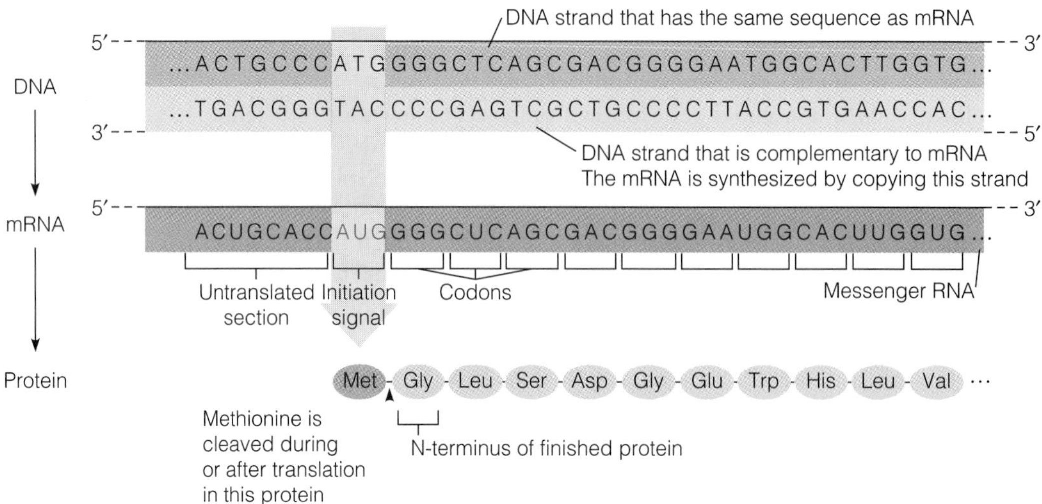

FIGURE **5.18**

Relationships of DNA to mRNA to polypeptide chain. These relationships are shown for the first 10 residues of seal myoglobin. Note that the DNA strand that is transcribed *from* is complementary to the mRNA.

From Blanchetot, *Nature* (1983) 301:732–734. © 1983 Macmillan Magazines Ltd.

during or immediately after translation. Figure 5.18 shows the relationship between DNA, mRNA, and polypeptide sequences for the N-terminal portion of seal myoglobin. In this case the N-terminal methionine is removed.

TRANSLATION

Translation requires activation of amino acids by coupling them to specific tRNAs.

We are now in a position to discuss the process of translation of mRNA into protein in a bit more detail than was presented in Chapter 4. We noted earlier in this chapter that the thermodynamics of peptide bond formation requires that amino acids be activated before they can be added to a polypeptide chain. This activation is accomplished by coupling each amino acid to the 3′ end of an appropriate transfer RNA to yield an **aminoacyl-tRNA,** as shown in Figure 5.19. (The structure of tRNA is shown in Figure 4.20.) This coupling is catalyzed by specific enzymes called **aminoacyl-tRNA synthetases,** each of which recognizes both a particular amino acid and its appropriate tRNA. The free energy source for the reaction is the hydrolysis of ATP to AMP and pyrophosphate.

Aminoacyl tRNAs bind to ribosomes by pairing the tRNA anticodon with a specific codon on mRNA.

The synthesis of a polypeptide is outlined in Figure 5.20. Each tRNA contains, in a region known as the **anticodon loop,** a nucleotide sequence called the **anticodon** that is complementary to the appropriate codon in the message. The messenger RNA has been bound to a ribosome, as mentioned in Chapter 4. The aminoacyl tRNAs also bind here, one by one, matching their anticodons to the codons on the message, as shown in Figure 5.20, step 1. The amino acid carried by each incoming tRNA is transferred to the growing peptide chain (step 2). The first tRNA is then released, and the ribosome moves one codon length along the message, allowing the next tRNA to come into place, carrying *its* amino acid (step 3). Again,

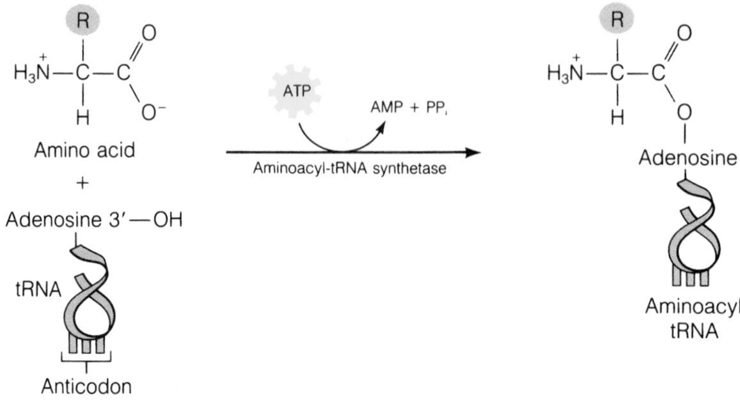

FIGURE **5.19**

Activation of amino acids for incorporation into proteins. The enzyme aminoacyl-tRNA synthetase recognizes both a particular amino acid and the tRNA carrying the corresponding anticodon. This synthetase catalyzes the formation of an aminoacyl tRNA, with accompanying hydrolysis of one ATP.

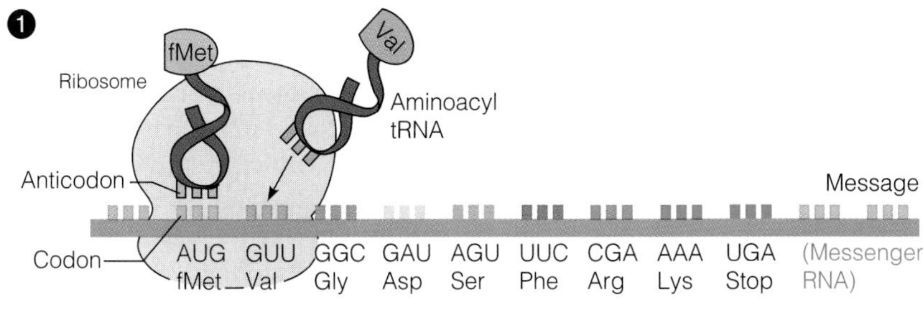

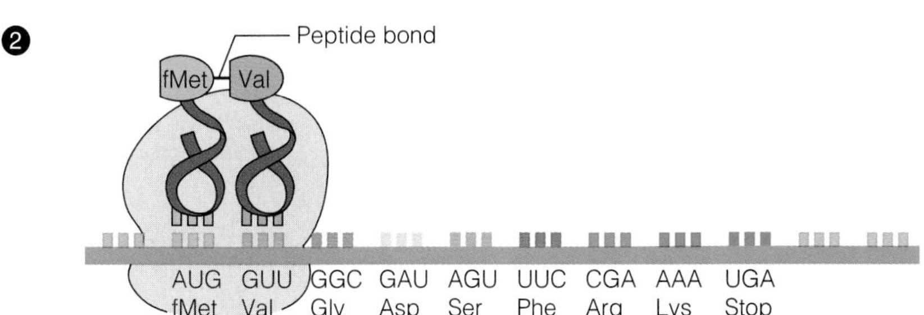

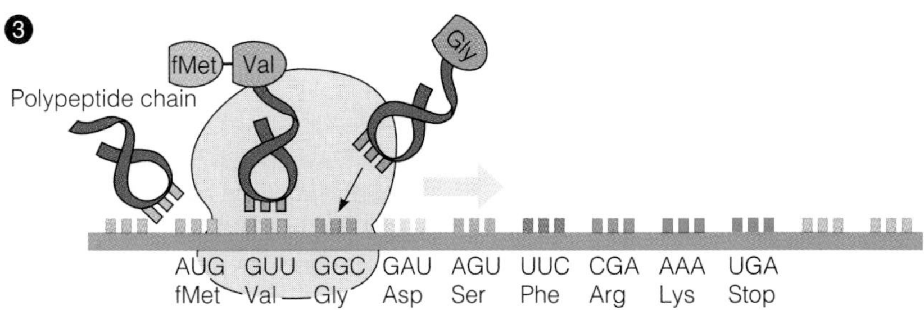

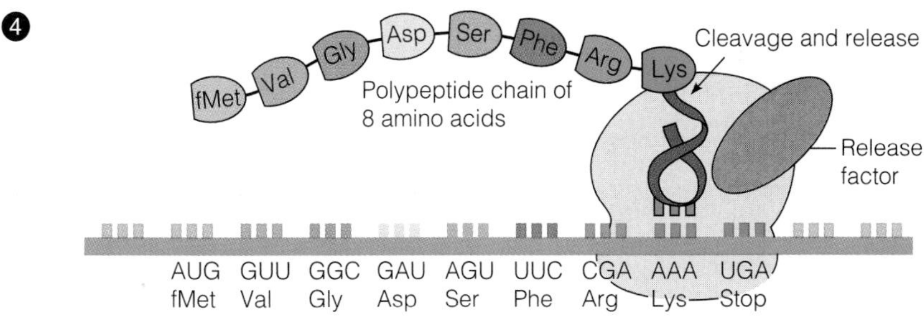

FIGURE **5.20**

Translation of an RNA message into a protein. As the ribosome moves along the message, it accepts specific aminoacyl tRNAs in succession, selecting them by matching the anticodon on the tRNA to the codon on the RNA message (step 1). The amino acid (in this case, the second one of the chain, Val) is transferred to the growing polypeptide chain (step 2), and the ribosome moves on to the next codon to repeat the process, while releasing the next-to-last-added tRNA (step 3). The preceding steps are repeated, adding more amino acids to the chain, until a stop signal is read (step 4), whereupon a protein release factor causes both the polypeptide and the mRNA to be released. The polypeptide shown here is unrealistically short, to illustrate both starting and stopping. Many details of translation have been omitted here; they are presented in Chapters 27 and 28.

expenditure of energy from high-energy phosphate hydrolysis is required at each step in the movement. As the ribosome moves along the messenger RNA, it eventually encounters a "stop" codon. At this point, the polypeptide chain is released. Step 4 shows a completed (although unrealistically short) protein. In every cell, of every kind of organism, this remarkable machinery translates the information coded in thousands of different genes into thousands of different proteins. In Chapters 27 and 28 we discuss details of how the translation process starts, proceeds, stops, and is energized.

It is not only cells that synthesize polypeptides. Biochemists do too, in order to make specific peptides needed as reagents in the laboratory. How this is done is described in Tools of Biochemistry 5D.

POSTTRANSLATIONAL PROCESSING OF PROTEINS

After being translated from mRNA, a protein may be modified in many ways.

When a polypeptide chain is released from a ribosome, it is not necessarily finished. It must fold into its correct three-dimensional structure, and in some cases disulfide bonds must form. Certain amino acid residues may be acted upon by

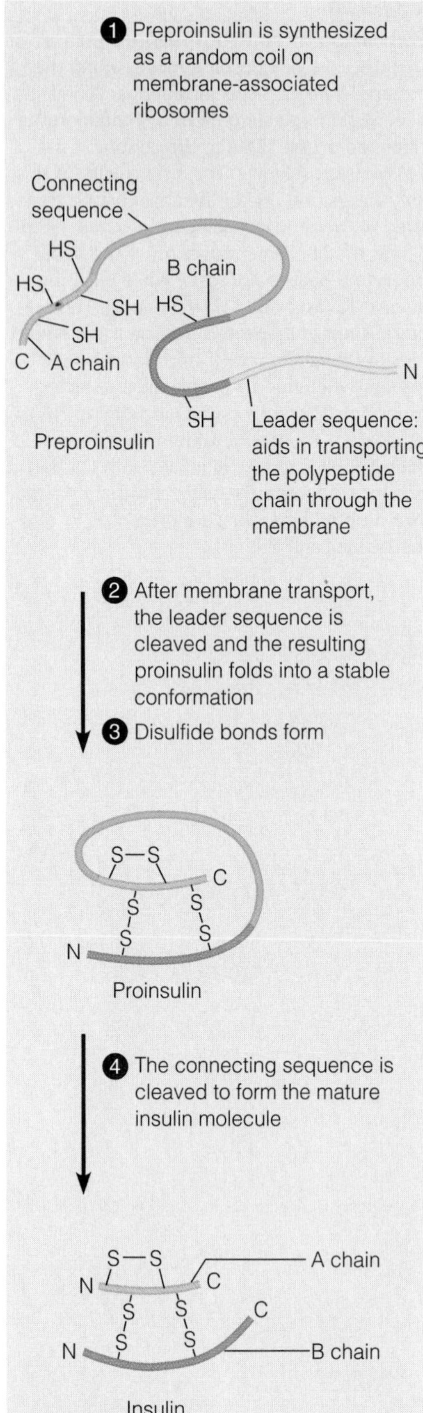

FIGURE **5.21**

Structure of preproinsulin and its conversion to insulin.

enzymes in the cell to produce, for example, the kinds of modifications shown on page 134.

Many proteins are further modified by specific proteolytic cleavage which shortens the chain length. A remarkable example is found in the synthesis of insulin (Figure 5.21). We have encountered insulin as a two-chain protein held together by disulfide bonds (Figure 5.15). Insulin is actually synthesized as a single, much longer polypeptide chain, called **preproinsulin** (Figure 5.21, step 1). The residues at the N-terminus of the molecule (the exact number varies with the species) serve as a "signal peptide" (also called a **leader sequence**) to help the preproinsulin molecule be transported through the hydrophobic cell membranes. This transport is essential, for insulin is one of a class of proteins that function outside the cells in which they are synthesized. The leader sequence is then cut off by a specific protease, leaving **proinsulin** (step 2). Proinsulin folds into a specific three-dimensional structure, which helps it form the correct disulfide bonds (step 3). The connecting sequence between the A chain and the B chain is then cut out by further protease action, yielding the finished insulin molecule (step 4). This achieves an important physiological advantage. Because proinsulin is not an active hormone, it can be present in tissues at high concentrations. Such levels of active insulin would be toxic. Yet the proinsulin can be converted to insulin rapidly when needed by the body.

The primary structure of a protein molecule is a sequence of *information*. The 20 kinds of amino acid side chains can be thought of as words in a long sentence. These words have been translated from another language, the language of nucleic acid sequences stored in the genes and copied into messenger RNA. After translation, the sentence has been edited, with certain words modified and others deleted in the posttranslational processing. In the next chapter we shall see that the information contained in the "sentence" of a protein sequence dictates how that protein folds in three dimensions. This folding, in turn, prescribes the function of the protein—how it interacts with small molecules and ions, with other proteins, and with substances like nucleic acids, carbohydrates, and lipids. The information expressed in protein sequences plays a primary role in determining how cells and organisms function.

SUMMARY

Proteins are polymers of L-α-amino acids. Twenty different amino acids are coded for in genes and incorporated into proteins. Other (nonprotein) amino acids exist, and there are also modifications of amino acids found in proteins. The variety of side chains—hydrophilic, hydrophobic, acidic, basic, neutral—allows much functional complexity in proteins. Additional variation is made possible by modification of some amino acids after they have been incorporated into proteins. The presence of both positive and negative charges on side chains makes proteins polyampholytes.

Oligopeptides and polypeptides are formed by polymerization of amino acids via peptide bonds. The peptide bond is nearly planar, and the *trans* form is favored. This bond is metastable and can be readily hydrolyzed in the presence of catalysts. The unique, defined sequence of amino acids in each protein constitutes its primary structure, dictated by its gene. Some proteins contain more than one polypeptide chain, held together by either covalent or noncovalent forces. Proteins are synthesized in the cell by an ATP-dependent process called translation. The messenger RNA for a protein attaches to a particle called a ribosome. Individual amino acids are coupled to specific tRNA molecules. A triplet anticodon on the tRNA matches with a codon triplet on the mRNA. As the ribosome moves along the mRNA, amino acids are transferred, one by one, to the growing polypeptide chain. Specific "start" and "stop" triplets in the genetic code limit the chain length.

Even after translation has been completed, the protein molecule can be further covalently modified, either by cleavage of particular portions or by modification of certain amino acid side chains.

REFERENCES

General

A number of excellent books provide more detailed or supplementary information on protein structure. We particularly recommend the following to supplement our Chapters 5, 6, and 7.

Branden, C., and J. Tooze (1991) *Introduction to Protein Structure.* Garland, New York, London. Lots of information and insights into all levels of structure. Excellent illustrations.

Creighton, T. E. (1993) *Proteins: Structure and Molecular Properties,* 2nd ed. Freeman, San Francisco. An elegant, thorough, contemporary exposition of all aspects of protein chemistry. Many good references throughout the text.

Dickerson, R. E., and I. Geis (1981) *The Structure and Action of Proteins,* 2nd ed. Benjamin/Cummings, Redwood City, Calif. Concise, well written, and well illustrated.

Hirs, C. H. W., and S. N. Timasheff, eds. (1983) *Enzyme Structure,* Part 1, in *Methods in Enzymology,* Vol. 91. Academic Press, New York. Contains a series of useful chapters on protein analytical methods.

Kyte, J. (1995) *Structure in Protein Chemistry.* Garland, New York. Together with Crieghton, the best contemporary treatises.

Reviews and Papers on Special Topics

Blake, C. C. F., and L. N. Johnson (1984) Protein structure. *Trends Biochem. Sci.* 9:147–151.

Dayhoff, M. O. (1972) *Atlas of Protein Sequence and Structure.* National Biomedical Research Foundation, Washington, D.C. (See also supplements to this volume published in subsequent years.)

Doolittle, R. (1985) Proteins. *Sci. Am.* 253(4):88–96.

Greenstein, J. P., and M. Winitz (1961) *Chemistry of the Amino Acids.* Wiley, New York.

Hegstrom, R. A., and D. K. Kondepudi (1990) The handedness of the universe. *Sci. Am.* January: 98–105. A clear discussion of theories of stereopreference.

Milton, R. C. deL., Milton, S. C. F., and Kent, S. B. H. (1992) Total chemical synthesis of a D-enzyme: The enantiomers of HIV-1 protease show demonstration of reciprocal chiral substrate specificity. *Science* 256:1445–1448.

Rose, G. D., A. R. Geselowitz, G. J. Lesser, R. H. Lee, and M. H. Zehfus (1985) Hydrophobicity of amino acid residues in globular proteins. *Science* 229:834–838.

Wilbur, P. J., and A. Allerhand (1977) Titration behavior and tautomeric states of individual histidine residues of myoglobin. *J. Biol. Chem.* 252:4968–4975.

PROBLEMS

1. Using the data in Table 5.1, calculate the *average* amino acid residue weight in a protein of typical composition. This is a useful number to know for approximate calculations.

2. The melanocyte-stimulating peptide hormone *α-melanotropin* has the following sequence:

   Ser–Tyr–Ser–Met–Glu–His–Phe–Arg–Trp–Gly–Lys–Pro–Val

   (a) Write the sequence using the one-letter abbreviations.
   (b) Calculate the molecular weight of α-melanotropin, using data in Table 5.1. Why is this result not *exactly* correct at neutral pH?

*3. (a) Sketch the titration curve you would expect for α-melanotropin (Problem 2). (Assume pK_a values in the middle of ranges given in Table 5.3.)
   (b) Approximately what charge would you expect at pH values of 11, 5, and 1?
   (c) Estimate the isoelectric point of α-melanotropin.

4. What peptides are expected to be produced when α-melanotropin (Problem 2) is cleaved by (a) trypsin, (b) cyanogen bromide, and (c) thermolysin? (Refer to Table 5.4.)

5. There is another melanocyte-stimulating hormone called *β-melanotropin.* Cleavage of β-melanotropin with trypsin produces the following peptides plus free aspartic acid.

   WGSPPK DSGPYK MEHFR

   If you assume maximum sequence homology between α-melanotropin and β-melanotropin, what must be the sequence of the latter?

6. Given the following peptide

   Ser–Glu–Pro–Ile–Met–Ala–Pro–Val–Glu–Tyr–Pro–Lys

   (a) Estimate the net charge at pH 7 and at pH 12.
   (b) How many peptides would result if this peptide were treated with (1) cyanogen bromide, (2) trypsin, and (3) chymotrypsin?
   (c) Suggest a method for separating the peptides produced by chymotrypsin treatment.

*7. A mutant form of polypeptide hormone angiotensin II has the amino acid *composition*

   (Asp, Arg, Ile, Met, Phe, Pro, Tyr, Val)

   The following observations are made:

   - Trypsin yields a dipeptide containing Asp and Arg, and a hexapeptide with all the rest.
   - Cyanogen bromide cleavage yields a dipeptide containing Phe and Pro, and a hexapeptide containing all the others.
   - Chymotrypsin cleaves the hormone into two tetrapeptides, of composition

     (Asp, Arg, Tyr, Val) and (Ile, Met, Phe, Pro)

   - The dipeptide of composition (Pro, Phe) cannot be cleaved by either chymotrypsin or carboxypeptidase.

   What is the sequence?

8. A protein has been sequenced after destruction of —S—S— bonds. It is known to contain 3 Cys residues, located as shown below. However, only one of these is a free —SH; two are involved in an S—S bond.

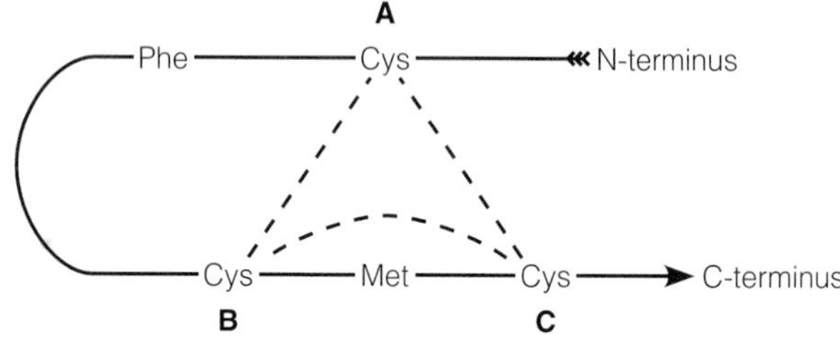

The only methionine and the only aromatic amino acid (Phe) in this protein are in the positions indicated. Cleavage of the *intact* protein (with —S—S— bridge intact) by either cyanogen bromide or chymotrypsin does *not* break the protein into two peptides. Where is the —S—S— bridge (AB, BC, or AC)?

9. *Apamine* is a small protein toxin present in the venom of the honeybee. It has the sequence

CNCKAPETALCARRCQQH

(a) It is known that apamine does not react with iodoacetate. How many disulfide bonds are present?
(b) Suppose trypsin cleavage gave two peptides. Where is(are) the S—S bond(s)?

10. (a) Write a possible sequence for an mRNA segment coding for apamine.
(b) Do you think apamine has been synthesized in the form shown in Problem 9, or is it more likely a product of proteolytic cleavage of a larger peptide? Explain.

11. Assume the following portion of an mRNA. Find a start signal, and write the amino acid sequence that is coded for.

5′··· GCCAUGUUUCCGAGUUAUCCCAAAGAUAAAAAAGAG··· 3′

12. A researcher has isolated an oligopeptide of unknown sequence and unknown amino acid composition. All attempts to sequence it by Edman degradation fail; the reaction just will not go.
(a) Suggest two quite different possible explanaitons for the problem.
(b) Suggest an experiment that will at least distinguish between these two.

*13. Suppose you had separated the A and B chains of insulin by disulfide reduction. What chromatographic method should allow the isolation of pure A and B chains? Describe your procedure in detail.

14. The ribosomes of *E. coli* are large particles, each of which is made up of two smaller particles. These are called the 30S subunit and the 50S subunit because of their sedimentation coefficients. The intact ribosome does not have a sedimentation coefficient of 80S; instead, the value found is 70S. Explain why this can be. (See Tools of Biochemistry 5A.)

TOOLS OF BIOCHEMISTRY 5A

Ways to Isolate and Purify Proteins and Other Macromolecules

The biochemist, in attempting to understand the chemical processes in cells and organisms, is faced with a formidable task. Even the simplest cell is a complex structure, made up of thousands of different kinds of compounds. Before their interactions and changes can be understood, it is necessary to *isolate* these various cellular constituents, identify them, and study their structures and properties.

To separate molecules, the biochemist seizes on differences among them that can be exploited. Such differences may be in solubility, size, mass, electrical charge, or affinity for other molecules. Electrophoresis, which uses electrical charge differences in separation, has been discussed in Tools of Biochemistry 2A. But electrophoresis is primarily an *analytical* tool, rather than a means of preparing samples. Though it is sometimes used to purify small quantities of proteins and nucleic acids, its capacity is limited; preparing the quantities of substances needed for characterization and study usually requires other methods. To go from cells or tissues to a purified material, we usually use a number of different techniques sequentially. We will describe some of these here.

Solubility

One of the oldest, simplest, and yet still effective ways to carry out a crude separation of a mixture of proteins is by taking advantage of differential solubility in concentrated salt solutions. In Chapter 2, it was mentioned that many proteins become insoluble in the presence of high concentrations of salts, especially polyvalent ones, which contribute strongly to ionic strength. What was not mentioned, however, is that the solubility varies widely between one protein and another and in concentrated solutions varies rapidly with ionic strength (see Figure 5A.1). It is clear from these data that in an ammonium sulfate solution of about 3 M, hemoglobin and serum albumin are almost completely insoluble whereas myoglobin remains quite soluble. Thus, one should be able to purify myoglobin from such contaminants

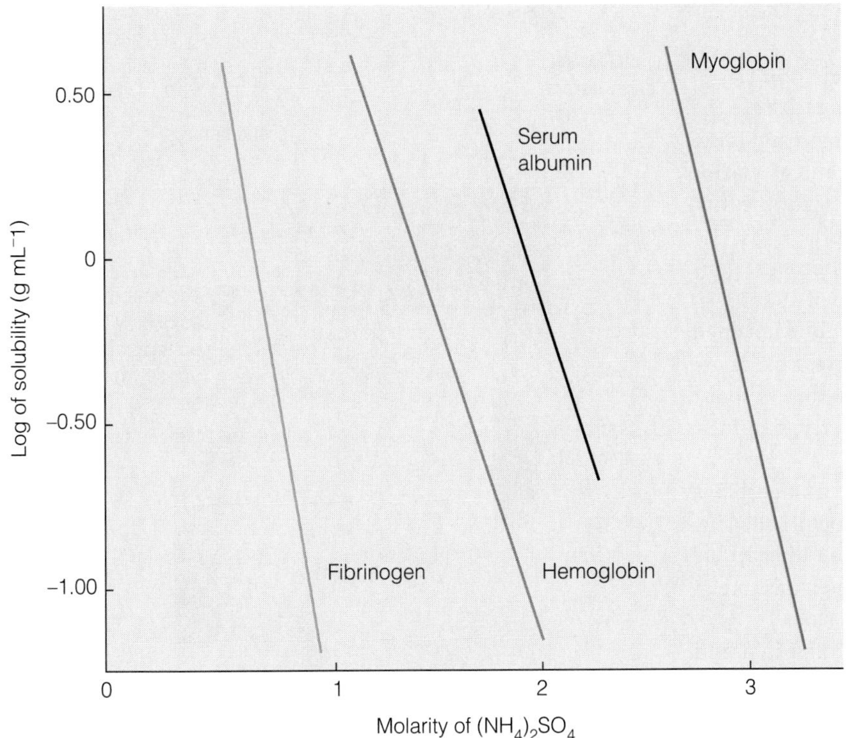

FIGURE **5A.1**

Solubility of some proteins in ammonium sulfate as a function of salt concentration.

by precipitating them from solution with ammonium sulfate. Ammonium sulfate is often used for such selective precipitations because quite high ionic strengths can be achieved without damage to the proteins.

Centrifugation

Any particle that is subjected to a centrifugal field by being spun in a centrifuge rotor can be said to be subject to a centrifugal force (Figure 5A.2). For a particle of mass m, this force is given by

$$f_c = m(1 - \bar{v}\rho)\omega^2 r \qquad (5A.1)$$

Here r is the distance of the particle from the center of the rotor, which is spinning at an angular velocity of ω radian/s ($\omega = (2\pi/60) \times$ RPM, where RPM is the number of revolutions per minute). The factor $(1 - \bar{v}\rho)$ is the **buoyancy factor,** which takes into account the fact that the particle is buoyed up by the surrounding solution. This factor contains the solution density ρ (g/mL) and the specific volume of the particle $\bar{v}$ (mL/g). The specific volume $\bar{v}$ can be considered equal to $1/\rho_P$, where ρ_P is the effective density of the particle. So the buoyancy factor $1 - \bar{v}\rho = 1 - \rho/\rho_P$. Obviously, if $\rho_P = \rho$, there is no net force on the particle, for it displaces its own mass of solution; such a particle would not sediment. As in electrophoresis, the particle's motion is resisted by a force $f_r = fv$, where v is the velocity and f is the **frictional coefficient.** A steady velocity of sedimentation is established, where $f_r = f_c$; therefore

$$fv = m(1 - \bar{v}\rho)\omega^2 r \qquad (5A.2)$$

As in electrophoresis, we divide velocity by field strength, in this case the centrifugal field strength $\omega^2 r$, and obtain

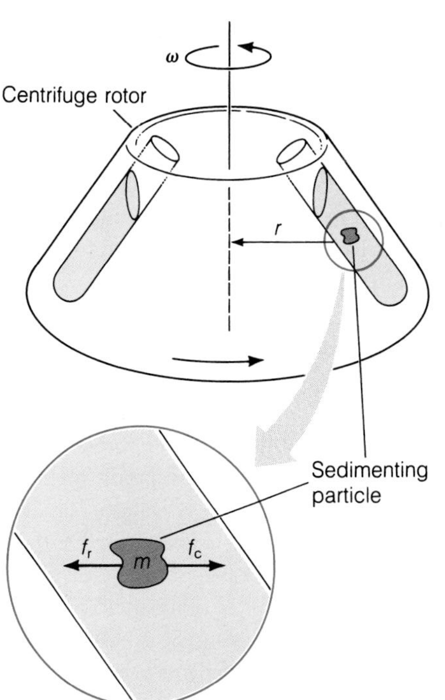

FIGURE **5A.2**

Forces on a particle in a centrifugal field in a fixed-angle rotor.

$$S = \frac{v}{\omega^2 r} = \frac{m(1 - \bar{v}\rho)}{f} = \frac{M(1 - \bar{v}\rho)}{Nf} \qquad (5A.3)$$

where M is molecular weight and N is Avogadro's number ($M = Nm$). *Molecular weight* (more correctly called *molar mass*) is the mass of 1 mole of substance, expressed in grams. The quantity S is the **sedimentation coefficient.**

Note that equation (5A.3) is analogous to equation (2A.2) on page 52 for the electrophoretic mobility. In each case the quantity (S or μ) is defined as velocity/field strength and is expressed in molecular terms (on the right) as the ratio of a factor describing the driving force to a factor accounting for the frictional resistance.

The sedimentation coefficient has the units of seconds. Typical values for molecules are in the neighborhood of 10^{-13} s, so this quantity has been designated 1 **Svedberg unit (S),** named after Theodor Svedberg, a pioneer in centrifuge research. For example, the sedimentation coefficient of hemoglobin is about 4×10^{-13} s, or 4S. Often cellular particles are referred to by their S value, for example, 70S ribosomes.

Clearly S increases with particle mass, but the relationship is neither linear nor simple, because the frictional coefficient, f, increases with the particle size and also depends on the particle shape. Although we can use sedimentation as a rough measure of molecular weight, the relationship is only approximate.

Since different-sized particles or molecules differ in S, and hence differ in sedimentation velocity, sedimentation is a useful tool for separation. A number of techniques are used, depending on what the researcher wants to separate. If the task is simply to remove large particles or aggregates from a solution of molecules, a low-speed centrifuge with a rotor carrying fixed-angle tubes (see Figure 5A.2) may suffice. For example, a first step in isolating proteins often involves breaking cells and separating the soluble proteins of the cytosol from cell nuclei, cell wall fragments, organelles, and other heavy debris. Sedimentation in a fixed-angle rotor at a few thousand RPM for 10–20 min usually accomplishes this separation. For more difficult cases, in which several kinds of large molecules or particles must be separated from one another, the **sucrose gradient** technique shown in Figure 5A.3 may be employed. The sucrose gradient method gives better separation than fixed-angle sedimentation, because components are separated into discrete bands. It could be used, for example, to obtain fractions enriched in individual proteins from the supernatant obtained from a low-speed centrifugation of broken cells. For separations of protein molecules, high rotor speeds (often up to 70,000 RPM) are required.

It is also possible to use sedimentation to analyze, rather than preparatively separate, mixtures. In such cases, an **analytical ultracentrifuge** is used, in which the progress of sedimentation of individual components can be followed during the experiment. By such experiments, the sedimentation coefficients of macromolecules can be determined with accuracy, using equation (5A.3). The use of sedimentation experiments for the actual determination of molecular weights is described in Tools of Biochemistry 6B.

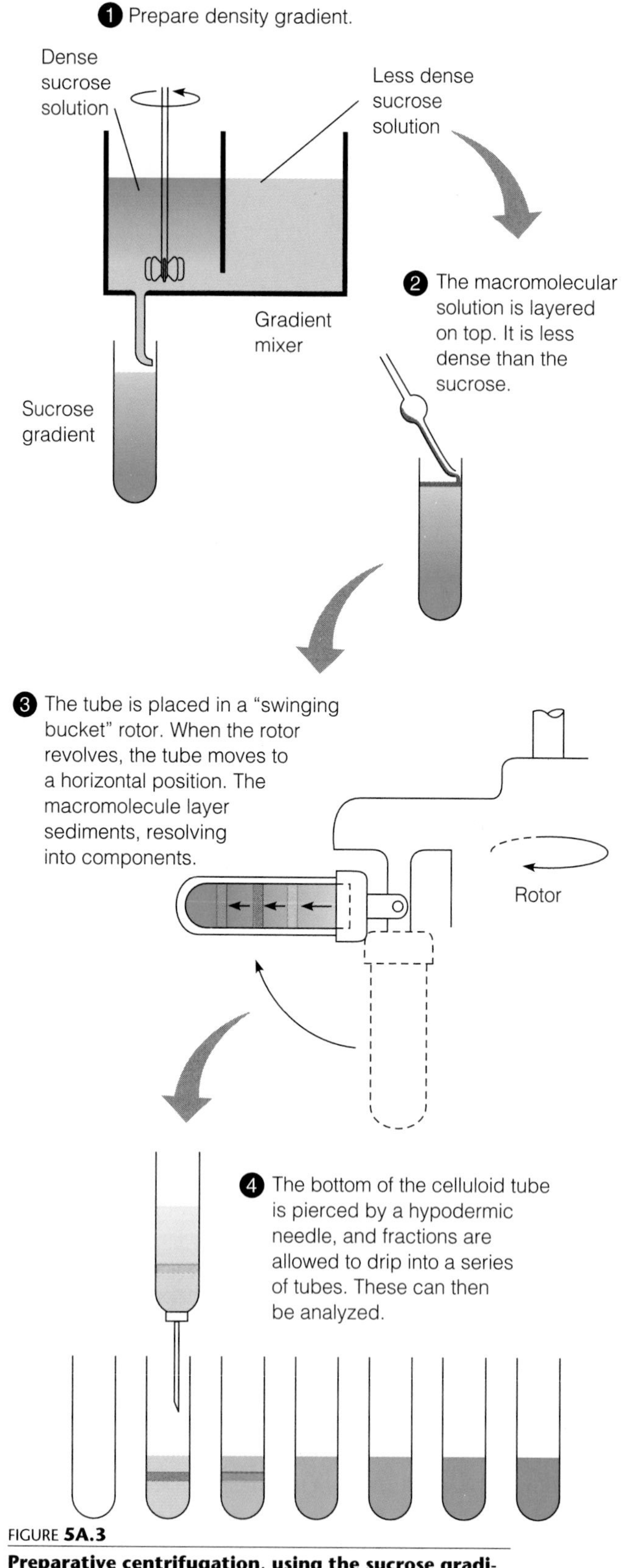

FIGURE **5A.3**

Preparative centrifugation, using the sucrose gradient method.

Chromatography

Much of modern biochemistry depends on the use of **column chromatographic** methods to separate molecules. Chromatographic methods involve passing a solution (the *mobile phase*) through a medium (the *immobile phase*) that shows selective absorption for different solute components. How rapidly any component in the mobile phase passes depends inversely on how strongly it interacts with the immobile phase. The general principles common to all chromatography are illustrated in Figure 5A.4. A column is packed with some material that can selectively adsorb molecules on the basis of some difference in their chemical structure. The column initially is wetted with the appropriate buffer solution. The mixture of the molecules to be separated is then placed on top of the column and slowly washed through the column with buffer. Fractions are taken, using an automatic fraction collector, as this process of **elution** (washing through the column) continues. Some kinds of molecules are adsorbed only weakly or not at all, and they are eluted first. The most strongly adsorbed are eluted last. Sometimes the composition of the buffer solution must be changed during the elution to remove the more tightly bound molecules. The column material used and the method of elution depend on the basis of separation desired. The following sections describe some important methods of column chromatography.

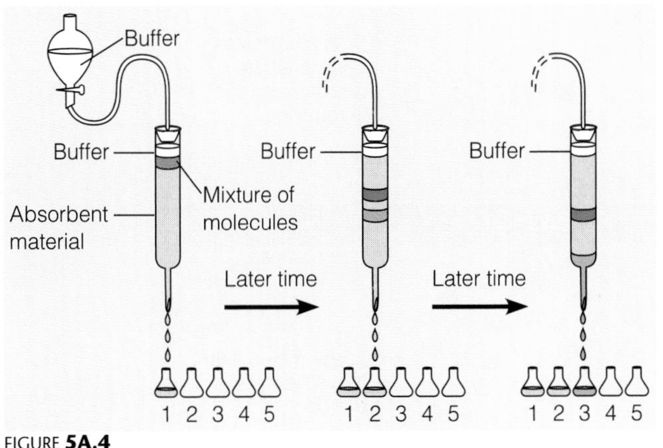

FIGURE **5A.4**
The principle of column chromatography.

Ion-Exchange Chromatography

Ion-exchange chromatography is used to separate molecules on the basis of their electrical charge. **Ion-exchange resins** are used, which are either *polyanions* or *polycations*. Suppose we wish to separate three kinds of molecules, one of which is negatively charged, one weakly positively charged, and one strongly positively charged. An *anionic* resin, carrying negatively charged groups, could be used for the column material. The negatively charged molecules in the mixture would pass through without adsorption and be found in early fractions. The two kinds of positively charged molecules would both be bound by the resin, but the weakly positive ones would be bound less tightly. Since increased salt concentrations tend to disrupt electrostatic interactions of this kind, a *salt gradient* might be applied after the first fractions were collected. That is, the eluting buffer would be made gradually more concentrated in salt (by using NaCl, for example). As salt concentration increases, the Na^+ ions can compete better and better for the negative sites on the resin. The weakly bound cations would be eluted next, and the tightly bound, strongly positive ones would be eluted only at a higher salt concentration.

Affinity Chromatography

Affinity chromatography is a more specific kind of chromatography, ideal for isolating one or a few proteins from a complex mixture. Many proteins exhibit quite strong interactions with other molecules. Examples are the interactions of enzymes with analogs of their substrates or with cofactors (see Chapter 11). The appropriate molecules, covalently coupled to an inert matrix material, will act as molecular "fishhooks" to pick up the desired protein. All remaining proteins will simply pass through the column. The captured protein molecules can then be released by eluting the column with a buffer solution containing free copies of the "fishhook" molecules, or some other reagent that can break the interaction. In a variant of this method, *antibodies* (see Chapter 7) to a particular substance are coupled to the resin and provide very specific retention of the desired material.

High-Performance Liquid Chromatography (HPLC)

Chromatography is often a slow process. In the usual procedures, only a small hydrostatic pressure is applied to force the fluids through the column, and elution takes many hours. This process is not only time-consuming but also sometimes deleterious to sensitive materials. Furthermore, the sample tends to spread out, because of diffusion, as it moves down the column. The longer the experiment takes, the more serious the spreading will be, and resolution of components will suffer. For all these reasons, biochemical researchers have welcomed the development of high-performance liquid chromatographic methods. In these techniques, pressures of 2000–3000 psi are used to force the solutions rapidly through the resin. This level of pressure has required the development of noncompressible resin materials and strong metal columns in which to carry out the process. In this way, separations that formerly required hours can now be done in minutes, yet with higher resolution.

Gel Filtration

Gel filtration is a form of chromatography in which the basis of separation is molecular size rather than chemical properties. Thus, it can provide an alternative to sedimentation. It is often a convenient method for separating different-sized macromolecules or for removing low-molecular-weight contaminants from solutions of large molecules.

The apparatus used for gel filtration (Figure 5A.5) is very similar to that described for column chromatography. A column is packed with porous gel beads, usually a cross-linked polysaccharide material. The porosity of the gel is chosen so that the smaller molecules in the mixture can penetrate the beads, whereas the larger ones cannot. The sample is applied to the top of the column, as in chromatography, and eluted with a buffer. As the sample moves down the column, the larger molecules move faster, for they cannot enter the gel beads and can only flow through the interstices between them. The small molecules can wander into the gel beads and therefore loiter behind. If fractions are collected as buffer is passed through the column, the earlier fractions will contain the larger molecules. By choosing the bead porosity

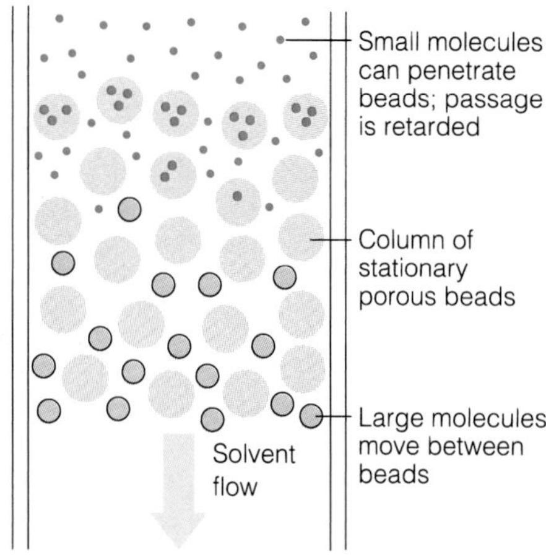

FIGURE **5A.5**
The principle of gel filtration.

properly, mixtures in different size ranges can be separated. The column can even be calibrated with known substances to allow a rough measure of molecular size from the point at which a particular component emerges.

Dialysis and Ultrafiltration

It is possible to obtain **semipermeable membranes** that have pores large enough to permit small molecules to pass freely but present a barrier to proteins and other macromolecules. Such membranes are used in both purification and concentration of biopolymers.

Dialysis is used routinely to remove small-molecule contaminants or to change buffer conditions gently. A solution of a protein, for example, is placed in a closed bag made from such a membrane and immersed in a much larger volume of buffer solution. Over many hours, the low-molecular-weight contaminants leak out and the original environment of the protein molecules is replaced by the outside buffer solution. (Often the outside solution is replaced several times during the process.) **Ultrafiltration** is sometimes used to concentrate solutions of macromolecules. A similar semipermeable membrane is used, but pressure is applied to force solvent and small molecules through the membrane, thereby concentrating the macromolecules.

An Example: How to Prepare Myoglobin

Having described a number of different separation and purification methods, it seems appropriate to give an example of how some of these might be used in an actual protein purification. The overall scheme for preparation of myoglobin is outlined in Table 5A.1, along with some of the reasons for each step. We chose myoglobin because the procedure is relatively simple and because we will use myoglobin frequently as an example of a typical protein.

The purity of the material obtained at the end of step 5 could be checked by gel electrophoresis or isoelectric focusing. A single band should be observed on the gel.

TABLE **5A.1**
HOW TO ISOLATE AND PURIFY MYOGLOBIN[a]

Steps	Rationale
1. Begin with frozen red muscle tissue.	A source rich in the protein desired should be used. It has been kept frozen to avoid degradation of the material.
2. Cut up, homogenize in blender with 65% saturated ammonium sulfate solution. Keep at 4°C.	The homogenization breaks the muscle cells, releasing the myoglobin and other cystolic proteins. The ammonium sulfate precipitates many of the soluble proteins, but *not* myoglobin.
3. Centrifuge at 8000 RPM in fixed-angle rotor for 30 min. Collect supernatant; discard pellet at bottom of tube.	This step removes nuclei, subcellular particles, broken cell walls, muscle fibers, and so forth, together with proteins precipitated by the $(NH_4)_2SO_4$. The supernatant is a crude preparation of myoglobin.
4. Dialyze at 4°C for 24 h against a buffer at pH 8.70, changing several times.	This step removes small molecules that were present in the cytosol and leaves the protein dissolved in the desired buffer. pH is chosen as one at which myoglobin is especially stable.
5. Chromatograph dialyzed solution over a DEAE[b]-cellulose cation-exchange resin. The main myoglobin component is the first heme-containing (red) fraction to emerge from the column.	This step provides a final separation from other proteins, including other heme proteins present in the mixture. The product should be a quite pure preparation of myoglobin.

[a]Adapted from T. E. Hugli and F. R. N. Gurd, *J. Biol. Chem.* (1970) 245:1930–1938.

[b]DEAE = diethylaminoethyl.

It should be emphasized that this example is a particularly simple protein preparation. Myoglobin is abundant in the kind of tissue used, and relatively few steps are required. Most purifications are much more difficult than this one.

References

Cantor, C. R., and P. R. Schimmel (1980) *Biophysical Chemistry,* Vol. 2. Freeman, New York. Detailed coverage of a number of techniques.

Freifelder, D. (1982) *Physical Biochemistry: Applications to Biochemistry and Molecular Biology.* Freeman, New York. More emphasis on practical separation techniques than in Cantor and Schimmel.

Scopes, R. (1982) *Protein Purification: Principles and Practice.* Springer-Verlag, New York.

van Holde, K. E., W. C. Johnson, and P.-S. Ho (1998) *Principles of Physical Biochemistry.* Prentice Hall, Upper Saddle River, N.J. Intermediate in level (for physical techniques like sedimentation) between Freifelder and Cantor and Schimmel.

TOOLS OF BIOCHEMISTRY 5B

Amino Acid Analysis of Proteins

Contemporary methods for determination of the amino acid composition of proteins involve three basic steps:

1. *Hydrolysis* of the protein to its constituent amino acids.
2. *Separation* of the amino acids in the mixture.
3. *Quantitation* of the individual amino acids.

What is described here is a technique used, with minor variations, in many laboratories. A small sample of the protein is first purified, perhaps by some combination of the methods described in Tools of Biochemistry 5A. The protein is dissolved in 6 M HCl, and the solution is sealed in an evacuated ampoule. It is then heated at 105–110°C for about 24 hours. Under these conditions, the metastable peptide bonds between the residues are completely hydrolyzed.

The hydrolyzed sample is then separated into the constituent amino acids on a cation-exchange column. The kinds of resin typically used are sulfonated polystyrenes:

$$-CH_2-CH-CH_2-CH-CH_2-CH-CH_2-CH-$$

(with phenyl, $C_6H_4SO_3^-$, $C_6H_4SO_3^-$, phenyl side groups)

Such a resin separates amino acids in two ways. First, because it is negatively charged, it tends to pass acidic amino acids first and retain basic ones. The pH of the eluting buffer is increased during elution to facilitate this separation. Second, the hydrophobic nature of the polystyrene itself tends to hold up the more hydrophobic amino acids such as leucine and phenylalanine. An example of such an analysis is shown in Figure 5B.1. Note the order of appearance of the amino acids, proceeding from the more acidic to the more basic. Modern amino acid analyzers are completely programmed and carry out both the chromatographic separation of the amino acids and their quantitation. Increasingly, researchers are turning to HPLC (see Tools of Biochemistry 5A) for separation of amino acids. The advantages are rapidity and even better resolution.

There are many methods for detection and quantitation of the amino acids eluting from the column. Older techniques employed the **ninhydrin** reaction (Figure 5B.2). Ninhydrin was mixed with and reacted at 100°C with the eluting amino acids in the analyzer; the absorbance of each peak (which is proportional to the amino acid concentration) was then detected spectrophotometrically in the same instrument. This technique can allow detection of a few nanomoles (10^{-9} mol) of an amino acid.

$R-CH(NH_3^+)-COO^-$

Amino acid

2 Ninhydrin → CO_2, R—CHO

Ninhydrin

Purple-blue compound

FIGURE **5B.2**
The ninhydrin reaction.

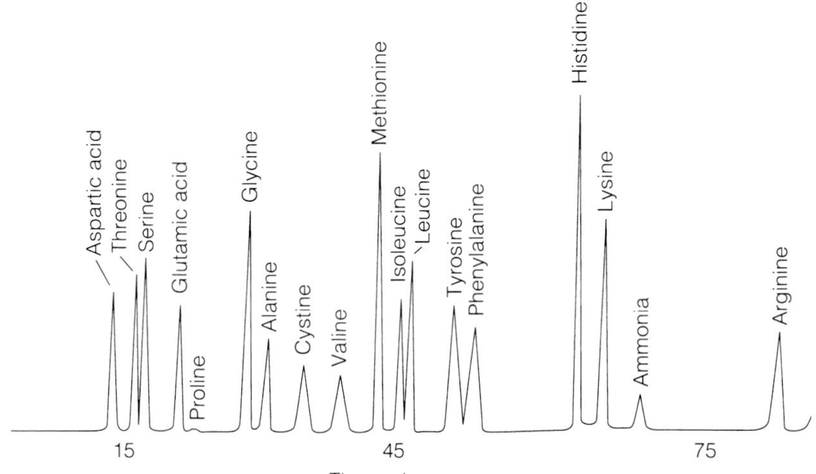

FIGURE **5B.1**
Analysis of a protein hydrolysate on a single-column amino acid analyzer.
Courtesy of Beckman Instruments, Inc.

More recently, the sensitivity of analysis has been greatly improved by the use of fluorescent reagents and fluorescence detection, which have almost entirely replaced spectrophotometric methods. For example, the amino acids may be reacted with *o*-phthalaldehyde to yield a fluorescent complex:

$$\text{o-Phthalaldehyde} + H_3\overset{+}{N}-\underset{R}{CH}-COO^- \;(\text{Amino acid}) + HS-CH_2-CH_2-OH \;(\beta\text{-Mercaptoethanol}) \longrightarrow \text{Isoindole derivative of amino acid} \;(S-CH_2-CH_2-OH,\; N-\underset{R}{CH}-COO^-) + 2H_2O + H^+$$

Such techniques easily give sensitivity to the picomole (pmol, or 10^{-12} mol) range. Recent advances using a new microelectrophoresis system and fluorescence detection have extended this sensitivity to the *attomole* (amol, or 10^{-18} mol) range. This amount corresponds to only a few thousand molecules. Indeed, amino acid analysis techniques have proceeded to the point that the amount of protein contained in one spot in two-dimensional gel electrophoresis can easily be analyzed.

Of course, these procedures are not as simple and trouble free as the foregoing discussion might imply. Some amino acids give problems in reaction with the compounds used for detection; proline in particular, since it is a cyclic amino acid, often reacts differently or not at all. Furthermore, some amino acids tend to be partially destroyed during the severe hydrolysis. Tryptophan is troublesome in this respect and must be determined by a separate analysis, usually based on its strong ultraviolet absorbance (see Figure 5.6). Serine, threonine, and tyrosine also tend to be degraded during long hydrolysis. To a considerable extent, these difficulties can be circumvented either by carrying out protective reactions first or by measuring the apparent content of the amino acid at different hydrolysis times and extrapolating to zero hydrolysis time. Asparagine and glutamine are invariably hydrolyzed to aspartic and glutamic acids, so that the total content of these acids observed includes the amides. This reaction, as well as the other degradation reactions mentioned above, can be avoided by using an enzymatic hydrolysis, with a mixture of proteolytic enzymes, in place of the acid hydrolysis. However, this method also has its drawbacks, because it is sometimes difficult to achieve complete hydrolysis and the enzymes themselves must be removed before analysis.

Despite such complications, amino acid analysis, using automated analyzers, has become a routine operation in many laboratories. One of the first analyses carried out on any newly discovered protein will invariably be for amino acid composition.

References

Chang, J.-Y., R. Knecht, and D. Braun (1981) Amino acid analysis at the picomole level. *Biochem. J.* 199:547–555.

Cheng, Y.-F., and N. Dovichi (1988) Subattomole amino acid analysis by capillary zone electrophoresis and laser-induced fluorescence. *Science* 242:562–564.

Liu, T.-Y. (1972) Determination of tryptophan. *Methods Enzymol.* 25:44–55.

TOOLS OF BIOCHEMISTRY 5C

How to Sequence a Protein

Determination of the primary structure of a protein, like similar analysis of nucleic acids, is commonly referred to as sequencing. There are a number of ways in which the amino acid sequence can be determined. As mentioned on page 142, determination of the gene sequence is rapidly becoming one of the easiest methods. Indeed, as the entire genomes of many organisms are now being determined, we are automatically being provided with information on the sequences of thousands of proteins, many of them of still unknown function. Nevertheless, as was pointed out, such sequences do not provide us with information concerning modification of amino acids or the existence of intramolecular cross-links such as disulfide bonds. To find these, we must sequence the protein itself. Today, virtually all direct sequencing is via the method devised by Pehr Edman. The Edman degradation utilizes the sequence of reactions shown in Figure 5C.1. The compound phenylisothiocyanate is reacted in alkali with the terminal amino group to yield a phenylthiocarbamyl derivative of the peptide (Figure 5C.1, step 1). This derivative is then treated with a strong anhydrous acid, which results in cleavage of the peptide bond between residues 1 and 2 (step 2).

The derivative of the N-terminal residue then rearranges to yield a phenylthiohydantoin (PTH) derivative of the amino acid (step 3). Two important things have been accomplished. First, the N-terminal residue has been marked with an identifiable label, but the rest of the polypeptide has not been destroyed; it has simply been shortened by one residue. The whole sequence of reactions can now be repeated and the second residue determined. By continued repetition, a long polypeptide can

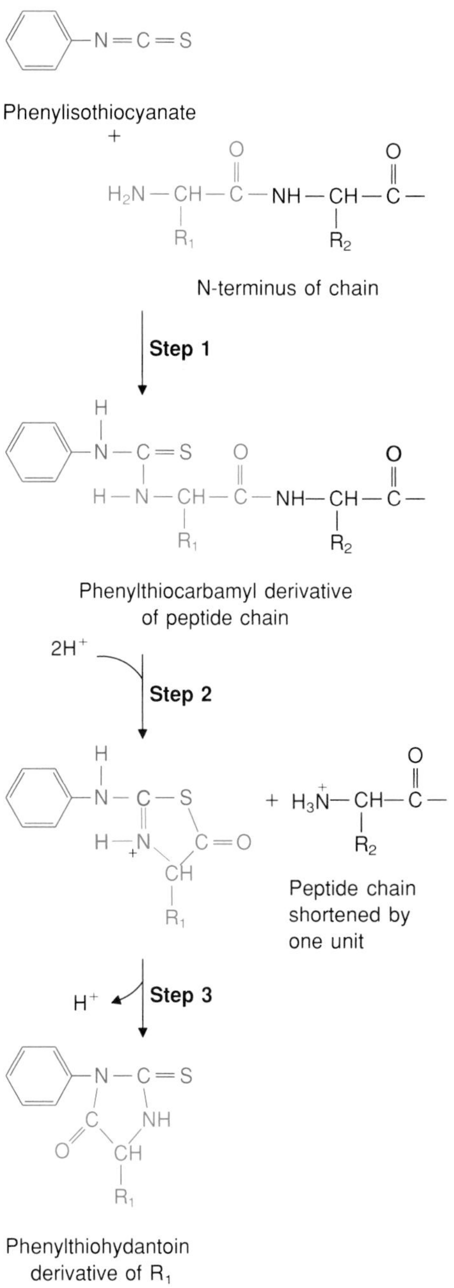

Phenylisothiocyanate

+

N-terminus of chain

Step 1

Phenylthiocarbamyl derivative of peptide chain

$2H^+$

Step 2

Peptide chain shortened by one unit

H^+

Step 3

Phenylthiohydantoin derivative of R_1

FIGURE **5C.1**
The Edman degradation.

be "read" starting from the N-terminal end. This procedure is the basis of the most important method for direct sequencing of polypeptide chains.

Nowadays it is done almost entirely automatically in instruments known as sequenators. Such a device is able to carry out the entire set of reactions shown in Figure 5C.1 over and over again. The sequenator will accumulate in a separate tube the phenylthiohydantoin (PTH) derivative of each amino acid residue in the polypeptide, starting with the N-terminal residue and proceeding for as many cycles as the operator desires or precision allows. The PTH derivatives are usually identified by high-performance liquid chromatography.

The protein we shall use as an example is bovine insulin. This choice is appropriate, for it was the first protein ever sequenced, by Frederick Sanger and his co-workers in the early 1950s. (Of course, the methods we describe are much more sophisticated than those available to Sanger in his pioneering work.) The example is also more complicated than most, for we must deal with two covalently connected chains, and locate S—S bonds. The steps of the procedure are outlined in Figure 5C.2.

The researcher intending to sequence a protein must first make sure that the material is pure. The protein can be separated from other proteins by some combination of the methods described in Tools of Biochemistry 5A and checked for purity by means of electrophoresis and/or isoelectric focusing. Next, it must be determined whether the material contains more than one polypeptide chain, for in some cases disulfide bridges covalently bond chains together. Sodium dodecyl sulfate (SDS) gel electrophoresis in the presence and absence of reducing agents can answer this question. In the insulin example, the investigator would find that there are two chains, A and B, as shown in Figure 5C.2. These chains must be separated and sequenced individually. To break disulfide bonds and thus separate the chains, several reactions are available. Descriptions of the two most important of these reactions follow.

Performic acid oxidation is the technique used in Figure 5C.2, step 1. The strong oxidizing agent performic acid will yield cysteic acid:

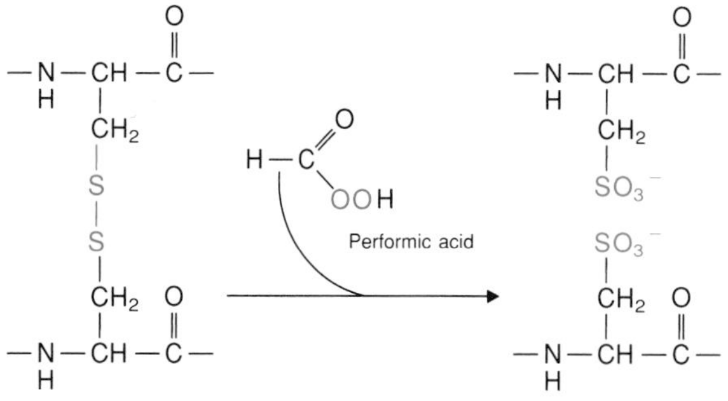

Performic acid

Reduction with β-mercaptoethanol is a milder technique.

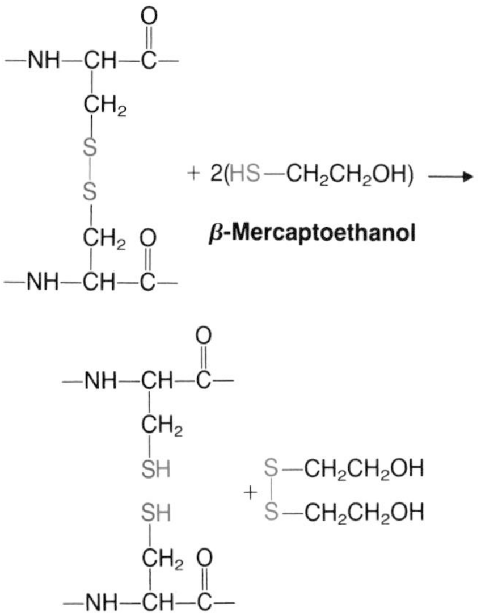

$+ 2(HS{-}CH_2CH_2OH) \longrightarrow$

β-Mercaptoethanol

Reduction leaves free sulfhydryl groups, often positioned so that reoxidation to re-form the disulfide bond is likely. Therefore, the sulfhydryls are usually blocked to prevent this. A common blocking reagent is iodoacetate:

$$\text{—NH—CH(CH}_2\text{SH)—C(=O)—} + \text{I—CH}_2\text{—COO}^- \longrightarrow \text{—NH—CH(CH}_2\text{—S—CH}_2\text{—COO}^-\text{)—C(=O)—} + H^+ + I^-$$

Cysteine residue **Iodoacetate** **δ-Carboxymethyl-cysteine residue**

FIGURE **5C.2**
Sequencing the B chain of insulin.

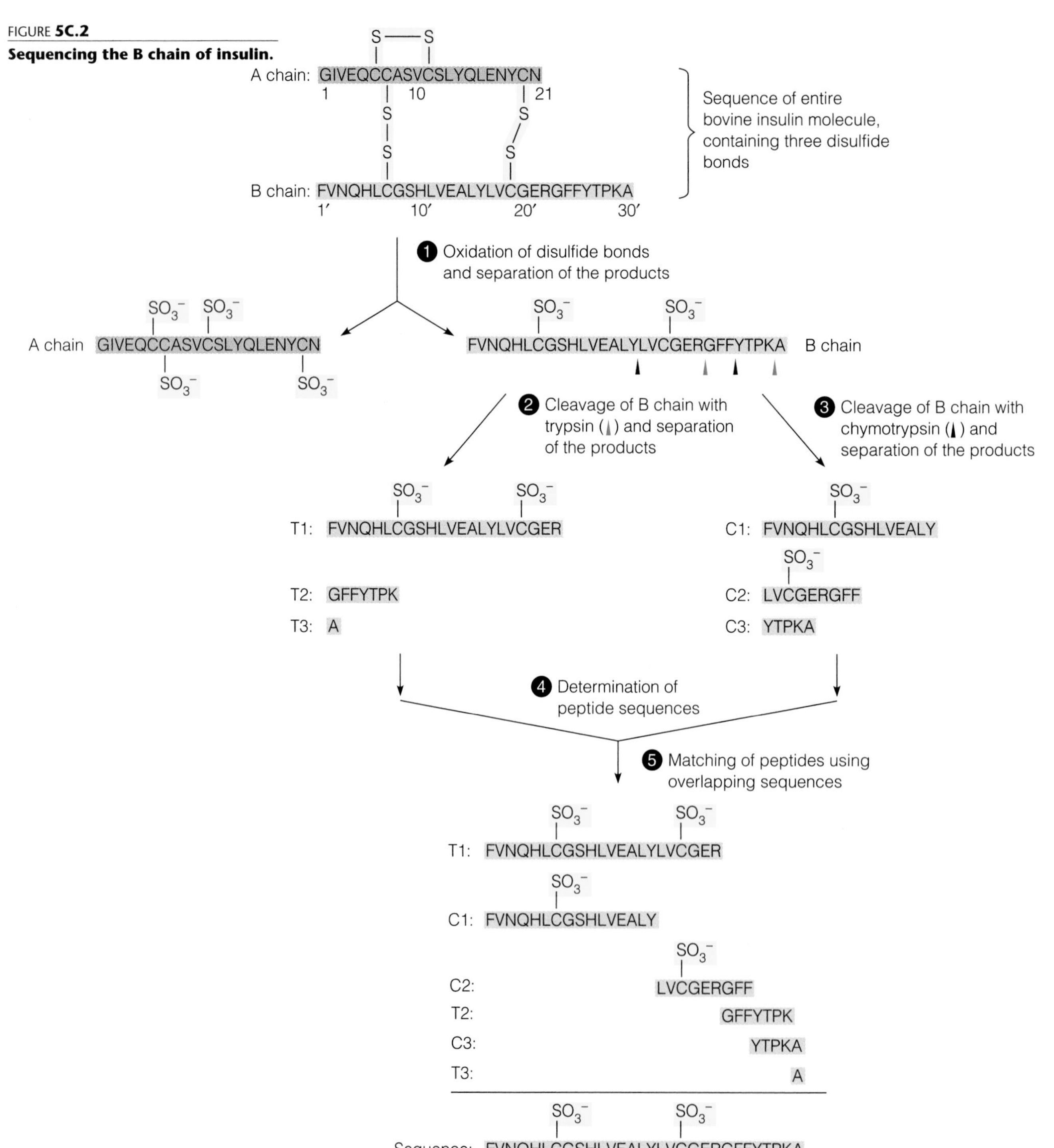

If either of these two methods is carried out with insulin, the whole protein is cleaved into A and B chains. These chains can then be separated by chromatographic methods.

Before sequencing of the individual chains is started, their amino acid composition is usually determined (see Tools of Biochemistry 5B). This determination may point to unusual compositions and thus warn the operator of potential problems. Furthermore, composition data will serve as a check on the sequencing results, because the sequence determined must be consistent with the composition.

Now sequencing can begin. The researcher must recognize that even the best Edman sequencing techniques cannot give reliable results beyond about 40–60 residues. Beyond this point, accumulated impurities and incompleted reactions make results ambiguous. If SDS gel electrophoresis has indicated that the chain is this long or longer, it is necessary first to cleave it into fragments of manageable size. In bovine insulin, the A and B chains are so short that modern techniques could sequence either in one sequenator run. But to demonstrate the methods needed for larger proteins, we assume that the investigator must cleave the insulin chains into small polypeptides. (This was indeed the case in Sanger's pioneering sequencing studies on insulin.) Suppose the insulin B chain is to be sequenced. A first step would be to cleave separate aliquots of the chain with two or more of the specific cleavage reagents described previously. Trypsin and chymotrypsin, for instance, would yield the sets of peptides shown in steps 2 and 3, respectively, of Figure 5C.2. The individual peptides would then be isolated from each of the two mixtures, using, for example, ion-exchange chromatography, and their sequences could be determined (Figure 5C.2, step 4).

A number of instrumental techniques are currently used to sequence peptides via the Edman technique. Descriptions of two of them follow.

In *solid-phase liquid-pulse sequenator* (Figure 5C.3), the polypeptide is covalently coupled, through its C-terminus, to a solid supporting matrix. This matrix is sometimes a thin polymer membrane, which is held in the reaction vessel, or it may be resin beads to which the protein is coupled. Reagents are added, reactions occur, and products are removed automatically. The phenylthiohydantoin derivatives are collected and analyzed.

In a variant technique, *gas-phase sequencing,* many of the reagents are added in the gaseous state. This minimizes losses and gives more rapid analysis. These solid-phase methods are capable of sequencing very small amounts of polypeptides; as little as 1 μg (10^{-3} mg) sometimes suffices.

Returning to the sequencing of bovine insulin: Suppose each of the peptides shown in Figure 5C.2 has been sequenced (step 4). Although the tryptic peptides alone cover the whole sequence, they are not sufficient to allow us to write down the sequence of the insulin B chain, because we do not know the order in which they appear in the chain. However, we also have the chymotryptic peptides, which overlap the tryptic peptides; therefore, all ambiguity is removed. Only one arrangement of the whole chain is consistent with the sequences of these two sets of peptides, as can be seen by matching overlapping sequences (step 5).

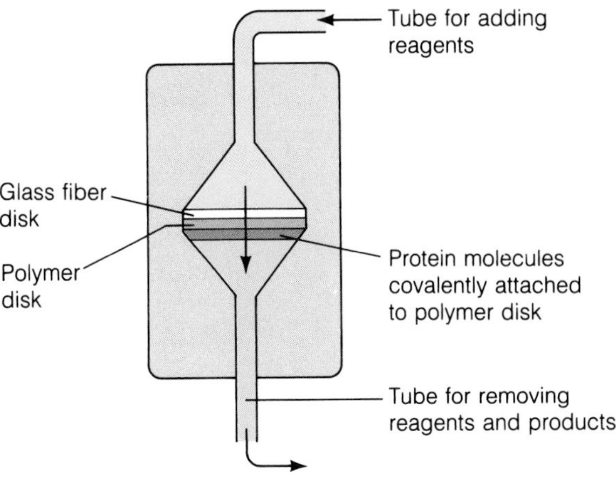

FIGURE **5C.3**
Reaction vessel for a solid-phase automated sequenator.

Finally, a complete characterization of the covalent structure of a protein requires that the positions of any disulfide bonds be located. In preparation for sequencing, these bonds would have been destroyed, but the positions of all cysteines, some of which *might* have been involved in bonding, would have been determined. The question is, Which cysteines are so linked in the native protein, and which is connected to which?

To determine the arrangement of disulfide bonds, the experimenter again starts with the native protein—insulin in the example shown in Figure 5C.4. Reaction with radioactive-labeled iodoacetate marks any free cysteine residues, and fragmentation of the protein into the same peptides used in sequencing allows the positions of these nonbonded cysteines to be identified (step 1). Then samples of the intact protein are cut with various cleavage reagents, but now without first cleaving disulfide bonds (steps 2 and 3). Some peptides, which are connected by these bonds, are attached to one another. These can then be isolated and their disulfide bonds cleaved to find out which of the original set of peptides each contains.

We have described how the entire amino acid sequence, or primary structure, of a protein can be determined. Such analyses have been carried out on several thousand different proteins in the years since Sanger first determined the sequence of insulin. The very fact that such sequencing can be done is convincing proof that each protein has a unique primary structure.

References

Edman, P., and G. Begg (1967) A protein sequenator. *Eur. J. Biochem.* 1:80–91. The first automated method.

Hunkapiller, M. W., J. E. Strickler, and K. J. Wilson (1984) Contemporary methodology for protein structure determination. *Science* 226:304–311.

Stein, S., and S. Udenfreund (1984) A picomole protein and peptide chemistry: Some applications to the opioid peptides. *Anal. Biochem.* 136:7–23.

Walsh, K. A., L. H. Ericsson, D. C. Parmelee, and K. Titani (1981) Advances in protein sequencing. *Annu. Rev. Biochem.* 50:261–284.

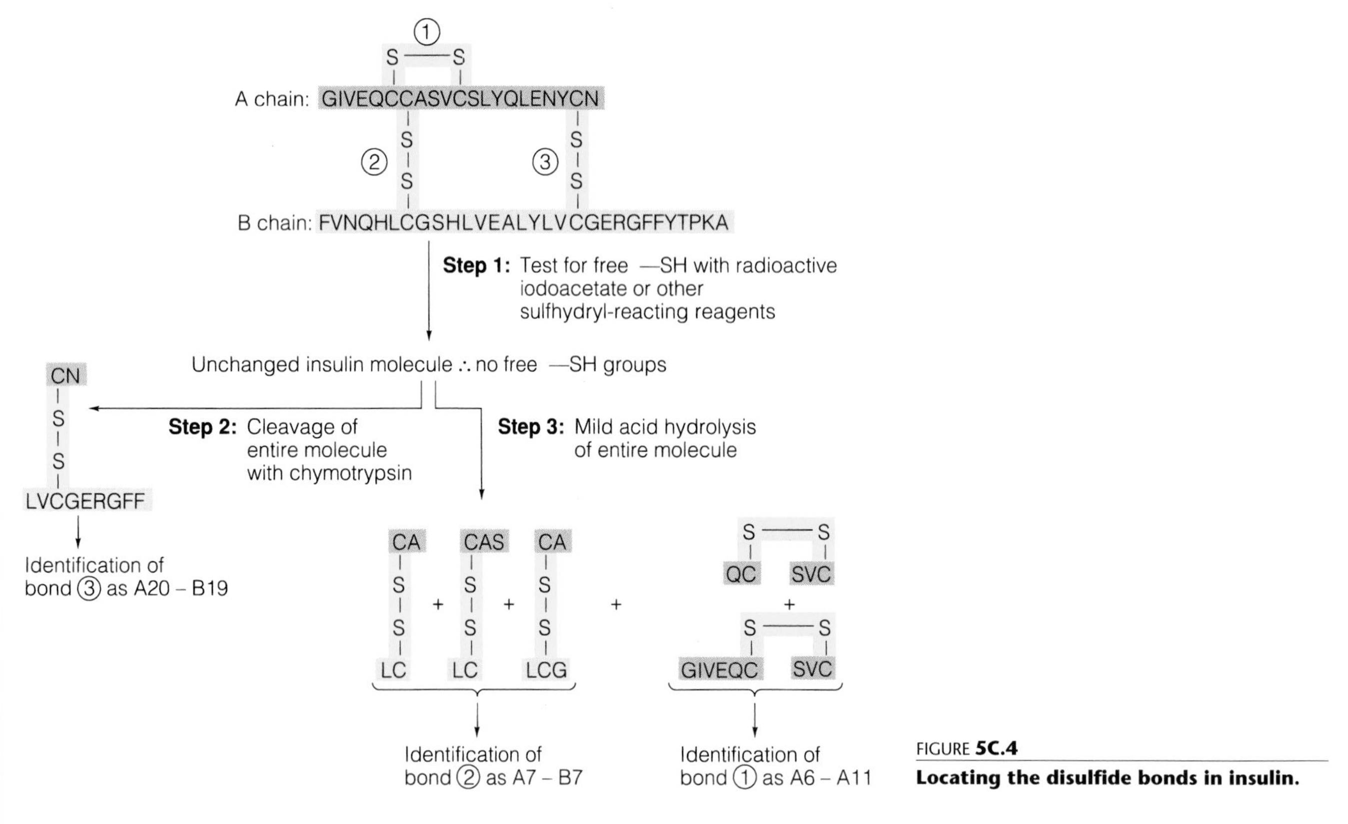

FIGURE **5C.4**
Locating the disulfide bonds in insulin.

TOOLS OF BIOCHEMISTRY 5D

How to Synthesize a Polypeptide

Chemical synthesis of peptides of defined sequence has become of great importance in medicine and molecular biology. Some synthetic hormones can be made with nonnatural modifications that make them more stable. Synthetic peptides can be used to efficiently elicit antibodies against portions of specific proteins; such antibodies are useful in studying the interaction of proteins with other molecules.

To synthesize a peptide of defined sequence, several criteria must be met:

1. It should be possible to add amino acids one at a time, preferably in an automated reactor.
2. Because of peptide bond metastability, the amino acids must be activated in some way.
3. To avoid side reactions, all reactive groups other than the carboxyl and amino groups adjacent to the α-carbon must be blocked. This is the function of the blocking agent.

The scheme that is used most commonly at present is shown in Figure 5D.1. It was devised by Bruce Merrifield.

The C-terminal amino acid of the desired peptide is covalently attached to a resin, with its $H_3\overset{+}{N}$— group exposed. The amino acid to be added next is activated by reaction at its carboxyl with a carbodiimide reagent, while its amino group is blocked with a *t*-butyloxycarbonyl (step 1). Other blocking groups can be used for other reactive groups on amino acid side chains. The coupling reaction yields a dipeptide, still carrying a blocked amino group. This group is deblocked using trifluoroacetic acid (step 2), and the next activated residue is then added (step 3). All reactions are carried out automatically, with the growing chains attached to the resin. In the final step (not shown), HF is added to cleave off the chain, as well as any blocking groups on lysine side chains and other reactive groups.

By using these methods, peptides approaching 100 residues in length can be routinely synthesized in good yield; even longer chains have been made in some instances. Merrifield has, for example, synthesized an active enzyme (ribonuclease) of 124 residues.

Light-Directed, Spatially Organized Combinatorial Peptide Arrays

Frequently it is necessary to simultaneously test a large number of different peptides for some kind of biological activity. One might want to know, for example, which member(s) of a large family of similar oligopeptides was the antigen reacting with a specific antibody. Formerly, this was an extremely laborious process, involving perhaps hundreds of separate syntheses.

Using techniques borrowed from photolithography, it is now possible to prepare microscopic, two-dimensional arrays containing many combinations of peptides grown on a solid surface. The basic technique is illustrated in Figure 5D.2. The amino acids to be used are each blocked on the N-terminus with a photolabile protective group and carry activated carboxyl groups. First, one class of amino acids (in this case Leu) is reacted with a surface coated with amino groups. The whole surface is then illuminated, which removes the protective groups. A second activated amino acid can then be added to each chain. In this example, after four rounds, the peptide GGFL has been grown on each site. Now the clever part begins: for the next round, a rectangular mask is placed over the surface, so that only half the squares in a checkerboard pattern are illuminated. This allows coupling Tyr residues in the illuminated portion. The other portion is then illuminated

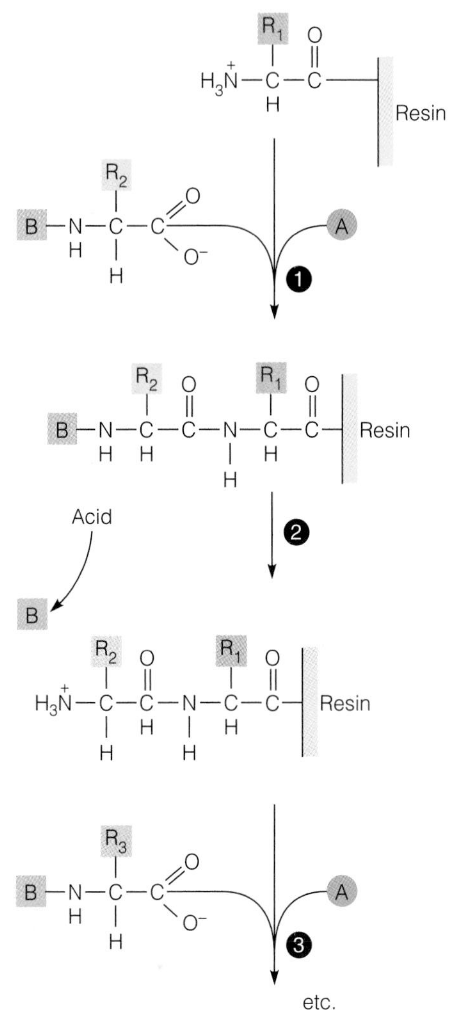

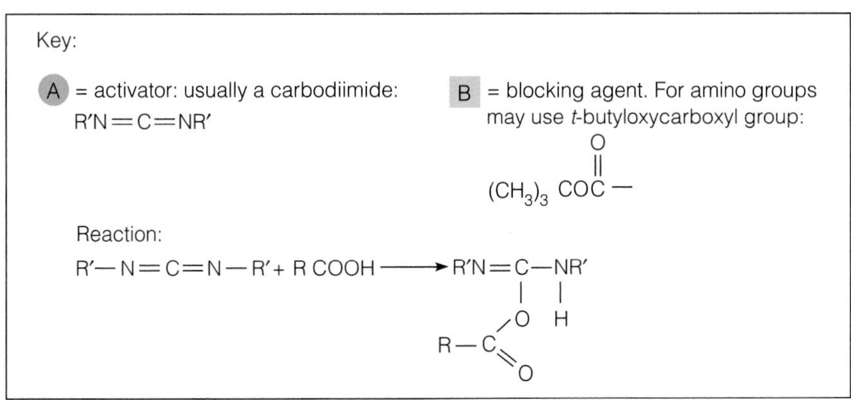

FIGURE **5D.1**

Reactions in solid-phase peptide synthesis.

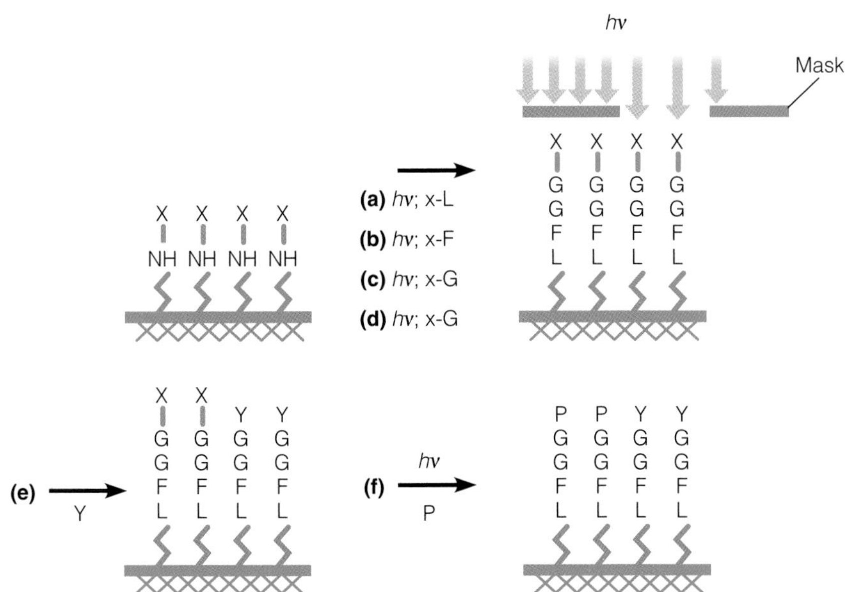

FIGURE **5D.2**

An example of light-directed, spatially patterned oligopeptide synthesis.
X indicates a photolabile blocking group attached to each amino acid residue added. In steps a–d, the tetrapeptide GGFL is built up on the surface. A mask is then used to illuminate and cleave blocking groups in defined areas, to allow addition of tyrosine (step e). An additional round of photo deprotection and coupling is used to add proline in the remaining areas.

From S. P. A. Fodor et al., *Science* (1991) 251:767–773.

and coupled with Pro. Thus, in this example a simple checkerboard pattern, with PGGFL and YGGFL alternating, is obtained. Figure 5D.3 shows the reaction of a fluorescent antibody reactive to YGGFL on such a surface. The example shown is simple: much more complex patterns can easily be generated by use of overlapping masks, allowing thousands of different peptides to be generated, in a prescribed pattern, on one surface.

References

Fodor, S. P. A., J. L. Reed, M. C. Pirrung, L. Stryer, A. T. Lu, and D. Solas (1991) Light-directed spatially addressable parallel chemical synthesis. *Science* 251:767–773.

Merrifield, B. (1986) Solid phase synthesis. *Science* 232:341–347.

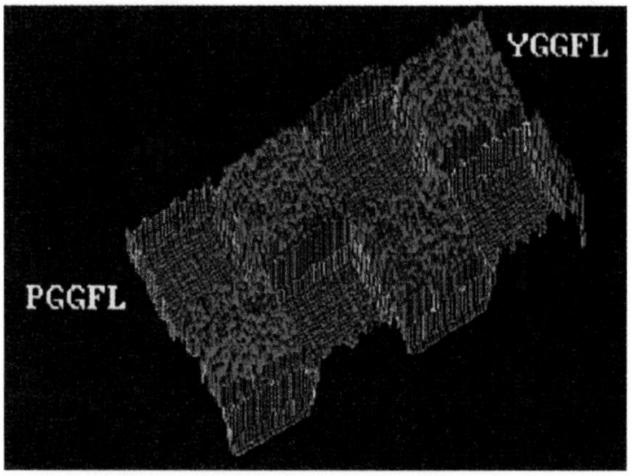

FIGURE **5D.3**

Three-dimensional representation of the checkerboard array of YGGFL and PGGFL. Fluorescence intensity data were converted into spike heights that are proportional to the number of counts detected from 2.5-μm square pixels. The spikes are also color coded.

Courtesy of S. Fodor, Affymetrix Research Institute, reprinted from S. Fodor et al., *Science* (1991) 251:767–773, with permission from *Science*.

CHAPTER 6

The Three-Dimensional Structure of Proteins

In Chapter 5 we introduced the concept of protein primary structure. We emphasized that this first level of organization, the amino acid sequence, is dictated by the DNA sequence in the gene for each protein. However, most proteins exhibit higher levels of structural organization as well. It is the specific three-dimensional structure of each protein that allows it to function in its particular biological role.

Figure 6.1 depicts another representation of the three-dimensional conformation of the myoglobin molecule we showed in Figure 5.1. A specific 3-D structure means that every one of the thousands of atoms in a protein molecule has a particular, well-defined spatial location within the molecule. This characteristic is emphasized in Figure 5.1. Figure 6.1, on the other hand, has been drawn to point out that there exist two distinguishable levels of three-dimensional folding of the polypeptide chain. First, the chain appears to be locally coiled into regions of helical structure (labeled A–H in the figure). Such local *regular* folding is called the **secondary structure** of the molecule. The helically coiled regions themselves are in turn folded into a specific compact structure for the entire polypeptide chain. We call this further level of folding the **tertiary structure** of the molecule. Later in this chapter we shall find that some proteins consist of several folded polypeptide chains, arranged in a regular manner. This arrangement we designate as the **quaternary** level of organization.

Protein molecules have four levels of structural organization: primary (sequence), secondary (local folding), tertiary (overall folding), and quaternary (multichain association).

This chapter is devoted to an examination of the several levels of protein structure—their geometry, how they are stabilized, and their importance in protein function.

Secondary Structure: Regular Ways to Fold the Polypeptide Chain

THE DISCOVERY OF REGULAR POLYPEPTIDE STRUCTURES

Our understanding of the protein secondary structure had its origins in the remarkable work of Linus Pauling, perhaps the greatest chemist of the twentieth

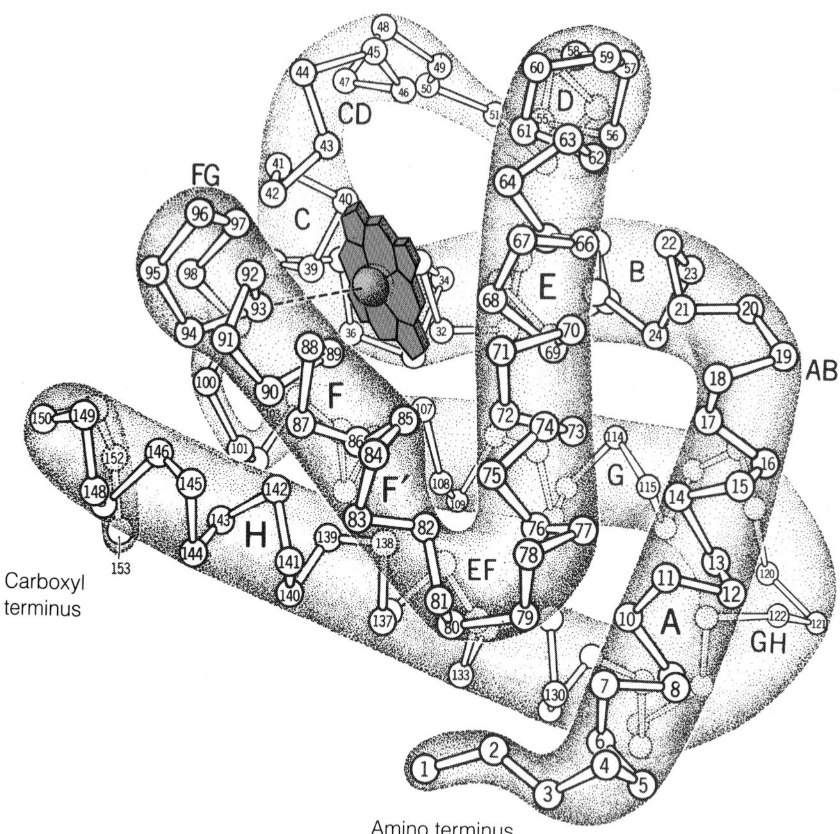

FIGURE **6.1**

Three-dimensional folding of the protein myoglobin. Each amino acid is indicated by a circle corresponding to its α carbon atom. Side chains are omitted, to emphasize how the polypeptide backbone is wrapped into helices and folded. Individual α-helical regions are labeled A–H, with turn regions designated by two letters (e.g., GH). This protein folds about a heme group (shown in purple), a planar heterocyclic structure that chelates iron and serves as the oxygen binding site.

century. As early as the 1930s, he had begun x-ray diffraction studies of amino acids and small peptides, with the aim of eventually analyzing protein structure. In the early 1950s, Pauling and his collaborators used these data together with unusual scientific intuition to begin a systematic analysis of the possible regular conformations of the polypeptide chain. They postulated several principles that any such structure must obey:

1. The bond lengths and bond angles should be distorted as little as possible from those found through x-ray diffraction studies of amino acids and peptides, as shown in Figure 5.12b (page 138).

2. No two atoms should approach one another more closely than is allowed by their van der Waals radii.

3. The amide group must remain planar and in the *trans* configuration, as shown in Figure 5.12b. (This feature had been recognized in the earlier x-ray diffraction studies of small peptides.) Consequently, rotation is possible only about the two bonds adjacent to the α-carbon in each amino acid residue, as shown in Figure 6.2.

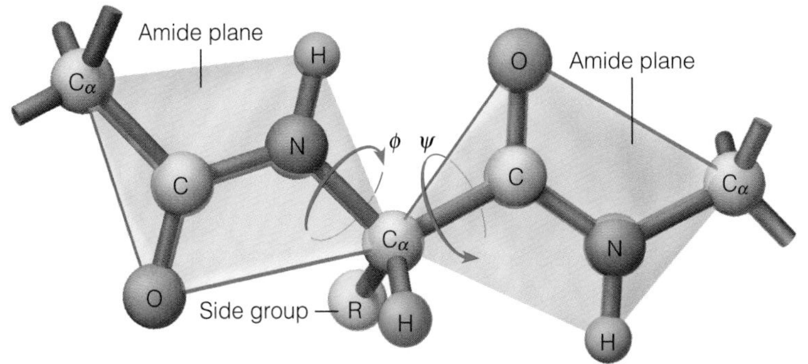

FIGURE **6.2**

Rotation around the bonds in a polypeptide chain. Two adjacent amide planes are shown in light green. Rotation is allowed only about the N—C_α and C_α—C(=O)— bonds. The angles of rotation about these bonds are defined as ϕ and ψ, respectively, with directions defined as positive rotation as shown by the arrows; positive rotation is clockwise as seen from the α-carbon. The extended conformation of the chain shown here corresponds to $\phi = +180°$, $\psi = +180°$.

4. Some kind of noncovalent bonding is necessary to stabilize a regular folding. The most obvious possibility is hydrogen bonding between amide protons and carbonyl oxygens:

$$\rangle\text{N—H}\cdots\text{O=C}\langle$$

Such a concept was natural to Pauling, who had had much to do with the development of the idea of H bonds. In summary, the preferred conformations must be those that allow a maximum amount of hydrogen bonding, yet satisfy criteria 1–3.

Working mainly with molecular models, Pauling and his associates were able to arrive at a small number of regular conformations that satisfied all of these criteria. Some were helical structures formed by a single polypeptide chain, and some were sheetlike structures formed by adjacent chains. The two structures they proposed as most likely—the *α helix* and *β sheet*—are shown in Figure 6.3a and b. These two structures turned out, in fact, to be the most common secondary structures in proteins. Figure 6.4 shows two other polypeptide helices that have since been defined. The 3_{10} helix is observed in some proteins but is not as common as the α helix. The π helix, though sterically possible, has not been observed, possibly because it has a hole down the middle too big to allow van der Waals interactions but too small to admit potentially stabilizing water molecules. All of the structures shown in Figures 6.3 and 6.4 satisfy the criteria listed earlier. In particular, in each structure the peptide group is planar, and every amide proton and every carbonyl oxygen (except a few near the ends of helices) is involved in hydrogen bonding. Each of these forms constitutes a possible kind of secondary structure in proteins.

Of the several possible secondary structures for polypeptides, the most frequently encountered are the α helix, the β sheet, and the 3_{10} helix.

DESCRIBING THE STRUCTURES: MOLECULAR HELICES AND PLEATED SHEETS

In Tools of Biochemistry 4A, we listed the distances that define a molecular helix: the crystallographic repeat (c), the pitch (p), and the rise (h). We also pointed out that helices may be either right-handed or left-handed and may contain either an integral number of residues per turn or a nonintegral number. We call the number of residues per turn n and the number of residues per repeat m. The latter number must always be an integer, because it defines an exact repeat of the structure. If there is an integral number of residues per turn, the pitch and the repeat will be equal, and $n = m$. Some helices of this kind are illustrated schematically in Figure 6.5. Note that as the number of residues per turn decreases, the structure changes progressively from a broad helix to a flat ribbon. Furthermore, progressive change in the same direction then leads to helices of the opposite twist. The

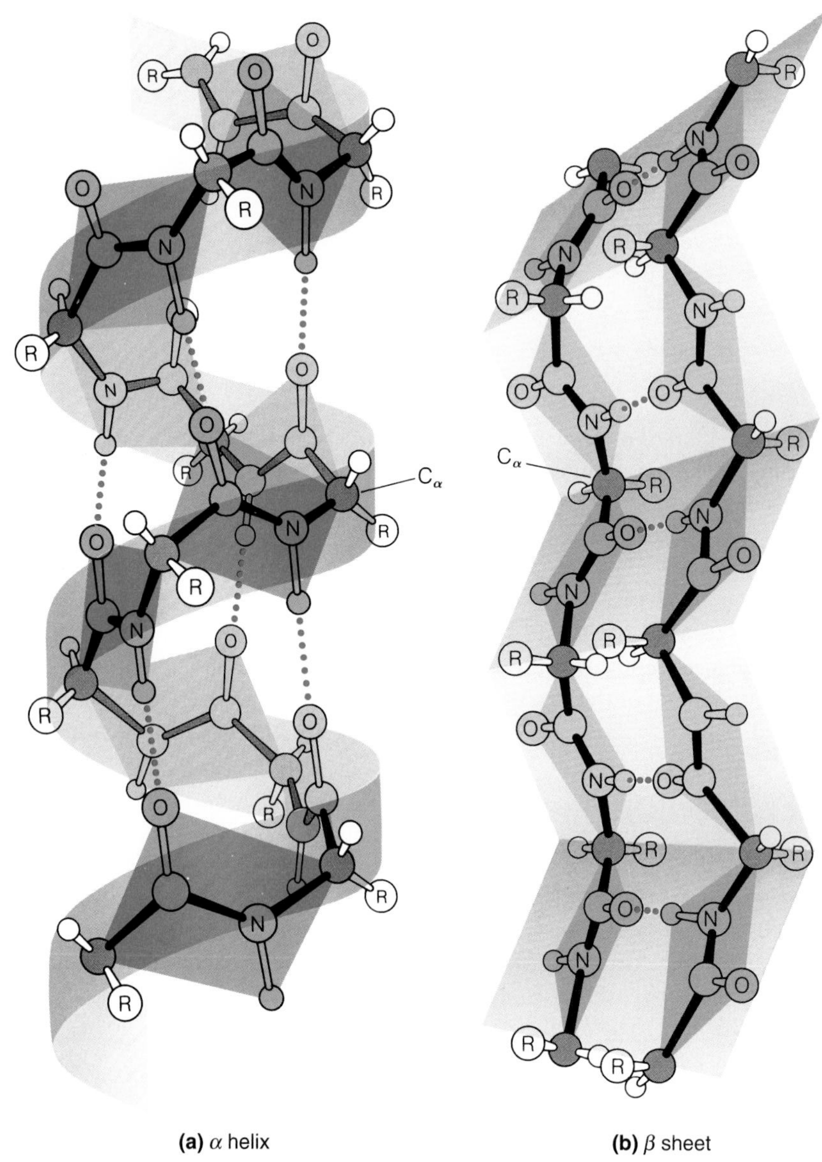

FIGURE **6.3**

The α helix and β sheet. These are the two most frequently encountered regular secondary structures of polypeptides. Their existence was predicted by Linus Pauling. **(a)** In the α helix the hydrogen bonds are within a single polypeptide chain and are almost parallel to the helix axis. **(b)** In the β sheet, the hydrogen bonds are between adjacent chains, of which only two are shown here. In this structure, the hydrogen bonds are nearly perpendicular to the chains.

same reversal of twist occurs when we attempt to make polypeptide helices with more than 5 residues/turn. The helix broadens to a closed ring. Not all of these structures are found in polypeptides. For example, the single-chain $n = 2$ structure shown in Figure 6.5 does not occur in nature, as we will explain shortly.

One of Pauling's major insights was to recognize that polypeptide helices do not have to have an integral number of residues per turn. For example, the α helix repeats after exactly 18 residues, which amounts to 5 turns. It has, therefore, 3.6 residues per turn. Because the pitch of a helix is given by $p = nh$, we have for the

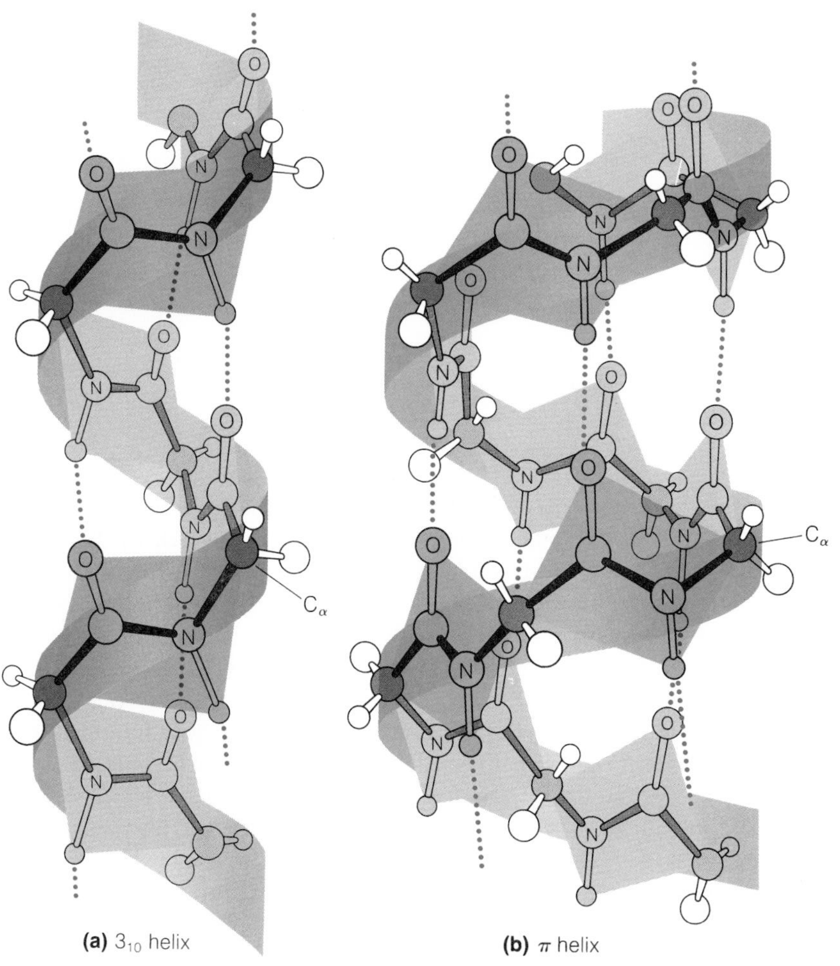

FIGURE **6.4**

Other possible secondary structures of polypeptides.
(a) The 3_{10} helix is found in proteins but is less common than the α helix shown in Figure 6.3a. **(b)** The π (or 4.4_{16}) helix is sterically possible but so far has not been observed in proteins.

α helix, with a rise of 0.15 nm/residue, $p = 3.6$ (res/turn) $\times$ 0.15 (nm/res) = 0.54 nm/turn. Parameters for the other helices shown in Figures 6.3 and 6.4 are listed in Table 6.1.

The parameters defined above describe most molecular helices. For polypeptide helices, which involve hydrogen bonding, there is an additional important quantity. If you examine the model for the α helix (Figure 6.3a), you will note that each carbonyl oxygen is hydrogen-bonded to the amido proton on the *fourth* residue up the helix. Thus, if we include the hydrogen bond, a loop of 13 atoms is formed. Figure 6.6 shows this schematically for each of the helices shown in Figures 6.3 and 6.4. Each helix type has a different number of atoms in such a hydrogen-bonded loop. We shall call this number *N*. A quick way to describe a polypeptide helix, then, is by the shorthand n_N, where *n* is the number of residues per turn. The 3_{10} helix fits this description; it has exactly 3.0 residues per turn and a 10-member hydrogen-bonded loop. The α helix could also be called a 3.6_{13} helix and the π helix a 4.4_{16} helix.

Because hydrogen bonds tend to be linear, the atoms N—H$\cdots$O in polypeptide helices should lie on a straight line. If you examine Figures 6.3 and 6.4, you will see that this requirement is at least approximately satisfied for the 3_{10}, α, and

TABLE **6.1** **Parameters of some polypeptide secondary structures**

Structure Type	Residues/ Turn	Rise (nm)	Number of Atoms in H-Bonded Ring	ϕ (°)	ψ (°)
Antiparallel β sheet	2.0	0.34	—[a]	−139	+135
Parallel β sheet	2.0	0.32	—[a]	−119	+113
3_{10} helix	3.0	0.20	10	−49	−26
α helix (3.6_{13})	3.6	0.15	13	−57	−47
π helix (4.4_{16})[b]	4.4	0.12	16	−57	−70

[a]Bonding is between polypeptide chains.
[b]Sterically permitted but not observed in protein.

π helices. However, it is very difficult to make helices with only two residues per turn *and* linear hydrogen bonds between residues in the same chain. Therefore, the only $n = 2$ structure that is found in proteins is *not* the flat ribbon shown in Figure 6.5 but the β *pleated sheet* structure shown in Figure 6.3b. In the β pleated sheet, each residue is rotated by 180° with respect to the preceding one, which makes each chain an $n = 2$ helix. If the chains are also folded in the accordion-like fashion shown in Figure 6.3b, linear hydrogen bonds can occur *between* adjacent chains. Forming interchain bonds allows correct bond angles with minimal strain when $n = 2$. There are two ways in which β sheets can be formed. The chains can have their N ⟶ C directions running parallel, to make the *parallel* β *sheet*, or they can be *antiparallel* as in Figure 6.3b. It is instructive to try to form such structures using molecular models. You will find that making an $n = 2$ helix with internal H bonds is awkward, but the pleated-sheet structures form naturally.

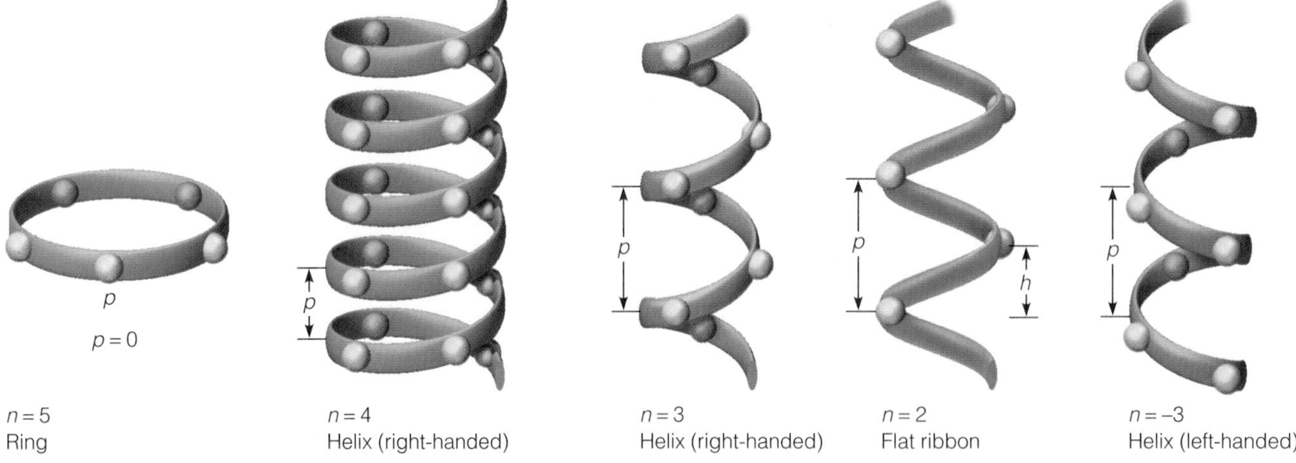

FIGURE **6.5**

Idealized helices. These hypothetical structures show the effect of varying the number (n) of polypeptide residues per turn of a helix. In each case the pitch (p) is indicated, and for $n = 2$ the rise (h) is also shown. Polypeptides can form helices ranging from a closed ring ($n = 5$, $p = 0$) to a 2-fold helix ($n = 2$), with each residue rotated by 180° with respect to the preceding one. All the integral positive values of n and one example of a negative value are shown here. The $n = 4$ and $n = 3$ helices are right-handed, the $n = -3$ helix is left-handed, and $n = 5$ (a ring) and $n = 2$ (a ribbon) have no handedness. The right-handed α helix (not shown here), with $n = 3.6$, is intermediate between the $n = 3$ and $n = 4$ structures.

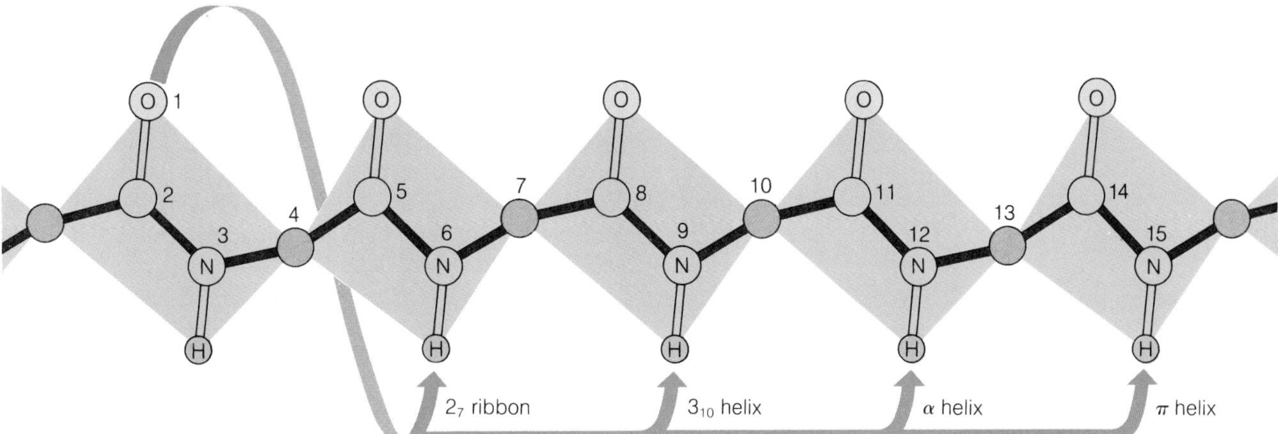

FIGURE **6.6**

Hydrogen bonding patterns for four helices. The structures are represented in a diagrammatic way to simplify counting the atoms in each H-bonded loop. For example, there are 13 atoms in the H loop corresponding to the α (3.6_{13}) helix.

FIGURE **6.7**

Segment of an α helix shown as a space-filling model. The segment illustrated is from the E helix in sperm whale myoglobin (see Figure 6.1).

Thus, possible secondary protein structures fall into two general classes: various helices and at least two types of pleated sheet. But just because a structure can be drawn and contains good H bonds does not mean that it necessarily exists. Many kinds of chain conformations can be imagined that are sterically impossible because atoms in the backbone and/or side chains would overlap. These steric restrictions can be fully appreciated only by examining space-filling models. For example, the α helix shown in Figure 6.7 is fully packed, with R groups on the outside. To examine steric crowding in a systematic way, we need a general procedure for describing polypeptide conformations.

RAMACHANDRAN PLOTS

As was shown in Figure 6.2, each residue in a polypeptide chain has two backbone bonds about which rotation is permitted—the bond between the nitrogen and the α-carbon, and the bond between the α-carbon and the carbonyl oxygen. The angles of rotation about these bonds, defined as ϕ (phi) and ψ (psi) respectively, may be adjusted to form different structures. Thus we could describe the backbone conformation of any particular residue in any protein by specifying these two angles. To make the definition meaningful, we must specify what we mean by a positive direction of rotation and the zero-angle conformation of each. The conventions chosen for directions of positive rotation about ϕ and ψ are given by the arrows in Figure 6.2—that is, clockwise when looking in either direction from the α-carbon. The conformation shown in that figure corresponds to $\phi = +180°$ and $\psi = +180°$, the fully extended form of the polypeptide chain.

With these conventions, the backbone conformation of any particular residue in a protein can be described by a point on a map (Figure 6.8) with coordinates ϕ and ψ. For particular simple types of secondary structure (α helices, for example), all residues are in equivalent conformations and have equivalent ϕ, ψ angles; thus the points for all residues would superimpose. A single point on such a map can therefore also describe a given secondary structure. Such maps are called **Ramachandran plots,** after the biochemist G. N. Ramachandran, who first made extensive use of them.

Many secondary structures can be imagined, but only a few are sterically allowed. The Ramachandran plot illustrates which conformations are allowed.

The Ramachandran plot in Figure 6.8 illustrates the positions of the various regular secondary structures previously discussed, as well as some additional conceivable ones. Note that all helices with a given integral number of residues per turn lie on one of the set of numbered lines drawn across the map. Particularly important are the line corresponding to the flat ribbon ($n = 2$) and the points corresponding to closed rings ($n = 5$). As these are passed, the handedness of the helix changes. This can easily be understood from the following thought experiment. Imagine unwinding a right-hand helix until it approaches a flat ribbon.

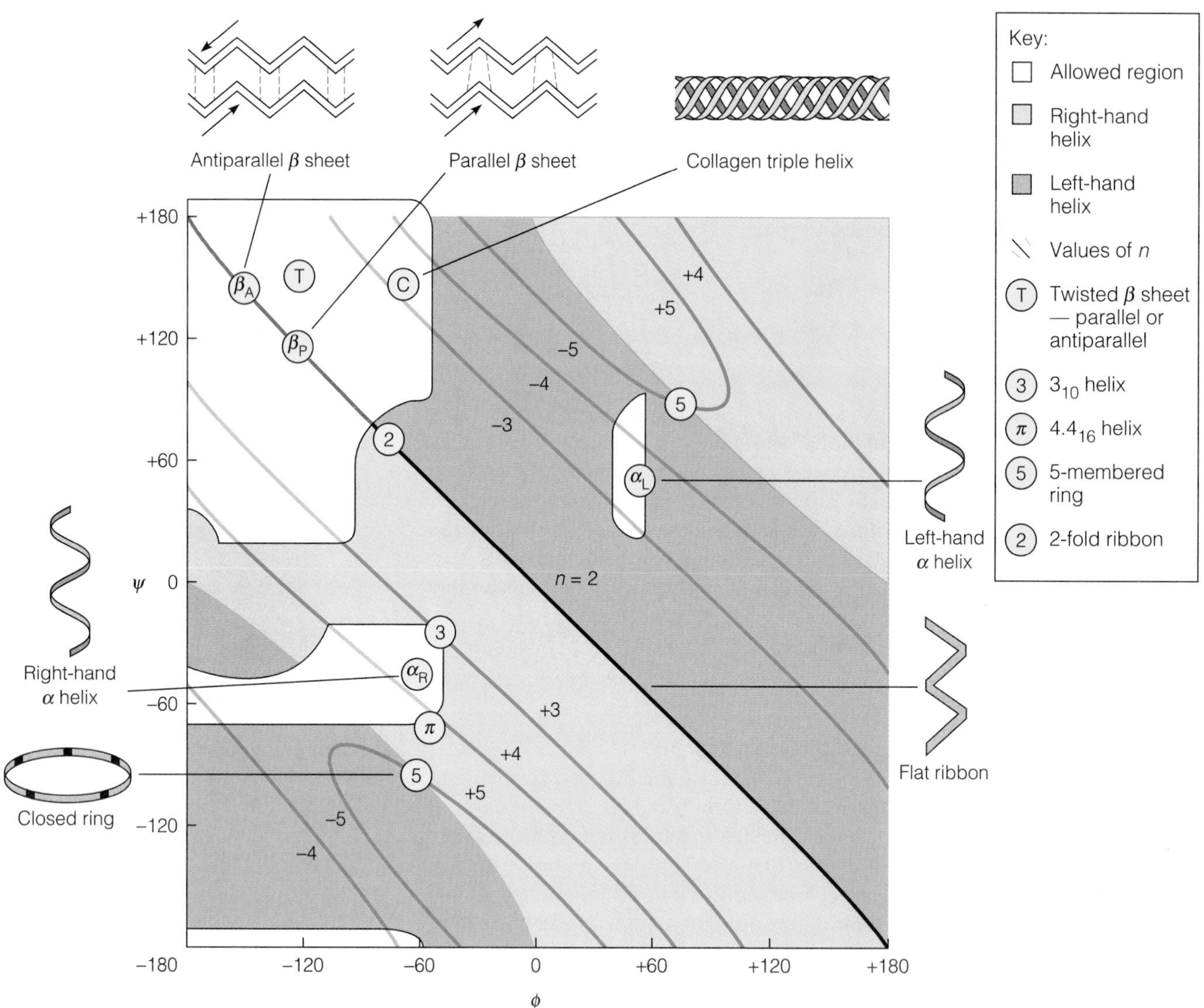

FIGURE **6.8**

A Ramachandran plot. A map of this type can be used to describe the backbone conformation of any polypeptide residue as well as some simple secondary structures of proteins. The coordinates are the bond angles ϕ and ψ, defined as in Figure 6.2. The white areas correspond to allowed conformations, which vary according to the bulkiness of the side chains; this plot assumes that all residues are L-alanines. The colored lines running across the graph correspond to various values of n (residues per turn); the line bisecting the graph corresponds to $n = 2$. The helix is right-handed when n is positive (tan regions), left-handed when n is negative (blue regions). The circles with symbols correspond to important secondary structures, as shown in margins.

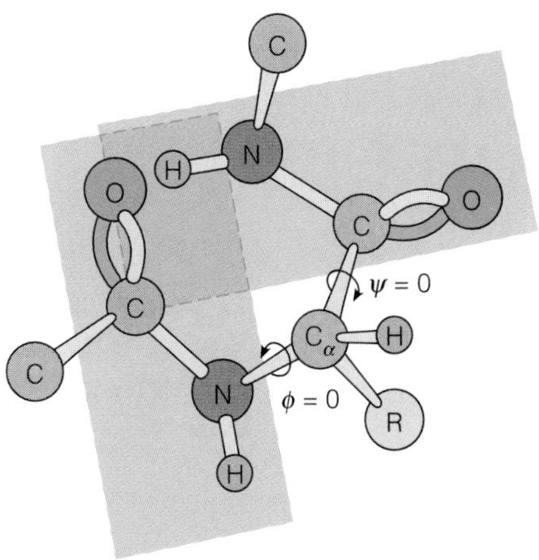

FIGURE **6.9**

A sterically nonallowed conformation. The conformation $\phi = 0$, $\psi = 0$ is not allowed in any polypeptide chain because of the steric clash between the carbonyl oxygen and amino proton.

The value of n will decrease until, at $n = 2$, the flat ribbon structure is reached. If winding is continued in the same direction, a left-hand helix begins to form. Similarly, in a progression to broader and broader helices, the pitch continually decreases. A polypeptide chain can form a five-membered ring with pitch = 0. This lies at the border between right and left helices.

One of the most useful features of Ramachandran maps is that they allow us to describe very simply which structures are sterically possible and which are not. Every point on the surface of the map corresponds to a pair of ϕ, ψ angles and thus to a *conceivable* secondary structure. But for many pairs of ϕ, ψ values some atoms in the chain would approach closer than their van der Waals radii would allow. An example is shown in Figure 6.9. Such conformations are sterically excluded. Ramachandran and other researchers have examined the entire map surface, using models and computers to determine which conformations are actually allowed. The allowed ϕ, ψ pairs lie in the white areas in Figure 6.8. Clearly, only a small fraction of the conceivable conformations is actually possible. All of the regular structures we have discussed fall into or very close to these regions.

Although Figure 6.8 shows the left-hand α helix lying on the edge of an allowed region, it is, in fact, not nearly as favored as the right-hand form. This difference is a consequence of the fact that all amino acids in proteins are of the L-form. With L-amino acids, steric interference between the side chains and the backbone of the helix is less with a right-hand helix than with a left-hand helix. This principle can be understood from a careful inspection of Figure 6.3a. Note that each R group is approximately *trans* to the adjacent carbonyl oxygen. If the amino acid were D instead of L, the orientation would be *cis*, with more likelihood of steric clash. Recall from Chapter 5 that chemists have recently synthesized proteins with all D-amino acids. These proteins have, as expected, left-hand α helices. The importance of such side-chain effects depends on the bulkiness of the side chain. The map shown in Figure 6.8 was drawn with the assumption that all residues are L-alanine (that is, all have CH_3 side chains). If bulkier side chains were considered, the "allowed" region would shrink. Conversely, glycine, with its —H side chain, allows more conformations than are shown in Figure 6.8.

The foregoing analysis of protein structure is quite abstract. How does it square with reality? Quite well, as is shown in Figure 6.10, which illustrates the individual ϕ, ψ angles for amino acid residues in a small protein. These quantities have

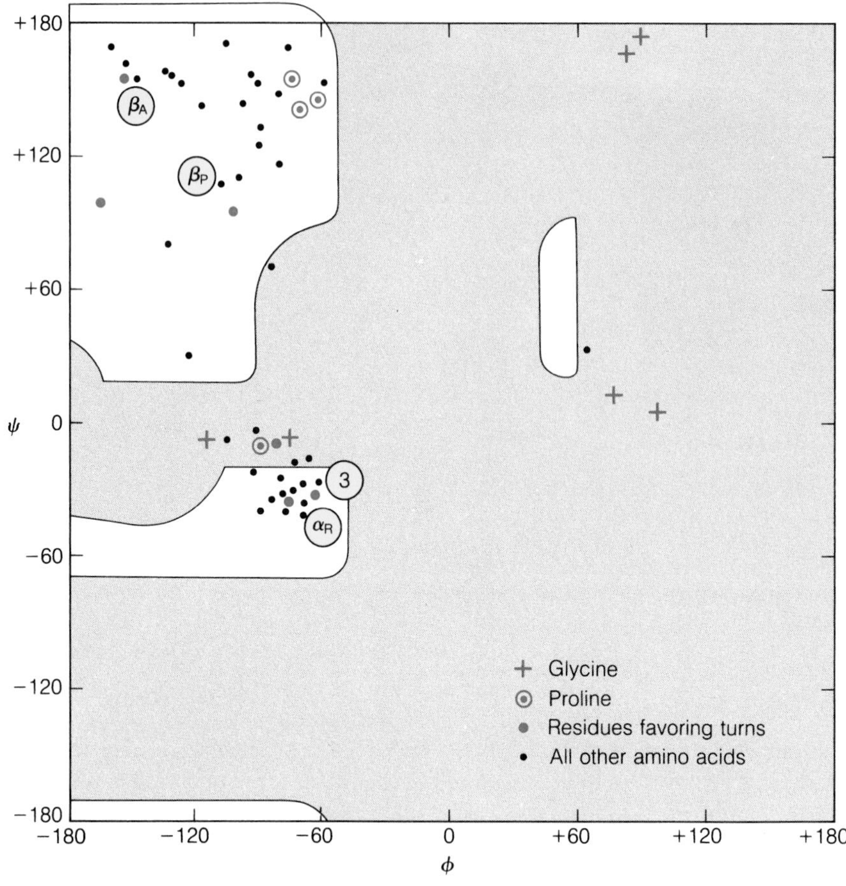

FIGURE **6.10**

Ramachandran plot of the residues in bovine pancreatic trypsin inhibitor (BPTI). The positions of major structures are indicated by yellow circles. Note that most ϕ, ψ pairs fall within or close to the allowed regions (white), and close to major structures. Most of the points far from allowed regions correspond to glycines.

been calculated from a detailed examination of the actual molecular structure, as determined by x-ray diffraction. Most of the points on this map fall close to the α_R position or the β sheet positions. But they do not correspond exactly to these points, testifying to the existence of distortions of these structures in a real protein and to the existence of regions of structure different from either β sheet or α helix. Although most of the points fall in "allowed" regions, a few lie in "nonallowed" regions. These are mainly glycines, for which a much wider range of ϕ, ψ angles is allowed because the side chain is so small.

Our discussion so far provides a background for understanding the basics of protein structure. It is now time to consider some specific cases. We begin with the observation that two major classes of proteins exist. These are called *fibrous* and *globular* proteins, and are distinguished by major structural differences. Let us first consider the fibrous proteins.

Fibrous Proteins: Structural Materials of Cells and Tissues

Fibrous proteins are elongated molecules with well-defined secondary structures. They usually play structural roles in the cell.

Fibrous proteins are distinguished from globular proteins by their filamentous, or elongated, form. Most of them play structural roles in animal cells and tissues—they hold things together. Fibrous proteins include the major proteins of

TABLE **6.2 Amino acid compositions of some fibrous proteins**

Amino Acid	α-Keratin (Wool)	Fibroin (Silk)	Collagen (Bovine Tendon)	Elastin (Pig Aorta)
Gly	8.1	44.6	32.7	32.3
Ala	5.0	29.4	12.0	23.0
Ser	10.2	12.2	3.4	1.3
Glu + Gln	12.1	1.0	7.7	2.1
Cys	11.2	0	0	—[e]
Pro	7.5	0.3	22.1[a]	10.7[c]
Arg	7.2	0.5	5.0	0.6
Leu	6.9	0.5	2.1	5.1
Thr	6.5	0.9	1.6	1.6
Asp + Asn	6.0	1.3	4.5	0.9
Val	5.1	2.2	1.8	12.1
Tyr	4.2	5.2	0.4	1.7
Ile	2.8	0.7	0.9	1.9
Phe	2.5	0.5	1.2	3.2
Lys	2.3	0.3	3.7[b]	3.6[d]
Trp	1.2	0.2	0	—[e]
His	0.7	0.2	0.3	—[e]
Met	0.5	0	0.7	—[e]

Note: The three most abundant amino acids in each protein are indicated in red. Values given are in mole percent.

[a]About 39% of this is hydroxyproline.

[b]About 14% of this is hydroxylysine.

[c]About 13% of this is hydroxyproline.

[d]Most (about 80%) is involved in cross-links.

[e]Essentially absent.

skin and connective tissue and of animal fibers like hair and silk. The amino acid sequence of each of these proteins favors a particular kind of secondary structure, which in turn confers a particular set of appropriate mechanical properties on the substance. Table 6.2 lists the amino acid composition of four specific examples, which we will now discuss.

THE KERATINS

Two important classes of proteins that have similar amino acid sequences and biological function are called α- and β-keratins. The **α-keratins** are the major proteins of hair and fingernails and compose a major fraction of animal skin. The α-keratins are members of a broad group of **intermediate filament proteins,** which play important structural roles in the nuclei, cytoplasm, and surfaces of many cell types. All of the intermediate filament proteins are predominantly α-helical in structure; in fact, it was the characteristic x-ray diffraction pattern of α-keratin that Pauling and his colleagues sought to explain by their α helix model.

The structure of a typical α-keratin, that of hair, is depicted in Figure 6.11. The individual molecules contain long sequences—over 300 residues in length—that are wholly α-helical. Pairs of these helices twine about one another in a left-hand **coiled-coil** structure. This pairing of α helices appears to be a consequence of a peculiarity of the amino acid sequence of α-keratin. Approximately every fourth amino acid has a nonpolar, hydrophobic side chain. Because the α helix has 3.6 residues/turn, this means there is a strip of hydrophobic residues along one side of each chain (actually, very slowly twining around the helix, since 4.0 does not exactly equal 3.6). As we noted in Chapter 2, hydrophobic groups tend to seek their own kind, so two α-keratin helices will stick together, bonded by their hydrophobic strips.

α-Keratin is built on a coiled-coil α-helical structure.

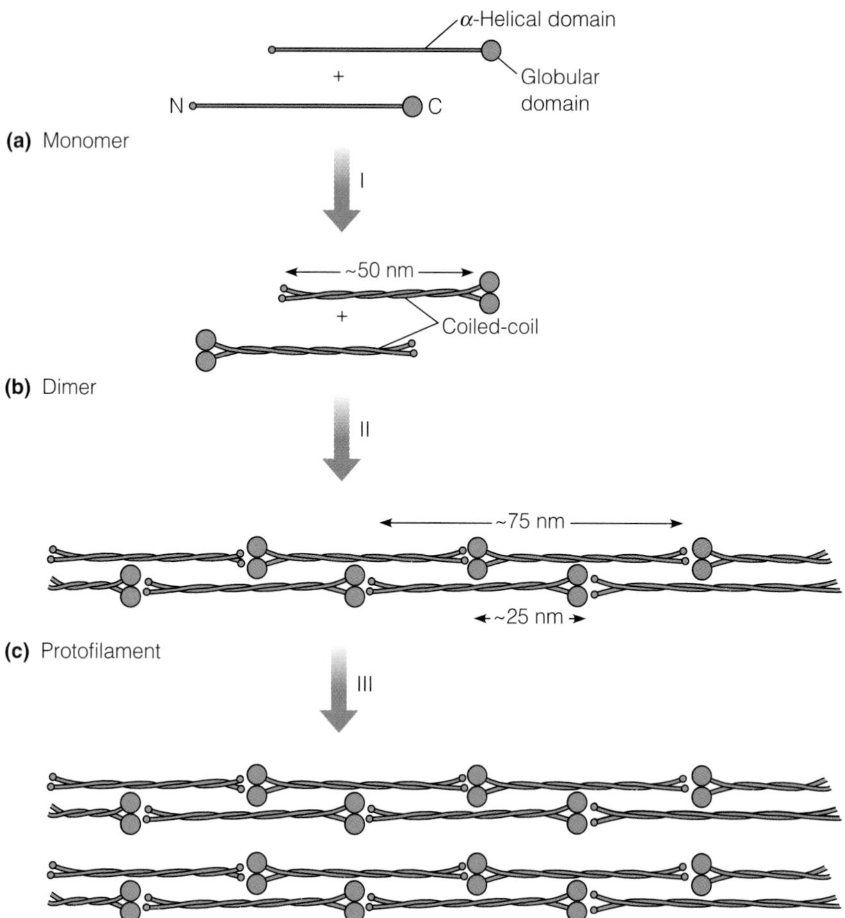

FIGURE **6.11**

Proposed structure for keratin-type intermediate filaments. Two monomers **(a)** pair via a parallel coiled-coil to form the 50-nm-long dimer **(b).** These then associate to form first a 4-strand protofilament **(c)** and then an 8-strand protofibril **(d).** The regular spacing of 25 nm along the fibers is accounted for by the overlap.

In intermediate filamants, pairs of coiled coils themselves tend to associate into a four-chain protofilament (Figure 6.11c), and two of these in turn pack together to form a protofibril (Figure 6.11d). The details of these higher levels of association are still unclear. Such twisted cables can be very stretchy and flexible, but in different tissues α-keratin is hardened, to differing degrees, by the introduction of disulfide cross-links within the several levels of fiber structure. (Note that α-keratin has an unusually high content of cysteine—see Table 6.2.) Fingernails have many cross-links in their α-keratin, whereas hair has relatively few. The process of introducing a "permanent wave" into human hair involves reduction of these disulfide bonds, rearrangement of the fibers, and reoxidation to "set" the waves thus introduced.

The β-keratins, as their name implies, contain much more β sheet structure. Indeed, they represented the second major structural class described by Pauling and co-workers. The β-keratins are found mostly in birds and reptiles, in structures like feathers and scales.

FIBROIN

Fibroin is a β sheet protein. Almost half of its residues are glycine.

The β sheet structure is most elegantly utilized in the fibers spun by silkworms and spiders. Silkworm fibroin (Figure 6.12) contains long regions of antiparallel β sheet, with the polypeptide chains running parallel to the fiber axis. The β sheet regions comprise almost exclusively multiple repetitions of the sequence

$$[\text{Gly–Ala–Gly–Ala–Gly–Ser–Gly–Ala–Ala–Gly–(Ser–Gly–Ala–Gly–Ala–Gly)}_8]$$

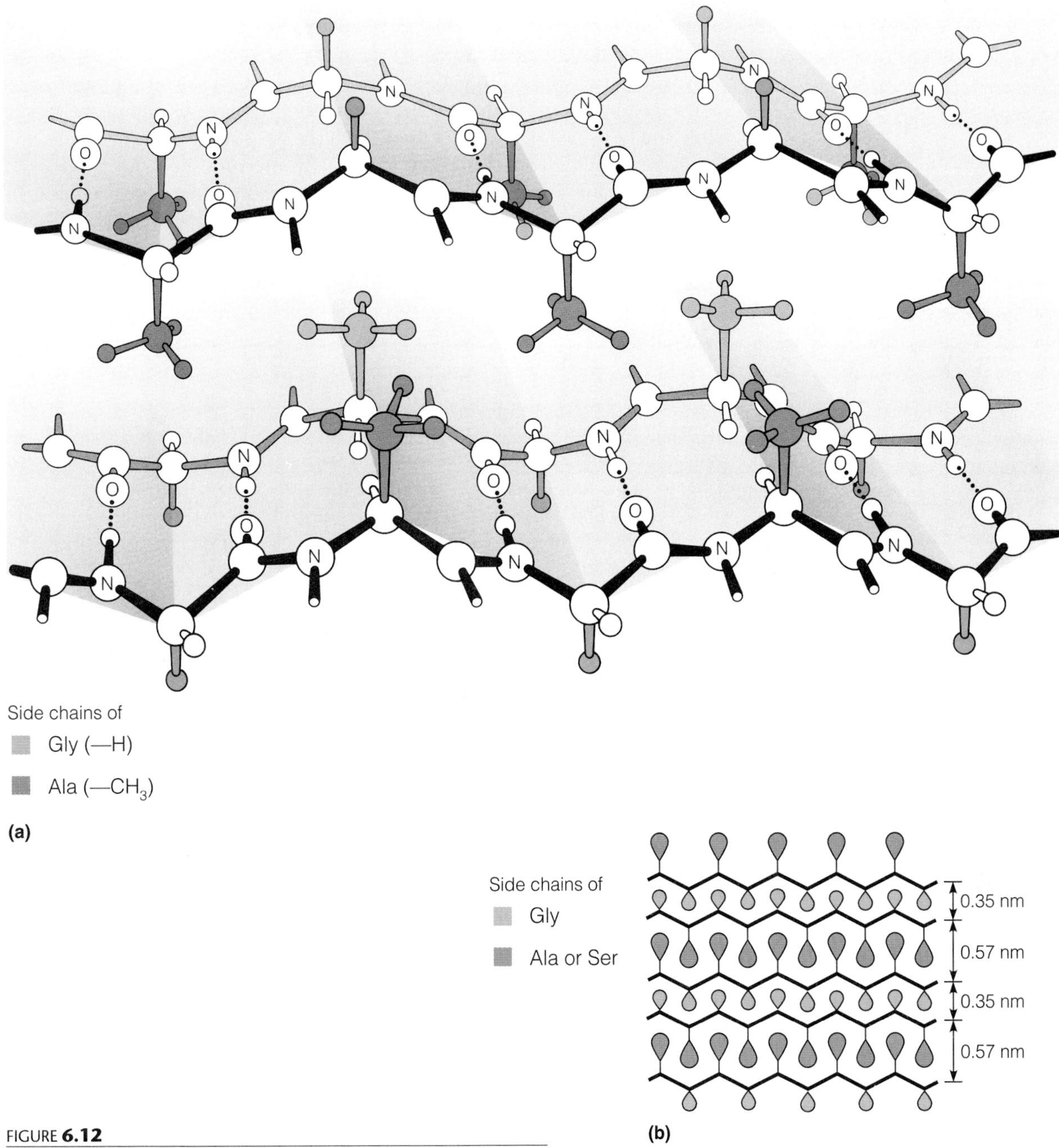

FIGURE **6.12**

The structure of silk fibroin. (a) A three-dimensional view of the stacked β sheets of fibroin, with the side chains shown in color. The region shown contains only alanine and glycine residues. **(b)** Interdigitation of alanine or serine side chains and glycine side chains in fibroin. The plane of the section is perpendicular to the folded sheets.

Examining this sequence, you will note that almost every other residue is Gly and that between them lie either Ala or Ser residues. This alternation allows the sheets to fit together and pack on top of one another in the manner shown in Figure 6.12. The arrangement results in a fiber that is strong and relatively inextensible, because the covalently bonded chains are stretched to nearly their maximum possible length. Yet the fibers are very flexible, because bonding between the sheets involves only the weak van der Waals interactions between the side chains, which provide little resistance to bending.

Not all of the fibroin protein is in β sheets. As the amino acid composition in Table 6.2 shows, fibroin contains small amounts of other, bulky amino acids like valine and tyrosine, which would not fit into the structure shown. These are carried in compact folded regions that periodically interrupt the β sheet segments, and they probably account for the amount of stretchiness that silk fibers have. In fact, different species of silkworms produce fibroins with different extents of such non–β sheet structure and corresponding differences in elasticity. The overall fibroin structure is a beautiful example of a protein molecule that has evolved to perform a particular function—to provide a tough, yet flexible fiber for the silkworm's cocoon or the spider's web.

COLLAGEN

Because it performs such a wide variety of functions, **collagen** is the most abundant single protein in most vertebrates. In large animals, it may make up a third of the total protein mass. Collagen fibers form the *matrix* material in bone, on which the mineral constituents precipitate; these fibers constitute the major portion of tendons; and a network of collagen fibers is an important constituent of skin. Basically, collagen holds most animals together.

Collagen Structure

Collagen fibers are built from triple helices of polypeptides rich in glycine and proline.

The basic unit of the collagen fiber is the **tropocollagen** molecule, a *triple helix* of three polypeptide chains, each about 1000 residues in length. This 3-fold helical structure, shown in Figure 6.13a and b, is unique to collagen. The individual chains are left-hand helices, with about 3.3 residues/turn. Three of these chains wrap around one another in a right-hand sense, with hydrogen bonds extending between the chains. Examination of the model reveals that every third residue, which must lie near the center of the triple helix, can be *only* glycine (see Figure 6.13a). Any side chain other than H would be too bulky. Formation of the individual helices of the collagen type is also favored by the presence of proline or hydroxyproline in the tropocollagen molecule. A repetitive theme in the sequence is of the form Gly–X–Y, where X is often proline and Y is proline or hydroxyproline. However, other residues are often tolerated in these positions. Like silk fibroin, collagen is a good example of how a particular kind of repetitive sequence dictates a particular structure. In order to properly serve the multiple functions it does, collagen exists in a larger number of genetic variants in higher organisms.

Collagen is also unusual in its widespread modification of proline to hydroxyproline. Most of the hydrogen bonds between chains in the triple helix are from amide protons to carbonyl oxygens, but the —OH groups of hydroxyproline also seem to participate in stabilizing the structure. Hydroxylation of lysine residues in collagen also occurs but is much less frequent. It plays a different role, serving to form attachment sites for polysaccharides.

Scurvy is caused by failure to hydroxylate prolines in collagen.

The enzyme catalyzing the hydroxylation of proline requires **vitamin C**, ascorbic acid (see Chapter 21). A symptom of extreme vitamin C deficiency, called **scurvy,** is the weakening of collagen fibers caused by the failure to hydroxylate proline. Consequences are as might be expected: Lesions develop in skin and gums, and blood vessels weaken. The condition quickly improves with administration of vitamin C.

The individual tropocollagen molecules pack together in a collagen fiber in a specific way (Figure 6.13c). Each molecule is about 300 nm long and overlaps its neighbor by about 64 nm, producing the characteristic banded appearance of the fibers shown in Figure 6.13d and e. This structure contributes remarkable strength: Collagen fibers in tendons have a strength comparable to that of hard-drawn copper wire.

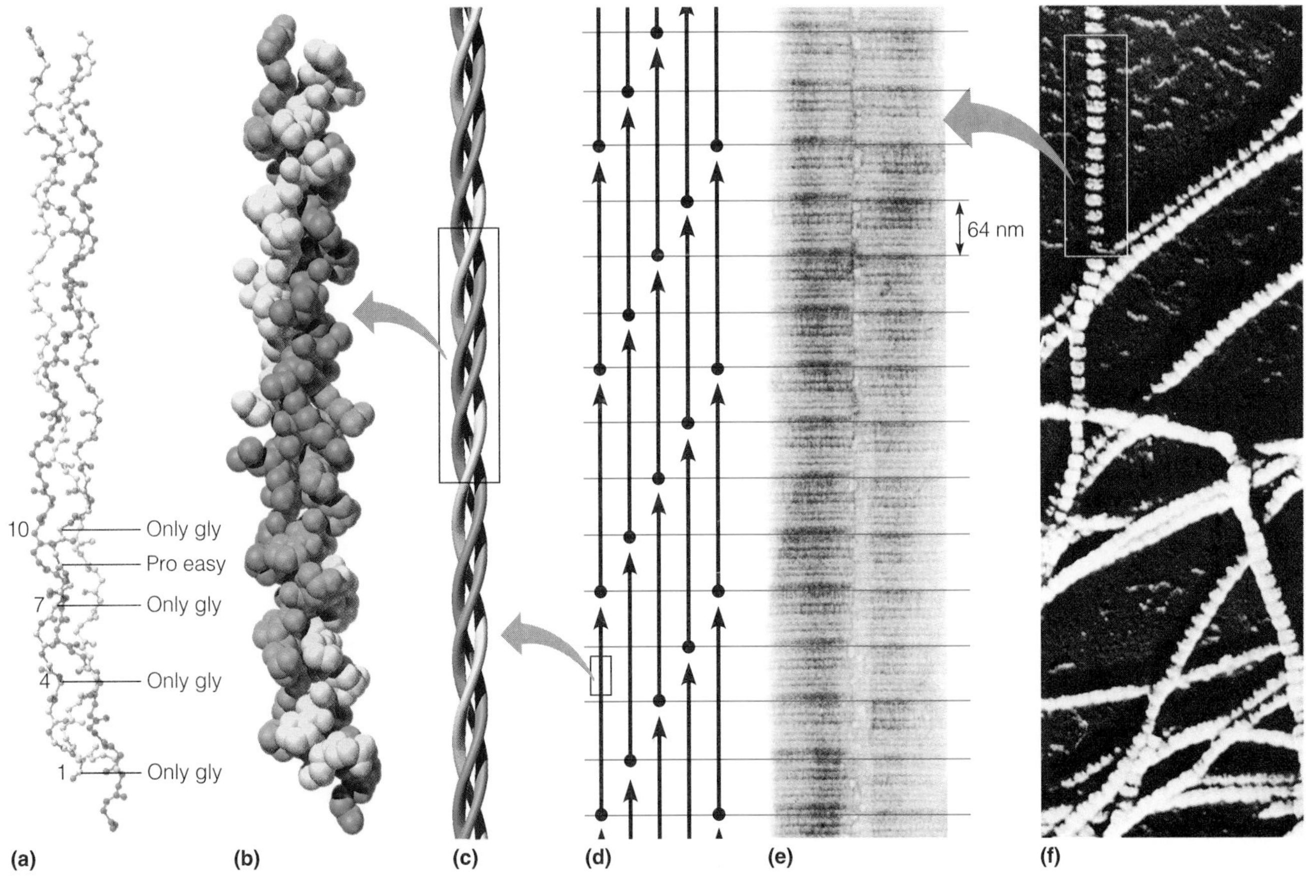

FIGURE **6.13**

The structure of collagen fibers. The protein collagen is made up of tropocollagen molecules packed together to form fibers. The tropocollagen molecule is a triple helix. **(a)** and **(b)** The primary and secondary structure of a tropocollagen molecule. **(c)** Lower magnification emphasizes the triple-helical secondary structure. **(d)** Tropocollagen molecules align side by side in a staggered fashion to form the collagen fiber. **(e)** The periodic pattern shown diagrammatically in (c) can be seen in an electron micrograph of a collagen fiber. **(f)** An electron micrograph of collagen at lower magnification shows the crisscrossing of many fibers, with the periodic pattern clearly visible in each.

(e,f) courtesy of Dr. Alan Hodges.

Part of the toughness of collagen is due to the cross-linking of tropocollagen molecules to one another via a reaction involving lysine side chains. Some of the lysine side chains are oxidized to aldehyde derivatives, which can then react with either a lysine residue or one another via an aldol condensation and dehydration to produce a cross-link:

$$-\text{N(H)}-\text{CH}(\text{C}{=}\text{O})-(\text{CH}_2)_3-\text{CH}{=}\text{C}(\text{CHO})-(\text{CH}_2)_2-\text{CH}(\text{C}{=}\text{O})-\text{N(H)}-$$

This process continues through life, and the accumulating cross-links make the collagen steadily less elastic and more brittle. As a result, bones and tendons in older individuals are more easily snapped, and the skin loses much of its elasticity. Many of the signs we associate with aging are consequences of this simple cross-linking process.

Collagen Synthesis

As you will have judged by now, collagen is a protein that undergoes extensive modification. Indeed, it can be considered an almost complete example of the posttranslational modification pathways we discussed at the end of Chapter 5. The tropocollagen triple helix that ends up cross-linked into an extracellular collagen fiber is very different from the molecule that is first synthesized on a ribosome. The steps in this transformation are shown in Figure 6.14, which begins with translation (step 1). The newly translated polypeptide is hydroxylated (step 2), and then sugars are attached (step 3) to yield **procollagen** (step 4). Procollagen contains

Collagen undergoes extensive posttranslational modification.

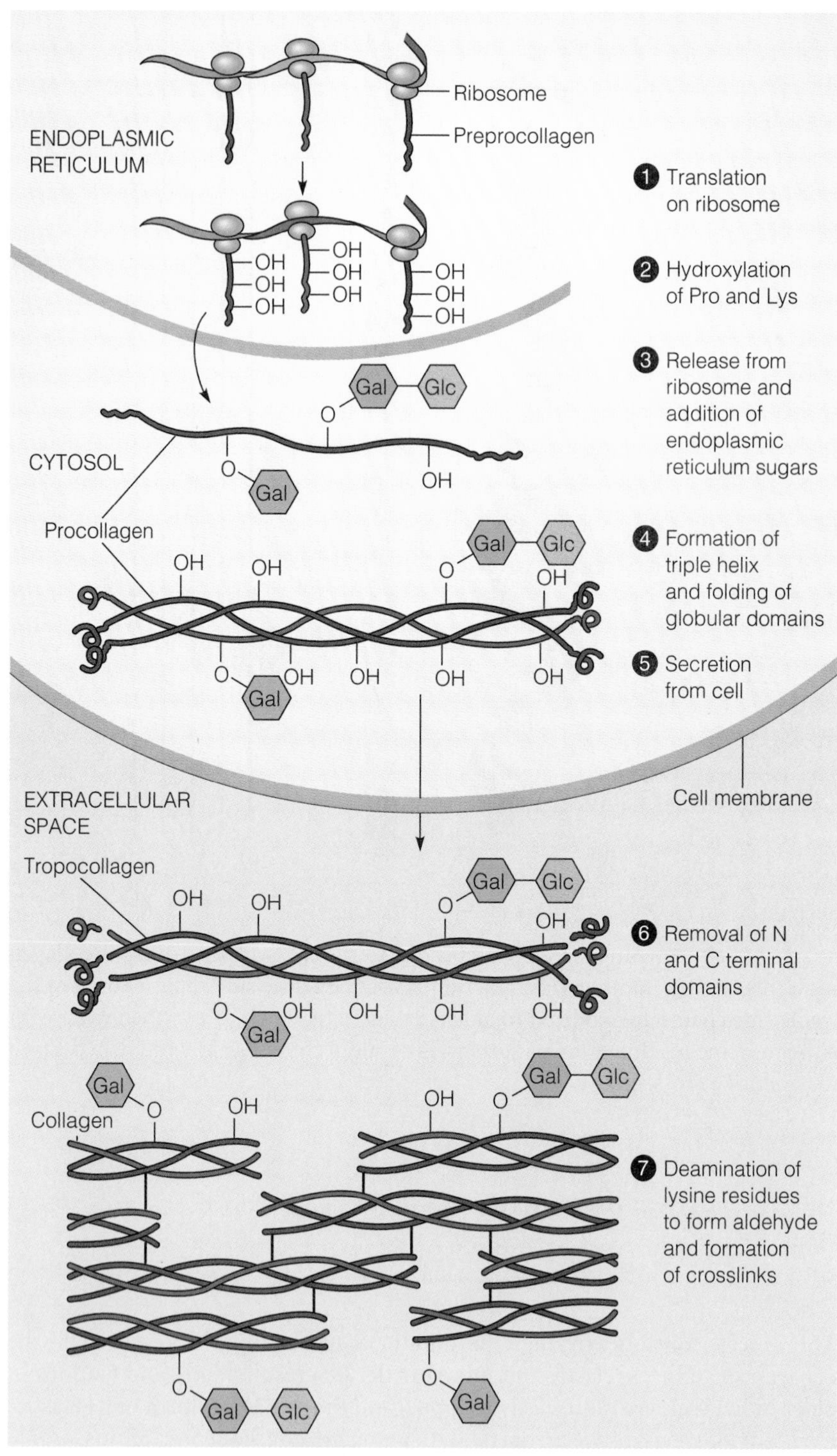

FIGURE **6.14**

Biosynthesis and assembly of collagen. The process can be visualized in several steps. Steps 1–4 occur in the endoplasmic reticulum and cytosol of collagen-synthesizing cells; steps 6 and 7 occur in the extracellular region. Gal = galactose, Glc = glucose.

almost 1500 residues, of which about 500 are in N-terminal and C-terminal regions that do not have the typical collagen fiber sequence described earlier. Three molecules of procollagen wrap their central regions into a triple helix, while the N- and C-terminal regions fold into globular protein structures. The procollagen triplexes are then exported into the extracellular space (step 5), at which point the N- and C-terminal regions are cleaved off by specific proteases, leaving only the tropocollagen triple helix, about 1000 residues long (step 6). These molecules then assemble into the staggered arrays shown in Figure 6.13c. Finally, cross-linking cements the molecules together into a tough collagen fiber.

ELASTIN

The protein elastin forms elastic fibers found in ligaments and blood vessels.

Collagen is found in tissues where strength or toughness is required, but some tissues, such as ligaments and arterial blood vessels, need highly elastic fibers. Such tissues contain large amounts of the fibrous protein **elastin.**

The polypeptide chain of elastin is rich in glycine, alanine, and valine and is very flexible and easily extended. In fact, its conformation probably approximates that of a random coil, with little secondary structure at all. However, the sequence also contains frequent lysine side chains, which can be involved in cross-links. These cross-links prevent the elastin fibers from being extended indefinitely, causing the fibers to snap back when tension is removed. Exactly the same principle accounts for the elasticity of vulcanized rubber, the flexible chains of which are also held by cross-links. The cross-links in elastin are rather different from those in collagen, for they are designed to hold several chains together. Four lysine side chains can be combined to yield a **desmosine** cross-link as shown in the margin. Because four separate chains are connected, only a small amount of such cross-linking is needed to convert elastin fibers into a highly interconnected, rubbery network.

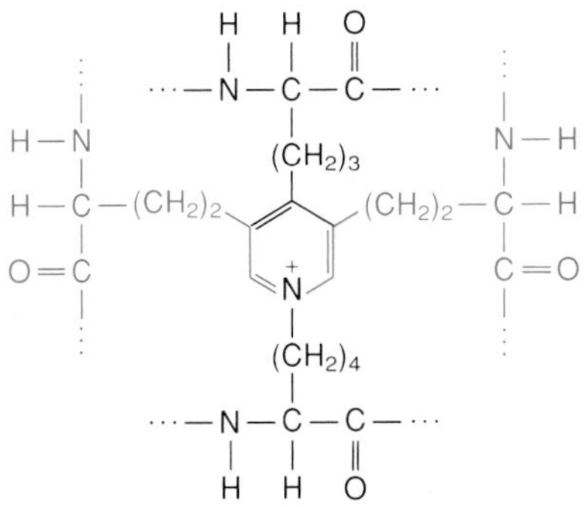

H H O
···—N—C—C—···
H—N (CH$_2$)$_3$ N—H
H—C—(CH$_2$)$_2$ (CH$_2$)$_2$—C—H
O=C N$^+$ C=O
(CH$_2$)$_4$
···—N—C—C—···
H H O

SUMMARY

This brief overview of a few of the structural proteins brings out several points. First, proteins can evolve to serve an almost infinite diversity of functions. Second, the structural fibrous proteins do this by taking advantage of the propensities of particular repetitive sequences of amino acid residues to favor one kind of secondary structure or another. Finally, posttranslational modification of proteins, including cross-linking, is an important adjunct in tailoring a protein to its function. We say more about the cellular sites for such modification in Chapters 27 and 28 when we consider the whole process of protein synthesis in detail.

These few examples do not exhaust the list of structural proteins. There are other, important ones, such as actin and myosin of muscle and tubulin in microtubules. But these proteins are constructed in a quite different way and are discussed in Chapter 8.

Globular Proteins: Tertiary Structure and Functional Diversity

DIFFERENT FOLDING FOR DIFFERENT FUNCTIONS

Abundant and essential as the structural proteins may be in any organism, they constitute only a small fraction of the *kinds* of proteins it possesses. Most of the chemical work of the cell—the synthesizing, transporting, and metabolizing—is carried out with the aid of an enormous class of **globular proteins.** These proteins are so named because their polypeptide chains are folded into compact structures very unlike the extended, filamentous forms of the fibrous proteins. Myoglobin (see Figure 6.1) is a typical globular protein. A glance at its three-dimensional structure, when compared with that of, say, collagen, immediately reveals this qualitative difference.

We now know a great deal about the structural details of many globular proteins, largely through the use of x-ray diffraction methods (see Tools of Biochemistry 4A). Often the resolution (the fineness of detail that can be discriminated) extends to less than 0.3 nm. This resolution is sufficient to identify individual amino acid residues, so the polypeptide chain can be traced through the molecule.

As it passes through the molecule, the polypeptide chain is often locally folded

Globular proteins not only possess secondary structures but also are folded into compact tertiary structures.

into one or another of the kinds of secondary structure (α helix, β sheet, and so forth) that we have already discussed. But to make the structure globular and compact, these regions must themselves be folded on one another. This folding is referred to as the tertiary structure of the protein; it is this folding that gives the molecule its overall three-dimensional shape. The distinction between secondary and tertiary structures can be clearly seen, for example, in the structure of myoglobin, as shown in Figure 6.1. About 70% of the myoglobin chain is α helix, and that helix is bent and folded to form a compact molecule. A pocket within this structure holds a **prosthetic group**, the heme. Many globular proteins carry prosthetic groups, small molecules that may be noncovalently or covalently bonded to the protein and enable the protein to fulfill special functions. In this case, the noncovalently bound heme group carries the oxygen binding site of myoglobin (see Chapter 7).

To show the great structural variation among globular proteins, let us consider another example. Figure 6.15 depicts one of the smallest and simplest globular proteins—bovine pancreatic trypsin inhibitor (BPTI). This protein, synthesized in the cow pancreas, is one of a number of proteins whose sole function is to bind to and inhibit proteolytic enzymes like trypsin. It is important to health, for it prevents autodigestion of the pancreas if trypsin is activated. We choose it as a beginning example because it is so simple that it has been studied in great detail. We shall use it as an example in many places.

BPTI is depicted in three ways in Figure 6.15. A skeletal model showing the positions of all atoms, which have been measured with high precision by x-ray diffraction, is given in Figure 6.15a. Figure 6.15b is a ribbon model of the chain backbone. In this model you can easily trace the path of the polypeptide chain and can clearly see both α helix and β sheet structures present in this molecule; also shown are the positions of the three disulfide bonds in BPTI. Finally, Figure 6.15c depicts a space-filling model of the molecule, illustrating the important point that globular proteins are tightly packed structures and are in fact often rather globular in shape.

Comparing Figure 6.15 with Figure 6.1, you can see that the structure of BPTI is wholly different from that of myoglobin. Whereas myoglobin is mostly α helix, BPTI has both helix and sheet regions, connected by bends in the chain. The point we wish to emphasize is this: Every globular protein has a unique tertiary struc-

FIGURE **6.15**

The tertiary structure of BPTI. Bovine pancreatic trypsin inhibitor, or BPTI, binds to trypsin and prevents it from catalyzing peptide hydrolysis. BPTI contains only 58 amino acid residues and is one of the most completely studied of all proteins. **(a)** A skeletal model, showing atomic positions from x-ray diffraction. **(b)** A ribbon model of the backbone. There are two short α helices, near the bottom of the molecule; much of the rest is made of antiparallel β sheet. **(c)** A space-filling model, with all atoms given their van der Waals radii. In (a) and (b), disulfide bonds are shown in yellow.

Courtesy of P.-S. Ho.

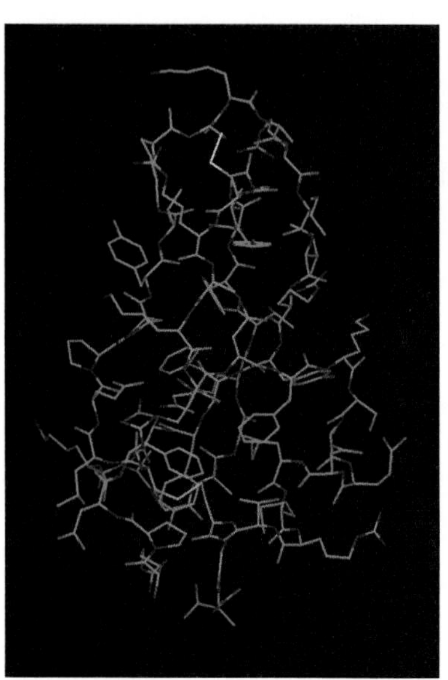

(a)

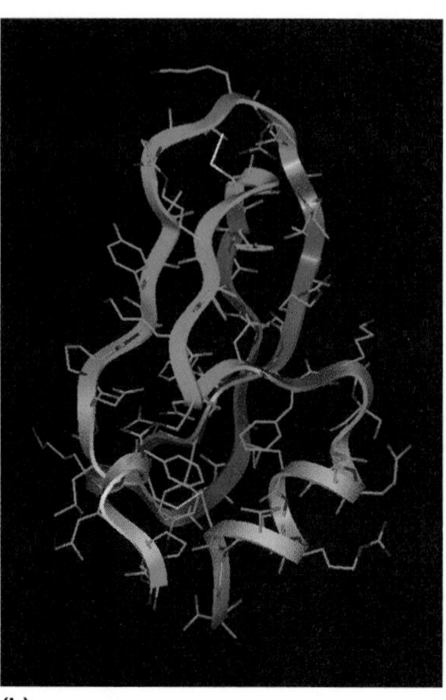

(b)

(c)

ture, made up of secondary structure elements (helices, β sheets, nonregular regions) folded in a specific way. As we examine proteins, we will find that each such conformation is suited to the particular functional role that the protein plays.

VARIETIES OF GLOBULAR PROTEIN STRUCTURE: PATTERNS OF FOLDING

At first glance, it might seem that there would be an almost infinite number of ways in which globular proteins could fold. If we examine every possible detail of folding, this is true. Yet when a large number of the known structures are examined, certain common motifs and principles emerge. The first principle is that many proteins are made up of more than one domain. A **domain** is a compact, locally folded region of tertiary structure. Domains are interconnected by the polypeptide strand that runs through the whole molecule. Multiple domains are especially common in the larger globular proteins, whereas very small proteins like BPTI tend to be single folded domains. As we shall see in later sections, different domains often perform differing functions, and a given domain type can sometimes be recognized in several different proteins.

Among the domain varieties, several distinct classes have been recognized by Jane Richardson (see References). The major folding patterns are of two kinds: those that are built about a packing of α helices, and those that are constructed on a framework of β sheet structures. A gallery of examples is shown in Figure 6.16. A detailed discussion of the differences between different classes of structures is

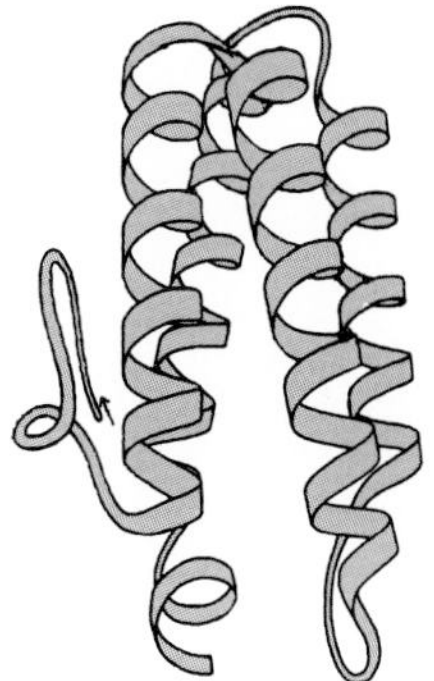

Myohemerythrin

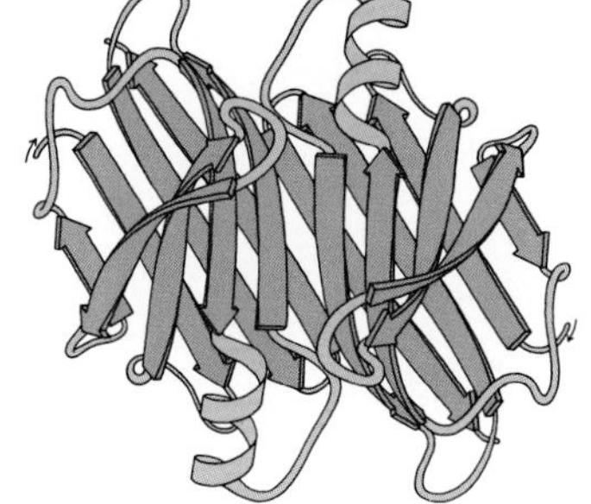

Prealbumin

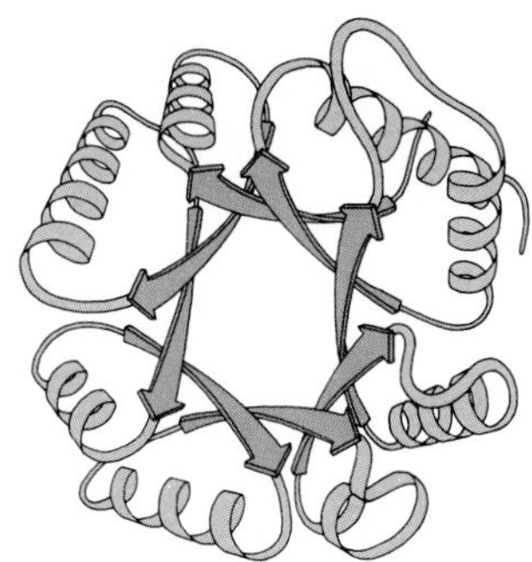

Pyruvate kinase, domain 1

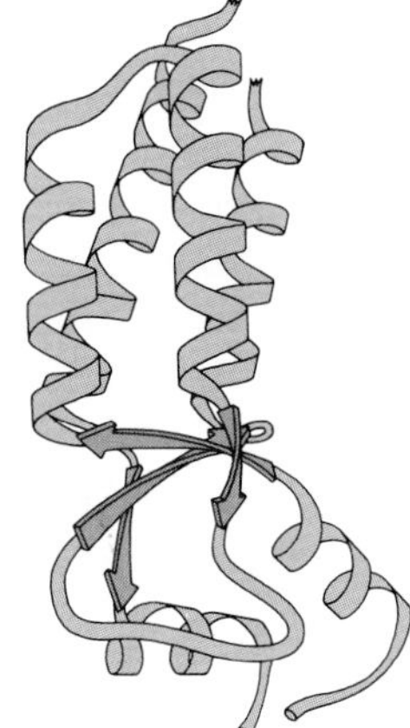

Tobacco mosaic coat protein

(a) Predominantly α helix

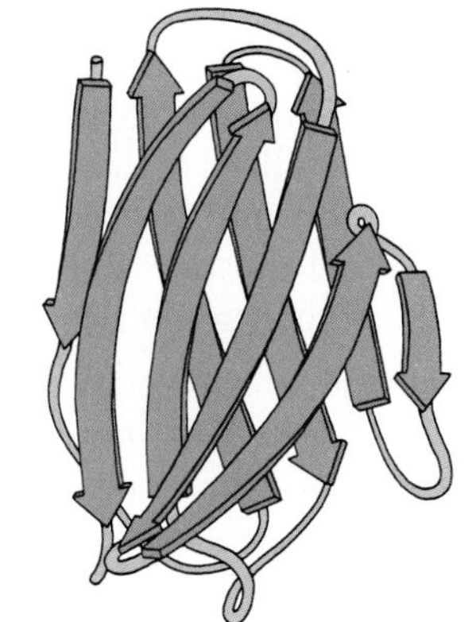

Immunoglobulin, V_2 domain

(b) Predominantly β sheet

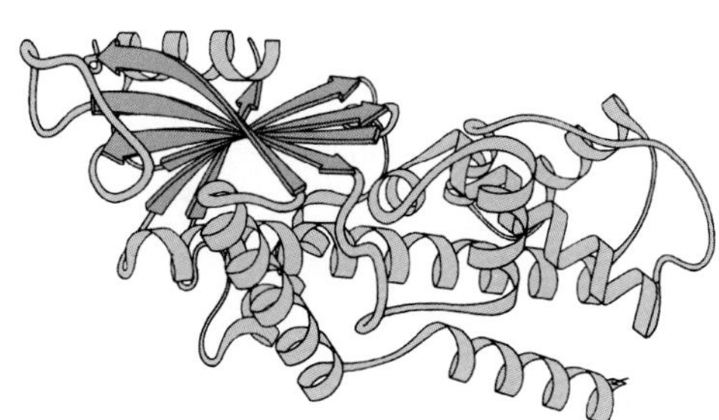

Hexokinase, domain 2

(c) Mixed α helix and β sheet

FIGURE **6.16**

Type structures of globular proteins. Blue indicates α helices; orange, β sheets; and green, irregularly structured regions. Arrowheads at the ends of β strands point in the N $\longrightarrow$ C direction. **(a)** Predominantly α helix (helix bundles). Top: Myohemerythrin. Bottom: Tobacco mosaic virus coat protein. **(b)** Predominantly β sheet. Top: Prealbumin (a dimer). Bottom: Immunoglobulin, V_2 domain. Prealbumin (also called transthyretin) could be described as a β sandwich, whereas the immunoglobulin V_2 domain is an antiparallel β barrel. **(c)** Mixed α helix and β sheet. Top: Pyruvate kinase, domain 1. This contains an excellent example of a parallel β barrel, seen here from the top. It is also a common sequence type termed α/β, in which α and β regions alternate. Bottom: Hexokinase, domain 2. This contains a clear example of a twisted β sheet, seen from the side.

Courtesy of Dr. Jane Richardson.

1 10 20 30 40 50
GCVEKGKKIFVQKCAQCHTVEKGGKHKTGPNLHGLFGRKTGQAPGFTYTD

ANKNKGITWKEETLMEYLENPKKYIPGTKMIFAGIKKKTEREDLIAYLKKATNE
60 70 80 90 100

(a)

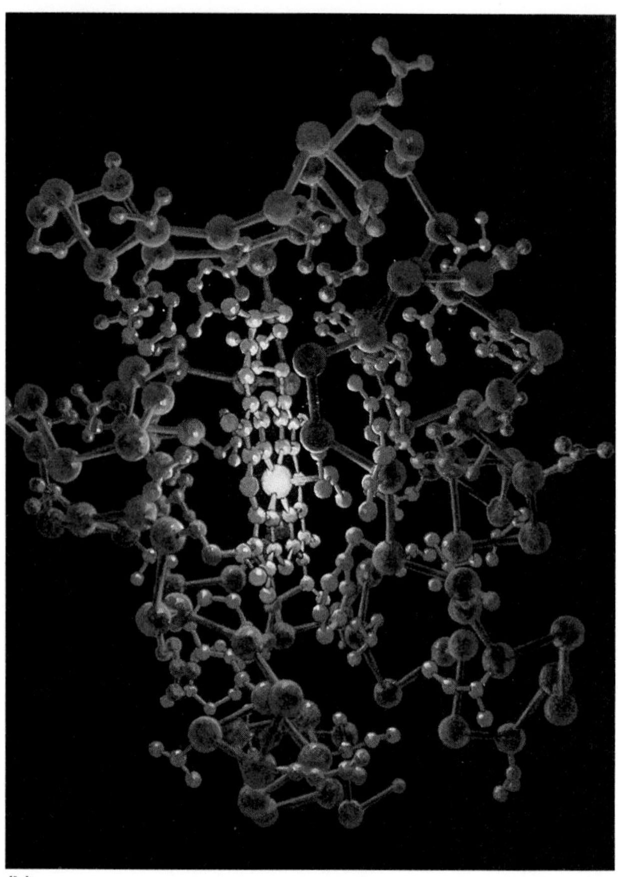

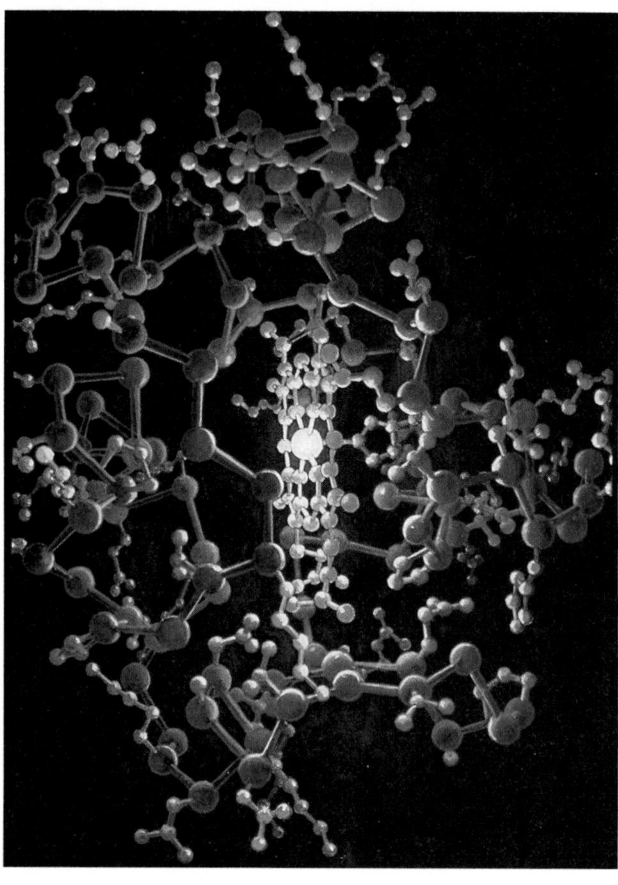

(b)

FIGURE **6.17**

The distribution of hydrophilic and hydrophobic residues in globular proteins. **(a)** The amino acid sequence of horse heart cytochrome *c*. Hydrophobic (red), hydrophilic (green), and ambivalent (white) residues appear to be scattered throughout the sequence. **(b)** The three-dimensional structure of the same protein. Left: Note how the hydrophobic amino acids (shown in red) cluster about the heme and on the inside of the molecule. Right: In a different view of the same molecule, the hydrophilic residues are shown in green. Note how they tend to lie on the molecular surface.

(b) © Irving Geis.

given by Branden and Tooze (see References). Analysis of the structures of hundreds of globular proteins has led to the formulation of certain general rules governing tertiary folding:

- *All globular proteins have a defined inside and outside.* If we examine the amino acid sequences of globular proteins, we find no particular distribution pattern of hydrophobic or hydrophilic residues (Figure 6.17a). But when we look at the positions of the amino acids in the three-dimensional structure, we invariably find that tertiary structure causes hydrophobic residues to be packed mostly on the inside, whereas the hydrophilic residues are on the surface, in contact with water (Figure 6.17b).
- *β sheets are usually twisted, or wrapped into barrel structures.* Clear examples can be seen in Figure 6.16. In most cases, twisted β sheets exhibit a twist that is left-handed, when viewed from a direction perpendicular to the chain axes. (For example, see hexokinase, domain 2.) It has been argued that this tendency of β sheets to left-hand twist is a consequence of the L configuration of the amino acid residues. It is probable that the structure of silk fibroin is not exactly planar, as depicted in Figure 6.12, but slightly twisted.
- *The polypeptide chain can turn corners in a number of ways,* to go from one β segment or α helix to the next. One kind of compact turn is called a β turn

(Figure 6.18). There are several varieties of β turn, each able to accomplish a complete reversal of the polypeptide chain direction in only four residues; in each case the carbonyl of residue i hydrogen-bonds to the amide hydrogen of residue $i + 3$. In the even tighter γ turn, bonding is to residue $i + 2$ (Figure 6.19). Proline often plays a role in turns, as in Figure 6.19, and also as a breaker of α helices, because this residue cannot be easily accommodated in the helix. Bends and turns most often occur at the surface of proteins.

- *Not all parts of globular proteins can be conveniently classified as helix, β sheet, or turns.* Examination of Figure 6.16, for example, reveals many strangely contorted loops and folds in the chains (the regions shown in green). These have sometimes been referred to as "random coil" regions, but this term is a misnomer, because such sections of the chain are not flexible in the same way that a true random coil is (see page 103 in Chapter 4). Rather, each such region has its own particular folding, exactly the same in each example of the particular protein molecule. We know of this regularity because x-ray diffraction studies show the same arrangement in all molecules in the crystal. We might call these *irregularly structured regions.* A few proteins also have true random-coil regions, which are usually found at the N- or C-terminus. When this happens, the random coil portion cannot be resolved by x-ray diffraction.

  In some proteins the folding is dominated by the need to bind a prosthetic group. Myog‚obin is an example. Although myoglobin might be roughly described as an α helix bundle, the tertiary structure of this protein has been distorted to form a hydrophobic cage about the heme group.

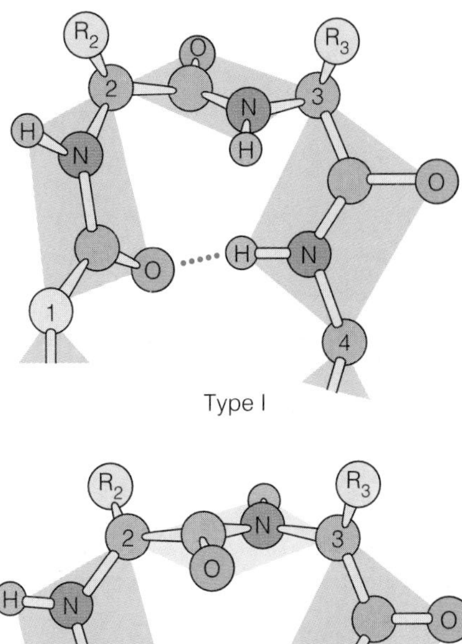

FIGURE **6.18**

Examples of β turns. Each of these turn types allows an abrupt change in polypeptide chain direction. In the type II turn, residue 3 is usually glycine, presumably because a bulky R group would clash with the carbonyl oxygen of residue 2.

Factors Determining Secondary and Tertiary Structure

THE INFORMATION FOR PROTEIN FOLDING

What ultimately determines the complex mixture of secondary and tertiary folding that characterizes each globular protein? Much evidence indicates that *most of the information for determining the three-dimensional structure of a protein is carried in the amino acid sequence of that protein.* This can be demonstrated by experiments in which the *native,* or natural, three-dimensional structure is broken up by changing the environmental conditions. If we raise the temperature sufficiently, or make the pH extremely acidic or alkaline, or add to the solvent certain kinds of organic molecules, such as alcohols or urea, the protein structure will unfold (Figure 6.20a). As with nucleic acids, this process is called **denaturation,** because the natural structure of the protein has been lost, along with many of its specific properties. The unfolded chain is a *random coil,* with freedom of rotation about bonds in both the polypeptide backbone and the side chains. It no longer has a single compact conformation but continually fluctuates among a large number of extended conformations. Differences between the native and random-coil conformations can be detected by several of the methods discussed in Tools of Biochemistry 5A.

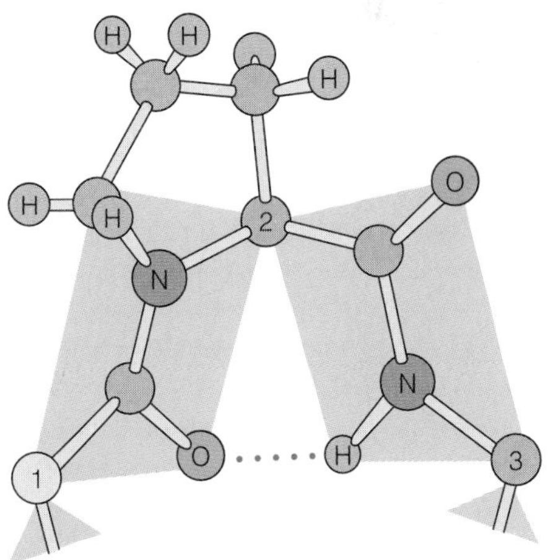

FIGURE **6.19**

A γ turn. Only one residue is out of the hydrogen bonding sequence. In this case it is a proline, which cannot make such a bond in any case.

In the classic experiment by C. Anfinsen depicted in Figure 6.20, the enzyme ribonuclease has been denatured by raising the temperature. Ribonuclease is a digestive enzyme that catalyzes the hydrolysis of ribonucleic acids. When ribonuclease is denatured, its tertiary and secondary structures are lost, and it can no longer act as a catalyst for the cleavage of RNA. The process can be tracked by various physical measurements, as shown in Figure 6.20b. Remarkably, this total scrambling of ribonuclease structure is fully reversible. If ribonuclease is restored to physiological conditions of temperature and solvent, it will spontaneously

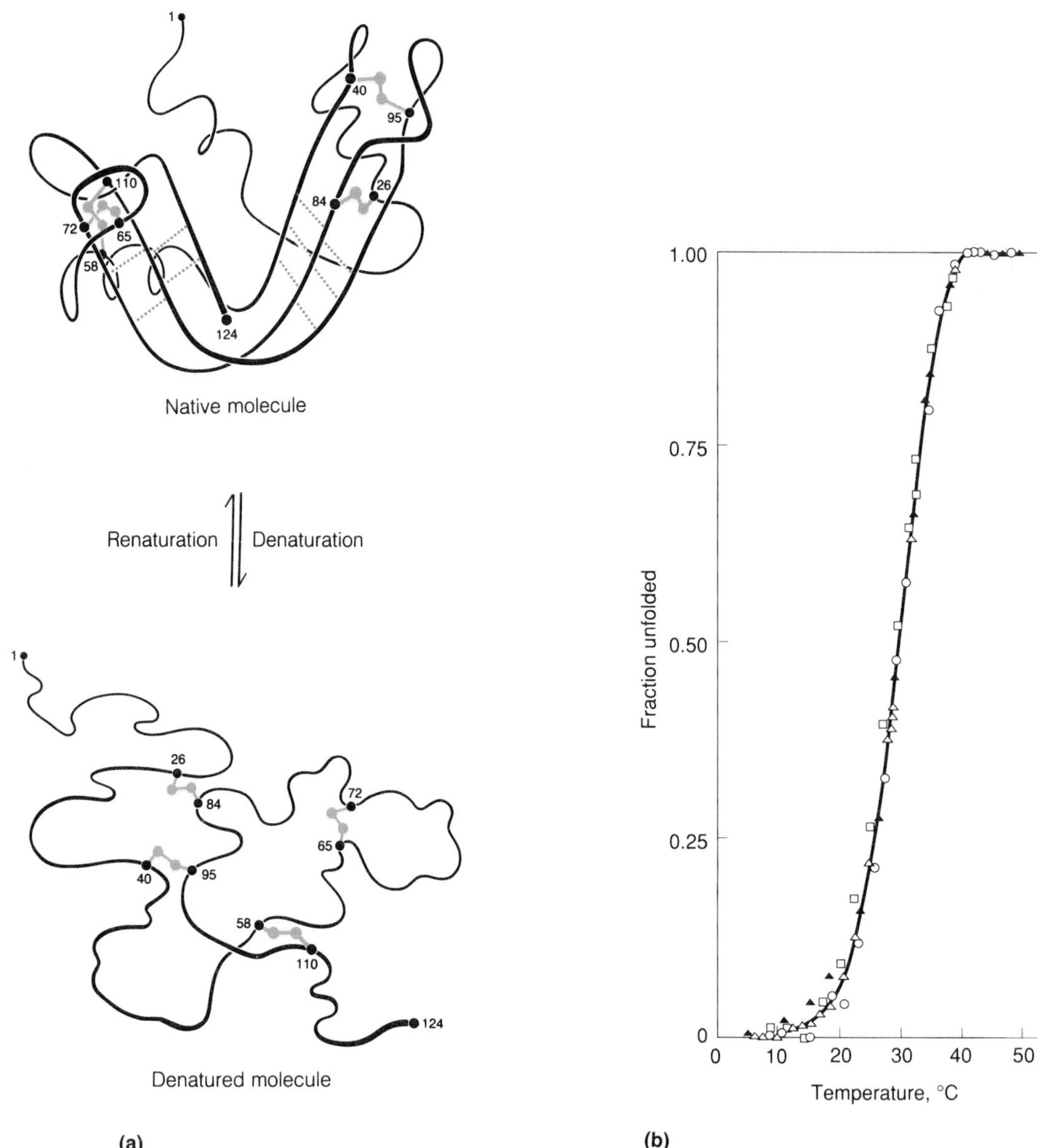

FIGURE **6.20**

The thermal denaturation of ribonuclease. (a) This schematic drawing depicts the conformational change to a disordered structure when ribonuclease is heated above its thermal denaturation temperature. Note that the disulfide bonds are still intact, providing some limitation to conformational flexibility. **(b)** Denaturation of ribonuclease followed by various physical methods. This graph shows the fraction denatured, as measured by the increase in solution viscosity (□), change in optical rotation at 365 nm (○), and change in UV absorbance at 287 nm (△). All three techniques indicate the same fraction unfolding. Measurements of a second denaturation after cooling (▲) produce the same curve, showing that the process is reversible. The experiments were conducted at pH 2.1, ionic strength 0.019 M. Under physiological conditions, ribonuclease is much more stable, not denaturing until about 70–80°C.

(b) Data from A. Ginsburg and W. R. Carroll, *Biochemistry* (1965) 4:2159–2174.

refold into its original structure. Furthermore, it will once again be a functional enzyme. Thus, the protein "knows" its own favored conformation; it needs no information to guide it other than that contained in its sequence. The same phenomenon has been observed experimentally for many other proteins. Thus, we might expect a newly synthesized polypeptide in a cell to spontaneously fold into the proper conformation, ready to take up its role. As we shall see later in this chapter, the actual processes in vivo may sometimes be more complicated and proteins may need "help" to avoid pitfalls of misfolding or aggregation. However, the basic principle of self-assembly of secondary and tertiary structure seems to be the general rule.

Amino acid sequence (primary structure) determines secondary and tertiary structure.

THE THERMODYNAMICS OF FOLDING

The folding of a globular protein is clearly a thermodynamically favored process under physiological conditions. In other words, the overall free energy change on folding must be negative. But this negative free energy change is achieved by a balance of several thermodynamic factors, which we will now describe.

Conformational Entropy

The folding process, which involves going from a multitude of random-coil conformations to a *single* folded structure, involves a decrease in randomness and thus a decrease in entropy. This change is termed the **conformational entropy** of folding. The free energy equation, $\Delta G = \Delta H - T\Delta S$, shows that this negative ΔS makes a *positive* contribution to ΔG. In other words, the conformational entropy change works *against* folding. To seek the explanation for an overall negative ΔG, we must seek features of protein folding that yield either a large negative ΔH or some other *increase* in entropy on folding. Both can be found.

The major source for a negative ΔH is energetically favorable interactions between groups within the folded molecule. These include many of the noncovalent interactions described in Chapter 2.

The decrease in conformational entropy when a protein folds disfavors folding. This is compensated in part by energy stabilization through internal noncovalent bonding.

Charge–Charge Interactions

Charge–charge interactions can occur between positively and negatively charged side chain groups. For example, a lysine side chain amino group may be placed close to the γ-carboxyl group of some glutamic acid residue. At neutral pH, one group will be charged positively and the other negatively, so an electrostatic attractive force exists between them. We could say that such a pair forms a kind of salt within the protein molecule; consequently, such interactions are sometimes called **salt bridges.** These ionic bonds are broken if the protein is taken to pH values high enough or low enough that either side chain loses its charge. This loss of salt bridges is a partial explanation for acid or base denaturation of proteins. The mutual repulsion between pairs of the numerous similarly charged groups that are present in proteins in very acidic or basic solutions contributes further to the instability of the folded structure under these conditions.

Internal Hydrogen Bonds

Many of the amino acid side chains carry groups that are either good hydrogen bond donors or good acceptors. Examples are the hydroxyls of serine or threonine, the amino groups and carbonyl oxygens of asparagine or glutamine, and the ring nitrogens in histidine. Furthermore, if amide protons or carbonyls in the polypeptide backbone are not involved in secondary structure formation, they are potential candidates for interaction with side chain groups. A network of several types of internal hydrogen bonds is seen in the portion of the molecule of the enzyme lysozyme shown in Figure 6.21. As we have seen before, hydrogen bonds are relatively weak in aqueous solution, but their large number can add a considerable contribution to stability.

van der Waals Interactions

The weak interactions between uncharged molecular groups can also make significant contributions to protein stability. As space-filling models of proteins like Figure 6.15c demonstrate, the interior is tightly packed, allowing maximum contact between side chain atoms.

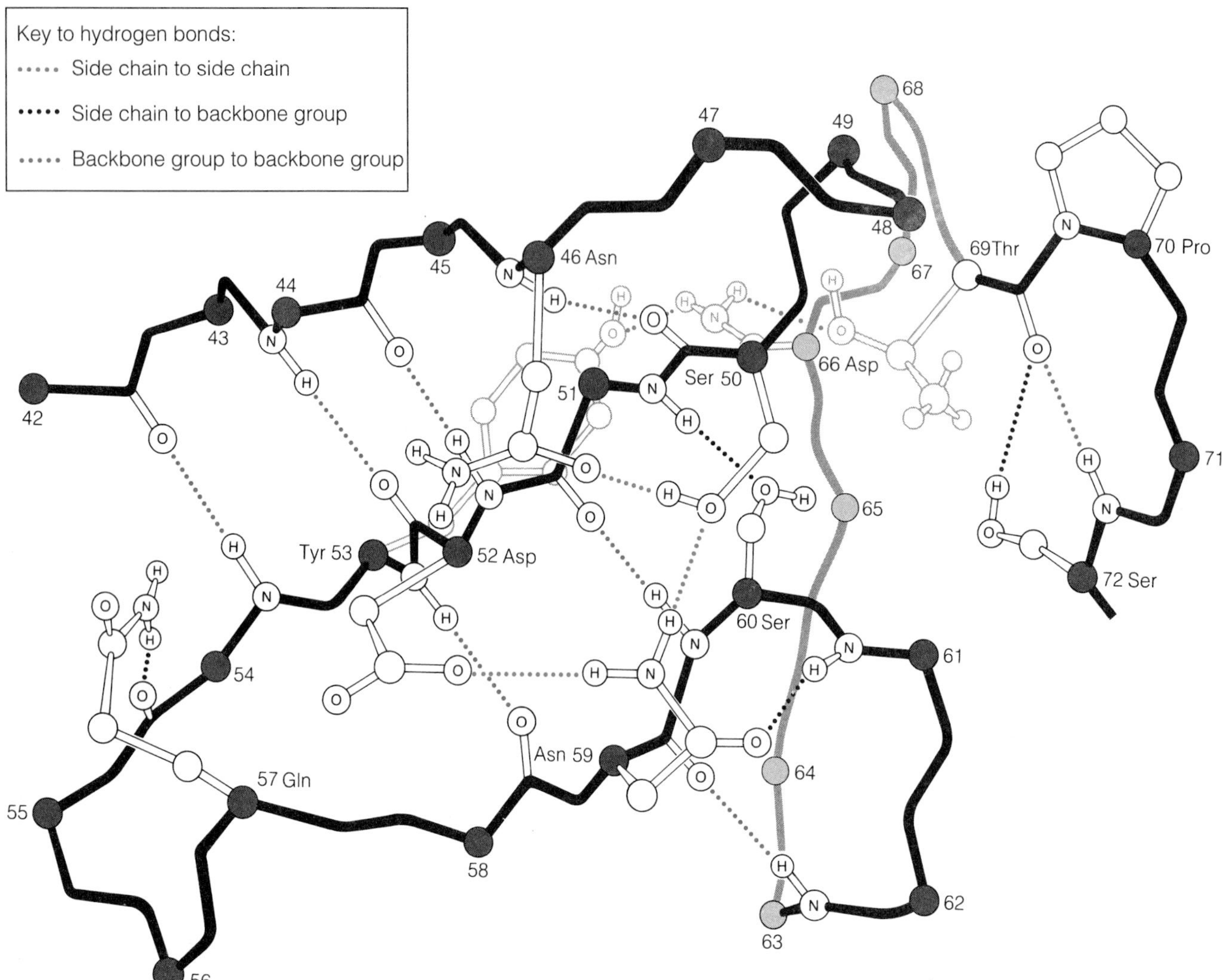

FIGURE **6.21**

Details of hydrogen bonding in a typical protein. A network of hydrogen bonds within the enzyme lysozyme is illustrated. Three kinds of H bonds are distinguished: bonds between side chain groups, bonds between backbone groups, and bonds between a side chain group and a backbone amide hydrogen or carbonyl oxygen.

The contribution of any of these interactions to the negative enthalpy of folding is diminished by the fact that when a protein molecule folds from an expanded random coil, it has to give up some favorable interactions with water. As we pointed out in Chapter 2, an unfolded α helix forms hydrogen bonds with water instead of forming internal bonds. It is not clear how much this exchange modifies the overall energy change. However, we do know that if internal H bonds were *not* formed when H-bond acceptors or donors were buried by folding, a substantial energy difference in favor of the unfolded state would result.

Each individual interaction can contribute only a small amount (at most only a few kilojoules) to the overall negative enthalpy of interaction. But the sum of the contributions of many interactions can yield powerful stabilization to the folded structure. Examples of the total enthalpy changes for folding are given for some representative proteins in Table 6.3. In many cases a favorable energy contribution from the sum of intramolecular interactions more than compensates for the unfavorable entropy of folding.

The burying of hydrophobic groups within a folded protein molecule produces a stabilizing entropy increase known as the hydrophobic effect.

The Hydrophobic Effect

Yet another factor makes a major contribution to the thermodynamic stability of many globular proteins. Recall from Chapter 2 that hydrophobic substances in contact with water cause the water molecules to form clathrate, or cagelike, structures around them. This ordering corresponds to a loss of randomness in the

TABLE 6.3 Thermodynamic parameters for folding of some globular proteins at 25°C in aqueous solution

Protein	ΔG (kJ/mol)	ΔH (kJ/mol)	ΔS (J/K·mol)
Ribonuclease	−46	−280	−790
Chymotrypsin	−55	−270	−720
Lysozyme	−62	−220	−530
Cytochrome *c*	−44	−52	−27
Myoglobin	−50	0	+170

Note: Data adapted from P. L. Privalov and N. N. Khechinashvili, *J. Mol. Biol.* (1974) 86:665–684. Each data set has been taken at the pH value where the protein is maximally stable; all are near physiological pH. Data are for the folding reaction: Denatured $\rightleftharpoons$ native.

system; the entropy is decreased. Suppose a protein contains, in its amino acid sequence, a substantial number of residues with hydrophobic side chains (for example, leucine, isoleucine, and phenylalanine). When the polypeptide chain is in an unfolded form, these residues are in contact with water and cause ordering of the surrounding water structure. But when the chain folds into a globular structure, the hydrophobic residues become buried within the molecule (see Figure 6.17b). The water molecules that were ordered by the protein when it was in the denatured state are now released, gaining freedom of motion. Thus, internalizing hydrophobic groups via folding increases the randomness of the whole system (protein plus water) and therefore yields an entropy *increase* on folding. This entropy increase makes a negative contribution to the free energy of folding and increases the stability of the protein structure.

The term *hydrophobic bonding* has sometimes been used to describe the stabilization resulting from the burying of hydrophobic groups, but this term is really a misnomer. The stabilization is not primarily the result of bonds forming between hydrophobic groups, although van der Waals interactions surely do occur. Rather, the overall system is stabilized by an entropy effect. The more appropriate term to describe this source of protein stabilization is **hydrophobic effect.** Examples of the importance of the hydrophobic effect can be seen in Table 6.3. The very small negative ΔS for cytochrome *c* and the positive value for myoglobin are a consequence of the hydrophobic effect in these proteins. Indeed, the stability of myoglobin comes mainly from the hydrophobic effect.

Different amino acid residues contribute very differently to the hydrophobic effect. Studies of the transfer of amino acids from water to organic solvents, as well as theoretical considerations, have led to the construction of various hydrophobicity scales. Two examples are shown in Table 6.4. The two scales, based on somewhat different premises, are not exactly the same. Nevertheless, such scales can be used to predict the importance of the hydrophobic effect in stabilizing any particular protein. In Chapter 10 we shall see how they can also be used to predict which portions of a protein will insert into a lipid membrane.

In summary, the stability of the folded structure of a globular protein depends on the interplay of three factors:

1. The unfavorable conformational entropy change, which favors random chains instead
2. The favorable enthalpy contribution arising from intramolecular side group interactions
3. The favorable entropy change arising from the burying of hydrophobic groups within the molecule

Thus, factor 1 works against folding, whereas 2 and 3 help stabilize folding. A schematic picture of the way these components contribute to the free energy of

TABLE 6.4 Two examples of hydrophobicity scales

Amino Acid	Scale of Engelman, Steitz, and Goldman[a]	Scale of Kyte and Doolittle[b]
Phe	3.7	2.8
Met	3.4	1.9
Ile	3.1	4.5
Leu	2.8	3.8
Val	2.6	4.2
Cys	2.0	2.5
Trp	1.9	−0.9
Ala	1.6	1.8
Thr	1.2	−0.7
Gly	1.0	−0.4
Ser	0.6	−0.8
Pro	−0.2	−1.6
Tyr	−0.7	−1.3
His	−3.0	−3.2
Gln	−4.1	−3.5
Asn	−4.8	−3.5
Glu	−8.2	−3.5
Lys	−8.8	−3.9
Asp	−9.2	−3.5
Arg	−12.3	−4.5

[a]Data from Engelman, D. M., T. A. Steitz, and A. Goldman, *Annu. Rev. Biophys. & Biophys. Chem.* (1986) 15:321–353.

[b]Data from Kyte, J., and R. F. Doolittle, *J. Mol. Biol.* (1982) 157:105–132.

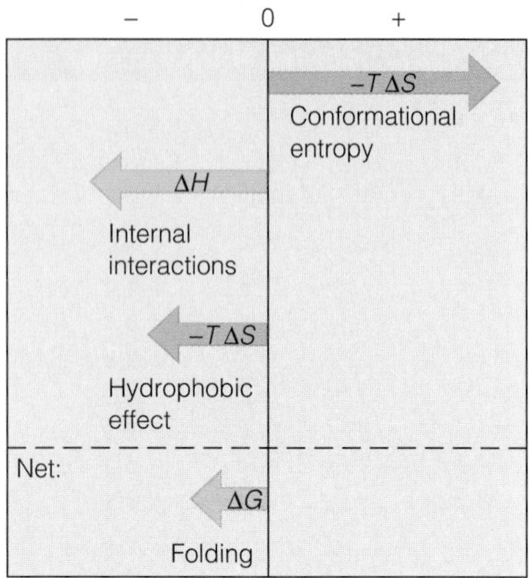

FIGURE **6.22**

Contributions to the free energy of folding of globular proteins. The conformational entropy change works against folding, but the enthalpy of internal interactions and the hydrophobic entropy change favor folding. Summing these quantities can make the total free energy of folding negative, so that the folded structure is stable.

folding is given in Figure 6.22. In different proteins, the apportionment of stabilization between side chain interactions and the hydrophobic effect differs (see Table 6.3), but the overall consequence is the same: Some particular folded structure corresponds to a free energy minimum for the polypeptide under physiological conditions. This is why the chains fold spontaneously.

Examination of the data in Table 6.3 reveals another important aspect of protein stabilization. The relatively small ΔG corresponding to the folding reaction is usually the difference between large ΔH and $T\Delta S$ terms. These large values arise because a protein tends to fold in a *cooperative* manner. A partially folded mixture contains mainly wholly unfolded and wholly folded molecules, with few intermediate structures. Thus, many folding interactions within a molecule form or break in concert. As a consequence of this cooperativity, the thermal denaturation of a protein usually occurs over a narrow temperature range, as is shown in Figure 6.20b. The data further illustrate the principle of **entropy–enthalpy compensation.** Although ΔH and ΔS values vary widely, they seem to be adjusted so that ΔG stays fairly constant. Although stability is required for protein function, it can be attained by many different combinations of contributions to ΔH and ΔS; evolution selects not for the values of these individual parameters but for some *combination* of values that yields stability at the temperature of the organism. Thermophilic organisms, such as those that live in hot springs, have proteins that are more stable at higher temperatures than those of organisms living in more temperate environments.

THE ROLE OF DISULFIDE BONDS

Some folded proteins are stabilized by internal disulfide bonds, in addition to noncovalent forces.

Once folding has occurred, the three-dimensional structure is in some cases further stabilized by the formation of disulfide bonds between cysteine residues. An extreme example of this bonding is found in the bovine pancreatic trypsin inhibitor (BPTI) protein, depicted in Figure 6.15. With three —S—S— bridges in 58 residues, this molecule is one of the stablest proteins known. It is quite inert to unfolding reagents like urea and exhibits thermal denaturation below 100°C only in very acidic solutions; the half-point for reversible denaturation is about 80°C at pH 2.1 (Figure 6.23). But if only *one* of the disulfide bonds (that between cysteine residues 14 and 38, Figure 6.15b) has been reduced and carboxymethylated, the midpoint is decreased to 59°C.

When all the disulfide bonds in BPTI are reduced, the protein is unfolded at room temperature. Yet upon reoxidation of the sulfhydryls, native protein with the three correct disulfide pairings is efficiently formed. This re-formation is not what would be expected by chance. Suppose a BPTI molecule has been reduced, yielding 6 cysteine residues, and we now randomly reoxidize the —SH groups. The first —SH group to pick a partner will have 5 choices, the second group 3, and the last only 1, so there are 5 × 3, or 15, equally probable combinations. Thus, we would expect only about 7% of reduced BPTI to refold successfully *by chance.* But many studies of this and other proteins containing disulfide bonds indicate that correct pairing is regained in almost 100% of the molecules if sufficient time is allowed. This finding must mean that it is the preferred folding of the protein that places the —SH groups in position for correct pairing. The corollary of this statement is that the —S—S— bridges are not themselves essential for correct refolding. They do, however, contribute to the stability of the structure once it is folded. A molecule containing —S—S— bridges has a smaller number of conformations available in the unfolded form than does a comparable protein without the bridges. Consequently it shows a smaller entropy gain on unfolding and is therefore stabilized.

This advantage raises a question: Why don't *most* proteins have disulfide bonds? In fact, such bonds are relatively rare and are found primarily in proteins that are

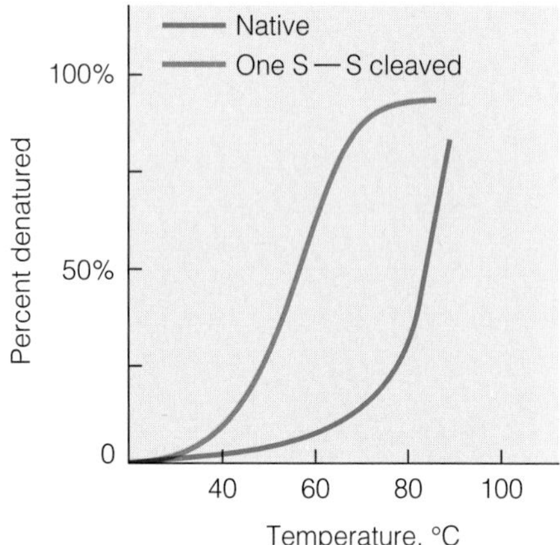

FIGURE **6.23**

Thermal denaturation of BPTI. The percent denaturation as a function of temperature at pH 2.1 is indicated for the native protein and for the protein in which the Cys 14–Cys 38 disulfide bond has been reduced and carboxymethylated.

Data from J. P. Vincent, R. Chicheportiche, and M. Lazdunski, *Eur. J. Biochem.* (1971) 23:401–411.

exported from cells, such as ribonuclease, BPTI, and insulin. One explanation is that the environment inside most cells is reducing and tends to keep sulfhydryl groups in the reduced state. External environments, for the most part, are oxidizing and stabilize —S—S— bridges.

A powerful new method for studying the sources of protein structural stabilization has been developed from recombinant DNA techniques. Now that the genes for proteins can be cloned and expressed in bacterial carriers, it is possible to make specific changes at desired positions. This method of **site-directed mutagenesis** (see Tools of Biochemistry 25E) allows us to test the effect of changing one or more amino acid residues or adding or removing disulfide bonds. The method has already been used, for example, to delete specific disulfide bonds in BPTI by replacing cysteines with serines or alanines.

Dynamics of Globular Protein Structure

The structural and thermodynamic descriptions we have developed so far tend to give too static a picture of globular proteins. The precise pictures produced by x-ray diffraction give the impression of a rigid, inflexible structure. Likewise, the thermodynamic analysis of folding concentrates on the initial (unfolded) and final (folded) states. But it is now recognized that globular proteins fold via complex *kinetic* pathways and that even the folded structure, once attained, is a dynamic structure. This section explores some of these dynamic aspects of globular proteins.

KINETICS OF PROTEIN FOLDING

Protein folding can be rapid but seems to involve well-defined intermediate states.

The folding of globular proteins from their denatured conformations is a remarkably rapid process, often complete in less than a second. This observation has been a continual source of surprise and interest to biochemists, for at first glance, the attainment of the very precise structure typical of globular proteins would seem to be very difficult. This point of view was dramatically expressed in "Levinthal's paradox," first enunciated by Cyrus Levinthal in 1968: A rough estimate would say that about 10^{50} conformations are possible for a polypeptide chain such as ribonuclease (124 residues). Even if the molecule could try a new conformation every 10^{-13} second, it would still take about 10^{30} years to try a significant fraction of them! Yet ribonuclease is experimentally observed to fold, in vitro, in about 1 minute. Clearly, something is wrong with such a calculation.

Years of experimental study and theoretical speculation have begun to resolve the paradox. First, it became clear from rapid kinetic studies, using a variety of physical techniques to monitor different aspects of protein structure, that folding takes place through a series of intermediate states; these observations led to the "pathway" model of folding, depicted as follows:

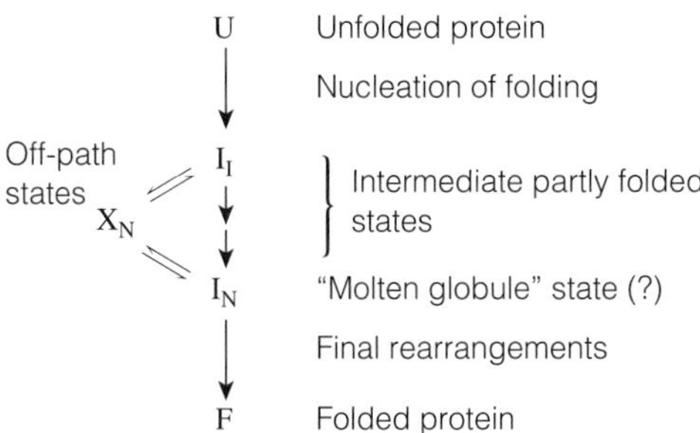

The nucleation step is critical, because it is much more difficult to begin an α helix, for example, than to extend it (note that at least four residues must fold

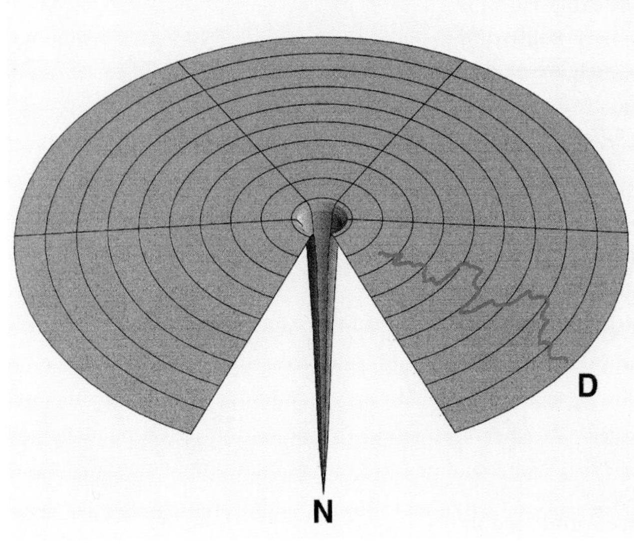

(a)

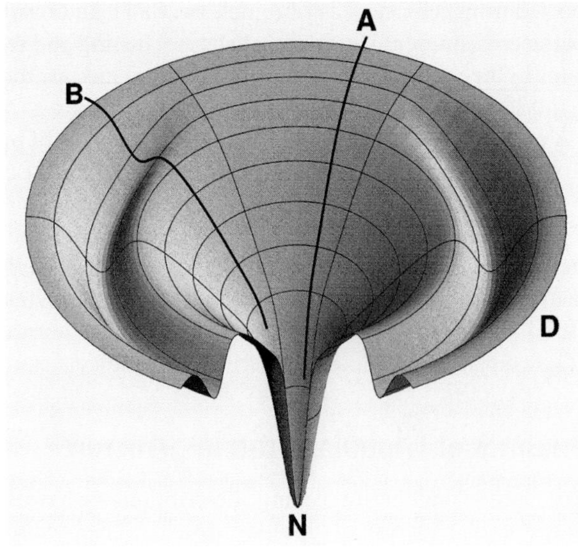

(b)

FIGURE **6.24**

Energy surfaces to visualize protein conformations. (a) The Levinthal "golf-course" energy landscape. **N** is the native conformation and **D** the denatured state. The chain searches for **N** randomly, that is, on a level playing field of energies, and has very little chance of finding the tiny hole leading to the native state. **(b)** A more realistic energy landscape, which tends to "funnel" the protein toward the native state. Many paths are possible, some of which lead straight "downhill" like **A,** and give rapid folding. Others, like **B,** may lead to secondary energy minima, which may delay folding. Real energy landscapes are probably much more complex funnels.

Courtesy of K. Dill and H. S. Chan, *Nature Struct. Biol.* (1997) 4:10–19.

properly to make the first stabilizing H bond.) On the other hand, it is now recognized that nucleation may start at a number of points and that all of these partially folded structures will be "funneled" by energy minimizations toward the final state. The funneling model proposes that there is not just one but *many* possible paths from the denatured state to the folded state, and each path leads downhill in energy. Thus, any one molecule need sample only an infinitesimal fraction of the total configurations possible, and Levinthal's paradox is averted (see Figure 6.24).

During the descent toward the free energy minimum, there may be pauses corresponding to metastable intermediates, as incorporated in the pathway model. An important folding intermediary for many proteins appears to be what is called the "molten globule" state, a compact structure in which much of the secondary and tertiary folding has occurred, but the internalized hydrophobic residues have not yet settled into their final packing.

Folding can be delayed by trapping of molecules in "off-path" states.

There is also evidence for "off-path" states—those in which some key element is incorrectly folded. Such states can correspond to local free energy minima in the funnel and may temporarily trap the protein (see Figure 6.24b). As we shall see later in this chapter, the cell has ways to assist incorrectly folded proteins to find their proper conformations. Larger, complex proteins may never achieve 100% folding success in vivo, but the cell can recognize those which are incorrect and dispose of them. One of the most common of these folding errors occurs via the incorrect *cis–trans* isomerization of the amide bond adjacent to a proline residue:

cis ⇌ **trans**

Unlike other peptide bonds in proteins, for which the *trans* isomer is highly favored (by a factor of about 1000), proline residues favor the *trans* form in the preceding peptide bond by a factor of only about 4. Therefore, there is a significant chance that the "wrong" isomer (in terms of the functional protein conformation) will form first. Conversion to the correct configuration may involve a significant chain rearrangement and hence may be slow in vitro. There is now evidence

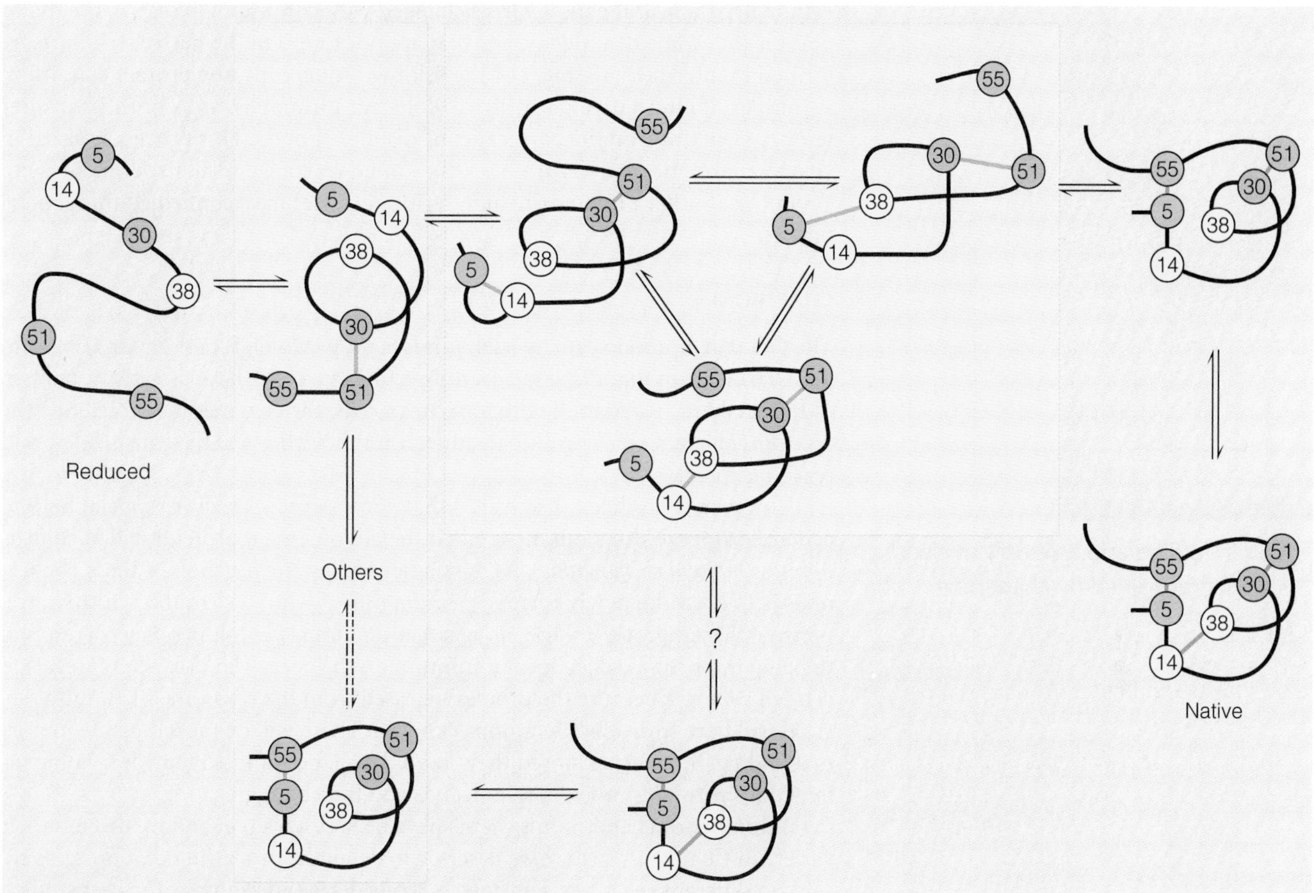

FIGURE **6.25**

Refolding and disulfide bond formation. This schematic drawing shows the postulated pathway for refolding and disulfide-bond formation in BPTI. Pairs of residues that form disulfide bonds with each other in the native structure are shown in the same color. Note that the intermediate structures may contain both native and nonnative disulfide bonds. Intermediates grouped together in colored boxes interconvert rapidly; the 5–55 intermediate interconverts more slowly than the other one-disulfide intermediates.

Adapted from D. P. Goldenberg, *Trends Biochem. Sci.* (1992) 17:257–261. © 1992 with permission of Elsevier Science.

that cells have enzymes to catalyze this *cis–trans* isomerization, and thus speed in vivo folding.

KINETICS OF DISULFIDE BOND FORMATION

If a protein is folded from a state in which disulfide bonds have been cleaved and must therefore be re-formed, the process is more complicated and slower—often taking many minutes. More detailed analysis is possible in these cases, because chemical methods can trap intermediates in which only some of the disulfide bonds have formed. These intermediates can then be separated by methods like gel electrophoresis or high-resolution chromatography. Trapping reactions are simple and fast; the blocking of unreacted sulfhydryls with iodocetate (see Tools of Biochemistry 5C) is one example.

The refolding process has been studied in great detail by using BPTI, which is suitable because it is small and contains three —S—S— bridges. Furthermore, when the bridges are reduced, BPTI is wholly unfolded at room temperature. Experiments begun in the laboratories of T. E. Creighton have provided a fascinating and surprising picture of this process, which is diagrammed in Figure 6.25.

Formation of the correct disulfide bonds in a protein is sometimes delayed by incorrect intermediates.

Although details of the model have changed with continuing research, the major point is that some disulfide bonds that are not found in the native structure are formed in intermediate stages of the folding. Apparently, the protein can utilize a number of alternative pathways to fold but ultimately finds both its proper tertiary structure and the correct set of disulfide bonds. This process is aided in vivo by enzymatic catalysis of —S—S— bond rearrangement. Experiments with a number of other proteins (including ribonuclease) have confirmed this general picture.

CHAPERONINS

The fact that a protein can, by itself, find its proper folded state in vitro does not necessarily mean that the same events occur in vivo. We have already noted two of the catalytic aids to folding that occur in cells, the enzymes that accelerate the *cis–trans* conversion at proline residues, and those that catalyze disulfide bond rearrangement.

Protein folding and assembly in vivo is sometimes aided by chaperone proteins.

However, it has been discovered that some proteins require special help to achieve proper folding and must enlist the aid of special proteins called *chaperonins* or *molecular chaperones.* As the name implies, the function of these chaperones is to keep the newly formed protein out of trouble. Trouble, in this case, means either improper folding or aggregation. Improper folding may correspond to being trapped in a deep local minimum on the energy landscape. Aggregation is often a danger because the protein, released from the ribosome in an unfolded state, will have hydrophobic groups exposed. These will be tucked inside in normal folding, but when exposed they stand the chance of making hydrophobic interactions with *other* polypeptide strands and thereby aggregating.

A number of chaperonin systems have been discovered, and we discuss their variety in Chapter 27, where we describe in detail the processing that proteins receive in different cell types. For now, we consider, as an example, the best studied of all chaperonins: the GroEL-ES complex from *E. coli.* The structure of this enormous and remarkable complex has recently been elucidated by x-ray diffraction (see Figure 6.26a). It consists of two basic portions—GroEL is made of two rings each consisting of seven protein molecules. The center of each ring is an open cavity, accessible to the solvent at the ends. Either cavity can be "capped" with GroES, again a seven-membered ring of smaller subunits.

Basically, the cavities provide "shelters" in which new protein chains can be "incubated" until they have folded properly. The GroEL-ES complex does not stipulate the folding pattern; that is up to the protein itself to do, as we have seen. But insulation from the environment prevents chances for aggregation or misfolding. The cycle experienced by a protein molecule is schematically shown in Figure 6.26b. The unfolded protein enters into a form of the cavity lined with hydrophobic residues. Then the cavity changes, presenting a hydrophilic lining. This releases the protein from the walls, and it folds and is then released. Note that ATP is required, presumably to drive the process in one direction.

Only a fraction of the proteins made in *E. coli* (or any cell) are processed via chaperonins. Some are too large to be accommodated within the cavity. Others can seemingly fold safely on their own. But the discovery of chaperonins tells us once again how wonderfully complex and subtle the molecular processes of life can be.

MOTIONS WITHIN GLOBULAR PROTEIN MOLECULES

Globular proteins are not static but continually undergo a wide variety of internal motions.

Much evidence indicates that various kinds of motions are continually occurring within folded protein molecules. A protein molecule undergoes continued, rapid fluctuations in its energy, as a consequence of interactions with its environment. The resulting motions can be roughly grouped into several classes, as shown in Table 6.5. Class 1 motions occur even within protein molecules in crystals and

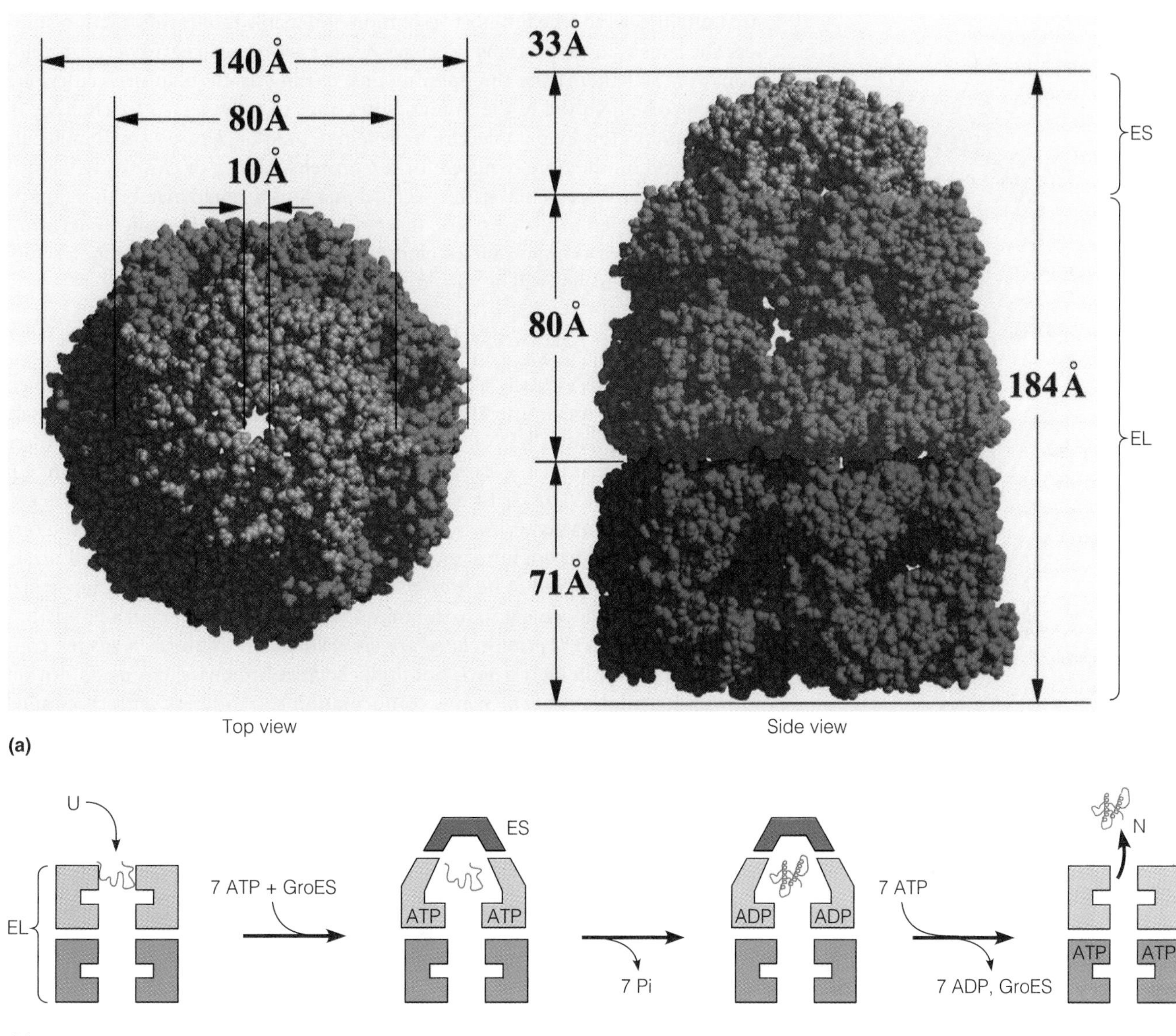

FIGURE **6.26**

The GroEL-GroES chaperonin. (a) X-ray diffraction structure of the EL·ES $(ADP)_7$ complex. **(b)** A schematic of function: U = unfolded protein, N = native protein.

(a) Courtesy of P. B. Sigler from Z. Xu, A. L. Horwich, and P. B. Sigler, *Nature* (1997) 388: 741–750. © 1997 Macmillan Magazines, Ltd.; **(b)** adapted from W. S. Netzer and F. U. Hartl, *Trends Biochem. Sci.* (1998) 23:68–73. © 1998 with permission from Elsevier Science.

TABLE **6.5 Motions in protein molecules**

		Approximate Range	
Class	Type of Motion	**Amplitude (nm)**	**Time (s)**
1	Vibrations and oscillations of individual atoms and groups	0.2	10^{-15}–10^{-12}
2	Concerted motions of structural elements, like α helices and groups of residues	0.2–1	10^{-12}–10^{-8}
3	Motions of whole domains; opening and closing of clefts	1–10	$\geq 10^{-8}$

account, in part, for the limits of resolution obtainable in x-ray diffraction studies. The larger, slower motions in classes 2 and 3 are more likely to occur in solution. Some of them, like the opening and closing of clefts in molecules, are thought to be involved in the catalytic functions of enzymes. As we shall see in later chapters, how long it takes for a protein to bind or release a small molecule may depend on the time required for the protein to open or close a cleft. Similarly, the protein "gates" that pass molecules and ions through membranes rapidly change from open to closed states. It seems likely that the dynamic behavior of proteins is at least as important in their function as the static details of their structure are. This behavior will be a fruitful field for future investigation.

PRIONS: PROTEIN FOLDING AND MAD COW DISEASE

Until very recently, virtually all researchers believed that the only ways in which diseases could be transmitted from one organism to another were via viruses or microorganisms. After all, DNA and RNA were the information carriers. But there is now evidence that there exists a class of diseases that is transmitted by a protein and nothing more. The best known of these diseases is *bovine spongiform encephalopathy,* or "mad cow disease," but they also include *scrapie* in sheep and certain neuropathologies in humans. The infectious agent has been termed *prion,* and the protein believed to be responsible is called *prion-related protein,* or *PrP.* The PrP is normally present in many animals (including humans) in a nonpathological form call PrP^c (*prion-related protein cellular*). In this form it has the conformation shown in Figure 6.27. But under certain circumstances, as yet not yet fully understood, PrP^c can change conformation to a different structure called PrP^{sc}, or *prion-related protein scrapie.* It is this form, in which the disordered N-terminal portion appears to at least partially fold into a β sheet, that wreaks havoc

FIGURE **6.27**

The prion protein. This model depicts NMR results for the structure of most of PrP^c. The globular, α-helical region lies at the C-terminus. Most of the N-terminal region appears to be a disordered random coil. It is believed that it is this portion of the molecule that folds, to produce the increase in β sheet observed in PrP^{sc}.

Courtesy of D. G. Donne, *Proc. Natl. Acad. Sci.* (1997) 94:13452–13457. With permission of PNAS.

with the nervous system. Even more remarkable is the fact that ingesting PrP^{sc} can induce conversion of PrP^{c} in the recipient to PrP^{sc}; thus the condition is transmitted. How this conversion is catalyzed is unknown, but it strongly suggests that PrP^{c} represents an especially stable "off-path" folding of the type hypothesized in the preceding section. The recognition of the relationship of PrP to these diseases won Stanley Prusiner the Nobel Prize in Physiology or Medicine in 1997.

Prediction of Secondary and Tertiary Protein Structure

Protein secondary structure can now be predicted with moderate accuracy. The more complex tertiary structure still cannot.

Can protein structure be predicted? In one sense, the answer to this question must surely be yes. We know that the molecular information necessary to determine the secondary and tertiary structures is carried in the amino acid sequence itself; thus, the gene "predicts" the structure. The implication of this fact is that if we fully understood the rules of folding, we could describe the entire three-dimensional conformation of any protein, starting with nothing more than a knowledge of its sequence. This kind of prediction cannot yet be done completely. We can predict secondary structure fairly well, but prediction of tertiary folding is a much more complicated problem, and only very recently has progress been made.

PREDICTION OF SECONDARY STRUCTURE

Although a number of approaches have been applied to the problem of predicting secondary structure, the most successful method is entirely empirical. From analysis of the *known* structures of a number of proteins, tables have been compiled to show the relative frequency (P_{α}, P_{β}, P_t) with which a particular kind of amino acid residue lies in α helices, β sheets, or turns. An example is given in Table 6.6. From these data, certain clear distinctions can be made. For example, Leu, Met, and Glu are all strong helix formers; Gly and Pro do not favor helices. Similarly, Ile, Val, and Phe are strong β sheet formers, whereas Pro does not fit well into β sheets. Gly is frequently found in β turns, whereas Val is not. We have already mentioned that Pro tends to lie in turns. Why other residues are generally found in one structure or another is not so clear.

Various rules have been proposed to use P values to predict structures; those developed by P. Y. Chou and G. D. Fasman are listed in Table 6.6. The structures predicted by applying these rules often agree fairly well, although never exactly, with secondary structures determined experimentally by x-ray diffraction (Figure 6.28). Differences are most frequently found in definition of the precise location of the ends of α and β regions.

TERTIARY STRUCTURE: COMPUTER SIMULATION OF FOLDING

Attempts to predict tertiary structure have proved much more difficult, probably because the higher-order folding depends so critically on specific side chain interactions, often between residues far removed from one another in the sequence. However, recent recognition of overall patterns in tertiary folding has allowed some successes—for example, predictions of secondary structure have been used to predict an α/β barrel structure for the enzyme tryptophan synthase that is in excellent agreement with the x-ray result.

In a few cases, attempts have been made to predict the overall three-dimensional structures of small globular proteins in a more direct way. The method depends on the fact that, in their spontaneous folding, proteins are seeking a free energy minimum. A random-coil chain is allowed, in computer simulation, to undergo a very large number of small permutations in its conformation, through rotation about individual bonds. The computer program keeps track of the total energy, in terms of possible interactions, and seeks an energy minimum. If one attempts simply to search at random through possible conformations, the task is impossible; one is, in essence, confronting the Levinthal problem (see page 187). Somehow, a search for a minimum must be more directed. Recently, a number of researchers have developed much more efficient algorithms, which break the folding process down into reasonable steps. An example is the program LINUS (Local

TABLE **6.6 Correspondence of amino acid residues to protein secondary structure**

A. Relative probabilities of amino acid residue occurrence in different globular protein secondary structures[a]

Amino Acid	α Helix (P_α)	β Sheet (P_β)	Turn (P_t)	
Ala	1.29	0.90	0.78	Favor α helices
Cys	1.11	0.74	0.80	
Leu	1.30	1.02	0.59	
Met	1.47	0.97	0.39	
Glu	1.44	0.75	1.00	
Gln	1.27	0.80	0.97	
His	1.22	1.08	0.69	
Lys	1.23	0.77	0.96	
Val	0.91	1.49	0.47	Favor β sheets
Ile	0.97	1.45	0.51	
Phe	1.07	1.32	0.58	
Tyr	0.72	1.25	1.05	
Trp	0.99	1.14	0.75	
Thr	0.82	1.21	1.03	
Gly	0.56	0.92	1.64	Favor turns
Ser	0.82	0.95	1.33	
Asp	1.04	0.72	1.41	
Asn	0.90	0.76	1.23	
Pro	0.52	0.64	1.91	
Arg	0.96	0.99	0.88	

B. Chou–Fasman rules for prediction[b]

1. Any segment of six residues or more, with $\langle P_\alpha \rangle \geq 1.03$, as well as $\langle P_\alpha \rangle > \langle P_\beta \rangle$, and not including Pro, is predicted to be α helix.
2. Any segment of five residues or more, with $\langle P_\beta \rangle \geq 1.05$, and $\langle P_\beta \rangle > \langle P_\alpha \rangle$, is predicted to be β sheet.
3. Examine the sequence for tetrapeptides with $\langle P_\alpha \rangle < 0.9$, $\langle P_t \rangle > \langle P_\beta \rangle$. They have a good chance of being turns. The actual rules for predicting β turns are more complex, but this method will work in most cases.

[a]Data adapted from M. Levitt, *Biochemistry* (1978) 17:4277–4285.

[b]Adapted with permission from P. Y. Chou and G. D. Fasman, *Biochemistry* (1974) 13:222–245. © 1974 American Chemical Society. The symbols $\langle P_\alpha \rangle$, $\langle P_\beta \rangle$, and $\langle P_t \rangle$ denote *average* values of these quantities in a region of the sequence.

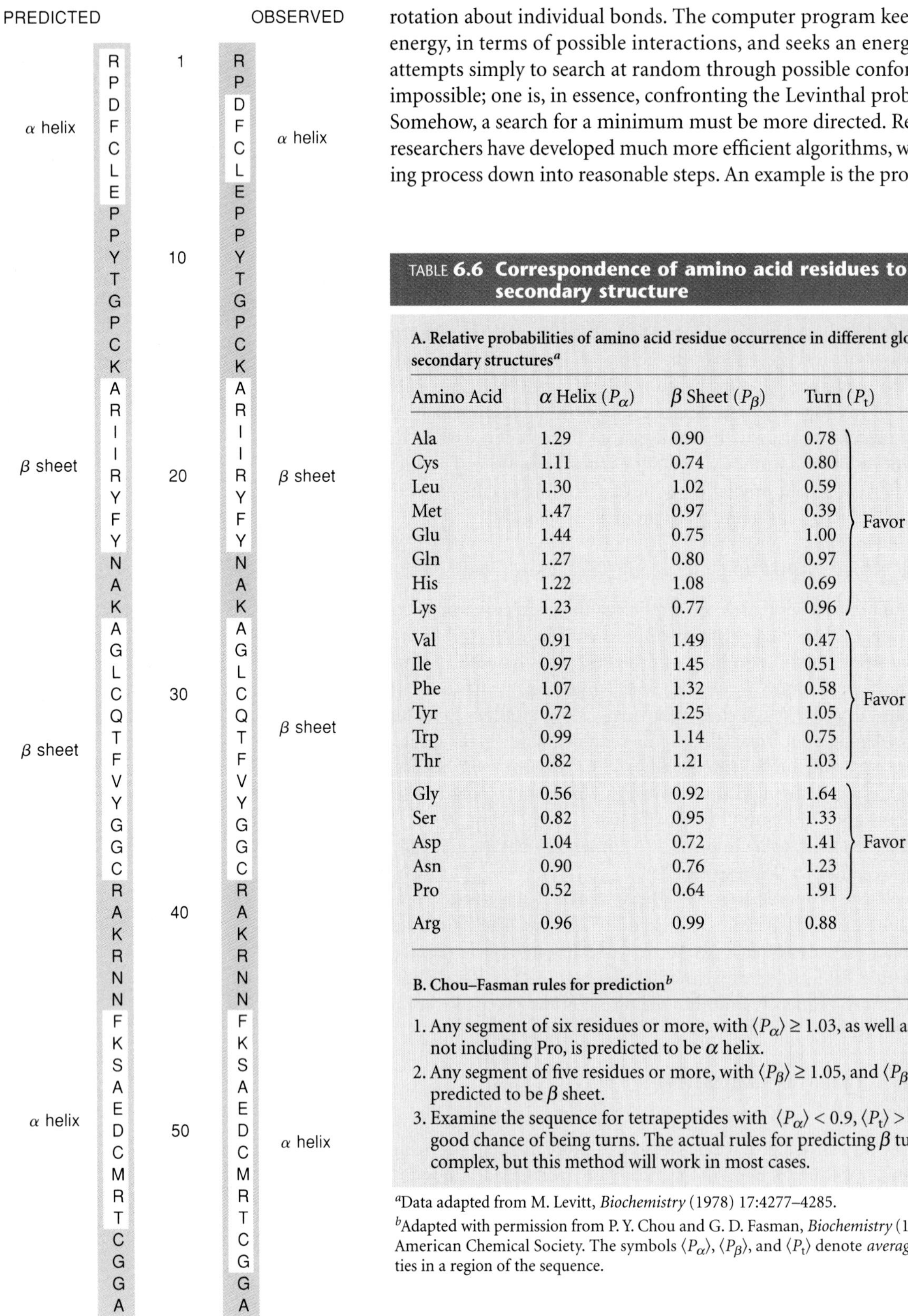

FIGURE **6.28**

Prediction of the secondary structure of BPTI. The sequence on the left shows the secondary structural elements predicted by P. Y. Chou and G. D. Fasman for bovine pancreatic trypsin inhibitor. The sequence on the right shows the results of x-ray diffraction studies of the same protein. The exceptionally good agreement found between the predicted and observed structures is somewhat better for BPTI than for most proteins.

Independently Nucleated Units of Structure) (see Srinivasan and Rose in the References). LINUS first finds local regions likely to have particular secondary structure, and then asks how these can fold together to minimize the energy. This is the way we believe real proteins fold, so it is perhaps not surprising that LINUS has made a number of successful predictions. This approach is still in its infancy, yet it gives the hope that we may ultimately be able to predict not only the folding patterns of proteins but also the pathways that lead to the final structures. A quite different approach comes from the growing recognition that certain domain types have been strongly conserved in their folding throughout evolution. If they can be recognized by sequence, we may make good first guesses as to how they ought to fold.

Quaternary Structure of Proteins

In Chapter 5 and in this chapter, we have explored increasingly complex levels of protein structure, from primary to secondary to tertiary (Figure 6.29a–c). Functional protein organization can reach at least one more level—**quaternary structure** (Figure 6.29d). Many proteins exist in the cell (and in solution, under physiological conditions) as specific aggregates of two or more folded polypeptide chains, or *subunits.* Methods for determining whether or not a protein is composed of multiple subunits are described in Tools of Biochemistry 6B. Such quaternary organization can be of two kinds—association between identical or nearly identical polypeptide chains (**homotypic**) or interactions between subunits of very different structures (**heterotypic**). In either case, *multisubunit proteins* are formed.

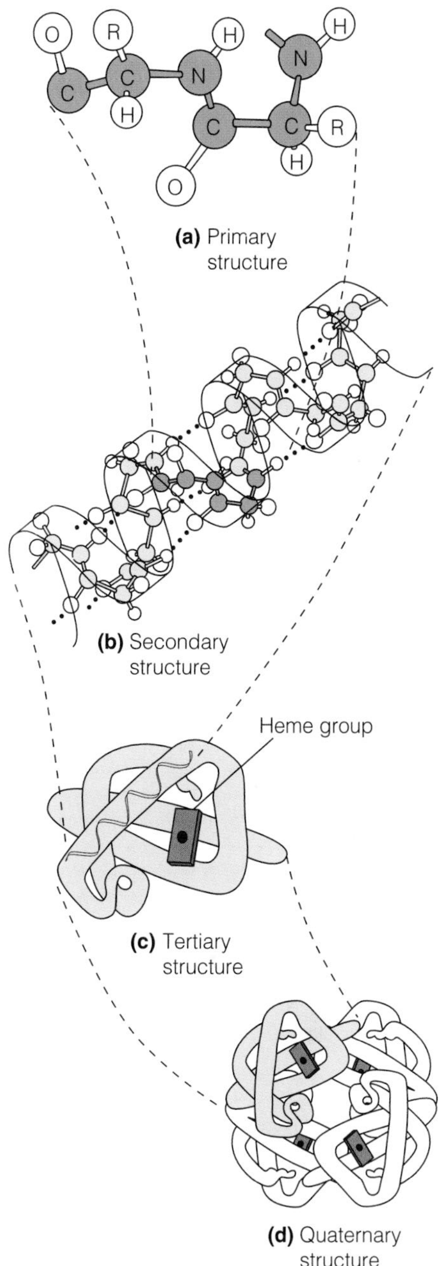

FIGURE **6.29**

The four levels of protein structure. This summary of the structural levels of protein uses the hemoglobin molecule, a tetramer of myoglobin-like chains, as an example.

(a, b) From W. M. Becker, L. J. Kleinsmith, and J. Hardin, *The World of the Cell,* 4th ed. (San Francisco, CA: Addison Wesley Longman, 2000) © Addison Wesley Longman, Inc.; **(c, d)** © Irving Geis.

MULTISUBUNIT PROTEINS: HOMOTYPIC PROTEIN–PROTEIN INTERACTIONS

The interactions between the folded polypeptide chains in multisubunit proteins are of the same kinds that stabilize tertiary structure—salt bridges, hydrogen bonding, van der Waals forces, hydrophobic interactions, and sometimes disulfide bonding. These interactions provide the energy to stabilize the multisubunit structure.

Each polypeptide chain is an asymmetric unit in the aggregate, but the overall quaternary structure may exhibit a wide variety of symmetries, depending on the geometry of the interactions. For purposes of illustration, we shall use an asymmetric object familiar to everyone—a right shoe. Think of this shoe as a polypeptide chain folded into a compact three-dimensional form. We can stick shoes together in many ways. If the interacting surfaces (A and B) were at toe and heel, a linear aggregate could form:

Because the two interacting groups lie in entirely different regions of the subunit, we would call this interaction **heterologous.** Heterologous interactions must be specially oriented to give a truly linear aggregate. More often, the interaction is such that each unit is twisted through some angle with respect to the preceding one. This twisting gives rise to a helical structure. Figure 6.30a shows an arrangement of shoes that forms a right helix with n units per turn. The top of the toe of each shoe is attached to the sole of the toe of the next, with a rotation of $360/n$ degrees. Two biological examples, both of **helical symmetry,** are shown in Figure 6.31: the helix of the muscle protein actin and the helical coat of tobacco mosaic virus. Note that both linear and helical arrays are potentially capable of indefinite

Association of polypeptide chains to form specific multisubunit structures is the quaternary level of protein organization.

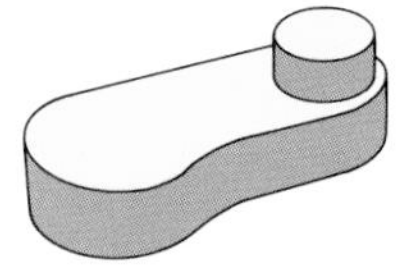
The asymmetric motif (a right shoe)

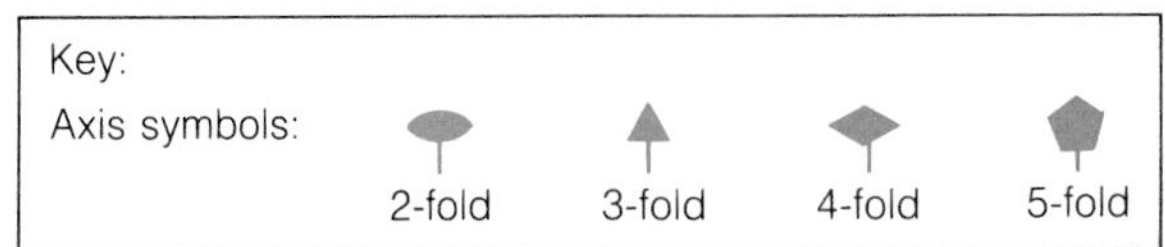

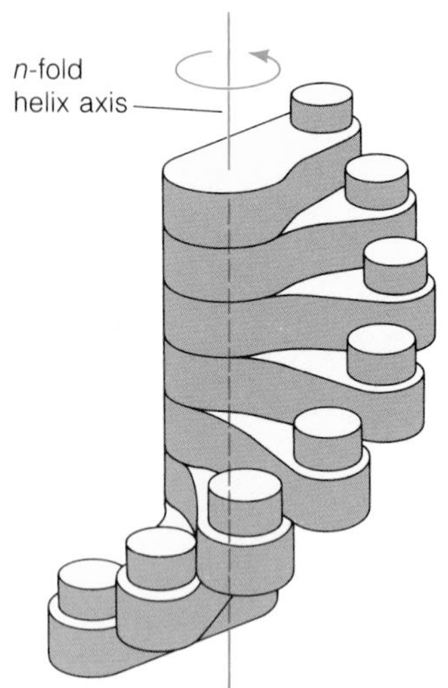

(a) Helical symmetry

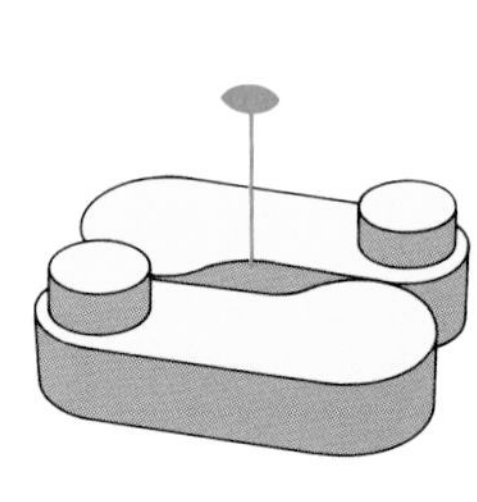
(b) C_2 symmetry

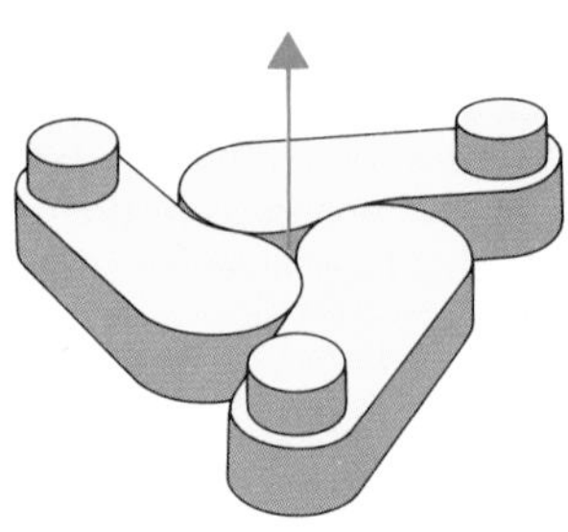
(c) C_3 symmetry

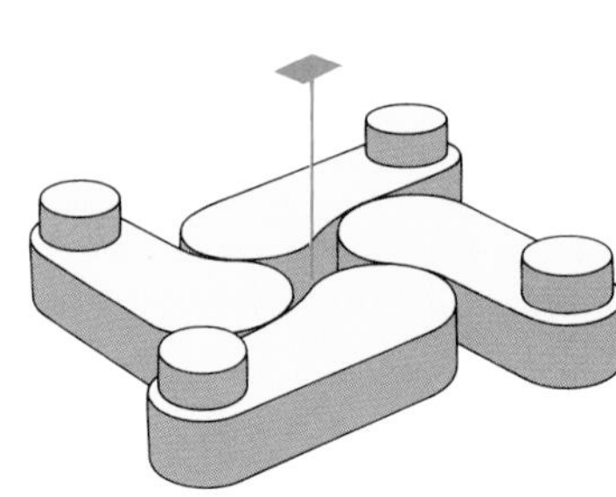
(d) C_4 symmetry

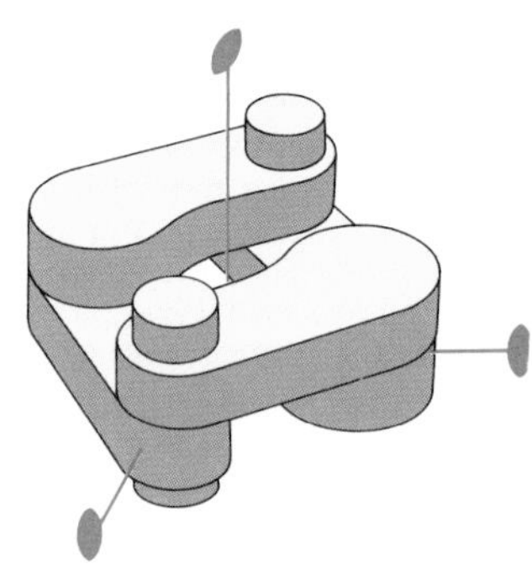
(e) D_2 symmetry

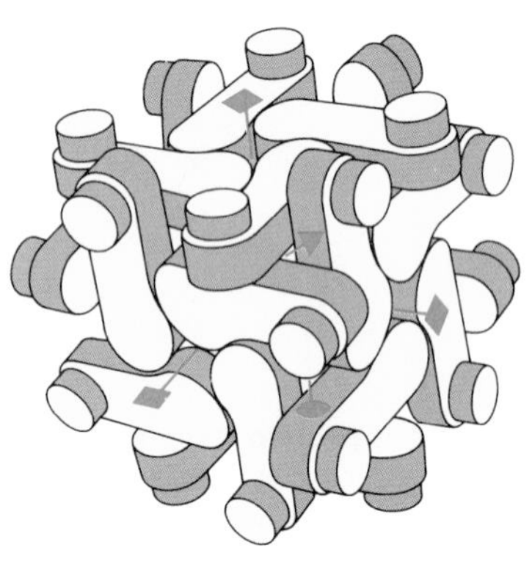
(f) Cubic symmetry

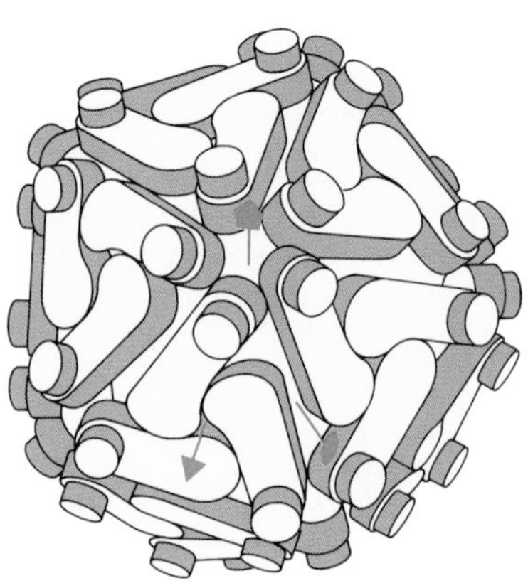
(g) Icosahedral symmetry

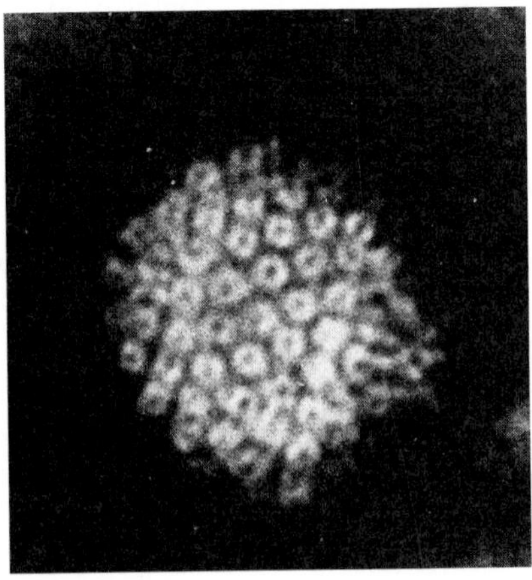
(h)

FIGURE **6.30**

Symmetries of protein quaternary structures. Although composed of asymmetric polypeptides, proteins adopt many symmetrical patterns in forming quaternary structures. In this figure, a right shoe represents the asymmetric structural unit. **(a)** A helix formed by rotation of each unit by 360/n degrees with respect to the preceding one. Such rotation produces an n-unit-per-turn helix of indefinite length. **(b)** A dimer with C_2 symmetry: one 2-fold axis. **(c)** A trimer with C_3 symmetry: one 3-fold axis. **(d)** A tetramer with C_4 symmetry: one 4-fold axis. **(e)** A tetramer with D_2 symmetry: three 2-fold axes. **(f)** A 24-mer exhibiting cubic symmetry. This structure has 4-fold, 3-fold, and 2-fold axes. **(g)** An icosahedral 60-mer, the kind of structure found in the protein coat of a number of viruses. This structure has 5-fold, 3-fold, and 2-fold axes. **(h)** An electron micrograph of a herpesvirus with icosahedral symmetry.

(h) Courtesy of B. Roizman, University of Chicago/BPS.

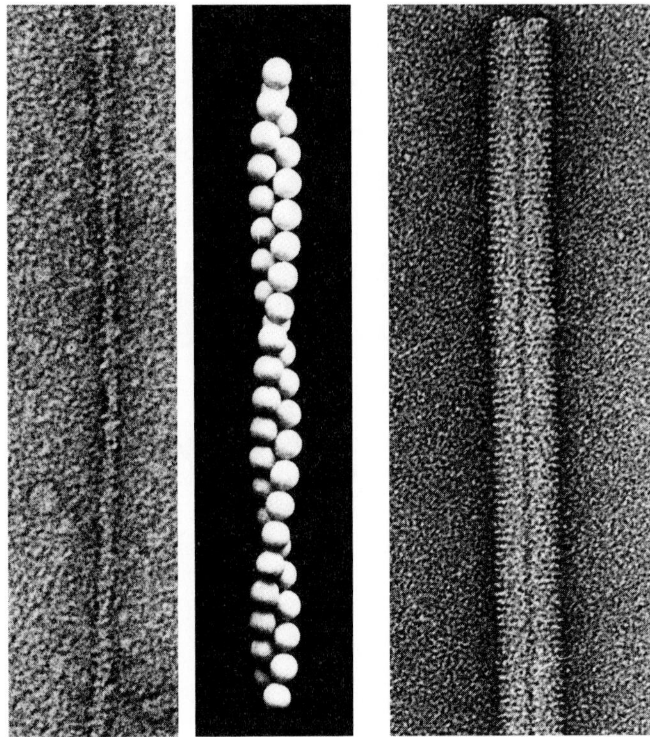

(a) Actin (b) Tobacco mosaic virus

FIGURE **6.31**

Two helical proteins. Beside each electron micrograph is a diagrammatic representation of the helical aggregate structure. **(a)** Actin. **(b)** Tobacco mosaic virus. In the virus, protein subunits form a helical array about a helically coiled RNA (red).

(a) Courtesy of T. Pollard, reproduced from *J. Cell Biol.* (1981) 91:156 by copyright permission of Rockefeller University Press; **(b)** courtesy of R. C. Williams, University of California, Berkeley.

growth by addition of more subunits. Actin filaments can be thousands of units in length.

Most assemblies of protein subunits are based not on helical symmetry but on one of the classes of **point-group symmetry** (Figure 6.30b–g). The classes of point-group symmetry involve a defined number of subunits arranged about one or more **axes of symmetry.** An n-fold axis of symmetry corresponds to rotation of each subunit by $360/n$ degrees with respect to its neighbor; thus, a 2-fold axis corresponds to 180° rotation. The simplest kinds of point-group symmetry are the cyclic symmetries C_n, shown in Figure 6.30b–d. These rings of subunits involve heterologous interactions where $n = 3$ or greater.

Two general classes of symmetry—helical and point-group—characterize most quaternary structures.

A special situation of great importance arises when two subunits are related to one another by a 2-fold axis (also called a **dyad** axis) to give C_2 symmetry. That is, each subunit is rotated by 180° about this axis with respect to the other:

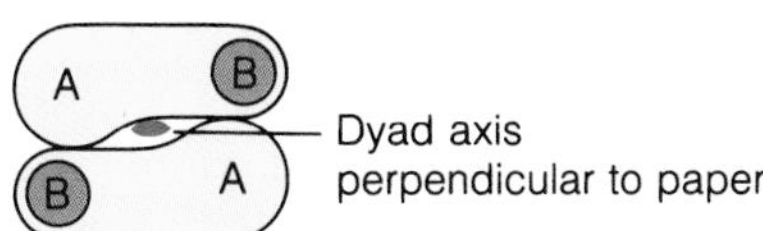

(This arrangement is visualized in three dimensions in Figure 6.30b.) Imagine that there are interacting groups at A and B. For example, A could be a hydrogen bond donor, B an acceptor. Note that in this case the 2-fold symmetry means that *two* identical interactions occur, symmetrically placed about the dyad axis. Such a symmetric interaction is called **isologous,** as opposed to the asymmetric heterologous interaction. Isologous interactions will be found whenever 2-fold axes are present. Dimers are the most common of all quaternary structures, and they are almost always bound together in this way. An example is shown in Figure 6.32. Further isologous interactions can easily give rise to more complex quaternary structures of higher symmetry. An example is **dihedral** symmetry (Figure 6.30e), the most common structure for tetrameric proteins. It has three mutually

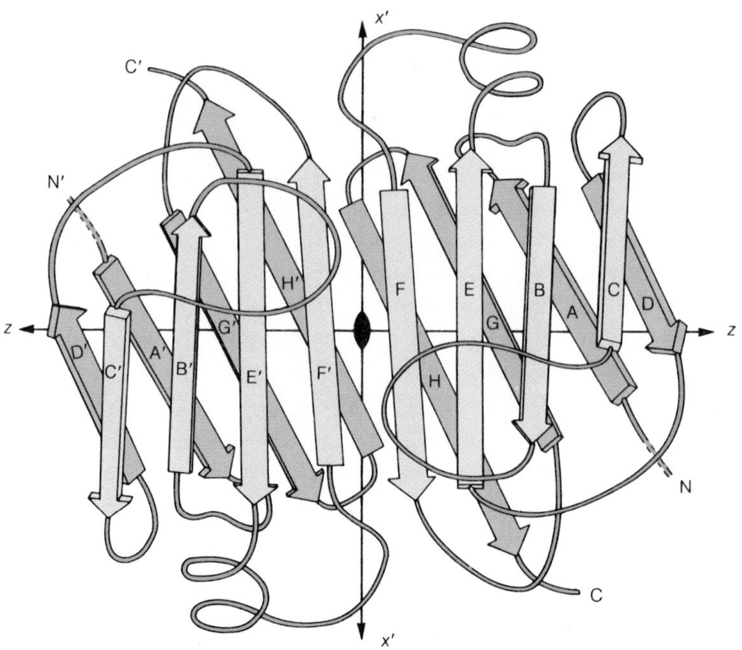

FIGURE **6.32**

The prealbumin dimer. In the prealbumin dimer, also seen in Figure 6.16b, the two monomers combine to form a complete β sandwich, or flattened β barrel. The dimer has 2-fold symmetry about an axis perpendicular to the paper. The isologous interactions are mostly hydrogen bonds between specific β sheet strands: F to F′ and H to H′. Prealbumin can also form a tetramer from two of these dimers by a second set of isologous interactions.

C. C. F. Blake et al., *J. Mol. Biol.* (1974) 88:1–12. © 1974 Academic Press.

perpendicular 2-fold axes and, therefore, involves three pairs of isologous interactions. In fact, any D_n symmetry can be generated by isologous interactions alone.

Other, more complex point-group symmetries exist in some protein quaternary structures; examples are shown in Figure 6.30f and g. Note that the more complex structures involve both 2-fold axes and axes with $n > 2$. Such molecules exhibit both isologous and heterologous interactions. Most important is that molecules exhibiting any of the point-group symmetries are always constrained to a definite number of subunits. Most multisubunit proteins exhibit this kind of association geometry, rather than the linear or helical aggregation that can lead to indefinite growth. This is why most protein molecules, even if they contain multiple subunits, have a well-defined size, shape, and mass.

Two exceptions to this simple characterization of possible structures must be noted. First, although most dimers have 2-fold symmetry and isologous binding, it is possible to construct dimers in which the binding is heterologous but indefinite association is still sterically blocked. Such dimers do not have 2-fold symmetry (Figure 6.33). A second complication is encountered whenever more than one type of polypeptide chain is incorporated into a specific multisubunit structure. In such cases the symmetry is reduced from the level you might expect from the total number of polypeptide chains, because two or more different chains will now form one asymmetric unit. We consider one such example, *hemoglobin*, in Chapter 7.

Note that each level of protein structure is built on the lower levels, as summarized in Figure 6.29. Tertiary structure can be thought of as a folding of elements of secondary structure, and quaternary structure is established by combining folded subunits. *All* of this higher-level structuring is dictated by primary struc-

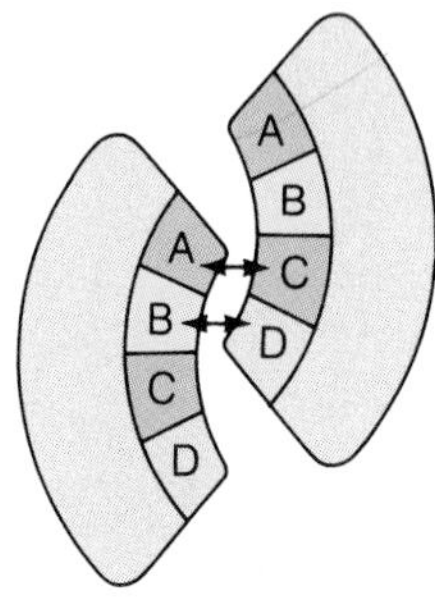

FIGURE **6.33**

A dimer without symmetry. This schematic picture shows how two subunits can associate by heterologous interactions and still not yield an indefinite chain. The interaction sites (A, B, C, D) are blocked from reaction with more monomers by the close fit of the surfaces.

ture and ultimately by the gene. This understanding of protein structure is one of the most important ideas in molecular biology.

All higher levels of structure of a protein are dictated by its genes.

HETEROTYPIC PROTEIN–PROTEIN INTERACTIONS

The preceding section has centered on the association of identical or near-identical protein subunits. However, the range of protein–protein interactions is much vaster; specific associations between entirely different protein molecules are common. Sometimes these associations lead to organized structures containing a dozen or more different subunit types.

The interactions that form these assemblies are of the same kind we described earlier—noncovalent forces at complementary protein surfaces. A simple example involves BPTI, a protein we have described in detail in this chapter. Bovine pancreatic trypsin inhibitor is so named because it forms a tight, specific complex with the enzyme trypsin, thereby inhibiting trypsin proteolytic activity in the pancreas. Figure 6.34, a computer simulation of this interaction, shows that the two protein surfaces fit one another closely. The "docking" of one protein on another predicted by the computer model closely matches the actual structure of the complex as determined by x-ray diffraction. The accuracy of the prediction suggests that it may be easier to predict some aspects of quaternary structure than tertiary structure.

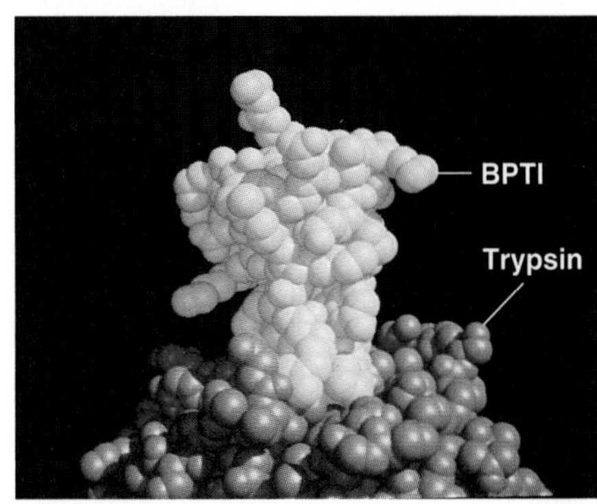

FIGURE **6.34**

Interaction of BPTI with trypsin. The BPTI molecule (top) fits snugly onto the surface of the trypsin molecule (bottom), blocking the active site of trypsin. This figure is from a computer "docking" simulation, which closely (within 0.05 nm) matches the x-ray structure of the complex.

Courtesy of B. K. Soichet and I. D. Kuntz, *J. Mol. Biol.* (1991) 221:327–346. © 1991 Academic Press.

SUMMARY

Protein molecules typically have several levels of organization. The first, or primary, level is the amino acid sequence, dictated by the gene. This sequence in turn dictates local folding (secondary structure), global folding (tertiary structure), and organization into multisubunit structures (quaternary structure).

Of many conceivable secondary structures, only a limited number are sterically allowed and can be stabilized by hydrogen bonds. These include the α helix, the β sheet, and the 3_{10} helix. There are also specific structures that allow a polypeptide chain to make sharp turns. Ramachandran diagrams provide a way to visualize the possibilities.

Proteins can be grouped into two broad categories—fibrous and globular. Fibrous proteins are elongated, usually of regular secondary structure, and perform structural roles in the cell and organism. Important examples include the keratins (α helix), the fibroins (β sheet), collagen (triple helix), and elastin (cross-linked random coils). Globular proteins have more complex tertiary structures and fold into compact shapes that often contain defined domains. A number of classes of tertiary folding have been recognized, such as α helix bundles, twisted β sheets, and β barrels.

A number of factors determine globular protein stability—conformational entropy, enthalpy from internal noncovalent bonding, the hydrophobic effect, and disulfide bonds. The folding of many globular proteins occurs spontaneously and rapidly under "native" conditions. In the cell, proteins called chaperonins help prevent formation of incorrect structures or undesired interactions. Even when folded, globular proteins are dynamic structures undergoing several kinds of internal motions. We can now predict protein secondary structure with some degree of accuracy, but prediction of tertiary structure is much more difficult.

Many (perhaps most) globular proteins exist and function as multisubunit assemblies forming a quaternary level of structure. A few of these proteins are elongated structures with helical symmetry; most have a small number of subunits (often 2, 4, or 6) and exhibit point-group symmetry. All of the levels of protein structure are determined by the gene sequence.

REFERENCES

General

Branden, C. and J. Tooze (1991) *Introduction to Protein Structure,* Garland, New York. A very well illustrated primer on structural principles.

Creighton, T. E. (1993) *Proteins: Structures and Molecular Properties,* 2nd ed. Freeman, San Francisco. A detailed, authoritative treatment of protein structure and function.

Kyte, J. (1995) *Structure in Protein Chemistry.* Garland, New York. Complements well the 2nd edition of Creighton; primary emphasis on structure and methods.

Oxender, D. L., and C. F. Fox (1987) *Protein Engineering.* Alan R. Liss, New York. An excellent collection of contemporary papers on protein design and modification.

Historical

Anfinsen, C. B. (1973) Principles that govern the folding of protein chains. *Science* 181:223–230. Describes some of the important early experiments that demonstrated the spontaneity of folding.

Pauling, L., R. B. Corey, and H. R. Branson (1951) The structure of proteins: Two hydrogen bonded helical conformations of the polypeptide chain. *Proc. Natl. Acad. Sci. USA* 37:205–211. The first description of the α helix.

Ramachandran, G. N., and V. Sassiekharan (1968) Conformation of polypeptides and proteins. *Adv. Protein Chem.* 28:283–437. Introduction of Ramachandran plots.

Fibrous Proteins

Fuchs, E., and P. A. Coulombe (1992) Of mice and men: Genetic skin diseases of keratin. *Cell* 69:899–902.

Kaplan. D., W. W. Adams, B. Farmer, and C. Viney (1994) *Silk Polypeptides.* American Chemical Society Press, New York.

Piez, K. A., and A. H. Reddi (1984) *Extracellular Matrix Biochemistry.* Elsevier Science Publishing, New York.

vanderRest, M., and P. Bruckner (1993) Collagens: Diversity at the molecular and supramolecular levels. *Curr. Opin. Struct. Biol.* 3:430–436.

Globular Proteins: Secondary and Tertiary Structure

Matthews, B. W. (1987) Structural basis of protein stability and DNA–protein interactions. In: *The Harvey Lectures,* ser. 81, pp. 33–51. Alan R. Liss, New York. Describes use of site-directed mutagenesis to analyze protein stability.

Richardson, J. S. (1981) The anatomy and taxonomy of protein structure. *Adv. Protein Chem.* 34:167–339. A seminal analysis, in which the concept of structural motifs is introduced.

Rose, G. (1979) Hierarchic organization of domains in globular proteins. *J. Mol. Biol.* 134:447–470. An important paper in our understanding of protein organization.

Wetzel, R. (1988) Harnessing disulfide bonds using protein engineering. *Trends Biochem. Sci.* 12:478–482. A brief discussion of the importance of —S—S— bonds in protein function and stability.

Protein Folding

Baldwin, R. L., and G. Rose (1999) Is protein folding hierarchic? I. Local structure and peptide folding. *Trends Biochem. Sci.* 24:26–33. This and the following paper provide a current, critical view of the state of the folding problem.

Baldwin, R. L., and G. Rose (1999) Is protein folding hierarchic? II. Folding intermediates and transition states. *Trends. Biochem. Sci.* 24:77–83.

Dill, K. A., and H. S. Chan (1997) From Levinthal to pathways to funnels. *Nature Struct. Biol.* 4:10–19. A wonderfully clear and interesting discussion of the "Levinthal paradox" and its resolution.

Gierasch, L. M., and J. King (1990) *Protein Folding.* American Association for Advancement of Science, Washington, D.C. An excellent collection of papers on many aspects of the folding problem.

Goldenberg, D. P. (1992) Native and non-native intermediates in the BPTI folding pathway. *Trends Biochem. Sci.* 17:257–261.

Kim, P. S. and Baldwin, R. L. (1980) Structural intermediates trapped during the folding of ribonuclease A by amide exchange. *Biochemistry* 19:6124–6129.

Chaperonins

Braig, K., Z. Otwinowski, R. Hedge, D. C. Boisvert, A. Joachimiak, A. L. Horwich, and P. B. Sigler (1994) The crystal structure of the bacterial chaperonin GnoEL at 2.8Å. *Nature* 371:578–586.

Georgopoulos, C. (1992) The emergence of the chaperone machine. *Trends Biochem. Sci.* 17:295–299.

Xu, Z., A. L. Horwich, and P. B. Sigler (1997) The crystal structure of the asymmetric GroEL-GroES-$(ADP)_7$ chaperonin complex. *Nature* 388:741–750.

Prediction of Protein Structure

Chou, P. Y., and G. D. Fasman (1978) Empirical predictions of protein structure. *Annu. Rev. Biochem.* 47:251–276.

Kabsch, W., and C. Sander (1983) How good are predictions of protein secondary structure? *FEBS (Fed. Eur. Biochem. Soc.) Lett.* 155:179–182.

Srinivasan, R., and G. Rose (1995) *Proteins Struct. Funct. Genet.* 22:81–99.

Protein Dynamics

Karplus, M., and J. A. McCannon (1986) The dynamics of proteins. *Sci. Am.* 254(4):42–51. An excellent and readable introduction.

Ringe, D., and G. R. Petsko (1985) Mapping protein dynamics by x-ray diffraction. *Prog. Biophys. Mol. Biol.* 45:197–235.

Prions

Pan, K.-M., M. Baldwin, J. Nguyen, M. Gasset, A. Serban, D. Groth, I. Mehlhorn, Z. Huang, R. J. Fletterick, F. E. Cohen, and S. B. Prusiner (1993) Conversion of α-helices into β-sheets features in the formation of the scrapie prion proteins. *Proc. Natl. Acad. Sci. USA* 90:10962–10966.

Prusiner, S. B. (1997) Prion diseases and the BSE crisis. *Science* 278:245–251. An excellent overview of prions and their pathological effects.

Quaternary Structure

Caspar, D. L. P., and A. Klug (1962) Physical principles in the construction of regular viruses. *Cold Spring Harbor Symp. Quant. Biol.* 27:1–24. A classic early paper on symmetry.

Klotz, I. M., N. R. Langerman, and D. W. Darnall (1970) Quaternary structure of proteins. *Annu. Rev. Biochem.* 39:25–62. A good overall view of structures and bonding.

Matthews, B. W., and S. A. Bernhard (1973) Structure and symmetry of oligomeric enzymes. *Annu. Rev. Biophys. Bioeng.* 2:257–317.

PROBLEMS

1. Polyglycine, a simple polypeptide, can form a helix with $\phi = -80°$, $\psi = +150°$. From the Ramachandran plot (see Figure 6.8), describe this helix with respect to (a) handedness and (b) number of residues per turn.
2. Certain polypeptide sequences that show a pronounced tendency to dimerize in an antiparallel coiled-coil structure exhibit an exact repetition of leucine every 7 residues. If you also know that the α helix is distorted somewhat from its usual 3.6 residues/turn in coiled coils, propose a mechanism for the dimerization. What does this suggest that the residue/turn value might be in a coiled coil?
3. A schematic structure of the subunit of hemerythrin (an oxygen-binding protein from invertebrate animals) follows.
(a) It has been found that in some of the α-helical regions of hemerythrin, about every third or fourth amino acid residue is a hydrophobic one. Suggest a structural reason for this finding.
(b) What would be the effect of a mutation that placed a proline residue at point A in the structure?
*(c) Sketch a reasonable mechanism (that is, series of steps) for the folding of hemerythrin from a random coil to the structure shown.

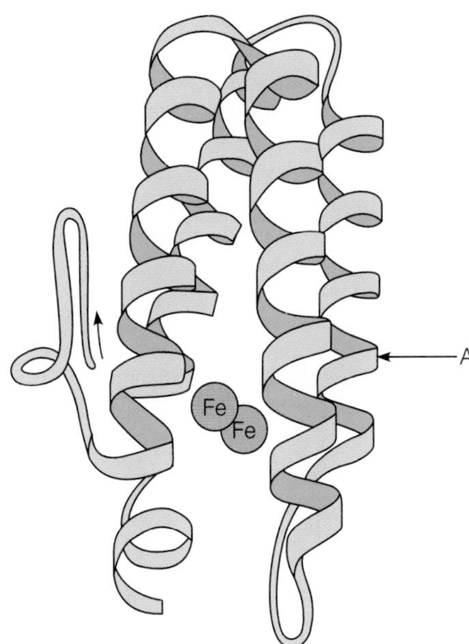

4. In the protein *adenylate kinase,* the C-terminal region is α-helical, with the sequence

Val–Asp–Asp–**Val**–**Phe**–Ser–Gln–**Val**–Cys–
Thr–His–**Leu**–Asp–Thr–**Leu**–Lys–

The hydrophobic residues in this sequence are presented in boldface type. Suggest a possible reason for the periodicity in their spacing.
5. Although the bond energy for the hydrogen bond in a vacuum is estimated to be about 20 kJ/mol, we find that each hydrogen bond in a folded protein contributes much less—probably less than 5 kJ/mol—to the enthalpy of protein stabilization. Suggest an explanation for this difference.
6. Consider a small protein containing 101 amino acid residues. The protein will have 200 bonds about which rotation can occur. Assume that three orientations are possible about each of these bonds.
(a) Based on these assumptions, about how many *random-coil* conformations will be possible for this protein?
(b) The estimate obtained in (a) is surely too large. Give one reason why.

*7. (a) Based on a more conservative answer to Problem 6 (2.7×10^{92} conformations), estimate the conformational entropy change on folding a mole of this protein into a structure with only one conformation.
(b) If the protein folds *entirely* into α helix with H bonds as the only source of enthalpy of stabilization, and each mole of H bonds contributes -5 kJ/mol to the enthalpy, estimate $\Delta H_{\text{folding}}$. Note that you cannot form 4 H bonds at one end.
(c) From your answers to (a) and (b), estimate $\Delta G_{\text{folding}}$ for this protein at 25°C. Is the folded form of the protein stable at 25°C?

*8. The following sequence is part of a globular protein. Using Table 6.6 and the Chou–Fasman rules, predict the secondary structure in this region.

···RRPVVLMAACLRPVVFITYGDGGTYYHWYH···

9. (a) A protein is found to be a tetramer of identical subunits. Name two symmetries possible for such a molecule. What kinds of interactions (isologous or heterologous) would stabilize each?
(b) Suppose a tetramer, like hemoglobin, consists of two each of two types of chains, α and β. What is the highest symmetry now possible?

*10. Under physiological conditions, the protein hemerythrin exists as an octamer of eight chains of the kind shown in Problem 3.
(a) Name two symmetries possible for this molecule.
(b) Which do you think is more likely? Explain.
(c) For the more likely symmetry, what kinds of interactions (isologous, heterologous, or both) would you expect? Why?

*11. A researcher studies the thermal denaturation of hemerythrin by two methods: (1) using circular dichroism, which measures the α helix content (see Tools of Biochemistry 6A) and (2) using differential scanning calorimetry (see Tools of Biochemistry 10A). He observes a considerably larger ΔH by calorimetry than he finds using circular dichroism. Suggest a reason for this.

12. The peptide hormone *vasopressin* is used in the regulation of salt-water balance in many vertebrates. Porcine (pig) vasopressin has the sequence

Asp–Tyr–Phe–Glu–Asn–Cys–Pro–Lys–Gly

(a) Using the data in Figure 5.6 and Table 5.1, estimate the extinction coefficient ε (in units of cm^2/mg) for vasopressin, using radiation with $\lambda = 280$ nm.
(b) A solution of vasopressin is placed in a 0.5-cm-thick cuvette. Its absorbance at 280 nm is found to be 1.3. What is the concentration of vasopressin, in mg/cm^3?
(c) What fraction of the incident light is passed through the cuvette in (b)?
(See Tools of Biochemistry 6A.)

*13. A protein gives, under conditions of buffer composition, pH, and temperature that are close to physiological conditions, a molecular weight by sedimentation equilibrium measurements of 140,000 g/mol. When the same protein is studied by SDS gel electrophoresis in the absence or presence of the reducing agent β-mercaptoethanol (BME), the patterns seen in lanes A and B respectively are observed. Lane C contains standards of molecular weight indicated. From these data, describe the native protein, in terms of the kinds of subunits present, the stoichiometry of subunits, and the kinds of bonding (covalent, noncovalent) existing between subunits. (See Tools of Biochemistry 6B.)

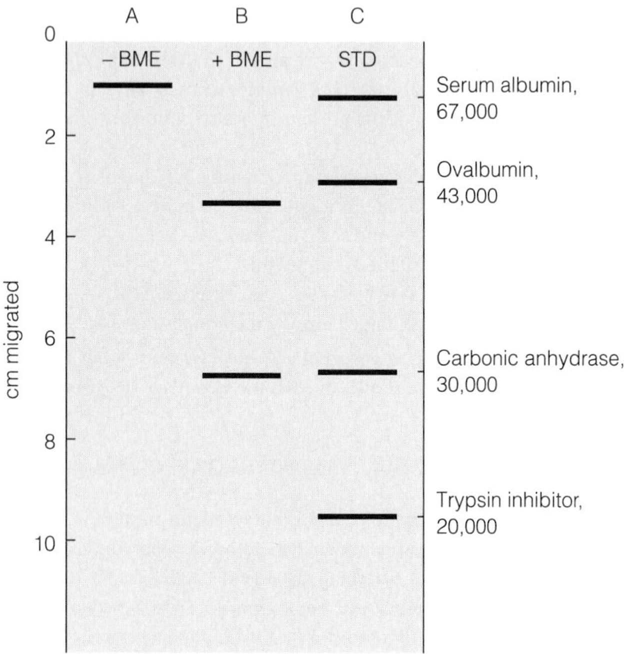

14. The following data were obtained for human γ-globulin in the analytical ultracentrifuge at 20°C:

$S = 7.12 \times 10^{-13}$ s $\bar{v} = 0.718$ mL/g

$D = 4.00 \times 10^{-7}$ cm^2 s^{-1} $\rho = 1.00$ g/cm^3

What is the molecular weight of γ-globulin? Note that the gas constant R, in cgs units, is 8.314×10^7 erg/K·mol. (See Tools of Biochemistry 6B.)

*15. A protein gives a single band on SDS gel electrophoresis, as shown in lanes 1 and 2 below. There is little if any effect from adding β-mercaptoethanol (BME) in the gel; if anything, the protein runs a little bit slower. When treated with the proteolytic enzyme thrombin (see Chapter 5) and electrophoresis in the absence of BME, the protein migrates a bit more rapidly (lane 3). But if BME is present, two much more rapidly migrating bands are found (lane 4). Explain these results in terms of a model for the protein.

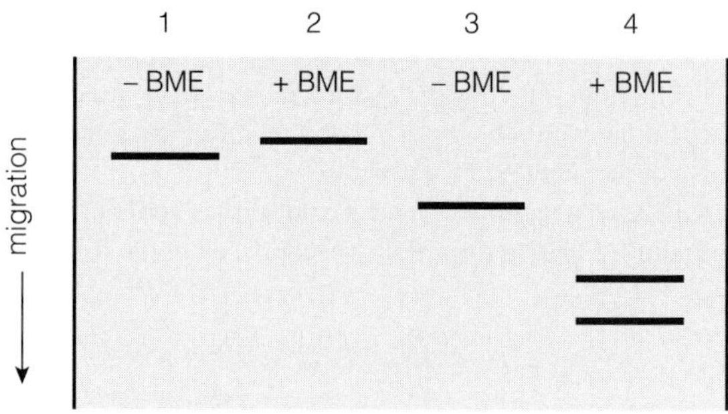

16. It has been postulated that the normal (noninfectious) form of prion differs from the infectious form only in secondary/tertiary structure (see page 192).
(a) How might you show that changes in secondary structure occur?
(b) How might you check for changes in quaternary structure?
(c) If this model is correct, what are the implications for structural prediction schemes like that of Chou and Fasman (page 194)?

TOOLS OF BIOCHEMISTRY 6A

Spectroscopic Methods for Studying Macromolecular Conformation in Solution

X-ray diffraction (see Tools of Biochemistry 4A) is a very powerful method for determining the details of the three-dimensional structure of globular proteins and other biopolymers. Yet this technique has the fundamental limitation that it can be employed only when the molecules are crystallized, and crystallization is not always easy or even possible. Furthermore, x-ray diffraction cannot easily be used to study conformational changes in response to changes in the molecules' environment. Other methods, however, allow us to study molecules in the dissolved state. A number of these methods can be grouped in the category of **spectroscopic techniques.**

Absorption Spectroscopy

Proteins, carbohydrates, and nucleic acids are complex molecules and can absorb radiation over a wide spectral range. Yet the basic principles of their absorption can be explained in terms of the simplest kind of molecule, a diatomic molecule.

When two atoms interact to form a molecule, the potential energy curve for the lowest-energy electronic state (the **ground state**) will look like the lower curve in Figure 6A.1a. **Excited electronic states** will have similar curves for energy versus interatomic distance, but at higher energies. For each electronic state of the molecule, there will be a series of allowed **vibrational states,** with energy levels indicated by horizontal lines in the figure. The basics of molecular spectroscopy can be understood by two simple rules: (1) Transitions are possible only between allowed energy states of the molecule (energy levels are **quantized**), and (2) the energy (ΔE) that has to be absorbed in any transition determines the wavelength (λ) of the radiation that is absorbed to accomplish that transition. The energy in a **quantum of radiation** is inversely proportional to λ:

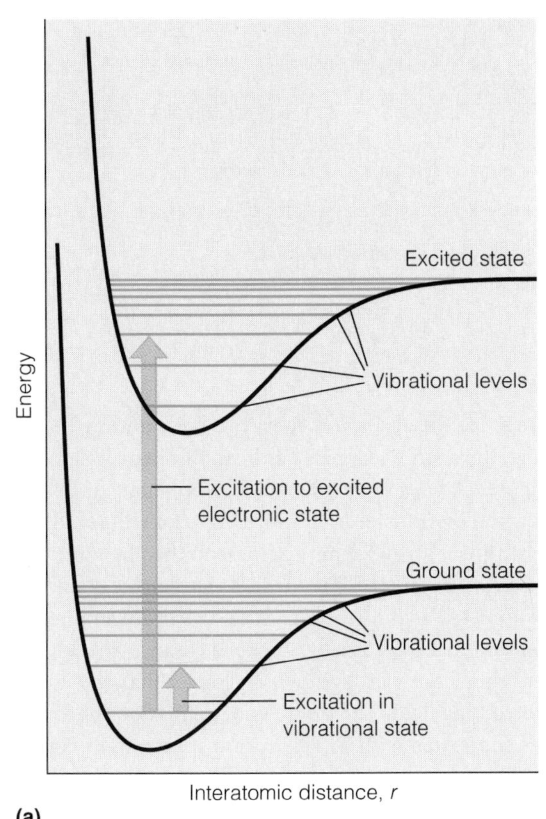

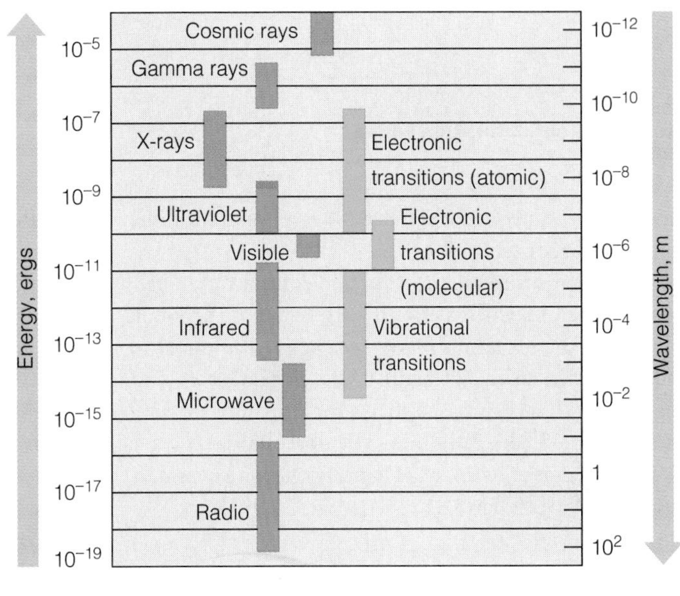

FIGURE **6A.1**

The principles of absorption spectroscopy. (a) Electronic and vibrational transitions in a diatomic molecule. **(b)** The electromagnetic spectrum.

$$\Delta E = hc/\lambda \qquad \Delta E = E_{\text{final state}} - E_{\text{initial state}} \qquad (6A.1)$$

Here h is Planck's constant (6.626×10^{-34} J s), and c is the velocity of light (3×10^8 m/s). According to equation (6A.1), small energy differences correspond to long wavelengths and large differences to short wavelengths. This relationship is in accord with Figure 6A.1b, which indicates that the high-energy transitions between electronic states of a molecule lead to absorption in the visible or ultraviolet region of the spectrum, whereas the low-energy transitions between different vibrational energy levels correspond to absorption of infrared energy.

Complex biopolymers like proteins and nucleic acids can undergo a multitude of kinds of molecular vibrations and oscillations. **Infrared spectroscopy** can therefore provide direct information concerning macromolecular structure. For example, the exact positions of infrared bands corresponding to vibrations in the polypeptide backbone are sensitive to the conformational state (α helix, β sheet, and so forth) of the chain. Thus, studies in this region of the spectrum are often used to investigate the conformations of protein molecules.

Most biopolymers do not absorb visible light to a significant extent. Some proteins are colored, but they invariably contain prosthetic groups (such as the heme in myoglobin) or metal ions (such as copper) that confer the absorption. Blood and red meat owe their color to the heme groups carried by hemoglobin, myoglobin, and other heme proteins. Such absorption can often be exploited to investigate changes in the molecular environment of the prosthetic group. An example is the use of absorption spectroscopy in the visible spectrum to follow the oxygenation of myoglobin or hemoglobin (see page 217 in Chapter 7). However, by far the most common uses of spectroscopic techniques in biochemistry involve *ultraviolet spectroscopy.* In the ultraviolet region both proteins and nucleic acids absorb strongly (Figure 6A.2). The strongest protein absorption is found in two wavelength ranges within the ultraviolet regions, in the neighborhoods of 280 and 200 nm. In the range 270–290 nm, we see absorption by the aromatic side chains of phenylalanine, tyrosine, and tryptophan (see also Figure 5.6). Because this region of the

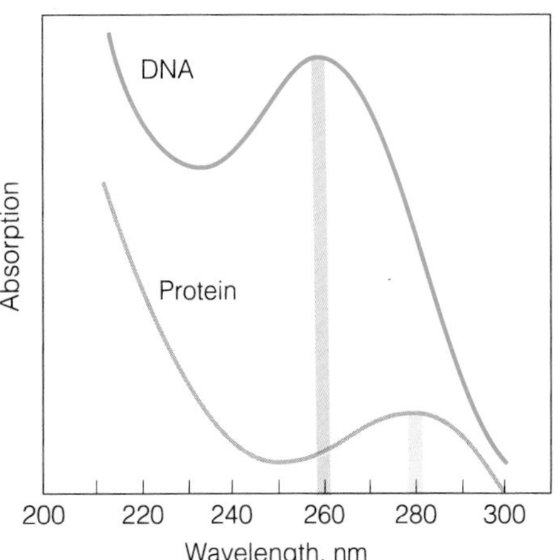

FIGURE **6A.2**

The near-ultraviolet absorption spectra of a typical protein and of DNA. Absorption at 280 nm is commonly used to measure protein concentrations, whereas absorption at 260 nm is more sensitive for nucleic acids.

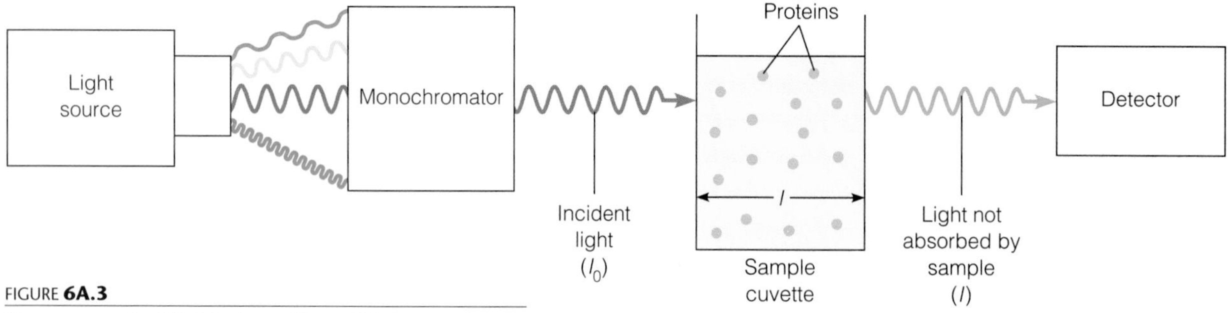

FIGURE **6A.3**
Measurement of light absorption with a spectrophotometer.

spectrum is easy to study, absorption at 280 nm is used routinely to measure protein concentrations.

Spectroscopic measurements of protein concentration use a **spectrophotometer,** in which a cuvette with light path l containing a solution of the protein is placed in a beam of monochromatic radiation of intensity I_0 (Figure 6A.3). The intensity of the emerging beam will be decreased to a value I because the solution absorbs some of the radiation. The **absorbance** at wavelength λ is defined as $A_\lambda = \log(I_0/I)$ and is related to l and the concentration c by **Beer's law:**

$$A_\lambda = \varepsilon_\lambda lc \tag{6A.2}$$

Here ε_λ is the **extinction coefficient** at wavelength λ for the particular substance being studied (see Problem 12). Its dimensions depend on the concentration units employed. If protein concentration is measured in mg/cm^3 and l in cm, then ε_λ must have the dimensions cm^2/mg, because A is a dimensionless quantity. On the other hand, when concentrations are expressed in molarity (M), the units of ε become M^{-1}cm^{-1}. Once the extinction coefficient for a particular protein has been determined (for example, by measuring the absorbance of a solution containing a known weight of the protein), the concentration of any other solution of that protein can be calculated from a simple absorbance measurement, using equation (6A.2). The same method is routinely used with nucleic acids, but in that case a wavelength of 260 nm is usually employed, because nucleic acids absorb most strongly in that spectral region.

The second region of strong absorption in the protein spectrum lies in the range 180–220 nm. Absorption at such wavelengths arises from electronic transitions in the polypeptide backbone itself and is therefore sensitive to the backbone conformation. Actually, *both* regions of the protein absorption spectrum are somewhat affected by conformational state. Extinction coefficients change slightly when a molecule goes from a globular to a random-coil form, because the local environment of all groups, including aromatic side chains, is modified. Because spectroscopy can be performed with great accuracy, even small changes can be measured and used to follow protein denaturation. An example is given in Figure 6.20.

Fluorescence

In most cases, molecules raised to an excited electronic state by absorption of radiant energy return to the ground state by **radiationless transfer** of the excitation energy to surrounding molecules. In short, the energy reappears as heat. But occasionally, as shown in Figure 6A.4a, a molecule will lose only part of its energy of excitation by transfer (yellow arrow) and will reradiate the larger part (green arrow). This gives rise to the phenomenon called **fluorescence.** Because, as Figure 6A.4a shows, the quantum of energy re-emitted as fluorescence is always of lower energy than the quantum that was initially absorbed (orange arrow), the wavelength of the fluorescent light will be longer than the wavelength of the exciting light. The **fluorescence emission spectrum** of tyrosine is contrasted with its absorption spectrum in Figure 6A.4b. In proteins, tyrosine and tryptophan are the major fluorescent groups. The environment of these residues can greatly modify the intensity of their fluorescence, and this technique can then be used to monitor changes in protein conformation. Further, excitation of fluorescence by plane polarized light (see the next section) provides a way of studying the dynamics of protein structure. If the excited residues are able to move or rotate appreciably before the fluorescent light is re-emitted, the fluorescence will be depolarized to some extent. Measurement of the extent of this depolarization provides a measure of group or molecule rotational mobility.

Recently, fluorescence has become widely used as a tool for precisely locating proteins in cells or subcellular organelles. Confocal microscopy (see Tools of Biochemistry 1A) allows such location, if the protein can be specifically labeled. Sometimes this can be done by covalently attaching fluorescent dyes, but this is hard to do in vivo. A powerful new technique uses a highly fluorescent protein found in some jellyfish, called green fluorescent protein (GFP). The intense fluorescence is due to an unusual chromophore, generated by oxidation of the amino acid sequence Ser-Tyr-Gly. GFP is most effectively used as a fusion protein; the gene for GFP is fused to the gene for the protein being studied, and the fused product is expressed in the organism of interest. The fusion protein usually functions and localizes like the native protein and provides a brilliant marker in microscopy.

Circular Dichroism

Although absorption spectroscopy and fluorescence can be helpful in following molecular changes, such measurements are difficult to interpret directly in terms of changes of secondary structure. For this purpose, techniques involving polarized light have become important.

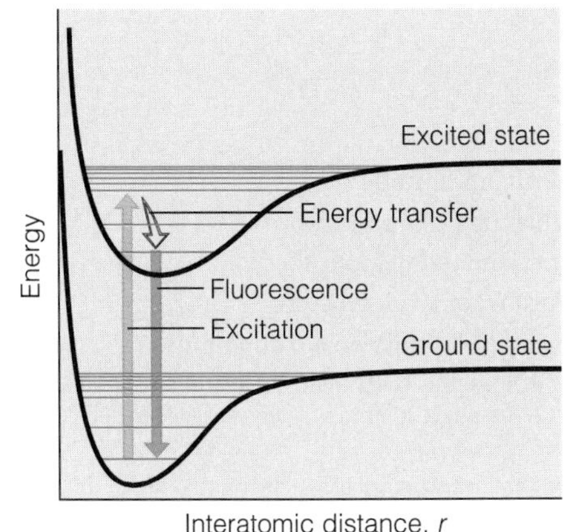

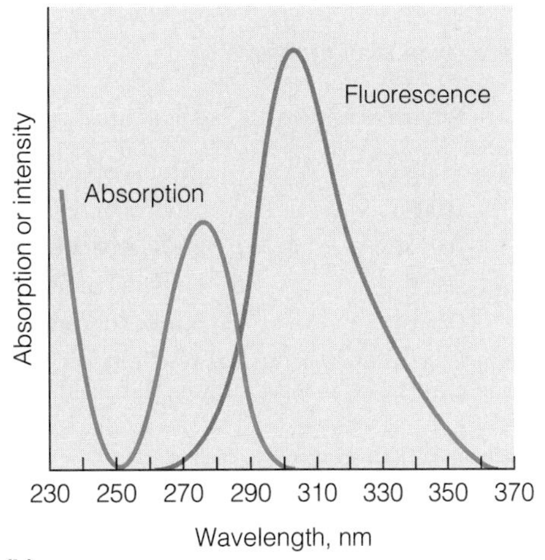

FIGURE **6A.4**

Fluorescence. (a) The principle of fluorescence. **(b)** Absorption and fluorescence emission spectra of tyrosine.

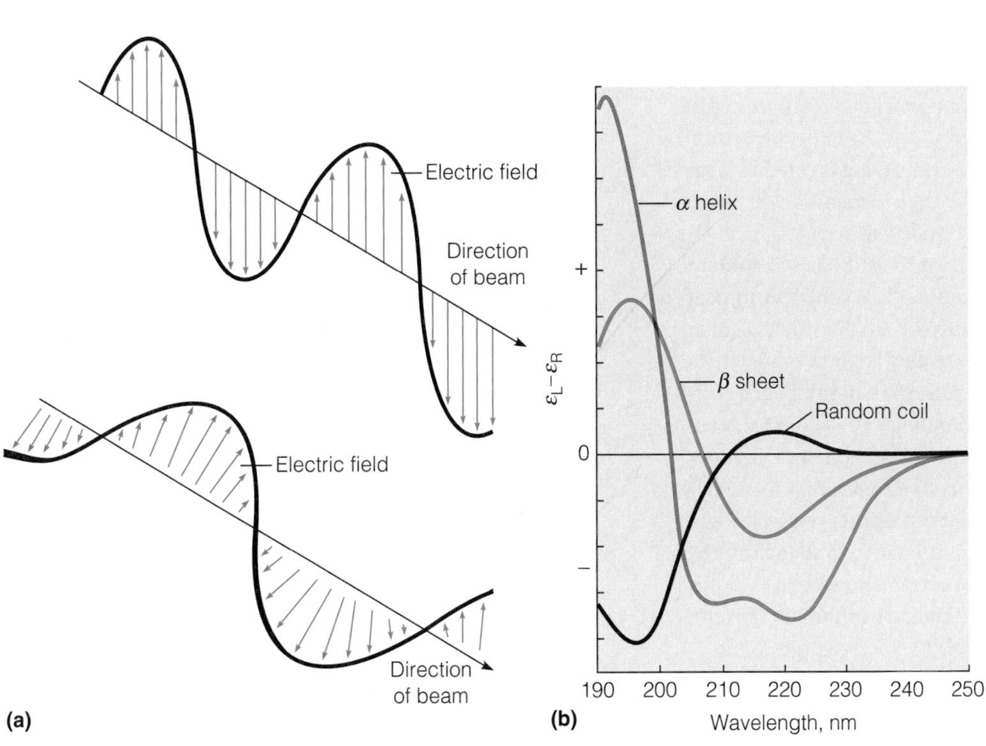

FIGURE **6A.5**

Circular dichroism. (a) Polarization of light. Above: Plane polarized light. Below: Circularly polarized light. **(b)** Circular dichroism spectra for polypeptides in various conformations.

Light can be polarized in various ways. Most familiar is **plane polarization** (Figure 6A.5a, top), in which the varying electric field of the radiation has a fixed orientation. Whereas *unpolarized* light consists of waves vibrating in *all* planes perpendicular to the direction of travel, plane polarized light has waves vibrating in a single plane. Less familiar, but equally important, is **circular polarization,** in which the direction of polarization *rotates* with the frequency of the radiation (Figure 6A.5a, bottom). If you observe a circularly polarized beam coming toward you, the electric field can be rotating in either a clockwise or counterclockwise direction. The former is called *right circularly polarized light,* the latter *left circularly polarized light.*

Most of the molecules studied by biochemists are asymmetric—for example, L- and D-amino acids, right- and left-hand protein helices, and right- and left-hand nucleic acid helices. Such molecules exhibit a preference for the absorption of either left or right circularly polarized light. For example, a right circularly polarized beam interacts differently with a right-hand α helix than does a left circularly polarized beam. This difference in absorption, called **circular dichroism,** is defined as

$$\Delta A = \frac{A_L - A_R}{A} \tag{6A.3}$$

where A_L is the absorbance for left circularly polarized light, A_R is the corresponding quantity for right circular polarization, and A is the absorbance for unpolarized light. Because ΔA can be either positive or negative, a **circular dichroism spectrum** (or **CD spectrum**) is unlike a normal absorption spectrum in that both + and − values are allowed.

TABLE **6A.1**

NUCLEI MOST OFTEN USED IN BIOCHEMICAL NMR EXPERIMENTS

Isotope	Spin	Natural Abundance[a] (%)	Relative Sensitivity[b]	Applications
^{1}H	½	99.98	(1.000)	Almost every kind of biochemical study
^{2}H	1	0.02	0.0096	Studies of selectively deuterated compounds
^{13}C	½	1.11	0.0159	Studies of specific carbon-containing groups
^{19}F	½	100.00	0.834	Substituted for H as a "probe" of local structure
^{31}P	½	100.00	0.0664	Studies of nucleic acids and phosphorylated compounds

[a]The number represents the percentage of this isotope in the naturally occurring mix of isotopes of each element. Isotopes with figures that are close to 100% can be studied directly in the naturally occurring biopolymers. Rare isotopes, such as ^{2}H (deuterium) and ^{13}C, often have to be artificially enriched in substances to be studied.

[b]Indicates the sensitivity (relative to ^{1}H) of conventional NMR instruments to each isotope. Low values mean that the experiment will be more difficult or time-consuming.

Figure 6A.5b shows CD spectra for polypeptides in the α helix, β sheet, and random-coil conformations. The three spectra in the figure are very different, so circular dichroism can be a powerful tool for following conformational changes in proteins. For example, if a protein is denatured so that its native structure, containing α helix and β sheet regions, is transformed into an unfolded, random-coil structure, this transformation will be reflected in a dramatic change in its CD spectrum.

Circular dichroism can also be used to estimate the content of α helix and β sheet in native proteins. The contributions of these different secondary structures to the circular dichroism at different wavelengths are known, so we may attempt to match an observed spectrum for a protein by a combination of such contributions. This kind of analysis frequently turns out to agree with the secondary structure composition as determined by x-ray studies, and it has given support to the idea that the structures of globular proteins observed in crystals are preserved when the crystals are dissolved in buffer solutions at physiological pH.

Although circular dichroism is an extremely useful technique, it is not a very discriminating one. That is, it cannot, at present, tell us what is happening at a particular *point* in a protein molecule. A method that has the potential to do so is nuclear magnetic resonance.

Nuclear Magnetic Resonance

General Principles: One-Dimensional NMR

The nuclei of certain isotopes of some elements have a property referred to as **spin,** which makes the nuclei behave like minute magnets. Only a limited number of isotopes have this property; some that are of use to biochemists are listed in Table 6A.1. If an external magnetic field is applied to a sample containing such nuclei, different orientations of the nuclear spin will have different energies. Microwave radiation can flip these nuclei from one energy state to another, a phenomenon called **nuclear magnetic resonance (NMR).** NMR is really a kind of spectroscopy. As Figure 6A.6a shows, at a given magnetic field, a particular energy corresponding to a particular wavelength in the microwave region of the spectrum will correspond exactly to the energy difference between the spin states. So the experimenter could use a fixed magnetic field and change the microwave wavelength until "resonance" was obtained. Equally well, the experiment can be

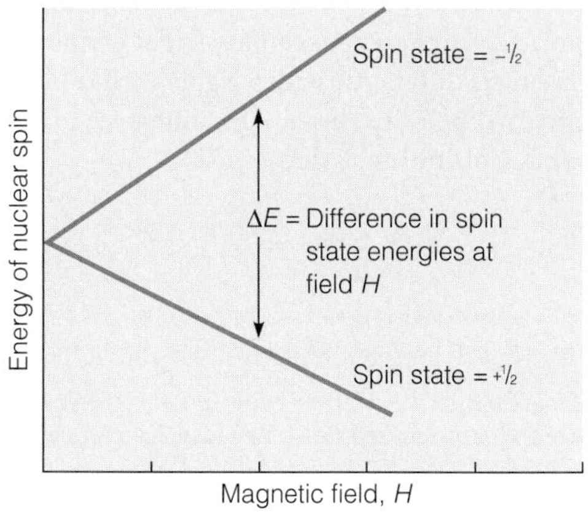

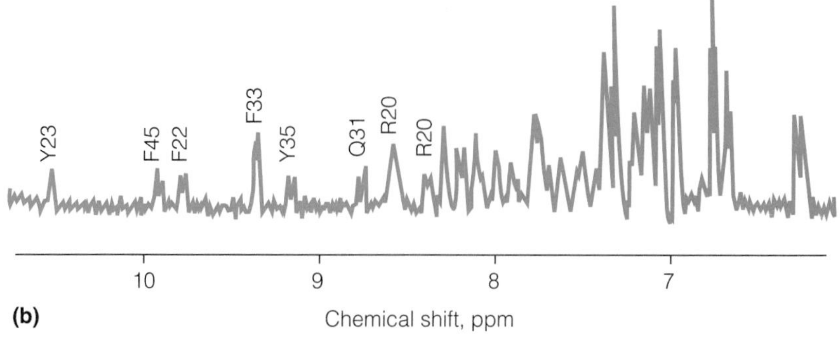

FIGURE **6A.6**

Nuclear magnetic resonance (NMR). (a) The principle of NMR. **(b)** A portion of the ^{1}H NMR spectrum of bovine pancreatic trypsin inhibitor. The axis is the chemical shift, δ, in parts per million (ppm). Letters refer to resonances assigned to specific groups of atoms.

K. Wüthricht and G. Wagner, *J. Mol. Biol.* (1979) 130:1–18. © 1979 Academic Press.

done the other way around: The wavelength of the radiation is held fixed, and the magnetic field is varied to achieve resonance.

The energy levels of a spinning nucleus in a magnetic field are very sensitive to the environment surrounding the atom in question. Different hydrogens in a compound, for example, will reach resonance at different magnetic field strengths. These differences are usually expressed in terms of *chemical shifts* (δ) defined with respect to a reference material added in the sample:

$$\delta = \frac{H_{ref} - H}{H_{ref}} \times 10^6 \qquad (6A.4)$$

Here H is the field strength for resonance for the nucleus in question, and H_{ref} is that for a reference nucleus.

With modern NMR instruments, it is possible to resolve resonances for most or all of the protons in even a large molecule; an NMR spectrum for the protein bovine pancreatic trypsin inhibitor, for example, is shown in Figure 6A.6b. If these resonances can be identified, a task of some difficulty, it becomes possible to ask what is happening to a particular group or atom in a molecule as large as a protein. An example of such discrimination is given in Figure 6A.7, which uses NMR to trace the titration curves of individual histidine residues in the protein ribonuclease. The figure also graphically demonstrates a principle alluded to in Chapter 5: Individual groups of a given type can show quite different pK_a values because of their different environments within the complex protein molecule.

Examples of the discriminatory power of NMR are legion. As another example, we can use ^{31}P NMR to show what the phosphate groups in a nucleic acid molecule are doing. The ^{13}C NMR spectra in Figure 6A.8 resolve the individual types of carbon atoms in the synthetic polypeptide poly(γ-benzyl)-L-glutamate. Figure 6A.8 demonstrates another important application of NMR. When atoms can move only slowly in solution, NMR lines broaden. Note the contrast in sharpness of the peaks between the helix and random-coil forms of this polypeptide. In the random coil, the backbone and side chains are free to swing about in random motion, but in the α helix each is locked in place and can move only with the sluggish motion of the whole large rodlike molecule. Thus, in this case NMR allows us to study in a most direct manner the helix ⟶ random-coil transition.

Multidimensional NMR

The kind of one-dimensional spectrum shown in Figure 6A.6b can tell us much about the behavior of individual atoms in a protein, but it does not reveal all of the information potentially available in NMR. The development of "two-dimensional NMR" of proteins, over the past several years, has opened whole new ranges of applications.

Basically, the 2-D NMR techniques of *correlation spectroscopy* (COSY) and *nuclear Overhauser effect spectroscopy* (NOESY) depend on the fact that spins on different protons interact or "couple" with one another. By the use of special techniques, involving

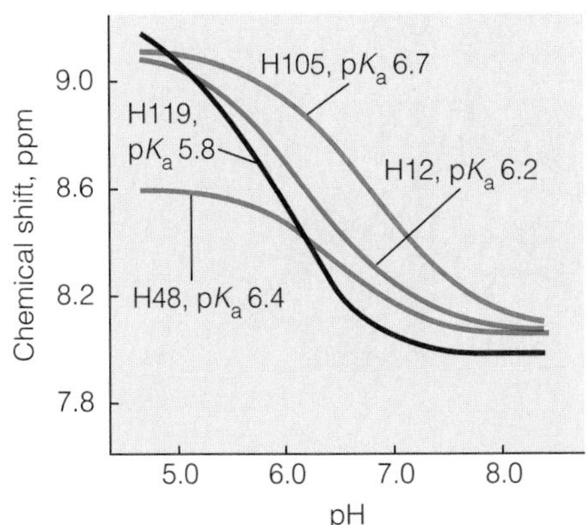

FIGURE **6A.7**

Titration of the four histidine residues in ribonuclease by NMR. Each curve follows the titration of an individual histidine group. The labels such as H12 and H48 indicate the positions of the groups in the chain. The two histidines with lowest pK_a (H12 and H119) are involved in the catalytic process.

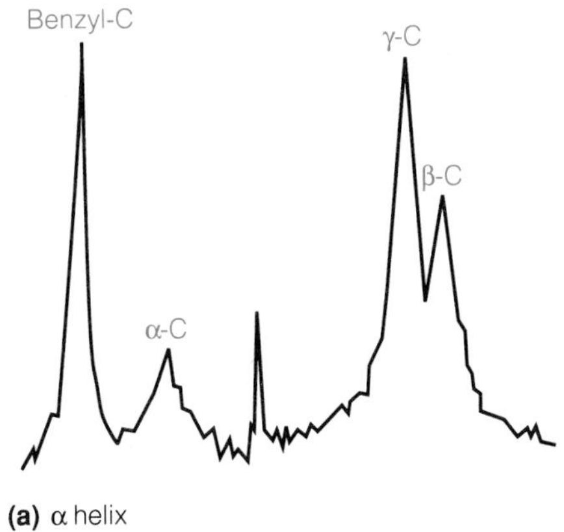

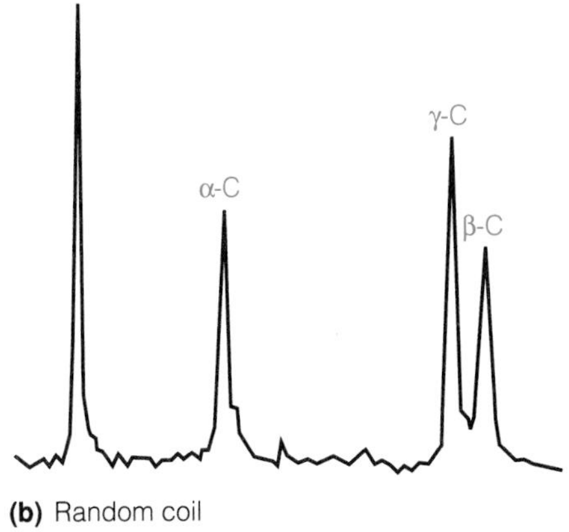

FIGURE **6A.8**

The ^{13}C NMR spectra of two conformations of a polypeptide. Individual peaks for the α, β, γ, and benzyl carbons of the side chains are clearly resolved. **(a)** Spectrum of poly(γ-benzyl)-L-glutamate in the α-helical conformation. **(b)** Spectrum of the same polypeptide in the random-coil conformation. Note how the peaks narrow in going from helix to random coil.

multiple pulses of energy to align nuclear spins, it is possible to perturb the spin of one nucleus and detect its effect on the spin state of another. Protons that are attached to adjacent atoms, and hence can be directly spin-coupled, can be studied by the COSY method. This method allows the researcher to assign NMR frequencies to particular protons by "tracking" from one atom to the next. Even more important is the NOESY spectrum, which depends on the fact that two protons closer than about 0.5 nm will perturb one another's spins *even if they are not closely coupled in the primary structure.* Thus, NOESY allows us to ask which protons are close to which *in space* and provides a route to three-dimensional structure determination. A typical NOESY 2-D spectrum is shown in Figure 6A.9. The spots on the diagonal correspond to the conventional 1-D spectrum, but the spots off the diagonal show close interactions between pairs of protons corresponding to pairs of diagonal spots. For example, that protons are within 0.5 nm is shown by the spot so labeled, and by its counterpart across the diagonal. When combined with proton assignments and bond constraints, such data can provide quite accurate models for small proteins in solution. An example is shown in Figure 6A.10, where the x-ray backbone structure of BPTI is compared with five possible models generated from 2-D NMR spectra.

Recently, the power of these methods has been extended through expansion to other "dimensions," by looking at interactions between, for example, proton spins and ^{13}C spins. The power of NMR to study protein structures is rapidly increasing.

(a) YGRGDSP
1234567

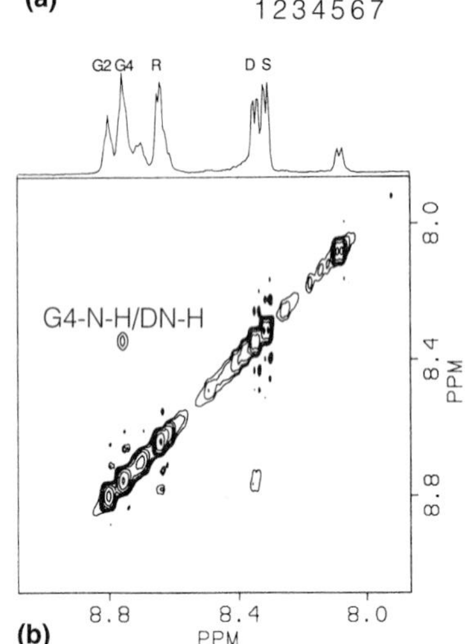

(b)

FIGURE **6A.9**

The use of NOESY NMR spectra to determine atom proximity. (a) The peptide YGRGDSP. **(b)** The NOESY spectrum of the peptide (amide-H region). The presence of a cross-peak corresponding to the interaction of the proton on glycine 4 (G4N-H) with the N—H proton on aspartate (DN—H) shows that these are close in space.

From W. Curtis Johnson, Jr. et al., *Biochem.* (1993) 32:268, fig. 4. © 1993 American Chemical Society.

References

Campbell, I. D., and R. A. Dwek (1984) *Biological Spectroscopy.* Benjamin/Cummings, Menlo Park, Calif. An excellent practical description of many techniques; numerous fine examples and problems.

Clore, G. M., and A. M. Gronenborn (1989) Determination of three-dimensional structures of proteins and nucleic acids in solution by nuclear magnetic resonance spectroscopy. *CRC Crit. Rev. Biochem.* 24:479–564.

Johnson, W. C., Jr. (1990) Protein secondary structure and circular dichroism: A practical guide. *Proteins Struct. Funct. Genet.* 7:205–214.

van Holde, K. E., W. C. Johnson, and P.-S. Ho (1998) *Principles of Physical Biochemistry.* Prentice Hall, Upper Saddle River, N.J. More detailed treatment of these topics than the presentation given here.

Wagner, G., W. Braun, T. F. Havel, T. Schaumann, G. Nobuhiro, and K. Wüthrich (1987) Protein structures in solution by nuclear magnetic resonance and distance geometry. The polypeptide fold of the bovine pancreatic trypsin inhibitor determined using two different algorithms, DISGEO and DISMAN. *J. Mol. Biol.* 196:611–639.

FIGURE **6A.10**

A comparison of the NMR and x-ray structures of BPTI. The heavy line represents the backbone conformation of the x-ray structure. The light lines represent five possible structures calculated from NMR data.

Courtesy of K. Wüthrich from *J. Mol. Biol.* (1987) 196:611. © 1987 Academic Press.

TOOLS OF BIOCHEMISTRY 6B

Determining Molecular Weights and the Number of Subunits in a Protein Molecule

When a new protein has been identified and purified, two questions immediately arise:

1. Does the protein exist under physiological conditions as a single polypeptide chain, or is it made up of multiple subunits?
2. If the functional protein has more than one subunit, are the subunits identical, or are there several kinds?

The answers to these questions can usually be obtained by first determining the molecular weight of the native protein and then subjecting it to conditions under which dissociation into subunits should occur. If subunits are held together by noncovalent interactions, changing the solvent environment will often effect dissociation. For example, the pH might be raised or lowered well outside the physiological range. Alternatively, denaturing solvents like concentrated solutions of urea or guanidinium hydrochloride might be used. These compounds, which are excellent hydrogen bonders, destroy the regular water structure. For this reason they are sometimes called **chaotropic** ("chaos-forming") agents. Destruction of the water structure decreases the hydrophobic effect and thereby promotes the unfolding and dissociation of protein molecules. Even more effective are detergents like sodium dodecyl sulfate (SDS; see bottom of page), which form micelle-like structures about individual polypeptide chains. By determining the molecular weights of the dissociated products in such a solvent and comparing them with the native molecular weight, we can tell how many subunits are involved.

Determining the Molecular Weight of the Native Structure

To determine the molecular weights of proteins in their physiological states, several techniques are available. Recall from Tools of Biochemistry 5A that the *sedimentation coefficient, S,* is related to the molecular weight, *M,* as follows:

$$S = \frac{M(1 - \bar{v}\rho)}{Nf} \tag{6B.1}$$

You will recall from Tools of Biochemistry 5A that $\bar{v}$ is the specific volume of the protein and ρ is the density of the solution. This equation contains the frictional coefficient, *f,* which also depends on molecular size. To eliminate *f,* a separate determination of the **diffusion coefficient,** ***D,*** of the protein may be made. This quantity measures how fast molecules diffuse in solution because of thermal Brownian movement. The diffusion coefficient depends on the temperature, *T,* and the frictional coefficient:

$$D = \frac{RT}{Nf} \tag{6B.2}$$

where *R* is the gas constant, 8.31×10^7 erg/K·mol. The diffusion coefficient can be measured by determining how rapidly a boundary between a solution of the molecule and pure solvent is blurred, or by laser light-scattering techniques (see van Holde et al., Chapter 4). Combining equation (6B.1) with (6B.2), we obtain the *Svedberg equation* for *M:*

$$M = \frac{RTS}{D(1 - \bar{v}\rho)} \tag{6B.3}$$

A more direct way to measure *M* uses the technique of **sedimentation equilibrium.** If a protein solution is sedimented for many hours at low rotor speed, an equilibrium will be established between the tendency of the molecules to sediment and their tendency to diffuse back into the solution. A steady concentration gradient will be set up (Figure 6B.1a). It can be demonstrated that a homogeneous protein will give a gradient described by the equation

$$\frac{c(r)}{c(r_0)} = e^{M(1 - \bar{v}\rho)\omega^2(r^2 - r_0^2)/2RT} \tag{6B.4}$$

where $c(r)$ is the concentration at a distance r from the center of rotation, $c(r_0)$ is the concentration at the meniscus position (r_0), and other quantities are as defined in Tools of Biochemistry 5A. According to equation (6B.4), if $\ln c(r)$ is graphed against $r^2 - r_0^2$, a straight line should be obtained (Figure 6B.1b):

$$\ln c(r) = \ln c(r_0) + \frac{M(1 - \bar{v}\rho)\omega^2(r^2 - r_0^2)}{2RT} \tag{6B.5}$$

The slope of the line is $M(1 - \bar{v}\rho)\omega^2/2RT$, allowing the determination of *M.* An analytical ultracentrifuge allows quite exact measurement of *M* by this technique. Details of sedimentation equilibrium and other physical techniques that can be used to determine the molecular weights of native proteins are given in van Holde et al. (see References).

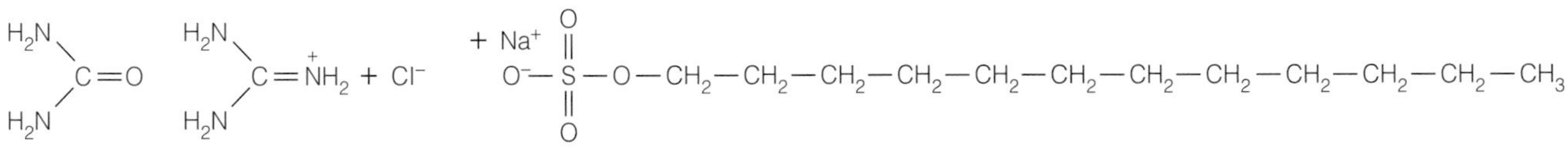

$(H_2N)_2C{=}O$ — **Urea**

$(H_2N)_2C{=}\overset{+}{N}H_2 + Cl^-$ — **Guanidine hydrochloride (guanidinium chloride)**

$Na^+\ ^-O{-}SO_2{-}O{-}CH_2{-}CH_2{-}CH_2{-}CH_2{-}CH_2{-}CH_2{-}CH_2{-}CH_2{-}CH_2{-}CH_2{-}CH_2{-}CH_3$ — **Sodium dodecyl sulfate (SDS)**

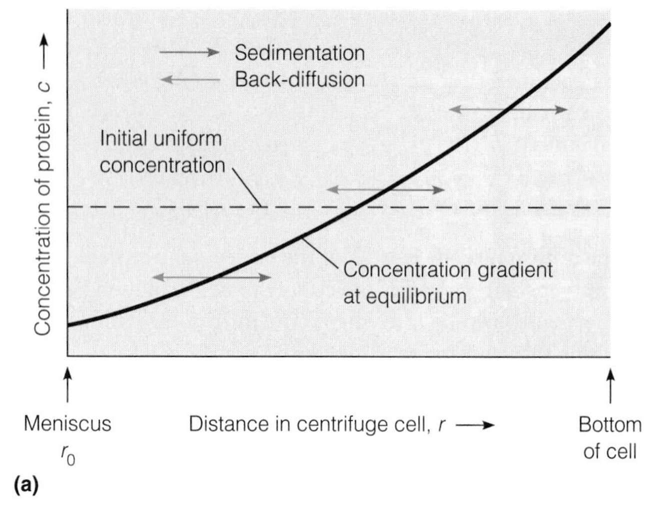

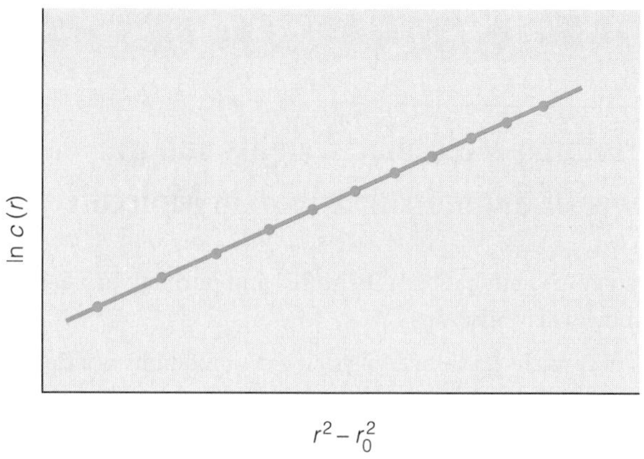

FIGURE **6B.1**

Using sedimentation equilibrium to measure molecular weight. (a) Sedimentation equilibrium is attained when the outward flow of solute caused by its sedimentation is balanced by backflow due to diffusion. **(b)** At the equilibrium state, the concentration gradient is such that a graph of ln c versus r^2 is a straight line, with slope given by $M(1 - \bar{v}\rho)\omega^2/2RT$.

Determining the Number and Approximate Weights of Subunits: SDS Gel Electrophoresis

Once the native molecular weight has been determined, the easiest way to estimate the molecular weight(s) of the subunits is to use gel electrophoresis in the presence of SDS. Under these conditions, quaternary, tertiary, and secondary structures of proteins are all broken down. The chain is unfolded and surrounded by SDS molecules. The numerous negative charges carried by the many SDS molecules bound to the protein make the charge carried by the protein insignificant. The folded polypeptide chain is therefore transformed into an elongated object, the length and charge of which are each proportional to the length (and hence molecular weight) of the chain. As pointed out in Tools of Biochemistry 2A, such particles will migrate in gel electrophoresis with relative mobilities depending only on their lengths. This phenomenon is demonstrated by the graph shown in Figure 6B.2. If electrophoresis of an unknown protein chain is carried out on the same gel as such a set of standards, the molecular weight of the unknown can be measured by interpolation in a graph like Figure 6B.2.

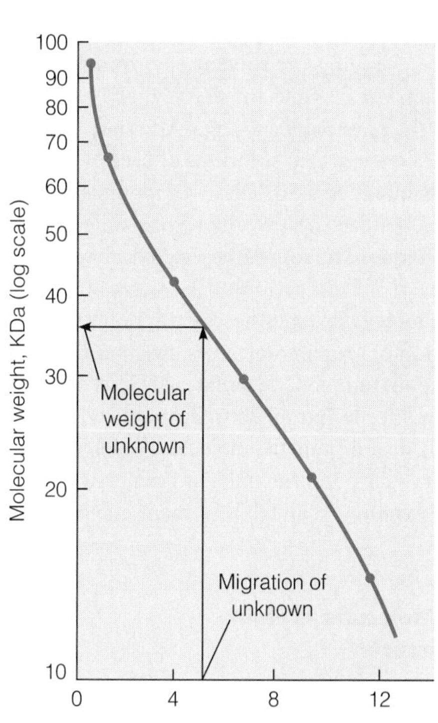

FIGURE **6B.2**

SDS gel electrophoresis. The graph plots log M versus relative electrophoretic mobility for a series of proteins dissolved in a solution containing the detergent SDS. The resulting curve is used to interpolate data for an unknown protein.

In investigating the subunits of a protein by this technique, it is advisable to do two experiments: one in the presence of a disulfide bond–reducing agent like β-mercaptoethanol ($HSCH_2CH_2OH$) and one in its absence. This will distinguish between subunits that are held to one another by —S—S— bridges and those that are held together only by noncovalent forces. If a single band is found on each of these SDS gels, corresponding in molecular weight to that of the native protein, we may conclude that the protein exists under physiological conditions as a single polypeptide chain. If the band or bands observed are of much lower molecular weight, a multisubunit structure is indicated. Even though the molecular weight values obtained from gel electrophoresis may be approximate, they should be sufficiently accurate for a good guess as to the number of subunits (see Problem 13).

Assuming that multiple subunits are indicated, is there only one kind or are there several? More than one band on the SDS gel is a clear indication of multiple types of subunits. But finding only one band does not prove that subunits are identical. There may, in fact, be several kinds of subunits with distinct amino acid sequences but nearly identical molecular weights; these different subunits usually cannot be resolved on SDS gels. To be satisfied that only one type of chain is present, the researcher must turn to other methods. If a way can be found to dissociate the protein without the use of detergents or chaotropic agents, isoelectric focusing can be a very sensitive technique (see Tools of Biochemistry 2A).

Determining Exact Molecular Weights of Protein Subunits: Mass Spectrometry

In recent years an old physical technique (mass spectrometry) has been applied to the study of proteins, with remarkable results. In mass spectrometers, molecules are ionized and then accelerated through a vacuum by an electric field. Particles of different mass/charge ratio are then separated by either deflecting them in a magnetic field, or by simply measuring their "time of flight" to a detector. (See Figure 6B.3.)

Each molecule receives a given charge (often only a few electron units, + or −) during its ionization. The electric field between the anode (the source) and a cathode containing a small aperture gives each molecule a kinetic energy proportional to its charge. Consider two macromolecules of different mass, each of which receives one unit of positive charge. The kinetic energy given to each will be the same, but their velocities will differ because $E = ½\, mv^2$. The bigger molecule will move more slowly. Thus, by timing the arrival at the detector through the "drift" region in the mass spectrometer, one can measure molecular mass.

The results can be extraordinarily accurate—to one part in 10,000 for molecules the size of myoglobin. This is even accurate enough to detect small changes in amino acid sequence. The technique has now been extended to allow the study of molecules as large as that of immunoblogulin (about 150,000 Da).

References

Chait, B. T., and S. B. H. Kent (1992) Weighing naked proteins: Practical, high-accuracy mass measurement of peptides and proteins. *Science* 257:1885–1893.

Hames, B. D., and D. Rickwood, eds. (1990) *Gel Electrophoresis of Proteins,* 2nd ed. IRL Press, Oxford, Washington, D.C.

Senko, M. W., and F. W. McLafferty (1994) Mass spectrometry of macromolecules. *Annu. Rev. Biophys. Biomol. Struct.* 23:763–785.

van Holde, K. E., W. C. Johnson, and P.-S. Ho (1998) *Principles of Physical Biochemistry,* Chapters 4–6. Prentice Hall, Upper Saddle River, N.J.

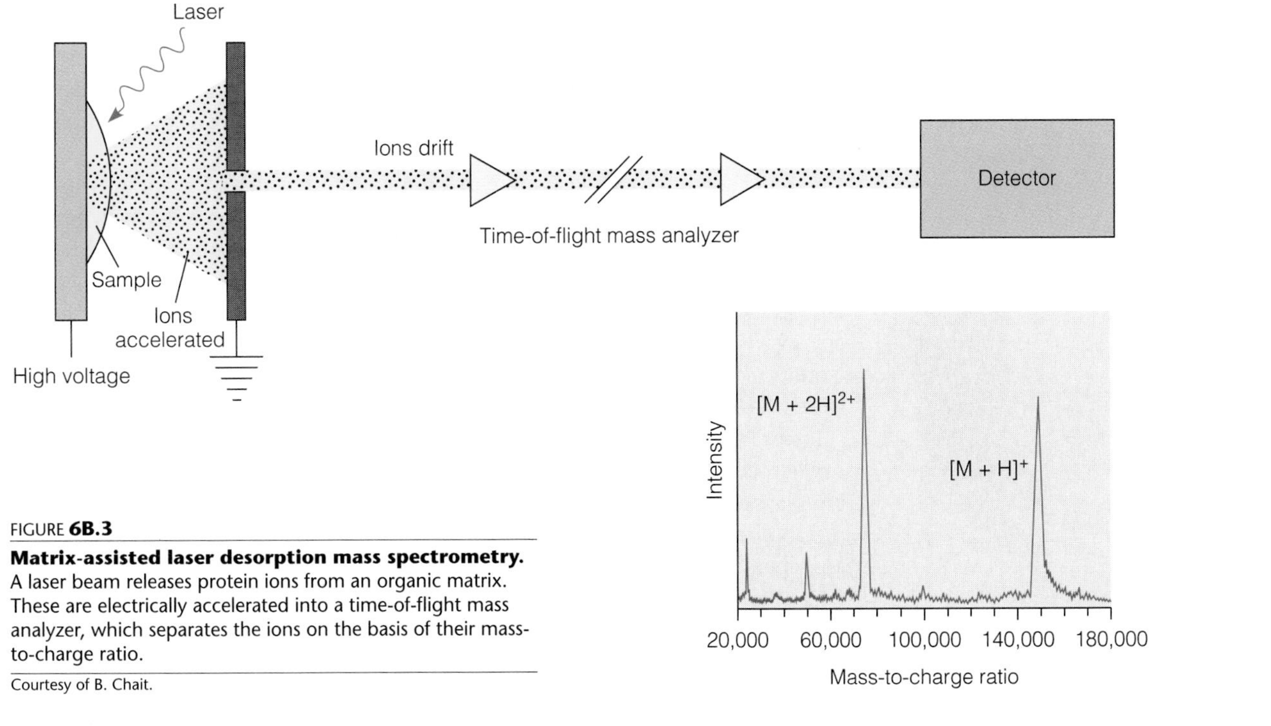

FIGURE **6B.3**

Matrix-assisted laser desorption mass spectrometry. A laser beam releases protein ions from an organic matrix. These are electrically accelerated into a time-of-flight mass analyzer, which separates the ions on the basis of their mass-to-charge ratio.

Courtesy of B. Chait.

CHAPTER 7

Protein Function and Evolution

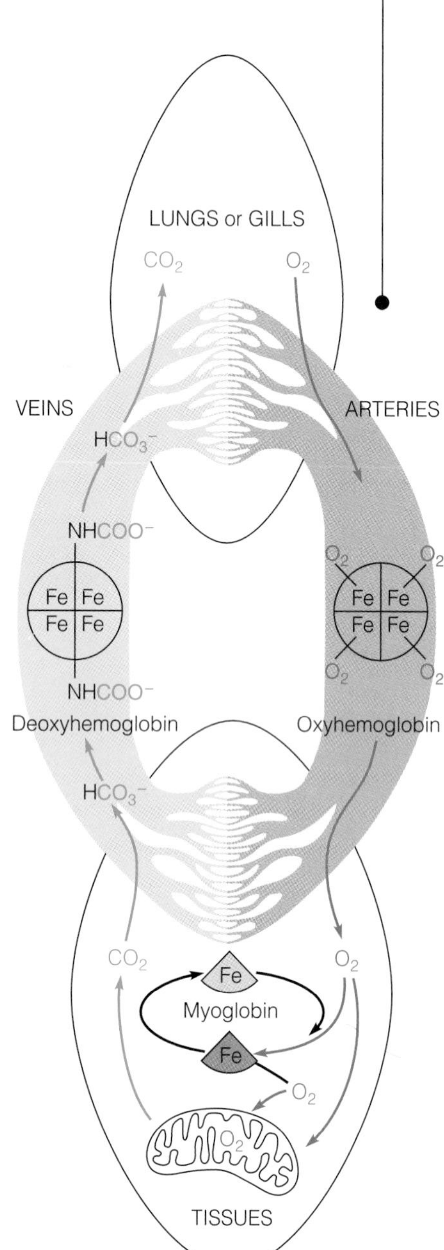

NOW THAT WE HAVE DEVELOPED AN UNDERSTANDING OF THE COMPLEX, FOLDED structures of globular proteins, we are in a position to look more closely at how such structures are related to the molecules' functions and how they may have evolved to fulfill those functions. We will take as our examples in this chapter two groups of proteins, the globins and immunoglobulins, whose main function is binding other molecules.

To begin with, we will examine **myoglobin** (abbreviated **Mb**) and its molecular relative **hemoglobin (Hb)**, members of a family of proteins collectively termed **globins.** We have chosen these examples for a number of reasons. First, the hemoglobins and myoglobins play vital roles in one of the most important aspects of animal metabolism—the acquisition and utilization of oxygen. The most efficient energy-generating mechanisms in animal cells require molecular oxygen for the oxidation of foodstuffs. Therefore, proteins that can deliver oxygen to cells and store it until needed are essential for any higher organism. Myoglobin is the oxygen storage protein used in virtually all animal species; hemoglobin is used for oxygen transport in all vertebrates and some invertebrates. In addition, hemoglobin plays a second role in removing CO_2 from tissues. CO_2 is a major product of metabolite oxidation, and must be continually removed and exhaled. These roles are schematically illustrated in Figure 7.1. Because oxygen delivery and CO_2 uptake must be carefully regulated to meet tissue needs, these examples can teach much about regulation of protein function. Finally, the relationship between he-

FIGURE **7.1**

Role of the globins in oxygen transport and storage. Vertebrate animals use hemoglobin and myoglobin to provide their tissues with a continuous oxygen supply. Hemoglobin transports oxygen from the lungs or gills to the tissues, where part of it may be used directly for metabolism in the mitochondria. The remaining oxygen may be stored by binding to myoglobin and then becomes available when oxygen demand is heavy. CO_2 released by oxidative processes in the tissues is carried back to the lungs or gills by hemoglobin and expired.

moglobin and myoglobin gives some important insights into how protein function evolves. We probably know more about the structure, function, and evolution of the globin family than about any other group of proteins.

Whereas the hemoglobins and myoglobins are devoted primarily to the binding of a single kind of molecule (oxygen), the **immunoglobulins** (or antibody molecules) are protein structures that can be produced in a multitude of variations to bind many different kinds of molecules in the immune response. Our primary defenses against disease and infection depend on the versatility of these proteins. Structure–function relationships for another very large and important group of proteins, the enzymes, are discussed in Chapter 11.

Oxygen Transport and Storage: The Roles of Hemoglobin and Myoglobin

Hemoglobin and myoglobin are proteins that have evolved to carry out the specialized functions of oxygen transport and storage in animals. Any animal larger than a few millimeters in diameter faces a serious problem in carrying out **aerobic** (oxygen-requiring) metabolism. It must ensure a steady supply of oxygen to cells throughout its body and remove metabolic waste products such as carbon dioxide. These gases will diffuse through tissues, but transport by diffusion becomes very slow if appreciable distances must be crossed. Insects solve the problem by having **tracheae,** tubular networks that lead from the body surface down into the tissues. They have, in effect, increased their body surface area to the extent that diffusion is practicable. This mechanism works because insects are small. (Alternatively, we might conclude that insects are small because they rely on this mechanism for obtaining oxygen.*)

Almost all other animals pick up oxygen in lungs or gills and pump it in the blood through arteries to the tissues (see Figure 7.1). Carbon dioxide is returned in the venous blood and released in the lungs or gills. In some primitive organisms, the gases are simply dissolved in the blood, but this mechanism is very inefficient, for oxygen has low solubility. Much blood must be pumped, at great metabolic expense, to deliver even a little oxygen in this way. Evolution of all higher organisms has been accompanied by the development of **oxygen transport proteins,** which allow the blood to carry a much greater load of oxygen than would be permitted by solubility alone. Oxygen transport proteins may be either dissolved in the blood (as in some invertebrates) or concentrated in specialized cells, like the human erythrocytes shown in Figure 7.2. In all vertebrates the oxygen transport protein is hemoglobin, a protein that can pick up oxygen in lungs or gills and deliver it to tissues.

All but the smallest animals need a protein—hemoglobin in vertebrates—to transport O_2 from gills or lungs to tissues. The oxygen storage protein used by most animals is myoglobin.

Once the oxygen is transported to the tissues, it must be released for utilization. Some tissues, such as muscle, require large oxygen reserves for periods when energy demands are high. To ensure a large oxygen reserve, most animals use myoglobin to store oxygen. Even some one-cell organisms (*Paramecium,* for example) have such proteins, probably to store oxygen for periods of oxygen deficit.

Myoglobin and hemoglobin are built on a common structural motif, as shown in Figure 7.3. In myoglobin, a single polypeptide chain is folded about a prosthetic group, the **heme,** which contains the oxygen binding site. Hemoglobin is a tetrameric protein, made up of four polypeptide chains each of which closely resembles myoglobin. We consider now the question, How do these structures serve

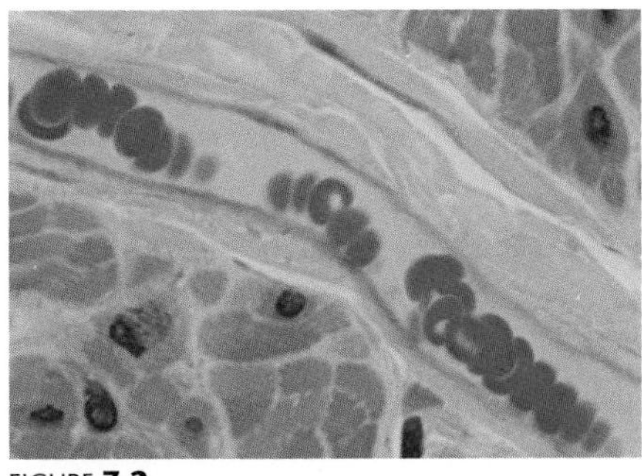

FIGURE **7.2**

Human erythrocytes. These red blood cells, or erythrocytes, are shown moving in a capillary. Each erythrocyte contains about 300 million hemoglobin molecules.

*This hypothesis raises an intriguing problem: How do we explain the giant dragonflies of the Carboniferous period, which had wingspreads up to 2 feet? Was the oxygen content of the atmosphere higher then (about 300 million years ago), or did they use some other mechanism for oxygen uptake? Recent evidence suggests that the first explanation may be correct.

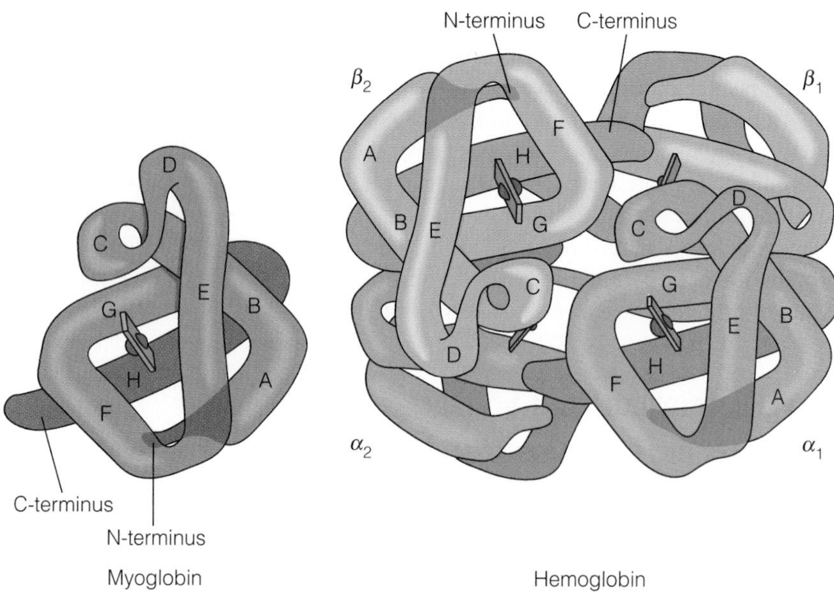

FIGURE **7.3**

Comparison of myoglobin and hemoglobin. These drawings show the structures of the two oxygen-binding molecules as revealed by x-ray crystallography. Each of the four chains in hemoglobin has a folded structure similar to that of myoglobin, and each carries a heme (shown in red). Hemoglobin contains two identical α chains and two identical β chains. The α and β chains are very similar but are distinguishable in both primary structures and folding. The letters A–H indicate α-helical regions.

the similar, but different, functions of oxygen storage and transport? We begin with the simpler protein, myoglobin, asking first how oxygen is bound.

The Mechanism of Oxygen Binding by Heme Proteins

An oxygen-storing or transport molecule must satisfy strict requirements. Any such molecule must be able to bind O_2, not allow it to oxidize any other substance (which would reduce the O_2), and then give up the O_2 on demand. How can a protein molecule play this role? The organic structure of a protein is wholly unsuited in itself to binding of oxygen. But certain transition metals in their lower oxidation states—particularly Fe(II) and Cu(I)—have a strong tendency to bind oxygen. The evolution of the myoglobin–hemoglobin family of proteins has produced a way in which Fe(II) can be bound to the proteins so as to produce an O_2 binding site.

THE OXYGEN BINDING SITE

Various iron-containing proteins can hold Fe(II) in a number of possible ways. Throughout the myoglobin–hemoglobin family, the iron is chelated by a tetrapyrrole ring system called **protoporphyrin IX,** one of a large class of **porphyrin** compounds. The basic ring structure of a porphyrin is shown in Figure 7.4a, and protoporphyrin IX is shown in Figure 7.4b. We will encounter other porphyrins in chlorophyll (Chapter 17), the cytochrome proteins (Chapter 15), and some natural pigments. Like most compounds with large conjugated ring systems, the porphyrins are strongly colored. The iron-porphyrin in hemoglobin accounts for

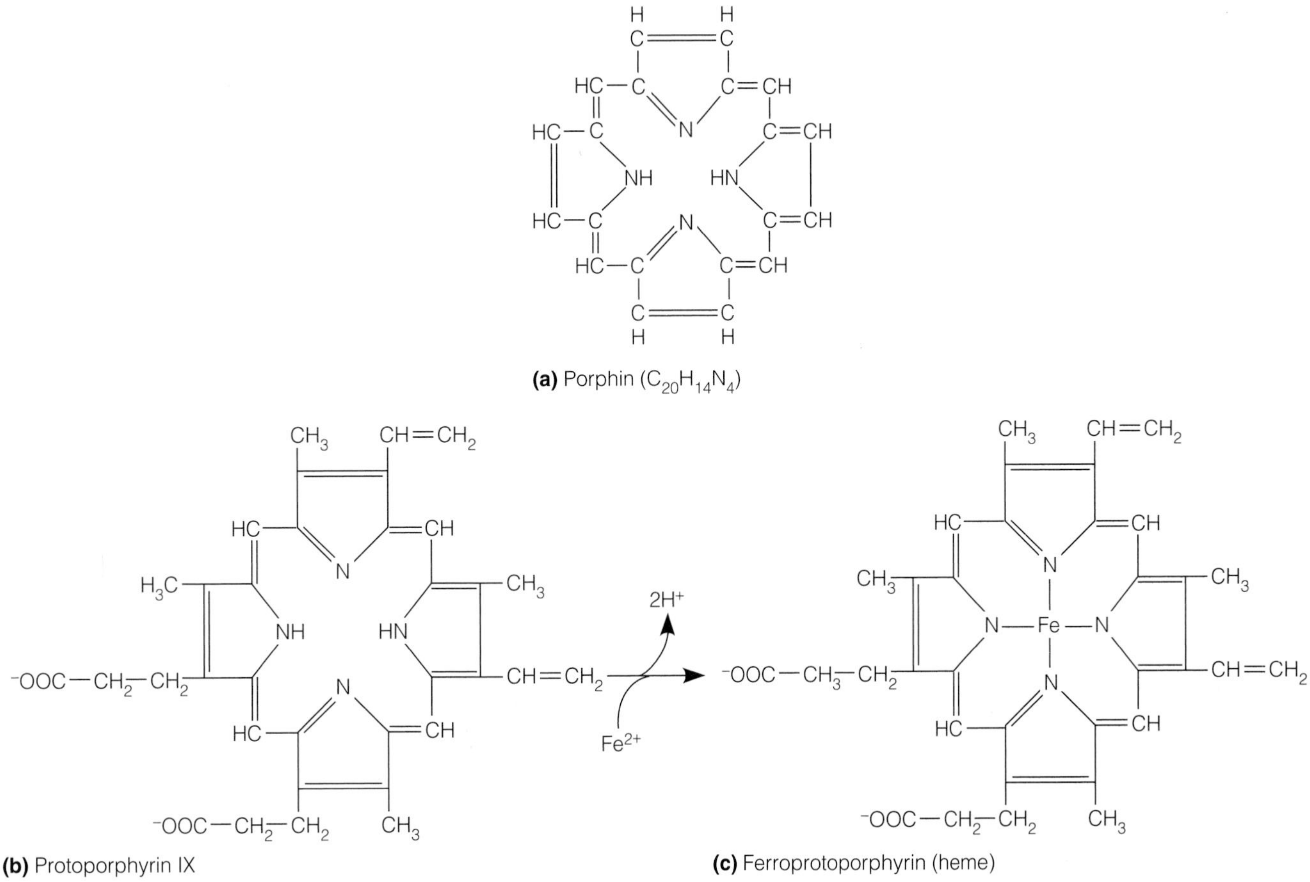

(a) Porphin ($C_{20}H_{14}N_4$)

(b) Protoporphyrin IX

(c) Ferroprotoporphyrin (heme)

FIGURE **7.4**

The structures of porphyrins. (a) The porphyrin ring structure is seen clearly in porphin, the simplest of the porphyrin molecules. **(b)** The more complex protoporphyrin IX is the organic portion of the heme molecule. **(c)** Heme, which is protoporphyrin IX complexed with Fe(II), is the prosthetic group of hemoglobin and myoglobin. Because of resonance delocalization of the electrons in the porphyrin ring, all bonds to the protons in protoporphyrin IX or to the iron atom in heme are equivalent.

Adapted from A. White et al., *Principles of Biochemistry,* 6th ed. (New York: McGraw-Hill, 1978). © 1978 McGraw-Hill.

the red color of blood, and the magnesium-porphyrin in chlorophyll is responsible for the green of plants.

The complex of protoporphyrin IX with Fe(II) is called heme (Figure 7.4c). This prosthetic group is noncovalently bonded in a hydrophobic crevice in the myoglobin or hemoglobin molecule (see Figure 7.3). The binding of oxygen to heme is illustrated in Figure 7.5, which shows the oxygenated form of myoglobin. Ferrous iron (Fe^{2+}) is normally octahedrally coordinated, which means it should have six **ligands,** or binding groups, attached to it. As shown in Figure 7.5a, the nitrogen atoms of the porphyrin ring account for only four of these ligands. Two remaining coordination sites are available, and they lie along an axis perpendicular to the plane of the ring. In both the deoxygenated and the oxygenated forms of myoglobin, one of these remaining coordination sites is occupied by the nitrogen of histidine residue number 93. The eight helical segments in the globins are called A through H (as shown in Figure 6.1 and Figure 7.3), and residue 93 is located in the F helix (Figure 7.5b). In the nomenclature often employed, this residue is called histidine F8. Because it is in direct contact with the Fe, it is also called the **proximal** histidine. In **deoxymyoglobin** the remaining coordination

Coordination of Fe(II) in a porphyrin (heme) within a hydrophobic globin pocket allows O_2 binding without iron oxidation.

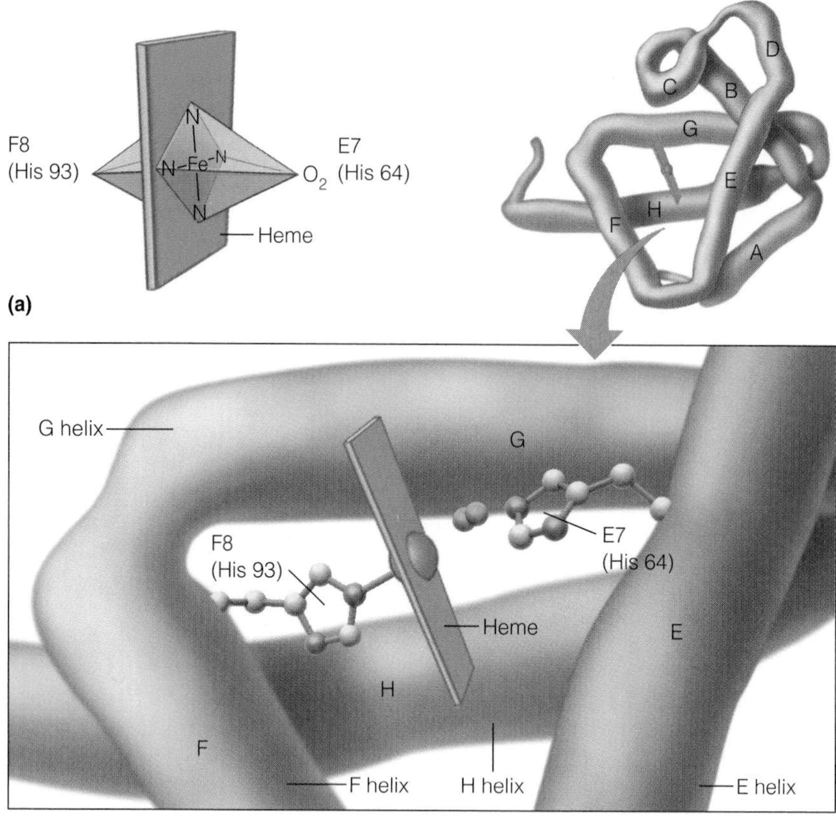

FIGURE **7.5**

The geometry of iron coordination in oxymyoglobin. (a) The octahedral coordination of the iron atom. The iron and the four nitrogens from protoporphyrin IX lie nearly in a plane. A histidine (F8, or His 93) occupies one of the axial positions, and O_2, held on the other side by another histidine (E7, or His 64), the other. **(b)** Close-up of the heme pocket, showing the proximal (F8) and distal (E7) histidine side chains.

site, on the other side of the iron, is occupied by a water molecule. When oxygen is bound, making **oxymyoglobin,** the O_2 molecule occupies this site. On the other side of the bound O_2 lies the **distal** histidine, number 64, or E7. Thus, the O_2 molecule is sandwiched between the ring nitrogen of the distal histidine and the iron atom of heme. An almost identical mode of oxygen binding is found in each subunit of hemoglobin.

Normally, an oxygen molecule in such close contact with a ferrous ion would oxidize the latter to the ferric [Fe(III)] state. The heme alone does not protect the iron, for heme dissolved free in solution is readily oxidized by O_2. But in the hydrophobic, protected environment provided by the interior of the myoglobin or hemoglobin molecule, the iron does not easily become oxidized. The oxygen is bound, and a temporary electron rearrangement occurs. When the oxygen is released, the iron remains in the ferrous state, able to bind another O_2. When myoglobin or hemoglobin is stored in air, outside the cellular environment, the iron does slowly become oxidized, to form what is called **metmyoglobin** or **methemoglobin.** When this oxidation occurs, the binding site is inactivated. The met-proteins will not bind O_2; a water molecule occupies the O_2 site instead.

The protection of an oxygen-binding metal from irreversible oxidation is the functional reason for the existence of myoglobin and hemoglobin. These two molecules provide environments in which the first step of an oxidation reaction (the binding of oxygen) is permitted, but the final step (oxidation) is blocked.

Although the heme pocket is ideally adapted to hold an O_2 molecule, it will accept some other small molecules. The most physiologically important of these others is carbon monoxide, which is approximately the same size and electronic configuration as O_2. However, CO is bound with much greater affinity to myoglobin and hemoglobin than is O_2, and the binding is not readily reversible. This property explains why CO is such a toxic gas; it ties up oxygen binding sites and thereby blocks respiration.

ANALYSIS OF OXYGEN BINDING BY MYOGLOBIN

The binding of oxygen by myoglobin must meet certain physiological requirements. As Figure 7.1 shows, myoglobin in tissues—especially in muscle—accepts oxygen from hemoglobin in the circulating arterial blood. It then delivers the oxygen to the oxygen-consuming organelles of the cells (the mitochondria) when their oxygen needs are sufficiently great. To understand these functions on a quantitative basis, we must examine how the binding of a ligand like oxygen depends on its concentration in the surroundings.

First, we need a way to measure the concentration of dissolved oxygen. According to Henry's law, the concentration of any gas dissolved in a fluid is proportional to the *partial pressure* of that gas above the fluid. Therefore, we can conveniently regulate (and measure) the concentration of dissolved O_2 by regulating the partial O_2 pressure above the myoglobin solution being studied. In fact, we can express oxygen concentration as this partial pressure—P_{O_2}.

To describe or study binding, we must have a way of measuring the fraction of myoglobin molecules carrying oxygen. When myoglobin is oxygenated, the absorption spectrum changes because of electron displacements in the porphyrin ring. This change allows a spectrophotometric determination of the fraction of myoglobin molecules that are oxygenated. The results of such analysis, using myoglobin in solution at neutral pH, are shown in Figure 7.6. Such a graph is called a *binding curve*, because it describes how the fraction of the myoglobin sites that have oxygen bound to them (θ) depends on the concentration (partial pressure) of free oxygen.

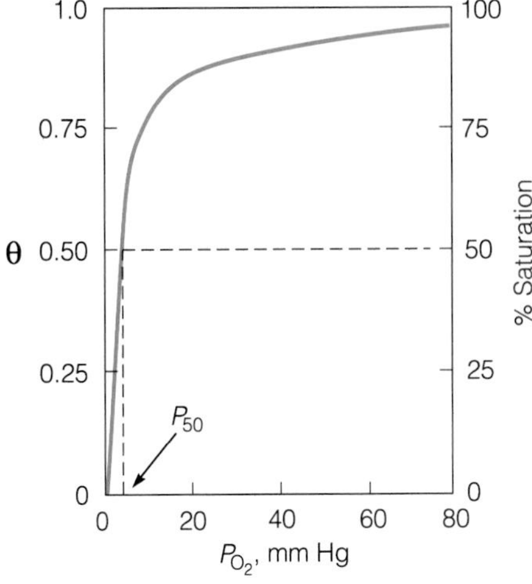

FIGURE **7.6**

Oxygen binding curve for myoglobin. The free oxygen concentration is expressed as P_{O_2}, the partial pressure of oxygen. The proportion of myoglobin binding sites that are occupied is expressed as a fraction (θ, on the left) or as percent saturation (on the right). As P_{O_2} becomes large, saturation is approached asymptotically, as described by equation (7.6). The value of P_{50}, the partial pressure of oxygen at 50% saturation, is indicated on the graph.

The oxygen binding curve for myoglobin has a hyperbolic shape which can be explained as follows. Because the binding equilibrium is described by the reaction

$$\text{Mb} + \text{O}_2 \rightleftharpoons \text{MbO}_2$$

we must have

$$K = \frac{[\text{MbO}_2]}{[\text{Mb}][\text{O}_2]} \tag{7.1}$$

where the equilibrium constant K is called an **association constant** or **affinity constant.** The quantities in brackets denote molar concentrations of oxygenated myoglobin $[MbO_2]$, nonoxygenated myoglobin [Mb], and free oxygen $[O_2]$. The fraction of myoglobin sites occupied is defined as follows:

Binding of a ligand (like O_2) to a single site on a protein (like Mb) is described by a hyperbolic binding curve.

$$\theta = \frac{\text{sites occupied}}{\text{total available sites}} \tag{7.2}$$

Each myoglobin molecule has only one site, so the total number of potentially available sites is proportional to the total molar concentration of myoglobin, $[MbO_2] + [Mb]$. Therefore

$$\theta = \frac{[\text{MbO}_2]}{[\text{Mb}] + [\text{MbO}_2]} = \frac{K[\text{Mb}][\text{O}_2]}{[\text{Mb}] + K[\text{Mb}][\text{O}_2]} \tag{7.3}$$

where we have used $[MbO_2] = K[Mb][O_2]$ from equation (7.1) to obtain the expression on the right. The concentration of myoglobin without bound oxygen, [Mb], can be factored out of the top and bottom and then cancels, to give

$$\theta = \frac{K[\text{O}_2]}{1 + K[\text{O}_2]} = \frac{[\text{O}_2]}{1/K + [\text{O}_2]} \tag{7.4}$$

Equation (7.4) can be rewritten as

$$\theta = \frac{[\text{O}_2]}{[\text{O}_2]_{1/2} + [\text{O}_2]} \tag{7.5}$$

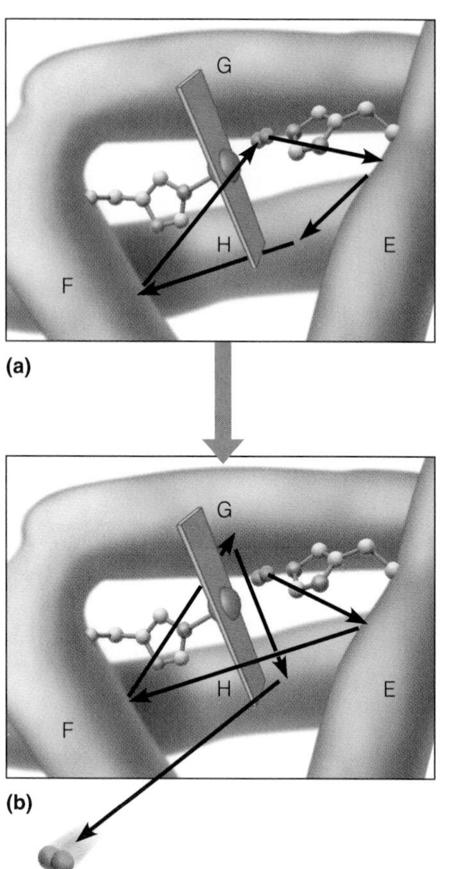

FIGURE **7.7**

Dynamics of oxygen release by myoglobin. (a) The O_2 molecule has been released from the binding site, but it bounces around inside the heme pocket and may rebind. **(b)** The released O_2 escapes if fluctuations in the myoglobin structure open a pathway to the outside.

because $1/K = [O_2]_{1/2}$, the oxygen concentration when half of the myoglobin molecules have oxygen bound to them. You can check this relationship by setting $\theta = ½$ in equation (7.4). Because oxygen concentration is proportional to oxygen partial pressure, equation (7.5) can equally well be written as

$$\theta = \frac{P_{O_2}}{P_{50} + P_{O_2}} \tag{7.6}$$

where P_{50} is the oxygen partial pressure for half-saturation.

We see that equation (7.6) describes the *hyperbolic* binding curve shown in Figure 7.6; θ starts at zero at $P_{O_2} = 0$ and approaches one as P_{O_2} gets very large. The P_{50} for myoglobin is very low (about 4 mm Hg), signifying that myoglobin has a high affinity for oxygen. This characteristic is appropriate for a protein that must extract oxygen from the blood. At the oxygen concentration existing in the capillaries (about 30 mm Hg), the myoglobin in adjacent tissues is nearly saturated. When cells are metabolically active, their internal P_{O_2} falls to much lower levels. Under these conditions, myoglobin will deliver its oxygen.

Thus, the myoglobin molecule must not only provide an environment for the reversible binding of oxygen but also ensure that the affinity constant K (or P_{50}) is of just the right magnitude. We can gain some insight into how this function is accomplished if we recall that the affinity constant is an equilibrium constant and therefore must be the ratio of two rate constants, k_1 for the *binding reaction* and k_{-1} for the *release reaction.* That is,

$$K = \frac{k_1}{k_{-1}} \tag{7.7}$$

Thus, the oxygen affinity could be controlled at the molecular level by regulating either binding or release. We do not yet know much about the binding reaction, but rapid kinetic studies have revealed that much of the regulation involves oxygen release. Computer simulations of the behavior of oxymyoglobin suggest that the rate-limiting process in oxygen release is the opening of a pathway for the O_2 molecule to escape from the heme pocket. In fact, the oxygen may spend some time rattling in its cage—and perhaps being recaptured—before the tertiary structure of the myoglobin shifts enough to let it escape (Figure 7.7). This process is an explicit example of a principle set forth in the preceding chapter—the dynamic internal motions of globular protein molecules play important roles in regulating the processes proteins mediate. The key to just how *strongly* myoglobin binds oxygen may lie in the flexibility of the myoglobin molecule.

In summary, what we observe in myoglobin is a complicated molecular structure that has been selected through evolution to produce exactly the right environment to allow the binding and release of oxygen under the appropriate conditions.

Oxygen Transport: Hemoglobin

All higher animals contain some kind of an oxygen transport protein. In vertebrates and some invertebrates this protein is hemoglobin. All hemoglobins are found to be multisubunit proteins, in contrast to the single-subunit myoglobins. Why should this be so? Investigating this question reveals new aspects of protein function.

COOPERATIVE BINDING AND ALLOSTERY

Consider the peculiar demands placed on an oxygen transport protein. To be useful, it must accept oxygen efficiently at the partial pressure found in lungs or gills (approximately 100 mm Hg) and then unload an appreciable fraction of the O_2

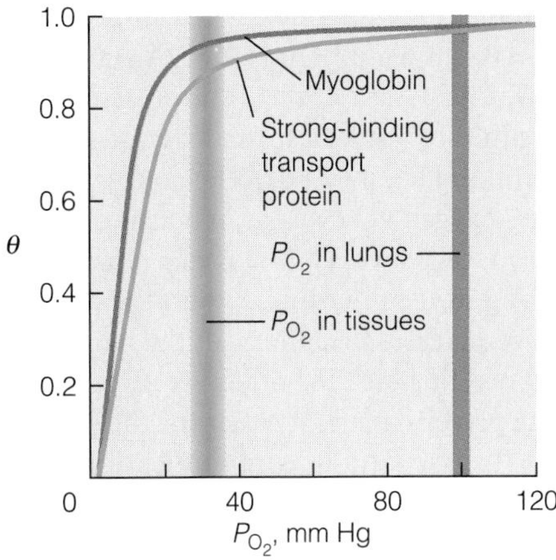

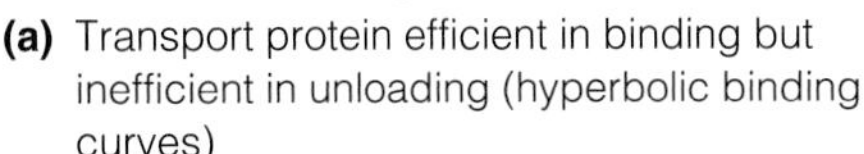
(a) Transport protein efficient in binding but inefficient in unloading (hyperbolic binding curves)

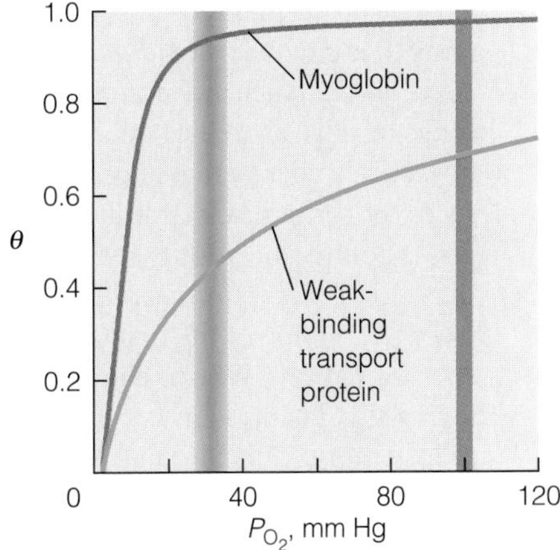

(b) Transport protein efficient in unloading but inefficient in binding (hyperbolic binding curves)

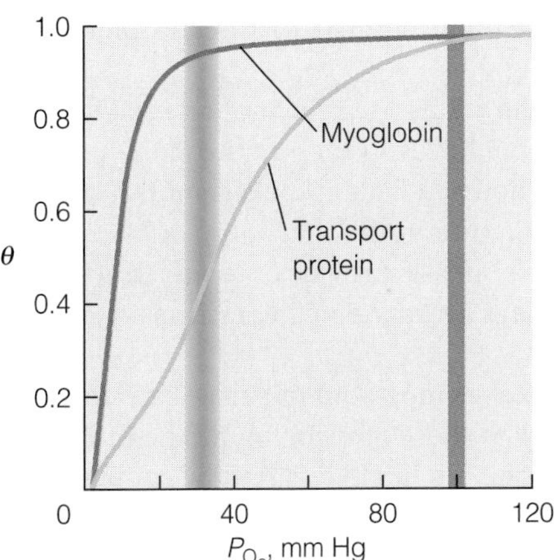

(c) Transport protein efficient in both binding and unloading, because it has a sigmoidal binding curve.

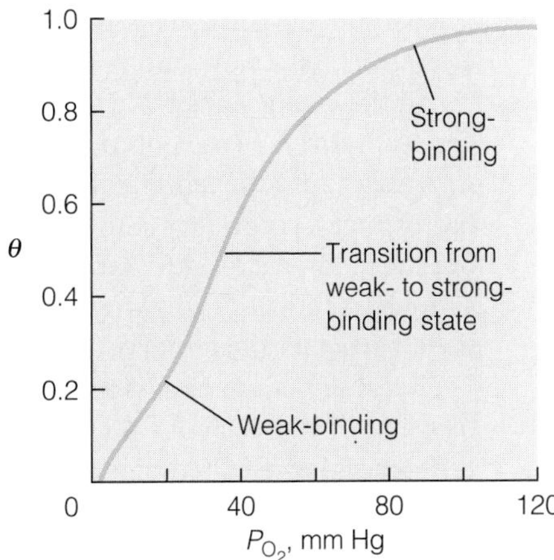

(d) Switch from weak– to strong–binding state explains the sigmoidal curve

FIGURE **7.8**

The binding curve required for a transport protein. These graphs show why an oxygen transport protein such as hemoglobin must have a sigmoidal binding curve. Graphs (a) and (b) show what would happen if the transport protein had a hyperbolic curve like that of myoglobin, with efficiency in either binding (a) or unloading (b). Graphs (c) and (d) show hemoglobin's actual binding curve. **(a)** If the transport protein had a high O_2 affinity, ensuring saturation in the lungs, unloading of O_2 to myoglobin in the tissues would be inefficient. **(b)** If the transport protein had a lower O_2 affinity, it could unload to myoglobin, but then it would not be saturated in the lungs. **(c)** A transport protein that binds efficiently in the lungs and unloads efficiently in the tissues requires a sigmoidal binding curve. **(d)** The sigmoidal binding curve represents the transport protein's switch from a weak-binding state at low oxygen pressures to a strong-binding state at high oxygen pressures.

to tissues, at a pressure of about 30 mm Hg. In other words, an ideal oxygen transport protein would be nearly saturated at 100 mm Hg and unsaturated at about 20–40 mm Hg. In that way, each transport protein molecule could deliver a significant fraction of its oxygen load. If the transport protein had a hyperbolic binding curve like that of myoglobin, such behavior would be impossible to achieve. With such a binding curve, the protein would have to be inefficient either in oxygen uptake or in delivery, as shown in Figure 7.8a and b.

Efficiency in O_2 transport is achieved by cooperative binding in multisite proteins, described by a sigmoidal binding curve.

The problem has been solved through the evolution of oxygen transport proteins that have the *sigmoidal* binding curve shown in Figure 7.8c. A sigmoidal curve is very efficient, because it allows nearly full saturation of the protein in the lungs or gills, as well as maximal release in the capillaries. You can understand how such a curve can be possible by examining Figure 7.8d. At low oxygen pressures the protein acts as if it were binding oxygen very weakly, but as more oxygen is bound, the affinity for oxygen becomes greater. This behavior must mean that a *cooperative interaction* exists among oxygen binding sites in the protein molecule. Filling of the first sites somehow increases the affinity of the remaining sites for oxygen. We can equally well express this idea the other way around, by saying that losing a few oxygens from the protein makes it easier for the protein to lose more. This idea can be true only if some kind of communication takes place among binding sites. A single-site protein, such as myoglobin, cannot accomplish this sort of communication, for one myoglobin molecule is completely ignorant of the state of another. They are independent entities.

Cooperativity in binding requires communication between binding sites.

By contrast, such communication *is* possible between the subunits of a multisubunit protein. This is exactly the evolutionary stratagem that has been adopted. Virtually all oxygen transport proteins are multisubunit structures, exhibiting cooperative interaction among their binding sites.

In the evolutionary line that led to the vertebrates, the protein used for oxygen transport is hemoglobin.* Hemoglobin has evolved from the more primitive myoglobin into the *tetrameric structure* shown in Figure 7.3. Each of the subunits has primary, secondary, and tertiary structures rather like those of myoglobin, but the amino acid side chains in hemoglobin also provide other necessary interactions—salt bridges, hydrogen bonds, and hydrophobic interactions—to stabilize a particular quaternary structure.

Because hemoglobin is a tetramer, each hemoglobin molecule can bind four oxygens, one in each of four sites similar to those of myoglobin. The functional difference between hemoglobin and myoglobin lies in the binding cooperativity exhibited by the hemoglobin sites. This cooperativity is possible because the oxygenation state (filled or empty) of one site can be communicated to another.

We have emphasized that this cooperativity results from a conformational "switch" from a weak-binding state to a strong-binding state. This switching is not easily visualized when binding curves are represented as in Figure 7.8c or d, nor do such curves give an easy way to measure the degree of cooperativity. A rearrangement of equation (7.6) allows a better visualization of cooperativity. If we calculate the quantity $\theta/(1-\theta)$, we obtain

$$\frac{\theta}{1-\theta} = \frac{P_{O_2}}{P_{50}} \tag{7.8}$$

Or, taking logarithms of both sides,

$$\log \frac{\theta}{1-\theta} = \log P_{O_2} - \log P_{50} \tag{7.9}$$

The Hill coefficient is a measure of binding cooperativity.

Graphing $\log [\theta/(1-\theta)]$ versus $\log P_{O_2}$ produces what is called a **Hill plot** (Figure 7.9). The Hill plot for noncooperative binding will, according to equation (7.9), be a straight line, with a slope = 1. In such a plot, the abscissa value (that is, the value of $\log P_{O_2}$) corresponding to $\log[\theta/(1-\theta)] = 0$ will equal $\log P_{50}$. Now

*Hemoglobin is not, however, the only oxygen binding protein used by animals. Most mollusks and some arthropods have a quite different protein, hemocyanin, which contains copper. Still other invertebrates use a wholly unrelated iron-containing protein called hemerythrin. This variety shows that the same function can often be arrived at by several independent evolutionary routes.

consider the Hill plot for a cooperatively binding protein such as hemoglobin. When hemoglobin first begins binding (at low P_{O_2}), its Hill plot has a slope $\cong 1$ and corresponds to the weak-binding state (large P_{50}). As binding progresses, the curve switches over to approach another, parallel straight line that describes the strong-binding state (small P_{50}).

On a Hill plot, the transition between binding states is clear, and the behavior is unmistakably different for cooperative and noncooperative systems. Furthermore, the Hill plot gives a direct numerical measure of the degree of cooperativity from its maximum slope, n_H, which is called the **Hill coefficient.** Three cases may be considered for a molecule with n binding sites:

1. $n_H = 1$: The molecule binds noncooperatively. This situation may happen even with a multisite protein if the sites do not communicate with one another.
2. $1 < n_H < n$: This situation is the usual one for a cooperatively binding protein, as depicted in Figure 7.9. The Hill coefficient must be greater than unity in order for the curve to switch over from the weak-binding line to the strong-binding line.
3. $n_H = n$: In this hypothetical situation the molecule is *wholly* cooperative. In such a situation, one molecule would fill up its sites before any others had taken oxygen, so that only wholly unliganded and wholly liganded molecules would be present at any point in the binding process. If this were the case, the binding curve would have the form

$$\theta = \frac{P_{O_2}{}^n}{P_{50}{}^n + P_{O_2}{}^n} \tag{7.10}$$

and the Hill equation would become

$$\log \frac{\theta}{1-\theta} = n \log P_{O_2} - n \log P_{50} \tag{7.11}$$

which is a straight line with slope = n. This case is never seen in reality. For example, the Hill coefficient of hemoglobin ($n = 4$) is about 3 under physiological conditions.

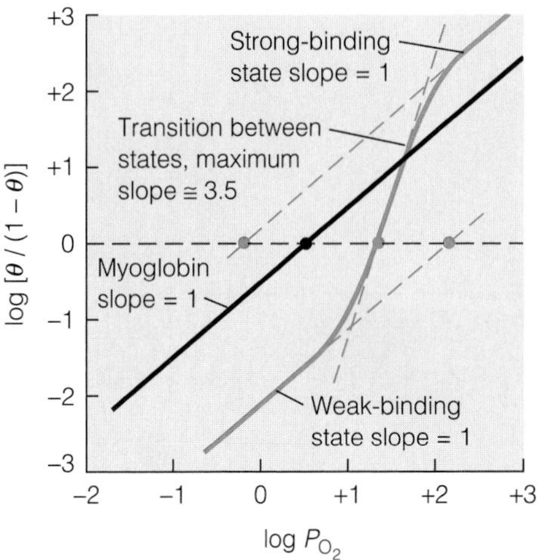

FIGURE **7.9**

Hill plots of oxygen binding for myoglobin and hemoglobin. The circles show the P_{50} of each protein and binding state. The plot for myoglobin, which binds oxygen noncooperatively, is a solid black line with a slope of 1. The plot for hemoglobin (in red), which binds cooperatively, shows the switch from a weak-binding state (large P_{50}) to a strong-binding state (small P_{50}) and has a maximum slope (Hill coefficient) of about 3.5. These data are for proteins under near-physiological conditions.

The cooperative binding of oxygen by hemoglobin is one example of what is referred to as an **allosteric** effect. In allosteric binding, the uptake of one ligand by a protein influences the affinities of remaining unfilled binding sites. The ligands may be of the same kind, as in this case, or they may be different. As discussed in Chapter 11, allostery is also an important mechanism for regulating the activity of enzymes. In that case, allostery allows one kind of small molecule to regulate the action of a protein on another kind of molecule. The ability of multisubunit proteins to permit allosteric regulation is one of the reasons why these proteins are so often involved in chains of reactions which must be carefully controlled.

MODELS FOR THE ALLOSTERIC CHANGE IN HEMOGLOBIN

How do allosteric transitions from weak-binding states to strong–binding states actually occur? A number of theories have been developed to describe allosteric transitions. They may be generally grouped into three classes. Although we shall discuss these models in terms of hemoglobin, they are generally applicable to any multisubunit allosteric proteins.

1. *Sequential models:* The prototype for such models is that of Koshland, Nemethy, and Filmer (Figure 7.10a). The KNF model assumes that the subunits can change their tertiary conformation one at a time in response to binding of oxygen. Cooperativity arises because the presence of some subunits carrying oxygen favors the strong-binding state in adjacent subunits

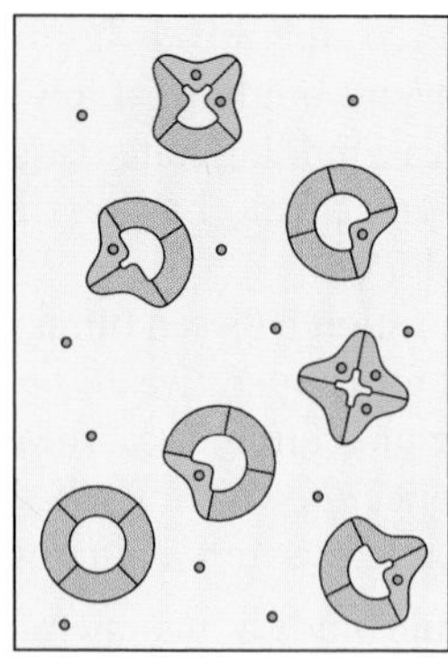

Key:

Hb subunits, weak-binding state

Hb subunits, strong-binding state

Hb tetramer, T state

Hb tetramer, R state

Oxygen

No oxygen bound

Oxygen bound

1 No oxygen bound. Almost all subunits in all molecules are in weak-binding state. Only a few happen to be in the strong-binding state.

2 Some oxygen bound. Each binding of an oxygen molecule favors the transition of adjacent subunits to the strong-binding state and promotes their binding of oxygen.

3 More oxygen bound. More and more subunits next to oxygen-occupied sites are switching to the strong-binding state.

4 Approaching saturation. Almost all sites are filled, and almost all subunits are now in the strong-binding state.

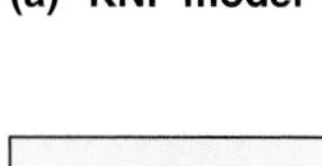

(a) KNF model

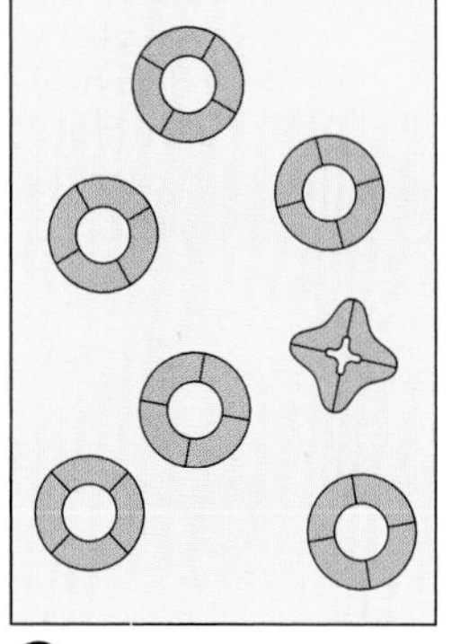
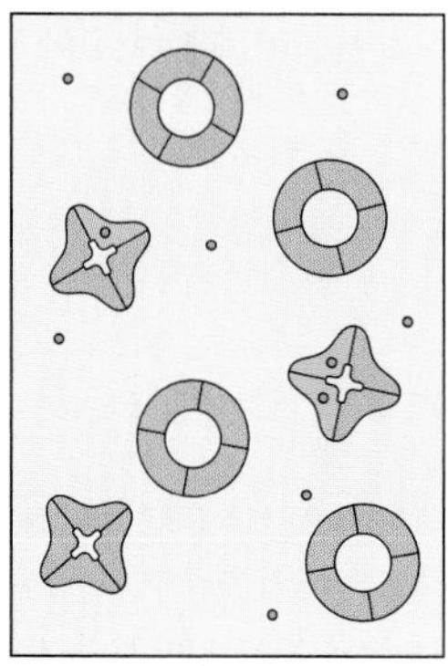
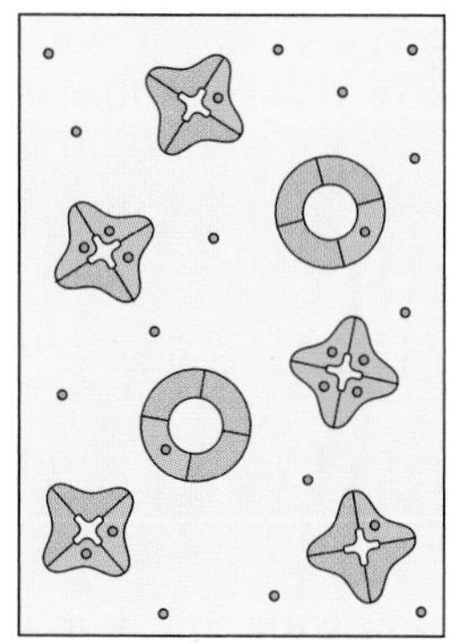
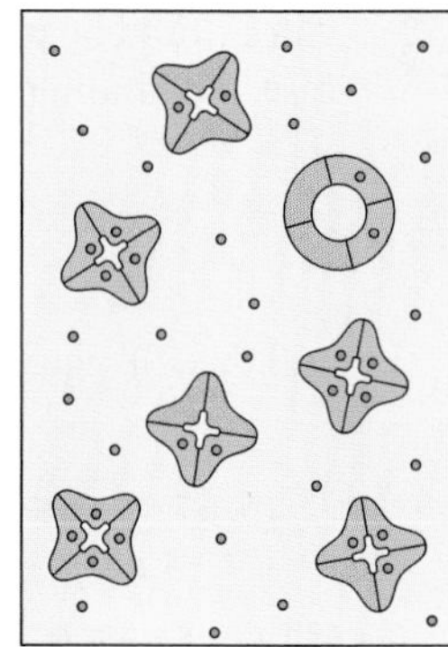

1 No oxygen bound. Most tetramers are in the T state, with only a few in the R state.

2 Some oxygen bound. Preference is for binding to molecules in R state so T ⇌ R equilibrium is shifted toward R.

3 More oxygen bound. Now most molecules are in R state. Note that T also binds oxygen, but more weakly.

4 Approaching saturation. Almost all sites are filled. Almost all molecules have shifted to R state.

(b) MWC model

FIGURE **7.10**

Two models for the cooperative transition of hemoglobin. Both models can explain cooperative binding by some, but certainly not all, real proteins. **(a)** The Koshland, Nemethy, and Filmer (KNF) model. As each subunit binds a ligand, it promotes a conformational change in an adjacent subunit, which thereby becomes strong-binding. **(b)** The Monod, Wyman, and Changeux (MWC) model. The entire molecule has two different states—tense (T) and relaxed (R)—which are in equilibrium. Binding of ligands shifts the equilibrium toward the strong-binding (R) state.

whose sites are not yet filled. Thus, as oxygenation progresses, almost all the sites become strong-binding. Such models are characterized by the existence of molecules with some subunits in the weak state, some in the strong.

2. *Concerted models:* At the opposite extreme lies the theory of Monod, Wyman, and Changeux (Figure 7.10b). According to the MWC model, the entire hemoglobin tetramer exists in an equilibrium between two forms. In the deoxy (T) state, all subunits in each molecule are in the weak-binding con-

formation, and in the oxy (R) state, all are in the strong-binding form. (The symbols T and R stand for "tense" and "relaxed"; the significance of this will be seen in the next section.) An equilibrium between these states is presumed to exist, and partial oxygenation shifts that equilibrium toward the R state. The shift is a *concerted* one, so that mixed molecules with some subunits in the weak-binding state and some in the strong-binding state are specifically excluded.

3. *Multistate models:* It has become clear in recent years that neither the KNF nor the MWC model can *exactly* explain the allosteric behavior of proteins, including hemoglobin. Consequently, more complex models have been devised. Most such models retain the MWC concept of a concerted switch in conformation but involve more than two states for the entire molecule. We shall see that a more complex model is needed for hemoglobin.

CHANGES IN HEMOGLOBIN STRUCTURE ACCOMPANYING OXYGEN BINDING

To understand the allosteric behavior of hemoglobin, it is necessary to examine the protein in more detail. The hemoglobin of higher vertebrates is made up of two types of chains, referred to as α and β. Their primary structures are compared with that of myoglobin in Figure 7.11. As you can see, the α and β sequences have considerable similarity to one another and some similarity to the sequence of myoglobin. Essential residues, like the proximal and distal histidines (F8 and E7, respectively), are conserved, and apparently those critical to the tertiary structure are conserved as well, for the hemoglobin chains and myoglobin all have very similar tertiary structure. The hemoglobin molecule contains two of each kind of chain, so the whole molecule can be called an $\alpha_2\beta_2$ tetramer. The chains are placed in a roughly tetrahedral arrangement as shown schematically in Figure 7.3. When hemoglobin is dissolved in concentrated urea solutions, it dissociates into $\alpha\beta$ dimers, suggesting that the closest and strongest contacts are between α and β chains, rather than α–α or β–β. In other words, the molecule could be thought of as a dimer of $\alpha\beta$ dimers. Figure 7.3 also shows that the heme groups, with their O_2 binding sites, are all close to the surface but *not* close to one another. Therefore, we cannot seek the source of cooperative binding in anything so unsubtle as direct heme–heme interaction.

A key to what is actually happening during oxygenation can be seen in Figure 7.12a and b, in which two viewpoints are given of the structural change on going from the deoxygenated state to the fully oxygenated molecule carrying four oxygens. X-ray diffraction studies reveal that what occurs is mostly a change in the quaternary structure, accompanied by much smaller tertiary structure changes. One $\alpha\beta$ dimer pair rotates and slides with respect to the other, as seen in Figure 7.12a. This movement brings the β chains closer together and narrows a central cavity in the molecule, as can be seen in Figure 7.12b. To a first approximation, then, we can regard the hemoglobin molecule as having two states of quaternary structure, one characteristic of the deoxy form and the other favored by the oxy form. The oxy structure has the higher affinity for O_2, and the switch to this state is what accounts for the cooperativity in binding.

We can now interpret the Hill plot shown in Figure 7.9 in terms of such an allosteric shift between two molecular conformations. Wholly deoxygenated hemoglobin molecules are in the deoxy conformation, so as we add oxygen to a solution of such molecules, the binding initially occurs along the line corresponding to the weak-binding state. But partial oxygenation favors transition to the strong-binding oxy state. As oxygen is bound, more and more of the remaining available sites are in hemoglobin molecules that have this conformation. Therefore, the binding curve passes over to that for the strong-binding state. As the last few sites are

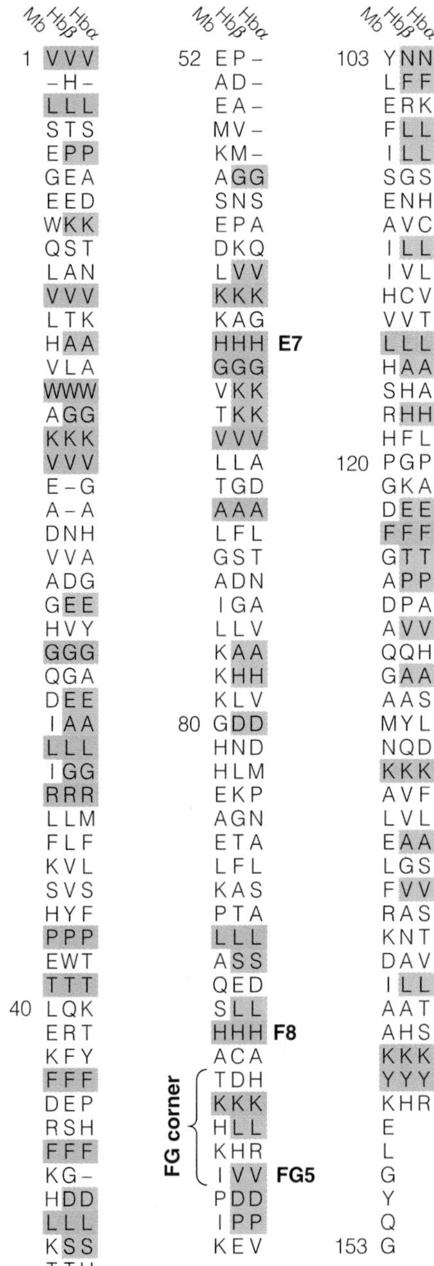

FIGURE **7.11**

Comparison of sequences of myoglobin and the α and β chains of hemoglobin. The aligned sequences are those of whale myoglobin and the two human hemoglobin chains. Gaps (indicated by dashes) have been inserted where necessary to provide maximum alignment of the sequences; the residue numbers to the left of the chains are for the myoglobin sequence. A residue critical to the functioning of these proteins is indicated to the right of the chains; F8 and E7 are the proximal and distal histidines, respectively (see Figure 7.5). Brown indicates the residues that are common to all three chains, and purple the residues common to both hemoglobin chains.

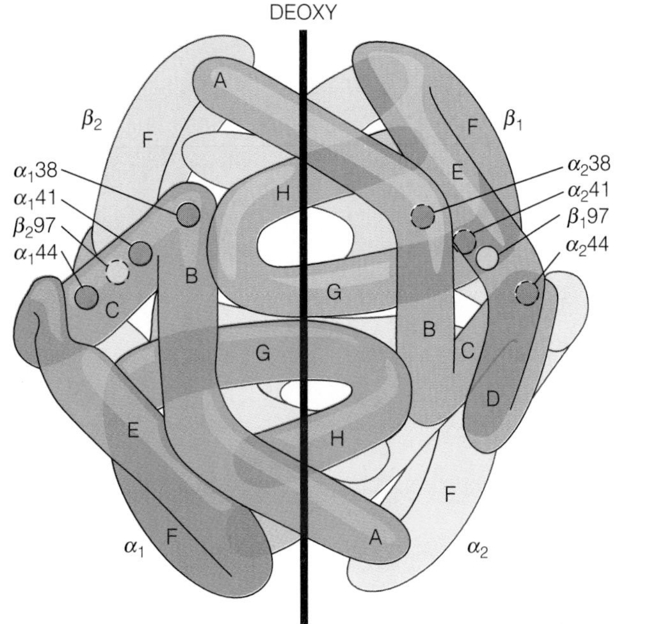

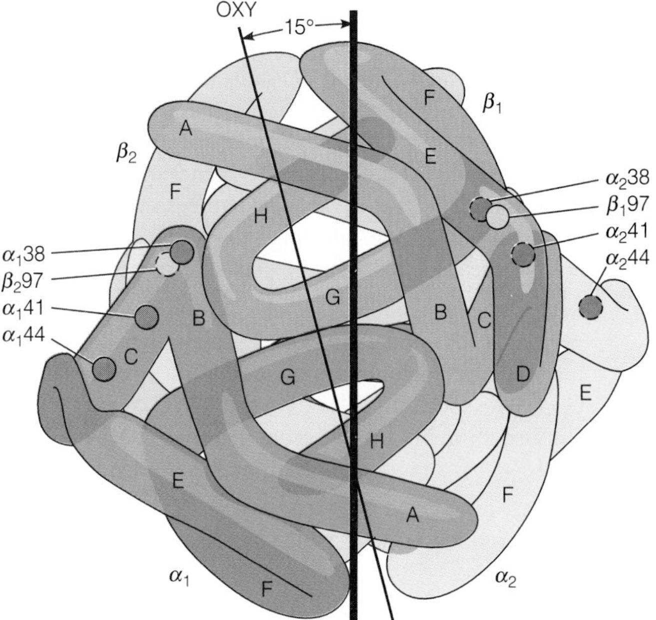

(a)

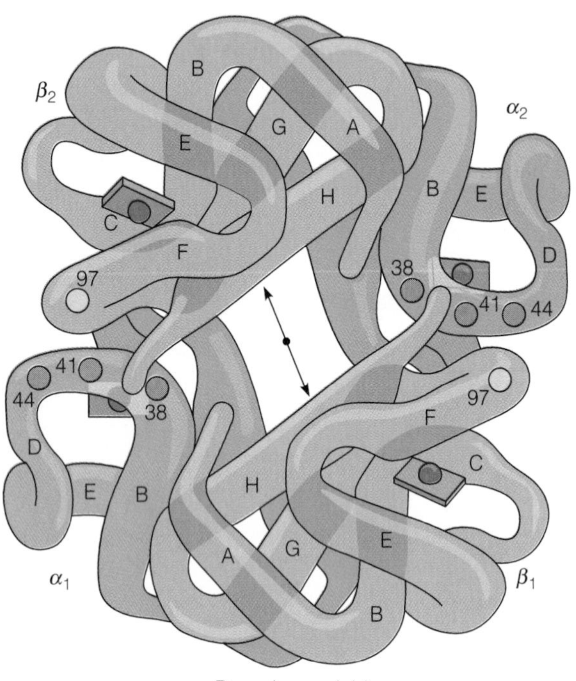

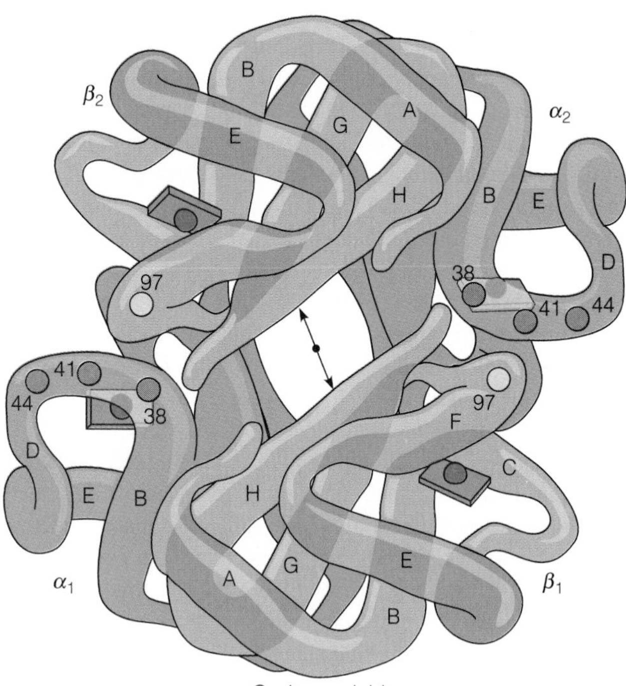

(b)

FIGURE **7.12**

The change in hemoglobin quaternary structure during oxygenation. (a) The transition viewed along an axis perpendicular to the 2-fold axis, with the $\alpha_1\beta_1$ dimer (darker blue and red areas) in front of the $\alpha_2\beta_2$ dimer. Deoxyhemoglobin is shown on the left, and oxyhemoglobin on the right. Note the rotation of $\alpha_1\beta_1$ with respect to $\alpha_2\beta_2$ and the shift of β97 with respect to α41 and α44. The rotation of about 15° is accompanied by sliding, because the center of rotation is not centrally located. **(b)** Top views of hemoglobin, looking down the 2-fold axis (dot in center). The two β subunits are in the foreground; α subunits are in the background. Note in deoxyhemoglobin that the central cavity is broad and that residue 97 on one β chain lies between residues 41 and 44 on the adjacent α chain. The shift from the deoxy to the oxy state is evident by the shrinkage of the central cavity and the shift in the contact of residue β97 with the α chain.

filled, all the molecules have adopted the strong-binding form. This leaves, however, many important questions unanswered. What triggers the change? Will either KNF or MWC models suffice to explain the results?

Vertebrate hemoglobins are tetramers ($\alpha_2\beta_2$) made up of two kinds of myoglobin-like chains.

A CLOSER LOOK AT THE ALLOSTERIC CHANGE IN HEMOGLOBIN

Because x-ray diffraction has made it possible to determine details of both the deoxy and the oxy states of hemoglobin, it is now possible to formulate a quite complete description of the overall change and to speculate concerning its mechanism.

Oxygenation causes hemoglobin quaternary structure to change: One $\alpha\beta$ pair rotates and slides with respect to the other.

The transition from the deoxy to the oxy conformation involves major changes in the details of subunit–subunit interaction. Some idea of this process can be gained from a closer study of Figure 7.12b. Note the region, to the lower left, where the β_2 subunit interacts with the α_1 chain. In the deoxy form, the C-terminus of β_2 (residue 146) lies atop the C helix of α_1 (residues 36–42) and is held in this position by a network of hydrogen bonds and salt bridges. His 97 in the FG corner of β_2 is pushed against the CD corner of α_1, between Thr 41 and Pro 44. In the oxy form, rotation and sliding of the subunits have pulled the C-termini of β chains away from α contacts (Figure 7.12b). The salt bridges and hydrogen bonds holding the C-terminus have been broken, and His 97 of β_2 now lies between Thr 38 and Thr 41 of α_1. Because of the symmetry of the structure, an exactly equivalent set of changes occurs at the $\alpha_2\beta_1$ interface. The molecule has, as it were, "switched" and clicked into a new set of interactions. In the process a number of strong interactions (those involving the C-termini in particular) have been broken. In terms of the MWC model, this looser conformation would be called *relaxed* (R). The energy price for this change is paid by the binding of O_2 to the molecule. Once the O_2 has departed, the molecule will naturally fall back into its lower-energy deoxy conformation. This tighter conformation is, in MWC nomenclature, the *tense* (T) state.

The oxy form of hemoglobin has higher affinity for O_2, accounting for binding cooperativity.

Exactly how is the energy of O_2 binding communicated to effect this molecular switching? Again, the details are complicated, but a partial idea can be gained from Figure 7.13, which shows the relationship of His F8 and the neighboring Val (FG5) to the heme in deoxyhemoglobin. The figure includes an important fact not mentioned previously: Not only is the iron atom in the deoxy conformation a bit above the heme plane, but also the heme itself is not quite flat; it is distorted into a dome shape. Furthermore, in both deoxymyoglobin and deoxyhemoglobin, the axis of His F8 is not exactly perpendicular to the heme but is tilted by about 8°. When oxygen binds to the other side, it pulls the iron atom a short distance down into the heme and flattens the heme (Figure 7.13b,c). This change cannot happen without molecular rearrangement, for such motion would bring both the ε-hydrogen of His F8 and the side chain of Val FG5 too close to the heme.

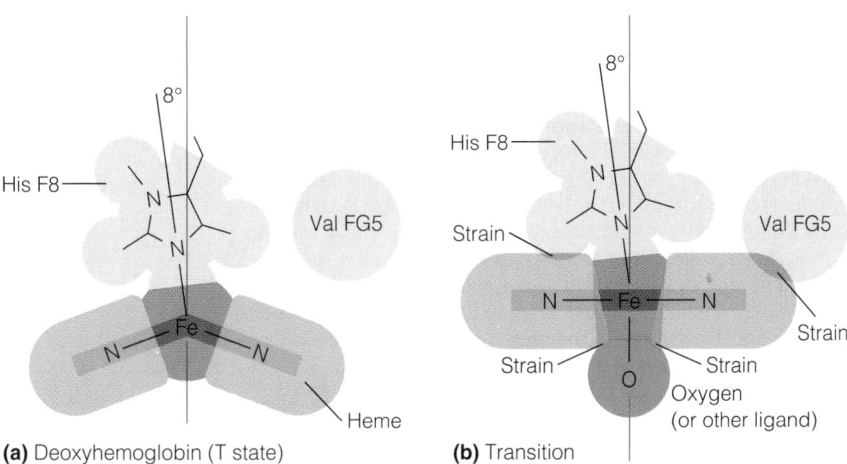

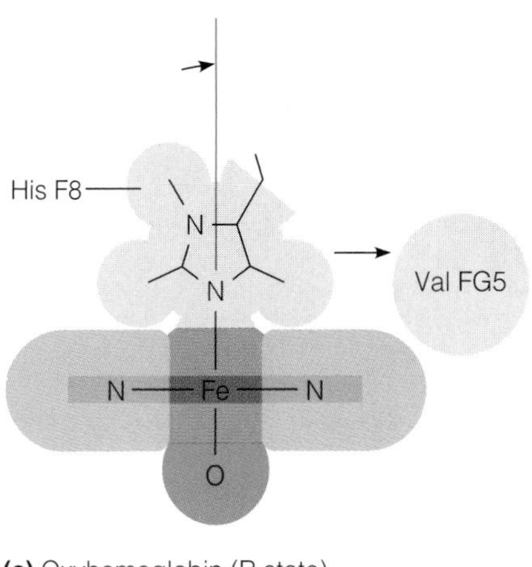

FIGURE **7.13**

Mechanism of the T ⟶ R transition in hemoglobin. The binding of oxygen to deoxyhemoglobin causes conformational changes in the heme. **(a)** In the deoxy state, heme has a dome shape, exaggerated in this figure. **(b)** Binding of the O_2 ligand pulls the iron into the heme plane, flattening the heme and causing strain. **(c)** A shift in the orientation of His F8 relieves the strain, partly because Val FG5 is pushed to the right. In this way, the tertiary change in heme is communicated to the FG corner.

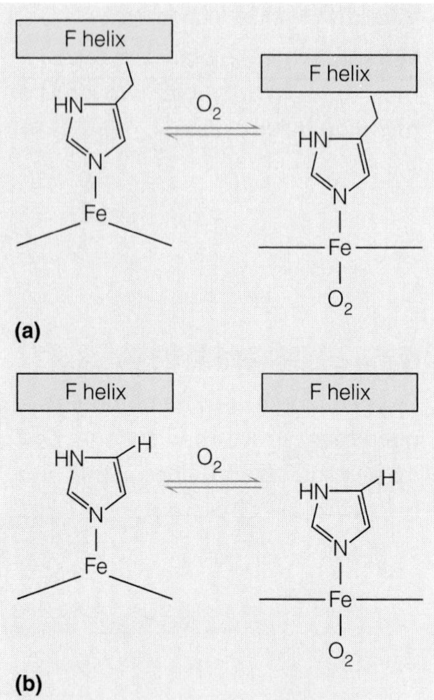

FIGURE **7.14**

The effect of replacing the proximal histidine in hemoglobin with a glycine residue and adding a noncovalently bonded imidazole. (a) The effect of O_2 binding according to the Perutz model: the F helix is drawn toward the heme. **(b)** Now lacking a connection to the heme, the F helix is not disturbed by O_2 binding and there is no cooperativity.

Adapted from D. Barrick et al., *Nature Struct. Biol.* (1997) 4:78–83. © 1997 Macmillan Magazines, Ltd.

What happens is that the histidine shifts its orientation toward the perpendicular, pulling on the F helix and the FG corner as it does so. This movement in turn distorts and weakens the whole complex of H bonds and salt bridges that connect FG corners of one subunit with C helices of another. Consequently, the rearrangement shown in Figure 7.12 occurs.

In the simplest terms, what has happened is that the binding of O_2, by pulling the iron a fraction of a nanometer into the heme, has produced by a lever effect a much larger shift in the surrounding structure, particularly at the critical α–β interfaces.

This mechanism to explain the cooperativity in oxygen binding was proposed in 1970 by M.F. Perutz, one of the pioneers in x-ray diffraction studies of proteins. But does it correspond to reality? Recent ingenious experiments indicate that it is at least a reasonable approximation. Barrick et al. (see References) have utilized the technique of site-directed mutagenesis to replace the proximal histidine residues in α and β chains with glycines. The protein was then studied in the presence of 10 mM imidazole; the small imidazole molecule can substitute for the histidine residue *but is not linked to the F helix* (see Figure 7.14). As a consequence, although oxygen binding can still flatten the heme, it does not move the F helix. What is observed is that cooperativity in binding is largely lost, as would be predicted by the Perutz model.

The changes described in the Perutz model constitute a rearrangement of the *tertiary* structure of each subunit upon oxygen binding. We know that a major rearrangement of the quaternary structure also occurs between the fully deoxy and fully oxy (T and R) states of the whole tetramer. How are the tertiary and quaternary structural changes connected? Much of the answer has been provided by research in the laboratory of Gary Ackers and is summarized in Figure 7.15. The tertiary structure changes that accompany oxygen binding can be tolerated up to a certain point before the T ⟶ R switch occurs. Specifically, whenever one site

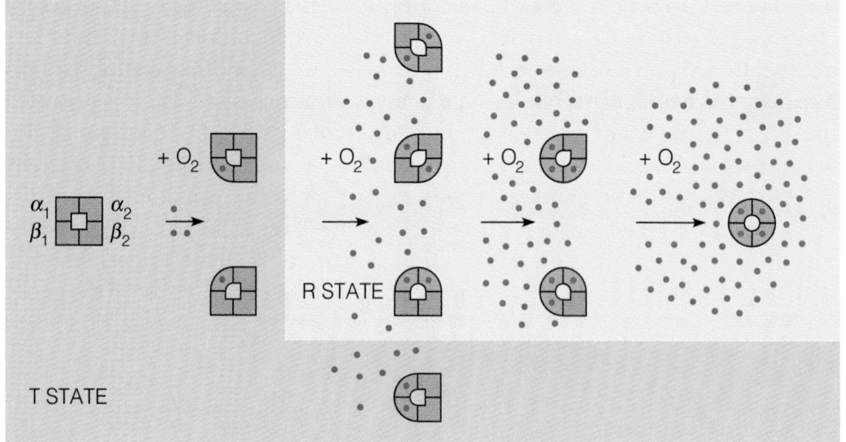

FIGURE **7.15**

A recent model for the cooperative transition of hemoglobin. This model, based on recent studies, is more general than either the KNF or the MWC model (see Figure 7.10) and contains features of both. Subunits are depicted here with $\alpha_1\beta_1$ dimers on the left, $\alpha_2\beta_2$ on the right. Deoxygenated subunits have squared corners, and oxygenated subunits have curved corners. As oxygenation progresses, tertiary strain accumulates. Whenever both $\alpha\beta$ dimers contain one or more oxy subunits, the T ⟶ R transition occurs. Thus, forms in the green area are all T, whereas those in the yellow area are all R. Not all forms are shown.

Adapted from G. K. Ackers et al., *Science* (1992) 255:54–63. © 1992 American Association for the Advancement of Science.

is occupied on *each* of the two $\alpha\beta$ dimers, the molecule as a whole adopts the R quaternary structure. Thus, hemoglobin obeys *neither* the KNF nor the MWC model completely but follows a novel path containing features of both models. This more recent model does not mean that the earlier models are incorrect in general. We shall show later, for example, that allosteric proteins exist that appear to follow the MWC model almost exactly.

Effects of Other Ligands on the Allosteric Behavior of Hemoglobin

Cooperative binding and transport of oxygen are only part of the allosteric behavior of hemoglobin. The realities of animal physiology impose further demands. First, as oxygen is utilized in tissues, carbon dioxide is produced and must be transported back to the lungs or gills. Accumulation of CO_2 also lowers the pH in erythrocytes through the *bicarbonate reaction,*

$$CO_2 + H_2O \rightleftharpoons HCO_3^- + H^+$$

This reaction in erythrocytes is catalyzed by the enzyme *carbonic anhydrase.* At the same time, the high demand for oxygen, especially in muscle involved in vigorous activity, can result in oxygen deficit. As shown in Chapter 13, a consequence of this deficit is the production of lactic acid, which also lowers the pH. The falling pH in tissue and venous blood signals a demand for more oxygen delivery.

Hemoglobin functions efficiently to meet these requirements. It does so through its allosteric transition between structurally different high-affinity and low-affinity states. Carbon dioxide, protons, and other substances that promote these changes are called **allosteric effectors.**

RESPONSE TO pH CHANGES: THE BOHR EFFECT

A lower pH in tissues, signaling O_2 demand, facilitates O_2 unloading. This response is the Bohr effect.

A pH drop in the capillaries has the effect of lowering the oxygen affinity of hemoglobin, allowing even more efficient release of the last traces of oxygen. This response of hemoglobin to pH change is called the **Bohr effect.** The overall reaction may be written

$$\text{Hb} \cdot 4O_2 + nH^+ \rightleftharpoons \text{Hb} \cdot nH^+ + 4O_2$$

where n has a value somewhat greater than 2. Physiologically, this reaction has two consequences. First, in the capillaries, hydrogen ions promote the release of O_2 by driving the reaction to the right. Then, when the venous blood recirculates to the lungs or gills, the oxygenation has the effect of releasing the H^+ by shifting the equilibrium to the left. This, in turn, tends to release CO_2 from bicarbonate dissolved in the blood by the reversal of the bicarbonate reaction. The free CO_2 can then be exhaled.

The mechanism of the Bohr effect can be explained by the model that was used to explain cooperative binding of O_2. Certain proton binding sites in hemoglobin are of higher affinity in the deoxy form than in the oxy form. A major contribution comes from histidine residue 146, which lies at the C-terminus of each β chain. In the deoxy form, this residue can make a salt bridge with Asp 94 in the same chain, *if the histidine is protonated.* This histidine residue has an abnormally high pK_a, because the salt bridge stabilizes the proton against dissociation. But in the oxy form, this salt bridge simply cannot be formed, so the pK_a falls to its normal value of about 6.5. Consequently, at blood pH ($\sim$7.4), His 146 is largely unprotonated in oxyhemoglobin. Therefore, a high concentration of protons, which favors protonation, also favors the *deoxy* form and thus promotes the release of oxygen.

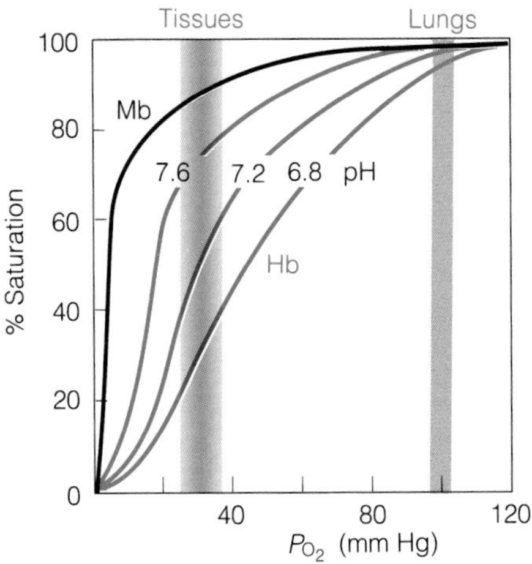

FIGURE **7.16**

The Bohr effect in hemoglobin. Oxygen binding curves for hemoglobin (Hb) are shown at pH 7.6, 7.2, and 6.8. Note that the efficiency of oxygen unloading, as measured by the differences in the curves at $P_{O_2} = 30$ mm Hg, increases greatly as the pH drops. As the blood circulates from lungs to tissues, the lower pH makes oxygen binding shift to the lower-affinity curves. Myoglobin (Mb) displays little Bohr effect, so its oxygen binding curve is approximately the same at all three pH values.

Other residues are also involved in the Bohr effect, including the N-terminal amino groups of the α chains. The basic mechanism is the same as for His 146: Protons associated with the residues are allosteric effectors favoring the deoxy conformation. The effect of lowering pH on the oxygen affinity of hemoglobin is illustrated in Figure 7.16. Note that a decrease in pH of only 0.8 unit shifts the P_{50} from less than 20 mm Hg to over 40 mm Hg, greatly increasing the amount of oxygen unloaded to myoglobin.

CARBON DIOXIDE TRANSPORT

Hemoglobin also transports CO_2 from tissues to gills or lungs. CO_2 acts as an allosteric effector of O_2 binding.

Release of carbon dioxide from respiring tissues lowers the oxygen affinity of hemoglobin in two ways. First, as mentioned above, some of the carbon dioxide becomes bicarbonate, releasing protons that contribute to the Bohr effect. Some of this bicarbonate is transported out of the erythrocytes and is carried dissolved in the blood serum. A portion reacts directly with hemoglobin, binding to the N-terminal amino groups of the chains to form **carbamates:**

$$-\overset{+}{N}H_3 + HCO_3^- \rightleftharpoons -\overset{\displaystyle H}{\overset{|}{N}}-COO^- + H^+ + H_2O$$

This *carbamation reaction* allows hemoglobin to aid in the transport of CO_2 from tissues to lungs or gills. It has two additional effects. First, the protons released on binding of HCO_3^- contribute to the Bohr effect. Second, a negatively charged group is introduced at the N-terminus of the chains, stabilizing salt bridge formation between α and β chains, which is characteristic of the deoxy state. Both the latter effect and the lower pH promote oxygen release when CO_2 is abundant. The reverse of this reaction, which occurs in lungs or gills, is equally important. Here, the high O_2 concentration favors oxygenation and hence the oxy form of the molecule. When this switch occurs, stabilization of the carbamated N-termini is decreased, and CO_2 is expelled and exhaled.

We may summarize the effects of H^+ and CO_2 in terms of the respiratory cycle shown in Figure 7.1: In the lungs or gills of an animal O_2 is abundant. Oxygenation favors the oxy conformation of hemoglobin, which stimulates the release of CO_2. As the blood then travels via arteries into the tissue capillaries, the lower pH and high CO_2 content favor the deoxy form, promoting O_2 release and binding of CO_2. Carbon dioxide, both in forming bicarbonate and in reacting with hemoglobin, causes the release of more protons, further stimulating O_2 release and CO_2 binding.

A consequence of the lack of CO_2 stimulation of O_2 release is seen in hyperventilation. If a person breathes too rapidly, CO_2 is effectively purged from the tissues, and consequently release of oxygen into the tissues is impaired. This condition leads to dizziness and, in extreme cases, unconsciousness. Hyperventilation can be easily corrected by breathing into a paper bag—this brings exhaled CO_2 back into the blood.

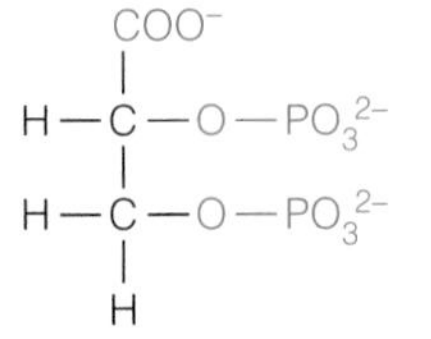

BISPHOSPHOGLYCERATE

H^+ and CO_2 are the effectors that function rapidly to facilitate the exchange of O_2 and CO_2 in the respiratory cycle. One other major effector operates over longer periods to permit organisms like humans to adapt to gradual changes in oxygen availability. It is a common observation that people who move to high altitudes at first experience some distress but gradually acclimate to the lower oxygen pressure. In part, this acclimation results from increased synthesis of hemoglobin, but another effect is due to changes in the amount of an allosteric effector called **2,3-bisphosphoglycerate** (**BPG**), or glycerate-2,3-bisphosphate (Figure 7.17a). Bisphosphoglycerate was formerly known as diphosphoglycerate (DPG); you will sometimes find the older term in the literature.

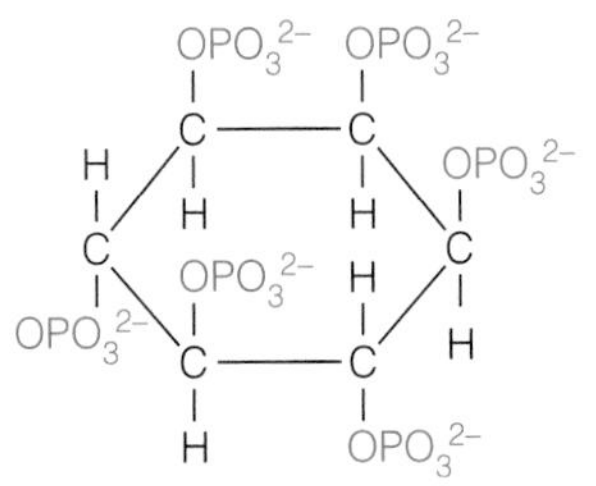

COO^- / $H-C-O-PO_3^{2-}$ / $H-C-O-PO_3^{2-}$ / H

(a) 2,3-Bisphosphoglycerate

Six-membered ring of CH carbons, each bearing OPO_3^{2-}

(b) Inositol hexaphosphate

FIGURE **7.17**

Two anionic compounds that bind to deoxyhemoglobin. (a) 2,3-Bisphosphoglycerate (BPG), found in mammals. **(b)** Inositol hexaphosphate (IHP), found in birds.

Like the effects of H^+ and CO_2, binding of BPG acts to lower the oxygen affinity of hemoglobin. At first glance this may seem like a strange way to adapt to lower O_2 pressure, but in fact, the more efficient unloading of oxygen in the tissues more than compensates for the slight decrease in loading efficiency in the lungs. The action of BPG is illustrated in Figure 7.18. BPG binds in the cavity between the β chains, making electrostatic interactions with positively charged groups surrounding this opening. Comparison of the two hemoglobin conformations shown in Figure 7.12b will show you that this opening is much narrower in oxyhemoglobin than in deoxyhemoglobin. In fact, BPG cannot be accommodated in the oxy form. The higher the BPG content in red blood cells, the more stable the deoxy structure will be. Once again, a decrease in O_2 affinity is explained by stabilization of the deoxy structure. Increased BPG levels are also found in the blood of smokers, who because of the carbon monoxide in smoke also suffer from limitation in oxygen supply.

Bisphosphoglycerate (BPG) is an allosteric effector of hemoglobin that regulates long-term changes in O_2 affinity.

BPG plays one other subtle, but important, role in the respiration of humans and other mammals. Consider the problem faced by a fetus, which must obtain oxygen from the mother's blood by exchange through the placenta. For this exchange to work well, fetal blood must have a higher O_2 affinity than the mother's blood. In fact, the human fetus has a hemoglobin different from the adult form. Whereas adult hemoglobin (HbA) has two α and two β chains ($\alpha_2\beta_2$), in the fetus the β chains are replaced by similar, but distinctly different, polypeptides. These are called γ chains, so fetal hemoglobin (HbF) is written as an $\alpha_2\gamma_2$ structure. The intrinsic oxygen affinity of HbF is very similar to that of HbA, but HbF has a much lower affinity for BPG than does HbA. This difference is largely due to the replacement of His 143 in the adult β chain by a serine in the fetal γ chain. As Figure 7.18 shows, the positively charged His 143 helps to bind the negative BPG molecule. The concentration of this small, diffusible molecule is about the same in the circulatory systems of mother and fetus. Under these conditions, HbF will have less BPG bound than will HbA, and therefore HbF will have a higher oxygen affinity.

The use of effectors that facilitate oxygen release is not restricted to mammals. The blood of birds contains **inositol hexaphosphate** (see Figure 7.17b), and fish use ATP for a similar purpose. All of these molecules have a strong negative charge and bind in the central cleft of deoxyhemoglobin. All of these allosteric effectors, including H^+, CO_2, and BPG, act in the same general manner—by biasing the

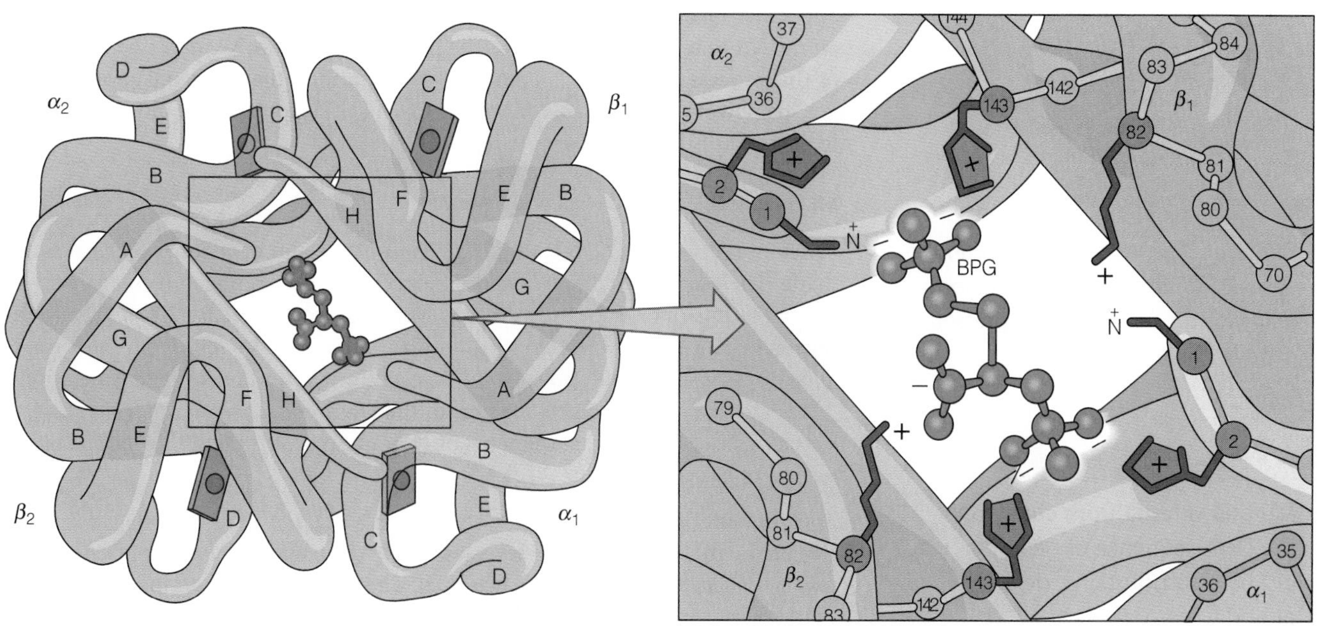

FIGURE **7.18**

Binding of 2,3-bisphosphoglycerate to deoxyhemoglobin. The binding site, in the central cavity of the hemoglobin tetramer (see Figure 7.12b), is lined with eight positively charged groups that help bind the negatively charged BPG molecule. Note the histidine residues (β143) that are replaced by serine in fetal hemoglobin.

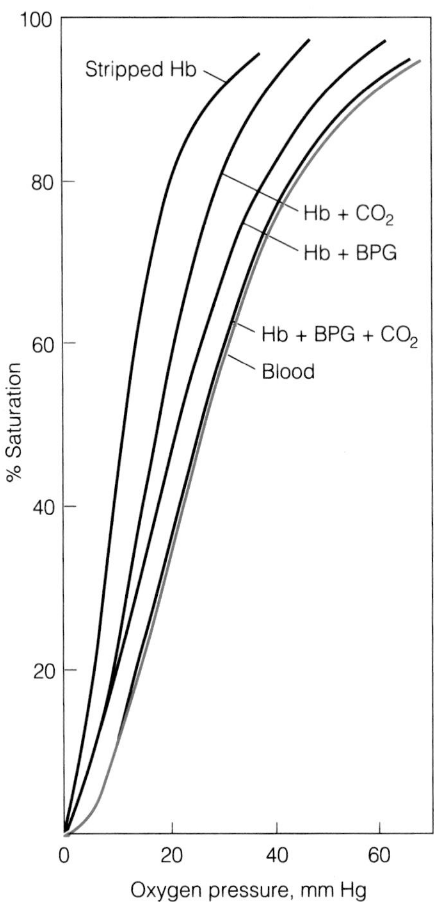

FIGURE **7.19**

Combined effects of CO_2 and BPG on oxygen binding by hemoglobin. Hemoglobin that has been stripped of both CO_2 and BPG has a high oxygen affinity. When both substances are added to hemoglobin at the levels found in blood emerging from the capillaries, the hemoglobin displays almost exactly the same binding curve as observed for whole blood.

J. V. Kilmartin, *Br. Med. Bull.* (1976) 32:209–212.

conformational equilibrium in hemoglobin toward the deoxy form. However, they interact at distinctly different sites, and therefore their effects can be additive, as illustrated for CO_2 and BPG in Figure 7.19.

Myoglobin and hemoglobin represent sophisticated molecular machines, each finely tuned for its function. In the following section we explore how these structures might have evolved.

Protein Evolution: Myoglobin and Hemoglobin as Examples

We have emphasized that for each polypeptide chain that an organism produces, a corresponding gene exists. The nucleotide sequence in that gene dictates the amino acid sequence of the protein, which in turn defines the protein's secondary, tertiary, and quaternary structures. Evolution of proteins occurs through accumulated changes in the nucleotide sequences of genes. To explore this process, we will use as an example the evolutionary development of the myoglobin–hemoglobin family of proteins. First, however, we must examine in a bit more detail the structure of eukaryotic genes and the mechanisms through which mutation can occur.

THE STRUCTURE OF EUKARYOTIC GENES: EXONS AND INTRONS

In previous chapters we implied that a direct correspondence exists between the nucleotide sequence in a gene and (via mRNA) the amino acid sequence of the polypeptide chain it codes for. For most genes in prokaryotic organisms, this concept is true. But investigation of the genomes of higher organisms has produced a surprising result: Within most eukaryotic genes are DNA sequences that are never expressed in the polypeptide chain. These noncoding regions, called **introns,** alternate with regions called **exons** that *are* expressed in the polypeptide sequence. Figure 7.20 shows how the exon–intron structure of the β globin gene is related to the structure of β globin. Only the parts of the gene shown in color correspond to portions of the polypeptide chain.

Eukaryotic genes are discontinuous, containing both expressed regions (exons) and regions not expressed as protein sequence (introns).

Clearly, this remarkable situation means that mRNA production in eukaryotes must be a more complex process than had been assumed at first. As Figure 7.20 shows, what actually happens is that transcription first produces a primary transcript, or **pre-mRNA,** corresponding to the whole gene—exons, introns, and portions of flanking regions. The pre-mRNA, while still in the cell nucleus, is cut and spliced to remove the regions corresponding to introns, thereby producing an mRNA that codes correctly for the polypeptide chain. We describe the details of this process in Chapter 28. For now, keep in mind that most eukaryotic genes are "patchwork" structures, containing extensive regions that do not correspond to any part of the protein sequence.

MECHANISMS OF PROTEIN MUTATION

When organisms reproduce, they copy their DNA, and occasionally mistakes are made. These mistakes may be random errors that occur during copying, or they may be results of damage the DNA has sustained from radiation or chemical **mutagens,** substances that produce mutations (Table 7.1). In any event, these alterations will appear as **mutations** in the DNA of the next and subsequent generations. There are two basic kinds of changes in the DNA sequence that may give rise to mutations in proteins: replacements of DNA bases by others and deletions or insertion of bases in the gene.

Mutations result from changes in the DNA sequence of genes, including base substitutions, deletions, or additions.

Replacement of DNA Bases

Replacement of bases can have several possible consequences. First, the base change may not affect the protein sequence at all. The change may occur in an intron, for example. But even if it is in a protein-coding region (exon), the replacement will

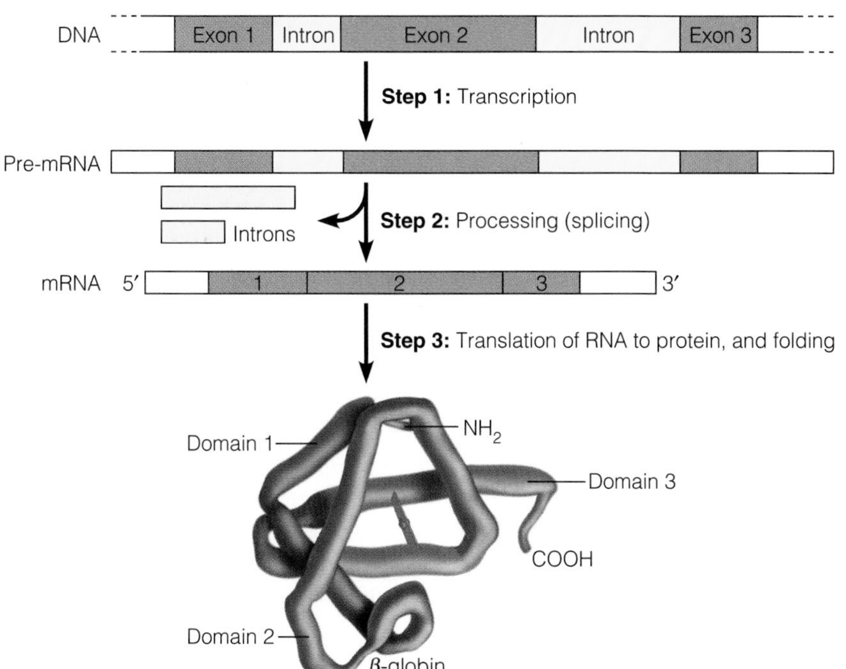

FIGURE **7.20**

Coding and noncoding regions of the β hemoglobin gene. The gene for the human β hemoglobin chain has coding regions, or exons, alternating with noncoding regions, or introns. This figure follows the transcription and translation of the gene to yield the final β hemoglobin chain. **Step 1, transcription:** A primary transcript (pre-mRNA) containing complementary copies of the exons and introns is produced from the gene. **Step 2, splicing:** The intron sequences are removed and the exons spliced together to yield the final mRNA. **Step 3, translation:** Coding regions of the spliced mRNA produce a β chain, which adopts its favored three-dimensional structure and incorporates a heme group. Note that the entire heme-binding region (domain 2) is coded for by one exon, as is the C-terminal region (domain 3), which is so important for allosteric interaction.

TABLE **7.1 Some mutagenic agents**

Substance	Formula	Source	Effects
Benzo[a]pyrene		Smoke	Intercalates in DNA, causes frameshifts; also is metabolized to a product which binds to residues, causing base substitutions.
Nitrous acid	HO—N=O	Some prepared foods	Deaminates C to U, leads to missense
Dimethyl nitrosamine	$(H_3C)_2N{-}N{=}O$	Some prepared foods	Methylating agent, modifies bases, yields missense (see Chapter 25)
Aflatoxin		Moldy nuts or grains	Alkylating agent, modifies guanine, causes missense
Ultraviolet light		Sunlight	Causes the formation of pyrimidine dimers (see Chapter 25)

make no difference in the protein sequence if the new codon codes for the same amino acid as the original one. The redundancy of the genetic code (see Figure 5.16, page 143) is such that fairly frequently a base change does not alter the protein product. On the other hand, an amino acid residue in the original protein may be replaced by a different one in the mutated protein; this type of replacement is called a **missense mutation** (Figure 7.21a). Occasionally, the codon for an amino acid residue within the original protein will be changed to a *stop* codon. We call this a **nonsense mutation,** for the protein will be terminated prematurely

Residue number		1	2	3	4	5	6	7	8	9	10
Normal β gene	...A T G	G T G	C A C	C T G	A C **T**	C C T	G **A** G	G A G	**A** A G	T C T	G C C...
		Val	His	Leu	Thr	Pro	Glu	Glu	Lys	Ser	Ala
(a) Missense mutation		G T G	C A C	C T G	A C T	C C T	G T G	G A G	A A G	T C T	G C C...
		Val	His	Leu	Thr	Pro	Val	Glu	Lys	Ser	Ala
(b) Nonsense mutation		G T G	C A C	C T G	A C T	C C T	G A G	G A G	T A G	T C T	G C C...
		Val	His	Leu	Thr	Pro	Glu	Glu	Stop		

(c) Frameshift mutation by deletion

G T G C A C C T G A C □ C C T G A G G A G A A G T C T G C C...

Val His Leu Thr Leu Arg Arg Ser Leu

FIGURE **7.21**

Mutation types. Some of the ways in which mutations can occur in the β hemoglobin chain are shown here. The first 10 residues of the normal human β chain, together with their DNA codons, are shown at top. **(a)** A missense mutation has occurred in residue 6. This is the sickle-cell mutation. **(b)** A nonsense mutation has introduced a stop signal after residue 7, terminating the chain prematurely. **(c)** A frameshift mutation has occurred by deletion of a single T residue. The rest of the chain, with a completely altered sequence, will continue to be produced until a stop signal is encountered in the new frame. Both (b) and (c) would result in β-thalassemia (see page 240).

and usually be nonfunctional (Figure 7.21b). Sometimes the opposite happens—a stop codon mutates into a codon for an amino acid residue. In this case translation continues, elongating the chain.

Nucleotide Deletions or Insertions

Deletions or insertions in the gene may be large or small. Such mutations outside the coding regions will generally have no effect, unless they modify sites of transcriptional control. Large insertions or deletions in coding regions almost invariably prevent the production of useful protein; sometimes even whole protein genes may be deleted. The effect of short deletions or insertions depends on whether they involve multiples of three bases. If one, two, or more *whole codons* are removed or added, the consequence is the deletion or addition of a corresponding number of amino acid residues. However, a deletion or insertion in a coding region of any number of bases *other* than a multiple of three has a much more profound effect: It causes a shift in the reading frame during translation. Such **frameshift mutations** are sometimes called *gibberish mutations,* because they result in a complete and meaningless change in the amino acid sequence in the C-terminal direction from the point of mutation (Figure 7.21c).

The effects of these kinds of mutations on the functionality of the protein product, and therefore on the organism itself, can be quite varied. Base substitutions may, in some cases, be neutral in effect, either not changing the amino acid coded for or changing it to another that functions equally well at that position in the protein. More often, the result is deleterious. Occasionally, such mutations increase the efficiency of a protein, and the mutated organisms may be selected for in future generations. Nonsense mutations and frameshift mutations, by contrast, almost always result in destruction of protein function. If the protein is important to the life of the organism, such mutations are strongly selected against in the course of evolution—those who inherit them die before they can reproduce.

Genomes can also be modified by gene duplication, gene fusion, or exon recombination.

Gene Duplications and Rearrangements

By accumulating many small mutational changes over millions of years, proteins gradually evolve. The diversity of functions that they can collectively perform is increased by two other phenomena: **gene duplication** and **exon recombination.**

Very occasionally, replication of the genome occurs in such a way that some DNA sequence, containing a particular gene, is copied twice. Initially, the only result of such duplication is that the descendants of the organism have two copies of the same gene. This mutation may be advantageous if the protein is needed in large amounts, because the efficiency of its production will be increased. In such cases, there will be selective pressures to maintain two or even more copies of the same gene. Alternatively, the two copies may evolve independently. One copy may continue to express the protein fulfilling the original function, but the other may evolve through mutations into an entirely different protein with a new function. Another way in which the diversity of proteins can increase is through the *fusion* of two or more initially independent genes. Such fusion may lead to the production of multidomain proteins exhibiting new combinations of functions.

The intervening sequences in eukaryotic genes (introns) offer a further possibility for diversification of protein structure and function. Because these regions are not used for coding, they represent positions where genes can be safely cut and recombined in the process of **genetic recombination.** The mechanisms of recombination are described in Chapter 25; at this point we are concerned only with its consequences. Suppose that an exon from one gene, which codes for a protein region with physiological function B, is inserted into an intron region in a gene for a protein carrying function A. The new hybrid protein is now capable of both functions A and B and may serve a new physiological function.

Through the combined effects of mutations, gene duplication, and genetic rearrangement, organisms can develop new abilities, adapt to new environments, and become new species. The process of organismal evolution, which we see exhibited in the fossil record and in the incredible variety of existing plants, animals, and microorganisms, is largely a consequence of this molecular evolution of proteins.

EVOLUTION OF THE MYOGLOBIN–HEMOGLOBIN FAMILY OF PROTEINS

Myoglobin and hemoglobin evolved from an ancestral myoglobin-like protein.

We have already seen an example of the process of protein evolution. If we compare the sequences of sperm whale and human myoglobin (see Figure 5.14), we find 25 amino acid changes. Since fossil evidence indicates that the evolutionary lines that led to sperm whales and humans diverged from a common mammalian ancestor about 100 million years ago, we can gain an idea of the rate of this process. If the rate was uniform, there has been an average of one replacement every 4 million years.

If we compare human myoglobin with that of the shark, we find about 88 differences. Because these evolutionary lines diverged about 400 million years ago, the accumulated differences are about what we would expect from the preceding example. In other words, the number of amino acid substitutions in two related proteins is roughly proportional to the evolutionary time that has elapsed since the proteins (and the species) had a common ancestor. Using this principle, we can compare the sequences of both hemoglobins and myoglobins and attempt to construct a "family tree" of globin proteins. The tree is complicated by the fact that higher eukaryotes, including humans, carry genes for both myoglobin and several *different* hemoglobin chains. These different genes are expressed at different times in human development (Figure 7.22). The α and β chains, as mentioned earlier, are normally present in adults. But in the early embryo, the hemoglobin genes expressed are those for the embryonic chains, ζ and ε. As the fetus develops, these chains are replaced by α and γ chains, and finally, at about the time of birth, the γ chains are replaced by β chains. In addition, after birth a small amount of a δ chain is produced. These developmental types of hemoglobin chain are slightly different, and each is coded for by a separate gene in the human genome.

Comparison of the sequences of many hemoglobins from many different species yields the evolutionary tree shown in Figure 7.23. According to these results, very primitive animals had only a myoglobin-like, single-chain ancestral globin for oxygen storage. Most of these animals, like protozoans and flatworms, were so small that they did not require a transport protein. Roughly 500 million years ago, an important event occurred: The ancestral myoglobin gene was duplicated.

FIGURE **7.22**

Expression of human globin genes at different stages in development. The human ζ and ε genes make the $\zeta_2\varepsilon_2$ hemoglobin found in the very early embryo. This is soon supplanted by the $\alpha_2\gamma_2$ hemoglobin of the fetus. At about the time of birth, transcription of the γ gene ceases, and the β gene begins to be transcribed. By six months, the infant will have almost all $\alpha_2\beta_2$ (adult) hemoglobin. The δ gene is never transcribed at high rates. There are two copies of the α gene: α_1 and α_2. Both contribute to the production of α chains.

W. G. Wood, *Br. Med. Bull.* (1976) 32:282–287.

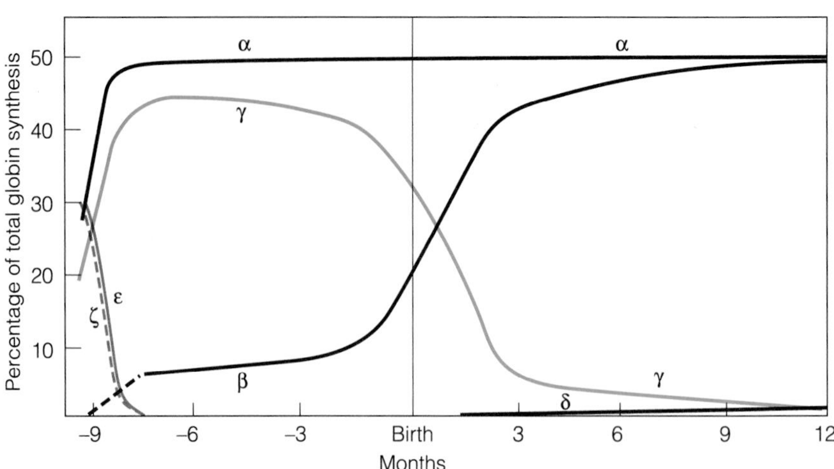

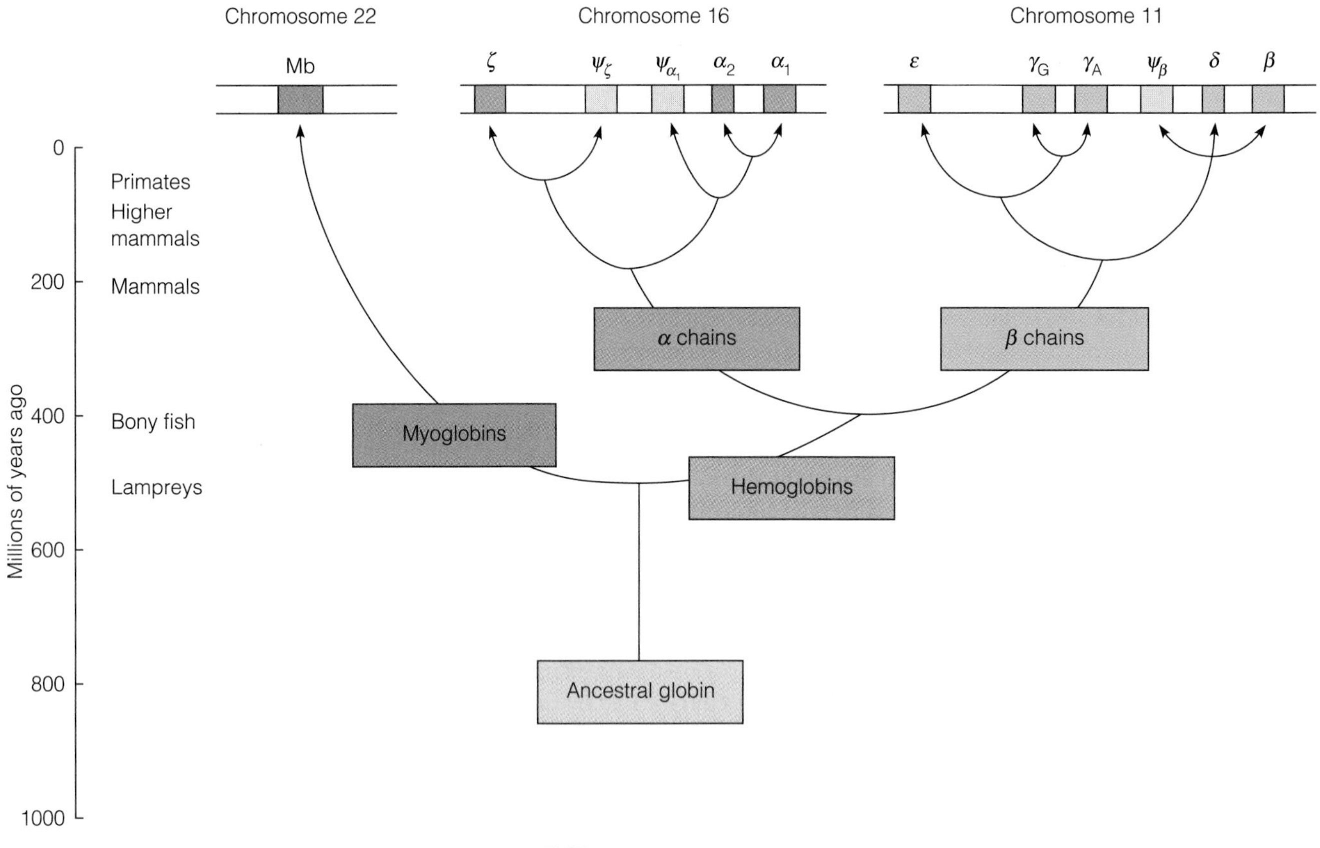

Chromosome 22
Mb
Chromosome 16
ζ ψ_ζ ψ_{α_1} α_2 α_1
Chromosome 11
ε γ_G γ_A ψ_β δ β

Millions of years ago
0
Primates
Higher mammals
200
Mammals
400
Bony fish
Lampreys
600
800
1000

α chains
β chains
Myoglobins
Hemoglobins
Ancestral globin

FIGURE **7.23**

Evolution of the globin genes. The arrangement of human globin genes is shown at the top. Note that they are found in three different chromosomes. Functional genes are shown in color; pseudogenes, which are nontranscribed variants of a gene, are in gray. The diagram underneath shows the probable evolution of the globin gene family, based on sequence differences among the various globin genes in humans and other animals. The times at which gene duplications occurred are inferred from a combination of sequence and fossil evidence and are only approximate. The two α genes and the two γ genes are too similar in sequence to allow us to judge the time of their divergence. We know only that it must have happened quite recently.

One of the copies became the ancestor of the myoglobin genes of all higher organisms; the other evolved into the gene for an oxygen transport protein and gave rise to the hemoglobins.

Along the evolutionary line leading to vertebrates and mammals, the most primitive animals to possess hemoglobin are the lampreys. Lamprey hemoglobin can form dimers but not tetramers and is only weakly cooperative; it represents a first step toward allosteric binding. But subsequently a *second* gene duplication occurred, giving rise to the ancestors of the present-day α and β hemoglobin chain families. Reconstruction, from sequence comparison, indicates that this must have happened about 400 million years ago, at about the time of divergence of the sharks and bony fish. The evolutionary line of the latter led to the reptiles and eventually to the mammals, all carrying genes for both α and β globins and capable of forming tetrameric $\alpha_2\beta_2$ hemoglobins. Further gene duplications have occurred in the hemoglobin line, leading to the embryonic forms ζ and ε and the fetal γ. As Figure 7.23 shows, the duplications that led to a distinction between adult and embryonic subtypes coincide fairly well with the development of placental mammals, about 200 million years ago. This concurrence is functionally appropriate, because in these mammals the later stages of embryo development occur within the mother, and a special hemoglobin, adapted to promote oxygen transfer through the placenta from mother to fetus, is essential (see page 229).

During the long evolution of the myoglobin–hemoglobin family of proteins, only a few amino acid residues have remained invariant. These *conserved residues* may mark the truly essential positions in the molecule. As Figure 7.11 shows, they include the histidines proximal and distal to the heme iron (F8 and E7; see Figure 7.5b). Interestingly, Val FG5, which has been implicated in the hemoglobin deoxy–oxy conformation change described earlier, is invariant in hemoglobins, replacing the isoleucine found at this position in most myoglobins. Other regions highly conserved in hemoglobins are those near the α_1–β_2 and α_2–β_1 contacts. These contacts are most directly involved in the allosteric conformational change.

Despite the major changes that have occurred in the primary structure of the myoglobin–hemoglobin family over hundreds of millions of years, the secondary and tertiary structures of these proteins have remained surprisingly unchanged. Figure 7.24 shows the backbone structure of members of this family, ranging from insect to horse. All are recognizable as the same basic folding, and the similarity is particularly strong in the region that binds the heme. At first glance this similarity seems inconsistent with our earlier statements that primary structure determines secondary and tertiary structure. However, careful examination of many sequences shows that many of the replacements have been *conservative*—that is, an amino acid has been replaced by another of the same general class. Obviously,

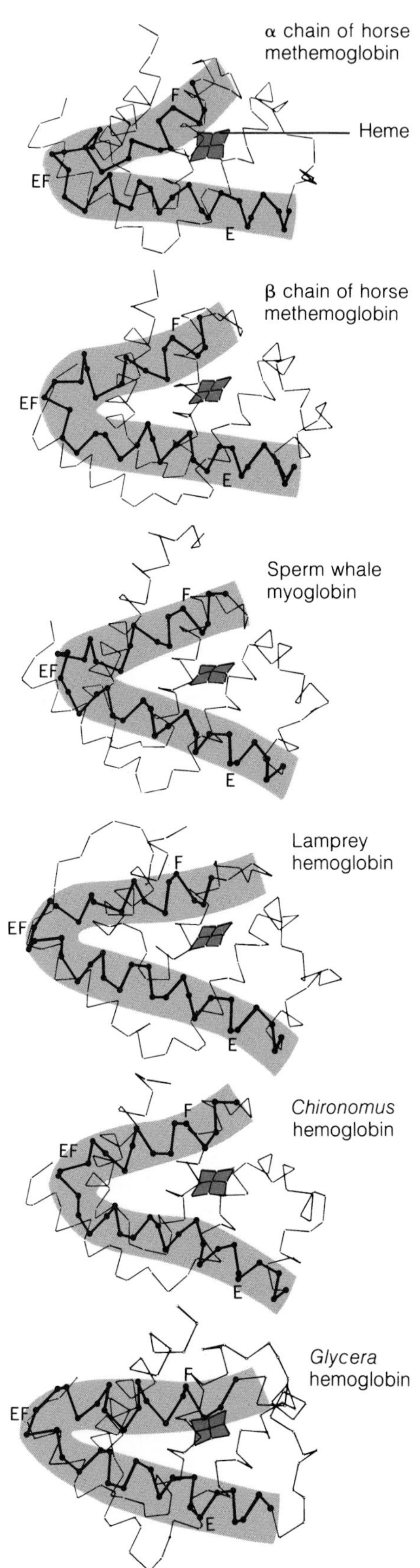

FIGURE **7.24**

Evolutionary conservation of the globin folding pattern. As these drawings emphasize, the overall tertiary structure of myoglobin and hemoglobin chains has remained nearly constant despite extensive changes in the primary structure. The shaded regions delineate the E and F helices, which surround the heme. Note that they are almost invariant and that changes tend to be concentrated near the ends of the chains. The most primitive proteins shown are the single-chain "hemoglobins" of the marine worm *Glycera* and the fly *Chironomus.* The lamprey is the most primitive creature to have distinct myoglobin and hemoglobin. Lamprey hemoglobin forms dimers and exhibits some cooperativity in binding. The α and β chains of horse hemoglobin are almost identical to those of all other mammals.

Evolution of globins has retained the common "globin fold" that holds the heme.

evolution of these proteins has proceeded not at random, but under the constraint of maintaining a physiologically functional structure. Survival of mutant proteins in the globin family has been restricted to those that maintain the basic "globin fold."

Hemoglobin Variants: Evolution in Progress

VARIANTS AND THEIR INHERITANCE

Evidence for continuing evolution is found in the many variant proteins in existing species.

Evidence for the ongoing evolution of hemoglobin genes can be seen in the existence of hemoglobin variants or, as they are often called, abnormal hemoglobins. Today several hundred recognized mutant hemoglobins exist within the human population. A number of mutation positions on the tetramer are shown in Figure 7.25. Most proteins in existing plants and animals probably show comparable diversity, but few of them have been as thoroughly studied as human hemoglobins. Each of the mutant forms of hemoglobin exists in only a small fraction of the total human population; some forms have been recognized in only a few individuals. Some of these mutant forms are deleterious and give rise to recognized pathologies; under conditions of natural selection they would eventually disappear. Most are, as far as we can tell, harmless, and are often referred to as neutral mutations. A very few may have as yet unrecognized advantages and therefore may come, in time, to dominate in the population.

We shall consider only a few of these abnormal hemoglobins. First, it is necessary to review a bit of genetics. All human cells, except for the germ cells (sperm and ova), are **diploid;** that is, they carry two copies of each chromosome. Therefore, they carry two copies of each gene, one on each of the paired chromosomes. Suppose we consider a gene such as the adult β hemoglobin gene, which can exist in two forms—the "normal" type, β, and a variant (mutant) type, β^*. An individual can have three possible combinations of these genes in his or her paired chromosomes:

A. $\beta + \beta$: **homozygous** (same genes) in the normal type

B. $\beta + \beta^*$: **heterozygous** (mixed genes)

C. $\beta^* + \beta^*$: homozygous in the variant type

Having genes for only the normal β hemoglobin, individual A will produce only normal β hemoglobin chains. Individual C, who has genes for only the variant type, will produce only variant hemoglobin chains. Individual B, with genes for both types, will produce both. If the mutation is deleterious, C will be in serious

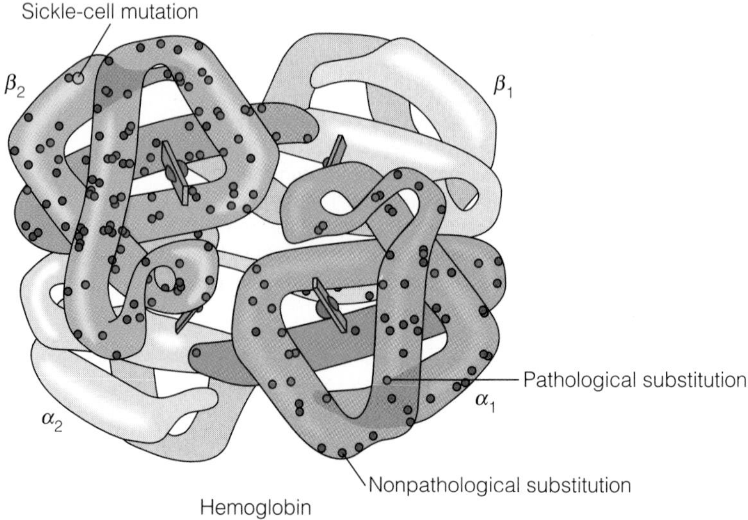

FIGURE **7.25**

Distribution of mutations in human hemoglobins. The blue and red dots represent all positions at which amino acid substitutions have been found in the α and β chains (only one pair is illustrated, for clarity). Those substitutions that have known pathological effects are shown in red. At many of these positions, more than one substitution has been observed. Position 6 in the β chain, at which the sickle-cell mutation occurs, is shown in yellow.

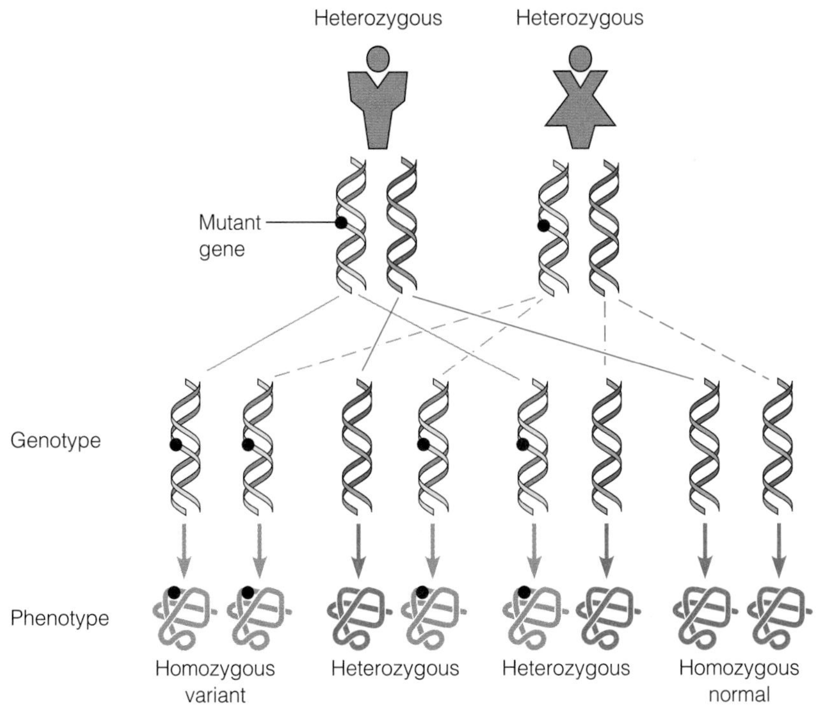

FIGURE **7.26**

Inheritance of normal and variant proteins in a heterozygous cross. Diploid organisms can exist as one of three types with respect to any gene: homozygous normal, homozygous variant, or heterozygous. The offspring of a heterozygous pair may be any of the three types, as shown here: homozygous normal, heterozygous, or homozygous variant, with a probability ratio of 1:2:1. We leave it as an exercise for you to work out other possibilities—for instance, the offspring of one homozygous normal parent and one homozygous variant parent. This pattern is referred to as classical Mendelian inheritance.

trouble. B, on the other hand, may do fairly well, because he or she will make normal protein chains along with the variant ones.

When two individuals produce offspring, each parent donates to a child one copy of the β hemoglobin gene, the selection of which will be random. If both parents carry only the normal gene, the child must receive two copies the same. If both carry only the variant gene, the child must also be homozygous for that gene. If both parents are heterozygous for the gene, Figure 7.26 shows that the child has one chance in four of being homozygous normal, one in four of being homozygous for the variant gene, and two in four of being heterozygous. Because most variant hemoglobin genes are rare in the human population, only occasionally do we find an individual homozygous for the variant type.

PATHOLOGICAL EFFECTS OF VARIANT HEMOGLOBINS

Although most hemoglobin mutations appear to be neutral, some are deleterious.

Of the large number of hemoglobin mutations, a significant fraction have deleterious effects. As Figure 7.25 shows, the known deleterious mutations are mostly clustered about the heme pockets and in the vicinity of the α–β contact region that is so important in the allosteric transition. A few of the well-studied pathological missense mutations are listed in Table 7.2. For example, a class of variants known as *hemoglobins M* tend to be readily oxidized to methemoglobin, which cannot bind O_2. Many of these mutations involve replacement of either the proximal or the distal histidine by other residues. Individuals carrying such mutations experience difficulty in transporting enough O_2 to tissues. Other variants that involve changes at the subunit interfaces can have several kinds of effects. Some, like *hemoglobin St. Lukes,* destabilize the hemoglobin tetramer, whereas others (e.g., *hemoglobin Suresnes*) tend to stabilize either the oxy or the deoxy conformation, inhibiting the allosteric switch. Finally, there are those like *hemoglobin Hammersmith* in which the tertiary structure of the molecule is so unstable that some of these variants cannot hold the heme effectively.

The most infamous of all variant hemoglobins, *sickle-cell hemoglobin,* is a source of misery and early death to many humans. The variant has gained its name because it causes red blood cells to adopt an elongated, sickle shape at low oxygen concentrations (Figure 7.27a). This "sickling" is a consequence of the tendency of

TABLE 7.2 Selected list of missense mutations in human hemoglobins

Effect	Residue Changed	Change	Name	Consequences of Mutation	Explanation
Sickling	β6 (A3)	Glu → Val	S	Sickling	Val fits into EF pocket in chain of another hemoglobin molecule.
	β6 (A3)	Glu → Ala	G Makassar	Not significant	Ala probably does not fit the pocket as well.
	β121 (GH4)	Glu → Lys	O Arab, Egypt	Enhances sickling in S/O heterozygote	β121 lies close to residue β6; Lys increases interaction between molecules.
Change in O_2 affinity	α87 (F8)	His → Tyr	M Iwate	Forms methemoglobin, decreased O_2 affinity	The His normally ligated to Fe has been replaced by Tyr.
	α141 (HC3)	Arg → His	Suresnes	Increases O_2 affinity by favoring R state	Replacement eliminates bond between Arg 141 and Asn 126 in deoxy state.
	β74 (E18)	Gly → Asp	Shepherds Bush	Increases O_2 affinity by decrease in BPG binding	The negative charge at this point decreases BPG binding.
	β146 (HC3)	His → Asp	Hiroshima	Increases O_2 affinity, reduced Bohr effect	Disrupts salt bridge in deoxy state and removes a His that binds a Bohr-effect proton.
	β92 (F8)	His → Gln	St. Etienne	Loss of heme	The normal bond from F8 to Fe is lost, and the polar glutamine tends to open the heme pocket.
Heme loss	β42 (CD1)	Phe → Ser	Hammersmith	Unstable, loses heme	Replacement of hydrophobic Phe with Ser attracts water into heme pocket.
Dissociation of tetramer	α95 (G2)	Pro → Arg	St. Lukes	Dissociation	Chain geometry is altered in subunit contact region.
	α136 (H19)	Leu → Pro	Bibba	Dissociation	Pro interrupts helix H.

Sickle-cell disease results from a single base substitution in the β chain.

the mutant hemoglobin, in its deoxygenated state, to aggregate into long, rodlike structures (Figure 7.28). The elongated cells tend to block capillaries, causing inflammation and considerable pain. Even more serious is that the sickled cells are fragile (Figure 7.27b); their breakdown leads to an anemia that leaves the victim susceptible to infections and diseases. Individuals who are homozygous for the sickle-cell mutation often do not survive into adulthood, and those who do are seriously debilitated. Heterozygous individuals, who can still produce some normal hemoglobin, usually suffer distress only under conditions of severe oxygen deprivation. For example, flying may be dangerous for HbS heterozygotes because of the lower oxygen level.

Linus Pauling, in one of his many inspired insights, first suggested in 1949 (see References) that sickle-cell disease was a "molecular disease" resulting from a mutation in the hemoglobin molecule. Remarkably, sickling stems from what we might expect to be an innocuous mutation in a part of the molecule far from the critical regions mentioned earlier. The glutamic acid residue normally found at position 6 in β chains is replaced by a valine (see Figure 7.21a). This hydrophobic valine can fit into a pocket at the EF corner of a β chain in another hemoglobin molecule, and thus, as shown in Figure 7.28c, adjacent hemoglobin molecules can fit together into a long, rodlike helical fiber. Why sickling occurs with deoxyhemoglobin, but not with the oxygenated form, is simply explained: In the oxy form the rearrangement of subunits makes the EF pocket inaccessible to Val 6.

Sickle-cell disease is confined largely to populations originating in tropical areas of the world. At first glance, this distribution seems unexpected. Why should a *genetic* disease be climate-related? The answer tells us something about the persistence of what seem to be unfavorable traits. A high incidence of sickle-cell disease in a population generally coincides with a high incidence of malaria, a parasitic disease carried by a tropical mosquito. Individuals *heterozygous* for sickle-cell hemoglobin have a higher resistance to malaria than those who do not carry the

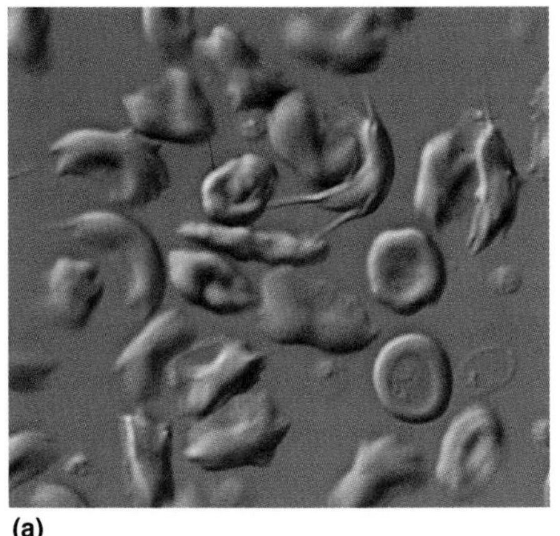

(a)

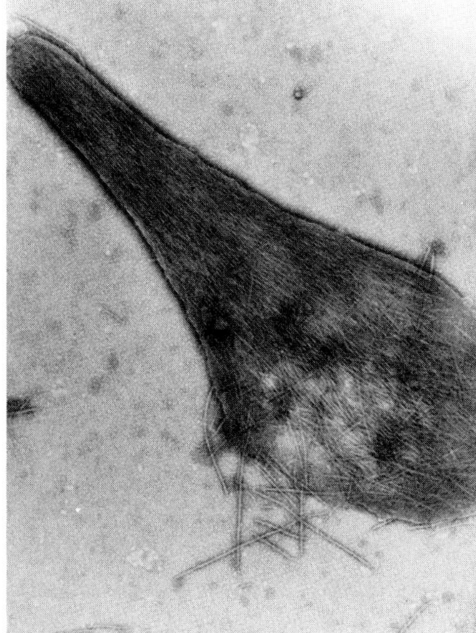

(b)

FIGURE **7.27**

Erythrocytes in sickle-cell disease. **(a)** Typical sickled cells, together with some normal, rounded red blood cells. **(b)** Scanning electron micrograph of a sickled cell that has ruptured, with hemoglobin fibers spilling out.

(a) © G. W. Willis, M.D./BPS; (b) Courtesy of T. Wellems and R. Josephs.

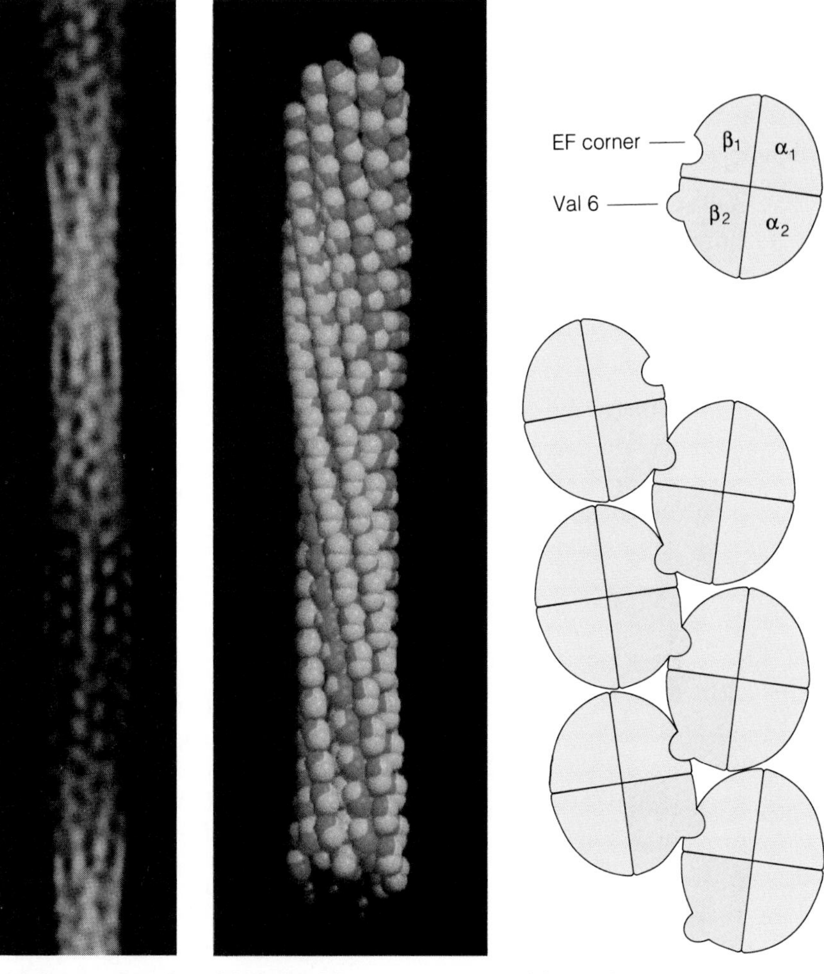

(a) Sickle-cell Hb fiber **(b)** Model of fiber **(c)** Fiber formation

FIGURE **7.28**

Sickle-cell hemoglobin. Molecules of sickle-cell hemoglobin tend to aggregate, forming long fibers. **(a)** An electron micrograph of one sickle-cell fiber. **(b)** A computer-graphic depiction of one fiber. **(c)** A schematic model of fiber formation. Hemoglobin molecules lock together to form a two-stranded cluster, because Val 6 in the β chain of one hemoglobin molecule fits into a pocket in an adjacent molecule. Interaction of these two-stranded structures with one another produces the multistrand fibers shown in (a) and (b).

Courtesy of B. Carragher, D. Bluemke, M. Potell, and R. Josephs.

sickle-cell mutation. The malarial parasite spends a portion of its life cycle in human red cells, and the increased fragility of the sickled cells, even in heterozygous individuals, tends to interrupt this cycle. In addition, the distortion of the cell membrane of intact sickled cells leads to a loss of potassium ions from these cells, providing a less favorable environment for the parasite. Heterozygous individuals have a higher survival rate—and therefore a better chance of passing on their genes—in malaria-infested regions. However, the high incidence of these genes in the population leads to the birth of many people who are homozygous for the mutant trait.

Sickle-cell disease is also the kind of genetic disease that many scientists hope will eventually be treatable by recombinant gene therapy. If a way could be found to introduce functional β-globin genes into an individual homozygous for the sickle-cell mutation, he or she would be rendered effectively heterozygous, with greatly increased chances for a long and productive life.

THALASSEMIAS: EFFECTS OF MISFUNCTIONING HEMOGLOBIN GENES

Thalassemias are hemoglobin mutations in which one or more genes are wholly or partially nonfunctional.

The human hemoglobin variants we have mentioned so far are all consequences of missense mutations. Because of a base substitution in a gene coding for one of the chains, one amino acid is substituted for another. There are, however, other genetic defects involving hemoglobin in which one or more of the chains are simply not produced or are produced in an insufficient amount. The pathological condition that arises is called **thalassemia.** The condition of thalassemia can arise in several ways:

1. One or more of the genes coding for hemoglobin chains may have been deleted.
2. All genes may be present, but one or more may have undergone a nonsense mutation that produces a shortened chain or a frameshift mutation that produces a nonfunctional chain (see Figure 7.21b and c).
3. All genes may be present, but a mutation may have occurred outside the coding regions, leading to a block in transcription or to improper processing of the pre-mRNA, so that the protein is not produced or is not functional.

In case 1 or 2, the gene produces no functional protein. In case 3 limited transcription and translation of the correct polypeptide sequence may occur.

The human genome contains a number of globin genes, corresponding to the protein chains used at different developmental stages, so there are many varieties of thalassemia. We describe here only two major classes—those involving loss or misfunction of genes for the adult β and α chains.

β-Thalassemia

If the β globin gene is lost or cannot be expressed, a most serious condition arises in individuals homozygous for this defect. They can make *no* β chains and must rely on continued production of the fetal γ chains to make a functional hemoglobin, $\alpha_2\gamma_2$ (see Figure 7.22). Such individuals may produce γ chains well into childhood, but they usually die before reaching maturity. Much less serious is the heterozygous state, in which one β gene is still functioning. There are also milder thalassemias (called β^+) in which transcription or processing of the β genes is partially inhibited; thus β globin production is limited but not entirely blocked.

α-Thalassemias

Thalassemias involving the α chain present a more complicated situation. Two copies of the gene (α_1 and α_2) are next to each other on the human chromosome (Figure 7.23). Their α_1 and α_2 chains differ by only one amino acid, and one can replace the other in the assembled hemoglobin tetramer. An individual can have 4, 3, 2, 1, or 0 copies of an α gene. Only if three or more genes are nonfunctional are serious effects observed. Individuals with only one α gene are anemic, because their total hemoglobin production is low. The low level of α hemoglobin is partially compensated for by formation of β_4 tetramers *(hemoglobin H)* and γ_4 tetramers *(hemoglobin Bart's)*. These tetramers can bind and carry oxygen, but they do not exhibit the allosteric transition (they remain always in the R state), nor do they exhibit a Bohr effect. So unloading of oxygen to tissues is inefficient. In the condition known as *hydrops fetalis,* all four α gene copies are missing. Individuals with this condition are inevitably stillborn. They can form only a γ_4 hemoglobin, and because the supply of γ chains falls near birth, not enough hemoglobin is available to support the near-term fetus.

Because there are two copies of the α gene but only one of the β gene, most of the deleterious mutations in mammalian hemoglobins occur in the β chains (see Figure 7.25). This phenomenon may suggest a functional role for gene duplication: If two or more copies of a gene are present, the species is somewhat protected from the harmful effects of mutations.

We have concentrated our discussion of protein mutation on hemoglobin, but it must be understood that the same principles apply to all other proteins. Although our knowledge of hemoglobin mutations is the most complete, missense mutations and deletions are found in many other proteins as well—those that are deleterious give rise to the wide class of *genetic diseases,* of which hemoglobin pathologies are only one subclass. Many other examples will be encountered in the chapters on metabolism.

Immunoglobulins: Variability in Structure Yields Versatility in Binding

In the remainder of this chapter, we turn to a group of proteins whose primary function, like that of myoglobin and hemoglobin, is binding of other substances, but whose biological role is entirely different. These proteins are the **immunoglobulins,** proteins which exhibit very specific binding to other substances.

THE IMMUNE RESPONSE

When a foreign substance—a virus, a bacterium, or even a foreign protein—invades the tissues of a higher vertebrate (like a human), the organism defends itself by what is called the **immune response.** This defense has two facets. In the **humoral immune response,** lymphatic cells called **B lymphocytes** synthesize specific immunoglobulin molecules that are secreted from the cell and bind to the invading substance. This binding aggregates the foreign substance and marks it for destruction by cells called **macrophages.** In the **cellular immune response,** lymphatic cells called **T lymphocytes,** bearing immunoglobulin-like molecules on their surfaces, recognize and kill foreign or aberrant cells. In this section we shall be mainly concerned with the humoral immune response.

The immune response involves the defense of the body against foreign substances or pathogens.

The substance that elicits an immune response is called the **antigen,** and a specific immunoglobulin that binds to this substance is called the **antibody.** If the invading particle is large, like a cell, a virus, or a protein, many different antibodies may be elicited, each type binding specifically to a given **antigenic determinant** (or **epitope**) on the surface of the particle (Figure 7.29a). Such antigenic determinants may be, for example, groups of amino acids on a protein surface or groups of sugar residues in a carbohydrate; an example is shown in Figure 7.29b.

In the humoral immune response, B lymphocytes secrete antibodies (immunoglobulins) that react with specific antigens.

The immune response is our major line of defense against infection and probably against cancer cells as well. It is the crippling of the immune system by the human immunodeficiency virus (HIV) that makes **AIDS (acquired immune deficiency syndrome)** a disease that has proved so devastating. Victims of AIDS do not die of the direct effects of the virus—they perish from infectious diseases or cancers that their immune system is no longer able to defend against. (See page 248.)

The immune response has some remarkable features. First, it is incredibly versatile, being able to respond to an enormous number of different foreign substances. These foreign substances range from cells of another individual of the same species (the basis of tissue graft or organ transplant rejection) to synthetic molecules that could never have been encountered in nature. Second, the immune response has a so-called *memory:* After an initial exposure to a given antigen, a second exposure at a later date will result in rapid and much more massive production of the specific antibodies.

Scientists were long perplexed by the immune response. In particular, it seemed difficult to explain how specific antibodies could be generated against millions of different antigens, some of which no organism had ever encountered before. Early theories, called **instructive theories,** suggested that an antigenic determinant could somehow induce an antibody molecule to take up a particular tertiary

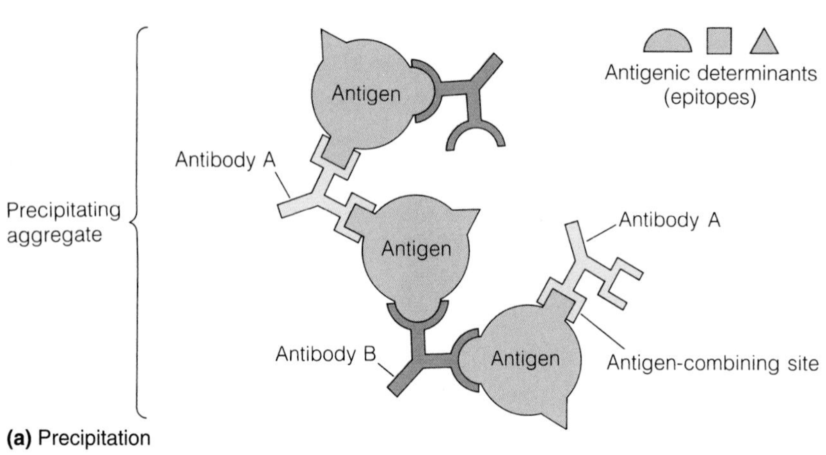

(a) Precipitation

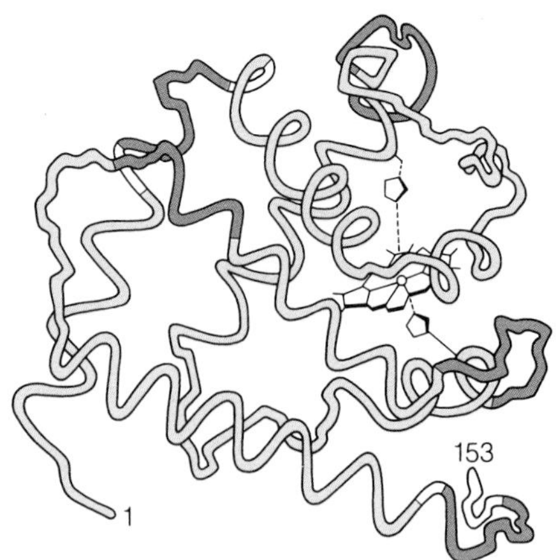

(b) Antigenic determinants, sperm whale myoglobin

FIGURE **7.29**

Antigenic determinants. (a) A foreign object, or antigen (such as a virus, a bacterial cell, or a foreign protein molecule), may elicit the production of antibodies to several different antigenic determinants on its surface. When the antigen is mixed with this collection of antibodies, precipitation occurs because each antibody molecule has two binding sites for its determinant. Thus, a network is formed. **(b)** The antigenic determinants of sperm whale myoglobin. The purple portions represent segments of the polypeptide chain that act as determinants. Some determinants involve portions of chain that are far apart in the primary sequence but close together in the tertiary structure. The white portions form part of the antigenic determinant with only some antibodies.

(b) M. Z. Atassi, *Immunochem.* (1975) 12:435. © 1975 with permission of Elsevier Science.

folding, which would then serve as a binding site for that determinant. However, models of this kind violate the principle we discussed in Chapter 6: Higher-order protein structure is dictated entirely by the primary structure. Evidence that the same principle holds for antibody binding sites came from experiments in which an antibody to a specific determinant was denatured and then allowed to renature. The capacity to bind the antigen, which had been lost on denaturation, was regained even if refolding took place in the absence of antigen.

Modern explanations for the immune response are based on the clonal selection theory.

Thus, the evidence suggested that the body has an inherent ability to produce an immense diversity of antibodies with different amino acid sequences that are able to bind an enormous range of antigens. How such a system can work is described in what is called the **clonal selection theory,** which is now supported by overwhelming evidence. The basic postulates of the clonal selection theory, illustrated in Figure 7.30, are as follows:

1. **B stem cells** in the bone marrow differentiate to become B lymphocytes, each producing a single type of immunoglobulin molecule, each type with a binding site that will recognize a specific molecular shape. (We will discuss the generation of this variation later in the chapter.) These immunoglobulins, or antibodies, are attached to the cell membrane and exposed on the outer surfaces of the B lymphocytes.
2. Binding of an antigen to one of these antibodies stimulates the cell carrying it to replicate, generating a **clone** (a collection of cells with identical genetic information). This *primary response* is aided by a special class of T cells called **helper T cells.** If a helper T cell recognizes a bound antigen, it binds to the appropriate B lymphocyte and transmits to it a signal protein (**interleukin-2**) that stimulates B-cell reproduction. Thus, only those clones of B cells that recognize antigens are stimulated to continued cell division.
3. As shown in Figures 7.30 and 7.31, two classes of cloned B cells are produced. **Effector B cells,** or **plasma cells,** now produce *soluble* antibodies,

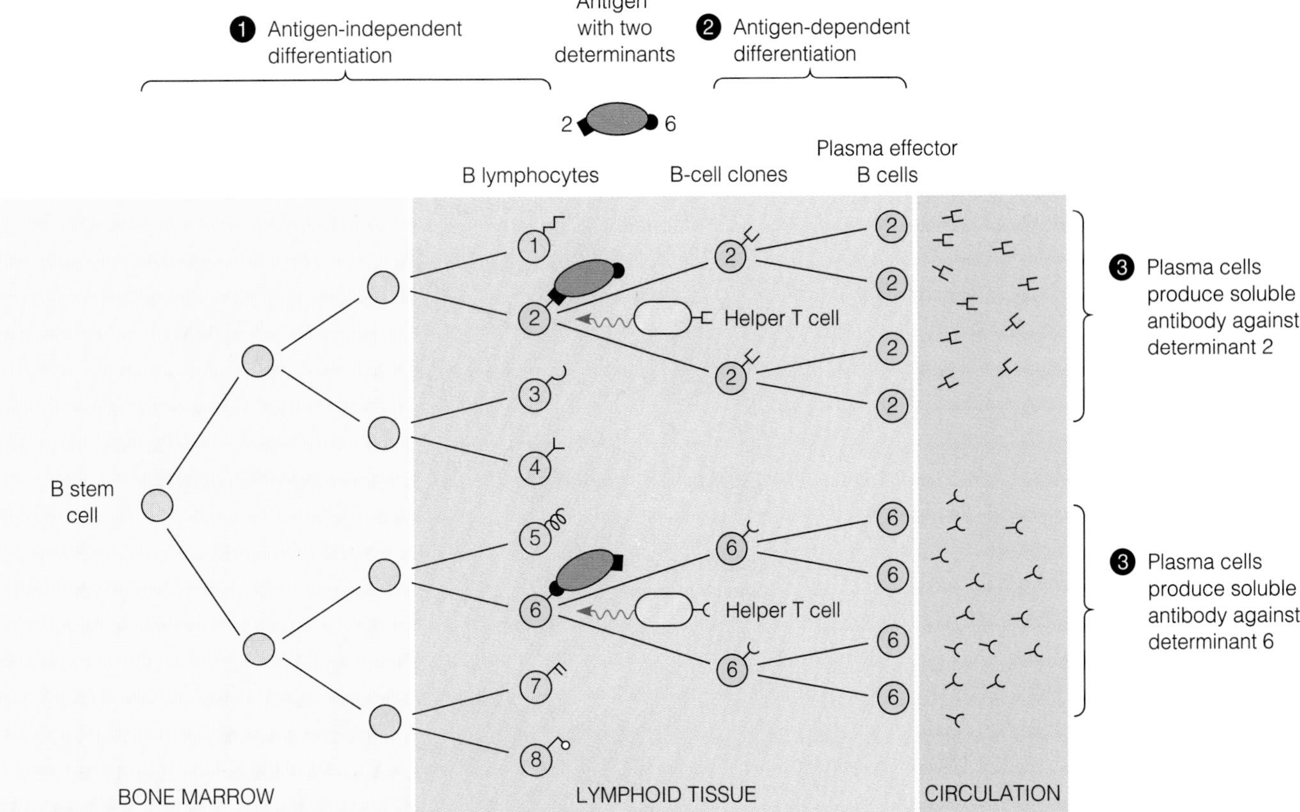

FIGURE **7.30**

The clonal selection theory of the immune response. Step 1: Stem cells in the bone marrow (B stem cells, left) differentiate and migrate to the lymphoid tissue. Each of the differentiated cells (B lymphocytes) synthesizes a unique kind of antibody, which it carries on its surface. The B lymphocytes in this figure carry antibodies numbered 1–8. **Step 2:** When the B lymphocytes encounter antigens, the B cells that carry antibodies to the antigenic determinants are stimulated by helper T cells to multiply, forming clones of B cells. In this case, the cells selected to form clones are those carrying antibodies 2 and 6. **Step 3:** Some of the cloned B cells, called plasma cells, produce soluble antibodies, with each clone producing antibody against a single determinant. Other cloned B cells are called memory cells; their role is illustrated in Figure 7.31.

Adapted with permission from J. Darnell, H. Lodish, and D. Baltimore, *Molecular Cell Biology;* ©1986 Scientific American Books, Inc.

which are secreted into the circulatory system. These antibodies have the same antigen binding sites as the surface antibodies of the B lymphocyte from which the effector cells arose, but they lack the hydrophobic tail that bound the surface antibodies to the lymphocyte membrane. The other class of cells in the clone—**memory cells**—will persist for some time, even after antigen is no longer present. This persistence constitutes the immune memory: It allows a rapid *secondary response* to a second stimulation by the same antigen, as shown in Figure 7.31.

The clonal selection theory explains many features of the immune response and does so in accord with what we know about the determinants of protein folding. But a critical question may have occurred to you: Why do we not find clones producing antibodies against *our own* proteins and tissues? The answer is a fascinating one that tells much about how biochemical "self" is established. When immature B cells in the fetus encounter substances that bind to their surface antibodies, they are *not* stimulated to replicate. Rather, these fetal B cells are destroyed. Thus, B cells producing antibodies against all of the potential "self" antigens to

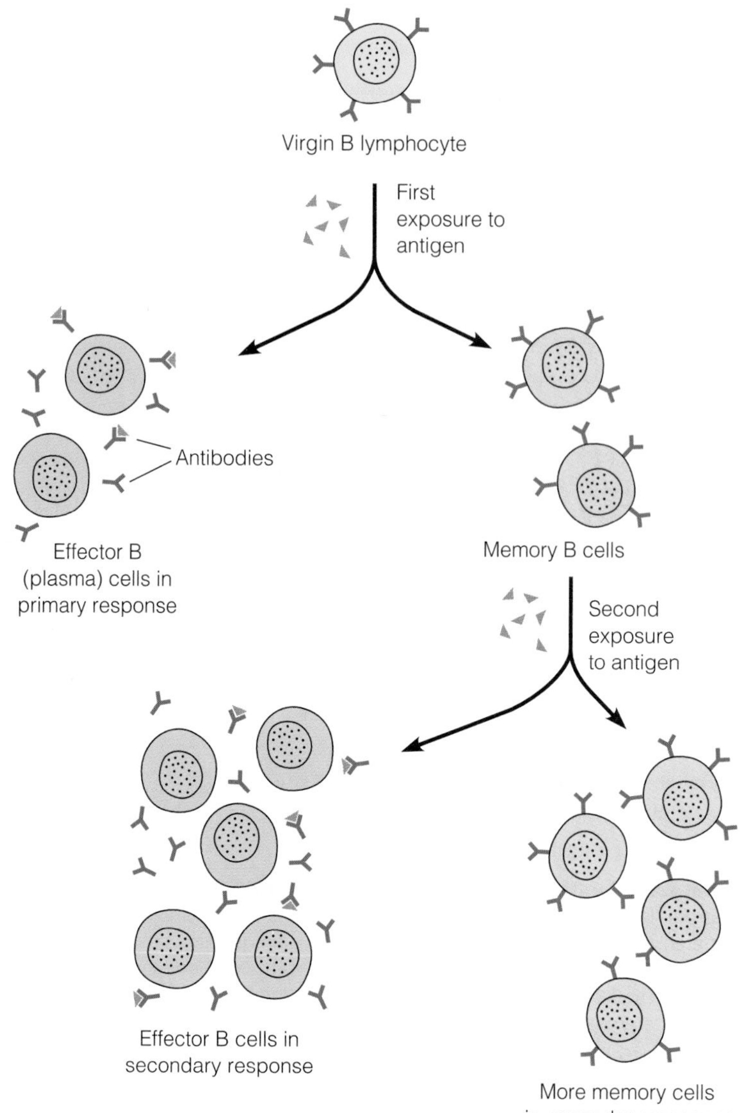

FIGURE **7.31**

Two developmental paths for stimulated B lymphocytes. Exposure to antigen causes two kinds of cells to develop from B lymphocytes. Cells of one type (effector B cells, or plasma cells) synthesize soluble antibody. Cells of the second class (memory cells) carry membrane-bound antibody to allow a rapid response to a second exposure.

which we might react are eliminated before birth. The only B cells that mature are those that produce antibodies against "nonself" substances.

Occasionally, the immune system goes awry and produces antibodies against the normal tissues of an adult. The reasons for such **autoimmunity** are not wholly understood, but the resulting diseases can be devastating. In *lupus erythematosus,* for example, the individual's own nucleic acids become the object of attack.

THE STRUCTURE OF ANTIBODIES

To see how clonal selection actually works at the molecular level, we must explore the structure of the immunoglobulin molecules that constitute the antibody arsenal. There are five classes of immunoglobulin molecules, which carry out various functions in the immune system (Table 7.3). However, all are built from the same basic immunoglobulin pattern, which is shown in schematic molecular detail in Figure 7.32. Different kinds of antibodies may contain from one to five immunoglobulin molecules; when more than one is present, the monomers are linked by a second type of polypeptide, called a J chain (see Table 7.3).

Immunoglobulin molecules contain both constant and variable regions. The variable regions are the antigen binding sites.

Each immunoglobulin monomer consists of four chains, two heavy chains ($M = 53{,}000$ each) and two light chains ($M = 23{,}000$ each), held together by disulfide bonds. In each chain are **constant domains** (identical in all antibodies of a

TABLE **7.3 The five classes of immunoglobulins**

Description	Structure
IgM is produced during the early response to an invading microorganism. It is the largest immunoglobulin, containing five Y-shaped units of two light and two heavy chains each. The units are held together by a component called a J chain. The relatively large size of IgM restricts it to the bloodstream. It is also effective in triggering an important mechanism for foreign cell destruction, called the complement system.	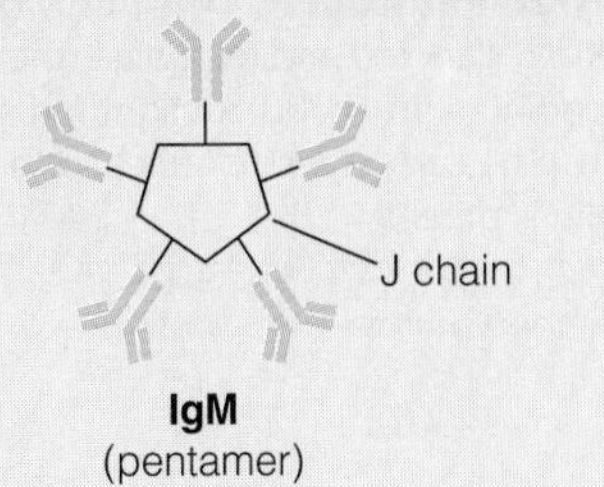 **IgM** (pentamer)
IgG molecules, also known as *γ-globulin,* are the most abundant of circulating antibodies. A variant is attached to B-cell surfaces. IgG molecules consist of a single Y-shaped unit and can traverse blood vessel walls rather readily; they also cross the placenta to carry some of the mother's immune protection to the developing fetus. Specific receptors allow such passage. IgG also triggers the complement system.	**IgG** (monomer)
IgA is found in body secretions, including saliva, sweat, and tears, and along the walls of the intestines. It is the major antibody of colostrum, the initial secretion from a mother's breasts after birth, and of milk. IgA occurs as a monomer or as double-unit aggregates of the Y-shaped protein molecule. IgA molecules tend to be arranged along the surface of body cells and to combine there with antigens, such as those on a bacterium, thus preventing the foreign substance from directly attaching to the body cell. The invading substance can then be swept out of the body together with the IgA molecule.	J chain **IgA** (monomer or dimer)
Less is known about the IgD and IgE immunoglobulins. IgD molecules are found on the surface of B cells, though little is known about their function. IgE is associated with some of the body's allergic responses, and its levels are elevated in individuals who have allergies. The constant regions of IgE molecules can bind tightly to mast cells, a type of epithelial and connective tissue cell that releases histamines as part of the allergic response. Both IgD and IgE consist of single Y-shaped units.	**IgD** (monomer) **IgE** (monomer)

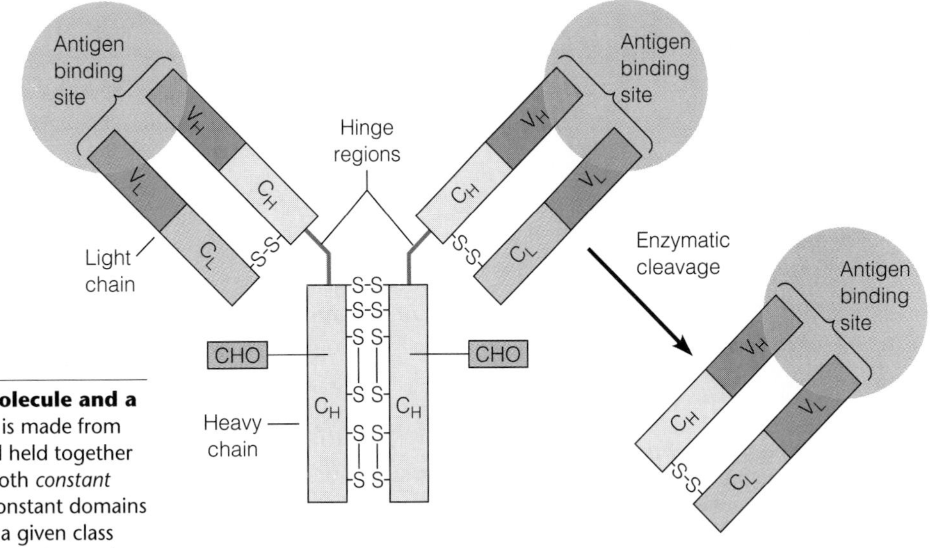

FIGURE **7.32**

Schematic models of an antibody molecule and a Fab fragment. The antibody molecule is made from two heavy chains and two light chains, all held together by disulfide bonds. Each chain contains both *constant* domains (C) and *variable* domains (V). Constant domains are the same in all antibody molecules of a given class (see Table 7.3), whereas variable domains confer specificity to a given antigenic determinant. Cleavage by certain proteolytic enzymes such as papain at the *hinge region* allows production of monovalent *Fab fragments.* The carbohydrate (CHO) attached to the heavy chains aids in determining the destinations of antibodies in the tissues and in stimulating secondary responses such as phagocytosis. A molecular model of an immunoglobulin molecule, derived from x-ray diffraction studies, is shown in Figure 7.33.

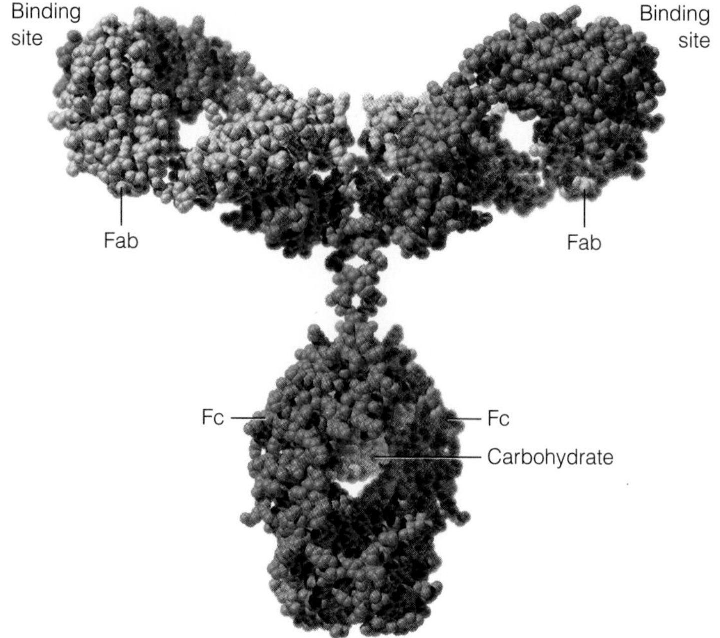

FIGURE **7.33**

A model of the IgG molecule from x-ray diffraction studies. The heavy chains are colored dark red and dark blue; the corresponding light chains are light red and light blue. The carbohydrate attached to the heavy chain is green.

Adapted from E. A. Padlan *Mol. Immunol.* (1994) 31:169–217. © 1975 with permission of Elsevier Science.

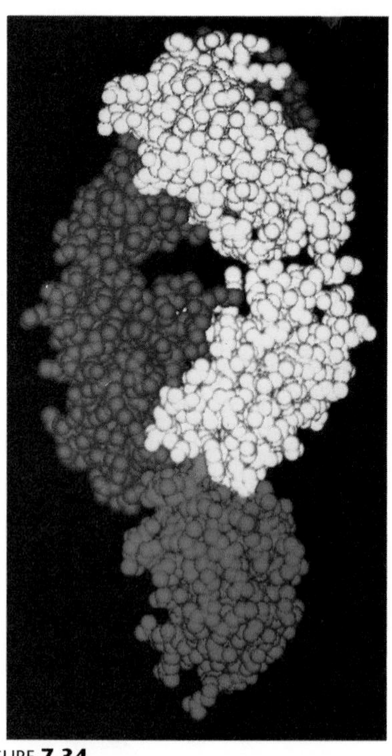

FIGURE **7.34**

Antigen binding by a Fab fragment. The binding of a Fab fragment to a specific antigen, the protein lysozyme, shows the close contact that occurs between antigen and antibody surfaces. The antibody light chain is shown in yellow, the heavy chain fragment in blue, and the lysozyme molecule in green. The red dot corresponds to Glu 121 of lysozyme, which fits snugly into a cleft between the light and heavy chains. The surfaces of the lysozyme and the antibody binding site fit closely together, and a number of specific hydrogen bonds are made across this surface.

Courtesy of A. G. Amit, R. A. Mariuzza, S. E. Phillips, and R. J. Poljiak, *Science* (1986) 233:747–753; © 1986 AAAS.

given class) and a **variable domain.** It is variations in the amino acid sequence (and therefore the tertiary structure) of the variable domains of the light and heavy chains that confers the multitudinous specificities of antigens to different determinants. Note that the four variable domains are carried at the ends of the Y-like fork of the molecule, where they form two binding sites for antigenic determinants.

A large protein, a virus, or a bacterial cell has on its surface many different potential antigenic determinants. Antibodies may be generated to several of these determinants, binding many antigen molecules together and thereby aggregating the antigen (see Figure 7.29a). If the antigen is so small that it has only one determinant, binding will occur but aggregation will not. Precipitation also requires the antibody to be *bivalent* (to have two binding sites). By careful proteolysis, it is possible to cleave antibodies at the hinge region (see Figure 7.32) to produce **Fab fragments** with only one binding site each. Such fragments will bind but not precipitate antigen.

The antigen binding site lies at the extreme end of the variable domains and involves amino acid residues from the variable regions of both heavy and light chains. Different sequences in these variable regions give rise to different local secondary and tertiary structure and can thereby define binding sites to fit different antigens. Figure 7.34 shows the results of x-ray diffraction studies of the interaction of a Fab fragment with a protein antigen, lysozyme. Close contacts are made between 16 lysozyme residues and 10 heavy chain and 7 light chain residues. The antigen and antibody surfaces fit together in a highly complementary fashion.

The constant domains of the heavy chains in the base of the Y-shaped molecule serve not only to hold the chains together. These regions also function as effectors, to signal macrophages in the circulatory system to attack particles or cells that have been labeled by antibody binding. Macrophages are large white blood cells that are specially adapted to engulf and digest foreign particles. In addition, differences in heavy chains identify immunoglobulin types for delivery to different tissues or for secretion (see Table 7.3).

GENERATION OF ANTIBODY DIVERSITY

How can the enormous diversity of immunoglobulin molecules be generated so that antibodies to an almost unlimited range of antigens are provided? The human genome simply does not have enough room to encode a gene for each of the millions of different immunoglobulin molecules occurring in B stem cells. Instead, two special processes occur in these cells.

The major source of antibody diversity is *recombination of exons.* The genomes of higher vertebrates contain "libraries" of exons corresponding to different portions of the immunoglobulin molecule and mechanisms for rearranging these exons to create different combinations in both the heavy and light chains. We have already mentioned that such rearrangements, when they occur in sperm- or ovum-producing cells, play a role in protein evolution. The same process, when it occurs in B cells, creates new immunoglobulins in individual cells. The details and mechanism of this process are described in Chapter 25.

Through somatic recombination and rapid mutation, a human can generate over 100 million different antibodies.

An additional source of antibody diversity is *somatic mutations,* mutations that are not inherited, because they occur in somatic cells (cells that are not germline). In the cells that generate antibodies, certain portions of the variable regions in the immunoglobulin genes mutate at an unusually high rate. The reason for these localized high rates of mutation is still obscure, but this process, together with recombination of gene fragments, can account for the generation of an immense diversity of immunoglobulin molecules. It has been calculated that about 100 million combinations can be made from the library of immunoglobulin gene fragments available in the human genome.

T CELLS AND THE CELLULAR RESPONSE

Whereas the humoral immune response is based on antibody-produced aggregation, usually followed by digestion by macrophages, the cellular immune response involves a quite different mechanism for killing of foreign cells. The cellular response plays a major role in tissue rejection and in destroying virus-infected cells. It may also destroy potential cancer cells before they have a chance to propagate. Although the mechanisms of the humoral and cellular processes are quite different, similar immunoglobulin molecules are involved in both cases (Figure 7.35), pointing to a common evolutionary origin for the humoral and cellular responses.

The major participants in the cellular immune response are **cytotoxic T cells,** also referred to as **killer T cells.** These cells carry on their surfaces receptor molecules that are structurally similar to the Fab fragments of antibody molecules. Like antibodies, these fragments have a wide range of binding specificities, mostly directed toward short oligopeptide sequences. Such oligopeptides might be produced, for example, by a virus-infected cell when it partially digests virus particles within it. The T-cell receptor does not recognize free oligopeptides. Instead, the oligopeptides must be presented on the surface of the infected cell and bound to another class of immunoglobulin-like molecules, proteins of the **major histocompatibility complex** (**MHC** proteins; see Figure 7.35). When a killer T cell identifies (via its receptor) an appropriate antigen carried on the surface of another cell by an MHC protein, it releases a protein called **perforin.** This protein forms pores in the plasma membrane of the cell being attacked, allowing critical ions to diffuse out and thereby killing the cell.

The cellular immune response uses killer T cells to destroy foreign or infected cells.

As Figure 7.35 emphasizes, both the antibodies of the humoral response and the molecules involved in the cellular response contain elements of common structure. The similarity is even closer than the figure suggests. The domains in these molecules are built on a common motif—the *immunoglobulin fold,* in which two antiparallel β sheets lie face to face (Figure 7.36). This structure probably represents the primitive structural element in the evolution of the immune response. Indeed, the immunoglobulin fold is also found in a number of other proteins that are involved in cell recognition.

It is instructive to compare the immunoglobulin family of proteins with the myoglobin–hemoglobin family. In both cases the primary function of the proteins is binding. In the myoglobin–hemoglobin family we see evidence for the progressive evolution of more and more sophisticated methods for regulating the binding of a particular molecule—oxygen—and for coupling oxygen binding to

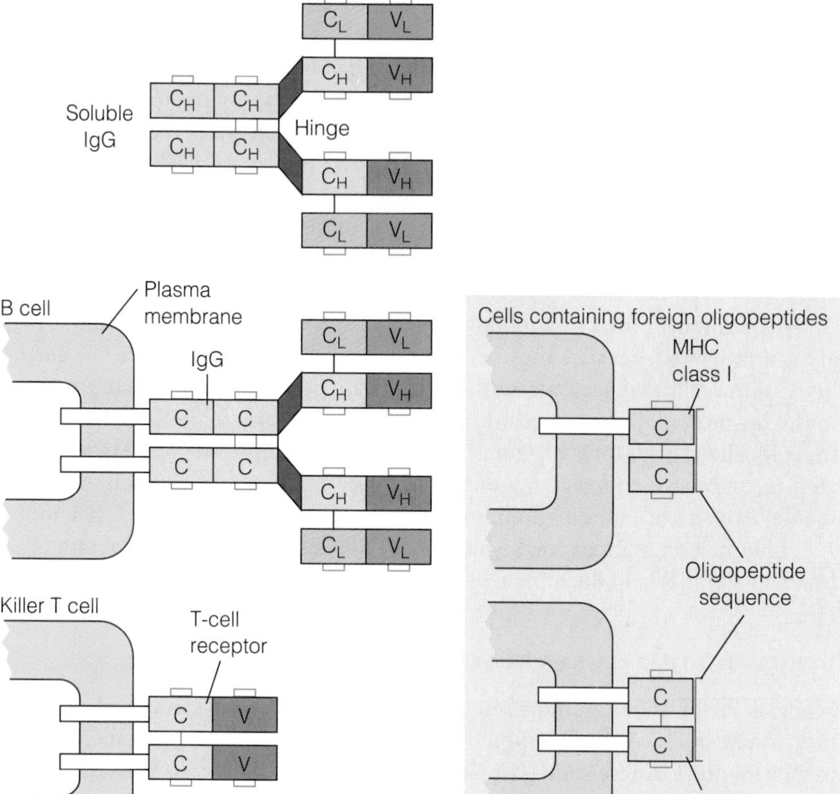

FIGURE **7.35**

Comparison of proteins of the humoral and cellular immune responses. This schematic figure shows the structural similarity of members of the immunoglobulin superfamily of proteins, which includes not only the immunoglobulin family but many related families of proteins as well. C and V indicate the constant and variable domains, respectively, each with a structure similar to that shown in Figure 7.36. MHC indicates the major histocompatibility complex. Connecting disulfide bridges are shown in blue. White rectangles represent membrane-binding domains.

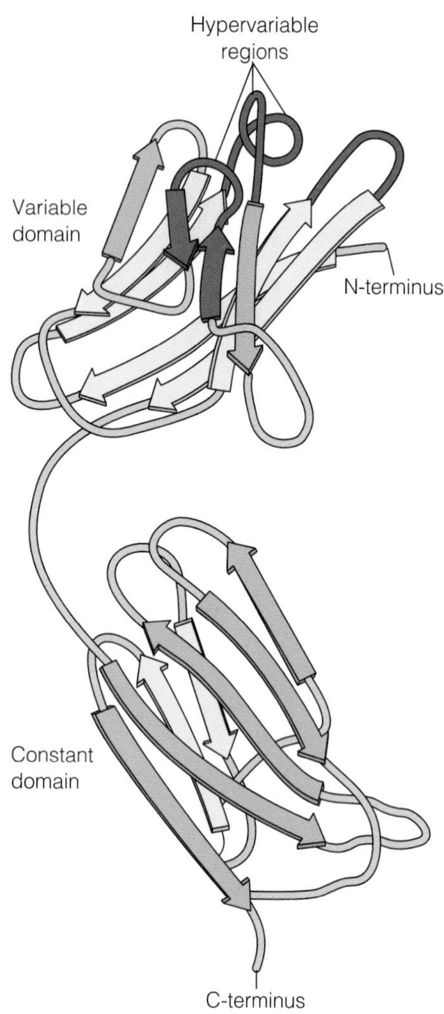

FIGURE **7.36**

The immunoglobulin fold. The immunoglobulin fold is a common structure in domains of many proteins in the immunoglobulin superfamily. Two layers of antiparallel β sheet are stacked face to face. In this model of an immunoglobulin light chain, the fold occurs twice—once in the constant region and once in the variable region.

From C. Branden and J. Tooze, *Introduction to Protein Structure* (New York: Garland Publishing 1991), p. 187. © 1991 Garland Publishing.

the binding of CO_2. In the immunoglobulin family, evolution from a simple motif has led to an enormous diversification in binding function. A mechanism has evolved that allows the production of an immense range of molecules with specific binding capacities. Different tasks require different tools.

AIDS and the Immune Response

AIDS (acquired immune deficiency syndrome) is a disease of the immune system. It is caused by the human immunodeficiency virus, or HIV (Figure 7.37), which attacks a number of kinds of cells but is particularly virulent toward a class of helper T cells. The virus wages a long battle with rapidly replicating T cells, but eventually the rate of cell destruction exceeds the rate of replication. The consequence is a deterioration of the whole immune response, in particular the ability of B cells to proliferate in response to antigen stimulation. In addition, there is a general failure in T-cell activation. Most AIDS patients succumb either to diseases they could have easily resisted before contracting AIDS or to certain kinds of cancer. AIDS is so deadly because it attacks our most fundamental defenses against all disease.

Because AIDS poses such a grave threat to world health, searches for a vaccine are being intensely pursued. Such searches entail unusual problems, for the AIDS virus has an unparalleled capacity to mutate and thus develop strains resistant to any vaccine. Mutations occur in the HIV genome at a rate many times higher than in the human genome. The genetic variation is increased by the many virus replication cycles and by the fact that the reverse transcriptase (Figure 7.37) is very error-prone. The magnitude of the problem can be grasped by considering our experience with the influenza virus. We have never been able to produce a

FIGURE **7.37**

(a) The human immunodeficiency virus (HIV) that is responsible for AIDS. **(b)** A schematic model of HIV. The surface protein gp160 is composed of two fragments, gp41 and gp120. The RNA genome is transcribed into DNA by a reverse transcriptase. This DNA integrates into the host cell genome, and is then retranscribed to produce new viral RNA.

(a) Courtesy of Hans Gelderblom; **(b)** From Hoth, Jr., Myers, and Stein, *Hospital Practice* 27:9, p. 154. Illustration © Alan D. Iselin.

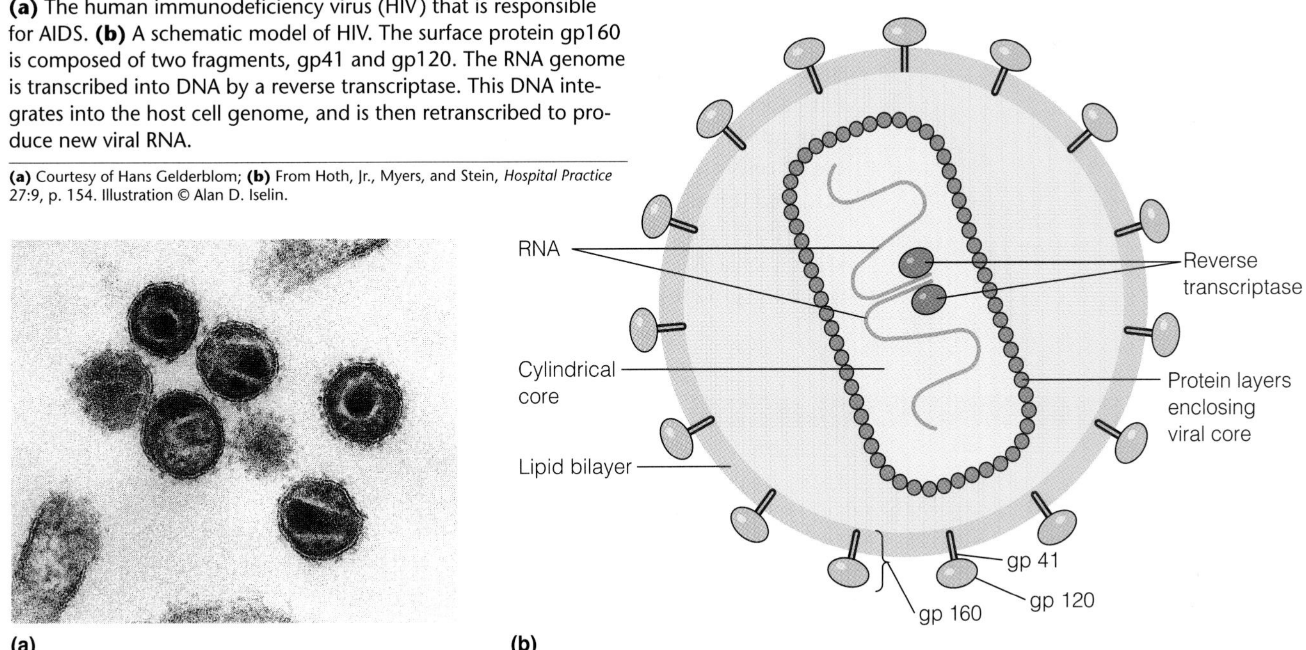

lifelong "flu" vaccine because of the great variability of the influenza virus. HIV mutates about 60 times faster than the influenza virus.

Because of the difficulty in producing vaccines and the fact that the HIV infection has already spread so widely, equally active efforts continue to be made to produce a therapy. The efforts which have been most successful have used a combination of approaches. On the one hand, the replication of the virus can be slowed by specific inhibitors of the reverse transcriptase. A quite different approach uses a protease inhibitor to block a proteolytic step essential in the maturation of new viruses within the infected cells. One may hope that combinations of such therapies, together with an eventual vaccine, will halt this plague.

In AIDS, the causal virus attacks helper T cells, destroying the body's immunological defense system.

SUMMARY

Most organisms need oxygen. Vertebrates use hemoglobin for oxygen transport and myoglobin for oxygen storage. In these proteins O_2 is bound at an Fe(II)-porphyrin (heme); the heme is carried in a hydrophobic pocket, inhibiting oxidation of the iron. Myoglobin carries a single oxygen binding site and consequently exhibits a noncooperative, hyperbolic binding curve. Hemoglobin binds O_2 cooperatively, with a sigmoidal binding curve that leads to more efficient transfer. It accomplishes this cooperativity by being a tetrameric, allosteric protein. Binding O_2 to hemoglobin sites causes tertiary structure changes. When strain from these changes accumulates, a quaternary ($T \longrightarrow R$) transition occurs, shifting the molecule from the weak-binding form to the strong-binding form. The allosteric transition also allows allosteric effectors (H^+, CO_2, BPG) to modify oxygen binding, leading to more efficient O_2 and CO_2 transport.

Myoglobin and hemoglobin, like other proteins, are evolving via mutations, duplications, and recombinations in their genes. Both types of globin evolved from a myoglobin-like ancestral protein, with the development of a true hemoglobin coinciding approximately with the emergence of vertebrates. Evolution of these proteins continues, as evidenced by the existence of a multitude of variant hemoglobins in the human population. Most base substitution (missense) mutations are neutral, but some, like the sickle-cell hemoglobin mutation, are deleterious. Thalassemias are hemoglobin pathologies that involve either deletion or faulty expression of whole genes or sets of genes.

The immune response is the body's main defense against infection. In the humoral response, antibodies (specific immunoglobulin molecules) that will bind with specific antigens are generated and secreted. This process occurs because recognition of the antigen by a few cells leads to clonal selection of a large number of cells producing the appropriate antibody. Immense antibody diversity is achieved through multiple somatic recombinations and rapid mutation in antibody-producing cells. The cellular immune response involves receptor-bearing killer T cells. AIDS is a disease of the immune system; HIV attacks T cells essential to the growth of B-cell clones.

REFERENCES

General

Dickerson, R. E., and I. Geis (1983) *Hemoglobin: Structure, Function, Evolution, and Pathology.* Benjamin/Cummings, Redwood City, Calif. A wealth of detail on myoglobin and hemoglobin, including much on hemoglobin evolution and variants.

van Holde, K. E., W. C. Johnson, and P.-S. Ho (1998) *Principles of Physical Biochemistry.* Prentice Hall, Upper Saddle River, N.J. Chapter 15 contains a more detailed discussion of binding equilibrium than is presented here.

Allosteric Models

The papers by Koshland et al. and Monod et al. introduced the major classes of models for allostery. The other papers describe more recent studies.

Ackers, G. (1998) Deciphering the molecular code of hemoglobin allostery. *Adv. Protein Chem.* 51:185–253.

Barrick, D., N. T. Ho, V. Simplaceanu, F. Dahlquist, and C. Ho (1997) A test of the role of the proximal histidines in the Perutz model for cooperativity in haemoglobin. *Nature Struct. Biol.* 4:78–83.

Gelin, B. R., A. W.-M. Lee, and M. Karplus (1983) Hemoglobin tertiary structural change on ligand binding. *J. Mol. Biol.* 171:489–559. Detailed and difficult.

Koshland, D. E., G. Nemethy, and D. Filmer (1966) Comparison of experimental binding data and theoretical models in proteins containing subunits. *Biochemistry* 5:365–385.

Monod, J., J. Wyman, and J.-P. Changeux (1965) On the nature of allosteric transitions: A plausible model. *J. Mol. Biol.* 12:88–118.

Dynamics of Oxygen Binding and Release

Case, D. A., and M. Karplus (1979) Dynamics of ligand binding to heme proteins. *J. Mol. Biol.* 132:343–368. A pioneering paper on protein dynamics.

Evolution of Hemoglobin

Hardison, R. (1991) Evolution of globin gene families. In: *Evolution at the Molecular Level,* edited by R. K. Selander, A. G. Clark, and T. S. Whittam, pp. 272–295. Sinauer Associates, Sunderland, Mass.

Variant Hemoglobins and Hemoglobin Pathologies

Embury, S. H. (1986) The clinical pathophysiology of sickle cell disease. *Annu. Rev. Med.* 37:361–376.

Honig, G. R., and J. G. Adams (1986) *Human Hemoglobin Genetics.* Springer-Verlag, Berlin, New York. A comprehensive treatment.

Ingram, V. M. (1957) Gene mutation in human haemoglobin: The chemical difference between normal and sickle cell haemoglobin. *Nature* 180:326–328. A classic paper, first identifying the amino acid locus and genetic change in a genetic disease.

Orkin, S. H., and D. G. Nathan (1981) The molecular genetics of thalassemia. *Adv. Hum. Genet.* 11:233–280.

Pauling, L., H. A. Itano, S. J. Singer, and I. C. Wells (1949) Sickle cell anemia: A molecular disease. *Science* 110:543–548. The introduction of the concept of molecular diseases.

Antibodies

Darnell, J., H. Lodish, and D. Baltimore (1986) *Molecular Cell Biology,* Chapter 24. Scientific American Publishing Co., New York. An elegant, detailed presentation.

Kuby, J. (1992) *Immunology.* W. H. Freeman, New York. A lucid and comprehensive text.

Stanfield, R. L., T. M. Fieser, R. A. Lerner, and I. A. Wilson (1990) Crystal structures of an antibody to a peptide and its complex with peptide antigen at 2.8Å. *Science* 248:712–719. Details of the interaction.

Tonegawa, S. (1985) The molecules of the immune system. *Sci. Am.* 253(4):104–115. Readable and concise description.

AIDS

The June 19, 1998 issue of *Science* has a series of papers on AIDS and HIV.

Coffin, J. M. (1995) HIV population dynamics in vivo: Implications for genetic variation, pathogenesis, and therapy. *Science* 267:483–489.

PROBLEMS

1. The following data describe the binding of oxygen to human myoglobin at 37°C.

P_{O_2} (mm Hg)	θ	P_{O_2} (mm Hg)	θ
0.5	0.161	6	0.697
1	0.277	8	0.754
2	0.434	12	0.821
3	0.535	20	0.885
4	0.605		

From these data, estimate (a) P_{50} and (b) the fraction saturation of myoglobin at 30 mm Hg, the partial pressure of O_2 in venous blood.

2. What qualitative effect would you expect each of the following to have on the P_{50} of hemoglobin?
(a) Increase in pH from 7.2 to 7.4
(b) Increase in P_{CO_2} from 20 to 40 mm Hg
(c) Dissociation into monomer polypeptide chains

*3. Measurements of oxygen binding by whole human blood, at 37°C, at pH 7.4, and in the presence of 40 mm Hg of CO_2 and normal physiological levels of BPG (5 mmol/L of cells), give the following:

P_{O_2} (mm Hg)	% Saturation (= 100 × θ)
10.6	10
19.5	30
27.4	50
37.5	70
50.4	85
77.3	96
92.3	98

(a) From these data, construct a binding curve, and estimate the percent oxygen saturation of blood at (1) 100 mm Hg, the approximate partial pressure of O_2 in the lungs, and (2) 30 mm Hg, the approximate partial pressure of O_2 in venous blood.
(b) Under these conditions, what percentage of the oxygen bound in the lungs is delivered to the tissues?
(c) Using the data in Figure 7.16, repeat the calculation of part (b) if the pH drops to 6.8 in capillaries but goes back to 7.4 as CO_2 is unloaded in the lungs.

4. It is observed that chloride ion acts as a negative allosteric effector for hemoglobin. Suggest a possible explanation for why this should be so.

5. Precise data have been obtained for the oxygen binding of stripped human hemoglobin at 25°C:

P_{O_2} (mm Hg)	% Saturation (= 100 × θ)	P_{O_2} (mm Hg)	% Saturation (= 100 × θ)
0.10	0.315	5.75	76.0
0.350	0.990	7.94	90.9
0.794	3.06	12.88	96.9
1.748	9.09	29.51	99.0
2.884	24.0	67.60	99.7
4.467	50.0		

Use a Hill plot to determine (a) P_{50}, (b) n_H (maximum slope), and (c) the P_{50} values corresponding to the T and R states.

*6. G. Ackers, M. L. Johnson, F. C. Mills, et al., in *Biochemistry* (1975) 14:5128–5134, have observed that the P_{50} of purified hemoglobin decreases as the concentration of hemoglobin in solution is decreased. Suggest an explanation.

7. Oxygen binding by the hemocyanin of the shrimp *Callianassa* has been measured. Using the following data, prepare a Hill plot and determine (a) P_{50}, (b) n_H (the Hill coefficient), and (c) the *minimum* number of oxygen binding sites on the protein molecule.

P_{O_2} (mm Hg)	θ	P_{O_2} (mm Hg)	θ
1.1	0.003	136.7	0.557
7.7	0.019	166.8	0.673
10.7	0.035	203.2	0.734
31.7	0.084	262.2	0.794
71.9	0.190	327.0	0.834
100.5	0.329	452.0	0.875
123.3	0.487	736.7	0.913

8. Suggest probable consequences of the following real or possible hemoglobin mutations. [Note: Consult Figures 7.12 and 7.18.]
(a) At β 146 (HC3) His → Asp
(b) At β 92 (F8) His → Leu
(c) At β 2 (NA2) His → Asp
In each case, tell whether a single base change is sufficient for the mutation.

9. Suppose each of the mutants listed in Problem 8 were electrophoresed in comparison with native hemoglobin (pI = 7.0) at pH 8.0. Which would move faster toward the anode than native protein, and which would move more slowly?

*10. In principle, an allosteric molecule could exhibit *negative* cooperativity; that is, binding of the first ligands could decrease affinity for additional ones.
(a) What would a Hill plot look like for negative cooperativity?
(b) It has been noted that the KNF theory allows negative cooperativity but the MWC theory doesn't. Explain.

11. In the experiments of Barrick et al. (see page 226) it was observed that replacement of histidine by a noncovalently bonded imidazole not only reduced cooperativity but also increased the oxygen affinity of the hemoglobin. Suggest an explanation.

12. Assume that a new oxygen-transport protein has been discovered in certain invertebrate animals. X-ray diffraction of the deoxy protein reveals that it has the dimeric structure shown here in part (a) with a salt bridge between residues histidine 13 and aspartic acid 85. The two monomers interact by salt bridges between carboxyl termini and N-termini. The O_2 site lies *between* the two iron atoms shown, which are rigidly linked to α helices A and C (see part (b)). In the deoxy form, the space between the iron atoms is too small to hold O_2, and so the Fe atoms must be forced apart when O_2 is bound.

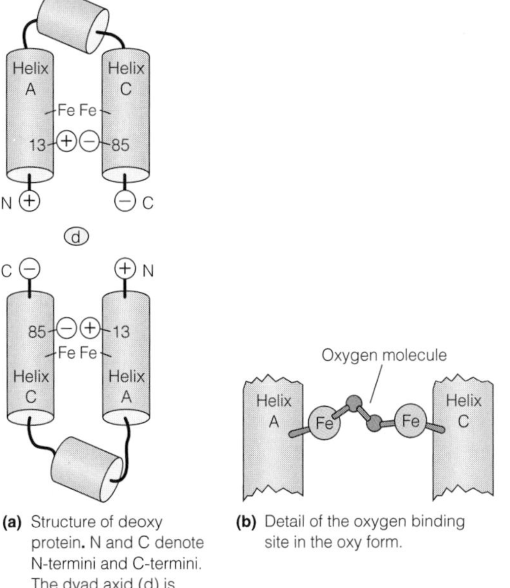

(a) Structure of deoxy protein. N and C denote N-termini and C-termini. The dyad axid (d) is perpendicular to this page.

(b) Detail of the oxygen binding site in the oxy form.

Answer the following questions, explaining your answer in each case in terms of the structure.
(a) Will this molecule show cooperative oxygen binding?
(b) Will this molecule exhibit a Bohr effect?
(c) What would be the effect of a mutation which replaced aspartic acid 85 by a lysine residue?

*13. The binding of antigens to antibodies can be thought of in ways similar to those we have used for discussing oxygen binding. Assuming we have monovalent antigens and n-valent antibodies, we can write the following equation for the number (r) of antigen molecules bound to an antibody at a concentration (c) of free antigen:

$$r = \frac{nKc}{1 + Kc}$$

where K is the equilibrium constant for the binding (the affinity constant).

(a) Show that this equation can be rearranged to give

$$\frac{r}{c} = Kn - Kr$$

This is called the *Scatchard equation*. It predicts that a graph of r/c versus r will be a straight line.

(b) Use the following data and a graph according to the Scatchard equation to obtain n and K for an antibody–antigen reaction.

c(M)	r
1.43×10^{-5}	0.50
2.57×10^{-5}	0.77
6.00×10^{-5}	1.20
1.68×10^{-4}	1.68
3.70×10^{-4}	1.85

14. What physiological effect would you predict from a mutation which replaced by serine that cysteine in the constant part of the immunoglobulin light chain which is involved in disulfide bond formation with the heavy chain? (See Figure 7.32.)

TOOLS OF BIOCHEMISTRY 7A

Immunological Methods

Because of the ease with which antibodies against biological materials can be prepared in the laboratory, and because of their great specificity, antibodies form the core of many important analytical and preparative biochemical procedures. We shall discuss here some of the methods most important to biochemists. Other immunological techniques are described in Tools of Biochemistry 23A.

Experiments in which an antigen is injected into a rabbit show that one antigen can elicit formation of several different antibodies. Each of these antibodies recognizes one particular portion of the antigen molecule, called an *antigenic determinant,* or *epitope.* Figure 7.29b shows the epitopes that have been identified in sperm whale myoglobin. Each epitope is a region encompassing five or six residues in the myoglobin sequence. Thus, a myoglobin **antiserum** (that is, serum from an animal immunized against myoglobin) carries at least five different antimyoglobin antibodies, each directed against one of the five epitopes shown in the figure.

Generally speaking, immunogenic substances (substances that elicit an immune reaction, such as synthesis of antibodies) are macromolecules. However, some of the most useful immunological reagents are antibodies against low-molecular-weight substances. For example, radioimmunoassay (Tools of Biochemistry 23A) is used to test for levels of steroid hormones or drugs. These compounds do not themselves stimulate antibody synthesis, but when one of them is coupled to an antigenic protein, the resultant conjugate is immunogenic. Some of the antibodies produced in response to the conjugate are directed against the low-molecular-weight, nonprotein constituent, which is called a **hapten.**

Most antibodies that are useful in biochemistry are of the IgG type (see Table 7.3). Each of these Y-shaped monomers has two antigen-combining sites. In an antigen–antibody reaction, each site usually binds to a different antigen molecule if sufficient antigen is present. If the antigen has two or more antibody-combining sites, as most do, a giant insoluble network of linked antigen and antibody molecules, called a **precipitin reaction,** is formed (see Figure 7.29a). The insoluble nature of this complex is at the heart of some of the older analytical techniques, such as immunodiffusion and immunoelectrophoresis.

In **immunodiffusion,** wells in an agar plate are filled with antigen and antibody, which diffuse toward each other in the agar. Where the substances meet, a precipitin reaction forms an insoluble precipitate, which can be visualized either directly or after staining with Coomassie blue or similar protein stain. This technique can be used to compare different antigens (Figure 7A.1).

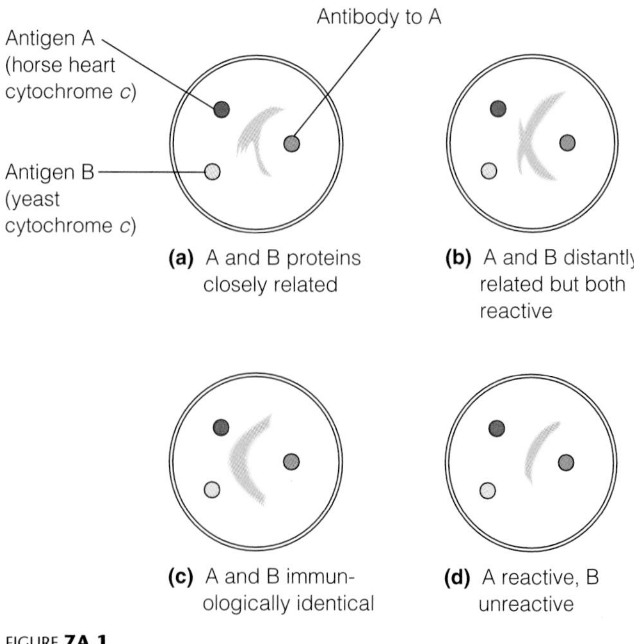

FIGURE **7A.1**
Gel diffusion precipitin test.

For example, suppose you wanted to know whether the protein cytochrome *c* from horse heart was related to cytochrome *c* from yeast. You could place horse cytochrome *c* in one well, the yeast protein in another, and antibody to the horse protein in a third. If the proteins are closely related, the antibody may react with both of the proteins. In that case the precipitin lines will intersect, as shown in Figure 7A.1a, usually with a "spur" pointing toward the less reactive antigen (in this experiment, the yeast protein). If the two antigens are unrelated, but are similar enough that each reacts with an antibody in the antiserum that was placed in the well, two precipitin lines are formed that simply cross each other (Figure 7A.1b). If the proteins are immunologically identical, a single, unbroken precipitin line is found (Figure 7A.1c). Finally, if the antibody binds to one protein but not at all to the other, only a short line is seen (Figure 7A.1d).

Immunodiffusion can be used to characterize mutant organisms that may be defective in synthesis of a biologically active protein. The presence of a cross-reacting antigen, such as seen in the pattern of Figure 7A.1a, suggests that a complete protein is made but that one of the residues essential for biological activity has been changed by the mutation. Thus, the proteins are closely related, so as to share epitopes, but are not identical.

Immunoelectrophoresis allows analysis of more complex mixtures of antigens and antibodies. Proteins in a complex mixture, such as serum, are subjected to electrophoresis, after which antibody is applied in a trough, as shown in Figure 7A.2. This method provides many different precipitin reactions in one experiment, and with relatively simple variations it can be adapted to quantify antigens in a mixture.

Much more widely used now to quantify antigen–antibody reactions is the **enzyme-linked immunosorbent assay (ELISA)**. A cross-linking reagent is used to covalently couple an enzyme to an antibody that is being used as an assay for a particular substance. The enzyme chosen is one whose activity can be easily assayed spectrophotometrically. Although this method has many variations, the principle is to assay for bound antibody by analyzing for the activity of the conjugated enzyme. This technique forms the basis of many clinical diagnostic tests, such as the most widely used current test for HIV infection. An enzyme-linked antibody to one of the surface proteins of the human immunodeficiency virus is used to detect the presence of the viral antigen in human blood samples. The presence of the enzyme activity on antigen–antibody complexes can be detected with great sensitivity.

Another analytical technique useful to characterize proteins is **Western blotting,** so called because of its superficial resemblance to a nucleic acid analytical technique called Southern blotting (described in Tools of Biochemistry 25D). Western blotting (more recently called **immunoblotting**) is used to detect, in a mixture of proteins or fragments of proteins, those that react with the same antibody. It can be used, for example, in studying the posttranslational cleavage of proteins that often occurs as part of protein maturation. In this technique the antibody-reactive proteins in a mixture are analyzed by first resolving the proteins in that mixture by denaturing gel electrophoresis. Often, 2-D gels are used. After electrophoresis, the gel is placed in contact with a sheet of nitrocellulose, and the proteins are transferred (or "blotted") to the nitrocellulose by an electric current. The proteins are bound irreversibly to the nitrocellulose sheet, so the antigen–antibody reactions can be visualized after treatment of the sheet with antibody. Visualization involves either the use of antibody radiolabeled with iodine-125 followed by radioautography, or a form of the ELISA technique. An example is shown in Figure 7A.3.

Because of their high specificity in protein binding, antibodies can also be used to purify proteins. In this technique, called **immunoaffinity chromatography,** the antibody is coupled to a chromatographic support and a column of this material is used to adsorb selectively the protein being purified. The protein is then desorbed, usually by a pH adjustment in the eluting solution and often in a state close to homogeneity.

Biochemists, like cell biologists, must be concerned with intracellular organization and with the location of enzymes that catalyze reactions of interest. An array of techniques with the generic term **immunocytochemistry** uses antibodies to help localize particular antigens in cytological preparations. In the simplest form, an antibody is conjugated with a fluorescent dye such as fluorescein. A thin section of cell or tissue is then immersed in a solution of the fluorescent antibody. After the excess is washed off, the bound antibody can be visualized by fluorescence microscopy. In the upper panel of Figure 7A.4, a fluorescent antibody has been made against a cytoplasmic protein; in the lower panel a nuclear protein is probed. Alternatively, the antibody can be linked to the iron-binding protein ferritin, and the bound iron can be visualized from its high electron density in the electron microscope. Another technique involves the binding of antibody linked to horseradish peroxidase, an enzyme that can be visualized by either light microscopic or electron microscopic techniques.

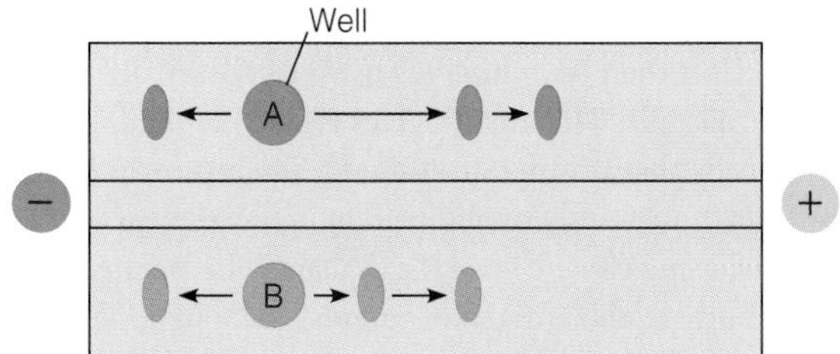

Antigen mixtures A and B are placed in wells cut in a gel and are subjected to an electrical field. Proteins, indicated by bands, migrate from the wells.

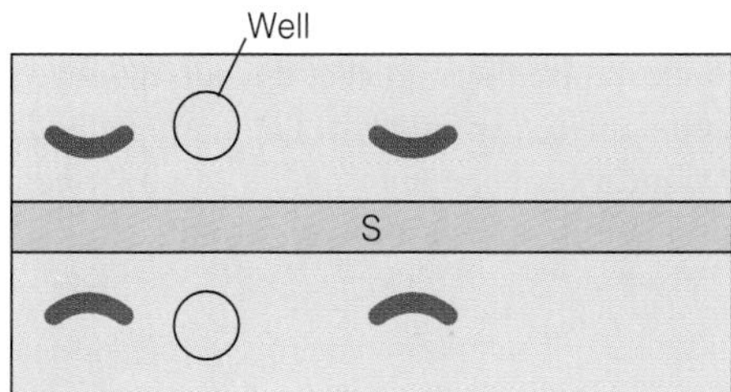

Antiserum (S), placed in the trough between A and B, diffuses into the gel. Precipitin arcs are formed with any components that react with the antiserum.

FIGURE **7A.2**

Immunoelectrophoresis.

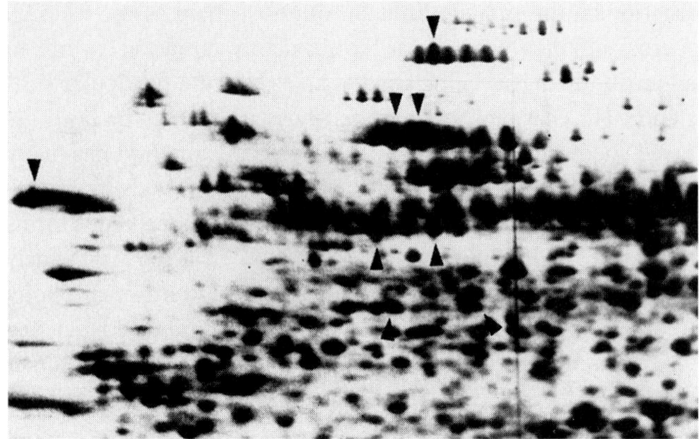

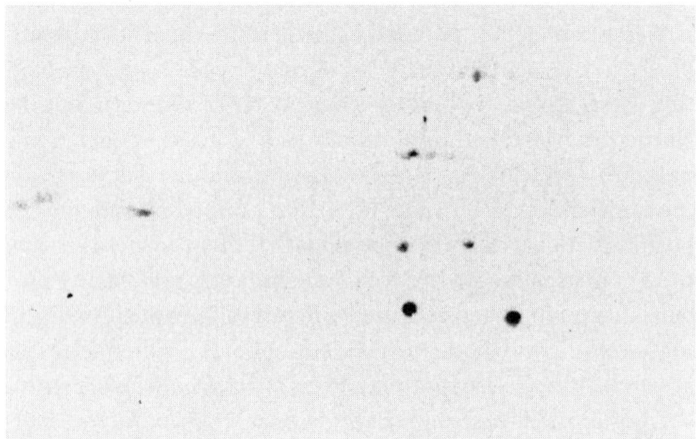

FIGURE **7A.3**

Western blotting. On the top is a 2-D gel of total protein from tobacco leaf. On the bottom is the same gel, blotted against antibody to proteins phosphorylated at threonine in mitosis.

Courtesy of J. A. Traas, A. F. Bevan, J. H. Doonan, J. Cordewener, and P. J. Shaw, *Plant J.* (1992) 2:723–732, by permission of Blackwell Scientific Ltd. and Bios Scientific Publishers.

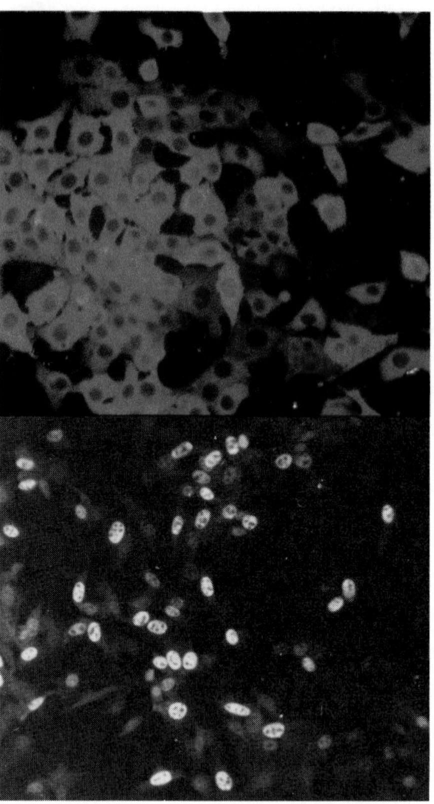

FIGURE **7A.4**

Immunofluorescence microscopy. This technique is used to differentiate between cytoplasmic (top) and nuclear (bottom) proteins in cultured 3T6 mouse cells.

Courtesy of Y. Engstöm et al., *EMBO J.* (1988) 7:1617.

As noted earlier, an antiserum prepared against a pure antigen, such as a protein, usually contains several different antibodies against that antigen. Moreover, the serum also contains all other antibodies that the animal carried in its bloodstream at the time that it was immunized against the protein of interest. Clearly the specificity, sensitivity, and reproducibility of all of the foregoing techniques would be increased greatly if they could be carried out with pure antibodies. Because of the chemical similarities among different IgGs, their purification by standard protein fractionation techniques is virtually impossible. Fortunately, the use of **monoclonal antibodies** is an alternative way to attain pure antibody in unlimited quantity.

Each antibody-forming B lymphocyte is specialized for the synthesis and secretion of one and only one antibody. If such cells could be grown in cell culture, it would be possible in principle to isolate a clone, or population derived from one cell, that synthesized one antibody directed against the protein of interest. Although B lymphocytes themselves do not grow in culture, Georges Kohler and Cesar Milstein in 1975 discovered a way to propagate antibody-forming lymphocytes with cultured cells from a mouse with multiple myeloma, a cancerous proliferation of white cells (Figure 7A.5). A mouse is immunized with the antigen of interest, and then lymphocytes from its spleen are fused with the myeloma cells. This fusion gives rise to cell lines called **hybridomas**—cells that proliferate endlessly in culture, like cancer cells, but synthesize only one antibody. By screening a large number of hybridoma clones, several can usually be isolated that synthesize antibody to the antigen of interest, each making antibody to a different antigenic determinant. Antibodies can easily be purified from cultures of the appropriate hybridomas, and they have many uses in addition to the techniques already described. Note that this method does not require a homogeneous antigen. An investigator who has the patience to examine many hybridomas can immunize the animal at the outset with a crude preparation of the antigen of interest. Later, after the desired monoclonal antibody has been obtained, it can be used as the basis of an immunoaffinity purification of the antigen.

References

Harlow, E., and D. Lane (1988) *Antibodies: A Laboratory Manual.* Cold Spring Harbor Laboratory, Cold Spring Harbor, New York.

Weir, D. M., ed. (1986) *Handbook of Experimental Immunology.* Oxford University Press, London, New York. A good description of modern techniques.

FIGURE **7A.5**

Production of monoclonal antibodies.

From G. J. Tortora, B. R. Funke, C. L. Case, *Microbiology: An Introduction*, 5th ed. (Redwood City, Calif.: Benjamin/Cummings, 1995.) © 1995 Benjamin/Cummings Publishing Co.

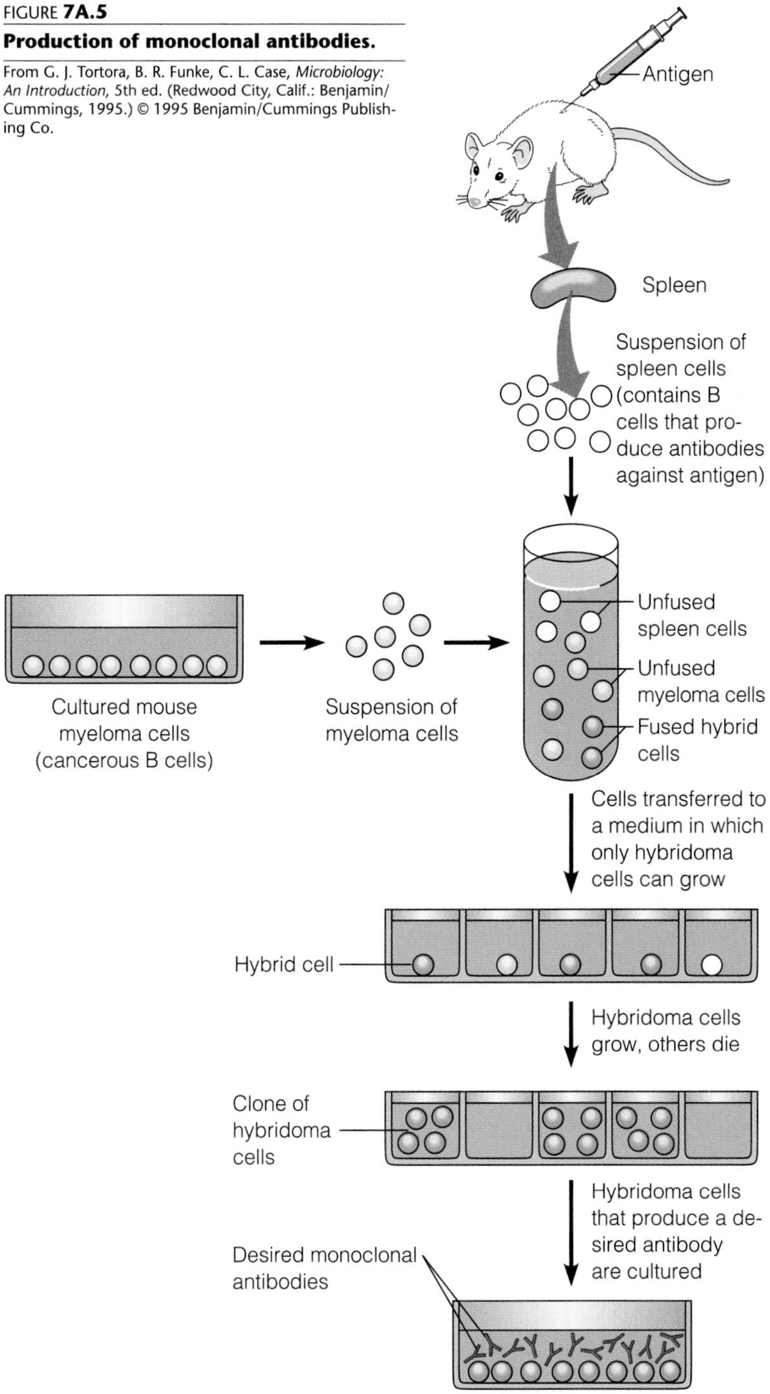

CHAPTER 8

Proteins in Motion: Contractile Systems and Molecular Motors

IN THE PREVIOUS TWO CHAPTERS, WE HAVE SEEN HOW PROTEIN MOLECULES, either singly or in defined multisubunit structures, can carry out a variety of functions. To continue to explore this idea, we now turn to examples in which protein molecules organize into large, complex structures involving many kinds of polypeptide chains. Such supramolecular structures perform many cellular functions; the one we shall consider here is the production of movement. This motion may involve the whole organism, parts thereof, cells, or subcellular constituents.

Of the many kinds of motion that living systems exhibit, the one we are most aware of is the muscle contraction required for bodily movement. However, muscle contraction accomplishes a remarkable variety of other things as well. Even the emission of sound is a muscular action, as is the injection of venom by an insect or a snake. Equally important muscular motions maintain the animal's internal world, including the beating of its heart, the breathing of its lungs or gills, and the peristaltic motions of its digestive system. Each of these kinds of movements is produced by a specific muscular tissue.

All muscles, as well as some other contractile systems we will encounter, are based on the interaction of two major proteins, **actin** and **myosin.** We often refer to these systems as **actin–myosin contractile systems.** However, certain kinds of directed motions exist—motions of individual cells and parts of cells—that do not depend on the actin–myosin system at all but use other protein mechanisms. For example, the beating of cilia and flagella and the movement of chromosomes and organelles within cells are accomplished by the interactions of a number of proteins with **microtubules,** filamentous structures made of a protein called **tubulin.** Even more remarkable are the variety of "molecular motors" that have been discovered in recent years. Some of these serve to carry molecules and vesicles along microtubules and other filaments; others produce rotation of flagella and are true microscopic motors. Also, some very well-known molecules like RNA polymerases (Chapters 26 and 28) are now recognized as molecular motors, moving along and untwisting a DNA template, fueled by phosphate ester hydrolysis.

Certain proteins act as energy transducers, using free energy from ATP hydrolysis to do mechanical work.

All of these biological systems that produce movement share one common feature: The energy released by the hydrolysis of ATP is converted into work by producing motions in parts of protein molecules. Thus, proteins can act as **en-**

ergy transducers. That is, some proteins can convert the chemical energy of ATP hydrolysis into mechanical work. When the motions of proteins are properly coordinated, directed macroscopic motion occurs.

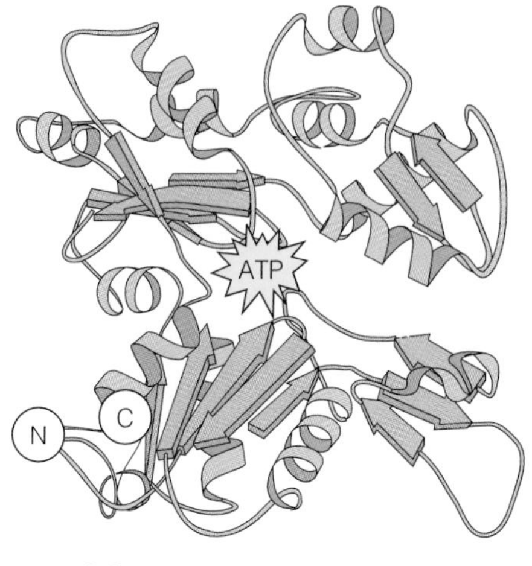

FIGURE **8.1**

G-actin. The structure of the actin monomer, as deduced by x-ray diffraction, shows the two-domain conformation clearly. The position of the bound nucleotide is also shown.

Adapted from Dr. W. Kabsch.

Muscles and Other Actin–Myosin Contractile Systems

The major proteins in muscle are actin and myosin. The best known function of these proteins is in muscle cells. However, actin and myosin are also found in many other kinds of cells and are involved in several kinds of cellular and intracellular motions. To understand how muscles and other actin–myosin systems work, we must consider the properties of these two proteins.

The major muscle systems of most animals are based on the proteins actin and myosin.

ACTIN AND MYOSIN

Actin

Under physiological conditions, actin exists as a long, helical polymer (fibrous actin, or F-actin) of a globular protein monomer (G-actin). The G-actin monomer, shown in Figure 8.1, is a two-domain molecule with a mass of 42,000 daltons. The binding of ATP by a G-actin monomer leads to polymerization; the ATP is subsequently hydrolyzed, but the ADP is held in the actin filament. In F-actin filaments, the G-actin monomers are arranged in a two-strand helix (Figure 8.2; see also Figure 6.31a). Because of the asymmetry of the subunits, the F-actin filament has a defined directionality, and the two ends have been called the **plus end** and the **minus end.** The polymerization reaction exhibits a preferred direction, and the plus end is defined as that which grows more rapidly. The actin filament carries sites on each subunit that can bind to myosin.

Myosin

The functional myosin molecule (Figure 8.3) is composed of six polypeptide chains: two identical heavy chains ($M = 230{,}000$) and two each of two kinds of

FIGURE **8.2**

F-actin. A model for F-actin filaments. G-actin monomers are distinguished by colors. Green residues bind to myosin.

Courtesy of Dr. Kenneth Holmes from *Nature* (1990) 347:44–49.

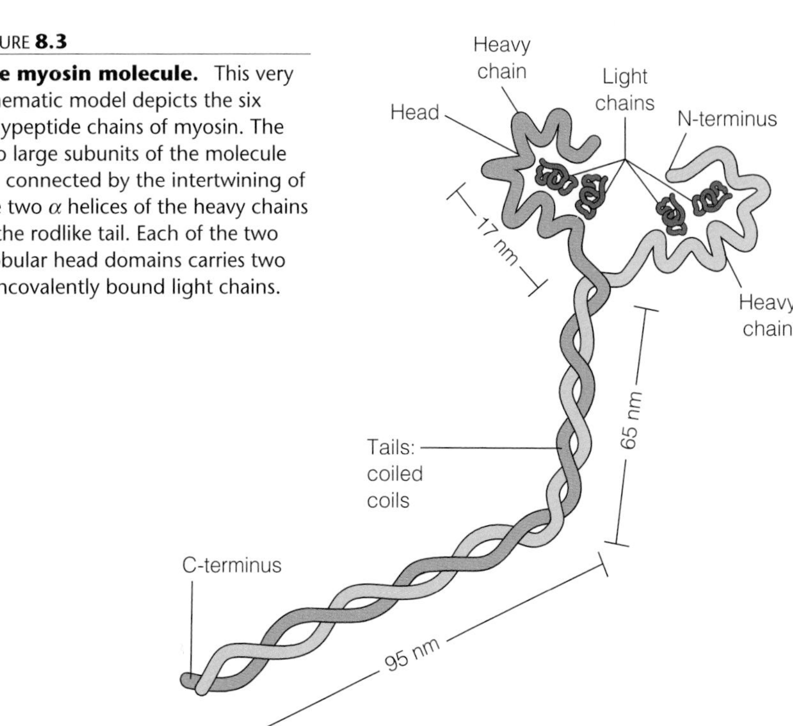

FIGURE **8.3**

The myosin molecule. This very schematic model depicts the six polypeptide chains of myosin. The two large subunits of the molecule are connected by the intertwining of the two α helices of the heavy chains in the rodlike tail. Each of the two globular head domains carries two noncovalently bound light chains.

light chains ($M = 20{,}000$). Together, they form a complex of molecular weight 540,000. The heavy chains have long α-helical tails, which are interwound into a two-strand coiled coil, and globular head domains to which the light chains are bound. Between each head domain and tail domain the heavy chain acts as a flexible stalk. The coiled-coil structure of the tails is reminiscent of the structure of α-keratin (see Figure 6.11).

The myosin molecule can be cleaved by proteases, as shown in Figure 8.4. The tail domain can be cleaved at a specific point by trypsin to yield fragments called **light meromyosin** and **heavy meromyosin.** Further cleavage of heavy meromyosin by papain cuts the stalks to yield two **S1 fragments,** each consisting of a head domain carrying the light chains. The stalk removed by papain is called an S2 fragment. The ability to cleave the myosin molecule in these specific ways has helped researchers understand the functions of its several parts. Myosin exhibits aspects of both fibrous and globular proteins, and its functional domains play quite different roles. The tail domains have a pronounced tendency to aggregate, causing myosin molecules to form the kind of thick bipolar filaments shown in Figure 8.5. The head domains with their attached light chains are often called **headpieces;** they have a strong tendency to bind to actin. The S1 fragments have now been crystallized and their structure determined by x-ray diffraction (Figure 8.6).

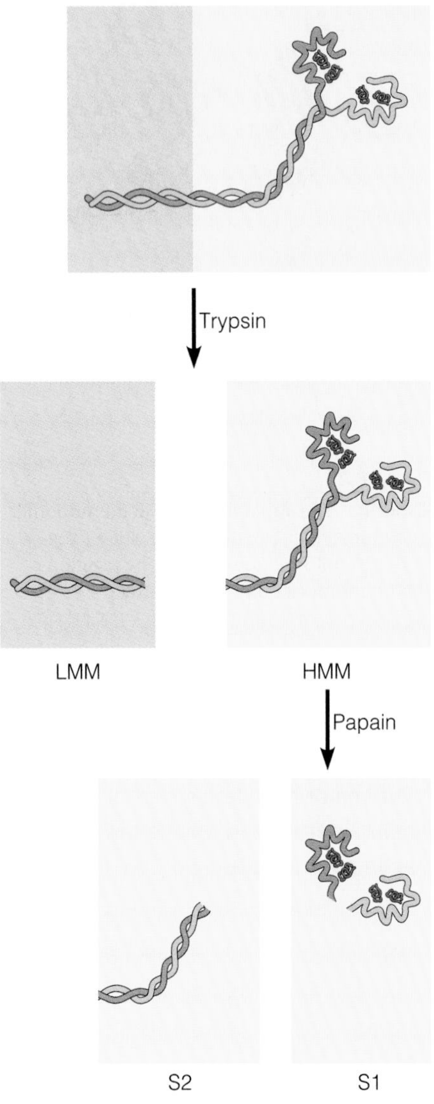

FIGURE **8.4**
Dissection of myosin by proteases.
Trypsin cleavage cuts through the myosin tail to produce light meromyosin (LMM) and heavy meromyosin (HMM). Treatment of heavy meromyosin with the protease papain digests part of the stalk structure, allowing separation of the two headpieces (head domains with their bound light chains). These separated headpieces are called S1 fragments, and the remainder of the tail is called an S2 fragment.

The Reaction of Myosin and Actin

If an actin filament is allowed to react with isolated S1 fragments, the filament will become "decorated" with these myosin headpieces, producing an asymmetric "arrowhead" pattern that demonstrates the polarity of the actin filament (Figure 8.7). In the presence of actin, whole myosin molecules or individual S1 fragments have ATPase activity, and the hydrolysis of ATP disrupts the binding. As we shall see, these observations together with the detailed structure of the S1 fragment have strong implications with respect to the mechanisms of muscle contraction.

THE STRUCTURE OF MUSCLE

In muscle tissue the actin and myosin filaments interact to produce the contractile structure. Vertebrates like ourselves have three morphologically distinct kinds of muscle. *Striated muscle* is the kind we most often associate with the term *mus-*

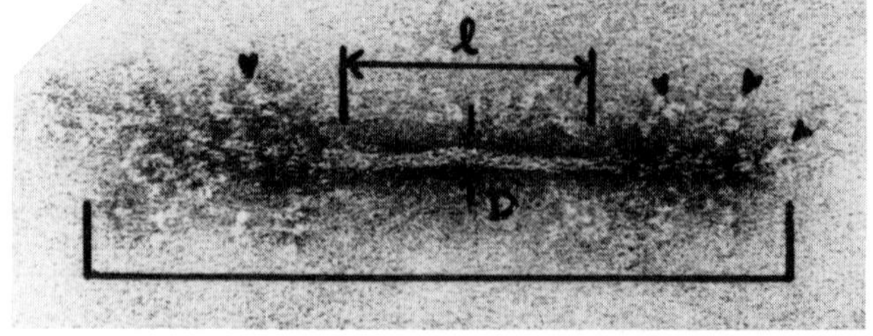

(a)

14.3 nm

15 nm

160 nm
Bare zone (ℓ)

325 nm

(b)

FIGURE **8.5**
A thick filament of myosin molecules. (a) An electron micrograph. The zone that is bare of headpieces is indicated by ℓ; some myosin headpieces are indicated by arrowheads. **(b)** A drawing of the filament structure, showing dimensions in nanometers. The projections are the pairs of headpieces on each myosin molecule. Note that the bare zone is just the length of the myosin tail, as shown in Figure 8.3.

Courtesy of T. Pollard.

cle, for it is the striated muscles in arms, legs, eyelids, and so forth that make voluntary motions possible. *Smooth muscle* surrounds internal organs such as blood vessels, intestines, and gallbladder, which are capable of slow, sustained contractions that are not under voluntary control. *Cardiac muscle* can be considered a specialized form of striated muscle, adapted for the repetitive, involuntary beating of the heart. In this discussion we will consider only the structure of striated muscle.

Figure 8.8 shows successive levels of organization of a typical vertebrate striated muscle. The individual muscle fibers, or **myofibers,** are actually very long (1–40 mm) multinucleate cells, formed by the fusion of muscle precursor cells. Each myofiber contains a bundle of protein structures called **myofibrils.** A myofibril exhibits a periodic structure when examined under the light microscope. Dark **A bands** alternate with lighter **I bands.** The I bands are divided by thin lines called **Z disks** (or sometimes **Z lines**). At the center of the A band is found a lighter region called the **H zone.** The repeating unit of muscle structure can be taken as extending from one Z disk to the next. It is called the **sarcomere** and is about 2.3 μm long in relaxed muscle.

The molecular basis for this periodic structure of the myofibril can be seen by electron microscopic studies of thin sections of muscle, as shown in Figure 8.9. **Thin filaments** of actin extend in both directions from the Z disks, interdigitating with myosin **thick filaments.** The regions in which the thick and thin filaments overlap form the dark areas of the A band. The I bands contain only thin filaments, which extend to the edges of the H zones. Within the H zones, only thick filaments are found. The dark line at the center of the H zone (sometimes called the M band) is believed to indicate positions where thick filaments associate with one another.

The composition of the thick and thin filaments has been demonstrated by extracting myofibrils with appropriate salt or detergent solutions to remove virtually all of the myosin. Because the A bands are abolished in this process, the thick A-band filaments must be composed of myosin. The myosin thick filaments are bipolar structures of the kind shown in Figure 8.5, in which the helical tails of the myosin molecules join together, with the headpieces projecting with a regular

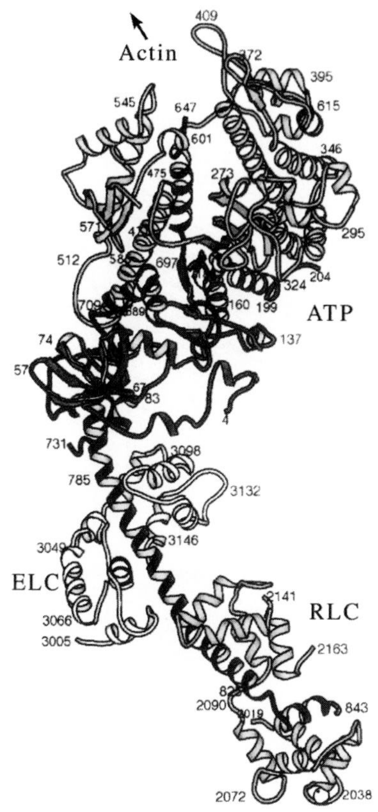

FIGURE **8.6**

The structure of an S1 fragment of myosin. The heavy chain occupies most of the structure; the two light chains are indicated by ELC and RLC. The position of ATP binding is shown, as well as the point of contact with actin.

Courtesy of S. Lowey, *Biophys. J.* (1995) 68:120s–127s, modified by M. K. Reedy from Rayment et al. (1993). With permission from Elsevier Science.

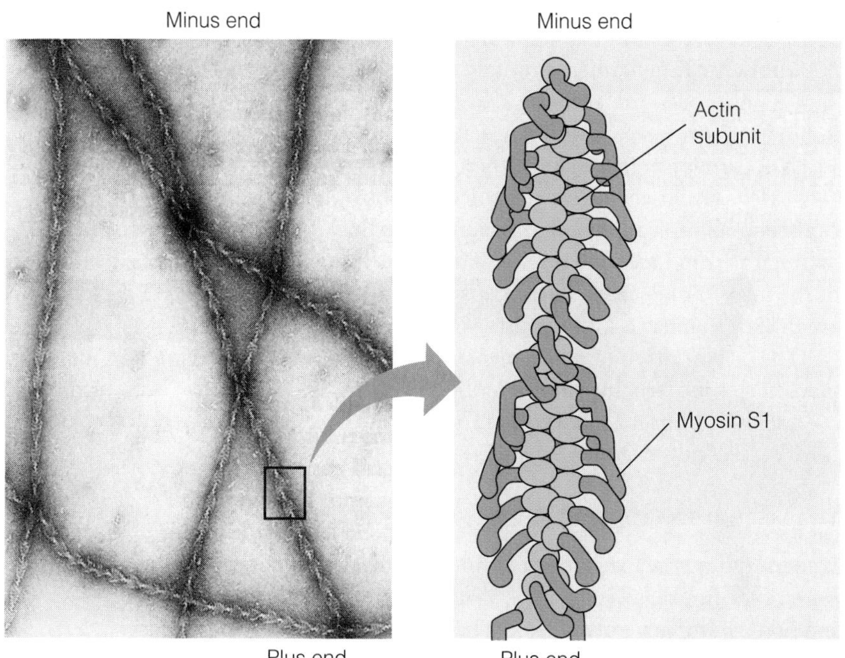

FIGURE **8.7**

An actin filament decorated with myosin headpieces. (left) An electron micrograph. **(right)** A schematic at higher magnification. The headpieces, indicated in the schematic, produce the arrowhead pattern visible in the micrograph.

Photo courtesy of Roger Craig.

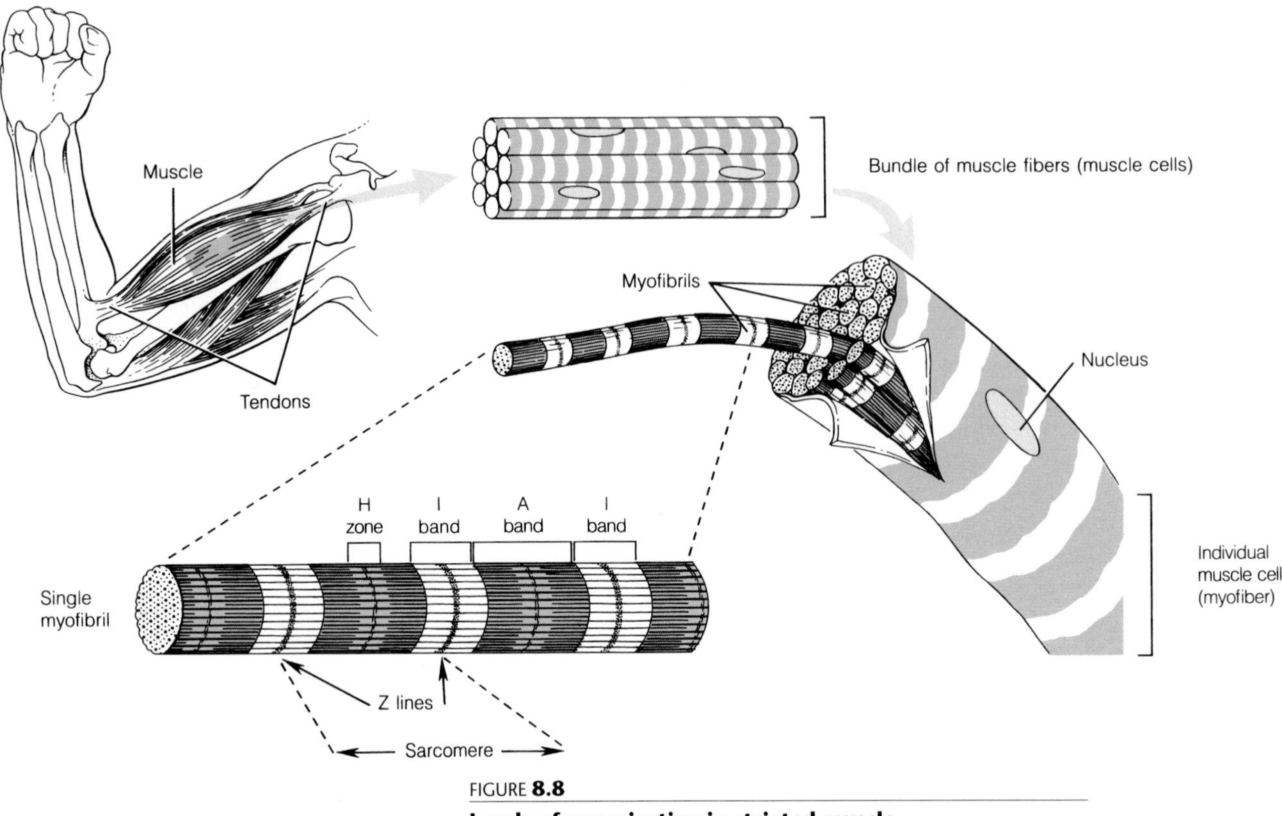

FIGURE **8.8**

Levels of organization in striated muscle.

From W. M. Becker, L. J. Kleinsmith, and J. Hardin, *The World of the Cell,* 4th ed. (San Francisco, CA: Addison Wesley Longman, 2000) © Addison Wesley Longman, Inc.

The sarcomere is the basic repeating unit of a muscle myofibril.

Thin filaments are mainly actin; thick filaments are mainly myosin. They are connected by breakable cross-bridges.

14.3-nm spacing at either end. This spacing appears to be generated by a 14.3-nm periodicity in the amino acid sequence in the myosin tails.

That the thin filaments contain actin can be shown in the following way. If myofibrils from which myosin has been removed are perfused with a solution of S1 fragments, the thin filaments are decorated in the arrowhead pattern (Figure 8.7); thus, the thin filaments contain actin. Furthermore, the arrowhead patterns always point outward from the Z disks, demonstrating the polarity of these thin filaments. However, the thin filaments are not composed entirely of F-actin. As we shall see later in this chapter, they contain other important proteins as well.

If we look closely at electron micrographs of myofibrils (see Figure 8.9d), we can see small projections extending from the thick (myosin) filaments, often contacting the thin (actin) filaments. The projections correspond to the headpieces of the myosin molecules. These *cross-bridges* between myosin and actin filaments are the key to muscle contraction.

The organization of actin, myosin, and other muscle proteins into the complex yet specific structure found in the sarcomere is a remarkable example of how several kinds of protein can combine in a specific way to form a functional structure. We will now examine how this structure works.

THE MECHANISM OF CONTRACTION: THE SLIDING FILAMENT MODEL

Understanding the mechanism of muscle contraction has come from observation both of the fine details of muscle structure and of changes in the sarcomere banding pattern during contraction. The muscle sections shown in Figures 8.8 and 8.9, and the top of Figure 8.10, are in the relaxed, or extended, state. In a full contraction of the muscle, each sarcomere shortens from a length of about 2.3 nm to

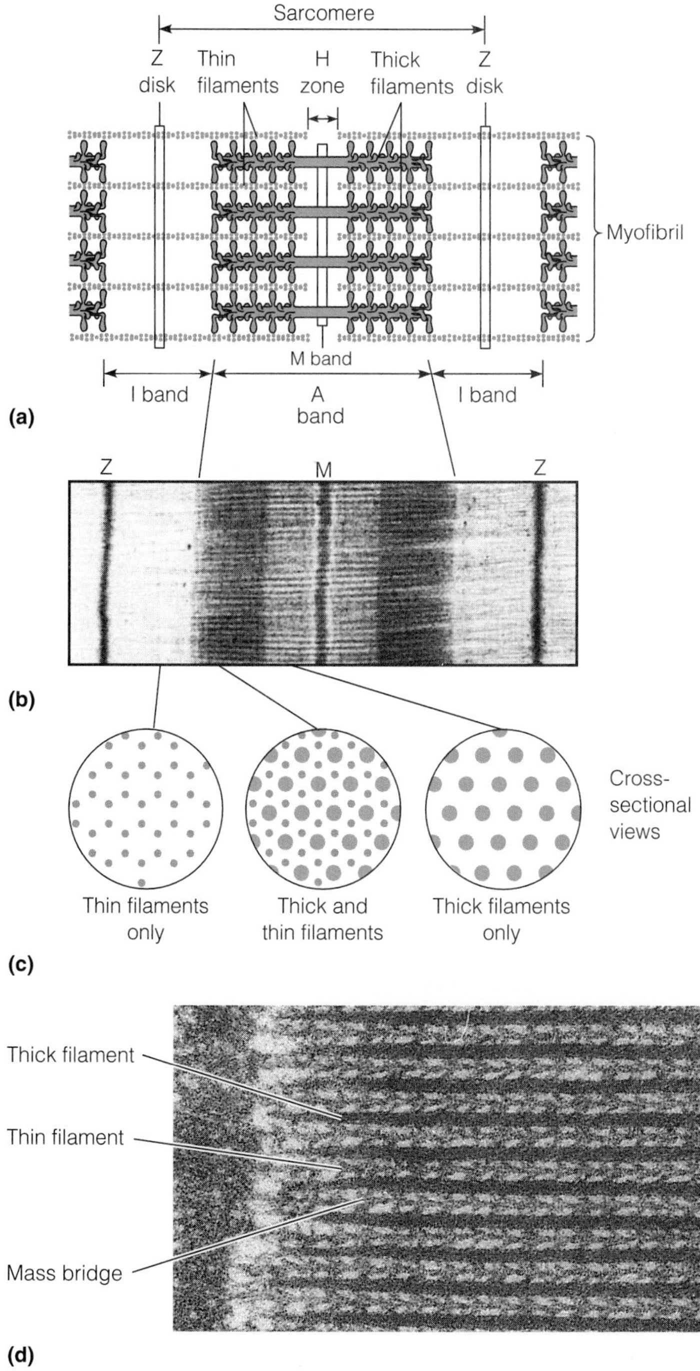

FIGURE **8.9**

Muscle structure seen at the EM level. (a) A model of the sarcomere, the repeating unit in striated muscle. The I bands, A bands, and Z disks shown in Figure 8.8 are identified, and structural elements of the sarcomere are indicated. **(b)** An electron micrograph showing the same features. **(c)** A schematic drawing of cross sections of a sarcomere in the various regions shown in (a) and (b). Thick filaments are indicated by heavy brown dots, thin filaments by small purple dots. **(d)** A higher magnification within an A band showing cross-bridges between actin and myosin filaments.

(b) Courtesy of H. E. Huxley; **(d)** courtesy of Mary Reedy.

1.0 μm. During this process, the I bands and H zones disappear, and the Z disks move right up against the A bands (bottom of Figure 8.10). Such observations led two independent (and unrelated) investigators, Hugh Huxley and Andrew Huxley, to propose in the 1950s the **sliding filament model** of muscle contraction depicted in Figure 8.10. According to this model, which is now supported by strong evidence, the myosin headpieces are presumed to "walk" along the interdigitated actin filaments, pulling them past, and thereby shortening the sarcomere.

To produce such directed motion against an opposing force on the muscle, energy must be expended. You might expect that energy would be derived in some way from ATP hydrolysis, and our previous mention of the ATPase activity of the

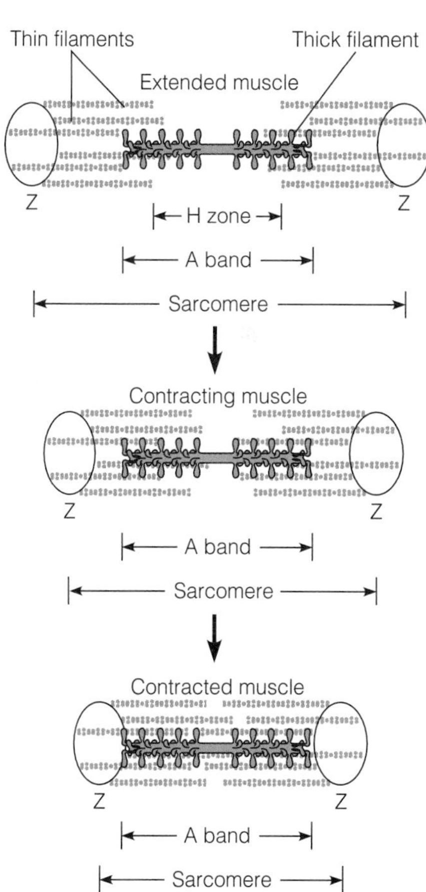

FIGURE **8.10**

The sliding filament model of muscle contraction. Contraction of striated muscle occurs when the myosin headpieces pull the actin filaments toward the center of the sarcomere.

actin–myosin complex hints at how this energy could be gained. Recall that ATP binding and hydrolysis cause the *release* of the actin–myosin interaction. According to the currently accepted mechanism, each myosin headpiece takes part in a repetitive cycle of making and breaking cross-bridges to an adjacent thin filament. We imagine the cycle starting with the myosin attached to actin, as shown at the top of Figure 8.11. Binding of ATP leads to release of the myosin cross-bridge (step 1). Hydrolysis of ATP then causes a conformational change, "cocking" the headpiece (step 2). Binding of Ca^{2+} causes the headpiece to rebind the thin filament (step 3), but at a site upstream of its previous position, because it has now cocked. The myosin molecule now forms the strongly bound complex (step 3). Phosphate release (step 4) prepares for the power stroke (step 5), which pulls the thin filament toward the center of the sarcomere. Binding a new ATP will then restart the cycle and ready the headpiece for another stroke.

In the sliding filament model, the periodic attachment and release of cross-bridges, with a cross-bridge conformational change, slide the thin and thick filaments past one another.

At the end of each cycle, the actin filament has been moved with respect to the myosin, so that each headpiece makes successive steps along the thin filament. The "walking" is rather like that of a millipede—some legs are always in contact with the thin filament, so that it cannot slip back during contraction. It should be noted that some details of head motion remain controversial, and the process may be more complex than what we have described.

Recently, several novel experimental methods have made it possible to measure both the force developed and the distance moved in each power stroke (see Figure 8.12). The force is found to depend upon the load placed on the myosin–actin filament couple but averages about 5 pN (piconewtons) at high load. This corresponds to an energy expenditure of about 10^{-20} J per power stroke—approximately one fifth of the energy released when a single ATP molecule is hydrolyzed. At lower loadings, there is evidence that several power strokes can be accomplished per ATP cycle. The distance the actin filament is moved per power

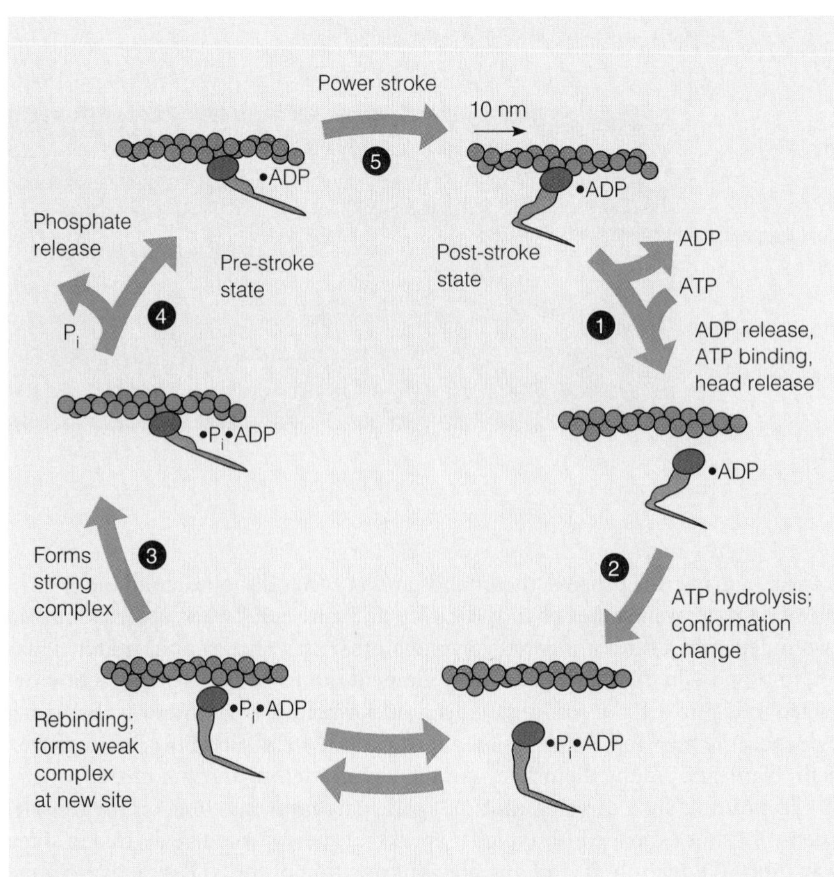

FIGURE **8.11**

The current view of the ATP cycle in muscle contraction. The slow step in the process is (3), the conversion from the "weak" complex to the "strong" complex. Exactly at what point in the cycle all conformational change in the headpiece occurs is unclear.

Adapted from J. Spudich, *Nature* (1994) 372:515–518. © 1994 Macmillan Magazines, Ltd.

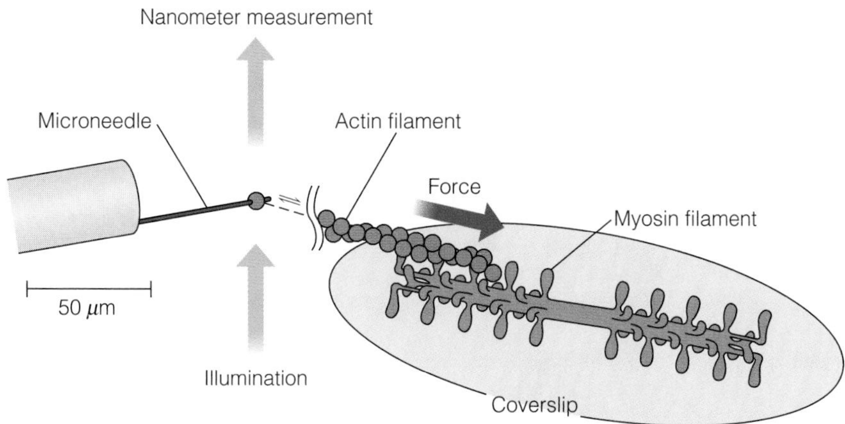

FIGURE **8.12**

Measurement of motion and force generated by interaction of a single actin filament with a single myosin filament. The myosin is attached to a coverslip, and the actin is attached to a microneedle of known stiffness. The actin and myosin were visualized by fluorescence. Motion of the actin filament as well as force generated upon addition of ATP is determined by the needle deflection.

From T. Yanagida and A. Ishijima, *Biophy. J.* (1995) 68: 312s–320s.

stroke is about 10–20 nm, and this is what can be accomplished per ATP cycle at high loading. When working against weaker resistance, the multiple power strokes can propel the actin for as much as 100 nm per ATP.

STIMULATION OF CONTRACTION: THE ROLE OF CALCIUM

The critical substance in stimulating contraction is not ATP, which is generally available in the myofibril, but the Ca^{2+} that enters at step 3 of Figure 8.11. To understand how calcium regulates muscle contraction, we must examine the molecular structure of the thin filament in a bit more detail.

A thin filament, as found in striated muscle, is more than just an F-actin polymer. Four other proteins, shown in Figure 8.13, are essential to the thin filament's contractile function. One of these proteins is **tropomyosin,** a fibrous protein that exists as elongated dimers lying along, or close to, the groove in the F-actin helix. Bound to each tropomyosin molecule are three small proteins called **troponins I, C,** and **T.** The presence of tropomyosin and the troponins inhibits the binding of myosin heads to actin *unless calcium is present* at a concentration of about 10^{-5} M. In resting muscle, Ca^{2+} concentrations are in the neighborhood of 10^{-7} M, so new cross-bridges cannot be formed. An influx of Ca^{2+} stimulates contraction, because the ion is bound by troponin C, causing a rearrangement of the troponin–tropomyosin complex. This shift makes new sites on actin available for binding by the myosin headpieces. The postulated mechanism shown in Figure 8.14 permits step 3 and the subsequent steps in the cycle of Figure 8.11 to take place.

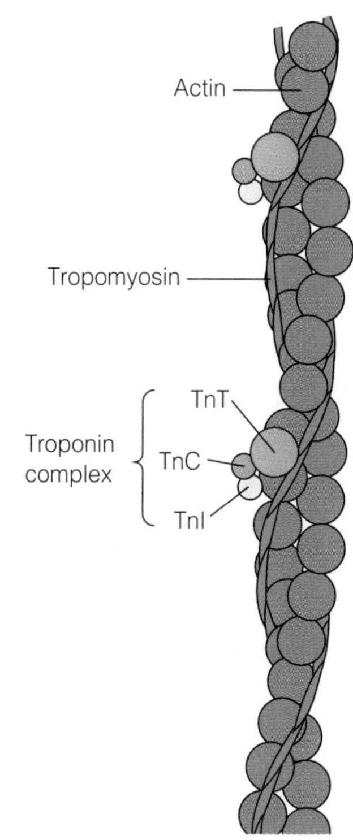

FIGURE **8.13**

F-actin and its associated proteins. This schematic drawing shows the proteins present in the thin filaments of striated muscle: F-actin, tropomyosin, and troponins (Tn) I, C, and T.

We have now traced the activation of muscle contraction to the influx of calcium into the myofibrils. But why does this influx occur? In particular, how can it be brought about by the nerve impulses that excite muscles to contract? The answer can be found from a closer examination of the myofiber, or muscle cell (Figure 8.15). Within the cell, each myofibril is surrounded by a structure called the **sarcoplasmic reticulum,** formed of membranous tubules. In resting muscles the level of Ca^{2+} in the myofibrils is maintained at about 10^{-7} M, whereas the Ca^{2+} level within the lumen of the sarcoplasmic reticulum may be 10,000-fold higher. Impulses from motor nerves depolarize the membrane of the sarcoplasmic reticulum, opening gated Ca^{2+} channels (see Chapter 10), and Ca^{2+} pours out of the lumen into the myofibrils, stimulating contraction. The signal is rapidly transmitted to the entire sarcoplasmic reticulum of a myofiber via **transverse tubules,** invaginations of the plasma membrane that connect at periodic intervals with the reticulum.

Muscle contraction is stimulated by an influx of Ca^{2+} into the sarcomere. Ca^{2+} binding by troponin C causes a rearrangement of the troponin–tropomyosin–actin complex, allowing actin–myosin crossbridges to form.

Although abrupt change in Ca^{2+} level is the universal *signal* for muscle contraction, it obviously cannot in itself provide the necessary energy. From where does the energy required for muscular work come?

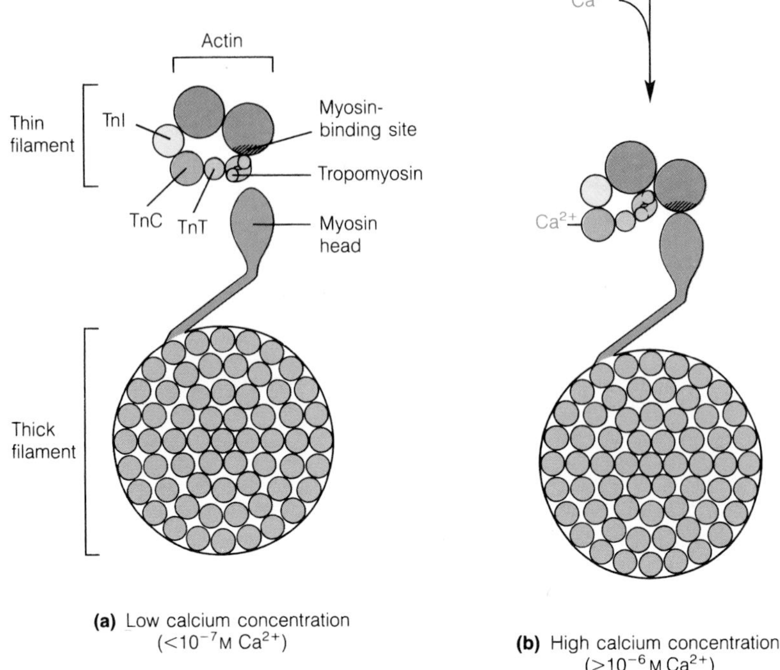

FIGURE **8.14**

The regulation of muscle contraction by calcium. A single myosin headpiece is shown next to a thin filament, in cross-sectional view. **(a)** Relaxed muscle. At low Ca^{2+} levels, the configuration of actin, tropomyosin, and the troponin complex in the thin filament blocks most myosin headpieces from contacting the thin filament. **(b)** The binding of Ca^{2+} to troponin C causes a rearrangement in the thin-filament components so that myosin-binding sites on the actin are made available. Cross-bridge formation (step 3 of Figure 8.11) can then occur, and the muscle contracts.

From W. M. Becker, L. J. Kleinsmith, and J. Hardin, *The World of the Cell*, 4th ed. (San Francisco, CA: Addison Wesley Longman, 2000) © Addison Wesley Longman, Inc.

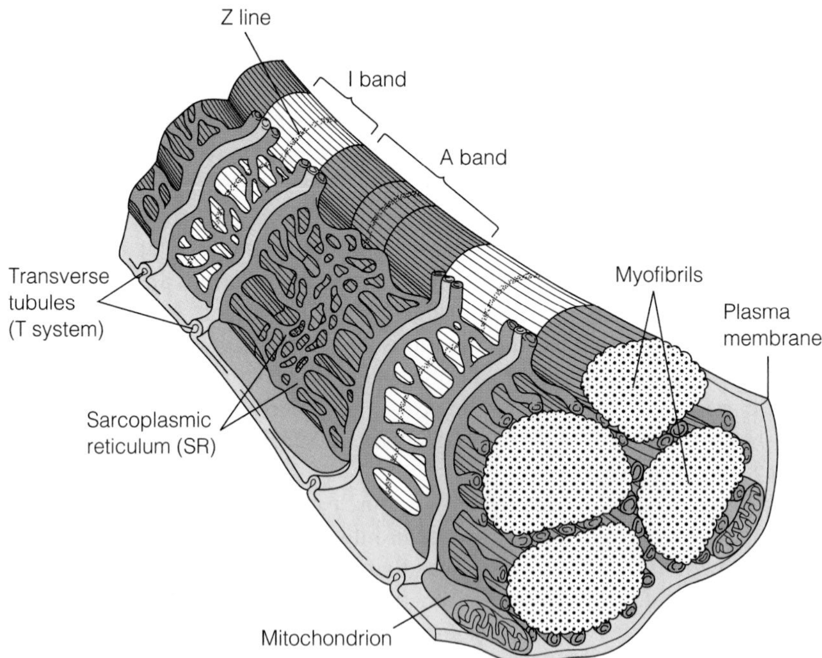

FIGURE **8.15**

Structure of a myofiber (muscle cell). The sarcoplasmic reticulum (SR) is a network of specialized endoplasmic reticulum tubules that surrounds the myofibrils within the myofiber. In resting muscle, the SR accumulates Ca^{2+}, which it discharges into the myofibrils when a neural signal reaches the plasma membrane. The transverse tubules (T system) are invaginations of the plasma membrane that make contact with the SR at many points, ensuring uniform response to the signal.

From W. M. Becker, L. J. Kleinsmith, and J. Hardin, *The World of the Cell*, 4th ed. (San Francisco, CA: Addison Wesley Longman, 2000) © Addison Wesley Longman, Inc.

ENERGETICS AND ENERGY SUPPLIES IN MUSCLE

Basically, muscle is a mechanism for converting the chemical free energy released in ATP hydrolysis into mechanical work. The conversion can be remarkably efficient, approaching values of 80% under optimal circumstances. This efficiency is much higher than can be attained by artificial chemical engines.

How is the ATP generated? Even in striated muscle the answer can vary, depending on the particular *kind* of muscle and its function. Striated muscles can be divided into two categories—*red muscle,* designed for relatively continuous use, and *white muscle,* employed for occasional, often rapid motions. Red muscle

TABLE **8.1 Comparison of red and white striated muscle**

	Red	White
Relative fiber size	Small	Large
Mode of contraction	Slow twitch	Fast twitch (about 5 times faster)
Vascularization	Heavy	Lighter
Mitochondria	Many	Few
Myoglobin	Much	Little
Major stored fuel	Fat cells	Glycogen in muscle
Main source of ATP	Fatty acid oxidation	Glycolysis

owes its dark color to its abundant heme proteins: It is well supplied with blood vessels and therefore hemoglobin, it has many mitochondria with cytochromes, and it has large stores of myoglobin. Red muscle depends heavily on aerobic metabolism in mitochondria, so the primary energy source in red muscle is the oxidation of fat. White muscle, on the other hand, relies on glycogen as a primary energy source. Glycogen is excellent for quick energy production but cannot sustain activity for long periods. (See Table 8.1 for a more detailed comparison of red and white muscle and Chapters 13 and 23 for a more extensive discussion of fuel metabolism.)

The functional differences between the two types of striated muscle are clearly revealed in birds. In the domestic chicken, the flight muscles of the breast, used only for brief fluttering or short flights, are white, whereas the heavily used leg muscles are red. Wild flying birds, which make sustained flights but rarely walk, have just the opposite distribution of light and dark meat.

Careful observation of ATP levels in red striated muscle has shown that the provision of energy is more complicated than it might appear at first. The amount of ATP needed for a single contraction may be greater than all the ATP immediately available to a sarcomere. Yet even after relatively long exercise, ATP levels in the sarcomeres remain essentially constant. Only after extreme exhaustion do ATP levels begin to fall. This finding suggests that ATP is an intermediary, and not the ultimate, energy storage compound in these muscles. Indeed, it has been known for many years that the high-energy compound steadily depleted during muscular activity is *creatine phosphate*. As its high phosphate transfer potential suggests (see Figure 3.7), this compound is capable of phosphorylating ADP very efficiently. The reaction is catalyzed by the enzyme *creatine kinase*.

The energy source in red muscle is creatine phosphate, which regenerates ATP continually as it is depleted by muscle contraction.

$$^{-}OOC-CH_2-N(CH_3)-C(=\overset{+}{N}H_2)-NH-Ⓟ + ADP \underset{\text{Creatine kinase}}{\rightleftharpoons} {}^{-}OOC-CH_2-N(CH_3)-C(=\overset{+}{N}H_2)-NH_2 + ATP \qquad \Delta G^{\circ\prime} = -12\ \text{kJ/mol}$$

Creatine phosphate **Creatine**

Because the equilibrium lies well to the right, virtually all of the muscle adenylate is maintained in the ATP form, rather than as ADP or AMP, as long as creatine phosphate is available. The consumption of creatine phosphate during exercise, while an almost constant level of ATP is maintained, is clearly demonstrated by the NMR studies of human muscle shown in Figure 12.14.

NONMUSCLE ACTIN AND MYOSIN

Although actin and myosin have traditionally been associated with muscle, members of the actin and myosin families are, in fact, found in most eukaryotic cells, even those that are in no way involved in muscular tissues. Actin and myosin

Motion and shape changes of many types of cells are produced by a nonmuscle actin–myosin system.

appear to play major roles in cell motility and changes of cell shape. Actin is a major component of the **cytoskeleton**—the fibrous array that pervades almost every kind of cell and gives it a specific shape (Figure 8.16a). Staining with fluorescent antibodies reveals that myosin is also associated with this network (Figure 8.16b). The myosin in such intracellular networks is different in sequence from muscle myosin. Rather than forming thick filaments, nonmuscle myosin tends to form dimers, interacting with cytoplasmic actin to form the kind of loose contractile network schematized in Figure 8.16c. Organized contraction and relaxation of such networks can lead to a wide variety of cell movements and responses, including amoeboid crawling.

Another intracellular process in which an intracellular actin–myosin contractile complex seems to be involved is **cytokinesis,** the division of cells in the last stages of mitosis (Chapter 28). The process can be seen clearly in the sea urchin egg, a favorite model system for such studies (Figure 8.17). Toward the end of mitosis, when the daughter nuclei are clearly separated at the two poles of the cell, a ring of indentation is observed in the cell surface, defining a plane perpendicular to the mitotic spindle. This ring contracts, forming the **cleavage furrow** and eventually cutting the cell in two. Electron microscopy shows that the ring consists of fibers, and staining with fluorescent antibodies shows that the fibers contain both actin and myosin.

The involvement of myosin in cytokinesis has been most elegantly demonstrated by S. Inoue and his colleagues in the experiment depicted in Figure 8.18. After the sea urchin egg had undergone one division, the daughter cell on the right was microinjected with antimyosin antibodies, which should make myosin nonfunctional. After 10 h, the control cell (left) had undergone many divisions, to construct half of an embryo. In the treated cell, mitosis continued, as evidenced

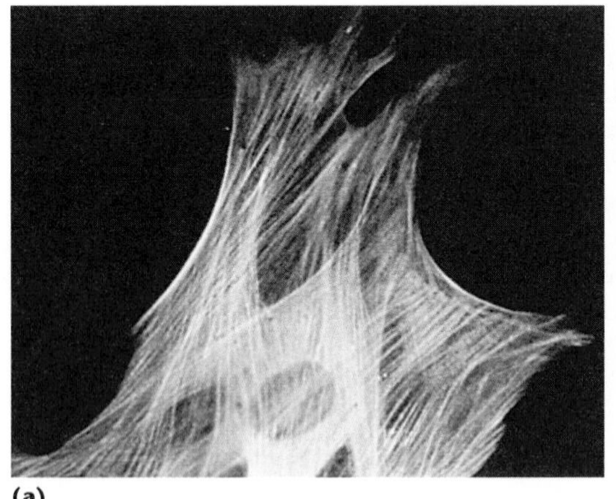
(a)

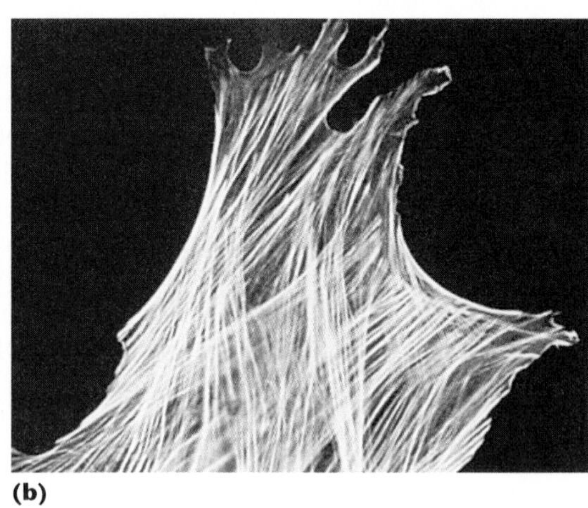
(b)

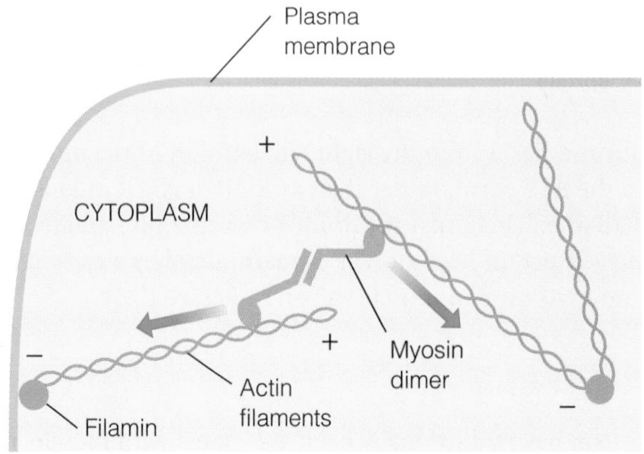

(c)

FIGURE **8.16**

Cytoskeletal actin and myosin in fibroblasts. (a) Actin fibers as detected by fluorescent phalloidin, which binds specifically to actin. **(b)** Myosin in the same cell detected by an injected fluorescent antibody. **(c)** A possible mechanism for contractility in the cytoskeleton. The globular proteins (green) linking actin molecules to one another and to the cell membrane are called *filamin.*

(a, b) From B. M. Jockusch et al., in *J. Cell Sci., Suppl. 14,* edited by R. A. Cross and J. Kendrick-Jones, pp. 41–47 (Cambridge: The Company of Biologists, Ltd., 1991).

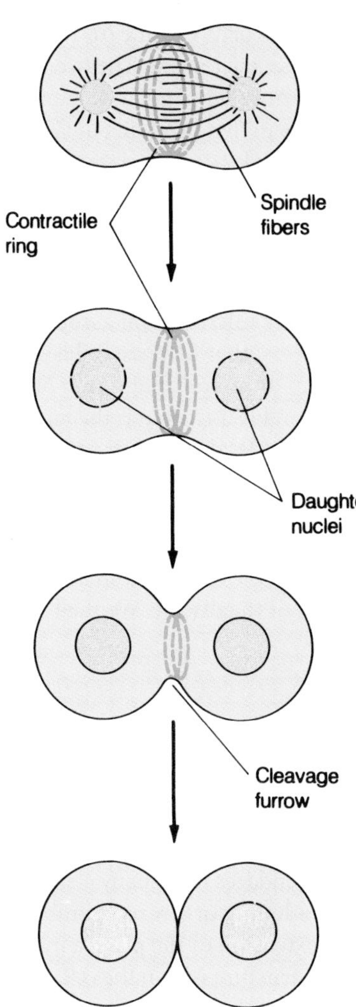

FIGURE **8.17**

Actin and myosin in cytokinesis. This drawing shows cytokinesis in a fertilized sea urchin egg. The contractile ring that cleaves the cell in two is an actin–myosin complex located just beneath the plasma membrane.

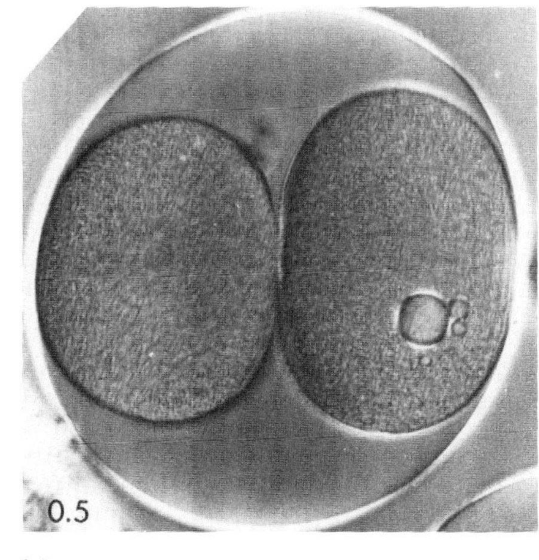

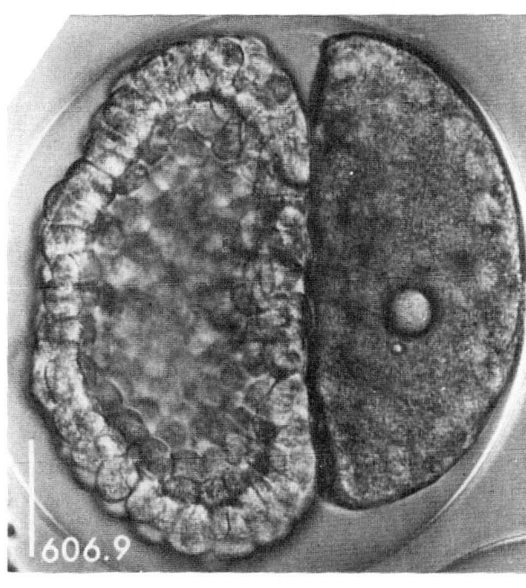

(a) (b)

FIGURE **8.18**

Myosin is essential to cytokinesis but not to mitosis. **(a)** A sea urchin embryo at the two-cell stage. The cell to the right has been injected with antimyosin antibodies; the oil droplet shows the site of injection. **(b)** After 10 h, the cell on the left has undergone many divisions, forming half of a normal embryo. In the cell on the right cytokinesis has been completely blocked, but the many new nuclei show that mitosis has continued.

Courtesy of D. P. Kiehart, I. Mabuchi, and S. Inoue, *J. Cell Biol.* (1982) 94:165, by copyright permission of Rockefeller University Press.

by the many nuclei present, but *cytokinesis had been completely blocked.* Therefore, myosin is essential for cytokinesis.

The experiment demonstrates one other important fact. Because mitosis continued even in the cell treated with antimyosin antibodies, the contractile process in the mitotic spindle apparently does *not* require the participation of myosin. There have been many reports of the presence of actin and myosin in the spindle, and it was long believed that they were essential for chromosome separation. But these and other experiments show that some other contractile system must be involved. It is to this second general class of motility-generating systems that we now turn.

Microtubule Systems for Motility

Microtubules are helical, tubular polymers made from two kinds of tubulin subunits.

A class of motile systems completely different from and unrelated to the actin–myosin contractile systems is used in places as diverse as the mitotic spindle, protozoan and sperm flagella, and nerve axons, to name only a few. These systems are constructed from **microtubules,** very long, tubular structures built from a helical wrapping of the protein **tubulin** (Figure 8.19). There are two kinds of tubulin subunits, α and β, each of molecular weight 55,000. They are present in equimolar quantities in the microtubule, which can be considered a helical array

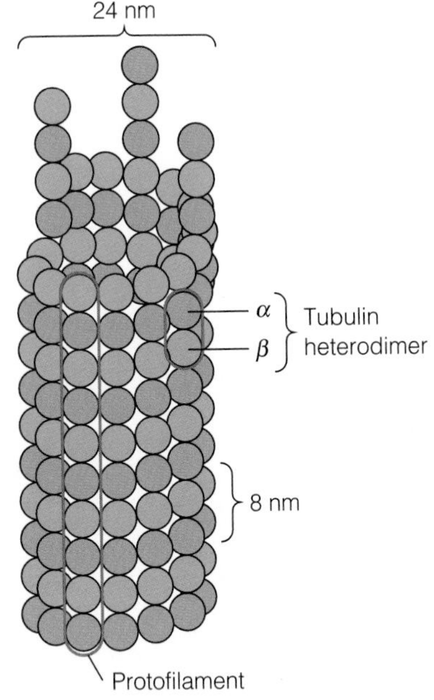

FIGURE **8.19**

Portion of a microtubule. The α and β tubulin subunits are shown in brown and teal, respectively. The microtubule can be thought of either as a helical array of αβ dimers or as 13 parallel rows of αβ dimers. These rows are referred to as protofilaments.

of αβ dimers. Alternatively, we can view the microtubule as consisting of 13 rows, or **protofilaments**, of alternating α and β subunits. Because the α and β units are asymmetrical proteins, with a defined and reproducible orientation in the fiber, the microtubule has a definite sense of direction. The structure of the αβ tubulin dimer has recently been solved (see Nogales et al. in the References). The two subunits are found to be very similar, as was expected from their similar sequences.

The assembly of microtubules bears a resemblance to that of actin, but GTP is required rather than ATP. The αβ dimers bind GTP and then associate to form oligomers. These oligomers form nucleation sites for the growth of microtubules (Figure 8.20). One end, called the plus end, grows more rapidly than the other, minus end. As in actin polymerization, the nucleotide is hydrolyzed but is held in the filament. The final assembly of a functional microtubule usually involves the binding of other proteins to its surface. Some of these **microtubule-associated proteins (MAPs)** play functional roles, as we shall see. Other MAPs stabilize microtubule structure and/or promote the association of microtubules into bundles. A member of this family that has drawn considerable attention is the tau (τ) MAP found in neuronal tissue. Phosphorylation of tau results in its dissociation from microtubules and their destabilization. Hyperphosphorylation has a much more dramatic effect, resulting in the formation of tangles of τ-filaments in neural axons, one of the major cellular symptoms of Alzheimer's disease. This in turn may imply that inappropriate synthesis or activation of the enzymes (kinases) involved in tau phosphorylation may trigger the onset of the disease. We shall now consider some of the things that microtubules do.

MOTIONS OF CILIA AND FLAGELLA

Many kinds of eukaryotic cells are propelled by the beating of cilia or flagella. Cilia are shorter than flagella and exert a coordinated rowing motion to move a microorganism through solution (Figure 8.21a). Eukaryotic flagella, such as sperm tails, are longer and propel the cell by an undulatory motion (Figure 8.21b). The structures of the two types of motile appendages have many elements in common (Figure 8.22). Each of them contains a highly organized bundle of microtubules called an **axoneme**, enveloped by an extension of the plasma membrane and connected to a **basal body**, an anchoring structure within the cell.

The internal structure of the axoneme is truly remarkable. As the cross section in Figure 8.22c shows, the most obvious feature is the arrangement of

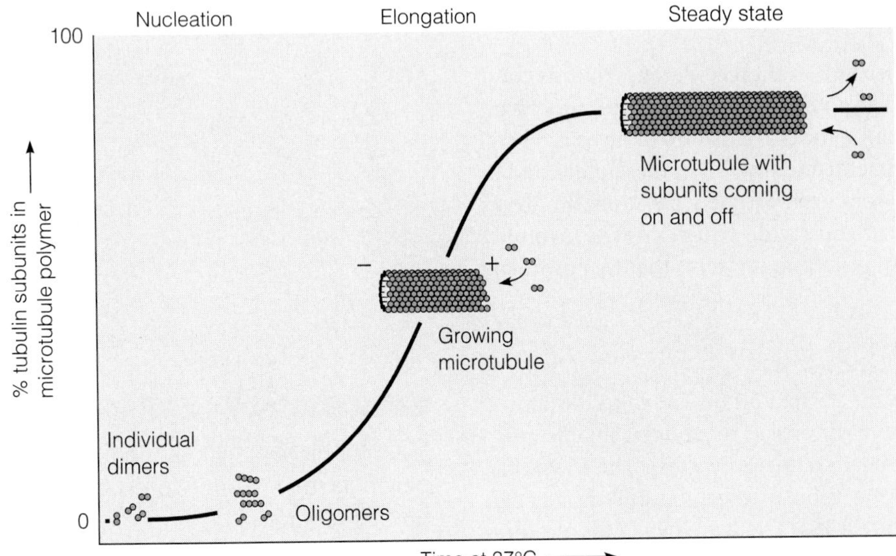

FIGURE **8.20**

Assembly of microtubules. During an initial lag period, αβ dimers form oligomers large enough to nucleate fiber formation. The microtubules then grow until most of the free dimers are used up, and equilibrium between growth and dissociation is reached. For simplicity, we show growth only at one end, because growth is more rapid at the plus end.

From Alberts et al., *Molecular Biology of the Cell* (New York: Garland Publishing, 1994), fig. 16.23. Copyright © 1994 Garland Publishing.

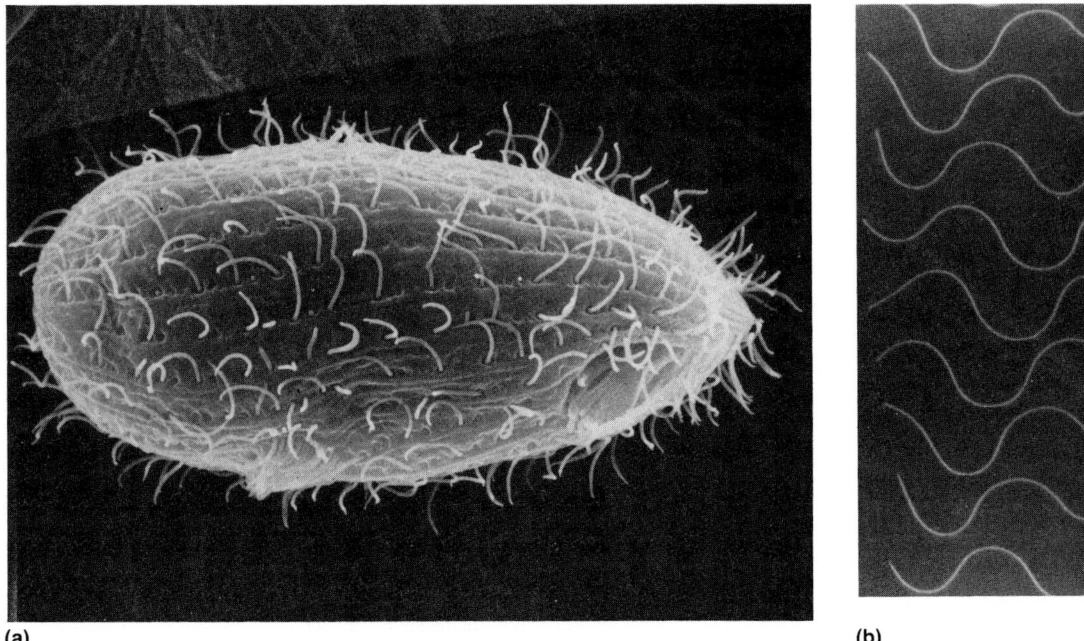

FIGURE **8.21**

Cilia and flagella. (a) The protozoan *Tetrahymena* is covered with rows of cilia. **(b)** A sperm of the tunicate *Ciona.* A series of time-lapse photographs shows how undulations of the flagellum propel the sperm.

(a) Courtesy of U. W. Goodenough, *J. Cell Biol.* (1983) 96:1610, by copyright permission of Rockefeller University Press; **(b)** courtesy of C. J. Brokaw, California Institute of Technology.

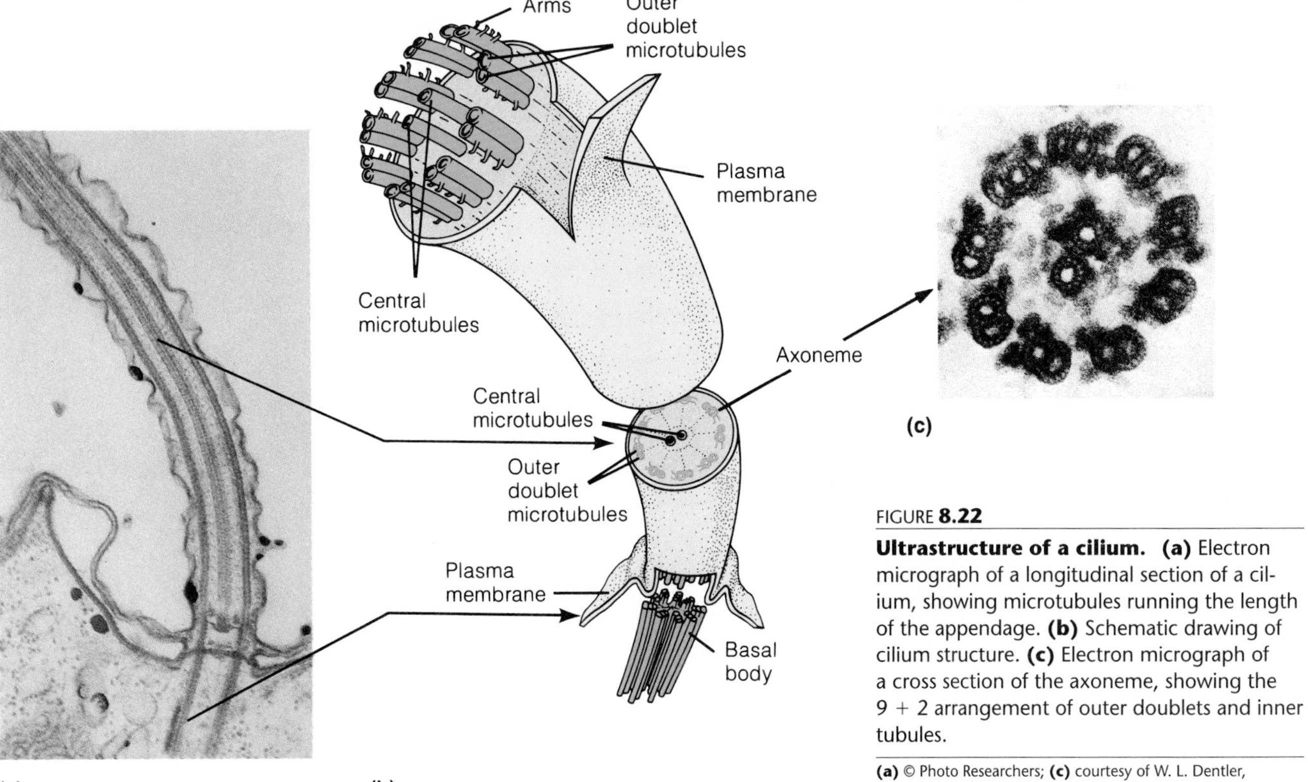

FIGURE **8.22**

Ultrastructure of a cilium. (a) Electron micrograph of a longitudinal section of a cilium, showing microtubules running the length of the appendage. **(b)** Schematic drawing of cilium structure. **(c)** Electron micrograph of a cross section of the axoneme, showing the 9 + 2 arrangement of outer doublets and inner tubules.

(a) © Photo Researchers; **(c)** courtesy of W. L. Dentler, University of Kansas/BPS.

Many kinds of cells are propelled by cilia or flagella containing microtubules.

microtubules known as a 9 + 2 array: two central microtubules ringed by nine microtubule doublets. The single microtubules in the center are complete, each having 13 protofilaments of $\alpha\beta$ tubulin dimers. By contrast, each of the nine surrounding doublets is composed of one complete microtubule (the **A fiber**) to which is fused an incomplete microtubule, carrying only 10 or 11 protofilaments (the **B fiber**). Closer inspection of electron micrographs reveals even greater complexity, as diagrammed in Figure 8.23. The outer doublets are periodically interconnected by a protein called *nexin* and carry at regular intervals *sidearms* composed of the protein **dynein.** In addition, *radial spokes,* each consisting of a head and an arm, project from the outer doublets to connect with the central pair.

Bending of cilia and flagella is accomplished by the dynein-driven sliding of microtubules past one another.

The full complexity of axoneme structure is revealed by gel electrophoresis studies of isolated axonemes. About 200 polypeptides can be resolved. Analysis indicates at least 6 proteins in the spoke heads and 11 others in the arms of the spokes. Much of this apparatus seems to be directly involved in the beating motions of cilia and flagella. If ATP is added to isolated axonemes, adjacent doublets can be seen to slide past one another. The best current model holds that this slid-

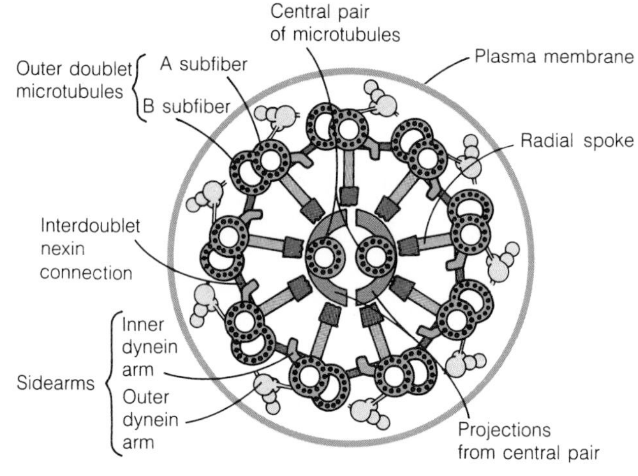

FIGURE **8.23**
Diagram of the cross section of an axoneme. The dynein arms have ATPase activity and can cause adjacent doublets to slide with respect to each other. The nexin connections between doublets and the radial spoke system give stability to the whole assembly.

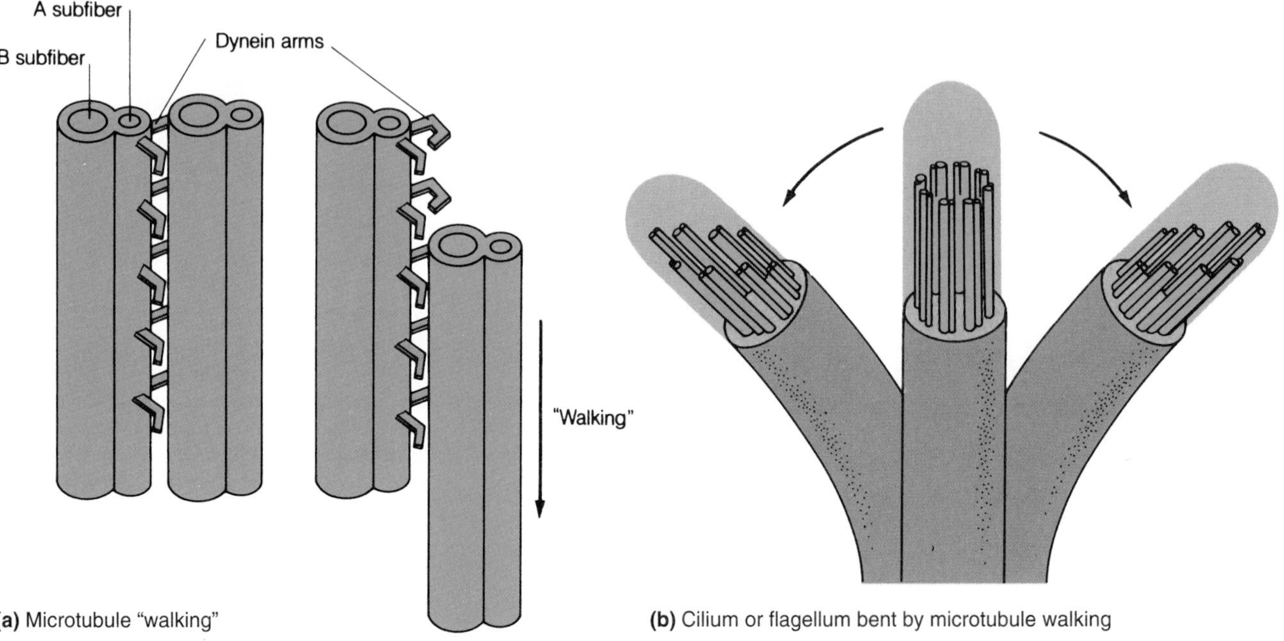

FIGURE **8.24**
Model for the bending of cilia and flagella. (a) Isolated microtubule doublets can "walk" past one another by making and breaking dynein arm contacts, if ATP is present. **(b)** The back-and-forth bending of a cilium or flagellum is produced by synchronized short "walks" by microtubules on opposite sides of the appendage. Long walks are prevented by the cross-connections between microtubules.

ing occurs by "walking" of the dynein sidearms along the adjacent doublet (Figure 8.24a). Doublets slide past each other first on one side of the axoneme and then the other, with the length of the slide limited by the central spokes and nexin connectors. In this way, the sliding of doublets is transformed into back-and-forth bending of the whole cilium or flagellum (Figure 8.24b). If connections within the axoneme are removed by careful proteolysis, ATP simply causes axonemes to extend and thin, as the outer doublets slide past one another with no stopping point.

It has been demonstrated that dynein has ATPase activity, with binding of ATP associated with the *breaking* of dynein cross-bridges. Thus, an obvious similarity exists between the mechanisms of beating of cilia and flagella and the ATP-driven walking of myosin heads along the actin fiber, and there appears to be similarity between the two systems at the protein structure level. Dynein resembles, in fact, the S1 head of the myosin molecule.

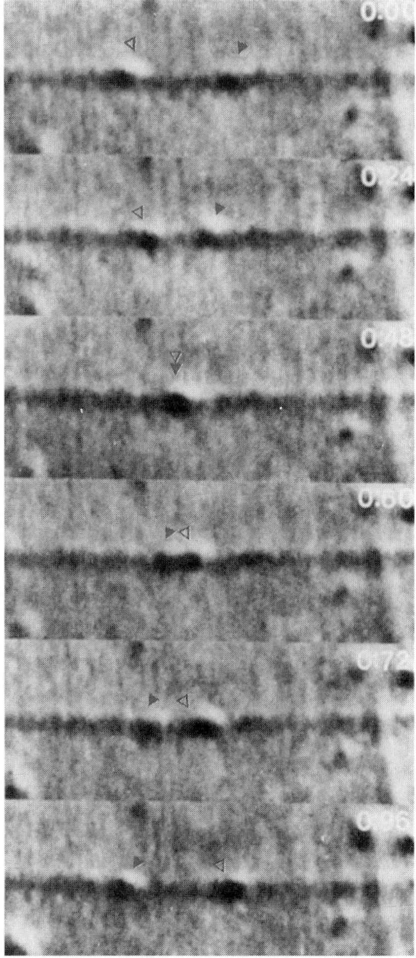

FIGURE **8.25**

Movement on microtubule tracks. A series of time-lapse video micrographs shows the bidirectional motion of organelle vesicles on a single microtubule filament from squid giant axon. The two organelles (pointed to by the open and solid red triangles) move in opposite directions and pass each other. Elapsed time (s) is given in the upper right corner of each frame.

B. J. Schnapp, R. D. Vale, M. Sheetz, and T. S. Reese, *Cell* (1985) 40:455–462, © 1985 Cell Press.

INTRACELLULAR TRANSPORT

At one time, all transport of materials within the cytoplasm of cells was thought to occur through simple diffusion. It is now known that some proteins and organelles are rapidly transported over long distances along microtubules, which serve as tracks that direct and facilitate the motion.

The clearest evidence comes from studies of transport in **axons,** the long projections that allow one nerve cell to contact another. Because nerve axons can be many centimeters in length, sufficiently rapid movement of materials between the cell body and the end of the axon cannot be accomplished by diffusion. The problem can be studied directly by using the mammalian sciatic nerve, which has very long axons extending from the cell body in the spinal cord. If radiolabeled amino acids are injected into the cell body, they are incorporated into proteins by the cell's ribosomes. After some time, the axon can be sectioned, and the location of newly synthesized, radiolabeled proteins can be determined. This method reveals that although transport rates vary greatly, some proteins, especially those associated with lipid vesicles, move as rapidly as 40 cm/day, much faster than could be accounted for by diffusion.

Recent advances in television-enhanced microscopy allow us to study such dynamic processes in new detail. As shown in Figure 8.25, small vesicles or whole organelles can actually be seen moving along microtubule bundles in an axon. Transport along the microtubules occurs in both directions, in each case by the attachment of "molecular motors" to the objects to be transported. These motors are of two kinds. One, called **cytoplasmic dynein,** resembles the dynein involved in the motion of cilia and flagella (see previous section) and is responsible for transport from the plus end of the microtubule toward the minus end. The other, called **kinesin,** is used to transport objects in the opposite direction. Kinesin and cytoplasmic dynein are now known to represent families of proteins, with similar but distinct transport functions in a wide variety of cell types. The two proteins are similar in structure and bear some similarity to the myosin family. This similarity can be seen in Figure 8.26, which depicts the structure of a kinesin dimer. As in myosin, there are two head groups, connected by an extended coiled-coil tail. There are even two associated light chains, of unknown function (not shown in Figure 8.26). Careful studies of the motion of kinesin and dynein on microtubules indicate that they "walk" along the microtubule track, with a step size of about 8 nm. This is exactly the distance from one tubulin dimer to the next. A proposed model for the step cycle for kinesin is illustrated in Figure 8.27. The molecule is believed to pivot, engaging one head group and then the other to the tubulin monomers. The helical tail of the molecule connects it to whatever "burden" the dynein or kinesin is carrying, perhaps through associated proteins (which may include the light chains).

Some organelles and other objects are transported within cells along molecular "tracks" of microtubules or actin fragments.

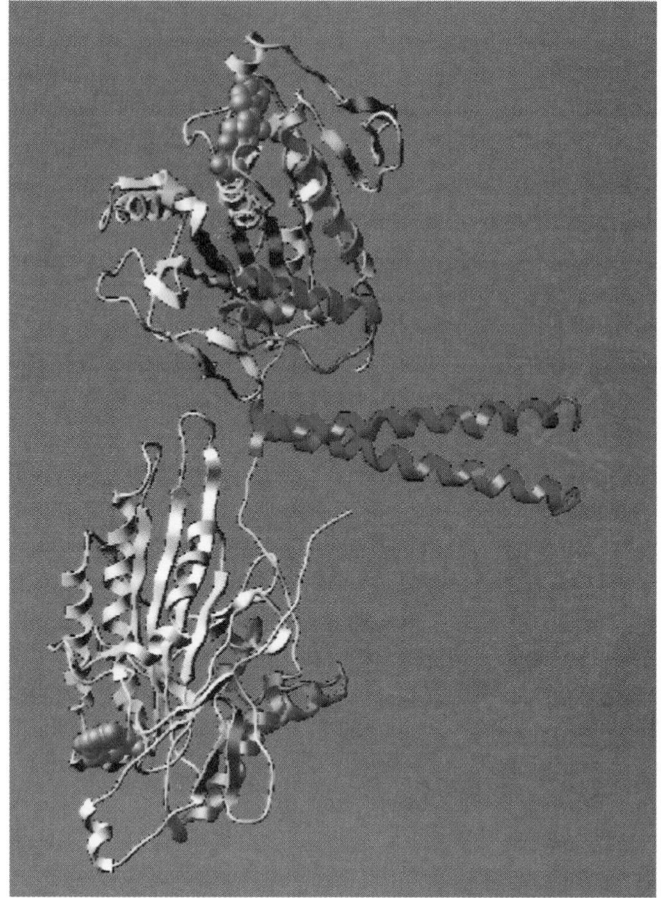

FIGURE **8.26**

The kinesin dimer. The globular regions are connected by an α-helical coiled-coil (red). The green portion of each globular region is involved in binding to tubulin. Bound ADP molecules are shown in orange.

Adapted from E. Mandelkow and A. Hoenger, *Curr. Opin. Cell. Biol.* (1999) 11:34–44.

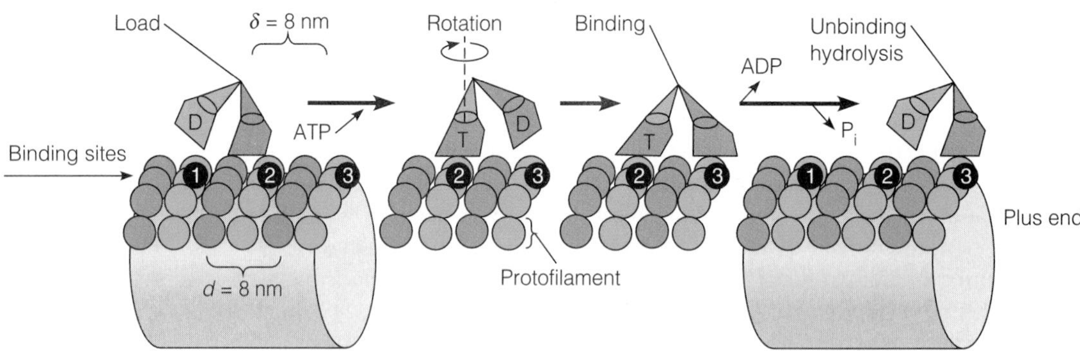

FIGURE **8.27**

A model for the motion of kinesin on a microtubule. The α and β subunits of the microtubule are shown as brown and green spheres, respectively. The symbols T and D indicate the presence of ATP and ADP, respectively. Binding of ATP to one kinesin head is presumed to cause a rotation of the dimer, bringing the ADP-bearing head (D) to where it can make contact with the next tubulin dimer. Binding results in the loss of ADP from this subunit. Hydrolysis of the ATP on the other head allows it to release from the tubulin, beginning the cycle all over again. One ATP is hydrolyzed for each step of 8 nm. Note the similarity in mechanism to the action of myosin.

Adapted from J. Howard, *Nature* (1997) 389:561–567. © 1997 Macmillan Magazines, Ltd.

There is also clear evidence for yet another mechanism for transport in the cytoplasm. Careful observation of motions of organelles on microtubules in squid axoplasm has revealed that some organelles periodically make short linear movements, with abrupt changes in direction, in regions where *no microtubules can be observed.* This research has led to the conclusion that motion can also occur on some or all of the actin fibers common in this and other cells. Further studies have shown that the motors here are special myosins, and that an organelle can have both kinesin and myosin. The observation raises the interesting idea that whereas microtubules may represent the "superhighways" for intracellular transport, actin fibers may constitute the "country roads." Clearly, much is happening in cells that we are only now beginning to understand.

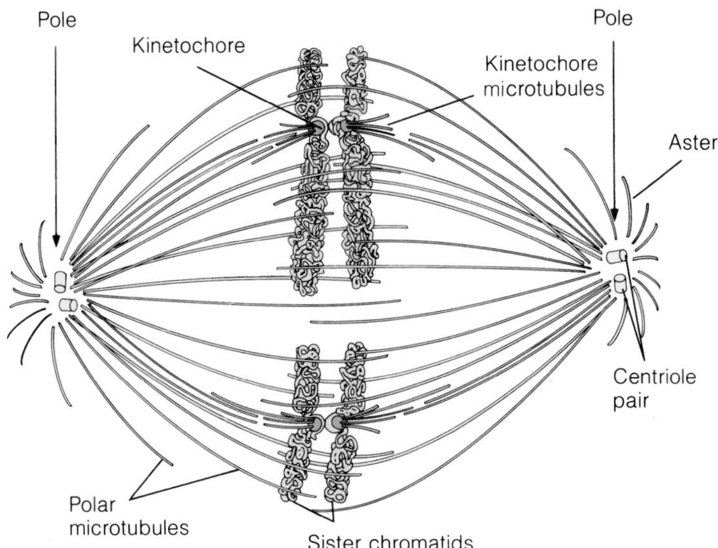

FIGURE **8.28**

Microtubules in the mitotic spindle. The separation of both the centrioles and the chromatids during mitosis appears to be accomplished by spindle microtubules. Polar microtubules radiate from the centrioles and seem to push them apart. Kinetochore microtubules emerge from points near the kinetochore of each chromosome and are thought to pull the chromatids toward the centrioles.

MICROTUBULES AND MITOSIS

Studies indicate that the mitotic spindle (see Chapter 28) is composed mainly of microtubules. Specifically, various conditions that block the formation of microtubules (certain plant alkaloids, low temperature, high pressure) also prevent the formation of the spindle and the completion of mitosis. The mitotic spindle contains microtubules that have a variety of functions (Figure 8.28). Some, called **polar microtubules,** extend between the centrioles, apparently pushing them apart. Others, the **kinetochore microtubules,** are attached to the kinetochores of the individual chromosomes and seem to pull the chromatids to the poles in telophase.

The mechanism by which microtubule motion is generated to carry out the complex processes of mitosis is still poorly understood but is the object of intense study. It has become clear that a number of members of both the kinesin and cytoplasmic dynein families are involved and that microtubules are both pulled and pushed during the process.

The separation of chromosomes in mitosis is accomplished by microtubules.

Bacterial Motility: Rotating Proteins

It is appropriate to close this chapter by examining a system that is almost without rival in its elegance and simplicity. In bacterial motility we find a mechanism that would never have been believed, had the evidence not become indisputable. The bacterial flagellum is a right-hand helical fiber, composed almost entirely of one fibrous protein, **flagellin.** It does not contain microtubules, actin, myosin, or any contractile system. Yet for many years it was assumed that the bacterial flagellum underwent in-plane bending motions, like those of sperm tails. Thus, researchers were surprised to learn that, in fact, it *rotates.* This mechanism was demonstrated most simply when the flagellum of a bacterium was stuck to a glass plate by antiflagellin antibodies. Because the flagellum could no longer rotate, the bacterium did.

The remarkable structure that attaches the flagellum to the bacterium and generates the rotation is shown in Figure 8.29. The fiber of the flagellum is attached through a hook structure to a rod that passes through a "bushing" in the outer bacterial membrane and into the inner membrane. There it terminates in a multisubunit "rotor," which is surrounded by a "stator" ring. Each of these components is made up of protein molecules, and most have been characterized. In other words, the flagellum is made to rotate by an ultramicroscopic motor,

Some bacteria move by the rotation of flagella, using molecular rotating motors set in the cell membrane.

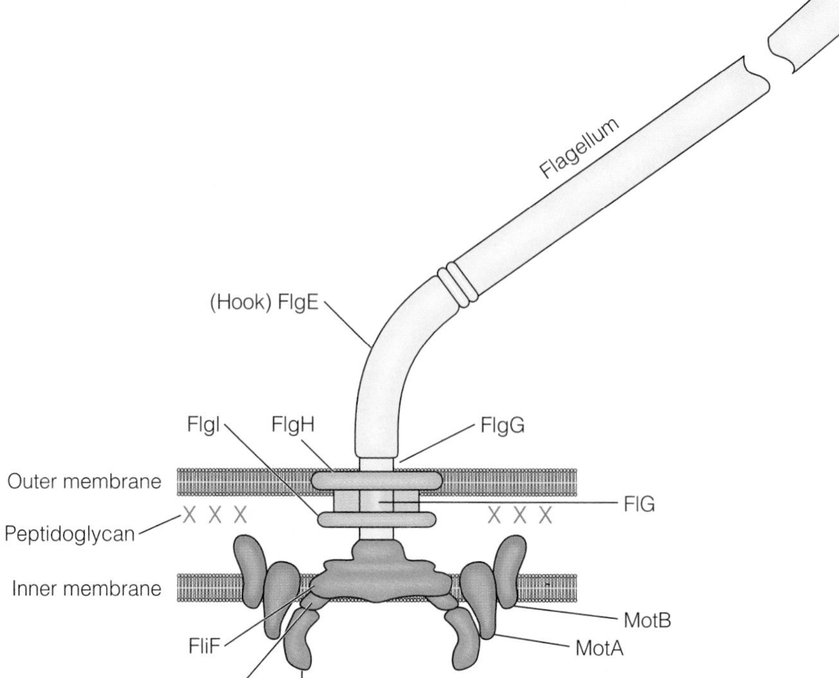

FIGURE **8.29**

Structure of the bacterial flagellar motor. The components corresponding to specific proteins are designated as follows: FlgE and FlgG form the rod and hook, respectively. The rod passes through two "bushings" (FlgH and FlgI), which connect to the outer membrane and peptidoglycan layer (see Chapter 9), respectively. The rod then connects to the rotor (FliF, FliG), which turns within a stator composed of MotA and MotB. Protons pass through MotA, into the cytoplasm, to drive the motor. The element FliM,N constitutes the "switch" that can reverse the motor. Moving parts are shown in red or pink, stationary parts in blue.

Adapted from H.C. Berg, *Biophys. J.* (1995) 68:163s–167s.

consisting entirely of protein subunits. In a sense it is an electric motor, because the driving force comes from protons moving across the bacterial inner membrane. The proton gradient is generated by ATP hydrolysis (see Chapter 15). In vivo the motor runs at about 100 revolutions per second and requires the passage of about 1000 protons per revolution. Such rotary motors are turning out to be fairly common in biology and exhibit a wide variety of functions and mechanisms. Certain marine bacteria propel themselves using motors driven by sodium ion fluxes rather than proton fluxes. On the other hand, proton flux through a rotary device is used to generate ATP from oxidative metabolism in most organisms, as we describe in Chapter 15.

The flagellar motor has still one more remarkable property—it can be reversed. That is, it can rotate the flagellum in either a clockwise or a counterclockwise direction. This ability is important to the bacterium, for it allows for both steady, rectilinear motion and changes in direction. If the multiple flagella are all rotating counterclockwise (Figure 8.30a), their right-hand helical sense makes them *push* together. They tend to be drawn together in a bundle, and propel the bacterium in a straight line, a movement known as *running.* But if the flagella rotate clockwise (Figure 8.30b), they fly out from the surface and *pull* in all directions. The result is that the bacterium *tumbles.*

Escherichia coli and a number of other flagellated bacteria demonstrate a response to chemicals, called **chemotaxis.** (The general phenomenon of **taxis,** widespread throughout the animal and plant worlds, comprises movements in response to external stimuli.) Chemotactic bacteria move preferentially toward attractants, such as nutrients, and away from repellents, such as poisons. We can now describe bacterial chemotaxis in terms of running and tumbling motions (Figure 8.31). In a neutral and uniform environment, periods of running lasting a few seconds alternate with brief periods of tumbling, and the bacterium wanders about randomly (Figure 8.31a). The presence of a gradient of either a nutrient or a noxious repellent biases this distribution of running and tumbling. If a bacterium is moving up a gradient of nutrient, tumbling is delayed, resulting in

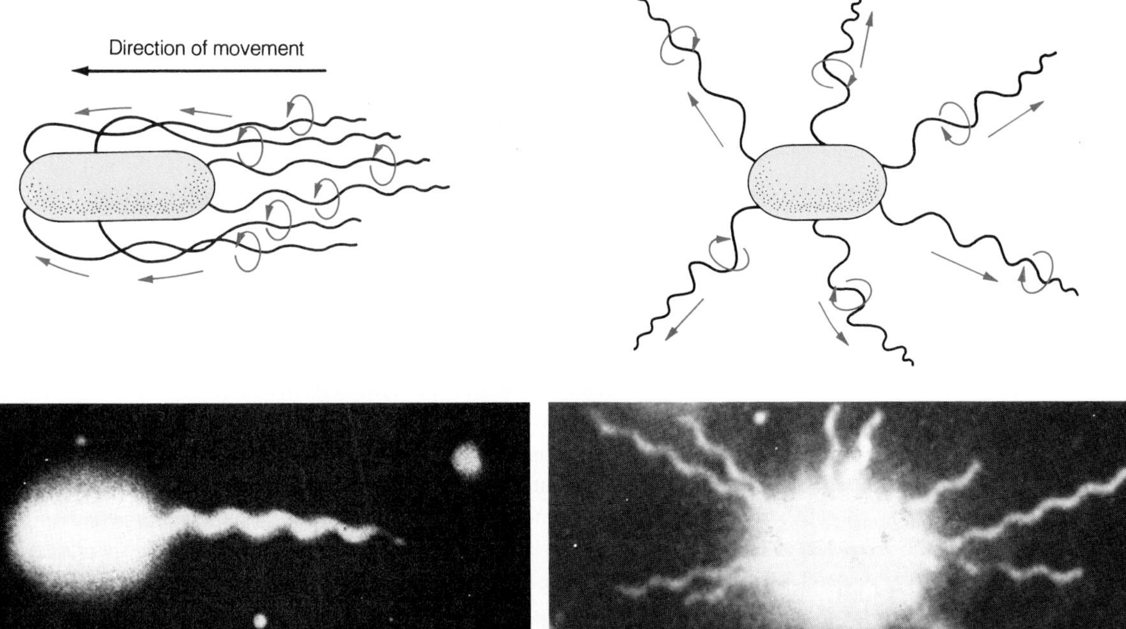

FIGURE **8.30**

Effect of the direction of flagellar rotation. These electron micrographs and accompanying diagrams show how the direction of flagellar rotation affects a bacterium with several flagella. **(a)** If the flagella rotate counterclockwise, their right-helical structure makes them be pulled together into a bundle and drive the bacterium in a straight line (running). **(b)** When rotation is clockwise, the flagella fly out in all directions, and the bacterium tumbles randomly.

Electron micrographs courtesy of R. M. McNab and M. K. Ornston, *J. Mol. Biol.* (1977) 112:1–30, © 1977 Academic Press.

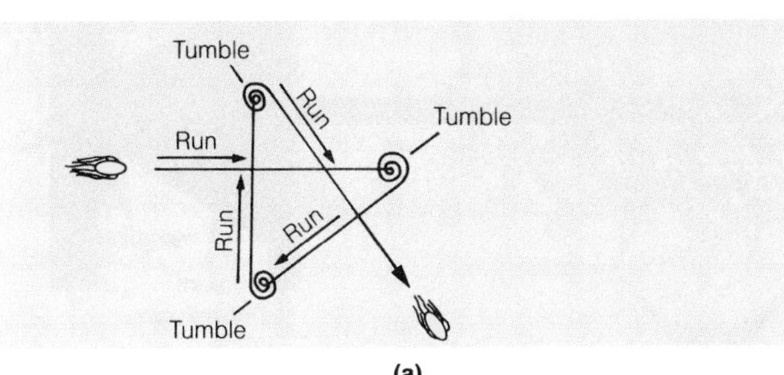

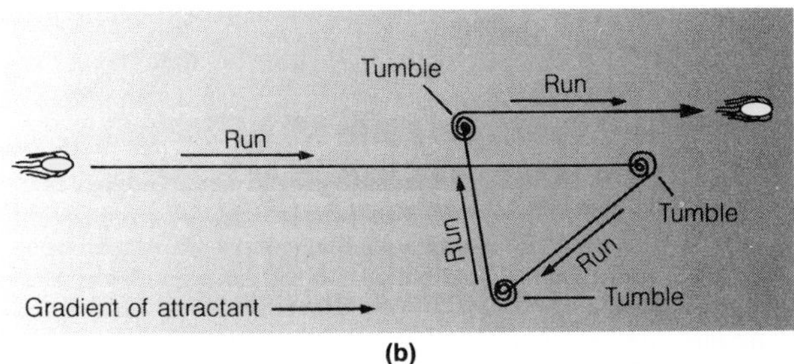

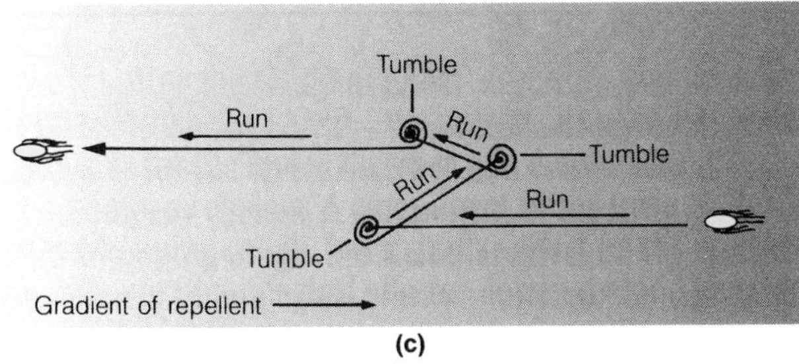

FIGURE **8.31**

Chemotactic motion of bacteria. (a) In the absence of either attractants or repellents, the bacterium stops and tumbles frequently, each time starting out in a new, random direction. **(b)** When a gradient of an attractant is present, a bacterium heading toward the attractant tends to run for longer periods without tumbling. **(c)** A gradient of a repellent has the opposite effect, favoring long runs away from the repellent source.

From W. M. Becker, L. J. Kleinsmith, and J. Hardin, *The World of the Cell,* 4th ed. (San Francisco, CA: Addison Wesley Longman, 2000) © Addison Wesley Longman, Inc.

net motion toward the source of nutrition (Figure 8.31b). Conversely, a bacterium moving away from a repellent continues to do so for longer than usual before tumbling, resulting in an avoidance reaction (Figure 8.31c).

These observations imply that flagellated bacteria must have some mechanism for sensing gradients of attractants or repellents and relaying this information to the motors of their flagella. In fact, they do. We do not have space to describe this remarkable mechanism here. The interested reader should consult the references at the end of the chapter.

SUMMARY

A number of macromolecular protein systems exist in animals to convert ATP energy into physical work. A major example is the actin–myosin system of muscle. In muscle, interdigitating filaments of actin and myosin are driven past one another by attachment, motion, and detachment of myosin cross-bridges. Muscle contraction is stimulated by the influx of calcium, which causes rearrangement of actin-associated proteins. The direct source of contractile energy is ATP, with creatine phosphate being the ultimate energy store.

There also exist many other nonmuscle systems for producing motion and doing work. Many kinds of cell motion, including amoeboid crawling and cytokinesis, use nonmuscle actin and myosin. On the other hand, flagella and cilia are driven by the ATP-driven sliding of microtubules, filaments formed by the polymerization of tubulin. Microtubules do many other things, including acting as "roadways" for transport of organelles and proteins within cells and accomplishing the separation of chromosomes in mitosis.

A remarkable molecular motor drives the rotation of bacterial flagella. This motor, by reversing, can cause the bacterium to tumble and thus seek new directions in which to find nutrients or avoid toxic substances.

REFERENCES

General

Cook, R. (Discussion Chairman) (1995) *The Seventh Biophysical Discussion: Molecular Motors, Structure, Mechanics, and Energy Transduction.* Supplement to Volume 68 of the *Biophysical Journal.* An excellent collection of short papers and abstracts, complete with discussion by members of the meeting.

Cross, R. A., and J. Kendrick-Jones, eds. (1991) *Motor Proteins: Supplement 14 of the Journal of Cell Science.* The Company of Biologists, Ltd., Cambridge. A splendid collection of papers. About half are on actin–myosin systems, about half on microtubule transport.

Muscle

Rayment, I., and H. M. Holden (1994) The three-dimensional structure of a molecular motor. *Trends Biochem. Sci.* 19:129–134.

Spudich, J. (1994) How molecular motors work. *Nature* 372:515–518. A short synopsis (mostly on muscle) by an outstanding authority.

Stroud, R. M. (1996) Balancing ATP in the cell. *Nature Struct. Biol.* 3:567–569. The newly determined structure of creatine kinase clarifies the way in which compounds involved in energy production are channeled from mitochondria to cytoplasm.

Yanagida, T., and A. Ishijima (1995) Forces and steps generated by single myosin molecules. *Biophys. J.* 68:312s–320s.

Microtubules, Dynein, and Kinesin

Amos, L. W., R. W. Linck, and A. Klug (1976) Molecular structure of flagellar microtubules. In: *Cell Motility,* edited by R. Goldman, T. Pollard, and J. Rosenbaum, pp. 847–868. Cold Spring Harbor Press, Cold Spring Harbor, N.Y. A classic early paper on microtubule structure.

Goodenough, U., and J. E. Heuser (1985) Substructure of inner dynein arms, radial spokes, and the central pair/projection complex of cilia and flagella. *J. Cell Biol.* 100:2008–2018.

Hirokawa, N., Y. Noda, and Y. Okada (1998) Kinesin and dynein superfamily proteins in organelle transport and cell division. *Curr. Opin. Cell Biol.* 10:60–73.

Nogales, E., S. C. Wolf, and K. H. Downing (1998) Structure of the $\alpha\beta$ tubulin dimer by electron crystallography. *Nature* 391:199–203.

Schnapp, B. J., R. D. Vale, M. P. Sheetz, and T. S. Reese (1985) Single microtubules from squid axoplasm support bidirectional movement of organelles. *Cell* 40:455–462. A pioneering study.

The Bacterial Motor and Chemotaxis

Berg, H. C. (1995) Torque generation by the flagellar rotary motor. *Biophys. J.* 68:163s–167s.

Meister, M., G. Lowe, and H. C. Berg (1987) The proton flux through the bacterial flagellar motor. *Cell* 49:643–650.

Schuster, S. C., and S. Kahn (1994) The bacterial flagellar motor. *Annu. Rev. Biophys. Biomol. Struct.* 23:509–539.

Segall, J. E., S. M. Block, and H. C. Berg (1986) Temporal comparisons in bacterial chemotaxis. *Proc. Natl. Acad. Sci. USA* 83:8987–8991.

Wolfe, A. J., M. P. Conley, T. J. Kramer, and H. C. Berg (1987) Reconstitution of signaling in bacterial chemotaxis. *J. Bacteriol.* 169:1878–1885.

PROBLEMS

1. A typical relaxed sarcomere is about 2.3 μm in length and contracts to about 2 μm in length. Within the sarcomere, the thin filaments are about 1 μm long and the thick filaments are about 1.5 μm long.
(a) Describe the overlap of thick and thin filaments in the relaxed and contracted sarcomere.
(b) An individual "step" by a myosin head in one cycle pulls the thin filament about 15 nm. How many steps must each actin fiber make in one contraction?

2. Each gram of mammalian skeletal muscle consumes ATP at a rate of about 1×10^{-3} mol/min during contraction. Concentrations of ATP and creatine phosphate in muscle are about 4 mM and 25 mM, respectively, and the density of muscle tissue can be taken to be about 1.2 g/cm^3.
(a) How long could contraction continue on ATP alone?
(b) If all creatine phosphate were converted into ATP and utilized as well, how long could contraction continue?
(c) What do these answers tell you?

3. The drug cytocholasin is known to bind to the ends of actin fibers. It does not affect striated muscle contraction, but it completely inhibits motility and changes in cell shape in eukaryotic cells. What do these findings suggest?

*4. Tubulin binds GTP, and the polymerization of a microtubule is greatly stimulated when a tubulin molecule at the growing end carries GTP. Tubulin also has a GTPase activity, with low turnover number. Microtubules carrying GDP at their ends are more likely to lose tubulin monomers. Show how these facts can explain the remarkable observation that at certain GTP levels some microtubules in a mixture will grow while others simultaneously shrink.

*5. Suppose that a spherical organelle 0.1 μm in diameter is being transported along a microtubule by kinesin at a velocity of 10 μm/min. Calculate the maximum distance it can be moved by the hydrolysis of one ATP molecule, assuming $\Delta G = -40$ kJ/mol in cellular conditions. [Hint: Go back to Tools of Biochemistry 5A and note that the resulting force on a moving particle is given by fv, where v is the velocity and f the frictional coefficient. For spheres, $f = 6\pi\eta r$ where r is the radius and η is the viscosity of the medium. You may take the viscosity in the cytoplasm to be about 0.5 kg/s · m, which is 5 times the viscosity of water. Finally, remember that work (energy) = force × distance.]

*6. At the present time, there is controversy concerning the distance actin and myosin filaments can slide past one another as a consequence of hydrolysis of one ATP molecule. At high loads, the result appears to be consistent with the stroke length (about 10 nm), but at low loads much greater distances have been claimed by some workers. Discuss this problem in terms of the model proposed in Figure 8.11.

CHAPTER 9

Carbohydrates

WE TURN NOW TO THE THIRD GREAT CLASS OF BIOLOGICAL MOLECULES, THE **carbohydrates,** or **saccharides.** Many of these substances are already familiar to you. The simplest carbohydrates are small, monomeric molecules—the **monosaccharides,** which include simple sugars such as *glucose* (Figure 9.1a). Other important carbohydrates are formed by linking such monosaccharides together. If only a few monomer units are involved, we call the molecule an **oligosaccharide.** An example is *maltose* (Figure 9.1b), a **disaccharide** made by linking two glucose molecules together. Long polymers of the monosaccharides, like the starch *amylose* (Figure 9.1c), are called **polysaccharides.** Many kinds of polysaccharides exist, some of which are complex polymers made from many types of sugar monomers.

Carbohydrates are compounds with the stoichiometric formula $(CH^2O)_n$ or are derivatives of such compounds.

Saccharides are often called by the more familiar name *carbohydrates* because many of them can be represented by the simple stoichiometric formula $(CH_2O)_n$. The name was first given when chemists knew only the stoichiometry of saccharides and thought of them as "hydrated carbon." The formula is an oversimplification, however, because many saccharides are modified, and some contain amino, sulfate, and phosphate groups. Nevertheless, all of the compounds described in this chapter either have this formula or can be derived from substances that do.

Carbohydrate formation in photosynthesis and oxidation in metabolism together constitute the major energy cycle in life.

The saccharides play an enormous variety of roles in living organisms. Indeed, the major energy cycle of the biosphere depends largely on carbohydrate metabolism. Before we turn to carbohydrate structure, let us look briefly at this cycle, as illustrated in Figure 9.2. In *photosynthesis,* plants take up CO_2 from the atmosphere and "fix" it into carbohydrates. The basic reaction can be described (in a vastly simplified way) as the light-driven reduction of CO_2 to carbohydrate, here represented as glucose. Much of this carbohydrate is stored in the plants as starch or cellulose. Animals obtain the carbohydrates by eating plants or plant-eating animals. Thus, plant-synthesized carbohydrates ultimately become the principal sources of carbon in all animal tissues. In the other half of the cycle, both plants and animals carry out, via oxidative metabolism, a reaction that is essentially the reverse of photosynthesis, to yield once again CO_2 and H_2O (Figure 9.2). This oxidation of carbohydrates is the primary energy-generating process in metabolism. The central role of carbohydrates is obvious when we consider that

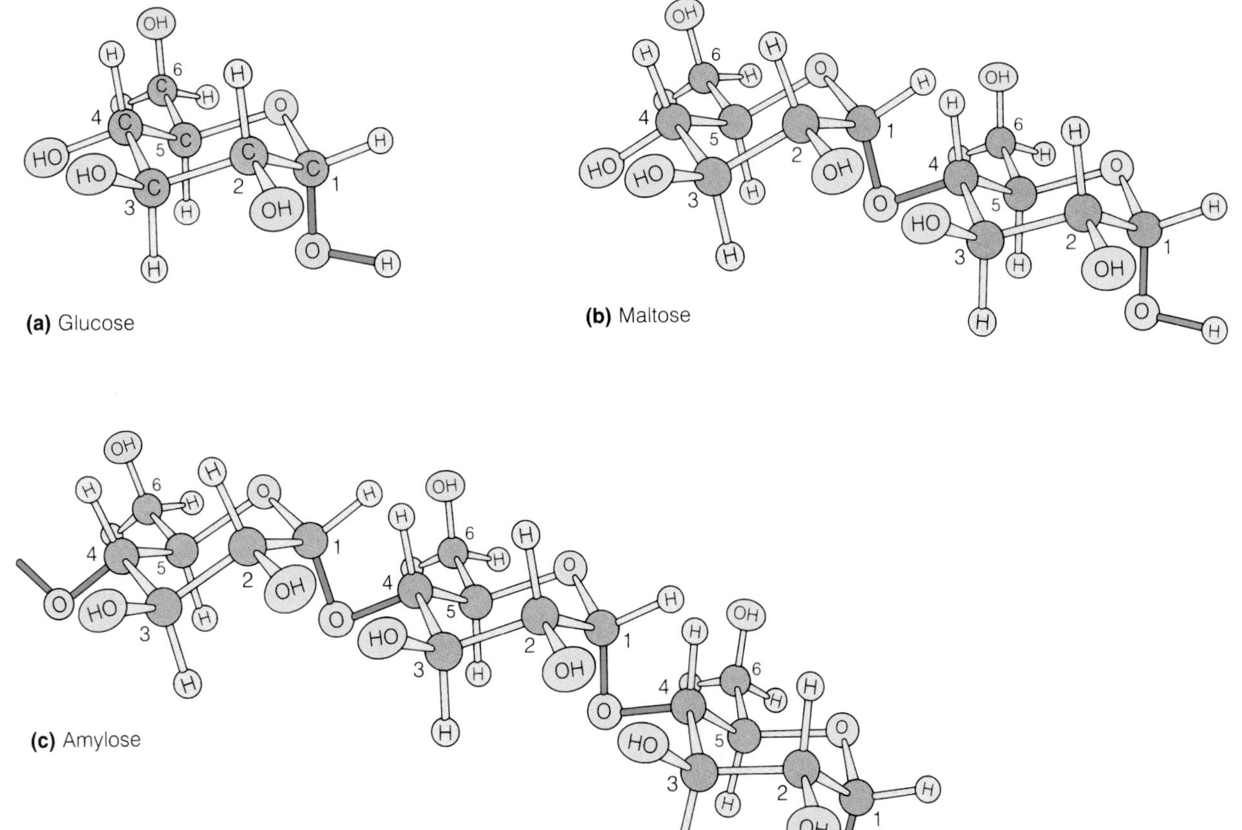

FIGURE **9.1**

Representative carbohydrates. The three compounds shown here are composed entirely of C, H, and O, with glucose forming the monomer for the oligomer and the polymer. **(a)** Glucose, a monosaccharide. **(b)** Maltose, a disaccharide containing two glucose units. **(c)** A portion of a molecule of amylose, a glucose polymer.

the basic foodstuff of most humans is the starch in such plant foods as rice, wheat, or potatoes. Even the meat we consume ultimately can be traced in large part to the carbohydrates eaten by grazing animals.

As critical as energy storage and generation are, they are not the only functions of carbohydrates. Many biological structural materials are polysaccharide in whole or in part. Important examples are the cellulose of woody plants, the cell walls of bacteria, and the exoskeletons of insects and other arthropods. Additionally, polysaccharides on cell surfaces or attached to proteins aid in molecular recognition. Examples include highly specific processes such as the binding of viruses or antibodies on particular cells. Thus, like proteins, carbohydrates are extremely versatile molecules, essential to every kind of living organism.

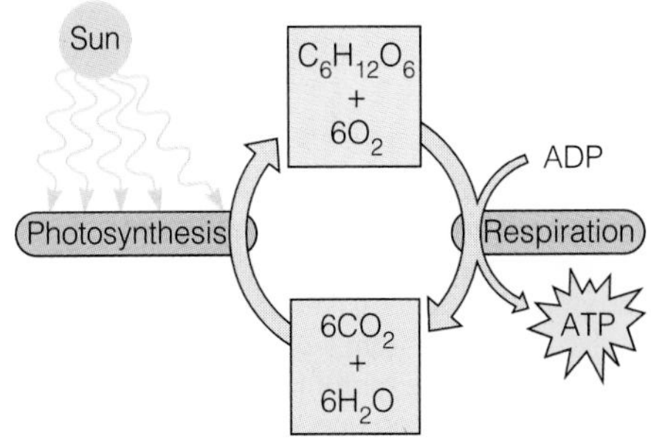

FIGURE **9.2**

The energy cycle of life. In photosynthesis, plants use the energy of sunlight to combine carbon dioxide and water into carbohydrates, releasing oxygen in the process. In respiration, both plants and animals oxidize the carbohydrates made by plants, releasing energy and re-forming CO_2 and H_2O.

Monosaccharides

We begin our discussion of carbohydrates with the simple, monomeric sugars—the monosaccharides. The simplest compound with the empirical formula of the class $(CH_2O)_n$ is found when $n = 1$. However, *formaldehyde,* $H_2C{=}O$, has little in common with our usual concept of sugars; indeed, it is a noxious, poisonous gas.

```
    H
    |
    C1=O
    |
H — C2 — OH
    |
H — C3 — OH
    |
    H
```

D-Glyceraldehyde (an aldose)

```
    H
    |
H — C1 — OH
    |
    C2=O
    |
H — C3 — OH
    |
    H
```

Dihydroxyacetone (a ketose)

FIGURE **9.3**

Trioses, the simplest monosaccharides. The two triose tautomers illustrate the difference between aldose and ketose monosaccharides. Carbon numbering begins in all aldoses with the aldehyde carbon and in ketoses with the end carbon closest to the ketone group. (Because dihydroxyacetone has only three carbons, the two end carbons are equivalent and either of them could be designated number one.)

```
    H                    H                      H
    |                    |                      |
    C1=O                 C1—OH              H — C1 — OH
    |                    ||                     |
H — C2 — OH   ⇌          C2—OH       ⇌          C2=O
    |                    |                      |
H — C3 — OH          H — C3 — OH            H — C3 — OH
    |                    |                      |
    H                    H                      H
```

D-Glyceraldehyde (aldotriose) | Enediol intermediate | Dihydroxyacetone (ketotriose)

FIGURE **9.4**

Aldose–ketose interconversion via an enediol intermediate. The intermediate is unstable and cannot be isolated.

The smallest molecules usually regarded as monosaccharides are the **trioses,** with $n = 3$. (The suffix *ose* is commonly used to designate compounds as saccharides.)

ALDOSES AND KETOSES

There are two trioses: *glyceraldehyde* and *dihydroxyacetone* (Figure 9.3). These molecules, as simple as they are, exhibit certain features that we shall encounter again and again in discussing sugars. In fact, they represent the two major classes of monosaccharides. Glyceraldehyde is an aldehyde, one of a class of monosaccharides called **aldoses.** Dihydroxyacetone is a ketone; such monosaccharides are called **ketoses.** Note that glyceraldehyde and dihydroxyacetone have the same atomic composition. They are **tautomers** (structural isomers differing in the location of their hydrogens and double bonds) and can interconvert via an unstable *enediol* intermediate, as shown in Figure 9.4. Such tautomeric interconversions occur to a certain extent between all such pairs of aldose and ketose monosaccharides, but the reactions are usually very slow unless catalyzed. Thus, glyceraldehyde and dihydroxyacetone can each exist as a quite stable compound.

The two major classes of monosaccharides are aldoses and ketoses.

ENANTIOMERS

An important feature of monosaccharide structure can be seen by examining the formula for glyceraldehyde a bit more carefully. The second carbon atom carries four different substituents, so it is a *chiral* carbon, like the α-carbon in most α-amino acids. Therefore, glyceraldehyde has two stereoisomers of the type called **enantiomers,** which are nonsuperimposable mirror images of one another. Three-dimensional drawings of the two forms, designated as D- and L-glyceraldehyde, are shown in Figure 9.5. Such three-dimensional orientation about the

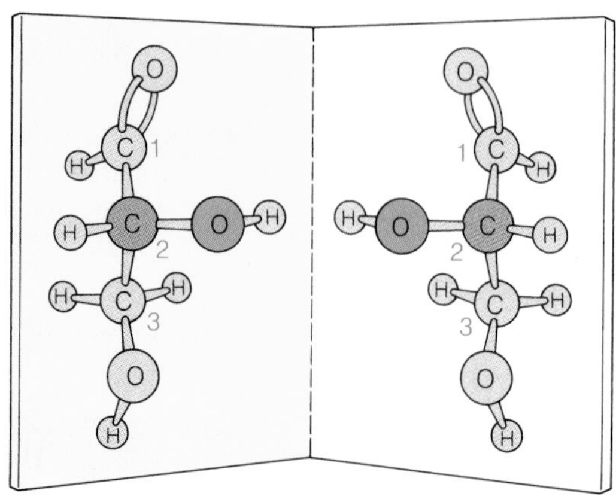

FIGURE **9.5**

The enantiomers of glyceraldehyde. The configuration of groups around the chiral carbon 2 (shown in darker gray) distinguishes D-glyceraldehyde from L-glyceraldehyde. The two molecules are mirror images and cannot be superimposed on one another.

asymmetric carbon can also be represented by the bond convention we used in Chapter 5:

CHO
H ► C ◄ OH
CH₂OH
D-Glyceraldehyde

CHO
HO ► C ◄ H
CH₂OH
L-Glyceraldehyde

D and L forms of a monosaccharide are nonsuperimposable mirror images and are called enantiomers.

Note that we do not need to draw the spatial orientation of atoms about carbons 1 or 3 because these carbons are not chiral centers.

The most compact way to represent enantiomers is to use a **Fischer projection.** In a Fischer projection the bonds that are drawn horizontally are imagined as coming toward you; those drawn vertically are receding. Thus, for D-glyceraldehyde and L-glyceraldehyde, we have

D-Glyceraldehyde	L-Glyceraldehyde	Carbon number:
CHO	CHO	1
H—C—OH	HO—C—H	2
CH_2OH	CH_2OH	3

D-Glyceraldehyde **L-Glyceraldehyde**

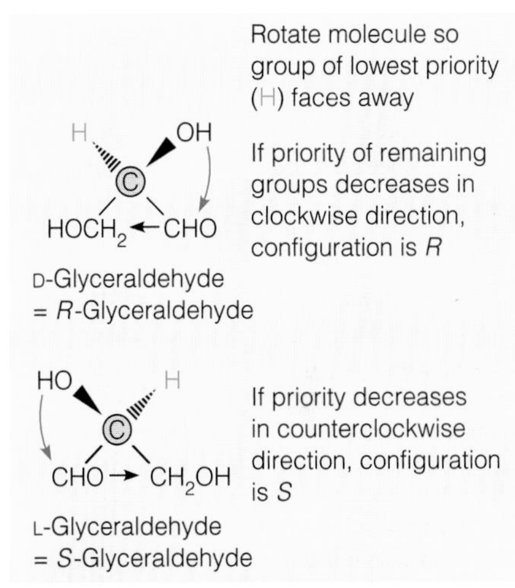

FIGURE **9.6**

***R–S* nomenclature.** The *R–S* system describes absolute stereochemical configuration, as shown in this example. Each type of group attached to a chiral carbon (gray) is given a priority, according to a set of defined rules. Priorities for groups common in carbohydrate chemistry are OR > OH > NH_2 > CO_2H > CHO > CH_2OH > CH_3 > H. We view the molecule with the group of lowest priority away from us (H in our example). If the priority of the remaining three groups *decreases clockwise,* the absolute configuration is called *R* (from Latin *rectus,* "right"). If priority *decreases counterclockwise,* the configuration is *S* (from Latin *sinister,* "left"). In this notation, D-glyceraldehyde is *R*-glyceraldehyde, and L-glyceraldehyde is *S*-glyceraldehyde.

Alternative Designations for Enantiomers: D–L and *R–S*

Originally, the terms D and L were meant to indicate the direction of rotation of the plane of polarization of polarized light: D for right (dextro), L for left (laevo). It is true that a solution of D-glyceraldehyde does rotate the plane of polarization to the right, as do many other D-monosaccharides, but this correspondence does not always hold, because the magnitude and even the direction of optical rotation are a complicated function of the electronic structure surrounding the chiral center. Another disadvantage of the D–L nomenclature is that it is not absolute; the designation is always with respect to some reference compound. Accordingly, an *absolute* convention has been developed that allows us to assign a stereochemical designation to any compound from examination of its three-dimensional structure. An example of how this so-called *R–S* convention works is shown in Figure 9.6. Although the *R–S* convention is more general, it is little used by biochemists. It becomes difficult to apply in the common situation in which a molecule contains more than one asymmetric carbon atom.

Monosaccharide Enantiomers in Nature

Just as in the case of amino acids, one enantiomeric form of monosaccharides dominates in living organisms. In proteins it was the L-amino acids; in carbohydrates it is the D-monosaccharides. Again, there is no obvious reason why this preference was established in nature. But once fixed in early evolution, it has persisted, for most of the cellular machinery has become geared to operate with D-sugars. However, just as D-amino acids are sometimes found in living organisms, so are L-monosaccharides. Like the "abnormal" D-amino acids, the L-monosaccharides play rather specialized roles. Table 9.1 includes, along with D-monosaccharides, the most commonly occurring L-monosaccharides, with examples of their occurrence and functions.

The most important naturally occurring saccharides are the D-enantiomers.

DIASTEREOMERS

When we consider monosaccharides with more than three carbons, a further structural complication appears. Such a monosaccharide may have more than one chiral carbon, which results in its having two types of stereoisomers. These

TABLE 9.1 Examples of occurrence and biochemical roles of monosaccharides

Monosaccharides	Natural Occurrence	Physiological Role[a]
Trioses		
Glyceraldehyde	Widespread (as phosphate)	The 3-phosphate is an intermediate in glycolysis
Dihydroxyacetone	Widespread (as phosphate)	The 1-phosphate is an intermediate in glycolysis
Tetroses		
D-Erythrose	Widespread	The 4-phosphate is an intermediate in carbohydrate metabolism
Pentoses		
D-Arabinose	Some plants, tuberculosis bacilli	Plant glycosides, cell walls
L-Arabinose	Widely distributed in plants, bacterial cell walls	Constituent of cell walls, plant glycoproteins
D-Ribose	Widespread, in all organisms	Constituent of ribonucleic acid
2-D-Deoxyribose	Widespread, in all organisms	Constituent of deoxyribonucleic acid
D-Xylose	Woody materials	Constituent of plant polysaccharides
Hexoses		
D-Galactose	Widespread	Milk (as part of lactose); structural polysaccharides
L-Galactose	Agar, other polysaccharides	Polysaccharide structures
D-Glucose	Widespread	A major energy source for animal metabolism; structural role in cellulose
D-Mannose	Plant polysaccharides, animal glycoproteins	Polysaccharide structures
D-Fructose	A major plant sugar; part of sucrose	Intermediate in glycolysis (phosphate esters)
Heptoses		
D-Sedoheptulose	Many plants	Intermediate in Calvin cycle in photosynthesis and pentose phosphate pathway

[a]Some of these monosaccharides have additional roles that are not listed.

types are *enantiomers* (mirror image isomers), which we have already discussed, and *diastereomers*, which we first encounter in the tetrose monosaccharides.

When monosaccharides contain more than one chiral carbon, the prefix D or L designates the configuration about the carbon farthest from the carbonyl group. Isomers differing in orientation about other carbons are called diastereomers and given different names.

Tetrose Diastereomers

Tetroses, with the empirical formula $(CH_2O)_4$, have two chiral carbons in the aldose forms. Therefore, an aldotetrose will have four stereoisomers, as shown in Figure 9.7. In general, a molecule with n chiral centers will have 2^n stereoisomers, because there are two possibilities at each chiral center. The following convention attempts to give a rational method for naming and distinguishing the stereoisomers of such a molecule: The prefix D or L is used to designate the orientation about the chiral carbon *farthest* from the carbonyl group—carbon number 3 in this case. Molecules with different orientations about the carbons preceding this reference carbon are given separate names. Thus, *threose* and *erythrose* are two aldotetroses with opposite orientations about carbon 2. Stereoisomers of this kind, which are *not* mirror images, are called **diastereomers.** Threose and erythrose are diastereomers, and each has two enantiomers (D and L) that are nonsuperimposable mirror images. Unfortunately, there is no general logical rule for forming the specific names (such as threose and erythrose); they must simply be learned, like the names of the amino acids.

The four-carbon *ketose*, which is called *erythrulose*, has only one pair of enantiomers, because this monosaccharide has only one chiral carbon (Figure 9.8). Another naming convention appears at this point: Usually the ketose name is derived from the corresponding aldose name by insertion of the letters *ul*. Thus *erythrose* becomes *erythrulose*. As with glyceraldehyde (and other monosaccharides),

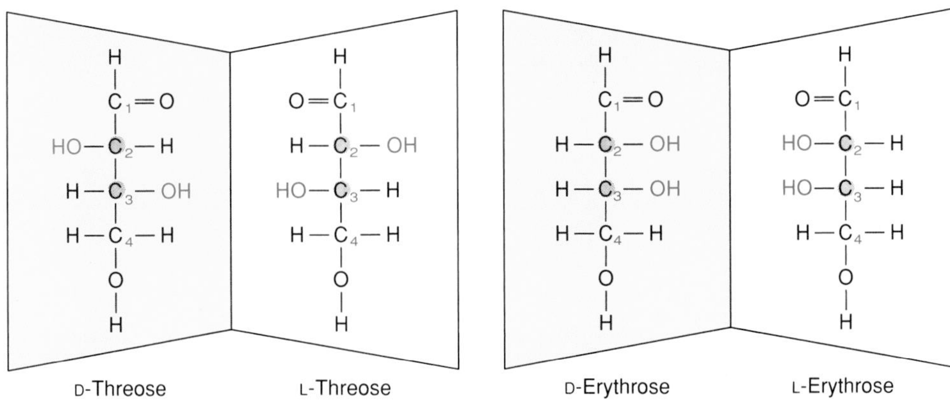

FIGURE **9.7**

Stereochemistry of aldotetroses. These molecules have two chiral carbons (2 and 3) and thus have two diastereomeric forms, threose and erythrose, each with a pair of enantiomers. Note that the threose enantiomers have *opposite* configuration about carbons 2 and 3, whereas erythrose enantiomers have the *same* configuration about these two carbons.

the ketose and aldose forms are interconvertible via tautomerization in dilute alkali. The aldose–ketose conversion also provides a route for interconversion of aldose diastereomers, using the ketose as an intermediate.

Pentose Diastereomers

Adding one more carbon, we obtain the **pentoses.** The **aldopentoses** have three chiral centers; therefore we expect 2^3, or eight, stereoisomers—in four pairs of enantiomers. The D forms of the pentoses are shown in Figure 9.9a, which provides a summary of the aldoses containing from three to six carbons. Note that each of the aldopentoses shown has the D orientation about carbon 4 and that all possible combinations of orientations about carbons 2 and 3 are included. (From here on, in our illustration of carbohydrate structure we will show only the D forms; you can easily draw the L forms from the rules given above.) **Ketopentoses,** as shown in Figure 9.9b, have two chiral carbons, so four isomers (two pairs of enantiomers) must exist. The D diastereomers are called D-*ribulose* and D-*xylulose.*

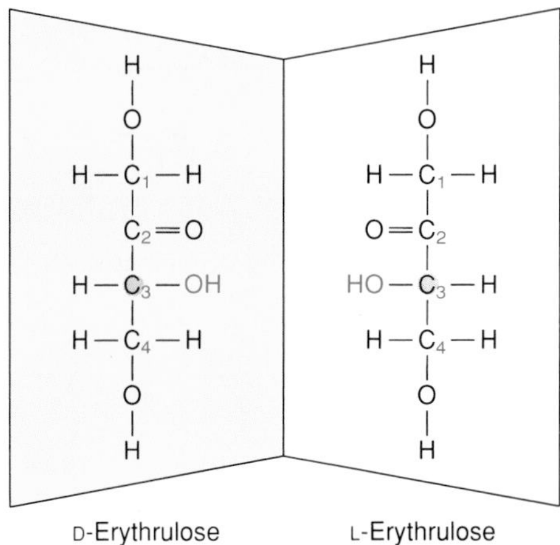

FIGURE **9.8**

The two enantiomers of erythrulose. Unlike the four-carbon aldoses (see Figure 9.7), the four-carbon ketose has only one chiral carbon (C_3) and only one pair of enantiomers.

Hexose Diastereomers

Monosaccharides containing six carbon atoms are called **hexoses.** As you might imagine, there is a large number of possible hexoses. To keep their structures in mind, it is useful to relate them to the simpler pentoses, tetroses, and trioses. Figure 9.9 provides a compact summary of these relationships. The hexoses we will most frequently encounter are *glucose* and *fructose.* However, *mannose* and *galactose* are also widespread in nature (see Table 9.1). In fact, almost all of the hexoses play some significant biological role.

RING STRUCTURES

With the pentoses and hexoses, another feature of monosaccharide chemistry assumes major importance. Having five or six carbons in the chain gives these compounds the potential to form very stable ring structures via internal *hemiacetal* formation. The bond angles characteristic of carbon and oxygen bonding are such that rings containing fewer than five atoms are strained to a considerable extent, whereas five- or six-membered rings are easily formed. In principle, aldotetroses can also form five-membered ring structures, but they rarely do.

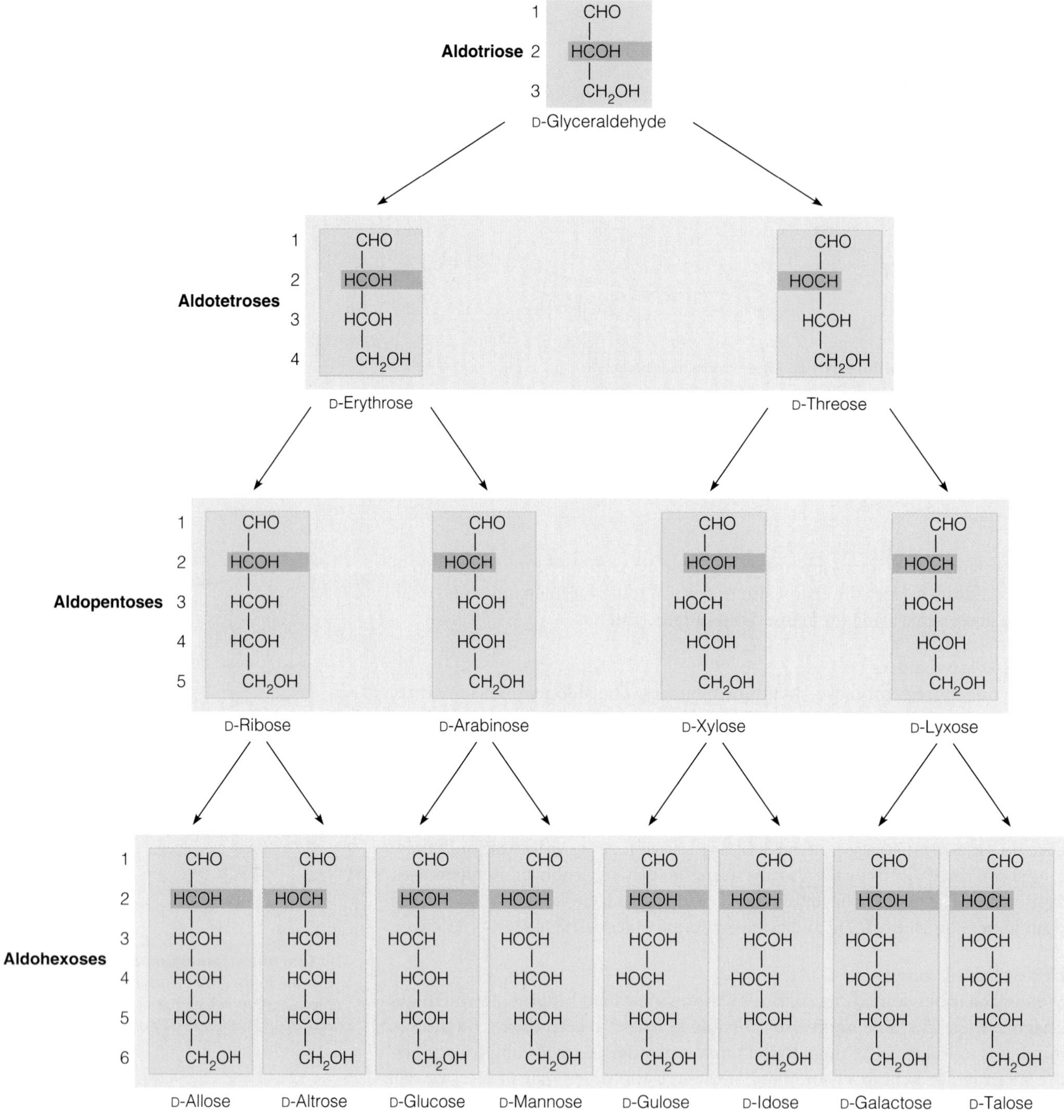

(a) D-Aldoses

FIGURE **9.9**

Stereochemical relationships of the D-aldoses and D-ketoses. This figure shows the relationships between pairs of diastereomers in the D-aldose series **(a)** and the D-ketose series. **(b)** Each series is generated by successive additions of one CHOH group (shaded) just below the carbonyl carbon. In each case, the two possible orientations of the added group generate a pair of diastereomers. The L forms are not shown; they are just the mirror images of the D forms.

Pentose Rings

Consider this hemiacetal ring formation in an aldopentose, such as D-ribose (see Figure 9.9a). Two modes of ring closure are possible, as shown in Figure 9.10. Reaction of the C-1 oxygen of D-ribose with the C-4 hydroxyl produces a five-membered ring structure called a **furanose**; the name reflects its structural similarity to the heterocyclic compound furan. Alternatively, a six-membered ring is obtained if the reaction occurs with the C-5 hydroxyl. Such a six-membered ring is called a **pyranose,** to indicate its relation to the heterocyclic compound pyran.

Both of the reactions shown in Figure 9.10 have equilibria that highly favor the cyclic structures for pentoses or larger sugars. Under physiological condi-

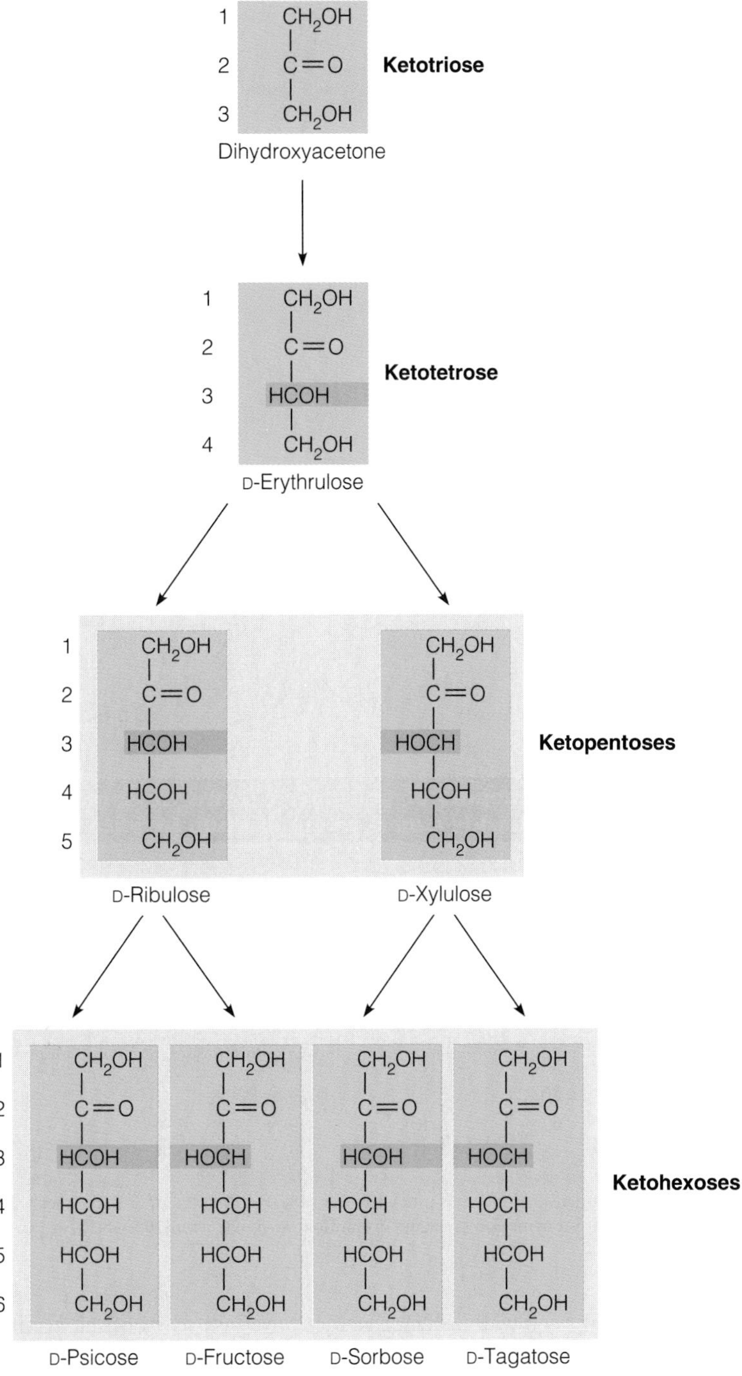

(b) D-Ketoses

tions in solution, monosaccharides with five or more carbons are typically more than 99% in the ring forms. The distribution between pyranose and furanose forms depends on the particular sugar structure, the pH, the solvent composition, and the temperature. Representative data obtained from nuclear magnetic resonance studies are shown in Table 9.2. When the monomers are incorporated into polysaccharides, the structure of the polymer may also influence the ring form chosen. For example, as Table 9.2 shows, D-ribose exists in solution as a mixture of the two ring forms. But in biological polysaccharides, specific forms are stabilized. Ribonucleic acid, for example, contains exclusively ribofuranose,

Monosaccharides with five or more carbons exist preferentially in five- or six-membered ring structures, resulting from internal hemiacetal formation.

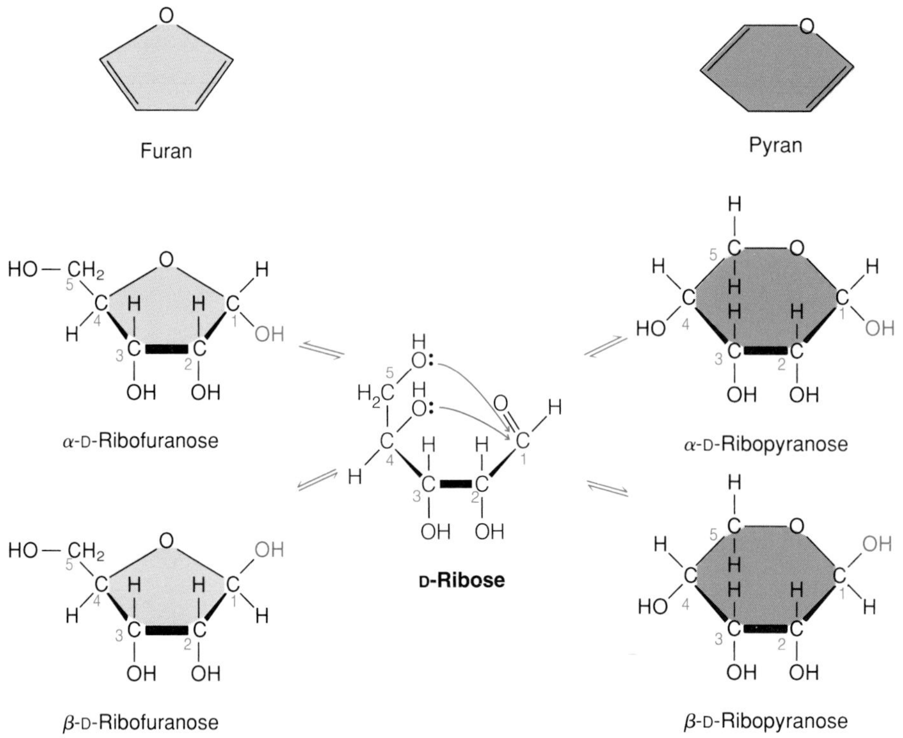

FIGURE **9.10**

Formation of ring structures by pentoses. The example shown here is D-ribose, which can form either a five-membered furanose ring or a six-membered pyranose ring. The reactions involve formation of hemiacetals from the aldehyde group. In each case, two anomeric forms, α and β, are possible. (Anomers differ in conformation only at carbon 1.) The sugar rings are depicted here as *Haworth projections,* with bonds closer to the viewer drawn more darkly to suggest perspective.

Furan

Pyran

α-D-Ribofuranose

D-Ribose

α-D-Ribopyranose

β-D-Ribofuranose

β-D-Ribopyranose

TABLE **9.2 Relative amounts of tautomeric forms for some monosaccharide sugars at equilibrium in water at 40°C**

	Relative Amount (%)				
Monosaccharide	α-Pyranose	β-Pyranose	α-Furanose	β-Furanose	Total Furanose
Ribose	20	56	6	18	24
Lyxose	71	29	—[a]	—[a]	<1
Altrose	27	40	20	13	33
Glucose	36	64	—[a]	—[a]	<1
Mannose	67	33	—[a]	—[a]	<1
Fructose	3	57	9	31	40

Note: In all cases, the open-chain form is much less than 1%. For data on other sugars, see S. J. Angyal, The composition and conformation of sugars in solution, *Angew. Chem.* (1969) 8:157–226.

[a]Much less than 1%.

whereas some plant cell wall polysaccharides have pentoses entirely in the pyranose form.

It is instructive to look a bit more closely at the ring structures shown in Figure 9.10. Cyclization has created a new asymmetric center at carbon 1. That is why we have had to draw two stereoisomers of D-ribofuranose, referred to as α-D-ribofuranose and β-D-ribofuranose, as well as a corresponding pair of ribopyranoses. Like other kinds of stereoisomers, these α and β forms rotate the plane of polarized light differently and can be distinguished in that way. Such isomers, differing in configuration only at carbon 1, are called **anomers,** and carbon 1 is often referred to as the *anomeric carbon atom.* The monosaccharides can undergo interconversion between the α and β forms, using the open-chain structure as an intermediate. This process is referred to as **mutarotation.** A purified anomer, dissolved in aqueous solution, will approach the equilibrium mixture, with an accompanying change in the optical rotation of the solution. Enzymes called **mutarotases** catalyze this process in vivo.

The representation of a cyclic sugar structure we have used in Figure 9.10 is called a **Haworth projection.** You are to imagine that you are seeing the ring in perspective, and the groups attached to the ring carbons (H, OH, CH_2OH) are pictured as being above or below the ring. In all D-monosaccharides, the —CH_2OH is above the ring. The relationship between hydroxyl orientations in a Fischer and a Haworth projection is straightforward. Those represented to the right of the chain in a Fischer projection are shown below the ring in a Haworth. For example, Fischer projections of α-D-ribofuranose and β-D-ribofuranose would look like this:

```
H—C—OH ┐            HO—C—H ┐
  |    |               |    |
H—C—OH O             H—C—OH O
  |    |               |    |
H—C—OH |             H—C—OH |
  |    |               |    |
H—C————┘             H—C————┘
  |                    |
 CH2OH                CH2OH
```

α-D-Ribofuranose **β-D-Ribofuranose**

Even Haworth projections do not accurately depict the three-dimensional structure of molecules like ribofuranose or ribopyranose. Saturated five- and six-membered rings cannot be planar, because the C—C—C bond angles are about 109° and the C—O—C angle is about 118°. Furthermore, the ring can pucker out of plane in many different ways. The different ring conformations produced by slightly different bond angles are called **conformational isomers.** Ball-and-stick models of two of the several possible conformational isomers of β-D-ribofuranose are shown in Figure 9.11.

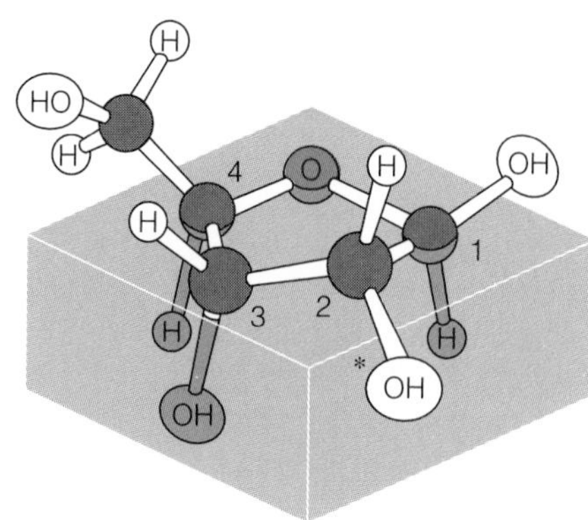

(a) β-D-Ribofuranose, C-2 endo

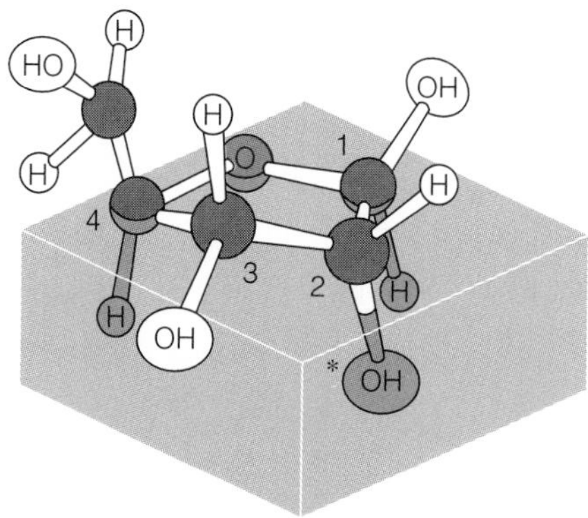

(b) β-D-Ribofuranose, C-3 endo

FIGURE **9.11**

Conformational isomers. These models show two of the possible ring conformations for β-D-ribofuranose. In both of them, C-1, O, and C-4 define a plane. In the C-2 endo conformation **(a)**, C-2 is above the plane. In the C-3 endo conformation **(b)**, C-3 is above the plane. These isomers are the two most common conformations for ribose and deoxyribose in nucleic acids. (In DNA the hydroxyl at carbon 2, indicated here by *, is replaced by hydrogen.) A C-3 exo conformation would look like the figure in (b), but C-3 would be flipped below the plane.

We have already encountered β-D-ribofuranose (and its close relative β-D-2-deoxyribofuranose) in Chapter 4. These sugars play a major role in biochemistry, for they are part of the backbone structures of ribonucleic acid and deoxyribonucleic acid, respectively. Only the β anomers are involved in nucleic acid structure, and the 2-endo and 3-endo conformations shown in Figure 9.11 are favored. However, there is some variation in ring conformation, even locally, along DNA and RNA chains with resulting changes in secondary structure. This flexibility points up a fundamental difference between *conformation* and *configuration.* Conformational isomers can interchange by a simple deformation of the molecule. But *configurational* isomers, such as the various kinds of stereoisomers described earlier, can interconvert only through the breaking and re-formation of covalent bonds.

Like the aldopentoses, the ketopentoses exist almost entirely in the ring form under physiological conditions. However, only the furanose form is possible for ketopentoses. An example is *α-D-ribulose,* which is an intermediate in the carbon fixation processes in photosynthesis.

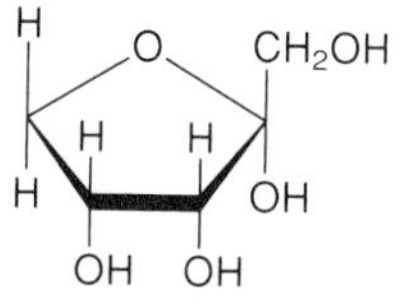

α-D-Ribulose

Hexose Rings

The hexoses also exist primarily in ring forms under physiological conditions. As with the aldopentoses, two kinds of rings are found: five-membered furanoses and six-membered pyranoses. In each case, α and β anomers are possible. An example, illustrated by Haworth projections, follows:

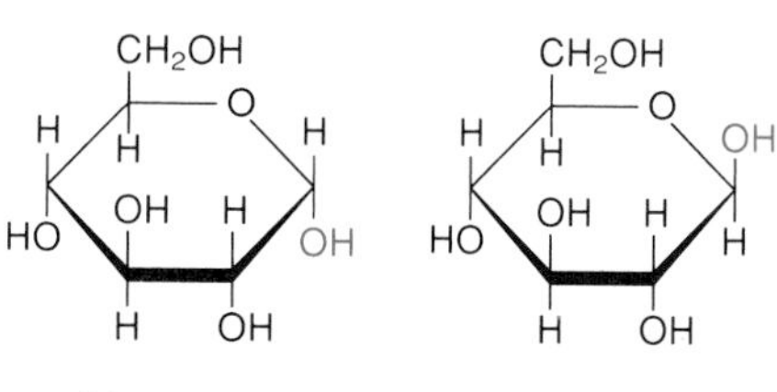

α-D-Glucopyranose **β-D-Glucopyranose**

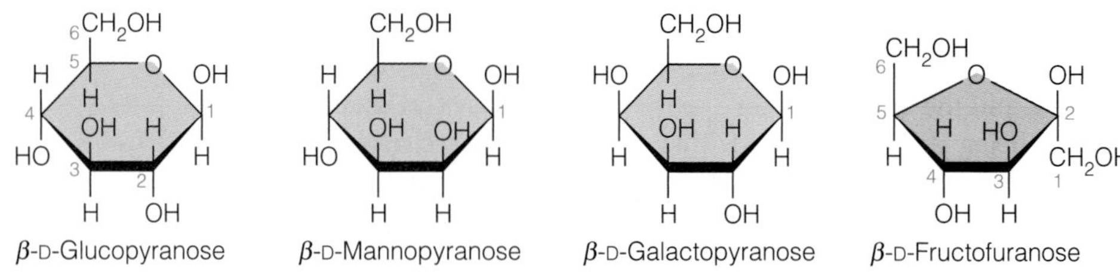

FIGURE **9.12**

The four most common hexoses. These Haworth projections represent the D enantiomers; only the β anomers are shown.

Figure 9.12 shows Haworth projections of the structures of the four most common hexoses in their usual configurations. Table 9.2 gives the fractions of the furanose and pyranose forms found at equilibrium for a number of hexoses. Clearly, which forms are favored depends greatly on the structure of the particular sugar and its environment, although we can make the generalization that hexoses prefer the pyranose ring structure when in aqueous solution. This preference is also true of fructose, but we have depicted D-fructose in Figure 9.12 in its furanose configuration, because that is how it is found in its most common biological source, the disaccharide sucrose. The elucidation of the distribution of anomeric and tautomeric forms of the sugars existing in solutions has been greatly facilitated by the technique of nuclear magnetic resonance spectroscopy (see Tools of Biochemistry 6A). Only this technique has the exquisite sensitivity to structure of molecules in solution to allow such analyses.

We have already shown that Haworth projections of the furanoses do not depict the actual three-dimensional structure correctly. The same is true for the pyranoses. Two major classes of pyranose conformations exist for the 6-carbon sugars—the more stable "chair" form and the less favored "boat" form. These two conformations are depicted as ball-and-stick models in Figure 9.13a. We will fre-

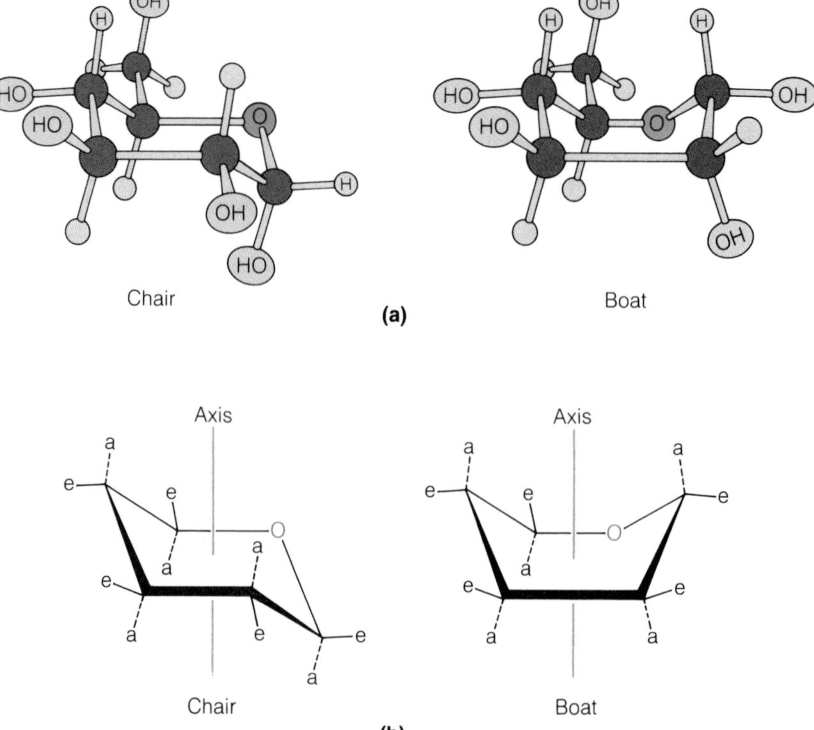

FIGURE **9.13**

The pyranose ring in chair and boat conformations. Three-dimensional representations of α-D-glucopyranose in the chair form (left) and the boat form (right). **(a)** Ball-and-stick models. **(b)** Skeletal diagrams of the bonding. Axial bonds (a) and equatorial bonds (e) are indicated.

quently depict them by skeletal diagrams in the ways shown in Figure 9.13b. For both the boat and chair forms of pyranose rings, a molecular axis can be defined perpendicular to the central plane of the molecule. Bonds to substituents on ring carbons can then be classed as *axial (a)* or *equatorial (e)*, depending on whether they are approximately parallel or perpendicular to the axis (Figure 9.13b). For most sugars, the chair form is more stable, because substituents on axial bonds tend to be more crowded in the boat form.

Hexoses can exist in boat and chair conformations. Usually the chair is more stable.

Sugars with More Than Six Carbons

Monosaccharides with seven or even more carbons exist in nature, but most are of minor importance. However, one heptose, called *sedoheptulose*, plays a major role in the fixation of CO_2 in photosynthesis (see Table 9.1 and Chapter 17).

By this point, you may have become confused by all of the terms used to describe the structures of sugar molecules—enantiomers, diastereomers, anomers, and ring conformations. For review, this terminology is summarized in Figure 9.14.

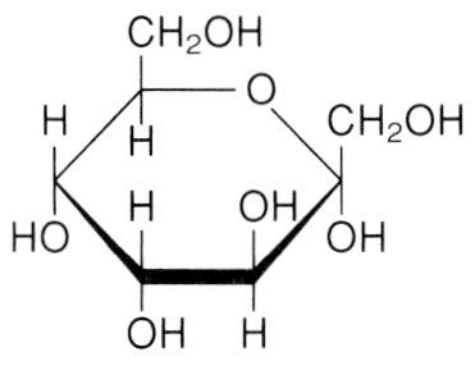

α-D-Sedoheptulopyranose

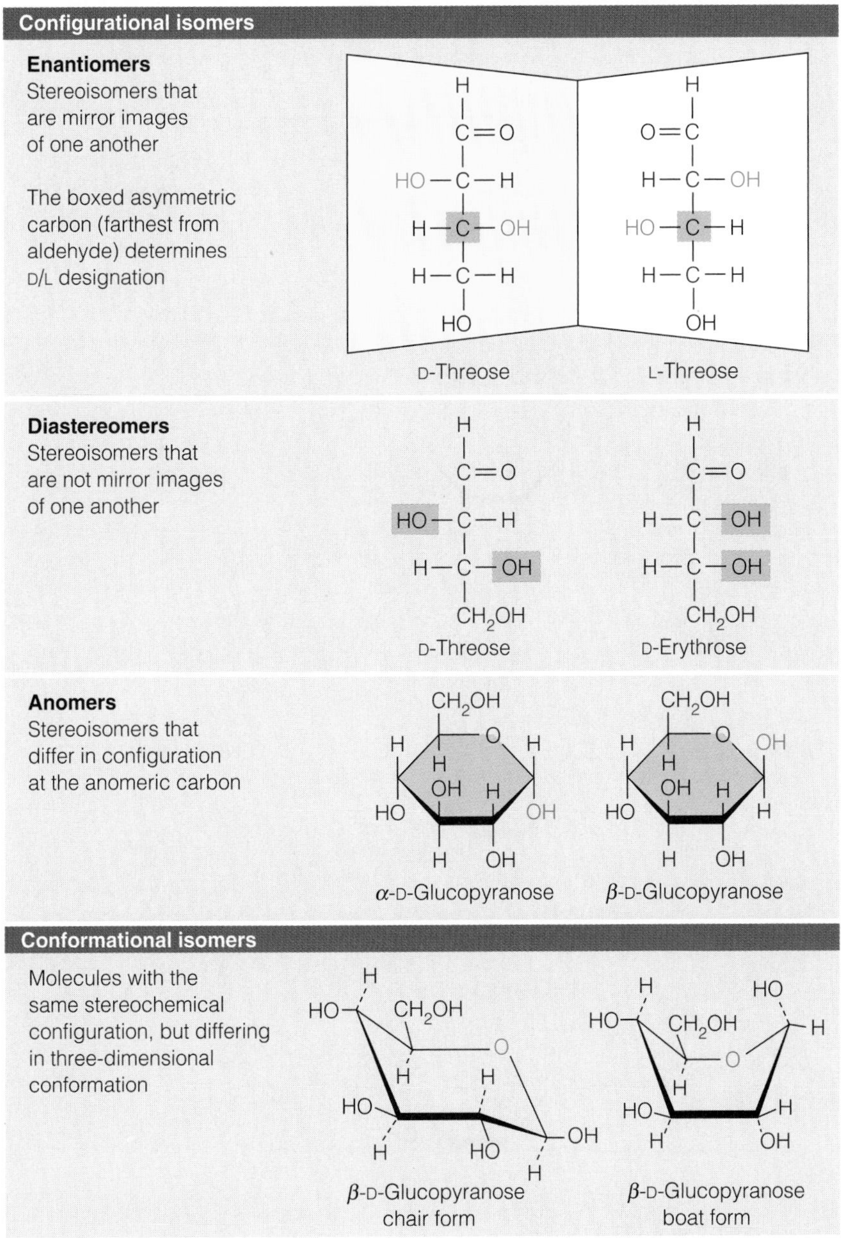

FIGURE **9.14**

Terminology describing the structure of sugar molecules. Conformational isomers are distinguished from configurational isomers in that the former can interconvert without breaking and re-forming bonds.

Derivatives of the Monosaccharides

The monosaccharides each carry a number of hydroxyl groups to which substituents might be attached or which could be replaced by other functional groups. An enormous number of sugars are in fact modified in this way. We shall describe here only a small number of them—primarily those that play biologically important roles.

PHOSPHATE ESTERS

Sugar phosphates are important intermediates in metabolism, functioning as activated compounds in syntheses.

We have already encountered sugar phosphorylation, in compounds like AMP, ATP, and the nucleic acids. As we shall see in later chapters, the phosphate esters of the monosaccharides themselves are major participants in many metabolic pathways. Table 9.3 illustrates a number of the more important phosphate esters and in-

TABLE 9.3 A number of biochemically important phosphate esters of monosaccharides

Name	Structure	$\Delta G^{\circ\prime a}$ (kJ/mol)	pK_{a1}	pK_{a2}
D-Glyceraldehyde-3-phosphate	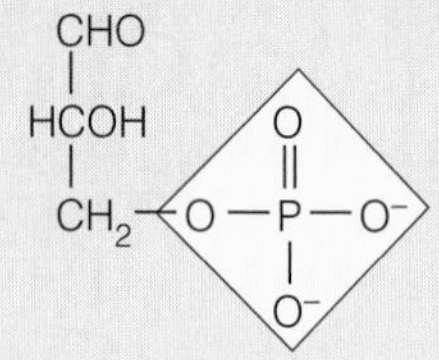 CHO HCOH CH2 O—P—O⁻ O O⁻	~−12	2.10	6.75
β-D-Glucose-1-phosphate	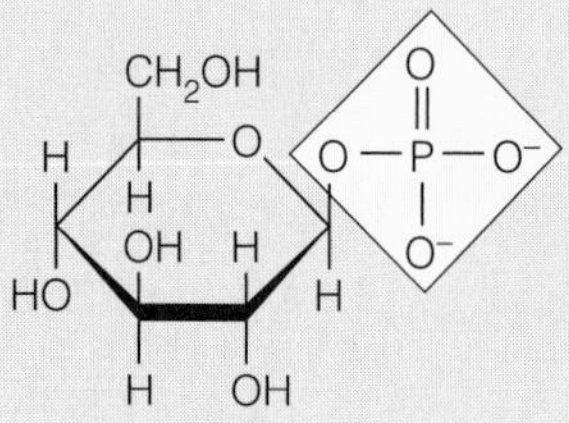 CH2OH H O O—P—O⁻ O O⁻ H OH H HO H H OH	−20.9	1.10	6.13
β-D-Glucose-6-phosphate	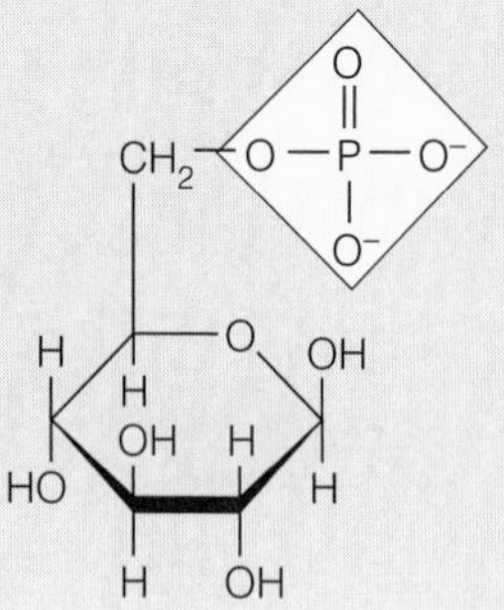 CH2 O—P—O⁻ O O⁻ H O OH H OH H HO H H OH	−13.8	0.94	6.11
α-D-Fructose-6-phosphate	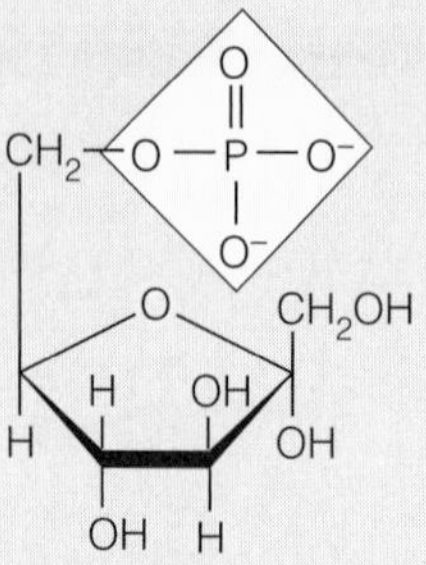 CH2 O—P—O⁻ O O⁻ O CH2OH H OH H OH OH H	−13.8	0.97	6.11

[a]Free energy of hydrolysis at pH 7.0 and 37°C.

cludes values for the standard state free energies of hydrolysis. In all cases, these values are less negative than the free energy of hydrolysis of ATP (−31 kJ/mol); thus ATP can act as a phosphate donor to monosaccharides. On the other hand, because hydrolysis of the phosphate esters of sugars is thermodynamically favorable, these derivatives can behave as "activated" compounds in many metabolic reactions.

Sugar phosphate esters are quite acidic, with pK_a values for the two stages of phosphate ionization of about 1–2 and 6–7, respectively (see Table 9.3).

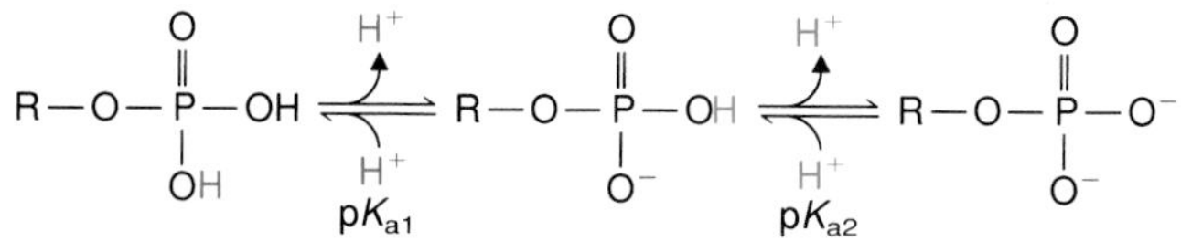

$$\mathrm{R{-}O{-}P(=O)(OH){-}OH} \underset{H^+}{\overset{H^+}{\rightleftharpoons}} \mathrm{R{-}O{-}P(=O)(O^-){-}OH} \underset{H^+}{\overset{H^+}{\rightleftharpoons}} \mathrm{R{-}O{-}P(=O)(O^-){-}O^-}$$

$$pK_{a1} \qquad pK_{a2}$$

Consequently, these compounds exist under physiological conditions as a mixture of the monoanions and dianions.

In additon to the sugar phosphates, a large number of other derivatives of monosaccharides play varied and important roles in biochemistry. We consider here a few of them and the reactions by which they may be generated from monosaccharides.

ACIDS AND LACTONES

Oxidation of monosaccharides can proceed in a variety of ways, depending on the oxidizing agent used. For example, mild oxidation of an aldose with alkaline Cu(II) (Fehling's solution) produces the **aldonic acids,** as in the following example:

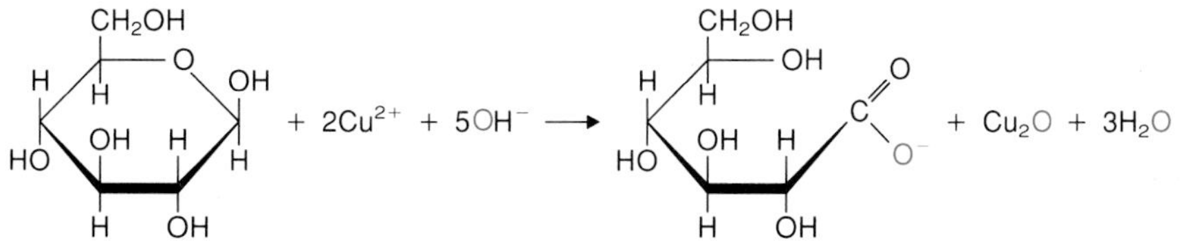

$$\beta\text{-D-Glucopyranose} + 2Cu^{2+} + 5OH^- \longrightarrow \text{D-Gluconic acid} + Cu_2O + 3H_2O$$

β-D-Glucopyranose **D-Gluconic acid**

The production of a red precipitate of Cu_2O is a classic sugar test and was used formerly to test for excess sugar in the urine of persons thought to have diabetes. Another, similar reaction involves the use of Ag^+ ion as an oxidant; its reduction to metallic silver leaves a characteristic "mirror" on the glassware. These older methods have now been replaced by more specific enzyme assays. Free aldonic acids, such as gluconic acid, are in equilibrium in solution with **lactones.**

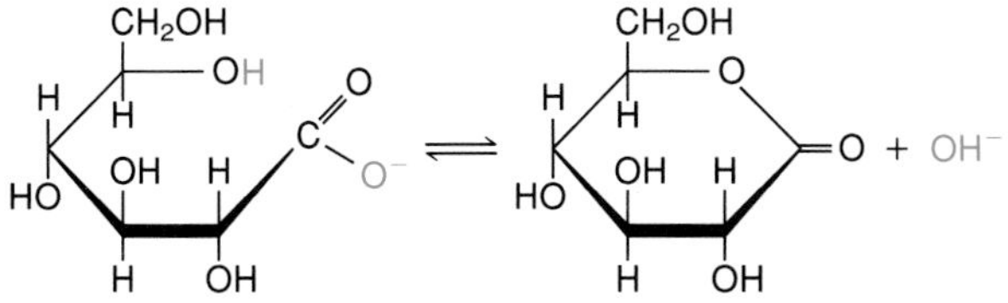

$$\text{D-Gluconic acid} \rightleftharpoons \text{D-}\delta\text{-Gluconolactone} + OH^-$$

D-Gluconic acid **D-δ-Gluconolactone**

Enzyme-catalyzed oxidation of monosaccharides gives a variety of specific products, including lactones and **uronic acids** such as **glucuronic acid,** in which oxidation has occurred at carbon 6. Uronic acids are, as we shall see later in this chapter, important constituents of certain natural polysaccharides.

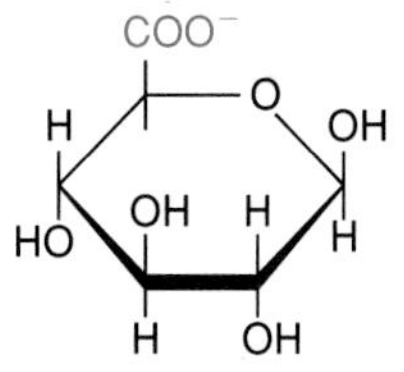

β-D-Glucuronic acid

ALDITOLS

Reduction of the carbonyl group on a sugar gives rise to the class of polyhydroxy compounds called **alditols.** Important naturally occurring ones are *erythritol,* D-*mannitol,* and D-*glucitol,* often called *sorbitol.*

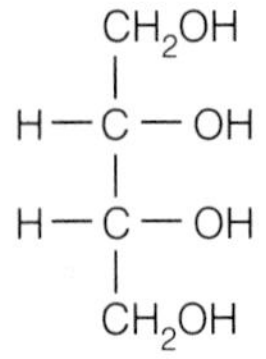

Erythritol* 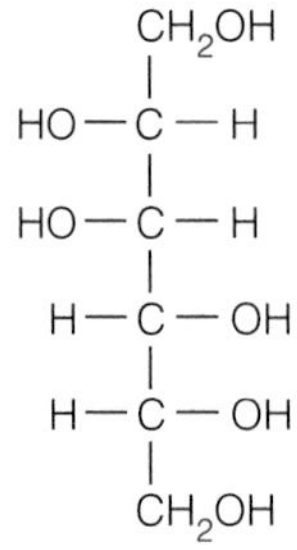 **D-Mannitol** 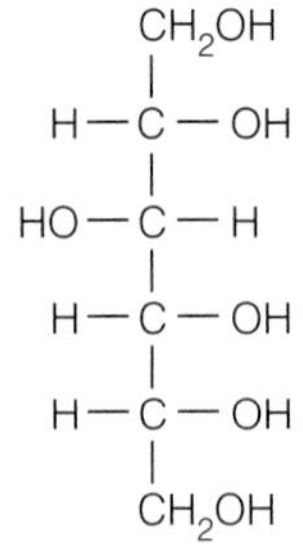 **D-Glucitol (sorbitol)**

Each is named from the corresponding monosaccharide. When sorbitol accumulates in the lens of a diabetic's eye, it can lead to the formation of cataracts.

AMINO SUGARS

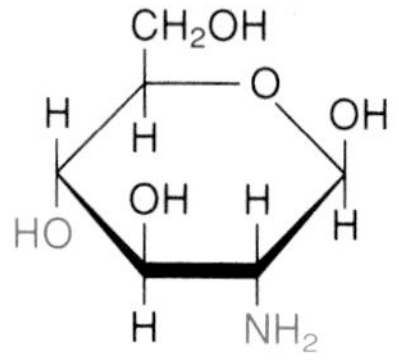

Two amino derivatives of simple sugars are widely distributed in natural polysaccharides: *glucosamine* and *galactosamine,* derived from glucose and galactose, respectively. Further modifications of these amino sugars are common. For example, the following compounds are derived from β-D-glucosamine:

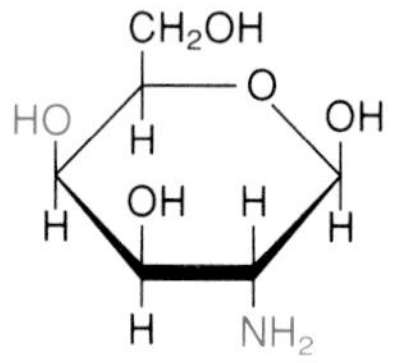

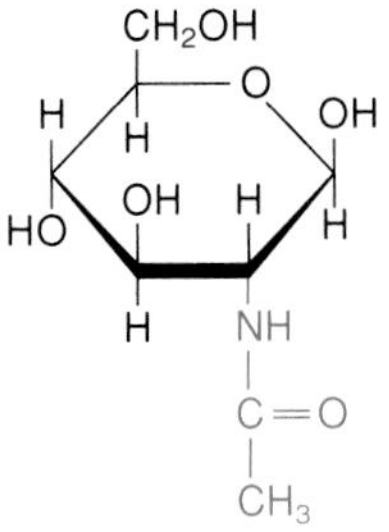

β-D-Glucosamine **β-D-Galactosamine**

β-D-*N*-Acetylglucosamine 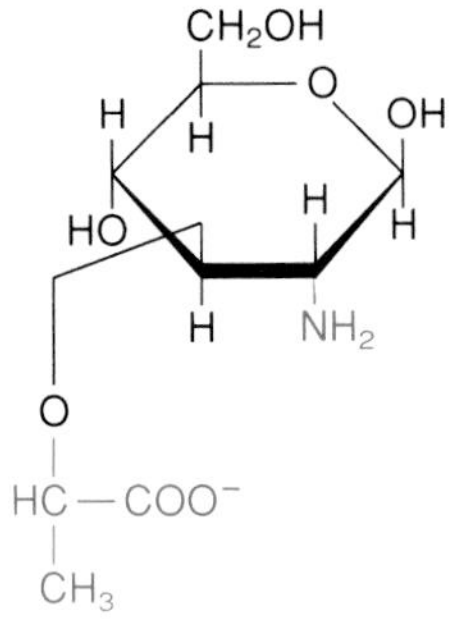 **Muramic acid** 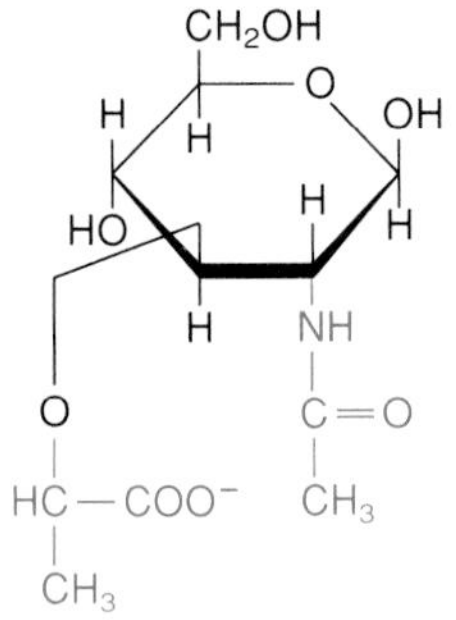 ***N*-Acetylmuramic acid**

These sugar derivatives are important constituents of many natural polysaccharides. Two others we shall encounter are the following:

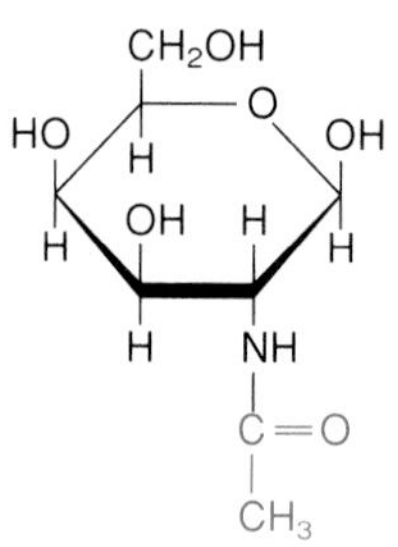

Amino sugars are found in many polysaccharides.

β-D-*N*-Acetylgalactosamine 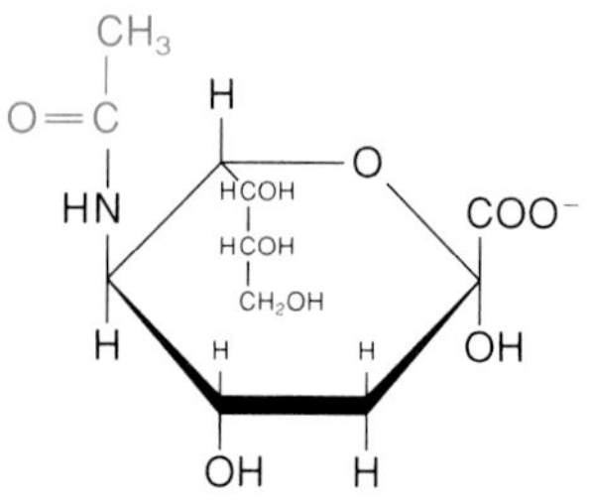 ***N*-Acetylneuraminic acid (sialic acid)**

*Erythritol, although it contains chiral carbons, is not optically active, because it has a plane of symmetry, between C-2 and C-3.

The modified sugars—especially the amino sugars—are most often found as monomer residues in complex oligosaccharides and polysaccharides. To aid in writing the structures of such molecules, it is useful to have a shorthand notation, as is used in describing nucleic acid and protein structure. Therefore, a set of abbreviations has been defined for the simple sugars and their derivatives. A number of the most important ones are listed in Table 9.4.

TABLE **9.4 Abbreviations for some common monosaccharide residues**

Monosaccharides	
Arabinose	Ara
Fructose	Fru
Fucose	Fuc
Galactose	Gal
Glucose	Glc
Lyxose	Lyx
Mannose	Man
Ribose	Rib
Xylose	Xyl
Monosaccharide Derivatives	
Gluconic acid	GlcA
Glucuronic acid	GlcUA
Galactosamine	GalN
Glucosamine	GlcN
N-Acetylgalactosamine	GalNAc
N-Acetylglucosamine	GlcNAc (or NAG)
Muramic acid	Mur
N-Acetylmuramic acid	MurNAc (or NAM)
N-Acetylneuraminic acid (or sialic acid)	NeuNAc (or Sia)

GLYCOSIDES

Elimination of water between the anomeric hydroxyl of a cyclic monosaccharide and the hydroxyl group of another compound yields an **O-glycoside** (the O signifying attachment at a hydroxyl). The ether bond formed is referred to as a **glycosidic bond.** A simple example is the formation of methyl-α-D-glucopyranoside:

α-D-Glucopyranose + CH_3OH $\xrightarrow{\text{Acidic solution}}$ Methyl-α-D-glucopyranoside + H_2O

Unlike the anomers of the sugars themselves, the anomeric glycosides (i.e., methyl-α-D-glucopyranoside and methyl-β-D-glucopyranoside) do not interconvert by mutarotation in the absence of an acid catalyst, a property that makes them useful in the determination of sugar configurations.

Many glycosides are found in plant and animal tissues. Some are very toxic substances, in most cases because they act as inhibitors of enzymes involved in ATP utilization. Two toxic glycosides, ouabain and amygdalin, are shown in Figure 9.15. *Ouabain* inhibits the action of the enzymes that pump Na^+ and K^+ ions across cell membranes to maintain necessary electrolyte balance. It comes from an African shrub and was discovered when it was observed that Somali hunters dipped arrowheads in an extract from the plant. Ouabain now finds use in treatment of some cardiac conditions (see page 342). *Amygdalin* is toxic for a quite different reason. Found in the seeds of bitter almonds, this glycoside yields hydrogen cyanide (HCN) on hydrolysis. It is for this reason that HCN gas is said to have the odor of bitter almonds.

O-glycosides are formed by elimination of a water molecule between a hydroxyl group on a saccharide and a hydroxyl on another compound.

Ouabain

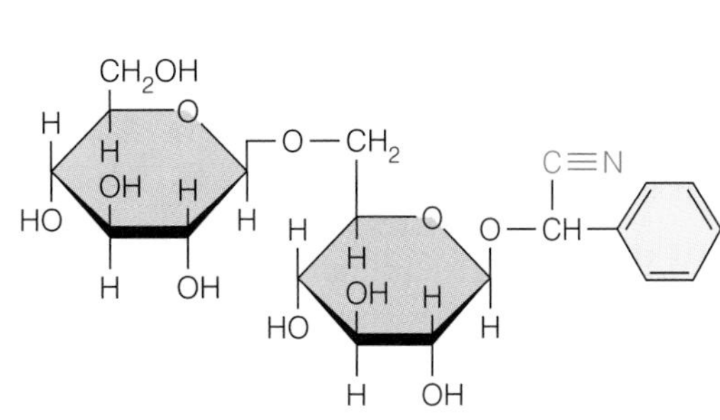
Amygdalin

FIGURE **9.15**

Two naturally occurring glycosides. Ouabain and amygdalin are very toxic glycosides produced by plants.

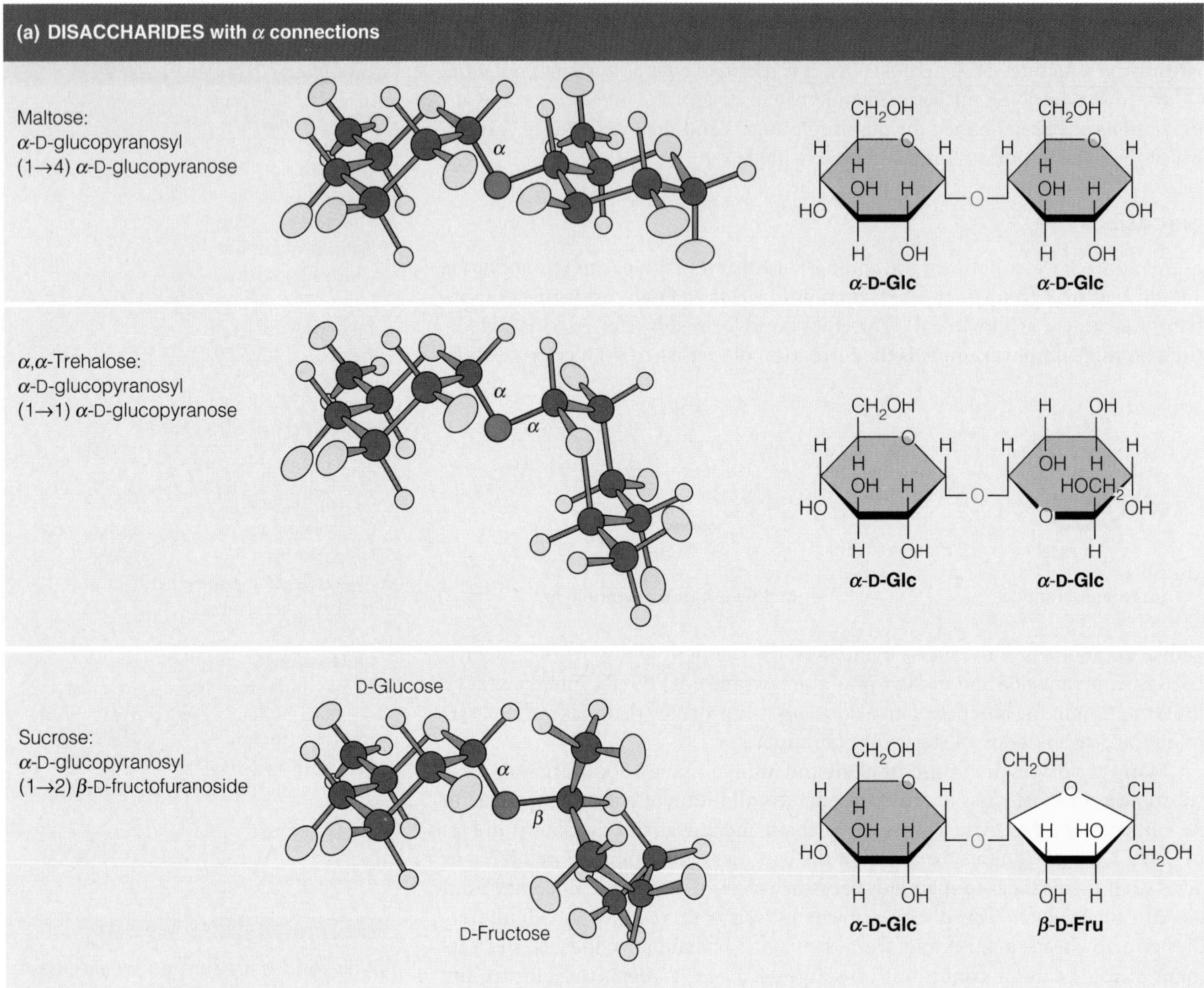

FIGURE **9.16**

Structures of some important disaccharides. Ball-and-stick models are shown on the left, with anomeric oxygens in red. On the right are Haworth projections of the same molecules, with color-coded monomers: blue = glucose, yellow = fructose, teal = galactose. **(a)** Disaccharides linked through the C-1 of the α anomer: maltose, trehalose, and sucrose. **(b)** Disaccharides with β linkage: cellobiose, lactose, and gentiobiose. Note the convention used to draw glycosidic bonds between monomers in disaccharides. The "bent bonds" allow the Haworth projections of the monomers to be drawn in parallel. The "corners" do *not* imply extra carbon atoms, as they often do in organic structure representations.

Oligosaccharides

Just as monosaccharides can form glycosidic bonds with other kinds of hydroxyl-containing compounds, they can do so with one another. Such bonding gives rise to the **oligosaccharides** and **polysaccharides.**

OLIGOSACCHARIDE STRUCTURES

The simplest and biologically most important oligosaccharides are the *disaccharides,* made up of two residues. As Table 9.5 shows, the disaccharides play many roles in living organisms. Some, like *sucrose, lactose,* and *trehalose,* are soluble energy stores in plants and animals. Others, like *maltose* and *cellobiose,* can be regarded primarily as intermediate products in the degradation of much longer polysaccharides. Still others, like *gentiobiose,* are found principally as constituents of more complex, naturally occurring substances. The structures of these important disaccharides are depicted in Figure 9.16.

Distinguishing Features of Different Disaccharides

Four major features distinguish disaccharides from one another:

1. *The two specific sugar monomers involved, and their stereoconfigurations.* The monomers may be of the same kind, as the two D-glucopyranose

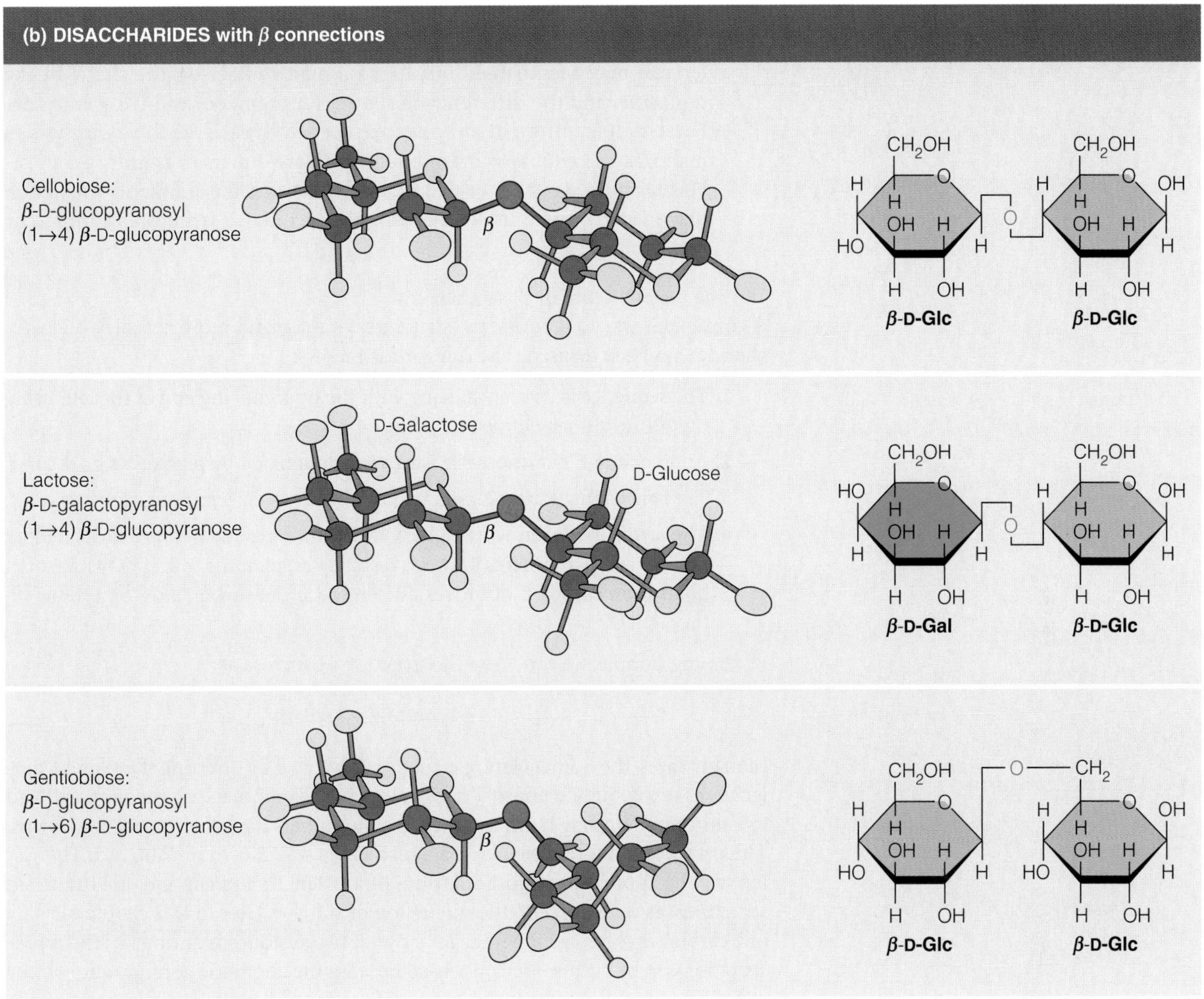

(b) DISACCHARIDES with β connections

Cellobiose:
β-D-glucopyranosyl
(1→4) β-D-glucopyranose

β-D-Glc β-D-Glc

Lactose:
β-D-galactopyranosyl
(1→4) β-D-glucopyranose

D-Galactose D-Glucose

β-D-Gal β-D-Glc

Gentiobiose:
β-D-glucopyranosyl
(1→6) β-D-glucopyranose

β-D-Glc β-D-Glc

residues in maltose, or they may be different, as the D-glucopyranose and D-fructofuranose residues in sucrose.

2. *The carbons involved in the linkage.* Although many possibilities exist, the most common linkages are 1→1 (as in trehalose), 1→2 (as in sucrose), 1→4 (as in lactose, maltose, and cellobiose), and 1→6 (as in gentiobiose). Note that all of these disaccharides involve the anomeric hydroxyl of at least one sugar as a participant in the bond.
3. *The order of the two monomer units, if they are different kinds.* The glycosidic linkage involves the anomeric carbon on one sugar, but in most cases the other is free. Thus, the two ends of the molecule can be distinguished by their chemical reactivity. For example, the glucose residue in lactose, having a free anomeric carbon and thus a potential free aldehyde group, could be oxidized by Fehling's solution; the galactose residue could not be. Lactose is therefore a reducing sugar, and the glucose residue is at its *reducing end.* The other end is called the *nonreducing end.* In sucrose, neither residue has a potential free aldehyde group; both anomeric carbons are involved in the glycosidic bond. Therefore sucrose is a nonreducing sugar.
4. *The anomeric configuration of the hydroxyl group on carbon 1 of each residue.* This feature is especially important for the anomeric carbon(s) involved

in the glycosidic bond. The configuration may be either α (as in the disaccharides shown in Figure 9.16a) or β (as in those in Figure 9.16b). This difference may seem small, but it has a major effect on the shape of the molecule, and the difference in shape is recognized readily by enzymes. For example, different enzymes are needed to catalyze the hydrolysis of maltose and cellobiose, even though both are dimers of D-glucopyranose. Furthermore, we shall see that in polysaccharides the anomeric orientation plays a critical role in determining the secondary structures adopted by these polymers.

Writing the Structure of Disaccharides

A convenient way to describe the structures of these and more complex oligosaccharides has been devised. The rules are as follows:

1. The sequence is written starting with the nonreducing end at the left, using the abbreviations defined in Table 9.4.
2. Anomeric and enantiomeric forms are designated by prefixes (e.g., α-, D-).
3. The ring configuration is indicated by a suffix (*p* for pyranose, *f* for furanose).
4. The atoms between which glycosidic bonds are formed are indicated by numbers in parentheses between residue designations (e.g., (1→4) means a bond from carbon 1 of the residue on the left to carbon 4 of the residue on the right).

As an example, we can write the structure of sucrose as

$$\alpha\text{-D-Glc}p(1\rightarrow2)\text{-}\beta\text{-D-Fru}f$$

In many cases, the nomenclature is further shortened by omitting the D and L designations (except in the unusual cases in which L enantiomers are encountered) and by omitting the *p* and *f* suffixes when the monomers have their usual ring forms. Thus, we would more likely write sucrose in the way shown in Table 9.5. The system can be applied to oligosaccharides of any length and can include branched structures, as we will see in the discussion of starches later in this chapter. If only one carbon involved in the linkage between two residues is anomeric, the representation can be even more condensed because the anomeric configuration at the reducing end will equilibrate in solution. For example, maltose can be represented as Glcα(1→4)Glc.

The list of biologically important oligosaccharides is by no means restricted to dimeric structures. Many trimers, tetramers, and even larger, yet specifically constructed, molecules are known. Examples of these compounds will be encountered later in this chapter, when we examine the oligosaccharides attached to certain proteins and to cell surfaces. Tools of Biochemistry 9A describes techniques used to sequence oligosaccharides.

TABLE 9.5 Occurrence and biochemical roles of some representative disaccharides

Disaccharide	Structure	Natural Occurrence	Physiological Role
Sucrose	Glcα(1→2)Fruβ	Many fruits, seeds, roots, honey	A final product of photosynthesis; used as primary energy source in many organisms
Lactose	Galβ(1→4)Glc	Milk, some plant sources	A major animal energy source
α,α-Trehalose	Glcα(1→1)Glcα	Yeast, other fungi, insect blood	A major circulatory sugar in insects; used for energy
Maltose	Glcα(1→4)Glc	Plants (starch) and animals (glycogen)	The dimer derived from the starch and glycogen polymers
Cellobiose	Glcβ(1→4)Glc	Plants (cellulose)	The dimer of the cellulose polymer
Gentiobiose	Glcβ(1→6)Glc	Some plants (e.g., gentians)	Constituent of plant glycosides and some polysaccharides

STABILITY AND FORMATION OF THE GLYCOSIDIC BOND

Formation of the glycosidic bond between two monomers in an oligosaccharide involves the elimination of a molecule of water. Thus, we might expect the synthesis of lactose to proceed as follows:

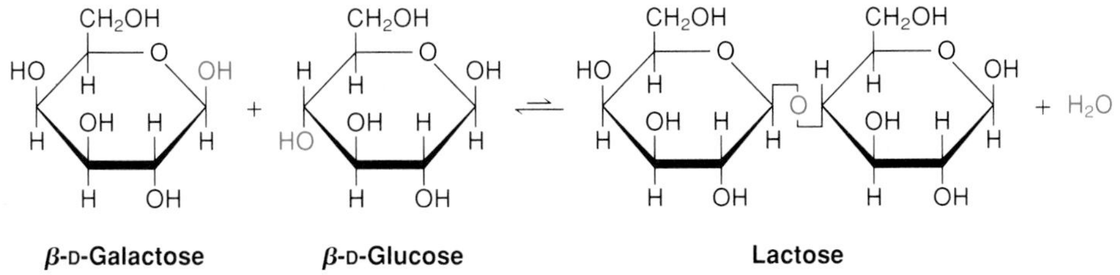

β-D-Galactose + β-D-Glucose ⇌ Lactose + H_2O

This reaction is analogous to the elimination of water between amino acids in the formation of polypeptides or between nucleotides in the formation of nucleic acids. As in those cases, the reaction as written is thermodynamically unfavored. Instead, the hydrolysis of oligosaccharides and polysaccharides is favored under physiological conditions by a standard state free energy change of about −15 kJ/mol, corresponding to an equilibrium constant of about 800 in favor of the hydrolysis products. Nevertheless, like peptides and oligonucleotides, saccharide polymers are sufficiently metastable to persist for long periods unless their hydrolysis is catalyzed by enzymes or acid. So the situation is the same as we have encountered with the other important biopolymers: The breakdown of oligosaccharides and polysaccharides in vivo is controlled by the presence of specific enzymes. Furthermore, synthesis of these sugar polymers never proceeds in living organisms by reactions like the one we have just shown. As in the case of protein and nucleic acid synthesis, activated monomers are required. For example, the formation of lactose in mammary tissue proceeds through the reactions shown in Figure 9.17. A phosphate bond between UDP and galactose is hydrolyzed to drive the coupling. A specific enzyme, *lactose synthase,* is the catalyst. The reaction is further favored by the hydrolysis of the inorganic pyrophosphate produced.

Like the phosphodiester bond in nucleic acid and amide bond in proteins, the glycosidic bond is metastable. Enzymes control its hydrolysis.

Because different disaccharides (and oligosaccharides and polysaccharides) are distinguished by both the kinds of monomers involved and the precise glycosidic linkages between them, the enzymes needed for their synthesis and breakdown must be very specific. For example, hydrolysis of the common nutritional disaccharides maltose, lactose and sucrose, which takes place in cells lining the wall of the small intestine, requires three different and specific enzymes (see page 469). None will substitute for another.

Likewise, the synthesis of disaccharides, as exemplified in Figure 9.17, requires enzymes that are specific not only for the pair of sugars to be joined, but also for the specific glycosidic linkage to be formed between them. Note that each galactose molecule has five different hydroxyl groups, each of which is potentially capable of forming a glycosidic linkage with each of the five hydroxyls on a glucose partner. Even allowing for steric restriction, this still means that many different galactose-glucose disaccharides could conceivably be formed. Yet the enzymes involved in the reactions shown in Figure 9.17 are so specific that only a single product is ever made.

As we shall see in later sections of this chapter, the larger oligosaccharides and the polysaccharides can exhibit extremely complex structures. There is one important way in which oligosaccharide and polysaccharide synthesis differs from synthesis of nucleic acids and proteins. These sugar polymers are never copied from template molecules. Instead, in the formation of oligosaccharides or polysaccharides, a different enzyme is employed to catalyze the addition of each kind of monomer unit. Details of the mechanisms of synthesis of oligosaccharides and polysaccharides are given in Chapter 16. Clearly, a vast array of plant and animal enzymes must be devoted to the synthesis and degradation of saccharide polymers.

β-D-Galactose

ATP → ADP

β-D-Galactose-1-phosphate

UTP → PP → 2 P_i

UDP-Galactose

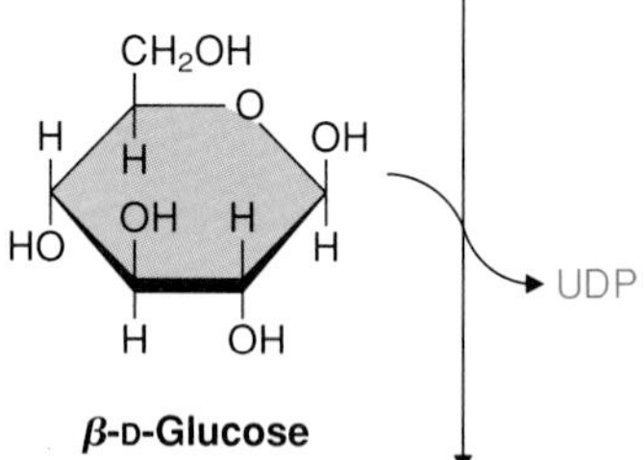

β-D-Glucose

UDP

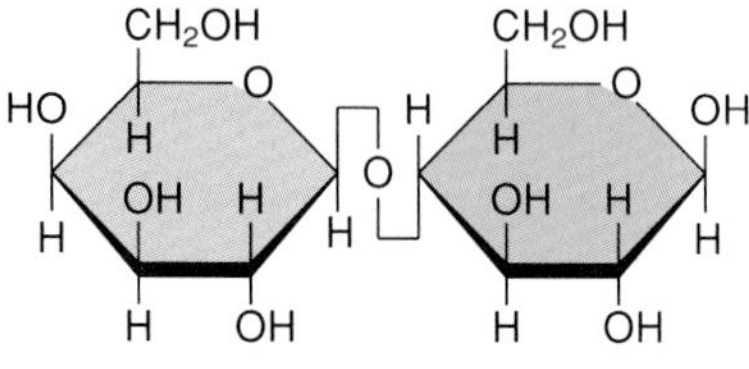

Lactose

FIGURE **9.17**

Formation of lactose in vivo. The reaction shown occurs in the formation of milk in mammary tissue. Galactose is phosphorylated by ATP, then transferred to uridine diphosphate (UDP). UDP-galactose transfers galactose to glucose, with the accompanying cleavage of a phosphate bond. The reaction is catalyzed by the enzyme lactose synthase.

Polysaccharides

Polysaccharides fulfill a wide variety of functions in living organisms. Some, like starch and glycogen (sometimes called animal starch), serve mainly to store sugars for energy in plants and animals. Others, like **cellulose, chitin,** and the polysaccharides of bacterial cell walls, are structural materials analogous to the structural proteins. It is simplest to consider these molecules in terms of their functional categories.

As with polypeptides and polynucleotides, the sequence of monomer residues in a polysaccharide defines its primary structure. Whereas proteins usually have complicated sequences, polysaccharides often have rather simple primary structures. In some cases (e.g., cellulose), the polymer is made from only one kind of monomer residue (β-D-glucose for cellulose); these kinds of polymers are referred to as **homopolysaccharides.** If two or more kinds of residues are involved, the polymer is called a **heteropolysaccharide.** Even those storage and structural polysaccharides that are heteropolymers are rarely complex; usually no more than two kinds of residues are involved. In further contrast to protein and nucleic acid molecules, which are almost always of defined length, polysaccharide chains grow to random lengths.

The functional reasons for these differences are not hard to find. A storage material, such as starch, needs neither to convey information nor to adopt a complicated three-dimensional form. It is simply a bin in which to put away glucose molecules for future use. Many structural polysaccharides (like structural proteins) form extended, regular secondary structures, well suited to the formation of fibers or sheets. Often a regular repetition of some simple monosaccharide or disaccharide motif will serve this function. (Recall, for comparison, the simple and repetitive amino acid sequences of collagen and silk fibroin described in Chapter 6.) The only saccharide polymers in which well-defined and complex sequences are found are some of the oligosaccharides attached to cell surfaces or those attached to specific glycoproteins. Because these oligomers serve to identify cells or molecules, they must convey information. This function requires precisely defined "words" in the polysaccharide language, just as nucleic acid sequences spell out information in their own language.

STORAGE POLYSACCHARIDES

The principal storage polysaccharides are **amylose** and **amylopectin,** which together constitute starch in plants, and **glycogen,** which is stored in animal and microbial cells. Both starch and glycogen are stored in granules within cells (Figure 9.18). Starch is found in almost every kind of plant cell, but grain seeds, tubers, and unripe fruits are especially rich in this material. Glycogen is deposited in the liver, which acts as a central energy storage organ in many animals. Glycogen is also abundant in muscle tissue, where it is more immediately available for energy release.

Amylose, amylopectin, and glycogen are all polymers of α-D-glucopyranose. They are homopolysaccharides of the class called **glucans,** the polymers of glucose. The three polymers differ only in the kinds of linkages between glucose residues. Amylose is a linear polymer, involving exclusively $\alpha(1\to4)$ links between adjacent glucose residues. Amylopectin (Figure 9.19) and glycogen are both branched polymers, because they contain, in addition to the $\alpha(1\to4)$ links, some $\alpha(1\to6)$ links as well. The branches in glycogen are somewhat more frequent and shorter than those in amylopectin, and glycogen is usually of higher molecular

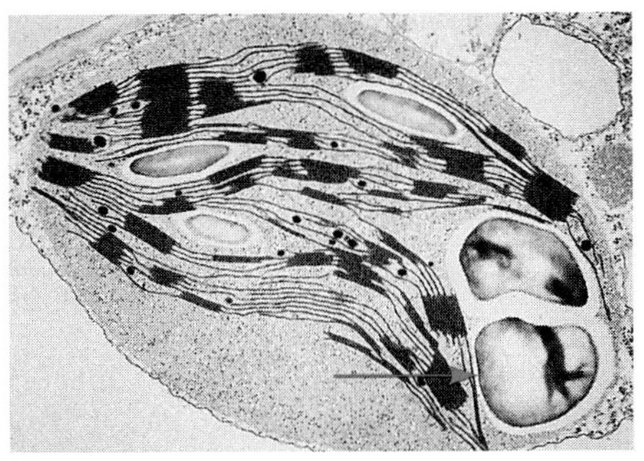

(a) Chloroplast granules

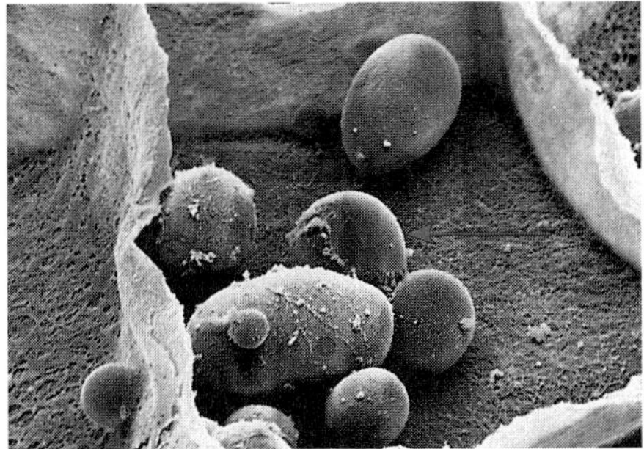

(b) Tuber cell granules

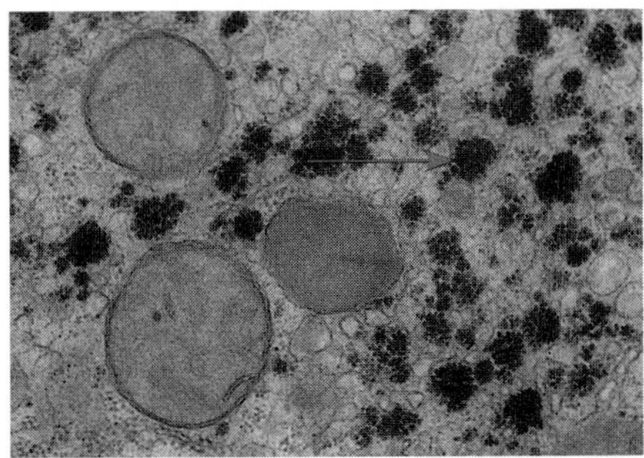

(c) Liver granules

FIGURE **9.18**

Storage of starch and glycogen in granules. In each case a representative granule is indicated by an arrow. **(a)** Starch granules in a plant leaf chloroplast. **(b)** Starch granules in potato tuber cells. **(c)** Glycogen granules in liver.

(a) From Science Source, © Biophoto Associates/Photo Researchers, Inc.; **(b)** Courtesy of Dr. L. M. Beidler; **(c)** Micrograph courtesy of Don Fawcett, M.D.

FIGURE **9.19**

Amylopectin, a branched glucan. **(a)** The primary structure of amylopectin. Nonreducing ends (N) and reducing ends (R) are indicated. **(b)** Detailed structure of a branch point.

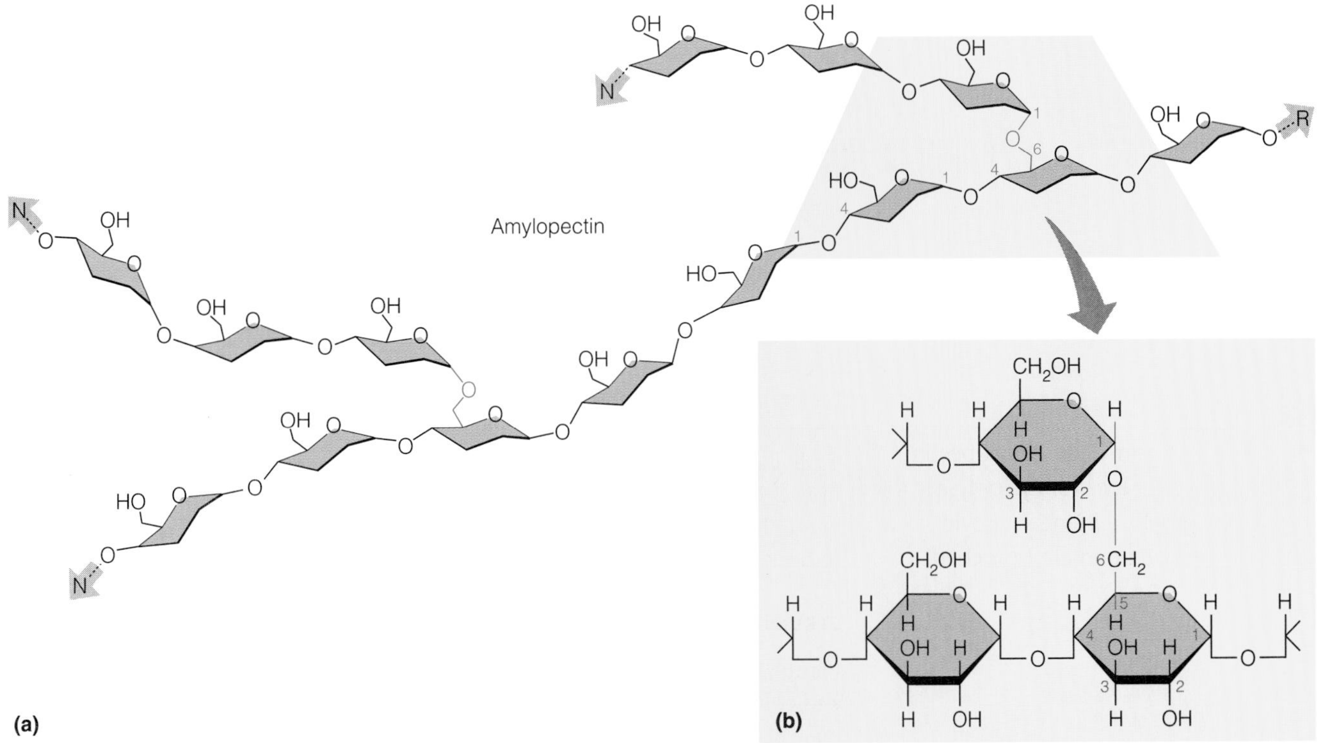

Amylopectin

(a)

(b)

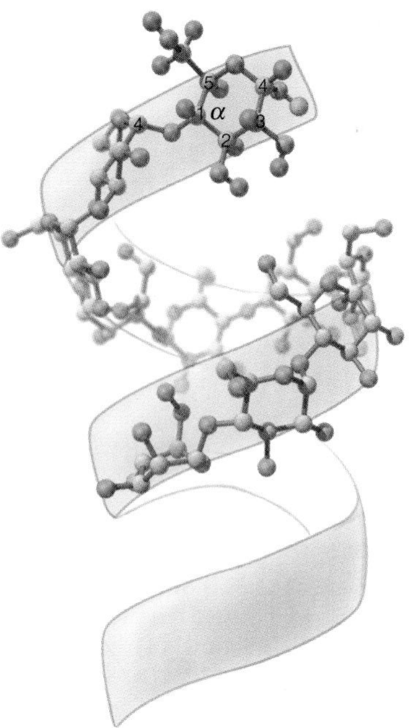

FIGURE **9.20**

The secondary structure of amylose. The orientation of successive glucose residues favors helix generation. Note the large core. Hydrogen bonds stabilize the helix.

weight, but in most respects the structures of these two polysaccharides are very similar.

The regular and simple primary structure of amylose allows a regular secondary structure for this molecule. As with polynucleotides and polypeptides, the details of this structure have come from x-ray diffraction studies. In fact, amylose was the first biopolymer whose structure was elucidated by this method. Because of the $\alpha(1\rightarrow4)$ link, each residue is angled with respect to the preceding residue, favoring a regular helical conformation (Figure 9.20). The branched nature of amylopectin and glycogen inhibits the formation of helices, because the helix requires 6 residues for each turn; there is a branch point about every 10–20 residues in amylopectin and about every 8 in glycogen.

The storage polysaccharides are admirably designed to serve their function. Glucose and even maltose are small, rapidly diffusing molecules, which are difficult to store. Were such small molecules present in large quantities in a cell, they would give rise to a very large cell osmotic pressure, which would be deleterious in most cases. Therefore, most cells build the glucose into long polymers, so that large quantities can be stored in a manner that prevents its diffusion and loss. Whenever glucose is needed, it can be obtained by selective degradation of the polymers by specific enzymes. These processes are discussed in detail in Chapter 13, but one aspect should be mentioned now. Most of the enzymes employed attack the chains at their nonreducing ends, releasing one glucose residue at a time. Such "end-nibbling" (as opposed to internal cutting) prevents the continual breakup of the long polymers, which would lead to their complete solubilization. The branched structure of both amylopectin and glycogen is such that each molecule has *many* nonreducing ends that can be attached simultaneously (see Figure 9.19), allowing rapid mobilization of glucose when it is needed. On the other hand, the linear chain of amylose with its single nonreducing end is used mainly for long-term storage of glucose.

The starches—amylose, amylopectin, and glycogen—are storage polysaccharides. Amylose is linear; amylopectin and glycogen are branched.

STRUCTURAL POLYSACCHARIDES

Plants do not seem to synthesize or use fibrous structural proteins (like keratin and collagen) but instead rely entirely on special polysaccharides. Animals use

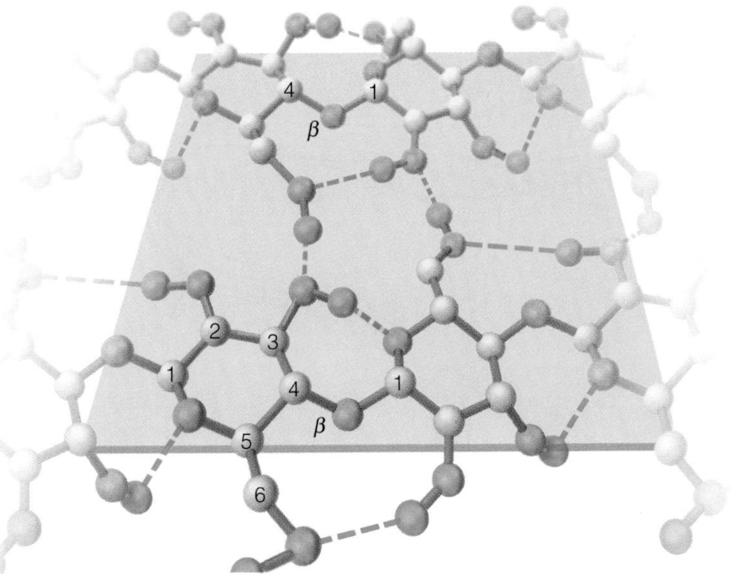

FIGURE **9.21**

Cellulose structure. The $\beta(1\rightarrow4)$ linkages of cellulose generate a planar structure. The parallel cellulose chains are linked together by a network of hydrogen bonds. Hydrogens involved in such bonds are shown in blue. For clarity, all of the hydrogens are shown in only one glucose residue (with carbons numbered).

both kinds of materials. Because each structural use requires different properties, a great variety of structural polysaccharides exists. We shall begin by considering those from plants.

Cellulose

The major polysaccharide in woody and fibrous plants (like trees and grasses), cellulose is the most abundant single polymer in the biosphere. Like amylose, cellulose is a linear polymer of D-glucose (and hence is also a glucan), but in cellulose the sugar residues are connected by $\beta(1\rightarrow4)$ linkages (Figure 9.21). This seemingly small difference from starch (e.g., amylose) has remarkable structural consequences. Cellulose can exist as fully extended chains, with each glucose residue flipped by 180° with respect to its neighbor in the chain. In this extended form, the chains can form ribbons that pack side by side with a network of hydrogen bonds within and between them. This arrangement is reminiscent of the β-sheet structure in silk fibroin, and as in fibroin, the fibrils of cellulose have great mechanical strength but limited extensibility.

The same small difference between cellulose and starch has another important consequence: Animal enzymes that are able to catalyze the cleavage of the $\alpha(1\rightarrow4)$ link in starch cannot cleave cellulose. For this reason, humans, even if starving, are unable to utilize the enormous quantities of glucose all around them in the form of cellulose. Ruminants such as cows can digest cellulose only because their digestive tracts contain symbiotic bacteria that produce the necessary **cellulases.** Termites manage to eat woody substances in a somewhat more complicated fashion—their guts harbor protozoans capable of cellulose digestion but their salivary glands also produce a cellulase. Many fungi also produce such enzymes, which is why some mushrooms can live on wood as a carbon source.

It should not be supposed that the fibrous parts of plants are made exclusively from cellulose. A variety of other polysaccharides are present in plant cell walls. These include the **xylans,** which are polymers with $\beta(1\rightarrow4)$-linked

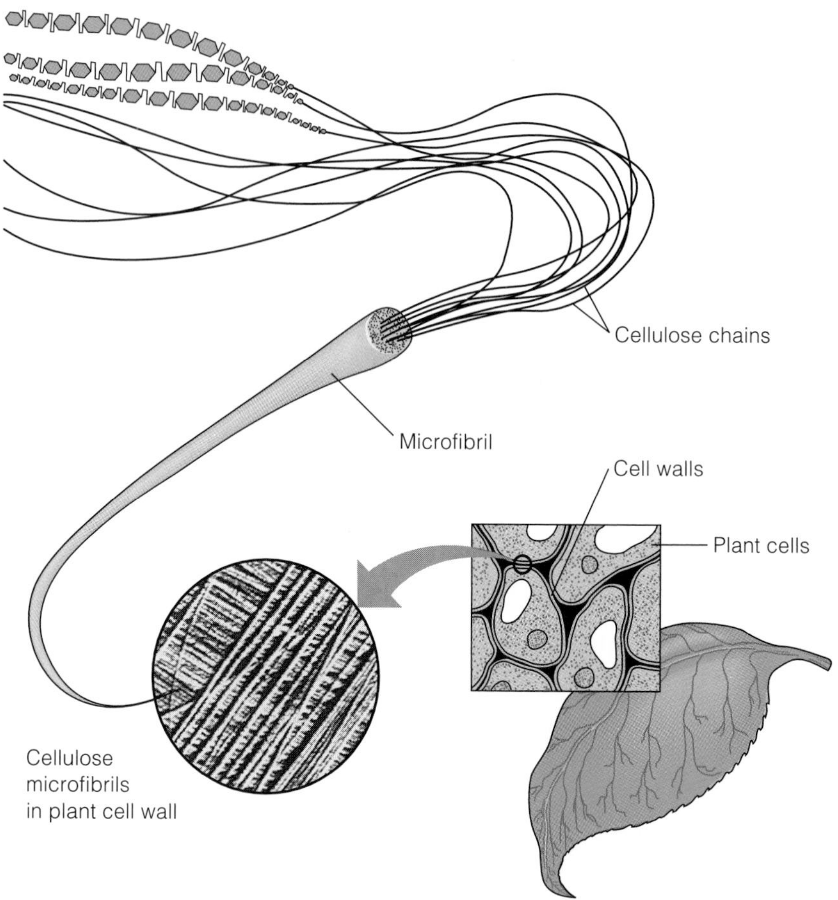

FIGURE **9.22**

Organization of plant cell walls. Microfibrils of cellulose are embedded in a matrix of hemicellulose. Note that the fibers are laid down in a crosshatched pattern to give strength in all directions.

From W. M. Becker, L. J. Kleinsmith, and J. Hardin, *The World of the Cell*, 4th ed. (San Francisco, CA: Addison Wesley Longman, 2000) © Addison Wesley Longman, Inc.; photo courtesy of E. Frei and R. D. Preston.

D-xylopyranose, often with substituent groups attached; the **glucomannans;** and many other polymers. Often these polysaccharides are grouped together under the term **hemicellulose.**

$$\cdots\beta\text{-D-Xyl}p(1\rightarrow4)[\beta\text{-D-Xyl}p(1\rightarrow4)]_7\text{-}\beta\text{-D-Xyl}p(1\rightarrow4)\text{-}\beta\text{-D-Xyl}p(1\rightarrow4)\cdots$$

(Acetyl at C-2 or C-3 on the second residue; 4-O-Me-α-D-Glc$p(1\rightarrow2)$ on the fourth residue)

A typical xylan structure

$$\cdots\beta\text{-D-Glc}p(1\rightarrow4)\text{-}\beta\text{-D-Man}p(1\rightarrow4)\text{-}\beta\text{-D-Man}p(1\rightarrow4)\text{-}\beta\text{-D-Man}p(1\rightarrow4)\cdots$$

(β-D-Gal$p(1\rightarrow6)$ on the third residue; Acetyl at C-2 or C-3 on the fourth residue)

A typical glucomannan structure

The cell wall of a plant is a complex structure, made up of several layers. Microfibrils of cellulose are laid down in a crosshatched pattern (Figure 9.22) and impregnated with a matrix of the other polysaccharides and some proteins. The same principle is used when glass fibers are embedded in a tough resin to produce strong, durable sheets of fiberglass.

Cellulose is not confined exclusively to the plant kingdom. The marine invertebrates called *tunicates* contain considerable quantities of cellulose in their hard outer mantle. There are even reports of small amounts of cellulose in human connective tissue. However, as a structural material, cellulose seems to have been largely passed over in animal evolution. In the fungi, extensive use is made of other glucans, with $\beta(1\rightarrow3)$ or $\beta(1\rightarrow6)$ linkages between glucose residues, as structural polysaccharides.

Chitin

A homopolymer of *N*-acetyl-β-D-glucosamine, chitin has a structure basically similar to that of cellulose, except that the hydroxyl on carbon 2 of each residue is replaced by an acetylated amino group.

Cellulose and chitin are examples of structural polysaccharides. Unlike starches, which have $\alpha(1\rightarrow4)$ links, these fibrous polymers have $\beta(1\rightarrow4)$ linkages.

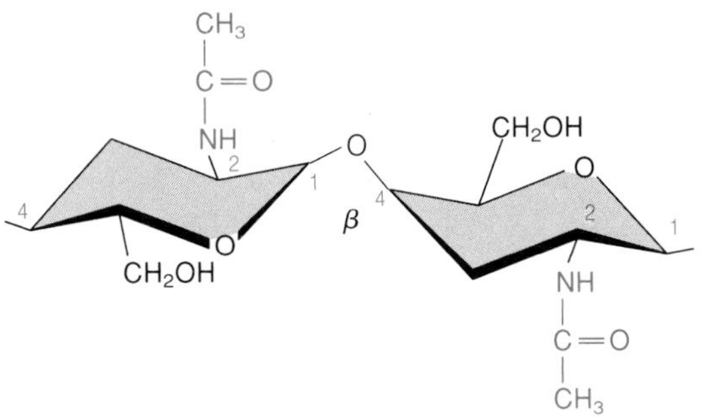

CH_3 C=O NH 2 O 1 β 4 4 CH_2OH O CH_2OH O 2 1 NH C=O CH_3

Chitin

Chitin is widely distributed among the kingdoms of organisms. It is a minor constituent in most fungi and some algae, where it often substitutes for cellulose or other glucans. In dividing yeast cells, chitin is found in the septum that forms between the separating cells. The best known role of chitin, however, is in invertebrate animals; it constitutes a major structural material in the exoskeletons of many arthropods and mollusks. In many of these exoskeletons, chitin forms a matrix on which mineralization takes place, much as collagen acts as a matrix for mineral deposition in vertebrate bones. The evolutionary implications are interesting. As animals evolved to the size that made rigid body parts essential, quite different paths were taken. The ancestors of the vertebrates developed a mineral skeleton on a collagen matrix. Annelids such as earthworms also use collagen, but in a segmented exoskeleton. The arthropods and mollusks also developed exoskeletons, but theirs were built on chitin—a carbohydrate rather than a protein matrix.

GLYCOSAMINOGLYCANS

One group of polysaccharides is of major structural importance in vertebrate animals—the **glycosaminoglycans,** formerly called *mucopolysaccharides.* Important examples are the *chondroitin sulfates* and *keratan sulfates* of connective tissue, the *dermatan sulfates* of skin, and *hyaluronic acid.* All are polymers of repeating disaccharide units, in which one of the sugars is either *N*-acetylgalactosamine or *N*-acetylglucosamine or one of their derivatives. All are acidic, through the presence of either sulfate or carboxylate groups. Representative structures of glycosaminoglycans are shown in Figure 9.23.

Glycosaminoglycans are negatively charged heteropolysaccharides that serve a number of structural functions in animals.

The Proteoglycan Complex

A major function of the glycosaminoglycans is the formation of a matrix to hold together the protein components of skin and connective tissue. An example is given in Figure 9.24, which illustrates the protein–carbohydrate, or **proteoglycan,** complex in cartilage. The filamentous structure is built on a single long hyaluronic acid molecule, to which extended core proteins are attached noncovalently. The core proteins, in turn, have chondroitin sulfate and keratan sulfate chains covalently bound to them through serine side chains. In cartilage, this kind of structure binds collagen (see Chapter 6) and helps hold the collagen fibers in a tight, strong network. The binding apparently involves electrostatic interactions between the sulfate and/or carboxylate groups of the proteoglycan complex and the basic side chains in collagen.

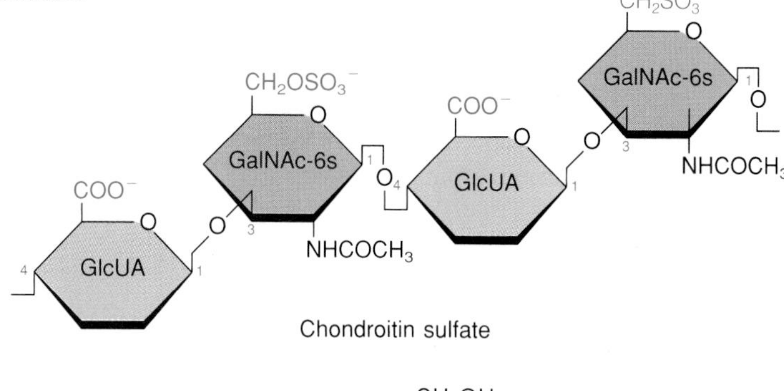

FIGURE **9.23**

Repeating structures of some glycosaminoglycans. In each case, the repeating unit is a disaccharide, of which two are shown for each structure. Abbreviations of residues (−6s means sulfonated at carbon 6) are in Table 9.4. To simplify the figure, hydrogens and nonreacted hydroxyls are not shown.

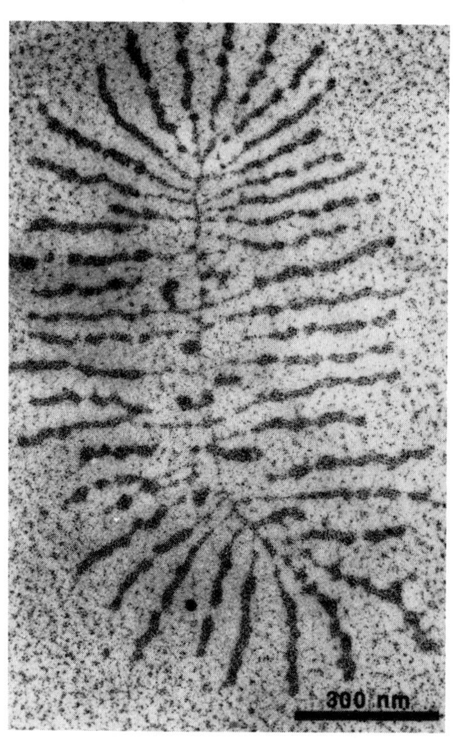

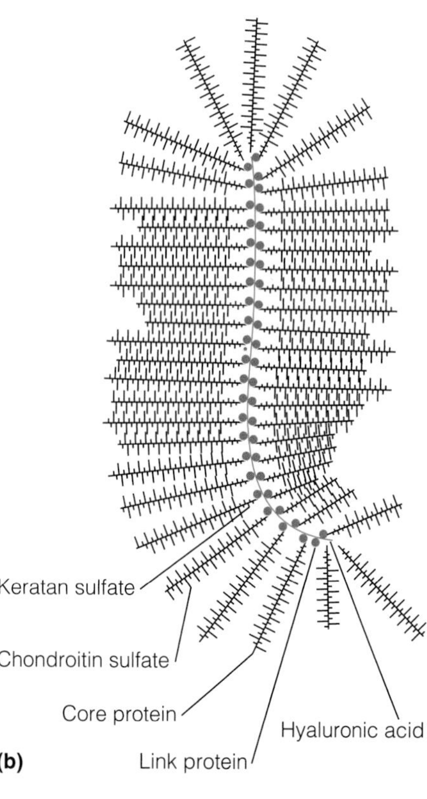

FIGURE **9.24**

Proteoglycan structure in bovine cartilage. (a) An electron micrograph of a proteoglycan aggregate. **(b)** A schematic drawing of the same structure. Keratan sulfate and chondroitin sulfate are covalently linked to extended core protein molecules. The core proteins are noncovalently attached to a long hyaluronic acid molecule with the aid of a link protein.

(a) Courtesy of J. A. Buckwalter and L. Rosenberg from *Collagen Relat. Res.* (1983) 3:489–504, Amsterdam: North Holland, 1975.

The proteoglycan complexes of connective tissue constitute one of the few examples in which the element silicon enters into biology. Some of the carbohydrate chains are cross-linked by bridges of the type

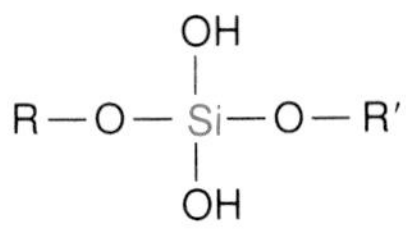

$$R-O-\underset{\displaystyle OH}{\overset{\displaystyle OH}{Si}}-O-R'$$

where R and R′ are sugar monomers of adjacent chains. There is about one silicon atom for every 100 sugar monomers.

Nonstructural Roles of Glycosaminoglycans

Hyaluronic acid has other functions in the body besides being a structural component. The polymer is very soluble in water and is present in synovial fluid of joints and in the vitreous humor of the eye. It appears to act as a viscosity-increasing agent or lubricating agent in these fluids.

Another highly sulfated glycosaminoglycan is **heparin.** One fragment of its complex chain is shown below. Heparin appears to be a natural anticoagulant and is found in many body tissues. It binds strongly to a blood protein, antiprothrombin III, and the complex inhibits enzymes of the blood clotting process (see Chapter 11). Therefore, heparin is used medicinally to inhibit clotting in blood vessels.

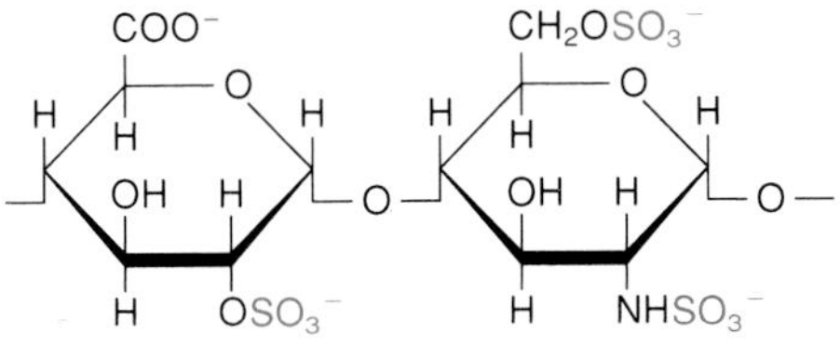

COO^- $CH_2OSO_3^-$ H H H H H H OH H O OH H O H OSO_3^- H $NHSO_3^-$

Heparin

The glycosaminoglycans are interesting examples of how sugar residues can be modified to provide polymers with a wide variety of properties and functions.

BACTERIAL CELL WALL POLYSACCHARIDES

In Chapter 1 we noted that bacteria and most other unicellular organisms possess a *cell wall.* The nature of this cell wall is the basis for categorizing bacteria into two major classes: those that retain the Gram stain (a dye–iodine complex), which are called *Gram-positive* bacteria, and those that do not, which are termed *Gram-negative* (Figure 9.25). Gram-positive bacteria have a cell wall with a cross-linked, multilayered polysaccharide–peptide complex called **peptidoglycan** at the surface, outside the lipid cell membrane (Figure 9.25a). Gram-negative bacterial cell walls also contain peptidoglycan, but it is single-layered and covered by an outer lipid membrane layer (Figure 9.25b). This difference allows the Gram stain to be washed from Gram-negative bacteria.

The cell walls of many bacteria are constructed of peptidoglycans, composite polymers of polysaccharides and oligopeptides.

The chemical structure of the peptidoglycan of a Gram-positive bacterium is shown in Figure 9.26. Long polysaccharide chains, which are strictly alternating copolymers of *N*-acetylglucosamine (NAG) and *N*-acetylmuramic acid (NAM), are cross-linked through short peptides (see Figure 9.25a). These peptides have unusual structures. Attached to the lactic acid moiety of the *N*-acetylmuramic acid is a tetrapeptide with the sequence

(L-Ala)–(D-Glu)–(L-Lys)–(D-Ala)

This peptide is unusual in two respects: It contains some D-amino acids, and the glutamic acid residue is linked into the chain through its γ-carboxyl instead of the

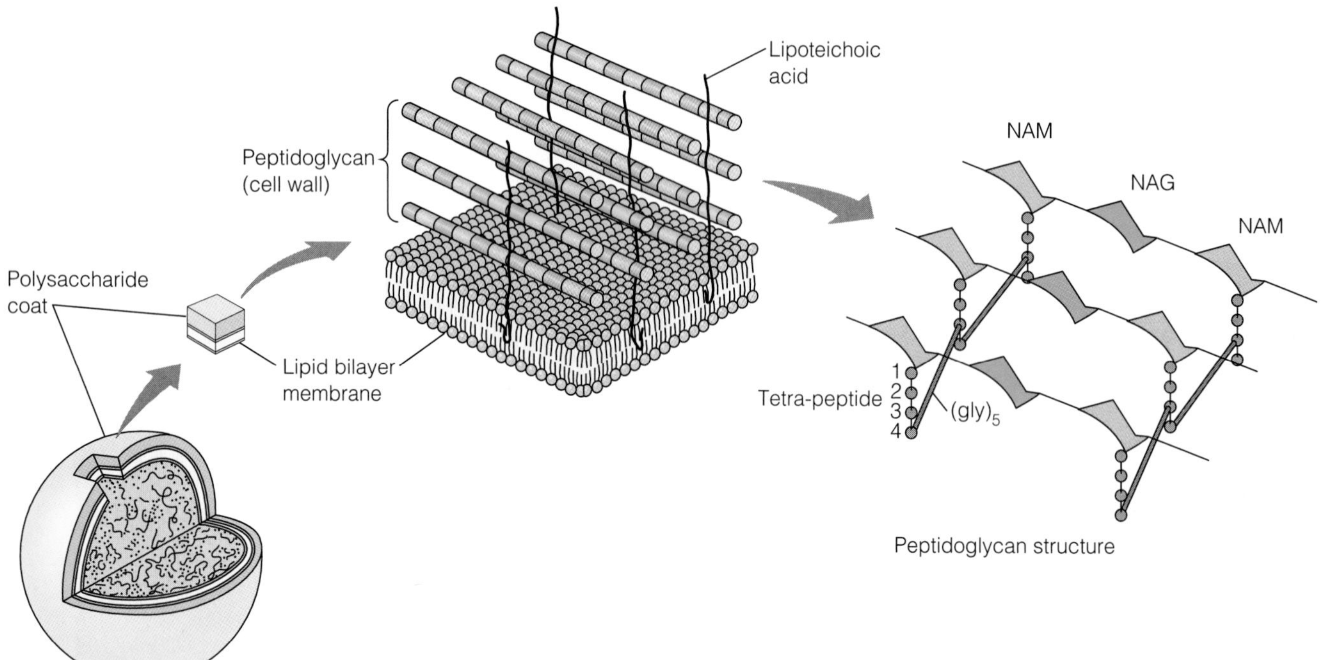

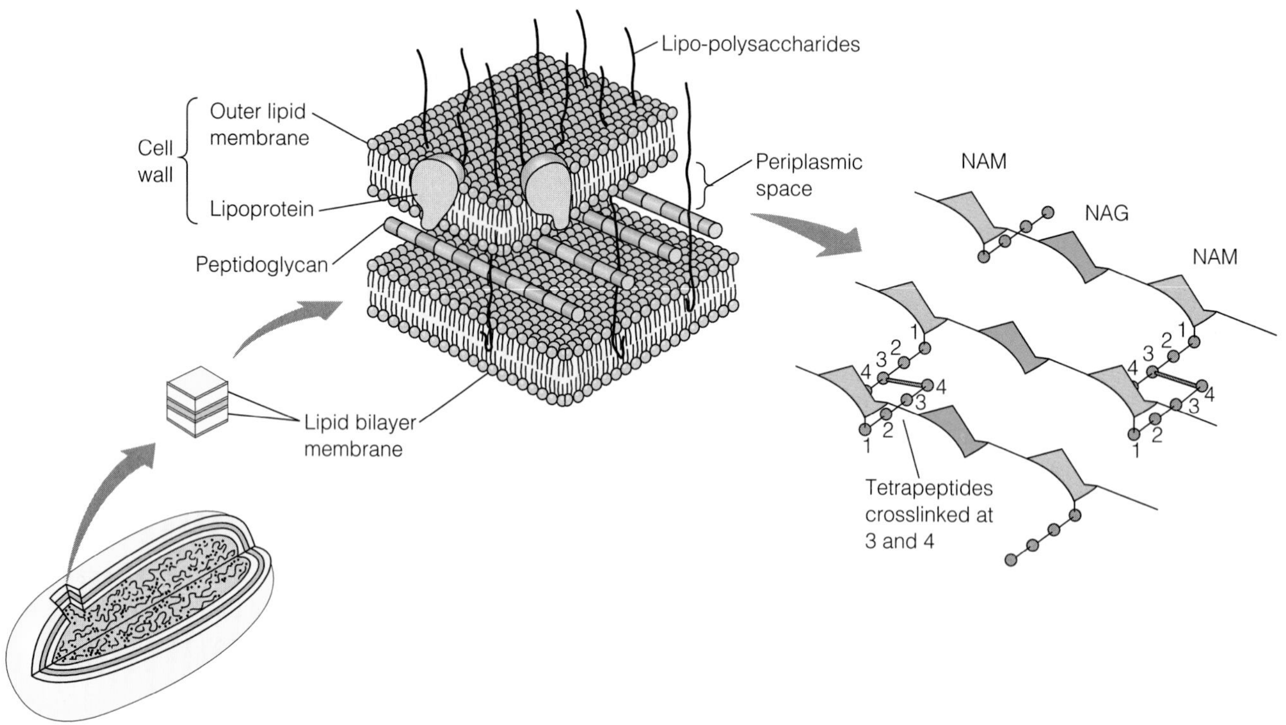

FIGURE **9.25**

Bacterial cell walls. These schematic drawings show the structures of the two types of bacterial cell wall. **(a)** The cell wall of a representative Gram-positive bacterium, *Staphylococcus aureus,* consists of a thick peptidoglycan layer made up of polysaccharide chains and short peptides. The peptides are linked by glycine pentapeptides. **(b)** The cell wall of a representative Gram-negative bacterium, *Escherichia coli,* has a thin peptidoglycan layer and an outer lipid membrane. The cross-links here are between tetrapeptides attached to the NAM residues in adjacent chains.

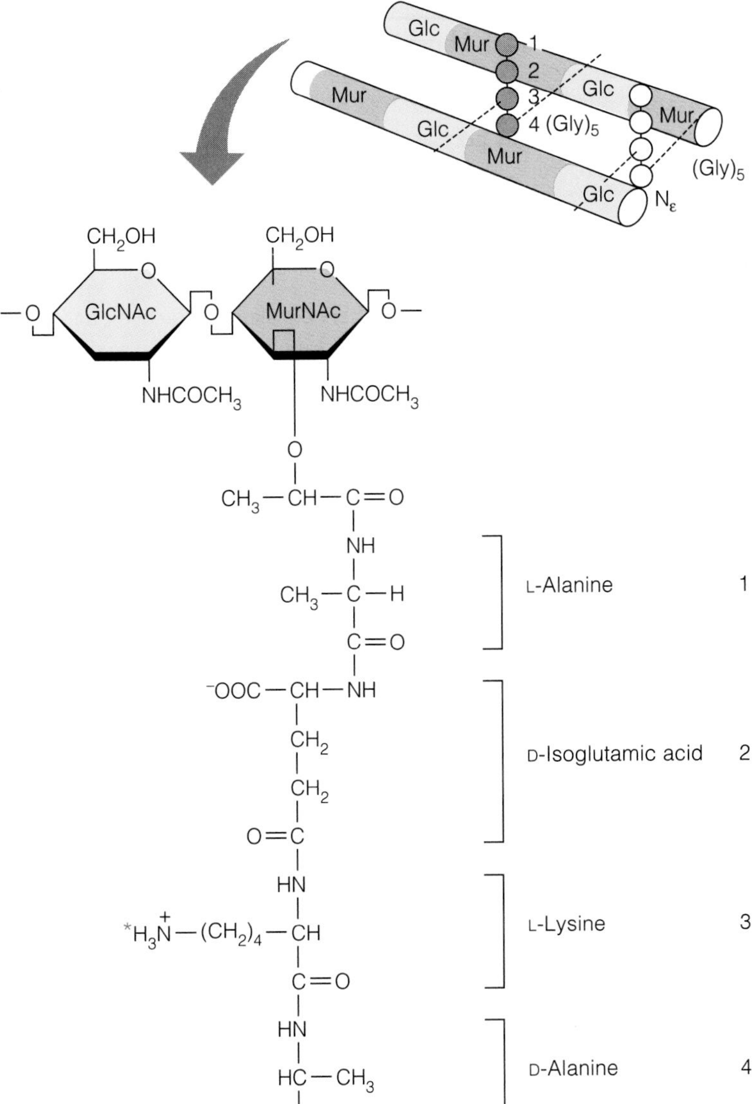

FIGURE **9.26**

The peptidoglycan layer of Gram-positive bacteria. Cross-links between the peptides are formed by pentaglycine chains between the ε-amino group of the lysine (*) on one chain and the C-terminal carboxyl group of the alanine (**) on an adjacent chain.

usual α-carboxyl linkage. To the ε-amino group of each lysine residue is attached a glycine pentapeptide, which is bonded at its other end to the terminal D-Ala residue of an adjacent chain. The result is the formation of a covalently cross-linked structure that envelops the bacterial cell. The entire cell wall can be regarded as a single enormous molecule made up of multiple layers of cross-linked peptidoglycan strands. In addition to the components mentioned above, elongated lipid–oligosaccharide complexes called **lipotechoic acids** (Figure 9.27) protrude from the membrane through the peptidoglycan wall. The cell wall protects bacteria from lysis when they are in the blood of host animals.

In the Gram-negative bacteria, the peptidoglycan layer is much thinner. Although the same basic polysaccharide structure is present, the peptide chains and their linkage are somewhat different (see Figure 9.25b).

Clearly, assembly of a structure as complex as the bacterial cell wall requires a battery of enzymes and reactions. A number of antibiotics (for example, penicillin) inhibit bacterial growth by interfering with the formation of the peptidoglycan layer. It is appropriate to defer discussion of these antibiotics until we consider the carbohydrate synthetic pathways in Chapter 16. However, we should note that one class of naturally occurring antibiotic substances acts not by interfering with cell wall synthesis but by attacking the peptidoglycan layer itself. These

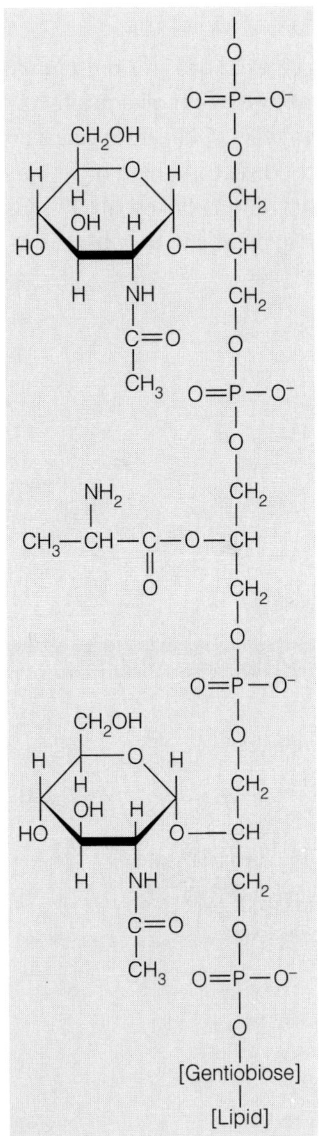

FIGURE **9.27**

The structure of a lipotechoic acid. D-Alanyl and NAG groups are arranged irregularly on the chain, which is anchored in the membrane by lipid.

substances are the **lysozymes,** enzymes with wide distribution—they are found in bacteriophages, egg white, and human tears, for example. In egg and tears they help maintain asepsis; in bacteriophage they help the phage to exist in infected bacteria. Lysozymes catalyze the hydrolysis of the glycosidic links between GlcNAc and MurNAc residues in the polysaccharide. Thus, they dissolve the cell wall, resulting in lysis and bacterial death.

Glycoproteins

More than half of all eukaryotic proteins carry covalently attached oligosaccharide or polysaccharide chains. There is an astonishing variety of these modified proteins, which are known as **glycoproteins,** and they serve many different functions. A few representative examples are listed in Table 9.6.

N-LINKED AND O-LINKED GLYCOPROTEINS

Oligosaccharides and proteins can be linked to form glycoproteins in two ways: O-linked glycans are attached via threonine or serine hydroxyls, and N-linked glycans via asparagine amino groups.

The saccharide chains (**glycans**) can be linked to proteins in two major ways (Figure 9.28). **N-linked glycans** are attached, usually through *N*-acetylglucosamine, or sometimes through *N*-acetylgalactosamine, to the side chain amino group in an asparagine residue. A common sequence surrounding the asparagine is –Asn–X–Ser/Thr–, where X may be any amino acid residue. **O-linked glycans** are usually attached by an O-glycosidic bond between *N*-acetylgalactosamine and the hydroxyl group of a threonine or serine residue, although in a few cases—collagen, for example—hydroxylysine or hydroxyproline is employed. Of the examples given in Table 9.6, the first two are O-linked, the remainder N-linked.

N-Linked Glycans

Careful study of many glycoproteins has revealed an enormous variety of N-linked oligosaccharide side chains, often exhibiting a complex branched structure. However, a common motif is often seen. The following structure often serves as a foundation for further elaboration:

$$\begin{matrix}\text{Man}\alpha(1\rightarrow6) \searrow & \\ & \text{Man}\beta(1\rightarrow4)\text{GlcNAc}\beta(1\rightarrow4)\text{GlcNAc}\beta(1\rightarrow\text{N})\text{Asn} \\ \text{Man}\alpha(1\rightarrow3) \nearrow & \end{matrix}$$

TABLE 9.6 Carbohydrate moieties of some glycoproteins

Glycoprotein	Oligosaccharide[a] and Attachment Site (in red)	No. of Chains in Proteins	Function of Protein
Fish antifreeze protein	Gal–GalNAc–Thr	From 4 to 50 in different proteins	Lowering of body fluid freezing point
Sheep submaxillary mucin	Sia–GalNAc–Ser (or Thr)	Many	Lubrication
Ribonuclease B	$(\text{Man})_6$–GlcNAc–GlcNAc–Asn	1	Enzyme
Hen ovalbumin	Man–Man↘ Man–GlcNAc–GlcNAc–Asn; Man–Man↗ (with Man branch on Man) (only one of many variants)	1	Storage protein in egg white
Human IgG	Sia–Gal–GlcNAc–Man↘ Man–GlcNAc–GlcNAc(–Fuc)–Asn; Sia–Gal–GlcNAc–Man↗ (many other variants in antibodies)	2 or more	Antibody molecule

[a]For compactness, the anomeric forms and linkages have been eliminated. For details, see R. C. Hughes, *Glycoproteins* (London: Chapman and Hall, 1983).

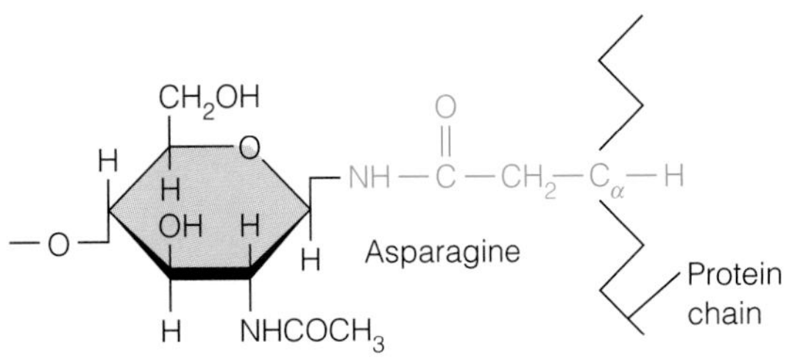

(a) *N*-Acetylglucosamine

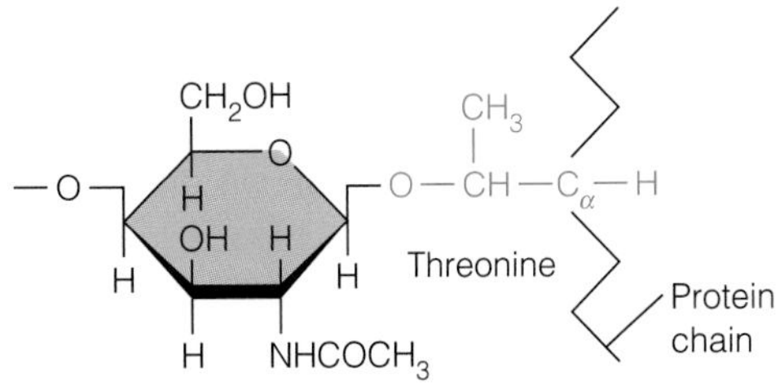

(b) *N*-Acetylgalactosamine

FIGURE **9.28**

Glycoprotein bonds. Two ways of linking oligosaccharide chains (glycans) with proteins to form glycoproteins are shown. **(a)** Bonding of an N-linked glycan. The bond is between *N*-acetylglucosamine and the amide side chain of an asparagine residue. In N-linked chains, *N*-acetylglucosamine is almost always the first sugar residue. **(b)** Bonding of an O-linked glycan. An O-glycosidic bond links *N*-acetylgalactosamine to the hydroxyl of threonine or serine.

This motif can be seen, for example, in the glycan moieties of ovalbumin and the immunoglobulins. The structure of the oligosaccharide found attached to a human immunoglobulin G (IgG) is shown in Table 9.6. The residue denoted Fuc is α-L-fucose, a residue often attached near the protein connection of N-linked glycans. The immunoglobulins represent an important example of the informational function of the glycan chains on glycoproteins. Recall from Chapter 7 that every immunoglobulin has carbohydrate attached to the constant domain of each heavy chain. The different types of immunoglobulins must be recognized, both for proper tissue distribution and for interaction with phagocytic cells, which will destroy the antigen–immunoglobin complex. A part, at least, of this recognition is based on differences in the oligosaccharide chains.

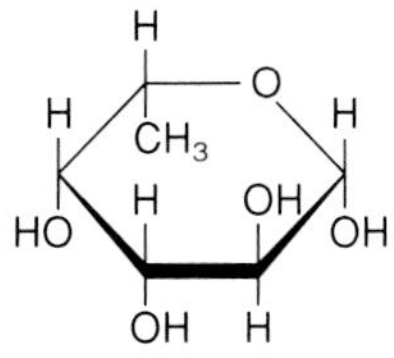

α-L-Fucose

A very important further use of N-linked oligosaccharides is in intracellular targeting in eukaryotic organisms. In Chapters 16 and 28 we see how proteins destined for certain organelles or for excretion from the cell are marked specifically by oligosaccharides during posttranslational processing. This marking ensures that they arrive at their proper destinations.

O-Linked Glycans

Many proteins carry O-linked oligosaccharides that serve a variety of functions. Antarctic fish contain a glycoprotein that serves as an "antifreeze," preventing the freezing of body fluids even in extremely cold water. The **mucins,** glycoproteins found extensively in salivary secretions, contain many short O-linked glycans. The highly extended and highly hydrated mucins increase the viscosity of the fluids in which they are dissolved. Some O-linked glycans also appear to function in intracellular targeting and molecular and cellular identification. An example of import to all of us is found in the blood group antigens.

BLOOD GROUP ANTIGENS

A most important group of oligosaccharides is the **blood group antigens.** On some cells these antigens are attached as O-linked glycans to membrane proteins. Alternatively, the oligosaccharide may be linked to a lipid molecule to form a **glycolipid** (see Chapter 10). The lipid portion of the molecule helps anchor the antigen to the outside surface of erythrocyte membranes. It is these oligosaccharides

The blood group substances are a set of antigenic oligosaccharides attached to the surfaces of red cells.

Sia
Gal
GalNAc
R
NHCOCH$_3$
Fuc
Type O

GalNAc
Sia
Gal
GalNAc
R
NHCOCH$_3$
NHCOCH$_3$
Fuc
Type A

Gal
Sia
Gal
GalNAc
R
NHCOCH$_3$
Fuc
Type B

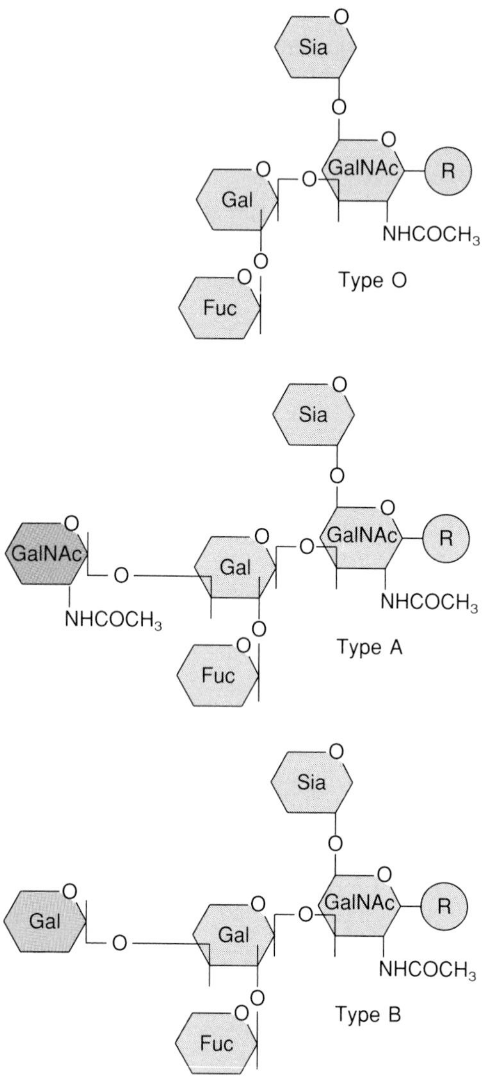

FIGURE **9.29**

The ABO blood group antigens. The O oligosaccharide (top) does not elicit antibodies in most humans. The A and B antigens are formed by addition of GalNAc or Gal, respectively, to the O oligosaccharide. Each of the A and B antigens can elicit a specific antibody. In this figure, R can represent either a protein molecule or a lipid molecule.

TABLE **9.7 Transfusion relationships among ABO blood types**

Person Has Blood Type:	Makes Antibodies Against:	Can Safely Receive Blood from:	Can Safely Donate Blood to:
O	A, B	O	O, A, B, AB
A	B	O, A	A, AB
B	A	O, B	B, AB
AB	None	O, A, B, AB[a]	AB

[a]In principle, this relationship is true. However, ABs are never given donations from other types, because the donor's antibodies could react with the recipient's antigens.

that determine the blood group types in humans. Their presence in a blood sample is detected by blood typing—determining whether antibodies to a particular antigen cause the red cells of that blood sample to clump, or agglutinate. Although the system consisting of blood types A, B, AB, and O is probably most familiar to you, it is just one of 14 genetically characterized blood group systems, with more than 100 different blood group antigens. These substances are also present in many cells and tissues other than blood, but we often focus on blood because of the widespread use of typing in establishing familial relationships and selecting blood for transfusion.

For simplicity, we will take the ABO system as an example. Figure 9.29 depicts cell surface oligosaccharides corresponding to each of these blood types. Almost all humans can produce the type O saccharide, but addition of either galactose (to make type B) or *N*-acetylgalactosamine (to make type A) requires an additional enzyme (see Chapter 16 for details of this synthesis). Some individuals possess one of these enzymes, some possess the other, and a few are heterozygous and can produce both. The heterozygous individuals have type AB blood, with both A and B oligosaccharides present on cell surfaces.

Humans can produce antibodies against the A and B oligosaccharides, but the O type are nonantigenic. Normally, a person does not produce antibodies against his or her own antigen but does produce them against the other antigen type. Thus, an individual with type A blood carries antibodies directed against the B polysaccharide. If he or she accepts blood from a type B donor, these antibodies will cause clumping and precipitation of the donated blood cells. Nor can a type B individual safely accept type A blood. People with type O blood normally have antibodies against both A and B and thus can receive from neither. Those with AB type, since they carry both A and B antigens themselves, have antibodies against neither.

In donating blood, an inverse relationship holds. Those with type O blood, which carries no antigenic determinants, can safely donate to any other person—they are the "universal donors." Type AB individuals can donate *only* to other ABs; a person of any other type will carry antibodies to A, or B, or both. These relationships are summarized in Table 9.7.

Oligosaccharides as Cell Markers

Biologists are beginning to realize that such molecules as the blood group antigens represent only a special case of a much more general phenomenon—cell marking by oligosaccharides. In a multicellular organism, it is essential that different kinds of cells be marked on their surfaces so that they can interact properly with other cells and molecules and so that an organism can recognize its own cells as immunologically distinct from foreign cells. In accord with this view is the growing appreciation that the surfaces of many cells are nearly covered with poly-

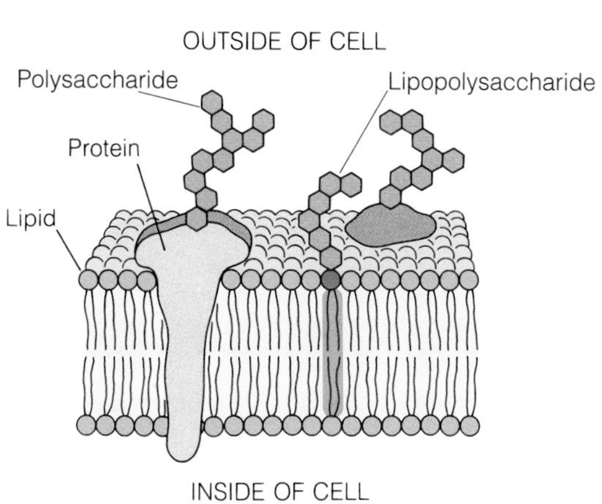

(a) Cell surface oligosaccharides

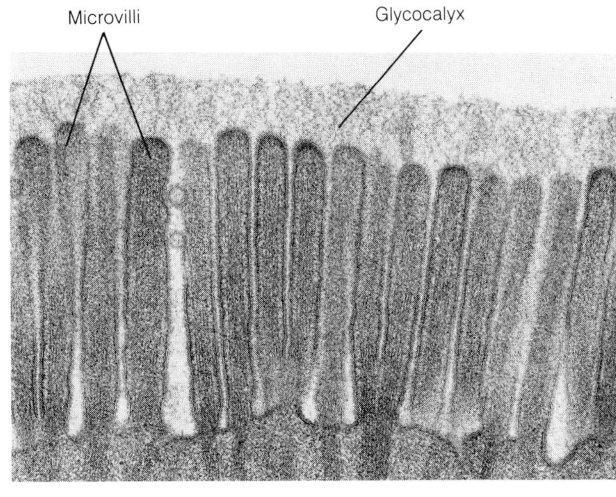

(b) Polysaccharide glycocalyx

FIGURE **9.30**

Cell surface recognition factors. **(a)** Schematic view of a lipid membrane. Oligosaccharides are attached to the outer surface, through either membrane-embedded proteins or special lipid molecules. **(b)** Electron micrograph of the surface of an intestinal epithelial cell. The cellular projections, called microvilli, are covered on their outer surface by a layer of branched polysaccharide chains attached to proteins in the cell membrane. This carbohydrate layer, called the glycocalyx, is found on many animal cell surfaces.

(b) Courtesy of Dr. Susumu Ito.

saccharides, which are attached to either proteins or lipids in the cell membrane (Figure 9.30a). Some animal cells have an extremely thick coating of polysaccharides called a **glycocalyx** (literally "sugar coat"); Figure 9.30b shows the glycocalyx of an intestinal cell. Glycocalyx oligosaccharides interact with other substances: with bacteria in the intestine, and with collagen of the intercellular matrix in some other tissues.

Many cells carry a complex layer of polysaccharides—the glycocalyx—on their surfaces.

For oligosaccharides or polysaccharides to serve as recognition signals, there must be proteins that bind to them specifically. One such class is the immunoglobulins. Another, very diverse group of saccharide-binding proteins is the **lectins.** The lectins were first recognized in plant tissues, where they appear to play defensive roles and to aid in adhering nitrogen-fixing bacteria to roots. It is now known that lectins are widely distributed and play a great variety of roles in animals as well. For example, lectins seem to be involved in interactions between cells and proteins of the intercellular matrix, such as collagen, and help to maintain tissue and organ structure. Lectins in the walls of intestinal bacteria help to bond the bacteria to the glycocalyx of the intestinal epithelium. The fact that cell surface polysaccharides are important in determining cell–cell interactions (including both adhesion and avoidance) has much medical significance. For example, it is now known that the polysaccharides on the surfaces of many cancer cells are abnormal. This may account in part for the loss in tissue specificity that such cells commonly exhibit.

Why do oligosaccharides so often play the role of cellular markers? We do not know, but certain possibilities suggest themselves. First, oligosaccharides can present a remarkable variety of structures in relatively short chains. The multiple choices of monomers (including modified sugars), linkages, and branching patterns allow a vast but specific vocabulary. Second, oligosaccharides are especially potent antigens, which means that specific antibodies can be elicited swiftly against them (see Chapter 7). Whether this interaction is the result of some intrinsic property of sugar molecules or of the antibody molecules is unclear. It is possible that antibodies evolved as a defense against bacteria, which have polysaccharide-rich walls, and thus have always favored saccharides as targets.

It seems likely that carbohydrate biochemistry, which has long been considered a rather prosaic part of the field, is about to enter an exciting new phase. As biochemists turn their attention more and more to the relationships between cells and tissues that make for an integrated organism, they are finding that many of these relationships are mediated by carbohydrates.

SUMMARY

Carbohydrates (or saccharides) are compounds with the stoichiometric formula $(CH_2O)_n$ or derivatives of such compounds. They are a major product of photosynthesis, and their oxidation provides a major energy source for both plants and animals. Because most monosaccharides have multiple chiral centers, these saccharides exist as enantiomeric pairs (D and L mirror images) of multiple diastereomers. Monosaccharides may be either aldoses or ketoses. Most of the important ones are D-aldoses. Those containing five or more carbons exist mainly in the form of rings of five (furanose) or six (pyranose) atoms, resulting from internal hemiacetal formation. Such rings exist as α or β anomers and exhibit multiple conformations (e.g., boat and chair) as well.

Important derivatives of monosaccharides include phosphate esters, acids and lactones, alditols, amino sugars, and glycosides. Phosphate esters are important as metabolic intermediates; glycosides represent a large class of compounds formed by elimination of water between a sugar and another hydroxy compound. Oligosaccharides and polysaccharides are formed by making glycosidic links between monosaccharides; the glycosidic linkage is metastable, so enzymes control its hydrolysis in vivo. Polysaccharides serve multiple kinds of functions—sugar storage (starch and glycogen), structural roles (cellulose, xylans, chitin, glycosaminoglycans, cell wall polysaccharides), and identification tags (the oligosaccharides and polysaccharides on glycoproteins and cell surfaces). The blood group antigens are important examples of the identification function.

REFERENCES

General

Binkley, R. W. (1988) *Modern Carbohydrate Chemistry.* Marcel Dekker, New York. A comprehensive contemporary survey.

Pigman, W., and D. Horton, eds. (1972) *The Carbohydrates,* 2nd ed. Academic Press, New York. A valuable sourcebook for details of carbohydrate structure and chemistry.

Sugar Conformations

Barker, R., and A. S. Serianni (1986) Carbohydrates in solution: Studies with stable isotopes. *Acc. Chem. Res.* 19:307–313. A brief review of ^{13}C NMR work that establishes conformations.

Carver, J. P. (1991) Experimental structural determination of oligosaccharides. *Curr. Opin. Struct. Biol.* 1:716–720.

Storage Saccharides

Goodwin, T. W., and E. I. Mercer (1983) *Introduction to Plant Biochemistry.* Pergamon, Oxford.

Meléndez-Hevia, E., T. G. Waddell, and E. D. Shelton (1993) Organization of molecular design in the evolution of metabolism: The glycogen molecule. *Biochem. J.* 295:477–483.

Bacterial and Plant Cell Walls

Albersheim, P. (1975) The walls of growing plants. *Sci. Am.* 232(4):80–95.

Kjellen, L., and U. Lindahl (1991) Proteoglycans: Structure and interactions. *Annu. Rev. Biochem.* 60:443–475.

Loomis, W. D., and R.W. Durst (1991) Boron and cell walls. *Curr. Top. Plant Biochem. Physiol.* 10:149–178. An unusual interaction of boron with polysaccharides, involved in plant cell growth.

Schockman, G. D., and J. F. Barnett (1983) Structure, function, and assembly of cell walls of Gram-positive bacteria. *Annu. Rev. Microbiol.* 37:501–527.

Glycoproteins

Hughes, R. C. (1983) *Glycoproteins.* Chapman and Hall, London. A small book, packed with information on glycoprotein structure and biosynthesis.

Rudd, P. M., and R. A. Dwek (1997) Glycosylation: Heterogeneity and the 3D structure of proteins. *Crit. Rev. Biochem. Mol. Biol.* 32:1–100.

van den Steen, P., P. M. Rudd, R. A. Dwek, and G. Opdenakker (1998) Concepts and principles of O-linked glycosylation. *Crit. Rev. Biochem. Mol. Biol.* 33:151–208.

Oligosaccharides and Cell Recognition

Barondes, S. H. (1984) Soluble lectins: A new class of extracellular proteins. *Science* 223:1259–1264.

Labat-Robert, J., R. Timpl, and R. Ladiglas, eds. (1986) *Structural Glycoproteins in Cell–Matrix Interaction.* Karger, New York.

PROBLEMS

1. Draw Haworth projections for the following:

   (a) 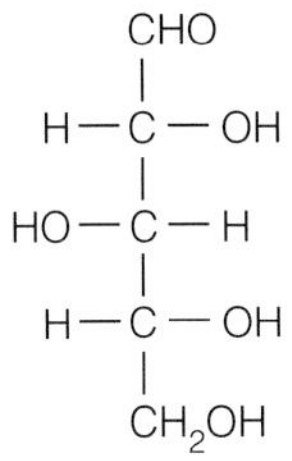 CHO–(H–C–OH)–(HO–C–H)–(H–C–OH)–CH_2OH in α-furanose form. Name the sugar.

   (b) The L isomer of (a)

   (c) α-D-GlcNAc

   (d) α-D-Fructofuranose

2. α-D-Galactopyranose rotates the plane of polarized light, but the product of its reduction with sodium borohydride (galactitol) does not. Explain the difference.

3. Provide an explanation for the fact that α-D-mannose is more stable than β-D-mannose, whereas the opposite is true for glucose.

*4. Using data in Table 9.2, calculate the standard state free energy change for the conversion of D-glucose from the α to the β anomer at 40°C. Can you provide a qualitative explanation for this? Do you think that ΔG arises primarily from an enthalpy or entropy contribution?

5. The disaccharide α,β-trehalose differs from the α,α structure in Figure 9.16a by having an $(\alpha 1 \rightarrow \beta 1)$ linkage. Draw its structure.

6. A *reducing* sugar will undergo the Fehling reaction (see page 291), which requires a (potential) free aldehyde group. Which of the disaccharides shown in Figure 9.16 are reducing, and which are nonreducing?

7. *Dextrans* are polysaccharides produced by certain species of bacteria. They are glucans, with primarily $\alpha(1 \rightarrow 6)$ linkages and with frequent $\alpha(1 \rightarrow 3)$ branching. Draw a Haworth projection of a portion of a dextran, including one $(1 \rightarrow 3)$ branch point.

8. What is the natural polysaccharide whose repeating structure can be symbolized by GlcUA$\beta(1 \rightarrow 3)$GlcNAc, with these units connected by $\beta(1 \rightarrow 4)$ links?

*9. Decide whether the structures shown are *R* or *S* in the absolute system.

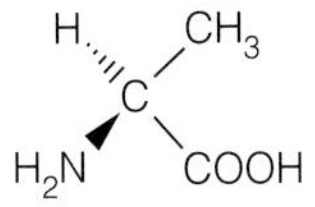

   (a) C with H, CH_3, H_2N, COOH

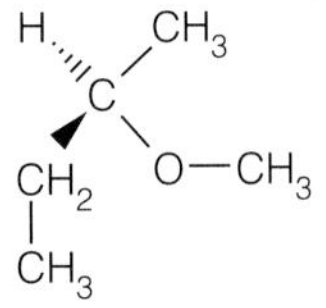

   (b) C with H, CH_3, CH_2–CH_3, O–CH_3

*10. The reagent periodate (IO_4^-) oxidatively cleaves the carbon–carbon bonds between two carbons carrying hydroxyl groups. Explain how periodate oxidation might be used to distinguish between methyl glycosides of glucose in the pyranose and furanose forms.

11. Draw (using Haworth projections) the fragments of xylan and glucomannan structure shown on page 302.

12. A research student is attempting to sequence an oligosaccharide attached to an orosomucoid (see Figure 9A.1) from a mutant cell line. She finds from analysis the presence of Sia, Gal, and GlcNAc. Neuraminidase cleavage succeeds, but β-galactosidase is without effect. Suggest an explanation.

13. One or more of the compounds shown below will satisfy each of the following statements. Not all compounds may be used; some may be used twice. Put the letter(s) in the blank.

    (a) Found in chitin. ______

    (b) An L-saccharide. ______

    (c) The first residue attached to asparagine in N-linked glycans. ______

    (d) A uronic acid. ______

    (e) A ketose. ______

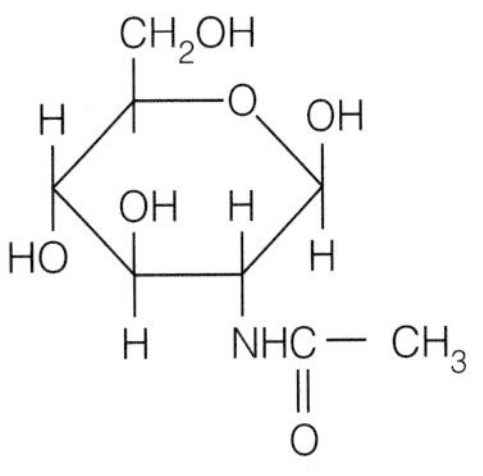

    (a) ring: CH_2OH, H, O, OH, OH, H, HO, H, H, NHC(=O)–CH_3

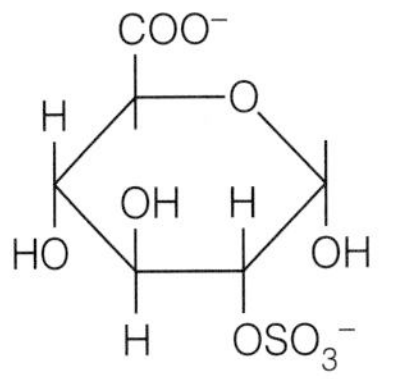

    (b) ring: COO^-, H, O, OH, H, HO, OH, H, OSO_3^-

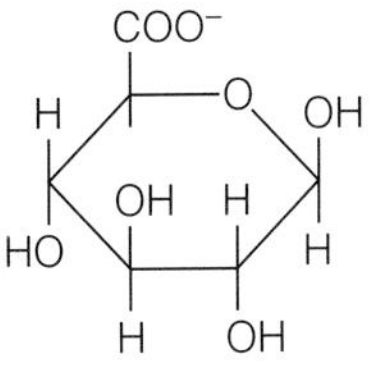

    (c) ring: COO^-, H, O, OH, OH, H, HO, H, H, OH

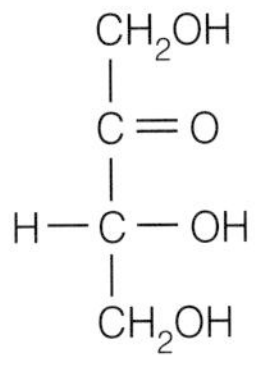

    (d) CH_2OH–C=O–(H–C–OH)–CH_2OH

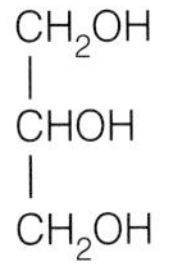

    (e) CH_2OH–CHOH–CH_2OH

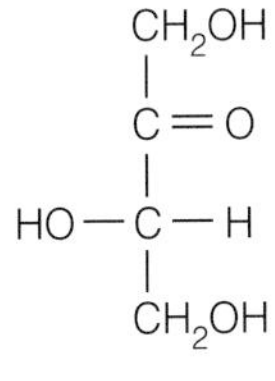

    (f) CH_2OH–C=O–(HO–C–H)–CH_2OH

TOOLS OF BIOCHEMISTRY 9A

Sequencing Oligosaccharides

Determining oligosaccharide sequences presents problems similar to, but more difficult than, those encountered in protein sequencing. Because of the many types of monomers that may be encountered and the variety of linkages between them, no single method like the Edman degradation of polypeptides has been devised.

TABLE **9A.1**

SOME SPECIFIC GLYCOSIDASES USED IN SEQUENCING OLIGOSACCHARIDES

Enzyme Name	Source	Specificity
Exoglycosidases		
Neuraminidase	*Streptococcus pneumoniae*	Siaα(2→3 or 6)Gal or Siaα(2→6)GlcNAc
β-Galactosidase	*Streptococcus pneumoniae*	Galβ(1→4)GlcNAc
α-Fucosidase	*Clostridium perfringens*	Fucα(1→2)Gal
Endoglycosidases		
Endo-β-galactosidase	*Escherichia freundii*	··· GlcNAcβ(1→3)Galβ(1→4)Glc(GlcNAc) ···
Almond emulsion	Bitter almond seeds	Cleaves bond to Asn in many N-linked oligosaccharides

The first step in any sequence analysis is, as in polypeptide analysis, the determination of composition. The oligomer is hydrolyzed in acidic solution, yielding a mixture of monosaccharides. At present, these monosaccharides are almost always separated, identified, and quantified by gas–liquid chromatography.

Determination of the sequence itself is much more difficult. In the past, chemical methods were extensively used, but they have been largely supplanted by enzymatic cleavage of the oligomer followed by sophisticated methods for identification of the fragments. Researchers are now familiar with a large number of enzymes (**glycosidases**) that catalyze the cleavage of glycosidic bonds between sugar moieties. Some of these enzymes are very specific in their action. They may be divided into two groups: *exoglycosidases,* which remove the terminal residue from an oligosaccharide chain, and *endoglycosidases,* which catalyze cleavage within the chain. A few examples of these enzymes are listed in Table 9A.1.

A simple example of the application of specific glycosidases to sequence determination is shown in Figure 9A.1. The oligosaccharide shown is a portion of one of several oligosaccharides attached to the blood serum protein *orosomucoid.* The residue at the nonreducing end of the chain can be removed by neuraminidase. According to Table 9A.1, the terminal residue must be sialic acid, attached to either Gal or GlcNAc. Sialic acid is found, and subsequent release of Gal by *Streptococcus* β-galactosidase indicates that the next residue is Gal, attached by a 1→4 linkage to GlcNAc. The latter residue is confirmed by its release by β-*N*-acetylglucosaminidase. Using modern techniques for analysis, sequences may be determined in this manner at the picomole level.

Although such procedures can often provide definitive information, they are laborious and not always effective. In recent years there has been remarkable development of high-resolution NMR and mass spectrometry as techniques for identification of complex oligosaccharides. The use of these methods is described in the reviews by Jones and by Wait (see References, following) but is too technical to be described here.

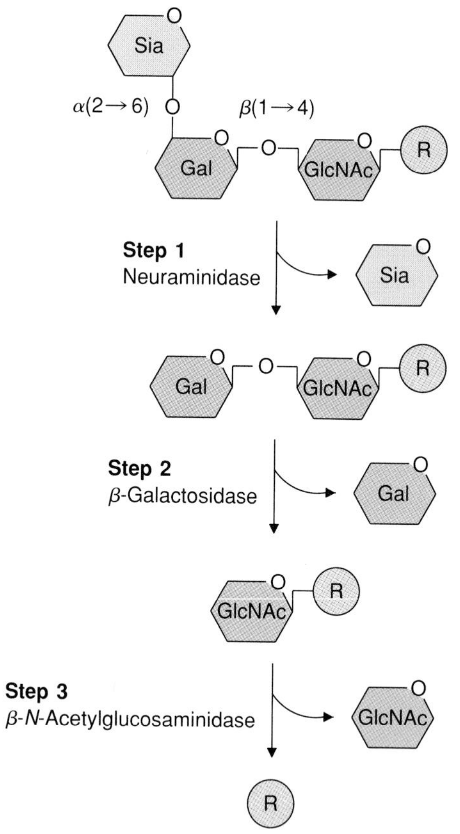

FIGURE **9A.1**

Cleavage of an oligosaccharide by glycosidases. The oligosaccharide is attached to the protein orosomucoid (R).

References

Jones, C. (1991) Nuclear magnetic resonance spectroscopy methods for the analysis of polysaccharides and glycoprotein carbohydrate chains. *Adv. Carbohydr. Anal.* 1:145–184. Includes a survey of useful techniques.

McCleary, B. V., and N. K. Matheson (1986) Enzymatic analysis of polysaccharide structures. *Adv. Carbohydr. Chem. Biochem.* 44: 147–276.

Wait, R. (1991) Structural analysis of carbohydrates by mass spectrometry. *Adv. Carbohydr. Anal.* 1:335–440.

CHAPTER 10

Lipids, Membranes, and Cellular Transport

THE MOLECULES WE ENCOUNTER IN THIS CHAPTER, THE **LIPIDS**, CARRY OUT MULtiple functions. Some—the fats—are used for energy storage and heat insulation. However, the largest fraction of the lipids in most cells are used to form **membranes,** the partitions that divide compartments from one another and separate the cell from its surroundings. The membranes are much more than passive walls, for they contain highly selective gates that promote the passage of certain materials in certain directions and block others altogether. It is this property of *selective membrane permeability* that allows each of the different parts of the cell to carry out its specific operations.

Lipids are an extremely diverse group of molecules, and we shall not discuss all of them in this chapter. In addition to lipids that serve the functions already described, some lipids function as vitamins (vitamin E, for example), and others make up the large classes of steroid hormones and prostaglandins. We describe some of them here, but we reserve discussion of lipids with more specialized functions and chemistry for Chapters 18 and 19. In those chapters, we also discuss lipid transport and the lipid–protein complexes or *lipoproteins* involved in that process.

The Molecular Structure and Behavior of Lipids

Unlike the proteins, nucleic acids, and polysaccharides, lipids are not polymers. Rather, they are quite small molecules that have a strong tendency to associate through noncovalent forces. Lipids are usually characterized by the kind of structure shown in the margin: a polar, hydrophilic "head" connected to a nonpolar, hydrophobic hydrocarbon "tail." Lipid molecules in an aqueous environment tend to clump together in noncovalent association for two fundamental reasons. Just as nonpolar groups in proteins associate via an entropy-driven hydrophobic effect, so do the nonpolar tails of lipids. A second stabilizing force comes from the van der Waals interaction between the hydrocarbon regions of the molecules.

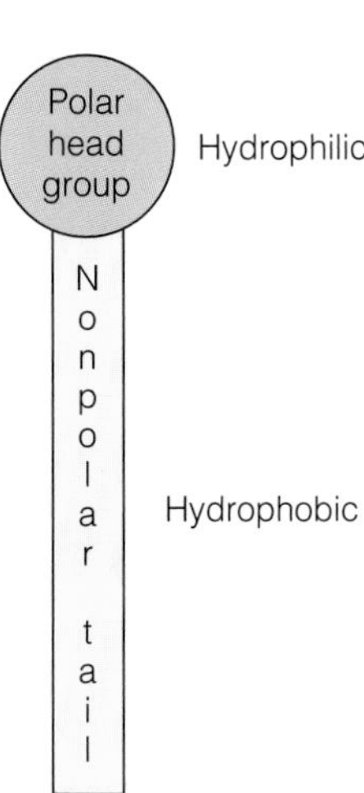

General lipid structure

The polar, hydrophilic head groups of lipid molecules, on the other hand, tend to associate with water. Lipids are, then, prime examples of the kind of

Lipid molecules are amphipathic. They tend to form surface monolayers, bilayers, micelles, or vesicles when in contact with water.

amphipathic substance described in Chapter 2. This "molecular schizophrenia" has a number of consequences, including the formation of surface monolayers, bilayers, micelles, and vesicles by lipids in contact with water (see Figure 2.15). From a biological point of view, the most important of these consequences is the tendency of lipids to form micelles and membrane bilayers. Exactly what kind of structure is formed when a lipid is in contact with water depends on the specific molecular structure of the hydrophilic and hydrophobic parts of that lipid molecule. Thus, it is appropriate that we now examine the structure of some of the major types of lipids.

FATTY ACIDS

The simplest lipids are the **fatty acids,** which are also constituents of many more complex lipids. Their basic structure exemplifies the general lipid model described above: A hydrophilic carboxylate group is attached to one end of an (often long) hydrocarbon chain. An example is *stearic acid,* which is widely distributed in organisms. We show it in Figure 10.1a and c as the ionized form, the *stearate* ion. Stearic acid is an example of a **saturated** fatty acid, one in which the carbons of the tail are all saturated with hydrogen atoms. A number of the more biologically important saturated fatty acids are listed in Table 10.1. Note that each has a common name (such as stearic acid) and a systematic name (in this case, *n*-octadecanoic acid).

Many important naturally occurring fatty acids are **unsaturated**—that is, they contain one or more double bonds (see Table 10.1). *Oleic acid,* which is found in many animal fats, yields the *oleate* ion (Figure 10.1b and c). In most of the naturally occurring unsaturated fatty acids, the orientation about double bonds is *cis* rather than *trans.* This orientation has an important effect on molecular structure, for each *cis* double bond inserts a bend into the hydrocarbon chain. Keep in mind, however, that although Figure 10.1 depicts the molecules as rigid structures, there is freedom of rotation about each single bond in the hydrocarbon chain. Thus, many conformations are possible.

Most naturally occurring fatty acids contain an even number of carbon atoms. If double bonds are present (unsaturation), they are usually *cis.*

Most of the naturally occurring fatty acids have an even number of carbon atoms. Although the hydrocarbon chains are linear in most fatty acids, some fatty acids (found primarily in bacteria) contain branches or even cyclic structures (see Table 10.1).

FIGURE **10.1**

Structures of the ionized forms of some representative fatty acids. Hydrophilic portions (head groups) of the molecules are indicated by a pale blue background in the models, hydrophobic portions (tails) by yellow. **(a)** The anion of stearic acid, a saturated fatty acid. **(b)** The anion of oleic acid, an unsaturated fatty acid with one *cis* double bond. **(c)** Formulas for (a) and (b).

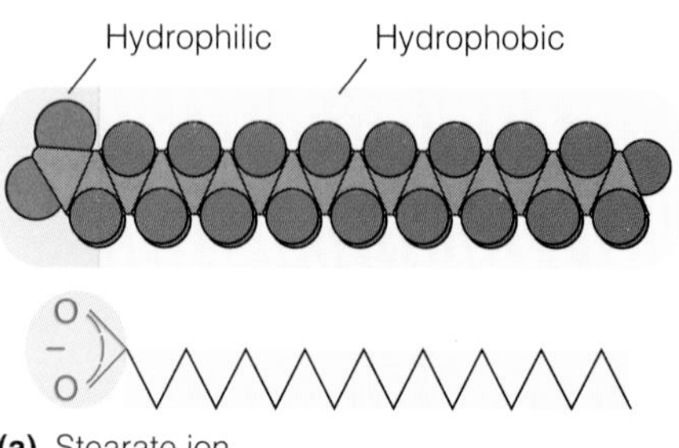

(a) Stearate ion

(b) Oleate ion

$^{-}O(O=)CCH_2CH_2CH_2CH_2CH_2CH_2CH_2CH_2CH_2CH_2CH_2CH_2CH_2CH_2CH_2CH_2CH_3$

Polar head group — Hydrocarbon tail

Stearate ion

$^{-}O(O=)CCH_2CH_2CH_2CH_2CH_2CH_2CH_2CH=CHCH_2CH_2CH_2CH_2CH_2CH_2CH_2CH_3$

Oleate ion

(c) Formulas

TABLE 10.1 Some biologically important fatty acids

Common Name	Systematic Name	Abbreviation	Structure	Melting Point (°C)
Saturated Fatty Acids				
Capric	*n*-Decanoic	10:0	$CH_3(CH_2)_8COOH$	31.6
Lauric	*n*-Dodecanoic	12:0	$CH_3(CH_2)_{10}COOH$	44.2
Myristic	*n*-Tetradecanoic	14:0	$CH_3(CH_2)_{12}COOH$	53.9
Palmitic	*n*-Hexadecanoic	16:0	$CH_3(CH_2)_{14}COOH$	63.1
Stearic	*n*-Octadecanoic	18:0	$CH_3(CH_2)_{16}COOH$	69.6
Arachidic	*n*-Eicosanoic	20:0	$CH_3(CH_2)_{18}COOH$	76.5
Behenic	*n*-Docosanoic	22:0	$CH_3(CH_2)_{20}COOH$	81.5
Lignoceric	*n*-Tetracosanoic	24:0	$CH_3(CH_2)_{22}COOH$	86.0
Cerotic	*n*-Hexacosanoic	26:0	$CH_3(CH_2)_{24}COOH$	88.5
Unsaturated Fatty Acids				
Palmitoleic	*cis*-9-Hexadecenoic	16:1cΔ9	$CH_3(CH_2)_5CH{=}CH(CH_2)_7COOH$	0
Oleic	*cis*-9-Octadecenoic	18:1cΔ9	$CH_3(CH_2)_7CH{=}CH(CH_2)_7COOH$	16
Linoleic	*cis,cis*-9,12-Octadecadienoic	18:2cΔ9,12	$CH_3(CH_2)_4CH{=}CHCH_2CH{=}CH(CH_2)_7COOH$	5
Linolenic	all-*cis*-9,12,15-Octadecatrienoic	18:3cΔ9,12,15	$CH_3CH_2CH{=}CHCH_2CH{=}CHCH_2CH{=}CH(CH_2)_7COOH$	−11
Arachidonic	all-*cis*-5,8,11,14-Eicosatetraenoic	20:4cΔ5,8,11,14	$CH_3(CH_2)_4CH{=}CHCH_2CH{=}CHCH_2CH{=}CHCH_2CH{=}CH(CH_2)_3COOH$	−50
Branched and Cyclic Acids				
Tuberculostearic	*l*-D-10-Methyloctadecanoic		$CH_3(CH_2)_7CH(CH_3)(CH_2)_8COOH$ (CH₃ branch on C-10)	13.2
Lactobacillic	ω-(2-*n*-Octylcyclopropyl)-octadecanoic		$CH_3(CH_2)_5CH{-}CH(CH_2)_9COOH$ (the two CH joined through a bridging CH_2, forming a cyclopropane ring)	29

To provide a more convenient and definitive way of referring to fatty acids, a system of abbreviations has been developed; it is illustrated in Table 10.1. The rules are simple: The number before the colon gives the total number of carbons, and the number after the colon gives the count of double bonds. The configurations and positions of double bonds are indicated by c *(cis)* or t *(trans)* followed by Δ and one or more numbers. These numbers denote the carbon atom (counting from the carboxyl) where each double bond starts. Thus, oleic acid is designated by 18:1cΔ9 and linolenic by 18:3cΔ9,12,15.

The fatty acids are weak acids, with pK_a values averaging about 4.5:

$$\mathrm{RCOOH} \underset{pK_a \cong 4.5}{\rightleftharpoons} \mathrm{RCOO^- + H^+}$$

Thus, these acids exist in the anionic form ($RCOO^-$) at physiological pH, and we should more properly speak of stearate and oleate, rather than stearic and oleic acids, under these conditions. The charge on the carboxyl group makes it extremely hydrophilic, whereas the long hydrocarbon tails are very hydrophobic. As a result, the fatty acids behave as typical amphipathic substances when we attempt to dissolve them in water. As shown in Figure 2.15, they tend to form **monolayers** at the air–water interface, with the hydrophilic carboxyl groups immersed in water and the hydrocarbon tails out of water.

If fatty acids are shaken with water, they will make **micelles**, in which the hydrocarbon tails cluster together within the structure and the carboxylate heads

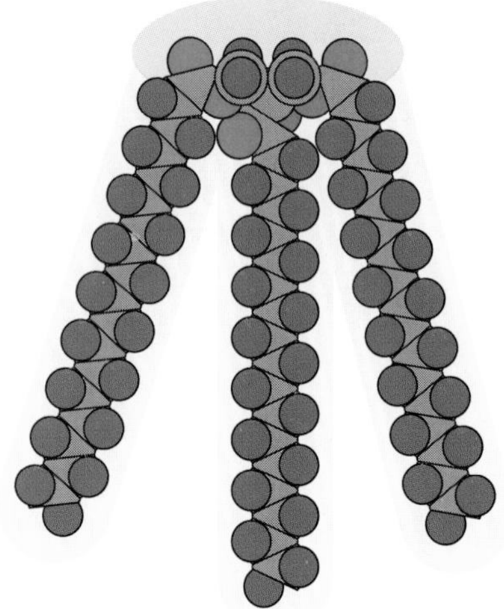

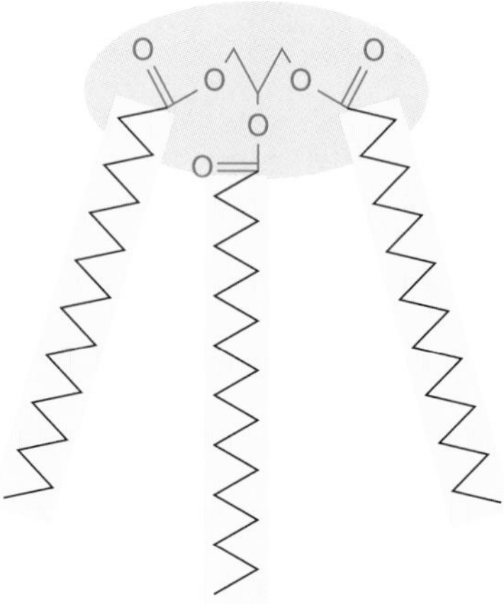

FIGURE **10.2**

The structure of tristearin, a fat. Tristearin is a triacylglycerol (fat) composed of glycerol and three stearate molecules. The hydrophilic portion of tristearin consists of glycerol and the head groups of the stearates; the hydrophobic portion consists of the hydrocarbon tails of the stearates.

are in contact with the surrounding water. If fatty acids are mixed with water *and* an oily or greasy substance (for example, a hydrocarbon), the micelles will form about droplets of the oil, emulsifying it. This behavior is the basis of the action of soaps and synthetic detergents.

Although the fatty acids play important roles in metabolism, large quantities of the free acids or their anions are never found in living cells. Instead, these compounds almost always occur as constituents of more complex lipids. We now turn to consideration of some of these classes of biologically important lipid molecules.

TRIACYLGLYCEROLS: FATS

The long hydrocarbon chains of fatty acids are extraordinarily efficient for energy storage, because they contain carbon in a fully reduced form and will therefore yield a maximum amount of energy on oxidation. They are, in fact, much more efficient energy stores than carbohydrates are. (Explicit analysis of this difference is made in Chapter 18.) For this reason, lipids are used by many organisms, including humans, for energy storage.

Storage of fatty acids in organisms is largely in the form of **triacylglycerols,** or **fats.** These substances are *triesters* of fatty acids and **glycerol;** the general formula is shown here.

$$\begin{array}{l} H_2C-O-\overset{\displaystyle O}{\overset{\|}{C}}-R_1 \\ \;\;| \\ H-C-O-\overset{\displaystyle O}{\overset{\|}{C}}-R_2 \\ \;\;| \\ H_2C-O-\overset{\displaystyle O}{\overset{\|}{C}}-R_3 \end{array}$$

Triacylglycerol

Here R_1, R_2, and R_3 correspond to the hydrocarbon tails of various fatty acids. We have depicted the structure with the hydrophobic chains to the right, according to our convention. This convention does not indicate stereochemical configuration. (The correct stereochemical configuration is depicted in Figure 10.6a.) As a particular example, if $R_1 = R_2 = R_3 = (CH_2)_{16}CH_3$, the hydrocarbon tail of stearic acid, the molecule is *tristearin* (Figure 10.2). Most triacylglycerols, however, contain a mixture of fatty acids, often including unsaturated ones. Table 10.2 lists the fatty acid composition of some naturally occurring fats. Comparison of common experience with these fats and data in the table reveals an interesting correlation. Fats rich in unsaturated fatty acids (like olive oil) are liquid at room temperature, whereas those with a higher content of saturated fatty acids (like butter) are more solid. Indeed, a wholly saturated fat is a quite firm solid, especially if the hydrocarbon chains are long. This is made clear from the melting point data in Table 10.1. The reason is simple: long saturated chains can pack closely together, to form regular, semicrystalline structures. In contrast, the kind of bend imposed by one or more *cis* double bonds (see Figure 10.1b and c) makes molecular packing more difficult. Indeed, partial **hydrogenation** of unsaturated fat oils (like corn oil) is used commercially to convert them to firmer fats, which can be used as butter substitutes such as oleomargarine.

Esterification with glycerol greatly diminishes the hydrophilic character of the head groups of the fatty acids. As a consequence, not only are triacylglycerols water-insoluble, but they do not even form micelles very effectively. Fats accumulated in plant and animal cells therefore form as oily droplets in the cytoplasm. In **adipocytes,** animal cells specialized for fat storage, almost the entire volume of

TABLE **10.2 Composition of some natural fats in percent of total fatty acids**

Number of C Atoms in Chain	Percent Present in: Olive Oil	Butter[a]	Beef Fat
Saturated			
4–12	2	11	2
14	2	10	2
16	13	26	29
18	3	11	21
Unsaturated			
16–18	80	40	46

[a]Numbers do not total 100% because the substance contains small amounts of other fatty acids.

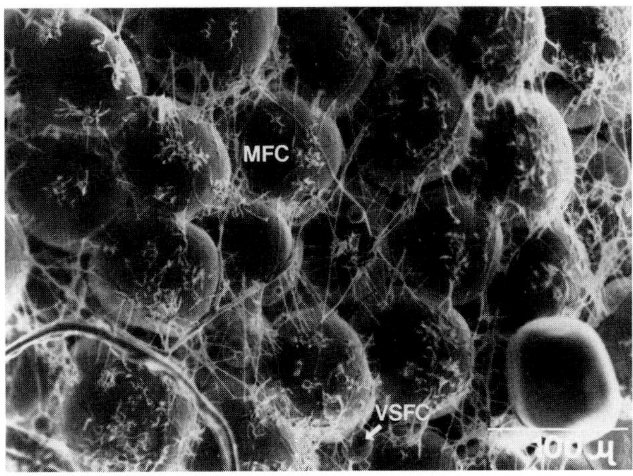

FIGURE **10.3**

Adipocytes. Adipocytes, or animal fat storage cells, make up a large part of adipose tissue. The designations MFC and VSFC correspond to "mature fat cell" and "very small fat cell," respectively.

From P. Julien, J.-P. Despres, and A. Angel, *J. Lipid Res.* (1989) 30:293–299. Used by permission.

Fats, or triacylglycerols, are triesters of fatty acids and glycerol. They are the major mode for energy storage in many organisms.

each cell is filled by a fat droplet (Figure 10.3). Such cells make up most of the adipose (fatty) tissue of animals.

Fat storage in animals serves three distinct functions:

1. *Energy production.* Most fat in most animals is oxidized for the generation of ATP, to drive metabolic processes.
2. *Heat production.* Some specialized cells (in "brown fat" of warm-blooded animals, for example) oxidize triacylglycerols for heat production, rather than to make ATP.
3. *Insulation.* In animals that live in a cold environment, layers of fat cells under the skin serve as thermal insulation. The blubber of whales is one obvious example.

SOAPS AND DETERGENTS

If fats are hydrolyzed with alkalis such as NaOH or KOH (in earlier times, wood ashes were used), a *soap* is produced. This process is called **saponification.** The fatty acids are released as either sodium or potassium salts, which are fully ionized. However, as cleansers, soaps have the disadvantage that the fatty acids are precipitated by the calcium or magnesium ions present in "hard" water, forming a scum and destroying the emulsifying action. Synthetic detergents have been devised that do not have this defect. One class is exemplified by *sodium dodecyl sulfate (SDS):*

$$^{-}O_3SO\ (CH_2)_{11}CH_3\ + Na^+$$

The salts of dodecyl sulfate with divalent cations (i.e., Ca^{2+} and Mg^{2+}) are more soluble. We have already encountered SDS, for it is widely used in forming micelles about proteins for gel electrophoresis. There are also synthetic nonionic detergents, like *Triton X-100:*

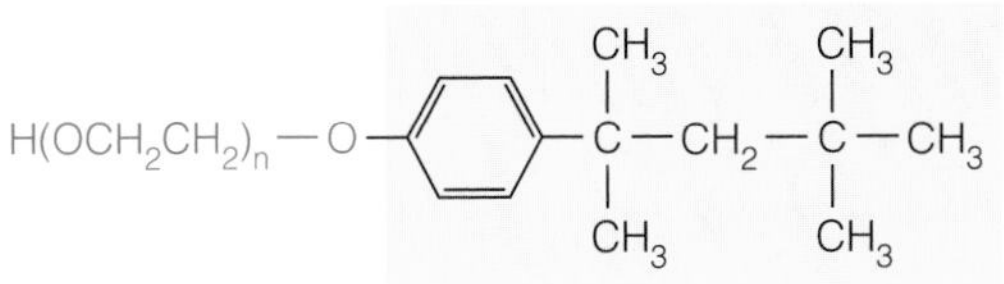

$$H(OCH_2CH_2)_n-O-C_6H_4-C(CH_3)_2-CH_2-C(CH_3)_2-CH_3$$

The hydrophilic group here is the polyoxyethylene head group (shown in blue), which in the commercial product averages about 9.5 residues in length.

WAXES

In the natural **waxes,** a long-chain fatty acid is esterified to a long-chain alcohol (Figure 10.4). This yields a head group that is only weakly hydrophilic, attached

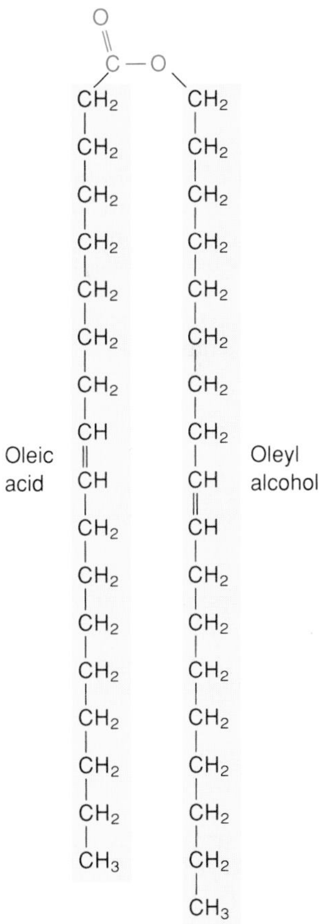

Oleic acid: $O{=}C(-O-)-(CH_2)_7-CH{=}CH-(CH_2)_7-CH_3$

Oleyl alcohol: $-O-(CH_2)_8-CH{=}CH-(CH_2)_7-CH_3$

FIGURE **10.4**

Structure of a typical wax. Waxes are formed by esterification of fatty acids and long-chain alcohols. The small head group can contribute little hydrophilicity, in contrast to the significant hydrophobic contribution of the two long tails.

to two hydrocarbon chains. As a consequence, the waxes are completely water-insoluble. In fact, they are so hydrophobic that they often serve as water repellents, as in the feathers of some birds and the leaves of some plants. In some marine microorganisms, waxes are used instead of other lipids for energy storage. In beeswax, they serve a structural function. As with the triacylglycerols, the hardness of waxes increases with chain length and degree of hydrocarbon saturation.

The Lipid Constituents of Biological Membranes

The major lipid components of biological membranes are glycerophospholipids, sphingolipids, glycosphingolipids, and glycoglycerolipids.

All biological membranes contain lipids as major constituents. The molecules that play the dominant roles in membrane formation all have highly polar head groups and, in most cases, two hydrocarbon tails. This composition makes molecular sense: If a large head group is attached to a single hydrocarbon chain, the molecule is wedge-shaped and will tend to form spherical micelles (Figure 10.5a). A double tail yields a roughly cylindrical molecule (Figure 10.5b); such cylindrical molecules can easily pack in parallel to form extended sheets of **bilayer** membranes with the hydrophilic head groups facing outward into the aqueous regions on either side (Figure 10.5c). The four major classes of membrane-forming lipids—glycerophospholipids, sphingolipids, glycosphingolipids, and glycoglycerolipids—share this type of cylindrical molecular structure. They differ principally in the nature of the head group. We shall describe a few examples of each class.

GLYCEROPHOSPHOLIPIDS

Glycerophospholipids (also called *phosphoglycerides*) are the major class of naturally occurring **phospholipids,** lipids with phosphate-containing head groups.

FIGURE **10.5**

Phospholipids and membrane structure. Whereas fatty acids **(a)** are wedge-shaped and tend to form spherical micelles, phospholipids **(b)** are more cylindrical and pack together better in the kind of lipid bilayer structure shown in **(c)**, a computer simulation of a phospholipid bilayer.

(c) From R. M. Venable, Y. Zhang, B. J. Hardy, and R. W. Pastor, *Science* (1993) 262:223–228. © 1993 by AAAS.

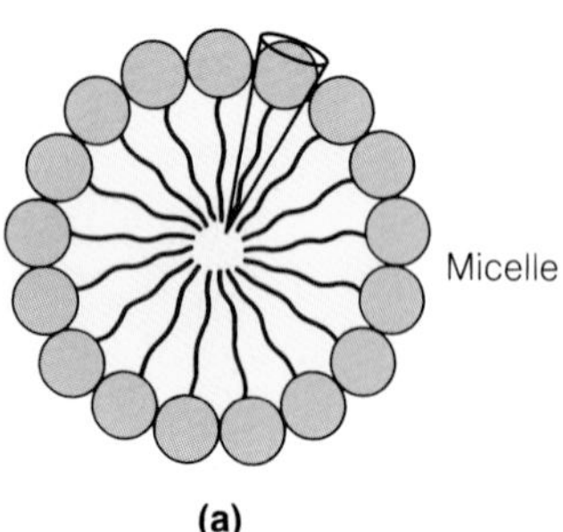

(a)

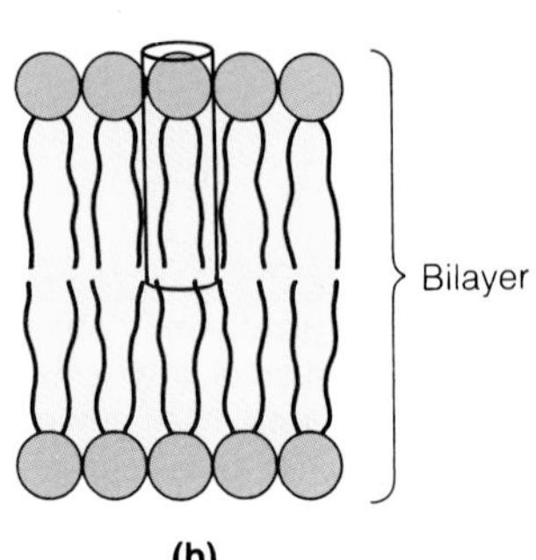

(b)

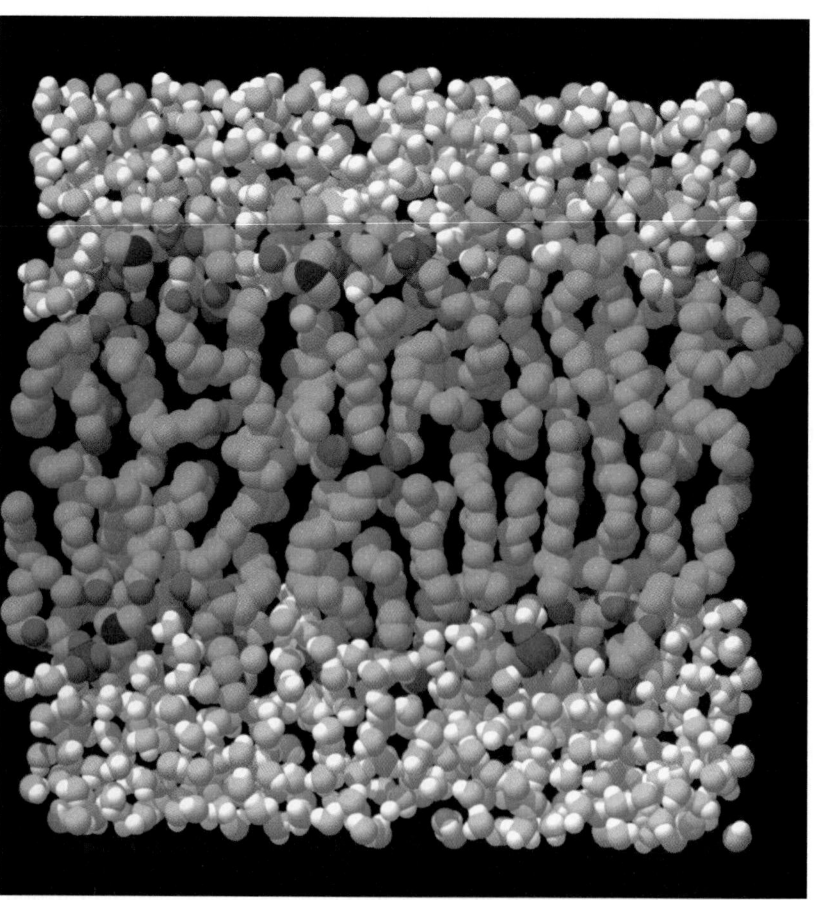

(c)

These compounds make up a significant fraction of the membrane lipids throughout the bacterial, plant, and animal kingdoms. All glycerophospholipids can be considered to be derivatives of glycerol-3-phosphate. Carbon 2 in glycerol-3-phosphate is a chiral center, and the naturally occurring glycerophospholipids are derivatives of the L enantiomer. The stereochemical configuration of the general structure of glycerophospholipids is shown in Figure 10.6a. Figure 10.6b shows the molecule in the manner we will generally use to represent membrane lipids, with the hydrophobic tails drawn to the right and the hydrophilic head group to the left. Usually, R_1 and R_2 are acyl side chains derived from the fatty acids; often one is saturated, the other unsaturated. The hydrophilic R_3 group varies greatly, and it is this that confers the greatest variation in properties among the glycerophospholipids. A gallery of the most common glycerophospholipids is shown in Figure 10.7, and their relative abundances in some membranes are given in Table 10.3. The simplest member of the group, **phosphatidic acid,** is only a minor membrane constituent; its principal role is as an intermediate in the synthesis of other glycerophospholipids (described in Chapter 19). The names of glycerophospholipids are derived from phosphatidic acid: *phosphatidylcholine, phosphatidylethanolamine,* and so on. As Figure 10.7 shows, the glycerophospholipids have very polar head groups, all carrying some charge. Because the hydrocarbon tails are derived from the naturally occurring fatty acids in various combinations, an enormous variety of glycerophospholipids exists. For example, the erythrocyte membrane contains molecules with hydrocarbon chains of 16 to 24 carbons, with 0 to 6 double bonds. Such variation in membrane composition allows "fine-tuning" of membrane properties for the diverse functions that different membranes must perform.

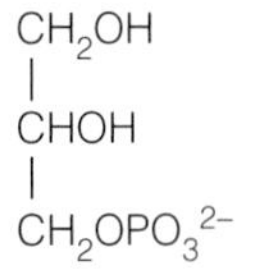

CH_2OH — $CHOH$ — $CH_2OPO_3^{2-}$

Glycerol-3-phosphate

SPHINGOLIPIDS AND GLYCOSPHINGOLIPIDS

A second major class of membrane constituents is built on the long-chain amino alcohol **sphingosine,** rather than on glycerol. If a fatty acid is linked via an amide

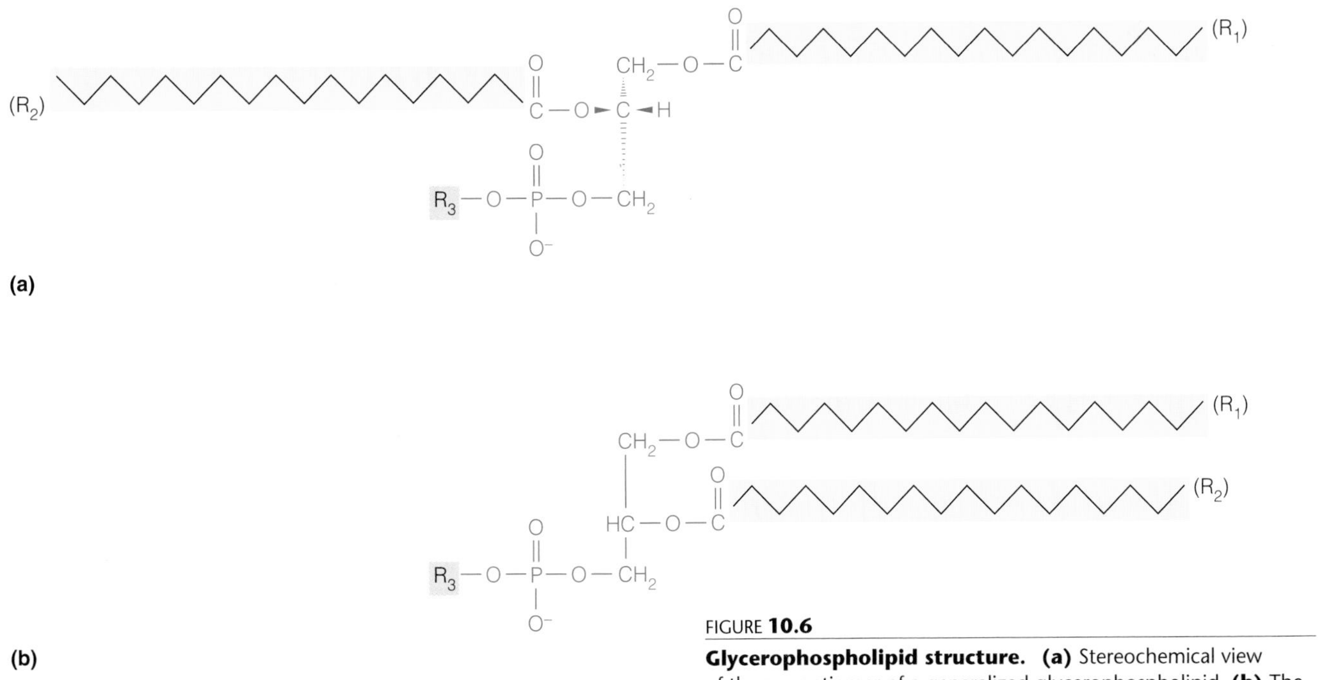

(R₂) C—O—C—H, CH₂—O—C (R₁), R₃—O—P—O—CH₂, O⁻

(a)

(R₁), (R₂), CH₂—O—C, HC—O—C, R₃—O—P—O—CH₂, O⁻

(b)

FIGURE **10.6**

Glycerophospholipid structure. (a) Stereochemical view of the L enantiomer of a generalized glycerophospholipid. **(b)** The same structure represented in our convention, with hydrophobic groups to the right, hydrophilic to the left. R_3 is a hydrophilic group. (See Figure 10.7.)

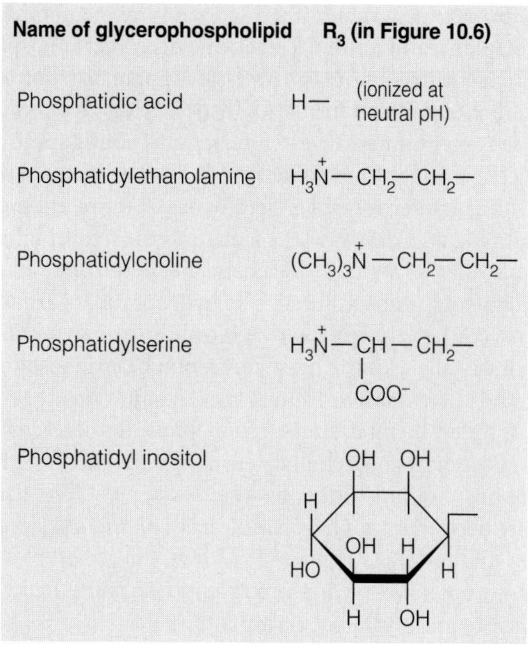

FIGURE **10.7**

The hydrophilic groups (R_3 in Figure 10.6) that distinguish common glycerophospholipids. In addition to this variation, there is also a great deal of variation in the hydrocarbon tails (R_1, R_2) in the structures shown in Figure 10.6.

Name of glycerophospholipid	R_3 (in Figure 10.6)
Phosphatidic acid	H— (ionized at neutral pH)
Phosphatidylethanolamine	$H_3\overset{+}{N}-CH_2-CH_2-$
Phosphatidylcholine	$(CH_3)_3\overset{+}{N}-CH_2-CH_2-$
Phosphatidylserine	$H_3\overset{+}{N}-CH(COO^-)-CH_2-$
Phosphatidyl inositol	(inositol ring: OH, OH, H, HO, OH, H, H, H, OH)

TABLE 10.3 Lipid composition of some biological membranes

Lipid	Percentage of Total Composition in: Human Erythrocyte Plasma Membrane	Human Myelin	Beef Heart Mitochondria	*E. coli* Cell Membrane
Phosphatidic acid	1.5	0.5	0	0
Phosphatidylcholine	19	10	39	0
Phosphatidylethanolamine	18	20	27	65
Phosphatidylglycerol	0	0	0	18
Phosphatidylinositol	1	1	7	0
Phosphatidylserine	8.0	8.0	0.5	0
Sphingomyelin	17.5	8.5	0	0
Glycolipids	10	26	0	0
Cholesterol	25	26	3	0
Others	0	0	23.5	17

Source: Data from C. Tanford, *The Hydrophobic Effect* (New York: Wiley, 1973).

A broad class of sphingolipids is built upon a sphingosine core.

bond to the $-NH_2$ group, the class of **sphingolipids** referred to as **ceramides** is obtained:

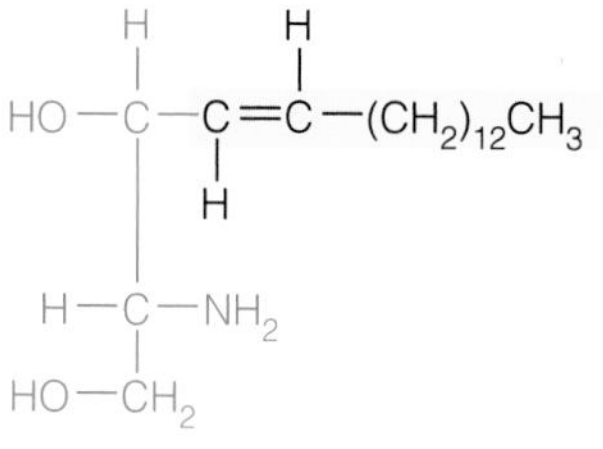

$$HO-CH(-CH(NH_2)-CH_2OH)-CH=CH-(CH_2)_{12}CH_3$$

Sphingosine = D-4-sphinganine

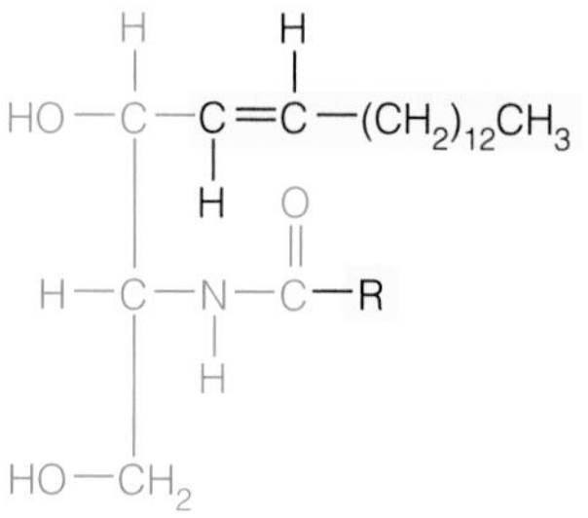

$$HO-CH(-CH(NH-CO-R)-CH_2OH)-CH=CH-(CH_2)_{12}CH_3$$

General structure of a ceramide (R = hydrocarbon)

Ceramides consist *only* of sphingosine and a fatty acid. Further modification, by addition of groups to the C-1 hydroxyl of sphingosine, leads to a variety of other sphingolipids. An especially important example is *sphingomyelin,* in which a *phosphocholine* group is attached to the C-3 hydroxyl.

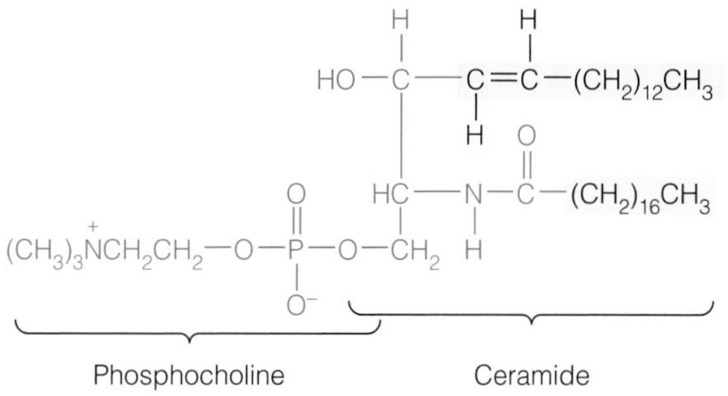

Phosphocholine Ceramide

Sphingomyelin

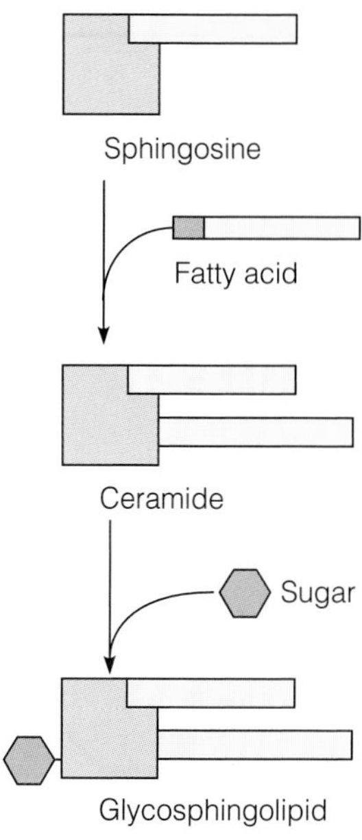

Composition of generalized glycosphingolipid

In some of the membrane lipids built on sphingosine, the head group contains saccharides. Lipids containing saccharide groups go under the general name of **glycolipids.** The **glycosphingolipids** constitute the third major class of membrane lipids. Their general structure is shown in the margin. They include such molecules as the **cerebrosides** (monoglycosyl ceramides) and **gangliosides,** anionic glycosphingolipids containing one or more sialic acid residues. Examples of these are shown in Figure 10.8. As the names of these compounds suggest, they are especially common in the membranes of brain and nerve cells.

GLYCOGLYCEROLIPIDS

Another class of lipids, less common in animal membranes but widespread in plant and bacterial membranes, are the **glycoglycerolipids,** exemplified by **monogalactosyl diglyceride:**

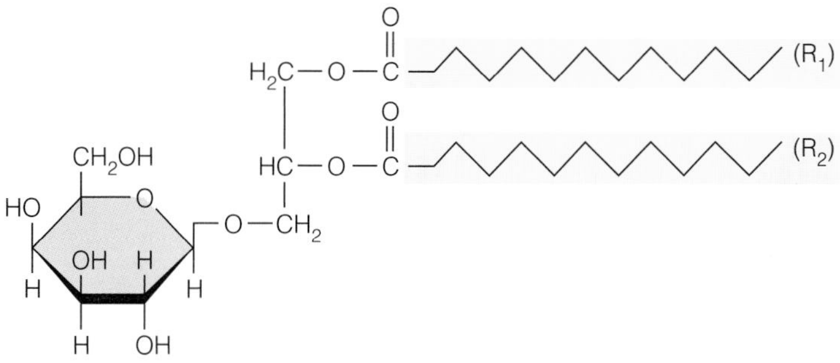

This compound may actually be the most abundant of all polar lipids, for it constitutes about half the lipid in chloroplast membranes (see Chapter 17). Such lipids are also abundant in archaebacteria, where they are the major membrane components.

CHOLESTEROL

One important lipid constituent of many membranes bears little superficial resemblance to the compounds we have studied so far. This substance is **cholesterol,** the structure of which is shown in Figure 10.9. Cholesterol is a member of a large group of substances called **steroids,** which include a number of important hormones, among them the sex hormones of higher animals. In fact, cholesterol is the precursor for the synthesis of many of these substances. Its role in these syntheses is discussed in Chapter 19, along with a detailed description of other

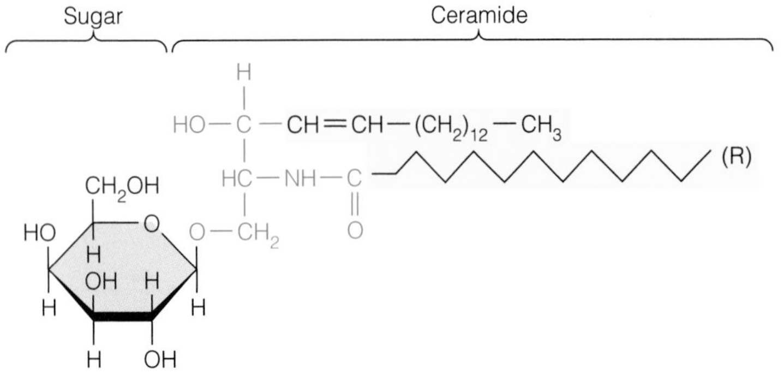

Sugar Ceramide

(a) Galactosylceramide

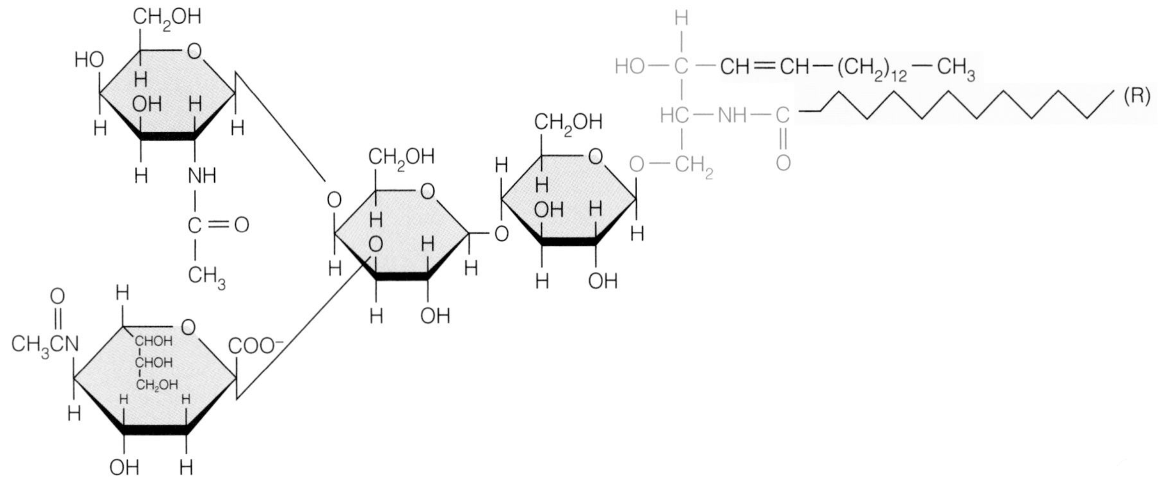

(b) GalNAcβ(1→4)Galβ(1→4)Glcβ(1→1)ceramide (3 ↑ α2 Sia)

FIGURE **10.8**

Examples of glycosphingolipids. (a) A cerebroside, an important constituent of brain cell membranes. **(b)** A ganglioside. This particular ganglioside, called GM_2 or the Tay-Sachs ganglioside, accumulates in neural tissue of infants with Tay-Sachs disease. The defect responsible for this inherited condition is the lack of an enzyme that normally cleaves the terminal GalNAc. (See Chapter 19.)

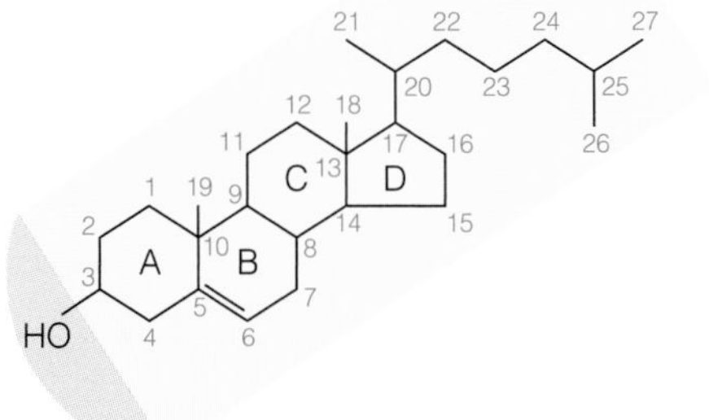

(a)

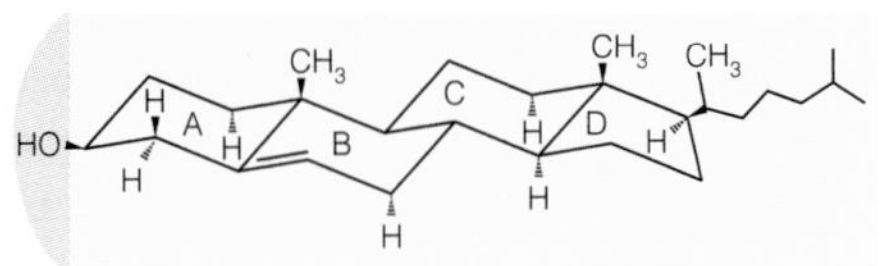

(b)

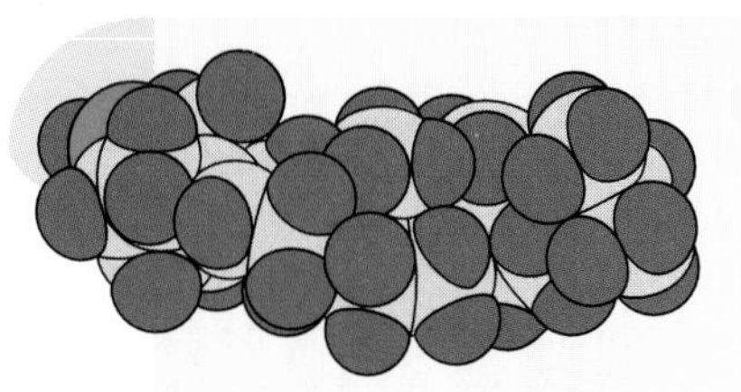

(c)

FIGURE **10.9**

Cholesterol. (a) Structural formula. **(b)** Skeletal model. **(c)** Space-filling model.

steroids and their functions. Other steroids are also found in membranes; for example, *lanosterol* (page 691) is prominent in plant cell membranes.

Cholesterol is a weakly amphipathic substance, because of the hydroxyl group at one end of the molecule. However, some of the cholesterol present in membranes has been rendered more hydrophobic by esterification of the hydroxyl group to a fatty acid. As the conformational structure in Figure 10.9b shows, the fused cyclohexane rings in cholesterol are all in the chair conformation. This makes cholesterol a bulky, rigid structure as compared with other hydrophobic membrane components such as the fatty acid tails. The cholesterol molecule fits awkwardly into membrane lipids and tends to disrupt regularity in membrane structure. This property can have a major effect, because cholesterol constitutes 25% or more of the lipid content in some membranes (see Table 10.3). Changes in membrane regularity can have, as we shall see, profound effects on such properties as membrane stiffness and permeability.

Cholesterol, a component of many animal membranes, influences membrane fluidity by its bulky structure.

The molecules described above constitute the major portion of membrane lipids in most organisms. However, one of the "three kingdoms" of organisms—the archaebacteria—are unique in having glycoglycerolipids as their major membrane lipids.

The Structure and Properties of Membranes and Membrane Proteins

The membranes of living cells are remarkable bits of molecular architecture, with many and varied functions. To say that a membrane is essentially a phospholipid bilayer is a gross oversimplification. To be sure, the phospholipid bilayer, as depicted in Figure 10.5b and c, forms the basic structure, but there is much more to the plasma membranes found in living cells. A more realistic representation of a typical eukaryotic cell membrane is shown in Figure 10.10. A wide variety of

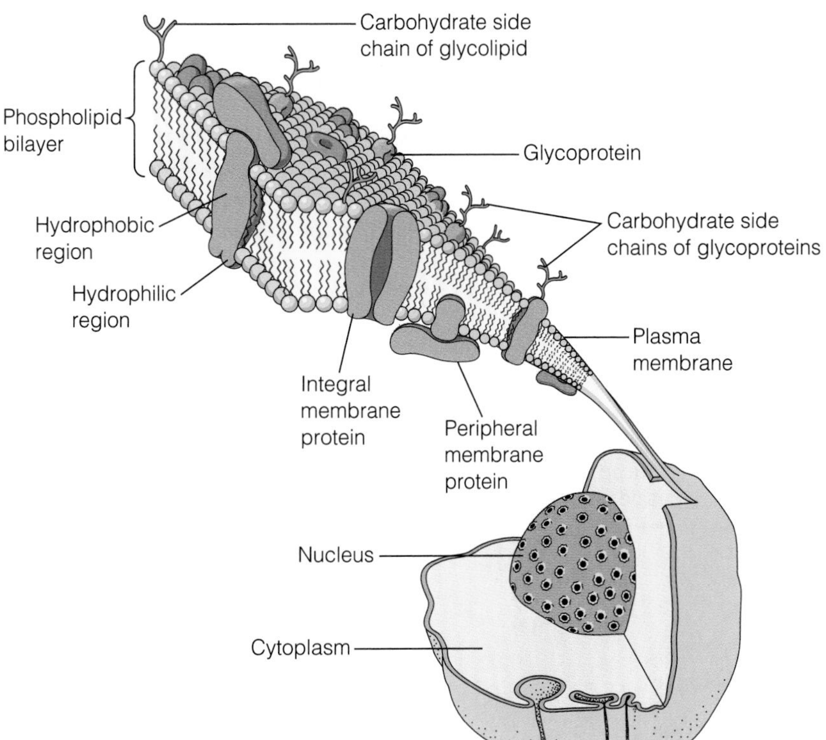

FIGURE **10.10**

Structure of a typical cell membrane. In this schematic view, a strip of the plasma membrane of a eukaryotic cell has been peeled off. Proteins are embedded in and on the phospholipid bilayer; some of them are glycoproteins, carrying oligosaccharide chains. The membrane is about 6 nm thick.

According to the fluid mosaic model, a membrane is a fluid mixture of lipids and proteins.

specific proteins are contained in the membrane or are bound to its surface. Many of these proteins carry oligosaccharide groups that project into the surrounding aqueous medium. Other oligosaccharides are carried by glycolipids, with the lipid portions inserted in the membrane. The two sides of the bilayer are usually different, both in lipid composition and in the placement and orientation of proteins and oligosaccharides.

Much of our current understanding concerning biological membranes is based upon the **fluid mosaic model** proposed by S. J. Singer and G. L. Nicholson in 1972. This is the model depicted in Figure 10.10. The fluid, asymmetric lipid bilayer carries within it a host of proteins. Some of them, called **peripheral membrane proteins,** are exposed at only one membrane face or the other. They are held to the membrane by interaction with lipid heads or integral membrane proteins. The **integral membrane proteins** are largely buried within the membrane but are usually exposed on both faces. Integral proteins are frequently involved in transmitting either specific substances or chemical signals through the membrane. The whole membrane is a mosaic of lipids and proteins.

Some membrane proteins are peripheral and confined to one face. Others, the integral proteins, extend through the membrane.

MOTION IN MEMBRANES

A functioning biological membrane is not a rigid, frozen structure. In fact, many of the lipid and protein components are in constant motion. This motion can be demonstrated in a direct and dramatic way. If human and mouse cells, each carrying a distinctive fluorescent marker in its plasma membrane, are fused together, the two kinds of markers gradually become completely intermixed (Figure 10.11). This demonstrates that *lateral diffusion* (parallel to the membrane surface) can occur in the membrane. The rapidity with which such two-dimensional diffusion can occur depends on the membrane fluidity, which in turn depends on temperature and lipid composition. Under physiological conditions, the average time required for a phospholipid molecule to wander completely around a cell is on the order of seconds to minutes; membrane proteins also move, but more slowly.

Motion in Synthetic Membranes

The effects of temperature and composition on fluidity can be most simply studied using artificial membranes containing only one or a few kinds of lipids and no proteins (see Tools of Biochemistry 10A). Figure 10.12a depicts the behavior of a membrane made entirely from phosphatidylcholine carrying 16-carbon saturated chains (PC-16:0/16:0, in shorthand). At low temperatures the hydrocarbon tails pack together closely to form a nearly solid *gel* state. If the temperature is raised above 41°C, this regular order is lost, and the hydrocarbon tails become free to move about. The membrane "melts" to adopt a semifluid *liquid crystalline*

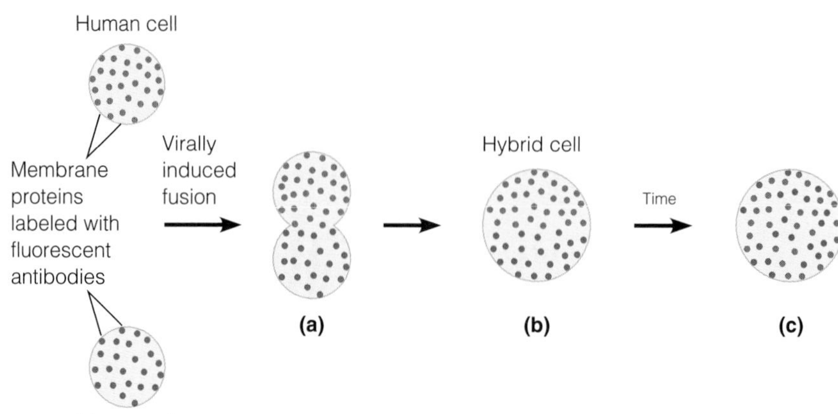

FIGURE **10.11**

Experimental demonstration of membrane fluidity. When cells with surface membrane protein marked by fluorescent tags are induced to fuse, the proteins gradually mix over the fused surface.

From W. M. Becker, L. J. Kleinsmith, and J. Hardin, *The World of the Cell*, 4th ed. (San Francisco, CA: Addison Wesley Longman, 2000) © Addison Wesley Longman, Inc.

state. The temperature at which this happens is called the *transition temperature.* This abrupt change in synthetic membrane properties can be detected by a number of the techniques described in Tools of Biochemistry 10A. Figure 10.12b shows the transition as detected by calorimetry.

The transition temperature is very sensitive to the nature of the hydrocarbon tails. If a PC-14:0/14:0 membrane is used (with tails only two carbons shorter than those just described), the transition temperature drops to 23°C. If a single *cis* double bond is incorporated into each 16-carbon tail (PC-16:1/16:1), melting occurs at −36°C. As explained earlier, *cis* double bonds put bends in the chains, inhibiting their close packing; thus, such chains must be cooled to a lower temperature to produce the rigid gel. Changing the head group can also make a big difference: If phosphatidylethanolamine is substituted for phosphatidylcholine in PC-16:0/16:0, the thermal transition is raised to 63°C. The sensitivity of the transition to lipid composition is shown dramatically by the fact that the small changes described above can change the transition temperature over a range of 100°C.

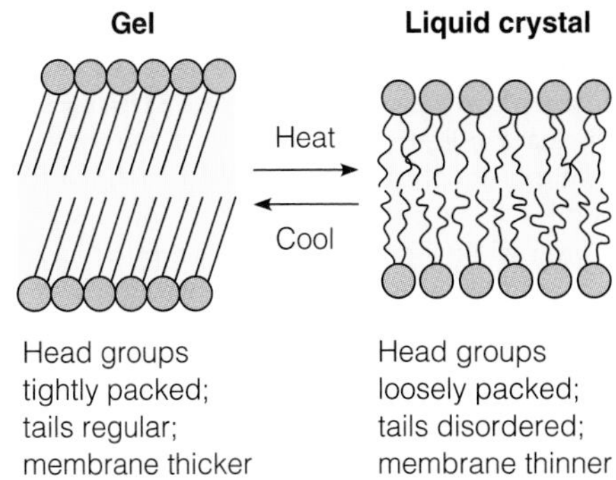

(a) Transition

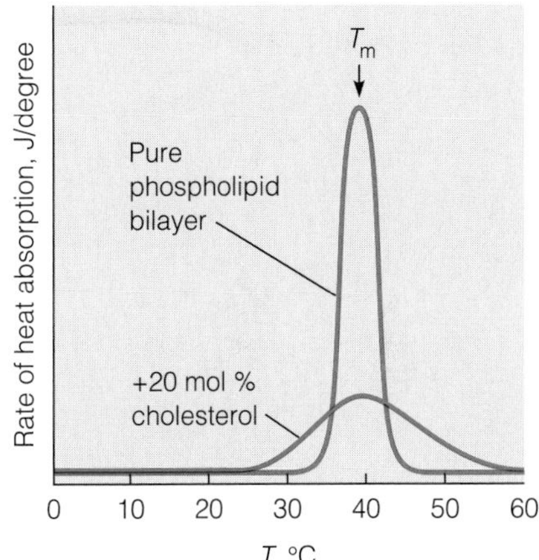

(b) Transition with and without cholesterol

FIGURE **10.12**

The gel–liquid crystalline transition in a synthetic lipid bilayer. (a) A schematic view of the change at the transition temperature. Below this temperature the hydrocarbon tails are packed together in a nearly rigid, nearly crystalline gel state (left). Above this temperature the chains become free to move about, and the interior of the membrane resembles a liquid hydrocarbon (right). **(b)** Detection of the transition by calorimetry. Measurement of the heat absorbed by a membrane as the temperature is raised each degree shows a sharp spike at the transition temperature (T_m) for a pure dipalmitoylphosphatidylcholine bilayer. This well-defined transition from gel to liquid is called melting of the membrane. When 20 mol % cholesterol is mixed into the bilayer, the transition temperature is not changed, but the transition is broadened.

Motion in Biological Membranes

Biological membranes, which contain complex mixtures of lipid components plus protein, exhibit much broader and more complex phase transitions than those observed for synthetic bilayers of the kind described above. Indeed, there is now evidence for the existence of quite stable "domains" of different composition in different parts of a cell membrane. This explains the broader transition observed. Because it is essential that the membranes in living cells be fluid, the membrane composition is regulated so as to keep the transition temperature below the body temperature of the organism. One example is found in bacteria, which will alter the saturated/unsaturated fatty acid ratio in their membranes in response to a change in the temperature at which they are grown. A remarkable case in the animal kingdom is that of the reindeer's leg. Its cell membranes show an increase in relative amount of unsaturated fatty acids near the hoof, which is usually cooler than the rest of the body.

Cholesterol has a specific and complex effect on membrane fluidity. As Figure 10.12b shows for a synthetic membrane, cholesterol does not influence the transition temperature markedly, but it does broaden the transition. It has been hypothesized that this broadening occurs because cholesterol can both stiffen the membrane above the transition temperature and inhibit regularity in structure formation below the transition temperature. Thus, it blurs the distinction between the gel and the fluid state. There is evidence that variations in cholesterol content are used to regulate membrane behavior in some organisms.

The effects of cholesterol on membrane structure are strongly dependent on its concentration in the membrane. Recent x-ray scattering studies have shown that at moderate concentrations cholesterol fits into the bilayer but thickens it (Figure 10.13), whereas at high concentrations "islands" of cholesterol bilayers form. These, it is speculated, may provide nuclei for the formation of cholesterol plaques in the circulatory system. The formation of such structures has been demonstrated to be an in vivo effect by experiments in which rats were fed high levels of cholesterol for many weeks.

In contrast to the ease of lateral movement, the "flip-flop" of lipid molecules *across* synthetic lipid bilayers, from one side to the other, is much slower. The reason is not hard to see: When a phospholipid molecule turns from one face to the other, it must duck its very hydrophilic head into the inhospitable medium of the hydrocarbon tails and pass it through that region. Such an event is very unfavorable from an energetic point of view, and hence the process is slow. There exist a class of enzymes (*translocases*) that catalyze flipping. In contrast to phospholipids, the transport of fatty acids across membranes is much more rapid, for a simple reason. Unlike the phosphate groups on phospholipids, the carboxylates on fatty

Under physiological conditions, biological membranes exist in a semifluid liquid crystalline state.

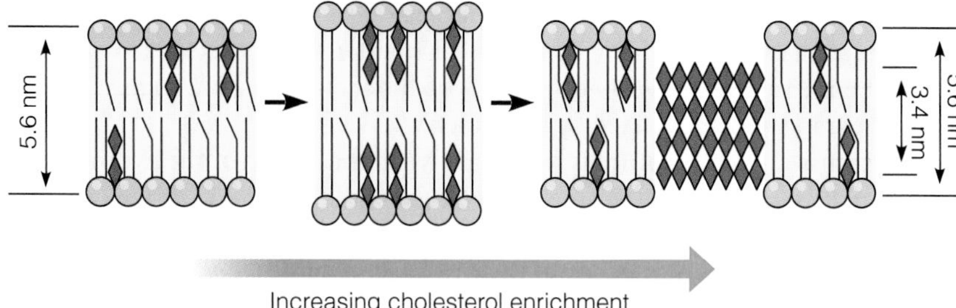

FIGURE **10.13**

A schematic model of the effects of cholesterol on plasma membrane structure. The data indicate an initial effect of the cholesterol/phospholipid ratio, C:PL,on membrane width. At C:PL mode ratios at or below 0.8:1, bilayer width increases with increasing C:PL. At C:PL mole ratios above 0.9:1, two separate lamellar phases form, one representing a liquid crystalline lipid bilayer and another representing an immiscible cholesterol phase. The immiscible cholesterol phase forms by phase separation as the membrane "saturates" with cholesterol.

Adapted from T. Tulenko, M. Chen, P. E. Mason, and R. P. Mason, *J. Lipid Res.* (1998) 29:947–956.

acids are protonated, and therefore uncharged, in the very nonpolar environment of the lipid bilayer. Thus, they can be better accommodated in the membrane than an ionized molecule and can flip-flop in less than a second.

THE ASYMMETRY OF MEMBRANES

Every biological membrane has two distinct faces, each encountering a different environment. Examples can be seen in the illustrations of cell ultrastructure shown in Chapter 1: The plasma membrane of a cell faces the external environment on the outside and the cytoplasm on the inside, whereas the membrane around a chloroplast faces the photosynthetic apparatus on the inside and the cytoplasm on the outside. Because the two faces of a membrane must deal with different surroundings, the faces are usually quite different in composition and structure.

This difference extends even to the level of phospholipid composition. Recall that all phospholipid membranes are bilayers; the two individual layers are often called *leaflets.* The compositions of the two leaflets in the plasma membranes of several kinds of cells are shown in Figure 10.14. Not only are the individual lipids distributed very asymmetrically, but the distribution also varies considerably among cell types.

The two leaflets of a membrane usually differ in lipid composition.

The consequences of such differences in phospholipid composition are numerous. Fluidity may be different on one side of the membrane or the other. The difference in charged groups on the two surfaces contributes to the membrane potential (discussed later in this chapter). Glycoproteins and glycolipids carried in the outer leaflet of a plasma membrane contribute, via their oligosaccharide chains, to identification of cells (see Chapter 9).

Much of our knowledge of membrane asymmetry comes from studies of **vesicles,** fragments of membrane that have resealed to form hollow shells, with an inside and an outside. Reagents can be either captured inside the vesicle or added only to the surrounding solution, so that they can react specifically with either outward-facing or inward-facing proteins or lipids. A membrane protein in a vesicle may be reacted covalently with a radioactively labeled reagent, isolated, and cleaved into peptides by proteases. Identification of which peptides are labeled by "inside" or "outside" reactants can reveal which portions of the protein were on the inner face and which were on the outer face. In a similar way, lipids can be tested, using enzymes or other reagents that cleave off or otherwise modify the head groups. Experiments of this kind, performed inside or outside vesicles, have provided much of the kind of information shown in Figure 10.14.

Biological membranes are highly dynamic structures. Not only must they continually expand as cells grow and divide, but also, even in resting cells, a continual turnover and renewal of the membrane components apparently occurs. Indeed, this dynamic, nonequilibrium state is a necessity if asymmetry is to be maintained. From the second law of thermodynamics (see Chapter 3) we know that the equilibrium state of a two-layer membrane would require an equal distribution of every component on either side. Like so many other biological sys-

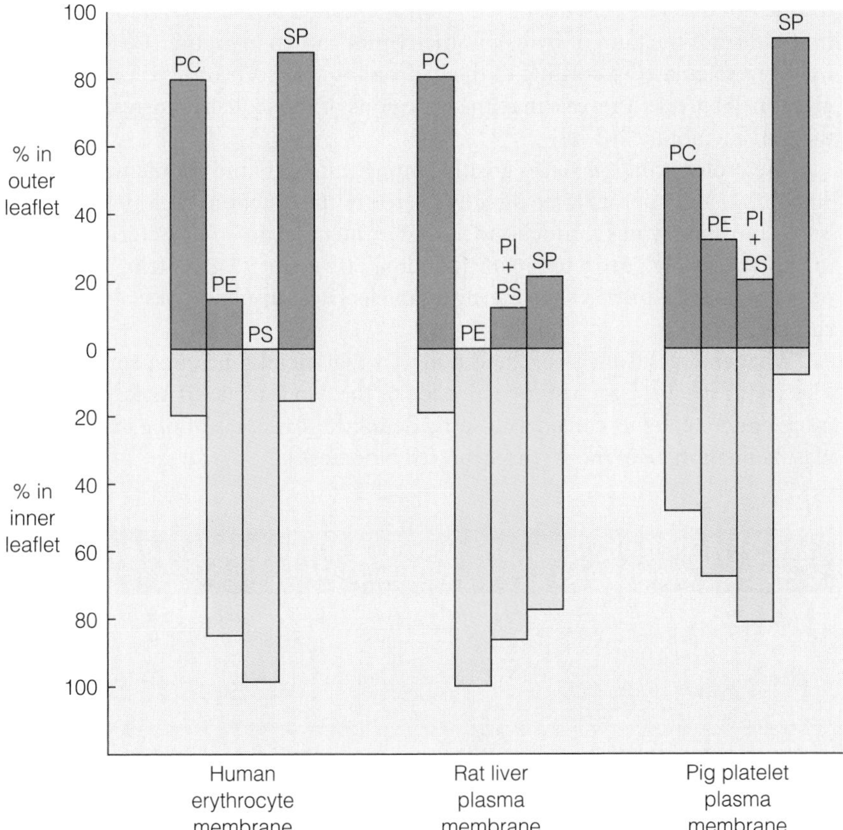

FIGURE **10.14**

Phospholipid asymmetry in plasma membranes. Lipid composition in the outer leaflet (green) and inner leaflet (gold) of the plasma membrane is graphed for three cell types. PC = phosphatidylcholine; PE = phosphatidylethanolamine; PS = phosphatidylserine; PI = phosphatidylinositol; SP = sphingomyelin.

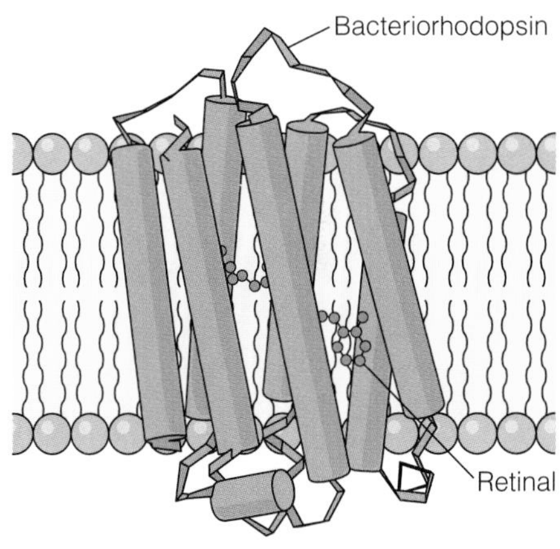

FIGURE **10.15**

An integral membrane protein. The protein shown is bacteriorhodopsin, which is used as a light-driven proton pump in certain bacteria. Seven helices span the membrane and hold a molecule of the pigment retinal.

Adapted from R. Henderson, J. M. Baldwin, J. A. Ceska, F. Zemlin, E. Beckman, and K. H. Downing, *J. Mol. Biol.* (1990) 213:899–929. © 1990 Academic Press.

tems, membranes exist as they do because they are *not* in equilibrium but rather represent dynamic, steady-state structures.

CHARACTERISTICS OF MEMBRANE PROTEINS

Membrane proteins possess special characteristics that distinguish them from other globular proteins. They often contain a high proportion of hydrophobic amino acids, particularly in the parts of the protein molecules that are embedded in the membrane. The segments of proteins that span membranes are often α-helical. Figure 10.15 depicts bacteriorhodopsin, an integral membrane protein whose structure has been solved to high resolution. Like many such proteins, it contains a bundle of seven α-helical segments that pass back and forth through

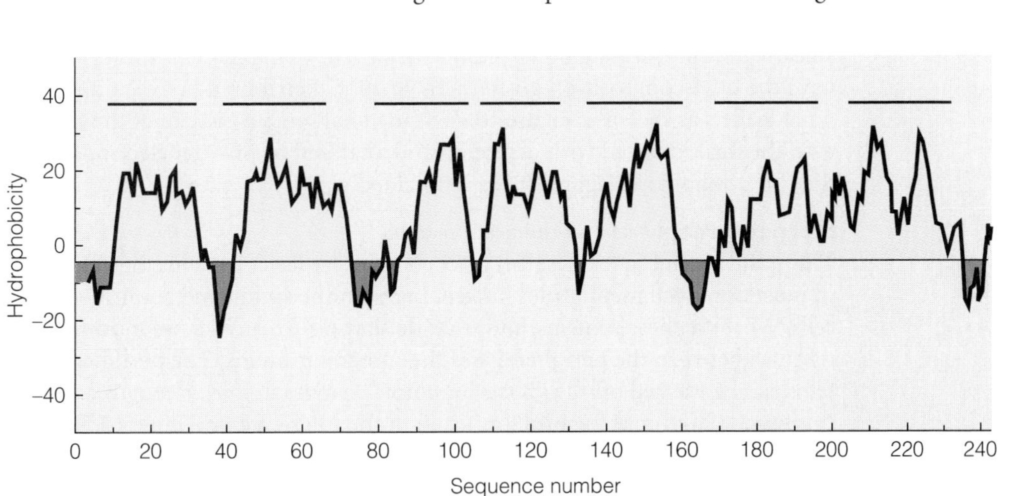

FIGURE **10.16**

Hydrophobicity plot for the bacteriorhodopsin molecule depicted in Figure 10.15. The hydrophobicity index has been calculated at each residue by the method of Kyte and Doolittle. The black bars show the approximate positions of the transmembrane helices shown in Figure 10.15.

Adapted from J. Kyte and R. F. Doolittle, *J. Mol. Biol.* (1982) 157:105–132.

the membrane. The presence of such transmembrane segments can sometimes be inferred from the kind of **hydrophobicity plot** shown in Figure 10.16. This plot has been calculated according to the hydrophobicity scale of Kyte and Doolittle given in Table 6.4. It reveals maxima in regions of the sequence corresponding to the transmembrane helices.

The protein content varies greatly among different kinds of membranes (see Table 10.4) and appears to be directly related to the functions each type of membrane must carry out. Mitochondrial inner membranes and bacterial cell wall membranes, which carry out many functions, are about 75% protein. The myelin of nerve fibers, which acts primarily as an electrical insulator, has a low protein content.

What emerges from all of these studies is that membranes are complex structures, with specific compositions for each of the two leaflets. To make the picture more concrete, let us consider in some detail the structure of one example, the plasma membrane of the erythrocyte (red blood cell).

TABLE **10.4 Protein, lipid, and carbohydrate content of some membranes**

	Percent by Weight		
Membrane	Protein	Lipid	Carbohydrate
Myelin	18	79	3
Human erythrocyte (plasma membrane)	49	43	8
Bovine retinal rod	51	49	0
Mitochondria (outer membrane)	52	48	0
Amoeba (plasma membrane)	54	42	4
Sarcoplasmic reticulum (muscle cells)	67	33	0
Chloroplast lamellae	70	30	0
Gram-positive bacteria	75	25	0
Mitochondria (inner membrane)	76	24	0

Source: Adapted from G. Guidotti, *Annu. Rev. Biochem.* (1972) 41:731.

THE ERYTHROCYTE MEMBRANE: AN EXAMPLE OF MEMBRANE STRUCTURE

The erythrocytes of mammals are among the simplest of all cells. In their mature state in the circulating blood, they have lost nucleus, mitochondria, and internal membranes and are essentially bags of hemoglobin. They can easily be lysed to release their contents and produce membrane *ghosts,* large vesicles that represent a nearly pure preparation of the plasma membrane of the cells. Erythrocyte ghosts have the lipid composition given in Table 10.3, distributed between inner and outer leaflets as shown in Figure 10.14. If the total protein content of the erythrocyte ghost is extracted with detergent and analyzed by SDS gel electrophoresis, the pattern shown in Figure 10.17a is obtained.

Separating Peripheral and Integral Proteins

The plasma membrane of erythrocytes contains far fewer proteins than are found in most other cell membranes, in keeping with the simple metabolism of these cells. What are these proteins, and what do they do? To answer, we must first distinguish between the peripheral and the integral proteins. The peripheral proteins can be washed off the ghosts by simple changes in ionic strength or pH. In this way, we find that the proteins listed in the top section of Table 10.5 are peripheral. Furthermore, all of these proteins turn out to be attached to the inside (cytoplasmic face) of the erythrocyte membrane. Electron microscopy after freeze-etching (see Tools of Biochemistry 10A) reveals a nearly smooth outer surface,

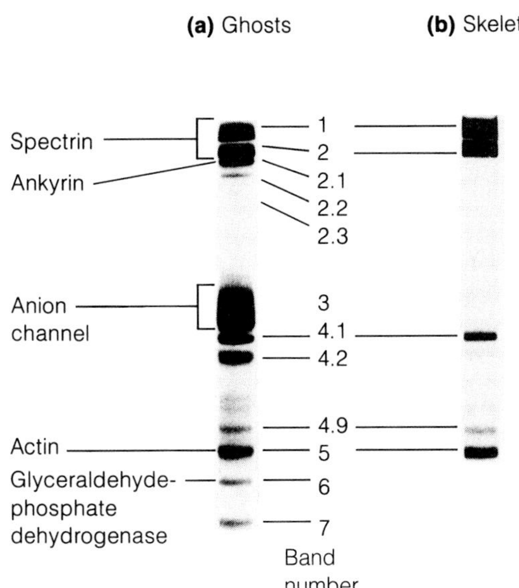

FIGURE **10.17**

Gel electrophoretic analysis of erythrocyte membrane proteins. (a) Proteins of the total erythrocyte ghost. Details concerning these proteins are given in Table 10.5. Glycophorin and band 4.5 protein do not show up here because they do not stain with the dye that was used. **(b)** Proteins of the erythrocyte skeleton. The skeleton consists of the peripheral membrane proteins that remain after integral proteins have been extracted with the detergent Triton X-100.

From D. Branton et al., *Cell* (1981) 24:24–32. © Cell Press.

TABLE **10.5 The major proteins of the human erythrocyte membrane**

Band No.[a]	Protein Name	Subunit Molecular Weight	Probable State of Assembly	Number of Copies per Cell	Role
Peripheral Proteins					
1	α-Spectrin	260,000	$\alpha_2\beta_2$ tetramers	10^5 tetramers	Membrane skeleton
2	β-Spectrin	225,000			
2.1	Ankyrin	215,000	Monomer	10^5	Links skeleton to band 3
*	Adducin	105,000 / 100,000	Heterodimer	3×10^4	
4.1	—	78,000	Monomer	2×10^5	Involved in spectrin junctions
4.2	Palladin	72,000	?	2×10^5	?
4.9	Demantin	48,000	Trimer?	5×10^4	Involved in spectrin–actin interaction
5	Actin	43,000	Oligomers of 12–17 units	5×10^5	Involved in spectrin junctions
*	Tropomyosin binding protein	43,000	Monomer	3×10^4	Binds tropomyosin
6	Glyceraldehyde-3-phosphate dehydrogenase	35,000	Tetramer	5×10^5	Glycolytic enzyme
*	Tropomyosin	29,000 / 27,000	Heterodimer	7×10^4	Binds to actin
7	—	29,000	?	5×10^5	?
8	—	23,000	?	10^5	?
Integral Proteins					
3	—	89,000	Dimer + tetramers	10^6 dimers	Anion channel
4.5		55,000	?	1.5×10^6	Glucose transport
	Glycophorin A	31,000	Dimer	4×10^5	Cell recognition
	Glycophorin B	23,000	?	$\sim 10^5$	Cell recognition
	Glycophorin C	29,000	?	$\sim 10^5$	Linkage to 4.1?

Source: Most of the data are from V. Bennett, *Annu. Rev. Biochem.* (1985) 54:273–304.

[a]Band numbers correspond to those in Figure 10.17. The glycophorins do not stain well with protein stains but can be detected by carbohydrate-specific stains.

*Components that do not constitute major bands on gels but have demonstrable roles in membranes.

with only integral membrane proteins and attached carbohydrates decorating the surface, whereas the inside of the membrane and the inner surface are rich in protein particles.

The major integral proteins of this membrane are the remainder of the proteins shown in Table 10.5: the anion channel (band 3), band 4.5, and the glycophorins. (We will identify these proteins later. Note that glycophorin and band 4.5 do not appear in Figure 10.17 because of the particular dye used in the analysis; some others are not seen because of their low abundance.) The integral proteins, along with much of the lipid material, can be extracted from intact membranes by use of the nonionic detergent Triton X-100.

The Protein Skeleton

The erythrocyte, like many other cells, has a complex "skeleton" of proteins underlying and attached to its plasma membrane.

Surprisingly, treatment with Triton X-100 leaves intact a protein **skeleton,** which retains the shape of the membrane ghost. The proteins of this membrane skeleton are identified in Figure 10.17b, and a schematic picture is shown in Figure 10.18. The skeleton is a two-dimensional network of some of the peripheral proteins, the major constituents being **spectrins,** actin, and band 4.1. The 200-nm-long fibers in this network are made up of $\alpha_2\beta_2$ tetramers of spectrin molecules. These very elongated molecules contain a large fraction of α helix and appear to be linked at their ends through short chains of actin molecules, together with the band 4.1 protein and adducin (see Table 10.5). The actin is also complexed with an erythrocyte-specific tropomyosin (see Chapter 8). The skeleton is anchored to the membrane itself by the protein **ankyrin,** which binds to both spectrin and the band 3 integral protein. Band 4.1 adds to the structure by also banding to glycophorin, another integral protein.

What purpose does this elaborate underpinning to the erythrocyte membrane serve? An obvious suggestion is that it helps to maintain the shape of the erythrocyte, despite the squeezings and buffetings a cell suffers in passing through the circulatory system. The erythrocyte is durable, typically surviving for about 120 days, or 10 million heartbeats. The discoid shape of the cell allows efficient exchange of O_2 and CO_2 to the hemoglobin inside, and even if that shape is momentarily deformed, the skeleton helps to regain it. Indeed, some forms of anemia, in which the symptoms are easy lysis of erythrocytes, can be traced to deficiencies in skeletal proteins. The kind of structure described here is not confined to erythrocytes; many other cell types contain similar but distinct membrane skeletons. It seems likely that connections exist between the membrane skeleton and the cytoskeleton (see Chapter 8), so that the membrane is linked to intracellular structure.

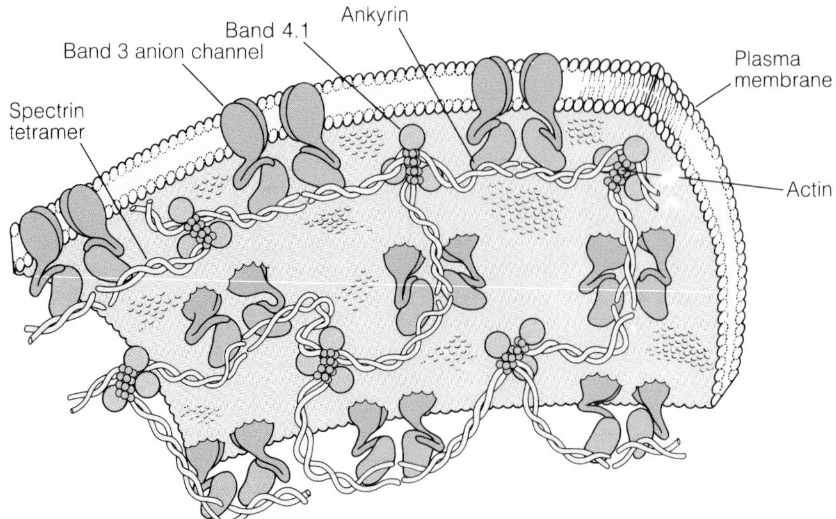

FIGURE **10.18**

Model of the postulated structure of the erythrocyte membrane skeleton. The proteins are identified in Figure 10.17 and Table 10.5. Note that ankyrin anchors the membrane to the skeleton by interacting with both spectrin and the integral band 3 protein (anion channel).

The Major Integral Membrane Proteins

The most abundant protein in the erythrocyte membrane is the band 3 protein, which is an anion carrier, facilitating the exchange of HCO_3^- for Cl^- across the erythrocyte membrane. (Recall from Chapter 7 the importance of bringing HCO_3^- into the erythrocytes to accomplish CO_2 transport; exchange for Cl^- keeps ionic balance.) Band 3 protein functions as a complex of 2 or 4 subunits. Each subunit chain crosses and recrosses the membrane a number of times, to create the channel through which the ions can be exchanged. The N-terminal portion of the protein extends into the cytosol, where it makes a number of interesting interactions. As mentioned previously, it contacts the skeletal protein ankyrin, providing a major attachment between the skeleton and the membrane. In addition, however, band 3 protein contacts a number of cytosolic proteins, including some glycolytic enzymes (glyceraldehyde-3-phosphate dehydrogenase and aldolase, for examples; see Chapter 13) and hemoglobin. The significance of these interactions is not wholly clear, but the fact that these enzymes are involved in bisphosphoglycerate (BPG) production and that the contact between band 3 and hemoglobin is in the BPG binding site (Chapter 7) is suggestive.

The other major integral red cell membrane proteins are the **glycophorins,** which apparently serve a variety of functions. Each of these proteins has an external carbohydrate-carrying domain, a single transmembrane helix, and a cytosolic C-terminal domain (see Figure 10.19). The O-linked carbohydrate groups on the external N-terminal domain carry sialic acid residues, giving a negative charge to

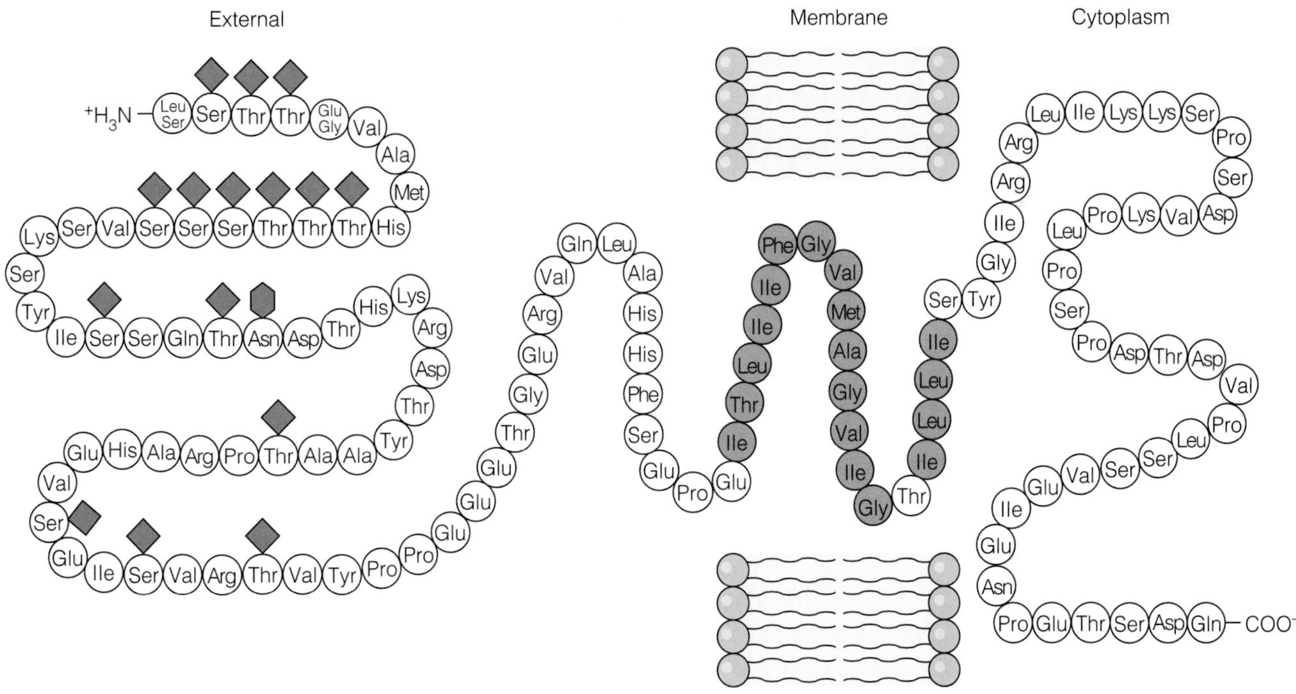

FIGURE **10.19**

The sequence and postulated structure of glycophorin A. This protein was the first integral membrane protein to be sequenced. The external (N-terminal) domain carries 15 O-linked and one N-linked oligosaccharides; together these constitute about 60% of the total protein mass. The single transmembrane helix is highly hydrophobic, whereas the cytosolic C-terminal domain is quite hydrophilic.

Adapted from V. T. Marchesi, *Semiu. Hemat.* (1979) 16:8.

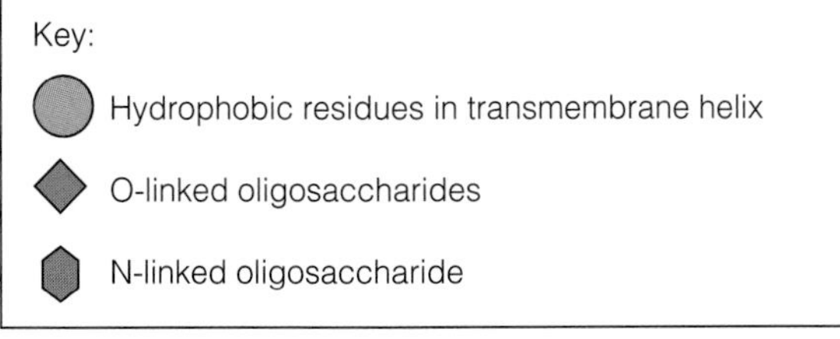

the exterior of the red blood cell. This may minimize the chances of such cells sticking to capillary walls during circulation. The role of the cytosolic domain is less clear, but in some glycophorins it appears to interact with band 4.1 protein, stabilizing skeleton–membrane adhesion.

The asymmetry of the orientation of integral proteins must be emphasized. Just as the erythrocyte membrane is asymmetric in its distribution of lipid components, it is equally asymmetric in its distribution of proteins, because each kind of protein is oriented in a particular direction. Just how these remarkable membrane structures are assembled is only now beginning to be understood. As discussed in Chapter 27, there appear to be special modes of protein synthesis that direct the placement of proteins in membranes and ensure their asymmetric orientation.

Transport Across Membranes

A cell or an organelle cannot be either wholly open or wholly closed to its surroundings. Its interior must be protected from certain toxic compounds, and yet metabolites must be taken in and waste products removed. Because the cell must contend with thousands of substances, it is not surprising that much of the complex structure of membranes is devoted to the regulation of transport.

In this section we consider the various ways in which molecules are transported across membranes. The three categories of transport—passive, facilitated, and active—are quite different and generally serve different purposes in the cell.

THE THERMODYNAMICS OF TRANSPORT

Before considering specific mechanisms of transport, it is useful to review some ideas that were first presented in Chapter 3. There we discussed the general thermodynamic principles governing the transfer of substances across membranes or surfaces. It was shown that the free energy change, ΔG, for transporting 1 mole of a substance from a region in which its concentration is C_1 to a place where its concentration is C_2 is given by

$$\Delta G = RT \ln \frac{C_2}{C_1} \tag{10.1}$$

The normal state of equilibrium for a substance that can pass through a membrane is equalization of the concentration on both sides.

According to this equation, if C_2 is less than C_1, ΔG is negative, and the process is thermodynamically favorable. As more and more substance is transferred (between two finite compartments), C_1 decreases and C_2 increases, until $C_2 = C_1$. At this point $\Delta G = 0$, and the system is at equilibrium. *Unless other factors are involved,* this equilibrium is the ultimate state approached by transport across any membrane: A substance that can traverse the membrane will eventually reach the same concentration on both sides. We can describe the same process in kinetic terms. If the molecules are wandering into the membrane at random, the number entering from each side will be proportional to the concentration on that side. When the concentrations become equal, the rates of transport in the two directions will be the same, and no net transport will occur.

There are three circumstances under which this equalization can be circumvented, and each is important in the behavior of real membranes:

1. A substance may be preferentially bound by macromolecules confined to one side of the membrane or may be chemically modified once it crosses. We may find that compound A is more concentrated inside a cell (in terms of total moles of A per unit volume) than outside. But much of A may be bound to some cellular macromolecules or may have been modified; that portion does not really count in equation (10.1), which simply states that the concentrations of *free* A on the two sides must be equal at equilibrium.

An appropriate example is oxygen in erythrocytes. If we were to measure the *total* oxygen concentration in an erythrocyte, we would find it higher than the concentration of O_2 in the surrounding blood plasma. But the total concentration inside the cell includes oxygen bound to hemoglobin. The *free* oxygen concentration in the fluids inside and outside an erythrocyte is the same at equilibrium.

2. A **membrane electrical potential** may be maintained across a membrane that influences the distribution of ions. This tendency can be expressed quantitatively in the following way. For an ion of charge Z, the free energy change for transport across a cell or organelle membrane now involves two contributions: the normal concentration term, as given in equation (10.1), plus a second term describing the energy change (or work involved) in moving a mole of ions across the potential difference. We consider a process in which ions are transported from *outside* to inside.

$$\Delta G = RT \ln \frac{C_{in}}{C_{out}} + ZF\Delta\psi \tag{10.2}$$

Here F is the Faraday constant (96.5 kJ mol^{-1}V^{-1}), and $\Delta\psi$ is the membrane potential in volts. We define the direction for $\Delta\psi$ as we did for transport (out→in). If $\Delta\psi$ is negative (in going from outside to inside, for example) and Z is positive, the $ZF\Delta\psi$ term in equation (10.2) makes a negative contribution to ΔG. That is, the transport of cations *into* the cell is favored. For anions, of course, the opposite is true; they will be driven out. If the potential difference is somehow maintained, the equilibrium state ($\Delta G = 0$) will *not* correspond to the same concentration of ions on the two sides of the membrane. However, energy must be expended continually to keep up the potential difference; otherwise migration of ions would neutralize it. Conversely, equation (10.2) may be interpreted to mean that if a difference in ionic concentration is maintained, a potential will be produced across the membranes (see Problem 9).

3. If some thermodynamically favored process is *coupled* to the transport, then the ΔG for this process must be included in the free energy equation. This is the general case of *active transport*, for which we can write

$$\Delta G = RT \ln \frac{C_{in}}{C_{out}} + \Delta G' \tag{10.3}$$

The quantity $\Delta G'$ could correspond to some thermodynamically favored reaction (like ATP hydrolysis) that was somehow coupled to the process of transport. This equation is a generalization of equation (10.2), now allowing a variety of processes—not just those that maintain an electrical potential difference—to participate in the transport.

Equalization across a membrane can be circumvented by binding of the substance to macromolecules, by maintaining a membrane potential, or by coupling transport to an exergonic process.

With this background, we turn now to the mechanisms whereby substances are passed through membranes. We may introduce the problem by asking two questions: (1) Does the transport occur against unfavorable concentration gradients, or is it eventually limited by equation (10.1)? That is, are equal free concentrations on both sides approached? (2) How fast does the transport occur? Some molecules that are not actively transported against a concentration gradient can still traverse some membranes very rapidly, whereas others are transported so slowly as to be effectively excluded.

PASSIVE TRANSPORT: DIFFUSION

Passive transport, or **passive diffusion,** is accomplished by the random wandering of molecules through membranes. The process is the same as the Brownian

The rate of passive transport, as measured by membrane permeability, is proportional to the diffusion and partition coefficients and inversely proportional to membrane thickness.

motion of molecules in any fluid, which is termed **molecular diffusion.** Passive transport ultimately results in the free concentration of the diffusing substance being the same on both sides of the membrane. The net rate of transport, J (in moles per square centimeter per second), is, as you might expect, proportional to the concentration difference $(C_2 - C_1)$ across the membrane:

$$J = -\frac{KD_1(C_2 - C_1)}{l} \quad (10.4)$$

where l is the thickness of the membrane, D_l is the **diffusion coefficient** of the diffusing substance in the membrane, and K is the **partition coefficient** for the diffusing material between membrane lipid and water (the ratio of solubilities of the material in membrane lipid and water). For ions and other hydrophilic substances, K is a very small number, with the result that diffusion of such substances through lipid membranes is extremely slow. There is simply not enough of a hydrophilic substance dissolved in the bilayer to provide rapid transport. In agreement with equation (10.1), equation (10.4) says that net transport will stop when $C_2 = C_1$. If C_1 and C_2 are expressed in mol/cm^3 and l in cm, then D_l has the units cm^2/s. D_l is not the same as the diffusion coefficient (D) that the same molecule would have in aqueous solution, for it depends not only on the size and shape of the molecule but also on the viscosity of the membrane lipid.

We usually do not know either K, D_l, or the exact thickness of the membranes involved, so we often describe the rate of passive transport in terms of a **permeability coefficient, *P*,** which can be measured by direct experiment:

$$J = -P(C_2 - C_1) \quad (10.5)$$

By comparing equations (10.5) and (10.4), we see that P is given by

$$P = \frac{KD_1}{l} \quad (10.6)$$

with units of cm/s.

Table 10.6 lists permeability coefficients for a number of small molecules and ions in membranes. The low P values of the ions are what we would expect, because ions, as already mentioned, have low values of K. However, the relatively large permeability value for water is surprising. Despite their hydrophobicity, biological membranes are not, in fact, very good barriers against water. Although the reasons for this are not entirely clear, it is probably fortunate for life, for it allows cells to exchange water readily with their surroundings. When water loss is

TABLE 10.6 Permeability coefficients from some ions and molecules through membranes

Permeability Coefficient (cm/s) for	Synthetic Membrane (Phosphatidylserine)	Biological Membrane (Human Erythrocyte)
K^+	$<9 \times 10^{-13}$	2.4×10^{-10}
Na^+	$<1.6 \times 10^{-13}$	10^{-10}
Cl^-	1.5×10^{-11}	1.4×10^{-4}*
Glucose	4×10^{-10}	2×10^{-5}*
Water	5×10^{-3}	5×10^{-3}

Source: Data from M. K. Jain and R. C. Wagner, *Introduction to Biological Membranes* (New York: Wiley, 1980).

*Facilitated transport. Note that whenever facilitated transport is encountered, the permeability coefficient rises dramatically.

to be strenuously avoided, as in the leaves of desert plants, waxy substances with their much more hydrophobic structures provide a nearly impermeable barrier. It seems likely that the conformational flexibility of membrane lipids above the gel–liquid crystal transition creates "holes" or "pockets" that allow the diffusion of small neutral molecules like H_2O or O_2.

FACILITATED TRANSPORT: ACCELERATED DIFFUSION

For many substances, the slow transport provided by passive diffusion is simply insufficient for the functional and metabolic needs of cells, and means must be found to increase transport rates. For example, exchange of Cl^- and HCO_3^- is essential to erythrocyte function. If we examine the permeability of erythrocyte membranes to chloride or bicarbonate ions, we find permeability coefficients of about 10^{-4} cm/s. This value is about 10 million times greater than the permeability coefficient for ions in pure lipid bilayers like the artificial phosphatidylserine membrane listed in Table 10.6. Clearly, some special mechanism is required to account for this difference. Two general types of **facilitated transport,** or **facilitated diffusion,** are known to occur. One type involves pores formed by transmembrane proteins (Figure 10.20a). The other type is mediated by transmembrane carrier molecules (Figure 10.20b).

Facilitated transport, via pores or carriers, can increase the rate of diffusion across a membrane by many orders of magnitude.

Pore-Facilitated Transport

We have already encountered the apparatus responsible for facilitated transport of Cl^- and HCO_3^- in erythrocyte membranes: it is the transmembrane protein called the band 3 protein, or anion channel (see page 333). The band 3 protein passes through the membrane, forming a channel or *ion pore* through which Cl^- and HCO_3^- can pass. As explained in Chapter 7, much of the CO_2 generated in tissues is transported as HCO_3^-, which is formed in erythrocytes by the action of carbonic anhydrase on CO_2. Exit of the HCO_3^- is balanced by influx of Cl^-; this both maintains charge balance and facilitates O_2 release (see Chapter 7). The band 3 protein doesn't just form a hole in the membrane for the passage of ions. Rather, the channel is very selective, exchanging HCO_3^- and Cl^- on a 1:1 basis. By contrast, such facilitated transport is not necessary for O_2. This tiny nonpolar molecule can move rapidly through the membrane by passive diffusion.

$$CO_2 + H_2O \xrightleftharpoons{\text{Carbonic anhydrase}} HCO_3^- + H^+$$

Another facilitated diffusion process in the erythrocyte membrane is essential for the cell's survival. The small energy demands of an erythrocyte are met by

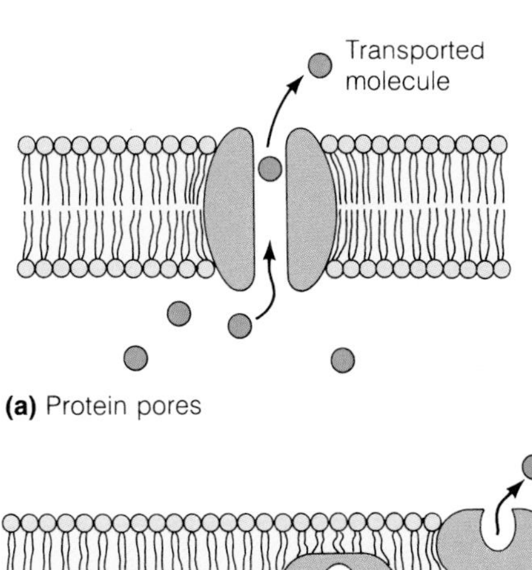

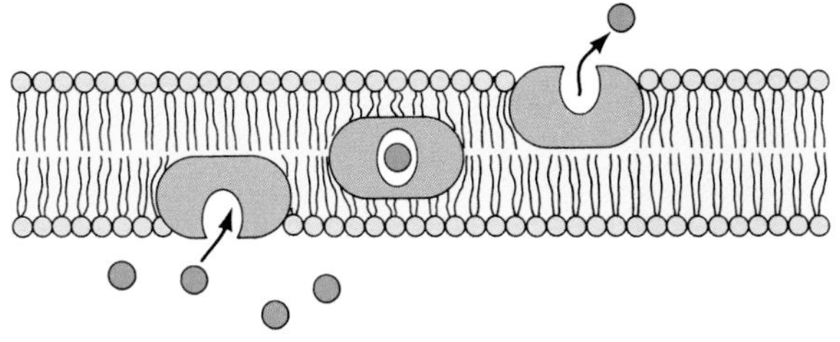

FIGURE **10.20**

The two major mechanisms for facilitated transport. (a) Protein pores. **(b)** Carrier molecules.

glucose, which is readily available in the surrounding blood plasma. But as Table 10.6 shows, the passive transport of glucose through artificial phospholipid membranes is agonizingly slow: $P = 4 \times 10^{-10}$ cm/s. Therefore, erythrocytes and other cells need a mechanism for the facilitated diffusion of glucose. This facilitation, like facilitated ion exchange, appears to be accomplished by a transmembrane protein that functions as a pore, increasing the diffusion rate 50,000-fold. This *glucose transport protein* (band 4.5; see Table 10.5) is a 492 residue molecule with 12 membrane-spanning helices. The transport protein is quite discriminating; for example, D-glucose is transported orders of magnitude more rapidly than L-glucose. Facilitated transport of metabolites like glucose appears to be a common feature in cells.

Ion channels can be weapons as well as necessities. Many pathogenic bacteria synthesize and excrete protein toxins that act by punching holes in the plasma membranes of cells of the host organism. Such channels (called **ionophores**) allow the indiscriminate efflux and influx of ions, destroying the precise ion balance that cells must maintain to live. An example, shown in Figure 10.21, is the α-toxin (α-hemolysin) from *Staphylococcus aureus.* This protein is made up of seven subunits, which combine when they encounter an appropriate cell surface receptor to produce the cylindrical structure shown. The stem β-barrel is very hydrophobic and forms a hollow tube passing through the cell membrane.

In defense against bacteria, many organisms have evolved *antibiotic* substances. A considerable number of these substances are also ionophores, directed specifically against bacterial membranes. They disrupt the carefully maintained differences in ion concentrations across bacterial membranes, thereby killing the microorganisms. For example, *gramicidin A* produced by the bacterium *Bacillus brevis* acts as a cation-specific ion pore, allowing a breakdown in the unequal ratio of K^+ and Na^+ normally maintained between the inside and outside of living cells (see below). Gramicidin A is a 15-residue polypeptide, containing alternating L- and D-amino acids (Figure 10.22a). Gramicidin adopts an open helical conformation when dissolved in the membrane, but one molecule of the antibiotic is long enough to traverse only half the thickness of the membrane. An open pore forms only when two gramicidin molecules line up to form an end-to-end dimer (Figure 10.22b). Potassium ions (and, to a lesser extent, sodium ions) can then pass through the channel. When we study the ionic conductance of membranes at very low gramicidin concentrations, we see a remarkable phenomenon: The conductance changes up or down in quantized steps. Each step corresponds to the opening or closing of an individual channel, probably by the diffusing together or separating of two gramicidin molecules, one in each membrane leaflet. It has been found that under some enviromental circumstances gramicidin can

FIGURE **10.21**

The channel-forming hemolysin from *Staphylococcus aureus.* Ribbon drawings of the α-hemolysin heptamer, viewed **(a)** looking down the sevenfold axis and **(b)** perpendicular to the sevenfold axis. **(c)** One protomer extracted from the heptamer structure. Structural features are labeled. The heptamer is 10 nm in diameter and 10 nm in length, as measured along the sevenfold axis. The stem, which penetrates the membrane, is about 6 nm long.

Structure determined by M. Hobaugh, L. Song, and E. Gouaux. Reprinted from E. Gouaux, *Curr. Opin. Struct. Biol.* (1997) 7:556.

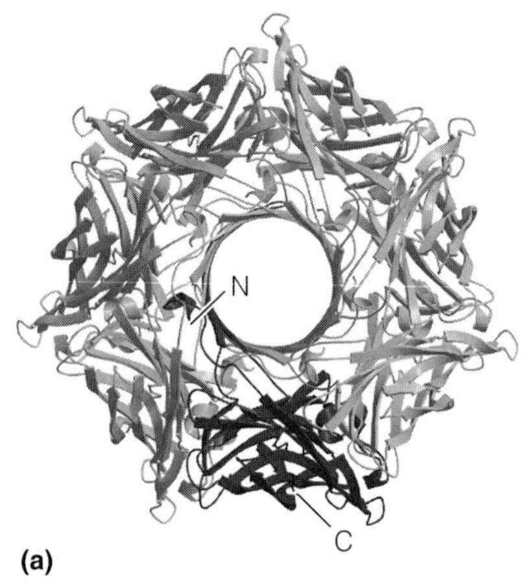

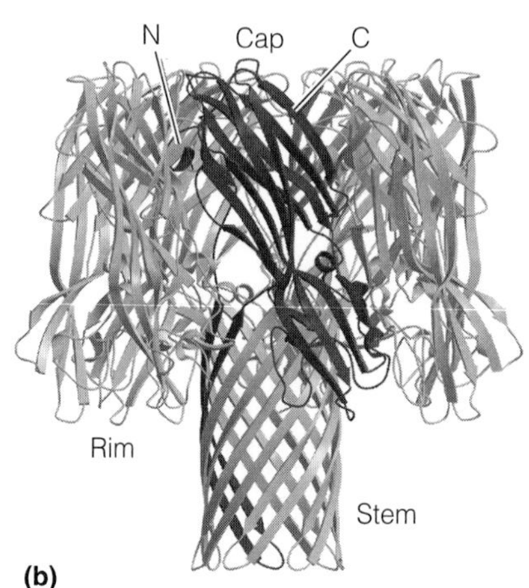

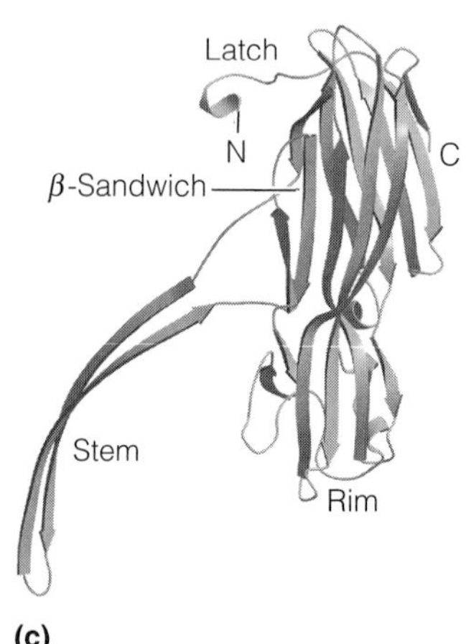

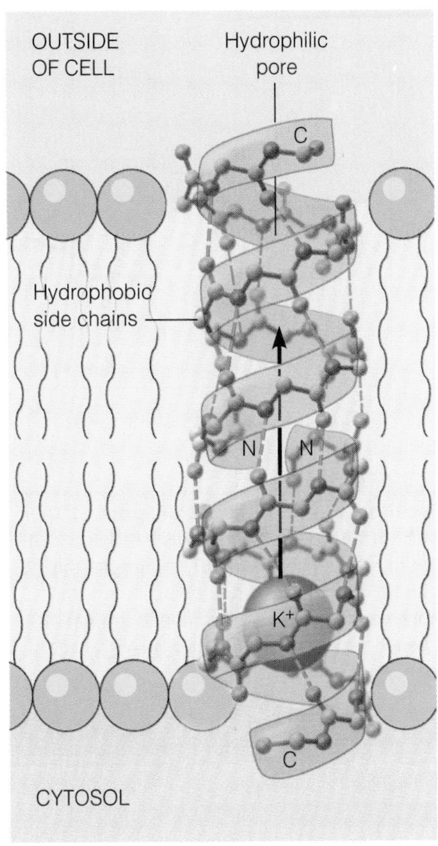

FIGURE **10.22**

Gramicidin A, an antibiotic that acts as an ion pore. Two molecules of gramicidin A form a pore through the membrane by adopting a helical conformation, with their hydrophobic side chains in contact with the lipid. The inside of the helix forms the hydrophilic pore. The hydrogen bonding in this open helical structure resembles that in β-sheet polypeptides. This is possible because of the alternating D and L residues.

adopt different, double-helical structures. However, that shown in Figure 10.22 is believed to predominate in bilayers.

Carrier-Facilitated Transport

Antibiotic ionophores also provide the best known examples of **carrier-facilitated transport** (see Figure 10.20b). For example, *valinomycin*, produced by a *Streptomyces*, has the structure shown in Figure 10.23. When complexed with K^+ it is a cyclic polypeptidelike molecule, involving three repeats of the sequence (D-valine)–(L-lactate)–(L-valine)–(D-hydroxyisovalerate). Its folded conformation presents an outside surface rich in $—CH_3$ groups and an interior cluster of nitrogens and oxygens that is well suited to chelating cations. The dimensions of the interior cavity nicely accommodate a K^+ ion but do not fit other cations as well. This structure is exactly what is needed for a cation carrier: The outer surface is hydrophobic, making the molecule soluble in the lipid bilayer, whereas the inside mimics in some ways the hydration shell that the cation would have in aqueous solution. A molecule like valinomycin can diffuse to one surface of a membrane, pick up an ion, and then diffuse to the other surface and release it. There is no *directed* flow, but the carrier in effect increases the solubility of the ion in the membrane. We could say that it increases the factor K in equation (10.6).

A number of other ion-carrier antibiotics have the same kind of structure. These molecules are either cyclic molecules or linear chains that can fold into cyclic structures. Their relative affinities for different ions vary greatly. For example, valinomycin has nearly a 20,000-fold preference for K^+ over Na^+, whereas the antibiotic *monensin* prefers Na^+ by 10-fold.

Distinguishing Different Types of Diffusion

How can facilitated diffusion be distinguished from passive diffusion? Aside from the generally much higher transport rate, a simple test is that facilitated transport systems are *saturable*. Any membrane has a limited number of carriers or pores. Each carrier can handle only one molecule or ion at a time; each pore can accommodate only one or a few ions or molecules at any moment. Thus, if we measure the rate of transport at increasingly high concentration differences of a substance transported across the membrane, a limiting rate is approached when all carriers or pores are busy (Figure 10.24). The rate of passive transport, on the other hand, increases linearly with the concentration difference, as expressed in equation (10.4) or (10.5) and Figure 10.24. In passive transport there are no sites to saturate, at least at any concentrations of transported molecules usually studied.

There is also an easy way to distinguish between pore-facilitated and carrier-facilitated diffusion. The latter should be extremely sensitive to membrane fluidity,

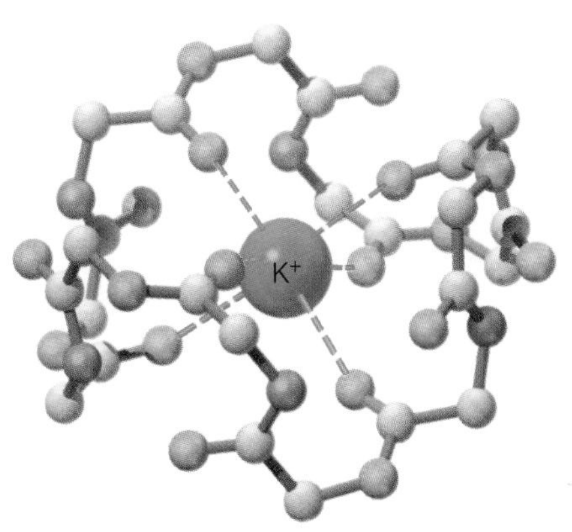

FIGURE **10.23**

Valinomycin, an antibiotic that acts as an ion carrier. The outside of this roughly spherical cyclic polymer is hydrophobic. The central cavity surrounded by oxygens (red) complexes a K^+ ion. Nitrogens are shown in blue, oxygens in the polypeptide backbone are red, and carbons are gray. The surface is covered with CH_3 groups (not shown).

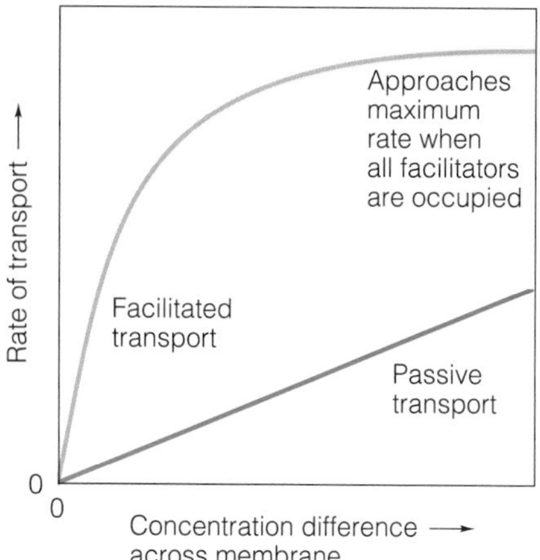

FIGURE **10.24**

Facilitated and passive transport. If the rate of transport is plotted against the concentration difference driving the transport, the graphs show that the rate of facilitated transport reaches a limit at high concentration difference, whereas the rate of passive transport increases linearly.

because the carrier must actually move in the membrane. If the temperature of a membrane is lowered below its fluid–gel transition temperature, transport by a carrier like valinomycin virtually ceases. Transport by a pore structure like gramicidin A, on the other hand, is affected little by temperature changes. A simple analogy can be made to a ferry and a bridge: If the river freezes, the ferry is stopped, but the bridge can continue to transport.

In conclusion, we must emphasize that even though facilitated transport is sometimes very fast and very selective, it is still only a special form of diffusion. Pores transport faster because they provide open gates; carriers effectively increase the solubility of the substance in the membrane. The equilibrium state for a system exhibiting facilitated transport is the same as for passive transport—the substance transported will eventually end up at equal free concentrations on the two sides of the membrane.

ACTIVE TRANSPORT: TRANSPORT AGAINST A CONCENTRATION GRADIENT

In active transport, substances are moved across a membrane against a concentration gradient. Coupling ATP hydrolysis to the transport usually provides the required free energy.

Facilitated diffusion is useful in many biochemical processes, but in some situations it is imperative that cells or cellular compartments be able to transport substances *against* concentration gradients, even very unfavorable ones. To take an extreme example, under some circumstances a calcium ion ratio of 30,000 must be established across membranes of the sarcoplasmic reticulum in muscle fibers (see Chapter 8). According to equation (10.1), this ratio corresponds to $\Delta G = +26.6$ kJ/mol, a formidable barrier. Nevertheless, this ratio is built up and maintained in living cells. Such transport against a concentration gradient is called **active transport.** Clearly, to pump ions against a gradient requires a free energy source of some kind. As you might expect, this energy usually comes from the hydrolysis of ATP. Altogether, it is estimated that most cells spend about 25% of their ATP just on active transport. However, the hydrolysis of ATP can be coupled to transport in a number of different ways, some of them rather indirect. To give an idea of the range of these mechanisms, we shall now consider some specific examples.

Ion Pumps: Direct Coupling of ATP Hydrolysis to Transport

The Na^+–K^+ pump acts in all cells to maintain higher concentrations of K^+ inside and Na^+ outside.

The best known physiological example of active transport is the maintenance of sodium and potassium gradients across the plasma membranes of cells. The fluid surrounding cells in most animals is about 140 mM in Na^+ and 5 mM in K^+. Yet animal cells maintain an Na^+ concentration of about 10 mM and a K^+ concentration of about 100 mM in their cytosol.

Even though Na^+ and K^+ pass very slowly through membranes by passive diffusion, such inequalities would ultimately vanish unless something were done to keep K^+ moving in and Na^+ out. This movement is accomplished by the action of the **sodium–potassium pump** (Figure 10.25). This molecular machine consists of large chains (α) of 113 kilodaltons each and smaller (β) subunits of 55 kilodaltons each. The α subunit is directly involved in the transport process and is an enzyme that hydrolyzes ATP. The free energy change in that reaction is used to drive the transport. The α subunit traverses the membrane 10 times, forming a multihelix channel. The site for ATP binding and phosphorylation lies on the cytoplasmic side (see Figure 10.25). Facing the outside are multiple sites for binding of **cardiotonic steroids,** including ouabain and **digitoxin** (digitalis). The medical importance of these is discussed below. The β subunit has a single membrane-traversing helix and carries a large (20 kDa) polysaccharide on the outer surface. The role of the polysaccharide or the β-protein is unknown. Although the single $\alpha\beta$ pair will function in reconstituted systems, the actual quaternary structure in cell membranes remains uncertain. Many believe that $(\alpha\beta)_2$ dimers predominate. The sodium–potassium pump is only one member of a large class of ATPases that function in active transport.

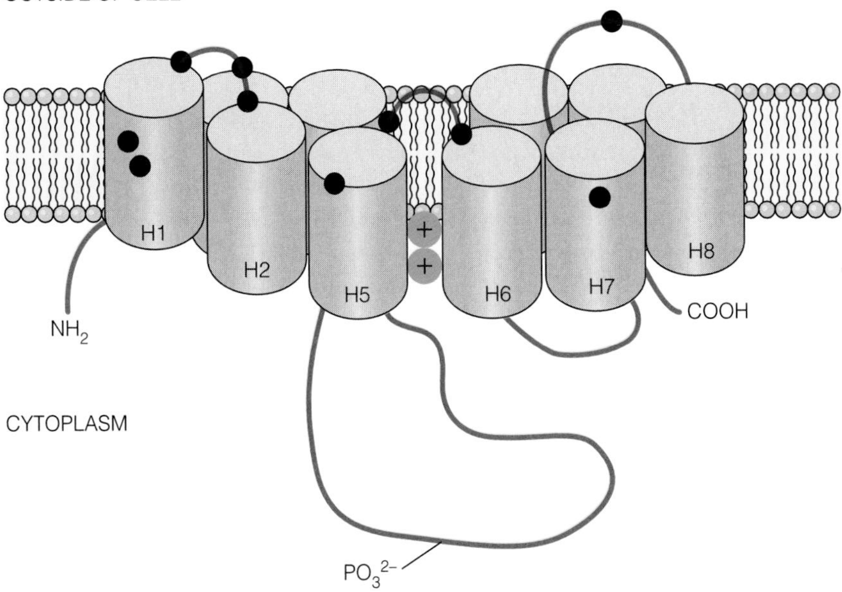

FIGURE **10.25**

A model for α subunit of the Na⁺, K⁺-ATPase. The molecule is shown in the open-to-cytoplasm conformation, with two K^+ (green) bound in the channel. The phosphorylation site is indicated, as well as the multiple sites for ouabain binding (black dots). Locations of the transmembrane helices and overall conformation are hypothetical. The β subunit is not shown, but passes through the membrane adjacent to the α subunit.

Adapted from J. B. Lingrel, J. Argüello, J. von Huysse, and J. A. Kuntzweiler, *Ann. N. Y. Acad. Sci.* (1997) 834:196–206.

Before we look at how the pump works, let us consider what it must do. Present estimates indicate that about two K^+ ions are pumped into the cell and three Na^+ ions are pumped out for every ATP hydrolyzed. Is this estimate reasonable from a thermodynamic point of view? To answer this, we calculate the free energy required to take 3 moles of Na^+ from 10 mM to 140 mM and 2 moles of K^+ from 5 mM to 100 mM at 37°C. First, let us use equation (10.2) to calculate the free energy required to transport 3 moles of Na^+ from within the cell to outside. We must take into account the membrane potential of about 0.07 volt. The inside of the membrane is more negative than the outside, so this potential opposes the flow. Per mole of Na^+, we have

$$\begin{aligned}\Delta G &= RT \ln \frac{C_{Na^+}\,(\text{out})}{C_{Na^+}\,(\text{in})} + Z_{Na^+} F\,\Delta\psi_{\text{in}\rightarrow\text{out}} \\ &= 8.314\ \text{J/K}\cdot\text{mol} \times 310\ \text{K} \times \ln \frac{140}{10} + 1 \times 96{,}480\ \text{J/volt}\cdot\text{mol} \times 0.07\ \text{volt} \qquad (10.7) \\ &= 6800\ \text{J/mol} + 6750\ \text{J/mol} = 13{,}550\ \text{J/mol} = 13.55\ \text{kJ/mol}\end{aligned}$$

For 3 moles, then, we have $\Delta G = 3 \times 13.55$ kJ = 40.65 kJ.

When K^+ is transported inward, the membrane potential is working in favor of the flow. Per mole of K^+, we have

$$\begin{aligned}\Delta G &= 8.314\ \text{J/K}\cdot\text{mol} \times 310\ \text{K} \times \ln \frac{100}{5} + 1 \times 96{,}480\ \text{J/volt}\cdot\text{mol} \times (-0.07\ \text{volt}) \\ &= 7720\ \text{J/mol} - 6750\ \text{J/mol} = 970\ \text{J/mol} = 0.97\ \text{kJ/mol}\end{aligned}$$

or, for 2 moles, $\Delta G = 1.94$ kJ. Note that the membrane potential is nearly sufficient in itself to maintain the K^+ gradient. The total free energy requirement for the outward transport of 3 moles of Na^+ and the inward transport of 2 moles of K^+ is then

$$\Delta G_{total} = 40.65 \text{ kJ} + 1.94 \text{ kJ} = 42.59 \text{ kJ}$$

At first glance, it would appear that the hydrolysis of 1 mole of ATP would not provide the necessary energy, for we have stated that $\Delta G°'$, the standard state free energy change for ATP hydrolysis under physiological conditions, is about −30 kJ/mol. In most cells, however, ATP is in much higher concentration than ADP or P_i, so the actual free energy change per mole is more like −45 to −50 kJ/mol. Thus, ATP hydrolysis is sufficient to maintain these concentration gradients under the observed stoichiometry of transport, but it could not transport any more than 3 Na^+ and 2 K^+ for each ATP. Furthermore, if the 3 + 2 estimate is correct, it means that the pump is very efficient.

Despite the transport against strong gradients, the sodium–potassium pump involves no violation of thermodynamic principles. The only requirement is that ATP hydrolysis and transport be *coupled.* This coupling is apparently accomplished in a multistep process. A current model for the entire process is diagrammed as a cycle in Figure 10.26. Extensive study has led to a model for the mechanism of the Na^+–K^+ pump (and others like it). It is proposed that the pump can exist in two kinds of conformations, one open to cytosol, the other open to the cell's surroundings. (See Figure 10.26.) Transition to the cytosol-open conformation, which allows K^+ release and Na^+ uptake, is triggered by binding of ATP and release of phosphate. Transition to the outside-open state, which permits Na^+ release and K^+ uptake, occurs upon phosphorylation of the α subunit and release of ADP.

The outside-open state designated E-P in Figure 10.26 has an especially high affinity for cardiotonic steroids, like digitoxin and ouabain. These agents inhibit the Na^+–K^+ pump by locking it in this conformation. Such inhibition has major effects on muscles, especially in the heart. The accumulation of Na^+ in cells leads to measures to reduce it, including a Ca^{2+}–Na^+ exchange process catalyzed by another pump. The resulting increase in Ca^{2+} in the sarcoplasmic reticulum of heart muscle cells leads to much stronger contractions (see page 263). This is why substances like digitoxin and ouabain are used as heart stimulants.

An important point should be noted for future reference: *A pump driven backward can act as an energy generator.* That is, the same kind of molecular mechanism described above, if allowed to pass a substance down a gradient, can sometimes be used to *generate* ATP. In fact, this mechanism is probably the major way in which ATP is produced in living organisms (see Chapter 15).

Cotransport Systems

In cotransport, the movement of a substance through the membrane is coupled to the spontaneous transport of another substance.

There are other kinds of active transport that do not depend directly on ATP as an energy source but employ ATP hydrolysis in an indirect way. You can imagine how this might occur, if you consider that the kind of ATP-driven ion pumps described above can generate gross inequalities of ion concentrations across membranes. These ion gradients are far from equilibrium and hence represent in themselves a potential source of free energy. The **sodium–glucose cotransport system** of the small intestine (Figure 10.27) is an example of how an ion gradient is used in driving transport. The transport of each glucose molecule from within the intestine into the cells of the intestine wall is accompanied by the simultaneous movement of one Na^+ ion in the same direction. Because a favorable Na^+ gradient is maintained by the ATP-driven Na^+–K^+ pump of these cells, glucose can be transported against an unfavorable gradient in glucose concentration. Glucose "piggybacks"

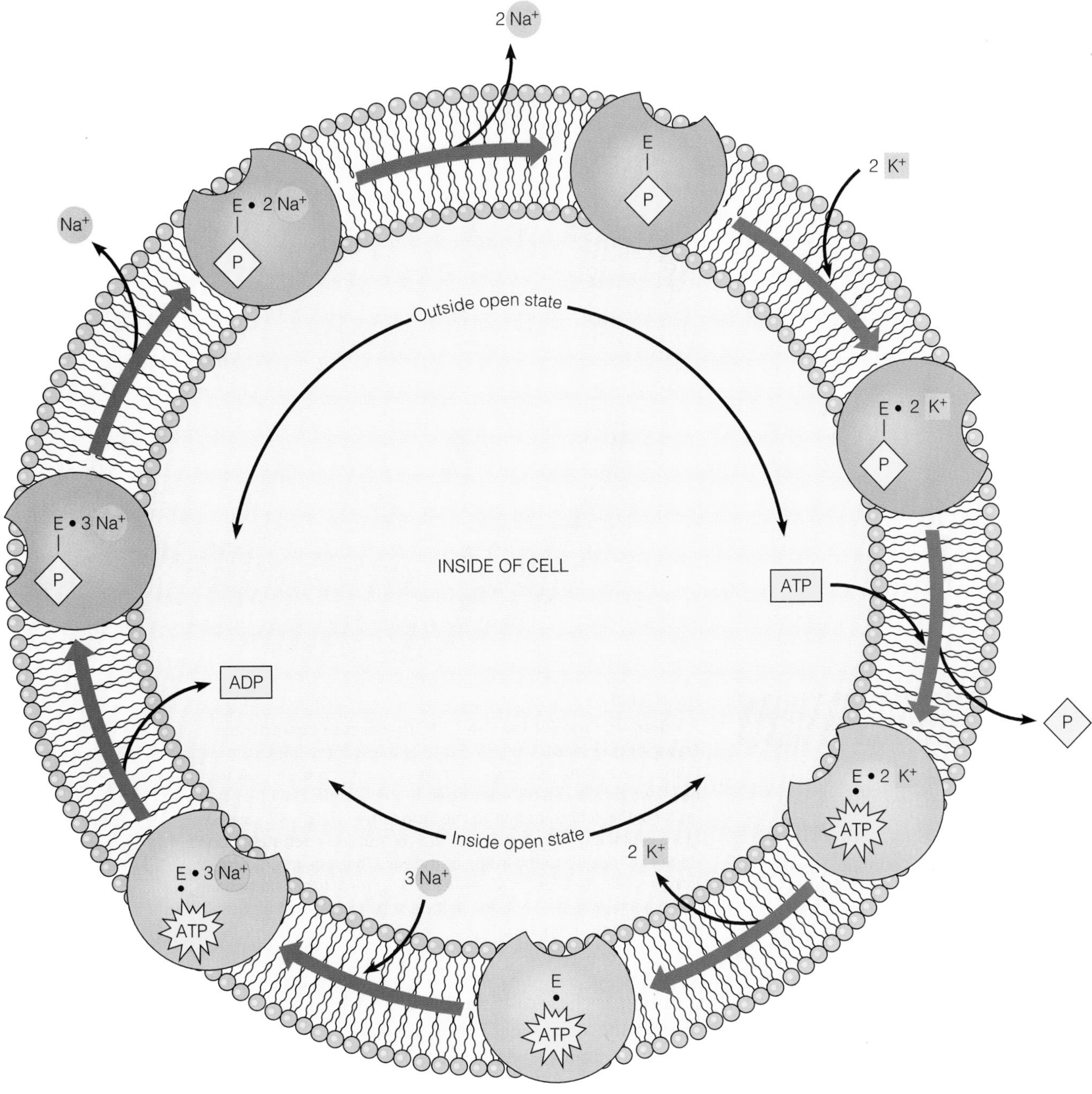

FIGURE **10.26**

A schematic diagram of the functional cycle of the Na^+–K^+ pump. The α subunit is believed to have two states, one open to the outside (dark brown), the other to the inside (blue). A dot (·) between two symbols indicates noncovalent binding, a line (|) indicates covalent attachment (as in phosphorylation).

on the thermodynamically favored Na^+ transport. When transport of the two molecules is in the same direction, as in this instance, it is referred to as **symport;** when in the opposite direction, it is called **antiport.**

A large number of such cotransport systems are known, many of them utilized to move nutrients into cells. A few examples are listed in Table 10.7. Many use the Na^+ gradient as a driving force, but some, like the *lactose permease system* in *E. coli,* depend on an H^+ gradient. As we shall see in later chapters, generation of H^+ gradients is a central step in energy production by most cells.

Transport by Modification

One other method cells have of achieving transport against a gradient uses the following trick: Suppose a molecule, on moving into a cell by passive or facilitated

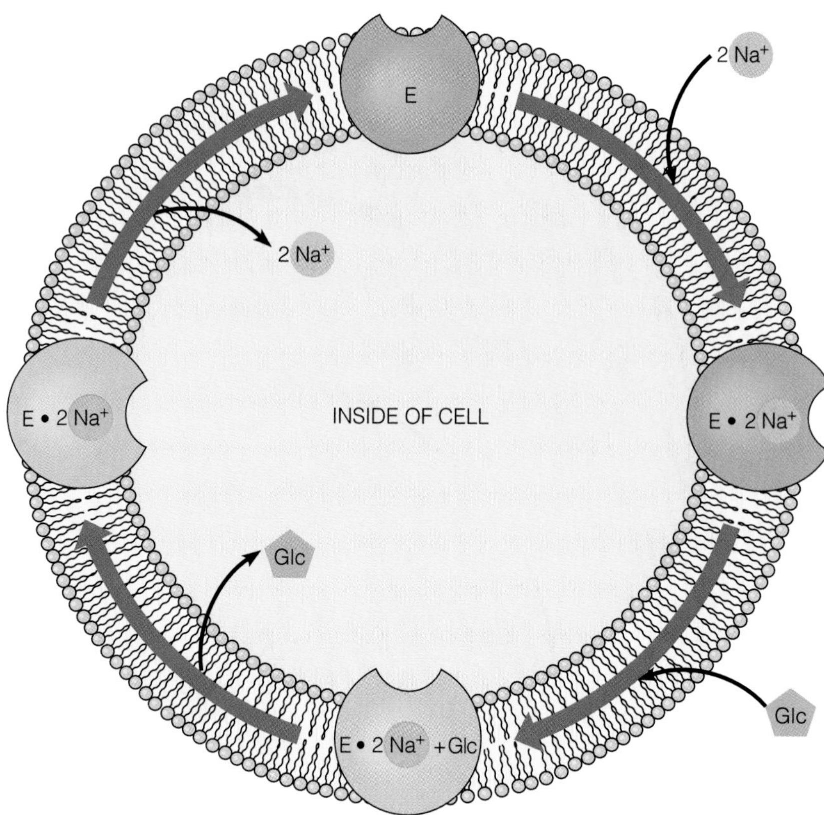

FIGURE **10.27**

A schematic model for the sodium–glucose cotransport (symport) system. As in the case of the sodium–potassium pump, the sodium–glucose cotransport channel is presumed to have two possible states—one open to the outside, the other to the inside of the cell. Transition to the inside-open state is stimulated by glucose binding to E · Na^+. Return to the outside-open state occurs upon Na^+ release to the inside of the cell. The sodium gradient from inside to outside provides the driving force for this otherwise unfavorable process. That gradient must be maintained by the sodium–potassium pump.

TABLE **10.7 Some cotransport systems**

Molecule Transported	Ion Gradient Used	Organism or Tissue
Glucose	Na^+	Intestine, kidney of many animals
Amino acids	Na^+	Mouse tumor cells
Glycine	Na^+	Pigeon erythrocytes
Alanine	Na^+	Mouse intestine
Lactose	H^+	*E. coli*

In transport by modification, a substance that has diffused through a membrane is modified so that it cannot return.

diffusion, is chemically modified in such a way that it can no longer pass back through the membrane. The net result is that quantities of the modified molecule steadily accumulate within the cell. This method is used by many bacteria for the uptake of sugars. The sugars are phosphorylated, either during their diffusion through the membrane or as soon as they emerge into the cytosol. Membranes are impermeable to the charged, phosphorylated monosaccharides, and thus these products remain in the cell. In the best-studied example, the **phosphotransferase system** of *E. coli,* transport is facilitated by a transmembrane protein, and the sugar molecule is apparently phosphorylated by ATP while in the pore. The process has the added convenience that phosphorylation of monosaccharides is, as discussed in Chapter 13, the first step in their metabolic utilization. Thus, the sugars taken up by *E. coli* are already primed for metabolism. Although this transport mechanism looks very different from the direct coupling in ion pumps, they are basically the same. In both cases a high-energy phosphate compound (ATP) has been hydrolyzed to accomplish the directed transport of a molecule across the membrane.

In this section we have described only a few examples of specific membrane transport. You will continue to encounter this phenomenon in later chapters on metabolism. To give an idea of how important specific transport is, Figure 10.28

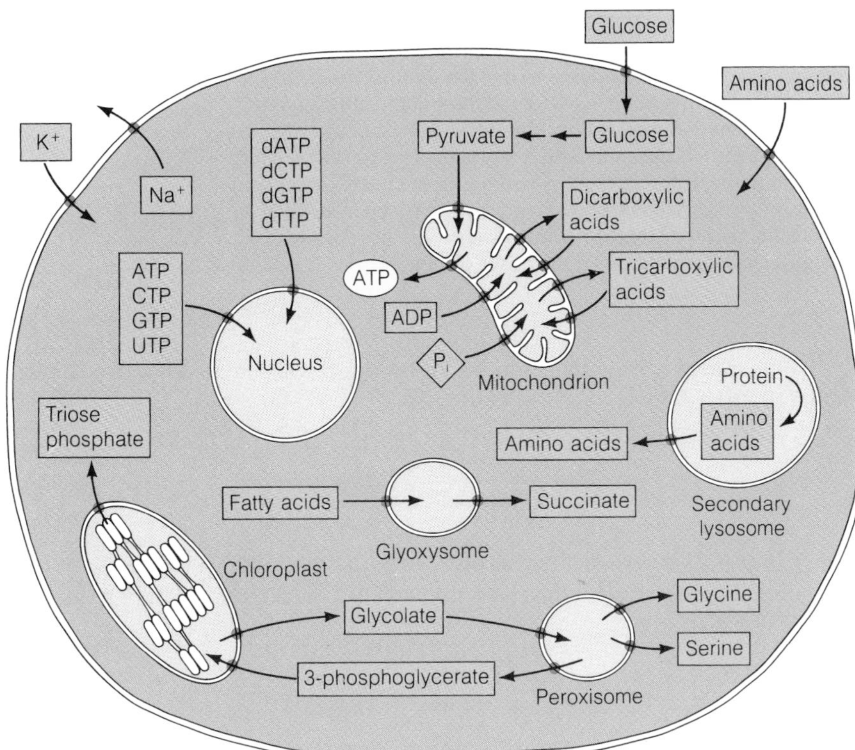

FIGURE **10.28**

Specific transport processes. This composite plant–animal cell illustrates some of the most important specific transport processes. All of the substances shown here, and many more, are transported in specific directions across cellular membranes. Red dots signify known transport proteins.

illustrates some of the known examples, using a hypothetical cell containing some characteristics of both plant and animal cells.

Excitable Membranes, Action Potentials, and Neurotransmission

We close with an example that demonstrates the enormous variety of properties that membranes can exhibit, through their ability to regulate ion transport. The conduction of neural impulses in animals is a remarkable process, but it depends on very simple physical principles.

Neurons, the nerve cells responsible for conduction of impulses, have filamentous extensions called dendrites and axons that act as the "wires" of the nervous system (Figure 10.29). Neurons are truly remarkable cells, which must meet unusual requirements. They must be able to conduct impulses without significant signal loss, yet rapidly enough to allow an animal to run with agility. Nerve conduction is accomplished not by electron flow, as in wires, but by waves in the membrane electrical potential on the surface of the axon. When such a wave passes a point on an axon or a dendrite, the **resting potential** of the membrane is momentarily changed to a moving **action potential** that constitutes the impulse. To understand how this occurs, we must examine how membrane potentials are generated and how they can be changed. In the limited space here, we can give only the briefest introduction to this vast and complex field.

In neural conduction, a wave of depolarization in the membrane potential moves along the fiber. This is called the action potential.

THE RESTING POTENTIAL

First, we must understand the source and nature of the resting potential. We begin with an oversimplified model, which draws on our earlier discussion of the electrochemical potential difference across a semipermeable membrane (see equation (10.2)). Suppose we have an ion (M^Z) of charge Z that is present outside the membrane at concentration $[M^Z]_{out}$ and inside at concentration $[M^Z]_{in}$.

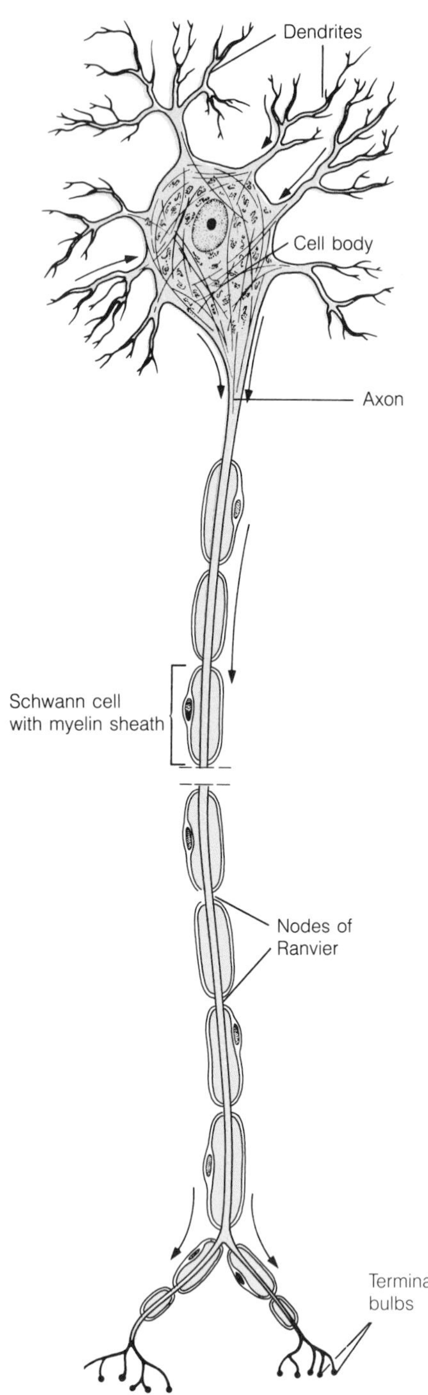

FIGURE **10.29**

Structure of a typical mammalian motor neuron. A motor neuron transmits nerve impulses to muscles or other nerve cells. The cell body contains the nucleus and most of the other organelles. The dendrites receive signals from the axons of other neurons; the axon transmits signals to other cells via the terminal bulbs, which connect to the dendrites of other neurons or to muscle cells. Along the axon are Schwann cells, which envelop the axon in layers of an insulating myelin membrane. The Schwann cells are separated by nonmyelinated regions called the nodes of Ranvier.

If the system is at equilibrium, ΔG for transport will be zero. Then, from equation (10.2), we find

$$\frac{RT}{ZF} \ln \frac{[M^Z]_{out}}{[M^Z]_{in}} = \Delta\psi \tag{10.8}$$

where $\Delta\psi$ is, as before, defined as out→in. Equation (10.8) is called the **Nernst equation.** For monovalent ions ($Z = \pm 1$) at 20°C, the Nernst equation reduces to

$$\Delta\psi = \pm 59 \log_{10} \frac{[M]_{out}}{[M]_{in}} \tag{10.9}$$

when $\Delta\psi$ is expressed in millivolts (mV).

According to equation (10.9), if we somehow maintain an ionic concentration difference across a membrane, an electrical potential difference will be produced. For example, if an ion such as K^+ ($Z = +1$) was kept 10 times as concentrated inside as outside, $\Delta\psi$ would be -59 mV. The membrane would be polarized, with a potential 58 mV lower on the inside. If the single ion that was unevenly distributed in this way was chloride, the potential would be $+59$ mV.

The major mechanisms creating ionic imbalance across cellular membranes are the specific ion pumps that continually act to concentrate certain ions on one side or the other. This imbalance gives rise to the resting potential across the membrane of a nerve axon. A much-studied example is the giant axon of the squid. This is a favorite experimental tool, because squids are unusual in having axons as large as 1 mm in diameter. It is possible, as shown in Figure 10.30, to insert recording electrodes into such an axon and measure the potential difference across the membrane. (The figure also shows stimulating electrodes, which are used in an experiment we will describe shortly.)

In the squid axon, we have a situation more complex than can be described by the Nernst equation. Several ions are involved, each of which can pass through the membrane, at least to some degree, and which are maintained in unequal concentrations on the two sides. If we used the Nernst equation to calculate the potential from the distribution of K^+ alone, we would predict a value of -75 mV (which we will call $\Delta\psi_{K^+}$). On the other hand, using the Nernst equation with the Na^+ concentration inside and outside, we would find $\Delta\psi_{Na^+} = +55$ mV. But when we measure the potential across a resting squid axonal membrane, we find a value of about -60 mV. What determines the actual potential in this case?

The resting potential of a nerve fiber is determined by the permeabilities of the membrane to different ions and the concentration of these ions on the two sides.

The key is that when several kinds of ions can pass through the membrane, the *permeability* of the membrane to different ions is important in determining $\Delta\psi$. The various ions are not at true equilibrium across the axonal membrane but are in a steady state, the position of which is determined in part by the individual permeabilities. This steady state can be described quantitatively by the **Goldman equation.** For the membrane potential determined by a collection of monovalent anions and cations the Goldman equation gives

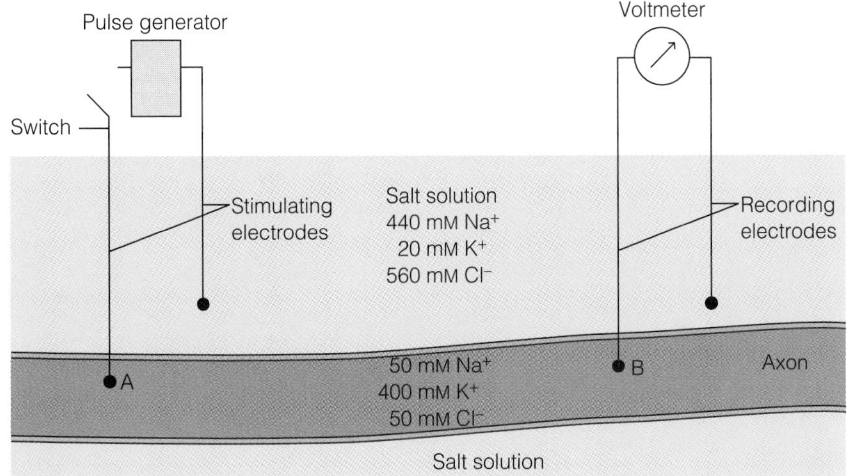

FIGURE **10.30**

Use of squid giant axons for studies of neural transmission. Electrodes attached to a voltmeter record the potential across the axonal membrane. At the resting axon ion concentrations shown here, the voltmeter would read about −60 mV. If the axon is stimulated at point A by a depolarizing pulse, the traveling action potential will shortly pass point B, where it can be recorded.

$$\Delta\psi = \frac{RT}{F} \ln\left(\frac{\Sigma_+ P_i[M_i^+]_{out} + \Sigma_- P_j[X_j^-]_{in}}{\Sigma_+ P_i[M_i^+]_{in} + \Sigma_- P_j[X_j^-]_{out}}\right) \tag{10.10}$$

Here, the sums (denoted by Σ) are taken over all cations (Σ_+) and anions (Σ_-) with significant permeability, and the *P*'s are the relative membrane permeabilities for these ions. Note that if any one ion were to have a much greater permeability than any of the others, it would dominate in the Goldman equation, and equation (10.10) would reduce to the Nernst equation for that ion.

Now let us apply the Goldman equation to calculate the resting potential in the squid giant axon. The only ions that contribute appreciably to the membrane potential in this case are K^+, Na^+, and Cl^-, and their relative permeabilities are $P_{K^+} = 1.0$, $P_{Na^+} = 0.04$, and $P_{Cl^-} = 0.45$. If these values are inserted into equation (10.10), together with the ion concentrations given in Figure 10.30, we find $\Delta\psi = -61$ mV, in agreement with experimental observation.

THE ACTION POTENTIAL

The resting potential value of −61 mV has a vital significance. Because −61 mV lies much closer to −75 mV than to +55 mV, it means that at the potential existing across the squid axon membrane, K^+ is much closer to its equilibrium distribution than is Na^+. If the membrane were to become fully permeable to ions, the major event would be a massive influx of sodium ions, with an accompanying shift in the membrane potentials toward ψ_{Na^+}.

This is just what happens when an action potential is transmitted along a nerve (Figure 10.31). Axonal fibers have specific, voltage-sensitive **gated channels** for the facilitated transport of Na^+ and K^+ through the membrane. Gated channels are protein ionophores that are open or closed, depending on the membrane potential; they are closed in the resting state (Figure 10.31a). Suppose we perform an experiment using electrodes placed in a squid giant axon as shown in Figure 10.30. At some distance from the recording electrodes, we place stimulating electrodes connected to a voltage source. If we apply at this electrode a pulse sufficient to depolarize the membrane locally by about 20 mV—that is, to a potential of −40 mV (the threshold for opening the Na^+ channels)—the Na^+ channel **activation gates** are opened (Figure 10.31b). The permeability to Na^+ increases about 100-fold and a flood of sodium ions rushes in, bringing the membrane potential up to about +40 mV within less than a millisecond (Figure 10.31c). In terms of the Goldman equation (10.10), the sudden high permeability of Na^+ ions makes them dominate, and the membrane potential *approaches* $\Delta\psi_{Na^+}$ (+55 mV).

The action potential is generated and propagated because a small depolarization of the nerve cell membrane opens gated channels, allowing ions to flow through.

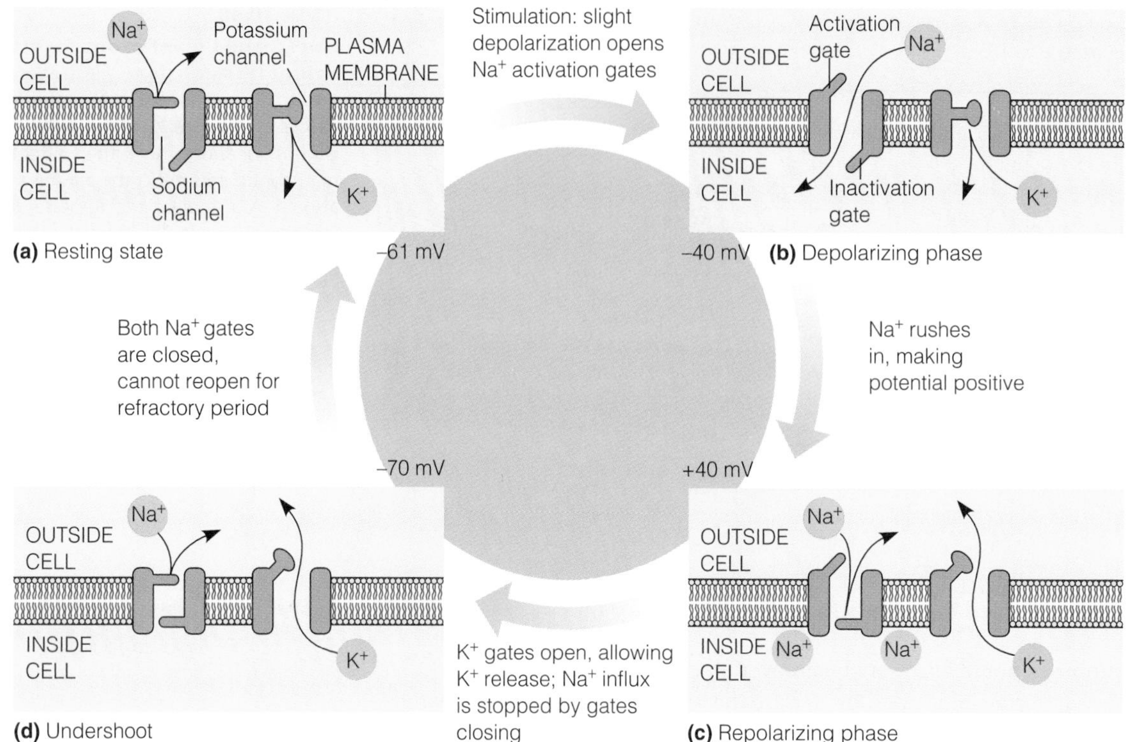

FIGURE **10.31**

The action potential. (a) In the resting state, K^+ is concentrated on the inside of the axon membrane and Na^+ on the outside. **(b)** An electrical stimulus causes local, partial depolarization of the membrane, which in turn causes the sodium channel activation gates to open, allowing a large inward rush of Na^+. The inrush of Na^+ causes the membrane potential to become positive. **(c)** As Na^+ approaches equilibrium, the inrush slows, and Na^+ diffuses along inside the axon. Meanwhile, K^+ gates open, and K^+ flows outward. **(d)** The outward flow of K^+ causes an extreme decrease in the membrane potential, which becomes temporarily more negative than the resting potential. This very negative potential causes the Na^+ gates to close and become resistant to opening. Thus, the resting potential is regained, but the region of membrane is temporarily inactivated (refractory).

From Neil Campbell, Jane Reece, and Larry Mitchell, *Biology*, 5th ed. (Menlo Park, CA: Addison Wesley Longman, 1999). © Addison Wesley Longman, Inc.

It does not reach this value, however, for further changes now occur. As seen in Figure 10.31b, the stimulus also causes K^+ channel gates to open, but more slowly, and these pour forth K^+ ions into the surrounding medium. This event reverses the potential again, overshooting to about −70 V. This change closes the Na^+ channel **inactivation gates** (which are separate from the activation gates), temporarily making the Na^+ channels resistant to opening (the refractory period—Figure 10.31d). The potential and permeability changes shown in Figure 10.31b–d occur in just a few milliseconds. The same changes are depicted graphically in Figure 10.32.

All of this activity would be a dramatic but localized effect were it not for a phenomenon shown in Figure 10.31c and d. As Na^+ ions pour in, they diffuse ahead of the region of stimulus and trigger the same round of depolarization in an adjacent section of the fiber. Thus a wave of depolarization proceeds down the axon. Following the wave front are the reverse polarization due to K^+ efflux, the refractory zone where K^+ channels are open and the Na^+ channels are inactivated, and the final return to the resting condition (Figures 10.31a and 10.33). Thus, the recording electrode in Figure 10.30 will see traveling past it, at some time after the stimulating pulse, exactly the same pattern of depolarization and reversal depicted in Figure 10.31b–d and Figure 10.32. This traveling pulse is called the *action potential.* The movement is depicted in Figure 10.33. The time required for the impulse to pass from stimulating electrode to recording electrode is proportional to the distance between the two electrodes and is inversely proportional to the velocity of propagation of the pulse. Typical values for propagation of the action potential range from 1 to 100 meters per second.

Several features of the action potential arise directly from the fact that it represents a wave of membrane depolarization. These features distinguish it clearly from the kind of conductance by electron flow seen in common electrical circuits.

1. The action potential does not appreciably decrease with distance transmitted, because it is continually renewed at each point along the axon.

2. The action potential is an all-or-none phenomenon. If the stimulus is sufficient to activate it, it occurs, and its magnitude is independent of the stimulus voltage as long as the stimulus is above the threshold value required for activation.
3. After an impulse has passed, the region of axon immediately behind it is unable to transmit another impulse for a period of some milliseconds because of the refractory period.

These three features account for some of the peculiar properties of the neural systems of animals. Because of feature 2, neural networks are more like digital

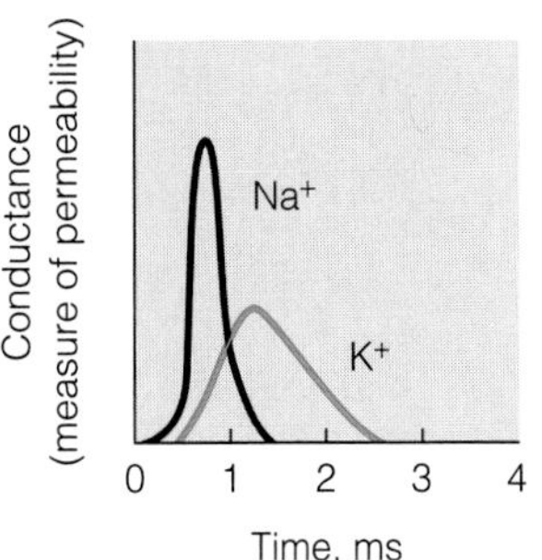

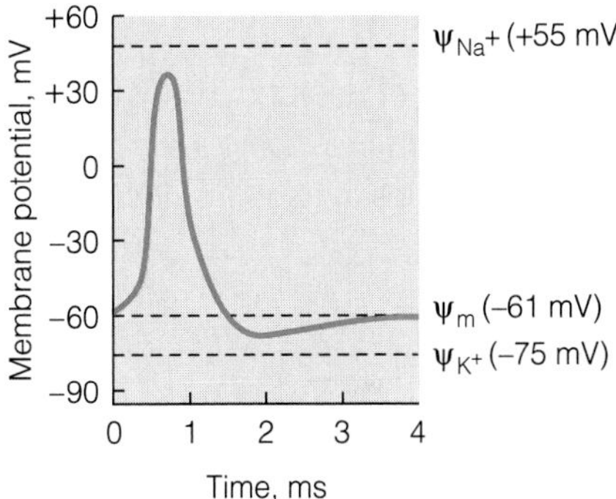

FIGURE **10.32**

The action potential. (a) Changes in membrane conductance at a point on an axon as a neural impulse passes. The membrane first becomes permeable to sodium ions, allowing a large inward rush of Na^+. A decrease in the Na^+ permeability results and is in turn followed by an outward flow of K^+. **(b)** Changes in membrane potential accompanying the permeability changes shown in (a). As Na^+ rushes in, the potential increases and becomes positive. As K^+ influx increases, the potential decreases, undershooting the resting potential, ψ_m, before leveling off. The potentials ψ_{Na^+} and ψ_{K^+} are the theoretical values for a membrane potential determined solely by either Na^+ or K^+, respectively.

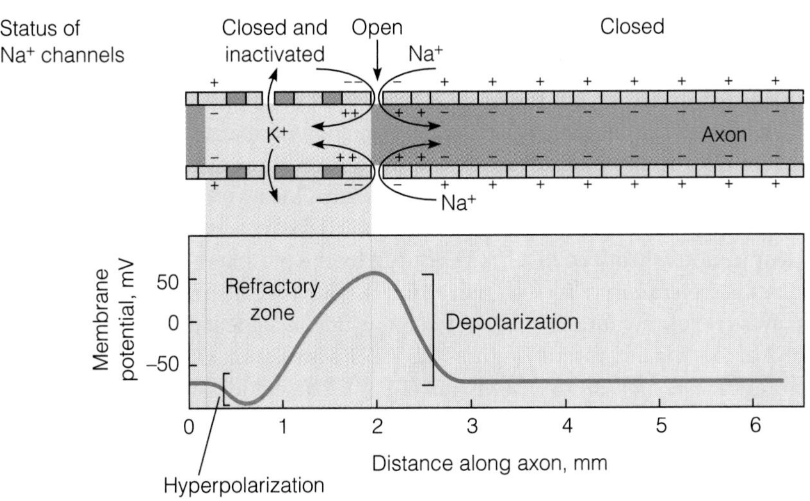

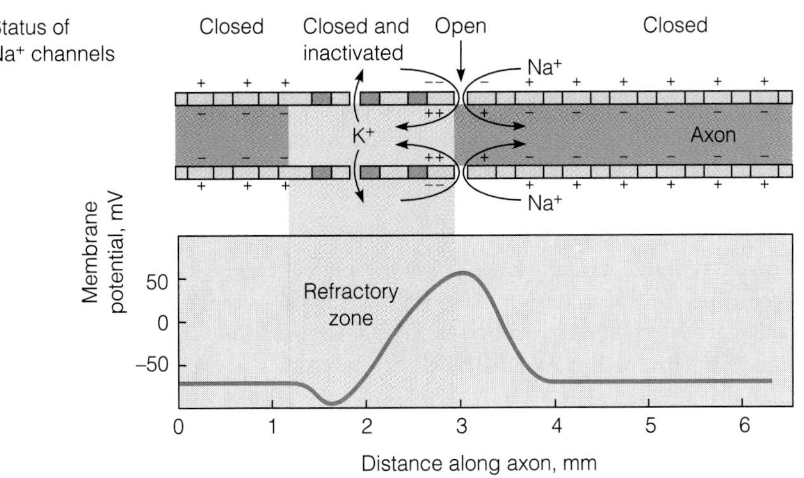

FIGURE **10.33**

Transmission of the action potential. Shown are two "snapshots," taken 1 ms apart, of potential along the axon. Red indicates inactivated channels. Arrows show Na^+ influx and K^+ efflux. **(a)** At time = 0, an action potential is occurring at the 2-mm position. The depolarization spreads down the axon, triggering development of the action potential downstream. **(b)** At time = 1 ms, the action potential peak has moved to the 3-mm position. The potential can move in only one direction, because after it has passed, the region behind the potential becomes refractory for a few milliseconds.

Adapted from J. Darnell et al., *Molecular Cell Biology*.

circuits than analog circuits. A neuron fires, or it does not. Furthermore, more intense stimuli do not give larger action potentials—they simply result in more frequent impulses. But because of the refractory period (feature 3), a neural system can become "saturated"; it can handle only so many impulses per second.

THE VELOCITY OF NEURAL TRANSMISSION

The velocity of neural transmission increases with increasing axon size or amount of axon insulation by myelin.

The properties of the axon itself have a major effect on how rapidly a single pulse can pass along it. Basically, the velocity depends on how rapidly the region of the axon ahead of the pulse can be depolarized to the threshold value necessary to open the Na^+ channel gates. If Na^+ ions can spread rapidly ahead of their point of entry, the pulse will move quickly. The resistance to Na^+ movement can be decreased by using a big axon. This approach to the problem is seen in the squid axons we have been talking about. Most invertebrates, and particularly mollusks, are sluggish creatures, but the squid is an exception. To put on a burst of speed, it must be able to expel water quickly by contraction of its mantle cavity. The axons that control the appropriate muscles are very large, allowing a fast response.

Vertebrates have evolved a different approach. Most vertebrate axons, particularly in the peripheral nervous system, are covered over much of their length by a discontinuous sheath of **myelin,** produced by the plasma membranes of special cells called *Schwann cells* (see Figure 10.29). The myelin sheath is a wrapping of many layers of membrane about the axon, which insulate it and prevent conductance across the membrane (Figure 10.34). The insulation is interrupted periodically at the *nodes of Ranvier* (see Figure 10.29). These nodes are the only points in a myelinated axon at which exchange of ions with the surroundings can occur; the Na^+ and K^+ channels are all concentrated here. In such an axon, the action potential jumps rapidly from one node to another, in a process known as *saltatory* (i.e., jumping) conduction. The efficiency of myelination can be judged from the fact that a 12-μm-diameter myelinated axon conducts as rapidly as a 600-μm-diameter, nonmyelinated squid axon.

Certain neurological diseases are associated with loss of myelin, the best-known being *multiple sclerosis.* In this condition a gradual destruction of myelin from certain areas of the brain and spinal cord occurs. Conduction of action potentials is slowed, with an accompanying loss of motor coordination.

TOXINS AND NEUROTRANSMISSION

Neurotoxins act by blocking gates in the axonal membrane in closed or open states.

Many extremely toxic substances have their effect by blocking the action of the specific ion gates necessary for development of the action potential. These substances are often called **neurotoxins.** *Tetrodotoxin* is found in some organs of the puffer fish. This fish is considered a delicacy in Japan, where special chefs are trained

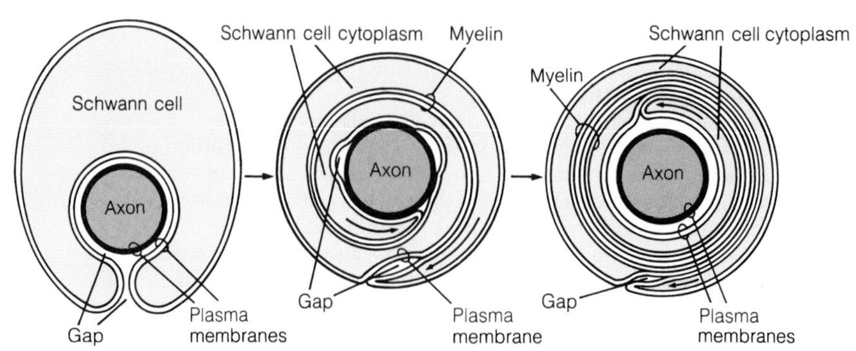

FIGURE **10.34**

How an axon is myelinated. During neural development, Schwann cells envelop the axon and grow in a spiral, wrapping many layers of the myelin membrane about the axon. These membranes are unusually low in protein content and correspondingly high in lipids. Therefore they make excellent insulators. The Schwann cells are spaced regularly along the axon, separated by the nodes of Ranvier, as shown in Figure 10.29.

and certified for their ability to remove the toxin-containing organs. Tetrodotoxin binds specifically to the Na^+ channel, blocking all ion movement. The same effect is produced by *saxitoxin,* contained in the marine dinoflagellates responsible for "red tide." These microscopic algae, along with their toxin, are ingested by shellfish and can in turn be consumed by humans. These two toxins, which attack a fundamental process of the nervous system, are among the most poisonous substances known, and their accidental ingestion leads to many deaths every year. A third very poisonous substance, *veratridine,* is found in the seeds of a plant of the lily family, *Schoenocaulon officinale.* This toxin also binds to the Na^+ channels but blocks them in the "open" configuration.

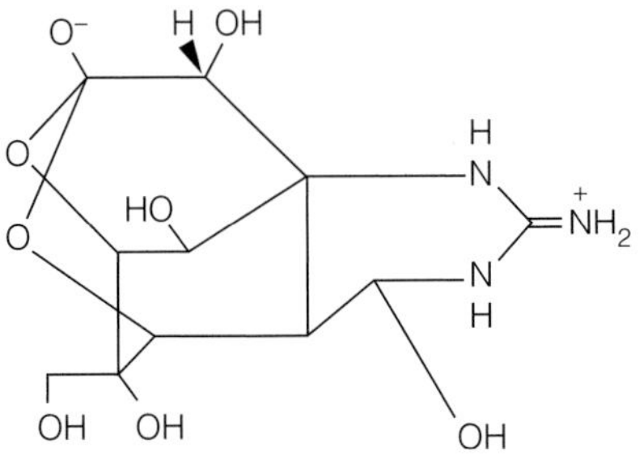

Tetrodotoxin

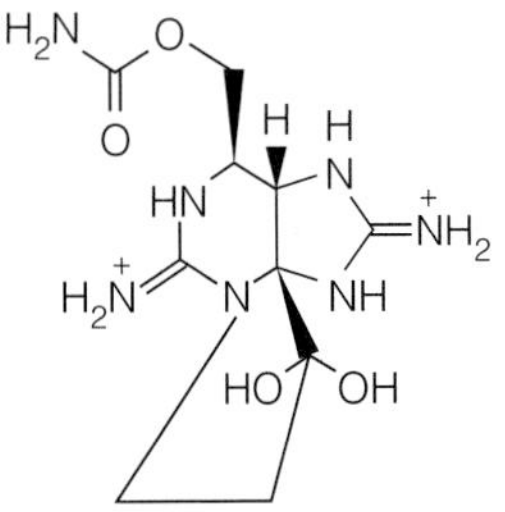

Saxitoxin

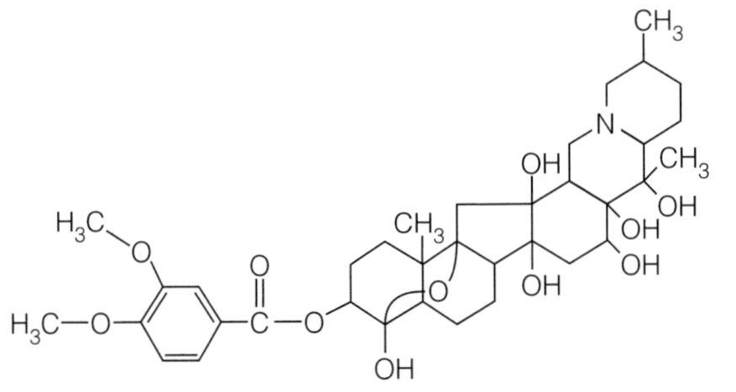

Veratridine

These toxins have proved to be useful in studies of axonal structure and conduction, for their tight binding makes them excellent affinity labels for the Na^+ channel. Their use has enabled researchers to determine that nonmyelinated axons contain about 100 channels per square micrometer, whereas the nodes of Ranvier in myelinated fibers have a channel density about 200 times greater.

Such studies have revealed that the Na^+ channels are typically composed of one large (α) subunit of about 260 to 300 kilodaltons, plus one or more smaller (β) subunits of 30 to 40 kilodaltons each. In some cases the β subunits are missing, and it is clear that most of the functional properties of the channel are associated with the α subunit. The amino acid sequence of the α subunit has been determined. The protein contains four hydrophobic domains, which probably form membrane-spanning walls about a central pore. Some segments of these domains contain an unusual repeating sequence in which every third residue is lysine or arginine, separated by two hydrophobic residues. Such sequences are suspected to be part of the gating mechanism, susceptible to conformational change when the membrane potential is altered.

We have described here only one part of the whole phenomenon of the transmission of neural impulses—the conduction along a single nerve fiber. The equally important problem of how these impulses are transmitted from one cell to another is discussed in Chapter 21, where the *neurotransmitter* substances are described.

SUMMARY

Many of the important properties of lipids stem from the fact that these substances are amphipathic, containing both hydrophobic and hydrophilic regions. Most naturally occurring fatty acids contain an even number of carbon atoms. When they are unsaturated, the double bonds are usually *cis*. Fatty acids are present in fats (triacylglycerols), where they serve for energy storage and insulation, and in membranes, where they are constituents of phospholipids, sphingolipids, glycosphingolipids, and glycoglycerolipids. Membranes are bilayer structures, containing proteins and lipids in a fluid mosaic. The two leaflets often differ in protein and lipid composition. Peripheral proteins are confined to one face or the other, whereas integral proteins extend through the membrane, with hydrophobic α helices common in the transmembrane region.

Transport through membranes may be achieved by passive diffusion, may be facilitated by pores or carriers, or actively driven by exergonic reactions. Only in the last case can transport against a concentration gradient occur. An example is the Na^+–K^+ pump, which maintains the ionic imbalance and membrane potential found between cells and their surroundings. Active transport may be indirect, as in cotransport or transport by modification.

Neural conduction of impulses depends upon a moving wave of depolarization (an action potential) in the membrane potential of a neural cell. This depolarization is produced by the flow of ions through gated channels in the membrane. The rate of neural transmission depends upon the dimensions of the axons and whether or not they are myelinated.

REFERENCES

General

Gennis, R. B. (1989) *Biomembranes.* Springer-Verlag, New York. A readable, comprehensive treatise.

Gurr, A. I., and J. L. Harwood (1991) *Lipid Biochemistry: An Introduction,* 4th ed. Chapman & Hall, New York. A valuable source for general information concerning lipids.

Lipowsky, R., and E. Sackmann (1995) *The Structure and Dynamics of Membranes.* Elsevier Science, Amsterdam. An up-to-date collection of papers.

Science (1992) 258:917–969. This issue of *Science* (6 November 1992) contains a collection of eight review papers on various aspects of membrane chemistry.

Vance, D. E., and J. E. Vance, eds. (1996) *Biochemistry of Lipids, Lipoproteins, and Membranes.* Elsevier Science, Amsterdam. A collection of chapters, each written by an expert, providing information on a wide variety of topics.

Membrane Asymmetry and Assembly

On den Kamp, J. A. F. (1979) Lipid asymmetry in membranes. *Annu. Rev. Biochem.* 48:47–71.

Voelker, D. R. (1996) Lipid assembly into cell membranes. In *Biochemistry of Lipids, Lipoproteins, and Membranes,* D. E. Vance and J. E. Vance, eds. Elsevier Science, Amsterdam.

Membrane Proteins

Branden, C., and J. Tooze (1991) *Introduction to Protein Structure,* Chapter 13. Garland, New York. Includes a brief, well visualized review of membrane proteins.

Lemmon, M. A., and D. Engelman (1992) Helix–helix interactions inside lipid bilayers. *Curr. Opin. Struct. Biol.* 2:511–518.

Singer, S. J., and G. L. Nicholson (1972) The fluid mosaic model of the structure of membranes. *Science* 175:720–731. The classic paper presenting this model.

Unwin, N., and R. Henderson (1984) The structure of proteins in biological membranes. *Sci. Am.* 250(2):78–94. Describes a pioneering structural study.

The Membrane Skeleton

Bennett, V. (1985) The membrane skeleton of human erythrocytes and its implication for more complex cells. *Annu. Rev. Biochem.* 54:273–304.

Coleman, T. R., D. J. Fishkind, M. E. Mooseker, and J. S. Morrow (1989) Functional diversity among spectrin isoforms. *Cell Motility Cytoskeleton* 12:225–247.

Liu, S.-C., and L. H. Derick (1992) Molecular anatomy of the red blood cell membrane skeleton: Structure–function relationships. *Semin. Hematol.* 29:231–243.

Transport

Gouaux, E. (1997) Channel-forming toxins: Tales of transformation. *Curr. Opin. Struct. Biol.* 7:566–573. A clear résumé of some interesting channel formers.

Graves, J. S., ed. (1985) *Regulation and Development of Membrane Transport Processes.* Wiley, New York. A good collection of papers.

Kühlbrandt, W., M. Auer, and G. A. Scarborough (1998) Structure of the P-type ATPases. *Curr. Opin. Struct. Biol.* 8:510–516. Structures to 8 Å resolution of two members of this class of ion pumps.

Lingrel, J. B., and T. Kuntzweiler (1994) Na^+, K^+-ATPase. *J Biol. Chem.* 269:19659–19662. A clear and informative minireview.

Reithmeier, R A. F. (1993) The erythrocyte anion transporter (band 3). *Curr. Opin. Struct. Biol.* 3:515–523.

Taniguchi, K., K. Suzuki, D. Kai, I. Matsuoka, K. Tomita, and S. Iida (1984) Conformational change of sodium and potassium-dependent adenosine triphosphatase. *J. Biol. Chem.* 259:15228–15233.

Neural Transmission

Bradford, H. F. (1986) *Chemical Neurobiology*. Freeman, San Francisco.

Hille, B. (1994) *Ionic Channels of Excitable Membranes*. Sinauer Associates, Sunderland, Mass.

PROBLEMS

1. Give structures for the following, based on the data in Table 10.1.
(a) *cis*-9-Dodecenoic acid
(b) 18:1cΔ11
(c) A saturated fatty acid that should melt below 30°C

2. Given these molecular components—glycerol, fatty acid, phosphate, long-chain alcohol, and carbohydrate—answer the following:
(a) Which two are present in both waxes and sphingomyelin?
(b) Which two are present in both fats and phosphatidylcholine?
(c) Which are present in a ganglioside but not in a fat?

3. The classic demonstration that cell plasma membranes are composed of bilayers depends on the following kinds of data:
   - The membrane lipids from 4.74×10^9 erythrocytes will form a monolayer of area 0.89 m^2 when spread on a water surface.
   - The surface of one erythrocyte is approximately 100 $(\mu m)^2$ in area.

   Show that these data can be accounted for only if the erythrocyte membrane is a bilayer.

4. The lipid portion of a typical bilayer is about 3 nm thick.
(a) Calculate the number of residues in an α helix that will just span this distance.
(b) The epidermal growth factor receptor has a single transmembrane helix. Find it in this partial sequence:

   ···RGPKIPSIATGMVGALLLLVVALGIGILFMRRRH···

5. In the situations described below, what is the free energy change if 1 mole of Na^+ is transported across a membrane from a region where the concentration is 1 μM to a region where it is 100 μM? (Assume T = 37°C.)
(a) In the absence of a membrane potential.
(b) When the transport is opposed by a membrane potential of 70 mV.
In each case, will hydrolysis of 1 mole of ATP suffice to drive the transport of 1 mole of ion? (Assume ΔG for ATP hydrolysis is about -50 kJ/mol under these conditions.)

*6. Consider passive diffusion of ions across the erythrocyte membrane, as measured by the permeability coefficients in Table 10.6. Calculate the number of moles of K^+ that would diffuse across a single erythrocyte membrane in 1 min, given the following data:

$$C_{K^+}\text{ (inside)} = 100\text{ mM}$$
$$C_{K^+}\text{ (outside)} = 15\text{ mM}$$

The surface area of one erythrocyte = 100 μm^2. [Note: Be careful of units.]

7. If the volume of the erythrocyte in Problem 6 is about 100 μm^3, what percentage of the K^+ ions would escape by passive diffusion in 1 minute?

*8. Assuming that the flow, J, in carrier-facilitated transport is proportional to the fraction of carriers occupied, derive an equation for J as a function of the concentration [A] of the substance transported.

9. Suppose calcium ion is maintained within an organelle at a concentration 1000 times greater than outside the organelle (T = 37°C). What is the contribution of Ca^{2+} to the membrane potential? Which side of the organelle membrane is positive, and which is negative?

10. Calculate the equilibrium membrane potentials to be expected across a membrane at 37°C, with an NaCl concentration of 0.10 M on the right and 0.01 M on the left, given the following conditions. In each case, state which side is + and which is −.
(a) Membrane permeable only to Na^+.
(b) Membrane permeable only to Cl^-.
(c) Membrane equally permeable to both ions.

11. In each of a, b, and c of Problem 10, will any appreciable transport of material take place in establishing the membrane potential?

*12. The rod cells in the retina contain membranes that are depolarized when a photon of light is absorbed. Suppose the following concentrations exist inside and outside a retinal rod cell:

	Concentration (mM)	
	Inside	Outside
K^+	100	5
Na^+	10	140
Cl^-	10	100

(a) If the relative permeabilities of K^+ and Cl^- are 1.0 and 0.45, respectively, what must be the relative permeability for Na^+ to give a potential of −30 mV, the value found across the resting rod cell membrane? Assume T = 37°C. [Hint: You must rearrange and solve the Goldman equation.]
(b) If the Na^+ gates were to close completely on stimulation by a photon, what value of the membrane potential would be reached?

13. Many transmembrane proteins are oligomeric, with several identical subunits. The oligomers are usually found to have some form of C_n symmetry, rather than D_n or any higher order. Suggest a reason for this selectivity.

14. The average human generates approximately his or her weight in ATP every day. A resting person uses about 25% of this in ion transport—mostly via the Na^+, K^+-ATPase. About how many grams of Na^+ and K^+ will a sedentary 70-kg person pump across membranes in a day?

TOOLS OF BIOCHEMISTRY 10A

Techniques for the Study of Membranes

A battery of special techniques exists for the study of membrane structure and function. Without them, our current level of knowledge concerning membranes could not have been attained.

Electron Microscopic Methods

Examinations of membrane structure as it exists within cells depends heavily on electron microscopy, and virtually all of the variants of that method mentioned in Tools of Biochemistry 1A have been employed at one time or another. For example, transmission EM of thin sections of cells embedded in a plastic matrix reveals cellular membranes in cross section (see Figures 1.11 and 1.12 for examples), whereas scanning EM can show surface details. An especially useful variant of this method is the **freeze fracture** technique. If a membrane is frozen quickly and then broken by a sharp blow from a microtome knife, it frequently splits along the plane between the bilayer leaflets (Figure 10A.1). One layer is thus peeled back, revealing the internal structure. The sample can then be metal-shadowed and studied by scanning EM. In a variant called **freeze-etching,** some of the ice is sublimed off before shadowing, revealing surface and subsurface details.

Although electron microscopy has revealed much of the elaborate architecture of natural membranes, investigators often need to use simplified systems to study specific membrane properties. For these purposes, synthetic bilayers and vesicles are frequently used.

Preparation of Bilayers and Vesicles

Membranes from individual types of cells or purified organelles can usually be obtained by lysis of the cell or organelle, followed by differential centrifugation.

If, as shown in Figure 10A.2, a membrane is extracted with organic solvents (a chloroform–ethanol mixture, for example), the soluble lipid constituents can be separated from insoluble protein and oligosaccharides. The lipid mixture can then be fractionated by methods such as high-pressure liquid chromatography to yield pure lipid components and an analysis of the lipid content. Alternatively, the investigator may wish to use the entire lipid mixture from a membrane.

If the organic solvent in such a preparation is removed by evaporation, and the membrane lipids are dispersed in aqueous solution, they will form vesicles (also called *liposomes*)—small spherical bilayer structures. Alternatively, a bilayer may be spread across a small hole in a partition between two compartments. Such preparations are often used to study permeability across membranes.

The vesicles can be used for many kinds of studies. For example, it is possible to reconstitute specific transport systems by iso-

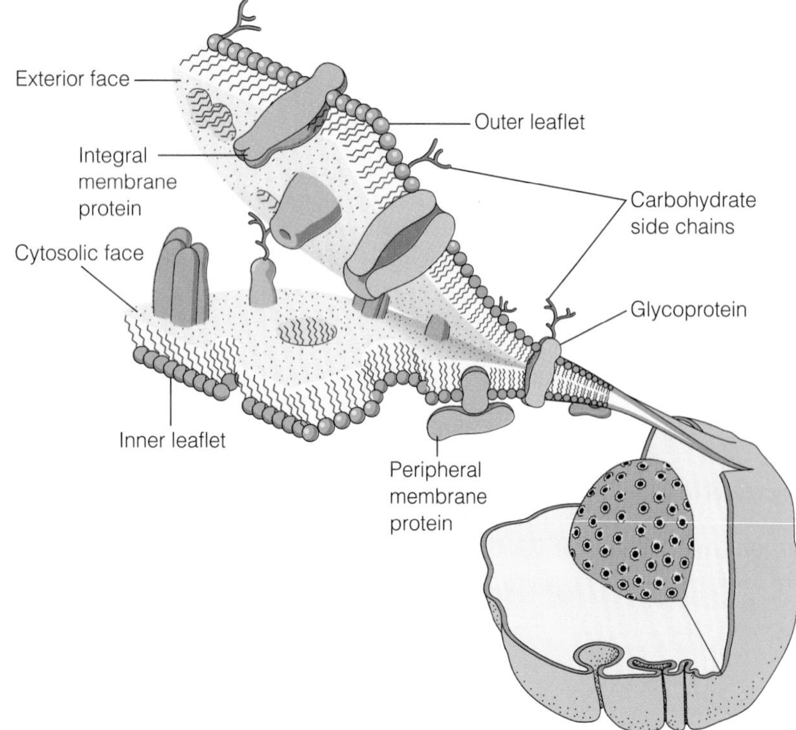

FIGURE **10A.1**

Freeze fracture. A schematic view of a freeze-fractured membrane.

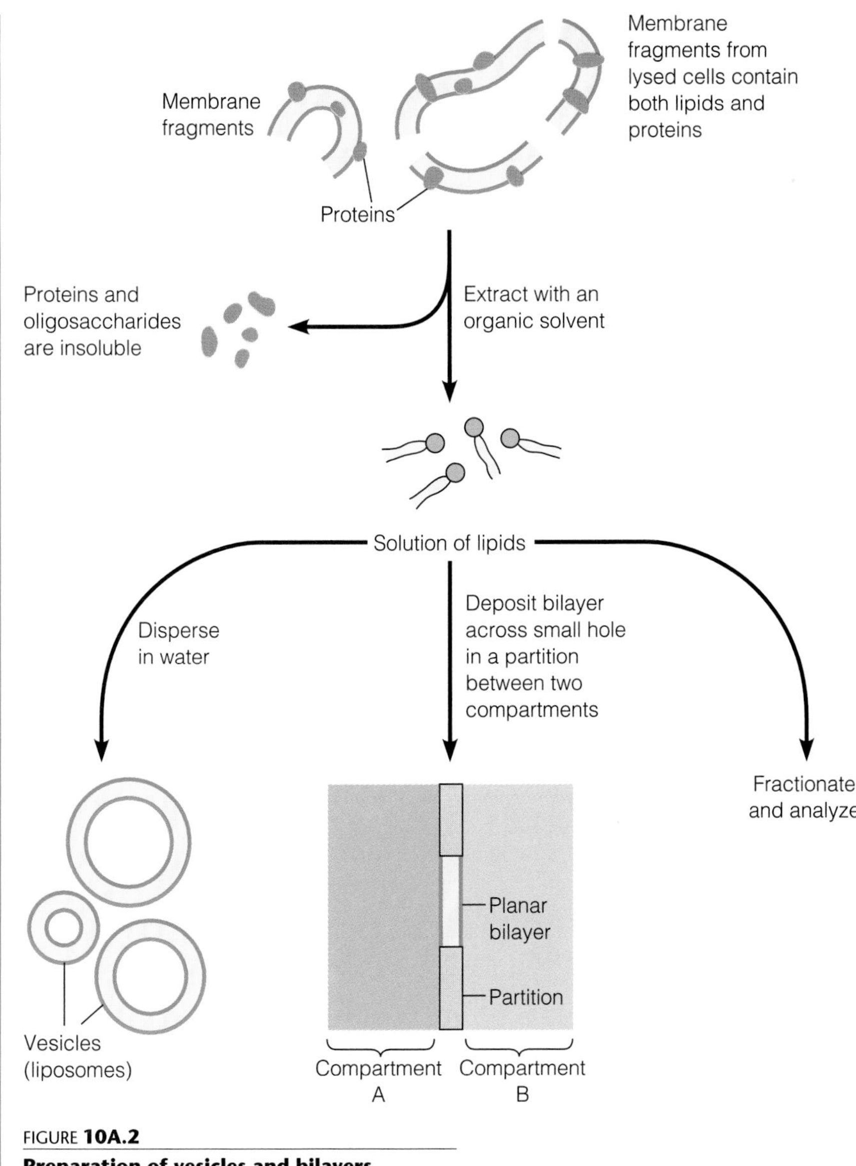

FIGURE **10A.2**
Preparation of vesicles and bilayers.

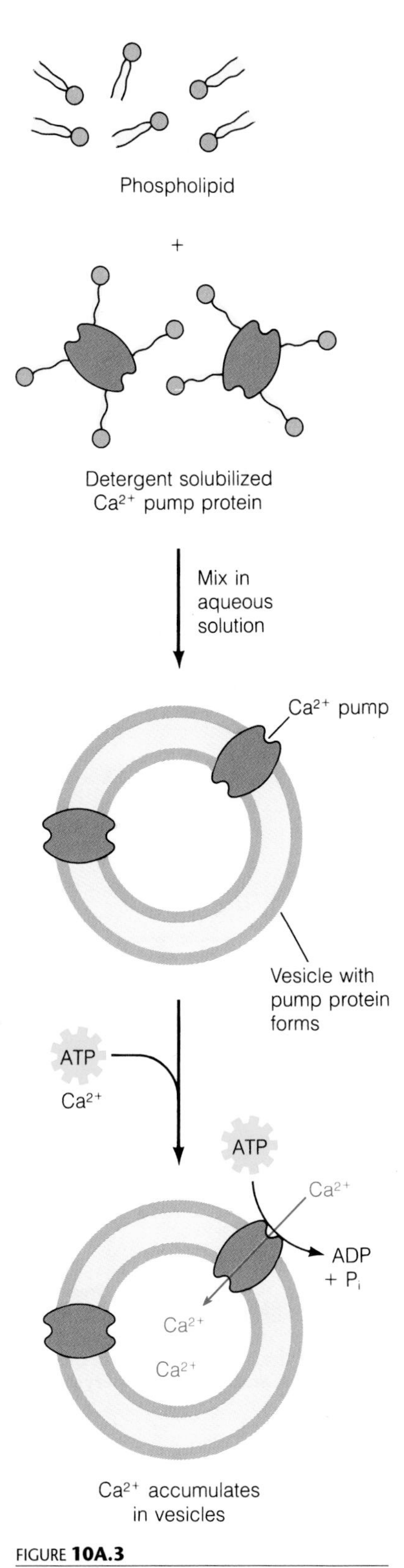

FIGURE **10A.3**
Reconstitution of the Ca^{2+} pump.

lating the transport proteins through detergent solubilization and then adding them to a vesicle-forming system. In the presence of ATP, properly reconstituted systems will display active transport. Figure 10A.3 demonstrates reconstitution of the Ca^{2+} pump of muscle cells. Vesicles made with specific mixtures of lipids and other components are also excellent objects for the study of such processes as the phase transition or diffusion in membranes.

In other kinds of experiments, it is desirable to retain all of the natural components in the membrane. This can often be done by carefully lysing cells or organelles, isolating the intact membranes, and then dispersing them in a solution in which they will reseal to form vesicles. In some cases, as shown with an erythrocyte in Figure 10A.4, it is possible to adjust conditions so that the vesicles reseal preferentially either right side out or inside out. Such preparations have provided much of our information concerning membrane asymmetry.

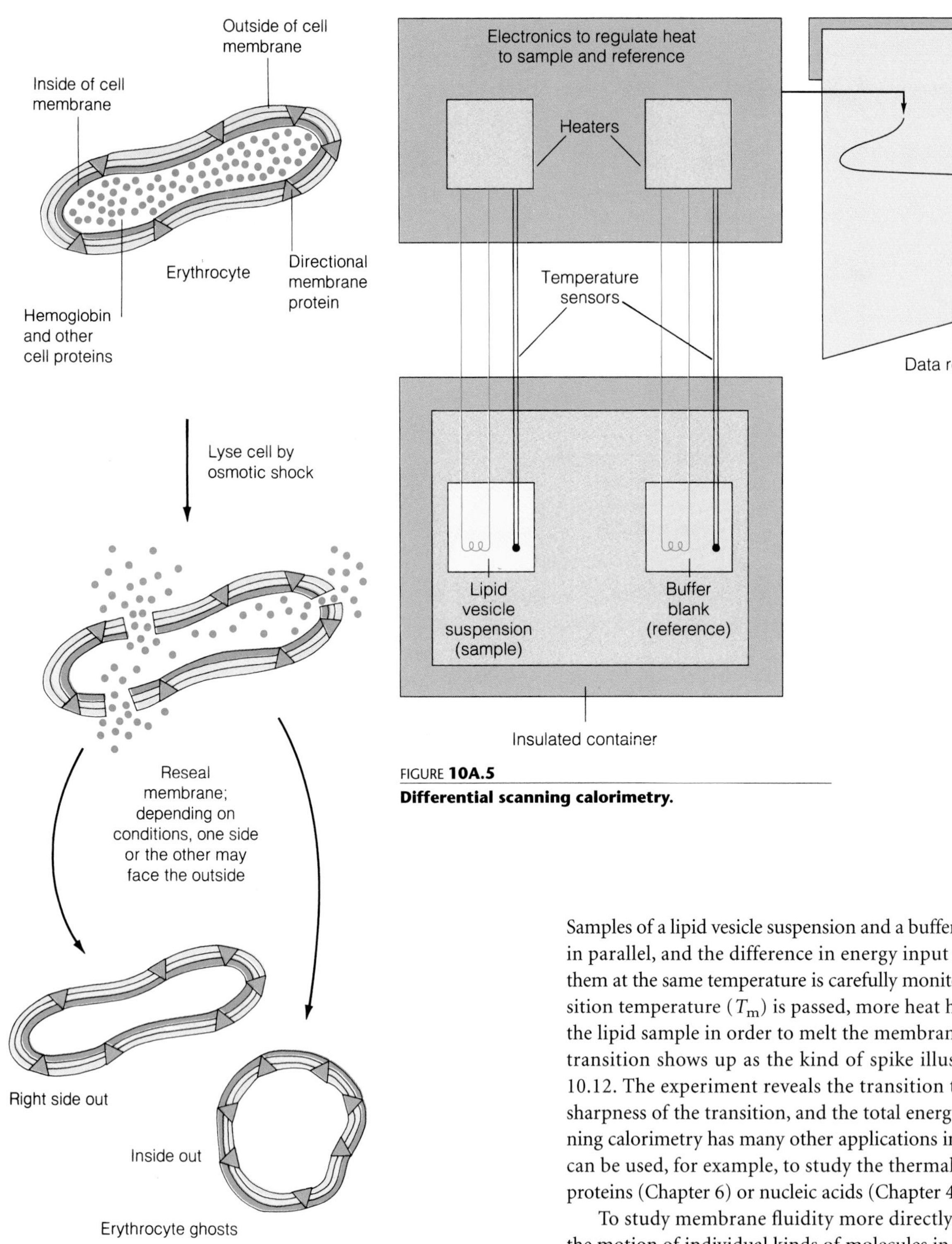

FIGURE **10A.4**
Preparation and resealing of erythrocyte ghosts.

FIGURE **10A.5**
Differential scanning calorimetry.

Physical Techniques

Much of our information concerning structural transitions in membranes has come from **scanning calorimetry.** A simplified schematic of a scanning calorimeter is shown in Figure 10A.5. Samples of a lipid vesicle suspension and a buffer blank are heated in parallel, and the difference in energy input required to keep them at the same temperature is carefully monitored. As the transition temperature (T_m) is passed, more heat has to be put into the lipid sample in order to melt the membrane structure. This transition shows up as the kind of spike illustrated in Figure 10.12. The experiment reveals the transition temperature, the sharpness of the transition, and the total energy required. Scanning calorimetry has many other applications in biochemistry. It can be used, for example, to study the thermal denaturation of proteins (Chapter 6) or nucleic acids (Chapter 4).

To study membrane fluidity more directly, and to examine the motion of individual kinds of molecules in the membrane, a number of other techniques are employed. **Electron spin resonance** (ESR) has been of major importance. Electron spin resonance bears similarities to nuclear magnetic resonance but involves changes in the spin of unpaired electrons rather than nuclei. The resonance spectrum is sensitive, in both absorption line spacing and sharpness, to the environment of the unpaired electron. As in NMR, narrow spectral lines are characteristic of a fluid environment with rapid molecular motion, and broadened lines are observed when molecular motion is sluggish.

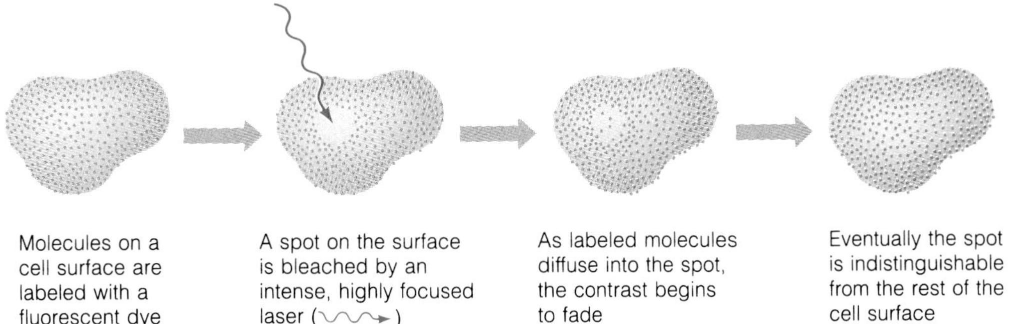

FIGURE **10A.6**
Fluorescence photobleaching recovery.

Most compounds do not have unpaired electrons, but certain nitroxide compounds, such as *tempocholine,* contain an unpaired electron in an N—O bond. Tempocholine has been substituted for choline in phosphatidylcholine to act as a "reporter group," sensing the freedom of motion of the head groups in membranes. Other reporter groups or molecules can probe the fluidity of the inside of membranes, or the mobility of membrane proteins.

$HO-CH_2-CH_2-N^+(CH_3)_2-$ (ring with two $C(CH_3)_2$ groups and $N-O^{\cdot}$)

Tempocholine

A new and powerful technique is called **fluorescence photobleaching recovery** (Figure 10A.6). Selected molecules in cell membranes are given a fluorescent label. Then a tiny spot on the membrane surface is exposed to a high-intensity laser beam. This "bleaches" the fluorescent label, forming a nonfluorescent spot on the cell surface. The cell is then observed under a fluorescence microscope. As bleached molecules diffuse out of the spot and unbleached molecules diffuse in, the spot gradually recovers its original fluorescence intensity. This procedure provides a direct way to measure the lateral movement of selected molecules in membranes.

References

Elson, E. L. (1986) Membrane dynamics studied by fluorescence correlation spectroscopy and photobleaching recovery. *Soc. Gen. Phys. Ser.* 40:367–383.

Kornberg, R. D., and H. M. McConnell (1971) Inside–outside transitions of phospholipids in vesicle membranes. *Biochemistry* 10:1111–1120. Describes use of ESR methods to measure motion across membranes.

Madden, T. D. (1986) Current concepts in membrane protein reconstitution. *Chem. Phys. Lipids* 40:207–222.

Prasad, R., ed. (1996) *Manual on Membrane Lipids.* Springer-Verlag, New York. This little manual contains a wealth of detailed, up-to-date techniques for a wide variety of lipid and membrane problems.

PART III

Dynamics of Life: Catalysis and Control of Biochemical Reactions

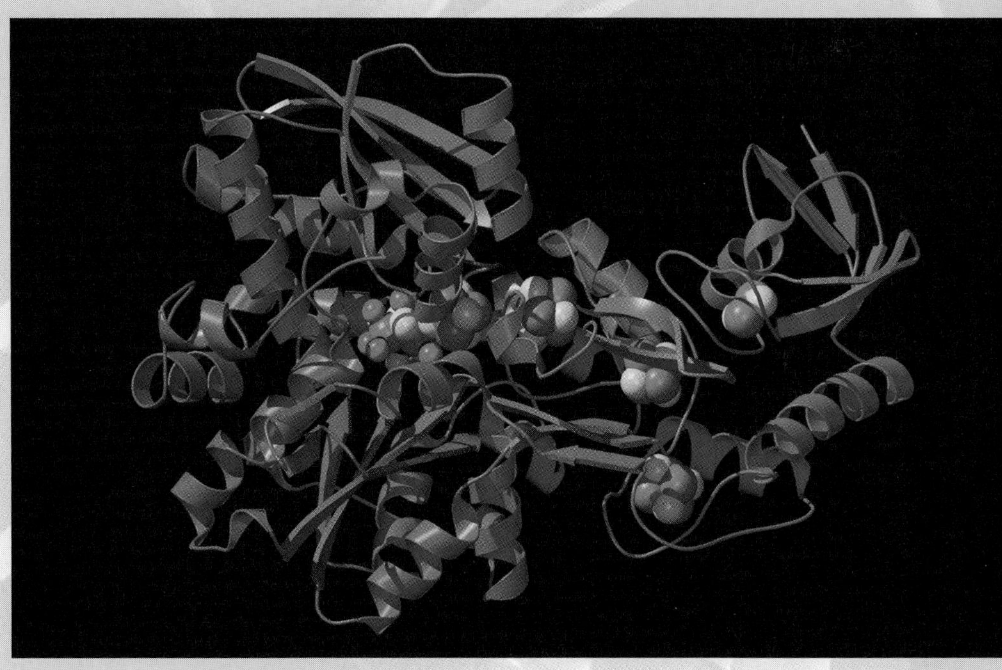

CHAPTER 11

Enzymes: Biological Catalysts

AGAIN AND AGAIN WE HAVE ALLUDED TO THE IMPORTANCE OF THE SPECIFIC catalysts called **enzymes** in regulating the chemistry of cells and organisms. Catalysis is essential to make many critical biochemical reactions proceed at useful rates under physiological conditions. A reaction that takes many hours to approach completion cannot be metabolically useful to a bacterium that must reproduce in 20 minutes or to a human nerve cell that must respond instantly to a stimulus.

In fact, life takes a clever advantage of the fact that most reactions must be catalyzed. In the complex milieu of the cell, countless reactions are *possible* among the molecules. The cell seizes on specific catalysis to channel substances into *useful* pathways rather than into wasteful side reactions. Furthermore, the enzymes the cell employs are unusual catalysts—in almost all cases, the efficiency of their operation can be controlled so as to modulate the production of different substances in response to cellular and organismal needs.

The rest of this book elaborates on this theme. Each of the hundreds of reactions we will encounter is catalyzed by a specific enzyme, tailored by evolution to do just its required task. What are the special properties of enzymes that make them such efficient catalysts? That is what we will explore in this chapter.

The Role of Enzymes

Catalysts increase the velocity of chemical reactions. Enzymes are biological catalysts.

A **catalyst** is a substance that increases the rate, or velocity, of a chemical reaction without itself being changed in the overall process. Most (but not all) biological catalysts are proteins; these we call enzymes. We have already encountered enzymes. For example, the protein trypsin catalyzes hydrolysis of peptide bonds in proteins and polypeptides. The substance that is acted on by an enzyme is called the **substrate** of that enzyme. Thus, polypeptides are appropriate substrates for trypsin.

We can see the power of enzyme catalysis in an example familiar to all—the decomposition of hydrogen peroxide into water and oxygen:

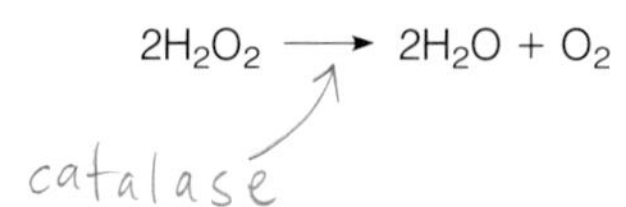

This reaction, although strongly favored thermodynamically, is very slow unless catalyzed. You can purchase a bottle of H_2O_2 solution and keep it on the shelf for many months before it breaks down. If, however, you were to add a bit of ferric ion (as $FeCl_3$, for example), you would find that the reaction rate increased about 1000-fold. The iron-containing protein hemoglobin is even better at increasing the rate of this reaction. If you apply hydrogen peroxide solution to a cut finger, you will notice an immediate bubbling from released O_2—the reaction is now proceeding about 1 million times faster than the uncatalyzed process. But even higher rates can be achieved. *Catalase,* an enzyme carried in many cells, increases the uncatalyzed rate of H_2O_2 decomposition about 1 billion–fold. Hydrogen peroxide is produced in some cellular reactions and is a dangerous oxidant (see Chapter 15), so catalase has been evolved to defend against it. This example shows that the rate of a favorable reaction depends greatly on whether a catalyst is present and upon the nature of the catalyst. Biological catalysts, including enzymes, are among the most efficient and specific catalysts known.

Two important facts must be emphasized. First, a true catalyst, although it participates in the reaction process, is unchanged by it. For example, after catalyzing the decomposition of an H_2O_2 molecule, catalase is found again in exactly the same state as before, ready for another round. Second, catalysts change *rates* of processes but do not affect the position of equilibrium of a reaction. A thermodynamically favorable process is not made more favorable, nor is an unfavorable process made favorable, by the presence of a catalyst. The equilibrium state is just approached more quickly.

Chemical Reaction Rates and the Effects of Catalysts

Before analyzing enzyme behavior in detail, we must review a bit of elementary chemical kinetics to provide precise definitions of reaction rates and a clear understanding of what catalysts do.

REACTION RATES AND REACTION ORDER

First-Order Reactions: The Rate Constant

To understand what is meant by a reaction rate and how it might be measured, let us first consider the simplest possible reaction, the *irreversible* conversion of substance A to substance B:

$$A \longrightarrow B$$

The single arrow here means that the reverse reaction ($B \longrightarrow A$) proceeds to only an infinitesimal extent; that is, the equilibrium state lies far to the right.

We can define the **reaction rate,** or **velocity** (V), at any instant as the rate of formation of the product, in this case B:

$$V = \frac{d[B]}{dt} \qquad (11.1)$$

The units of V are moles per liter per second—$(mol/L)s^{-1}$—if [B] symbolizes molar concentration of B. If we note that, for every B molecule formed, an A molecule must disappear, it is clear that V can equally well be written as

$$V = -\frac{d[A]}{dt} \qquad (11.2)$$

The change of each molecule of A into B is an independent event. Therefore, as molecules of A are consumed, the number of molecules left to change is diminished, and the rate decreases as the reaction proceeds (see Figure 11.1a). Mathematically, we state this by saying that the rate is proportional to [A]:

$$V = \frac{d[\mathrm{B}]}{dt} = -\frac{d[\mathrm{A}]}{dt} = k_1[\mathrm{A}] \tag{11.3}$$

A first-order reaction is one whose rate is directly proportional to the first power of the reactant concentration.

The constant k_1 is called the **rate constant** and for this reaction has units of (seconds)$^{-1}$. The rate constant provides a direct measure of how fast this reaction is. A large k_1 means the reaction is rapid, and a small k_1 means it is a slow reaction. This reaction is called a **first-order reaction,** because its rate depends on the first power of the reactant concentration.

If we want to prove that a reaction is first-order or we want to measure the rate constant, it is more convenient to have an equation that describes how the concentration of A changes with time during the reaction. This description can be obtained by integrating equation (11.3):

$$\int_{[\mathrm{A}]_0}^{[\mathrm{A}]} d\ln[\mathrm{A}] = -k_1 \int_0^t dt \tag{11.4a}$$

$$\ln \frac{[\mathrm{A}]}{[\mathrm{A}]_0} = -k_1 t \tag{11.4b}$$

or

$$\frac{[\mathrm{A}]}{[\mathrm{A}]_0} = e^{-k_1 t} \tag{11.4c}$$

A first-order reaction is characterized by exponential decay of the reactant.

where $[\mathrm{A}]_0$ is the starting concentration, when $t = 0$. Equation (11.4c) tells us that the concentration of A decreases exponentially with time, as shown in Figure 11.1a. A characteristic of such an exponential decay is the **half-life** ($t_{1/2}$), the time for half of whatever amount is left to be lost (Figure 11.1a). The half-life is inversely proportional to k_1 (see Problem 1). To test whether a reaction is truly first-order, we need only make a graph of $\ln[\mathrm{A}]$ versus t, as shown in Figure 11.1b. A straight line confirms equation (11.4b); the slope gives $(-k_1)$. An alternative procedure is shown in Figure 11.1c and d: We measure the *initial rate* of the reaction at different values of the initial concentration, $[\mathrm{A}]_0$, and plot these initial rates against $[\mathrm{A}]_0$. If the initial rate is proportional to $[\mathrm{A}]_0$, the reaction is first-order. The rate constant k_1 can be determined from the slope of a graph of initial rate versus $[\mathrm{A}]_0$. The most common example of a first-order reaction is the decay of radioactive elements.

Most biochemical processes cannot be described over their full course by equations as simple as (11.4). One reason is that many of the reactions and processes we encounter are *reversible*—that is, equilibrium does not lie far to one side—and as product accumulates, the reverse reaction becomes important. For example, we may have a reaction like the following:

$$[\mathrm{A}] \underset{k_{-1}}{\overset{k_1}{\rightleftharpoons}} [\mathrm{B}]$$

Because A is being consumed in the reaction to the right and formed by the reaction to the left, the corresponding rate equation is

$$-V = \frac{d[\mathrm{A}]}{dt} = -k_1[\mathrm{A}] + k_{-1}[\mathrm{B}] \tag{11.5}$$

Here, k_1 and k_{-1} are the rate constants for the first-order forward and reverse reactions. Such a reaction approaches a state of equilibrium, at which point the

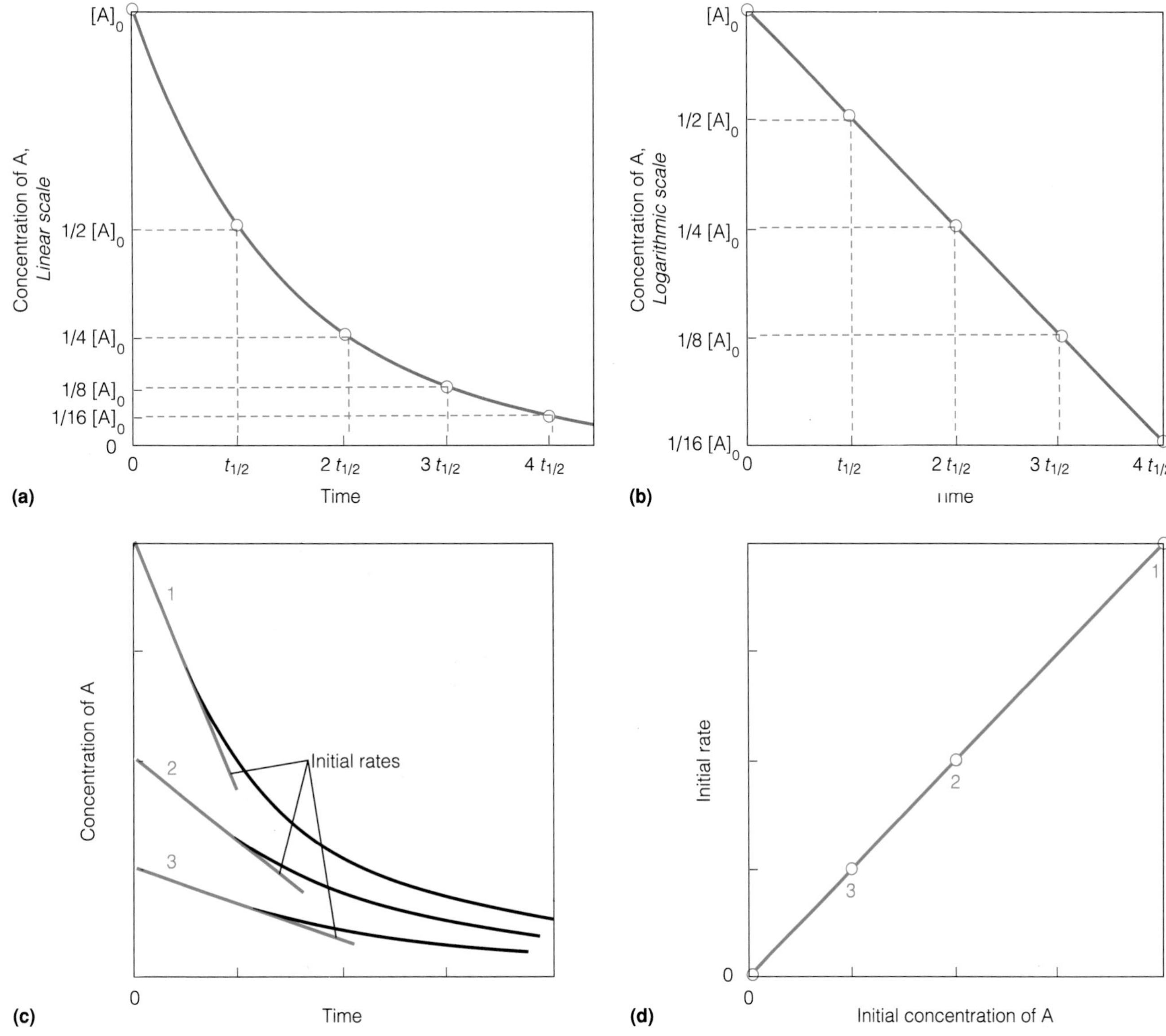

FIGURE **11.1**

Determining the order and rate constant of an irreversible first-order reaction. Graphs (a) and (b) analyze the rate of a single reaction over its entire course, with time expressed as multiples of the half-life ($t_{1/2}$) of the reactant. Note that for each interval of $t_{1/2}$ the reactant concentration is halved. Graphs (c) and (d) analyze the *initial* rates of a reaction at different starting concentrations. **(a)** A graph of [A] versus *t* shows that the rate, defined as the slope of the line, decreases as the reaction continues. **(b)** A graph of ln[A] versus *t*, when linear, indicates that the reaction follows equation (11.4b) and hence is first-order. The slope of this line ($d\ln[A]/dt$) is equal to $-k_1$. **(c)** The initial reaction rate is determined for three values of $[A]_0$. **(d)** The three initial rates determined in (c) are plotted against $[A]_0$. If the initial rates are proportional to $[A]_0$, yielding a straight-line graph, the reaction is first-order. The slope of the line is k_1.

rates of the forward and reverse reactions become equal, and so the overall rate becomes zero. Then

$$0 = -k_1[A]_{eq} + k_{-1}[B]_{eq} \tag{11.6a}$$

or

$$\frac{[B]_{eq}}{[A]_{eq}} = \frac{k_1}{k_{-1}} = K \tag{11.6b}$$

where *K* is the equilibrium constant. For a reversible reaction that is first-order in both directions, the equilibrium constant is always the ratio of the forward and reverse rate constants.

Second-Order Reactions

The reactions we have described so far are first-order—they involve changes happening in individual molecules. But many biochemical reactions are more complex, involving encounters between molecules. A **second-order reaction** occurs typically when two molecules must come together to form products. A simple example is

$$2A \xrightarrow{k_2} A_2$$

The rate of a second-order reaction is proportional to the second power of the reactant concentration.

The rate of such a reaction is proportional to the *second* power of the concentration of reactant, for the reaction can occur only when two molecules collide:

$$V = \frac{-d[A]}{dt} = -k_2[A]^2 \qquad (11.7)$$

Here k_2 is the **second-order rate constant.** It has dimensions of $(\text{mol/L})^{-1}\,\text{s}^{-1}$. Obtaining the integrated expression for concentration as a function of time from equation (11.7) is not difficult; it is left for an example in Problem 7.

Many more complicated kinds of reactions occur, including complex, multistep processes. We shall not be concerned with their kinetics at this point, although we shall see that enzyme-catalyzed reactions, when analyzed in detail, are generally more complicated than those described above. Often, however, the analysis of complex reaction schemes can be simplified by the recognition of a **rate-limiting step.** This concept is familiar to anyone who has encountered repair work on the highway or tried to check in at the airport gate for a crowded flight—the slowest step in a multistep process may determine how long the whole process takes.

TRANSITION STATES AND REACTION RATES

Our primary question is, What determines the rate of a chemical reaction? That is, what makes a rate constant large or small? Thermodynamics gives no information concerning rates. If we draw a free energy diagram for a favorable reaction on the basis of thermodynamics alone, it will look like Figure 11.2a. Here we plot the free energy of the system versus the **reaction coordinate,** a generalized measure of the progress of the reaction through intermediate states. Its physical meaning will differ from one reaction to another (which we will discuss further). We can see that the standard state free energy of the products of this reaction is lower than that of the reactants. However, what may be most important in determining the reaction rate is what happens in the *transition* from reactants to product. Equilibrium measurements, which pertain to final and initial states, do not reveal any information about the transition or the barrier which the transition presents.

Barriers to chemical reactions occur because a reactant molecule must pass through a high-energy transition state to form products. This free energy barrier is called the activation energy.

A molecule in a first-order reaction must only occasionally reach an energy state in which the process can occur; otherwise, all molecules would already have reacted. This observation suggests that only a fraction of the molecules, which are sufficiently *energetic,* can undergo reaction. Similarly, in a second-order reaction not all encounters between reactants can be productive, because some collisions may not be sufficiently energetic, or the colliding molecules may not be properly oriented with respect to each other to react. Such considerations have given rise to the idea of a **free energy barrier** to reaction and the concept of an **activated state,** or **transition state** (symbolized by ‡). The transition state is thought of as a stage through which the reacting molecule or molecules must pass, often one in which a molecule is strained or distorted or has a particular electronic structure, or in which molecules collide properly. To make concrete the idea of a transition state, we diagram in Figure 11.2b the free energy as a function of the reaction coordinate. This concept is somewhat abstract, but we give a simple, concrete example in Figure 11.2c, the boat $\longrightarrow$ chair conversion of a pyranose ring (see Chapter 9, pages 288–289). The reaction coordinate in this case is just the angle θ. Both the initial state (boat) and the final state (chair) are of lower energy than the most strained, flattened state (the transition state). The ring must go through the half-chair state, which is such a high-energy transition state, to make the conversion.

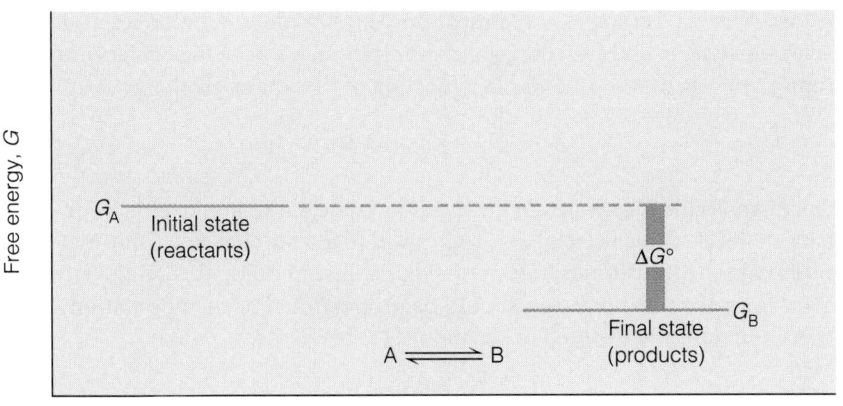

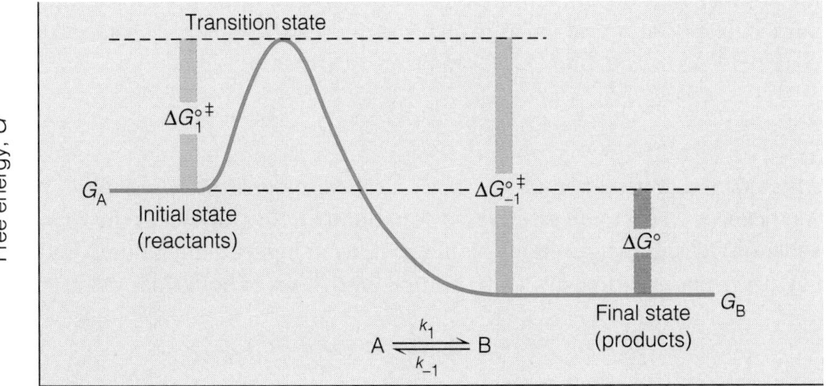

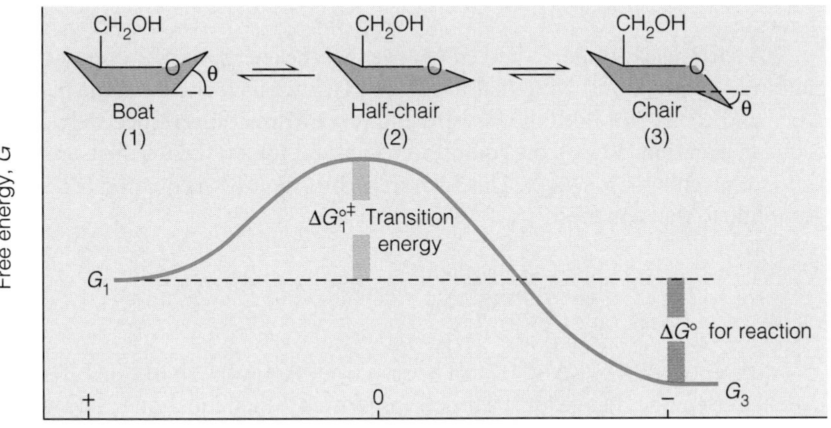

FIGURE **11.2**

Free energy diagrams for the simple reaction A ⇌ B. **(a)** Information provided by thermodynamic studies of the equilibrium: Only the free energy difference between the initial state and the final state is revealed. G_A and G_B represent the *average* free energies per mole of A and B molecules. $\Delta G°$ is the standard state free energy change for the reaction. **(b)** Free energy diagram filled in to include the transition state (‡) through which the molecule must pass to go from A to B or vice versa. $\Delta G_1^{\circ\ddagger}$ marks the energy of activation for the A ⟶ B transition, $\Delta G_{-1}^{\circ\ddagger}$ is for the B ⟶ A transition. **(c)** A reasonable path for the transition of a pyranose (such as glucose) from boat (1) to chair (3) conformation. The highest energy state—the transition state—will look something like (2).

The **standard free energy of activation,** $\Delta G^{\circ\ddagger}$, represents the additional free energy (above the average free energy of reactant molecules) that molecules must have to attain the transition state. We know that in any sample or solution, not all molecules have the same energy at any instant. If the barrier to reaction is high, only a small fraction of the molecules will have enough energy to surmount it, or only a small fraction of collisions will be energetic enough. If we assume that molecules of the reactant A are in equilibrium between the initial state and the activated state, the concentration of activated molecules at any instant will be given by equation (3.25), introduced on page 71, for a chemical equilibrium

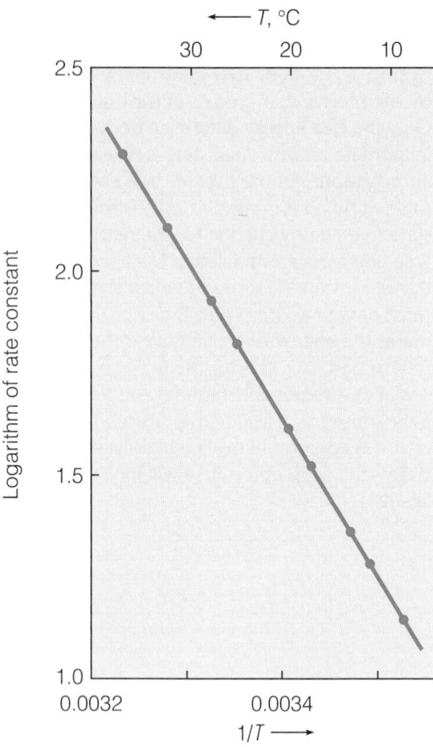

FIGURE **11.3**

Test of the Arrhenius equation. The logarithm of the rate constant for the reaction L-malate ⟶ fumarate + H_2O is plotted versus the reciprocal of the absolute temperature (in kelvins). A straight line is obtained, as predicted by equation (11.11). The temperature in degrees Centigrade is shown on the top scale. Note the strong dependence of rate on temperature: Increasing the temperature by 10 degrees produces almost a 3-fold increase in rate.

($K = e^{-\Delta G°/RT}$). Letting $[A]^\ddagger$ represent the concentration of molecules having the activation energy and $[A]_o$ the total concentration, we write the equilibrium constant $[A]^\ddagger/[A]_o$ in the same way as a function of the activation energy, $\Delta G^{\circ\ddagger}$:

$$[A]^\ddagger = [A]_o e^{-\Delta G^{\circ\ddagger}/RT} \tag{11.8}$$

Once a molecule is activated to the transition state, it can hop to either side of the barrier—back to the reactant side or forward to the product side. Only molecules that attain the transition state have the possibility of doing this, so the rate constant for a chemical reaction should be proportional to the population of the transition state, as expressed in equation (11.8):

$$k = Qe^{-\Delta G^{\circ\ddagger}/RT} \tag{11.9}$$

The factor Q, which has units of s^{-1}, represents the frequency of hopping toward the product side. It is sometimes more convenient to think in terms of the standard state enthalpy and entropy of activation ($\Delta H^{\circ\ddagger}$, $\Delta S^{\circ\ddagger}$). Because $\Delta G^{\circ\ddagger} = \Delta H^{\circ\ddagger} - T\Delta S^{\circ\ddagger}$, equation (11.9) can be rewritten as

$$k = Qe^{\Delta S^{\circ\ddagger}/R} \cdot e^{-\Delta H^{\circ\ddagger}/RT} = Q'e^{-\Delta H^{\circ\ddagger}/RT} \tag{11.10}$$

where Q' is a constant equal to $Qe^{\Delta S^{\circ\ddagger}/R}$ called the *preexponential term.* We always expect $\Delta H^{\circ\ddagger}$ to be positive, so equation (11.10) (known as the *Arrhenius* equation) tells us that reactions should go faster at higher temperatures. Equation (11.10) can be rewritten, by taking natural logarithms of both sides, as

$$\ln k = \ln Q' - \Delta H^{\circ\ddagger}/RT \tag{11.11}$$

Accordingly, a graph of $\ln k$ versus $1/T$ should be a straight line, and its slope should allow experimental determination of $\Delta H^{\circ\ddagger}$. Figure 11.3 is typical of the results for many reactions and is evidence that equation (11.10) is correct, at least in form.

It is particularly important to emphasize that the energy barrier opposes reaction in *both* directions. Note that in Figure 11.2b we have included the barriers $\Delta G_1^{\circ\ddagger}$ and $\Delta G_{-1}^{\circ\ddagger}$ for both the forward and reverse directions respectively. As we stated in equation (11.6b), the equilibrium constant for a reversible first-order reaction is given by $K = k_1/k_{-1}$. This is entirely consistent with equation (11.9), for according to that equation

$$K = \frac{Qe^{-\Delta G_1^{\circ\ddagger}/RT}}{Qe^{-\Delta G_{-1}^{\circ\ddagger}/RT}} = e^{-(\Delta G_1^{\circ\ddagger} - \Delta G_{-1}^{\circ\ddagger})/RT} \tag{11.12}$$

The difference ($\Delta G_1^{\circ\ddagger} - \Delta G_{-1}^{\circ\ddagger}$) can be seen from Figure 11.2b to equal $\Delta G°$, the overall free energy change for the process as written. Thus, equation (11.12) yields the relation between K and $\Delta G°$ given in equation (3.25) (page 71):

$$K = e^{-\Delta G°/RT} \tag{11.13}$$

These results show directly why the equilibrium constant K itself can tell us nothing about rates of processes—it depends only on the free energy *difference* between final and initial states and carries no information about the height of the barrier between these states.

WHAT A CATALYST DOES

Catalysts function by lowering the activation energy.

A catalyst lowers the energy barrier to a reaction, thereby increasing the fraction of molecules that have enough energy to attain the transition state and making the reaction go faster in both directions. However, the presence of a catalyst has no effect on the position of equilibrium, because the *difference* in barrier heights in the two directions is exactly the same whether a catalyst is present or not (Figure 11.4). Thus $\Delta G°$ is unchanged by a catalyst, and so is K, which equals k_1/k_{-1}, even though k_1 and k_{-1} may each be thousands or millions of times larger than they were for the uncatalyzed process.

The presence of a catalyst increases forward and reverse rates, but does not affect the equilibrium reached.

We now ask *how* a catalyst lowers the free energy barrier. To answer this question, let us examine $\Delta G°^{\ddagger}$ a bit more closely. Since $\Delta G°^{\ddagger} = \Delta H°^{\ddagger} - T\Delta S°^{\ddagger}$, either a requirement for a great deal of energy (a large positive $\Delta H°^{\ddagger}$) or a very improbable conformation in the transition state (a large negative $\Delta S°^{\ddagger}$) will make $\Delta G°^{\ddagger}$ large and positive and thus make a reaction slow. The energy contribution is easy to visualize. Very often, the reactive molecule or molecules must go through energy-demanding strained and distorted states in order for the reaction to occur (see Figure 11.2c for an example). Sometimes the catalyst can reduce this energy requirement by forcing the reactant molecule into an **intermediate** state that resembles the transition state but is of lower energy because of the favorable energetics of the binding to the catalyst (Figure 11.5). The result is that two very low transition energy barriers replace the single higher barrier. We distinguish such an intermediate state from a transition state by the fact that the former corresponds to a free energy minimum, and the latter to a free energy maximum. An intermediate state is a metastable state of the molecule.

The entropy term in the transition free energy reflects the fact that a particular *orientation* between reactants or parts of a molecule may be necessary to enter the transition state. For example, when collisions between molecules occur (as in

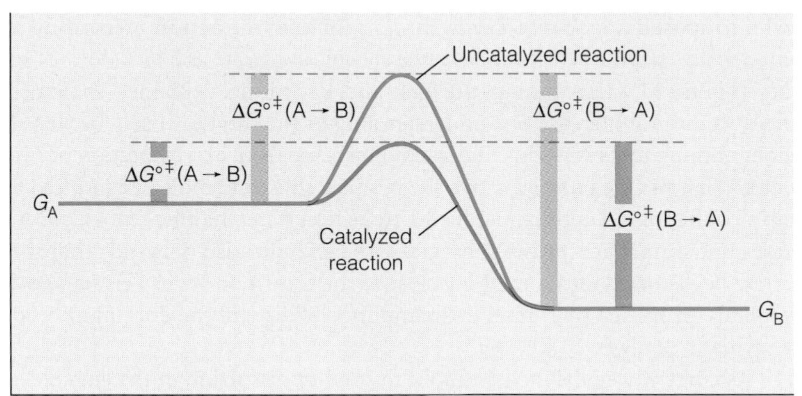

FIGURE **11.4**

Effect of a catalyst on activation energy. Here we have drawn the same free energy diagram as in Figure 11.2b but have provided an alternative *catalyzed* path (red) for the reaction. The catalyst lowers the standard free energy of activation, $\Delta G°^{\ddagger}$, so more of the reactant molecules have the energy needed to reach the lowered transition state.

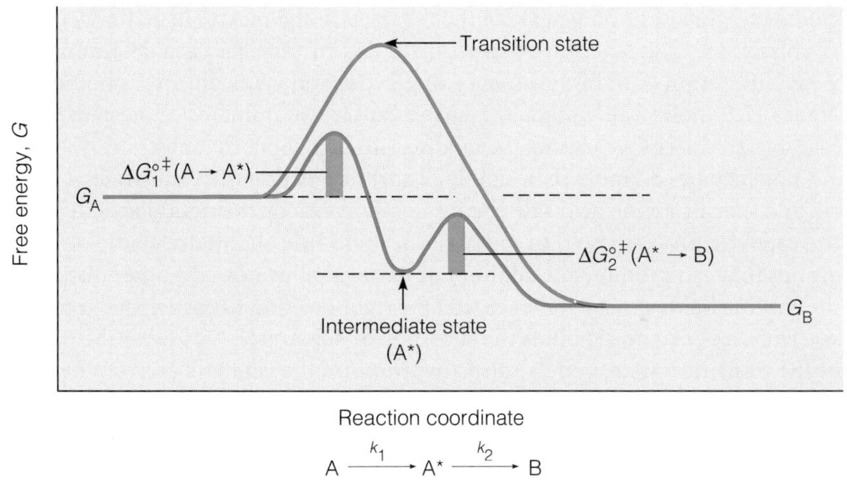

$$A \xrightarrow{k_1} A^* \xrightarrow{k_2} B$$

FIGURE **11.5**

Importance of intermediate states. Often a catalyst (such as an enzyme) binds the substrate in an intermediate conformation that resembles the transition state but has a lower energy. The activation energies for formation of the intermediate state and for conversion of the intermediate to product ($\Delta G_1^{\circ\ddagger}$ and $\Delta G_2^{\circ\ddagger}$, respectively) are lower than the activation energy for the uncatalyzed reaction (see Figure 11.2). In this figure, only activation energies for the forward reaction (A $\longrightarrow$ B) are shown.

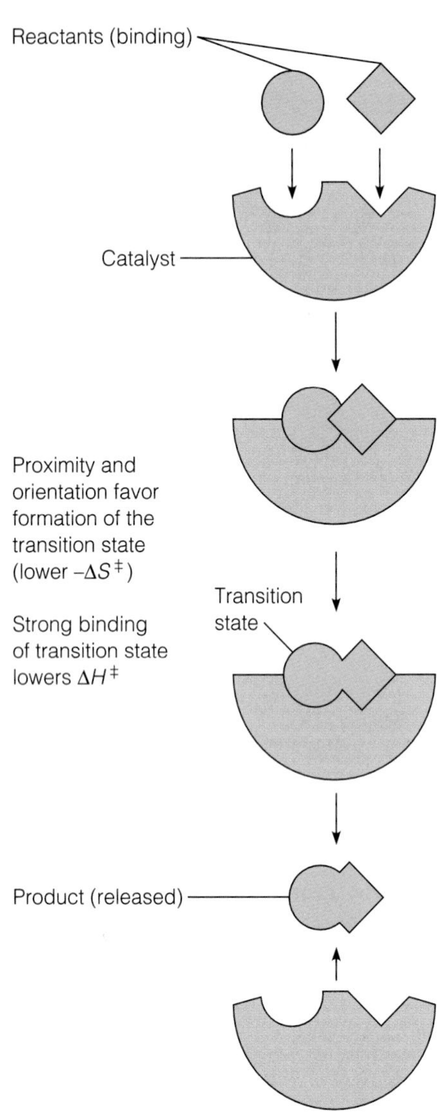

FIGURE **11.6**

Entropic and enthalpic factors in catalysis. In this example, two reactants are bound to an enzyme, which ensures their correct mutual orientation and proximity and binds them most strongly when they are forced into the transition state.

a second-order reaction), most encounters are unproductive just because the molecules happen to be pointed the wrong way when they hit. A catalyst that can bind two reacting molecules in proper mutual orientation will increase their reactivity by making the entropy of activation less negative (Figure 11.6). In a first-order reaction, which involves something happening *within* a single molecule, parts of the molecule must often be oriented properly to allow the transition state to be reached. This orientation is part of what a catalyst achieves. As an external matrix, it can force molecules or parts of molecules into conformations that favor reaction. We shall now examine how biological catalysts, or enzymes, carry out these functions.

How Enzymes Act as Catalysts: Principles and Examples

GENERAL PRINCIPLES: THE INDUCED FIT MODEL

We have seen that the role of a catalyst is to decrease $\Delta G^{\circ\ddagger}$ by facilitating the formation of the transition state or a transition-like state. An enzyme binds a molecule of substrate (or in some cases several substrates) into a region of the enzyme called the **active site,** as shown schematically in Figure 11.7. The active site is often a pocket or cleft surrounded by amino acid side chains that help bind the substrate and by other side chains that play a role in catalysis. The complex tertiary structure of enzymes makes it possible for this pocket to fit the substrate quite closely, which explains the extraordinary specificity of enzyme catalysis. This possibility was realized as early as 1894 by the great German biochemist Emil Fischer, who proposed a *lock-and-key hypothesis* for enzyme action. According to this model, the enzyme accommodates the specific substrate as a lock does its specific key (Figure 11.7a). Although the lock-and-key model explained enzyme specificity, it did not increase our understanding of the catalysis itself, because a lock does nothing to its key. This understanding came from an elaboration of Fischer's idea: What fits the enzyme active site best is a substrate molecule induced to take up a configuration approximating the transition state. In other words, the enzyme does not simply accept the substrate—the enzyme also demands that the substrate be distorted into something close to the transition state. This *induced fit hypothesis,* proposed by Daniel Koshland in 1958, is still the dominant model for enzymatic catalysis.

The induced fit hypothesis states that an enzyme induces a bound substrate molecule to adapt a conformation resembling the transition state.

As currently envisioned, induced fit implies distortion of the enzyme as well as the substrate (Figure 11.7b). This distortion may be local, or it may involve a major change in enzyme conformation. A graphic visualization of such distortion can be seen with the enzyme *hexokinase,* which catalyzes the phosphorylation of glucose to glucose-6-phosphate in the first step in the metabolic pathway called glycolysis (see Chapter 13). The structure of this enzyme has been determined by x-ray diffraction both in the absence of glucose and when glucose is bound. As Figure 11.8 shows, the binding of glucose causes two domains of the enzyme to fold toward each other, closing the binding site cleft about the substrate.

An enzyme binds substrate(s), promotes the reaction, and releases product(s).

But enzymes do more than simply distort or position their substrates. Often, we find specific amino acid side chains poised in exactly the right places to aid in the catalytic process itself. In many cases, these side chains are acidic or basic groups that can promote the addition or removal of protons. In other instances, the enzyme holds a metal ion in exactly the right position to participate in catalysis. Thus, an enzyme (1) binds the substrate or substrates, (2) lowers the energy of the transition state, and (3) directly promotes the catalytic event. When the catalytic process has been completed, the enzyme must be able to release the product or products and return to its original state, ready for another round of cataly-

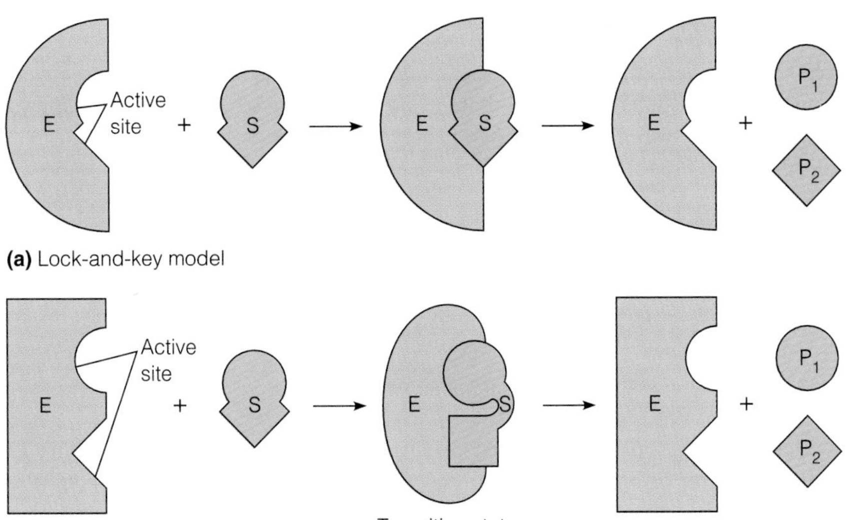

FIGURE **11.7**

Two models for enzyme–substrate interaction. In this example, the enzyme catalyzes a cleavage reaction. **(a)** The lock-and-key model. In this early model, the active site of the enzyme fits the substrate as a lock does a key. **(b)** The induced fit model. In this elaboration of the lock-and-key model, both enzyme and substrate are distorted on binding. The substrate is forced into a conformation approximating the transition state; the enzyme keeps the substrate under stress.

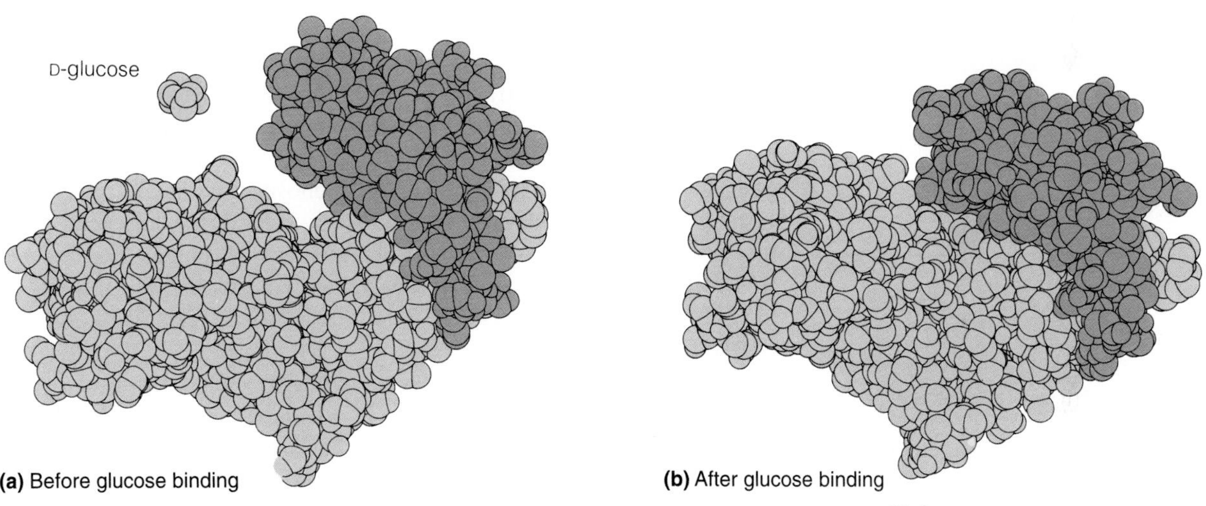

FIGURE **11.8**

The induced conformational change in hexokinase. The binding of glucose to hexokinase induces a major conformational change in the enzyme. The enzyme is a single polypeptide chain, but its two major domains are shaded differently to distinguish them. Notice how the cleft between the domains closes around the glucose molecule.

sis. For an enzyme (E) that catalyzes the conversion of a single substrate (S) into a single product (P), the simplest way to write the overall reaction is in two steps:

$$S + E \rightleftharpoons ES$$

$$ES \longrightarrow E + P$$

Here ES represents the **enzyme–substrate complex,** in which we envision the substrate as being bound to the enzyme and in or near the transition state. We have assumed for the sake of simplicity that although the first reaction (the binding step) is reversible, the second (or catalytic step) is irreversible. We shall see later that much more complex representations are possible, with one or more intermediate states, but the equations above give the essence of the process and emphasize the catalytic role of the enzyme.

To illustrate these general principles in a more concrete way, we turn to two specific enzyme-catalyzed reactions, in which details of the catalysis are well understood. They are also reactions that you have already encountered and will encounter again.

TRIOSE PHOSPHATE ISOMERASE

The well-studied enzyme triose phosphate isomerase catalyzes the interconversion of glyceraldehyde-3-phosphate (G3P) and dihydroxyacetone phosphate (DHAP), a reaction that is known to proceed via an enediol intermediate:

G3P ⇌ *cis*-enediol ⇌ DHAP

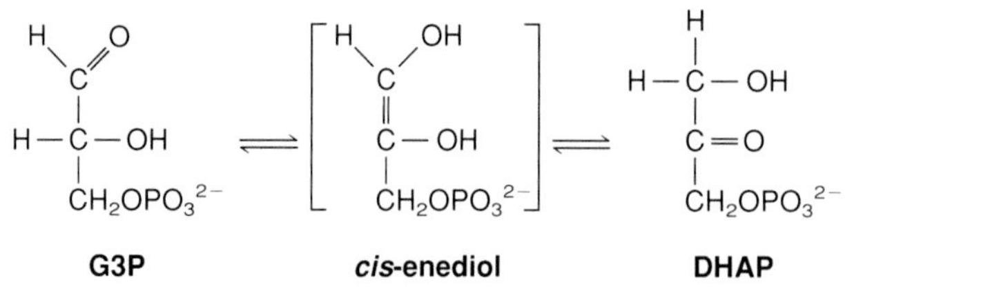

We have already considered this reaction in Chapter 9, and we find in Chapter 13 that it is an important step in glycolysis. As we have written the reaction, G3P is considered the substrate and DHAP the product. But we could equally well have written the reaction the other way around, with DHAP as the substrate. The enediol intermediate is unstable, in that it normally has a much more positive free energy than either G3P or DHAP. Thus, to catalyze the reaction, the enzyme must both promote the proton exchange that forms the enediol and stabilize this intermediate.

Catalysis by triose phosphate isomerase involves promotion and stabilization of an enediol intermediate.

The name given to this enzyme, *triose phosphate isomerase* (TPI, or TIM in the older literature), illustrates a general principle in enzyme naming. Although more complex and specific rules will be given later, note that the name tells what the enzyme does: It isomerizes triose phosphates. The suffix *-ase* is widely used to designate enzymes. The structures of triose phosphate isomerase and its complex with DHAP have been determined in detail by x-ray diffraction methods. The active enzyme is a dimer of two identical subunits. Each has the conformation shown in Figure 11.9a: a parallel β barrel with α helices in the interconnecting loops. The active site lies near the top of the barrel and can accommodate either G3P or DHAP (Figure 11.9b).

The active site of triose phosphate isomerase includes a glutamate residue (Glu 165) and a histidine (His 95) that have been found essential for the function of this enzyme. The glutamate is negatively charged (therefore basic) at physiological pH and helps to extract the proton from carbon 2 of G3P. At the same time histidine acts as an acidic proton-donating group to shuttle protons between the substrate's carbonyl group and the —OH on C-2. Thus, the reaction seems to proceed as follows:

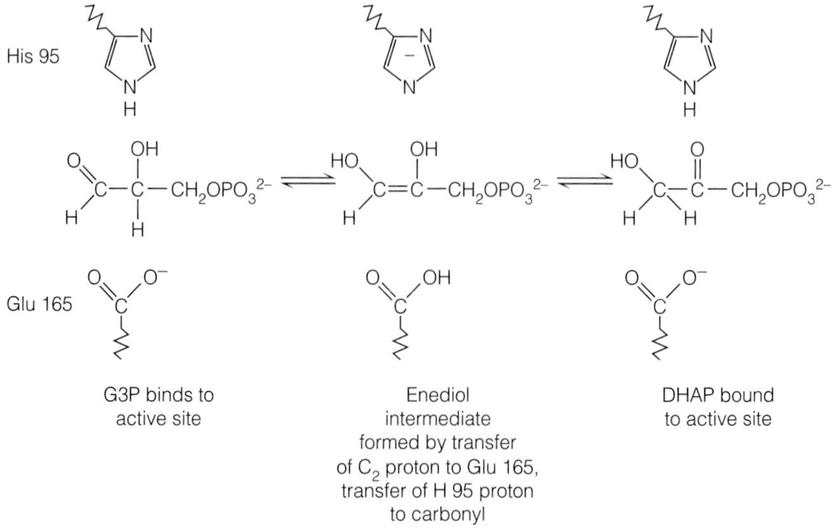

His 95

Glu 165

G3P binds to active site

Enediol intermediate formed by transfer of C_2 proton to Glu 165, transfer of H 95 proton to carbonyl

DHAP bound to active site

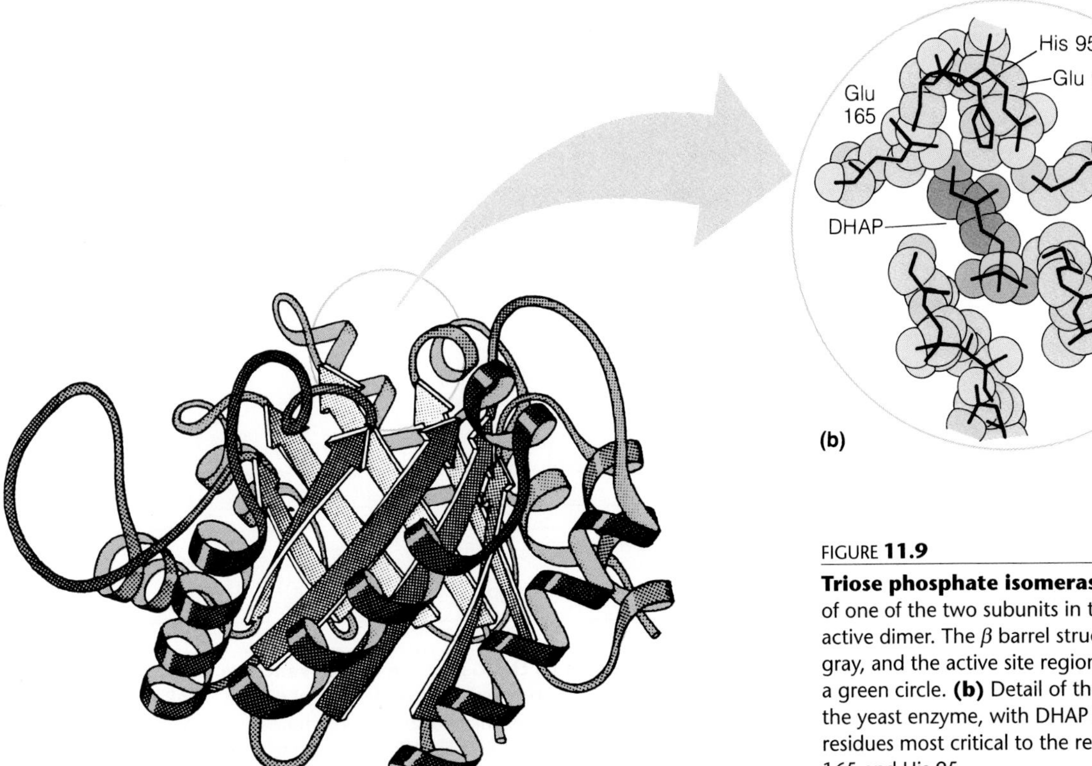

FIGURE **11.9**

Triose phosphate isomerase. (a) Structure of one of the two subunits in the enzymatically active dimer. The β barrel structure is shown in gray, and the active site region is indicated by a green circle. **(b)** Detail of the active site in the yeast enzyme, with DHAP bound. The two residues most critical to the reaction are Glu 165 and His 95.

(a) Courtesy of Dr. J. S. Richardson; **(b)** T. C. Alber et al., *Cold Spring Harbor Symp. Quant. Biol.* (1987) 52:603.

Note what happens. First, the enzyme provides a "cage," or binding site, in which the substrate is held. Indeed, a loop of 11 residues (166 to 176) forms a "lid" that closes down upon the substrate to trap and protect the enediol intermediate. If this loop is removed by mutation, the mutant enzyme tends to lose the enediol, which then dephosphorylates to produce the very toxic methylglyoxal. Thus, the loop is essential to safe functioning of the enzyme in the body. The catalytic efficiency of this mutant is decreased by a factor of 10^5. Second, the presence of Glu 165 and His 95 in exactly the right positions favors the exchange of protons essential for the formation of the intermediate state. The importance of correct positioning has also been demonstrated by site-directed mutagenesis; Glu 165 has been replaced by Asp, which moves the carboxyl only 0.1 nm farther away from the substrate. This change reduces the catalytic rate by a factor of 1000. Both entropy (positioning) and enthalpy (stabilization) factors are involved in reducing the free energy barrier to reaction. Thus, we see that the catalysis of the G3P-to-DHAP conversion is a bit more complicated than the simplest scheme for enzyme catalysis described above. A minimal description of this reaction requires that we write

$$\mathrm{E + G3P \rightleftharpoons E \cdot G3P} \quad \text{Binding of G3P}$$

$$\mathrm{E \cdot G3P \rightleftharpoons E \cdot ed} \quad \text{Conversion to enediol}$$

$$\mathrm{E \cdot ed \rightleftharpoons E \cdot DHAP} \quad \text{Conversion to DHAP}$$

$$\mathrm{E \cdot DHAP \rightleftharpoons E + DHAP} \quad \text{Release of DHAP}$$

The reaction has been studied in great detail, and the free energy barriers to the individual steps have been determined. As Figure 11.10 shows, the presence of the enzyme leads to several low-energy barriers instead of the much higher barrier that would exist in its absence.

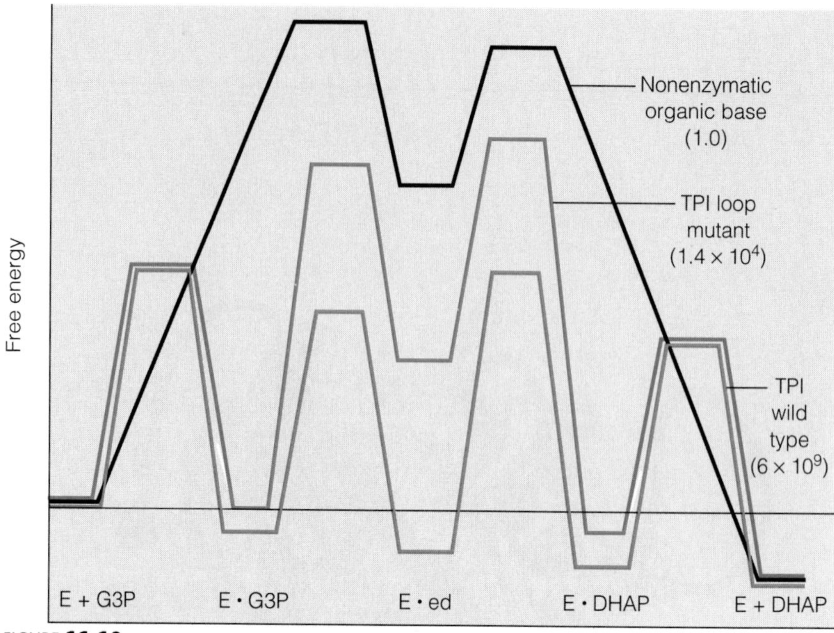

FIGURE **11.10**

Free energy barriers to the G3P ⟶ DHAP reaction. The diagram compares the free energy barriers to the conversion of G3P to DHAP with three different catalysts. The barriers have been drawn schematically, because we don't know their exact shapes. Relative values of the rate constant are shown in parentheses. Black line: The reaction as catalyzed at pH 7 by a simple organic base. Red line: The reaction as catalyzed by wild-type (normal) triose phosphate isomerase. Blue line: The reaction as catalyzed by a triose phosphate isomerase mutant in which the "lid" has been partly removed. Note the decreased stabilization of the enzyme enediol complex in the case of the mutant enzyme.

Adapted from J. Knowles, *Nature* (1991) 350:121–124. © 1991 Macmillan Magazines, Ltd.

SERINE PROTEASES

As a second example, let us consider the catalysis of peptide bond hydrolysis by one of the **serine proteases.** This important class of enzymes includes trypsin and chymotrypsin, which we first encountered in Chapter 5. These enzymes are called proteases because they catalyze the hydrolysis of peptide bonds in polypeptides and proteins. Many kinds of proteases exist, exhibiting a wide range of substrate specificities and utilizing a variety of catalytic mechanisms. The *serine* proteases are distinct because they all have a serine residue that plays a critical role in the catalytic process.

Each of the serine proteases preferentially cuts a polypeptide chain just to the carboxyl side of specific kinds of amino acids. Examples of sites of preferential cutting are shown in Table 5.4 (see page 139). For example, trypsin cuts preferentially to the carboxylate side of basic amino acid residues like lysine or arginine, whereas chymotrypsin acts most strongly if a hydrophobic residue (like phenylalanine) is in this position. The serine proteases also hydrolyze a wide variety of esters—a fact of little physiological importance, but one that biochemists make use of in kinetic studies.

Most of the serine proteases have similar three-dimensional structures and are obviously evolutionarily related. The active site regions of all of the serine proteases have a number of common factors. In particular, an aspartate residue, a histidine residue, and a serine residue are always clustered about the active site depression. These residues are Asp 102, His 57, and Ser 195 in the structure of chymotrypsin shown in Figure 11.11. A fourth feature of the active site differs from one serine protease to another. This is a "pocket," always located close to the active site serine (Figure 11.12). In trypsin the pocket is deep and narrow, with a

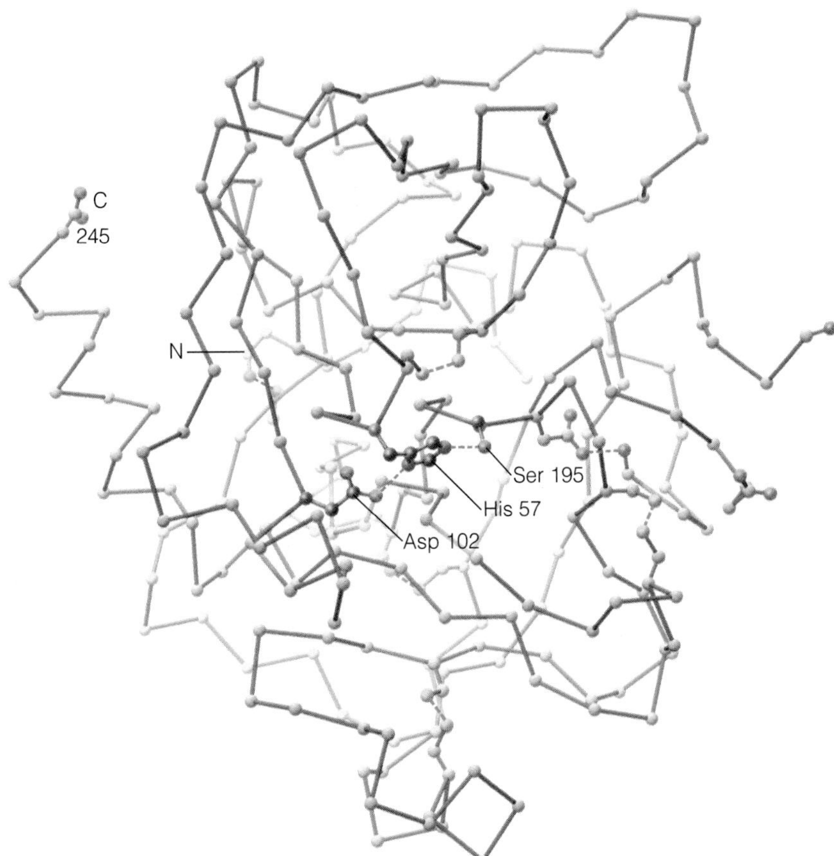

FIGURE **11.11**

The structure of chymotrypsin. Critical residues in the active site are shown in red and blue. The active site pocket lies just below and to right of Ser 195.

negatively charged carboxylate at its bottom—just the thing to catch and hold a long, positively charged side chain like that of lysine or arginine. In chymotrypsin the pocket is wider and is lined with hydrophobic residues to accommodate a hydrophobic side chain, such as that of phenylalanine. It is the nature of this pocket that gives each of the serine proteases its specificity.

Catalysis of peptide bond hydrolysis by a serine protease (chymotrypsin, in this example) proceeds as shown in Figure 11.13. First, the polypeptide chain to be cleaved is bound to the enzyme surface (step 1). Most of the polypeptide binds nonspecifically, but the side chain of the residue to the N-terminal side of the peptide bond to be cleaved must fit in the active site pocket. This pocket defines not only the position of cutting but also the *stereospecificity* of serine proteases. If a D-amino acid residue were present at this position, the side chain would project in the wrong direction, *away* from the pocket. Serine proteases would therefore not cut next to D-amino acids.

This very specific binding of a particular type of amino acid places the active site serine very close to the carbonyl group of the bond to be cleaved. Serine residues are not usually reactive, but this one is in an unusual environment—it sits very close to His 57. The serine proton is transferred to the histidine ring (step 2), leaving a negative charge on the serine. Normally, this transfer would be unlikely due to the high pK_a of alcoholic —OH groups, but it appears to be facilitated by Asp 102, which, by its negative charge, stabilizes the protonation of the adjacent histidine ring. The activated serine is a strong nucleophile and can attack the carbonyl of the substrate, forming the *tetrahedral transition state* shown in step 2 of the figure. Cleavage of the peptide bond (step 3) occurs in this activated state, yielding an *acyl-enzyme intermediate* in which the N-terminal part of the polypeptide substrate is left covalently bound to the enzyme. The C-terminal portion of the polypeptide extracts the proton (which was originally the serine proton)

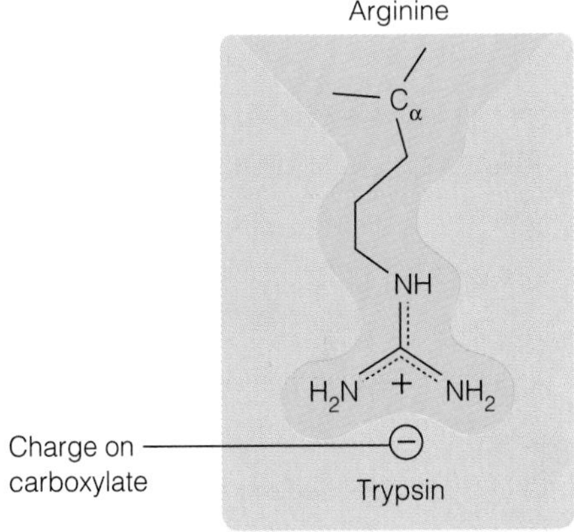

FIGURE **11.12**

Active site pockets of two serine proteases. Different characteristics of the active site pockets account for the different specificities of serine proteases such as trypsin and chymotrypsin.

Hydrophobic pocket
(N-terminus)
Ser 195
Polypeptide substrate
His 57
(C-terminus)
Asp 102
ENZYME–SUBSTRATE COMPLEX

1 Polypeptide substrate binds noncovalently with side chains of hydrophobic pocket.

Ser 195
Cleavage site
His 57
Asp 102
FIRST TRANSITION STATE

2 H⁺ is transferred from Ser to His. The substrate forms a tetrahedral transition state with the enzyme.

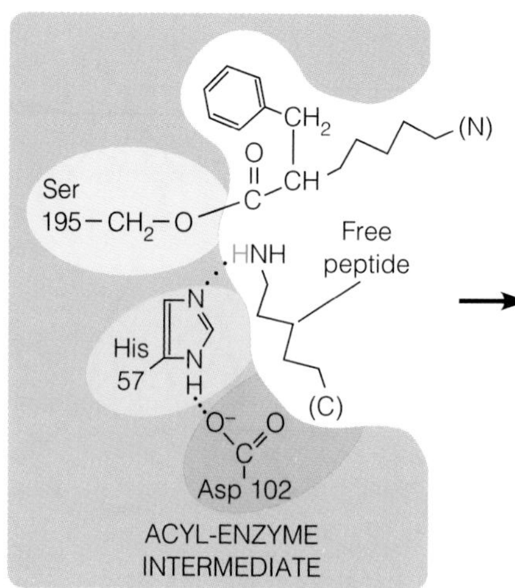

3 H⁺ is transferred to the C-terminal fragment, which is released by cleavage of the C—N bond. The N-terminal peptide is bound through acyl linkage to serine.

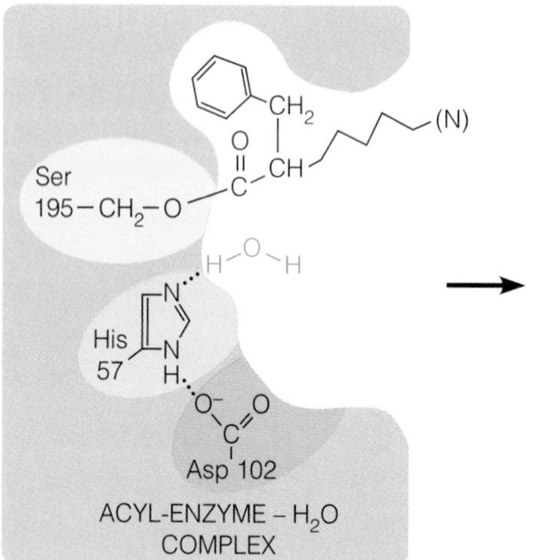

4 A water molecule binds to the enzyme in place of the departed polypeptide.

Ser 195
His 57
Asp 102
SECOND TRANSITION STATE

5 The water molecule transfers its proton to His 57 and its —OH to the remaining substrate fragment. Again a tetrahedral transition state is formed.

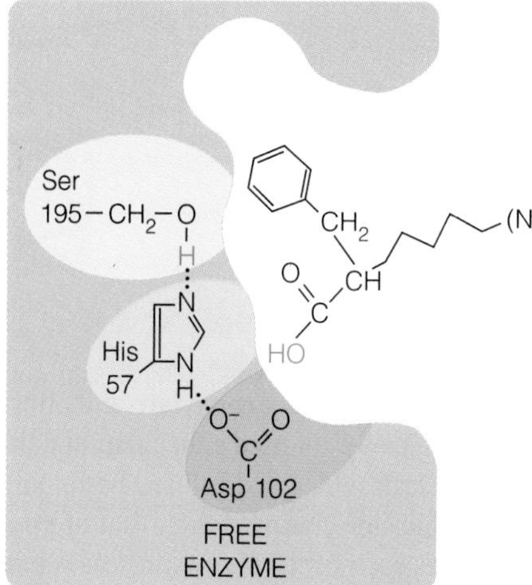

6 The second peptide fragment is released: The acyl bond is cleaved, the proton is transferred from His back to Ser, and the enzyme returns to its initial state.

FIGURE **11.13**

Catalysis of peptide bond hydrolysis by chymotrypsin. The figure shows the steps in the cleavage of a polypeptide chain as catalyzed by chymotrypsin. The brown area represents the enzyme. The figure is highly schematic and does not represent the actual spatial arrangement of atoms.

from the histidine to form a new terminal amino group. This portion of the substrate chain is not by itself bound strongly to the enzyme.

Until this point H_2O has not entered the picture, but it comes in now to first displace the C-terminal portion of the chain, and then to cleave the acyl intermediate. In step 4, a water molecule positions itself between the acyl group and His 57; in step 5, it transfers one proton to His 57 and then links to the acyl intermediate to form a second tetrahedral transition state. Note that this process is essentially a reversal of the formation of the initial acyl intermediate, with the water molecule playing the role of the released portion of the polypeptide chain. Finally (step 6), the proton is transferred from the histidine back to Ser 195, and the rest of the polypeptide chain is released. The enzyme is back in its original state, ready to catalyze the hydrolysis of another chain.

A key to the mechanism of serine protease catalysis lies in the stability of the two tetrahedral intermediate states, which are very similar to the essential transition states. Why are they so stable? A likely possibility has emerged from careful study of the x-ray structures of trapped intermediates. These indicate that backbone amino protons from residues Ser 195 and Gly 193 can form hydrogen bonds to one of the oxygens in the tetrahedral complex. This is the oxygen that was the carbonyl oxygen in the substrate, and the hydrogen bonding can occur only with formation of the tetrahedral state. Thus, this hydrogen bonding stabilizes such intermediates.

The catalysis of peptide bond cleavage by serine proteases involves stabilization of tetrahedral intermediate states.

Chymotrypsin is an excellent example of how enzymes operate. Note that both energetic stabilization of activated states and correct positioning of reactants are important in lowering the free energy of activation.

You may have noted that the two enzymes we have described in detail, triose phospate isomerase and chymotrypsin, both have histidine and an acidic residue in their active site. This is by no means unusual. Histidine is very common in active sites, because of its facility to accept or donate protons at physiological pH. Residues like Glu, Asp, Lys, and Arg are also common participants in proton transfers, as well as frequently serving to make electrostatic bonds with substrate molecules. A number of other residues, such as serine and cysteine, are also occasionally found to play important roles in the active sites of enzymes.

The Kinetics of Enzymatic Catalysis

In the preceding section, we have described details of the mechanisms of examples of enzyme-catalyzed reactions. Very many more are known, and such knowledge is essential in many aspects of modern medicine. It helps in the understanding of pathological conditions and in the design of drugs and therapies to deal with these. Where does this mechanistic knowledge come from? In large part, it comes from the careful mathematical analysis of enzyme kinetics. Thus, it is to that analysis that we must now turn.

REACTION RATE FOR A SIMPLE ENZYME-CATALYZED REACTION: MICHAELIS–MENTEN KINETICS

We have described mechanisms of two enzyme-catalyzed reactions in detail. These examples demonstrate that the *minimal* equation to describe a simple one-substrate, one-product reaction catalyzed by an enzyme is the following:

$$E + S \underset{k_{-1}}{\overset{k_1}{\rightleftharpoons}} ES \xrightarrow{k_2} E + P$$

We have assumed that conditions are such that the reverse reaction between E and P is negligible. The catalytic formation of the product, with enzyme regeneration, will then be a simple first-order reaction, and its rate will be determined solely by the concentration of ES and the value of k_2. Therefore, we can express the reaction rate or velocity, defined as the rate of formation of products, just as we did in equation (11.3):

$$V = k_2[ES] \tag{11.14}$$

If we could measure V and [ES] for a specific enzyme and substrate, they would give us the rate constant, k_2, for that particular reaction. However, [ES] is usually not a measurable concentration. What we *can* measure are substrate (or product) concentrations and the *total* concentration of enzyme, which must be the sum of free enzyme and occupied enzyme:

$$\underset{\text{Total enzyme}}{[E]_t} = \underset{\text{Free enzyme}}{[E]} + \underset{\text{Enzyme in ES complex}}{[ES]} \tag{11.15}$$

Thus, we would like to express the rate, V, in terms of the substrate concentration [S] and the total enzyme concentration $[E]_t$.

You might think, from the way we have written the reaction, that E and S should be in equilibrium with ES, with an equilibrium dissociation constant K_S:

$$K_S = \frac{k_{-1}}{k_1} = \frac{[E][S]}{[ES]} \tag{11.16}$$

This is usually an incorrect assumption, but under certain circumstances (k_2 much less than k_{-1}), this approximation is valid. It was used in early attempts to solve the problem of expressing the reaction rate. But it won't work in general, for E, S, and ES are not truly in equilibrium; some ES is continually being drained off to make P. An analysis that avoids the assumption of equilibrium was presented by G. E. Briggs and J. B. S. Haldane in 1925. The Briggs–Haldane model is based on the following argument: The more ES that is present, the faster ES will dissociate either to products or back to reactants. Therefore, when the reaction is started by mixing enzymes and substrates, the ES concentration builds up at first, but quickly reaches a **steady state,** in which it remains almost constant. This steady state will persist until almost all of the substrate has been consumed (Figure 11.14). Because the steady state accounts for nearly all the reaction time, we can calculate the reaction velocity by assuming steady state conditions. Normally, we measure rates only after the steady state has been established and before [ES] has changed much. We can then express the velocity as follows.

The steady state hypothesis states that the concentration of enzyme–substrate complex remains nearly constant through much of the reaction.

In the steady state, the rates of formation and breakdown of ES are equal. Therefore,

$$\underset{\text{Formation from enzyme and substrate}}{k_1[E][S]} = \underset{\text{Breakdown into enzyme and substrate}}{k_{-1}[ES]} + \underset{\text{Breakdown into enzyme and products}}{k_2[ES]} \tag{11.17}$$

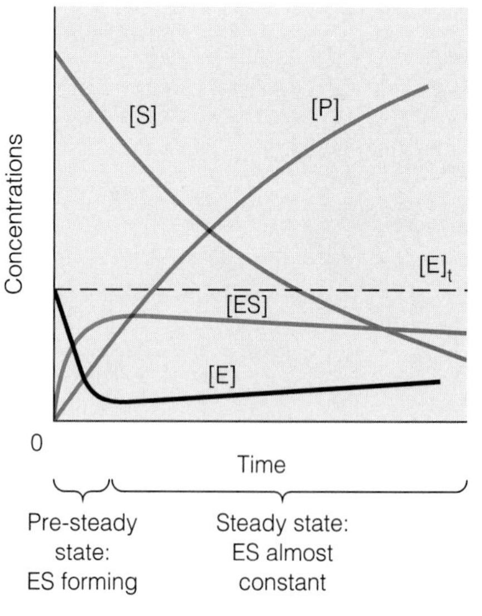

FIGURE **11.14**

The steady state in enzyme kinetics. For a simple enzyme-catalyzed reaction ($E + S \rightleftharpoons ES \longrightarrow E + P$), the graph shows how the concentrations of substrate [S], free enzyme [E], enzyme–substrate complex [ES], and product [P] vary with time. After a very brief initial period, [ES] reaches a steady state in which ES is consumed approximately as rapidly as it is formed, so $d[ES]/dt \cong 0$. The amounts of E and ES are greatly exaggerated for clarity. Note that $[E] + [ES] = [E]_t$, or total enzyme concentration, and that [ES] actually falls very slowly as substrate is consumed, while [E] accordingly rises.

which we can rearrange to give

$$[ES] = \left(\frac{k_1}{k_{-1} + k_2}\right)[E][S] \tag{11.18}$$

We combine the ratio of rate constants in equation (11.18) into a single constant, K_M:

$$K_M = \frac{k_{-1} + k_2}{k_1} \tag{11.19}$$

Equation (11.18) can now be rewritten as

$$K_M[ES] = [E][S] \tag{11.20}$$

At this point, we have [ES] in terms of [E] and [S]. To get $[E]_t$ into the equation, rather than [E], recall from equation (11.15) that $[E] = [E]_t - [ES]$. Putting this into equation (11.20) yields

$$K_M[ES] = [E]_t[S] - [ES][S] \tag{11.21}$$

If we transfer the second term on the right side of the equation to the left side and then factor, we get

$$[ES] \cdot (K_M + [S]) = [E]_t[S] \tag{11.22}$$

This rearranges to

$$[ES] = \frac{[E]_t[S]}{K_M + [S]} \tag{11.23}$$

Finally, if we insert this result into equation (11.14), we achieve what we sought—an expression for V in terms of $[E]_t$ and [S]:

$$V = \frac{k_2[E]_t[S]}{K_M + [S]} \tag{11.24}$$

Equation (11.24) is called the **Michaelis–Menten equation,** and K_M the **Michaelis constant,** honoring two pioneers in the analysis of enzyme kinetics, Leonor Michaelis and Maude Menten. We will discuss the meaning of K_M shortly; in the meantime, there are two important points to keep in mind. First, because K_M is a ratio of the rate constants for a specific reaction (see equation (11.19)), it is a characteristic of that reaction. Thus, a given enzyme acting upon a given substrate has a defined K_M. Second, you can see from equation (11.20) that K_M has units of concentration.

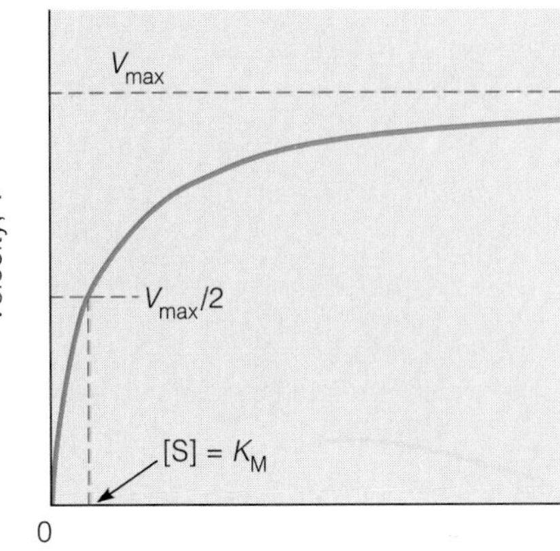

FIGURE **11.15**

The effect of substrate concentration on reaction velocity. This graph, a plot of equation (11.26), shows the variation of reaction velocity with substrate concentration according to Michaelis–Menten kinetics. At the point where $[S] = K_M$, the reaction has exactly half its maximum velocity. Note that it is difficult to estimate maximum velocity from the data, using this graph, because V_{max} is approached asymptotically.

Now consider the graph of V versus [S] shown in Figure 11.15. At high substrate concentrations, where [S] is much greater than K_M, the reaction approaches a **maximum velocity, V_{max},** because the enzyme molecules are *saturated;* every enzyme molecule is occupied by substrate and is busy carrying out the catalytic step. When we express this mathematically by making [S] much larger than K_M in equation (11.24), so that $K_m + [S] \cong [S]$ we obtain the expression for V_{max}:

$$V_{max} = k_2[E]_t \tag{11.25}$$

Thus, we can express $k_2[E]_t$ in equation (11.24) as V_{max}, and rewrite the Michaelis–Menten rate equation as

$$V = \frac{V_{max}[S]}{K_M + [S]} \tag{11.26}$$

This is the form in which we shall mostly use it.

EXPRESSING REACTION RATES FOR MULTISTEP REACTIONS

We have derived the Michaelis–Menten equation for the simple two-step enzymatic reaction given on page 375. However, this equation can also be used to describe more complicated processes under certain circumstances. Consider the reaction

$$E + S \underset{k_{-1}}{\overset{k_1}{\rightleftharpoons}} ES \xrightarrow{k_2} EP \xrightarrow{k_3} E + P$$

Such a multistep reaction can be described by an equation of the form of the Michaelis–Menten equation, but k_2 in equation (11.24) must be replaced by a more general constant, which we call k_{cat}:

$$V = \frac{k_{cat}[E]_t[S]}{K_M + [S]} \tag{11.27}$$

The quantity k_{cat} incorporates the rate constants for all the reactions between ES and E + P. For the two-step reactions we considered earlier, $k_{cat} = k_2$. For more complex reactions, k_{cat} is a combination of other rate constants; its value depends on which steps in the process are rate-limiting. (See page 382 for an example.)

In any event, equation (11.26) remains unchanged in form, if we define $V_{max} = k_{cat}[E]_t$. Because the definition of k_{cat} includes k_2 as a special case, we shall employ k_{cat} in all subsequent equations, even for the simplest reactions.

THE SIGNIFICANCE OF K_M, k_{cat}, AND k_{cat}/K_M

The Michaelis constant, K_M, measures the substrate concentration at which the reaction rate is $V_{max}/2$.

The turnover number, k_{cat}, measures the rate of the catalytic process.

The ratio k_{cat}/K_M is a convenient measure of enzyme efficiency.

The two quantities that characterize an enzyme obeying Michaelis–Menten kinetics are K_M and k_{cat}. What do they mean? The Michaelis constant, K_M, is often associated with the affinity of enzyme for substrate. This relationship is certainly true in the limiting case, a two-step reaction in which k_2 is much smaller than k_{-1}, for then equation (11.19) yields $K_M \cong k_{-1}/k_1 = K_S$, the equilibrium constant defined in equation (11.16). Under these circumstances, a large K_M means that k_{-1} is much greater than k_1, and the enzyme binds substrate very weakly. Thus, when the rate of product formation is low, we can think of K_M as an inverse measure of substrate binding strength. But a very large value of k_2 can also lead to a large K_M, as equation (11.19) shows. We might say that K_M measures the propensity of ES to dissociate in either direction, relative to its propensity to be formed. This definition is not very useful, however, because a high K_M can mean either that product is formed rapidly (high k_2) or that the complex dissociates rapidly (high k_{-1}). Perhaps the most useful way to think of K_M is shown in Figure 11.15, which is a plot of equation (11.26). This plot shows that, for any reaction that follows the Michaelis–Menten equation, K_M is numerically equal to the substrate concentration at which the reaction velocity has attained *half* of its maximum value. Thus, K_M is a measure of the substrate concentration required for effective catalysis to occur. *An enzyme with a high K_M requires a higher substrate concentration to achieve a given reaction velocity than an enzyme with a low K_M* but the same k_{cat}. Table 11.1 lists K_M values for a number of important enzymes.

The second constant, k_{cat}, gives a direct measure of the catalytic production of product under optimum conditions (saturated enzyme). The units of k_{cat} are s^{-1}, so the reciprocal of k_{cat} can be thought of as a time—the time required by an enzyme molecule to "turn over" one substrate molecule. Alternatively, k_{cat} measures the number of substrate molecules turned over per enzyme molecule per second. Thus, k_{cat} is sometimes called the **turnover number.** Some typical turnover numbers are listed in Table 11.1.

The enzymes listed in Table 11.1 are arranged in increasing order of the ratio k_{cat}/K_M. This ratio is often thought of as a measure of enzyme efficiency. Note that either a large value of k_{cat} (which means rapid turnover) or a small value of K_m (which means high affinity for substrate) will make k_{cat}/K_m large. We can gain another insight into the meaning of k_{cat}/K_m by considering the situation at very low substrate concentrations. In this case, $[S] << K_M$, and most of the enzyme is free, so $[E]_t \cong [E]$. Then equation (11.27) becomes

$$V \cong \frac{k_{cat}}{K_M}[E][S] \tag{11.28}$$

Therefore, under these circumstances, the ratio k_{cat}/K_M behaves as a second-order rate constant for the reaction between substrate and free enzyme. This ratio is important, for it provides a direct measure of enzyme efficiency and specificity. It shows what the enzyme and substrate can accomplish when abundant enzyme sites are available, and it allows direct comparison of the effectiveness of an enzyme toward different substrates. Suppose an enzyme has a choice of two substrates, A or B, present at equal concentrations. Then under conditions in which both substrates are dilute and are competing for the enzyme, we find

$$\frac{V_A}{V_B} = \frac{(k_{cat}/K_M)_A[E][A]}{(k_{cat}/K_M)_B[E][B]} = \frac{(k_{cat}/K_M)_A}{(k_{cat}/K_M)_B} \tag{11.29}$$

TABLE **11.1 Michaelis–Menten parameters for selected enzymes, arranged in order of increasing efficiency as measured by k_{cat}/K_M**

Enzyme	Reaction Catalyzed	K_M (mol/L)	$k_{cat}(s^{-1})$	k_{cat}/K_M [(mol/L)$^{-1}$ s^{-1}]
Chymotrypsin	Ac–Phe–Ala $\xrightarrow{H_2O}$ Ac–Phe + Ala	1.5×10^{-2}	0.14	9.3
Pepsin	Phe–Gly $\xrightarrow{H_2O}$ Phe + Gly	3×10^{-4}	0.5	1.7×10^3
Tyrosyl-tRNA synthetase	Tyrosine + tRNA $\longrightarrow$ tyrosyl-tRNA	9×10^{-4}	7.6	8.4×10^3
Ribonuclease	Cytidine 2′, 3′ cyclic phosphate $\xrightarrow{H_2O}$ cytidine 3′-phosphate	7.9×10^{-3}	7.9×10^2	1.0×10^5
Carbonic anhydrase	$HCO_3^- + H^+ \longrightarrow H_2O + CO_2$	2.6×10^{-2}	4×10^5	1.5×10^7
Fumarase	Fumarate $\xrightarrow{H_2O}$ malate	5×10^{-6}	8×10^2	1.6×10^8

Table 11.2 lists values of k_{cat}/K_M for chymotrypsin with various substrates. Within the group shown, k_{cat}/K_M varies 1 million–fold, showing the range of preference the enzyme has for even quite similar substrates. The preference to cleave next to the most hydrophobic residues is quite clear from these data.

We have just seen that the ratio k_{cat}/K_M corresponds to the second-order rate constant for enzyme–substrate combination under circumstances of low substrate concentration. Such a rate constant has a maximum possible value, which is determined by the frequency with which enzyme and substrate molecules can collide. A reaction which attains such a velocity is said to be "diffusion-limited"; every encounter leads to reaction, so nothing but the rate of molecular encounters limits the velocity. If *every* collision results in formation of an enzyme–substrate complex, diffusion theory predicts that k_{cat}/K_M will attain a value of about 10^8 to 10^9 (mol/L)$^{-1}$ s^{-1} (see Problem 4). Thus, an enzyme that approaches maximum possible efficiency will demonstrate this by having a value of k_{cat}/K_M in this range. As Table 11.1 shows, enzymes such as carbonic anhydrase and fumarase actually approach this limit. Triose phosphate isomerase, which we have already studied, is another example, having $k_{cat}/K_M = 2.4 \times 10^8$ (mol/L)$^{-1}$ s^{-1}. In fact, it has been argued that triose phosphate isomerase is an almost perfect enzyme, having evolved to nearly maximum efficiency. In support of this idea is the observation that triose phosphate isomerases from organisms as evolutionarily distant as yeasts and vertebrates show very little change in structure. Apparently, this vital

TABLE **11.2 Preferences of chymotrypsin in the hydrolysis of several *N*-acetyl amino acid methyl esters, as measured by k_{cat}/K_M**

Amino Acid in Ester	Amino Acid Side Chain	k_{cat}/K_M [(mol/L)$^{-1}$ s^{-1}]
Glycine	—H	1.3×10^{-1}
Norvaline	$—CH_2CH_2CH_3$	3.6×10^2
Norleucine	$—CH_2CH_2CH_2CH_3$	3.0×10^3
Phenylalanine	$—CH_2$–(phenyl ring)	1.0×10^5

N-acetyl amino acid methyl ester

enzyme reached near perfection early in evolution and has changed little since that time.

ANALYSIS OF KINETIC DATA: TESTING THE MICHAELIS–MENTEN EQUATION

Suppose we wish to see whether an enzyme-catalyzed reaction follows the Michaelis–Menten law (equation (11.27)) and to determine the constants K_M and k_{cat}. How do we proceed? First, we will need some way to follow the formation of product or the consumption of substrate, in order to measure reaction velocity.

A number of analytical methods for the measurement of rates are described in Tools of Biochemistry 11A. One general point should be noted: In principle, we could simply mix enzyme and substrate and follow the changes in both substrate concentration and reaction velocity with time, as shown in Figure 11.1a. As substrate is consumed, the velocity decreases, until equilibrium is eventually reached. But measuring the instantaneous velocity at specific times during the reaction is difficult and usually inaccurate. It is usually easier to set up a series of experiments, all at the same enzyme concentration but at different substrate concentrations, and measure the *initial* rates (Figure 11.1c). Because we know the initial [S] precisely, and the change in [S] versus t is almost linear in the initial stages, accurate data for V as a function of [S] can thus be obtained.

The most convenient ways to determine K_M and k_{cat} are via Lineweaver–Burk plots or Eadie–Hofstee plots.

Given the data for concentrations and initial rates, how do we find K_M and k_{cat}? The best way is to rearrange equation (11.26) in such a way that it produces a linear graph. Several kinds of graphs are possible, but it is most common to use a **double reciprocal plot,** also called a **Lineweaver–Burk plot** (Figure 11.16). If we simply invert both sides of equation (11.26), we find

$$\frac{1}{V} = \frac{K_M + [S]}{V_{max}[S]} = \frac{K_M}{V_{max}[S]} + \frac{[S]}{V_{max}[S]} \tag{11.30a}$$

or

$$\frac{1}{V} = \left(\frac{K_M}{V_{max}}\right)\left(\frac{1}{[S]}\right) + \frac{1}{V_{max}} \tag{11.30b}$$

So, plotting $1/V$ versus $1/[S]$, we should get a straight line. Where $1/[S] = 0$, [S] is infinitely large and the reaction velocity is at its maximum. Therefore, we can find $1/V_{max}$ from the intercept of the line with the $1/V$ axis. Having V_{max} and knowing $[E]_t$ from the weight concentration and molecular weight of the enzyme, we can calculate k_{cat} from $V_{max} = k_{cat}[E]_t$. To obtain K_M, we extrapolate the Lineweaver–Burk plot to the hypothetical point where $1/V = 0$, which gives

$$0 = \frac{K_M}{V_{max}[S]_0} + \frac{1}{V_{max}} \tag{11.31}$$

where $[S]_0$ denotes the value of [S] at $1/V = 0$. We then obtain from equation (11.31):

$$1/[S]_0 = -1/K_M \tag{11.32}$$

Thus, the intercept of the Lineweaver–Burk plot with the $1/[S]$ axis gives $-1/K_M$. (See Figure 11.16.) Alternatively, we can obtain K_m/V_{max} from the slope.

A Lineweaver–Burk plot provides a quick test for adherence to Michaelis–Menten kinetics and allows easy evaluation of the critical constants. As we shall see, it also allows discrimination between different kinds of enzyme inhibition and regulation. A disadvantage of a Lineweaver–Burk plot is that a long extrapolation is often required to determine K_M, with corresponding uncertainty in the

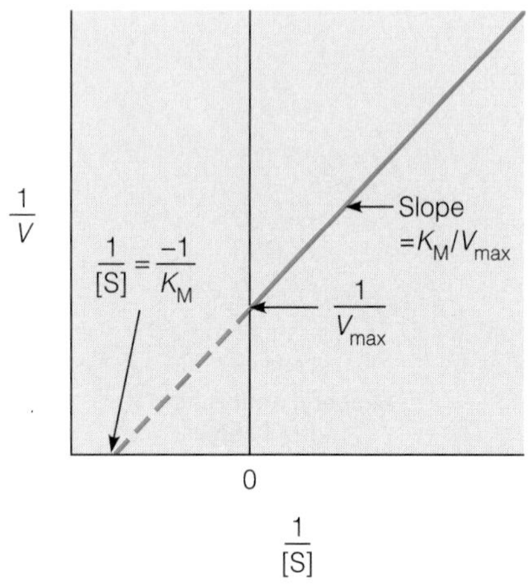

FIGURE **11.16**

A Lineweaver–Burk plot. In this double reciprocal plot, $1/V$ is graphed versus $1/[S]$, according to equation (11.30b). Note that a linear extrapolation of the data gives both V_{max} and K_M.

result. Consequently, other ways of plotting the data are sometimes used. One alternative is to rearrange equation (11.26) into the form

$$V = V_{max} - K_M \frac{V}{[S]} \tag{11.33}$$

and graph V versus $V/[S]$. This yields what is called an **Eadie–Hofstee plot** (Figure 11.17).

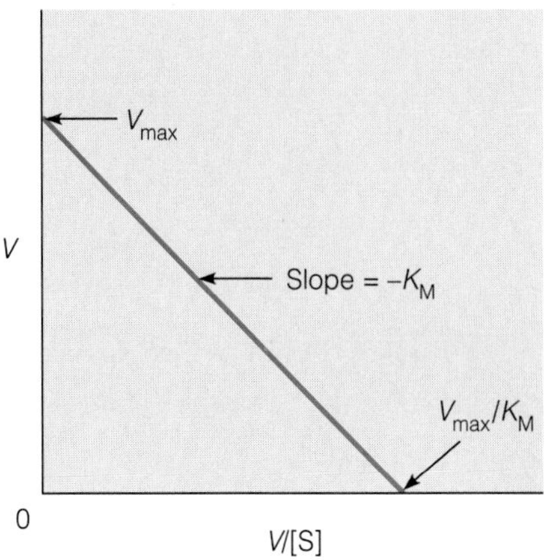

FIGURE **11.17**

An Eadie–Hofstee plot. Graphing V versus $V/[S]$, we obtain V_{max} at $(V/[S]) = 0$ and K_M from the slope of the line.

MULTISUBSTRATE REACTIONS

Our discussions of enzyme kinetics have, to this point, centered on simple reactions in which one substrate molecule is bound to an enzyme and undergoes reaction there. In fact, such reactions are in the minority. Most biochemical reactions involve two or more substrates, often resulting in multiple products. An example we have already discussed is proteolysis, which involves two substrates (the polypeptide and water) and two products (the two fragments of the cleaved polypeptide chain). Phosphorylation of glucose, as catalyzed by hexokinase, is another such case: The two substrates are glucose and ATP, and the products are glucose-6-phosphate and ADP.

When an enzyme binds two or more substrates and releases multiple products, the order of the steps becomes an important feature of the enzyme mechanism. Several major classes of mechanisms for multisubstrate reactions are recognized. We shall illustrate them with examples using two substrates, S1 and S2, and two products, P1 and P2.

Multisubstrate reactions fall into several classes, depending on the order of substrate binding: random, ordered, or ping-pong.

Random Substrate Binding

In random substrate binding, either substrate can be bound first, although in many cases one substrate will be favored for initial binding, and its binding may promote the binding of the other. The general pathway is

either S1: $E \xrightarrow{S1} E \cdot S1 \xrightarrow{S2} E \cdot S1 \cdot S2 \longrightarrow E + P1 + P2$

or S2: $E \xrightarrow{S2} E \cdot S2 \xrightarrow{S1} E \cdot S1 \cdot S2 \longrightarrow E + P1 + P2$

The phosphorylation of glucose by ATP, with hexokinase as the enzyme, appears to follow such a mechanism, although there is some tendency for glucose to bind first.

Ordered Substrate Binding

In some cases, one substrate *must* bind before a second substrate can bind significantly. We then have

$$E \xrightarrow{S1} E \cdot S1 \xrightarrow{S2} E \cdot S1 \cdot S2 \longrightarrow E + P1 + P2$$

This mechanism is often observed in oxidations of substrates by the coenzyme nicotinamide adenine dinucleotide (NAD^+), which will be discussed later in this chapter.

The "Ping-Pong" Mechanism

Sometimes the sequence of events in catalysis goes like this: One substrate is bound, one product is released, a second substrate comes in, and a second product is released. This is called a ping-pong reaction:

$$E \xrightarrow{S1} E \cdot S1 \xrightarrow{P1} E^* \xrightarrow{S2} E^* \cdot S2 \xrightarrow{P2} E$$

Here E* is a modified form of the enzyme, often carrying a fragment of S1. A good example is the cleavage of a polypeptide chain by a serine protease such as trypsin or chymotrypsin. We describe the polypeptide as S = A · B, where A and B designate the C-terminal and N-terminal portions of the chain from the point of cleavage:

$$\mathrm{E} \xrightarrow{\mathrm{S}} \mathrm{E\cdot S} \xrightarrow{\mathrm{A}} \mathrm{E^{*}\cdot B} \xrightarrow{\mathrm{H_2O}} \mathrm{E^{*}\cdot B\cdot H_2O} \xrightarrow{\mathrm{B}} \mathrm{E}$$

Here E* · B and E* · B · H_2O indicate the covalent intermediates described in our earlier discussion (see Figure 11.13).

A CLOSER LOOK AT SOME COMPLEX REACTIONS

How do we actually analyze the mechanism of a complex enzyme-catalyzed reaction and determine the rate constants for different steps? As an example, let us consider the cleavage of a substrate by a serine protease such as chymotrypsin.

In the first place, note that we cannot analyze the step E* · B + H_2O ⟶ E* · B · H_2O. The concentration of water is essentially fixed in aqueous solution and is not a variable. Therefore, it will suffice to write the reaction as

$$\mathrm{E+S} \underset{k_{-1}}{\overset{k_1}{\rightleftharpoons}} \mathrm{E\cdot S} \xrightarrow[k_2]{\mathrm{A}} \mathrm{E^{*}\cdot B} \xrightarrow[k_3]{} \mathrm{E+B}$$

We have, then, a number of constants to determine. Steady state measurements will not in themselves be sufficient. It can be shown that the steady state velocity for this reaction is given by the equation

$$V = \frac{\left(\dfrac{k_2k_3}{k_2+k_3}\right)[\mathrm{E}]_t[\mathrm{S}]}{[\mathrm{S}] + \left(\dfrac{K_Sk_3}{k_2+k_3}\right)} \tag{11.34}$$

In other words, the enzyme obeys Michaelis–Menten kinetics, but k_{cat} and K_M are now defined as follows:

$$k_{cat} = \frac{k_2k_3}{k_2+k_3} \tag{11.35a}$$

$$K_M = \frac{K_Sk_3}{(k_2+k_3)} \tag{11.35b}$$

$$K_S = k_{-1}/k_1 \tag{11.35c}$$

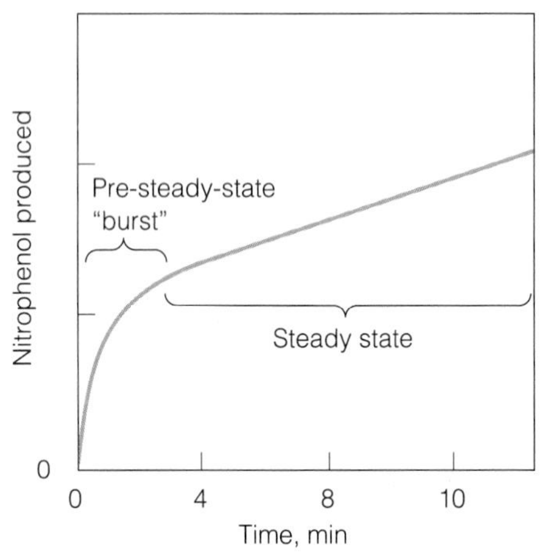

FIGURE **11.18**

The pre–steady state. This plot shows the kinetics of chymotrypsin-catalyzed hydrolysis of *p*-nitrophenylacetate. Production of the first product (*p*-nitrophenol) is followed spectrophotometrically after the enzyme and substrate are mixed. The initial burst of product formation ceases when enzyme is almost all bound in the acyl intermediate.

These results emphasize that the appropriate expressions for k_{cat} and K_M depend upon the reaction mechanism, even in cases where the Michaelis–Menten equation describes the velocity. To obtain the individual rate constants in such a case, measurements outside the steady state range must be employed. One of the first indications that an intermediate might be involved in hydrolysis came from rapid kinetic studies of the early stages in the hydrolysis of esters by chymotrypsin. If we follow the release of product A in our example reaction, we find that its concentration increases quickly for a few minutes, until about 1 mole has been produced per mole of enzyme. After this time, we observe steady state production (Figure 11.18).

The initial burst of production of the first product, called *pre-steady-state production,* has been explained in the following way. For ester hydrolysis, k_3 is much smaller than k_2. Thus, the acyl intermediate forms quickly on each enzyme molecule, with accompanying release of 1 mole of product A. But after this period, more A can be formed only after each acyl intermediate breaks down and the en-

TABLE **11.3 Rate constants for the hydrolysis of two *N*-acyl amino acid esters by chymotrypsin**

Substrate	k_{cat} (s^{-1})	k_2 (s^{-1})	k_3 (s^{-1})	K_M (mM)	K_S (mM)
C_6H_5–CO–NH–CH(CH_3)–CO–O–C_2H_5	0.069	0.069	0.6	5.87	5.97
CH_3–CO–NH–CH(CH_2–C_6H_4–OH)–CO–O–C_2H_5	192	5000	200	0.663	17.2

zyme becomes available again. The dissociation of the acyl intermediate is the rate-limiting step.

Still faster measurements, using stopped-flow techniques (see Tools of Biochemistry 11A), allow measurement of the rate of formation of the enzyme–substrate complex (ES). Measurements of the decay of the acyl intermediates after substrate is exhausted provide k_3. By using a combination of such methods, together with steady state studies, we can obtain all of the constants in equation (11.34). Two examples of such detailed kinetic data are given in Table 11.3 for hydrolysis of *N*-acyl amino acid esters. In example 1, we find $k_{cat} \cong k_2$, and $K_M \cong K_S$. These are the results expected when the acylation reaction (k_2) is rate-limiting, with $k_2 << k_3$. In the second instance, deacylation is rate-limiting ($k_2 >> k_3$) and $k_{cat} \cong k_3$. In this situation, $K_M = K_S(k_3/k_2)$. You can check each of these statements by examination of how equations (11.35) behave in the specific cases.

This example demonstrates that a steady state analysis is only a first step in the study of any enzyme and that a variety of techniques must be employed to unravel mechanisms.

Enzyme Inhibition

We can learn more about how enzymes act by studies of inhibition. Many different kinds of molecules inhibit enzymes, and they act in a variety of ways. A major distinction must be made between **reversible inhibition** and **irreversible inhibition.** The former involves *noncovalent* binding of the inhibitor and can always be reversed, at least in principle, by removal of the inhibitor. In some cases, noncovalent binding may be so strong as to *appear* irreversible under physiological conditions. Trypsin inhibitor binding to trypsin is one such example. In irreversible inhibition, on the other hand, a molecule is *covalently* bound to the enzyme and truly incapacitates it. Irreversible inhibition is frequently encountered in the action of specific toxins and poisons, many of which kill by incapacitating key enzymes (see p. 386). On the other hand, the therapeutic action of many drugs depends on their acting as enzyme inhibitors, as we shall see in many examples.

Inhibition of enzymes can be either reversible or irreversible.

REVERSIBLE INHIBITION

The various modes of reversible inhibition all involve the noncovalent binding of an inhibitor to the enzyme, but they differ in the mechanisms by which they decrease the enzyme's activity and in how they affect the kinetics of the reaction.

Competitive Inhibition

Suppose a molecule exists that so closely resembles the substrate for an enzyme-catalyzed reaction that the enzyme will accept this molecule in its binding site. If this molecule can also be processed by the enzyme, it is merely a competing alternative substrate. However, if the molecule binds to the active site but *cannot* undergo the catalytic step, it simply wastes the enzyme's time. Such a molecule is called a **competitive inhibitor** (Figure 11.19).

A competitive inhibitor competes with substrate for the binding site. It increases the apparent K_M.

For whatever fraction of time a competitive inhibitor molecule is occupying the active site, the enzyme is unavailable for catalysis. The overall effect is as if the enzyme cannot bind substrate as well when the inhibitor is present. Thus, we expect that the enzyme would act as if its K_M were increased by the presence of the inhibitor. Mathematically, we express these ideas by writing the reaction scheme as

$$\begin{array}{l} E + S \underset{k_{-1}}{\overset{k_1}{\rightleftharpoons}} ES \xrightarrow{k_{cat}} E + P \\ + \\ \text{I} \\ K_I \updownarrow \\ EI \end{array}$$

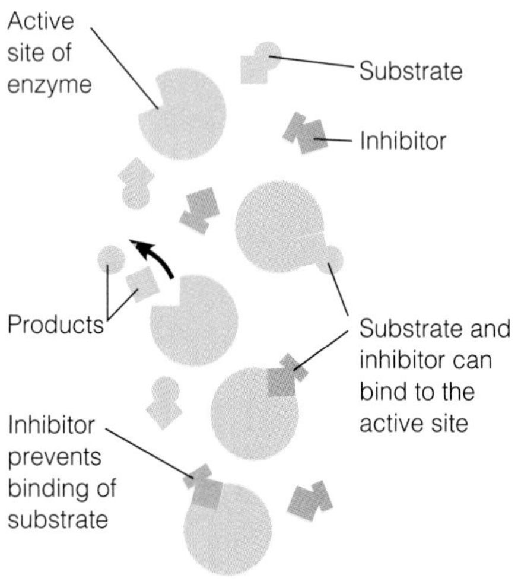

FIGURE **11.19**

Competitive inhibition. Both substrate and inhibitor can fit the active site. Substrate can be processed by the enzyme, but inhibitor cannot.

Here I stands for the inhibitory substance and K_I is a dissociation constant for inhibitor binding, defined as $K_I = [E][I]/[EI]$, where $[I]$ = concentration of free inhibitor. We can solve the rate equations just as we did in the previous section, but noting that now

$$[E]_t = [E] + [ES] + [EI] \tag{11.36}$$

$[E]_t$		$[E]$		$[ES]$		$[EI]$
Total enzyme		Free enzyme		Enzyme bound to substrate		Enzyme bound to inhibitor

Upon analysis of this case, the expression for V is found to be

$$V = \frac{k_{cat}[E]_t[S]}{K_M(1 + [I]/K_I) + [S]} \tag{11.37a}$$

which may be rewritten as

$$V = \frac{k_{cat}[E]_t[S]}{K_M^{app} + [S]} = \frac{V_{max}[S]}{K_M^{app} + [S]} \tag{11.37b}$$

This expression looks just like the Michaelis–Menten equation, with an "apparent" K_M given by $K_M^{app} = K_M(1 + [I]/K_I)$. As we predicted, increasing $[I]$ causes an increase in the apparent K_M. Note that the maximum velocity is unchanged, for as $[S]$ becomes very large, V approaches V_{max}, just as in the absence of inhibition, and $V_{max} = k_{cat}[E]_t$. Physically, this simply means that if we make $[S]$ very large at a given $[I]$, the numerous substrate molecules will outcompete the inhibitor. The effect of competitive inhibition on a graph of V versus $[S]$ is shown in Figure 11.20a.

Because the system, at a given $[I]$, still obeys an equation of the Michaelis–Menten form, we must expect that Lineweaver–Burk plots and Eadie–Hofstee plots will still be linear graphs, with K_M (but not V_{max}) changed by the presence of inhibitor. As Figure 11.20b shows, this is exactly what happens. If we graph the K_M^{app} values obtained from the Lineweaver–Burk intercepts obtained at different inhibitor concentrations versus $[I]$, we can determine both the true K_M and K_I, as shown in Figure 11.20c.

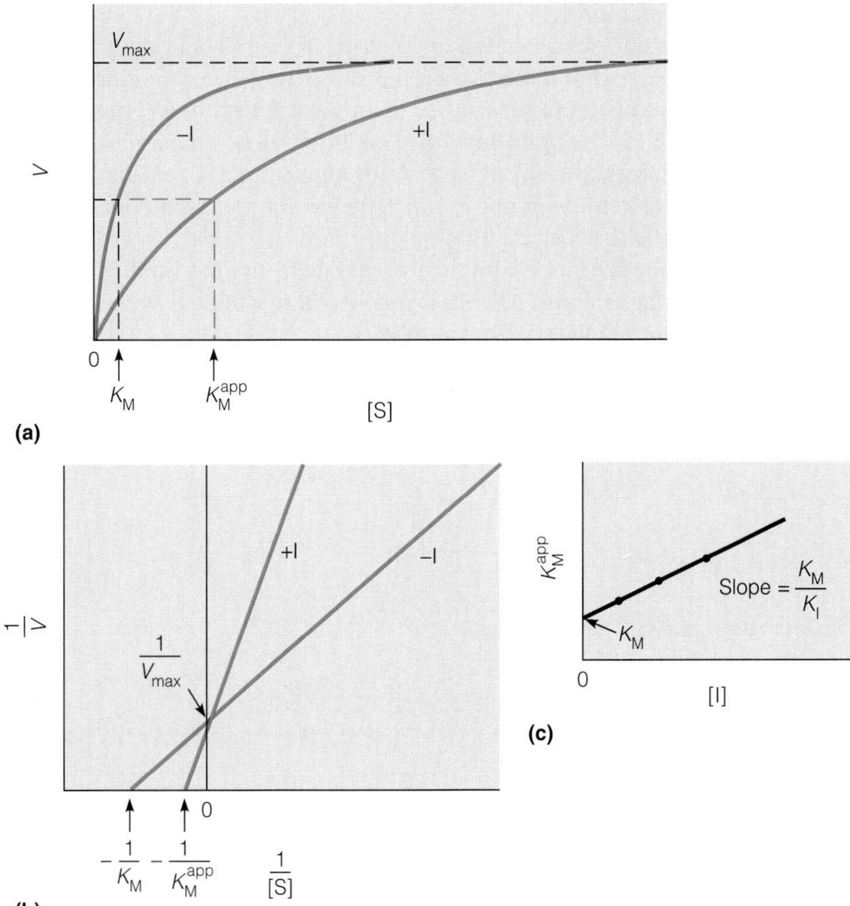

FIGURE **11.20**

Effects of competitive inhibition on enzyme kinetics. (a) The effect of a competitive inhibitor (I) on reaction velocity at different substrate concentrations. Addition of the inhibitor decreases the velocity but not the V_{max}. The apparent K_M is higher in the presence of inhibitor. **(b)** Lineweaver–Burk plots of the reactions shown in (a). The lines cross the $1/V$ axis at the same V_{max}, showing that I is a competitive inhibitor. **(c)** Determination of K_M and K_I. If the measurement of K_M^{app} is repeated at different concentrations of I, K_I can be determined from the slope of the line, and the true K_M from the line's intercept where [I] = 0.

A clear example of a competitive inhibitor is shown in Figure 11.21. The dinucleotide UpA is an excellent substrate for the enzyme ribonuclease, which catalyzes hydrolysis of the phosphodiester bond between the two nucleotides. But if the oxygen atom at the cleavage site in UpA is replaced by a CH_2 group, the molecule becomes the *phosphonate* analog UpcA, a strong competitive inhibitor. Ribonuclease binds the analog strongly enough in the active site to allow x-ray diffraction studies of the complex, but it cannot cleave the phosphonate bond.

A variant of competitive inhibition is **nonproductive binding.** Sometimes a substrate molecule has an extra way of fitting into the binding site, a way in which the normal catalytic event cannot occur:

$$\begin{array}{l} E + S \underset{k_{-1}}{\overset{k_1}{\rightleftharpoons}} ES \xrightarrow{k_{cat}} E + P \\ + \\ S \\ K'_S \updownarrow \\ ES' \end{array}$$

Here ES′ is the enzyme–substrate complex that cannot lead to product. In such cases both K_M and k_{cat} are modified, for even at saturating concentrations of substrates a fraction of the substrate molecules will be bound in the nonproductive mode. An example is found in the action of carboxypeptidase on dipeptides such as glycyl-L-tyrosine. Recall from Chapter 5 that carboxypeptidase cleaves the C-terminal residues from polypeptide chains. Dipeptides are cleaved only slowly, presumably because they can be bound in a way in which the N-terminal amino group interferes with the catalytic site.

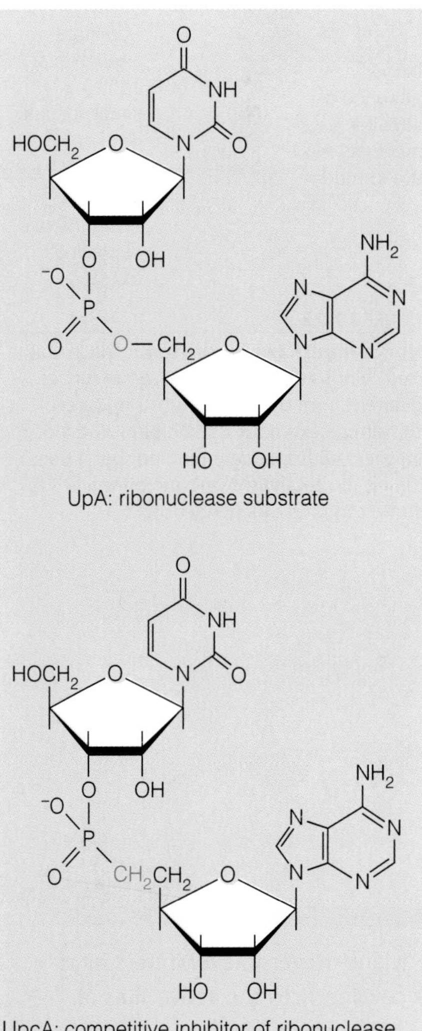

FIGURE **11.21**

A substrate and its competitive inhibitor. The substrate UpA and the very similar molecule UpcA are competitors for the enzyme ribonuclease. The single difference between the substrate and the inhibitor is shown in red.

A noncompetitive inhibitor does not compete for the active site but affects the catalytic event. It changes the apparent k_{cat}.

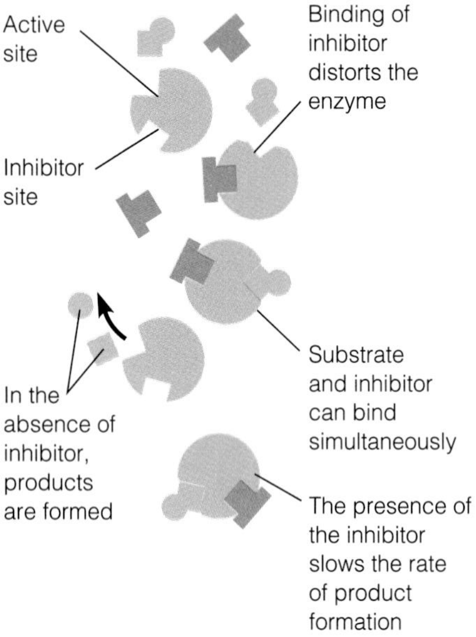

FIGURE **11.22**

Noncompetitive inhibition. The inhibitor (red) binds at a site on the enzyme surface different from that of the substrate (green). In this simple example, the inhibitor does not interfere with substrate binding, but it does inhibit the catalytic event. Inhibitor and substrate can bind independently.

Noncompetitive Inhibition

This form of inhibition occurs when a molecule or an ion can bind to a *second* site on an enzyme surface (not the active site) in such a way that it modifies k_{cat}. It might, for example, distort the enzyme so that the catalytic process is not as efficient (Figure 11.22). Such a **noncompetitive inhibitor** can be a molecule that does not resemble the substrate at all but has a strong affinity for a second binding site. The simplest case to consider is one in which the inhibitor molecule does not interfere in any way with substrate binding but completely prevents the catalytic step. Since the I binding site is completely separate from the S binding site, we may assume that the inhibitor will bind equally well to E and ES, with the same K_I for both. We diagram the reactions as follows:

$$\begin{array}{ccccc} E + S & \underset{k_{-1}}{\overset{k_1}{\rightleftharpoons}} & ES & \xrightarrow{k_{cat}} & E + P \\ + & & + & & \\ I & & I & & \\ K_I \updownarrow & & \updownarrow K_I & & \\ EI + S & \underset{k_{-1}}{\overset{k_1}{\rightleftharpoons}} & EIS & & \end{array}$$

Mathematical analysis yields

$$V = \frac{\{k_{cat}/(1 + [I]/K_I)\}[E]_t[S]}{K_M + [S]} = \frac{k_{cat}^{app}[E]_t[S]}{K_M + [S]} \tag{11.38}$$

The result is exactly what we would expect. The apparent K_M is uninfluenced by inhibitor, but the apparent k_{cat}, given by $k_{cat}/([1 + [I]/K_I])$, *decreases* with increasing [I]. Therefore V_{max} is changed in this case (Figure 11.23a), for at high [S] we find

$$V \longrightarrow V_{max}^{app} = k_{cat}^{app}[E]_t = \frac{k_{cat}[E]_t}{1 + [I]/K_I} \tag{11.39}$$

The effect of noncompetitive inhibition on a Lineweaver–Burk plot is shown in Figure 11.23b. Both the true k_{cat} and K_I may be determined by graphing $1/V_{max}^{app}$ versus [I] (Figure 11.23c). An example is found in the inhibition of acetylcholinesterase (see Chapter 21) by tertiary amines (R_3N). Such compounds bind to the enzyme in its various forms, but the acyl–intermediate–amine complex cannot break down into enzyme plus product.

In reality, the situation is usually more complex than the simple case we have described. For example, the complex ESI may also be able to undergo the catalytic process slowly, or the binding of inhibitor may modify both k_{cat} and K_M. The latter case is called, for obvious reasons, **mixed inhibition.** The kinetic equations in such circumstances become more complicated and will not be considered here.

IRREVERSIBLE INHIBITION

Many irreversible inhibitors bind covalently to the active sites of enzymes.

Some substances combine *covalently* with enzymes so as to inactivate them irreversibly. Almost all **irreversible enzyme inhibitors** are toxic substances, either natural or synthetic. Table 11.4 (page 388) lists a number of them. In most cases, such substances react with some functional group in the active site to block the site from substrate or to leave it catalytically inactive.

A typical example of an irreversible competitive inhibitor is found in *diisopropyl fluorophosphate (DFP).* This compound reacts rapidly and irreversibly with serine hydroxyl groups to form a covalent *adduct,* as shown in Figure 11.24. Therefore, DFP acts as an irreversible inhibitor of enzymes that contain an essential serine in their active site. These enzymes include, among others, the serine

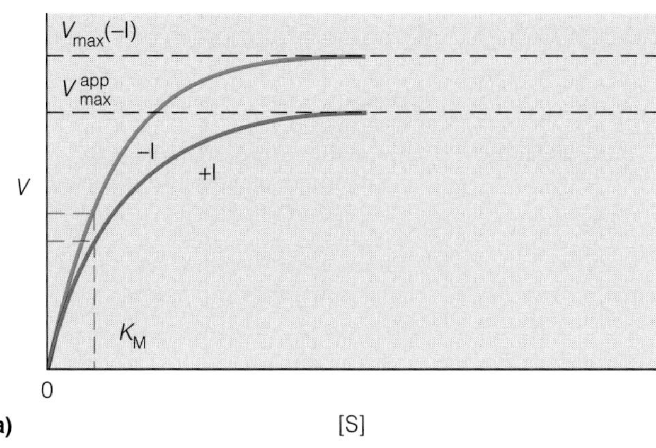

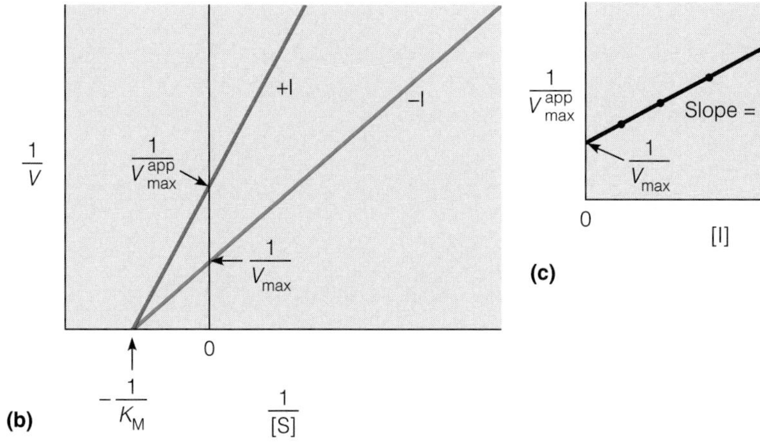

FIGURE **11.23**

Effects of noncompetitive inhibition on enzyme kinetics. (a) The effect of a noncompetitive inhibitor (I) on reaction velocity at different substrate concentrations. In this simple example, K_M is not affected, but V_{max} is decreased because the enzyme is not as catalytically efficient in the presence of the inhibitor. **(b)** Lineweaver–Burk plots of the reactions shown in (a). The lines cross the $1/V$ axis at different points, clearly distinguishing this situation from competitive inhibition (see Figure 11.20b). **(c)** Determination of K_I and the uninhibited V_{max} (which will give k_{cat}).

proteases and the enzyme *acetylcholinesterase.* It is the inhibition of acetylcholinesterase that makes DFP such an exceedingly toxic substance to animals. Acetylcholinesterase is essential for nerve conduction (see Chapter 21), and its inhibition causes rapid paralysis of vital functions. Many insecticides and nerve gases resemble DFP and are potent acetylcholinesterase inhibitors. Examples are given in Table 11.4.

In order for such irreversible inhibitors to react selectively with a critical residue, they must bind strongly to the active site. Many do so because they contain a group of atoms in a configuration resembling the transition state. Examples of such **transition state analogs** include DFP and *sarin* (see Table 11.4), which have a tetrahedral structure surrounding the phosphorus atom very similar to the tetrahedral transition state of the substrate in many hydrolytic enzymes.

In other cases, irreversible inhibitors may be extremely selective because they strongly resemble the substrate rather than its transition state. An example is *N-tosyl-L-phenylalaninechloromethyl ketone (TPCK)* (see Table 11.4). TPCK is an excellent inhibitor for chymotrypsin, because the phenyl group fits nicely into the active site pocket, positioning the chlorine for nucleophilic displacement by a nitrogen of the imidazole ring of His 57. A large number of such specific irreversible inhibitors have been synthesized to aid in the analysis of enzyme mechanisms and to control enzyme activity. For example, a biochemist who is using chymotrypsin to hydrolyze a protein can stop the reaction instantly at any point by simply adding TPCK. Another use for such substances is to label active site residues of an enzyme specifically to aid in their identification. When irreversible inhibitors are used in this way, they are often called **affinity labels.** In some cases an affinity label is unreactive until it is acted on by the enzyme, at which point it binds irreversibly. Such substances are called **suicide inhibitors,** because the enzyme "kills" itself by processing the inhibitor.

Active site serine

DFP

+ HF

FIGURE **11.24**

Irreversible inhibition by adduct formation. Diisopropyl fluorophosphate (DFP) reacts with a serine group on a protein to form a covalent adduct. The covalent bond renders the catalytically important serine ineffective in catalysis. The adduct also may block the site to substrate.

TABLE 11.4 Irreversible enzyme inhibitors

Name	Formula[a]	Source	Mode of Action
Cyanide	CN^-	Bitter almonds	Reacts with enzyme metal ions (i.e., Fe, Zn, Cu); respiratory chain enzymes are primary targets (see Chapter 15)
Diisopropyl fluorophosphate (DFP)	$(H_3C)_2CH-O-P(=O)(F)-O-CH(CH_3)_2$	Synthetic	Inhibits enzymes with active site serine, including acetylcholinesterase
Sarin	$(H_3C)_2CH-O-P(=O)(F)-CH_3$	Synthetic (nerve gas)	Like DFP
Physostigmine	$CH_3-NH-C(=O)-O-$ (tricyclic ring system with CH_3, $N-CH_3$, $N-CH_3$)	Calabar beans	Like DFP
Parathion	$(C_2H_5O)_2P(=S)-O-C_6H_4-NO_2$	Synthetic (insecticide)	Like DFP, but especially inhibitory to insect acetylcholinesterase
N-Tosyl-L-phenylalaninechloromethyl ketone (TPCK)	$C_6H_5-CH_2-CH(NH-SO_2-C_6H_4-CH_3)-C(=O)-CH_2-Cl$	Synthetic	Reacts with His 57 of chymotrypsin
Penicillin	$R^a-C(=O)-NH-HC-CH-S-C(CH_3)_2-CH(COO^-)-N-C(=O)$ (β-lactam/thiazolidine rings)	From *Penicillium* fungus	Inhibits enzymes in bacterial cell wall synthesis (see Chapter 16)

[a]R = variable group; differs on different penicillins.

Many natural toxins are irreversible enzyme inhibitors. The alkaloid *physostigmine* (see Table 11.4), which is contained in calabar beans, is toxic because it is a potent inhibitor of acetylcholinesterase. The *penicillin* antibiotics also act as irreversible inhibitors of serine-containing enzymes used in bacterial cell wall synthesis (see Chapter 16).

Coenzymes, Vitamins, and Essential Metals

The complexity of globular protein structure and the variety of side chain residues available in a protein allow the formation of many kinds of catalytic sites. This variability, in turn, allows proteins themselves to act as efficient enzyme catalysts for many reactions. However, for some kinds of biological processes, the molecular vocabulary of protein side chains alone is not sufficient. A protein may require the help of some other small molecule or ion to carry out the reaction. The molecules that are bound to enzymes for this purpose are called **coenzymes.** Like enzymes, coenzymes are not irreversibly changed during catalysis; they are either unmodified or regenerated.

Many enzymes use small bound molecules called coenzymes to aid in catalysis.

COENZYMES AND WHAT THEY DO

Coenzymes often have complex organic structures that cannot be synthesized by some organisms—mammals in particular. The water-soluble vitamins, those usually referred to as the vitamin-B complex, are metabolic precursors of a number of coenzymes. This is why such vitamins are so important in metabolism. Our approach in this book is to introduce the detailed biochemistry of each coenzyme as we first encounter that coenzyme in discussions of metabolic pathways. Table 11.5 lists a number of important coenzymes, together with their related vitamins, the kinds of reactions they are associated with, and where you will find detailed descriptions and discussion. At this point, just to give a more concrete idea of how coenzymes function, we will describe one class in some detail. These are the nicotinamide nucleotides, a major example being **nicotinamide adenine dinucleotide** (**NAD$^+$**) derived from the vitamin **niacin.**

Many essential vitamins are constituents of coenzymes.

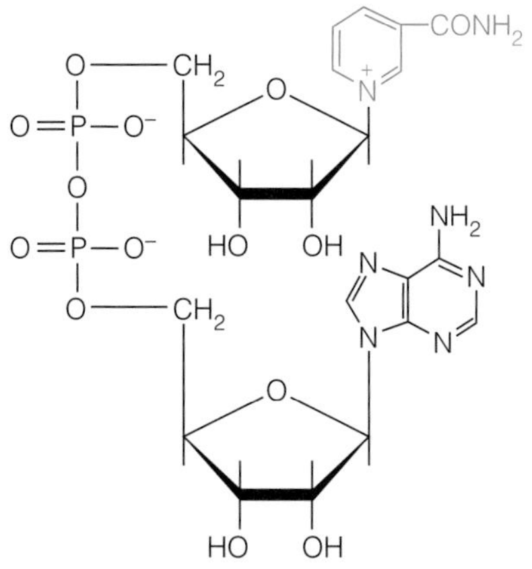

NAD$^+$

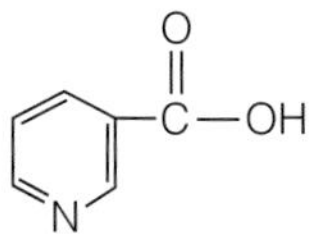

Niacin

TABLE **11.5 Some important coenzymes and related vitamins**

Vitamin	Coenzyme	Reactions Involving These Coenzymes	Page Where Coenzyme Is Introduced
Thiamine (vitamin B_1)	Thiamine pyrophosphate	Activation and transfer of aldehydes	490
Riboflavin (vitamin B_2)	Flavin mononucleotide; flavin adenine dinucleotide	Oxidation–reduction	492
Niacin	Nicotinamide adenine dinucleotide; nicotinamide adenine dinucleotide phosphate	Oxidation–reduction	389, 423
Pantothenic acid	Coenzyme A	Acyl group activation and transfer	494
Pyridoxine	Pyridoxal phosphate	Various reactions involving amino acid activation	731
Biotin	Biotin	CO_2 activation and transfer	507
Lipoic acid	Lipoamide	Acyl group activation; oxidation–reduction	492
Folic acid	Tetrahydrofolate	Activation and transfer of single-carbon functional groups	732
Vitamin B_{12}	Adenosyl cobalamin; methyl cobalamin	Isomerizations and methyl group transfers	738

The nicotinamide portion is the "business end" of NAD^+, for it is capable of being reduced and thus can serve as an oxidizing agent.

$$\text{NAD}^+ \;(\text{nicotinamide ring: } -C(=O)NH_2,\ \overset{+}{N}-R) + 2e^- + H^+ \rightleftharpoons \text{NADH}\;(\text{ring with } CH_2,\ N-R,\ -C(=O)NH_2)$$

NAD⁺ **NADH**

Here R stands for the remainder of the molecule. In reduction, two electrons and a proton are added to the nicotinamide ring. The reaction can probably be considered most accurately as a **hydride ion** transfer: $NAD^+ + H^- \rightleftharpoons NADH$.

A typical reaction in which NAD^+ acts as an oxidizing agent is the conversion of alcohols to aldehydes or ketones (for example, by the *alcohol dehydrogenase* of liver):

$$CH_3-CH_2-OH \rightleftharpoons CH_3CHO + 2e^- + 2H^+$$

$$NAD^+ + 2e^- + H^+ \rightleftharpoons NADH$$

$$\underset{\textbf{Ethanol}}{CH_3CH_2OH} + NAD^+ \rightleftharpoons \underset{\textbf{Acetaldehyde}}{CH_3CHO} + H^+ + NADH$$

It is the C-linked H, not the O-linked H, that is transferred to NAD^+, as can be demonstrated by studies using deuterated compounds. Furthermore, these reactions have a strong stereospecificity. Even when the hydroxyl carbon has *two* hydrogens attached (as with ethanol), a particular one of the hydrogens is transferred to NAD^+. This specificity may seem surprising, since the hydroxyl carbon of ethanol is not a chiral center. How can a particular H be favored when the substrate molecule has a plane of symmetry? The answer lies in the asymmetric nature of the enzyme surface, to which both NAD^+ and the alcohol are bound. If even a symmetrical molecule like ethanol is bound by *three* points to an asymmetric object, the two H atoms are no longer equivalent; they are said to be *prochiral* (Figure 11.25). Furthermore, although the nicotinamide ring is planar, the transfer of hydrogen in a particular reaction is always to a specific face of the ring, for the two faces are *not* equivalent in the asymmetric environment of an enzyme. Such considerations lie behind the high stereospecificity of many enzyme-catalyzed reactions, in contrast to nonenzymatic catalysis.

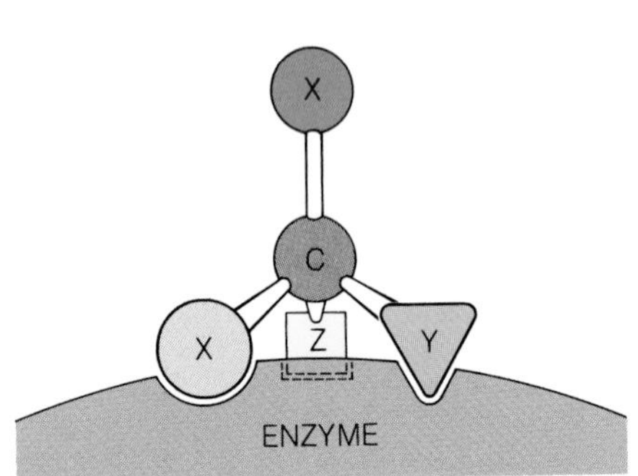

FIGURE **11.25**

Stereospecificity conferred by an enzyme. This figure shows how the asymmetric surface of an enzyme can confer stereospecificity in the reaction of a symmetric substrate. If the substrate molecule X_2CYZ must make three contacts with unique complementary groups on the enzyme, its two X atoms are no longer equivalent. Only a specific one of the two X atoms can contact the surface properly.

Sometimes it is difficult to make a clear distinction between a true coenzyme and a second substrate in a reaction. The reactions we have just discussed are good examples of this problem. The dehydrogenase enzymes, such as alcohol dehydrogenase, each have a strong binding site for the oxidized form, NAD^+. After oxidation of the substrate, the reduced form, NADH, leaves the enzyme and is reoxidized by other electron-acceptor systems in the cells. The NAD^+ so formed can now bind to another enzyme molecule and repeat the cycle. In such cases, NAD^+ is acting more like a second substrate than a true coenzyme. Yet NAD^+ and NADH differ from most substrates in that they are continually recycled in the cell and are used over and over again. Because of this behavior, we consider them coenzymes.

An example of NAD^+ behaving unambiguously as a coenzyme is found in the reaction involving *UDP-galactose 4-epimerase* shown in Figure 11.26. This enzyme facilitates synthesis of complex polysaccharides by interconverting UDP-glucose and UDP-galactose (see Chapters 9 and 13). The mechanism by which the hydroxyl at position 4 is changed in stereochemical orientation involves oxidation of the hydroxyl to a carbonyl as an intermediate state. In this case, NAD^+ and NADH never leave the enzyme; they are reduced and reoxidized in a cyclic fashion, providing a temporary resting place for electrons and the hydrogen from the substrate. This reaction provides a good example of what coenzymes do and why they are necessary. The carbonyl intermediate provides an excellent intermediate state for interconversion of the sugars, but none of the normal amino acid side chains of a protein are really well suited to promote this kind of oxidation and reduction. By binding NAD^+, the enzyme can carry out this function.

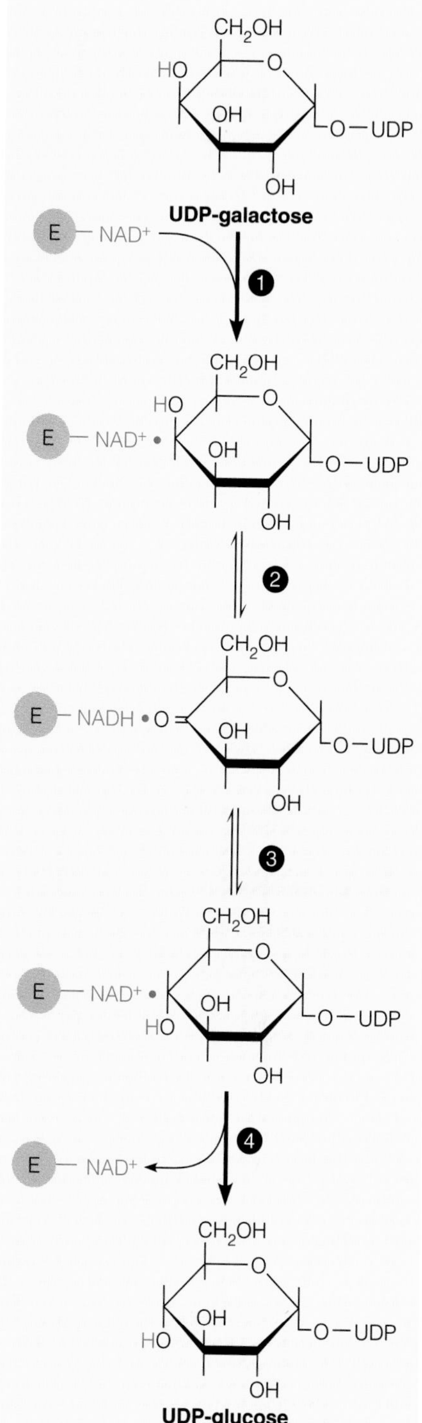

FIGURE **11.26**

Proposed mechanism for UDP-galactose epimerase. UDP-galactose is bound to the enzyme, which carries the coenzyme NAD^+ (step 1). A proton is extracted to produce the carbonyl intermediate (step 2) and then reinserted with opposite enantiomeric orientation (step 3). The product, UDP-glucose, is then released (step 4).

METAL IONS IN ENZYMES

Many enzymes contain metal ions, usually held by coordinate covalent bonds from amino acid side chains, but sometimes bound by a prosthetic group like heme. Such enzymes are called **metalloenzymes.** The ions act in much the same way as coenzymes, conferring on the metalloenzyme a property it would not possess in their absence. As Table 11.6 shows, the roles these ions play are diverse. Some, like the zinc ion in *carboxypeptidase A,* act as metal catalysts for hydrolytic reactions (Figure 11.27). The zinc ion is involved in stabilizing the intermediate state for hydrolysis in much the same way as serine is in chymotrypsin (see Figure 11.13 for comparison).

In other cases, the metal in a metalloenzyme serves as a redox reagent. We have mentioned the example of the heme-iron-containing enzyme *catalase,* which catalyzes the breakdown of hydrogen peroxide, a potentially destructive agent in cells. Because the reaction involves both reduction and oxidation of H_2O_2, the Fe^{2+} is reversibly oxidized and reduced, acting as an electron exchanger. As noted previously, catalase is a very efficient enzyme; its k_{cat}/K_M value of 4×10^7 approaches the theoretical limit. Hemoglobin, which also contains Fe^{2+}, has weaker catalase activity because oxidation to Fe^{3+} is largely irreversible in the different protein environment. Such redox activity requires metals like Fe or Cu with multiple oxidation states.

In many other enzymatic reactions, certain ions are necessary for catalytic efficiency, even though they may not remain permanently attached to the protein

TABLE **11.6 Metals and trace elements important as enzymatic cofactors**

Metal	Example of Enzyme	Role of Metal
Fe	Cytochrome oxidase	Oxidation–reduction
Cu	Ascorbic acid oxidase	Oxidation–reduction
Zn	Alcohol dehydrogenase	Helps bind NAD^+
Mn	Histidine ammonia-lyase	Aids in catalysis by electron withdrawal
Co	Glutamate mutase	Co is part of cobalamin coenzyme
Ni	Urease	Catalytic site
Mo	Xanthine oxidase	Oxidation–reduction
V	Nitrate reductase	Oxidation–reduction
Se	Glutathione peroxidase	Replaces S in one cysteine in active site

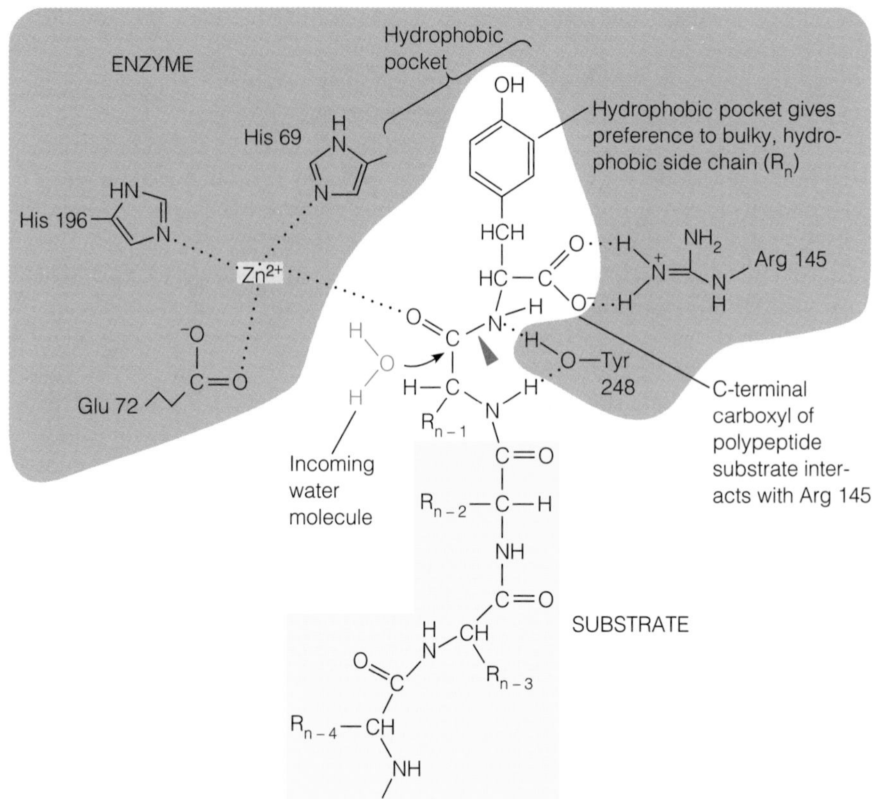

FIGURE **11.27**

Active site of the protease carboxypeptidase A. The zinc atom serves as a metal ion catalyst to promote hydrolysis. It does so by stabilizing the negative charge on the oxygen in the tetrahedral transition state.The bond cleaved is indicated by the red wedge.

Some enzymes require metal ions for their catalytic function.

nor play a direct role in the catalytic process. For example, a number of enzymes that couple ATP hydrolysis to other processes require Mg^{2+} for efficient function. In most cases Mg^{2+} is necessary because the Mg–ATP complex (see Chapter 3) is a better substrate than ATP itself.

The Diversity of Enzymatic Function

CLASSIFICATION OF PROTEIN ENZYMES

By this point, it should be clear that an enormous number of different proteins act as enzymes. Many of these enzymes were given common names, especially during the earlier years of enzymology. Some enzyme names, like *triose phosphate isomerase,* are descriptive of the enzyme's function; others, like *trypsin,* are not. To reduce confusion, a rational naming and numbering system has been devised by the Enzyme Commission of the International Union of Biochemistry and Molecular Biology (IUBMB). Enzymes are divided into six major classes, with subgroups and sub-subgroups to define their functions more precisely. The major classes are as follows:

1. *Oxidoreductases* catalyze oxidation–reduction reactions.
2. *Transferases* catalyze transfer of functional groups from one molecule to another.
3. *Hydrolases* catalyze hydrolytic cleavage.
4. *Lyases* catalyze removal of a group from or addition of a group to a double bond, or other cleavages involving electron rearrangement.
5. *Isomerases* catalyze intramolecular rearrangement.
6. *Ligases* catalyze reactions in which two molecules are joined.

TABLE 11.7 Examples of each of the major classes of enzymes

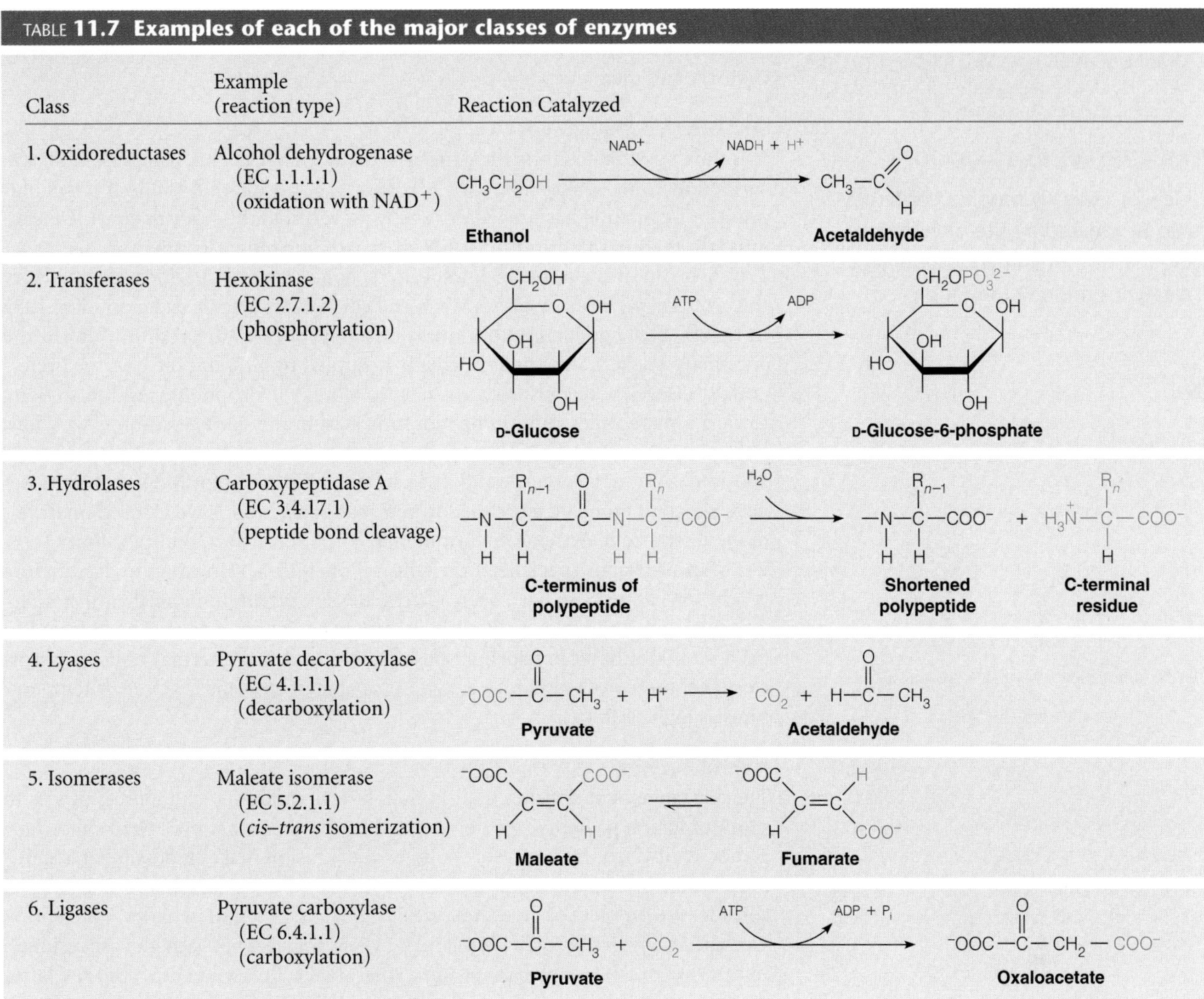

Class	Example (reaction type)	Reaction Catalyzed
1. Oxidoreductases	Alcohol dehydrogenase (EC 1.1.1.1) (oxidation with NAD^+)	CH_3CH_2OH (Ethanol) → CH_3CHO (Acetaldehyde); NAD^+ → $NADH + H^+$
2. Transferases	Hexokinase (EC 2.7.1.2) (phosphorylation)	D-Glucose → D-Glucose-6-phosphate; ATP → ADP
3. Hydrolases	Carboxypeptidase A (EC 3.4.17.1) (peptide bond cleavage)	C-terminus of polypeptide + H_2O → Shortened polypeptide + C-terminal residue
4. Lyases	Pyruvate decarboxylase (EC 4.1.1.1) (decarboxylation)	$^-OOC{-}CO{-}CH_3 + H^+ \rightarrow CO_2 + H{-}CO{-}CH_3$ (Pyruvate → Acetaldehyde)
5. Isomerases	Maleate isomerase (EC 5.2.1.1) (*cis–trans* isomerization)	Maleate ⇌ Fumarate
6. Ligases	Pyruvate carboxylase (EC 6.4.1.1) (carboxylation)	$^-OOC{-}CO{-}CH_3 + CO_2 \rightarrow {}^-OOC{-}CO{-}CH_2{-}COO^-$ (Pyruvate → Oxaloacetate); ATP → ADP + P_i

The IUBMB Enzyme Commission (EC) has given each enzyme a number with four parts, like EC 3.4.21.5. The first three numbers define major class, subclass, and sub-subclass, respectively. The last is a serial number in the sub-subclass, indicating the order in which each enzyme is added to the list, which is continually growing. Listings for almost all currently recognized enzymes, together with information on each and literature references, are given in the book *Enzyme Nomenclature* (Academic Press, 1992). The 3196 entries in this edition surely do not include all enzymes; more are being discovered all the time. Indeed, it has been estimated that the typical cell contains many thousands of different kinds of enzymes. Table 11.7 lists one example enzyme and reaction from each of the major classes. We will discuss each of these reactions later in this book. The main point we wish to emphasize here is the enormous diversity of enzymatic functions and how their nomenclature has been rationalized.

MOLECULAR ENGINEERING OF NEW AND MODIFIED ENZYMES

Despite the variety of enzymatic functions available in nature, modern biotechnology continually faces needs for substances with new catalytic abilities, or enzymes that function with different specificities or under unusual conditions. These needs have generated a field, still in its infancy, of enzyme design and engineering. Several approaches are being taken in this quest, which has enormous potential in industry and medicine.

New or radically modified enzymes can be created by site-directed mutagenesis, protein hybridization, or catalytic antibody formation.

Site-Directed Mutagenesis

It is now possible to clone the genes for many enzymes and, using methods described in Tools of Biochemistry 25E, to make specific mutations at particular points. The method has proved extremely powerful in the study of enzyme mechanisms, as illustrated by the study of triose phosphate isomerase (see pages 370–371). But it is also being employed to change enzyme specificities. A study of the protease *subtilisin* by James Wells and co-workers (see Oxender and Fox 1987 in References) has focused on mutations at a specific site (residue 166) in the specificity pocket. This site is normally occupied by Gly, and the enzyme preferentially cleaves polypeptide chains next to a bulky hydrophobic residue. Activity toward polypeptides containing glutamic acid in the same position is very low. Replacing Gly 166 by Lys increases the frequency of cutting next to glutamic acid 500-fold. Although some attempts at protein engineering for industrial purposes have aimed at tailoring enzymes for new reactions, more have focused on developing tolerance to extreme environmental conditions (heat, acidity, salinity) typical of industrial or specialized agricultural operations. However, to date, nature (evolution) appears to have been a better designer than humans; the most interesting enzymes capable of tolerating extreme environments have been found in bacteria that inhabit hot springs, deserts, and undersea thermal vents. It is now clear that there exist in nature enzymes capable of efficient function at temperatures as high as 100°C.

Hybrid Enzymes

The rearrangement of genes that is possible with modern techniques allows us to produce **fusion proteins**—proteins coded for by genes that have been spliced together in vitro from two or more sources. Such fusion makes it possible to recombine binding sites and catalytic sites in novel ways. To take just one example, consider the problem of digesting cereal β-glucans, as found in barley. These polysaccharides involve both β (1$\rightarrow$4) and β (1$\rightarrow$3) linkages between glucose residues. In brewing practice, a glucanase and a microbial cellulase are used, but the latter is often ineffective in industrial conditions, leading to viscous products that interfere with beer manufacture. It was found possible to construct a hybrid enzyme that has the capability of cleaving both kinds of bonds and thereby digesting the glucan to glucose in one step.

Another kind of hybrid enzyme is depicted in Figure 11.28. Here a synthetic oligonucleotide of defined sequence has been grafted onto the enzyme *staphylococcal nuclease* through a disulfide bond. Staphylococcal nuclease is efficient at cutting single-strand RNA or DNA, but it normally does so with very little specificity. When it has been attached to an oligonucleotide, however, the oligonucleotide binds to a specific complementary sequence in a single-strand nucleic

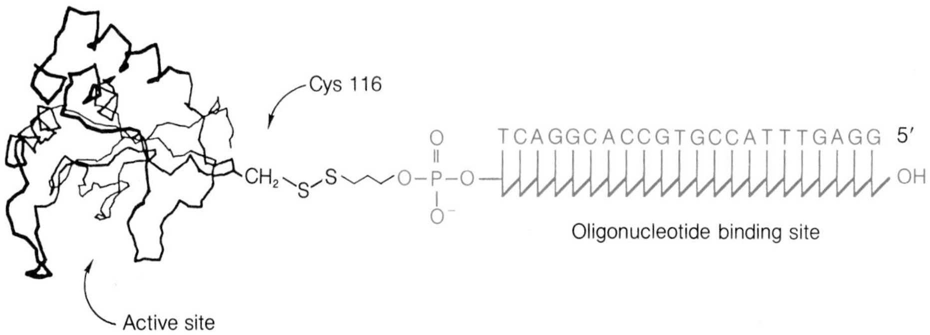

FIGURE **11.28**

A hybrid enzyme. The enzyme staphylococcal nuclease (black) has been attached through a disulfide bridge at cysteine 116 to an oligonucleotide of defined sequence (blue). The oligonucleotide base-pairs with complementary sequences on single-strand polynucleotides, providing site specificity for the nuclease.

P. G. Schultz, *Science* (1988) 240:425–432, © 1988 by the AAAS.

acid, and the enzyme cuts next to that sequence. In this way, high specificity can be designed into a normally nonspecific enzyme.

Catalytic Antibodies

Recall from Chapter 7 that antibodies show remarkably high specificity in binding to their antigens. Enzymes bind most strongly to the transition state in a reaction. What happens if we make antibodies against molecules that are structurally analogous to the transition states of particular substrates? The answer is that these antibodies act like enzymes, so they are now often called **abzymes.** Suppose, for example, we wish to make an antibody that will function as a catalyst for hydrolysis of esters. Just as in the hydrolysis of amides, a tetrahedral transition state (see page 373) is required in ester hydrolysis. A kind of molecule that mimics such a structure is the following:

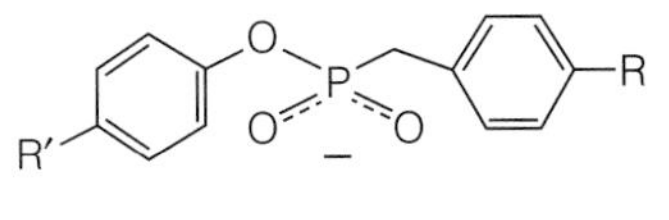

By using various kinds of molecules as antigens, it has been possible to produce abzymes catalyzing a number of the classes of reactions shown in Table 11.7. In some cases, reaction rates as much as 10^7 times the uncatalyzed rate have been obtained.

For many years, the major difficulty in generating catalytic antibodies directed toward specific compounds or functional groups was the necessity of using the immune system of some animal to make the selection. More recently, however, selection systems have been developed that circumvent this requirement. The basic idea is that randomly rearranged Fab fragments (see Chapter 7) are cloned and the mixture subjected to selection by chemical affinity to the desired molecule or structure. Such techniques have allowed the development of abzymes directed toward synthetic molecule substrates that would be very difficult to present as antigens to an in vivo system because of toxic effects.

Catalytic antibodies are becoming of considerable importance in synthetic organic chemistry. A major problem in the in vitro synthesis of complex organic molecules is that of obtaining the correct stereochemistry. The remarkable stereospecificity exhibited by enzymes (including abzymes) has enormously aided in some such syntheses. It seems at this point that biochemists have only begun to explore the possibilities of engineering enzymes for specific purposes.

Nonprotein Biocatalysts: Ribozymes

Ribozymes, a class of ribonucleic acids, function as biological catalysts.

Throughout this chapter, we have described how the proteins called enzymes function as biocatalysts. Indeed, until recently, it was assumed that *all* biochemical catalysis was carried out by proteins. But biochemistry is full of surprises, and research has revealed something wholly unexpected: Some RNA molecules can act as enzymes. These molecules are called **ribozymes.**

The first hint of this phenomenon came from studies of **ribonuclease P,** an enzyme that cleaves the precursors of tRNAs to yield the functional tRNAs (Figure 11.29). It had been known for some time that active ribonuclease P contained both a protein portion and an RNA "cofactor," but it was widely assumed that the active site resided on the protein portion. However, careful studies of the isolated components by Sidney Altman and co-workers in 1983 revealed an astonishing fact: Whereas the protein component alone was wholly inactive, the RNA by itself, if provided with either a sufficiently high concentration of magnesium or a small amount of magnesium plus the small basic molecule spermine, was capable of catalyzing the specific cleavage of pre-tRNAs. Furthermore, the RNA acted like a true enzyme, being unchanged in the process and obeying Michaelis–Menten kinet-

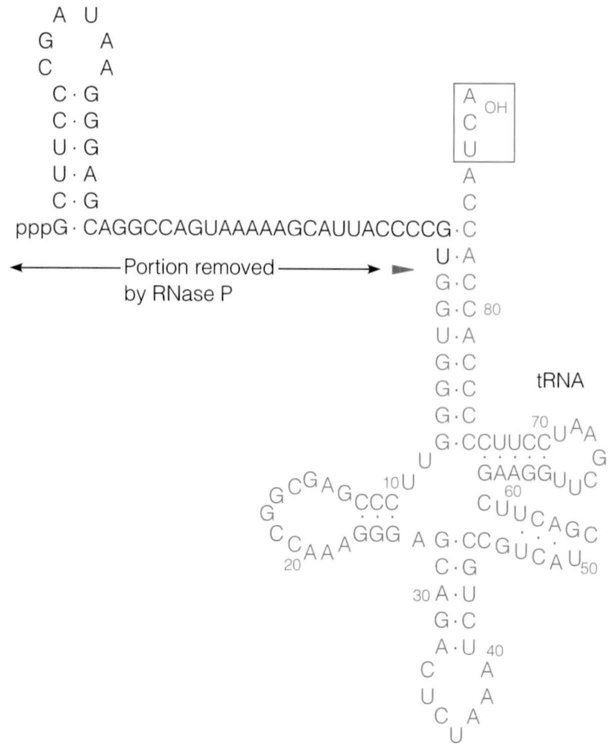

FIGURE **11.29**

Cleavage of a typical pre-tRNA by ribonuclease P. The production of tRNA from pre-tRNA is catalyzed by an RNA–protein complex called ribonuclease P. The portion removed from tRNA is shown in black, and the resulting tRNA is in blue. The RNA portion of ribonuclease P can by itself catalyze the hydrolysis of the specific phosphodiester bond indicated by the red wedge.

ics. Addition of the protein portion of ribonuclease P does enhance the activity (k_{cat} is markedly increased) but is in no way essential for either substrate binding or cleavage. At high salt concentrations the RNA itself becomes a very efficient catalyst; K_M becomes very low, and k_{cat}/K_M approaches 10^7 $(\text{mol/L})^{-1}\,\text{s}^{-1}$.

At about the same time, another remarkable class of RNA reactions was discovered by Thomas Cech and his colleagues. Examining the removal of an intron (intervening sequence, or IVS) from the preribosomal RNAs of the protist *Tetrahymena*, they found that the rRNA *itself* carried out the excision of its 413-nucleotide intron and the necessary resplicing as well (Figure 11.30a). Furthermore, the excised IVS went through a further series of site-specific reactions. The final product is a molecule called L-19 IVS—the intervening sequence with 19 more nucleotides removed. This activity is not regarded as true catalysis, because the “catalyst” itself is modified in the reaction. However, L-19 IVS does have true catalytic abilities: It is capable of either lengthening or shortening small oligonucleotides, in the manner shown in Figure 11.30b. This example is by no means unique; we discuss other RNA-catalyzed reactions in Chapter 28.

Although many biochemists were astounded by these discoveries, in retrospect we can see that there is no reason why RNA molecules should not have catalytic functions. As discussed in Chapter 4, RNA molecules can adopt complex tertiary structures, just as proteins do, and this is the kind of structure that appears to be essential for enzymatic activity.

The fact that RNA molecules possess the potential for both self-replication and catalysis has led some scientists to suggest that these molecules may have well been the primordial substances in the evolution of life. Such theorists envision an “RNA world” before proteins and DNA had evolved, and only self-replicating RNA molecules existed, capable of catalyzing a simple metabolism. Support for the idea that ribozymes could be self-replicating comes from remarkable experiments. Wright and Joyce (see References) have shown that the ribozyme depicted in Figure 11.31 is capable of self-catalyzed replication (with a little help from some protein enzymes). The structure labeled RNA is capable of catalyzing the ligation of another RNA segment (in this case carrying a bacteriophage T7 pro-

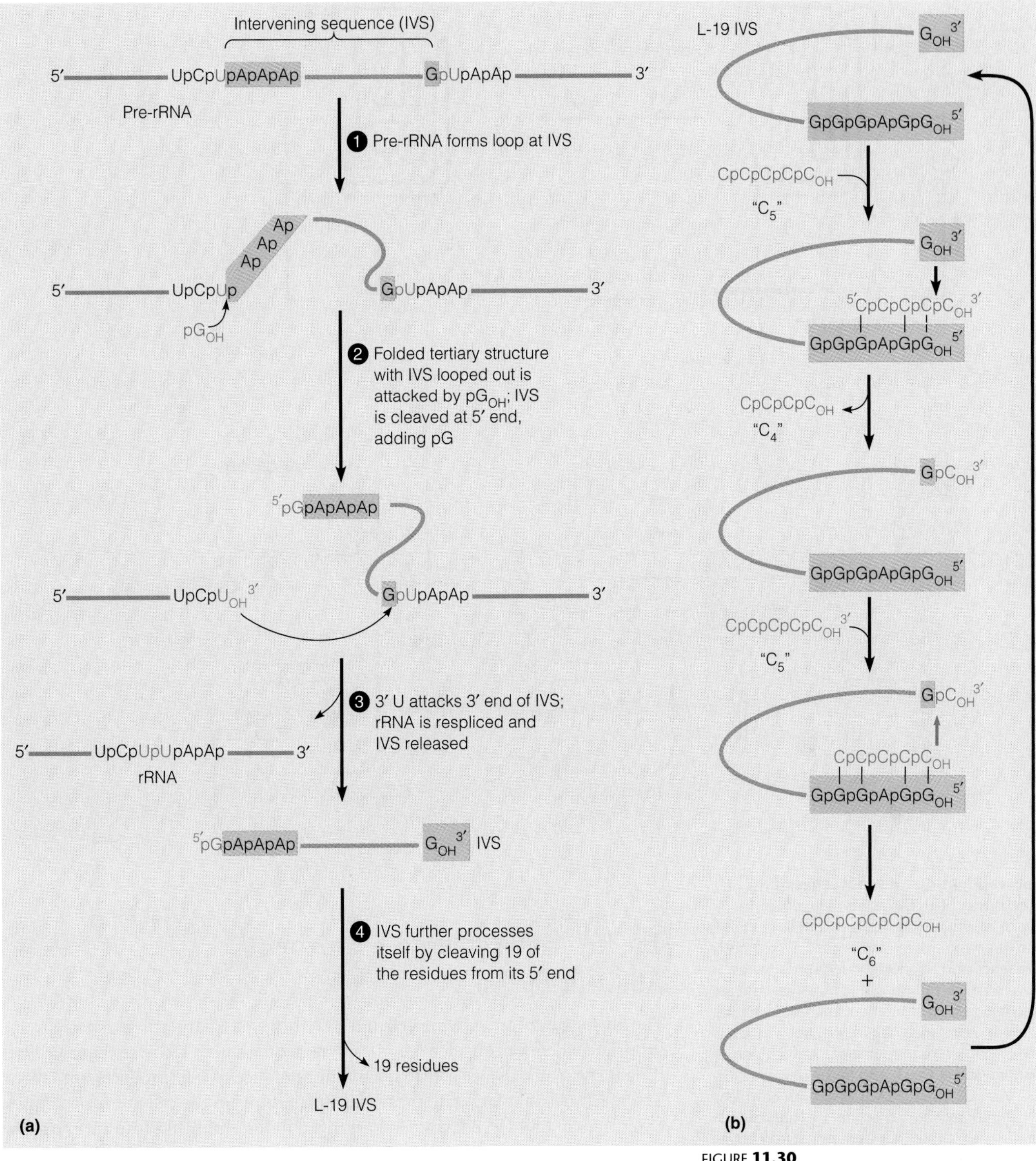

FIGURE **11.30**

Catalysis by the intervening sequence in *Tetrahymena* preribosomal RNA. **(a)** Self-excision and splicing of the intervening sequence (IVS). Note that a pG_{OH} is added in the reaction. A series of further steps reduces the IVS to L-19 IVS. **(b)** Conversion of $2C_5$ to $C_4 + C_6$ by L-19 IVS. This oligonucleotide can itself either shorten or elongate small oligonucleotides, acting here as a true ribozyme catalyst.

moter) to itself. In the presence of the enzymes reverse transcriptase and RNA polymerase (see Chapter 26), a new copy of the RNA is generated. If this is "fed" with the T7 promoter fragment, ligation occurs and a new copy of the ribozyme is formed. In one such experiment, the system was led through over 1000 doublings in a period of 2 days. During this time, the sequence evolved to a more efficient ribozyme. Although no experiments to date have managed to dispense with proteins completely, such studies provide support for the "RNA-world" model for life's origin.

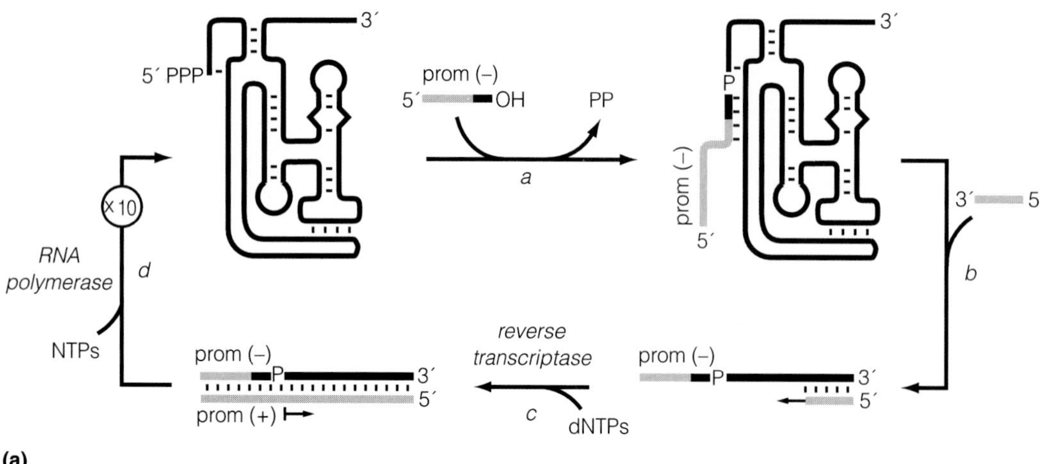

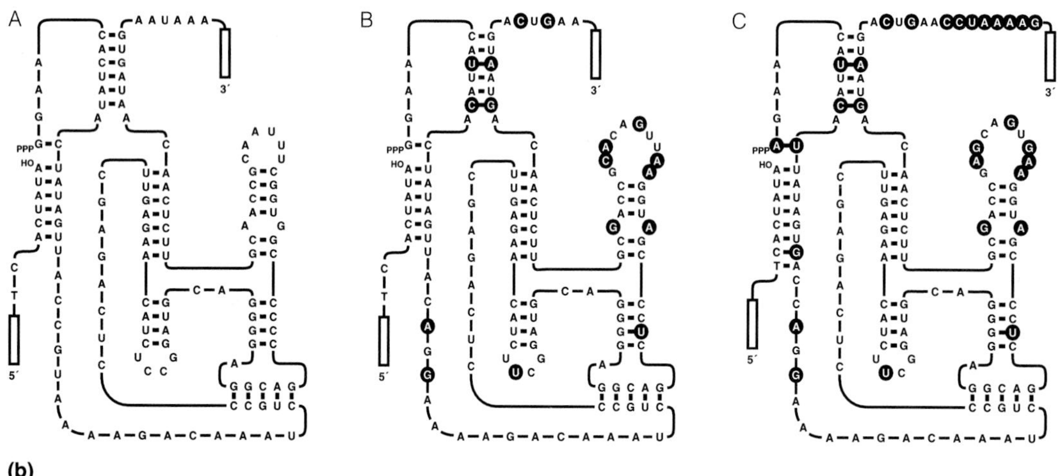

FIGURE **11.31**

Self-replication and mutation of a ribozyme. (a) The ribozyme replication cycle, which involves repeated replication of the ribozyme, when "fed" with a T7 promoter sequence and ribonucleotide triphosphates. Reverse transcriptase and RNA polymerase are also needed. **(b)** Mutations that arose during ribozyme evolution. Open rectangles indicate the 5′ portion of the substrate and the primer binding site at the 3′ end of the ribozyme (5′-CCAAUCGCAGGCUCAGC-3′), both of which are immutable during evolution. Highlighted residues are those that were mutated relative to the starting ribozyme. **(A)** Prototype ribozyme used to construct the initial pool. **(B)** An individual ribozyme isolated before beginning continuous evolution. **(C)** An individual ribozyme isolated after 52 hours of continuous evolution.

Courtesy of G. F. Joyce from M. C. Wright and G. F. Joyce, *Science* (1997) 276:614–616. © 1997 AAAS, with permission from *Science*.

The Regulation of Enzyme Activity: Allosteric Enzymes

The analogy between a living cell and a factory (see Chapter 1) is especially appropriate when we consider the roles that enzymes play. We note that a cell has certain raw materials available to it and must produce specific products from them. The machines that facilitate these transformations in the cell are the enzymes. Often, as we shall see, enzymes are arranged in "assembly lines" to carry out the necessary sequential steps in a metabolic pathway.

No factory operates efficiently if every machine is operating at its maximum rate. The capabilities of machines vary greatly, and if all of them were running at top speed, massive problems would soon arise. Intermediate products would pile up in some assembly lines, and certain parts of the finished product would be produced in vast excess. Different assembly lines might draw on the same raw material, and the faster ones could deplete the supplies so completely that other, equally important lines would have to shut down. Obviously, *coordination* and *regulation* are required to run a large factory efficiently.

The same kinds of problems could occur if the enzymatic machinery of the cell were not regulated precisely. The efficiencies with which individual enzymes operate must be controlled in a manner that reflects the availability of substrates, the utilization of products, and the overall needs of the cell. Thus we will see, in the following chapters, many examples of regulation.

Regulation of enzyme activity is essential for the ordered flow of metabolism.

SUBSTRATE-LEVEL CONTROL

Some enzyme regulation occurs in a simple way, through direct interaction of the substrates and products of each enzyme-catalyzed reaction with the enzyme itself. This is referred to as **substrate-level control.** As our analysis of kinetics has shown, the higher a substrate concentration is, the more rapidly a reaction occurs, at least until saturation of the enzyme is approached. Conversely, high levels of product, which can also bind to the enzyme, tend to inhibit the conversion of substrate to product. Insofar as the metabolically desired reaction is concerned, product can act as a competitive inhibitor. As an example, consider the first step in glycolysis (see Chapter 13)—the phosphorylation of glucose to yield glucose-6-phosphate:

$$\text{Glucose} + \text{ATP} \xrightarrow{\text{Hexokinase}} \text{glucose-6-phosphate} + \text{ADP}$$

The enzyme hexokinase, which catalyzes this reaction, is inhibited by its product, glucose-6-phosphate. If glycolysis is blocked for any reason, glucose-6-phosphate will accumulate. This accumulation will inhibit hexokinase and slow down further entry of glucose into the pathway.

However, substrate-level control is not sufficient for the regulation of many metabolic pathways. In many instances, it is essential to have an enzyme regulated by some substance quite different from the substrate or immediate product. This requires quite different mechanisms, as we shall shortly see.

FEEDBACK CONTROL

We have emphasized that most metabolic pathways resemble assembly lines. The simplest metabolic assembly line looks like this:

$$A \xrightarrow{\text{Enzyme } 1} B \xrightarrow{\text{Enzyme } 2} C \xrightarrow{\text{Enzyme } 3} D \xrightarrow{\text{Enzyme } 4} E$$

where A is the initial reactant or raw material; B, C, and D are intermediate products; and E is the final product.

The final product of this pathway will probably be used in some other pathway. Similarly, the "raw material" may also participate in some other set of processes. Suppose the utilization of E suddenly slows down. If everything kept going as before, E would accumulate, and consumption of A would continue. But this process is inefficient. A modern, automated assembly line would handle the problem by keeping an inventory on E and, as E accumulated, sending a signal back to slow the line. The cell can control generation of the final product through activation (⊕) or inhibition (⊘) of a step in the pathway. It would be most efficient to slow the *first step*—the conversion of A to B. So the A ⟶ B "machine" should be regulated by the amount of E.

$$A \xrightarrow{⊘} B \longrightarrow C \longrightarrow D \longrightarrow E$$

(dashed feedback arrow from E back to the A ⟶ B step)

This process is called **feedback control** or, more precisely, **negative feedback control** because an *increase* in the concentration of E leads to a *decrease* in its rate of production. Note that by inhibiting the first step, we prevent both unwanted

Feedback control is important in the efficient regulation of complex metabolic pathways.

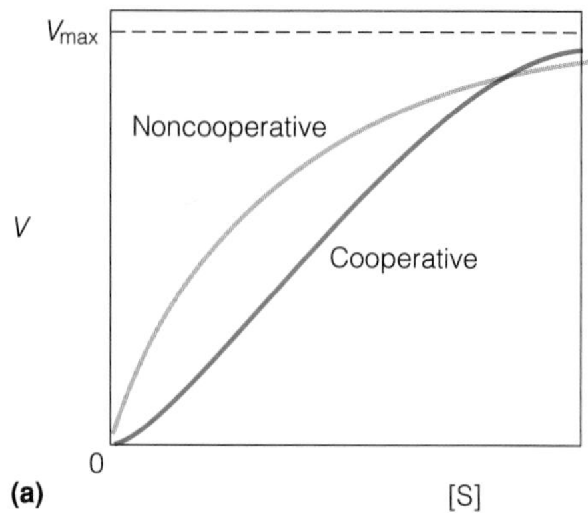

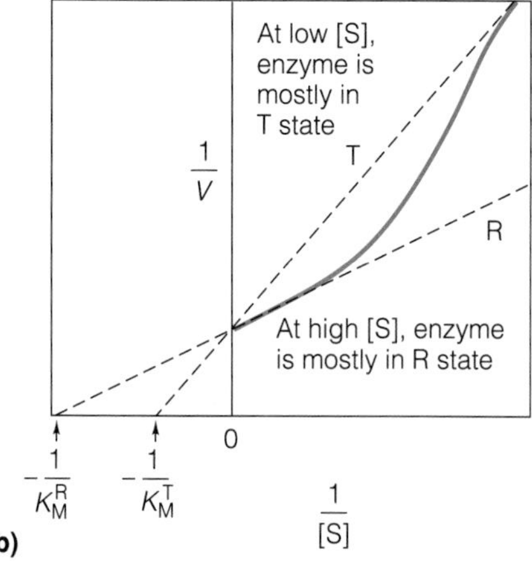

FIGURE **11.32**

Effect of cooperative substrate binding on enzyme kinetics. (a) Comparison of V-versus-[S] curves for a noncooperative enzyme and an allosteric enzyme with cooperative binding. The two enzymes are assumed to have the same V_{max}. Compare this plot with the myoglobin and hemoglobin oxygen binding curves shown in Figure 7.8 (page 219). **(b)** Lineweaver–Burk plot corresponding to the cooperative binding curve shown in (a). The T state has a high K_M (binds S weakly). As more S is bound, the T ⟶ R equilibrium shifts toward R, with stronger binding and a lower K_M.

utilization of A and accumulation of E. Furthermore, because most biochemical processes are reversible to some extent, generation of a large quantity of E will tend to build up the concentration of intermediate products. The feedback control mechanism visualized above prevents accumulation of any intermediates, which might have undesired effects on metabolism.

Other metabolic situations require more complicated patterns, in which **activation** as well as inhibition may be useful. For example, consider a slightly more complex case, in which A is fed into two pathways, which lead to two products needed in roughly equivalent amounts. Then a scheme like the following emerges:

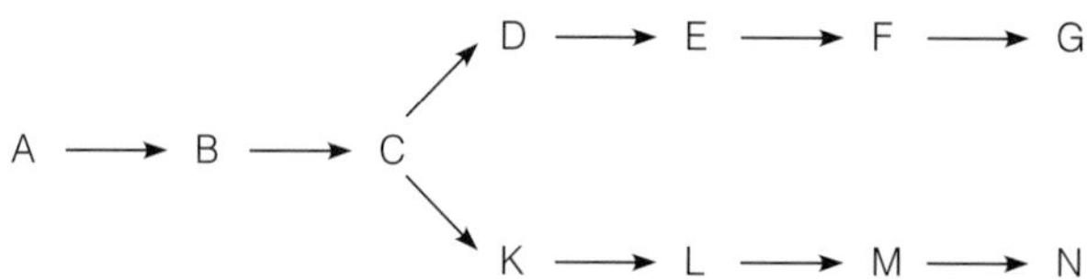

To control the pathways so that G and N keep in balance, high concentrations of G might *inhibit* the C ⟶ D enzyme and/or *activate* the C ⟶ K enzyme. Conversely, N might inhibit the C ⟶ K enzyme and/or activate the C ⟶ D enzyme. Finally, it might be useful to have both G and N inhibit the A ⟶ B enzyme, to provide overall regulation. An example of this kind of control is found in the synthesis of the purine and pyrimidine monomers that go into making DNA, because approximately equal quantities of all four deoxyribonucleotides are required for DNA replication.

It is important to note that both inhibition and activation of enzymes are essential to regulated metabolism. Furthermore, control of pathways by their end products means that the necessary inhibitions and activations *must* be produced by molecules that come from far down the assembly line and therefore bear no resemblance to either the substrates or the direct products of the enzymes to be regulated. None of the kinds of regulation we have discussed up to this point will satisfy these needs. To attain this kind of control, organisms have evolved a special class of enzymes, capable of **allosteric regulation.** The term *allosteric* is derived from Greek words meaning "other structure," emphasizing that structures of regulators need not resemble substrate or direct product.

ALLOSTERIC ENZYMES

Allosteric enzymes are invariably multisubunit proteins, with multiple active sites. They exhibit cooperativity in substrate binding (**homoallostery**) and regulation of their activity by other, effector molecules (**heteroallostery**).

We have already studied an example of allosteric control of protein function. Hemoglobin (see Chapter 7) is a four-subunit protein that has four binding sites for its "substrate," oxygen. The binding of oxygen is cooperative and is influenced by other molecules and ions. The basic ideas that were presented in the analysis of hemoglobin function apply equally well to allosteric enzymes.

Homoallostery

Let us first consider the homoallosteric effects (cooperative substrate binding). In Chapter 7 we contrasted O_2 binding by the single-subunit protein myoglobin with binding by the multisubunit hemoglobin. Myoglobin gave a hyperbolic binding curve (Figure 7.6); hemoglobin, with its cooperative binding, gave a sigmoidal curve (Figure 7.19). We find *exactly the same contrast* when we compare the V-versus-[S] curve of a single-site enzyme obeying Michaelis–Menten kinetics with that of a multisite enzyme showing cooperative binding (Figure 11.32a). The same kind of reasoning applies: An enzyme that binds substrate cooperatively will behave, at low substrate concentration, as if it were poor at substrate binding (that is, as if it had a large K_M). But as the substrate levels are increased and more substrate is bound, the enzyme becomes more and more effective, for it binds substrate more avidly in the last sites to be filled (see Figure 11.32b). We imagine this happening, as with hemoglobin, because the enzyme undergoes, as substrate is bound, a transition from a weak-binding state (T state) to a strong-binding state (R state). The kinds of models that have been used to describe O_2 binding by hemoglobin (i.e., the MWC model and KNF model, pages 221–223) can equally well account for the kinetics exhibited by enzymes that show cooperative substrate binding.

What physiological function does sigmoidal kinetics fulfill? In extreme cases, enzymes obeying sigmoidal kinetics can regulate substrate levels to quite constant values. Consider a substrate that is being supplied constantly by other reactions and is acted on by an enzyme that exhibits the extreme cooperativity shown in Figure 11.33. Substrate can easily accumulate up to the critical level $[S]_c$; the enzyme is essentially inactive at lower [S]. But any further increase leads to a greatly increased enzyme activity, so that the substrate level will be maintained near the value $[S]_c$. Although real allosteric enzymes rarely if ever exhibit curves as extremely sigmoidal as that in Figure 11.33, the principle remains: Multisubunit enzymes may help to maintain the homeostasis of a dynamic system. Put another way, homoallostery sharpens substrate-level control.

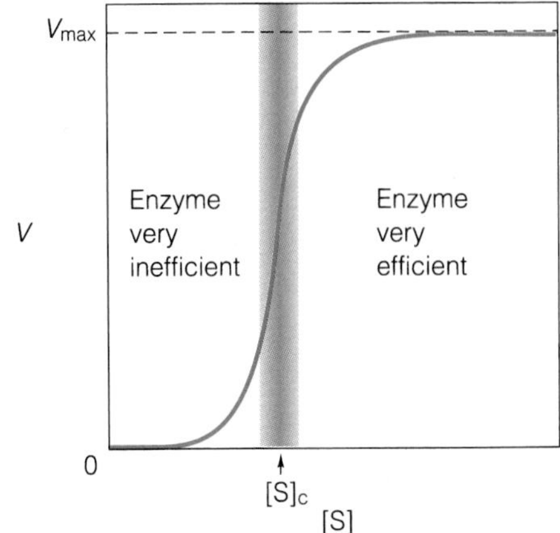

FIGURE **11.33**
Effect of extreme homoallostery. The *V*-versus-[S] curve is shown for a hypothetical enzyme with *very* high cooperativity in substrate binding. At concentrations below $[S]_c$, the enzyme is almost inactive; above this concentration, it is very active. Substrate can easily accumulate to the level $[S]_c$, but at higher concentrations it will be processed rapidly.

Heteroallostery

The major advantage of allosteric control is found in the role of **heteroallosteric effectors,** which may be either inhibitors or activators. These effectors are the analogs, in enzyme kinetics, of the CO_2, BPG, and H^+ that so elegantly regulate O_2 binding by hemoglobin. The activation and inhibition of enzymes by allosteric effectors are the key to the kind of complex feedback control described on pages 399–400. If an enzyme molecule can exist in two conformational states (T and R) that differ dramatically in the strength with which substrate is bound or in the catalytic rate, then its kinetics can be controlled by *any* other substance that, in binding to the protein, shifts the T $\rightleftharpoons$ R equilibrium. Allosteric *inhibitors* shift the equilibrium toward T, and *activators* shift it toward R (Figure 11.34). Some enzymes are regulated by multiple inhibitors and activators, allowing extremely subtle and complex patterns of metabolic control.

Allosteric enzymes show cooperative substrate binding and can respond to a variety of inhibitors and activators.

ASPARTATE CARBAMOYLTRANSFERASE: AN EXAMPLE OF AN ALLOSTERIC ENZYME

An excellent example of allosteric regulation is provided by the enzyme **aspartate carbamoyltransferase** (also known as aspartate transcarbamoylase, or ATCase), a key enzyme in pyrimidine synthesis. As can be seen from Figure 11.35, ATCase stands at a crossroads in biosynthetic pathways. Glutamine, glutamate, and aspartate are also used in protein synthesis; but once aspartate has been carbamoylated to form ***N*-carbamoyl-L-aspartate** (CAA), the molecule is committed to pyrimidine synthesis. Thus, the enzyme that controls this step must be sensitive to pyrimidine need. In bacteria like *E. coli*, the activity of ATCase is regulated to respond to this need. This enzyme, as shown in Figure 11.36, is inhibited by cytidine triphosphate (CTP) and activated by ATP. Both responses make physiological sense; when CTP levels are already high, more pyrimidines are not needed. On the other hand, high ATP signals both a purine-rich state (needing accompanying pyrimidines) and an energy-rich cell condition under which DNA and RNA synthesis will be active.

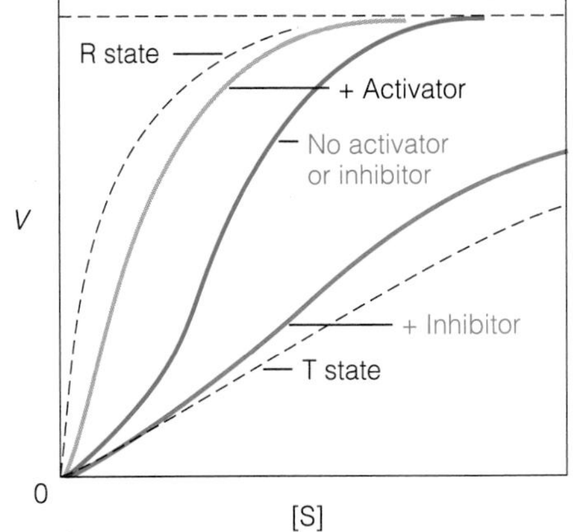

FIGURE **11.34**
Heteroallosteric control of an enzyme. In the absence of activation or inhibitors, the *V*-versus-[S] curve is sigmoidal. Activators shift the system toward the R state; inhibitors stabilize the T state.

ATCase is, as you would expect, a multisubunit protein. Its quaternary structure has been examined in some detail and is depicted schematically in Figure 11.37a. There are six *catalytic* subunits, in two tiers of three, held together by six *regulatory* subunits. Pairs of regulatory subunits appear to connect catalytic subunits in the two tiers. The three-dimensional structure of ATCase has been solved to high resolution, and a detailed representation of one catalytic subunit and one

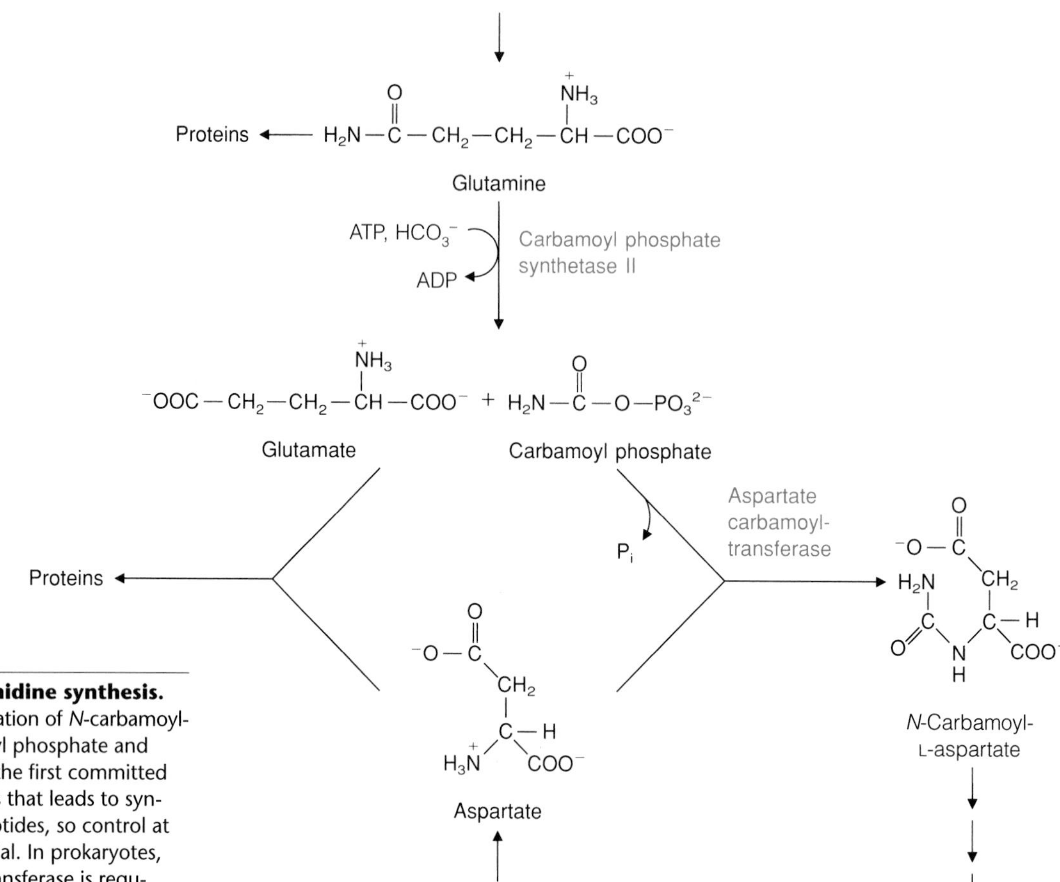

FIGURE **11.35**

Control points in pyrimidine synthesis. This figure shows the formation of *N*-carbamoyl-L-aspartate from carbamoyl phosphate and aspartate. This reaction is the first committed step in a series of reactions that leads to synthesis of pyrimidine nucleotides, so control at or near this point is essential. In prokaryotes, the aspartate carbamoyltransferase is regulated; in most eukaryotes, regulation is on the carbamoyl phosphate synthetase II.

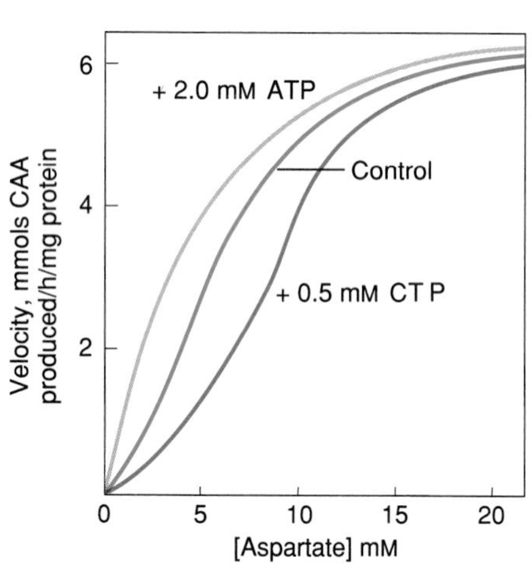

FIGURE **11.36**

Regulation of aspartate carbamoyltransferase by ATP and CTP. ATP is an activator of aspartate carbamoyltransferase, and CTP is an inhibitor. The curve marked as "control" shows the behavior of the enzyme in the absence of both regulators. *N*-Carbamoyl-L-aspartate (CAA) is the product of the reaction.

regulatory subunit is shown in Figure 11.38. The catalytic subunit comprises two domains, one binding aspartate and the other carbamoyl phosphate: the active site lies between them. The regulatory subunit is likewise of two parts; the so-called zinc domain and the allosteric domain. The former binds a structurally necessary zinc atom; the latter contains the ATP/CTP binding site. ATP and CTP thus compete for the same site, so that the activity of ATCase is regulated by the *ratio* of ATP to CTP in the cell.

As in the case of hemoglobin, the allosteric regulation of ATCase involves changes in the quaternary structure of the molecule. Conformations of the R and T states have been determined by x-ray diffraction. As Figure 11.37b shows, a major rearrangement of subunit positions occurs in the T $\longrightarrow$ R transition.

Virtually every metabolic pathway we shall encounter in the following chapters is subject to complex feedback control, and in almost all cases multisubunit, allosteric enzymes are employed. The pattern of control, even in a given pathway, is not the same in every organism. To take a relevant example, whereas ATCase is the major control point in the pyrimidine pathway in bacteria, eukaryotes regulate at the preceding step—the synthesis of carbamoyl phosphate (see Figure 11.35). In mammals, the **carbamoyl phosphate synthetase II** is inhibited by UDP, UTP, CTP, dUDP, and UDP-glucose. These compounds all inhibit binding of the ATP substrate. In addition, glycine acts as a competitive inhibitor for glutamine.

It should be clear at this point that organisms can regulate metabolism in complex and subtle ways through allosteric enzymes. But this kind of regulation is not sufficient for all needs. We turn now to covalent modification, an entirely different kind of regulatory mechanism.

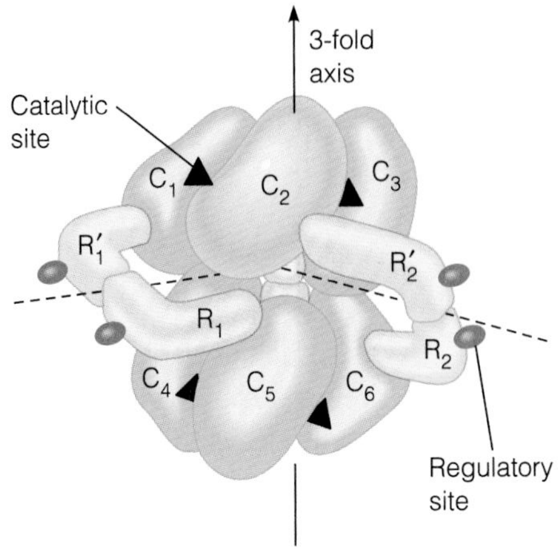

(a) ATCase: T state

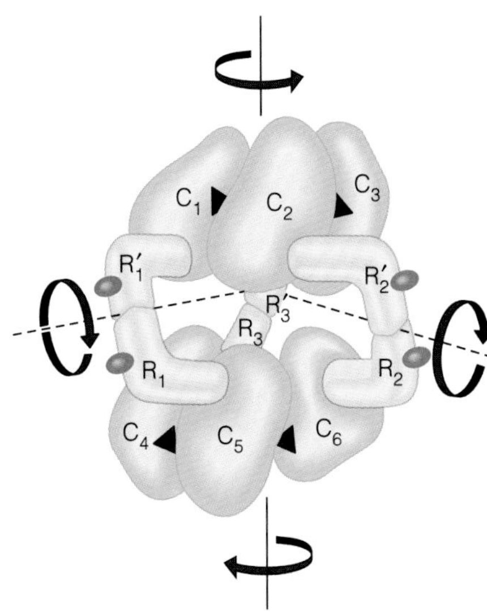

(b) ATCase: R state

FIGURE **11.37**

Quaternary structure of aspartate carbamoyltransferase (ATCase). **(a)** Quaternary structure of ATCase in the T state. This schematic view of the enzyme shows the six catalytic subunits (C) and six regulatory subunits (R). Six catalytic sites (black triangles) lie in or near the grooves between the catalytic subunits. Regulatory sites (red ellipses) lie on the outer surfaces of the regulatory subunits. The molecule has one 3-fold axis and three 2-fold axes (D_3 symmetry). This is a side view of the molecule with the 3-fold axis in the plane of the paper. **(b)** Transition of ATCase to the R state. The transition involves a rotation of the regulatory subunits, which pushes the two tiers of catalytic subunits apart and rotates them slightly about the 3-fold axis.

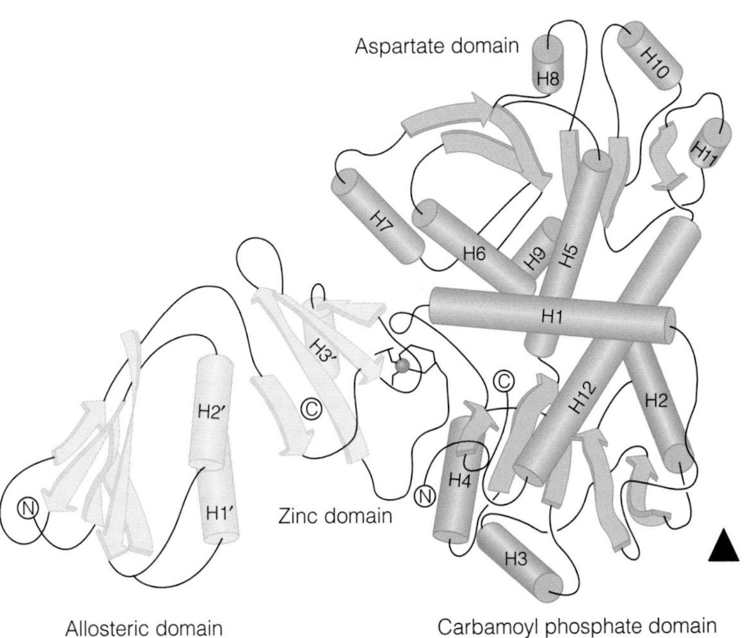

FIGURE **11.38**

The detailed structure of one catalytic subunit (green) and adjacent regulatory subunit (yellow) of ATCase. The view is down the 3-fold axis, which is at the lower right. The regulatory subunit lies mostly below the plane, the catalytic subunit mostly above.

From W. N. Lipscomb, *Advances In Enzymology and Related Areas of Molecular Biology,* (1994) 73:677–151.

Covalent Modifications Used to Regulate Enzyme Activity

In the factory analogy, allosteric regulation can be thought of as the feedback control of continuously running machines. But any large factory also has machinery that is used only from time to time and is left on standby until needed. The same is true for the cell. In this section, we discuss enzymes that are wholly inactive until they are changed by a **covalent modification** and then begin to function. In some cases such modification acts in the opposite direction, to inactivate otherwise active enzymes. Some such modifications can be reversed; others cannot.

A number of kinds of covalent modification are commonly used to regulate enzyme activity. One of the most widespread is phosphorylation or dephosphorylation of various amino acid side chains (serine, threonine, tyrosine, and histidine, for example). However, these kinds of regulation are most often a part of complex regulatory pathways, frequently under hormonal control. Therefore, we shall defer discussion of this mechanism until later chapters, where its significance will be clearer. At this point, we shall concentrate instead on another and simpler kind of covalent modification—the activation of certain enzymes by proteolytic cleavage of precursor forms.

PANCREATIC PROTEASES: ACTIVATION BY CLEAVAGE

Some enzymes, such as pancreatic proteases, are irreversibly switched on by proteolytic cleavage.

An important example of covalent enzyme activation, **proteolytic cleavage,** is found in the **pancreatic proteases.** They include a number of enzymes—for example, trypsin, chymotrypsin, elastase, and carboxypeptidase—some of which we have already discussed. All are synthesized in the pancreas. They are secreted through the pancreatic duct into the duodenum of the small intestine in response to a hormone signal generated when food passes from the stomach. They are not, however, synthesized in their final, active form, because a battery of potent proteases free in the pancreas would digest the pancreatic tissue. Rather, they are made as slightly longer, catalytically inactive molecules, called **zymogens.** The names given to the zymogens of the enzymes mentioned above are *trypsinogen, chymotrypsinogen, proelastase,* and *procarboxypeptidase,* respectively. The zymogens must be cleaved proteolytically in the intestine to yield the active enzymes. These enzymes are broken down after serving their purpose, so they do not endanger the intestinal tissue, which is also somewhat protected by its glycosylated surface. The cleavage of zymogens to active enzymes is diagrammed in Figure 11.39.

The first step is the activation of trypsin in the duodenum. A hexapeptide is removed from the N-terminal end of trypsinogen by *enteropeptidase,* a protease secreted by duodenal cells. This action yields the active trypsin, which then activates the other zymogens by specific proteolytic cleavages. In fact, once some active trypsin is present, it will activate other trypsinogen molecules to make more trypsin; thus its activation is *autocatalytic.* This is an example of the kind of cascade process frequently observed when enzymes are activated by covalent modification. The production of just a few trypsin molecules leads quickly to many more, for each enzyme molecule, when activated, can process many more every minute. These molecules can in turn activate the other zymogens. A most spectacular example of such a cascade process occurs in blood clotting, which we describe in the next section.

The activation of chymotrypsinogen to chymotrypsin is one of the most complex and best studied examples of proteolytic activation of an enzyme; it is illustrated in Figure 11.40. In the first step, trypsin cleaves the bond between arginine 15 and isoleucine 16. The N-terminal peptide remains attached to the rest of the

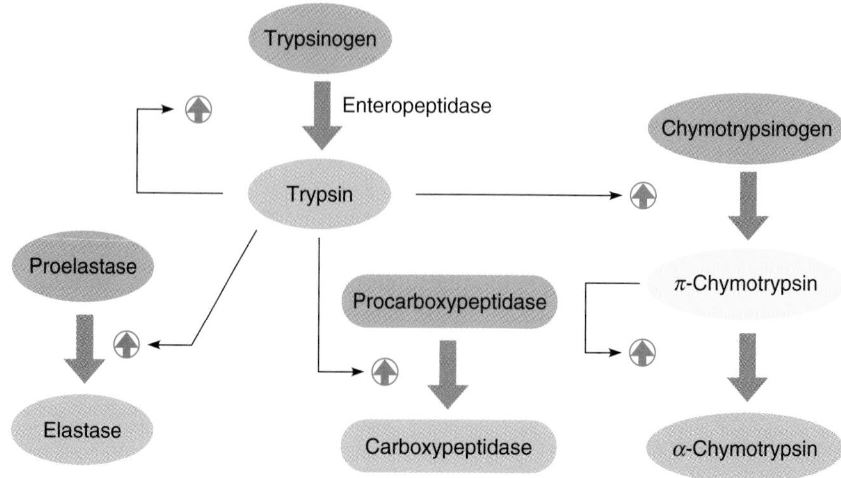

FIGURE **11.39**

Zymogen activation by proteolytic cleavage. This schematic view shows the activation of pancreatic zymogens, molecules that become catalytically active when cleaved. Zymogens are shown in red, and active proteases in yellow or green. The difference between π-chymotrypsin and α-chymotrypsin is shown in Figure 11.40.

molecule because of the disulfide bond between residues 1 and 122. The product, called *π-chymotrypsin,* is an active enzyme.

Just how the cleavage of one peptide bond transforms a wholly inactive protein into an active one can now be understood as a result of detailed x-ray diffraction studies of the zymogen and the enzyme. Cleavage of the peptide bond between residues 15 and 16 creates a new, positively charged N-terminal residue at Ile 16. This residue shifts its position and forms a salt bridge with Asp 194, the neighbor of the active site Ser 195 (see Figure 11.11). This change in turn triggers further conformational rearrangements in the neighborhood of the active site. These changes include the correct modeling of the active site pocket and movement of main chain amino groups of residues 193 and 195 to where they can hydrogen-bond to the substrate oxyanion in the tetrahedral transition state. Thus, both the binding pocket and the catalytic site are formed correctly only after the peptide bond between Arg 15 and Ile 16 has been cleaved.

Chymotrypsin is not quite "finished" by this cleavage. More autocatalytic cleavages remove residues 14–15 and 147–148 from the molecule, to produce the final *α-chymotrypsin,* which is the principal and fully active form found in the digestive tract.

This battery of enzymes, trypsin, chymotrypsin, elastase, and carboxypeptidase, together with the *pepsin* of the stomach and other proteases secreted by the intestinal wall cells, is capable of ultimately digesting most ingested proteins into free amino acids, which can be absorbed by the intestinal epithelium. The enzymes themselves are continually subjected to mutual digestion and autodigestion, so that high levels of these enzymes never accumulate in the intestine.

Even inactive zymogens are a potential source of danger to the pancreas. Because trypsin activation can be autocatalytic, the presence of even a single active trypsin molecule could set the whole chain in motion prematurely. Therefore the pancreas protects itself further by synthesizing a protein called the *secretory pancreatic trypsin inhibitor* (to be distinguished from the pancreatic trypsin inhibitor described in Chapter 6, which is an intracellular protein found only in ruminants). This competitive inhibitor binds so tightly to the active site of trypsin that it effectively inactivates it even at very low concentration. The bonding between trypsin and its inhibitor is among the strongest noncovalent associations known in biochemistry. Only a tiny amount of trypsin inhibitor is present—far less than needed to inhibit all of the potential trypsin in the pancreas. Thus, only a fraction of the trypsin generated in the duodenum is inhibited, and the rest can be activated. Because protection is limited, zymogen activation can sometimes be triggered in the pancreas—for example, if the pancreatic duct is blocked. The active enzymes then begin to digest the pancreatic tissue itself. This condition, called *acute pancreatitis,* is extremely painful and sometimes fatal.

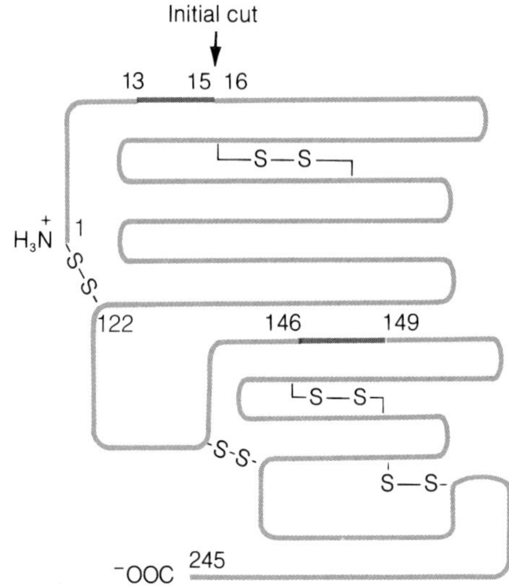

FIGURE **11.40**

Activation of chymotrypsinogen. The figure is a schematic rendition of the chymotrypsinogen molecule. A series of cleavages produces the enzyme chymotrypsin, with the disulfide bonds continuing to hold the structure together. The initial cleavage between amino acids 15 and 16 (arrow) results in the formation of π-chymotrypsin. Subsequent removal of the segments shown in black yields α-chymotrypsin.

A FURTHER LOOK AT ACTIVATION BY CLEAVAGE: BLOOD CLOTTING

Activation of zymogens is the key to another biologically important process—the clotting of vertebrate blood. If a blood clot is examined in the electron microscope, it is found to be composed of striated fibers of a protein called *fibrin* (Figure 11.41a). The fibrin monomers are elongated molecules, about 46 nm long, that stick together in a staggered array as shown in Figure 11.41b. Fibrin monomers are derived from a precursor, *fibrinogen* (Figure 11.41c), by proteolytic cleavages that release small *fibrinopeptides* (A and B in Figure 11.41c). Loss of these peptides uncovers positions at which the fibrin molecules can stick together. After the clot is formed, it is further stabilized by covalent cross-links between glutamine and lysine residues.

Blood clotting involves a cascade of proteolytic activation of specific proteases, culminating in the transition of fibrinogen to fibrin.

The proteolysis of fibrinogen to fibrin is catalyzed by the serine protease *thrombin.* Thrombin has sequence and structural similarities to trypsin, but as a protease with a very specific function it cleaves only a few types of bonds, mainly

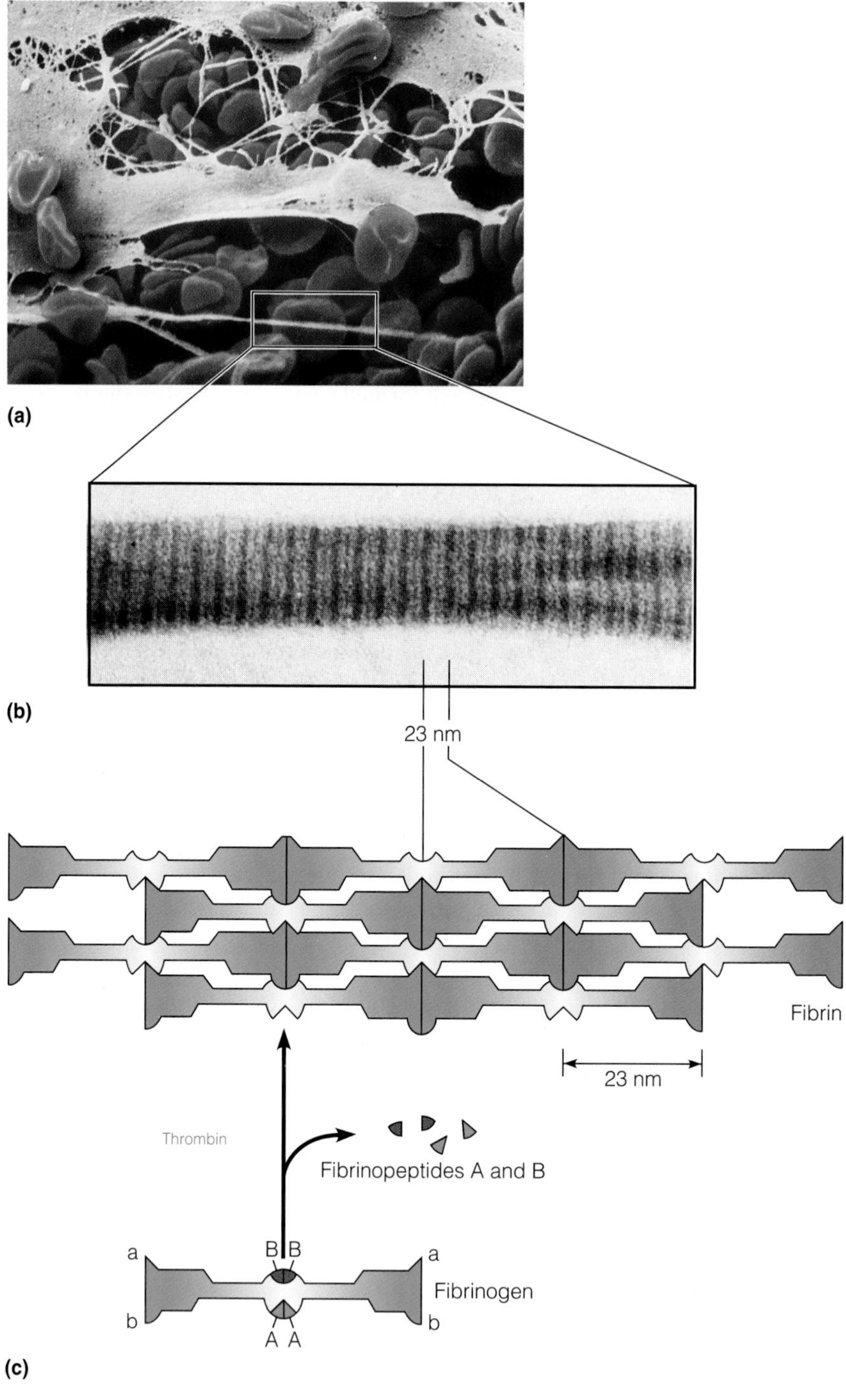

FIGURE **11.41**

Formation of a blood clot. (a) Red blood cells enmeshed in the insoluble strands of a fibrin clot. **(b)** Electron micrograph of part of a fibrin fiber. **(c)** Schematic view of how fibrin monomers are thought to associate to form a fiber. Removal of fibrinopeptides A and B from fibrinogen by thrombin makes sites accessible for association with complementary sites a and b on adjacent monomers. The molecules are believed to overlap as shown because the striations seen in the fibers are 23 nm wide, exactly half the length of the fibrinogen molecule.

(a) Lennart Nilsson/Albert Bonniers Forlag AB; **(b)** courtesy of Dr. Henry Slayter.

Arg–Gly. Thrombin itself is produced from *prothrombin* by another specific protease; in fact, as Figure 11.42 shows, a whole cascade of proteolytic activation reactions leads ultimately to the formation of a fibrin clot. Involved are a series of proteases referred to as *factors.* In damaged tissues, the proteins *kininogen* and *kallikrein* activate factor XII (also called *Hageman factor*), which in turn activates factor XI—and the cascade of reactions proceeds as shown. This set of initial reactions is called the **intrinsic pathway.** Alternatively, damage to blood vessels leads to the release of *tissue factor* and activation of factor VII, starting the **extrinsic pathway.** The two pathways merge in the activation of factor X, which will proteolyze and thereby activate prothrombin.

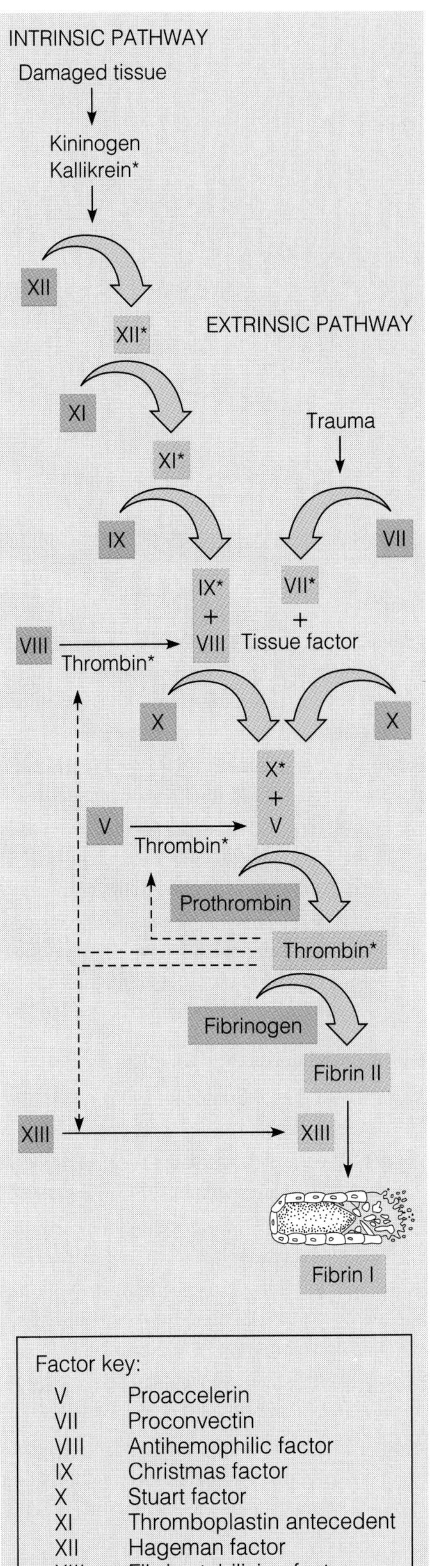

FIGURE **11.42**

The cascade process in blood clotting. Each factor (protease) in the pathway can exist in an inactive form (red) or an active form (green). The cascade of proteolytic activations can start from exposure of blood at damaged tissue surfaces (intrinsic pathway) or from internal trauma to blood vessels (extrinsic pathway). The common result is activation of fibrinogen to clotting fibrin. Auxiliary factors that aid some steps are also shown. Asterisk (*) denotes serine proteases.

Some of the activation steps require auxiliary proteins. For example, activation of factor X in the intrinsic pathway by factor IX (*Christmas factor*) requires a 330-kilodalton protein called *antihemophilic factor* (factor VIII). The partial or complete absence of factor VIII activity is the cause of classic **hemophilia.** The gene for factor VIII is carried on the X chromosome, so women, who have two copies of this chromosome, can be heterozygous carriers of the trait but will exhibit the symptoms only if they are homozygous. However, a male descendent who receives on his single X chromosome the damaged copy of the factor VIII gene will experience more or less severe difficulty in blood clotting. The condition can now be treated by frequent transfusions of a blood serum fraction concentrated in factor VIII. The gene for this protein has recently been cloned and expressed in bacteria, and the availability of this synthetic factor VIII may allow such patients to avoid the dangers of regular transfusions.

As wounds heal, or tissue damage is repaired, it is essential that blood clots be dissolved. The principal agent for clot dissolution is an enzyme called **plasmin,** which cleaves fibrin. Plasmin itself is derived by proteolytic cleavage of an inactive precursor, **plasminogen.** Activation of plasminogen is catalyzed by a number of proteases, the most important being tissue-type **plasminogen activator (t-PA).** In addition to its normal function, t-PA can be extremely effective in initiating the cascade to dissolve the unwanted blood clot involved in stroke or heart attack. Recent development of methods to produce t-PA by recombinant DNA techniques promises new and effective therapy.

The mechanisms of regulation we have mentioned here by no means describe the cell's whole repertoire. In addition to regulation of enzyme *performance,* cells and organisms can regulate both the synthesis and degradation of enzymes. But description of these processes is more appropriate in the broader framework of Chapter 12.

SUMMARY

The rate of a chemical reaction is determined by reactant concentrations and by the rate constant. The rate constant in turn depends upon the activation energy needed to reach the transition state. All catalysts function by lowering the activation energy for a reaction. In doing so, they do not affect chemical equilibrium but only increase rates. Enzymes are biological catalysts that increase the rates of biochemical processes but are themselves unchanged. Most (but not all) enzymes are proteins. In enzyme catalysis, one or more substrates are bound at the active site of an enzyme, to form the enzyme–substrate complex; products are then released. The induced fit hypothesis states that enzymes induce bound substrates to adopt conformations close to the transition state. Most simple enzymatic reactions can be described by the Michaelis–Menten equation, with two parameters, the Michaelis constant, K_M, and the turnover number, k_{cat}. Enzymes can be inhibited reversibly or irreversibly. Reversible inhibition can be competitive or noncompetitive. Competitive inhibition affects the apparent K_M; noncompetitive inhibition affects the apparent k_{cat}. Irreversible inhibition usually involves covalent binding to the active site. Many enzymes utilize coenzymes in their function; others require specific metal ions. A number of coenzymes are closely related to vitamins required in the human diet.

It is now possible to create new or modified enzymes by molecular engineering or by the generation of antibodies with enzymatic capabilities. In addition, it has been found that some ribonucleic acid molecules function as enzymes; they are called ribozymes.

Regulation of enzymatic activity takes many forms. Substrate-level regulation simply depends on ambient concentrations of reactants and products. Allosteric regulation provides sensitive feedback control for complex metabolic pathways. For more drastic changes in activity, some enzymes are switched on or off (or both) by covalent modification.

REFERENCES

General

Creighton, T. E. (1993) *Proteins,* 2nd ed. W. H. Freeman, New York. Chapter 9 of this excellent book contains a succinct summary of enzyme catalysis, together with many references to the appropriate literature.

Fersht, A. (1985) *Enzyme Structure and Mechanism,* 2nd ed. W. H. Freeman, New York. A fine treatise on almost all aspects of enzymology.

Gutfreund, H. (1995) *Kinetics for the Life Sciences.* Cambridge University Press, Cambridge, UK. A thoughtful analysis and exposition of kinetic principles, as applied to biological problems.

Webb, E. (1992) *Enzyme Nomenclature.* Academic Press, Orlando, Fla. This is the latest authoritative listing of enzymes prepared by the Nomenclature Committee of the Enzyme Commission of the International Union of Biochemistry and Molecular Biology.

Enzyme Mechanisms and Kinetics

Åqvist, J., and M. Fothergill (1996) Computer simulation of the triosephosphate isomerase reaction. *J. Biol. Chem.* 271: 10010–10016.

Bennet, W. S., and T. A. Steitz (1978) Glucose-induced conformational change in yeast hexokinase. *Proc. Natl. Acad. Sci. USA* 75:4848–4852.

Cleland, W. W. (1977) Determining the chemical mechanisms of enzyme catalyzed reactions by kinetic studies. *Adv. Enzymol.* 45:273–387.

Knowles, J. R. (1991) Enzyme catalysis: Not different, just better. *Nature* 350:121–124. Mostly about triose phosphate isomerase, but with implications for enzymes in general.

Kraut, J. (1988) How do enzymes work? *Science* 242:533–540. A thoughtful reanalysis of transition state theory in terms of current information.

Walsh, C. T. (1977) *Enzymatic Reaction Mechanisms.* W. H. Freeman, New York. A comprehensive work, including a thorough treatment of coenzymes.

Proteases

Corey, D. R., and C. S. Craik (1992) An investigation into the minimum requirements for peptide hydrolysis by mutation of the catalytic triad of trypsin. *J. Am. Chem. Soc.* 114:1784–1790.

Kraut, J. (1977) Serine proteases: Structure and mechanism of catalysis. *Annu. Rev. Biochem.* 46:331–358.

Steitz, T. A., and R. G. Shulman (1982) Crystallographic and NMR studies of the serine proteases. *Annu. Rev. Biochem. Biophys.* 11:419–444.

Irreversible Inhibition

Abeles, R. H. (1983) Suicide enzyme inactivators. *Chem. Eng. News* 61:48–56. A brief, clear discussion.

Rando, R. R. (1975) Mechanisms of action of naturally occurring irreversible enzyme inhibitors. *Acc. Chem. Res.* 8:281–288.

Molecular Engineering of Enzymes

Olsen, O., K. K. Thomsen, J. Weber, J. Duus, I. Svendsen, C. Wegener, and D. von Wettstein (1996) Transplanting two unique β-gluconase catalytic activities into one multienzyme which forms glucose. *Nat. Biotechnol.* 14:71–76.

Oxender, D. L., and C. F. Fox, eds. (1987) *Protein Engineering.* Alan R. Liss, New York. A fascinating compilation of papers on techniques and results. Chapter 25, by J. A. Wells and 12 collaborators, describes the work on subtilisin referred to in the text.

Schultz, P. G. (1988) The interplay between chemistry and biology in the design of enzymatic catalysts. *Science* 240:426–433. A short, clearly written exposition of methods.

Catalytic Antibodies

Benkovic, S. J. (1992) Catalytic antibodies. *Annu. Rev. Biochem.* 61:29–54. A comprehensive review by one of the leaders in the field.

Janda, K. D., L.-C. Lo, C.-H. Lo, M.-M. Sim, R. Wang, C.-H. Wong, and R. A. Lerner (1997) Chemical selection for catalysis in combinatorial antibody libraries. *Science* 275:945–948.

Powell, M. J., and D. E. Hansen (1989) Catalytic antibodies—a new direction in enzyme design. *Protein Eng.* 3:69–75.

Ribozymes

Cech, T. R. (1987) The chemistry of self-splicing RNA and RNA enzymes. *Science* 236:1532–1539.

Gesteland, R. F., T. R. Cech, and J. F. Atkins (1999) *The RNA World.* Cold Spring Harbor Press, Cold Spring Harbor, New York.

McCorkle, G. M., and S. Altman (1987) RNA's as catalysts. *Concepts Biochem.* 64:221–226.

Waugh, D. S., C. J. Green, and N. R. Pace (1989) The design and catalytic properties of a simplified ribonuclease P RNA. *Science* 244:1569–1571. Ribozyme bioengineering.

Wright, M. C., and G. F. Joyce (1997) Continuous in vitro evolution of catalytic function. *Science* 276:614–616.

Allosteric Regulation

Lipscomb, W. N. (1994) Aspartate transcarbamylase from *Escherichia coli:* Activity and regulation. *Adv. Enzymol.* 73:67–151. A very complete description of our current knowledge of this important example.

Monod, J., J.-P. Changeux, and F. Jacob (1963) Allosteric proteins and cellular control systems. *J. Mol. Biol.* 6:306–329. This paper introduced the concept of allosteric control.

Zymogen Activation

Bode, W., and R. Huber (1986) Crystal structure of pancreatic serine endopeptidases. In: *Molecular and Cellular Basis of Digestion,* edited by P. Desnuelle, H. Sjorstrom, and O. Noren, pp. 213–234. Elsevier, New York.

Neurath, H. (1986) The versatility of proteolytic enzymes. *J. Cell. Biochem.* 32:35–49. A useful survey of proteases.

Blood Clotting

Davie, E. W. (1986) Introduction to the blood coagulation cascade and the cloning of blood coagulation factors. *J. Protein Chem.* 5:247–253.

Doolittle, R. F. (1984) Fibrinogen and fibrin. *Annu. Rev. Biochem.* 53:195–229.

Lawn, R. M., and G. A. Vehar (1986) The molecular genetics of hemophilia. *Sci. Am.* 254:48–65. A readable overall view.

PROBLEMS

1. Show that the half-life for a first-order reaction is inversely proportional to the rate constant, and determine the constant of proportionality.
2. A substance A is consumed by a reaction of unknown order. The initial concentration is 2 mM, and concentrations at later times are as shown:

Time (min)	[A] (mM)
1	1.66
2	1.44
4	1.12
8	0.76
16	0.48

Test whether or not a first-order reaction fits the data.

3. The enzyme urease catalyzes the hydrolysis of urea to ammonia plus carbon dioxide. At 21°C the uncatalyzed reaction has an activation energy of about 125 kJ/mol, whereas in the presence of urease the activation energy is lowered to about 46 kJ/mol. By what factor does urease increase the velocity of the reaction?
4. The maximum possible rate for a reaction between two molecules occurs when every collision results in reaction. In this case, the second-order rate constant is predicted to be

$$k = \frac{4\pi N}{1000}(D_1 + D_2)r_{12}$$

with units of $(\text{mol/L})^{-1}\,\text{s}^{-1}$. Here N is Avogadro's number (6.02×10^{23} molecules/mol), r_{12} is the critical distance for reaction, and D_1 and D_2 are the diffusion coefficients for the two participants. In a typical enzyme-catalyzed reaction, we might expect r_{12} to be about 10^{-7} cm, $D_{enzyme} = 5 \times 10^{-7}$ cm²/s, and $D_{substrate} = 5 \times 10^{-6}$ cm²/s. What does this predict for the value of k? Compare with values of k_{cat}/K_M in Table 11.1. [Hint: The equation is written for the units used above. Do not convert.]

*5. In some reactions, in which a protein molecule is binding to a specific site on DNA, a rate *greater* than that predicted by the equation in Problem 4 is observed. Suggest an explanation. [Hint: The protein molecule can also bind weakly and nonspecifically to *any* DNA site.]

*6. Derive an expression for the change of concentration of reactant with time for a reversible first-order reaction (equation (11.5)), assuming [B] = 0 at $t = 0$. (Warning: This is much more difficult than it looks.)

7. (a) Derive an expression for the concentration of reactant as a function of time for the second-order reaction 2A ⟶ B.
(b) On the basis of your answer to (a), show how a linear graph could be obtained for a second-order reaction.
(c) Test the data given in Problem 2 with your method.
8. The initial rate for an enzyme-catalyzed reaction has been determined at a number of substrate concentrations. Data are as follows:

[S] (μmol/L)	V [(μmol/L) min^{-1}]
5	22
10	39
20	65
50	102
100	120
200	135

(a) Estimate V_{max} and K_M from a direct graph of V versus [S]. Do you find difficulties in getting clear answers?
(b) Now use a Lineweaver–Burk plot to analyze the same data. Does this work better?
(c) Finally, try an Eadie–Hofstee plot of the same data.

9. (a) If the total enzyme concentration in Problem 8 was 1 nmol/L, how many molecules of substrate can a molecule of enzyme process in each minute?
(b) Calculate k_{cat}/K_M for the enzyme reaction in Problem 8. Is this a fairly efficient enzyme? (See Table 11.1.)

*10. (a) If we write the Michaelis–Menten equation as

$$\frac{d[S]}{dt} = -\frac{d[P]}{dt} = -\frac{V_{max}[S]}{K_M + [S]}$$

it should be possible to integrate it and obtain an expression for substrate concentration as a function of time. Do so, calling $[S]_0$ the initial substrate concentration at $t = 0$.
(b) Show from the result in (a) that under conditions where $[S]_0 >> K_M$, the decrease in substrate concentration with time is approximately linear. Interpret this result.

*11. Devise a graphical method for using the equation obtained in Problem 10a to determine V_{max} and K_M.

12. The catalytic efficiency of many enzymes depends on pH. Chymotrypsin shows a maximum value of k_{cat}/K_M at pH 8. Detailed analysis shows that k_{cat} increases rapidly between pH 6 and 7 and remains constant at higher pH. K_M increases rapidly between pH 8 and 10. Suggest explanations for these observations.

13. The following data describe the catalysis of cleavage of peptide bonds in small peptides by the enzyme elastase.

Substrate	K_M (mM)	k_{cat} (s^{-1})
P A P A↓G	4.0	26
P A P A↓A	1.5	37
P A P A↓F	0.64	18

The arrow indicates the peptide bond cleaved in each case.
(a) If a mixture of these three substrates was presented to elastase with the concentration of each peptide equal to 0.5 mM, which would be digested most rapidly? Which most slowly? (Assume enzyme is present in excess.)
(b) On the basis of these data, suggest what features of amino acid sequence dictate the specificity of proteolytic cleavage by elastase.
(c) Elastase is closely related to chymotrypsin. Suggest two kinds of amino acid residues you might expect to find in or near the active site.

14. The serine protease, subtilisin, is used in some laundry detergents to help remove protein-type stains.
(a) What unusual kind of stability does this suggest for subtilisin?
(b) Subtilisin does have a problem, in that it becomes inactivated by oxidation of a methionine close to the active site. Suggest a way to make a better subtilisin.

15. The steady state kinetics of an enzyme are studied in the absence and presence of an inhibitor (inhibitor A). The initial rate is given as a function of substrate concentration in the following table:

	V [(mmol/L)min^{-1}]	
[S] (mmol/L)	No inhibitor	Inhibitor A
1.25	1.72	0.98
1.67	2.04	1.17
2.50	2.63	1.47
5.00	3.33	1.96
10.00	4.17	2.38

(a) What kind of inhibition (competitive or noncompetitive) is involved?
(b) Determine V_{max} and K_M in the absence and presence of inhibitor.

16. The same enzyme as in Problem 15 is studied in the presence of a different inhibitor (inhibitor B). In this case, two different concentrations of inhibitor are used. Data are as follows:

	V [(mmol/L)min^{-1}]		
[S] (mmol/L)	No inhibitor	3 mM inhibitor B	5 mM inhibitor B
1.25	1.72	1.25	1.01
1.67	2.04	1.54	1.26
2.50	2.63	2.00	1.72
5.00	3.33	2.86	2.56
10.00	4.17	3.70	3.49

(a) What kind of inhibitor is inhibitor B?
(b) Determine the apparent V_{max} and K_M at each inhibitor concentration.
(c) Estimate K_I from these data.

17. We have mentioned Eadie–Hofstee plots as an alternative to Lineweaver–Burk plots for expression of kinetic data. Sketch what Eadie–Hofstee plots would look like for a series of experiments at different concentrations of
(a) A competitive inhibitor
(b) A noncompetitive inhibitor

18. We have discussed the role of TPCK as an irreversible inhibitor of chymotrypsin. Design a comparable inhibitor for trypsin.

*19. In another kind of reversible inhibition, called *uncompetitive*, the inhibitor binds only to the ES complex (not to the free enzyme) but blocks catalysis when it does. Devise the kinetic equation for this case, and sketch Lineweaver–Burk plots.

20. The following data are found for the steady state kinetics of a multisubunit enzyme:

[S] (mmol/L)	Initial rate [(mmol/L)s^{-1}]
0.25	0.26
0.33	0.45
0.50	0.92
0.75	1.80
1.00	2.50
2.00	4.10
4.00	4.80

(a) Does this enzyme follow Michaelis–Menten kinetics?
(b) Estimate V_{max}.

*21. Derive an equation expressing $\log\left(\frac{V/V_{max}}{1 - V/V_{max}}\right)$ as a function of log[S] for an enzyme that obeys the Michaelis–Menten equation.
(a) For an *allosteric* enzyme, show how you could use this equation to obtain K_M^T and K_M^R (the K_M values for T and R states) from a Hill plot of kinetic data (see Figure 7.9).
(b) Apply your result to the data in Problem 20.

*22. In a few instances, multisubunit enzymes have been demonstrated to exhibit *negative cooperativity;* that is, the binding of the first substrate molecule markedly decreases the affinity for binding of subsequent molecules. Show what a Hill plot would look like for an enzyme exhibiting negative cooperativity. [Hint: Consult Problem 21a.]

23. The MWC theory (see pages 222–223 in Chapter 7) cannot account for negative cooperativity, but the KNF theory can. Explain.

*24. Although the MWC theory describing binding to a multisite protein is quite complicated, we can easily derive equations for a simple case. Assume a dimeric enzyme, with two sites, with states T and R. For simplicity, assume that the R state can bind one or two molecules of substrate, but the T state cannot bind. We then have the equations

$$T \rightleftharpoons R \qquad L = \frac{[T]}{[R]}$$

$$R + S \rightleftharpoons RS \qquad K = \frac{2[R][S]}{[RS]}$$

$$RS + S \rightleftharpoons RS_2 \qquad 2K = \frac{[RS][S]}{[RS_2]}$$

The factor 2 appears because there are two equivalent sites on each R to bind S.

(a) From this information, show that the fraction of sites occupied at substrate concentration [S] is

$$\theta = \frac{[S]}{K}\left(\frac{1 + [S]/K}{L + (1 + [S]/K)^2}\right) = \frac{V}{V_{max}}$$

(b) Picking a large value of L (say 10^5), and with $K = 10^{-6}$ M, show that a sigmoidal curve is obtained for V/V_{max} versus [S].

25. Suggest the effects of each of the following mutations on the physiological role of chymotrypsinogen:
(a) Arg 15 ⟶ Ser
(b) Cys 1 ⟶ Ser
(c) Thr 147 ⟶ Ser

26. Suppose you had available a sample of the ribozyme IVS-19 from *Tetrahymena*. Describe the protocol for a simple experiment that would demonstrate the reaction shown in Figure 11.30b.

27. From the example shown on page 395, suggest the design of an antigen that might produce a catalytic antibody capable of amide bond hydrolysis.

TOOLS OF BIOCHEMISTRY 11A

How to Measure the Rates of Enzyme-Catalyzed Reactions

There are essentially two approaches to enzyme kinetic analysis. The first and simplest is to make measurements of rates under conditions in which the steady state approximation holds (see page 376). Under these conditions, the Michaelis–Menten equation is often applicable, and determination of the reaction velocity as a function of substrate and enzyme concentrations will yield K_M and k_{cat}. Almost all enzymatic studies at least start in this way. But if the experimenter wishes to learn more of the details of the mechanism, it is often important to carry out studies before the steady state has been attained. Such *pre-steady-state* experiments require the use of special fast techniques. On pages 382–383 we described how a combination of such approaches can be used to dissect a complex enzymatic process and to understand it in detail. Here we describe some of the experimental techniques that can be employed.

Analysis in the Steady State

The steady state in most enzymatic reactions is established within seconds or a few minutes and persists for many minutes or even hours thereafter. Therefore, extreme rapidity of measurement is not important, and many techniques are available to the experimenter wishing to follow the reaction. Descriptions of the most commonly used techniques follow.

Spectrophotometry

Spectrophotometric methods are simple and accurate (see Tools of Biochemistry 6A). However, an obvious requirement is that either a substrate or a product of the reaction must absorb light in a spectral region where other substrates or products do not. Classic examples are reactions that generate or consume NADH. NADH absorbs quite strongly at 340 nm, but NAD^+ does not absorb in this region. Thus we could, for example, follow the oxidation of ethanol to acetaldehyde, as catalyzed by alcohol dehydrogenase, by measuring the formation of NADH spectrophotometrically. Even if the reaction being studied does not involve a light-absorbing substance, it may be possible to *couple* this reaction to another, very rapid reaction that does.

Fluorescence

The applications of fluorescence are similar to those of spectrophotometry, and the problems are similar: A substrate or a product must have a distinctive fluorescence emission spectrum (see Tools of Biochemistry 6A). However, fluorescence often has the advantage of high sensitivity, so extremely dilute solutions may be employed, enabling an experimenter to greatly extend the concentration range over which studies are practicable.

Automatic Titration

If the reaction produces or consumes acid or base, it can be followed by using a device called a *pH-stat.* A glass electrode senses the pH of the solution, and its signal is used to actuate a motor-driven syringe that titrates acid or base into the reaction vessel to keep the pH constant. The record of acid or base consumed is then a record of the progress of the enzymatically catalyzed reaction.

Radioactivity Assays

If a substrate is labeled with a radioactive isotope that will be lost or transferred during the reaction to be studied, measurement of changes in radioactivity can be an extremely sensitive kinetic method. This procedure requires that the labeled compound can be separated quickly at different, precisely defined times during the reaction. An example is a method often used with radioactive ATP. The ATP can be adsorbed on charcoal-impregnated filter disks by very fast filtration of aliquots from the reaction mixture. The radioactivity can then be measured in a scintillation counter (see Tools of Biochemistry 12A).

Analysis of Very Fast Reactions

Reactions that are extremely rapid require special techniques in order to investigate the pre-steady-state processes. Three major methods are currently employed, to cover the time scales shown in Figure 11A.1.

Stopped Flow

Figure 11A.2 shows a stopped-flow apparatus. Enzyme and substrate are initially in separate syringes. The syringes are rapidly

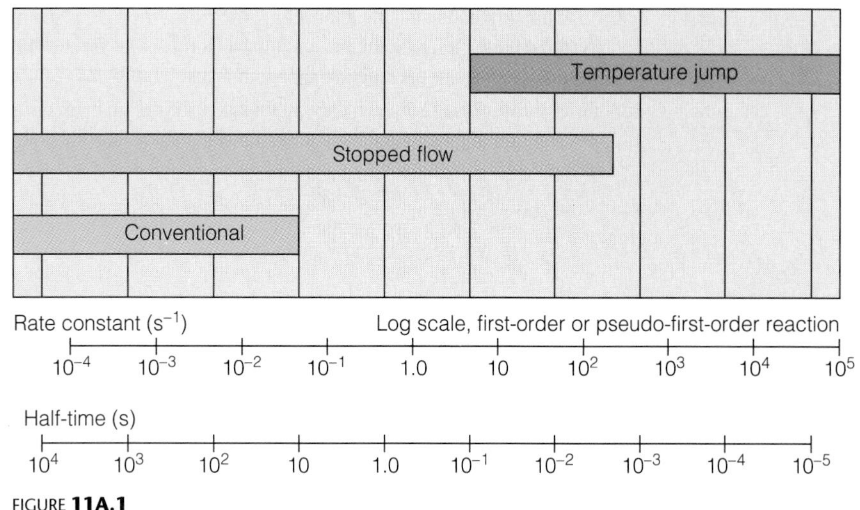

FIGURE **11A.1**

Time scales for kinetic techniques described here.

Courtesy of Dionex Instruments.

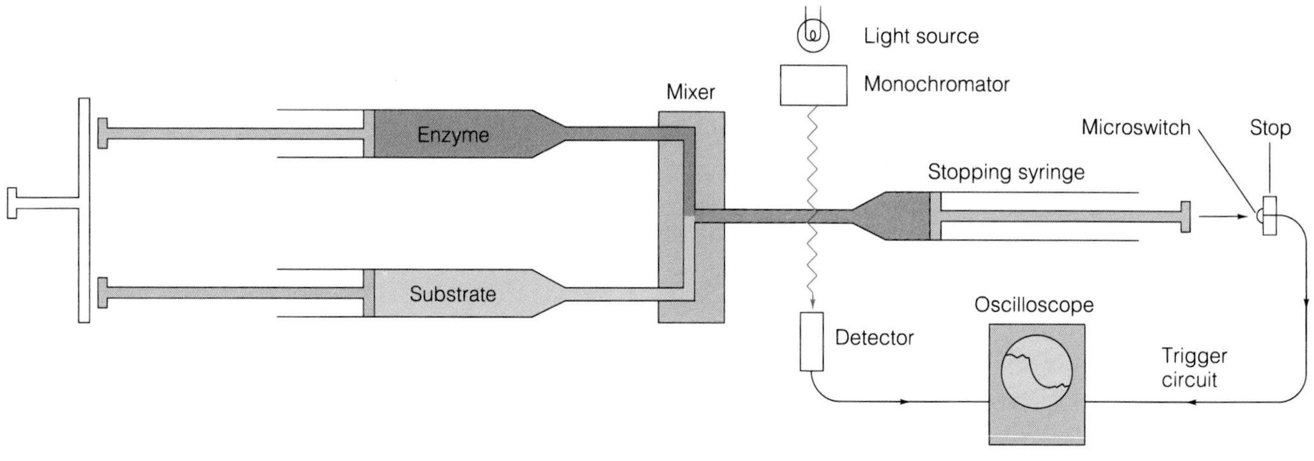

FIGURE **11A.2**

Typical stopped-flow apparatus.

driven to deliver their contents through a mixing chamber and into a third, "stopping" syringe. This step triggers a detector to begin observing (for example, by light absorption or fluorimetry) the solution in the tube connecting the mixer to the stopping syringe. Flow rates can easily be made as high as 1000 cm/s. If the mixture was moving at this rate when the flow was stopped, and if the observation point is 1 cm from the mixer, the detection system first sees a mixture that is 1 ms "old." The reaction can then be followed for as long as desired—often for a period of only a few seconds. The limitations of the method are imposed only by the initial "dead time" (in the example above, 1 ms) and the rapidity of the detection system.

Temperature Jump

Some processes are so fast that they are essentially completed in the dead time of a stopped-flow apparatus. The experimenter may then turn to temperature jump (*T*-jump) methods. The basic apparatus and principle of the method are shown in Figure 11A.3a and b, respectively. A reaction mixture that is at equilibrium at a temperature T_1 is suddenly jumped to a temperature T_2. The position of equilibrium will shift, and reaction must occur to attain this new equilibrium. A rapid jump in temperature (5–10°C in 1 μs) can be obtained by passing a large burst of electrical current between electrodes immersed in the reaction mixture. Even more rapid jumps (10–100 ns) can be obtained if a pulsed infrared laser is used to heat the mixture. The relaxation (approach) to a new equilibrium, monitored by absorption or fluorescence measurements, is an exponential process. For a simple reaction, the change in reactant concentration is given by

$$\Delta[\mathrm{A}] = (\Delta[\mathrm{A}])_{\text{total}} e^{-t/\tau} \qquad (11\mathrm{A}.1)$$

where τ is called the *relaxation time* and can be related to the rate constants for the reaction. For example, for the simple reversible isomerization

$$\mathrm{A} \underset{k_{-1}}{\overset{k_1}{\rightleftharpoons}} \mathrm{B}$$

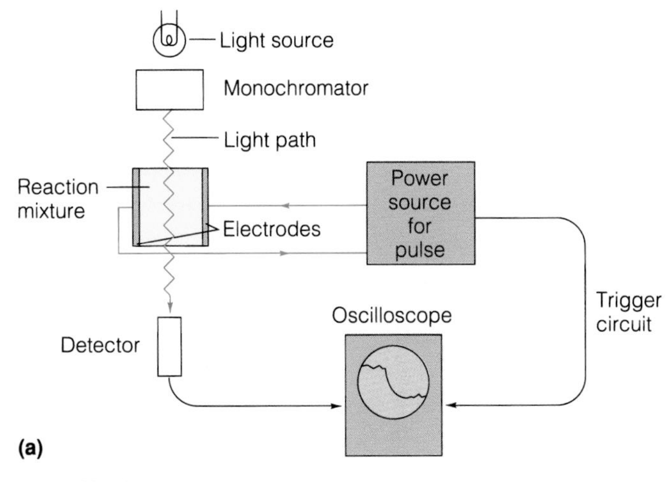

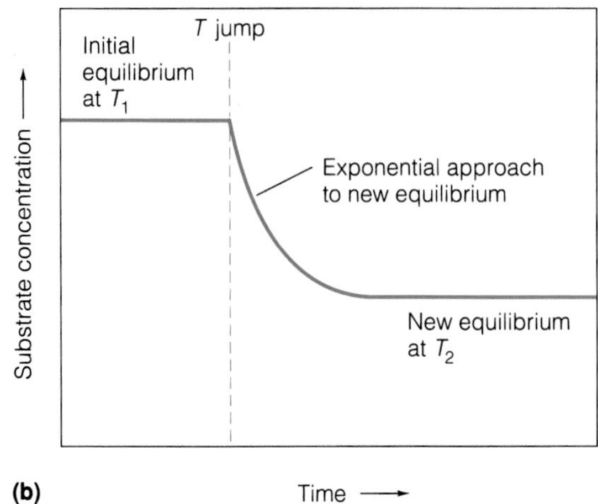

FIGURE **11A.3**
The temperature jump method.

we have

$$\frac{1}{\tau} = k_1 + k_{-1} \tag{11A.2}$$

More complex reactions involve multiple relaxation times and more complex curves than expressed by equation (11A.1). *T*-jump experiments can handle reactions with τ values as low as about 10^{-5} s.

Although a number of other techniques are employed for even faster reactions, including some newly developed NMR methods and pulsed laser techniques, the methods described here are the ones most widely used. If we consider the variety of techniques available to the experimenter, we can see that they cover a wide time range. Altogether, times from nanoseconds to hours can be studied.

References

Fersht, A. (1985) *Enzyme Structure and Mechanism,* 2nd ed. W. H. Freeman, New York. Chapters 4, 6, and 7 contain much information about techniques and references.

Himori, K. (1979) *Kinetics of Fast Enzyme Reactions.* Halstead, New York.

CHAPTER 12

Introduction to Metabolism

A CHEMIST CARRYING OUT AN ORGANIC SYNTHESIS RARELY RUNS MORE THAN one reaction in a single reaction vessel. This procedure is essential to prevent side reactions and to optimize the yield of the desired product. Yet a living cell carries out thousands of reactions simultaneously, with each reaction sequence controlled so that unwanted accumulations or deficiencies of intermediates and products do not occur. Reactions of great mechanistic complexity and stereochemical selectivity proceed smoothly under mild conditions—1 atm pressure, moderate temperature, and a pH near neutrality. A goal of the next several chapters is to understand how cells carry out and regulate these complex reaction sequences.

A First Look at Metabolism

As noted above, a principal task of the biochemist is to understand how a cell regulates its myriad reaction sequences and, in so doing, controls its internal environment. In Chapter 11 we discussed the properties of individual enzymes and control mechanisms that affect their activity. Here we consider specific reaction sequences, or **pathways;** the relationship between each pathway and cellular architecture; the biological importance of each pathway; control mechanisms that regulate **flux,** or intracellular reaction rate; and experimental methods used to investigate metabolism. Figure 12.1, a simplified view of the processes we shall consider, illustrates two important principles: (1) Metabolism can be subdivided into two major categories—**catabolism,** those processes related to degradation of complex substances, and **anabolism,** those processes concerned primarily with the synthesis of complex organic molecules. (2) Both catabolic and anabolic pathways occur in three stages of complexity—stage 1, the interconversion of polymers and complex lipids with monomeric intermediates; stage 2, the interconversion of monomeric sugars, amino acids, and lipids with still simpler organic compounds; and stage 3, the ultimate degradation to, or synthesis from, inorganic compounds, including CO_2, H_2O, and NH_3. As we proceed through this chapter, we shall add detail to this figure, introducing you thereby to each major metabolic process and identifying the functions of each.

During our discussion we shall see that energy-yielding pathways also generate intermediates used in biosynthetic processes. Thus, although we will focus at first upon degradation of organic compounds to provide energy, you should be aware that metabolism is really a continuum, with many of the same reactions playing roles in both degradative and biosynthetic processes.

We shall also use the terms **intermediary metabolism, energy metabolism,** and **central pathways.** Intermediary metabolism comprises all reactions concerned with storing and generating metabolic energy and with using that energy in biosynthesis of low-molecular-weight compounds and energy storage compounds. Not included are nucleic acid and protein biosynthesis from monomeric precursors. The reactions of intermediary metabolism can be thought of as those that do not involve a nucleic acid template, because the information needed to specify each reaction is provided within the structure of the enzyme catalyzing that reaction. Energy metabolism is that part of intermediary metabolism consisting of pathways that store or generate metabolic energy. Chapters 12 through 23 present intermediary metabolism, and a major focus in Chapters 13 through 18 is on its energetic aspects. The central pathways of metabolism are substantially the same in many different organisms, and they account for relatively large amounts of mass transfer and energy generation within a cell; they are the quantitatively major pathways. Later we will identify and further discuss each central pathway, but for now let us focus on the overall process.

Intermediary metabolism refers primarily to the biosynthesis, utilization, and degradation of low-molecular-weight compounds.

Most organisms derive both the raw materials and the energy for biosynthesis from organic fuel molecules such as glucose. The central pathways involve the oxidation of fuel molecules and the synthesis of small biomolecules from the resulting fragments; these pathways are found in all aerobic organisms. But a fundamental distinction among these organisms lies in the source of their fuel molecules. **Autotrophs** (from Greek, "self-feeding") synthesize glucose and all of their other organic compounds from inorganic carbon, supplied as CO_2. In contrast, **heterotrophs** ("feeding on others") can synthesize their organic metabolites only from other organic compounds, which they must therefore consume. A primary difference between plants and animals is that plants are autotrophs and animals are heterotrophs. With the exception of rare insect-eating plants, such as the Venus flytrap, green plants obtain all of their organic carbon through photosynthetic fixation of carbon dioxide. Animals feed on plants or other animals and synthesize their metabolites by transforming the organic molecules they consume. Microorganisms exhibit a wide range of biosynthetic capabilities and sources of metabolic energy.

Microorganisms also show adaptability with respect to their ability to survive in the absence of oxygen. Virtually all multicellular organisms and many bacteria are strictly **aerobic** organisms; they depend absolutely upon **respiration,** the coupling of energy generation to the oxidation of nutrients by oxygen. By contrast, many microorganisms either can or must grow in **anaerobic** environments, deriving their metabolic energy from processes that do not involve molecular oxygen.

To the extent that biological molecules are synthesized ultimately from CO_2 that undergoes photosynthetic carbon fixation, the sun can be considered as the ultimate source of biological energy. However, this concept is not quite accurate, because of the existence of extremely **thermophilic** organisms, which live at temperatures as high as 100°C or more, either in hydrothermal vents deep in the ocean, or in geothermal vents in active volcanic regions. Although there is much to learn about metabolism in these thermophiles, it is clear that most of their metabolic energy does not come from sunlight.

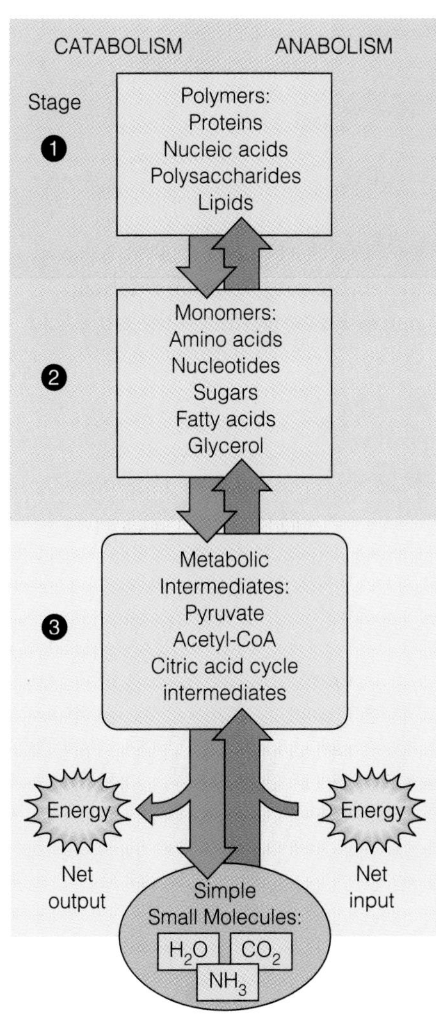

FIGURE **12.1**

A brief overview of metabolism.

Freeways on the Metabolic Road Map

Now let us return to the central pathways and their identification. Probably you have seen metabolic charts—those wall hangings like giant road maps that adorn biochemistry laboratories and offices. Figure 12.1 is a highly simplified metabolic chart. Figure 12.2 presents metabolism in more detail and is the basic road map for this section of the book. This figure will reappear in subsequent chapters, with the pathways presented in those chapters highlighted.

The central pathways account for most of the mass transformations in metabolism.

Faced with the thousands of individual reactions that constitute metabolism, how do we approach this vast topic? Our first concerns are with central pathways and with energy metabolism. Therefore, in the next several chapters we consider the degradative processes that are most important in energy generation—the catabolism of carbohydrates and lipids. We shall also consider how these substances are biosynthesized. These reactions are located in the middle of the metabolic charts and are illustrated with the biggest arrows—freeways, so to speak, on the metabolic road map.

The road map analogy is also useful when we consider directional flow in metabolism. Just as traffic flows from the suburbs to downtown in the morning and in the reverse direction in the evening, so also will we see that some conditions favor biosynthesis whereas others favor catabolism, and that parts of the same highways are used in both sets of processes.

CENTRAL PATHWAYS OF ENERGY METABOLISM

Recall from our discussion of Figure 12.1 that metabolism can be subdivided into three stages of complexity of the metabolites involved. The first pathway that we present in detail (in Chapter 13) is **glycolysis,** a stage 2 pathway for degradation of carbohydrates, in either aerobic or anaerobic cells. As schematized in Figure 12.3, the major input to glycolysis is glucose, usually derived from either energy storage polysaccharides or dietary carbohydrates. This pathway leads to pyruvate, a three-carbon keto acid. Anaerobic organisms reduce pyruvate to a variety of products—for example, ethanol and carbon dioxide. These processes are called *fermentations* (see page 422). In oxidative metabolism (respiration) the major fate of pyruvate is its oxidation to a metabolically activated two-carbon fragment, **acetyl-coenzyme A,** or acetyl-CoA (see page 494). The two carbons in the acetyl group then undergo oxidation in the **citric acid cycle** (Figure 12.4). In aerobic organisms the citric acid cycle, presented in Chapter 14, is the principal stage 3 pathway. This cyclic pathway accepts simple carbon compounds, derived not only from carbohydrate but also from lipid or protein, and oxidizes them to CO_2. Using the freeway analogy again, we will see that numerous on-ramps from the highways and byways of stage 1 and stage 2 metabolism lead to the citric acid cycle. In fact, all catabolic pathways converge at this point.

In aerobic organisms all catabolic pathways converge on the citric acid cycle.

Oxidative reactions of the citric acid cycle generate reduced electron carriers whose reoxidation drives ATP biosynthesis, primarily through processes in the mitochondrial respiratory chain—**electron transport** and **oxidative phosphorylation,** shown also in Figure 12.4. As described in Chapter 15, the mitochondrial membrane uses oxidative energy to maintain a transmembrane gradient of hydrogen ion concentration, and discharge of this osmotic energy is coupled to the synthesis of ATP from ADP.

Stage 2 pathways other than glycolysis also deliver fuel to the citric acid cycle. Acetyl-CoA comes not only from pyruvate oxidation but also from the breakdown of fatty acids by **β-oxidation** (Figure 12.5; presented in Chapter 18) and from some amino acid oxidation pathways (Figure 12.6; presented in Chapters 20 and 21). If the two carbons of acetyl-CoA are not oxidized in the citric acid cycle, they can go in the anabolic direction, providing substrates for synthesis of fatty acids and steroids (Figure 12.5; presented in Chapter 18). These and other biosynthetic processes use a reduced electron carrier, NADPH, which is structurally very similar to NADH (page 423).

Several important biosynthetic processes for carbohydrates will concern us (Figure 12.7). Chapter 16 presents **gluconeogenesis,** the synthesis of glucose from

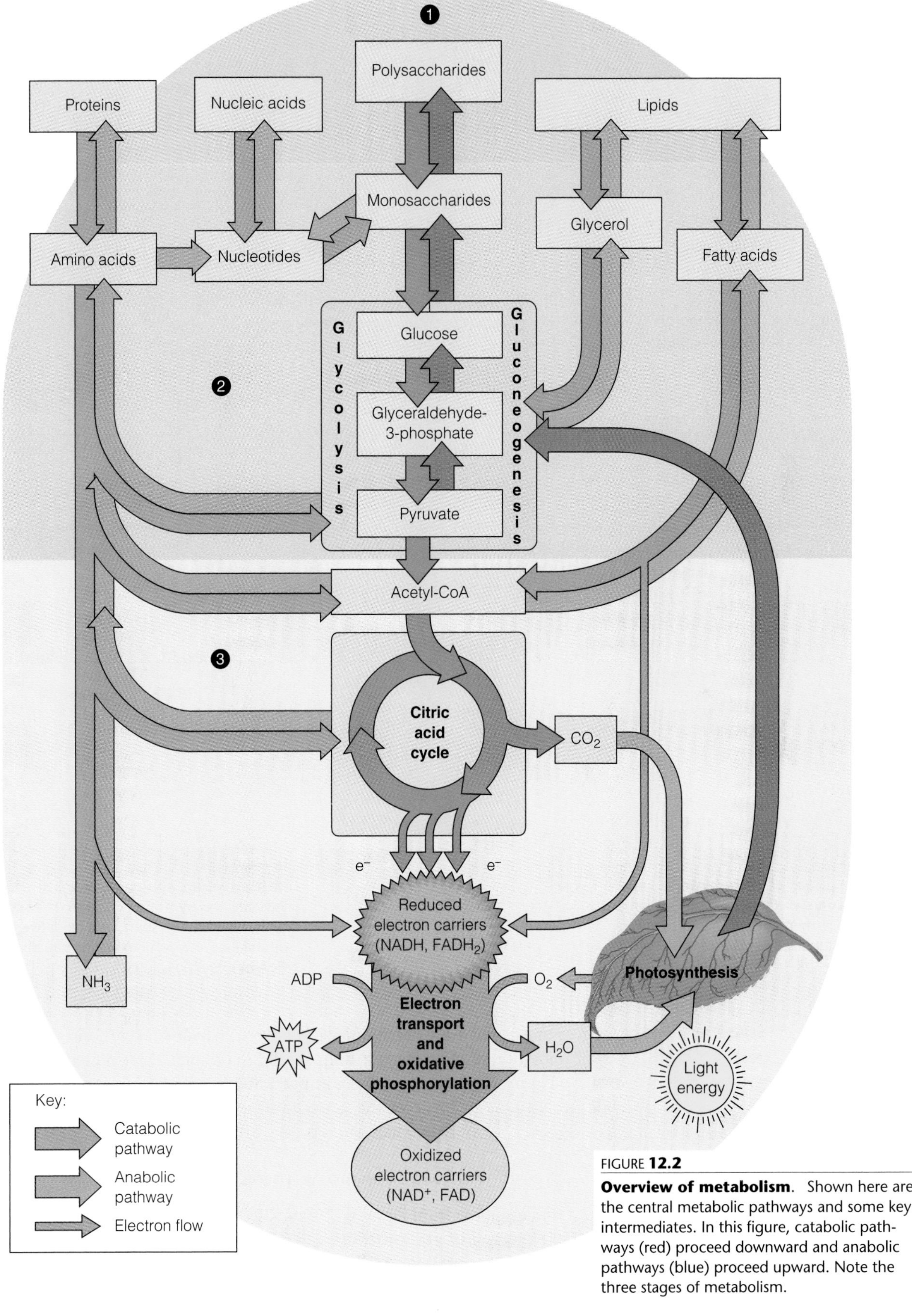

FIGURE **12.2**

Overview of metabolism. Shown here are the central metabolic pathways and some key intermediates. In this figure, catabolic pathways (red) proceed downward and anabolic pathways (blue) proceed upward. Note the three stages of metabolism.

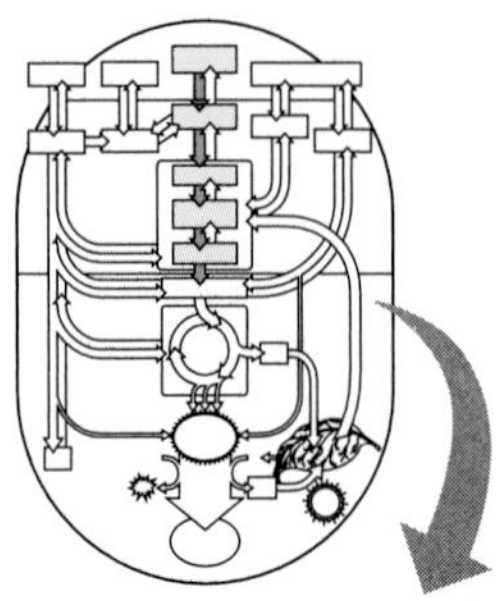

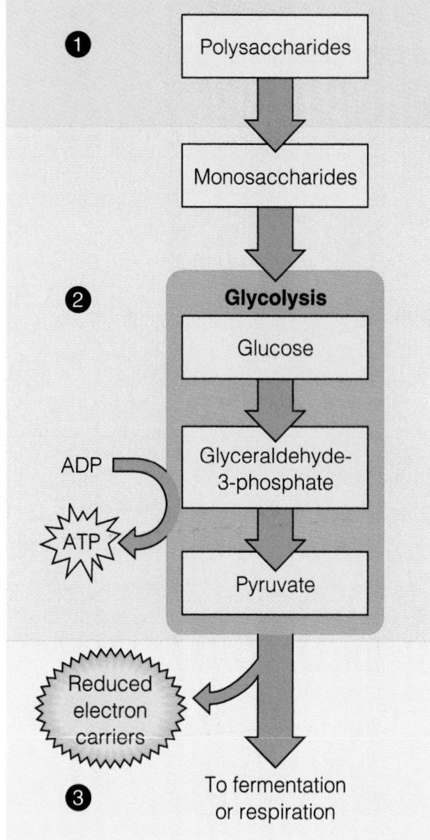

FIGURE **12.3**

The initial phase of carbohydrate catabolism: glycolysis. Pyruvate either undergoes reduction in fermentation reactions or enters oxidative metabolism (respiration) via conversion to acetyl-CoA.

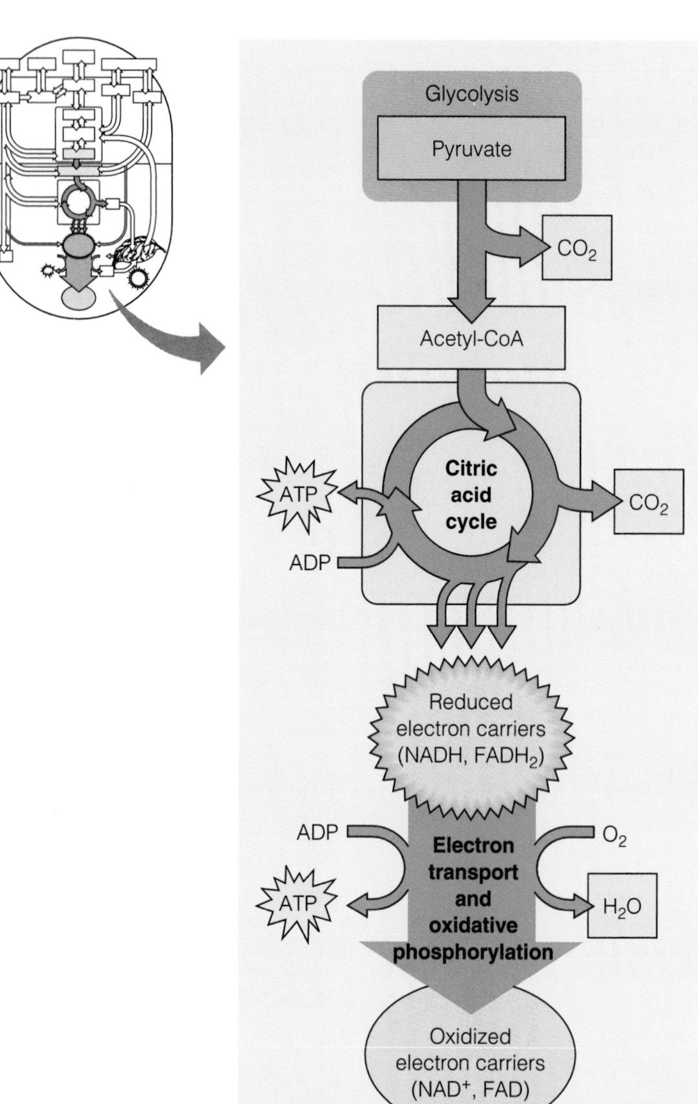

FIGURE **12.4**

Oxidative metabolism. Oxidative metabolism includes pyruvate oxidation, the citric acid cycle, electron transport, and oxidative phosphorylation. Pyruvate oxidation supplies acetyl-CoA to the citric acid cycle.

noncarbohydrate precursors, and polysaccharide biosynthesis, particularly the biosynthesis of glycogen in animal cells. In Chapter 17 we present **photosynthesis** (Figure 12.8), the supremely important process by which green plants capture light energy to drive the generation of energy (ATP) and reducing power (NADPH), both of which are used for carbohydrate synthesis.

DISTINCT PATHWAYS FOR BIOSYNTHESIS AND DEGRADATION

It may appear from Figures 12.1 and 12.2 that some pathways operate simply as the reversal of other pathways. For example, fatty acids are synthesized from acetyl-CoA, but they are also converted to acetyl-CoA by β-oxidation. Similarly, glucose-6-phosphate is synthesized from pyruvate in gluconeogenesis, which looks at first glance like a simple reversal of glycolysis. It is important to realize

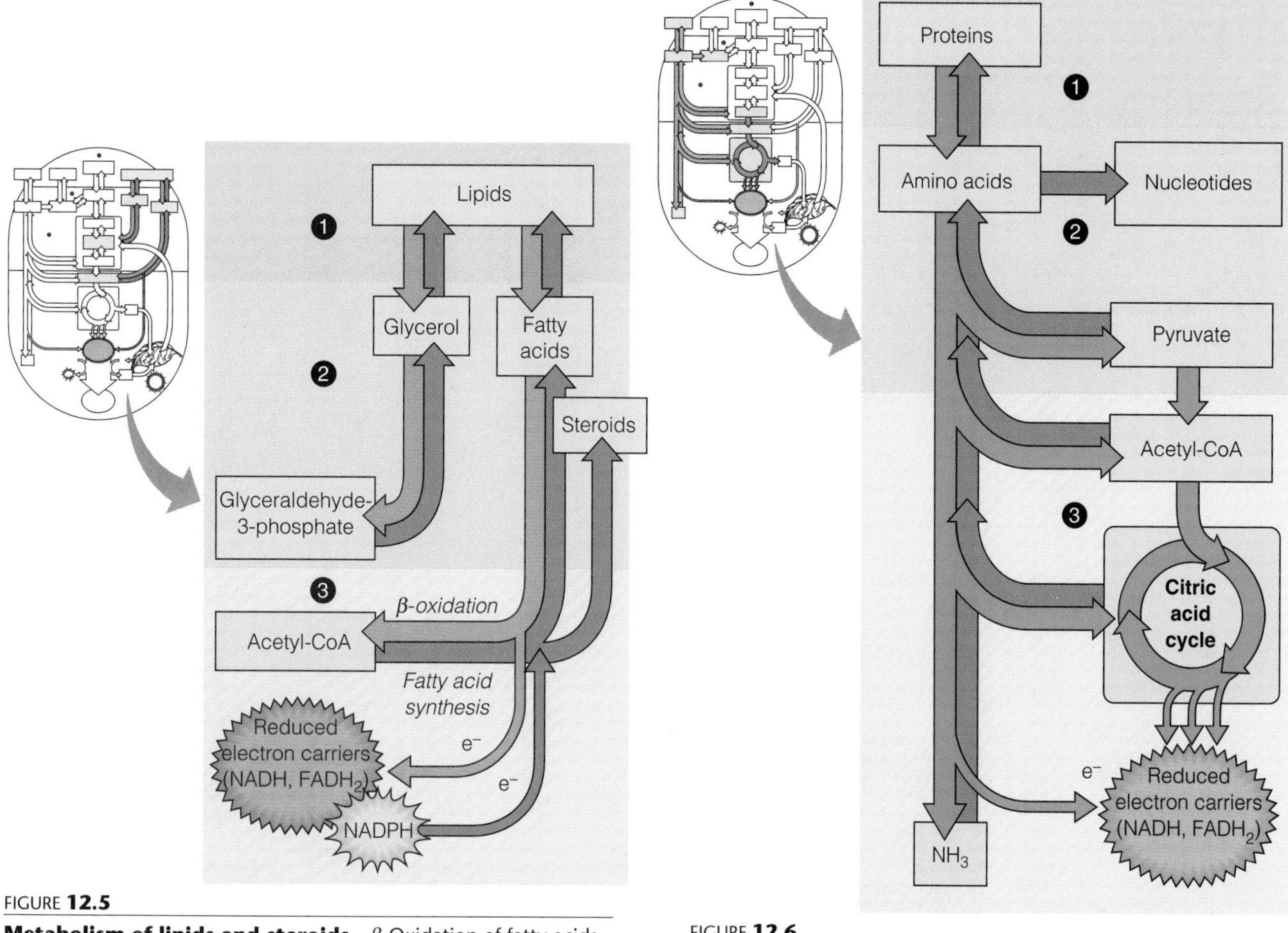

FIGURE **12.5**

Metabolism of lipids and steroids. β-Oxidation of fatty acids supplies acetyl-CoA to the citric acid cycle. Reductive biosynthetic reactions use NADPH as the coenzyme (see page 423). Pathways leading to membrane phospholipids are not shown here.

FIGURE **12.6**

Metabolism of amino acids. Amino acids supply fuel to the citric acid cycle via several pathways and are also used in biosynthesis of nucleotides and many other metabolites.

that in these cases the opposed pathways are quite distinct from one another. They may share some common intermediates or enzymatic reactions, but they are separate reaction sequences, regulated by distinct mechanisms and with different enzymes catalyzing their regulated reactions. They may even occur in separate cellular compartments. For example, fatty acid synthesis takes place in cytosol, whereas fatty acid oxidation takes place in mitochondria.

Biosynthetic and degradative pathways are rarely, if ever, simple reversals of one another, even though they often begin and end with the same metabolites. The existence of separate pathways is important for two reasons. First, to proceed in a particular direction, a pathway must be exergonic in that direction. If a pathway is strongly exergonic, then reversal of that pathway is just as strongly endergonic under the same conditions.

Degradative and biosynthetic pathways are distinct for two reasons: A pathway can be exergonic in only one direction, and pathways must be separately regulated to avoid futile cycles.

Second and equally important is the need to control the flow of metabolites in relation to the bioenergetic status of a cell. When ATP levels are high, there is less need for carbon to be oxidized in the citric acid cycle. At such times the cell can store carbon as fats and carbohydrates, so fatty acid synthesis, gluconeogenesis, and related pathways come into play. When ATP levels are low, the cell must mobilize stored carbon to generate substrates for the citric acid cycle, so carbohydrate and fat breakdown must occur. Using separate pathways for the biosynthetic and degradative processes is crucial for control, so conditions that activate one pathway tend to inhibit the opposed pathway and vice versa.

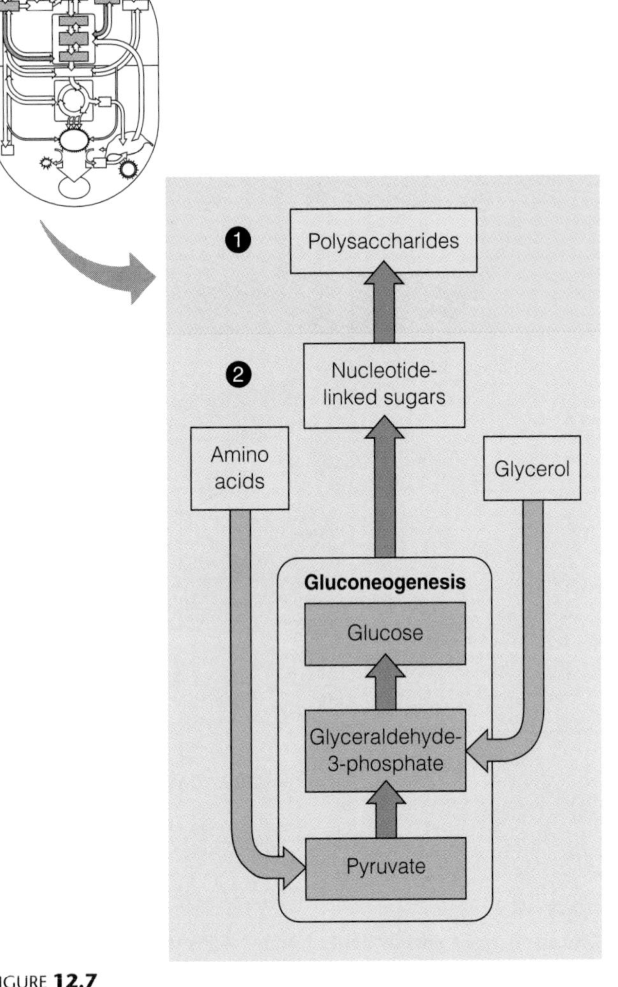

FIGURE **12.7**

Carbohydrate anabolism. Biosynthesis of carbohydrates includes gluconeogenesis and polysaccharide synthesis.

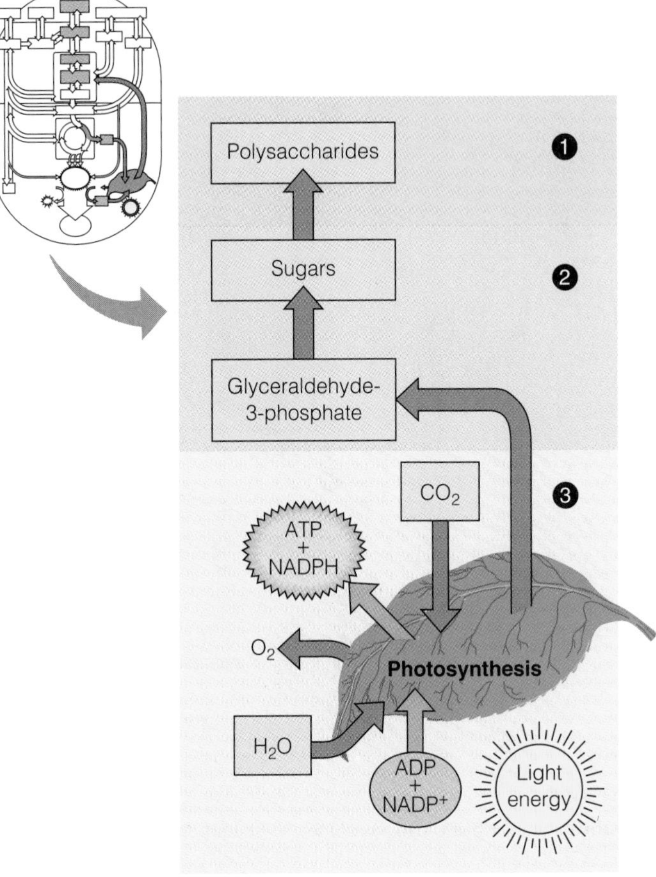

FIGURE **12.8**

Photosynthesis.

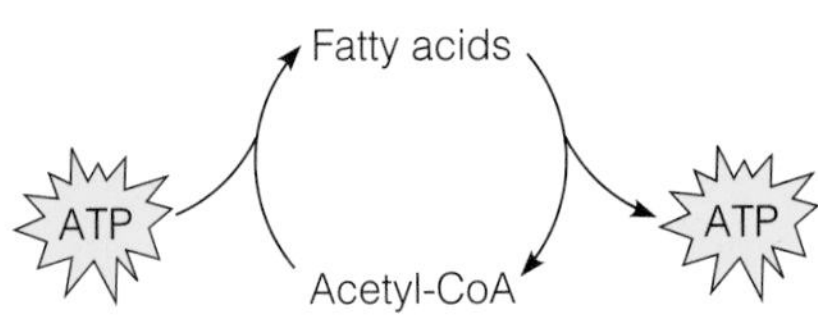

Compartmentation and allosteric control of anabolic and catabolic processes prevent futile cycles, which simply waste energy.

Consider what would happen, for example, if fatty acid synthesis and oxidation took place in the same cell compartment and in an uncontrolled fashion. Two-carbon fragments released by oxidation would be immediately used for resynthesis, a situation called a **futile cycle.** No useful work is done, and the net result is simply consumption of the ATP used in the endergonic reactions of fatty acid synthesis. A similar futile cycle could result from the interconversion of fructose-6-phosphate with fructose-1,6-bisphosphate in carbohydrate metabolism.

$$\text{Fructose-6-phosphate} + \text{ATP} \longrightarrow \text{fructose-1,6-bisphosphate} + \text{ADP}$$

$$\text{Fructose-1,6-bisphosphate} + H_2O \longrightarrow \text{fructose-6-phosphate} + P_i$$

$$\text{Net:} \quad \text{ATP} + H_2O \longrightarrow \text{ADP} + P_i$$

The first reaction occurs in glycolysis, and the second participates in a biosynthetic pathway, gluconeogenesis. Both processes occur in the cytosol. The net effect of carrying out both reactions simultaneously would be the wasteful hydrolysis of ATP to ADP and P_i. However, enzymes catalyzing both of the above reactions respond to allosteric effectors, such that one enzyme is inhibited by conditions that activate the other. Thus, effective control prevents the futile cycle from occurring, even though the two enzymes occupy the same cell compartment. Therefore, it is more appropriate to call this situation—two seemingly opposed cellular reactions that are independently controlled—a **substrate cycle.**

Studies on metabolic control suggest that a substrate cycle represents an efficient regulatory mechanism, because a small change in the activity of either or both enzymes can have a much larger effect on the flux of metabolites in one direction or the other. You can test this idea in Problem 6 at the end of this chapter.

Some Bioenergetic Considerations

OXIDATION AS A METABOLIC ENERGY SOURCE

Because the primary focus of the next few chapters will be on energy metabolism, we will now consider briefly how metabolic energy is generated. As we saw in Chapter 3, a thermodynamically unfavorable, or endergonic, reaction will proceed smoothly in the unfavored direction only if it can be coupled to a thermodynamically favorable, or exergonic, reaction. In principle, any exergonic reaction can serve this purpose, provided that it releases sufficient free energy. In living systems most of the energy needed to drive biosynthetic reactions is derived from the *oxidation* of organic substrates. Oxygen, the ultimate electron acceptor for aerobic organisms, is a strong oxidant; it has a marked tendency to attract electrons, becoming reduced in the process. Given this tendency and the abundance of oxygen in our atmosphere, it is not surprising that living systems have gained the ability to derive energy from the oxidation of organic substrates.

Biological Oxidations: Energy Release in Small Increments

In a thermodynamic sense, the biological oxidation of organic substrates is comparable to nonbiological oxidations such as the burning of wood. The free energy release is the same whether we are talking about oxidation of the glucose polymer cellulose in a wood fire, combustion of glucose in a calorimeter, or the metabolic oxidation of glucose:

Most biological energy derives from oxidation of reduced metabolites in a series of reactions, with oxygen as the final electron acceptor.

$$C_6H_{12}O_6 + 6O_2 \longrightarrow 6CO_2 + 6H_2O \qquad \Delta G^{\circ\prime} = -2870 \text{ kJ/mol}$$

Biological oxidations, however, are far more complex processes than combustion. When wood is burned, all of the energy is released as heat; useful work cannot be performed, except through the action of a device such as a steam engine. In biological oxidations, by contrast, oxidation reactions occur without a large increase in temperature and with capture of some of the free energy as chemical energy. This energy capture occurs largely through the synthesis of ATP, and, as discussed in Chapter 3, the hydrolysis of ATP can be coupled to many processes to provide energy for biological work. In catabolism of glucose, about 40% of the released energy is used to drive the synthesis of ATP from ADP and P_i.

Unlike the oxidation of glucose by oxygen shown in the previous equation, most biological oxidations do not involve direct transfer of electrons from a reduced substrate to oxygen (Figure 12.4). Rather, a series of coupled oxidation–reduction reactions occurs, with the electrons passed to intermediate electron carriers such as NAD^+ and finally transferred to oxygen. The reaction sequence is called the **electron transport chain**, or **respiratory chain**, and oxygen is called the **terminal electron acceptor.** Because the potential energy stored in the organic substrate is released in small increments, it is easier to control oxidation and capture some of the energy as it is released—small energy transfers waste less energy than a single large transfer does. This situation is somewhat analogous to the generation of hydroelectric power, in which a series of small dams on a river generates more electricity than a single large dam, even though the total distance through which the water drops is the same in both cases.

Not all metabolic energy comes from oxidation by oxygen. Substances other than oxygen can serve as terminal electron acceptors. Many microorganisms either can or must live **anaerobically** (in the absence of oxygen). Most of these organisms derive their energy from **fermentations,** which are energy-yielding catabolic pathways that proceed with no net change in the oxidation state of the products as compared with that of the substrates. A good example is the production of ethanol and CO_2 from glucose, presented in Chapter 13. Other anaerobic energy-yielding pathways are seen in some deep-sea hydrothermal vent bacteria, which reduce sulfur to sulfide as the terminal electron transfer reaction, and in other bacteria that reduce nitrite to ammonia. These organisms oxidize the substrates that sustain them, but they use electron acceptors other than oxygen.

Energy Yields, Respiratory Quotients, and Reducing Equivalents

If metabolic energy comes primarily from oxidative reactions, it follows that the more highly reduced a substrate, the higher its potential for generating biological energy. We can use a calorimeter to measure the heat output (enthalpy) from oxidation of fat, carbohydrate, or protein. The combustion of fat provides more heat energy than the combustion of an equivalent amount of carbohydrate. In other words, fat has a higher **caloric content** than carbohydrate. For illustration, compare the oxidation of glucose with the oxidation of a typical saturated fatty acid, palmitic acid:

$$C_6H_{12}O_6 + 6O_2 \longrightarrow 6CO_2 + 6H_2O \qquad \Delta G^{\circ\prime} = -3.74 \text{ kcal/g}$$

$$C_{16}H_{32}O_2 + 23O_2 \longrightarrow 16CO_2 + 16H_2O \qquad \Delta G^{\circ\prime} = -9.30 \text{ kcal/g}$$

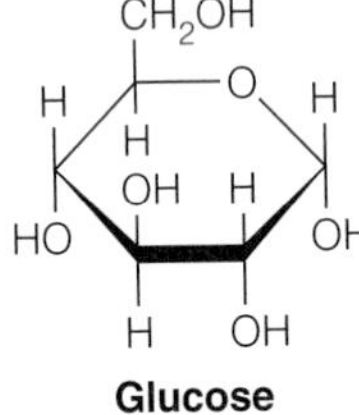

Glucose

$CH_3(CH_2)_{14}COOH$

Palmitic acid

Converting calories (the units of nutrition) to joules (the units of biochemistry), we see that the oxidation of glucose yields 15.64 kJ/g, and the oxidation of palmitic acid yields 38.90 kJ/g. The carbons in fat are in general more highly reduced than those in carbohydrate; thus, they contain more protons and electrons to combine with oxygen on the path to CO_2 than do the carbons in sugar. We can see this by counting oxygen atoms. Glucose has more oxygens per carbon than does palmitic acid; each carbon in glucose is linked to at least one oxygen atom.

We can also tell that glucose is the more highly oxidized substance, because its oxidation produces more moles of CO_2 per mole of O_2 consumed during oxidation, a ratio called the **respiratory quotient,** or **RQ.** The above equations show RQ for glucose to be 1.0 ($6CO_2/6O_2$), whereas that for palmitic acid is 0.70 ($16CO_2/23O_2$). In general, the lower the RQ for a substrate, the more oxygen consumed per carbon oxidized and the greater the potential per mole of substrate for generating ATP.

Another way to express the degree of substrate oxidation is to say that more **reducing equivalents** are derived from oxidation of fat than from oxidation of carbohydrate. A reducing equivalent can be defined as 1 mole of hydrogen atoms (one proton and one electron per H atom). For example, two reducing equivalents are used in the reduction of one-half mole of oxygen to water:

$$\tfrac{1}{2}O_2 + 2e^- + 2H^+ \longrightarrow H_2O$$

Whereas the breakdown of complex organic compounds yields both energy and reducing equivalents, the biosynthesis of such compounds utilizes both. For example, we know that both carbons of acetate are used for fatty acid biosynthesis:

$$\underset{\textbf{Acetate}}{8CH_3COO^-} \longrightarrow \longrightarrow \longrightarrow \underset{\textbf{Palmitate}}{CH_3(CH_2)_{14}COO^-}$$

Fifteen of the 16 carbon atoms of palmitate are highly reduced—14 at the methylene level and 1 at the methyl level. Therefore, many reducing equivalents are required to complete this biosynthesis.

The major source of electrons for reductive biosynthesis is **NADPH, nicotinamide adenine dinucleotide phosphate (reduced).** $NADP^+$ and NADPH are identical to NAD^+ and NADH, respectively, except that the former have an additional phosphate esterified at C-2′ on the adenylate moiety. NAD^+ and $NADP^+$ are equivalent in their thermodynamic tendency to accept electrons; they have equal standard reduction potentials (discussed further in Chapter 15). However, for reasons still not clear, nicotinamide nucleotide–linked enzymes that act primarily in a catabolic direction usually use the NAD^+–NADH pair, whereas those acting primarily in anabolic pathways use $NADP^+$ and NADPH. In other words, as shown in Figure 12.9, nicotinamide nucleotide–linked enzymes that oxidize substrates (dehydrogenases) usually use NAD^+, and those enzymes that reduce substrates (reductases) usually use NADPH. An exception is two dehydrogenases in the **pentose phosphate pathway** (see Chapter 14), which convert $NADP^+$ to NADPH and represent the principal route for synthesis of the reduced nucleotide.

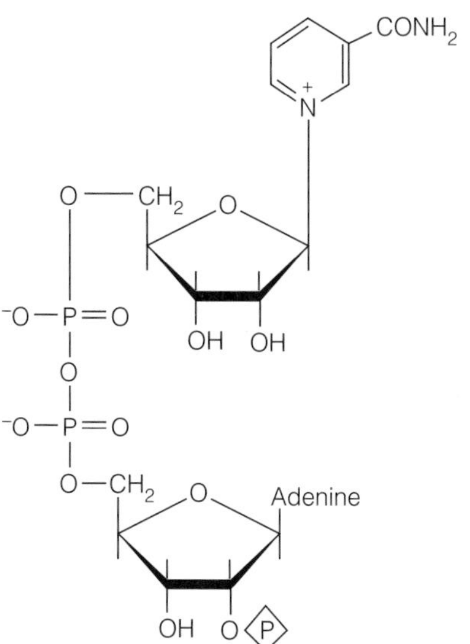

NADP⁺
Nicotinamide adenine dinucleotide phosphate (oxidized)

NAD^+ is the cofactor for most dehydrogenases that oxidize metabolites. NADPH is the cofactor for most reductases.

ATP AS A FREE ENERGY CURRENCY

ATP is commonly referred to as a "free energy currency." What does this mean? Currency is a medium of exchange. A twenty-dollar bill has a generally recognized value that can be readily exchanged for a variety of goods or services—for example, four pairs of socks, or dinner at a moderately priced restaurant, or about a half-hour's labor by a skilled auto mechanic. In the same sense cells exchange the energy released from the breakdown of ATP to carry out essential functions, often converting the chemical energy released in ATP hydrolysis to other forms of energy—mechanical energy, for example, in muscle contraction, or electrical

FIGURE **12.9**

Nicotinamide nucleotides in catabolism and biosynthesis. NAD^+ is the cofactor for most enzymes that act in the direction of substrate oxidation (dehydrogenases), whereas NADPH usually functions as a cofactor for reductases, enzymes that catalyze substrate reduction. NADPH is synthesized either from $NADP^+$ in the pentose phosphate pathway (see Chapter 14) or from NADH through the action of mitochondrial energy-linked transhydrogenase (see Chapter 15). $NADP^+$ is synthesized from NAD^+ by an ATP-dependent kinase reaction.

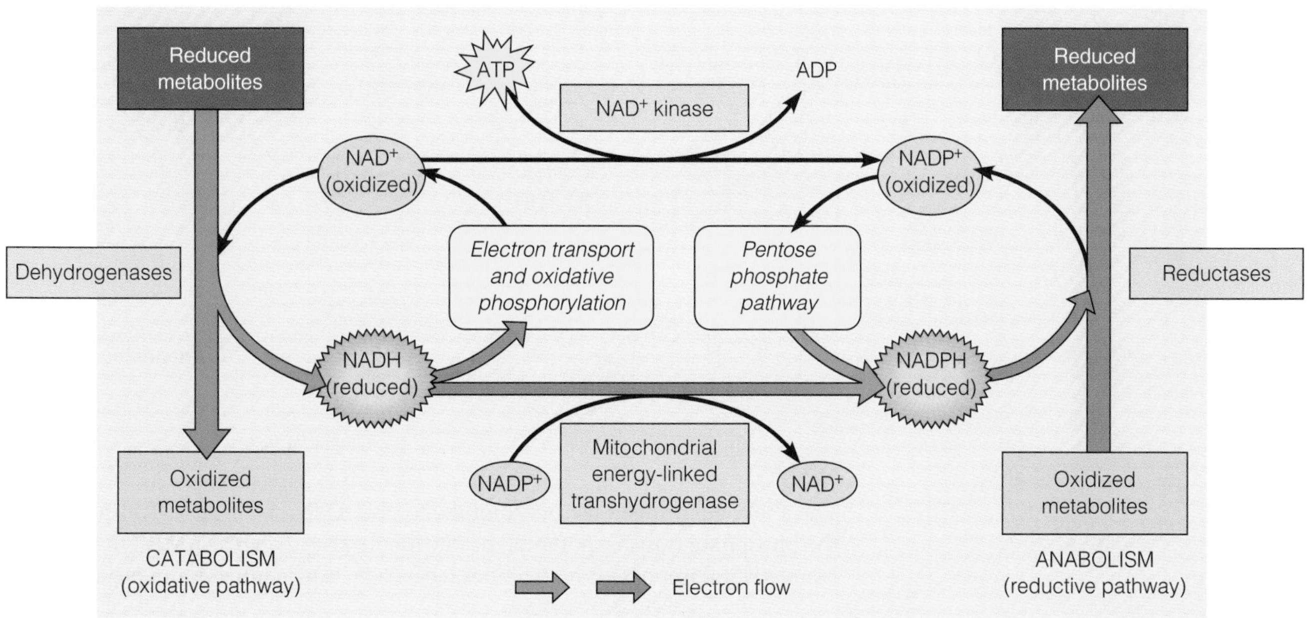

energy in conducting nerve impulses, or osmotic energy in transporting substances across membranes against a concentration gradient.

Thermodynamic Properties of ATP

What factors equip ATP for its special role as energy currency? First, there is nothing unique about the chemistry of ATP; the phosphoanhydride bonds whose breakdown can be coupled to drive endergonic reactions are shared by all other nucleoside di- and triphosphates and by several other metabolites. When we call ATP a "high-energy compound," as we did in Chapter 3, we use that term within a defined context; i.e., a high-energy compound is one containing at least one bond with a high $\Delta G^{\circ\prime}$ of hydrolysis. ATP has two phosphoanhydride bonds. Cleavage of one yields adenosine diphosphate (ADP) and inorganic phosphate (P_i), and cleavage of the other gives adenosine monophosphate (AMP) and pyrophosphate (PP_i). Either reaction proceeds with $\Delta G^{\circ\prime}$ of hydrolysis of about −31 kJ/mol. Note, however, that calling a substance a high-energy compound does not mean that it is chemically unstable or unusually reactive. In fact, ATP is a relatively stable compound; its spontaneous hydrolysis is slow, but when hydrolysis does occur, whether spontaneously or enzyme-catalyzed, substantial free energy is released. Be aware, however, that utilization of that energy to drive endergonic reactions usually does *not* involve hydrolysis. Rather, ATP breakdown is usually coupled with a thermodynamically unfavorable reaction, such as the synthesis of glucose-6-phosphate from glucose. In this case the phosphate released from ATP does not become P_i but instead is transferred to glucose, forming the esterified phosphate of glucose-6-phosphate. Thus, it is probably more accurate to say that ATP has a high "phosphoryl group transfer potential" than to call it a high-energy compound. So long as you understand the context in which the term is used, the concept of a high-energy compound is quite useful.

Recall from Chapter 3 that several factors contribute to the thermodynamic stability of a hydrolyzable bond and determine whether $\Delta G^{\circ\prime}$ is relatively high, as in the phosphate anhydride bonds of ATP, ADP, or pyrophosphate, or relatively low, as in the phosphate ester bonds of glucose-6-phosphate or AMP. These factors include electrostatic repulsion among the negative charges in the molecule before hydrolysis, resonance stabilization of the products of hydrolysis, and the tendency of the hydrolysis products to deprotonate (see page 77). By contrast, the hydrolysis of a phosphate ester, such as AMP, generates an alcohol (the sugar 5′ hydroxyl), which has almost no tendency to lose a proton.

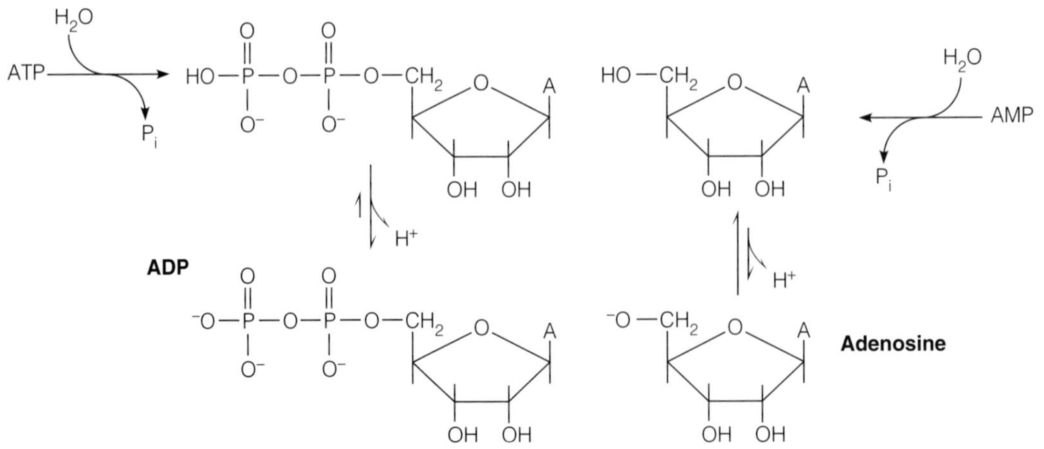

The above-mentioned factors combine to give ATP a $\Delta G^{\circ\prime}$ of −31 kJ/mol, twice the phosphate transfer potential of phosphate esters such as AMP. On the other hand, several important metabolites have $\Delta G^{\circ\prime}$ values much more negative than that of ATP. Examples include phosphoenolpyruvate, 1,3-bisphosphoglycerate, and creatine phosphate (see Figure 3.7), with $\Delta G^{\circ\prime}$ values of -62, -49, and -43 kJ/mol, respectively. This means that ATP is actually intermediate on the scale of "energy wealth." This is important also, because it means that the breakdown of a "super-high-energy" compound such as phosphoenolpyruvate can be coupled to drive the synthesis of ATP itself from ADP and P_i. In fact, such coupled reactions, called **substrate-level phosphorylation** reactions, represent the process by which ATP is synthesized in glycolysis, as we discuss in Chapter 13.

The Importance of Differences Between $\Delta G'$ and $\Delta G^{\circ\prime}$

But what provides the energy for synthesis of compounds with a much higher phosphate transfer potential than that of ATP itself? Much of the answer lies in the fact that $\Delta G'$ values under intracellular conditions are quite different from standard ($\Delta G^{\circ\prime}$) values. This is mainly because intracellular concentrations are far different from the 1 M concentrations used to compute standard free energies. If you work Problem 5 at the end of this chapter, you will see that ATP hydrolysis has a considerably higher $\Delta G'$ value at intracellular concentrations of ATP, ADP, and AMP than it has under standard conditions. This allows ATP to drive the synthesis of the more energy-rich compound creatine phosphate. This compound shuttles phosphate bond energy from ATP in mitochondria to myofibrils, where that bond energy is transduced to the mechanical energy of muscle contraction.

Creatine phosphate is synthesized within the mitochondrion by the enzyme **creatine kinase:**

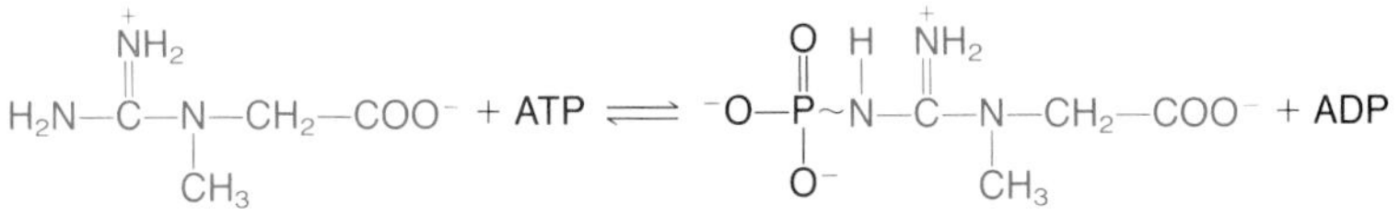

$$H_2N{-}\overset{\overset{+}{N}H_2}{\overset{\|}{C}}{-}\underset{CH_3}{\underset{|}{N}}{-}CH_2{-}COO^- + ATP \rightleftharpoons {}^-O{-}\underset{O^-}{\underset{|}{\overset{O}{\overset{\|}{P}}}}{\sim}\overset{H}{\overset{|}{N}}{-}\overset{\overset{+}{N}H_2}{\overset{\|}{C}}{-}\underset{CH_3}{\underset{|}{N}}{-}CH_2{-}COO^- + ADP$$

Creatine **Creatine phosphate**

From the respective $\Delta G^{\circ\prime}$ values for creatine phosphate and ATP, you can calculate that this reaction is endergonic under standard conditions, with a $\Delta G^{\circ\prime}$ value of $+12$ kJ/mol. However, because ATP levels are very high within mitochondria, and creatine phosphate levels are relatively low, the reaction is exergonic as written and proceeds to the right. Creatine phosphate then diffuses from mitochondria to the myofibrils, where it provides the energy for muscle contraction. However, the direct energy source for that contraction is ATP hydrolysis once again. High levels of ADP formed during contraction favor the reverse reaction—resynthesis of ATP—at the expense of creatine phosphate cleavage to creatine, which can then return to mitochondria for resynthesis of creatine phosphate. The popularity of creatine as a dietary supplement for athletes, to increase muscle strength, suggests that the biosynthesis of creatine itself (see Chapter 20 for the pathway) may be a limiting factor in operating this intracellular energy shuttle.

ATP can drive the synthesis of higher-energy compounds, if nonequilibrium intracellular concentrations make such reactions exergonic.

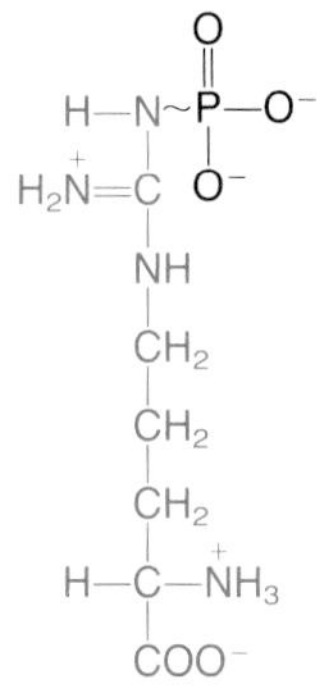

In some invertebrate animals **arginine phosphate,** instead of creatine phosphate, plays a similar role in storing high-energy phosphate for rapid production, as needed, of ATP. In both instances, the varying concentrations of adenine nucleotides in the two environments—mitochondria and myofibrils—both at levels far from standard values, are critical in understanding how "super-high-energy" compounds can be synthesized and utilized.

$$\begin{array}{c} H{-}N{\sim}\overset{O}{\overset{\|}{P}}{-}O^- \\ H_2\overset{+}{N}{=}C\quad \underset{|}{O^-} \\ | \\ NH \\ | \\ CH_2 \\ | \\ CH_2 \\ | \\ CH_2 \\ | \\ H{-}C{-}\overset{+}{N}H_3 \\ | \\ COO^- \end{array}$$

Arginine phosphate

But factors in addition to concentration combine to make biological ΔGs quite different from standard values. For example, as pH increases, the negative charge on the ATP molecule increases. This in turn increases the electrostatic repulsion between oxygen atoms linked to adjacent phosphorus atoms, and this in turn promotes hydrolysis, thus making ΔG more negative.

Also significant is the fact that most ATP is chelated within cells, as a complex with Mg^{2+}:

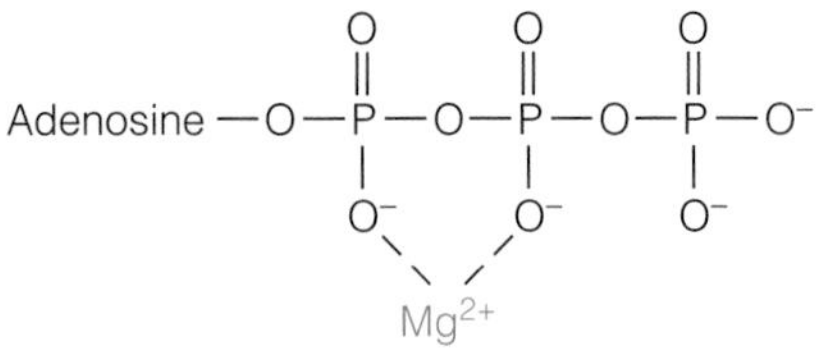

$$\text{Adenosine}-O-\overset{\overset{\displaystyle O}{\|}}{\underset{\underset{\displaystyle O^-}{|}}{P}}-O-\overset{\overset{\displaystyle O}{\|}}{\underset{\underset{\displaystyle O^-}{|}}{P}}-O-\overset{\overset{\displaystyle O}{\|}}{\underset{\underset{\displaystyle O^-}{|}}{P}}-O^- \quad (\alpha\text{- and }\beta\text{-}O^- \text{ chelating } Mg^{2+})$$

ADP is also complexed with Mg^{2+}, but ADP has a different affinity for Mg^{2+} from that of ATP. Varying levels of Mg^{2+} will change ΔG in complicated ways, depending upon relative affinities of reactants and products for the magnesium ion. (For a detailed discussion of the effects of pH, magnesium, and other ionic conditions, see the article by R. Alberty cited in the Chapter 3 references.)

Other High-Energy Nucleotides

As discussed earlier, there is nothing unique about the properties endowing ATP with its special role as energy currency. All other nucleoside triphosphates, as well as more complex nucleotides, such as NAD^+, have $\Delta G^{\circ\prime}$ values close to -31 kJ/mol and could have been selected for the role played by ATP. However, evolution has created an array of enzymes that preferentially bind ATP and use its free energy of hydrolysis to drive endergonic reactions. There are exceptions, such as the use of GTP as the primary energy-providing nucleotide in protein synthesis. But phosphate bond energy is created almost exclusively at the adenine nucleotide level, through oxidative phosphorylation in aerobic cells, photosynthesis in plants, and glycolysis in virtually all organisms. As a result, ATP is usually the most abundant nucleotide.

In most cells, ATP levels, at 2–8 mM, are severalfold higher than those of the other nucleoside triphosphates and also severalfold higher than the levels of ADP or AMP. These factors give ATP a strong tendency to distribute its γ (outermost) phosphate in the synthesis of other nucleoside triphosphates. This is accomplished through the action of **nucleoside diphosphate kinase,** which synthesizes CTP from CDP in the following example.

$$\text{ATP} + \text{CDP} \rightleftharpoons \text{ADP} + \text{CTP}$$

Nucleoside diphosphate kinase is active with a wide variety of phosphate donors and acceptors. Because its equilibrium constant is close to unity and because ATP is the most abundant nucleotide within cells, the enzyme normally uses ATP to drive the synthesis of the other common ribo- and deoxyribonucleoside triphosphates from their respective diphosphates.

Some metabolic reactions, such as the activation of amino acids for protein synthesis, cleave ATP, not to ADP and P_i, but to AMP and PP_i. The conversion of AMP to ATP, allowing reuse of the nucleotide, involves another enzyme, **adenylate kinase** (also called myokinase because of its abundance in muscle).

$$\text{AMP} + \text{ATP} \rightleftharpoons 2\text{ADP}$$

ADP is reconverted to ATP by substrate-level phosphorylation, oxidative phosphorylation, or (in plants) photosynthetic energy. Because the reaction is readily reversible, it can also be used for resynthesis of ATP when ADP levels rise, for example, after a burst of energy consumption. This function is particularly important in muscle metabolism.

Adenylate Energy Charge

As noted above, ATP levels are normally much higher than the levels of ADP and AMP in well-nourished, energy-sufficient cells. Many enzymes that participate in regulating energy-generating or storage pathways are acutely sensitive to concentrations of adenine nucleotides. In general, energy-generating pathways, such as glycolysis and the citric acid cycle, are activated at low energy states, when levels of ATP are relatively low and those of ADP and AMP are relatively high. It is useful to be able to describe the energy status of a cell in quantitative terms. Daniel Atkinson has likened the cell to a battery. When the cellular battery is fully charged, all of the adenine ribonucleotides are present in the form of ATP. When fully discharged, all of the ATP has been broken down to AMP. Atkinson has proposed the term **adenylate energy charge,** which is defined in terms of the intracellular concentrations of ATP, ADP, and AMP.

$$\text{Adenylate energy charge} = \frac{[\text{ATP}] + 0.5[\text{ADP}]}{[\text{ATP}] + [\text{ADP}] + [\text{AMP}]} \qquad (12.1)$$

The term is the proportion of total energy-rich bonds that could be present in the adenine nucleotides in a cell to those that are actually present. Note that ADP has one energy-rich bond compared with two in ATP, so it carries one-half the weight of ATP in the numerator of equation (12.1). Well-nourished aerobic cells normally have adenylate energy charge values close to 0.9, as expected if ATP levels are far higher than those of ADP or AMP.

Major Metabolic Control Mechanisms

The living cell uses a marvelous array of regulatory devices to control its functions. These mechanisms include those that act primarily to control enzyme *activity,* such as substrate concentration and allosteric control, as discussed in Chapter 11. Control of enzyme *concentration,* through regulation of enzyme synthesis and degradation, is the major focus of Chapters 26–28. In eukaryotic cells, *compartmentation* represents another regulatory mechanism, with the fate of a metabolite being controlled by the flow of that metabolite through a membrane. Overlying all of these mechanisms are the actions of *hormones,* chemical messengers that act at all levels of regulation.

CONTROL OF ENZYME LEVELS

If you were to prepare a cell-free extract of a particular tissue and determine intracellular concentrations of several different enzymes, you would find tremendous variations. Enzymes of the central energy-generating pathways are present at many thousands of molecules per cell, whereas enzymes that have limited or specialized functions might be present at fewer than a dozen molecules per cell. Two-dimensional gel electrophoresis of a cell extract (see Tools of Biochemistry 12A) gives an impression, from the varying spot intensities, of the wide variations in amounts of individual proteins in a particular cell.

The level of a single enzyme can also vary widely under different environmental conditions, in large part because of variation in the enzyme's rate of synthesis. For example, when a usable substrate is added to a bacterial culture, the abundance of the enzymes needed to process the substrate may increase, through synthesis of new protein, from less than one molecule per cell to many thousands of molecules per cell. This phenomenon is called enzyme **induction.** Similarly, the presence of the end product of a pathway may turn off the synthesis of enzymes needed to generate that end product, a process called **repression.**

Enzyme levels in a cell may change in response to changes in metabolic needs.

For some time it was thought that controlling the intracellular level of a protein was primarily a matter of controlling the *synthesis* of that protein—in other words, through genetic regulation. We now know that intracellular protein *degradation* is also important in determining enzyme levels.

CONTROL OF ENZYME ACTIVITY

Enzyme activity is regulated by interaction with substrates, products, and allosteric effectors and by covalent modification of enzyme protein.

The *catalytic activity* of an enzyme molecule can be controlled in two ways: by reversible interaction with ligands (substrates, products, or allosteric modifiers) and by covalent modification of the protein molecule.

Ligands that control enzyme activity can be polymers. For example, protein–protein interactions can affect enzyme activity, and several enzymes of nucleic acid metabolism are activated by binding to DNA. Here our concern is with low-molecular-weight ligands, principally substrates and allosteric effectors. Substrates are usually present within cells at concentrations lower than the K_M values for the enzymes that act on them, but within an order of magnitude of these values. In other words, substrate concentrations usually lie within the first-order ranges of substrate concentration–velocity curves for the enzymes that act on them. Therefore, enzyme activities respond to small changes in substrate concentration.

We saw in Chapter 11 that allosteric activation or inhibition usually acts on committed steps of a metabolic pathway, often initial reactions. The effectors function by binding at specific regulatory sites, thereby affecting subunit–subunit interactions in the enzyme protein. This effect in turn either facilitates or hinders the binding of substrates. Such a mechanism operates in an obvious way to control product formation if a pathway is unidirectional and unbranched. However, some substrates are involved in numerous pathways, so many branch points exist. Therefore, some of the allosteric enzymes that we will describe display somewhat more complicated regulation than the examples presented in Chapter 11.

Covalent modification of enzyme structure represents another efficient way to control enzyme activity. Three types of covalent modification are shown in Figure 12.10. In Chapter 11 we introduced the regulation of glycogen phosphorylase activity, effected through the ATP-dependent **phosphorylation** of a specific serine residue. Other proteins can undergo phosphorylation of threonine or tyrosine residues. More complex reactions involve **adenylylation,** the transfer of an adenylate moiety from ATP, and **ADP-ribosylation,** the transfer of an ADP-ribosyl moiety from NAD^+. However, phosphorylation is by far the most widespread covalent modification used to control enzyme activity.

Control through covalent modification is often associated with regulatory cascades. Modification activates an enzyme, which in turn acts on a second enzyme, which may activate yet a third enzyme, which finally acts on the substrate. Because enzymes act catalytically, this cascading provides an efficient way to *amplify* the original biological signal. Suppose that the original signal modifying enzyme A activates it 10-fold, that modified enzyme A then activates enzyme B 100-fold, and that B activates enzyme C by 1000-fold. Thus, with the involvement of relatively few molecules of enzyme, a pathway is activated by 1000-fold.

The first well-understood regulatory cascade was the one controlling glycogen breakdown in animal cells, a critical process that provides carbohydrate substrates for energy generation. This regulatory cascade, involving enzyme phosphorylation and dephosphorylation, was introduced in Chapter 11 and is described in detail in Chapter 13. Blood clotting, also described in Chapter 11, is another well-understood regulatory cascade.

Intense interest in protein modification reactions has focused on the protein products of **oncogenes** (cancer-causing genes), many of which have been found to be **protein kinases.** Aberrant activities of these enzymes, which phosphorylate other proteins within a cell and alter their activities, are involved in the transformation of a normal cell to a cancer cell, as discussed in Chapter 23.

COMPARTMENTATION

We have already described the physical division of labor that exists in a eukaryotic cell, in the sense that enzymes participating in the same process are localized to a particular *compartment* within the cell. For example, RNA polymerases are

Enzyme
OH
ATP
ADP
Enzyme
O—P—O⁻
Phosphate moiety

(a) Phosphorylation

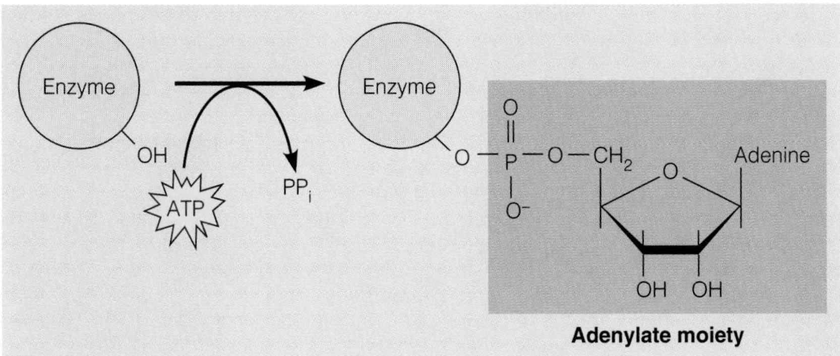

(b) Adenylylation

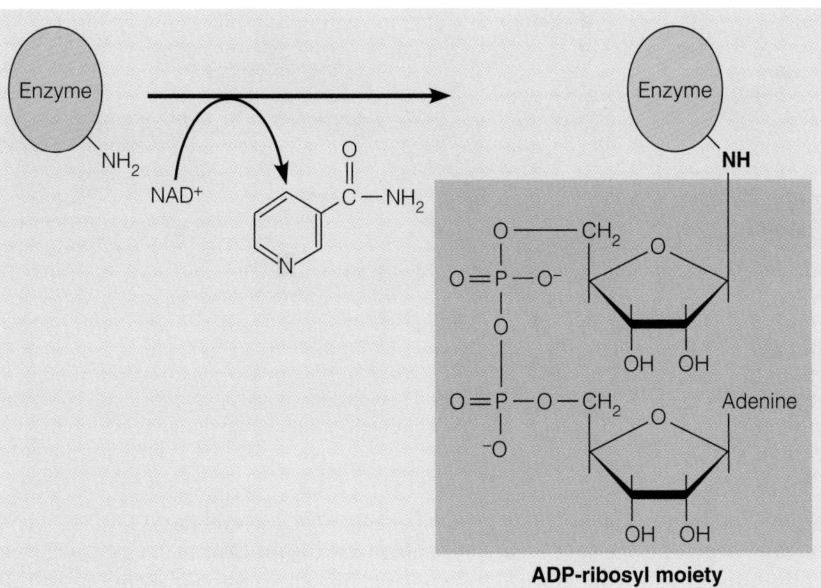

(c) ADP-ribosylation

FIGURE **12.10**

Three types of covalent modifications that control the activities of enzymes. The target residue for phosphorylation or adenylylation is usually serine, threonine, or tyrosine, whereas ADP-ribosylation can involve arginine or a modified histidine residue.

found in the nucleus and nucleolus, where DNA transcription occurs, and the enzymes of the citric acid cycle are all found in mitochondria. Figure 12.11 presents the locations of a number of metabolic pathways within eukaryotic cells.

Compartmentation creates a division of labor within a cell, which increases the efficiency of cell function. In addition, compartmentation has an important regulatory function. This function derives largely from the selective permeability of membranes to different metabolites, thereby controlling the passage of intermediates from one compartment into another. Typically, intermediates of a pathway remain trapped within an organelle, while specific carriers allow substrates to enter and products to exit. The flux through a pathway, therefore, can be

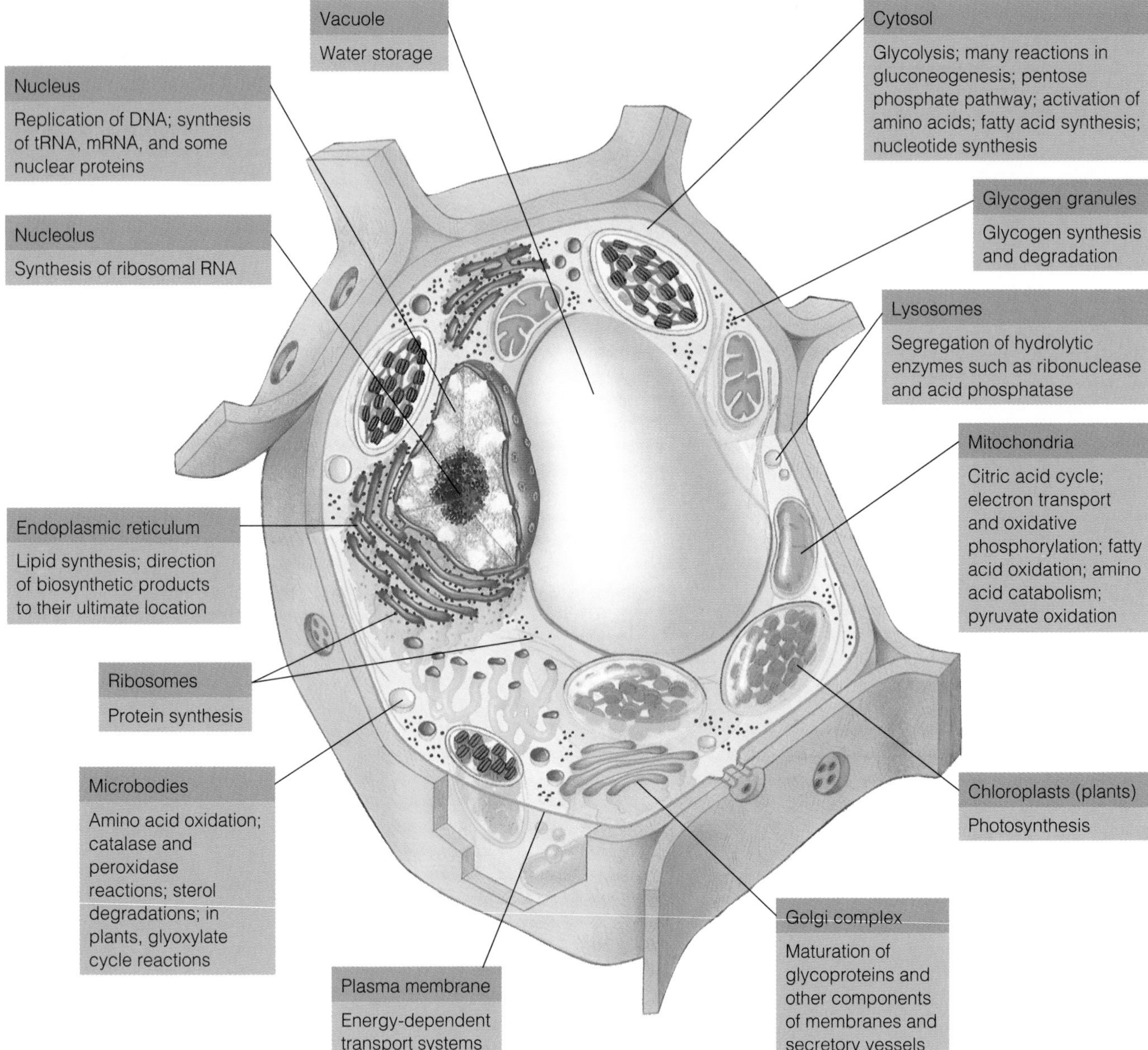

FIGURE **12.11**

Locations of major metabolic pathways within a eukaryotic cell. This hypothetical cell combines features of a plant cell and an animal cell.

From Neil Campbell, Jane Reece, and Larry Mitchell, *Biology,* 5th ed. (Menlo Park, CA: Addison Wesley Longman, 1999). © Addison Wesley Longman, Inc.

Enzymes catalyzing sequential reactions are often associated, even in the cytosol, where organized structures are difficult to visualize.

regulated by controlling the rate at which a substrate enters the compartment. For example, one of the ways in which the hormone insulin stimulates carbohydrate utilization is by moving glucose transporters into the plasma membrane, so that glucose is more readily taken into cells for catabolism or for synthesis of glycogen.

Compartmentation is more than a matter of tucking enzymes into the proper organelles. Juxtaposition of enzymes that catalyze sequential reactions localizes substrates even in the absence of membrane-bound organelles. Opportunities for diffusion are reduced, because the product of one reaction is released close to the active site of the next enzyme in a pathway. The enzymes may be bound to one another in a membrane, as are the enzymes of mitochondrial electron transport. Alternatively, they may be part of a highly organized multiprotein complex, such as the pyruvate dehydrogenase complex, a major entry point to the citric acid cycle.

Compartmentation can also result from weak interactions among enzymes that do not remain complexed when they are isolated. For example, conversion of glucose to pyruvate by glycolysis is catalyzed by enzymes that interact quite weakly in solution. However, there is evidence that these enzymes interact within the cy-

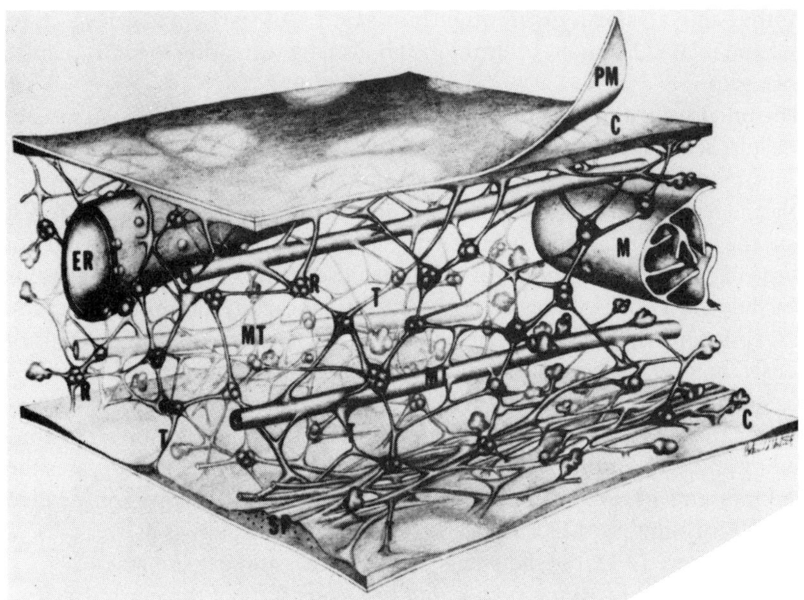

FIGURE **12.12**

Model of the structural organization of the cytomatrix. The model is based on high-voltage electron microscopic examination of cultured mammalian fibroblast cells. PM = plasma membrane; C = cortex (part of the cell surface); ER = endoplasmic reticulum; MT = microtubules; R = free ribosomes; M = mitochondrion. Approximate magnification, ×150,000.

Reproduced from J. J. Wolosewick and K. R. Porter, *J. Cell Biol.* (1979) 82:114–139, by copyright permission of Rockefeller University Press.

tosol, forming a supramolecular structure that facilitates the multistep glycolytic pathway. The concept of intracellular interactions among readily solubilized enzymes developed as scientists began to realize that the cytosol is much more highly structured than was formerly thought. High-resolution electron micrographs of mammalian cytosol reveal the outlines of an organized structure that has been termed the **cytomatrix,** a model of which is shown in Figure 12.12. It is likely that such structures form as a result of the extremely high concentrations of proteins inside cells, which decrease the concentration of water and drive weakly interacting proteins to associate. It has been proposed that soluble enzymes are bound within the cell to the structural elements of the cytomatrix.

Whether highly structured or loosely associated, multienzyme complexes allow for efficient control of reaction pathways. Enzyme complexes restrict diffusion of intermediates, thereby keeping the average concentrations of intermediates low (but their local concentrations high, at enzyme catalytic sites). This complexing reduces *transient time*, the average time needed for a molecule to traverse a pathway. Thus, the flux through a pathway can change quickly in response to a change in the concentration of the first substrate for that pathway.

HORMONAL REGULATION

Overlaid and interspersed with the regulatory mechanisms operating within a eukaryotic cell are messages dispatched from other tissues and organs. The process of transmitting these messages and bringing about metabolic changes is called **signal transduction.** The extracellular messengers include hormones, growth factors, neurotransmitters, and pheromones. A **hormone** is a substance synthesized in specialized cells and carried via the circulation to remote target cells. There it interacts with specific receptors, resulting in specific metabolic changes in the target cell; an example is the rapid generation of energy that results from secretion of **adrenaline** (now called **epinephrine**). Other modes of signal transduction, discussed in Chapters 21 and 23, have similar mechanisms.

Two types of metabolic response to hormones are particularly well understood. The first type, seen with steroid hormones, involves changes in gene expression. A steroid hormone, such as estrogen, enters a target cell, where it interacts with an intracellular protein receptor, and the resultant hormone–receptor complex then migrates to the nucleus. There the complex binds to specific sites in the genome and either activates or represses the transcription of particular genes. Thus, steroid and related hormones control metabolism by regulating the intracellular levels of proteins.

CH2 O Adenine
O O OH
P
O O⁻

Adenosine-3′, 5′-monophosphate

The other well-understood response to hormones involves the synthesis of intracellular **second messengers** that control metabolic reactions (see Figure 12.13). The hormone (the first messenger) acts extracellularly, by binding to receptors in the plasma membrane. Because the receptor protein traverses the membrane, it can, on the intracellular side, stimulate the formation of a second messenger in response to extracellular binding of the first messenger. Hormones acting in this way include most polypeptide hormones, such as **glucagon,** and some hormones, such as epinephrine, that are related to amino acids. As an example of this type of response, consider the events leading to the breakdown of intracellular glycogen stores to glucose-1-phosphate. We have already introduced these events as an example of a metabolic cascade. Glycogen mobilization, or **glycogenolysis,** is controlled by glucagon in the liver and by epinephrine in muscle. In both cases the second messenger is **adenosine 3′,5′-monophosphate,** more commonly called **cyclic AMP,** or simply **cAMP.** Here we will consider glycogenolysis in liver cells, as outlined in Figure 12.13, but the events in muscle cells are very similar.

The signal transduction system is modular, consisting of three protein components—a **receptor,** a **transducer,** and an **effector.** The receptor is a membrane-spanning protein that recognizes and binds a specific hormone, in this case, glucagon. The transducer is a **G protein,** so-called because of its high affinity for guanine nucleotides. Interaction of the hormone–receptor complex with the transducer stimulates an exchange reaction, in which GDP bound to the G protein is replaced by GTP. This exchange activates the G protein, which then interacts with the effector, the enzyme **adenylate cyclase.** In the response of liver cells to glucagon, that interaction stimulates adenylate cyclase, which catalyzes the conversion of ATP to cyclic AMP, the intracellular second messenger. As discussed in Chapter

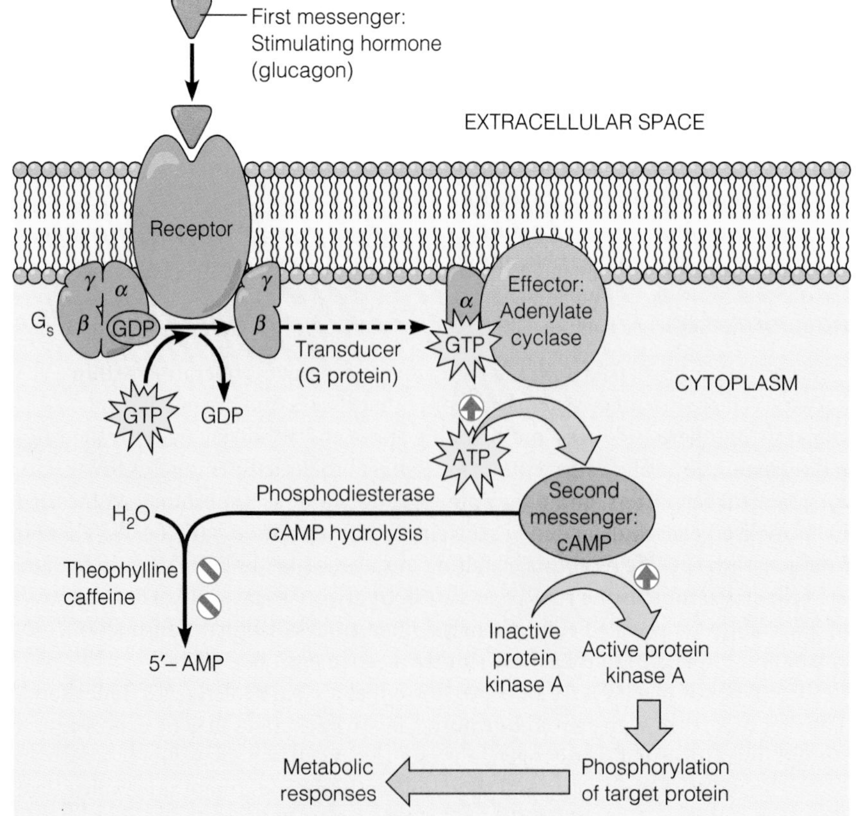

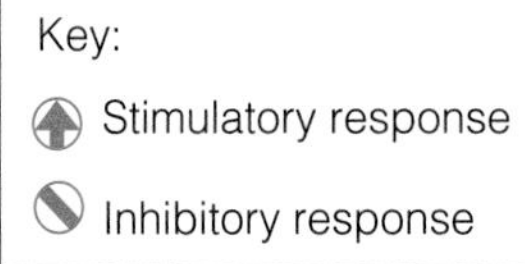

FIGURE **12.13**

Signal transduction pathway involving adenylate cyclase. The example shown is initiation of glycogenolysis by the hormone glucagon. Interaction between a glucagon–receptor complex and a G protein activates the G protein (G_s in the figure). The G protein (the transducer) in turn stimulates cAMP synthesis by activating adenylate cyclase (the effector). cAMP (the second, or intracellular, messenger) activates protein kinase A (cAMP-dependent protein kinase), and the resultant phosphorylation of target proteins leads to metabolic changes. Other G proteins are activated by different hormone–receptor complexes, and they act to *inhibit* adenylate cyclase. Theophylline and caffeine (components of tea and coffee, respectively) inhibit the catabolism of cAMP and thereby promote cAMP-stimulated pathways.

13, cAMP activates a protein kinase. Some enzymes are activated by phosphorylation, whereas others are inhibited; reactions in the metabolic cascade leading to glycogenolysis are activated. Thus, binding of the hormone at the cell surface stimulates synthesis of the second messenger inside the cell, which in turn effects a desirable metabolic response.

The modular nature of this hormonal control system allows a diversity of metabolic responses to be based on the same operating principles. First, the use of different receptors—for glucagon in liver and for epinephrine in muscle—allows the same metabolic response in either tissue, but only upon specific demand, occasioned by the release of either one hormone or the other. Second, the transducer protein involved in glycogenolysis is but one member of a large family of G proteins. Some G proteins activate adenylate cyclase, and others inhibit it. Inhibition lowers the level of cAMP and turns off responses controlled by that messenger. Third, there are effector proteins other than adenylate cyclase, each catalyzing the production of a different second messenger. For example, in Chapter 23 we discuss the **phosphoinositide system,** a pair of second messengers derived from the hormonally stimulated hydrolysis of **phosphatidylinositol 4,5-bisphosphate.** Thus, secretion of one hormone can have quite diverse effects in different tissues, depending upon the nature of the receptor and transducer proteins and upon the ensembles of effector proteins and second messenger systems in different target cells. Moreover, a single second messenger may have diverse effects within a single cell. For example, cyclic AMP, which activates the glycogenolytic cascade, also activates a cascade that inhibits the synthesis of glycogen. This dual effect is an example of a coordinated metabolic response: Glycogen synthesis is *inhibited* under the same physiological conditions that *promote* glycogen breakdown. We consider hormonal regulation further in Chapter 23.

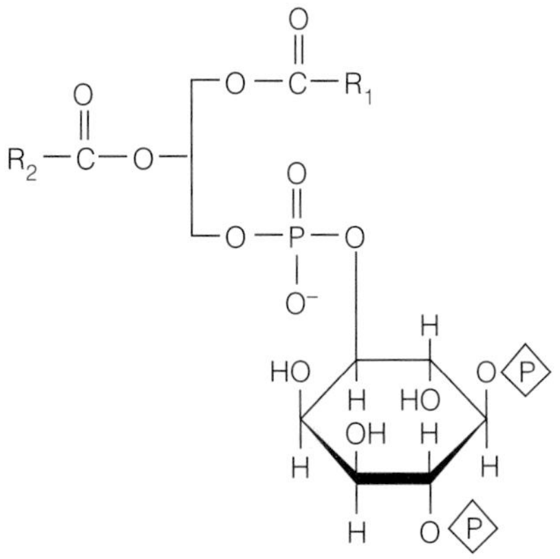

Second messengers transmit information from hormones bound at the cell surface, thereby controlling intracellular metabolic processes.

Phosphatidylinositol 4,5-bisphosphate

DISTRIBUTIVE CONTROL OF METABOLISM

In recent years, an important principle of metabolic regulation has emerged. With the discovery in the 1950s and 1960s of allosterically regulated enzymes, the concept arose that metabolic pathway flux is regulated primarily through control of the intracellular activity of one or a few key enzymes in that pathway. When it was realized that allosteric enzymes often catalyze committed reactions, i.e., the first reaction in a pathway that leads to an intermediate with no other known function, the concept arose that these enzymes catalyze the "rate-limiting reactions" in metabolic pathways. Taken literally, the term *rate-limiting reaction* implies that the flux rate of the pathways is identical to the intracellular activity of the rate-limiting step. However, because in a metabolic steady state all of the steps along a linear pathway are proceeding at the same rate, it is difficult or impossible to establish that the rate is controlled by a single enzymatic step. In other words, the concept of a rate-limiting enzyme is seriously flawed.

We now realize that metabolic regulation is more complex and that all of the enzymes in a pathway contribute toward control of pathway flux. An approach called **metabolic control analysis** assigns to each enzyme in a pathway a **flux control coefficient**, a value that can vary between zero and one. For a given enzyme, the flux control coefficient is the relative increase in flux, divided by the relative increase in enzyme activity that brought about that flux increase. For a true rate-limiting enzyme, the flux control coefficient is one; a 20% increase in the activity of that enzyme would increase that flux rate by 20%. But metabolic control theory predicts that all enzymes in a pathway contribute toward regulation, meaning that all enzymes have flux control coefficients greater than zero (but none has a value as high as one, which would be the case if flux were truly controlled solely by one rate-limiting enzyme.)

The predictions of metabolic control theory can be tested, for example, by using mutations affecting a specific enzyme to bring about defined changes in the activity of that enzyme in vivo, and then measuring the change in flux rate of a pathway in which that enzyme is involved. Such analyses confirm that enzymes catalyzing committed reactions do play large roles in regulation (i.e., they have high flux control coefficients). More important, however, these analyses confirm that *every* enzyme in a pathway contributes toward control of that pathway; in other words, every enzyme has a flux control coefficient greater than zero. Thus, regulation of a pathway is *distributed* among all of the enzymes involved in the pathway, giving rise to the concept of **distributive control of metabolism.** In retrospect, this concept should have been predicted simply from the long-understood complexity of metabolism. Many intermediates participate in more than one pathway, making different pathways interdependent and interlocking. Regulatory schemes that depend on control of just one or two enzymes in each pathway lack the flexibility and subtlety to account for the ability of cells to maintain homeostasis under widely varying nutritional and energetic conditions. To be sure, for each pathway or process, one or a few regulatory enzymes of primary importance have been identified, and these will be pointed out as we present the individual pathways involved.

Experimental Analysis of Metabolism

GOALS OF THE STUDY OF METABOLISM

Given that metabolism consists of all the chemical reactions in living matter, how does a biochemist approach metabolism in the laboratory? To subdivide a particular metabolic process into experimentally attainable goals, the biochemist seeks (1) to identify reactants, products, and cofactors, plus the stoichiometry, for each reaction involved; (2) to understand how the rate of each reaction is controlled in the tissue of origin; and (3) to identify the physiological function of each reaction and control mechanism. These goals necessitate isolating and characterizing the enzyme catalyzing each reaction in a pathway. This latter task—extrapolating from test tube biochemistry to the intact cell—is especially challenging. For instance, given that most enzymes catalyze reactions that can proceed in either direction, what is the direction of an enzymatic reaction in vivo? Many reactions originally found to proceed one way in vitro have been shown to proceed in the opposite direction in vivo. The mitochondrial enzyme that synthesizes ATP from ADP, for example, was originally isolated as an ATPase—an enzyme that hydrolyzes ATP to ADP and P_i. Therefore, it is not enough to isolate an enzyme and to demonstrate that it catalyzes a particular reaction in the test tube. We must show also that the same enzyme catalyzes the same reaction in intact tissue, usually a more difficult task.

To achieve the stated goals, a biochemist must perform analyses at several levels of biological organization, from living organisms and intact cells to broken-cell preparations and ultimately, to purified components. Cell-free, or in vitro, preparations can be manipulated in ways that intact cells cannot—for example, by addition of substrates and cofactors that will not pass through cell membranes. The researcher attempts to duplicate in vitro the process that is known to occur in vivo.

Preparation of cell-free components usually destroys biological organization. Cells can be broken open by sonic oscillation, shear forces, or enzymatic digestion of cell walls. These harsh treatments lead to mixing of components that were in separate compartments in the intact cell, making it possible to misinterpret data obtained from in vitro systems. An example comes from studies of protein biosynthesis. The existence of messenger RNA as a template for protein synthesis was predicted in 1961 from the behavior of bacterial mutants altered in genetic regulation. However, it was difficult to demonstrate the existence of messenger

RNA in vitro, because the putative template was present in very small amounts and was rapidly degraded by enzymes in cell-free extracts. Only when investigators learned how to prevent this degradation could they prove the existence of messenger RNA.

The foregoing discussion reveals the necessity of studying metabolism at various levels of biological organization, from intact organism to purified chemical components. Here we discuss what can be learned at each level.

LEVELS OF ORGANIZATION AT WHICH METABOLISM IS STUDIED

Whole Organism

Biochemists must investigate metabolism in whole organisms, because our ultimate aim is to understand chemical processes in intact living systems. Radioisotopic tracers are widely used to characterize metabolic pathways, as described in Tools of Biochemistry 12A. A classic example is the elucidation of cholesterol synthesis in the 1940s. Konrad Bloch injected ^{14}C-labeled acetate into rats and followed the flow of label into intermediates by killing rats at intervals and analyzing the radioactive compounds in their livers. In designing experiments of this type, the investigator must pay attention to the efficiency of transport of the labeled precursor to the organs of interest, uptake of the precursor into cells, and competition of exogenous precursors with preexisting pools of unlabeled intermediates.

Many diagnostic tests in clinical medicine are in vivo metabolic experiments. Instead of using radioisotopes, we sample tissue at intervals and carry out biochemical assays. In the **glucose tolerance test,** for example, a subject consumes a large oral dose of glucose, and its level in the blood is then determined at intervals over several hours. The glucose tolerance test is used to diagnose diabetes and other disorders of carbohydrate metabolism.

In recent years, *nuclear magnetic resonance* (NMR) spectroscopy has become widely available for noninvasive monitoring of intact cells and organs. As explained in Tools of Biochemistry 6A, compounds containing certain atomic nuclei can be identified from an NMR spectrum, which measures shifts in the frequency of absorbed electromagnetic radiation. A researcher can determine an NMR spectrum of whole cells, or of organs or tissues in an intact plant or animal. For the most part, macromolecular components do not contribute to the spectrum, nor do compounds that are present at less than about 0.5 mM. The nuclei most commonly used in this in vivo technique are ^{1}H, ^{31}P, and ^{13}C. Figure 12.14 shows ^{31}P NMR spectra that represent components in the human forearm muscle. The five major peaks correspond to the phosphorus nuclei in orthophosphate (P_i), creatine phosphate, and the three phosphates of ATP. Because peak area is proportional to concentration, the energy status of intact cells can be determined. For example, an energy-rich muscle has lots of creatine phosphate, whereas a fatigued muscle uses up most of its creatine phosphate in order to maintain ATP levels (note also the accumulation of AMP—peak 6—in the third scan). NMR is finding wide applicability in monitoring recovery from heart attacks, in which

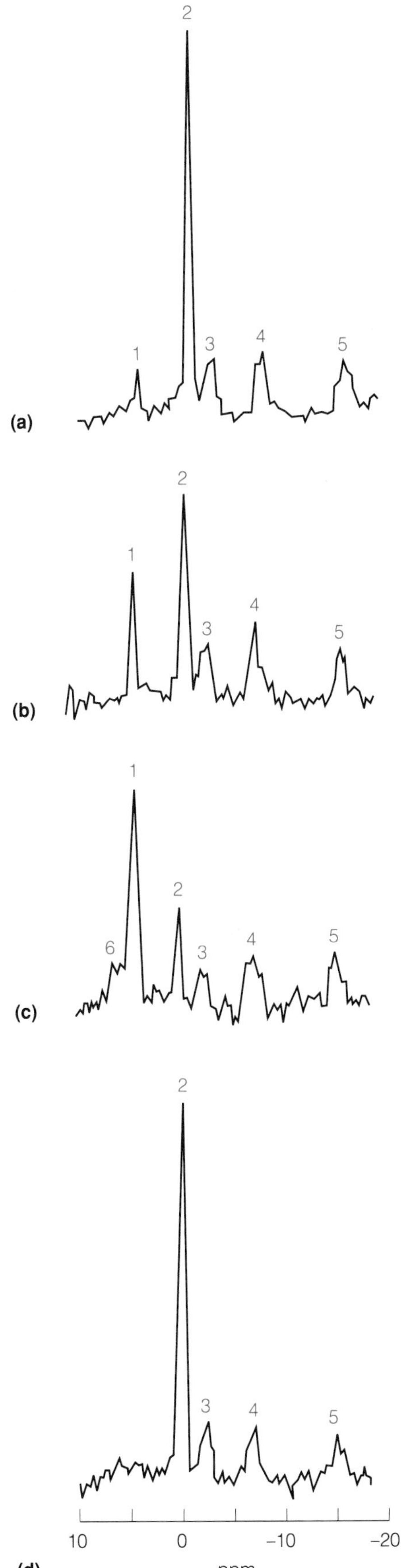

FIGURE **12.14**

Effect of anaerobic exercise on ^{31}P NMR spectra of human forearm muscle. **(a)** Before exercise. **(b)** One minute into a 19-minute exercise period. **(c)** The 19th minute. **(d)** Ten minutes after exercise. Peak areas are proportional to intracellular concentrations. Peak 1 = P_i; 2 = creatine phosphate; 3 = ATP γ-phosphate; 4 = ATP α-phosphate; 5 = ATP β-phosphate; 6 = phosphomonoesters. See Tools of Biochemistry 6A for interpretation of NMR spectra.

cellular ischemia (insufficient oxygenation) damages cells by reducing ATP content. NMR can also be used to study metabolite compartmentation, flux rates through major metabolic pathways, and intracellular pH.

Isolated or Perfused Organ

Some of the difficulties in transporting a precursor or inhibitor to the desired organ can be circumvented by using an isolated organ. A researcher usually **perfuses** the isolated organ during the experimental manipulations. This procedure involves pumping a buffered isotonic solution containing nutrients, drugs, or hormones through the organ. The solution partly takes the place of the normal circulation, delivering nutrients and removing waste products. The researcher can also perfuse an organ within a living animal, following appropriate surgical procedures. Of course, perfusion is much less efficient than circulation; thus, experiments at this level must be of limited duration.

The circulation problem can be partly overcome by cutting the tissue into thin slices before the experimental manipulations begin. The structural integrity of the organ is lost, but most of the cells remain intact, and they are in better contact with the fluid that bathes the tissue. The cells may be better oxygenated and supplied with substrates than in a whole organ. The elucidation of the citric acid cycle resulted largely from experiments with slices of liver and heart.

Whole Cells

Tissue slices are not now widely used, partly because methods are now available for disaggregating an organ or a tissue into its component cells. Liver, kidney, and heart cells can be prepared by treatment of the organ with trypsin or collagenase, to break down the extracellular matrix that holds the organ together. For plant cells, enzymes such as cellulase or pectinase, which attack the cell wall, can be used to make comparable preparations.

Any plant or animal organ contains a complex mixture of different cell types. Several means of fractionating the cells after disaggregation of an organ are used to obtain preparations enriched in one cell type. One method is centrifugation to separate cells on the basis of size. In recent years the **fluorescence-activated cell sorter** has come into widespread use. In a typical application a cell suspension is treated with a fluorescent-tagged antibody to a cell surface antigen, which is present in varying amounts among different cell types. Cells pass in single file through a laser beam and are physically separated according to the amount of fluorescence recorded from each cell. Such machines, which can sort several thousand cells per second, result in fractionation based on the abundance of the selected surface antigen.

Uniformity in a cell population is often achieved by growth of cells in **tissue culture.** Disaggregated cells of an organ or a tissue can, with special care, be induced to grow in a medium containing cell nutrients and protein growth factors. The cells grow and divide independently of one another, much like the cells in a bacterial culture. Although animal cells usually cease growth after a certain number of divisions in culture, variant lines arise that are capable of indefinite growth, as long as they are adequately nourished. In such cultures, **clonal** cell lines can be generated in which all of the cells in a line are derived from a single cell, so that they are genetically and metabolically uniform. This uniformity is a boon for many biochemical investigations. For example, much of our understanding of virus replication depends upon the ability to infect a large number of identical cells in culture simultaneously and then follow the metabolic changes by sampling the cell culture at various times after infection.

One problem with tissue culture is that cells adapted to long-term growth in culture take on characteristics different from those of their parent cells, which were originally embedded in plant or animal tissue. Maintaining specialized cell characteristics in culture always presents a challenge.

Cell-Free Systems

Problems of transport through membranes are obviated by working with broken-cell preparations. Animal cells are easily lysed (ruptured) by mild shear forces, suspension in hypotonic medium, or freezing and thawing. Bacterial cells have a rigid cell wall that requires vigorous treatment such as sonic oscillation. Enzymatic digestion with lysozyme is often used to open bacterial cells under relatively mild conditions. Breaking the especially tough cell walls of yeasts and plants usually requires combinations of enzymatic and mechanical treatments.

Initial metabolic experiments are usually carried out in unfractionated cell-free homogenates. However, localizing a metabolic pathway within a particular cell compartment requires fractionating the homogenate to separate the organelles, usually by **differential centrifugation.** Lysis is carried out in isotonic sucrose solutions and generally yields morphologically intact organelles. These components can be precipitated by centrifugation at different speeds and for different lengths of time. Typically, nuclei, mitochondria, chloroplasts, lysosomes, and **microsomes** (membranous aggregates of ribosomes) can all be at least partially separated from each other. The contents of the cytosol remain in the supernatant after the final centrifugation step. Much of our understanding of DNA replication and transcription in eukaryotic cells comes from investigations with isolated nuclei, whereas purified mitochondria have yielded much of our understanding of respiratory electron transport and oxidative phosphorylation.

Purified Components

To understand a biological process at the molecular level, the investigator must purify to homogeneity all of the factors thought to be involved and determine their interactions. Often, as with the citric acid cycle, this process is simply a matter of purifying the individual enzymes involved, determining the substrate and cofactor requirements of each, recombining the purified enzymes, and showing that the entire process can be catalyzed by purified components. This process is called **reconstitution.** Some pathways require cell constituents other than enzymes, such as the ribosomes and transfer RNAs needed for protein synthesis.

In purifying individual components, the biochemist continually risks losing factors that are essential for normal control or for some other aspect of the process under study. Avoiding such pitfalls requires painstaking experiments in which the researcher defines criteria for biological activity and continually examines fractions to ensure that each activity is retained through each fractionation step. A good example of this approach is presented in Chapter 24, in which we discuss the several enzymes and proteins that must function at a DNA replication fork.

METABOLIC PROBES

An invaluable biochemical aid is the use of metabolic probes, agents that allow a researcher to interfere specifically with one or a small number of reactions in a pathway. The consequences of such interference can be extremely informative. Two kinds of probes are most widely used—*metabolic inhibitors* and *mutations.* By blocking a specific reaction in vivo and determining the results of the blockade, such probes help identify the metabolic role of a reaction. For example, respiratory poisons such as carbon monoxide and cyanide block specific steps in respiration, and metabolic inhibitors helped identify the order of electron carriers in the respiratory electron transport chain (see Chapter 15).

By inactivating individual enzymes, mutations and enzyme inhibitors help identify the metabolic roles of enzymes.

Inhibitors can present difficulties for the researcher, such as poor transport into cells or multiple sites of action. It is often easier to interfere with a pathway by selecting mutant strains deficient in the enzyme of interest. In the 1940s George Beadle and Edward Tatum were the first to use mutations as biochemical probes, in their work with the bread mold *Neurospora crassa.* Beadle and Tatum isolated a number of x-ray-induced mutants that required arginine for growth, in

FIGURE **12.15**

Using mutations as biochemical probes. The steps of a hypothetical metabolic pathway are identified by analysis of mutants defective in individual steps of the pathway. For example, we can identify metabolite C as the substrate for enzyme III by the absence of this enzyme in mutants that accumulate C. We know that D and E follow C in the pathway, because feeding either D or E to mutants defective in enzyme III bypasses the genetic block and allows the cells to grow.

A mutant defective in enzyme	accumulates metabolite	requires an external source of	allows the growth of another mutant, defective in enzyme
I	A	B, C, D, or E	—
II	B	C, D, or E	I
III	C	D, or E	I or II
IV	D	E	I, II, or III

Analysis of mutants

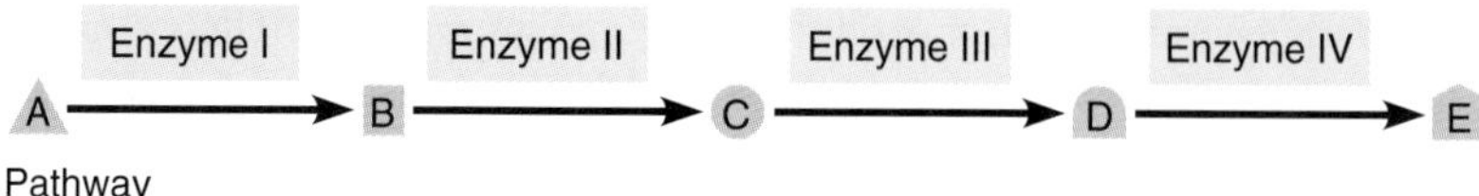

Pathway

addition to the constituents of minimal medium. Furthermore, different mutations affected different enzymes in the arginine biosynthetic pathway, with each mutant accumulating the intermediate that was the substrate for the deficient enzyme. This observation allowed Beadle and Tatum to order the enzymes according to the reactions they controlled, by the rationale illustrated in Figure 12.15. If a culture filtrate from one mutant allowed a second mutant to grow without arginine, the researchers concluded that the first mutation blocked an enzymatic step *later* in the pathway than the step blocked by the second mutation. Ultimately, the accumulating intermediates were identified and the pathway was elucidated.

In addition to identifying pathways, mutants have been used to elucidate genetic regulatory mechanisms. The earliest successes came from studies in Paris by François Jacob and Jacques Monod, who isolated dozens of *Escherichia coli* mutants with defects in regulation of lactose catabolism or with abnormalities in virus–host relationships. These data led ultimately to the discovery of mRNA and the repressor–operator mechanism of genetic regulation, which we discuss in Chapter 26.

A single investigation can use both metabolic inhibitors and mutations. This combined approach helped to elucidate the function of DNA gyrase, one of the DNA topoisomerases mentioned in Chapter 4. This enzyme is inhibited by nalidixic acid. When nalidixic acid is administered to bacteria, DNA replication is inhibited, suggesting that DNA gyrase plays an essential role in DNA replication. However, because nalidixic acid *could* inhibit DNA replication by blocking some other enzyme, stronger evidence is needed. Such evidence was obtained when mutants resistant to nalidixic acid were found to contain an altered form of DNA gyrase that was resistant to nalidixic acid. Thus, a single mutation abolished nalidixic acid sensitivity for both the enzyme and the ability of the cells to replicate their DNA, strongly supporting an essential role for DNA gyrase in DNA replication.

In this chapter we have described the general strategy of metabolism, identified the major pathways, described how pathways are regulated, and identified experimental approaches to understanding metabolism. We are now prepared for detailed descriptions of metabolic pathways, which we begin in Chapter 13 with carbohydrates.

SUMMARY

Metabolism is the totality of chemical reactions occurring within a cell. Catabolic pathways break down substrates to provide energy, largely through oxidative reactions, whereas anabolic pathways synthesize complex biomolecules from small molecules, often from intermediates in catabolic pathways. Catabolic and anabolic pathways with the same end points are actually different pathways, not simple reversals of each other, so that both pathways can function exergonically. Most metabolic energy comes from oxidation of substrates, with energy release coming in a series of small steps as the electrons released are transferred ultimately to oxygen. The more highly reduced a substrate, the more energy is released through its catabolism.

Flux through metabolic pathways is controlled by regulation of enzyme concentration (through control of enzyme synthesis and degradation), enzyme activity (through concentrations of substrates, products, and allosteric effectors, and covalent modification of enzyme proteins), compartmentation, and hormonal control. Hormonal regulation may involve control of enzyme synthesis at the genetic level or regulation of enzyme activity. In the latter case, intracellular second messengers are formed in response to hormonal signals.

The understanding of metabolic processes requires identification of each reaction in a pathway and a knowledge of the reaction's function and control. This understanding requires experimentation at all levels of biological organization, from living organism to purified enzyme. The ability to block specific enzymes, either with inhibitors or mutations, is of great value in identifying the functions of those enzymes.

REFERENCES

Experimental Techniques in the Study of Metabolism

Davey, H. M., and D. B. Kell (1996) Flow cytometry and cell sorting of heterogeneous microbial populations: The importance of single-cell analyses. *Mirobiol. Rev.* 60: 641–696. The power of fluorescence-activated cell sorting.

Radda, G. K. (1992) Control, bioenergetics, and adaptation in health and disease—Noninvasive biochemistry from nuclear magnetic resonance. *Science* 233:640–645. A contemporary review.

Shulman, R. G., D. L. Rothman, and A. M. Blamire (1994) NMR studies of human brain function. *Trends Biochem. Sci.* 19:522–526. Imaging and metabolic analysis by whole-organ NMR.

Tsien, R. Y. (1994) Fluorescence imaging creates a window on the cell. *Chem. Eng. News,* July 18, pp. 34–44. This review describes another emerging technique for noninvasive metabolic monitoring of individual cells.

Compartmentation and Intracellular Enzyme Organization

Bessman, S. P., and C. L. Carpenter (1985) The creatine–creatine phosphate energy shuttle. *Annu. Rev. Biochem.* 54:831–862. A contemporary account of the bioenergetic role of creatine phosphate, emphasizing the importance of compartmentation as a metabolic control phenomenon.

Goodsell, D. S. (1991) Inside a living cell. *Trends Biochem. Sci.* 16:203–206. Drawings of the interior of a bacterial cell, based upon physical information about the sizes, shapes, and distribution of cellular constituents.

Srere, P. A., M. E. Jones, and C. K. Mathews, eds. (1990) *Structural and Organizational Aspects of Metabolic Regulation.* Proceedings of a UCLA Conference. W. J. Wiley, New York. Papers presented at a conference dealing with structural organization of sequential metabolic pathways, including both membranous complexes and complexes involving soluble enzymes.

Enzyme Control and Metabolic Regulation

Lefkowitz, R. J., and M. G. Caron (1988) Adrenergic receptors: Models for the study of receptors coupled to guanine nucleotide regulatory proteins. *J. Biol. Chem.* 263:4993–4996. A minireview of membrane receptor structure and function.

Newsholme, E. A., R. A. J. Challiss, and B. Crabtree (1984) Substrate cycles: Their role in improving sensitivity in metabolic control. *Trends Biochem. Sci.* 9:277–280. A brief but lucid discussion of substrate cycle control, with several examples.

Taussig, R., and A. G. Gilman (1995) Membrane-bound adenylyl cyclases. *J. Biol. Chem.* 270:1–4. Discussion of cyclic AMP synthesis coauthored by the discoverer of G proteins.

Taylor, S. S., J. A. Buechler, and W. Yonemoto (1990) cAMP-dependent protein kinase: Framework for a diverse family of regulatory enzymes. *Annu. Rev. Biochem.* 59:971–1006. General insights into metabolic regulation from structural studies of a key enzyme.

Distributive Control of Metabolism

Fell, D., R. Heinrich, J.-P. Mazat, R. Small, and H. Westerhoff (1995) Henrik Kacser, 1918–1995. *Trends Biochem. Sci.* 20:297–298. This obituary of one of the founders of metabolic control analysis details his roles in developing its methods and concepts.

Ovádi, J., and P. A. Srere (1996) Metabolic consequences of enzyme interactions. *Cell Biochem. Funct.* 14:249–258. One of a collection of review articles, organized by Srere, which also includes articles on noninvasive metabolic monitoring and molecular genetic approaches to understanding pathway control.

Veech, R. L., and D. A. Fell (1996) Distribution control of metabolic flux. *Cell Biochem. Funct.* 14:229–236. A readable review, studded with recent references.

PROBLEMS

1. Write a balanced equation for the complete oxidation of each of the following, and calculate the respiratory quotient for each substance.
   (a) Ethanol
   (b) Acetic acid
   (c) Stearic acid
   (d) Oleic acid
   (e) Linoleic acid
2. Given what you know about the involvement of nicotinamide nucleotides in oxidative and reductive metabolic reactions, predict whether the following intracellular concentration ratios should be (1) unity, (2) greater than unity, or (3) less than unity. Explain your answers.
   (a) $[NAD^+]/[NADH]$
   (b) $[NADP^+]/[NADPH]$
   Since NAD^+ and $NADP^+$ are essentially equivalent in their tendency to attract electrons, discuss how the two concentration ratios might be maintained inside cells at greatly differing values.
3. (a) NAD^+ kinase catalyzes the ATP-dependent conversion of NAD^+ to $NADP^+$. How many reducing equivalents are involved in this reaction?
   (b) How many reducing equivalents are involved in the conversion of ferric to ferrous ion?
   (c) How many reducing equivalents are involved in reducing one molecule of oxygen gas to water?
4. On page 422 we showed that the oxidation of glucose and palmitic acid yields 15.64 kJ/g and 38.90 kJ/g, respectively. Calculate these values in terms of kJ/mol and kJ per carbon atom oxidized for both glucose and palmitic acid.
5. Free energy changes under intracellular conditions differ markedly from those determined under standard conditions. $\Delta G°'$ for ATP hydrolysis to ADP and P_i is -30.5 kJ/mol. Calculate $\Delta G'$ for ATP hydrolysis in a cell at 37°C that contains ATP at 3 mM, ADP at 0.2 mM, and P_i at 50 mM.
6. Consider a substrate cycle operating with enzymes X and Y in the following hypothetical metabolic pathway:

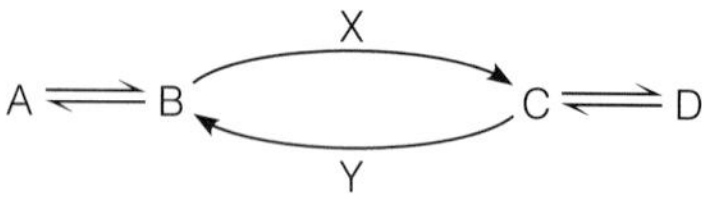

   (a) Under intracellular conditions the activity of enzyme X is 100 pmol/10^6 cells/s, and that of enzyme Y is 80. What are the direction and rate of metabolic flux between B and C?
   (b) Calculate the effect on metabolic flux rate and direction of (1) an inhibitor that reduces the activity of X by 10%, (2) an activator that increases the activity of X by 10%, and (3) doubling the activity of enzyme Y.
   (c) Discuss possible experimental approaches to the question of whether this substrate cycle plays a significant role in regulating the pathway from A to D.

Read Tools of Biochemistry 12A before attempting to work Problems 7–9.

*7. Two-dimensional gel electrophoresis of proteins in a cell extract provides a qualitative way to compare proteins with respect to intracellular abundance. Describe a quantitative approach to the determination of number of molecules of an enzyme per cell.

*8. Mammalian cells growing in culture were labeled for a long time in [^{3}H]thymidine to estimate the rate of DNA synthesis. The thymidine administered had a specific activity of 3000 cpm/pmol. At intervals, samples of culture were taken and acidified to precipitate nucleic acids. The rate of incorporation of isotope into DNA was 1500 cpm/10^6 cells/min. A portion of culture was taken for determination of specific activity of the intracellular dTTP pool, which was found to be 600 cpm/pmol.
   (a) What fraction of the intracellular dTTP is synthesized from the exogenous precursor?
   (b) What is the rate of DNA synthesis, in molecules per minute per cell of thymine nucleotides incorporated into DNA?
   (c) How might you determine the specific activity of the dTTP pool?

*9. Suppliers of radioisotopically labeled compounds usually provide each product as a mixture of labeled and unlabeled material. Unlabeled material is added deliberately as a **carrier**, partly because the specific activity of the carrier-free product is too high to be useful and partly because the product is more stable at lower specific activities. Using the radioactive decay law, calculate the following.
   (a) The specific activity of carrier-free [^{32}P]orthophosphate, in mCi/μmol.
   (b) In a preparation of uniform-label [^{3}H]leucine, provided at 10 mCi/μmol, the fraction of H atoms that are radioactive.

TOOLS OF BIOCHEMISTRY 12A

Radioisotopes and the Liquid Scintillation Counter

Radioisotopes revolutionized biochemistry when they became available to investigators shortly after World War II. Radioisotopes extend by several orders of magnitude the sensitivity with which chemical species can be detected. Traditional chemical analysis can detect and quantify molecules in the micromole or nanomole range (i.e., 10^{-6} to 10^{-9} mole). A compound that is "labeled," containing one or more atoms of a radioisotope, can be detected in picomole or even femtomole amounts (i.e., 10^{-12} or 10^{-15} mole). Radiolabeled compounds are called **tracers** because they allow an investigator to follow specific chemical or biochemical transformations in the presence of a huge excess of nonradioactive material.

Recall that isotopes are different forms of the same element. They have different atomic weights but the same atomic number.

Thus, the chemical properties of the different isotopes of a particular element are virtually identical. Isotopic forms of an element exist naturally, and substances enriched for rare isotopes can be isolated and purified from natural sources. Most of the isotopes used in biochemistry, however, are produced in nuclear reactors. Simple chemical compounds produced in such reactors are then converted to radiolabeled biochemicals by chemical and enzymatic synthesis.

Stable Isotopes

Although radioisotopes are widely used in biochemistry, stable isotopes are also used as tracers, particularly when convenient radioisotopes are not available. For example, the two rare isotopes of hydrogen include a stable isotope ($^2H_1$, or **deuterium**) and a radioactive isotope ($^3H_1$, or **tritium**). Of the many uses of stable isotopes in biochemical research, we mention three applications here. First, incorporation of a stable isotope often increases the density of a material, because the rare isotopes usually have higher atomic weights than their more abundant counterparts. This difference presents a way to separate labeled from nonlabeled compounds physically, as in the Meselson–Stahl experiment on DNA replication (see Chapter 4). Second, compounds labeled with stable isotopes, particularly ^{13}C, are widely used in nuclear magnetic resonance studies of molecular structure and reaction mechanisms (see Tools of Biochemistry 6A). Third, stable isotopes can be used as tracers in the absence of suitable radioisotopes, as already noted. For example, there are no radioisotopes of oxygen or nitrogen, so ^{18}O and ^{15}N are useful tracers of these respective elements. Stable isotopes are detected and quantified in the mass spectrometer, a somewhat laborious methodology compared to counting radioactivity.

Table 12A.1 gives information about the isotopes, both stable and radioactive, that have found the greatest use in biochemistry.

TABLE **12A.1**

SOME USEFUL ISOTOPES IN BIOCHEMISTRY

				Maximum Energy
Isotope	Stable or Radioactive	Emission	Half-life	(mev[a])
^2H	Stable			
^3H	Radioactive	β	12.1 years	0.018
^{13}C	Stable			
^{14}C	Radioactive	β	5568 years	0.155
^{15}N	Stable			
^{18}O	Stable			
^{24}Na	Radioactive	β (and γ)	15 hours	1.39
^{32}P	Radioactive	β	14.2 days	1.71
^{35}S	Radioactive	β	87 days	0.167
^{45}Ca	Radioactive	β	164 days	0.254
^{59}Fe	Radioactive	β (and γ)	45 days	0.46, 0.27
^{131}I	Radioactive	β (and γ)	8.1 days	0.335, 0.608

[a]mev = million electron volts.

The Nature of Radioactive Decay

The atomic nucleus of an unstable element can decay, giving rise to one or more of the three types of ionizing radiation: α-, β-, or γ-rays. Only β- and γ-emitting radioisotopes are useful in biochemical research; the most useful are shown in Table 12A.1. A β-ray is an emitted electron, and a γ-ray is a high-energy photon. γ-Ray detectors have found wide use in immunological research, because γ-emitting isotopes of iodine are available, and antibodies, like many proteins, can easily be iodinated without substantial changes in their biological properties. Most biochemical uses of radioisotopes, however, involve β-emitters.

Radioactive decay is a first-order kinetic process. The probability that a given atomic nucleus will decay is affected neither by the number of preceding decay events that have occurred nor by interaction with other radioactive nuclei. Rather, it is an intrinsic property of that nucleus. Thus, the number of decay events occurring in a given time interval is related only to the number of radioactive atoms present. This phenomenon gives rise to the **law of radioactive decay:**

$$N = N_0 e^{-\lambda t}$$

where N_0 is the number of radioactive atoms at time 0, N is the number remaining at time t, and λ is a radioactive decay constant for a particular isotope, related to the intrinsic instability of that isotope. This equation states that the *fraction* of nuclei in a population that decays within a given time interval is constant. For this reason, a more convenient parameter than the decay constant λ is the **half-life**, $t_{1/2}$, the time required for half of the nuclei in a sample to decay. The half-life is equal to $-\ln 0.5/\lambda$, or $+0.693/\lambda$. The half-life, like λ, is an intrinsic property of a given radioisotope (see Table 12A.1).

The basic unit of radioactive decay is the **curie** (Ci). This unit is defined as an amount of radioactivity equivalent to that in 1 g of radium, namely 2.22×10^{12} disintegrations per minute (dpm). Biochemists usually work with far smaller amounts of radioactive material—the **millicurie** (mCi) and the **microcurie** (μCi), which correspond, respectively, to 2.2×10^9 and 2.2×10^6 dpm. Because detectors of radioactivity rarely record every decay event in a sample—that is, they are not 100% efficient—we often speak of radioactivity in terms of the decay events actually recorded: counts per minute, or cpm. Counting efficiency is the percentage of decay events actually recorded, as determined, for example, by reference to standards. A counter that is 50% efficient for a given isotope would show a count rate of 1.1×10^5 cpm for a 0.1-μCi sample.

Detection of Radioactivity: The Liquid Scintillation Counter

Geiger counters were in widespread use four decades ago, but virtually all work with β-emitting isotopes is now done with the **liquid scintillation counter,** for two reasons. First, the scintillation counter has much higher efficiency than the Geiger counter, and second, radioactivities of two different isotopes in the same sample can be simultaneously determined.

In scintillation counting the sample is usually dissolved or suspended in an organic solvent, although aqueous mixtures are available as well. Also present are one or two fluorescent organic compounds, or **fluors.** A β-particle emitted from the sample has a high probability of hitting a molecule of the solvent; this contact excites the solvent molecule, driving an electron into a higher-energy orbital. When that electron returns to the ground state, a photon of light is emitted. The photon is absorbed by a molecule of the fluor, which in turn becomes excited. Fluorescence involves the absorption of light at a given energy, followed by emission of that light at lower energy, or longer wavelength. A photomultiplier detects that tiny flash of light and for each disintegration converts it to an electrical signal, which is recorded and counted.

Whereas γ-rays are emitted at distinct and characteristic energy values, a β-emitting isotope will display a range of energy values among its emissions. Each β-emitter shows a characteristic **energy spectrum**—that is, a plot of energy (in millions of electron volts, or mev) against the probability of an individual emission having that energy. Figure 12A.1 shows the energy spectra for three widely used radioisotopes—^3H, ^{14}C, and ^{32}P.

These differences in emission energies are exploited in the liquid scintillation counter, so that two isotopes in the same sample can be simultaneously quantified (a dual-label experiment). A strong β-emission excites more fluor molecules than a weak emission does, so a brighter light flash is produced. This flash level can be detected by setting electronic discriminators so that energy values falling only within a desired range are recorded—that is, setting a "window" that is optimized for detection of a particular isotope. This restriction reduces counting efficiency, because only a fraction of the emissions from the higher-energy isotope are counted, but it does permit great selectivity. Most scintillation counters have three different sets of discriminators, or **counting channels.** Thus, isotopes can be counted simultaneously in three different windows.

Because radioactive decay is a random process, the more counts observed during the analysis of a sample, the more closely the measured radioactivity will approach the true decay rate. In practice, therefore, a researcher wants to count a sample for enough time to accumulate several thousand counts. Also, the researcher must consider the signal-to-noise ratio and design an experiment so that the measured radioactivity of the samples of interest will be many times higher than **background radioactivity,** or the counts recorded in the absence of a radioactive sample.

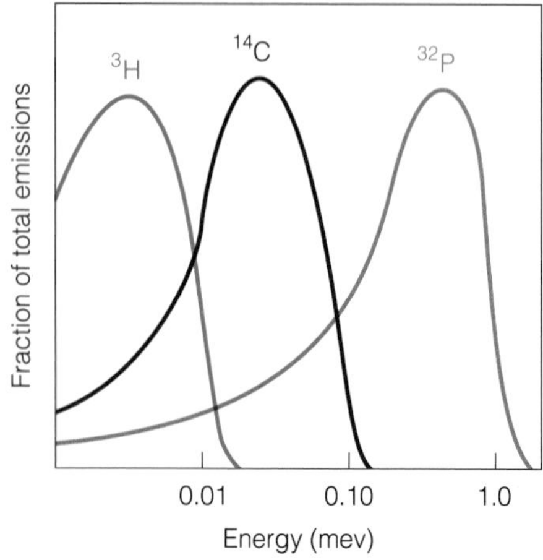

FIGURE **12A.1**

Energy spectra for some β-emitting isotopes. Shown here are spectra for the three β-emitters most widely used in biochemistry—^3H, ^{14}C, and ^{32}P.

Some Uses of Radioisotopes in Biochemistry

Radioisotopes have many uses in metabolic investigations. Here we describe a few examples. Tracer experiments can be categorized in terms of the time of exposure of the biological system to the radioisotope and include (1) equilibrium labeling, (2) pulse labeling, and (3) pulse-chase labeling.

In **equilibrium labeling** the exposure to the tracer is relatively long, so that each labeled species reaches a constant **specific radioactivity,** or **specific activity.** The specific activity is a measure of the relative abundance of radioactive molecules in a labeled sample and is reported as radioactivity per unit mass—for example, cpm/μmol.

Experiments to identify metabolic precursors often involve equilibrium labeling conditions. The investigator administers a radiolabeled precursor so as to maximize the chance of detecting label in the product. A good example, mentioned earlier in this chapter, is Konrad Bloch's study of cholesterol biosynthesis, in which ^{14}C-labeled acetate was administered to rats. Bloch then isolated cholesterol from the liver and determined, by chemical degradation, the specific carbon atoms in cholesterol that had incorporated radioactivity.

Equilibrium labeling can be used as part of a procedure for purifying biological species of interest. An example is Mark Ptashne's purification of a genetic repressor (see Chapter 26). Ptashne had two bacterial strains, one designed to overproduce the putative repressor, the other identical except for its inability to synthesize repressor. The first strain was grown for several generations in the presence of [^{14}C]leucine, and the second was grown similarly in [^3H]leucine. The cultures were mixed and the proteins fractionated, with the expectation that repressor would be the only protein present that was labeled with ^{14}C but not ^3H. One protein with a high $^{14}C/^3H$ ratio was isolated and subsequently shown to be the repressor.

Another use of equilibrium labeling is to measure rates of biological processes from the rates of labeling by low-molecular-weight precursors. An example is the widespread use of radiolabeled thymidine to follow rates of DNA synthesis, either in cell cultures or in intact organisms. Incorporation of thymidine into macromolecules other than DNA is negligible, so the investigator merely samples the population with time and observes the incorporation into acid-insoluble material. However, this measurement alone does not give the true rate of DNA replication. The investigator must also consider the metabolism of the labeled precursor en route to its ultimate destination. As shown in Figure 12A.2, thymidine is converted first to thymidine monophosphate, then to the diphosphate, and finally to the triphosphate, the actual substrate for DNA synthesis. At each stage the labeled

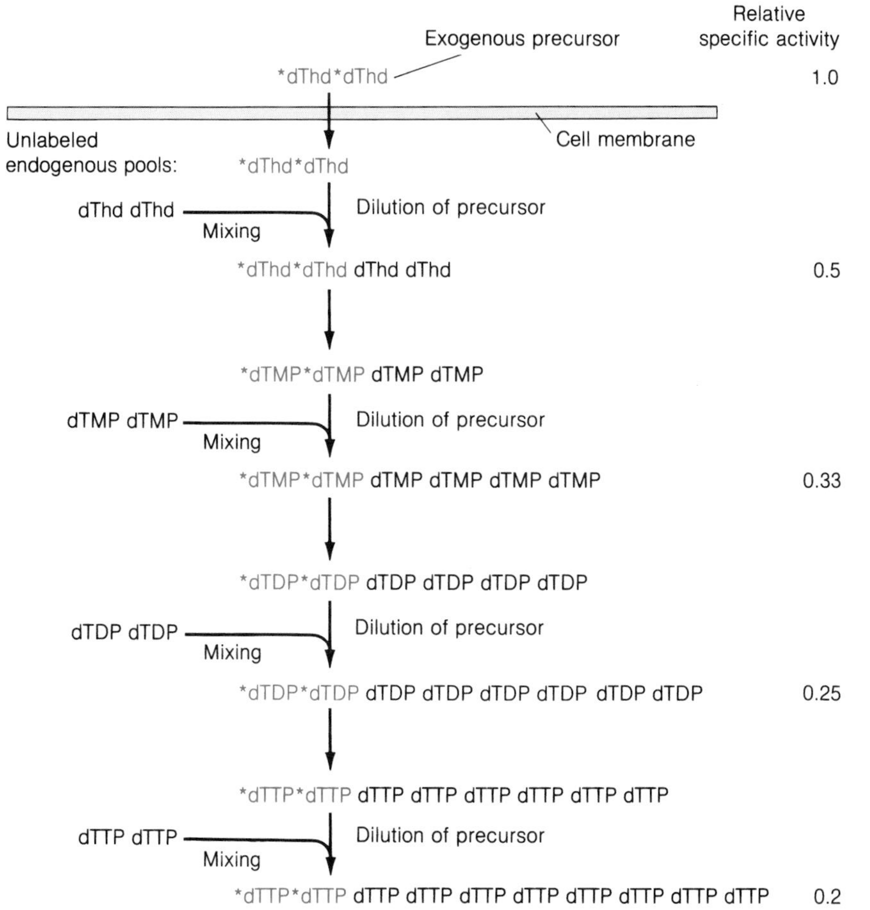

FIGURE **12A.2**

Dilution of an exogenous metabolic precursor (radiolabeled thymidine, *dThd) by unlabeled endogenous pools. In the example shown, the specific activity of thymidine triphosphate is reduced to 20% of that of the exogenous thymidine. Thus, the rate of incorporation of radioactivity into DNA would underestimate the true rate of DNA synthesis by 5-fold.

precursor mixes with intracellular pools of unlabeled precursors, which dilute its specific radioactivity. The labeling experiment gives only a rate of DNA labeling, in cpm incorporated per cell per unit time. To get a true rate, in molecules incorporated per cell per unit time, the initial result must be divided by the specific activity of the immediate precursor, in this case, dTTP. Calculating that specific activity requires isolation of dTTP in sufficient amount and purity to allow determination of both its mass and radioactivity.

Overlaid on these difficulties is the possibility of compartmentation. In this case, a cell may have two or more pools of dTTP, each of which becomes labeled at different rates but only one of which is used for DNA replication. There are various ways to deal with this situation, the best being to use a mutant strain that is unable to synthesize thymidine nucleotides. All thymidine compounds must then be supplied exogenously, and all intracellular dTTP pools will reach the same specific activity, namely that of the exogenous precursor.

Pulse labeling involves administration of an isotopic precursor for an interval that is short relative to the process under study. Radiolabel accumulates preferentially in the shortest-lived species—that is, the earliest intermediates in a metabolic pathway—because the shorter the labeling interval is, the less time is available for loss of radioactivity from a labeled pool through breakdown of rapidly labeled and short-lived metabolites. Melvin Calvin identified the pathway of photosynthetic carbon fixation by labeling green algae with $^{14}CO_2$ for just a few seconds. If labeling was carried out for 10 s, a dozen or more radioactive compounds could be detected. After a 5-s pulse of radioactivity, only a single compound was labeled, namely 3-phosphoglycerate. This finding led ultimately to the discovery of ribulose-1,5-bisphosphate carboxylase as the first enzyme in the photosynthetic carbon fixation pathway (see Chapter 17).

In a **pulse-chase** experiment, the investigator administers label for a short time and then rapidly reduces the specific activity of the isotopic precursor to prevent further incorporation. This reduction can be done by adding unlabeled precursor at a molar excess of about 1000-fold, which greatly dilutes the radioisotope still present. The investigator samples at various times afterward, to determine the metabolic fate of the material labeled during the pulse. Messenger RNA was originally detected by pulse-labeling bacterial cultures (see Chapter 26). Traditional analytical methods could not detect mRNA because of its low abundance and its metabolic instability in bacterial cells. When label incorporated into mRNA by a pulse of [^{32}P]orthophosphate or labeled uridine was chased out, the label was ultimately found to be distributed uniformly in all cellular RNA species. This finding showed that the metabolic instability of mRNA involves its degradation to nucleotides, which can then be used for synthesis of other RNA species.

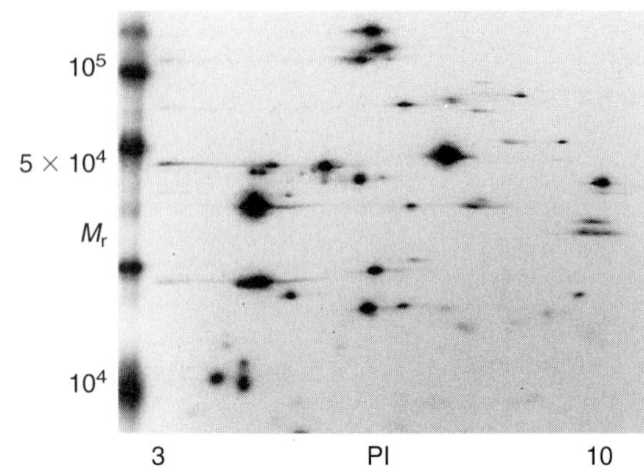

FIGURE **12A.3**

Two-dimensional electrophoresis of an extract of T4 bacteriophage proteins, labeled from 3 to 8 minutes after infection of *E. coli*. The radioactive bands at the left identify molecular weight standards.

Courtesy of Linda J. Wheeler.

A final application of radioisotopes is **radioautography**, in which the investigator incorporates an isotopic precursor into a biomolecule and prepares an image of the radiolabeled molecule on a sheet of photographic film. Chapter 24 shows several radioautographs of individual DNA molecules made radioactive by growth of *E. coli* in the presence of $[^3H]$thymidine. Of interest here is **two-dimensional gel electrophoresis,** a technique that allows visualization of virtually all the proteins in a cell or tissue. In the example shown in Figure 12A.3, a T4 bacteriophage-infected *E. coli* culture was pulse-labeled with $[^{35}S]$methionine. An extract of the labeled proteins was resolved first by isoelectric focusing (see Tools of Biochemistry 2A), which separates proteins from left to right as shown, and then by SDS polyacrylamide gel electrophoresis, which separates on the basis of molecular weight. The gel was dried and used to expose a sheet of x-ray film. Each spot represents one protein, whose rate of synthesis is related to intensity of the spot. The molecular weight and isoelectric point given on the radioautograph, along with other information (such as the properties of mutants affecting synthesis of a particular protein), allow identification of many of the protein spots.

Reference

Freifelder, D. (1982) *Physical Biochemistry,* 2nd ed. W. H. Freeman, San Francisco. Chapter 5 of this book presents a clear description of techniques in radioactive labeling and counting.

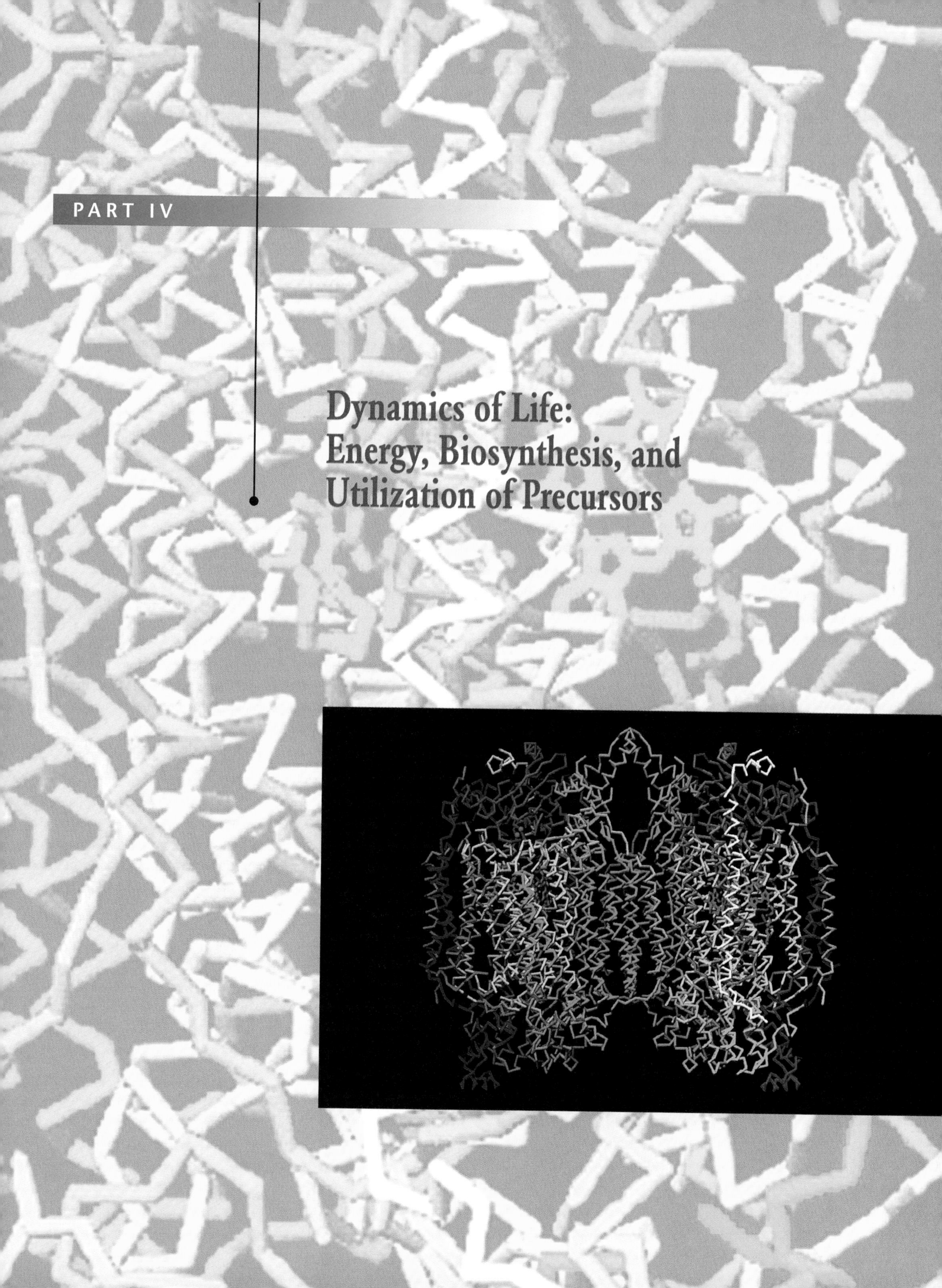

PART IV

Dynamics of Life: Energy, Biosynthesis, and Utilization of Precursors

CHAPTER 13

Carbohydrate Metabolism I: Anaerobic Processes in Generating Metabolic Energy

OUR DETAILED STUDY OF METABOLISM BEGINS WITH THE ANAEROBIC PHASES OF carbohydrate metabolism (Figure 13.1). Most of this chapter is devoted to glycolysis, the initial pathway in the catabolism of carbohydrates. The term *glycolysis* is derived from Greek words meaning "sweet" and "splitting." These words are literally correct terms, for glycolysis is the pathway by which six-carbon sugars (which are sweet) are split, yielding a three-carbon compound, pyruvate. During glycolysis, some of the potential energy stored in the hexose structure is released and used to drive the synthesis of ATP from ADP. Glycolysis can proceed under anaerobic conditions, with no net oxidation of the sugar substrates taking place. **Anaerobes,** microorganisms that live in oxygen-free environments, can derive all of their metabolic energy from this process. However, aerobic cells also use glycolysis. In these cells, glycolysis is the initial, anaerobic part of an overall degradation pathway that involves considerable oxygen consumption and the complete oxidation of carbohydrates.

Glycolysis is an appropriate point to begin a detailed study of metabolism, for several reasons. First, it was the earliest metabolic pathway to be understood in detail. Second, the pathway is nearly universal in living cells. Third, the regulation of glycolysis is particularly well understood. Last but not least is the central metabolic role this pathway plays in generating both energy and metabolic intermediates for other pathways. It is one of the busiest freeways on the metabolic road map, but it is also connected to many less traveled roads.

Although cells can metabolize a variety of hexose sugars via glycolysis, glucose is the major carbohydrate fuel for most cells. Indeed, some animal tissues,

FIGURE **13.1** (facing page)

Anaerobic processes in the generation of metabolic energy. The purple portions of the metabolic map show the glycolytic pathway and the breakdown of polysaccharides that supply this pathway. Glycolysis generates ATP anaerobically and provides fuel for the aerobic energy-generating pathways. The numbers 1, 2, and 3 identify the three stages of metabolism (see Chapter 12).

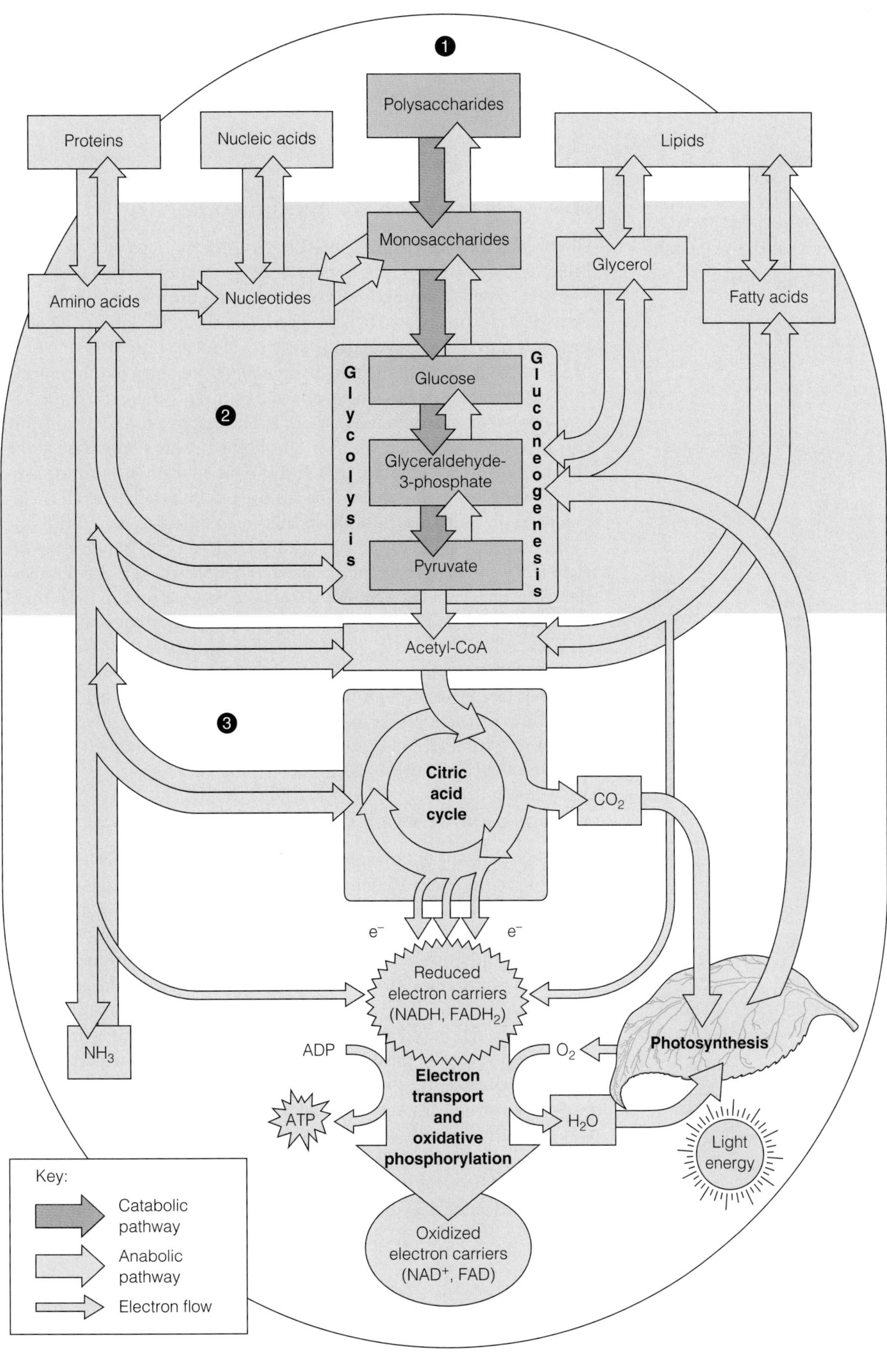

1
Polysaccharides
Proteins
Nucleic acids
Lipids
Monosaccharides
Glycerol
Amino acids
Nucleotides
Fatty acids
Glycolysis
Glucose
Gluconeogenesis
2
Glyceraldehyde-3-phosphate
Pyruvate
Acetyl-CoA
3
Citric acid cycle
CO_2
e^-
e^-
Reduced electron carriers ($NADH$, $FADH_2$)
ADP
O_2
Photosynthesis
NH_3
Electron transport and oxidative phosphorylation
ATP
H_2O
Light energy
Key:
Catabolic pathway
Anabolic pathway
Electron flow
Oxidized electron carriers (NAD^+, FAD)

such as brain, normally use glucose as the sole energy source, and all energy generation in such cells begins with glycolysis. Most cells, however, can utilize other sugars, and we shall explore how those sugars are converted to intermediates in glycolysis. In addition we will consider processes by which stored carbohydrate in the form of polysaccharides is made available for use in glycolysis.

Glycolysis: An Overview

The ten reactions of glycolysis occur in two phases: energy investment (first five reactions) and energy generation (last five reactions).

RELATION OF GLYCOLYSIS TO OTHER PATHWAYS

Glycolysis is a 10-step pathway that converts one molecule of glucose to two molecules of pyruvate, with the concomitant generation of two molecules of ATP. The breakdown of storage polysaccharides and the metabolism of oligosaccharides yield glucose, related hexose sugars, and sugar phosphates, all of which find their way into the glycolytic pathway. We will focus initially on the pathway as it begins with glucose and then discuss the routes for entry of other carbohydrates.

The 10 reactions between glucose and pyruvate can be considered as two distinct phases, schematized in Figure 13.2. The first five reactions constitute an **energy investment phase,** in which sugar phosphates are synthesized at the expense of 2 moles of ATP (converted to ADP), and the six-carbon substrate is split into 2 three-carbon sugar phosphates. The last five reactions represent an **energy generation phase,** in which the triose phosphates are converted to energy-rich compounds. These transfer 4 moles of phosphate to ADP, leading to 4 moles of ATP. The net yield, per mole of glucose metabolized, is 2 moles of ATP and 2 moles of pyruvate. Note that 2 reducing equivalents are generated as well, in the form of NADH.

In aerobic organisms, glycolysis is the first step in the complete oxidation of glucose to CO_2 and water. The second step is oxidation of pyruvate to acetyl-CoA, and the final process is oxidation of the acetyl group carbons in the citric acid cycle (see Figure 13.1). Chapter 14 presents the latter processes in detail. Glycolysis also provides biosynthetic intermediates. Thus, glycolysis is both an anabolic and a catabolic pathway, with an importance that extends beyond the synthesis of ATP and substrates for the citric acid cycle.

ENERGY INVESTMENT PHASE

Glucose

2ADP 2ATP

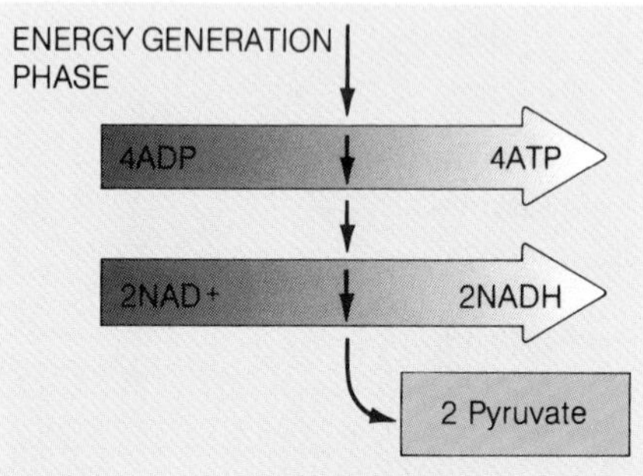

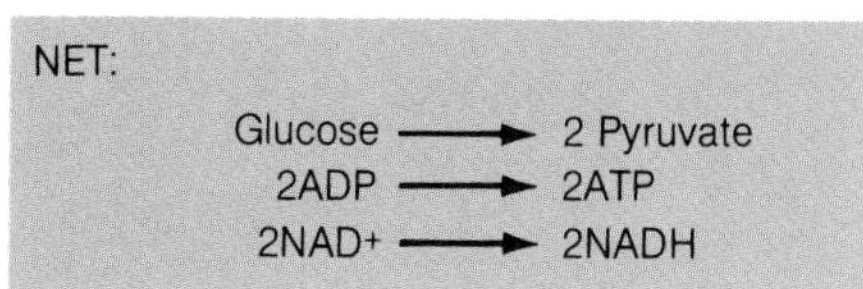

FIGURE **13.2**
The two phases of glycolysis and the products of glycolysis.

ANAEROBIC AND AEROBIC GLYCOLYSIS

Glycolysis is an ancient metabolic pathway that was probably used by the earliest known bacteria, some 3.5 billion years ago. Because that time was about 1 billion years before the earliest known photosynthetic organisms began contributing O_2 to the earth's atmosphere, glycolysis had to function initially under completely anaerobic conditions—with no net change in the oxidation state as substrates are converted to products. However, note in Figure 13.2 that the conversion of glucose to pyruvate, which does oxidize the carbons of glucose, involves the concomitant reduction of 2 moles of NAD^+ to NADH. For the pathway to operate anaerobically, NADH must be reoxidized to NAD^+ by transferring its electrons to an **electron acceptor** so that *a steady state is maintained.* Some microorganisms growing anaerobically can generate additional energy by transferring the electrons to inorganic substances such as sulfate ion or nitrate ion, and some microorganisms reduce organic substrates. Most straightforward is the route used by lactic acid bacteria, which simply use NADH to reduce pyruvate to lactate, via the enzyme **lactate dehydrogenase.** This reaction occurs when milk sours.

$$\underset{\textbf{Pyruvate}}{\begin{matrix} COO^- \\ | \\ C{=}O \\ | \\ CH_3 \end{matrix}} + NADH + H^+ \rightleftharpoons \underset{\textbf{L-Lactate}}{\begin{matrix} COO^- \\ | \\ HO{-}C{-}H \\ | \\ CH_3 \end{matrix}} + NAD^+ \qquad \Delta G^{\circ\prime} = -25.1 \text{ kJ/mol}$$

Glycolysis is therefore part of a fermentation, which is defined as an energy-yielding metabolic pathway that involves no net change in oxidation state. The lactic acid fermentation (conversion of glucose to lactate) is important in the manufacture of cheese. Another important fermentation involves cleavage of pyruvate to acetaldehyde and CO_2 (see page 461), with the acetaldehyde then reduced to ethanol by **alcohol dehydrogenase:**

A fermentation is an energy-yielding metabolic pathway with no net change in the oxidation state of products compared to substrates.

$$CH_3CHO + NADH + H^+ \rightleftharpoons CH_3CH_2OH + NAD^+$$

As carried out by yeasts, this fermentation generates the ethanol in alcoholic beverages. Yeasts used in baking also carry out the alcoholic fermentation; the CO_2 produced by pyruvate decarboxylation causes bread to rise, and the ethanol produced evaporates during baking. Among the dozens of other useful fermentations are those leading to acetic acid (manufacture of vinegar) and propionic acid (manufacture of Swiss cheese).

Animal cells, like lactic acid bacteria, can reduce pyruvate to lactate, and they do so when pyruvate is produced faster than it can be oxidized through the citric acid cycle. During strenuous exertion, skeletal muscle cells derive most of their energy from this **anaerobic glycolysis**—glycolysis occurring under anaerobic conditions.

Anaerobic glycolysis (like aerobic glycolysis) leads to pyruvate, but the pyruvate is then reduced, so no net oxidation of glucose occurs.

By contrast, consider a cell undergoing active *respiration,* the oxidative breakdown and release of energy from nutrient molecules by reaction with oxygen. In these cells, pyruvate is oxidized to acetyl-CoA, which enters the citric acid cycle. The NADH produced during glycolysis is reoxidized through the mitochondrial electron transport chain for additional energy production (see Chapter 15), with the electrons transferred ultimately to O_2, the terminal electron acceptor. The conversion of glucose to pyruvate in a respiring cell is called **aerobic glycolysis.**

THE CRUCIAL EARLY EXPERIMENTS

For as long as humans have used yeasts in baking and in brewing, glycolysis has been exploited, even though it was not understood until this century. (The early definition of *fermentation* was "a chemical change with effervescence.") Louis Pasteur's demonstration in 1856 that fermentations are carried out by microorganisms ranks as a milestone in the history of science. The dominant viewpoint of the time was that a process such as the fermentation of glucose to ethanol was so complex that it could not be reproduced outside a living cell. As we saw in Chapter 1, however, Eduard and Hans Büchner showed in 1897 that fermentation could occur under cell-free conditions.

In 1905 Arthur Harden and William Young found that inorganic phosphate, when added to yeast extract, stimulated and prolonged the fermentation of glucose. During fermentation the inorganic phosphate disappeared from the reaction medium, which led Harden and Young to suggest that fermentation was functioning via the formation of one or more sugar phosphate esters.

This observation opened the door to dissection of the individual chemical reactions involved in fermentation, a feat accomplished in Germany in the 1930s, largely by G. Embden, O. Meyerhof, and O. Warburg. In fact, glycolysis is often referred to as the **Embden–Meyerhof pathway.** These scientists identified 10 different reactions, virtually identical in a wide range of organisms, leading from glucose to pyruvate. Glycolysis is the first metabolic pathway to have been elucidated as a series of defined chemical reactions. Extensive information is now available about the structure and mechanism of action of each enzyme involved.

STRATEGY OF GLYCOLYSIS

Glycolysis is such an important pathway that we shall examine each of its 10 reactions in some detail. Before doing so, let us look at the pathway as a whole. First,

ATP is synthesized by three major routes—substrate-level phosphorylation, oxidative phosphorylation, and photophosphorylation.

recall from Chapter 12 that in eukaryotic cells glycolysis occurs in the cytosol, and the further oxidation of pyruvate occurs in mitochondria. (Certain trypanosomes, the parasitic protozoans that cause African sleeping sickness, present an interesting exception. They carry out the first seven reactions of glycolysis in an organized cytoplasmic organelle called the **glycosome.**)

Figure 13.3 presents an abbreviated look at the conversion of glucose to pyruvate. In the energy investment phase (the first five reactions), the sugar is metabolically activated by phosphorylation. This process yields a six-carbon phosphorylated sugar, **fructose-1,6-bisphosphate,** which undergoes cleavage to yield 2 moles of triose phosphate: **glyceraldehyde-3-phosphate** and **dihydroxyacetone phosphate.**

In the energy generation phase (reactions 6 through 10), the triose phosphates undergo further activation to yield two compounds containing energy-rich phosphate bonds—first **1,3-bisphosphoglycerate** and then **phosphoenolpyruvate.** Recall from Figure 3.7 (page 75) that each of these compounds has a higher $\Delta G^{\circ\prime}$ of hydrolysis than ATP; they can be considered as super-high-energy compounds. During the energy generation phase, each of these compounds transfers its high-energy phosphate to ADP, yielding ATP. This process is called **substrate-level phosphorylation**—the transfer of a phosphoryl group from a super-high-energy compound to ADP, yielding ATP. Substrate-level phosphorylation is distinguished from oxidative phosphorylation, the synthesis of ATP driven by electron transport (see Chapter 15), and **photophosphorylation,** the utilization of photosynthetic light energy to drive ATP synthesis (see Chapter 17).

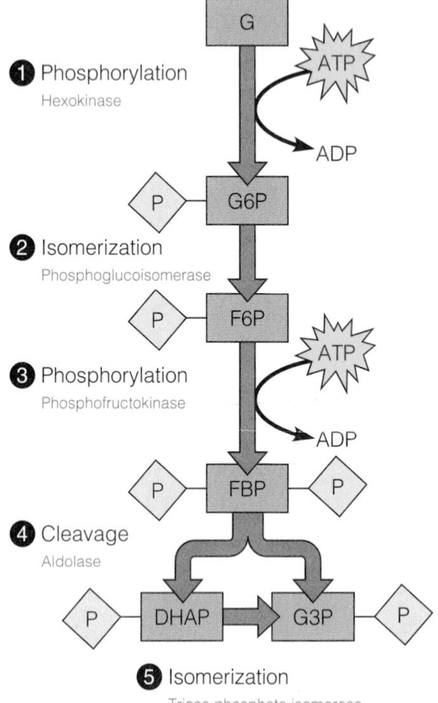

Because 2 moles of triose phosphate are metabolized per mole of glucose, the yield from the two substrate-level phosphorylations of glycolysis is 4 moles of ATP per mole of glucose. Subtracting the 2 moles of ATP invested in the first phase (reactions 1–5), we see a net gain of two ATP molecules synthesized per molecule of glucose converted to pyruvate (see Figure 13.2).

Reactions of Glycolysis

Now let us consider in sequence the 10 reactions leading from glucose to pyruvate, numbering each reaction as indicated in Figure 13.3. The complete names of substrates and products are given when each reaction is presented, but in the text these names are shortened for simplicity. Thus, glucose-6-phosphate is the same as α-D-glucose-6-phosphate.

REACTIONS 1–5: THE ENERGY INVESTMENT PHASE

The first five reactions, which constitute the energy investment phase, are summarized in the margin.

Reaction 1: The First ATP Investment

We begin with the ATP-dependent phosphorylation of glucose, catalyzed by **hexokinase.**

α-D-Glucose + ATP $\xrightarrow{Mg^{2+}}$ α-D-Glucose-6-phosphate + ADP + H^+ $\quad \Delta G^{\circ\prime} = -16.7$ kJ/mol

Magnesium ion is required, because the reactive form of ATP is its chelated complex with Mg^{2+}(see page 426). This is true for virtually all ATP-requiring enzymes. Hexokinase exists in various forms in different organisms but is generally characterized by broad specificity for sugars and low K_M for the sugar substrate (0.01 to

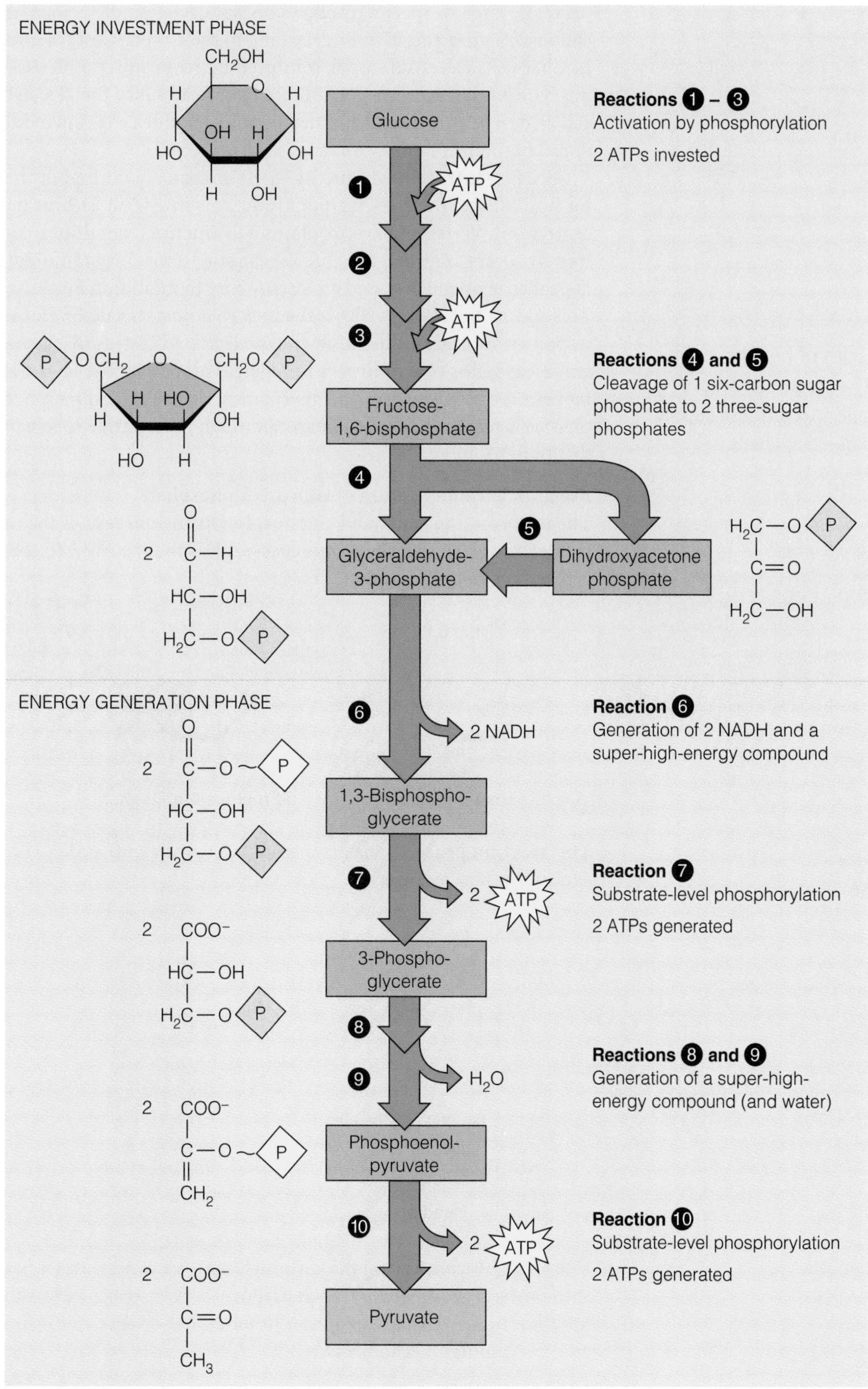

FIGURE **13.3**

An overview of glycolysis. This condensed view of glycolysis shows the key intermediates and reactions in each of the two major phases. In the energy-generating phase, two ATPs are produced for each ATP utilized in the energy-investment phase. Here and elsewhere, energy-rich phosphate bonds are shown by the yellow color of the phosphate group.

0.1 mM). The low specificity allows phosphorylation of various hexose sugars, including fructose and mannose, leading to their utilization via glycolysis. As noted in Chapter 11, hexokinase is inhibited by its product, glucose-6-phosphate, a mechanism that controls the influx of substrates into the glycolytic pathway. Recall also that the structure of hexokinase provides striking evidence for the induced fit model of enzyme catalysis (see page 369).

Because intracellular glucose levels are usually far higher than the K_M value for hexokinase, the enzyme often functions in vivo at saturating substrate concentrations. Vertebrate liver contains a distinctive form of hexokinase, characterized by a very high K_M for glucose (about 10 mM), a sigmoidal concentration dependence on glucose, and an insensitivity to inhibition by glucose-6-phosphate. This special hexokinase allows the liver to adjust its rate of glucose utilization in response to variations in blood glucose levels. In fact, as discussed in Chapters 16 and 23, a major role of liver is to regulate blood glucose levels, and this enzyme represents one of the principal mechanisms by which it does so. This form of hexokinase is often called **glucokinase,** although its substrate specificity is identical to that of hexokinase.

A high-K_M species of hexokinase in liver permits that organ to adjust glucose utilization to glucose supply at high blood glucose levels.

Reaction 2: Isomerization of Glucose-6-phosphate

The next reaction, catalyzed by **phosphoglucoisomerase,** is the readily reversible isomerization of the aldose, glucose-6-phosphate (G6P), to the corresponding ketose, **fructose-6-phosphate** (F6P).

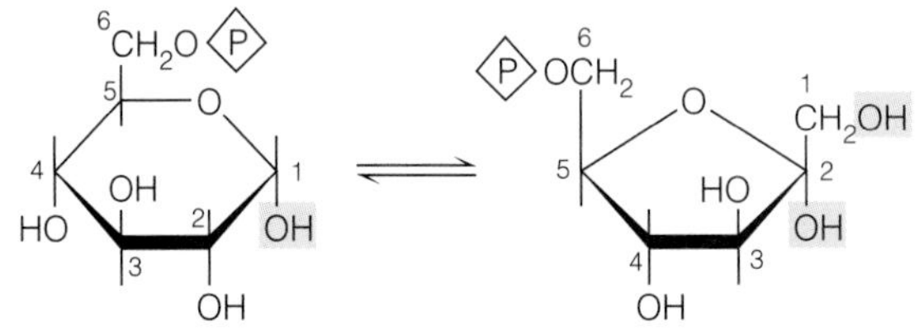

$\Delta G^{\circ\prime} = +1.7$ kJ/mol

α-D-Glucose-6-phosphate **D-Fructose-6-phosphate**

This reaction proceeds via an enediol intermediate: B and B-H represent active-site amino acid residues.

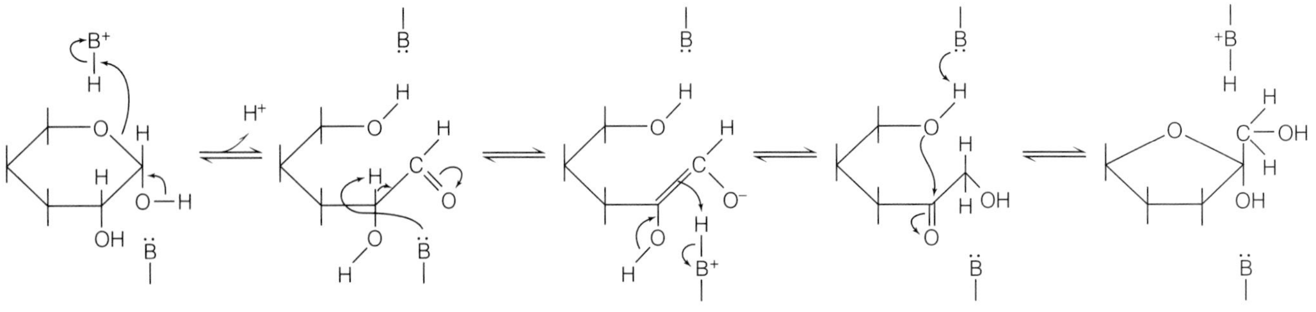

G6P **Enediol** **F6P**

The effect of transferring the carbonyl oxygen from carbon 1 to carbon 2 is that the hydroxyl group generated at carbon 1 can be readily phosphorylated in the next reaction. We shall encounter other aldose–ketose isomerizations that proceed by a similar mechanism.

Reaction 3: The Second Investment of ATP

In reaction 3, **phosphofructokinase** of ATP carries out a second ATP-dependent phosphorylation, to give a hexose derivative phosphorylated at both carbons 1 and 6. The product, fructose-1,6-bisphosphate (FBP), was formerly called fructose-1,6-diphosphate. Renaming was done to show that the two phosphates are separate, rather than linked as in ADP.

The phosphofructokinase reaction is the primary step at which glycolysis is regulated.

$$\text{D-Fructose-6-phosphate} + \text{ATP} \xrightarrow{Mg^{2+}} \text{D-Fructose-1,6-bisphosphate} + \text{ADP} + H^+ \quad \Delta G^{\circ\prime} = -14.2 \text{ kJ/mol}$$

Like the phosphorylation of glucose, this reaction is sufficiently exergonic to be essentially irreversible in vivo. Irreversibility is important, because phosphofructokinase (PFK) represents the primary site for regulation of the flow of carbon through glycolysis. PFK is an allosteric enzyme whose activity is acutely sensitive to the energy status of the cell, as well as to the levels of various other intermediates, particularly citrate and fatty acids. Interactions with allosteric effectors, which are discussed later in this chapter, activate PFK. This activation increases carbon flux through glycolysis when there is a need to generate more ATP and inhibits it when the cell contains ample stores of ATP or oxidizable substrates.

Higher plants contain two different PFKs—the ATP-dependent enzyme and a unique form, which uses pyrophosphate instead of ATP as the phosphorylating agent.

$$\text{Fructose-6-phosphate} + PP_i \longrightarrow \text{fructose-1,6-bisphosphate} + P_i$$

This enzyme, which is comparable in activity to ATP-dependent phosphofructokinase, seems to represent an alternative route to catalysis of the third step in glycolysis.

Reaction 4: Cleavage to Two Triose Phosphates

Reaction 4 is catalyzed by **fructose-1,6-bisphosphate aldolase,** usually called **aldolase,** because its reaction is similar to the reverse of an aldol condensation. In this reaction the "splitting of sugar" that is connoted by the term *glycolysis* occurs, because the six-carbon compound fructose-1,6-bisphosphate is cleaved to give 2 three-carbon intermediates, glyceraldehyde-3-phosphate and dihydroxyacetone phosphate.

$$\text{D-Fructose-1,6-bisphosphate} \rightleftharpoons \text{Dihydroxyacetone phosphate} + \text{D-glyceraldehyde-3-phosphate} \quad \Delta G^{\circ\prime} = +23.9 \text{ kJ/mol}$$

This reaction illustrates an important metabolic principle. Note that the reaction is strongly endergonic under standard conditions, such that formation of fructose-1,6-bisphosphate is highly favored. However, from the actual intracellular concentrations of the reactant and products, as determined in rabbit skeletal muscle, a ΔG of -1.3 kJ/mol is calculated, consistent with the observation that the reaction runs rightward in vivo. This example illustrates the importance of considering conditions *in the cell,* and *not* standard state conditions, in deciding which direction of a reaction is favored.

Aldolase cleaves fructose-1,6-bisphosphate under intracellular conditions, even though one might predict it to act oppositely under standard conditions.

Aldolase from most vertebrate sources is a tetrameric protein. The enzyme activates the substrate for cleavage by condensing the keto carbon at position 2 with a lysine ε-amino group in the active site to give a **Schiff base** intermediate, as shown in Figure 13.4. A Schiff base is a condensation product between an amino group and a carbonyl group. The activated substrate undergoes abstraction of a

Glyceraldehyde-3-phosphate

$K_{eq} = 10^{-4}$

Fructose-1,6-bisphosphate

Schiff base

Dihydroxyacetone phosphate

Aldolase

FIGURE **13.4**

Reaction mechanism for fructose-1,6-bisphosphate aldolase. The figure shows the Schiff base intermediate between the substrate and the active site's lysine residue. B, a cysteine residue on the enzyme, accepts a proton from the hydroxyl on C-4 and returns it to a histidine residue after cleavage between C-3 and C-4.

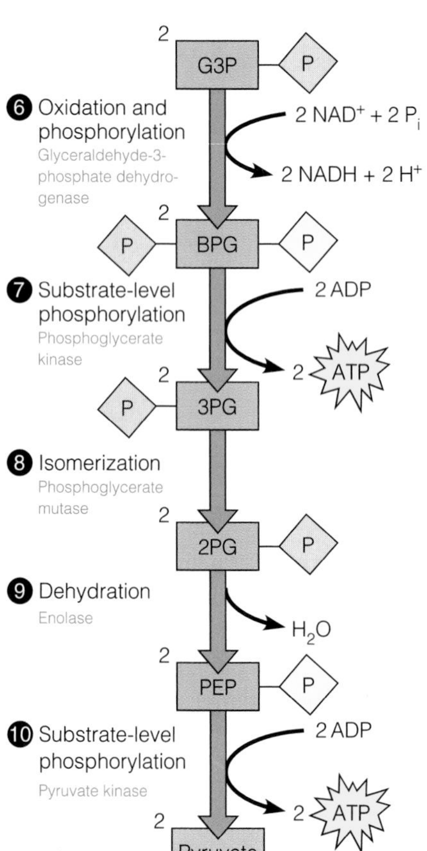

proton from the hydroxyl group at carbon 4, followed by an elimination of the resulting enolate ion, which results in splitting the bond between C-3 and C-4.

Reaction 5: Isomerization of Dihydroxyacetone Phosphate

As noted above, the aldolase reaction yields 2 three-carbon sugar phosphates. The function of reaction 5, catalyzed by **triose phosphate isomerase,** is conversion of one of these products, dihydroxyacetone phosphate (DHAP), to the other, glyceraldehyde-3-phosphate (G3P). Because G3P is the substrate for the next glycolytic reaction, this reaction permits use of all six carbon atoms of glucose.

Dihydroxyacetone phosphate ⇌ **D-Glyceraldehyde-3-phosphate** $\Delta G^{\circ\prime} = +7.6$ kJ/mol

This reaction is also weakly endergonic under standard conditions, but the intracellular concentration of glyceraldehyde-3-phosphate is low, drawing the reaction toward the right. As was described in Chapter 11 (page 370), the isomerization of dihydroxyacetone phosphate proceeds via an enediol intermediate.

At this point, glycolysis has expended two ATP molecules and converted one hexose sugar to two molecules of glyceraldehyde-3-phosphate, each of which will next be metabolized to give high-energy compounds that can drive the synthesis of ATP. The energy investment phase of the cycle is complete, and the energy generation phase is about to begin.

REACTIONS 6–10: THE ENERGY GENERATION PHASE

The five reactions of the energy generation phase are summarized in the margin.

Reaction 6: Generation of the First Energy-Rich Compound

Mechanistically, this reaction, catalyzed by **glyceraldehyde-3-phosphate dehydrogenase,** is among the most important in glycolysis, partly because it generates the first high-energy intermediate and partly because it generates a pair of reducing equivalents (see Figure 13.5). The overall reaction is as follows:

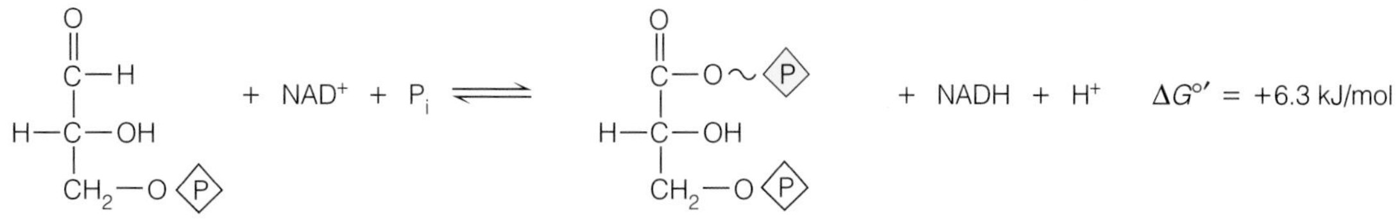

$$\text{D-Glyceraldehyde-3-phosphate} + NAD^+ + P_i \rightleftharpoons \text{1,3-Bisphosphoglycerate} + NADH + H^+ \qquad \Delta G^{\circ\prime} = +6.3 \text{ kJ/mol}$$

D-Glyceraldehyde-3-phosphate **1,3-Bisphosphoglycerate**

Reaction 6 involves a two-electron oxidation of the carbonyl carbon of glyceraldehyde-3-phosphate to the carboxyl level, a reaction that is normally quite exergonic. However, the overall reaction is slightly *endergonic* (under *standard* conditions), because the enzyme utilizes most of the energy released to drive the synthesis of a super-high-energy compound, 1,3-bisphosphoglycerate (BPG). This compound contains a carboxylic-phosphoric acid anhydride, or an **acyl-phosphate group,** at position 1, a functional group with a very high standard free energy of hydrolysis, −49.4 kJ/mol. This enzyme also requires a coenzyme, NAD^+, to accept electrons from the substrate being oxidized.

Glyceraldehyde-3-phosphate dehydrogenase creates a high-energy compound and generates a pair of reducing equivalents.

Because the acyl-phosphate group is much more energy rich than the phosphate anhydride of ATP, 1,3-bisphosphoglycerate can drive the synthesis of ATP from ADP. Indeed, it does so in the next reaction in the sequence, the first of two substrate-level phosphorylations in glycolysis. Because of the importance of understanding how ATP is synthesized, much attention has focused on understanding how the super-high-energy compounds in substrate-level phosphorylation are synthesized.

For glyceraldehyde-3-phosphate dehydrogenase, that understanding derived in large part from an old observation that glycolysis is inhibited by iodoacetate and by heavy metals such as mercury. Both compounds react with free sulfhydryl groups, as the following shows for iodoacetate:

$$RSH + ICH_2COO^- \longrightarrow RS{-}CH_2COO^- + HI$$

The finding that these compounds inhibit glycolysis specifically by inhibition of glyceraldehyde-3-phosphate dehydrogenase strongly implied that the enzyme contains one or more essential thiol groups. We now know that the reaction proceeds as outlined in Figure 13.5, starting with formation of a **thiohemiacetal**

FIGURE **13.5**

Reaction pathway for glyceraldehyde-3-phosphate dehydrogenase. Step 1: Formation of the initial thiohemiacetal intermediate between glyceraldehyde-3-phosphate and the enzyme. **Step 2:** Oxidation of the initial intermediate by NAD^+ to give an acyl-enzyme intermediate. **Step 3:** Phosphorolytic cleavage of the thioester bond in the acyl-enzyme intermediate.

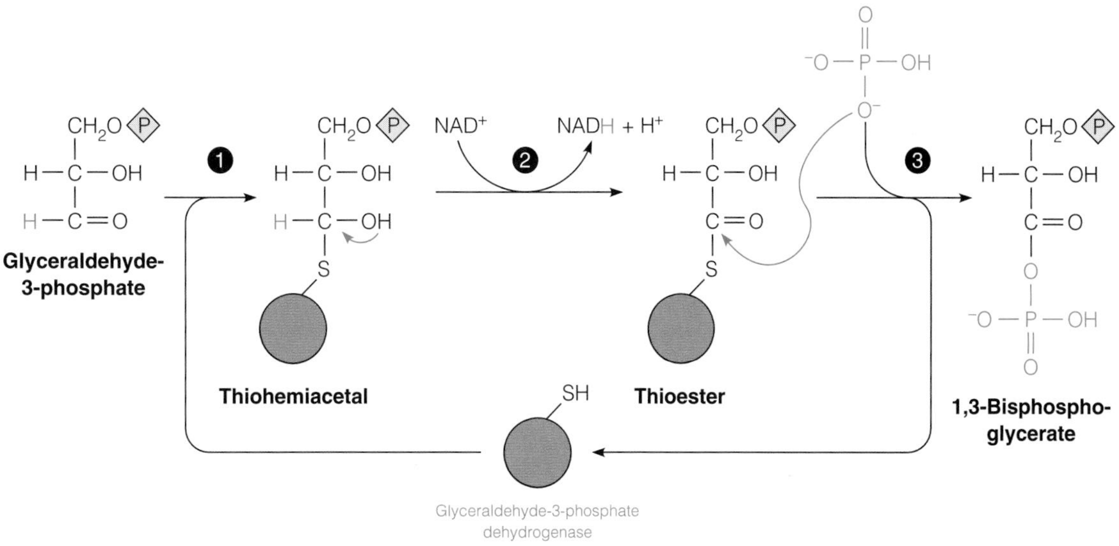

group involving the substrate carbonyl group and a cysteine thiol group on the enzyme. The thiohemiacetal is next oxidized by NAD^+ to give an acyl-enzyme intermediate, or thioester. Thioesters are high-energy compounds; phosphorolysis of this thioester by P_i preserves much of the energy as the acyl phosphate, which is the product.

The overall stoichiometry of the reaction involves reduction of 1 mole of NAD^+ to NADH + H^+. This reaction is the source of the NADH formed in glycolysis, which was first identified in Figure 13.2.

Reaction 7: The First Substrate-Level Phosphorylation

As noted previously, 1,3-bisphosphoglycerate, because of its high group transfer potential, has a strong tendency to transfer its acyl-phosphate group to ADP, with resultant formation of ATP. This substrate-level phosphorylation reaction is catalyzed by **phosphoglycerate kinase,** as follows:

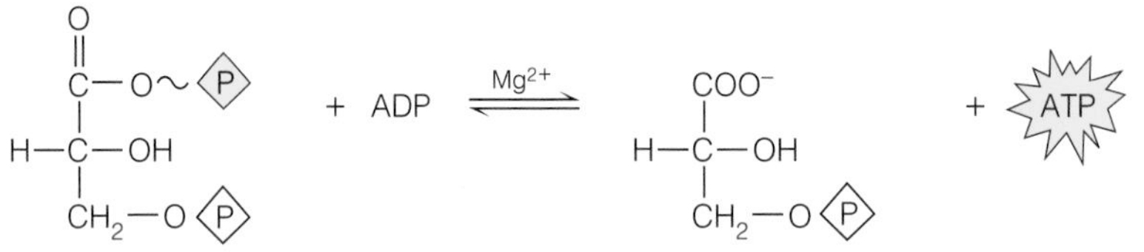

1,3-Bisphosphoglycerate + ADP $\xrightleftharpoons{Mg^{2+}}$ 3-Phosphoglycerate + ATP $\quad \Delta G^{\circ\prime} = -18.8$ kJ/mol

1,3-Bisphosphoglycerate **3-Phosphoglycerate**

Phosphoglycerate kinase catalyzes the first glycolytic reaction that forms ATP.

At this stage the net ATP yield from the glycolytic pathway is zero. Recall that two ATPs per mole of glucose were invested to generate 2 moles of triose phosphate. The reaction shown here generates one ATP from each mole of triose phosphate, or two ATPs per mole of glucose. The pathway as a whole becomes exergonic in the remaining three reactions. This involves activation of the remaining phosphate, which, in 3-phosphoglycerate (3PG), has a relatively low phosphate transfer potential.

Reaction 8: Preparing for Synthesis of the Next High-Energy Compound

Activation of 3-phosphoglycerate begins with an isomerization catalyzed by **phosphoglycerate mutase.** The enzyme transfers phosphate from position 3 to position 2 of the substrate to yield 2-phosphoglycerate. Mg^{2+} is required.

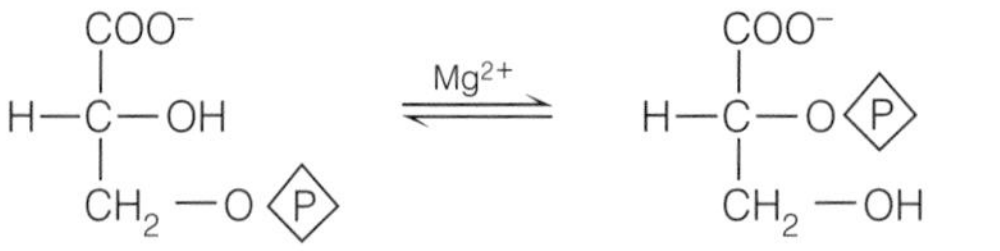

3-Phosphoglycerate $\xrightleftharpoons{Mg^{2+}}$ 2-Phosphoglycerate $\quad \Delta G^{\circ\prime} = +4.4$ kJ/mol

3-Phosphoglycerate **2-Phosphoglycerate**

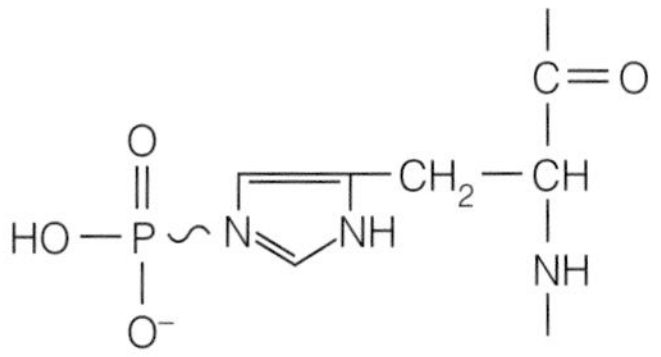

***N*-Phosphohistidine residue**

The reaction is slightly endergonic under standard conditions. Again, the intracellular level of 3-phosphoglycerate is high relative to that of 2-phosphoglycerate (2PG), so that in vivo the reaction proceeds to the right without difficulty. The enzyme contains a phosphohistidine residue in the active site. In the first step of the reaction, the phosphate is transferred from the enzyme to the substrate to give an intermediate, 2,3-bisphosphoglycerate. Breakdown of the enzyme-bound intermediate regenerates the phosphorylated enzyme and forms the product, which is released.

Enzyme–P + 3-P-glycerate ⟶ [Enzyme–2,3-bis-P-glycerate] ⟶ Enzyme–P + 2-P-glycerate

Reaction 9: Synthesis of the Second High-Energy Compound

Reaction 9, catalyzed by **enolase,** generates phosphoenolphyruvate (PEP), another super-high-energy compound. PEP participates in the second substrate-level phosphorylation of glycolysis.

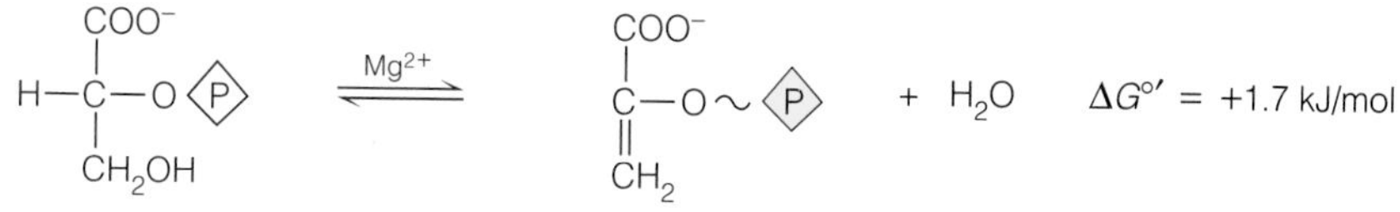

$$\text{2-Phosphoglycerate} \xrightleftharpoons{Mg^{2+}} \text{Phosphoenolpyruvate} + H_2O \qquad \Delta G^{\circ\prime} = +1.7 \text{ kJ/mol}$$

2-Phosphoglycerate **Phosphoenolpyruvate**

The reaction involves a simple dehydration, or α,β-elimination, and the overall free energy change is small. However, the effect is to increase enormously the free energy of hydrolysis of the phosphate bond—from –15.6 kJ/mol for 2-phosphoglycerate to −61.9 kJ/mol for phosphoenolpyruvate. Carbon 2 of phosphoenolpyruvate is "locked into" the unfavored enol configuration and, as discussed in Chapter 3, the great thermodynamic instability of enolpyruvate is chiefly responsible for the large negative free energy of hydrolysis of phosphoenolpyruvate.

Reaction 10: The Second Substrate-Level Phosphorylation

In the last reaction, catalyzed by **pyruvate kinase,** phosphoenolpyruvate transfers its phosphoryl group to ADP in another substrate-level phosphorylation. Note that the enzyme is named as if it were acting leftward in the reaction shown below, even though it is strongly exergonic in the rightward direction. Many enzymes were named before the function or direction of intracellular catalysis had been identified.

Pyruvate kinase catalyzes the second glycolytic reaction that forms ATP.

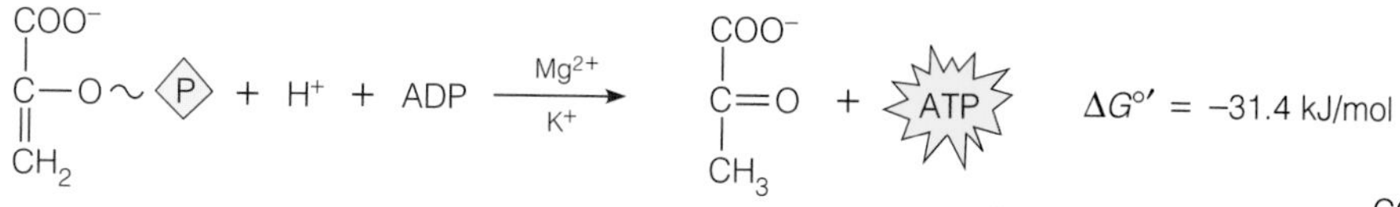

$$\text{Phosphoenolpyruvate} + H^+ + ADP \xrightarrow[K^+]{Mg^{2+}} \text{Pyruvate} + ATP \qquad \Delta G^{\circ\prime} = -31.4 \text{ kJ/mol}$$

Phosphoenolpyruvate **Pyruvate**

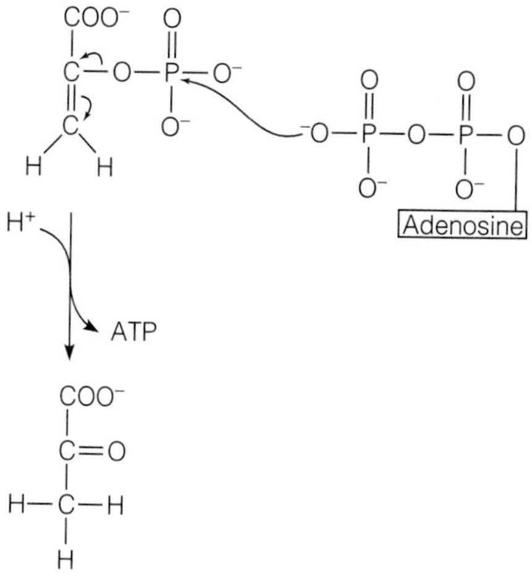

The enzyme requires Mg^{2+} and K^+. Even though the reaction involves the endergonic synthesis of ATP, the overall reaction is strongly exergonic because, as noted in Chapter 3, the spontaneous tautomerization of the product, enolpyruvate, to the highly favored keto form provides a strong thermodynamic drive in the forward direction.

The pyruvate kinase reaction is another site for metabolic regulation. In vertebrate liver the enzyme, a tetramer of M_r (molecular weight) about 250,000, is allosterically inhibited at high ATP concentrations and activated by fructose-1,6-bisphosphate. The synthesis of the liver enzyme is under dietary control; intracellular activity may increase as much as 10-fold from increased enzyme synthesis, or induction, as a result of high carbohydrate ingestion. Whatever the genetic regulatory mechanisms involved, this induction may contribute to the efficacy of "carbohydrate loading," the practice of eating a great deal of carbohydrate before an athletic event requiring great endurance, such as a marathon run. Augmented pyruvate kinase levels increase the rate at which energy can be generated by glycolysis.

Pyruvate kinase activity in the liver is also regulated by phosphorylation and dephosphorylation of the enzyme protein. The dephosphorylated form is far more active than the phosphorylated form. Phosphorylation, which is under hormonal control, diverts phosphoenolpyruvate to gluconeogenesis (see Chapter 16) when fatty acid oxidation and the citric acid cycle are already operating at rates sufficient to meet the energy needs of the cell. By contrast, virtually all of the phosphoenolpyruvate produced in muscle is converted to pyruvate.

Dietary carbohydrate induces pyruvate kinase and increases the ability of the body to obtain energy from glycolysis.

Human genetic deficiencies of erythrocyte pyruvate kinase have been studied. Accumulation of phosphoenolpyruvate leads to excessive levels in blood of other glycolytic intermediates. Of major clinical importance is the accumulation of 2,3-bisphosphoglycerate, which was introduced in Chapter 7 as an allosteric inhibitor of oxygen binding to hemoglobin. This accumulation leads to impaired oxygen uptake in the lungs and impaired transport through the bloodstream to tissues.

TABLE 13.1 Summary of glycolysis

Reaction	Enzyme	ATP Yield	$\Delta G°'$ (kJ/mol)	ΔG (kJ/mol)
ENERGY INVESTMENT PHASE				
Glucose (G)				
1 (ATP → ADP)	Hexokinase (HK)	−1	−16.7	−33.5
Glucose-6-phosphate (G6P)				
2	Phosphoglucoisomerase (PGI)		+1.7	−2.5
Fructose-6-phosphate (F6P)				
3 (ATP → ADP)	Phosphofructokinase (PFK)	−1	−14.2	−22.2
Fructose-1,6-bisphosphate (FBP)				
4	Aldolase (ALD)		+23.9	−1.3
Glyceraldehyde-3-phosphate (G3P) + dihydroxyacetone phosphate (DHAP)				
5	Triose phosphate isomerase (TPI)		+7.6	+2.5
Two glyceraldehyde-3-phosphate				
ENERGY GENERATION PHASE				
6 ($2NAD^+ + 2P_i$ → $2NADH + 2H^+$)	Glyceraldehyde-3-phosphate dehydrogenase (G3PDH)		+12.6	−3.4
Two 1,3-bisphosphoglycerate (BPG)				
7 (2ADP → 2 ATP)	Phosphoglycerate kinase (PGK)	+2	−37.6	+2.6
Two 3-phosphoglycerate (3PG)				
8	Phosphoglycerate mutase (PGM)		+8.8	+1.6
Two 2-phosphoglycerate (2PG)				
9 (→ $2H_2O$)	Enolase (ENO)		+3.4	−6.6
Two phosphoenolpyruvate (PEP)				
10 (2ADP → 2 ATP)	Pyruvate kinase (PK)	+2	−62.8	−33.4
Two pyruvate (Pyr)				
Net: Glucose + 2ADP + $2P_i$ + $2NAD^+$ ⟶ 2 pyruvate + 2ATP + 2NADH + $2H^+$ + $2H_2O$		+2	−73.3	−96.2

Note: ΔG values are estimated from the approximate intracellular concentrations of glycolytic intermediates in rabbit skeletal muscle. All ΔG values past reaction 5 are doubled, because each reaction involves 2 three-carbon substrates per glucose molecule.

The pyruvate kinase reaction converts the overall glycolytic pathway from an energy-neutral process to one that involves net synthesis of ATP. Two high-energy phosphates per mole of hexose are generated here, to go with the two generated by phosphoglycerate kinase. Subtracting the two ATPs invested at hexokinase and phosphofructokinase gives a net yield of two high-energy phosphates per mole of glucose—not a high yield, to be sure, but the process can meet the energy requirements of many anaerobes. Moreover, subsequent metabolism of pyruvate through aerobic pathways generates much additional high-energy phosphate.

Table 13.1 summarizes the reactions of glycolysis, showing free energy changes and ATP yields at each step.

Metabolic Fates of Pyruvate

Pyruvate represents a central metabolic branch point. Its fate depends crucially on the oxidation state of the cell, which is related to the reaction catalyzed by glyceraldehyde-3-phosphate dehydrogenase (reaction 6). Recall that this reaction converts 1 mole of NAD^+ per mole of triose phosphate to NADH. *This NADH must be reoxidized to* NAD^+ for glycolysis to continue. During aerobic glycolysis, as noted earlier, this NADH is oxidized by the mitochondrial electron transport chain, with the electrons transferred ultimately to oxygen. This oxidation of NADH, which we consider in detail in Chapter 15, yields additional energy, with about 3 moles of ATP synthesized from ADP per mole of NADH oxidized. Because 2 moles of NADH are produced per mole of glucose entering the pathway, aerobic glycolysis yields considerably more ATP than anaerobic glycolysis. In addition, the oxidation of pyruvate through the citric acid cycle generates much more energy.

LACTATE METABOLISM

In aerobic cells that are undergoing very high rates of glycolysis, the NADH generated in glycolysis cannot all be reoxidized at comparable rates in the mitochondrion. In such cases or in anaerobic cells, which lack mitochondria, NADH must be used to drive the reduction of an organic substrate in order to maintain homeostasis. As noted earlier, that substrate is pyruvate itself, both in eukaryotic cells and in lactic acid bacteria, and the product is lactate. The enzyme catalyzing this reaction is lactate dehydrogenase (see page 448). The equilibrium for this reaction lies far to the right. Figure 13.6, which depicts the energy profile of anaerobic glycolysis, shows that NADH produced in the oxidation of glyceraldehyde-3-phosphate is used to reduce pyruvate to lactate. Thus, during anaerobic glycolysis or **lactic acid fermentation**, an overall electron balance is maintained.

Pyruvate must be reduced to lactate when tissues are insufficiently aerobic to oxidize all of the NADH formed in glycolysis.

In vertebrates some tissues, such as red blood cells, derive most of their energy from anaerobic metabolism. Skeletal muscle, which derives most of its energy from respiration when at rest, relies heavily on glycolysis during exertion, when glycogen stores are rapidly broken down, or mobilized, to provide substrates for glycolysis. Normally the lactate produced diffuses from the tissue and is transported through the bloodstream to highly aerobic tissues, such as heart and liver. The aerobic tissue can catabolize lactate further, through respiration, or can convert it back to glucose, through gluconeogenesis. However, if lactate is produced in large quantities, it cannot be readily consumed. Then, as we discussed in Chapter 7, the blood pH falls and the Bohr effect functions to increase oxygen supplies to the tissues.

Until recently it was thought that lactate accumulation in skeletal muscle was largely a consequence of anaerobic metabolism, which occurs when the need for tissues to generate energy exceeds their capacity to oxidize the pyruvate produced in glycolysis. Metabolic studies, including ^{31}P NMR analyses of the levels of phosphorylated intermediates in living muscle cells during exercise, suggest that

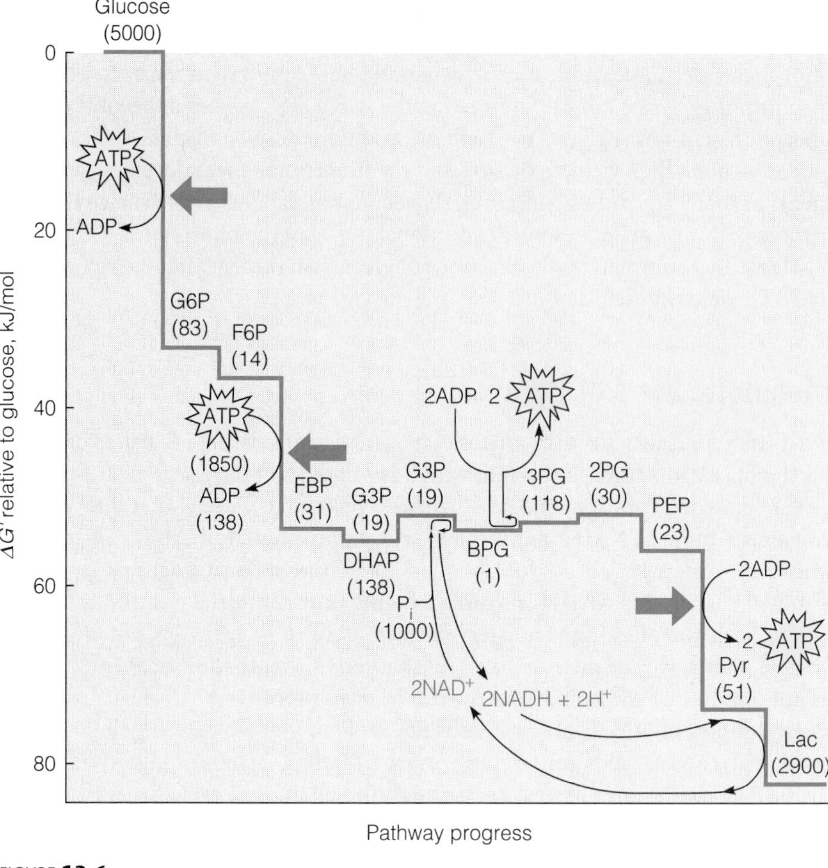

FIGURE **13.6**

Energy and electron profile of anaerobic glycolysis. The graph shows the $\Delta G'$ for each reaction, calculated from $\Delta G^{\circ\prime}$ values and the estimated concentration of each intermediate in the human erythrocyte. (The numbers in parentheses are the approximate micromolar concentrations.) Note these points: (1) Two of the four ATPs generated are used to repay the initial ATP investment, (2) the reducing equivalents generated by glyceraldehyde-3-phosphate dehydrogenase must be used to reduce an organic substrate in anaerobiosis, (3) enzymes subject to allosteric control are those that catalyze reactions so highly exergonic as to be virtually irreversible (arrows), and (4) because a reaction must have a negative $\Delta G'$ to proceed at all, imprecision in measuring metabolite concentrations is probably responsible for the recording of positive $\Delta G'$ values for some reactions.

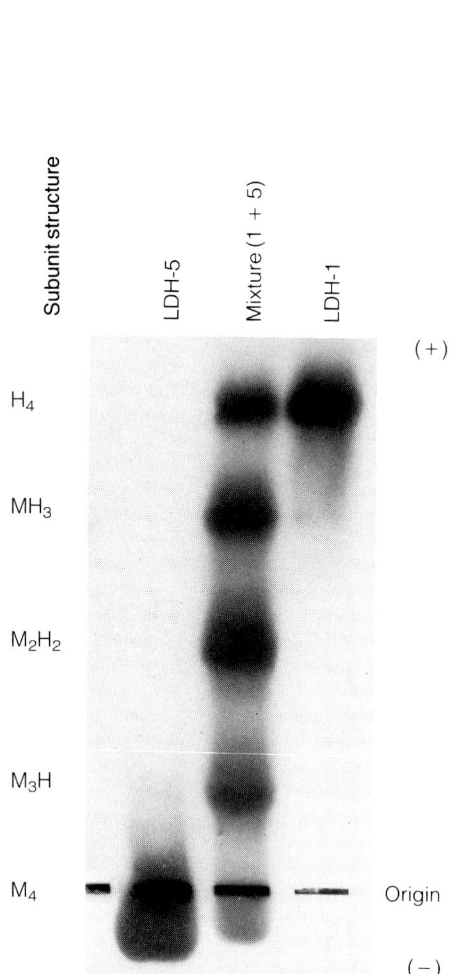

FIGURE **13.7**

Structural basis for the existence of isoenzymes of lactate dehydrogenase. Preparations were subjected to electrophoresis in a starch gel, which was then treated to reveal bands containing enzymatically active protein. LDH-1 is a tetramer containing only the H subunit, whereas LDH-5 contains only M subunits. The middle lane depicts an experiment in which equal amounts of LDH-1 and LDH-5 were mixed. The subunits were dissociated and then allowed to reassociate. The presence of five different enzyme forms and their relative amounts show that individual M and H subunits can associate randomly to form tetramers of mixed subunit composition.

lactate is actually an intermediate and not a metabolic "dead end," whose only fate is reconversion to pyruvate. These studies show that even in fully oxygenated muscle tissue, as much as 50% of the glucose metabolized is converted to lactate. This may represent a means for coordinating energy-storing and energy-generating pathways in different tissues, but the mechanisms involved are not yet clear.

ISOENZYMES OF LACTATE DEHYDROGENASE

Lactate dehydrogenase, like many enzymes, exists in animal tissues in multiple molecular forms. Different molecular forms of an enzyme catalyzing the same reaction are called **isoenzymes,** or **isozymes.** Lactate dehydrogenase was the first enzyme that established the structural basis for the existence of isoenzymes. Most tissues contain five isoenzymes of lactate dehydrogenase. They can be resolved electrophoretically, as shown in Figure 13.7.

Lactate dehydrogenase (LDH) is a tetrameric protein consisting of two types of subunits, called M and H, which have small differences in amino acid sequence. M subunits predominate in skeletal muscle and liver, and H subunits predominate in heart. M and H subunits combine randomly with each other, so that the five major isoenzymes have the compositions M_4, M_3H, M_2H_2, MH_3, and H_4. Because of random subunit reassortment, the isoenzymic composition of a tissue is determined primarily by the activities of the genes specifying the two subunits.

The physiological need for the existence of different forms of this enzyme is not yet clear. However, the tissue specificity of isoenzyme patterns is useful in clinical medicine. Such pathological conditions as myocardial infarction, infectious hepatitis, and muscle diseases involve cell death of affected tissue, with release of cell contents to the blood. The pattern of LDH isoenzymes in the blood serum is representative of the tissue that released the isoenzymes. This information can be used to diagnose such conditions and to monitor the progress of treatment.

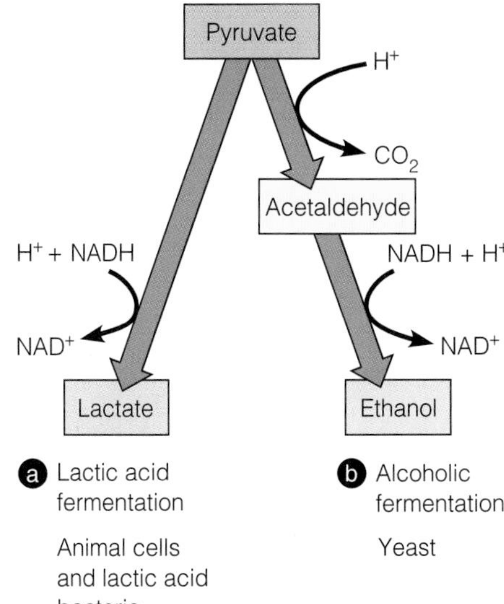

(a) Lactic acid fermentation

Animal cells and lactic acid bacteria

(b) Alcoholic fermentation

Yeast

ETHANOL METABOLISM

Pyruvate has numerous alternative fates in anaerobic microorganisms. As we have seen, lactic acid bacteria reduce pyruvate to lactate in a single step (see margin). By contrast, yeasts convert pyruvate to ethanol in a two-step pathway. This **alcoholic fermentation** starts with the nonoxidative decarboxylation of pyruvate to acetaldehyde, catalyzed by **pyruvate decarboxylase.** This reaction is followed by the NADH-dependent reduction of acetaldehyde to ethanol, catalyzed by alcohol dehydrogenase.

$$\underset{\textbf{Pyruvate}}{CH_3-\overset{\overset{\displaystyle O}{\|}}{C}-COO^-} \xrightarrow[\text{Pyruvate decarboxylase}]{H^+ \quad CO_2} \underset{\textbf{Acetaldehyde}}{CH_3-\overset{\overset{\displaystyle O}{\|}}{C}-H} \underset{\text{Alcohol dehydrogenase}}{\overset{NADH + H^+ \quad NAD^+}{\rightleftharpoons}} \underset{\textbf{Ethanol}}{CH_3-CH_2-OH}$$

The first reaction requires **thiamine pyrophosphate** as a coenzyme. This coenzyme, derived from vitamin B_1, participates in a number of group transfer reactions involving an activated aldehyde moiety (see Chapter 14).

Industrial production of ethanol has assumed immense importance as humanity attempts to deal with two serious problems—(1) replacement of nonrenewable petroleum by renewable fuel sources and (2) utilization of biological waste materials. Efforts in these arenas involve bioengineering, to generate bacterial strains that can convert materials such as the cellulose in wood waste or straw or more complex materials in human and animal wastes to hexose sugars, and analysis of glycolytic regulation, aimed at maximizing ethanol production once suitable substrates for glycolysis have been produced.

Animal tissues also contain alcohol dehydrogenase, even though ethanol is not a major metabolic product in animal cells. Some of the major metabolic consequences of ethanol intoxication result from ethanol oxidation by this enzyme in the liver. First, there is massive reduction of NAD^+ to NADH, which depletes NAD^+ levels, thereby decreasing flux through glyceraldehyde-3-phosphate dehydrogenase, with consequent inhibition of energy generation. Second, acetaldehyde is quite toxic, and many of the unpleasant effects of hangovers result from actions of acetaldehyde and its subsequent metabolites.

Energy and Electron Balance Sheets

By writing a balanced chemical equation for glycolysis, we can compute the energy yield accompanying conversion of 1 mole of glucose. For anaerobic glycolysis or for a lactic acid fermentation, we can write the following balanced equation:

$$\text{Glucose} + 2\text{ADP} + 2\text{P}_i \longrightarrow 2\text{ lactate} + 2\text{ATP} + 2\text{H}_2\text{O}$$

Similarly, we can write a balanced equation for alcoholic fermentation:

$$\text{Glucose} + 2\text{ADP} + 2\text{P}_i + 2\text{H}^+ \longrightarrow 2\text{ ethanol} + 2\text{CO}_2 + 2\text{ATP} + 2\text{H}_2\text{O}$$

Note first that both processes involve no net change in oxidation state; NAD^+ and NADH, both of which participate in the reaction pathways, do not appear in the overall reactions. This is also shown in Figure 13.6.

During aerobic glycolysis, the nicotinamide nucleotides do appear in the overall equation:

$$\text{Glucose} + 2\text{ADP} + 2\text{P}_i + 2\text{NAD}^+ \longrightarrow 2\text{ pyruvate} + 2\text{ATP} + 2\text{NADH} + 2\text{H}^+ + 2\text{H}_2\text{O}$$

The NADH generates reducing equivalents within the mitochondrion by an energy-requiring transport process, as described in Chapter 15. We know empirically that reoxidation of 1 mole of NADH in the mitochondrion generates about 3 moles of ATP.

$$2\text{NADH} + 8\text{H}^+ + \text{O}_2 + 6\text{ADP} + 6\text{P}_i \longrightarrow 2\text{NAD}^+ + 8\text{H}_2\text{O} + 6\text{ATP}$$

Summing these last two equations, you can see that glycolysis occurring during respiration yields 8 moles of ATP per mole of glucose.

$$\text{Glucose} + 8\text{ADP} + 6\text{H}^+ + 8\text{P}_i + \text{O}_2 \longrightarrow 2\text{ pyruvate} + 8\text{ATP} + 10\text{H}_2\text{O}$$

Metabolism of glucose to either lactate or ethanol represents a nonoxidative process, as you can see by comparing the empirical formulas for glucose ($C_6H_{12}O_6$) and lactate ($C_3H_6O_3$). Clearly, there is no change in the overall oxidation state of the carbons, because the numbers of hydrogens and oxygens bound per carbon atom are identical for glucose and lactate. The same is true for ethanol plus CO_2, when one counts the atoms in both. However, some individual carbon atoms of lactate and ethanol plus CO_2 undergo oxidation, and some become reduced. By contrast, pyruvate is more highly oxidized than glucose, as seen from its empirical formula ($C_3H_4O_3$).

Glycolysis, which yields 2 ATP per glucose anaerobically or 8 ATP aerobically, releases but a small fraction of the energy available from glucose.

Note also that glycolysis, whether aerobic or anaerobic, releases but a small fraction of the potential energy stored in the glucose molecule. As noted earlier (see Chapters 3 and 12), the complete combustion of glucose to CO_2 and H_2O releases 2870 kJ/mol of free energy under standard conditions. As we shall see in the next chapter, about 38 moles of ATP are synthesized from ADP per mole of glucose carried completely through glycolysis and the citric acid cycle. The free energy needed to drive the synthesis of these 38 ATPs represents about 40% of the potential energy released in glucose combustion. Since catabolism of glucose to lactate or pyruvate yields but 2 or 8 moles of ATP, respectively, you can see that most of the potential energy originally present in glucose is still waiting to be released after glycolysis. Aerobic metabolism yields more energy from glucose; therefore, aerobic organisms in general are more successful and widespread than anaerobic organisms. The early evolution of aerobic metabolism made possible the large, active animals that exist today. Nevertheless, many large animals still derive a large fraction of their metabolic energy from glycolysis, under certain physiological circumstances. A good example is the crocodile—torpid (and aerobic) for much of its life, yet capable of short bursts of intensely rapid movement. In the latter circumstance glycolysis, coupled with the breakdown of carbohydrate energy stores, represents a rapid, though inefficient, way to mobilize energy.

Regulation of Glycolysis

Glycolysis is closely coordinated with other major pathways of energy generation and utilization, notably synthesis and breakdown of glycogen (or starch), gluconeogenesis, the pentose phosphate pathway, and the citric acid cycle. Metabolic factors that control glycolysis tend to regulate other processes in a coordinated fashion. Thus, it is difficult to consider regulation of glycolysis in isolation from these other processes, and we return to this topic after we have presented the other major pathways in energy metabolism (see Chapter 23). However, it is important here to describe two key glycolytic enzymes that serve as regulatory targets—phosphofructokinase (the major target) and pyruvate kinase. Note that hexokinase also catalyzes a regulated step (see page 368 and Figure 11.8). Control of hexokinase by its product, glucose-6-phosphate, is also involved in other processes, such as glycogen synthesis and the maintenance of blood glucose levels.

Phosphofructokinase and pyruvate kinase are the major control sites for glycolysis.

THE PASTEUR EFFECT

The recognition that glycolysis is controlled primarily by the activity of phosphofructokinase developed largely from a discovery made over a century ago by Louis Pasteur: When anaerobic yeast cultures metabolizing glucose were exposed to air, the rate of glucose utilization decreased dramatically. It became clear that this phenomenon, the **Pasteur effect,** involves the inhibition of glycolysis by oxygen. This effect makes biological sense, because far more energy is derived from complete oxidation of glucose than from glycolysis alone. What is the mechanism of this effect if oxygen is not an active participant in glycolysis? The needed insight came from analyses of the intracellular contents of glycolytic intermediates in aerobic and anaerobic cells. These analyses required techniques for the rapid interruption of metabolism and extraction of metabolites. One such technique is **freeze-clamping,** in which tissue is rapidly compressed between metal plates cooled to liquid nitrogen temperatures. The solid tissue can then be powdered and extracted for analysis.

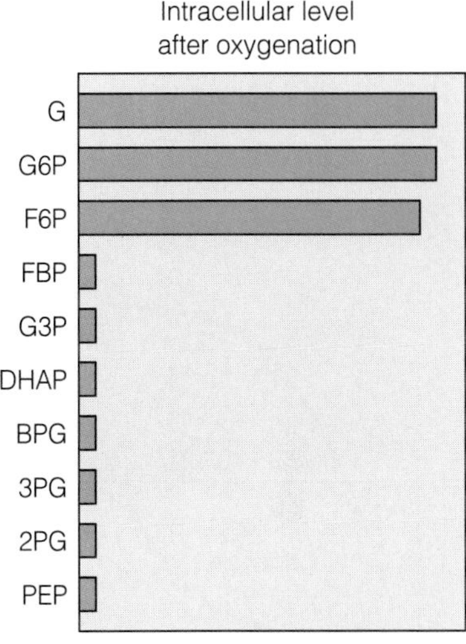

Experiments of this type revealed that when oxygen is introduced to anaerobic cells, the levels of all the glycolytic intermediates from fructose-1,6-bisphosphate onward *decrease*, while all of the *earlier* intermediates accumulate at higher levels. This finding is consistent with the idea that the metabolic flux through phosphofructokinase is specifically decreased in the presence of O_2.

OSCILLATIONS OF GLYCOLYTIC INTERMEDIATES

Other important conclusions emerged from the discovery that the intracellular levels of glycolytic intermediates are not constant under many conditions but undergo periodic variations, or oscillations, as shown in Figure 13.8. These variations can be visualized most readily by following fluorescence of a yeast cell suspension at 450 nm. The major contributor to this fluorescence is NADH, so this type of experiment monitors changes in the intracellular NADH pool with time. Oscillations are a common feature of feedback-controlled systems, and the cyclic variations in levels of glycolytic intermediates provide important clues to regulatory mechanisms affecting glycolysis.

During the time that the fluorescence of a yeast cell suspension is increasing, NADH is accumulating, through reduction of intracellular NAD^+. Under these conditions, glycolysis is turned on and NADH is being produced by glyceraldehyde-3-phosphate dehydrogenase faster than it can be used to reduce pyruvate. Presumably, during this period one or more regulatory substances are also accumulating. Once they have accumulated sufficiently to inhibit glycolysis, the NADH level falls—until the supply of regulators is depleted to the point that the pathway is inhibited again. This cycle occurs repeatedly.

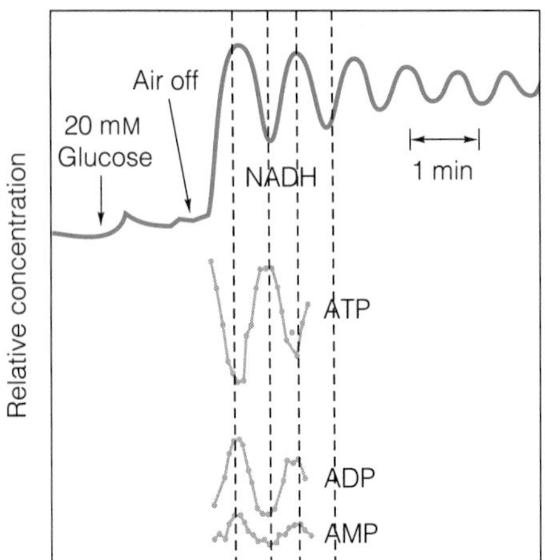

FIGURE **13.8**

Periodic oscillations of the levels of glycolytic intermediates in yeast cells undergoing glycolysis. The culture was shifted from aerobic to anaerobic conditions at the point labeled "Air off." The upper tracing (blue) depicts continuous monitoring of fluorescence of the cell suspension, which is related to the intracellular concentration of NADH. The nucleotide levels (orange) were determined in a parallel experiment in which samples of culture were removed at various times, extracted, and assayed for their contents of ATP, ADP, and AMP.

Once investigators realized that the intracellular levels of NADH were varying periodically, they began sampling extracts of oscillating cells to analyze the intracellular levels of other substrates for glycolytic enzymes. These intermediates were found to rise and fall periodically as well. Note from Figure 13.8 that the levels of ADP and AMP rise and fall precisely in phase with NADH, whereas the level of ATP is 180° out of phase. This pattern suggests that the activity of glycolysis depends in some way on the adenylate energy charge (see Chapter 12): When the charge is high, the pathway is turned off; when it is low, the pathway is activated. These and other observations suggested an enzyme regulated by energy charge as the major point of regulation. Phosphofructokinase is such an enzyme.

ALLOSTERIC REGULATION OF PHOSPHOFRUCTOKINASE

Phosphofructokinase is a complex multisubunit enzyme. In muscle it exists as a homotetramer with M_r of about 360,000 that undergoes reversible dissociation to a dimeric form. Some of the allosteric activators and inhibitors of the enzyme act by influencing the interconversion of the dimer and the tetramer. *Activators* of phosphofructokinase include AMP, ADP, and a more recently discovered compound, **fructose-2,6-bisphosphate.**

Fructose-2,6-bisphosphate

As shown in Figure 13.9a, a very low concentration of fructose-2,6-bisphosphate activates phosphofructokinase. Fructose-2,6-bisphosphate is considered to be the major regulator controlling carbon flux through glycolysis and gluconeogenesis in the liver; its synthesis is catalyzed by another form of phosphofructokinase, called PFK-2. (The glycolytic enzyme is now called PFK-1.) PFK-2 activity in turn is regulated by reversible phosphorylation and dephosphorylation, controlled ultimately by cyclic AMP, in response to hormonal stimuli. The details of these interactions are presented when we discuss the control of gluconeogenesis in Chapter 16.

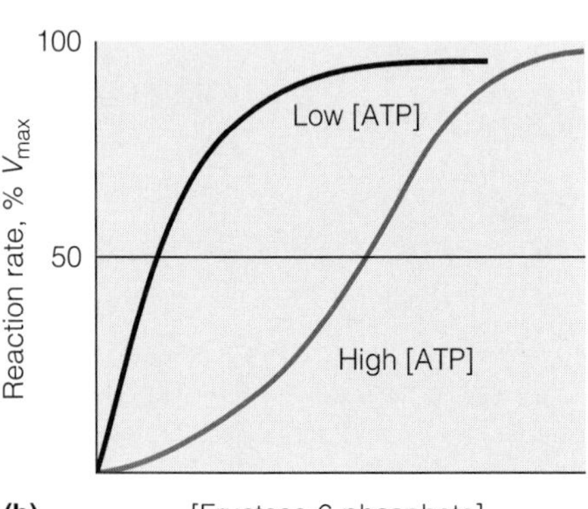

FIGURE **13.9**

Allosteric control of liver phosphofructokinase. (a) Activation by fructose-2,6-bisphosphate. **(b)** How ATP increases the apparent K_M for substrate fructose-6-phosphate.

The most significant *inhibitors* of phosphofructokinase, from a biological standpoint, are ATP (Figure 13.9b) and citrate. The effect of ATP may seem anomalous, because ATP is a substrate and hence essential for the reaction. As an inhibitor, ATP binds to a site on the enzyme separate from the catalytic site, and with lower affinity. At low ATP concentrations the substrate saturation curve for fructose-6-phosphate is nearly hyperbolic, because the regulatory site is not occupied, and the enzyme is almost all in the R state (recall the discussion of R and T states in Chapter 11, page 402). At high ATP levels the T state predominates, causing the curve to become sigmoidal and shift far to the right (Figure 13.9b). Thus, inhibition is achieved because the apparent affinity for fructose-6-phosphate is greatly reduced.

The control of PFK by adenine nucleotides represents a way in which energy metabolism responds to the adenylate energy charge. At high energy charge, the relative abundance of ATP signals that the energy-yielding glycolytic pathway should diminish in activity; the signal involves inhibition of PFK. Conversely, a high AMP or ADP level signals that energy charge is low and that flux through glycolysis should increase. Inhibition by citrate represents another energy level sensor. At high energy charge, flux through the citric acid cycle diminishes, via mechanisms that are discussed in Chapter 14. Under these conditions, citrate accumulates and is transported out of mitochondria. Interaction with PFK in the cytosol can signal that energy generation is adequate, and hence the production of citric acid cycle precursors via glycolysis can be diminished.

CONTROL OF PYRUVATE KINASE

Earlier we identified pyruvate kinase as a control point for glycolysis. Like PFK, pyruvate kinase is a multisubunit enzyme that is inhibited by ATP, in a kinetically similar fashion: High ATP levels reduce the apparent affinity of pyruvate kinase for its other substrate, phosphoenolpyruvate. A second allosteric effect is the **feed-forward activation** of pyruvate kinase by fructose-1,6-bisphosphate. This effect, the converse of feedback inhibition, ensures that carbon passing the first regulated step in the pathway (PFK) will be able to complete its passage through glycolysis and that undesirable accumulation of intermediates will not occur. A third feedback control effect is inhibition of pyruvate kinase by acetyl-CoA, the major product of fatty acid oxidation. This inhibition allows the cell to reduce glycolytic flux when ample substrates are available from fat breakdown.

Phosphofructokinase and pyruvate kinase regulate the flow of intermediates from glucose-6-phosphate to pyruvate. Glycolysis is regulated also at the points of entry of carbon into the pathway. We have already mentioned hexokinase inhibition by glucose-6-phosphate as one such control site. The other major control point, at least in animal metabolism, is the breakdown of glycogen, catalyzed by *glycogen phosphorylase.* This extremely important process is discussed in detail later in this chapter.

GLYCOLYSIS AS BOTH A CATABOLIC AND AN ANABOLIC PATHWAY

Why should the cell regulate glycolysis at more than one point? Any answer to this question should take into account that glycolysis not only generates ATP and provides pyruvate for oxidation via the citric acid cycle but also is a biosynthetic pathway. Intermediates in glycolysis are precursors to a number of compounds, particularly lipids and amino acids. These processes are discussed throughout the book. At this point, Figure 13.10 identifies some of the major biosynthetic roles of glycolytic intermediates. This figure illustrates why glycolysis is considered a major metabolic thoroughfare. Many pathways lead into glycolysis, and many pathways diverge from it, creating a substantial flux through the pathway. Thus, even though the amount of fuel for oxidative degradation is ample, the regulatory demands placed on the cell are too complex to meet with a single rate-controlling reaction. Figure 13.11 summarizes the regulatory relationships between glycolysis and other metabolic pathways.

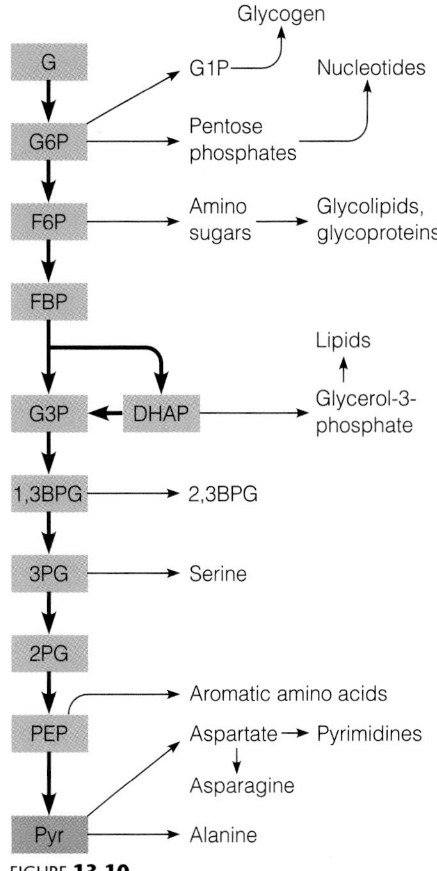

FIGURE **13.10**

Alternative fates of glycolytic intermediates in biosynthetic pathways. G1P is glucose-1-phosphate.

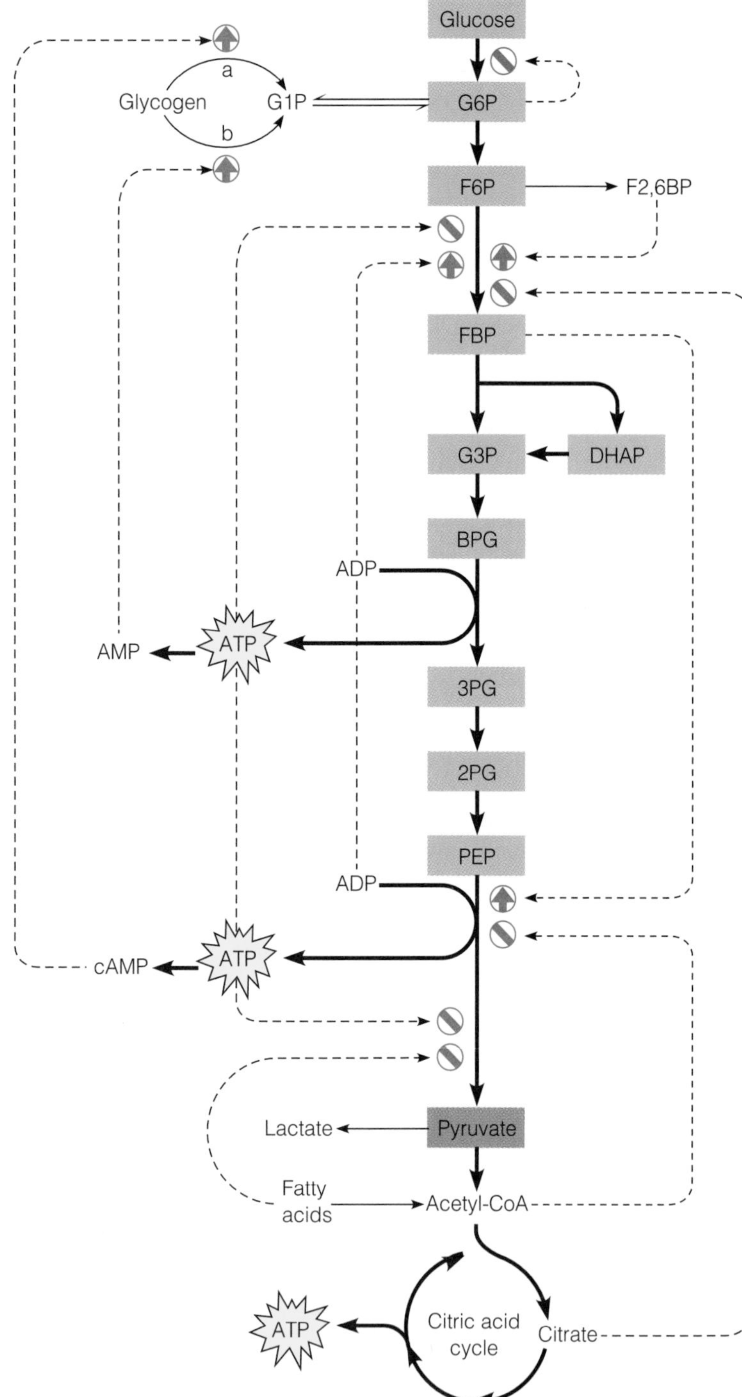

FIGURE **13.11**

Overview of the regulation of glycolysis. The diagram shows points of coordination with other metabolic pathways. F2,6BP = fructose-2,6-bisphosphate; other abbreviations are as in Table 13.1. The circled upward arrows identify activation, and the circled downward slashes represent inhibition. The control of glycogen breakdown is discussed later in this chapter.

Entry of Other Sugars into the Glycolytic Pathway

Thus far our discussion of glycolysis has focused on glucose as a source of carbon for this pathway. Many other sources of carbohydrate energy are available, whether through digestion of foodstuffs or utilization of endogenous metabolites. This section focuses on the utilization of monosaccharides other than glucose, of disaccharides, and of glycerol derived from fat metabolism. These pathways are summarized in Figure 13.12.

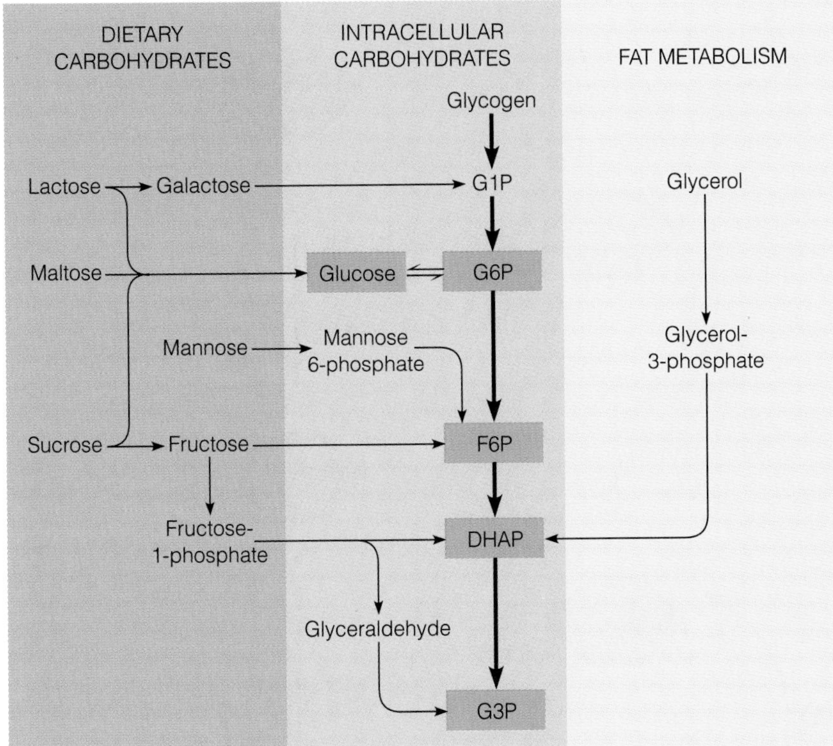

FIGURE **13.12**

Routes for utilizing substrates other than glucose in glycolysis. In animals most of the carbohydrate other than glucose and glycogen comes from the diet, and most of the glycerol is derived from lipid catabolism.

MONOSACCHARIDE METABOLISM

As stated earlier, hexokinase has a broad substrate specificity. Thus, it can participate in utilization of hexoses other than glucose, particularly fructose and mannose. A separate enzyme, galactokinase, converts galactose to galactose-1-phosphate.

Galactose Utilization

D-Galactose is derived principally from hydrolysis of the disaccharide lactose, which is particularly abundant in milk. The main route for galactose utilization is conversion to glucose-6-phosphate, as shown in Figure 13.13. This pathway begins with the ATP-dependent conversion of galactose to galactose-1-phosphate, catalyzed by **galactokinase** (Figure 13.13, reaction 1). Glucose-1-phosphate, which also participates, is formed largely from glycogen breakdown (page 471). Transformation of galactose-1-phosphate to glucose-1-phosphate involves epimerization at carbon 4. However, before epimerization can occur, galactose-1-phosphate must be metabolically activated (reaction 2), by a transferase reaction with a nucleotide-linked sugar, **uridine diphosphate glucose,** also called UDP-glucose, UDP-Glc, or UDPG. This reaction produces another nucleotide-linked sugar, **uridine diphosphate galactose,** abbreviated UDP-galactose or UDP-Gal.

Uridine diphosphate glucose

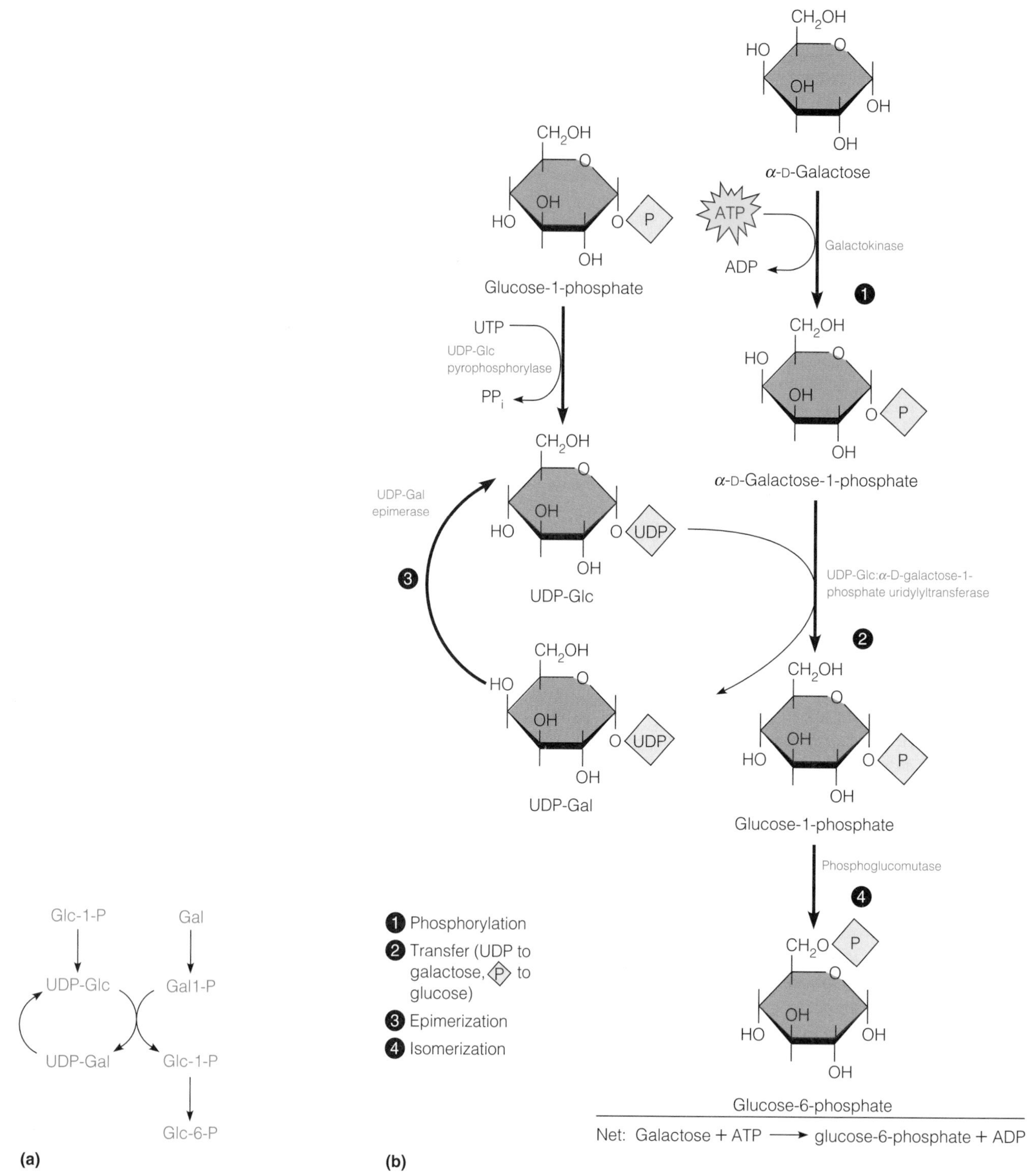

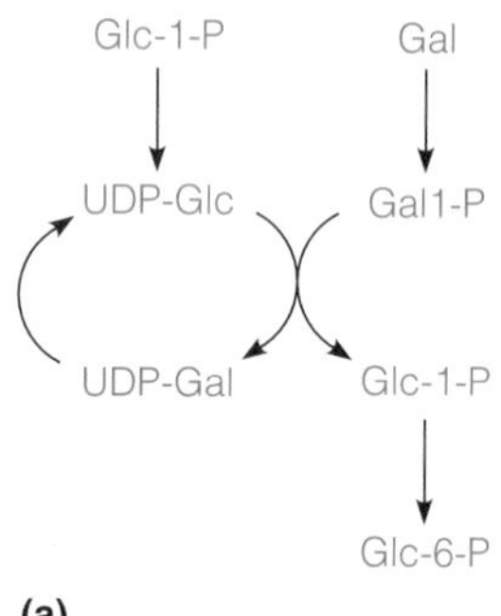

(a)

α-D-Galactose
ATP
ADP
Galactokinase
1
α-D-Galactose-1-phosphate
Glucose-1-phosphate
UTP
UDP-Glc pyrophosphorylase
PP_i
UDP-Gal epimerase
3
UDP-Glc
UDP-Gal
UDP-Glc:α-D-galactose-1-phosphate uridylyltransferase
2
Glucose-1-phosphate
Phosphoglucomutase
4
Glucose-6-phosphate

1 Phosphorylation
2 Transfer (UDP to galactose, P to glucose)
3 Epimerization
4 Isomerization

Net: Galactose + ATP ⟶ glucose-6-phosphate + ADP

(b)

FIGURE **13.13**

Pathway for utilizing galactose by converting it to glucose-6-phosphate. **(a)** An overview of the pathway and **(b)** a detailed diagram.

The NAD^+-linked enzyme **UDP-galactose 4-epimerase** can then convert UDP-Gal to UDP-Glc (reaction 3). Details of this epimerization reaction are shown in Figure 13.14. Returning to Figure 13.13, reaction 3, note that UDP-Glc is formed from glucose-1-phosphate and UTP by **UDP-Glc pyrophosphorylase.** This enzyme is named from its reverse reaction, involving cleavage of the phosphoric acid anhydride bond in UDP-Glc by addition across that bond of the elements of pyrophosphoric acid. The glucose-1-phosphate formed in reaction 2 is then converted to glucose-6-phosphate by **phosphoglucomutase** (reaction 4), an enzyme involved also in glycogen synthesis. In our discussion of glycogen biosynthesis in Chapter 16, we see that nucleoside diphosphate sugars such as UDP-Glc are widely used intermediates in polysaccharide biosynthesis.

The enzymes shown in Figure 13.13 also participate in mammary gland in the synthesis of lactose in milk. Lactose is formed from UDP-Gal plus glucose by **lactose synthase,** in the presence of the protein **α-lactalbumin.** The endergonic synthesis of UDP-Gal in this tissue proceeds smoothly because of its high rate of conversion to lactose. As discussed in Chapters 16 and 19, UDP-Gal is also used in glycoprotein and glycolipid biosynthesis.

A variety of genetic disorders in humans go by the generic name **galactosemia.** They all involve a failure to metabolize galactose, so that galactose, galactose-1-phosphate, or both accumulate in the blood and tissues. Clinical consequences include mental retardation, visual cataracts, and enlargement of the liver and other organs. These disorders result from hereditary deficiency of any one of three enzymes involved in galactose utilization. The most common form is caused by a deficiency of UDP-glucose:α-D-galactose-1-phosphate uridylyltransferase (see Figure 13.13). Rarer forms involve deficiencies of galactokinase or UDP-galactose 4-epimerase. Because the major dietary source of galactose is lactose in milk, the symptoms usually occur in infants. The condition can be alleviated by withholding milk and milk products from the diet.

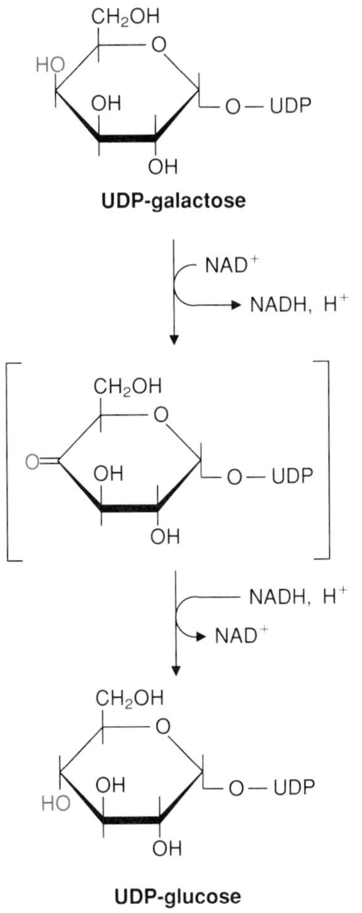

FIGURE **13.14**

Reaction pathway for UDP-galactose 4-epimerase. This detailed view of reaction 3 in Figure 13.13 shows the catalytic role of NAD^+ in oxidizing and then reducing the substrate.

Fructose Utilization

Fructose is present as the free sugar in many fruits, and it is also derived from hydrolysis of sucrose (see Figure 13.12). Phosphorylation of fructose in most tissues yields fructose-6-phosphate, a glycolytic intermediate. A different pathway is involved in vertebrate liver, where the enzyme **fructokinase** phosphorylates fructose to **fructose-1-phosphate** (F1P). This intermediate is then cleaved by a specific enzyme, **aldolase B.** Cleavage products are dihydroxyacetone phosphate, a glycolytic intermediate, and D-glyceraldehyde. The latter is then phosphorylated in an ATP-dependent reaction to give the glycolytic intermediate glyceraldehyde-3-phosphate. This pathway of utilization bypasses phosphofructokinase regulation and may account for the ease with which dietary sucrose is converted to fat (i.e., F1P→G3P+DHAP→glycerol-3 phosphate→triacylglycerols; see Chapter 18).

Mannose Utilization

Finally, among the major hexoses, mannose arises through digestion of foods containing certain polysaccharides or glycoproteins. The hexokinase-catalyzed phosphorylation of mannose to mannose-6-phosphate is followed by isomerization of the latter to fructose-6-phosphate (see Figure 13.12).

DISACCHARIDE METABOLISM

The three disaccharides most abundant in foods are maltose, lactose, and sucrose. Maltose is available primarily as an artificial sweetener, derived from starch, while lactose and sucrose are abundant natural products. In animal metabolism they are hydrolyzed in cells lining the small intestine, to give the constituent hexose sugars:

$$\text{Maltose} + H_2O \xrightarrow{\text{Maltase}} 2\ \text{D-glucose}$$

$$\text{Lactose} + H_2O \xrightarrow{\text{Lactase}} \text{D-galactose} + \text{D-glucose}$$

$$\text{Sucrose} + H_2O \xrightarrow{\text{Sucrase}} \text{D-fructose} + \text{D-glucose}$$

The hexose sugars pass via the portal vein to the liver, where they are catabolized as described in the previous section.

In some humans the enzyme lactase disappears from the intestinal mucosal cells after age 4 to 6, when milk drinking usually decreases. This causes **lactose**

intolerance, a condition in which ingestion of milk or lactose-containing milk products causes intestinal distress, because of bacterial action on the lactose that accumulates.

Plants and microorganisms have different pathways for metabolizing disaccharides. Bacteria metabolize sucrose through the action of **sucrose phosphorylase:**

$$\text{Sucrose} + P_i \rightleftharpoons \text{D-glucose-1-phosphate} + \text{D-fructose}$$

GLYCEROL METABOLISM

The digestion of neutral fat (triacylglycerols) and most phospholipids generates glycerol as one product. In animals, glycerol enters the glycolytic pathway first by the action in liver of **glycerol kinase.**

$$\underset{\textbf{Glycerol}}{HO{-}CH(CH_2OH)_2} + ATP \longrightarrow \underset{\textbf{Glycerol-3-phosphate}}{CH_2OH{-}C(H)(OH){-}CH_2O\text{Ⓟ}} + ADP + H^+$$

The product is then oxidized by **glycerol-3-phosphate dehydrogenase,** to yield dihydroxyacetone phosphate, which is catabolized by glycolysis (see Figure 13.12).

$$\underset{\textbf{Glycerol-3-phosphate}}{CH_2OH{-}C(H)(OH){-}CH_2O\text{Ⓟ}} + NAD^+ \longrightarrow \underset{\textbf{Dihydroxyacetone phosphate}}{CH_2OH{-}C(=O){-}CH_2O\text{Ⓟ}} + NADH + H^+$$

Catabolism of Polysaccharides

In animal metabolism, two primary sources of glucose are derived from polysaccharides: (1) digestion of dietary polysaccharides, chiefly starch from plant foodstuffs and glycogen from meat, and (2) mobilization of the animal's own glycogen reserves. Recall from Chapter 9 that starch, the major nutrient polysaccharide of plants, consists of the unbranched glucose polymer amylose and the branched polymer amylopectin. Glucose residues in both polymers are linked by $\alpha(1\rightarrow4)$ glycosidic bonds, but amylopectin also has $\alpha(1\rightarrow6)$ linkages, which provide branch points in the otherwise linear polymer. Glycogen is chemically similar to amylopectin, except that it is more highly branched and is of higher molecular weight. Many microorganisms, like animals, store carbohydrate as glycogen.

HYDROLYTIC AND PHOSPHOROLYTIC CLEAVAGES

Dietary polysaccharides are metabolized by hydrolysis to monosaccharides. Intracellular carbohydrate stores, as glycogen, are mobilized by phosphorolysis.

Polysaccharide digestion and glycogen mobilization both involve sequential cleavage of monosaccharide units from nonreducing ends of glucose polymers. The first of these processes occurs via *hydrolysis* and the second via *phosphorolysis.* These processes are chemically similar, as illustrated in Figure 13.15. Hydrolysis is the cleavage of a bond by addition across that bond of the elements of water, and a phosphorolytic cleavage occurs by addition of the elements of phosphoric acid. An enzyme catalyzing a phosphorolysis is called a **phosphorylase,** to be distinguished from a *phosphatase* (or, more precisely, a *phosphohydrolase*), which catalyzes the hydrolytic cleavage of a phosphate ester bond.

Energetically speaking, the advantage of a phosphorolytic mechanism is that mobilization of glycogen yields most of its monosaccharide units in the form of sugar phosphates. These units can be converted to glycolytic intermediates directly, without the investment of additional ATP. By contrast, starch digestion yields glucose plus some maltose; ATP and the hexokinase reaction are necessary to initiate glycolytic breakdown of these sugars.

The hydrolytic mechanism is useful, however, for the digestion of dietary carbohydrate, which occurs largely in the intestine. Digestion products must be absorbed and transported to the liver, where they are converted into glucose. Because sugar phosphates, like other charged compounds, are inefficiently transported across cell membranes, the hydrolytic digestion of polysaccharides to yield hexose sugars makes good metabolic sense.

In plant metabolism we are not concerned with digestion, because with few exceptions plants synthesize both monosaccharides and energy storage polysaccharides via photosynthesis. However, the same enzymatic mechanisms are used to mobilize stored carbohydrate in plant metabolism as in animals—both hydrolysis and phosphorolysis of starch, with hydrolysis predominating. In the brewing of beer, the controlled germination of cereal seeds such as barley develops hydrolytic enzymes that break starch down to mono- and disaccharides for later fermentation by yeast. This process is called malting.

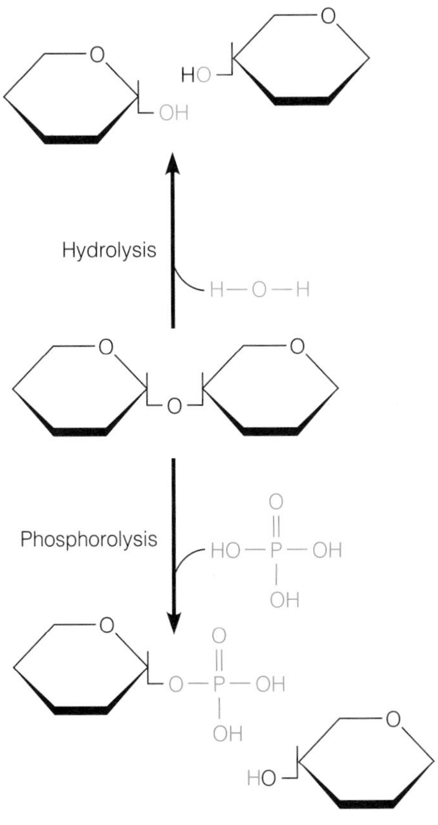

FIGURE **13.15**

Cleavage of a glycosidic bond by hydrolysis or phosphorolysis. This formal diagram shows how the elements of water or phosphoric acid, respectively, are added across a glycosidic bond.

STARCH AND GLYCOGEN DIGESTION

In animals the digestion of starch and glycogen begins in the mouth, with the action of **α-amylase** secreted in saliva. This enzyme cleaves internal $\alpha(1\rightarrow4)$ linkages of both polymers. In the intestine, digestion continues, aided by α-amylase secreted by the pancreas. α-Amylase degrades amylose to maltose and a little glucose. However, it only partially degrades amylopectin and glycogen, as shown in Figure 13.16, because it cannot cleave the $\alpha(1\rightarrow6)$ linkages found at branch points. The product of exhaustive digestion of amylopectin or glycogen by α-amylase is called a **limit dextrin;** its continued degradation requires the action of a "debranching enzyme," **$\alpha(1\rightarrow6)$-glucosidase** (also called isomaltase). This action exposes a new group of $\alpha(1\rightarrow4)$-linked branches, which can be attacked by α-amylase until a new set of $\alpha(1\rightarrow6)$-linked branches is reached. The end result of the sequential action of these two enzymes is the complete breakdown of starch or glycogen to maltose and some glucose. Maltose is cleaved hydrolytically by **maltase,** yielding 2 moles of glucose, which is then absorbed into the bloodstream and transported to various tissues for utilization.

GLYCOGEN MOBILIZATION

The principal glycogen stores in vertebrates are in skeletal muscle and liver. Breakdown of these stores into usable energy, or **mobilization** of glycogen, involves sequential phosphorolytic cleavages of $\alpha(1\rightarrow4)$ bonds, catalyzed by **glycogen phosphorylase.** In plants, starch is similarly mobilized by the action of **starch phosphorylase.** Both reactions release glucose-1-phosphate from nonreducing ends of the glucose polymer:

$$\text{Glucose}\alpha(1\rightarrow4)\text{glucose}\alpha(1\rightarrow4)\text{glucose}\alpha(1\rightarrow4)\text{glucose}\cdots$$

$$\xrightarrow[\text{Phosphorylase}]{P_i}$$

$$\alpha\text{-D-Glucose-1-}\langle P\rangle + \text{glucose}\alpha(1\rightarrow4)\text{glucose}\alpha(1\rightarrow4)\text{glucose}\cdots$$

The cleavage reaction is slightly disfavored under standard conditions ($\Delta G^{\circ\prime} = +3.1$ kJ/mol), but the relatively high intracellular levels of inorganic phosphate cause this reaction to operate in vivo almost exclusively in the degradative, rather than the synthetic, direction. Note that the reaction proceeds with retention of configuration at carbon 1.

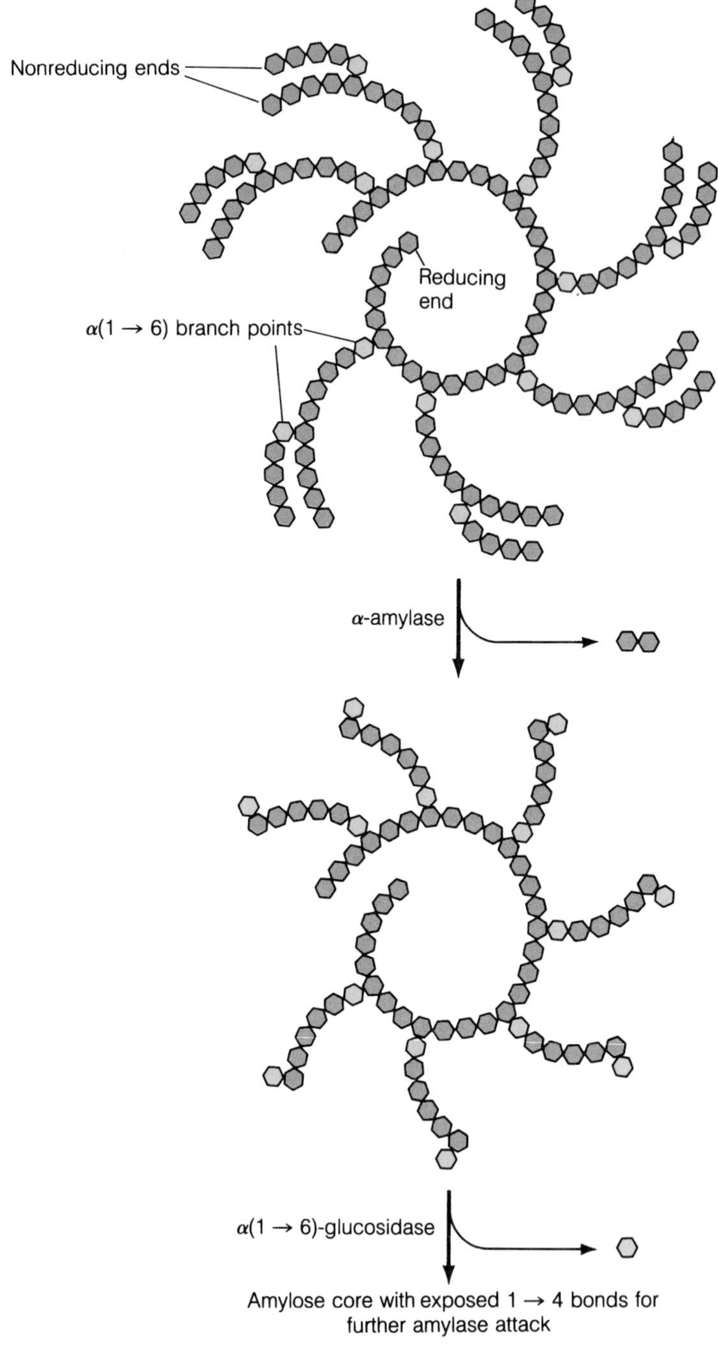

FIGURE **13.16**

Sequential digestion of amylopectin or glycogen by α-amylase and α(1→6)-glucosidase. (top) α-Amylase in saliva cleaves 1→4 bonds between the maltose units of amylopectin (or glycogen). However, it cannot cleave 1→6 glycosidic bonds in the branched polymer, and a limit dextrin (gray) accumulates unless α(1→6)-glucosidase is present. **(bottom)** α(1→6)-Glucosidase in the intestine cleaves the branch points, exposing the amylose core to further digestion by amylase.

Like α-amylase, phosphorylases cannot cleave past α(1→6) branch points. In fact, cleavage stops four glucose residues from a branch point. The debranching process involves the action of a second enzyme, as shown in Figure 13.17. This "debranching enzyme," **(α1,4→α1,4)glucantransferase,** catalyzes two reactions. First is the transferase activity, in which the enzyme removes three of the remaining glucose residues and transfers this trisaccharide moiety intact to the end of some other outer branch. Next, the remaining glucose residue, which is still attached to the chain by an α(1→6) bond, is cleaved by the α(1→6)-glucosidase activity of the same debranching enzyme. This yields one molecule of free glucose and a branch of three α(1→4)-linked glucose residues. This newly exposed

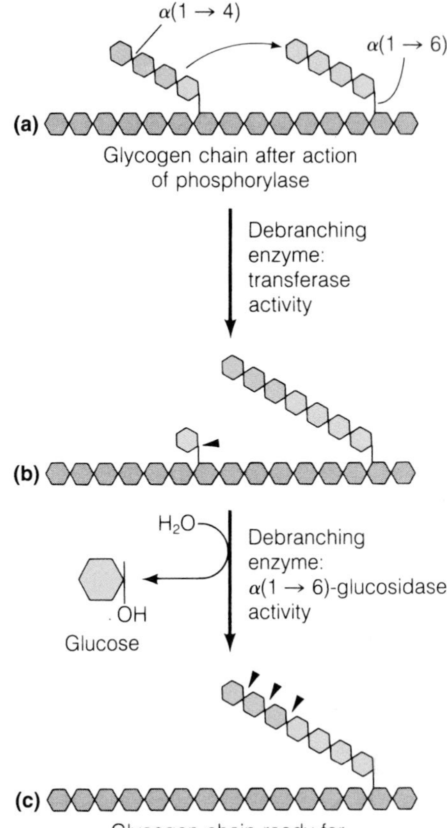

FIGURE **13.17**

The debranching process in glycogen catabolism. (a) A glycogen chain following activity by phosphorylase, which cleaves off glucose residues to within four residues of the branch point. **(b)** The glycogen chain following transferase activity by the debranching enzyme. The three remaining glucose residues with $\alpha(1\rightarrow4)$ linkage have been transferred to a nearby nonreducing end. **(c)** The glycogen chain following $\alpha(1\rightarrow6)$-glucosidase activity by the debranching enzyme, which has removed the last remaining glucose residue of the branch. Phosphorylase will cleave off all but four glucose units of the newly elongated branch, beginning the debranching process again. The new cleavage points are indicated by wedges.

branch is now available for further attack by phosphorylase. The end result of the action of these two enzymes is the complete breakdown of glycogen to glucose-1-phosphate (the major product) and glucose.

At this point you might wonder why the glycogen breakdown scheme has evolved to include this complex debranching process. The importance of storing carbohydrate energy in the form of a highly branched polymer may well lie in an animal's need to generate energy very quickly following appropriate stimuli. Glycogen phosphorylase attacks *exoglycosidic* bonds—it cleaves sequentially from nonreducing ends. The more such ends that exist in a polymer, the faster the polymer can be mobilized.

To be metabolized via glycolysis, the glucose-1-phosphate produced by phosphorylase action must be converted to glucose-6-phosphate. This isomerization is accomplished by phosphoglucomutase. This reaction is also important in glycogen synthesis (Chapter 16). The reaction is mechanistically similar to that of phosphoglycerate mutase (page 456), except that in phosphoglucomutase a phosphoserine residue on the enzyme reacts with substrate, instead of phosphohistidine:

Enzyme-phosphate + glucose-1-phosphate	$\rightleftharpoons$	enzyme + glucose-1,6-bisphosphate	
Glucose-1,6-bisphosphate + enzyme	$\rightleftharpoons$	enzyme-phosphate + glucose-6-phosphate	
Net: Glucose-1-phosphate	$\rightleftharpoons$	glucose-6-phosphate	$\Delta G^{\circ\prime} = -7.3$ kJ/mol

The serine residue that carries the phosphate group is unusually reactive, as shown by the fact that phosphoglucomutase, like chymotrypsin and other serine proteases, is irreversibly inhibited by diisopropylfluorophosphate. The inhibition, like that of chymotrypsin (see Chapter 11), involves acylation of only the active site serine.

Most of the glycogen in vertebrate animals is stored as granules in cells of liver and skeletal muscle. A major function of the liver is to provide glucose for metabolism by other tissues. This function is accomplished both through glycogen mobilization and through gluconeogenesis. Both processes yield phosphorylated forms of glucose, which cannot exit from liver cells. Conversion to free glucose requires the action of **glucose-6-phosphatase,** which hydrolyzes glucose-6-phosphate to glucose and orthophosphate. This enzyme is also present in kidney and intestine. By contrast, muscle glycogen serves primarily as a source of glucose-6-phosphate for catabolism within muscle cells. Accordingly, glucose-6-phosphatase is absent from muscle, as it is from brain, which depends almost exclusively on glucose from the blood as its primary energy source. This ensures that glucose-6-phosphate formed from glycogen cannot diffuse out of these cells, because, as mentioned earlier, sugar phosphates do not readily traverse cell membranes.

REGULATION OF GLYCOGEN BREAKDOWN

In Chapter 11 we mentioned the control of glycogen breakdown, or glycogenolysis, as a particularly well-understood example of a regulatory cascade, a process in which the intensity of an initial regulatory signal is amplified manyfold through a series of enzyme activations. This amplification is particularly important in the case of glycogenolysis, because fright, for example, or the need to catch prey, can trigger an instantaneous requirement for increased energy generation and utilization. Glycogen represents the most immediately available *large-scale* source of metabolic energy, and hence it is important that animals be able to activate glycogen mobilization rapidly. Moreover, glycogen breakdown is the first hormone-controlled process for which the molecular action of the hormone was understood in detail. Therefore, although the process was introduced in Chapters 11 and 12, we present it here in more detail (Figure 13.18).

Structure of Glycogen Phosphorylase

In skeletal muscle, glycogen phosphorylase is a dimer containing two identical polypeptide chains, each of 97,400 daltons. The enzyme exists in two interconvertible forms—the relatively *active* phosphorylase *a* and the relatively *inactive* phosphorylase *b*. Phosphorylation of serine 14 converts the relatively inactive phosphorylase *b* to the active phosphorylase *a*. As shown in Figure 13.18, activation is catalyzed by a specific **phosphorylase *b* kinase**, which transfers phosphate from ATP to the two serine residues. Deactivation is brought about by a specific **phosphorylase phosphatase.**

Control of Phosphorylase Activity

Glycogen mobilization is controlled hormonally by a metabolic cascade that is activated by cAMP formation and involves successive phosphorylations of enzyme proteins.

Phosphorylase *b* kinase is also activated by phosphorylation from an inactive to an active form. This reaction is catalyzed by cyclic AMP–dependent protein kinase. As pointed out in Chapter 12, cyclic AMP (cAMP) plays numerous roles in regulating metabolism, both in eukaryotes and in prokaryotes. In eukaryotes its role is as a second messenger, receiving hormonal messages originating outside the cell and transmitting them within the cell. This transmission involves the activation of some metabolic processes and the inhibition of others. In glycogenolysis, cyclic AMP exerts a rapid and efficient activation. At the same time, it inhibits glycogen synthesis through a separate regulatory cascade, discussed in Chapter 16.

The primary hormone promoting glycogenolysis in muscle is epinephrine (formerly called adrenaline), which is secreted from the adrenal medulla and binds to specific receptors on muscle cell membranes. Liver glycogen mobilization is stimulated largely by the pancreatic peptide hormone glucagon. In both cases, binding of the hormone at the membrane stimulates the synthesis of cyclic AMP by membrane-bound adenylate cyclase, through the action of a G protein, G_s. Cyclic AMP in turn activates protein kinase, which catalyzes the phosphorylation of phosphorylase *b* kinase. This kinase in turn catalyzes the phosphorylation of phosphorylase *b* to *a* and, hence, the activation of glycogen breakdown, through the action of phosphorylase *a*. These events explain how the secretion of relatively few molecules of hormone, such as epinephrine, can, within just a few moments, trigger a massive conversion of glycogen to glucose-1-phosphate.

The rapid mobilization of muscle glycogen triggered by epinephrine is one of several components of the "fight or flight" response.

Epinephrine, also known as adrenaline, is the principal hormone governing the "fight or flight" response to various stimuli. In addition to stimulating glycogenolysis, the hormone triggers a variety of physiological events, such as increasing depth and frequency of heartbeats. These effects are also mediated via cyclic AMP, as discussed further in Chapter 23. Cyclic AMP also regulates other metabolic processes, including the stimulation of fat breakdown and the inhibition of glycogen synthesis. We shall return to these effects as we proceed through metabolism.

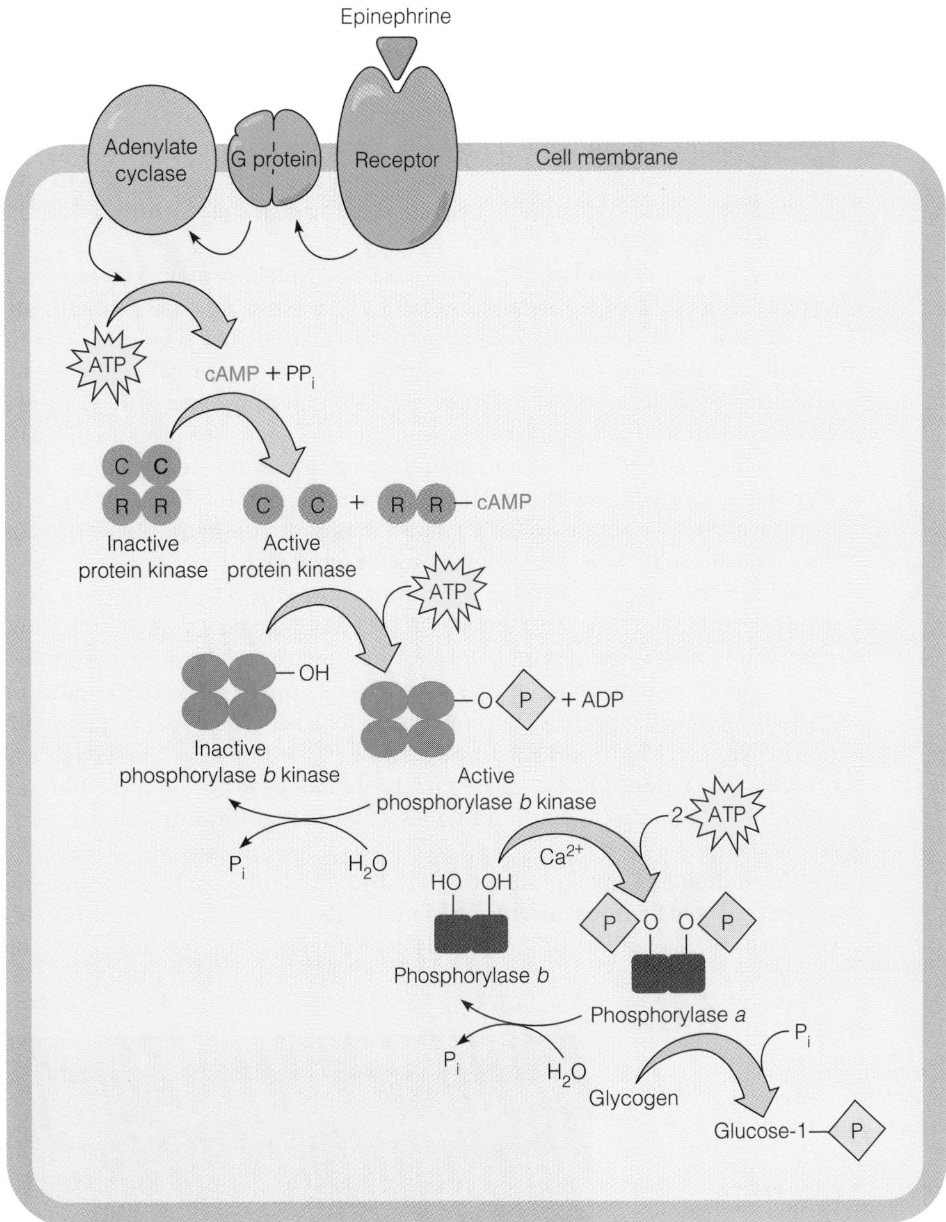

FIGURE **13.18**

The regulatory cascade controlling glycogen breakdown. The hormonal regulation of glycogen breakdown is shown here as it might occur in a muscle cell after epinephrine stimulation. Hormone binding to its receptor triggers an interaction with a G protein, which in turn interacts with and activates adenylate cyclase (see Figure 12.13). Binding of cyclic AMP to the R (regulatory) subunits of the protein kinase A R_2C_2 tetramer causes their dissociation from the C (catalytic) subunits. The active C monomer catalyzes phosphorylation of specific serine residues on inactive phosphorylase *b* kinase, activating phosphorylase *b* in the process. The active kinase phosphorylates a serine residue on each of the two subunits of the homodimeric phosphorylase *b*. This converts the inactive phosphorylase *b* to the active phosphorylase *a*, which then catalyzes glycogen breakdown. Each reaction in the regulatory cascade amplifies the hormonal signal, so that binding of very few epinephrine molecules at the cell surface triggers enormous release of glucose-1-phosphate from intracellular glycogen stores. Inactivation of the pathway involves the actions of specific phosphatases, which remove the phosphates from phosphorylase *b* kinase and phosphorylase *a*. Actions of these phosphatases are also subject to regulation (Chapter 16).

Proteins in the Glycogenolytic Cascade

Our presentation of the glycogenolytic cascade started with the phosphorylase reaction and then worked backward, to the initial hormonal signal. Now let us start with the hormone and work forward, with emphasis on the proteins involved (again, refer to Figure 13.18). The hormone binds to a specific receptor located on the outside of the cytoplasmic membrane. This binding leads to activation of adenylate cyclase, which is bound to the inside of the membrane; the activation is mediated by a G protein, G_s.

Cyclic AMP–dependent protein kinase, also called protein kinase A, is a tetramer consisting of two *catalytic* subunits, C, and two *regulatory* subunits, R. The tetramer, R_2C_2, is catalytically inactive. Structural analysis of the catalytic subunit reveals a core structure that is similar in all known protein kinases. Figure 13.19 shows the crystal structure of the C subunit complexed with ATP and a peptide inhibitor that is similar to the inhibitory domain of the R subunit. The catalytic mechanism involves nucleophilic attack by a serine residue in the protein substrate upon the γ-phosphate of ATP. The effect of inhibitor binding is to increase greatly the affinity for ATP (shown in green in the figure), essentially by burying it and preventing access to the protein substrate.

Binding of cyclic AMP to the R subunits causes the tetramer to dissociate, giving the catalytically active monomer (C). This protein now catalyzes the phosphorylation of phosphorylase *b* kinase. Phosphorylase *b* kinase is a complex multisubunit protein, one of whose subunits is a protein called **calmodulin,** or calcium-modulating protein. Calcium ion has long been known as an important physiological regulator, particularly of processes related to nerve conduction and muscle contraction. Most of these effects are mediated through the binding of Ca^{2+} to calmodulin, which amplifies small changes in intracellular Ca^{2+} concentration.

Calmodulin is a small protein (M_r ~17,000) of highly conserved amino acid sequence. It contains four calcium ion binding sites (Figure 13.20). Each site binds Ca^{2+} with a K_D of about 10^{-6} M, consistent with observations that calcium can

FIGURE **13.19**

Protein kinase A. Space-filling model of the catalytic subunit bound to ATP (shown in green) and a peptide inhibitor similar to the R subunit inhibitory domain (shown in red). The amino terminus of the protein is shown in dark purple.

Courtesy of E. Radzio-Andzeim and S. S. Taylor. Coordinates taken from the ternary complex of the Catalytic Subunit: ATP:PKI (5-24). Reproduced with permission from J. Zheng et al., *Biochem.* (1993) 32:2154, fig. 2, Protein Data Bank (Reference 1ATP). © American Chemical Society.

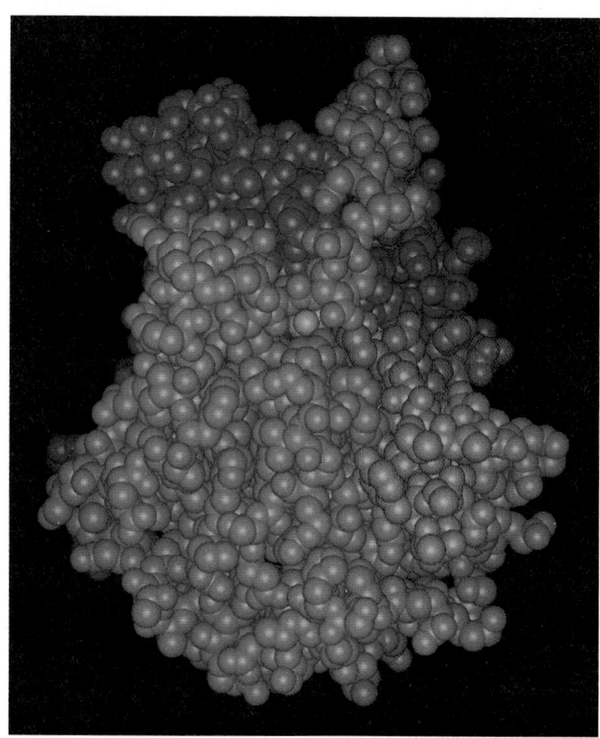

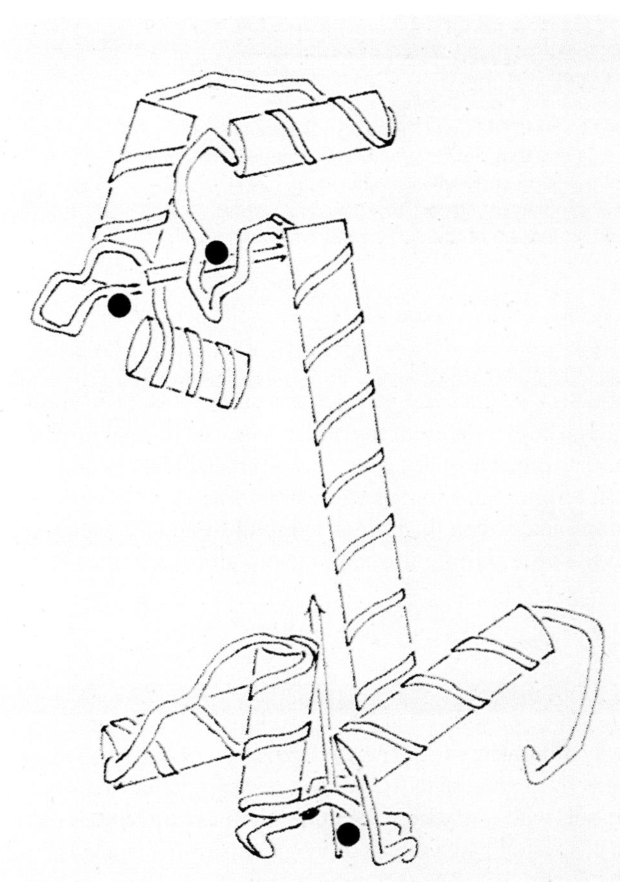

FIGURE **13.20**

Calmodulin. This model shows the structure of bovine brain calmodulin, as determined by x-ray crystallography. The four calcium-binding sites are depicted as black circles. Each end of the dumbbell-shaped molecule consists of three α-helical regions and binds two Ca^{2+} ions. The helix joining the two ends probably undergoes length changes as a result of calcium binding. These conformational changes are responsible for changes in calmodulin's affinity for calcium-regulated targets.

Reprinted from R. H. Kretsinger, S. E. Rudwick, and L. J. Weissman, *J. Inorg. Chem.* (1986) 28:293.

effect intracellular metabolic changes in concentrations as low as 1 μM. Binding stimulates a major conformational change in the protein, leading to a more compact and more highly helical structure, which augments the affinity of calmodulin for a number of regulatory target proteins. In the case of phosphorylase *b* kinase, calmodulin plays a special role as an integral subunit of the enzyme. Hence, the glycogenolysis cascade depends on intracellular calcium concentration as well as on cyclic AMP levels. This dependence is particularly important in muscle, where contraction is stimulated by calcium release. Thus, Ca^{2+} plays a dual role, in provision of the energy substrates needed to support muscle contraction and in contraction itself.

Nonhormonal Control of Glycogenolysis

Glycogen breakdown is under nonhormonal, as well as hormonal, control. Recall that phosphorylase *b* is relatively inactive. This form of the enzyme is activated allosterically by 5′-AMP (but not by cyclic AMP). Usually this activation does not occur in the cell, because ATP, which is far more abundant and does not activate phosphorylase *b*, competes with AMP for binding to the enzyme. However, under energy-deprived conditions, AMP may accumulate at the expense of ATP breakdown, to the extent that phosphorylase *b* and hence glycogenolysis are activated. Recent crystallographic studies show that the structural changes induced in phosphorylase *b* by AMP are remarkably similar to those induced by phosphorylation of phosphorylase *b* to *a*, even though AMP binding and phosphorylation occur at quite distant sites (Figure 13.21). In any case the mobilization of energy reserves from glycogen can be brought about either by hormonal stimulation, reflecting a physiological need for increased ATP production, or by an allosteric mechanism

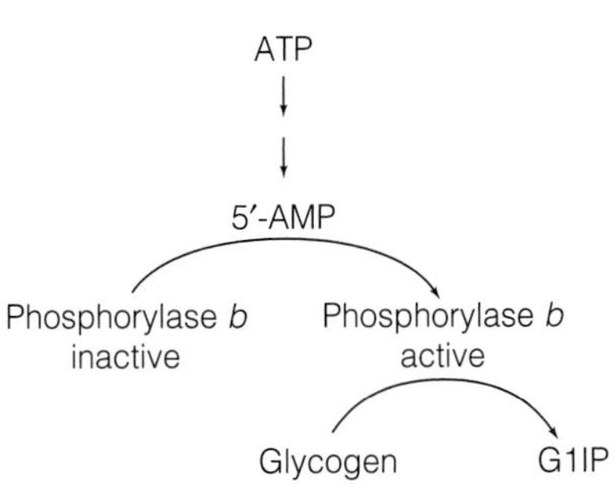

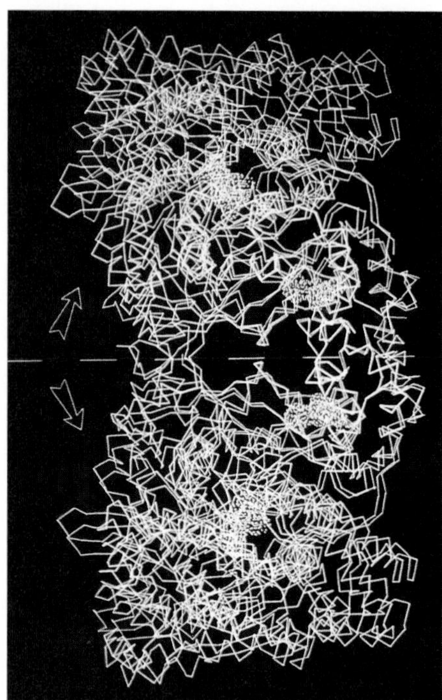

FIGURE **13.21**

Structural basis for activation of glycogen phosphorylase. The C_α atom tracings represent the structures of the two active forms of glycogen phosphorylase: phosphorylase *a* (blue) and phosphorylase *b* activated by AMP (yellow). The two structures are nearly superimposable, showing that similar structural transitions result from *b* to *a* phosphorylation and AMP activation of *b*. Catalytic sites and sites of AMP binding are shown in yellow. The arrows show the direction of relative motion of the dimer subunits upon activation.

Courtesy of S. R. Sprang, S. G. Withers, E. J. Goldsmith, R. J. Fletterick, and N. B. Madsen, *Science* (1991) 254:1367–1371.

triggered when the energy level is deficient for maintenance of normal functions. The nonhormonal mechanism, which does not involve a metabolic cascade, stimulates glycogenolysis in response to a low energy charge, whereas the hormonally induced cascade predominates when the need is to rapidly augment energy generation. In both cases the phosphorolysis of glycogen to glucose-1-phosphate is enhanced.

SUMMARY

Glycolysis is the central pathway by which energy is extracted from carbohydrates. A 10-step pathway leads from glucose to pyruvate in respiring cells. In anaerobic microorganisms or in cells with impaired respiration, pyruvate undergoes reductive reactions, so that the overall pathway can proceed with no net change in oxidation state. Glycolysis can be viewed as occurring in two phases—first, an energy investment phase in which ATP is used to synthesize a six-carbon sugar phosphate that is split to yield two triose phosphates, and second, an energy generation phase, in which the energy of two super-high-energy compounds is used to drive ATP synthesis from ADP. Phosphofructokinase and pyruvate kinase are the two major sites for control of the pathway. Much of the control is related to the energy needs of a cell, with conditions of low energy charge stimulating the pathway, and conditions of energy abundance retarding the pathway. Intracellular polysaccharide stores in animals are mobilized by a hormonally controlled metabolic cascade, in which cyclic AMP transmits the hormonal signal and sets in motion events that activate the breakdown of glycogen to glucose-1-phosphate.

REFERENCES

Intracellular Organization of Glycolytic Enzymes

Ovádi, J. ed. (1991) Physiological significance of metabolite channeling. *J. Theor. Biol.* 152:1–141. A special issue of this journal, containing reviews on the organization of metabolic enzymes, with special emphasis on glycolysis.

Regulation of Carbohydrate Metabolism

Beitner, R., ed. (1985) *Regulation of Carbohydrate Metabolism,* Vols. 1 and 2. CRC Press, Boca Raton, Fla. Review chapters deal with fructose-2,6-bisphosphate as a biological regulator, structural organization of glycolytic enzymes, control of glycogen metabolism, and aberrant carbohydrate metabolism in disease states, among other topics.

Boyer, P. D., and E. G. Krebs, eds. (1986) *The Enzymes,* 3rd ed., Vols. 17 and 18. Academic Press, Orlando, Fla. Discusses enzyme systems whose activity is controlled by phosphorylation, including protein kinases, glycogen phosphorylase, pyruvate kinase, the enzymes of fructose-2,6-bisphosphate synthesis, and phosphoprotein phosphatases.

Dell'Acqua, M. L., and J. D. Scott (1997) Protein kinase A anchoring. *J. Biol. Chem.* 272:12881–12884. Many hormonal signals involve cAMP-dependent protein kinase A activation, and location of protein kinase within the cell helps to establish specificity of particular signaling pathways.

Knighton, D. R., J. Zheng, L. F. TenEyck, V. A. Ashford, N. Xuong, S. S. Taylor, and J. M. Sowadski (1991) Crystal structure of the catalytic subunit of cyclic adenosine monophosphate–dependent protein kinase. *Science* 253:407–420. Gives the structural basis for the activation of protein kinase upon dissociation into its subunits.

Lynd, L. R., J. H. Cushman, R. J. Nichols, and C. E. Wyman (1991) Fuel ethanol from cellulosic biomass. *Science* 251:1318–1323. Optimization of fuel production from biomass will necessitate thorough understanding of glycolytic control mechanisms.

McCoy, M. (1998) Biomass ethanol inches forward. *Chem. Eng. News* December 7:29–32. A news article that describes processes under development for converting waste products to usable ethanol.

Meador, W. E., A. R. Means, and F. A. Quiocho (1992) Target enzyme recognition by calmodulin: 2.4 Å structure of a calmodulin–peptide complex. *Science* 257:1251–1254. A high-resolution crystallographic picture of calmodulin action.

Pilkis, S. J., I. T. Weber, R. W. Harrison, and G. I. Bell (1994) Glucokinase: Structural analysis of a protein involved in susceptibility to diabetes. *J. Biol. Chem.* 269:21925–21928. A recent minireview.

Rhoads, R., and F. Friedberg (1997) Sequence motifs for calmodulin recognition. *Fed. Proc.* 11:331–340. A review of the calcium-controlled processes that are mediated by calmodulin, with discussion of the structural basis for the interactions of target proteins with calmodulin.

Sirover, M. A. (1997) Role of the glycolytic protein, glyceraldehyde-3-phosphate dehydrogenase, in normal cell function and in cell pathology. *J. Cell. Biochem.* 66:133–140. G3P dehydrogenase has a surprising range of functions in addition to its catalytic role in glycolysis.

Sprang, S. R., S. G. Withers, E. J. Goldsmith, R. J. Fletterick, and N. B. Madsen (1991) Structural basis for the activation of glycogen phosphorylase *b* by adenosine monophosphate. *Science* 254:1367–1371. One of a series of reports describing the crystal structure of glycogen phosphorylase in activated and unactivated states.

Su, Y., R. G. Dostmann, F. W. Herberg, K. Durick, N.-H. Xuong, L. Ten Eyck, S. S. Taylor, and K. I. Varughese (1995) Regulatory subunit of protein kinase A: Structure of deletion mutant with cAMP binding domains. A structural analysis of the cyclic AMP binding site reveals its mode of cooperative binding to regulatory subunits. *Science* 269:807–813.

Analysis of Carbohydrate Metabolism by In Vivo NMR

Chance, B., and K. Wasserman (1986) Anaerobiosis, lactate, and gas exchange during exercise, I and II. *Fed. Proc.* 45:2904–2957. Reports from a symposium in which modern metabolic techniques, including ^{31}P NMR, were used to relate muscular exertion to the control of carbohydrate metabolism.

Shulman, R. G. (1988) High resolution NMR in vivo. *Trends Biochem. Sci.* 13:37–39. A minireview that summarizes how NMR studies have changed some of our concepts about energy metabolism and its regulation.

Oscillations of Glycolytic Intermediates

Bier, M., B. Teusink, B. N. Kholodenko, and H. V. Westerhoff (1996) Control analysis of glycolytic oscillations. *Biophys. Chem.* 62:15–24. Mathematical modeling of oscillations helps in understanding control of glycolysis.

Richter, P. H., and J. Ross (1981) Concentration oscillations and efficiency: Glycolysis. *Science* 211:715–716. A theoretical discussion of the energetic advantages to a living system of the oscillations observed in levels of glycolytic intermediates.

PROBLEMS

1. Intracellular concentrations in resting muscle are as follows: fructose-6-phosphate, 1.0 mM; fructose-1,6-bisphosphate, 10 mM; AMP, 0.1 mM; ADP, 0.5 mM; ATP, 5 mM; and P_i, 10 mM. Is the phosphofructokinase reaction in muscle *more* or *less* exergonic than under standard conditions? By how much?

2. Methanol is highly toxic, not because of its own biological activity but because it is converted metabolically to formaldehyde, through action of alcohol dehydrogenase. Part of the medical treatment for methanol poisoning involves administration of large doses of ethanol. Explain why this treatment is effective.

3. Refer to Figure 13.6, which indicates $\Delta G'$ for each glycolytic reaction under intracellular conditions. Assume that glyceraldehyde-3-phosphate dehydrogenase was inhibited with iodoacetic acid. Which glycolytic intermediate would you expect to accumulate most rapidly, and why?

4. In different organisms sucrose can be cleaved either by hydrolysis or by phosphorolysis. Calculate the ATP yield per mole of sucrose metabolized by anaerobic glycolysis starting with (a) hydrolytic cleavage and (b) phosphorolytic cleavage.

5. Suppose it were possible to label glucose with ^{14}C at any position or combination of positions. For yeast fermenting glucose to ethanol, which form or forms of labeled glucose would give the *most* radioactivity in CO_2 and the *least* in ethanol?

6. Write balanced chemical equations for each of the following: (a) anaerobic glycolysis of 1 mole of sucrose, cleaved initially by sucrose phosphorylase; (b) aerobic glycolysis of 1 mole of maltose; (c) fermentation of one glucose residue in starch to ethanol, with the initial cleavage involving α-amylase.

7. Because of the position of arsenic in the periodic table, arsenate (AsO_4^{3-}) is chemically similar to inorganic phosphate and is used by phosphate-requiring enzymes as an alternative substrate. However, organic arsenates are quite unstable and spontaneously hydrolyze. Arsenate is known to inhibit ATP production in glycolysis. Identify the target enzyme, and explain the mechanism of inhibition.

8. Suppose that you made some wine whose alcohol content was 10% w/v (i.e., 10 g of ethanol per 100 mL of wine). The initial fermentation mixture would have had to contain what molar concentration of glucose or its equivalent to generate this much ethanol? Is it likely that an initial fermentation mixture would contain that much glucose? In what other forms might the fermentable carbon appear?

9. Briefly discuss why each of the three common forms of galactosemia involves impaired utilization of galactose. Which metabolic process is blocked in each condition?

*10. Not all anaerobic bacteria ferment glucose to lactate or ethanol. Some ferment glucose to acetate, and others ferment it to propionate. Propose a straightforward pathway leading from pyruvate to acetate. The pathway should proceed with no net change in the oxidation state of the cell. For this and for more complex fermentation pathways, consult a microbiology textbook.

11. Write a pathway leading from glucose to lactose in mammary gland, and write a balanced equation for the overall pathway.

12. Sketch a curve that would describe the expected behavior of phosphofructokinase.

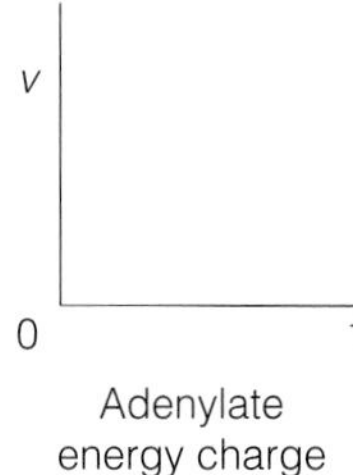

13. Explain the basis for the following statement. For efficient conversion of galactose to glucose-1-phosphate, UDP-glucose need be present in catalytic amounts only.
14. The muscle isozyme of lactate dehydrogenase is inhibited by lactate. Steady state kinetic analysis yielded the following data, with lactate either absent or present at a fixed concentration.

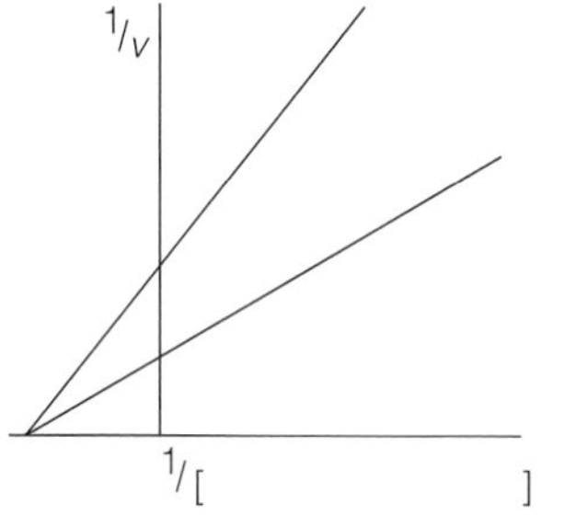

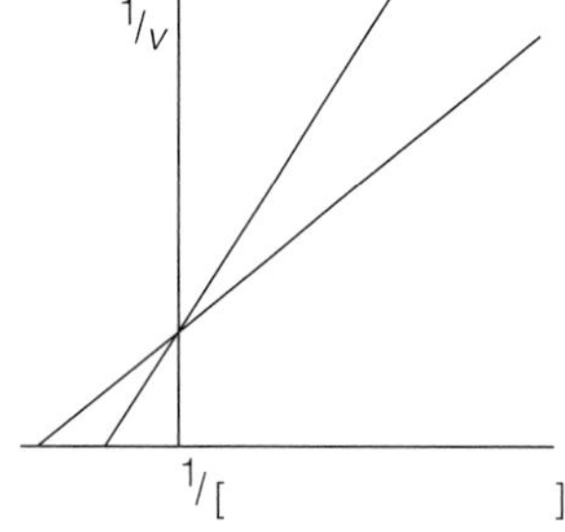

(a) Pyruvate is the substrate whose concentration is varied in one plot, NADH in the other. Identify each. Use an arrow and the appropriate letter (b, c, d, or e) to identify each of the following.
(b) Reciprocal of V_{max} for the uninhibited enzyme.
(c) The line representing data obtained in the presence of lactate acting as a competitive inhibitor with respect to the variable substrate.
(d) The line representing data obtained in the presence of lactate acting as a noncompetitive inhibitor with respect to the variable substrate.
(e) Reciprocal of K_M in the presence of lactate acting as a competitive inhibitor.
(f) If K_M for NADH is 2×10^{-5} M, which is the most appropriate NADH concentration to use when determining K_M for pyruvate? 10^{-7} M, 10^{-6} M, 10^{-5} M, 10^{-4} M, or 10^{-3} M

TOOLS OF BIOCHEMISTRY 13A

Detecting and Analyzing Protein–Protein Interactions

The enzymes of glycolysis are readily isolated as soluble proteins. However, several lines of evidence support the notion that they are physically associated within living cells. For years, biochemistry students have been told, "A cell is not a bag of enzymes," implying that enzymes are organized into functional supramolecular units in intact cells. Often these organized units are stabilized by weak, noncovalent forces that are easily disrupted when cells are broken open, as must occur if the enzymes within are to be isolated and characterized. Even when cells are gently lysed, most protein extraction processes dilute intracellular contents by several orders of magnitude, and that alone can disrupt associations that are highly concentration-dependent. Biochemists are trying to define how the organization of functionally related enzymes facilitates the flow of metabolites and the control and coordination of metabolic pathways.

An early indication that glycolytic enzymes might interact within cells came from observations that the intracellular molar concentrations of glycolytic intermediates are actually lower than the concentrations of the enzymes that act upon those intermediates. This finding suggested that most of the supply of an intermediate is enzyme-bound within cells and this, in turn, led to the idea that intermediates are passed directly from enzyme to enzyme, without release to the surrounding milieu, as if the glycolytic enzymes functioned as part of a multienzyme complex. If such a complex could be isolated intact, its properties could be explored by the methods for molecular weight determination, described in Tools of Biochemistry 6B. However, because of difficulties often encountered in isolating enzyme complexes held together by ephemeral forces, scientists usually use multiple approaches to demonstrate and characterize the protein–protein interactions involved. A few of those techniques are described here.

Bifunctional Cross-Linking Reagents

These are reagents containing two functional groups capable of forming covalent bonds with specific amino acid residues in closely associated proteins. For instance, *dimethylsuberimidate* reacts with lysine ε–amino groups and N-terminal amino groups, linking two proteins in a form

$$H_3CO{-}\overset{\overset{\displaystyle NH}{\|}}{C}{-}CH_2{-}CH_2{-}CH_2{-}CH_2{-}CH_2{-}CH_2{-}\overset{\overset{\displaystyle NH}{\|}}{C}{-}OCH_3$$

that can be detected by gel electrophoresis because of the increase in molecular weight. Some reagents have cleavable cross-links, such as a disulfide bond that can be reductively cleaved, allowing analysis of the separate cross-linked partners. Although the technique can be very informative, experimentation with many reagents is often required to find the right combination of functional groups and distance between the reactive partners for the cross-linking reaction to proceed to a measurable extent. Also, care must be taken not to overinterpret results, for even transient contacts between molecules, which may occur nonspecifically, sometimes lead to cross-linking.

Affinity Chromatography

In this technique, one protein is immobilized on a chromatographic support, and a mixture of proteins is passed through a column of this material. Proteins that are retained can be identified after elution, by biological activity or by electrophoretic techniques, such as immunoblotting or two-dimensional electrophoretic analysis. The chief limitations of this technique are the need to have one of the test proteins available in pure form for immobilization and the fact that interactions occur in a rather artificial environment. Again, controls are essential because of nonspecific retention of some proteins on affinity columns.

Immunoprecipitation

Antibody to a purified protein can be added to a protein mixture, often with immunoprecipitation of both the antigenic protein and any interactive proteins bound to it. Although this technique is qualitative, like the approaches described above, it is simple to do, and it needs only small amounts of material. Because multiple assays can be run simultaneously, immunoprecipitation can be used, for example, to study the effects of the binding of small molecules (substrates or effectors) upon protein associations.

Kinetic Analysis

If enzymes catalyzing sequential reactions interact, the interactions can facilitate the flow of metabolites through multistep pathways (metabolic channeling), and this can be detected in vitro in several ways. Generally, a channeled pathway will display one or more of the following characteristics: (1) reduced *transient time,* the interval after initiating a multistep pathway needed for the formation of final product to reach its maximal rate; (2) steady-state levels of intermediates much lower than expected if they must seek the next enzyme acting on them by diffusion rather than by direct or facilitated transfer to a nearby enzyme molecule; and (3) restricted ability of an exogenous intermediate to equilibrate with the same intermediate in a channeled pathway, as determined usually by radioisotope experiments.

Fluorescence Polarization

As indicated in Tools of Biochemistry 6A, rotation of a macromolecule will tend to depolarize the emitted fluorescent light when the fluorescence of that macromolecule is excited by plane-polarized light, and the rate of depolarization is dependent upon the speed of rotation. Because a protein–protein complex is larger, and often more asymmetric, than the individual protein molecules, complex formation slows down rotation, and this can be measured by following the rate at which fluorescent light becomes depolarized. In practice, this involves measurement over very short times of fluorescent light intensity in planes parallel to and perpendicular to the plane of incident light.

Library-Based Methods

These are methods that allow the screening of a large number, or library, of cloned genes (see Chapter 25). These methods allow tentative identification of interacting partners without a prior requirement for purification and identification of one of the partners. A popular method called the two-hybrid system uses a transcriptional activation system in yeast that requires two proteins to interact in order for transcription at an appropriate gene site to be initiated (see Chapter 28). One of these proteins binds at the DNA site, and the other activates transcription. Two hybrid, or fusion, proteins are generated by recombinant DNA techniques (Chapter 25); the gene for one test protein (X) is fused to the DNA-binding protein gene, and the gene for another test protein (Y), or a library of cloned genes, is fused to the gene for the transcriptional activation domain. The recombinant genes are transferred into yeast cells, where the interaction of proteins X and Y can form a fully functional transcriptional activator (assuming that the functional domains of the fusion proteins fold as they do in their native state). Transcription of the target gene is then monitored by assays for the activity of a reporter gene, a gene cloned downstream of the promoter and whose biological activity is easily assayed. Once a specific protein association has been detected, it becomes essential to isolate the interactive partners as full-length proteins and to ascertain that the interactions detected by this somewhat qualitative method are indeed biologically significant.

Biosensor Analysis

In recent years, a new kind of instrumentation has been developed that allows both qualitative and quantitative analysis of protein–protein interactions, using rather small quantities of purified proteins. One such instrument, a BIACORE, measures an optical property called *surface plasmon resonance,* which is related to minute changes in refractive index that occur when a protein in solution interacts with a protein immobilized on

FIGURE **13A.1**

BIACORE analysis of a protein–protein association. Test protein flows past the immobilized protein in the association phase and is replaced by buffer in the dissociation phase. The height of the plateau response, as compared with standards, is related to the stoichiometry of the association.

Courtesy of Biacore.

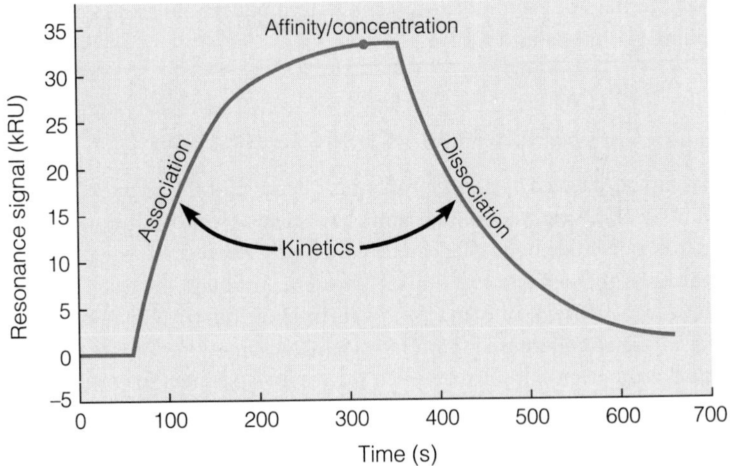

a chip. The signal measured is proportional to total protein concentration, over a wide range. Thus, the kinetics of a protein association reaction can be monitored by following the increase in signal as a protein in solution passes over a chip containing immobilized protein. The amount of protein bound at equilibrium gives the affinity constant for the interaction, and the kinetics of dissociation can then be followed by passing buffer over the chip and following the decrease in signal, as indicated in Figure 13A.1. Limitations of this useful technique, which can be controlled for, include the possibility that immobilization alters the protein in a way that affects the interaction and the fact that the two interacting proteins are in different phases (solid, or immobilized, and liquid).

References

Phizicky, E. M., and S. Fields (1995) Protein–protein interactions: Methods for detection and analysis. *Microbiol. Rev.* 59:94–123. A comprehensive review by the originator of the widely used two-hybrid technique.

Szabo, A., L. Stolz, and R. Granzow (1995) Surface plasmon resonance and its use in biomolecular interaction analysis (BIA). *Curr. Opin. Struct. Biol.* 5:699–705.

CHAPTER 14

Oxidative Processes: Citric Acid Cycle and Pentose Phosphate Pathway

IN CHAPTER 13 WE EXPLORED THE INITIAL, ANAEROBIC, PHASE OF CARBOHYDRATE degradation. Here we follow the subsequent aerobic reactions by which carbohydrates are ultimately oxidized to carbon dioxide and water (Figure 14.1). However, this chapter is not primarily about carbohydrate metabolism. That is because the pathway involved, the citric acid cycle, is the central oxidative pathway in respiration, the process by which *all* metabolic fuels—carbohydrate, lipid, and protein—are catabolized in aerobic organisms and tissues.

The citric acid cycle is a pathway for oxidizing all metabolic fuels.

Although respiration and glycolysis each conserve about 40% of energy released in the form of ATP, much less total energy is generated, per mole of substrate catabolized, in glycolysis than in respiration. Lactate, like other end products of anaerobic carbohydrate catabolism, is at the same oxidation level as the starting material, glucose. Therefore, relatively little of the potential energy stored in the glucose molecule is released by its conversion to lactate.

Far more energy is generated in the subsequent reactions by which lactate or pyruvate becomes completely oxidized to CO_2. That energy release involves dehydrogenation reactions that generate reduced electron carriers, primarily NADH. These carriers are next reoxidized in the mitochondrial respiratory (electron transport) chain. These reactions provide the energy that drives ATP synthesis, through oxidative phosphorylation. The electrons released are ultimately transferred to oxygen, which becomes reduced to water. This chapter focuses on fates of the oxidizable substrates, and the next chapter focuses on the chain of electron carriers and the synthesis of ATP. In short, if we look at the end products of respiration—CO_2, H_2O, and ATP—our emphasis in this chapter is on the generation of CO_2 and in Chapter 15 on the generation of H_2O and ATP.

Most of the energy yield from substrate oxidation in the citric acid cycle comes from subsequent reoxidation of reduced electron carriers.

Overview of Pyruvate Oxidation and the Citric Acid Cycle

THE THREE STAGES OF RESPIRATION

It is convenient to think of the metabolic oxidation of organic substrates as a three-stage process, schematized in Figure 14.2. Stage 1 is the generation of an activated

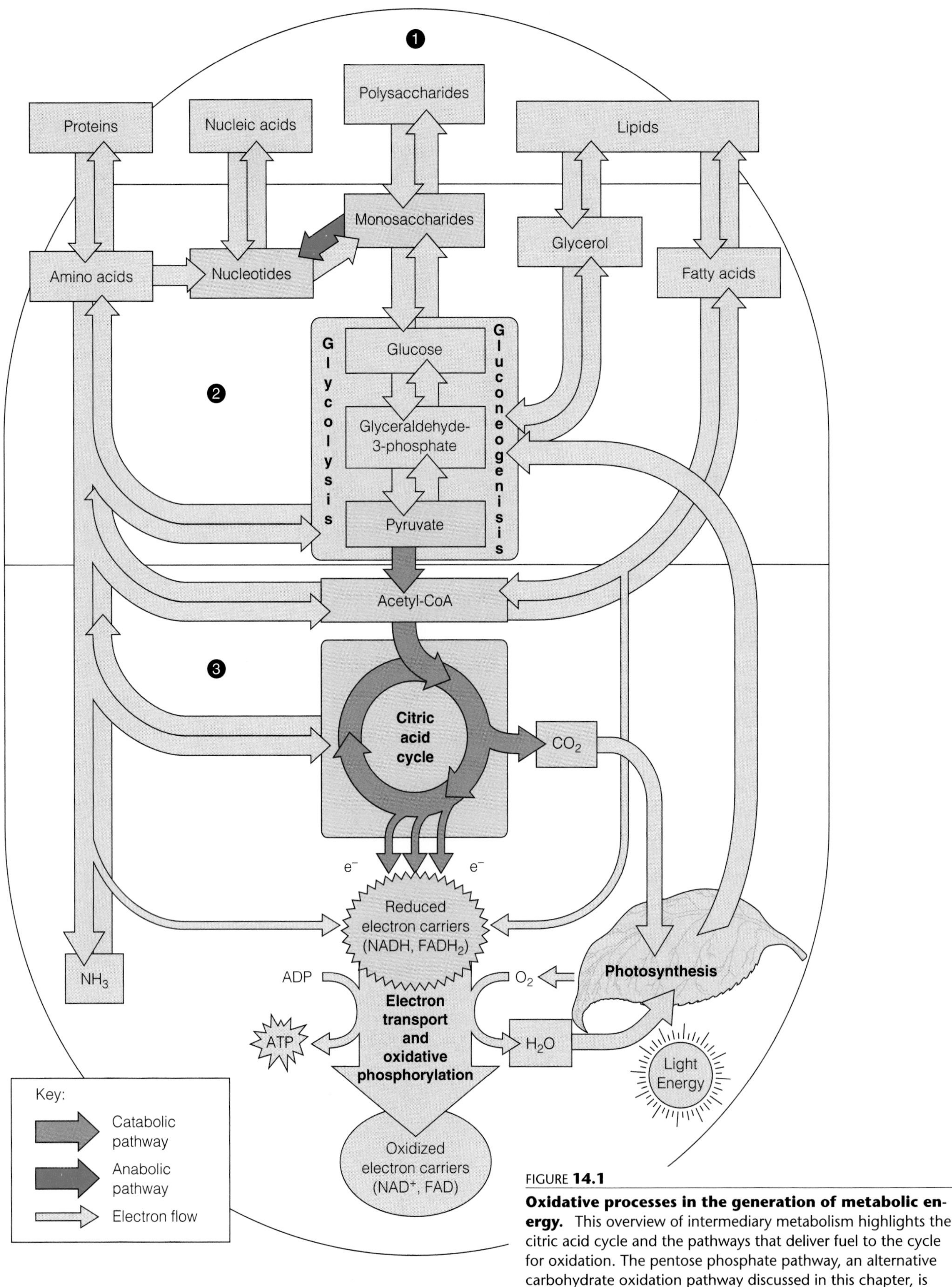

FIGURE **14.1**

Oxidative processes in the generation of metabolic energy. This overview of intermediary metabolism highlights the citric acid cycle and the pathways that deliver fuel to the cycle for oxidation. The pentose phosphate pathway, an alternative carbohydrate oxidation pathway discussed in this chapter, is needed for nucleotide synthesis.

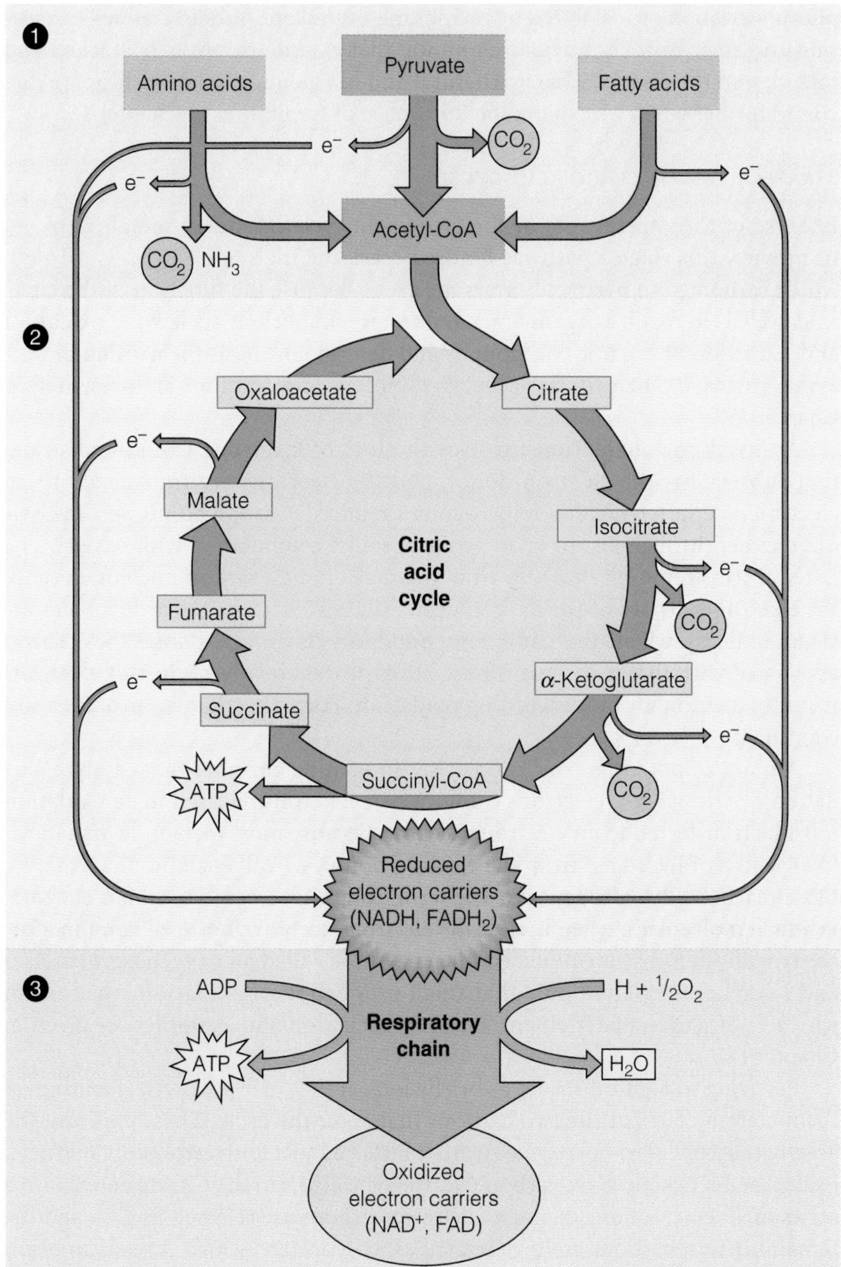

FIGURE **14.2**

The three stages of respiration. In stage 1, carbon from metabolic fuels is incorporated into acetyl-CoA. In stage 2, the citric acid cycle, oxidation of carbon produces CO_2, reduced electron carriers, and a small amount of ATP. In stage 3, the reduced electron carriers are reoxidized, providing energy for the synthesis of additional ATP.

two-carbon fragment—the acetyl group of **acetyl-coenzyme A, or acetyl-CoA.** (Recall from Chapter 12 that coenzyme A activates and transfers acyl groups; more about this on pages 494–495). Stage 2 is oxidation of those two carbon atoms in the citric acid cycle. Stage 3 is electron transport and oxidative phosphorylation, in which the reduced electron carriers generated in the citric acid cycle become reoxidized, with concomitant synthesis of ATP. Stage 1 is a family of pathways, which operate separately on carbohydrate, fat, and protein. Carbon from carbohydrate enters stage 1 as pyruvate; oxidation of pyruvate to acetyl-CoA is described in this chapter. Fat breakdown generates acetyl-CoA primarily through β-oxidation of fatty acids (see Chapter 18), whereas several different pathways generate acetyl-CoA and citric acid cycle intermediates from amino acid catabolism (see Chapter 21).

Reactions of the first two stages of respiration occur within the interior, gel-like matrix of the mitochondrion, and electron transport and oxidative phosphorylation are catalyzed by membrane-bound enzymes in **cristae**—projections of inner mitochondrial membrane that extend, extensively stacked and folded, into the matrix. These structural and biochemical relationships are discussed further when we explore the third stage of respiration in Chapter 15.

STRATEGY OF THE CITRIC ACID CYCLE

Because of the central role of the citric acid cycle in energy metabolism, let us preview this stage 2 pathway before describing the stage 1 process through which carbon from pyruvate enters the cycle. Because the function of the citric acid cycle is to oxidize organic metabolites, we first briefly review the oxidation and reduction of organic compounds and describe the basis for naming the enzymes involved. Quantitative aspects of biological oxidations are presented in Chapter 15.

As you know, oxidation involves the loss of electrons from a substrate; that substrate is the electron donor, and the electrons are transferred to an electron acceptor, which thereby becomes reduced. Carbon atoms become oxidized either through loss of hydrogen or through combination with oxygen. The latter process removes electrons from the shell about a carbon nucleus, because the electronegativity of the oxygen draws shared electrons toward its own nucleus. Similarly, when an organic compound loses hydrogen, it loses the electron associated with that hydrogen. Thus, either process involves a loss of electrons from the carbon atom undergoing oxidation. Formally, the two processes are equivalent.

A point of potential confusion arises in naming enzymes that catalyze oxidation reactions. Free electrons cannot exist; electrons released in an oxidation must be transferred to an electron acceptor. Because most metabolic oxidations involve loss of hydrogen from the electron donor, we call enzymes that catalyze those reactions **dehydrogenases.** The term *oxidase* is reserved for those enzymes in which molecular oxygen itself is the electron acceptor. If oxygen combines directly with the substrate oxidized, the enzyme is called an **oxygenase.** Oxidases and oxygenases catalyze only that small proportion of oxidation reactions in which O_2 is a direct participant. Further discussion and examples are given in Chapter 15.

Dehydrogenases catalyze substrate oxidations. Oxidases catalyze the subset of oxidations in which O_2 is the electron acceptor.

Referring to Figure 14.3, let us briefly look at the citric acid cycle, focusing on the metabolic fates of the two carbons that enter the cycle. These carbons, the acetyl group of acetyl-coenzyme A, are transferred to a four-carbon organic acid, **oxaloacetate,** to yield a six-carbon tricarboxylic acid, citrate. Citrate enters into a series of seven reactions during which two carbons are released as CO_2 and the remaining four carbons are regenerated as oxaloacetate, which is ready to begin the process again. Hence, the cyclic nature of the pathway: Oxaloacetate is present at the beginning, to react with an activated two-carbon fragment, and it is present at the end, after two carbons have been oxidized to CO_2. Note also that of the eight reactions shown in Figure 14.3, four are dehydrogenations, which together generate reduced cofactors: NADH and a reduced flavin coenzyme.

Just as we considered two phases of glycolysis, so also is it convenient to consider two distinct phases to the citric acid cycle. The first phase (reactions 1 through 4) serves primarily to oxidize two carbons to CO_2, whereas the second phase (reactions 5 through 8) serves primarily to regenerate oxaloacetate.

DISCOVERY OF THE CITRIC ACID CYCLE

The idea that organic fuels are oxidized via a cyclic pathway was proposed in 1937 by Hans Krebs, on the basis of studies for which he later shared a Nobel Prize with Fritz Lipmann (the discoverer of coenzyme A). Krebs was studying oxygen consumption in minced pigeon breast muscle, a tissue with a very high rate of

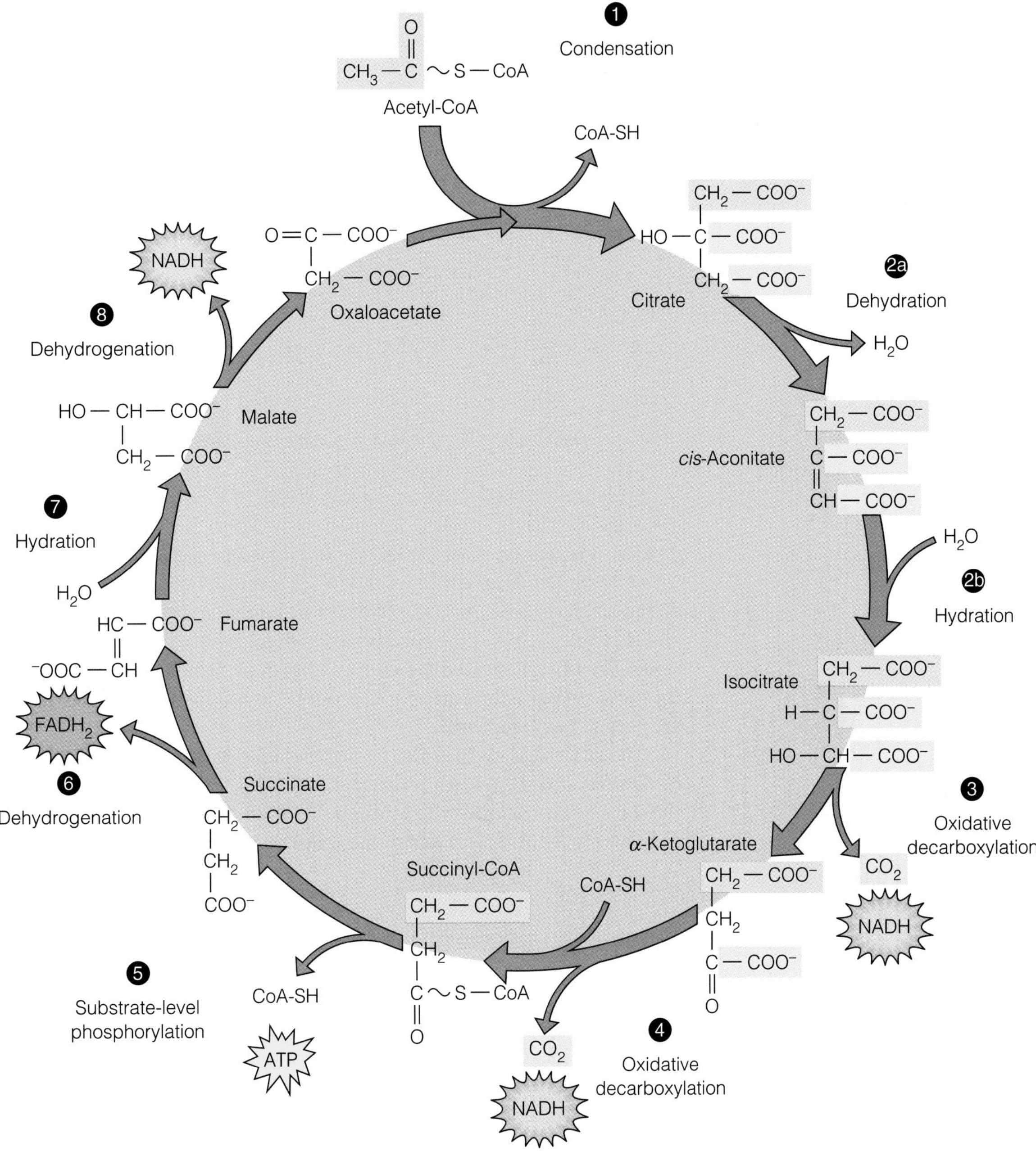

FIGURE **14.3**

The fate of carbon in the citric acid cycle. Acetyl-CoA entering the citric acid cycle is highlighted (in blue) to show the fate of its two carbons through reaction 4. After reaction 5, the carbon atoms most recently entered are no longer highlighted, because succinate is a symmetrical molecule. Carboxyl groups that leave the cycle as CO_2 in reactions 3 and 4 are shown in green. Note that these departing groups contain carbons incorporated as acetyl-CoA in earlier turns of the cycle.

respiration, and he made several fundamental observations. First, the anions of three tricarboxylic acids, namely, citrate, isocitrate, and *cis*-aconitate, stimulated both oxygen consumption and pyruvate oxidation out of proportion to the amounts added, suggesting that they were somehow acting catalytically, rather than as substrates in a linear pathway.

$$\begin{array}{c} CH_2-COO^- \\ | \\ HO-C-COO^- \\ | \\ CH_2-COO^- \end{array} \qquad \begin{array}{c} CH_2-COO^- \\ | \\ H-C-COO^- \\ | \\ H-C-COO^- \\ | \\ OH \end{array} \qquad \begin{array}{c} CH_2-COO^- \\ | \\ C-COO^- \\ \| \\ H-C-COO^- \end{array}$$

Citrate **Isocitrate** ***cis*-Aconitate**

Second, malonate, an analog of succinate and a known inhibitor of succinate dehydrogenase, blocked the oxidation of pyruvate, pointing to a role for succinate dehydrogenase in pyruvate oxidation. Third, malonate-inhibited cells accumulated citrate, α-ketoglutarate, and succinate, suggesting that citrate and α-ketoglutarate are both normal precursors of succinate. Finally, addition of pyruvate plus oxaloacetate to this minced muscle preparation led to accumulation of citrate in the medium, suggesting that these two acids are precursors to citrate.

Malonate	Succinate	α-Ketoglutarate	Oxaloacetate
COO^-–CH_2–COO^-	COO^-–CH_2–CH_2–COO^-	COO^-–CH_2–CH_2–C=O–COO^-	COO^-–CH_2–C=O–COO^-

From these observations and a knowledge of the structures and reactivities of the organic acids that could stimulate respiration, Krebs proposed both the sequence of reactions involved and the cyclic nature of the pathway. He postulated that a cyclic sequence could result from a reaction between pyruvate and oxaloacetate to give citrate plus CO_2. We now know that pyruvate must first be oxidized and that acetyl-CoA is the species that subsequently reacts with oxaloacetate. We know also of an activated intermediate form of succinate, **succinyl-CoA.** Except for these changes, the pathway as proposed by Krebs was correct. The reactions are outlined in Figure 14.3.

The citric acid cycle is also known by other names: the Krebs cycle, after its discoverer; and the tricarboxylic acid (TCA) cycle, because it was apparent from the outset that tricarboxylic acids are involved as intermediates. It became clear only later that citrate was one of those intermediates.

Pyruvate Oxidation: A Major Entry Route for Carbon into the Citric Acid Cycle

As noted earlier, pyruvate derived from carbohydrate oxidation is but one of the major suppliers of acetyl-CoA for oxidation in the citric acid cycle. Chapters 18 and 20 focus upon oxidation of fatty acids and amino acids, respectively, to give this central metabolite. The conversion of pyruvate to acetyl-CoA, catalyzed by **pyruvate dehydrogenase,** is an oxidative decarboxylation. In the overall reaction the carboxyl group of pyruvate is lost as CO_2, while the remaining two carbons form the acetyl moiety of acetyl-CoA.

$$CH_3{-}\overset{O}{\overset{\|}{C}}{-}COO^- + NAD^+ + \text{coenzyme A} \longrightarrow CH_3{-}\overset{O}{\overset{\|}{C}}{-}\text{coenzyme A} + NADH + CO_2$$

$$\Delta G^{\circ\prime} = -33.5 \text{ kJ/mol}$$

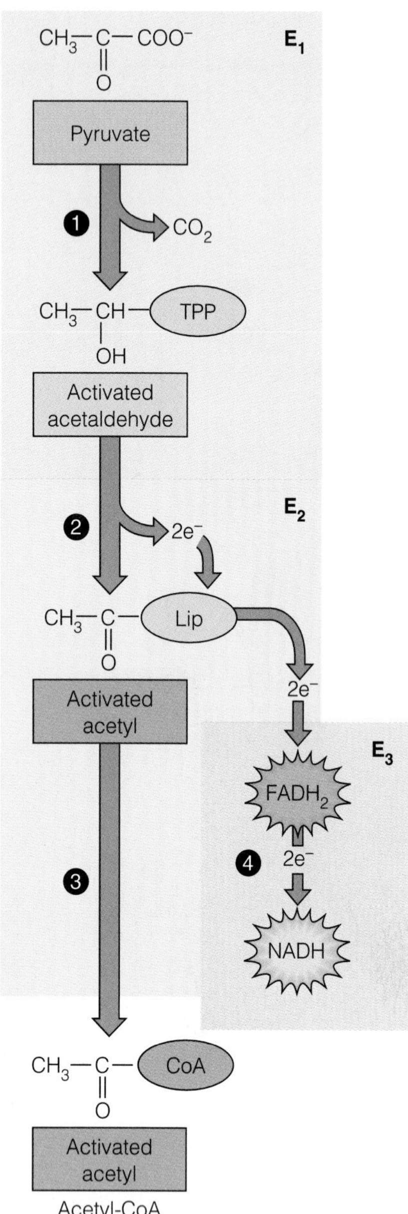

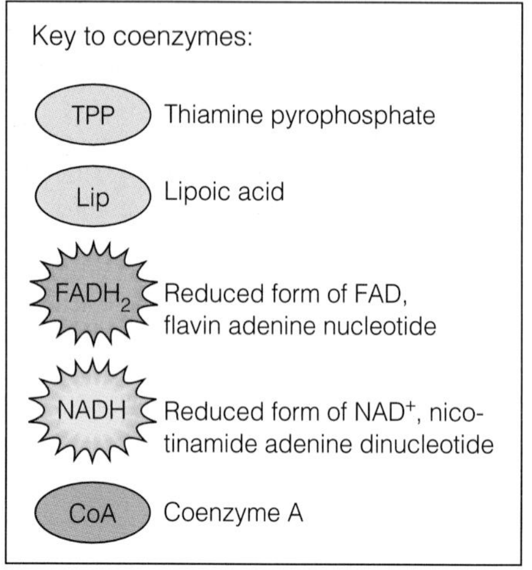

FIGURE **14.4**

Overview of the reactions of the pyruvate dehydrogenase complex. This simplified diagram shows oxidation of pyruvate to acetyl-CoA by the pyruvate dehydrogenase complex. The oxidation involves three enzymes and five coenzymes. Three of the coenzymes—TPP, FAD, and lipoic acid—are bound to enzymes in the complex.

Although the overall reaction may look straightforward, it is rather complicated, involving generation of a reduced electron carrier (NADH), decarboxylation of pyruvate, and metabolic activation of the remaining two carbons of pyruvate. The reaction is highly exergonic and is essentially irreversible in vivo. Three enzymes are involved, plus five coenzymes, including the two coenzymes—NAD^+ and coenzyme A—that appear in the overall reaction. The three enzymes involved are assembled into a highly organized multienzyme assembly called the **pyruvate dehydrogenase complex.** Figure 14.4 schematizes the reactions catalyzed by the three individual enzymes—**pyruvate dehydrogenase** (E_1), **dihydrolipoamide transacetylase** (E_2), and **dihydrolipoamide dehydrogenase** (E_3).

Pyruvate oxidation to acetyl-CoA is a virtually irreversible reaction that involves three enzymes and five coenzymes.

Figure 14.5 shows an electron micrograph of the complex as purified from *E. coli* and a model of the *Azotobacter* complex based upon x-ray crystallographic analysis of one of its components. The complex contains 24 polypeptide chains of E_1, 24 chains of E_2, and 12 chains of E_3. To E_1 is linked the coenzyme **thiamine pyrophosphate (TPP)**, E_2 contains tightly bound *lipoic acid,* and E_3 contains *flavin adenine dinucleotide (FAD).*

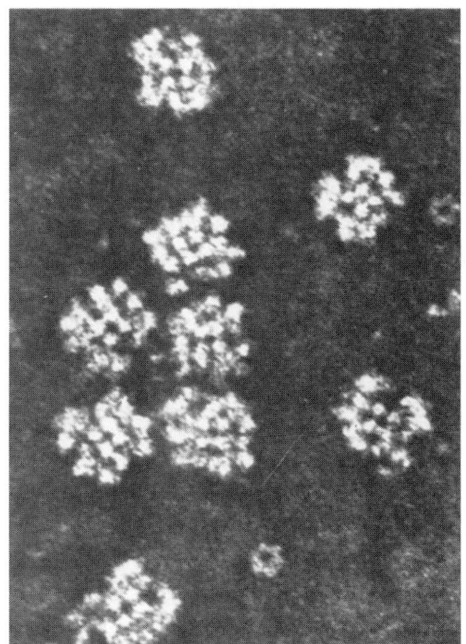

(a)

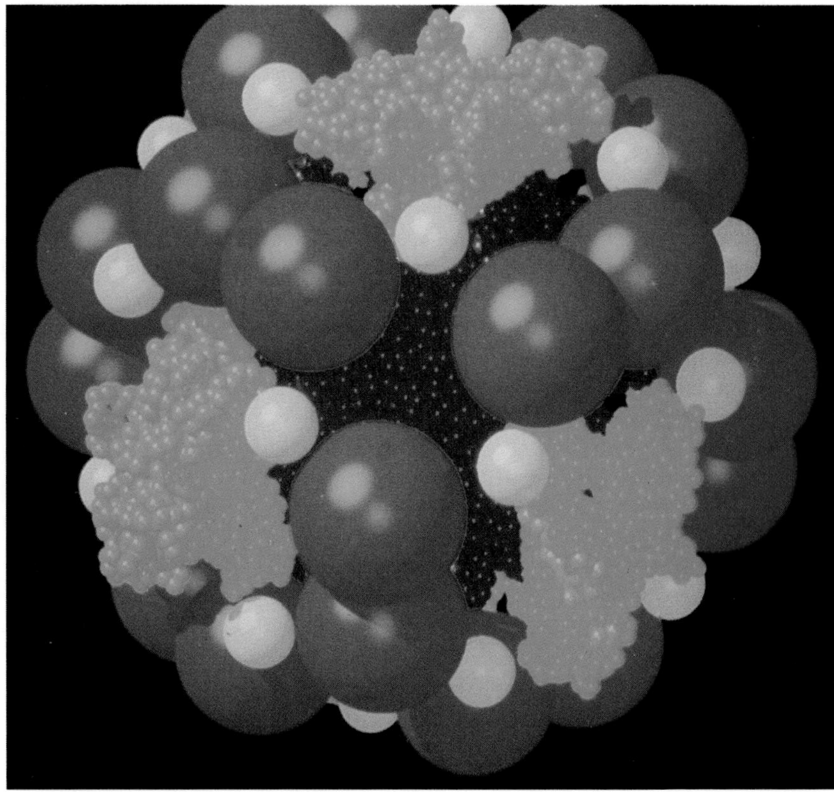

(b)

FIGURE **14.5**

Structure of the pyruvate dehydrogenase complex. (a) Electron micrograph of the purified pyruvate dehydrogenase complex from *E. coli.* **(b)** A model for the complex, based upon the crystal structure of E_2 from *Azotobacter vinelandii.* The complex is viewed along a 3-fold axis. E_2 and E_3 are shown in purple and green, respectively. E_1 is shown in blue, and each lipoamide binding site swinging arm (see page 496) is shown in yellow.

(a) R. M. Oliver and L. J. Reed, in *Electron Microscopy of Proteins,* Vol. 2, edited by J. R. Harris (Orlando, Fla.: Academic Press, 1982); **(b)** Courtesy of Mattevi et al., *Science* (1992) 255:1544–1550. Photograph courtesy of Wim Hol.

The pyruvate dehydrogenase complex as isolated from *E. coli* has a mass of about 4.6 million daltons, slightly larger than a ribosome. In mammalian mitochondria the complex is about twice that size. The eukaryotic complex contains small amounts of two regulatory enzymes as well—a kinase that phosphorylates three serine residues in E_1 and a phosphatase that removes those phosphates. Regulation of the activities of the complex will concern us later in this chapter.

To understand how the three enzymes in the pyruvate dehydrogenase complex interact, we must understand the functions of the five coenzymes. Although we introduced all five coenzymes in Chapter 11, we have discussed only the nicotinamide nucleotides in detail. Therefore, we digress here to describe reactions of the other four.

Coenzymes Involved in Pyruvate Oxidation and the Citric Acid Cycle

The story of these coenzymes is in large part the story of the vitamins from which they are derived. It began with observations in Asia, a century ago, regarding the paralytic disease *beriberi.* Polished rice (with the husks removed) was a dietary staple in the region. It was found that the husks, or bran, from rice contained a substance that relieved the paralysis. Casimir Funk, a Polish biochemist, reasoned that beriberi and three other nutritional diseases—*pellagra, rickets,* and *scurvy*—all resulted from lack of a dietary trace substance, which he called vitamine (literally, amine of life). This led to adoption of the generic term, **vitamin.**

The idea that vitamins could prevent disease was not widely accepted, because of the dominance of Pasteur's idea that diseases were caused only by infection. Now, however, we recognize the existence of about a dozen vitamins. The first known example is thiamine, the component of rice bran that is nutritionally deficient in beriberi. Later the vitamin deficiencies associated with pellagra, rickets, and scurvy were related to vitamins B_6, D, and C, respectively.

THIAMINE PYROPHOSPHATE

Because thiamine was the first of the B vitamins to be identified, it is also called vitamin B_1. The vitamin is structurally complex, but its conversion to the coenzyme form, thiamine pyrophosphate, or TPP, involves simply an ATP-dependent pyrophosphorylation.

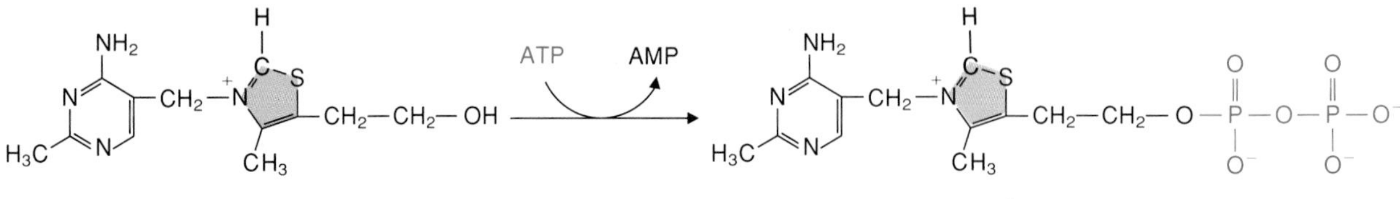

NH2 H CH2 N C S CH2—CH2—OH H3C N CH3 —ATP→AMP— NH2 H CH2 N C S CH2—CH2—O—P(O)(O⁻)—O—P(O)(O⁻)—O⁻ CH3

Thiamine **Thiamine pyrophosphate**

The thiazole ring of TPP is the functional part of the coenzyme, allowing it to bind and transfer activated aldehydes.

Thiamine pyrophosphate is the coenzyme for all decarboxylations of α-keto acids. The mechanism shown in Figure 14.6 for pyruvate decarboxylation is involved in all of these reactions. Note that TPP contains two heterocyclic rings: a substituted pyrimidine and a thiazole. Recent NMR studies have shown that both rings participate in the formation of a reactive carbanion at C-2 of the thiazole ring—the carbon atom between the nitrogen and the sulfur. As shown in the following diagram, a glutamate carboxyl group in the enzyme attracts a proton

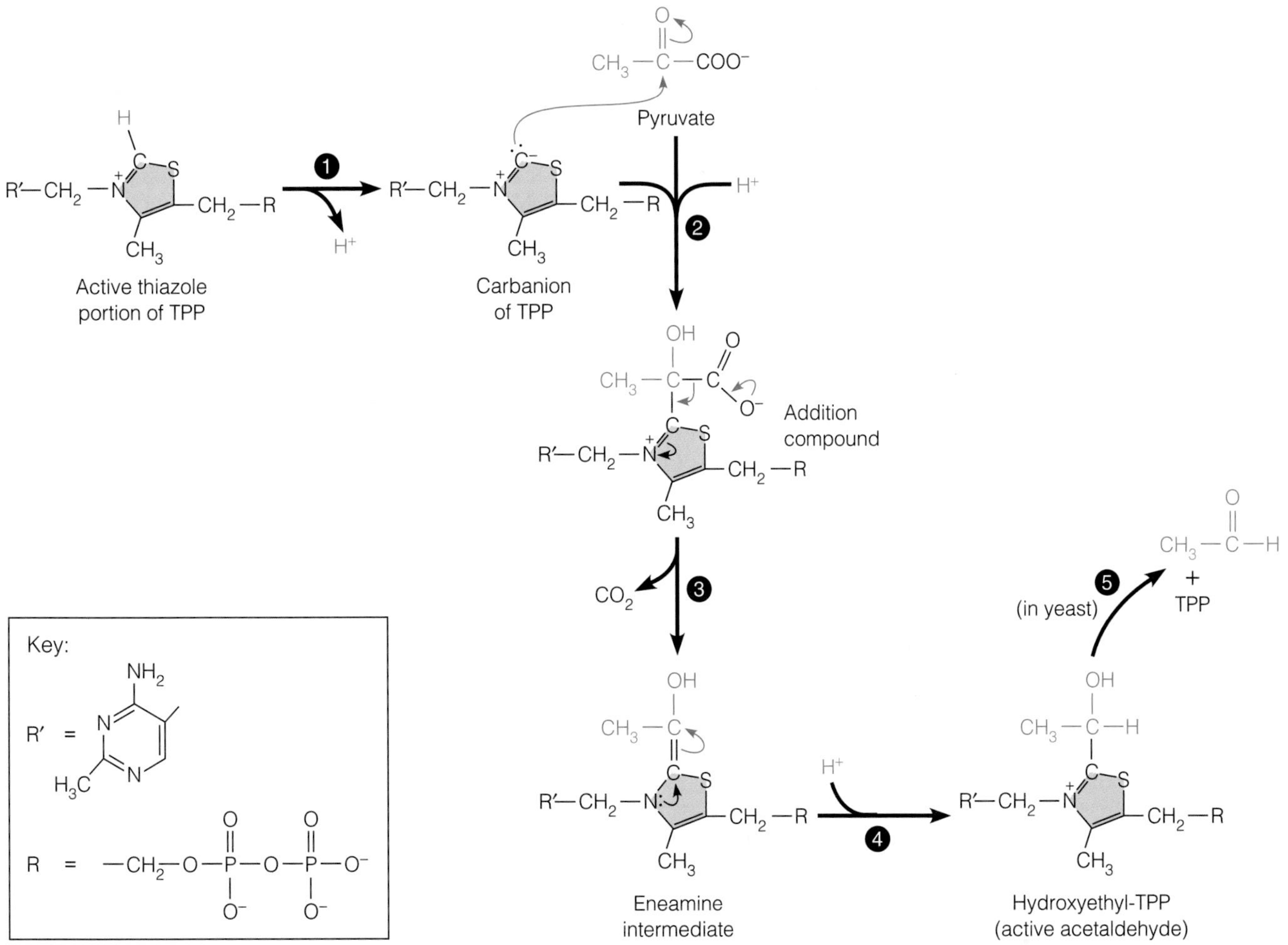

FIGURE **14.6**

Thiamine pyrophosphate in the pyruvate dehydrogenase reaction. Thiamine pyrophosphate (TPP) is the coenzyme for the pyruvate dehydrogenase reaction and other nonoxidative decarboxylations of α-keto acids. The key reaction (step 2) is attack by the carbanion of TPP on the carbonyl carbon of pyruvate and is followed by nonoxidative decarboxylation of the coenzyme-bound pyruvate (step 3); the electron pair remains with the ring nitrogen. In step 4, the two-carbon fragment (red) bound to TPP extracts a proton from pyruvate decarboxylase (not shown), generating a hydroxyethyl group. This fragment remains at the aldehyde oxidation level. In fermentation of glucose to ethanol, hydroxyethyl-TPP is converted to acetaldehyde and TPP by yeast pyruvate decarboxylase (step 5).

linked to N-1 of the pyrimidine, which in turn increases the basicity of the amino group, facilitating the deprotonation of C-2 of the thiazole ring.

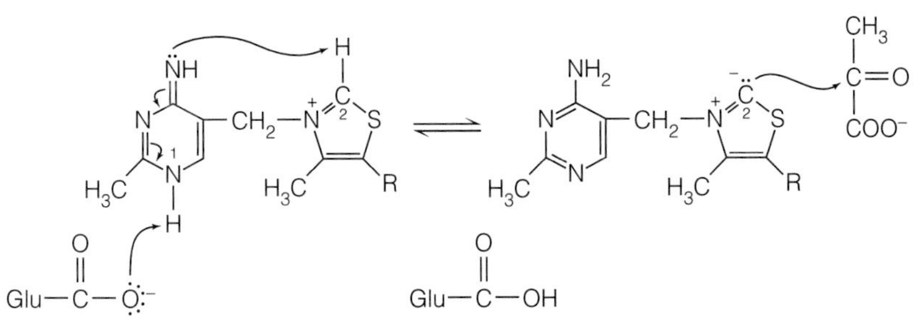

This carbon forms a carbanion (step 1 in Figure 14.6), which in turn can attack the carbonyl carbon of α-keto acids, such as pyruvate, giving an addition compound (step 2). The addition compound undergoes nonoxidative decarboxylation (step 3), with the thiazole ring acting as an electron sink in forming a resonance-stabilized eneamine. Protonation (step 4) gives a species called **active acetaldehyde** or, more accurately, **hydroxyethyl-TPP.** In the fermentation of glucose to ethanol in yeast (step 5), this intermediate undergoes an elimination reaction to yield acetaldehyde and the TPP carbanion. In the pyruvate dehydrogenase reaction (not shown here), the activated two-carbon fragment is simultaneously oxidized and transferred to another enzyme, as discussed in more detail shortly. Thus, in general terms, TPP functions in the generation of an activated aldehyde species, which may or may not undergo oxidation as it is transferred to an acceptor.

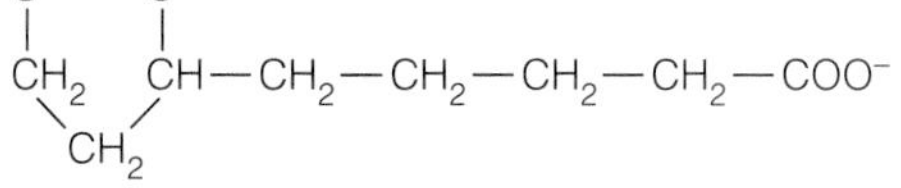

LIPOIC ACID (LIPOAMIDE)

In pyruvate oxidation, the next acceptor of the aldehyde generated by TPP is **lipoic acid**, which is the internal disulfide of 6,8-dithiooctanoic acid. The coenzyme is joined to its apoenzyme via an amide bond linking the carboxyl group of lipoic acid to a lysine ε-amino group. Thus, the reactive species is an amide, called **lipoamide.**

Lipoic acid

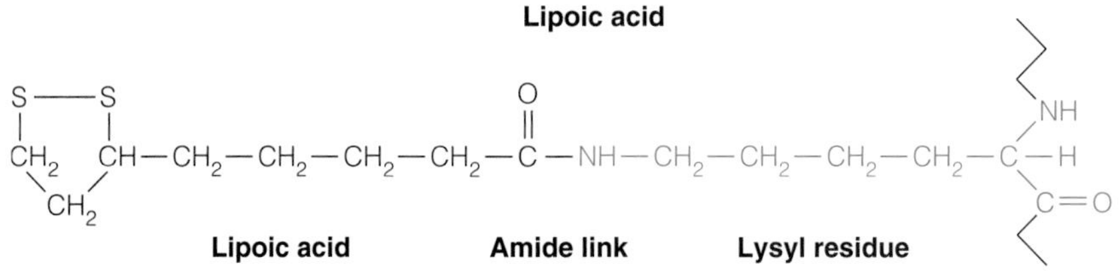

Lipoic acid   Amide link   Lysyl residue

Lipoamide is a carrier of both electrons and acyl groups.

Transfer of the active aldehyde moiety from TPP to the sulfur on carbon 6 of lipoamide involves simultaneous oxidation of the aldehyde, coupled to reduction of the disulfide. This generates an acyl group which, in pyruvate dehydrogenase, is transferred next to coenzyme A. Thus, lipoamide is both an electron carrier and an acyl group carrier.

FLAVIN COENZYMES

Flavin adenine dinucleotide, or **FAD,** is one of two coenzymes derived from vitamin B_2, or **riboflavin.** The other is the simpler **flavin mononucleotide** (FMN), or riboflavin phosphate (Figure 14.7). The functional part of both coenzymes is the **isoalloxazine ring** system, which serves as a two-electron acceptor. Compounds containing such a ring system are called **flavins.** In riboflavin and its derivatives the ring system is attached to **ribitol,** an open-chain version of ribose with the aldehyde carbon reduced to the alcohol level. The 5′ carbon of ribitol is linked to phosphate in FMN, and FAD is an adenylylated derivative of FMN. Thus, these compounds are somewhat analogous to nicotinamide mononucleotide and NAD^+, respectively.

Enzymes that use a flavin coenzyme are called **flavoproteins.** FMN and FAD undergo virtually identical electron transfer reactions. Flavoprotein enzymes preferentially bind either FMN or FAD. In a few cases, that binding is covalent. However, in most cases the flavin is bound tightly but noncovalently, such that the coenzyme can be reversibly dissociated from the holoenzyme.

Flavin coenzymes participate in two-electron oxidoreduction reactions that can proceed in 2 one-electron steps.

Like the nicotinamide coenzymes, the flavins undergo two-electron oxidation and reduction reactions. The flavins, however, are distinctive in having a stable one-electron-reduced species, a **semiquinone** free radical, as shown in Figure 14.8. This free radical can be detected spectrophotometrically; whereas oxidized FAD and FMN are bright yellow, and fully reduced flavins are colorless, the semiquinone intermediate is either red or blue, depending on pH. The stability of the

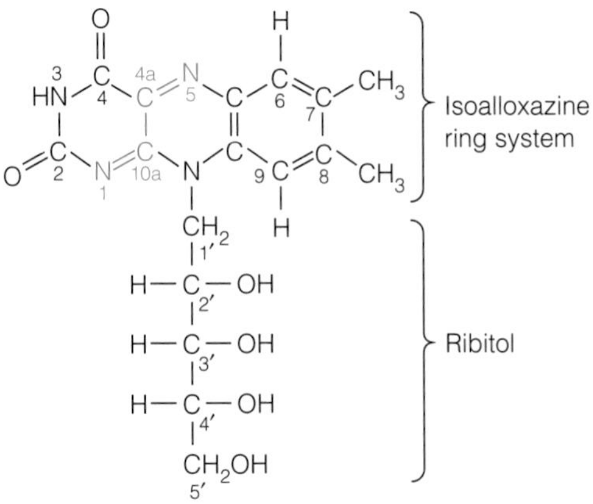

Riboflavin

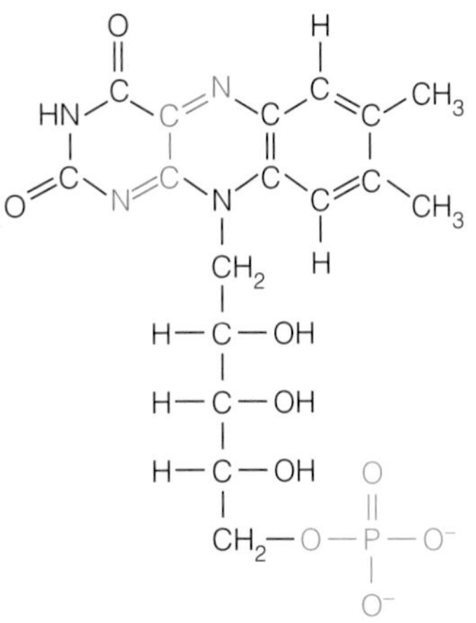

Flavin mononucleotide (FMN)
(also called riboflavin phosphate)

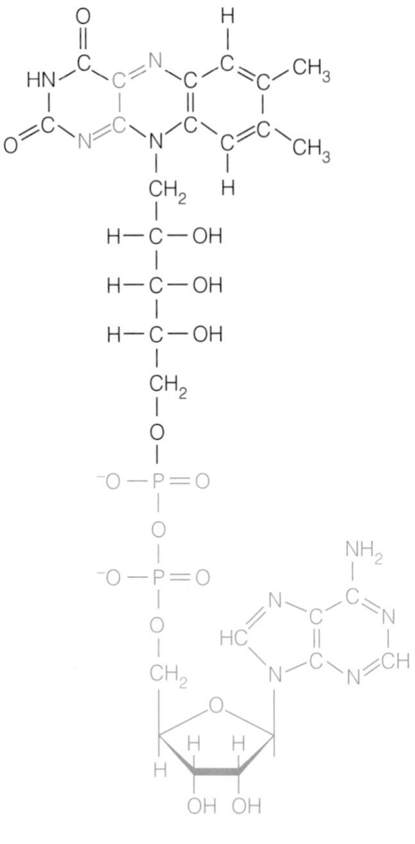

Flavin adenine dinucleotide (FAD)

FIGURE **14.7**

Structures of riboflavin and the flavin coenzymes. Riboflavin and its coenzyme derivatives, FMN and FAD, all contain an isoalloxazine ring system and ribitol. The figure identifies (in red) the cluster of two carbon and two nitrogen atoms within the ring system that participates in the oxidation–reduction reactions of the flavin coenzymes.

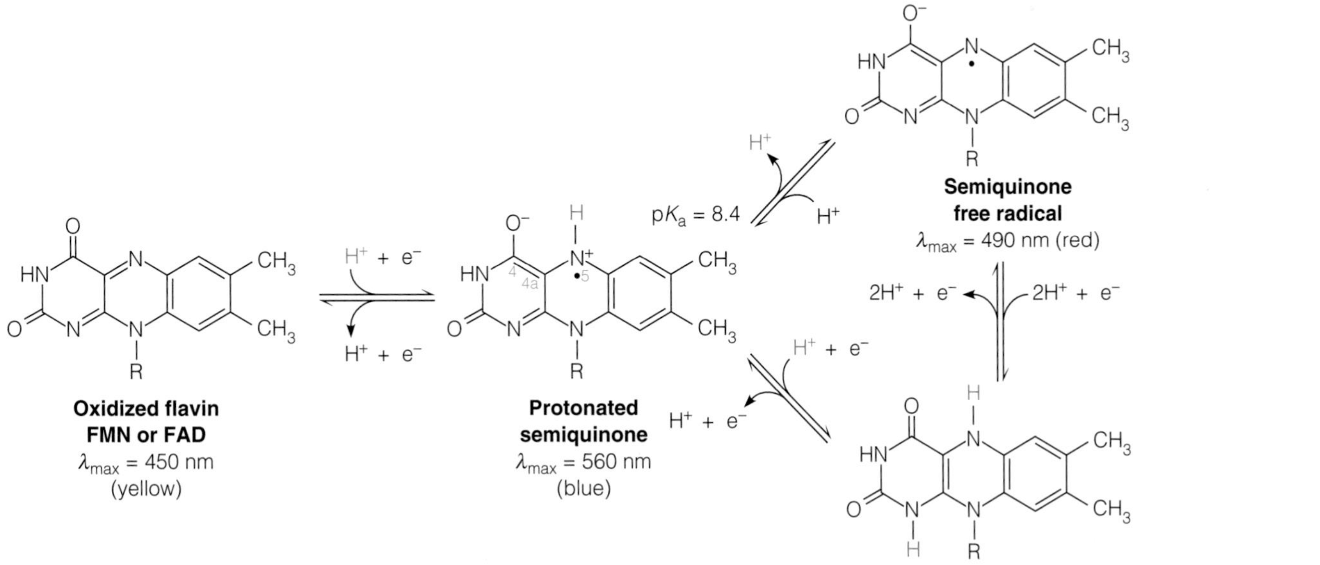

Reduced flavin
$FMNH_2$ or $FADH_2$
(colorless)

FIGURE **14.8**

Oxidation and reduction reactions involving flavin coenzymes. Flavins participate in two-electron reactions, but the existence of the semiquinone free radical intermediate allows these reactions to proceed one electron at a time. Thus, reduced flavins can readily be oxidized by one-electron acceptors. Spectral maxima (λ_{max}) are indicated for the oxidized flavin and the protonated and deprotonated forms of the semiquinone intermediate. In both semiquinone forms the unpaired electron is delocalized between N-5 and C-4a.

intermediate gives flavins a catalytic versatility not shared by nicotinamide coenzymes, in that flavins can interact with either two-electron or one-electron donor–acceptor pairs. Also, flavoproteins can interact directly with oxygen. Thus, some, but not all, flavoproteins are oxidases.

COENZYME A AND THE ACTIVATION OF ACYL GROUPS

Coenzyme A (A for acyl) participates in activation of acyl groups in general, including the acetyl group derived from pyruvate. The coenzyme is derived metabolically from ATP, the vitamin **pantothenic acid,** and β-mercaptoethylamine.

$$HS-CH_2-CH_2-NH-\overset{O}{\overset{\|}{C}}-CH_2-CH_2-NH-\overset{O}{\overset{\|}{C}}-\underset{OH}{\overset{H}{C}}-\underset{CH_3}{\overset{CH_3}{C}}-CH_2-O-\underset{O^-}{\overset{O}{\overset{\|}{P}}}-O-\underset{O^-}{\overset{O}{\overset{\|}{P}}}-O-CH_2\text{-(ribose 3'-}ⓟ\text{, Adenine)}$$

β-Mercaptoethylamine | **Pantothenic acid** | **Adenosine 3′-phosphate 5′-diphosphate**

A free thiol on the β-mercaptoethylamine moiety is the functionally significant part of the coenzyme molecule; the rest of the molecule provides enzyme binding sites. In acylated derivatives, such as acetyl-coenzyme A, the acyl group is linked to the thiol group to form an energy-rich thioester.

$$\underset{\textbf{Coenzyme A}}{\text{CoA-SH}} + \underset{\textbf{Acetyl group}}{-\overset{O}{\overset{\|}{C}}-CH_3} \longrightarrow \underset{\textbf{Acetyl-CoA}}{\text{CoA}-S-\overset{O}{\overset{\|}{C}}-CH_3}$$

We designate the acylated forms of coenzyme A as $R-\overset{O}{\overset{\|}{C}}-S-\text{CoA}$, or acyl-CoA, and the unacylated form as CoA-SH.

The energy-rich nature of thioesters, as compared with ordinary esters, is related primarily to resonance stabilization (Figure 14.9). Most esters have two resonance forms (Figure 14.9). Stabilization involves π-electron overlap, giving

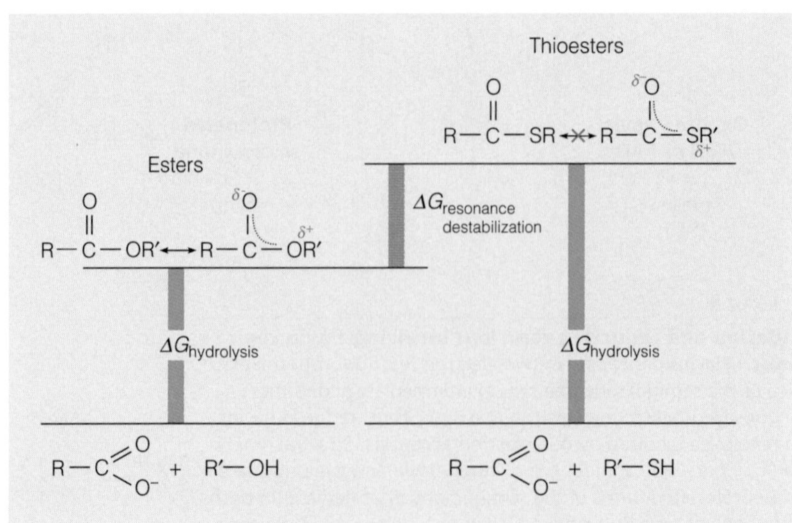

FIGURE **14.9**

Comparison of free energies of hydrolysis of thioesters and other esters. Lack of resonance stabilization (shown by the red X through the resonance arrow) is the basis for the higher ΔG of hydrolysis of thioesters, relative to that of ordinary esters. The free energies of the hydrolysis products are similar for the two classes of compounds.

partial double-bond character to the C—O link. In thioesters, the larger atomic size of S (as compared with O) reduces the π-electron overlap between C and S, so that the C-S structure does not contribute significantly to resonance stabilization. Thus, the thioester is *destabilized* relative to an ester, so that its ΔG of hydrolysis is increased.

The lack of double-bond character in the C—S bond of acyl-CoAs makes this bond weaker than the corresponding C—O bond in ordinary esters, in turn making the thioalkoxide ion ($R—S^-$) a good leaving group in nucleophilic displacement reactions. Thus, the acyl group is readily transferred to other metabolites, which is what occurs in the first reaction of the citric acid cycle.

Thioesters such as acetyl-CoA are energy rich because the C—S bond destabilizes it relative to a C—O bond.

Action of the Pyruvate Dehydrogenase Complex

As noted earlier, the oxidation of pyruvate to acetyl-CoA involves the coenzymes TPP, lipoic acid, FAD, NAD^+, and CoA-SH, acting in concert with three enzymes, in the pyruvate dehydrogenase complex. Now we are ready to see how all of these components function together to effect the conversion of pyruvate to acetyl-CoA. The reactions catalyzed by each enzyme are shown here, and the overall process is summarized in Figure 14.10.

1. Pyruvate + TPP + $H^+ \xrightarrow{E_1}$ hydroxyethyl-TPP + CO_2
2. Hydroxyethyl-TPP + $\mathrm{Lip}_1\!<\!\begin{matrix}\mathrm{S}\\ |\\ \mathrm{S}\end{matrix} \xrightarrow{E_1}$ TPP + $\mathrm{Lip}_1\!<\!\begin{matrix}\text{S-acetyl}\\ \mathrm{SH}\end{matrix}$
3. $\mathrm{Lip}_1\!<\!\begin{matrix}\text{S-acetyl}\\ \mathrm{SH}\end{matrix} + \mathrm{Lip}_2\!<\!\begin{matrix}\mathrm{S}\\ |\\ \mathrm{S}\end{matrix} \xrightarrow{E_2} \mathrm{Lip}_1\!<\!\begin{matrix}\mathrm{S}\\ |\\ \mathrm{S}\end{matrix} + \mathrm{Lip}_2\!<\!\begin{matrix}\text{S-acetyl}\\ \mathrm{SH}\end{matrix}$
4. $\mathrm{Lip}_2\!<\!\begin{matrix}\text{S-acetyl}\\ \mathrm{SH}\end{matrix}$ + CoA-SH $\xrightarrow{E_2}$ acetyl-CoA + $\mathrm{Lip}_2\!<\!\begin{matrix}\mathrm{SH}\\ \mathrm{SH}\end{matrix}$
5. $\mathrm{Lip}_2\!<\!\begin{matrix}\mathrm{SH}\\ \mathrm{SH}\end{matrix}$ + FAD $\xrightarrow{E_3} \mathrm{Lip}_2\!<\!\begin{matrix}\mathrm{S}\\ |\\ \mathrm{S}\end{matrix}$ + $FADH_2$
6. $FADH_2$ + $NAD^+ \xrightarrow{E_3}$ FAD + NADH + H^+

Sum: Pyruvate + CoA-SH + $NAD^+ \longrightarrow$ acetyl-CoA + CO_2 + NADH

A central feature of this reaction sequence is the interaction of two lipoamide moieties on E_2 first with hydroxyethyl-TPP bound to E_1, next with CoA-SH on E_2 to yield acetyl-CoA, and finally with bound FAD on E_3 to regenerate the disulfide species of lipoamide. Transfer and concomitant oxidation of the hydroxyethyl group probably occur as follows, where B denotes basic residues on the E_1 protein:

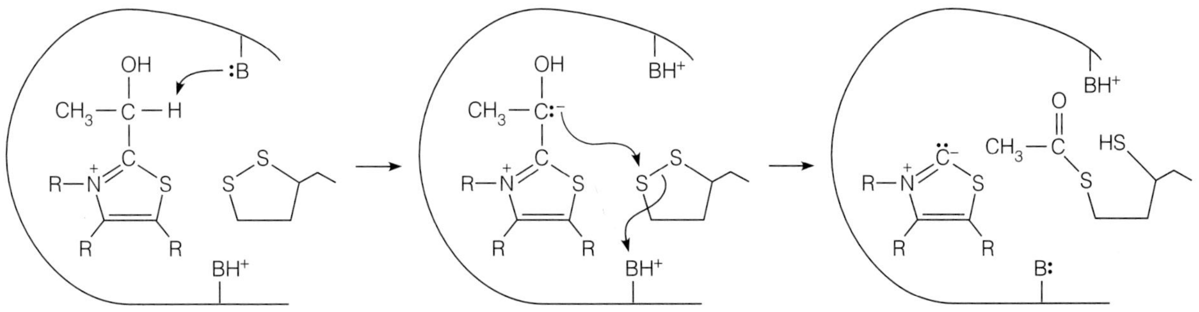

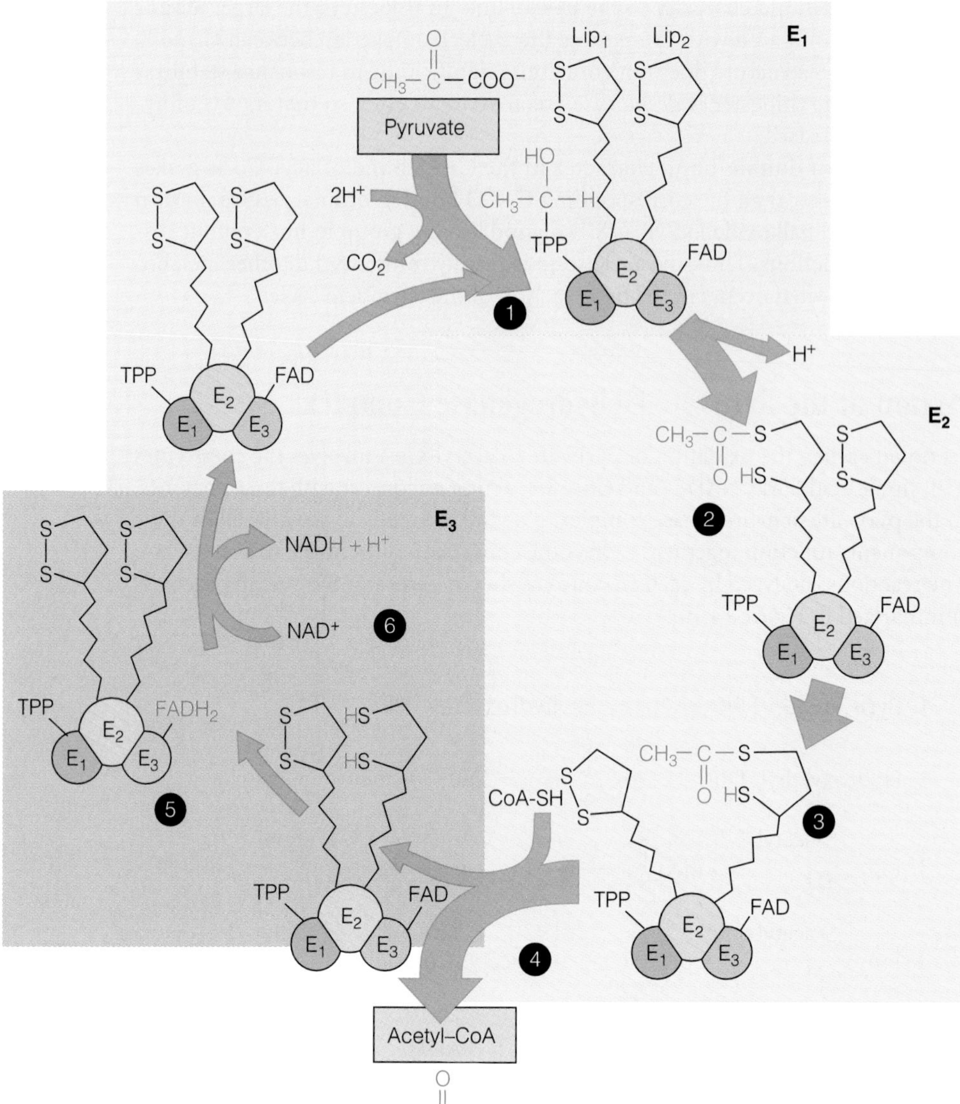

FIGURE **14.10**

Mechanisms of the pyruvate dehydrogenase complex. This detailed diagram of the oxidation of pyruvate to acetyl-CoA shows the role of the swinging arms of lipoamide in the functioning of the pyruvate dehydrogenase complex. The number identifying each reaction is colored the same as the enzyme catalyzing that reaction. **Reaction 1:** E_1 accepts a two-carbon aldehyde group from pyruvate and binds it to TPP, forming hydroxyethyl-TPP. **Reaction 2:** The aldehyde group is transferred by E_1 to the first lipoamide swinging arm on E_2 and simultaneously oxidized to an acetyl group. **Reaction 3:** The acetyl group is transferred to the second swinging arm, which positions it for transfer to CoA-SH. **Reaction 4:** The acetyl group is transferred to CoA-SH, producing acetyl-CoA. **Reaction 5:** E_3 oxidizes the reduced lipoamide swinging arm by transferring two hydrogens to FAD. **Reaction 6:** The reduced flavin ($FADH_2$) is oxidized by NAD^+. E_1: pyruvate dehydrogenase. E_2: dihydrolipoamide transacetylase. E_3: dihydrolipoamide dehydrogenase.

Lipoamide is tethered to one enzyme (E_2) in the pyruvate dehydrogenase complex, but it interacts with all three enzymes via a flexible swinging arm.

Attachment of the eight-carbon lipoic acid chain with a six-carbon lysine residue gives a flexible chain that, when fully extended, is about 1.4 nm in length. However, structural studies indicate that bound TPP on E_1 is about 4.5 to 6.0 nm away from bound FAD on E_3. Thus, a single lipoamide arm on E_2 cannot swing all the way between participating sites on E_1 and E_3. The problem is solved through the participation of at least two lipoamide arms on each E_2 chain, with transfer of the acetyl group from one arm to the next. Because each 1.4-nm arm can arc about twice that distance, two arms can connect functional groups up to 5.6 nm apart. Operation of these swinging arms is shown diagrammatically in Figure 14.10. The first arm accepts the two-carbon fragment from E_1 and then transfers it as an acetyl group to the second arm. From there it is transferred directly to coenzyme A. The reduced lipoamide arm, with two thiol groups, is then reoxidized to the cyclic disulfide by E_3.

Physical juxtaposition of the enzymes of the complex allows the overall reaction to proceed smoothly, without unwanted side reactions or diffusion of intermediates from catalytic sites. The pyruvate dehydrogenase complex represents one of the best-understood examples of how cells can achieve economy of function by juxtaposition of enzymes catalyzing sequential reactions.

The Citric Acid Cycle

Figure 14.3 presented the entire citric acid cycle, showing the structure of each intermediate. The importance of this pathway merits examination of each reaction. First, recall that the cycle proceeds in two phases: (1) addition of a two-carbon moiety (acetyl-CoA) to a four-carbon compound (oxaloacetate) to give a six-carbon organic anion, citrate, followed by loss of two carbons as CO_2, and (2) the regeneration of oxaloacetate.

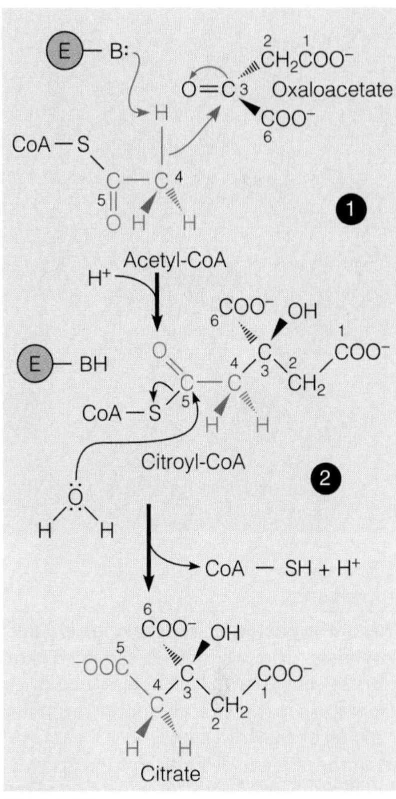

FIGURE **14.11**

Mechanism of the citrate synthase reaction. Step 1: A basic residue (B:) on the enzyme extracts a proton from the methyl group of acetyl-CoA, creating a nucleophilic carbanion that attacks the keto carbon of oxaloacetate. **Step 2:** The citroyl-CoA formed in step 1 spontaneously hydrolyzes.

PHASE 1: INTRODUCTION AND LOSS OF TWO CARBON ATOMS

Step 1: Introduction of Two Carbon Atoms as Acetyl-CoA

The initial reaction, catalyzed by **citrate synthase,** is akin to an aldol condensation.

$$CH_3{-}\overset{O}{\overset{\|}{C}}{-}S{-}CoA + \underset{CH_2-COO^-}{\overset{O}{\overset{\|}{C}}{-}COO^-} + H_2O \longrightarrow HO{-}\underset{CH_2-COO^-}{\overset{CH_2-COO^-}{C}}{-}COO^- + CoA{-}SH + H^+$$

Acetyl-S-CoA **Oxaloacetate** **Citrate** $\Delta G°' = -32.2$ kJ/mol

As shown in Figure 14.11, the methyl carbon of the activated acetyl moiety on acetyl-CoA loses a proton, with nucleophilic attack of the resultant carbanion on the carbonyl carbon of oxaloacetate. This reaction generates the highly unstable **citroyl-CoA,** which spontaneously hydrolyzes while enzyme-bound, to yield the products. As expected for the first committed step in a pathway, the reaction is highly exergonic and has been considered to be a site of regulation for the overall pathway (see page 504). Crystallographic analysis of citrate synthase (Figure 14.12) gives excellent evidence for the induced fit model of enzyme catalysis (see Chapter 11).

Step 2: Isomerization of Citrate

Substrates for decarboxylation reactions are generally α- or β-keto acids. The tertiary alcoholic hydroxyl group of citrate must migrate in order to undergo oxidation to a carbonyl. Isomerization, catalyzed by **aconitase,** generates the secondary alcoholic compound, isocitrate, which can be oxidized. The reaction involves successive dehydration and hydration, through *cis*-aconitate as a dehydrated intermediate, which remains enzyme-bound.

$$\begin{array}{c} CH_2-COO^- \\ | \\ HO-C-COO^- \\ | \\ H-C-COO^- \\ | \\ H \end{array} \underset{H_2O}{\overset{H_2O}{\rightleftharpoons}} \left[\begin{array}{c} CH_2-COO^- \\ | \\ C-COO^- \\ \| \\ CH-COO^- \end{array}\right] \underset{H_2O}{\overset{H_2O}{\rightleftharpoons}} \begin{array}{c} CH_2-COO^- \\ | \\ H-C-COO^- \\ | \\ H-C-COO^- \\ | \\ OH \end{array} \qquad \Delta G°' = +6.3 \text{ kJ/mol}$$

Citrate ***cis*-Aconitate** **D-Isocitrate**

An equilibrium mixture of these three acids at 25°C contains about 90% citrate, 4% *cis*-aconitate, and 6% isocitrate, but the exergonic nature of the next reaction draws the reaction to the right as written. The enzyme contains nonheme iron and acid-labile sulfur in a cluster called a 4Fe–4S *iron–sulfur center,* normally associated with oxidoreductases (see Chapter 15). In aconitase the iron forms a complex with citrate, through two carboxyl groups and one hydroxyl group on the citrate molecule.

Aconitase is the target site for the toxic action of **fluoroacetate,** a plant product that has been used as a pesticide. Its use by ranchers in the West to control coyote populations has led also to the death of eagles. Fluoroacetate blocks the

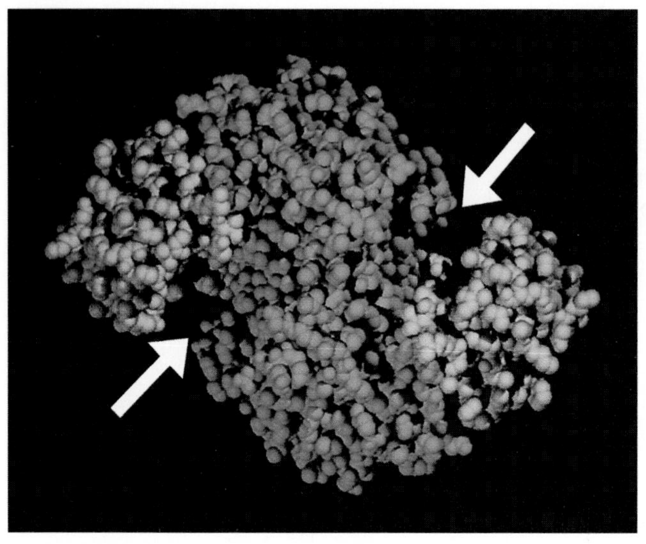

(a)

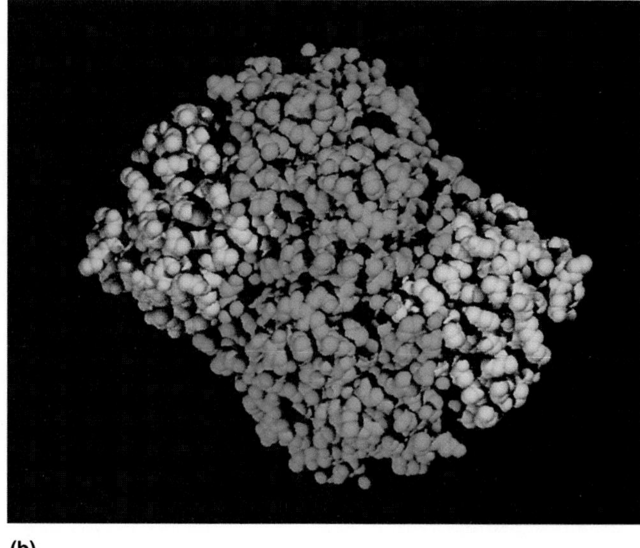

(b)

FIGURE **14.12**

Three-dimensional structure of citrate synthase. The two forms of pig heart citrate synthase shown here were determined by crystallographic methods and support the induced fit model of enzyme catalysis (see Chapter 11). **(a)** In the absence of substrates the enzyme crystallizes in an "open" form, with large clefts (at arrows) in both catalytic domains of the homodimeric protein. **(b)** Binding of substrates or CoA-SH causes the enzyme to adopt a "closed" conformation, with the clefts essentially filled.

Reproduced with permission from G. Wiegand and S. J. Remington, *Annu. Rev. Biophys. & Biophys. Chem.*, Vol. 15, © 1986 by Annual Reviews, Inc.

citric acid cycle by its metabolic conversion to **fluorocitrate,** which is a potent inhibitor of aconitase. Fluoroacetate is an example of a **suicide substrate.** As noted in Chapter 11, a suicide substrate by itself is not toxic to cells, but it resembles a normal metabolite closely enough that it undergoes metabolic transformation to a product that does inhibit a crucial enzyme. The cell "commits suicide" by transforming the analog to a toxic product. In this case fluoroacetate is converted to fluoroacetyl-CoA, which serves as a substrate for citrate synthase. However, the resultant fluorocitrate cannot be acted on by aconitase.

Fluoroacetate acts as a poison by being converted, via citrate synthase, to the aconitase inhibitor fluorocitrate.

$$\underset{\textbf{Fluoroacetate}}{F-CH_2-COO^-} \xrightarrow{CoA-SH} \underset{\textbf{Fluoroacetyl-CoA}}{F-CH_2-\overset{O}{\overset{\|}{C}}\sim S-CoA} \longrightarrow \underset{\textbf{Fluorocitrate}}{H-C(F)(COO^-)-C(OH)(COO^-)-CH_2-COO^-}$$

The aconitase reaction is *stereospecific:* Of the four possible diastereomers of isocitrate, only one is produced. Aconitase is one of the earliest enzymes for which stereospecificity was demonstrated and elucidated. To describe those developments, we must first present the next reaction in the citric acid cycle.

Step 3: Generation of CO_2 by an NAD^+-Linked Dehydrogenase

The first of two oxidative decarboxylations in the cycle is catalyzed by **isocitrate dehydrogenase.**

$$\underset{\textbf{Isocitrate}}{CH_2(COO^-)-CH(COO^-)-CH(OH)-COO^-} + \underset{(NADP^+)}{NAD^+} \rightleftharpoons \underset{\alpha\textbf{-Ketoglutarate}}{CH_2(COO^-)-CH_2-C(=O)-COO^-} + \underset{(NADPH)}{NADH} + CO_2 \qquad \Delta G^{\circ\prime} = -20.9\ \text{kJ/mol}$$

The reaction involves dehydrogenation to **oxalosuccinate,** an unstable enzyme-bound intermediate that spontaneously decarboxylates before release of the product.

$$\text{Isocitrate} \xrightarrow{NAD^+ \rightarrow NADH + H^+} [\text{Oxalosuccinate}] \xrightarrow{H^+ \rightarrow CO_2} \alpha\text{-Ketoglutarate}$$

Isocitrate **Oxalosuccinate** **α-Ketoglutarate**

A mitochondrial form of isocitrate dehydrogenase that is specific for NAD^+ is probably the chief participant in the citric acid cycle. However, an $NADP^+$-specific form of the enzyme is also present in both cytosol and mitochondria.

As noted earlier, the stereospecificity of enzyme–substrate interactions was first shown in an experiment involving aconitase. That experiment, carried out in the late 1940s, at first cast into question whether citrate was a participant in the overall cycle. Carboxyl-labeled acetate was administered to animal tissue preparations; this labels citrate via conversion of acetate to acetyl-CoA. Because citrate is a symmetrical compound, it was expected that aconitase could transfer the alcoholic hydroxyl in either direction (Figure 14.13). Thus, α-ketoglutarate was expected to be labeled equally in both carboxyl groups. However, only the γ-labeled form was detected, suggesting that only asymmetric tricarboxylic acids could be involved in these early steps of the cycle. This suggestion would rule out the symmetrical citrate molecule. However, Alexander Ogston pointed out that if the enzyme bound the substrate at three points or more, then the binding site itself was asymmetric and could bind the substrate in only one way (Figure 14.13). Thus, citrate was the first substance to be recognized as **prochiral,** or symmetrical, but becoming asymmetric upon binding to the asymmetric surface of an enzyme, or upon some similar change in one of two equivalent groups.

Step 4: Generation of a Second CO_2 by a Multienzyme Complex

The fourth reaction of the citric acid cycle is a multistep reaction entirely comparable to the pyruvate dehydrogenase reaction. An α-keto acid substrate undergoes oxidative decarboxylation, with concomitant formation of an acyl-CoA.

$$\alpha\text{-Ketoglutarate} + NAD^+ + \text{CoA-SH} \longrightarrow \text{Succinyl-CoA} + CO_2 + NADH \qquad \Delta G^{\circ\prime} = -33.5 \text{ kJ/mol}$$

α-Ketoglutarate **Succinyl-CoA**

This reaction is catalyzed by the **α-ketoglutarate dehydrogenase complex,** an enzyme cluster similar to the pyruvate dehydrogenase complex, with three analogous enzyme activities and the same five coenzymes—TPP, NAD^+, FAD, lipoic acid, and CoA-SH. The first reaction of the series decarboxylates the substrate, as shown in Figure 14.14. Subsequent transfer of the remaining four-carbon unit to lipoic acid and oxidation by FAD and NAD^+ are analogous to the reactions shown in Figure 14.10 for the pyruvate dehydrogenase complex. One important distinction between these two multienzyme complexes is that the regulatory activities associated with the pyruvate dehydrogenase complex are not present in the α-ketoglutarate dehydrogenase complex.

PHASE 2: REGENERATION OF OXALOACETATE

At this point in the cycle, two carbon atoms have been introduced as acetyl-CoA (by citrate synthase), and two have been lost as CO_2. Because of the stereochemistry of the aconitase reaction, the two carbon atoms lost are not the same as the two atoms introduced at the beginning of the cycle. In the remaining reactions

Two carbon atoms enter the citric acid cycle as acetyl-CoA, and two are lost as CO_2 in reactions 3 and 4 of the cycle.

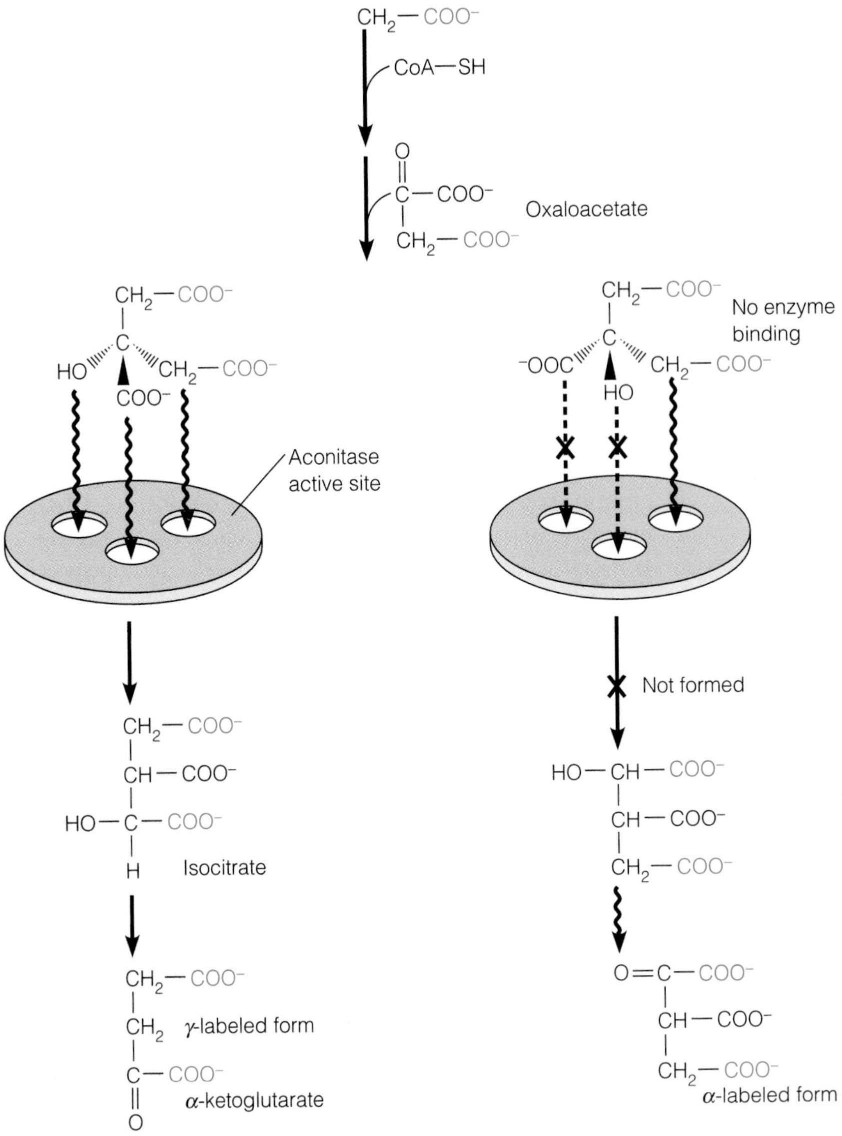

FIGURE **14.13**

Asymmetric binding of citrate to the active site of aconitase. The radiolabeled carboxyl introduced as acetate is shown in red. Binding of citrate to aconitase at three sites or more makes the two carboxyl groups (red and blue) nonequivalent, even though citrate is a symmetrical molecule. Thus, only the labeling pathway shown at the left is stereochemically allowed.

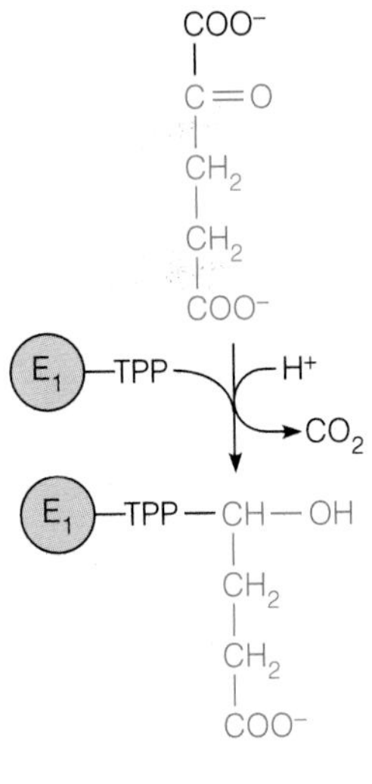

FIGURE **14.14**

Decarboxylation of α-ketoglutarate. The first reaction carried out by the α-ketoglutarate complex is a decarboxylation catalyzed by α-ketoglutarate decarboxylase (E_1 of the complex).

the four-carbon intermediate, succinyl-CoA, is converted to the four-carbon product, oxaloacetate, with two of the four steps involving dehydrogenation reactions.

Step 5: A Substrate-Level Phosphorylation

Succinyl-CoA is an energy-rich compound, and its potential energy is used to drive the formation of a nucleoside triphosphate from a diphosphate. This reaction, catalyzed by **succinyl-CoA synthetase,** is comparable to the two substrate-level phosphorylation reactions that we encountered in glycolysis, except that in animal cells the energy-rich nucleotide product is not always ATP but, in some tissues, GTP.

$$\text{Succinyl-CoA} + P_i + \underset{\text{or ADP}}{\text{GDP}} \rightleftharpoons \text{succinate} + \underset{\text{or ATP}}{\text{GTP}} + \text{CoA-SH} \qquad \Delta G^{\circ\prime} = -2.9 \text{ kJ/mol}$$

In animals, liver cells contain primarily the GTP-linked succinyl-CoA synthetase, while in brain and heart the ATP-linked enzyme predominates. Much of the GTP formed drives the ultimate synthesis of ATP, through the action of nucleoside diphosphate kinase.

$$\text{GTP} + \text{ADP} \rightleftharpoons \text{GDP} + \text{ATP} \qquad \Delta G^{\circ\prime} = 0.0 \text{ kJ/mol}$$

In plants and bacteria, ATP is formed directly in step 5, as shown in Figure 14.3, through reaction of succinyl-CoA with ADP. In either case, step 5 involves formation of the anhydride succinyl phosphate (Figure 14.15), which then activates succinyl-CoA synthetase by phosphorylating a specific histidine residue. The resulting *N*-phosphohistidine residue then transfers its phosphate to the nucleoside diphosphate substrate.

Step 6: A Flavin-Dependent Dehydrogenation

Completion of the cycle involves conversion of the four-carbon succinate to the four-carbon oxaloacetate. The first of the three reactions involved, catalyzed by **succinate dehydrogenase,** is the FAD-dependent dehydrogenation of two saturated carbons to a double bond.

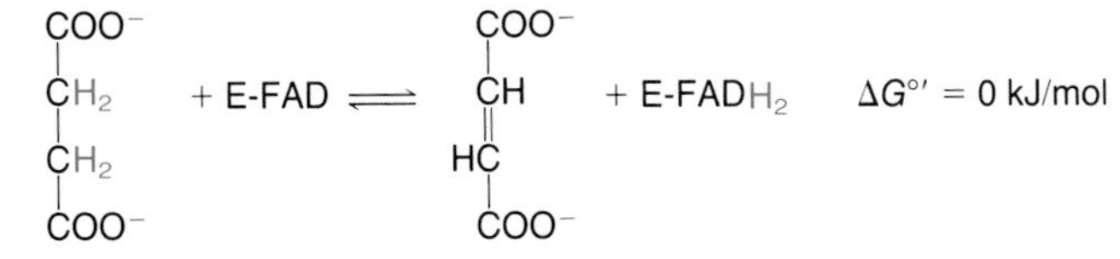

$$\underset{\text{Succinate}}{COO^-\text{—}CH_2\text{—}CH_2\text{—}COO^-} + \text{E-FAD} \rightleftharpoons \underset{\text{Fumarate}}{COO^-\text{—}CH\text{=}HC\text{—}COO^-} + \text{E-FADH}_2 \qquad \Delta G^{\circ\prime} = 0 \text{ kJ/mol}$$

This reaction is followed in the next two steps by hydration of that double bond and dehydrogenation of the resulting α-hydroxy acid to the α-keto acid oxaloacetate.

A C—C single bond is more difficult to oxidize than a C—O bond. Therefore, the redox coenzyme for succinate dehydrogenase is not NAD^+ but the more powerful oxidant FAD, which in turn transfers its electrons to an iron–sulfur center in the enzyme molecule. FAD is bound covalently to the enzyme protein, designated E, through a specific histidine residue. The enzyme is tightly bound to the mitochondrial inner membrane (see Chapter 15). The importance of this binding is that the reduced flavin, which must be reoxidized for the enzyme to act

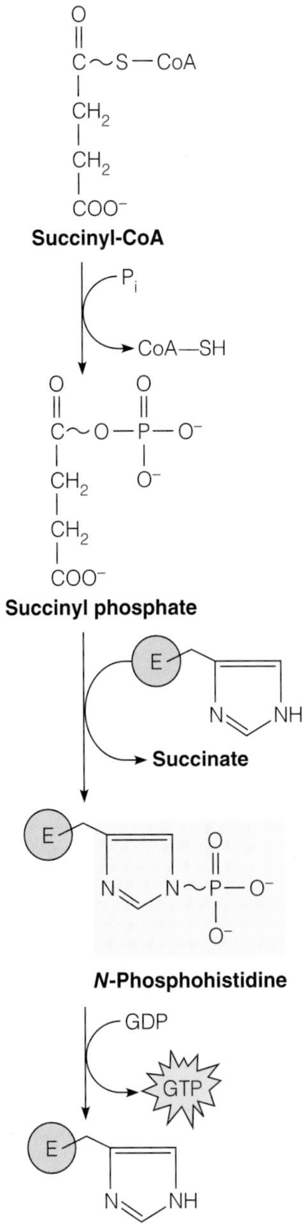

Succinyl-CoA
P_i
CoA—SH
Succinyl phosphate
E
Succinate
E
N-Phosphohistidine
GDP
GTP
E

FIGURE **14.15**

Production of energy-rich nucleotides by the succinyl-CoA synthetase reaction. In animal cells, an *N*-phosphohistidine residue of succinyl-CoA synthetase serves as an intermediate in the production of GTP. In plants and bacteria, the energy-rich nucleotide produced in this step is ATP.

again, becomes reoxidized through interaction with the mitochondrial electron transport system, also bound to the membrane.

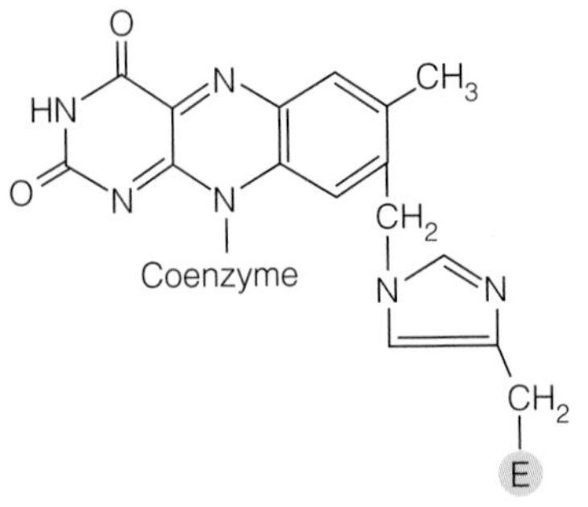

Note that action of succinate dehydrogenase is also stereoselective, with only the *trans* isomer, fumarate, formed. The *cis* isomer, maleate, is not formed.

Step 7: Hydration of a Carbon–Carbon Double Bond
The stereospecific *trans* hydration of the carbon–carbon double bond is catalyzed by **fumarate hydratase,** more commonly called **fumarase.**

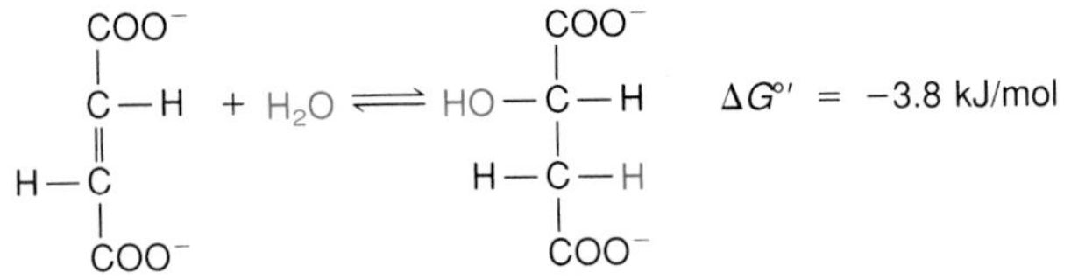

$$\text{Fumarate} + H_2O \rightleftharpoons \text{L-Malate} \qquad \Delta G^{\circ\prime} = -3.8\ \text{kJ/mol}$$

Fumarate **L-Malate**

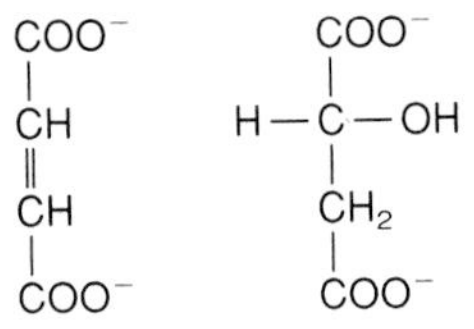

Maleate **D-Malate**

The *cis* isomer of fumarate, namely maleate, is not a substrate for the forward reaction, nor can the enzyme act on D-malate in the reverse direction.

Step 8: A Dehydrogenation That Regenerates Oxaloacetate
Finally, the cycle is completed with the NAD^+-dependent dehydrogenation of malate to oxaloacetate, catalyzed by **malate dehydrogenase.**

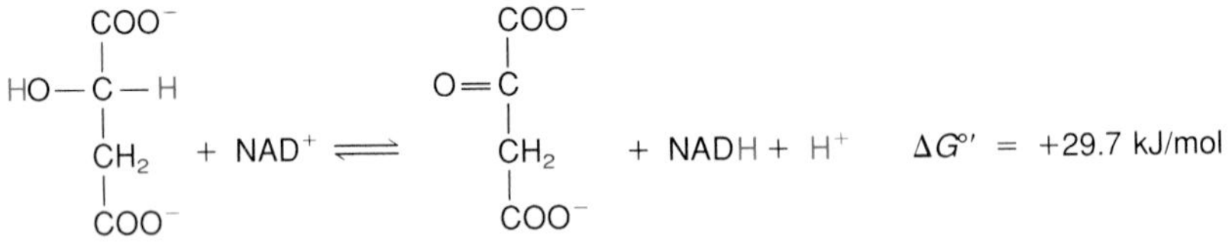

$$\text{L-Malate} + NAD^+ \rightleftharpoons \text{Oxaloacetate} + NADH + H^+ \qquad \Delta G^{\circ\prime} = +29.7\ \text{kJ/mol}$$

L-Malate **Oxaloacetate**

This highly endergonic reaction proceeds to the right as written because the highly exergonic citrate synthase reaction keeps intramitochondrial oxaloacetate levels exceedingly low (below 10^{-6} M).

Stoichiometry and Energetics of the Citric Acid Cycle

Now let us review what has been accomplished in one turn of the citric acid cycle, summarized in Table 14.1. The cycle started when a two-carbon fragment (acetyl-CoA) combined with a four-carbon acceptor (oxaloacetate). Then, two carbons were removed as CO_2 as the resultant citrate was further metabolized. Four oxidation reactions occurred during the cycle, with NAD^+ serving as coenzyme for

TABLE **14.1 Reactions of the citric acid cycle**

Reaction	Enzyme	$\Delta G°'$ (kJ/mol)
1. Acetyl-CoA + oxaloacetate + H_2O ⟶ citrate + CoA-SH + H^+	Citrate synthase	−32.2
2a. Citrate ⇌ *cis*-aconitate + H_2O	Aconitase	+6.3
2b. *cis*-Aconitate + H_2O ⇌ isocitrate	Aconitase	
3. Isocitrate + NAD^+ ⇌ α-ketoglutarate + CO_2 + NADH	Isocitrate dehydrogenase	−20.9
4. α-Ketoglutarate + NAD^+ + CoA-SH ⇌ succinyl-CoA + CO_2 + NADH	α-Ketoglutarate dehydrogenase complex	−33.5
5. Succinyl-CoA + P_i + GDP ⇌ succinate + GTP + CoA-SH	Succinyl-CoA synthetase	−2.9
6. Succinate + FAD (enzyme-bound) ⇌ fumarate + $FADH_2$ (enzyme-bound)	Succinate dehydrogenase	0
7. Fumarate + H_2O ⇌ L-malate	Fumarase	−3.8
8. L-Malate + NAD^+ ⇌ oxaloacetate + NADH + H^+	Malate dehydrogenase	+29.7
	Net	−57.3

three and FAD as coenzyme for the fourth. High-energy phosphate was generated directly at only one reaction (catalyzed by succinyl-CoA synthetase). Finally, oxaloacetate was regenerated; it is now ready to start the cycle again by condensation with another molecule of acetyl-CoA.

We can write a balanced chemical equation representing the sum of the eight reactions involved in one turn of the cycle:

$$\text{Acetyl-CoA} + 2H_2O + 3NAD^+ + FAD + GDP + P_i \longrightarrow 2CO_2 + 3NADH + FADH_2 + \text{CoA-SH} + GTP$$

The GTP formed in the animal succinyl-CoA synthetase reaction is energetically equivalent to ATP, because nucleoside diphosphate kinase can convert the GTP that is formed to ATP at no net free energy cost. Therefore, we substitute ATP for GTP in subsequent discussion.

One turn of the citric acid cycle generates one high-energy phosphate through substrate-level phosphorylation, plus three NADH and one $FADH_2$ for subsequent reoxidation.

Now if we take into account the pyruvate dehydrogenase reaction, and if we recall that each molecule of glucose generates two molecules of pyruvate, we can write the following equation for catabolism of glucose through glycolysis and the citric acid cycle.

$$\text{Glucose} + 2H_2O + 10NAD^+ + 2FAD + 4ADP + 4P_i \longrightarrow 6CO_2 + 10NADH + 6H^+ + 2FADH_2 + 4ATP$$

At this point the ATP yield per mole of glucose metabolized has not increased greatly over the yield from glycolysis: 2 moles of ATP per glucose in glycolysis alone to 4 moles here. Most of the ATP generated during glucose oxidation is not formed directly from reactions of glycolysis and the citric acid cycle but is formed from the reoxidation of reduced electron carriers in the respiratory chain. These compounds, NADH and $FADH_2$, are themselves energy rich, in the sense that their oxidation is highly exergonic. As electrons are transferred from these reduced carriers stepwise toward molecular oxygen, the coupled synthesis of ATP from ADP takes place, with about 3 moles of ATP per mole of NADH reoxidized and about 2 moles of ATP per mole of $FADH_2$ reoxidized. As we see in Chapter

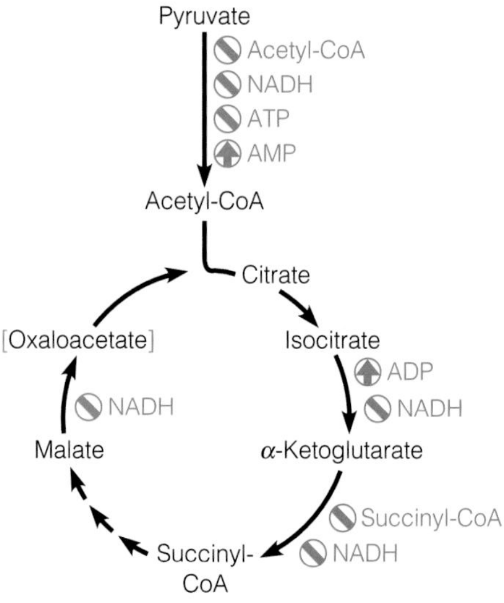

FIGURE **14.16**

Major regulatory factors controlling pyruvate dehydrogenase and the citric acid cycle. Red brackets indicate concentration dependence. NADH can inhibit through allosteric interactions, but apparent NADH inhibition can also be a reflection of reduced NAD^+ availability.

15, this coupled synthesis generates about 38 moles of ATP per mole of glucose oxidized to CO_2 and water.

Regulation of Pyruvate Dehydrogenase and the Citric Acid Cycle

Because the citric acid cycle is a source of biosynthetic intermediates, as well as a route for generating metabolic energy, regulation of the cycle is somewhat more complex than if it were just an energy-generating pathway. As in glycolysis, regulation occurs both at the level of entry of fuel into the cycle and at the level of control of key reactions within the cycle. Figure 14.16 summarizes the major factors involved in regulation at both levels.

CONTROL OF PYRUVATE OXIDATION

Fuel enters the cycle primarily as acetyl-CoA, which arises from carbohydrate via pyruvate dehydrogenase and from lipid primarily from the β-oxidation of fatty acids. Because fatty acid oxidation is discussed in Chapter 18, we concentrate here on control of pyruvate dehydrogenase. Activity of this complex is controlled by allosteric inhibition and, as noted earlier, by a covalent modification that is in turn controlled by the energy state of the cell. E_2, the transacetylase component (see Figure 14.10), is inhibited by acetyl-CoA and activated by CoA-SH. E_3, the dihydrolipoamide dehydrogenase component, is inhibited by NADH and activated by NAD^+. ATP is an allosteric inhibitor of the complex, and AMP is an activator. Thus, the activity of this key reaction is coordinated with the energy charge, the $[NAD^+]/[NADH]$ ratio, and the ratio of acetylated to free coenzyme A.

Activity of the pyruvate dehydrogenase complex is regulated by phosphorylation and dephosphorylation of the E_1 subunit.

In the pyruvate dehydrogenase complex from mammalian sources, regulation occurs also by covalent modification of E_1, the pyruvate decarboxylase component of the complex. As shown in Figure 14.17, this involves phosphorylation and dephosphorylation of serine residues in E_1. NADH and acetyl-CoA both activate another component of the complex, **pyruvate dehydrogenase kinase,** which phosphorylates three specific E_1 serine residues, resulting in loss of activity of pyruvate dehydrogenase. A specific **pyruvate dehydrogenase phosphatase** hydrolytically removes the bound phosphate and reactivates the complex. The phosphatase is activated by Ca^{2+} and Mg^{2+}. Because ATP binds Mg^{2+} more tightly than does ADP, the concentration of free Mg^{2+} reflects the ATP/ADP ratio within the mito-

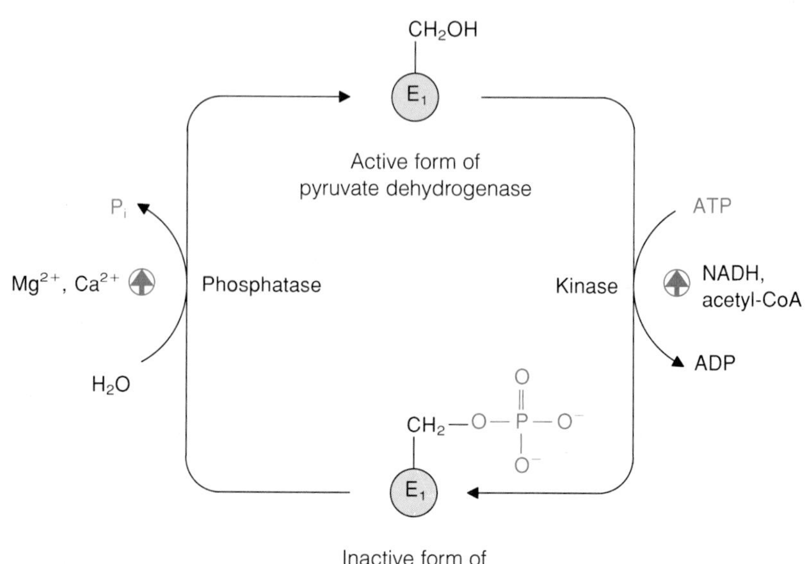

FIGURE **14.17**

Regulation of the pyruvate dehydrogenase complex by modification of E_1. A kinase and a phosphatase inactivate and activate the first component (E_1) of the pyruvate dehydrogenase complex by phosphorylating and dephosphorylating, respectively, three specific serine residues (depicted as —CH_2OH).

chondrion. That is, free Mg^{2+} accumulates at low [ATP]/[ADP] ratios. Thus, pyruvate dehydrogenase responds to ATP levels by being turned off when ATP is abundant and further energy production is unneeded. In mammalian tissues at rest, much less than half of the total pyruvate dehydrogenase is in the active, nonphosphorylated form. The complex can be turned on when low ATP levels signal a need to generate more ATP. The kinase protein is an integral part of the pyruvate dehydrogenase complex, whereas the phosphatase is but loosely bound.

CONTROL OF THE CITRIC ACID CYCLE

Flux through the citric acid cycle is controlled by allosteric interactions, but the concentrations of substrates also play a critical role. Though details of regulation vary among different cells and tissues, the major effects are as summarized in Figure 14.16.

The most important factor controlling citric acid cycle activity is the intramitochondrial ratio of [NAD^+] to [NADH]. NAD^+ is a substrate for three cycle enzymes (Figure 14.3), as well as for pyruvate dehydrogenase. Under conditions that *decrease* the [NAD^+]/[NADH] ratio, such as limitation of the oxygen supply, the low concentration of NAD^+ can limit the activities of these dehydrogenases.

The citric acid cycle is controlled primarily by the relative intramitochondrial concentrations of NAD^+ and NADH.

In some mammalian tissues, notably liver, the levels of citrate vary as much as 10-fold, and at low levels flux through the citrate synthase reaction is limited by substrate availability. Recall that in some animal tissues citrate is also a prime regulator of flux through *glycolysis* via allosteric regulation of phosphofructokinase (PFK). That is not true for all tissues. Heart cells, for example, cannot transport citrate out of mitochondria, so interaction with cytosolic PFK probably does not occur to a significant extent. However, citrate levels can still control the citric acid cycle in heart.

Key sites for allosteric regulation are the reactions catalyzed by isocitrate dehydrogenase and α-ketoglutarate dehydrogenase. In many cells, isocitrate dehydrogenase is activated by ADP and is inhibited directly by NADH; this control is in addition to the indirect diminution of activity seen at low [NAD^+]/[NADH] ratios. Isocitrate dehydrogenase is also inactivated by phosphorylation at one serine residue, which prevents binding of isocitrate. α-Ketoglutarate dehydrogenase activity is inhibited by succinyl-CoA and by NADH. The mechanisms are comparable to the mechanisms by which levels of acetyl-CoA and NADH control pyruvate dehydrogenase activity.

For some time it was thought that citrate synthase represented another site for allosteric control. The enzyme is subject to inhibition by NADH, NADPH, or succinyl-CoA. However, measurements of intramitochondrial levels of oxaloacetate, acetyl-CoA, and citrate show that the enzyme is operating close to equilibrium conditions; that is, the ratio of [citrate] to ([acetyl-CoA] × [oxaloacetate]) is close to K_{eq} for the citrate synthase reaction. On the other hand, it is evident that intramitochondrial levels of oxaloacetate, which are quite low, can exercise substrate-level control over flux through citrate synthase.

To summarize, citric acid cycle flux is responsive to the energy state of the cell, through allosteric activation of isocitrate dehydrogenase by ADP; to the redox state of the cell, through flux rate limitation caused when intramitochondrial [NAD^+] decreases; and to the availability of energy-rich compounds, through inhibition of relevant enzymes by acetyl-CoA or succinyl-CoA.

Anaplerotic Sequences: The Need to Replace Cycle Intermediates

So far, our discussion of the citric acid cycle has focused on its role in catabolism and energy generation. The cycle also serves as an important source of biosynthetic intermediates. Figure 14.18 summarizes the most important anabolic

Citric acid cycle intermediates used in biosynthetic pathways must be replenished to maintain flux through the cycle. Anaplerotic pathways serve this purpose.

pathways involved. These pathways tend to draw carbon from the cycle by utilizing intermediates in the cycle. Succinyl-CoA is used in the synthesis of heme and other porphyrins. Oxaloacetate and α-ketoglutarate are the keto acid analogs of the amino acids aspartate and glutamate, respectively, and are used in the synthesis of these and other amino acids by transamination (see page 508). In some tissues citrate is transported from mitochondria to the cytosol, where it is cleaved to provide acetyl-CoA for fatty acid biosynthesis. Because these and other reactions tend to deplete citric acid cycle intermediates by drawing carbon away, operation of the cycle would be impaired were it not for other processes that replenish the stores of citric acid cycle intermediates. These processes are called **anaplerotic** pathways, from a Greek word that means "filling up." In most cells the flow of carbon out of the cycle is balanced by these anaplerotic reactions, so that the intramitochondrial concentrations of citric acid cycle intermediates remain constant with time. The anaplerotic processes are also summarized in Figure 14.18.

REACTIONS THAT REPLENISH OXALOACETATE

In animals the most important anaplerotic reaction, particularly in liver and kidney, is the reversible, ATP-dependent carboxylation of pyruvate to give oxaloacetate. This reaction is catalyzed by **pyruvate carboxylase.**

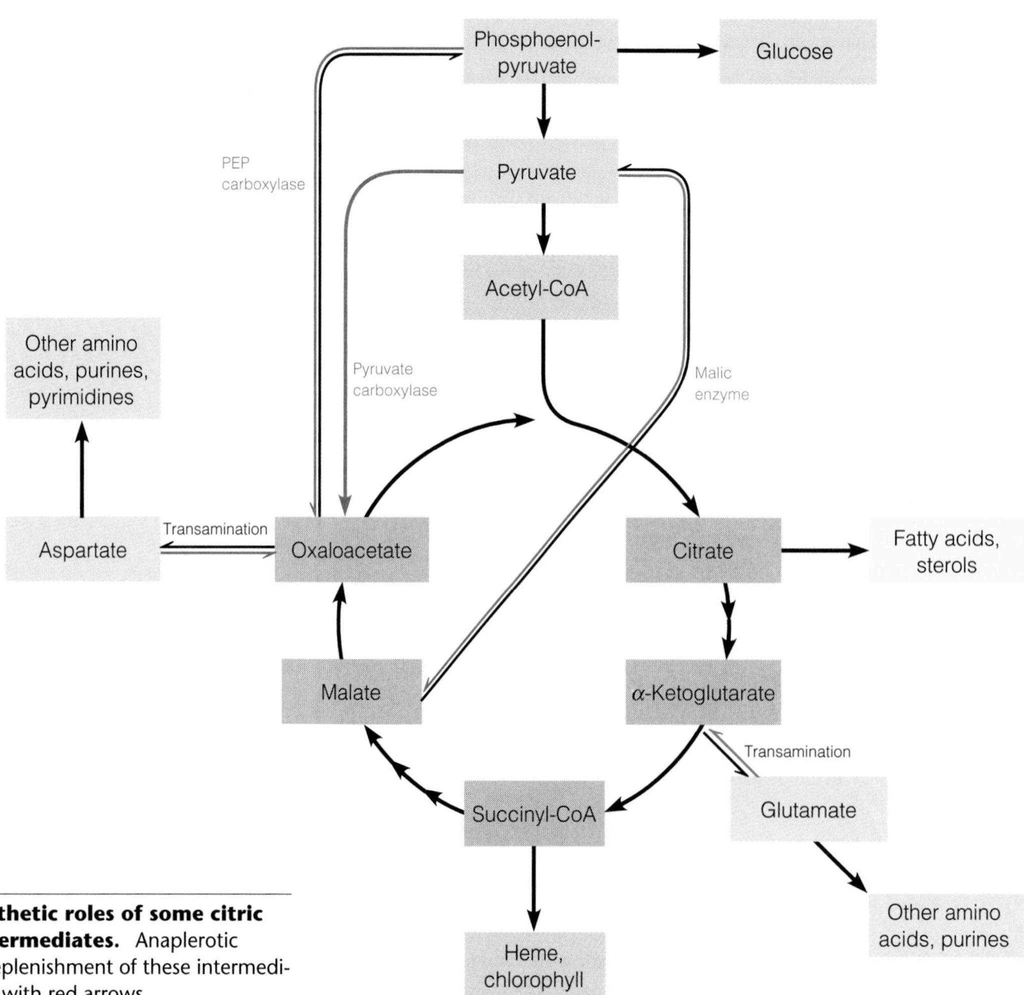

FIGURE **14.18**

Major biosynthetic roles of some citric acid cycle intermediates. Anaplerotic pathways for replenishment of these intermediates are shown with red arrows.

$$\underset{\textbf{Pyruvate}}{\begin{matrix} CH_3 \\ | \\ C{=}O \\ | \\ COO^- \end{matrix}} + HCO_3^- + ATP \rightleftharpoons \underset{\textbf{Oxaloacetate}}{\begin{matrix} COO^- \\ | \\ CH_2 \\ | \\ C{=}O \\ | \\ COO^- \end{matrix}} + ADP + P_i + H^+ \qquad \Delta G^{\circ\prime} = -2.1 \text{ kJ/mol}$$

The enzyme is activated allosterically by acetyl-CoA; in fact, it is all but inactive in the absence of this effector. This process represents a feedforward activation, because the effect of acetyl-CoA accumulation is to promote its own utilization by stimulating the synthesis of oxaloacetate. The latter compound in turn reacts with acetyl-CoA, via the citrate synthase reaction. Alternatively, oxaloacetate can be used for carbohydrate synthesis, via gluconeogenesis, and acetyl-CoA accumulation can be seen as a signal that adequate carbon is available for some of it to be stored as carbohydrate.

In Chapter 11 we identified biotin as a cofactor in most carboxylation reactions involving CO_2. Pyruvate carboxylase is a tetrameric protein carrying four molecules of biotin, each bound covalently through an amide bond involving the ε-amino group of a lysine residue. Pyruvate carboxylate operates by a two-step mechanism (Figure 14.19), starting with an ATP-dependent carboxylation of the cofactor to give *N*-carboxybiotin. This activated derivative then transfers the carboxyl group directly to pyruvate.

In plants and bacteria an alternative route leads directly from phosphoenolpyruvate to oxaloacetate. Because phosphoenolpyruvate is such an energy-rich compound, this reaction, catalyzed by **phosphoenolpyruvate carboxylase,**

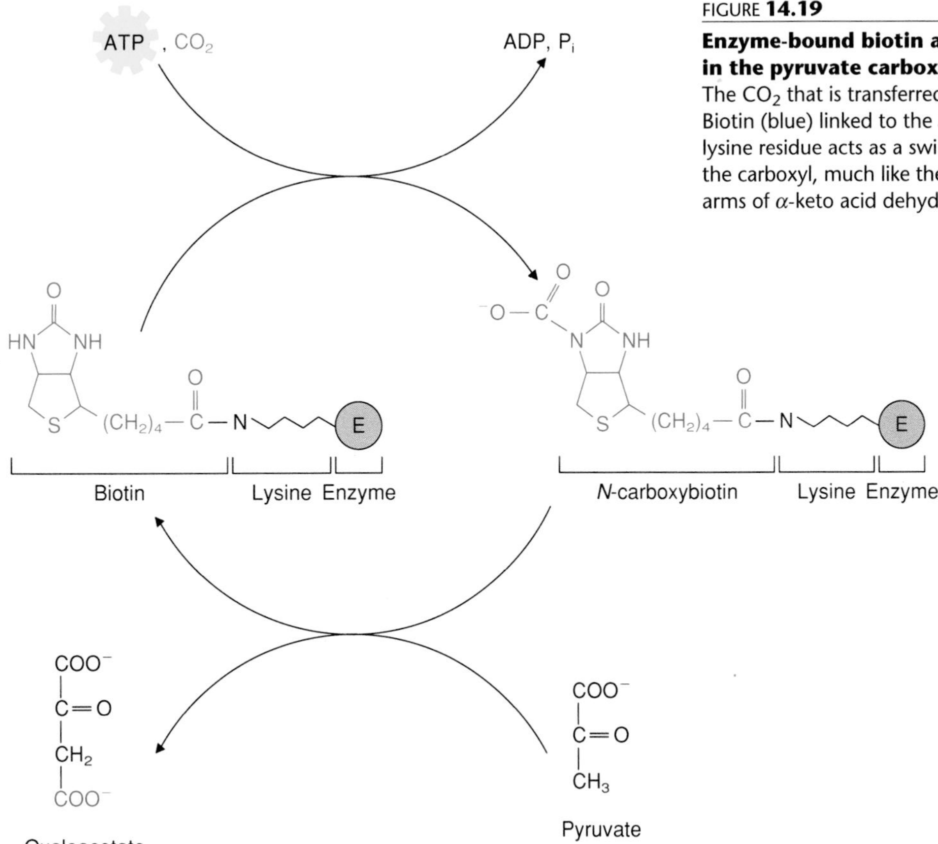

FIGURE **14.19**

Enzyme-bound biotin as a carrier for CO_2 in the pyruvate carboxylase reaction. The CO_2 that is transferred is shown in red. Biotin (blue) linked to the ε-amino group of a lysine residue acts as a swinging arm to transfer the carboxyl, much like the lipoamide swinging arms of α-keto acid dehydrogenase complexes.

requires neither an energy cofactor nor biotin. This reaction is important in the C_4 pathway of photosynthetic CO_2 fixation (Chapter 17).

$$CH_2{=}C(O\text{Ⓟ}){-}COO^- + HCO_3^- \longrightarrow {}^-OOC{-}CH_2{-}C(=O){-}COO^- + P_i$$

Phosphoenolpyruvate **Oxaloacetate**

A related enzyme, phosphoenolpyruvate carboxykinase, also converts phosphoenolpyruvate to oxaloacetate. Because the major function of this enzyme is its role in gluconeogenesis, we discuss it in detail in Chapter 16.

THE MALIC ENZYME

In addition to pyruvate carboxylase and phosphoenolpyruvate carboxylase, a third anaplerotic process is provided by an enzyme commonly known as **malic enzyme** but more officially as **malate dehydrogenase (decarboxylating:NADP$^+$)**. The malic enzyme catalyzes the reductive carboxylation of pyruvate to give malate.

$$CH_3{-}C(=O){-}COO^- + HCO_3^- + NADPH + H^+ \rightleftharpoons {}^-OOC{-}CH_2{-}CH(OH){-}COO^- + NADP^+ + H_2O$$

Pyruvate **L-Malate**

This reaction is formally equivalent to the isocitrate dehydrogenase reaction operating in reverse, with pyruvate and malate being analogous, respectively, to α-ketoglutarate and isocitrate.

REACTIONS INVOLVING AMINO ACIDS

Although not usually classified as anaplerotic pathways, **transamination** reactions can be regarded that way also, because they are reversible reactions that can yield citric acid cycle intermediates. In transamination an amino acid transfers its amino group to a keto acid and is thereby converted itself to a keto acid. The mechanism is discussed in Chapter 20.

$$R_1{-}CH(\overset{+}{N}H_3){-}COO^- + R_2{-}C(=O){-}COO^- \rightleftharpoons R_1{-}C(=O){-}COO^- + R_2{-}CH(\overset{+}{N}H_3){-}COO^-$$

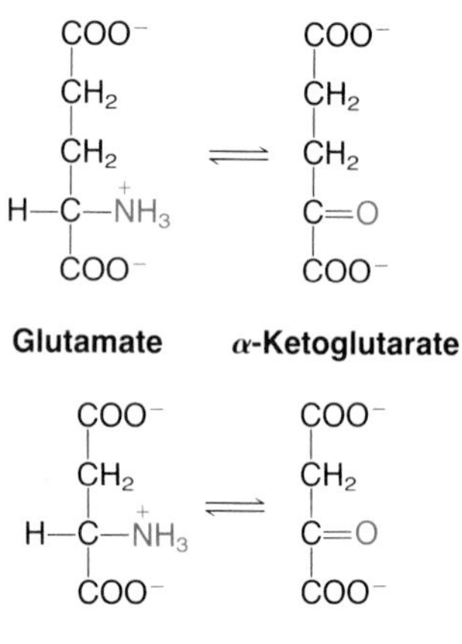

Glutamate and aspartate undergo transamination to generate the citric acid cycle intermediates α-ketoglutarate and oxaloacetate, respectively. Hence, cells containing amino acids in abundance can convert them via transamination to citric acid cycle intermediates. Another enzyme, **glutamate dehydrogenase,** presents an additional route for synthesis of α-ketoglutarate from glutamate.

$$\text{Glutamate} + NAD(P)^+ + H_2O \rightleftharpoons \alpha\text{-ketoglutarate} + NAD(P)H + NH_4^+$$

Glutamate dehydrogenase, which we discuss in more detail in Chapter 20, uses either NAD$^+$ or NADP$^+$. Being reversible, transaminations and the glutamate dehydrogenase reaction can be used either for amino acid synthesis or for replenishment of citric acid cycle intermediates, depending upon the needs of the cell.

Finally, many plants and bacteria can convert two-carbon fragments to four-carbon citric acid cycle intermediates via the glyoxylate cycle, as described in the next section.

Glyoxylate Cycle: An Anabolic Variant of the Citric Acid Cycle

Metabolically, plant and animal cells differ in many important respects. Of particular concern here is that plant cells, along with some microorganisms, can carry out the net synthesis of carbohydrate from fat. This conversion is crucial to the development of seeds, in which a great deal of energy is stored in the form of triacylglycerols. (In fact, most vegetable oils available in grocery stores are mixtures of triacylglycerols derived from seeds.) When the seeds germinate, triacylglycerol is broken down and converted to sugars, which provide energy and raw material needed for growth of the plant. By contrast, animal cells cannot carry out the net synthesis of carbohydrate from fat.

Plants synthesize sugars by using the **glyoxylate cycle,** which can be considered an anabolic variant form of the citric acid cycle. To understand the importance of this cycle, consider first the two primary fates of acetyl-CoA in animal metabolism—oxidation through the citric acid cycle, and the synthesis of fatty acids. Because of the virtual irreversibility of the pyruvate dehydrogenase reaction, acetyl-CoA cannot undergo net conversion to pyruvate and hence cannot participate in the *net* synthesis of carbohydrate. To be sure, the two carbons of acetyl-CoA can be incorporated into oxaloacetate, which is an efficient gluconeogenic precursor. However, because two carbons are lost in this part of the citric acid cycle, there is no net accumulation of carbon in carbohydrate. The glyoxylate cycle permits the net synthesis of oxaloacetate.

O
‖
C—H
|
COO^-
Glyoxylate

The glyoxylate cycle (Figure 14.20) is a cyclic pathway that converts two acetyl units, as acetyl-CoA, to one molecule of succinate. The pathway uses some of the same enzymes as the citric acid cycle, but it bypasses the reactions in which carbon is lost during the citric acid cycle. The second mole of acetyl-CoA is brought in during this bypass (Figure 14.20). Thus, each turn of the cycle involves incorporation of 2 two-carbon fragments and results in the net synthesis of a four-carbon molecule. This process occurs in the **glyoxysome,** a specialized organelle that carries out both β-oxidation of fatty acids to acetyl-CoA and utilization of that acetyl-CoA in the glyoxylate cycle (Figure 14.21). The succinate generated is transported from the glyoxysome to the mitochondrion, where it is converted, via reactions 6, 7, and 8 of the citric acid cycle (see Figure 14.3), to oxaloacetate. The oxaloacetate is readily utilized for carbohydrate synthesis via gluconeogenesis.

The glyoxylate cycle allows plants and bacteria to carry out net conversion of fat to carbohydrate, bypassing CO_2-generating reactions of the citric acid cycle.

The glyoxylate cycle also allows many microorganisms to metabolize two-carbon substrates, such as acetate. *E. coli,* for example, can grow in a medium that provides acetate as the sole carbon source, as can many fungi, protozoans, and algae. These cells synthesize acetyl-CoA, which is used both for energy production, via the citric acid cycle, and for synthesis of gluconeogenic precursors, via the glyoxylate cycle.

Now let us examine the individual reactions of the glyoxylate cycle. As noted, acetyl-CoA is provided from fatty acid oxidation. Alternatively, acetate itself is converted to acetyl-CoA by **acetate thiokinase.**

$$\text{Acetate} + \text{CoA-SH} + \text{ATP} \rightleftharpoons \text{acetyl-CoA} + \text{AMP} + \text{PP}_i$$

Next, acetyl-CoA condenses with oxaloacetate to give citrate, just as in the citric acid cycle, and citrate reacts with aconitase to give isocitrate. At this point the

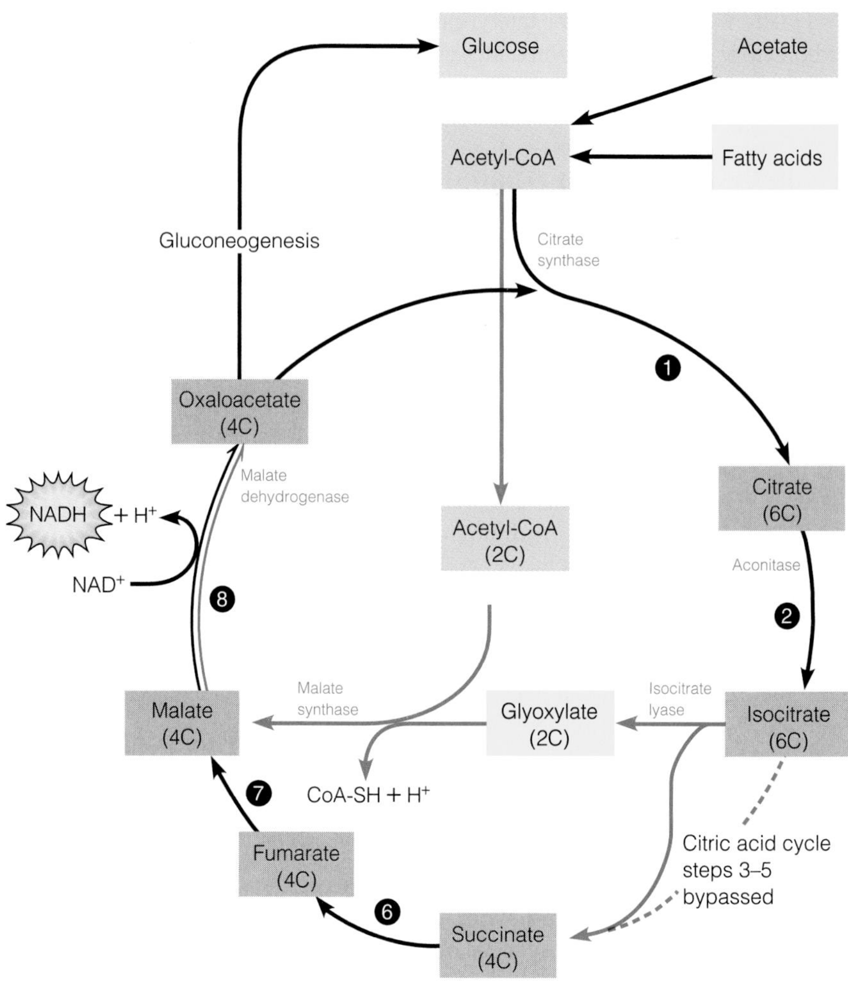

FIGURE **14.20**

Reactions of the glyoxylate cycle. The reactions catalyzed by isocitrate lyase and malate synthase (red arrows) bypass the three citric acid cycle steps between isocitrate and succinate (blue arrows), so that the two carbons lost in the citric acid cycle are saved, resulting in the net synthesis of oxaloacetate.

glyoxylate cycle diverges from the citric acid cycle. The next reaction, catalyzed by **isocitrate lyase,** cleaves isocitrate to glyoxylate and succinate.

$$\begin{array}{c} CH_2{-}COO^- \\ | \\ H{-}C{-}COO^- \\ | \\ HO{-}C{-}COO^- \\ | \\ H \end{array} \longrightarrow \begin{array}{c} CH_2{-}COO^- \\ | \\ CH_2{-}COO^- \end{array} + \begin{array}{c} O{=}C{-}COO^- \\ | \\ H \end{array}$$

Isocitrate **Succinate** **Glyoxylate**

Glyoxylate then accepts acetate from another acetyl-CoA, in a reaction catalyzed by **malate synthase.**

$$\begin{array}{c} O{=}C{-}COO^- \\ | \\ H \end{array} + CH_3{-}\overset{\overset{O}{\|}}{C}{\sim}S{-}CoA + H_2O \longrightarrow \begin{array}{c} COO^- \\ | \\ HO{-}C{-}H \\ | \\ CH_2 \\ | \\ COO^- \end{array} + CoA{-}SH + H^+$$

Glyoxylate **Acetyl-CoA** **Malate**

Mechanistically, this reaction is comparable to that catalyzed by citrate synthase, involving nucleophilic attack of the carbanion form of acetyl-CoA on a carbonyl carbon, in this case the aldehyde carbon of glyoxylate. The malate is

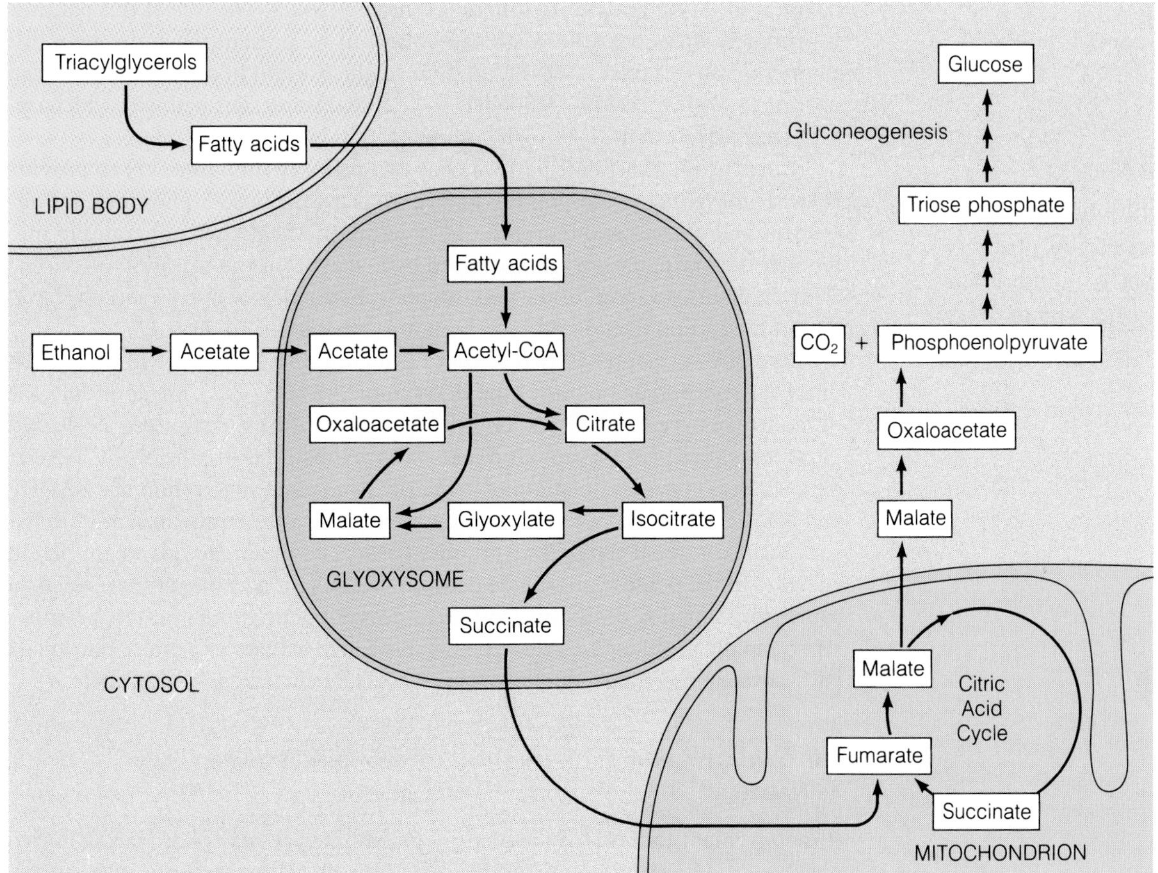

FIGURE **14.21**

Intracellular relationships involving the glyoxylate cycle in plant cells. Fatty acids released in lipid bodies are oxidized in glyoxysomes to acetyl-CoA, which can also come directly from acetate. Acetyl-CoA is then converted to succinate in the glyoxylate cycle, and the succinate is transported to mitochondria. There it is converted in the citric acid cycle to oxaloacetate, which is readily converted to sugars by gluconeogenesis.

then dehydrogenated to regenerate oxaloacetate. The enzyme involved, malate dehydrogenase, is localized in glyoxysomes and is distinct from the mitochondrial form of the enzyme, which is involved in the citric acid cycle. The same is true for citrate synthase and aconitase.

As noted earlier, the glyoxylate cycle results in the net conversion of 2 two-carbon fragments, acetyl-CoA, to a four-carbon compound, succinate, as shown by the following balanced equation.

$$2\ \text{Acetyl-CoA} + \text{NAD}^+ + 2\text{H}_2\text{O} \longrightarrow \text{succinate} + \text{NADH} + 3\text{H}^+ + 2\text{CoA-SH}$$

The primary fate of succinate is its entry into gluconeogenesis via its conversion to oxaloacetate.

A Biosynthetic Pathway That Oxidizes Glucose: The Pentose Phosphate Pathway

The predominant pathway for glucose catabolism is glycolysis to yield pyruvate, followed by oxidation to CO_2 in the citric acid cycle. An alternative process, the **pentose phosphate pathway,** is a remarkable, multipurpose pathway that operates to varying extents in different cells and tissues. The role of this pathway is primarily anabolic rather than catabolic, but we present the pathway in this chapter because it does involve oxidation of glucose and in certain modes it can operate to oxidize glucose completely to CO_2 and water. The pathway, which operates exclusively in the cytosol, is summarized in Figure 14.22.

The pentose phosphate pathway primarily generates NADPH for reductive biosynthesis and ribose-5-phosphate for nucleotide biosynthesis.

The pentose phosphate pathway has two primary functions: (1) to provide NADPH for reductive biosynthesis and (2) to provide ribose-5-phosphate for nucleotide and nucleic acid biosynthesis. In addition, the pathway operates to metabolize dietary pentose sugars, derived primarily from the digestion of nucleic acids. In plants, a variant of the pentose phosphate pathway operates in reverse as part of the carbon fixation process of photosynthesis (see Chapter 17).

Recall from Chapter 12 that $NADP^+$ is identical to NAD^+ except for the additional 2′ phosphate on one of the ribose moieties of $NADP^+$. Metabolically, the difference between NAD^+ and $NADP^+$ is that nicotinamide nucleotide–linked enzymes whose primary function is to *oxidize* substrates use the NAD^+/NADH pair, whereas enzymes functioning primarily in a *reductive* direction use $NADP^+$ and NADPH. Because NADPH is used for fatty acid and steroid biosynthesis, tissues such as adrenal gland, liver, adipose tissue, and mammary gland are rich in enzymes of the pentose phosphate pathway. NADPH is also the ultimate electron source for reduction of ribonucleotides to deoxyribonucleotides for DNA synthesis, so rapidly proliferating cells generally have high activity of pentose phosphate pathway enzymes, for production of both NADPH and ribose-5-phosphate.

THE OXIDATIVE PHASE: GENERATION OF REDUCING POWER AS NADPH

It is convenient to think of the pentose phosphate pathway as operating in two phases—oxidative and nonoxidative. Two of the first three reactions in this pathway are oxidative, each involving reduction of one $NADP^+$ to NADPH. As shown

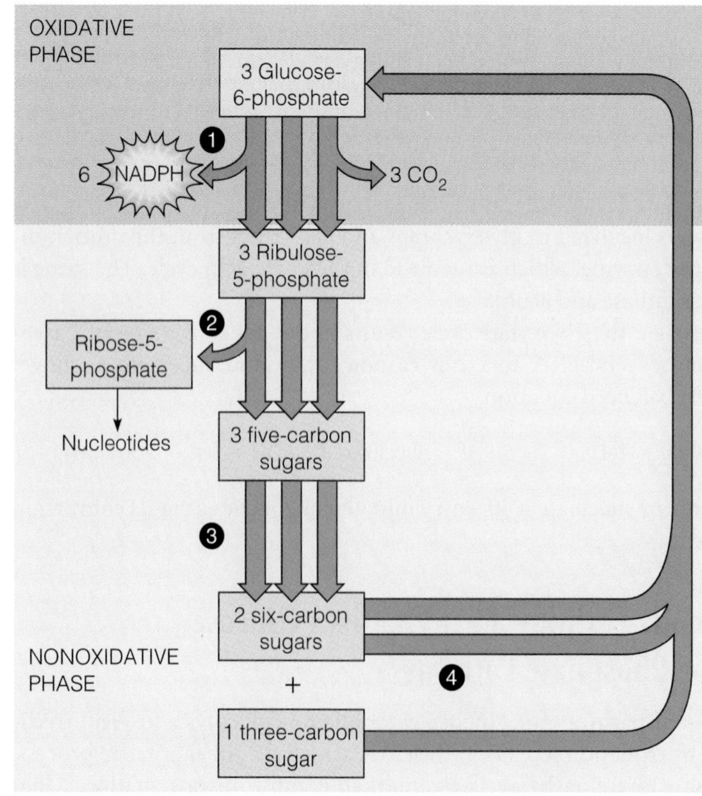

FIGURE **14.22**

Overall strategy of the pentose phosphate pathway. The pentose phosphate pathway converts glucose to various other sugars, which can be used for energy. Its most important products, however, are NADPH and ribose-5-phosphate. In **step 1,** the oxidative phase, glucose-6-phosphate is oxidized to ribulose-5-phosphate and CO_2, with production of NADPH. (The three reactions involved in this step are shown in Figure 14.23.) The remaining steps constitute the nonoxidative phase of the pathway. In **step 2,** some ribulose-5-phosphate is converted to other five-carbon sugars, including ribose-5-phosphate. The ribose phosphate may be used in nucleotide synthesis (its primary use) or in the next step of the pentose phosphate pathway. In **step 3,** a series of reactions converts three molecules of five-carbon sugar to two molecules of six-carbon sugar and one of three-carbon sugar. In **step 4,** some of these sugars are converted to glucose 6-phosphate, and the cycle repeats. Figure 14.25 presents a more detailed overview of the pathway and shows how it varies under different metabolic conditions.

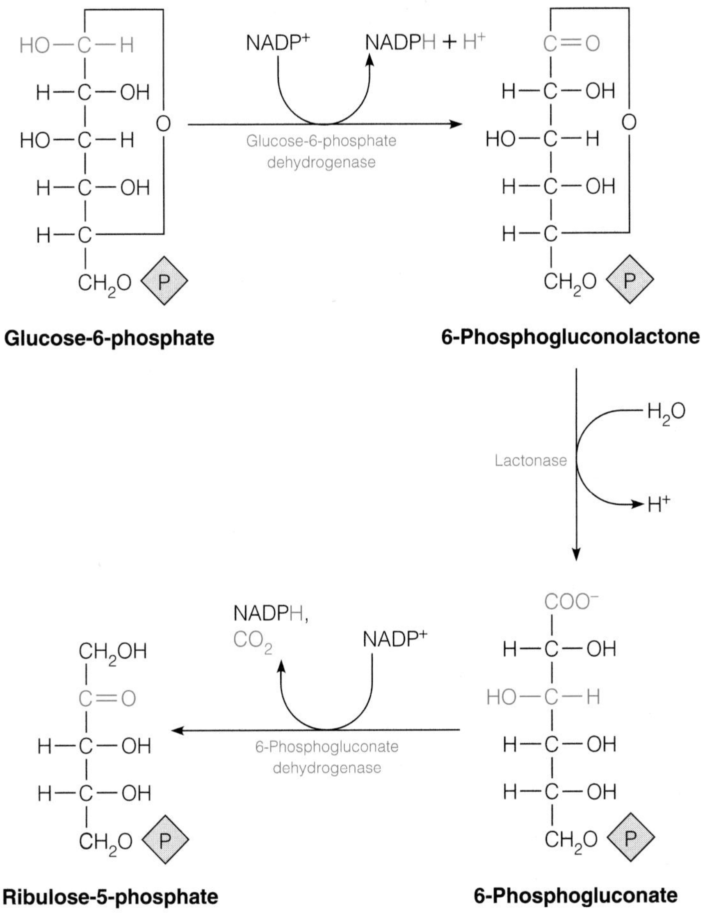

Glucose-6-phosphate → (NADP+ → NADPH + H+; Glucose-6-phosphate dehydrogenase) → 6-Phosphogluconolactone → (H2O → H+; Lactonase) → 6-Phosphogluconate → (NADP+ → NADPH, CO2; 6-Phosphogluconate dehydrogenase) → Ribulose-5-phosphate

FIGURE **14.23**

Oxidative phase of the pentose phosphate pathway. The three reactions of the oxidative phase include two oxidations, which produce NADPH.

in Figure 14.23, the first reaction, catalyzed by **glucose-6-phosphate dehydrogenase,** oxidizes glucose-6-phosphate to 6-phosphogluconolactone, the corresponding **lactone** (an internal ester linking carbons 1 and 5). 6-Phosphogluconolactone is hydrolyzed by a specific **lactonase** to **6-phosphogluconate,** which undergoes an oxidative decarboxylation to yield CO_2, another NADPH, and **ribulose-5-phosphate** (a pentose phosphate). The net result of the oxidative phase is generation of 2 moles of NADPH, oxidation of one carbon to CO_2, and synthesis of 1 mole of pentose phosphate.

THE NONOXIDATIVE PHASE: ALTERNATIVE FATES OF PENTOSE PHOSPHATES

In the nonoxidative phase, some of the ribulose-5-phosphate produced in the oxidative phase is converted to ribose-5-phosphate by **phosphopentose isomerase.**

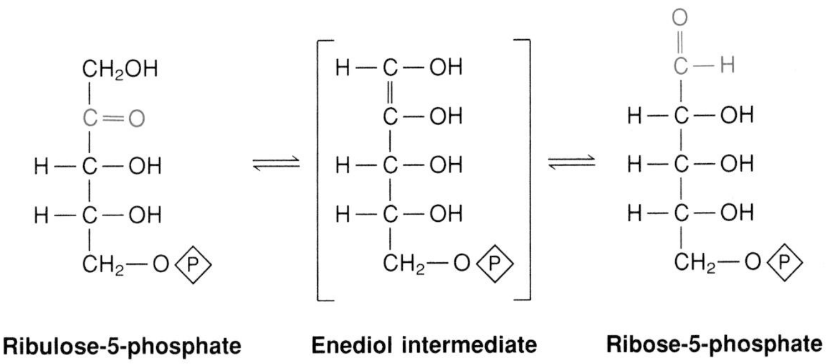

Ribulose-5-phosphate ⇌ [Enediol intermediate] ⇌ Ribose-5-phosphate

The reaction proceeds via an enediol intermediate, just as in two different reactions of glycolysis—those catalyzed by triose phosphate isomerase (see page 454) and phosphoglucoisomerase (see page 452).

Production of Six-Carbon and Three-Carbon Sugar Phosphates

At this stage the primary functions of the pathway have been fulfilled, namely the generation of NADPH and ribose-5-phosphate. We can write a balanced equation for what has transpired thus far:

$$\text{Glucose-6-phosphate} + 2\text{NADP}^+ \longrightarrow \text{ribose-5-phosphate} + CO_2 + 2\text{NADPH} + 2\text{H}^+$$

Many cells need the NADPH for reductive biosynthesis but do not need the ribose-5-phosphate in such large quantities. How, then, is this ribose-5-phosphate catabolized? The process involves a series of sugar phosphate transformations that may look complicated but that have a simple result. *The reaction sequence converts 3 five-carbon sugar phosphates to 2 six-carbon sugar phosphates and 1 three-carbon sugar phosphate.* The hexose phosphates formed can be catabolized either by recycling through the pentose phosphate pathway or by glycolysis. The triose phosphate is glyceraldehyde-3-phosphate, a glycolytic intermediate. Three enzymes are involved: **phosphopentose epimerase, transketolase,** and **transaldolase.**

The pathway begins with both ribulose-5-phosphate and ribose-5-phosphate, the latter having been formed by phosphopentose isomerase. Phosphopentose epimerase converts ribulose-5-phosphate to its epimer, xylulose-5-phosphate.

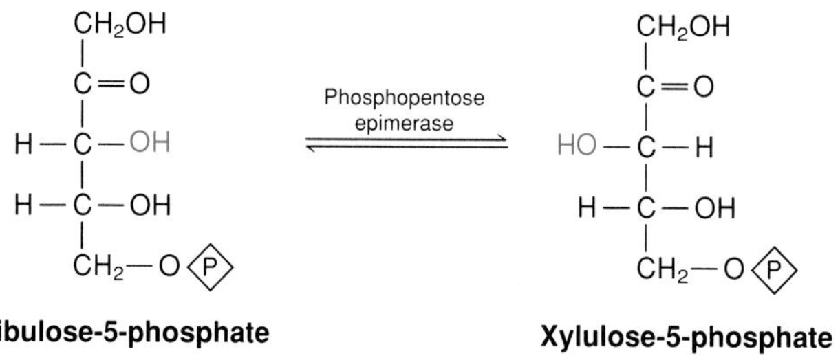

Phosphopentose epimerase

Ribulose-5-phosphate **Xylulose-5-phosphate**

One mole of xylulose-5-phosphate then reacts with one mole of ribose-5-phosphate. The reaction is catalyzed by **transketolase,** which transfers a two-carbon fragment from xylulose-5-phosphate to ribose-5-phosphate to give a triose phosphate, glyceraldehyde-3-phosphate, and a seven-carbon sugar, **sedoheptulose-7-phosphate.**

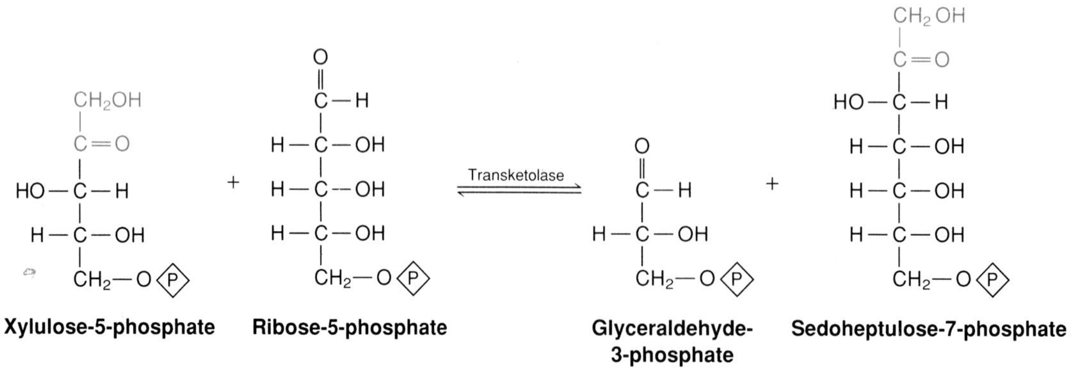

Transketolase

Xylulose-5-phosphate **Ribose-5-phosphate** **Glyceraldehyde-3-phosphate** **Sedoheptulose-7-phosphate**

The two-carbon fragment transferred is an activated **glycolaldehyde** fragment. Recall that pyruvate dehydrogenase transfers an active *acetaldehyde* fragment, with the aid of thiamine pyrophosphate (TPP). Transketolase also requires TPP as a cofactor, with the two-carbon fragment bound transiently to carbon 2 of the thiazole ring of TPP. The mechanism of activation and transfer of two-carbon fragments is very similar in the reactions catalyzed by these two enzymes.

Glycolaldehyde

Glycolaldehyde fragment

Thiazole ring of TPP

Next, transaldolase acts on the two products of the transketolase reaction, with transfer of a three-carbon **dihydroxyacetone** unit from the seven-carbon substrate to the three-carbon substrate. The products are a four-carbon sugar phosphate and a six-carbon sugar phosphate: **erythrose-4-phosphate** and fructose-6-phosphate, respectively. The combined actions of transketolase and transaldolase convert 2 five-carbon sugar phosphates to a four-carbon sugar phosphate and a six-carbon sugar phosphate.

Transaldolase

Sedoheptulose-7-phosphate + Glyceraldehyde-3-phosphate ⇌ Erythrose-4-phosphate + Fructose-6-phosphate

Figure 14.24 shows the transaldolase reaction in more detail. The enzyme activates the ketose substrate by forming a Schiff base with a lysine residue on the

FIGURE **14.24**

Mechanism of the transaldolase reaction.

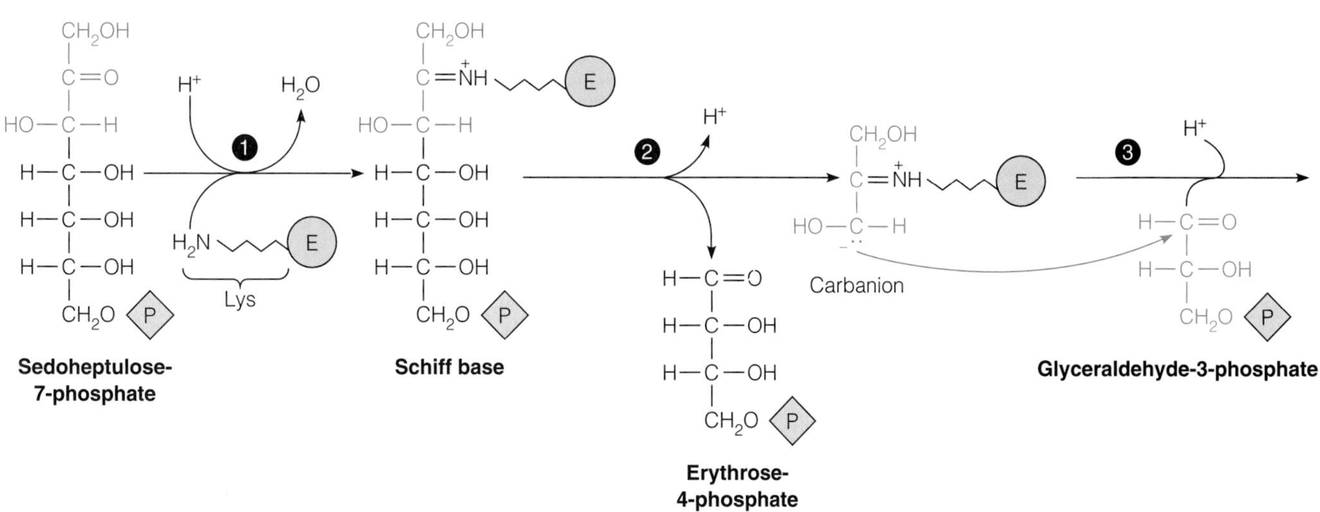

Schiff base

Fructose-6-phosphate

enzyme (step 1). Protonation of the Schiff base leads to carbon–carbon bond cleavage, much as occurs in the fructose-bisphosphate aldolase reaction of glycolysis (see page 454), with release of a four-carbon aldose phosphate (step 2). The dihydroxyacetone unit remains bound as a resonance-stabilized carbanion, which then attacks the carbonyl carbon of glyceraldehyde-3-phosphate, in an aldol condensation reaction (step 3). Hydrolysis of the Schiff base yields the six-carbon product, fructose-6-phosphate (step 4). In the final reaction of pentose phosphate catabolism, transketolase acts on another molecule of xylulose-5-phosphate, transferring a glycolaldehyde fragment to erythrose-4-phosphate and generating a three-carbon product and a six-carbon product—glyceraldehyde-3-phosphate and fructose-6-phosphate, respectively.

Xylulose-5-phosphate + Erythrose-4-phosphate $\xrightleftharpoons{\text{Transketolase}}$ Glyceraldehyde-3-phosphate + Fructose-6-phosphate

So far, the pathway has required input of three pentose phosphate molecules—two for the first transketolase reaction and one for the second. Thus, to summarize the pathway to this point, we must consider three molecules of glucose-6-phosphate passing through the oxidative phase.

$$3\ \text{Glucose-6-phosphate} + 6\text{NADP}^+ + 3H_2O \longrightarrow 3\ \text{pentose-5-phosphate} + 6\text{NADPH} + 6H^+ + 3CO_2$$

Now the rearrangements of the nonoxidative phase result in conversion of three pentose phosphates to 2 six-carbon and 1 three-carbon sugar phosphates.

$$2\ \text{Xylulose-5-phosphate} + \text{ribose-5-phosphate} \longrightarrow 2\ \text{fructose-6-phosphate} + \text{glyceraldehyde-3-phosphate}$$

Thus, we can write a balanced equation for the entire pathway as follows.

$$3\ \text{Glucose-6-phosphate} + 6\text{NADP}^+ + 3H_2O \longrightarrow 2\ \text{fructose-6-phosphate} + \text{glyceraldehyde-3-phosphate} + 6\text{NADPH} + 6H^+ + 3CO_2$$

Tailoring the Pentose Phosphate Pathway to Specific Needs

In the equation for the overall reaction, three hexose phosphates yield two hexose phosphates, one triose phosphate, and three molecules of CO_2, for a net oxidation of three carbons to CO_2. In a formal sense, therefore, the pathway can be seen as a means to oxidize the six carbons of glucose-6-phosphate to CO_2, just as occurs in glycolysis plus the citric acid cycle.

However, as noted above, the pentose phosphate pathway is not primarily an energy-generating pathway. The actual fate of the sugar phosphates depends on the metabolic needs of the cell in which the pathway is occurring. If the primary need is for nucleotide and nucleic acid synthesis, the major product is ribose-5-phosphate, and most of the rearrangements of the nonoxidative phase do not take place (Figure 14.25a). If the primary need is for NADPH generation (for fatty acid or steroid synthesis), the nonoxidative phase generates compounds that can easily be reconverted to glucose-6-phosphate, for subsequent passage through the oxidative phase (Figure 14.25b). In this mode, repeated turns of the cycle

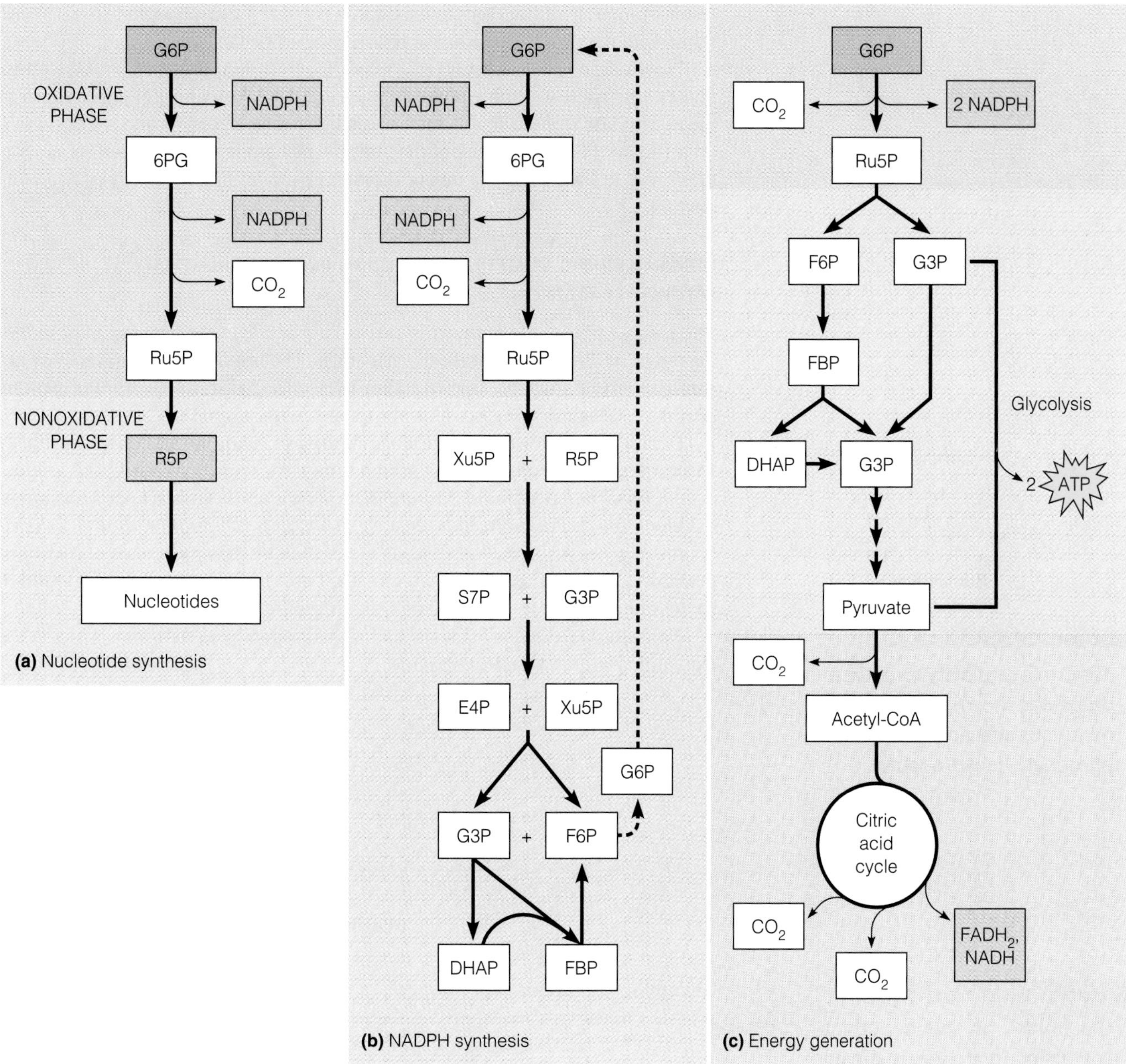

FIGURE **14.25**

Alternative pentose phosphate pathways. The pentose phosphate pathway has different modes of operation to meet varying metabolic needs. **(a)** When the primary need is for nucleotide biosynthesis, the primary product is ribose-5-phosphate. Reducing equivalents from NADPH are used to reduce ribonucleotides to deoxyribonucleotides (Chapter 22). **(b)** When the primary need is for reducing power (NADPH), fructose phosphates are reconverted to glucose-6-phosphate for reoxidation in the oxidative phase. **(c)** When only moderate quantities of pentose phosphates and NADPH are needed, the pathway can also be used to supply energy, with the reaction products being oxidized through glycolysis and the citric acid cycle. Abbreviations: G6P = glucose-6-phosphate; 6PG = 6-phosphogluconate; R5P = ribose-5-phosphate; Ru5P = ribulose-5-phosphate; Xu5P = xylulose-5-phosphate; S7P = sedoheptulose-7-phosphate; G3P = glyceraldehyde-3-phosphate; E4P = erythrose-4-phosphate; F6P = fructose-6-phosphate; DHAP = dihydroxyacetone phosphate; FBP = fructose-1,6-bisphosphate.

result ultimately in the complete oxidation of glucose-6-phosphate to CO_2 and water, with maximal generation of reducing equivalents.

Finally, in a cell with moderate needs for both NADPH and pentose phosphates, the fructose-6-phosphate and glyceraldehyde-3-phosphate produced in the nonoxidative phase can be further catabolized by glycolysis and the citric acid cycle (Figure 14.25c). Because of the multiple metabolic needs of a cell for biosynthesis, it is unlikely that any one of these three modes operates exclusively in any one cell.

HUMAN GENETIC DISORDERS INVOLVING PENTOSE PHOSPHATE PATHWAY ENZYMES

The pentose phosphate pathway is particularly active in the generation of reducing power in the red blood cells of vertebrates. The importance of this activity became apparent through investigation of a fairly widespread human genetic disorder, a deficiency of glucose-6-phosphate dehydrogenase.

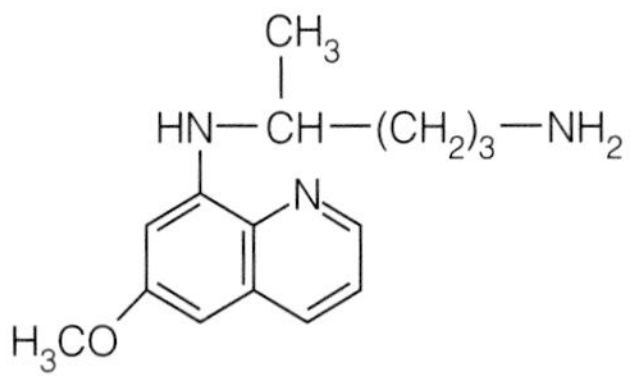

CH_3 — HN—CH—$(CH_2)_3$—NH_2 — H_3CO

Primaquine

Abnormal sensitivity to antimalarial drugs was shown to result from mutations affecting glucose-6-phosphate dehydrogenase.

During World War II the antimalarial drug primaquine was prophylactically administered to members of the armed forces. As a result, a significant proportion of servicemen suffered a severe hemolytic anemia (massive destruction of red blood cells). They were also sensitive to a variety of compounds that, like primaquine, generate oxidative stress, as manifested by the appearance of hydrogen peroxide and organic peroxides in red cells. These individuals were later found to be deficient in glucose-6-phosphate dehydrogenase.

Normally, peroxides are inactivated via reduction by **glutathione,** which is the tripeptide γ-glutamylcysteinylglycine.

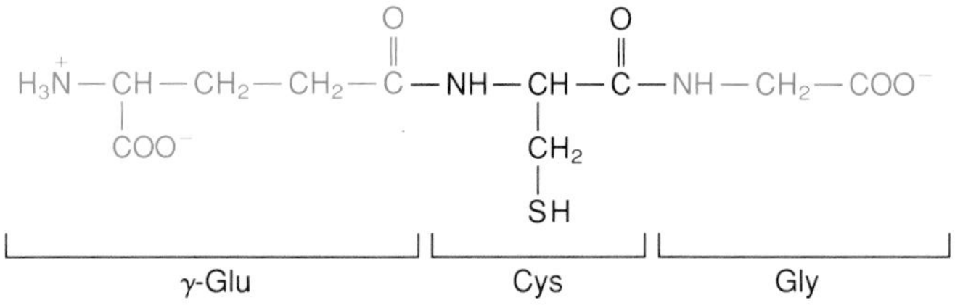

$$H_3\overset{+}{N}-\underset{COO^-}{CH}-CH_2-CH_2-\overset{O}{\overset{\|}{C}}-NH-\underset{\underset{SH}{CH_2}}{CH}-\overset{O}{\overset{\|}{C}}-NH-CH_2-COO^-$$

γ-Glu | Cys | Gly

Glutathione

Glutathione, an abundant thiol-containing tripeptide, is a major intracellular reductant.

Glutathione is abundant in most cells, and because of its free thiol group, it represents a major protective mechanism against oxidative stress. For example, it helps keep cysteine thiol groups in proteins in the reduced state. If two thiol groups become oxidized, they can be reduced nonenzymatically by glutathione.

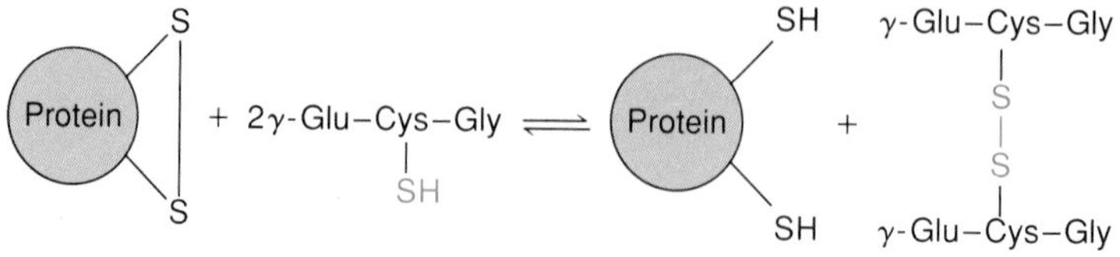

$$\text{Protein}(S\text{—}S) + 2\gamma\text{-Glu-}\underset{SH}{\text{Cys}}\text{-Gly} \rightleftharpoons \text{Protein}(SH)(SH) + \gamma\text{-Glu-}\underset{\underset{\gamma\text{-Glu-Cys-Gly}}{S-S}}{\text{Cys}}\text{-Gly}$$

And, as noted, glutathione also carries out the reduction of peroxides; this is an enzymatic reaction, catalyzed by **glutathione peroxidase** (see page 555).

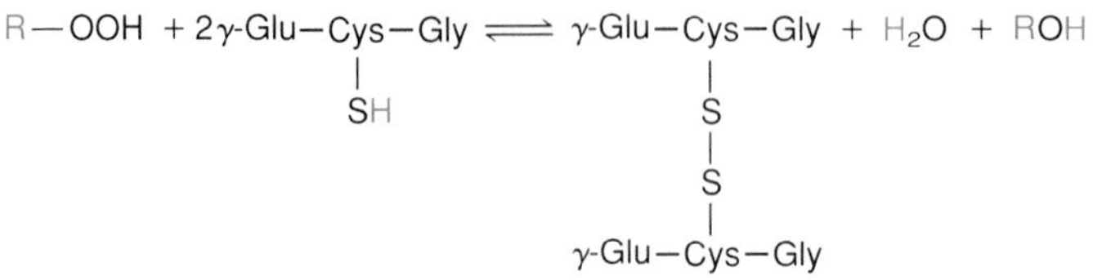

$$R\text{—}OOH + 2\gamma\text{-Glu-}\underset{SH}{\text{Cys}}\text{-Gly} \rightleftharpoons \gamma\text{-Glu-}\underset{\underset{\gamma\text{-Glu-Cys-Gly}}{S-S}}{\text{Cys}}\text{-Gly} + H_2O + ROH$$

Oxidized glutathione (GSSG) is reduced by the NADPH-dependent enzyme **glutathione reductase.**

$$\text{GSSG} + \text{NADPH} + \text{H}^+ \longrightarrow 2\text{GSH} + \text{NADP}^+$$

This FAD-dependent enzyme acts essentially unidirectionally, so that the ratio of reduced glutathione (GSH) to oxidized glutathione in most cells is about 500 to 1.

In the erythrocyte, a particularly important role of glutathione is to maintain hemoglobin in the reduced (Fe^{2+}) state; recall that methemoglobin (Fe^{3+}) is nonfunctional (see page 216). Therefore, the erythrocyte is especially sensitive to depletion of reduced glutathione. And because the pentose phosphate pathway is virtually the only pathway for generation of NADPH, the erythrocyte is especially vulnerable to conditions that impair flux through this pathway and thereby lower intracellular NADPH levels. Thus, the individuals who were deficient in glucose-6-phosphate dehydrogenase were the ones most sensitive to primaquine.

In most cases of glucose-6-phosphate dehydrogenase deficiency the enzyme in red cells is not totally inactive but instead is decreased in activity by about 10-fold. Individuals with this deficiency are asymptomatic until stressed. That is, they are asymptomatic until primaquine or a related agent generates enough peroxides that the available GSH becomes depleted. Reduction of the resultant GSSG back to GSH is impaired because NADPH levels are inadequate to allow glutathione reductase to function. This causes methemoglobin (Fe^{3+}) to accumulate at the expense of hemoglobin, which in turn changes the structure of the cell, weakening the membrane and rendering it sensitive to rupture, or hemolysis.

Interestingly, glucose-6-phosphate dehydrogenase deficiency, like sickle-cell trait, confers resistance to malaria caused by *Plasmodium falciparum* (see Chapter 7). Thus, the deficiency has a positive survival value in tropical and subtropical regions of the world, where malaria is common. This explains the observation that glucose-6-phosphate dehydrogenase deficiency is seen most frequently among individuals of African or Mediterranean extraction.

Another genetic disorder related to the pentose phosphate pathway is the **Wernicke–Korsakoff syndrome.** This mental disorder is coupled with loss of memory and partial paralysis and develops when affected individuals suffer a moderate thiamine deficiency. The symptoms often become manifest in alcoholics, whose diets are apt to be vitamin deficient.

A mutation in transketolase that increases its K_M for TPP is responsible for the neurological difficulties of Wernicke–Korsakoff syndrome.

The basis for the Wernicke–Korsakoff syndrome is an alteration of transketolase that reduces its affinity for thiamine pyrophosphate by about 10-fold. Other TPP-dependent enzymes, notably pyruvate dehydrogenase and α-ketoglutarate dehydrogenase, are not affected. Symptoms of the disease become manifest when TPP levels drop below the values needed to saturate the abnormal transketolase. Normal individuals contain a transketolase that binds TPP strongly enough that no change in enzyme function occurs as a result of these slight to moderate thiamine deficiencies.

Both glucose-6-phosphate dehydrogenase deficiency and the Wernicke–Korsakoff syndrome, like sickle-cell disease (Chapter 7), illustrate the interdependence of genetic and environmental factors in the onset of clinical disease. Symptoms of the hereditary change express themselves only after some kind of moderate stress that does not affect normal individuals.

SUMMARY

The citric acid cycle is a central pathway for oxidation of carbohydrates, lipids, and proteins. A principal entrant to this cyclic pathway is pyruvate produced in glycolysis, which undergoes oxidation to acetyl-CoA by the pyruvate dehydrogenase complex. This three-enzyme system uses five coenzymes: NAD^+, CoA-SH,

FAD, lipoamide, and thiamine pyrophosphate. Each turn of the citric acid cycle involves entry of two carbons as the acetyl group of acetyl-coenzyme A and loss of two carbons as CO_2. Acetyl-CoA condenses with oxaloacetate to form citrate. After one turn of the cycle, oxaloacetate is regenerated, to begin the process anew. During the cycle, reduced electron carriers, primarily NADH, are generated, and their reoxidation in mitochondria provides the energy for ATP synthesis. Anaplerotic reactions replace citric acid cycle intermediates that are consumed in biosynthetic pathways. In plants and bacteria the glyoxylate cycle bypasses the decarboxylation reactions of the citric acid cycle, allowing acetyl-CoA to undergo net conversion to carbohydrate. An alternative carbohydrate oxidative pathway, the pentose phosphate pathway, generates NADPH for reductive biosynthesis and pentose phosphates for nucleotide biosynthesis.

REFERENCES

Regulation of the Citric Acid Cycle

Atkinson, D. E. (1977) *Cellular Energy Metabolism and Its Regulation.* Academic Press, New York. Provocative remarks by the person who originated the concept of adenylate energy charge.

Hurley, J. H., A. M. Dean, J. L. Sohl, D. E. Koshland, Jr., and R. M. Stroud (1990) Regulation of an enzyme by phsophorylation at the active site. *Science* 249:1012–1046. Isocitrate dehydrogenase changes its structure very little upon phosphorylation, but the active site becomes inaccessible to substrate.

McCormack, J. G., and R. M. Denton (1986) Ca^{2+} as a second messenger within mitochondria. *Trends Biochem. Sci.* 11:258–262. Brief review of the relationship between intramitochondrial calcium levels and the demand for energy generation.

Enzymes of the Citric Acid Cycle and Related Pathways

Gruer, M. J., P. J. Artymiuk, and J. R. Guest (1997) The aconitase family: Three structural variations on a common theme. *Trends Biochem. Sci.* 22:3–6. The iron–sulfur protein aconitase has several evolutionary relatives, which participate in catalysis and regulation of iron metabolism.

Kern, D., G. Kern, H. Neef, K. Tittmann, M. Killenberg-Jabs, C. Wilkner, G. Schneider, and G. Hübner (1997) How thiamine diphosphate is activated in enzymes. *Science* 275:67–70. This paper describes NMR experiments that establish how the reactive carbanion is formed in thiamine pyrophosphate–dependent reactions.

Knowles, J. R. (1989) The mechanism of biotin-dependent enzymes. *Annu. Rev. Biochem.* 58:195–221. Evolutionary relationships and mechanisms, with pyruvate carboxylase as the primary focus of interest.

Mattevi, A., G. Obmolova, E. Schulze, K. H. Kalk, A. H. Westphal, A. de Kok, and W. G. J. Hol (1992) Atomic structure of the cubic core of the pyruvate dehydrogenase multienzyme complex. *Science* 255:1544–1550. An x-ray crystallographic analysis of the E_2 component.

Patel, M. S., and T. E. Roche (1990) Molecular biology and biochemistry of pyruvate dehydrogenase complexes. *FASEB J.* 4:3224–3232. A contemporary discussion of the regulation of pyruvate oxidation vis-à-vis structures of enzymes in the complex.

Popják, G. (1970) Stereospecificity of enzymic reactions. In: *The Enzymes,* 3rd ed., Vol. 2, edited by P. D. Boyer, pp. 115–215. Academic Press, New York. A general review of sterochemical issues with specific reference to aconitase and other citric acid cycle enzymes.

Velot, C., M. B. Mixon, M. Teige, and P. A. Srere (1997) Model of a quinary structure between Krebs TCA cycle enzymes: A model for the metabolon. *Biochemistry* 36:14271–14276. This paper describes a novel experimental approach to understanding how enzymes that catalyze sequential reactions interact with each other to facilitate catalysis of multistep pathways.

Experimental Background of the Citric Acid Cycle

Krebs, H. A. (1970) The history of the tricarboxylic acid cycle. *Perspect. Biol. Med.* 14:154–170. A historical account by the man responsible for most of the history.

Snell, E. E. (1993) From bacterial nutrition to enzyme structure: A personal odyssey. *Annu. Rev. Biochem.* 62:1–28. A memoir by one of the scientists most intimately involved in discoveries of vitamins and coenzymes.

The Glyoxylate Cycle

Tolbert, N. E. (1981) Metabolic pathways in peroxisomes and glyoxysomes. *Annu. Rev. Biochem.* 50:133–158. Reviews the cell biology and biochemistry of the glyoxylate cycle, primarily in plants.

Protection Against Intracellular Oxidation

Meister, A., and M. E. Anderson (1983) Glutathione. *Annu. Rev. Biochem.* 52:711–760. Reviews the chemistry and biochemistry of this important biological reductant.

Ortiz de Montellano, P., ed. (1986) *Cytochrome P-450: Structure, Mechanism, and Biochemistry.* Plenum, New York. A multiauthored treatise on this diverse and important group of enzymes.

PROBLEMS

1. Design a radiotracer experiment that would allow you to determine which proportion of glucose catabolism in a given tissue preparation occurs through the pentose phosphate pathway and which proportion through glycolysis and the citric acid cycle. Assume that you can synthesize glucose labeled with ^{14}C in any desired position or combination of positions. Assume also that you can trap CO_2 after administration of labeled glucose and determine its radioactivity.

2. Write a balanced chemical equation for the pentose phosphate pathway in each of the three modes depicted in Figure 14.25: (a) where ribose-5-phosphate synthesis is maximized; (b) where NADPH production is maximized, by conversion of the sugar phosphate products to glucose-6-phosphate for repeated operations of the pathway; (c) where the fructose-6-phosphate and glyceraldehyde-3-phosphate generated by each passage through the pathway are catabolized via glycolysis and the citric acid cycle.
3. Consider the fate of pyruvate labeled with ^{14}C in each of the following positions: carbon 1 (carboxyl), carbon 2 (carbonyl), and carbon 3 (methyl). Predict the fate of each labeled carbon during one turn of the citric acid cycle.
4. Suppose that aconitase did *not* bind its substrate asymmetrically. What fraction of the carbon atoms introduced as acetyl-CoA would be released in one turn of the cycle? In two turns?
5. [methyl-^{14}C]Pyruvate was administered to isolated liver cells in the presence of sufficient malonate to block succinate dehydrogenase completely. After a time, isocitrate was isolated and found to contain label in both carbon 2 and carbon 5:

$$
\begin{array}{l}
{}^{14}CH_2-COO^- \\
\quad | \\
HC-COO^- \\
\quad | \\
H^{14}C-COO^- \\
\quad | \\
\quad OH
\end{array}
$$

How do you explain this result?
6. Considering the evidence that led Krebs to propose a cyclic pathway for oxidation of pyruvate, discuss the type of experimental evidence that might have led to realization of the cyclic nature of the glyoxylate pathway.
7. Which carbon or carbons of glucose, if metabolized via glycolysis and the citric acid cycle, would be most rapidly lost as CO_2?
*8. [1-^{14}C]Ribose-5-phosphate is incubated with a mixture of purified transketolase, transaldolase, phosphopentose isomerase, phosphopentose epimerase, and glyceraldehyde-3-phosphate. Predict the distribution of radioactivity in the erythrose-4-phosphate and fructose-6-phosphate that are formed in this mixture.
9. Would you expect NAD^+ or CoA-SH to affect the activity of pyruvate dehydrogenase kinase? Briefly explain your answer.
*10. In deciding which form of isocitrate dehydrogenase plays the more important role in the citric acid cycle—the NAD^+-dependent or the $NADP^+$-dependent form—what kinds of information would help you?
11. Referring to Figure 14.10, write a balanced equation for the overall reaction catalyzed by each of the three enzymes in the pyruvate dehydrogenase complex—E_1, E_2, and E_3.
12. Briefly describe the biological rationale for each of the following allosteric phenomena: (a) activation of pyruvate carboxylase by acetyl-CoA; (b) activation of pyruvate dehydrogenase kinase by NADH; (c) inhibition of isocitrate dehydrogenase by NADH; (d) activation of isocitrate dehydrogenase by ADP; (e) inhibition of α-ketoglutarate dehydrogenase by succinyl-CoA; (f) activation of pyruvate dehydrogenase complex by AMP.
13. Given what you know about the function of the glyoxylate cycle and the regulation of the citric acid cycle, propose control mechansims that might regulate the glyoxylate cycle.
14. Consider the labeling experiment described in Figure 14.13; then predict the fate of carboxyl-labeled acetate in the glyoxylate cycle.
15. In Figure 14.6, the product of step 3 is identified as a resonance-stabilized eneamine. Draw the structures of the two resonance forms.
16. Write a balanced equation for the conversion in the glyoxylate cycle of two acetyl units, as acetyl-CoA, to oxaloacetate.
17. As discussed in Chapter 15, FAD is a stronger oxidant than NAD^+; FAD has a higher standard reduction potential than NAD^+. Yet in the last reaction of the pyruvate dehydrogenase complex, $FADH_2$ bound to E_3 is oxidized by NAD^+. Explain this apparent paradox.
18. Pyruvate carboxylase is thought to activate CO_2 by ATP, through formation of carboxyl phosphate as an intermediate. Propose a mechanism for formation of this intermediate.
19. Given the roles of NAD^+/NADH in dehydrogenation reactions and NADPH/$NADP^+$ in reductions, as discussed on page 423, would you expect the intracellular ratio of NAD^+ to NADH to be high or low? What about the ratio of $NADP^+$ to NADPH? Explain your answers.

CHAPTER 15

Electron Transport, Oxidative Phosphorylation, and Oxygen Metabolism

THE AVERAGE ADULT HUMAN GENERATES ENOUGH METABOLIC ENERGY TO SYNthesize his or her own weight in ATP *every day.* How is this massive amount of energy mobilized? Glycolysis and the citric acid cycle by themselves generate relatively little energy as ATP. However, six dehydrogenation steps—one in glycolysis, another in the pyruvate dehydrogenase reaction, and four more in the citric acid cycle—collectively reduce 10 moles of NAD^+ to NADH and 2 moles of FAD to $FADH_2$ per mole of glucose. Reoxidation of these reduced electron carriers generates most of the energy needed for ATP synthesis, the process that will concern us for much of this chapter. *Anaerobic* cells maintain an electronic steady state by transferring electrons from reduced carriers to electron acceptors, as in the reduction of pyruvate to lactate in anaerobic glycolysis. By contrast, *aerobic* cells and tissues transfer electrons in a stepwise process, from reduced carriers to molecular oxygen.

The aerobic metabolism of pyruvate generates many more reducing equivalents than does anaerobic catabolism. In respiration those reducing equivalents take the form of NADH and $FADH_2$, which in eukaryotic cells become reoxidized by electron transport proteins bound to the inner mitochondrial membrane. A series of coupled oxidation and reduction reactions occurs, with electrons being passed along a series of electron carriers—the electron transport chain, or respiratory chain (Figure 15.1). The final step is reduction of O_2 to water. The overall electron transport sequence is quite exergonic. One pair of reducing equivalents, generated from 1 mole of NADH, suffices to drive the coupled synthesis of about 3 moles of ATP from ADP and P_i by oxidative phosphorylation. How is the energy released from the oxidative reactions of the respiratory chain harnessed, or **coupled,** to drive the synthesis of ATP? The mechanism of this coupling will concern us throughout this chapter. In addition, we will consider a number of other important metabolic roles oxygen plays in aerobic cells.

Oxidation of 1 mole of NADH provides sufficient energy for synthesis of about 3 moles of ATP from ADP.

The Mitochondrion: Scene of the Action

Our comprehension of biological oxidations requires an understanding of both the chemistry of oxidation–reduction reactions and the cell biology of the mitochondrion. Before reviewing the chemistry, let us describe the intracellular sites where these reactions occur. Cellular metabolism generates reduced compounds in all of the major compartments of a eukaryotic cell. As noted earlier, glycolysis takes place in the cytosol of eukaryotic cells, whereas pyruvate oxidation, fatty acid β-oxidation, amino acid oxidation, and the citric acid cycle occur within the mitochondrial matrix. Individual cells vary widely in the abundance and structure of their mitochondria. Most vertebrate cells contain several hundred mitochondria, but the number can be as low as 1 or as high as 100,000.

The mitochondrion consists of four distinct subregions, shown in Figure 15.2a and b—the outer membrane, the inner membrane, the intermembrane space, and the matrix, located within the inner membrane. The inner membrane is highly folded into **cristae** that project into, and often nearly through, the interior of the mitochondrion. Because respiratory proteins are bound to the inner membrane, the density of cristae is related to the respiratory activity of a cell. For example, heart muscle cells, which have high rates of respiration, contain mitochondria with densely packed cristae. By contrast, liver cells have much lower respiration rates and mitochondria with more sparsely distributed cristae.

Whatever the compartment in which biological oxidations occur, all of these processes generate reduced electron carriers, primarily NADH. Most of this NADH is reoxidized, with concomitant ATP production, by the enzymes of the respiratory chain, firmly embedded in the inner membrane. The inner membrane itself consists of about 70% protein and 30% lipid, making it perhaps the most protein rich of all biological membranes. About half of the inner membrane protein in bovine heart mitochondria consists of proteins directly involved in electron transport and oxidative phosphorylation. Most of the remaining proteins are involved in transport of substances into and out of mitochondria. By contrast, a completely different set of proteins is bound to the outer membrane, including enzymes of amino acid oxidation, fatty acid elongation, membrane phospholipid biosynthesis, and enzymatic hydroxylations.

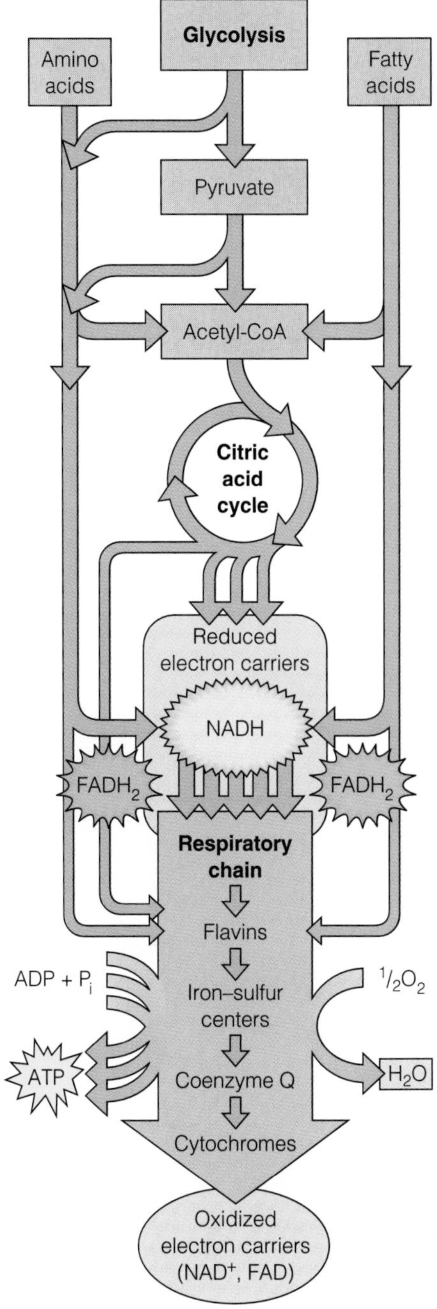

FIGURE **15.1**

Overview of oxidative energy generation.

Embedded within the inner membrane are the protein carriers, primarily cytochromes, which constitute the respiratory chain. They are assembled in the form of five multiprotein complexes, named I, II, III, IV, and V (Figure 15.2b). Complex I and complex II receive electrons from the oxidation of NADH and succinate, respectively, and pass them along to a lipid electron carrier, coenzyme Q (see pages 530–531), which moves freely through the membrane. Complex III oxidizes the reduced form of coenzyme Q and in turn reduces cytochrome *c*, a protein electron carrier that is also mobile within the inner membrane. Finally, complex IV couples the oxidation of cytochrome *c* to the reduction of O_2 to water. The energy released by these exergonic reactions creates a proton gradient across the inner membrane, with protons being pumped into the intermembrane space. Protons then reenter the membrane through a specific channel in complex V. The energy released by this exergonic process drives the endergonic synthesis of ATP from ADP and inorganic phosphate. Throughout this chapter we develop the structural and functional basis for our understanding of these energy coupling processes.

Critical to comprehension of these processes was the isolation of physiologically intact mitochondria, using differential centrifugation of cell homogenates. This feat was accomplished in the late 1940s by Eugene Kennedy and Albert Lehninger, who demonstrated that isolated mitochondria could synthesize ATP from ADP and P_i in vitro, but only if an oxidizable substrate was present as well. Much of our understanding of the sequence of electron carriers, and of the localization of specific enzymes within mitochondria, has come from fractionation and analysis of the complexes described earlier, yielding snapshots of individual parts of the overall reaction sequence of electron transport and oxidative phosphorylation.

Most electron carriers in the respiratory chain are embedded in the mitochondrial inner membrane.

The situation in prokaryotic cells is comparable, although different electron carriers are involved. However, because prokaryotic cells lack organelles, all of the electron carriers and enzymes of oxidative phosphorylation are bound to the inner

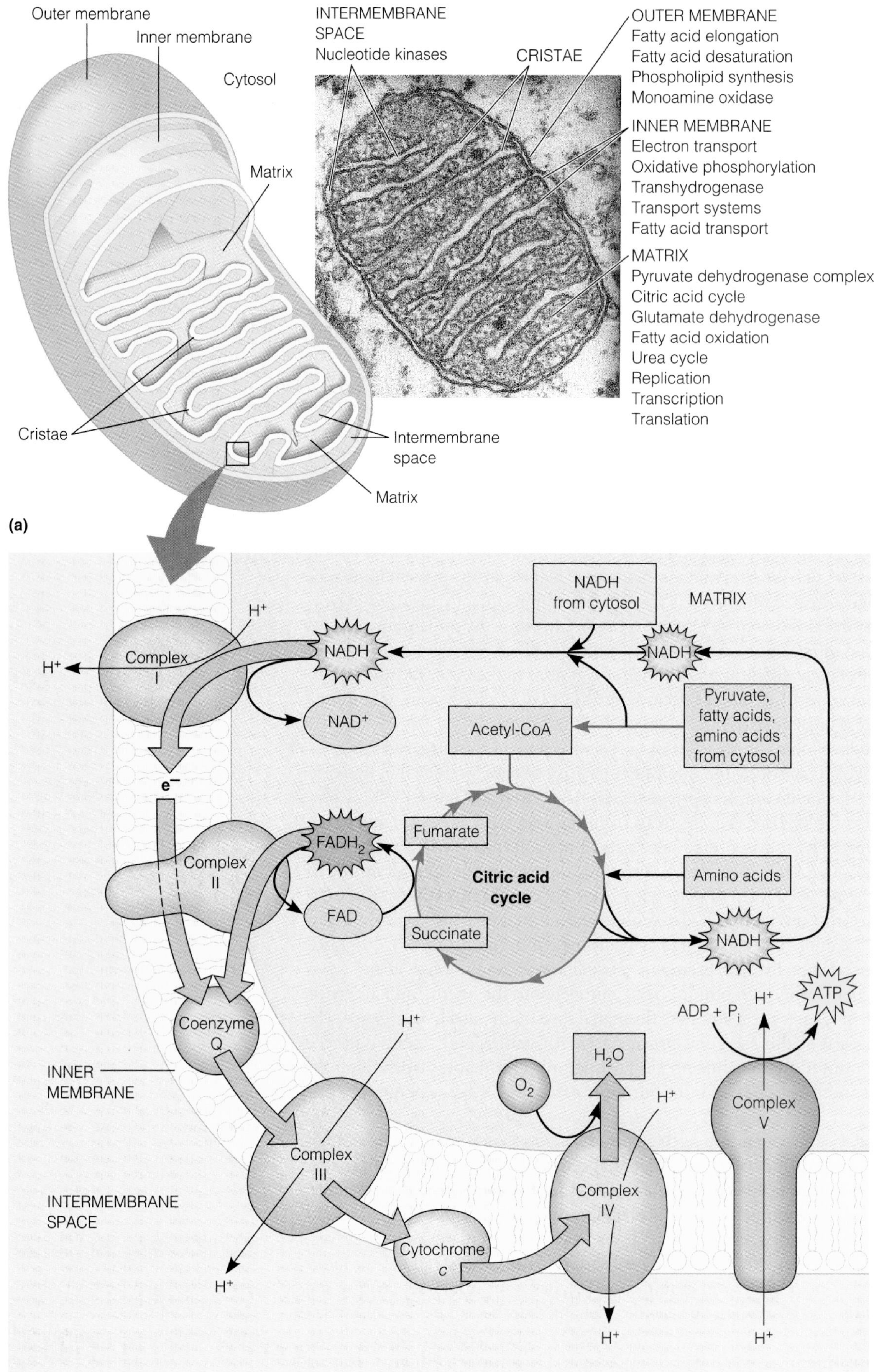
Outer membrane
Inner membrane
Cytosol
Matrix
Cristae
Intermembrane space
Matrix
INTERMEMBRANE SPACE
Nucleotide kinases
CRISTAE
OUTER MEMBRANE
Fatty acid elongation
Fatty acid desaturation
Phospholipid synthesis
Monoamine oxidase
INNER MEMBRANE
Electron transport
Oxidative phosphorylation
Transhydrogenase
Transport systems
Fatty acid transport
MATRIX
Pyruvate dehydrogenase complex
Citric acid cycle
Glutamate dehydrogenase
Fatty acid oxidation
Urea cycle
Replication
Transcription
Translation
(a)
NADH from cytosol
MATRIX
H^+
Complex I
NADH
NAD$^+$
Acetyl-CoA
Pyruvate, fatty acids, amino acids from cytosol
e$^-$
Complex II
FADH$_2$
FAD
Fumarate
Succinate
Citric acid cycle
Amino acids
NADH
Coenzyme Q
INNER MEMBRANE
Complex III
Cytochrome c
O_2
H_2O
Complex IV
ADP + P$_i$
ATP
Complex V
INTERMEMBRANE SPACE
(b)

FIGURE **15.2** (facing page)

Localization of respiratory processes in the mitochondrion. (a) A mitochondrion from a pancreatic cell, shown as a thin section in the electron micrograph. The major intramitochondrial compartments are shown, along with principal enzymes and pathways localized to each compartment. Magnification, ×155,000. **(b)** Overview of oxidative phosphorylation. Reduced electron carriers, produced by cytosolic dehydrogenases and mitochondrial oxidative pathways, become reoxidized by enzyme complexes bound in the inner membrane. These complexes actively pump protons outward, creating an energy gradient whose discharge through complex V drives ATP synthesis.

(a) Photo courtesy of A. Tzagoloff, *Mitochondria* (New York: Plenum, 1982).

membrane of the cell surface. Therefore, electron transport and oxidative phosphorylation occur at the cell periphery. As discussed in Chapter 17, there is reason to believe that mitochondria and chloroplasts, both of which contain genes and the means to express them, are descended from free-living primitive prokaryotic cells.

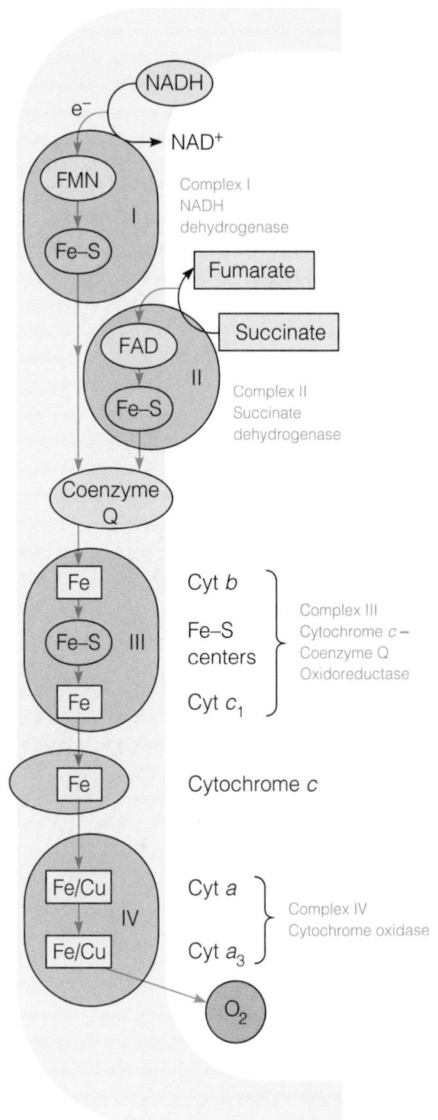

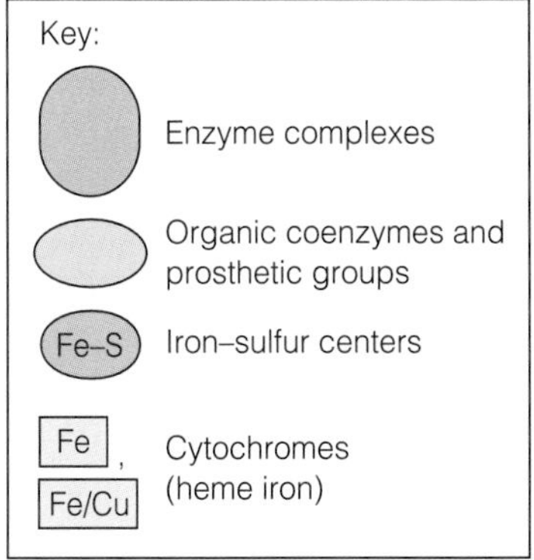

FIGURE **15.3**

Respiratory electron carriers in the mitochondrion. This figure shows the sequence of electron carriers that oxidize succinate and NAD^+-linked substrates in the inner membrane.

Oxidations and Energy Generation

Biological electron transport consists of a series of linked oxidations and reductions, or redox reactions. To understand the logic behind the sequence of reactions in the respiratory chain, and the mechanisms by which metabolic energy is generated from these reactions, you must understand how to calculate the free energy available from a redox reaction. Figure 15.3 identifies the major respiratory electron carriers; we shall return to this scheme throughout this chapter. Although the redox chemistry of these biological electron carriers may seem more complicated than that governing simple inorganic reactions, such as oxidation of Fe^{2+} to Fe^{3+}, the basic principles are identical.

QUANTITATION OF REDUCING POWER: STANDARD REDUCTION POTENTIAL

Redox chemistry is comparable in many ways to acid–base chemistry, which we discussed in Chapter 2. In a protonic equilibrium we have an acid and its conjugate base, which represent a proton donor and a proton acceptor, respectively.

$$\text{Acid (proton donor)} \rightleftharpoons \text{conjugate base} + H^+ \qquad (\text{e.g., HOAc} \rightleftharpoons OAc^- + H^+)$$

Similarly, in a redox reaction we have a donor and acceptor of *electrons.*

$$\text{Reduced compound (e}^-\text{ donor)} \rightleftharpoons \text{oxidized compound} + e^-$$

$$(\text{e.g., } Fe^{2+} \rightleftharpoons Fe^{3+} + e^-)$$

As you know, free protons and free electrons exist at negligible concentrations in aqueous media, so these equilibrium expressions are merely half-reactions in an overall acid–base or redox equilibrium. A complete redox reaction must show as one reactant an electron acceptor, which becomes reduced by gaining electrons. Of the two substrates the electron donor is the **reductant,** which becomes oxidized while transferring electrons to the other substrate, the **oxidant.**

$$\text{Reductant} + \text{oxidant} \rightleftharpoons \text{oxidized reductant} + \text{reduced oxidant}$$

A simple example is the oxidation of Cu^+ by Fe^{3+}.

$$\underset{\textbf{Reductant}}{Cu^+} + \underset{\textbf{Oxidant}}{Fe^{3+}} \rightleftharpoons \underset{\textbf{Oxidized reductant}}{Cu^{2+}} + \underset{\textbf{Reduced oxidant}}{Fe^{2+}}$$

Cu^+ is the reductant in this reaction, because it is the electron donor. The reductant is thus comparable to the acid in a protonic equilibrium.

Critical to our understanding of acid–base chemistry is the concept of pK_a, which represents a quantitative measure of the tendency of an acid to lose a proton. In the same sense, our understanding of biological oxidations demands a comparable measure of the tendency of a reductant to lose electrons (or of an oxidant to gain electrons). Such an index is provided by the *standard reduction potential,* or E_0. In acid–base equilibria we arbitrarily define water, with a pK_a of 7.0, as neutral. Anything with a pK_a below 7, which tends to protonate water, is called an acid, and compounds that tend to be protonated by water are called bases. Redox chemistry also employs a reference standard, namely, the standard hydrogen electrode in an electrochemical cell.

E_o is the tendency of a reductant to lose an electron, in the same sense that pK_a is the tendency of an acid to lose a proton.

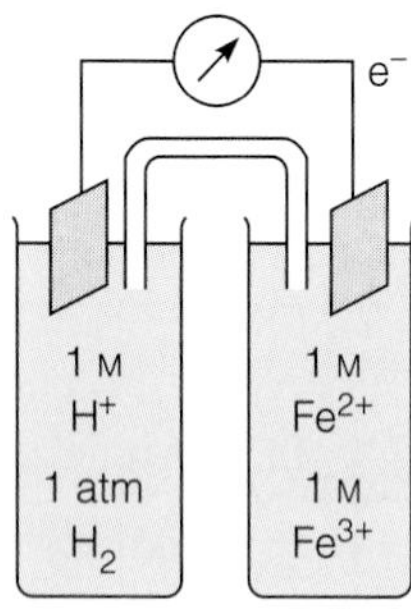

An electrochemical cell consists of two **half-cells,** each containing an electron donor and its conjugate acceptor (see margin). In the diagram the left-hand beaker constitutes the reference half-cell, a standard hydrogen electrode. The right-hand beaker contains the test half-cell, with the solution containing the test electron donor and its conjugate acceptor, each at 1 M concentration. In this example, the solution contains Fe^{2+} and Fe^{3+}, each at 1 M. Electrical neutrality is maintained with an agar salt bridge. The galvanometer that links the two half-cells measures the **electromotive force,** or **emf,** in volts. Electromotive force is a measure of the potential, or "pressure," for electrons to flow from one half-cell to the other. The electrons may flow either toward or away from the reference half-cell, depending on whether H_2 or the test electron donor has the greater tendency to lose electrons. Because H_2 loses electrons more readily than Fe^{2+}, the electrons in our example will flow from the reference half-cell to the test half-cell, oxidizing H_2, reducing Fe^{3+}, and causing the galvanometer to record a positive emf. However, if the test electron donor loses electrons more readily than H_2, then electrons flow in the reverse direction, reducing H^+ to H_2 in the reference half-cell and causing a negative emf to be recorded. The stronger oxidant, whether H^+ or the test electron acceptor, will draw electrons away from the other half-cell and become reduced.

The tendency of an electron donor to reduce its conjugate acceptor is called the **reduction potential,** ***E***. Under standard conditions (25°C, donor and acceptor each at 1 M), this term becomes the **standard reduction potential,** $\boldsymbol{E_0}$. The emf recorded in an electrochemical cell as described above is the *difference* between the reduction potentials in the reference and test half-cells. If both half-cells are at standard conditions, we measure the difference in E_0 values.

The higher the standard reduction potential, the higher the tendency of the oxidized member of a redox couple to attract electrons.

By convention, E_0 for the hydrogen electrode is set at 0.00 volts. Any redox couple that tends to donate electrons to the standard hydrogen electrode has a negative value of E_0, which can be determined directly as the recorded emf in a half-cell linked to a standard hydrogen half-cell. A positive E_0 means that electrons from H_2 are flowing toward the test cell and reducing the electron acceptor, or, in other words, that the test acceptor is oxidizing H_2. *The higher the value of E_0 for a redox couple, the stronger an oxidant is the electron acceptor of that couple.* For example, we know from experience that Fe^{3+} is a strong oxidant. This fact is borne out when we measure the reduction potential of a standard Fe^{3+}/Fe^{2+} half-cell with reference to a standard hydrogen half-cell. That value is +0.77 V. By contrast, a high negative value of E_0 connotes a ready tendency to lose electrons and, hence, strong reducing power.

Standard conditions for biochemists include a pH value of 7.0, a condition far from that seen in the standard hydrogen electrode, which contains H^+ at 1.0 M. Therefore, biochemists use a modified term, E'_0, which is the standard reduction potential measured at pH 7.0 (10^{-7} M H^+). These are the values used in this book and most other biochemical references. E'_0 values for a number of biochemically important redox pairs are recorded in Table 15.1. Note that for the hydrogen ion itself, E'_0 is −0.42 V, significantly different from the standard value determined at pH 0 (i.e., at 1 M H^+). You should realize that all of the E'_0 values in Table 15.1 were recorded with reference to a standard hydrogen electrode; pH was 7.0 in the *test* electrode but 0 in the *reference* electrode. Thus, the NADH + H^+/NAD^+ couple, with an E'_0 of −0.32 V, will lose electrons to a *standard* hydrogen electrode (where $[H^+] = 1$ M) but will tend to gain electrons from a hydrogen electrode at pH 7.0 (where $[H^+] = 10^{-7}$ M). This observation again illustrates the principle that, *in a spontaneous reaction, electrons flow from the half-cell of lower potential to the half-cell of higher potential.* Related to this, the O_2/H_2O couple, with a high E'_0 of +0.82 V, has a strong tendency to oxidize other substances. Conversely, there is very little tendency for water to become oxidized to O_2, because none of the common biological oxidants has a higher E'_0 than O_2/H_2O. Photosynthesis, which does oxidize H_2O to O_2, requires considerable energy, as light, to accomplish this feat. (More about this in Chapter 17.)

Test Electrode	Reference Electrode	Measured E
H^+/H_2 (pH 7) ($E'_0 = -0.42$)	H^+/H_2 (pH 0) ($E'_0 = 0$)	−0.42 V
NAD^+/NADH (pH 7) ($E'_0 = -0.32$)	H^+/H_2 (pH 0) ($E'_0 = 0$)	−0.32 V
NAD^+/NADH (pH 7) ($E'_0 = -0.32$)	H^+/H_2 (pH 7) ($E'_0 = -0.42$)	+0.10 V

FREE ENERGY CHANGES FROM OXIDATION–REDUCTION REACTIONS

To recapitulate, the higher the value of E'_0 for a redox couple, the greater is the tendency for that couple to participate in oxidation of another substrate. We can

TABLE 15.1 Standard reduction potentials of interest in biochemistry

Oxidant	Reductant	n	E'_0, V
Acetate + CO_2 + $2H^+$	Pyruvate + H_2O	2	−0.70
Succinate + CO_2 + $2H^+$	α-Ketoglutarate + H_2O	2	−0.67
Acetate + $3H^+$	Acetaldehyde + H_2O	2	−0.60
O_2	O_2^-	1	−0.45
Ferredoxin (oxidized)	Ferredoxin (reduced)	1	−0.43
$2H^+$	H_2	2	−0.42
Acetoacetate + $2H^+$	β-Hydroxybutyrate	2	−0.35
Pyruvate + CO_2 + H^+	Malate	2	−0.33
NAD^+ + H^+	NADH	2	−0.32
$NADP^+$ + H^+	NADPH	2	−0.32
FMN (enzyme-bound) + $2H^+$	$FMNH_2$ (enzyme-bound)	2	−0.30
Lipoate (oxidized) + $2H^+$	Lipoate (reduced)	2	−0.29
1,3-Bisphosphoglycerate + $2H^+$	Glyceraldehyde-3-phosphate + P_i	2	−0.29
Glutathione (oxidized) + $2H^+$	2 Glutathione (reduced)	2	−0.23
FAD + $2H^+$	$FADH_2$	2	−0.22
Acetaldehyde + $2H^+$	Ethanol	2	−0.20
Pyruvate + $2H^+$	Lactate	2	−0.19
Oxaloacetate + $2H^+$	Malate	2	−0.17
α-Ketoglutarate + $\overset{+}{N}H_4$ + $2H^+$	Glutamate + H_2O	2	−0.14
Methylene blue (oxidized) + $2H^+$	Methylene blue (reduced)	2	0.01
Fumarate + $2H^+$	Succinate	2	0.03
CoQ + $2H^+$	$CoQH_2$	2	0.04
Cytochrome *b* (+3)	Cytochrome *b* (+2)	1	0.07
Dehydroascorbate + $2H^+$	Ascorbate	2	0.08
Cytochrome c_1 (+3)	Cytochrome c_1 (+2)	1	0.23
Cytochrome *c* (+3)	Cytochrome *c* (+2)	1	0.25
Cytochrome *a* (+3)	Cytochrome *a* (+2)	1	0.29
½O_2 + H_2O	H_2O_2	2	0.30
Ferricyanide	Ferrocyanide	2	0.36
Nitrate + $2H^+$	Nitrite + H_2O	1	0.42
Cytochrome a_3 (+3)	Cytochrome a_3 (+2)	1	0.55
Fe (+3)	Fe (+2)	1	0.77
½O_2 + $2H^+$	H_2O	2	0.82

Note: E'_0 is the standard reduction potential at pH 7 and 25°C, n is the number of electrons transferred, and each potential is for the partial reaction written as follows: Oxidant + ne^- ⟶ reductant.

describe this tendency in quantitative terms, because free energy changes (Chapter 3) are directly related to differences in reduction potential:

$$\Delta G^{\circ\prime} = -nF\,\Delta E_0' = -nF[E_0'\,(\text{acceptor}) - E_0'\,(\text{donor})] \tag{15.1}$$

where n is the number of electrons transferred in the half-reactions, F is Faraday's constant (96.5 kJ $mol^{-1}V^{-1}$), and $\Delta E_0'$ is the difference in standard reduction potentials between the two redox couples. Note that $\Delta E_0'$ is to ΔE_0 as $\Delta G^{\circ\prime}$ is to ΔG°; the "prime" has the same significance in both cases: each reactant and product (except H^+) at 1M and a pH of 7.0.

Free Energy Changes Under Standard Conditions

As an example, consider the oxidation of ethanol catalyzed by alcohol dehydrogenase:

$$\text{Ethanol} + NAD^+ \rightleftharpoons \text{acetaldehyde} + NADH + H^+$$

The two half-reactions are as follows, both written in the direction of reduction:

(a) $$NAD^+ + H^+ + 2e^- \rightleftharpoons NADH \qquad E_0' = -0.320\text{ V}$$

(b) $$\text{Acetaldehyde} + 2H^+ + 2e^- \rightleftharpoons \text{ethanol} \qquad E_0' = -0.197\text{ V}$$

But because ethanol becomes oxidized in the reaction, we reverse the second half-reaction *and* the sign of E_0' for this half-reaction.

(c) $$\text{Ethanol} \rightleftharpoons \text{acetaldehyde} + 2H^+ + 2e^- \qquad E_0' = +0.197\text{ V}$$

Now the overall reaction is the sum of half-reactions (a) and (c), and $\Delta E_0'$ is the algebraic sum of the two individual E_0' values:

$$\Delta E_0' = -0.320\text{ V} + 0.197\text{ V} = -0.123\text{ V} \tag{15.2}$$

Note that $\Delta E_0'$ calculated this way is equal to E_0' (acceptor) − E_0' (donor) as shown in equation (15.1).

The standard free energy change is given, then, by

$$\Delta G^{\circ\prime} = -nF\,\Delta E_0' = -2(96.5)(-0.123)\text{ kJ/mol} = +23.74\text{ kJ/mol} \tag{15.3}$$

Note from this example that a negative $\Delta E_0'$ value gives a positive $\Delta G^{\circ\prime}$ and hence corresponds to a reaction that is *not* favored in the direction written. Note also that if we were to calculate $\Delta G^{\circ\prime}$ for the reverse reaction (reduction of acetaldehyde by NADH), E_0' (acceptor) − E_0' (donor) would be +0.123 V, and $\Delta G^{\circ\prime}$ would have the same numerical value but the opposite sign.

Free Energy Changes Under Nonstandard Conditions

The values given in Table 15.1 allow calculation of free energy changes only under standard conditions (including, by convention, a pH of 7.0). To calculate reduction potentials under nonstandard conditions, we must use the *Nernst equation,*

The Nernst equation makes it possible to compute reduction potentials under nonstandard conditions.

$$E' = E_0' + \frac{2.303RT}{nF}\log\frac{[\text{electron acceptor}]}{[\text{electron donor}]} \tag{15.4}$$

where R is the gas constant (8.314 J K^{-1} mol^{-1}), T is the absolute temperature in kelvins, and 2.303 is the conversion factor from natural to common logarithms. At 25°C the term $2.303RT/nF$ has the value 0.059 for a one-electron transfer and 0.0295 (rounded to 0.03) for a two-electron transfer (where $n = 2$). Thus, for a two-electron transfer we can use a simplified version of equation (15.4):

$$E' = E'_0 + 0.03 \log \frac{[\text{electron acceptor}]}{[\text{electron donor}]} \quad (15.5)$$

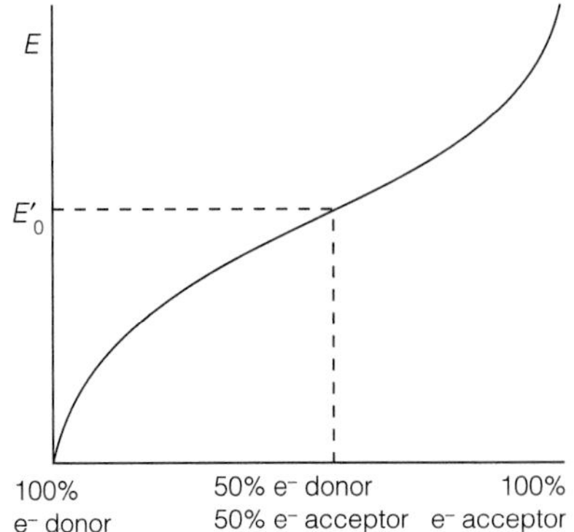

Note that equation 15.5 is of the same form as the Henderson–Hasselbalch equation: $pH = pK_a + \log([\text{proton acceptor}]/[\text{proton donor}])$. In the same sense, then, that pK_a is defined by the midpoint of an acid titration curve, E'_0 is defined by the midpoint of an electrochemical titration, where electron acceptor and electron donor are present in equal concentrations.

Free Energy Changes in Biological Oxidations

Each of the coupled redox reactions in biological electron transport involves the transfer of electrons from one redox couple to another couple of higher reduction potential. Thus, each individual redox reaction in the sequence is exergonic under standard conditions. For electrons entering the respiratory chain as NADH, the overall reaction sequence is given by the following equation:

$$\text{NADH} + \text{H}^+ + \tfrac{1}{2}\text{O}_2 \rightleftharpoons \text{NAD}^+ + \text{H}_2\text{O}$$

This sequence is strongly exergonic:

$$\Delta G^{\circ\prime} = -nF\Delta E'_0 = -2(96.5)[0.82 - (-0.32)] \text{ kJ/mol} = -220 \text{ kJ/mol} \quad (15.6)$$

As discussed later in this chapter, the oxidation of 1 mole of NADH in the respiratory chain proceeds concomitantly with synthesis of about 3 moles of ATP from ADP and P_i. Because $\Delta G^{\circ\prime}$ for ATP hydrolysis is -31 kJ/mol, synthesis of three ATPs requires 93 kJ under standard conditions, giving an efficiency for oxidative phosphorylation of about 40%. However, $\Delta G'$ for ATP hydrolysis under intracellular conditions is significantly higher (-40 kJ/mol or more), so the intracellular efficiency is probably somewhat higher than 40%.

Electron Transport

ELECTRON CARRIERS IN THE RESPIRATORY CHAIN

If you compare the sequence of respiratory electron carriers (Figure 15.3) with the standard reduction potentials of those carriers (Table 15.1), you will see that E'_0 for each carrier increases in the same order as the sequence of their use in electron transport. This order suggests that each individual oxidoreduction reaction in electron transport is exergonic under standard conditions and that electrons flow in continuous fashion from low-potential to high-potential carriers. Very neat, but is it real? After all, we have seen that glycolysis and the citric acid cycle both proceed smoothly despite the involvement of some reactions with large positive $\Delta G^{\circ\prime}$ values. We shall explore the lines of evidence by which the currently accepted pathway of electron transport was determined. First, however, let us become better acquainted with the participants—the electron carriers and the enzymes involved.

NADH and NADH Dehydrogenase

Numerous dehydrogenases in the cell generate NADH. The NADH becomes oxidized in the first step of electron transport by mitochondrial complex I, or **NADH dehydrogenase,** which contains flavin mononucleotide (FMN) as a tightly bound prosthetic group and catalyzes the following reaction.

$$\text{NADH} + \text{H}^+ + \text{FMN} \rightleftharpoons \text{NAD}^+ + \text{FMNH}_2$$

NADH dehydrogenase is a large, multisubunit complex with about 25 separate polypeptide chains. The complex also contains a number of **iron–sulfur**

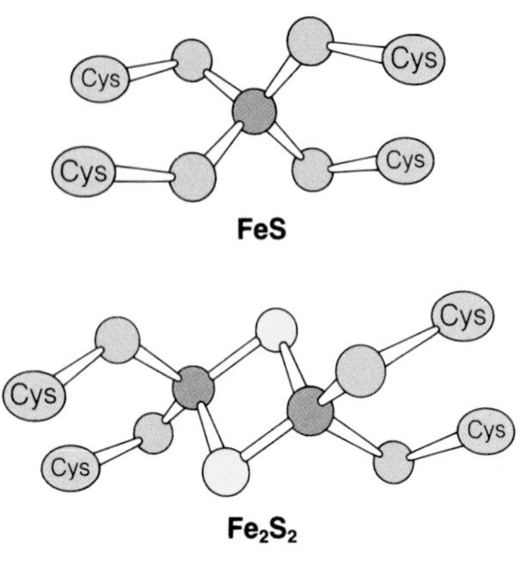

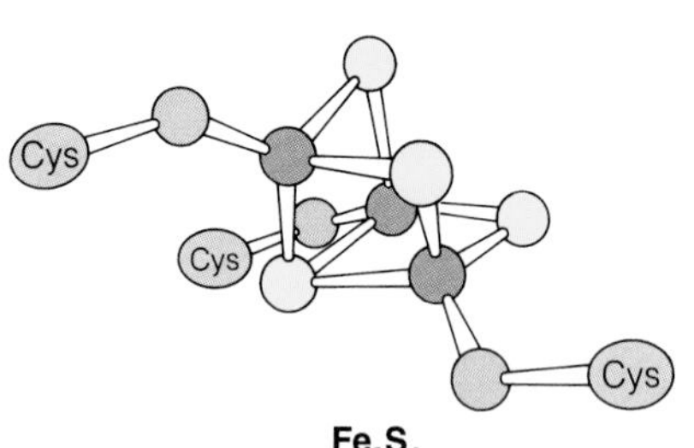

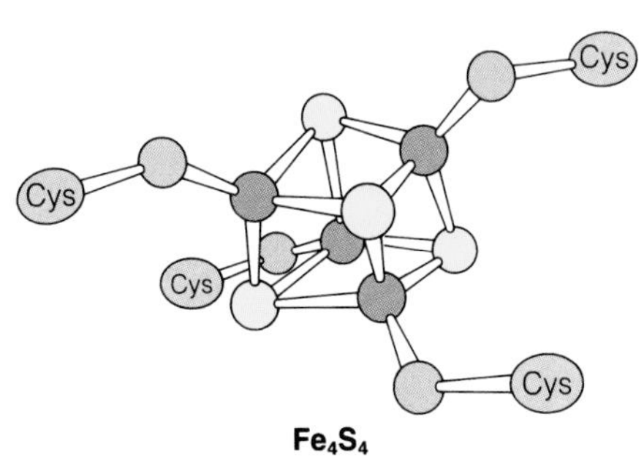

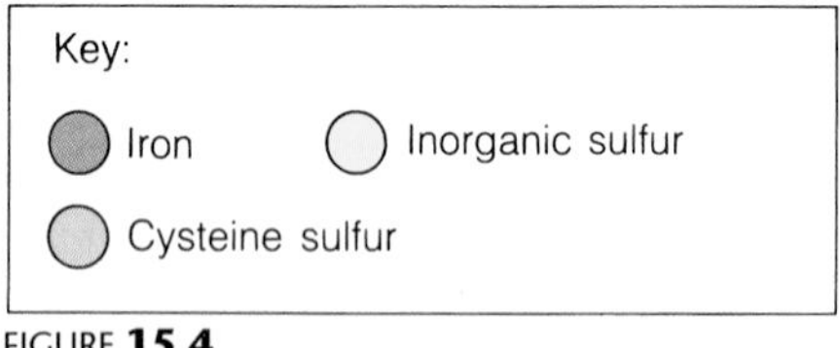

FIGURE **15.4**
Structures of iron–sulfur centers.

The respiratory electron carriers include NAD^+, flavins, iron–sulfur centers, coenzyme Q, and cytochromes.

centers, which transfer electrons from reduced flavin to another respiratory carrier, coenzyme Q. Iron–sulfur centers consist of nonheme iron complexed to sulfur in four known ways (Figure 15.4). The simplest form, designated **FeS,** involves one iron tetrahedrally complexed with the thiol sulfurs of four cysteine residues. The second form, denoted $\mathbf{Fe_2S_2}$, contains two irons, each complexed with two cysteine residues and two inorganic sulfides. The third and most complicated form ($\mathbf{Fe_4S_4}$) contains four irons, four sulfides, and four cysteine residues. NADH dehydrogenase contains both Fe_2S_2 and Fe_4S_4 centers. A fourth, less common form called Fe_3S_4 contains three irons, four sulfides, and four cysteine residues in a cubelike structure with one vertex missing. In all of these centers the iron can undergo cyclic oxidoreduction between ferrous and ferric states, as shown below for the NADH dehydrogenase complex. Because the electrons are then used to reduce coenzyme Q, a more descriptive name for this complex is **NADH–coenzyme Q reductase.**

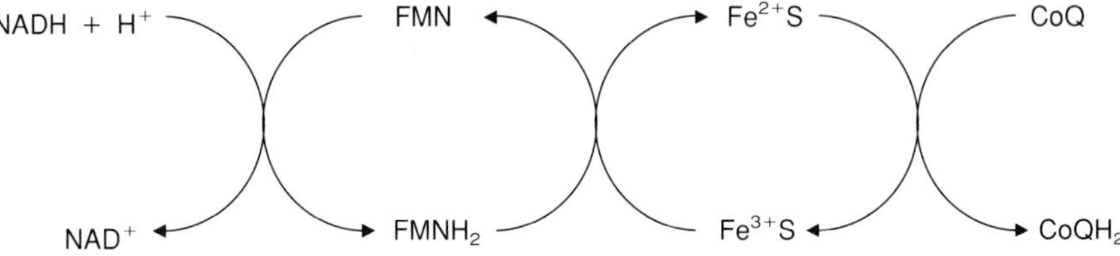

Several important redox proteins contain iron–sulfur centers, including the cytochromes—other proteins we will encounter in the respiratory chain. This group goes by the collective term **iron–sulfur proteins.** Not all iron–sulfur proteins are involved in redox reactions. For example, aconitase (see page 497) has an iron–sulfur center that is involved in substrate binding and activation.

NADH dehydrogenase can also accept electrons from NADPH, although rather inefficiently. A more widely used route for NADPH oxidation, when that becomes necessary, is the transfer of two reducing equivalents from NADPH to NAD^+ via an enzyme called **transhydrogenase:**

$$\text{NADPH} + \text{NAD}^+ \rightleftharpoons \text{NADP}^+ + \text{NADH}$$

Coenzyme Q

The respiratory electron carrier coenzyme Q was discovered when treatment of isolated mitochondria with an organic solvent such as isooctane was observed to completely abolish the ability of the mitochondria to oxidize substrates. Addition of the material extracted by isooctane completely restored the oxidative capacities of the mitochondria, suggesting the presence of an extremely lipophilic electron carrier that was but loosely linked to protein. This carrier was found to be a benzoquinone linked to a number of isoprene units, usually 10 in mammalian cells and 6 in bacteria. Because the substance is ubiquitous in living cells, one group of researchers named it *ubiquinone,* while another called it **coenzyme Q,** or **CoQ.** The term Q_{10} is used to specify the form of CoQ containing 10 isoprene units. The isoprenoid tail gives the molecule its apolar character, which allows CoQ to diffuse rapidly through the inner mitochondrial membrane.

CoQ draws electrons into the respiratory chain, not only from NADH but also from succinate, as shown in Figure 15.3, and from intermediates in fatty acid oxidation. Succinate dehydrogenase uses an FAD coenzyme, as noted in Chapter 14. Unlike the other citric acid cycle enzymes, succinate dehydrogenase is an inner membrane protein. The enzyme can thus transfer electrons directly from its bound $FADH_2$ to the other membrane-bound respiratory carriers. Like NADH dehydrogenase, succinate dehydrogenase transfers electrons via iron–sulfur centers to coenzyme Q, and it is more completely named **succinate–coenzyme Q reductase** (it is also called succinate dehydrogenase or complex II—see Figures 15.2 and 15.3).

Other flavoprotein dehydrogenases, such as the one involved in fatty acid oxidation (see Chapter 18), also transport electrons to electron acceptors via CoQ.

Because CoQ is subsequently oxidized by cytochromes, it can be seen as a collection point, gathering electrons from several flavoprotein dehydrogenases and passing them along to cytochromes for ultimate transport to O_2. Further, because CoQ oxidoreduction proceeds one electron at a time through a *semiquinone* intermediate, CoQ provides an interface between two-electron carriers and the one-electron cytochromes.

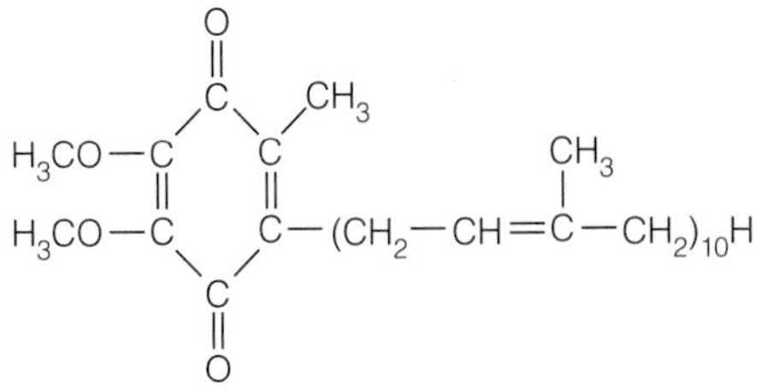

Oxidized coenzyme Q_{10} (CoQ)

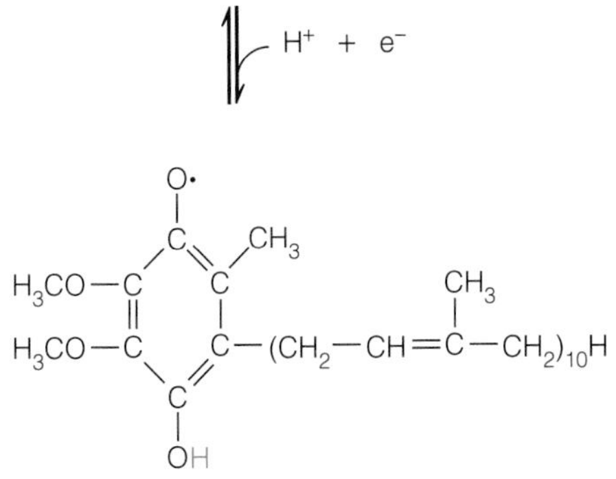

$\rightleftharpoons H^+ + e^-$

Semiquinone form of coenzyme Q

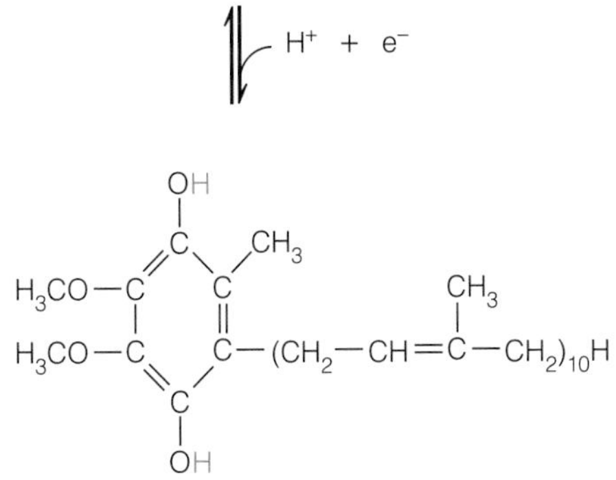

$\rightleftharpoons H^+ + e^-$

Reduced coenzyme Q_{10} ($CoQH_2$)

Cytochromes

Finally, we come to the cytochromes, a group of red or brown heme proteins having distinctive visible-light spectra. These proteins were first characterized and their role in respiration demonstrated by the Englishman David Keilin. Using a hand spectroscope, Keilin observed red-brown pigments in the flight muscles of insects. During muscular exertion (when an immobilized fly tried to free itself), the spectra of these pigments underwent marked changes. This observation led Keilin to postulate a role for these substances in carrying electrons from biological fuels to oxygen.

The major respiratory cytochromes are classified as *b*, *c*, or *a*, depending on the wavelengths of the spectral absorption peaks. Figure 15.5 shows the spectral characteristics of typical *b*-, *c*-, and *a*-type cytochromes. Within each class (*b*, *c*, or *a*), the cytochromes are distinguished by smaller spectral differences. For example, cytochrome c_1 has a spectrum similar to that of cytochrome *c*, but the α and γ absorption peaks are shifted slightly toward the red.

Among the respiratory electron carriers are three *b*-type cytochromes, cytochromes *c* and c_1, and cytochromes *a* and a_3. Cytochromes *b*, *c*, and c_1 all contain the same heme found in hemoglobin and myoglobin—iron complexed with protoporphyrin IX. In cytochromes *c* and c_1, but not *b*, this heme is linked covalently to the protein component, via thioether bonds formed between two of the vinyl side chains and two cysteine residues (Figure 15.6a). Cytochromes *a* and a_3 contain a modified form of heme, called heme A, in which two of the side chains are modified (Figure 15.6b). Cytochromes *a* and a_3 represent two identical heme A moieties, attached to the same polypeptide chain, but within different environments in the inner membrane and, hence, having different reduction potentials. Each of the hemes is associated with a copper ion, located close to the heme iron. Cytochromes undergo oxidoreduction through the complexed metal, which cycles between +2 and +3 states of heme iron and +1 and +2 states for the copper in cytochromes *a* and a_3. Thus, the cytochromes are one-electron carriers.

Cytochrome *c* is a small protein (M_r = 13,000), which is associated with the inner membrane but is readily extracted in soluble form. Because it is small and relatively abundant, detailed structural studies have been carried out with this protein. The amino acid sequence of cytochrome *c* has been highly conserved in evolution, with nearly 50% identity between residues at corresponding positions of cytochromes *c* in organisms as diverse as yeast and human. The other cytochromes are integral membrane proteins and are exceedingly difficult to dissociate from the membrane. Accordingly, we know less about their structures. Cytochromes *a* and a_3 form part of mitochondrial complex IV (see Figures 15.2 and 15.3), a multiprotein complex also called **cytochrome oxidase.** This complex catalyzes the ultimate step in electron transport—the reduction of oxygen to water. In eukaryotes the cytochrome oxidase complex contains up to 13 polypeptide chains, 3 of which are known to be encoded by the mitochondrial genome and synthesized within the mitochondrion. The rest are encoded by nuclear genes.

DETERMINING THE SEQUENCE OF RESPIRATORY ELECTRON CARRIERS

To comprehend the mechanism by which energy from biological oxidations is captured to drive ATP synthesis, we must understand the oxidation reactions

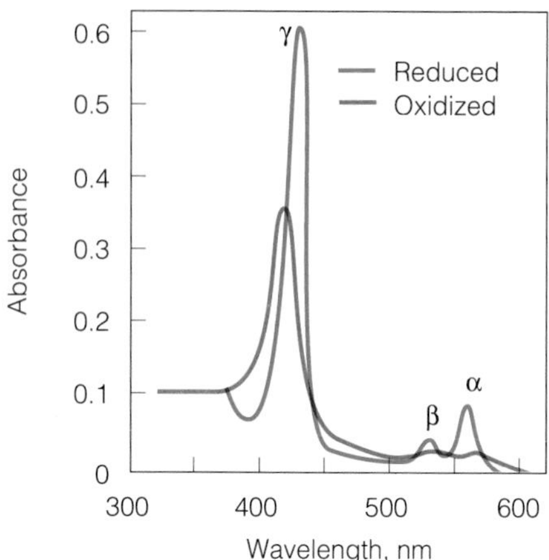

(a) Cytochrome *b*

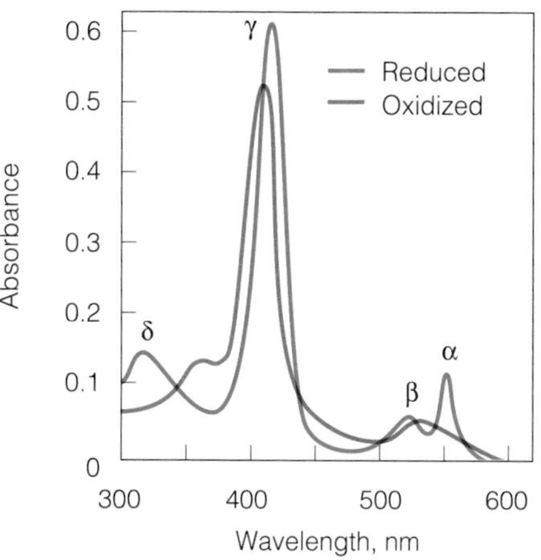

(b) Cytochrome *c*

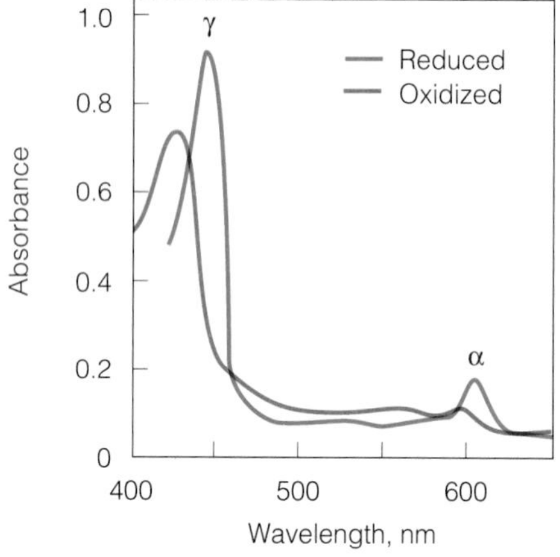

(c) Cytochromes *a* and a_3

FIGURE **15.5**

Absorption spectra of cytochromes. The plots show the absorption spectra of cytochromes *b*, *c*, and *a* in their oxidized (red) and reduced (blue) states. **(a)** Cytochrome *b* from *Neurospora*; **(b)** cytochrome *c* from horse heart; **(c)** beef heart cytochrome oxidase (which contains both cytochromes *a* and a_3).

Adapted from A. Tzagoloff, *Mitochondria* (New York: Plenum, 1982). Reprinted by permission.

(a) General structure of cytochromes *c* and c_1

(b) Heme A in cytochromes *a* and a_3

FIGURE **15.6**

The hemes found in cytochromes. (a) The covalent bond formed between heme and the protein component in cytochromes *c* and c_1. The vinyl groups on heme are linked to the thiol groups of two cysteine residues (red). **(b)** Heme A, the form found in cytochromes *a* and a_3. Note the modified side chains—a formyl group (red) and a 17-carbon side chain (blue).

of electron transport, both the sequence in which electrons are carried from reduced substrates to oxygen and the energetics of individual reactions. Figure 15.7 shows E_0' values for the major respiratory electron carriers. If this figure accurately represents the sequence, then we can visualize respiratory electron transport as a series of coupled exergonic reactions in which the total energy available from oxidation of NADH by O_2 is released in a series of small steps, some of which are sufficiently exergonic to generate the 31 kJ/mol needed to drive ATP synthesis.

However, the data of Figure 15.7 involve standard conditions, and conditions are much different within the mitochondrion. In particular, the hydrophobic nature of the membrane environment changes *E* values in ways that are difficult to predict or to measure. We shall outline here three experimental techniques that have been used to identify both the actual order of electron carriers and the spe-

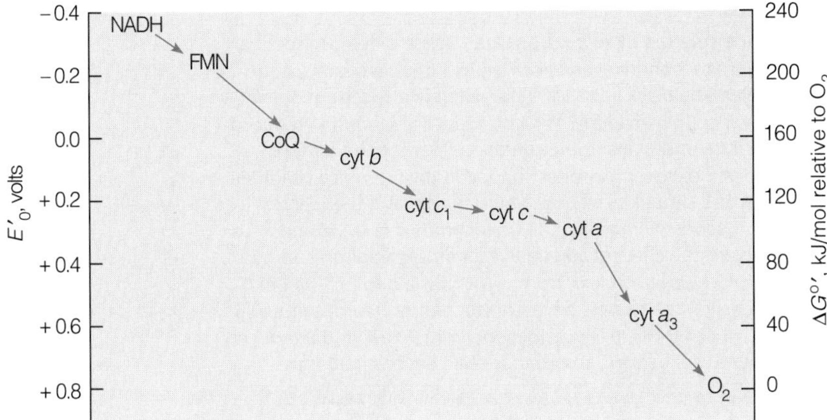

FIGURE **15.7**

Standard reduction potentials of the major respiratory electron carriers. Three reactions in the respiratory chain have $\Delta G^{\circ\prime}$ values greater than 31 kJ/mol, the $\Delta G^{\circ\prime}$ for ATP hydrolysis: FMN $\longrightarrow$ CoQ, cyt *b* $\longrightarrow$ cyt c_1, and cyt *a* $\longrightarrow$ O_2.

cific reactions that drive ATP synthesis. We focus first on three approaches used to identify the order of electron carriers: (1) spectrophotometric techniques to measure the redox status of electron carriers in intact mitochondria; (2) use of specific respiratory inhibitors and artificial electron acceptors; and (3) fractionation of mitochondria into respiratory subassemblies, each capable of catalyzing specific portions of the overall sequence.

Difference Spectra

For nicotinamide nucleotides, flavin nucleotides, and cytochromes, the absorption spectrum for the reduced carrier differs from that of its oxidized counterpart. We should, therefore, be able to scan the absorption spectrum of a mixture of these carriers and ascertain the proportions of each in the oxidized and reduced states. The sensitivity of the technique is increased if a **difference spectrum** is obtained. Here the sample cuvette contains the mixture of electron carriers under study, and the reference cuvette contains not a blank but an equimolar mixture of carriers in a known state, for instance, entirely oxidized. Thus, any small absorbance changes, either positive or negative, result from reduction of a portion of the carriers in the test sample.

When these carriers are embedded in the mitochondrion, however, the task becomes quite difficult. Mitochondrial suspensions are turbid, and the resultant light scattering makes it impossible to measure difference spectra with ordinary spectrophotometers. Britton Chance greatly improved this technique when he developed a dual-wavelength, double-beam spectrophotometer that allowed him to obtain difference spectra with intact mitochondria. In the example shown by the black lines in Figure 15.8, the reference cuvette contains mitochondria saturated with oxygen, so that all carriers are oxidized, while the sample cuvette contains anaerobic mitochondria plus an oxidizable substrate, so that all of the carriers are reduced. The difference spectrum identifies wavelengths of maximal and minimal absorbance differences. Absorbance readings at these wavelengths allow determinations of the concentrations of the reduced and oxidized forms of an electron carrier that absorbs light in these ranges. For example, the higher the absorbance at 340 nm, the greater the proportion of the NAD^+/NADH couple that is present as NADH; and the lower the negative absorbance at 460 nm, the greater the proportion of flavin nucleotide that is in the oxidized form.

Two important observations were made soon after the introduction of this technique in the mid-1950s. First, in actively respiring mitochondria, NADH predominated over NAD^+, whereas cytochrome a_3 was largely oxidized. For the intermediate carriers, the proportion in the oxidized state increased in the same order as the order of their presumed function in respiration. Second, when oxygen was added to anaerobic mitochondria, and difference spectra were obtained at intervals after oxygen addition, it was possible to determine the order in which

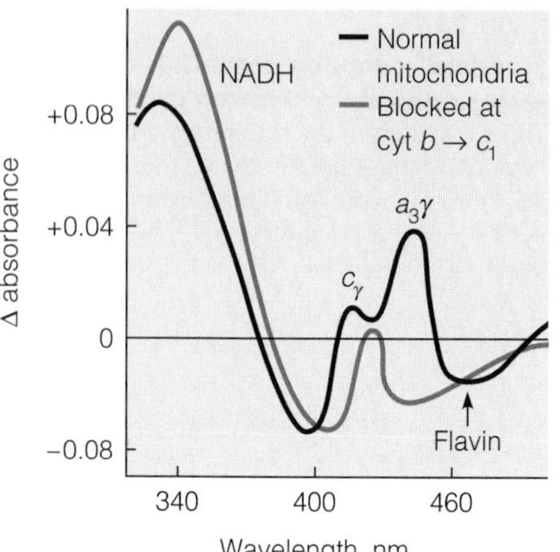

(a) Difference spectra for wavelengths below 500 nm

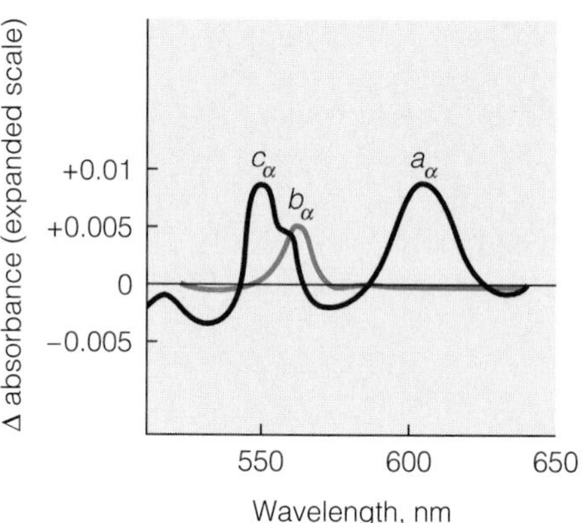

(b) Difference spectra continued with extended scale

FIGURE **15.8**

Difference spectra of mitochondria. These difference spectra of rat liver mitochondria were recorded in a double-beam spectrophotometer. The black line shows the difference spectrum of fully reduced versus fully oxidized mitochondria. Mitochondria reduced with substrate under anaerobic conditions were in the sample chamber, and oxygenated mitochondria in the reference chamber. The peaks and shoulders refer to NADH, flavin, and the α and γ absorption bands, as indicated, for cytochromes *a*, a_3, *b*, and *c*. The red line shows the effect of adding a respiratory inhibitor, antimycin A, which blocks electron flow from cytochrome *b* to c_1. The inhibitor causes all of the carriers beyond cytochrome *b* to become fully oxidized, while NADH, flavin, and cytochrome *b* are in the reduced state. Note the expanded absorbance scale beyond 500 nm.

Adapted from A. Tzagoloff, *Mitochondria* (New York: Plenum, 1982). Reprinted by permission.

each carrier went from fully reduced to partially oxidized. That order was the same as the presumed order of function in respiration.

Inhibitors and Artificial Electron Acceptors

Further information was obtained from difference spectrophotometry in conjunction with exogenous compounds that functioned either as respiratory inhibitors or as artificial electron donors or acceptors. The sites of action of several important inhibitors are shown in Figure 15.9. These inhibitors include (1) **rotenone,** a plant product from South America that is used as an insecticide and blocks electron flow from NADH to coenzyme Q; (2) **amytal,** a barbiturate drug that acts at the same site; (3) **antimycin A,** a *Streptomyces* antibiotic that blocks electron flow from cytochrome *b* to c_1; and (4) the cytochrome oxidase inhibitors cyanide, azide, and carbon monoxide. Cyanide and azide react with the oxidized form of the cytochrome target, and CO reacts with the reduced form.

To see the utility of respiratory inhibitors, consider what happens in actively respiring mitochondria to which antimycin A is added (refer to the red lines in Figure 15.8). Because electrons cannot flow from cytochrome *b* to c_1, all of the carriers before cytochrome *b* become reduced, while all of the subsequent carriers become fully oxidized, analogous to the accumulation of water upstream of a dam in a river. This site of inhibition is called a **crossover point**—a specific target of inhibition when an overall pathway is blocked. Crossover points in the respiratory chain can be detected by dual-beam spectrophotometry, which allows identification of all the carriers preceding a site of inhibition (NAD, flavin, CoQ, and cytochrome *b* after antimycin treatment) and all those following that site (cytochromes c_1, *c*, *a*, and a_3).

Artificial electron donors and acceptors are compounds that can either feed in or draw off electrons from the respiratory chain in spontaneous nonenzymatic reactions. For example, 2,6-dichlorophenol-indophenol (DCIP) can spontaneously oxidize cytochrome *b*, but probably not c_1, because of the E_0' values involved (see Figure 15.9). This demonstrated that cytochrome *b* lies beyond the entry point for electrons from succinate, as well as from NADH. The key observation was that addition of DCIP to cyanide-inhibited mitochondria allows these mitochondria to oxidize both NAD^+-linked substrates and succinate. In this system, electrons flow from substrate to cytochrome *b* and then to the exogenous electron carrier, DCIP, which becomes reduced.

Difference spectrophotometry, analysis of respiratory inhibitors, and properties of membrane complexes establish the order of action of electron carriers in the respiratory chain.

Respiratory Complexes

Mitochondria can be disrupted by mechanical treatment, such as sonic oscillation, or by nonionic detergents such as *digitonin,* which preferentially solubilizes the outer membrane but tends not to denature enzymes. By combinations of these techniques, one can fractionate mitochondria into four separate enzyme complexes (complexes I, II, III, and IV—see Figure 15.2b), each of which contains part of the entire respiratory sequence, plus a fifth (complex V), which catalyzes ATP synthesis

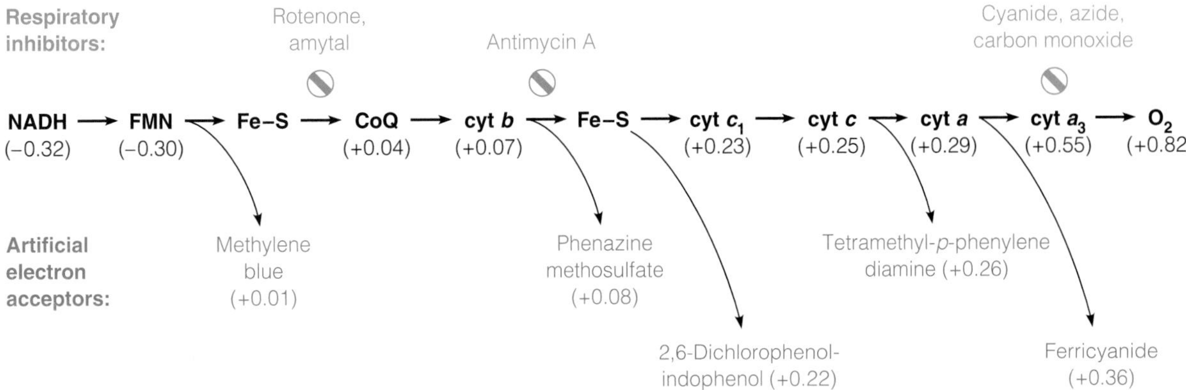

FIGURE **15.9**

Sites of action of some respiratory inhibitors and artificial electron acceptors. This schematic of the respiratory chain from NADH to O_2 shows the sites of action of some useful inhibitors (red) and some artificial electron acceptors (blue). Each acceptor is positioned according to its E_0' value (in parentheses), identifying the most likely site at which an acceptor will withdraw electrons from the respiratory chain when added to mitochondria.

from ADP. Analysis of each complex for the presence of electron carriers, as well as for reactions catalyzed, has helped to establish the currently accepted sequence of carriers. Figure 15.10 provides more detailed information about the protein composition and catalytic activities of each complex. As noted earlier, complexes I, II, III, and IV each accept electrons from a relatively mobile electron carrier (one that is not tightly membrane-bound) and pass electrons along to another mobile carrier. These mobile carriers are NADH, succinate, coenzyme Q, cytochrome *c*, and oxygen. Energy released by the actions of complexes I, III, and IV drives the synthesis of ATP by complex V, which is also called the ATP synthase complex.

The oxidation of reduced coenzyme Q by complex III is more complicated than indicated in Figure 15.10, because at this point, a two-electron donor, $CoQH_2$, is transferring electrons to one-electron acceptors, the cytochromes. In this process, called the **Q cycle**, $CoQH_2$ first transfers one electron to an iron–sulfur center, which is transferred directly to cytochrome c_1 and then to cytochrome *c*. The resultant CoQ semiquinone then transfers another electron to the heme b_{566} component of cytochrome *b* (this heme has a light absorption maximum at 566 nm), and this electron next passes to the heme b_{562} component and then reduces a molecule of oxidized CoQ to the semiquinone. This process is repeated with a second molecule of $CoQH_2$, but this time the electron from the heme b_{562} reduces the CoQ semiquinone to $CoQH_2$ (see margin). The result is that two molecules of $CoQH_2$ become oxidized and one molecule of CoQ becomes reduced, for a net transfer of two electrons being used to reduce two molecules of cytochrome *c*. Because the proton-consuming reactions occur within the matrix, while proton release takes place in the intermembrane space, the Q cycle helps to maintain the proton gradient needed to drive ATP synthesis (see page 540).

QH₂ → QH• → Q; e⁻ → Fe•S → c_1 → *c*; e⁻ → b_{566} → b_{562} → e⁻ + Q → QH• → QH_2

$$2\ CoQH_2 \rightarrow 2\ CoQ + 4H^+ + 4e^-$$
$$2e^- + 2H^+ + CoQ \rightarrow CoQH_2$$
$$\text{Net: } CoQH_2 \rightarrow CoQ + 2H^+ + 2e^-$$

Q cycle

SHUTTLING ELECTRON CARRIERS INTO MITOCHONDRIA

So far we have discussed NADH oxidation within the mitochondrion, without describing how reducing equivalents are generated there. Some NADH is generated within the mitochondrial matrix by enzymes such as malate dehydrogenase. However, NADH generated in glycolysis and by other cytosolic dehydrogenases must transfer reducing equivalents into the mitochondrion, for reoxidation by the respiratory chain. Specific transport systems are required, because the NADH generated by a cytosolic enzyme, such as glyceraldehyde-3-phosphate

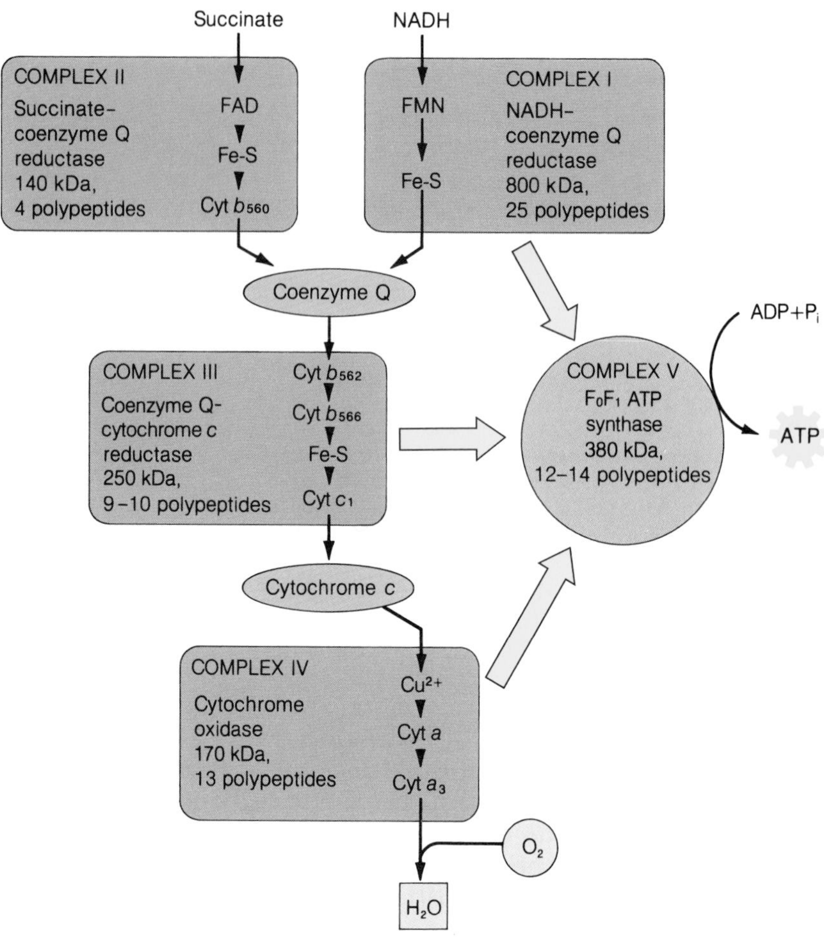

FIGURE **15.10**

Multiprotein complexes in the respiratory assembly. The subscripts for the *b* cytochromes denote their spectral maxima. The two *b* hemes in complex III, identified as cyt b_{562} and cyt b_{566}, are bound to the same polypeptide chain.

Electrons are transported into mitochondria by metabolic shuttles.

dehydrogenase, cannot itself traverse the mitochondrial membrane to be oxidized by the respiratory chain. Therefore, the reducing equivalents must be *shuttled* to respiratory assemblies in the inner mitochondrial membrane, without physical movement of the coenzyme. This process involves the reduction of a substrate by NADH in the cytoplasm, passage of the reduced substrate into the mitochondrial matrix via a specific transport system, reoxidation of that compound inside the matrix, and passage of the oxidized substrate back to the cytoplasm, where it can undergo the same cycle again.

The earliest known shuttle system is the dihydroxyacetone phosphate/glycerol-3-phosphate shuttle, which is particularly active in brain (and in the flight muscle of insects). As shown in Figure 15.11a, dihydroxyacetone phosphate (DHAP) is reduced by NADH in the cytosol, followed by passage of the resultant glycerol-3-phosphate into the mitochondrion, where it is reoxidized by a flavoprotein, glycerol-3-phosphate dehydrogenase, bound at the outer face of the inner membrane. This process involves reduction of FAD, followed by transfer of an electron pair from $FADH_2$ to coenzyme Q, just as intramitochondrial NADH transfers electrons to CoQ. Once dihydroxyacetone phosphate has returned to the cytosol, the net effect has been to transfer two reducing equivalents from cytosolic NADH to mitochondrial $FADH_2$ and from there on up the respiratory chain.

A different shuttle system, particularly active in liver and heart, is the malate/aspartate shuttle, shown in Figure 15.11b. Here a cytosolic isozyme of malate dehydrogenase, together with NADH, reduces oxaloacetate to malate, which passes into the matrix via a specific active transport system in the inner mitochondrial membrane. The malate is then reoxidized by the malate dehydrogenase of the citric acid cycle, which also uses NAD^+. Because oxaloacetate cannot cross the inner

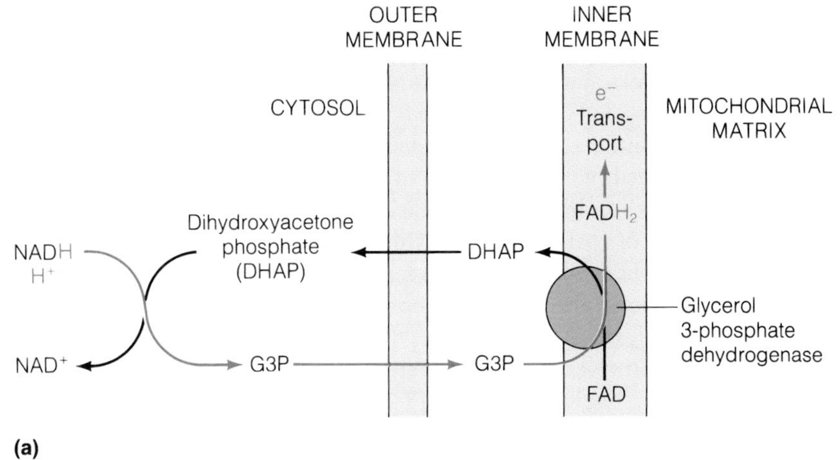

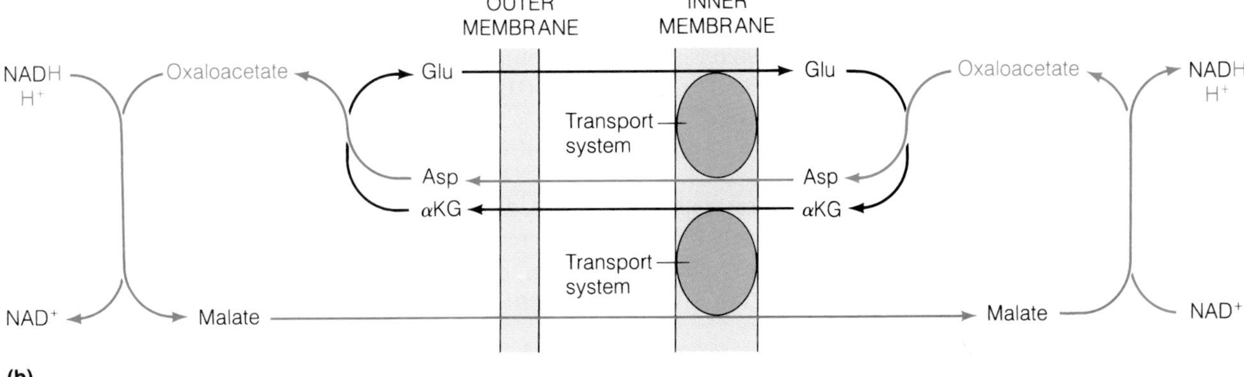

FIGURE **15.11**

Shuttles for transfer of reducing equivalents from cytosol into mitochondria. **(a)** The dihydroxyacetone phosphate/glycerol-3-phosphate shuttle. **(b)** The malate/aspartate shuttle. Glu = glutamate; Asp = aspartate; αKG = α-ketoglutarate.

membrane, it is transaminated to aspartate, which is then transported out for reconversion to oxaloacetate, to begin the cycle anew. Because of the transamination involved, this process requires that α-ketoglutarate be continuously transported out of mitochondria and that glutamate be continuously transported in.

Oxidative Phosphorylation

Having discussed how energy is generated from the mitochondrial oxidation of reduced substrates, we turn now to the question of how that energy is made available for ATP synthesis—in short, the mechanism of oxidative phosphorylation. Mechanistically, oxidative phosphorylation is far more complex than the substrate-level phosphorylation reactions in glycolysis and the citric acid cycle, and until recently this field of inquiry has been one of the most contentious biochemical research arenas. As the late Efraim Racker stated, "Anyone who is not confused about oxidative phosphorylation just doesn't understand the situation." As we shall see, a great deal of the needed understanding has developed in the three decades since Racker wrote these words.

THE P/O RATIO: EFFICIENCY OF OXIDATIVE PHOSPHORYLATION

To measure the efficiency of oxidative phosphorylation, we must determine energy captured in the form of ATP as a fraction of total energy released in the oxidation of a substrate. This measurement became possible once we could measure the quantity of ATP synthesized per mole of substrate oxidized in isolated mitochondria. What we usually measure is the **P/O ratio,** which is the number of molecules of ATP synthesized per pair of electrons carried through electron transport. ATP synthesis is quantitated as phosphate incorporation into ATP, and electron pairs are quantitated as oxygen uptake, in μatoms reduced to water (a μatom of O_2 is 0.5 μmole). Oxygen uptake is determined with a recording oxygen electrode.

Until recently it was widely agreed that the mitochondrial oxidation of NADH proceeds with a P/O ratio of 3 and, hence, that any substrate metabolized via a mitochondrial NAD^+-linked dehydrogenase should yield 3 moles of ATP per mole of NADH oxidized. This belief was consistent with older theories of oxidative phosphorylation, which postulated a direct coupling of ATP synthesis to an exergonic reaction, as seen in the substrate-level phosphorylation reactions of glycolysis.

However, it is now apparent that phosphorylation and oxidation are not directly coupled. As Figure 15.2 suggests, coupling occurs indirectly. This mechanism does not require an integral stoichiometric relationship between reducing equivalents consumed and ATP synthesized. On the other hand, most measurements of P/O ratios do give values close to 3 for oxidation of NAD^+-linked substrates and 2 for succinate, and we shall use those values in our subsequent discussions. With this in mind, we can write a balanced equation for the mitochondrial oxidation of NADH.

$$\text{NADH} + 4\text{H}^+ + \tfrac{1}{2}\text{O}_2 + 3\text{ADP} + 3\text{P}_i \rightleftharpoons \text{NAD}^+ + 4\text{H}_2\text{O} + 3\text{ATP}$$

As discussed earlier, the oxidation of NADH by O_2 has a $\Delta G°'$ of -220 kJ/mol. Reversal of ATP hydrolysis requires 31 kJ/mol under standard conditions. Therefore, coupling the synthesis of three ATPs to oxidation of one NADH traps 93 kJ (3×31), or about 42%, of the energy released (under standard conditions).

OXIDATIVE REACTIONS THAT DRIVE ATP SYNTHESIS

A glance at Figure 15.3, 15.7, 15.9, or 15.10 reveals that the transfer of reducing equivalents from an NAD^+-oxidizable substrate to O_2 involves about a dozen consecutive, linked oxidoreduction reactions. Which of these reactions actually drive ATP synthesis? This question was of paramount concern in the early days of bioenergetics research, when it was thought that ATP synthesis was directly coupled to individual exergonic reactions, as it is in substrate-level phosphorylation. The most straightforward interpretation of the P/O ratio of 3 observed for NADH oxidation is that three of the individual reactions of the respiratory chain are sufficiently exergonic to drive the synthesis of one ATP molecule each. Indeed, three of these reactions have $\Delta G°'$ values exceeding -31 kJ/mol, the minimum barrier that must be overcome (under standard conditions) to make the synthesis of each ATP exergonic (see Figure 15.7). Those three reactions are the oxidation of $FMNH_2$ by coenzyme Q, the oxidation of cytochrome *b* by cytochrome c_1, and the cytochrome oxidase reaction. Each of these reactions was considered to be a **coupling site** for ATP synthesis; that is, each was considered to be a reaction in which ATP synthesis is driven directly by the energy released from that reaction.

Three reactions in the respiratory chain are sufficiently exergonic to drive the synthesis of ATP from ADP.

Because we now know that coupling between oxidation and ATP synthesis is indirect, the concept of coupling sites is an oversimplification. Nevertheless, the concept was useful, because it provided a framework for experiments identifying each of the above three reactions as individually capable of energizing the membrane for ATP synthesis, even when the entire electron transport chain is not operating. With that in mind, let us examine this evidence, which is summarized in Figure 15.12.

First, succinate is oxidized with a P/O ratio of 2, not 3, suggesting the existence of a coupling site associated with NADH dehydrogenase. This was confirmed by blocking electron transport past cytochrome *b* with antimycin A. Ferricyanide was added as an artificial electron acceptor, so that electrons could continue to flow. Under these conditions NAD^+-linked substrates, such as β-

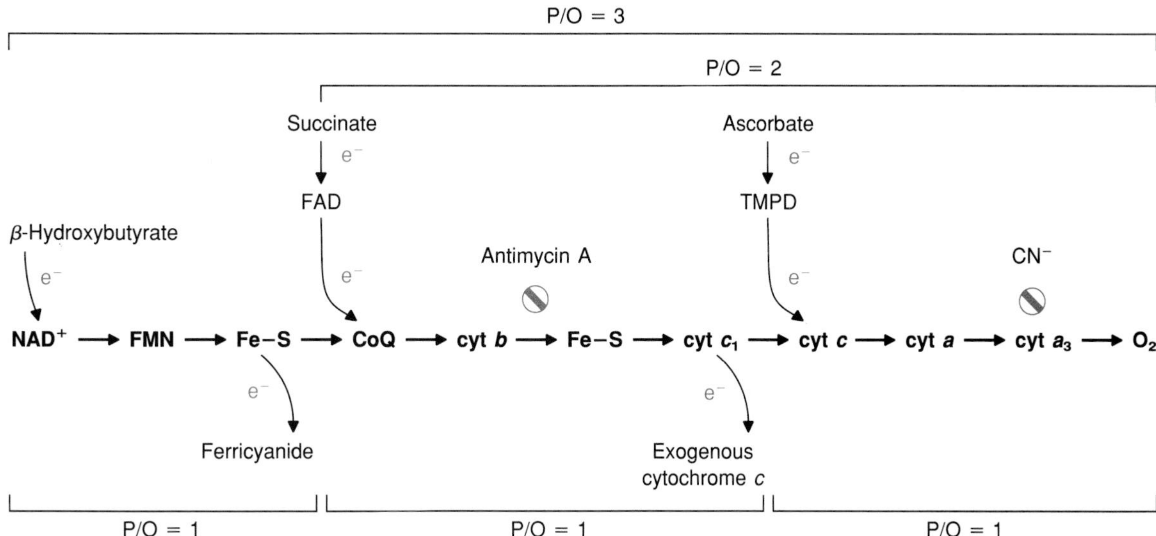

FIGURE **15.12**

Experimental identification of "coupling sites." Electron transport was restricted to particular parts of the chain by use of selected electron donors, electron acceptors, and respiratory inhibitors, as indicated. For each segment of the chain that was thus isolated, P/O ratios were determined, as shown on the figure. Antimycin A was added in the experiment that identified the first coupling site, to block electron flow past cytochrome *b*.

hydroxybutyrate, were oxidized with a P/O ratio of 1, confirming the existence of one site before cytochrome *b*.

Another approach involved an artificial electron donor, ascorbate. In the presence of another electron acceptor, tetramethyl-*p*-phenylene diamine (TMPD—see Figure 15.9), electrons could be supplied to the respiratory chain at cytochrome *c*. These electron carriers reduced cytochrome *c* nonenzymatically, and its oxidation via cytochrome oxidase proceeded with a P/O ratio of 1, thus localizing one coupling site beyond cytochrome *c*. Finally, purified cytochrome *c*, when added to mitochondria, can act as an electron acceptor, withdrawing electrons from the electron transport chain. The further addition of a cytochrome oxidase inhibitor, such as cyanide, forces electrons to exit from the respiratory chain at cytochrome *c*. Under these conditions succinate is oxidized with a P/O ratio of 1, which localizes a site between cytochromes *b* and *c*. In sum, these experiments demonstrate that complexes I, III, and IV are each capable of driving ATP synthesis, but that complex II is not.

THE ENZYME SYSTEM FOR ATP SYNTHESIS

That complexes I, III, and IV could individually drive ATP synthesis was confirmed in elegant reconstitution studies involving complex V, which catalyzes the actual synthesis of ATP from ADP. Let us review the discovery and nature of complex V and then discuss the reconstitution experiments.

When mitochondria are negatively stained with phosphotungstate, electron microscopy reveals that the cristae are covered with knoblike projections on the matrix side, each attached to the inner membrane by a short stalk (Figure 15.13). The knobs are known as **F_1 spheres.** Disruption of mitochondria by sonic oscillation generates fragments of inner membrane, which reseal in the form of closed vesicles. The membrane closes on itself inside out, so the knoblike projections are on the outside. These submitochondrial particles respire and synthesize ATP, just as intact mitochondria do. Efraim Racker and his colleagues showed that treatment

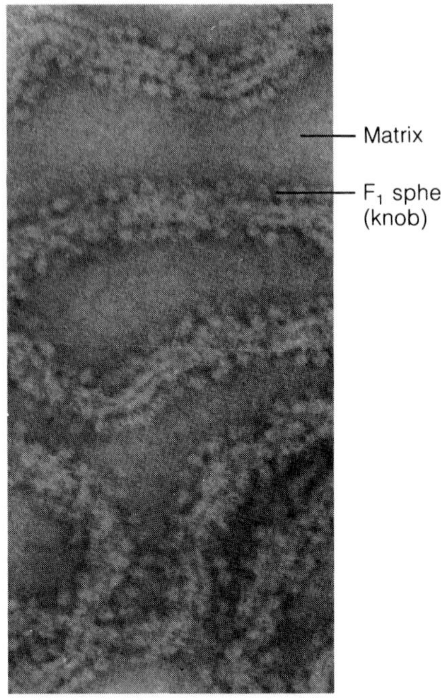

(a)

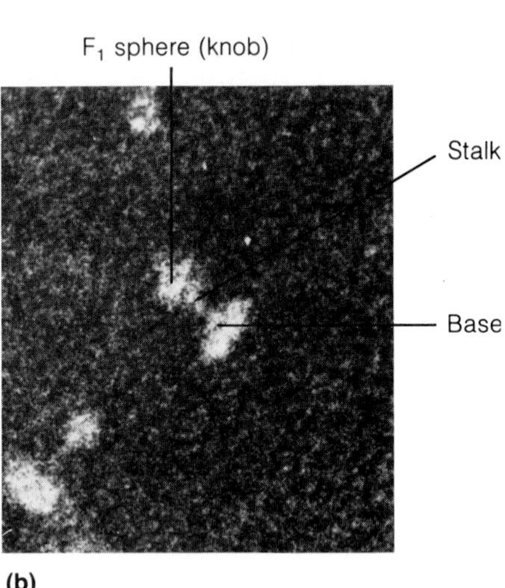

(b)

FIGURE **15.13**

Fine structure of mitochondrial cristae. Negatively stained preparations of the inner membrane show the F_1 spheres as "knobs" projecting from the cristae. **(a)** A portion of bovine heart mitochondrial inner membrane, showing knoblike projections along the matrix side of the membrane. **(b)** Purified preparations of knobs from rat liver mitochondria. The knob is attached by a short stalk to a base, which is embedded in the inner membrane of intact mitochondria.

Courtesy of A. Tzagoloff, *Mitochondria* (New York: Plenum, 1982).

of these vesicles with trypsin or urea caused the knobs to dissociate from the vesicles. After centrifugation to separate the "stripped" vesicles from the knobs, the vesicles could still oxidize substrates and reduce oxygen, but no ATP was synthesized. When knobs were added back to the vesicles, there was substantial reconstitution of particles that could then catalyze ATP synthesis as a result of the oxidation of exogenous substrates.

Because readdition of the knobs recoupled ATP synthesis to electron transport, the knobs were originally called **coupling factors.** Purification of the knobs attached to the underlying stalks revealed a large multiprotein aggregate consisting of more than a dozen polypeptide chains, as schematized in Figure 15.14. The entire structure, called the $\mathbf{F_0F_1}$ **complex,** consists of the knob, the stalk to which it is attached, and an attached complex that is embedded in the inner membrane. The stalked knob is called F_1, and the base is called F_0. The F_1 complex consists of five proteins, designated α, β, γ, δ, and ε, with a subunit stoichiometry of $\alpha_3\beta_3\gamma\delta\varepsilon$. The F_0 complex contains at least three protein subunits, a, b, and c, with a stoichiometry of ab_2c_{12}.

The entire F_0F_1 complex has an ATPase activity in vitro, as does factor F_1 alone; the ATPase activity was assumed to represent the reverse of the true physiological reaction, namely, ATP synthesis. One of the proteins in the F_0 complex is the site of binding of the ATP synthesis inhibitor **oligomycin** (discussed further on page 543). For this reason, some workers call this structure F_o (the letter o stands for oligomycin), rather than F_0. The ATPase activity, sensitivity to oligomycin, and results of the reconstitution experiments confirm that the role of the F_0F_1 complex (also called complex V) is to synthesize ATP. The action of this amazing structure, a molecular "rotary engine," is discussed in more detail on page 546.

Continuing with his reconstitution experiments, Racker found that isolated submitochondrial respiratory complexes (I, II, and so on) could be reconstituted by sonic oscillation into artificial membranes (liposomes) containing purified phospholipids. When F_0F_1 complex was included in the sonication mixture, it was also incorporated into the vesicles. In this case each preparation had the electron transport properties of the original complex, plus a phosphorylation activity. For reconstituted complexes I, III, and IV, the P/O ratio was 1 in each case. For example, reconstituted complex III could transfer electrons from coenzyme Q to cytochrome *c*, with concomitant synthesis of 1 mole of ATP per pair of reducing equivalents. This type of evidence showed that respiratory complexes I, III, and IV each contain one coupling site. By contrast, complex II (succinate dehydrogenase) showed no ATP synthesis, confirming the absence of a coupling site in this complex.

The F_0F_1 complex contains a proton channel and the enzyme that synthesizes ATP.

MECHANISM OF OXIDATIVE PHOSPHORYLATION: CHEMIOSMOTIC COUPLING

What is the actual mechanism by which energy released from respiration is harnessed to drive the synthesis of ATP? Even though most biochemists now agree on the overall outlines of the answer, important questions remain.

Initial mechanistic interest focused on a **chemical coupling** mechanism in which the energy released would be used directly for the synthesis of an energy-rich intermediate, as seen in substrate-level phosphorylations. However, such activated intermediates have never been demonstrated in oxidative phosphorylation.

Although other models have been considered, there is now widespread acceptance of a model involving **chemiosmotic coupling,** proposed in 1961 by British biochemist Peter Mitchell. Although this model was resisted at first, overwhelming evidence has now accumulated in its support, and Mitchell's achievements were recognized in 1978 with a Nobel Prize. In its most basic form, this model proposes that energy from electron transport drives an active transport system, which somehow pumps protons out of the mitochondrial matrix into the intermembrane space. This action generates an electrochemical gradient for protons, with a lower pH value outside the inner mitochondrial membrane than inside. The protons on the outside have a thermodynamic tendency to flow back in, so as to equalize pH on both sides of the membrane. Another way to state this is that free energy must be expended to maintain the proton gradient. When protons do flow back into the matrix, that energy is dissipated, some of it being used to drive the synthesis of ATP.

To understand the chemiosmotic theory in more detail, recall that some but not all of the reactions of electron transport transfer hydrogen ions (protons) as well as electrons. These reactions include the dehydrogenations of NADH, $FMNH_2$, $FADH_2$, and reduced coenzyme Q. Mitchell proposed that the enzymes catalyzing these dehydrogenations are asymmetrically oriented in the inner membrane, so that protons are always taken up from inside the matrix and released in the intermembrane space. Figure 15.15 shows how this process might occur. This **proton pumping** by respiratory proteins results in conversion of the energy of respiration to osmotic energy, in the form of an **electrochemical gradient**—or a gradient of chemical concentration that establishes an electrical potential. The energy released from discharging this gradient can be coupled with the phosphorylation of ADP to ATP, with no isolatable intermediates being formed. This process involves the F_0F_1 complex (complex V). The F_0 portion of the complex spans the inner membrane and is thought to contain a specific channel for return of protons to the mitochondrial matrix. The free energy released as H^+ traverses this channel to return to the matrix is harnessed to drive the synthesis of ATP, catalyzed by the F_1 component of the complex.

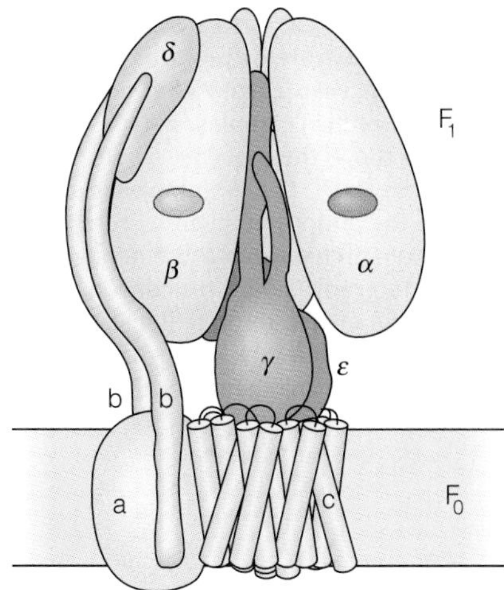

FIGURE **15.14**

Structure of the F_0F_1 complex. The F_0F_1 complex, also called ATP synthase or complex V, contains an F_1 knob projecting into the mitochondrial matrix and connected by a stalk to the F_0 base. The F_1 knob contains three $\alpha\beta$ dimers, forming what appear to be six segments of an orange spaced about the stalk, which is made up of γ and ε proteins; δ is attached to protein b of the F_0 base, which also contains proteins a and c plus others. This abc complex is called a "stator," a structure that opposes the rotation of the $\alpha\beta\gamma\delta\varepsilon$ "rotor." This model is based upon the crystal structure of the bovine mitochondrial F_1 complex.

Reprinted with permission from W. Junge, H. Lill, and S. Engelbrecht, *Trends Biochem. Sci.* (1997) 22:420. © 1997 with permission of Elsevier Science.

Chemiosmotic coupling refers to the establishment of a transmembrane proton gradient to drive endergonic processes.

A CLOSER LOOK AT CHEMIOSMOTIC COUPLING: THE EXPERIMENTAL EVIDENCE

Let us now explore the experimental evidence for chemiosmotic coupling in some detail—partly because of its importance to an understanding of oxidative phosphorylation and partly because it provides insight into other biological processes, including active transport and photosynthesis.

Membranes Can Establish Proton Gradients

When it became possible to measure changes in pH and electrical potential across mitochondrial membranes, it became clear that mitochondria can pump protons from the matrix to the intermembrane space. In fact, the pH value outside an actively respiring mitochondrion is about 1.4 units lower than in the matrix. The pH gradient also generates an electrical potential of about 0.14 V across the membrane, because of the net movement of positively charged protons outward across the inner membrane. The pH gradient and the membrane potential both contribute to an *electrochemical H^+ gradient,* or **proton motive force (pmf),** as shown here and also in Chapter 10 [equation (10.2)].

$$\underset{\text{Electrochemical H}^+\text{ gradient}}{\Delta\mu_H} = \underset{\text{Membrane potential}}{\Delta\psi} - \underset{\text{pH gradient}}{2.3RT\,\Delta pH/F} \qquad (15.7)$$

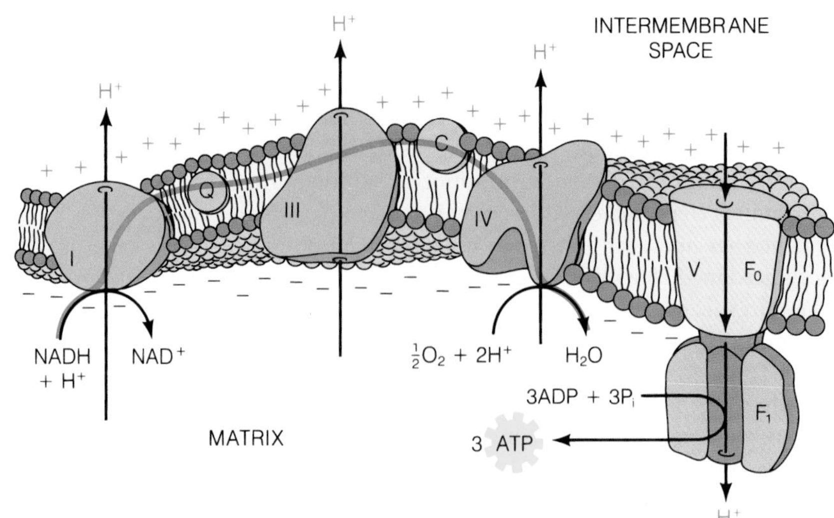

FIGURE **15.15**

Vectorial transport of protons by complexes of the respiratory chain. This depiction of protein complexes in the inner membrane shows the sequence of electron carriers from NAD^+ to O_2. Complex II is not included. The protons pumped by complexes I, III, and IV are probably not those removed from electron carriers during oxidation. Proton reentry to the matrix, through the F_0 channel (complex V), provides the energy to drive ATP synthesis. Stoichiometric relationships between oxidation reactions and proton pumping have not been established.

$\Delta\mu_H$ is also called Δp, the proton motive force, or pmf. ΔpH has a negative value (−1.4) because it is measured from inside to outside. Thus, the contribution of the pH gradient is positive: $-2.3RT(-1.4)/F$, which equals +0.084 V, and the $\Delta\mu_H$ is 0.14 V + 0.084 V, or +0.224 V. From the relationship between standard reduction potential and $\Delta G^{\circ\prime}$ [see equation (15.1)], we see that this value corresponds to a standard free energy change of about 21 kJ per mole of protons, which would generate about 42 kJ per pair of reducing equivalents. This amount of energy is in the right range to drive the synthesis of 1 mole of ATP, although it is not universally accepted that only two protons are pumped at each coupling site. In fact, recent evidence suggests that the number of protons translocated per pair of reducing equivalents is as high as 12.

Comparable experiments have shown that electrochemical proton gradients are used in energy transactions other than oxidative phosphorylation. Bacterial membranes use proton pumping to transduce energy both for oxidative phosphorylation and for driving flagellar motors that allow movement of the cell. Proton pumping across the chloroplast thylakoid membrane drives ATP synthesis in photophosphorylation (see Chapter 17). Proton gradients also drive active transport (see Chapter 10), as well as heat production and the synthesis of NADPH by transhydrogenase.

An Intact Inner Membrane Is Required for Oxidative Phosphorylation

When the physical continuity of the membrane is interrupted—for example, by sonic oscillation—the resultant particles can carry out electron transport but not ATP synthesis. The necessity of a structurally intact membrane for maintenance of a membrane potential is consistent with the idea that a proton gradient is essential for oxidative phosphorylation.

Key Electron Transport Proteins Span the Inner Membrane

If the respiratory proteins are to serve as proton pumps, then the electron carriers that carry protons, such as NADH dehydrogenase, should be in contact with both the inner and the outer sides of the membrane. Moreover, these carriers should be asymmetrically oriented in the membrane, to account for transport in one direction—outward. Asymmetric orientation has been demonstrated by the use of agents that react with respiratory proteins but cannot themselves traverse the membrane, such as antibodies, proteolytic enzymes, or labeling reagents. Treatment of intact mitochondria with such reagents allows detection of proteins located at the outer surface of the inner membrane, whereas reaction with membrane vesicles allows access to the inner, or matrix, side. Such approaches have shown, for example, that the cytochrome oxidase complex (complex IV) binds to cytochrome *c* only on the cytosolic side. Moreover, 9 of the 13 subunits of the complex can be labeled from only one side or the other, indicating asymmetric placement of the complex in the membrane. Similar findings have now been made for the proteins of complexes I, II, and III.

NO_2, NO_2, OH

2,4-Dinitrophenol (DNP)

F_3CO—C₆H₄—NH—N=C(CN)(CN)

Trifluorocarbonylcyanide phenylhydrazone (FCCP)

Uncouplers Act by Dissipating the Proton Gradient

A class of compounds, exemplified by **2,4-dinitrophenol (DNP)** and **trifluorocarbonylcyanide phenylhydrazone (FCCP)** are called uncoupling agents, or **uncouplers.** Uncoupling agents, when added to mitochondria, block ATP synthesis, while permitting continued electron transport along the respiratory chain to O_2. The pK_a of the phenolic hydroxyl group in DNP is such that it is normally dissociated at intracellular pH. However, a DNP molecule that approaches the inner membrane from the outside becomes protonated, because of the lower pH value in this vicinity. This protonation increases the hydrophobicity of DNP, allowing it to diffuse into the membrane and by mass action to pass through. Once inside the matrix, the higher pH causes the phenolic hydroxyl to deprotonate. Thus, the uncoupler has the effect of transporting H^+ back into the matrix, bypassing the F_0 proton channel and thereby preventing ATP synthesis.

The action of uncouplers provides strong support for chemiosmotic coupling.

Another phosphorylation inhibitor, the antibiotic **oligomycin,** acts quite differently. By binding to a specific protein of the F_0 complex, it blocks the flow of protons through the F_0 proton channel and hence inhibits oxidative phosphorylation directly. Proton pumping by the respiratory carriers is inhibited as well, but this is a secondary consequence of the inhibition of ATP synthesis.

In summary, an inhibitor such as DNP dissipates the proton gradient by transporting protons into mitochondria at sites other than the F_0 proton channel, whereas oligomycin blocks the F_0 channel so that the pH gradient cannot be dissipated at all.

Extensive data on the transport of ions other than H^+ confirm that a functionally intact membrane is essential to oxidative phosphorylation. The antibiotic *valinomycin* (see page 339 in Chapter 10) is an example of an **ionophore** ("ion carrier"). This lipid-soluble compound forms a specific complex with potassium ion. Because the complex is lipophilic and can diffuse into the membrane, just as protonated DNP does, valinomycin brings about the transport of K^+ through the inner membrane in much the same sense that DNP transports protons. Valinomycin acts by decreasing the $\Delta\psi$ (membrane potential) component of the pmf, without a direct effect on the pH gradient. Another antibiotic, **nigericin,** acts as a K^+/H^+ **antiport;** it carries H^+ in one direction, coupled with the reverse transport of K^+. Thus, nigericin dissipates the ΔpH component of the pmf, with little effect on $\Delta\psi$. Neither compound alone is a particularly effective uncoupler of oxidative phosphorylation, but in combination they collapse both elements of the pmf, and ATP synthesis is effectively inhibited.

Antibiotics such as valinomycin and nigericin interfere with oxidative phosphorylation or with maintenance of the proton gradient.

Generation of a Proton Gradient Permits ATP Synthesis Without Electron Transport

Andre Jagendorf, while studying photosynthetic ATP production, provided important evidence for chemiosmotic coupling in the chloroplast. As presented in Chapter 17, the chloroplast couples light energy to ATP synthesis. Jagendorf showed that ATP synthesis can proceed in the chloroplast in the absence of electron transport, so long as a proton gradient is present. Chloroplasts were incubated at pH 4 for several hours and then quickly transferred to a buffer at pH 8. Thus, like the situation in intact cells, the inside of the organelle was at a lower pH than the outside (chloroplast membranes pump protons *inward,* not outward). Addition of ADP and P_i to these chloroplasts generated a burst of ATP synthesis, simultaneous with dissipation of the pH gradient. Similar results have now been observed with mitochondria. These experiments show that the

Chloroplasts use light energy to drive protons inward, establishing a proton gradient for ATP synthesis.

establishment of a proton gradient, even without a corresponding energy input, suffices to drive the synthesis of ATP.

A dramatic variation on Jagendorf's experiment involved a membrane protein, *bacteriorhodopsin,* from the photosynthetic bacterium *Halobacterium halobium.* Bacteriorhodopsin pumps protons when the bacteria are supplied with light. Bacteriorhodopsin was isolated in a membrane-free form and then incorporated into synthetic vesicles, along with isolated liver mitochondrial F_0F_1 ATPase in the right orientation. When these vesicles were illuminated, ATP synthesis occurred, showing that phosphorylation can occur as a direct consequence of the formation of a proton gradient.

STRUCTURAL INSIGHTS INTO OXIDATIVE PHOSPHORYLATION

The mid-1990s saw three heroic achievements in x-ray crystallography, which enormously expanded our mechanistic understanding of electron transport and oxidative phosphorylation—the structures of complexes III, IV, and V from bovine heart mitochondria. The structure of complex III (11 subunits, 248 kDa) provided strong support for the model of the Q cycle, shown on page 535. As shown in Figure 15.16, one of the two hemes of cytrochrome *b* (b_L in the figure, identical to b_{566}) lies close enough to the iron–sulfur center that binding of $CoQH_2$ between both sites can lead to its transfer of one electron to each of the two carriers. Once b_L has passed its electron on to the nearby b_H (identical to b_{562}), a molecule of CoQ can bind at a second site and be reduced by b_H.

In 1996 the structure of bovine complex IV (cytochrome oxidase) was described. This 13-subunit, 200-kDa complex, modeled in Figure 15.17, consists largely of transmembrane helices. The two hemes, *a* and a_3, lie at nearby sites in the interior of the membrane, with associated coppers (Cu_A and Cu_B). Both the coppers and the heme irons undergo cyclic oxidation and reduction during the

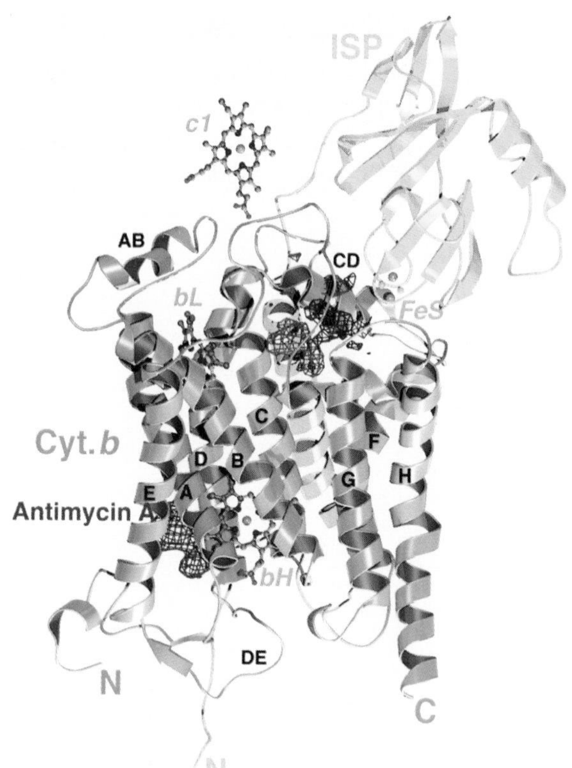

FIGURE **15.16**

Structure of the redox components of complex III—cytochrome *b*, the iron–sulfur protein (ISP), and the heme of cytochrome c_1. The figure shows one monomer of a dimeric complex. Transmembrane helices in cytochrome *b* (green) are identified with letters A through H, and some of the connecting loops (e.g., AB) are also labeled. Binding of the $CoQH_2$ that undergoes the initial oxidation occurs between b_L and the iron–sulfur center (FeS) on ISP (shown in gold). In the complex, cytochrome c_1, which accepts electrons from FeS, binds near the cytosolic face. Not shown is a connection between the ISP transmembrane helix and cytochrome *b* of the second monomer. The antimycin A binding site is shown in purple.

Courtesy of J. Deisenhofer from H. Kim, D. Xia, and J. Deisenhofer, *Proc. Natl. Acad. Sci.* (1995) 95:8026–8033, with permission of PNAS.

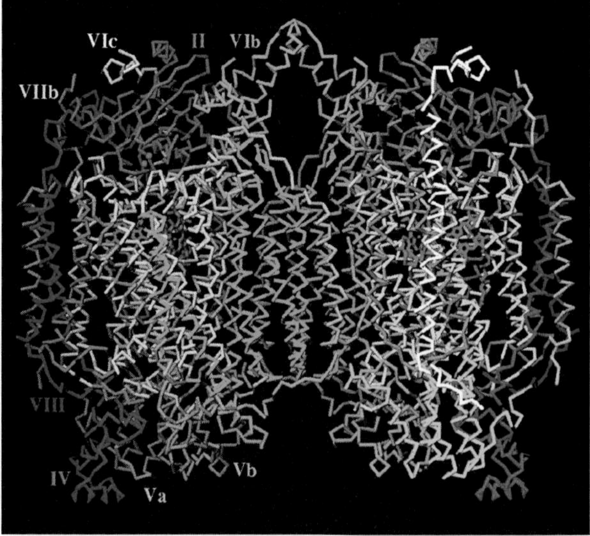

FIGURE **15.17**

The Cα-backbone trace of a dimer of bovine heart cytochrome oxidase. Each subunit is shown in a different color. The two hemes, with associated coppers, are shown in red.

Courtesy of S. Yoshikawa from T. Tsukihara et al., *Science* (1996) 272:1136, with permission from *Science.*

passage of electrons from cytochrome *c* to oxygen. The initial oxidation of cytochrome *c* is carried out by Cu_A, with the electron transferred to heme *a*, and then to Cu_B and the iron of heme a_3, which are so close together (0.5 nm) as to constitute a "binuclear center," functioning as a single unit in electron transfer. At this site O_2 undergoes its reduction to water. The overall reaction catalyzed by complex IV is as follows.

$$4 \text{ cyt } c\text{–}Fe^{2+} + 8H^+ \text{ (inside)} + O_2 \longrightarrow 4 \text{ cyt } c\text{–}Fe^{3+} + 2H_2O + 4H^+ \text{ (outside)}$$

Clearly, the most fundamental question about the function of this complex is the mechanism by which four protons per molecule of oxygen reduced are pumped from the matrix to the intermembrane space. At least two processes seem to be involved. A recent analysis of complex IV in both oxidized and reduced forms provided a clue to one of these processes when it revealed significant movement of an aspartate residue, D51, accompanying oxidation. As schematized in Figure 15.18, the final stage in ejection of a proton to the intermembrane space (top in the figure) involves rotation of the protonated carboxyl group of D51 and its deprotonation, concomitant with binding of a water molecule, produced by

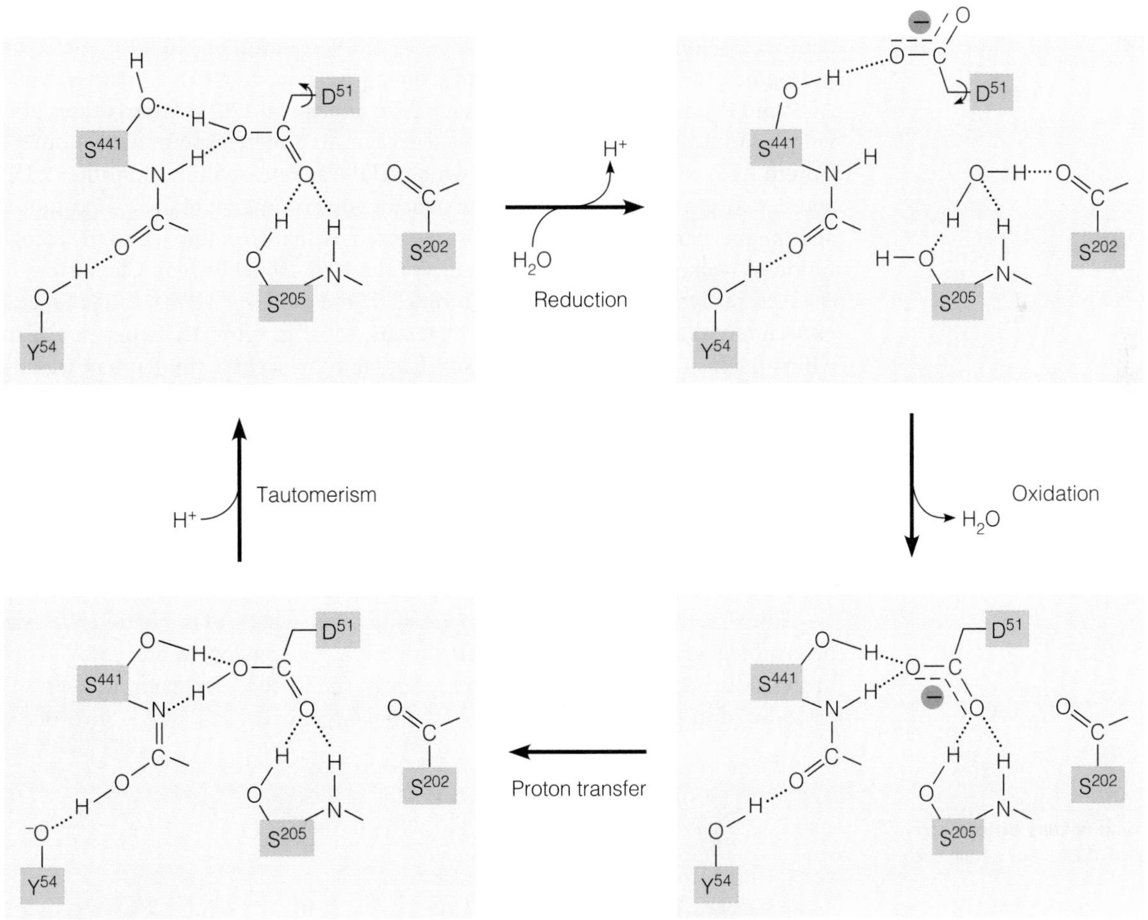

FIGURE **15.18**

A model for the final stages in proton pumping by cytochrome oxidase. This model is suggested by structural analysis of reduced and oxidized cytochrome oxidase. For details see text.

From S. Yoshikawa et al., *Science* (1998) 280:1727, fig. 4. Reprinted with permission.

reduction of O_2, to serines 202 and 205. Release of that water causes the aspartate carboxyl to rotate back and undergo protonation from a tyrosine, Y54, that is also part of the proton transfer network.

In 1994 the publication by John Walker's group of the crystal structure of the 371-kDa F_1 component of complex V clarified the mechanism by which the passage of protons through the F_0F_1 complex drives the synthesis of ATP. As was suggested in Figure 15.14, the knob of the F_1 complex contains three identical $\alpha\beta$ dimers. However, the crystal structure showed significant structural differences among the three. Most significant, the nucleotide-binding site of one contained ADP, while another contained AMP-PNP (a nonhydrolyzable ATP analog added during crystallization), and the third was empty.

The N-terminal regions of each dimer formed β barrels whose relationship to the interior γ subunit suggested bearings in a rotating engine. All of this supported a mechanism proposed by Paul Boyer some years earlier, in which rotation of the γ subunit causes sequential conformational changes in the $\alpha\beta$ assemblies. As shown in Figure 15.19, the three conformations of the nucleotide-binding site are loose (L), tight (T), and open (O). In this model, ATP is formed spontaneously, by condensation of ADP and P_i. The energy-dependent step is not ATP *synthesis* but rather its *release* from a tight-binding site. This release is brought about by the energy-dependent rotation of γ, which drives the conformational changes in the $\alpha\beta$ assemblies. This release occurs simultaneously with binding of ADP and P_i to another assembly. Thus, in step 1 of Figure 15.19, rotation of γ by 120°, fueled by proton passage through channels in F_0, opens site T, leading to ATP release, while ADP and P_i bind to an open site. In step 2, an additional 120° rotation causes previously bound ADP + P_i, now bound at a T site, to undergo spontaneous conversion to ATP, while the O site from which ATP was released binds another ADP and P_i to begin the process anew. The structure determination of the F_0F_1 complex and the mechanism of ATP synthesis suggested by that structure led to the recognition of Walker and Boyer as recipients of the 1997 Nobel Prize in Chemistry.

Is F_1 really a rotating engine, as implied by the model of Figure 15.19? Several experimental approaches indicate that it is. Most graphic are experiments in which the β subunits were immobilized and a fluorescent actin filament was attached to the γ protein (see Figure 15.20). When this structure was examined by fluorescence microscopy, the addition of ATP could be seen to stimulate the rotation of the fluorescent probe. Analysis showed the rotation to be counterclockwise (when viewed from the membrane) and to occur in discrete steps of 120°, just as predicted from the model. By contrast, when the complex operates as it does in vivo, in the direction of ATP synthesis, rotation is assumed to be in the opposite (clockwise) direction.

Note that the crystal structure supporting the mechanism in Figure 15.19 was determined with F_1 bound to a nonhydrolyzable ATP analog, to prevent ATP hydrolysis during crystallization. More recently, the F_1 complex from rat liver was crystallized in the presence of ATP, but with Mg^{2+} absent so that ATP hydrolysis

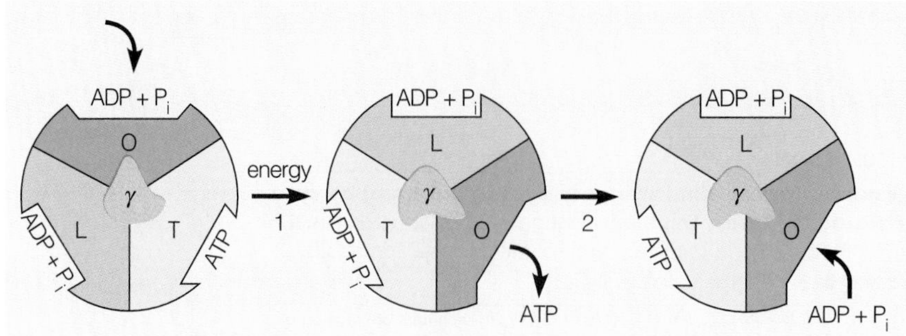

FIGURE **15.19**

F_1 ATP synthase as a rotary engine driving the synthesis of ATP. For details see text.

From Y. Zhou, T. M. Duncan, and R. L. Cross, *Proc. Natl. Acad. Sci. USA* (1997) 94:10583. Reprinted with permission of the PNAS.

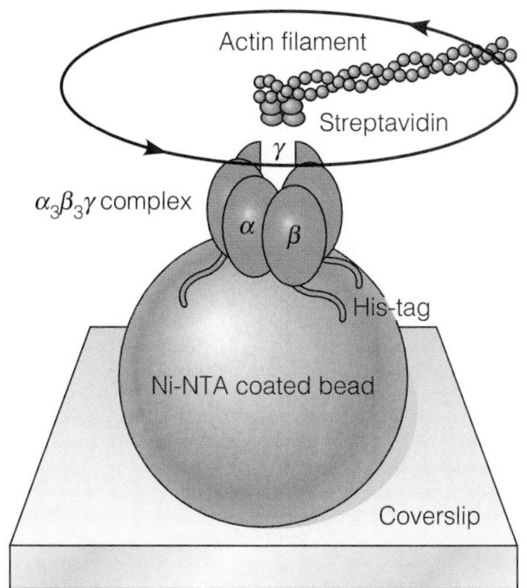

FIGURE **15.20**

The experimental system that permits observation of rotation in the F_1 component of F_0F_1 ATP synthase. The cloned gene encoding the F_1 β subunit was modified by adding a sequence that encodes an oligohistidine sequence that binds to a nickel-coated bead ("Ni-NTA coated bead") in the conformation shown. Thus, the F_1 complex was immobilized on the bead. Streptavidin is a protein used to couple fluorescent-tagged actin to the γ subunit. Fluorescence microscopic examination showed that the actin molecule was rotating, which proved that the γ subunit itself was rotating following addition of ATP and its subsequent hydrolysis.

Reprinted with permission from R. Yasuda et al., *Cell* (1998) 93:1118.

would not occur. This structure modified the above model somewhat, by showing that each of the three $\alpha\beta$ assemblies binds a nucleotide in each of the three 120° phases of a complete rotation. Thus, ATP release is followed immediately by ADP binding, before the γ subunit has rotated to its next site.

RESPIRATORY STATES AND RESPIRATORY CONTROL

Like any metabolic process, oxidative phosphorylation can occur only in the presence of adequate quantities of its substrates. It is controlled not by allosteric mechanisms but simply by substrate availability. Those substrates include ADP, P_i, O_2, and an oxidizable metabolite that can generate reduced electron carriers—NADH and/or $FADH_2$. Under different metabolic conditions any one of these four substrates can limit the rate of oxidative phosphorylation.

The dependence of oxidative phosphorylation on ADP reveals an important general feature of this process: *Respiration is tightly coupled to the synthesis of ATP.* Not only is ATP synthesis absolutely dependent on continued electron flow from substrates to oxygen, but electron flow in normal mitochondria occurs only when ATP is being synthesized as well. This regulatory phenomenon, called **respiratory control,** makes biological sense, because it ensures that substrates will not be oxidized wastefully. Their utilization is controlled by the physiological need for ATP.

The rate of electron transport is limited by the availability of ADP for conversion to ATP.

In most aerobic cells the level of ATP exceeds that of ADP by 4- to 10-fold. Thus, it is convenient to think of respiratory control as a dependence of respiration on ADP as a substrate for phosphorylation. If the energy demands on a cell cause ATP to be consumed at high rates, the resultant accumulation of ADP will stimulate respiration, with concomitant activation of ATP resynthesis. Conversely, in a relaxed and well-nourished cell, ATP accumulates at the expense of ADP, and the depletion of ADP limits the rate of both electron transport and its own phosphorylation to ATP. Thus, the energy-generating capacity of the cell is closely attuned to its energy demands. The energy state of the cell also has profound effects on the structure of the mitochondrion. In actively respiring mitochondria the inner membrane seems to fold on itself, leaving a greatly enlarged intermembrane space (Figure 15.21).

Experimentally, respiratory control is demonstrated by following oxygen utilization in isolated mitochondria (Figure 15.22). In the absence of added substrate or ADP, oxygen uptake, caused by oxidation of endogenous substrates, is slow. Addition of an oxidizable substrate, such as glutamate or malate, has but a small effect on the respiration rate. If ADP is then added, however, oxygen uptake

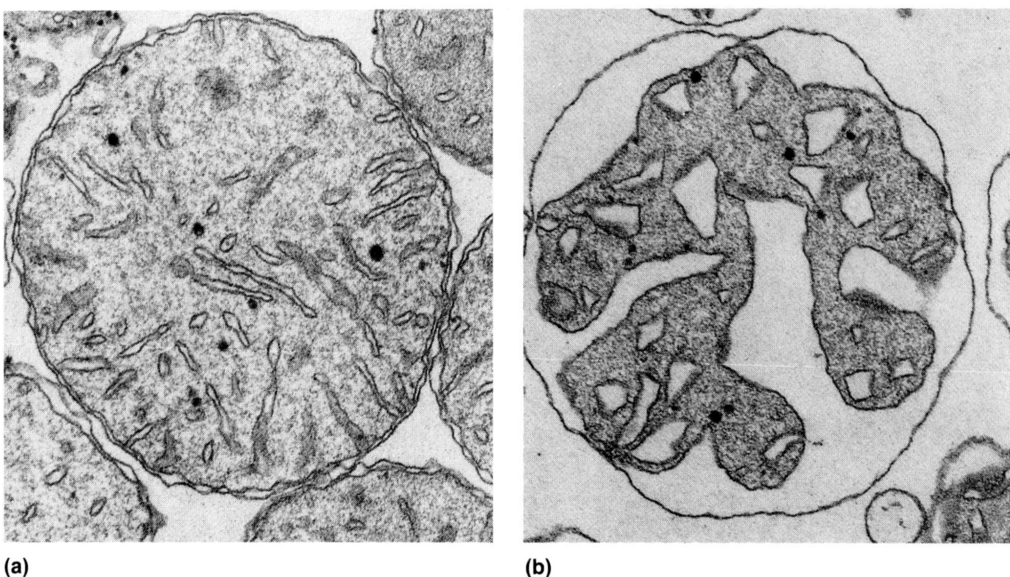

FIGURE 15.21

Morphological differences between resting and active mitochondria. These electron micrographs show the morphological changes in rat liver mitochondria as they go from a resting state to a state of active respiration. **(a)** In resting mitochondria most of the internal space is occupied by matrix. **(b)** In the high-energy state the intermembrane space expands as the matrix volume decreases.

Courtesy of A. Tzagoloff, *Mitochondria* (New York: Plenum, 1982).

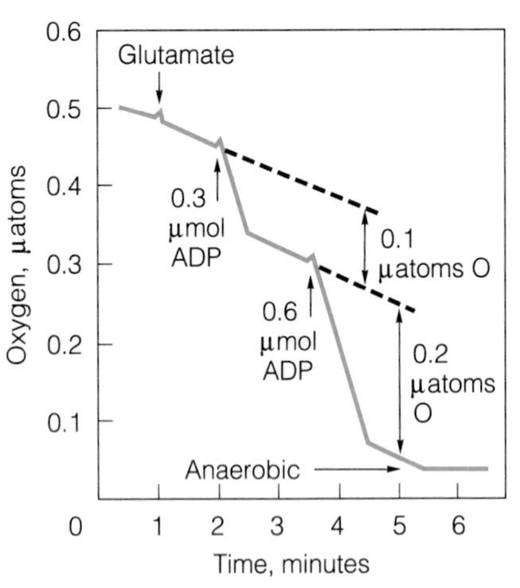

FIGURE 15.22

Experimental demonstration of respiratory control. Oxygen uptake is monitored in carefully prepared mitochondria. The addition of an exogenous oxidizable substrate (glutamate) stimulates respiration only slightly, unless ADP is added as well. Both ADP additions represent limiting amounts; the second addition is twice the amount of the first, to show that the magnitude of oxygen uptake is stoichiometric. The slow oxygen uptake at the beginning results from endogenous substrates in the mitochondrion. ADP stimulates respiration only until all of the ADP has been converted to ATP. Oxygen uptake is recorded in μatoms, because one pair of electrons reduces one atom of O, not one molecule of O_2.

proceeds at an enhanced rate until all of the added ADP has been converted to ATP, and then oxygen uptake returns to the basal rate. This stimulation of respiration is stoichiometric; that is, addition of twice as much ADP causes twice the amount of oxygen uptake at the enhanced rate. If excess ADP is present instead of oxidizable substrate, the addition of substrate in limiting amounts will stimulate oxygen uptake until the substrate is exhausted.

Maintenance of respiratory control depends on the structural integrity of the mitochondrion. Disruption of the organelle causes electron transport to become uncoupled from ATP synthesis. Under these conditions, oxygen uptake proceeds at high rates even in the absence of added ADP. ATP synthesis is inhibited, even though electrons are being passed along the respiratory chain and used to reduce O_2 to water. Before carrying out experiments with freshly isolated mitochondria, biochemists usually ascertain that their mitochondria are tightly **coupled,** by determining rates of oxygen uptake in the presence and absence of added ADP. In carefully prepared mitochondria the ratio of these two O_2 uptake rates may be as high as 10. By contrast, aged or disrupted mitochondria may yield ratios as low as 1, showing an absence of coupling. Uncoupling of respiration from phosphorylation can also be achieved chemically. As noted earlier, uncouplers such as DNP or FCCP act by dissipating the proton gradient. Addition of an uncoupler to mitochondria stimulates oxygen utilization even in the absence of added ADP. No phosphorylation occurs under these conditions because there is no ADP to be phosphorylated.

Another group of compounds, exemplified by oligomycin, act as inhibitors of oxidative phosphorylation. Addition of oligomycin to actively respiring, well-coupled mitochondria inhibits both oxygen uptake and ATP synthesis, as shown in Figure 15.23. However, no direct inhibition of electron transport occurs, as shown by the fact that subsequent addition of an uncoupler such as DNP greatly stimulates oxygen uptake. As mentioned earlier, oligomycin acts by direct inhibition of the F_0F_1 ATP synthase.

Under some natural conditions the ability to uncouple respiration from phosphorylation is highly desirable. Many mammals, particularly those that are born hairless, those that hibernate, and those that are cold-adapted, have special needs for maintenance of body temperature. Such animals have a special tissue, called brown fat, in the neck and upper back. Mitochondria in this tissue are specialized to generate heat from fat oxidation, uncoupled from phosphorylation. These mitochondria are especially rich in respiratory electron carriers, particularly cytochromes, which give this tissue its brown color. A comparable phenomenon is seen in the plant world among species that emerge in early spring, often when the ground is still covered with snow. The floral spike of the skunk cabbage is a particularly dramatic example; this tissue can maintain a temperature some 10° to 25° above ambient temperature, by uncoupling oxidation from phosphorylation.

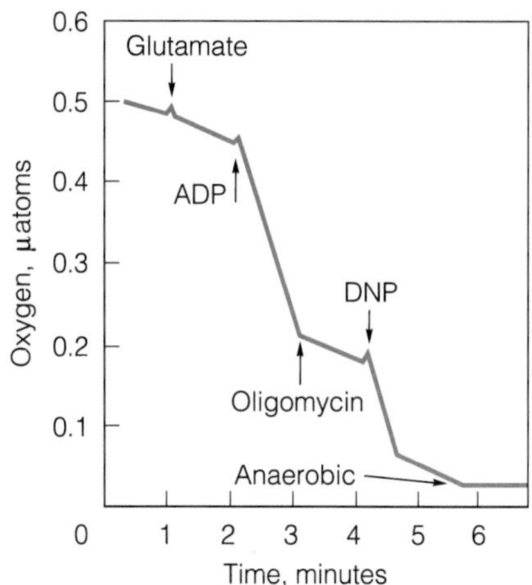

FIGURE **15.23**

Effects of an inhibitor and an uncoupler on oxygen uptake. The plot shows the results of an experiment in which an inhibitor and an uncoupler of oxidative phosphorylation were added to a mixture of isolated mitochondria, an oxidizable substrate (glutamate), and excess ADP. The addition of oligomycin inhibits phosphorylation and consequently slows respiration. Dinitrophenol (DNP) uncouples respiration from phosphorylation, so that O_2 uptake is stimulated even in the presence of oligomycin.

MITOCHONDRIAL TRANSPORT SYSTEMS

Whereas the mitochondrial outer membrane is freely permeable to many substances, the permeability of the inner membrane is severely limited. The importance of this selective permeability can be seen from our discussions of electrochemical gradients and the shuttle systems used to transport reducing equivalents into the mitochondrion. We must also consider substrate transport, including the inward transport of intermediates for oxidation in the citric acid cycle, the export of intermediates used for biosynthesis in other cell compartments, and the exit of newly synthesized ATP. Properties of the principal transport systems are outlined in Figure 15.24. (Note: Additional systems transport fatty acids and amino acids into mitochondria for subsequent oxidation.)

First let us consider ATP, ADP, and P_i, the participants in oxidative phosphorylation. Two systems are involved, an **adenine nucleotide translocase** and a **phosphate translocase.**

The adenine nucleotide translocase spans the inner membrane, and it binds ADP to a specific site on the outer surface of the inner membrane. The protein couples the efflux of ATP from the matrix to the influx of an equivalent amount of ADP from the intermembrane space. Because this system exchanges ATP, with a charge of −4, for ADP, with a charge of −3, its action is driven by the proton gradient established through action of the F_0F_1 ATP synthase. It is generally true of the mitochondrial transport systems that at least one of the participants is moving down a concentration gradient, so that no further energy source is required.

The phosphate translocase acts in either the *antiport* or the *symport* mode (see Chapter 10). As an antiporter, it transports $H_2PO_4^-$ into the matrix, coupled with the efflux of a hydroxide ion. In the alternative symport mode, it transports HPO_4^{2-} into the matrix, along with two protons. Both modes of transport maintain electrical neutrality. The net effect of the adenine nucleotide and phosphate transport systems is to couple the inward transport of the substrates of oxidative phosphorylation, ADP and P_i, to the efflux of the product, ATP.

Next, let us consider the substrates for oxidation. The major substrate from carbohydrate catabolism is pyruvate, which, like phosphate, is exchanged for OH^-. Dicarboxylic acid substrates—namely succinate, fumarate, and malate—can be exchanged for each other or for orthophosphate in the dicarboxylate transport system. Similarly, the tricarboxylate transport system carries either citrate or isocitrate, coupled each with the other or with a dicarboxylic acid. The influx of fatty acids for β-oxidation involves an additional transport system, which is presented in Chapter 18.

In addition to the transport systems shown in Figure 15.24, there is an important transport system for calcium ion. Ca^{2+} regulates a number of metabolic

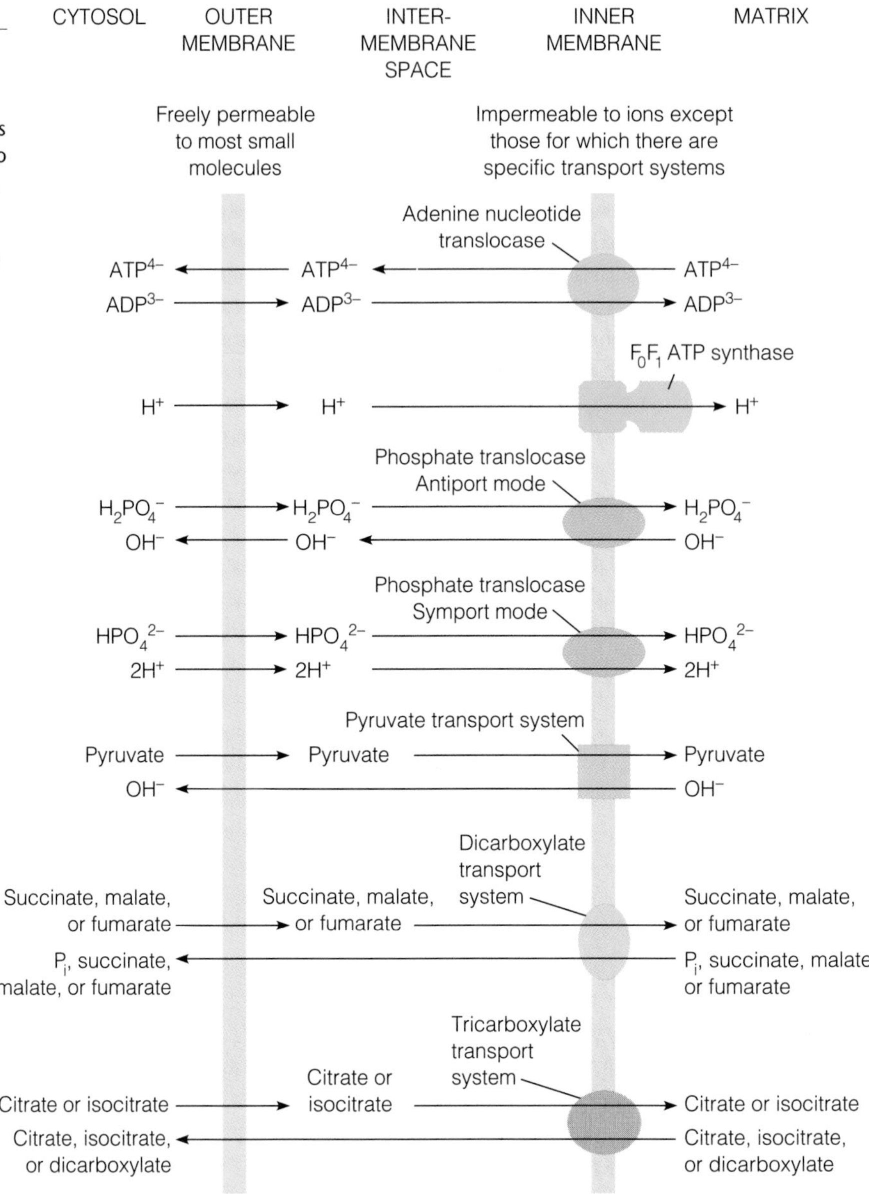

FIGURE **15.24**

Major inner membrane transport systems for respiratory substrates and products. Adenine nucleotide translocase and the phosphate translocase move substrates for oxidative phosphorylation (ADP and P_i) into the mitochondrion and the product (ATP) out. Other transport systems move substrates and products for citric acid cycle oxidation into or out of the matrix, as dictated by the metabolic needs of the cell.

processes in the cytosol, and its release from mitochondrial stores may represent one way to change the cytosolic concentration of Ca^{2+} (see Chapter 23). Calcium is transported inward by the membrane potential, which is more negative inside. It is effluxed by exchange with H^+ from the outside or, in heart mitochondria, for cytosolic H^+.

Energy Yields from Oxidative Metabolism

Much of the past three chapters has been devoted to the pathways by which carbohydrates are oxidized to CO_2 and water. Finally, we are in a position to calculate the total energy yield and metabolic efficiency of these combined pathways. Let us review how much energy is recovered in the form of ATP from the entire oxidative catabolism of glucose. First, we present a balanced equation for each of the three pathways involved, and then we estimate the amount of ATP that can be derived through oxidative phosphorylation from the reduced electron carriers.

Glycolysis:
$$\text{Glucose} + 2\text{ADP} + 2\text{P}_i + 2\text{NAD}^+ \longrightarrow 2\text{ pyruvate} + 2\text{ATP} + 2\text{NADH} + 2\text{H}_2\text{O} + 4\text{H}^+$$

Pyruvate dehydrogenase complex:
$$2\text{ Pyruvate} + 2\text{NAD}^+ + 2\text{CoA-SH} \longrightarrow 2\text{ acetyl-CoA} + 2\text{NADH} + 2\text{CO}_2$$

Citric acid cycle (including conversion of GTP to ATP):
$$2\text{ Acetyl-CoA} + 6\text{H}_2\text{O} + 6\text{NAD}^+ + 2\text{FAD} + 2\text{ADP} + 2\text{P}_i \longrightarrow 4\text{CO}_2 + 6\text{NADH} + 2\text{FADH}_2 + 2\text{CoA-SH} + 2\text{ATP}$$

Net:
$$\text{Glucose} + 10\text{NAD}^+ + 2\text{FAD} + 4\text{H}_2\text{O} + 4\text{ADP} + 4\text{P}_i \longrightarrow 6\text{CO}_2 + 10\text{NADH} + 4\text{H}^+ + 2\text{FADH}_2 + 4\text{ATP}$$

These three processes generate 4 moles of ATP directly, plus 10 moles of NADH and 2 moles of $FADH_2$. Although, as noted earlier, P/O ratios for oxidation of NADH and $FADH_2$ may not have integral values, we shall use the values 3 and 2, respectively. Therefore, the total realizable ATP yield is about 38 per mole of glucose oxidized ($4 + 3 \times 10 + 2 \times 2$). In prokaryotes and in cells using the malate/aspartate shuttle (see Figure 15.11b), these reducing equivalents are carried into the mitochondrion with no energy cost. However, cells using the glycerol phosphate shuttle (see Figure 15.11a) must pay an energy cost. As described on page 536, the electrons from cytosolic NADH enter the respiratory chain as $FADH_2$. Therefore, the ATP yield from each of these two NADHs is 2, not 3. This decreases the overall ATP yield to 36 per mole of glucose. For the following discussion, we shall use 38 as the best estimate of total moles of ATP derived per mole of glucose oxidized. Recalling that $\Delta G^{\circ\prime}$ for glucose oxidation is -2870 kJ/mol (see page 421) and that $\Delta G^{\circ\prime}$ for ATP hydrolysis is -31 kJ/mol, we can calculate the efficiency for operation of this biochemical machine: $(38 \times 31)/2870$, or 40%, under standard conditions. As noted earlier, the efficiency in vivo is probably higher.

The complete oxidation of 1 mole of glucose generates about 38 moles of ATP synthesized from ADP.

Oxygen as a Substrate for Other Metabolic Reactions

In most cells at least 90% of the molecular oxygen consumed is utilized for oxidative phosphorylation. The remaining O_2 is used in a wide variety of specialized metabolic reactions. At least 200 known enzymes use O_2 as a substrate. Because O_2 is rather unreactive, virtually all of these 200 enzymes use a metal ion to enhance the reactivity of oxygen, just as cytochrome oxidase does. In this section we briefly categorize these enzymes, and we consider the metabolism of partially reduced forms of oxygen, which arise continually in all cells and are highly toxic because of their great reactivity.

OXIDASES AND OXYGENASES

The term **oxidase** is applied to enzymes that catalyze the oxidation of a substrate without incorporation of oxygen from O_2 into the product. A two-electron oxidation is usually involved, so the oxygen is converted to H_2O_2. Most oxidases utilize either a metal or a flavin coenzyme. D-Amino acid oxidases, for example, use FAD as a cofactor.

Oxidases and oxygenases are enzymes that use O_2 as a substrate.

$$\text{R—CH(}\overset{+}{\text{N}}\text{H}_3\text{)—COO}^- + \text{H}_2\text{O} + \text{FAD} \longrightarrow \text{R—C(=O)—COO}^- + \overset{+}{\text{N}}\text{H}_4 + \text{FADH}_2$$

$$\text{FADH}_2 + \text{O}_2 \longrightarrow \text{FAD} + \text{H}_2\text{O}_2$$

Oxygenases are enzymes that incorporate oxygen atoms from O_2 into the oxidized products; there are two classes—monooxygenases and dioxygenases. **Dioxygenases,** which incorporate both atoms of O_2 into one substrate, are of limited distribution. An example is tryptophan 2,3-dioxygenase, which contains a heme cofactor and catalyzes the first reaction in tryptophan catabolism:

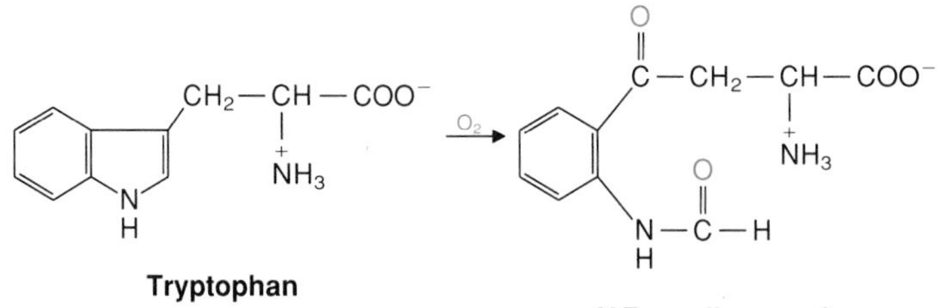

Tryptophan

N-Formylkynurenine

Far more widely distributed are **monooxygenases,** which incorporate one atom from O_2 into a product and reduce the other atom to water. A monooxygenase has one substrate that accepts oxygen and another that furnishes the two H atoms that reduce the other oxygen to water. Because two substrates are oxidized, enzymes of this class are also called **mixed-function oxidases.** The general reaction catalyzed by monooxygenases is as follows:

$$AH + BH_2 + O{=}O \longrightarrow A{-}OH + B + H_2O$$

Because the substrate AH usually becomes hydroxylated by this class of enzymes, the term **hydroxylase** is also used. An example of this type of reaction is the hydroxylation of steroids. Here NADPH is the reductive cofactor, BH_2.

$$RH + NADPH + H^+ + O_2 \longrightarrow R{-}OH + NADP^+ + H_2O$$

Several compounds other than NADPH function as BH_2 in monooxygenase reactions, including α-ketoglutarate, in the hydroxylation of collagen proline residues to hydroxyproline. This and other examples are discussed in Chapter 21.

Cytochromes P450 catalyze hydroxylations of numerous unreactive substrates, making them easier to metabolize.

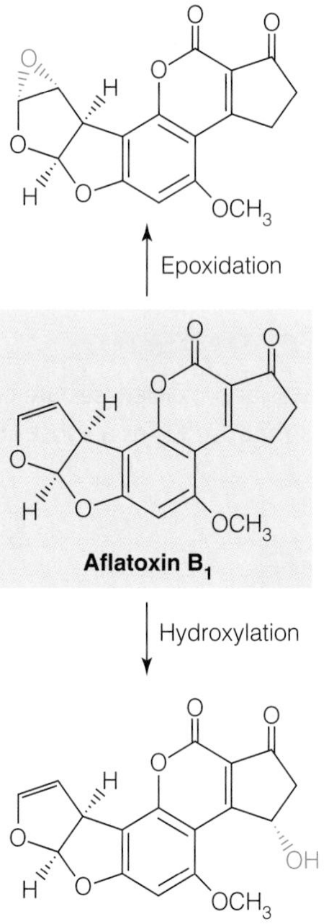

Epoxidation

Aflatoxin B_1

Hydroxylation

CYTOCHROME P450

The most numerous hydroxylation reactions involve a family of heme proteins with the collective name **cytochrome P450.** Most vertebrate genomes contain more than 40 different structural genes for cytochromes P450, making this a large and diverse protein family. These proteins resemble hemoglobin and mitochondrial cytochrome oxidase in being able to bind both O_2 and carbon monoxide. Cytochromes P450 are distinctive, however, in that the reduced form of the heme, when complexed with carbon monoxide, absorbs light strongly at 450 nm. A common structural feature of cytochromes P450 is a cysteine thiolate ion, in which the sulfur occupies one of the six coordination positions with the heme iron. The other ligands are the four pyrrole nitrogens of the protoporphyrin IX in heme and the bound oxygen. Not all proteins with the cysteinyl-heme configuration, however, are classified as cytochromes P450. This same structural motif is used, for example, by thromboxane synthase (page 702) and nitric oxide synthase (page 749). Cytochromes P450 are usually found in the endoplasmic reticulum of eukaryotic cells, not in the mitochondria.

Cytochromes P450 are involved in hydroxylating a large variety of compounds. These reactions include the hydroxylation reactions in steroid hormone biosynthesis (see Chapter 19) and the synthesis of hydroxylated fatty acids and fatty acid epoxides. In addition, cytochromes P450 act upon thousands of **xenobiotics** (foreign compounds), including drugs such as phenobarbital and environmental carcinogens such as benzpyrene, a constituent of tobacco smoke, or aflatoxin B, a carcinogenic compound produced by a mold and found in peanuts that have not been properly screened. Hydroxylation of foreign substances usually increases their solubility and is a step in their detoxification, or metabolism and excretion. However, some of these reactions result in activation of potentially carcinogenic substances to more reactive species as shown for aflatoxin B_1, which is converted to more reactive species either by hydroxylation or epoxidation (see margin).

A key to the reactivity of cytochrome P450 is its ability to split O_2, with one oxygen atom binding to the cytochrome's heme iron. This bond forms a **perferryl ion,** which can be represented as FeO^{3+}. This highly reactive group can abstract a hydrogen atom, even from an unreactive substrate such as a hydrocarbon. In such a hydroxylation, reducing equivalents are typically transferred to the cytochrome from NADPH, the usual electron donor in hydroxylation reactions. A general mechanism is shown in Figure 15.25. In this mechanism, substrate binding is followed by O_2 binding. Transfer of two electrons reduces one oxygen atom, such that splitting of the oxygen molecule generates water plus the perferryl ion, which then oxygenates the substrate.

Cytochrome P450 systems participate in a wide variety of additional reactions, including epoxidation, peroxygenation, desulfuration, dealkylation, deamination, and dehalogenation. These reactions are particularly active in liver, where a number of cytochromes P450 are inducible; that is, their synthesis is stimulated by substrates that are metabolized by these enzymes. Inducers include drugs such as phenobarbital and other barbiturates.

REACTIVE OXYGEN SPECIES, ANTIOXIDANT DEFENSES, AND HUMAN DISEASE

Partially reduced oxygen species—superoxide, peroxide, and hydroxyl radical—are extremely toxic. Their toxicity is counteracted by both enzymatic and nonenzymatic mechanisms.

Formation of Reactive Oxygen Species

As we have seen, the terminal step in electron transport is the four-electron reduction of O_2 to water. Cytochrome oxidase, like most oxidases, transfers electrons to oxygen from metal ions that change their valence states by one electron at a time—such as the heme iron and copper in cytochrome oxidase. Because the

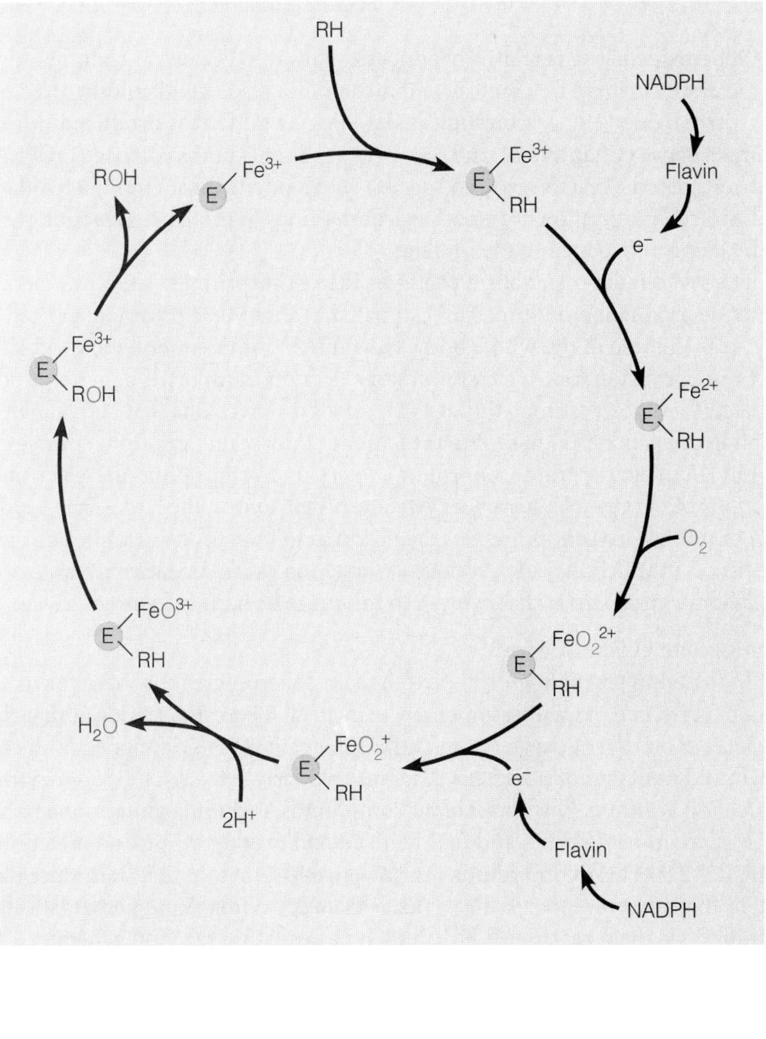

FIGURE **15.25**

Enzymatic hydroxylation involving NADPH and cytochrome P450. In this depiction of a hydroxylation mechanism, the substrate (RH) is converted to the hydroxylated product (ROH) by cytochrome P450 (E). A flavoprotein enzyme (flavin) delivers electrons one at a time from NADPH for the two single-electron reductions depicted.

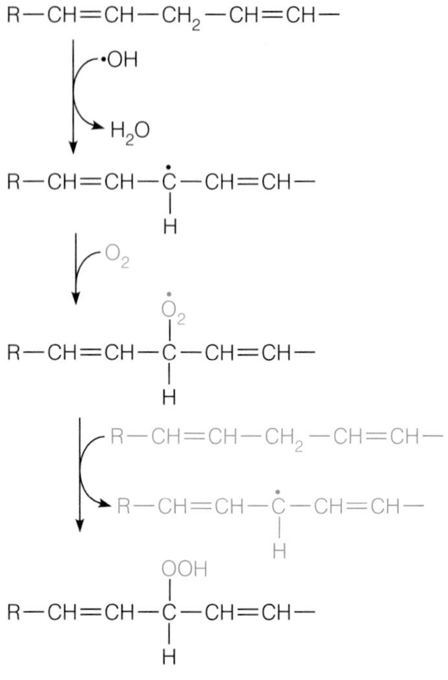

Lipid peroxidation

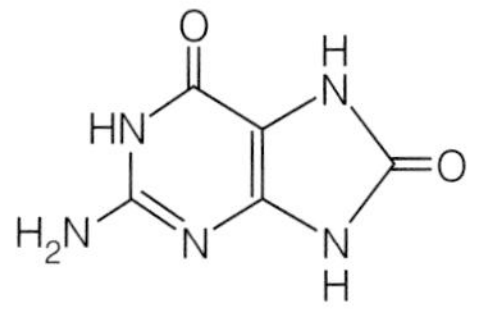

8-Oxoguanine

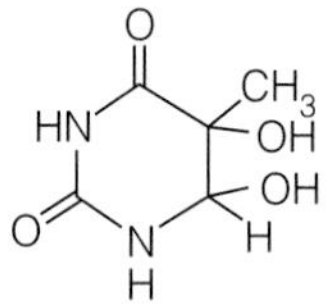

Thymine glycol

interactions of one-electron carriers with two-electron carriers are rarely 100% efficient, oxidases often generate incompletely reduced oxygen species—**superoxide** ($O_2\cdot^-$), formed from a one-electron reduction of O_2; hydrogen peroxide (H_2O_2), formed from a two-electron reduction; and **hydroxyl radical** ($OH\cdot$), formed via a three-electron reduction. In addition, some enzymes, such as xanthine oxidase (Chapter 22) and amino acid oxidases (page 551), generate hydrogen peroxide as their ordinary products. Superoxide, peroxide, and hydroxyl radical are all more reactive than O_2 and are referred to collectively as **reactive oxygen species** (ROS). Hydroxyl radical is particularly reactive and is responsible for damage to other biological molecules. Hydroxyl radical damages proteins in various ways and damages membranes by initiating the oxidation of fatty acids in membrane lipids, a process termed **lipid peroxidation**. As shown in the margin, lipid peroxidation is a chain reaction, because each fatty acyl moiety that undergoes peroxidation generates a radical that can initiate another peroxidation reaction. Hydroxyl radical also damages nucleic acids, both by causing polynucleotide strand breakage (double-strand DNA breaks are lethal) and by changing the structure of DNA bases. About 20 different base changes, or DNA lesions, are known to result from reactions of hydroxyl radical with DNA. Some lesions are mutagenic, because the altered base created (such as **8-oxoguanine**, see margin) forms non-Watson–Crick base pairs during DNA replication. Other lesions, such as **thymine glycol**, are potentially lethal, because, unless the lesion is repaired (Chapter 25), their occurrence in DNA blocks replication past that site. Hydroxyl radical is produced as a result of ionizing radiation and represents the most active mutagen derived from ionizing radiation. Hydroxyl radical is also produced from H_2O_2 in the Fenton reaction:

$$H_2O_2 + Fe^{2+} \text{ (or } Cu^{+}) \longrightarrow Fe^{3+} \text{ (or } Cu^{2+}) + OH\cdot + OH^-$$

Superoxide per se is relatively nontoxic. However, because it contains an unpaired electron, it is a free radical, and it combines readily with another free radical, **nitric oxide** ($NO\cdot$), a biological signaling agent that is produced in many animal tissues (Chapters 21 and 23). The product is **peroxynitrite** ($OONO^-$), also considered a reactive oxygen species. Peroxynitrite causes lipid peroxidation and also causes nitration of tyrosyl hydroxyl groups in proteins, a reaction particularly damaging to membrane proteins.

Normal cellular metabolism produces ROS in surprisingly large amounts; superoxide production in humans is estimated at about 0.02 picomole per cell per day, or 0.15 mole in the whole body. How much of this superoxide production occurs in mitochondria, during respiratory electron transport, is a matter of some debate. It has been estimated that 1–2% of all the electrons that start down the respiratory chain never make it to cytochrome oxidase but leak from complexes I, II, and III, primarily from iron–sulfur centers, and bring about one-electron reductions of oxygen to superoxide. Whatever their source, the large-scale production of reactive oxygen species has the potential to inflict considerable damage on the tissues in which they are produced, a situation called **oxidative stress.** A series of elaborate mechanisms has evolved to minimize its harmful consequences.

Dealing with Oxidative Stress

As discussed in Chapter 17, the earth had an anaerobic atmosphere for its first billion years, and oxygen was intensely toxic to all life forms existing at that time. With the evolution of oxygen in our atmosphere, life forms developed both enzymatic and nonenzymatic defenses against oxidative stress. The nonenzymatic protection is afforded by **antioxidant** compounds, including glutathione (Chapter 14), vitamins C and E, and **uric acid**, an end product of purine metabolism (Chapter 22). These compounds can scavenge ROS before they can cause damage, or they can prevent oxidative damage from spreading. Lipid peroxidation, for example, is a chain reaction in which each peroxidation reaction generates a free radical that can initiate another peroxidation. Thus, trapping one radical can break a chain in which multiple fatty acyl moieties would otherwise be damaged. Vitamin E, a family of compounds in which α-tocopherol is most common, is the principal lipid-soluble antioxidant compound and plays an important role in preventing membrane damage. β-Carotene and other carotenoid compounds related to vitamin A (page 696) are lipid-soluble antioxidants that also play roles in free radical trapping. Glutathione, as noted in Chapter 14, is abundant within cells and plays a particularly important role in cellular antioxidant protection. Vitamin C, or **ascorbic acid**, is an important antioxidant by virtue of its water solubility and its ready oxidation to dehydroascorbic acid. In extracellular fluids, ascorbate levels are far higher than those of glutathione, and ascorbate probably plays the predominant role in extracellular antioxidant protection.Recent evidence suggests that a major antioxidant role of uric acid is its ability to bind and inactivate peroxynitrite.

γ-Glu – Cys – Gly

Glutathione

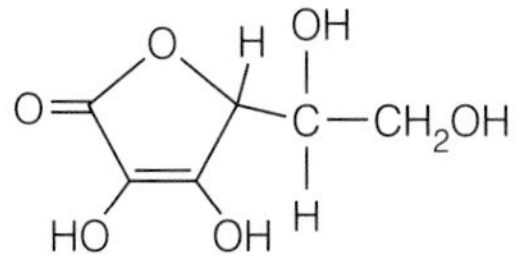

L-Ascorbic acid (vitamin C)

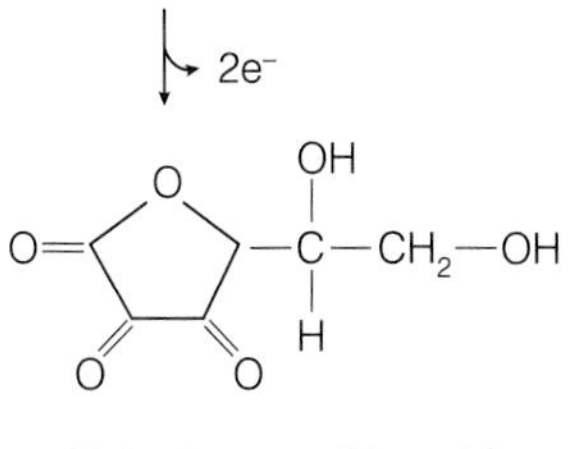

2e⁻

Dehydroascorbic acid

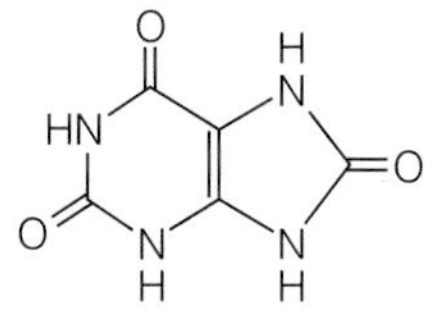

Uric acid

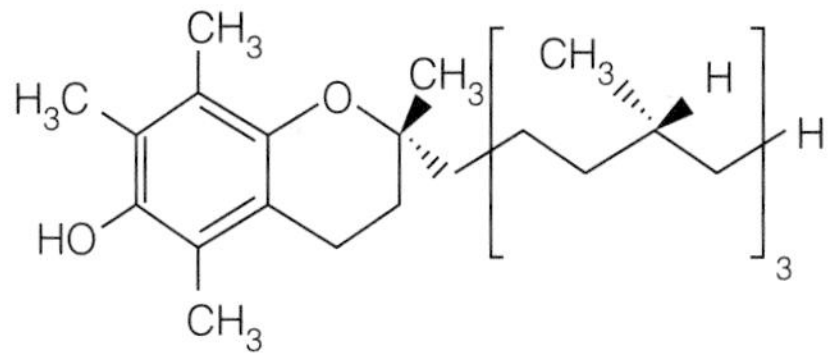

α-Tocopherol (vitamin E)

Among enzymatic mechanisms the first line of defense is **superoxide dismutase (SOD)**, a family of metalloenzymes that catalyze a **dismutation** (a reaction in which two identical substrate molecules have different fates). Here one molecule of superoxide is oxidized and one is reduced.

$$O_2^- + O_2^- + 2H^+ \longrightarrow H_2O_2 + O_2$$

A copper- and zinc-containing form of this enzyme is found in the cytosol of eukaryotic cells; a manganese-containing form is found in both mitochondria and bacterial cells; and a related iron-containing form is found in bacteria, cyanobacteria, and some plants. A nickel-containing bacterial SOD has recently been described.

Hydrogen peroxide is metabolized either by *catalase,* a widely distributed enzyme, or by a more limited family of **peroxidases.** Catalase is a heme protein with an extremely high turnover rate (>40,000 molecules per second). It catalyzes the following reaction:

$$2H_2O_2 \longrightarrow 2H_2O + O_2$$

Peroxidases, which are widely distributed in plants, reduce H_2O_2 to water at the expense of oxidation of an organic substrate. An example of a peroxidase is found in erythrocytes, which are especially sensitive to peroxide accumulation. (See pages 518 through 519 in Chapter 14 for a discussion of the consequences of peroxide accumulation in glucose-6-phosphate dehydrogenase deficiency.) Within erythrocytes is *glutathione peroxidase,* an enzyme that reduces H_2O_2 to water, along with the oxidation of glutathione.

$$2GSH + H_2O_2 \longrightarrow GSSG + 2H_2O$$

Glutathione peroxidase is interesting in that it contains one residue per mole of an unusual amino acid, *selenocysteine,* an analog of cysteine that contains selenium in place of sulfur. This may be related to the current interest in dietary supplements of selenium to prevent cancer.

In some cases the production of reactive oxygen species is not undesirable but is a normal part of the functioning of a cell. For example, certain white blood cells contribute to defense against infectious agents by **phagocytosis** (Greek, "cell eating"). Such cells can engulf a bacterial cell. This event is followed by a **respiratory burst,** a rapid increase in oxygen uptake, stimulated by mechanisms still under study. Much of this oxygen is reduced to superoxide ion and to H_2O_2, which help to kill the engulfed bacterium. Thus, whereas most cells contain elaborate mechanisms to protect against toxicity of reactive oxygen species, the respiratory burst involves a deliberate and controlled production of these species.

Oxygen Metabolism and Human Disease

Because oxidative stress can damage many biomolecules—lipids, proteins, and nucleic acids—the tissue injury that results can, in principle, lead to a variety of disease states. Oxidative damage has been implicated in as many as one hundred disease states, including cardiovascular disease, cancer, stroke, neurodegenerative diseases, and chronic inflammatory diseases. Determining precise cause-and-effect relationships is difficult at this stage, and it is clear that in some states, oxidative stress is not a cause but rather a result of tissue injury from some other factor, which exacerbates the original problem. However, epidemiological evidence points strongly to the value of adequate dietary intake of antioxidant compounds in preventing many of these diseases. The evident health-promoting effects of diets rich in fresh fruits and vegetables, particularly with respect to incidence of cardiovascular disease and cancer, probably result in large part from their high content of antioxidant compounds, particularly vitamins C and E. Many people now take dietary supplements of vitamins C and E as a preventive measure.

Whatever the cause-and-effect relationship between oxidative damage and disease, there is little doubt that the DNA damage caused by ionizing radiation, which is known to be carcinogenic, is mediated through the mutagenic effects of hydroxyl radicals. In addition, hydroxyl radicals are generated by oxidizing agents independent of radiation effects. As discussed in Chapters 23, 25, and 28, cancer is clearly a genetic disease, resulting from an accumulation within a precancerous cell of mutations that ultimately destroy the cell's ability to regulate its own growth and its scheduled death (apoptosis, see page 1110). The generation of altered bases in DNA, such as 8-oxoguanine or 5-hydroxycytosine, is intensely mutagenic. Even though such lesions are usually removed by DNA repair systems (Chapter 25), the repair systems are not 100% effective, and with time the accumulation of unrepaired DNA damage can contribute to the burden of mutations that ultimately transform a normal cell to a cancer cell. Partly because cancer incidence is strongly correlated with age, many scientists also attribute normal aging to the accumulation of unrepaired mutagenic DNA lesions, and oxidative stress is implicated in what has been called the "free radical theory of aging."

Other intriguing evidence, both circumstantial and genetic, points strongly to the roles of oxygen metabolism in human disease; a few examples will be given. Human mutations in the gene encoding the copper/zinc form of superoxide dismutase are associated with amyotrophic lateral sclerosis, a neurodegenerative disorder better known as Lou Gehrig's disease, although the precise causal relationship is not yet well understood. The role of peroxynitrite in the nerve degeneration causing multiple sclerosis (MS) has been brought into focus with the observation that patients with gout hardly ever develop MS, suggesting that the chronic uric acid elevation causing gout (Chapter 22) also prevents the development of MS. Finally, a growing number of mutations in mitochondrial genes that encode respiratory proteins have been shown to cause disease, including optic nerve degeneration and a variety of muscle diseases. Mutations affecting the mitochondrially encoded subunits of cytochrome oxidase have been associated with Alzheimer's disease. Clearly, the relationship between oxidative metabolism and human disease represents one of the most exciting research frontiers in biochemistry.

SUMMARY

Most of the energy captured for ATP synthesis from oxidative reactions in cells is generated in mitochondrial oxidative phosphorylation. Reduced electron carriers, both NADH and $FADH_2$, shuttle their reducing equivalents into the mitochondrial matrix. Enzyme complexes bound to the inner mitochondrial membrane pass these electrons through the respiratory chain, a series of electron carriers of ever-increasing reduction potential. The complexes are numbered I (NADH dehydrogenase), II (succinate–coenzyme Q oxidoreductase), III (coenzyme Q–cytochrome *c* oxidoreductase), IV (cytochrome oxidase), and V (ATP synthase). Electrons are eventually transferred to O_2, which is reduced to water. The reactions of complexes I, III, and IV provide energy to pump protons through the inner membrane, making the outer surface much more acidic than the matrix. Discharge of the resultant proton gradient, when protons pass back into the matrix through a specific ion channel, generates energy that is used to drive ATP synthesis. Although respiration accounts for about 90% of the total oxygen uptake in most cells, in numerous other reactions dozens of enzymes use O_2 as a substrate—oxygenases, oxidases, and hydroxylases. Some reactions generate partially reduced oxygen species—hydroxyl radical, superoxide, and peroxide—which are toxic and mutagenic. Cells possess numerous mechanisms for detoxification of these reactive oxygen species.

REFERENCES

Historical Background

Lehninger, A. L. (1965) *The Mitochondrion: Molecular Basis of Structure and Function.* Benjamin, New York. An account of the earlier work by one who contributed much to it.

Saier, M. H., Jr. (1997) Peter Mitchell and his chemiosmotic theories. *ASM News* 63:13–21. A short scientific biography of the biochemist who proposed the proton gradient as the driving force for ATP synthesis.

Mitochondrial Structure and Function

Kiberstis, P. A. (1999) Mitochondria make a comeback. *Science* 283: 1475. An introductory essay to a special section of *Science,* with four contemporary reviews—of mitochondrial evolution, mitochondrial diseases, oxidative phosphorylation, and mitochondrial genetics.

Tzagoloff, A. (1982) *Mitochondria.* Plenum, New York. A concise, well-illustrated book-length review of mitochondrial structure and function.

Mechanisms in Electron Transport

Beinert, H., R. H. Holm, and E. Münck (1997) Iron–sulfur clusters: Nature's modular, multipurpose structures. *Science* 277:653–659. A review of the roles of these structures in oxidative enzymes and their numerous other roles.

Gennis, R. B. (1998) How does cytochrome oxidase pump protons? *Proc. Natl. Acad. Sci. USA* 95:12747–12749. A minireview and commentary that concisely summarizes cytochrome oxidase structure and function.

Tsukihara, T., and eight coauthors (1996) The whole structure of the 13-subunit oxidized cytochrome *c* oxidase at 2.8 Å. *Science* 272:1136–1144. The crystal structure of complex IV, a landmark achievement.

Xia, D., and seven coauthors (1997) Crystal structure of the cytochrome bc_1 complex from bovine heart mitochondria. *Science* 277:60–66. Insights into the Q cycle from the structure of complex III, also a landmark achievement.

Mechanisms in Oxidative Phosphorylation

Abrahams, J. P., R. Lutter, A. G. W. Leslie, and J. E. Walker (1994) Structure at 2.8 Å resolution of F_1 ATPase from bovine heart mitochondria. *Nature* 370:621–626. A structure that confirms essential features of the mechanism of oxidative phosphorylation.

Bianchet, M. A., J. Hullihen, P. L. Pedersen, and L. M. Amzel (1998) The 2.80Å structure of rat liver F_1-ATPase: Configuration of a critical intermediate in ATP synthesis–hydrolysis. *Proc. Natl. Acad. Sci. USA* 95:11065–11070. The structure of the F_1-ATP complex led to a modification of the now-accepted model for ATP synthesis.

Boyer, P. D. (1997) The ATP synthase—A splendid molecular machine. *Annu. Rev. Biochem.* 66:717–750. A mechanistic analysis of the function of F_0F_1 ATP synthase by the person who predicted the correct mechanism of ATP synthesis and did the crucial early experiments.

Hinkle, P. C., M. A. Kumar, A. Resetar, and D. L. Harris (1991) Mechanistic stoichiometry of mitochondrial oxidative phosphorylation. *Biochemistry* 30:3576–3582. A careful conceptual and experimental analysis that questions whether P/O ratios need be integral.

Lee, C. P., Q. Gu, Y. Xiong, R. A. Mitchell, and L. Ernster (1996) P/O ratios reassessed. Mitochondrial P/O ratios consistently exceed 1.5 with succinate and 2.5 with NAD-linked substrates. *FASEB J.* 10:345–350. Even though values need not be integral, the highest experimental values recently achieved correspond closely to historical values.

Trumpower, B. L. (1990) The protonmotive Q cycle: Energy transduction by coupling of proton translocation by the cytochrome bc_1 complex. *J. Biol. Chem.* 285:11409–11412. A minireview of the mechanism of energy coupling in complex III.

Yasuda, R., H. Noji, K. Kinosita, Jr., and M. Yoshida (1998) F_1 ATPase is a highly efficient molecular motor that rotates with discrete 120° steps. *Cell* 93:1117–1124. Direct evidence for rotation from fluorescence miscroscopy.

Yoshikawa, S., and 12 coauthors (1998) Redox-coupled crystal structure changes in bovine heart cytochrome *c* oxidase. *Science* 280: 1723–1729. Insight into proton pumping from the structural changes resulting from a change in redox state of complex IV.

Zhou, Y., T. M. Duncan, and R. L. Cross (1997) Subunit rotation in *Escherichia coli* F_0F_1-ATP synthase during oxidative phosphorylation. *Proc. Natl. Acad. Sci. USA* 94:10583–10587. The use of protein cross-linking reagents to provide evidence for physical rotation.

Oxygen Metabolism

Ames, B. N. (1983) Dietary carcinogens and anticarcinogens. *Science* 221:1256–1264. One of the earliest explicit proposals that reactive oxygen species are involved in mutation, cancer, and aging and that antioxidants in natural foods might counteract these effects.

Beckman, K. B., and B. N. Ames (1998) The free radical theory of aging. *Physiol. Rev.* 78:547–581. A comprehensive review from Bruce Ames, an early proponent of the idea that oxidative damage causes cancer and aging.

Chanock, S. J., J. E. Benna, R. M. Smith, and B. M. Babior (1994) The respiratory burst oxidase. *J. Biol. Chem.* 269:24519–24522. Superoxide production as an antibacterial defense mechanism.

Estabrook, R. W. (1996) The remarkable P450s: A historical overview of these versatile hemeprotein catalysts. *FASEB J.* 10:202–204. The introduction to a series of review articles on cytochromes P450 by a pioneer in the field.

Frei, B. (1994) Reactive oxygen species and antioxidant vitamins: Mechanisms of action. *Am. J. Med.* 97(suppl 3A):5S–13S. An informative review of the origins of reactive oxygen species and the actions of antioxidants.

Fridovich, I. (1995) Superoxide radical and superoxide dismutases. *Annu. Rev. Biochem.* 64:97–112. A review by the discoverer of superoxide dismutase.

Halliwell, B. (1994) Free radicals, antioxidants, and human disease: Curiosity, cause, or consequence? *Lancet* 344:721–724. A brief but informative review with references to earlier literature.

Malmström, B. (1982) Enzymology of oxygen. *Rev. Biochem.* 51:21–59. A review of enzymes that use oxygen as a substrate.

Vaughan, M. (1997) Oxidative modification of macromolecules minireview series. *J. Biol. Chem.* 272:18513. The first in a coordinated series of six JBC minireviews, providing recent information on oxidative modification of nucleic acids, proteins, and lipoproteins.

PROBLEMS

1. Referring to Table 15.1 for E_0' values, calculate $\Delta G°'$ for oxidation of malate by malate dehydrogenase.

2. When pure reduced cytochrome *c* is added to carefully prepared mitochondria along with ADP, P_i, antimycin A, and oxygen, the cytochrome *c* becomes oxidized, and ATP is formed, with a P/O ratio approaching 1.0.
   (a) Indicate the probable flow of electrons in this system.
   (b) Why was antimycin A added?
   (c) What does this experiment tell you about the location of coupling sites for oxidative phosphorylation?
   (d) Write a balanced equation for the overall reaction.
   (e) Calculate $\Delta G°'$ for the above reaction, using E_0' values from Table 15.1 and a $\Delta G°'$ value for ATP hydrolysis of about −31 kJ/mol.

3. Freshly prepared mitochondria were incubated with β-hydroxybutyrate, oxidized cytochrome *c*, ADP, P_i, and cyanide. β-Hydroxybutyrate is oxidized by an NAD^+-dependent dehydrogenase.

$$\mathrm{COO^-{-}CH_2{-}CH(OH){-}CH_3} + \mathrm{NAD^+} \longrightarrow \mathrm{COO^-{-}CH_2{-}C(=O){-}CH_3} + \mathrm{NADH} + \mathrm{H^+}$$

   The experimenter measured the rate of oxidation of β-hydroxybutyrate and the rate of formation of ATP.
   (a) Indicate the probabie flow of electrons in this system.
   (b) How many moles of ATP would you expect to be formed per mole of β-hydroxybutyrate oxidized in this system?
   (c) Why is β-hydroxybutyrate added rather than NADH?
   (d) What is the function of the cyanide?
   (e) Write a balanced equation for the overall reaction occurring in this system.
   (f) Calculate the net free energy change ($\Delta G°'$) in this system, using E_0' values from Table 15.1 and $\Delta G°'$ for hydrolysis of ATP of about −31 kJ/mol.

4. If you were to determine the P/O ratio for oxidation of α-ketoglutarate, you would probably include some malonate in your reaction system. Why? Under these conditions, what P/O ratio would you expect to observe?

5. Of the various oxidation reactions in glycolysis and the citric acid cycle, the only one that does not involve NAD^+ is the succinate dehydrogenase reaction. What would $\Delta G°'$ be for an enzyme that oxidizes succinate with NAD^+ instead of FAD? If the intramitochondrial concentration of succinate was 10-fold higher than that of fumarate, what minimum $[NAD^+]/[NADH]$ ratio in mitochondria would be needed to make this reaction exergonic?

6. Intramitochondrial ATP concentrations are about 5 mM, and phosphate concentration is about 10 mM. If ADP is five times more abundant than AMP, calculate the molar concentrations of ADP and AMP at an energy charge of 0.85. Calculate $\Delta G'$ for ATP hydrolysis under these conditions. The energy charge is the concentration of ATP plus half the concentration of ADP divided by the total adenine nucleotide concentration:

$$\frac{[\mathrm{ATP}] + \tfrac{1}{2}[\mathrm{ADP}]}{[\mathrm{ATP}] + [\mathrm{ADP}] + [\mathrm{AMP}]}$$

7. From E_0' values in Table 15.1, calculate the equilibrium constant for the glutathione peroxidase reaction.

*8. In the early days of "mitochondriology," P/O ratios were determined from measurements of volume of O_2 taken up by respiring mitochondria and chemical assays for disappearance of inorganic phosphate. Now, however, it is possible to measure P/O ratios simply with a recording oxygen electrode. How might this be done?

9. Years ago there was interest in using uncouplers such as dinitrophenol as weight control agents. Presumably, fat could be oxidized without concomitant ATP synthesis for re-formation of fat or carbohydrate. Why was this a bad idea?

10. Biochemists working with isolated mitochondria recognize five energy "states" of mitochondria, depending on the presence or absence of essential substrates for respiration—O_2, ADP, oxidizable substrates, and so forth. The characteristics of each state are given in the accompanying table (see table and graph on the next page).
    (a) On the graph, identify a state that might predominate under each set of conditions indicated with a letter.
    (b) To determine whether isolated mitochondria exhibit respiratory control, one determines the ratio of rates of oxygen uptake in two different states. Which states?
    (c) Which state probably predominates in vivo in skeletal muscle fatigued from a long and strenuous workout?
    (d) Which state probably predominates in resting skeletal muscle of a well-nourished animal?

	State 1	State 2	State 3	State 4	State 5
O_2 availability	Aerobic	Aerobic	Aerobic	Aerobic	Anaerobic
ADP level	Low	High	High	Low	High
Oxidizable substrate	Endogenous (low)	Near zero	High	High	High
Respiration rate	Slow	Slow	Fast	Slow	Zero
Rate-limiting component	ADP	Oxidizable substrate	Electron transport	ADP	O_2

(e) Which state probably predominates in heart muscle most of the time?

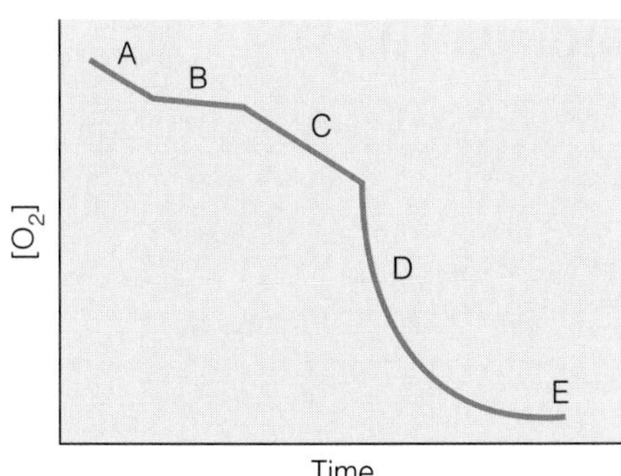

11. Referring to Figure 15.12, predict the P/O ratio for oxidation of ascorbate by isolated mitochondria.

*12. Cytochromes P450 catalyze reactions other than hydroxylations—deaminations, dehalogenations, desulfurations, peroxidations, dealkylations, and epoxidations. Write balanced equations for any two of these reactions.

*13. $NAD^+/NADH$ and $NADP^+/NADPH$ have virtually identical standard reduction potentials. Why, then, does the transhydrogenase reaction require energy?

14. As a representation of the respiratory chain, what is wrong with this picture? There are four deliberate errors.

Malate — NAD⁺ — $FADH_2$ — $CoQH_2$ — cyt b^{2+}

Oxalo-acetate — NADH — FAD — CoQ — cyt b^{3+}

cyt c^{3+} — cyt c_1^{2+} — cyt a-a_3^{3+} — H_2O_2

cyt c^{2+} — cyt c_1^{3+} — cyt a-a_3^{2+} — $1/2O_2$

15.
$$GSSG + NADPH + H^+ \longrightarrow 2GSH + NADP^+$$
$$E'_0\ (GSSG \longrightarrow 2GSH) = -0.23\ V$$
$$E'_0\ (NADP^+/NADPH) = -0.32\ V \qquad F = 96.5\ kJ/mol\text{-}V$$

(a) Calculate $\Delta G°'$ for the glutathione reductase reaction in the direction shown.
(b) Suppose that a cell contained an isoform of glutathione reductase that used NADH instead of NADPH as the reductive coenzyme. Would you expect $\Delta G°'$ for this enzyme to be higher, lower, or the same as the corresponding value for the real glutathione reductase? Briefly explain your answer.
(c) Given what you know about the metabolic roles and/or intracellular concentration ratios of $NAD^+/NADH$ and $NADP^+/NADPH$, would you expect $\Delta G'$ (not $\Delta G°'$) for this enzyme to be higher, lower, or the same as $\Delta G'$ for the real enzyme under intracellular conditions? Briefly explain your answer.

16. To carefully prepared mitochondria were added succinate, cytochrome *c*, ADP, orthophosphate, and sodium cyanide. Referring to Figures 15.9 and 15.12, answer the following.
(a) List the sequence of electron carriers in this system.
(b) Write a balanced equation for the overall reaction occurring in this system, showing oxidation of the initial electron donor, reduction of the final acceptor, and synthesis of ATP.
(c) Calculate $\Delta G°'$ for the overall reaction. $\Delta G°'$ for ATP hydrolysis is -31 kJ/mol.
(d) Why was cyanide added in this experiment?
(e) What would the P/O ratio be if the same experiment were run with addition to the mitochondria of 2,4-dinitrophenol?

17. In order to function as an oxidative phosphorylation uncoupler (page 543), 2,4-dinitrophenol must act catalytically, not stoichiometrically. What does this mean? Identify and discuss an important implication of this conclusion.

CHAPTER 16

Carbohydrate Metabolism II: Biosynthesis

IN THIS CHAPTER OUR ATTENTION TURNS FROM ENERGY-YIELDING DEGRADATIVE pathways to energy-requiring biosynthetic processes. The pathways of carbohydrate biosynthesis that receive emphasis in this chapter (Figure 16.1) include gluconeogenesis, or the synthesis of glucose from noncarbohydrate precursors; the synthesis of polysaccharides, primarily glycogen in animals; and the synthesis of **glycoconjugates,** molecules containing carbohydrate linked to a protein or lipid molecule. A supremely important carbohydrate biosynthetic process—photosynthesis in plants and some microorganisms—is presented separately in Chapter 17.

Here we encounter the first instances of a principle enunciated in Chapter 12: *Biosynthetic processes are never simply the reversal of corresponding catabolic pathways.* Superficially, gluconeogenesis looks very much like glycolysis in reverse, but different enzymatic reactions are used at crucial sites. These sites are strongly exergonic reactions that are controlled largely in reciprocal fashion, so that physiological conditions that activate glycolysis inhibit gluconeogenesis and vice versa. Much the same picture will emerge from our discussion of glycogen synthesis as compared with glycogen mobilization, later in this chapter.

Gluconeogenesis

PHYSIOLOGICAL NEED FOR GLUCOSE SYNTHESIS IN ANIMALS

Most animal organs can metabolize a variety of carbon sources to generate energy—triacylglycerols, various sugars, pyruvate, amino acids, and so forth. However, the brain and central nervous system require glucose as the sole or primary carbon source. The same is true for some other tissues, such as kidney medulla, testes, and erythrocytes (Figure 16.2). Consequently, animal cells must be able to synthesize glucose from other precursors and also to maintain blood glucose levels within narrow limits—both for proper functioning of the brain and central nervous system and also for providing precursors for glycogen storage in other tissues. The glucose requirements of the human brain are relatively enormous—120 grams per day, out of about 160 grams needed by the entire body. The amount of glucose that can be generated from the body's glycogen reserves at any time is

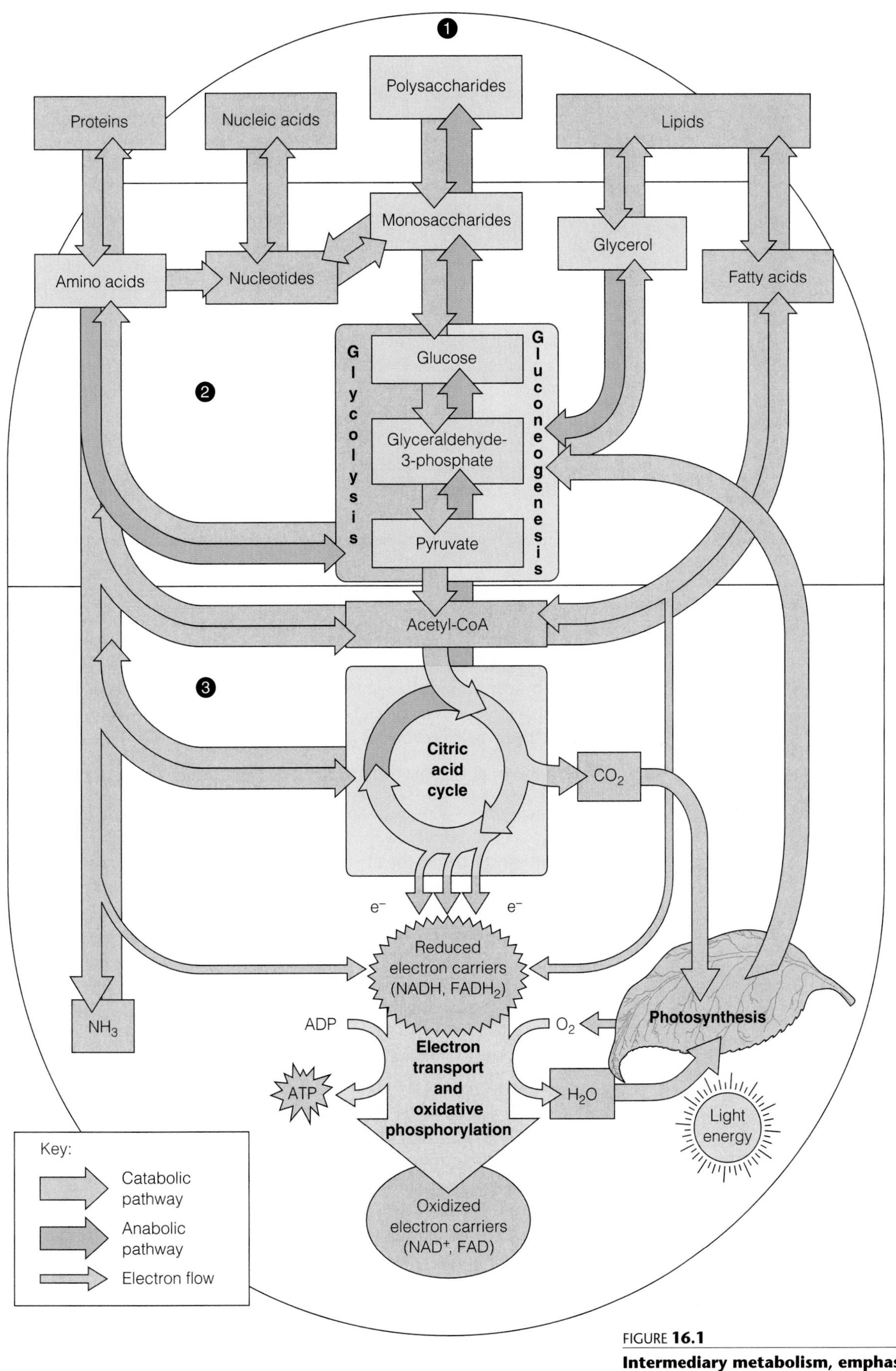

FIGURE **16.1**

Intermediary metabolism, emphasizing pathways in carbohydrate biosynthesis.

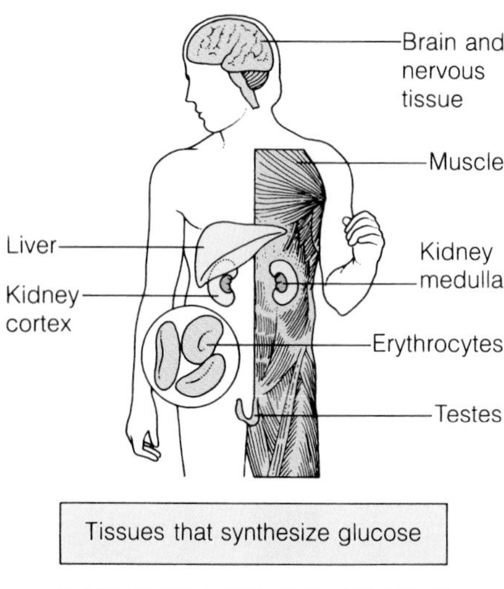

FIGURE **16.2**

Synthesis and use of glucose in the human body. Liver and kidney cortex are the primary gluconeogenic tissues. Brain, skeletal muscle, kidney medulla, erythrocytes, and testes use glucose as their sole or primary energy source, but they lack the enzymatic machinery to synthesize it.

about 190 grams, and the total amount of glucose in body fluids is little more than 20 grams. Thus, the readily available glucose reserves amount to about one day's supply. During periods of fasting for more than one day, glucose must be formed from other precursors. The same is true during intense exertion, for example, during a marathon run, when the glucose reserves are rapidly depleted.

The synthetic process is called **gluconeogenesis**—literally, the production of new glucose. Gluconeogenesis is defined as the biosynthesis of carbohydrate from three-carbon and four-carbon precursors, generally noncarbohydrate in nature. The principal substrates for gluconeogenesis are *lactate,* produced primarily from glycolysis in skeletal muscle and erythrocytes; *amino acids,* generated from dietary protein or from the breakdown of muscle protein during starvation; the specific amino acid *alanine,* produced in muscle through the glucose–alanine cycle (see Chapter 20); *propionate,* derived from the breakdown of some fatty acids and amino acids; and *glycerol,* derived from the catabolism of fats. The fatty acids released during lipid breakdown are mostly converted to acetyl-CoA and cannot be used for carbohydrate synthesis, except in organisms that have a functioning glyoxylate cycle (see Figure 14.20, page 510).

Synthesis of glucose from noncarbohydrate precursors is essential for maintenance of blood glucose levels within acceptable limits.

Gluconeogenesis occurs primarily in the cytosol, although some precursors are generated in mitochondria and must be transported to the cytosol to be utilized. The primary gluconeogenic organ in animals is the liver, with kidney cortex contributing in a lesser but still significant way (see Figure 16.2). The major fates of glucose formed by gluconeogenesis are catabolism by nervous tissue and utilization by skeletal muscle. In addition, glucose is the primary precursor for all other carbohydrates, including amino sugars, complex polysaccharides, and the carbohydrate components of glycoproteins and glycolipids. The need for glucose as a biosynthetic intermediate means that gluconeogenesis is an important pathway in plants and microorganisms, as well as in animals, and the pathway is essentially identical in all organisms. However, the wealth of information on control of gluconeogenesis in animals leads us to concentrate on animal metabolism for the first part of this chapter.

ENZYMATIC RELATIONSHIP OF GLUCONEOGENESIS TO GLYCOLYSIS

Gluconeogenesis should be an easy pathway to learn, because it closely resembles glycolysis in reverse. However, there are some important differences, which allow the pathway to run in the direction of glucose *synthesis* in the cell.

Recall that a metabolic pathway can proceed smoothly only if $\Delta G'$ is strongly negative for the overall pathway in the direction written. In Chapter 13 you learned that glycolysis from glucose to pyruvate is strongly exergonic; under typical intracellular conditions $\Delta G'$ is about -96 kJ/mol. How, then, can the conversion of pyruvate to glucose be made exergonic? Recall that three reactions of the glycolytic pathway are so strongly exergonic as to be essentially irreversible—those catalyzed by hexokinase, phosphofructokinase, and pyruvate kinase. In gluconeogenesis, different enzymes are used at each of these steps, so that, for example, the conversion of fructose-1,6-bisphosphate to fructose-6-phosphate is not simply a reversal of the phosphofructokinase reaction. In essence, the three irreversible reactions of glycolysis are bypassed by enzymes specific to gluconeogenesis, which catalyze quite different reactions that run strongly in the direction of glucose synthesis. This biosynthetic process involves a substantial energy cost, which must be paid if the overall process is to be thermodynamically favored (see page 564). The remaining seven reactions of gluconeogenesis are catalyzed by glycolytic enzymes that catalyze reversible reactions and are driven in either direction by mass action. Another way to relate glycolysis to gluconeogenesis is to say that they differ only at three steps controlled by *substrate cycles* (see page 421).

Gluconeogenesis uses specific enzymes to bypass three irreversible reactions of glycolysis.

The entire gluconeogenic pathway, from pyruvate to glucose, is summarized in Figure 16.3. We focus here on the reactions that bypass the three irreversible steps in glycolysis.

Step 1: Conversion of Pyruvate to Phosphoenolpyruvate

The bypass of pyruvate kinase involves two reactions, the first of which we encountered in Chapter 14. **Pyruvate carboxylase** catalyzes the ATP- and biotin-dependent conversion of pyruvate to oxaloacetate (see Figure 14.19, page 507). The enzyme requires acetyl-CoA as an allosteric activator:

$$\text{Pyruvate} + CO_2 + H_2O + \text{ATP} \longrightarrow \text{oxaloacetate} + \text{ADP} + P_i + 2H^+ \quad \Delta G^{\circ\prime} = -2.1 \text{ kJ/mol}$$

This is one of the *anaplerotic* reactions used to maintain levels of citric acid cycle intermediates (see page 505). Pyruvate carboxylase generates oxaloacetate in the mitochondrial matrix, where it can be oxidized in the citric acid cycle. To be used for gluconeogenesis, oxaloacetate must move out of the mitochondrion to the cytosol, where the remainder of the pathway occurs. However, the mitochondrial membrane does not have an effective transporter for oxaloacetate. Therefore, oxaloacetate is reduced by mitochondrial malate dehydrogenase to malate, which is transported into the cytosol by exchange for orthophosphate and then reoxidized by cytosolic malate dehydrogenase. This is one of the mitochondrial transport processes presented in Chapter 15.

Once in the cytosol, oxaloacetate is acted on by **phosphoenolpyruvate carboxykinase** (abbreviated **PEPCK**) to give phosphoenolpyruvate:

$$\text{Oxaloacetate} + \text{GTP} \rightleftharpoons \text{phosphoenolpyruvate} + CO_2 + \text{GDP} \quad \Delta G^{\circ\prime} = +2.9 \text{ kJ/mol}$$

Note the use of GTP, rather than ATP, as an energy donor. Note also that the CO_2 that was fixed by pyruvate carboxylase is released in this reaction, so that no net fixation of CO_2 occurs. The PEPCK reaction requires Mg^{2+} or Mn^{2+} and is readily reversible. In the reaction the carboxyl group formed from the transferred CO_2 provides electrons to facilitate O−P bond formation:

GDP–O–P(OH)(O⁻)(⁻O) + ⁻OOC–C(=O)–CH₂–C(=O)–O⁻ → GDP–O⁻ + (HO)(⁻O)P(O⁻)–O–C(COO⁻)=CH₂ + O=C=O

PEPCK is localized to the cytosol of most animal cells, but in some species it is found in both cytosol and mitochondria. The cytosolic form is the more important player in gluconeogenesis, as shown by the fact that its intracellular concentration is regulated by hormonal conditions known to control gluconeogenesis.

The overall reaction for the bypass of pyruvate kinase is as follows:

$$\text{Pyruvate} + \text{ATP} + \text{GTP} + H_2O \longrightarrow \text{phosphoenolpyruvate} + \text{ADP} + \text{GDP} + P_i + 2H^+$$
$$\Delta G^{\circ\prime} = +0.8 \text{ kJ/mol}$$

$\Delta G^{\circ\prime}$ for the two reactions combined is slightly positive. However, under intracellular conditions the sequence is quite exergonic, with $\Delta G'$ of about −25 kJ/mol. As shown in the summary reaction, two high-energy phosphates must be invested for the synthesis of one super-high-energy phosphoenolpyruvate. After this bypass, phosphoenolpyruvate is converted to fructose-1,6-bisphosphate by glycolytic enzymes acting in reverse.

Step 2: Conversion of Fructose-1,6-bisphosphate to Fructose-6-phosphate

The phosphofructokinase reaction of glycolysis is essentially irreversible but only because it is driven by phosphate transfer from ATP. A bypass reaction in

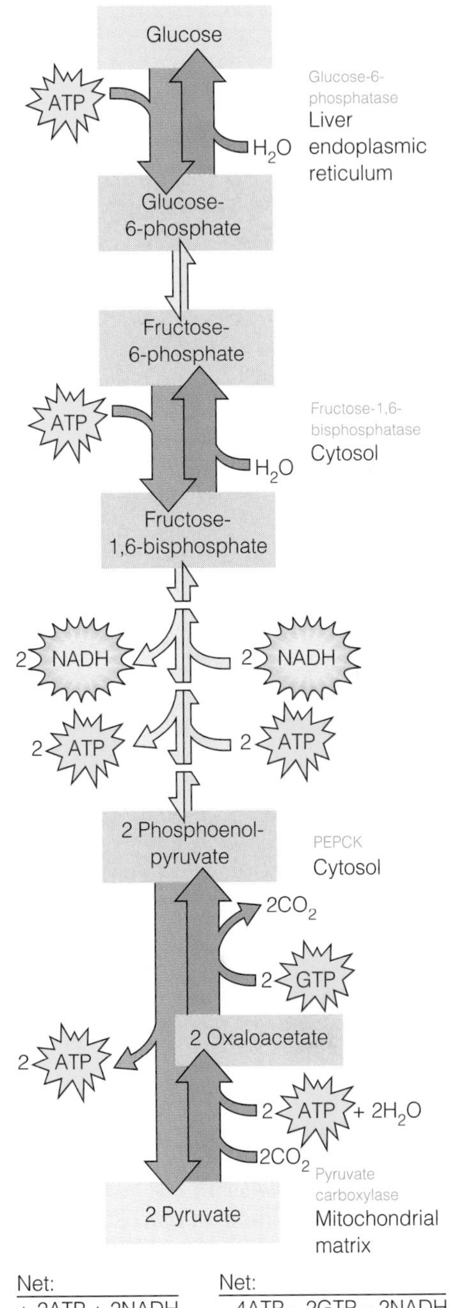

FIGURE **16.3**

Reactions of glycolysis and gluconeogenesis. Irreversible reactions of glycolysis are shown in dark purple. The opposed reactions in gluconeogenesis, which bypass these steps, are shown in dark blue. Pale arrows identify reversible reactions used in both pathways.

gluconeogenesis involves a simple hydrolytic reaction, catalyzed by **fructose-1,6-bisphosphatase.**

$$\text{Fructose-1,6-bisphosphate} + H_2O \longrightarrow \text{fructose-6-phosphate} + P_i \qquad \Delta G^{\circ\prime} = -16.3 \text{ kJ/mol}$$

The negative $\Delta G^{\circ\prime}$ for this reaction favors the rightward reaction. The multisubunit enzyme requires Mg^{2+} for activity and represents one of the major control sites regulating the overall gluconeogenic pathway. The fructose-6-phosphate formed in this reaction then undergoes isomerization by phosphoglucoisomerase to glucose-6-phosphate.

Step 3: Conversion of Glucose-6-phosphate to Glucose
Glucose-6-phosphate cannot be converted to glucose by reverse action of hexokinase or glucokinase, because of the high positive $\Delta G^{\circ\prime}$ of that reaction; phosphate transfer from ATP makes that reaction virtually irreversible. Another enzyme specific to gluconeogenesis, **glucose-6-phosphatase,** comes into play instead. This bypass reaction also involves a simple hydrolysis.

$$\text{Glucose-6-phosphate} + H_2O \longrightarrow \text{glucose} + P_i \qquad \Delta G^{\circ\prime} = -12.1 \text{ kJ/mol}$$

Glucose-6-phosphatase, which also requires Mg^{2+}, is found primarily in the endoplasmic reticulum of the liver, with its active site on the cytosolic side. The significance of its location in liver is that a unique function of liver is to synthesize glucose for export to the tissues via the bloodstream. Recent evidence suggests that glucose-6-phosphatase also plays a role in glucose-6-phosphate synthesis from alternate phosphate donors such as pyrophosphate, and the ER-bound form of the enzyme might be part of a multiprotein system that regulates G6P levels. Because most tissues, notably brain and skeletal muscle, lack this enzyme, they cannot release free glucose into the bloodstream, although they do synthesize glucose-6-phosphate for their own use.

STOICHIOMETRY AND ENERGY BALANCE OF GLUCONEOGENESIS

Twelve high-energy phosphates are consumed per mole of glucose synthesized by gluconeogenesis.

We have emphasized that catabolic pathways generate energy, whereas anabolic pathways carry an energy cost. For gluconeogenesis we can estimate that cost. The overall conversion of 2 moles of pyruvate to 1 mole of glucose is quite exergonic, as shown in Table 16.1. $\Delta G^{\circ\prime}$ for the overall process is about -47.6 kJ/mol. But the synthesis of glucose is costly to the cell in an energetic sense. Six high-energy phosphate groups are consumed (four ATPs and two GTPs), as well as 2 moles of NADH, which is the energetic equivalent of six more ATPs (because mitochondrial oxidation of 1 mole of NADH generates 3 moles of ATP).

By contrast, if glycolysis could operate in reverse, the net equation would show far less energy input—2 moles of NADH and 2 moles of high-energy phosphate:

$$2\,\text{Pyruvate} + 2\text{ATP} + 2\text{NADH} + 2H^+ + 2H_2O \longrightarrow \text{glucose} + 2\text{ADP} + 2P_i + 2\text{NAD}^+$$

This process would be highly endergonic, however, with a $\Delta G^{\circ\prime}$ of $+73.3$ kJ/mol. Therefore, it is clear that the investment of four extra high-energy phosphate bonds is essential if the net synthesis of glucose is to occur as an irreversible process.

SUBSTRATES FOR GLUCONEOGENESIS

The most important gluconeogenic precursors are lactate, alanine, glycerol, and propionate.

As indicated earlier, gluconeogenesis draws precursors from diverse sources, including lactate, amino acids, glycerol, and propionate. The pathways by which these substrates enter gluconeogenesis are shown in Figure 16.4 and are discussed in this section.

TABLE 16.1 Summary of gluconeogenesis, from pyruvate to glucose

Reaction		$\Delta G^{\circ\prime}$ (kJ/mol)
Pyruvate + CO_2 + ATP + H_2O ⟶ oxaloacetate + ADP + P_i + $2H^+$	× 2	−4.2
Oxaloacetate + GTP ⇌ phosphoenolpyruvate + CO_2 + GDP	× 2	+5.8
Phosphoenolpyruvate + H_2O ⇌ 2-phosphoglycerate	× 2	−3.4
2-Phosphoglycerate ⇌ 3-phosphoglycerate	× 2	−9.2
3-Phosphoglycerate + ATP ⇌ 1,3-bisphosphoglycerate + ADP	× 2	+37.6
1,3-Bisphosphoglycerate + NADH + H^+ ⇌ glyceraldehyde-3-phosphate + NAD^+ + P_i	× 2	−12.6
Glyceraldehyde-3-phosphate ⇌ dihydroxyacetone phosphate		−7.6
Glyceraldehyde-3-phosphate + dihydroxyacetone phosphate ⇌ fructose-1,6-bisphosphate		−23.9
Fructose-1,6-bisphosphate + H_2O ⟶ fructose-6-phosphate + P_i		−16.3
Fructose-6-phosphate ⇌ glucose-6-phosphate		−1.7
Glucose-6-phosphate + H_2O ⟶ glucose + P_i		−12.1
Net: 2 Pyruvate + 4ATP + 2GTP + 2NADH + $6H_2O$ ⟶ glucose + 4ADP + 2GDP + $6P_i$ + $2NAD^+$ + $2H^+$		−47.6

Note: The reactions in boldface type are those that bypass irreversible glycolytic reactions; the remaining reactions are reversible reactions of glycolysis. The first six reactions are multiplied by 2, because 2 three-carbon precursors are required to make one molecule of glucose.

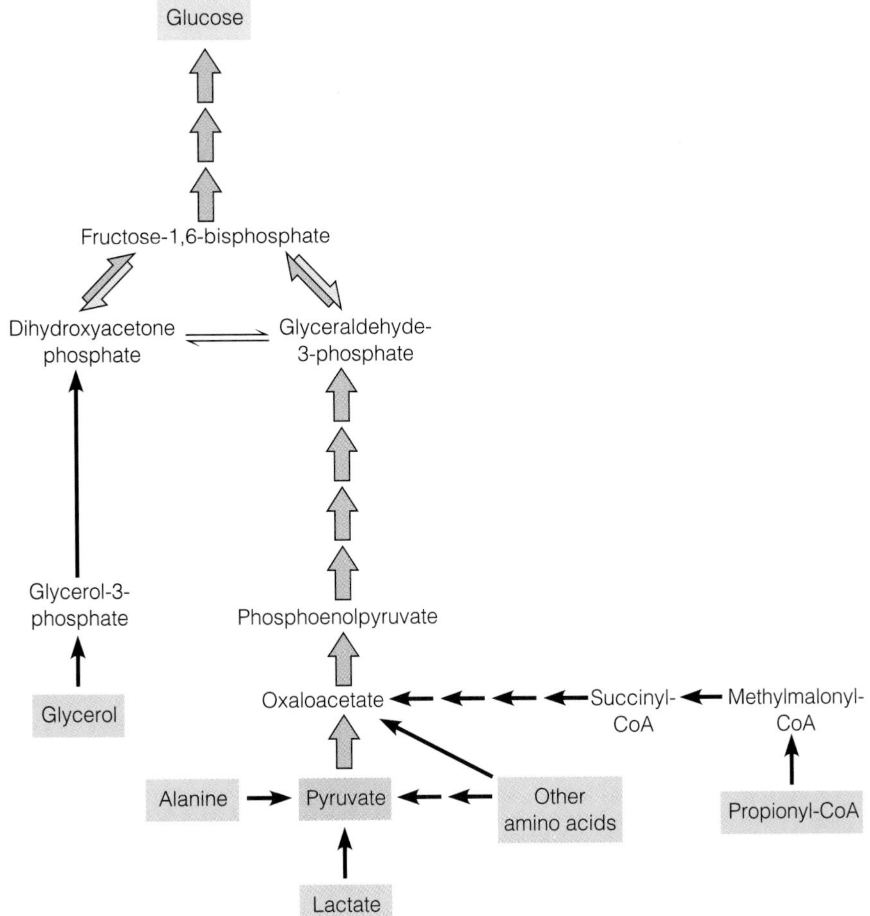

FIGURE **16.4**

Outline of pathways for glucose synthesis from the major gluconeogenic precursors. Note that both glucose and lactate are carried in the blood.

Lactate

In quantitative terms, lactate is the most significant gluconeogenic precursor. Recall from Chapter 13 that skeletal muscle derives much of its energy from glycolysis, particularly during intense exertion, when respiration cannot deliver sufficient oxygen to the tissues for complete oxidation of glucose. Under these conditions, glycogen stores are mobilized, and glucose-6-phosphate is converted to pyruvate more rapidly than it can be further metabolized via pyruvate dehydrogenase and the citric acid cycle. Lactate dehydrogenase is abundant in muscle, and the equilibrium strongly favors pyruvate reduction to lactate. Thus, lactate is released to the blood, whence it is readily taken up by the heart and oxidized as fuel. The accumulation of lactate during prolonged exertion is a significant factor limiting athletic performance.

Some of the lactate produced in muscle enters the liver and is reoxidized to pyruvate. This pyruvate can then undergo gluconeogenesis to give glucose, which is returned to the bloodstream and taken up by muscle to regenerate the glycogen stores. This process, described originally by Carl and Gerti Cori and appropriately called the **Cori cycle,** is schematized in Figure 16.5. The pathway is particularly active during recovery from intense muscular exercise. During this time the breathing rate is elevated, and the increased oxidative metabolism generates more ATP, much of which is used to rebuild glycogen stores via gluconeogenesis.

In a parallel process, the **glucose–alanine cycle,** pyruvate in peripheral tissues undergoes transamination to alanine, which is returned to the liver and used for gluconeogenesis. This pathway, which is presented in detail in Chapter 20, helps tissues to dispose of toxic ammonia formed during protein degradation.

Amino Acids

Like alanine, many other amino acids can readily be converted to glucose, primarily through degradative pathways that generate citric acid cycle intermediates,

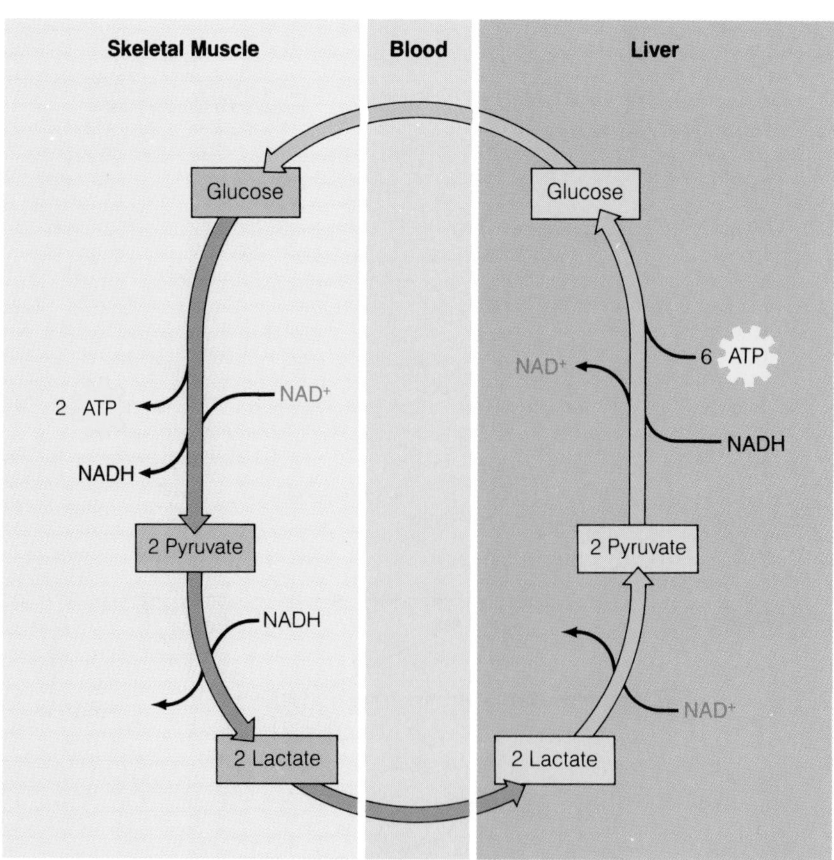

FIGURE **16.5**

The Cori cycle. Lactate produced in glycolysis during muscle exertion is transported to the liver, for resynthesis of glucose by gluconeogenesis.

which can be converted to oxaloacetate (see Figure 16.4). Such amino acids are called **glucogenic** (that is, able to be converted to glucose), although *gluconeogenic* is probably a more accurate term. Among the 20 amino acids found in proteins, only the catabolic pathways for leucine and lysine do not generate gluconeogenic precursors. During fasting, when insufficient carbohydrate is ingested, the catabolism of muscle proteins is the major source of intermediates needed to maintain normal blood glucose concentrations. The same is true in the disease diabetes mellitus, as discussed further in Chapter 23.

Glycerol

In general, lipids are poor gluconeogenic precursors. Catabolism of triacylglycerols yields fatty acids and glycerol. Fatty acids undergo β-oxidation to yield acetyl-CoA. In plants and bacteria, acetyl-CoA can be incorporated into carbohydrate via the glyoxylate cycle. In animals, however, acetyl-CoA cannot be converted to pyruvate or to any other gluconeogenic precursor. Hence, *fatty acids cannot undergo net conversion to carbohydrates.* Although it is true that two-carbon units from acetyl-CoA can proceed to oxaloacetate in the citric acid cycle, there is no net conversion, because two carbons are lost in each turn of the cycle. Therefore, aside from a minor contribution by odd-chain fatty acids (explained in the next section), the only fat breakdown product that can enter gluconeogenesis is glycerol. Its utilization involves phosphorylation, followed by dehydrogenation to dihydroxyacetone phosphate (see Figure 16.4).

Because animals lack a glyoxylate cycle, they cannot undergo net conversion of fat to carbohydrate.

Propionate

In all organisms a three-carbon acyl-CoA, **propionyl-CoA,** is generated either from the breakdown of some amino acids or from the oxidation of fatty acids with odd numbers of carbon atoms. Propionyl-CoA enters gluconeogenesis via its conversion to succinyl-CoA and thence to oxaloacetate. The process, which is detailed in Chapter 18, involves a coenzyme derived from vitamin B_{12}.

$$\underset{\textbf{Propionyl-CoA}}{CH_3{-}CH_2{-}\overset{O}{\overset{\|}{C}}{-}S{-}CoA} \xrightarrow{CO_2} \underset{\textbf{Methylmalonyl-CoA}}{CH_3{-}\underset{COO^-}{CH}{-}\overset{O}{\overset{\|}{C}}{-}S{-}CoA} \longrightarrow \underset{\textbf{Succinyl-CoA}}{{}^-OOC{-}CH_2{-}CH_2{-}\overset{O}{\overset{\|}{C}}{-}S{-}CoA}$$

Although all organisms use propionate as a gluconeogenic substrate, it is particularly important in the metabolism of ruminant animals such as cattle. These animals display enormous rates of gluconeogenesis from a variety of substrates, owing to the great amount of bacterial fermentation occurring in their several stomach chambers. In cattle the four chambers of the stomach have a total volume as high as 70 liters. The action of various bacteria degrades plant materials, particularly cellulose, to glucose. But before the glucose can be absorbed into the bloodstream, as in human digestion, it is fermented further to various products, notably lactate and propionate. Propionate is converted to propionyl-CoA and thence to succinyl-CoA, and lactate is simply dehydrogenated to pyruvate.

ETHANOL CONSUMPTION AND GLUCONEOGENESIS

Although it is possible to visualize pathways by which ethanol could be converted to glucose, ethanol is actually a poor gluconeogenic precursor. In fact, ethanol strongly inhibits gluconeogenesis and can bring about **hypoglycemia,** a potentially dangerous decrease in blood glucose levels.

Ethanol is metabolized primarily in the liver, by alcohol dehydrogenase:

$$\text{Ethanol} + NAD^+ \rightleftharpoons \text{acetaldehyde} + NADH + H^+$$

This reaction elevates the [NADH]/[NAD$^+$] ratio in liver cytosol, which in turn shifts the equilibrium of the lactate dehydrogenase and glyceraldehyde 3-phosphate dehydrogenase, inhibiting glycolysis. The same mechanism shifts the equilibrium of cytosolic malate dehydrogenase, so that oxaloacetate tends to be reduced to malate and hence becomes unavailable for gluconeogenesis. The resultant hypoglycemia can affect the parts of the brain concerned with temperature regulation. This response, in turn, can lower the body temperature by as much as 2°C. Therefore, the time-honored practice of feeding brandy or whiskey to those rescued from cold or wet conditions is counterproductive. To be sure, alcohol creates a sense of warming through vasodilation, but this peripheral vasodilation causes further heat loss. Metabolically speaking, glucose would be far more effective in raising body temperature.

ROLES OF EXTRAHEPATIC PHOSPHOENOLPYRUVATE CARBOXYKINASE

In addition to its role in liver, the predominant gluconeogenic tissue, PEPCK is important in the metabolism of two other tissues. First, in kidney cortex, the other major gluconeogenic tissue, PEPCK participates in acid–base regulation. The cortex produces ammonia, to titrate the acidity of urine. The primary source of ammonia is glutamine, which generates two moles of ammonia:

$$\text{Glutamine} + H_2O \longrightarrow \text{glutamate} + NH_3$$

$$\text{Glutamate} + NAD^+ + H_2O \longrightarrow \alpha\text{-ketoglutarate} + NH_3 + NADH + H^+$$

The α-ketoglutarate produced in the second reaction (glutamate dehydrogenase, page 508) is converted via the citric acid cycle to oxaloacetate, which generates phosphoenolypyruvate via the PEPCK reaction, for ultimate synthesis of glucose.

Second, in adipose (fatty) tissue, PEPCK participates in a process that has been called **glyceroneogenesis**, which evidently acts to produce sufficient glycerol-3-phosphate for triacylglycerol formation, particularly during starvation, when fat breakdown has been activated (Chapters 18 and 23) and there is a need to balance breakdown with resynthesis. In this tissue, the phosphoenolpyruvate produced by pyruvate carboxylase and PEPCK is *not* converted to glucose. Instead, it is diverted at dihydroxyacetone phosphate for reduction to glycerol-3-phosphate, which combines with the coenzyme A derivatives of free fatty acids (Chapter 18) to produce triacylglyerols.

Regulation of Gluconeogenesis

Regulation of gluconeogenesis is crucial for many physiological functions, but particularly so for proper functioning of nervous tissue. Although other organs can use a variety of energy sources, the well-being of the central nervous system demands maintenance of blood glucose levels within narrow limits. Gluconeogenic control is important also as an animal adjusts to muscular exertion or to cycles of feeding and fasting. Flux through the pathway must increase and decrease, based on the availability of lactate produced by the muscles, of glucose from the diet, or of other gluconeogenic precursors.

Gluconeogenic flux rates are inversely related to the carbohydrate content of the diet. This effect is mediated hormonally.

Gluconeogenesis is controlled in large part by the diet. Animals fed a high-carbohydrate diet show low rates of gluconeogenesis, whereas fasted animals or those fed carbohydrate-poor diets show high flux through this pathway. These hormonal effects, mediated primarily through insulin and glucagon, involve both control of the synthesis of phosphoenolpyruvate carboxykinase and regulation effected through control of cyclic AMP levels. Our discussion here focuses upon these cAMP-mediated effects as well as other mechanisms affecting enzyme activities. We return to discussion of hormonal effects upon enzyme synthesis in Chapter 23, when we present hormone action in detail.

RECIPROCAL REGULATION OF GLYCOLYSIS AND GLUCONEOGENESIS

Conditions that promote glycolysis inhibit gluconeogenesis, and vice versa.

Gluconeogenesis and glycolysis both proceed largely in the cytosol. Because gluconeogenesis synthesizes glucose and glycolysis catabolizes glucose, it is evident that *gluconeogenesis and glycolysis must be controlled in reciprocal fashion.* In other words, intracellular conditions that activate one pathway tend to inhibit the other. If it were not for reciprocal control, glycolysis and gluconeogenesis would operate together as a giant futile cycle. Reciprocal regulation is related in large part to the adenylate energy charge. Conditions of low energy charge tend to activate the rate-controlling steps in glycolysis while inhibiting carbon flux through gluconeogenesis. Conversely, gluconeogenesis is stimulated at high energy charge, under conditions where catabolic flux rates are adequate to maintain sufficient ATP levels.

Recall from Chapter 13 that glycolysis is controlled primarily by regulation of the three strongly exergonic reactions of the pathway—those catalyzed by hexokinase (or glucokinase in liver), phosphofructokinase, and pyruvate kinase. The opposed reactions in gluconeogenesis—those catalyzed by glucose-6-phosphatase, fructose-1,6-bisphosphatase, and the combination of pyruvate carboxylase and phosphoenolpyruvate carboxykinase—are also strongly exergonic and represent the chief targets for control of this pathway. In other words, the three substrate cycles that differentiate glycolysis from gluconeogenesis (see Figure 16.3) represent the primary sites for reciprocal regulation of these pathways. Figure 16.6 identifies the major allosteric activators and inhibitors of the key exergonic reactions in glycolysis and gluconeogenesis.

Glucose-6-phosphatase is not known to be allosterically controlled, but its K_M for glucose-6-phosphate is far higher than intracellular concentrations of this metabolite. Thus, intracellular activity is largely controlled in first-order fashion by the concentration of this substrate.

Energy charge affects the control of glycolysis and gluconeogenesis, by regulating the interconversion of fructose-6-phosphate and fructose-1,6-bisphosphate. Phosphofructokinase is activated by AMP, and the opposed enzyme, fructose-1,6-bisphosphatase, is inhibited by AMP. Thus, as energy charge decreases, glycolysis is activated and gluconeogenesis inhibited by the opposed effects of AMP on phosphofructokinase and fructose-1,6-bisphosphatase, respectively. However, although intracellular adenine nucleotide levels do vary in parallel with changes in flux through glycolysis and gluconeogenesis, as expected if energy charge is a major regulatory factor, the correspondence is not absolute. These observations suggested additional control mechanisms and led to the discovery of the much more important physiological regulator, fructose-2,6-bisphosphate, which we introduced in Chapter 13 and describe in detail later in this chapter.

Energy charge, however, does control one of the two isoenzyme forms of pyruvate kinase. In liver and in other gluconeogenic tissues, the so-called L form of pyruvate kinase predominates, whereas the M form is found largely in muscle. The L form is inhibited both by ATP and by some amino acids, particularly alanine, the major gluconeogenic precursor among the amino acids. This relationship allows inhibition of glycolysis, with consequent activation of gluconeogenesis, specifically in gluconeogenic tissues, when ample energy and substrates are available. Control at the pyruvate kinase step allows conservation of high-energy phosphate in the phosphoenolpyruvate molecule.

Acetyl-CoA can also be seen as a reciprocal regulator of glycolysis and gluconeogenesis, acting on the enzymes that interconvert pyruvate and phosphoenolpyruvate. Acetyl-CoA is a required activator of pyruvate carboxylase and an inhibitor of pyruvate kinase and of pyruvate dehydrogenase complex. It can thus signal, when its levels rise, that adequate substrates are available to provide energy through the citric acid cycle and that more carbon can be shuttled into gluconeogenesis and ultimately stored as glycogen. However, some biochemists have questioned whether the activation of pyruvate carboxylase by acetyl-CoA represents

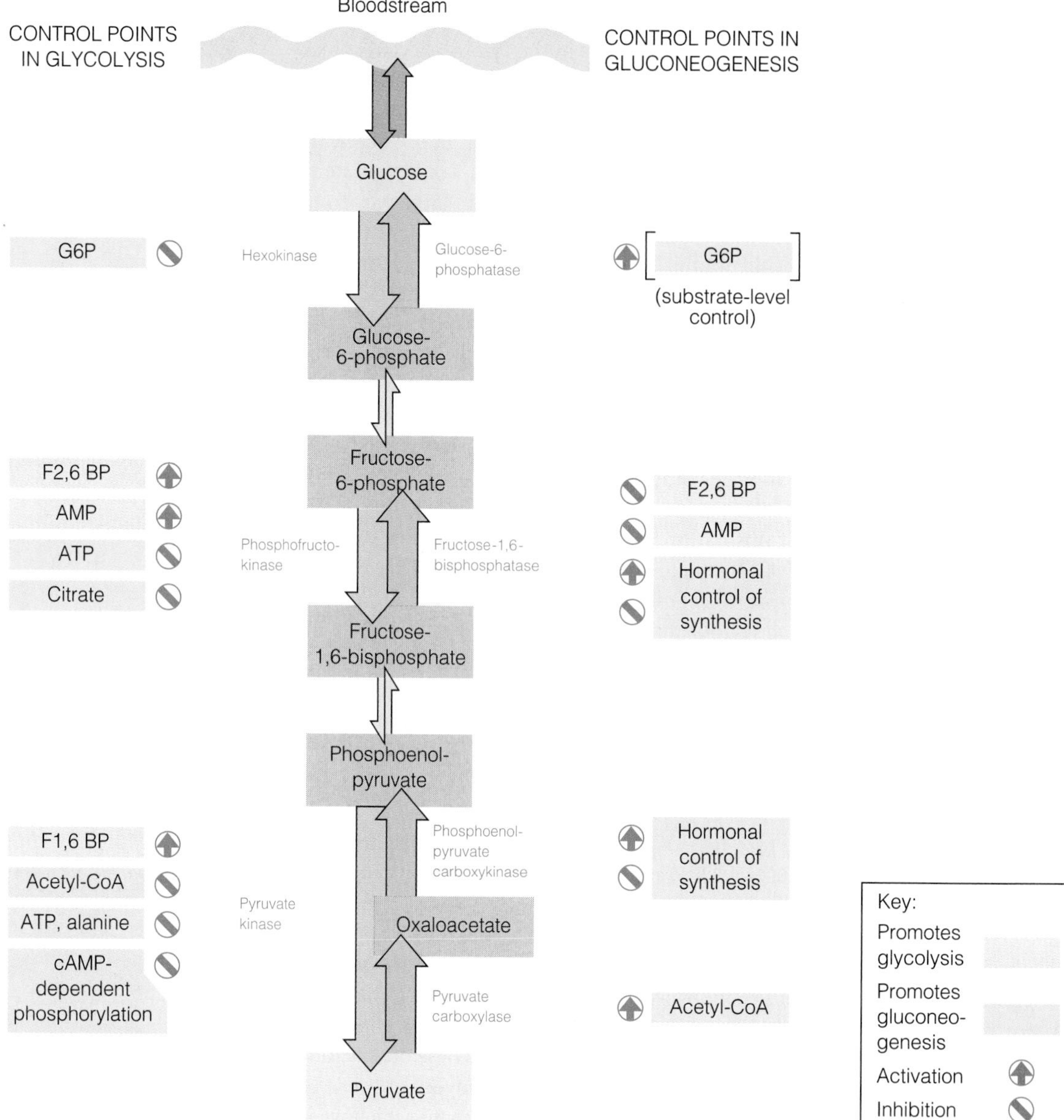

FIGURE **16.6**

Major control mechanisms affecting glycolysis and gluconeogenesis. The figure shows the strongly exergonic reactions of glycolysis and gluconeogenesis and the major activators and inhibitors of these reactions.

an important regulatory mechanism, because its intramitochondrial levels under most conditions are far higher than the concentration giving half-maximal stimulation. Thus, the activity of pyruvate carboxylase in vivo might not vary in response to changing acetyl-CoA levels.

FRUCTOSE-2,6-BISPHOSPHATE AND THE CONTROL OF GLUCONEOGENESIS

Fructose-2,6-bisphosphate, the most important regulator of gluconeogenesis, is synthesized and degraded by different forms of the same enzyme.

Although the effects discussed above are all significant for controlling glycolysis and gluconeogenesis, developments since 1980 have identified **fructose-2,6-bisphosphate** as the most important regulator, through its effects on the interconversion of fructose-6-phosphate and fructose-1,6-bisphosphate. Fructose-2,6-bisphosphate is active at much lower concentrations than the other physiological regulators we have discussed. For example, fructose-2,6-bisphosphate inhibits fructose-1,6-bisphosphatase with a K_i of 2.5 μM, whereas AMP has a K_i for the same enzyme of 25 μM. Recall from Chapter 13 that fructose-2,6-bisphosphate is also an allosteric activator of phosphofructokinase. Thus, accumulation of this regulator has the effect of activating glycolysis and inhibiting gluconeogenesis.

The level of fructose-2,6-bisphosphate itself is controlled ultimately by cyclic AMP, through the action of cAMP-dependent protein kinase.

As shown in Figure 16.7, fructose-2,6-bisphosphate is formed from fructose-6-phosphate by a distinct form of phosphofructokinase, called PFK-2 to distinguish it from the well-known PFK of glycolysis, which we can call PFK-1 for clarity. The activity of PFK-2 is in turn controlled by cyclic AMP. A 49-kilodalton subunit of PFK-2 is subject to phosphorylation by cAMP-dependent protein kinase. This phosphorylation *decreases* the activity of PFK-2, by increasing its K_M for fructose-6-phosphate. The resultant drop in fructose-2,6-bisphosphate levels causes it to dissociate from PFK-1, thereby increasing the sensitivity of PFK-1 to allosteric inhibitors—citrate, ATP, and phosphoenolpyruvate.

Another enzyme activity, called **fructose-2,6-bisphosphatase,** cleaves fructose-2,6-bisphosphate back to fructose-6-phosphate. Control of this activity can contribute to regulation of the level of fructose-2,6-bisphosphate. Fructose-2,6-bisphosphatase is strongly inhibited by fructose-6-phosphate. More important, this enzyme is also regulated by cAMP-stimulated phosphorylation of a 49-kilodalton subunit, except that in this case phosphorylation *enhances* the activity of the enzyme.

The finding that both activities are regulated by phosphorylation of a 49-kilodalton protein led to the discovery that the same protein is involved in both effects. In fact, PFK-2 and fructose-2,6-bisphosphatase constitute a bifunctional enzyme. When the 49-kilodalton protein is in the *nonphosphorylated* form, the bifunctional enzyme catalyzes the synthesis of fructose-2,6-bisphosphate, and when the 49-kilodalton protein is phosphorylated, the phosphatase activity is activated. Thus, cAMP has a dual effect on fructose-2,6-bisphosphate levels, by inactivating PFK-2 and by stimulating fructose-2,6-bisphosphatase activities, and

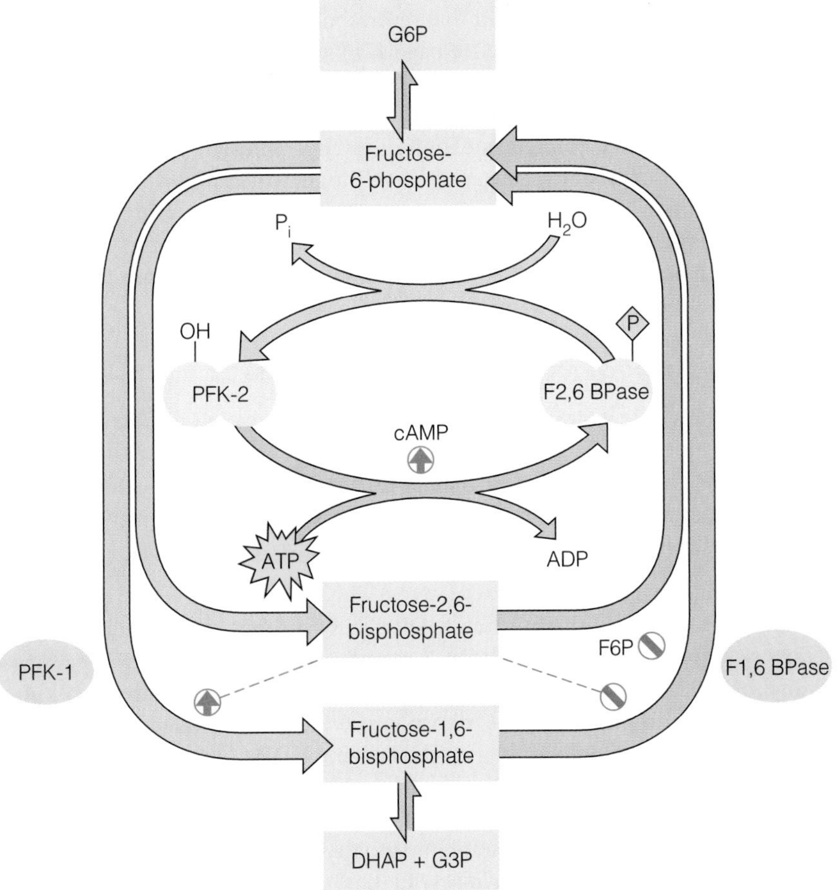

FIGURE **16.7**

Biosynthesis and degradation of fructose-2,6-bisphosphate by action of a bifunctional enzyme. The phosphorylation of one subunit of PFK-2 is carried out by cAMP-dependent protein kinase. Fructose-2,6-bisphosphate activates phosphofructokinase-1 and inhibits fructose-1,6-bisphosphatase. Reactions promoting glycolysis are shown with purple arrows, those favoring gluconeogenesis with blue arrows.

both effects tend to *decrease* the levels of fructose-2,6-bisphosphate. This in turn reduces flux through glycolysis and *stimulates* gluconeogenesis by (1) diminishing the stimulation of PFK-1 and (2) relieving the inhibition of fructose-1,6-bisphosphatase.

Recall from Chapter 13 that the primary hormone whose action raises cAMP levels in liver is the pancreatic hormone *glucagon.* Glucagon elevates blood glucose levels by two distinct mechanisms: (1) the cAMP stimulation of the regulatory cascade that causes glycogen breakdown and (2) as described here, the cAMP depression of fructose-2,6-bisphosphate levels, which stimulates gluconeogenesis. In addition, glucagon controls levels of the key enzyme phosphoenolpyruvate carboxykinase (PEPCK), by activating transcription of the structural gene for PEPCK. Insulin has the converse effect. By inhibiting PEPCK gene transcription, insulin tends to depress gluconeogenic flux rates. Glucagon has an additional action at the genetic level: It represses synthesis of pyruvate kinase, the glycolytic enzyme converting phosphoenolpyruvate to pyruvate, thereby contributing to increased flux from pyruvate to phosphoenolpyruvate.

Fructose-2,6-bisphosphate also plays a regulatory role in plants. By inhibiting a cytosolic fructose-1,6-bisphosphatase, it controls the flow of three-carbon sugars, produced by photosynthesis, out of chloroplasts and into the pathway for sucrose synthesis in the cytosol.

Glycogen Biosynthesis

A major fate of glucose in animals is the synthesis of glycogen, the highly branched $\alpha(1\rightarrow4)$-linked glucose polymer (Chapter 9). This is a good place to begin a discussion of polysaccharide synthesis, because glycogen synthesis illustrates general mechanisms used in synthesizing glycosidic bonds. In addition, the reciprocal regulation of glycogen synthesis and breakdown is of great metabolic importance.

For many years it was thought that reversal of the glycogen phosphorylase reaction described in Chapter 13 (see page 471) was the major route for glycogen synthesis. However, three observations could not be reconciled with this notion. First, epinephrine secretion activates glycogen metabolism only in the direction of breakdown; in fact, epinephrine inhibits glycogen biosynthesis. Second, intracellular levels of orthophosphate are relatively high, which would make it difficult on equilibrium grounds for phosphorylase to catalyze glycogen synthesis in vivo. Third, although phosphorylase can synthesize glycogen in vitro, the product is of much lower molecular weight than natural glycogen. All of these factors suggested that glycogen is synthesized by a different enzyme.

GLYCOGEN SYNTHASE AND THE BRANCHING PROCESS

UDP-glucose is the metabolically activated form of glucose for glycogen synthesis.

In the late 1950s, Argentine biochemist Luis Leloir discovered *uridine diphosphate glucose,* commonly called UDP-glucose or UDP-Glc. Although we discussed the role of UDP-glucose in galactose metabolism in Chapter 13, its first known role was as a substrate for glycogen biosynthesis. The enzyme involved, **glycogen synthase,** is a tetramer, bound tightly to intracellular glycogen granules.

Biosynthesis of UDP-Glucose

Let us first review how the substrate UDP-Glc is synthesized from blood glucose. Glucose is transported into cells by a membrane-bound protein, the **glucose transporter.** As shown in Figure 16.8, it is then phosphorylated by hexokinase or glucokinase to give glucose-6-phosphate, which is isomerized to glucose-1-phosphate by phosphoglucomutase. The enzyme **UDP-glucose pyrophosphorylase,** which was introduced in Chapter 13 (see Figure 13.13b), then catalyzes the synthesis of UDP-glucose. The reaction is drawn to the right by rapid enzymatic cleavage of pyrophosphate to orthophosphate, catalyzed by pyrophosphatase. Hydrolysis of pyrophosphate yields about 30 kJ/mol under standard conditions.

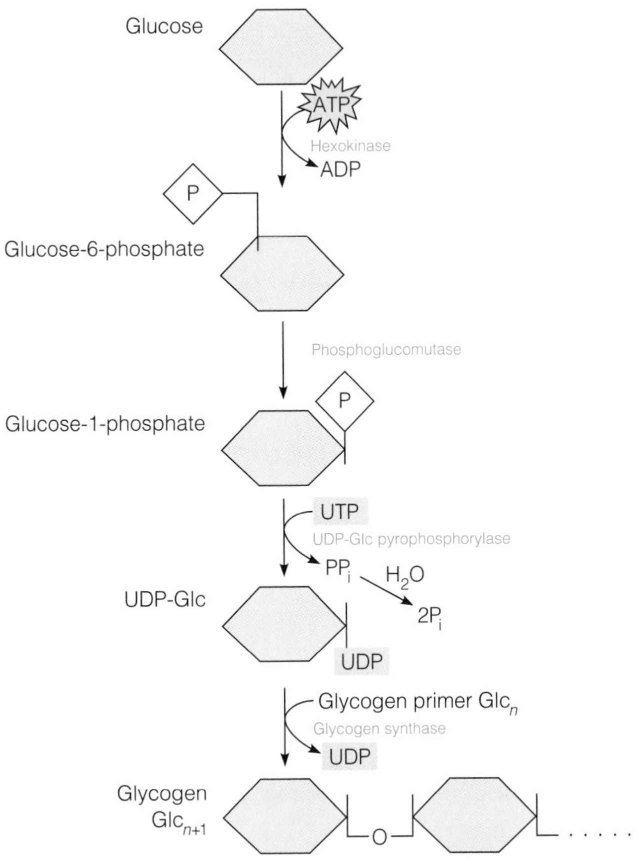

FIGURE **16.8**

Pathway for conversion of glucose monomers to polymeric glycogen.

The Glycogen Synthase Reaction

UDP-glucose is the immediate donor of a glucosyl residue to the nonreducing end of a glycogen branch, which must be at least four glucose residues in length. Glycogen synthase is a **glycosyltransferase**—an enzyme that transfers an activated sugar unit to a nonreducing sugar hydroxyl group. The reaction, depicted in Figure 16.9, generates an $\alpha(1\rightarrow4)$ glycosidic linkage between carbon 1 of the incoming glucosyl moiety and carbon 4 of the glucose residue at the terminus of the glycogen chain. The enzyme continues to add glucose residues successively to the 4-hydroxyl group at the nonreducing end. Because UDP-Glc is a high-energy compound, the glycogen synthase reaction is exergonic, with a $\Delta G^{\circ\prime}$ of about -13.4 kJ/mol.

The primer for glycogen synthase is a short chain of glucose residues assembled by a protein called **glycogenin**, which transfers glucose from UDP-Glc to a tyrosine residue on the protein itself. Glycogenin then transfers additional glucosyl units from UDP-Glc, to give $\alpha(1\rightarrow4)$-linked primers up to eight residues long. These primers are then extended by glycogen synthase.

Formation of Branches

Glycogen synthesis involves both polymerization of glucose units and branching from $\alpha(1\rightarrow6)$ linkages. These branches are important because they increase the solubility of the polymer and also increase the number of nonreducing ends from which glucose-1-phosphate can be derived during glycogen mobilization. However, these branches cannot be introduced by glycogen synthase. Another enzyme, called **branching enzyme** but more accurately termed **amylo-(1,4→1,6)-transglycosylase,** comes into play, as shown in Figure 16.10. This branching enzyme transfers a terminal fragment, some 6 or 7 residues long, from a branch terminus at least 11 residues in length to a hydroxyl group at the 6-position of a glucose residue in the interior of the polymer. The reaction involves nucleophilic attack

Glycogen biosynthesis requires glycogen synthase for polymerization and a transglycosylase to create branches.

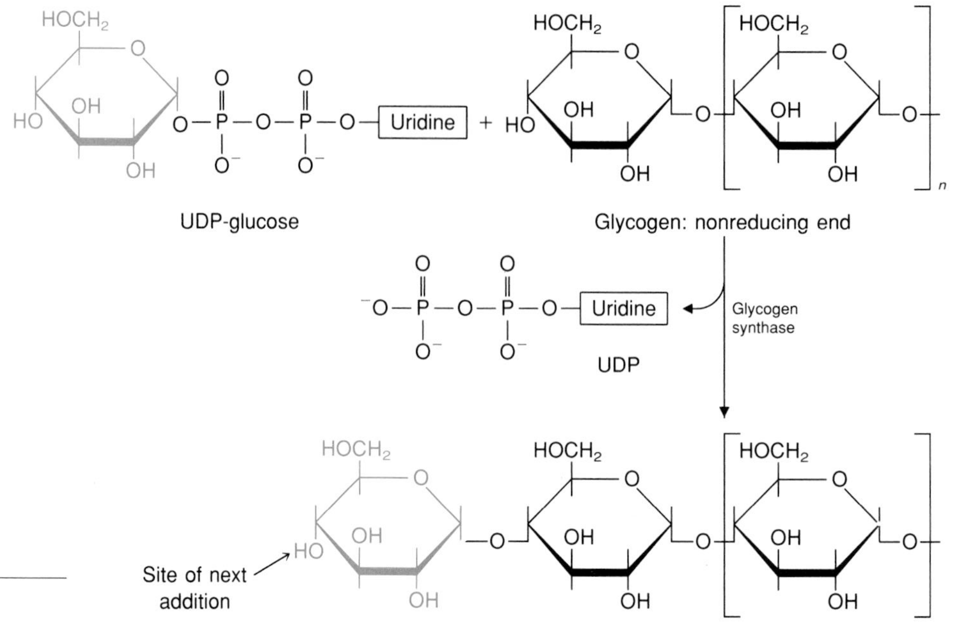

FIGURE **16.9**

The glycogen synthase reaction.

of the C-6 hydroxyl on C-1 of the oligosaccharide that will form the branch. The reaction thus creates two termini for continued action by glycogen synthase, whereas just one existed before. The branching process does not involve a large free energy change, because of the chemical similarity of (1→4) and (1→6) linkages.

RECIPROCAL RELATIONSHIP BETWEEN GLYCOGEN SYNTHESIS AND MOBILIZATION

Conditions that activate glycogen breakdown inhibit glycogen synthesis, and vice versa.

Earlier we noted that epinephrine secretion inhibits glycogen synthesis in muscle at the same time that it promotes glycogen mobilization. Glucagon has similar effects in liver. Control of both synthesis and degradation of glycogen is mediated by distinct regulatory cascades involving cyclic AMP–dependent protein kinase and reversible protein phosphorylations. However, whereas the cascade controlling glycogenolysis leads to *activation* of glycogen phosphorylase (see Figure 13.18), the cascade controlling glycogen synthesis leads to *inhibition* of glycogen synthase (Figure 16.11).

Phosphorylation of Glycogen Synthase

Glycogen synthase from vertebrate tissues is a tetrameric protein comprising four identical subunits and having a total molecular weight of about 350,000 daltons.

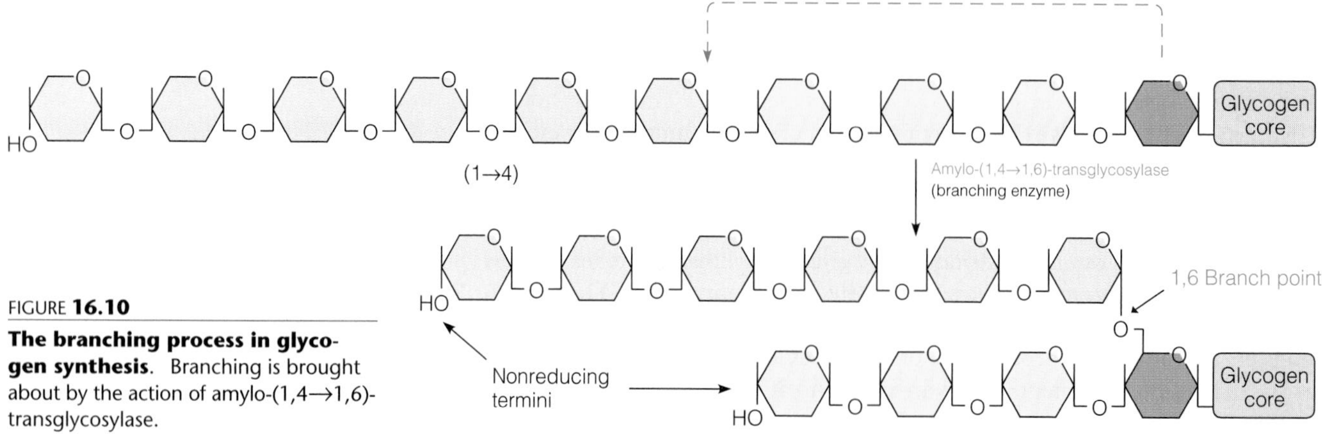

FIGURE **16.10**

The branching process in glycogen synthesis. Branching is brought about by the action of amylo-(1,4→1,6)-transglycosylase.

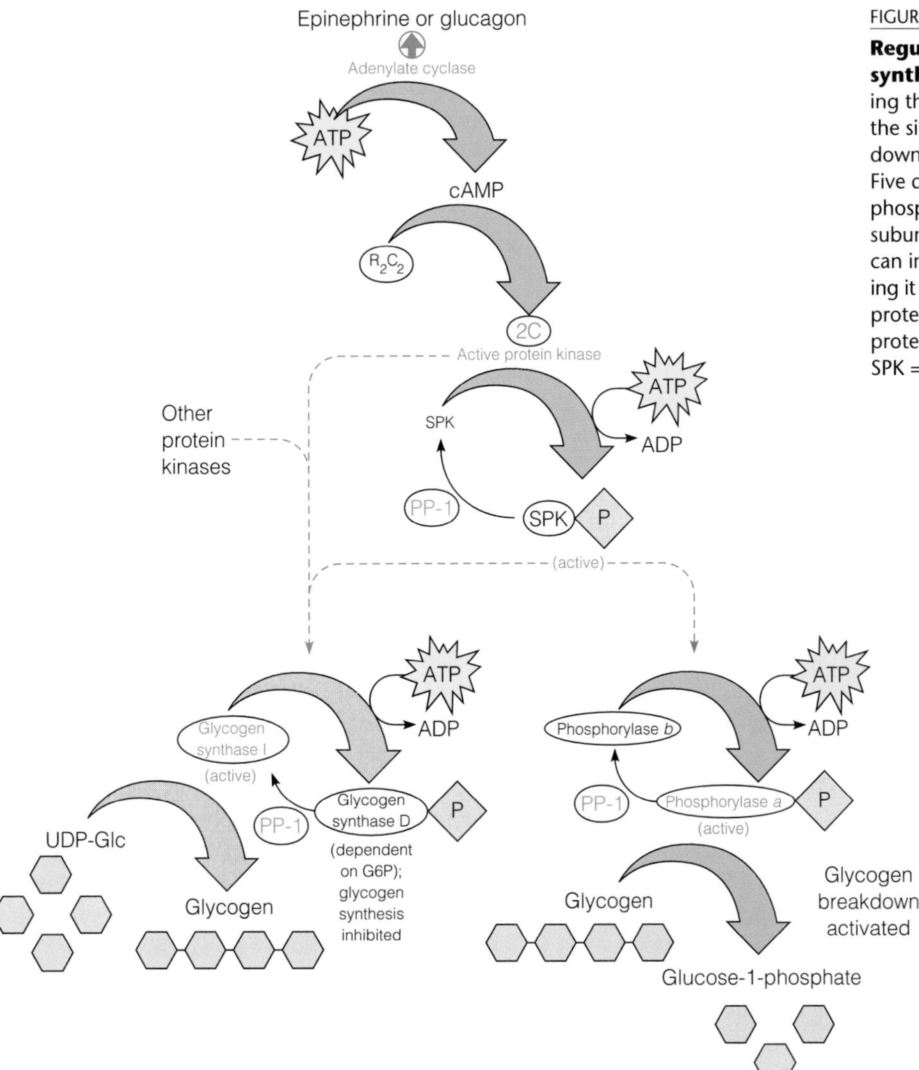

FIGURE **16.11**

Regulatory cascades affecting glycogen synthesis and mobilization. Events following the activation of adenylate cyclase promote the simultaneous activation of glycogen breakdown and inhibition of glycogen synthesis. Five different protein kinases, including the phosphorylated form of SPK and the catalytic subunit of cAMP-dependent protein kinase, can inactivate glycogen synthase I by converting it to the less active D form. PP-1 = phosphoprotein phosphatase; R_2C_2 = cAMP-dependent protein kinase; C = catalytic subunit of R_2C_2; SPK = synthase-phosphorylase kinase.

Like phosphorylase, glycogen synthase exists in phosphorylated and dephosphorylated states, with a phosphate reversibly bound to a serine residue on each subunit. Some of the phosphorylations and dephosphorylations are catalyzed by the same protein kinases and phosphatases that regulate glycogenolysis. In fact, the enzyme introduced in Chapter 13 as phosphorylase *b* kinase is also called **synthase-phosphorylase kinase (SPK),** to emphasize its dual specificity.

Glycogen synthase activity is controlled by phosphorylation, through mechanisms comparable to those controlling glycogen breakdown by phosphorylase but having reciprocal effects on enzyme activity.

The dephosphorylated form of glycogen synthase, known as **glycogen synthase I,** is the active form. Protein kinases phosphorylate serine residues and generate the less active **glycogen synthase D.** In kinetic terms, the phosphorylation of glycogen synthase from the I to the D form increases the K_M for UDP-glucose, *but only when the reactions occur in the absence of glucose-6-phosphate.* Glucose-6-phosphate is an allosteric effector, and when it is present at sufficient concentrations (> 1 mM), there is no effect on the K_M for UDP-Glc. Practically, this means that the activity of glycogen synthase D is *dependent* on the presence of glucose-6-phosphate, which is the significance of the letter D to designate the less active form of glycogen synthase. By contrast, the dephosphorylated I form of the enzyme is *independent* of the presence of glucose-6-phosphate for its activity. In resting muscle the I form predominates; during muscle contraction the major species is the D form, which is active only at high glucose-6-phosphate concentrations.

The relationship of glycogen synthase D activity to glucose-6-phosphate concentration makes good metabolic sense. When glucose-6-phosphate levels are high, activation of glycogen synthase can serve as a signal to convert some of that glucose-6-phosphate into glycogen (via glucose-1-phosphate and UDP-Glc). This effect can come into play in the absence of a hormonal stimulus—comparable to the activation of glycogen phosphorylase *b* by 5′-AMP. These effects of glucose-6-phosphate and AMP represent nonhormonal mechanisms that regulate glycogen metabolism in metabolically reasonable ways.

5′-AMP and glucose-6-phosphate regulate glycogenolysis and gluconeogenesis, respectively, in nonhormonal ways.

Cyclic AMP and Glycogen Synthase Regulation

Now let us examine the consequences of hormone release upon glycogen synthase (see Figure 16.11). Just as depicted in Figure 13.18, the activation of adenylate cyclase by epinephrine or glucagon promotes the dissociation of cAMP-dependent protein kinase (R_2C_2) to give free catalytic monomers (C). C can phosphorylate glycogen synthase I to D directly, or it can phosphorylate SPK, which in turn activates SPK to phosphorylate glycogen synthase I.

To complicate the picture somewhat, three additional protein kinases, less well understood, act on glycogen synthase I. Each of the five enzymes phosphorylates different serine residues, so that there are several different forms of glycogen synthase D, and it is an oversimplification to speak of just two forms. In general, as more sites are phosphorylated, the activity of the enzyme progressively decreases because of the following progressive changes: (1) decreased affinity for the substrate, UDP-glucose; (2) decreased affinity for the allosteric activator, glucose-6-phosphate; and (3) increased affinity for ATP and P_i, both of which tend to antagonize the activation by glucose-6-phosphate. Thus, there is a graded series of responses to changing metabolic conditions, involving a series of different protein kinases. The control mechanisms affecting the three kinases other than cAMP-dependent protein kinase and SPK are not yet well understood. Whichever kinase is used, the net effect of phosphorylating glycogen synthase is inhibition of the enzyme, with consequent inhibition of glycogen synthesis.

Note that the glycogen *synthesis* cascade can have one less cycle than the glycogen *breakdown* cascade, because C can phosphorylate glycogen synthase directly, whereas it can act on glycogen phosphorylase only through its action on SPK. The extra cycle allows for more sensitive regulation of glycogen breakdown than of its synthesis, which is consonant with needs of animals for exceedingly rapid changes in demand for energy generation in muscle. In fact, experimental observations show that the maximum rate of muscle glycogen breakdown is some 300-fold higher than that of glycogen synthesis.

A Closer Look at Glycogen Synthase Regulation: Dephosphorylation of the D Form

A number of different phosphatases act to regenerate the dephosphorylated forms of glycogen synthase and SPK. Of primary physiological importance is **phosphoprotein phosphatase-1,** also called **PP-1.** The activity of this enzyme in turn is controlled by a protein called **phosphoprotein phosphatase inhibitor, PI-1.** A phosphorylated form of PI-1 is active as an inhibitor, and the dephosphorylated form is inactive. Phosphorylation is carried out by C, the cAMP-dependent protein kinase. Thus, as shown in Figure 16.12, cAMP exerts two effects in inhibiting glycogen synthesis: (1) phosphorylation of glycogen synthase, causing its inactivation, and (2) inhibition of phosphoprotein phosphatase (PP-1), whose activity would tend to restore activity of glycogen synthase. PP-1 and other phosphoprotein phosphatases play converse roles in glycogenolysis, in which *dephosphorylation* of glycogen phosphorylase kinase (SPK) causes its inactivation.

Phosphoprotein phosphatase is also one of several sites of action of insulin. Secretion of this hormone stimulates the phosphatase activity, which tends to oppose the effects of glucagon and epinephrine.

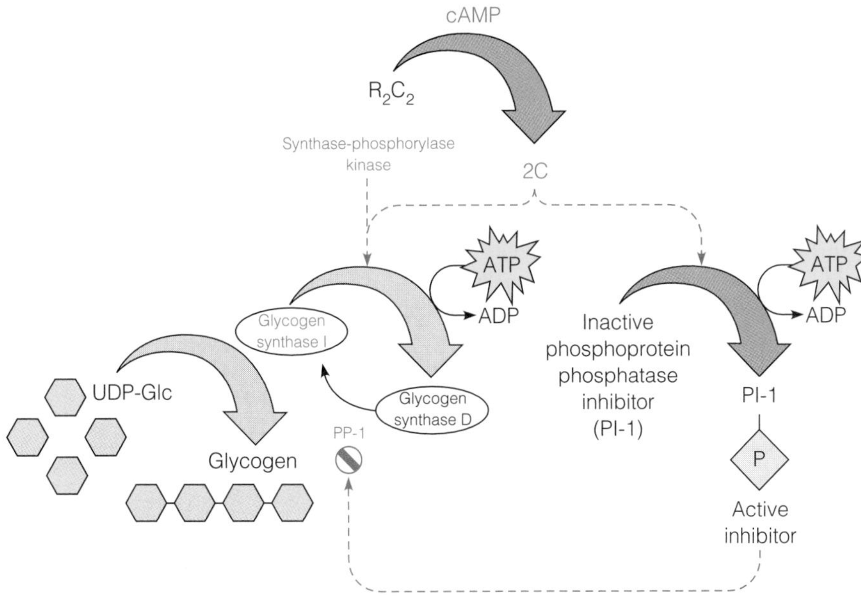

FIGURE **16.12**

Regulation of glycogen synthase activity through cAMP-mediated control of phosphoprotein phosphatase (PP-1) activity.

FUNCTIONS OF GLYCOGEN STORES IN MUSCLE AND LIVER

Glycogen is the major energy source for contraction of skeletal muscle. Because liver derives most of its own metabolic energy from fatty acid oxidation, liver glycogen has quite a different function—as a source for blood glucose, to be transported to other tissues for catabolism. Liver serves primarily as a "glucostat," sensing the blood glucose levels and adjusting synthesis and breakdown of glycogen accordingly; much of this regulation involves control of glycogen synthase and phosphorylase. As befits this role, the liver contains relatively large glycogen stores, from 2% to 8% of the weight of the organ. In liver the maximal rates of glycogen synthesis and degradation are about equal, whereas in muscle the maximal rate of glycogenolysis exceeds that of glycogen synthesis by about 300-fold. Although the enzymology of glycogen synthesis and breakdown is similar in liver and muscle, the endocrine control in liver is quite different, as we discuss in Chapter 23. The enzymes differ structurally as well.

The liver regulates blood glucose levels partly by control of its glycogen synthase and phosphorylase.

CONGENITAL DEFECTS OF GLYCOGEN METABOLISM IN HUMANS

Several of the enzymes of glycogen metabolism can undergo inactivation by mutation, though not necessarily with lethal consequences. The clinical symptoms of these conditions, called **glycogen storage diseases,** can be quite severe and usually result from storage of abnormal quantities of glycogen or storage of glycogen with abnormal properties. Accumulation of abnormal glycogen results from its failure to be broken down. Studies on these conditions have helped identify the roles of the enzymes involved in glycogen metabolism.

Human mutations affecting enzymes of glycogen metabolism can have mild or profound clinical consequences.

Among the earliest glycogen storage diseases to be described was *von Gierke's disease,* named for a German physician who studied an 8-year-old girl with chronically enlarged liver. After her death in 1929 from influenza, her liver was found to contain 40% glycogen. The glycogen appeared normal but could not be degraded by extracts of the girl's liver, only by extracts of other livers. Today it is recognized that these symptoms can result from deficiency of either glucose-6-phosphatase or the debranching enzyme. When the debranching enzyme is deficient, phosphorylase can degrade glycogen only until branch points are reached and no farther.

Table 16.2 provides information on several of the glycogen storage diseases that have been characterized. Among the most serious clinically is the type I disease, resulting from functional lack of glucose-6-phosphatase. Individuals with this condition can break down glycogen normally, but because they cannot cleave

TABLE **16.2 Human congenital defects of glycogen metabolism**

Type	Common Name	Enzyme Deficiency	Glycogen Structure	Organ Affected
Ia	von Gierke's disease	Glucose-6-phosphatase	Normal	Liver, kidney, intestine
Ib		Glucose-6-phosphatase translocase	Normal	Liver
II	Pompe's disease	$\alpha(1\rightarrow4)$Glucosidase	Normal	Generalized
III	Cori's disease	Debranching enzyme	Short outer chains	Liver, heart, muscle
IV	Andersen's disease	Branching enzyme	Abnormally long unbranched chains	Liver and other organs
V	McArdle–Schmidt–Pearson disease	Muscle glycogen phosphorylase	Normal	Skeletal muscle
VI	Hers' disease	Liver glycogen phosphorylase	Normal	Liver, leukocytes
VII		Muscle phosphofructokinase	Normal	Muscle
VIII		Liver phosphorylase kinase	Normal	Liver

G6P to glucose for release from liver to the bloodstream, they are chronically hypoglycemic. In a less severe form of this disease, blood glucose levels are normal except after stress, when the normal hyperglycemic response is inhibited. One form of this disease (type Ia) results from deficiency of glucose-6-phosphatase itself. The type Ib disease involves deficiency of a specific **translocase,** a protein that helps transport glucose-6-phosphatase to its site of activity in the lumen of the endoplasmic reticulum.

Other forms of glycogen storage diseases involve abnormalities that can be easily understood in terms of the known enzymatic defect. In type III individuals, who have a defective debranching enzyme, glycogen with very short outer branches accumulates, which leads to enlargement of the liver. By contrast, type IV disease, which is associated with a defective branching enzyme, involves accumulation of glycogen with very long outer branches. Early death from liver failure is often observed in type IV individuals. Type III, V, VI, VII, and VIII diseases have less severe symptoms. For instance, individuals with type V disease, with a deficiency of muscle glycogen phosphorylase, usually show no symptoms until about age 20. Once symptoms have appeared, the principal ones are severe muscle cramps upon exercising and failure of lactate to accumulate in blood after exercise.

Biosynthesis of Other Polysaccharides

Polysaccharide biosynthesis in general involves nucleotide-activated sugar intermediates and glycosyltransferase enzymes.

Synthesis of other polysaccharides involves the same mechanisms just presented for glycogen, particularly the use of nucleotide-linked sugars as activated biosynthetic intermediates and glycosyltransferase enzymes. In this section we consider briefly the synthesis of the most abundant and widely distributed polysaccharides—cellulose, starch, dextran, and several heteropolysaccharides.

UDP-glucose is used in some plant species for the synthesis of cellulose, a straight-chain glucose homopolymer with $\beta(1\rightarrow4)$ linkages (page 301). The mechanism is identical to that of glycogen synthesis, except for the stereochemistry of glycosidic bond formation. Other nucleotide-linked sugars are also active in polysaccharide synthesis. *Adenosine* diphosphate glucose and *cytidine* diphosphate glucose are the substrates for cellulose biosynthesis in some plants. Because of the importance of cellulose in textiles and other fiber-based products, there is great interest in the mechanism of cellulose biosynthesis. 1998 saw the first cloning of a plant gene for a cellulose synthesizing enzyme.

ADP-glucose is also the intermediate used in plant starch synthesis and in the synthesis of glycogen in bacterial cells. *Chitin,* a structural polysaccharide in the exoskeletons of arthropods and mollusks, is synthesized in insects from UDP-*N*-acetylglucosamine, the product being a $\beta(1\to4)$-linked polymer of the sugar (see Chapter 9). The synthesis of the biological lubricant *hyaluronic acid* in animals involves two enzymes, because the product is a heteropolymer with a strictly alternating sequence of glucuronic acid and *N*-acetylglucosamine. The two enzymes use as substrates UDP-glucuronate and UDP-*N*-acetylglucosamine, respectively.

An intriguing exception to the use of nucleoside diphosphate sugars as activated intermediates is the biosynthesis in some bacteria of **dextran,** an $\alpha(1\to6)$-linked polymer of glucose with $\alpha(1\to2)$, $\alpha(1\to3)$, or $\alpha(1\to4)$ branch points. The polymerization, catalyzed by **dextran sucrase,** involves a transglycosylation reaction with sucrose as the substrate:

$$n \text{ Sucrose} \longrightarrow \text{glucose}_n \text{ (dextran)} + n \text{ fructose}$$

Several bacteria growing in the human oral cavity synthesize large quantities of dextran, which contributes to formation of dental plaque—hence one concern nutritionists have about sucrose consumption.

The synthesis of sucrose itself in plants occurs via a nucleotide-linked sugar:

$$\text{UDP-glucose} + \text{fructose-6-phosphate} \longrightarrow \text{sucrose-6-phosphate} + \text{UDP}$$

$$\text{sucrose-6-phosphate} \xrightarrow{H_2O \;\to\; P_i} \text{sucrose}$$

The exergonic hydrolysis of sucrose-6-phosphate drives the process in the direction of sucrose synthesis.

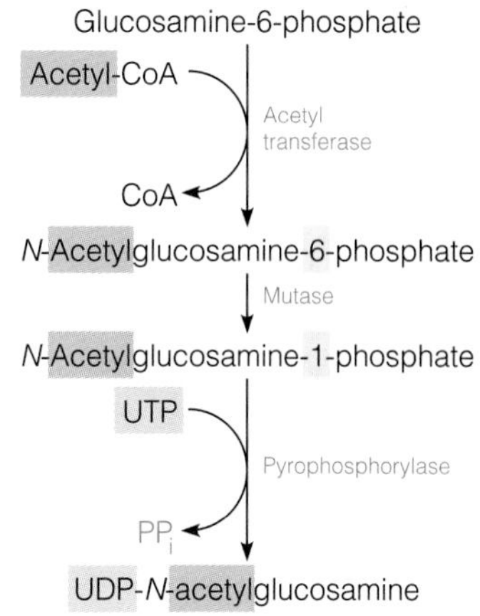

Glucosamine-6-phosphate
Acetyl-CoA → CoA (Acetyl transferase)
N-Acetylglucosamine-6-phosphate
(Mutase)
N-Acetylglucosamine-1-phosphate
UTP → PP_i (Pyrophosphorylase)
UDP-*N*-acetylglucosamine

FIGURE **16.13**

Pathway for biosynthesis of UDP-*N*-acetylglucosamine from glucosamine-6-phosphate.

Biosynthesis of Amino Sugars

Amino sugars are major constituents of *glycoconjugates*—macromolecules containing covalently bound oligosaccharide chains. As described in Chapter 9 and the next section of this chapter, the glycoconjugates include glycoproteins and glycolipids. Here we consider the biosynthesis of some of the building blocks for glycoconjugates, the amino sugars themselves.

Glucose is the metabolic precursor for all other sugars, although in most animals other sugars are available in the diet as well. The earliest-formed amino sugar, which leads to all the others, is formed not from glucose but from fructose-6-phosphate. The nitrogen comes from the amide group of glutamine, in an essentially irreversible reaction catalyzed by **glutamine:fructose-6-phosphate amidotransferase:**

Glucose is the metabolic precursor for virtually all sugars, including amino sugars.

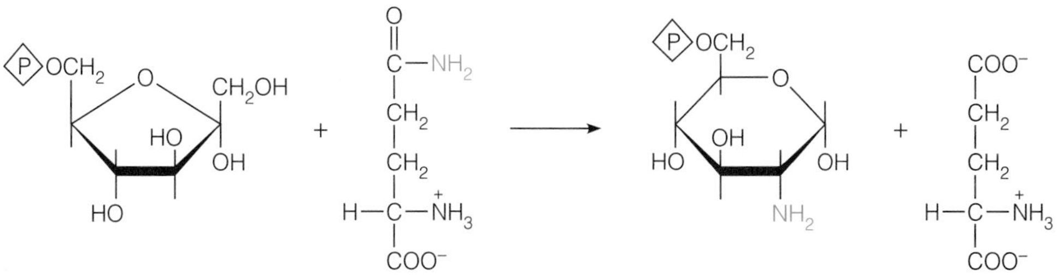

Fructose-6-phosphate + **Glutamine** ⟶ **Glucosamine-6-phosphate** + **Glutamate**

Note that the enzyme catalyzes both amide group transfer and an internal oxidoreduction of the sugar, with oxidation of C-1 coupled to reduction of C-2. Glucosamine-6-phosphate is then converted to UDP-*N*-acetylglucosamine by a three-step sequence, shown in Figure 16.13.

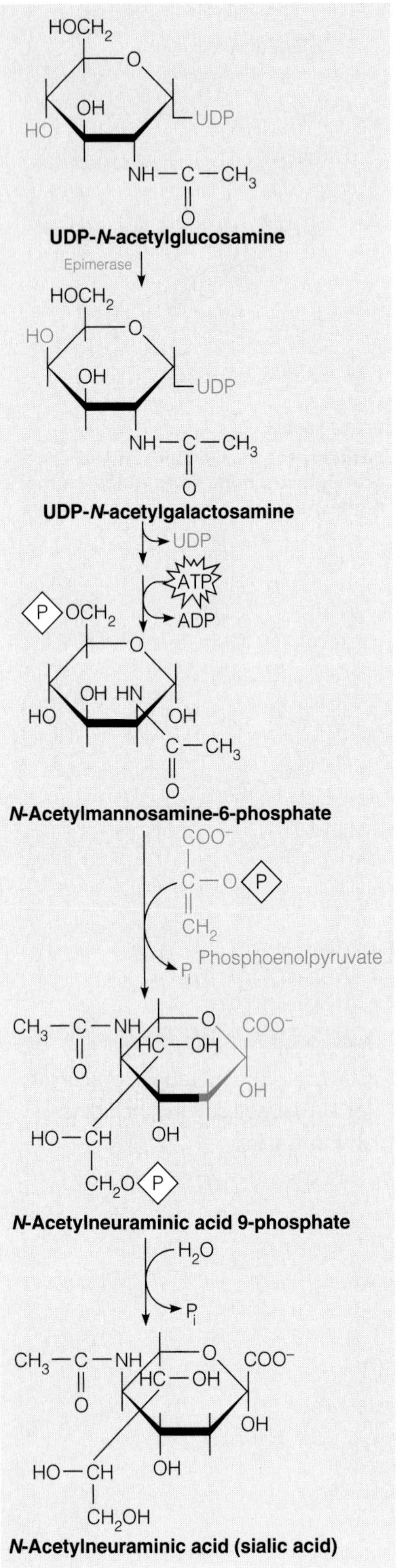

FIGURE **16.14**

Biosynthesis of *N*-acetylneuraminic acid (sialic acid) from UDP-*N*-acetylglucosamine.

Sialic acid (***N*****-acetylneuraminic acid**) is an important constituent of glycoproteins, as discussed shortly. Its synthesis (Figure 16.14) involves an epimerization of UDP-*N*-acetylglucosamine, comparable to the interchange of UDP-glucose and UDP-galactose (see pages 467–469 and Figure 13.14). The product, UDP-*N*-acetylgalactosamine, is converted in a short sequence involving epimerization at C-2 and C-4 to ***N*****-acetylmannosamine-6-phosphate.** The latter undergoes a reaction with phosphoenolpyruvate, somewhat akin to an aldol condensation. The product, ***N*****-acetylneuraminic acid 9-phosphate,** is cleaved to give *N*-acetylneuraminic acid, or sialic acid. Metabolic activation of sialic acid for oligosaccharide biosynthesis involves formation not of a nucleoside diphosphate sugar but of a nucleoside *monophosphate* sugar, **cytidine monophosphate–sialic acid,** or CMP-sialic acid, from cytidine triphosphate (CTP).

$$\text{CTP} + \text{sialic acid} \longrightarrow \text{CMP-sialic acid} + \text{PP}_i$$

Interestingly, CMP-sialic acid is synthesized in the nucleus of animal cells, whereas all other known nucleotide-linked sugars are synthesized in the cytosol.

An intriguing aspect of amino sugar metabolism involves the biosynthesis of ***N*****-acetyl-β-lactosamine,** a constituent of the carbohydrate portion of glycoproteins.

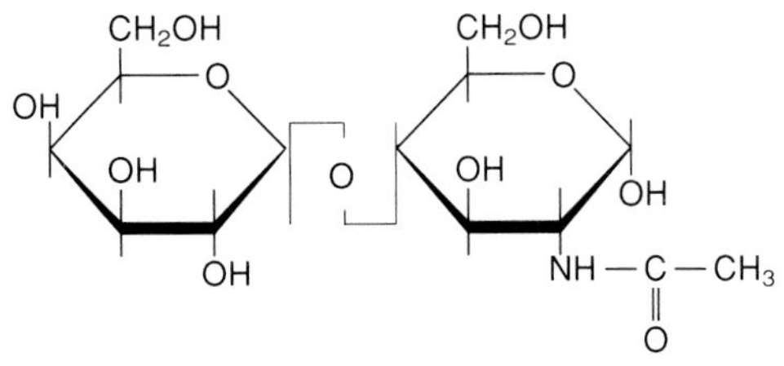

***N*-Acetyl-β-lactosamine**

The synthetic reaction is catalyzed by **galactosyltransferase.**

$$\text{UDP-galactose} + N\text{-acetylglucosamine} \longrightarrow \text{UDP} + N\text{-acetyllactosamine}$$

The common form of the enzyme contains a single polypeptide subunit. The presence of an additional subunit changes the specificity of the enzyme so that lactose is synthesized instead.

$$\text{UDP-galactose} + \text{glucose} \longrightarrow \text{UDP} + \text{lactose}$$

This modified enzyme, termed **lactose synthase,** is found in animals only in mammary gland, where it synthesizes the major sugar of milk. The polypeptide that modifies the specificity of the enzyme is the mammary gland protein *α-lactalbumin.* Synthesis of α-lactalbumin is activated hormonally in mothers shortly after giving birth. The protein combines with preexisting galactosyltransferase, changes its specificity, and activates the large amount of lactose synthesis needed for milk production.

Biosynthesis of Glycoconjugates

As indicated in Chapter 9, the covalent binding of carbohydrate to protein or lipid brings about large changes in the physical properties of these substances that allow them to serve specialized biochemical functions. For example, in some glycoproteins rich in carbohydrate (50% or more) the presence of sulfated polysaccharides or uronic acids provides a net negative charge that makes them effective biological lubricants. In glycolipids, furthermore, the carbohydrate portion provides a polar head group, allowing these components to be inserted into membranes. As a result, glycolipids occur abundantly in the outer layer of the plasma membrane of most cells. The biosynthesis of glycolipids is presented in Chapter 19.

As noted in Chapter 9, the vast structural diversity possible among carbohydrates allows them to be used for both intracellular and extracellular molecular recognition in glycoproteins, even when present in very small amounts. The carbohydrate components of cell surface glycoproteins direct the interaction of the cells with other cells and with their environment. Such interactions involve processes as diverse as cell movements in development, cell adhesion, cell motility, cell growth control, oncogenic transformation, and **endocytosis** (internalization of material from the extracellular milieu—see Chapter 18).

In this section we consider the biosynthesis of the carbohydrate components of some selected glycoconjugates—the oligosaccharide components of glycoproteins and the peptidoglycan and O-antigen portions of bacterial cell walls. The examples chosen have general biological interest, and they illustrate mechanisms used in synthesizing the vast known array of glycoconjugates. In general, those mechanisms involve the action of specific glycosyltransferases that, like glycogen synthase, transfer a monosaccharide unit from a nucleotide-linked sugar to a nonreducing end of an oligosaccharide chain or to an appropriate functional group on the protein component. As stated earlier, the nucleotide-linked sugars are synthesized in the cytosol, with the exception of CMP-sialic acid, which is formed in the nucleus. The glycosyltransferases are bound to membranes of the smooth or rough endoplasmic reticulum or the Golgi apparatus.

A particularly fascinating part of glycoprotein synthesis is the protein sorting, or "traffic," that occurs as oligosaccharide chains are growing. In some cases the structure of the oligosaccharide chain or chains on a protein serves as a molecular recognition determinant, directing that protein to the proper intracellular location for the next step in oligosaccharide synthesis and, ultimately, to the site where the mature protein will reside.

O-LINKED OLIGOSACCHARIDES: BLOOD GROUP ANTIGENS

Mammalian glycoproteins are classified as **O-linked** or **N-linked.** As described in Chapter 9, N-linked glycoproteins contain an *N*-acetylglucosamine residue linked to the amide nitrogen of an asparagine residue in the protein. The most common O linkage involves a terminal *N*-acetylgalactosamine residue in the oligosaccharide linked to a serine or threonine residue. The A, B, and O blood group substances found on erythrocyte surfaces are the best-known examples of O-linked glycoproteins.

Modifications of the carbohydrate chains of N-linked glycoproteins help target these proteins to their intracellular or extracellular destinations.

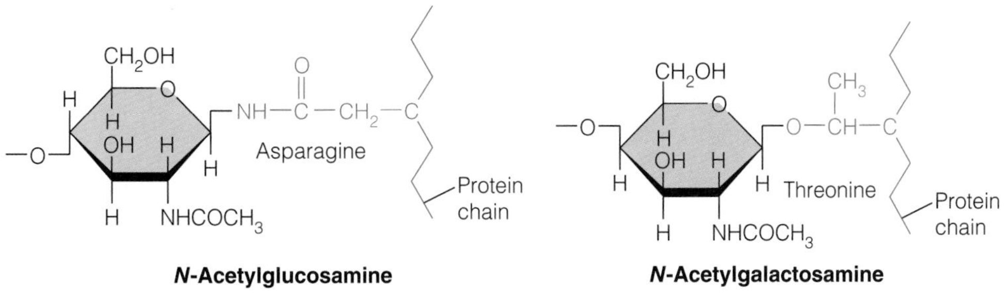

CH_2OH, O, H, H, OH, H, H, H, $NHCOCH_3$, —O—, NH—C(=O)—CH_2—C, Asparagine, Protein chain

***N*-Acetylglucosamine**

CH_2OH, O, H, OH, H, H, H, H, $NHCOCH_3$, —O—, O—CH(CH_3)—C, Threonine, Protein chain

***N*-Acetylgalactosamine**

The sugars in these blood group oligosaccharides are fucose (Fuc), galactose (Gal), *N*-acetylgalactosamine (GalNAc), and sialic acid (Sia). As shown in Figure 16.15, the synthetic pathway starts with a protein carrier (R). A series of

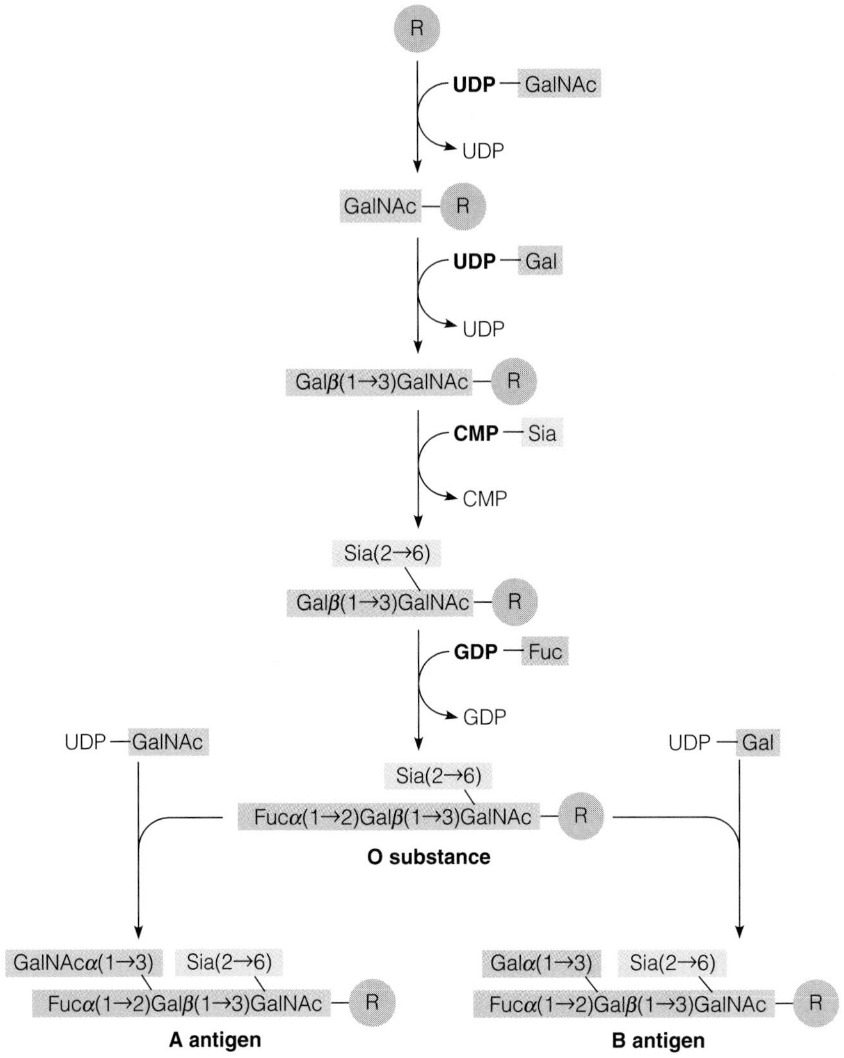

FIGURE **16.15**

Biosynthesis of O-linked oligosaccharide units on glycoproteins of the O, A, and B blood group substances.

The presence or absence of one or two specific glycosyltransferases involved in O-linked oligosaccharide synthesis determines blood type A, B, AB, or O.

membrane-bound glycosyltransferase reactions follows, each of which uses a nucleotide-linked sugar substrate. The pathway leads to two different pentasaccharide products, the A antigen and the B antigen. The last enzyme in the synthesis of the A antigen is a glycosyltransferase involving UDP-GalNAc; in synthesis of the B antigen the last reaction involves UDP-Gal. Individuals with type A blood have the former glycosyltransferase and synthesize the A antigen, whereas type B individuals have the galactosyltransferase and synthesize the B antigen. AB individuals are heterozygotes; that is, they have both enzymes and carry both blood group substances on their erythrocytes. Type O individuals carry neither glycosyltransferase and hence synthesize neither antigen; they carry the tetrasaccharide identified as O substance in Figure 16.15.

N-LINKED OLIGOSACCHARIDES: GLYCOPROTEINS

Quite a different process is involved in the synthesis of N-linked oligosaccharides, common to most glycoproteins. Here, oligosaccharide assembly occurs not on the polypeptide chain but on a *lipid-linked intermediate.* A precursor oligosaccharide is then transferred to a polypeptide chain, which is itself still in the midst of being synthesized. This type of reaction is called **cotranslational.** Finally, the transferred oligosaccharide is subject to further processing steps as it passes from the rough and smooth endoplasmic reticulum through the Golgi apparatus.

The many N-linked glycoproteins of known structure can be categorized according to three basic oligosaccharide structures, as summarized in Figure 16.16—**complex, hybrid,** and **high-mannose.** All of the known N-linked oligosaccharides have a common core pentasaccharide structure (the boxed areas in the figure; see also page 308 in Chapter 9).

$$\begin{array}{l}\text{Man}\alpha(1\rightarrow6)\searrow \\ \qquad\qquad\qquad \text{Man}\beta(1\rightarrow4)\text{GlcNAc}\beta(1\rightarrow4)\text{GlcNAc-Asn} \\ \text{Man}\alpha(1\rightarrow3)\nearrow\end{array}$$

That core is assembled as part of a larger oligosaccharide intermediate linked to the isoprenoid lipid compound **dolichol phosphate.**

$$\text{H(CH}_2\text{—}\overset{\overset{\text{CH}_3}{|}}{\text{C}}\text{=CH—CH}_2)_{18-20}\text{—CH}_2\text{—}\overset{\overset{\text{CH}_3}{|}}{\text{CH}}\text{—CH}_2\text{—CH}_2\text{—O—}\underset{\underset{\text{O}^-}{|}}{\overset{\overset{\text{O}}{\|}}{\text{P}}}\text{—O}^-$$

Dolichol phosphate

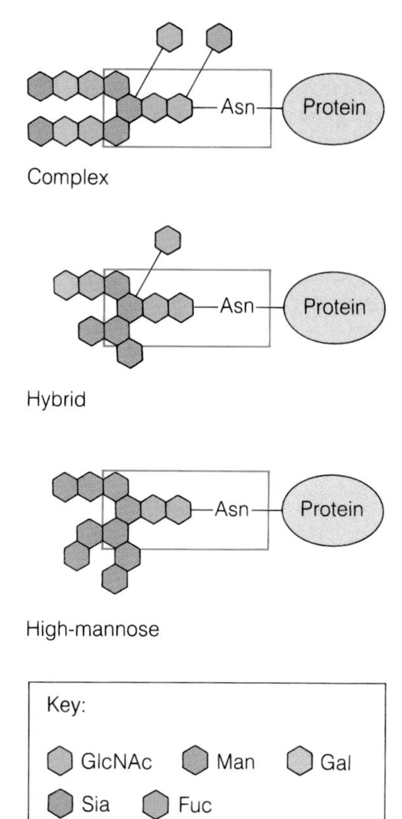

FIGURE **16.16**
Structures of the major types of asparagine-linked (N-linked) oligosaccharides. The red boxes contain the core common to all known N-linked structures.

In vertebrate tissues, dolichol contains 18 to 20 isoprenoid units with two *trans* double bonds, the remainder being *cis*, except for a saturated terminal isoprene unit. Dolichol is synthesized from the same pathway that yields cholesterol, other sterols, and other isoprenoid compounds (see Chapter 19) and is then converted to dolichol phosphate (Dol-P) by a CTP-requiring kinase.

Synthesis of the Lipid-Linked Intermediate

The first step in glycoprotein synthesis is assembly of a lipid-linked oligosaccharide intermediate, which serves as precursor to all known N-linked oligosaccharides. That process takes place in the endoplasmic reticulum (ER). Figure 16.17 summarizes the biosynthetic pathway leading to this intermediate. The first seven sugars are transferred to dolichol phosphate from nucleoside diphosphate sugars, UDP-GlcNAc and GDP-Man. Each reaction is catalyzed by a separate glycosyltransferase. The first of these enzymes is specifically inhibited by the antibiotic *tunicamycin,* which blocks the reaction of UDP-GlcNAc and Dol-P and hence inhibits the synthesis of all N-linked glycoproteins.

The next seven glycosyltransferases utilize as substrates not nucleotide-linked sugars but dolichol-linked sugars, Dol-P-Man and Dol-P-Glc. These in turn are synthesized from Dol-P plus GDP-Man or UDP-Glc, respectively. During this stage the lipid-linked intermediate ($\text{Man}_5\text{GlcNAc}_2$-P-P-Dol, the third intermediate in Figure 16.17) is somehow translocated from the exterior surface of the ER membrane to the luminal, or interior, side. This translocation explains the need for dolichol phosphate–linked sugars for the subsequent glycosylations, because the nucleoside diphosphate–linked sugars cannot penetrate to the luminal side of the membrane.

The polysaccharide chains of N-linked glycoproteins are formed while linked to a lipid compound, dolichol phosphate.

The next step, transfer of the oligosaccharide unit to a polypeptide acceptor, also occurs in the lumen, catalyzed by a specific **oligosaccharyltransferase.** The acceptor site is an asparagine residue in the sequence Asn–X–Ser/Thr, where X is any amino acid. The acceptor site must be accessible in a loop or a bend in the polypeptide chain, which is probably why the transfer occurs simultaneously with translation of the acceptor polypeptide. The three glucosyl residues on the transferred oligosaccharyl unit somehow facilitate the transfer but are not absolutely required. For virtually all glycoproteins, these glucosyl residues are removed in subsequent processing steps.

Processing of the Oligosaccharides

Processing of the oligosaccharide-linked polypeptides begins in the lumen of the rough endoplasmic reticulum and continues as the nascent glycoprotein moves into the smooth ER and ultimately through various cisternae of the Golgi apparatus. A great variety of oligosaccharide structures is generated during this

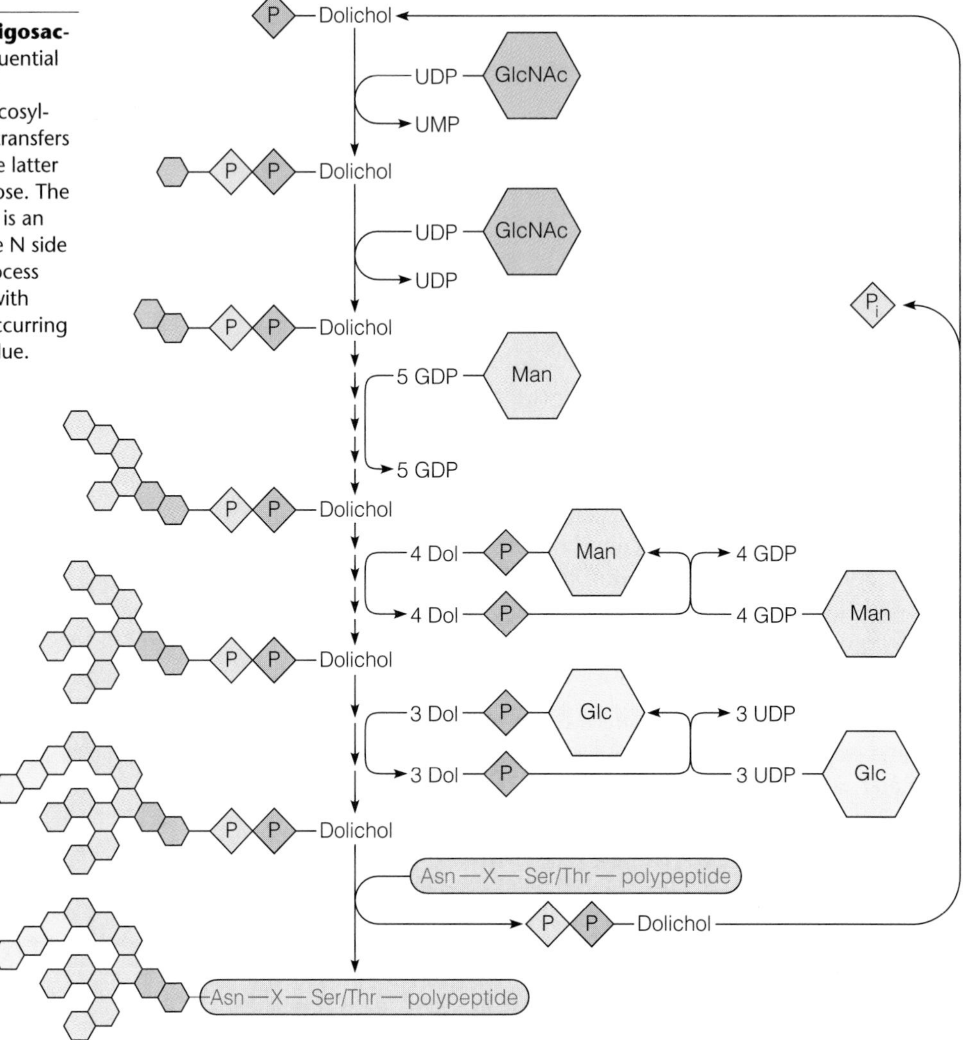

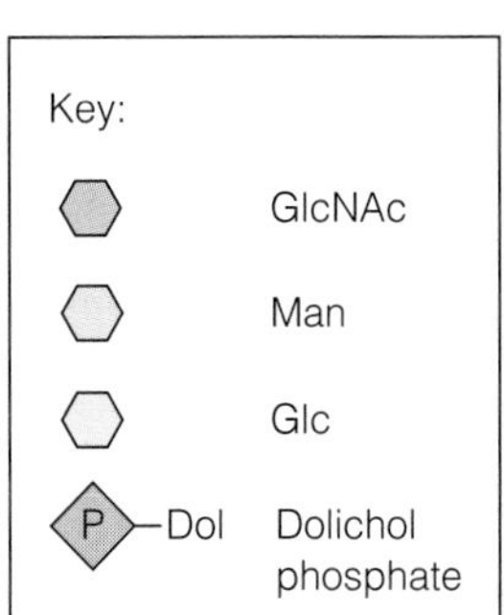

FIGURE **16.17**

Biosynthesis of the lipid-linked oligosaccharide intermediate. The five sequential mannosyl transfer reactions from GDP-mannose are catalyzed by separate glycosyltransferases, as are the four mannosyl transfers from dolichol phosphate mannose. The latter is synthesized in turn from GDP-mannose. The acceptor site on the polypeptide chain is an asparagine residue two positions to the N side of a serine or threonine. The whole process occurs in the endoplasmic reticulum, with translocation to the lumen of the ER occurring after transfer of the fifth mannose residue.

processing. Part of the diversity arises from differences in conformation of the protein moiety in the vicinity of a carbohydrate chain, which affect the accessibility of the chain to glycosidases and glycosyltransferases. Oligosaccharide chains probably generate recognition sites for targeting each processed compound to different sites both inside and later outside the Golgi, where specific membrane-bound processing enzymes are found (Figure 16.18). In virtually all cases, processing begins (after transfer to a polypeptide chain) with removal of the three glucosyl residues in the rough ER, followed by removal of some of the mannosyl residues in the Golgi apparatus. These processess are critical to the proper folding of some proteins in the ER. Those glycoproteins destined to be of the "complex" type are further processed by addition of *N*-acetylglucosamine, followed by further trimming of mannosyl residues. Fucosyl, galactosyl, and sialyl residues are added from appropriate nucleotide-linked sugars by specific glycosyltransferases. Comparable pathways lead to the other classes of glycoproteins.

Processing and Intracellular Protein Traffic

As noted earlier, oligosaccharide chains help direct glycoproteins to their ultimate intracellular destinations. A graphic demonstration is provided by the generation of mannose-6-phosphate residues during processing of the glycoproteins known as lysosomal acid hydrolases. All known enzymes of this type contain between one and five mannose-6-phosphate units, which evidently help target proteins to, and through, the lysosomal membrane. The role of mannose-6-phosphate in di-

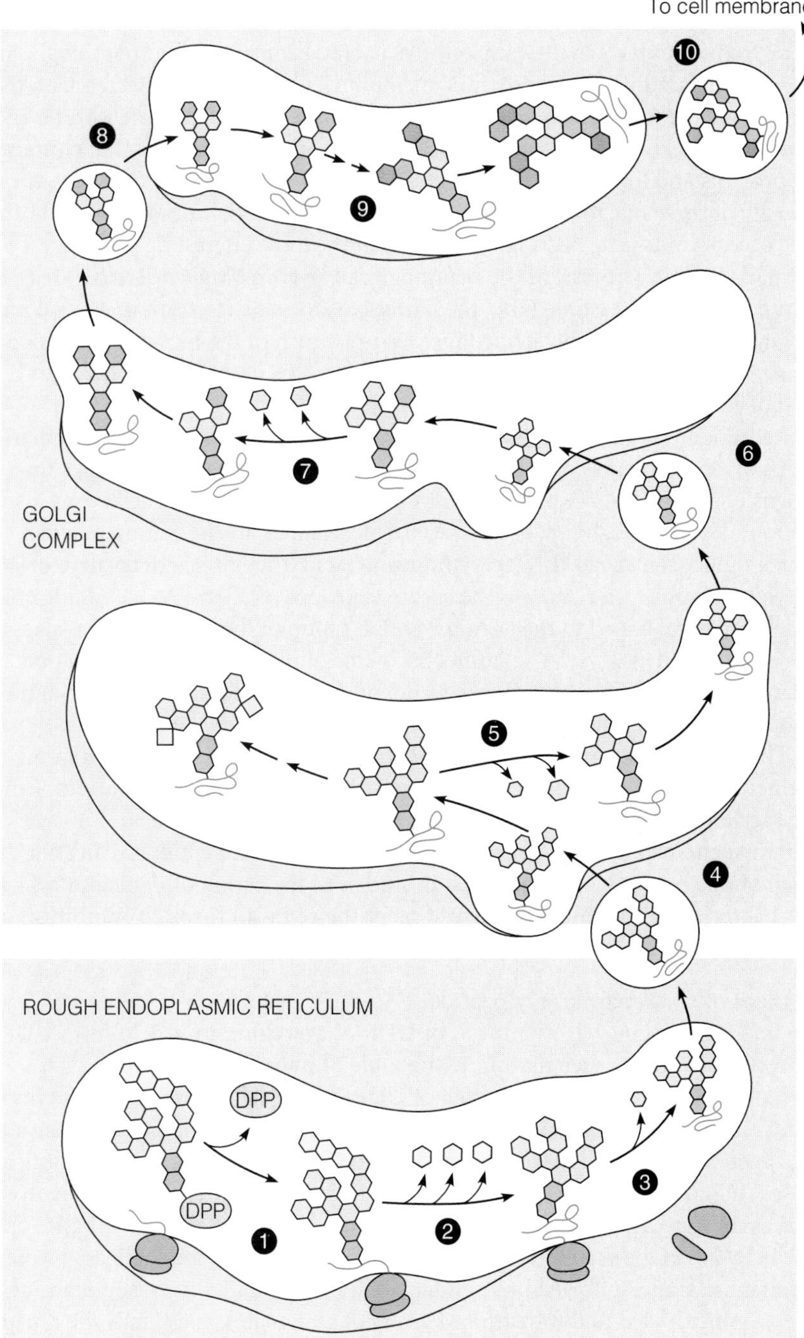

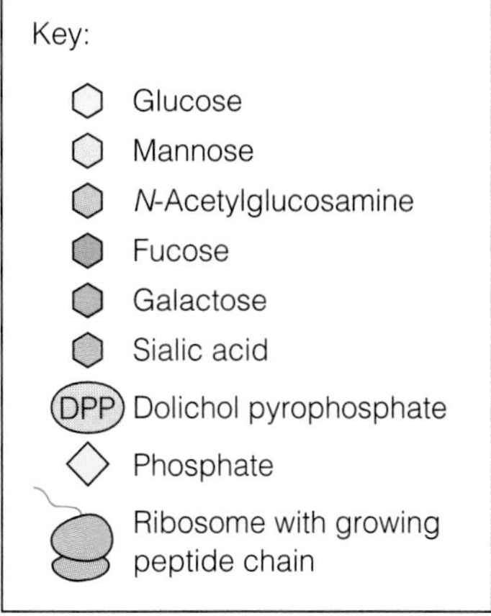

FIGURE **16.18**

Schematic pathway of oligosaccharide processing on newly synthesized glycoproteins. Starting at the bottom of the figure, the oligosaccharide is first transferred from its dolichol phosphate carrier to a polypeptide chain (red) while the latter is still being synthesized on the ribosome (step 1). As the peptide chain grows within the endoplasmic reticulum, monosaccharides may be cleaved from the nonreducing ends of the oligosaccharide (step 2). After completion of the polypeptide chain synthesis (step 3), the nascent glycoprotein is carried in a transport vesicle (step 4) to the Golgi apparatus, where further modification of the oligosaccharide occurs. New monosaccharides may be added and others removed in a multistep process involving several transfers to different parts of the Golgi (steps 5–9). The completed glycoprotein is transported to its ultimate destination in a membrane (step 10) or is secreted.

recting protein traffic is confirmed by the existence of a rare and fatal congenital abnormality called **I-cell disease.** In this condition, lysosomes cannot perform their normal function of intracellular digestion, because a deficiency of the first glycosyltransferase that generates a mannose-6-phosphate residue causes the acid hydrolase content of lysosomes to be very low. Lysosomal enzymes, manufactured in the ER, cannot be targeted to their ultimate destinations but instead are secreted into the extracellular milieu. When purified and characterized, these enzymes are also found to lack mannose-6-phosphate in their oligosaccharide chains. Our understanding of the molecular recognition involved in targeting these proteins has been advanced by the recent cloning of the gene for the lysosomal membrane mannose-6-phosphate receptor.

MICROBIAL CELL WALL POLYSACCHARIDES: PEPTIDOGLYCAN

Recall from Chapter 9 that Gram-positive bacteria contain a rigid peptidoglycan cell wall surrounding the cytoplasmic membrane. Gram-negative bacteria, on the other hand, contain a third layer, in addition to the cytoplasmic membrane and peptidoglycan layer. This outer membrane is a complex structure that contains lipoproteins and lipopolysaccharides (see Figure 9.25, page 306). The structural diversity of these macromolecules among different bacterial species is enormous, and space permits introduction of only two of the most interesting pathways. The first pathway, biosynthesis of the peptidoglycan layer in *Staphylococcus aureus,* is of interest for two reasons. First, the pathway is the site of action of several important antibiotics, notably penicillins. Second, much of the biosynthetic process occurs on the outside of the cell, where there is no ready supply of energy in the form of ATP.

Recall from Figure 9.25 that bacterial peptidoglycans consist of polymeric chains of amino sugars cross-linked by oligopeptide chains, to form a huge three-dimensional network in which the entire peptidoglycan layer of a cell is one giant macromolecule. The glycan, or polysaccharide, chain is an alternating polymer of *N*-acetylglucosamine and *N*-acetylmuramic acid, the latter a derivative of *N*-acetylglucosamine. In *S. aureus* the carboxyl groups of all the *N*-acetylmuramic acid residues are linked to the terminal amino group of the tetrapeptide L-alanyl-D-γ-isoglutaminyl-L-lysyl-D-alanine. Each cross-link takes the form of a pentaglycine chain that joins the carboxyl group of a D-alanine residue to the ε-amino group of a lysine residue in an adjacent oligopeptide.

The biosynthesis of the *S. aureus* peptidoglycan can be considered in three distinct stages: (1) synthesis of *N*-acetylmuramylpeptide, (2) formation of the polysaccharide chain by polymerization of *N*-acetylglucosamine and *N*-acetylmuramylpentapeptide, and (3) cross-linking of individual peptidoglycan strands. Much of this pathway was elucidated in studies of the action of penicillin, which kills bacterial cells by blocking synthesis of the cell wall through inhibition of peptidoglycan synthesis.

HOCH2 O OH HO UDP NHCOCH3

UDP-*N*-acetylglucosamine

COO⁻ C—O—Ⓟ CH2 Phosphoenolpyruvate → P_i

HOCH2 O O HO UDP NHCOCH3 $CH_2{=}C—COO^-$

H^+ + NADPH → NAD^+

HOCH2 O O HO UDP NHCOCH3 $CH_3—CH(H)—COO^-$

UDP-*N*-acetylmuramic acid

Synthesis of *N*-Acetylmuramylpeptide

The first stage begins with synthesis of UDP-*N*-acetylmuramic acid from UDP-*N*-acetylglucosamine (see margin.) One mole of phosphoenolpyruvate is transferred, to give the 3-enolpyruvyl ether of UDP-*N*-acetylglucosamine. The pyruvyl group is reduced by NADPH to give the 3-O-D-lactyl ether of *N*-acetylglucosamine, or *N*-acetylmuramic acid. Then a pentapeptide is built up in stepwise fashion, as shown in Figure 16.19. No messenger RNA template or ribosomes are involved here, as they are in the synthesis of polypeptide chains in protein synthesis. The specificity lies in the sequential actions of a series of ATP-dependent ligases, which add first L-alanine, followed by D-glutamate (later amidated to D-isoglutamine), then L-lysine (linked to the γ-carboxyl group of glutamate), and finally the dipeptide D-alanyl-D-alanine. One of these two D-alanyl residues will be removed at a later step (see Figure 16.21).

Formation of the Peptidoglycan Chain

The next stage is polymerization of *N*-acetylglucosamine and *N*-acetylmuramylpentapeptide to give a linear peptidoglycan chain. This process involves a lipid carrier, **undecaprenol phosphate,** comparable to dolichol phosphate, which we just encountered in the synthesis of N-linked oligosaccharides.

$$H(CH_2—\overset{CH_3}{\overset{|}{C}}{=}CH—CH_2)_{10}—CH_2—\overset{CH_3}{\overset{|}{C}}{=}CH—CH_2—O—\overset{O}{\overset{\|}{\underset{O^-}{\underset{|}{P}}}}—O^-$$

Undecaprenol phosphate

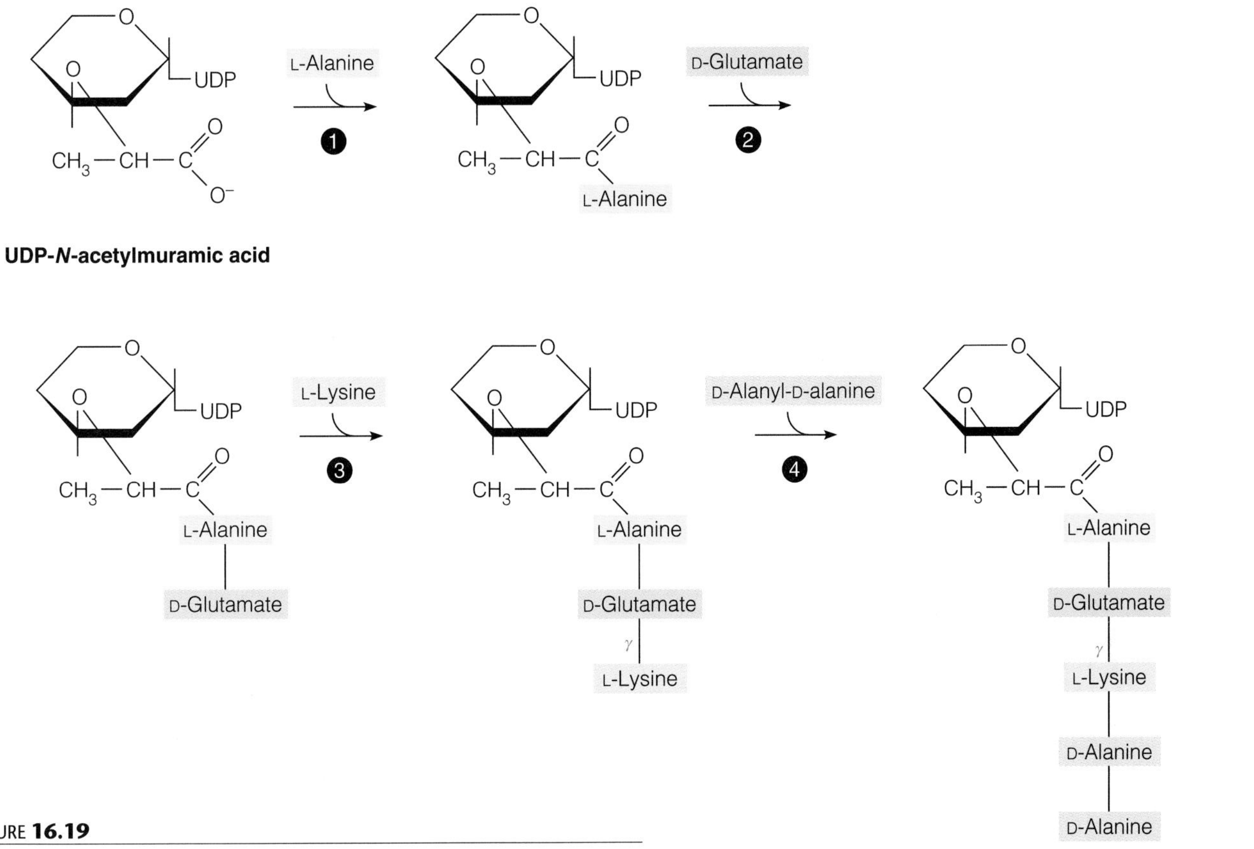

UDP-*N*-acetylmuramic acid

L-Alanine ❶ D-Glutamate ❷ L-Lysine ❸ D-Alanyl-D-alanine ❹

UDP-*N*-acetylmuramylpentapeptide

FIGURE **16.19**

Biosynthesis of UDP-*N*-acetylmuramylpentapeptide from UDP-*N*-acetylmuramic acid. The sugar structure is shown in outline form.

Undecaprenol phosphate is a 55-carbon compound containing 11 isoprenoid units, with phosphate linked at the terminus. To this phosphate is transferred the *N*-acetylmuramylpentapeptide moiety from UDP-*N*-acetylmuramylpentapeptide (Figure 16.20, step 1). This compound then accepts *N*-acetylglucosamine from UDP-*N*-acetylglucosamine (step 2), followed by the sequential addition of five glycyl residues, from glycyl transfer RNA (step 3). It seems likely that the phospholipid carrier then transports the peptidodisaccharide unit through the membrane at this stage (step 4), because polymerization—addition to the reducing end of a preexisting peptidoglycan chain—occurs on the outside of the cell wall (step 5). The antibiotics *bacitracin* and *vancomycin* inhibit specific steps in this process at the sites shown in Figure 16.20.

Cross-Linking of Peptidoglycan Strands

Finally, the cross-linking occurs between adjacent chains, also outside the cell. This involves a **transpeptidation** reaction, with the cleavage of one peptide bond providing the energy needed to drive the formation of another peptide bond. This means that the free energy needed to drive cross-link formation was built into the structures while they were still accessible to ATP, inside the cell. As shown in Figure 16.21, the transpeptidation involves nucleophilic attack by the free terminal amino group of the pentaglycine chain on the amide carbon linking the terminal D-alanines in an adjacent chain.

Much of the biosynthesis of peptidoglycan oligosaccharide chains occurs outside the cell wall, using activated intermediates synthesized inside the cell.

The cross-linking reaction is the target for the action of two important classes of antibiotics, the *penicillins* and the *cephalosporins.* Penicillin is thought to react irreversibly with the transpeptidase that catalyzes cross-linking. That enzyme normally forms an acyl-enzyme intermediate, via the penultimate D-alanine of the pentapeptide chain (Figure 16.21, right). Penicillin evidently resembles the

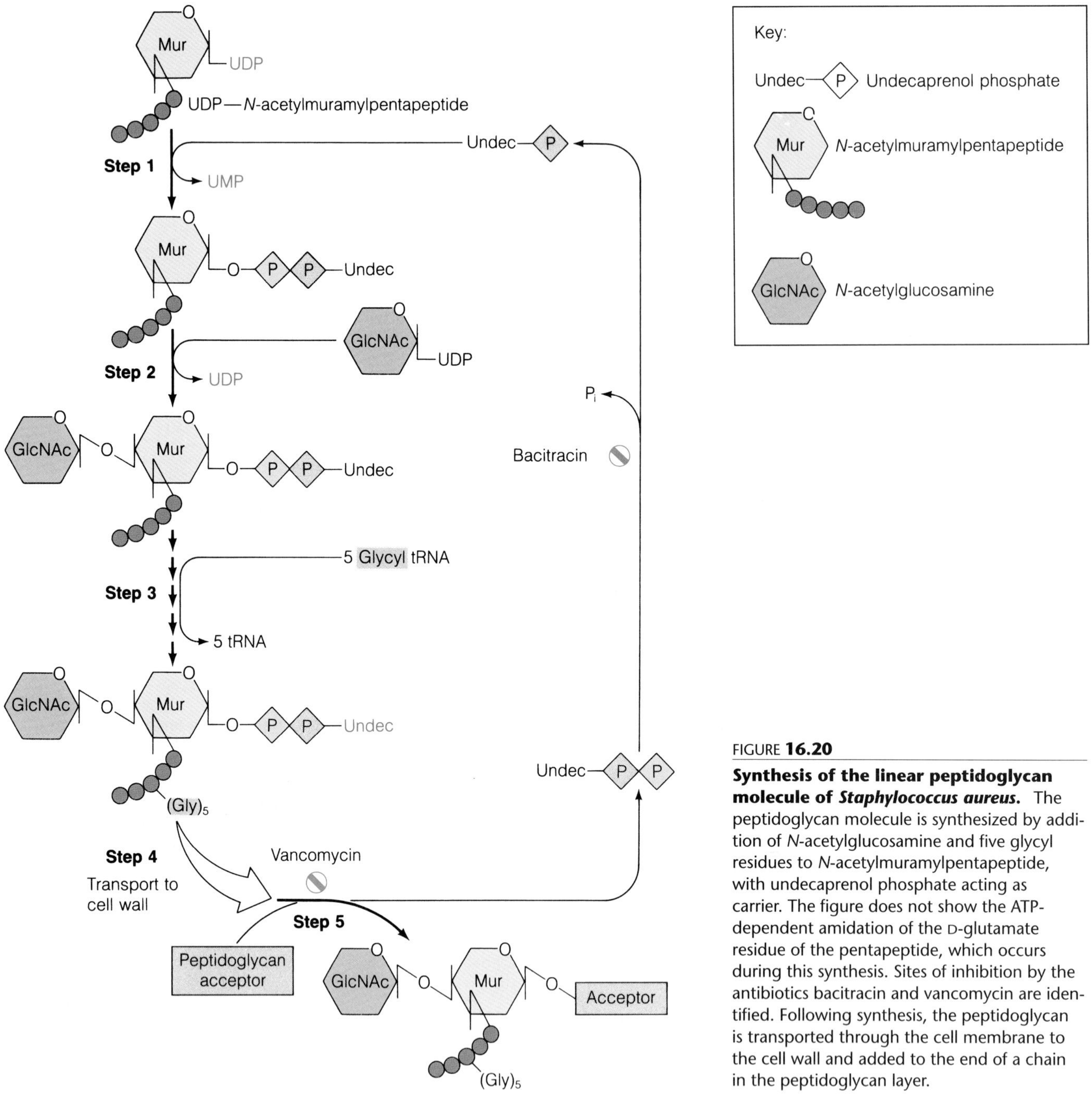

Mur
UDP
UDP—*N*-acetylmuramylpentapeptide
Step 1
Undec—P
UMP
Mur—O—P—P—Undec
GlcNAc—UDP
Step 2
UDP
GlcNAc—O—Mur—O—P—P—Undec
5 Glycyl tRNA
Step 3
5 tRNA
GlcNAc—O—Mur—O—P—P—Undec
(Gly)$_5$
Step 4
Transport to cell wall
Vancomycin
Step 5
Peptidoglycan acceptor
GlcNAc—O—Mur—O—Acceptor
(Gly)$_5$
Undec—P—P
P_i
Bacitracin

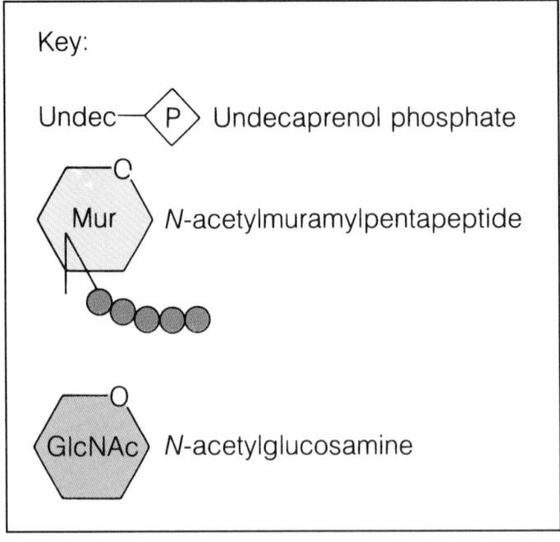

FIGURE **16.20**

Synthesis of the linear peptidoglycan molecule of *Staphylococcus aureus*. The peptidoglycan molecule is synthesized by addition of *N*-acetylglucosamine and five glycyl residues to *N*-acetylmuramylpentapeptide, with undecaprenol phosphate acting as carrier. The figure does not show the ATP-dependent amidation of the D-glutamate residue of the pentapeptide, which occurs during this synthesis. Sites of inhibition by the antibiotics bacitracin and vancomycin are identified. Following synthesis, the peptidoglycan is transported through the cell membrane to the cell wall and added to the end of a chain in the peptidoglycan layer.

The antibiotic activity of penicillin derives from its interference with extracellular peptidoglycan synthesis.

terminal dipeptide of this structure to the point that it can also react with the transpeptidase. The reaction is driven in part by the strain built into the four-membered **lactam ring** of penicillin, for that ring opens during the reaction. Penicillin has been widely studied as an "ideal" antibiotic, because the cross-linking reaction has no counterpart in animal metabolism. Because the bacterial cell must continue to synthesize cell wall in order to grow and divide, inhibition of a step in this process provides a completely specific way to interfere with the growth of bacterial pathogens. Unfortunately, resistance to penicillin can be acquired. This resistance usually involves the synthesis, directed by an extrachromosomal gene, of **lactamase,** an enzyme that hydrolyzes the lactam ring of penicillin and destroys its ability to interfere with peptidoglycan synthesis.

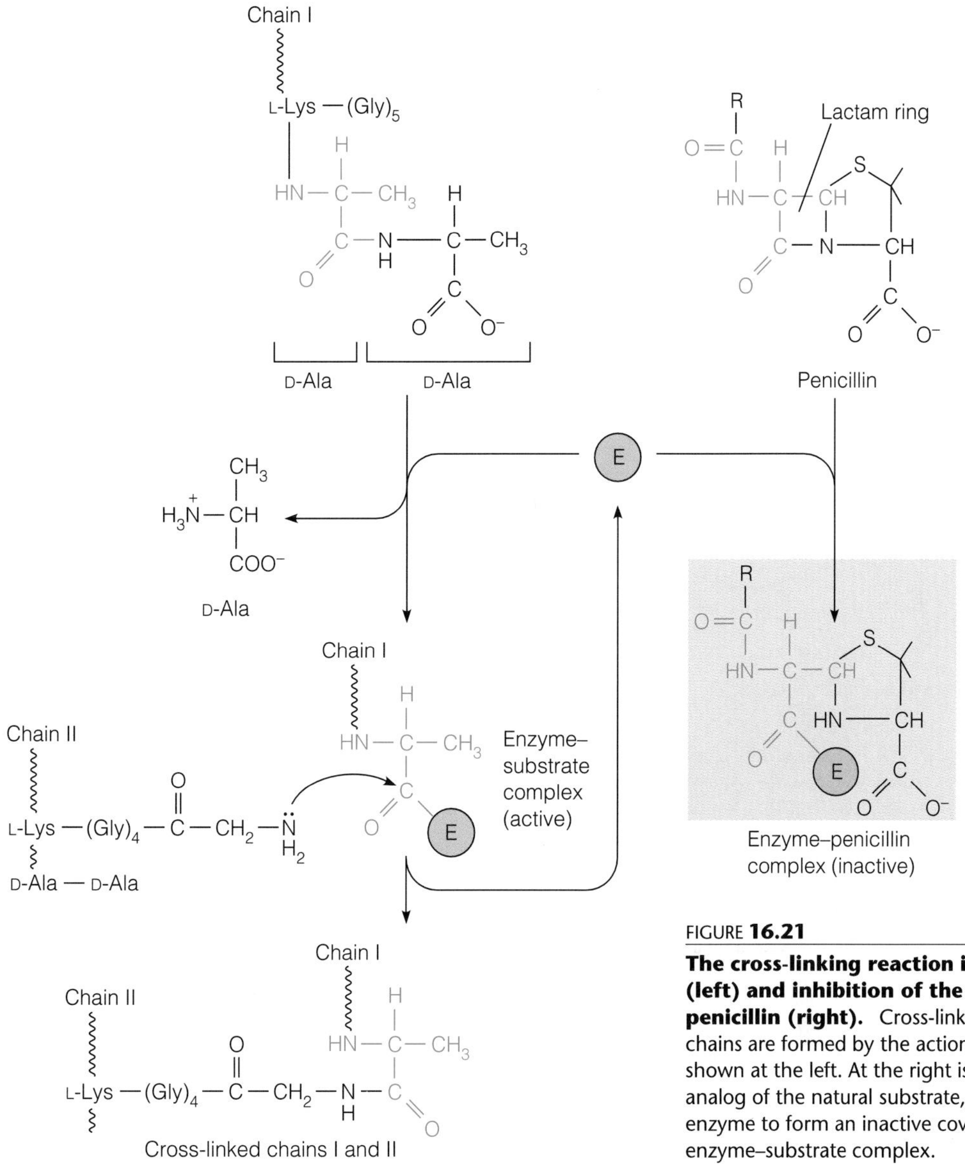

FIGURE **16.21**

The cross-linking reaction in peptidoglycan synthesis (left) and inhibition of the transpeptidase enzyme, E, by penicillin (right). Cross-links between adjacent peptidoglycan chains are formed by the action of a transpeptidase enzyme, as shown at the left. At the right is shown how penicillin, a structural analog of the natural substrate, reacts with the active form of the enzyme to form an inactive covalent complex that resembles the enzyme–substrate complex.

MICROBIAL CELL WALL POLYSACCHARIDES: O ANTIGENS

The second example we consider is the biosynthesis of the O antigen of the Gram-negative *Salmonella typhimurium.* The O antigen (not to be confused with the type O blood group substance of human erythrocytes) is the major lipopolysaccharide component of the outer membrane. Lipopolysaccharides contain repeating oligosaccharide units that are attached to a basal core polysaccharide. The latter is, in turn, attached to a complex called lipid A. The repeating oligosaccharide units protrude like minute fibers from the outer membrane surface. Because they represent the outer surface of the cell and are composed of specific carbohydrate structures, these fibers provoke strong immune reactions—hence the term *O antigen* that is applied to the fibers. Production of antibodies directed against O antigens represents a primary defense mechanism used by vertebrates against bacterial infection. Bacteria have responded evolutionarily by being able to change O-antigen structure through extremely rapid genetic change, the mechanism of which is not fully understood. Consequently, hundreds of different serotypes (immunologically distinct strains) of bacteria such as *S. typhimurium* exist, each with a different O antigen repeating unit.

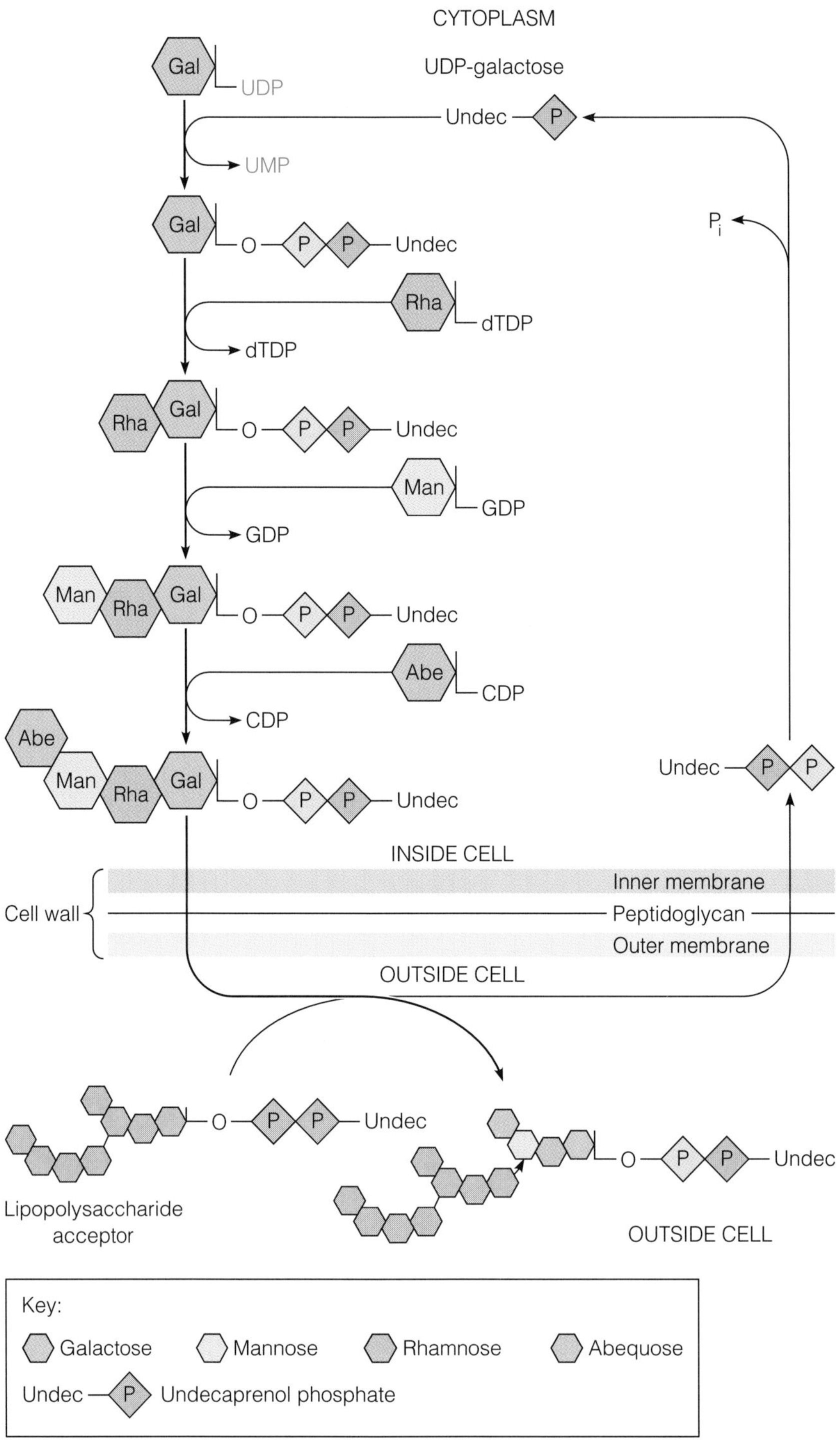

FIGURE **16.22**

Biosynthesis of the repeating oligosaccharide unit of the O antigen of *Salmonella typhimurium*. The first four reactions occur within the inner membrane. Transfer of the activated tetrasaccharide unit to the unactivated terminus of a growing polysaccharide unit occurs on the outside of the outer membrane.

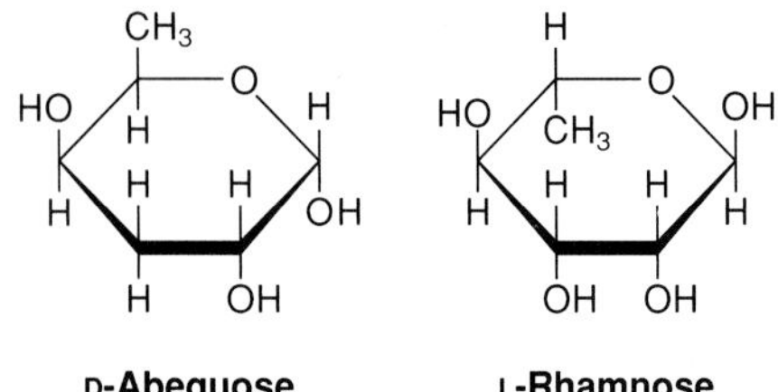

D-Abequose **L-Rhamnose**

In wild strains of *S. typhimurium* the O-antigen repeating unit has the structure abequose($\alpha 1 \rightarrow 3$)mannose($\alpha 1 \rightarrow 4$)rhamnose($\beta 1 \rightarrow 3$)galactose. Note from the margin that abequose and rhamnose are both deoxy sugars. The oligosaccharide unit is assembled on a lipid carrier, undecaprenol phosphate, the same carrier we encountered in peptidoglycan synthesis. A lipid-linked tetrasaccharide is assembled within the inner membrane as shown in Figure 16.22 and then passes outward through the outer membrane, where the activated galactose of a lipid-linked polymer attacks the mannose of the activated tetrasaccharide. As noted earlier, this general scheme can lead to tremendous structural diversity among the lipopolysaccharides.

SUMMARY

Biosynthesis of carbohydrates involves distinct pathways leading to monosaccharides, oligosaccharides, and polysaccharides. All organisms carry out gluconeogenesis, the synthesis of carbohydrate from noncarbohydrate three-carbon and four-carbon compounds. Gluconeogenesis uses seven glycolytic enzymes and four specific gluconeogenic enzymes, the latter to bypass the three irreversible steps in glycolysis. The four enzymes specific to gluconeogenesis are pyruvate carboxylase, phosphoenolpyruvate carboxykinase, fructose-1,6-bisphosphatase, and glucose-6-phosphatase. Regulation occurs at the sites of these three substrate cycles. Control is supremely important in animal metabolism, which requires that blood glucose levels be maintained within narrow limits. Hormonal and allosteric mechanisms are involved, with fructuose-2,6-bisphosphate being a key regulator.

All oligosaccharide and polysaccharide synthesis involves glycosyltransferases, enzymes that transfer the sugar unit from a nucleotide-linked or otherwise activated sugar to an acceptor sugar, at a nonreducing end. Glycogen synthase uses uridine diphosphate glucose as its glucosyl donor. The enzyme is regulated by hormonal and nonhormonal processes that are complementary and opposed to those that regulate glycogen breakdown by phosphorylase.

The carbohydrate portions of glycoproteins are also synthesized by glycosyltransferases, which synthesize oligosaccharide chains and link them to proteins. Carbohydrates added to the surfaces of proteins provide molecular recognition determinants that help move them to appropriate intracellular and extracellular locations. Steps in synthesis and processing of complex glycoproteins are sites of action for numerous antibiotics.

REFERENCES

Gluconeogenesis

Foster, J. D., B. A. Pederson, and R. C. Nordlie (1997) Glucose-6-phosphatase structure, regulation, and function: An update. *Proc. Soc. Exp. Biol. Med.* 207:314–332. This minireview discusses models to account for the membranous location of this gluconeogenic enzyme.

Granner, D., and S. Pilkis (1990) The genes of hepatic glucose metabolism. *J. Biol. Chem.* 265:10173–10176. A minireview describing progress toward cloning the genes for regulatory enzymes in glycolysis and gluconeogenesis, and analysis of the control of those genes.

Hanson, R. W., and L. Reshef (1997) Regulation of phosphoenolpyruvate carboxykinase (GTP) gene expression. *Annu. Rev. Biochem.* 66: 581–611. A great deal is known about hormonal and dietary regulation of the synthesis of PEPCK.

Hanson, R. W., and Y. M. Patel (1994) Phosphoenolpyruvate carboxykinase (GTP): The gene and the enzyme. *Adv. Enzymol. Relat. Areas* 69:203–281. Structure, functions, and regulation of this important gluconeogenic enzyme.

Pilkis, S. J., M. R. El-Maghrabi, and T. H. Claus (1988) Hormonal regulation of hepatic gluconeogenesis and glycolysis. *Annu. Rev. Biochem.* 57:755–784. A review of fructose-2,6-bisphosphate and the hormonal control of levels of this gluconeogenic regulator.

Glycogen Metabolism

Chen, Y.-T., and A. Burchell (1995) Glycogen storage diseases. In: *The Metabolic and Molecular Bases of Inherited Disease,* 7th ed., edited by C. R. Scriver, A. L. Beaudet, W. S. Sly, and D. Valle, pp. 935–966. McGraw-Hill, New York. A chapter in the three-volume treatise considered the most authoritative reference on heritable metabolic human diseases.

Krebs, E. G. (1994) The growth of research on protein phosphorylation. *Trends Biochem. Sci.* 19:439–440. The opening article in a special TIBS issue devoted to protein phosphorylation.

Leloir, L. F. (1983) Long ago and far away. *Annu. Rev. Biochem.* 52:1–16. A personal reminiscence, describing the author's Nobel Prize–winning role in the discovery of nucleotide-linked sugars and the mechanism of glycogen synthesis.

Millward, T. A., S. Zolnierowicz, and B. A. Hemmings (1999) Regulation of protein kinase cascades by protein phosphatase 2A. *Trends Biochem. Sci.* 24:186–191. Recent information about the control of protein phosphorylation and dephosphorylation.

Segal, S., and G. T. Berry (1995) Disorders of galactose metabolism. In: *The Metabolic and Molecular Bases of Inherited Disease,* 7th ed., edited by C. R. Scriver, A. L. Beaudet, W. S. Sly, and D. Valle, pp. 967–1000. McGraw-Hill, New York. A comprehensive review of galactosemias and related disorders.

Shulman, R. G., and D. L. Rothman (1996) Enzymatic phosphorylation of muscle glycogen synthase: A mechanism for maintenance of metabolic homeostasis. *Proc. Natl. Sci. USA* 93:7491–7495. This paper applies metabolic control analysis (see Chapter 12) to the complexities of controlling glycogen synthase at multiple phosphorylation sites.

Silverman, M. (1991) Structure and function of glucose transporters. *Annu. Rev. Biochem.* 60:757–794. The five different glucose transporter proteins, each with its own tissue distribution, metabolic roles, and modes of regulation, are described.

Glycoprotein Synthesis

Abeijon, C., and C. B. Hirschberg (1992) Topography of glycosylation reactions in the endoplasmic reticulum. *Trends Biochem. Sci.* 17:32–37. A good synthesis of the cell biology and biochemistry needed to understand glycoprotein processing.

Allan, B. B., and W. E. Balch (1999) Protein sorting by directed maturation of Golgi compartments. *Science* 285:63–66. A recent review.

Elbein, A. D. (1987) Inhibitors of the biosynthesis and processing of N-linked oligosaccharide chains. *Annu. Rev. Biochem.* 56:497–534. The complexity of the synthetic and processing steps has made inhibitors especially valuable in determining the roles of individual reactions.

Featherstone, C. (1998) Coming to grips with the Golgi. *Science* 282: 2172–2174. A news-type article written 100 years after Camillo Golgi's first description of this structure.

Gahmberg, C. G., and M. Tolvanen (1996) Why mammalian cell surface proteins are glycoproteins. *Trends Biochem. Sci.* 21:308–311. How the diversity of possible carbohydrate structures adapts them for use as recognition determinants on cell surfaces.

Kornfeld, S. (1992) Structure and function of the mannose 6-phosphate/insulinlike growth factor II receptors. *Annu. Rev. Biochem.* 61: 307–330. A review describing the lysosomal membrane targeting system and the unexpected discovery that the receptor is identical to a cell growth factor discovered independently.

Rothman, J. E., and F. T. Wieland (1996) Protein sorting by transport vesicles. *Science* 272:227–233. An up-tp-date review of the sorting that occurs during protein and glycoprotein maturation.

Plant and Bacterial Polysaccharide Synthesis

Arioli, T., and 13 coauthors (1998) Molecular analysis of cellulose biosynthesis in *Arabidopsis*. *Science* 279:717–720. This article and an accompanying news story describe the first cloning of a gene controlling a step in cellulose synthesis.

Tipper, D. J., and A. Wright (1979) The structure and biosynthesis of bacterial cell walls. In: *Mechanisms of Adaptation*, edited by J. R. Sokatch and L. N. Ornston, Vol. 7 of *The Bacteria*, pp. 291–426. Academic Press, New York. A detailed review including pathways of complex carbohydrate synthetic processes.

Walsh, C. T. (1989) Enzymes in the D-alanine branch of bacterial cell wall peptidoglycan assembly. *J. Biol. Chem.* 264:2393–2396. A minireview describing mechanisms in this pathway, which is the site of action of penicillin and other antibiotics.

Bacterial Cell Wall Polysaccharide Synthesis

Stinson, S. C. (1996) Drug firms restock antibacterial arsenal. *Chem. Eng. News*, September 23 issue, pp. 75–100. The proliferation of antibiotic-resistant bacterial strains has spurred the search for new antibiotics and new drug targets, which are discussed in this "product review."

PROBLEMS

1. How many ATPs are consumed in the conversion of each of the following to a glucosyl residue in glycogen?
   (a) Dihydroxyacetone phosphate
   (b) Fructose-1,6-bisphosphate
   (c) Pyruvate
   (d) Glucose-6-phosphate
2. How many high-energy phosphates are generated in (a) converting 1 mole of glucose to lactate? (b) converting 2 moles of lactate to glucose?
3. Avidin is a protein that binds extremely tightly to biotin. Therefore, it is a potent inhibitor of biotin-requiring enzyme reactions. Consider glucose biosynthesis from each of the following substrates and predict which of these pathways would be inhibited by avidin.
   (a) Lactate
   (b) Oxaloacetate
   (c) Fumarate
   (d) Fructose-6-phosphate
   (e) Phosphoenolpyruvate
4. $^{14}CO_2$ was bubbled through a suspension of liver cells that was undergoing gluconeogenesis from lactate to glucose. Which carbons in the glucose molecule would become radioactive?

*5. Propose synthetic pathways for the following nucleotide sugar intermediates involved in the biosynthesis of *Salmonella typhimurium* O antigen: UDP-Gal, GDP-Man, CDP-abequose, and dTDP-L-rhamnose.

6. Write a balanced equation for each of the following reactions or reaction sequences.
   (a) The reaction catalyzed by PFK-2
   (b) The conversion of 2 moles of oxaloacetate to glucose
   (c) The conversion of glucose to UDP-Glc
   (d) The conversion of 2 moles of glycerol to glucose
   (e) The conversion of 2 moles of malate to glucose-6-phosphate
7. Sketch curves for reaction velocity versus [fructose-6-phosphate] for the phosphorylated *and* nonphosphorylated forms of PFK-2.
8. Based on information presented on pages 575–576, sketch curves relating glycogen synthase reaction velocity to [UDP-glucose], for both the I and D forms of the enzyme, in the presence and absence of glucose-6-phosphate.
9. Glycogen synthesis and breakdown are regulated primarily at the hormonal level. However, important *nonhormonal* mechanisms also control the rates of synthesis and mobilization. Describe these nonhormonal regulatory processes.
10. For a bacterial cell that contains enzymes of the glyoxylate cycle, calculate the number of ATPs consumed in the biosynthesis of 1 mole of glucose from acetyl-CoA. How many moles of acetyl-CoA are required?
11. Why does it make good metabolic sense for phosphoenolpyruvate carboxykinase, rather than pyruvate carboxylase, to be the primary target for regulation of gluconeogenesis at the level of control of enzyme synthesis?
12. What factors determine whether pyruvate carboxylase acts primarily as an anaplerotic or a gluconeogenic enzyme?
13. What is the metabolic significance of the following observations? (1) Only the liver form of pyruvate kinase is inhibited by alanine, and (2) only gluconeogenic tissues contain appreciable levels of glucose-6-phosphatase.

*14. Predict how phosphoprotein phosphatase-1 (PP-1) and phosphoprotein phosphatase inhibitor (PI-1) might interact with components of the glycogenolytic cascade to effect regulation reciprocal to their effects upon glycogen synthesis.

15. Write a one-sentence explanation for each of the following statements.
(a) In liver, glucagon stimulates glycogen breakdown via cyclic AMP. Although one might expect glucagon also to stimulate catabolism of the glucose formed, glucagon *inhibits* glycolysis and stimulates gluconeogenesis.
(b) An individual with a glucose-6-phosphatase deficiency suffers from chronic hypoglycemia.
(c) The action of synthase phosphorylase kinase simultaneously activates glycogen breakdown and inhibits glycogen synthesis.
(d) An individual with type AB blood cannot be a donor for individuals with type A, B, or O blood.
(e) The presence in liver of glucose-6-phosphatase is essential to the function of the liver in synthesizing glucose for use by other tissues.

CHAPTER 17

Photosynthesis

IN EARLIER CHAPTERS, WE HAVE DESCRIBED IN CONSIDERABLE DETAIL THE WAYS in which organisms extract, and store in ATP, a substantial portion of the energy available from the oxidation of carbohydrates. Using glucose as an example, we wrote for the overall reaction:

$$C_6H_{12}O_6 + 6O_2 \longrightarrow 6CO_2 + 6H_2O \qquad \Delta G^\circ = -2870 \text{ kJ/mol}$$

We noted that as much as 40% of this energy could be recovered for useful biochemical work.

But life cannot depend on oxidative metabolism as its ultimate source of energy, and it cannot continue indefinitely returning organic carbon to the atmosphere as CO_2. The reaction above is only half of the great energy–carbon cycle of nature (Figure 17.1). The reverse of the carbohydrate oxidation reaction is accomplished by plants, algae, and some microorganisms, using the energy from sunlight to provide the enormous amount of free energy required.

$$6CO_2 + 6H_2O \xrightarrow{\text{Light energy}} C_6H_{12}O_6 + 6O_2 \qquad \Delta G^\circ = +2870 \text{ kJ/mol}$$

Photosynthesis provides carbohydrates for energy production, fixes CO_2, and is the major source of atmospheric O_2.

This process is called **photosynthesis.** Not only does it provide carbohydrates for energy production in plants and animals, but it is also the major path through which carbon reenters the biosphere—that is, the principal means of carbon fixation. Furthermore, photosynthesis is the major source of oxygen in the earth's atmosphere.

Prior to the evolution of photosynthetic organisms, the earth's atmosphere was probably devoid of oxygen (though rich in carbon dioxide). Prephotosynthetic organisms must have used abiotically synthesized energy-rich molecules for their metabolism. Without the advent of photosynthesis, these energy sources would have been wholly consumed, and life would have perished. The fossil record suggests that photosynthetic organisms first appeared about 3.5 billion years ago. Their gradual conversion of the primitive, nonoxidizing atmosphere of the earth to an oxidizing atmosphere paved the way for aerobic metabolism and the

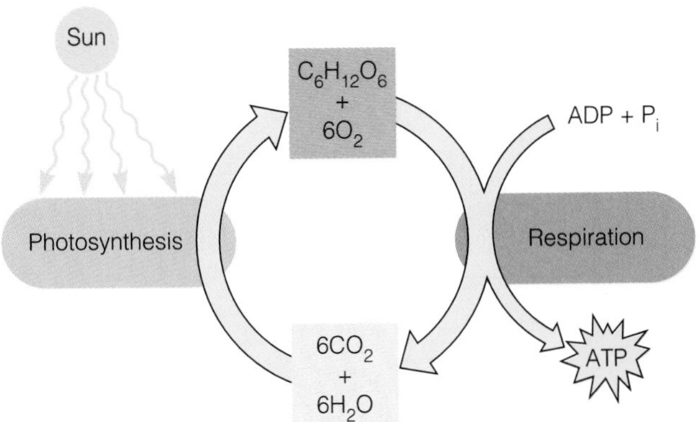

FIGURE **17.1**

The carbon cycle in nature. Carbon dioxide and water are combined through photosynthesis in plants to form carbohydrates. In both plants and animals, these carbohydrates can be reoxidized to regenerate CO_2 and H_2O. Part of the energy obtained from photosynthesis or oxidation is trapped in ATP.

evolution of animals. Today, photosynthesis represents the ultimate source of energy for almost all life.* It is used by plants, algae, and a wide variety of prokaryotes—all of which are food sources for other organisms. A comprehensive view of the relationship of photosynthesis to other pathways we have studied is shown in Figure 17.2.

The Basic Processes of Photosynthesis

The equation we have just shown for the photosynthetic reaction is, of course, a great oversimplification. As you might expect, the actual process of photosynthesis involves many intermediate steps. Furthermore, a hexose itself is not the primary carbohydrate product. Therefore, the photosynthetic reaction is usually written in this more general form:

$$CO_2 + H_2O \xrightarrow{\text{Light energy}} [CH_2O] + O_2$$

where $[CH_2O]$ represents a general carbohydrate.

Because the burning of carbohydrates to form CO_2 is an oxidative process, converting CO_2 into carbohydrate must involve a *reduction* of the carbon. The preceding reaction statement shows H_2O as the ultimate reducing agent, which is the case in plants, most algae, and cyanobacteria. However, there are photosynthetic processes in many bacteria that use other reductants. Thus, an even more general reaction can be written:

$$CO_2 + 2H_2A \xrightarrow{\text{Light energy}} [CH_2O] + H_2O + 2A$$

where H_2A is a general reductant and A is the oxidized product. Examples of photosynthetic reactions are given in Table 17.1. Comparison of the reactions shown in the table suggests that the source of the oxygen released in photosynthesis by plants, algae, and cyanobacteria must be H_2O, rather than CO_2. This source was predicted in the early 1930s by C. B. van Niel, one of the pioneers in photosynthesis studies, and was confirmed in 1941 by isotope labeling experiments using

Photosynthesis requires a reductant, usually H_2O, to reduce CO_2 to the carbohydrate level.

*Recent developments make it necessary to qualify the statement that *all* life derives its energy, directly, from photosynthesis. It has been found that some bacteria, such as those associated with submarine biothermal vents, use the oxidation of substances like H_2S or H_2 as an alternate energy source, in the complete absence of light. This energy cycle represents, however, only a small fraction of the energy flow in the biosphere.

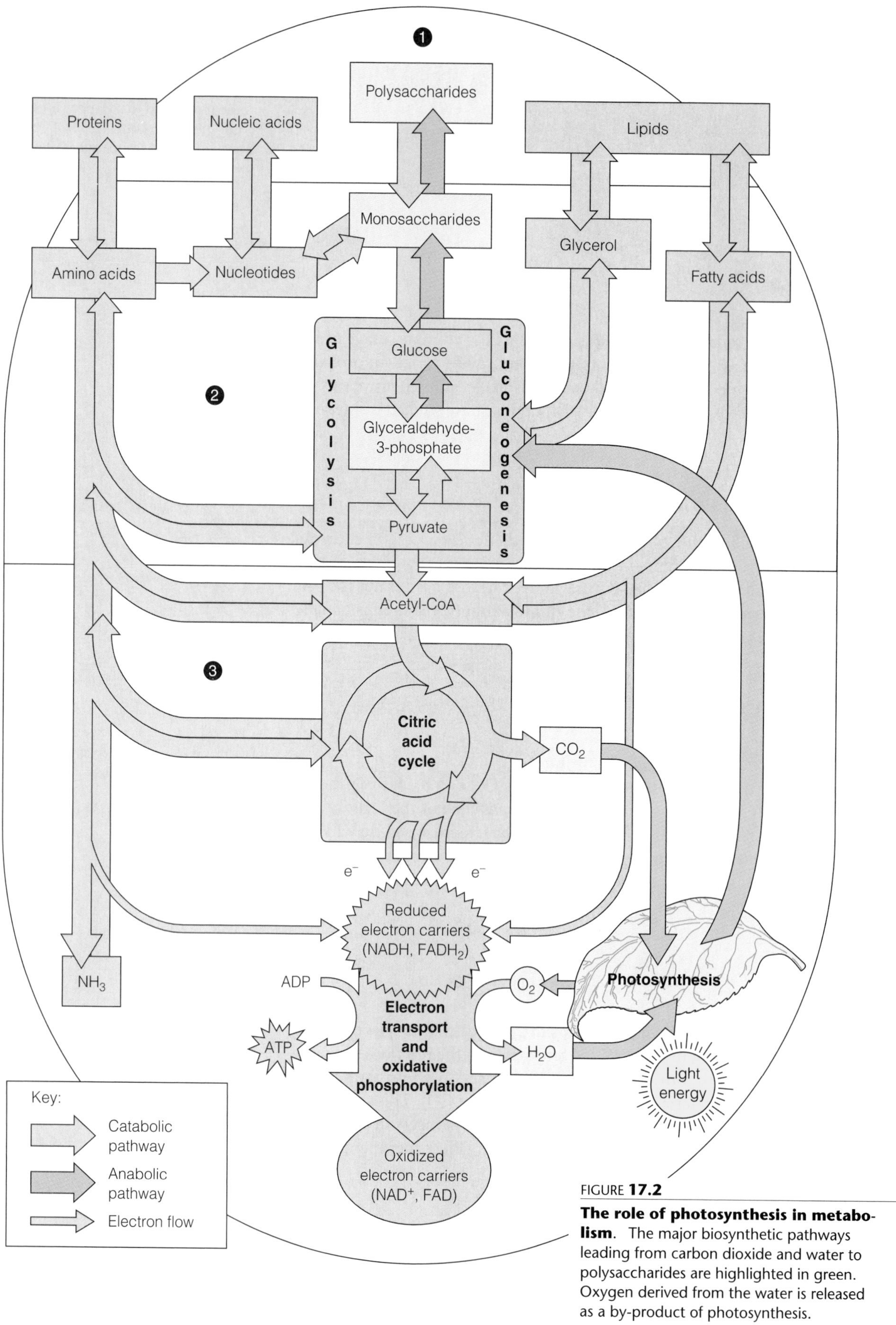

FIGURE **17.2**

The role of photosynthesis in metabolism. The major biosynthetic pathways leading from carbon dioxide and water to polysaccharides are highlighted in green. Oxygen derived from the water is released as a by-product of photosynthesis.

TABLE **17.1** **Examples of some photosynthetic reactions**

Organisms	Reductant	Reaction
Plants, algae, cyanobacteria	H_2O	$CO_2 + 2H_2O \longrightarrow [CH_2O] + H_2O + O_2$
Green sulfur bacteria	H_2S	$CO_2 + 2H_2S \longrightarrow [CH_2O] + H_2O + 2S$
Purple sulfur bacteria	$[HSO_3^-]$	$CO_2 + H_2O + 2[HSO_3^-] \longrightarrow [CH_2O] + 2[HSO_4^-]$
Nonsulfur photosynthetic bacteria	H_2 or many other reductants, such as lactate	$CO_2 + 2H_2 \longrightarrow [CH_2O] + H_2O$ $CO_2 + 2\begin{pmatrix} CH_3 \\ HC{-}OH \\ COO^- \end{pmatrix} \longrightarrow [CH_2O] + H_2O + 2\begin{pmatrix} CH_3 \\ C{=}O \\ COO^- \end{pmatrix}$ Lactate ... Pyruvate

^{18}O-labeled water and unlabeled CO_2. These experiments showed that neither of the oxygen atoms in O_2 comes from CO_2. Therefore, it is more correct to write the photosynthetic reactions in the fashion shown in Table 17.1, which makes it clear that one of the oxygens from CO_2 ends up in carbohydrate, the other in water:

$$CO_2 + 2H_2O \xrightarrow{\text{Light energy}} [CH_2O] + H_2O + O_2$$

Light energy cannot be used *directly* to drive this reaction; in fact, H_2O does not reduce CO_2 *directly* under any known circumstances. The overall process we have just described is actually separated, both chemically and physically, into two subprocesses in all photosynthetic organisms. A slightly more sophisticated version of what actually happens is shown in Figure 17.3. In the first subprocess, in a series of steps called the **light reactions,** energy from sunlight is used to carry out the photochemical oxidation of H_2O. Two things are accomplished by this oxidation. First, the oxidizing agent $NADP^+$ is reduced to NADPH, producing reducing equivalents, and O_2 is released. Second, part of the energy from sunlight is captured by phosphorylating ADP to produce ATP. This is called **photophosphorylation.** In the second subprocess, the so-called **dark reactions** of photosynthesis, the NADPH and ATP produced by the light reactions are used in the reductive synthesis of carbohydrate from CO_2 and water. These reactions were originally termed *dark* to emphasize that they do not require the direct participation of light energy. Though this is true, the term carries the unfortunate implication that these synthetic reactions occur only in the dark. Nothing could be further from the truth. Rather, these reactions occur at all times and are actually accelerated by light. Because the term *dark reactions* is well established, we shall retain it, but you should not be misled by it.

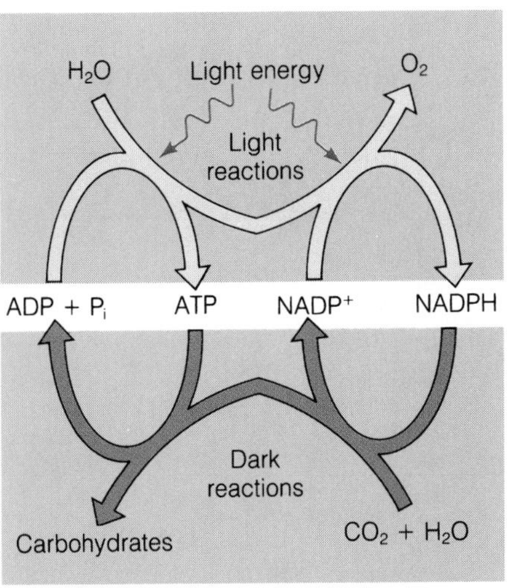

FIGURE **17.3**
The two subprocesses of photosynthesis. The overall process of photosynthesis is divided into light reactions and dark reactions. The light reactions, which require visible light as an energy source, produce reducing power (in the form of NADPH), ATP, and O_2. The NADPH and ATP drive the so-called dark reactions, which occur in both the presence and the absence of light and fix CO_2 into carbohydrates.

Before considering details of either the light reactions or the dark reactions, it is appropriate to consider the sites of photosynthesis. Just as all eukaryotic cells have organelles (mitochondria) specialized for oxidative metabolism, plants and algae have organelles specialized for photosynthesis.

Photosynthesis can be divided into light reactions, which use sunlight energy to produce NADPH and ATP, releasing O_2 in the process, and dark reactions, which use NADPH and ATP to fix CO_2.

The Chloroplast

In all higher plants and algae, photosynthetic processes are localized in organelles called **chloroplasts.** In plants, most of the chloroplasts are found in cells just under the leaf surface (mesophyll cells). Each cell may contain 20 to 50 of these organelles (Figure 17.4). The eukaryotic algae also have chloroplasts, but often only one very large one is found in each cell.

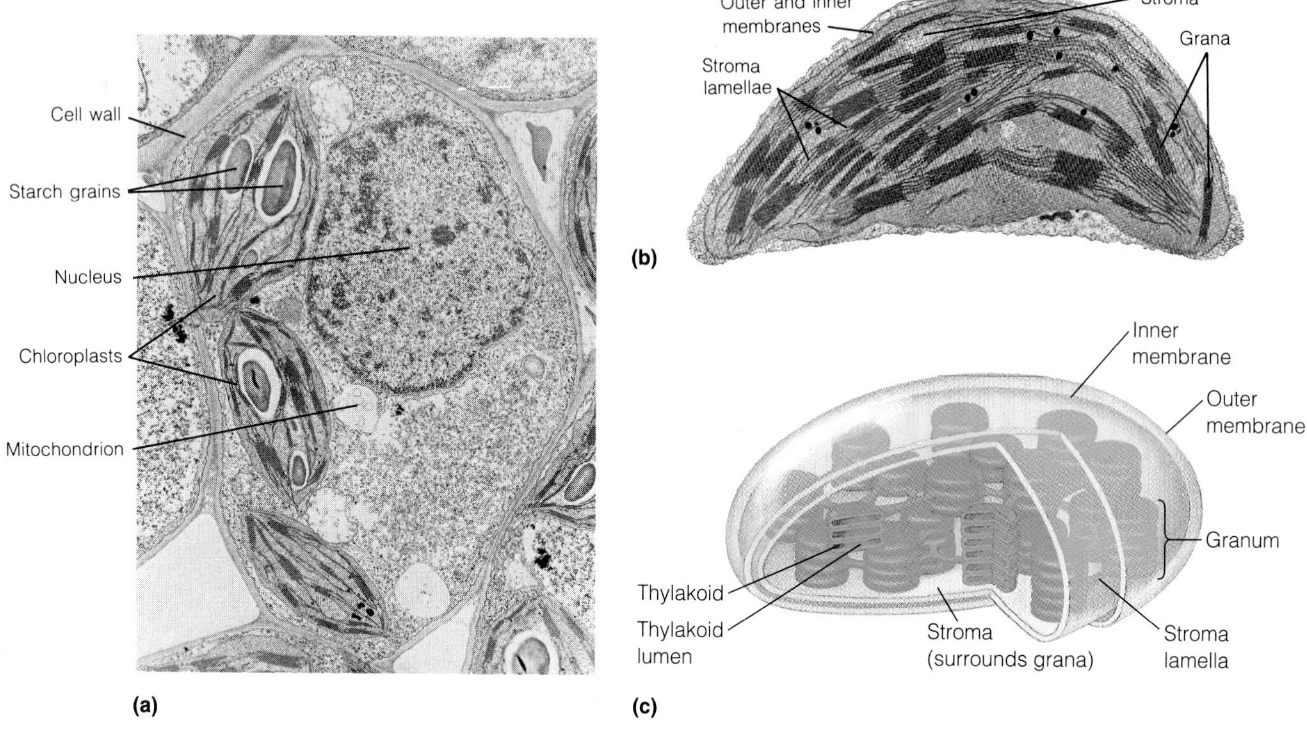

FIGURE **17.4**

Chloroplasts, the photosynthetic organelles of green plants and algae. **(a)** Several chloroplasts are shown in a cross section of a cell from a *Coleus* leaf. **(b)** Enlarged view of a single chloroplast from a leaf of timothy grass. **(c)** Schematic rendering of a chloroplast.

(a) Micrograph by M. W. Steer, photo provided by E. H. Newcomb; **(b)** micrograph by K. P. Wergin, photo provided by E. H. Newcomb/BPS; **(c)** From Neil Campbell, Jane Reece, and Larry Mitchell, *Biology,* 5th ed. (Menlo Park, CA: Addison Wesley Longman, 1999). © Addison Wesley Longman, Inc.

Photosynthesis in plants and algae occurs in organelles called chloroplasts.

Like mitochondria, chloroplasts are semiautonomous, carrying their own DNA to code for some of their proteins, as well as the ribosomes necessary for translation of the appropriate messenger RNAs. There is now much evidence that chloroplasts evolved from unicellular organisms similar to cyanobacteria (blue-green algae). Such prokaryotic photosynthesizers do not contain chloroplasts but have membrane structures that play the same roles as chloroplast membranes (Figure 17.5). To a certain extent, the cyanobacteria resemble free-living chloroplasts. It is believed that, early in evolution, primitive unicellular organisms took up cyanobacteria-like prokaryotes and that eventually the relationship became symbiotic: The photosynthetic organelles were no longer capable of independent life, and the algae depended upon them as energy sources. Today, some chloroplast genes are coded in the organelle genome, and some are in the cell nucleus.

The internal structure of a chloroplast, as shown in Figure 17.4b and c, bears some resemblance to that of a mitochondrion (see Figure 15.2a). There is an outer, freely permeable membrane and an inner membrane that is selectively permeable. The inner membrane encloses a material called the **stroma** that is analogous to the mitochondrial matrix. Immersed in the stroma are many flat, saclike membrane structures called **thylakoids,** which are often stacked like coins to form units called *grana* (see Figure 17.4c). Individual grana are irregularly interconnected by thylakoid extensions called *stroma lamellae.* The thylakoid membrane encloses an interior space, the **lumen** of the thylakoid.

Absorption of light and the light reactions occur in the chloroplast membranes. The dark reactions occur in the stroma.

The division of labor within a chloroplast is simple. Absorption of light and all of the light reactions occur within or on the thylakoid membranes. The ATP and NADPH produced by these reactions are released into the surrounding stroma, where all of the synthetic dark reactions occur. Thus, there are analogies in structure and role between mitochondrial matrix and chloroplast stroma and between the inner membrane of the mitochondrion and the thylakoid membrane of the chloroplast. Indeed, we shall find that a very similar kind of chemiosmotic ATP generation is carried out across these membranes in both mitochondria and chloroplasts. To see how this ATP generation occurs, we must first examine the light reactions in detail, beginning with the process of light absorption.

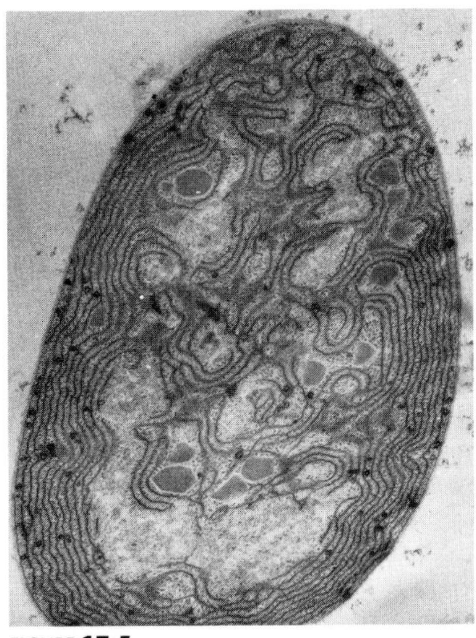

FIGURE **17.5**

A photosynthetic prokaryote. This electron micrograph of a thin section of the cyanobacterium *Anabaena azollae* shows the folded membranes, which resemble the thylakoids of eukaryotic chloroplasts.

Courtesy of N. Lang, University of California, Davis/BPS.

The Light Reactions

ABSORPTION OF LIGHT: THE LIGHT-HARVESTING SYSTEM

The Energy of Light

To understand how energy from sunlight can be captured and utilized, we must first review the nature of electromagnetic radiation. The quantum mechanical theory of radiation states that light (and all other electromagnetic radiation) has two aspects: wavelike and particle-like. We can characterize a particular kind of radiation by its wavelength (λ) or frequency (ν); these parameters characterize the *wave* aspects of the light. If waves with a length of λ are passing an observer at a velocity c, the number of waves passing per second is the frequency, ν. Thus

$$\nu = c/\lambda \tag{17.1}$$

where c is the velocity of light, 2.99×10^8 m/s. The red light from a neon laser has a wavelength of 632.8 nm, or 6.328×10^{-7} m. Thus, its frequency is $4.73 \times 10^{14}\ s^{-1}$. But to see how *energy* might be obtained from light, it is necessary to consider the particulate aspect of radiation. We must think of a light beam as a stream of light particles, or **photons.** Each photon has an associated unit of energy called a **quantum.** The energy value of a quantum—that is, the energy per photon—is related to the frequency of the light by one of the most basic equations in physics, Planck's law:

$$E = h\nu \tag{17.2}$$

where h is Planck's constant, 6.626×10^{-34} J s. Thus, the neon laser in our example can deliver light energy only in packets, or quanta, of 3.14×10^{-19} J (or $[6.626 \times 10^{-34}\text{ J s}] \times [4.73 \times 10^{14}\ s^{-1}]$). However, biochemists rarely deal with single photons. Because we are interested in how radiation can promote chemical or biochemical processes, which are usually expressed on a molar basis, the more appropriate quantity for our purposes is the energy of a *mole* (6.02×10^{23}) of photons. For the neon laser light, multiplying the energy per photon by 6.02×10^{23} gives 189 kJ. A mole of photons is called one **einstein.**

Figure 17.6 shows a graph of energy per mole of photons as a function of wavelength, through the infrared, visible, and ultraviolet parts of the spectrum. For comparison, the energies associated with molecular vibrations and various covalent bonds are indicated. When photons of infrared radiation are absorbed by a molecule, they can do little except stimulate molecular vibrations, which we perceive as heat. Photons of far-ultraviolet radiation, on the other hand, have energies quite capable of breaking covalent bonds. Far-ultraviolet radiation is chemically destructive to humans and to other organisms, but fortunately most of it is screened from the earth's surface by the ozone layer. This is why depletion of the ozone layer is of such serious concern.

Photosynthesis depends primarily on light in the visible and near-infrared regions of the spectrum, lying between the extremes of covalent bond–breaking and stimulating molecular vibrations. Photons in the visible and near infrared

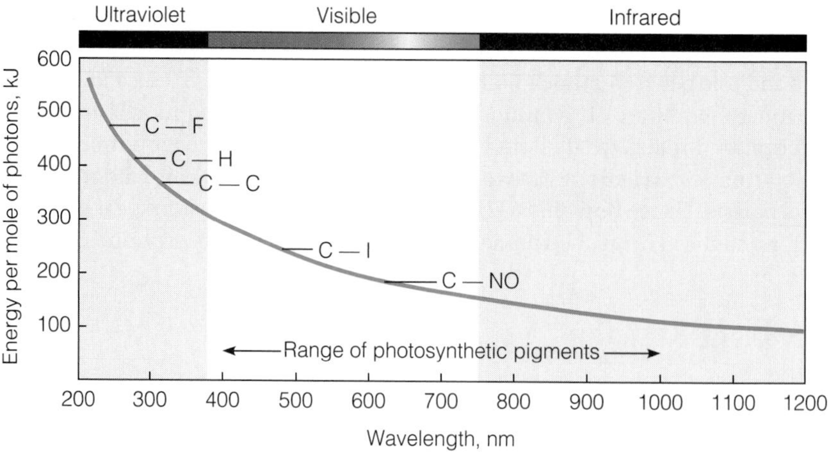

FIGURE **17.6**

The energy of photons. The graph shows energy per mole of photons as a function of wavelength, compared with energies of several chemical bonds. Light in the ultraviolet range has enough energy to break many chemical bonds directly. Visible light can break some weak bonds. Light in the long-wavelength portion of infrared region of the spectrum causes only heat-producing molecular vibrations.

are not highly destructive, but can cause transitions in the electronic states of organic molecules that can drive reactions and thus capture the energy in a chemical form. The ability to use radiation in this range has had clear evolutionary advantages for photosynthetic organisms. Most of the sun's energy that reaches the earth's surface lies in this spectral range. The small amount of ultraviolet radiation that does get through can penetrate only a very short distance into water and thus would have been unavailable to primitive photosynthetic organisms living in the sea. The photons of far-infrared radiation have energies too low to be useful for any photochemical processes.

The Light-Absorbing Pigments

To capture the useful portion of the light energy, photosynthetic organisms have evolved a set of pigments that efficiently absorb visible and near-infrared light. The light-absorbing portions of these pigments are sometimes referred to as **chromophores.** Structures of a few of the most important photosynthetic chromophores are shown in Figure 17.7. In Figure 17.8 the absorption spectra of these photosynthetic pigments are compared with the distribution of solar radiation in the spectrum. Together, the chromophores "blanket" the visible spectrum; scarcely a photon can come through that cannot be absorbed by one chromophore or another. The most abundant pigments in higher plants are chlorophyll *a* and chlorophyll *b*. As you can see by comparing Figure 17.7a with Figure 7.4b (page 215), these molecules are related to the protoporphyrin IX found in hemoglobin and myoglobin. However, the bound metal in the chlorophylls is Mg^{2+} rather than Fe^{2+}. In Figure 17.7b and c, two accessory pigments are also shown. All of these molecules absorb light in the visible region of the spectrum because they have large conjugated double-bond systems. Because chlorophylls *a* and *b* absorb strongly in both the deep blue and red, the light that is *not* absorbed but *reflected* from chloroplasts is green, the color we associate with most growing plants. The other observed colors, such as the red, brown, or purple of algae and photosynthetic bacteria, are accounted for by differing amounts of accessory pigments. Loss of chlorophylls in autumn leaves allows the colors of the accessory pigments, as well as nonphotosynthetic pigments, to become evident. Some photosynthetic bacteria use pigments that absorb wavelengths up to about 1000 nm, in the near infrared.

The Light-Gathering Structures

Chlorophyll and some of the accessory pigments are contained in the **thylakoid membranes** of the chloroplast. The composition of these membranes is rather unusual. They contain only a small fraction of the common phospholipids but are rich in glycolipids. They also contain much protein, and some of the photosynthetic pigments are attached to certain of these proteins. Other photosynthetic pigments, including chlorophylls *a* and *b*, are not covalently bound but

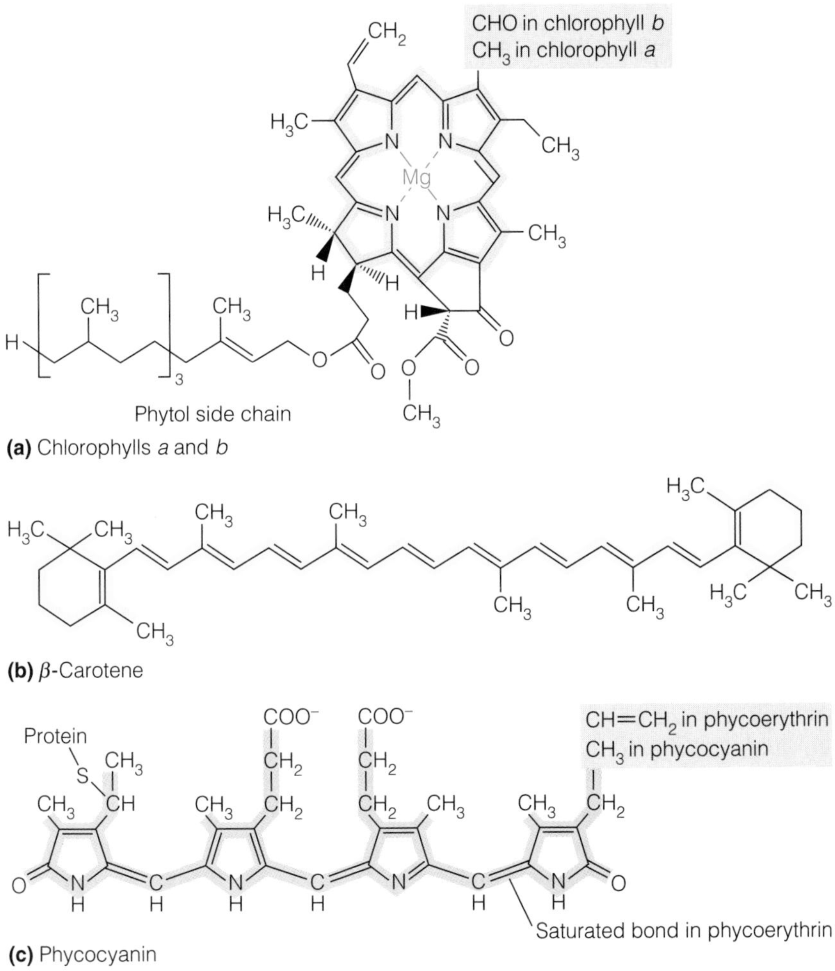

CHO in chlorophyll *b*
CH_3 in chlorophyll *a*

Phytol side chain

(a) Chlorophylls *a* and *b*

(b) β-Carotene

$CH{=}CH_2$ in phycoerythrin
CH_3 in phycocyanin

Protein

Saturated bond in phycoerythrin

(c) Phycocyanin

FIGURE **17.7**

Some photosynthetic pigments.
Chlorophylls *a* and *b* are the most abundant plant pigments, whereas β-carotene and phycocyanin are examples of accessory pigments. Phycocyanin is covalently attached to a protein via a sulfhydryl group. The related phycoerythrins also form pigment–protein complexes. The chlorophylls *a* and *b* are found in higher plants and algae. There are also bacteriochlorophylls, which differ slightly in structure.

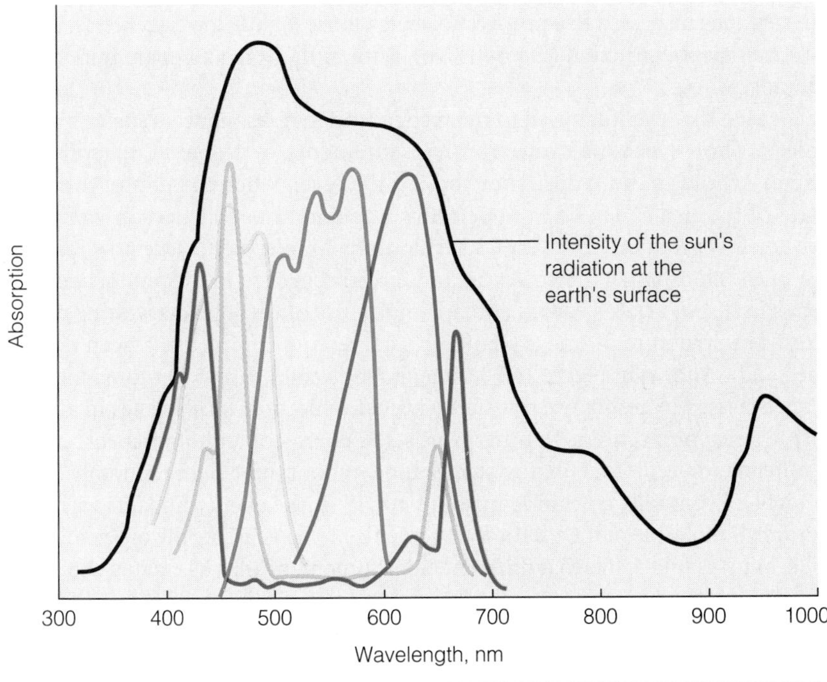

FIGURE **17.8**

Absorption spectra and light energy.
The absorption spectra of various plant pigments are compared with the spectral distribution of the sunlight that reaches the earth's surface.

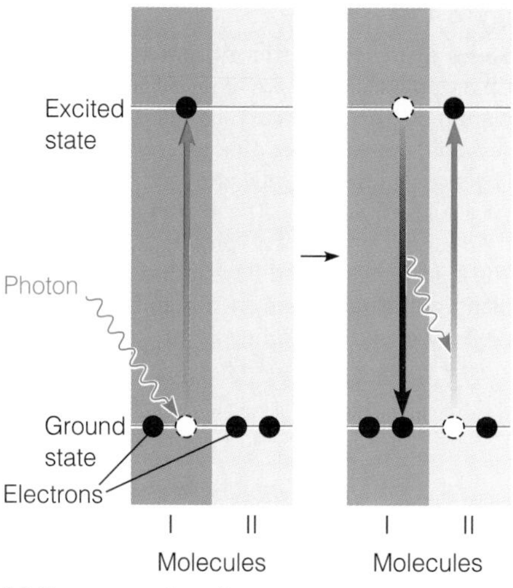

(a) Resonance transfer

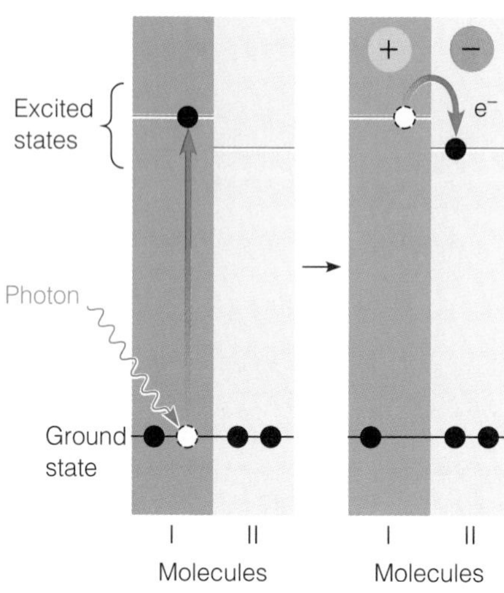

(b) Electron transfer

FIGURE **17.9**

Two modes of energy transfer following photoexcitation. For each of the two types of energy transfer that occur in a photosystem, the left-hand illustration shows a molecule being excited to a higher energy state by absorption of a photon of radiation. The right-hand illustrations show how the energy is transferred to an adjacent molecule. **(a)** In *resonance transfer* molecule I transfers its excitation energy to an identical molecule II, which rises to its higher energy state as molecule I falls back to the ground state. **(b)** In *electron transfer* an excited electron in molecule I is transferred to the slightly lower excited state of molecule II, making molecule I a cation and molecule II an anion.

interact with both proteins and membrane lipids. These pigments interact with membrane lipids through their hydrophobic phytol tails (see Figure 17.7a).

The assemblies of light-harvesting pigments in the thylakoid membrane, together with their associated proteins, are organized into well-defined **photosystems,** structural units dedicated to the task of absorbing light photons and recovering some of their energy in a chemical form. The first part of this process takes place in what are referred to as **light-harvesting complexes.** Each is a multisubunit protein complex containing multiple **antenna** pigment molecules (chlorophylls and some accessory pigments) and a pair of chlorophyll molecules that act as the **reaction center,** trapping energy quanta excited by the absorption of light.

To understand how this system functions, we must look a bit more closely into what can happen when a molecule absorbs a quantum of radiant energy. Recall from Tools of Biochemistry 6A that absorption in the visible region of the spectrum excites the molecule from the ground state to a higher electronic state. In the case of the photosynthetic pigments, the electron so excited is one occupying a π orbital in the conjugated bond system. In Tools of Biochemistry 6A we described two ways in which the energy could be lost to return the molecule to its ground state: radiationless dissipation of the energy as heat, or reradiation as fluorescence. However, when similar absorbing molecules are packed tightly together, as in a photosystem, two other possibilities arise. First, the excitation energy may be passed from one molecule to an adjacent one—a process called *resonance transfer* or *exciton transfer* (Figure 17.9a). Alternatively, the excited electron itself may be passed to a nearby molecule with a slightly lower excited state—an *electron transfer* reaction (Figure 17.9b). Both of these processes are important in photosynthesis.

Most chlorophyll molecules are used as antennae to catch photons and pass their energy on to reaction centers.

The clue that eventually led to the recognition that resonance transfer played a role in photosynthesis came from measurements by Robert Emerson and William Arnold in the 1930s. They showed that even when the photosynthetic system of the alga *Chlorella* was operating at maximum efficiency, only one O_2 molecule was produced for every 2500 chlorophyll molecules. As we now realize, most of the chlorophyll molecules are not directly engaged in the photochemical process itself but act, instead, as antenna molecules of the light-harvesting complexes. The structure of one type of light-harvesting complex has been determined and is shown in Figure 17.10. Antenna molecules absorb photons, and the energy is passed by resonance transfer to specific chlorophyll molecules in a relatively few reaction centers. In other words, the energy of a photon absorbed by any antenna molecule in a photosystem wanders about the system randomly (Figure 17.11). Eventually (meaning in about 10^{-10} s), the energy finds its way to a chlorophyll molecule in the reaction center. This molecule is like the other chlorophylls, but it is in a somewhat different environment, so that its excited state energy level is a bit lower. Thus, it acts as a trap for quanta of energy absorbed by any of the other pigment molecules. It is the excitation of this reaction center that begins the actual photochemistry of the light reactions, for it starts a series of electron transfers.

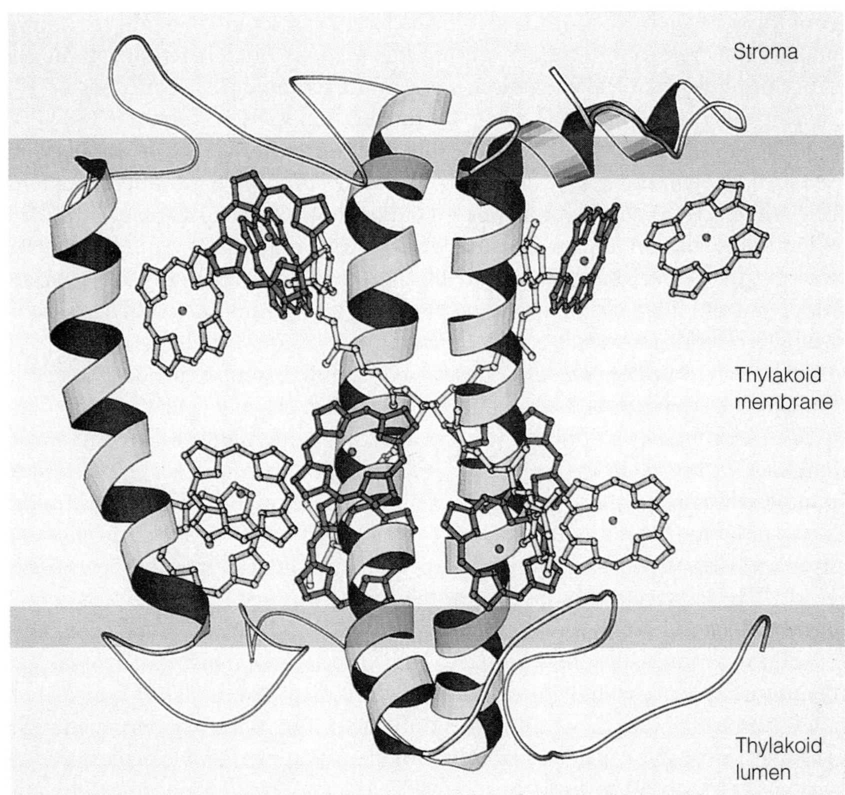

FIGURE **17.10**

Three-dimensional structure of a monomer of the trimeric light-harvesting complex of plants. This side view shows the light-harvesting monomer buried in the thylakoid membrane. Chlorophyll molecules and accessory pigments are clearly resolved within the structure.

Courtesy of W. Kühlbrandt, et al., *Nature* (1994) 367:614–621. © 1994 Macmillan Magazines, Ltd.

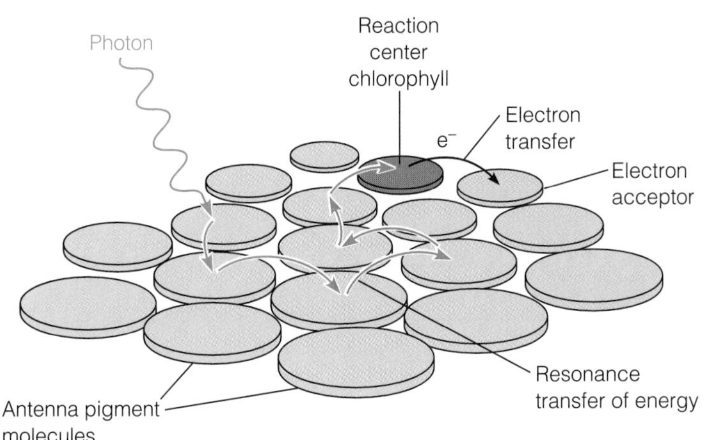

FIGURE **17.11**

Resonance transfer of energy in a light-harvesting complex. The excitation energy originating in a photon of light wanders from one antenna molecule to another until it reaches a reaction center. There an electron is transferred to a primary electron acceptor molecule, and the energy is trapped.

PHOTOCHEMISTRY IN PLANTS AND ALGAE: TWO PHOTOSYSTEMS IN SERIES

Our understanding of the photochemical light reactions has developed from many elegant experiments in many different laboratories. A pioneering study in 1939, by Robert Hill at the University of Cambridge, made the seminal observation that isolated chloroplasts could promote reduction when illuminated in the presence of any of a variety of electron acceptors. For example, when ferric ion was used, the following reaction proceeded efficiently:

$$4Fe^{3+} + 2H_2O \xrightarrow{\text{Light energy}} 4Fe^{2+} + 4H^+ + O_2$$

A number of such reactions involving different inorganic oxidants are known and are now referred to collectively as *Hill reactions*. Such reactions, in the absence of photochemical activation, are very unfavorable. Fe^{3+}, for example, is a much weaker oxidant than O_2; $\Delta G°$ for the reaction written is about +180 kJ; so equilibrium should lie far to the left. Hill's discoveries showed that *chloroplasts irradiated by light are capable of driving thermodynamically unfavorable reactions.* The Hill reactions also demonstrated that the photosynthetic system can oxidize water to O_2 without any involvement of CO_2 (see Figure 17.3). This observation was the first clear indication that the light and dark reactions are separate processes, and it led ultimately to the discovery that the final electron acceptor of the light reactions in vivo is $NADP^+$, yielding NADPH.

Further studies revealed that *two* kinds of photosystems must be involved in photosynthesis in plants. The first hint came from experiments that measured the quantum efficiency of photosynthesis in algae, using light of different wavelengths. The *quantum efficiency (Q)* is the ratio of oxygen molecules released to photons absorbed. As the wavelength of the monochromatic light used was raised above 680 nm (far red), an abrupt drop in Q was noted. This "red drop" was a strange observation, for chlorophylls in plants still show appreciable absorbance even at higher wavelengths. Somehow the energy was just not being used as efficiently above 680 nm. Even more remarkable was the observation that simultaneous illumination with yellow light (650 nm) produced a marked increase in the quantum efficiency from light at 700 nm. Even if the yellow light was switched off a few minutes before the measurement, the quantum efficiency remained high. The only reasonable explanation for these results is that two complementary photosystems exist, one absorbing most strongly at wavelengths around 700 nm and the other at shorter wavelengths. The action of *both* must be required for photosynthesis to proceed with maximal efficiency.

The two photosystems predicted by early experimenters have now been identified and characterized. They are both localized in the thylakoid membrane. Each photosystem is a multisubunit, transmembrane protein complex, carrying antenna and reaction center chlorophyll molecules and electron transport agents. The photosystems have been named according to the order in which they were discovered. The one showing absorbance up to 700 nm is called **photosystem I (PSI)**, and the one that absorbs only to a wavelength of about 680 nm is called **photosystem II (PSII)**. In algae, cyanobacteria, and all higher plants, these two photosystems are linked in series to carry out the complete sequence of the light reactions. The basic sequence is illustrated in Figure 17.12a, which depicts the path of electrons through the two systems. Figure 17.12b emphasizes the energetics of the electron flow and places the major participants in the light reactions on a scale of reduction potential.

Two photosystems, linked in series, are involved in the photosynthetic light reactions in algae, cyanobacteria, and higher plants.

In each of the two photosystems, the primary step is transfer of a light-excited electron from a reaction center (P680 or P700) into an electron transport chain. The ultimate source of the electrons is the water molecules shown at the left in both parts of Figure 17.12. The final destination of the electrons is the molecule of $NADP^+$ at the right, which is thereby reduced to NADPH. At two stages in the electron transport process, protons are released into the thylakoid lumen. Some of the protons come from the H_2O that is broken down, some come from the stroma. This transfer of protons into the lumen produces a pH gradient across the thylakoid membrane. It should not surprise you to find that this proton gradient drives ATP production, just as a proton gradient produces ATP in mitochondria (see Chapter 15). Thus, ATP and reducing power in the form of NADPH are the products of the light reactions. These compounds are exactly what is needed to drive the syntheses carried out in the dark reactions. To examine their generation in detail, we begin with photosystem II, for it is where electrons enter the scheme.

FIGURE **17.12**

The two-photosystem light reactions. In the two-photosystem mode of photosynthesis, the light reactions are carried out by two photosystems linked in series. **(a)** A schematic view of the path of electrons through the two photosystems. The two systems and the cytochrome complex are embedded in the thylakoid membrane. Electrons taken from water in photosystem II are transferred to photosystem I via quinones (Q), the cytochrome b_6f complex, and plastocyanin (PC) . In photosystem I the electrons are excited by light again, for transfer via a series of intermediates to ferredoxin. Reduced ferredoxin reduces $NADP^+$. **(b)** Energetics of the two-photosystem light reactions. In each of the two reaction centers, P680 and P700, electrons are raised to an excited state by absorption of photons. In each photosystem the excited electrons are passed through an electron transport chain, which drives the pumping of hydrogen ions into the thylakoid lumen. The two-photosystem mode is sometimes called the Z-scheme because of the pattern of energy changes shown here. **Key:** MnC = manganese center; Z = donor to P680; P680 = photosystem II reaction center chlorophyll; Ph = pheophytin acceptor; Q_A, Q_B = protein-bound plastoquinones; QH_2 = plastoquinol (reduced plastoquinone) in membrane; Cyt b_6f = cytochrome b_6f complex; PC = plastocyanin; P700 = photosystem I reaction center chlorophyll; A_0 = chlorophyll acceptor; A_1 = protein-bound phylloquinone; F_A, F_B, F_X = iron–sulfur clusters; Fd = ferredoxin; FNR = ferredoxin:$NADP^+$ oxidoreductase.

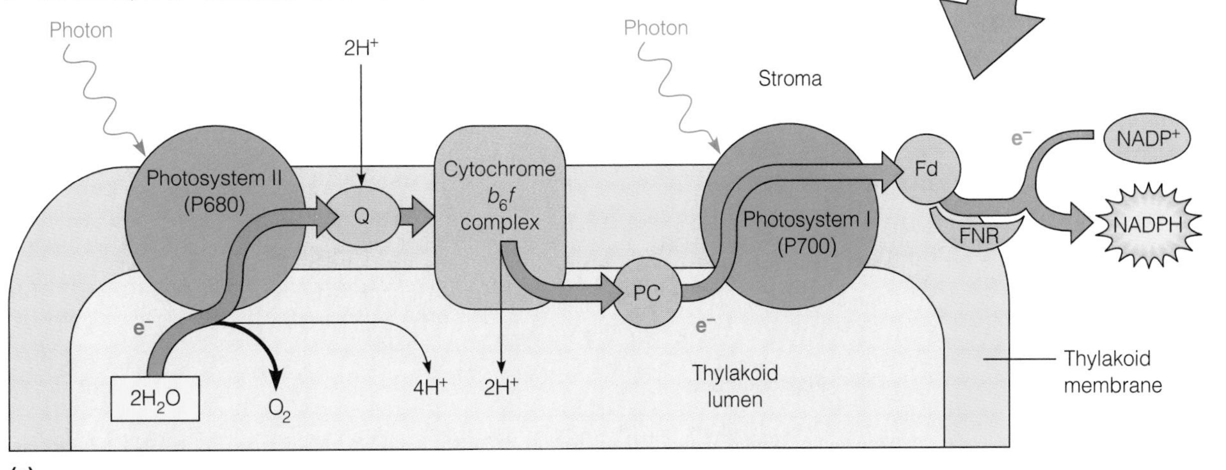

(a)

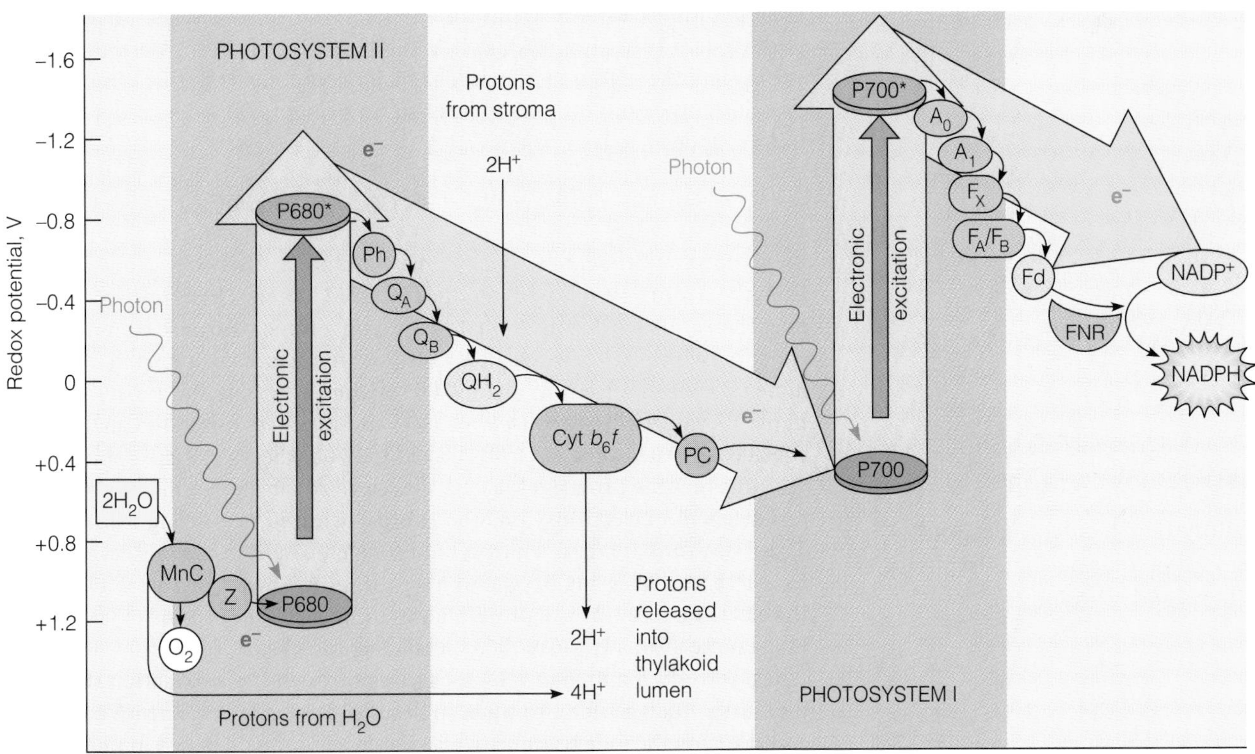

(b)

Photosystem II: The Splitting of Water

Each of the photosystems contains an electron transport chain, which extracts energy when an excited electron loses its energy of excitation in a stepwise fashion. The photosystem carries out a series of oxidation–reduction reactions. It is easiest to follow the events in a photosystem by starting with the absorption of a photon picked up by the light-harvesting system of photosystem II. The photon is funneled to a reaction center chlorophyll, designated P680 in Figure 17.12. Excitation of P680 raises the molecule from the ground state to an excited state at −0.8 volt. Thus, the excited P680 has become an excellent reducing agent, able to transfer quickly an electron from P680 to a lower-energy primary electron acceptor—*pheophytin a (Ph),* as shown in Figure 17.12b. The pheophytins are molecules identical to chlorophylls, except that two protons substitute for the centrally bound magnesium ion. We can consider this excited electron as a low-redox-potential electron (see Chapter 15).

The electron is then transferred to a series of *plastoquinone* molecules (Q_A and Q_B) associated with PSII proteins. Ultimately, two electrons and two protons are picked up by the plastoquinone Q_B; the protons come from the stroma. The reduced plastoquinone, QH_2 (*plastoquinol*), is then released into the lipid portion of the thylakoid membrane. The overall reduction of plastoquinone can be written as follows:

$$\text{Plastoquinone} \; (CH_3)_2C_6O_2\text{—}(CH_2\text{—}CH{=}C(CH_3)\text{—}CH_2)_n\text{—}H + 2e^- + 2H^+ \longrightarrow \text{Plastoquinol} \; (CH_3)_2C_6(OH)_2\text{—}(CH_2\text{—}CH{=}C(CH_3)\text{—}CH_2)_n\text{—}H$$

$n = 6$ to 10

Plastoquinone **Plastoquinol**

Plastoquinol then interacts with a membrane-bound complex of cytochromes and iron–sulfur proteins, the cytochrome b_6f complex. This complex catalyzes the transfer of the electrons to a copper protein, *plastocyanin (PC).* In doing so, the b_6f complex serves two purposes. First, it transmits activated electrons from photosystem II to photosystem I. At the same time, it pumps protons from the stroma into the thylakoid lumen. The major components of this complex are two cytochromes (f and b_6) and an iron–sulfur protein (ISF). The order of electron transfer appears to be

$$\text{Plastoquinol} \longrightarrow \text{cytochrome } b_6 \longrightarrow \text{ISF} \longrightarrow \text{cytochrome } f \longrightarrow \text{plastocyanin}$$

As plastoquinol is oxidized back to plastoquinone, the two protons it has taken from the stroma are released into the thylakoid lumen. The cytochrome b_6f complex plays a role analogous to that of the cytochrome reductase complex in mitochondria, which it closely resembles. Plastocyanin, a mobile protein in the thylakoid lumen, passes the electrons on to P700 reaction centers. In this process the copper in plastocyanin is first reduced to Cu(I) and then reoxidized to Cu(II). We shall consider the fate of the electrons passed to P700 when we discuss photosystem I.

Photosystem II extracts electrons from water, passing them to photosystem I and releasing O_2.

Note that the processes we have described so far have left P680 reaction centers deficient in electrons—in other words, oxidized to a strong oxidant, $P680^+$. These electrons are regained from water, which is split in the presence of an electron acceptor, releasing oxygen in the process. As shown in Figure 17.12b (left), the electron acceptor is a protein containing a cluster of four oxygen-bridged manganese atoms (MnC). This metal cluster can exist in a series of oxidation states, as indicated in Figure 17.13; cycling through these oxidation states allows the cluster to dismantle two water molecules, passing four electrons back to P680 and releasing the four accompanying protons into the thylakoid lumen. Exactly at which points in the cycle individual electrons and protons are released is still a

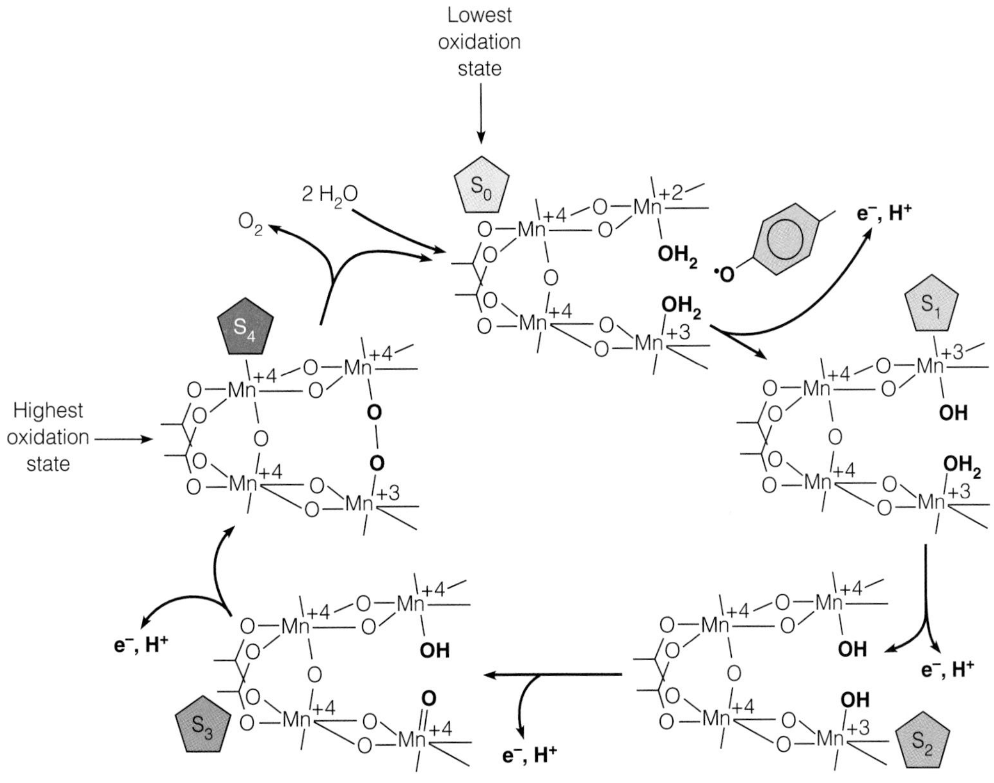

FIGURE **17.13**

A model for the function of the MnC cluster in PSII. The e⁻, H⁺ pairs are successively transfered to the tyrosyl radical as H atoms. For clarity, the radical is shown only once.

Reprinted with permission from C. W. Hoganson and G. T. Babcock, *Science* (1997) 277:1953–1955. © 1997 AAAS.

matter of debate, with several models proposed. In the version shown in Figure 17.13, electrons and protons are released in pairs, which amounts to a hydrogen extraction. This idea is consistent with the observation that the electron donor returning electrons to oxidized P680 is a tyrosine in one of the PSII proteins. This yields a tyrosine radical, which has been observed by electron spin resonance (see Tools of Biochemistry 10A), and releases a proton. It is proposed, then, that each of the steps in Figure 17.13 in which a hydrogen atom ($H^+ + e^-$) is extracted involves the following cycle:

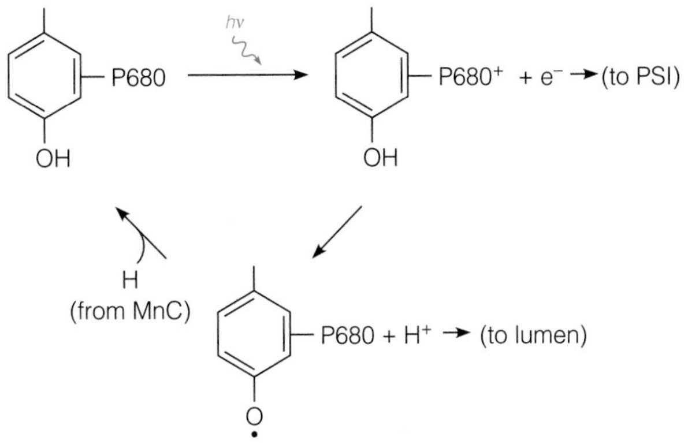

The system has in effect "dismembered" two water molecules. The oxygen produced is released and diffuses out of the chloroplast. The four protons that are produced from $2H_2O$ are released into the thylakoid lumen, helping to generate a pH difference between the lumen and stroma. We may summarize the reaction carried out by photosystem II as follows:

$$2H_2O \xrightarrow{4h\nu} 4H^+ + 4e^- + O_2$$

The electrons produced have traveled through the transport chain of photosystem II and will be passed on to photosystem I, through the b_6f complex.

Photosystem I: Production of NADPH

We have seen that in plants, which utilize two photosystems, photosystem II accomplishes the splitting of water with evolution of O_2 and helps generate a proton gradient across the thylakoid membrane. However, the electrons from the water molecules have not yet reached their final destination in NADPH. This process is the task of photosystem I, in which electrons are again released from a reaction center by light excitation and passed through a second electron transport chain. These electrons are replaced by those passed on from photosystem II.

Photosystem I is a multiprotein complex, containing at least 11 polypeptide chains. It also contains many antenna chlorophylls, and a reaction center chlorophyll, P700, which can absorb light of up to 700 nm. As shown in Figure 17.12b, excitation by a photon absorbed by antenna chlorophylls raises electrons in P700 from a ground state to an excited state at about -1.3 V. Each excited electron then passes through a transport chain. It is first taken up by a chlorophyll-like acceptor (designated A_0), then transferred to a molecule of *phylloquinone* (A_1, also known as vitamin K_1—see page 698), and finally passed through a series of three iron–sulfur proteins (F_X, F_B, and F_A). These proteins contain iron–sulfur clusters of the kinds depicted in Figure 15.4. Finally, the electron is transferred to another iron–sulfur protein, *soluble ferredoxin* (Fd), which is present in the stroma. The enzyme ferredoxin:NADP$^+$ oxidoreductase catalyzes the transfer of electrons to NADP$^+$, after ferredoxin has been reduced by photosystem I:

Photosystem I receives electrons from photosystem II and transfers them to NADP$^+$ to make NADPH.

$$2\text{Fd (red)} + H^+ + NADP^+ \xrightarrow[\text{Reductase}]{} 2\text{Fd (ox)} + NADPH$$

In a sense ferredoxin, rather than NADP$^+$, can be considered the *direct* recipient of electrons from the pathway. Although much of the reduced ferredoxin is used to reduce NADP$^+$, some is used for other reductive reactions, which we discuss later. In fact, we may consider reduced ferredoxin a source of low-potential electrons for many reductive processes. The NADPH produced by ferredoxin oxidation is released into the stroma, where it will be used in dark reactions.

The electrons that have been driven through photosystem I originated in electron transfer from P700 reaction centers. The oxidized reaction centers (P700$^+$) so produced must be resupplied with electrons for photosynthesis to continue. In two-system photosynthesis, these electrons are provided from photosystem II via plastocyanin.

Recently, x-ray diffraction studies have revealed, in high resolution, the structure of the entire photosystem I complex from a cyanobacterium (see References). The complex exists as a trimer, in which each unit consists of several polypeptide chains and contains about 100 chlorophylls (most of which are antenna molecules) and all components of the electron transport chain except for ferredoxin (Figure 17.14a). The complex spans the thylakoid membrane. The approximate locations of the electron transport components with respect to the lumen and stroma faces are diagrammed in Figure 17.14b. Recent studies of the structure of photosystem II reveal surprising similarities, suggesting a common evolutionary origin.

Summation of the Two Systems: The Overall Reaction and ATP Generation

We can now summarize the electron flow through the two-system light reactions. As shown in Figure 17.12, electrons are taken from water and end up in NADPH. We wrote for the overall reaction in photosystem II:

$$2H_2O \xrightarrow{4h\nu} 4H^+ + 4e^- + O_2$$

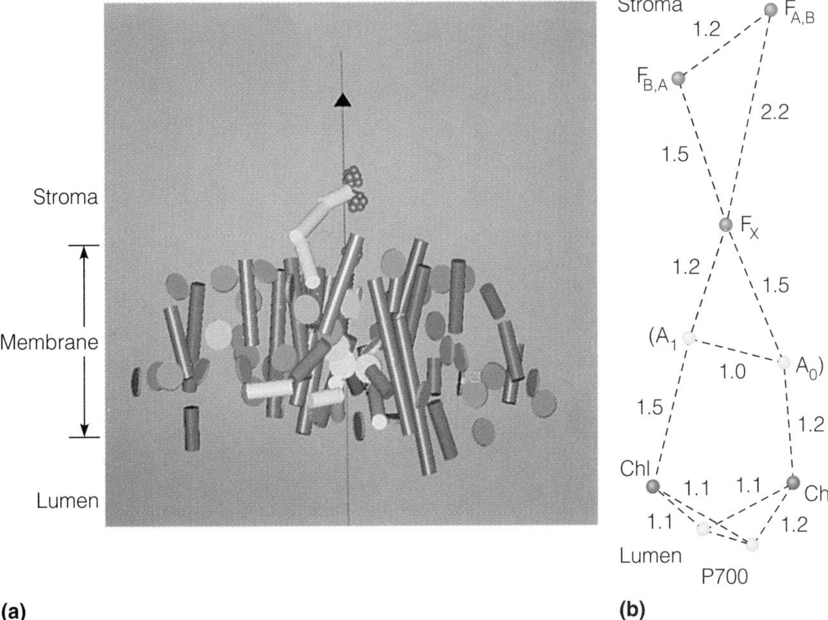

FIGURE **17.14**

The structure of photosystem I. (a) A view of one monomer of the trimeric structure as seen from the side (the membrane plane is perpendicular to the page; the 3-fold axis of the trimer, designated by a line with a black triangle, is perpendicular to the membrane plane). The stroma is to the top, the lumen to the bottom. Several polypeptide chains are contained in the complex. Their individual domains cannot be resolved at the 0.6 nm resolution of this study. **(b)** A schematic view showing the positions of the elements of the electron transport chain. The locations of A_0 and A_1 are not certain. Distances in nm are indicated.

The reactions of photosystem I, if written for four electrons and with all intermediates eliminated, are

$$4e^- + 2H^+ + 2NADP^+ \longrightarrow 2NADPH$$

Adding these two reactions gives us the following summation of the light reactions:

$$2H_2O + 2NADP^+ \xrightarrow{4h\nu} 2H^+ + O_2 + 2NADPH$$

with the understanding that additional protons have been pumped from the stroma into the thylakoid lumen during the passage of each electron through the electron transport chain. Current estimates of the total number of protons are somewhat uncertain because the number transported per electron by the b_6f complex is not known exactly. However, a good estimate would range from 8 to 12 per oxygen released. The net result from the combined function of system I and II is the reduction of $NADP^+$ and the generation of a proton gradient across the thylakoid membrane, with the lumen becoming more acidic than the stroma.

The pH difference produced across the thylakoid membrane can become very large—as much as 3.5 pH units in brightly illuminated chloroplasts. As in the case of ATP generation in mitochondria, these protons can pass back through the thylakoid membrane only through membrane-bound ATP synthase complexes. In chloroplasts these complexes are called CF_0–CF_1 complexes, and they bear considerable resemblance to the F_0–F_1 complexes of mitochondria (see Chapter 15). The pH gradient across the membrane corresponds to a ΔG of about -20 kJ/mol for the passage of one proton. It has been estimated that one ATP is produced for

Both photosystems transport protons from the stroma into the thylakoid lumen. The return of protons, through CF_0–CF_1 complexes, is used to generate ATP.

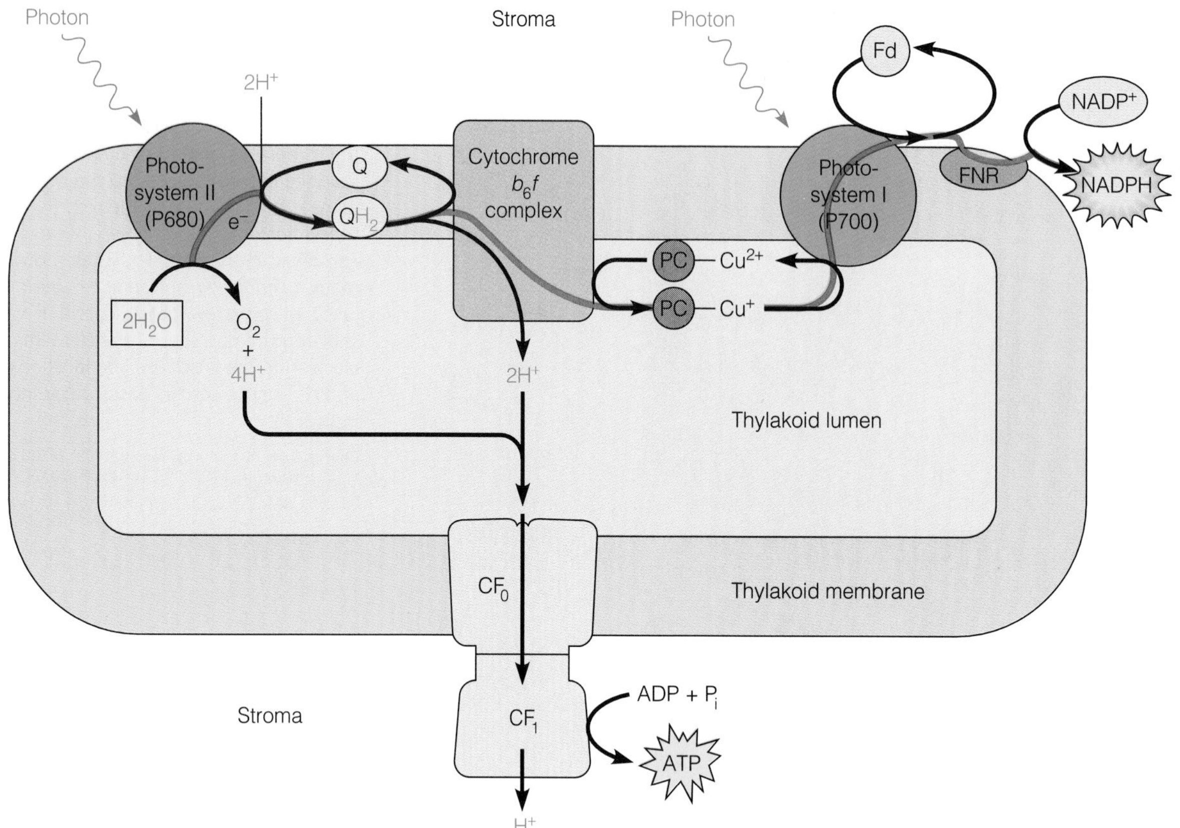

FIGURE **17.15**

Summary view of the light reactions as they occur in the thylakoid. Photosystems I and II and the cytochrome b_6f complex are physically separate protein complexes embedded in the thylakoid membrane. Electron transfer from PSII to cytochrome b_6f is by diffusion of reduced plastoquinone (QH_2) in the membrane lipid. Transfer from b_6f to PSI is mediated by plastocyanin (PC), soluble in the lumen. The protons added to the thylakoid lumen during the light reactions pass through the thylakoid membrane via ATP synthase complexes (CF_0–CF_1). The ATP synthase particles are in the CF_1 subunit and face into the stroma, so that ATP is generated in that compartment. Reduction of $NADP^+$ also occurs on or near the stromal surface of the membrane.

each three protons passing through the CF_0–CF_1 complex, a thermodynamically reasonable result. Because two or three H^+ are transported per electron, up to one ATP is generated for each electron that passes through the chain.

A summary view of the whole set of light reactions is shown in Figure 17.15. It should be noted that photosystems I and II, the cytochrome b_6f complex, and ATP synthase (CF_0–CF_1) are all individual entities embedded in the thylakoid membrane but are not necessarily contiguous. The components that link the photosystems and the b_6f complex are mobile—plastoquinone in the lipid phase of the membrane and plastocyanin in the thylakoid lumen. Thus, electrons can be moved over long distances in this system. Such long-range transport is necessary because of the arrangement of components in the thylakoid membrane. Careful analysis of the composition of grana indicates that the interior membrane layers of the grana are rich in photosystem II; by contrast, the stroma lamellae are rich in photosystem I (see Figures 17.4b and 17.16).

AN ALTERNATE LIGHT REACTION MECHANISM: CYCLIC ELECTRON FLOW

In the two-system light reactions just described, the electrons displaced from photosystem I by excitation are replaced by photosystem II, which receives them from water. The entire process is called **noncyclic electron flow,** and generation of ATP by this process is called **noncyclic photophosphorylation.** An alternative pathway for the light reactions, called **cyclic electron flow,** utilizes the components of photosystem I, plus plastocyanin and the cytochrome b_6f complex (Figure 17.17). Whether this pathway is used or not depends on the levels of $NADP^+$ in the chloroplast stroma. When $NADP^+$ is present in only small amounts, electrons excited in the P700 center are not transferred to $NADP^+$. Instead, they are

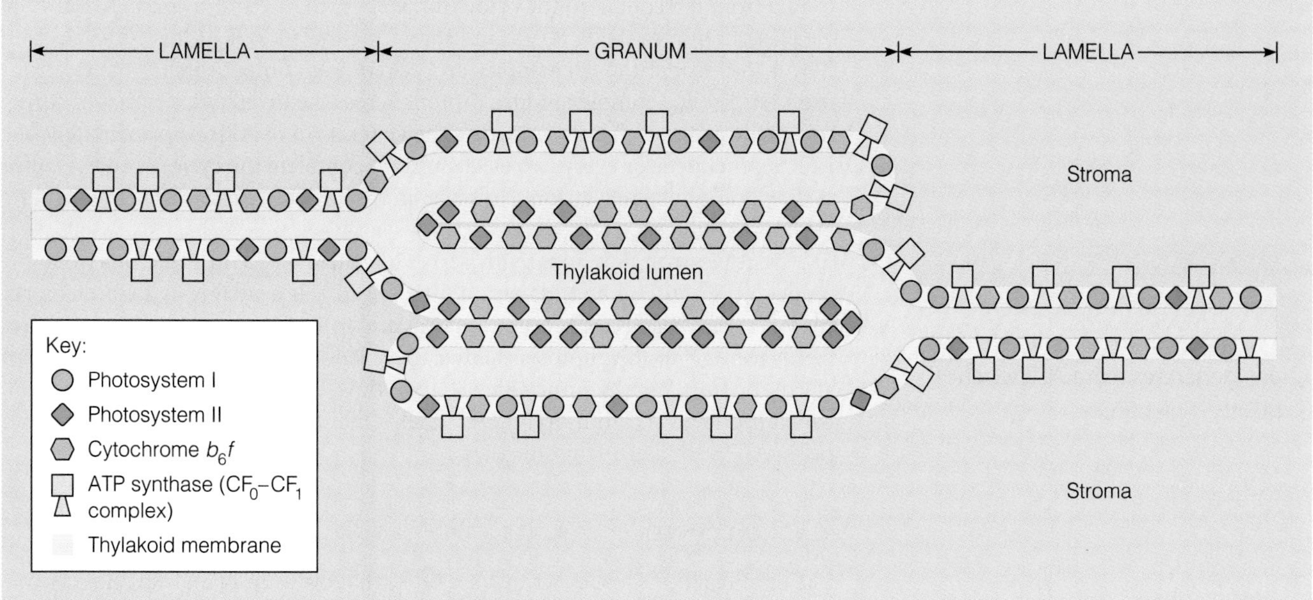

FIGURE **17.16**

Arrangement of components of the two photosystems on the thylakoid membrane. The membrane layers in the interior of the granum are rich in photosystem II. The stroma lamellae and the top and bottom surfaces of the granum are rich in photosystem I and ATP synthase particles, allowing $NADP^+$ reduction and ATP generation to occur at or near these stroma-facing surfaces.

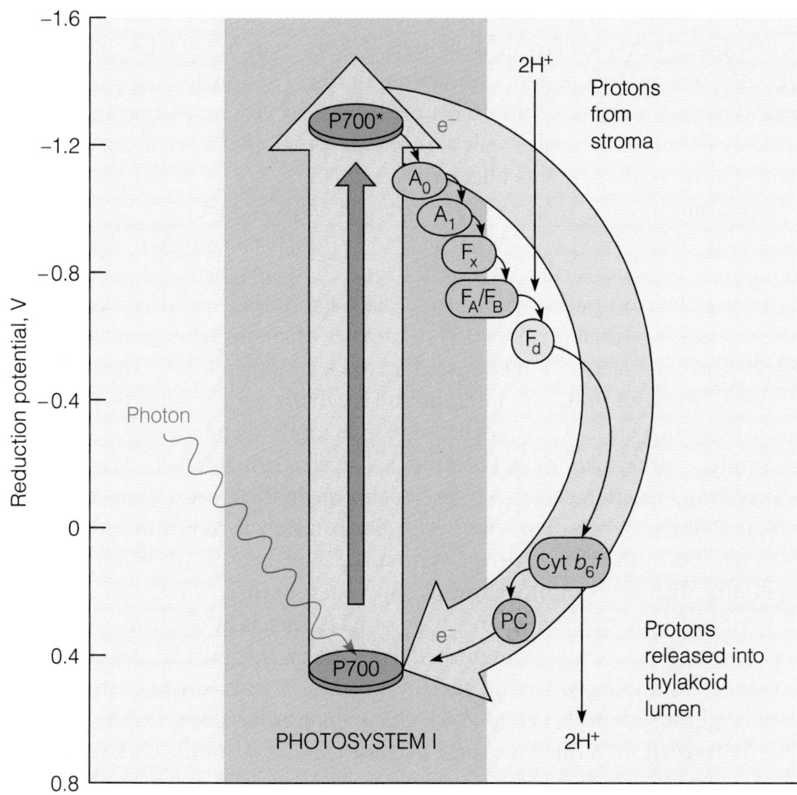

FIGURE **17.17**

Cyclic electron flow. When levels of $NADP^+$ are low and levels of NADPH are high, electrons from the P700 center are returned to it via the cytochrome b_6f complex. There is no $NADP^+$ reduction, but protons are pumped across the membrane and therefore ATP is generated. Symbols are as in Figure 17.12.

passed from ferredoxin to the cytochrome b_6f complex and from there returned via plastocyanin to the P700 ground state. One way to look at this cyclic electron flow is to consider the b_6f complex and $NADP^+$ as competitors for the electrons from Fd. The b_6f complex pumps protons across the thylakoid membrane during this cyclic process, thereby ensuring the generation of ATP. Approximately one ATP is generated for every two electrons that complete the cycle, a process called **cyclic photophosphorylation.** However, in this process, no O_2 is released and no $NADP^+$ is reduced.

Cyclic electron flow, an alternative to two-system (noncyclic) electron flow, generates extra ATP when NADPH is plentiful.

Cyclic electron flow apparently serves to generate ATP in situations when the reductant NADPH is abundant and little $NADP^+$ is available as an electron acceptor. It may also play a more fundamental role. As we shall see, the requirements for ATP in the photosynthetic dark reactions are substantial and may not always be fully met by noncyclic electron flow. Cyclic photophosphorylation, which produces ATP but no NADPH, helps maintain the necessary balance between ATP and NADPH production.

REACTION CENTER COMPLEXES IN PHOTOSYNTHETIC BACTERIA

The light reactions just described are those that occur in plants and algae. However, some of our most precise information concerning functioning of the light reactions has come from studies of photosynthesizing bacteria. Pioneering studies by Roderick Clayton showed that reaction centers from these organisms could be isolated in pure form. Later, the entire molecular structure of a crystallized reaction center complex from the purple sulfur bacterium *Rhodobacter viridis* was determined by Johann Deisenhofer, Hartmut Michel, and Robert Huber using x-ray diffraction. This work received the Nobel Prize in chemistry in 1988.

A model of the reaction center complex is shown in Figure 17.18. It is a transmembrane protein, consisting of four polypeptides. The portion of the complex that lies outside the bacterial plasma membrane, in the periplasmic space, is a cytochrome carrying four heme groups. Subunit H, on the other hand, lies largely on the cytosolic face of the membrane. Buried within the membrane are two subunits (L and M) that are largely α-helical. They carry four bacteriochlorophyll *b* molecules, two bacteriopheophytins, two quinones (designated Q_A and Q_B), and a bound iron atom. Two of the chlorophylls lie very close together; they constitute the reaction center itself. Light absorbance is maximal in the near infrared, at about 870 nm, so the center is referred to as P870.

Some photosynthetic bacteria use a photosystem that generates ATP in a manner analogous to that of photosystem II.

Chemically, the *Rhodobacter* center complex most closely resembles photosystem II in plants, given that it contains pheophytins (bacteriopheophytin, or BPh) and quinones. Studies of the kinetics of the reactions in isolated centers have elucidated the electron pathway (Figure 17.19). Excitation of the reaction center leads very quickly (in about 10^{-12} s) to transfer of an electron to one of the two pheophytins. The electron is then passed on to Q_A and then to Q_B. These quinones are normally bound in the complex, but on receiving a second electron (and two protons), Q_B dissociates. QH_2 is believed to then move to and reduce a cytochrome bc_1 complex (rather like the b_6f complex). The electron is returned from the bc_1 complex to the reaction center via the cytochrome in the reaction center complex.

Note that the net result of this cyclic electron flow is to pump protons from the bacterial cytosol into the periplasmic space. The cytosols of such bacteria become quite alkaline as photosynthesis continues. Return of protons is through ATP synthase complexes, with generation of ATP. The cyclic electron flow in these bacteria must be carefully distinguished from the cyclic flow which can occur via photosystem I in plants. The bacterial system is, in its electron carriers, much more like photosystem II. Though no reducing power has been directly generated by the light reaction in this case, these bacteria can carry out the dark reactions of photosynthesis by using ATP energy to transfer electrons from various substrates to $NADP^+$.

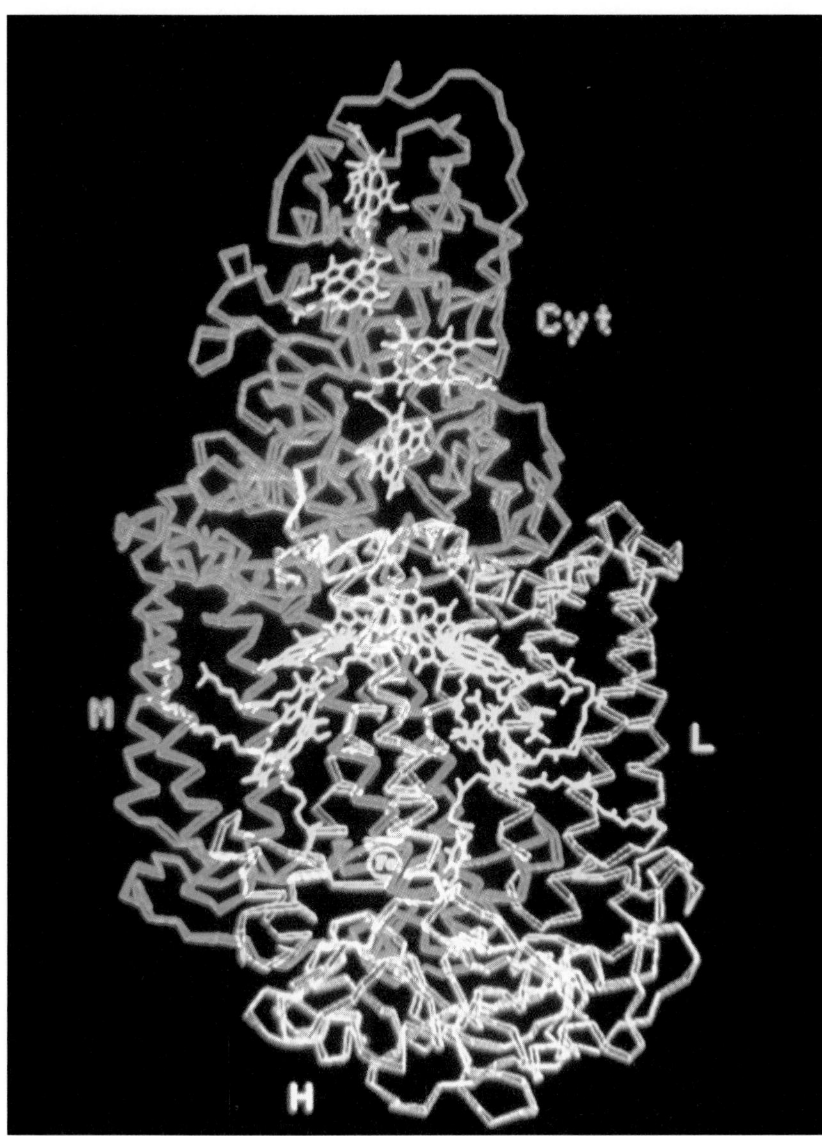

FIGURE **17.18**

Model of a bacterial reaction center complex. This model of the enzyme complex containing the reaction center of the photosynthetic bacterium *Rhodobacter viridis* was deduced from x-ray diffraction measurements. The cytochrome (green), carrying four heme groups (yellow), lies in the periplasmic space between the bacterial inner and outer membranes. Subunits M and L, each with five transmembrane α helices, span the membrane. These subunits carry four bacteriochlorophylls, two bacteriopheophytins, two quinones, and an iron atom—all involved in photon harvesting and electron transfer. Two of the bacteriochlorophylls constitute the reaction center. Subunit H lies mostly on the cytosolic side of the membrane but has one membrane-spanning α helix.

Courtesy of H. Michel and J. Deisenhofer, Max Planck Institut für Biochemie, Munich.

ARTIFICIAL PHOTOSYNTHESIS

The great incentive to devise new, environmentally safe energy sources has led a number of scientists to attempt to mimic the most efficient solar energy converters known—photosynthesizing plants. Although many kinds of molecules can capture photons, in most cases the energy is simply degraded to heat at low temperatures and cannot produce useful work. The trick is to produce, by capture of radiation, a long-lived excited state that can be coupled to a desired energy-saving process. Recently, it has been discovered that compounds in which a porphyrin pigment is covalently sandwiched between a carotenoid and a quinone can be photochemically excited to a long-lived diradical state, with electron transfer from the carotenoid to the quinone:

$$\text{C—P—Q} \xrightarrow{h\nu} \text{C}^{+}\text{—P—Q}^{-}$$

When such molecules are placed in lipid vesicle membranes, together with a lipid-soluble quinone, irradiation with visible light will cause protons to be pumped from outside the vesicle to the inside. If the membranes also contain the ATP

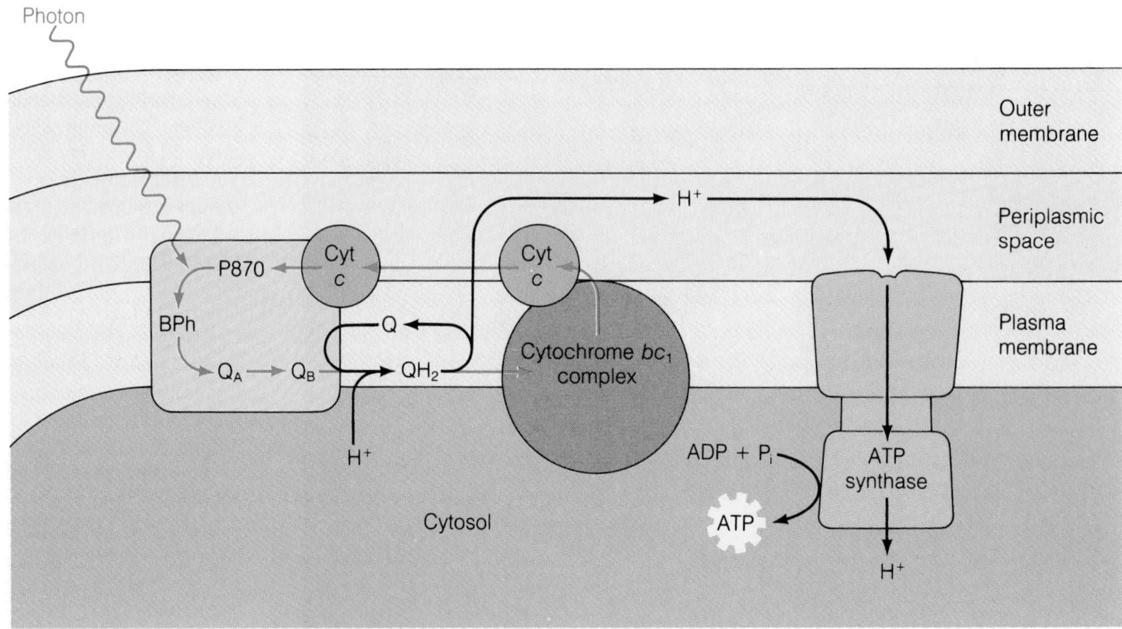

FIGURE **17.19**

Postulated mechanism for bacterial photosynthesis. This process somewhat resembles the light reactions in the thylakoid (see Figure 17.15), with a reaction center and a membrane-bound cytochrome complex. However, there is only one kind of reaction center, and water is not split, nor is $NADP^+$ directly reduced.

synthase complex, ATP will be synthesized. This system is analogous to the photosynthetic systems we have described that are found in plants, algae, and photosynthesizing bacteria.

A wide variety of other model systems have been devised or proposed, in what is a rapidly developing field. It may well turn out that a major energy source in the future will be patterned on that developed by nature several billion years ago.

The Dark Reactions: The Calvin Cycle

The Calvin cycle uses the ATP and NADPH generated in the light reactions to fix atmospheric CO_2 into carbohydrate.

The dark reactions occur in the stroma of the chloroplast. Their function is to fix atmospheric carbon dioxide into carbohydrates, utilizing ATP energy and reducing power (NADPH) generated by the light reactions. As noted earlier, the dark reactions can occur without light but are accelerated in the presence of light.

Carbon dioxide fixation is accomplished by adding one CO_2 at a time to an acceptor molecule and passing the molecule through a cyclic series of reactions, shown schematically in Figure 17.20. The whole series is called the **Calvin cycle,** after the American biochemist Melvin Calvin, who in 1961 received the Nobel Prize for his work in this field. The cycle results in the formation of hexoses and in the regeneration of the acceptor molecule. The Calvin cycle can be envisioned as divided into two stages. In stage I, the carbon dioxide is trapped as a carboxylate and reduced to the aldehyde–ketone level found in sugars, resulting in net carbohydrate synthesis. Stage II is dedicated to regenerating the acceptor molecule. Let us examine each stage in turn.

STAGE I: CARBON DIOXIDE FIXATION AND SUGAR PRODUCTION

Incorporation of CO_2 into a Three-Carbon Sugar

Carbon dioxide is incorporated into glyceraldehyde-3-phosphate (G3P) via the intermediates shown in Figure 17.20. The acceptor molecule for CO_2 is **ribulose-**

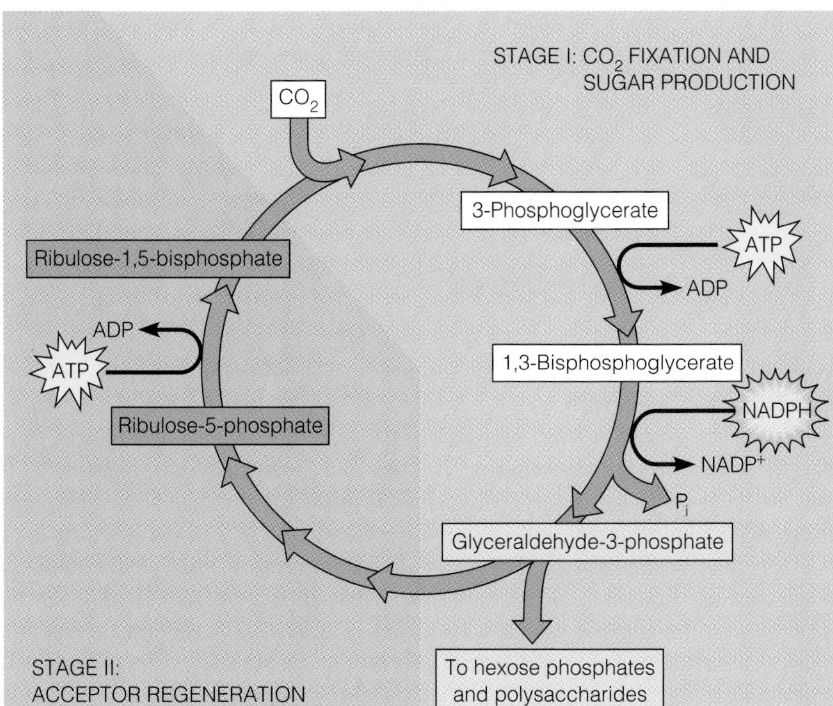

FIGURE **17.20**

Schematic view of the Calvin cycle. The cycle may be divided into two stages. In stage I, CO_2 is fixed and glyceraldehyde-3-phosphate is produced. Part of this G3P is used to make hexose phosphates and eventually polysaccharides. Another fraction of the G3P is used in stage II to regenerate the acceptor molecule, ribulose-1,5-bisphosphate.

1,5-bisphosphate (RuBP). Carbon dioxide from the air diffuses into the stroma of the chloroplast, where it is added at the carbonyl carbon of RuBP. The reaction is catalyzed by the enzyme **ribulose-1,5-bisphosphate carboxylase,** also known as ribulose-1,5-bisphosphate carboxylase/oxygenase (or **rubisco**). This enzyme is one of the most important in the biosphere and certainly the most abundant. It makes up about 15% of all chloroplast proteins, and there are an estimated 40 million tons of it in the world—about 20 pounds for every living person. As its full name implies, the enzyme also has an alternative oxygenase activity. We shall see the consequences of this other activity later. For the moment we will concentrate on its CO_2-fixing (carboxylase) function. The true substrate is the five-carbon enediol intermediate:

Ribulose-1,5-bisphosphate ($CH_2OPO_3^{2-}$–C=O–H–C–OH–H–C–OH–$CH_2OPO_3^{2-}$) → Enediol intermediate ($CH_2OPO_3^{2-}$–C–OH=C–OH–H–C–OH–$CH_2OPO_3^{2-}$) —(CO_2, H^+)→ 2-Carboxy-3-keto-D-arabinitol-1,5-bisphosphate ($CH_2OPO_3^{2-}$–HO–C–COO^-–C=O–H–C–OH–$CH_2OPO_3^{2-}$) —(H_2O)→ Hydrated intermediate ($CH_2OPO_3^{2-}$–HO–C–COO^-–HO–C–OH–H–C–OH–$CH_2OPO_3^{2-}$) —(H^+)→ 2 Molecules of 3-phosphoglycerate ($CH_2OPO_3^{2-}$–H–C–OH–COO^- + COO^-–H–C–OH–$CH_2OPO_3^{2-}$)

The enediol intermediate is carboxylated, and the product is hydrated and then cleaved to yield two molecules of 3-phosphoglycerate (3PG). The reaction is essentially irreversible, with $\Delta G° = -51.9$ kJ/mol. At this point CO_2 has already been fixed into a carbohydrate. The remainder of the Calvin cycle reactions are dedicated to producing hexoses from the triose and regenerating RuBP.

Each molecule of 3PG is phosphorylated by ATP, in a reaction catalyzed by *phosphoglycerate kinase*. The 1,3-bisphosphoglycerate so produced is then reduced to glyceraldehyde-3-phosphate (G3P), with accompanying loss of one

The Calvin cycle has two stages. First, CO_2 is fixed by addition to ribulose-1,5-bisphosphate (RuBP) and hexoses are formed. In the second stage RuBP is regenerated.

phosphate. The reducing agent is NADPH, produced in the light reaction, and the reaction is catalyzed by the enzyme *glyceraldehyde-3-phosphate dehydrogenase:*

$$\underset{\textbf{3-Phosphoglycerate}}{\mathrm{{}^{-}O{-}C({=}O){-}CH(OH){-}CH_2OPO_3^{2-}}} \xrightarrow[\text{Phosphoglycerate kinase}]{\text{ATP} \;\to\; \text{ADP}} \underset{\textbf{1,3-Bisphosphoglycerate}}{\mathrm{{}^{2-}O_3PO{-}C({=}O){-}CH(OH){-}CH_2OPO_3^{2-}}} \xrightarrow[\text{Glyceraldehyde-3-phosphate dehydrogenase}]{\text{NADPH} \;\to\; \text{NADP}^+ + \text{P}_i} \underset{\textbf{Glyceraldehyde-3-phosphate}}{\mathrm{H{-}C({=}O){-}CH(OH){-}CH_2OPO_3^{2-}}}$$

We have encountered similar enzymes earlier, in connection with their roles in glycolysis (see Chapter 13).

At this stage of the cycle a molecule of CO_2 has already been fixed into a simple (three-carbon) monosaccharide. It is useful to note the requirements in ATP and NADPH up to this point. For each CO_2 molecule that has passed through these steps, two molecules of ATP have been hydrolyzed and two molecules of NADPH have been oxidized. However, it is more appropriate to keep accounts on a "per glucose" basis because we want to see what must happen to account for the generation of one hexose molecule from CO_2. Figure 17.21 provides a schematic view of the stoichiometry of the whole Calvin cycle. Six molecules of CO_2 will have to enter the cycle to provide the six carbons needed for every new molecule of hexose produced. That will require formation of 12 G3P, and therefore 12 ATP and 12 NADPH will be needed.

At this point the pathway splits, so as to satisfy the two essential goals—to make hexoses and to regenerate the acceptor. Of the 12 molecules of G3P that have been produced, 2 will be used to make a molecule of a hexose. The remaining 10 will be utilized to regenerate the 6 molecules of ribulose bisphosphate nec-

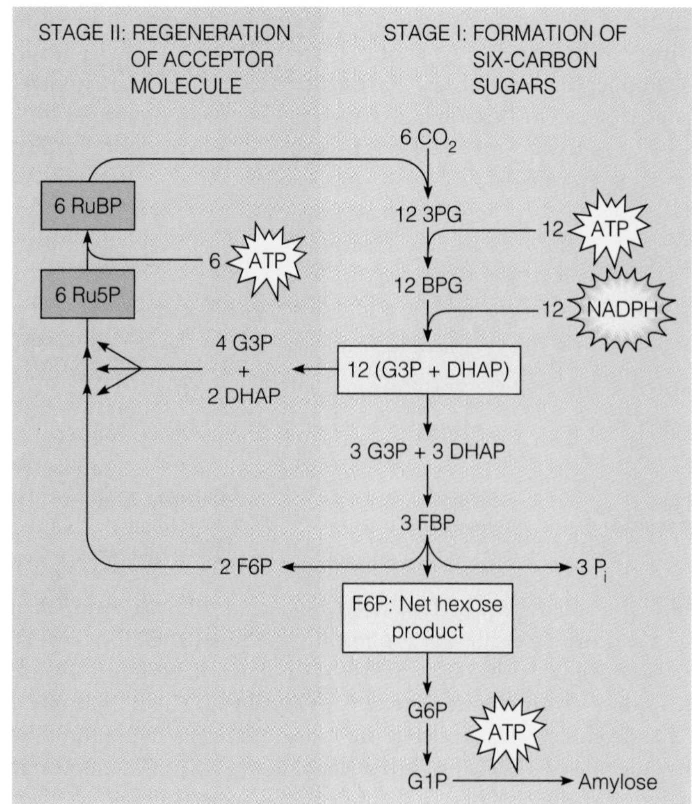

FIGURE **17.21**

Stoichiometry of the Calvin cycle. In six turns of the Calvin cycle, six CO_2 molecules will have entered and bound to six molecules of ribulose-1,5-bisphosphate (RuBP) to yield 12 molecules of glyceraldehyde-3-phosphate (G3P). Because G3P is in isomeric equilibrium with dihydroxyacetone phosphate (DHAP), the 12 G3P may be considered an interconvertible stock of 12 molecules of (G3P + DHAP). Of these, six are used to make three molecules of fructose-1,6-bisphosphate (FBP), of which *one* constitutes the net hexose product of the six turns. The other two FBPs are used, together with the six remaining molecules of (G3P + DHAP), to form six molecules of ribulose-5-phosphate (Ru5P), which are then phosphorylated to regenerate the required six molecules of RuBP.

essary to maintain the cycle. That is, 10 three-carbon molecules will be converted to 6 five-carbon molecules.

Formation of Hexose Sugars

Let us consider the hexose formation first. This is actually familiar ground, for it follows a portion of the gluconeogenic pathway described in Chapter 16. The reactions are shown schematically in Figure 17.21. Recall that glyceraldehyde-3-phosphate can be isomerized to dihydroxyacetone phosphate (DHAP) by triose phosphate isomerase (pages 370–371). Thus, the 12 molecules of G3P produced can be considered to be an interconvertible pool of G3P and DHAP. A molecule of G3P and a molecule of DHAP can be combined, via the enzyme *fructose bisphosphate aldolase,* to yield fructose-1,6-bisphosphate (FBP). As Figure 17.21 shows, 6 of the G3P molecules follow this path, to yield 3 molecules of FBP. The FBP is dephosphorylated to yield three molecules of fructose-6-phosphate (F6P). Of these, two will be employed in the regeneration pathway, but one is available as a net product of the Calvin cycle; it is then isomerized to glucose-6-phosphate (G6P) and finally to glucose-1-phosphate (G1P).

Glucose-1-phosphate is, in plants as in animals, the precursor to oligosaccharide and polysaccharide formation. Formation of plant starch (amylose) follows a path similar to that used by animals in glycogen synthesis. However, instead of using UTP to activate the glucose monomer, as in glycogen formation, ATP is employed in the polymerization of amylose:

$$\text{Glucose-1-phosphate} + \text{ATP} \longrightarrow \text{ADP-glucose} + \text{PP}_i$$

$$\text{ADP-glucose} + (\text{glucose})_n \longrightarrow (\text{glucose})_{n+1} + \text{ADP}$$

Amylose, which is not very soluble, is a storage carbohydrate. However, much of the saccharide synthesized in plant leaves is exported to other parts of the plant, mostly in the form of sucrose. Sucrose is synthesized in the cytosol of plant leaves by the following sequence of reactions:

$$\text{UTP} + \text{glucose-1-P} \longrightarrow \text{UDP-glucose} + \text{PP}_i$$

$$\text{UDP-glucose} + \text{fructose-6-P} \longrightarrow \text{UDP} + \text{sucrose-6-P}$$

$$\text{Sucrose-6-P} + \text{H}_2\text{O} \longrightarrow \text{sucrose} + \text{P}_i$$

The UDP produced is then reconverted to UTP by phosphate transfer from ATP.

STAGE II: REGENERATION OF THE ACCEPTOR

The reactions we have considered to this point can account for the introduction of one carbon into one molecule of hexose, with subsequent formation of oligosaccharides or polysaccharides. But to complete the Calvin cycle, it is necessary to regenerate enough ribulose-1,5-bisphosphate to keep the cycle going. This means we will need to regenerate 6 moles of RuBP for every 6 moles of CO_2 taken up. This is accomplished by the set of reactions shown in Figure 17.22, which constitute the regenerative phase of the cycle schematized in Figures 17.20 and 17.21. Note that the *input* molecules in this somewhat complex reaction pathway are as follows:

1. Two molecules of DHAP and four molecules of G3P, from the six G3P that were diverted to the regeneration pathway in Figure 17.21.
2. Two of the three molecules of fructose-6-phosphate (F6P) that were produced from the remaining three G3P and three DHAP.

In order to make five-carbon molecules from six-carbon and three-carbon molecules, several rearrangements are required. They are accomplished by *transketolases*

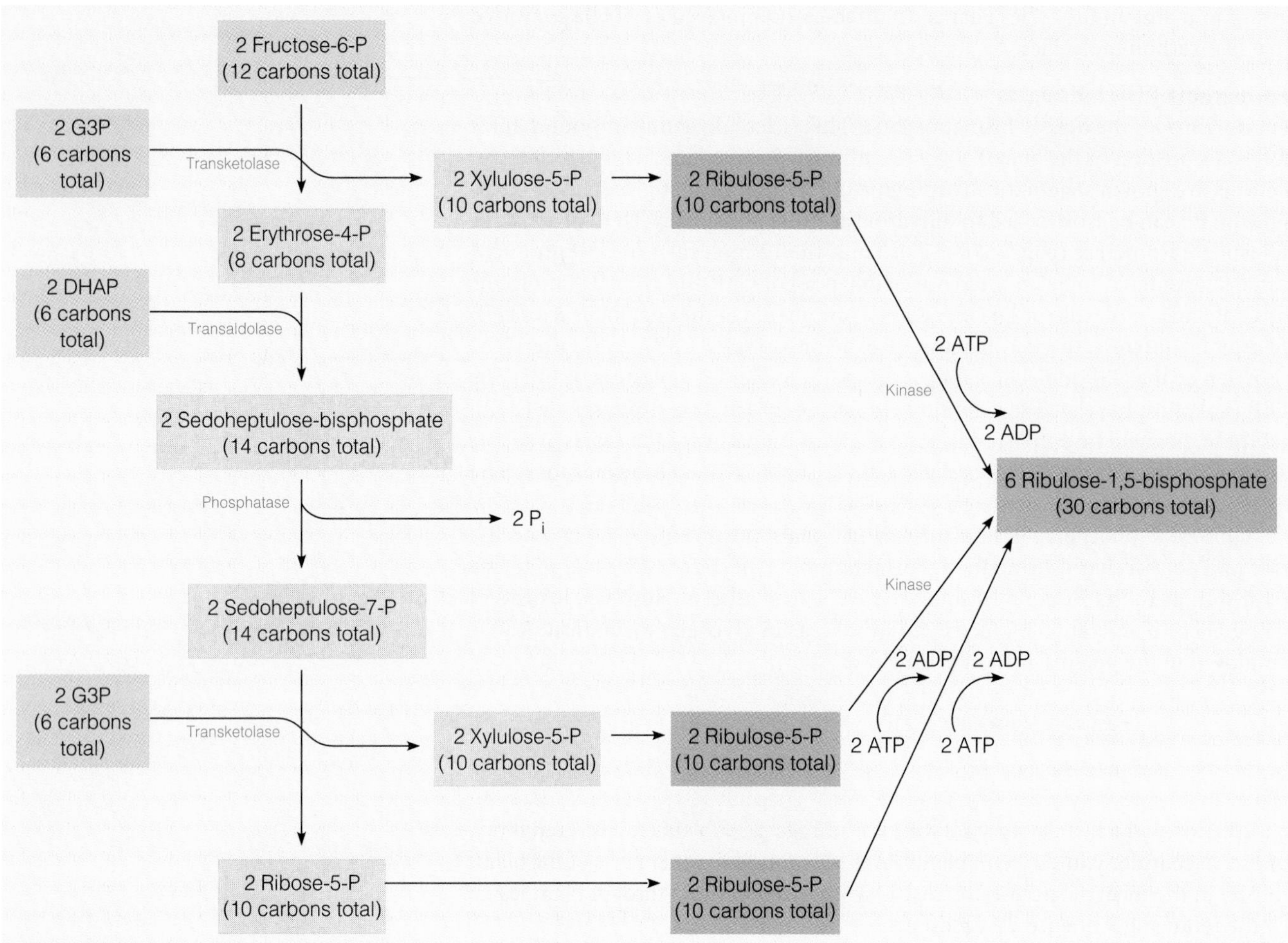

FIGURE **17.22**

Regeneration phase of the Calvin cycle. The stoichiometry here follows that in Figure 17.21. The two molecules of fructose-6-phosphate entering at the top are combined with four molecules of G3P and two molecules of DHAP to yield the required six molecules of ribulose-5-phosphate. These are then phosphorylated to produce the required RuBP. Note the similarity of this pathway to parts of the pentose phosphate pathway running in reverse (see Figure 14.22, page 512).

and *transaldolases.* The structures of the sugars involved in these reactions are all given in Chapter 9. What is important here is the way in which two hexoses and six trioses have been rearranged and recombined to form six pentoses.

The final step in the regeneration of ribulose-1,5-bisphosphate is a phosphorylation, catalyzed by the enzyme *ribulose-5-phosphate kinase* and utilizing ATP. For six rounds of the cycle, this step will require 6 ATPs in addition to the 12 already accounted for. Therefore, the requirements for synthesizing 1 mole of hexose from CO_2 are 12 moles of NADPH and 18 moles of ATP. The overall dark reaction may then be written as

$$6CO_2 + 18ATP + 12NADPH + 12H_2O \longrightarrow C_6H_{12}O_6 + 18ADP + 18P_i + 12NADP^+ + 6H^+$$

A Summary of the Light and Dark Reactions in Two-System Photosynthesis

THE OVERALL REACTION AND THE EFFICIENCY OF PHOTOSYNTHESIS

The ATP and NADPH needed for the dark reactions are released into the stroma by the light reactions of photosynthesis. If we recall that two photons are required for every electron to pass through photosystems I and II, and that two electrons are required to reduce each $NADP^+$, then four photons are necessary for the production of each NADPH molecule. This corresponds to eight photons per O_2, a number in agreement with the quantum efficiency experimentally observed when both photosystems are operating—about 0.12 O_2 per photon. For the 12 NADPH needed in the dark reaction, as summarized in the previous section, 48 photons must be absorbed. If we assume that these photons will also pump enough protons across the thylakoid membrane to yield the 18 ATP required, we may, as an approximation, write the light reactions as

$$12NADP^+ + 18ADP + 18P_i + 6H^+ \xrightarrow{\text{48 photons}} 6O_2 + 12NADPH + 18ATP + 6H_2O$$

This equation differs from the second equation on page 609 because we now include the ATP generation from the proton gradient. Adding this equation for the light reaction to the overall dark reaction, we obtain

$$6H_2O + 6CO_2 \xrightarrow{\text{48 photons}} C_6H_{12}O_6 + 6O_2$$

This estimate of 48 photons assumes that noncyclic photophosphorylation provides enough ATP for the dark reactions. If, as many workers in the field believe, additional ATP from cyclic photophosphorylation is required, the number of photons needed will be greater.

The overall energy efficiency of photosynthesis can approach 35%.

We can, on the basis of these calculations, estimate the energy efficiency of photosynthesis. Forming a mole of hexose from CO_2 and water requires, as we have seen, 2870 kJ. The energy input per photon depends on the wavelength of light used. Assuming that light of 650 nm wavelength is used, 48 einsteins of such light correspond to about 8000 kJ (see Figure 17.6). From this figure, we estimate a theoretical efficiency of approximately 35%. Direct experimental measurements of the efficiency under optimal conditions give results in the same range or slightly lower. At high levels of illumination, when not all photons absorbed by the chloroplasts can be used for reaction center excitation, efficiency is much lower.

REGULATION OF PHOTOSYNTHESIS

The photosynthetic dark reactions are regulated by the amount of light available to the organism.

It should be evident that the so-called dark reactions of photosynthesis, which result in the production of sugars, require careful regulation. Because the dark reactions depend on the reductive power and ATP supplied by the light reactions, it is not surprising that they are stimulated by the light reactions. There are two major ways in which this stimulation is accomplished. A central enzyme in the dark reactions, ribulose-1,5-bisphosphate carboxylase, is stimulated by high pH and by both CO_2 and Mg^{2+}. The pumping of protons from the stroma into the thylakoid lumen by the light reactions increases the stromal pH; at the same time, Mg^{2+} ions enter the stroma to compensate for the positive charge of the H^+ ions that have been lost. Recent experiments have revealed that directly light-dependent pathways also exist for the stimulation of this enzyme.

Three other enzymes of the Calvin cycle are specifically activated by another light-dependent mechanism. They are sedoheptulose-1,7-bisphosphatase (the phosphatase shown in Figure 17.22), glyceraldehyde-3-phosphate dehydrogenase (see page 616), and ribulose-5-phosphate kinase (see Figure 17.22). This activation is dependent on reduction of disulfides to sulfhydryls in the enzymes, which is promoted by a disulfide exchange reaction with the protein *thioredoxin* (Figure 17.23). Thioredoxin, a small protein carrying two reversibly oxidizable —SH groups, is used in a wide variety of redox reactions (see page 812). The reduction of thioredoxin, in turn, is promoted by oxidation of the reduced form of ferredoxin via a reaction catalyzed by the enzyme *ferredoxin–thioredoxin reductase.* In strongly irradiated chloroplasts, in which $NADP^+$ stores are depleted, reduced ferredoxin accumulates. High levels of reduced ferredoxin thereby lead to activation of the Calvin cycle enzymes, stimulating the Calvin cycle reactions when the

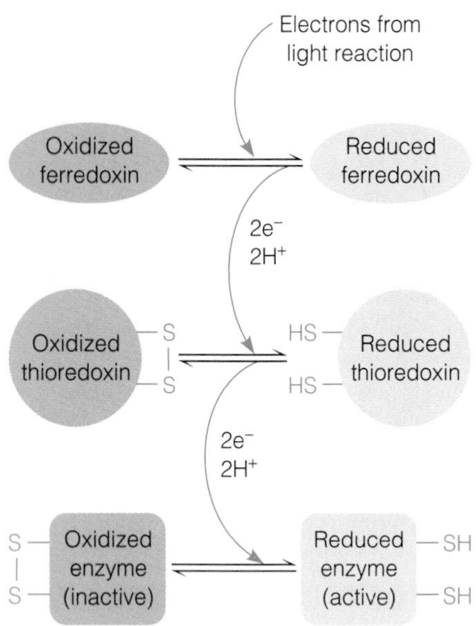

FIGURE **17.23**

Light-dependent activation of dark-reaction enzymes. Some enzymes of the Calvin cycle are activated by disulfide reduction, which is mediated by reduced thioredoxin. Thioredoxin is reduced by reduced ferredoxin, which accumulates in irradiated chloroplasts. Thioredoxin-mediated activation and other light-dependent mechanisms cause the dark reaction to speed up in the presence of light. Additional functions of thioredoxin are presented in Chapter 22.

light reactions are very active. The same compound, reduced thioredoxin, also stimulates the CF_0–CF_1 complexes, ensuring a high rate of ATP generation when illumination is intense.

In the dark, the plant "turns into an animal" in terms of its biochemistry. Although the dark reactions of photosynthesis may continue for some time, using photosynthesized ATP and NADPH, the plant must ultimately begin to draw on its energy reserves, using pathways familiar from our studies of animal catabolism: glycolysis, the citric acid cycle, and the pentose phosphate pathway. In general, these pathways are inhibited in plants in the presence of sunlight and become more active in the dark. The key light-inhibited enzymes are phosphofructokinase (in glycolysis) and glucose-6-phosphate dehydrogenase (in the pentose phosphate pathway). The latter is inhibited by the same reduced form of thioredoxin that *activates* Calvin cycle enzymes.

Finally, it should be mentioned that there is now evidence for regulation of chloroplast genes at the transcriptional level. Sucrose and glucose can act as repressors.

Photorespiration and the C_4 Cycle

Ribulose bisphosphate carboxylase is a peculiar enzyme. Under some environmental conditions, it can behave as an *oxygenase* instead of a carboxylase:

Ribulose-1,5-bisphosphate → **Enediol intermediate** →(O_2) **Hydroperoxide intermediate** →(H_2O) **Hydrated intermediate** →(H^+) **Phosphoglycolate** + **3-Phosphoglycerate** + H_2O + $2H^+$

Glyoxylate

This reaction happens mostly under conditions of high O_2 and low CO_2 concentrations, for the K_M for O_2 is about 10-fold higher than for CO_2. When the oxygenase reaction becomes significant, it initiates a reaction pathway known as **photorespiration,** with production of 3-phosphoglycerate and *phosphoglycolate* in the chloroplast. As Figure 17.24 shows, the phosphoglycolate is then dephosphorylated and passed into organelles called **peroxisomes.** Here, it is further oxidized, yielding hydrogen peroxide and glyoxylate. The toxic H_2O_2 is broken down by catalase, and the glyoxylate is aminated, producing glycine. The glycine enters mitochondria, where *two* molecules are converted into *one* molecule of serine, plus one molecule each of CO_2 and NH_3. This process, involving the multipro-

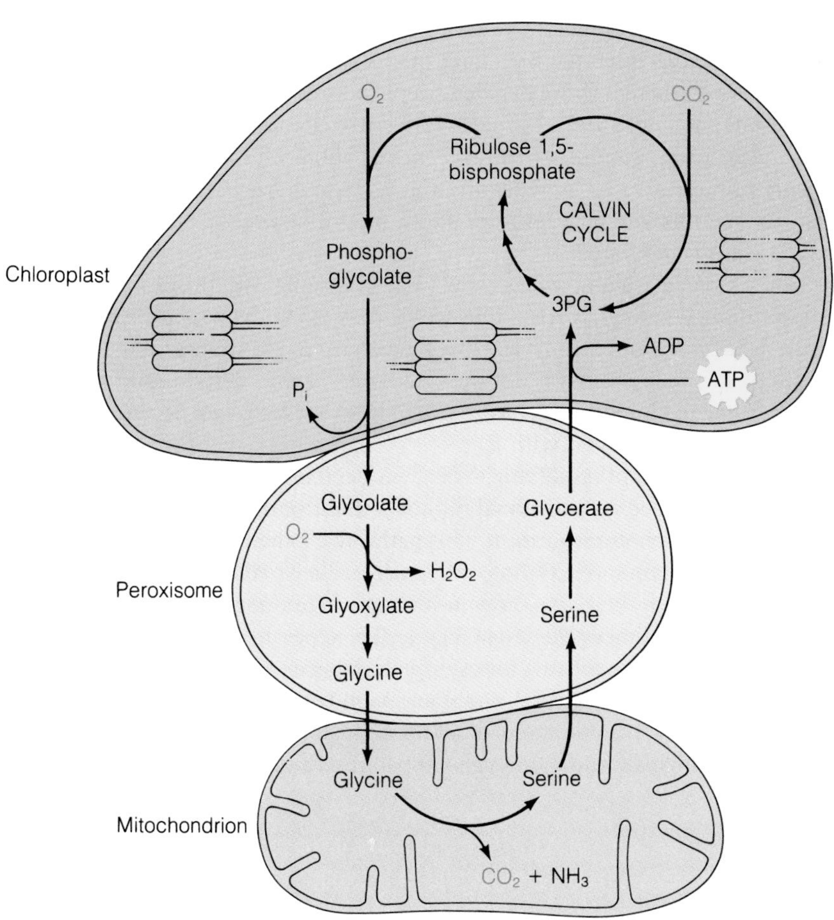

FIGURE **17.24**

Photorespiration. Ribulose-1,5-bisphosphate can be diverted from the Calvin cycle, especially when the concentration of CO_2 is low. RuBP carboxylase/oxygenase catalyzes the oxidation of RuBP to form phosphoglycolate. In the reactions that follow, more O_2 is used, CO_2 is generated, and ATP is hydrolyzed, as metabolites pass from the chloroplast to nearby peroxisomes and mitochondria and then back into the chloroplast.

tein glycine cleavage system, is described in detail in Chapter 20. The gases, CO_2 and ammonia, are released. The serine passes back into the peroxisome, where a series of reactions convert it to glycerate. Returning to the chloroplast, the glycerate is rephosphorylated (using ATP) to yield 3-phosphoglycerate.

Under conditions of low CO_2 and high O_2, plants exhibit photorespiration, in which O_2 is consumed and CO_2 released.

However it is looked at, photorespiration appears to be a losing process. Note the following:

1. Ribulose-1,5-bisphosphate is lost from the Calvin cycle.
2. The fixation of CO_2 is reversed: O_2 is consumed, and CO_2 is released.
3. Only a part of the carbon is returned to the chloroplast.
4. ATP is needlessly expended.

It is difficult to see any useful function for photorespiration. It essentially works against the entire photosynthetic process, undoing what has been accomplished. The fact that photorespiration becomes important only under conditions where CO_2 levels are very low (as in plants exposed to high levels of illumination so that they consume most of the CO_2 in the surroundings) suggests one possible explanation. Continued photosynthesis under such conditions would lead to the production of reactive oxygen species like O_2^- (Chapter 15), which can produce cellular damage. Under these circumstances, photorespiration will lower oxygen levels and inhibit the light reactions. Support for this idea comes from recent experiments in which the capacity of tobacco plants to carry out photorespiration was modified by genetic engineering. The researchers observed that the rate-limiting process in photorespiration was the utilization of the ammonia generated (Figure 17.24); they therefore constructed plants that produced higher or lower levels of the enzyme glutamine synthetase, which combines ammonia with glutamate to give glutamine (see pages 715–718). Those with high levels (and therefore efficient photorespiration) tolerated high light intensity with less damage to leaves than the wild type, whereas those deficient in photorespiration were more sensitive.

Despite such potential advantages in adverse circumstances, the fact remains that under normal conditions the existence of photorespiration decreases the efficiency of most plants. Consequently, much research has been devoted in recent years to attempts to reduce levels of photorespiration in crop plants. So far, success has been very limited.

Some plants (called C_4 plants) minimize the wastefulness of photorespiration by utilizing an alternative to the Calvin cycle.

Some plants have evolved their own way of dealing with the problem of photorespiration. You might expect they would have done so by modification of the ribulose bisphosphate carboxylase/oxygenase enzyme to suppress the oxygenase function. Surprisingly, that is not the case. This enzyme, despite (or perhaps because of) its vital importance, has changed little over long ages. It remains a relatively inefficient catalyst (with $k_{cat} \cong 2\ s^{-1}$) and has never lost its oxygenase function. So instead of modifying the enzyme, certain plants, which are called **C_4 plants,** have evolved an additional photosynthetic pathway that helps conserve CO_2 released by photorespiration. This pathway is called the **C_4 cycle** because it involves incorporation of CO_2 into a C_4 intermediate (oxaloacetate). This cycle is distinguished from the Calvin cycle, which utilizes a three-carbon intermediate, and is hence sometimes called the C_3 cycle. The C_4 cycle is found in several crop species (maize and sugarcane, for example) and is important in tropical plants, which are exposed to intense sunlight and high temperatures. Although photorespiration occurs to some extent at all times in all plants, it is most active under conditions of high illumination, high temperature, and CO_2 depletion.

FIGURE **17.25**

Structural differences between leaves of C_3 and C_4 plants. The figure shows structural differences in mesophyll and bundle sheath cells associated with their different uses in C_3 and C_4 plants. Cells that carry out the Calvin cycle (C_3 cycle) are shown in dark green; those that carry out the C_4 cycle are shown in light green. The stomata (plural of stoma) are orifices through which the leaf exchanges gases and water vapor with the surroundings. **(a)** C_3 plants. In C_3 plants, which have no C_4 cycle, Calvin cycle photosynthesis occurs mostly in the mesophyll cells. **(b)** C_4 plants. In C_4 plants the mesophyll cells carry out the C_4 cycle. The C_4 compounds are then delivered to the bundle sheath cells, where most of the Calvin cycle photosynthesis takes place.

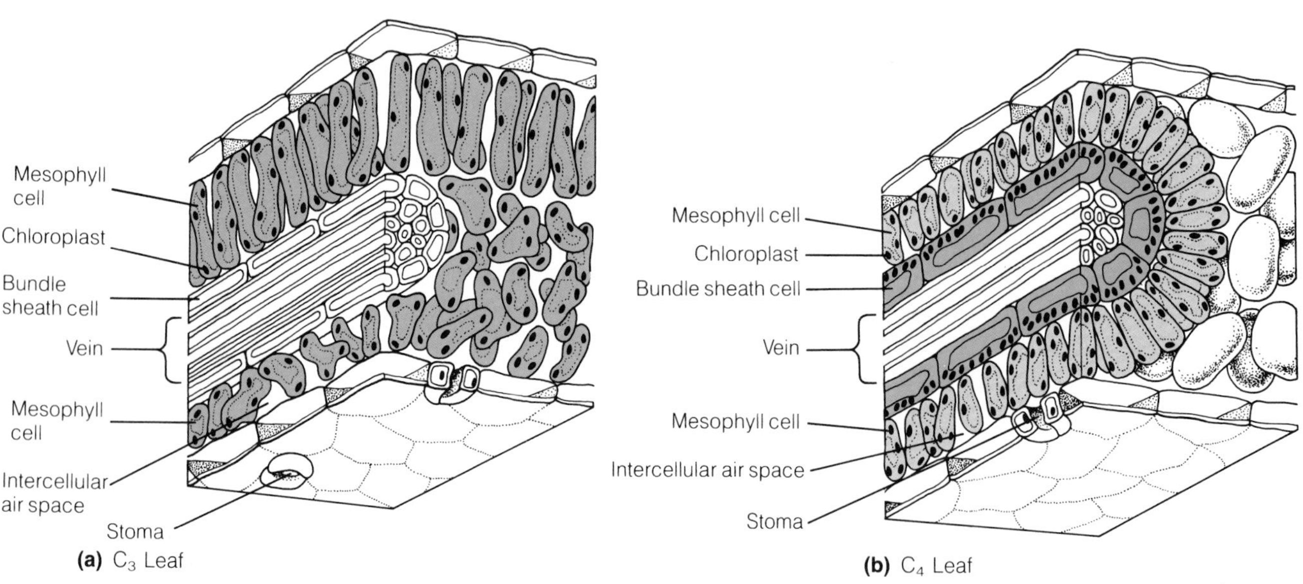

C_4 plants concentrate their Calvin cycle (C_3) photosynthesis in specialized *bundle sheath cells,* which lie below a layer of mesophyll cells (Figure 17.25). The mesophyll cells on the other hand, which are most directly exposed to external CO_2, contain the enzymes for the C_4 cycle. This pathway, as it operates in most C_4 plants, is shown in Figure 17.26. It is essentially a mechanism for trapping CO_2 into the four-carbon compound oxaloacetate and passing it on to bundle sheath cells for use in their Calvin (C_3) cycle.

The key to the efficiency of C_4 plants is that the CO_2-fixing enzyme used in this pathway, *phosphoenolpyruvate carboxylase,* lacks the oxygenase activity shown by ribulose bisphosphate carboxylase, and has a much lower K_M for CO_2. Thus, even under conditions of high O_2 concentration and low CO_2 concentration, the mesophyll cells continue to pump CO_2 to the photosynthesizing bundle sheath cells. This process helps maintain high enough CO_2 levels in the bundle sheath cells that fixation, rather than photorespiration, is favored. Furthermore, if photorespiration *does* occur, the CO_2 that is released in that process can be largely salvaged in the surrounding mesophyll cells and returned to the Calvin cycle.

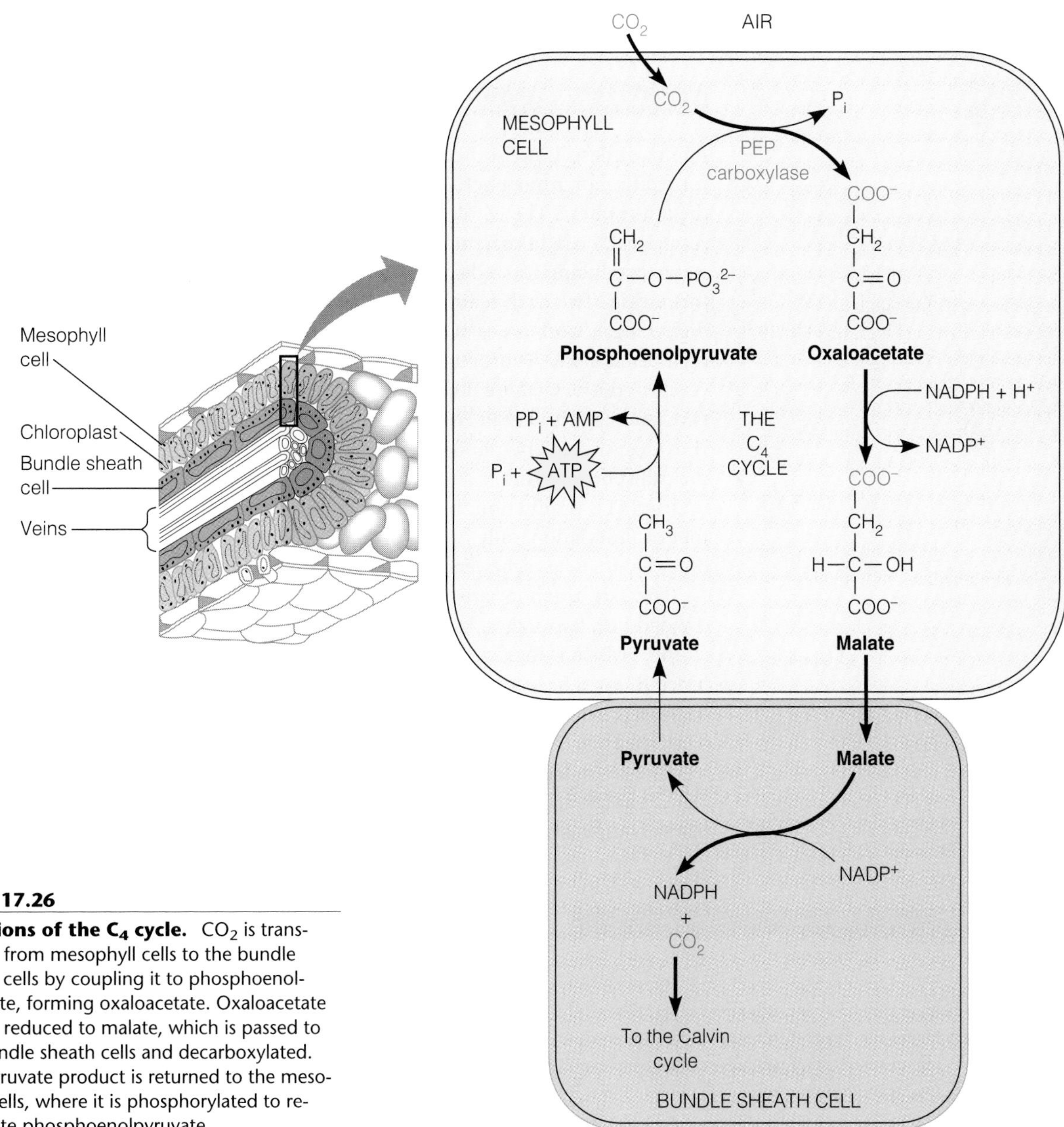

FIGURE **17.26**

Reactions of the C_4 cycle. CO_2 is transported from mesophyll cells to the bundle sheath cells by coupling it to phosphoenolpyruvate, forming oxaloacetate. Oxaloacetate is then reduced to malate, which is passed to the bundle sheath cells and decarboxylated. The pyruvate product is returned to the mesophyll cells, where it is phosphorylated to regenerate phosphoenolpyruvate.

As Figure 17.26 shows, the C_4 cycle costs the plant energy in the form of ATP. In fact, because ATP is hydrolyzed to AMP and inorganic phosphate in regenerating phosphoenolpyruvate, the expense is equivalent to *two* extra ATPs for every CO_2 molecule fixed. Nevertheless, the price appears to be worth paying under circumstances when photorespiration would dominate.

The inefficiency of rubisco as an enzyme and its willing participation in photorespiration greatly reduce the efficiency of plants as food producers. Not only is energy wasted in photorespiration, but also the very large amounts of rubisco that must be synthesized place seemingly unnecessary demands on plant metabolism. If a more efficient enzyme could be developed, crop yields could be greatly increased and nitrogen demands reduced. Therefore, there are intensive efforts in progress to engineer a more efficient rubisco into crop plants. These have taken two directions, neither of which has been successful to date. On the one hand, attempts are made to modify the rubisco of higher plants. This effort is complicated by the complexity of the protein (16 subunits) and our incomplete understanding of its enzymology. A quite different approach makes use of the observation that the rubisco enzyme in some red algae is considerably more efficient. Attempts are being made to incorporate this enzyme into crop plants.

SUMMARY

Photosynthesis is the source of most of the energy in the biosphere and accounts for fixation of atmospheric CO_2 and the production of most or all of the O_2 in the atmosphere. The whole process can be divided into light reactions and dark reactions. The light reactions use the energy of sunlight to extract electrons from water, producing O_2, reductive potential, and a proton gradient that drives ATP formation. The dark reactions reduce CO_2 into carbohydrates. In plants and higher algae, both types of reactions take place in chloroplasts. Photons for the light reaction are absorbed by antenna pigments, and the energy is transferred to reaction centers, where it enters either photosystem I or photosystem II. These two systems, working in conjunction, carry out the light reactions. Photosystem II oxidizes water, and photosystem I reduces $NADP^+$. Together, the systems drive the transport of protons across chloroplast membranes to provide a pH gradient to drive ATP production. In a "short circuit" of photosystem I called cyclic photophosphorylation, only ATP is produced. Some photosynthetic bacteria, on the other hand, use a cyclic version of photosystem II to generate ATP.

The dark reactions are largely summarized in the Calvin cycle, which may be divided into two stages. In the first, CO_2 is added to ribulose-1,5-bisphosphate (RuBP), which is then cleaved and reduced to form trioses that can then be combined to form hexose. The second stage of the cycle uses most of the trioses and hexoses to regenerate RuBP. A number of these dark reactions are regulated by light intensity.

Under conditions of low CO_2 and high O_2, plants undergo an oxidative process called photorespiration. This process is essentially inefficient, and some tropical plants compensate for it via the C_4 cycle, which is less sensitive to high O_2 levels.

REFERENCES

General

Clayton, R. K. (1980) *Photosynthesis: Physical Mechanisms and Chemical Patterns.* Cambridge University Press, Cambridge. Although somewhat superseded by recent work, this remains an excellent summary of the more physical aspects of photosynthesis.

Diesenhofer, J., and J. R. Norris, eds. (1993) *The Photosynthetic Reaction Center.* Academic Press, San Diego. This two-volume set contains a wealth of information on both structure and function.

Lawlor, D. W. (1987) *Photosynthesis: Metabolism, Control, and Physiology.* Longman, Essex, England. This little book is a readable source of information.

Youvan, D. C., and B. L. Marrs (1987) Molecular mechanisms of photosynthesis. *Sci. Am.* 256(6):42–48. A brief but lucid account of the field.

Chloroplasts

Bogorad, L., and I. K. Vasil, eds. (1991) *The Molecular Biology of Plastids.* Academic Press, San Diego.

Hoober, J. K. (1984) *Chloroplasts.* Plenum, New York. A detailed, extensive survey.

Light Reactions

Barber, J., ed. (1992) *The Photosystems: Structure, Function, and Molecular Biology.* Elsevier, Amsterdam.

Blankenship, R. E., and R. C. Prince (1985) Excited-state redox potentials and the Z-scheme of photosynthesis. *Trends Biochem. Sci.* 10:382–383.

Freemantle, M. (1998) Mimicking natural photosynthesis. *Chem & Eng. News*, October, pp. 37–46. An up-to-date overview of some of the proposed methods for "artificial photosynthesis."

Hoganson, C. W., and G. T. Babcock (1997) A metalloradical mechanism for the generation of oxygen from water in photosynthesis. *Science* 277:1953–1956. A recent proposal for how the Mn cluster functions to release O_2.

Metz, J. G., P. J. Nixon, M. Rögner, G. Brudvig, and B. A. Diner (1989) Directed alteration of the D1 polypeptide of photosystem II: Evidence that tyrosine-161 is the redox component Z, connecting an oxygen-evolving complex to the primary electron donor, P680. *Biochemistry* 28:6960–6969.

Woodbury, N. W., M. Becker, D. Middendorf, and W. W. Parson (1985) Picosecond kinetics of the initial photochemical electron transfer reaction in bacterial photosynthetic reaction centers. *Biochemistry* 24:7516–7521. An example of the power of fast kinetic methods.

Structures

Cramer, W. A., S. E. Martinez, P. N. Furbacher, D. Huang, and J. L. Smith (1994) The cytochrome b_6f complex. *Curr. Opin. Struct. Biol.* 4: 536–544.

Deisenhofer, J., and H. Michel (1991) High resolution structure of photosynthetic reaction centers. *Annu. Rev. Biophys. & Biophys. Chem.* 20:247–266. An overview by two of the pioneers of the new structural studies.

Golbeck, J. (1993) The structure of photosystem I. *Curr. Opin. Struct. Biol.* 3:508–514. Includes information on the minor peptides and their functions in the complex.

Krauss, N., W. Hinrichs, I. Witt, P. Fromme, W. Pritzkow, Z. Dauter, C. Betzel, K. Wilson, H. Witt, and W. Saenger (1993) Three dimensional structure of system I of photosynthesis at 6Å resolution. *Nature* 361:326–331.

Kühlbrandt, W., and D. N. Wang (1991) Three-dimensional structure of plant light-harvesting complex determined by electron crystallography. *Nature* 350:130–134. The first structure determined for such a complex.

Rhee, K.-H., E. P. Morris, J. Barber, and W. Kühlbrandt (1998) Three-dimensional structure of photosystem II reaction center at 8 Å resolution. *Nature* 396:283–286.

Synthetic Reactions and Photorespiration

Chapman, M. S., S. W. Suh, P. Curmi, D. Cascio, W. W. Smith, and D. Eisenberg (1988) Tertiary structure of plant RuBisCo: Domains and their contacts. *Science* 241:71–74.

Ellis, R. J., and J. C. Gray, eds. (1986) Ribulose bisphosphate carboxylase–oxygenase. *Philos. Trans. R. Soc. London Ser. B* 313:303–469. A volume containing a number of excellent papers on this important enzyme.

Hartman, F. C., and M. R. Harpel (1993) Chemical and genetic probes of the active site of D-ribulose-1,5-bisphosphate carboxylase/oxygenase: A retrospective based on the three-dimensional structure. In *Advances in Enzymology*, Vol. 67, edited by A. Meister, pp. 1–75. Described by its title, this carries much information about this important enzyme.

Kozaki, A., and G. Taheba (1996) Photorespiration protects C3 plants from photo-oxidation. *Nature* 384:557–560.

Sheen, J. (1990) Metabolic repression of transcription in higher plants. *Plant Cell* 2:1027–1038.

PROBLEMS

1. According to Figure 17.12b, the P700 center is raised in potential from about +0.4 to −1.3 volts upon excitation. To what value of $\Delta G°$ does this correspond? How does it compare with the energy in an einstein of 700 nm photons?
2. In cyclic photophosphorylation, it is estimated that two electrons must be passed through the cycle to pump enough protons to generate one ATP. Assuming that the ΔG for hydrolysis of ATP under conditions existing in the chloroplast is about −50 kJ/mol, what is the corresponding percent efficiency of cyclic photophosphorylation, using light of 700 nm?
3. Assume a pH gradient of 4.0 units across a thylakoid membrane, with the lumen more acidic than the stroma. What is the *longest* wavelength of light that could provide enough energy per photon to pump one proton against this gradient, assuming a 20% efficiency in photosynthesis and $T = 25°C$?

*4. Suppose a brief pulse of $^{14}CO_2$ is taken up by a green plant.
   (a) Trace the ^{14}C label through the steps leading to fructose-1,6-bisphosphate synthesis, showing which carbon atoms in each compound should carry the label during the first cycle.
   (b) Will all molecules of fructose-1,6-bisphosphate carry two ^{14}C atoms? Explain.

5. The flux of solar energy reaching the earth's surface is approximately 7 J/cm^2 s. Assume that *all* of this energy is used by a green leaf (10 cm^2 in area), with the maximal efficiency of 35%. How many moles of hexose could the leaf theoretically generate in an hour? You may use 600 nm for an average wavelength.

*6. The substance dichlorophenyldimethylurea (DCMU) is an herbicide that inhibits photosynthesis by blocking electron transfer between plastoquinones in photosystem II.
   (a) Would you expect DCMU to interfere with cyclic photophosphorylation?
   (b) Normally, DCMU blocks O_2 evolution, but addition of ferricyanide to chloroplasts allows O_2 evolution in the presence of DCMU. Explain.

7. Suppose a researcher is carrying out studies in which she adds a nonphysiological electron donor to a suspension of chloroplasts. Illumination of the chloroplasts yields oxidation of the donor. How could she tell whether photosystem I, II, or both are involved?
8. Suppose ribulose-5-phosphate, labeled with ^{14}C in carbon 1, is used as the substrate in dark reactions. In which carbon of 3PG will the label appear?

9. The following data, presented by G. Bowes and W. L. Ogre in *J. Biol. Chem.* (1972) 247:2171–2176, describe the relative rates of incorporation of CO_2 by rubisco under N_2 and under pure O_2. Decide whether O_2 is a competitive or uncompetitive inhibitor.

$[CO_2]$ (mM)	Under N_2	Under O_2
0.20	16.7	10
0.10	12.5	5.6
0.067	8.3	4.2
0.050	7.1	3.2

*10. J. C. Servaites, in *Plant Physiol.* (1985) 78:839–843, observed that rubisco from tobacco leaves collected before dawn had a much lower specific activity than the enzyme collected at noon. This difference persisted despite extensive dialysis, gel filtration, or heat treatment. However, precipitation of the predawn enzyme by 50% $(NH_4)_2SO_4$ restored the specific activity to the level of noon-collected enzyme. Suggest an explanation.

11. It is believed that the ratio of cyclic photophosphorylation to noncyclic photophosphorylation changes in response to metabolic demands. In each of the following situations, would you expect the ratio to increase, decrease, or remain unchanged?
(a) Chloroplasts carrying out both the Calvin cycle and the reduction of nitrite (NO_2^-) to ammonia (This process does not require ATP.)
(b) Chloroplasts carrying out not only the Calvin cycle but also extensive active transport
(c) Chloroplasts using both the Calvin cycle and the C_4 pathway

12. If a photosynthetic organism is illuminated in a closed, sealed environment, it is observed that the CO_2 and O_2 levels in the surrounding atmosphere reach a constant ratio.
(a) Suggest an explanation.
(b) What factor would you think primarily determines the value of this ratio?

13. If algae are exposed to $^{14}CO_2$ for a brief period while illuminated, the labeled carbon is initially found almost entirely in the carboxyl group of 3-phosphoglycerate. However, if illumination is continued after the label pulse, other carbon atoms become labeled. Explain.

CHAPTER 18

Lipid Metabolism I: Fatty Acids, Triacylglycerols, and Lipoproteins

LIKE THE CARBOHYDRATES WE HAVE DISCUSSED IN PREVIOUS CHAPTERS, LIPIDS play roles in energy metabolism and in a variety of other processes. For lipids, those other processes include their roles as membrane constituents, hormones, fat-soluble vitamins, thermal insulators, and biological regulators such as the prostaglandins. This chapter focuses on bioenergetic aspects of lipid metabolism. We discuss the synthesis and breakdown of energy storage lipids, as well as fatty acid oxidation and biosynthesis (Figure 18.1), processes that are quite similar among plants, animals, and microorganisms. We also present some topics related more directly to animal metabolism—fat digestion, absorption, storage, and mobilization. Metabolism of membrane lipids and lipids of more specialized metabolic functions is covered in Chapter 19.

Utilization and Transport of Fat and Cholesterol

As discussed in Chapter 10, the great bulk of the lipid in most organisms is in the form of *triacylglycerols* (formerly called *triglycerides*). The term *fat,* or *neutral fat,* refers to this most abundant class of lipids. Fat utilization in animals is intertwined with lipoprotein metabolism, as is the metabolism of cholesterol. Therefore, we will consider fat and cholesterol metabolism together although cholesterol biosynthesis is presented in Chapter 19.

A mammal contains 5% to 25% or more of its body weight as lipid, with as much as 90% of this lipid in the form of triacylglycerols. Most of this fat, which is stored in adipose tissue, constitutes the primary energy reserve. In animal systems, fat is stored in specialized cells, the *adipocytes,* where giant fat globules occupy most of the intracellular space (see Figure 10.3). Plant seeds store great quantities of fat to provide energy to the developing plant embryo (Figure 18.2). Because plant lipids contain mostly unsaturated fatty acids, the triacylglycerols of seeds are largely in the form of liquid oils.

Triacylglycerols play roles other than in energy storage. Fat serves to cushion organs against shock, and it provides an efficient thermal insulator, particularly in marine mammals, which must maintain a body temperature far higher than that of the seawater in which they live.

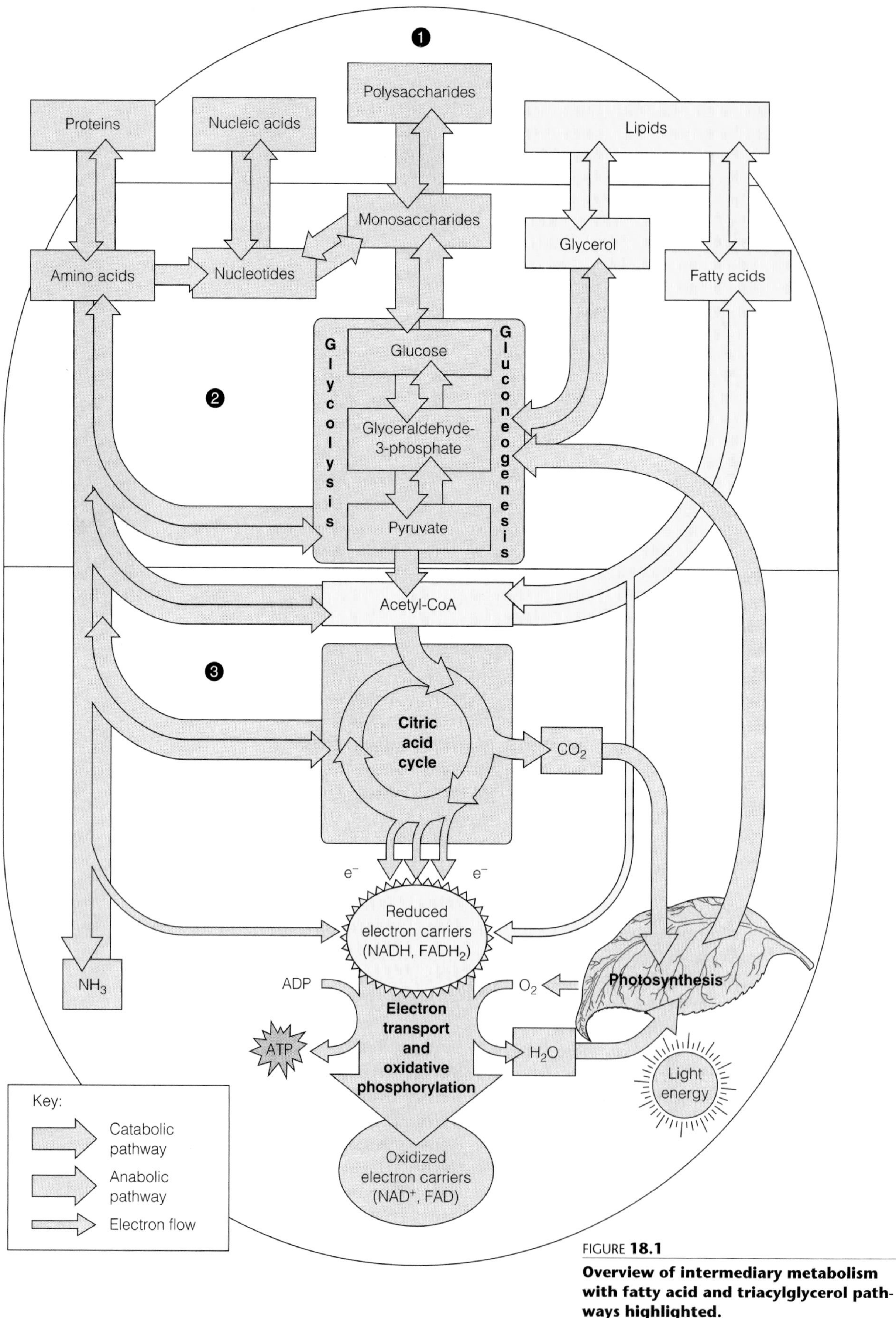

FIGURE **18.1**

Overview of intermediary metabolism with fatty acid and triacylglycerol pathways highlighted.

FATS AS ENERGY RESERVE

Recall that most of the carbon in triacylglycerols is more highly reduced than the carbon in carbohydrates. To be sure, the carboxyl carbons of fatty acids are highly oxidized, but most of the fatty acid carbons are at the reduced methyl or methylene level. Thus, metabolic oxidation of fat consumes more oxygen, on a weight basis, than oxidation of carbohydrate, with correspondingly larger metabolic energy release. The complete metabolic oxidation of triacylglycerols yields 37 kJ/g or more, whereas that of carbohydrates and proteins yields about 17 kJ/g. Adding to this difference between fat and carbohydrate is the hydrophilic nature of glucose polymers. Glycogen binds about 2 grams of water per gram of carbohydrate. Fat, being extremely nonpolar, is anhydrous. Thus, because 1 gram of intracellular glycogen contains but ⅓ gram of anhydrous glucose polymer, intracellular fat contains about six times as much potential metabolic energy, on a mass basis, as intracellular glycogen. This is an obvious advantage in many situations, such as in hibernating animals, which must store several months' worth of food, or in the flight muscles of small birds, in which weight is at a premium. Incredibly, some small land birds prepare for migration by increasing their body weight about 15% per day, with all of the weight gain being triacylglycerols. Such obese birds can then fly nonstop for as long as 60 hours or more. In addition, the insolubility of fat allows it to be stored in cells without affecting intracellular osmotic pressure.

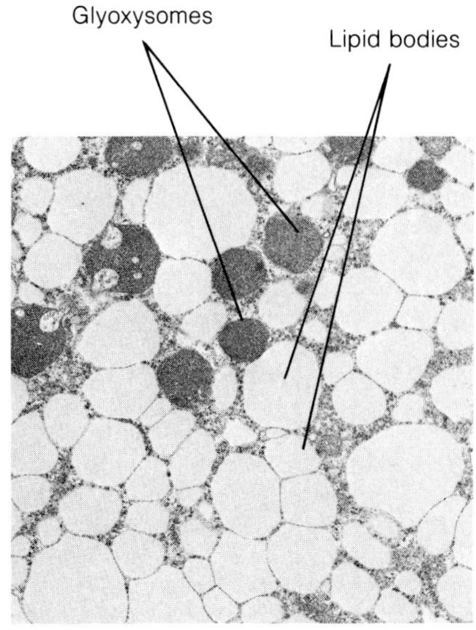

FIGURE **18.2**

Fat storage in a plant seedling. The electron micrograph (×6500) shows a cell from a cucumber cotyledon (seed leaf) a few days after germination. Fat stored in lipid bodies is degraded, oxidized, and converted to carbohydrate in neighboring glyoxysomes (or microbodies) to support the growth of the plant.

Courtesy of R. N. Trelease, P. J. Gruber, W. M. Becker, and E. H. Newcomb, *Plant Physiol.* (1971) 48:461.

Little wonder, then, that fat is the major energy storage form in most cells. A typical 70-kg human may have fuel reserves of 400,000 kJ in total body fat and about 100,000 kJ in total protein (mostly muscle protein). By contrast, the glycogen stores amount to just 2500 kJ of available energy, and the total glucose to about 170 kJ. Fat stores are maintained from the diet; about 40% of the caloric value of Western diets comes from fat. Most nutritionists recommend that this value be closer to 25–30%, for cardiovascular health. In addition, carbohydrate ingested in excess of its ability to be catabolized or stored as glycogen is readily converted to fat.

Most of the energy derived from fat breakdown comes from oxidation of the constituent fatty acids. Fatty acid oxidation provides the major energy source for many animal tissues. Brain is distinctive in being unable to use fatty acids as a significant energy supply; brain has a highly specific requirement for glucose. However, under conditions of starvation, when blood glucose levels decrease, brain can adjust to use a class of lipid-related compounds called *ketone bodies,* as discussed later in this chapter.

Fat has six times more caloric content by weight than carbohydrate, because fat is more highly reduced and is anhydrous.

FAT DIGESTION AND ABSORPTION

Triacylglycerols are derived from three primary sources: (1) the diet; (2) de novo biosynthesis, particularly in liver; and (3) storage depots in adipocytes. Processes by which these sources are utilized in animals are summarized in Figure 18.3. The major problem that animals must cope with in the digestion, absorption, and transport of dietary lipids is their insolubility in aqueous media. The action of **bile salts,** detergent substances synthesized in liver and stored in the gallbladder, is essential to the digestion of lipids and their absorption through the intestinal mucosa. The problem of transport through the blood and lymph is dealt with in part by the complexing of lipids with proteins to form soluble aggregates called **lipoproteins.**

A bile salt molecule is made up of a **bile acid,** such as cholic acid, and an associated cation. Bile acids (discussed further in Chapter 19) are derived from cholesterol. As shown in Figure 18.4, the bile salt molecule has both hydrophobic and hydrophilic surfaces. This amphipathic character allows bile salts to orient at an oil–water interface, with the hydrophobic surface in contact with the apolar phase and the hydrophilic surface in contact with the aqueous phase. This detergent action emulsifies lipids and yields micelles (see Chapter 10), allowing digestive

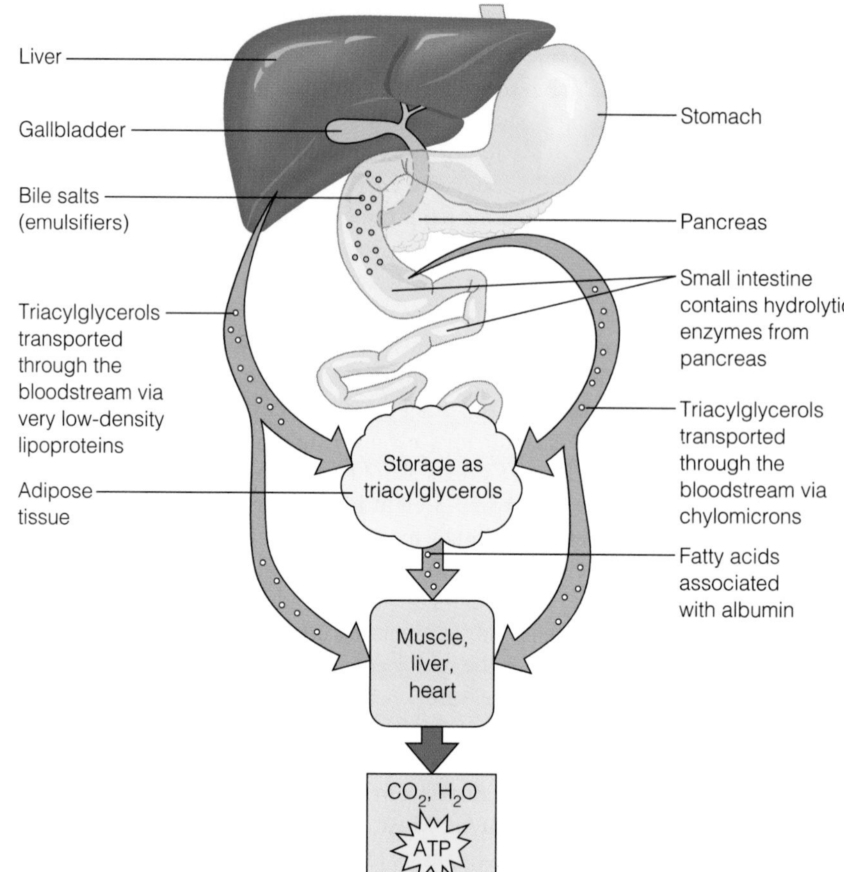

FIGURE **18.3**

Overview of fat digestion, absorption, storage, and mobilization in the human. Triacylglycerols (fats) are ingested, synthesized in the liver, or mobilized from storage. Ingested triacylglycerols are hydrolyzed in the lumen of the small intestine by pancreatic lipase and other enzymes. Hydrolysis products absorbed by the intestinal mucosa are recombined into triacylglycerols, which combine with apoproteins to form the lipoproteins called chylomicrons. This process solubilizes the lipids and permits their transport through blood and lymph. Triacylglycerols synthesized in liver are combined with other apoproteins to form very low-density lipoproteins (VLDLs) for transport. Lipoproteins transported to peripheral tissues are hydrolyzed at the inner surfaces of capillaries. Hydrolysis products entering cells are either catabolized for energy or recombined into triacylglycerols for storage. Mobilization of stored triacylglycerols is hormonally regulated.

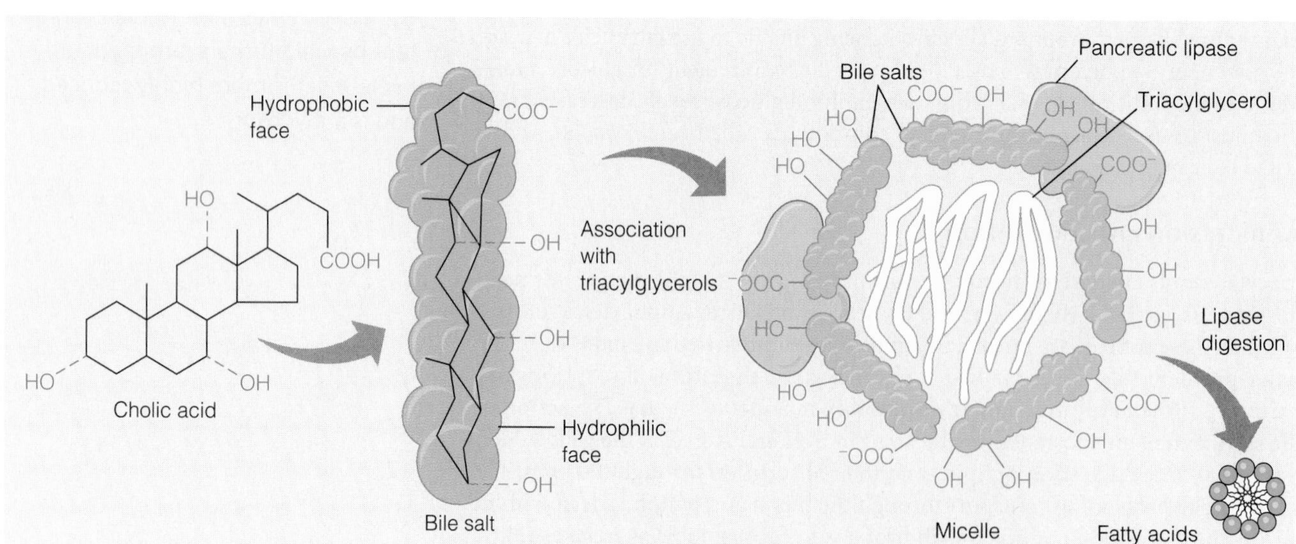

FIGURE **18.4**

Action of bile salts in emulsifying fats in the intestine. Cholic acid, a typical bile acid, ionizes to give its cognate bile salt.The hydrophobic surface of the bile salt molecule associates with triacylglycerol, and several such complexes aggregate to form a micelle. The polar surface of the bile salts faces outward, allowing the micelle to associate with pancreatic lipase. Action of this enzyme frees the fatty acids to associate in a much smaller micelle that can be absorbed through the intestinal mucosa.

attack by water-soluble enzymes and facilitating the absorption of lipid through intestinal mucosal cells. Most of the digestion occurs through action of **pancreatic lipase,** an unusual calcium-requiring enzyme that catalyzes a reaction at an oil–water interface. The substrate being cleaved is in an apolar phase, and the other substrate, of course, is water. Structural studies on pancreatic lipase show that the catalytic center is exposed as the result of a conformational change that occurs only at an oil–water interface. Despite its unusual physical properties, pancreatic lipase acts via an active site serine and an acyl-enzyme intermediate, like serine proteases such as chymotrypsin.

Bile salts emulsify fats, thereby promoting their hydrolysis in digestion.

The products of fat digestion comprise a mixture of glycerol, free fatty acids, monoacylglycerols, and diacylglycerols. Less than 10% of the original triacylglycerol remains unhydrolyzed. During absorption through intestinal mucosal cells, much resynthesis of triacylglycerols occurs from the hydrolysis products. This resynthesis occurs in the endoplasmic reticulum and Golgi complex of mucosal cells. Triacylglycerols emerge into the lymph system complexed with protein to form the lipoproteins called **chylomicrons.** The chylomicron is essentially an oil droplet coated with more polar lipids and a skin of protein, which help disperse and partially solubilize the fat for transport to tissues. The chylomicron is also a transport vessel for dietary cholesterol.

In general, fats containing substantial amounts of unsaturated fatty acids, such as oleic and linoleic acids, which tend to be liquid at body temperature, are relatively easily absorbed, whereas lipids containing mostly saturated fatty acids, such as palmitic and stearic acids, are digested and absorbed more slowly.

$CH_3—(CH_2)_{14}—COOH$

Palmitic acid

$CH_3(CH_2)_{16}—COOH$

Stearic acid

$CH_3—(CH_2)_4—CH{=}CH—CH_2—CH{=}CH—(CH_2)_7—COOH$

Linoleic acid

$CH_3—(CH_2)_7—CH{=}CH—(CH_2)_7—COOH$

Oleic acid

TRANSPORT OF FAT TO TISSUES: LIPOPROTEINS

Chylomicrons constitute just one class of lipoproteins found in the bloodstream. These complexes play essential roles in the transport of lipids to tissues, either for energy storage or for oxidation. Free lipids are all but undetectable in blood. The **apoprotein** or apolipoprotein components, or polypeptide chains, of lipoproteins are synthesized mainly in the liver, though about 20% are produced in intestinal mucosal cells.

Classification and Functions of Lipoproteins

Distinct families of lipoproteins have been described, each of which plays defined roles in lipid transport. These families are classified in terms of their density, as determined by centrifugation (Table 18.1). Lipoproteins in each class contain

TABLE **18.1 Properties of major human plasma lipoprotein classes**

	Chylomicron	VLDL	IDL	LDL	HDL
Density (g/mL)	<0.95	0.950–1.006	1.006–1.019	1.019–1.063	1.063–1.210
Components (% dry weight)					
Protein	2	8	15	22	40–55
Triacylglycerol	86	55	31	6	4
Free cholesterol	2	7	7	8	4
Cholesterol esters	3	12	23	42	12–20
Phospholipids	7	18	22	22	25–30
Apoprotein composition	A-I, A-II, B-48, C-I, C-II, C-III	B-100, C-I, C-II, C-III, E	B-100, C-I, C-II, C-III, E	B-100	A-I, A-II, C-I, C-II, C-III, D, E

Sources: Data from D. E. Vance and J. E. Vance (eds.), *Biochemistry of Lipids and Membranes* (Redwood City, Calif.: Benjamin/Cummings, 1985); and J. F. Mead, R. B. Alfin-Slater, D. R. Howton, and G. Popják, *Lipids* (New York: Plenum, 1986).

TABLE **18.2 Apoproteins of the human plasma lipoproteins**

Apoprotein	Molecular Weight	Characteristics
A-I	28,300	Major protein in HDL; activates LCAT
A-II	17,400	Major protein in HDL
B-48	241,000	Found exclusively in chylomicrons
B-100	513,000	Major protein in LDL
C-I	7,000	Found in chylomicrons; activates LCAT and LPL
C-II	10,000	Found primarily in VLDL; activates LPL
C-III	9,300	Found primarily in chylomicrons, VLDL, and HDL; inhibits LPL
D	35,000	HDL protein, also called cholesterol ester transfer protein
E	33,000	Found in VLDL, LDL, and HDL

Sources: Data from D. E. Vance and J. E. Vance, eds., *Biochemistry of Lipids and Membranes* (Redwood City, Calif.: Benjamin/Cummings, 1985), and J. F. Mead, R. B. Alfin-Slater, D. R. Howton, and G. Popják, *Lipids* (New York: Plenum, 1986).

Note: LCAT = lecithin:cholesterol acyltransferase, LPL = lipoprotein lipase.

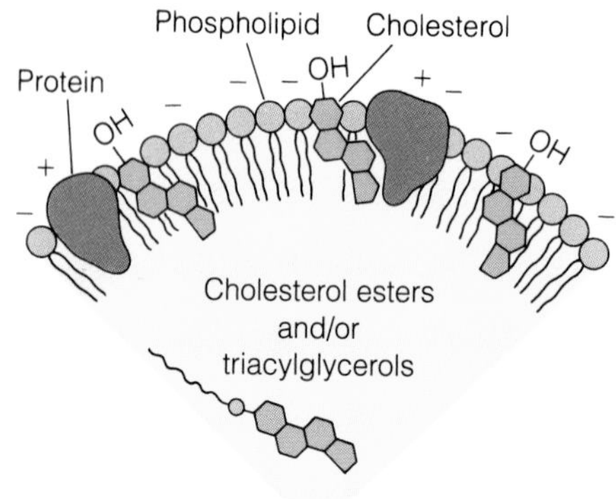

FIGURE **18.5**

Generalized structure of a plasma lipoprotein. The spherical particle, part of which is shown, contains neutral lipids in the interior and phospholipids, cholesterol, and protein at the surface.

Lipoproteins are lipid–protein complexes that allow movement of apolar lipids through aqueous environments.

characteristic apoproteins and have distinctive lipid compositions. A total of nine major **apolipoproteins** are found in human lipoproteins. Their properties are summarized in Table 18.2. Because lipids are of much lower density than proteins, the lipid content of a lipoprotein class is inversely related to its density: The higher the lipid abundance, the lower the density. The standard lipoprotein classification includes, in increasing order of density: *chylomicrons,* **very low-density lipoprotein (VLDL), intermediate-density lipoprotein (IDL), low-density lipoprotein (LDL),** and **high-density lipoprotein (HDL).** Some classification schemes recognize two classes of HDL, and in addition there is a quantitatively minor lipoprotein called very high-density lipoprotein (VHDL).

Despite their differences in lipid and protein composition, all lipoproteins share common structural features, notably a spherical shape that can be detected by electron microscopy. As shown in Figure 18.5, the hydrophobic parts, both lipid and apolar amino acid residues, form an inner core, and hydrophilic protein structures and polar head groups of phospholipids are on the outside.

Some apolipoproteins have specific biochemical activities other than their roles as passive carriers of lipid from one tissue to another. For instance, apo C-II is an activator of triacylglycerol hydrolysis by **lipoprotein lipase,** a cell surface enzyme that hydrolyzes triacylglycerols in lipoproteins. A human deficiency of apo C-II is associated with elevated triacylglycerol levels in blood. Other apoproteins target specific lipoproteins to specific cells, by being recognized by receptors in the plasma membranes of these cells. Of great interest is an association of a variant form of apo E with increased risk for developing Alzheimer's disease. There are three common allelic forms of apo E, and individuals homozygous for the relevant variant invariably develop Alzheimer's disease by age 80.

Together, the lipoproteins help maintain in emulsified form some 500 mg of total lipid per 100 mL of human blood in the postabsorptive state, after the contents of a meal have been digested and absorbed into the bloodstream. Of this 500 mg, typically about 120 mg is triacylglycerol, 220 mg is cholesterol (two-thirds esterified with fatty acids, one-third free), and 160 mg is phospholipids, principally phosphatidylcholine and phosphatidylethanolamine.

Transport and Utilization of Lipoproteins

As noted previously, chylomicrons represent the form in which dietary fat is transported from the intestine to peripheral tissues, notably heart, muscle, and adipose tissue (see Figure 18.3). VLDL plays a comparable role for triacylglycerols

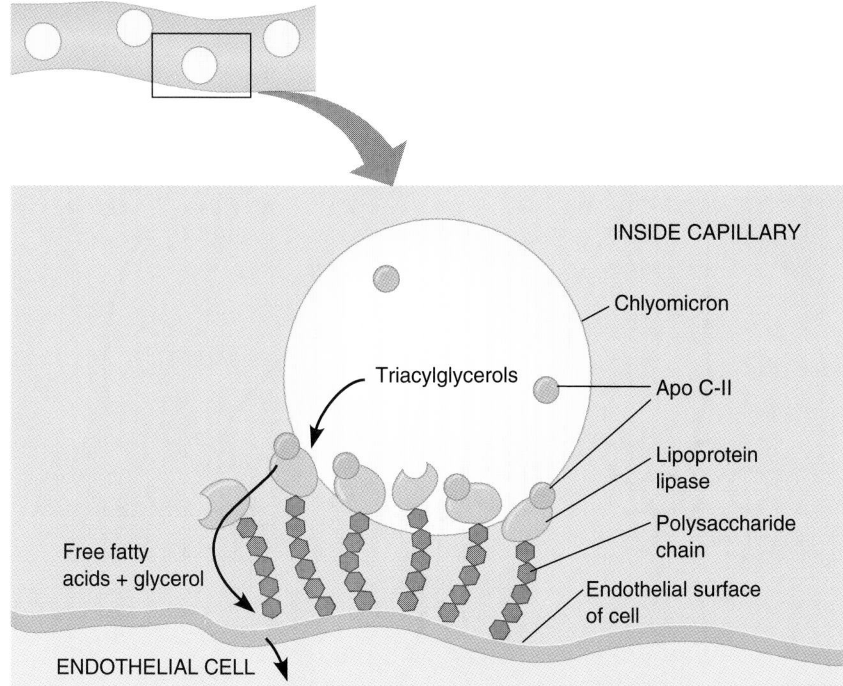

FIGURE **18.6**

Binding of a chylomicron to lipoprotein lipase on the inner surface of a capillary. The chylomicron is anchored by lipoprotein lipase, which is linked by a polysaccharide chain to the endothelial cell surface. When activated by apoprotein C-II, the lipase hydrolyzes the triacylglycerols in the chylomicron, allowing uptake into the cell of the glycerol and free fatty acids.

synthesized in liver. The triacylglycerols in both lipoproteins are hydrolyzed to glycerol and fatty acids at the inner surfaces of capillaries in the peripheral tissues. This hydrolysis involves activation of the extracellular enzyme lipoprotein lipase by apoprotein C-II (Figure 18.6). Some of the released fatty acids are absorbed by nearby cells, while others, still rather insoluble, become complexed with serum albumin for transport to more distant cells. After absorption into the cell, the fatty acids derived from lipoprotein lipase action can be either catabolized to generate energy or, in adipose cells, used to resynthesize triacylglycerols. However, because adipocytes lack glycerol kinase, glycerol-3-phosphate for resynthesis of triacylglycerols must come from glycolysis. Glycerol is returned from adipocytes to liver, for resynthesis of glucose by gluconeogenesis. Figure 18.7 summarizes overall aspects of lipoprotein metabolism and transport.

A major consequence of liver dysfunction is an inability to synthesize apolipoproteins and, hence, to transport fat out of the liver.

As a consequence of triacylglycerol hydrolysis in the capillaries, both chylomicrons and VLDL are degraded to protein-rich remnants. The IDL class of lipoprotein is derived from VLDL, and chylomicrons are degraded to what are simply called chylomicron remnants. Both classes of remnants are taken up by the liver through interaction with specific receptors and further degraded in liver lysosomes. Apoprotein B-100 is reused for synthesis of LDL (via IDL). As described in the next section, LDL is the principal form in which cholesterol is transported to tissues, and HDL plays the primary role in returning excess cholesterol from tissues to the liver for metabolism or excretion. The importance of lipoproteins as transport vehicles is evident from the fact that a major consequence of chronic liver cirrhosis is fatty liver degeneration, where the liver becomes engorged with fat. Because the liver is the major site of apolipoprotein synthesis, damage to this organ causes endogenously synthesized fat to accumulate there because it cannot be transported to peripheral tissues.

CHOLESTEROL TRANSPORT AND UTILIZATION IN ANIMALS

Cholesterol accumulation in the blood is correlated with development of atherosclerotic plaque. Precise mechanisms of atherogenesis are unknown at present.

As you undoubtedly know, a primary risk factor predisposing to heart disease is an abnormally elevated level of cholesterol in the blood. Prolonged cholesterol accumulation contributes to the development of **atherosclerotic plaques,** fatty deposits that line the inner surfaces of coronary arteries. Because of the medical

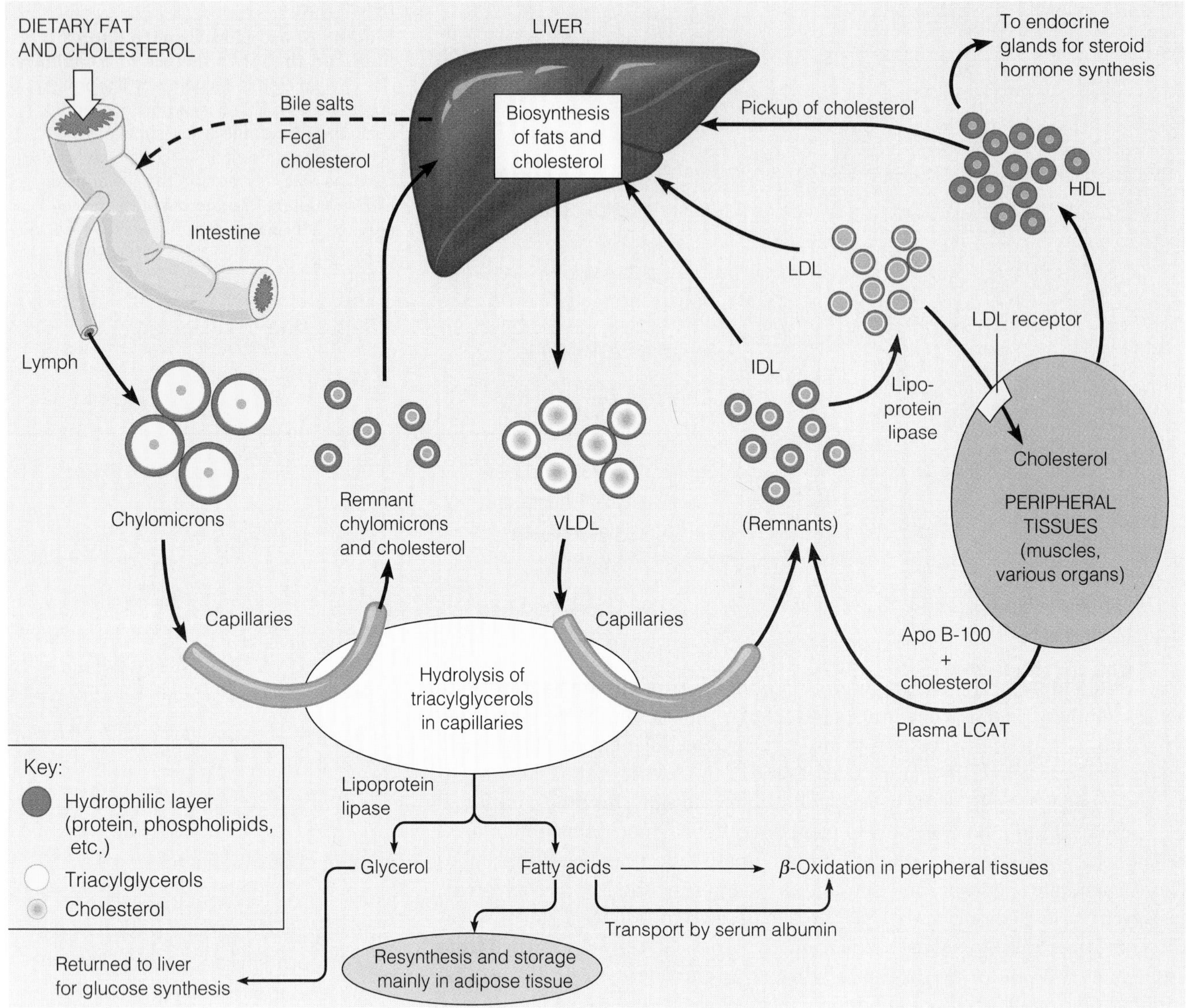

FIGURE **18.7**

Overview of lipoprotein transport pathways and fates.

Cholesterol

Cholesterol esters

importance of knowing how cholesterol contributes to **atherogenesis,** or plaque development, most of what we know about cholesterol metabolism and its regulation comes from studies on mammals.

Recall from Table 18.1 that cholesterol in plasma lipoproteins exists both as the free sterol and as cholesterol esters. Esterification occurs at the cholesterol hydroxyl position with a long-chain fatty acid, usually unsaturated. Cholesterol esters are synthesized in plasma from cholesterol and an acyl chain on phosphatidylcholine, through the action of **lecithin:cholesterol acyltransferase (LCAT),** an enzyme that is secreted from liver into the bloodstream:

$$\text{Phosphatidylcholine} + \text{cholesterol} \rightleftharpoons \text{lysolecithin} + \text{cholesterol ester}$$

Cholesterol esters are considerably more hydrophobic than cholesterol itself.

Of the five lipoprotein classes, LDL is by far the richest in cholesterol. The amounts of cholesterol and cholesterol esters associated with LDL are typically about two-thirds of the total plasma cholesterol (total plasma cholesterol ranges from 130 to 260 mg/100 mL of human plasma, with the most desirable levels between 160 and 200). More than 40% of the weight of the LDL particle is cholesterol esters, and the total of esterified and free cholesterol amounts to well over half the total weight. The LDL particle contains a single molecule of apoprotein B-100 (M_r = 513,000) as its primary protein component. Because cholesterol biosynthesis is confined primarily to the liver with some occurring also in intestine, LDL plays an important role in delivering cholesterol to other tissues.

The LDL Receptor and Cholesterol Homeostasis

The importance of understanding cholesterol homeostasis can be seen by reviewing the consequences of prolonged high plasma cholesterol levels. Cholesterol is highly insoluble, and it accumulates in white blood cells that become deposited at sites of injury or damage on inner arterial walls. If cholesterol levels are too high for its subsequent removal into the bloodstream, these cells become engorged with fatty deposits, which then harden into plaque; this condition, called atherosclerosis, ultimately blocks key blood vessels and causes myocardial infarctions, or heart attacks.

To understand the relationship between elevated cholesterol levels and atherogenesis, we must know how cholesterol is taken up from LDL into cells, because cholesterol esters are too hydrophobic to traverse cell membranes by themselves. The answer to this question came from the research of Michael Brown and Joseph Goldstein, who showed in the mid-1970s that cholesterol uptake by cells is a receptor-mediated process and that the quantity of receptors themselves is subject to regulation.

In 1972 Brown and Goldstein began to study a hereditary condition called **familial hypercholesterolemia,** or FH. Individuals with the rare homozygous form of this disease (about 1 in 1 million) have grossly elevated levels of serum cholesterol, from 650 to 1000 mg/100 mL (about 5-fold over normal levels). They develop atherosclerosis early in life and usually die of heart disease before age 20. The more common heterozygous condition, characterized by one defective allele instead of two, affects about one individual in 500. These individuals have less severely elevated cholesterol levels, in the range of 250 to 500 mg/100 mL. They are at high risk to have heart attacks in their thirties and forties, although many enjoy a normal life span.

Crucial to the success of Brown and Goldstein was their ability to demonstrate the defective FH phenotype in cell culture. Fibroblasts from FH patients synthesized cholesterol at abnormally high rates in culture, whereas normal cells showed low rates of synthesis. When cultured in the presence of LDL, normal cells showed low activity of hydroxymethylglutaryl-CoA reductase (HMG-CoA reductase, the major regulated enzyme in cholesterol synthesis—see page 689). In the absence of LDL the same cells showed reductase activities some 50- to 100-fold higher. These results suggested that cholesterol is normally transported into the cell, where it regulates its own synthesis by suppressing the activity of the rate-limiting enzyme. By contrast, cells from FH individuals showed high levels of reductase activity, whether cultured in the presence or absence of LDL, suggesting that they were deficient in ability to take up cholesterol from the medium.

These observations suggested that cholesterol is taken into cells through the action of a specific receptor, which is deficient or defective in FH patients. In short order Brown and Goldstein and their colleagues demonstrated the existence of this receptor, the **LDL receptor** (see Figure 18.7), and demonstrated a new mechanism by which cells can interact with their environment—**receptor-mediated endocytosis.** By conjugating LDL with an electron-dense material and binding this material to cells, the investigators were able to visualize the LDL receptor on cell surfaces (Figure 18.8). These experiments showed that the receptors are clustered in a structure called a **coated pit,** an invagination whose most abundant protein is **clathrin,** a self-interacting protein capable of forming a cagelike structure (Figure 18.9).

Endocytosis is a process by which cells take up large molecules from the extracellular environment. Although LDL uptake involves a cell surface receptor, the

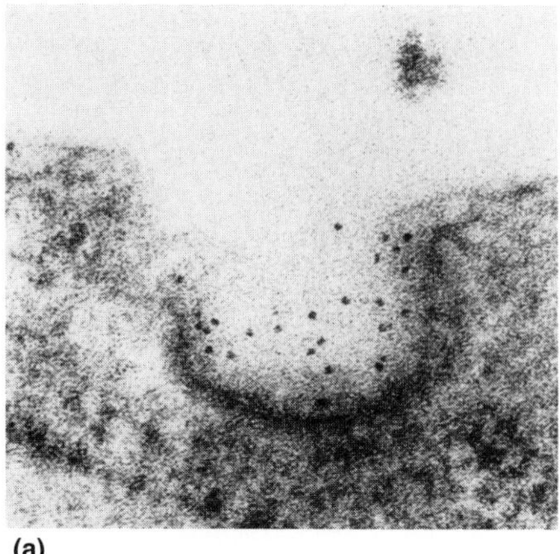

(a)

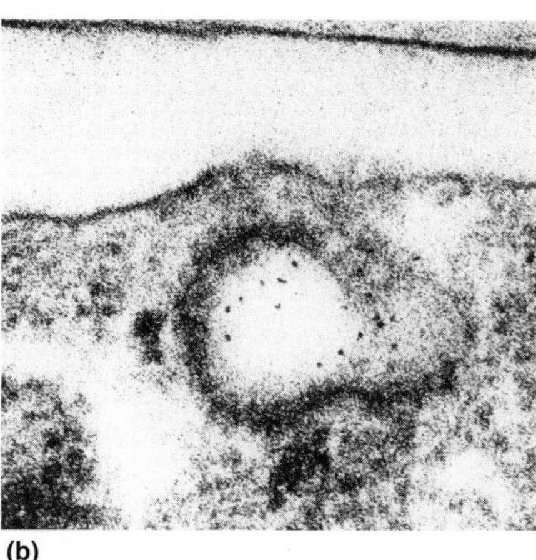

(b)

FIGURE **18.8**

Receptor-mediated endocytosis of LDL. Low-density lipoprotein (LDL) was conjugated with ferritin to permit electron microscopic visualization. **(a)** The LDL–ferritin (dark dots) binds to a coated pit on the surface of a cultured human fibroblast (a type of connective tissue cell). **(b)** The plasma membrane closes over the coated pit, forming an endocytotic vesicle.

Uptake of cholesterol from the blood occurs at the LDL receptor via receptor-mediated endocytosis.

interaction of LDL with its receptor is unlike the interaction of hormones such as epinephrine with their receptors. As discussed in Chapters 12 and 13, the binding of epinephrine at its receptor in the plasma membrane triggers intracellular metabolic changes, but the hormone itself does not enter the cell. By contrast, when LDL binds to its receptor, through recognition of the B-100 apoprotein by the receptor, the entire LDL molecule is engulfed and taken into the cell, as schematized in Figure 18.10. The plasma membrane fuses in the vicinity of the LDL–receptor complex, and the coated pit becomes an endocytotic vesicle. Several of these clathrin-lined vesicles fuse to form an **endosome.** The endosome then fuses with a lysosome, putting the LDL–receptor complex in contact with the hydrolytic enzymes of the lysosome. The LDL apoprotein is hydrolyzed to amino acids, and

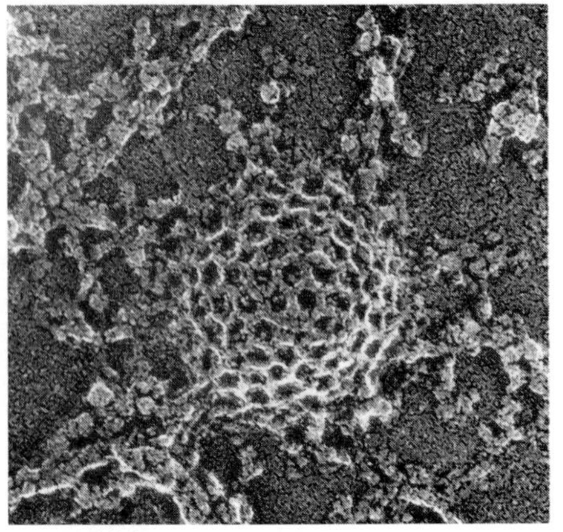

FIGURE **18.9**

Structure of a coated pit. A coated pit on the inner surface of the plasma membrane of a cultured mammalian cell is visualized by freeze-fracture electron microscopy. The cagelike structure of the pit is due to the structure of clathrin, the major protein in such pits.

Micrograph produced by John Heuser, Washington University School of Medicine.

FIGURE **18.10**

Involvement of LDL receptors in cholesterol uptake and metabolism. LDL receptors are synthesized in the endoplasmic reticulum and mature in the Golgi complex. They then migrate to the cell surface, where they cluster in clathrin-coated pits. LDL, made up of cholesterol esters and apoprotein, binds to the LDL receptors and is internalized in endocytotic vesicles. Several such vesicles fuse to form an organelle called an endosome. Proton pumping in the endosome membrane causes the pH to drop, which in turn causes LDL to dissociate from the receptors. The endosome fuses with a lysosome, and the receptor-bearing clathrin coat dissociates and returns to the membrane. The receptor–LDL complex is degraded in the lysosomes, and cholesterol has various fates. Regulatory actions of cholesterol are shown in red.

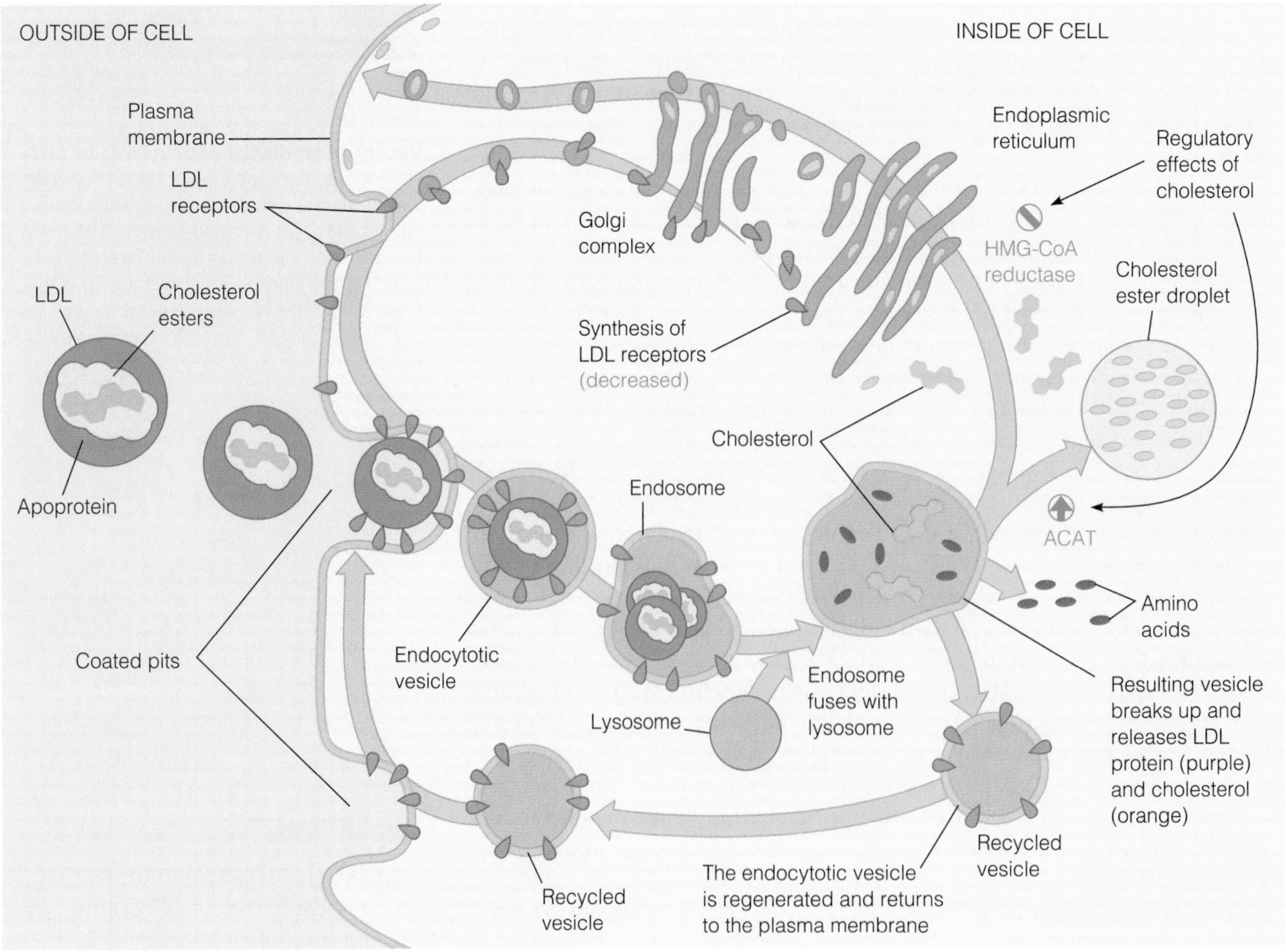

the cholesterol esters are hydrolyzed to give free cholesterol. The receptor itself is recycled, moving back to the plasma membrane to pick up more LDL. About 10 minutes is required for each round trip.

Much of the cholesterol released moves to the endoplasmic reticulum, where it is used for membrane synthesis. The internalized cholesterol exerts three regulatory effects. (1) As mentioned earlier, it suppresses endogenous cholesterol synthesis, by inhibiting HMG-CoA reductase and also by suppressing transcription of the gene for this enzyme and accelerating degradation of the enzyme protein. (2) It activates **acyl-CoA:cholesterol acyltransferase** (**ACAT**), an intracellular enzyme that synthesizes cholesterol esters from cholesterol and a long-chain acyl-CoA. This promotes the storage of excess cholesterol in the form of droplets of cholesterol esters. (3) It regulates the synthesis of the LDL receptor itself, by somehow lowering the content of mRNA for the receptor. Decreased synthesis of the receptor ensures that cholesterol will not be taken into the cell in excess of the cell's needs, even when extracellular levels are very high. This regulatory mechanism explains why excessive dietary cholesterol leads directly to elevations of blood cholesterol levels. With intracellular cholesterol levels so well regulated, the extracellular cholesterol accumulates because it has nowhere else to go.

Intracellular cholesterol regulates its own level by controlling (1) de novo cholesterol biosynthesis, (2) formation and storage of cholesterol esters, and (3) LDL receptor density.

Current therapeutic approaches to lowering cholesterol levels are focused on developing inhibitors of HMG-CoA reductase, on the assumption that they will depress de novo cholesterol biosynthesis and, hence, intracellular cholesterol levels; consequently, the production of LDL receptors for clearance of extracellular cholesterol from the blood will increase. One such inhibitor, called **lovastatin,** received approval in 1987 for treatment of patients with very high cholesterol levels. Several other HMG-CoA reductase inhibitors are now available.

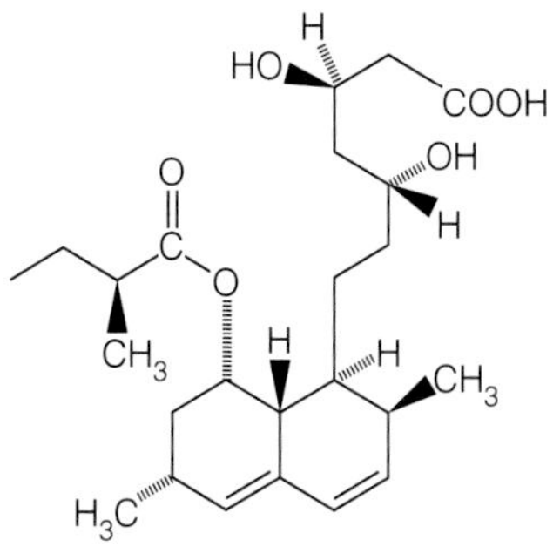

HO H COOH OH H O C O CH3 H H CH3 H3C

Lovastatin

Receptor-mediated endocytosis is now known to be a widely used pathway for internalization of extracellular substances, including other lipoproteins, cell growth factors, and some viruses. Also, studies of many individuals with FH have revealed four types of mutation that generate the FH phenotype, and these studies are giving a remarkably detailed picture of the receptor and its action. The LDL receptor is a glycoprotein with an 839-residue polypeptide chain and 18 O-linked oligosaccharide chains. The receptor binds specifically the apolipoproteins B-100 and E. Gene cloning and DNA sequence analysis have allowed identification of the four types of mutations affecting the receptor and its metabolism in humans.

First, and most common, are mutations that lead to insufficient receptor synthesis. Second are mutations in which receptor is synthesized but fails to migrate from the endoplasmic reticulum to the Golgi complex, for transport to the cytoplasmic membrane. Third are mutations in which receptor is synthesized and processed normally and reaches the cell surface, but fails to bind LDL. Finally, there is a class of mutant receptors that reach the cell surface and bind LDL but fail to cluster in coated pits.

Cholesterol, LDL, and Atherosclerosis

Thanks principally to Brown and Goldstein, we now know a great deal about the genetic and biochemical factors that control serum cholesterol levels, and overwhelming epidemiological evidence now links prolonged hypercholesterolemia to the development of atherosclerotic plaque. Much, however, remains to be learned. We don't know, for example, why diets rich in saturated fatty acids tend to elevate serum cholesterol levels. Nor do we know why a particular class of polyunsaturated fatty acids called **ω-3 fatty acids** tends to depress levels of both serum cholesterol and triacylglycerols. But nutritionists have found that adding to a Western diet fish or fish oils, which are abundant in this class of fatty acids, does indeed have this effect, and that is why we are being urged to substitute fish for red meat, which tends to be rich in both saturated fatty acids and cholesterol. The most prominent ω-3 fatty acid is linolenic acid, which is an 18:3cΔ9,12,15 fatty acid (nutritionists number fatty acids backward from biochemists, so the term

Polyunsaturated fat ingestion is correlated with low plasma cholesterol levels. The mechanisms involved are not understood.

$CH_3—CH_2—CH{=}CH—CH_2—CH{=}$
$CH—CH_2—CH{=}CH—(CH_2)_7—COOH$

Linolenic acid

$CH_3—CH{=}CH—CH_2—CH{=}CH—$
$CH_2—CH{=}CH—CH_2—CH{=}$
$CH—(CH_2)_7—COOH$

Arachidonic acid

ω-3 refers to a double bond on the third carbon from the terminal methyl group, what we would call the bond between C-15 and C-16 in this 18-carbon molecule). Current evidence suggests that linolenic and other ω-3 fatty acids act by somehow competing with the metabolism of arachidonic acid (20:4cΔ9,12,15,18), the principal precursor of a class of biological regulators called **prostaglandins** and **thromboxanes,** which we discuss in Chapter 19; however, the biochemical relationships involved are still obscure.

Uptake of oxidized LDL by the scavenger receptor is a key event in atherogenesis.

Progress is being made, however, in learning how elevated cholesterol levels lead to atherogenesis (plaque formation). LDL undergoes rather ready oxidation, both in cells and in plasma, to a mixture of molecules collectively called **oxidized LDL.** Although the specific oxidation reactions are not well defined, they include peroxidation of unsaturated fatty acids (page 554), hydroxylation of cholesterol itself, and oxidation of amino acid residues in the apoprotein. LDL undergoes uptake into a class of white blood cells that accumulate at sites of arterial injury. Uptake occurs through a receptor called the **scavenger receptor;** the normal ligands for this receptor have not been identified, but it takes up many substances in addition to oxidized LDL. Unlike the LDL receptor, the scavenger receptor is not down-regulated by cholesterol, so cholesterol uptake into these cells is virtually unlimited, which converts these cells to a cholesterol-engorged species called a **foam cell.** These events have a chemotactic effect, causing more white cells to migrate to the site and leading them to accumulate more cholesterol, which ultimately becomes one of the chief chemical constituents of the plaque that forms at such a site.

You have probably heard about "bad cholesterol" and "good cholesterol." These are actually inappropriate terms, because cholesterol itself is a natural metabolite, an essential component of all membranes, and the precursor to all steroid hormones and bile acids. However, cholesterol present in LDL is considered "bad," because prolonged elevation of LDL levels is what leads to atherosclerosis. By contrast, cholesterol in HDL is called "good," because high levels of HDL counteract atherogenesis. This is because cholesterol cannot be metabolically degraded, and excess cholesterol is returned from peripheral cells to the liver, for passage through the bile to the intestine, for ultimate excretion. As the agent for this transport back to the liver, HDL plays a role in lowering total serum cholesterol levels, which is "good." Recent studies have shown that cholesterol from HDL is taken up into cells not by endocytosis, like LDL, but instead by a "docking" reaction, in which HDL interacts with a cell surface receptor, deposits its cholesterol for uptake by still unknown mechanisms, and departs as a remnant without itself being incorporated in the cell's interior.

MOBILIZATION OF STORED FAT

In general, the capacity of storage depots to store fat is virtually unlimited. Whatever appears in the body from the diet is absorbed, and most of it is transported to adipose tissue for storage. The lack of control of this process is sadly evident from the prevalence of obesity among humans, in whom fat is stored in excess of its need to supply energy. By contrast, the release of fat from storage depots in adipose tissue is controlled hormonally, to meet the needs of the organism for energy generation.

Fat mobilization in adipose cells is hormonally controlled, via the cyclic AMP–dependent phosphorylation of the lipase involved.

The catabolism of fat begins with the hydrolysis of triacylglycerol to yield glycerol plus free fatty acid (often abbreviated FFA). About 95% of the energy derived from subsequent oxidation of the fat comes from the fatty acids, with only 5% coming from glycerol. All of the carbons from fatty acids are catabolized to two-carbon fragments, as acetyl-coenzyme A, except for the small proportion of fatty acids that contain odd-numbered chains.

The release of metabolic energy stored in triacylglycerols is comparable to the mobilization of carbohydrate energy stored in animal glycogen, in that the first step of fat breakdown—its hydrolysis to glycerol and fatty acids—is hormonally regulated. The first reaction is controlled by a cascade involving cyclic AMP (Figure 18.11). Depending on the physiological state, glucagon or epinephrine or β-corticotropin binds to a plasma membrane receptor, leading to activation of adenylate cyclase, as described in Chapter 13. This, in turn, activates protein kinase A, by releasing active catalytic subunits (C) from the inactive R_2C_2 tetramer. Protein kinase A phosphorylates and thereby activates an enzyme called **triacylglycerol lipase,** also called **hormone-sensitive lipase.** This enzyme catalyzes the hydrolytic release of fatty acid from carbon 1 or 3 of the glycerol moiety. The fatty acid release is followed by the action of a diacylglycerol lipase and a monoacylglycerol lipase, in turn. Together, these three enzymes degrade the original molecule to glycerol and three unesterified fatty acids. In adipose tissue the primary hormonal effects are mediated by epinephrine in stress situations and by glucagon during fasting.

The triacylglycerol hydrolysis products exit the adipocyte by passive diffusion and find their way to the blood plasma, where the fatty acids become bound to **albumin.** This is the most abundant plasma protein, about 50% of the total plasma protein in humans. The protein, with an M_r of 66,200, contains 17 disulfide bridges. Each molecule of albumin can bind up to 10 molecules of free fatty acid, although the actual amount bound is usually far lower. Fatty acids are released from albumin and taken up by tissues largely by passive diffusion, so that fatty

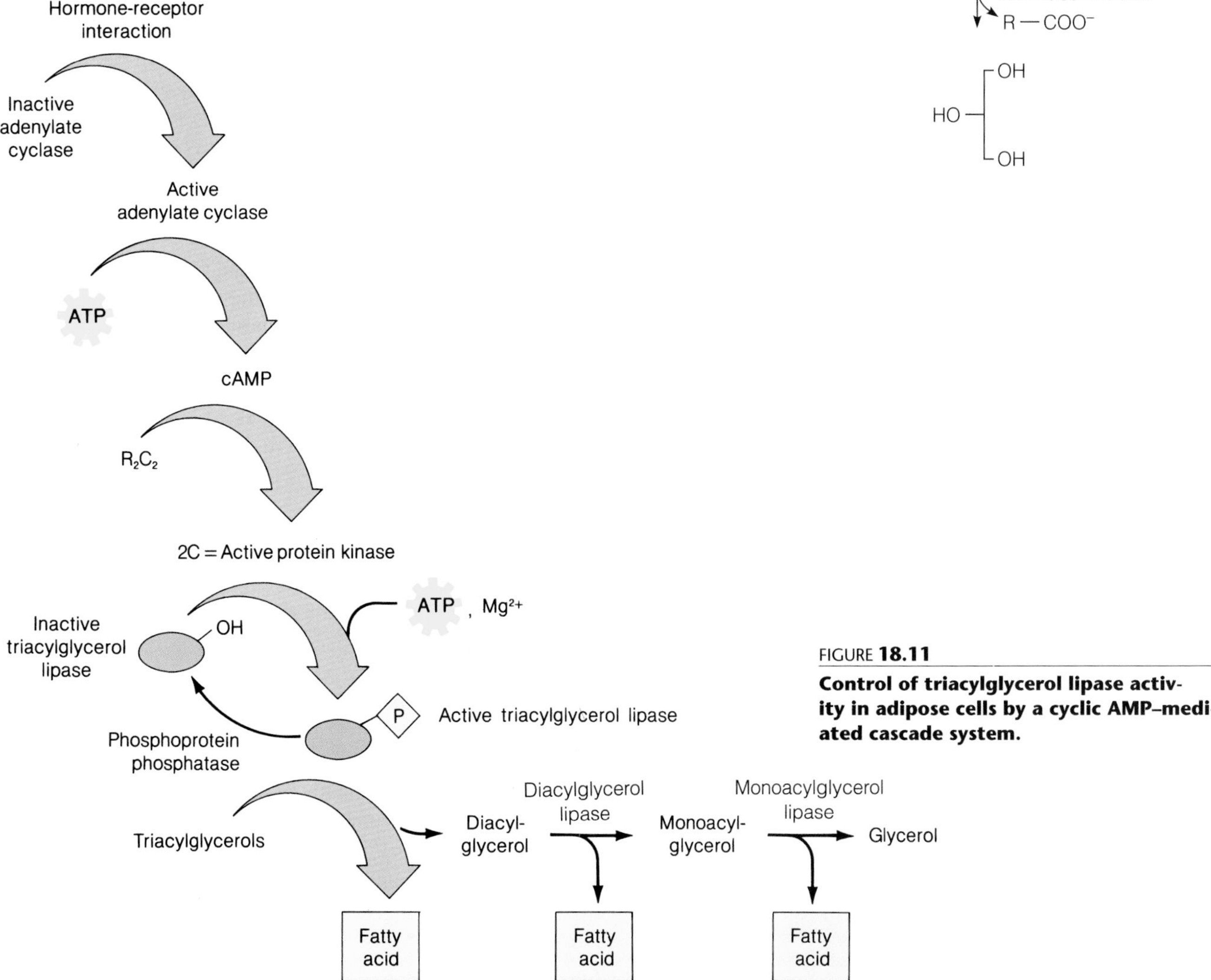

FIGURE **18.11**

Control of triacylglycerol lipase activity in adipose cells by a cyclic AMP–mediated cascade system.

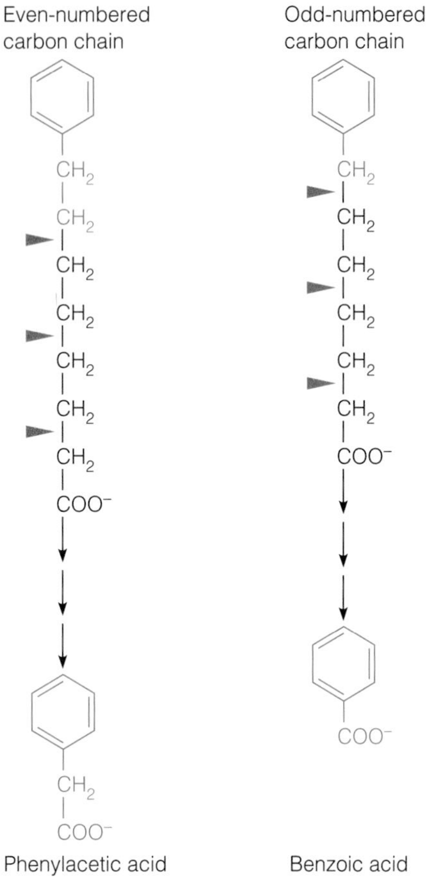

Even-numbered carbon chain

Odd-numbered carbon chain

Phenylacetic acid

Benzoic acid

FIGURE **18.12**

Oxidation of phenyl derivatives of fatty acids in Knoop's experiment. Red triangles represent presumed sites of cleavage of these model fatty acids.

acid uptake into cells is driven primarily by concentration. Most of the glycerol released to the bloodstream is taken up by liver cells, where it serves as a gluconeogenic substrate, leading to glucose.

Fatty Acid Oxidation

EARLY EXPERIMENTS

The nature of the pathway by which fatty acids are oxidized was revealed as early as 1904, in a brilliant series of experiments by a German chemist named Franz Knoop. The experiments were inspired, because they involved the first known use of metabolic tracers, more than 40 years before radioactive tracers became available. Knoop fed dogs a series of fatty acids in which the terminal methyl group was derivatized with a phenyl group. The expectation was that these analogs followed metabolic pathways similar to those used for oxidizing normal fatty acids. Knoop found that when the fed fatty acid had an even-numbered carbon chain, the final breakdown product, recovered from urine, was phenylacetic acid. When the fed fatty acid had an odd-numbered chain, the product was benzoic acid (Figure 18.12).

These results led Knoop to propose that fatty acids are oxidized in a stepwise fashion, with initial attack on carbon 3 (the β-carbon with respect to the carboxyl group). This attack would release the terminal two carbons, and the remainder of the fatty acid molecule could undergo another oxidation. Release of a two-carbon fragment would occur at each step in the oxidation. With the analogs, the process would be repeated until the remaining acid, either phenylacetic or benzoic acid, could not be further metabolized and would be excreted in the urine.

The next major development came in the 1940s, when Luis Leloir and Albert Lehninger independently demonstrated fatty acid oxidation in cell-free liver homogenates. Lehninger showed that ATP was essential for this process, suggesting that ATP somehow activates the carboxyl group of the fatty acid. Working with Eugene Kennedy, Lehninger also showed that the process occurs in mitochondria and that it releases two-carbon fragments that are oxidized in the citric acid cycle. In Munich, Feodor Lynen demonstrated that the ATP-dependent activation esterifies the fatty acid carboxyl group with the thiol group of coenzyme A, and it was later shown that all of the intermediates in the subsequent oxidative reactions are fatty acyl-CoA thioesters. Thus, by the mid-1950s the basic outlines of the fatty acid oxidation pathway were clear. As shown in Figure 18.13, the pathway consists of activation of the carboxyl group, transport into the mitchondrial matrix, and stepwise oxidation of the carbon chain, two carbons at a time, from the end containing the carboxyl group.

FATTY ACID ACTIVATION AND TRANSPORT INTO MITOCHONDRIA

For the most part, fatty acids arise in the cytosol, either through biosynthesis (as discussed later in this chapter) or through triacylglycerol or fatty acid transport from fat depots outside the cell. These fatty acids must be transported into the mitochondrial matrix for oxidation. Because the inner membrane is impermeable to free long-chain fatty acids and acyl-CoAs, a specific transport system comes into play. That transport system operates hand in hand with the metabolic activation needed to initiate the β-oxidation pathway.

A series of **fatty acyl-CoA ligases**, specific for short-chain, medium-chain, or long-chain fatty acids, catalyzes formation of the fatty acyl thioester conjugate with coenzyme A (steps 1 and 1′ in the upper part of Figure 18.13):

Fatty acids are activated for oxidation by ATP-dependent acylation of coenzyme A.

$$\text{R—COO}^- + \text{ATP} + \text{CoA—SH} \rightleftharpoons \text{R—}\overset{\text{O}}{\overset{\|}{\text{C}}}\text{—S—CoA} + \text{AMP} + \text{PP}_i \qquad \Delta G^{\circ\prime} = -0.8\ \text{kJ/mol}$$

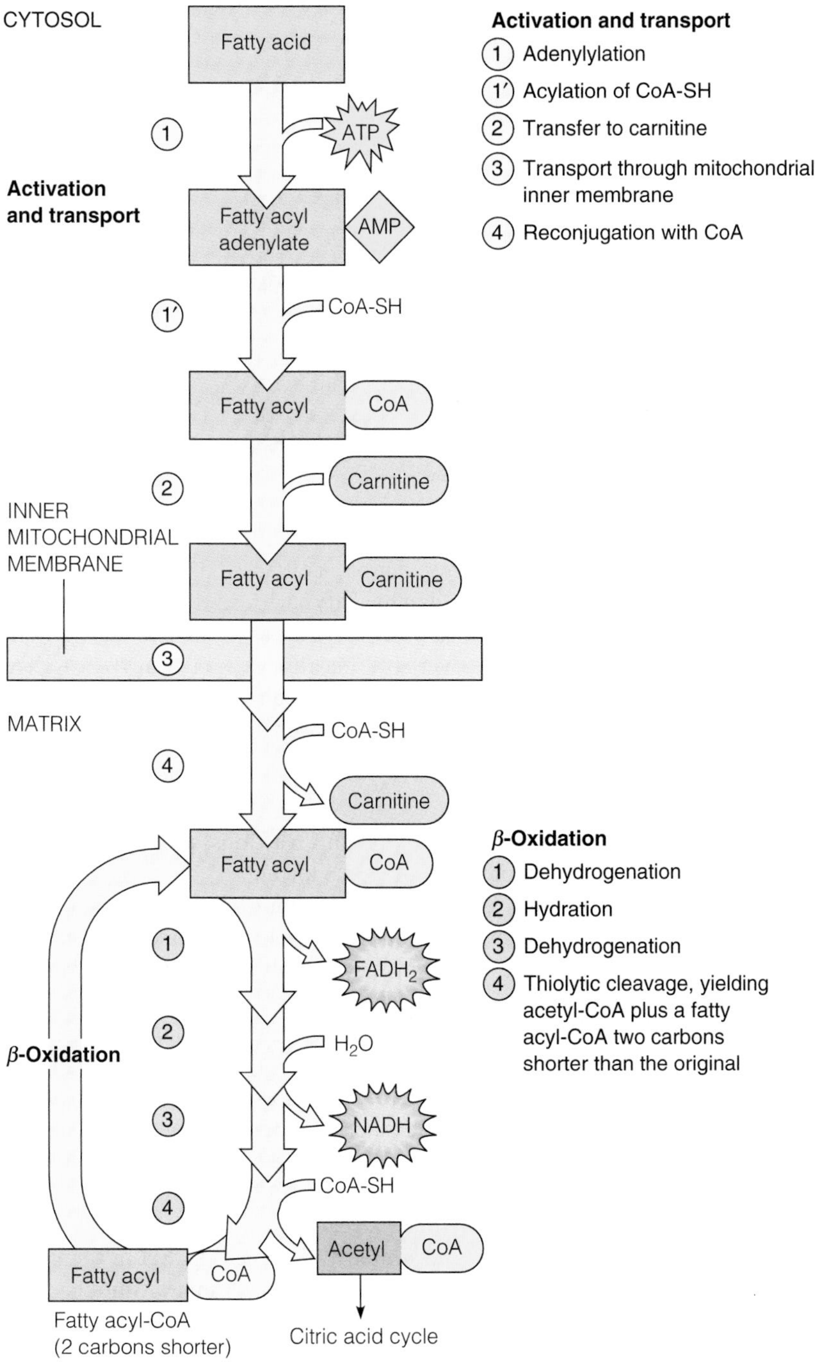

CYTOSOL

Fatty acid

① ATP

Activation and transport

Fatty acyl adenylate AMP

①′ CoA-SH

Fatty acyl CoA

② Carnitine

INNER MITOCHONDRIAL MEMBRANE

Fatty acyl Carnitine

③

MATRIX

④ CoA-SH Carnitine

Fatty acyl CoA

① $FADH_2$

β-Oxidation

② H_2O

③ NADH

④ CoA-SH

Fatty acyl CoA

Fatty acyl-CoA (2 carbons shorter)

Acetyl CoA

Citric acid cycle

Activation and transport

① Adenylylation

①′ Acylation of CoA-SH

② Transfer to carnitine

③ Transport through mitochondrial inner membrane

④ Reconjugation with CoA

β-Oxidation

① Dehydrogenation

② Hydration

③ Dehydrogenation

④ Thiolytic cleavage, yielding acetyl-CoA plus a fatty acyl-CoA two carbons shorter than the original

FIGURE **18.13**

Overview of the fatty acid oxidation pathway.

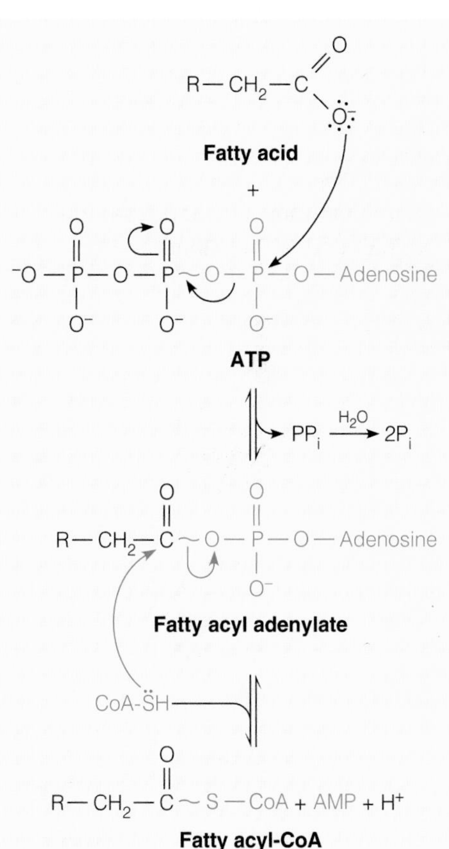

FIGURE **18.14**

Mechanism of fatty acyl-CoA ligase reactions. The figure shows reversible formation of the activated fatty acyl adenylate, nucleophilic attack by the thiol sulfur of CoA-SH on the activated carboxyl group, and the quasi-irreversible pyrophosphatase reaction, which draws the overall reaction toward fatty acyl-CoA.

The ligase specific for long-chain acids is a membrane-bound enzyme, found in both the endoplasmic reticulum and the outer mitochondrial membrane; the short-chain and medium-chain enzymes are found primarily in the mitochondrial matrix. The long-chain enzyme, which plays the predominant role in initiating fatty acid oxidation, acts on fatty acids with chain lengths of 10 to 20 carbons; the medium-chain enzyme acts on 4- to 12-carbon chains; and the short-chain enzyme prefers acetate and propionate.

Chemically, the energy-rich thioester link in long-chain fatty acyl-CoAs is identical to that of acetyl-CoA (see Chapter 14). Recall that pyruvate oxidation provides the energy to drive acetyl-CoA formation. The fatty acyl-CoA ligases, on the other hand, use a two-step mechanism involving cleavage of ATP to drive the endergonic thioester formation (Figure 18.14). First comes activation of the

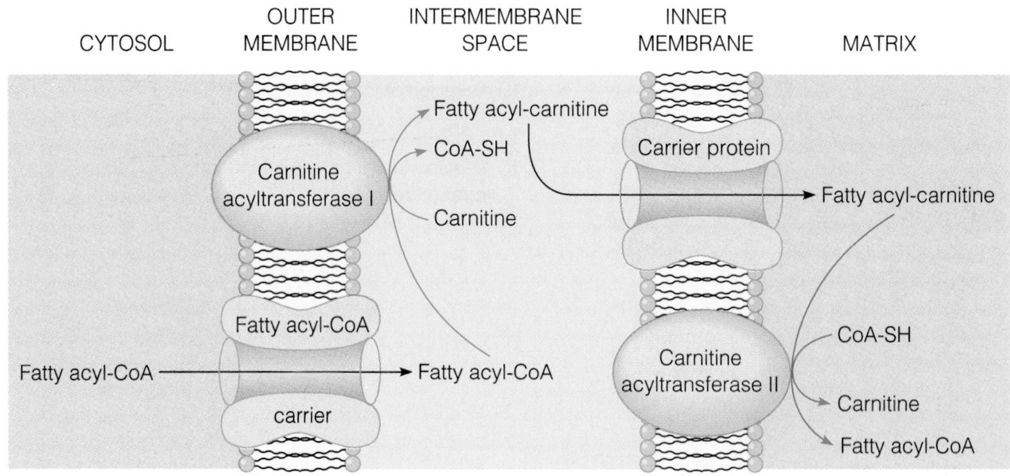

FIGURE **18.15**
The carnitine cycle, for transport of fatty acyl-CoAs into mitochondria.

carboxyl group by ATP to give a **fatty acyl adenylate,** with concomitant release of pyrophosphate. Next, the activated carboxyl group is attacked by the thiol group of CoA, thereby displacing AMP and forming the fatty acyl-CoA derivative. Carboxyl groups of amino acids are activated for protein synthesis in very similar fashion.

Fatty acid activation is readily reversible, because each fatty acyl-CoA, like ATP itself, is an energy-rich compound. However, the direction of this reaction in cells is far to the right, because of the active pyrophosphatase present in most cells:

$$PP_i + H_2O \longrightarrow 2P_i \qquad \Delta G^{\circ\prime} = -33 \text{ kJ/mol}$$

Thus, the overall reaction (sum of the two previous reactions) proceeds far in the direction of completion, with a net $\Delta G^{\circ\prime}$ of -33.8 kJ/mol.

Carnitine transports acyl-CoAs into mitochondria for oxidation.

Fatty acyl-CoAs are formed on the outer mitochondrial membrane. Hence, they must move through the inner mitochondrial membrane to become oxidized. This movement involves transfer of the fatty acyl moiety to a carrier called **carnitine** (step 2, upper part of Figure 18.13). The reaction is catalyzed by **carnitine acyltransferase I,** located in the outer mitochondrial membrane, and it yields a derivative, **fatty acyl-carnitine,** that can traverse the inner membrane (Figure 18.15 and step 3, upper part of Figure 18.13). The enzyme **carnitine acyltransferase II,** located on the matrix side of the inner membrane, completes the transfer process by exchanging fatty acyl-carnitine for free carnitine and producing fatty acyl-CoA within the matrix (step 4, upper part of Figure 18.13).

$$R{-}CH_2{-}\overset{O}{\overset{\|}{C}}{\sim}S{-}CoA \; + \; H{-}C(OH)(CH_2\overset{+}{N}(CH_3)_3)(CH_2COO^-)$$

Fatty acyl-CoA **Carnitine**

$$\rightleftharpoons$$

$$H{-}C(CH_2\overset{+}{N}(CH_3)_3)(CH_2COO^-){-}O{\sim}\overset{O}{\overset{\|}{C}}{-}CH_2{-}R \; + \; CoA{-}SH$$

Fatty acyl-carnitine **CoA-SH**

The free carnitine formed in the matrix readily returns to the intermembrane space via the carrier protein, and similarly, the free CoA-SH in the intermembrane space returns to the cytosol to begin the process anew. Although fatty acyl-carnitines are ordinary esters, the ester bond in these compounds is somewhat activated, as shown by the ready reversibility of the carnitine acyltransferase reactions.

What is the point of this rather complex shuttling process? Probably it exists to regulate fatty acid oxidation, preventing the futile cycle that would occur if oxidation and resynthesis were taking place in the same cell. Carnitine acyltransferase I is strongly inhibited by **malonyl-CoA,** the first intermediate in fatty acid synthesis (page 652). Thus, conditions in the cell that favor fatty acid synthesis prevent the transfer of fatty acyl moieties to their intracellular sites of oxidation and, hence, prevent that oxidation.

THE β-OXIDATION PATHWAY

Once inside the mitochondrial matrix, fatty acyl-CoAs are oxidized as predicted by Knoop, with initial oxidation of the β-carbon and a series of steps that each releases a two-carbon fragment, in the form of acetyl-CoA, from the fatty acid

FIGURE **18.16**

Outline of the β-oxidation of fatty acids. In the diagram a 16-carbon saturated fatty acyl-CoA (palmitoyl-CoA) undergoes seven cycles of oxidation to yield eight molecules of acetyl-CoA. These reactions correspond to steps 1–4, lower part of Figure 18.13.

$CH_3(CH_2)_{11}CH_2-CH_2-CH_2-C(=O)SCoA$

C_{16} acyl-CoA

1 Dehydrogenation (FAD → $FADH_2$)

$CH_3(CH_2)_{11}CH_2-CH=CH-C(=O)SCoA$

***trans*-Δ2-Enoyl-CoA**

2 Hydration (H_2O)

$CH_3(CH_2)_{11}CH_2-CH(OH)-CH_2-C(=O)SCoA$

L-3-Hydroxyacyl-CoA

3 Dehydrogenation (NAD^+ → $NADH + H^+$)

$CH_3(CH_2)_{11}CH_2-C(=O)-CH_2-C(=O)SCoA$

3-Ketoacyl-CoA

4 Thiolytic cleavage (CoA-SH)

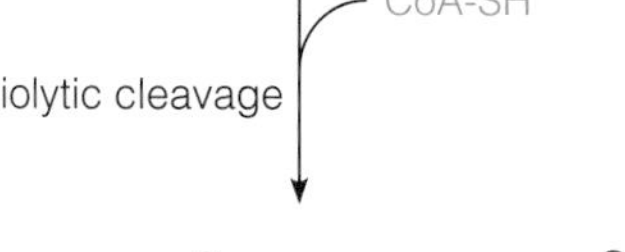

C_{14} acyl-CoA **Acetyl-CoA**

6 additional cycles of reactions 1–4

$7\ CH_3-C(=O)-S\ CoA$

Acetyl-CoA

undergoing oxidation. Each step involves four reactions (Figure 18.16 and steps 1–4, lower part of Figure 18.13). The pathway is cyclic in that each step ends with formation of an acyl-CoA, shortened by two carbons, which undergoes the same process in the next step, or cycle. For example, 1 mole of palmitoyl-CoA, derived from a 16-carbon fatty acid, undergoes seven cycles of oxidation to give 8 moles of acetyl-CoA. Each cycle releases 1 two-carbon unit, concomitant with 2 two-electron oxidation–reduction reactions. Because each step begins with oxidation of the β-carbon, the pathway is called **β-oxidation.**

Mechanistically, this pathway is remarkably similar to that used to oxidize succinate in the citric acid cycle. As shown in Figure 18.16, each cycle in the oxidation of a saturated fatty acyl-CoA involves the following reactions: (1) dehydrogenation to give an enoyl derivative; (2) hydration of the resultant double bond, with the β-carbon undergoing hydroxylation; (3) dehydrogenation of the hydroxyl group; and (4) cleavage by attack of a second molecule of coenzyme A on the β-carbon, to release acetyl-CoA and a fatty acyl-CoA two carbons shorter than the original substrate. Oxidation of unsaturated fatty acyl-CoAs is slightly different, as discussed on page 645.

Acetyl-CoA from β-oxidation enters the citric acid cycle, where it is oxidized to CO_2 in the same fashion as the acetyl-CoA derived from the oxidation of pyruvate. Like the citric acid cycle, β-oxidation generates reduced electron carriers, whose reoxidation in the mitochondria generates ATP via oxidative phosphorylation from ADP. Now let us describe the individual reactions in detail.

Reaction 1: The Initial Dehydrogenation

The first reaction is catalyzed by a **fatty acyl-CoA dehydrogenase,** which dehydrogenates between the α-carbon and β-carbon to give a *trans*-Δ2-enoyl-CoA as the product.

$$\text{Fatty acyl-S-CoA} + \text{E–FAD} \longrightarrow \textit{trans}\text{-}\Delta 2\text{-enoyl-S-CoA} + \text{E–FADH}_2$$

Three forms of this enzyme exist and are specific for short-, medium-, or long-chain fatty acyl-CoAs, respectively. Each enzyme carries a tightly bound FAD prosthetic group. As shown in Figure 18.17, the enzyme-bound $FADH_2$ contributes a pair of electrons to a shuttle protein, the **electron-transferring flavoprotein (ETFP).** These electrons are passed in turn to coenzyme Q via ETF-Q oxidoreductase, an integral membrane protein, and are then shuttled along the respiratory chain, yielding two ATPs via oxidative phosphorylation (Figure 18.17). In this respect, ETF-Q oxidoreductase is comparable to NADH dehydrogenase and succinate dehydrogenase. All three are flavoproteins that transfer electrons to the mobile electron carrier coenzyme Q.

Reactions 2 and 3: Hydration and Dehydrogenation

Like succinate oxidation, an initial FAD-dependent fatty acyl-CoA oxidation is followed by hydration and an NAD^+-dependent dehydrogenation. In β-oxidation the latter two reactions are catalyzed by **enoyl-CoA hydratase** and **3-hydroxyacyl-CoA dehydrogenase,** respectively. Both reactions are stereospecific.

$$\textit{trans}\text{-}\Delta 2\text{-Enoyl-S-CoA} + H_2O \longrightarrow \text{L-3-hydroxyacyl-S-CoA}$$

$$\text{L-3-Hydroxyacyl-S-CoA} + NAD^+ \longrightarrow \text{3-ketoacyl-S-CoA} + NADH + H^+$$

Because carbon 3 is β with respect to the carboxyl carbon, the products of these two reactions are sometimes called L-β-hydroxyacyl-CoA and β-ketoacyl-CoA, respectively. Hence the term *β-oxidation.*

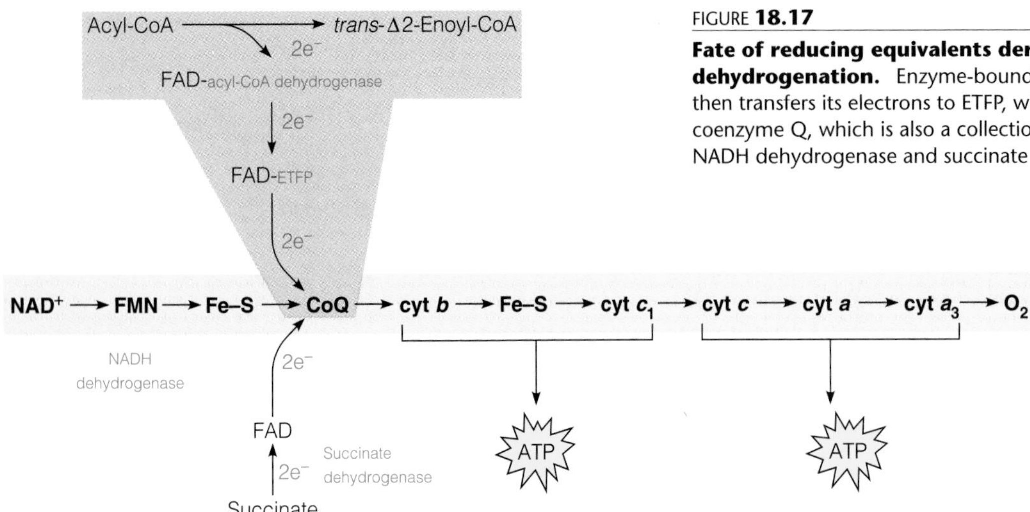

FIGURE **18.17**

Fate of reducing equivalents derived from fatty acyl-CoA dehydrogenation. Enzyme-bound FAD becomes reduced and then transfers its electrons to ETFP, which in turn passes them to coenzyme Q, which is also a collection point for electrons from NADH dehydrogenase and succinate dehydrogenase.

Fatty acids are oxidized by repeated cycles of dehydrogenation, hydration, dehydrogenation, and thiolytic cleavage, with each cycle yielding acetyl-CoA and a fatty acyl-CoA shorter by two carbons than the input acyl-CoA.

Mechanism of thiolytic cleavage

Reaction 4: Thiolytic Cleavage

The fourth and last reaction in each cycle of the β-oxidation pathway involves attack of the nucleophilic thiol sulfur of coenzyme A on the electron-poor keto carbon of 3-ketoacyl-CoA, with cleavage of the α—β bond and release of acetyl-CoA. The other product is a shortened fatty acyl-CoA, ready to begin a new cycle of oxidation:

$$\underset{\textbf{3-Ketoacyl-CoA}}{R{-}\overset{O}{\overset{\|}{C}}{-}CH_2{-}\overset{O}{\overset{\|}{C}}{-}S{-}CoA} + CoA\text{-}SH \longrightarrow \underset{\textbf{Acyl-CoA}}{R{-}\overset{O}{\overset{\|}{C}}{-}S{-}CoA} + \underset{\textbf{Acetyl-CoA}}{CH_3{-}\overset{O}{\overset{\|}{C}}{-}S{-}CoA}$$

Because this reaction involves cleavage by a thiol, it is referred to as a **thiolytic cleavage,** by analogy with hydrolysis, which involves cleavage by water. The enzyme is commonly called **β-ketothiolase,** or simply **thiolase.** An essential thiol group on the enzyme (E-SH) attacks the substrate, with formation of an acyl-enzyme intermediate and acetyl-CoA. Free CoA-SH then attacks the intermediate.

As noted earlier, the overall oxidation pathway as just described is applicable to the most abundant fatty acids—those that contain even numbers of carbon atoms and are fully saturated. Presently, we shall describe the variations in this pathway that permit oxidation of other fatty acids. For the saturated, even-chain fatty acyl-CoAs, oxidation simply proceeds stepwise, with two carbons lost as acetyl-CoA after each cycle. For the C_{16} palmitoyl-CoA, the example shown in Figure 18.16, the first cycle yields acetyl-CoA plus the C_{14} myristoyl-CoA. A second cycle, acting on the latter substrate, yields acetyl-CoA plus the C_{12} lauroyl-CoA. In the seventh and last cycle, the 3-hydroxyacyl-CoA dehydrogenase reaction yields acetoacetyl-CoA. Thiolytic cleavage of this substrate yields 2 moles of acetyl-CoA. Thus, the oxidation of 1 mole of palmitic acid involves six successive cycles, each of which yields 1 mole of acetyl-CoA, and a seventh cycle, which yields 2 moles. Other saturated even-chain fatty acids are degraded identically. For example, stearic acid oxidation involves eight steps, with two acetyl-CoAs resulting from the last cycle.

ENERGY YIELD FROM FATTY ACID OXIDATION

We can now write a balanced equation for the overall degradation of palmitoyl-CoA to 8 moles of acetyl-CoA.

$$\text{Palmitoyl-CoA} + 7\text{CoA-SH} + 7\text{FAD} + 7\text{NAD}^+ + 7H_2O \longrightarrow 8\text{ Acetyl-CoA} + 7\text{FADH}_2 + 7\text{NADH} + 7H^+$$

Each of the products is metabolized exactly as described earlier for oxidation of carbohydrates. Acetyl-CoA is catabolized via the citric acid cycle, and $FADH_2$ and NADH transfer electrons to the respiratory chain through ETFP (page 644) and NADH dehydrogenase, respectively. Thus, we can easily compute the metabolic energy yield from fatty acid oxidation in terms of moles of ATP synthesized from ADP. Recall from Chapter 15 that oxidation of acetyl-CoA in one turn of the citric acid cycle yields 12 ATPs and that the P/O ratios for oxidation of flavoproteins and NADH are 2 and 3, respectively. The following summation, using palmitate as an example, gives the total energy yield:

Reaction	ATP Yield
Activation of palmitate to palmitoyl-CoA	−2
Oxidation of 8 acetyl-CoA	8 × 12 = 96
Oxidation of 7 $FADH_2$	7 × 2 = 14
Oxidation of 7 NADH	7 × 3 = 21
Net: Palmitate ⟶ $CO_2 + H_2O$	129

From this, you can calculate the ATP yield per carbon oxidized to CO_2 as 129/16, or about 8.2. The corresponding value for glucose is 6.3 (38 ATPs formed per 6 carbons oxidized). Thus, the energy yield from fat oxidation is higher than that from oxidation of the less highly reduced carbohydrate, whether measured on a weight basis (see page 422) or a molar basis.

CH_3–C(=O)–CH_2–C(=O)–S–CoA

Thiolase
CoA-SH

CH_3–C(=O)–S–CoA + CH_3–C(=O)–S–CoA

The final thiolytic cleavage

OXIDATION OF UNSATURATED FATTY ACIDS

Two enzymes, enoyl-CoA isomerase and 2,4-dienoyl-CoA reductase, play essential roles in the oxidation of unsaturated fatty acids.

Recall from Chapter 10 that many fatty acids in natural lipids are unsaturated; that is, they contain one or more double bonds (see Table 10.1, page 317). Because these bonds are in the *cis* configuration, they cannot be simply acted on by enoyl-CoA hydratase, which acts only on *trans* compounds. Two additional enzymes, **enoyl-CoA isomerase** and **2,4-dienoyl-CoA reductase,** must come into play for these fatty acids to be oxidized. The isomerase acts upon monounsaturated fatty acids, such as the 18-carbon Δ9 compound, oleic acid, which contains a *cis* double bond between carbons 9 and 10. Oleic acid is activated, transported into mitochondria, and carried through three cycles of β-oxidation just as are the saturated fatty acids. The product of the third cycle is the CoA ester of a 12-carbon fatty acid with a *cis* double bond between carbons 3 and 4. Not only is the double bond in the wrong configuration to be hydrated, but it is also in the wrong position. The enoyl-CoA isomerase enzyme converts this *cis*-Δ3-enoyl-CoA to the corresponding *trans*-Δ2-enoyl-CoA, which can then be acted on by enoyl-CoA hydratase (see margin). This hydration and all subsequent reactions are identical to those already described for saturated fatty acids.

The other auxiliary enzyme, 2,4-dienoyl-CoA reductase, comes into play during the oxidation of polyunsaturated fatty acids, such as linoleic acid. This 18-carbon fatty acid contains *cis* double bonds between carbons 9 and 10 and between carbons 12 and 13. As shown in Figure 18.18, linoleyl-CoA undergoes three cycles of β-oxidation, just as does oleyl-CoA, to give a C_{12} acyl-CoA with *cis* double bonds between carbons 3 and 4 and between 6 and 7. The isomerase converts the Δ3 double bond from *cis* to *trans.* There follow hydration, dehydrogenation, and thiolytic cleavage to give acetyl-CoA plus a 10-carbon enoyl-CoA, unsaturated between carbons 4 and 5. Action of acyl-CoA dehydrogenase yields a dienoyl-CoA, unsaturated at C-4—C-5 and at C-2—C-3. The NADPH-dependent 2,4-dienoyl-CoA reductase converts this to a C_{10} *cis*-Δ3-enoyl-CoA. The isomerase comes into play once more, generating a *trans*-Δ2-enoyl-CoA, which undergoes the remaining cycles of β-oxidation normally.

By the pathways described here, both monounsaturated and diunsaturated 18-carbon fatty acids can undergo degradation to 9 moles of acetyl-CoA. There

FIGURE **18.18**

***β*-Oxidation pathway for polyunsaturated fatty acids.** This example, using linoleyl-CoA, shows points for the action of enoyl-CoA isomerase and 2,4-dienoyl-CoA reductase. Enzymes specific to unsaturated fatty acid oxidation are identified with red type.

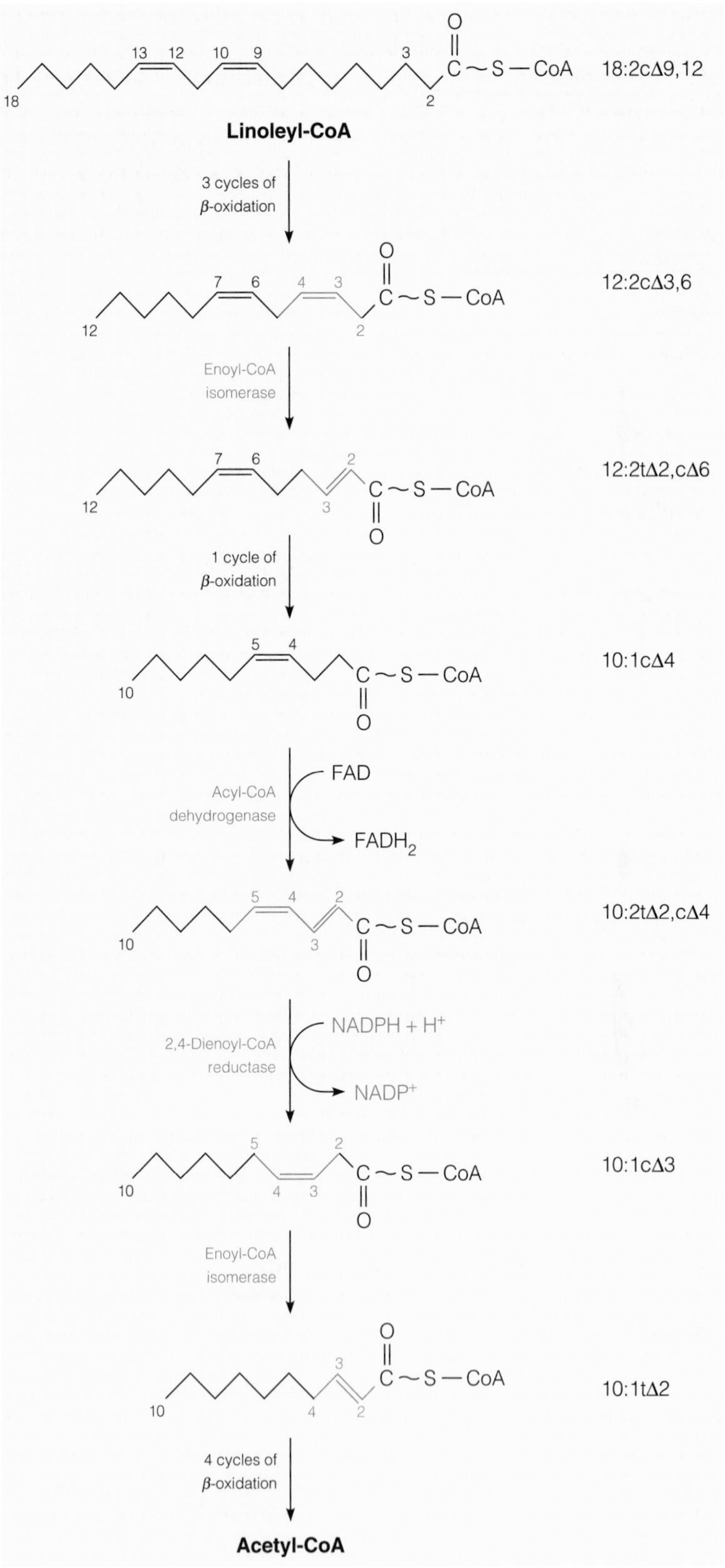

is, of course, a slight reduction in the overall energy yield because each double bond in the original fatty acid means one less FAD reduction step in the overall process. The two auxiliary enzymes allow all of the even-chain polyunsaturated fatty acids to be similarly degraded, with the following exception. A significant portion of dietary lipid contains unsaturated fatty acids with double bonds in the *trans* configuration. These fatty acids arise through microbial action in the digestive systems of ruminant mammals and also chemically, through partial hydrogenation of fats and oils. Thus, they are relatively abundant in dairy products and in margarine. Because there is some evidence implicating *trans* fatty acids in coronary artery disease, margarine manufacturers are beginning to remove them from their product.

OXIDATION OF FATTY ACIDS WITH ODD-NUMBERED CARBON CHAINS

Odd-numbered fatty acid chains yield upon oxidation 1 mole of propionyl-CoA, whose conversion to succinyl-CoA involves a biotin-dependent carboxylation and a coenzyme B_{12}–dependent rearrangement.

Though most of the fatty acids in natural lipids contain even-numbered carbon chains, a small proportion have odd-numbered carbon chains. The latter group presents a special metabolic problem, which is solved in a novel way. The substrate for the last cycle of β-oxidation of an odd-chain acyl-CoA is the five-carbon homolog of acetoacetyl-CoA. Thiolytic cleavage of this substrate yields 1 mole each of acetyl-CoA and **propionyl-CoA.**

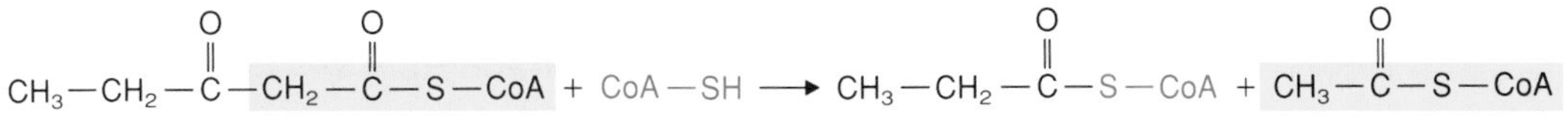

$$CH_3-CH_2-\overset{O}{\overset{\|}{C}}-CH_2-\overset{O}{\overset{\|}{C}}-S-CoA + CoA-SH \longrightarrow CH_3-CH_2-\overset{O}{\overset{\|}{C}}-S-CoA + CH_3-\overset{O}{\overset{\|}{C}}-S-CoA$$

Acetoacetyl-CoA homolog **Propionyl-CoA** **Acetyl-CoA**

Unlike acetyl-CoA, which is catabolized via the citric acid cycle, propionyl-CoA must be further metabolized before its carbon atoms can enter the citric acid cycle for complete oxidation to CO_2. That further metabolism (Figure 18.19) involves first the ATP-dependent carboxylation of propionyl-CoA, catalyzed by the biotin-containing enzyme **propionyl-CoA carboxylase.** The product, D-methylmalonyl-CoA, then undergoes epimerization to its L stereoisomer by action of methylmalonyl-CoA epimerase. Next, this branched-chain acyl-CoA derivative is converted to the corresponding straight-chain compound, which happens to be succinyl-CoA, by an unusual reaction. The enzyme, L-methylmalonyl-CoA mutase, requires a cofactor called **adenosylcobalamin,** derived from vitamin B_{12}. Mechanistically, the side-chain migration makes this reaction quite interesting. However, because B_{12} coenzymes are also involved in amino acid metabolism, we reserve study of this and other B_{12}-dependent reactions for Chapter 20.

Inability to catabolize propionyl-CoA properly has severe consequences in humans. If there is defective activity of L-methylmalonyl-CoA mutase or of the synthesis of the adenosylcobalamin coenzyme, L-methylmalonyl-CoA accumulates and exits from cells as methylmalonic acid. This process causes a severe acidosis (lowering of blood pH) and also damages the central nervous system. This rare condition, called **methylmalonic acidemia,** is usually fatal in early life. The disease can sometimes be treated by administering large doses of vitamin B_{12}. In these cases the mutation decreases affinity of the mutase for its B_{12} coenzyme, and the enzyme can be induced to function if the coenzyme concentration can be increased substantially.

CONTROL OF FATTY ACID OXIDATION

In most cells, fatty acid oxidation is controlled by availability of substrates for oxidation, the fatty acids themselves. In animals this availability is controlled in turn by the hormonal control of fat mobilization in adipocytes. Because the function of adipose tissue is to store fat for use in other cells, it makes good metabolic sense for breakdown and release of this stored fat to be regulated by hormones, which are extracellular messengers. Recall from page 639 that triacylglycerol lipase activity

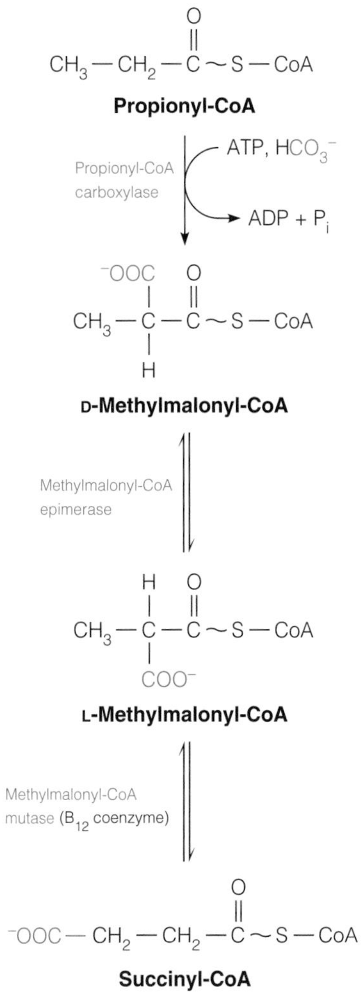

$$CH_3-CH_2-\overset{O}{\overset{\|}{C}}\sim S-CoA$$

Propionyl-CoA

Propionyl-CoA carboxylase: ATP, HCO_3^- → ADP + P_i

$$CH_3-\underset{H}{\underset{|}{\overset{^-OOC}{\overset{|}{C}}}}-\overset{O}{\overset{\|}{C}}\sim S-CoA$$

D-Methylmalonyl-CoA

⇅ Methylmalonyl-CoA epimerase

$$CH_3-\underset{COO^-}{\underset{|}{\overset{H}{\overset{|}{C}}}}-\overset{O}{\overset{\|}{C}}\sim S-CoA$$

L-Methylmalonyl-CoA

⇅ Methylmalonyl-CoA mutase (B_{12} coenzyme)

$$^-OOC-CH_2-CH_2-\overset{O}{\overset{\|}{C}}\sim S-CoA$$

Succinyl-CoA

FIGURE **18.19**
Pathway for catabolism of propionyl-CoA.

is regulated by hormonally initiated regulatory cascades involving cyclic AMP. The action of glucagon or epinephrine causes fat breakdown and release, which leads ultimately to fatty acid accumulation in other cells. Also, as noted on page 642, malonyl-CoA provides another important regulatory mechanism, by inhibiting fatty acyl-CoA movement into mitochondria by the acylcarnitine shuttle.

PEROXISOMAL β-OXIDATION OF FATTY ACIDS

A modified version of the β-oxidation pathway occurs in **peroxisomes,** organelles that are present in most eukaryotic cells. Peroxisomes are quite similar to plant cell glyoxysomes, except that peroxisomes lack the enzymes of the glyoxylate pathway (see Chapter 14). Both peroxisomes and glyoxysomes carry out a β-oxidation pathway in which the FAD-linked acyl-CoA dehydrogenase transfers electrons not to the respiratory electron transport chain but directly to oxygen. The latter is reduced to hydrogen peroxide, which in turn is acted on by catalase.

$$\text{E—FAD} + \text{R—CH}_2\text{—CH}_2\text{—}\overset{\text{O}}{\overset{\|}{\text{C}}}\text{—S—CoA} \longrightarrow \text{R—CH=CH—}\overset{\text{O}}{\overset{\|}{\text{C}}}\text{—S—CoA} + \text{E—FADH}_2$$

$$\text{E—FADH}_2 + \text{O}_2 \longrightarrow \text{E—FAD} + \text{H}_2\text{O}_2$$

$$\text{H}_2\text{O}_2 \longrightarrow \text{H}_2\text{O} + \tfrac{1}{2}\,\text{O}_2$$

Because electrons are not shuttled into the respiratory chain, the peroxisomal pathway is not coupled to energy production, but it does generate heat. In animal peroxisomes the pathway proceeds only as far as C_4 and C_6 acyl-CoAs. However, the acyl groups can be transferred to carnitine for transport into mitochondria, where oxidation can be completed. By contrast, plant glyoxysomes carry the oxidation to acetyl-CoA, which is used for carbohydrate synthesis via the glyoxylate pathway. The function of the peroxisomal pathway is not yet clear, but it probably involves initial stages in oxidizing very long-chain fatty acids and other lipids.

α-OXIDATION OF FATTY ACIDS

Although β-oxidation is the major pathway for fatty acid degradation, a minor pathway oxidizes certain fatty acids with initial oxidation occurring on the α-carbon rather than the β-carbon. This pathway came to light through analysis of a rare and severe congenital neurological disorder called **Refsum's disease.** Patients with this condition accumulate large amounts of an unusual fatty acid, **phytanic acid,** which is derived from **phytol,** a constituent of chlorophyll (Figure 18.20). The methyl group on carbon 3 of phytol prevents β-oxidation of this substrate. However, the α-carbon can evidently undergo oxidation, to give **pristanic acid,** a substrate that can be degraded by β-oxidation. In Refsum's disease the α-oxidation pathway is defective, and the phytanic acid is not converted to a compound that can be degraded. The only known treatment is to feed a diet containing little or no chlorophyll. This treatment is difficult because it rules out both leafy green vegetables and meat and milk that come from herbivorous animals—all of these foods contain substantial amounts of phytanic acid.

KETOGENESIS

Thus far we have written about acetyl-CoA as though it had only two major metabolic fates, either oxidation to CO_2 in the citric acid cycle or biosynthesis of fatty acids. Another major pathway comes into play in mitochondria (primarily in liver) when acetyl-CoA accumulates beyond its capacity to be oxidized or used for fatty acid synthesis. That pathway is called **ketogenesis,** and it leads to a class of compounds called **ketone bodies.**

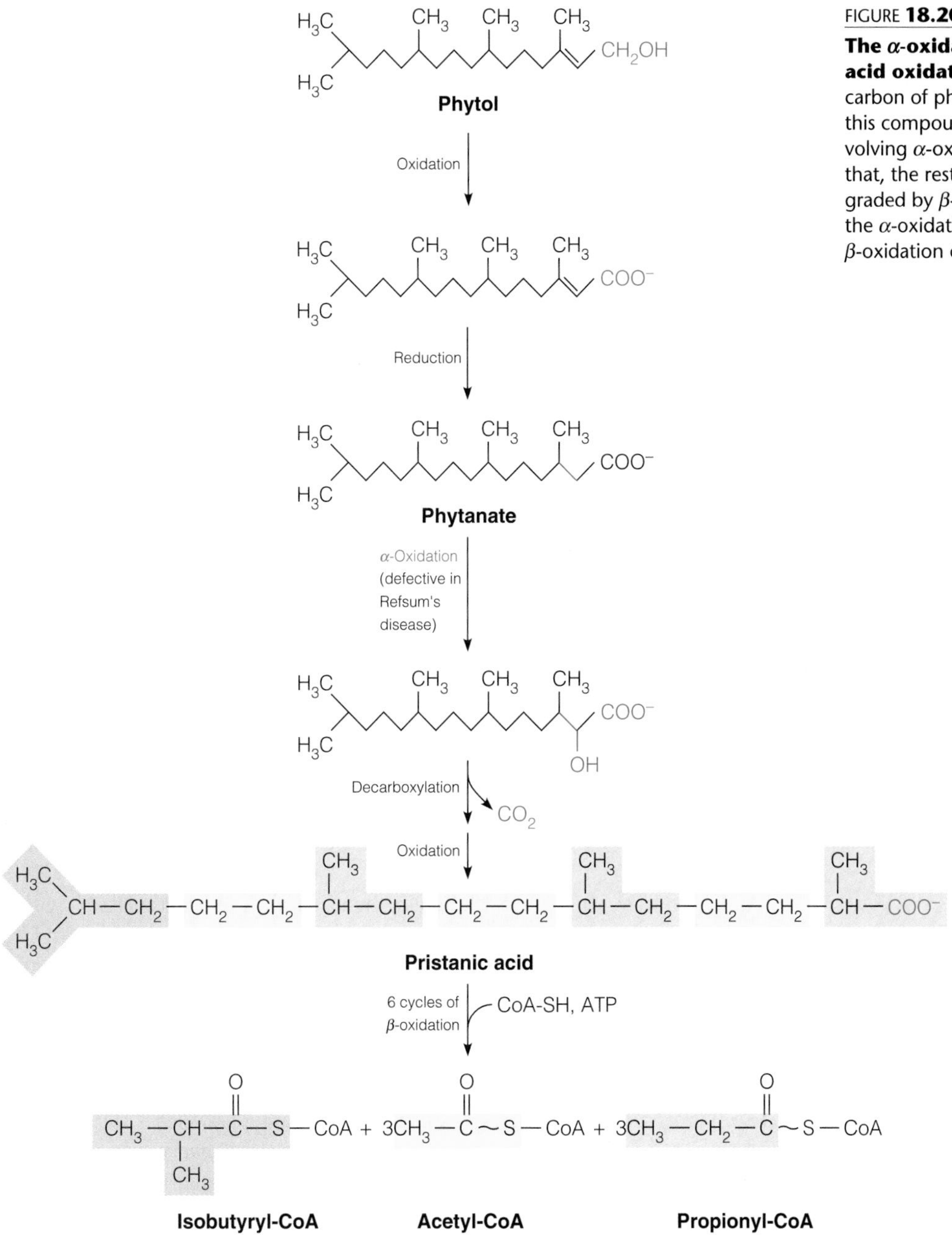

FIGURE **18.20**

The α-oxidation pathway for phytanic acid oxidation. The methyl group on the β-carbon of phytanic acid prevents β-oxidation of this compound, so an additional pathway, involving α-oxidation, comes into play. After that, the rest of the molecule can be degraded by β-oxidation. In Refsum's disease the α-oxidation pathway is defective, so β-oxidation cannot proceed.

When acetyl-CoA levels are high, 2 moles of acetyl-CoA undergo a reversal of the thiolase reaction to give acetoacetyl-CoA (Figure 18.21). This occurs particularly when levels of oxaloacetate are low, so that flux through citrate synthase is impaired. Acetoacetyl-CoA can react in turn with a third mole of acetyl-CoA to give **β-hydroxy-β-methylglutaryl-CoA (HMG-CoA),** catalyzed by **HMG-CoA synthase.** When formed in cytosol, HMG-CoA is an early intermediate in cholesterol biosynthesis (see Chapter 19). In mitochondria, however, HMG-CoA is acted on by **HMG-CoA lyase** to yield **acetoacetate** plus acetyl-CoA. Acetoacetate undergoes either NADH-dependent reduction to give **D-β-hydroxybutyrate** or, in very small amounts, spontaneous decarboxylation to acetone. Collectively, acetoacetate, acetone, and β-hydroxybutyrate are called ketone bodies, even though the last compound does not contain a keto carbonyl group.

In some circumstances, ketogenesis can be considered an "overflow pathway." As noted above, it is stimulated when acetyl-CoA accumulates because of

When carbohydrate catabolism is limited, acetyl-CoA is converted to ketone bodies, mainly acetoacetate and β-hydroxybutyrate—important metabolic fuels in some circumstances.

deficient carbohydrate utilization. Ketogenesis occurs primarily in liver, because of the high levels of HMG-CoA synthase in that tissue. Ketone bodies are transported from liver to other tissues, where acetoacetate and β-hydroxybutyrate can be reconverted to acetyl-CoA for energy generation. The reconversion involves enzymatic transfer of a CoA moiety from succinyl-CoA to acetoacetate, yielding acetoacetyl-CoA and succinate.

$$\beta\text{-Hydroxybutyrate} \xrightarrow{NAD^+ \;\rightarrow\; NADH + H^+} \text{acetoacetate} \xrightarrow{\text{Succinyl-CoA} \;\rightarrow\; \text{succinate}} \text{acetoacetyl-CoA}$$

As discussed further in Chapter 23, ketogenesis becomes extremely important in starvation, when the brain, which normally uses glucose as its main fuel, undergoes metabolic adaptation to the use of ketone bodies. Under normal conditions, certain other tissues, particularly heart, derive much of their energy by metabolizing ketone bodies produced in the liver.

Fatty Acid Biosynthesis

RELATIONSHIP OF FATTY ACID SYNTHESIS TO CARBOHYDRATE METABOLISM

We have noted that the vast majority of the stored fuel in most animal cells is in the form of fat. However, a large proportion of the caloric intake of many animal diets—certainly most human diets—is carbohydrate. Because carbohydrate storage reserves are strictly limited, there must be efficient mechanisms for conversion of carbohydrate to fat. In this section our primary focus is on fatty acid synthesis.

As schematized in Figure 18.22, a central metabolite is acetyl-CoA, which comes both from pyruvate in the pyruvate dehydrogenase reaction and from fatty acid β-oxidation. Acetyl-CoA is, in turn, converted in the cytosol to fatty acids. Thus, acetyl-CoA is derived from both fat breakdown and carbohydrate breakdown and is also the major fat precursor. However, in animals *acetyl-CoA cannot undergo net conversion to carbohydrate.* This is because of the virtual irreversibility of the pyruvate dehydrogenase reaction. As noted in Chapter 14, the glyoxylate cycle in plants and some microorganisms permits a bypass of this step, with net conversion of acetyl-CoA to gluconeogenic precursors. However, *in animals the conversion of carbohydrate to fat is unidirectional.* Moreover, although fatty acid synthesis is regulated, the total capacity for fat storage is not.

EARLY STUDIES OF FATTY ACID SYNTHESIS

Early in the twentieth century, when it became evident that most fatty acids in lipids contain even-numbered chains, it was reasonable to expect that the biosynthetic process would involve some stepwise addition of activated two-carbon fragments, in the same sense that oxidation proceeds two carbons at a time. Indeed, this process was demonstrated experimentally in the 1940s, in one of the first metabolic experiments using isotopic tracers. David Rittenberg and Konrad

$2CH_3-C(=O)\sim S-CoA$

Acetyl-CoA

β-Ketothiolase → CoA-SH

$CH_3-C(=O)-CH_2-C(=O)\sim S-CoA$

Acetoacetyl-CoA

HMG-CoA synthase: $CH_3-C(=O)-S-CoA$ in, CoA-SH out

$^-OOC-CH_2-C(OH)(CH_3)-CH_2-C(=O)\sim S-CoA$

β-Hydroxy-β-methylglutaryl-CoA (HMG-CoA)

HMG-CoA lyase → $CH_3-C(=O)\sim S-CoA$

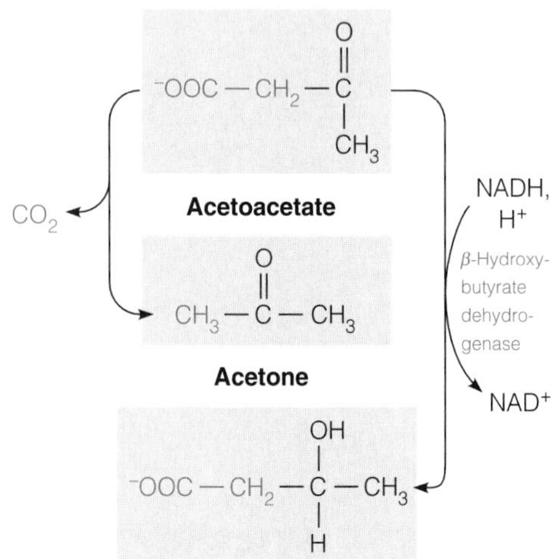

FIGURE **18.21**

Biosynthesis of ketone bodies in the liver. The three compounds commonly called ketone bodies are boxed. Acetone is formed in very small quantities, possibly by nonenzymatic decarboxylation of acetoacetate.

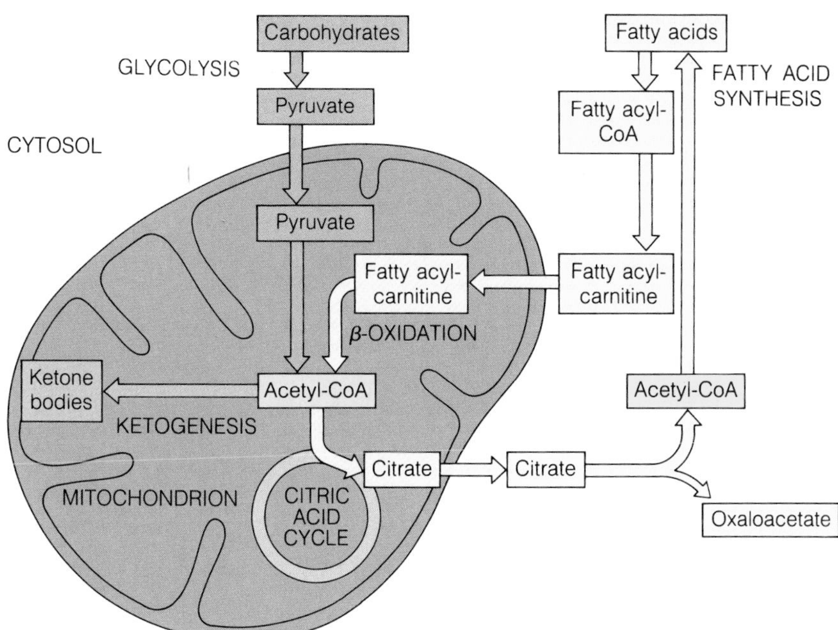

FIGURE **18.22**

Acetyl-CoA as a key intermediate between fat and carbohydrate metabolism. Arrows identify major routes of formation or utilization of acetyl-CoA. Citrate serves as a carrier to transport acetyl units from the mitochondrion to the cytosol for fatty acid synthesis.

Bloch fed to mice acetate labeled with the stable isotopes ^{13}C and deuterium ($^{13}C^{2}H_3{}^{13}COO^-$) and found both isotopes incorporated into fatty acids.

Once the β-oxidation pathway had been discovered, it was generally thought that fatty acid synthesis would proceed simply by a reversal of its degradation pathway. However, when biochemists began to fractionate enzyme systems capable of synthesizing fatty acids, they found that the activities of β-oxidation were lacking from their purified fractions. The key discovery that established fatty acid synthesis as an entirely different pathway was Salih Wakil's observation in the late 1950s that fatty acid synthesis has an absolute requirement for bicarbonate. The carbon from that bicarbonate was not incorporated in the final product, however. These observations led to the discovery of a three-carbon compound, *malonyl-CoA*, as the first committed intermediate in fatty acid biosynthesis. Today we know that, although the chemistries of fatty acid synthesis and degradation are similar, the pathways differ in the enzymes involved, acyl group carriers, stereochemistry of the intermediates, electron carriers, intracellular location, and regulation. Indeed, fatty acid metabolism is one of the best examples of the statement that anabolic pathways are never the simple reversal of catabolic pathways.

Carbohydrate is readily converted to fat, but animals cannot carry out net conversion of fat to carbohydrate.

$$^{-}OOC{-}CH_2{-}\overset{\displaystyle O}{\overset{\|}{C}}{-}S{-}CoA$$

Malonyl-CoA

The overall process of fatty acid synthesis is similar in all prokaryotic and eukaryotic systems analyzed to date. Three separate enzyme systems catalyze, respectively, (1) biosynthesis of palmitate from acetyl-CoA, (2) chain elongation starting from palmitate, and (3) desaturation. In eukaryotic cells the first pathway occurs in the cytosol, chain elongation occurs both in mitochondria and in the endoplasmic reticulum, and desaturation occurs in the endoplasmic reticulum.

Fatty acid synthesis occurs through intermediates similar to those of fatty acid oxidation, but with differences in electron carriers, carboxyl group activation, stereochemistry, and cellular location.

BIOSYNTHESIS OF PALMITATE FROM ACETYL-COA

As noted earlier and outlined in Figure 18.23, the chemistry of palmitate synthesis is remarkably similar to that of palmitate oxidation run in reverse. The synthetic process comprises stepwise additions of two-carbon units, with each step proceeding via condensation, reduction, dehydration, and another reduction. The major distinctions are the need for an activated intermediate, malonyl-CoA, at each two-carbon addition step, the nature of the acyl group carrier, and the use of NADPH-requiring enzymes in the reductive reactions. Details of these and the other reactions follow.

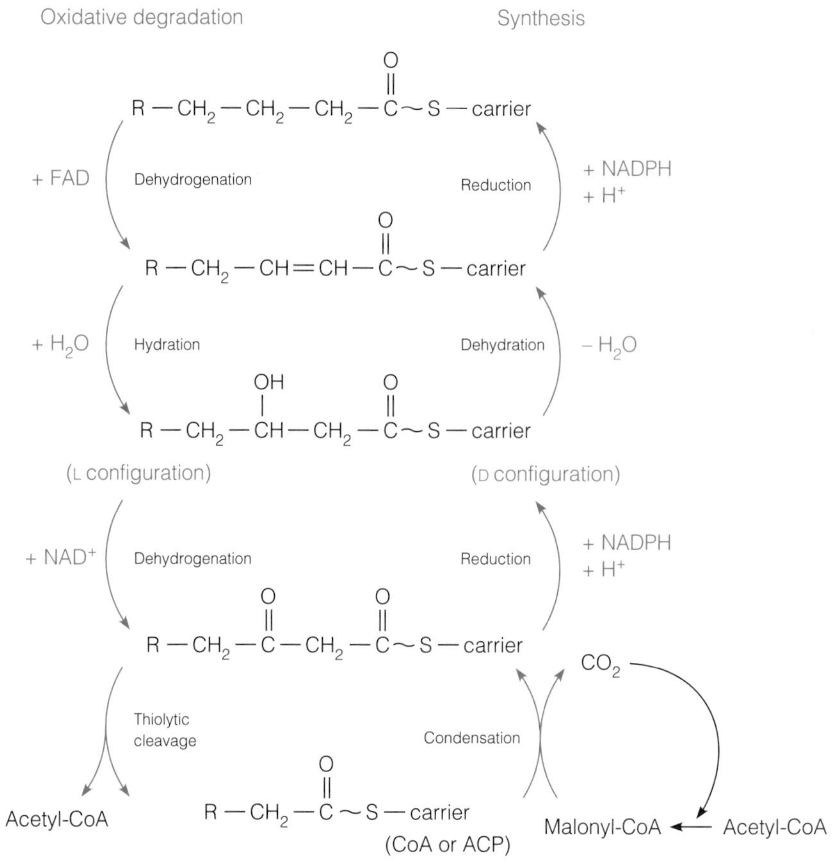

FIGURE **18.23**

Chemical similarities between oxidation and synthesis of a fatty acid. The figure shows a single cycle of oxidation (down) or addition (up) of a two-carbon fragment. Coenzyme A is the acyl group carrier for oxidation, and acyl carrier protein (ACP) is the carrier for synthesis.

Synthesis of Malonyl-CoA

The first committed step in fatty acid biosynthesis is the formation of malonyl-CoA from acetyl-CoA and bicarbonate, catalyzed by **acetyl-CoA carboxylase** (Figure 18.24, reaction 1).

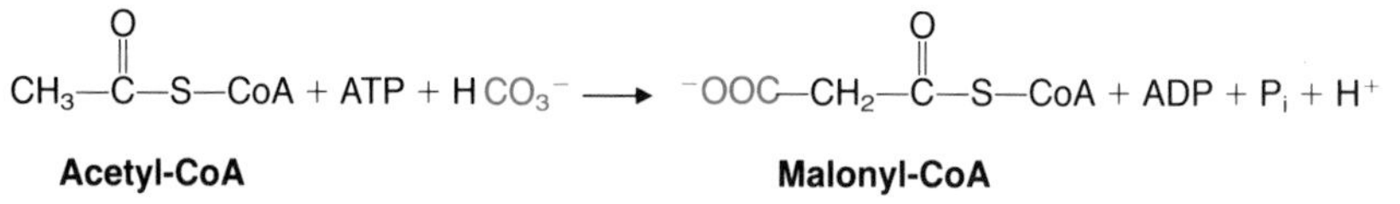

$$\underset{\textbf{Acetyl-CoA}}{CH_3{-}\overset{O}{\overset{\|}{C}}{-}S{-}CoA} + ATP + HCO_3^- \longrightarrow \underset{\textbf{Malonyl-CoA}}{^-OOC{-}CH_2{-}\overset{O}{\overset{\|}{C}}{-}S{-}CoA} + ADP + P_i + H^+$$

Like other committed steps in biosynthetic pathways, this reaction is so exergonic as to be virtually irreversible. Like other enzymes catalyzing carboxylation reactions (see page 507), acetyl-CoA carboxylase has a biotin cofactor, covalently

FIGURE **18.24**

The first three reactions in each addition cycle of fatty acid synthesis. The reactions shown here produce malonyl-ACP and acetyl-ACP, which are used in the remaining reactions of the cycle.

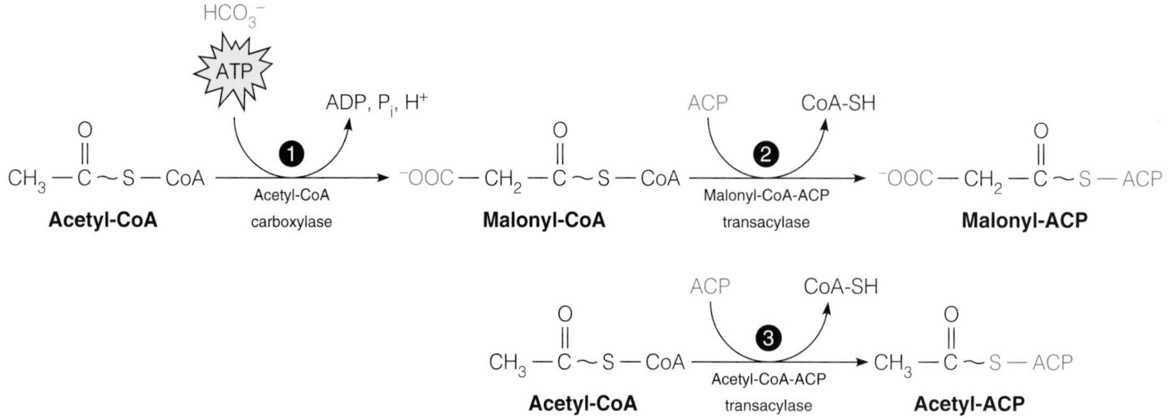

bound via a lysine ε-amino group. The reaction proceeds via a covalently bound *N*-carboxybiotin intermediate.

$$\text{E-biotin} + \text{ATP} + \text{HCO}_3^- \longrightarrow \text{E-}N\text{-carboxybiotin} + \text{ADP} + \text{P}_i$$

$$\text{E-}N\text{-carboxybiotin} + \text{acetyl-CoA} \longrightarrow \text{malonyl-CoA} + \text{E-biotin}$$

The prokaryotic form of this enzyme, exemplified by the enzyme purified from *E. coli,* consists of three separate proteins: (1) a small carrier protein that contains the bound biotin, (2) a **biotin carboxylase,** which catalyzes the formation of *N*-carboxybiotin, and (3) a **transcarboxylase,** which transfers the activated carboxyl group from *N*-carboxybiotin to acetyl-CoA. The hydrocarbon chains in both biotin and its associated lysine residue act as a flexible swinging arm, which allows the biotin to interact with the catalytic sites of both catalytic subunits.

By contrast, acetyl-CoA carboxylase in eukaryotes consists of a single protein containing two identical polypeptide chains, each with M_r of about 230,000. The dimeric protein itself is inactive, but in the presence of citrate it polymerizes to a novel filamentous form, with M_r of $4–8 \times 10^6$, that can readily be visualized in the electron microscope (Figure 18.25). The equilibrium between inactive protein dimers and the active filamentous form, and its control by metabolic intermediates, probably represents a mechanism for regulating fatty acid biosynthesis. The primary physiological regulator appears to be not citrate but long-chain fatty acyl-CoAs, which promote depolymerization of the active form.

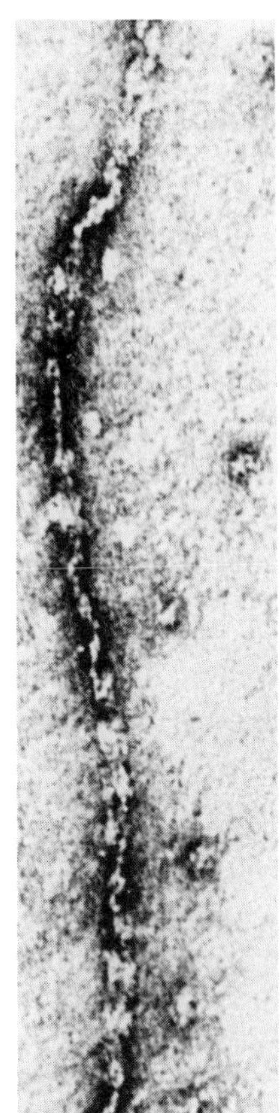

FIGURE **18.25**

The active filamentous form of eukaryotic acetyl-CoA carboxylase.

Reprinted with permission from M. D. Lane, J. Moss, and S. E. Polakis, *Curr. Top. Cell. Regul.* (1974) 15:144–145. Academic Press, Inc.

Acyl Carrier Protein

All of the intermediates in fatty acid oxidation are activated via their linkage to a carrier molecule, coenzyme A. A similar activation is involved in fatty acid synthesis, but the carrier is different. It is a small protein (77 residues in *E. coli*) called **acyl carrier protein (ACP).** The chemistry of activation is identical to that in acyl-CoAs. Recall from Chapter 14 that the reactive sulfhydryl group in CoA is part of a phosphopantetheine moiety, derived from pantothenic acid. Similarly, in ACP, a phosphopantetheine moiety is linked to a serine group in the polypeptide (Figure 18.26). It was not immediately clear why the synthetic pathway should use a protein carrier, but later research showed that a swinging arm formed from part of a protein is involved in the subsequent steps of fatty acid synthesis, as well as in the formation of malonyl-CoA.

ACP becomes involved in fatty acid synthesis through the actions of **malonyl-CoA-ACP transacylase** and **acetyl-CoA-ACP transacylase** (see Figure 18.24, reactions 2 and 3). In each case the acyl group is transferred from acyl-CoA to ACP.

FIGURE **18.26**

Phosphopantetheine as the reactive unit in ACP and CoA.

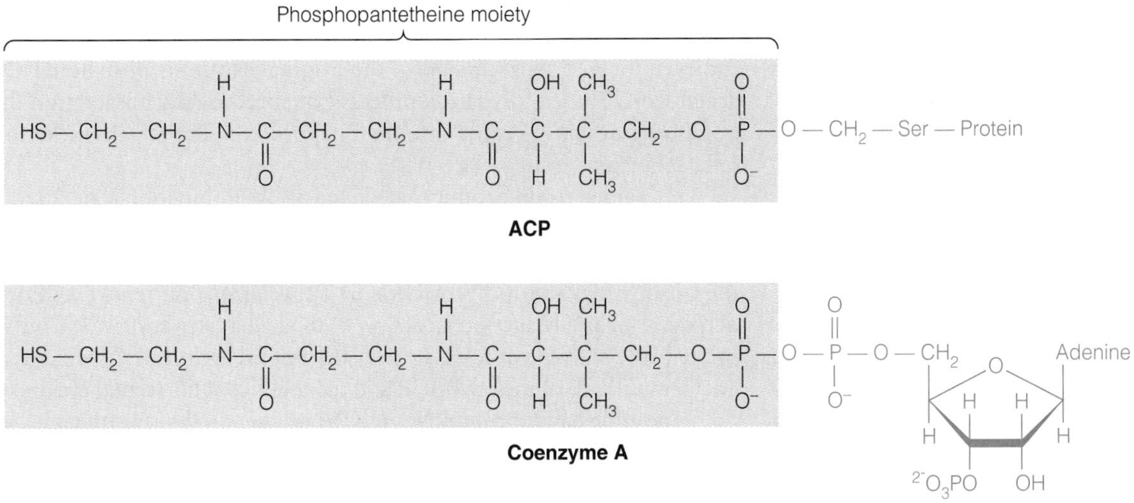

Phosphopantetheine moiety

HS—CH$_2$—CH$_2$—N(H)—C(=O)—CH$_2$—CH$_2$—N(H)—C(=O)—C(OH)(H)—C(CH$_3$)(CH$_3$)—CH$_2$—O—P(=O)(O$^-$)—O—CH$_2$—Ser—Protein

ACP

HS—CH$_2$—CH$_2$—N(H)—C(=O)—CH$_2$—CH$_2$—N(H)—C(=O)—C(OH)(H)—C(CH$_3$)(CH$_3$)—CH$_2$—O—P(=O)(O$^-$)—O—P(=O)(O$^-$)—O—CH$_2$—[ribose: H, H, H, H, $^{2-}$O$_3$PO, OH]—Adenine

Coenzyme A

Because the energy-rich bonds in acyl-CoAs and acyl-ACPs are identical, these reactions are readily reversible.

$$\text{Acetyl-CoA} + \text{ACP} \rightleftharpoons \text{acetyl-ACP} + \text{CoA-SH}$$

$$\text{Malonyl-CoA} + \text{ACP} \rightleftharpoons \text{malonyl-ACP} + \text{CoA-SH}$$

Although the malonyl transacylase is highly specific, the acetyl transacylase can react to some extent with other acyl-CoA substrates. In fact, this is how the synthesis of odd-chain fatty acids begins, with propionyl-CoA as the substrate, instead of acetyl-CoA.

From Malonyl-ACP to Palmitate

As noted earlier, the fatty acid chain is built up by successive additions of two-carbon units. Each cycle of addition consists of seven reactions, starting with acetyl-CoA carboxylase. The reaction pathway is identical in all known organisms, but the protein chemistry involved is startlingly variable. In *E. coli,* in other bacteria, and in plants the reactions are catalyzed by seven distinct enzymes, which can be separately purified. By contrast, in animals and in lower eukaryotes all of the activities are associated in a highly structured multienzyme complex called **fatty acid synthase.** Let us first focus on the seven reactions and then consider the nature of the fatty acid synthase complex.

The first three reactions we have already presented—those catalyzed by acetyl-CoA carboxylase, malonyl-CoA-ACP transacylase, and acetyl-CoA-ACP transacylase (see Figure 18.24). These reactions are identical in each two-carbon addition cycle. For the first cycle of synthesis (Figure 18.27, reactions 4–7), we start with 1 mole each of malonyl-ACP and acetyl-ACP, and in four reactions we generate 1 mole of butyryl-ACP. These are the reactions that resemble the reactions (in reverse) of fatty acid oxidation (see Figure 18.23). The synthetic cycle proceeds via *condensation, reduction, dehydration, reduction,* whereas oxidation (in reverse) involves *thiolytic cleavage, dehydrogenation, hydration, dehydrogenation.* Some of the major differences are evident from Figure 18.23. ACP is the acyl carrier for synthesis, and NADPH is the electron carrier for both reductive steps. Also, the use of a malonyl group has no counterpart in fatty acid oxidation.

Malonyl-CoA represents an activated source of two-carbon fragments for fatty acid biosynthesis, with the loss of CO_2 driving C—C bond formation.

What is the molecular logic of using malonyl-ACP as a donor of an acetyl unit? Probably it is that the condensation of two activated acetyl units is quite endergonic. A comparable reaction in reverse—namely the thiolytic cleavage of acetoacetyl-CoA—is strongly exergonic. However, the carboxyl group of malonyl-ACP is a good leaving group. In the β-ketoacyl-ACP synthase reaction (Figure 18.27, reaction 4), the acetyl group is first transferred from ACP to a cysteine thiol on the enzyme. Next, malonyl-ACP decarboxylates, activating its methylene carbon as a nucleophile, which attacks the electrophilic keto carbon of the acetyl group. The involvement of ATP to drive this endergonic reaction is apparent but indirect, for ATP participated in the original synthesis of malonyl-CoA from acetyl-CoA. This condensation process explains the early observation that bicarbonate is not incorporated into the final product. Rather, all of the carbons in fatty acids come from acetate.

The condensation product, a β-ketoacyl-ACP thioester, is next reduced to a D-3-hydroxyacyl-ACP (Figure 18.27, reaction 5). By contrast, the 3-hydroxyacyl-CoAs produced in fatty acid *oxidation* have the L configuration. Dehydration of the D-3-hydroxyacyl-ACP (reaction 6) yields a *trans*-$\Delta 2$ -enoyl-ACP, which undergoes a second reduction (reaction 7) to yield a fatty acyl-ACP—butyryl-ACP in the first cycle of synthesis. To start the second cycle, butyryl-ACP reacts with another molecule of malonyl-ACP, and the product of the second cycle is hexanoyl-ACP. The same pattern continues until the product of the seventh cycle, palmitoyl-ACP, undergoes hydrolysis to yield palmitate and free ACP.

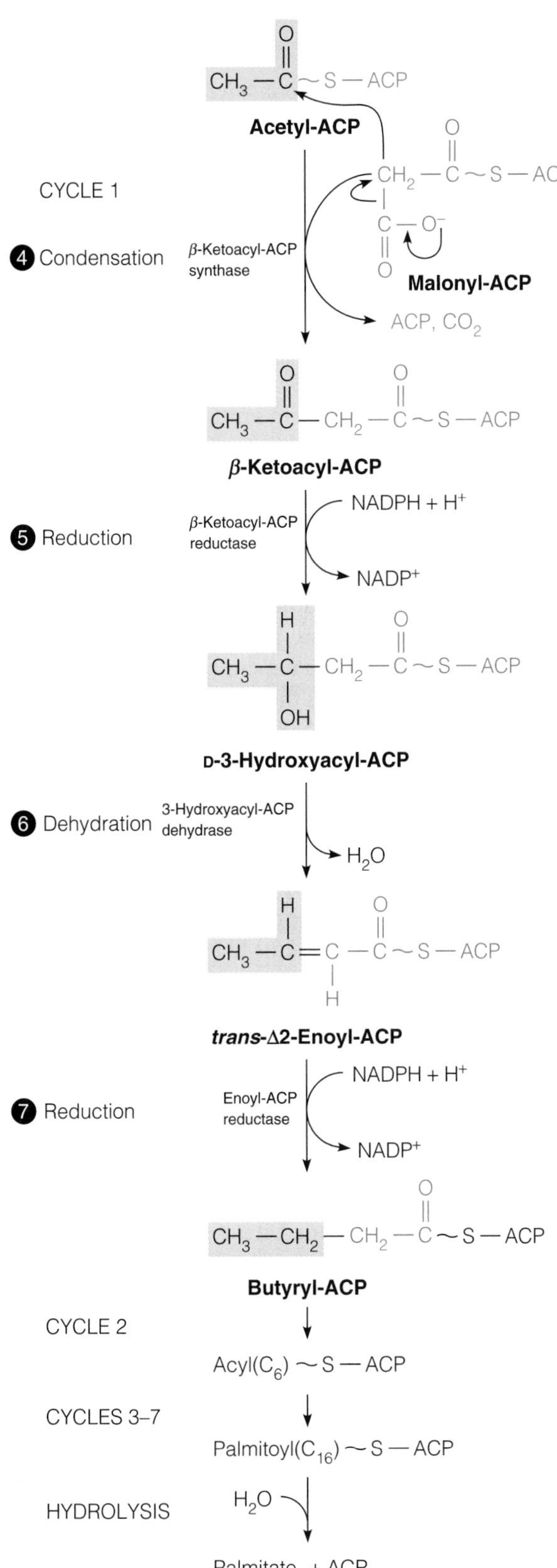

FIGURE **18.27**

Synthesis of palmitate, starting with malonyl-ACP and acetyl-ACP. The first cycle of four reactions generates butyryl-ACP, which reacts with a second mole of malonyl-ACP, leading to a second cycle of two-carbon addition. A total of seven such cycles generates palmitoyl-ACP. Hydrolysis of this product releases palmitate. When fatty acid synthesis is carried out by a multienzyme complex, as it is in eukaryotes, the acyl group (acetyl in cycle 1, butyryl in cycle 2, etc.) is carried not on ACP, as shown here, but on the ketoacyl-ACP synthase domain of the complex.

Like most biosynthetic pathways, this one requires both *energy* (as ATP) and *reducing equivalents* (as NADPH). The quantitative requirements can be seen from the stoichiometry of the complete seven-cycle process:

$$\text{Acetyl-CoA} + 7\ \text{malonyl-CoA} + 14\text{NADPH} + 20\text{H}^+ \longrightarrow \text{palmitate} + 7\text{CO}_2 + 14\text{NADP}^+ + 8\text{CoA-SH} + 6\text{H}_2\text{O}$$

To see the ATP requirement, we must consider the synthesis of the 7 moles of malonyl-CoA:

$$7\ \text{Acetyl-CoA} + 7\text{CO}_2 + 7\text{ATP} \longrightarrow 7\ \text{malonyl-CoA} + 7\text{ADP} + 7\text{P}_i + 7\text{H}^+$$

Hence, the following equation describes the overall process.

$$8\ \text{Acetyl-CoA} + 7\text{ATP} + 14\text{NADPH} + 13\text{H}^+ \longrightarrow \text{palmitate} + 14\text{NADP}^+ + 8\text{CoA-SH} + 6\text{H}_2\text{O} + 7\text{ADP} + 7\text{P}_i$$

In eukaryotes, fatty acid synthesis is carried out by an organized multienzyme complex that contains multifunctional proteins.

Multifunctional Proteins in Fatty Acid Synthesis

As noted above, the enzymes of fatty acid synthesis constitute a tightly coupled multienzyme complex in eukaryotic cells. In the mid-1970s both biochemical and genetic studies in yeast revealed that the complex actually contained two multifunctional polypeptide chains. This complex has a mass of about 2.2 million daltons and can readily be visualized by electron microscopy (Figure 18.28). The complex contains six molecules each of two polypeptide chains, which are called subunit A and subunit B. Subunit A ($M_r = 185{,}000$) contains the acyl carrier protein, the condensing enzyme, and the β-ketothioester reductase, and subunit B ($M_r = 175{,}000$) contains the remaining four activities—truly, a *multifunctional* protein. In vertebrate tissues the complex is smaller, with two identical subunits of about 240,000 daltons each. Evidently each subunit contains an ACP region plus all of the enzyme activities involved. An activity catalyzing the final release of palmitate is present as well. The yeast complex lacks the latter activity, because the final product of fatty acid synthase activity in yeast is not palmitate but palmitoyl-CoA.

It is interesting that in eukaryotic complexes the ACP is not a small protein, as it is in bacteria, but apparently a specific domain of the vastly larger polypeptide chain that makes up the complex. It seems likely that the bound phosphopantetheine serves as a swinging arm to bring the acyl moiety into contact with the several active sites, just as we discussed for acetyl-CoA carboxylase and for the pyruvate dehydrogenase complex (Figure 18.29). There is already considerable information available on the architecture of the multifunctional protein. Limited proteolysis of chicken liver fatty acid synthase yields products that represent globular domains of the multifunctional protein, each with specific catalytic activities. These plus more recent physical studies yield a model of the fatty acid synthase, as shown in Figure 18.30.

It has long been known that intermediates between acetyl-CoA and palmitate do not accumulate in cells that are synthesizing fatty acids. The basis for this is clear, if the intermediates are covalently bound to the ACP domain of a multifunctional protein, as they are in eukaryotic cells. This arrangement ensures that substrates need not seek catalytic sites by simple diffusion. Moreover, because the genetic apparatus needs to regulate the synthesis of only one or two polypeptide chains instead of seven, this type of organization is attractive from the standpoint of coordinating the activities involved. Multifunctional proteins have now been described in most major metabolic processes, although few other enzymes carry as many as four activities on one polypeptide chain.

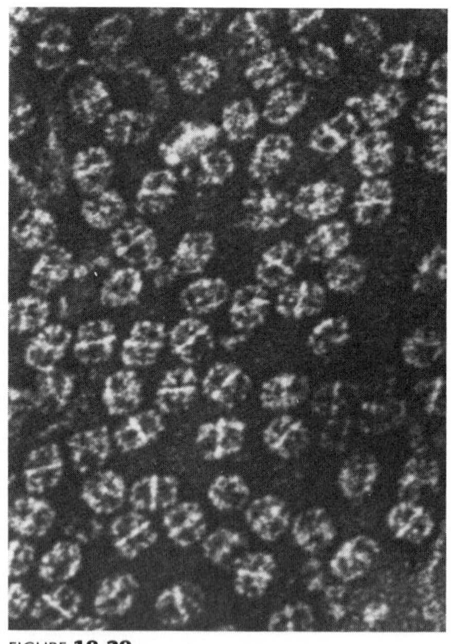

FIGURE **18.28**

The fatty acid synthase complex from yeast. The electron micrograph is magnified ×200,000.

Courtesy of S. J. Wakil, J. K. Stoops, and the late V. C. Joshi, reproduced with permission from *Annu. Rev. Biochem.* (1983) 52:537–580.

Transport of Acetyl Units and Reducing Equivalents into the Cytosol

Because acetyl-CoA is generated in the mitochondrial matrix, it must be transported to the cytosol for use in fatty acid synthesis. Like longer-chain acyl-CoAs, acetyl-CoA cannot penetrate the inner membrane. A shuttle system is used, which is interesting both because it provides a control mechanism for fatty acid synthesis and because it generates much of the NADPH needed for the process. The shuttle involves citrate, which is formed in mitochondria from acetyl-CoA and

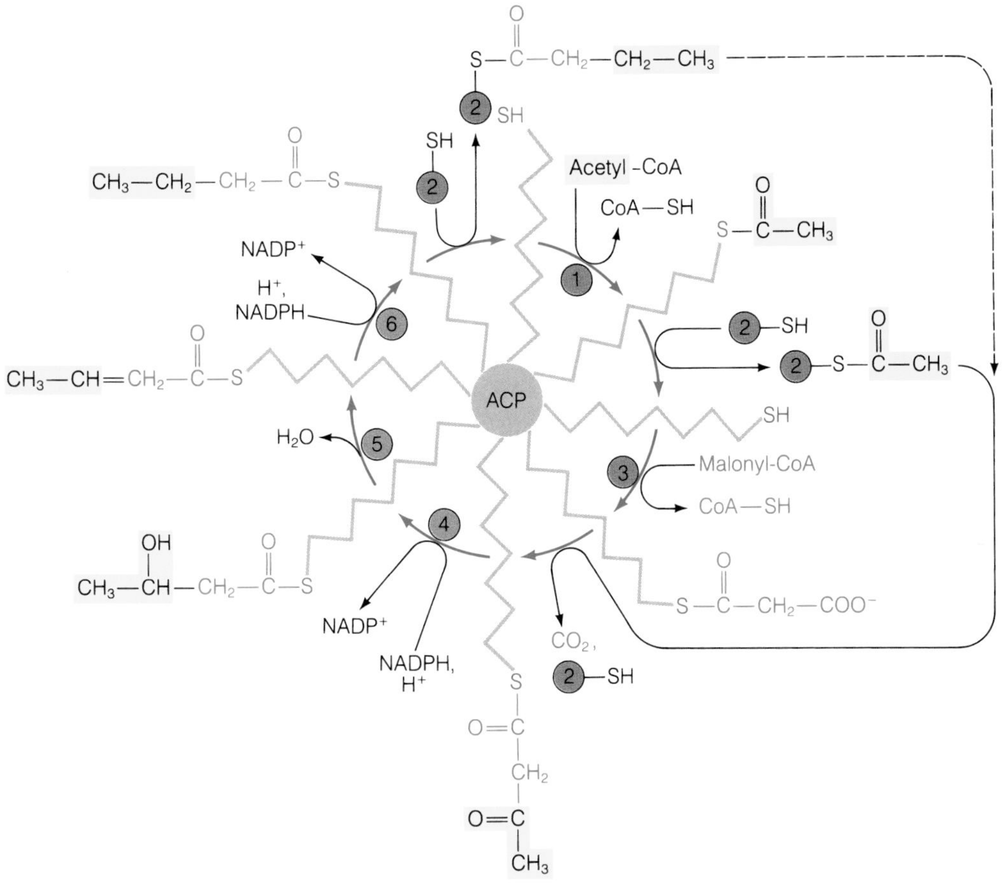

FIGURE **18.29**

A swinging arm mechanism in the eukaryotic fatty acid synthase complex. A phosphopantetheine swinging arm serves as a mechanism for bringing acyl groups into contact with all of the active sites of fatty acid synthesis. The individual active sites (numbered circles) are (1) acetyl-CoA-ACP transacylase, (2) β-ketoacyl-ACP synthase, (3) malonyl-CoA-ACP transacylase, (4) β-ketoacyl-ACP reductase, (5) β-hydroxyacyl-ACP dehydrase, and (6) enoyl-ACP reductase. The cycle begins with transfer of the acetyl group from acetyl-CoA to the phosphopantetheine swinging arm (site 1). The acetyl group is then transferred to a cysteine thiol group on β-ketoacyl-ACP synthase (site 2). At the end of the first cycle, the resultant butyryl group is transferred to the same cysteine thiol group, so that another round (dashed line) can begin by transfer of a malonyl group from malonyl-CoA to the phosphopantetheine swinging arm.

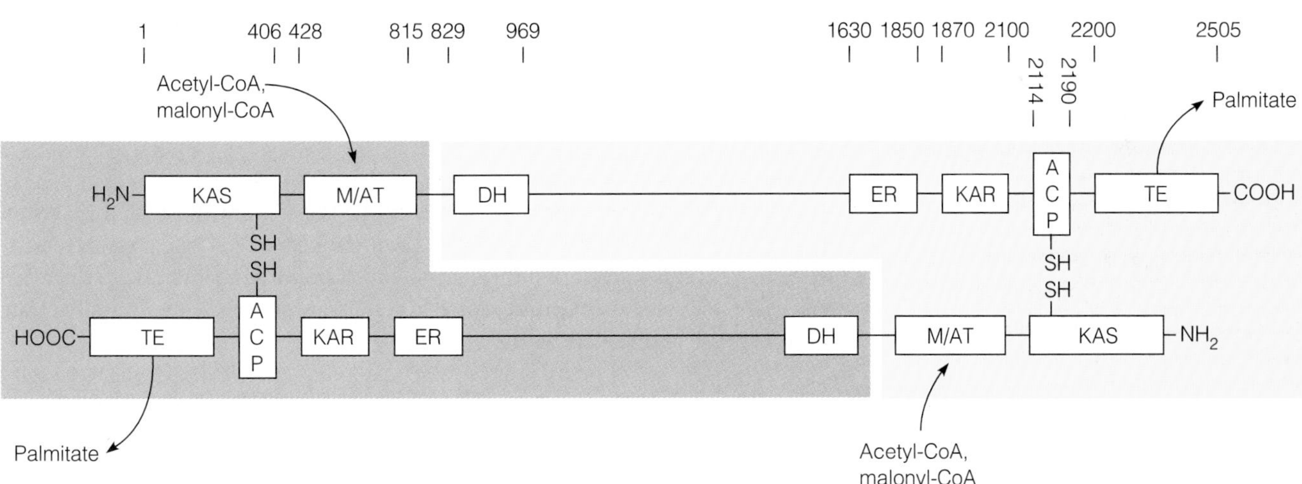

FIGURE **18.30**

Domain structure of the vertebrate fatty acid synthase complex. In the rat enzyme, two identical 2505-residue subunits are lined up head to tail. The residue numbers encompassing each functional domain are shown at the top of the figure. Abbreviations (listed in order of the reaction sequence in the synthetic pathway): M/AT = malonyl/acetyl transacylase; ACP = acyl carrier protein; KAS = ketoacyl-ACP synthase; KAR = ketoacyl-ACP reductase; DH = hydroxyacyl-ACP dehydrase; ER = enoyl-ACP reductase; TE = thioesterase (the hydrolytic enzyme that converts palmitoyl-CoA to palmitate). The complex includes two functional units, each comprising domains, as indicated by the colored lines surrounding each functional unit, on each of the two polypeptide chains. Note that the three-dimensional structure of the complex is quite different from the simple representation in this diagram, because each N-terminal dehydrase domain (DH) must interact with intermediates bound more than 1000 residues away.

Adapted from A. K. Joshi, A. Witkowski, and S. Smith, *Biochemistry* (1997) 36:2316–2322. © 1997 American Chemical Society.

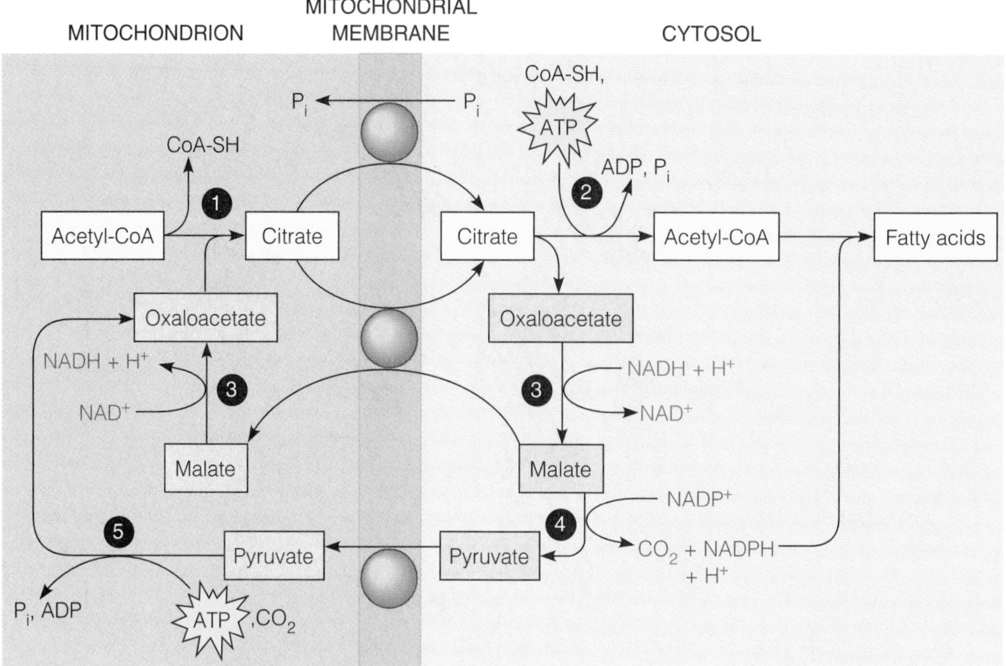

FIGURE **18.31**

Transport of acetyl units and reducing equivalents used in fatty acid synthesis. This diagram shows the shuttle mechanism for transferring acetyl units and reducing equivalents from mitochondria to the cytosol, for use in fatty acid synthesis. Citrate must be exchanged for a carrier as it moves out of the mitochondrion. Some citrate is evidently exchanged for orthophosphate and some for malate. The malate that is not exchanged generates some of the NADPH for fatty acid synthesis, through action of the malic enzyme. Purple circles represent transport systems located in the mitochondrial membrane. 1 = citrate synthase; 2 = citrate lyase; 3 = malate dehydrogenase; 4 = malic enzyme; 5 = pyruvate carboxylase.

Citrate serves as a carrier of two-carbon fragments from mitochondria to cytosol for fatty acid biosynthesis.

oxaloacetate in the first step of the citric acid cycle (Figure 18.31, step 1). When citrate is being generated in excess of the amount needed for oxidation in the citric acid cycle, it is transported through the mitochondrial membrane to the cytosol. There it is acted on by **citrate lyase** to regenerate acetyl-CoA and oxaloacetate at the expense of one ATP (step 2):

$$\text{Citrate} + \text{ATP} + \text{CoA-SH} \longrightarrow \text{acetyl-CoA} + \text{ADP} + \text{P}_i + \text{oxaloacetate}$$

Oxaloacetate cannot directly return to the mitochondrial matrix, because the inner membrane lacks a transporter for this compound. First it is reduced by a cytosolic malate dehydrogenase to malate (step 3), and some malate is oxidatively decarboxylated by the malic enzyme to give pyruvate (step 4; the malic enzyme is working here opposite to the direction shown in Chapter 14, page 508). Some of the malate formed returns to the mitochondrion and is exchanged for citrate, however.

$$\text{Oxaloacetate} + \text{NADH} + \text{H}^+ \longrightarrow \text{malate} + \text{NAD}^+$$

$$\text{Malate} + \text{NADP}^+ \longrightarrow \text{pyruvate} + \text{CO}_2 + \text{NADPH} + \text{H}^+$$

The resultant pyruvate is transported back into mitochondria, where it is reconverted to oxaloacetate by pyruvate carboxylase (step 5).

$$\text{Pyruvate} + \text{CO}_2 + \text{ATP} + \text{H}_2\text{O} \longrightarrow \text{oxaloacetate} + \text{ADP} + \text{P}_i + \text{H}^+$$

The net reaction catalyzed by these three enzymes is as follows:

$$NADP^+ + NADH + ATP + H_2O \longrightarrow NADPH + NAD^+ + ADP + P_i + H^+$$

For each mole of malate remaining in the cytosol, 1 mole of NADPH is generated. Most of the remainder of the 14 moles of NADPH required to synthesize 1 mole of palmitate is generated in the cytosol via the pentose phosphate pathway.

ELONGATION OF FATTY ACID CHAINS

Because fatty acid synthase action leads primarily to palmitate, we must consider the processes that lead from palmitate to give the variations observed among fatty acids in both chain length and degree of unsaturation. In eukaryotic cells, elongation occurs in both mitochondria and endoplasmic reticulum. The latter, so-called microsomal, system has far greater activity and is the one described here. It is similar to the fatty acid synthase sequence that leads to palmitate, but it involves acyl-CoA derivatives and separate enzymes. The first reaction is a condensation between malonyl-CoA and a long-chain fatty acyl-CoA substrate.

$$\underset{\textbf{Acyl-CoA}}{R-\overset{O}{\overset{\|}{C}}-S-CoA} + \underset{\textbf{Malonyl-CoA}}{^-OOC-CH_2-\overset{O}{\overset{\|}{C}}-S-CoA} \longrightarrow \underset{\boldsymbol{\beta}\textbf{-Ketoacyl-CoA}}{R-\overset{O}{\overset{\|}{C}}-CH_2-\overset{O}{\overset{\|}{C}}-S-CoA} + CoA-SH + CO_2$$

The resultant β-ketoacyl-CoA undergoes NADPH-dependent reduction, dehydration of the resultant hydroxyacyl-CoA, and another NADPH-dependent reduction to give a saturated fatty acyl-CoA two carbons longer than the original substrate.

$$\underset{\boldsymbol{\beta}\textbf{-Ketoacyl-CoA}}{R-\overset{O}{\overset{\|}{C}}-CH_2-\overset{O}{\overset{\|}{C}}\sim S-CoA} \xrightarrow{NADPH + H^+ \;\searrow\nearrow\; NADP^+} \underset{\boldsymbol{\beta}\textbf{-Hydroxyacyl-CoA}}{R-\overset{OH}{\overset{|}{\underset{H}{\underset{|}{C}}}}-CH_2-\overset{O}{\overset{\|}{C}}\sim S-CoA} \xrightarrow{\nearrow H_2O}$$

$$\underset{\textbf{Enoyl-CoA}}{R-CH{=}CH-\overset{O}{\overset{\|}{C}}\sim S-CoA} \xrightarrow{NADPH + H^+ \;\searrow\nearrow\; NADP^+} \underset{\textbf{Acyl-CoA}}{R-CH_2-CH_2-\overset{O}{\overset{\|}{C}}\sim S-CoA}$$

At least two different condensing enzymes are present in endoplasmic reticulum, one of which acts on unsaturated fatty acyl-CoAs. Apparently, a single set of enzymes carries out the remaining three reactions.

FATTY ACID DESATURATION

The most common monounsaturated fatty acids in animal lipids are oleic acid, an 18:1cΔ9 acid, and palmitoleic acid, a 16:1cΔ9 compound (see Table 10.1, page 317). These compounds are synthesized from stearate and palmitate, respectively, by a microsomal system called **fatty acyl-CoA desaturase** (Figure 18.32). The overall reaction for stearoyl-CoA desaturation is as follows:

$$CH_3-(CH_2)_7-CH{=}CH-(CH_2)_7-COOH$$
Oleic acid

$$CH_3-(CH_2)_5-CH{=}CH-(CH_2)_7-COOH$$
Palmitoleic acid

$$\text{Stearoyl-CoA} + NADH + H^+ + O_2 \longrightarrow \text{oleyl-CoA} + NAD^+ + 2H_2O$$

Note that both substrates undergo two-electron oxidations in this reaction. The overall electron transfer in this reaction involves another enzyme, the flavin-dependent **cytochrome b_5 reductase.**

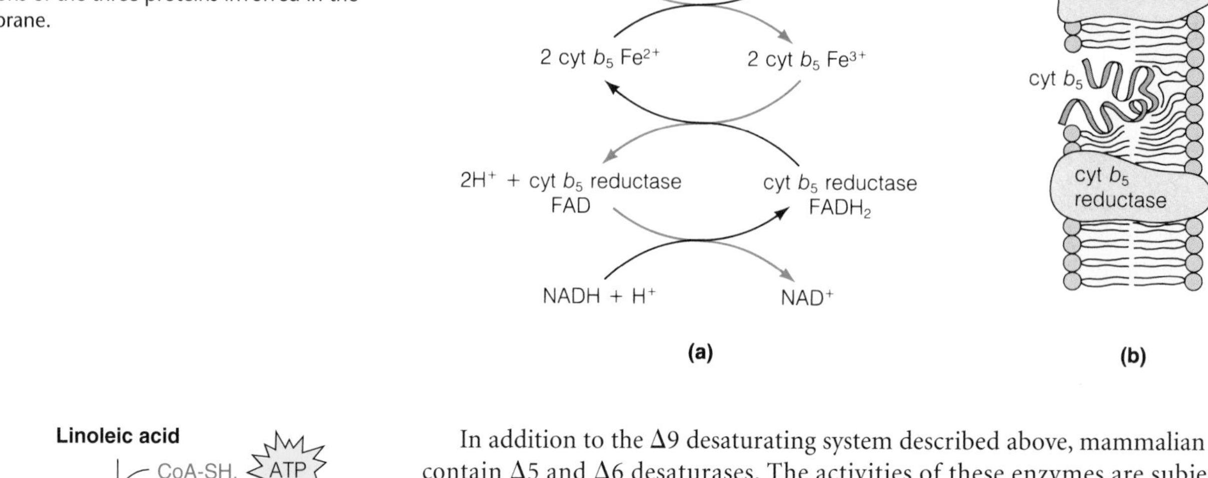

FIGURE **18.32**

Fatty acid desaturation system. (a) The redox reactions involved. **(b)** A model for the locations of the three proteins involved in the membrane.

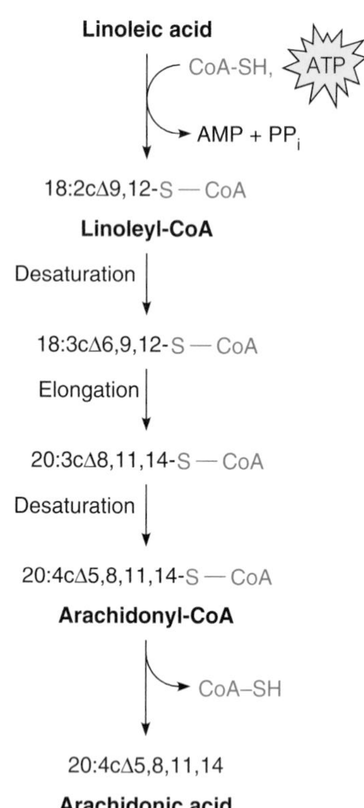

FIGURE **18.33**

Pathway for conversion of linoleic acid to arachidonic acid in mammals. Although arachidonyl-CoA can be cleaved as shown to give arachidonic acid, most of the arachidonyl-CoA produced in this pathway is used for phospholipid synthesis. The arachidonic acid for eicosanoid biosynthesis is derived from hydrolysis of these phospholipids.

In addition to the Δ9 desaturating system described above, mammalian cells contain Δ5 and Δ6 desaturases. The activities of these enzymes are subject to complex hormonal control. All three activities are enhanced by insulin, but other hormones have differential effects—activating one desaturase and inhibiting others. The significance of these effects remains a focus of active investigation.

Mammals are unable to introduce double bonds beyond Δ9 in the fatty acid chain. Hence, they cannot synthesize either linoleic acid (18:2cΔ9,12) or linolenic acid (18:3cΔ9,12,15). These are called **essential fatty acids** because they are essential lipid components that must be provided in the diet. After ingestion in mammals, they are, in turn, substrates for further desaturation and elongation reactions. Particularly important is the pathway summarized in Figure 18.33, which leads from linoleic acid to **arachidonic acid** (20:4cΔ5,8,11,14). Arachidonic acid is the precursor to a class of compounds called the *eicosanoids.* As discussed in Chapter 19, eicosanoids include two important classes of metabolic regulators, the prostaglandins and the thromboxanes.

CONTROL OF FATTY ACID SYNTHESIS

To a large extent, fatty acid biosynthesis is controlled by hormonal mechanisms. Much of the fatty acid synthesis in animals takes place in adipose tissue, where fat is being stored for release and transport to other tissues on demand, to help meet their energy needs. As extracellular messengers, hormones are well suited to these interorgan regulatory roles.

Figure 18.34 summarizes the major effects in regulation of fatty acid synthesis in animal cells. Insulin acts in several ways to stimulate fatty acid synthesis. One of its effects is to stimulate glucose entry into cells. This effect increases flux through glycolysis and the pyruvate dehydrogenase reaction, which provides acetyl-CoA for fatty acid synthesis. Insulin also activates the pyruvate dehydrogenase complex, by stimulating its dephosphorylation to the active form (see page 490). The mechanism of this effect of insulin is not clear, but it may involve increases in levels of Ca^{2+}, which stimulates pyruvate dehydrogenase phosphatase.

Another site for regulation (not shown in the figure) is the transfer of acetyl units from the mitochondrial matrix to the cytosol, where fatty acid synthesis occurs. There is evidence that citrate lyase, a key enzyme in this process (page 658), is controlled by phosphorylation and dephosphorylation of the enzyme protein, but the physiological mechanisms involved are not yet known.

The first enzyme whose action is committed to fatty acid synthesis is acetyl-CoA carboxylase. Activities of this enzyme are quite low in starved animals, probably reflecting hormonal control mediated via cyclic AMP. Phosphorylation of acetyl-CoA carboxylase by cyclic AMP–dependent protein kinase tends to inacti-

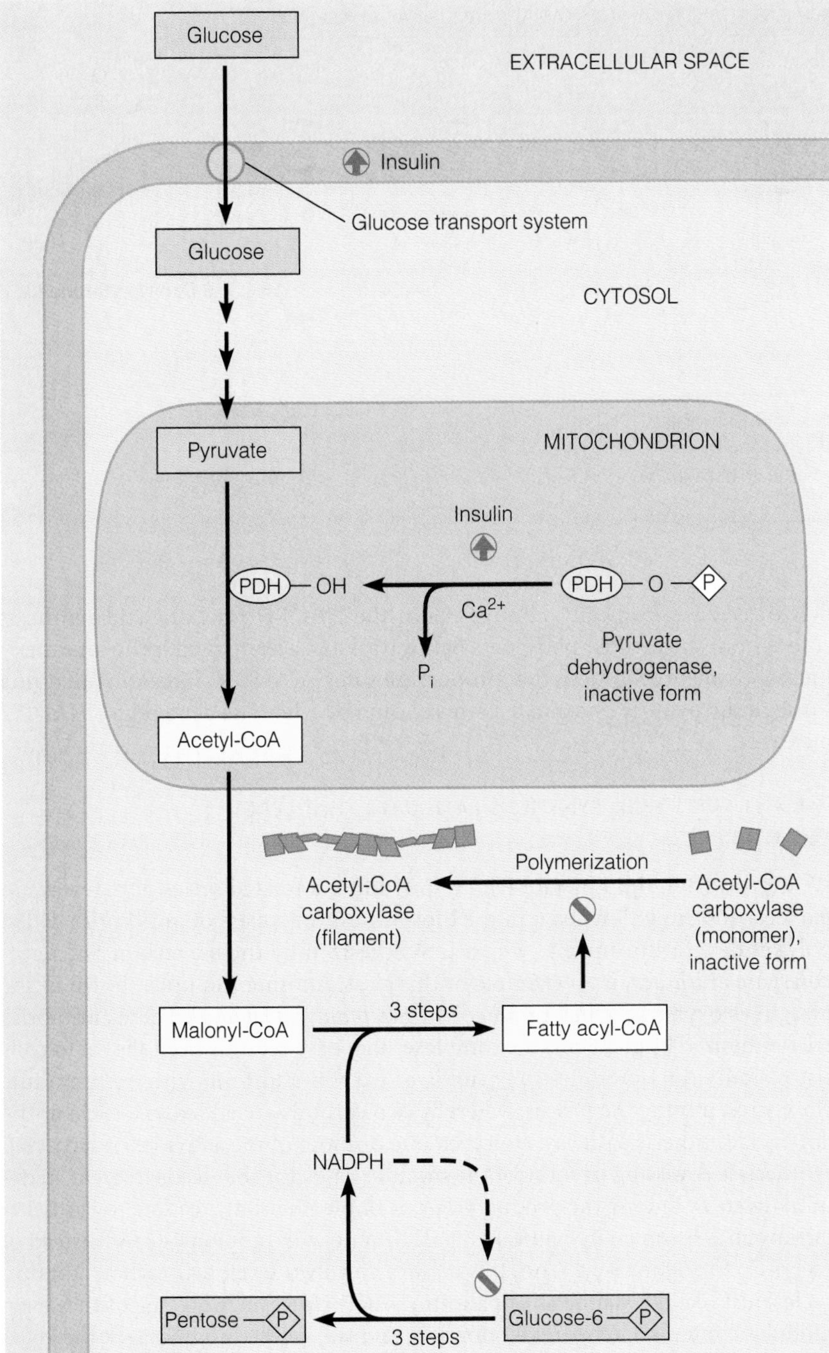

FIGURE **18.34**

Regulation of fatty acid synthesis in animal cells, such as liver cells. The dephosphorylated form of pyruvate dehydrogenase is the enzymatically active form. In adipocytes, the lack of glycerol kinase makes essential the insulin-controlled uptake of glucose, in order to generate glycerol-3-phosphate for triacylglycerol synthesis.

vate the enzyme. Also, as noted earlier, acetyl-CoA carboxylase must undergo a reversible polymerization in order to be active. Long-chain fatty acyl-CoAs at low levels prevent polymerization, thereby inactivating the enzyme and providing an apparent feedback inhibition of the pathway. The levels of fatty acyl-CoAs are lowered by insulin, another mechanism by which insulin stimulates fatty acid synthesis.

Acetyl-CoA carboxylase is the committed enzyme and major control point for fatty acid synthesis.

Finally, there is evidence that fatty acid synthesis is controlled by the availability of reducing equivalents. Recall that NADPH comes from both the transport of citrate out of mitochondria (see page 658) and the pentose phosphate pathway (see Chapter 14). The pentose phosphate pathway in turn is controlled through inhibition by NADPH of glucose-6-phosphate dehydrogenase and 6-phosphogluconate

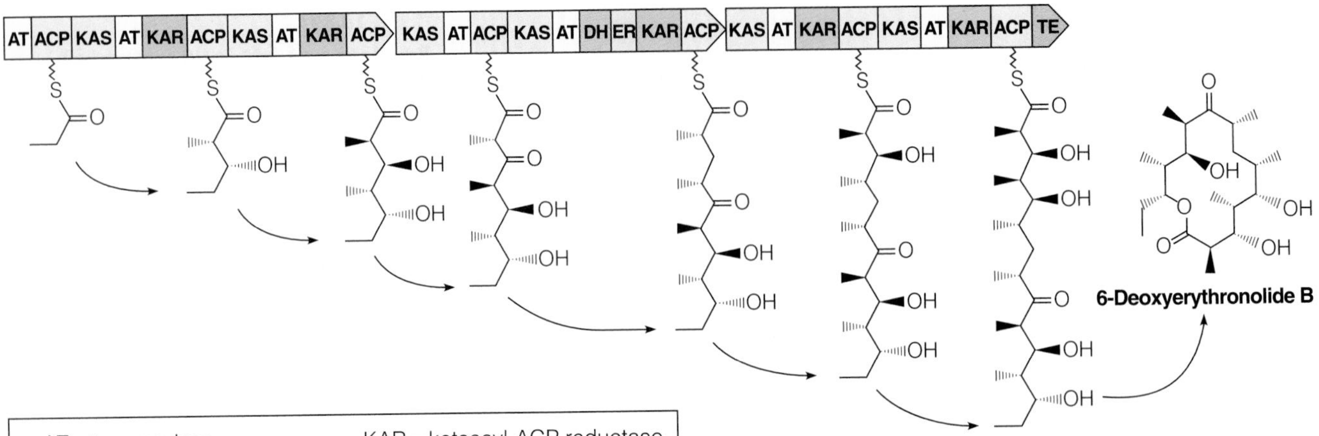

AT ACP KAS AT KAR ACP KAS AT KAR ACP KAS AT ACP KAS AT DH ER KAR ACP KAS AT KAR ACP KAS AT KAR ACP TE

6-Deoxyerythronolide B

AT = transacylase
ACP = acyl carrier protein
KAS = ketoacyl-ACP synthase
KAR = ketoacyl-ACP reductase
DH = dehydrase
ER = enoyl-ACP reductase
TE = thioesterase

FIGURE **18.35**

Biosynthetic route leading to erythromycin and related antibiotics. This process, as it occurs in *Saccharopolyspora erythraea,* involves three multifunctional proteins, each of which is shown with a separate arrow and with the sequential enzyme activities involved. Abbreviations are the same as in Figure 18.30, except for the final activity. This **thioesterase** (TE) activity cyclizes the product in the process of removing it from the ACP domain to which it is bound. Conversion of the product, 6-deoxyerythronolide B, to erythromycin A involves two glycosylation steps.

Modified from R. L. Rawls, *Chem. Eng. News* (1998) 76(10), pp. 29–32. © 1998 American Chemical Society.

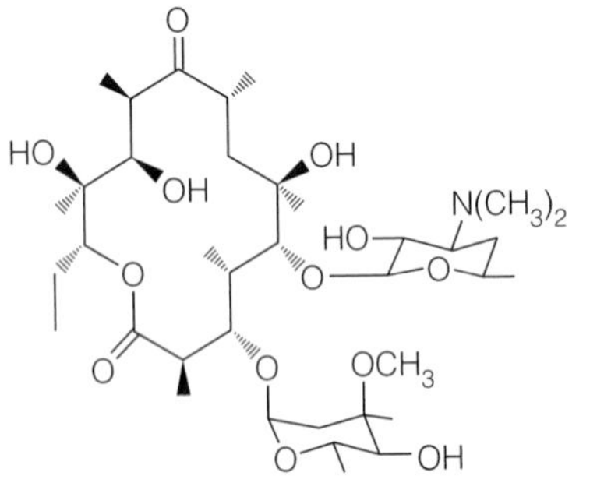

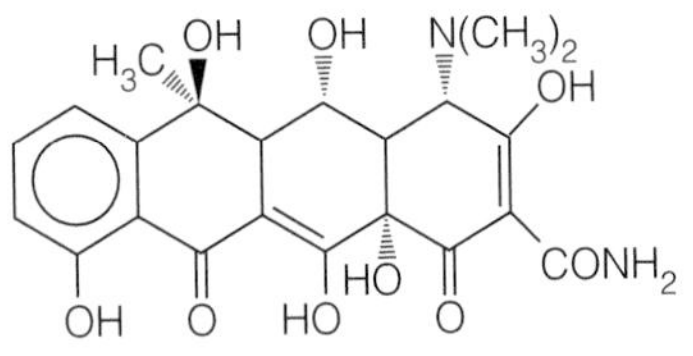

dehydrogenase. Typically, about 60% of the NADPH for fatty acid synthesis comes from the pentose phosphate pathway. When acetate is added to an experimental tissue preparation, that proportion can rise to 80%, indicating that flux through the pentose phosphate pathway can rise when the demand for NADPH increases.

VARIANT FATTY ACID SYNTHESIS PATHWAYS THAT LEAD TO ANTIBIOTICS

We digress from lipid metabolism to introduce a related series of pathways in bacteria and fungi, involved in the biosynthesis of a class of antibiotics called **polyketides.** Erythromycin, which is synthesized by the bacterium *Saccharopolyspora erythraea,* is an example of this class. Another example, shown in the margin, is oxytetracycline, produced by *Streptomyces rimosus.* These antibiotics are synthesized by giant enzyme complexes that have recently been shown to consist of individual modules for rounds of carbon addition, with each module closely resembling the process whereby two carbons are added in a cycle of the fatty acid synthesis pathway. However, one or more of the activities of fatty acid synthesis are missing in some of the modules, a factor that leads to great structural diversity among the products. For example, the route leading toward erythromycin A is shown in Figure 18.35. Beginning with propionyl-CoA (instead of acetyl-CoA) and malonyl-CoA, this pathway involves seven two-carbon addition cycles, just like the synthesis of palmitoyl-ACP. However, only one of the seven modules contains a dehydratase domain, and two of the modules lack the ketoacyl reductase domain. Thus, the product, unlike fatty acids, contains both keto and hydroxyl oxygen atoms. Much current excitement involves attempts to modify and rearrange the individual modules, by site-directed mutagenesis of the cloned genes, leading the further diversification of the pathways involved, and possibly leading to novel and useful antibiotics.

Erythromycin A

Oxytetracycline

Biosynthesis of Triacylglycerols

Fatty acyl-CoAs, along with glycerol-3-phosphate, serve as the major precursors to triacylglycerols. Glycerol-3-phosphate is derived either from the reduction of the glycolytic intermediate dihydroxyacetone phosphate, catalyzed by **glycerol phosphate dehydrogenase,** or from the ATP-dependent phosphorylation of glycerol by **glycerol kinase:**

Dihydroxyacetone phosphate + NADH + H^+ ⟶ L-glycerol-3-phosphate + NAD^+

Glycerol + ATP ⟶ L-glycerol-3-phosphate + ADP

The pathway involving DHAP predominates in adipose tissue, because adipocytes lack glycerol kinase. However formed, glycerol-3-phosphate undergoes two successive enzymatic esterifications with fatty acyl-CoAs to yield **diacylglycerol-3-phosphate:**

Fatty acyl-S-CoA + glycerol-3-phosphate ⟶ monoacylglycerol-3-phosphate + CoA-SH

Monoacylglycerol-3-phosphate + fatty acyl-S-CoA ⟶ diacylglycerol-3-phosphate + CoA-SH

Diacylglycerol-3-phosphate, also called **phosphatidic acid,** is a precursor both to phospholipids and to triacylglycerols. The pathway to triacylglycerols involves hydrolytic removal of the phosphate, followed by transfer of another fatty acyl moiety from an acyl-CoA:

Phosphatidic acid + H_2O ⟶ 1,2-diacylglycerol + P_i

1,2-Diacylglycerol + fatty acyl-CoA ⟶ triacylglycerol + CoA-SH

As we have noted, triacylglycerols represent the major form in which energy can be stored. Normally in an adult animal, synthesis and degradation are balanced, so that there is no net change in the total body amount of triacylglycerols. If dietary intake exceeds caloric needs, then proteins, carbohydrate, or fat can each readily provide acetyl-CoA to drive the synthesis of fatty acids and triacylglycerols. On the other hand, fat reserves allow animals to go for rather long times without eating and still maintain adequate energy levels. Such fasting does generate some metabolic stresses, as we describe further in Chapter 23.

Hibernating animals have adapted remarkably well to cope with such stresses. For example, bears store huge amounts of fat just before beginning a hibernation that may last as long as 7 months. During this period all of the bear's energy comes from breakdown of the stored fat. Moreover, the bear excretes so little water that the water released from fat oxidation meets the animal's needs. Similarly, the glycerol released from triacylglycerols provides a source of gluconeogenic precursors.

HO—CH₂(OH)—O—Ⓟ

Glycerol-3-phosphate

↓ R—CO—S—CoA

↓ R—CO—S—CoA

R—CO(=O)—O; —O—C(=O)—R; —O—Ⓟ

Diacylglycerol-3-phosphate (phosphatidic acid)

↓ H_2O

R—C(=O)—O; —O—C(=O)—R; —OH

1,2-Diacylglycerol

↓ R—CO—S—CoA

R—C(=O)—O; —O—C(=O)—R; —O—C(=O)—R

Triacylglycerol

Biochemical Insights into Obesity

About one-third of Americans are classified as seriously overweight, making obesity one of our most significant public health problems. Until recently, obesity was usually considered simply a consequence of overeating, leading to excessive triacylglycerol deposition in adipocytes. It is becoming clear that biochemical factors lead some individuals to be far more prone to obesity than others. In 1995 the product of the *ob* gene in mice (*ob*, short for *obese*) was identified. Mice bearing two defective alleles of the *ob* gene grow to body weights as much as three times normal. The *ob* gene specifies a 16-kDa protein called **leptin,** which is synthesized in adipocytes and acts as a hormone, binding to a specific receptor in the brain. Leptin evidently functions as a "lipostat," sensing the amount of fat stored in the adipocytes. When fat stores are adequate, leptin levels are high, and the signaling system controls feeding behavior to limit fat deposition. During starvation, leptin levels decline, which promotes feeding and fat storage within the adipocyte. *Obese* mice, lacking functional leptin, act as if perpetually starved, and their overeating makes them obese; injections of leptin lower their feeding rates and cause them to lose weight dramatically. Obese humans are different from obese mice, however, in that they contain high levels of leptin. Current research is aimed at the premise that these individuals are somehow unresponsive to normal leptin signaling.

Evidence is accumulating for other biochemical factors as significant elements in weight control. A number of hormones, including **serotonin** (page 782), control satiety, the feeling of fullness after eating. The antiobesity drug **fenfluramine,** which acts by increasing serotonin levels and affecting the appetite, was becoming quite popular until it was found to do serious damage to the heart and was withdrawn from the market. Current biochemical attention is focused upon uncoupling proteins, intramitochondrial proteins similar to those found in brown fat mitochondria (page 549), which evidently act as uncouplers of oxidative phosphorylation and, hence, decrease the amount of ATP synthesized during oxidative metabolism. Attention is also focused upon other hormonal factors and upon fatty acid–binding proteins, which evidently participate in the transport of fatty acids to different locations within the cell. Biochemical research on obesity is one of the most active current research frontiers.

SUMMARY

Triacylglycerols are the main form for storage of biological energy. In animals, dietary triacylglycerols are digested and then resynthesized as they are complexed with proteins to form chylomicrons, for transport to tissues. Triacylglycerols synthesized in liver are transported to peripheral tissues as very low-density lipoproteins. Low-density lipoproteins represent the major vehicle for transport of cholesterol to peripheral tissues. Cholesterol levels in blood are regulated through control of synthesis of LDL receptors, involved in cellular uptake of LDL by endocytosis. Faulty control of LDL levels contributes toward the development of atherosclerotic plaque.

Fat depots are mobilized by enzymatic hydrolysis of triacylglycerols to fatty acids plus glycerol. The process is hormonally controlled via cyclic AMP. Most fatty acid degradation occurs through β-oxidation, a mitochondrial process that involves stepwise oxidation and removal of two-carbon fragments as acetyl-CoA. Processing unsaturated fatty acids is a bit more complicated, because of the stereochemistry involved, but the pathways are straightforward. Under conditions in which oxidation of acetyl-CoA through the citric acid cycle is limited, acetyl-CoA is used to synthesize ketone bodies, which are excellent or satisfactory energy substrates for some organs.

Fatty acid biosynthesis occurs via the stepwise addition of two-carbon fragments, in a process that superficially resembles a reversal of β-oxidation. Metabolic activation involves acyl carrier protein, and the reductive power comes from NADPH. In eukaryotic cells the seven enzyme activities are linked covalently on multifunctional enzymes or multienzyme complexes.Fatty acid elongation beyond the C_{16} stage is mechanistically similar, but CoA derivatives are involved instead of acyl carrier protein. Unsaturated fatty acids are formed primarily by an ER-associated desaturating system. Triacylglycerols are synthesized by straightforward pathways in which acyl groups in fatty acyl-CoAs are transferred to the hydroxyl groups of glycerol-3-phosphate and diacylglycerol. Faulty hormone control of triacylglycerol deposition, along with excess dietary intake, can be responsible for obesity.

REFERENCES

General

Mead, J. F., R. B. Alfin-Slater, D. R. Howton, and G. Popják (1986) *Lipids: Chemistry, Biochemistry, and Nutrition.* Plenum, New York. A detailed and reasonably up-to-date textbook.

Töpfer, R., N. Martini, and J. Schell (1995) Modification of plant lipid synthesis. *Science* 268:681–686. Genetic engineering of plants is creating factories for production of nutritionally and industrially desirable fatty acids and oils.

Vance, D. E., and J. E. Vance, eds. (1985) *Biochemistry of Lipids and Membranes.* Benjamin/Cummings, Redwood City, Calif. This multi-authored text contains a number of detailed reviews. Particularly relevant to this chapter are the following articles: Oxidation of fatty acids, pp. 116–142; Fatty acid synthesis in eucaryotes, pp. 143–180; Fatty acid desaturation and chain elongation in eucaryotes, pp. 181–212; Metabolism of triacylglycerols, pp. 213–241; Metabolism of cholesterol and lipoproteins, pp. 404–474.

Lipid and Lipoprotein Metabolism in Animals

Brown, M., and J. L. Goldstein (1986) A receptor-mediated pathway for cholesterol homeostasis. *Science* 232:34–47. Brown and Goldstein's Nobel Prize address, chronicling the discovery and actions of the LDL receptor.

Derewenda, Z., and A. M. Sharp (1993) News from the interface: The molecular structures of triacylglyceride lipases. *Trends Biochem. Sci.* 18:20–25. Describes the conformational change that exposes a catalytic site at an oil–water interface.

Diaz, M. N., B. Frei, J. A. Vita, and J. F. Keaney, Jr. (1997) Antioxidants and atherosclerotic heart disease. *N. Engl. J. Med.* 337:408–416. A review of the processes leading to LDL oxidation and of the protective effects of antioxidants.

Havel, R. J., and J. P. Kane (1995) Introduction: Structure and metabolism of plasma lipoproteins. In: *The Metabolic and Molecular Bases of Inherited Disease,* 7th ed., edited by C. R. Scriver, A. L. Beaudet, W. S. Sly, and D. Valle, pp. 1841–1852. McGraw-Hill, New York. The first in a series of 10 chapters dealing with clinical disorders of lipid and lipoprotein metabolism.

Hodges, P., and J. Scott (1992) Apolipoprotein B mRNA editing: A new tier for the control of gene expression. *Trends Biochem. Sci.* 17:77–81. Both B apoproteins are products of the same gene, and the mRNA for the smaller protein is formed by a novel modification of the large-protein mRNA.

Nestel, P. J. (1990) Effects of n-3 fatty acids on lipid metabolism. *Annu. Rev. Nutrition* 10:149–167. A summary of the biochemical and nutritional aspects of ω-3 fatty acids and their effects on lipid metabolism.

Schmid, S. L. (1997) Clathrin-coated vesicle formation and protein sorting: An integrated process. *Annu. Rev. Biochem.* 66:511–548. A review of the biochemistry of endocytosis and protein sorting.

Steinberg, D. S. (1996) A docking receptor for HDL cholesterol esters. *Science* 271:460–461. A "Perspectives" article summarizing events in HDL interaction with cells.

Steinberg, D. S. (1997) Low density lipoprotein oxidation and its pathobiological significance. *J. Biol. Chem.* 272:20963–20966. A biochemical treatment of LDL oxidation.

Fatty Acid Metabolism

Joshi, A. K., A. Witkowski, and S. Smith (1997) Mapping of functional interactions between domains of the animal fatty acid synthase by mutant complementation *in vitro. Biochemistry* 36:2316–2322. An analysis of the functional relationships of the different domains of this multifunctional protein.

Reynolds, K. A. (1998) Combinatorial biosynthesis: Lesson learned from nature. *Proc. Natl. Acad. Sci. USA* 95:12744–12746. A minireview describing the modular nature of polyketide antibiotic synthesis and potential applications to production of novel antibiotics.

Somerville, C., and J. Browse (1991) Plant lipids: Metabolism, mutants, and membranes. *Science* 252:80–87. Genetic engineering is being used to direct plants to synthesize nutritionally or industrially important triacylglycerols and other lipids.

Tolbert, N. E. (1981) Metabolic pathways in peroxisomes and glyoxysomes. *Annu. Rev. Biochem.* 50:133–157. A complete description of the enzymology and functions of peroxisomal β-oxidation.

Wakil, S. J., J. K. Stoops, and V. C. Joshi (1983) Fatty acid synthesis and its regulation. *Annu. Rev. Biochem.* 52:537–579. A detailed discussion of the protein chemistry of the eukaryotic multifunctional proteins involved in fatty acid synthesis.

Biochemical Insights into Obesity

Campfield, L. A., F. J. Smith, and P. Burn (1998) Strategies and potential molecular targets for obesity treatment. *Science* 280:1383–1387. One of a series of articles on regulation of body weight in a special issue of *Science.*

Friedman, J. M., and J. L. Halaas (1998) Leptin and the regulation of body weight in mammals. *Nature* 395:763–770. A review of leptin and its role in preventing obesity.

Rawls, R. L. (1999) Weighing in on obesity. *Chem. Eng. News,* June 21 issue, pp. 35–44. A readable description of mechanisms and therapeutic strategies.

PROBLEMS

1. Calculate the metabolic energy yield from oxidation of palmitic acid, taking into account the energy needed to activate the fatty acid and transport it into mitochondria. Do the same for stearic acid, linoleic acid, and oleic acid.
2. If palmitic acid is subjected to complete combustion in a bomb calorimeter, one can calculate a standard free energy of combustion of 9788 kJ/mol. From the ATP yield of palmitate oxidation, what is the metabolic efficiency of this process, in terms of kilojoules saved as ATP per kilojoule released?
3. Calculate the number of ATPs generated by the complete metabolic oxidation of tripalmitin (tripalmitoylglycerol). Hydrolysis of the triacylglycerol occurs at the cell surface. Consider the energy yield from catabolism of glycerol, as well as from the fatty acids. Calculate the ATP yield per carbon atom oxidized, and compare it with the energy yield from glucose.
4. Write a balanced equation for the *complete* metabolic oxidation of each of the following. Include O_2, ADP, and P_i as reactants and ATP, CO_2, and H_2O as products.
   (a) Stearic acid
   (b) Oleic acid
   (c) Palmitic acid
   (d) Linoleic acid
5. Calculate the number of ATPs generated from the metabolic oxidation of the four carbons of acetoacetyl-CoA to CO_2. Now consider the homolog derived from oxidation of an odd-numbered carbon chain, namely propionoacetyl-CoA. Calculate the net ATP yield from oxidation of the five carbons of this compound to CO_2.
6. 2-Bromopalmitoyl-CoA inhibits the oxidation of palmitoyl-CoA by isolated mitochondria but has no effect on the oxidation of palmitoylcarnitine. What is the most likely site of inhibition by 2-bromopalmitoyl-CoA?

*7. When the identical subunits of chicken liver fatty acid synthase are dissociated in vitro, all of the activities can be detected in the separated subunits except for the β-ketoacyl synthase reaction and the overall synthesis of palmitate. Explain these observations.

8. Mammals cannot undergo *net* synthesis of carbohydrate from acetyl-CoA, but the carbons of acetyl-CoA can be incorporated into glucose and amino acids. Present pathways by which this could come about.
9. Describe a pathway whereby some of the carbon from a fatty acid with an odd-numbered carbon chain could undergo a net conversion to carbohydrate.
10. How many tritium atoms (^3H)are incorporated into palmitate when fatty acid synthesis is carried out in vitro with the following labeled substrate?

$$^{-}OOC-C\,^3H_2-\overset{O}{\overset{\|}{C}}-S-CoA$$

11. What would be the effect on fatty acid synthesis of an increase in intramitochondrial oxaloacetate level? Briefly explain your answer.
*12. Glucagon secretion causes inhibition of intracellular acetyl-CoA carboxylase activity by several mechanisms. Name all you can think of.
13. It has been observed that malonyl-CoA inhibits palmitic acid oxidation in isolated mitochondria, but that it has little or no effect on the oxidation of octanoic acid (C_8, saturated). Explain this observation.
14. Identify and briefly discuss each mechanism ensuring against simultaneous fatty acid synthesis and oxidation in the same cell.
15. Discuss the metabolic rationale for phosphorylation of acetyl-CoA carboxylase by cyclic AMP–dependent protein kinase.
16. Describe the probable effect in adipocytes of insulin-stimulated uptake of glucose into these cells.

CHAPTER 19

Lipid Metabolism II: Membrane Lipids, Steroids, Isoprenoids, and Eicosanoids

CHAPTER 18 WAS CONCERNED PRIMARILY WITH ENERGETIC ASPECTS OF LIPID metabolism—synthesis and oxidation of fatty acids, interorgan transport of lipoproteins, and metabolism of triacylglycerols. In addition to their roles in energy storage, lipids function as membrane components and as biological regulators. Our attention now shifts to the roles played by these more complex lipids, as well as pathways for their synthesis and degradation. The cell biology of some of these processes is schematized in Figure 19.1.

This chapter focuses on the following major classes: *glycerophospholipids* (also called *phosphoglycerides*), which are primarily membrane components but which play some specialized regulatory roles; *sphingolipids,* which in animals are found abundantly in nervous tissue; *steroids* and other *isoprenoid* compounds, which function as hormones, vitamins, and membrane constituents; and the *eicosanoids,* a class of biological regulators synthesized from arachidonic acid. Two major topics involving lipids and regulation are discussed later in this book: the actions of steroid hormones (in Chapters 23 and 28) and the second-messenger regulatory role of inositol phospholipids (in Chapter 23).

Metabolism of Glycerophospholipids

The most abundant phospholipids are those derived from glycerol. These **glycerophospholipids** are found primarily as components of membranes. Membrane phospholipids are also metabolic precursors to various regulatory elements of signal transduction pathways. In animals, phospholipids also participate in the transport of triacylglycerols and cholesterol, as discussed in Chapter 18, by forming the surface of lipoproteins. In addition, phospholipids play specific roles in processes as diverse as blood clotting and lung functioning. Pathways of glycerophospholipid synthesis are outlined in Figure 19.2 and are presented in more detail in the following sections. The major membrane lipids other than glycerophospholipids are the sphingolipids.

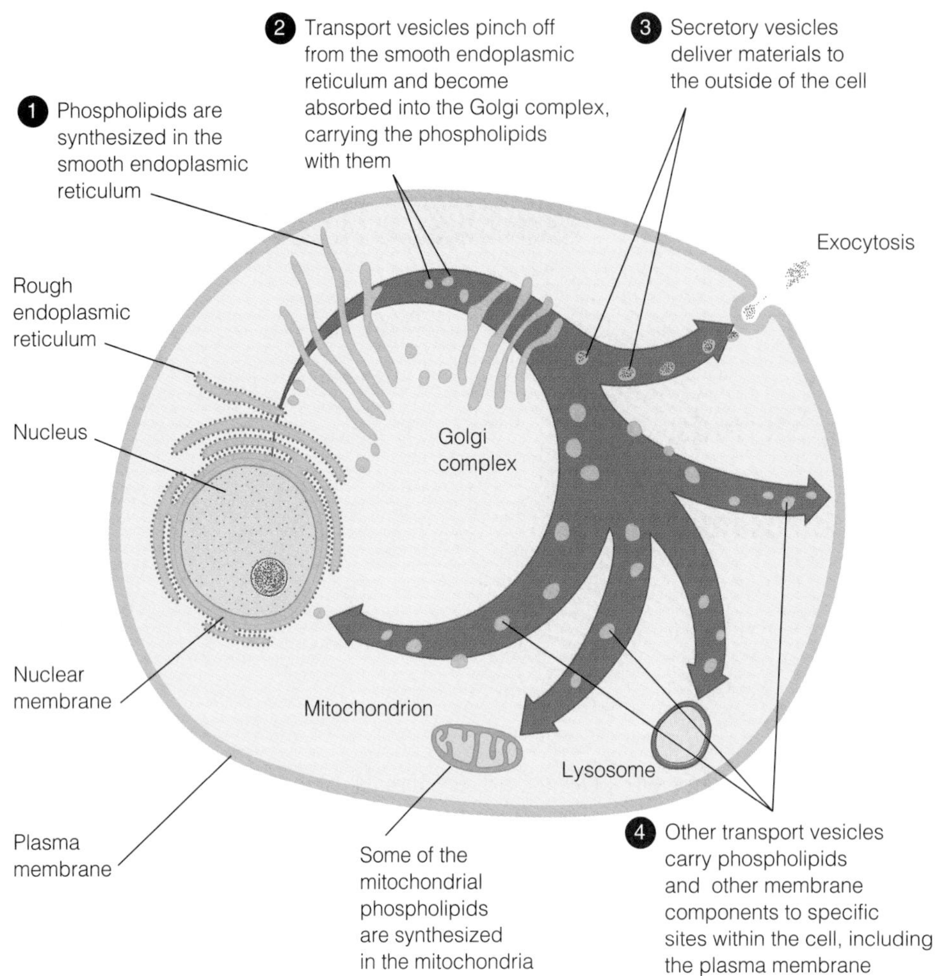

FIGURE **19.1**
Intracellular synthesis and transport of membrane phospholipids.

E. coli membranes contain just three different phospholipids, which contain predominantly three different fatty acids.

BIOSYNTHESIS OF GLYCEROPHOSPHOLIPIDS IN BACTERIA

We begin our discussion of phospholipid metabolism in the prokaryotic kingdom, partly because of the relative simplicity of the biological systems involved. In bacteria, phospholipids may constitute 10% of the dry weight of the cell, yet their only known role is as components of membranes. In *E. coli,* the most widely studied microorganism, the membranes contain but three phospholipids in significant amounts—phosphatidylethanolamine (75%–85%), phosphatidylglycerol (10%–20%), and cardiolipin (5%–15%). The fatty acid content of these lipids is also simple, with three dominating—palmitate (16:0), palmitoleate (16:1cΔ9), and *cis*-vaccenate (18:1cΔ11).

Because they are easy to grow in large quantities, bacteria provide abundant sources for large-scale isolation of the enzymes involved in lipid metabolism. Much of our earliest information on both phospholipid synthesis and fatty acid synthesis came from studies with *E. coli.* More recently, physiological studies on bacterial mutants have offered insight into control of membrane lipid synthesis and, in particular, the mechanisms used for temperature regulation of the fatty acid content of these lipids. Recall from Chapter 10 that, when grown at low temperature, bacteria increase the unsaturation of their membrane fatty acids to maintain optimal fluidity. One of the most intriguing areas of contemporary lipid biochemistry involves genetic analysis of mechanisms that maintain the optimal pattern of unsaturation at a given temperature.

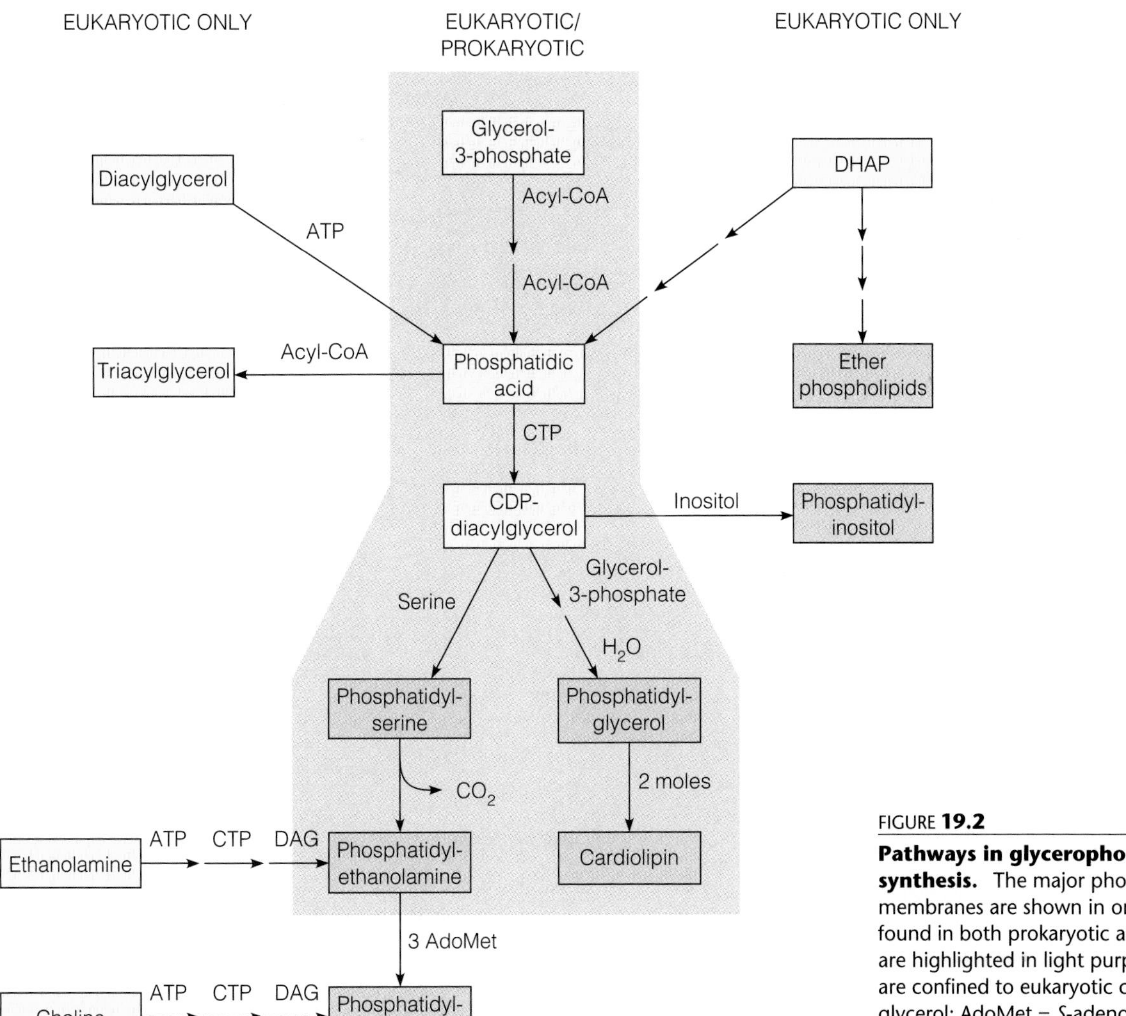

FIGURE **19.2**

Pathways in glycerophospholipid biosynthesis. The major phospholipids found in membranes are shown in orange. Pathways found in both prokaryotic and eukaryotic cells are highlighted in light purple. Other reactions are confined to eukaryotic cells. DAG = diacylglycerol; AdoMet = *S*-adenosylmethionine (see page 675).

Biosynthesis of Phosphatidic Acid and Polar Head Groups

In Chapter 18 we described the synthesis of phosphatidic acid (diacylglycerol-3-phosphate), starting from L-*sn*-glycerol-3-phosphate (see Figure 19.3 legend). As noted there and as shown in Figure 19.2, phosphatidic acid represents a branch point between the syntheses of triacylglycerols and phospholipids. The energy cofactor for phospholipid biosynthesis is cytidine triphosphate (CTP), whose role is similar to that of UTP in polysaccharide synthesis.

Phosphatidic acid is a branch metabolite between triacylglycerol biosynthesis and phospholipid biosynthesis.

Recall that phosphatidic acid is synthesized by two successive acylations of glycerol-3-phosphate. In bacteria the acyl groups are borne on acyl carrier protein (Figure 19.3). Two different acyltransferase enzymes are involved, as shown in part by their differences in specificity for fatty acyl-ACPs: About 90% of the acyl groups esterified at position 1 are saturated, whereas 90% at position 2 are unsaturated. Phosphatidic acid next becomes metabolically activated by reaction with CTP. This reaction yields **CDP-diacylglycerol,** which contains an anhydride bond linking the phosphates of cytidylic and phosphatidic acids. Mechanistically, this reaction is reminiscent of the activation of glucose-1-phosphate by UTP to yield UDP-glucose (Chapter 16).

Metabolic activation of phospholipid precursors is carried out by reaction with CTP.

CDP-diacylglycerol is now activated for the synthesis of polar head groups. In one reaction sequence (Figure 19.4, left side), CMP is exchanged for serine, giving

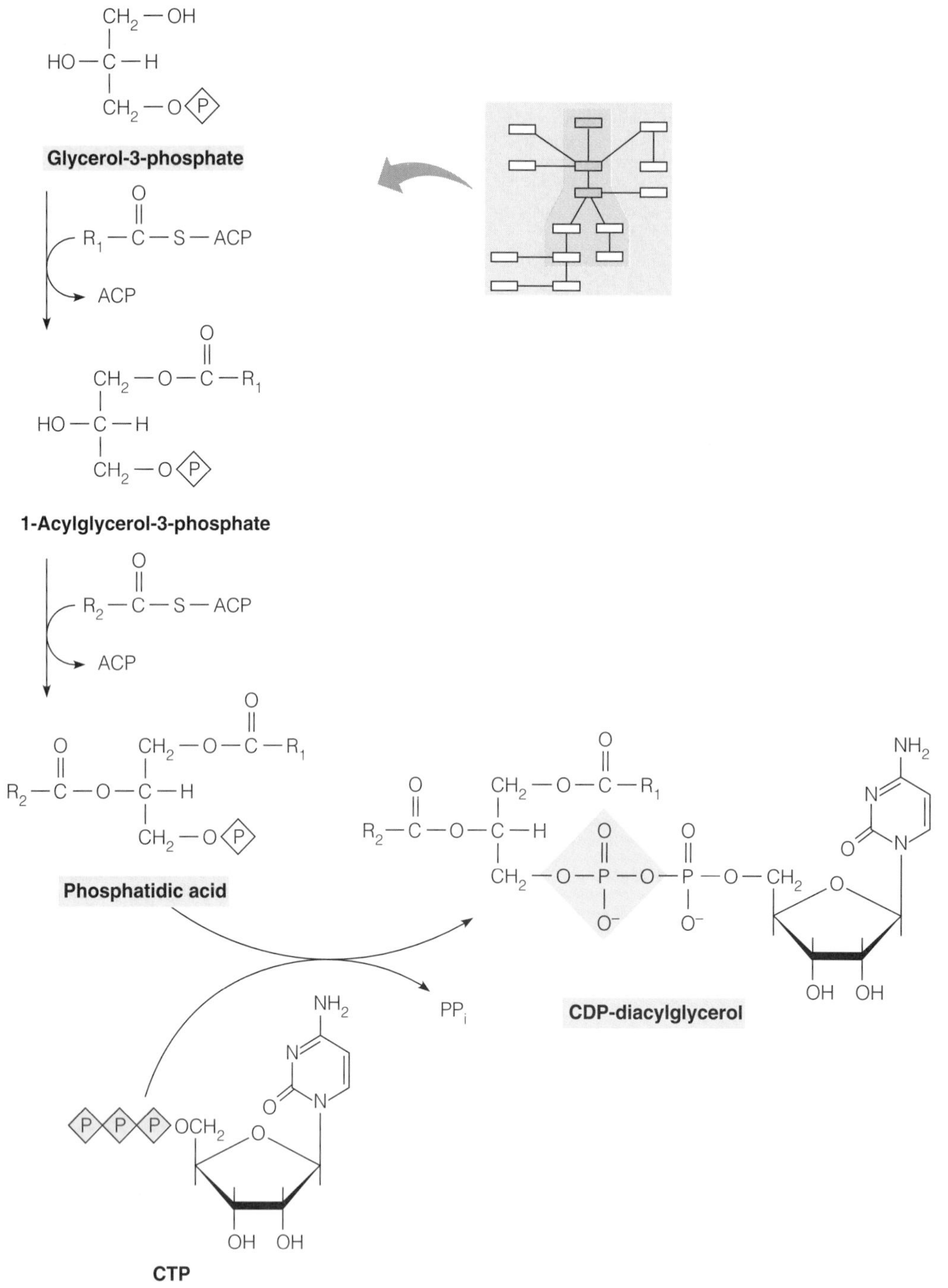

Glycerol-3-phosphate

1-Acylglycerol-3-phosphate

Phosphatidic acid

CDP-diacylglycerol

CTP

FIGURE **19.3**

Synthesis of phosphatidic acid and CDP-diacylglycerol in bacteria. Two separate acyltransferases are involved in phosphatidic acid synthesis. Although glycerol is a symmetrical molecule, most glycerol-containing metabolites have different substituents on each hydroxyl, making C-2 asymmetric. In the configuration drawn, glycerol is referred to as "*sn*-glycerol" (stereospecifically numbered), with C-1 on top and C-3 on the bottom (see also Figure 10.6). The reaction of CTP with phosphatidic acid is drawn to the right by the enzymatic hydrolysis of pyrophosphate.

phosphatidylserine, which immediately undergoes decarboxylation to phosphatidylethanolamine. Consequently, phosphatidylserine does not accumulate in bacteria. The other pathway (Figure 19.4, right side) transfers glycerol-3-phosphate to phosphatidic acid, followed by a phosphatase reaction to give phosphatidylglycerol. Reaction with another mole of phosphatidylglycerol gives **diphosphatidylglycerol,** or **cardiolipin.** Cardiolipin, which is particularly abundant in the membranes of spirochetes, is the principal antigenic component measured in the Wassermann test, formerly used for diagnosis of syphilis. In *E. coli* phosphatidylglycerol and cardiolipin play specfic roles in activating the protein product of the *dnaA* gene, involved in initiation of DNA replication at a membrane site (page 910, Chapter 24.)

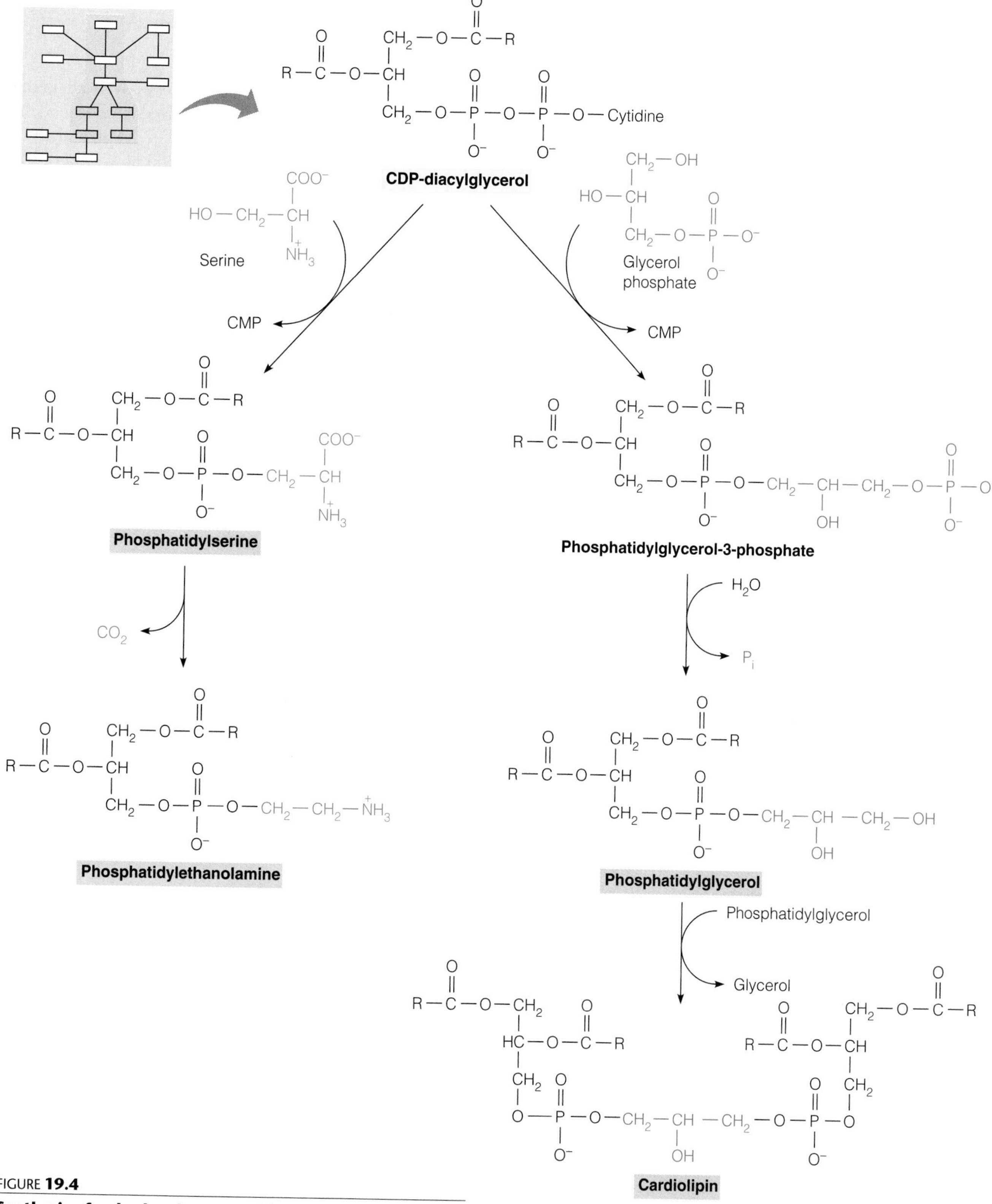

FIGURE **19.4**

Synthesis of polar head groups of bacterial phospholipids.

Not shown in Figure 19.2 is the fact that phosphatidylethanolamine and phosphatidylglycerol both turn over relatively rapidly, with transfer of the ethanolamine phosphate and glycerophosphate moieties, respectively, to membrane proteins and oligosaccharides. In both cases, the other product of turnover is diacylglycerol, which can then be converted to phosphatidic acid.

Control of Phospholipid Synthesis in Prokaryotes

The genetic analysis of *E. coli* phospholipid metabolism is fairly advanced, in the sense that the structural genes for most of the enzymes involved have been identified and mapped, mutant phenotypes have been analyzed in detail, and all of the genes have been cloned and sequenced. However, we still know rather little about how phospholipid synthesis is regulated. Current evidence suggests that the rate of phospholipid synthesis is controlled primarily at the level of fatty acid synthesis. This is suggested in part because cells that overproduce phospholipid biosynthetic enzymes do not overproduce phospholipids, suggesting that control lies at an earlier step. Because fatty acids in bacteria are used primarily for membrane synthesis, and not as energy substrates, it makes good metabolic sense to limit the synthesis of phospholipids at the earliest steps committed to membrane formation.

GLYCEROPHOSPHOLIPID METABOLISM IN EUKARYOTES

Most eukaryotic cells contain six classes of glycerophospholipids—the same phosphatidylethanolamine (PE), phosphatidylglycerol (PG), and cardiolipin (CL) found in bacteria, plus phosphatidylserine (PS), phosphatidylcholine (PC), and phosphatidylinositol (PI). As outlined in Figure 19.2, phosphatidic acid serves as a major precursor to all six compounds, and the pathways to PE, PS, PG, and CL are virtually identical to those already presented for bacteria. However, eukaryotic cells possess additional pathways that start with the free base—choline and ethanolamine, respectively, leading to PC and PE. These pathways are also outlined in Figure 19.2 and elaborated upon in a following section.

Synthesis of Phosphatidic Acid

In eukaryotes, phosphatidic acid has three different origins: glycerol-3-phosphate, dihydroxyacetone phosphate, and diacylglycerol.

Eukaryotic cells display three biosynthetic routes to phosphatidic acid. The major pathway, starting with glycerol-3-phosphate, is similar to that used by bacteria (see Figure 19.3), except that the acyltransferases use acyl-CoAs as substrates instead of acyl-ACPs. A second pathway starts with dihydroxyacetone phosphate (DHAP), which accepts a fatty acyl moiety at position 1 from an acyl-CoA, followed by reduction to 1-acylglycerol-3-phosphate and a second acylation.

$$\text{DHAP} \xrightarrow{R_1-\overset{O}{\overset{\|}{C}}-S-\text{CoA}} \text{1-Acyl-DHAP} \xrightarrow{\text{NADPH}} \text{1-Acylglycerol-3-phosphate} \xrightarrow{R_2-\overset{O}{\overset{\|}{C}}-S-\text{CoA}} \text{Phosphatidic acid}$$

The third route simply involves phosphorylation by a specific kinase of diacylglycerol, which arises from metabolic turnover of phospholipids (see page 854).

Regardless of which route is used, phosphatidic acid is converted to CDP-diacylglycerol as previously described for bacteria. CDP-diacylglycerol in turn serves as precursor to PS, PE, PG, and CL. Another set of pathways starts from free bases, as described in the next section.

Pathways to Phosphatidylcholine and Phosphatidylethanolamine

The most abundant phospholipids in most eukaryotic cells are phosphatidylcholine and phosphatidylethanolamine. Both can be synthesized from phosphatidylserine or through alternative pathways that start with free choline or ethanolamine, respectively. Because choline and ethanolamine arise largely through the turnover of preexisting phospholipids, the latter pathways can be considered "salvage pathways" for reutilization of these breakdown products. The significance of reutilizing choline lies in the fact that the three methyl groups of choline are derived from the amino acid methionine. As we discuss in Chapter 20, methionine is nutritionally essential for many animals, also making essential the reuse of scarce metabolites such as choline. In fact, choline itself is an essential component of most animal diets.

Salvage pathways to phospholipids, starting with choline or ethanolamine, are quantitatively important in eukaryotic cells.

The pathway for use of preformed choline, which predominates in most animal cells, is summarized in Figure 19.5. Choline is phosphorylated, and the resultant **phosphocholine** undergoes a reaction with CTP, similar to that described for bacterial diacylglycerol metabolism, to give **CDP-choline.** The phosphocholine moiety of this intermediate is transferred to diacylglycerol, yielding phosphatidylcholine.

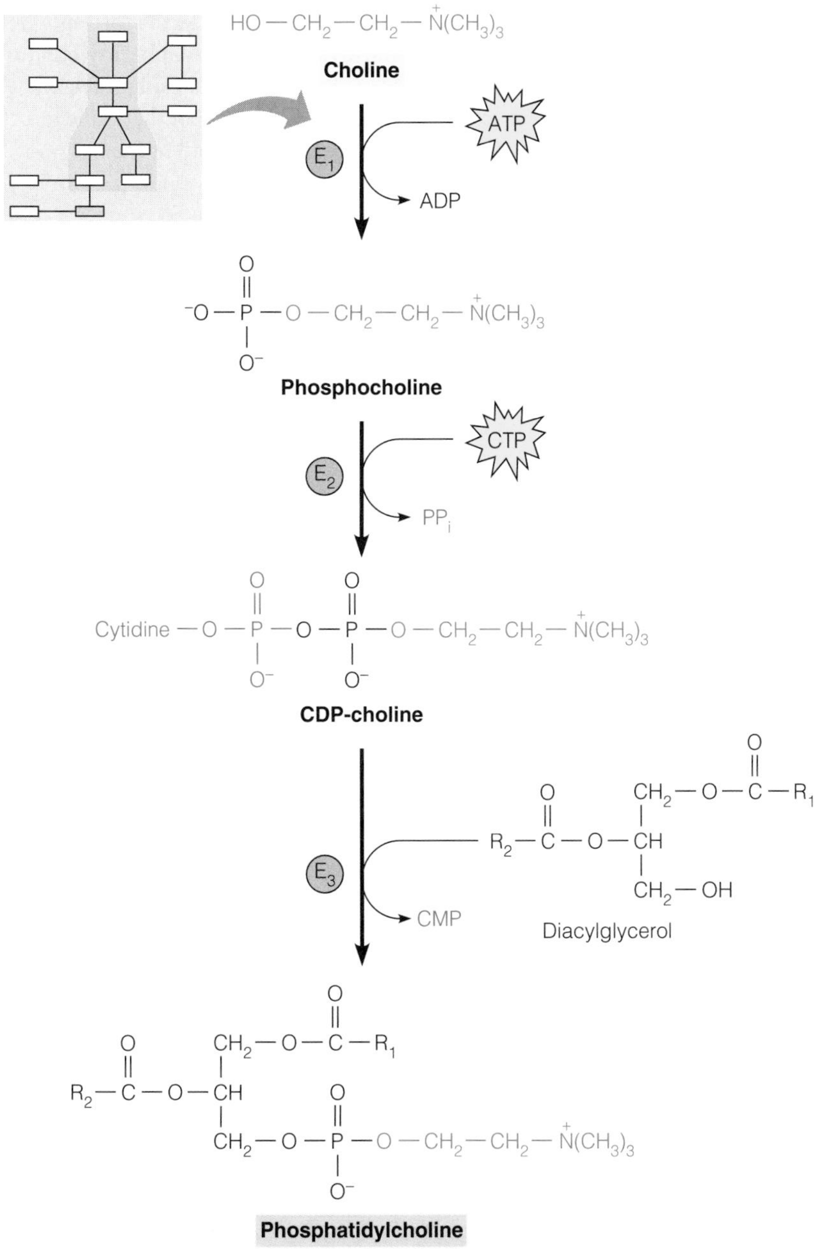

FIGURE **19.5**

Synthesis of phosphatidylcholine from choline. This pathway for reuse of choline for phosphatidylcholine synthesis occurs in animal cells. The same series of reactions converts ethanolamine to phosphatidylethanolamine, but the enzymes for the second and third reactions are different. E_1 = choline kinase; E_2 = CTP:phosphocholine cytidylyltransferase; E_3 = CDP-choline:1,2-diacylglycerol choline phosphotransferase.

The first enzyme in the pathway, **choline kinase** (E_1 in the figure) is cytosolic, whereas the second enzyme, **CTP:phosphocholine cytidylyltransferase** (E_2), is found in both cytosolic and microsomal fractions. The last enzyme, **CDP-choline:1,2-diacylglycerol choline phosphotransferase** (E_3), is membrane-bound in the endoplasmic reticulum. Recent evidence suggests that only the membrane-bound form of the cytidylyltransferase (E_2) is active and that the rate of phosphatidylcholine synthesis is controlled in part by translocation of this enzyme between cytosolic and membranous forms. Translocation is apparently controlled in turn by reversible phosphorylation and dephosphorylation of the enzyme, but the ultimate regulatory signals are not yet known.

The salvage route to phosphatidylethanolamine involves the same reactions, starting with ethanolamine instead of choline. The same enzyme carries out phosphorylation of both choline and ethanolamine, but the subsequent reactions are carried out by different enzymes in the two pathways.

The alternative pathway to PE and PC begins with the conversion of phosphatidylserine (PS) to PE, which is catalyzed by either of two different enzymes. **Phosphatidylserine decarboxylase** is a mitochondrial enzyme that, like the corresponding bacterial enzyme, decarboxylates phosphatidylserine to phosphatidylethanolamine. The second enzyme is a calcium-activated transferase, **phosphatidylethanolamine serinetransferase,** which exchanges free ethanolamine for the serine moiety of phosphatidylserine, yielding phosphatidylethanolamine and serine.

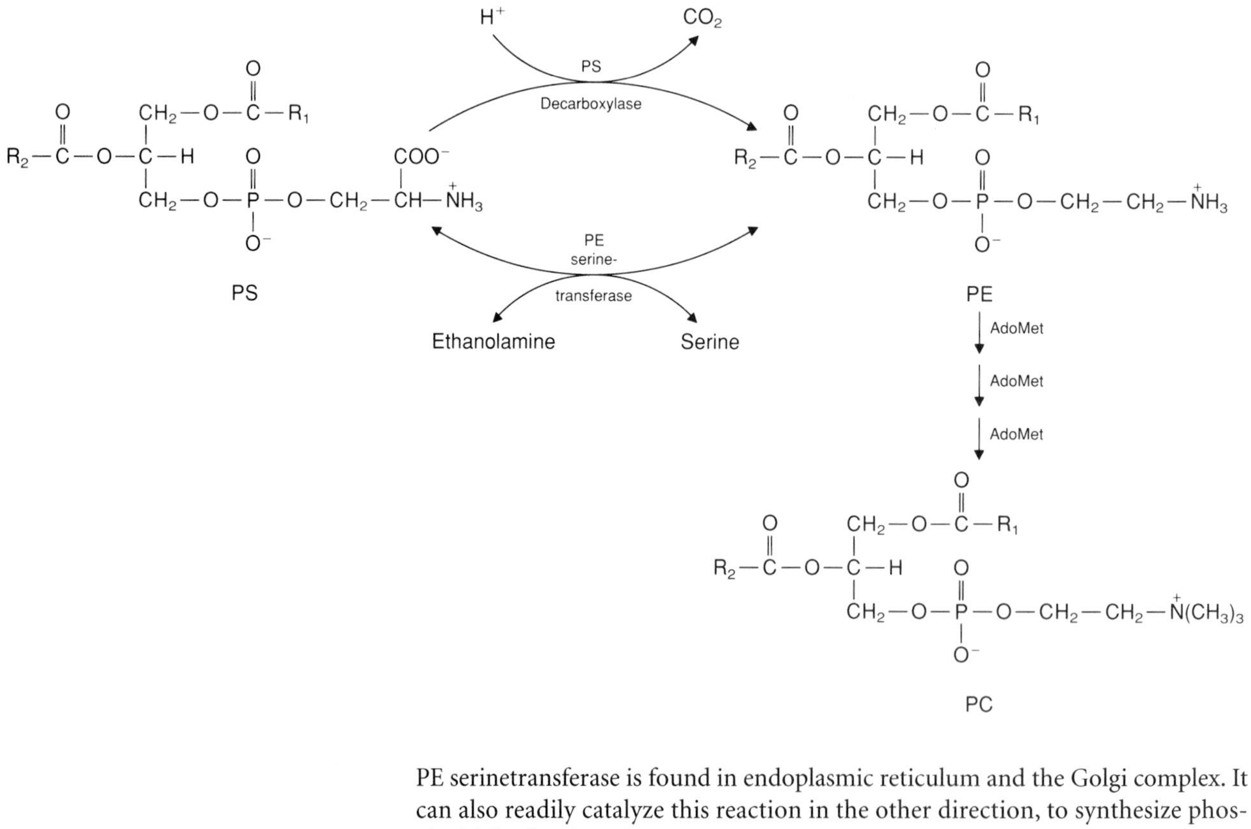

H^+ CO_2

PS Decarboxylase

$R_2{-}C(=O){-}O{-}CH(CH_2{-}O{-}C(=O){-}R_1){-}CH_2{-}O{-}P(=O)(O^-){-}O{-}CH_2{-}CH(COO^-){-}\overset{+}{N}H_3$

PS

PE serine-transferase

Ethanolamine Serine

$R_2{-}C(=O){-}O{-}CH(CH_2{-}O{-}C(=O){-}R_1){-}CH_2{-}O{-}P(=O)(O^-){-}O{-}CH_2{-}CH_2{-}\overset{+}{N}H_3$

PE

AdoMet

AdoMet

AdoMet

$R_2{-}C(=O){-}O{-}CH(CH_2{-}O{-}C(=O){-}R_1){-}CH_2{-}O{-}P(=O)(O^-){-}O{-}CH_2{-}CH_2{-}\overset{+}{N}(CH_3)_3$

PC

PE serinetransferase is found in endoplasmic reticulum and the Golgi complex. It can also readily catalyze this reaction in the other direction, to synthesize phosphatidylserine from PE.

S-Adenosylmethionine is the methyl group donor in synthesis of phosphatidylcholine and numerous other methylated metabolites.

However formed, phosphatidylethanolamine undergoes three successive methylations, probably all catalyzed by the same enzyme, to give phosphatidylcholine. In animals this pathway occurs primarily in the liver. The methyl group donor for these reactions is an activated derivative of methionine, ***S*-adenosylmethionine** (AdoMet). The product of methyl group transfer is ***S*-adenosylhomocysteine** (AdoHcy).

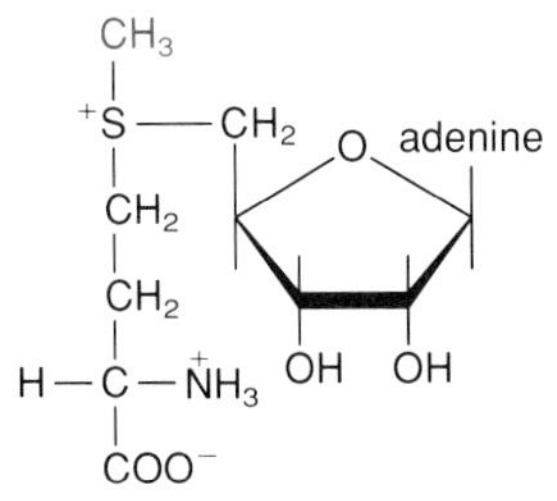

S-Adenosylmethionine (AdoMet)

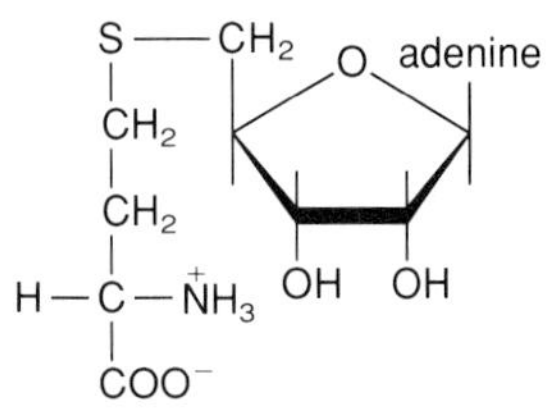

S-Adenosylhomocysteine (AdoHcy)

S-Adenosylmethionine is involved in numerous methyl group transfer reactions in lipid, protein, amino acid, and nucleic acid metabolism. AdoMet is formed from methionine and ATP in an unusual reaction, in which ATP is cleaved to yield inorganic triphosphate (PPP_i) plus an adenosyl moiety linked directly via C-5′ to the methionine sulfur. This creates an unstable **sulfonium ion,** which has a high thermodynamic tendency to transfer its methyl group and lose its charge. The triphosphate is hydrolyzed by the enzyme to pyrophosphate (PP_i) and orthophosphate (P_i).

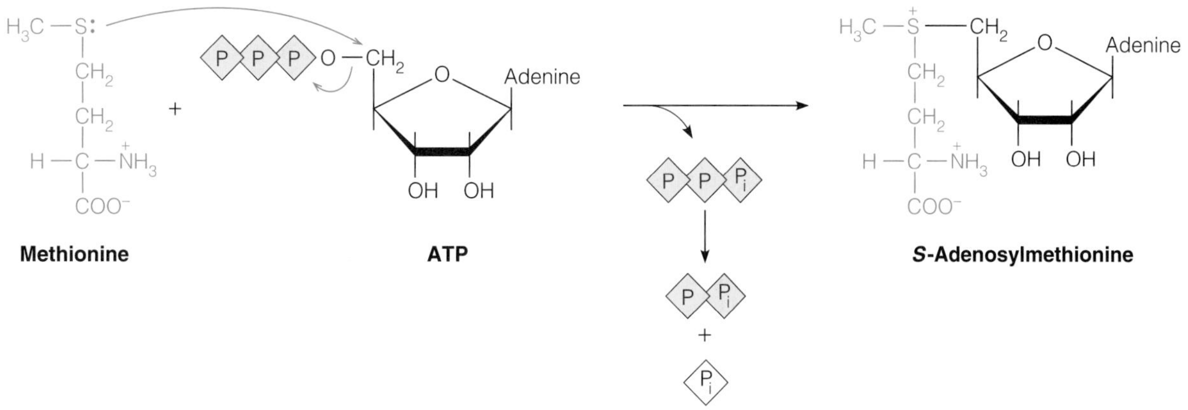

Methionine + **ATP** → **S-Adenosylmethionine** + PPP_i → PP_i + P_i

Redistribution of Phospholipid Fatty Acids: Lung Surfactant and Phospholipases

Having described the biosynthesis of polar head groups, let us now focus on the fatty acid constituents of phospholipids, after which we will return to one more important polar head group—the inositol of phosphoinositides. A variety of isotope labeling experiments show that phospholipids, even after insertion into membranes, are not metabolically inert. Specifically, the fatty acyl chains can change in response to varying environmental conditions or needs. Does this response involve modification of resident fatty acid chains or substitution of new fatty acids? To answer this question, investigators have explored the biosynthesis of the phospholipid component of **lung surfactant,** a lipid- and protein-containing substance that is secreted from lung and, by maintaining high surface tension, prevents collapse of the alveoli when air is expelled. Lung surfactant contains some 50% to 60% **dipalmitoylphosphatidylcholine,** a form of phosphatidylcholine in which palmitoyl chains occupy both positions 1 and 2. Infants afflicted with **respiratory distress syndrome** show defects in the metabolism of lung surfactant—deficiencies in either synthesis or secretion of this substance. Synthesis of this phospholipid in lung tissue is greatly activated after birth, as a result of translocation of CTP:phosphocholine cytidylyltransferase (E_2 in Figure 19.5) from cytosol to endoplasmic reticulum (ER) and consequent activation. This translocation is controlled by cyclic AMP–dependent phosphorylation of the cytidylyltransferase. The dephosphorylated form of the enzyme is preferentially

Phosphatidylcholine containing two palmitoyl groups plays an indispensable role in preventing collapse of the lungs after exhalation.

membrane-bound and active in the ER, whereas phosphorylation causes dissociation from the membrane. These observations point to the cytidylyltransferase as a major control point in eukaryotic phospholipid synthesis. Because this system is activated after birth, premature infants are often treated with a surfactant.

Because phospholipids are usually synthesized with an unsaturated chain at position 2, we can ask how the saturated chain arises at this position in surfactant. Although the question is not yet settled, current evidence rules out the possibility that an unsaturated fatty acid chain resident on position 2 becomes modified in situ to a palmitoyl chain. Two remaining possibilities are (1) transfer of palmitoyl chains from an acyl donor such as palmitoyl-CoA to glycerophosphorylcholine and (2) hydrolytic cleavage of acyl chains from phosphatidylcholine, followed by transfer of palmitoyl chains from a suitable acyl group donor.

Studies of lung surfactant biosynthesis illuminate mechanisms by which fatty acid chains in phospholipids are retailored to meet the needs of the organism.

Whatever the actual pathway for lung surfactant synthesis, it is apparent that phospholipid molecules in general can be retailored by cleavage and replacement of fatty acyl chains on carbons 1 and 2. Hydrolytic cleavage can occur through the action of **phospholipase A_2**, and reacylation would most probably occur from an acyl-CoA. Phospholipid remodeling by this mechanism occurs in membrane phospholipids, as well as in specialized lipid compounds such as surfactant.

Phospholipase A_2 is one of a class of four enzymes that hydrolyze specific bonds in phospholipids; the others are phospholipases A_1, C, and D (Figure 19.6). Phospholipases have been useful reagents in studies of both lipid and membrane structures. Structural studies on phospholipase A_2 have been of particular interest, partly because the enzyme is active at a membrane–water interface and serves as a probe of membrane structure, and partly because phospholipase actions lead to prostaglandins (page 702) and to second messengers (see Chapter 23). Phospholipase A_2 shows remarkable conservation of three-dimensional structure across evolutionary time, as shown in Figure 19.7. One form of the enzyme shown in the figure comes from cobra venom, which has the ability to cause hemolysis, or rupture of red blood cells. The release of one fatty acyl chain from phosphatidylcholine yields 1-acylglycerophosphorylcholine, more commonly known as **lysolecithin** (phosphatidylcholine itself is commonly called lecithin). Lysolecithin derives its name from the fact that, as an excellent detergent, it solubilizes membranes and hence causes cells to lyse; erythrocytes are particularly susceptible to this action.

Phospholipases play roles in signal transduction and in membrane repair.

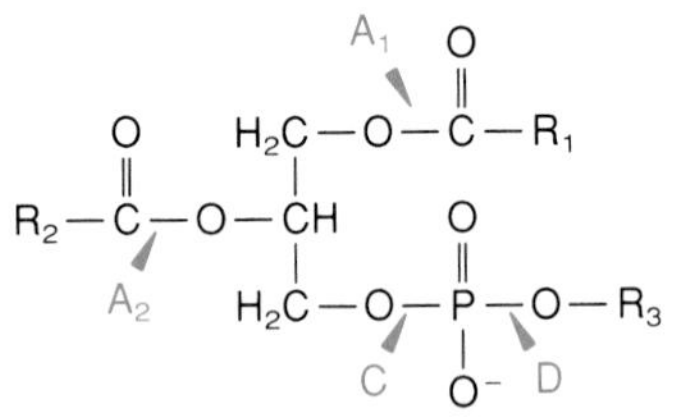

FIGURE **19.6**

Specificities of phospholipases A_1, A_2, C, and D.

Another postulated role for phospholipase A_2 is in repair of damaged membrane phospholipids. As discussed in Chapter 15, fatty acids are susceptible to nonenzymatic attack by oxygen or reactive oxygen species such as superoxide to give fatty acid hydroperoxides. When a membrane phospholipid undergoes peroxidation of a fatty acyl chain, the structure of the membrane is distorted, and the function of the membrane can be affected. Current evidence suggests that phospholipase A_2 can remove these abnormal fatty acids from phospholipids still resident in a lipid bilayer, which leads to replacement of the damaged acyl chains by normal fatty acids.

The simplest possible such "lysolipid" is **lysophosphatidic acid,** formed by action on phosphatidic acid of phospholipase A_2, which releases a fatty acid from position 2. Lysophosphatidic acid has recently been identified as a biological signaling agent. Its release from activated cells such as platelets stimulates the growth of other cell types, through interaction with specific receptors, evidently as part of the wound-healing process.

Biosynthesis of Other Acylated Glycerophospholipids

The remaining major pathways outlined in Figure 19.2 are for synthesis of phosphatidylserine, phosphatidylglycerol, cardiolipin, and phosphatidylinositol. In yeast the CDP-diacylglycerol pathway to phosphatidylserine predominates. In animals, phosphatidylserine is synthesized primarily by the calcium-activated exchange between phosphatidylethanolamine and serine. However, the ultimate source of ethanolamine for this process is not yet known. In animals the synthesis

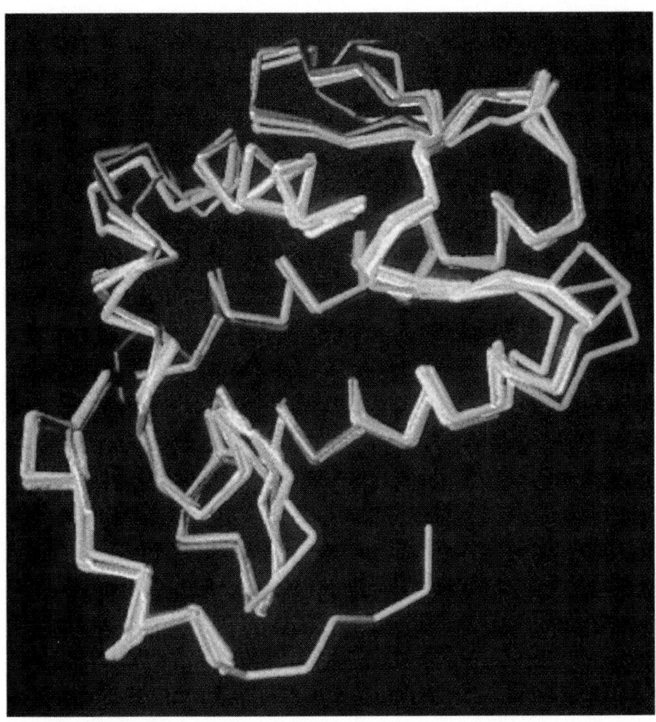

FIGURE **19.7**

Structure of phospholipase A_2, an enzyme that metabolizes membrane phospholipids. Catabolism of phospholipids at membrane–water interfaces is important both in modification of membrane structure and as a source of second messengers and other regulators. The crystal structures of phospholipase A_2 from sources as diverse as the Indian cobra (cyan), rattlesnake (green), bovine pancreas (red), and porcine pancreas (purple) show near-complete superimposability of their α-chain backbones. A loop in the porcine enzyme structure was removed for this comparison.

of phosphatidylglycerol, which is largely confined to mitochondria, is identical to the route used in bacteria. However, its conversion to cardiolipin involves CDP-diacylglycerol, rather than a second mole of phosphatidylglycerol, as the second substrate.

The biosynthesis of phosphatidylinositol, catalyzed by **phosphatidylinositol synthase,** involves CDP-diacylglycerol and L-*myo*-inositol (Figure 19.8). The latter is one of nine possible stereoisomers of hexahydroxycyclohexane; it is synthesized from D-glucose-6-phosphate. Phosphatidylinositol undergoes two successive phosphorylations to yield phosphatidylinositol-4-phosphate and phosphatidylinositol-4,5-bisphosphate, both of which are present in small but appreciable amounts. All three of these lipids, which are collectively termed **phosphoinositides,** are enriched in arachidonic acid at position 2. This enrichment evidently occurs via the deacylation–reacylation process we discussed earlier (see page 676).

It has long been known from ^{32}P labeling studies that the phosphoinositides are in a state of active metabolic flux. They are synthesized and degraded rapidly, particularly in nervous tissue and particularly in response to the binding of neurotransmitters. The phosphoinositides play important roles as precursors to second messengers in **transmembrane signaling,** the transmission of an extracellular signal to some element of the intracellular metabolic apparatus. Our current understanding of these events is presented in Chapter 23.

Phosphatidylinositol and its phosphorylated derivatives play important roles as precursors of second messengers.

Posttranslational Protein Modification by Lipids

Recently, another metabolic role for phosphatidylinositol has come to light through studies on variant forms of a cell surface glycoprotein of the protozoal parasite *Trypanosoma brucei,* which causes African sleeping sickness. The organism escapes immunological detection by rapidly changing the structure of this surface glycoprotein. The glycoprotein is linked, through its terminal carboxyl group, to a glycosylated form of phosphatidylinositol. This linkage provides both an anchor, binding the protein to the membrane, and a site for cleavage, when the organism replaces one glycoprotein with another at its surface. Figure 19.9 presents a model suggesting how this might occur. There is evidence that this mechanism is used by other eukaryotic cells to control the concentrations of particular

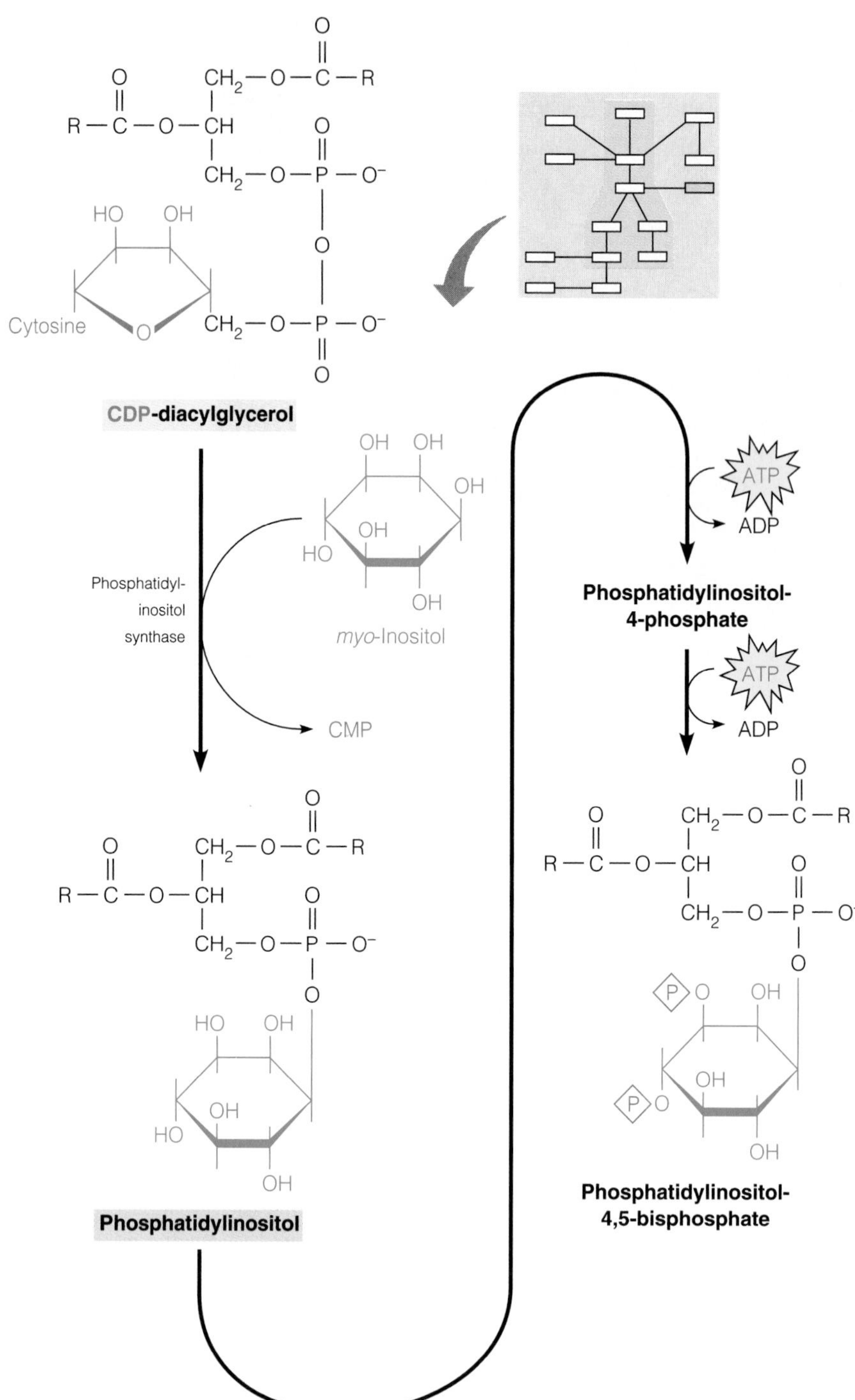

FIGURE **19.8**
Biosynthesis of phosphoinositides.

proteins bound at the cell surface. Glycolipids containing glucosamine, as does the trypanosome protein, have been described in several other eukaryotic cell membranes.

The glycosylphosphatidylinositol modification is but one of several recently described lipid modifications of proteins. Two of the others (which do not involve phosphoinositides) are *acylation* and *prenylation.* Acylation involves transfer of single saturated acyl groups—primarily myristate (C_{14}) linked as an amide to N-terminal glycine and palmitate (C_{16}) linked as a thioester to cysteine. A large number of viral proteins, membrane proteins, and proteins involved in signal

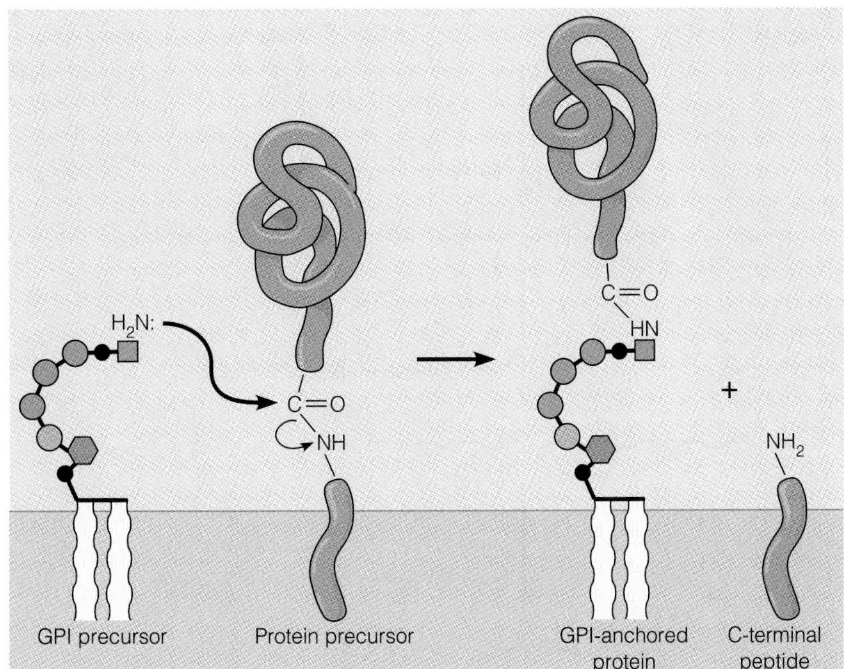

FIGURE **19.9**

Anchoring of proteins in membranes by a glycolipid containing phosphatidylinositol. Binding in this mode allows rapid exchange of bound glycoproteins. In the protozoan *Trypanosoma brucei* the membrane glycolipid (glycosylphosphatidylinositol, or GPI) contains inositol (purple), glucosamine (blue), mannose (green), phosphate (black), and ethanolamine (pink). The ethanolamine is linked via its amino group to the C-terminal residue in the protein.

Courtesy of T. L. Doering, W. J. Masterson, G. W. Hart, and P. T. Englund, *J. Biol. Chem.* (1990) 265:611–614, reprinted by permission of The American Society for Biochemistry & Molecular Biology.

transduction are acylated. Prenylation involves transfer of C_{15} or C_{20} groups from intermediates in cholesterol synthesis (see page 689) to cysteine residues four positions from the C-terminus. Subsequent modifications involve the removal of the three C-terminal amino acid residues and methylation of the carboxy terminus. This results in a prenylated and methylated C-terminal cysteine residue, which is very hydrophobic. The *ras* oncogene protein (see page 864) is the most prominent of the family of prenylated proteins. Recently, the enzyme **farnesyltransferase,** which transfers a C_{15} unit from **farnesyl pyrophosphate** to a substrate protein (Figure 19.10), has been recognized as an attractive target for inhibition in cancer chemotherapy. Inhibitors of this enzyme interfere with the growth of cancer cells in culture. Although the precise mechanisms of growth inhibition are not known, this is a promising area for molecular pharmacology.

In most cases, acylation and prenylation provide membrane anchors, like the GPI modification. However, some cytosolic proteins contain lipid modifications, and there the lipid may stabilize protein–protein interactions. Recently, the "hedgehog" family of proteins regulating differentiation has been shown to be esterified with cholesterol near the C-terminus, a novel lipid modification.

Ether Phospholipids

Ether lipids are lipids containing an alkyl group, rather than an acyl group, linked to one of the oxygen atoms of glycerol. Alkyl and alkenyl phospholipids are widely distributed, but their abundance in tissues varies greatly. For example, consider the **plasmalogens,** or **vinyl ethers,** phospholipids that contain an alkenyl ether at position *sn*-1 of glycerol. These compounds constitute some 50% of all choline phospholipids found in heart tissue but are virtually undetectable in many other tissues. So far, little is known about the functional significance of this class of lipids. However, a genetic deficiency of plasmalogen synthesis has serious consequences. Biosynthesis of ether lipids occurs in peroxisomes, organelles that also carry out a β-oxidation pathway (see page 648) and an α-oxidation pathway (see page 648). In a rare autosomal recessive disorder called **Zellweger syndrome,** peroxisomes are absent, and plasmalogen synthesis is severely deficient. Individuals with this condition suffer damage to brain, liver, and kidney, before reaching an early death.

$R_2-C(=O)-O-CH(CH_2-O-CH_2-CH_2-R_1)(CH_2-O-P(=O)(O^-)-O-X)$

Phospholipid with an alkyl ether at position *sn*-1

$R_2-C(=O)-O-CH(CH_2-O-CH{=}CH-R_1)(CH_2-O-P(=O)(O^-)-O-X)$

Phospholipid with an alkenyl ether at position *sn*-1

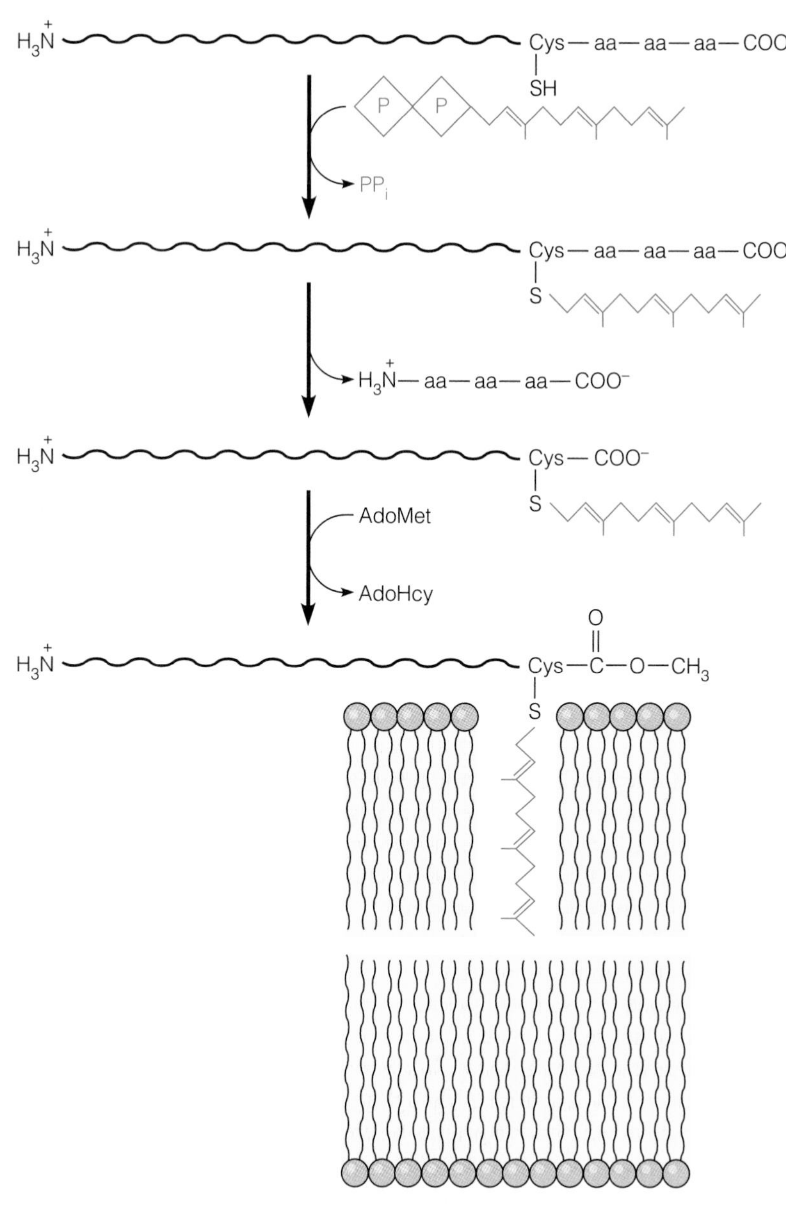

FIGURE **19.10**

The protein farnesyltransferase reaction. aa = amino acid residue.

1-Acyldihydroxyacetone phosphate

$R^*—C(=O)—S—CoA$

2NADPH

$R^*—OH$

Saturated fatty alcohol

$R_1—COO^-$

1-Alkyldihydroxyacetone phosphate

NADPH, H^+

$NADP^+$

1-Alkylglycerol-3-phosphate

$R_2—C(=O)—S—CoA$

CoA—SH

1-Alkyl-2-acylglycerol-3-phosphate

FIGURE **19.11**

Biosynthetic routes to alkyl ether phospholipids.

The biosynthesis of ether phospholipids (Figure 19.11) begins with 1-acyldihydroxyacetone phosphate (see page 672). This undergoes exchange of an alkyl group for the acyl group; the saturated fatty alcohol used in this reaction is derived from NADPH-dependent reduction of the corresponding fatty acyl-CoA. Carbon 2 is then reduced from the keto to the hydroxyl level and acylated. This gives the 1-alkyl analog of phosphatidic acid (last structure in the figure), which is converted to saturated ether phospholipids, or **glyceryl ethers,** by the pathways already presented in Figure 19.2 for phospholipid biosynthesis. The primary route leads to the glyceryl ether of phosphatidylethanolamine; the serine and choline analogs arise from the ethanolamine analog (as on page 674) by base exchange and methylation, respectively. The synthesis of plasmalogens from glyceryl ethers then involves desaturation of the alkyl group at position *sn*-1 (Figure 19.12). The microsomal enzyme system involved, like that used for desaturation of stearoyl-CoA (see page 659 in Chapter 18), requires O_2, NADH, and cytochrome b_5.

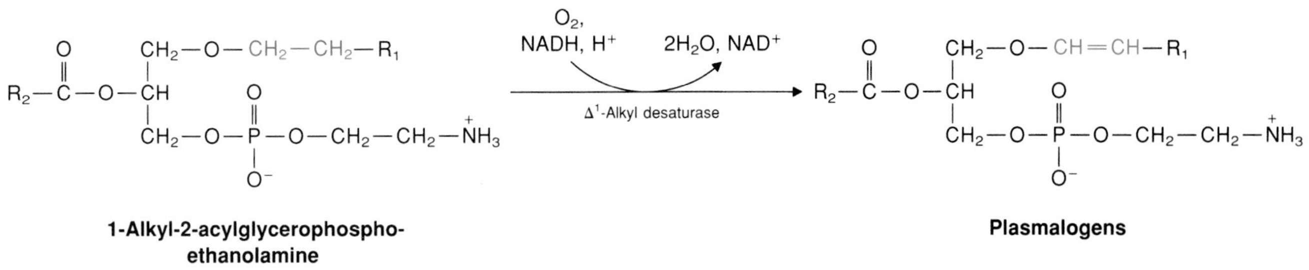

$$R_2-\overset{O}{\overset{\|}{C}}-O-CH(CH_2-O-CH_2-CH_2-R_1)(CH_2-O-\overset{O}{\overset{\|}{P}}(O^-)-O-CH_2-CH_2-\overset{+}{N}H_3) \xrightarrow[\Delta^1\text{-Alkyl desaturase}]{O_2,\ NADH,\ H^+ \ \to\ 2H_2O,\ NAD^+} R_2-\overset{O}{\overset{\|}{C}}-O-CH(CH_2-O-CH{=}CH-R_1)(CH_2-O-\overset{O}{\overset{\|}{P}}(O^-)-O-CH_2-CH_2-\overset{+}{N}H_3)$$

1-Alkyl-2-acylglycerophospho-ethanolamine → **Plasmalogens**

FIGURE **19.12**

Synthesis of a plasmalogen from a glyceryl ether. Desaturation of 1-alkyl-2-acylglycerophosphoethanolamine (the alkyl analog of phosphatidylethanolamine) yields the corresponding vinyl ether, or plasmalogen.

An unusual ether lipid, called **platelet-activating factor,** has the structure **1-alkyl-2-acetylglycerophosphocholine.** Physiologically, this compound is perhaps the most potent compound known. At concentrations as low as 1 *picomolar* (10^{-12} M), it has numerous effects, both in normal physiology and in inflammation reactions, including stimulation of blood platelet aggregation, reduction of blood pressure, activation of several white blood cell classes, decreased cardiac output, stimulation of glycogenolysis, and stimulation of uterine contraction. This lipid is synthesized via acetylation of the corresponding 1-alkylglycerophosphocholine by acetyl-CoA.

Platelet-activating factor is a glyceryl ether with potent physiological properties.

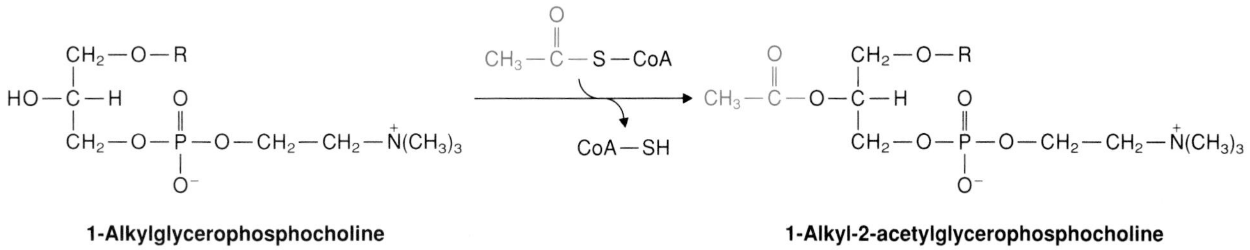

$$HO-C(H)(CH_2-O-R)(CH_2-O-\overset{O}{\overset{\|}{P}}(O^-)-O-CH_2-CH_2-\overset{+}{N}(CH_3)_3) \xrightarrow[CoA-SH]{CH_3-\overset{O}{\overset{\|}{C}}-S-CoA} CH_3-\overset{O}{\overset{\|}{C}}-O-C(H)(CH_2-O-R)(CH_2-O-\overset{O}{\overset{\|}{P}}(O^-)-O-CH_2-CH_2-\overset{+}{N}(CH_3)_3)$$

1-Alkylglycerophosphocholine → **1-Alkyl-2-acetylglycerophosphocholine**

The discovery of a phospholipid with such a striking biological activity was without precedent, and it has opened a fascinating new realm of biochemistry. The factor acts through binding to a high-affinity receptor in the membrane of susceptible cells. The receptors interact with signal transduction systems through G proteins (see Chapter 23).

Ether-containing lipids are quite abundant in the membranes of halophilic ("salt-loving") microorganisms. These bacteria and protozoans grow in media with NaCl concentrations as high as 4 M. Although we don't know the relationship between ether lipids and the ability to grow in a high-salt environment, the greater stability of alkyl ethers against hydrolysis, as compared with acyl esters, may be a factor.

Intracellular Transport of Membrane Phospholipids

Of the six major classes of glycerophospholipids in membrane lipids, phosphatidylglycerol and cardiolipin are found primarily in mitochondrial membranes and are synthesized in mitochondria. The remaining four classes are synthesized simultaneously with their insertion into the cytosolic side of membranes of the endoplasmic reticulum. From there they undergo translocation to the luminal side of the membrane and ultimately are transported to other membranes—the nuclear envelope, mitochondrial membranes, and the plasma membrane. Just how these events occur is the subject of one of the most active areas of contemporary cell biology. The three major questions are the following: (1) How do phospholipid molecules move from one side of a membrane to the other?

(2) How do phospholipid molecules move from one site to another within the cell? (3) How does phospholipid transport directed to specific organelles account for the differences in phospholipid composition of membranes within a single cell?

Investigations of transmembrane movement of phospholipids (question 1) use specific lipid probes that allow detection of a lipid on only one side of a bilayer. As mentioned in Tools of Biochemistry 10A, one such approach involves the use of a **spin label,** a lipid analog that is detectable from its electron paramagnetic resonance spectrum. Such measurements show that transbilayer movement, or "flip-flop," does occur spontaneously but is quite slow. Measurements in vivo show much faster transbilayer movement, so proteins or other factors may promote flip-flop in living cells.

Transport of phospholipids within the cell (question 2) involves largely the transfer of fragments of membranes of the ER into the Golgi complex, as was shown in Figure 19.1. Membrane vesicles are constantly pinched off from the Golgi, and these vesicles, containing secretory products, fuse with the plasma membrane for secretion of their contents via **exocytosis** (transport out of the cell). It seems likely that this route is used not only for extracellular secretion but also for transport of membrane lipids to the plasma membrane. Probably comparable processes transport membrane lipids to mitochondria, plant chloroplasts, and nuclei, although these processes are not as well understood.

Membranes are assembled by membrane vesicles' moving from synthesis sites in the endoplasmic reticulum and Golgi complex to existing membranes and fusing with them.

To explain the variability of membrane lipid composition within a given cell (question 3), we can postulate the existence in Golgi membranes of specific targeting proteins—proteins that preferentially associate with certain lipids and have an affinity for certain organelles. Another mechanism involves the action of **phospholipid exchange proteins**—cytosolic proteins that bind a phospholipid and can catalyze its exchange with a corresponding membrane lipid. The protein-bound lipid moves into the membrane, and the membrane lipid becomes bound to the cytosolic protein. This mechanism does not provide for net transfer of lipid to a membrane, but it does allow for modulation of the lipid composition of a particular membrane.

Metabolism of Sphingolipids

Interest in sphingolipids focuses largely on their important role in nervous tissue and, related to this role, a number of human genetic defects of sphingolipid metabolism. Sphingolipids are also widely distributed in the membranes of plant cells and in lower eukaryotes such as yeast.

$CH_3(CH_2)_{12}CH{=}CH{-}CH(OH){-}CH(NH_3^+){-}CH_2OH$

Sphingosine

$CH_3(CH_2)_{12}CH_2{-}CH(OH){-}CH(OH){-}CH(NH_3^+){-}CH_2OH$

Phytosphingosine

Recall from Chapter 10 that sphingolipids are derivatives of the base *sphingosine.* Plant sphingolipids contain a slightly different form of this compound, called **phytosphingosine.** The sphingolipids include *ceramide* (*N*-acylsphingosine), *sphingomyelin* (*N*-acylsphingosine phosphorylcholine), and a family of carbohydrate-containing sphingolipids called neutral and acidic **glycosphingolipids;** the latter substances include *cerebrosides* and *gangliosides* (which also contain sialic acid). Ceramide serves as the precursor to both sphingomyelin and the glycosphingolipids.

In animals the pathway to ceramide starts with the synthesis of a sphingosine derivative, **sphinganine,** from palmitoyl-CoA and serine (Figure 19.13). After reduction of the resulting keto group, the amino group of sphinganine is acylated to give a ceramide. The sphinganine unit of this compound is then desaturated to give a ceramide with a sphingosine base. Transfer of a phosphocholine unit from phosphatidylcholine yields sphingomyelin plus diacylglycerol.

The pathways leading to glycosphingolipids are more numerous, but the metabolic strategies are comparable to those we have encountered before in synthesis of the oligosaccharide chains of glycoproteins (see Chapter 16). The pathways involve the stepwise addition of monosaccharide units, using nucleotide-linked sugars as the activated biosynthetic substrates and with ceramide as the initial

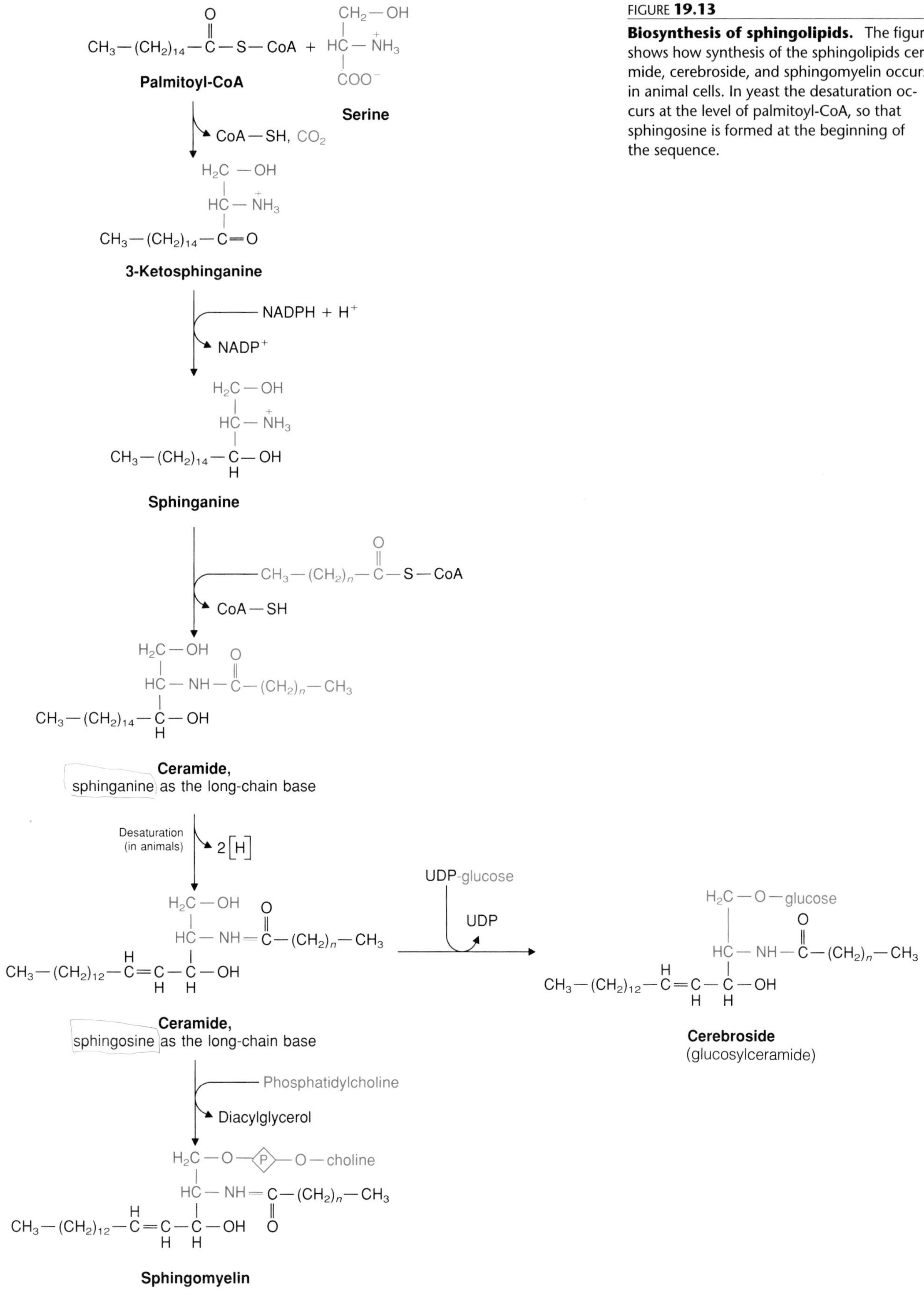

$CH_3-(CH_2)_{14}-C(=O)-S-CoA$ + $HC(CH_2-OH)(\overset{+}{N}H_3)(COO^-)$

Palmitoyl-CoA **Serine**

→ CoA—SH, CO_2

$CH_3-(CH_2)_{14}-C(=O)-CH(\overset{+}{N}H_3)-CH_2-OH$

3-Ketosphinganine

NADPH + H^+ → $NADP^+$

$CH_3-(CH_2)_{14}-CH(OH)-CH(\overset{+}{N}H_3)-CH_2-OH$

Sphinganine

$CH_3-(CH_2)_n-C(=O)-S-CoA$ → CoA—SH

$CH_3-(CH_2)_{14}-CH(OH)-CH(NH-C(=O)-(CH_2)_n-CH_3)-H_2C-OH$

Ceramide,
sphinganine as the long-chain base

Desaturation (in animals) → 2[H]

$CH_3-(CH_2)_{12}-CH=CH-CH(OH)-CH(NH-C(=O)-(CH_2)_n-CH_3)-H_2C-OH$

Ceramide,
sphingosine as the long-chain base

UDP-glucose → UDP

$CH_3-(CH_2)_{12}-CH=CH-CH(OH)-CH(NH-C(=O)-(CH_2)_n-CH_3)-H_2C-O-glucose$

Cerebroside
(glucosylceramide)

Phosphatidylcholine → Diacylglycerol

$CH_3-(CH_2)_{12}-CH=CH-CH(OH)-CH(NH-C(=O)-(CH_2)_n-CH_3)-H_2C-O-Ⓟ-O-choline$

Sphingomyelin

FIGURE **19.13**

Biosynthesis of sphingolipids. The figure shows how synthesis of the sphingolipids ceramide, cerebroside, and sphingomyelin occurs in animal cells. In yeast the desaturation occurs at the level of palmitoyl-CoA, so that sphingosine is formed at the beginning of the sequence.

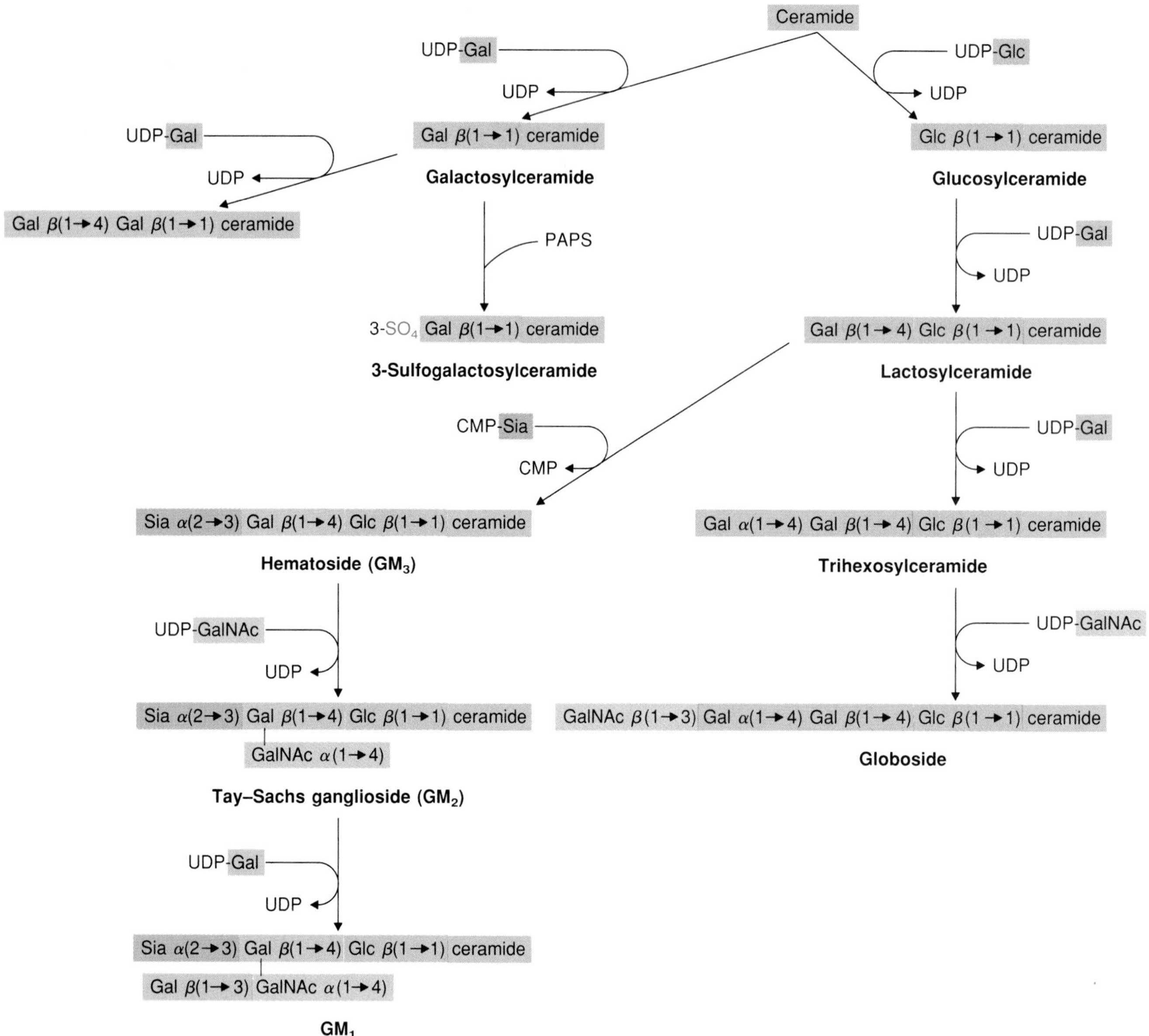

FIGURE **19.14**

Pathways of synthesis of glycosphingolipids. The common name of each compound is given. PAPS is a sulfate group donor (see Chapter 21).

Nucleotide-linked sugars and glycosyltransferases are involved in glycosphingolipid biosynthesis.

monosaccharide acceptor. The sugar nucleotides involved in glycosphingolipid synthesis include UDP-glucose (UDP-Glc), UDP-galactose (UDP-Gal), UDP-*N*-acetylgalactosamine (UDP-GalNAc), and CMP-*N*-acetylneuraminic acid (CMP-Sia, or CMP-sialic acid). Figure 19.14 shows pathways leading to some of the most abundant glycosphingolipids.

Sphingolipids, especially sphingomyelin, are abundant components of the *myelin sheath*, a multilayered structure that protects and insulates cells of the central nervous system (Figure 19.15; see also Figure 10.34). In human myelin, sphingolipids constitute some 25% of the total lipid. Sphingolipids are in a continuous state of metabolic turnover, both synthesis and degradation. Degradation occurs in the lysosomes, by a family of hydrolytic enzymes. These pathways are of great medical interest because of their relationship to a group of congenital diseases called **sphingolipidoses** (also known as **lipid storage diseases**). Each condition is characterized by deficiency of one of the degradative enzymes, with concomitant accumulation within the lysosome of the substrate for the deficient enzyme (Table 19.1). In fact, structural analysis of the abnormal metabolites that accumulate helped to establish the degradative pathways, which are depicted in Figure 19.16. Most of these diseases are autosomal recessive, which means that two defective

TABLE 19.1 Inherited diseases of sphingolipid catabolism

Disease	Defective Enzyme[a]	Accumulated Intermediate
GM_1 gangliosidosis	1 β-Galactosidase	GM_1 ganglioside
Tay–Sachs disease	2 β-*N*-Acetylhexosaminidase A	GM_2 (Tay–Sachs) ganglioside
Fabry's disease	3 α-Galactosidase A	Trihexosylceramide
Gaucher's disease	4 β-Glucosidase	Glucosylceramide
Niemann–Pick disease	5 Sphingomyelinase	Sphingomyelin
Farber's lipogranulomatosis	6 Ceramidase	Ceramide
Globoid cell leukodystrophy (Krabbe's disease)	7 β-Galactosidase	Galactosylceramide
Metachromatic leukodystrophy	8 Arylsulfatase A	3-Sulfogalactosylceramide
Sandhoff disease	9 *N*-Acetylhexosaminidases A and B	GM_1 ganglioside and globoside

[a]Numbers refer to enzymes shown in Figure 19.16.

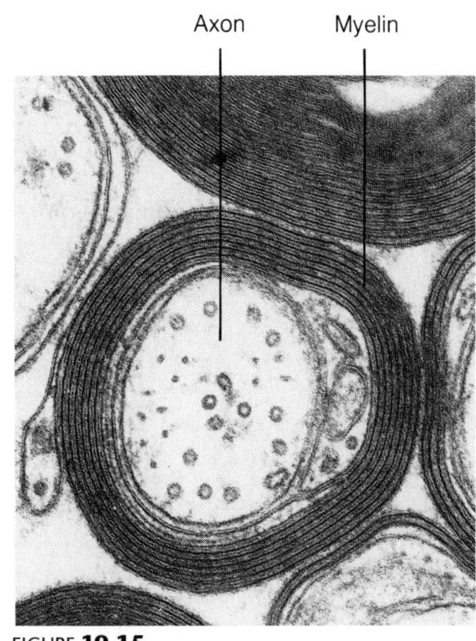

FIGURE **19.15**

A myelinated axon from the spinal cord. Myelin, an insulating layer wrapping about the axon, is rich in sphingomyelin.

Courtesy of Dr. Cedric Raine.

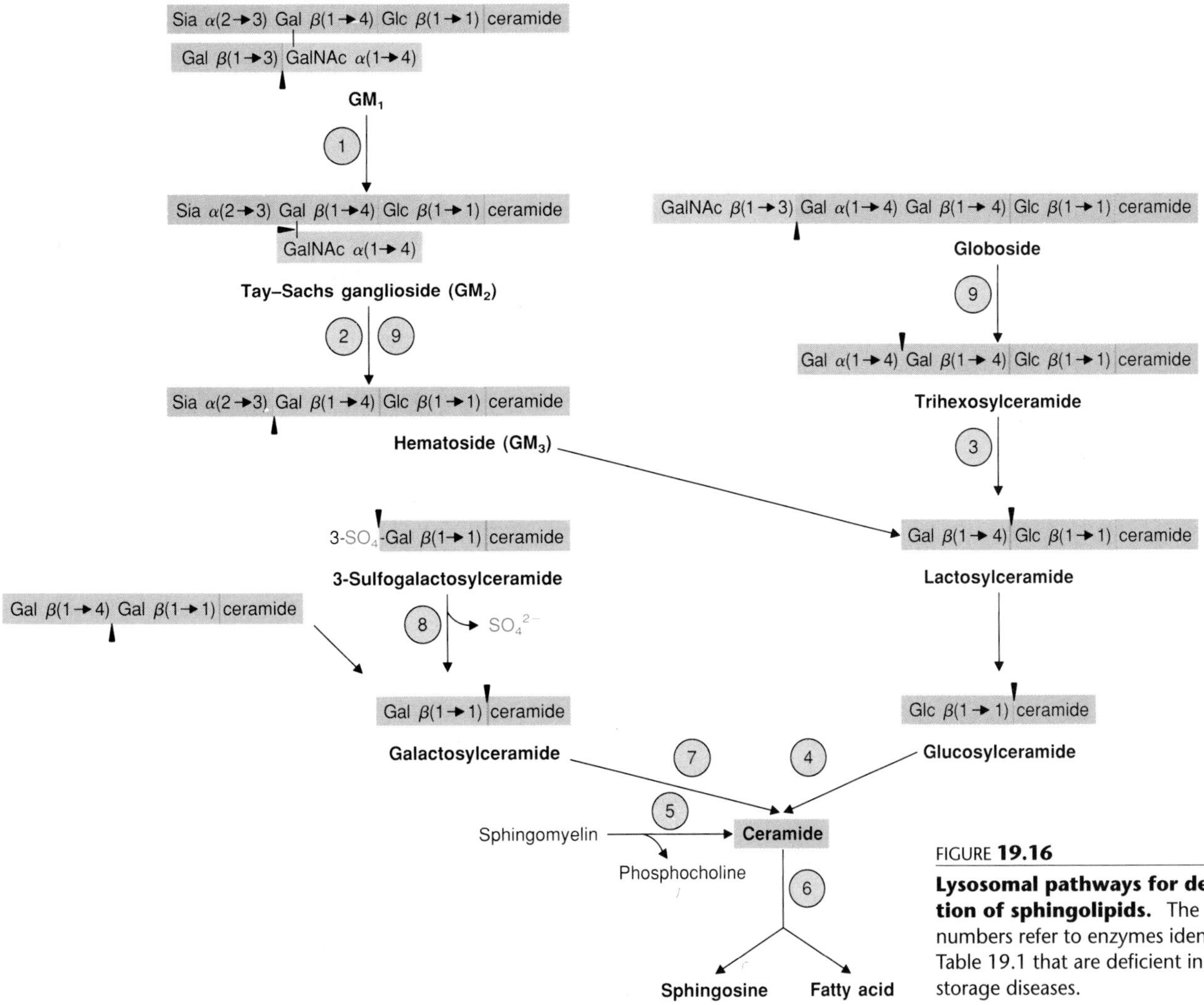

Sia α(2→3) Gal β(1→4) Glc β(1→1) ceramide
Gal β(1→3) GalNAc α(1→4)
GM_1
1
Sia α(2→3) Gal β(1→4) Glc β(1→1) ceramide
GalNAc α(1→4)
Tay–Sachs ganglioside (GM_2)
2 9
Sia α(2→3) Gal β(1→4) Glc β(1→1) ceramide
Hematoside (GM_3)
GalNAc β(1→3) Gal α(1→4) Gal β(1→4) Glc β(1→1) ceramide
Globoside
9
Gal α(1→4) Gal β(1→4) Glc β(1→1) ceramide
Trihexosylceramide
3
Gal β(1→4) Glc β(1→1) ceramide
Lactosylceramide
3-SO_4-Gal β(1→1) ceramide
3-Sulfogalactosylceramide
8
SO_4^{2-}
Gal β(1→4) Gal β(1→1) ceramide
Gal β(1→1) ceramide
Galactosylceramide
Glc β(1→1) ceramide
Glucosylceramide
7 4
5
Sphingomyelin
Phosphocholine
Ceramide
6
Sphingosine **Fatty acid**

FIGURE **19.16**

Lysosomal pathways for degradation of sphingolipids. The circled numbers refer to enzymes identified in Table 19.1 that are deficient in lipid storage diseases.

alleles of the gene encoding a particular enzyme must be present in an individual for disease symptoms to be manifest. Because of the large amounts of sphingolipids in nervous tissue, it is perhaps not surprising that most of the sphingolipidoses involve severely impaired central nervous system function.

Genetic defects in glycosphingolipid catabolism cause breakdown intermediates to accumulate in nervous tissue, with severe consequences.

The best known of the sphingolipidoses is **Tay–Sachs disease,** originally described in 1881, which is a deficiency of the lysosomal ***N*-acetylhexosaminidase A.** The enzyme deficiency causes accumulation of the ganglioside called GM_2, particularly in the brain (see Figure 19.16; shown also in Figure 10.8). The disease is devastating, causing nervous system degeneration, mental retardation, blindness, and death, usually by the age of four.

Although Tay–Sachs disease is rare in the general population, the defective gene is relatively common among Ashkenazic Jews (those of middle and eastern European extraction). Among American Jews, about 1 in 30 individuals carries the defective gene. Thus, two Jewish parents carry an appreciable risk of bearing a Tay–Sachs child. Because there is no known cure for the disease, attention has focused on prenatal detection. In fact, this was one of the first genetic diseases to be successfully diagnosed in conjunction with amniocentesis. Two prospective parents both shown to be heterozygous for this condition can be counseled that they have a 25% chance of conceiving a Tay–Sachs child.

Little is known about specific biochemical functions of sphingolipids, but their presence in the outer surface of plasma membranes of animal cells provides some tantalizing clues. Gangliosides are receptors for specific agents, such as cholera toxin, which binds to ganglioside GM_1, or influenza virus, which recognizes the sialic acid portion of certain gangliosides. Influenza virus encodes a neuraminidase, which cleaves these gangliosides as part of the viral entry process into cells. Inhibitors of this enzyme are being developed as possible drugs to treat or prevent influenza. Also of great interest is the observation that some gangliosides promote the growth of neural tissue in cell culture, suggesting that they might be used to promote regeneration of nerve tissue after spinal cord injury. In fact, New York Jets lineman Dennis Byrd was treated with GM_1 ganglioside, as well as with surgery, after breaking his neck in a 1992 football injury, and this may have been partially responsible for his remarkable recovery.

Lipids are being shown to have numerous roles as biological signaling agents. Examples that we have discussed include platelet-activating factor and lysophosphatidic acid. Among the sphingolipids, ceramide has been proposed, although not established, as participant in a pathway leading to **apoptosis,** or programmed cell death. This process, which we present in more detail in Chapter 28, involves the death of certain cells as part of normal development or after environmental damage to a cell so severe that the cell's survival would be harmful to the organism. Extracellular factors activate the enzymatic cleavage of sphingomyelin within membranes, and the ceramide that is released activates a number of protein kinases and phosphatases, a part of a signal transduction pathway not yet understood in detail.

Steroid Metabolism

$$CH_2{=}C(CH_3){-}C(H){=}CH_2$$ or

Isoprene

We turn now to an extraordinarily large and diverse group of lipids, the **isoprenoids,** or **terpenes.** These compounds are built up from one or more five-carbon units, activated derivatives of **isoprene.** The family includes steroids and bile acids; the lipid-soluble vitamins; the dolichol and undecaprenol phosphates we encountered in glycoprotein synthesis; phytol, the long-chain alcohol in chlorophyll; **gibberellins,** a family of plant growth hormones; insect juvenile hormones; the major components of rubber; coenzyme Q; and many more compounds.

Much of our discussion of isoprenoids focuses on a single steroid compound, cholesterol. As discussed in Chapter 10, this lipid is a major component of animal cell membranes, where it participates in modulation of membrane fluidity. In animals it also serves as precursor to all of the steroid hormones, to vitamin D, and to the bile acids, which aid in fat digestion. And as we discussed in Chapter 18, there is intense medical interest in cholesterol because of the relationships among diet, blood cholesterol levels, atherosclerosis, and heart disease. These biological relationships, coupled with the complex stereochemistry of its structure and the elegance of its biosynthetic pathway from a single low-molecular-weight precursor, have focused attention on this compound ever since its first isolation from gallstones, in 1784. Michael Brown and Joseph Goldstein have proclaimed cholesterol "the most highly decorated small molecule in biology," with 13 Nobel Prizes having been awarded to scientists, including themselves, who devoted major parts of their careers to cholesterol.

SOME STRUCTURAL CONSIDERATIONS

Steroids constitute a class of lipids that are derivatives of the saturated tetracyclic hydrocarbon **perhydrocyclopentanophenanthrene** (Figure 19.17; see also Chapter 10, page 324). Note the letters used to denote the four rings—A, B, C, and D, with D being the five-membered ring—and the carbon numbering system. Cholesterol differs from the basic ring system in having an aliphatic chain at C-17, axial methyl groups at C-10 and C-13, a double bond in ring B, and a hydroxyl group in ring A. The alcoholic functional group and the carbon chain at C-17 make cholesterol a **sterol,** which is the generic term used to identify steroid alcohols.

The cyclohexane rings of the steroids adopt puckered conformations, of which the more stable chair form (Chapter 10, Figure 10.9) predominates over the boat. This gives cholesterol a rigid molecular structure, with only the hydroxyl group generating a little polarity at one end (see Figure 10.9). Much of the cholesterol in lipoproteins and intracellular storage droplets is esterified at this position with a long-chain fatty acid, which makes the resultant cholesterol ester much more hydrophobic than cholesterol itself. The structure makes apparent how increasing concentrations of cholesterol in a membrane can reduce the fluidity of that membrane, by reducing the proportion of total lipid that can undergo a phase transition and by reducing the lateral mobility of polar lipids within the membrane.

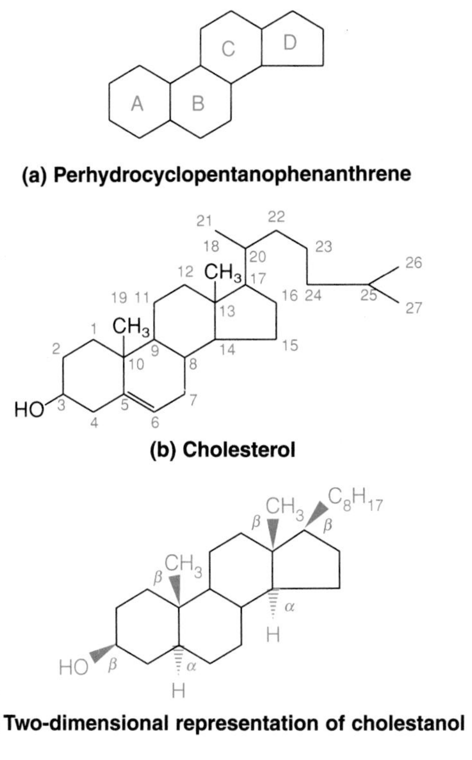

(a) Perhydrocyclopentanophenanthrene

(b) Cholesterol

(c) Two-dimensional representation of cholestanol

FIGURE **19.17**

Ring identification system (a) and carbon numbering system (b) used for steroids. (c) Structural conventions, with cholestanol as the example. α Substituents project below the plane of the steroid ring system (blue dashed wedge), and β substituents project above that plane (red solid wedge). The hydrogens at positions 5 and 14 have the α configuration, whereas the hydroxyl, the two methyl groups, and the aliphatic side chain at C-17 are all β substituents.

During most of our treatment of steroid metabolism, we shall use structural representations, as in Figure 19.17c, instead of three-dimensional configurational models. By convention, the methyl group at position 10 projects *above* the plane of the rings. This and all other substituents that project above the plane are denoted β and are drawn with a solid wedge. Substituents that project *below* the plane of the ring are called α and are denoted by a dashed wedge. These conventions are shown in Figure 19.17c for **cholestanol,** one of the two fully saturated derivatives of cholesterol.

BIOSYNTHESIS OF CHOLESTEROL

The pathway by which cholesterol is synthesized is worthy of study because of the diversity of metabolites synthesized by the pathway and the elegance of the pathway itself. Isotopic tracer studies have shown that all 27 carbons of cholesterol come from a two-carbon precursor—acetate. How could such a simple compound be built up to give a structure of the great complexity of cholesterol? That is what concerns us next.

Early Studies of Cholesterol Biosynthesis

Most of our early insights into cholesterol biosynthesis came from Konrad Bloch's laboratory in the 1940s. Taking note that cholesterol biosynthesis in vertebrates is confined largely to the liver, Bloch fed rats with acetate having ^{14}C either in the methyl group position or in the carboxyl group. After each administration, cholesterol was isolated from the liver and subjected to chemical degradation, with radioactive counting of the fragments. This procedure established the pattern shown in the margin, in which each carbon of cholesterol was found to originate from either the methyl carbon (blue) or the carboxyl carbon (red) of acetate (actually, acetyl-CoA).

HO ... $H_3C-C(=O)O^-$

Cholesterol, the precursor to all steroids, derives all of its carbon atoms from acetate.

Other early insights came from the realization that the five carbons of isoprene could be derived metabolically from three molecules of acetate, and the prediction that cholesterol was a product of the cyclization of the linear C_{30} hydrocarbon **squalene.** Squalene contains six isoprene units (delineated by red marks on the structures below), and its configuration makes it a plausible steroid precursor.

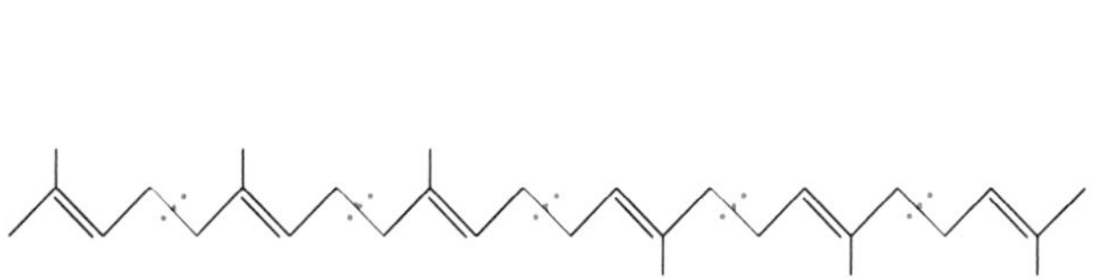

Squalene

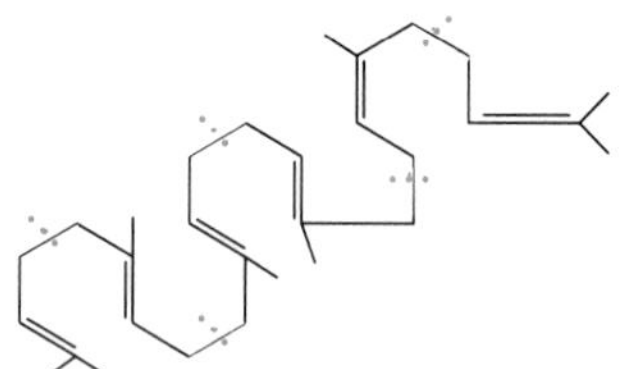

Postulated precyclization configuration of squalene

In 1956 another important development occurred, when Karl Folkers discovered that a C_6 organic acid, **mevalonic acid,** could permit the growth of certain acetate-requiring strains of *Lactobacillus.* Folkers showed that mevalonic acid was readily converted to an activated C_5 isoprenoid compound, **isopentenyl pyrophosphate.** Interestingly, *Lactobacillus* does not synthesize steroids, but it uses the first several steps of the pathway to synthesize other isoprenoid compounds. In animals, mevalonate is readily converted to squalene. Once this had been established, the stage was set for considering cholesterol biosynthesis as three distinct processes.

Hydroxymethylglutaryl-CoA reductase, which catalyzes an early reaction in cholesterol biosynthesis, is the major control point for the overall process.

1. Conversion of C_2 fragments (acetate) to a C_6 isoprenoid precursor (mevalonate)
2. Conversion of six C_6 mevalonates, via activated C_5 intermediates, to the C_{30} squalene
3. Cyclization of squalene and its transformation to the C_{27} cholesterol

Now let us consider these three processes in detail.

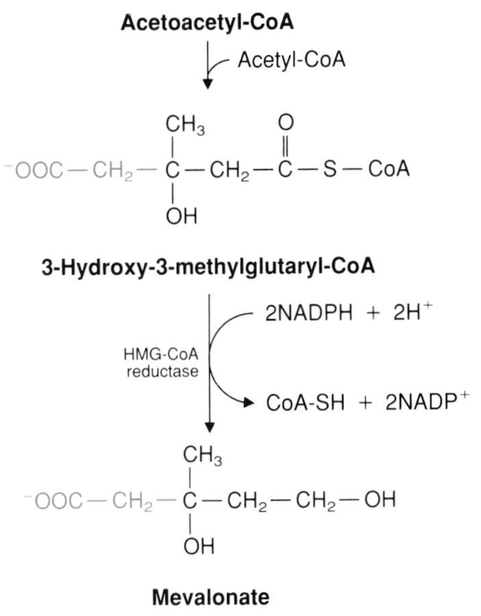

Acetoacetyl-CoA → (Acetyl-CoA) → 3-Hydroxy-3-methylglutaryl-CoA → (HMG-CoA reductase; 2NADPH + 2H⁺ → CoA-SH + 2NADP⁺) → Mevalonate

FIGURE **19.18**

Biosynthesis of mevalonate.

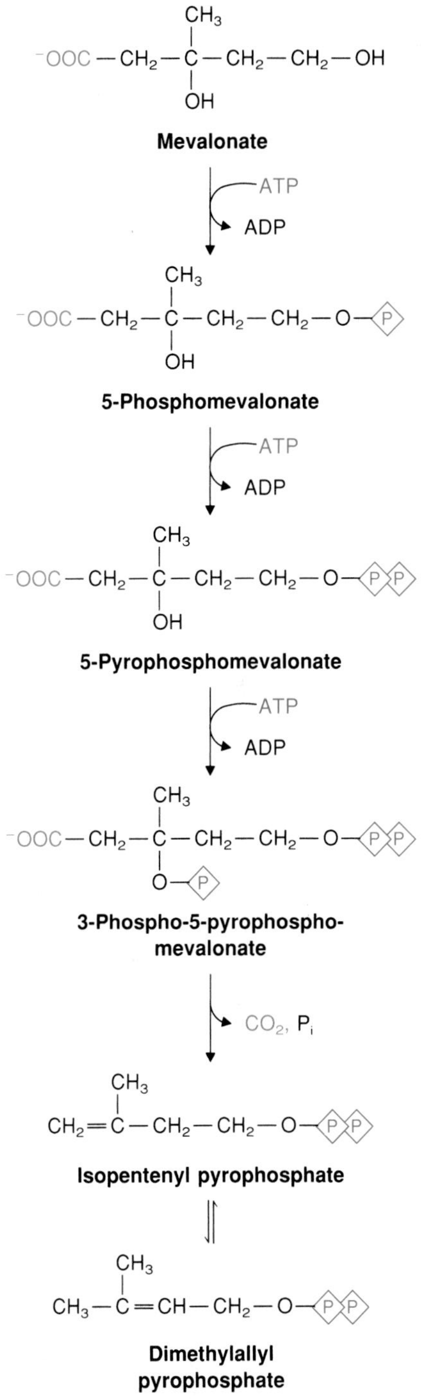

Mevalonate → (ATP → ADP) → 5-Phosphomevalonate → (ATP → ADP) → 5-Pyrophosphomevalonate → (ATP → ADP) → 3-Phospho-5-pyrophospho-mevalonate → (CO_2, P_i) → Isopentenyl pyrophosphate ⇌ Dimethylallyl pyrophosphate

FIGURE **19.19**

Conversion of mevalonate to isopentenyl pyrophosphate and dimethylallyl pyrophosphate. The carboxyl carbon of mevalonate (red) is lost as CO_2. There is evidence that an ATP-dependent decarboxylation occurs as a converted reaction, without the 3-phospho intermediate.

Stage 1: Formation of Mevalonate

The first part of the pathway is identical to reactions used in ketogenesis (see Chapter 18), although it occurs in a different cell compartment. Ketogenesis occurs in mitochondria, whereas cholesterol biosynthesis occurs in the cytosol and the endoplasmic reticulum.

Stage 1 begins with condensation of two molecules of acetyl-CoA to give acetoacetyl-CoA. Figure 19.18 shows the rest of this stage. Acetoacetyl-CoA reacts with a third molecule of acetyl-CoA to give 3-hydroxy-3-methylglutaryl-CoA (HMG-CoA). Recall from Figure 18.21 that during ketogenesis, HMG-CoA cleaves to give acetoacetate plus acetyl-CoA in the mitochondrial matrix. However, the HMG-CoA lyase that accomplishes this cleavage is missing from the endoplasmic reticulum, where cholesterol biosynthesis begins. Instead, **HMG-CoA reductase** catalyzes the four-electron, NADPH-dependent reduction of the HMG-CoA to mevalonate. This is the major step that regulates the overall pathway.

Stage 2: Synthesis of Squalene from Mevalonate

The next several reactions, shown in Figures 19.19 and 19.20, occur in the cytosol. First, mevalonate is activated by three successive phosphorylations (Figure 19.19). The third phosphorylation, at position 3, sets the stage for a decarboxylation via *trans* elimination, to give isopentenyl pyrophosphate. One molecule of this isomerizes to the C_5 **dimethylallyl pyrophosphate.** The latter compound, as shown in Figure 19.20, reacts with a second molecule of isopentenyl pyrophosphate to give the C_{10} **geranyl pyrophosphate,** and still another molecule of isopentenyl pyrophosphate reacts with this product to give the C_{15} **farnesyl pyrophosphate.**

Figure 19.21 shows the final reactions of squalene synthesis. The first of these reactions is catalyzed by **farnesyl transferase** (also called squalene synthase), which is bound to membranes of the endoplasmic reticulum. This NADPH-dependent enzyme joins 2 molecules of farnesyl pyrophosphate to give **presqualene pyrophosphate,** which then undergoes pyrophosphate elimination and rearrangement via a cyclopropyl cation intermediate to yield squalene. All subsequent reactions occur in the endoplasmic reticulum.

A remarkable feature of this part of the pathway is its stereochemistry. In the 1960s two British scientists, George Popják and John Cornforth, identified 14 "stereochemical ambiguities"—that is, 14 steps in the overall process that could go in either of two ways. For example, the triphosphorylated mevalonate derivative shown in Figure 19.19 could undergo decarboxylation by either a *cis* or *trans* elimination of the carboxyl and phosphoryl groups. Thus, 2^{14}, or 16,384, *different*

Cyclization of squalene, a C_{30} hydrocarbon, creates the four-ring sterol nucleus.

FIGURE **19.20**

Conversion of isopentenyl pyrophosphate and dimethylallyl pyrophosphate to farnesyl pyrophosphate.

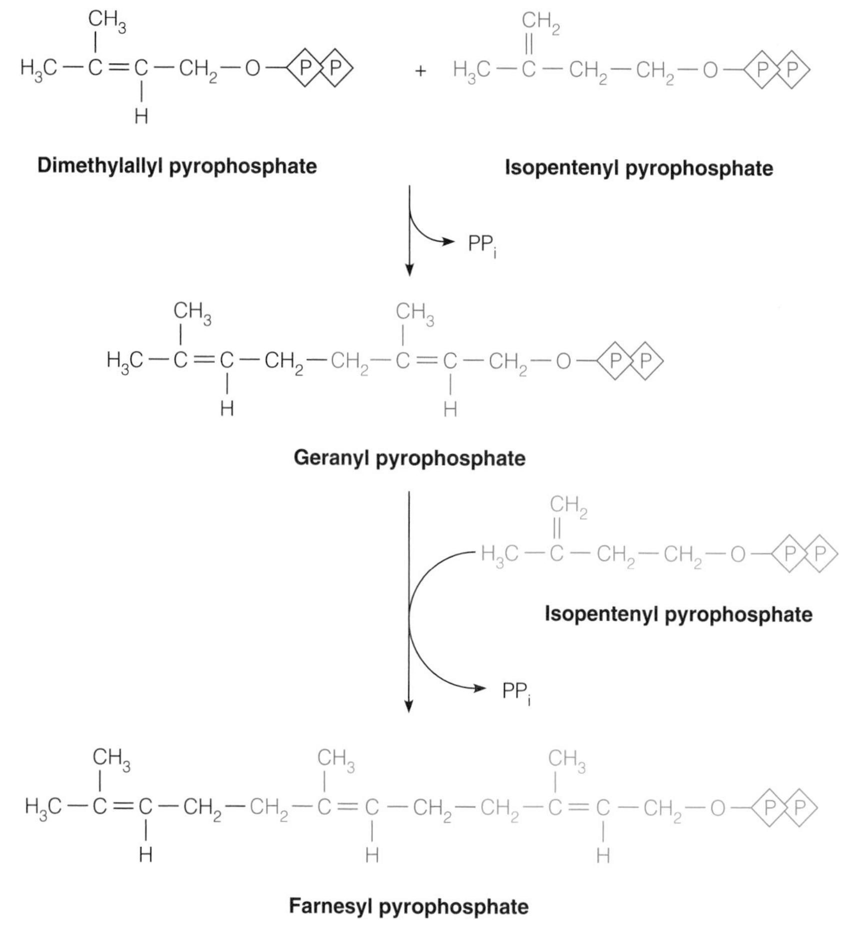

FIGURE **19.21**

Conversion of farnesyl pyrophosphate to squalene. Note the head-to-head reaction between two molecules of farnesyl pyrophosphate, which involves the formation of an activated cyclopropane intermediate, coupled with loss of one pyrophosphate.

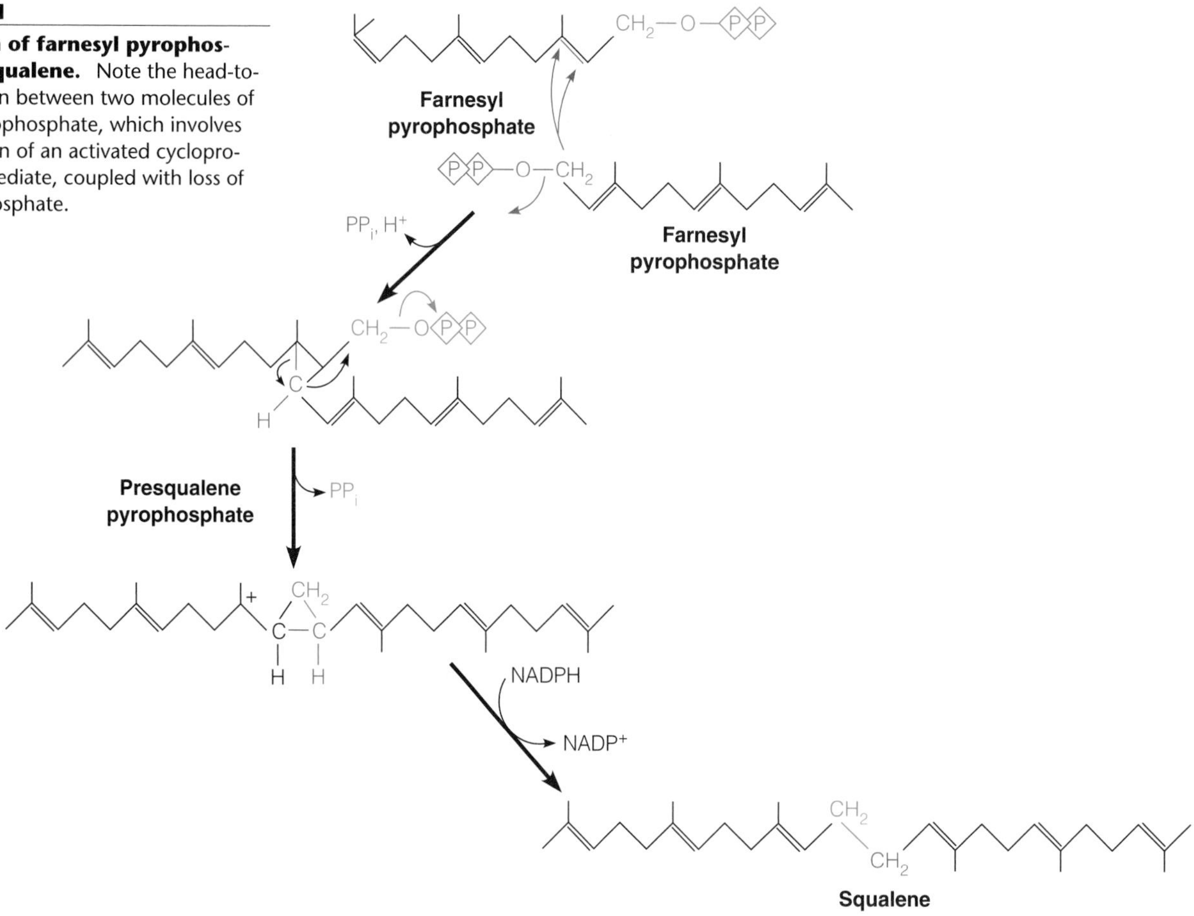

stereochemical routes were possible for conversion of mevalonate to squalene. Remarkably, these scientists and their colleagues were able to identify the single stereochemical pathway that actually takes place among these 16,384 possibilities.

Stage 3: Cyclization of Squalene to Lanosterol and Its Conversion to Cholesterol

The cyclization of squalene to lanosterol and the conversion of lanosterol to cholesterol are shown in Figure 19.22. The formation of lanosterol, which has the four-ring sterol nucleus, occurs in two steps. First, a mixed-function oxidase introduces an epoxide function at carbons 2 and 3. Protonation of this functional group initiates a series of *trans* 1,2 shifts of methyl groups and hydride ions, to produce lanosterol. A series of about 20 reactions follows, involving double-bond reductions and three demethylations; one methyl group is removed from C-14 and two from C-4. The penultimate product is 7-dehydrocholesterol, which undergoes a final reduction to yield cholesterol.

Control of Cholesterol Biosynthesis

As noted earlier, HMG-CoA reductase, which catalyzes the committed reaction in cholesterol biosynthesis, represents a major target for regulation of the overall pathway. It has long been known from feeding studies that dietary cholesterol efficiently suppresses the endogenous synthesis of cholesterol. This control occurs at both transcriptional and translational levels, with cholesterol itself or one of its derivatives playing an active role in translational control. In addition, the enzyme is regulated hormonally, by insulin and glucagon. Earlier evidence suggested that this control involved cyclic AMP–dependent phosphorylation of the enzyme protein, but more recent evidence supports the idea that these hormones are affecting synthesis of the enzyme at the transcriptional level. The mammalian enzyme is subject of reversible phosphorylation, but the regulatory significance of this reaction is not yet clear.

In vertebrates, cholesterol synthesis is controlled beautifully, through the rate at which cholesterol enters cells from the bloodstream. As we discussed in Chapter 18, homeostasis is maintained by a mechanism that coordinates dietary intake of cholesterol, rate of endogenous cholesterol synthesis in the liver (and to a lesser extent in the intestine), and rate of cholesterol use by cells. That mechanism involves the LDL receptor, the agent most responsible for transporting cholesterol in the bloodstream.

BILE ACIDS

Now let us turn to the use of cholesterol for synthesis of other important metabolites—bile acids and steroid hormones. As mentioned in Chapter 18, bile acids are steroid derivatives with detergent properties, which emulsify dietary lipids in the intestine and thereby promote fat digestion and absorption. They are secreted from the liver, stored in the gallbladder, and passed through the bile duct and into the intestine. Biosynthesis of bile acids represents the major metabolic fate of cholesterol, accounting for more than half of the 800 mg/day that is metabolized in the normal human adult. By contrast, steroid hormone synthesis accounts for only about 50 mg of cholesterol metabolized per day.

Although about 400 mg of bile acids is synthesized daily, much more than 400 mg is secreted into the intestine. Most of the bile acids secreted into the upper

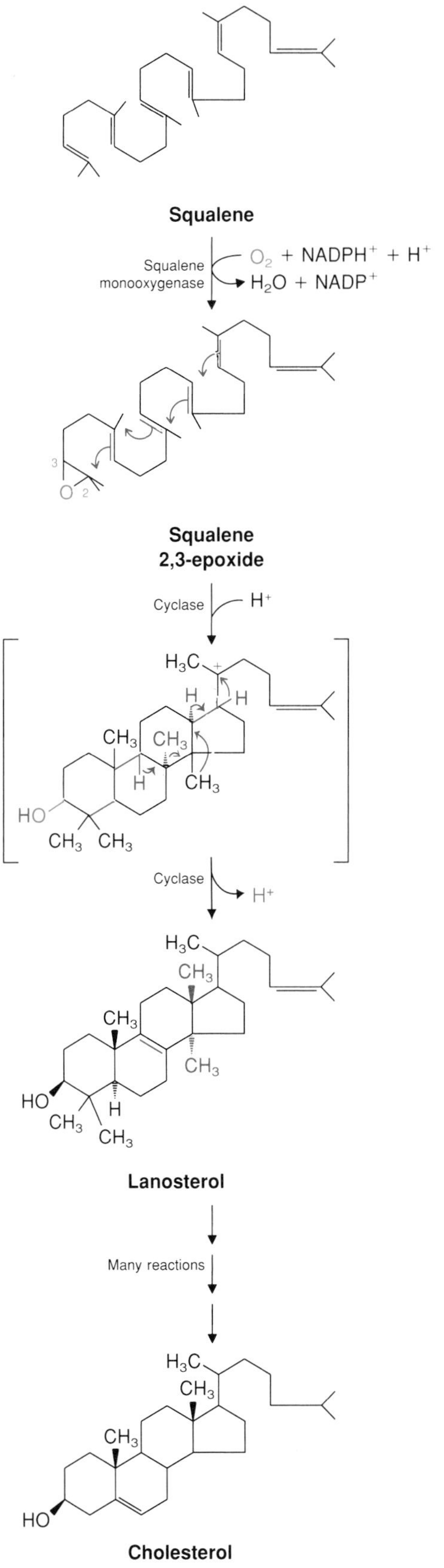

FIGURE **19.22**

Conversion of squalene to cholesterol. Formation of squalene epoxide leads to a series of double-bond electron shifts that close the four rings, and a migration of a carbon atom from C-14 to C-13 yields the first steroid intermediate, lanosterol. Many subsequent reactions lead to 7-dehydrocholesterol, which undergoes reduction to cholesterol.

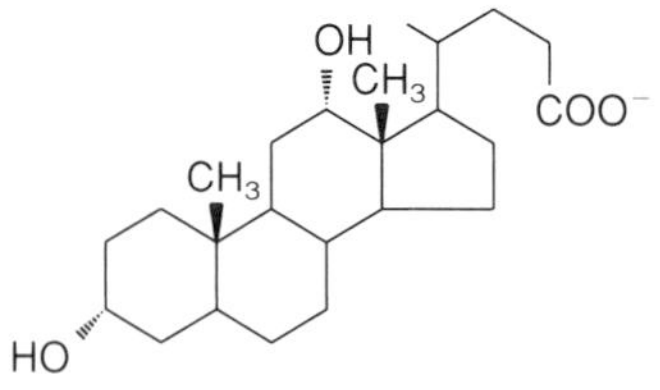

OH CH$_3$ COO$^-$ CH$_3$ HO

Deoxycholate

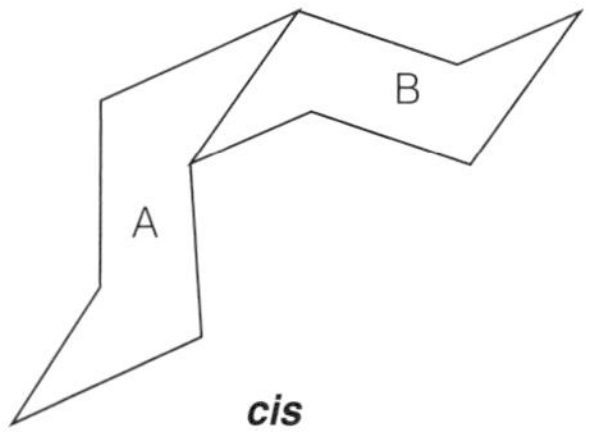

A B

cis

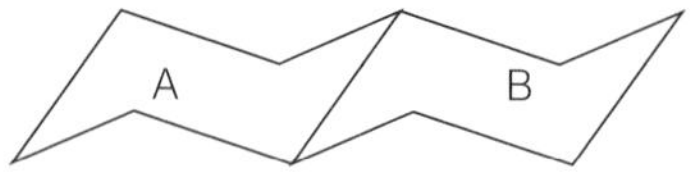

A B

trans

small intestine are absorbed in the lower small intestine and returned to the liver for reuse, through the portal blood. This process, which handles 20 to 30 g of bile acids per day, is called the **enterohepatic circulation.** Daily elimination of bile acids in the feces amounts to just 0.5 g/day or less.

The most abundant bile acids in humans are **cholic acid** and **chenodeoxycholic acid** (shown in Figure 19.23 as the respective bile salts, cholate and chenodeoxycholate). These are usually conjugated in amide linkage with the amino acid **glycine** or **taurine,** giving compounds called **bile salts.** The cholic acid conjugates with glycine and taurine are called **glycocholate** and **taurocholate,** respectively. Another bile acid, **deoxycholate,** is abundant in the bile of some other mammals. It is widely used as a laboratory reagent, to solubilize membrane proteins.

In the biosynthetic routes to cholate, glycocholate, and taurocholate outlined in Figure 19.23, a series of hydroxylations occurs, catalyzed by microsomal P450 mixed-function oxidases. The first of these, at C-7, is rate-determining and hence plays the major role in controlling the overall pathway. The dehydrogenation of the hydroxyl to a ketone at carbon 3, followed by its reduction back to a hydroxyl, inverts its configuration and joins the A and B rings in a *cis* configuration, yielding 3,7,12-trihydroxyprostanoate. At present we know little about the mechanism of the side chain modification, which converts the eight-carbon saturated hydrocarbon to a five-carbon chain with a terminal carboxylate.

STEROID HORMONES

Cholesterol is the biosynthetic source of all steroid hormones, the extracellular messengers elaborated by the gonads and the adrenal cortex, plus the placenta in pregnant females. In this chapter we review the biosynthetic pathways to steroid hormones; their actions are discussed in Chapter 23. In general, steroid hormones control metabolism at the gene level. They react with intracellular protein receptors, and the hormone–receptor complexes bind to specific sites on the genome and affect transcription of neighboring genes.

The principal categories of steroid hormones in vertebrates are progestins, glucocorticoids, mineralocorticoids, androgens, and estrogens.

Five major classes of hormone will concern us: (1) the **progestins** (progesterone), which regulate events during pregnancy and are the precursors to all other steroid hormones; (2) the **glucocorticoids** (cortisol and corticosterone), which promote gluconeogenesis and, in pharmacological doses, suppress inflammation reactions; (3) the **mineralocorticoids** (aldosterone), which regulate ion balance by promoting reabsorption of K^+, Na^+, Cl^-, and HCO_3^- in the kidney; (4) the **androgens** (androstenedione and testosterone), which promote male sexual development and maintain male sex characteristics; and (5) the **estrogens** (estrone and estradiol), or female sex hormones, which support female characteristics. Most of these hormones are shown in Figure 19.24, which also summarizes their routes of synthesis (described on page 695). In each case the side chain in cholesterol is either greatly shortened or nonexistent.

A general feature of steroid hormones is that they are not stored for release after synthesis. Therefore, the level of a circulating hormone is controlled primarily by its rate of synthesis, which is often controlled ultimately by signals from the brain. These signals usually act through intermediary hormones. For example, the neurohormone **corticotropin releasing factor** (**CRF**), is released from cells in the hypothalamus in response to central nervous system inputs (see Chapter 23). CRF stimulates release from the pituitary gland of **corticotropin,** or **adrenocorticotropic hormone** (**ACTH**), which in turn stimulates the synthesis of glucocorticoids in adrenal cortex.

Activation of steroid hormone synthesis involves stimulation of both hydrolysis of cholesterol esters and uptake of cholesterol into mitochondria of cells in the target organ. There a cytochrome P450 enzyme called **cholesterol desmolase** hydroxylates the side chain at C-20 and C-22 and cleaves it, to yield

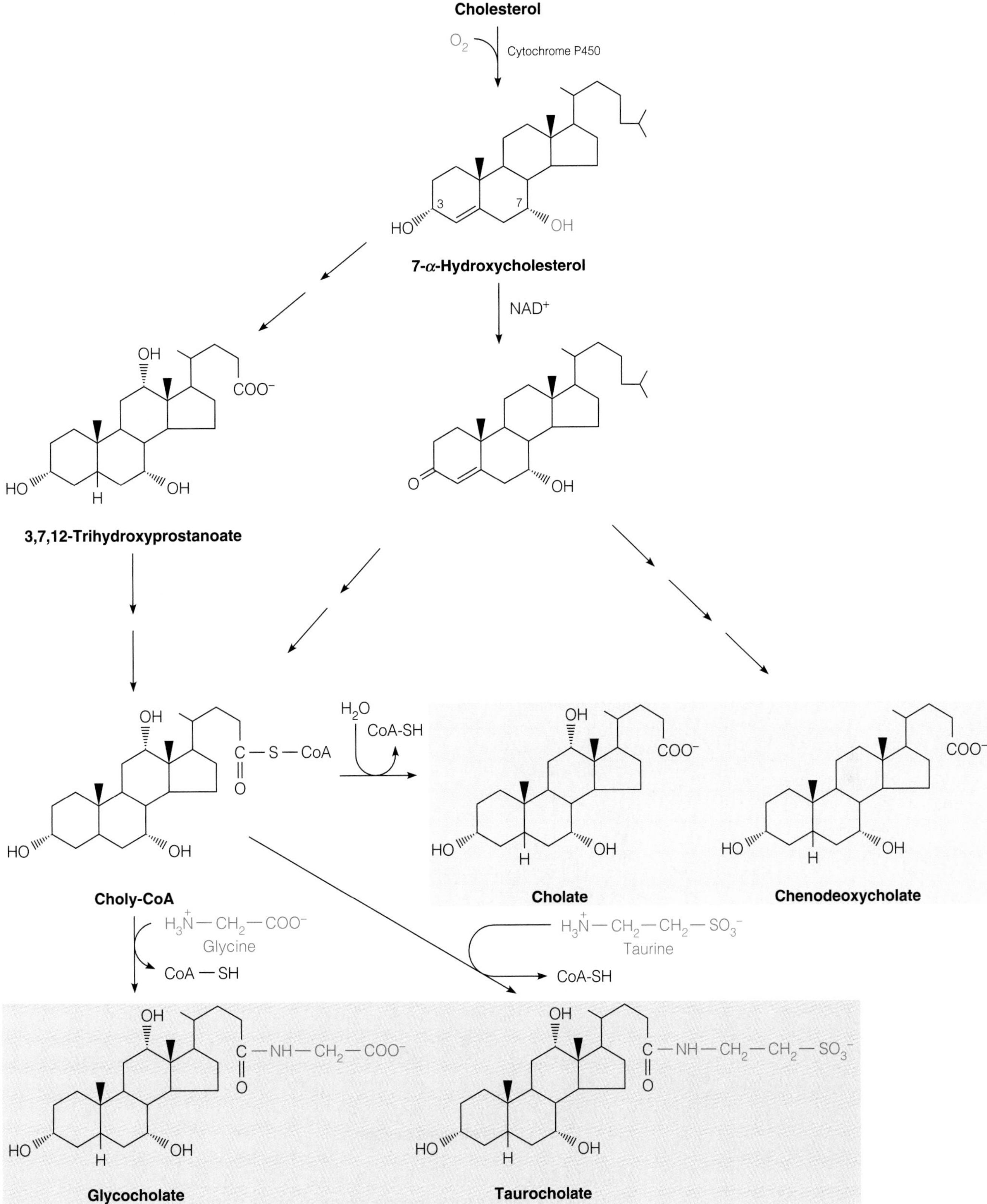

FIGURE **19.23**

Biosynthesis of bile acids and salts from cholesterol. The major pathway, beginning down the center of the figure, starts with hydroxylation of cholesterol by a mixed-function oxidase.

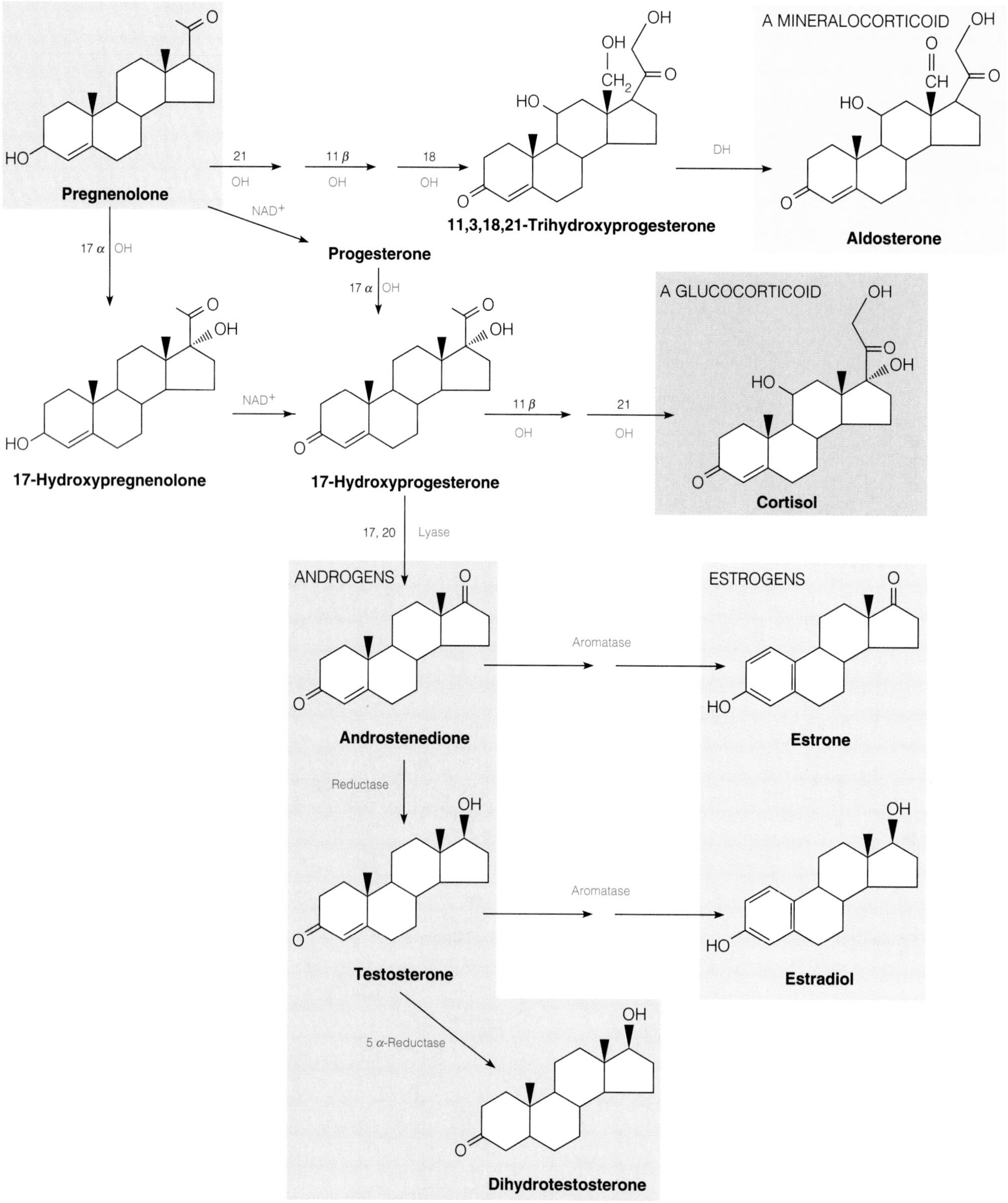

FIGURE **19.24**

Biosynthetic routes from pregnenolone to other steroid hormones. OH = hydroxylase; DH = dehydrogenase. An alternative pathway leads from pregnenolone through progesterone to 17-hydroxyprogesterone.

pregnenolone, the precursor to all other steroid hormones. Pregnenolone is converted to the steroid hormone progesterone by dehydrogenation and double-bond isomerization.

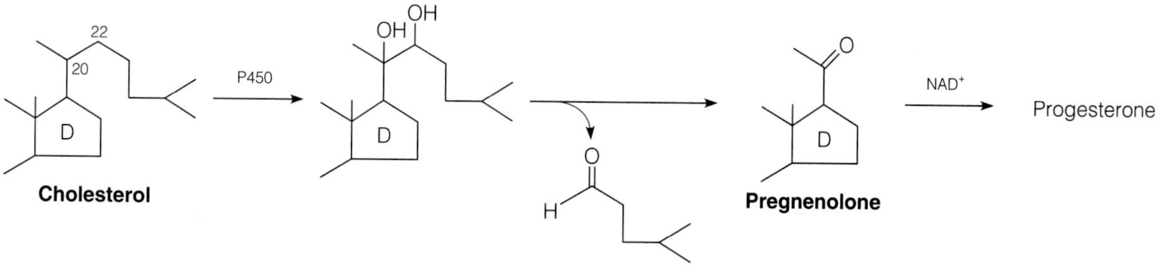

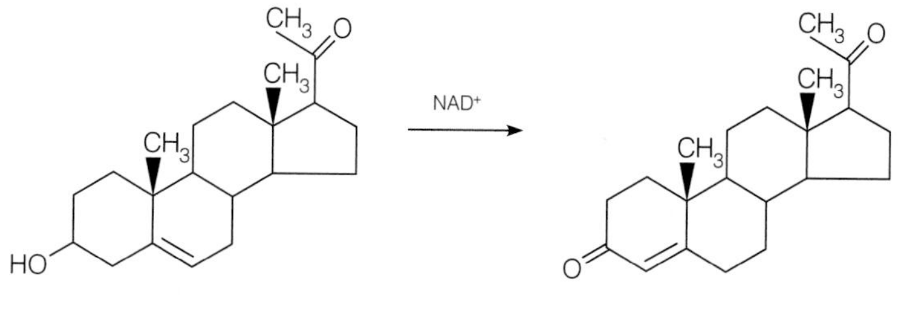

Cholesterol → P450 → Pregnenolone → NAD^+ → Progesterone

Pregnenolone → NAD^+ → Progesterone

The conversion of progesterone to other steroid hormones is shown in Figure 19.24. Hydroxylation at C-21 by an adrenal cortex enzyme, followed by two more hydroxylations and a dehydrogenation to form an aldehyde group, gives aldosterone, a mineralocorticoid. Hydroxylation of progesterone at C-17 gives 17 α-hydroxyprogesterone, the precursor to all other steroids. Two hydroxylations of this intermediate give cortisol (a glucocorticoid), primarily in the adrenal gland. An enzyme in adrenal cortex and in gonads cleaves the side chain of 17-hydroxyprogesterone at C-17, giving **androstenedione,** a precursor to androgens and estrogens (and a legal dietary supplement for professional baseball players, including Mark McGwire). The enzymes in these processes form a complex called **aromatase.** Note that the reactions catalyzed by this complex represent the only known route for synthesis of aromatic rings in animal cells. Testosterone undergoes reduction at C-5, giving 5α-dihydrotestosterone, a somewhat more potent androgen.

Pregnenolone is an intermediate en route from cholesterol to all other known steroid compounds.

Human enzyme deficiencies have been described for all of the above enzymes. A deficiency of the 17-hydroxylase decreases levels of cortisol, androgens, and estrogens, with severe effects upon sexual maturation. A deficiency of the 21-hydroxylase inhibits synthesis of glucocorticoids and mineralocorticoids, leading to overproduction of testosterone in the adrenal glands. At the same time, the underproduction of cortisol interferes with a feedback loop of hormonal control involving corticotropin releasing factor (Chapter 23) and ACTH; increased ACTH secretion stimulates the adrenals to grow and synthesize steroids, exacerbating the testosterone overproduction. This causes virilization (masculinization) of females. A deficiency of the 5α-reductase decreases effective androgen levels and leads to feminization of males. Fortunately, these and other steroid abnormalities can be treated with hormone replacement therapy, if detected early enough in life.

Hundreds of synthetic compounds with steroid hormone–like activity have been tested and used for various purposes. Widely used synthetic steroids are anti-inflammatory glucocorticoids. **Diethylstilbestrol,** a synthetic estrogen, was widely used to promote growth of beef cattle, until it was found to be potentially carcinogenic at the levels found in meat from treated cattle. Oral contraceptives are formulated with compounds containing progesterone and estrogen activities. Two widely used synthetic estrogens are **norethynodrel** and **mestranol.**

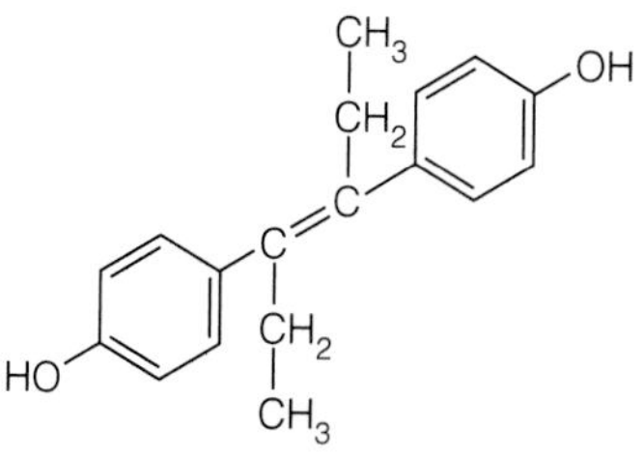

Diethylstilbestrol

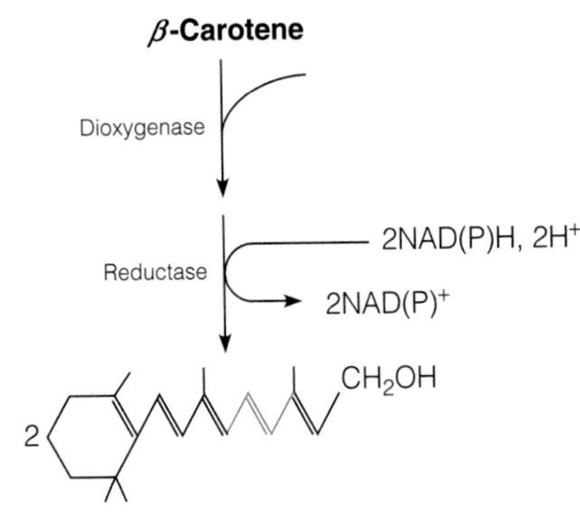

FIGURE **19.25**

Synthesis of vitamin A_1, all-*trans*-retinol. The 11-*cis* derivative of all-*trans*-retinol is a component of the visual pigment rhodopsin. Red color identifies the bond that undergoes *cis–trans* isomerization in the visual process.

C≡CH OH O **Norethynodrel**

C≡CH OH H_3CO **Mestranol**

Mixtures of either drug with progesterone inhibit pituitary secretion of hormones that control the female reproductive cycle. This inhibition suppresses maturation of the ovarian follicle and ovulation.

Recent concern has focused upon environmental substances produced through human activity, such as certain pesticides, which happen to have estrogen-like activity. Some of these have been shown to interact with estrogen receptors and to stimulate similar biochemical responses. Although solid evidence is difficult to obtain, these "endocrine disrupters" are thought to be responsible for declines in fertility of a number of animal species, possibly including humans.

Mammalian cells lack the capacity for complete degradation of steroid compounds. Although a number of catabolic reactions do occur, most steroids and their metabolites become conjugated, through their hydroxyl groups, to glucuronate or sulfate. Either modification greatly increases solubility of the steroid and facilitates its elimination in the urine.

Other Isoprenoid Compounds

LIPID-SOLUBLE VITAMINS

The four lipid-soluble vitamins—A, D, E, and K—are all isoprenoid compounds. They are made up, like steroids, of activated five-carbon units. Thus, as a group they have a structural relatedness not seen among the water-soluble vitamins. On the other hand, the water-soluble vitamins have a *functional* uniformity in that all are designed to carry mobile metabolic groups, whereas the lipid-soluble vitamins are diverse in their functions.

Vitamin A

Vitamin A, otherwise called ***trans*-retinol,** is an isoprenoid alcohol that plays a key role in vision. It also plays a role in controlling animal growth, somehow stimulating development of the nervous system. The vitamin can be either consumed in the diet or biosynthesized from **β-carotene,** an isoprenoid compound especially abundant in carrots. Vitamin A biosynthesis is shown in Figure 19.25.

Vitamin A plays an important role in the visual process in rod cells of the retina, the cells primarily responsible for low-light vision, with relatively little color detection (Figure 19.26). The rod cell outer segments contain lamellar protein disks that are rich in the protein **opsin.** *trans*-Retinol in the disks undergoes isomerization and dehydrogenation to give **11-*cis*-retinal.** The chemical changes in photoreception are shown in Figure 19.27. 11-*cis*-Retinal forms a Schiff base with a lysine residue in opsin, giving **rhodopsin** (step 1). Rhodopsin has very strong light absorption in the 400- to 600-nm range (the visible wavelength of

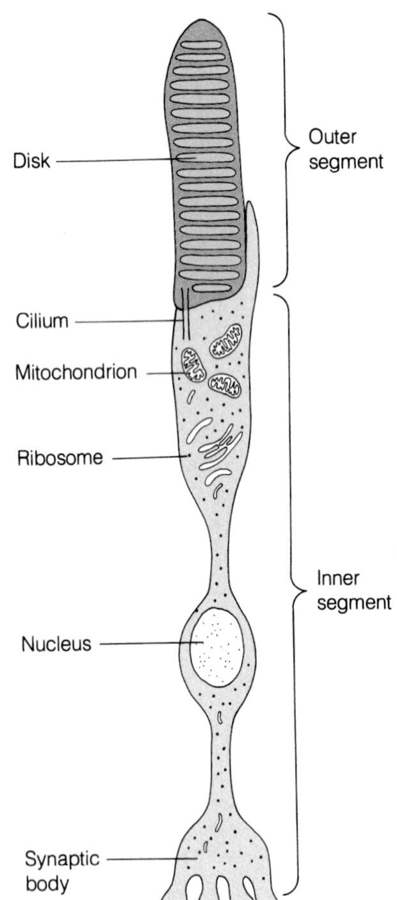

FIGURE **19.26**

Schematic drawing of a rod cell. The outer segment is a stack of membranous disks, which contain the photoreceptive pigments. This segment is connected by a thin cilium, to the inner segment, which contains the cell nucleus, cytosol, and synaptic body. The potential change produced in the outer segment travels to the synaptic body and is transmitted to one or more of the neurons of the retina.

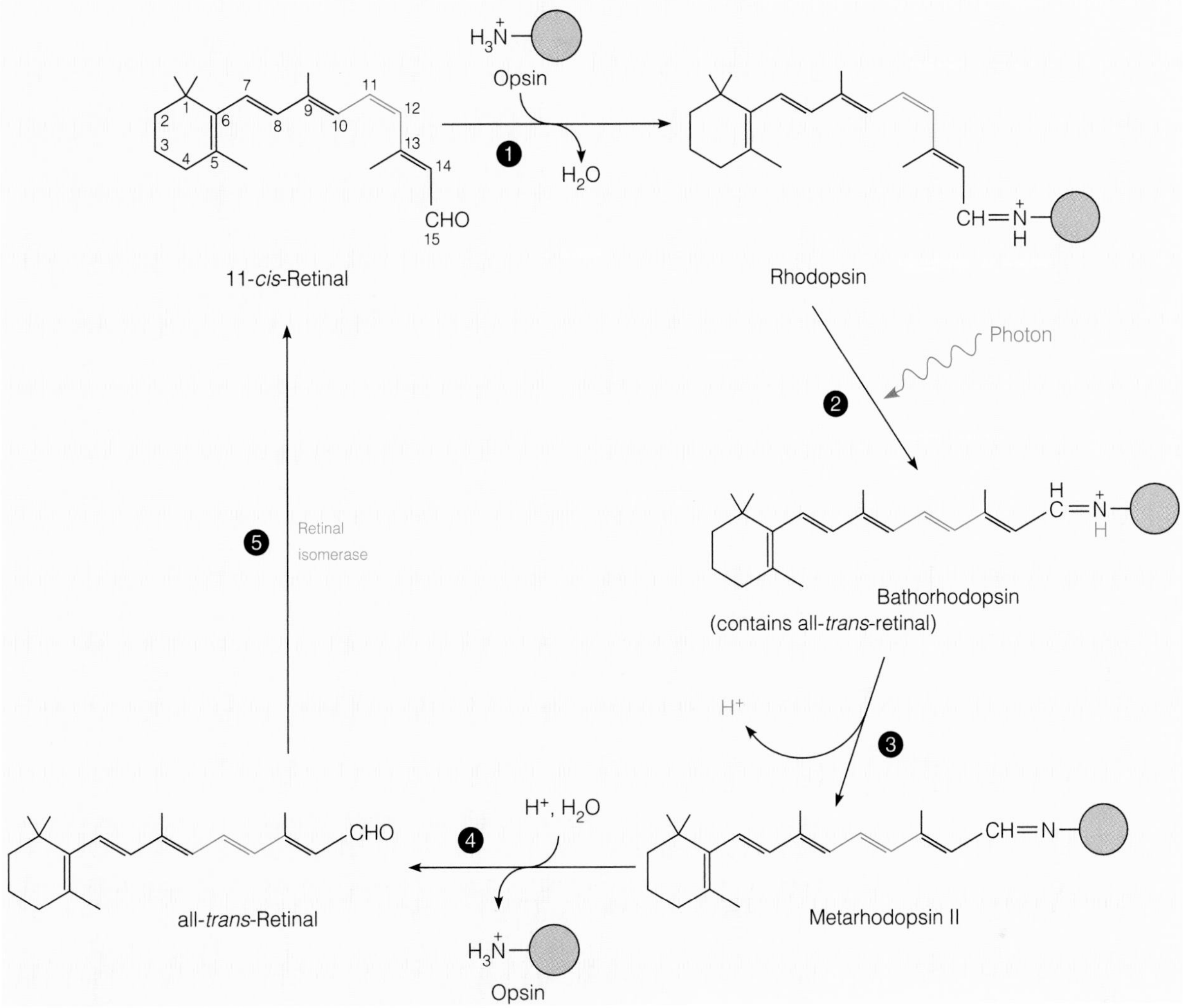

FIGURE **19.27**

The chemical changes in photoreception. 11-*cis*-Retinal and opsin in a rod cell combine to form rhodopsin. Absorption of a photon of light leads to the chemical changes shown in steps 2 and 3. Metarhodopsin II is the species that activates transducin (not shown) to initiate the visual cascade described in Chapter 23. After about 1 second, metarhodopsin II dissociates into all-*trans*-retinal, which isomerizes as the cycle begins again.

the spectrum). Absorption of a photon of light triggers a chain of events leading to neural excitation. The excited retinal in rhodopsin isomerizes to an all-*trans* form (step 2), followed by several conformational changes and release of all-*trans*-retinal (steps 3 and 4). After isomerization (step 5), the process can begin anew. Step 3 in Figure 19.27 is the key to the transduction of photon reception to a neural action potential. This process involves cyclic GMP and a G protein called **transducin.** Further details of this process are presented in Chapter 23.

Isomerization of a protein-bound form of vitamin A in the retina is the mechanism by which light energy is received in the eye.

As developmental regulators, retinoids (derivatives of retinol) act somewhat like steroid hormones (see page 858). They interact with specific receptor proteins in the cell nucleus. The ligand–receptor complexes bind to specific DNA sequences, where they control the transcription of particular genes. Identification of those genes is an active area of current research.

Vitamin D

1,25-Dihydroxycholecalciferol controls bone metabolism by regulating intestinal absorption of calcium.

The most abundant form of vitamin D is vitamin D_3, or **cholecalciferol.** This is not truly a vitamin, because it is not required in the diet. Instead, it arises by

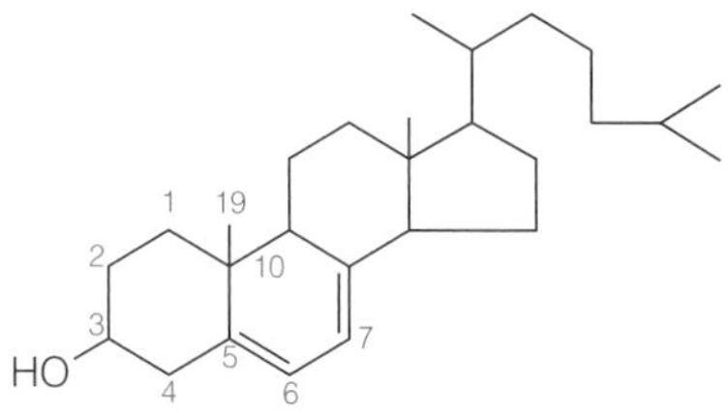

7-Dehydrocholesterol

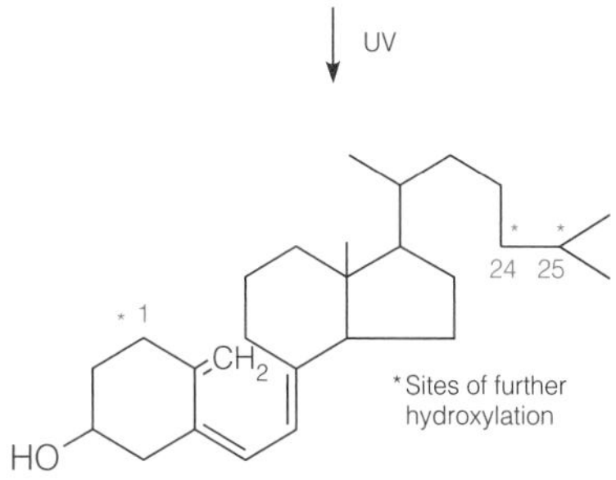

UV

Cholecalciferol (vitamin D_3)

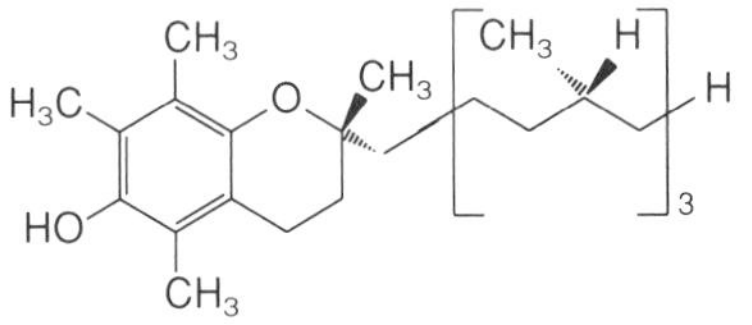

*Sites of further hydroxylation

synthesis from 7-dehydrocholesterol, an intermediate in cholesterol biosynthesis. It is more accurate to think of vitamin D_3 as a prohormone, because it is converted to a metabolite that acts analogously to a steroid hormone. Its action involves the regulation of calcium and phosphorus metabolism, particularly with respect to synthesis of the inorganic matrix of bone, which consists largely of calcium phosphate.

In skin cells, 7-dehydrocholesterol undergoes ultraviolet photolysis to give cholecalciferol. Because the UV rays come from sunlight, insufficient sunlight exposure can cause a deficiency of vitamin D_3 and result in the bone malformation known as **rickets.** Vitamin D_3 is often added to dairy products as a dietary supplement, because sunlight exposure is limited in many regions for much of the year.

Cholecalciferol undergoes two successive hydroxylations, each catalyzed by a mixed-function oxidase. The first, at carbon 25, involves a microsomal enzyme system in liver. 25-Hydroxycholecalciferol is then transported to the kidney, where a mitochondrial enzyme hydroxylates it at carbon 1. This reaction is activated by **parathyroid hormone,** which is secreted from the parathyroid gland when calcium levels are low. When calcium levels are adequate, the second hydroxylation occurs at C-24, instead of C-1, to give an inactive metabolite.

1,25-Dihydroxycholecalciferol, or 1,25(OH)D_3, is the hormonally active form of vitamin D. This compound migrates to target cells in the intestine and in osteoblasts (bone cells), where it binds to protein receptors that migrate to the cell nucleus. In intestine the hormone–receptor complex stimulates transcription, resulting in synthesis of a protein that stimulates calcium absorption into the bloodstream. In osteoblasts, 1,25(OH)D_3 stimulates calcium uptake for deposition as calcium phosphate.

α-Tocopherol (vitamin E)

Vitamin E

Vitamin E, also called **α-tocopherol,** was originally recognized in nutritional studies as an agent that prevented sterility in rats. The vitamin appears to play an antioxidant role, particularly in preventing attack of peroxides on unsaturated fatty acids in membrane lipids (see also page 554, Chapter 15). α-Tocopherol does prevent fatty acid peroxidation in vitro. However, vitamin E deficiency results in additional symptoms that are not relieved by other antioxidants. Therefore, additional biological roles seem likely for this vitamin.

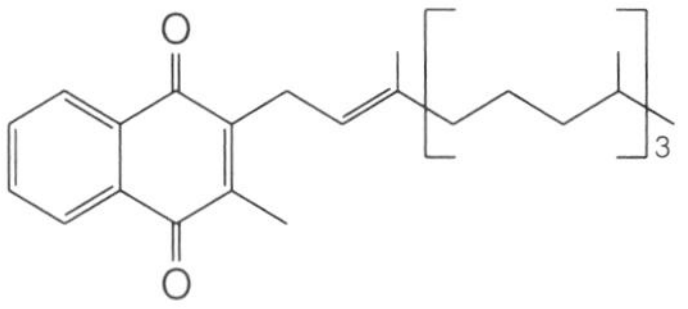

Phylloquinone (vitamin K_1)

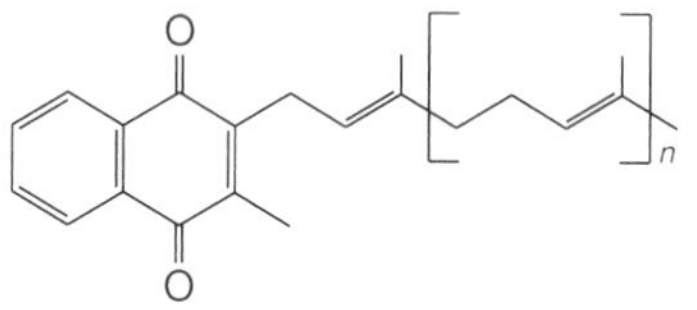

Menaquinone (vitamin K_2)

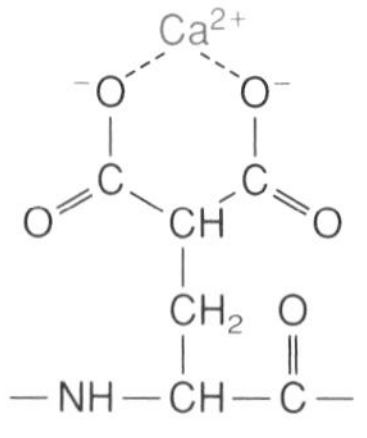

Vitamin K

Vitamin K was originally discovered as a lipid-soluble substance involved in blood coagulation. Vitamin K_1, or **phylloquinone,** is found in plants; the quinone portion of this molecule has a largely saturated side chain. Another form of the vitamin, vitamin K_2, or **menaquinone,** is found largely in animals and bacteria. Menaquinone has a partly unsaturated side chain. In animals, vitamin K_2 is essential for the carboxylation of glutamate residues in certain proteins, to give **γ-carboxyglutamate.** This modification allows the protein to bind calcium, an essential event in the blood clotting cascade (discussed in Chapter 11). Newborn children routinely receive vitamin K injections, because most of our vitamin K comes from intestinal bacteria, which have not yet colonized the guts of newborns.

Carboxylation of glutamate residues occurs in other proteins that are active in the mobilization or transport of calcium. The carboxylating enzyme uses the reduced, hydroquinone, form of vitamin K. During the reaction, the hydroquinone becomes oxygenated to a quinone epoxide, and this reaction facilitates the attack of CO_2 on C-4 of a glutamate residue by deprotonating that carbon.

A γ-carboxyglutamate residue complexed with calcium

OTHER TERPENES

Terpene is a generic term for all compounds that are biosynthesized from isoprene precursors. Therefore, the compounds we have been discussing—cholesterol, bile acids, steroids, and lipid-soluble vitamins—are terpenes; they receive special attention because of their importance in animal metabolism. Here we take

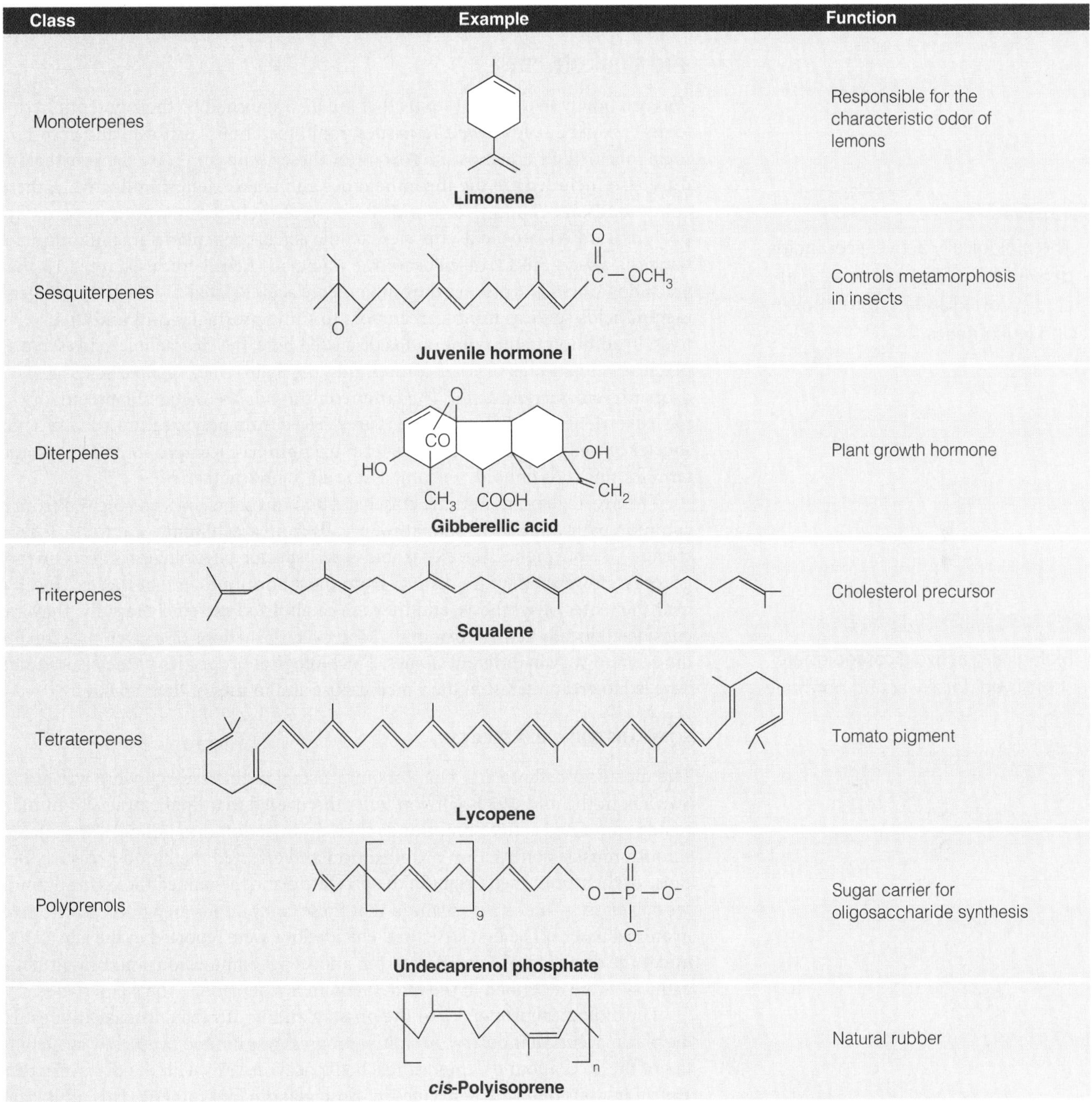

Class	Example	Function
Monoterpenes	**Limonene**	Responsible for the characteristic odor of lemons
Sesquiterpenes	**Juvenile hormone I**	Controls metamorphosis in insects
Diterpenes	**Gibberellic acid**	Plant growth hormone
Triterpenes	**Squalene**	Cholesterol precursor
Tetraterpenes	**Lycopene**	Tomato pigment
Polyprenols	**Undecaprenol phosphate**	Sugar carrier for oligosaccharide synthesis
	***cis*-Polyisoprene**	Natural rubber

FIGURE **19.28**

Some terpene compounds. These examples are representative of an enormous class of natural products.

a brief glimpse at the vast range of other terpene compounds. They include insect hormones and plant growth hormones, as well as the lipid-linked sugar carriers we encountered in Chapter 16.

Terpenes are biosynthesized ultimately from isopentenyl pyrophosphate (C_5) and dimethylallyl pyrophosphate (C_5). When these combine to yield geranyl pyrophosphate (C_{10}), any terpene formed thereby is called a **monoterpene.** When a compound is formed from 1 mole of farnesyl pyrophosphate (C_{15}), the product is called a **sesquiterpene. Triterpenes** (C_{30}) are formed from 2 moles of farnesyl pyrophosphate. Geranylgeranyl pyrophosphate (C_{20}) yields either **diterpenes** (C_{20}) or **tetraterpenes** (C_{40}). The dolichols and undecaprenol, introduced in Chapter 16, are examples of **polyprenols** (polyisoprenoid alcohols), which have more than 50 carbons. Structures of some common terpenes are given in Figure 19.28.

Eicosanoids: Prostaglandins, Thromboxanes, and Leukotrienes

The biologically active eicosanoids, derived from arachidonic acid, include prostaglandins, thromboxanes, and leukotrienes.

Biologically active eicosanoids are short-lived, locally acting hormones.

We turn finally to a class of lipids that are distinguished by their potent physiological properties, low levels in tissues, rapid metabolic turnover, and common metabolic origin. The most important of these compounds are the **prostaglandins;** also included are the **thromboxanes** and **leukotrienes.** Collectively, these compounds are called **eicosanoids** because of their common origin from C_{20} polyunsaturated fatty acids, the eicosaenoic acids, particularly arachidonic acid, which is all-*cis*-5,8,11,14-eicosatetraenoic acid. Recall from Chapter 18 that arachidonic acid is synthesized from linolenic acid. Related C_{20} trienoic and pentaenoic acids serve as minor precursors to some prostaglandins and their relatives. In addition to the compounds discussed here, the eicosaenoic acids serve as precursors to another class of compounds, the hydroxyeicosaenoic acids and hydroperoxyeicosaenoic acids. The latter compounds are metabolic precursors to the leukotrienes. Related compounds are derived from polyunsaturated fatty acids shorter or longer than C_{20}, and the term **oxylipin** has been proposed as a generic term for this class of lipids, encompassing all chain lengths.

The prostaglandins and the closely related thromboxanes are derived from a common pathway; a different pathway leads from arachidonic acid to the leukotrienes. Like hormones, the eicosanoids exert specific physiological effects on target cells. However, they are distinct from most hormones in that they act locally, near their sites of synthesis, and they are catabolized extremely rapidly; they are considered locally acting hormones. Moreover, the actions of a given prostaglandin seem to vary in different tissues. The biological properties of the eicosanoids have led to great interest in their medical use and in uses of their analogs.

SOME HISTORICAL ASPECTS

The most important early chapters in prostaglandin research were written in Sweden. In the mid-1930s Ulf von Euler discovered that lipid extracts of human semen contained active compounds that, on injection into animals, stimulated smooth muscle contraction or relaxation and affected the blood pressure. Because of their presumed origin in the prostate gland, he named these compounds *prostaglandins.* Later it was realized that these compounds are widely distributed in animal tissues. The first structural elucidations were reported in the late 1950s, under the direction of Sune Bergström and Bengt Samuelsson, and biosynthetic pathways were described in the mid-1960s, in Sweden and in the Netherlands.

The biological properties of the prostaglandins attracted intense interest in the pharmaceutical industry, but initial progress was limited by the low availabilities of these compounds. Interest reached a peak in 1971 with the discovery that aspirin inhibits one of the enzymes in prostaglandin biosynthesis. This inhibition is now known as the major site of action of aspirin and other nonsteroidal anti-inflammatory drugs. Later in the 1970s the thromboxanes and leukotrienes were discovered.

STRUCTURE

The first two prostaglandins to be isolated were called prostaglandins E and F, respectively, because of their preferential solubility in ether (E) or in phosphate buffer (F for *fosfat,* the Swedish word). We now denote these compounds by PGE and PGF, respectively; all other prostaglandins are denoted by a letter, for instance, PGA and PGH. Each prostaglandin has a cyclopentane ring and two side chains, with a carboxyl group in one side chain. A subscript numeral denotes the number of double bonds in the two chains. The most abundant prostaglandins, those synthesized from arachidonic acid, contain two double bonds; thus, PGE_2 would be the prostaglandin E derived from arachidonic acid. Finally, in the PGF se-

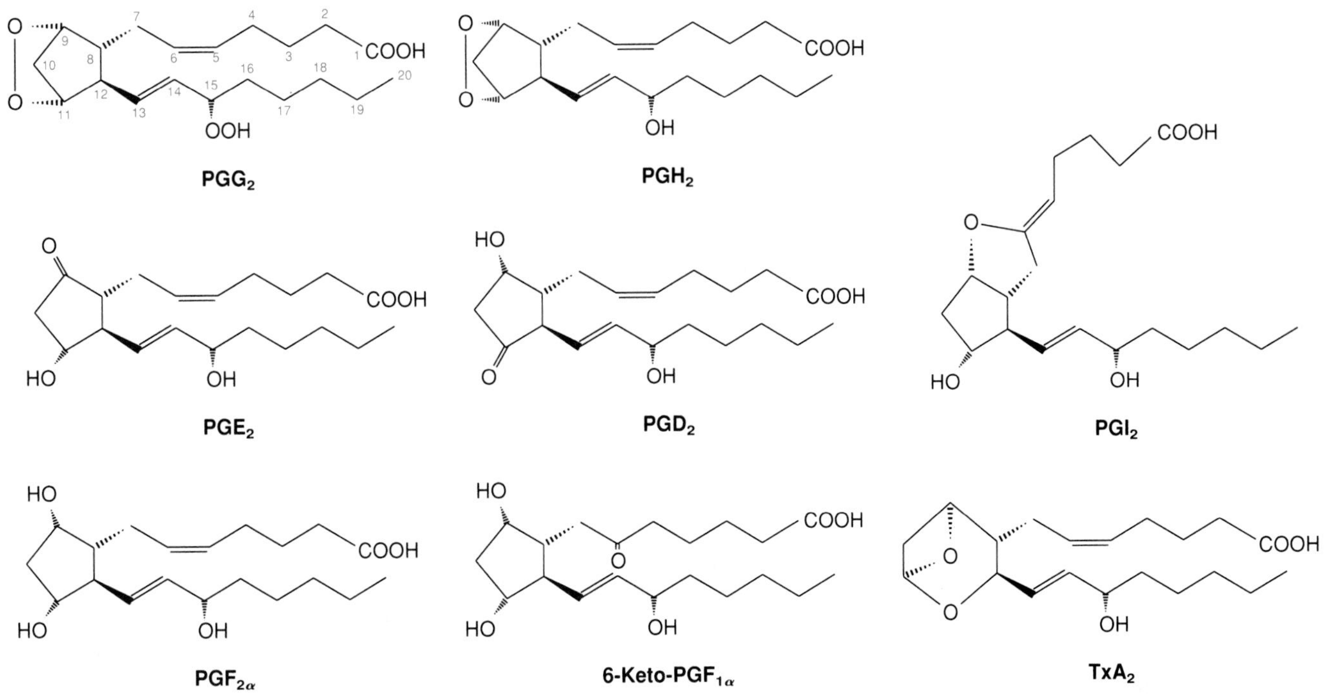

PGG$_2$ PGH$_2$ PGE$_2$ PGD$_2$ PGI$_2$ PGF$_{2\alpha}$ 6-Keto-PGF$_{1\alpha}$ TxA$_2$

FIGURE **19.29**

Structures of the major prostaglandins and thromboxane A$_2$. The figure shows the most abundant prostaglandins, those of the 2-series. They are derived from arachidonic acid, as is thromboxane A$_2$. Numbering of carbons begins with the carboxyl group as shown for the structure of PGG$_2$.

ries a subscript α indicates that the hydroxyl group at C-9 is *cis* to the 11-hydroxyl group, and β signifies a *trans* configuration. Structures of the most common prostaglandins are shown in Figure 19.29, along with that of thromboxane A$_2$ (TxA$_2$). TxA$_2$ was originally isolated from thrombocytes, or blood platelets, as a compound that stimulated platelet aggregation, an early step in blood clotting. Note its structural resemblance to PGE$_2$, except for the cyclic ether ring. Another thromboxane, TxB$_2$, is a hydrolysis product of TxA$_2$.

BIOSYNTHESIS AND CATABOLISM

Here we discuss only the biosynthesis of the 2-series of prostaglandins. The 1- and 3-prostaglandins are synthesized identically, from related C_{20} fatty acids. The biosynthetic pathways, which occur in endoplasmic reticulum, are shown in Figure 19.30. We can consider these pathways as occurring in three distinct stages: (1) release of arachidonic acid from membrane phospholipids; (2) oxygenation of arachidonate to yield PGH$_2$, a prostaglandin endoperoxide that serves as precursor to other prostaglandins; and (3) depending on the enzymes present in a cell, the conversion of PGH to other prostaglandins or to TxA$_2$.

The release of arachidonic acid in stage 1 occurs as a result of tissue-specific stimuli by hormones such as **bradykinin** or epinephrine or by proteases such as thrombin. Pathological release can occur if membranes are perturbed. For example, the inflammation caused by bee stings is probably due to arachidonate release stimulated by the venom protein **melittin.** Release evidently involves the action of a specific phospholipase A$_2$ on phosphatidylcholine or phosphatidylethanolamine, yielding arachidonate, or the action of a phospholipase C on phosphatidylinositol, yielding a diacylglycerol, which in turn undergoes cleavage to give free arachidonate.

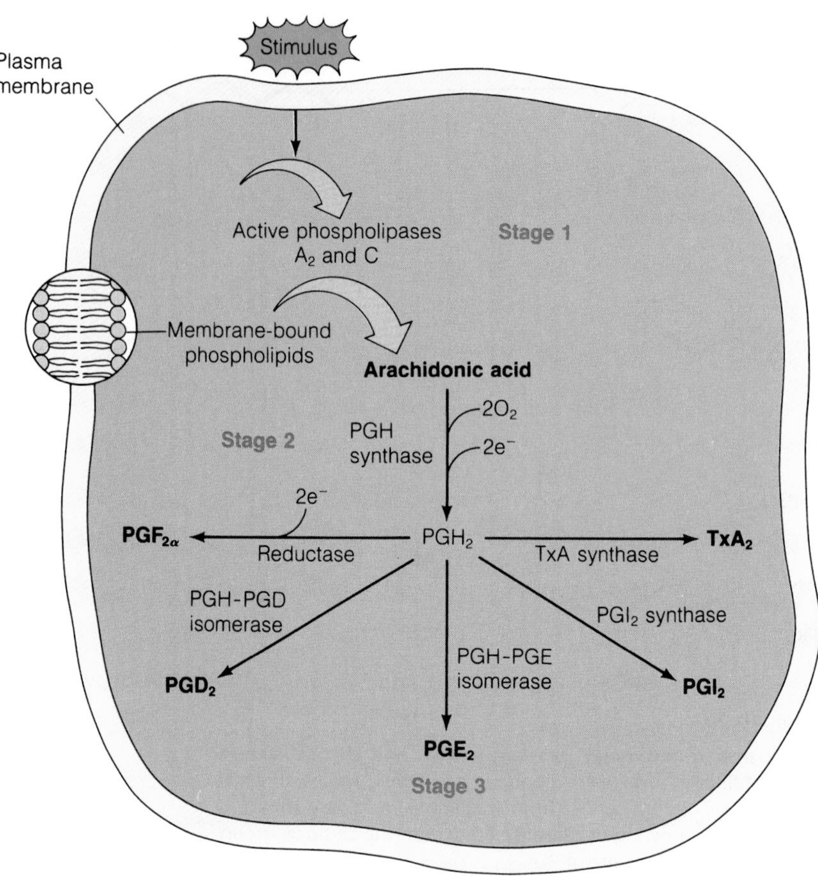

FIGURE **19.30**

Summary of biosynthetic routes to the major prostaglandins and thromboxane A_2.

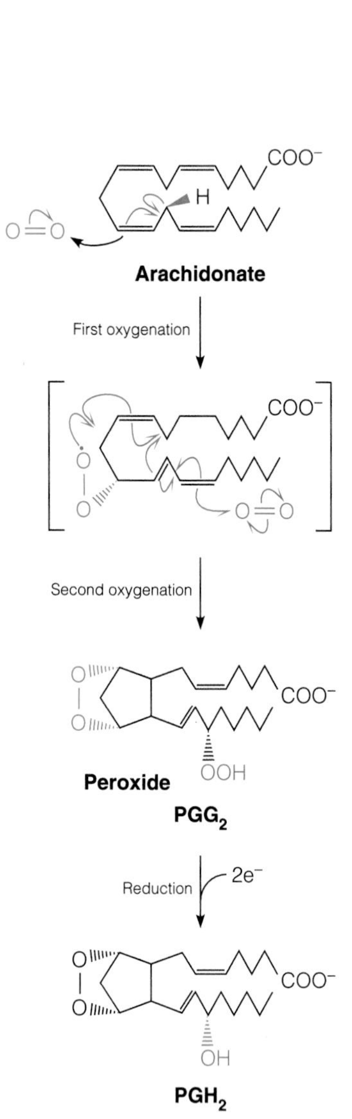

FIGURE **19.31**

Probable mechanism for the cyclooxygenation of arachidonic acid by PGH synthase. This free radical mechanism involves only one-electron shifts.

Free arachidonate is acted on, in stage 2, by **PGH synthase,** a bifunctional enzyme with two activities in a single heme-containing polypeptide chain. The first, a **cyclooxygenase,** introduces two molecules of O_2, one to form the ring and one to form a hydroperoxy group at C-15. The second activity involves a two-electron reduction of the peroxide, to give PGH_2, with a hydroxyl group at C-15, as shown in Figure 19.31. Mammalian cells contain two distinct forms of PGH synthetase, called PGHS-1 and PGHS-2 (or Cox-1 and Cox-2, where Cox stands for cyclooxygenase). Both isoforms are covalently modified, and hence inactivated, by reaction with aspirin (acetylsalicylic acid). As shown, aspirin acetylates a specific serine residue, which in turn blocks access of the fatty acid substrate to the active site.

Enzyme—Ser—CH_2OH + Acetylsalicylic acid ⟶ Enzyme—Ser—CH_2O—C(=O)—CH_3 + Salicylic acid

PGHS **Acetylsalicylic acid** **Acetylated PGHS** **Salicylic acid**

The anti-inflammatory and analgesic properties of aspirin derive from the inhibition of PGHS-2. However, the inhibition of PGHS-1 has undesirable side effects on the gastrointestinal system, including ulceration. Another widely used drug, **ibuprofen,** acts more specifically upon PGHS-2 but is a less effective inhibitor than aspirin. Using structural data on the two isoforms, molecular pharmacologists have recently designed aspirin analogs that selectively acylate PGHS-2,

Ibuprofen

and these have the desirable properties of aspirin without the side effects. A drug called Celebrex, released in late 1998 for relief of arthritis pain, binds 400-fold more tightly to PGHS-2 than to PGHS-1.

The cyclooxygenase reaction, one of the first steps in eicosanoid synthesis, is the target site for aspirin action.

In stage 3 a series of specific enzymes converts PGH_2 to other prostaglandins and to thromboxane A_2. Another pathway leads from arachidonate to the leukotrienes. Leukotriene C was originally discovered in the class of white blood cells called polymorphonuclear leukocytes and was named after the source (*leuko*cytes) and the *triene* structure (three double bonds). It is a potent muscle contractant that is thought to be involved in the pathogenesis of asthma, through constriction of the small airways in the lung. As shown in Figure 19.32, leukotrienes are formed from the initial attack on arachidonate of a **lipoxygenase,** which adds O_2 to C-5, giving 5-hydroperoxyeicosatetraenoic acid (5-HPETE). A dehydration to give the epoxide coupled with isomerization of double bonds gives leukotriene A_4. Hydrolysis of the epoxide ring yields leukotriene B_4. Transfer of the thiol group of glutathione yields leukotriene C_4. Subsequent modifications of the peptide chain (not shown) yield related compounds, leukotrienes D and E.

All of the eicosanoids are metabolized extremely rapidly, with most failing to survive a single pass through the circulatory system. The lung is a major site of prostaglandin catabolism. The many catabolic pathways all seem to start with conversion to 15-keto-13,14-dihydro derivatives.

BIOLOGICAL ACTIONS

As noted earlier, prostaglandins and their relatives can be considered as locally acting hormones. Evidently they act through binding to specific cellular receptors. We know relatively little about their subsequent effects at the molecular level, though there clearly are interactions with cyclic nucleotide metabolism. PGE stimulates adenylate cyclase in some cells, and PGF_2 has been reported to elevate levels of cyclic GMP in target cells. Although receptors for these compounds have not yet been characterized in detail, several compounds now under clinical study as anti-inflammatory agents evidently act by binding to receptors for leukotriene B_4. A number of the eicosanoids evidently play roles in the inflammation process, as evidenced by the fact that the anti-inflammatory effects of aspirin derive, at least in part, from the inhibition of cyclooxygenase.

Other biological effects include inhibition of platelet aggregation and relaxation of coronary arteries by PGI_2, effects that are counteracted by TxA_2. It seems likely that the action of PGI_2 could prevent platelet binding to arterial walls. In damaged areas the inhibition of PGI_2 synthesis could allow TxA_2 to bind and cause platelet aggregation, thereby promoting the formation of a clot.

Although little is known so far about molecular actions of the eicosanoids, knowledge of their physiology is being applied in useful ways. Recent studies indicate that long-term aspirin administration reduces the risk of heart attacks. Presumably this effect is related to the reduced synthesis of eicosanoids (particularly TxA_2) that induce platelet aggregation, an early step in the clot formation that participates in myocardial infarction. Because prostaglandin release is involved in the uterine muscle contraction that occurs in labor, $PGF_{2\alpha}$ is used when it is necessary to induce labor in mothers at term. Related to this effect is that $PGF_{2\alpha}$ inhibits progesterone secretion and regression of the corpus luteum. $PGF_{2\alpha}$ and PGE_2 are used as well to induce abortion in the second trimester or to induce delivery in case of the death of a fetus. Prostaglandin derivatives are used in animal husbandry, to bring a group of female animals into heat simultaneously. PGI_2 is used to reduce the risk of blood clotting during cardiopulmonary bypass operations. PGE_1, a vasodilator, is being tested against various circulatory disorders, and forms of PGE, which also inhibit gastric secretion, are being used in the experimental treatment of stomach ulcers. Efforts in the pharmaceutical industry are devoted to developing longer-lived prostaglandin analogs.

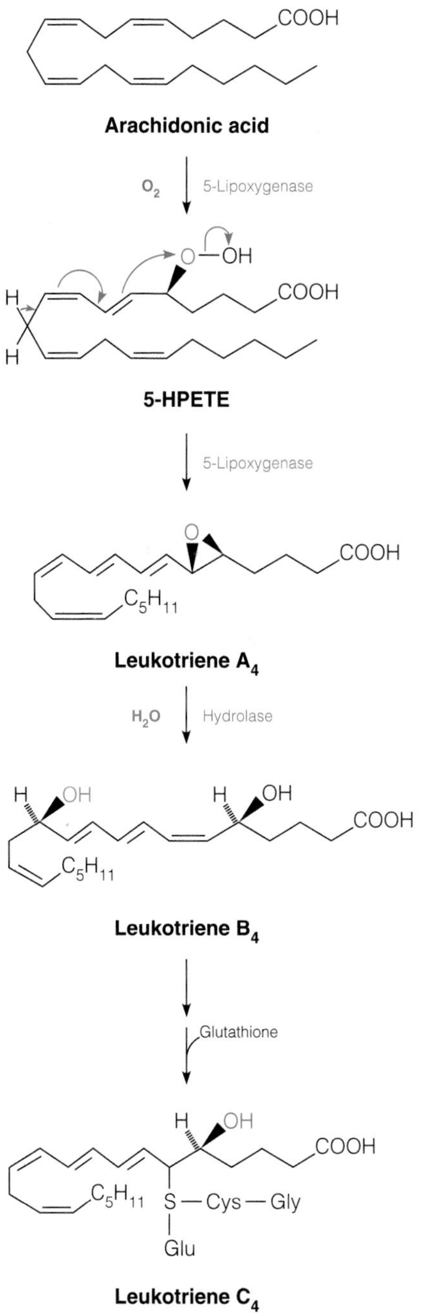

FIGURE **19.32**
Biosynthesis of leukotrienes.

SUMMARY

Glycerophospholipids, the predominant membrane lipids, are synthesized by routes that start from phosphatidic acid and intermediates activated by reaction with cytidine triphosphate. Retailoring of fatty acid side chains in phospholipids, exchange of polar head groups, and the actions of phospholipid exchange proteins, which insert phospholipids into membranes, all play roles in shaping the lipid composition of specific membranes. *S*-Adenosylmethionine is the methyl group donor for phosphatidylcholine synthesis.

Glycosphingolipids are assembled from ceramide and successive sugar additions involving glycosyltransferases and nucleotide-linked sugars. The pathway for turnover of these compounds in nervous tissue was elucidated from analysis of products that accumulated in cells of individuals with enzymatic defects in the pathway.

All steroid compounds—and indeed all isoprenoid compounds—are synthesized from acetate, by a pathway that proceeds through the six-carbon mevalonic acid and involves C_5, C_{10}, and C_{15} intermediates. Cyclization of the C_{30} hydrocarbon squalene leads to cholesterol, the precursor for all bile acids and steroid hormones. Synthesis of steroid hormones occurs in endocrine glands and involves hydroxylations, oxidoreduction reactions, and side chain cleavage reactions. All steroid hormone synthesis proceeds from cholesterol through pregnenolone.

Arachidonic acid and other C_{20} unsaturated fatty acids are precursors to physiologically potent, locally acting hormones that include prostaglandins, thromboxanes, and leukotrienes. Although actions of these biological regulators are not yet understood at the molecular level, metabolism of these eicosanoids presents therapeutic targets for drugs used to control inflammation, blood clotting, and gastric secretion and to manipulate reproductive processes in various ways.

REFERENCES

General References

Johnson, D. R., R. S. Bhatnagar, L. J. Knoll, and J. I. Gordon (1994) Genetic and biochemical studies of protein *N*-myristoylation. *Annu. Rev. Biochem.* 63:869–914. A contemporary review of an increasingly important posttranslational protein modification reaction.

Norman, A. W., and G. Litwack (1987) *Hormones.* Academic Press, Orlando, Fla. A clearly written textbook of biochemical endocrinology, with good sections on steroid metabolism.

Vance, D. E., and J. E. Vance, eds. (1985) *Biochemistry of Lipids and Membranes,* Benjamin/Cummings, Redwood City, Calif. This multi-authored textbook contains a number of detailed reviews. Particularly relevant to this chapter are the following articles: Lipid metabolism in prokaryotes, pp. 73–115; Phospholipid metabolism in eukaryotes, pp. 242–270; Metabolism, regulation, and function of ether-linked glycerolipids, pp. 271–298; Sphingolipids, pp. 361–403; The eicosanoids, pp. 325–360.

Vance, J. E. (1998) Eukaryotic lipid-biosynthetic enzymes: The same but not the same. *Trends Biochem. Sci.* 23:423–428. A thoughtful review article proposing that isoforms of some lipid-synthesizing enzymes are involved in compartmentation processes.

Phospholipid Metabolism

Chao, W., and M. S. Olson (1993) Platelet-activating factor: Receptors and signal transduction. *Biochem. J.* 292:617–629. A description of one of the most physiologically potent substances known.

Dowhan, W. (1997) Molecular basis for membrane phospholipid diversity: Why are there so many lipids? *Annu. Rev. Biochem.* 66:199–232. By using bacterial and yeast genetics, it is becoming possible to enumerate specific functions of lipids of very similar structure.

Downes, C.P (1998) Lipids in signaling and recognition. *Biochim. Biophys. Acta* 1436:1–261. A special issue of "BBA" devoted entirely to lipids and signal transduction.

Gelb, M. H., M. K. Jain, A. M. Hanel, and O. G. Berg (1995) Interfacial enzymology of glycerolipid hydrolases: Lessons from secreted phospholipase A_2. *Annu. Rev. Biochem.* 64:654–688. Enzymes that act upon lipid substrates in membranes operate at interfaces and follow rather different kinetic expressions.

Kuroki, Y., and D. R. Voelker (1994) Pulmonary surfactant proteins. *J. Biol. Chem.* 269:25943–25946. A description of the proteins that interact with dipalmitoylphosphatidylcholine in its lung surfactant role.

Moolenaar, W. H. (1995) Lysophosphatidic acid, a multifunctional phospholipid messenger. *J. Biol. Chem.* 270:12949–12952. A description of one of the most recently identified lipid signaling molecules.

Prescott, S. M. (1997) A thematic series on phospholipases. *J. Biol. Chem.* 272:15043. The introduction to a series of seven minireviews of the biology and biochemistry of phospholipases.

Snyder, F. (1995) Platelet-activating factor and its analogs: Metabolic pathways and related intracellular processes. *Biochim. Biophys. Acta* 1254:231–249. A contemporary review of this potent regulator.

Udenfriend, S., and K. Kodukula (1995) How glycosylphosphatidylinositol-anchored membrane proteins are made. *Annu. Rev. Biochem.* 64:563–591. A review of this recently discovered type of lipid modification of proteins.

van den Bosch, H., R. B. H. Schutgens, R. J. A. Wanders, and J. M. Tager (1992) Biochemistry of peroxisomes. *Annu. Rev. Biochem.* 62:157–197. These organelles carry out several lipid metabolic pathways, notably ether phospholipid synthesis.

Wilson, D. W., S. W. Whiteheart, L. Orci, and J. E. Rothman (1991) Intracellular membrane fusion. *Trends Biochem. Sci.* 16:334–337. How do membrane fusion events lead to specificity of membrane composition? This minireview describes early approaches to this question.

Wirtz, K. W. A. (1991) Phospholipid transfer proteins. *Annu. Rev. Biochem.* 60:73–99. A review describing a major process that shapes membranes.

Sphingolipids

Hoffmann, K., and V. M. Dixit (1998) Ceramide in apoptosis—does it really matter? *Trends Biochem. Sci.* 23:374–377. A brief article questioning the evidence for a signaling role for ceramide.

Moser, H. W. (1995) Ceramidase deficiency: Farber lipogranulomatosis. In: *The Metabolic and Molecular Bases of Inherited Desease,* 7th ed., edited by C. R. Scriver, A. L. Beaudet, W. S. Sly, and D. Valle, pp. 2589–2600. McGraw-Hill, New York. The first of 10 articles on lipid storage diseases in this three-volume series on inherited metabolic disorders.

van Echten, G., and K. Sandhoff (1993) Ganglioside metabolism: Enzymology, topology, and regulation. *J. Biol. Chem.* 268:5341–5344. A minireview of the cell biology and biochemistry of these processes.

Steroids and Isoprenoids

Gelb, M. H. (1997) Protein prenylation, et cetera: Signal transduction in two dimensions. *Science* 275:1750–1751. A "Perspectives" article in *Science* that discusses the significance of recent developments in this field.

Oliff, A. (1999) Farnesyltransferase inhibitors: Targeting the molecular basis of cancer. *Biochim. Biophys. Acta* 1423:C19–C30. Information about protein prenylation has important practical consequences.

Simons, S. S., Jr. (1996) Environmental estrogens: Can two "Alrights" make a wrong? *Science* 272:1451. A news article, with references, that discusses endocrine disrupters.

Tabernero, L., D. A. Bochar, V. W. Rodwell, and C. V. Stauffacher (1999) Substrate-induced closure of the flap domain in the ternary complex structures provides insight into the mechanism of catalysis by 3-hydroxy-3-methylglutaryl-CoA reductase. *Proc. Natl. Acad. Sci. USA* 96:7167–7171. Mechanistic understanding of this important enzyme comes from crystallography.

Lipid-Soluble Vitamins

Brigelius-Flohé, R., and M. G. Traber (1999) Vitamin E: Function and metabolism. *FASEB J.* 13:1145–1156. Reviews antioxidant and other properties of this vitamin.

Dowd, P., R. Hershline, S. W. Ham, and S. Naganathan (1995) Vitamin K and energy transduction: A base strength amplification mechanism. *Science* 269:1684–1691. A discussion of the surprisingly complex mechanism of carboxyglutamate synthesis.

Means, A. L., and L. J. Gudas (1995) The roles of retinoids in vertebrate development. *Annu. Rev. Biochem.* 64:201–233. A review of the activities of vitamin A–related compounds in gene regulation.

Eicosanoids

Needleman, P., J. Turk, B. A. Jakschik, A. R. Morrison, and J. B. Lefkowith (1986) Arachidonic acid metabolism. *Annu. Rev. Biochem.* 55:69–102. Several important regulatory molecules, including prostaglandins, thromboxanes, and leukotrienes, are derived metabolically from arachidonic acid. This review covers the field efficiently and completely.

Samuelsson, B., and C. D. Funk (1989) Enzymes involved in the biosynthesis of leukotriene B_4. *J. Biol. Chem.* 264:19469–19472. This minireview describes progress in cloning the genes for these enzymes.

Smith, W. L., R. M. Garavito, and D. L. DeWitt (1996) Prostaglandin endoperoxide H synthases (cyclooxygenases)-1 and -2. *J. Biol. Chem.* 271:33157–33160. A recent review of the structure and function of the first enzymes in eicosanoid synthesis.

PROBLEMS

1. Would you expect the reaction catalyzed by cardiolipin synthase to be strongly exergonic or strongly endergonic? Explain your reasoning.

*2. Phosphatidylserine (PS) is considered to be an intermediate in the biosynthesis of phosphatidylethanolamine (PE) in *E. coli,* yet PS is not found in appreciable amounts among *E. coli* membrane phospholipids. Because PS must be present in the membrane to serve as an intermediate, how might you explain its failure to accumulate to a significant extent? What kinds of experiments could test your proposed explanation?

3. Melittin is a protein in bee venom that activates phospholipase A_2. How might this effect contribute to the local inflammation that is caused by bee stings?

4. What would you expect to happen to levels of mevalonate in human plasma if an individual were to go from a meat-containing diet to a vegetarian diet?

5. Write a balanced equation for the synthesis of *sn*-1-stearoyl-2-oleylglycerophosphorylserine, starting with glycerol, the fatty acids involved, and serine.

6. If mevalonate labeled with ^{14}C in the carboxyl carbon were administered to rats, which carbons of cholesterol would become labeled?

7. Which step in lipid metabolism would you expect to be affected by 3,4-dihydroxybutyl-1-phosphonic acid (shown here)? Explain your answer.

```
       CH2OH
        |
  HO — C — H         O
        |            ||
       CH2 — CH2 — P — O⁻
                     |
                     O⁻
```

*8. Methyl-labeled [^{14}C]methionine at a specific activity of 2.0 millicuries per millimole was injected into rats. Six hours later the rats were killed. Phosphatidylcholine was isolated from the liver and found to have a specific activity of 1.5 millicuries per millimole. Calculate the proportions of phosphatidylcholine synthesized by the phosphatidylserine pathway and by the pathway starting from free choline. What further information would you need for your calculated values to reflect the true rates of these processes?

9. Identify a pathway for utilization of the four carbons of acetoacetate in cholesterol biosynthesis. Carry your pathway as far as the rate-determining reaction in cholesterol biosynthesis.

10. Explain why a deficiency of steroid 21-hydroxylase leads to excessive production of sex steroids (androgens and estrogens).

11. *cis*-Vaccenate is an 18-carbon unsaturated fatty acid abundant in *E. coli* membrane lipids. Propose a metabolic route for synthesis of this fatty acid, in light of the fact that stearic acid, the C_{18} saturated analogous fatty acid, is virtually absent from *E. coli* lipids.

12. The synthesis of phosphatidylcholine from phosphatidylethanolamine proceeds via three successive AdoMet-dependent methyla-

tion reactions. How might you determine experimentally whether these three reactions are catalyzed by the same enzyme or by three separate enzymes?

13. Briefly describe how cyclic AMP controls phospholipid synthesis.

*14. Assuming that all terpene compounds are synthesized by the HMG-CoA/mevalonate pathway, select a simple terpene (limonene, retinol, phylloquinone, or the like) and propose a route for its synthesis from isopentenyl pyrophosphate and dimethylallyl pyrophosphate.

15. Each structure in column A is correctly identified by one term in column B. Each description in column C is appropriate for one of the compounds listed in A and B, but with one deliberate error in each statement. Fill in the following box, matching each structure, name, and descriptive phrase, and correct each phrase by changing one or two words.

A	1	2	3	4	5	6
B						
C						

A

1. [Structure: steroid-derived ring system with CH_2 and HO]

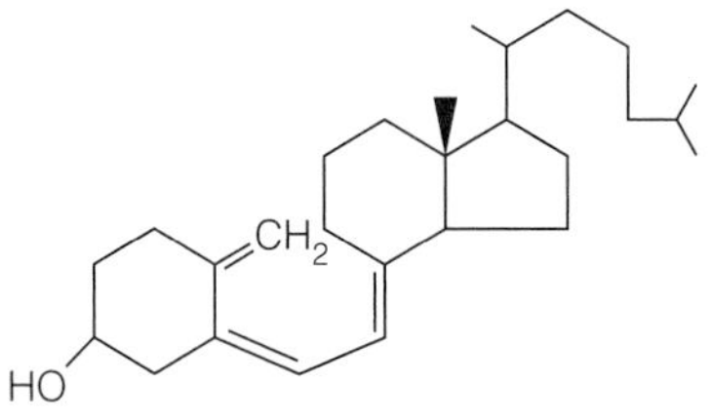

2. [Structure:] OH; H—C—CH=$CH(CH_2)_{12}$—CH_3; H—C—NH—C(=O)—$(CH_2)_n$—CH_3; Sia—Gal(—GalNAc)—Glc—O—CH_2

3. [Structure:] CH_2—O—P(=O)(O^-)—O^- on a ring with O, OH, HO, OH, OH

4. $^-OOC-CH_2-C(OH)(CH_3)-CH_2-C(=O)-S-CoA$

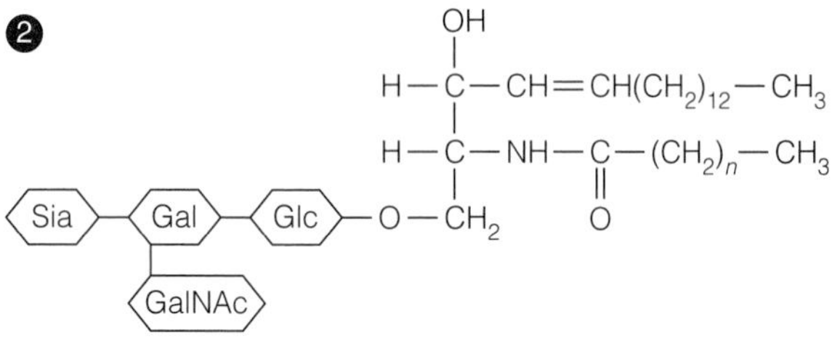

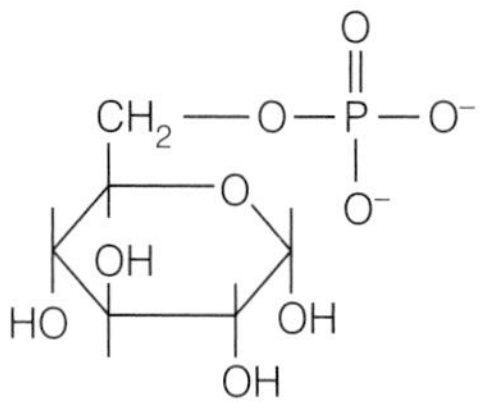

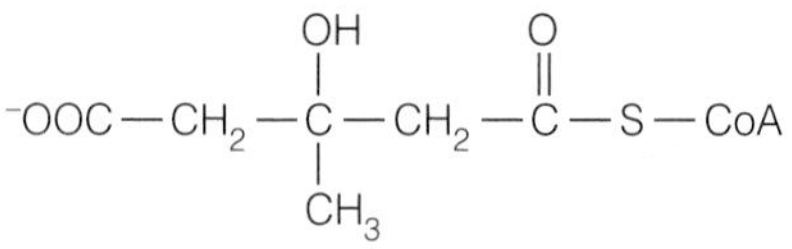

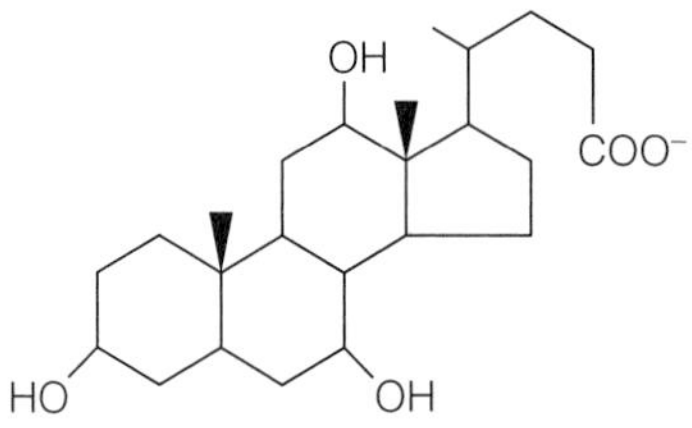

5. [Structure: steroid ring system with OH, HO, OH and side chain ending in COO^-]

6. HO—CH—CH=$CH(CH_2)_{12}$—CH_3; H—C—NH—C(=O)—$(CH_2)_n$—CH_3; CH_2—O—P(=O)(O^-)—O—CH_2—CH_2—$\overset{+}{N}(CH_3)$

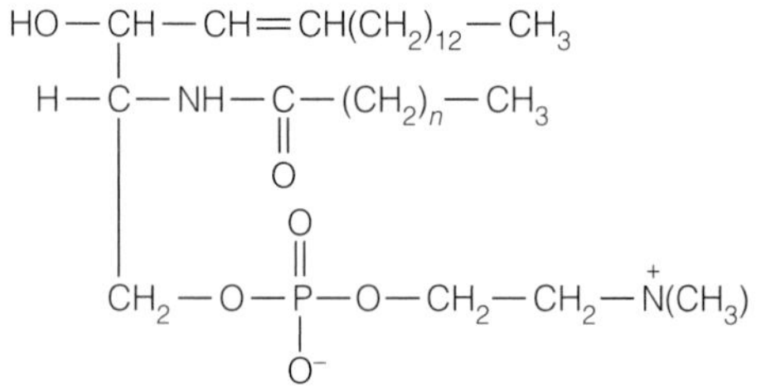

B

1. Glucose-6-phosphate
2. Squalene
3. Cholecalciferol
4. *N*-Acetylglucosamine
5. Sphingomyelin
6. Cholesterol
7. Tay–Sachs ganglioside
8. Acetoacetyl-CoA
9. 3-Hydroxy-3-methylglutaryl-CoA
10. Cholic acid

C

1. Glycerophospholipid abundant in myelin sheath surrounding nerve cells
2. Synthesized from cholesterol, it emulsifies triacylglycerols in chylomicrons
3. Product of the reaction that principally controls cholesterol biosynthesis
4. Accumulates in liver, causing a serious glycogen deficiency in von Gierke's disease, the first known glycogen storage disease
5. Accumulates in cell membranes because of a defect in glycolipid biosynthesis
6. Requires UV irradiation for its breakdown to 7-dehydrocholesterol

CHAPTER 20

Metabolism of Nitrogenous Compounds: Principles of Biosynthesis, Utilization, and Turnover

THUS FAR OUR STUDY OF METABOLISM HAS CONCERNED ITSELF PRIMARILY WITH compounds that can be degraded completely to carbon dioxide and water—in other words, compounds containing only carbon, hydrogen, and oxygen. We turn now to nitrogen-containing compounds—amino acids and their derivatives, nucleotides, and the polymeric nucleic acids and proteins. This chapter and the next two treat the metabolism of low-molecular-weight nitrogenous compounds (Figure 20.1). The metabolism of specific amino acids is presented in Chapter 21, and nucleotide metabolism is covered in Chapter 22. Chapters 24–28 consider how these compounds are utilized for protein and nucleic acid biosynthesis. Because the biological availability of nitrogen is limited, and because the breakdown of nitrogenous compounds often yields toxic products, we will encounter some new metabolic principles in this part of the biochemical scene.

The existence of 20 different amino acids in proteins implies the existence of 20 biosynthetic pathways and 20 degradative pathways. Although this might seem intimidating at first glance, the common features of pathways involving various amino acids should ease the task of learning. More important than many of the details of biosynthesis and degradation, however, are the numerous roles of amino acids other than as protein constituents, including their functions as precursors to hormones, vitamins, coenzymes, porphyrins, pigments, and neurotransmitters.

Chapters 20–22 also indicate how much we have learned from naturally occurring human mutations, as well as from mutations generated in the laboratory, in cultured cells or in bacteria. Whereas a mutation that inactivates an enzyme in one of the central energy-generating or energy-storing pathways is likely to be lethal and, hence, not expressed in living individuals, mutations that affect amino acid metabolism are often not lethal and *are* found in living humans. The clinical consequences of these mutations are often tragic, but these inherited metabolic diseases have greatly enhanced our understanding of human biochemistry.

Utilization of Inorganic Nitrogen: The Nitrogen Cycle

For many organisms, growth and reproduction are limited by the availability of utilizable nitrogen, which in turn is limited by the abilities of organisms to utilize

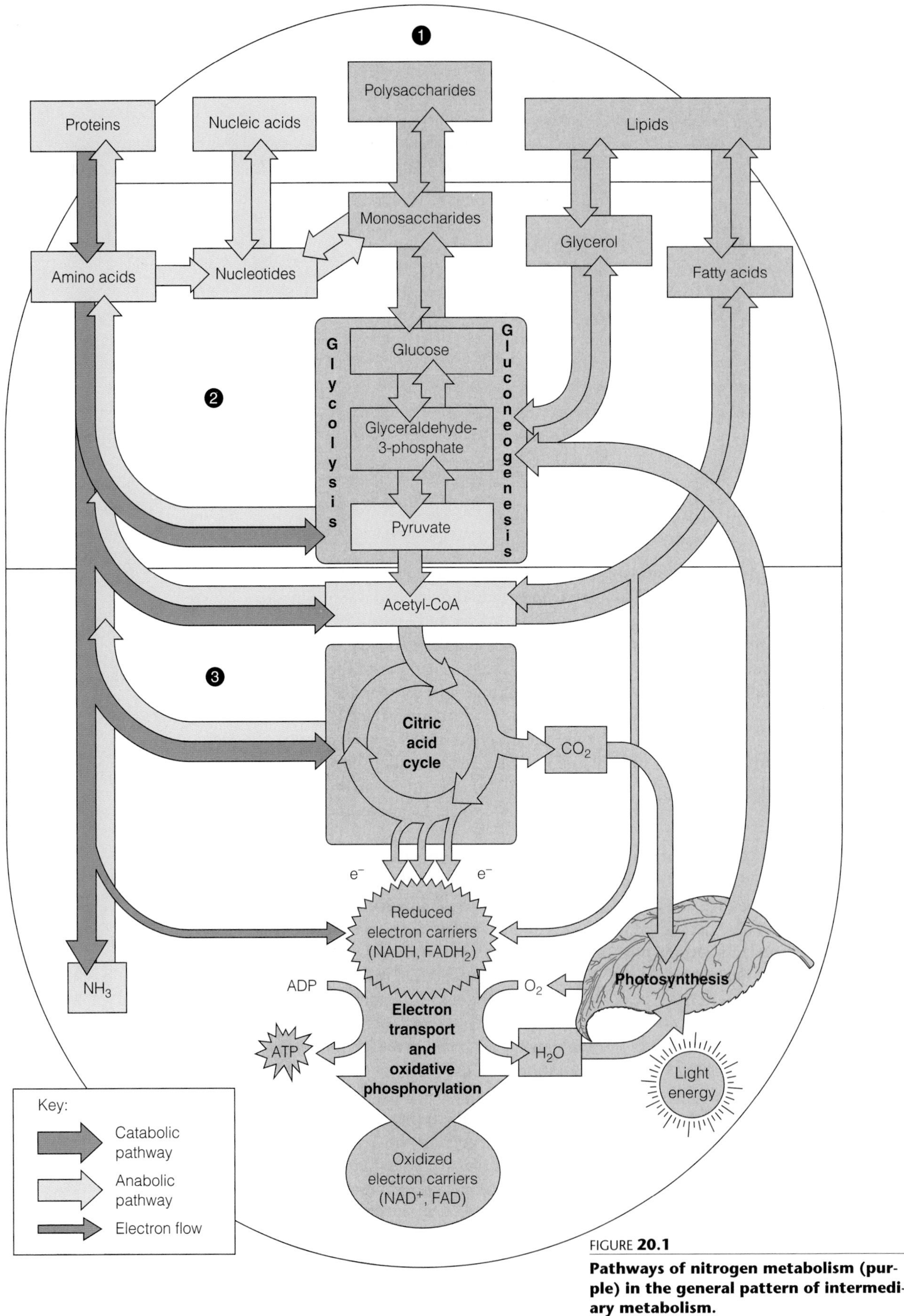

FIGURE **20.1**

Pathways of nitrogen metabolism (purple) in the general pattern of intermediary metabolism.

different inorganic forms of nitrogen. All organisms can convert ammonia (NH_3) to organic nitrogen compounds—that is, substances containing C—N bonds. However, not all organisms can synthesize ammonia from the far more abundant forms of inorganic nitrogen—dinitrogen gas (N_2), the most abundant component of the earth's atmosphere, and nitrate ion (NO_3^-), a soil constituent essential for the growth of most plants. The reduction of N_2 to NH_3, called **biological nitrogen fixation,** is carried out only by certain microorganisms, sometimes in symbiotic relationship with plants. The reduction of NO_3^- to NH_3, by contrast, is widespread among both plants and microorganisms.

Because few organisms can use the N_2 in air, and many soils are poor in nitrate, nitrogen bioavailability limits growth for most organisms.

As in the consideration of any limited resource, it is useful to think about nitrogen metabolism in terms of an economy—a **nitrogen economy**—that focuses on questions of supply, demand, turnover, reuse, growth, and maintenance of a steady state. Within the biosphere a balance is maintained between total inorganic and total organic forms of nitrogen. The conversion of inorganic to organic nitrogen, which starts with nitrogen fixation and nitrate reduction, is counterbalanced by catabolism, **denitrification,** and decay (Figure 20.2). Catabolism yields ammonia and various organic nitrogenous end products, which can in turn be metabolized by various bacteria: *Nitrosomonas* species oxidize ammonia to nitrite (NO_2^-), and *Nitrobacter* species oxidize nitrite to nitrate. These oxidations generate biological energy, just as other organisms derive energy from oxidation of carbohydrate or fat to CO_2. Other bacteria, the **denitrifying bacteria,** catabolize ammonia to N_2. Because of the toxicity of ammonia, there is much interest in using denitrifying bacteria and their enzymes in **bioremediation,** the use of living organisms to purify and detoxify environmental residues of human activity, such as manufacturing, or in waste disposal. Our concern here is with the use of nitrogen for amino acid and nucleotide biosynthesis, so our focus in the rest of this section will be on synthesis of ammonia from N_2 and from nitrate ion.

BIOLOGICAL NITROGEN FIXATION

Although nitrogen gas makes up about 80% of the earth's atmosphere, its reduction to ammonia occurs in relatively few living systems—some free-living soil bacteria, such as *Klebsiella* and *Azotobacter;* cyanobacteria (blue-green algae); and symbiotic nodules on the roots of leguminous plants, such as beans or alfalfa, that have been infected with certain bacteria, notably of the genus *Rhizobium* (Figure 20.3). The infecting bacterium assumes a modified form, called a

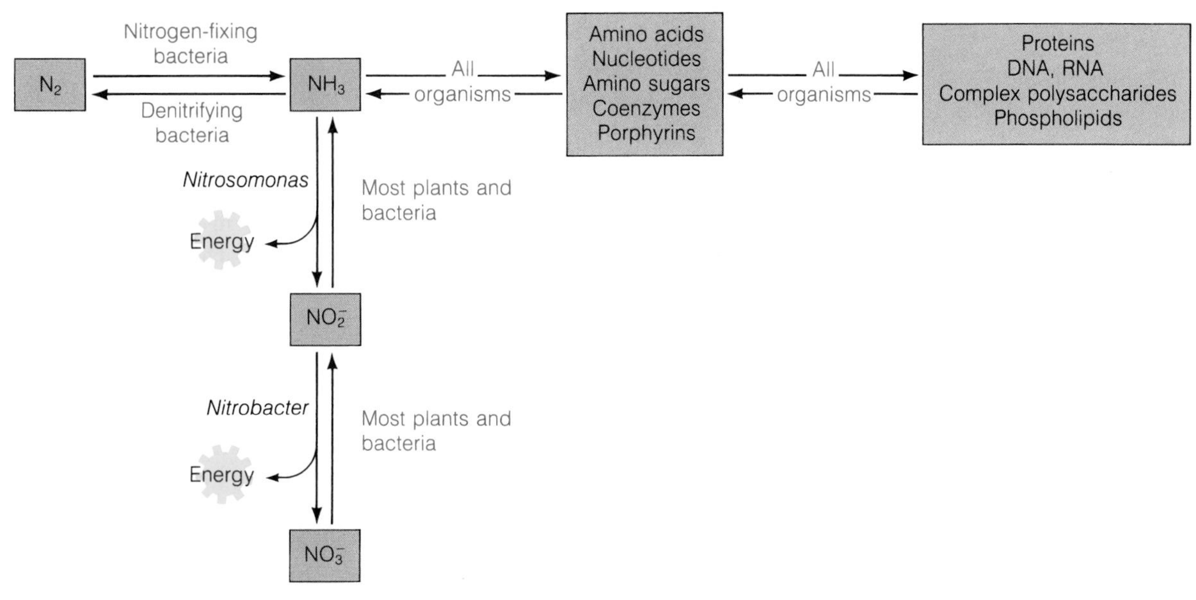

FIGURE **20.2**

Relationships between inorganic and organic nitrogen metabolism. Inorganic nitrogen compounds are shown in brown, and organic nitrogen compounds in purple. Interconversion of N_2, nitrate, and ammonia is limited in the biosphere, but steady state levels of these species are maintained. All organisms can use ammonia for biosynthesis, and it is a major metabolic end product.

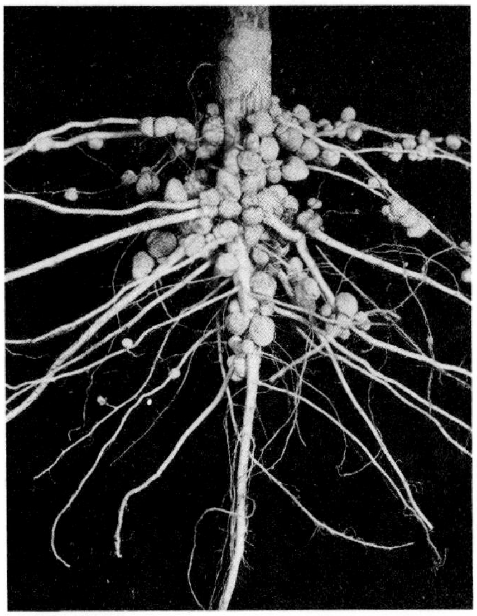

FIGURE **20.3**

The site of nitrogen fixation in symbiotic root nodules. This root of a soybean plant is infected by nitrogen-fixing bacteria of the genus *Rhizobium*.

Courtesy of the Nitragin Company.

$$N_2 + 3H_2 \xrightarrow[450°C,\ 270\ atm]{catalyst} 2NH_3$$

The Haber process

bacteroid, inside the cells of infected plants. Some trees, such as alder, also form nitrogen-fixing nodules and thus have the capacity to fix nitrogen.

Because nitrogen availability is the factor limiting the fertility of most soils, an understanding of biological nitrogen fixation is directly related to increasing the world's food supply. The triply bonded N_2 molecule, N≡N, with a bond energy of about 940 kJ/mol, is extraordinarily difficult to reduce. Industrially, the reduction is done by the Haber Process, a low-yield catalytic hydrogenation carried out at very high temperature and pressure. This process is used in the manufacture of ammonia-based fertilizers. Interest in the molecular details of biological nitrogen fixation has derived partly from hopes of supplanting this energy-intensive process with a means of ammonia production that can take place under milder conditions.

Formally, nitrogen fixation can be compared with photosynthesis. Both N_2 and CO_2 are stable inorganic compounds whose reduction requires both energy and low-potential electrons—electron carriers of very low E_0'. As we saw in Chapter 17, photosynthesis uses light to generate both energy (through photophosphorylation) and low-potential electrons (as ferredoxin). Comparable mechanisms used in nitrogen fixation are still not clear, because the enzymes involved are extremely sensitive to oxygen and can be studied only under anaerobic conditions. The major reason nitrogen can be fixed in root nodules of plants infected with *Rhizobium* is that the nodules contain an abundant protein called **leghemoglobin,** which maintains an anaerobic environment by binding any O_2 that finds its way into the nodule and presenting it to respiratory enzymes—in a manner somewhat akin to the behavior of myoglobin in animals.

Mechanistic details of nitrogen fixation appear quite similar among the species examined to date. As studied in *Klebsiella pneumoniae,* the stoichiometry of the overall reaction is as follows:

$$N_2 + 8e^- + 16ATP + 16H_2O \longrightarrow 2NH_3 + H_2 + 16ADP + 16P_i + 8H^+$$

Although we present a balanced equation, the precise number of ATP molecules in this process has not yet been established; in any case, a large amount of ATP is required. The ATP is generated through energy-yielding pathways of the organism, primarily carbohydrate catabolism. Electrons for N_2 reduction are derived from low-potential carriers, either ferredoxin or flavodoxin, a low-potential flavoprotein. Hydrogen is a by-product of nitrogen reduction. Some nitrogen-fixing species have the ability to "recycle" this hydrogen, to generate low-potential electrons for additional cycles of N_2 reduction.

The best-understood nitrogen fixation systems involve an iron protein, an iron–molybdenum protein, and hydrolysis of a large amount of ATP.

The enzyme system responsible for N_2 reduction, called the **nitrogenase complex,** consists of two separate proteins. As outlined in Figure 20.4, one protein—called **component I, nitrogenase,** or **molybdenum–iron protein**—catalyzes the reduction of N_2, and the other—called **component II, nitrogenase reductase,** or **iron protein**—transfers electrons from ferredoxin or flavodoxin to component I. Both proteins (component I and component II) contain iron–sulfur clusters, and component I also contains molybdenum, in the form of a tightly bound **iron–molybdenum cofactor** (FeMo-co). N_2 binds to this cofactor during its reduction, although the precise mode of binding is not yet known.

In 1992, crystal structures were described for both components I and II from *Azotobacter vinelandii* and, in 1997, a structure for the entire complex was reported, with a nonhydrolyzable ATP analog bound to component II (Figure 20.5). Binding of ATP evidently forces a conformational change in component II, which drives its docking to component I. As shown in Figure 20.5, component II contains a conventional Fe_4S_4 cluster, while component I contains two novel iron–sulfur complexes—the P cluster, which contains 7 sulfurs and 8 irons, and the iron–molybdenum cofactor, which, as shown, contains 9 sulfurs, 7 irons, and one molybdenum ion linked also to a molecule of **homocitrate.** Electrons flow from

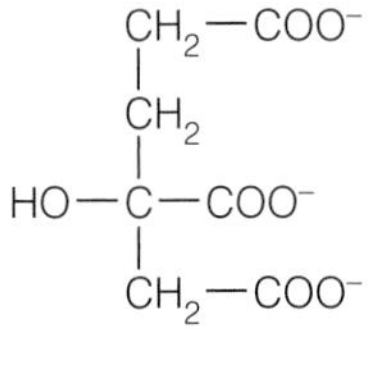

$CH_2—COO^-$
$|$
CH_2
$|$
$HO—C—COO^-$
$|$
$CH_2—COO^-$

Homocitrate

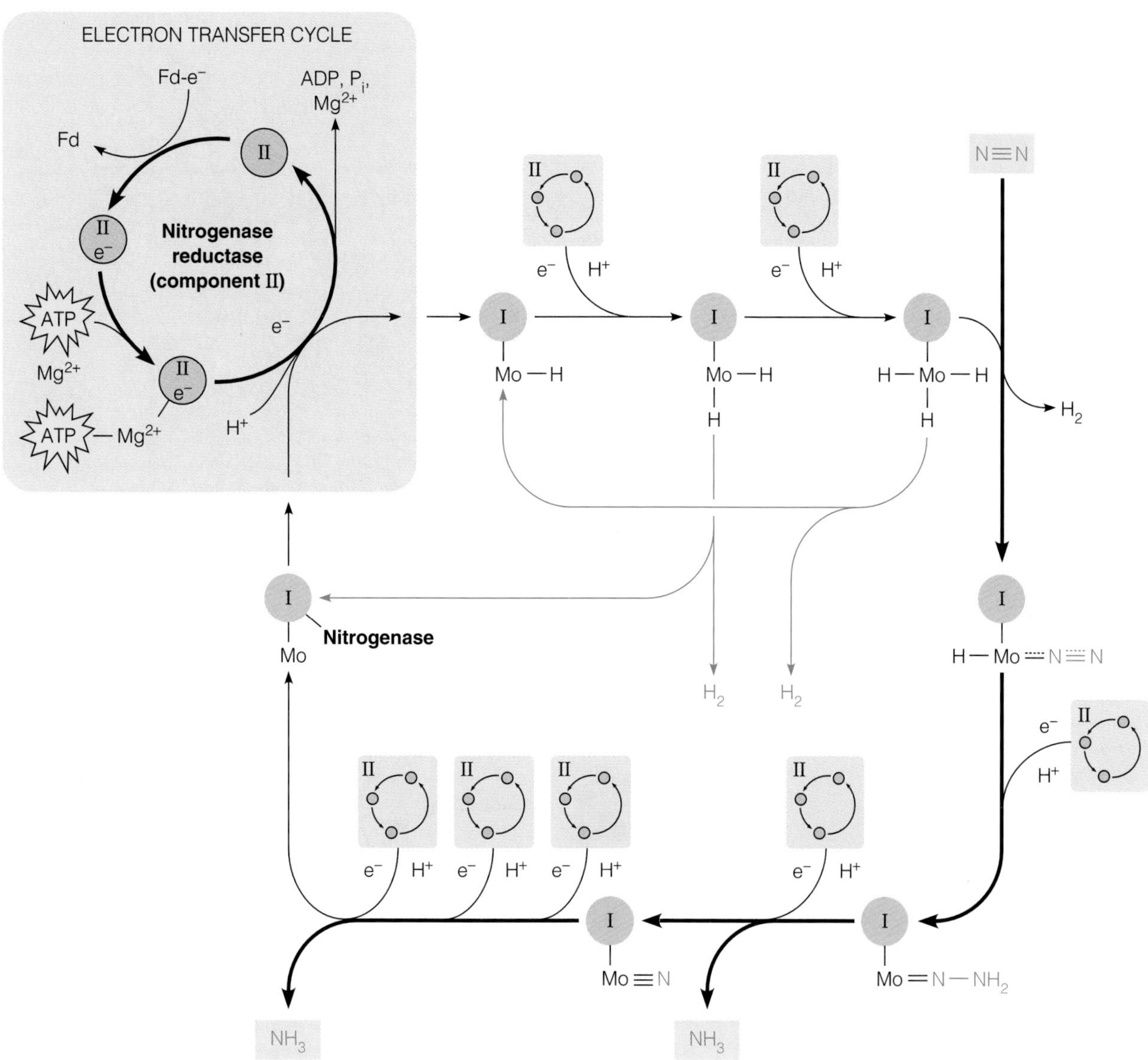

FIGURE **20.4**

Schematic view of nitrogen fixation. Component I (green) is nitrogenase, or molybdenum–iron protein. Component II (tan in light purple background) is nitrogenase reductase, or iron protein. Component I catalyzes the reduction of N_2 (shown before reduction at upper right) and is reduced in turn by component II. The transfer of H^+ and an electron from component II to component I (shown in detail in the enlarged light purple box) occurs eight times during the reduction of one N_2 molecule. Fd-e^-, the electron donor that reduces component II, is either ferredoxin (in *Rhizobium*) or flavodoxin (in *Klebsiella*). Binding of ATP to reduced component II is thought to generate an altered conformation of component II, with a very low reduction potential. Transfer of the electron from reduced component II to component I is accompanied by the return of component II to a relaxed state and splitting of bound ATP. These reactions generate mono-, di-, and trihydride forms of FeMo-co (possibly with the H not bound directly to the Mo) bound to component I. Binding of N_2 to Mo occurs concomitantly with release of two bound hydrogens as H_2. Hydrogen can also be released from the di- and trihydrides (blue arrows).

Concepts courtesy of Harold J. Evans.

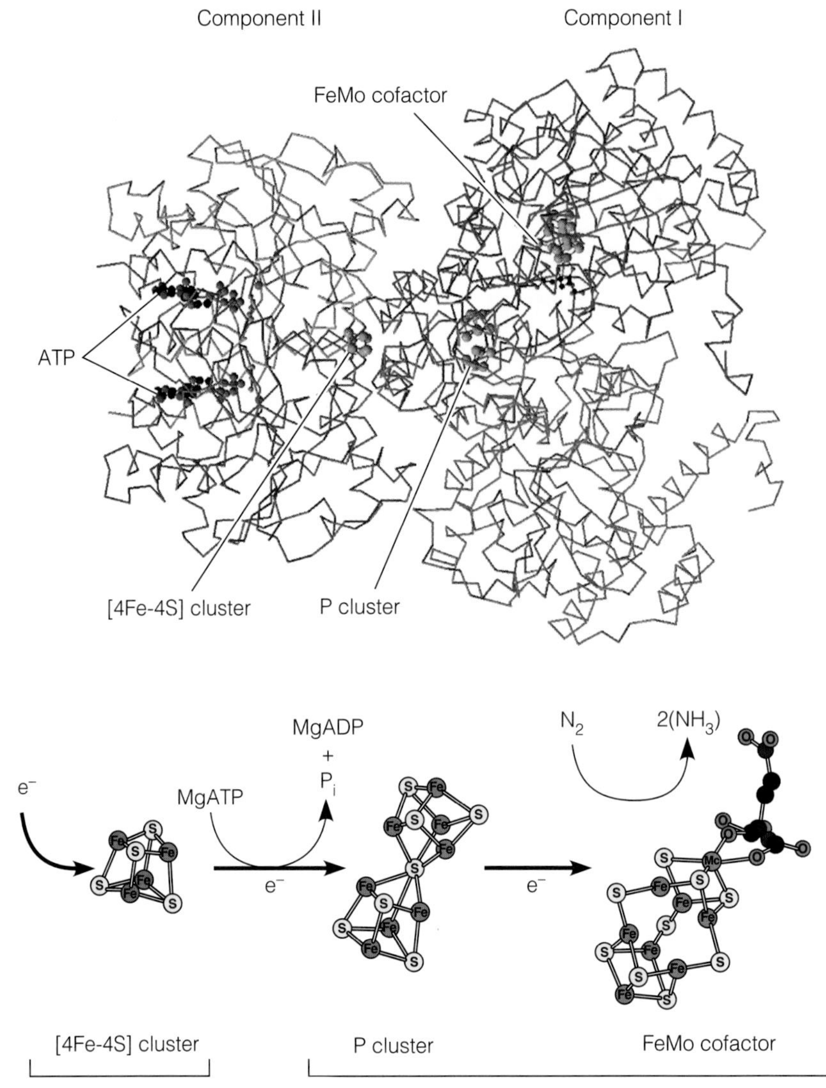

FIGURE **20.5**

Structure of nitrogenase complex. Top, skeletal model of the complex containing an ATP analog, with subunits of component II shown in yellow and green and those of component I in red and purple. Bottom, structures of the iron–sulfur clusters that participate in reductive electron transfer. Hydrolysis of bound ATP is thought both to drive the reduction of P cluster by component II and to trigger a conformational change in complex II that causes it to dissociate transiently from complex I, assuring unidirectional electron flow.

Courtesy of D. C. Rees and D. R. Dean, as reproduced in R. L. Rawls, *Chem. Eng. News* (1998), June 22 issue, pp. 29–32. Reprinted by permission of the author.

ferredoxin or flavodoxin to the Fe_4S_4 complex in component II, and the hydrolysis of bound ATP somehow drives the electrons to the P cluster in component I and then to FeMo-co. As shown in Figure 20.5, these three clusters are sufficiently close together in the complex to allow facile electron transfer. Almost certainly, the reduction of N_2 occurs at the bound coenzyme, but the mode of N_2 binding must be described in order for the mechanism of reduction to be understood.

Recently some bacteria have been found to contain more than one nitrogenase complex. *Azotobacter* contains three such systems, one of which uses vanadium instead of molybdenum, and one that has iron as the sole bound metal.

The genetics of nitrogen fixation is under intense study because of the desirability of transferring nitrogen-fixing capabilities to higher plants, thereby reducing the use of nitrogenous fertilizers. In *Klebsiella* 13 genes known to be essential to the process are linked within one region of DNA some 24,000 base pairs in length, the *nif* gene cluster. Seven interspersed genes participate but are not essential for nitrogen fixation. Products of the essential genes include the two polypeptide subunits of component I, the one subunit of component II, flavodoxin, and enzymes that synthesize FeMo-co.

NITRATE UTILIZATION

The ability to reduce nitrate to ammonia is common to virtually all plants, fungi, and bacteria. The first step, reduction of nitrate to nitrite (NO_2^-) is chemically difficult and involves a large and complex enzyme, **nitrate reductase.** This multisubunit enzyme, with M_r of about 800 kilodaltons, contains bound FAD, molybdenum, and a cytochrome called cytochrome 557 (which contains an Fe_4S_4 complex). The enzyme carries out the overall reaction:

$$NO_3^- + NAD(P)H + H^+ \longrightarrow NO_2^- + NAD(P)^+ + H_2O$$

Plants use NADH as the electron donor, whereas fungi and bacteria use NADPH. The electrons are transferred to enzyme-bound FAD, then to cytochrome 557, then to molybdenum, and finally to the substrate. The molybdenum is bound to a cofactor containing a *pteridine* ring (see page 732), which is quite distinct from the structure of FeMo-co. In fact, all known molybdenum-requiring enzymes except nitrogenase contain a structure similar to that of this **molybdopterin.**

Molybdopterin

Reduction of nitrite to ammonia is carried out in three steps ($NO_2^- \longrightarrow NO^- \longrightarrow NH_2OH \longrightarrow NH_3$) by one enzyme, **nitrite reductase.** This enzyme contains one Fe_2S_2 center and one molecule of **siroheme,** a partially reduced iron porphyrin. The electron donor for each step is ferredoxin.

Of considerable environmental interest is another molybdenum enzyme, **dimethylsulfoxide reductase.** As shown in Figure 20.6, metabolism of marine algae produces dimethyl sulfoniopropionate (DMSP) as an osmotic stabilizer and biological methylating agent. This compound readily hydrolyzes to dimethyl sulfide (DMS), which in turn undergoes oxidation to dimethylsulfoxide (DMSO). Bacterial reduction of DMSO regenerates DMS, the principal volatile form of sulfur in

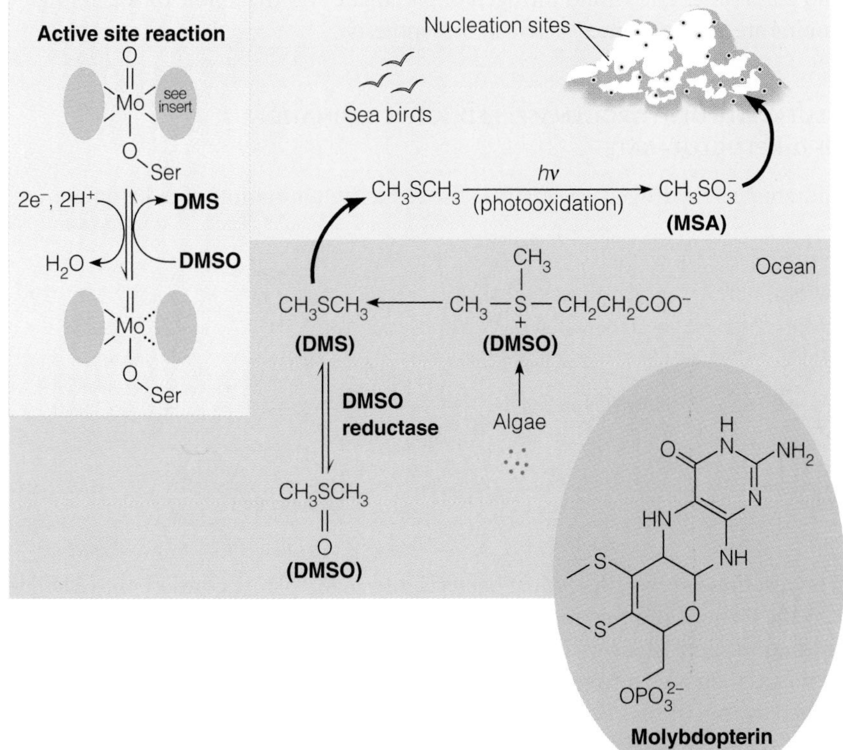

FIGURE **20.6**

Environmental significance of DMSO reductase.

Courtesy of E. I. Steifel, *Science* (1998) 272:1599–1600, reprinted with permission.

the oceans; its release to the atmosphere is probably responsible for the distinctive, and misnamed, "salt air fragrance" that we associate with the sea. This smell attracts sea birds, who recognize it as an indicator of biological productivity in the nearby ocean. Also, DMS in the atmosphere undergoes photooxidation to methyl sulfate (MSA), which serves as nucleation sites for the formation of clouds. Thus, the interplay between DMSO and DMS plays an important role in controlling climate, contributing toward modulating the earth's temperature. Recently, a structural determination of DMSO reductase from the bacterium *Rhodobacter spheroides* revealed its molybdenum cofactor to contain two molecules of bound molybdopterin, with the complex linked covalently to a serine residue on the enzyme.

Utilization of Ammonia: Biogenesis of Organic Nitrogen

Several ubiquitous enzymes use ammonia as substrate for synthesis of glutamate, glutamine, asparagine, or carbamoyl phosphate.

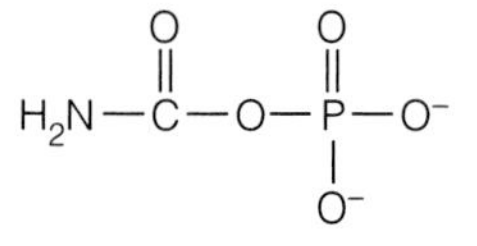

$$H_2N{-}\overset{O}{\overset{\|}{C}}{-}O{-}\overset{O}{\overset{\|}{\underset{O^-}{\underset{|}{P}}}}{-}O^-$$

Carbamoyl phosphate

Although plants, animals, and bacteria derive their nitrogen from different sources, virtually all organisms share a few common routes for utilization of inorganic nitrogen in the form of ammonia. Ammonia in high concentrations is quite toxic, but at lower levels it is a central metabolite, serving as substrate for five enzymes that convert it to various organic nitrogen compounds. At physiological pH the dominant ionic species is ammonium ion, $\overset{+}{N}H_4$. However, the five reactions involve the unshared electron pair of NH_3, which is therefore the reactive species.

All organisms assimilate ammonia via reactions leading to glutamate, glutamine, asparagine, and **carbamoyl phosphate** (Figure 20.7). Because carbamoyl phosphate is used only in the biosynthesis of arginine, urea, and the pyrimidine nucleotides, most of the nitrogen that finds its way from ammonia to amino acids and other nitrogenous compounds does so via the two amino acids glutamate and glutamine. The amino nitrogen of glutamate and the *amide* nitrogen of glutamine are both extremely active in biosynthesis.

GLUTAMATE DEHYDROGENASE: REDUCTIVE AMINATION OF α-KETOGLUTARATE

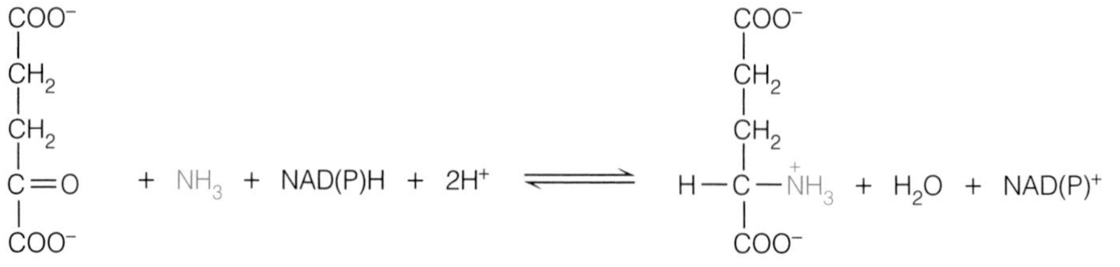

Glutamate dehydrogenase catalyzes the reductive amination of α-ketoglutarate:

$$\underset{\alpha\text{-Ketoglutarate}}{COO^-{-}CH_2{-}CH_2{-}C(=O){-}COO^-} + NH_3 + NAD(P)H + 2H^+ \rightleftharpoons \underset{\text{Glutamate}}{COO^-{-}CH_2{-}CH_2{-}CH(\overset{+}{N}H_3){-}COO^-} + H_2O + NAD(P)^+$$

The reaction is reversible. Most bacteria and many plants contain an NADPH-specific form of the enzyme, which acts primarily in the direction of glutamate formation. Consistent with this characteristic is that bacteria growing with ammonia as their sole nitrogen source use this reaction as the primary route for nitrogen assimilation. In animal cells the enzyme functions in either the synthetic direction or the catabolic direction, supplying α-ketoglutarate for the citric acid cycle; the catabolic role probably predominates. The animal enzyme uses NAD^+

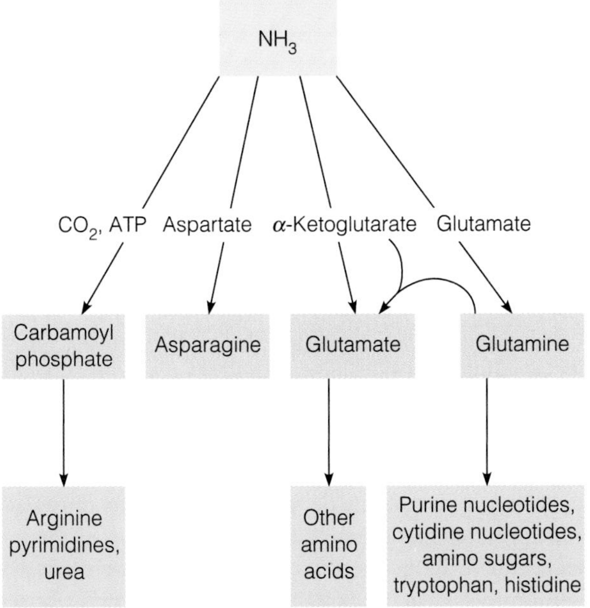

FIGURE **20.7**

Reactions in assimilation of ammonia and major fates of the fixed nitrogen. In animals, glutamine is the chief nitrogen source for pyrimidines.

as its principal cofactor, but it can also use $NADP^+$. In animals, glutamate dehydrogenase, a hexamer of identical subunits, is located in mitochondria, consistent with a primary role in energy generation. Moreover, the enzyme is allosterically controlled; α-ketogluturate synthesis is inhibited by ATP or GTP and stimulated by ADP or GDP. Thus, the enzyme is activated under conditions of low energy charge. Yeasts and fungi contain both types of glutamate dehydrogenase—each appropriately regulated, with one tailored for nitrogen assimilation and one functioning primarily in catabolism.

A related enzyme, **glutamate synthase,** catalyzes a reaction comparable to that catalyzed by glutamate dehydrogenase but functions primarily in glutamate biosynthesis:

$$\alpha\text{-Ketoglutarate} + \text{glutamine} + \text{NADPH} + \text{H}^+ \longrightarrow 2\ \text{glutamate} + \text{NADP}^+$$

Because glutamate dehydrogenase has a relatively high K_M for ammonia, glutamate synthase plays a larger role in glutamate synthesis in most cells (together with glutamine synthetase; see the next section). Glutamate synthase as isolated from several bacteria contains two types of subunits (α and β) in an 800-kilodalton holoenzyme; the $\alpha\beta$ protomer contains FAD, FMN, and several iron–sulfur centers. The enzyme from plants uses either NADPH, NADH, or ferredoxin, whereas glutamate synthase from other organisms uses NADH exclusively.

GLUTAMINE SYNTHETASE: GENERATION OF BIOLOGICALLY ACTIVE AMIDE NITROGEN

Whether formed by action of glutamate dehydrogenase or glutamate synthase, or by transamination (see Chapter 14), glutamate can accept a second ammonia moiety to form glutamine in the reaction catalyzed by **glutamine synthetase.** Mn^{2+} is required.

$$\underset{\textbf{Glutamate}}{{}^{-}OOC{-}CH(\overset{+}{N}H_3){-}CH_2{-}CH_2{-}C(=O){-}O^-} + NH_3 + ATP \longrightarrow \underset{\textbf{Glutamine}}{{}^{-}OOC{-}CH(\overset{+}{N}H_3){-}CH_2{-}CH_2{-}C(=O){-}NH_2} + ADP + P_i$$

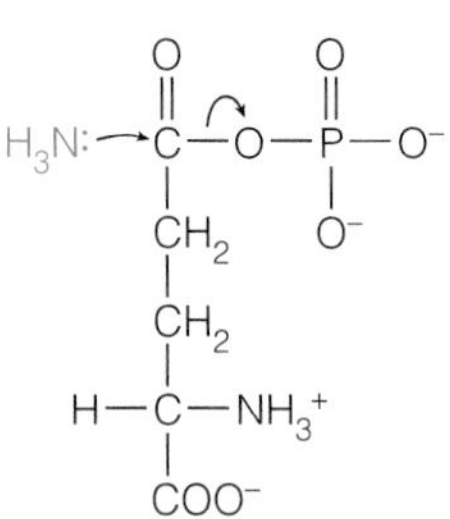

γ-**Glutamyl phosphate**

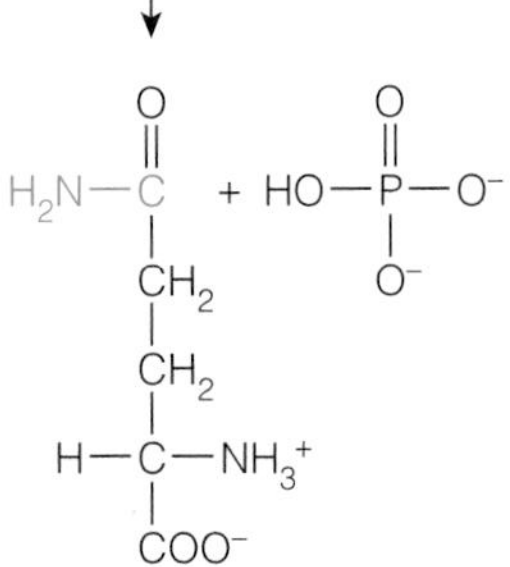

Glutamine

This enzyme is named a *synthetase*, rather than a *synthase*, because the reaction couples bond formation with the energy released from ATP hydrolysis. Both enzymes are classified as ligases (see page 392), but a synthase does not require ATP.

The glutamine synthetase reaction occurs via an acyl phosphate intermediate. ATP phosphorylates the δ-carbon of glutamate to give a carboxylic-phosphoric acid anhydride, which undergoes nucleophilic attack by the nitrogen of ammonia to give the amide product, glutamine.

As revealed originally by electron microscopy and more recently by x-ray crystallography, glutamine synthetase of *E. coli* is a dodecamer, whose 12 identical subunits form two facing hexagonal arrays (Figure 20.8). The holoenzyme has a molecular weight of about 600,000. Each catalytic site is formed at an interface between polypeptide subunits within a hexamer and is made up of residues from two adjacent subunits.

Regulation of Glutamine Synthetase

Glutamine occupies a central role in nitrogen metabolism (see Figure 20.7). The amide nitrogen is used in biosynthesis of several amino acids (including gluta-

FIGURE **20.8**

Structure of *E. coli* glutamine synthetase. The adenylylation site (Tyr 397) is shown in red and the two Mn^{2+} ions at the catalytic site are shown in orange. **(a)** Top view of the holoenzyme, as revealed from the crystal structure. Only 6 subunits can be seen from this viewpoint. **(b)** Schematic view of one of the 12 polypeptide chains.

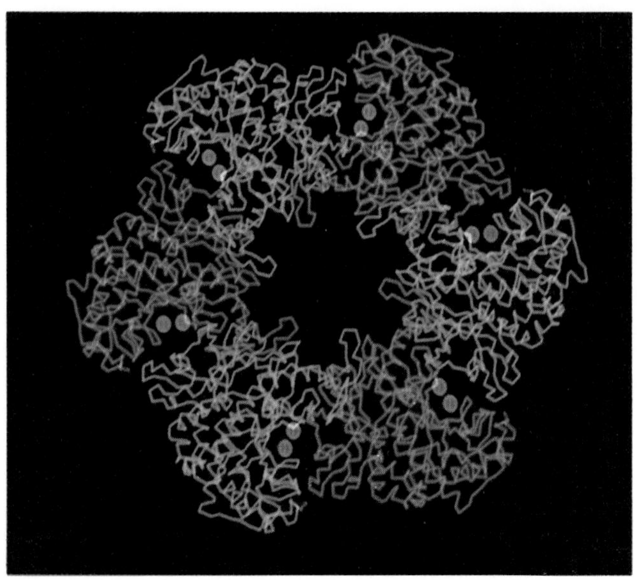

(a)

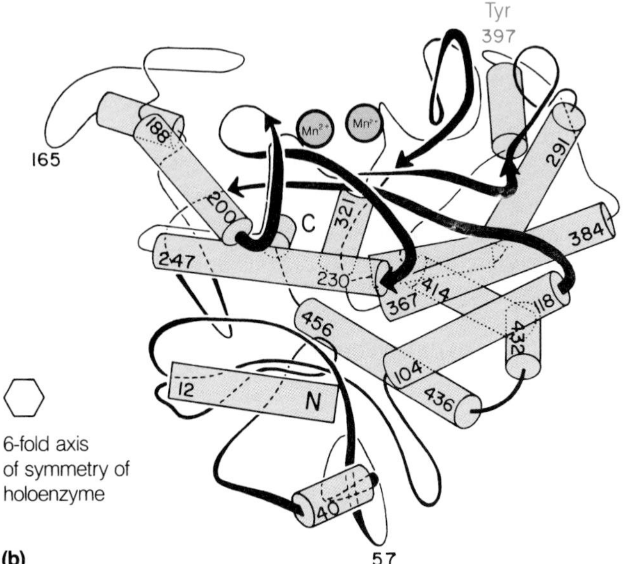

(b)

mate, tryptophan, and histidine), purine and pyrimidine nucleotides, and amino sugars. In animals, glutamine synthetase is a key participant in detoxifying ammonia formed from amino acid catabolism, particularly in brain. In fact, glutamate and glutamine are two of the most abundant free amino acids in brain cells; accumulation of these amino acids can deplete α-ketoglutarate, their principal precursor, thereby interfering with the citric acid cycle and energy generation. In addition, glutamine participates in ammonia excretion in the kidney, and it is a major fuel for cells in the immune system. It is not surprising, therefore, that the glutamine synthetase reaction is tightly regulated.

Prokaryotic glutamine synthetase is controlled by cumulative feedback inhibition and covalent modification.

As revealed primarily in *E. coli*, several remarkable control mechanisms for this reaction interact with one another in complex ways. The activity of glutamine synthetase is controlled by two distinct but interlocking mechanisms: (1) allosteric regulation by **cumulative feedback inhibition** and (2) covalent modification of the enzyme brought about by a regulatory cascade.

Cumulative feedback inhibition involves the action of eight specific feedback inhibitors. Those eight inhibitors are either metabolic end products of glutamine (tryptophan, histidine, glucosamine-6-phosphate, carbamoyl phosphate, CTP, and AMP; see Figure 20.7) or indicators in some other way of the general status of amino acid metabolism (alanine, glycine). Remarkably, each 50,000-dalton subunit of glutamine synthetase must contain binding sites for each of the eight inhibitors, as well as for substrates and products. Each of the eight compounds alone gives only partial inhibition, but in combination the degree of inhibition is increased until a mixture of all eight gives virtually complete blockage. This makes good metabolic sense, because it ensures that an accumulation of the end product of one pathway does not shut off the supply of a substrate (glutamine) needed for other pathways.

Superimposed on the cumulative feedback inhibition is a mode of regulation involving covalent modification of the enzyme. Glutamine synthetase is regulated by **adenylylation:** A specific tyrosine residue in the enzyme reacts with ATP to form an ester between the phenolic hydroxyl group and the phosphate of the resultant AMP. That tyrosine residue lies very close to a catalytic site. Adenylylation inactivates the adjacent catalytic site. An enzyme molecule with all 12 sites adenylylated is completely inactive, whereas partial adenylylation yields partial inactivation.

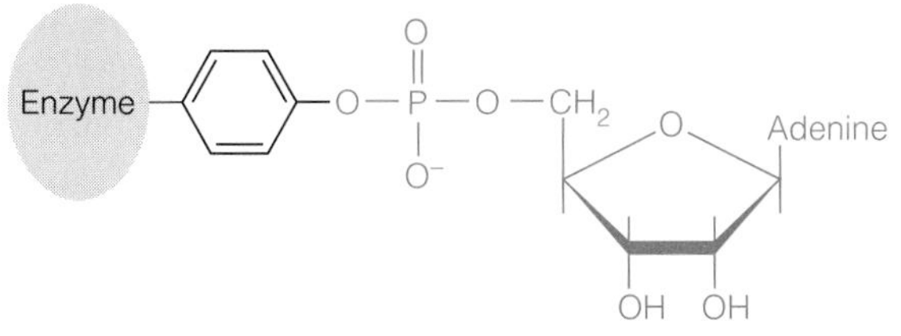

Adenylylated tyrosine residue

Adenylylation and deadenylylation of glutamine synthetase involve a complex series of regulatory cascades (Figure 20.9). Both reactions are catalyzed by the same enzyme: a complex of **adenylyl transferase** and a regulatory protein, $\mathbf{P_{II}}$. The molecular form of P_{II}—uridylylated or deuridylylated—determines whether the complex catalyzes adenylylation or deadenylylation. Uridylylation of P_{II} is catalyzed by yet another enzyme, **uridylyl transferase,** which transfers a UMP residue to a specific site on the P_{II} molecule. The product, $\mathbf{P_{II}}$**-UMP**, reacts with adenylyl transferase to stimulate its *de*adenylylation of glutamine synthetase. The non-UMP-containing form of P_{II} converts adenylyl transferase to an adenylylating enzyme. Activity of the uridylyl transferase is in turn stimulated by ATP and α-ketoglutarate and inhibited by glutamine. The [α-ketoglutarate]/[glutamine]

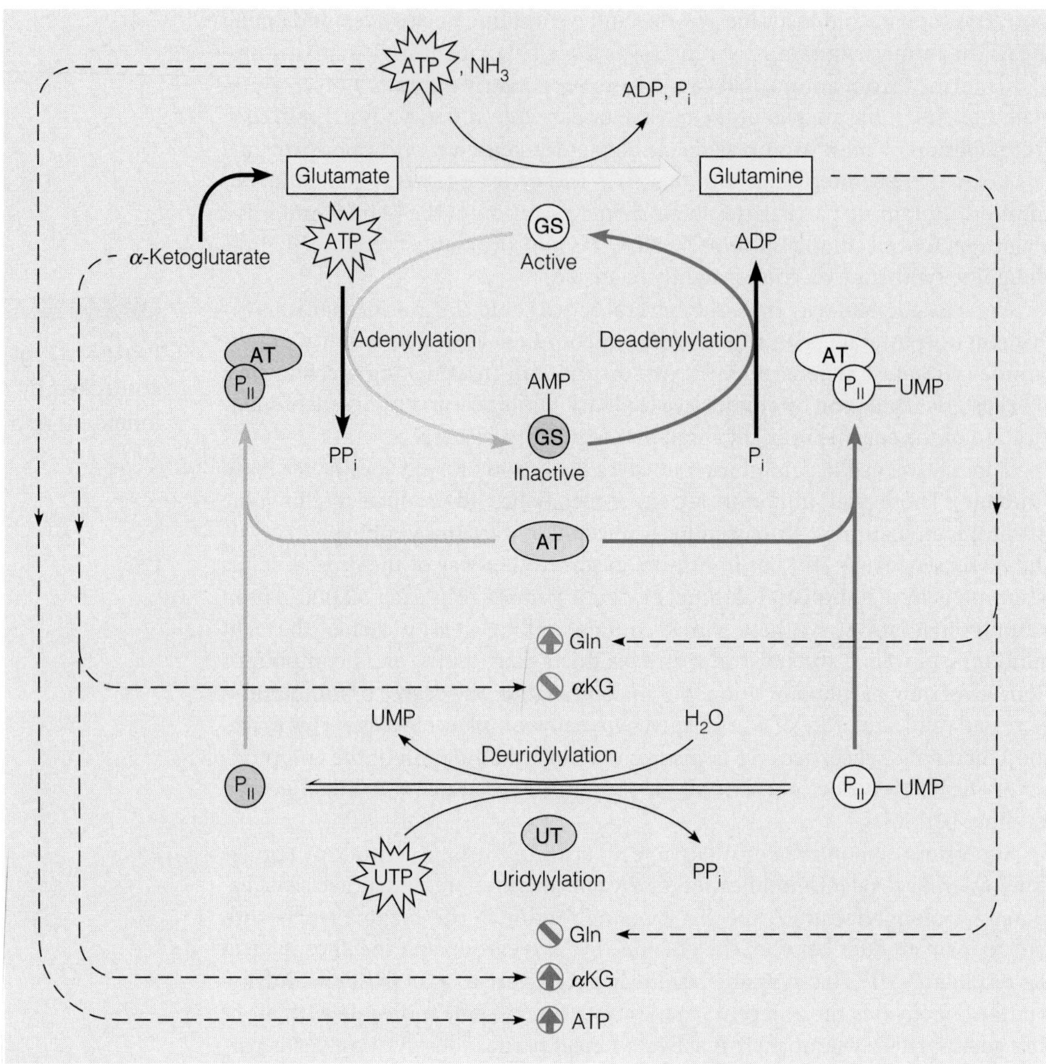

FIGURE **20.9**

Regulation of the activity of *E. coli* glutamine synthetase. The complex of AT (adenylyl transferase) and P_{II} (a regulatory protein) catalyzes both the adenylylation and deadenylylation of glutamine synthetase (GS), depending on whether P_{II} is deuridylylated (left) or uridylylated (right). Uridylylation of P_{II} is catalyzed by uridylyl transferase (UT). Components shown in blue tend to promote activity of glutamine synthetase, whereas those shown in orange are associated with enzyme inactivation.

ratio is crucial in determining whether uridylylation is favored or inhibited. Deuridylylation of P_{II}-UMP is catalyzed by a different form of the same enzyme.

These regulatory cascades provide a responsive mechanism ensuring that, when the supply of activated nitrogen (glutamine) is high, its further biosynthesis is shut down; the non-UMP-containing form of P_{II} accumulates and activates the adenylylation activity of adenylyl transferase. This causes the AMP-containing, or less active, form of glutamine synthetase to accumulate. Conversely, when activated nitrogen supplies are low, α-ketoglutarate accumulates and, provided that ATP is also abundant, stimulates the activity of glutamine synthetase by the converse mechanism.

At present we know little about how glutamine synthetase is regulated in animal cells. We do know that these exquisite control mechanisms are not involved. Recent evidence suggests that interconversions between octameric and tetrameric forms of the enzyme may be involved.

ASPARAGINE SYNTHETASE: A SIMILAR AMIDATION REACTION

An enzyme comparable to glutamine synthetase, namely **asparagine synthetase,** is also widespread, although the latter accounts for much less ammonia assimilation. Asparagine synthetase uses ammonia or glutamine in catalyzing the conversion of aspartate to asparagine.

$$\text{Aspartate} + \text{ATP} + NH_3\,(\text{Gln}) \longrightarrow \text{Asparagine} + \text{AMP} + PP_i + (\text{Glu})$$

Aspartate

Asparagine

Note that asparagine synthetase cleaves ATP to yield AMP and PP_i, whereas glutamine synthetase yields ADP and P_i. Asparagine synthetase also differs in that glutamine is strongly preferred as a substrate over ammonia. Probably the reactive species in bond formation is ammonia, which is generated at the active site by hydrolysis of the substrate glutamine.

CARBAMOYL PHOSPHATE SYNTHETASE: GENERATION OF AN INTERMEDIATE FOR ARGININE AND PYRIMIDINE SYNTHESIS

Recall from Figure 20.7 that the final route for assimilating ammonia first forms carbamoyl phosphate. The enzyme responsible is **carbamoyl phosphate synthetase.** Either ammonia or glutamine can serve as the nitrogen donor.

$$NH_3 + HCO_3^- + 2ATP \longrightarrow \text{carbamoyl phosphate} + 2ADP + P_i$$

$$\text{Glutamine} + HCO_3^- + 2ATP + H_2O \longrightarrow \text{carbamoyl phosphate} + 2ADP + P_i + \text{glutamate}$$

The bacterial enzyme can catalyze both reactions, although glutamine is the preferred substrate. Eukaryotic cells contain two forms of the enzyme. Form I, localized in mitochondria, has a preference for ammonia as substrate and is used in the arginine biosynthetic pathway and the urea cycle (see pages 727–729). Form II, present in the cytosol, has a strong preference for glutamine. It is inhibited by uridine triphosphate, consistent with its demonstrated involvement in pyrimidine nucleotide biosynthesis. As discussed in Chapter 22, the form II enzyme is part of a large protein with three distinct catalytic sites, which catalyze the first three reactions of pyrimidine nucleotide synthesis.

The Nitrogen Economy: Aspects of Amino Acid Synthesis and Degradation

METABOLIC CONSEQUENCES OF THE ABSENCE OF NITROGEN STORAGE COMPOUNDS

Protein and nucleic acid metabolism differs significantly from the metabolism of carbohydrates and lipids. Whereas carbohydrates and lipids can be stored for mobilization as needed by an organism for energy generation or for biosynthesis, most organisms past the embryonic state have no polymeric nitrogen compounds whose function is to be stored and released on demand. Although plants store some nitrogenous compounds, such as asparagine in asparagus, and some insects have storage proteins in their blood, such compounds do not represent widely used nitrogen storage depots. The lack of such depots imposes special requirements on organisms, particularly because of the limited availability of utilizable nitrogen. Animals must continually replenish nitrogen supplies through the diet to replace nitrogen lost through catabolism. In much of the world, protein-rich foods cannot be produced in sufficient quantity to meet the nutritional needs of humans and domestic animals. When dietary protein is insufficient, proteins manufactured for other purposes, mostly muscle proteins, are broken down and

Most organisms lack nitrogen storage depots.

not replaced. Such consequences occur even when the diet contains an adequate caloric content of protein, if that protein does not contain the needed amino acids (essential amino acids, see next section).

Just as we can think of a nitrogen economy for the biosphere, we can see it also in relation to individual organisms. Under optimal conditions, animals maintain nitrogen intake and excretion at equivalent rates. A well-nourished adult is said to be in **nitrogen equilibrium** or **normal nitrogen balance** if the daily intake of nitrogen through the diet is equal to that lost through excretion and other processes, such as perspiration. Positive nitrogen balance, in which normal nitrogen intake exceeds nitrogen loss, is seen during pregnancy, growth of a juvenile, or recovery from starvation. In negative nitrogen balance, more nitrogen is lost than is taken in, a situation found in senescence, starvation, and certain disease states. Plants and microorganisms commonly excrete very little nitrogen. Microorganisms often grow so rapidly that nitrogen released by catabolism is reassimilated, and in plants nitrogen is often available in such severely limited amounts that this factor itself limits cellular growth rates.

BIOSYNTHETIC CAPACITIES OF ORGANISMS

Organisms vary widely in their ability to synthesize amino acids. Many bacteria and most plants can synthesize all of their nitrogenous metabolites starting from a single nitrogen source, such as ammonia or nitrate. However, many microorganisms will use a preformed amino acid, when available, in preference to synthesizing that amino acid. Sometimes preformed amino acids are required. For example, over the course of evolution, *Lactobacillus* has lost many biosynthetic capacities because it grows in milk, a very nutrient-rich environment. This bacterium must therefore be provided with all 20 amino acids to be grown in the laboratory. Mammals are intermediate, being able to biosynthesize about half of the amino acids in quantities needed for growth and for maintenance of normal nitrogen balance.

Essential amino acids cannot be biosynthesized in adequate amounts and must be provided in the diet.

The amino acids that must be provided in the diet to meet an animal's metabolic needs are called **essential amino acids** (Table 20.1). Those that need not be provided because they can be biosynthesized in adequate amounts are called **nonessential amino acids.** In general, the essential amino acids include those with complex structures, including aromatic rings and hydrocarbon side chains. The nonessential amino acids include those that are readily synthesized from abundant metabolites, such as intermediates in glycolysis or the citric acid cycle.

Although dietitians recommend a protein intake of 50 to 100 grams per day or more, a human can do quite well on a diet containing as little as 20 grams per day, if that protein is of high nutritional quality—that is, if it contains adequate proportions of essential amino acids. In general, the more closely the amino acid composition of ingested protein resembles the amino acid composition of the animal eating the protein, the higher the nutritional quality of that protein. For humans, mammalian protein is of the highest nutritional quality, followed by fish and poultry, and then by fruits and vegetables. (In this context, nutritional quality refers only to the single criterion of essential amino acid content.) Plant proteins in particular are often deficient in lysine, methionine, or tryptophan. However, a vegetarian diet provides adequate protein if it contains a variety of protein sources, with a deficiency in one source being compensated for by excess in another source.

TABLE **20.1 Nutritional requirements for amino acids in mammals**

Essential
Arginine, histidine, isoleucine, leucine, lysine, methionine, phenylalanine, threonine, tryptophan, valine
Nonessential
Alanine, asparagine, aspartate, cysteine, glutamate, glutamine, glycine, proline, serine, tyrosine

Note: In both humans and rats, arginine and histidine are classified as essential amino acids, but nutritional studies show that they are required in the diet only during the growth of juveniles.

TRANSAMINATION

In Chapter 14 we introduced transamination, a process whereby amino acids can replenish citric acid cycle intermediates. Transamination plays a somewhat broader role in amino acid metabolism, in that it provides a route for redistribution of amino acid nitrogen. Because of the key role of glutamate in ammonia assimilation, it is a star player in transamination. In other words, glutamate is an abundant product of ammonia assimilation, and transamination uses glutamate nitrogen to synthesize other amino acids.

Transamination is the reversible transfer of an amino group from an amino acid to a keto acid, with pyridoxal phosphate as a coenzyme.

Transamination reactions are catalyzed by enzymes called **transaminases** or, more properly, **aminotransferases.** As mentioned in Chapter 14 and shown below, transamination involves transfer of the amino group, usually of glutamate, to an α-keto acid, with formation of the corresponding amino acid plus the α-keto derivative of glutamate, which is α-ketoglutarate.

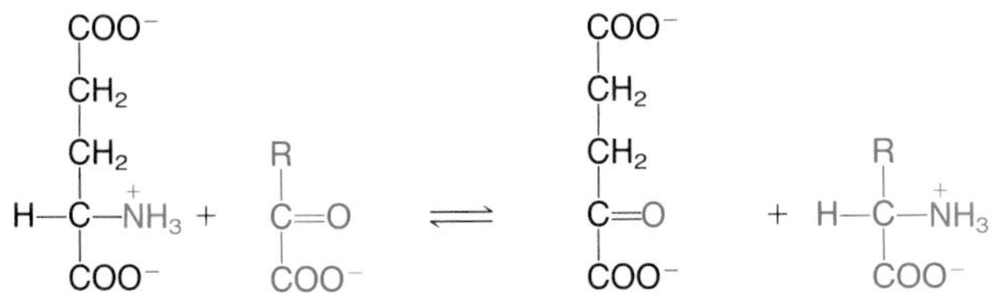

$$\text{Glutamate } (^-OOC{-}CH_2{-}CH_2{-}CH(\overset{+}{N}H_3){-}COO^-) + \alpha\text{-Keto acid } (R{-}C(=O){-}COO^-) \rightleftharpoons \alpha\text{-Ketoglutarate } (^-OOC{-}CH_2{-}CH_2{-}C(=O){-}COO^-) + \alpha\text{-Amino acid } (R{-}CH(\overset{+}{N}H_3){-}COO^-)$$

Glutamate **α-Keto acid** **α-Ketoglutarate** **α-Amino acid**

Specific aminotransferases exist in animal cells for the synthesis of all of the amino acids found in proteins except threonine and lysine, so long as the corresponding keto acids are available. Thus, the inability of animal cells to synthesize most of the essential amino acids results from an inability to synthesize the carbon skeletons in the form of α-keto acids.

Aminotransferases utilize a coenzyme, **pyridoxal phosphate,** that is derived from vitamin B_6. The functional part of the cofactor is an aldehyde functional group, —CHO, attached to a pyridine ring. Catalysis begins with condensation of this aldehyde with the amino group of an amino acid, to give a Schiff base, or aldimine, intermediate. Because pyridoxal phosphate participates in a diversity of reactions involving amino acids, we discuss catalytic mechanisms in a separate section, beginning on page 727.

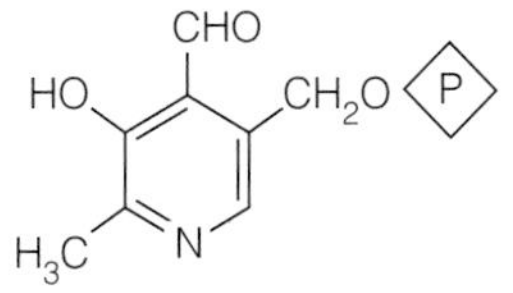

CHO, HO, CH_2O Ⓟ, H_3C, N

Pyridoxal phosphate

H, R—C—COO^-, N, CH, HO, CH_2O Ⓟ, H_3C, N

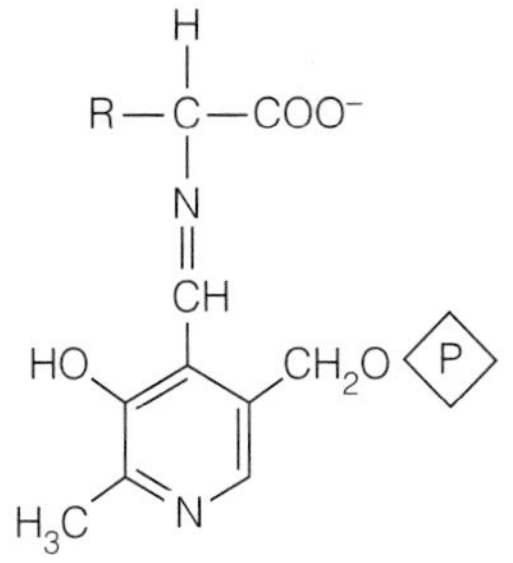

Schiff base between amino acid and pyridoxal phosphate

Transamination reactions have equilibrium constants close to unity. Therefore, the direction in which a particular transamination proceeds is controlled in large part by the intracellular concentrations of substrates and products. This means that transamination can be used not only for amino acid synthesis but also for degradation of amino acids that accumulate in excess of need. In degradation the transaminase works in concert with glutamate dehydrogenase, as exemplified by the degradation of alanine:

$$\text{Alanine} + \alpha\text{-ketoglutarate} \xrightarrow{\text{Aminotransferase}} \text{pyruvate} + \text{glutamate}$$

$$\text{Glutamate} + NAD^+ + H_2O \xrightarrow{\text{Glutamate dehydrogenase}} \alpha\text{-ketoglutarate} + NADH + \overset{+}{N}H_4$$

$$\text{Net: Alanine} + NAD^+ + H_2O \longrightarrow \text{pyruvate} + NADH + \overset{+}{N}H_4$$

Thus we see transamination as a mechanism for amino acid synthesis *or* degradation. Because the amino acids within a cell are rarely present in the proportions needed to synthesize the specific proteins of that cell, transamination plays an important role in bringing the amino acid composition into line with the organism's needs. It also participates in funneling excess amino acids toward catabolism and energy generation.

Most aminotransferases use glutamate/α-ketoglutarate as one of the two amino/keto acid pairs involved. Two such enzymes are important in the clinical diagnosis of human disease—serum glutamate-oxaloacetate transaminase (SGOT) and serum glutamate-pyruvate transaminase (SGPT):

$$\text{Glutamate} + \text{oxaloacetate} \xrightarrow{\text{SGOT}} \alpha\text{-ketoglutarate} + \text{aspartate}$$

$$\text{Glutamate} + \text{pyruvate} \xrightarrow{\text{SGPT}} \alpha\text{-ketoglutarate} + \text{alanine}$$

These enzymes, abundant in heart and in liver, are released from cells as part of the cell injury that occurs in myocardial infarction, infectious hepatitis, or other damage to either organ. Assays of these enzyme activities in blood serum can be used both in diagnosis and in monitoring the progress of a patient during treatment. Note the convention used in naming transaminases; the amino donor and the keto acid acceptor are named.

Protein Turnover

Proteins are like low-molecular-weight metabolic intermediates in that they are subject to continuous biosynthesis and degradation, a process called **protein turnover.** For an intracellular protein whose total concentration does not change with time, the steady state level is maintained by synthesis of the protein at a rate just sufficient to replenish protein lost by degradation. Many of the amino acids released during protein turnover are reutilized in the synthesis of new proteins.

QUANTITATIVE FEATURES OF PROTEIN TURNOVER

The macroscopic dimensions of protein turnover can be appreciated by considering a day in the life of a 70-kilogram person. That individual typically will consume 100 grams of protein during the day and, because he or she is in normal nitrogen balance, will excrete an equivalent amount of nitrogenous end products. Yet isotope labeling studies show that about 400 grams of protein are synthesized per day, and 400 grams broken down. About three-quarters of the released amino acids are reused in protein synthesis, with the remainder being degraded and the nitrogen excreted. Thus, the total amino acid pool consists of 500 g/day—100 ingested and 400 released via protein degradation. From this pool, 400 grams are used in protein synthesis and 100 grams are catabolized and excreted.

Individual proteins exhibit tremendous variability in their metabolic lifetimes, from a few minutes to many months. You should realize, however, that all of the proteins in the body are represented among the 400 grams broken down in a typical day. Extensive pulse-chase experiments in laboratory animals (see Tools of Biochemistry 12A) show that protein degradation follows first-order kinetics. For a particular protein, individual molecules are degraded at random, such that a semilogarithmic plot of isotope remaining in a protein versus time is linear. Thus, we can determine the metabolic half-life of a particular protein. In the rat the average protein has a half-life of 1 or 2 days. Table 20.2 gives information about the half-lives of specific proteins.

As you might expect, proteins that are secreted into an extracellular environment, such as digestive enzymes, polypeptide hormones, and antibodies, turn over quite rapidly, whereas proteins that play a predominantly structural role, such as collagen of connective tissue, are much more stable metabolically. Enzymes catalyzing rate-determining steps in metabolic pathways are also short-lived. Indeed, for many enzymes the rate of breakdown is an important regulatory factor in controlling intracellular enzyme levels. By contrast, proteins that do not represent metabolic control points turn over relatively slowly. In the rat, cytochrome *c* has a half-life of nearly a week; in the human, hemoglobin is as long-lived as the erythrocyte in which it resides (about 120 days). But why should such proteins turn over at all, if their degradation does not represent a metabolic control mechanism? Isn't such turnover wasteful of energy? Let us consider that point.

TABLE 20.2 **Half-lives and intracellular sites of degradation in protein turnover**

Half-life (hours)	Intracellular Location: Nucleus	Cytosol	Mitochondria	Endoplasmic Reticulum and Plasma Membrane
<2	Oncogene products	Ornithine decarboxylase, tyrosine aminotransferase, protein kinase C	δ-Aminolevulinic acid synthetase	HMG-CoA reductase
2–8	—	Tryptophan oxygenase, cAMP-dependent protein kinase	—	γ-Glutamyl transferase
9–40	Ubiquitin	Calmodulin, glucokinase	Acetyl-CoA carboxylase, alanine aminotransferase	LDL receptor, cytochrome P450
41–200	Histone H1	Lactate dehydrogenase, aldolase, dihydrofolate reductase, phytochrome P670	Cytochrome oxidase, pyruvate carboxylase, cytochrome *c*	Cytochrome b_5, cyt b_5 reductase
>200	Histones H2A, H2B, H3, H4	Hemoglobin, glycogen phosphorylase	—	Acetylcholine receptor

Source: From M. Rechsteiner, S. Rogers, and K. Rote, *Trends Biochem. Sci.* (1987) 12:390–394. © 1987 with permission from Elsevier Science.

Note: This table represents just a few examples of the many proteins whose half-lives have been determined in different organisms.

BIOLOGICAL IMPORTANCE OF PROTEIN TURNOVER

Like all other intracellular constituents, proteins are subjected to a barrage of environmental influences, primarily reactive oxygen species (see Chapter 15), which can affect their structure, conformation, and biological activity. The capacity of proteins to repair the resulting damage is limited. Protein turnover could be seen as an inefficient quality control system in which both normal and modified proteins are randomly degraded and replaced. Recent work, however, indicates that the process is nonrandom; protein molecules that have become chemically altered are preferentially degraded. A certain chemical change may mark a protein molecule, targeting it for degradation by a proteolytic enzyme that specifically recognizes the marker.

Though much remains to be learned about intracellular protein degradation, a great deal has been learned within the past decade. We know that in bacteria, mutant proteins are degraded much more rapidly than their wild-type counterparts. Evidently evolution has generated proteins whose conformation renders maximum stability in the intracellular environment, and most structural changes decrease this stability.

Protein turnover also represents a route for cellular adaptation to altered environmental conditions. For example, in many bacteria, extensive proteolysis is one of the metabolic events interlinked with **sporulation,** or spore formation. Spores represent a heat-stable form of the microorganism; they metabolize at negligible rates and can remain dormant for months or years. When metabolic conditions induce a growing cell to sporulate, extensive protein turnover occurs, with the amino acids released being used to synthesize proteins of the spore. This quasi-dormant state can be maintained indefinitely, with germination to vegetative cells occurring after improvement in the environmental conditions.

All proteins are in a constant state of turnover and replacement, for repair of damage and for biological regulation.

Except for specialized functions such as sporulation, it was thought until the early 1970s that protein turnover, meaning its complete degradation to amino acids, served just two major functions: (1) protein digestion, providing amino acids for synthesis of proteins and other metabolites derived from amino acids, and (2) ridding cells of defective proteins, including those that were mutationally or environmentally damaged. Accordingly, most intracellular proteases were thought to participate in protein turnover. Of course, it has long been recognized that specific endopeptidase protein cleavage reactions are involved in enzyme activation; an example is the blood clotting cascade, which we discussed in Chapter 11. Over the past two decades, it has become clear that this process of limited proteolysis—cleavage of a few specific peptide bonds in a protein—has a host of additional functions, including regulation of gene expression, response to environmental stress, and participation in cell signaling pathways. Of great current interest is the involvement of selective proteolytic reactions in signaling pathways leading to *apoptosis,* a process in normal development in which certain cells, having fulfilled their function in differentiation, undergo a preprogrammed death. These functions are presented in detail in Chapter 28. Here we concern ourselves with those aspects of protein turnover that relate specifically to amino acid metabolism, to identifying major classes of intracellular proteases, and to describing some of the structural features that predispose certain proteins to degradation.

INTRACELLULAR PROTEASES AND SITES OF TURNOVER

Because most proteins are used intracellularly, most turn over within the cell. The earliest intracellular proteases to be characterized were in lysosomes. It is clear, however, that proteins are degraded in all of the major cell compartments, inasmuch as proteolytic enzymes are found throughout the cell. In eukaryotic cells several major proteases have been found in the cytosol—two Ca^{2+}-activated proteases called **calpains,** a large (700-kilodalton) multisubunit neutral protease, and a still larger ATP-dependent protease called the **proteasome.** These enzymes are distinct from the lysosomal proteases, called **cathepsins,** which are designed to function in an acidic milieu. Although the specific roles of each protease are not yet defined, it seems likely that extracellular proteins taken up by a cell and long-lived cellular proteins are degraded in lysosomes, while selective protein turnover related to metabolic regulation occurs in other compartments.

Lysosomes, which form by budding from the Golgi complex, are bags of digestive enzymes, containing proteases, nucleases, lipases, and carbohydrate-cleaving enzymes. As discussed further in Chapter 28, lysosomes play various cellular roles: secretion of digestive enzymes, digestion of organelles destined for destruction, digestion of food particles or bacteria engulfed by phagocytosis, or intracellular release of enzymes followed by **autolysis**—the digestion and death of a cell as part of a normal morphogenetic process in development. For example, the webbing that is present between the toes and fingers of the early human fetus is destroyed by this kind of programmed cell death.

In contrast to lysosomal enzymes, which are usually safely sequestered in their vesicles, any protease activity free in normal cytosol must be under strict control, so as to attack only those proteins whose destruction is needed—damaged, mutant, or otherwise dispensable proteins. Also, although proteolysis is favored thermodynamically, much intracellular protein turnover has a substantial ATP requirement. Recent work shows that the proteasome has a tubular structure (Figure 20.10) and ATP energy is needed both to mark proteins for degradation and to move them into and through the tube. The identification, or marking, of those proteins whose degradation suits the interests of the cell involves some labeling schemes described in the next section.

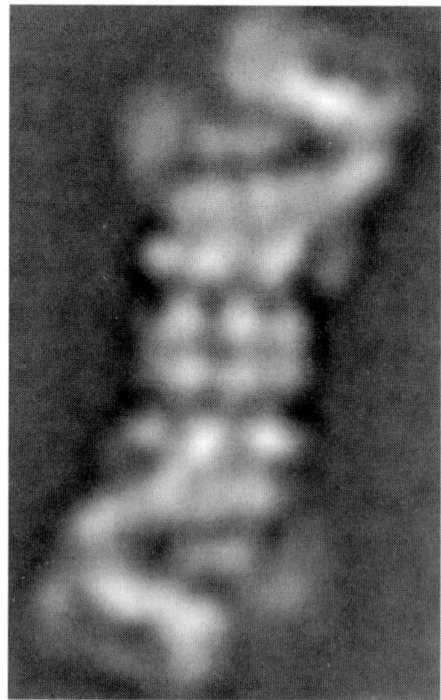

FIGURE **20.10**

Structure of the proteasome. This electron micrograph shows the 26S proteasome of the toad *Xenopus laevis.* This large protein in eukaryotes contains seven different α-type subunits (at the ends of the tube) and seven different β-type subunits. Proteins marked by reaction with the protein ubiquitin pass through the tube in an ATP-dependent fashion. A smaller entity, called the 20S proteasome, contains an ATP-independent proteolytic core. Prokaryotic cells contain only the 20S proteasome; the crystal structure of the 20S proteasome from an archaebacterium, *Thermoplasma acidophilum,* has been described.

Courtesy of W. Baumeister from A. Lupas, J. M. Flanagan, T. Tamura, and W. Baumeister, *Trends Biochem. Sci.* (1997) 22:399–404, with permission from Elsevier Science.

CHEMICAL SIGNALS FOR TURNOVER

The turnover rates for different proteins vary by as much as 1000-fold, whereas differences in protein stability, as measured by denaturation in vitro, may be much less. Four structural features are currently thought to be interrelated determinants of turnover rate: (1) **ubiquitination,** (2) metal-catalyzed oxidation of particular residues, (3) **PEST sequences,** and (4) particular N-terminal residues.

Ubiquitination

Ubiquitin is a small (76-residue), heat-stable protein found in all eukaryotic cells—it derives its name from its widespread distribution. Ubiquitin undergoes

FIGURE **20.11**

Programmed destruction of cytosolic proteins by the ubiquitin marking system. E_1 and E_2 are enzymes involved in the ubiquitin transfer to lysine residues on target proteins. **Step 1:** ATP-dependent formation of a thioester bond between ubiquitin C-terminus and a cysteine thiol on E_1. **Step 2:** Transfer of ubiquitin from E_1 to E_2. **Step 3:** Transfer of ubiquitin to lysine residues on target proteins. **Step 4:** ATP-dependent proteolysis of ubiquitin-tagged proteins.

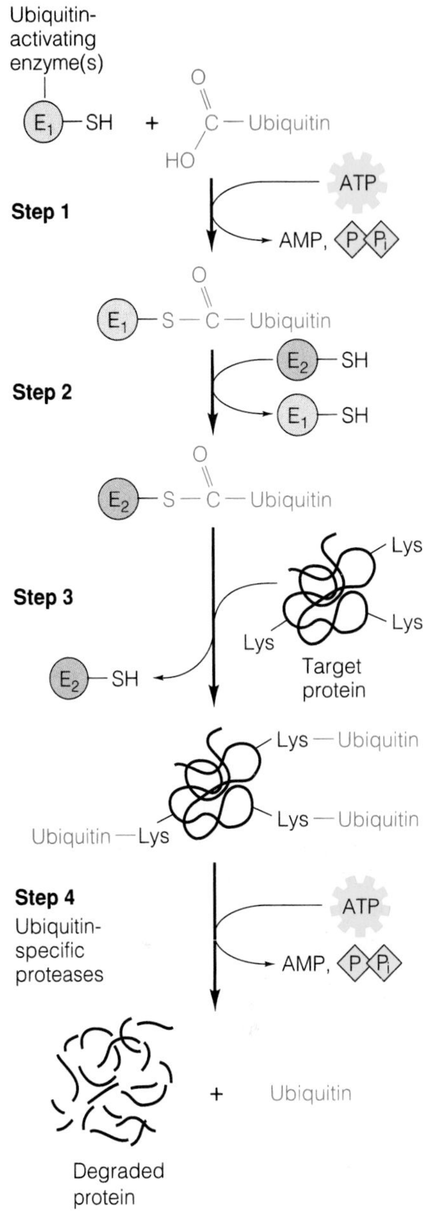

an ATP-dependent reaction with proteins, which condenses C-terminal glycine residues of ubiquitin with lysine amino groups on the target protein, as shown in Figure 20.11. Such modified proteins are degraded soon afterward, by the proteasome, which recognizes ubiquitinated proteins, and uses ATP energy to drive them through its tubular structure. It is still not clear how proteins are chosen to receive this "kiss of death," but it involves aspects of the amino acid sequence described in the following sections.

Oxidation of Amino Acid Residues

Earl Stadtman and co-workers have shown that many proteins undergo mixed-function oxidation of particular residues, promoted by conditions that generate oxygen radicals. Fe^{2+} and the hydroxyl radical, OH·, are essential reactants in this process, and lysine, arginine, and proline residues are the most susceptible to oxidation. These modifications evidently mark these proteins for subsequent degradation by cytosolic proteases. Consistent with this idea is the isolation of a protease, from both *E. coli* and rat liver, that cleaves oxidized glutamine synthetase in vitro but does not attack the native enzyme; presumably, other oxidized proteins are also targets for this enzyme. Note that oxidation is distinct from ubiquitination in that the latter is a specific signal designed to mark proteins for degradation. Recent evidence suggests that accumulation of oxidatively damaged proteins beyond the cell's capacity to degrade and replace them contributes importantly to cellular aging.

PEST Sequences

Examination of the amino acid sequences of short-lived proteins ($t_{1/2} < 2$ hours) shows that virtually all of these proteins contain one or more regions rich in proline, glutamate, serine, and threonine. From the one-letter designations for these amino acids (P, E, S, and T, respectively), these regions, some 12 to 60 residues long, have been called PEST sequences. Very few longer-lived proteins contain these regions. Although this evidence is largely circumstantial and says nothing about the biochemical function of the PEST sequences, the pattern does result from inspection of several dozen amino acid sequences. Moreover, creation of PEST sequences in long-lived proteins by site-directed mutagenesis increases their metabolic lability. Thus, it seems likely that the PEST region is part of a recognition scheme for the enzyme systems that degrade short-lived proteins, possibly including the ubiquitin marking system.

The metabolic lifetimes of proteins are determined by ubiquitination, which is determined by PEST sequences, particular N-terminal residues, and oxidation of particular residues.

N-Terminal Amino Acid Residue

Experiments with bacteria have revealed that the intracellular half-life of a particular protein varies considerably, depending on the identity of its N-terminal amino acid residue. An N-terminal residue of Phe, Leu, Tyr, Trp, Lys, or Arg is correlated with short metabolic lifetimes, whereas proteins with other amino termini are far longer-lived. These findings, originally made with natural proteins, are supported by experiments in which amino termini of a protein were altered by site-directed mutagenesis, with corresponding changes in metabolic half-lives of the mutant proteins. The same "N-end rule" has also been shown to apply to animal and plant cells.

These and other observations indicate that specific structural features on proteins convey information about the metabolic stability of the proteins. The molecular nature of that information processing and the identities of the enzymes involved remain to be determined, and, as discussed further in Chapter 28, proteolysis is becoming recognized as comparable to phosphorylation in its importance as a metabolic regulatory mechanism.

Amino Acid Degradation and Metabolism of Nitrogenous End Products

COMMON FEATURES OF AMINO ACID DEGRADATION PATHWAYS

In animals whose dietary protein intake exceeds the need for protein synthesis and other biosyntheses, the excess nitrogen is mostly degraded, with the carbon skeletons being metabolized in the citric acid cycle. Protein can thus be a significant contributor toward an animal's energetic requirements. In contrast, plants and bacteria generally can synthesize most of their own amino acids, and they regulate the anabolic pathways so that excesses rarely develop. Generally, microorganisms use preformed amino acids in preference to synthesizing their own, even though many bacteria can satisfy all of their requirements for nitrogen *and* carbon from a single amino acid. The degradative pathways active here are in general similar to those described in animals.

Amino acid degradation usually begins with conversion to the corresponding α-keto acid by transamination or oxidative deamination.

With a few exceptions, the first step in amino acid degradation involves removal of the α-amino group to give the corresponding α-keto acid. This modification is usually effected by transamination, with the concomitant synthesis of glutamate from α-ketoglutarate. This is followed by the glutamate dehydrogenase reaction, as was shown on page 721. Thus, the net process is the deamination of amino acid to the corresponding keto acid plus ammonia. The same net conversion can also be catalyzed by **L-amino acid oxidase,** a flavoprotein enzyme found in liver and kidney.

$$R{-}CH(NH_3^+){-}COO^- + FMN + H_2O \longrightarrow R{-}C({=}O){-}COO^- + FMNH_2 + NH_3$$

$$FMNH_2 + O_2 \longrightarrow FMN + H_2O_2$$

$$2H_2O_2 \xrightarrow{\text{Catalase}} 2H_2O + O_2$$

The peroxide formed is decomposed by catalase. Kidney and liver are also rich in FAD-containing **D-amino acid oxidase.** The function of this enzyme in animals is unknown, since the D isomers of amino acids are quite rare. However, bacterial cell walls contain D-amino acids (see Chapter 16), and a number of bacterial cells contain D-amino acid oxidases.

Once the nitrogen has been removed, the carbon skeleton can, depending on the physiological state of the organism, either proceed toward oxidation in the citric acid cycle or be used for biosynthesis of carbohydrate. Figure 20.12 shows the entry points into the citric acid cycle for the breakdown products of each of the amino acids. The individual pathways are presented in Chapter 21.

Amino acids whose skeletons generate pyruvate or oxaloacetate (examples are alanine and aspartate) are efficiently converted to carbohydrates via gluconeogenesis. Amino acids leading to acetyl-CoA or acetoacetyl-CoA, such as leucine, contribute heavily toward ketogenesis. The terms **glucogenic** and **ketogenic** have been used to classify amino acids as generators primarily of carbohydrate or ketone bodies, respectively.

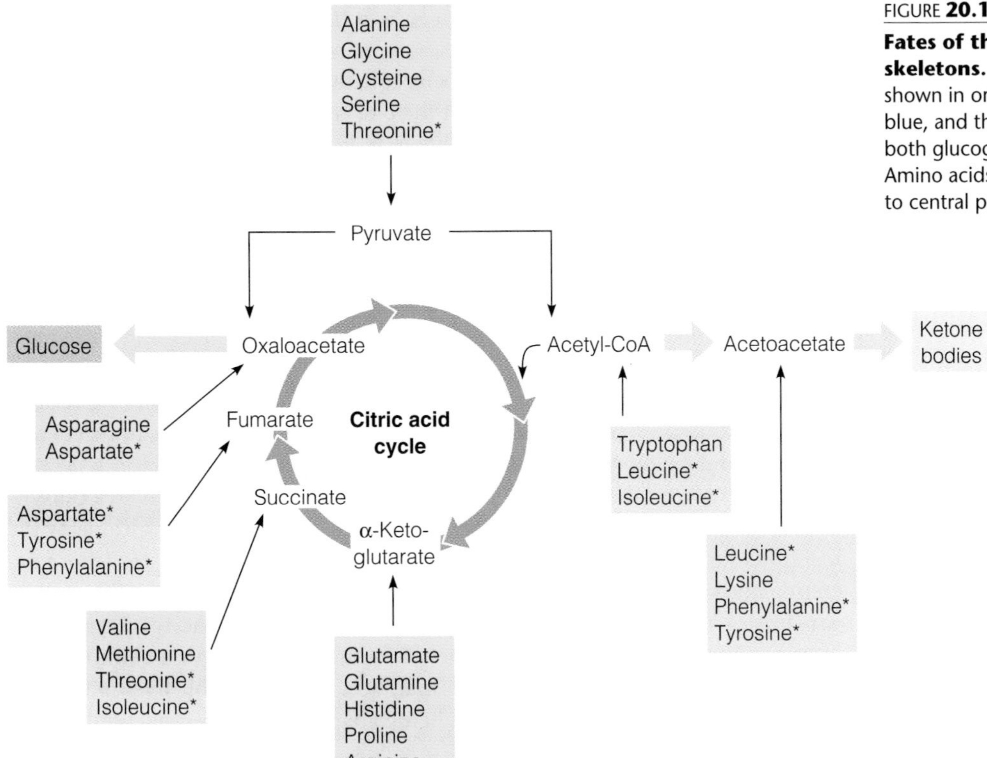

FIGURE **20.12**

Fates of the amino acid carbon skeletons. Glucogenic amino acids are shown in orange, ketogenic amino acids in blue, and the few amino acids that can be both glucogenic and ketogenic in purple. Amino acids with more than one route of entry to central pathways are marked by an asterisk.

DETOXIFICATION AND EXCRETION OF AMMONIA

Although ammonia is a universal participant in amino acid synthesis and degradation, its accumulation in abnormal concentrations has toxic consequences. Therefore, cells undergoing active amino acid catabolism must be able to detoxify and/or excrete ammonia as fast as it is generated. For most aquatic animals, which can take in and pass out unlimited quantities of water, ammonia simply dissolves in the water and diffuses away. Because terrestrial animals must conserve water, they convert ammonia to a form that can be excreted without large water losses. Birds, terrestrial reptiles, and insects convert most of their excess ammonia to **uric acid,** an oxidized purine. Because uric acid is quite insoluble, it precipitates and can be excreted without a large water loss and without building up osmotic pressure. This is particularly important during the part of each animal's lifetime that is spent in the egg. The biosynthesis of uric acid occurs by the route used to synthesize purine nucleotides, a pathway presented in Chapter 22. Most mammals excrete the bulk of their nitrogen in the form of **urea** (an interesting exception is the Dalmatian dog, which excretes most nitrogen as uric acid). Urea is highly soluble and, being electrically neutral, does not affect the pH when it accumulates, as does ammonia.

Uric acid

$$H_2N-\overset{O}{\overset{\|}{C}}-NH_2$$

Urea

Animals have evolved pathways, adapted to their lifestyles, for excretion of ammonia, uric acid, or urea as the major nitrogenous end product.

THE KREBS–HENSELEIT UREA CYCLE

Urea is synthesized almost exclusively in the liver and then transported to the kidneys for excretion. The synthetic pathway, which is cyclic, was discovered by Hans Krebs and Kurt Henseleit in 1932, five years before the other cycle for which Krebs is famous. Krebs and Henseleit were investigating the pathway by adding possible precursors to liver slices and then measuring the amount of urea produced. When arginine was added, urea was produced in 30-fold molar excess over the amount of arginine administered. Similar results were seen if either of two structurally related amino acids, **ornithine** or **citrulline,** was substituted for arginine. Because these three amino acids seemed to function catalytically to promote

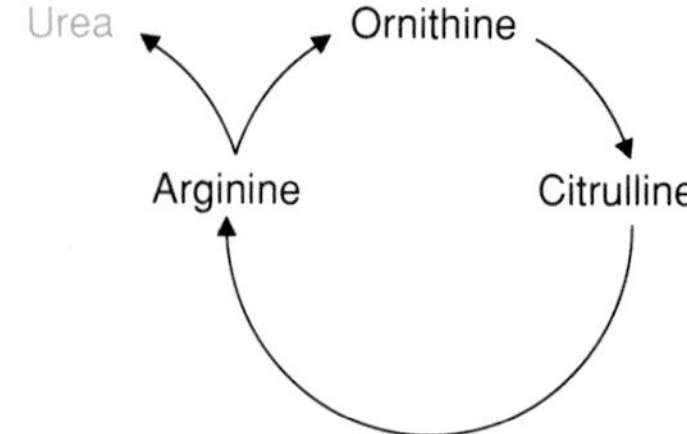

urea synthesis, Krebs and Henseleit proposed the existence of a cyclic pathway (see margin).

That proposal was correct, as was subsequently confirmed by isolation of the enzymes involved and identification of the biosynthetic route to ornithine, which begins the pathway. These details are shown in Figure 20.13. Ornithine serves as a "carrier," upon which are assembled the carbon and nitrogen atoms that will eventually constitute urea. Ornithine itself is synthesized from glutamate, by a pathway shown in Chapter 21 (see Figure 21.2). The source of the carbon and one nitrogen atom in urea is carbamoyl phosphate, which reacts with ornithine, via the enzyme **ornithine carbamoyltransferase,** to give citrulline. The second nitrogen comes from aspartate, which reacts with citrulline to form **argininosuccinate,** through the action of **argininosuccinate synthetase.** Next, **argininosuccinase** cleaves argininosuccinate in a nonhydrolytic, nonoxidative reaction to give arginine and fumarate. Arginine is cleaved hydrolytically by **arginase,** to regenerate ornithine and yield one molecule of urea.

The enzyme arginase is responsible for the cyclic nature of the urea biosynthetic pathway. Virtually all organisms synthesize arginine from ornithine by the reactions shown in Figure 20.13. However, only **ureotelic** organisms (those excreting most of their nitrogen as urea) contain arginase and, hence, only those organisms carry out the cyclic pathway. Interestingly, the capacity to synthesize arginase develops in frogs at the same time that they undergo metamorphosis from the tadpole stage to the adult animal. Because the tadpole lives in water, it can excrete ammonia. The adult frog, being adapted to a terrestrial lifestyle, develops the ability to synthesize urea.

As noted already, one nitrogen atom in urea comes from aspartate. This atom is derived from ammonia, which is transferred to glutamate via the glutamate dehydrogenase reaction, then to aspartate by transamination. Note from the bottom of Figure 20.13 that carbon balance is maintained in this process by conversion of the fumarate produced from argininosuccinate cleavage to oxalaoacetate in the citric acid cycle and then to aspartate by transamination.

The net reaction for one turn of the urea cycle is as follows:

$$CO_2 + \overset{+}{N}H_4 + 3ATP + \text{aspartate} + 2H_2O \longrightarrow \text{urea} + 2ADP + 2P_i + AMP + PP_i + \text{fumarate}$$

Urea is synthesized by an energy-requiring cyclic pathway that begins and ends with ornithine.

Two molecules of ATP are required to reconvert AMP to ATP, so really four (not three) high-energy phosphates are consumed in each turn of the cycle. Thus, the synthesis of this excretion product is energetically expensive. Extremely protein-rich diets are thus not necessarily beneficial. Because much energy is consumed both in fixing nitrogen and in excreting it, dietary intake of protein beyond its need for biosynthesis of nitrogenous metabolites is energetically wasteful.

The reactions of the urea cycle occur in both mitochondria and cytosol of liver cells. Glutamate dehydrogenase, the citric acid cycle enzymes, carbamoyl phosphate synthetase I, and ornithine carbamoyltransferase are localized in the mitochondrion, and the rest of the cycle occurs in the cytosol. This means that ornithine must be transported into mitochondria, and citrulline exported to the cytosol, in order for the cycle to proceed.

FIGURE **20.13**

The Krebs–Henseleit urea cycle. Urea (upper left) contains a carbon and a nitrogen (orange) derived from carbamoyl phosphate, and a nitrogen (purple) derived from aspartate. CO_2 and $\overset{+}{N}H_4$, the ultimate sources of these atoms, were incorporated in turn through the actions of carbamoyl phosphate synthetase (upper right) and glutamate dehydrogenase (lower right). Glutamate can serve directly as the source of some urea nitrogen. Yellow shading identifies reactions occurring in mitochondria. The rest of the pathway takes place in cytosol.

Following its synthesis, urea is transported in the bloodstream to the kidneys, which filter it for excretion. Measurements of blood urea nitrogen (BUN) levels provide a sensitive clinical test of kidney function, because filtration and removal of urea are impaired in cases of kidney malfunction. Analogously, blood ammonia

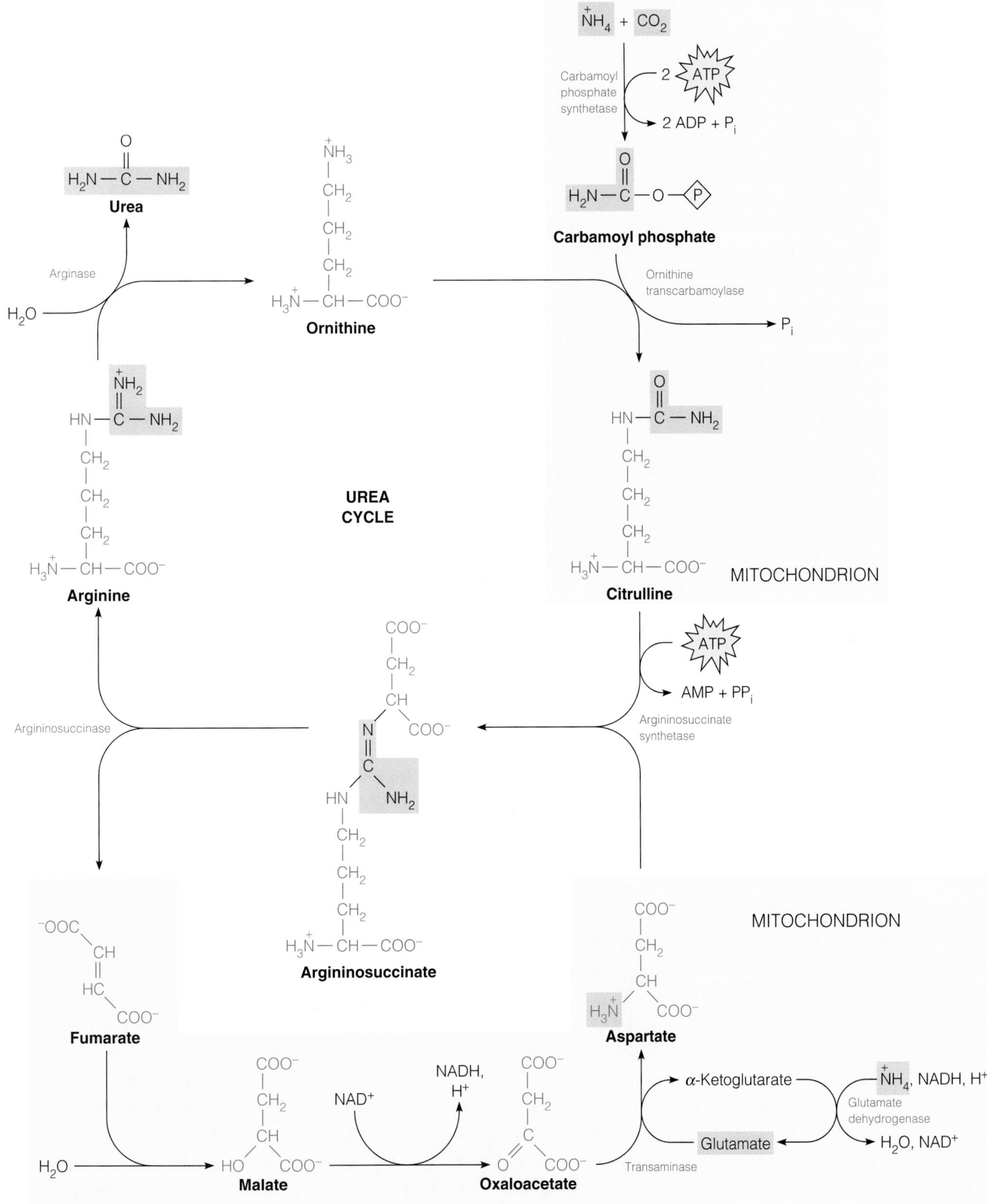

measurements are a sensitive test of liver function. Liver damage, whether acute (hepatitis, poisoning) or chronic (alcoholic cirrhosis), reduces activity of the urea cycle. The accumulation of ammonia is toxic to the brain and thus is related to the comatose condition seen in advanced cases of chronic alcoholism.

TRANSPORT OF AMMONIA TO THE LIVER

All animal organs degrade amino acids and produce ammonia. Two mechanisms are involved in transporting this ammonia from other tissues to liver for its eventual conversion to urea (Figure 20.14). Most tissues use glutamine synthetase to convert ammonia to the nontoxic, and electrically neutral, glutamine. The glutamine is transported in the blood to the liver, where, as noted also in Chapter 16, it is cleaved hydrolytically by **glutaminase.**

$$\text{Glutamine} + H_2O \longrightarrow \text{glutamate} + \text{ammonia}$$

The glucose–alanine cycle removes toxic ammonia from muscle. Glutamine synthetase and glutaminase do the same for most other tissues.

Muscle, which derives most of its energy from glycolysis, uses a different route, the **glucose–alanine cycle.** Glycolysis generates pyruvate, which undergoes transamination with glutamate to give alanine and α-ketoglutarate. The glutamate in turn has acquired its nitrogen from ammonia, via glutamate dehydrogenase. The resultant alanine is transported to the liver, where it loses its nitrogen by a reversal of the previous processes. This reversal yields ammonia for urea synthesis, plus pyruvate. The pyruvate undergoes gluconeogenesis to give glucose, which is released to the blood for transport back to the muscle or for nourishment of the brain. This cyclic process helps muscle get rid of ammonia, with the carbon from the pyruvate being returned to the liver for gluconeogenesis.

Although urea is usually regarded as an end product, studies on hibernating animals have shown that urea can be reutilized for amino acid synthesis in these animals. During the 6- to 7-month hibernation of the black bear, the animal does not urinate. Urea that accumulates in the bladder is somehow reabsorbed and returned to the tissues for amino acid synthesis. The pathways of utilization are not known, but presumably they involve hydrolytic cleavage of urea to ammonia. Current investigations focus upon whether bears can synthesize all needed amino acids under conditions of hibernation.

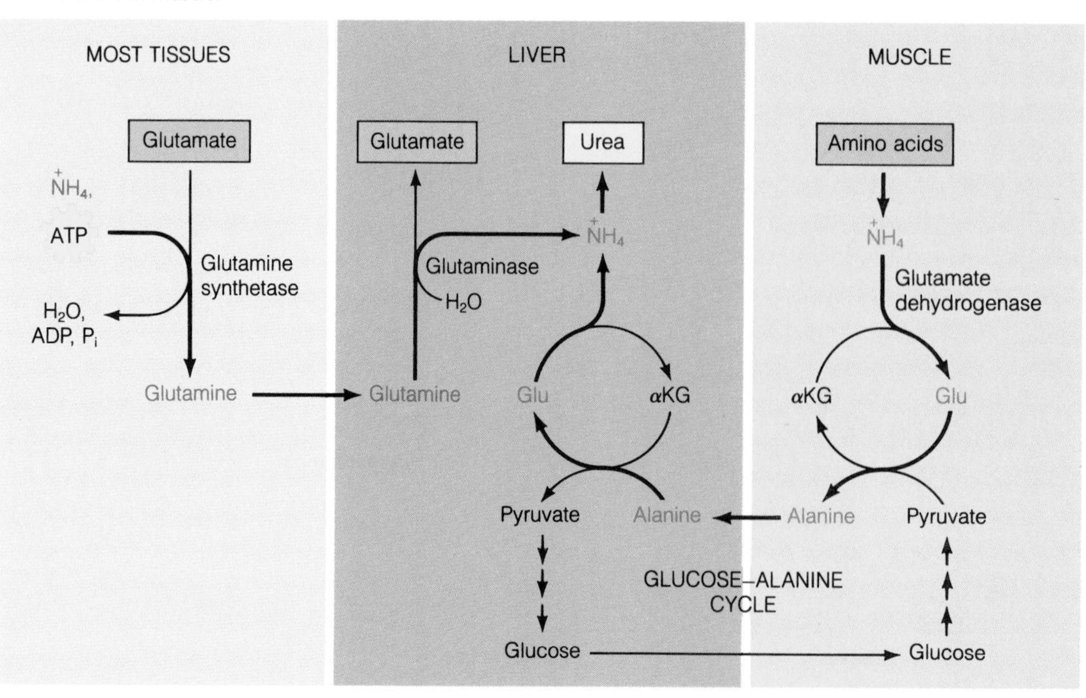

FIGURE **20.14**

Transport of ammonia to the liver for urea synthesis. The carrier is glutamine in most tissues but is alanine in muscle.

Coenzymes Involved Primarily in Nitrogen Metabolism

Before presenting in detail the metabolism of amino acids and nucleotides, as we do in the next two chapters, we must consider three families of coenzymes that function primarily in amino acid and/or nucleotide metabolism. Although all have been mentioned before, we shall consider their actions in detail here. These cofactors include (1) pyridoxal phosphate, the cofactor for transamination and many other reactions of amino acid metabolism; (2) the folic acid coenzymes, which transfer single-carbon functional groups in synthesizing nucleotides and certain amino acids; and (3) the B_{12}, or cobalamin, coenzymes, which participate in the synthesis of methionine and, as noted in Chapter 18, the catabolism of methylmalonyl-CoA.

PYRIDOXAL PHOSPHATE

Vitamin B_6 was discovered in the 1930s as the result of nutritional studies with rats fed vitamin-free diets. The vitamin as originally isolated is **pyridoxine,** named from its structural similarity to pyridine. Pyridoxine contains a hydroxymethyl group at position 4 of the pyridine ring. However, in the coenzyme this group has been oxidized to an aldehyde. In addition the hydroxymethyl group at position 5 is phosphorylated. Pyridoxal phosphate (which we shall abbreviate PLP) is the predominant coenzyme form, with pyridoxamine phosphate (PMP) being an intermediate in transamination reactions.

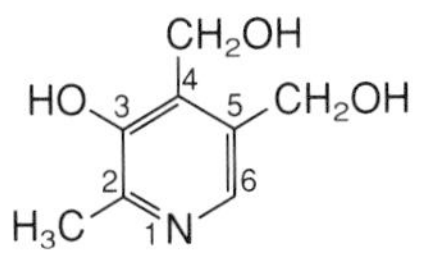

Nutritional requirements for vitamin B_6 are so low that dietary deficiency states are rarely observed in humans. However, many drugs and poisons *induce* deficiency states, usually by reacting with the aldehyde group and thereby sequestering the coenzyme. A well-understood situation is the induced B_6 deficiency arising during treatment of the mycobacterial infection known as tuberculosis. The antimycobacterial agent **isoniazid** (isonicotinic acid hydrazide) reacts covalently with pyridoxal to make it unavailable for phosphorylation by pyridoxal kinase. Because the mycobacterium contains low levels of the kinase, its growth is effectively blocked by the agent. Prolonged treatment with this drug can generate a B_6 deficiency in the patient by the same mechanism, unless the diet is supplemented with the vitamin.

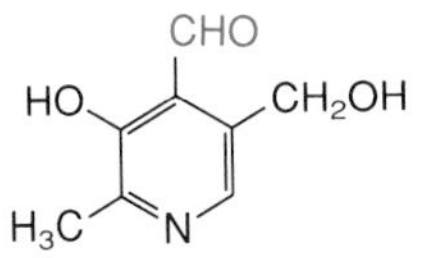

Pyridoxal phosphate is a remarkably versatile coenzyme. In addition to its involvement in transamination reactions, it participates in amino acid decarboxylations, racemizations, and numerous modifications of amino acid side chains. A key clue to the mechanism by which this coenzyme functions was the finding in Esmond Snell's laboratory in the 1940s that all of the known PLP-requiring enzyme reactions can be catalyzed, in the absence of any enzyme, by pyridoxal itself. Certain metal ions are required as well, such as Al^{3+} or Cu^{2+}. Although the reaction rates were much lower than those achieved by enzyme catalysis, the model studies permitted a detailed analysis that led to formulation of a unified mechanism for the action of PLP-requiring enzymes. The metal ion was postulated to stabilize a Schiff base or aldimine, formed between pyridoxal and the amino acid substrate, shown in the margin. Normally this role would be played by an amino acid residue in the active site of the enzyme.

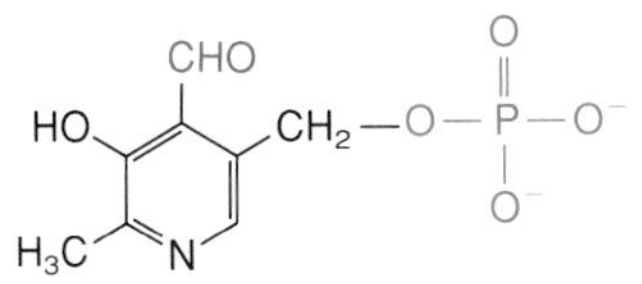

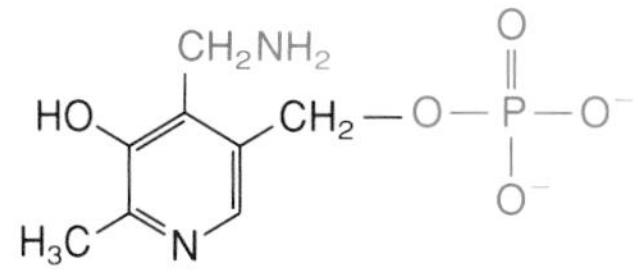

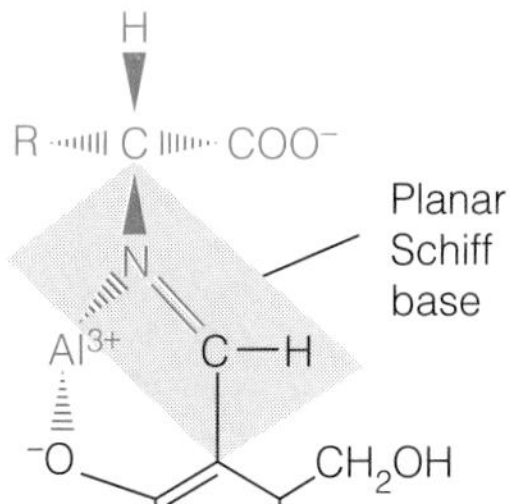

We now know that all pyridoxal phosphate–requiring enzymes act via the formation of a Schiff base between the amino acid and coenzyme (Figure 20.15). A cation, whether a metal (as in the nonenzymatic model system) or a proton (as in the enzymatic reaction), is essential to bridge the phenolate ion of the coenzyme and the imino nitrogen of the amino acid. This bridging maintains planarity of the structure, which is essential for catalysis. The most important catalytic feature of the coenzyme is the electrophilic nitrogen of the pyridine ring, which acts as an *electron sink,* drawing electrons away from the amino acid and stabilizing a carbanion intermediate. Interactions with the enzyme determine which bond is broken in the substrate and hence what specific reactions are catalyzed. *All of the*

Pyridoxine

Pyridoxal

Pyridoxal phosphate (PLP)

Pyridoxamine phosphate (PMP)

Planar Schiff base

All pyridoxal phosphate reactions involve initial Schiff base formation, followed by bond labilization caused by electron withdrawal to the coenzyme's pyridine ring.

known reactions of PLP enzymes can be described mechanistically in the same way: formation of a planar Schiff base or aldimine intermediate, followed by formation of a resonance-stabilized carbanion with a quinonoid structure, as shown in Figure 20.15. Depending on the bond labilized, formation of the aldimine can lead to a transamination (as shown), to decarboxylation, to racemization, or to numerous side chain modifications, such as β-elimination.

Although pyridoxal phosphate is the coenzyme for all of these reactions, we now know that the reactive species is not the aldehyde group but rather an aldimine, formed between the coenzyme and an ε-amino group of a lysine residue in the active site. This bond can be reduced by sodium borohydride, to give irreversible bonding of the coenzyme to the active lysine residue. This finding allowed identification of the catalytic site on the enzyme and the specific lysine residue involved in coenzyme binding.

Bond reduced by $NaBH_4$

Enzyme

Lysine residue

Enzyme-bound pyridoxal phosphate

For many years it was thought that PLP was the cofactor for all amino acid decarboxylases. However, a class of enzymes that use *pyruvic acid* as a cofactor is now known. Although structurally dissimilar to PLP, pyruvate evidently is mechanistically similar in this role; it is covalently bound to an amino group in the enzyme (although via an amide with its carboxyl group). Evidently during catalysis the keto oxygen of pyruvate plays a role comparable to that of the PLP ring nitrogen, as shown for the enzyme **histidine decarboxylase** of *Lactobacillus.*

Histidine substrate

Enzyme

Pyruvate cofactor

TETRAHYDROFOLATE COENZYMES AND ONE-CARBON METABOLISM

Discovery and Chemistry of Folic Acid

Coenzymes derived from the vitamin **folic acid** participate in the generation and utilization of single-carbon functional groups—methyl, methylene, and formyl. The vitamin itself was discovered in the 1930s, when it was found that people with a certain type of **megaloblastic anemia** could be cured by treatment with yeast or liver extracts. The condition is characterized, like all anemias, by reduced levels of erythrocytes. The cells that remain are characteristically large and immature, suggesting a role for the vitamin in cell proliferation and/or maturation. The active component in the extracts was also shown to be essential for growth of chicks and to be required in the growth media for certain bacteria, notably *Lactobacillus casei* and *Streptococcus faecium.* The latter findings allowed development of a rapid bioassay based on growth of these bacteria, and isolation and structural identification soon followed. The vitamin was found to be abundant in leafy green vegetables such as spinach, so it was named folic acid, from the same root as *foliage.*

Chemically, folic acid is formed from three distinct moieties: (1) a bicyclic, heterocyclic **pteridine** ring, 6-methylpterin; (2) ***p*-aminobenzoic acid (PABA),** which is itself required for the growth of many bacteria; and (3) glutamic acid. These three moieties are shown in the overall structure:

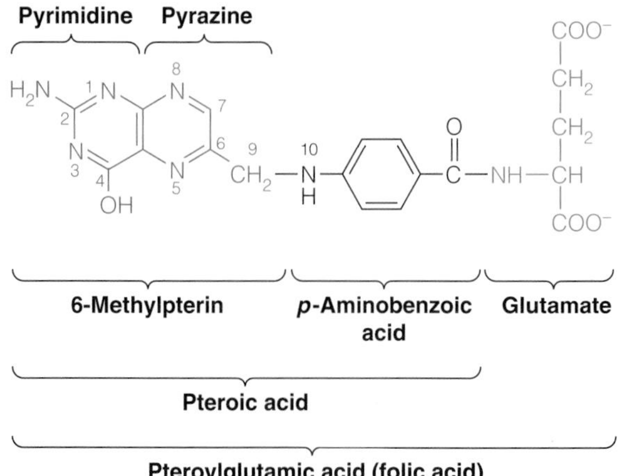

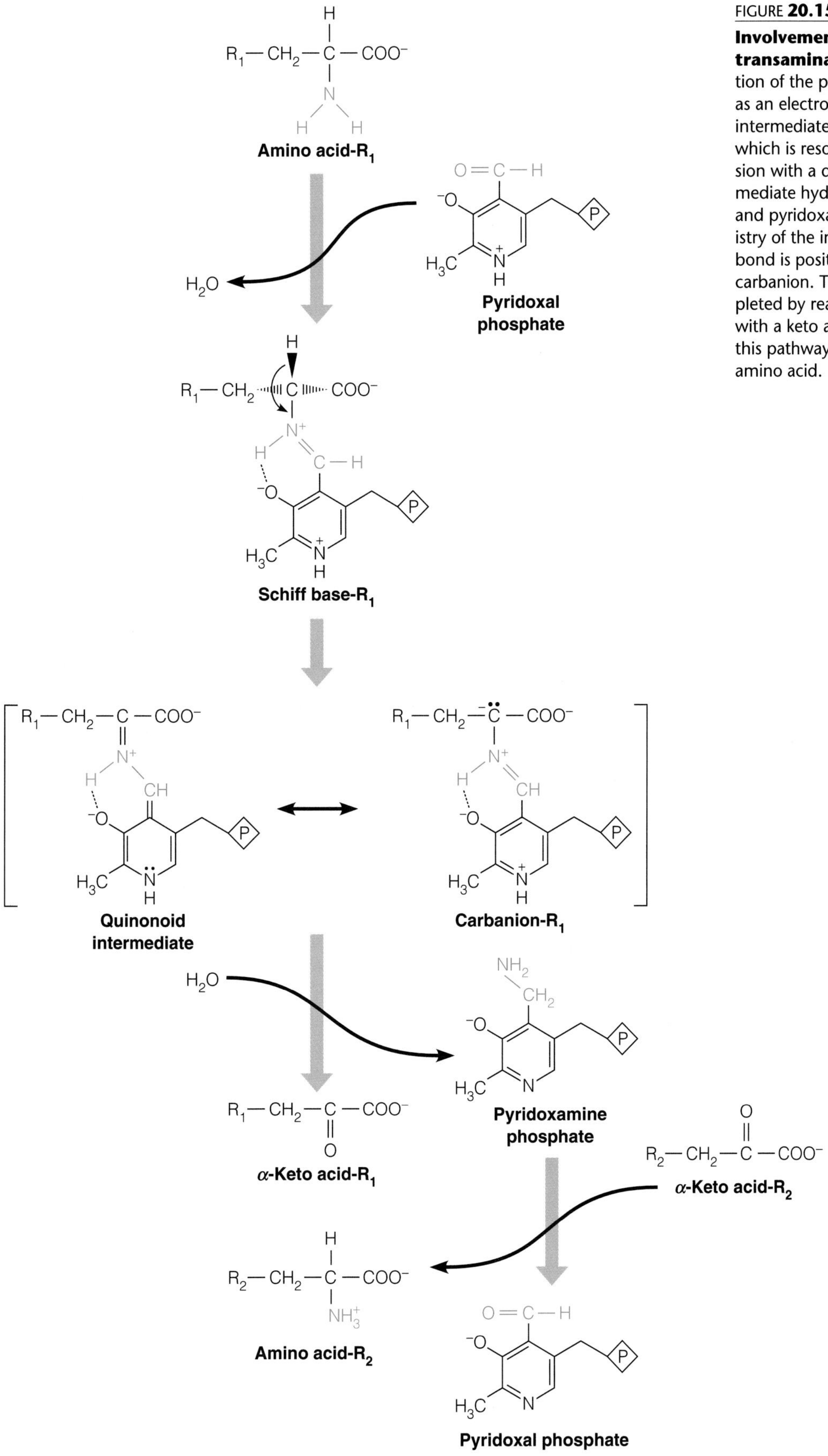

Amino acid-R_1

Pyridoxal phosphate

H_2O

Schiff base-R_1

Quinonoid intermediate

Carbanion-R_1

H_2O

α-Keto acid-R_1

Pyridoxamine phosphate

α-Keto acid-R_2

Amino acid-R_2

Pyridoxal phosphate

FIGURE **20.15**

Involvement of pyridoxal phosphate in transamination. The figure shows the action of the positively charged pyridinium ion as an electron sink. Formation of a Schiff base intermediate leads to formation of a carbanion, which is resonance stabilized by interconversion with a quinonoid intermediate. The intermediate hydrolyzes to yield a keto acid product and pyridoxamine phosphate. The stereochemistry of the initial Schiff base determines which bond is positioned to be broken and to form a carbanion. The transamination reaction is completed by reaction of pyridoxamine phosphate with a keto acid and conversion, by reversal of this pathway, to pyridoxal phosphate and an amino acid.

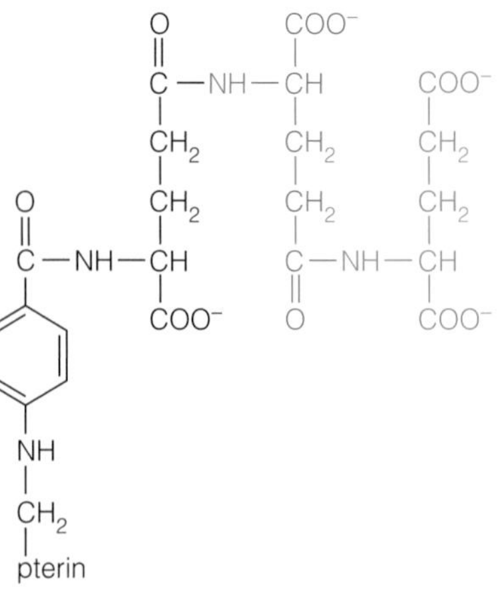

Pteroyl-γ-triglutamate

Folate coenzymes contain multiple glutamate residues, which evidently help them be retained within cells.

The pteridine ring was already known in nature, having been discovered in a large class of biological pigments. Insect wings and eyes contain pteridine pigments, as does the skin of amphibians and fish. Butterfly wings are particularly abundant in pteridines and were the first source from which any such compounds were identified structurally. These compounds are named after the Greek *pteron* ("wing").

In the structure of folic acid, 6-methylpterin is linked through the amino group of PABA to form **pteroic acid,** which is linked in turn via an amide to glutamate, to form **pteroylmonoglutamate.** Naturally occurring folates may differ from this compound in the number of glutamate residues per molecule of vitamin, which ranges from three to eight or more. These residues are linked to one another, not by the familiar peptide bond but rather by a modified peptide bond involving the α-amino group and the γ-carboxyl group. A typical representative of these folate polyglutamates is shown in the margin.

Most enzymes that use folate coenzymes bind more tightly to polyglutamated forms than to monoglutamates. The major need for the additional glutamate residues is probably for intracellular retention of folates. Animal cells take up folates by active transport, but only the monoglutamated form can be taken up. However, this form can also be transported out of cells, so conjugation with additional glutamate residues converts the folate to a form that cannot exit the cell.

Conversion of Folate to Tetrahydrofolate

Once inside a cell, folate is converted to active forms by two successive reductions of the pyrazine part of the pteridine ring. Both reactions are catalyzed by the NADPH-specific enzyme **dihydrofolate reductase.** The first reduction yields **7,8-dihydrofolate,** and the second reduction yields **5,6,7,8-tetrahydrofolate.**

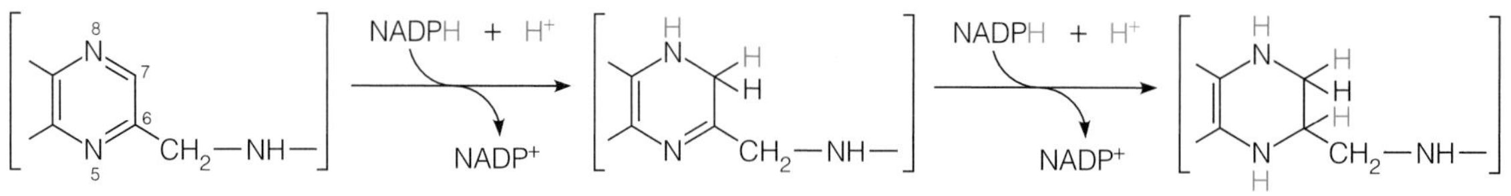

NADPH + H^+ → NADP$^+$; NADPH + H^+ → NADP$^+$

Folate (partial structure) **Dihydrofolate** **Tetrahydrofolate**

For reasons that will become clear later, dihydrofolate is the preferred substrate, and hence its name is given to the enzyme.

Dihydrofolate reductase has been thoroughly studied because it is the target for action of a number of clinically useful **antimetabolites.** An antimetabolite is a synthetic compound, usually a structural analog of a normal metabolite, that interferes with the utilization of the metabolite to which it is related structurally. As early as 1948, two analogs of folate—**aminopterin** and **amethopterin** (also known as **methotrexate**)—had been synthesized and found to induce remissions in acute leukemias.

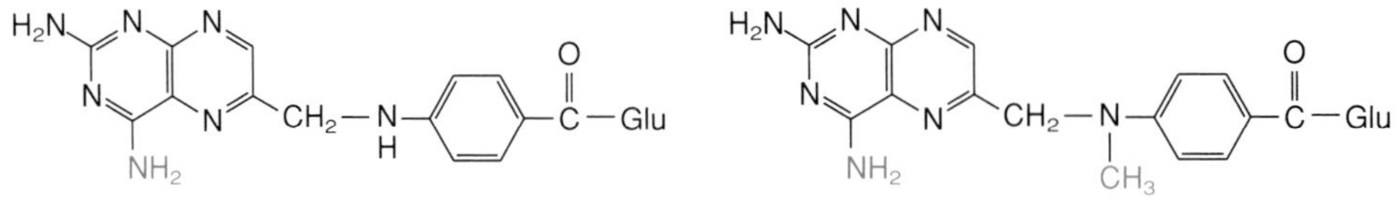

Aminopterin (4-aminofolate) **Amethopterin (4-amino-10-methylfolate)**

A decade later it was found that these compounds inhibit dihydrofolate reductase, binding to the enzyme at least 1000-fold more tightly than the normal substrates do. Thus, these analogs block the utilization of folate and dihydrofolate. We now know that their effectiveness derives from the involvement of dihydrofolate reductase in the biosynthesis of thymine nucleotides and, hence, of DNA. Inhibiting DNA synthesis blocks the proliferation of cancer cells, as discussed further in Chapter 22. Nearly two more decades passed before a detailed understanding of the mechanism of dihydrofolate reductase inhibition became available, through crystallization of enzyme–inhibitor complexes and determination of their three-dimensional structure (Figure 20.16).

Dihydrofolate reductase is the target for a number of useful anticancer, antibacterial, and antiparasitic drugs.

Folate analogs such as methotrexate have been used in treating many different cancers in addition to leukemia. Other clinically useful dihydrofolate reductase inhibitors show selectivity among species-specific forms of the enzyme. **Trimethoprim** specifically inhibits bacterial dihydrofolate reductases and is widely used to treat bacterial infections, and **pyrimethamine** shows similar specificity against the enzyme of protozoal origin.

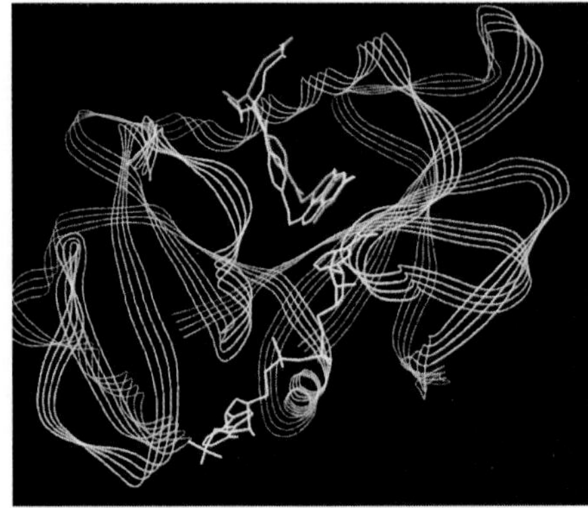

FIGURE **20.16**

Human dihydrofolate reductase complexed with ligands. This figure is based upon crystallography of the human protein and model building, using structures of enzyme–ligand complexes from other sources. NADPH is shown in yellow, methotrexate in orange, and dihydrofolate in yellow-green (partly superimposed on methotrexate). The additional amino group on methotrexate allows formation of an additional hydrogen bond to the enzyme, which increases its binding affinity and somewhat alters the conformation of the bound pteridine ring.

Courtesy of B. I. Schweitzer, A. P. Dicker, and J. R. Bertino, *FASEB J.* (1990) 4:2441–2452.

H_2N, N, N, NH_2, OCH_3, OCH_3, OCH_3

Trimethoprim

H_2N, N, N, $CH_2—CH_3$, NH_2, Cl

Pyrimethamine

The entire concept of antimetabolites as drugs arose from early work on folate metabolism. Before World War II, one of the few effective antibacterial drugs available was **sulfanilamide,** one of the class of sulfonamide drugs ("sulfa drugs"). A British biochemist, D. D. Woods, noted a structural similarity between sulfanilamide and *p*-aminobenzoate, which was known to be essential for bacterial growth. Before anything was known about the relation between PABA and folic acid, Woods proposed that sulfanilamide acts by blocking the normal utilization of PABA, and he coined the term *antimetabolite.* PABA is not required for growth of animal cells, so the drug is not toxic to human cells. Years later, when the pathway of folate biosynthesis had been established, it was learned that Woods was right; the enzyme incorporating PABA is inhibited by sulfonamides. Because animal cells do not carry out the synthetic pathway but instead take up fully formed folate from the diet, they are not harmed by the drug. The concept elucidated by Woods, of seeking a metabolic difference between normal cells and pathological cells—infecting parasites, virus-infected cells, or cancer cells—and of exploiting that difference chemically, has had an enormous impact on the field of pharmacology.

SO_2NH_2, NH_2

Sulfanilamide

COOH, NH_2

***p*-Aminobenzoic acid (PABA)**

Tetrahydrofolate in the Metabolism of Single-Carbon Units

The coenzymatic function of tetrahydrofolate is the mobilization and utilization of single-carbon functional groups. These reactions are involved in the metabolism of serine, glycine, methionine, and histidine, among the amino acids, and in the biosynthesis of purine nucleotides and the methyl group of thymine.

Tetrahydrofolate binds single-carbon units at the methyl, methylene, and formyl oxidation levels, equivalent in oxidation level to methanol, formaldehyde, and formic acid, respectively (Figure 20.17). Single-carbon groups on tetrahydrofolate can be carried on N-5 or N-10, or bridged between N-5 and N-10. Formation of a cyclic bridged adduct involves dehydration of a hydroxymethylated intermediate, so a methylene group ($—CH_2—$) is formally equivalent to a hydroxymethyl group ($—CH_2OH$) in an unbridged compound, and a formyl group ($—CHO$) cyclizes to become a **methenyl** group ($—CH{=}$). In addition, there is a single-carbon adduct of tetrahydrofolate in which the single-carbon unit contains a nitrogen atom as well. In this **formimino** group ($—C{=}NH$), the carbon atom is at the same oxidation level as a formyl group. The structure of one such adduct, **5,10-methylenetetrahydrofolate**, is shown:

Tetrahydrofolate coenzymes transfer and interconvert single-carbon units at the methyl, methylene, and formyl oxidation levels.

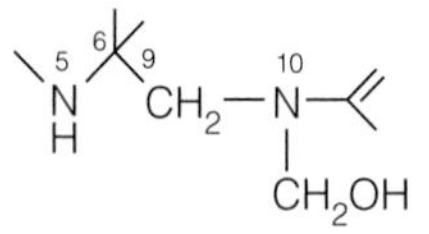

10-hydroxymethyl-THF

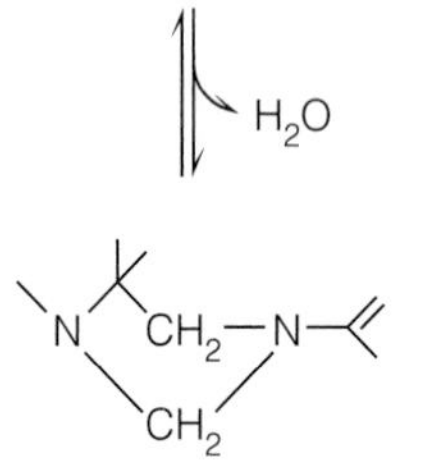

H_2O

5,10-methylene-THF

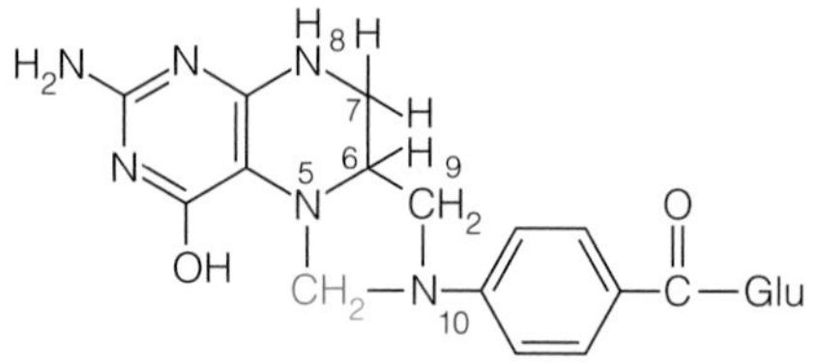

5,10-Methylenetetrahydrofolate

Tetrahydrofolate can acquire single-carbon units from diverse sources. For example, many cells carry out the ATP-dependent activation of formate to 10-formyltetrahydrofolate. The degradation of histidine, in both bacterial and animal cells, yields 5-formiminotetrahydrofolate, as does the bacterial fermentation of purines. However, most organisms derive most of their activated single-carbon units from the β-carbon of serine and the subsequent oxidation of glycine (reactions 10 and 11 in Figure 20.17). The first of these reactions is catalyzed by **serine transhydroxymethylase:**

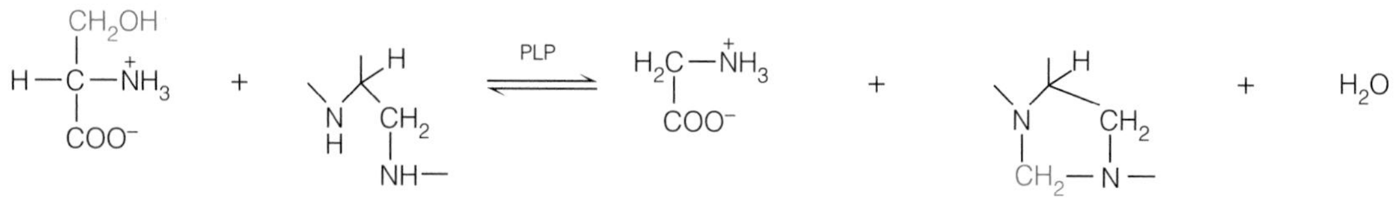

$$\text{Serine} + \text{Tetrahydrofolate} \xrightleftharpoons{\text{PLP}} \text{Glycine} + \text{5,10-Methylenetetrahydrofolate} + H_2O$$

This is a reversible reaction that, in the direction shown, yields glycine and 5,10-methylenetetrahydrofolate, but it can also be used for serine biosynthesis as needed. The enzyme also requires pyridoxal phosphate, so the actual substrate is the Schiff base formed from serine and PLP.

Glycine can yield an additional molecule of 5,10-methylenetetrahydrofolate through action of the **glycine cleavage system,** a multienzyme complex located in mitochondria:

$$\text{Glycine} + \text{tetrahydrofolate} + NAD^+ \longrightarrow \text{5,10-methylenetetrahydrofolate} + CO_2 + NH_3 + NADH + H^+$$

This reaction represents the chief catabolic route for glycine in most organisms. The overall reaction is as shown, but the pathway is mechanistically similar to that catalyzed by the pyruvate dehydrogenase complex (see Chapter 14). Four proteins are involved, along with pyridoxal phosphate, FAD, and lipoic acid. The serine transhydroxymethylase and glycine cleavage reactions comprise that part of the photorespiration pathway that occurs in plant mitochondria (Chapter 17, page 620).

Once a single-carbon unit has been activated via its attachment to tetrahydrofolate, it can undergo interconversions such as change in oxidation state, or it can be used directly in a biosynthetic reaction. Figure 20.17 shows most of the known reactions involving tetrahydrofolate coenzymes. Note the reactions that involve change in the oxidation level of the bound single-carbon unit: the reversible oxidation of 5,10-methylenetetrahydrofolate to 5,10-methenyltetrahydrofolate, catalyzed by **5,10-methylenetetrahydrofolate dehydrogenase** (reaction 7), and the irreversible reduction of 5,10-methylenetetrahydrofolate to the 5-methyl derivative, brought about by **5,10-methylenetetrahydrofolate reductase** (reaction 8). Many organisms contain multifunctional enzymes or complexes that help channel these scarce and/or unstable intermediates. For example, most eukaryotes combine activities 3, 4, and 7 into one trifunctional protein.

As shown in Figure 20.17, single-carbon units derived from tetrahydrofolate

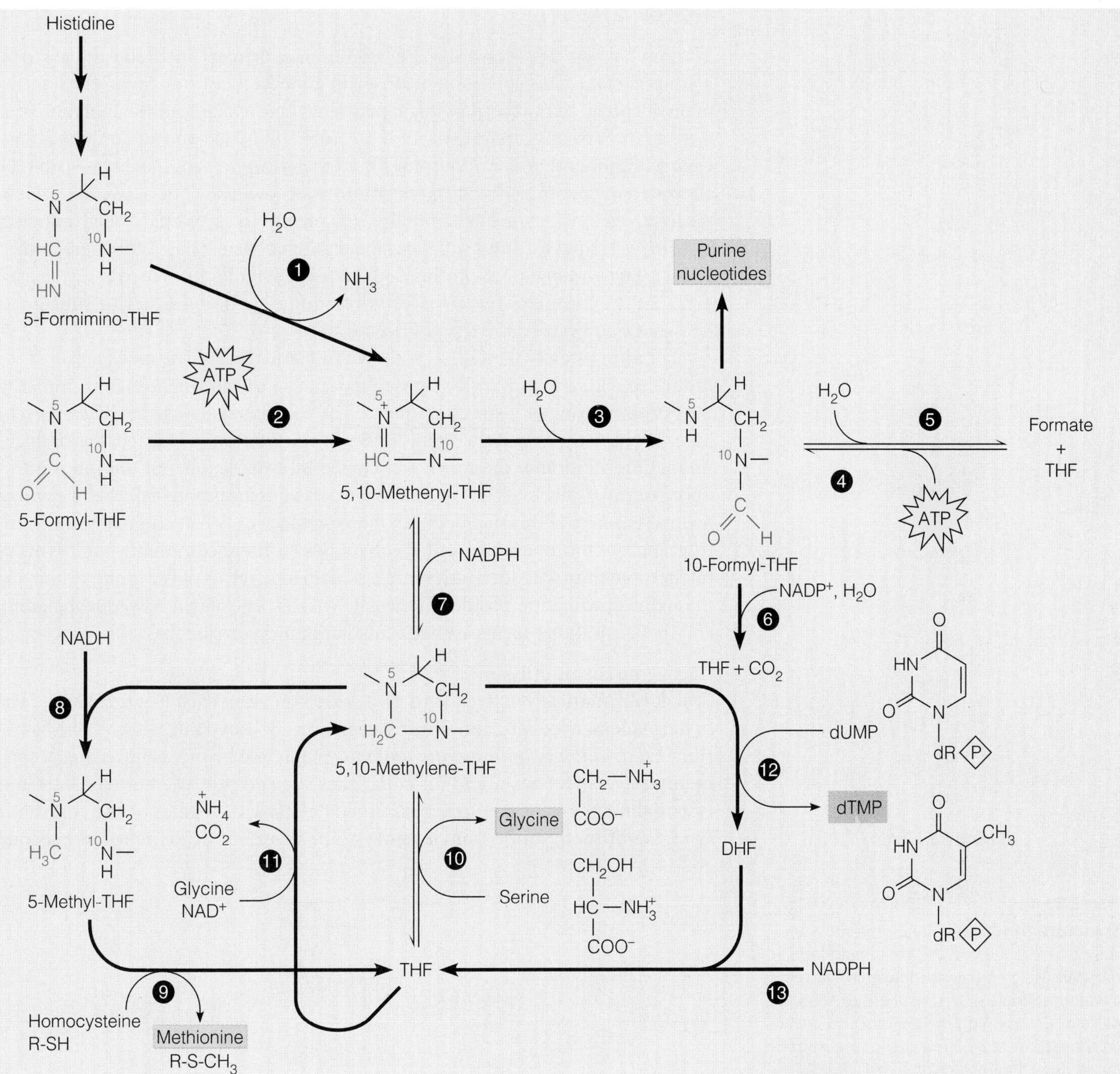

FIGURE **20.17**

Metabolic reactions involving synthesis, interconversion, and utilization of single-carbon adducts of tetrahydrofolate. Major end products of single-carbon metabolism are highlighted (purple). The enzymes involved are (1) cyclodeaminase, (2) methenyltetrahydrofolate synthetase, (3) methenyltetrahydrofolate cyclohydrolase, (4) formyltetrahydrofolate synthetase, (5) formyltetrahydrofolate hydrolase, (6) formyltetrahydrofolate dehydrogenase, (7) methylenetetrahydrofolate dehydrogenase, (8) methylenetetrahydrofolate reductase, (9) homocysteine methyltransferase (also called methionine synthase), (10) serine transhydroxymethylase, (11) glycine cleavage system, (12) thymidylate synthase, and (13) dihydrofolate reductase. THF = tetrahydrofolate, DHF = dihydrofolate.

coenzymes are used in synthesis of purine nucleotides, thymine nucleotides (dTMP), and methionine, in addition to the reactions we have discussed. Furthermore, in prokaryotes 10-formyltetrahydrofolate participates in synthesis of ***N*-formylmethionyl-tRNA**, which is involved in initiation of protein synthesis (discussed further in Chapter 27). In the synthesis of thymine nucleotides, catalyzed by **thymidylate synthase** (reaction 12), the tetrahydrofolate coenzyme serves both as a single-carbon donor and as a source of reducing power. Because this enzyme generates the methyl group of thymine from 5,10-methylenetetrahydrofolate, it catalyzes both a one-carbon transfer and a reduction. The electrons come from the reduced pteridine ring, to give dihydrofolate as a product. Although dihydrofolate reductase (reaction 13) can act on either folate or dihydrofolate, the reduction of dihydrofolate is more significant in vivo than that of folate because of the need for constant regeneration of tetrahydrofolate from dihydrofolate produced in the thymidylate synthase reaction.

B_{12} COENZYMES

Vitamin B_{12} was discovered through studies of a formerly incurable disease, pernicious anemia. This condition begins with a megaloblastic anemia, which is virtually identical to that seen in folate deficiency but which leads to an irreversible degeneration of the nervous system if untreated. In 1926 two Harvard physicians, George Minot and William Murphy, found that symptoms of the disease could be alleviated by feeding patients large amounts of raw liver. The active material in the liver, which was named vitamin B_{12}, was present in exceedingly small amounts, so many years passed until sufficient material had been isolated for characterization. In England in1964, Dorothy Hodgkin and her colleagues used x-ray crystallography to complete the structure determination for this active substance. Hodgkin was awarded the Nobel Prize for this work.

The structure of vitamin B_{12} is shown in Figure 20.18. The metal cobalt is coordinated with a tetrapyrrole ring system, called a **corrin** ring, which is similar to the porphyrin ring of heme compounds. The cobalt is also linked to a heterocyclic base, 5,6-dimethylbenzimidazole (DMB). In the vitamin as isolated, the sixth coordination position of the cobalt is occupied by cyanide ion, but this ion is introduced during isolation. The vitamin as it exists in tissues contains either water or a hydroxyl moiety at this site. Because of the presence of cobalt and many amide nitrogens, B_{12} compounds are called **cobamides** or, more commonly but less accurately, **cobalamins.** B_{12} derivatives are also named in terms of the group occupying the sixth coordination position. Thus, the vitamin as isolated is **cyanocobalamin,** and the intracellular forms are **aquocobalamin** or **hydroxocobalamin.**

Coenzyme Forms of B_{12}

Two coenzymatically active forms of B_{12} are known to exist (Figure 20.19). The first, **5′-adenosylcobalamin,** was discovered in 1964 by H. A. Barker. The bacterium *Clostridium cylindrosporum* ferments glutamate by first isomerizing it to β-methylaspartate. Barker and his colleagues showed that adenosylcobalamin is essential for this reaction to occur. Shortly afterward, studies of the enzymatic synthesis of methionine from **homocysteine** revealed the existence of a second

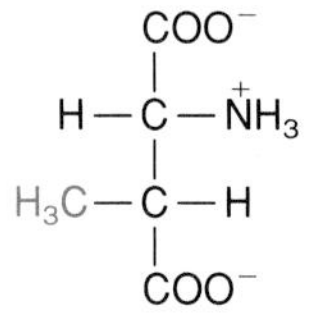

β-Methylaspartate

FIGURE **20.18**

Structure of vitamin B_{12}. The molecule shown here is the cyanide-containing form originally isolated (cyanocobalamin). In cells, a water molecule or hydroxyl group takes the place of CN, forming the precursor to the coenzyme forms of B_{12}. The corrin ring is shown in red. 5,6-Dimethylbenzimidazole (DMB), which is linked to the cobalt, is shown in blue.

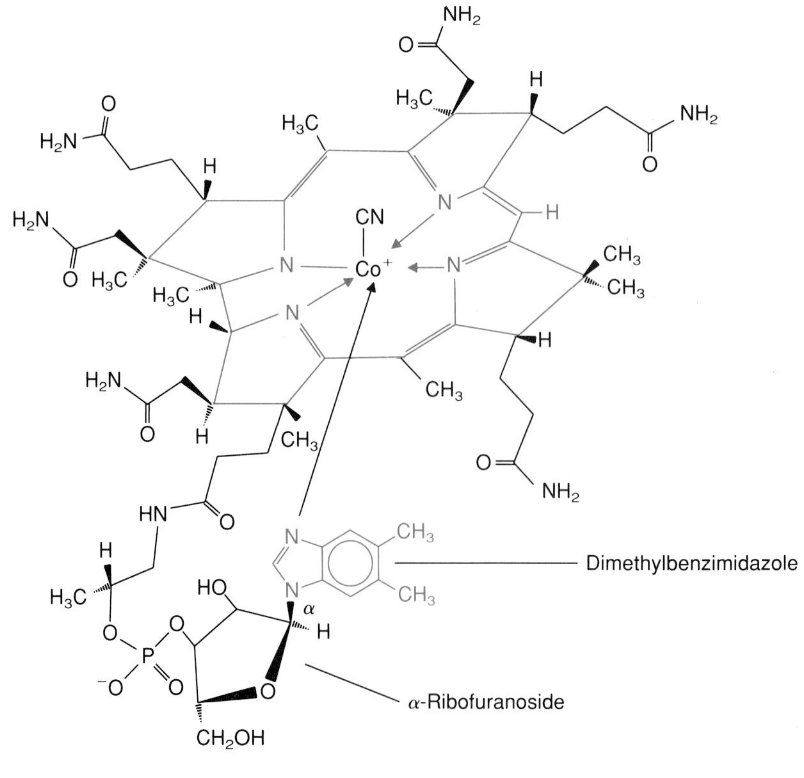

coenzymatically active form, **methylcobalamin,** or methyl-B_{12}. Both adenosylcobalamin and methylcobalamin contain a covalent carbon–cobalt bond, making them true organometallics. In methionine synthesis the coenzyme transfers its methyl group with a change in the oxidation state of the cobalt from +3 to +1.

B_{12} coenzymes have either a methyl group or a 5′-adenosyl moiety linked to Co^{+}, making them the first known organometallics in metabolism.

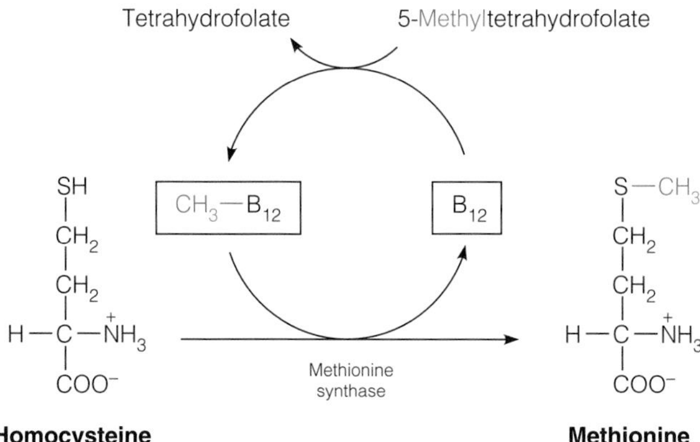

Note that the methyl group is derived ultimately from 5-methyltetrahydrofolate (reaction 9, Figure 20.17).

About 15 different B_{12}-requiring reactions are now known, most of which occur in only a few bacterial species that carry out specialized fermentations. Only two reactions occur to a significant extent in mammalian metabolism: the synthesis of methionine from homocysteine, which we have shown here, and the isomerization of methylmalonyl-CoA to succinyl-CoA, which we introduced in Chapter 18 as a key step in the oxidation of odd-chain fatty acids. Also of interest is the involvement of methyl-B_{12} in methane synthesis by methanogenic bacteria. With one exception, the known B_{12}-requiring reactions involve either (1) methyl group transfer, like the methionine synthase reaction, using methylcobalamin as the cofactor, or (2) adenosylcobalamin-dependent isomerizations involving exchange of a carbon-bound hydrogen with another carbon-bound functional group, as shown here; methylmalonyl-CoA mutase catalyzes a reaction of this type.

```
   b  c          b  c
   |  |          |  |
a—C—C—d      a—C—C—d
   |  |          |  |
   X  H          H  X
```

The one exception is an intermolecular transfer reaction catalyzed by a ribonucleotide reductase of *Lactobacillus* and a few other prokaryotes (see Chapter 22). This enzyme may be an evolutionary vestige, because all other known ribonucleotide reductases act by a mechanism that is independent of B_{12}, although similar overall.

Action of Adenosylcobalamin

Carbon–carbon bonds are generally difficult to break and to form. B_{12} coenzymes readily catalyze such reactions, and this activity, plus the novel cobalt–carbon bond, has focused attention on mechanisms of action of adenosylcobalamin. From studies on methylmalonyl-CoA mutase and one or two other enzymes, the following facts have emerged: (1) Hydrogen transfer is stereospecific; some of the reactions proceed with inversion of configuration, and others do not. (2) Hydrogen that is transferred does not exchange with the protons of water; isotopic hydrogen present in a substrate is conserved in the product. (3) The transferred hydrogen is transiently carried on the 5′ carbon of the adenosyl moiety; labeling of this position with tritium leads to incorporation of that label into product. (4) Spectral studies indicate that the cobalt changes its oxidation state during

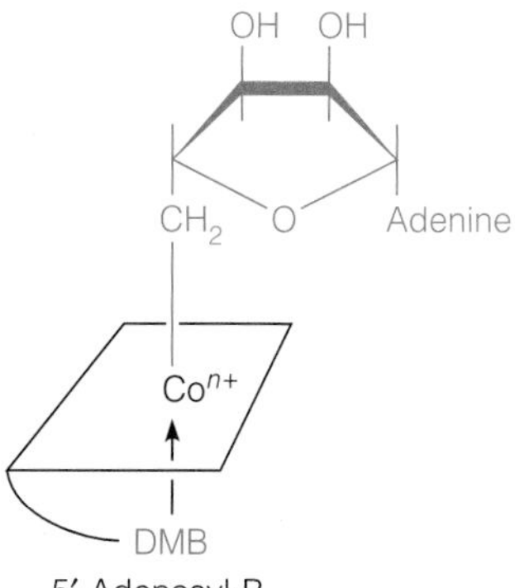

5′-Adenosyl-B_{12}
(5′-adenosylcobalamin)

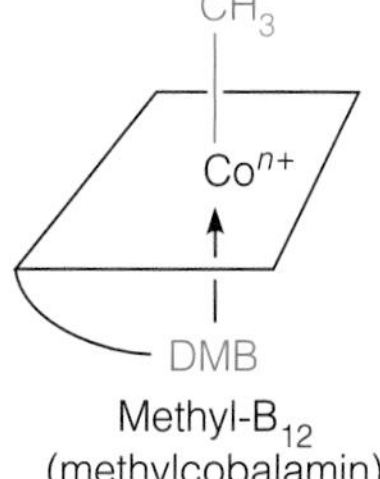

Methyl-B_{12}
(methylcobalamin)

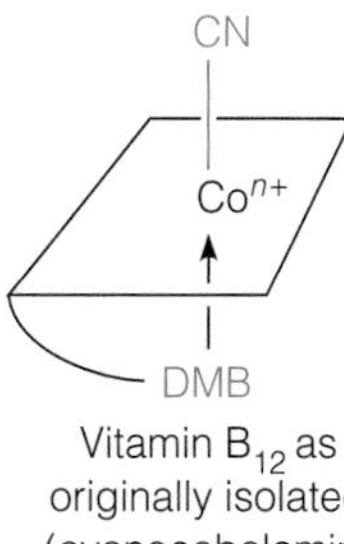

Vitamin B_{12} as originally isolated (cyanocobalamin)

FIGURE **20.19**

Coenzymes derived from vitamin B_{12}. The corrin ring, identical in all known forms of B_{12}, is indicated here schematically. The Co bears a positive charge (n = 1, 2, or 3), while each molecule is uncharged overall.

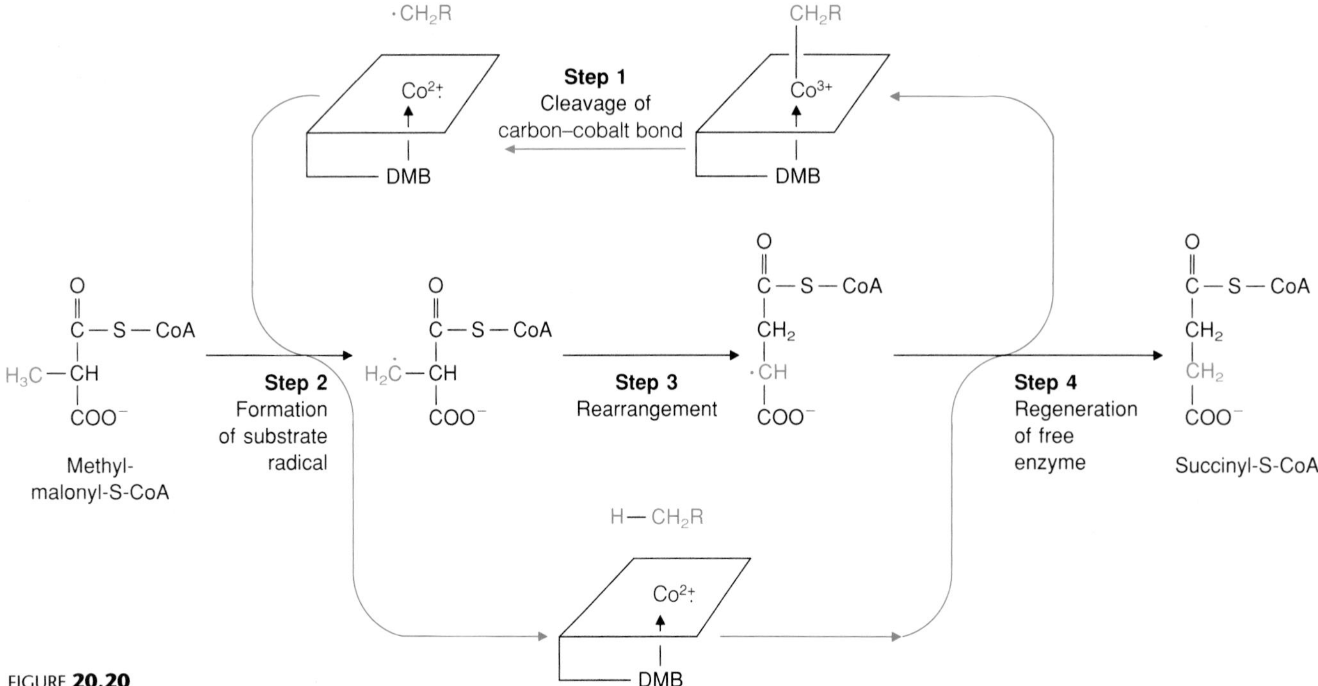

FIGURE **20.20**

The intramolecular rearrangement catalyzed by methylmalonyl-CoA mutase. This mechanism is consistent with experimental observations and may account for other B_{12}-dependent rearrangements. CH_2R is the adenosyl moiety. The precise mechanism by which the free radical form of the substrate rearranges is not known.

catalysis. (5) Electron spin resonance studies indicate that free radical intermediates are generated during catalysis.

All of these observations imply that the covalent carbon–cobalt bond on the coenzyme undergoes transient **homolytic** cleavage during catalysis. That is, the cobalt and the carbon each acquire one electron from the pair that formed the bond, creating a free radical at the adenosine C-5′. Interaction with the substrate then creates a substrate radical, as shown for methylmalonyl-CoA mutase in Figure 20.20, leading to rearrangement of the substrate.

One feature of the mechanism shown in Figure 20.20 may soon require modification. In late 1994 the crystal structure of methylcobalamin bound to methionine synthase was reported, and it showed that the coenzyme undergoes a major conformational change upon binding to the enzyme. As shown in Figure 20.21, the DMB-containing "tail" is no longer complexed to the cobalt. Instead, the "tail" has moved into a deep pocket, which helps bind the coenzyme tightly. Replacing the DMB in coordination with cobalt is a histidine residue on the enzyme, suggesting a role for the enzyme in stabilizing reaction intermediates. The relationship of this structure and the proposed mechanism to other cobalamin-dependent enzyme reactions remains to be seen.

B_{12} Coenzymes and Pernicious Anemia

Now let us return to the role of B_{12} coenzymes in mammalian metabolism. Recall that vitamin B_{12} was isolated as a factor that could cure pernicious anemia, which suggested that the disease is caused by a result of a B_{12} deficiency. In fact, pernicious anemia is a disease of the stomach. Gastric tissue secretes a glycoprotein called **intrinsic factor,** which complexes with ingested B_{12} in the digestive tract and promotes its absorption through the small intestine into the bloodstream. Pernicious anemia results from insufficient secretion of intrinsic factor. Indeed, patients who undergo surgical removal of the stomach for ulcers or other problems can also develop the symptoms of pernicious anemia. The uncomplexed vitamin can be absorbed, but so poorly that massive doses must be administered to cure or prevent the disease.

Pernicious anemia is caused by deficiency of a glycoprotein needed for intestinal absorption of vitamin B_{12}, leading to intracellular deficiencies of B_{12} coenzymes.

But what is the relationship between failure to absorb B_{12} and the deficiency of red blood cell formation that defines anemias? Given the hematologic similar-

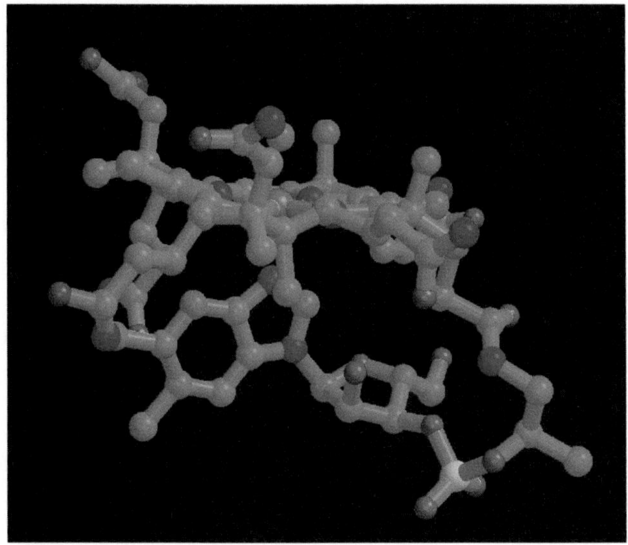

(a)

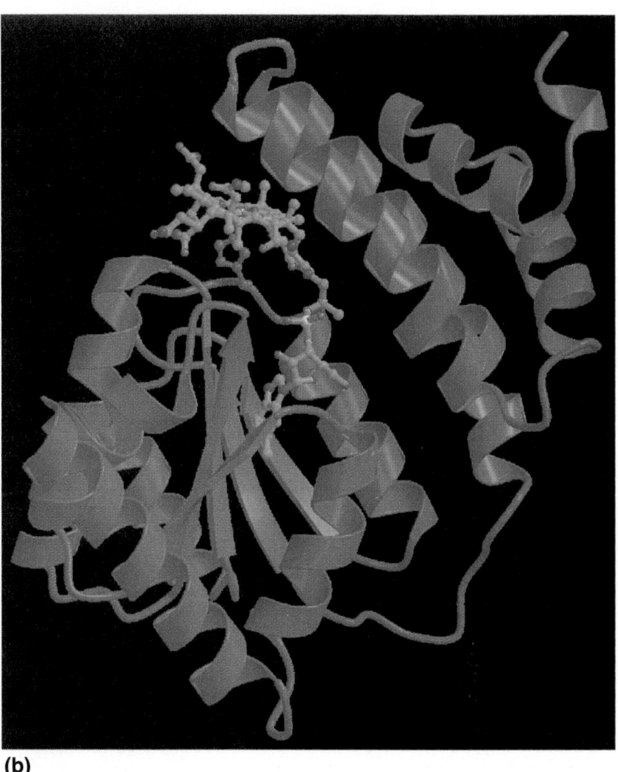

(b)

FIGURE **20.21**

Conformational change accompanying enzyme binding of B_{12}. (a) The crystal structure of methylcobalamin, with DMB complexed to the cobalt. **(b)** The B_{12}-binding domain of methionine synthase and bound methylcobalamin, showing the DMB buried in a deep crevice. A histidine residue on the enzyme contacts the lower face of the corrin ring, in place of DMB.

ity between anemias of folate deficiency and pernicious anemia, a relationship between folate and B_{12} metabolism has long been suspected. In fact, the megaloblastic anemia seen in early stages of pernicious anemia disappears if folic acid is administered. However, this treatment only hastens the onset of the far more severe neurological symptoms. Given that these neurological problems are never seen in simple folate deficiency, what is the metabolic relationship between these two vitamins?

Figure 20.22 outlines the probable explanation: (1) When B_{12} levels are low, flux through the methionine synthase reaction (reaction 9, Figure 20.17) decreases, but because adequate dietary methionine is usually available, there is no immediate disturbance of protein metabolism. (2) Reduction of 5,10-methylenetetrahydrofolate to 5-methyltetrahydrofolate (reaction 8, Figure 20.17) continues, because this reaction is virtually irreversible. (3) Because methionine synthase is

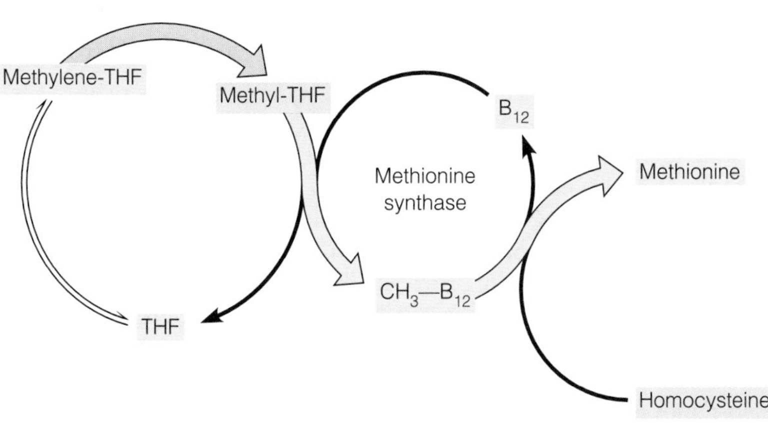

FIGURE **20.22**

A relationship between folate and B_{12} metabolism. This scheme is based on the apparent folate deficiency seen in early stages of B_{12} deficiency. The diagram identifies intermediates that either accumulate (blue) or are depleted (gold) as a result of decreased flux through the methionine synthase reaction.

B_{12} deficiency probably causes 5-methyltetrahydrofolate to accumulate, with concomitant depletion of other folate coenzymes.

the only mammalian enzyme known to act on 5-methyltetrahydrofolate, the decreased intracellular activity of this enzyme causes 5-methyltetrahydrofolate to accumulate, at the expense of depleted pools of the other tetrahydrofolate coenzymes. Thus, even though total folate levels may seem ample, there is a *functional* folate deficiency, with insufficient levels of the formyl and methylene derivatives needed for synthesis of nucleic acid precursors. From this discussion, it is apparent that methionine synthase plays a major role in maintaining tetrahydrofolate supplies by releasing THF from its "trapped" form, methyl-THF.

This does not explain why untreated pernicious anemia progresses to a neurological disease, because simple folate deficiency anemias show no such complications. Early observations suggested that abnormal fatty acid metabolism caused by inhibition of methylmalonyl-CoA mutase was responsible. The problem is extraordinarily difficult, because dietary B_{12} requirements are so low that generating animal models for B_{12} deficiency is almost impossible. Recently insight was gained through the observation that the commonly used anesthetic nitrous oxide (N_2O), or laughing gas, induces a pernicious anemia–like state by oxidizing the cobalt in B_{12} from the +1 to the +2 state, thereby inactivating the coenzyme and inhibiting enzymes that use B_{12} coenzymes. Data from N_2O-treated animals suggest that inhibition of methionine synthesis is responsible for the nervous system degeneration in pernicious anemia, although the mechanisms involved are yet to be elucidated.

Folic Acid, Vitamin B_{12}, and the Heart

In the mid-1990s, a series of clinical reports described correlations between folate deficiencies and increased risk of myocardial infarction. The same studies revealed that individuals at risk for heart attack also showed abnormally high levels of serum homocysteine. The simplest interpretation is that in folate-deficient individuals, decreased levels of tetrahydrofolate cofactors limit metabolic flux through the methionine synthase reaction (reaction 9, Figure 20.17), with consequent accumulation of homocysteine, the substrate for this enzyme. Homocysteine is presumed to be the metabolite responsible for damage to the heart, although toxic mechanisms are not known, and indeed, some studies have failed to show a correlation between folate status and heart disease. However, folate deficiencies have other recently recognized biological consequences, including abnormally high levels of uracil in DNA. As we discuss in Chapter 22, this phenomenon, which can lead to chromosome breakage, is a consequence of limitation of the biosynthesis of thymine nucleotides. By a probably similar mechanism, folate deficiencies during embryogenesis cause a significant proportion of neural tube defects, in which the nervous system fails to develop properly. Consequently, women are urged to take folic acid supplements throughout their pregnancies, but especially in the early stages, when the fetus's nervous system develops most rapidly.

SUMMARY

Although inorganic nitrogen is abundant, metabolism of most organisms is limited by nitrogen bioavailability. Reduction of N_2 in biological nitrogen fixation and reduction of nitrate in plant and bacterial metabolism generate ammonia, which all organisms can utilize. The capacity for amino acid synthesis varies greatly among organisms, with mammals requiring about half of the 20 common amino acids in the diet. Proteins are in a continual state of turnover and replacement, partly for replacement of damaged proteins and partly as the result of normal cellular regulatory mechanisms. Most amino acids released by protein turnover are reutilized for protein synthesis. When amino acids are degraded, either for catabolism of an oversupply or when needed for energy generation, the first step is usually removal of the α-amino group, either through transamination or oxidative deamination. The resultant ammonia is excreted directly (in fish), converted to uric acid (in most reptiles, insects, and birds), or converted to urea (in mammals). Urea synthesis is a cyclic pathway involving ornithine and arginine as intermediates. Transamination and numerous additional reactions undergone by amino acids use pyridoxal phosphate as a coenzyme. After condensation of the amino group of the amino acid with the aldehyde of the coenzyme to give a Schiff base, the pyridine ring of the coenzyme withdraws electrons transiently and destabilizes bonds that are broken in the reaction. Tetrahydrofolate binds single-carbon units at three different oxidation states, interconverts them, and transfers them in the synthesis of purine nucleotides, thymidine nucleotides, and several amino acids. B_{12} coenzymes include methylcobalamin, which participates in methionine biosynthesis, and adenosylcobalamin, the coenzyme for methylmalonyl-CoA mutase. Folate metabolism presents various chemotherapeutic targets, and folate and B_{12} deficiencies both have important clinical consequences.

REFERENCES

Inorganic Nitrogen Metabolism

Fischer, H-M. (1994) Genetic regulation of nitrogen fixation in rhizobia. *Microbiol. Rev.* 58:352–386. A contemporary view of the complexities of control of nitrogen fixation, with some perspective on practical use of this information.

Howard, J. B., and D. C. Rees (1994) Nitrogenase: A nucleotide-dependent molecular switch. *Annu. Rev. Biochem.* 63:235–264. A structural perspective on the mechanism of nitrogen fixation.

Kisker, C., H. Schindelin, and D. C. Rees (1997) Molybdenum-cofactor-containing enzymes: Structure and mechanism. *Annu. Rev. Biochem.* 66:233–267. Reviews all of the known molybdenum-requiring reactions, with emphasis upon structures of the proteins involved.

Rawls, R. L. (1998) Breaking up is hard to do. *Chem. Eng. News,* June 22 issue, pp. 29–34. A readable article, in news-style format, that nicely reviews recent developments in the chemistry and enzymology of biological nitrogen fixation.

General Aspects of Nitrogen Metabolism

Brusilow, S. W., and A. L. Horwich (1995) Urea cycle enzymes. In: *The Metabolic and Molecular Bases of Inherited Disease,* 7th ed., edited by C. R. Scriver, A. L. Beaudet, W. S. Sly, and D. Valle, pp. 1187–1232. McGraw-Hill, New York. Genetic disorders of urea cycle enzymes are reviewed in this chapter of the definitive work on inherited metabolic disorders.

Morris, S. M., Jr. (1992) Regulation of enzymes of urea and arginine synthesis. *Annu. Rev. Nutrition* 12:81–101. Recent studies, summarized here, yield insight into dietary and hormonal regulation of these enzymes in mammals.

Walsh, C. T. (1979) *Enzymatic Reaction Mechanisms.* Freeman, San Francisco. An excellent book, particularly valuable in the context of amino acid metabolism, one-carbon metabolism, cobalamin coenzymes, and oxygenases.

Woehl, E. U., and M. F. Dunn (1995) The roles of Na^+ and K^+ in pyridoxal phosphate enzyme catalysis. *Coordination Chem. Rev.* 144: 147–197. A recent review of one aspect of PLP function.

Protein Turnover

Goldberg, A. L. (1995) Functions of the proteasome: The lysis at the end of the tunnel. *Science* 268:522–523. A brief article with perspective on the crystal structure of an archaebacterial proteasome, described in detail in the same issue of *Science.*

Hopkin, K. (1997) Regulating cellular processes: The power of protein degradation. *J. NIH Res.* 9:36–42. A news-type article that presents a comprehensive review of intracellular proteolytic processes and their metabolic significance.

Lupas, A., J. M. Flanagan, T. Tamura, and W. Baumeister (1997) Self-compartmentalizing proteases. *Trends Biochem. Sci.* 21:399–404. This article, which reviews the structures of the proteasome and other proteolytic complexes, is one article in a special issue of this journal dedicated to the biology and biochemistry of controlled proteolysis.

Rechsteiner, M., L. Hoffman, and W. Dubiel (1993) The multicatalytic and 26 S proteases. *J. Biol. Chem.* 268:6065–6068. A minireview describing some recently characterized enzymes in protein turnover.

Rechsteiner, M., and S. W. Rogers (1996) PEST sequences and regulation by proteolysis. *Trends Biochem. Sci.* 21:267–271. This brief review summarizes experimental evidence that these sequences do indeed influence protein metabolic lifetimes.

Stadtman, E. R. (1992) Protein oxidation and aging. *Science* 257:1220–1224. One of Stadtman's several reviews of protein oxidation and turnover.

Folate and B_{12} Coenzymes

Appling, D. R. (1991) Compartmentation of folate-mediated one-carbon metabolism in eukaryotes. *FASEB J.* 5:2645–2651. Several of these reactions are catalyzed by multifunctional proteins or multienzyme complexes, and this has implications for optimal therapeutic use of folate antimetabolites.

Benkovic, S. J., and R. L. Blakley (1984) *Folates and Pterins,* Vol. 1. Academic Press, New York. A book-length review, which covers one-carbon metabolism and mechanisms of action of folate coenzymes.

Fenton, W. A., and L. E. Rosenberg (1995) Disorders of propionate and methylmalonate metabolism. In: *The Metabolic and Molecular Bases of Inherited Disease,* 7th ed., edited by C. R. Scriver, A. L. Beaudet, W. S. Sly, and D. Valle, pp. 1423–1450. McGraw-Hill, New York. Genetic disorders of B_{12} metabolism are reviewed in this chapter of the definitive work on inherited metabolic disorders.

Halpern, J. (1985) Mechanisms of coenzyme B_{12}-dependent rearrangements. *Science* 227:869–875. Discussion of experimental evidence for the mechanism outlined in this chapter.

Ludwig, M. L., and R. G. Matthews (1997) Structure-based perspectives on B_{12}-dependent enzymes. *Annu. Rev. Biochem.* 66:269–313. An excellent recent review of B_{12} mechanisms, focused upon the structures of the enzymes involved.

Metz, J. (1992) Cobalamin deficiency and the pathogenesis of nervous system disease. *Annu. Rev. Nutrition* 12:59–79. A discussion of the still unknown relationship between vitamin B_{12} and neurobiology.

Schweitzer, B. I., A. P. Dicker, and J. R. Bertino (1990) Dihydrofolate reductase as a therapeutic target. *FASEB J.* 4:2441–2452. Discusses a range of useful inhibitors from the structural standpoint of their interaction with the target enzyme.

Smaglik, P. (1997) Folic acid deficiency's role expands beyond birth defects. *The Scientist* October 13 issue, pp. 10–11. A highly readable article, in a news-style format, that nicely reviews a number of nutritional and clinical studies dealing with folic acid.

PROBLEMS

1. Identify the most likely additional substrates, products, and coenzymes for each reaction in the following imaginary pathway.

$$\mathrm{^-OOC{-}CH(NH{-}CHO){-}CH_2{-}CH_2{-}CH_2{-}CH(\overset{+}{N}H_3){-}COO^-} \longrightarrow \mathrm{^-OOC{-}CH(NH{-}CHO){-}CH_2{-}CH(CH_3){-}CH(\overset{+}{N}H_3){-}COO^-} \longrightarrow \mathrm{^-OOC{-}CH(\overset{+}{N}H_3){-}CH_2{-}CH(CH_3){-}CH(\overset{+}{N}H_3){-}COO^-} \longrightarrow$$

$$\mathrm{H_2N{-}C({=}O){-}CH(\overset{+}{N}H_3){-}CH_2{-}CH(CH_3){-}CH(\overset{+}{N}H_3){-}COO^-} \longrightarrow \mathrm{H_2N{-}C({=}O){-}CH(\overset{+}{N}H_3){-}CH_2{-}CH(CH_3){-}C({=}O){-}COO^-} \longrightarrow \mathrm{H_2N{-}C({=}O){-}CH(NH{-}CH_3){-}CH_2{-}CH(CH_3){-}C({=}O){-}COO^-}$$

2. The following diagram shows the biosynthesis of B_{12} coenzymes, starting with the vitamin. DMB is dimethylbenzimidazole.

(a) What one additional substrate or cofactor is required by enzyme B?

(b) Genetic deficiency in animals of enzyme C would result in excessive urinary excretion of what compound?

(c) Some forms of the condition described in (b) can be successfully treated by injection of rather massive doses of vitamin B_{12}. What kind of genetic alteration in the enzyme would be consistent with this result?

(d) Genetic deficiency in animals of enzyme B will result in excessive urinary excretion of what amino acid?

*3. Using the principles described in the text regarding pyridoxal phosphate mechanisms, propose a mechanism for the reaction catalyzed by serine transhydroxymethylase.

*4. A physician treating a patient with megaloblastic anemia might be tempted to treat the patient with folic acid, to see if a simple folate deficiency is involved. If the patient has pernicious anemia, this treatment hastens the onset of the far more serious neurological symptoms. Why?

5. Use numbers 1 to 5 to identify each carbon atom in the product of this reaction. What is the coenzyme?

$$\mathrm{\overset{5}{^-OOC}{-}\overset{4}{CH_2}{-}\overset{3}{CH_2}{-}\overset{2}{C}H(\overset{+}{N}H_3){-}\overset{1}{COO^-}} \longrightarrow \mathrm{^-OOC{-}CH(CH_3){-}CH(\overset{+}{N}H_3){-}COO^-}$$

*6. Based on the mechanism for methylmalonyl-CoA mutase shown in Figure 20.20, propose a mechanism for the diol dehydrase reaction:

$$\mathrm{HO{-}CH_2{-}CH_2{-}OH \longrightarrow CH_3CHO}$$

7. The precise mechanism of ammonia toxicity to the brain is not known. Speculate on a possible mechanism, based on possible effects of ammonia on levels of key intermediates in energy generation.

*8. Mutants of *Neurospora crassa* that lack carbamoyl phosphate synthetase I (CPS I) require arginine in the medium in order to grow, whereas mutants that lack carbamoyl-phosphate synthetase II (CPS II) require a pyrimidine, such as uracil. A priori, one would expect the active CPS II in the arginine mutants to provide sufficient carbamoyl phosphate for arginine synthesis, and the active CPS I in the pyrimidine mutants to "feed" the pyrimidine pathway. Explain these observations.

9. In some forms of leukemia the proliferating white blood cells contain very low levels of asparagine synthetase. Some years ago there was interest in treating these leukemias by purifying the enzyme asparaginase from *E. coli* and injecting it into the bloodstream of leukemic patients. Asparaginase catalyzes the hydrolysis of asparagine to aspartate plus ammonia. What is the rationale behind this mode of therapy, and why might you expect it not to work?

10. Indicate whether each of the following statements is true or false, and briefly explain your answer.

(a) In general, the metabolic oxidation of protein in mammals is less efficient, in terms of energy conserved, than the metabolic oxidation of carbohydrate or fat.

(b) Given that the nitrogen of glutamate can be redistributed by transamination, glutamate should be a good supplement for nutritionally poor proteins.

(c) Arginine is a nonessential amino acid for mammals, because the enzymes of arginine synthesis are abundant in liver.

(d) Alanine is an essential amino acid because it is a constituent of every protein.

11. Write a series of balanced equations and a summary equation for the reactions of the glucose–alanine cycle.

12. Consider the following questions about glutamate dehydrogenase.
(a) The reaction as shown on page 714 has NH_3 as a reactant, instead of $\overset{+}{N}H_4$, which is far more abundant at physiological pH. Why is NH_3 preferred?
(b) The equilibrium for the reaction greatly favors α-ketoglutarate reduction, yet in mitochondria the enzyme acts primarily to oxidize glutamate to α-ketoglutarate. Explain.
(c) Propose a reasonable mechanism for this reaction.

13. Explain the basis for the following statement: As a coenzyme, pyridoxal phosphate is covalently bound to enzymes with which it functions, yet during catalysis the coenzyme is not covalently bound.

14. Consider the regulation of *E. coli* glutamine synthetase, and explain the metabolic rationale for each of the following effects:
(a) Inhibition of glutamine synthetase by carbamoyl phosphate
(b) Inhibition of the deuridylylation of P_{II}-UMP by α-ketoglutarate
(c) Activation of the uridylylation of P_{II} by ATP

15. Suppose that you wanted to determine the metabolic half-life of glutamine synthetase in HeLa cells (a line of human tumor cells) growing in tissue culture. Describe how this could be done experimentally.

*16. Folic acid is synthesized in bacteria as dihydrofolate, in a pathway starting from guanosine triphosphate. In this pathway, C-8 is lost as formate. From the structural similarities between guanine and pterin, predict which carbon and nitrogen atoms of GTP are the precursors to N-1, C-2, C-4, N-5, C-7, N-8, and C-9, of dihydrofolate.

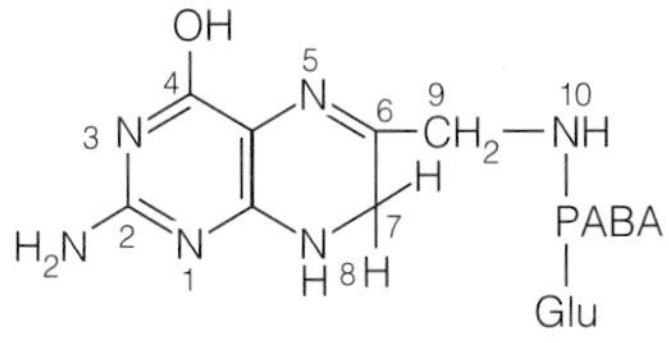

Dihydrofolate

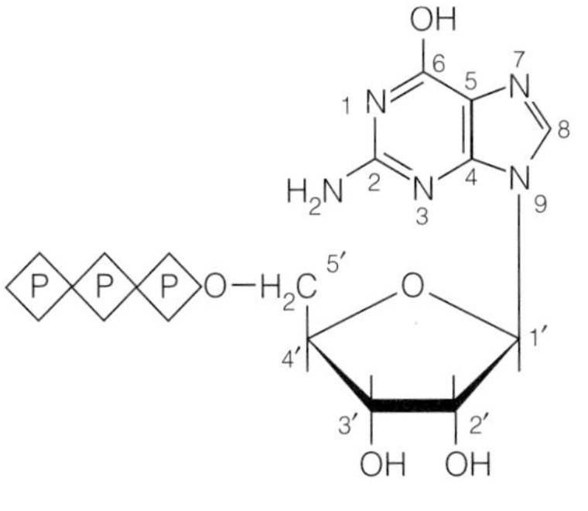

GTP

17. Which folate coenzyme
(a) Is the substrate for the enzyme that is inhibited by methotrexate and trimethoprim?
(b) Has the most highly oxidized single-carbon substituent?
(c) Is used in the conversion of serine to glycine?
(d) Transfers its single-carbon substituent to a B_{12} coenzyme? What amino acid is synthesized as the end result of this reaction?
(e) Is the coenzyme for the thymidylate synthase reaction?
(f) Is not known to exist in nature?
(g) Is used in purine nucleotide synthesis?

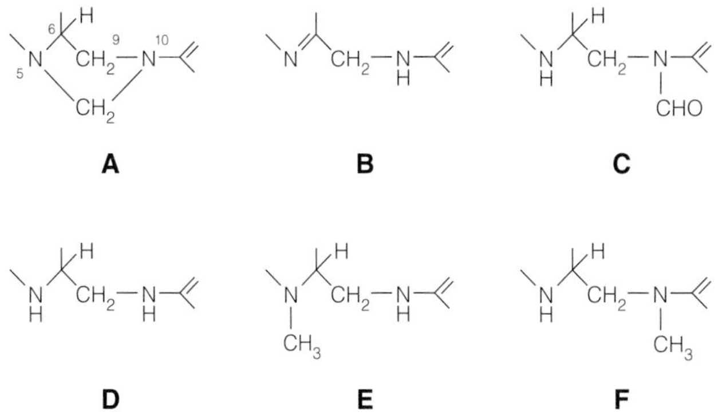

A **B** **C**

D **E** **F**

18. Glutamine affects the regulatory system for *E. coli* glutamine synthetase so as to promote the adenylylation of glutamine synthetase and inhibit the deadenylylation. Why do these effects make good metabolic sense?

19. Briefly discuss how a yeast cell might contain two glutamate dehydrogenases, one specialized for nitrogen assimilation and one for amino acid catabolism, and not dissipate energy in the futile cycle glutamate $\rightleftharpoons$ α-ketoglutarate.

CHAPTER 21

Metabolism of Nitrogenous Compounds: Amino Acids, Porphyrins, and Neurotransmitters

IN CHAPTER 20 WE PRESENTED UNIFYING PRINCIPLES OF AMINO ACID METABOLISM—general features of synthetic and degradative pathways, common routes for utilizing and excreting ammonia, and coenzymes used in nitrogen metabolism. Here we consider the metabolism of the 20 individual amino acids. Our approach is to organize these amino acids into families or groups that are related structurally or metabolically. Within each family, we shall discuss biosynthesis and degradation of each amino acid and the major roles of each as an intermediate en route to metabolites other than protein. We will pay particular attention to the processes and pathways that are most widespread in biology and to those that interrelate with other metabolic processes, illustrate important biological principles, or are mechanistically interesting. Some of the biological roles of amino acids—as neurotransmitters and neurotransmitter precursors and as porphyrin precursors—are so important that they are presented as separate major divisions within this chapter.

Amino Acids Related to Citric Acid Cycle Intermediates

About half of the 20 amino acids are biosynthesized more or less directly from intermediates in the citric acid cycle or from pyruvate, as shown in Figure 21.1. We include in this family glutamate, aspartate, and alanine, which can be formed by transamination from α-ketoglutarate, oxaloacetate, and pyruvate, respectively. The family also includes glutamine and asparagine, which are formed directly from glutamate and aspartate, respectively; and proline and arginine, which are formed in short pathways from glutamate. Several other amino acids—threonine, methionine, and isoleucine—are derived from aspartate, but we have assigned them to other families and will consider their metabolism later in the chapter.

SYNTHESIS AND CATABOLISM OF GLUTAMATE, ASPARTATE, ALANINE, GLUTAMINE, AND ASPARAGINE

Transamination provides major routes for both synthesis and degradation of glutamate, aspartate, and alanine. Reactions catalyzed by glutamate dehydrogenase

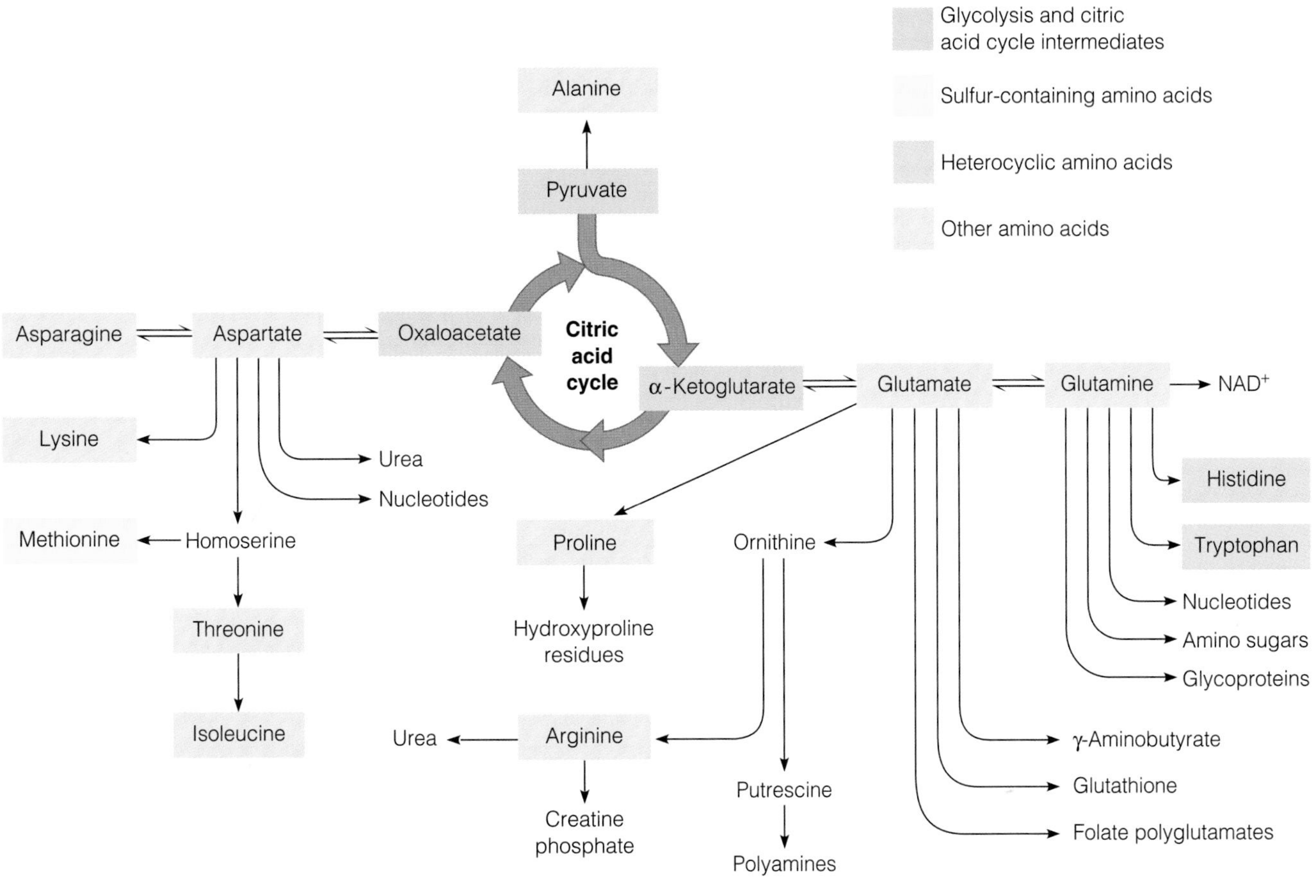

FIGURE **21.1**

Metabolic relationships among amino acids derived from citric acid cycle intermediates.

and glutamate synthase, introduced in Chapter 20, present additional routes for glutamate synthesis from α-ketoglutarate.

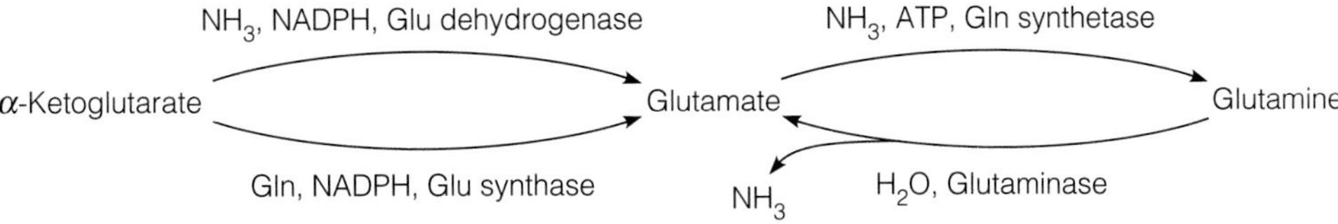

Glutamine is synthesized from glutamate by glutamine synthetase, and its major catabolic route is hydrolysis to glutamate by *glutaminase,* which in animals participates in transporting ammonia to the liver (see Figure 20.14). Similarly, asparagine is synthesized from aspartate by asparagine synthetase, and hydrolytic cleavage back to aspartate, catalyzed by **asparaginase,** is the principal degradative pathway. Aspartate is catabolized either by transamination or via deamination by **aspartase,** to give fumarate and ammonia.

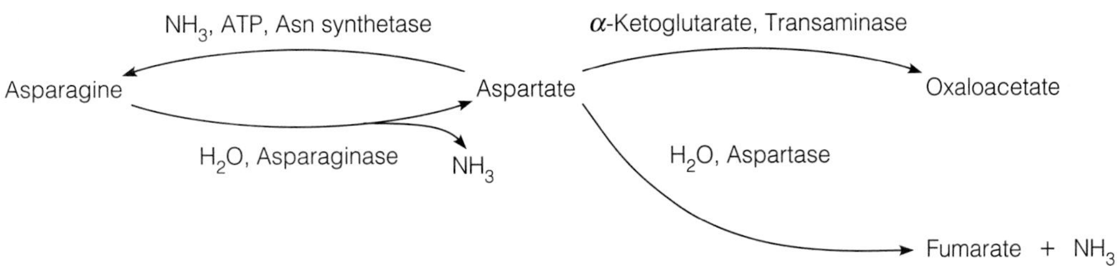

The amide nitrogen of glutamine is used in several reactions leading to purine and pyrimidine nucleotides, amino sugars, and nicotinamide nucleotides.

Asparagine has few metabolic functions aside from serving as a protein constituent.

In animals, a major function of alanine is its role in the glucose–alanine cycle as a carrier of carbon for gluconeogenesis from muscle to liver (see Figure 20.14). Other members of this family—glutamine, aspartate, and glutamate—all have numerous metabolic roles. Glutamine contributes its amide nitrogen in reactions leading to other amino acids, to nucleotides, and to amino sugars and glycoproteins. These reactions are catalyzed by a group of similar ATP-dependent ligases, the **amidotransferases.** One such reaction was presented in Chapter 16 (see page 579)—the biosynthesis of glucosamine-6-phosphate from fructose-6-phosphate, involved in the biosynthesis of amino sugars. Other amidotransferase reactions are presented later in this chapter and in Chapter 22.

Aspartate leads via homoserine to threonine, isoleucine, and methionine.

The nitrogen of aspartate is used in the biosynthesis of arginine and urea, as noted in Chapter 20. Similar reactions are involved in purine nucleotide synthesis, and the entire aspartate molecule is used in pyrimidine nucleotide biosynthesis; both of these processes are discussed in Chapter 22. Finally, in plants and bacteria, aspartate is a precursor to three other amino acids via its conversion to **homoserine,** as shown here. Separate pathways then lead from homoserine to methionine, threonine, and isoleucine. Also, in bacteria and plants, aspartic β-semialdehyde is a precursor to lysine.

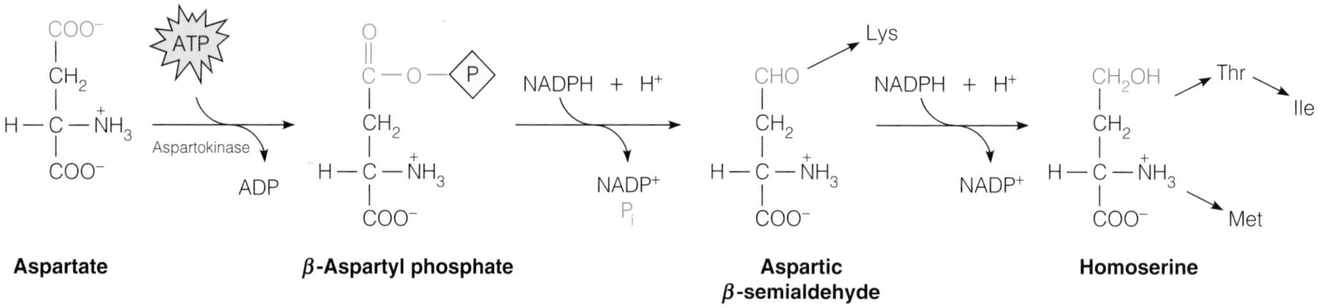

ATP
Aspartokinase
ADP
NADPH + H⁺
NADP⁺
P_i
Lys
NADPH + H⁺
NADP⁺
Thr
Ile
Met

Aspartate **β-Aspartyl phosphate** **Aspartic β-semialdehyde** **Homoserine**

The first enzyme in this pathway, **aspartokinase,** is a major site for regulation of each biosynthetic pathway. Bacteria contain three distinct forms of aspartokinase, each with its own mode of allosteric regulation. In *E. coli,* one form is specifically inhibited by threonine and one by lysine. The third enzyme is not feedback-inhibited by methionine, but synthesis of the enzyme is repressed specifically by methionine. These findings suggest that each enzyme is specialized to serve just one of the three pathways.

Homoserine has an interesting role in microbial metabolism, unrelated to its function in amino acid synthesis. Bacterial cultures carry out certain processes only after reaching a particular cell density. This phenomenon is called "quorum-sensing." Somehow bacteria sense when a certain population-density threshhold has been crossed, and this triggers physiological responses as diverse as luminescence, antibiotic synthesis, and conjugal gene transfer. For many bacteria the signaling molecule is one of several long-chain *N*-acyl derivatives of homoserine (see margin). *N*-Acylhomoserine lactone is synthesized and secreted at a low, constant rate, and it diffuses back into cells. At a sufficiently high cell density, the extracellular and, hence, the intracellular concentrations of the lactone have risen high enough for it to bind to genetic regulatory proteins, which in turn stimulate transcription of genes required to activate the particular process.

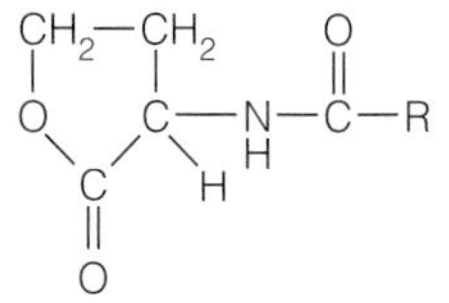

N-Acylhomoserine lactone

INTERMEDIARY METABOLISM OF GLUTAMATE

Biosynthesis of Ornithine, Arginine, and Creatine Phosphate

As suggested by Figure 21.1, glutamate is perhaps the most active of all the amino acids in terms of its number of metabolic roles. An important reaction of glutamate is the energy-requiring reduction of the γ-carboxyl group, to give **glutamic γ-semialdehyde** (Figure 21.2). This reaction, which leads toward both ornithine and proline, is comparable to the reduction of aspartate to aspartic semialdehyde, which was just presented. In plants and bacteria, glutamate is acetylated to **N-acetylglutamate** before reduction, with the acetyl group being removed a couple of steps later. Acetylation prevents cyclization of the molecule after reduction, by condensation between the resulting aldehyde and the α-amino group.

We shall discuss the role of glutamic γ-semialdehyde in proline synthesis shortly. Ornithine is formed directly from glutamic γ-semialdehyde by transamination at the aldehyde group. Recall from Chapter 20 that ornithine is a participant in the urea cycle, leading to arginine and urea. In muscle, arginine is the precursor to the energy storage compound creatine phosphate (Figure 21.3). The guanidino group is transferred to glycine, with regeneration of ornithine. The other product, **guanidinoacetic acid,** is methylated by S-adenosylmethionine to give creatine, followed by phosphorylation of the latter to creatine phosphate, catalyzed by creatine kinase.

Ornithine undergoes decarboxylation to give 1,4-diaminobutane, which has the trivial name **putrescine** (see margin) because of its original isolation from rotting meat. Putrescine is the precursor to a class of important and ubiquitous compounds, the **polyamines.** Synthesis of polyamines involves S-adenosylmethionine in a reaction that we present in the section on methionine metabolism (see Figure 21.10).

Glutamate, one of the most active amino acids, is a precursor to glutamine, arginine, creatine phosphate, proline, hydroxyproline, polyamines, glutathione, and γ-aminobutyric acid.

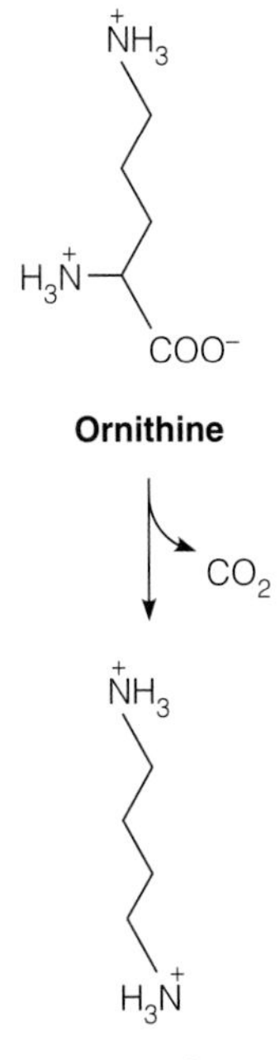

$\overset{+}{N}H_3$ / $H_3\overset{+}{N}$ / COO^-

Ornithine

CO_2

$\overset{+}{N}H_3$ / $H_3\overset{+}{N}$

Putrescine

Arginine as the Precursor to Nitric Oxide, a Novel Second Messenger

Beginning in the late 1980s an unexpected role for arginine was described as precursor to a novel second messenger and neurotransmitter. This novel regulator has been identified as a gas, **nitric oxide** (NO·), which is produced from arginine in an unusual reaction that also yields citrulline (Figure 21.3). The enzyme, NO· synthase, contains bound FMN, FAD, nonheme iron, and *tetrahydrobiopterin,* a relative of tetrahydrofolate that was originally identified as a cofactor for synthesis of tyrosine from phenylalanine (see page 768).

NO· was originally identified as a signal-transducing agent in vasodilation of endothelial vascular cells and underlying smooth muscle. A variety of signals that decrease blood pressure and inhibit platelet aggregation use NO· as an intermediary. In inflammatory and immune responses, an inducible form of nitric oxide synthase produces NO· at levels sufficient to be toxic to pathogenic organisms.

The NO· synthase in endothelial vascular cells is acutely sensitive to calcium ion concentration; activation of the enzyme by Ca^{2+} causes NO· accumulation. Because NO· is a gas, it can diffuse rapidly into neighboring cells and control their metabolism, primarily by stimulating cyclic GMP synthesis (see Chapter 23). Because it is unstable, with a chemical half-life of 1 to 5 seconds, its actions are short-lived. Studies of intracellular localization of NO· synthase and of NO· synthase inhibitors have identified additional roles for NO· as diverse as neurotransmission in the central nervous system and stimulation of erection of the penis. The drug Viagra, introduced in 1998 to counteract erectile dysfunction, acts by inhibiting cyclic GMP breakdown, thereby prolonging the effects of NO·.

Biosynthesis of Proline, Hydroxyproline, and Collagen

As noted in the following figure, glutamic γ-semialdehyde leads not only to ornithine but also to proline (see also Figure 21.1). First, spontaneous intramolecular Schiff base formation gives **Δ1-pyrroline carboxylic acid.** An NADPH-dependent reduction follows, to give proline.

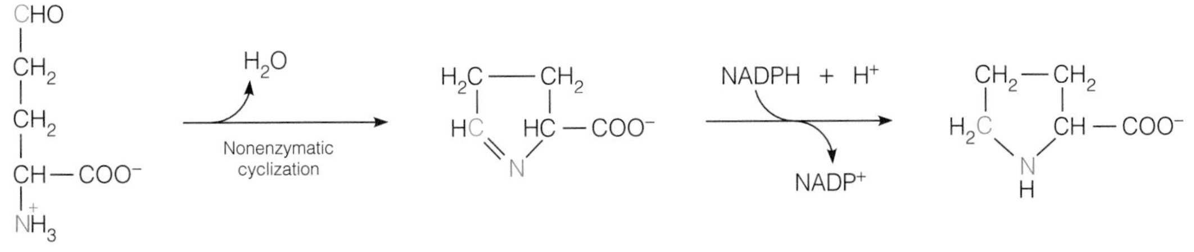

CHO — CH_2 — CH_2 — CH—COO^- — $\overset{+}{N}H_3$

H_2O

Nonenzymatic cyclization

H_2C—CH_2 / HC HC—COO^- / N

NADPH + H^+

$NADP^+$

CH_2—CH_2 / H_2C CH—COO^- / N / H

Glutamic γ-semialdehyde

Δ1-Pyrroline carboxylic acid

Proline

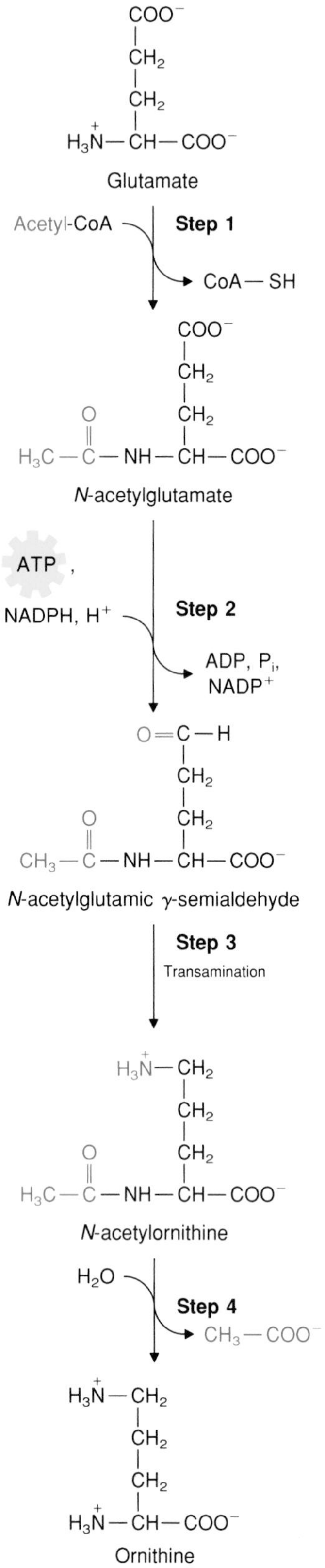

FIGURE **21.2**

Biosynthesis of ornithine from glutamate. The reduction of *N*-acetylglutamate (step 2) probably begins with phosphorylation of the carboxyl group by ATP, followed by NADPH-dependent reduction of the activated carboxyl group.

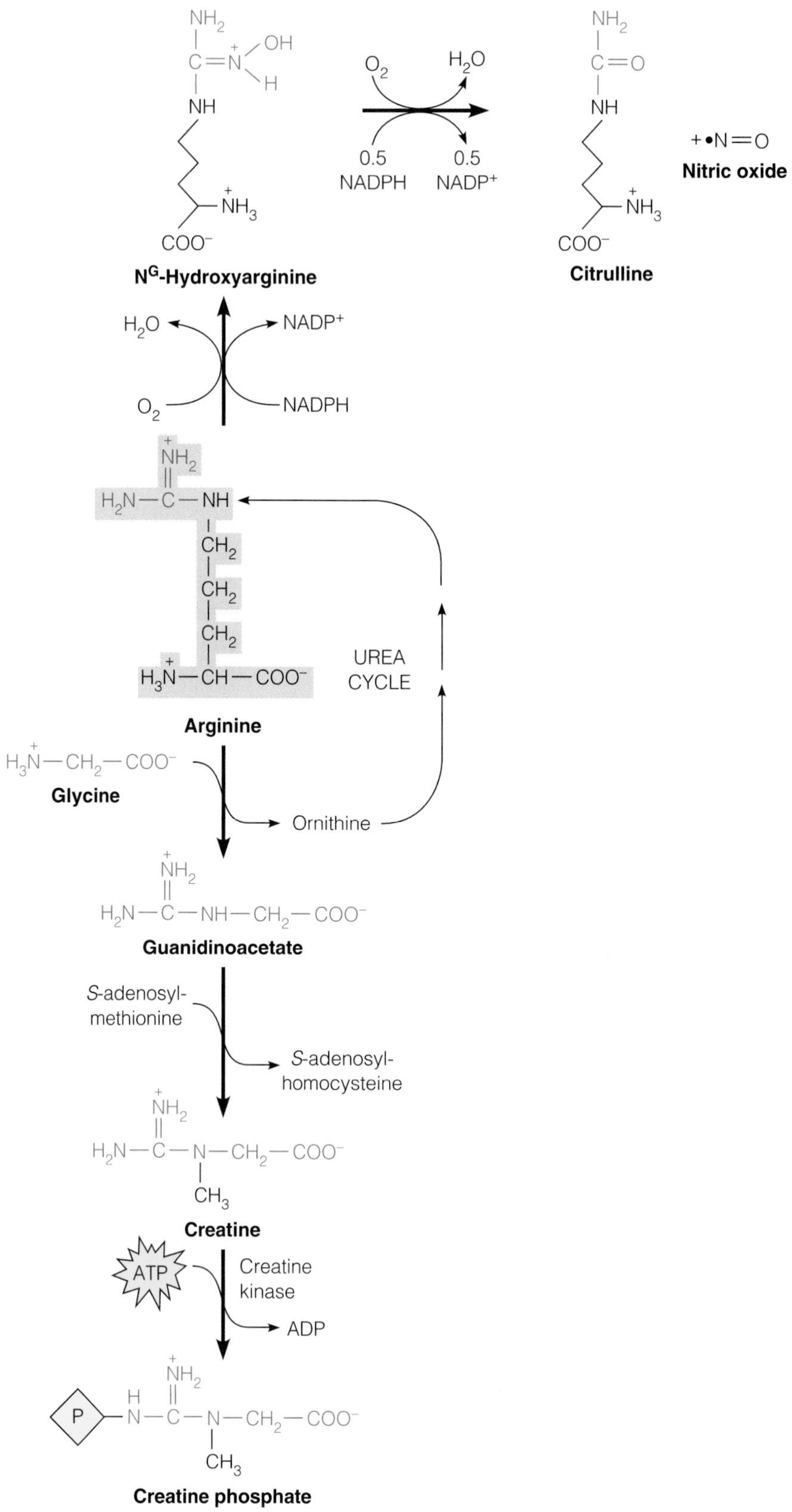

FIGURE **21.3**

Biosynthesis of creatine phosphate and nitric oxide from arginine. In N^G-hydroxyarginine, the G refers to the placement of the hydroxyl group on the guanidine group.

In plants, glutamic γ-semialdehyde is synthesized from *N*-acetylglutamic γ-semialdehyde by transfer of the acetyl group on the latter to glutamate.

N-Acetylglutamic γ-semialdehyde + glutamate $\rightleftharpoons$ *N*-acetylglutamate + glutamic γ-semialdehyde

An important metabolic role of proline is its incorporation into polypeptide precursors to collagen and other connective tissue proteins, where it serves as a precursor to **hydroxyproline.** As mentioned in Chapter 5 (see page 134), hydroxyproline residues are generated by posttranslational modification, following completion of the polypeptide chain. The nonhydroxylated collagen precursor is called **procollagen** (see Chapter 6). In this polypeptide, a proline residue two positions to the carboxyl side of a glycine residue is the preferred substrate for the action of **procollagen proline hydroxylase** (Figure 21.4). This unusual enzyme requires ferrous iron, ascorbic acid, molecular oxygen, and α-ketoglutarate. α-Ketoglutarate is oxidized during the reaction to succinate and CO_2. Ascorbic acid reduces O_2 with one atom incorporated into succinate, and the other ending up in the hydroxyproline hydroxyl group. This reaction is of particular interest because it represents one of the few well-defined roles for ascorbic acid, or vitamin C. A vitamin C deficiency, or scurvy, involves defects in connective tissue function, and it seems likely that these problems derive from defective synthesis or maturation of collagen in connective tissue.

γ-Aminobutyric Acid

Glutamate is one of several amino acids serving as precursors to compounds that function in transmission of nerve impulses. Decarboxylation of glutamate yields **γ-aminobutyric acid,** or **GABA.** In addition, glutamate itself is a neurotransmitter. Functions of both glutamate and GABA in neurotransmission will be discussed later in this chapter.

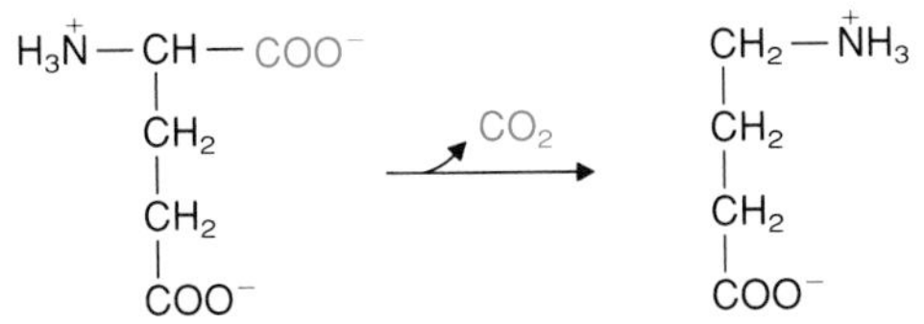

$$H_3\overset{+}{N}-CH(CH_2CH_2COO^-)-COO^- \xrightarrow{\ \ CO_2\ \nearrow\ \ } \overset{+}{N}H_3-CH_2-CH_2-CH_2-COO^-$$

Glutamate → **γ-Aminobutyric acid**

Other Functions of Glutamate

Glutamate has at least two other major and ubiquitous metabolic fates, each involving specific enzyme systems: (1) the synthesis of glutathione (see Chapter 14), along with cysteine and glycine, and (2) the synthesis, via an ATP-dependent

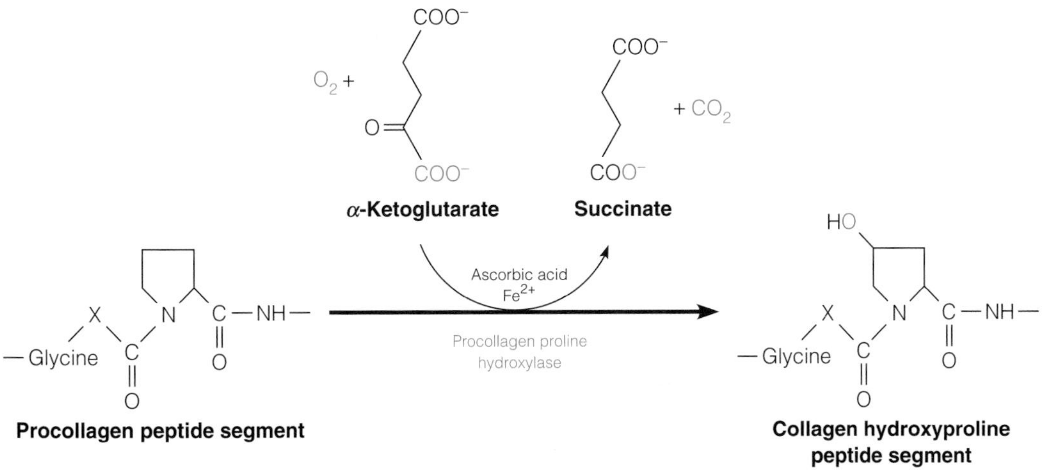

O_2 + COO⁻–CH₂–CH₂–C(=O)–COO⁻ (**α-Ketoglutarate**) → COO⁻–CH₂–CH₂–COO⁻ (**Succinate**) + CO_2

—Glycine–C(=O)–X–N(proline)–C(=O)–NH— (**Procollagen peptide segment**) —[Ascorbic acid, Fe^{2+}; Procollagen proline hydroxylase]→ —Glycine–C(=O)–X–N(HO-proline)–C(=O)–NH— (**Collagen hydroxyproline peptide segment**)

FIGURE **21.4**

Enzymatic hydroxylation of procollagen proline residues in the synthesis of collagen. The fates of the two atoms of O_2 are indicated in red. X is any amino acid residue.

conjugating system, of polyglutamate forms of folic acid and its coenzymes (see Chapter 20).

Metabolism of Sulfur-Containing Amino Acids

REDUCTION OF INORGANIC SULFUR

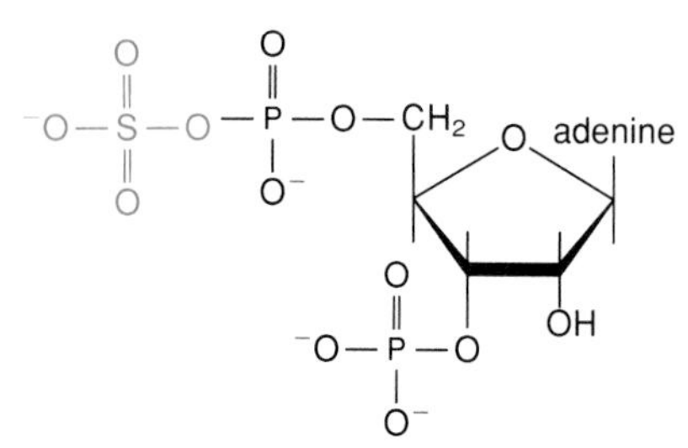

3′-Phosphoadenosine-5′-phosphosulfate

Like carbon and nitrogen, sulfur is made available to organisms largely in the form of inorganic compounds—principally sulfate, although some bacteria can synthesize organic compounds from elemental sulfur or from sulfite. Just as CO_2 and N_2 must undergo fixation to be utilized, the utilization of sulfate requires metabolic activation to a form that can readily undergo reduction. The process for sulfate is largely confined to plants and bacteria. The end product of reduction is H_2S and this is used for cysteine and methionine synthesis. The activated sulfate compound is **3′-phosphoadenosine-5′-phosphosulfate (PAPS).** This nucleotide is formed in two steps from ATP and sulfate ion.

$$\mathrm{SO_4^{2-}} + \mathrm{ATP} \xrightarrow{\nearrow \mathrm{PP_i}} \text{adenosine-5'-phosphosulfate} \xrightarrow{\mathrm{ATP} \searrow} \mathrm{PAPS} + \mathrm{ADP}$$

In all organisms, PAPS serves as an active agent for sulfate esterification, as in the synthesis of sulfated polysaccharides such as chondroitin sulfate.

Phosphoadenosine phosphosulfate is an activated form of sulfate used both for sulfation reactions and as a substrate for sulfate reduction.

In bacteria, PAPS also serves as the substrate for sulfate reduction. Reduction of the sulfate in PAPS to sulfite (SO_3^{2-}) involves **thioredoxin,** a small protein ($M_r \cong 12{,}000$) that contains two reversibly oxidizable cysteine thiol groups. Thioredoxin is involved in several other intracellular redox reactions, as described in Chapters 17 and 22. The sulfite is subsequently reduced by **sulfite reductase,** a large and complex enzyme that catalyzes a six-electron transfer. The electrons are shuttled along a pathway involving NADPH, FAD, FMN, an iron–sulfur center, and the porphyrin siroheme (see page 713). No intermediates accumulate, just the product H_2S. In plants, adenosine 5′-phosphosulfate, rather than PAPS, is the substrate for reduction.

SYNTHESIS OF CYSTEINE AND METHIONINE IN PLANTS AND BACTERIA

Our discussion of sulfur-containing amino acids deals only with cysteine and methionine. Bacteria and plants can synthesize both amino acids by incorporation of H_2S (Figure 21.5), whereas animals derive cysteine either from the diet or from dietary methionine. Although cystine is found in protein hydrolysates, recall from Chapter 5 that cystine disulfides are synthesized at the polypeptide level, by posttranslational oxidation of cysteine residues. Cystine normally does not exist in cells as the free amino acid; when it does accumulate, it is reduced nonenzymatically by glutathione.

Cysteine Biosynthesis

Plants and microorganisms use H_2S for the synthesis of cysteine, with serine providing the carbon skeleton. Some bacteria can condense these two substrates directly, via a pyridoxal phosphate–dependent enzyme.

$$\underset{\textbf{Serine}}{\mathrm{H{-}C(CH_2OH)(COO^-){-}\overset{+}{N}H_3}} + \mathrm{H_2S} \longrightarrow \underset{\textbf{Cysteine}}{\mathrm{H{-}C(CH_2SH)(COO^-){-}\overset{+}{N}H_3}} + \mathrm{H_2O}$$

However, plants and most microorganisms use β-*O*-acetylserine as the substrate reacting with H_2S.

$$\text{Serine} + \text{acetyl-CoA} \longrightarrow O\text{-acetylserine} + \text{CoA-SH}$$

$$O\text{-Acetylserine} + H_2S \longrightarrow \text{cysteine} + \text{acetate} + H_2O$$

Methionine Biosynthesis

In plants and bacteria, cysteine provides the sulfur for methionine synthesis, with the carbon coming from homoserine (Figure 21.6). Homoserine first reacts with succinyl-CoA to form **O-succinylhomoserine.** This reaction evidently is a control

FIGURE **21.5**

Outline of pathways for cysteine and methionine synthesis in plants and bacteria.

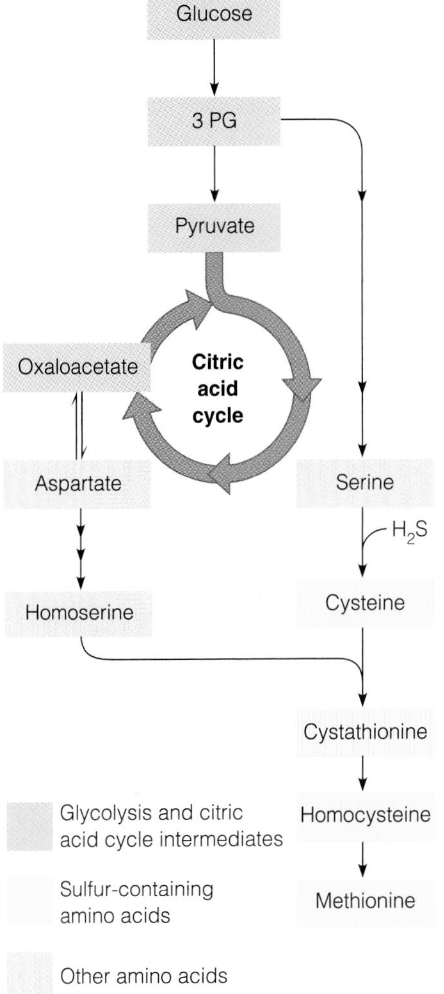

$CH_2{-}OH$ / CH_2 / $H_3\overset{+}{N}{-}CH{-}COO^-$

Homoserine

Succinyl—CoA → CoA (inhibited by Methionine)

$CH_2{-}O{-}\text{Succinate}$ / CH_2 / $H_3\overset{+}{N}{-}CH{-}COO^-$

O-Succinylhomoserine

Cysteine ($CH_2{-}SH$ / $H_3\overset{+}{N}{-}CH{-}COO^-$), B_6 → Succinate

$CH_2{-}S{-}CH_2$ / $H_3\overset{+}{N}{-}CH$ CH_2 / COO^- $CH{-}COO^-$ / $\overset{+}{N}H_3$

Cystathionine

→ NH_3, Pyruvate

$HS{-}CH_2$ / CH_2 / $H_3\overset{+}{N}{-}CH{-}COO^-$

Homocysteine

$CH_3{-}B_{12}$ → B_{12}, B_6

$H_3C{-}S{-}CH_2$ / CH_2 / $H_3\overset{+}{N}{-}CH{-}COO^-$

Methionine

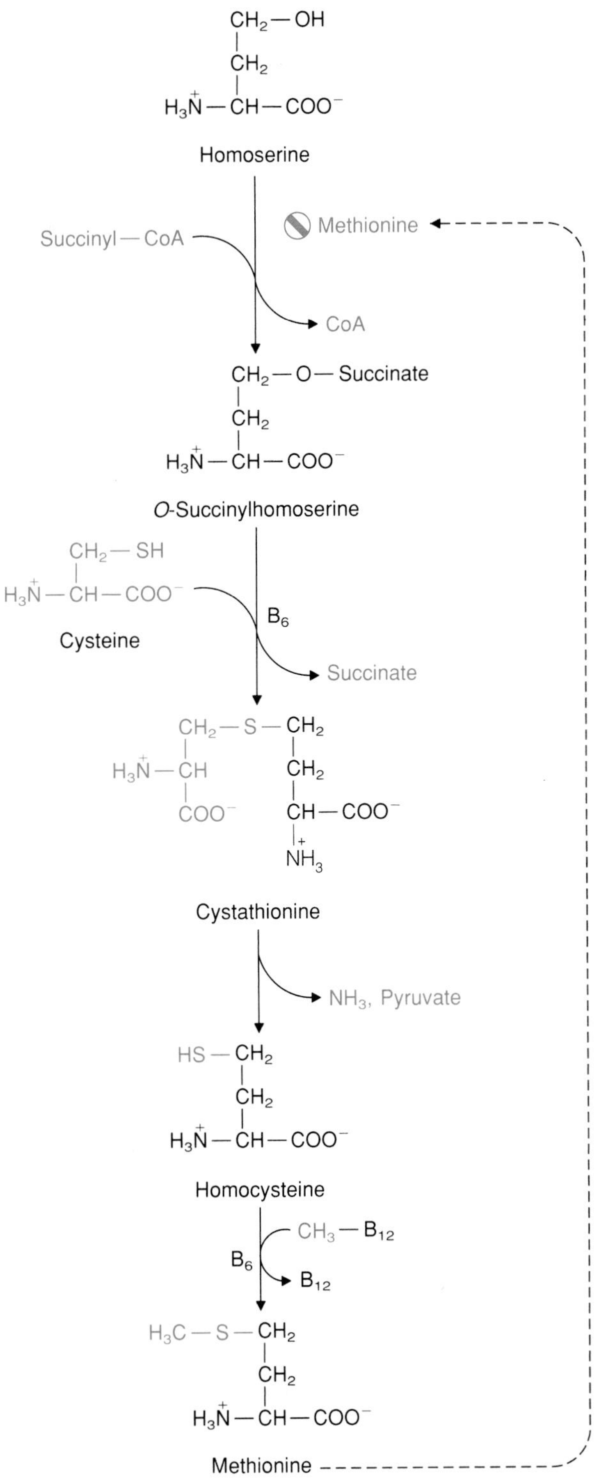

FIGURE **21.6**

Biosynthesis of methionine from homoserine, as it occurs in plants and bacteria.

point, because the enzyme catalyzing this reaction is feedback inhibited by methionine. *O*-Succinylhomoserine reacts with cysteine to give **cystathionine,** a thioether compound. Cleavage of cystathionine then occurs, with the sulfur thereby becoming linked to the four-carbon side chain that started as homoserine. The resulting amino acid, homocysteine, is the substrate for methionine synthase, which, as presented in Chapter 20, uses B_{12} to transfer a methyl group from 5-methyltetrahydrofolate, yielding methionine (see page 739).

METHIONINE AS THE SOURCE OF CYSTEINE SULFUR IN ANIMALS

Methionine is classified as an essential amino acid for mammals, and cysteine is considered nonessential. Actually, the biosynthetic route in animals proceeds from methionine to cysteine, as shown in Figure 21.7, so cysteine is nonessential only as long as the diet contains adequate methionine.

Plants and bacteria synthesize cysteine from inorganic sulfur and synthesize methionine from cysteine. Animals synthesize cysteine from dietary methionine.

The synthesis of cysteine in animals, shown in more detail in Figure 21.8, resembles the reverse of the methionine biosynthetic pathway we just discussed. Homocysteine condenses with serine, in a pyridoxal phosphate–dependent reaction catalyzed by **cystathionine synthase,** to give cystathionine. This compound then cleaves, also in a PLP-dependent reaction, to give cysteine. The four-carbon side chain deaminates, forming ammonia and α-ketobutyrate. Plants and prokaryotes carry out the same pathway, with α-ketobutyrate used also for isoleucine biosynthesis.

Note that cystathionine plays a central role both in the biosynthesis of methionine in plants and bacteria and in the biosynthesis of cysteine in animals. In methionine synthesis from cysteine, cystathionine cleavage accompanies the transfer of sulfur from a three-carbon to a four-carbon side chain. The converse occurs during the synthesis of cysteine from methionine.

In humans a genetic deficiency of cystathionine synthase leads to a condition called **homocystinuria,** in which homocysteine overaccumulates, as evidenced by excessive urinary excretion of homocystine (the disulfide derivative of homocysteine). The condition results in severe mental retardation, damage to blood vessels, and dislocation of the lens of the eye. After the cystathionine synthase deficiency was described, it was found that similar symptoms resulted from deficiencies of either of two related enzymes—methionine synthase or 5,10-methylenetetrahydrofolate reductase. From Figure 21.9, you can see how a deficiency in any one of these three enzymes would cause homocysteine accumulation.

METABOLISM OF GLUTATHIONE

As described in Chapter 14, cysteine plays an important metabolic role as a constituent of **glutathione,** or γ-glutamylcysteinylglycine. This tripeptide, abundant in all cells, protects against two kinds of metabolic stress. First, it can nonenzymatically reduce a number of substances, such as peroxides or free radicals, which accumulate in cells under oxidizing conditions. Glutathione maintains an intracellular reducing environment, which prevents intracellular protein thiols from oxidizing to disulfides. Second, through the action of widely distributed enzymes, the **glutathione *S*-transferases,** glutathione participates in detoxification of many substances, such as **xenobiotics** (foreign organic compounds not produced in metabolism) or electrophiles produced through the action of cytochrome P450–linked oxidases. Such compounds include organic halides, fatty acid peroxides derived from lipid oxidation, and products derived from radiation-damaged DNA. Glutathione

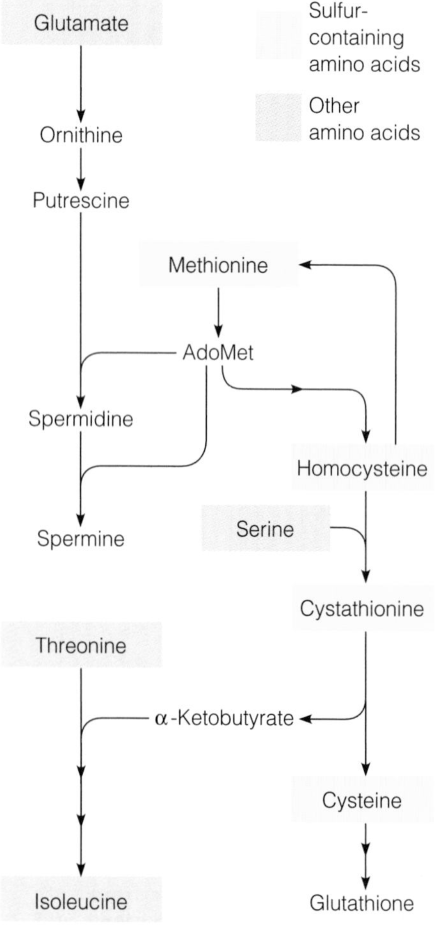

FIGURE **21.7**

Outline of methionine metabolism. Except for the synthesis of isoleucine from threonine (see page 775), which is limited to plants and bacteria, these pathways occur in virtually all organisms.

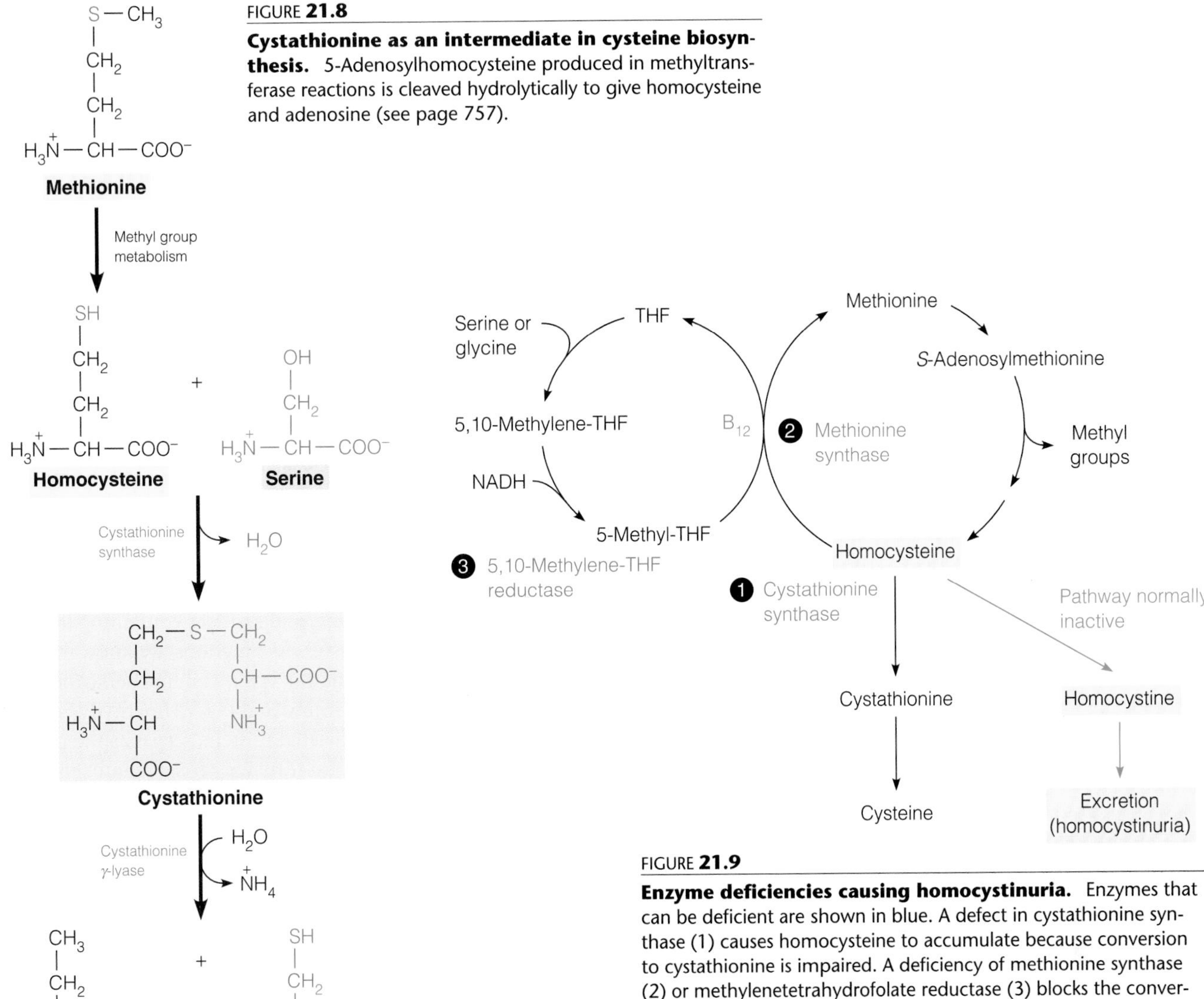

FIGURE **21.8**

Cystathionine as an intermediate in cysteine biosynthesis. 5-Adenosylhomocysteine produced in methyltransferase reactions is cleaved hydrolytically to give homocysteine and adenosine (see page 757).

FIGURE **21.9**

Enzyme deficiencies causing homocystinuria. Enzymes that can be deficient are shown in blue. A defect in cystathionine synthase (1) causes homocysteine to accumulate because conversion to cystathionine is impaired. A deficiency of methionine synthase (2) or methylenetetrahydrofolate reductase (3) blocks the conversion of homocysteine to methionine. In any of these cases the homocysteine that accumulates is oxidized to homocystine, the form in which it is excreted.

reacts with such a compound (denoted RX) as shown here, followed by cleavage of the γ-glutamyl and glycyl residues and then acetylation by acetyl-CoA to give a **mercapturic acid.** This more soluble, less toxic derivative of the original compound can then be excreted in the urine.

RX + HS—Cys(γ-Glu)(Gly) → (− X^-) R—S—Cys(γ-Glu)(Gly) → (− Glu + Gly) R—S—CH_2—CH($\overset{+}{N}H_3$)—COO^- → (Ac—S—CoA → CoA—SH) R—S—CH_2—CH(NH—Ac)—COO^-

Glutathione; R–cysteine conjugate; Mercapturic acid

Alternatively, detoxification may involve catabolism of the cysteine conjugate to a methylthio compound or a glucuronide.

Ovothiol

ε-*N*-Trimethyllysine

A recently described sulfur amino acid, **ovothiol,** is found in fertilized eggs, where it plays a role comparable to that of glutathione. Ovothiol, which is present at 5 mM concentration in sea urchin eggs, protects the egg against oxidative damage by peroxides produced at the egg surface early in fertilization. Ovothiol is in turn reduced by glutathione.

S-ADENOSYLMETHIONINE AND BIOLOGICAL METHYLATION

In Chapter 19 we introduced *S*-adenosylmethionine (AdoMet) as a metabolically activated form of methionine, when we described the biosynthesis of phosphatidylcholine from phosphatidylethanolamine. Recall that the synthesis of AdoMet from methionine and ATP generates a sulfonium compound with a high group transfer potential. Most, though not all, of the group transfer reactions involving AdoMet are **transmethylations,** in which the methyl group is transferred to an acceptor, with the other product being *S*-adenosylhomocysteine (AdoHcy). The syntheses of creatine (see Figure 21.3) and of phosphatidylcholine (see page 674) are good examples of transmethylations.

Table 21.1 lists a number of biologically important AdoMet-dependent transmethylations. Note that the substrates can be polymeric proteins or nucleic acids. We discuss the functions of DNA and RNA methylation in later chapters. However, a brief discussion of protein methylation is in order here. Methylatable residues include—in different proteins—lysine, arginine, and residues containing free carboxyl groups. Although we know little about specific functions, we know that histones become methylated, with specific arginine and lysine residues being modified at particular times in the cell cycle. We also know that ε-*N*-trimethyllysine, derived specifically from the hydrolysis of methylated protein, serves as the precursor to carnitine, whose role in fatty acyl group transfer across membranes was presented in Chapter 18. Furthermore, we know that, in bacteria, protein methylation plays an important role in **chemotaxis,** the process whereby bacteria sense a concentration gradient of a chemical substance in the medium and move either toward or away from it. Chemotaxis is being studied as a rudimentary model for sensory transduction; it involves the cyclic methylation and demethylation of a group of proteins called **MCPs,** or methylatable chemotactic proteins. Finally, there are indications that protein methylation somehow protects proteins, in at least two ways: (1) By blocking sites of ubiquitination (see Chapter 20), methylation evidently helps protect proteins from turnover. (2) Environmental damage to protein molecules modifies their aspartate residues, causing damage that can be repaired by a methylation reaction.

5-Methyltetrahydrofolate transfers a methyl group in methionine synthesis, but all other biological methyl transfers involve *S*-adenosylmethionine.

The central metabolic role of AdoMet can be appreciated if you keep in mind that, except for a few reactions in bacterial metabolism, *the only known methyl group transfer that does not involve AdoMet is the synthesis of methionine itself.* As noted in Chapter 20, a methyl group is generated de novo through the reduction of 5,10-methylenetetrahydrofolate to 5-methyltetrahydrofolate. The methyl group is then transferred to yield methionine via methyl-B_{12} and the action of methio-

TABLE **21.1 Some AdoMet-dependent transmethylations**

Methyl Group Acceptor	Methylated Product
Norepinephrine	Epinephrine
Guanidinoacetic acid	Creatine
Phosphatidylethanolamine	Phosphatidylcholine
DNA-adenine or -cytosine	DNA-*N*-methyladenine or 5-methylcytosine
tRNA bases	Methylated tRNA bases
Nicotinamide	N^1-Methylnicotinamide
Protein amino acid residues	Methylated amino acid residues

nine synthase. There is, however, an additional route for methionine synthesis that involves a different kind of transmethylation. *S*-Adenosylhomocysteine, formed from AdoMet-dependent transmethylations, cleaves hydrolytically to yield adenosine and homocysteine. Homocysteine is remethylated to methionine, either by methionine synthase or by transmethylation from **glycine betaine.** The latter is formed from the oxidation of choline.

$$CH_2OH{-}CH_2{-}\overset{+}{N}(CH_3)_3 \xrightarrow{\text{Oxidation}} COO^-{-}CH_2{-}\overset{+}{N}(CH_3)_3 \xrightarrow[COO^-{-}CH_2{-}N(CH_3)_2]{\text{Homocysteine}} S{-}CH_3{-}CH_2{-}CH_2{-}H{-}C{-}\overset{+}{N}H_3{-}COO^-$$

Choline **Glycine betaine** **Methionine**

The betaine, which is a quaternary amine derivative of an amino acid, has a positive charge and a high group transfer potential, as does AdoMet itself. Because the methyl groups on glycine betaine came originally from AdoMet, the existence of this transmethylation process is consistent with the statement that all methyl groups come from methionine except those used in the synthesis of methionine.

S-Adenosylmethionine plays an additional role in plant metabolism, as the precursor to the hydrocarbon hormone **ethylene.** Ethylene promotes plant growth and development and induces the ripening of fruit. Although mechanistic details are not yet clear, we know that the main carbon skeleton of methionine, rather than the methyl group, is split off in this process, to yield **1-aminocyclopropane-1-carboxylic acid.** This substance is fragmented as shown, to yield ethylene as one of the products.

$$CH_3{-}\overset{+}{S}(\text{adenine}){-}\text{ribose},\ {}_4CH_2{-}{}_3CH_2{-}{}_2HC{-}\overset{+}{N}H_3{-}{}_1COO^- \xrightarrow{CH_3{-}S(\text{adenine}){-}\text{ribose}} \text{cyclo}({}^3H_2C{-}{}^4H_2C{-}{}^2C)(\overset{+}{N}H_3)({}^1COO^-) \xrightarrow[\ ]{O_2 + 2H \text{ or } H_2O_2 \ \to\ \overset{+}{N}H_4,\ \overset{2}{H}COO^-} \overset{3}{H_2C}{=}\overset{4}{CH_2} + \overset{1}{C}O_2$$

AdoMet **1-Aminocyclopropane-1-carboxylate** **Ethylene**

Isotope labeling studies with methionine fed to plants have revealed the source of each carbon atom.

It is likely that a similar mechanism is involved in the synthesis of rare fatty acids that contain a cyclopropane ring in their hydrocarbon chains, such as lactobacillic acid, the example shown.

$$CH_3(CH_2)_5CH{-}CH(CH_2)_9COOH \text{ (with bridging } CH_2)$$

Lactobacillic acid

POLYAMINES

The synthesis of ethylene involves transfer, not of the methyl group but of the four-carbon moiety of AdoMet-bound methionine. A comparable process occurs during synthesis of the widely distributed polyamines **spermine** and **spermidine** (Figure 21.10). The names of these substances suggest the source of their original detection—human semen. These polyamines and the diamine putrescine are widely distributed cationic cell components, being especially abundant in rapidly proliferating cells. Although putrescine and its homolog **cadaverine** are classified among the polyamines, they are in fact diamines; they are synthesized by decarboxylation of ornithine and lysine, respectively. **Ornithine decarboxylase,** which synthesizes putrescine, is a highly regulated enzyme whose activity responds to

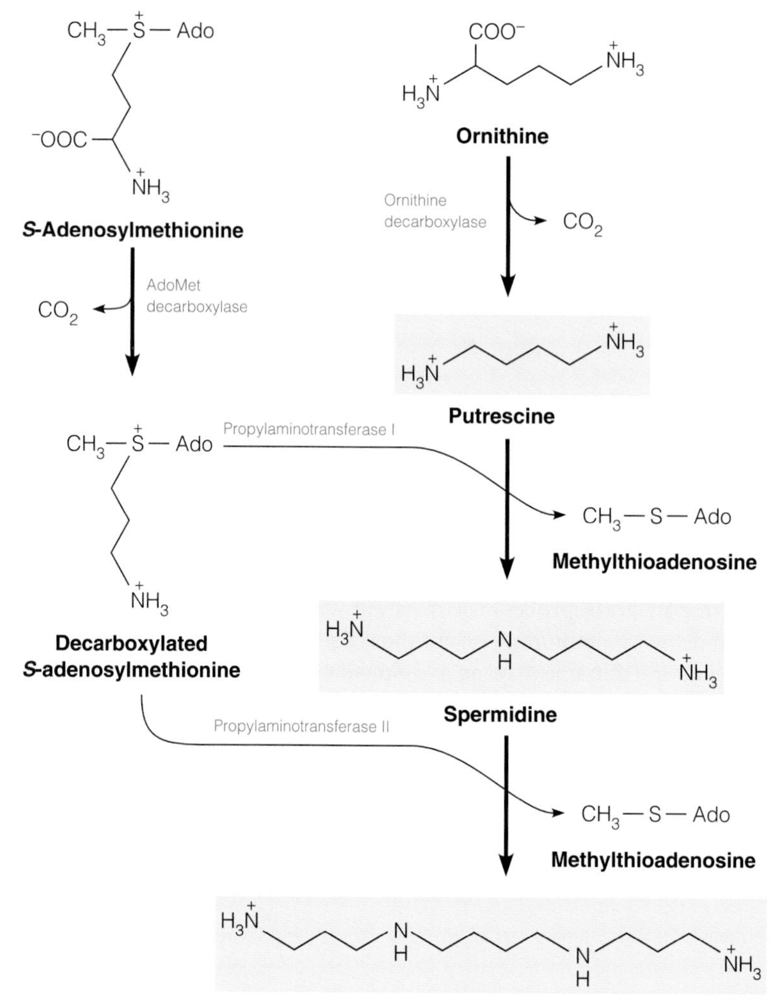

FIGURE **21.10**

Biosynthesis of putrescine, spermidine, and spermine.

many hormonal stimuli. It has an extremely short metabolic half-life (approximately 10 minutes), suggesting that its intracellular activity is controlled largely at the level of protein degradation.

Polyamines are required for cell proliferation, because of their roles in stabilizing duplex DNA structures.

As polycations, the polyamines play multiple roles in stabilizing intracellular conformations of negatively charged nucleic acids. A polyamine molecule can bind to phosphates on both strands of a duplex, thereby stabilizing double-stranded DNA or a duplex region of RNA. In bacteriophage T4, for instance, about 40% of the negative charge on the viral DNA is neutralized by polyamines. Transfer RNA from some sources contains two molecules of tightly bound spermine or spermidine per molecule of transfer RNA. Some proteins contain covalently bound polyamines, with the nitrogens linked to glutamate γ-carboxyl groups. In bacterial cells, polyamines also participate in regulating internal osmotic strength and they help to stabilize membranes. In animal cells, polyamines control the electrical properties of excitable membranes. Also, polyamine biosynthesis is closely related to the proliferative state of the cell; when nucleic acid synthesis is activated, so is polyamine synthesis. Consequently, investigators are asking whether activation of polyamine synthesis early in tumorigenesis could be a marker for early cancer diagnosis or a target for cancer treatment. Consistent with the latter, the antiparasitic drug **difluoromethylornithine** arrests progression through the cell division cycle (see Chapter 28) by inhibiting ornithine decarboxylase, suggesting a role for polyamines in cell cycle regulation.

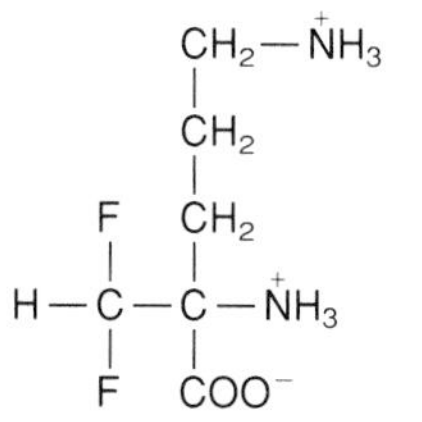

Difluoromethylornithine

Putrescine serves as the precursor to spermidine, then to spermine, through the AdoMet-mediated transfer of active propylamino groups (see Figure 21.10). First, AdoMet is decarboxylated by **AdoMet decarboxylase.** This enzyme is an example of a decarboxylase that contains as its cofactor not pyridoxal phosphate but covalently bound pyruvate (see page 732). The resulting propylamino group is then transferred to putrescine to give spermidine, and a second group is transferred to spermidine, by a different enzyme, to give spermine. The other product of these reactions is methylthioadenosine, which undergoes phosphorolytic cleavage to adenine and **5′-methylthioribose-1-phosphate.** The latter compound is used for resynthesis of methionine by another pathway, not shown here.

CATABOLISM OF CYSTEINE AND METHIONINE

The principal catabolic pathways for cysteine and methionine are outlined in Figure 21.11. β-Mercaptopyruvate, the transamination product of cysteine, can

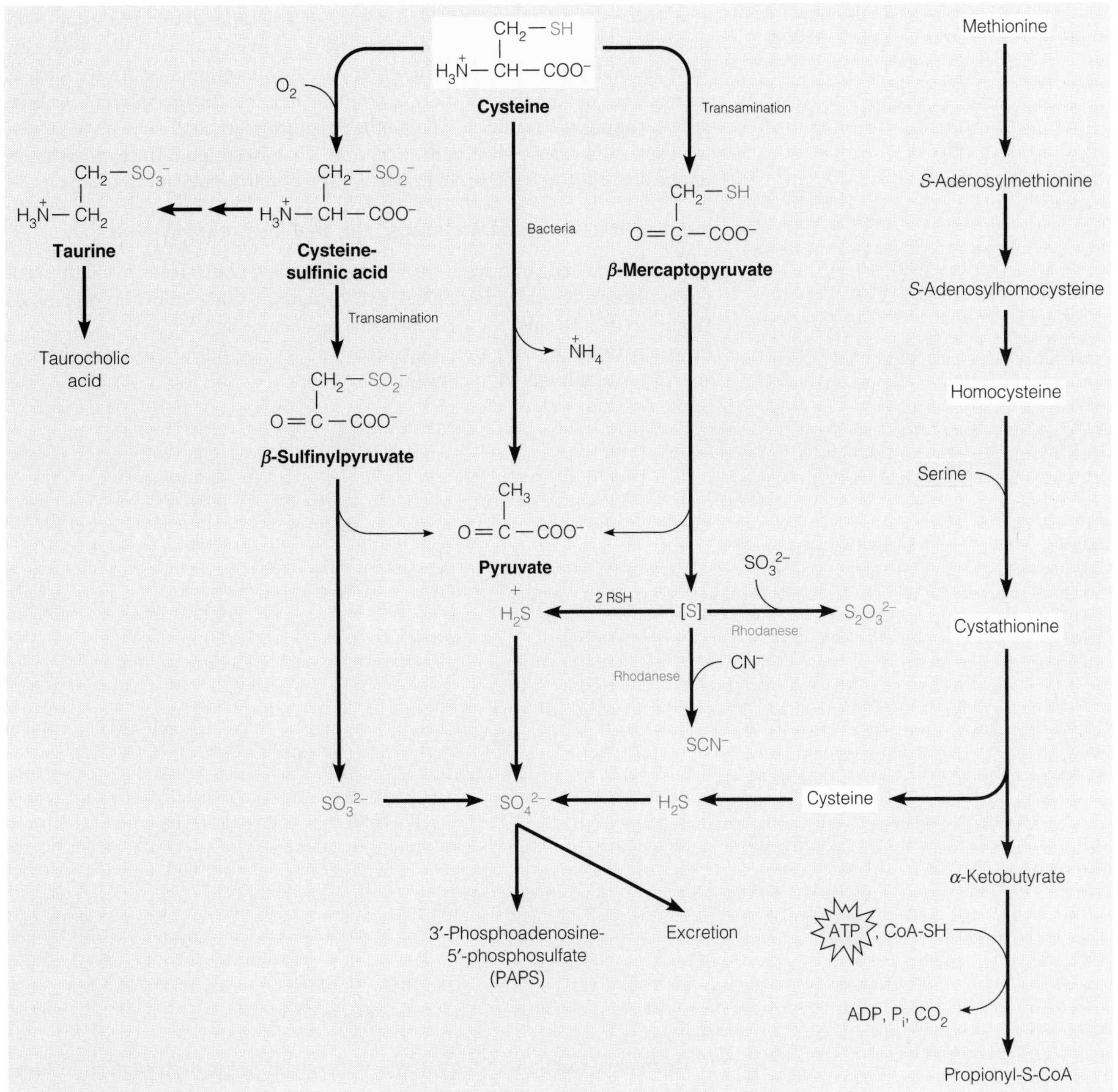

FIGURE **21.11**

Catabolic pathways for sulfur-containing amino acids. β-Mercaptopyruvate from cysteine degradation yields pyruvate. The various fates of the sulfur are shown, though the specific reactions involving the sulfur are not shown in detail.

undergo desulfuration by several routes, to give pyruvate and H_2S, sulfite (SO_3^{2-}), thiosulfate ($S_2O_3^{2-}$), or thiocyanate (SCN^-). A mitochondrial enzyme called **rhodanese** catalyzes several reactions, including the formation of thiocyanate from thiosulfate and cyanide ion. This reaction is involved in the detoxification of cyanide.

The catabolism of methionine proceeds largely through homocysteine, which is catabolized via cystathionine to give α-ketobutyrate. α-Ketobutyrate reacts with CoA and ATP to yield propionyl-CoA, which is then converted to succinyl-CoA by the B_{12}-requiring pathway we have discussed before. α-Ketobutyrate also has a biosynthetic role, in the synthesis of isoleucine.

Aromatic Amino Acids

In this section we consider the metabolism of phenylalanine, tyrosine, tryptophan, and histidine. Histidine is not usually classified among the aromatic amino acids, but because there are several similarities between its metabolism and that of the others, we shall consider it an "honorary aromatic" in our discussion. Synthesis of these heterocyclic and/or aromatic rings from noncyclic precursors involves complex chemistry. As with other lengthy biosynthetic pathways, such as the synthesis of vitamins, most of the aromatic biosynthetic capabilities have been lost during animal evolution. The synthetic pathways we shall discuss are limited to plants and bacteria—with one exception. That exception is the hydroxylation of phenylalanine to tyrosine, an important reaction in animal metabolism.

BIOSYNTHESIS OF AROMATIC RINGS: THE SHIKIMIC ACID PATHWAY

A single branched pathway in microorganisms and plants leads to synthesis of phenylalanine, tyrosine, tryptophan, and virtually all other aromatic compounds (Figure 21.12). Because these latter substances include **lignin,** a major constituent of woody tissue, it is apparent that a tremendous amount of carbon in the biosphere passes through this pathway.

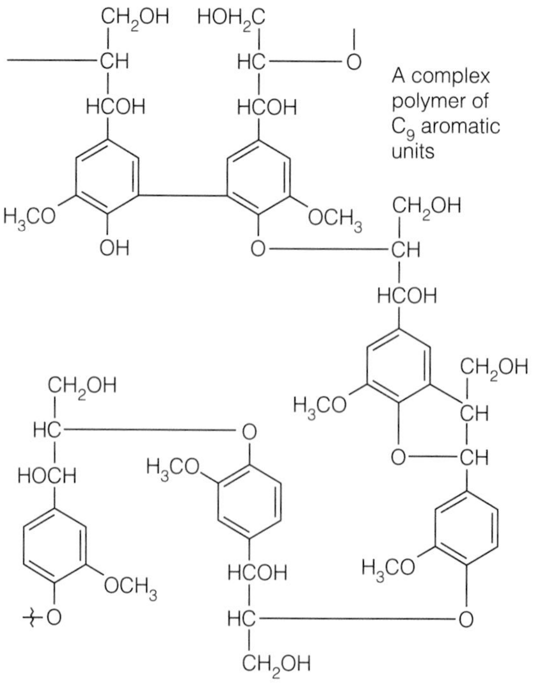

A complex polymer of C_9 aromatic units

Lignin

Individual reactions of the pathway were established in bacteria. A large number of auxotrophic mutants could be isolated and characterized genetically (map-

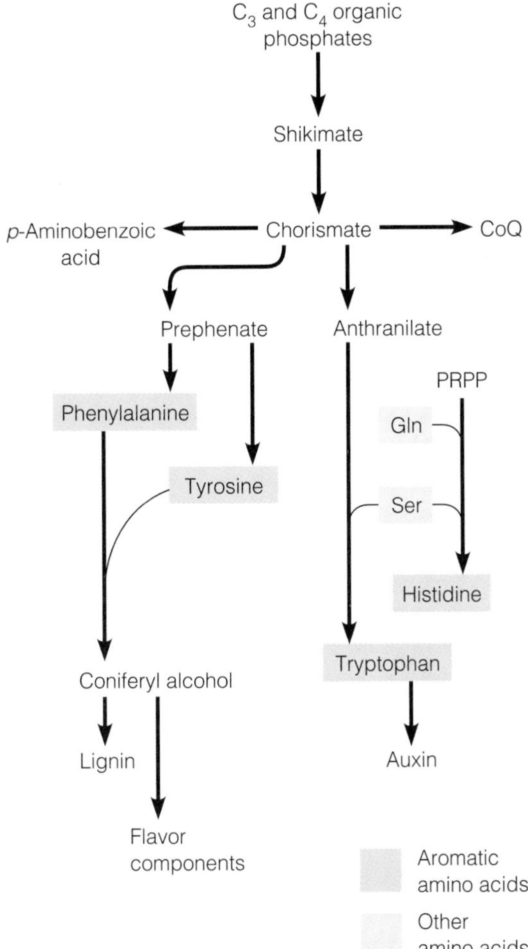

FIGURE **21.12**

Overview of the biosynthesis of aromatic amino acids and histidine. The central pathways leading to the four amino acids are essentially the same in plants and bacteria. Details of the shikimic acid pathway are presented in Figures 21.13 and 21.14, and histidine synthesis is shown in Figure 21.17. PRPP is 5-phosphoribosyl-1-pyrophosphate (Figure 21.14).

ping), physiologically (identification of compounds that could satisfy growth requirements), and biochemically (identification of intermediates accumulating when a given step is blocked). So far as is known, the processes are quite similar in plants.

The key findings were, first, that all of the carbon in phenylalanine and tyrosine was derived from erythrose-4-phosphate and phosphoenolpyruvate and, second, that a class of mutants requiring phenylalanine, tyrosine, tryptophan, *p*-aminobenzoic acid, and *p*-hydroxybenzoic acid for growth could have all five requirements met by a single compound—**shikimic acid.** We now know that these mutants are blocked in the formation of shikimate (the fourth reaction in the part of the pathway shown in Figure 21.13) and that, when it is provided, all of the subsequent steps in the pathway can occur. Note that an unbranched pathway leads through shikimic acid to **chorismic acid.** From chorismic acid one pathway leads to **prephenic acid,** with subsequent branches leading to phenylalanine and tyrosine. Another pathway leads through **anthranilic acid** to tryptophan, still another pathway leads to *p*-aminobenzoic acid, and a final pathway leads, via *p*-hydroxybenzoic acid, to coenzyme Q. Thus, the shikimic acid pathway is responsible for biosynthesis of virtually all aromatic compounds, because the products we have just mentioned serve in turn as precursors to other aromatic compounds.

The shikimic acid pathway leads to synthesis of nearly all aromatic compounds, including lignin, and is thus one of the most productive pathways in biology.

Great attention has focused in recent years on the sixth reaction of the shikimic acid pathway, catalyzed by **5-enoylpyruvylshikimate-3-phosphate synthase (EPSP synthase),** in higher plants. This enzyme is specifically inhibited by a compound called **glyphosate,** or glycine phosphonate (a *phosphonate* is a compound with a covalent bond linking carbon to phosphorus). The growth of most crop and weed

$$HO-\overset{\overset{O}{\|}}{\underset{\underset{O^-}{|}}{P}}-CH_2-NH-CH_2-COO^-$$

Glyphosate

FIGURE **21.13**

Details of the shikimic acid pathway, I. This figure depicts the initial unbranched pathway from erythrose-4-phosphate and phosphoenolpyruvate to chorismic acid. The first reaction is driven by loss of phosphate from phosphoenolpyruvate. In the second reaction an unusual cobalt-requiring enzyme effects ring closure with dehydrogenation and loss of the second phosphate. Dehydration in the third step yields dehydroshikimate, which is then reduced by NADPH to shikimate. A three-carbon side chain is then attached via phosphorylation of shikimate and reaction with a second molecule of phosphoenolpyruvate. Dephosphorylation of this intermediate gives chorismate, the branch point of the pathway.

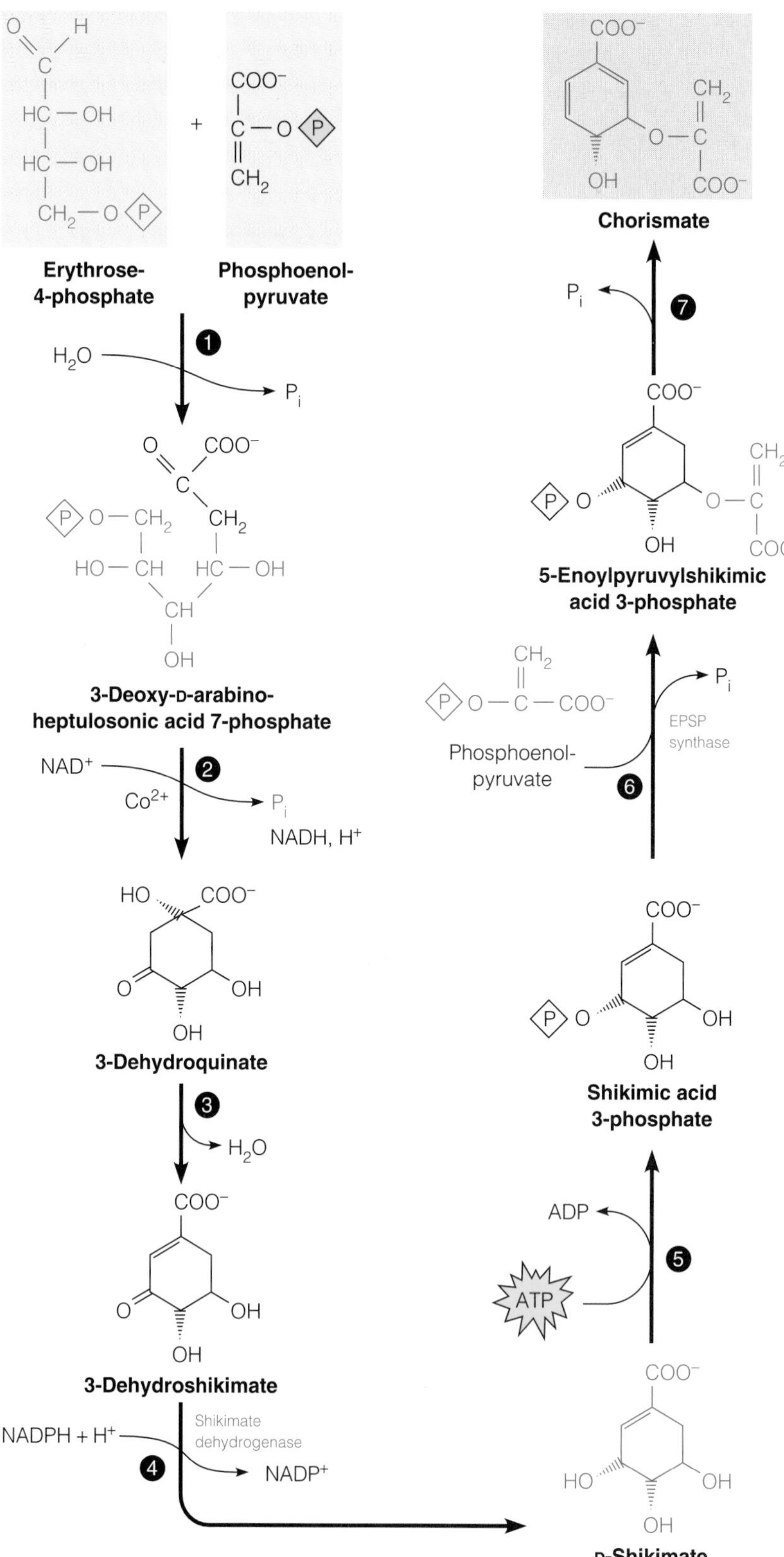

plants is inhibited by glyphosate, which is an effective broad-spectrum herbicide sold as Roundup. A recent achievement of biotechnology is the transfer of genes conferring resistance to glyphosate into crop plants. For example, "Roundup Ready" cotton seed was released to farmers in the mid-1990s, and other crop plants, similarly modified, are now on the market. This modification allows simplified and effective weed control. Spraying a field should eliminate all plants except the genetically engineered species.

Interesting control mechanisms are involved in regulating these pathways. Studies on the genetic control of tryptophan biosynthesis, described in Chapter 26, have provided some of our most important insights into transcriptional regulation. More recent studies on the reactions leading to chorismic acid have revealed the existence of several multifunctional enzymes—single polypeptide chains containing two or more active sites for catalysis of sequential reactions—obviously, an efficient way to control several reactions jointly.

The first reaction in the pathway from anthranilic acid to tryptophan (Figure 21.14) involves an activated sugar derivative, **5-phosphoribosyl-1-pyrophosphate (PRPP),** which plays its most widespread roles in nucleotide synthesis (see Chapter 22). In bacteria the genes encoding these enzymes are linked in a linear array, the **tryptophan operon.** As discussed in Chapter 26, an *operon* is a linked set of genes whose expression is regulated jointly at the level of transcription. In enteric bacteria the final enzyme in the pathway, **tryptophan synthase,** is an $\alpha_2\beta_2$ dimer. Isolated α and β subunits catalyze the following partial reactions, with the holoenzyme catalyzing a concerted reaction in which indole does not dissociate from the enzyme surface but immediately reacts with serine to give tryptophan.

$$\text{Indole-3-glycerol phosphate} \xrightarrow{\alpha \text{ subunit}} \text{indole} + \text{3-phosphoglyceraldehyde}$$

$$\text{Indole} + \text{serine} \xrightarrow{\beta \text{ subunit}} \text{tryptophan} + H_2O$$

Remarkably, x-ray crystallography shows that the intermediate, indole, is transported from the α subunit active site to the β active site, a distance of 2.5 nm, through a tunnel in the interior of the protein molecule (Figure 21.15). Kinetic analysis, complementing more detailed structural analysis, suggests that each active site is covered intermittently by a "lid" (Figure 21.16), which acts to keep indole within the channel.

BIOSYNTHESIS OF HISTIDINE

The biosynthesis of histidine presents several parallels with the shikimic acid pathway, regarding methods used for its elucidation, complexity of the reactions involved, elegance of genetic pathway regulation, and practical applications of knowledge of the pathway. However, the histidine pathway is distinctive in being unbranched. As established largely in the laboratories of Bruce Ames and Philip Hartman, the pathway is shown in Figure 21.17. Ten individual reactions are involved, starting with an unusual reaction that joins ATP and PRPP; both compounds contribute carbon to the product.

The 10 structural genes for the enzymes of histidine synthesis in enteric bacteria are linked to one another in the same order as the order of the reactions of the pathway. This set of genes, the **histidine operon,** is coordinately regulated at the transcriptional level, and all 10 genes are transcribed to give one large messenger RNA, which is translated to give the 10 enzymes. This highly organized gene arrangement may facilitate regulation of the pathway.

Once the genes and gene products had been identified, Bruce Ames used the mutant bacteria that had been generated for these investigations in a novel way—to search for mutagens in the environment. Using the **Ames test,** researchers can count mutations simply by measuring the rate at which mutants that cannot synthesize histidine (histidine **auxotrophs**) mutate to a form that can synthesize the amino acid (**prototrophs**). The researcher treats a culture of auxotrophs with a suspected mutagen, plates the bacteria on medium containing no histidine, and counts the colonies that appear as the result of reversion mutations. Using this system, Ames and colleagues reported a very high correlation between compounds known to be carcinogenic in animals and those found to be mutagens in this test.

Histidine auxotrophic mutations have been useful both for defining the biosynthetic pathway and for analyzing environmental mutagenesis.

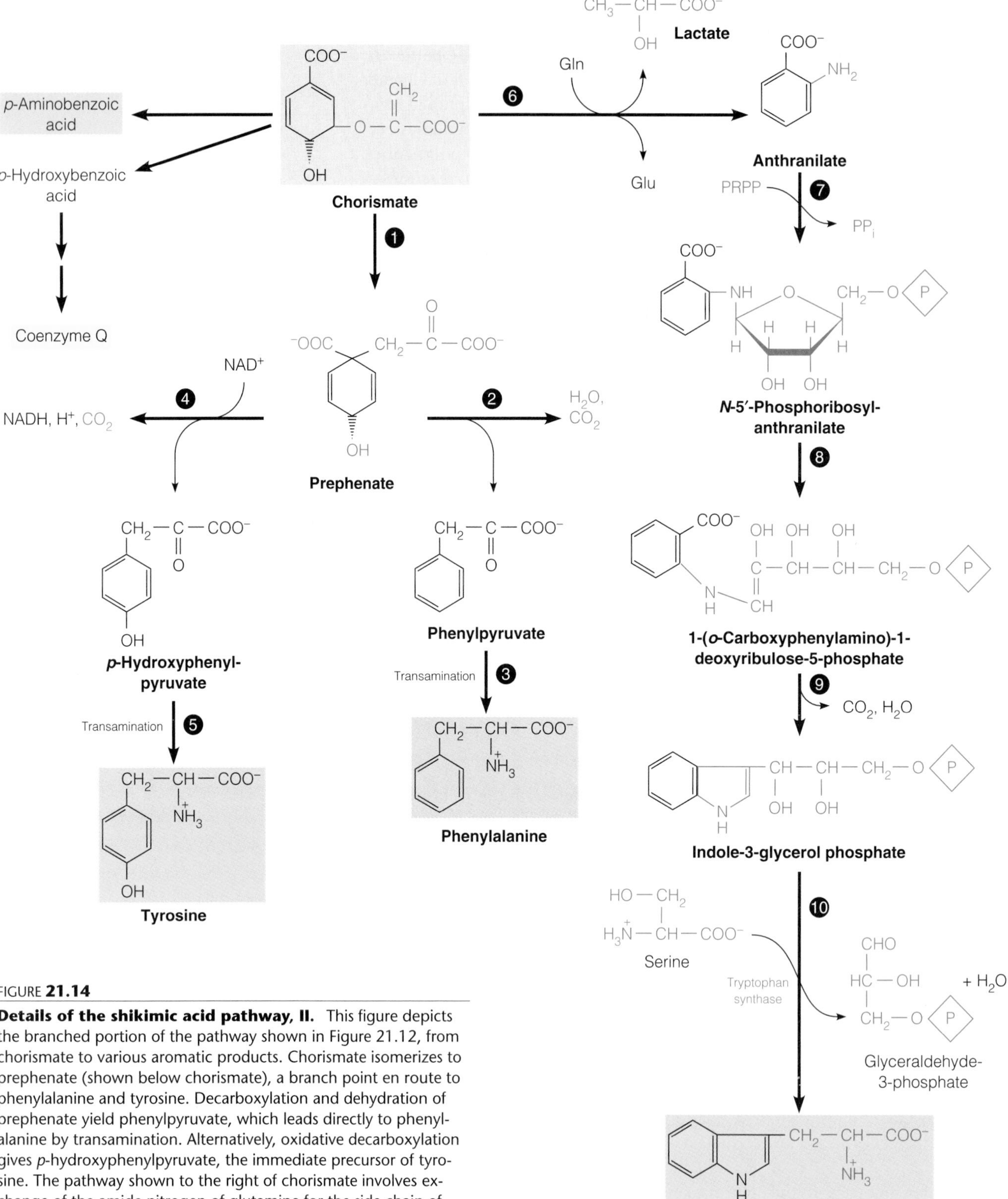

FIGURE **21.14**

Details of the shikimic acid pathway, II. This figure depicts the branched portion of the pathway shown in Figure 21.12, from chorismate to various aromatic products. Chorismate isomerizes to prephenate (shown below chorismate), a branch point en route to phenylalanine and tyrosine. Decarboxylation and dehydration of prephenate yield phenylpyruvate, which leads directly to phenylalanine by transamination. Alternatively, oxidative decarboxylation gives *p*-hydroxyphenylpyruvate, the immediate precursor of tyrosine. The pathway shown to the right of chorismate involves exchange of the amide nitrogen of glutamine for the side chain of chorismate. The enzyme, anthranilate synthetase, catalyzes the first committed step en route to tryptophan. In the next step, carbon 1 of 5-phosphoribosyl-1-pyrophosphate (PRPP) is linked to the nitrogen of anthranilate, driven by loss of pyrophosphate from carbon 1. Opening of the sugar ring is followed by decarboxylation and ring closure to give indole-3-glycerol phosphate. In the final step, the three-carbon side chain of the indole compound is exchanged for that of serine, yielding tryptophan.

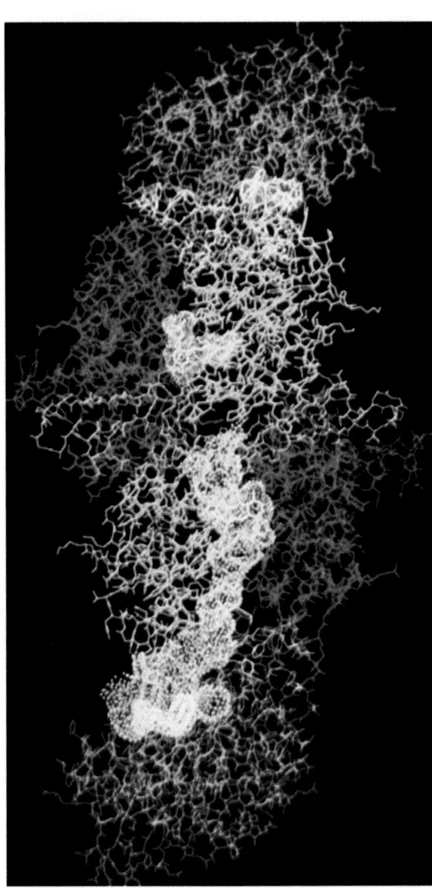

FIGURE **21.15**

Structure of tryptophan synthase. This model shows the structure of the $\alpha_2\beta_2$ tryptophan synthase complex from *Salmonella typhimurium*. The α subunits are shown in blue, the N-terminal domain of the β subunits in yellow, and the β C-terminal domains in red. Light blue shows an interior tunnel that connects the active sites on the lower α and β subunits. The active sites on the upper α and β subunits are shown in red and light blue, respectively, and the tunnel connecting these subunits is not highlighted.

Courtesy of C. C. Hyde et al., *J. Biol. Chem.* (1988) 263:17857–17871. Reprinted with permission from the American Society for Biochemistry and Molecular Biology.

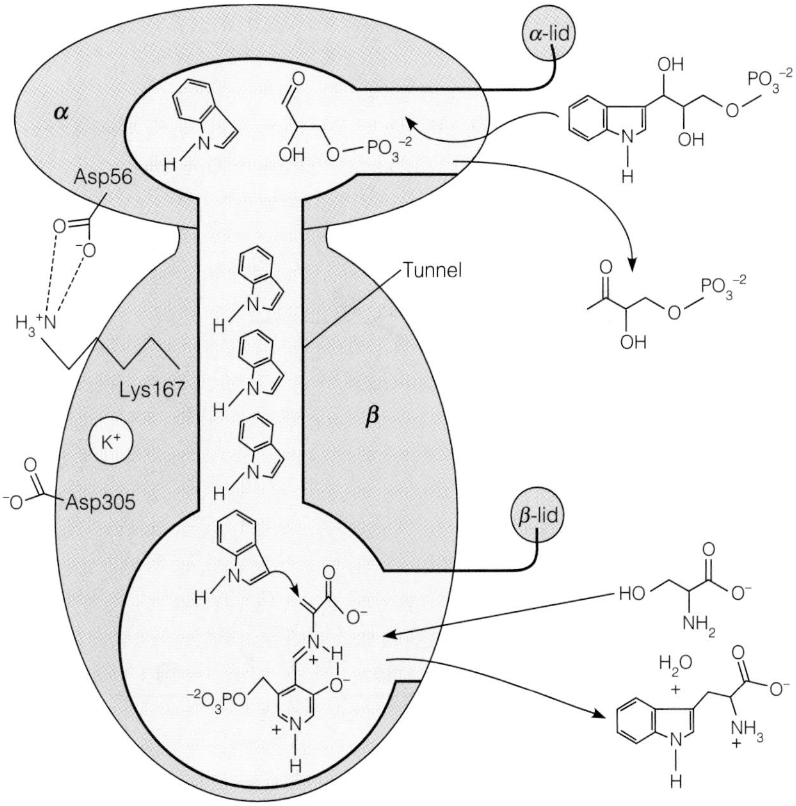

FIGURE **21.16**

Schematic depiction of the action of tryptophan synthase in one α–β subassembly. Shown are the active sites of the α (red) and β (orange) subunits, the interconnecting 2.5-nm tunnel filled with indole molecules, and the routes of substrate entry and product release for both subunits, with the covering "lids," whose alternate opening and closing restrain indole within the tunnel. Shown also are a binding site for the catalytically essential K^+ and a salt bridge known to provide allosteric linkage between the two subunits.

Courtesy of P. Pan, E. Woehl, and M. F. Dunn, *Trends in Biochem. Sci.* (1997) 22:22–27. © 1997 with permission of Elsevier Science.

Thus, the Ames test provides a quick and inexpensive way to search for suspected carcinogens in the environment. Moreover, these findings provided support for the idea that cancer arises as the result of a series of somatic cell mutations, an idea we discuss further in Chapter 23.

AROMATIC AMINO ACID UTILIZATION IN PLANTS

Phenylalanine and tyrosine serve as precursors to an enormous number of plant substances, ranging from the polymeric lignin to tannins, pigments, and many of the flavor components of spices. In fact, the role of these amino acids as precursors to such substances in cinnamon oil, wintergreen oil, bitter almond, nutmeg, cayenne pepper, vanilla bean, clove, and ginger is related to their designation as *aromatic* amino acids. These are derived from coniferyl alcohol, which is also the central intermediate in lignin synthesis.

Coniferyl alcohol

Phenylalanine also serves as the precursor to a large number of plant pigments and related polyphenolic compounds called **flavonoids.** These include many flower colorants, which serve in part as ultraviolet protectants, and also the respiratory inhibitor rotenone (see page 534). In the generic structure shown in

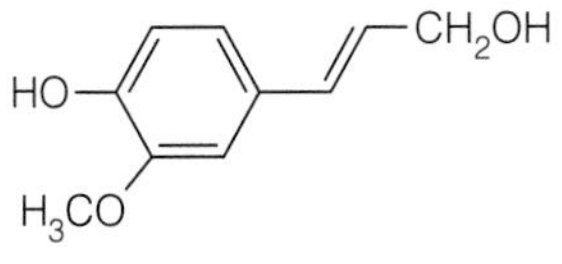

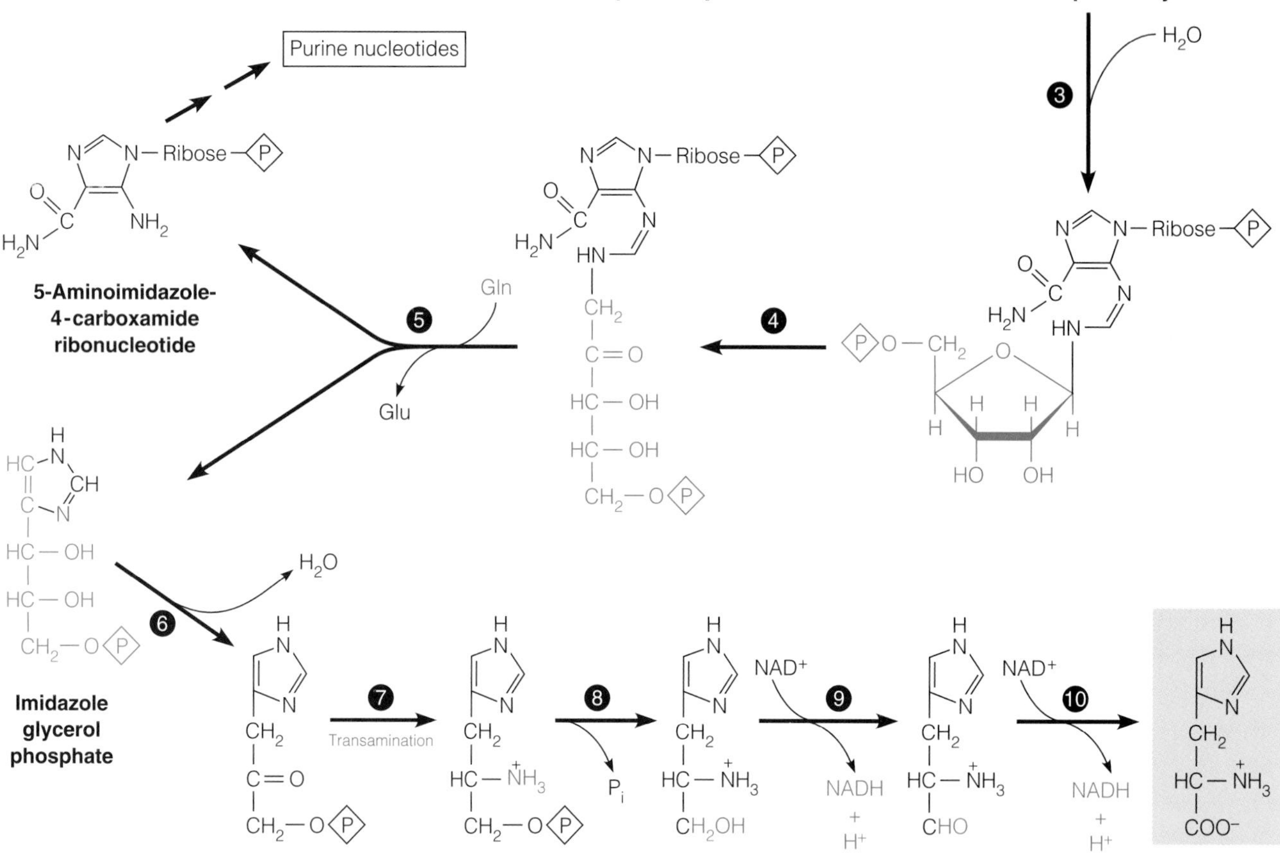

FIGURE **21.17**

The pathway for histidine biosynthesis. After activation of a purine ring by reaction with 5-phosphoribosyl-1-pyrophosphate (PRPP, step 1), the ring opens to give the third intermediate (steps 2 and 3). The ribose ring derived from PRPP then opens (step 4). The transfer of amide nitrogen from glutamine (step 5) is accompanied by cleavage and ring closure, giving the first imidazole compound, imidazole glycerol phosphate. The other product is an intermediate in the synthesis of purine nucleotides. Imidazole glycerol phosphate is transformed to histidine by a straightforward sequence involving dehydration (step 6), transamination (step 7), dephosphorylation (step 8), and dehydrogenation (steps 9 and 10).

the margin (this page), the aromatic ring to the right comes from phenylalanine via the scheme shown on the next page, and the ring to the left comes from malonyl-CoA, in a process akin to fatty acid and polyketide synthesis (Chapter 18). The substituent groups (R) are combinations of —H, —OH, and —OCH_3.

The biosynthetic scheme outlined at the end of the paragraph leads to a class of flavonoids called **anthocyanins,** which are common flower pigments. Substituents on the rings determine specific color, as shown. An offshoot from this pathway leads to synthesis of cocaine.

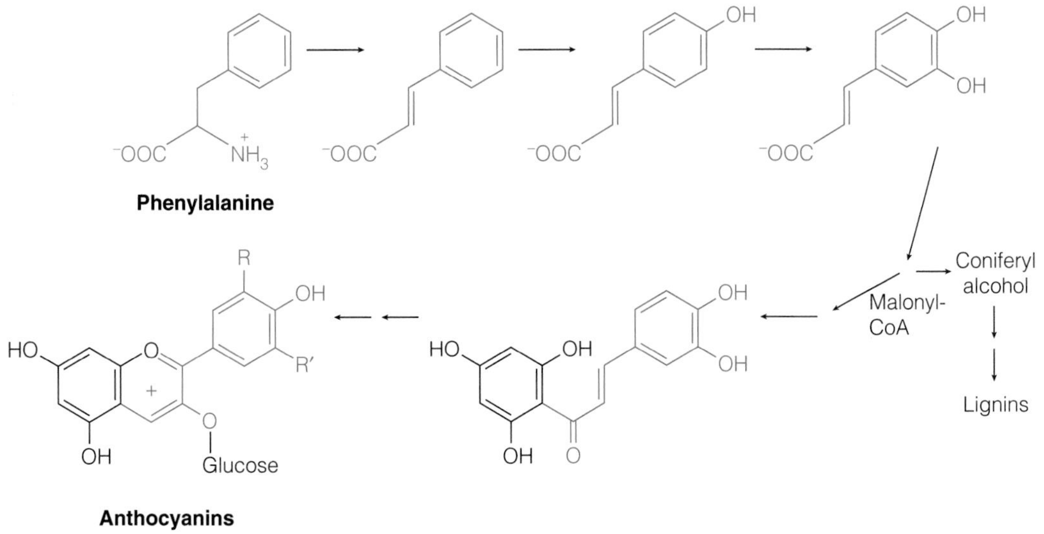

Phenylalanine

Coniferyl alcohol

Malonyl-CoA

Lignins

Glucose

Anthocyanins

If R and R′ are both H, color is orange-red
If R is H and R′ is OH, color is crimson-purple
If R and R′ are both OH, color is blue

As noted on page 760, coniferyl alcohol is the precursor to the complex and nearly inert lignins. Much of the effort in creating paper and textiles from wood involves degrading lignin to gain access to the cellulose fibers contained in woody tissues. An intriguing use of biotechnology involves the substitution of organisms for the pulping processes used to break down lignin, which release large quantities of sulfites to the environment. In these "biopulping" applications, fungi that produce lignin-degrading enzymes are being developed for production of cellulose without the accumulation of chemical pollutants.

Tryptophan is utilized for synthesis of a plant growth hormone. As shown here, the transamination product of tryptophan yields **indole-3-acetic acid,** or **auxin.**

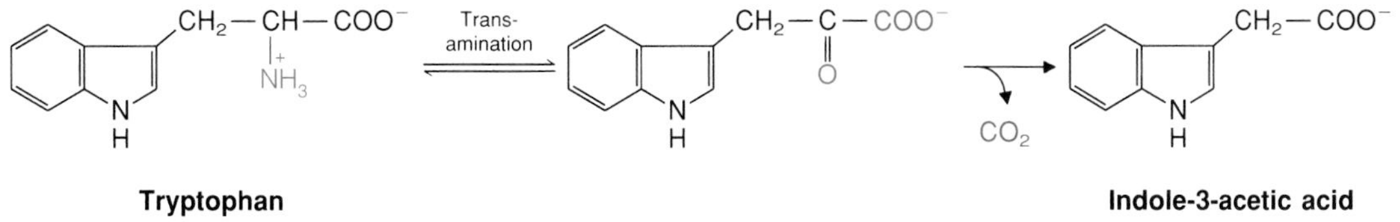

Trans-amination

CO_2

Tryptophan

Indole-3-acetic acid

AROMATIC AMINO ACID METABOLISM IN ANIMALS

Although animal cells do not synthesize aromatic rings de novo, the intermediary metabolism of aromatic amino acids in animals is extensive, and it involves many important biological processes. These include the biosynthesis of tyrosine from phenylalanine, the utilization of tyrosine in synthesis of pigments and hormones, and the utilization of tyrosine, tryptophan, and histidine in synthesis of **biogenic amines,** compounds that serve as hormones and as neurotransmitters. Aromatic amino acid metabolism also provides the backdrop for much of our early understanding of human biochemical genetics.

Tyrosine Biosynthesis

The only known reaction of aromatic amino acid biosynthesis in animals is the conversion of phenylalanine to tyrosine, catalyzed by **phenylalanine hydroxylase.** This interesting enzyme is a mixed-function oxygenase that uses a pteridine

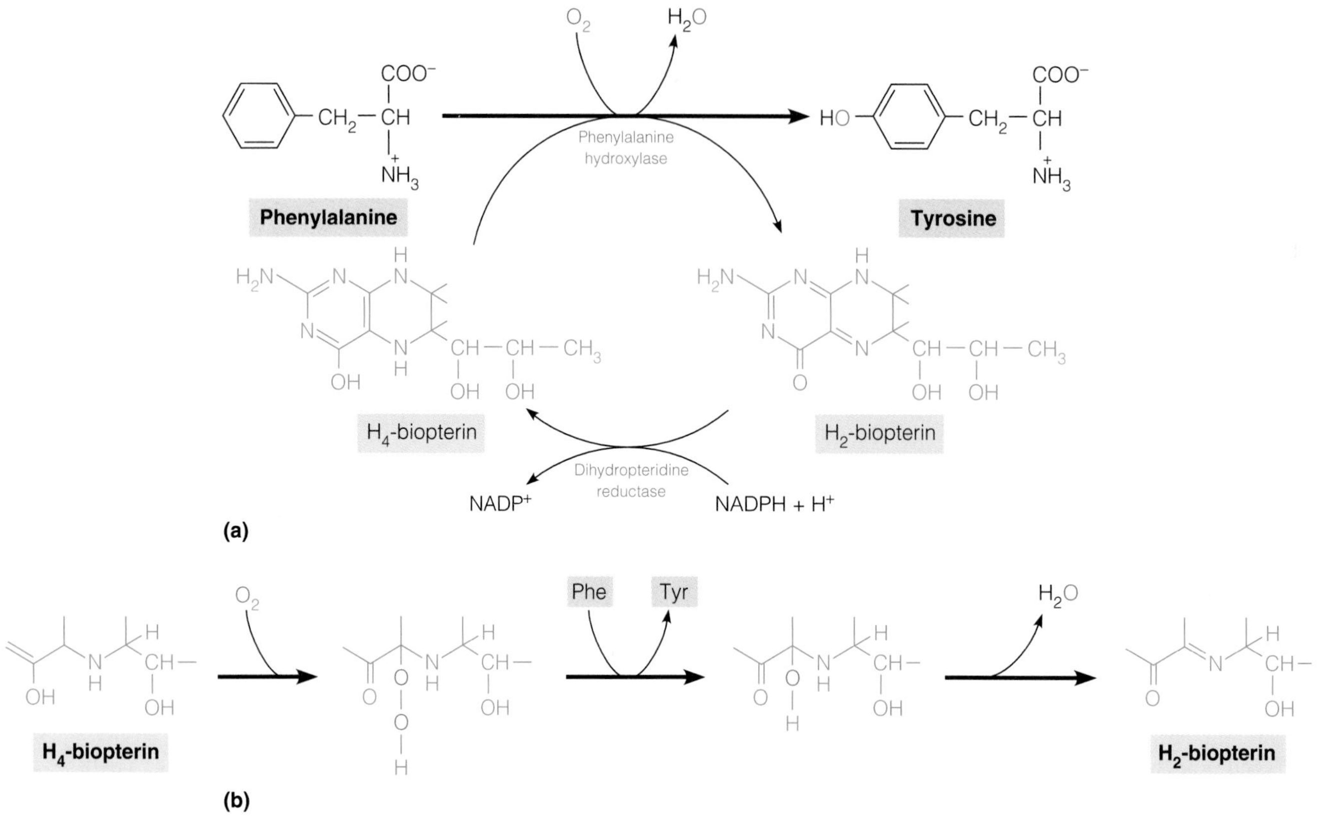

FIGURE **21.18**

The phenylalanine hydroxylase and dihydropteridine reductase reactions. **(a)** Conversion of phenylalanine to tyrosine is catalyzed by phenylalanine hydroxylase, accompanied by oxidation of the cofactor, tetrahydrobiopterin, to dihydrobiopterin. Tetrahydrobiopterin is regenerated by the enzyme dihydrobiopterin reductase. **(b)** A scheme for O_2 activation by tetrahydrobiopterin.

cofactor, **tetrahydrobiopterin** (which was introduced on page 749). The reaction oxidizes tetrahydrobiopterin to the quinonoid isomer of dihydrobiopterin, as shown in Figure 21.18. The coenzyme is regenerated through the action of the NADPH-requiring **dihydropteridine reductase** (analogous, but not identical, to dihydrofolate reductase). This enzyme system occurs almost entirely in the liver.

Phenylketonuria, a severe mental deficiency, arises from genetic insufficiency of phenylalanine hydroxylase or, rarely, dihydropteridine reductase.

A hereditary deficiency of phenylalanine hydroxylase is responsible for **phenylketonuria (PKU)**, a condition that afflicts about 1 in 10,000 newborn infants in western Europe and the United States. PKU is an autosomal recessive trait, meaning that two parents heterozygous for the trait have 1 chance in 4 of having a phenylketonuric child. From the incidence of the disease, we can estimate that about 2% of the population are carriers. In phenylketonuria, phenylalanine accumulates to very high levels because of the block in conversion to tyrosine, and much of this phenylalanine is metabolized via pathways that are normally little used—particularly transamination to phenylpyruvate, and also subsequent conversion of phenylpyruvate to phenyllactate and phenylacetate. These compounds are excreted in urine in enormous quantities (1 to 2 grams per day).

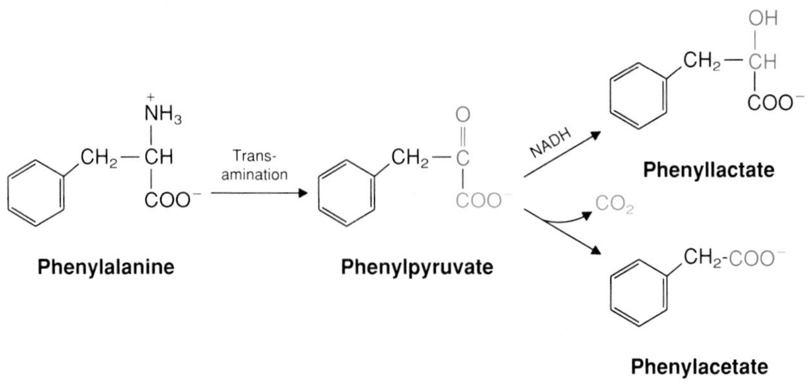

FIGURE **21.19**

Biosynthesis of thyroid hormones as residues in the protein thyroglobulin. The iodinated forms of tyrosine—triiodothyronine (T_3) and thyroxine (T_4)—are released from these proteins by proteolytic degradation.

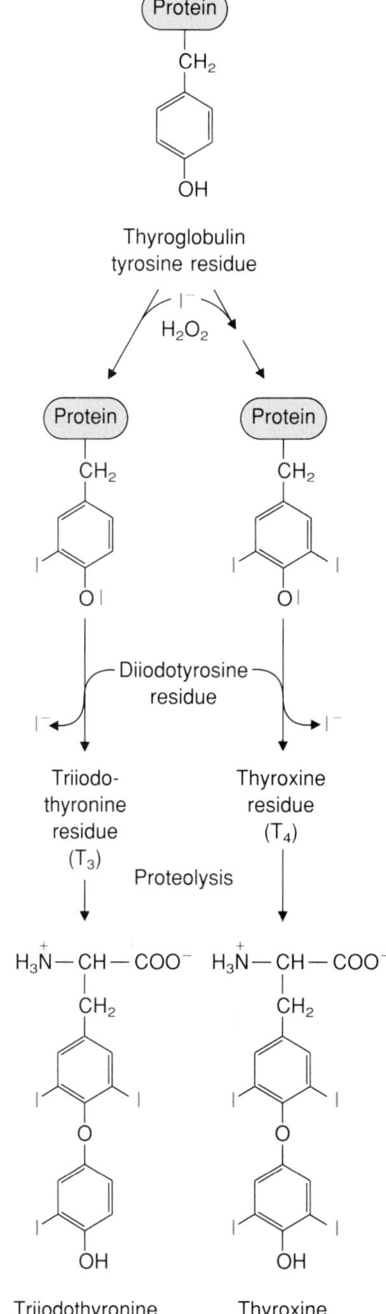

If undetected and untreated, PKU leads to profound mental retardation; the precise biochemical causation has not yet been identified. Fortunately, PKU can readily be detected at birth, and many hospitals carry out routine screening of newborns. If the condition is detected early, the onset of retardation can be prevented by feeding for several years a synthetic diet low in phenylalanine and rich in tyrosine, to allow normal development of the nervous system. Because the use of this synthetic diet is quite expensive, there has been much interest in prenatal diagnosis of PKU and in identification of heterozygous carriers. The human gene for phenylalanine hydroxylase has been cloned, providing the basis for a test, using nucleic acid hybridization, that can be applied in any cell.

In recent years a different form of PKU has been described that results from a hereditary deficiency of dihydropteridine reductase. This condition is much rarer than classical PKU and, because tetrahydrobiopterin is involved in other hydroxylations, much more severe in its symptoms.

Tyrosine Utilization and Catabolism

In animals, phenylalanine plays no significant role other than as a component of proteins and a precursor to tyrosine. However, tyrosine plays several important roles in animal metabolism—as precursor to thyroid hormones, to the biological pigments called **melanins,** and to the *catecholamines,* which serve both as hormones and as neurotransmitters (discussed later in this chapter).

Thyroid hormones stimulate a number of metabolic processes, through activation of the transcription of particular genes (see Chapter 23). The synthesis of thyroid hormones, principally **thyroxine** (T_4) and **triiodothyronine** (T_3), occurs by a novel pathway, at the level of tyrosine residues in a specific protein, **thyroglobulin.** As shown in Figure 21.19, two iodinated tyrosine residues on the same polypeptide chain react to give a residue of T_3 or T_4, which is released by proteolysis. This protein undergoes degradation to yield the free hormones, which are transported to their sites of action through the bloodstream. The synthesis of thyroid hormones involves iodination of the tyrosine ring. This process occurs in the thyroid gland, which concentrates iodide ion from the blood serum for this purpose. One result of iodine deficiency is **goiter,** a condition in which the thyroid gland grows abnormally large as it attempts to scavenge all available iodine. Before iodized salt came into widespread use, goiter was endemic in regions whose soil was deficient in iodine.

The synthesis of melanins (Figure 21.20) occurs in pigment-producing cells, the **melanocytes.** For some time it was thought that the entire pathway involves just one enzyme. However, in mammalian tissues, it is likely that the initial step, the hydroxylation of tyrosine to **3,4-dihydroxyphenylalanine** (usually called by the acronym **dopa**), involves a peroxidase reaction. Dopa, which also acts as a cofactor for this reaction, is the substrate for a subsequent oxidation to **dopaquinone,** catalyzed by a copper-containing oxygenase called tyrosinase. The subsequent reactions, leading to melanins, occur spontaneously in vitro. However, tyrosinase has been shown to catalyze another reaction, on the branch of the pathway leading to **black melanin**—the oxidation of 5,6-dihydroxyindole to indole-5,6-quinone. This reaction also requires dopa as a cofactor and is inhibited by tyrosine, apparently providing a means for regulation of pigment formation.

In another branch of the pathway, dopaquinone reacts with cysteine en route to a related series of polymers, the **red melanins.** An individual's pigmentation is determined by the relative amounts of red and black melanins in the skin. These in turn result from the distribution and density of melanocytes in the basal layers of the skin, as well as the activities of the pathways leading to the different melanins. A genetic deficiency of tyrosinase causes an individual to lack pigmentation, a condition called **albinism.**

The principal catabolic route for tyrosine involves first its transamination by **tyrosine aminotransferase,** an enzyme in the liver whose level is regulated hormonally (Figure 21.21). The product, *p*-hydroxyphenylpyruvate, is acted on by ***p*-hydroxyphenylpyruvate dioxygenase,** an unusual iron-containing enzyme, which catalyzes a ring hydroxylation, decarboxylation, and side chain migration, using ascorbate as a cofactor, in a similar fashion to its use with procollagen proline hydroxylase (see page 751). This reaction involves a mechanism called the **NIH shift,** after scientists at the National Institutes of Health, who described a ring hydroxylation that proceeds via formation of an epoxide intermediate.

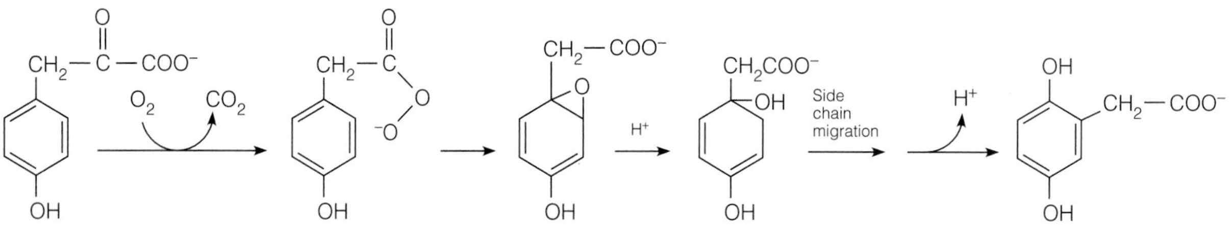

***p*-Hydroxyphenylpyruvate** → **Homogentisate**

The phenylalanine hydroxylase reaction also involves an NIH shift, with migration of the hydrogen at C-4 of phenylalanine to C-3 of tyrosine, without mixing with solvent.

The product of *p*-hydroxyphenylpyruvate oxidation, **homogentisic acid,** is oxidized by an iron-containing enzyme, **homogentisic acid dioxygenase,** that cleaves the ring to yield a straight-chain eight-carbon compound that isomerizes to **fumarylacetoacetate.** The latter ultimately cleaves to yield fumarate and acetoacetate, both of which are catabolized by standard energy-yielding pathways. In plants, homogentisate is the precursor to the aromatic ring portion of vitamin E (see page 698).

A hereditary deficiency of the enzyme homogentisic acid dioxygenase in humans causes a condition that was known for centuries as the "dark urine disease" but is now called **alkaptonuria.** Homogentisic acid accumulates and is excreted in large amounts in the urine; its oxidation on standing causes the urine to become dark. Although the clinical symptoms of the disease are not severe, it is of considerable historical interest. Early in the twentieth century Archibald Garrod examined pedigrees of the families of afflicted individuals, and in 1908 he proposed correctly that the deficiency was due to a gene controlling a normal reaction in the breakdown of aromatic compounds. In other words, he proposed that one gene encodes one enzyme, long before the chemical nature of either genes or enzymes was known.

Tryptophan

Tryptophan is transformed by many pathways, of which just two concern us here—the major catabolic route, proceeding via **kynurenine** to glutaryl-CoA (Figure 21.22), and the synthesis of nicotinamide nucleotides by a branch from this major catabolic pathway. The first reaction in the degradation of tryptophan is catalyzed by **tryptophan oxygenase,** an iron heme protein whose level is controlled by two mechanisms: (1) induction by certain hormones and (2) stabilization in vivo by its substrate, tryptophan, which increases enzyme levels by protecting the protein against intracellular degradation.

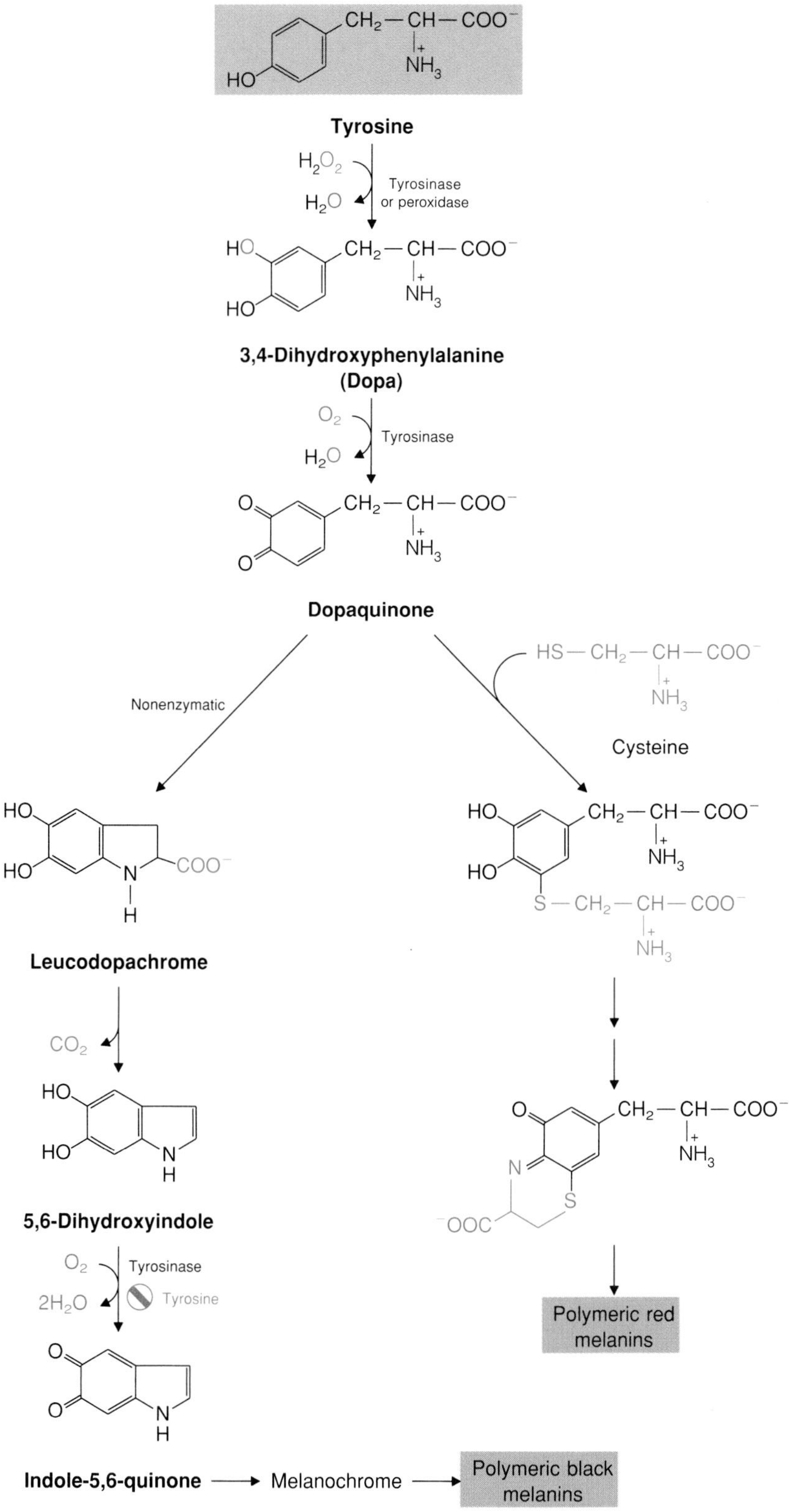

Tyrosine

H_2O_2 → H_2O (Tyrosinase or peroxidase)

3,4-Dihydroxyphenylalanine (Dopa)

O_2 → H_2O (Tyrosinase)

Dopaquinone

Nonenzymatic

Cysteine

Leucodopachrome

CO_2

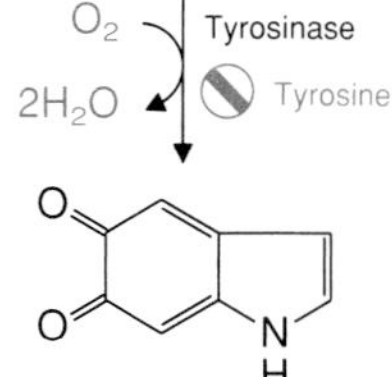

5,6-Dihydroxyindole

O_2 → $2H_2O$ (Tyrosinase; Tyrosine)

Indole-5,6-quinone → Melanochrome → Polymeric black melanins

Polymeric red melanins

FIGURE **21.20**

Biosynthetic pathways from tyrosine to melanins.

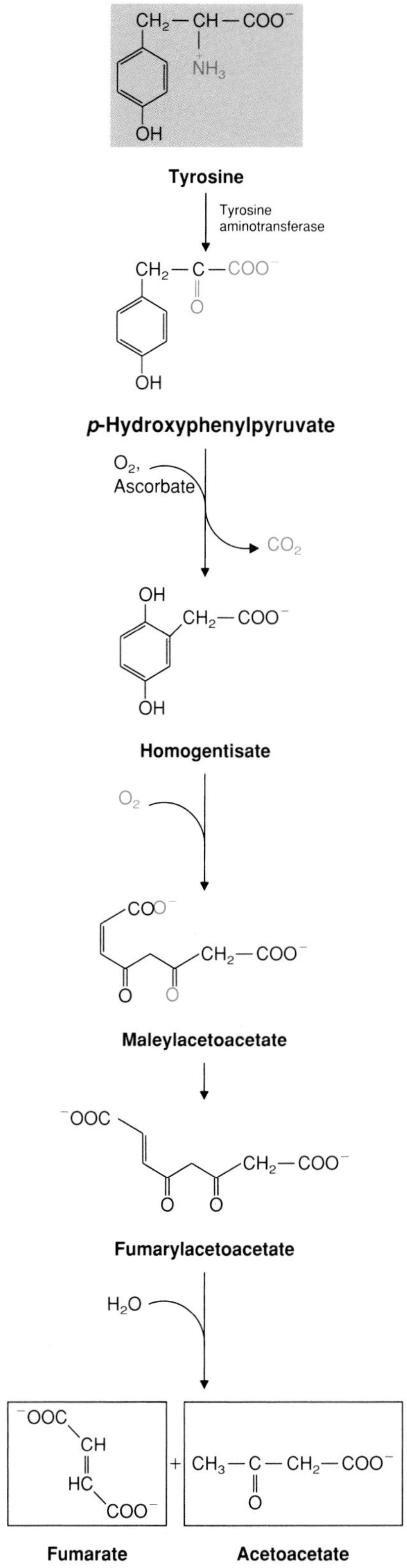

Tyrosine

Tyrosine aminotransferase

p-Hydroxyphenylpyruvate

O_2, Ascorbate → CO_2

Homogentisate

O_2

Maleylacetoacetate

Fumarylacetoacetate

H_2O

Fumarate + Acetoacetate

FIGURE **21.21**

Catabolism of tyrosine to fumarate and acetoacetate.

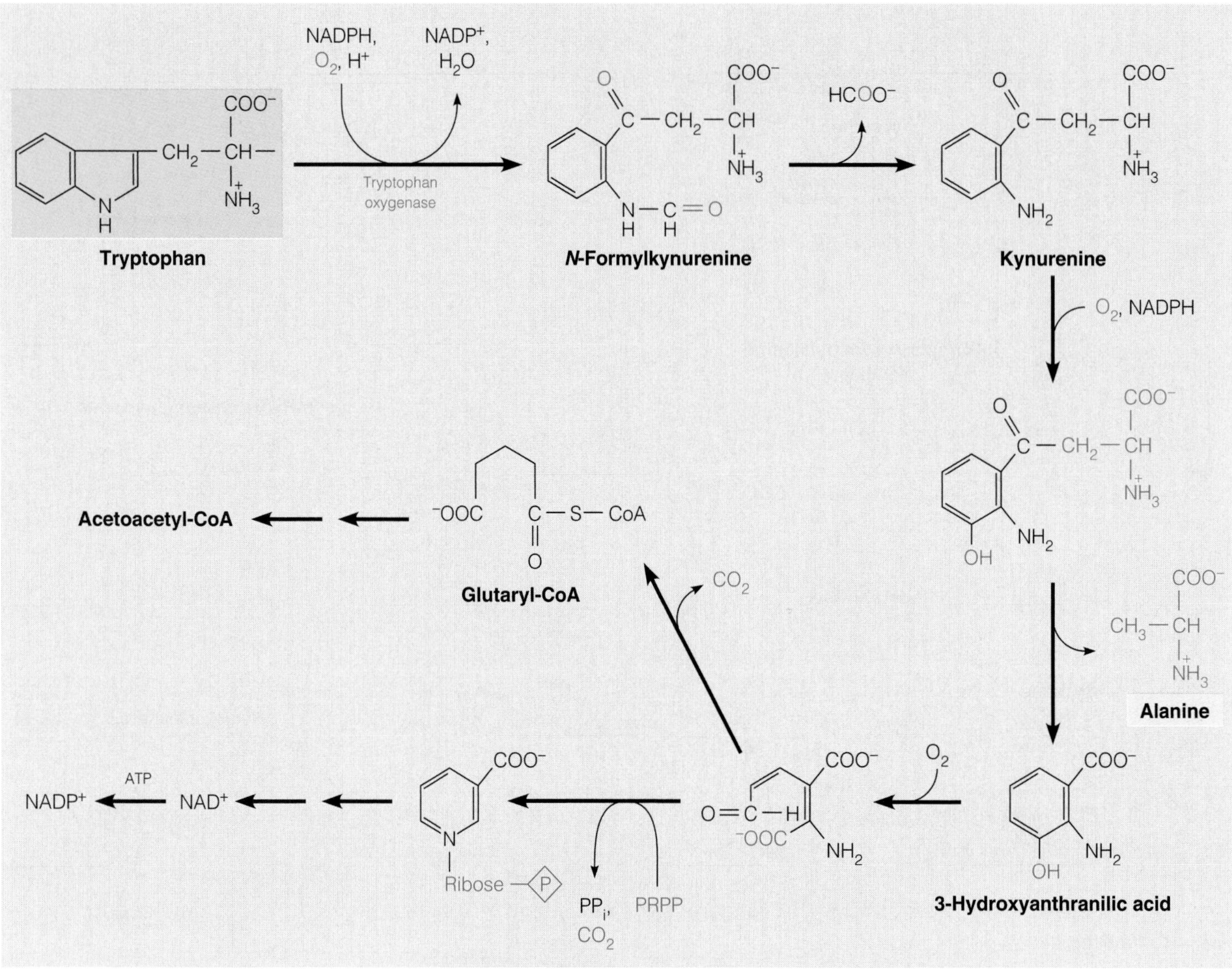

FIGURE **21.22**

Metabolic fates of tryptophan. The figure shows two major pathways—the synthetic pathway to the nicotinamide nucleotides and the major catabolic pathway that degrades most of the tryptophan molecule to acetoacetyl-CoA and CO_2.

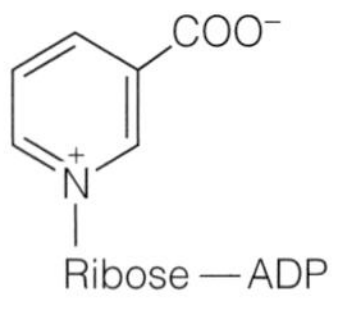

NAD^+ can be synthesized either from tryptophan or from the vitamin nicotinic acid (see Table 11.5 on page 389). For both pathways the last reaction is catalyzed by a glutamine-requiring amidotransferase (see margin). Both pathways probably contribute significantly to NAD^+ biosynthesis, as can be inferred from studies of the nicotinamide deficiency disease **pellagra.** Pellagra was formerly endemic in regions, such as the southern United States, where corn is a dietary staple. Because corn proteins contain little tryptophan, deficiencies of this amino acid were common. The symptoms of tryptophan deficiency are identical to those of nicotinamide deficiency, as expected if a major role of tryptophan is as a nicotinamide substitute.

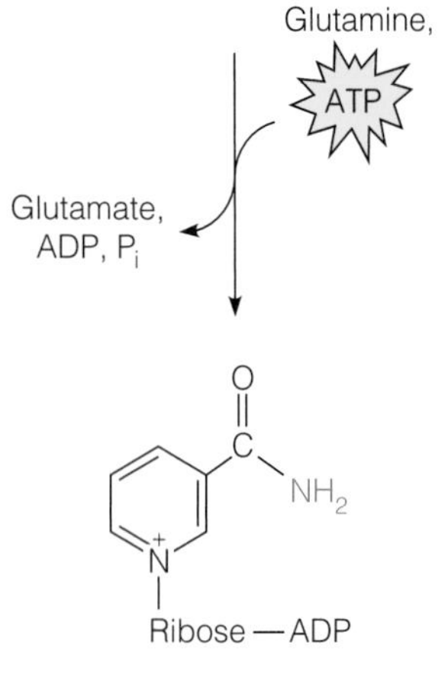

Histidine

Histidine undergoes decarboxylation to generate **histamine,** a substance with multiple biological actions. When secreted in the stomach, histamine promotes the secretion of hydrochloric acid and pepsin, both of which aid digestion. It is a potent vasodilator, released locally in sites of trauma, inflammation, or allergic reaction. The local enlargement of blood capillaries is the basis for the reddening

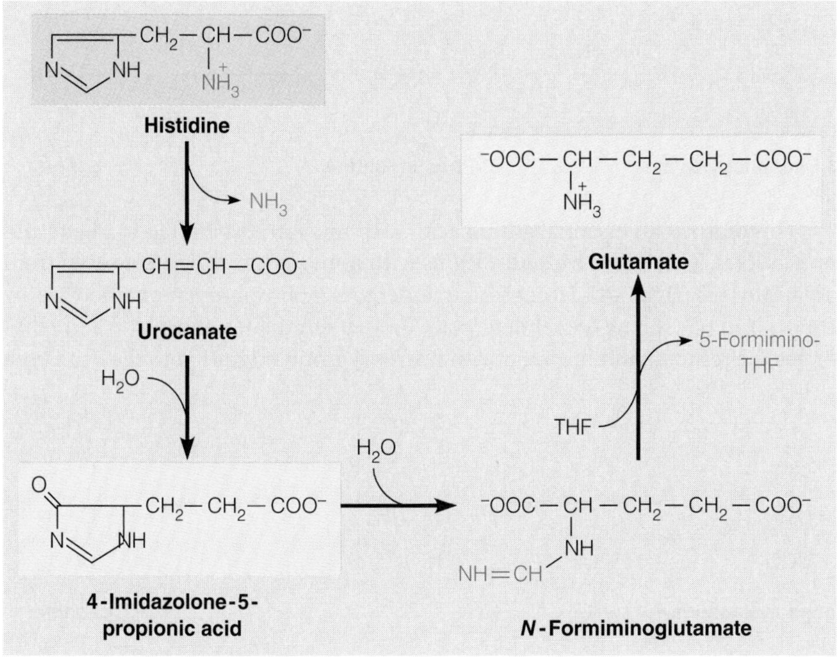

Histidine
NH_3
Urocanate
H_2O
4-Imidazolone-5-propionic acid
H_2O
N-Formiminoglutamate
THF
5-Formimino-THF
Glutamate

FIGURE **21.23**

Catabolism of histidine. The methene carbon of the imidazole ring is converted to a one-carbon tetrahydrofolate adduct.

that occurs in inflamed tissues. Release of histamine in trauma contributes to the dangerous lowering of blood pressure that can lead to shock. A large number of **antihistamines** are in use to treat allergies and other inflammations. Typically, these drugs prevent the binding of histamine to its receptors.

Figure 21.23 presents the major catabolic route for histidine. This amino acid, like tryptophan, does not undergo transamination at the start of its breakdown. Rather, a specific lyase cleaves out ammonia to give **urocanic acid.** Two subsequent steps bring about reduction and ring opening to yield **formiminoglutamic acid,** which is of interest because it serves as a donor of active one-carbon fragments. The formimino group is transferred to tetrahydrofolate, yielding 5-formiminotetrahydrofolate and glutamate.

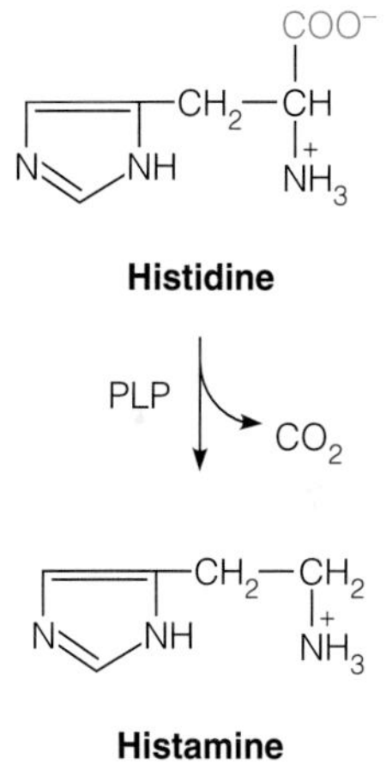

Histidine
PLP
CO_2
Histamine

Serine, Glycine, and Threonine

Although serine, glycine, and threonine do not form a natural grouping, we consider them together partly because of the hydroxyl group shared by serine and threonine and partly because of the close interconnection between serine and glycine via the serine transhydroxymethylase reaction (see page 736). Serine is quite active metabolically; we have already considered its roles in biosynthesis of phospholipids and cysteine, as well as its contribution of activated one-carbon units to the pool of tetrahydrofolate coenzymes. Glycine also plays multiple roles, including contributions to the one-carbon pool and as a precursor to glutathione, to purine nucleotides (see Chapter 22), and to porphyrins. Figure 21.24 summarizes the metabolic fates of glycine and serine. By contrast, threonine plays but one significant role other than as a constituent of proteins: It is a precursor to isoleucine in plants and microorganisms.

Serine is involved in glycine, phospholipid, and cysteine synthesis. Glycine is active in biosynthesis of purine nucleotides and porphyrins.

All three of these amino acids are synthesized and degraded by rather simple pathways. Serine can be synthesized from glycine, via the serine transhydroxymethylase reaction. However, this reaction probably proceeds more often in the reverse direction, as the principal biosynthetic route to glycine and to activated one-carbon compounds. Most serine biosynthesis occurs in a three-step sequence from the glycolytic intermediate 3-phosphoglycerate.

Serine and glycine are both major contributors to the pool of activated one-carbon groups, as 5,10-methylenetetrahydrofolate.

3-Phosphoglycerate $\xrightarrow{NAD^+}$ **3-Phosphopyruvate** $\xrightarrow{\text{Transamination}}$ **3-Phosphoserine** $\xrightarrow{H_2O}$ **Serine**

Threonine is an essential amino acid, with its synthesis limited to plants and prokaryotes. Threonine synthesis begins with homoserine, which is derived from aspartate (see page 748). Homoserine undergoes a phosphorylation, followed by a pyridoxal phosphate–dependent reaction that eliminates phosphate and rehydrates the resultant double bond with hydroxyl group migration to the β-carbon.

Homoserine $\xrightarrow{ATP \rightarrow ADP}$ ***O*-Phosphohomoserine** $\xrightarrow{P_i}$ [intermediate] $\xrightarrow{H_2O}$ **Threonine**

The major route for glycine degradation is via the mitochondrial glycine cleavage system (see Chapter 20), to yield NH_3, CO_2, and 5,10-methylenetetrahydrofolate. This reaction is particularly important in photorespiration in plants (see Chapter 17). Glycolate produced in photorespiration is oxidized to glyoxylate, which undergoes transamination to glycine. One molecule of glycine undergoes the glycine cleavage reaction, and the resultant 5,10-methylenetetrahydrofolate reacts with a second molecule of glycine, to give serine. Serine in turn is converted to 3-phosphoglycerate, which undergoes the normal photosynthetic carbon reduction cycle.

Serine is catabolized both by its conversion to glycine and by the action of **serine–threonine dehydratase.** This pyridoxal phosphate–dependent enzyme converts serine to pyruvate and threonine to α-ketobutyrate (Figure 21.25). α-Ketobutyrate is converted to propionyl-CoA and CO_2; the propionyl-CoA is then metabolized to succinyl-CoA by the same pathway used in odd-chain fatty acid oxidation (Chapter 18) and methionine catabolism. In plants and prokaryotes, α-ketobutyrate has a biosynthetic role as well, as the first intermediate en route to isoleucine biosynthesis.

An alternative route for threonine catabolism is its reversible cleavage to glycine and acetaldehyde, catalyzed by **threonine aldolase.** Though this reaction could

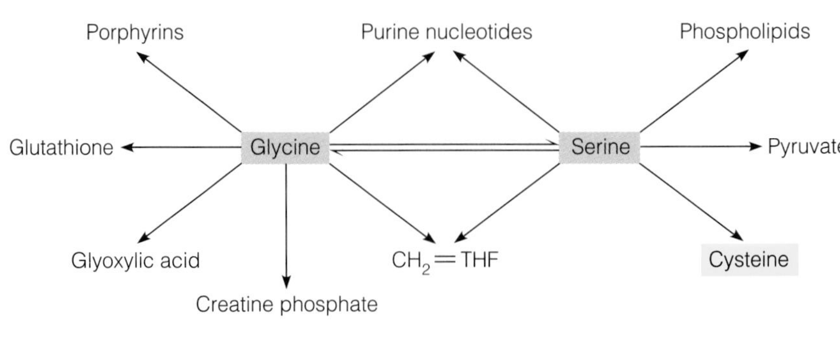

FIGURE **21.24**
Metabolic interconversions and fates of serine and glycine.

in principle lead to threonine biosynthesis in animals, acetaldehyde is highly toxic and cannot accumulate to the point that it would be a biosynthetic intermediate.

$$\underset{\textbf{Acetaldehyde}}{CH_3-CHO} + \underset{\textbf{Glycine}}{{}^{+}NH_3-CH_2-COO^-} \rightleftharpoons \underset{\textbf{Threonine}}{CH_3-CH(OH)-CH({}^{+}NH_3)-COO^-}$$

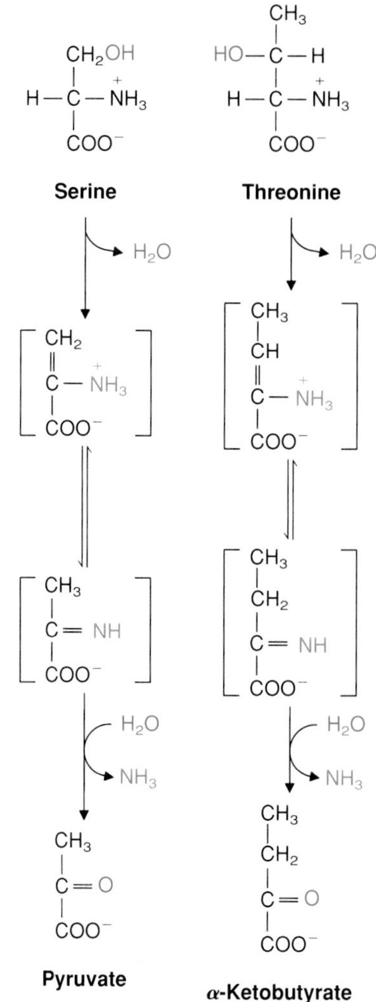

FIGURE **21.25**

Reactions catalyzed by serine–threonine dehydratase.

Valine, Leucine, Isoleucine, and Lysine

Valine, leucine, isoleucine, and lysine have in common the fact that they are essential amino acids for mammals, and they are synthesized primarily in plant and bacterial cells. Furthermore, none of these amino acids is known to play significant metabolic roles other than as protein constituents and as substrates for their own degradation. The pathways involved are complex, and they are shown here only in outline.

VALINE, LEUCINE, AND ISOLEUCINE

Valine, leucine, and isoleucine are structurally related, and they share certain reactions and enzymes in their biosynthetic pathways (Figure 21.26). The last four reactions in valine biosynthesis and in isoleucine biosynthesis are catalyzed by the same four enzymes. Valine biosynthesis begins with transfer of a two-carbon fragment from hydroxyethyl thiamine pyrophosphate to pyruvate. Similar transfer of a two-carbon unit to α-ketobutyrate begins the pathway to isoleucine. The keto acid analog of valine is the input for a four-step pathway to leucine. In bacteria each of these three amino acids controls its own synthesis by feedback inhibition of a different enzyme. In fact, the concept of allosteric control was developed largely in studies on the inhibition of threonine dehydratase by isoleucine.

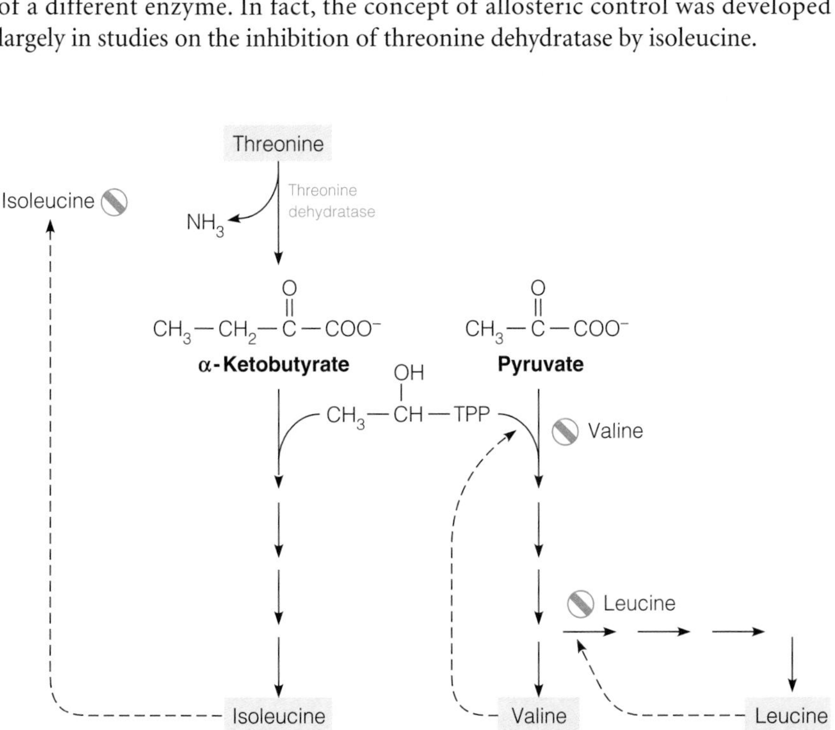

FIGURE **21.26**

Biosynthesis of valine and isoleucine. After the threonine dehydratase reaction, one set of enzymes catalyzes the comparable reactions in valine and isoleucine synthesis. In bacteria, each end product regulates its own synthesis by inhibiting a specific enzyme.

Degradation of leucine, isoleucine, and valine in animals starts with transamination followed by oxidative decarboxylation of the respective keto acids.

$$\underset{\textbf{Leu, Ile, or Val}}{R-\overset{\overset{+}{NH_3}}{\overset{|}{CH}}-COO^-} \xrightarrow{\text{Transamination}} R-\overset{\overset{O}{\|}}{C}-COO^- \xrightarrow[\text{Branched-chain } \alpha\text{-keto acid dehydrogenase}]{NAD^+,\ CoA-SH \quad NADH + H^+,\ CO_2} R-\overset{\overset{O}{\|}}{C}-S-CoA$$

A deficiency of branched-chain α-keto acid dehydrogenase complex, which metabolizes valine, leucine, and isoleucine in humans, creates a severe mental deficiency called maple syrup urine disease.

The latter reaction is carried out by a multienzyme complex, called the **branched-chain $\boldsymbol{\alpha}$-keto acid dehydrogenase complex,** which is similar in structure and mechanism to the pyruvate dehydrogenase and α-ketoglutarate dehydrogenase complexes. In a rare human disorder called **maple syrup urine disease,** this complex is defective. All three keto acids accumulate in the urine, and their characteristic odor gives the condition its name. The condition involves severe mental retardation.

The remainder of the degradative pathways are comparable to fatty acid oxidation, except that the intermediates must undergo a debranching not seen with β-oxidation.

LYSINE

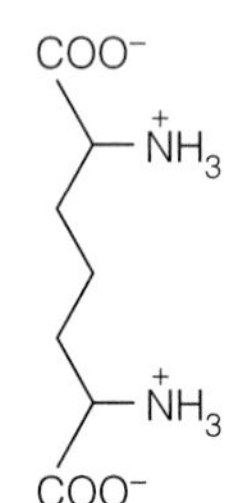

Diaminopimelate

$\downarrow$ CO_2

$\overset{+}{NH_3}$–(CH₂)₄–CH($\overset{+}{NH_3}$)–COO⁻

Lysine

Lysine metabolism is distinguished partly by the complexity of both its synthetic and degradative pathways and partly by the fact that there are two distinct biosynthetic pathways. The **diaminopimelic acid pathway** operates in bacteria, some lower fungi, algae, and higher plants. This pathway begins with condensation of pyruvate with aspartate β-semialdehyde and ends with decarboxylation of diaminopimelate. The principal intermediate, diaminopimelate, also serves an important function as a constituent of bacterial cell walls. The **$\boldsymbol{\alpha}$-aminoadipic acid pathway** is less widespread, functioning in other lower fungi, in higher fungi, and in the protist *Euglena.*

Porphyrin and Heme Metabolism

BIOSYNTHESIS OF TETRAPYRROLES: THE SUCCINATE–GLYCINE PATHWAY

A major metabolic fate of glycine is its utilization for tetrapyrrole biosynthesis. *Tetrapyrrole* is a generic term for compounds containing four linked pyrrole rings. Four such classes of compounds are widespread in biology: the widely distributed iron *porphyrin,* heme; the *chlorophylls* of plants and photosynthetic bacteria; the **phycobilins,** photosynthetic pigments of algae (Chapter 17); and the *cobalamins,* notably vitamin B_{12} and its derivatives. Structures of most of these compounds have been shown previously. All tetrapyrroles are synthesized from a common precursor, **$\boldsymbol{\delta}$-aminolevulinic acid** (ALA). Figure 21.27 illustrates the relationships among the various synthetic pathways.

We shall concentrate here on the well-understood porphyrin synthetic pathway, which leads to heme. This pathway is widespread in animal tissues and, so far as is known, is similar in all organisms containing heme proteins, such as cytochromes. Seven reactions are involved, and they occur in two different cell compartments. The first reaction occurs in mitochondria, followed by three reactions in the cytosol, and finally, three more mitochondrial reactions. As we shall see, this compartmentation provides the opportunity for a novel control mechanism for the pathway.

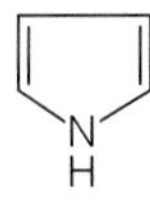

Pyrrole

Early labeling studies in animals revealed that *all of the nitrogen of heme is derived from glycine, and all of the carbon is derived from succinate and glycine.* Hence, this

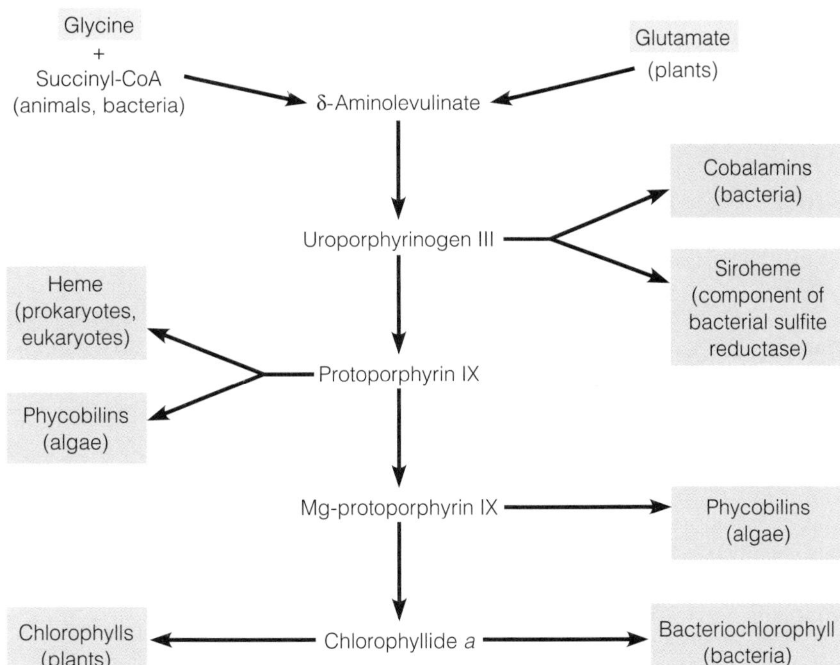

FIGURE **21.27**

Biosynthetic pathways to tetrapyrroles. Tetrapyrroles include heme, chlorophylls, phycobilins, and cobalamins. All are synthesized from δ-aminolevulinic acid, which is formed differently in plants from the way it is formed in bacterial and animal cells.

synthesis is often called the **succinate–glycine pathway.** The first reaction is catalyzed by a pyridoxal phosphate–dependent enzyme, **δ-aminolevulinic acid synthetase,** or ALA synthetase. As shown in Figure 21.28, the binding of glycine to pyridoxal phosphate probably activates the α-carbon of glycine for an attack on the thioester carbon of succinyl-CoA. Decarboxylation follows, to give the product ALA.

δ-Aminolevulinic acid synthetase brings together succinyl-CoA and glycine, which furnish all of the carbon and nitrogen in porphyrins, cobalamins, phycobilins, and chlorophylls.

In plants, ALA is formed by a completely different pathway, a three-step sequence beginning with glutamate (Figure 21.29). The first reaction of this novel pathway links glutamate, through its carboxyl group, to a specific transfer RNA, just as occurs in protein synthesis. The carboxyl group thus activated is then reduced by NADPH, giving glutamate 1-semialdehyde, which finally undergoes an internal transamination to give ALA. Because the major end product of the pathway in plants is chlorophyll, the synthesis of ALA is regulated by light. The identity of the specific light-regulated steps is now under active investigation.

Porphyrin biosynthesis involves (1) formation of a pyrrole ring; (2) condensation of four pyrrole moieties, giving a cyclic tetrapyrrole; and (3) side chain modifications and ring oxidations.

Whether in plants, animals, or microorganisms, the remainder of the porphyrin synthetic pathway involves three distinct processes: (1) synthesis of a substituted pyrrole compound, **porphobilinogen** from ALA; (2) condensation of four porphobilinogen molecules to yield a partly reduced precursor called a **porphyrinogen;** and (3) modification of the side chains, dehydrogenation of the ring system, and introduction of iron, to give the porphyrin product, heme. In the first stage, two molecules of ALA condense in the cytosol to form one molecule of porphobilinogen. The reaction is catalyzed by **ALA dehydratase:**

$$\underset{\delta\text{-ALA}}{H_2N{-}CH_2{-}C({=}O){-}CH_2{-}CH_2{-}COO^-} + \underset{\delta\text{-ALA}}{H_2N{-}CH_2{-}C({=}O){-}CH_2{-}CH_2{-}COO^-} \xrightarrow[\text{Dehydratase}]{2\,H_2O} \text{Porphobilinogen}$$

Next, four molecules of porphobilinogen combine in a PLP-requiring deaminase reaction to give the first tetrapyrrole compound, **uroporphyrinogen III** (Figure

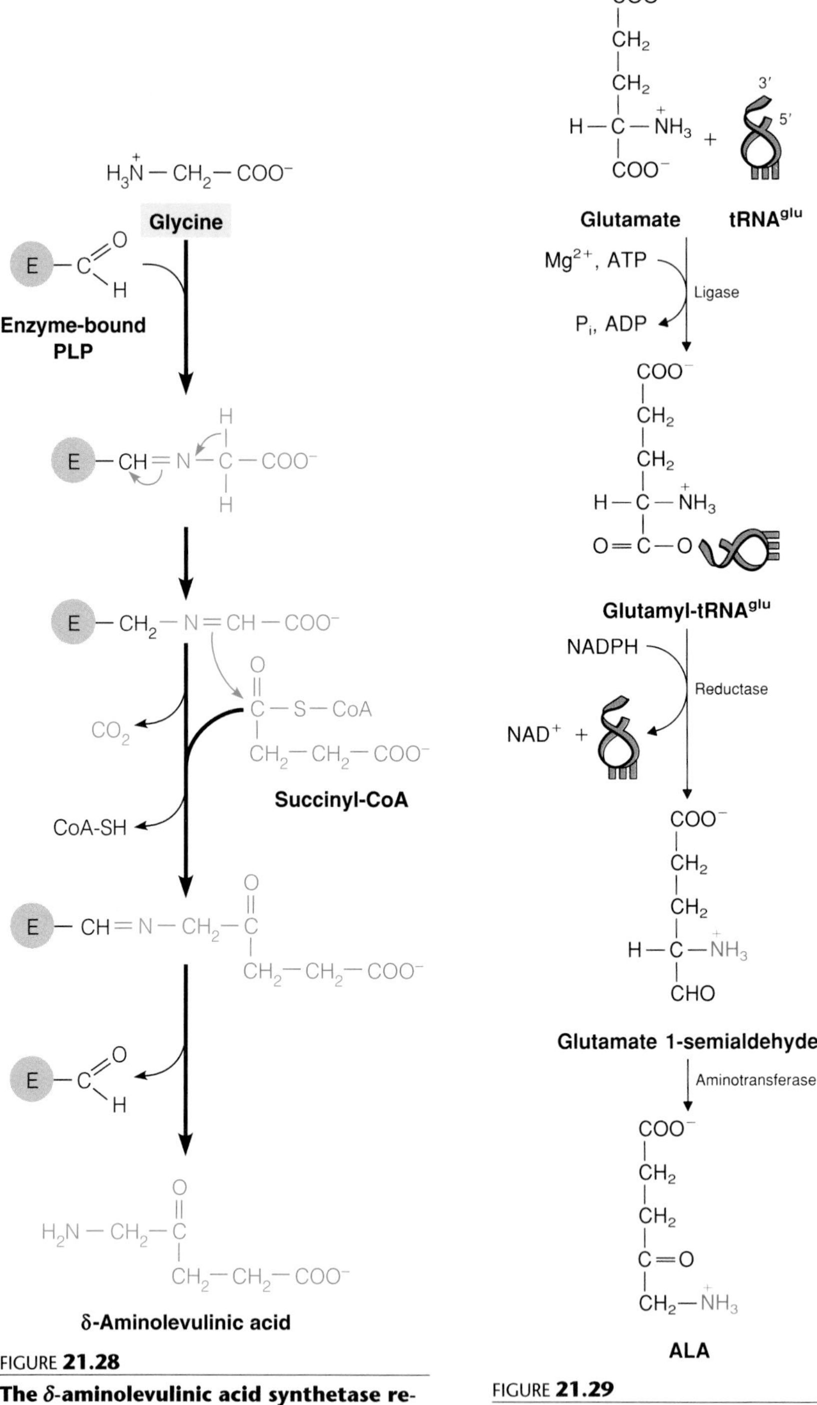

FIGURE **21.28**

The δ-aminolevulinic acid synthetase reaction. E-CHO is the enzyme–pyridoxal phosphate complex.

FIGURE **21.29**

Synthesis of δ-aminolevulinic acid in plants.

Uroporphyrinogen I

21.30; note the side chain abbreviations used in this figure). Two different proteins are involved in this reaction: **uroporphyrinogen I synthase** and **uroporphyrinogen III cosynthase.** The former protein is enzymatically active by itself, but when it acts alone, the product is the undesired symmetrical compound **uroporphyrinogen I.** Interaction between the synthase and cosynthase allows one of the rings to flip during the combining reaction, so that the product is the asymmetric uroporphyrinogen III. The symmetric compound and some metabolites derived from it are synthesized as nonfunctional side products, in low amounts.

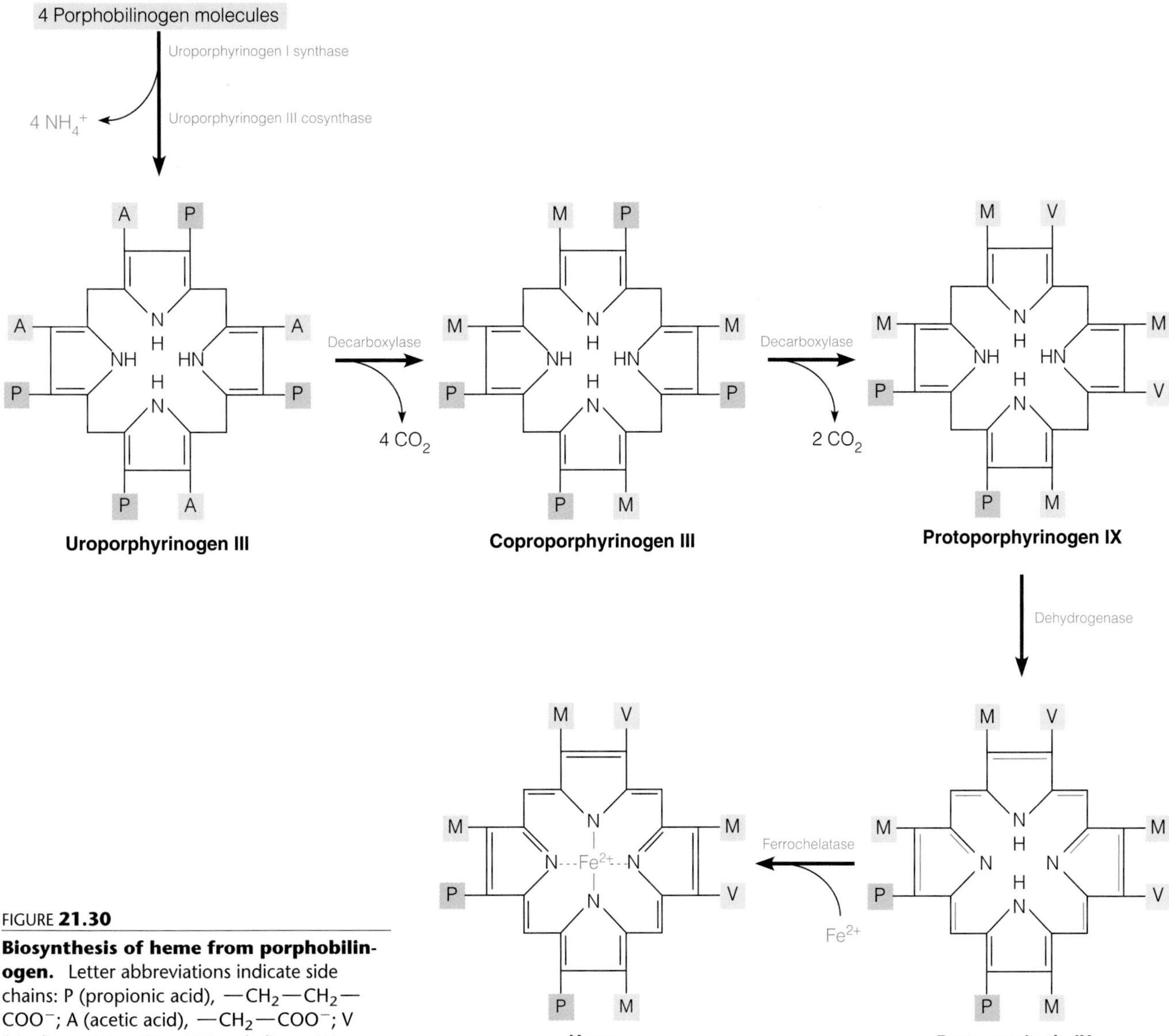

FIGURE **21.30**

Biosynthesis of heme from porphobilinogen. Letter abbreviations indicate side chains: P (propionic acid), $-CH_2-CH_2-COO^-$; A (acetic acid), $-CH_2-COO^-$; V (vinyl), $-CH{=}CH_2$; M (methyl), $-CH_3$.

Uroporphyrinogen III undergoes decarboxylation of its acetic acid side chains. The product then reenters the mitochondrion for further modifications: first, side chain modifications, then ring oxidation to yield a fully conjugated system, and finally the insertion of iron. The last reaction can proceed spontaneously, but it is catalyzed by **ferrochelatase,** an enzyme on the inner mitochondrial membrane that also requires a reducing agent. At this stage the completed heme combines with polypeptides to give completed heme proteins, including myoglobin and hemoglobin in vertebrates, and cytochromes and other heme proteins in all aerobic organisms.

In a hereditary condition called **congenital erythropoietic porphyria,** uroporphyrinogen III cosynthase is defective, and the symmetrical (and metabolically useless) type I porphyrins accumulate beyond the capacity of the body to excrete them. Their accumulation causes the urine to turn red, the skin to become acutely photosensitive, and the teeth to become fluorescent, all because of the deposition of the strongly light-absorbing porphyrins. In addition, erythrocytes are destroyed prematurely and insufficient heme is synthesized, making afflicted individuals quite anemic. It has been speculated that people labeled as vampires in medieval

Porphyrias involve abnormal accumulations of heme precursors, either from overproduction of the unnatural type I porphyrins or from abnormally high flux through δ-ALA synthetase.

folktales suffered from this condition, which would explain their preference for the dark, their bizarre appearance, and their propensity for drinking blood. In fact, individuals with congenital erythropoietic porphyria can be treated by injections of heme.

Quite distinct from the above condition is **acute intermittent porphyria,** which results from deficiency in uroporphyrinogen I synthase. This deficiency causes ALA and porphobilinogen to accumulate in the liver. The condition is accompanied by episodes of acute abdominal pain and neurological disorders. There is evidence that King George III of England suffered from this condition; evidence for this was put before the general public in the successful stage play and film *The Madness of King George.* Symptoms of porphyrias can also be acquired, most notably in lead poisoning. The crystal structure of ALA dehydratase shows that lead can displace zinc, the natural metal cofactor, and inhibit the enzyme, causing great accumulation of ALA.

Being the first committed step in heme synthesis, the ALA synthetase reaction (Figure 21.28) is the major control point. Heme and related compounds feedback-inhibit the enzyme. Heme also has two other important effects. At low concentrations, heme inhibits the *synthesis* of ALA synthetase at the translational level. At higher levels, heme somehow blocks the *translocation* of ALA synthetase from the cytosol, where it is synthesized on ribosomes, into the mitochondrion, where it acts. Heme also inhibits the ferrochelatase reaction (Figure 21.30). A number of drugs and poisons cause excessive heme synthesis. In some cases, this effect results from stimulation of the synthesis of cytochrome P450, which increases the demand for heme and hence activates ALA synthetase.

Tetrapyrrole biosynthesis is being exploited as a target for the action of weedkillers. The idea is to spray weeds in the dark with ALA. The pathway to chlorophyll begins, and when it becomes light, the pathway is completed, and chlorophyll is produced in such massive amounts that the plant weakens and dies.

DEGRADATION OF HEME IN ANIMALS

By far the most abundant porphyrin compound in vertebrates is the heme of hemoglobin. Therefore, the story of porphyrin degradation is largely the story of hemoglobin and heme degradation. Lacking nuclei, mammalian erythrocytes are incapable of renewal and self-destruct after characteristic intervals. In humans the average erythrocyte life span is 120 days. Aged erythrocytes are destroyed upon passage through the spleen or liver (Figure 21.31).

Heme protein degradation in animals releases amino acids and iron, which are reused, and bilirubin, which must be solubilized for excretion.

Amino acids released from the globin portion of the hemoglobin molecule are catabolized or reused for protein synthesis. The heme portion undergoes degradation, starting with a mixed-function oxidase reaction that opens the ring and converts one of the methene bridge carbons to carbon monoxide. Iron is released from the resulting linear tetrapyrrole, called **biliverdin,** and is transported to storage pools in bone marrow for reuse in erythrocyte production. The tetrapyrrole is next reduced to **bilirubin,** which is excreted. Bilirubin is quite insoluble, and its removal involves several organ systems. First, it complexes with serum albumin for transport to the liver. There, bilirubin is solubilized by conjugation with two molecules of **glucuronic acid.** The reaction is comparable to other glycosyltransferase reactions we have encountered (see Chapter 16), with the substrate being **UDP-glucuronate.** This solubilized compound, **bilirubin diglucuronide,** is secreted into the bile and ultimately excreted via the intestine.

Because several organ systems participate in the degradation of heme, there are numerous ways for things to go wrong. When heme catabolism is defective, bilirubin accumulates in the blood. This defect is first recognized because the distinctive color of bilirubin gives a yellow cast to the skin and the whites of the eyes. This condition, known as **jaundice,** is seen, for example, in acute or chronic liver disease, in which the glucuronate conjugating system is impaired and albumin

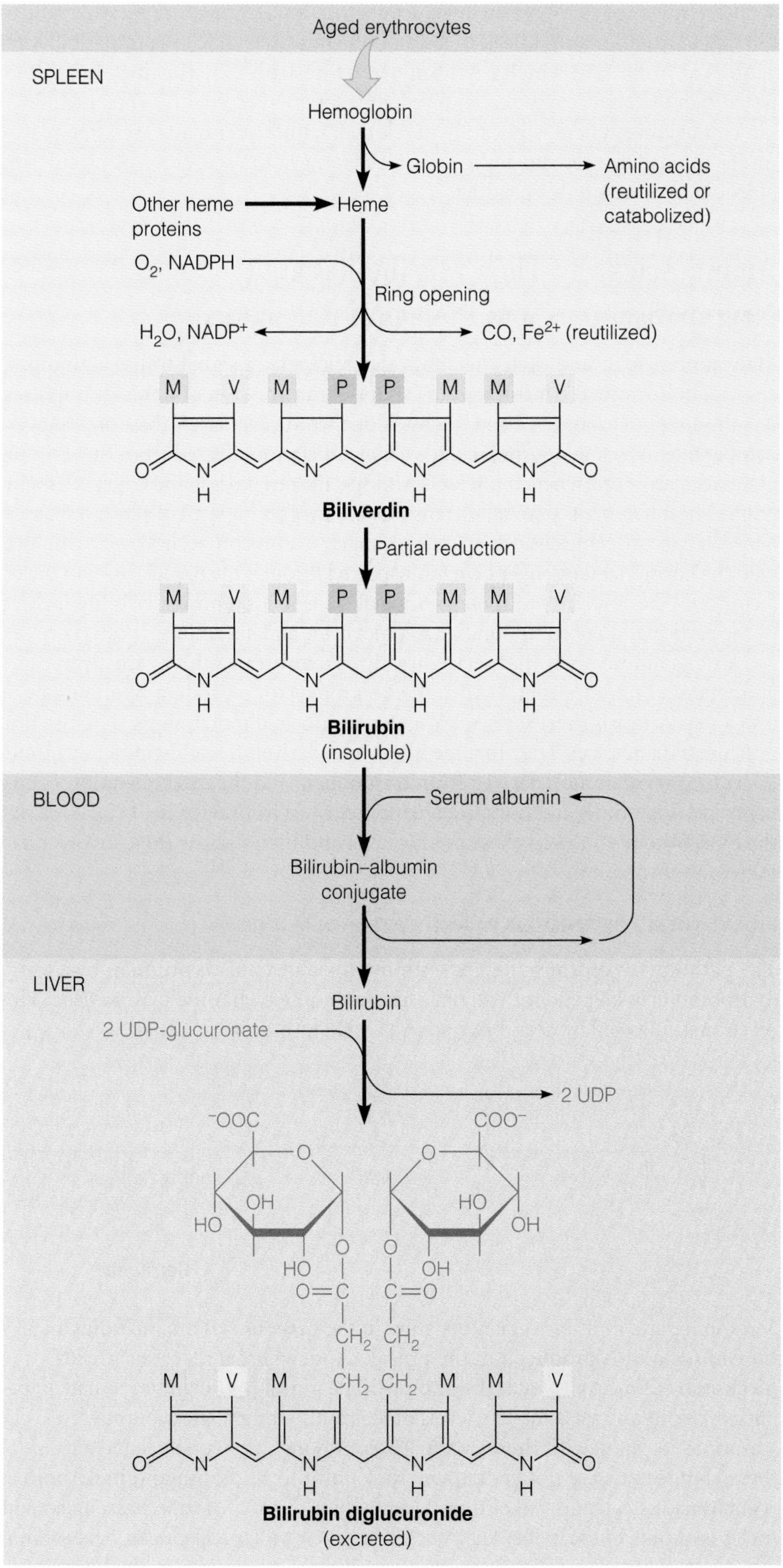

FIGURE **21.31**

Catabolism of heme. Most of the heme comes from breakdown of aged erythrocytes, but some comes from cytochromes and other heme proteins. Side chain designations are as in Figure 21.30.

synthesis might be defective; in bile duct obstruction (e.g., gallstone), when bilirubin diglucuronide cannot be secreted into the intestine; in Rh incompatibility reactions of infants, in which erythrocytes are destroyed by the immune system faster than the heme can be catabolized; or in premature infants, when the bilirubin conjugating system is not fully developed. Jaundiced infants are often placed under intense fluorescent light, which rearranges the structure of circulating bilirubin to more soluble products.

Amino Acids and Their Metabolites as Neurotransmitters and Biological Regulators

Many amino acids and their metabolites participate in signal transduction processes—in hormonal control and in synaptic transmission of nervous impulses. As introduced in Chapter 12 and discussed further in Chapter 23, these two roles are comparable in that a low-molecular-weight substance released from one cell migrates to a target cell, where it interacts with specific receptors in the target cell membrane. The difference is that neurotransmission involves movement across a synapse, between two adjacent cells, whereas hormonal transmission occurs over a distance, with the hormonal messenger being transported through the bloodstream to the effector cell. The similarity of these two signal transduction processes is highlighted by the participation of compounds like epinephrine and histamine in both processes.

Glutamate, tyrosine, glycine, and tryptophan serve as neurotransmitters or precursors to neurotransmitters.

Among amino acids that serve directly as neurotransmitters are glycine and glutamate. As noted earlier, GABA, the decarboxylation product of glutamate, is also a neurotransmitter. Several aromatic amino acid metabolites also function in neurotransmission. They include histamine, derived from histidine; **serotonin** (5-hydroxytryptamine), derived from tryptophan; and the **catecholamines**—epinephrine, **dopamine,** and **norepinephrine**—derived from tyrosine. Let us describe the biosynthetic routes to these compounds and then discuss their involvement in neurotransmission.

BIOSYNTHESIS OF SEROTONIN AND CATECHOLAMINES

The pathway to serotonin begins with hydroxylation of tryptophan by a tetrahydrobiopterin-dependent enzyme, similar to phenylalanine hydroxylase. This reaction is followed by decarboxylation to yield serotonin.

O_2 H_2O tetrahydro-biopterin dihydro-biopterin CO_2 PyP

Tryptophan **Serotonin**

Serotonin plays multiple regulatory roles in the nervous system, including neurotransmission. It is produced in the pineal gland, where it serves as precursor to **melatonin** (*O*-methyl-*N*-acetylserotonin). The pineal is known to regulate light–dark cycles in animals, and the levels of serotonin and melatonin undergo cyclic variations in phase with these cycles. Thus, although the cycle-related actions of these compounds are not yet known, they point to serotonin and melatonin as regulators of sleep and wakefulness. Many long-distance airline passengers take melatonin pills to escape jet lag by resetting their biological clocks. Serotonin is also secreted by cells in the small intestine, where it regulates intestinal peristalsis. Finally, serotonin is a potent vasoconstrictor, which helps regulate blood pressure. Several antiobesity treatments act by increasing serotonin levels, thereby creating a sense of satiety, or well-being with regard to food.

As shown in Figure 21.32, the pathway to catecholamines is similar, starting with a tetrahydrobiopterin-dependent hydroxylation (of tyrosine) followed by a decarboxylation. The hydroxylation product is dopa, which is formed by a quite different mechanism in melanin synthesis (see page 769). However, the latter pathway is localized to melanocytes, whereas most catecholamine synthesis occurs in the adrenal medulla and in the central nervous system.

Tyrosine is hydroxylated to dopa by two distinct mechanisms in the synthesis of catecholamines and melanins.

Once formed, dopa undergoes decarboxylation to give dopamine. Dopamine serves in turn as substrate for a copper-containing monooxygenase, **dopamine β-hydroxylase,** giving norepinephrine, which in turn is methylated by *S*-adenosylmethionine to give epinephrine. Although dopamine and norepinephrine are intermediates in epinephrine synthesis, each is a neurotransmitter in its own right, as discussed in the next section.

Decarboxylation reactions of tryptophan, dopa, and histidine lead to a series of potent biological regulators.

BIOCHEMISTRY OF NEUROTRANSMISSION

In Chapter 10 we noted that the transmission of nerve impulses involves two distinct components: (1) the transmission of an action potential within a neuron, by continuous polarization and depolarization of the membrane; and (2) transmission of the impulse across a synaptic junction, from one neuron to another neuron or to a muscle or gland cell. Transmission within a neuron is described in Chapter 10. Here we discuss transmission from cell to cell.

The Cholinergic Synapse

Neurotransmission usually involves the release of a chemical messenger, the **neurotransmitter,** from the presynaptic cell and its binding to receptors on the postsynaptic cell (nerve, muscle, or gland). Synapses involving **acetylcholine** as the neurotransmitter (**cholinergic synapses**) are the best understood. Recall from Chapter 19 that choline is synthesized principally as part of phosphatidylcholine, with the carbons of choline coming ultimately from serine.

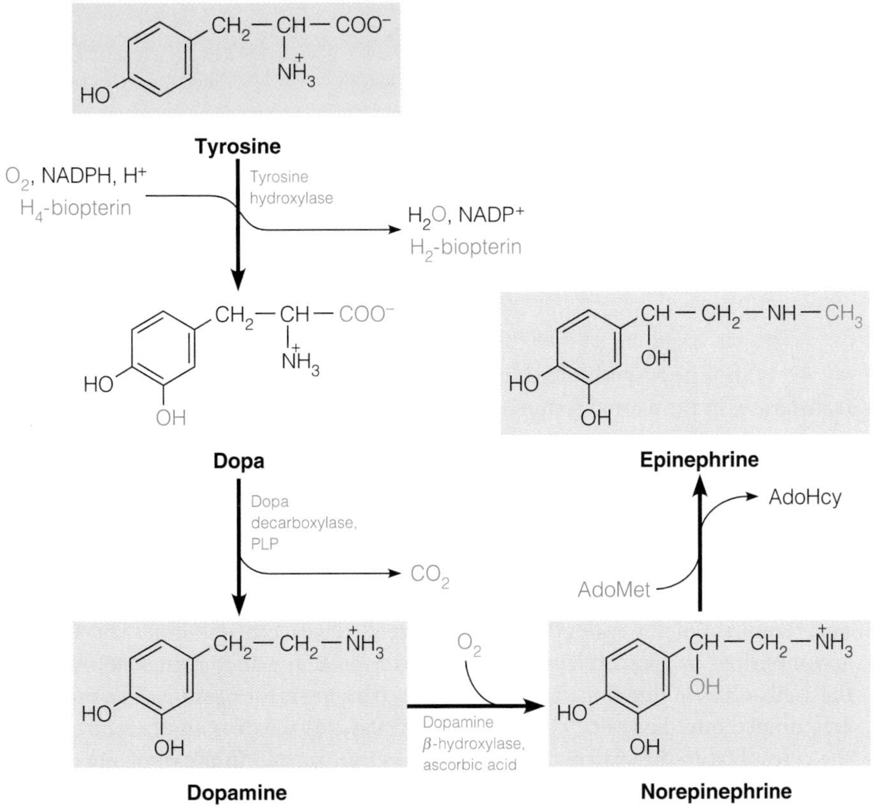

FIGURE **21.32**

Biosynthesis of the catecholamines—dopamine, norepinephrine, and epinephrine—from tyrosine.

Neurotransmission involves calcium release in the presynaptic nerve end, vesicle membrane fusion, neurotransmitter release, and binding of neurotransmitter to receptors in the postsynaptic cell membrane.

A cholinergic synapse is shown in Figure 21.33. The **terminal bulb** (also called the **synaptic knob**) of the **presynaptic axon** is separated from the **postsynaptic dendrite** by a **synaptic cleft** about 20 nm wide. The nerve impulse (action potential) moves down the presynaptic axon to the terminal bulb; the change in membrane potential in the bulb causes the opening of voltage-gated calcium channels, allowing Ca^{2+} ions to pass from the synaptic cleft into the axonal bulb (Figure 21.33a). Within the bulb are **synaptic vesicles,** each containing about 10^3 to 10^4 acetylcholine molecules. The increase in Ca^{2+} concentration causes these vesicles to fuse with the axonal membrane and open, spilling their contents into the synaptic cleft (Figure 21.33b). The postsynaptic membrane of the receptor dendrite has specific acetylcholine receptors, toward which the neurotransmitter diffuses (Figure 21.33c). Binding of acetylcholine triggers the opening of ion channels in the postsynaptic membrane (Figure 21.33d), initiating an action potential that can be passed on to the next axon (Figure 21.33e). This last process involves a wave of permeability changes leading to sodium influx and potassium efflux. The receptors shown are referred to as **nicotinic acetylcholine receptors,** because they can bind the alkaloid **nicotine;** we will discuss their relationship to the ion channels shortly. The other major type of acetylcholine receptor is called the *muscarinic acetylcholine receptor;* it participates in different synapses (see page 788).

$HO-CH_2CH_2\overset{+}{N}(CH_3)_3$

Choline

$CH_3-\overset{O}{\overset{\|}{C}}-S-CoA$ → $CoA-SH$ (Choline acetyltransferase)

CH_3COO^- ← H_2O (Acetylcholinesterase)

$H_3C-\overset{O}{\overset{\|}{C}}-O-CH_2CH_2\overset{+}{N}(CH_3)_3$

Acetylcholine

Acetylcholine is synthesized from choline and acetyl-CoA, by **choline acetyltransferase,** in the axonal terminal bulbs. After it has been released from vesicles and bound to the receptors, the neurotransmitter is rapidly hydrolyzed by an enzyme present in the synaptic cleft, **acetylcholinesterase,** yielding choline, which binds poorly to acetylcholine receptors. Degradation of acetylcholine restores the resting potential in the postsynaptic membrane.

In order to ready the synapse for another impulse (which must occur about 1000 times per second), the empty synaptic vesicles, which are returned to the axonal terminal bulb by endocytosis, must be refilled with acetylcholine. This task is accomplished by an **acetylcholine transporter** protein, which brings newly synthesized acetylcholine into the vesicles by exchanging it for protons. The protons are pumped into the vesicles by a **vacuolar ATPase.** As the protons are returned to the cytosol from the vesicles, acetylcholine is transported in the opposite direction. Similar events occur in neuron-to-neuron synapses.

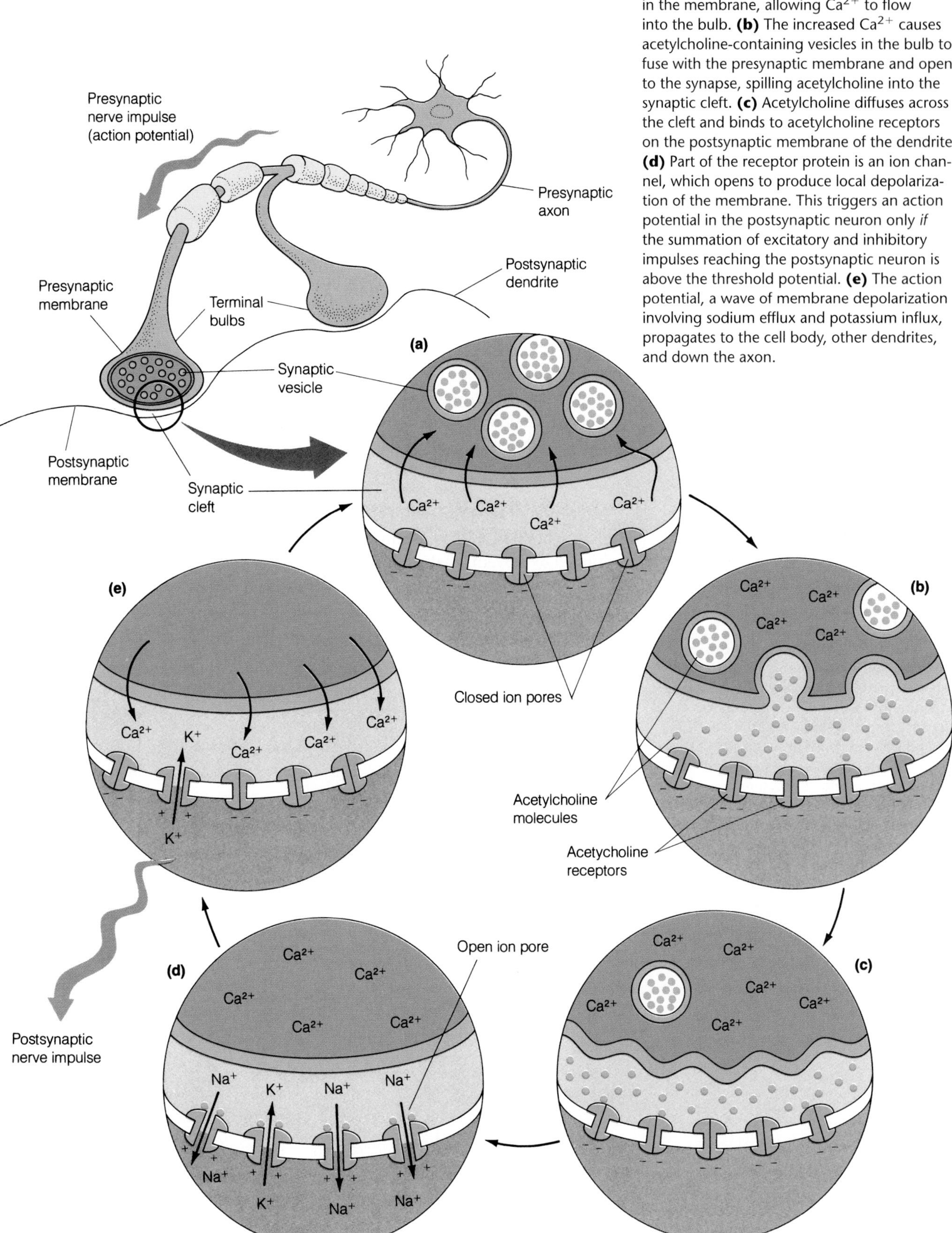

FIGURE **21.33**

Transmission of a neural impulse across a cholinergic synapse. (a) Movement of the incoming action potential into the bulb of the presynaptic axon opens Ca^{2+} channels in the membrane, allowing Ca^{2+} to flow into the bulb. **(b)** The increased Ca^{2+} causes acetylcholine-containing vesicles in the bulb to fuse with the presynaptic membrane and open to the synapse, spilling acetylcholine into the synaptic cleft. **(c)** Acetylcholine diffuses across the cleft and binds to acetylcholine receptors on the postsynaptic membrane of the dendrite. **(d)** Part of the receptor protein is an ion channel, which opens to produce local depolarization of the membrane. This triggers an action potential in the postsynaptic neuron only *if* the summation of excitatory and inhibitory impulses reaching the postsynaptic neuron is above the threshold potential. **(e)** The action potential, a wave of membrane depolarization involving sodium efflux and potassium influx, propagates to the cell body, other dendrites, and down the axon.

The Nicotinic Acetylcholine Receptor

Viewed directly from the surface, the nicotinic acetylcholine receptors appear in the electron microscope to be doughnutlike structures extending through the membrane and containing a central pore (Figure 21.34). Isolation of the receptors by affinity chromatography showed them to be pentamers made of four kinds of glycoprotein subunits (molecular masses 54, 56, 58, and 60 kilodaltons, in molar ratio 2:1:1:1). Reconstitution of receptors into lipid vesicles provides a system that can conduct ions when stimulated by acetylcholine. Thus, it appears that the receptor and ion channel are a single unit. The central pore presumably functions as the gated ion channel. Genes for many of the components of the synaptic system have now been cloned, including choline acetyltransferase, acetylcholinesterase, and components of the nicotinic receptors.

Although postsynaptic membranes are typically packed densely with receptors (of the order of 20,000 per square micrometer), these regions constitute only a small fraction of the cellular surface in a typical neural tissue. Therefore, biochemists often turn to more specialized tissues for study of synaptic processes and biochemistry. Favorite experimental subjects for such studies are the electric organs of the electric ray (*Torpedo*) and the electric eel (*Electrophorus*). These organs contain stacks of cells called **electroplaques,** which have a high density ($10^5/\mu m^2$) of nicotinic acetylcholine receptors extending over one whole face of the cell. Depolarization of the membrane on this face, while the other face remains at the resting potential, gives a $\Delta\psi$ of 130 mV across the cell. With thousands of cells stacked in series, potentials of several hundred volts are generated.

Inhibition of the Cholinergic Synaptic Transmission

Acetylcholinesterase is a serine esterase and consequently can be irreversibly inhibited by reagents that react with the active site serine, such as diisopropyl fluo-

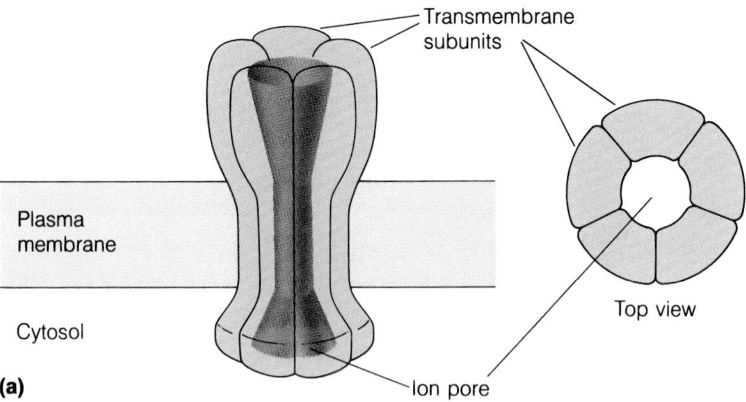

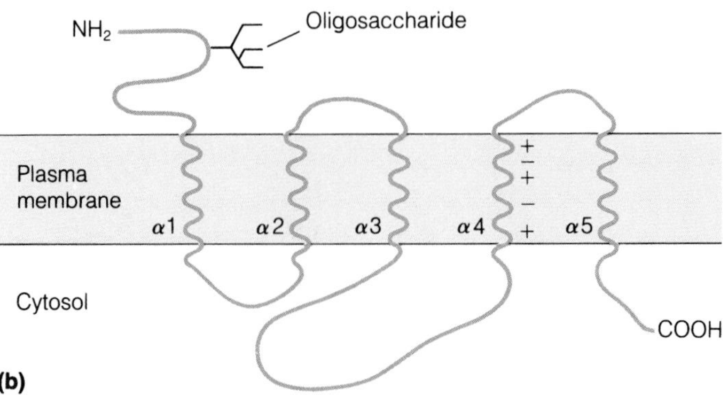

FIGURE **21.34**

The nicotinic acetylcholine receptor. **(a)** Schematic model of the receptor. Five subunits combine to form a transmembrane structure with an ion pore in the center. **(b)** Structure of an individual subunit. There are four different kinds of subunits, but their sequences are all similar, and each individual subunit has the kind of structure depicted here. Five α helixes ($\alpha 1$ to $\alpha 5$) in each subunit traverse the membrane. The charged residues on helix $\alpha 4$, which tend to be on one surface, probably line the wall of the pore.

rophosphate, sarin, physostigmine, and parathion (see Table 11.4). As you might expect, acetylcholinesterase inhibitors are extremely toxic substances, causing paralysis. Another class of toxins acts on the acetylcholine receptor itself either by blocking it (*d*-tubocurarine, from curare, and the small protein toxins in some snake venoms, like cobra toxin) or by locking the ion channels open (nicotine). The former class are called **antagonists,** and the latter **agonists.** Such compounds are useful, both as research tools and, with proper caution, as drugs (muscle relaxants).

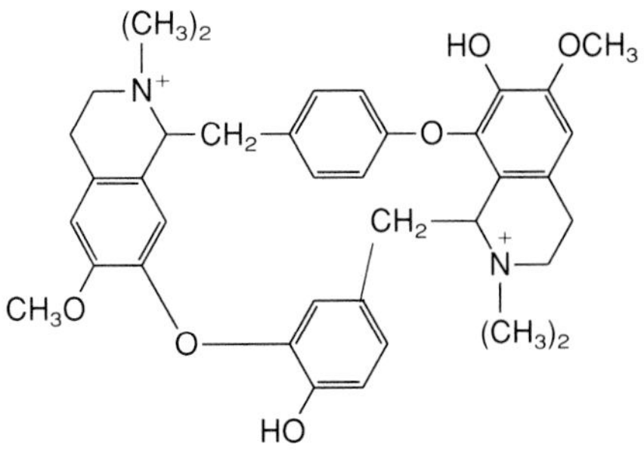

Tubocurarine, an antagonist

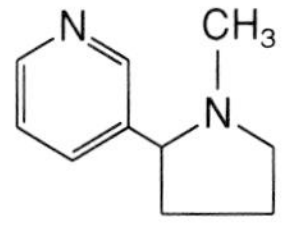

Nicotine, an agonist

Catecholamines and Adrenergic Neurons

We have concentrated on the cholinergic synapse because it is the best understood at present. But many other substances are known or suspected to be neurotransmitters, which function in different kinds of synapses. These include the catecholamines—dopamine, norepinephrine, and epinephrine. Because epinephrine is also an adrenal hormone, synapses that use catecholamines are called **adrenergic.**

The importance of dopamine in neural transmission is emphasized by the number of major neurological diseases that are associated with improper dopamine regulation. The earliest indication of this type of defect was the finding that dopamine levels are abnormally low in a particular region of the brain of patients with **parkinsonism,** a severe neurological disorder. Attempts to treat such patients with dopamine were futile, because this substance, after injection, does not cross the **blood–brain barrier,** a selective permeability blockade that prevents uptake of some substances by the brain. However, the dopamine precursor, dopa, does cross the blood–brain barrier. For many individuals with parkinsonism, daily doses of dopa have provided dramatic clinical improvement.

Circumstantial evidence links dopamine metabolism to schizophrenia, which may result partly from excess firing of **dopaminergic neurons**—neurons secreting dopamine. Part of the circumstantial evidence is the close structural relationship between dopamine (see Figure 21.32) and **mescaline,** a product of the peyote cactus known to induce a quasi-schizophrenic state. **Amphetamine** is another catecholamine analog with potent psychopharmacological properties. Other circumstantial evidence is that the drugs most useful for treating schizophrenia, notably **chlorpromazine,** are dopamine antagonists, which block dopamine binding to its receptors. In recent years the genes for four different dopamine receptors have been cloned, and intense interest is focused upon which, if any, of these receptors functions abnormally in schizophrenia.

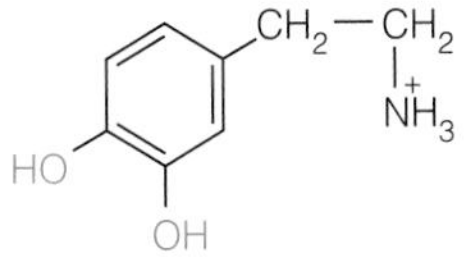

Dopamine

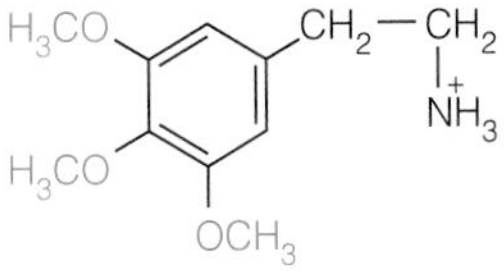

Mescaline

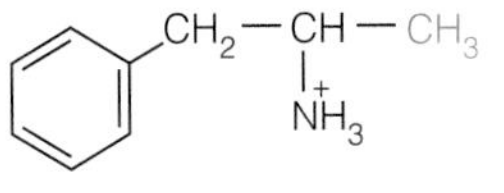

Amphetamine

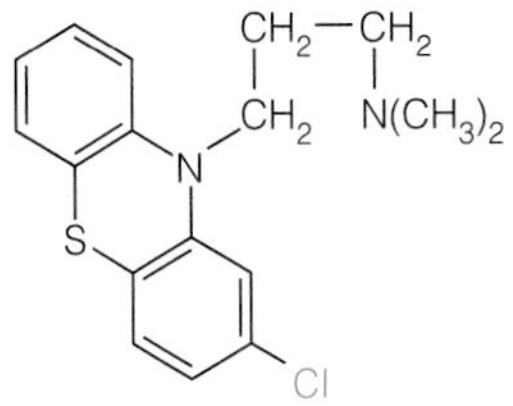

Chlorpromazine

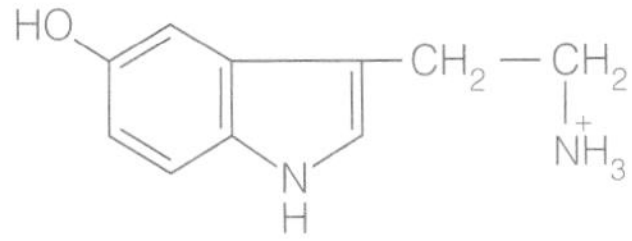

Serotonin

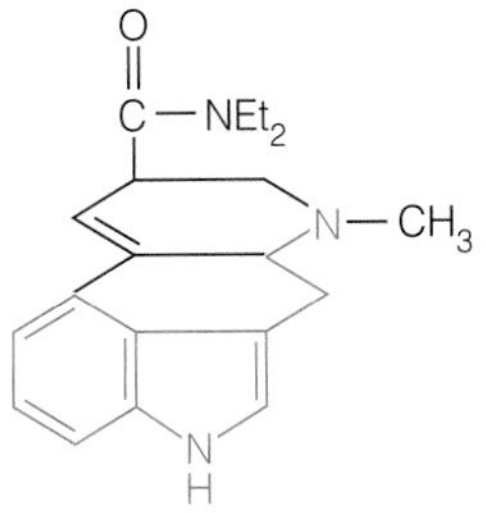

Lysergic acid diethylamide

Another psychotropic drug, whose action suggested a role for serotonin in neurotransmission, is **lysergic acid diethylamide (LSD)**, an indole derivative. LSD action is explained in part as mimicking the effects of serotonin at receptors in the central nervous system. More recently the drug **fluoxetine** (marketed as **Prozac**) has been shown to block the uptake of serotonin into presynaptic nerve terminals, with an increase in the amount available to bind to postsynaptic receptors. Originally marketed as an antidepressant drug, Prozac is now known to be active against a range of psychiatric disorders.

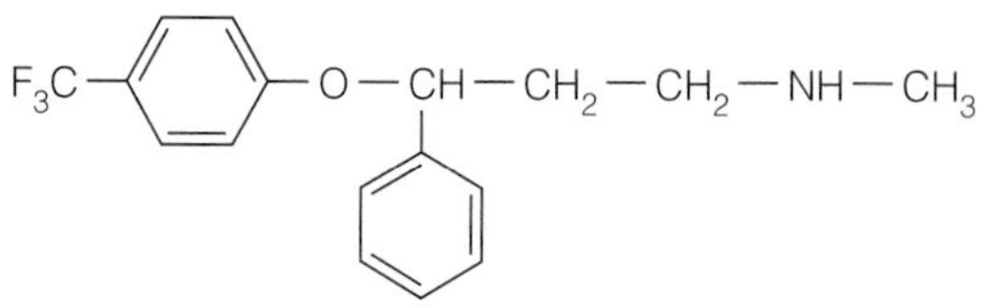

Prozac
(fluoxetine)

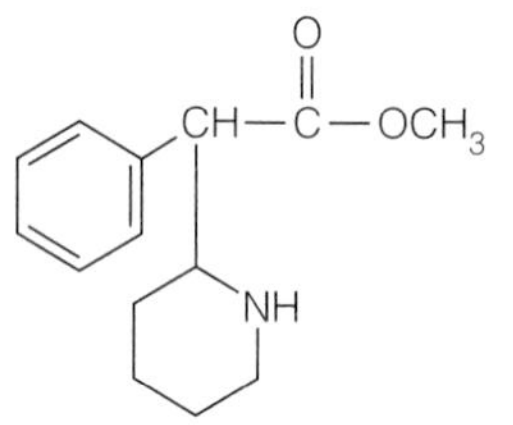

Also related to serotonin metabolism is another widely prescribed drug, **Ritalin,** used to treat children with hyperactivity–attention deficit disorder. Because Ritalin is a stimulant, known to act by raising dopamine levels, its effect in calming hyperactive children was paradoxical. Studies reported in early 1999, however, reported that the drug, in the low doses used with such children, actually exerts its calming effect by elevating levels of serotonin.

Ritalin
(methyl phenidylacetate)

Just as acetylcholinesterase limits the firing of cholinergic synapses, so also do two enzymes regulate the action of adrenergic neurons. These enzymes are **catecholamine *O*-methyltransferase** (COMT), which catalyzes an AdoMet-dependent transmethylation, and **monoamine oxidase** (MAO), a flavoprotein that oxidizes primary amines to aldehydes.

Excitatory and Inhibitory Neurotransmission

The necessity for many types of neurotransmitters and corresponding kinds of synapses probably derives from the complexity of vertebrate neural systems. Neurotransmitters and the synapses that they mediate vary widely in their properties. Some are rapid in their action, and some are slow. Some, such as the nicotinic cholinergic synapses and those involving glutamate, are stimulatory and promote an action potential in the postsynaptic cell. Others, such as those using γ-aminobutyric acid (GABA) as a transmitter, are inhibitory; an impulse received at these synapses will *discourage* the transmission of an action potential in the recipient neuron. This inhibition can occur, for example, by the opening of chloride channels. Whether the neuron fires depends on the net summation of stimulatory and inhibitory inputs. Given that GABA, a principal inhibitory neurotransmitter, is derived from glutamate, it is interesting that glutamate itself is a principal excitatory neurotransmitter. Glutamate ingested in large quantities, as the flavor enhancer monosodium glutamate (MSG), can damage the central nervous system by overstimulation. For this reason, MSG was removed from most infant formulas two decades ago, because the developing nervous system is particularly susceptible to damage. MSG is a major constituent of soy sauce, and some people experience sweatiness, flushing, and headaches from eating MSG-flavored foods.

A second type of receptor for acetylcholine, the **muscarinic acetylcholine receptor,** is quite different. This receptor, which has a different structure, can be inhibitory, though not by the chloride channel mechanism. Thus, whether acetylcholine is stimulatory or inhibitory depends on the type of receptor to which it binds. Inhibitory synapses play a major role in the regulation of neural transmission. A typical nerve cell, such as the motor neuron shown in Figure 21.35, receives input from the axons of many different neurons. Figure 21.36 is a schematic illustration of transmission and inhibition in a neural network.

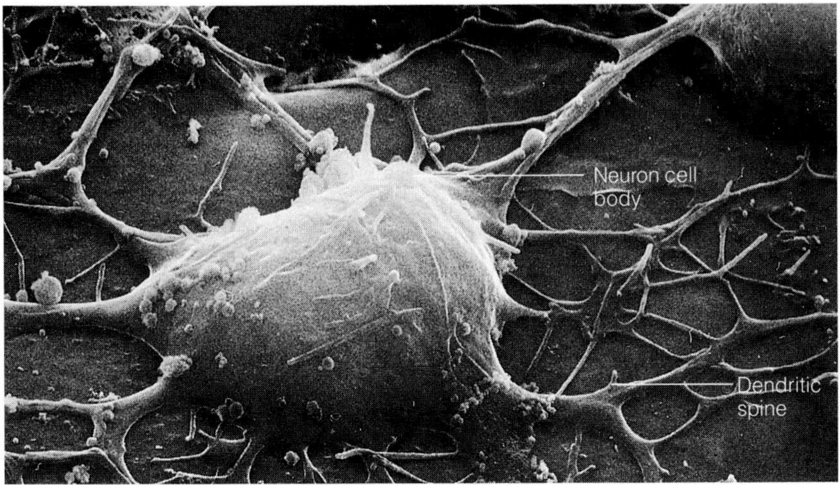

FIGURE **21.35**

Multiple synapses on the body of a single neuron. This scanning electron micrograph gives an idea of the complexity of interconnection in the nervous system. Some of these synapses will be stimulatory and others inhibitory.

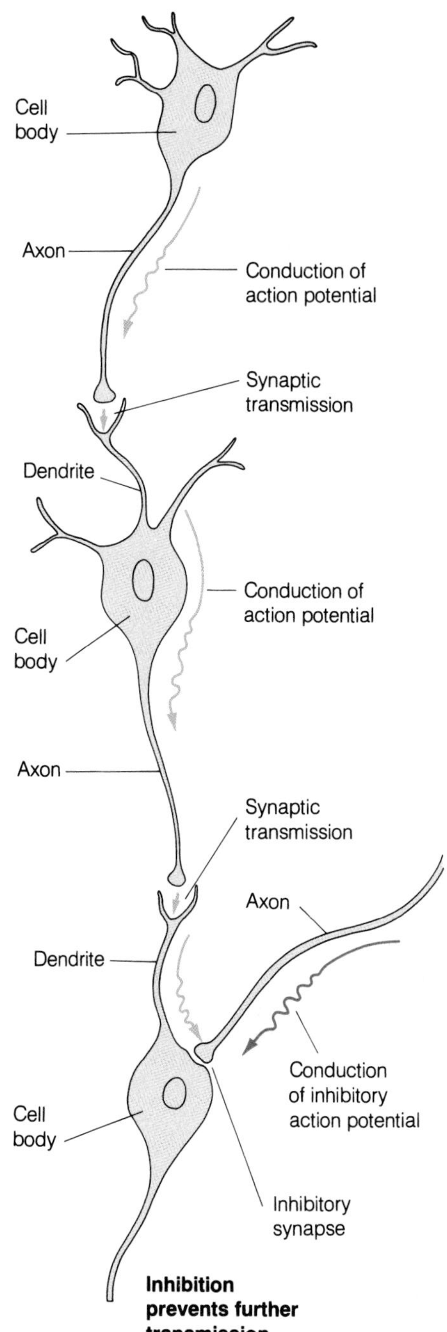

FIGURE **21.36**

Transmission and inhibition in neural networks. In the hypothetical example shown, an action potential passes from the axon at the top, through the neuron in the center, to reach the neuron at the bottom. However, another signal has generated an inhibitory response at a synapse on the lower cell body, blocking further transmission of the action potential.

Certain small peptides, such as **somatostatin, neurotensin,** and the **enkephalins,** also act as neurotransmitters (Table 21.2). In some cases, these compounds exhibit a second function as **neurohormones,** which are described in the last section.

Finally, there are even specialized synapses that do not use neurotransmitter substances at all. Although transmission through a synapse via neurotransmitters can be quite rapid (about a millisecond), some responses must be even quicker than this will allow. In such cases, there is direct electrical–ionic conduction between neural cells, using gap junctions (see Chapter 10). Such **electrical synapses** are frequently found in animals that live in cold environments yet need to make quick motions. The low body temperatures of such creatures would slow the chemical and diffusion processes involved in chemical synapses. Direct conduction provides a solution to this problem.

Neurotransmitter Receptors and Psychopharmacology

Recent years have seen explosive growth in our understanding of relationships between particular neurotransmitters and receptors, on the one hand, and specific behavior patterns and abnormalities, on the other—including aggressive behavior, anxiety, biochemical changes accompanying learning, and psychiatric disorders such as schizophrenia and drug addiction. Most of our enlarged understanding has emerged from the discovery of multiple receptor classes, such as the

TABLE **21.2 Some peptides that act as neurohormones (H) or neurotransmitters (T)**

Name	H/T	Sequence[a]
β-Endorphin	H	YGGFMTSFKSQTPLVTLFKNAIIKNAYKKGE
Met-enkephalin	H, T	YGGFM
Leu-enkephalin	H, T	YGGFL
Neurotensin	T	pELYENKPRRPYIL
Somatostatin	T	AGCKNFFWKTFTSC

[a]The subsequence YGGF, common to β-endorphin and the enkephalins, appears to be essential for their narcotic effects. The p at the N-terminal end of neurotensin signifies that the glutamate has been cyclized to the "pyro" form.

four distinct dopamine receptors mentioned earlier. This has permitted two lines of attack: (1) psychopharmacological approaches, in which studies on recombinant receptor proteins are allowing the development of agonists and antagonists specific to one or two receptor subtypes, for studies in both experimental animals and humans, and (2) molecular genetic approaches, in which the development of transgenic animal models (see Tools of Biochemistry 25B) allows the selective deletion of the gene for a particular subtype in an experimental animal, usually a "knockout mouse" (Tools of Biochemistry 25B).

Current research on the biochemistry of drug addiction provides good examples of what is being learned about molecular control of behavior. It has long been suspected that dopamine metabolism is intimately connected with addiction, because all addictive drugs have been shown to cause a surge in dopamine levels in the "reward" section of the brain, the **nucleus accumbens.** In one study, agonists for the closely related D1 and D2 dopamine receptors in rats were found to exert opposite effects upon the animals' tendency to dose themselves with cocaine after an initial drug experience. In another study a targeted deletion of the D4 receptor in mice caused them to become hypersensitive to ethanol, cocaine, and methamphetamine.

Recent work, however, has implicated glutamate receptors as well in control of addictive behavior. Inhibition of glutamate neurotransmission in rats has been shown to modulate the compulsive drug-seeking behavior after an initial drug experience, and neuroscientists are excited at the prospect of developing therapies involving glutamate antagonists, which might increase the likelihood that an addict would remain "clean" after treatment. Glutamate receptors are also strongly linked to the action of phencyclidine (PCP, or "angel dust"). This compound blocks glutamate binding to the *N*-methyl-*D*-aspartate (NMDA) class of glutamate receptors, thereby inducing a schizophrenia-like state thought until recently to result from decreased "glutamatergic" neurotransmission. However, neuroscientists have found that lowering brain glutamate levels with another drug greatly diminishes the effectiveness of PCP, suggesting a possible new approach to the treatment of schizophrenia.

Glutamate receptors are receiving greatly increased attention in other regards. A particular receptor subtype has been shown to be phosphorylated by a calmodulin-dependent kinase during "long-term potentiation," a neurophysiological response associated with learning, suggesting that this covalent receptor modification might be associated with long-term memory.

Neurohormones

Whereas transmitter substances carry out the actual passage of a signal across a synapse, other substances in nervous systems modify the way in which nerve cells respond to transmitters. These substances include peptides called *neurohormones,* which were discovered originally as a result of Solomon Snyder's studies on addiction to opiate drugs such as **morphine.** Snyder detected receptors in brain tissue specific for binding opiates. Because it seemed unlikely that vertebrate brains would contain specific receptors directed toward a product of the poppy plant, a search was made for the natural ligands that bound to these sites. The search led, in the 1970s, to the finding of several small peptides, called *enkephalins* and **endorphins,** which are natural analgesics (see Table 21.2). The modification of neural signals by these substances appears to be responsible for the insensitivity to pain that is experienced under conditions of great stress or shock. The effectiveness of opiate analgesics such as morphine is a consequence, perhaps accidental, of the recognition of these opiates by neurohormone receptors despite their structural differences from neurohormones.

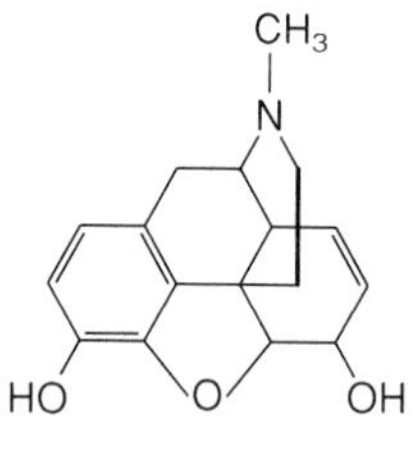

Morphine

The endorphins and enkephalins are synthesized as part of the much longer hormone precursor *prepro-opiomelanocortin.* As described in Chapter 23, this precursor is cleaved to release both the neurohormones and a number of other hormones with entirely different functions.

SUMMARY

Amino acids are synthesized from intermediates in the citric acid cycle, glycolysis, and the pentose phosphate pathway. Amino acids play numerous roles as intermediates in the biosynthesis of other metabolites, including purine nucleotides (glutamine, glycine, serine), pyrimidine nucleotides (aspartate, glutamine), polyamines (methionine), glutathione (glutamate, cysteine, glycine), creatine phosphate (arginine), neurotransmitters (tyrosine, tryptophan, glutamate, arginine), lignin, aromatic compounds and pigments (phenylalanine), hormones (tyrosine, histidine), porphyrins (glycine and glutamate in plants), and other amino acids. The roles of amino acids as neurotransmitters and neurotransmitter precursors are particularly important, as are their roles in porphyrin synthesis. Neurotransmission involves the diffusion of a small molecule from the end of a neuron to a target cell, either muscle or neuron, so that the action potential can be propagated. Agonists and antagonists of neurotransmitters are revealing the biochemical basis of learning and memory, mental illness, and addictive behavior. In porphyrin synthesis, glycine condenses with succinate to give a heterocyclic ring compound, porphobilinogen, which is the precursor to all four pyrrole rings in heme and other porphyrins. Degradation of porphyrins in animals yields iron, which is reutilized, and bilirubin, an insoluble tetrapyrrole that is excreted.

REFERENCES

The Citric Acid Cycle Superfamily

Cooper, J. B., J. A. Chen, G.-J. van Holst, and J. E. Varner (1987) Hydroxyproline-rich glycoproteins of plant cell walls. *Trends Biochem. Sci.* 12:24–27. Hydroxyproline was long thought to be present only in animal connective tissue proteins. This article summarizes its occurrence and roles in plant structural proteins.

Ecker, J. R. (1995) The ethylene signal transduction pathway in plants. *Science* 268:667–675. A timely review of the synthesis and actions of this important plant hormone.

Kivirikko, K. I., R. Myllylä, and T. Pihlajaniemi (1989) Protein hydroxylation: Prolyl 4-hydroxylase, an enzyme with four cosubstrates and a multifunctional subunit. *FASEB J.* 3:1609–1617. A review of the mechanism of proline hydroxylation and its role in collagen synthesis.

Mayer, B., and B. Hemmens (1997) Biosynthesis and action of nitric oxide in mammalian cells. *Trends Biochem. Sci.* 22:477–481. Surprisingly comprehensive for a four-page minireview.

Perry, J. M., and M. A. Marletta (1998) Effects of transition metals on nitric oxide synthase catalysis. *Proc. Natl. Acad. Sci USA* 95:11101–11106. A recent study of the role of tetrahydrobiopterin in this reaction, by a leader in mechanistic analysis of nitric oxide synthase.

Theologis, A. (1992) One rotten apple spoils the whole bushel: The role of ethylene in fruit ripening. *Cell* 70:181–184. A minireview describing ethylene biosynthesis and functions.

Wilson, E. K. (1998) Impotence drugs: Beyond Viagra. *Chem. Eng. News,* June 29 issue. This news article clearly explains the effects of Viagra on nitric oxide metabolism and the rationale behind development of related drugs.

Withers, H. L., and K. Nordström (1998) Quorum-sensing acts at initiation of chromosomal replication in *Escherichia coli. Proc. Natl. Acad. Sci. USA* 95:15694–15699. A recent research paper on quorum-sensing, with references to earlier literature.

Sulfur-Containing Amino Acids

Banerjee, R. V., and R. G. Matthews (1990) Cobalamin-dependent methionine synthase. *FASEB J.* 4:1450–1459. An integrated treatment of mechanism, genetics, and potential for therapeutic manipulation.

Cohen, S. S. (1998) *Biochemistry of the Polyamines.* Oxford University Press, New York. An all-encompassing review of the metabolism and functions of polyamines, written by a long-time leader in the field.

Cowley, G., and A. Underwood (1999) What is SAMe? *Newsweek,* July 5 issue, pp. 46–50. A news article touting *S*-adenosylmethionine as a wonder drug for depression, arthritis, and liver disease.

Pickett, C. B., and A. Y. H. Lu (1989) Glutathione S-transferases: Gene structure, regulation, and biological function. *Annu. Rev. Biochem.* 58:743–764. A discussion of this important class of detoxifying enzymes.

Aromatic Amino Acids

Brennan, M. M. (1998) New age paper and textiles. *Chem. Eng. News,* March 23 issue, pp. 39–47. A news-type article describing the use of biological reagents to degrade lignin and their applications in paper and textile production.

Goodwin, T. W., and E. I. Mercer (1983) *Introduction to Plant Biochemistry,* 2nd ed. Pergamon, Oxford. Different sections of this book treat in detail topics such as lignin biosynthesis, flavonoid metabolism, glyphosate and its action, and the conversion of homogentisic acid to vitamin E and its derivatives.

Kaufman, S. (1993) New tetrahydrobiopterin-dependent systems. *Annu. Rev. Nutr.* 13:261–286. Originally discovered as a phenylalanine hydroxylase cofactor, tetrahydrobiopterin has multiple coenzymatic functions.

Miles, E. W., S. Rhee, and D. R. Davies (1999) The molecular basis of substrate channeling. *J. Biol. Chem.* 274:12193–12196. This minireview discusses tryptophan synthase and other examples.

Okun, M. R. (1997) The role of peroxidase in neuromelanin synthesis: A review. *Physiol. Chem. Phys. Med. NMR* 29:15–22. This minireview summarizes the evidence that melanin synthesis, at least in some mammalian tissues, requires two enzymes.

Pan, P., E. Woehl, and M. F. Dunn (1997) Protein architecture, dynamics and allostery in tryptophan synthase channeling. *Trends Biochem. Sci.* 22:22–27. A nice summary relating structural and kinetic analysis in this well-understood example of metabolite channeling.

Scriver, C. R., S. Kaufman, R. C. Eisensmith, and S. L. C. Woo (1995) The hyperphenylalaninemias. In: *The Metabolic and Molecular Bases of Inherited Disease,* 7th ed., edited by Scriver, C. R., A. L. Beaudet, W. S. Sly, and D. Valle, pp. 1015–1076. McGraw-Hill, New York. This is the first of 12 articles in this compendium that treat heritable metabolic disorders of amino acid metabolism.

Valine, Leucine, and Isoleucine

Yeaman, S. J. (1986) The mammalian 2-oxoacid dehydrogenases: A complex family. *Trends Biochem. Sci.* 11:293–296. This minireview compares and contrasts the complexes involved in oxidizing pyruvate, α-ketoglutarate, and the branched-chain keto acids.

Porphyrin Metabolism

Jahn, D., E. Verkamp, and D. Söll (1992) Glutamyl-transfer RNA: A precursor of heme and chlorophyll biosynthesis. *Trends Biochem. Sci.* 17:215–218. This article discusses the unexpected role of RNA in δ-ALA synthetase.

Warren, M. J., J. B. Cooper, S. P. Wood, and P. M. Shoolingin-Jordan (1998) Lead poisoning, haem synthesis and 5-aminolaevulinic acid dehydratase. *Trends Biochem. Sci.* 23:217–221. The structural basis for the porphyria acquired in lead poisoning.

Warren, M. J., M. Jay, D. M. Hunt, G. H. Elder, and J. G. Rôhl (1996) The maddening business of King George III and porphyria. *Trends Biochem. Sci.* 21:229–234. A fascinating mixture of history and biochemistry, illustrated with scenes from the movie.

Neurotransmitters

Barondes, S. H. (1994) Thinking about Prozac. *Science* 263:1102–1103. A brief discussion of the wide range of illnesses treatable by this drug.

Barria, A., D. Muller, V. Derkach, L. C. Griffith, and T. R. Soderling (1997) Regulatory phosphorylation of AMPA-type glutamate receptors by CaM-KII during long-term potentiation. *Science* 276:2042–2045. This and an accompanying news article set out the case for receptor phosphorylation as a key event in learning and memory.

Hayaishi, O. (1994) Tryptophan, oxygen, and sleep. *Annu. Rev. Biochem.* 63:1–24. An autobiographical article by a leader in this field.

Kandel, E., and T. Abel (1995) Neuropeptides, adenylyl cyclase, and memory storage. *Science* 268:825–826. A timely and well-referenced minireview.

Moghaddam, B., and B. W. Adams (1998) Reversal of phencyclidine effects by a group II metabotropic glutamate receptor agonist in rats. *Science* 281:1349–1354. This article and an accompanying news story raise the likelihood that glutamate metabolism could present attractive targets for treating schizophrenia.

Rubinstein, M., and 14 coauthors (1997) Mice lacking dopamine D4 receptors are supersensitive to ethanol, cocaine, and methamphetamine. *Cell* 90:991–1001. An example of the use of knockout mice to assess the functions of specific receptor classes.

Snyder, S. H., and T. Narahashi, eds. (1990) Diseases of receptors and channels. *FASEB J.* 4:2707–2816. A special issue of this journal, devoted to parkinsonism, schizophrenia, and other disorders of receptor function.

Taubes, G. (1994) Will new dopamine receptors offer a key to schizophrenia? *Science* 265:1034–1035. A news article describing the excitement over discovery of new dopamine receptors.

Wickelgren, I. (1998) Teaching the brain to take drugs. *Science* 280: 2045–2047. A news-type article describing studies that relate glutamate receptors to drug addiction.

PROBLEMS

1. A clinical test sometimes used to diagnose folate deficiency or B_{12} deficiency is a histidine tolerance test, where one injects a large dose of histidine into the bloodstream and then carries out a series of biochemical determinations. What histidine metabolite would you expect to accumulate in a folate- or B_{12}-deficient patient, and why?

2. In bacteria much of the putrescine is synthesized, not from ornithine but from arginine, which decarboxylates to yield *agmatine.* Formulate a plausible pathway from arginine to putrescine, using this intermediate.

$$H_2N-\overset{\overset{+}{N}H_2}{\overset{\|}{C}}-NH-CH_2-CH_2-CH_2-CH_2-\overset{+}{N}H_3$$

Agmatine

3. The mitochondrial form of carbamoyl phosphate synthetase is allosterically activated by *N*-acetylglutamate. Briefly describe a rationale for this effect.

4. *Psilocybin* is a hallucinogenic compound found in some mushrooms. Present a straightforward pathway for its biosynthesis from one of the aromatic amino acids.

O(P)
CH₂—CH₂—N(CH₃)₂
N
H

Psilocybin

*5. One can identify phenylketonurics and PKU carriers (heterozygotes) by means of a phenylalanine tolerance test. One injects a large dose of phenylalanine into the bloodstream and measures its clearance from the blood by measuring serum phenylalanine levels at regular intervals. Sketch curves showing relative blood phenylalanine concentration versus time that you would expect to be displayed by (a) a PKU patient, (b) a heterozygote, and (c) a normal individual. What kind of tolerance test could you devise to distinguish between PKU resulting from either phenylalanine hydroxylase deficiency or dihydropteridine reductase deficiency?

6. (a) Formaldehyde reacts nonenzymatically with tetrahydrofolate to generate 5,10-methylenetetrahydrofolate. [^{14}C]Formaldehyde can be used to prepare serine labeled in the β-carbon. What else would be needed?

(b) [^{14}C]Serine, prepared as described above, is useful for many things, but you would probably not want to use it for studies on protein synthesis because it would label nucleic acids, carbohydrates, and lipids, as well as proteins. Indicate how each of these classes of compounds could become labeled by this precursor.

7. If oxidation of acetyl-CoA yields 12 ATPs per mole through the citric acid cycle, how many ATPs will be derived from the complete metabolic oxidation of 1 mole of alanine in a mammal? Would the corresponding energy yield in a fish be higher or lower? Why? How much energy would be derived from the metabolic oxidation of 1 mole of leucine to CO_2, H_2O, and NH_3? Of tyrosine?

8. Some bacteria contain three different forms of aspartokinase, each with its own mode of regulation. Based on the roles of aspartokinase, as discussed in the text, propose a regulatory scheme applicable to each form of aspartokinase.

9. Proline betaine is a putative osmoprotector in plants and bacteria, helping to prevent dehydration of cells.

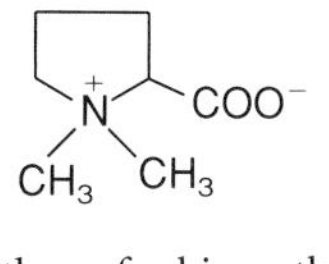

N^+ COO⁻
CH_3 CH_3

Propose a plausible pathway for biosynthesis of this compound.

10. Most bacterial mutants that require isoleucine for growth also require valine. Why? Which enzyme or reaction would be defective in a mutant requiring only isoleucine (not valine) for growth?

11. Describe a series of allosteric interactions that could adequately control the biosynthesis of valine, leucine, and isoleucine.

12. The structure shown below is an intermediate in the synthesis of which biogenic amine? Use arrows to show how the next intermediate in this reaction is formed, and draw the structure of that intermediate.

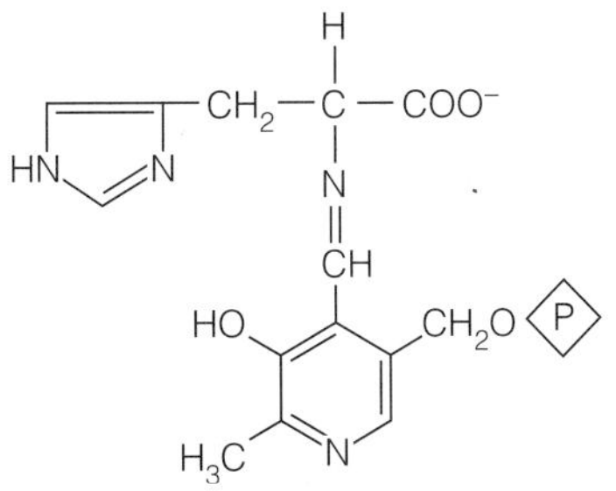

H
CH_2—C—COO^-
HN N
N
CH
HO CH_2O P
H_3C N

13. Identify carbon atoms, by number, that are incorporated from this structure into the compounds listed in a–f.

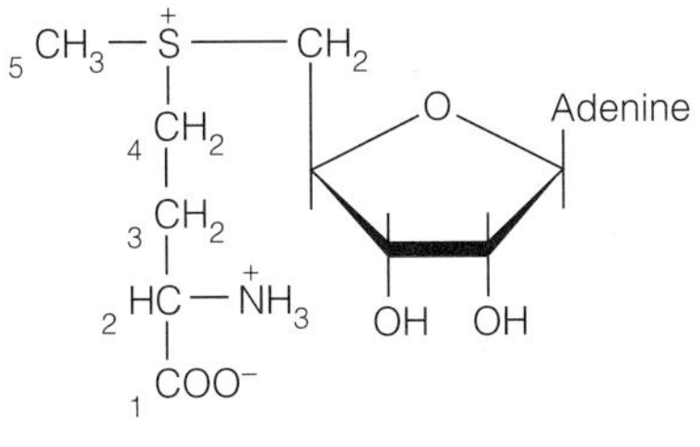

5 CH_3—$\overset{+}{S}$—CH_2
4 CH_2 O Adenine
3 CH_2
2 HC—$\overset{+}{N}H_3$ OH OH
1 COO^-

(a) Creatine phosphate ________
(b) Spermidine ________
(c) Ethylene ________
(d) Putrescine ________
(e) Glycine betaine ________
(f) Epinephrine ________

14. Why is phenylketonuria resulting from dihydropteridine reductase deficiency a more serious disorder than PKU resulting from phenylalanine hydroxylase deficiency?

*15. Propose a mechanism for the reaction catalyzed by dopamine β-hydroxylase. Assume that copper in the enzyme binds O_2 and that copper can change its oxidation state during the reaction.

16. Propose a plausible pathway for the biosynthesis of ovothiol from histidine, indicating the involvement of cofactors.

17. Glyphosate, which inhibits the EPSP synthase reaction of the shikimic acid pathway, is a phosphonic acid derivative of glycine; hence its name (glycine phosphonate). Based upon your knowledge of the EPSP synthase reaction (see pages 761–762), would you expect the inhibition to be competitive or noncompetitive with respect to each of the substrates? Briefly explain your answer.

18. In evaluating jaundiced patients, clinicians often determine whether the bilirubin that accumulates in the blood is primarily free bilirubin or the diglucuronide. Which form would accumulate primarily in chronic liver disease? In bile duct obstruction by a gallstone? In hemolytic anemia, such as that associated with glucose-6-phosphate dehydrogenase deficiency (see Chapter 14)? Briefly explain your answers.

19. Propose additional substrates and cofactors that might participate in each of the first three reactions in the flavonoid biosynthetic pathway shown on page 767.

20. Write a one-sentence answer to each question.
(a) Why does liver damage cause jaundice?
(b) Why do individuals with congenital erythropoietic porphyria become anemic?
(c) Why does a genetic deficiency of 5,10-methylenetetrahydrofolate reductase cause homocystinuria?

CHAPTER 22

Nucleotide Metabolism

WE HAVE ENCOUNTERED NUCLEOTIDES REPEATEDLY DURING OUR EXPLORATION of biochemistry. They serve as precursors to nucleic acids, as critical elements in energy metabolism, as carriers of activated metabolites for biosynthesis (such as nucleoside diphosphate sugars), as structural moieties of coenzymes, and, finally, as metabolic regulators and signal molecules (notably, cyclic AMP). In this chapter we discuss pathways of biosynthesis and degradation of purine and pyrimidine nucleotides, and we explore regulation of these processes—particularly critical in pathways leading to DNA replication. We discuss nucleotide biosynthetic enzymes as targets for the action of antimicrobial and anticancer drugs, and we describe the metabolic consequences of certain heritable alterations of nucleotide metabolism. The roles of nucleotides in metabolic or genetic regulation are discussed elsewhere in this book, in connection with the specific processes regulated.

Before starting this chapter, you may find it useful to review the information on nucleotide structure in Chapter 4. You should be aware also of the distinction between nucleosides and nucleotides. On complete hydrolysis, a *nucleoside* yields at least 1 mole each of a sugar and a heterocyclic base, whereas a *nucleotide* yields at least 1 mole each of a sugar, a base, and inorganic phosphate. A *mononucleotide* contains only 1 mole each of base and sugar, but it may contain more than one phosphate. If it contains, for example, three phosphates, it is called a *nucleoside triphosphate.* The *deoxyribonucleotides,* which are used in DNA synthesis, are formed from *ribonucleotides* (RNA constituents) by pathways discussed later in this chapter.

Outlines of Pathways in Nucleotide Metabolism

BIOSYNTHETIC ROUTES: DE NOVO AND SALVAGE PATHWAYS

Unlike the other classes of metabolites we have encountered, neither nucleotides nor the bases and nucleosides from which they are formed are required to meet nutritional requirements, with the exception of some protozoan parasites. Most organisms can synthesize purine and pyrimidine nucleotides from low-molecular-

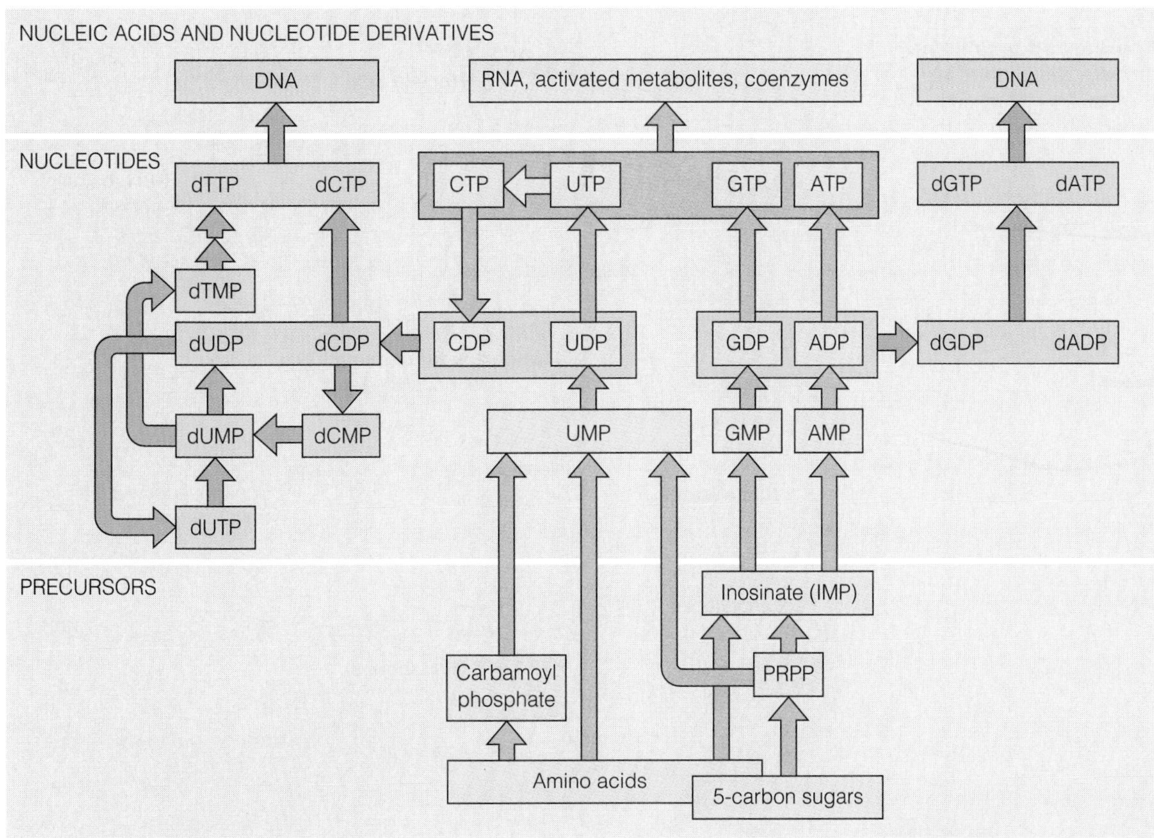

FIGURE **22.1**

Overview of nucleotide metabolism. De novo pathways for synthesis and utilization of ribonucleotides (orange) and deoxyribonucleotides (blue).

weight precursors in amounts sufficient for their needs. These so-called **de novo pathways** are essentially identical throughout the biological world (Figure 22.1). Most organisms can also synthesize nucleotides from nucleosides or bases that become available either in the diet or through enzymatic breakdown of nucleic acids. These processes are called **salvage pathways,** because they involve the utilization of preformed purine and pyrimidine compounds that would otherwise be lost to biodegradation. As we shall see, salvage pathways represent important targets for treatment of microbial or parasitic diseases, sites for manipulation of biological systems (for example, in studies of mutagenesis or in preparation of monoclonal antibodies), and biological processes in which genetic alterations have severe and far-reaching consequences.

Nucleotides arise through de novo synthesis from low-molecular-weight precursors or through salvage of nucleosides or bases.

NUCLEIC ACID DEGRADATION AND THE IMPORTANCE OF NUCLEOTIDE SALVAGE

Because salvage, or reuse, of purine and pyrimidine bases involves molecules released by nucleic acid degradation, let us begin by briefly considering these processes (Figure 22.2). Degradation can occur intracellularly (through the turnover of unstable messenger RNA species or through DNA repair pathways), as a result of cell death, or, in animals, through digestion of nucleic acids ingested in the diet.

In animals the extracellular hydrolysis of ingested nucleic acids represents the major route by which bases and nucleosides become available. The breakdown processes are comparable to those involved in protein digestion. Cleavage processes begin at internal linkages—in this case, phosphodiester bonds. Catalysis occurs via **endonucleases,** such as pancreatic ribonuclease or deoxyribonuclease, which function to digest nucleic acids in the small intestine. Endonucleolytic cleavages yield oligonucleotides, which are then cleaved **exonucleolytically** (at linkages

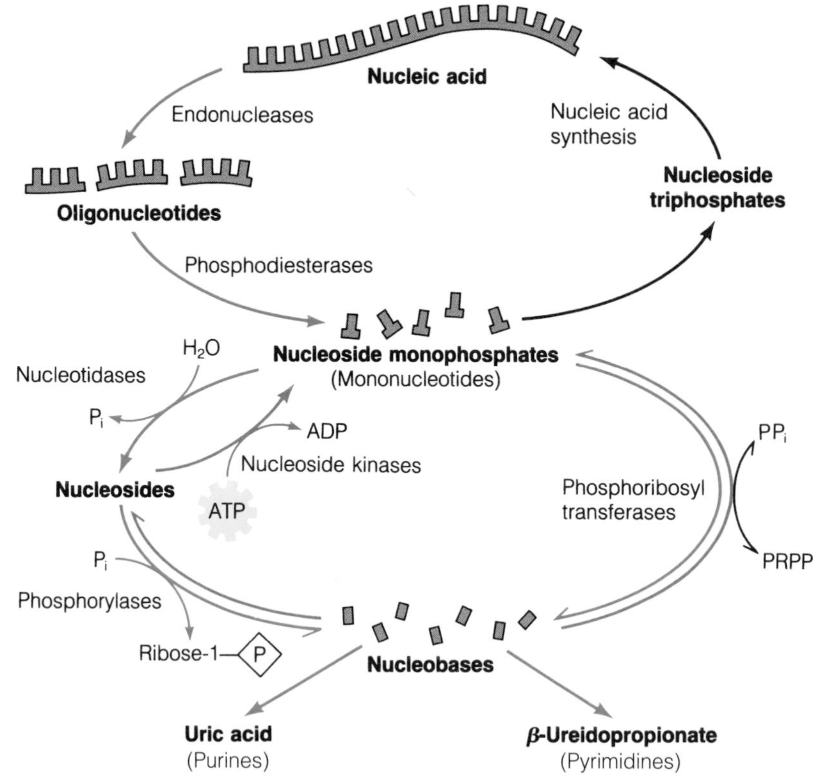

FIGURE **22.2**

Reutilization of purine and pyrimidine bases. The figure shows relationships between nucleic acid catabolism (blue) and resynthesis of nucleotides by salvage pathways (red).

near the ends of molecules) by nonspecific enzymes called **phosphodiesterases.** The products are mononucleotides—nucleoside 5′- or 3′-monophosphates, depending on the specificities of the enzymes involved. Nucleotides can then be cleaved hydrolytically, by a group of phosphomonoesterases called **nucleotidases,** to yield orthophosphate plus the corresponding nucleoside. Although hydrolytic cleavage of the resultant nucleoside does occur, the most common route for cleavage to the base involves the action of a **nucleoside phosphorylase.** Like glycogen phosphorylase, nucleoside phosphorylases cleave a glycosidic bond by adding across it the elements of inorganic phosphate, to yield the corresponding base plus ribose-1-phosphate (or deoxyribose-1-phosphate if the substrate is a deoxyribonucleoside):

$$\underset{\textbf{Guanosine}}{\text{(HOCH}_2\text{–ribose–guanine; OH OH)}} + {}^-\text{O–P(=O)(O}^-\text{)–OH} \rightleftharpoons \text{guanine} + \underset{\textbf{Ribose-1-phosphate}}{\text{(HOCH}_2\text{–ribose–O–P; OH OH)}}$$

These reactions are readily reversible, such that a nucleoside phosphorylase can also catalyze the first step in salvage synthesis of nucleotides from free nucleobases. When that occurs, the product nucleoside can be phosphorylated by ATP, through the action of a **nucleoside kinase.** Such enzymes are not universal. For example, animal cells contain neither a guanosine kinase nor a uridine phosphorylase, although these enzymes are found in other organisms.

If bases or nucleosides are not reused for nucleic acid synthesis via salvage pathways, the purine and pyrimidine bases are further degraded, to uric acid or β-ureidopropionate, respectively, as indicated in Figure 22.2. We consider these pathways later in this chapter.

PRPP: A CENTRAL METABOLITE IN DE NOVO AND SALVAGE PATHWAYS

An alternative salvage pathway, also shown in Figure 22.2, synthesizes nucleoside 5′-phosphates directly from free bases. This route involves a class of enzymes called **phosphoribosyltransferases** and an activated sugar phosphate, **5-phospho-α-D-ribosyl-1-pyrophosphate (PRPP).** Identified in Chapter 21 as an intermediate in histidine and tryptophan biosynthesis, PRPP is a key intermediate in the de novo synthesis of both purine and pyrimidine nucleotides. It is formed through the action of **PRPP synthetase,** which activates carbon 1 of ribose-5-phosphate by transferring to it the pyrophosphate moiety of ATP:

PRPP is an activated ribose-5-phosphate derivative used in both salvage and de novo pathways.

ATP + **Ribose-5-phosphate** ⇌ **5-Phospho-α-D-ribosyl-1-pyrophosphate (PRPP)** + AMP

A phosphoribosyltransferase reaction catalyzes the reversible transfer of a free base to the ribose of PRPP, producing a nucleoside monophosphate and pyrophosphate. Because the deoxyribose analog of PRPP is absent from most cells, these enzymes are not involved directly in deoxyribonucleotide metabolism.

PRPP + guanine ⇌ **GMP** + PP_i

In principle such a reaction could participate in nucleotide breakdown. However, in vivo, pyrophosphate is rapidly cleaved by pyrophosphatase to give inorganic phosphate. This dictates that phosphoribosyltransferases most commonly act in the direction of nucleotide biosynthesis.

De Novo Biosynthesis of Purine Nucleotides

EARLY STUDIES ON DE NOVO PURINE SYNTHESIS

The reactions of de novo purine nucleotide biosynthesis were identified in the 1950s, in the laboratories of John Buchanan and Robert Greenberg. Elucidation of the pathway began with the realization that birds excrete most of their excess nitrogen compounds in the form of *uric acid,* an oxidized purine (see Chapter 20). Thus, researchers were able to identify low-molecular-weight precursors to purines by administering isotopically labeled compounds to pigeons, crystallizing uric acid from the droppings, and, by selective chemical degradation, determining which positions were labeled by which precursors. This procedure yielded the pattern shown in Figure 22.3. At that time, 10-formyltetrahydrofolate was not known, but compounds such as formate or serine labeled in the hydroxymethyl carbon readily labeled C-2 and C-8 of uric acid.

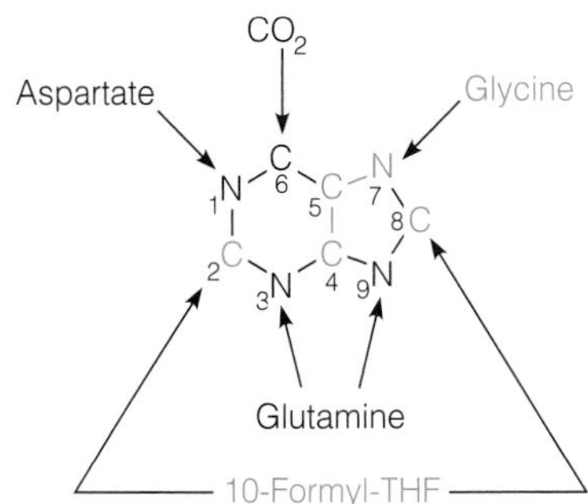

FIGURE **22.3**

Low-molecular-weight precursors to the purine ring. The source of each atom in the ring, as established with isotopic tracer studies of uric acid synthesis.

Next, two related antibiotics, **azaserine** and **6-diazo-5-oxonorleucine** (**DON**), were identified as potent inhibitors of purine nucleotide synthesis. Recognition that these compounds are structural analogs of glutamine led to the eventual realization that azaserine and 6-diazo-5-oxonorleucine are irreversible inhibitors of a class of enzymes called the **glutamine amidotransferases,** which catalyze the ATP-dependent transfer of the amido nitrogen of glutamine to an acceptor. Three such reactions occur in purine nucleotide synthesis (and one in pyrimidine nucleotide synthesis).

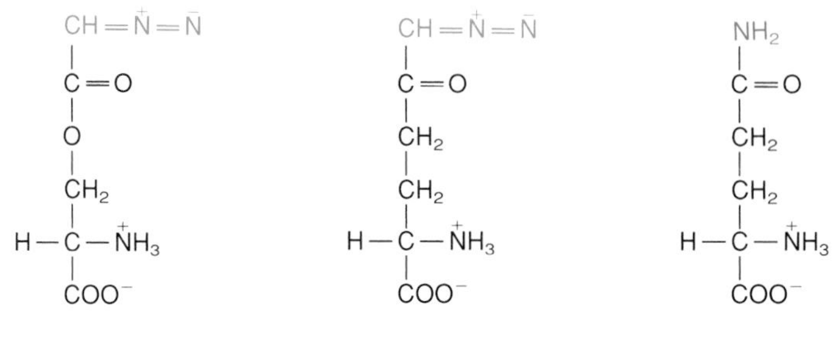

Azaserine **6-Diazo-5-oxonorleucine (DON)** **Glutamine**

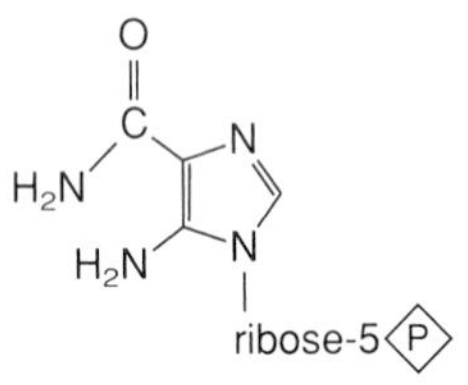

In later experiments, bacteria were treated with sulfonamide drugs, such as sulfanilamide. The bacteria excreted large quantities of a red compound identified as an oxidation product of **5-aminoimidazole-4-carboxamide ribonucleotide** (**AICAR**), which resembles an incomplete purine nucleotide. This finding suggested that AICAR is a biosynthetic intermediate whose use was somehow blocked by the drug. Because sulfonamides block the synthesis of folate coenzymes, the accumulation of AICAR suggested that a folate coenzyme participates in the next reaction. Moreover, these observations suggested that the pathway proceeds at the nucleotide level—in other words, that the purine ring is assembled while already attached to the ribose-5-phosphate moiety.

5-Aminoimidazole-4-carboxamide ribonucleotide (AICAR)

PURINE SYNTHESIS FROM PRPP TO INOSINIC ACID

Figure 22.4 summarizes the pathway leading from PRPP, the first intermediate, to the first fully formed purine nucleotide, **inosine 5′-monophosphate** (**IMP**), also called **inosinic acid.** This compound is the 5′-ribonucleotide of the purine base **hypoxanthine.** Note that there are two glutamine amidotransferase reactions in this process, reactions 1 and 4. They differ mechanistically in that PRPP amidotransferase (reaction 1) does not require ATP, because the substrate has been activated by ATP in the previous step. An inversion of configuration occurs in reaction 1, as the amido nitrogen displaces the pyrophosphate moiety. The latter is an excellent leaving group, giving a simple nucleotide (5-phosphoribosylamine), which carries the β-configuration at carbon 1 of the sugar, as do all the common nucleotides.

Purines are synthesized at the nucleotide level, starting with PRPP conversion to phosphoribosylamine and purine ring assembly on the amino group.

In reaction 2, a glycine molecule is transferred, with the aid of ATP, to the nitrogen of phosphoribosylamine. This is followed by a **transformylase** reaction, in which a formyl group is transferred from 10-formyltetrahydrofolate to the growing

FIGURE **22.4**

De novo biosynthesis of the purine ring, from PRPP to inosinic acid. In this and subsequent figures ribose—Ⓟ refers to a ribose-5-phosphate moiety in a nucleotide. The enzymes involved are (1) PRPP amidotransferase, (2) GAR synthetase, (3) GAR transformylase, (4) FGAR amidotransferase, (5) FGAM cyclase, (6) AIR carboxylase, (7) SAICAR synthetase, (8) SAICAR lyase, (9) AICAR transformylase, and (10) IMP synthase.

PRPP

Gln

Glu, PP_i

Reaction 1 PRPP amidotransferase

AMP, GMP

P—O—CH_2 O NH_2 HO OH

5-Phosphoribosylamine (PRA)

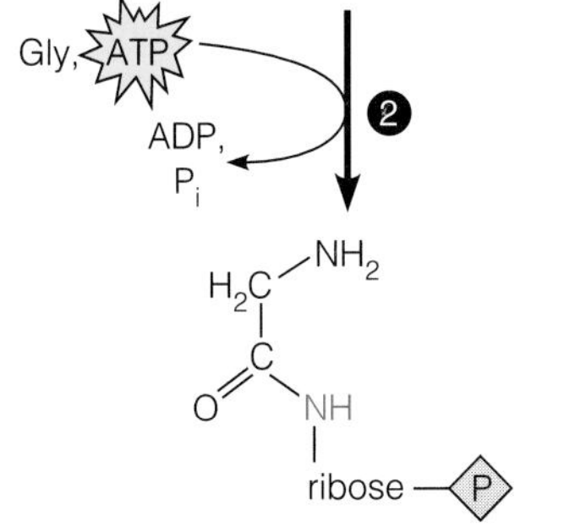

Glycinamide ribonucleotide (GAR)

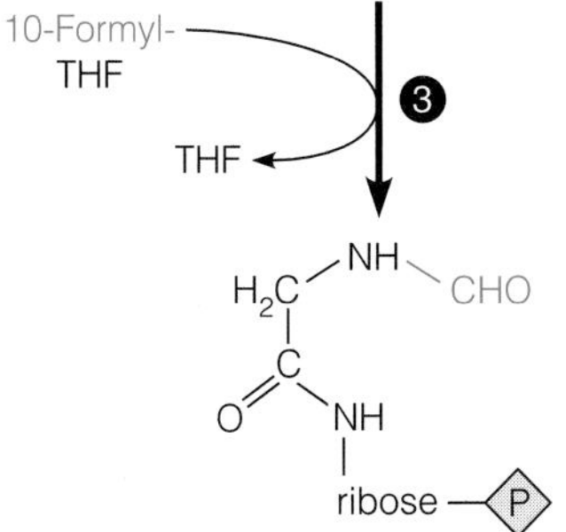

Formylglycinamide ribonucleotide (FGAR)

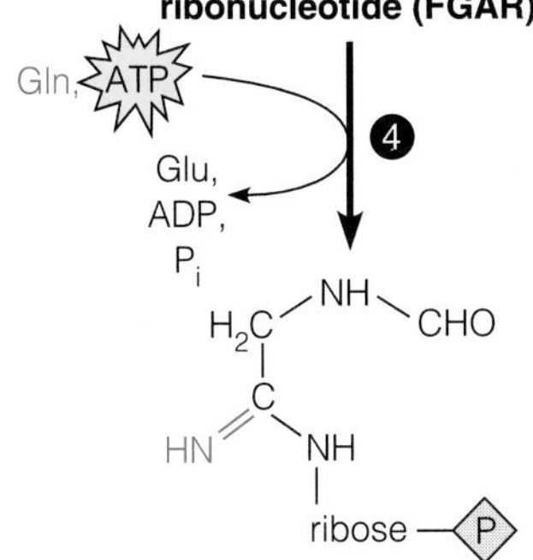

Formylglycinamidine ribonucleotide (FGAM)

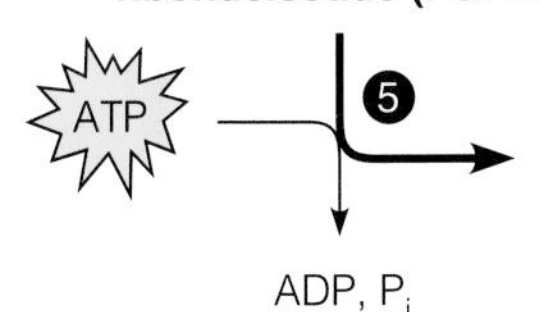

H_2N N CH N ribose P

5-Aminoimidazole ribonucleotide (AIR)

CO_2 6

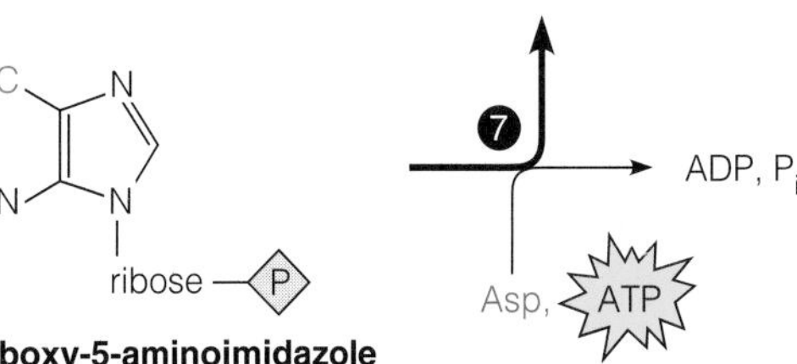

4-Carboxy-5-aminoimidazole ribonucleotide

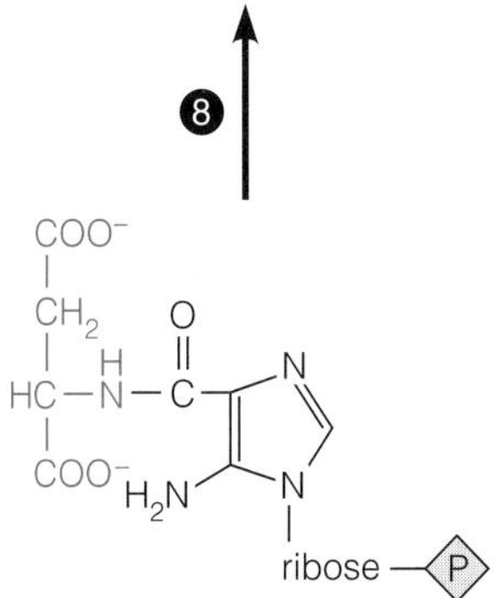

***N*-Succinylo-5-aminoimidazole-4-carboxamide ribonucleotide (SAICAR)**

COO^- HC CH COO^- **Fumarate** + H_2N C N H_2N N ribose P

5-Aminoimidazole-4-carboxamide ribonucleotide (AICAR)

9 THF 10-Formyl-THF

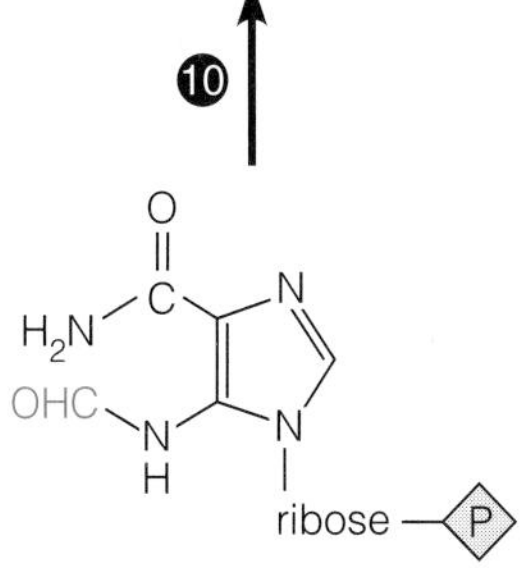

***N*-Formylaminoimidazole-4-carboxamide ribonucleotide (FAICAR)**

O HN N N N ribose P

Inosinic acid (IMP)

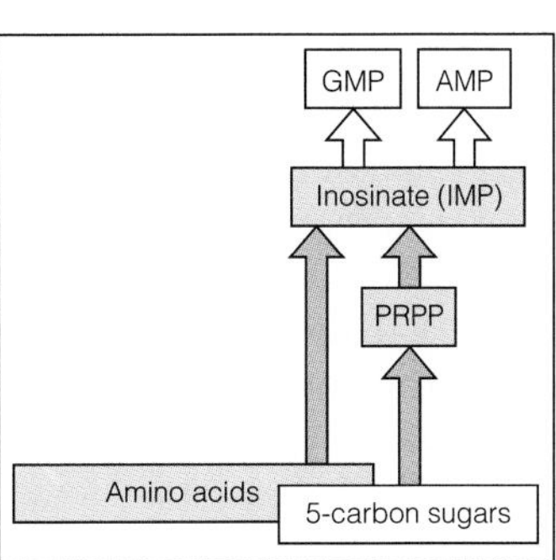

purine ring. As we have noted, reaction 4 is catalyzed by an ATP-dependent amidotransferase. Reaction 5 is an ATP-dependent ring closure, giving the imidazole portion of the purine ring. Reaction 6 is a reversible carboxylation reaction, which is noteworthy in that it does *not* require biotin. Reactions 7 and 8 result in the transfer of a nitrogen from aspartate, by a mechanism identical to that used to convert citrulline to arginine in the urea cycle (see Chapter 20). First, the entire aspartate molecule is transferred to the carboxyl group of 4-carboxy-5-aminoimidazole ribonucleotide (reaction 7). An α,β-elimination reaction follows (reaction 8), yielding AICAR, the intermediate shown to accumulate in sulfonamide-treated bacteria. Reaction 9 is another transformylase reaction, with a single-carbon group transferred from 10-formyltetrahydrofolate. Finally, an internal condensation reaction (reaction 10) yields the first purine compound, inosinic acid.

Vertebrate cells contain several of these activities in the form of multifunctional enzymes. This fact came to light when cloned genes for these enzymes were transferred into *E. coli,* and single cloned genes were found to complement (that is, to replace the function of) two or three different bacterial genes. For example, a single cloned cDNA (see Tools of Biochemistry 25B) allowed growth of bacterial purine auxotrophs defective in E_2, E_3, or E_5. Subsequent analysis showed that the cloned vertebrate DNA encoded a single polypeptide that catalyzed these three reactions. Similar observations confirmed that reactions 6 and 7 are catalyzed by a single bifunctional enzyme.

The two transformylase enzymes, which catalyze reactions 3 and 9, exist not as a single-chain bifunctional enzyme but as a multiprotein complex, in association with serine transhydroxymethylase and a trifunctional enzyme, formylmethenylmethylene-tetrahydrofolate synthetase (Figure 22.5). The latter protein carries out three activities of folate coenzyme synthesis (see page 735). Whether activities are carried on a multifunctional enzyme or a multienzyme complex, the advantages of juxtaposing catalytic sites are readily apparent—protecting labile tetrahydrofolate coenzymes, jointly regulating sequential enzyme activities, and "channeling" intermediates whose concentrations are low (that is, facilitating their direct transfer from one catalytic site to the next).

Control over the biosynthesis of inosinic acid is provided through feedback regulation of early steps in purine nucleotide synthesis. PRPP synthetase is inhibited by various purine nucleotides—particularly AMP, ADP, and GDP—and PRPP amidotransferase (reaction 1 of Figure 22.4) is inhibited allosterically by AMP, ADP, GMP, and GDP. In *E. coli* the expression of genes encoding these enzymes is controlled by a repressor protein (see Chapter 26), the product of the *purR* gene. This protein binds either hypoxanthine or guanine, and the resultant protein–

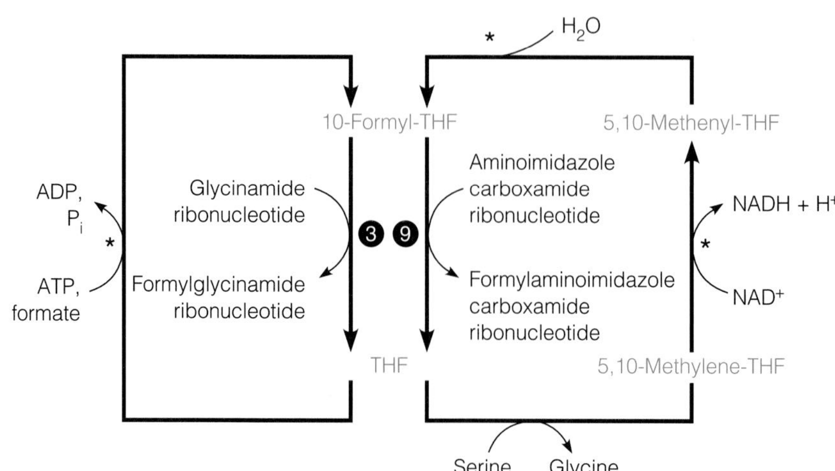

FIGURE **22.5**

Transformylation reactions in purine nucleotide synthesis. The transformylase enzymes shown in Figure 22.4 (reactions 3 and 9) are part of a multienzyme complex that catalyzes transformylation reactions in purine nucleotide synthesis. A trifunctional enzyme for tetrahydrofolate coenzyme synthesis is also part of the complex; its three activities are marked with an asterisk. The serine transhydroxymethylase reaction is shown at the lower right.

purine base complex binds to DNA sites upstream of several genes of purine (and pyrimidine) synthesis, thereby inhibiting their transcription.

SYNTHESIS OF ATP AND GTP FROM INOSINIC ACID

Inosinic acid represents a branch point in purine nucleotide synthesis. Figure 22.6 depicts the conversion of this compound both to adenosine 5′-monophosphate and to guanosine 5′-monophosphate. The pathway to guanine nucleotides begins with an NAD^+-dependent hydroxylation of the purine ring, yielding the nucleotide **xanthosine monophosphate (XMP)**; XMP contains the base xanthine. A glutamine-dependent amidotransferase reaction follows, yielding GMP. The route to AMP involves the transfer of nitrogen from aspartate to IMP, by a mechanism similar to that of reactions 7 and 8 in the de novo synthesis of the purine ring. First a succinylonucleotide intermediate is formed, and then an α,β-elimination reaction yields AMP plus fumarate. In fact, the same enzyme catalyzes both elimination reactions. Note from Figure 22.6 that the energy to drive the aspartate transfer reaction comes not from ATP but from GTP. This may represent a way to control the proportions of IMP that go to adenine and guanine nucleotide synthesis. GTP accumulation would tend to promote the pathway toward adenine nucleotides. Also, because the conversion of XMP to GMP is ATP-dependent, accumulation of ATP could promote guanine nucleotide synthesis.

IMP, the first fully formed purine nucleotide, is a branch point between adenine and guanine nucleotide biosynthesis.

Xanthine

Nucleotides are active in metabolism primarily as the nucleoside triphosphates. GMP and AMP are converted to their corresponding triphosphates through two

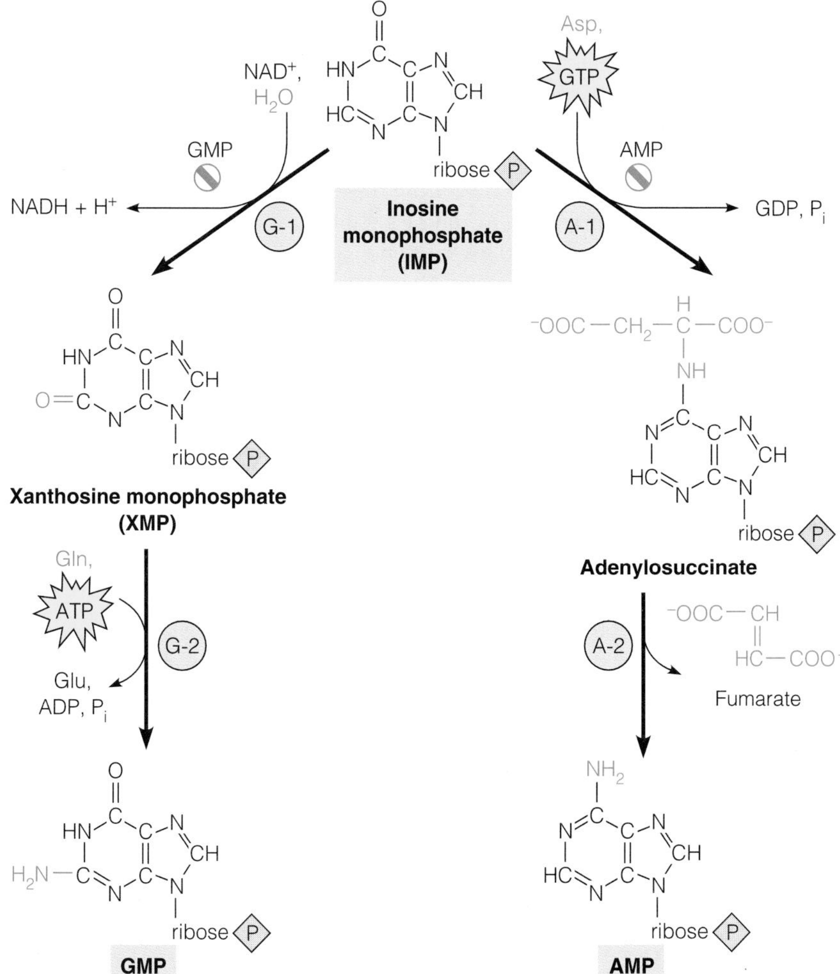

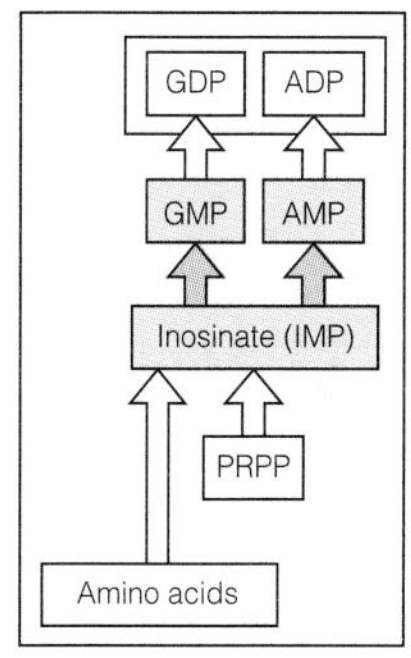

FIGURE **22.6**

Pathways from inosinic acid to GMP and AMP. Enzymes are as follows: G-1 = IMP dehydrogenase, G-2 = XMP aminase, A-1 = adenylosuccinate synthetase, and A-2 = adenylosuccinate lyase.

successive phosphorylation reactions. Conversion to the diphosphates involves specific ATP-dependent kinases.

$$\text{GMP} + \text{ATP} \xrightleftharpoons{\text{Guanylate kinase}} \text{GDP} + \text{ADP}$$

$$\text{AMP} + \text{ATP} \xrightleftharpoons{\text{Adenylate kinase}} \text{2ADP}$$

Phosphorylation of ADP to ATP occurs through energy metabolism—oxidative phosphorylation or substrate-level phosphorylations or (in plants) photophosphorylation. ATP can also be formed from ADP through the action of adenylate kinase, acting in the reverse of the direction shown here.

Nucleoside diphosphate kinase, an equilibrium-driven enzyme, transfers a phosphoryl group from ATP in the synthesis of all other nucleoside triphosphates.

ATP is the phosphate donor for conversion of GDP (and other nucleoside diphosphates) to the triphosphate level through the action of **nucleoside diphosphate kinase.** This enzyme is highly active but has broad specificity, with regard to both phosphoryl group donor and acceptor.

$$\text{GDP} + \text{ATP} \xrightleftharpoons{\text{NDP kinase}} \text{GTP} + \text{ADP} \qquad \Delta G^{\circ\prime} = 0$$

Because ATP is by far the most abundant nucleoside triphosphate in most cells, equilibrium and mass action considerations dictate that it is used most readily as the donor of the γ (outer) phosphate in synthesis of other nucleoside triphosphates.

In most organisms the biosynthesis of deoxyribonucleotides for DNA synthesis begins at the ribonucleoside diphosphate level, with reduction of the ribose moiety to 2′ deoxyribose. This process is presented in detail later. In purine metabolism, subsequent phosphorylation by nucleoside diphosphate kinase yields the deoxyribonucleoside triphosphates, dATP and dGTP.

Let us briefly summarize points of feedback regulation in de novo purine biosynthesis. We have already described the control of IMP synthesis, primarily through feedback regulation of PRPP amidotransferase. Beyond that point, GMP controls its own biosynthesis by inhibiting the conversion of IMP to XMP, and AMP controls its own formation by inhibiting the synthesis of adenylosuccinate. Additional control is exerted at the level of deoxyribonucleotide biosynthesis, as we discuss later.

UTILIZATION OF ADENINE NUCLEOTIDES IN COENZYME BIOSYNTHESIS

An important metabolic role of purine nucleotides is in the synthesis of coenzymes, primarily those containing an adenylate moiety. These include the flavin nucleotides, the nicotinamide nucleotides, and coenzyme A, as shown in the following summary:

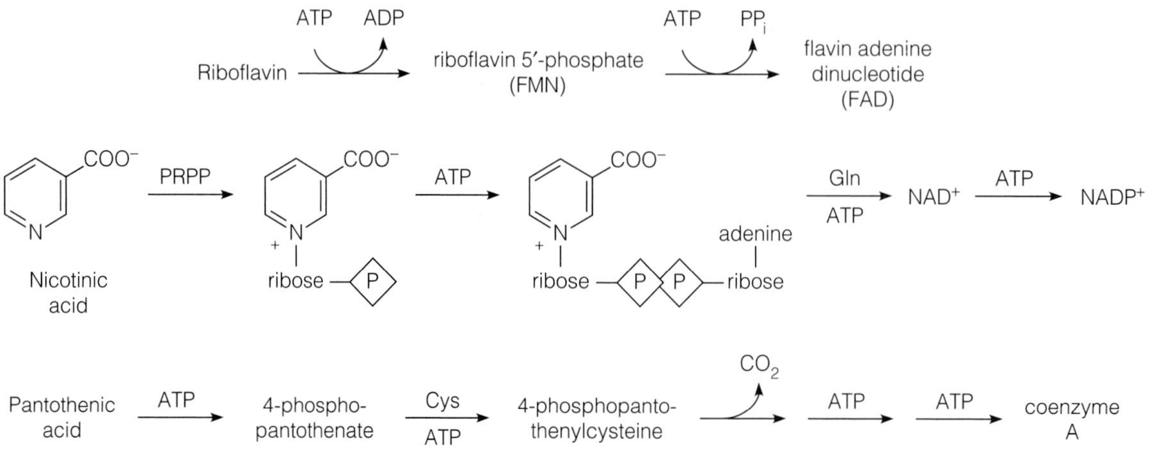

Riboflavin —(ATP → ADP)→ riboflavin 5′-phosphate (FMN) —(ATP → PP_i)→ flavin adenine dinucleotide (FAD)

Nicotinic acid (COO^-) —PRPP→ nicotinate ribose–P (COO^-, N^+) —ATP→ nicotinate ribose–P–P–ribose–adenine —(Gln, ATP)→ NAD^+ —ATP→ $NADP^+$

Pantothenic acid —ATP→ 4-phosphopantothenate —(Cys, ATP)→ 4-phosphopantothenylcysteine —(CO_2)→ —ATP→ —ATP→ coenzyme A

Purine Degradation and Clinical Disorders of Purine Metabolism

All purine degradation leads to uric acid. In some animals, additional degradation occurs.

FORMATION OF URIC ACID

Purine nucleotide catabolism yields uric acid, by routes shown in Figure 22.7. The specific pathways vary among organisms and among tissues of the same organisms. For example, AMP is either deaminated to yield inosinic acid (IMP) or hydrolyzed to yield adenosine. Deamination is particularly active in muscle, whereas hydrolysis predominates in most other animal tissues.

In degradative pathways, adenosine is deaminated by **adenosine deaminase (ADA)** to give inosine. Both inosine and guanosine can be formed by hydrolysis of the respective nucleoside monophosphates, which are acted on by **purine nucleoside phosphorylase (PNP)** to give hypoxanthine and guanine, respectively. Guanine is deaminated to xanthine by **guanine deaminase,** an enzyme abundant in mammalian brain and liver. Hypoxanthine is oxidized to xanthine, and xanthine to uric acid, by **xanthine oxidase.** This enzyme, which oxidizes several other heterocyclic nitrogen compounds, contains bound FAD, molybdenum, and nonheme iron. Electrons derived from the oxidation of substrates are passed to each of these carriers, ultimately reducing oxygen to H_2O_2, which is acted upon by catalase (see page 555).

Purine catabolism in primates ends with uric acid, which is excreted. However, most animals further oxidize the purine ring, to **allantoin** and then to **allantoic acid,** which is either excreted (in some fishes) or further catabolized to urea (in most fishes, some mollusks, and amphibians) or ammonia (in some marine invertebrates). Figure 22.8 shows the pathway from uric acid to CO_2.

EXCESSIVE ACCUMULATION OF URIC ACID: GOUT

Uric acid and its urate salts are quite insoluble. This property is advantageous to egg-laying animals, because it provides a route for disposition of excess nitrogen in a closed environment: The waste material simply precipitates in situ. However, the insolubility of urates can present difficulties in mammalian metabolism. In humans, about 3 individuals in 1000 suffer from **hyperuricemia**—chronic elevation of blood uric acid levels well beyond normal limits. Although the biochemical reasons for hyperuricemia vary, the condition goes by the single clinical name of

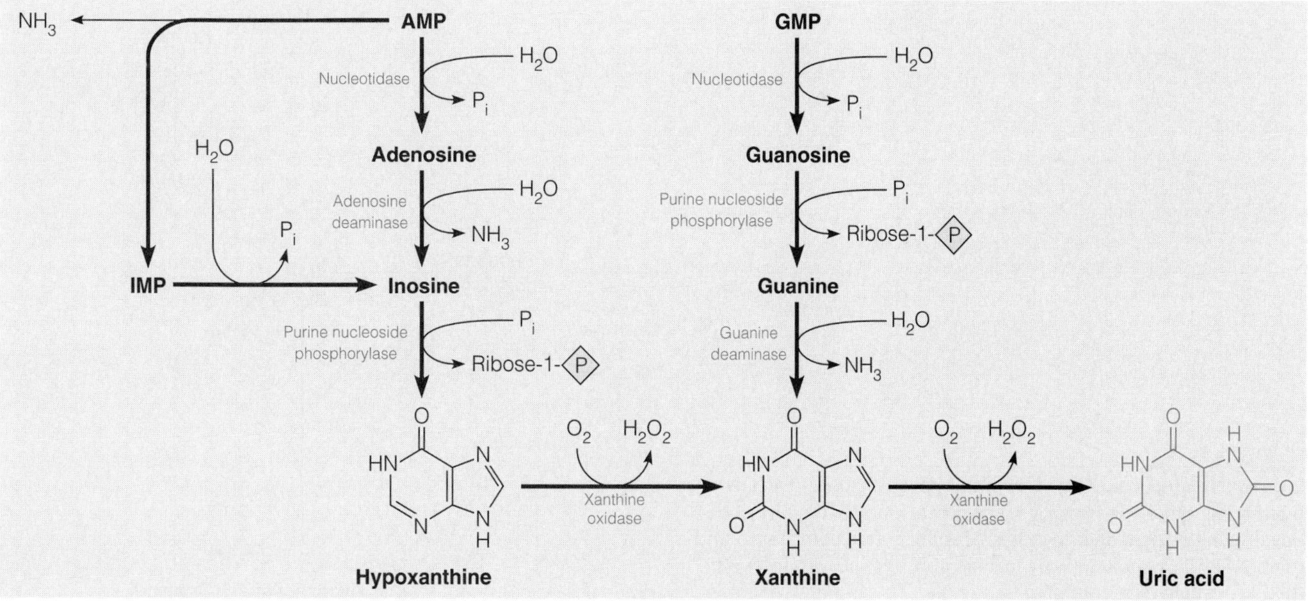

FIGURE **22.7**
Catabolism of purine nucleotides to uric acid.

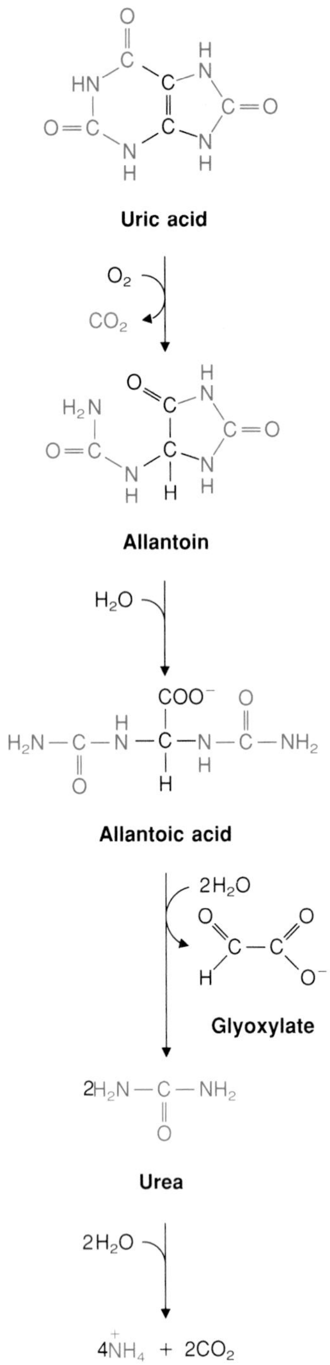

FIGURE **22.8**

Catabolism of uric acid to ammonia and CO_2.

gout. Prolonged or acute elevation of blood urate leads to its precipitation, as crystals of sodium urate, in the synovial fluid of joints. These precipitates cause inflammation, resulting in a painful arthritis, which, if untreated, leads ultimately to severe degeneration of the joints. Eating and drinking purine-rich foods are apt to stimulate acute gouty attacks in susceptible individuals. Because such foods include "rich" items such as liver, sweetbreads, anchovies, and wine, gout is historically associated with an excess of high living.

Gout results either from overproduction of purine nucleotides, leading to excessive uric acid synthesis, or from impaired uric acid excretion through the kidneys. Several specific enzymatic defects can lead to excessive purine synthesis, as shown in Figure 22.9. One form of gout is characterized by elevated activity of PRPP synthetase (defect 1), which may result from insensitivity to feedback inhibition by purine nucleotides. Because the activity of PRPP amidotransferase is probably controlled in part by the concentrations of substrates, an elevation of the steady state pool of PRPP increases flux through the amidotransferase reaction, which represents a major control point in de novo purine biosynthesis (see Figure 22.4). Gout can also result from mutations in PRPP amidotransferase that render it less sensitive to feedback inhibition by purine nucleotides (defect 2 in Figure 22.9). This loss of control also increases flux through the major regulated step. Another form of gout results from deficiency of the salvage enzyme **hypoxanthine–guanine phosphoribosyltransferase (HGPRT)** (defect 3). This is one of two phosphoribosyltransferases in animal purine metabolism; the other is specific for adenine.

We still don't know exactly why an HGPRT deficiency should increase the rate of purine nucleotide synthesis. A reasonable explanation is that the HGPRT reaction, when active, consumes PRPP. Decreased flux through this reaction, when the enzyme is deficient, could raise the steady state level of PRPP, thereby increasing the intracellular activity of PRPP amidotransferase through mass action.

As noted earlier, gout also results from impaired uric acid excretion. Patients with some forms of glycogen storage disease are gouty. Evidently, prolonged hypoglycemia causes accumulation of organic acids (lactate and the like), and this accumulation interferes with tubular secretion of uric acid in the kidney. Gout is also a consequence of cancer chemotherapy, presumably resulting from an overload of purines caused by nucleic acid degradation after death of tumor cells.

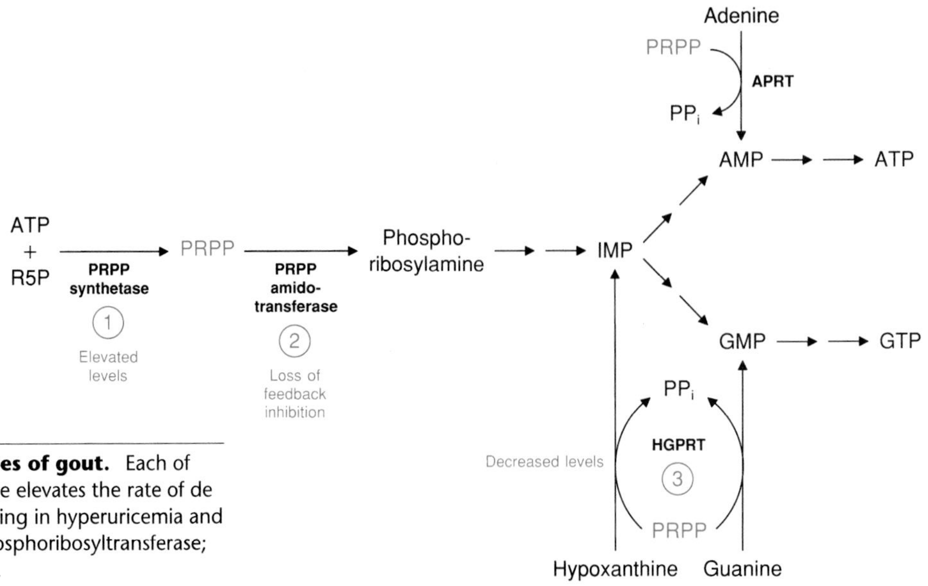

FIGURE **22.9**

Enzymatic abnormalities in three types of gout. Each of the three enzyme abnormalities shown here elevates the rate of de novo purine nucleotide biosynthesis, resulting in hyperuricemia and gout. HGPRT = hypoxanthine–guanine phosphoribosyltransferase; APRT = adenine phosphoribosyltransferase.

Many cases of gout are successfully treated by the antimetabolite **allopurinol,** a structural analog of hypoxanthine that strongly inhibits xanthine oxidase. This inhibition causes accumulation of hypoxanthine and xanthine, both of which are more soluble and, hence, more readily excreted than is uric acid.

Several known genetic alterations in purine metabolism can lead to purine oversynthesis, uric acid overproduction, and gout.

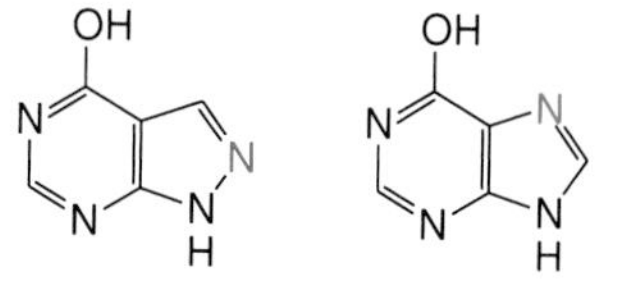

Allopurinol Hypoxanthine

DRAMATIC CONSEQUENCES OF A SEVERE HGPRT DEFICIENCY: LESCH–NYHAN SYNDROME

Careful analysis of patients with simple gout resulting from HGPRT deficiency reveals low but significant residual levels of the affected enzyme. Evidently the mutations involved alter the catalytic activity of the enzyme but do not abolish it completely. Far more serious consequences result from "null mutations," which result in total absence of the enzyme. This condition was first described in 1964 by medical student Michael Lesch and his faculty mentor, William Nyhan. Lesch–Nyhan syndrome is a sex-linked trait, because the structural gene for HGPRT is located on the X chromosome. Patients with this condition display a severe gouty arthritis, but they also have a dramatic malfunction of the nervous system, manifested as behavioral disorders, learning disability, and hostile or aggressive behavior, often self-directed. In the most extreme cases, patients nibble at their fingertips or, if restrained, their lips, causing severe self-mutilation. Nyhan has likened this behavior to "nailbiting, with the volume turned up." The biochemical reason for this bizarre behavioral pattern is unknown, but the condition, even though rare, is of great interest because all of the aberrations derive ultimately from the single well-characterized enzyme deficiency affecting HGPRT levels. At present there is no successful treatment, and afflicted individuals have such severe gouty arthritis that they rarely live beyond 20 years. However, the condition can be diagnosed prenatally through amniocentesis, because cells from Lesch–Nyhan fetuses are unable to incorporate radiolabeled hypoxanthine into nucleic acids.

UNEXPECTED CONSEQUENCES OF DEFECTIVE PURINE CATABOLISM: IMMUNODEFICIENCY

A surprising feature of human purine metabolism came to light in 1972, through studies on a hereditary condition called **severe combined immunodeficiency syndrome.** Patients with this condition are susceptible, often fatally, to infectious diseases because of a total inability to mount an immune response to antigenic challenge. In this condition, both B and T lymphocytes are affected; neither class of cells can proliferate as they must if antibodies are to be synthesized. In many such cases the immunodeficiency results from a heritable lack of the degradative enzyme adenosine deaminase (ADA).

What is the basis for this unexpected relationship? First, adenosine deaminase also acts on deoxyadenosine, which results from the degradation of DNA. Second, white blood cells have abundant levels of salvage enzymes, including nucleoside kinases; thus, adenosine and deoxyadenosine that accumulate are readily converted in white cells to their respective nucleotides. These nucleotides include dATP, which is known to be a potent inhibitor of DNA replication because it inhibits the synthesis of deoxyribonucleotides from ribonucleotides (see page 813). White cells must proliferate for an immune response to occur. In turn, proliferation requires ample synthesis of DNA and its precursors. Additional mechanisms are involved, because dATP has been found to kill white cells even when they are not proliferating. One such mechanism came to light in 1997, when dATP was reported to be a signaling agent that helped to trigger early metabolic events leading to apoptosis (Chapter 28).

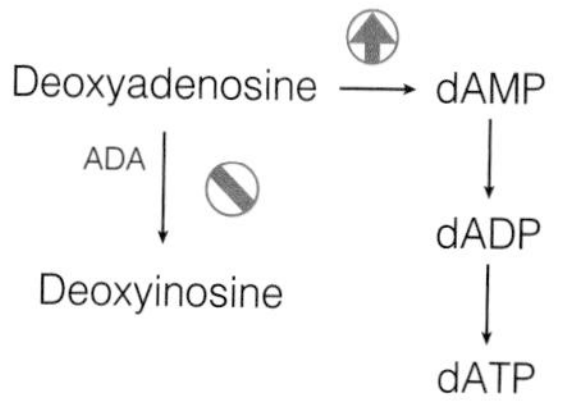

Deoxyadenosine → dAMP → dADP → dATP
ADA ↓ Deoxyinosine

Adenosine deaminase deficiency is the first condition to be treated by gene therapy. In 1995, two girls with the condition were treated with a viral vector into which the gene for adenosine deaminase had been spliced by recombinant DNA technology, in hopes that the engineered virus would establish itself in enough cells to yield sufficient enzyme to degrade the accumulated deoxyadenosine compounds. Three years after the experimental treatment, the two patients were reported to be in good health.

A less severe immunodeficiency results from the lack of another purine degradative enzyme, purine nucleoside phosphorylase (PNP). Decreased activity of this enzyme leads to accumulation primarily of dGTP. This accumulation also affects DNA replication, but less severely than does excessive dATP. Interestingly, the phosphorylase deficiency destroys only the T class of lymphocytes and not the B cells.

Pyrimidine Nucleotide Metabolism

DE NOVO BIOSYNTHESIS OF THE PYRIMIDINE RING

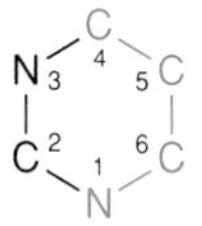

From carbamoyl phosphate: N3, C2 — From aspartate: C4, C5, C6, N1

Now we turn our attention to pyrimidine nucleotide biosynthesis, which is much simpler than formation of the structurally more complex purine nucleotides. Like purine biosynthesis, this pathway is identical in virtually all organisms that have been studied. As summarized in Figure 22.10, however, there are two major distinctions from the purine pathway. First, the pyrimidine ring is assembled as a free base, with conversion to a nucleotide occurring later in the pathway, when the base **orotic acid** is converted to orotidine monophosphate, or OMP. Second, the pyrimidine pathway is unbranched. Uridine triphosphate, one of the two common ribonucleoside triphosphates and hence an end product of the pathway, is also the substrate for formation of cytidine triphosphate, the other end product.

Pyrimidine nucleotide synthesis occurs primarily at the free base level, with conversion to a nucleotide occurring late in the unbranched pathway.

CONTROL OF PYRIMIDINE BIOSYNTHESIS IN BACTERIA

Pyrimidine synthesis begins with formation of carbamoyl phosphate, a reaction presented in Chapter 20 (see page 719 and reaction 1 in Figure 22.10). However, the first reaction committed solely to pyrimidine synthesis is the formation of carbamoyl aspartate from carbamoyl phosphate and aspartate, catalyzed by aspartate transcarbamoylase, or ATCase (reaction 2). In enteric bacteria, this enzyme represents a marvelous example of feedback control, as was discussed at length in Chapter 11. Recall that the enzyme is inhibited by the end product CTP and activated by ATP, the latter possibly representing a mechanism to keep purine and pyrimidine biosyntheses in balance. Recall also that the enzyme contains six each of two types of subunits, arranged as two catalytic trimers and three regulatory dimers.

Bacteria also regulate pyrimidine metabolism through control of the *synthesis* of ATCase and the other enzymes. The rate of transcription of an operon encoding both of the ATCase subunits can vary by as much as 150-fold, depending on the intracellular level of UTP. The higher the UTP concentration, the lower the rate of transcription of these genes.

MULTIFUNCTIONAL ENZYMES IN EUKARYOTIC PYRIMIDINE SYNTHESIS

In eukaryotes the first three reactions of pyrimidine synthesis are catalyzed by a multifunctional enzyme, the CAD protein.

Aspartate transcarbamoylase in eukaryotes is strikingly different from the *E. coli* enzyme. This came to light through analysis of ATCase inhibition by ***N*-phosphonoacetyl-L-aspartate** (PALA).

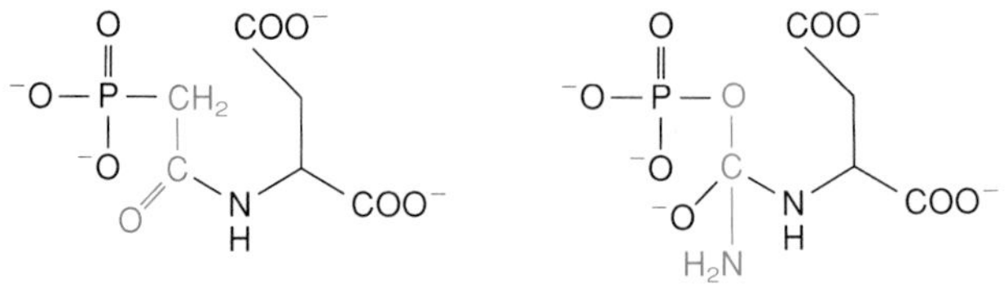

PALA **Putative transition state complex**

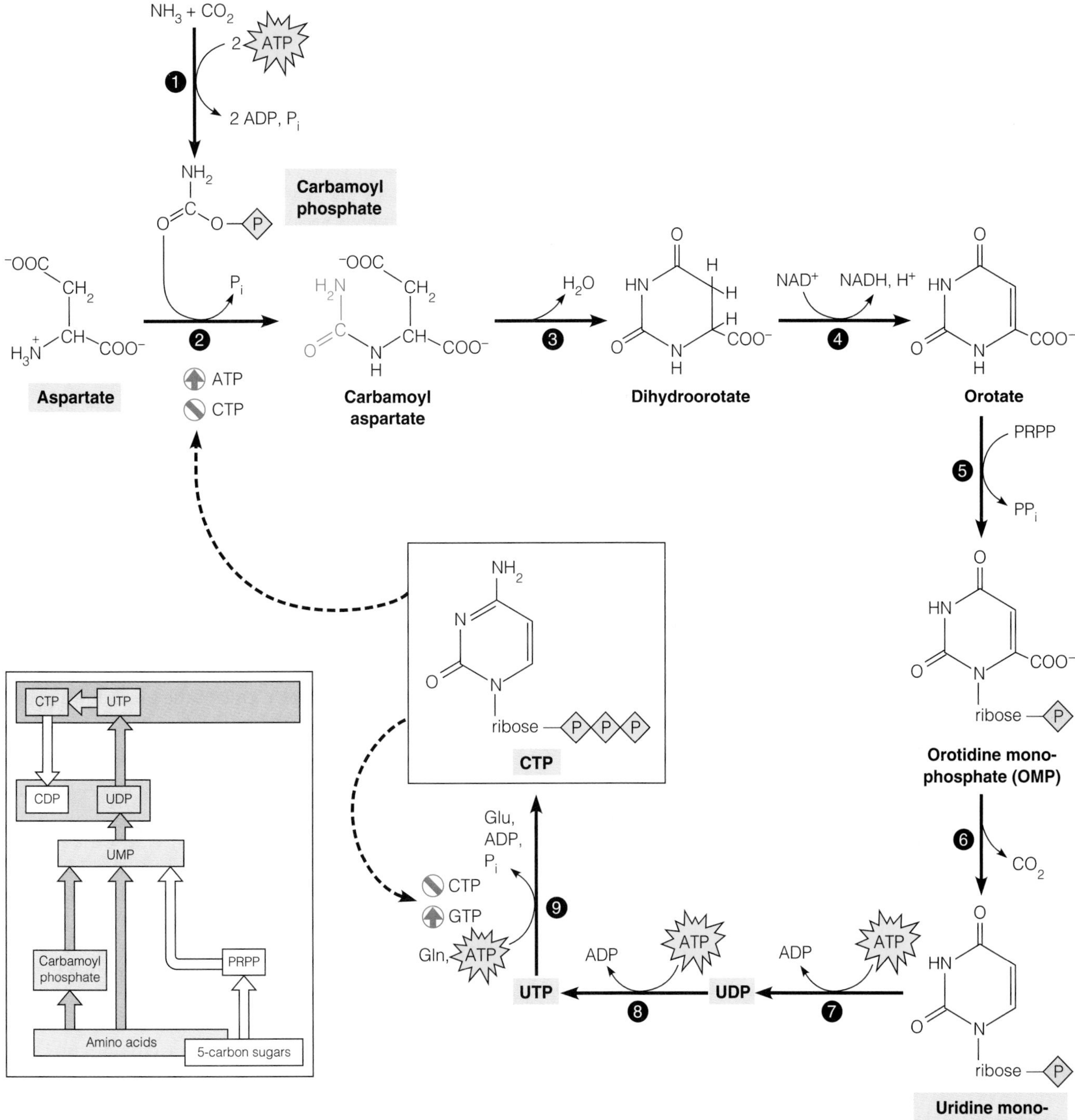

FIGURE **22.10**

De novo synthesis of pyrimidine nucleotides. Sites of allosteric control are indicated. Enzymes are as follows: (1) carbamoyl phosphate synthetase, (2) aspartate transcarbamoylase, (3) dihydroorotase, (4) dihydroorotate dehydrogenase, (5) orotate phosphoribosyltransferase, (6) orotidylate decarboxylase, (7) UMP kinase, (8) nucleoside diphosphate kinase, and (9) CTP synthetase.

This compound, synthesized as an analog of the putative transition state complex formed between the two substrates, inhibits pyrimidine synthesis in mammalian cells. However, cells eventually develop resistance to it, because levels of ATCase rise in these cells beyond the capacity of PALA to inhibit all of the activity. Surprisingly, these resistant cells contain similarly elevated levels of carbamoyl phosphate synthetase (see reaction 1 in Figure 22.10) and dihydroorotase (reaction 3). The explanation for this observation came with the discovery of a single protein containing three identical polypeptide chains, each with M_r of about 230,000, that catalyzes all three reactions.

George Stark has given this trifunctional enzyme the acronym **CAD** (from the first letter in the name of each enzyme). He showed that the protein accumulates

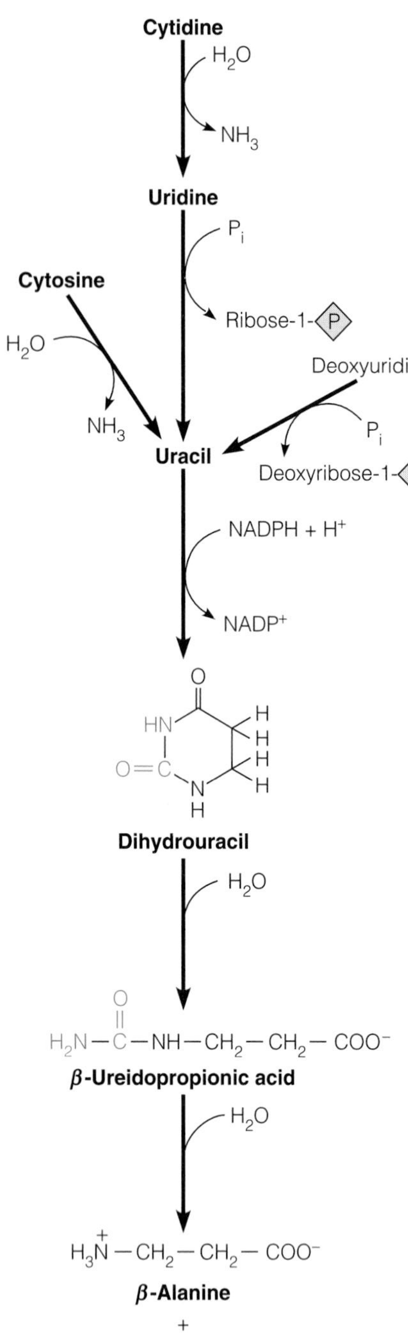

FIGURE **22.11**

Catabolic pathways in pyrimidine nucleotide metabolism.

in PALA-resistant cells because the gene encoding the protein becomes amplified as a consequence of the selective pressure exerted by PALA; resistant cells contain many more copies of the gene than the normal complement of two copies per diploid cell. This phenomenon of gene **amplification** has now been observed many times in eukaryotic cells exposed to prolonged and specific stresses. We discuss mechanisms involved in Chapter 25.

In mammalian cells, reactions 5 and 6 in Figure 22.10 are also catalyzed by a single protein, which has been called **UMP synthase.** We don't know much about how this affects regulation of pyrimidine biosynthesis, but the juxtaposition of active sites may allow channeling, as mentioned for the multifunctional proteins in purine synthesis. We still have much to learn about how, or whether, pyrimidine biosynthetic intermediates are channeled, because dihydroorotate dehydrogenase (see reaction 4 in Figure 22.10) is localized in the outer mitochondrial membrane, whereas the bifunctional and trifunctional enzymes are both cytosolic. Given that intermediates must travel into and then out of the mitochondria, the kinetic advantage provided by channeling of the first and last steps would seem to be negated. The metabolic logic of this incomplete metabolic channeling is not understood.

Another site for control of pyrimidine nucleotide synthesis is the amidotransferase, **CTP synthetase,** which converts UTP to CTP (reaction 9). This enzyme is inhibited allosterically by its product CTP and is activated by GTP.

SALVAGE SYNTHESIS AND PYRIMIDINE CATABOLISM

Pyrimidine nucleotides are also synthesized by salvage pathways involving phosphorylases and kinases, comparable to those already discussed for purines. The catabolic pathways for pyrimidines, summarized in Figure 22.11, are simpler than those for purines. Because the intermediates are relatively soluble, there are few known derangements of pyrimidine breakdown. One of the breakdown products, β-alanine, is used in the biosynthesis of coenzyme A.

Deoxyribonucleotide Biosynthesis and Metabolism

Most cells contain 5 to 10 times as much RNA as DNA. Moreover, as we have seen, ribonucleotides have multiple metabolic roles, while deoxyribonucleotides serve only as constituents of DNA. Therefore, most of the carbon that flows through nucleotide synthetic pathways goes into ribonucleoside triphosphate (rNTP) pools. However, the relatively small fraction that is diverted to the synthesis of deoxyribonucleoside triphosphates (dNTPs) is of paramount importance in the life of the cell. dNTPs are used almost exclusively in the biosynthesis of DNA. Consequently, there are especially close regulatory relationships between DNA synthesis and dNTP metabolism—more so than seen between other macromolecular biosyntheses and the pathways that provide their precursors. Overall pathways of dNTP biosynthesis are shown in Figure 22.12.

Recalling that DNA differs chemically from RNA in the nature of the sugar and in the identity of one of the pyrimidine bases, we can focus our discussion of deoxyribonucleotide biosynthesis on two specific processes—the conversion of ribose to deoxyribose, and the conversion of uracil to thymine. Both of these processes occur at the nucleotide level. Both processes are of great interest mechanistically, as target sites for chemotherapy for cancer or infectious diseases and from the standpoint of regulation. Accordingly, we shall present both processes in some detail.

REDUCTION OF RIBONUCLEOTIDES TO DEOXYRIBONUCLEOTIDES

Mechanistically, the reduction of ribose to deoxyribose involves replacement of the hydroxyl at C-2 by a hydride ion, with retention of configuration. Peter Reichard showed that this difficult reaction occurs at the nucleotide level, and this led to his discovery of the important enzyme **ribonucleotide reductase.** In all organ-

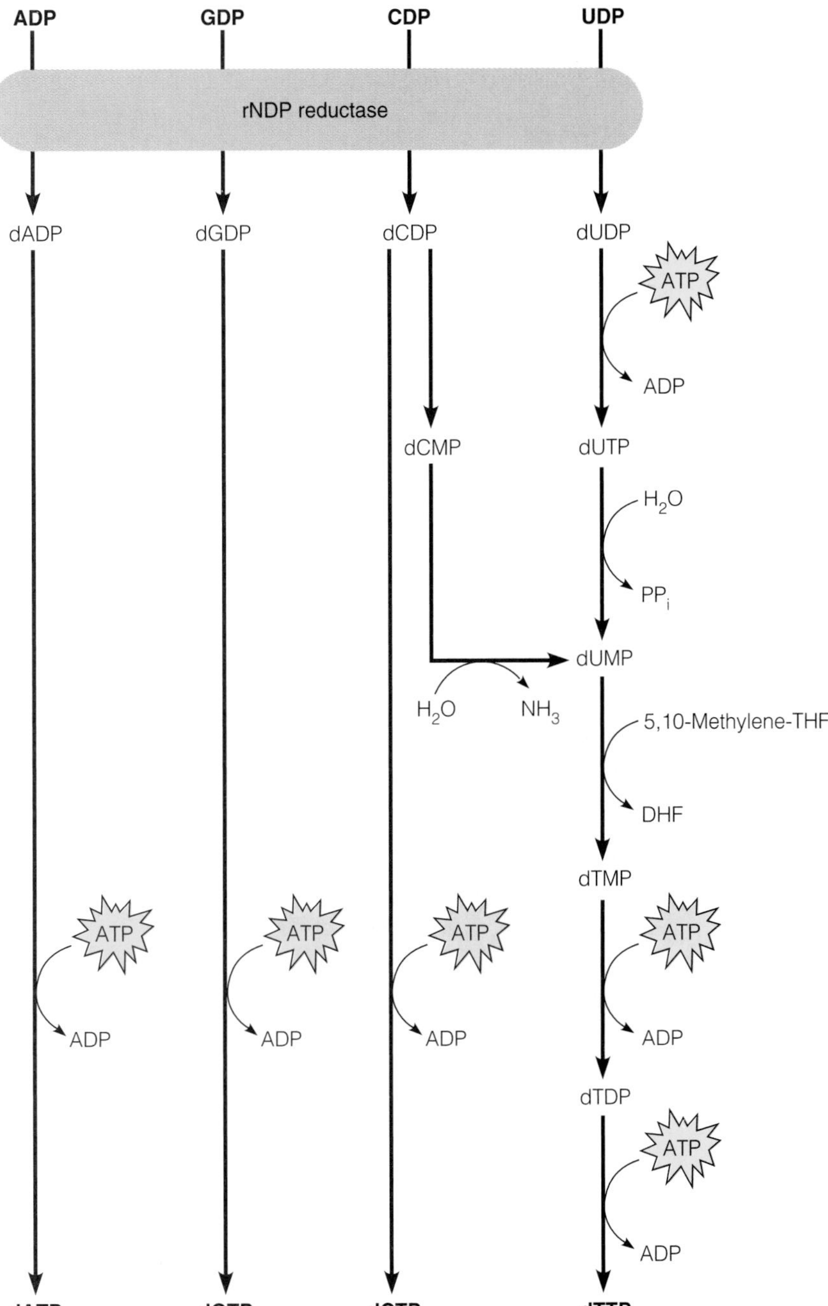

FIGURE **22.12**

Overview of deoxyribonucleoside triphosphate (dNTP) biosynthesis.

isms studied thus far, a single enzyme reduces all four common ribonucleotide substrates to the corresponding 2′-deoxyribonucleotides. A free radical mechanism is involved in the reaction. Although evolution has created three widely diverse mechanisms for generating a functional free radical, the three classes of ribonucleotide reductase evidently all use the same fundamental chemistry to reduce the substrates. The most widely distributed enzyme form, called class I ribonucleotide reductase, acts upon ribonucleoside diphosphate substrates, leading it to be also called **rNDP reductase.** This enzyme generates its radical on a specific tyrosine residue, with the aid of a diferric oxygen bridge (see page 812). The class II enzymes, found in cyanobacteria, some bacteria, and *Euglena,* act upon ribonucleoside triphosphate substrates and use adenosylcobalamin, a B_{12} coenzyme, to generate a free radical. The type III enzyme, found only in facultative or obligate anaerobes, also acts upon ribonucleoside triphosphate substrates. These enzymes

One enzyme, ribonucleotide reductase, reduces all four ribonucleotides to their deoxyribo derivatives.

use *S*-adenosylmethionine and an iron–sulfur center to generate the catalytically essential radical on a glycine residue. Our discussion will focus on class I, the most widespread form, whose action is shown in Figure 22.12.

Ribonucleotide reductase contains catalytic residues on each of its subunits—redox-active thiols and a tyrosine free radical stabilized by an iron–oxygen complex.

Structure of rNDP Reductase

As found in *E. coli* and mammalian cells, the class I rNDP reductase is an $\alpha_2\beta_2$ tetramer. The *E. coli* enzyme contains two proteins—R1, with two identical α polypeptide chains with M_r of 87,000 each, and R2, consisting of two 43,000-dalton β chains. The structure of the enzyme is illustrated in Figure 22.13. The catalytic site lies in the large (R1) protein. Within this site are three cysteine residues, which are conserved among different rNDP reductases. Two of the cysteine thiols are **redox-active,** so called because they undergo cyclic oxidation and reduction during the reaction. The third cysteine evidently functions as part of a free radical mechanism, as described in the next section.

The R2 dimer contains an unusual tyrosine free radical that is involved in the reaction and an oxygen atom bridging two ferric ions. This **dinuclear iron center** stabilizes the free radical. The R1 protein contains two classes of regulatory sites, which we shall discuss shortly. Finally, the R1 protein contains an additional pair of redox-active thiol groups, which interact with an external reductive cofactor. The mammalian enzyme is similar in structure.

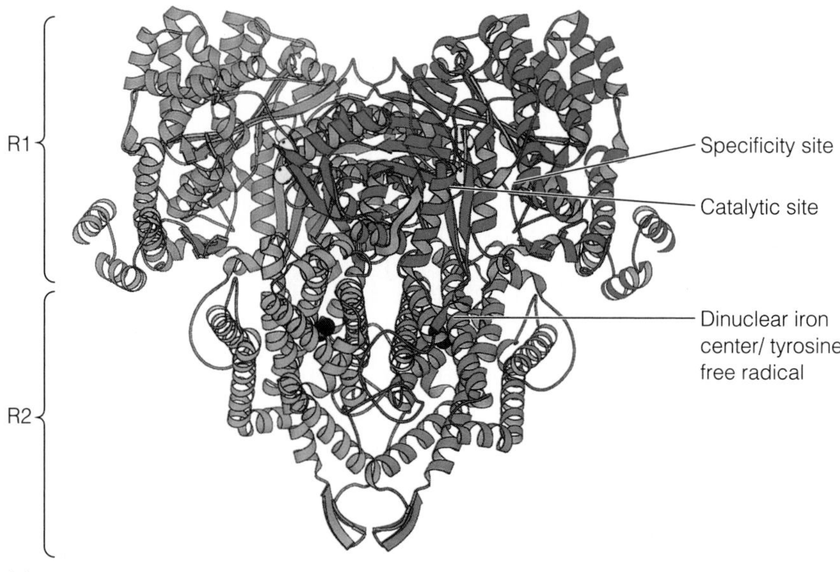

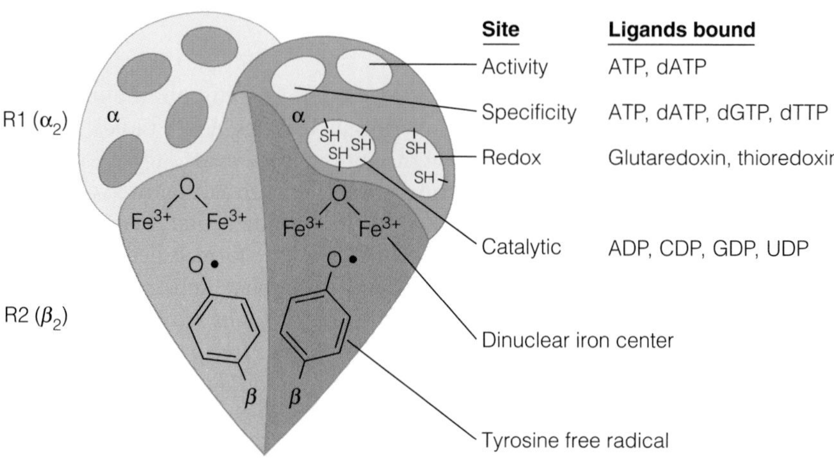

FIGURE **22.13**

Structure of *E. coli* ribonucleoside diphosphate reductase. (a) A model of the molecular structure of the R1R2 holoenzyme, based upon crystallographic analysis of the separate homodimeric R1 (blue) and R2 (red) proteins. Yellow spheres are active-site cysteines. **(b)** A schematic diagram identifying the functional center of R2—a tyrosine free radical and a dinuclear iron center with a bridged oxygen. Shown also are the ligand-binding sites in R1: a redox site for interaction with an external electron donor, the catalytic site, the activity site, and the substrate specificity site. The activity and specificity sites are allosterically controlled. The redox site contains two functional cysteine residues, and the catalytic site contains three.

(a) From U. Uhlin and H. Eklund, *Nature* (1994) 370:533–539.

Mechanism of Ribonucleotide Reduction

Although we do not yet know the complete mechanism of the rNDP reductase reaction, we can formulate a plausible mechanism based on the following observations. (1) Radiolabeling studies show that cleavage of the ribose C-3′—H bond occurs during the reaction. (2) The reaction proceeds with retention of configuration at C-2′, which rules out displacement of the hydroxyl group by a hydride ion in an S_N2 reaction. (3) The thiol groups undergo oxidation during the reaction. (4) The tyrosine free radical participates in the reaction. This was shown first by the fact that **hydroxyurea,** an inhibitor of rNDP reductase, reversibly destroys the free radical. A more elegant demonstration is depicted in Figure 22.14. The free radical gives a characteristic electron paramagnetic resonance (EPR) spectrum. Tyrosine 122, the residue thought to generate the radical in the *E. coli* enzyme, was changed to phenylalanine by site-directed mutagenesis of the cloned gene for R2. The modified protein was inactive and showed no EPR spectrum—hence, no evidence for the existence of a free radical. However, because the radical is located far from the catalytic site, as shown by crystallography (see Figure 22.13), one must postulate some sort of long-range process by which the unpaired electron in the tyrosine radical attracts an electron from an active site residue. Evidence indicates that (1) this residue is cysteine 439 in the *E. coli* enzyme and (2) a specific set of amino acid residues on R1 and R2 participates in the long-range electron transport process.

A plausible mechanism for rNDP reductase, based on the above observations, is shown in Figure 22.15. First, cysteine 439 in the active site is converted to a **thiyl**

HO—NH—C(=O)—NH_2

Hydroxyurea

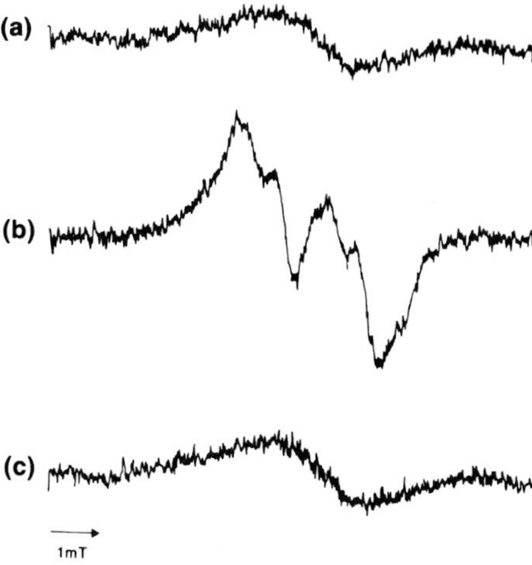

FIGURE **22.14**

Evidence that tyrosine 122 in *E. coli* R2 protein carries the essential free radical. The figure shows electron paramagnetic resonance spectra of *E. coli* cells containing a cloned gene that overexpresses R2. **(a)** Bacteria containing no clone (the normal level of R2 protein is too low to create a significant spectral signal). **(b)** Bacteria containing the cloned wild-type gene. **(c)** Bacteria containing a mutant cloned gene in which tyrosine 122 has been changed to phenylalanine. Treatment of the purified enzyme with hydroxyurea yields a spectrum like that shown in (c).

From Å. Larsson and B. M. Sjöberg, *EMBO J.* (1986) 3:2038. By permission of Oxford University Press.

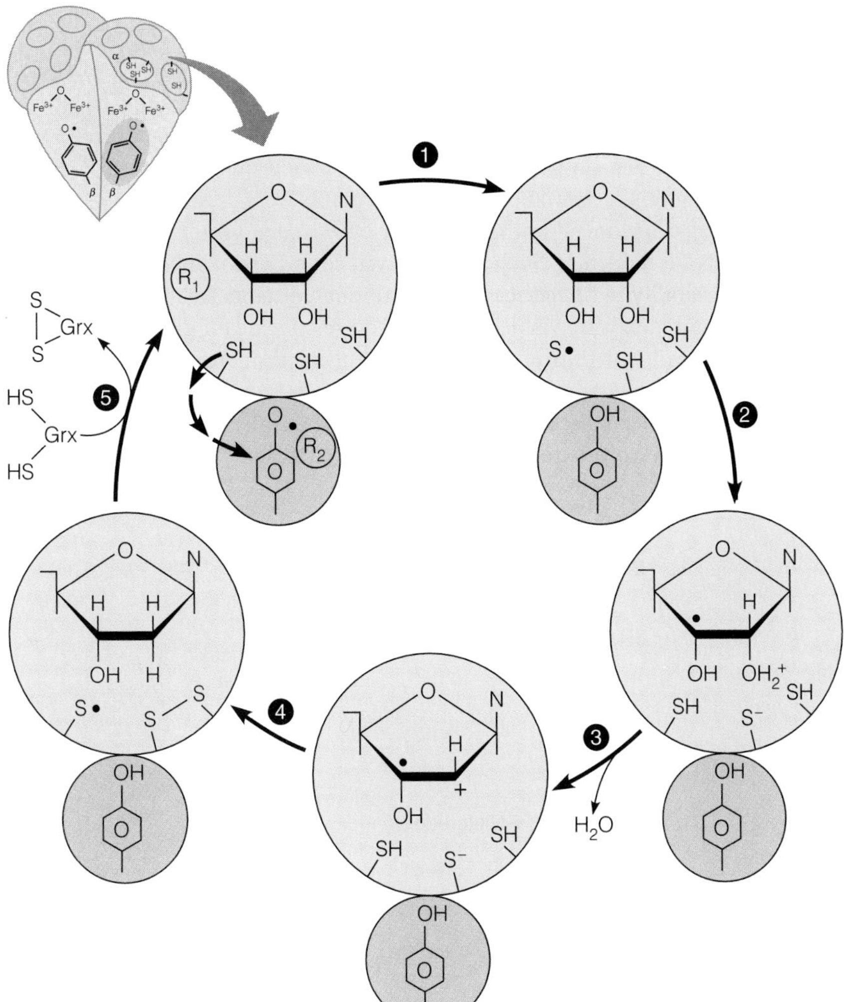

FIGURE **22.15**

Reduction of a ribonucleoside diphosphate by rNDP reductase. This partial mechanism is consistent with available evidence. N is adenine, cytosine, guanine, or uracil. **Step 1:** A cysteine in the R1 catalytic site (Cys 439 in *E. coli*) loses an electron to become a thiyl radical, and a reductive electron transport chain reduces the tyrosine free radical on R2 to a tyrosine residue. **Step 2:** The substrate nucleotide interacts with the radical, creating a radical at C-3′. **Steps 2 and 3:** Protonation and dehydration at C-2′ create a radical cation. **Step 4:** Reduction at C-2′ by the redox-active cysteines (225 and 462 in *E. coli*) and transfer of an electron to C-3′ produce the deoxyribonucleotide product and regenerate the thiyl radical. **Step 5:** The redox-active thiols are reduced by interaction with glutaredoxin (Grx), and dNDP dissociation (not shown) allows rNDP substrate binding.

$R2(Fe^{3+})_2$, Y·
↓ Iron chelator
R2–YH
↓ Fe^{2+}
$R2(Fe^{2+})_2$, YH
↓ O_2
$R2(Fe^{3+})(Fe^{3+}-OOH)$, YH
↓ H_2O
$R2(Fe^{4+})(Fe^{4+}=O)$, YH
↓ e^-
$R2(Fe^{3+}-O-Fe^{3+})$, Y·

free radical, by loss of an electron, in an electron transport chain that results in reduction of tyrosine 122 (step 1). Next, the thiyl radical participates in abstraction of a hydrogen atom from C-3′ of the substrate (step 2). This is followed by loss of a hydroxide ion from C-2′, giving a resonance-stabilized radical carbonium ion (steps 2 and 3). The enzyme transfers an electron pair from the redox-active cysteine thiols and a hydrogen atom from cysteine 439 (step 4), to give the nucleotide product. The resultant cystine residue is now reduced by disulfide exchange with another pair of redox-active thiols in the R1 subunit; this is not shown in the figure. The resulting disulfide is reduced by an external cofactor (shown as Grx—see the next section), regenerating the active form of the enzyme.

The mechanism shown in Figure 22.15 is incomplete, because several important questions remain. First, we don't know the role of the iron center in forming the free radical. One possible mechanism, reminiscent of cytochrome P450 activation, is shown in the margin. Also, as discussed in the next section, it is not clear how external reducing power is transferred through one pair of thiols at the redox site on R1, to the other pair in the catalytic site.

Source of Electrons for rNDP Reduction

Electrons for the reduction of ribonucleotides come ultimately from NADPH, but they are shuttled to rNDP reductase by a coenzyme that is unusual because it is itself a protein (Figure 22.16). The first known member of this class of redox-active proteins is **thioredoxin,** a small protein ($M_r \cong 12{,}000$) with two thiol groups in the sequence Cys–Gly–Pro–Cys. These thiols undergo reversible oxidation to the disulfide, thereby reducing sulfurs in the active site of rNDP reductase. Oxidized thioredoxin is reduced by NADPH via the action of a flavoprotein enzyme, **thioredoxin reductase.**

Since its discovery, thioredoxin has been found to have many activities in vitro, suggesting an astonishing range of biological functions. Some of them are listed in Table 22.1. Whether thioredoxin is the true intracellular cofactor for ribonucleotide reduction was brought into question by the isolation of *E. coli* mutants lacking this protein. Because these mutants were capable of DNA replication, investigators looked in these cells for other redox proteins that could interact with rNDP reductase. Such a protein was found and named **glutaredoxin,** because of its ability to be reduced by glutathione. Mutants lacking both thioredoxin and glutaredoxin are viable, but they require cysteine for growth. This finding confirms a role for thioredoxin in sulfate utilization (see page 752), and it suggests the existence of a third electron carrier. In fact, additional species of glutaredoxin have been discovered. Whichever carrier is the principal cofactor for rNDP reductase, the ultimate electron source is NADPH.

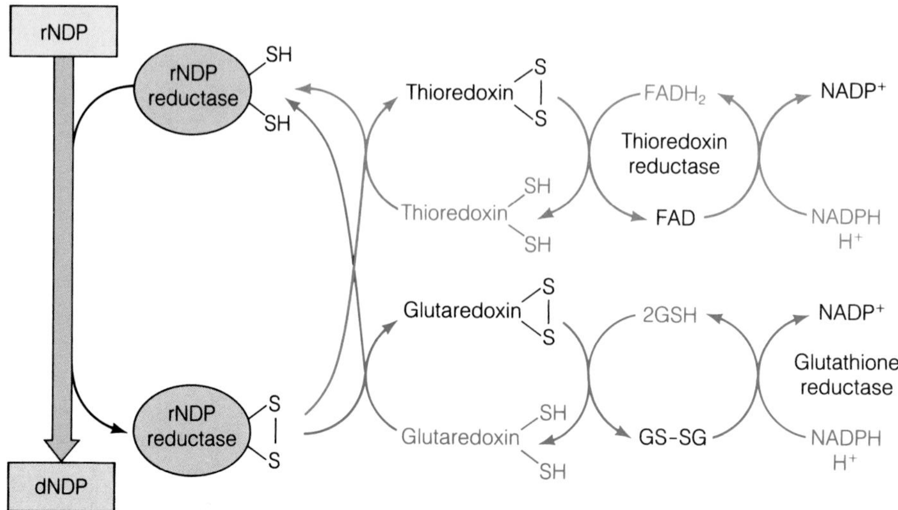

FIGURE **22.16**

Reductive electron transport sequences in the action of rNDP reductase. Either thioredoxin or glutaredoxin can reduce the oxidized form of the reductase.

TABLE **22.1 Biological activities of thioredoxin**

Activity	Organism
Cofactor for ribonucleotide reduction	All organisms
Protein folding (thioredoxin promotes correct disulfide bond formation)	All organisms
Possible control of insulin levels, through control of insulin reduction	Animals
Control of melanin formation (people with high levels of thioredoxin reductase tan easily)	Animals
Regulation of photosynthetic carbon fixation (see Chapter 17)	Plants
Sulfite reduction (see Chapter 21)	Plants, bacteria
Essential subunit of viral DNA polymerase	Bacteriophage T7
Maturation of filamentous phages by an unknown mechanism	Single-stranded DNA bacteriophages

Ribonucleotide reductase uses a protein cofactor—thioredoxin or glutaredoxin—to provide electrons for reduction of the ribonucleotide substrate. However, the ultimate electron donor is NADPH.

Regulation of Ribonucleotide Reductase Activity

Because deoxyribonucleotides are used only for DNA synthesis, and because one enzyme system is used for reduction of all four ribonucleotide substrates, regulation of both the *activity* and the *specificity* of ribonucleotide reductase is essential to maintain balanced pools of DNA precursors. This regulation is achieved through binding of nucleoside triphosphate effectors to two classes of regulatory sites on the R1 subunit (two of each site per molecule in the *E. coli* enzyme—see Figure 22.13). The **activity sites** bind either ATP or dATP, with relatively low affinity, whereas the **specificity sites** bind ATP, dATP, dGTP, or dTTP, all with relatively high affinity. Binding of ATP at the activity sites tends to increase the catalytic efficiency of rNDP reductase for all substrates, whereas dATP acts as a general inhibitor of all four reactions. Binding of nucleotides at the specificity sites modulates the activities of the enzyme toward different substrates, so as to maintain balanced rates of production of the four dNTPs. For example, binding of dTTP (with ATP bound in the activity site) activates the enzyme for reduction of GDP but decreases its ability to reduce either UDP or CDP. Table 22.2 summarizes the principal regulatory effects.

Ribonucleotide reductase has two classes of allosteric sites. Activity sites influence catalytic efficiency, and specificity sites determine specificity for one or more of the four substrates.

These effects are seen in vitro with purified enzymes. There is ample reason to conclude that similar regulatory effects also operate in intact cells. For example, either deoxyadenosine or thymidine will inhibit DNA synthesis when administered to intact cells. Measurements of intracellular pools of dNTPs show that in deoxyadenosine-treated cells the dATP pools expand (as expected from the effects of salvage pathways), while dTTP, dGTP, and dCTP pools shrink. This is probably why white blood cells cannot proliferate as needed in immunodeficiency states associated with a lack of adenosine deaminase (see page 805): dATP

TABLE **22.2 Regulation of the activities of mammalian ribonucleotide reductase**

Nucleotide Bound in			
Activity Site	**Specificity Site**	Activates Reduction of	Inhibits Reduction of
ATP	ATP or dATP	CDP, UDP	
ATP	dTTP	GDP	CDP, UDP
ATP	dGTP	ADP	CDP, UDP[a]
dATP	Any effector		ADP, GDP, CDP, UDP

[a]dGTP binding inhibits the reduction of pyrimidine nucleotides by the mammalian enzyme but not by the *E. coli* enzyme.

accumulation in these cells blocks deoxyribonucleotide synthesis and, hence, DNA replication.

Another example comes from cell biology. Researchers often **synchronize** cell cultures, that is, manipulate the cells so that all are brought to the same phase of the cell cycle. Synchrony can be accomplished by **thymidine block,** in which thymidine is added to the cells to inhibit DNA synthesis. This prevents the passage of cells from the G1 to the S phase of the cell cycle, and cells accumulate at this point, like car traffic stopped at a red light. Transferring the cells to medium containing no thymidine is like a green light, reversing the inhibition and allowing the cells to initiate DNA replication synchronously. dNTP pool measurements in thymidine-blocked cells show that dTTP accumulates, as expected from salvage synthesis, while there is a specific depletion of dCTP, as expected from the effects of dTTP on ribonucleotide reductase activity. Indeed, addition of deoxycytidine restores normal dCTP pools (by salvage synthesis) and relieves the thymidine block.

Inhibition of DNA synthesis by thymidine or deoxyadenosine involves allosteric inhibition of ribonucleotide reductase by dTTP or dATP, respectively.

Further support for the control of rNDP reductase activity in vivo comes from isolation of mammalian cell mutant lines whose growth is not inhibited by deoxyribonucleosides. rNDP reductase from these altered cells shows modifications in either activity or specificity sites, which render these enzymes less susceptible to inhibition by dNTP effectors. Some of these cell lines display both dNTP pool abnormalities and **mutator phenotypes.** That is, they show increased rates of spontaneous mutation at all genetic loci tested. Comparable observations have been made with mutant cells altered in either CTP synthetase or deoxycytidylate deaminase. These findings suggest that when dNTP concentrations are altered at DNA replication sites, the likelihood is increased for replication errors, which lead to mutations. This point is discussed further in Chapter 24.

The metabolic rationale for all of the effects shown in Table 22.2 is not immediately obvious. For example, why should dATP at the specificity site activate both CDP and UDP reduction? Part of the answer is that UDP reduction is a relatively minor pathway. Most dTTP comes from deoxycytidine nucleotides, via the dCMP deaminase reaction (see below).

BIOSYNTHESIS OF THYMINE DEOXYRIBONUCLEOTIDES

The previous section explored the first metabolic reaction committed to DNA synthesis—the formation of deoxyribonucleoside diphosphates through the action of rNDP reductase. Once formed, three of the diphosphates—dADP, dGDP, and dCDP—are converted directly to the corresponding triphosphates by nucleoside diphosphate kinase. Biosynthesis of deoxythymidine triphosphate occurs partly from the dUDP produced via the reductase and partly from deoxycytidine nucleotides; the ratio varies in different cells and organisms. Note that we use the terms *thymidine* and *deoxythymidine* interchangeably. That is because thymine *ribonucleotides* are not normal metabolites, so the nucleoside containing thymine and deoxyribose need not be specifically identified as a deoxyribonucleoside.

dUMP, the substrate for thymidylate synthesis, can arise either from UDP reduction and dephosphorylation or from deamination of a deoxycytidine nucleotide.

The pathways are summarized in Figures 22.12 and 22.17. Both of the de novo pathways shown lead to deoxyuridine monophosphate (dUMP), the substrate for synthesis of thymine nucleotides: (1) dUDP is phosphorylated to dUTP, which is then cleaved by a highly active diphosphohydrolase, **dUTPase.** (2) dCDP is dephosphorylated to dCMP, which then undergoes deamination to dUMP by an aminohydrolase called **dCMP deaminase.** The latter reaction represents a branch point for pyrimidine dNTP synthesis; the enzyme requires dCTP as an allosteric activator and is inhibited by dTTP. *E. coli* and some other bacteria use a different route to dUMP. Deamination is carried out at the triphosphate level by **dCTP deaminase,** and the resultant dUTP is cleaved by dUTPase to dUMP and PP_i.

However it is formed, dUMP serves as substrate for formation of thymidine monophosphate (dTMP), catalyzed by **thymidylate synthase.** This enzyme trans-

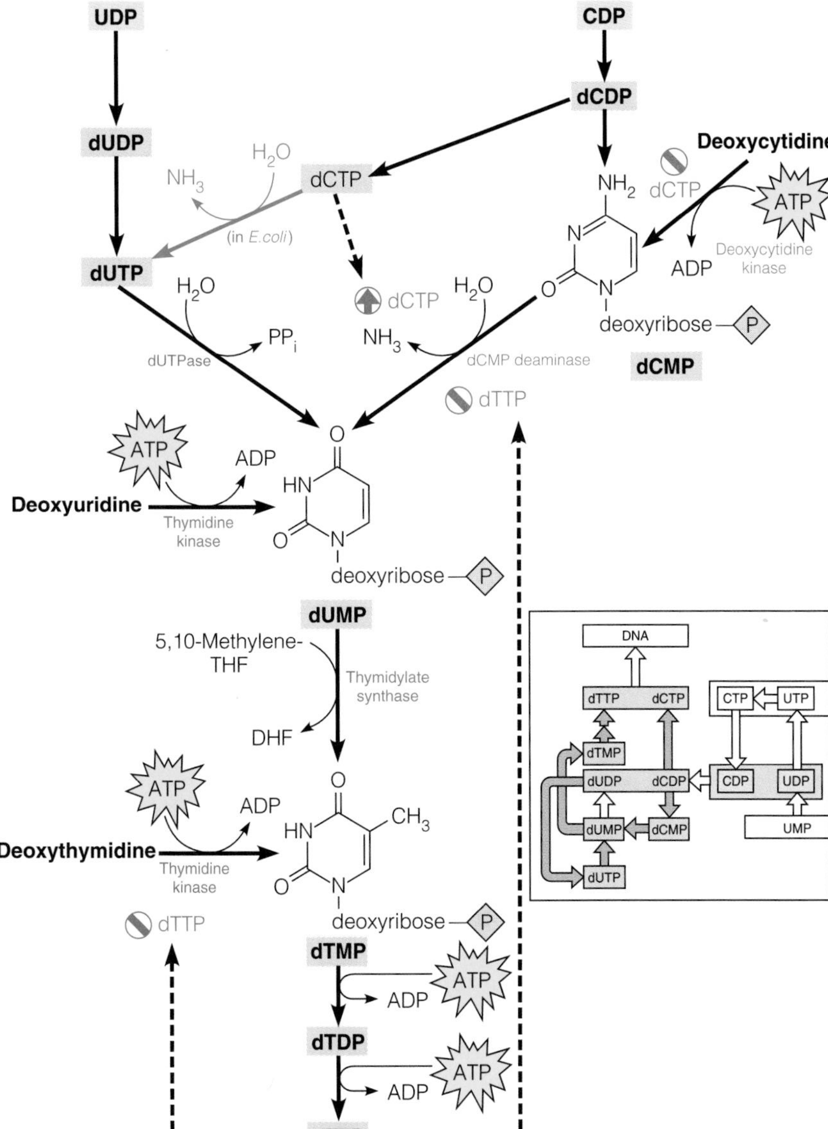

FIGURE **22.17**

Salvage and de novo synthetic pathways to thymine nucleotides. The de novo pathways start with UDP or CDP, shown at the top.

fers a one-carbon unit, at the methylene level of oxidation, and reduces it to the methyl level. The one-carbon donor is 5,10-methylenetetrahydrofolate, which in this unusual reaction also serves as a redox cofactor, to give dihydrofolate as the other reaction product (Figure 22.18). The cofactor must then be reduced, by dihydrofolate reductase, and it must acquire another methylene group, most commonly via serine transhydroxymethylase. Interruption of any of these steps of the cycle interferes with thymine nucleotide formation. dTMP, once formed, is converted to dTTP by two successive phosphorylations.

In the reaction catalyzed by thymidylate synthase, 5,10-methylenetetrahydrofolate donates both a single-carbon group and an electron pair to reduce that group to the methyl level.

DEOXYURIDINE NUCLEOTIDE METABOLISM

In addition to the biosynthetic function of dUTPase in forming dUMP for thymine nucleotide formation, the enzyme plays an important role in excluding uracil from DNA. Were dUTP not rapidly degraded, it could serve as a satisfactory substrate for DNA polymerases. In fact, as discussed in Chapter 24, cells have a rather elaborate mechanism to ensure that any dUMP residues that do find their way into DNA are efficiently excised. Because uracil is virtually identical

FIGURE **22.18**

Relationship between thymidylate synthase and enzymes of tetrahydrofolate metabolism. Only the active portion of the folate coenzyme is shown.

to thymine with respect to its base-pairing properties, why is it so important for the cell to allow only thymine to be stably incorporated into DNA? The answer probably lies in the fact that dUMP residues can arise in DNA not only by dUTP incorporation but also by the spontaneous deamination of dCMP residues. The latter process, which occurs at an appreciable rate, would change the sense of a genetic message, so it seems advantageous for the cell to maintain genetic stability by having a surveillance system that excises dUMP residues no matter how they arise. Why thymidine was selected originally instead of uracil as a DNA base is a related question. Thermodynamic studies indicate that the methyl group of thymidine contributes significantly to hydrophobic interactions that stabilize double-helical DNA.

dUTPase has an unusual structure (Figure 22.19). The *E. coli* enzyme is a homotrimer, with each of its three active sites comprising amino acid residues from two adjacent polypeptide subunits.

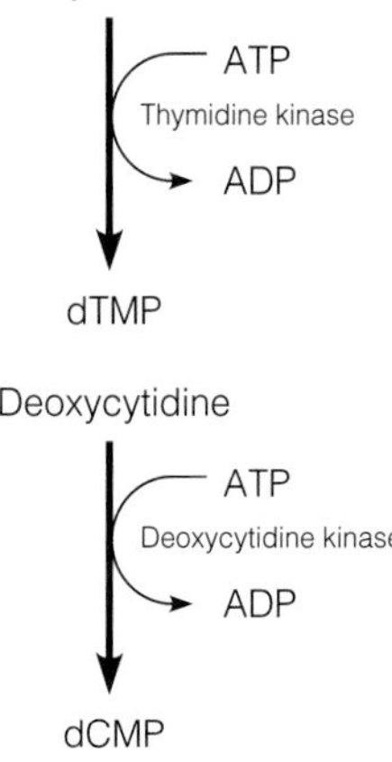

SALVAGE ROUTES TO DEOXYRIBONUCLEOTIDE SYNTHESIS

As noted previously, purine salvage usually involves phosphoribosyltransferase reactions, which generate ribonucleoside monophosphates from the purine bases and PRPP. After phosphorylation to the diphosphate level, these compounds enter deoxyribonucleotide metabolism via ribonucleotide reductase. However, deoxyribonucleoside kinases, leading directly to deoxyribonucleoside monophosphates, are widely distributed, and they involve both purines and pyrimidines. Cells and organisms vary widely in their contents of ribo- and deoxyribonucleoside kinases. Human cells contain four different deoxyribonucleoside kinases—(1) thymidine kinase, which is located in the cytosol; (2) deoxycytidine kinase, also a cytosolic enzyme, which phosphorylates deoxyadenosine and deoxyguanosine as well as deoxycytidine, but only at higher concentrations; (3) deoxyguanosine kinase, which is mitochondrial in origin; and (4) a mitochondrial isoform of thymidine kinase,

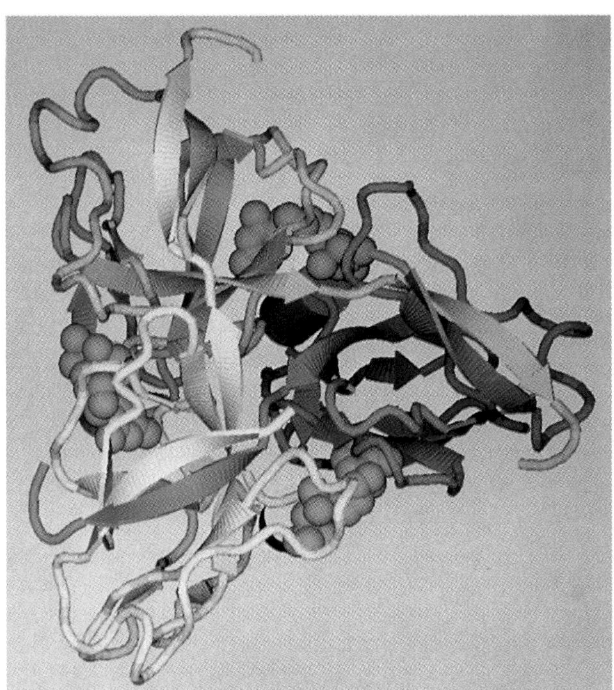

FIGURE **22.19**

Crystal structure of *E. coli* dUTPase. One of the three subunits of the homotrimer is shown in white, one in light blue, and one in dark blue. Each subunit contains a C-terminal arm that wraps about an adjacent subunit. A competitive inhibitor, dUDP (shown in orange), is bound at each of the three catalytic sites and each site is formed at a junction between subunits.

Courtesy of Elle Cedergren-Zeppenzauer and Gunilla Larsson.

which has a broader substrate specificity than the cytosolic enzyme, acting also upon deoxycytidine and deoxyuridine.

As discussed later in this chapter, several nucleoside analogs are being used or tested in the treatment of cancer and several viral diseases. Invariably, these drugs must be converted to deoxyribonucleotides in order to be effective, and this has focused attention upon the deoxyribonucleoside kinases. For example, a side effect of **3′-azido-2′,3′-dideoxythymidine** (AZT, page 823), the first drug to receive approval for treating human immunodeficiency virus (HIV) infections, is cardiotoxicity—damage to the heart muscle. The major route for azidothymidine utilization involves the mitochondrial isoform of thymidine kinase, also called TK2. Evidence suggests that deoxyribonucleotides of azidothymidine interfere with mitochondrial function, possibly by inhibiting mitochondrial DNA replication or transcription, and this is probably the basis for the cardiotoxicity. Therefore, current research is aimed at developing analogs whose metabolic activation does not occur in mitochondria.

Of the four deoxyribonucleoside kinases, three are synthesized constitutively, being produced at constant rates throughout the cell cycle. The exception is cytosolic thymidine kinase (TK1), the expression of which is highest when DNA is being replicated; in this respect TK1 resembles enzymes of de novo deoxyribonucleotide synthesis, such as ribonucleotide reductase. For reasons not yet understood, TK1 salvages exogenous thymidine extremely efficiently. Experiments with radiolabeled precursors show that dTTP derived from salvage synthesis is usually incorporated into DNA in preference to thymidine nucleotides generated by de novo synthesis. This is the basis for the widespread technique of estimating rates of DNA replication by measuring incorporation of radiolabeled thymidine into DNA. As discussed in Tools of Biochemistry 12A, however, an accurate measurement of the rate of DNA synthesis from thymidine incorporation data requires measuring the specific radioactivity of the labeled dTTP pool. Nevertheless, because of the great efficiency with which thymidine is utilized in most cells, simple measurements of thymidine incorporation often yield fairly accurate estimates of DNA replication rates.

3′-Azido-2′,3′-dideoxythymidine

Thymidylate Synthase: A Target Enzyme for Chemotherapy

A goal of **chemotherapy**—the treatment of diseases with chemical agents—is to exploit a biochemical difference between the disease process and the host tissue in order to interfere selectively with the disease process. Many chemotherapeutic agents were originally discovered by chance, through testing of analogs of normal metabolites. Most of these agents are limited in their effectiveness by unanticipated side effects, incomplete selectivity, and the development of resistance to the agent. One of the most exciting areas of modern biochemical pharmacology is drug architecture—the design of specific inhibitors based on knowledge of the molecular structure of the site to which the inhibitor will bind and the mechanism of action of the target molecule. For drugs whose target is an enzyme, it is necessary to know the three-dimensional structure of the enzyme and its mechanism of action. Obtaining this information requires a fusion of x-ray crystallography, classical bioorganic chemistry, site-directed mutagenesis, and computer-aided molecular graphics. Thymidylate synthase presents an excellent example of the utility of these approaches.

Inhibition of thymidylate synthase is an approach to cancer chemotherapy, by causing specific inhibition of DNA synthesis.

As noted here and in Chapter 20, the goal of chemotherapy is to attack selectively a metabolic process that is specific to the pathological condition. Because thymidylate synthase participates in the synthesis of a deoxyribonucleotide, any disease that involves uncontrolled cell proliferation can in principle be treated with inhibitors of thymidylate synthase: Blocking the production of an essential DNA precursor should inhibit DNA replication with minimal effects on other processes. Cells that are not undergoing rapid proliferation should be relatively immune to such agents. Thus, cancer and a wide range of infectious diseases should be amenable to treatment by this approach.

None of this was recognized in the mid-1950s. In fact, thymidylate synthase had not yet been discovered. It was known, however, that certain tumor cells took up and metabolized uracil much more rapidly than normal cells. Without knowing the metabolic fates of uracil in detail, Charles Heidelberger hoped to kill tumor cells selectively by treatment with analogs that would block uracil metabolism in tumor cells. To that end, he undertook the chemical synthesis of **5-fluorouracil (FUra)** and its deoxyribonucleoside, **5-fluorodeoxyuridine (FdUrd).** Both compounds were found to be potent inhibitors of DNA synthesis. Their action as inhibitors involves their intracellular conversion to **5-fluorodeoxyuridine monophosphate (FdUMP)**, a dUMP analog that acts as an irreversible inhibitor of thymidylate synthase.

Salvage pathways

ADP ATP

Thymidine kinase

HOCH$_2$

OH

5-Fluorouracil (FUra)

5-Fluorodeoxyuridine monophosphate (FdUMP)

5-Fluorodeoxyuridine (FdUrd)

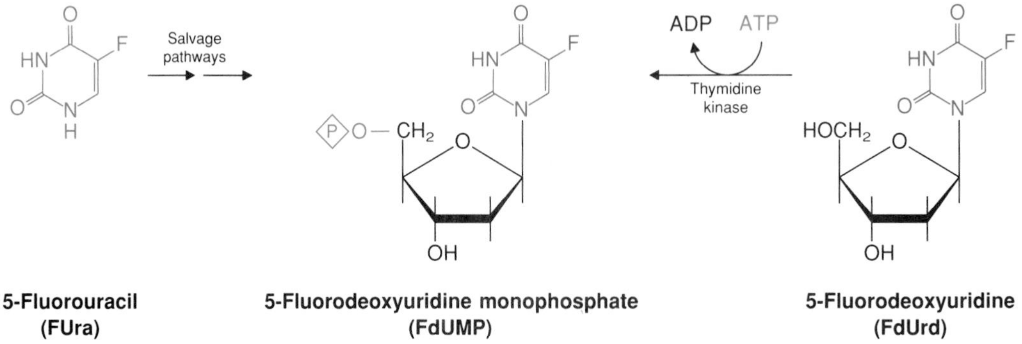

Both fluorouracil and fluorodeoxyuridine are used in cancer treatment. However, the fluorinated pyrimidines are not completely selective in their effects. For example, fluorouracil can be incorporated into RNA by salvage routes normally used for uracil, thereby interfering with the function of messenger RNA in both cancer and normal cells. Clearly, a detailed understanding of the active site of thymidylate synthase could lead to the design of completely specific enzyme inhibitors.

Analysis of the binding of 5-fluorodeoxyuridine monophosphate to thymidylate synthase has opened the door to understanding the enzyme's reaction mechanism and the structure of the active site. FdUMP is a true **mechanism-based inhibitor,** in that irreversible binding occurs only in the presence of 5,10-methylenetetrahydrofolate. Presumably, binding of the coenzyme induces a conformational change in the active site that duplicates early steps in the catalytic reaction and leads to irreversible FdUMP binding. Proteolytic digestion of the ternary complex containing FdUMP, methylenetetrahydrofolate, and enzyme led investigators in the laboratories of Charles Heidelberger and Daniel Santi to isolate a peptide fragment of the enzyme containing both the inhibitor and the coenzyme. Eventually it was shown that FdUMP was linked to the methylene carbon of the coenzyme through C-5 of the pyrimidine ring, and to the enzyme through a cysteine sulfur covalently bonded to C-6 of the pyrimidine. The structure of the complex suggested that the enzymatic reaction begins with nucleophilic attack by the cysteine thiol upon C-6 of the dUMP substrate.

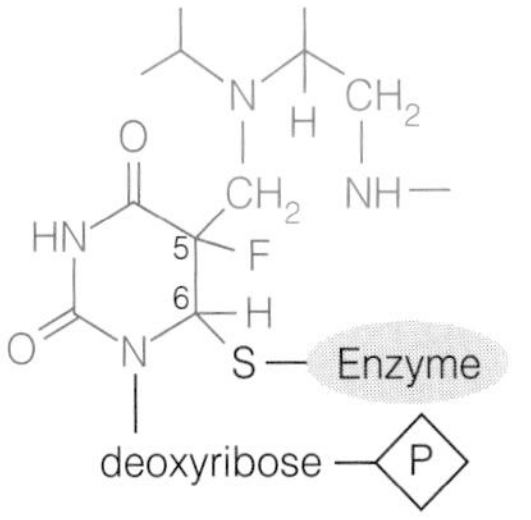

Methylene-THF (partial structure)

Complex between enzyme-bound FdUMP and 5,10-methylene-THF

The structure of the complex between enzyme-bound FdUMP and 5,10-methylene-THF suggested the mechanism outlined in Figure 22.20. As noted previously, a cysteine thiolate ion on the enzyme initiates a nucleophilic attack on C-6 of dUMP (step 1). This generates a resonance-stabilized enolate anion; C-5 now becomes a nucleophile, attacking the methylene carbon of the coenzyme (step 2). Loss of a proton from C-5 (step 3) initiates an electron shift that leads to transfer of the C-6 hydrogen of the cofactor to the pyrimidine (step 4), consistent with observations that this hydrogen is incorporated quantitatively into the thymidylate methyl group. This hydrogen transfer oxidizes the cofactor to dihydrofolate, which dissociates in step 5. In step 6 the covalent enzyme–substrate bond is broken, with re-formation of the double bond between C-5 and C-6 of the pyrimidine ring. Inhibition by FdUMP results from the electronegativity of fluorine, which generates a C—F bond at C-5 that cannot be broken. Thus, the reaction pathway cannot proceed to step 3.

Confirmation of this mechanism has come about with evidence that the substrate, dUMP, forms a covalent bond with the coenzyme during normal catalysis. More important evidence came from elucidation of a three-dimensional model of thymidylate synthase. In 1987 Robert Stroud, Daniel Santi, and their colleagues presented a model of the enzyme from *Lactobacillus casei* (Figure 22.21). Thymidylate synthase is highly conserved evolutionarily, with about 20% of its residues being invariant among mammalian, bacterial, viral, fungal, and protozoal sequences. The crystal structure shows a high percentage of these conserved residues in the cleft thought to represent the active site. The protein is a homodimer, and conserved residues from both subunits line the cleft and presumably contribute to catalysis. Cysteine 198, the residue that becomes linked to FdUMP, lies in this cleft, close to another conserved residue, Arg 218, which may lower the pK_a of the cysteine thiol and generate the reactive thiolate ion. Several conserved lysine residues lie nearby and are thought to represent binding sites for the polyglutamate tail on the folate coenzyme (thymidylate synthase binds folate polyglutamates about 100-fold more tightly than the monoglutamate).

Shortly after the enzyme model was presented, two laboratories crystallized *E. coli* thymidylate synthase as a complex with dUMP and an analog of 5,10-methylenetetrahydrofolate. Analysis of these complexes confirmed the conformations of substrate and cofactor bound in the active site (Figure 22.22) and identification of active site residues in contact with these ligands. Analysis of the binding interactions involved has led to design and synthesis of cofactor analogs that bear no obvious kinship with 5,10-methylenetetrahydrofolate but that compete with it quite effectively for binding to TS (K_i values as low as 30 nM). One such inhibitor, Thymitaq, is shown here. In 1991 the crystallization of the human enzyme was

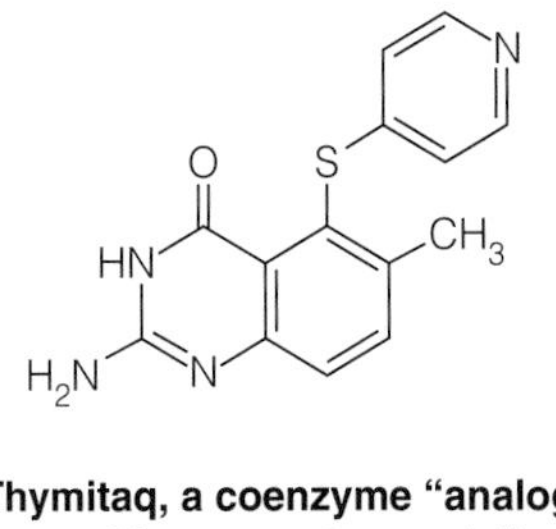

Thymitaq, a coenzyme "analog" created by computer modeling

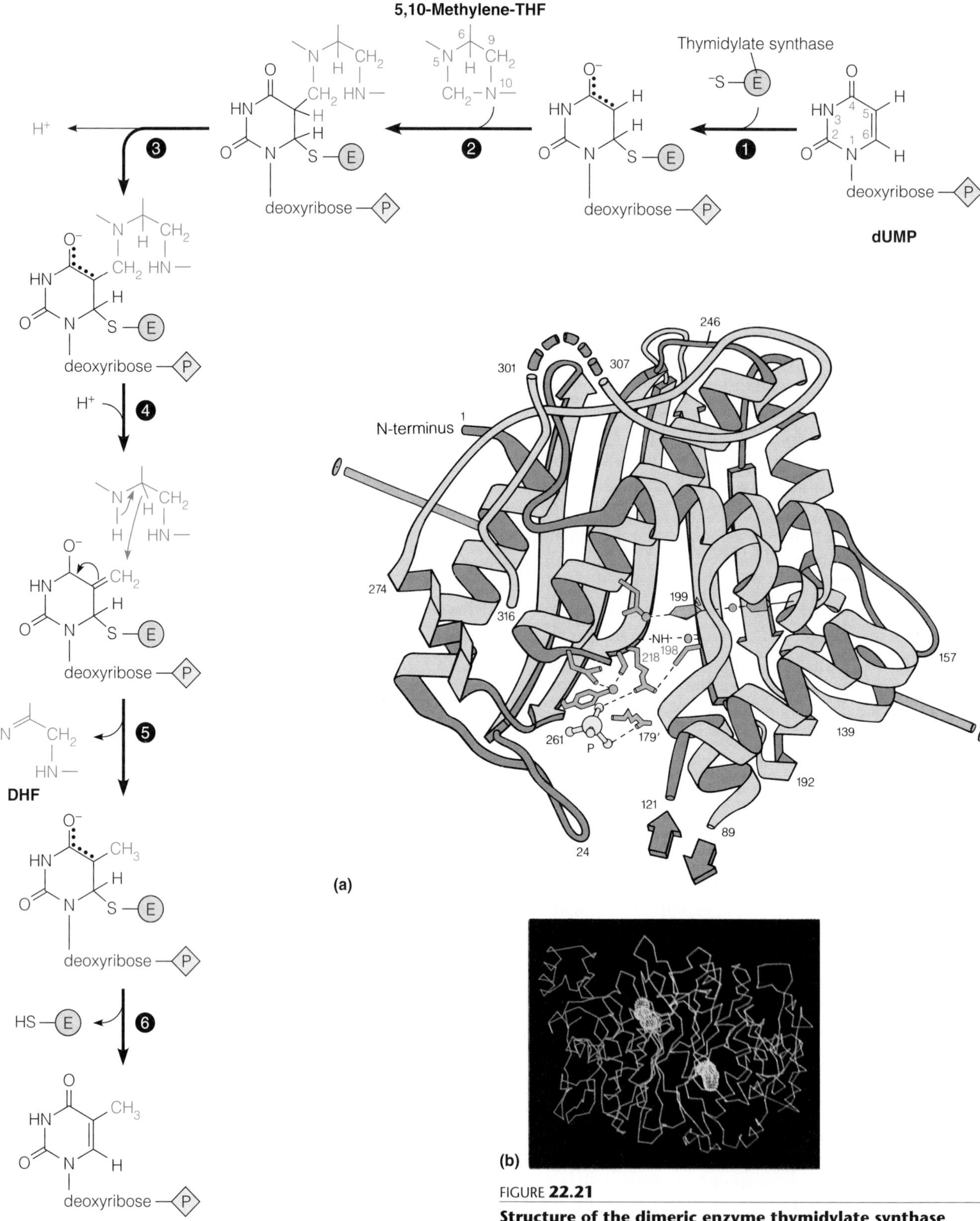

FIGURE 22.20
Mechanism for the reaction catalyzed by thymidylate synthase.

FIGURE 22.21
Structure of the dimeric enzyme thymidylate synthase from *Lactobacillus casei*. **(a)** Arrangement of helices and extended chain within one monomer. The amino terminus is at number 1. The broken chain at 301–307 and the break between 89 and 121 correspond to disordered regions in the structure. Conserved amino acid residues are shown in the active site cleft. Arg 179′ is from the adjacent monomer. **(b)** Structure of the homodimer.

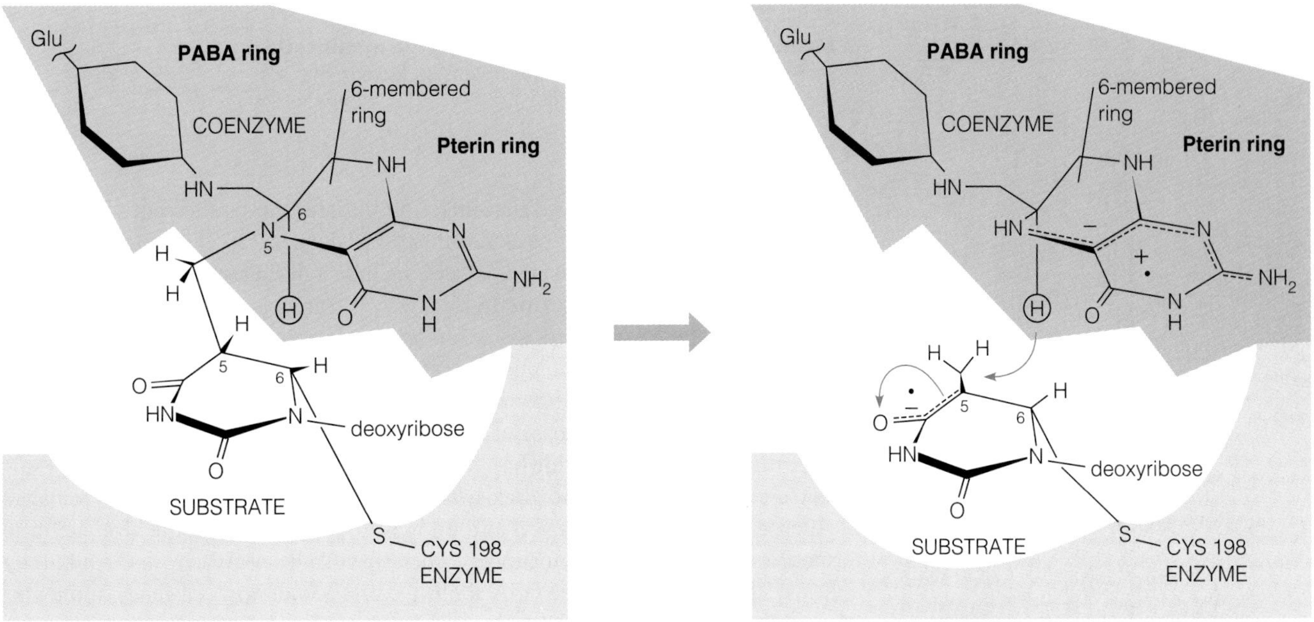

FIGURE **22.22**

Orientation of substrate and coenzyme in the active site of thymidylate synthase. The contacts shown in this figure were deduced from crystallographic analysis of a ternary enzyme–substrate–coenzyme analog complex. The pterin ring and PABA are parts of the 5,10-methylenetetrahydrofolate coenzyme.

Adapted with permission from J. S. Finer-Moore, W. R. Montfort, and R. M. Stroud, *Biochemistry* (1990) 29:6977–6986. © 1990 American Chemical Society.

reported, and this enzyme is the focus of further drug development. These efforts, plus mechanistic studies of variant enzymes produced by site-directed mutagenesis and structural determination of thymidylate synthases from other species, should pave the way for rational development of effective enzyme inhibitors that may be valuable anticancer drugs. The same type of approach is under way in dozens of laboratories, focused on drug receptors that include membrane-bound or intracellular receptor proteins and nucleic acids, as well as enzymes. The development of specific HIV protease inhibitors as anti-AIDS drugs is a striking example of the success of this approach.

Targets other than cancers are also susceptible to attack by inhibition of thymidylate synthase. For example, parasitic protozoans, such as those that cause malaria, synthesize an unusual form of thymidylate synthase—a bifunctional enzyme, with both thymidylate synthase and dihydrofolate reductase activities. Knowing the structure of this enzyme's active site should allow development of inhibitors that would block this specific enzyme but not thymidylate synthase of the animal or human host.

Virus-Directed Alterations of Nucleotide Metabolism

That viruses can redirect the metabolism of their host cells first came to light in 1957 through studies of nucleotide biosynthesis in *E. coli* bacteria infected by the T-even bacteriophages—T2, T4, and T6. G. R. Wyatt and Seymour Cohen had shown in 1952 that DNA of these viruses contains no cytosine but instead contains 5-hydroxymethylcytosine, with most of the hydroxymethyl groups further modified by being in glycosidic linkage with glucose moieties.

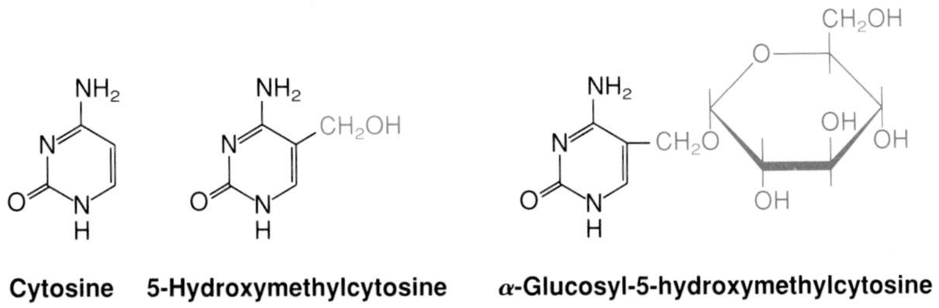

Cytosine **5-Hydroxymethylcytosine** **α-Glucosyl-5-hydroxymethylcytosine**

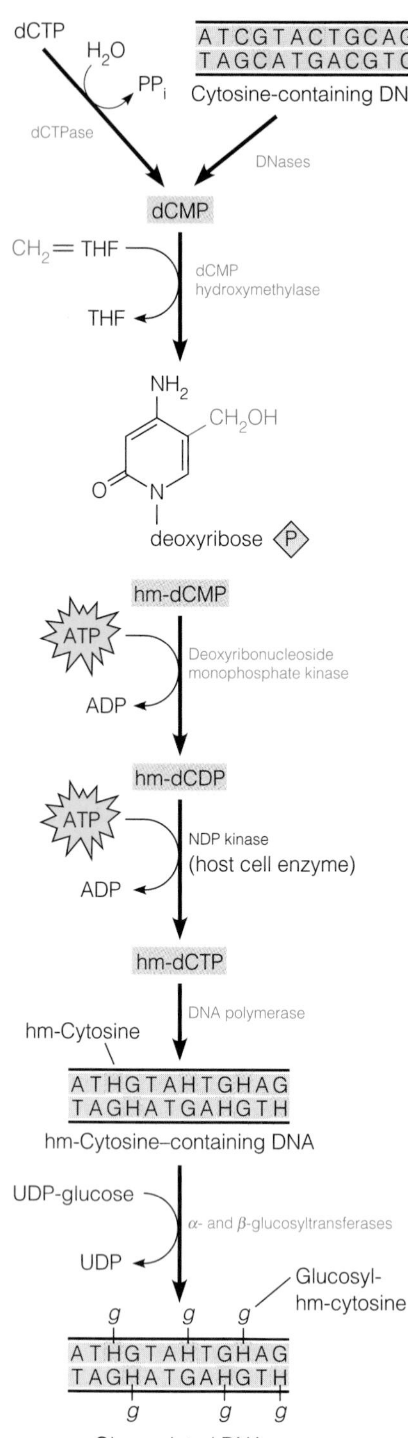

FIGURE **22.23**

Metabolic pathways leading to nucleotide modifications in T-even phage–infected *E. coli*. Virus-coded enzymes are shown in red. hm = hydroxymethyl.

Continuing work showed that infection initiates the synthesis of virus-coded enzymes, which carry out these modifications (Figure 22.23). The principal enzymes in T-even phage–infected bacteria include a **dCTPase,** which cleaves dCTP to dCMP; a **dCMP hydroxymethylase,** which transfers a one-carbon group to dCMP at the hydroxymethyl oxidation level; and a **deoxyribonucleoside monophosphate kinase,** which can phosphorylate the resultant 5-hydroxymethyl-dCMP. Phosphorylation of 5-hydroxymethyl-dCDP to the triphosphate is catalyzed by nucleoside diphosphate kinase of the host cell. The glucosylation reactions occur after the modified nucleotide has been incorporated into DNA. In T4 phage–infected bacteria there are two **glucosyltransferase** reactions, one of which transfers glucose in the α configuration and one in the β. In addition, the viral genome specifies several deoxyribonucleases, which specifically cleave cytosine-containing DNA. This process helps the virus abolish expression of host cell genes, and it also provides a source of precursors for viral DNA synthesis.

Although base substitutions in DNA are rather unusual, about a dozen instances have now come to light in which one of the four common deoxyribonucleotides is replaced, wholly or in part, by a chemically altered derivative that retains the base-pairing specificity of the replaced nucleotide. Some *Bacillus subtilis* phages, for example, substitute uracil for thymine in their DNA. Other *B. subtilis* phages contain 5-hydroxymethyluracil in place of thymine. A phage of *Xanthomonas oryzae* substitutes 5-methylcytosine for every one of its DNA cytosine residues. In each case investigated, the virus directs the nucleotide modifications through synthesis of virus-encoded enzymes that create novel metabolic pathways in the cells they infect.

Plant and animal viruses do not contain extensive nucleic acid base modifications of the type found in bacteriophages. (Most organisms contain a significant proportion of methylated nucleic acid bases, but these modifications occur after polymerization; see Chapter 25.) However, virus-coded enzymes are often produced to help the infected cell augment its synthesis of nucleic acid precursors. In some cases the virus-specified enzyme differs from its host cell counterpart sufficiently that scientists can design selective enzyme inhibitors and thereby achieve a specific chemotherapy directed against the virus-infected cell. The best current example is the use of **acyclovir** and its relative, **ganciclovir,** in treating herpesvirus infections. The herpesviruses are large DNA-containing viruses whose genomes encode several enzymes, including **deoxypyrimidine kinase.** Originally discovered as a thymidine kinase, the enzyme also phosphorylates dTMP. More important, however, is the extremely broad substrate specificity of the nucleoside kinase activity of this enzyme. The thymidine analog **5-iododeoxyuridine,** used to treat herpes simplex eye infections, is readily phosphorylated by deoxypyrimidine kinase and incorporated into DNA in place of thymidylate. More recently, acyclovir (also called **acycloguanosine**) and its relative ganciclovir have been shown also to be phosphorylated by viral deoxypyrimidine kinases, even though they are analogs of a purine nucleoside, deoxyguanosine.

Novel metabolic processes induced by viruses are attractive chemotherapeutic targets.

Both acyclovir and ganciclovir are now being used for topical and systemic treatment of herpes viral infections. All three analogs are converted ultimately to the 5′-triphosphate, which interferes with DNA replication. Uninfected cells do not efficiently phosphorylate acyclovir and ganciclovir, and they phosphorylate iododeoxyuridine weakly, so DNA replication and, hence, virus growth are inhibited selectively in the infected cells.

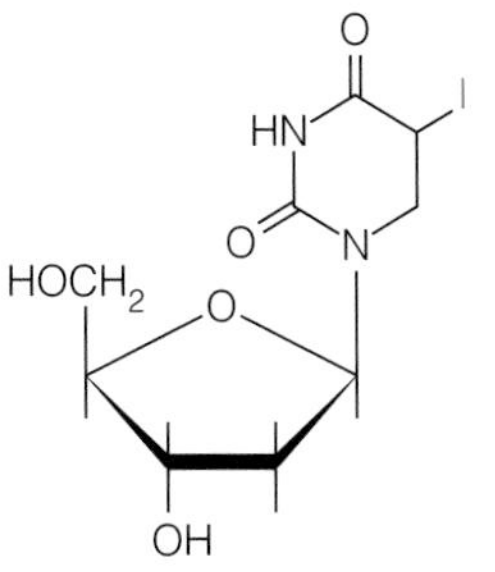

5-Iododeoxyuridine

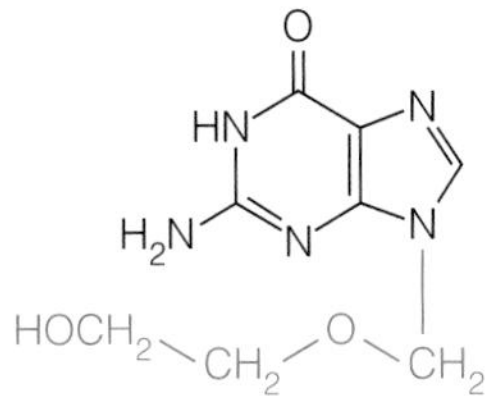

Acycloguanosine (acyclovir)

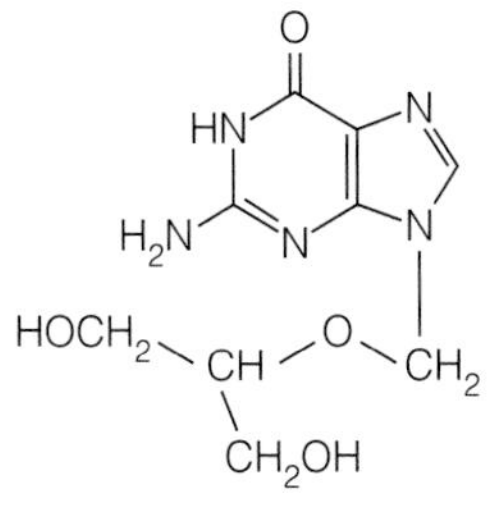

Ganciclovir

Biological and Medical Importance of Other Nucleotide Analogs

The previous examples have made clear the utility of nucleoside and nucleotide analogs, primarily as drugs. In this section we discuss additional analogs of medical importance, as well as several that are useful as research reagents. Nucleotides are poorly transported into cells because of the negative phosphate charge, so most of the compounds we shall discuss are introduced into cells as nucleosides or nucleoside derivatives, where they are acted upon initially by nucleoside kinases (see page 816). After uptake and conversion to nucleotides, these compounds interfere with metabolism in various ways. In this section we shall use the terms *nucleotide analogs* and *nucleoside analogs* interchangeably.

NUCLEOTIDE ANALOGS AS CHEMOTHERAPEUTIC AGENTS

The enzymes of nucleotide synthesis have been widely studied as target sites for the action of antiviral or antimicrobial drugs. As noted earlier, the aim is to identify a biochemical distinction between comparable processes of the uninfected host and the infected host.

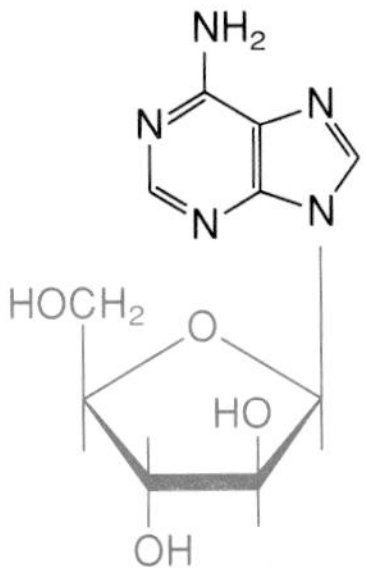

Arabinosyladenine (araA)

Arabinosylcytosine (araC)

3′-Thiacytidine

Antiviral Nucleoside Analogs

One of the earliest antiviral drugs approved for use in humans is **arabinosyladenine (araA)**. It is now being used to treat viral encephalitis, a neurological disease caused by another member of the herpesvirus family. Unlike acycloguanosine, araA is phosphorylated to the triphosphate level by cellular kinases. The triphosphate, araATP, is a selective inhibitor of DNA polymerases encoded by herpesviruses. Thus, araA selectively interferes with viral DNA replication, even though all cells, infected and uninfected, form the triphosphate. Because araA is susceptible to degradation by adenosine deaminase, its effectiveness can be increased when administered with an inhibitor of the latter enzyme. The arabinose analog of deoxycytidine, **arabinosylcytosine (araC)**, is being used in cancer chemotherapy. araCTP also interferes with DNA replication after conversion to the triphosphate.

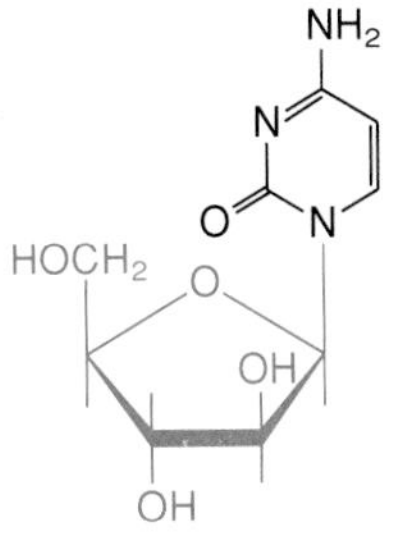

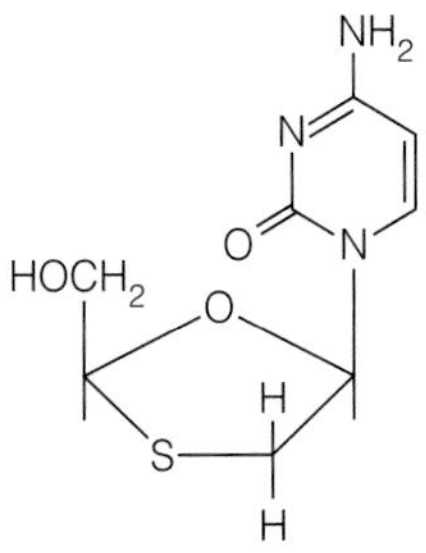

Other analogs receiving considerable attention are those being used to combat acquired immune deficiency syndrome (AIDS) caused by HIV. One such analog, 3′-azido-2′,3′-dideoxythymidine (AZT; see page 817) is anabolized to the corresponding 5′ triphosphate, which is an inhibitor of viral reverse transcriptase (the enzyme that makes a DNA copy of the viral RNA—see Chapter 24). Other nucleoside analogs—**2′,3′-dideoxycytidine (ddC)**, **2′,3′-dideoxyinosine (ddI)**, **3′-thiacytidine** (3TC), and **2′,3′-didehydro-3′-deoxythymidine** (d4T)—act by conversion to the corresponding triphosphate, which is incorporated into DNA but then blocks further replicative chain elongation because of the absence of a 3′ hydroxyl terminus. All four analogs have been approved for use in treating human HIV infections and both AZT and 3TC are components, along with HIV protease inhibitors, of the three-drug "cocktails" recently credited with long-term remission of HIV infections. In late 1994 two laboratories reported a synergistic effect of ddI and hydroxyurea against HIV infections, presumably starving the cell of dNTPs for reverse transcription (see Figure 24.45).

2′,3′-Dideoxycytidine **2′,3′-Dideoxyinosine** **2′,3′-Didehydro-3′-deoxythymidine**

Purine Salvage as a Target

An important biochemical anomaly is found in parasitic protozoans such as *Plasmodium,* which is responsible for malaria, and *Leishmania,* which causes a debilitating but usually nonfatal disease affecting the skin and visceral organs. Parasitic protozoans lack the capacity for de novo purine synthesis, and they depend entirely on salvage of nucleosides and bases provided by the host. Compounds such as allopurinol (see page 805) and **formycin B** inhibit the growth of these organisms in culture, partly through inhibition of salvage enzymes and partly through the ability of the salvage enzymes to anabolize the analog, an ability lacking in the corresponding host enzymes. For example, allopurinol is converted to an analog of inosinic acid and then to an AMP analog and is finally incorporated into RNA, where it interferes with messenger RNA coding in protein synthesis.

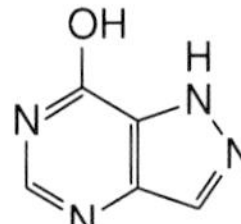

Formycin B

Inhibition of nucleoside salvage may be more effective if the analog is administered along with an inhibitor of nucleoside transport. In animal cells a single protein, the nucleoside transporter, appears to be responsible for uptake of a wide variety of nucleosides. Functioning of this protein is inhibited by **dipyridamole,** which blocks the uptake of most nucleosides.

Dipyridamole

Folate Antagonists

Recall from Chapter 20 that the folic acid analog methotrexate was found long ago to induce remissions in certain acute leukemias. What is the basis for this selectivity? Given that folate cofactors play essential roles in synthesizing precursors to DNA, RNA, protein, *and* phospholipids, one would expect inhibition of tetrahydrofolate synthesis to be toxic to all cells. However, there is a rationale for the selective toxicity of folate antagonists against proliferating cells. Recall that the thymidylate synthase reaction oxidizes methylenetetrahydrofolate to dihydrofolate; this is the only known tetrahydrofolate-requiring reaction that does not regenerate tetrahydrofolate. From the reactions shown in Figure 22.18, one can predict that inhibition of dihydrofolate reductase blocks the recycling of dihydrofolate back to tetrahydrofolate. Under these conditions the rate at which all of the intracellular reduced folates become oxidized is directly related to the intracellular activity of thymidylate synthase, which in turn is coordinated with the rate of DNA synthesis. Thus, proliferating cells, with rapid rates of DNA replication, will exhaust their tetrahydrofolate stores faster than nonproliferating cells.

Cell proliferation rate is a determinant of the effectiveness of chemotherapy with dihydrofolate reductase inhibitors, because thymidylate synthase reaction flux determines the rate at which tetrahydrofolate is oxidized to dihydrofolate.

Although antimetabolites such as fluorouracil or methotrexate do attack proliferating tissue selectively, they are also toxic to normal cells. Deleterious side effects are seen against tissues that proliferate as part of their normal function; such tissues include intestinal mucosa, hair cells, and components of the immune system. Equally serious is the development of drug-resistant cell variants. In such cases, levels of the target enzyme rise beyond the point that they can be controlled with inhibitors. Robert Schimke and his colleagues investigated the mechanism for the rise in dihydrofolate reductase activity, which can be several hundredfold in methotrexate-resistant cell lines. They found that prolonged incubation of cells in methotrexate often leads to selective amplification of the gene encoding dihydrofolate reductase; the number of DNA copies of this gene per cell increases many times, with a corresponding accumulation of the gene product, dihydrofolate reductase. Other resistance mutations involve more conventional mechanisms, such as altered transport of the drug into cells or alteration of the target enzyme, making it resistant to the antimetabolite.

Another class of dihydrofolate reductase inhibitors is exemplified by **trimethoprim.** This compound is a specific inhibitor of dihydrofolate reductases of prokaryotic origin. Trimethoprim and its relatives are widely used to treat both bacterial infections and certain forms of malaria. The success of these drugs derives from their being extremely weak inhibitors of vertebrate dihydrofolate reductases. Trimethoprim is often administered in conjunction with a sulfonamide drug to inhibit the *synthesis* of folate and hence to block sequential steps in the same pathway.

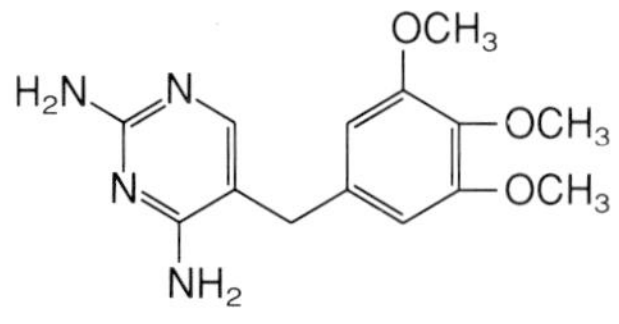

OCH_3 H_2N N N OCH_3 OCH_3 NH_2

Trimethoprim

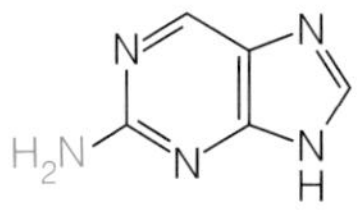

N N H_2N N N H

2-Aminopurine

O HN Br O N deoxyribose

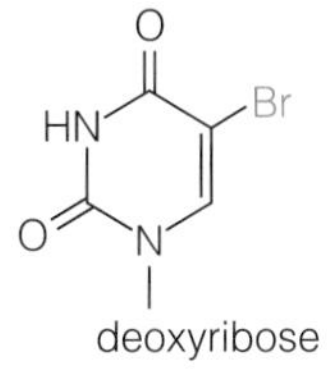

5-Bromodeoxyuridine

NUCLEOTIDE ANALOGS AND MUTAGENESIS

Some nucleotide analogs are excellent mutagens, useful both in the isolation of mutants and in studies on the mechanism of mutagenesis. Two such analogs are **2-aminopurine (2AP)** and **5-bromodeoxyuridine (BrdUrd).** 2-Aminopurine is incorporated into DNA in place of adenine, but when a 2AP-containing template replicates, the analog occasionally base-pairs with cytosine rather than thymine. Thus, incorporation of 2AP changes an A-T pair in DNA to a G-C pair (Figure 22.24).

Bromodeoxyuridine functions similarly, but it has other uses as well. It is an excellent thymidine analog, because the van der Waals radius of the bromine atom is close to that of the methyl group. Hence, it is efficiently incorporated into DNA. Mispairing of BrdUrd with a deoxyguanosine residue on replication of a BrdUrd-containing template can lead to mutagenesis by changing an A–T base pair to a G–C (see Figure 22.24). Alternatively, BrdUTP can compete with dCTP for incorporation opposite G in the template (not shown).

Because bromine is much heavier than a methyl group, it imparts an increased density to substituted DNA, thereby providing a physical basis for separating replicating from nonreplicating DNA (see Chapter 24). Finally, radiobiologists use BrdUrd as a radiosensitizing agent. Bromo-dUMP residues in DNA debrominate readily when the substituted DNA is irradiated with UV or near-UV light. This process generates free radicals, which cause various kinds of damage to the DNA structure.

Recent interest has focused upon a mutagenic nucleotide analog that is formed in normal metabolism. Recall from Chapter 15 that oxidative stress—generated, for example, by treating cells with hydrogen peroxide—damages DNA bases. The oxidation of guanine residues in DNA to 8-oxoguanine is strongly mutagenic, because 8-oxoguanine base-pairs during replication with adenine almost as efficiently as it pairs with cytosine. This mispairing event could cause mutation by changing a G–C base pair to a T–A. Recently an enzyme was discovered that specifically hydrolyzes the nucleotide 8-oxo-dGTP to 8-oxo-dGMP plus pyrophosphate (the *mutT* gene product; see Chapter 25). Mutations that inactivate this enzyme have a mutator phenotype, suggesting that a significant mutagenic pathway involves oxidation of guanine nucleotides to 8-oxoguanine nucleotides, followed by their incorporation into DNA, and that oxidative mutagenesis is minimized by the breakdown of 8-oxo-dGTP before this incorporation can occur.

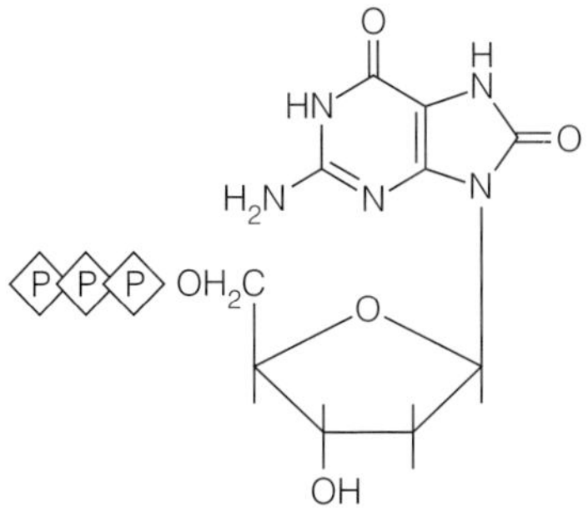

O H HN N O H_2N N N P P P OH_2C O OH

8-oxo-dGTP

dGTP
•OH
H_2O PP_i
8-oxo-dGTP → 8-oxo-dGMP
GC DNA
oxoGC + GC
oxoGA GC
TA oxoGC

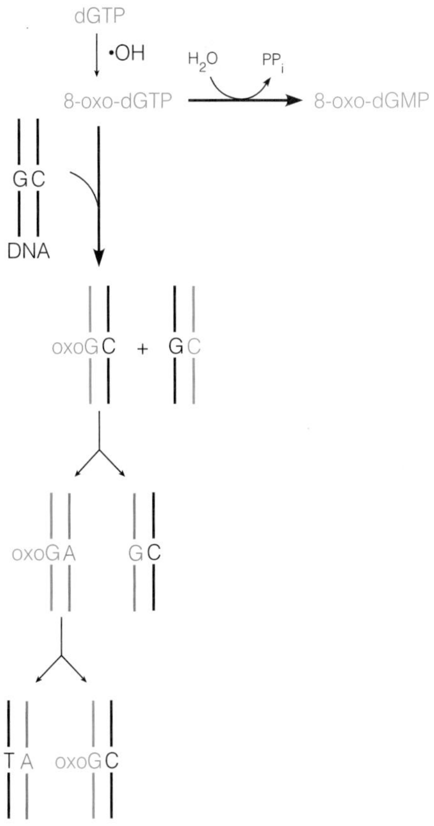

Nucleotide analogs with altered base-pairing properties are mutagenic, because non-Watson–Crick base pairs form when the analog is either in the template or in an incoming nucleoside triphosphate.

Genes for nucleotide-metabolizing enzymes are excellent selectable markers. Separate salvage and de novo pathways allow selection for survival or death of cells with particular metabolic traits.

NUCLEOTIDE-METABOLIZING ENZYMES AS SELECTABLE GENETIC MARKERS

Because most cells can synthesize nucleotides de novo, the enzymes of salvage synthesis are usually nonessential for cell viability. Moreover, as we have seen, many inhibitors of these enzymes are available. Consequently, nucleotide-metabolizing enzymes and the genes that encode them provide **selectable genetic markers,** which have a variety of uses. As the term implies, one can devise selective growth conditions such that only cells lacking a particular enzyme, or containing a particular enzyme, will grow. For example, **6-thioguanine** is a purine analog that is metabolized to a toxic intermediate by HGPRT. Culture of cells in thioguanine-containing medium allows growth of only the cells lacking active HGPRT.

SH, N, N, H_2N, N, N, H

6-Thioguanine

Similarly, one can isolate cells lacking thymidine kinase by selecting for a bromodeoxyuridine-resistant phenotype, because TK must be active in order to anabolize BrdUrd to a toxic metabolite. Thus, one can measure forward mutation rates by observing the appearance of these drug-resistant phenotypes.

By the same token, one can select for mutation in the reverse direction by adjusting culture conditions so that the capacity for salvage synthesis is essential to cell viability. A common technique, both in somatic cell genetic analysis and in preparation of monoclonal antibodies (see Tools of Biochemistry 7A), is **cell fusion.** Two cell lines of different origins are mixed under conditions in which some of them can physically fuse, resulting in two different nuclei in one cytoplasm. One can select for these cell hybrids by adjusting culture conditions so that only the hybrids will grow in **HAT medium** (normal cell culture medium augmented with hypoxanthine, aminopterin, and thymidine). Aminopterin inhibits dihydrofolate reductase and hence blocks de novo purine and thymidylate synthesis. Cells can survive only if they have active HGPRT, to utilize the hypoxanthine for purine synthesis, and TK, to utilize the thymidine for thymidylate synthesis.

Genes with selectable phenotypes represent an important adjunct to the use of recombinant DNA technology for introducing novel genetic material into cells—animal, plant, or microbial—because one can introduce recombinant DNA molecules that contain a selectable marker in tandem with a gene whose transfer is desired. Application of selective conditions then forces the growth of only those cells that have acquired the new pair of genes. An intriguing variant on this approach is described in a recent report of a novel treatment of brain tumors. A recombinant DNA containing the herpes simplex virus deoxypyrimidine kinase gene

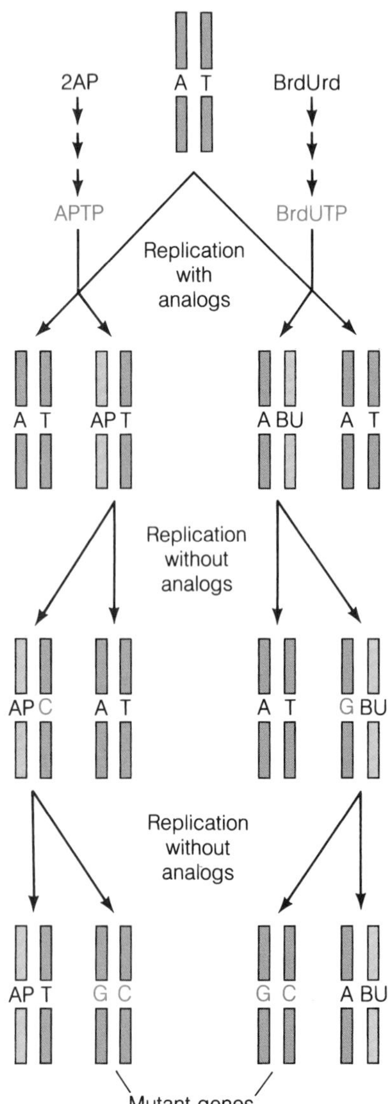

FIGURE **22.24**

Mechanisms of mutagenesis by nucleotide analogs. 2-Aminopurine (2AP) is converted by salvage pathways to dAPTP, the dNTP analog of dATP. BrdUrd is converted to BrdUTP, the dNTP analog of dTTP. The first round of replication occurs in the presence of the analog, and the second and third rounds occur in its absence. Replication of only the analog-containing duplex is shown in the second and third rounds. AP is a 2-aminopurine nucleotide residue in DNA, and BU is a bromodeoxyuridine nucleotide residue. In the pathways shown, both analogs change an A-T base pair to a G-C base pair (red letters). Other pathways can occur as well.

was injected into the tumor. In experimental animals the proliferating (tumor) cells took up and replicated this DNA. After several days the animals were treated with ganciclovir, which led to selective killing of the tumor cells because of selective phosphorylation of the drug in those cells.

SUMMARY

Nucleotides arise within cells from nucleic acid breakdown, from reuse (or salvage) of preformed nucleosides or nucleobases, or from de novo biosynthesis. Purine nucleotides are formed at the nucleotide level, in a 10-step pathway that leads from PRPP to inosinic acid. Beyond this branch point, separate pathways lead to adenine and guanine nucleotides. Purine catabolism yields uric acid, an insoluble compound that is formed in excess in a variety of disease states. Pyrimidines are synthesized at the base level, with conversion to a nucleotide occurring late in the pathway. An unbranched pathway leads to both UTP and CTP. In most organisms the ribonucleoside diphosphates are substrates for reduction of the ribose sugar in situ, yielding deoxyribonucleoside diphosphates, which in turn lead to the four dNTP DNA precursors. Ribonucleotide reductase is an important control site, inasmuch as it represents the first metabolic reaction committed to DNA synthesis. Biosynthesis of thymine nucleotides involves transfer of the methylene group of 5,10-methylenetetrahydrofolate to a deoxyuridine nucleotide, followed by reduction of the methylene group. Reactions of deoxyribonucleotide biosynthesis are target sites for enzyme inhibitors that have found use as anticancer, antimicrobial, antiviral, and antiparasitic drugs. Other nucleotide analogs have found use as research reagents—for example, in studies of mutagenesis or as DNA density labels.

REFERENCES

Enzymes of Nucleotide Metabolism

Blakley, R. L., and S. J. Benkovic (1984) *Folates and Pterins,* Vol. 1. Academic Press, New York. A multiauthored book containing reviews on dihydrofolate reductase, purine metabolism, and pyrimidine biosynthesis.

Elion, G. B. (1989) The purine path to chemotherapy. *Science* 244:41–47. Dr. Elion's Nobel Prize address, which described the development of allopurinol, acyclovir, 6-thioguanine, and other therapeutically valuable purine analogs.

Kornberg, A., and T. A. Baker (1992) *DNA Replication,* 2nd ed. Freeman, San Francisco. The first chapter contains an excellent summary of nucleotide metabolism.

Plagemann, P. G. W., R. M. Wohlhueter, and C. Woffendin (1988) Nucleoside and nucleobase transport in animal cells. *Biochim. Biophys. Acta* 947:405–443. For salvage pathways to occur, their substrates must get into cells. This review describes the transport processes involved.

Smith, J. L., E. J. Zaluzec, J.-P. Wery, L. Niu, R. L. Switzer, H. Zalkin, and Y. Satow (1994) Structure of the allosteric regulatory enzyme of purine biosynthesis. *Science* 264:1427–1433. Crystal structure of *B. subtilis* PRPP amidotransferase, the key regulatory enzyme of purine synthesis.

Webster, D. R., D. M. O. Becroft, and D. P. Suttle (1995) Hereditary orotic aciduria and other disorders of pyrimidine metabolism. In: *The Metabolic and Molecular Basis of Inherited Disease,* 7th ed., edited by C. R. Scrivener, A. L. Beaudet, W. S. Sly, and D. Valle, pp. 1799–1837. McGraw-Hill, New York. A thorough discussion of these rare diseases.

Zalkin, H., and J. E. Dixon (1992) De novo purine nucleotide biosynthesis. *Prog. Nucleic Acid Res. Mol. Biol.* 42:259–287. A review dealing mostly with genetic regulation in prokaryotes and structures of the enzymes and their genes in eukaryotes.

Deoxyribonucleotide Biosynthesis

Arnér, E. S. J., and S. Eriksson (1995) Mammalian deoxyribonucleoside kinases. *Pharmacol. Ther.* 67:155–186. Interest in these enzymes has increased because of their roles in metabolic activation of nucleoside drugs.

Carreras, C., and D. V. Santi (1995) The catalytic mechanism and structure of thymidylate synthase. *Annu. Rev. Biochem.* 64:721–762. A mechanistic analysis based upon recent structural information.

Danenberg, P. (1977) Thymidylate synthase—A target enzyme in chemotherapy. *Biochim. Biophys. Acta* 473:73–92. This older review nicely summarizes the insights gained into mechanism by studies of thymidylate synthase inhibition by FdUMP.

Elledge, S., Z. Zhou, and J. B. Allen (1992) Ribonucleotide reductase: Regulation, regulation, regulation. *Trends Biochem. Sci.* 17:119–123. In yeast this enzyme is both cell cycle–regulated and inducible by DNA damage, as discussed in this minireview.

Holmgren, A. (1989) Thioredoxin and glutaredoxin systems. *J. Biol. Chem.* 264:13963–13966. The two known electron carriers for rNDP reductase are discussed, plus the possible existence of another such carrier.

Jordan, A., and P. Reichard (1998) Ribonucleotide reductases. *Annu. Rev. Biochem.* 67:71–98. A comprehensive recent review, which focuses upon the evolutionary significance of the existence of widely divergent classes of this important enzyme.

Kunz, B. A., S. E. Kohalmi, T. A. Kunkel, C. K. Mathews, E. M. McIntosh, and J. A. Reidy (1994) Deoxyribonucleoside triphosphate levels: A critical factor in the maintenance of genetic stability. *Mutat. Res.* 318:1–64. This review describes genetic consequences of deoxyribonucleotide pool imbalances.

Li, P., D. Nijhawan, I. Budihardjo, S. M. Srinivasula, M. Ahmad, E. S. Alnemri, and X. Wang (1997) Cytochrome *c* and dATP-dependent formation of Apaf-1/Caspase 9 complex initiates an apoptotic protease cascade. *Cell* 91:479–489. Intriguing evidence that dATP accumulation helps to trigger apoptosis and kill cells even if they are not undergoing DNA replication.

Mathews, C. K. (1993) Enzyme organization in DNA precursor biosynthesis. *Prog. Nucleic Acid Res. Mol. Biol.* 44:167–203. This review summarizes evidence that dNTP biosynthetic enzymes are linked in multienzyme complexes, which may in turn be linked to DNA replication sites.

Stubbe, J. (1998) Ribonucleotide reductases in the twenty-first century. *Proc. Natl. Acad. Sci USA* 95:2723–2724. A brief but well-referenced commentary that focuses upon structural and mechanistic issues awaiting resolution.

Uhlin, U., and H. Eklund (1994) Structure of ribonucleotide reductase protein R1. *Nature* 370:533–539. A crystallographic analysis that presents a model for the reductase holoenzyme.

Nucleotide Analogs and Chemotherapy

Appelt, K., and 31 coauthors (1991) Design of enzyme inhibitors using iterative protein crystallographic analysis. *J. Med. Chem.* 34:1925–1934. The new era of drug design, applied to thymidylate synthase.

Culver, K. W., Z. Ram, S. Wallbridge, H. Ishii, E. H. Oldfield, and R. M. Blaese (1992) In vivo gene transfer with retroviral vector—Producer cells for treatment of experimental brain tumors. *Science* 256:1550–1552. An exciting way to use herpes viral deoxypyrimidine kinase as a selective agent for tumor cell killing.

Hardy, L. W., J. S. Finer-Moore, W. R. Montfort, M. O. Jones, D. V. Santi, and R. M. Stroud (1987) Atomic structure of thymidylate synthase: Target for rational drug design. *Science* 235:448–455. Describes determination of the crystal structure of this enzyme and its implications.

Jackson, R. C. (ed.) (1999) *Anti-folate drugs—Past and future perspectives* (Series: Cancer Drug Discovery and Development) Humana Press, Inc., Totowa, N.J. 451 pages. Contains 22 articles describing new approaches to drug development.

Knighton, D. R., C.-C. Kan, E. Howland, C. A. Janson, Z. Hostomska, K. M. Welsh, and D. A. Matthews (1994) Structure of and kinetic channelling in bifunctional thymidylate synthase–dihydrofolate reductase. *Struct. Biol.* 1:186–194. This distinctive enzyme is an attractive target for antiprotozoal chemotherapy.

Mitsuya, H. (ed.) (1997) *Anti-HIV Nucleosides: Past, Present, and Future.* R. G. Landes, Georgetown, Tex. This short book contains five articles by leading contributors to HIV drug development.

North, T. W. (1992) Chemotherapy, antiviral agents. In: *Encyclopedia of Human Biology,* Vol. 2, pp. 395–402. Academic Press, New York. An informative and readable review.

PROBLEMS

1. Identify each reaction catalyzed by (a) a nucleotidase; (b) a phosphorylase; (c) a phosphoribosyltransferase.

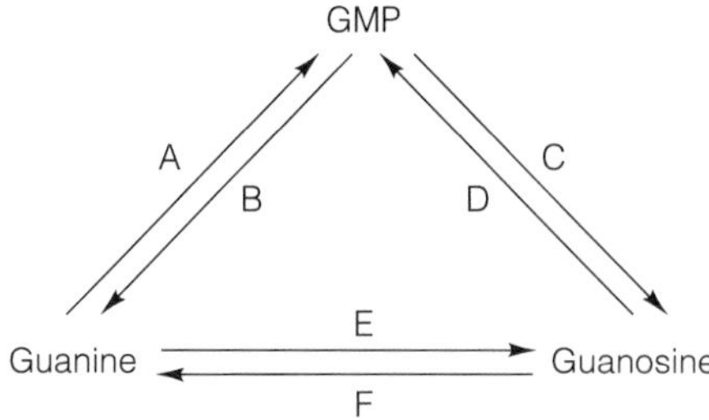

2. Describe the metabolic fate of 5-bromodeoxyuridine. Would you expect BrdUrd to *inhibit* DNA replication? Briefly explain your answer.

3. Predict the effects of the following compounds on intracellular nucleoside triphosphate levels. For each answer, plot percent initial nucleotide level as a function of time after administration of the agent, for each of the four nucleotides (semiquantitatively). Consider not only the primary effect of the compound but also any indirect effects on allosteric enzymes caused by nucleotide accumulation or depletion.
   (a) Effect of thymidine on dNTP levels
   (b) Effect of trimethoprim on bacterial rNTP levels
   (c) Effect of fluorodeoxyuridine on dNTP levels
   (d) Effect of hydroxyurea on dNTP levels
   (e) Effect of azaserine on rNTP levels

4. Radioactive uracil can be used to label all of the pyrimidine residues in DNA. Using either names or structures, present pathways for conversion of uracil to dTTP and to dCTP. For each reaction, show the involvement of cofactors, and identify sites of allosteric regulation.

5. Similarly, hypoxanthine (HX) can be used to label purine residues. As in Problem 4, write reactions showing the conversion of hypoxanthine to dATP and dGTP.

6. Leukemia is a neoplastic (cancerous) proliferation of white blood cells. Clinicians are currently testing deoxycoformycin, an adenosine deaminase inhibitor, as a possible antileukemic agent. Why might one expect this therapy to be effective?

*7. Under what conditions might one expect a deficiency of hypoxanthine–guanine phosphoribosyltransferase to affect the rate of *pyrimidine* nucleotide biosynthesis? How might one estimate the rate of pyrimidine nucleotide biosynthesis in living animals or people?

*8. A classic way to isolate thymidylate synthase–negative mutants of bacteria is to treat a growing culture with thymidine and trimethoprim. Most of the cells are killed, and the survivors are greatly enriched in thymidylate synthase–negative mutants.
   (a) What phenotype would allow you to identify these mutants?
   (b) What is the biochemical rationale for the selection? (That is, why are the mutants not killed under these conditions?)
   (c) How would the procedure need to be modified to select mammalian cell mutants defective in thymidylate synthase?

9. As stated in the text, mammalian cells can become resistant to the lethal action of methotrexate by the selective survival of cells containing increases in dihydrofolate reductase gene copy number, so that intracellular levels of the enzyme become very high. What other biochemical or genetic changes in cells could cause them to become resistant to methotrexate?

10. As stated in the text, bacteriophages have been discovered with the following base substitutions in their DNA:
(a) dUMP completely substituting for dTMP
(b) 5-hydroxymethyl-dUMP completely substituting for dTMP
(c) 5-methyl-dCMP completely substituting for dCMP.
For any one of these cases, formulate a set of virus-coded enzyme activities that could lead to the observed substitution. Write a balanced equation for each reaction you propose.

*11. Radioisotope "suicide techniques" are often used in the selection of mutants. In one such technique, cells are grown in the presence of [^{3}H]thymidine at very high specific activity. The cells are then stored frozen to allow for decay of some of the incorporated radioactivity. Decay of dTMP residues in DNA causes strand breakage and other potentially lethal events. Thus, cells that have incorporated thymidine into their DNA are very likely to be killed by this regimen.
(a) Identify an enzyme deficiency that would allow a mutant to survive such a regimen.
(b) Choose any enzyme of nucleotide metabolism, and devise a suicide procedure that could select for mutants deficient in that enzyme.

*12. For any of the multifunctional enzymes in purine or pyrimidine nucleotide biosynthesis, assume that the enzyme is available in purified form, and propose one or two experiments to determine whether the enzyme "channels" substrates through a multistep reaction sequence.

13. Write balanced equations for the three known reactions that transfer an amino group to a substrate by condensation with aspartate to give an intermediate that then undergoes α,β-elimination to give the product plus fumarate.

*14. CTP synthetase catalyzes the glutamine-dependent conversion of UTP to CTP. The enzyme is allosterically inhibited by the product, CTP. Mammalian cells defective in this allosteric inhibition are found to have a complex phenotype: They require thymidine in the growth medium, they have unbalanced nucleotide pools, and they have a mutator phenotype. Explain the basis for these observations.

15. If thymine nucleotides are degraded by the same enzymes as those catabolizing uridine nucleotides, give the structure of the thymine metabolite that corresponds to β-ureidopropionate.

16. (a) Explain the biochemical basis for the fact that one can synchronize cell populations by treating them with deoxythymidine.
(b) Explain the apparent paradox that dATP at low concentrations is an activator of ribonucleotide reductase, whereas at higher concentrations it becomes inhibitory.

17. The text states that ATP is synthesized primarily by energy metabolism, whereas other nucleoside triphosphates are formed from the action of nucleoside diphosphate kinase. What additional pathway exists for GTP synthesis?

CHAPTER 23

Metabolic Coordination, Metabolic Control, and Signal Transduction

OUR PRESENTATION OF INTERMEDIARY METABOLISM IN EARLIER CHAPTERS HAS placed primary emphasis on the cell, its reactions, its individual enzymes, and its other components. In this chapter we integrate these individual pathways in two ways. First, we review the metabolic profiles of the major organs in vertebrates: which fuels they use, which fuels they generate, and how the organs interact under stress to maintain appropriate energy balance. The interactions are controlled in large part by hormonal signals, some of which we have already discussed. In the second part of the chapter we describe molecular mechanisms in signal transduction, extending what we introduced in Chapter 12. Great insight has come in recent years, with discoveries of mutationally altered elements of signal transduction—growth factors, hormones, transcription factors, and receptors—in cancer cells. These discoveries have both established cancer as a genetic disease and illuminated normal cell control mechanisms, by showing the consequences of perturbed control.

Metabolite concentrations represent a significant intracellular control mechanism.

As we consider metabolic integration, keep in mind that metabolism is controlled to a great extent by the availability of substrates for specific metabolic pathways. In general, substrate concentrations within cells fall below saturating levels for the enzymes that metabolize them. Therefore, reaction fluxes through particular enzymes vary as the concentrations of the substrate vary. A good example is the metabolic adaptation that occurs during a marathon run. Once the glycogen stores in liver and muscle are exhausted, flux through glycolysis decreases in muscle, not for any hormonal reason but simply because glucose phosphates are less available. Hormonal adjustments then occur to allow increased use of fatty acids in muscle, but the primary factor in the metabolic decision made by the cell—which substrates to catabolize for energy—is the concentration of each of the usable substrates.

Interdependence of the Major Organs in Vertebrate Fuel Metabolism

In this section we look at metabolism not as the activities in just one cell but as the totality of chemical reactions in a complex multicellular animal. We emphasize the specialized roles that each of the major organs plays in fuel metabolism—

brain, muscle, liver, adipose tissue, and heart—and we describe the varying relationships among these organs as the animal encounters different physiological conditions.

FUEL INPUTS AND OUTPUTS

In a differentiated organism, each tissue must be provided with fuels that it can use, in amounts sufficient to meet its own energy needs and to perform its specialized roles. The kidney, for example, must generate ATP for the osmotic work of transporting solutes against a concentration gradient for excretion. Muscle must generate ATP for the mechanical work of contraction, and in heart muscle the energy supply must be continuous. The liver generates ATP for biosynthetic purposes, whether for plasma protein synthesis, cholesterol generation, fatty acid synthesis, gluconeogenesis, or the production of urea for nitrogen excretion. Energy production must meet needs that vary widely, depending on level of exertion, composition of fuel molecules in the diet, time since last feeding, and so forth. For example, in humans the daily caloric intake may vary by 4-fold, depending in part on the level of exertion—from 1500 to 6000 kcal/day in an average-sized human or, in the SI units used in this book, from 6000 to 25,000 kJ/day.

The major organs involved in fuel metabolism vary in their levels of specific enzymes, so that each organ is specialized for the storage, use, and generation of different fuels. The major fuel depots are *triacylglycerols,* stored primarily in adipose tissue; *protein,* most of which exists in skeletal muscle; and *glycogen,* which is stored in both liver and muscle. In general, an organ specialized to *produce* a particular fuel lacks the enzymes to *use* that fuel. For example, little catabolism of ketone bodies occurs in the liver. Now let us review how the mobilization of each depot is controlled, and how the organs involved communicate with each other, to meet the energy needs of the animal. This information is summarized in Figure 23.1 and Table 23.1.

The major fuel depots are triacylglycerols (adipose tissue), protein (muscle), and glycogen (muscle and liver).

METABOLIC DIVISION OF LABOR AMONG THE MAJOR ORGANS

Brain

The brain is the most fastidious, and one of the most voracious, of the organs. It must generate ATP in large quantities to maintain the membrane potentials essential for transmission of nerve impulses. Under normal conditions the brain uses only glucose to meet its prodigious energy requirement, which amounts to about 60% of the glucose utilization of a human at rest. The brain's need for about 120 grams of glucose per day is equivalent to 1760 kJ—about 15% of the total energy consumed by one person. The brain's quantitative requirement for glucose remains quite constant, even when an animal is at rest or asleep. The brain is a highly aerobic organ, and its metabolism demands some 20% of the total oxygen consumed by a human. Because the brain has no significant glycogen or other fuel reserves, the supply of both oxygen and glucose cannot be interrupted, even for a short time. Otherwise, irreversible brain damage results. However, the brain can adapt during fasting to use ketone bodies (see Chapter 18) instead of glucose as a major fuel.

Muscle

Muscle can utilize a variety of fuels—glucose, fatty acids, and ketone bodies. Skeletal muscle varies widely in its energy demands and the fuels it consumes, in line with its wide variations in activity. In resting muscle, fatty acids represent the major energy source; during exertion, glucose is the primary source. Early in a period of exertion, that glucose comes from mobilization of the muscle's glycogen reserves. Later, as glycogen reserves are depleted, fatty acids become the dominant fuel. Skeletal muscle stores about three-fourths of the total glycogen in humans, with most of the rest being stored in the liver. However, glucose released

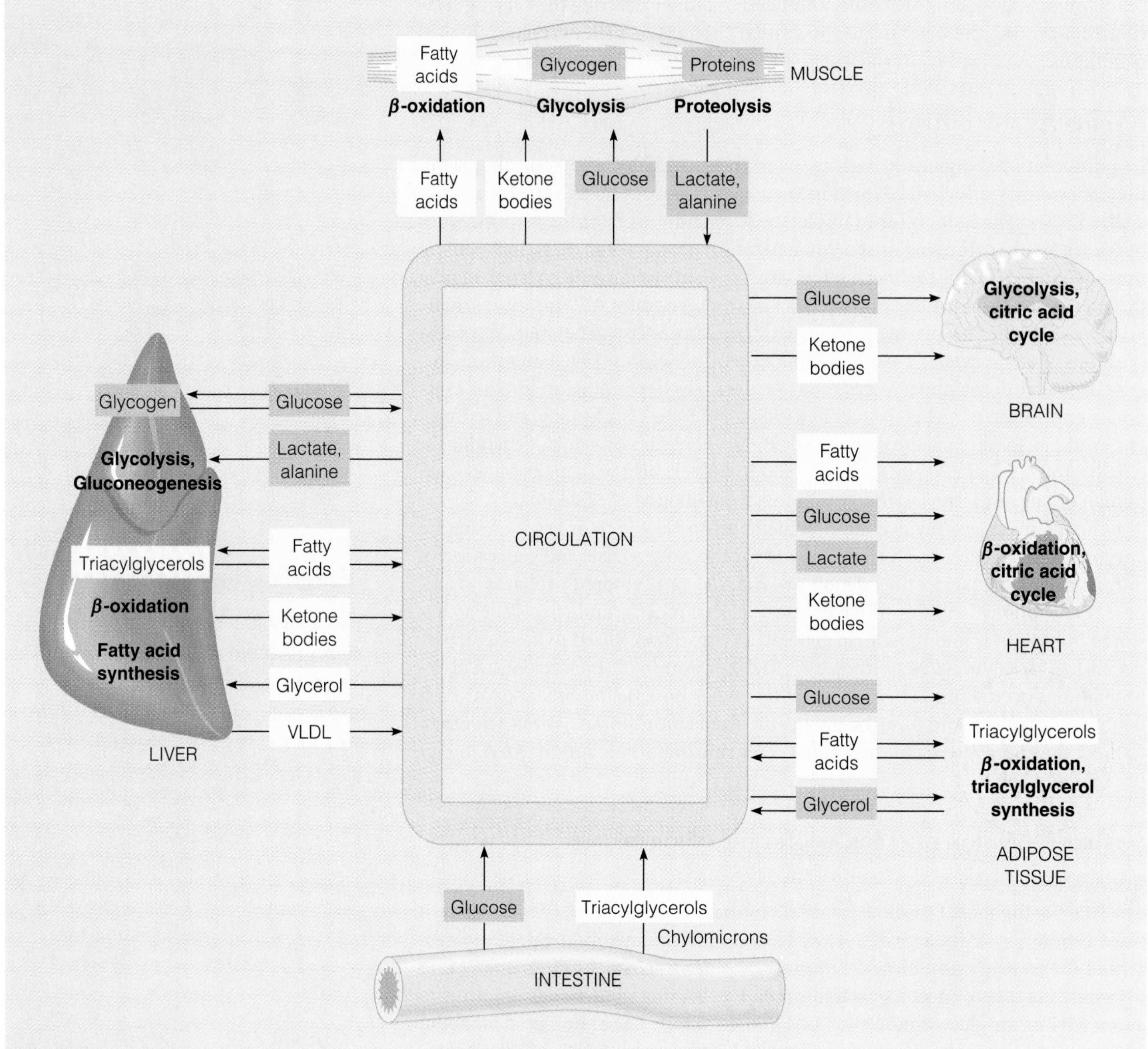

FIGURE **23.1**

Metabolic interactions among the major fuel-metabolizing organs. The major fuel metabolites imported and exported by each organ are shown, along with the major energy pathways and fuel reserves in each organ.

TABLE 23.1 Profiles of the major vertebrate organs in fuel metabolism

Tissue	Fuel Store	Preferred Fuel	Fuel Sources Exported
Brain	None	Glucose (ketone bodies during starvation)	None
Skeletal muscle (resting)	Glycogen	Fatty acids	None
Skeletal muscle (during exertion)	None	Glucose	Lactate, alanine
Heart muscle	None	Fatty acids	None
Adipose tissue	Triacylglycerols	Fatty acids	Fatty acids, glycerol
Liver	Glycogen, triacylglycerols	Amino acids, glucose, fatty acids	Fatty acids, glucose, ketone bodies

from muscle glycogen cannot be released from the cell for use by other tissues. Muscle lacks the enzyme glucose-6-phosphatase, so glucose phosphates derived from glycogen cannot be converted to glucose and released from the cell.

During exertion, the flux rate through glycolysis exceeds that through the citric acid cycle, so lactate accumulates and is released. Another metabolic product is alanine, produced via transamination from pyruvate in the glucose–alanine cycle (see page 730 in Chapter 20). Both lactate and alanine are transported through the bloodstream to the liver, where they are reconverted through gluconeogenesis to glucose, for return to the muscle and other tissues by the Cori cycle. However, the major fate of lactate is reoxidation in the heart (Chapter 16).

Muscle contains another readily mobilizable source of energy—its own protein. However, the breakdown of muscle protein to meet energy needs is both energetically wasteful and harmful to an animal, which must move about in order to survive. Protein breakdown is regulated so as to minimize amino acid catabolism except in starvation.

Finally, recall that muscle has an additional energy reserve in creatine phosphate, which generates ATP without the need for metabolizing fuels (see page 425 in Chapter 12). This reserve is exhausted early in a period of exertion and must be replenished, along with glycogen stores, as muscle rests after prolonged exertion.

Heart

The metabolism of heart muscle differs from that of skeletal muscle in three important respects. First, the variation in work output is far less than that seen in skeletal muscle. Second, the heart is a completely aerobic tissue, whereas skeletal muscle can function anaerobically for limited periods. Mitochondria are much more densely packed in heart than in other cells, making up nearly half the volume of a heart cell. Third, the heart contains negligible energy reserves as glycogen or lipid, although there is a small amount of creatine phosphate. Therefore, the supply of both oxygen and fuels from the blood must be continuous to meet the unending energy demands of the heart. The heart uses a variety of fuels—mainly fatty acids but also glucose, lactate, and ketone bodies.

Adipose Tissue

Adipose tissue represents the major fuel depot for an animal. The total stored triacylglycerols amount to some 565,000 kJ (135,000 kcal) in an average-sized human. This is enough fuel, metabolic complications aside, to sustain life for a couple of months in the absence of further caloric intake.

The adipocyte, or fat cell, is designed for continuous synthesis and breakdown of triacylglycerols, with breakdown controlled largely via the activation of hormone-sensitive lipase. Because adipocytes lack the enzyme glycerol kinase, some glucose catabolism must occur for triacylglycerol synthesis to take place—specifically, the formation of dihydroxyacetone phosphate, for reduction to glycerol-3-phosphate (see page 663 in Chapter 18). Glucose acts as a sensor in adipose tissue metabolism. When glucose levels are adequate, continuing production of dihydroxyacetone phosphate generates enough glycerol-3-phosphate for resynthesis of triacylglycerols from the released fatty acids. When intracellular glucose levels fall, the concentration of glycerol-3-phosphate falls also, and fatty acids are released from the adipocyte for export as the albumin complex to other tissues (see page 633 in Chapter 18).

Liver

A primary role of liver is the synthesis of fuel components for use by other organs. In fact, most of the low-molecular-weight metabolites that appear in the blood through digestion are taken up by the liver for this metabolic processing. The liver is a major site for fatty acid synthesis. It also produces glucose, both from its own glycogen stores and from gluconeogenesis, the latter using lactate and alanine from muscle, glycerol from adipose tissue, and the amino acids not needed for protein synthesis. Ketone bodies are also manufactured largely in the liver. In liver the level of malonyl-CoA, which is related to the energy status of the cell, is a determinant of the fate of fatty acyl-CoAs. When fuel is abundant, malonyl-CoA accumulates and inhibits carnitine acyltransferase I, preventing the transport of fatty acyl-CoAs into mitochondria for β-oxidation and ketogenesis. On the other hand, shrinking malonyl-CoA pools signal the cells to transport fatty acids into the mitochondria, for generation of energy and fuels.

One of the most important roles of liver is to serve as a "glucostat," monitoring and stabilizing blood glucose levels.

An important role of liver is to buffer the level of blood glucose. It does this largely through the action of glucokinase, an enzyme peculiar to liver, with a high K_M (about 10 mM) for glucose, and partly through a high-K_M transport protein, the **glucose transporter,** a membrane protein (actually, a family of proteins) that carries out facilitated diffusion of glucose. Thus, liver is unique in being able to respond to high blood glucose levels by increasing the uptake and phosphorylation of glucose, which results eventually in its deposition as glycogen. Glucose-6-phosphate accumulation activates the D form of glycogen synthase. In addition, glucose itself binds to glycogen phosphorylase *a*, increasing the susceptibility of phosphorylase *a* to dephosphorylation (see Figure 13.18, page 475), with consequent inactivation. Thus, in addition to hormonal effects, described shortly, liver senses the fed state and acts to store fuel derived from glucose. Liver also senses the fasted state and increases the synthesis and export of glucose when blood glucose levels are low. (Other organs also sense the fed state, notably the pancreas, which adjusts its glucagon and insulin outputs accordingly.)

To meet its internal energy needs, the liver can use a variety of fuel sources, including glucose, fatty acids, and amino acids.

Blood

All of the organs we have discussed are connected by the bloodstream, which transports what may be one organ's waste product but another organ's fuel (for example, alanine from muscle to liver). Blood also transports oxygen from lungs to tissues, enabling exergonic oxidative pathways to occur, followed by transport of the resultant CO_2 back to the lungs for exhalation, as described in Chapter 7. And, as described in Chapter 18, the lipoprotein components of blood plasma play indispensable roles in transporting lipids. Of course, blood is also the medium of transport of hormonal signals from one tissue to another, and of exit for metabolic end products, such as urea, via the kidneys.

In terms of the blood's own energy metabolism, the most prominent pathway is glycolysis in the erythrocyte. Blood cells constitute nearly half the volume of blood, and erythrocytes constitute more than 99% of blood cells. Mammalian erythrocytes contain no mitochondria and depend exclusively upon anaerobic glycolysis to meet their limited energy needs.

Hormonal Regulation of Fuel Metabolism

Maintenance of blood glucose within narrow limits is critical to brain function.

In animals it is supremely important to maintain blood glucose levels within rather narrow limits, particularly for proper functioning of the nervous system. Of course, blood glucose levels vary, depending on nutritional status. Several hours after a meal, the normal level in humans is about 80 mg per 100 mL of blood, or 4.4 mM. Shortly after a meal, that level might rise to 120 mg per 100 mL. In response, homeostatic mechanisms come into play to promote uptake of glucose into cells and its use by tissues. Similarly, when glucose levels fall, several hours after a meal, other mechanisms promote both glucose release, from intracellular glycogen stores, and gluconeogenesis, so that the normal level is maintained. Some of the homeostatic mechanisms were mentioned in the previous section; others involve hormonal regulation. Though we discuss molecular mechanisms of hormone action later in this chapter, it is appropriate here to discuss at the physiological level some of the hormones involved in fuel metabolism.

ACTIONS OF THE MAJOR HORMONES

The most important hormone promoting glucose uptake and use is insulin, whereas both glucagon and epinephrine have the opposite effect, to increase blood glucose levels. The major effects of these agents are summarized in Table 23.2. Figure 23.2 illustrates the interplay between the two pancreatic hormones, insulin and glucagon.

The key hormones regulating fuel metabolism are insulin, which promotes glucose use, and glucagon and epinephrine, which increase blood glucose.

Insulin

Insulin is a 5.8-kilodalton protein (see page 142) that is synthesized in the pancreas. The pancreas has both **endocrine cells,** which secrete hormones directly into the bloodstream, and **exocrine cells,** which secrete zymogen precursors of digestive enzymes into the upper small intestine. The endocrine tissue, which takes the form of cell clusters known as islets of Langerhans, contains at least four different cell types, each specialized for synthesis of one hormone. The A cells produce

TABLE **23.2 Major hormones controlling fuel metabolism in mammals**

Hormone	Biochemical Actions	Physiological Actions
Insulin	↑ Cell permeability to glucose (in muscle and adipose tissue)	Signals fed state
	↑ Glycolysis	↓ Blood glucose level
	↑ Glycogen synthesis	↑ Fuel storage
	↑ Triacylglycerol synthesis	↑ Cell growth and differentiation
	↓ Gluconeogenesis	
	↓ Lipolysis	
	↓ Protein degradation	
	↑ Protein, DNA, and RNA synthesis	
Glucagon	↑ cAMP level in liver and adipose tissue	↑ Glucose release from liver
	↑ Glycogenolysis	↑ Blood glucose level
	↓ Glycogen synthesis	
	↑ Triacylglycerol hydrolysis	
	↑ Gluconeogenesis	
	↓ Glycolysis	
Epinephrine	↑ cAMP level in muscle	↑ Glucose release from liver
	↑ Triacylglycerol mobilization	↓ Glucose use by muscle
	↑ Glycogenolysis	↑ Blood glucose level
	↓ Glycogen synthesis	

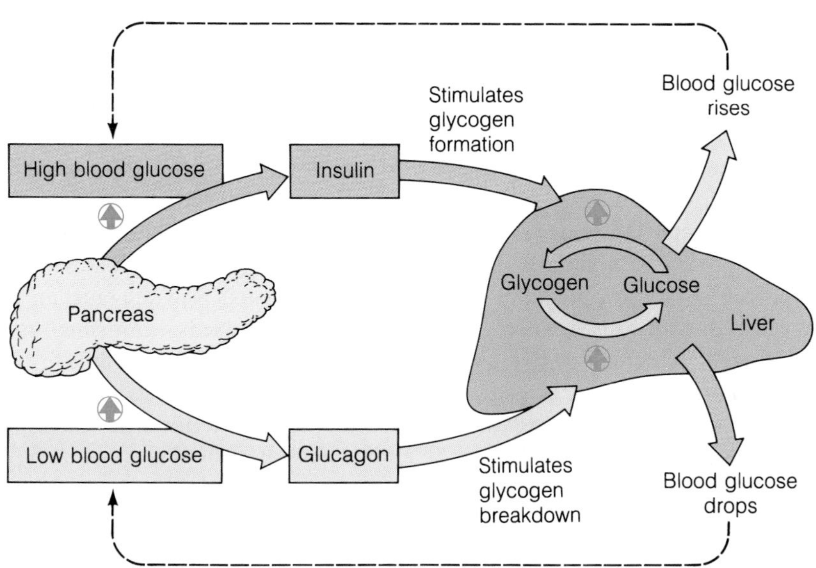

FIGURE **23.2**

Aspects of the control of blood glucose levels by pancreatic secretion of insulin and glucagon. Conditions resulting from high glucose levels are shown in blue, and those from low glucose levels in pink.

glucagon; the D cells, **somatostatin;** and the P cells, a recently discovered **pancreatic hormone.** Insulin is synthesized in the B cells, which sense glucose levels and secrete insulin in response to increased levels of blood glucose.

The simplest way to describe the several actions of insulin is to say that *insulin signals the fed state* and thereby promotes (1) uptake of fuel substrates into some cells, (2) storage of fuels (lipids and glycogen), and (3) biosynthesis of macromolecules (nucleic acids and protein). Specific effects include increased uptake of glucose in muscle and adipose tissue; activation of glycolysis in liver; increased synthesis of fatty acids and triacylglycerols in liver and adipose tissue; inhibition of gluconeogenesis in liver; increased glycogen synthesis in liver and muscle; increased uptake of amino acids into muscle with consequent activation of muscle protein synthesis; and inhibition of protein degradation. Because of its promotion of biosynthesis, it is appropriate to consider insulin a growth hormone.

The mechanism by which insulin stimulates glucose uptake into muscle and adipose cells is an area of intense investigation. One important action involves the glucose transporter. In cells not stimulated by insulin, members of this protein family are localized in the cytosol. The protein is translocated to the cell surface in response to insulin. An important consequence of glucose uptake in adipocytes is its conversion to glycerol-3-phosphate, which combines with fatty acids to stimulate triacylglycerol synthesis.

Glucagon

A 3.5-kilodalton polypeptide, glucagon is synthesized by A cells of the islets of Langerhans in the pancreas. These endocrine cells sense the blood glucose concentration and release the hormone in response to low levels (see Figure 23.2). Both synthesis and release of glucagon are controlled by insulin.

The primary target of glucagon is the liver, and its principal effect is to increase cyclic AMP levels in liver cells, as schematized in Figure 23.3. The resultant metabolic cascades, discussed in Chapters 13 and 16, promote glycogenolysis and inhibit glycogen synthesis. In addition, by activating the hydrolysis of fructose-2,6-bisphosphate, cAMP inhibits glycolysis and activates gluconeogenesis. Glucagon also brings about inhibition of pyruvate kinase (PK) in the liver, causing phosphoenolpyruvate (PEP) to accumulate. The level of pyruvate decreases, both because its synthesis from PEP is blocked and because it continues to be converted to PEP, via the pyruvate carboxylase and phosphoenolpyruvate carboxykinase reactions. Although accumulation of PEP is slight, it suffices to promote gluconeogenesis, while inhibition of pyruvate kinase diminishes the glycolytic flux rate.

Glucagon also raises cAMP levels in adipose tissue. There the chief effect of cAMP is to promote triacylglycerol mobilization via phosphorylation of hormone-sensitive lipase, yielding glycerol and fatty acids.

Epinephrine

The catecholamines epinephrine and norepinephrine, when released from presynaptic nerve endings, function as neurotransmitters (see Chapter 21). When released from adrenal medulla in response to low blood glucose levels, epinephrine interacts with second-messenger systems in many tissues, with varied effects. In muscle, epinephrine activates adenylate cyclase, with concomitant activation of glycogenolysis and inhibition of glycogen synthesis (see Chapters 13 and 16). Triacylglycerol breakdown in adipose tissue is also stimulated, providing fuel for the muscle tissue. In consequence, glucose uptake into muscle is diminished, contributing to an increase in blood glucose levels. Epinephrine also inhibits insulin secretion and stimulates glucagon secretion. These effects tend to increase glucose production and release by the liver. The net result is to increase blood glucose levels. Unlike glucagon, the catecholamines have short-lived metabolic effects. As discussed in Chapter 13, epinephrine action on skeletal and heart muscle cells is a crucial part of the "fight or flight" response.

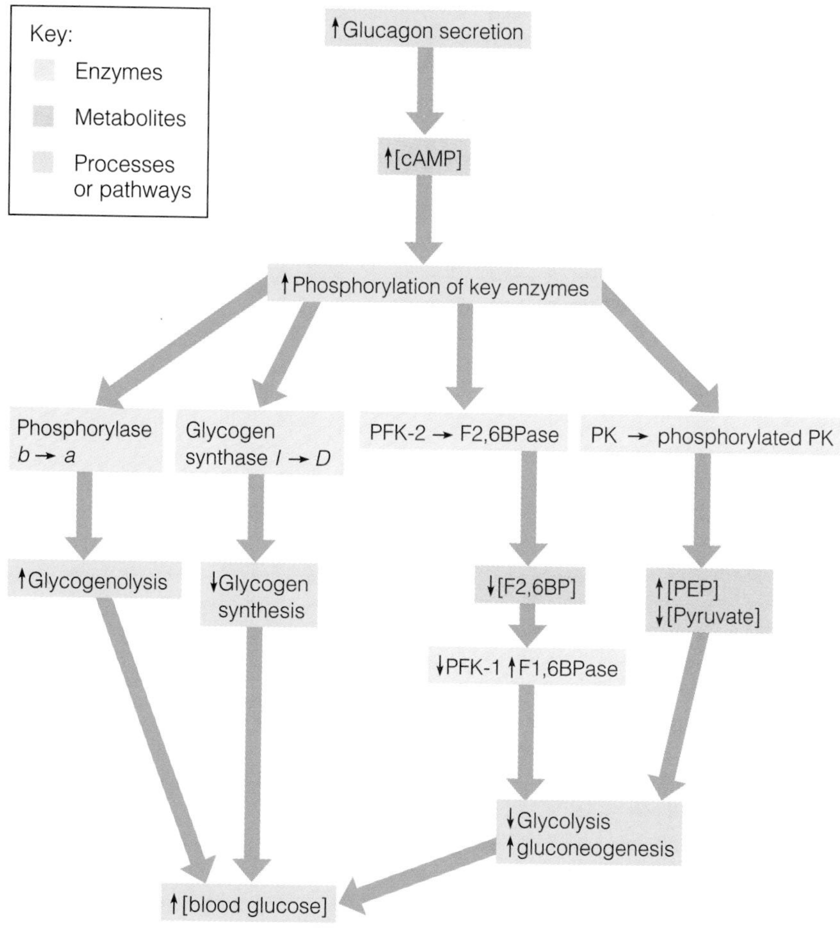

FIGURE **23.3**

Actions of glucagon in liver that lead to a rise in blood glucose. Brackets indicate concentration; ↑ and ↓ indicate increase or decrease, respectively, in enzyme activity, pathway flux, or metabolite level.

RESPONSES TO METABOLIC STRESS: STARVATION, DIABETES

An excellent way to understand how the interorgan and hormonal relationships we have discussed actually integrate fuel metabolism is to examine the effects of metabolic stress. In this section we consider two examples—prolonged fasting, in which the intake of fuel substrates is inadequate; and **diabetes mellitus,** in which a functional insufficiency of insulin impairs the ability of the body to use glucose, even when the sugar is present in abundance.

First let us review how glucose levels are maintained during normal feeding cycles (Figure 23.4). The blood glucose elevation occurring shortly after a carbohydrate-containing meal stimulates the secretion of insulin and suppresses the secretion of glucagon. Together these effects promote uptake of glucose into the liver, stimulate glycogen synthesis, and suppress glycogen breakdown. Flux through glucokinase increases in response to elevated glucose levels, providing substrates for glycogen synthesis. In addition, activation of acetyl-CoA carboxylase in the liver stimulates fatty acid synthesis, with subsequent transport to adipose tissue as triacylglycerols in very low-density lipoproteins. There, the increased levels of glycolytic intermediates and fatty acids stimulate triacylglycerol synthesis. Finally, increased glucose uptake into muscle increases levels of substrates for glycogen synthesis in that tissue as well.

Several hours later, when blood glucose levels begin to fall, the above events are reversed. Insulin secretion slows, and glucagon secretion increases. This promotes glycogen mobilization in liver via the cAMP-dependent cascade mechanisms that activate glycogen phosphorylase and inactivate glycogen synthase. Triacylglycerol breakdown in adipocytes is activated as well, via the action of hormone-sensitive

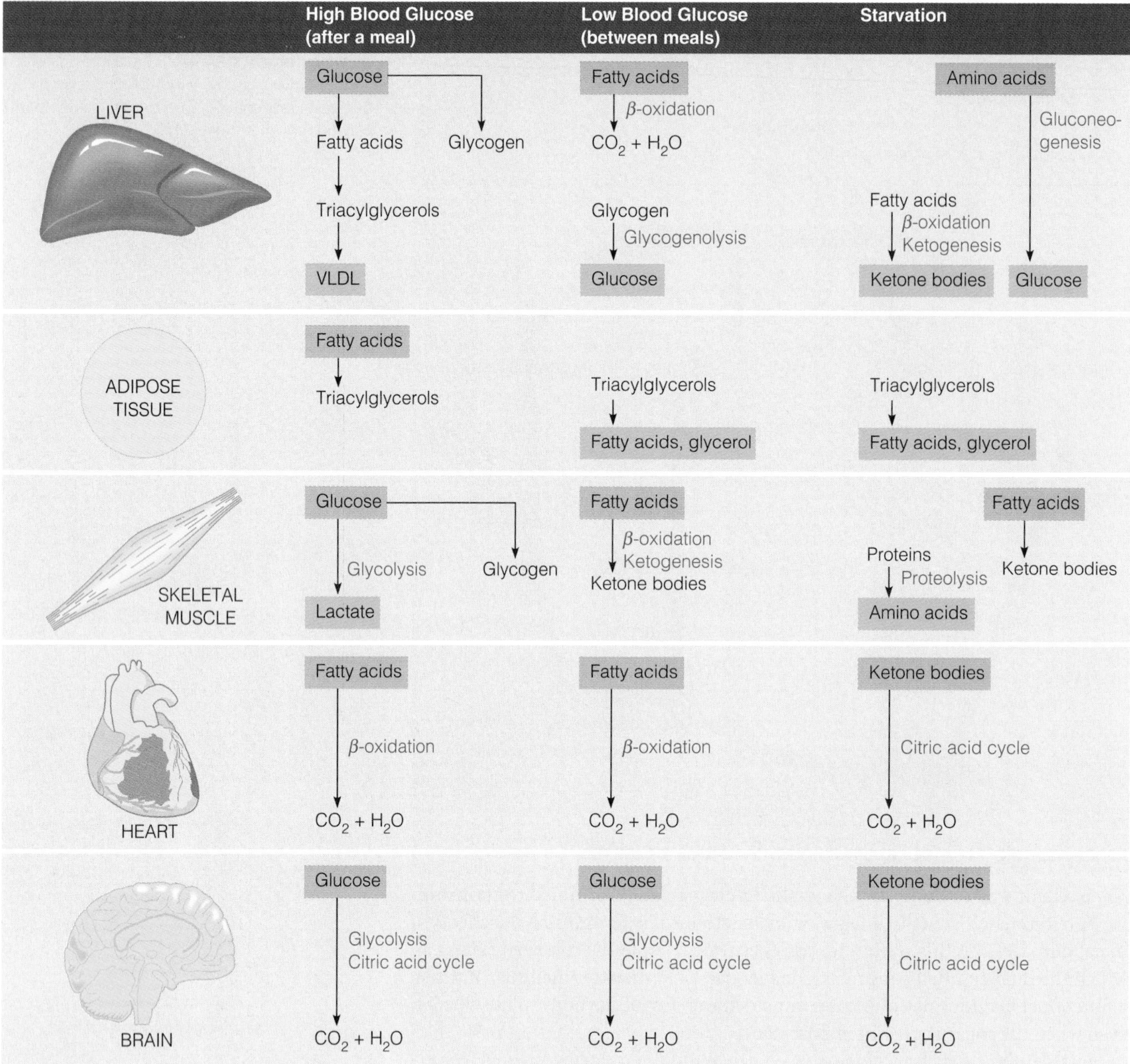

FIGURE **23.4**

Major events in the storage, retrieval, and use of fuels in the fed and unfed states and in early starvation. Purple indicates fuels imported into the tissue; green indicates fuels exported from tissue.

lipase, generating fatty acids for use as fuel by liver and muscle. At the same time, the decrease in insulin levels reduces glucose use by muscle, liver, and adipose tissue. Consequently, nearly all the glucose produced in the liver is exported to the blood and is available for use by the brain.

Starvation

Suppose that food intake is denied not just for a few hours, as just described, but for many days. Given that a 70-kg human can store at most the equivalent of 6700 kJ of energy as glycogen, this source of blood glucose will be exhausted in just a few hours. Because it is critical for brain function that blood glucose levels be maintained near 4.4 mM, the organism adapts metabolically to increase the use of fuels other than carbohydrate.

Before we discuss the metabolic adjustments involved, let us consider the other major energy stores: about 565,000 kJ as triacylglycerol, largely in adipose tissue, and 100,000 kJ as mobilizable proteins, largely in muscle. These stores provide sufficient energy to permit survival for up to several months. However, use of these stores presents problems. Triacylglycerol mobilization generates energy largely as acetyl-CoA, whose further oxidation in the citric acid cycle requires oxaloacetate. Recall from Chapter 14 that oxaloacetate and other citric acid cycle intermediates are used in other metabolic reactions and must be replenished via anaplerotic pathways. The most important of these processes is the pyruvate carboxylase reaction, with most of the pyruvate coming from carbohydrate catabolism. When carbohydrate availability is limited, the resupply of citric acid cycle intermediates is limited, and flux through the cycle may be reduced.

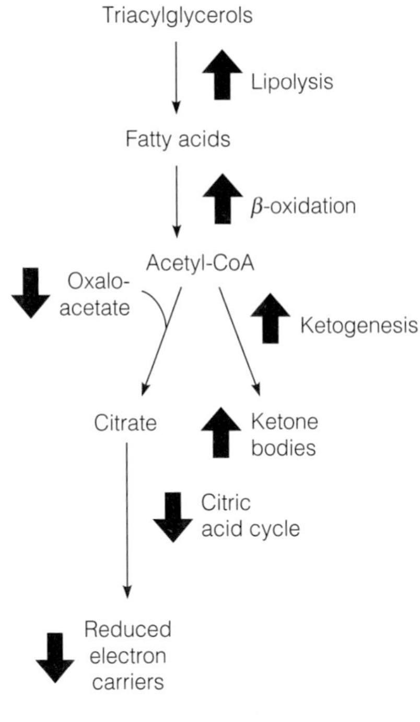

During carbohydrate limitation, citric acid cycle intermediates can be provided from other sources. For example, the glycerol released from lipolysis can be used, but it is not produced in amounts adequate to maintain levels of citric acid cycle intermediates. Alternatively, these intermediates can be produced from protein catabolism and transamination. However, this process is energetically wasteful and has the undesirable effect of wasting the muscle and weakening the fasting subject. Nevertheless, proteolysis is accelerated during the first few days of starvation, because amino acids for protein synthesis are not present in sufficient amounts to counterbalance protein breakdown, which continues at normal rates. A major fate of the released amino acids is gluconeogenesis, as the body attempts to cope with the absence of glycogen stores by synthesizing its own glucose. During this time, the liver and muscle are shifting to fatty acids as the dominant fuels for their own use.

Meanwhile, the increased use of carbon for gluconeogenesis diminishes the amount of oxaloacetate available to combine with acetyl-CoA in the citric acid cycle. Because fat breakdown has been activated, both acetyl-CoA and reduced electron carriers accumulate in the liver to the point that the acetyl-CoA cannot all be oxidized, and ketone bodies begin to accumulate. Accumulation of acetoacetate and β-hydroxybutyrate increases flux through the reactions that catabolize these ketone bodies. Thus, the brain adapts to reduced glucose levels by increasing the use of ketone bodies as alternative energy substrates. This trend continues for the duration of starvation. On the third day, the brain derives about one-third of its energy needs from ketone bodies; by day 40, that usage has increased to two-thirds. This adaptation reduces the need for gluconeogenesis and spares the mobilization of muscle protein. In fact, the loss of muscle protein *decreases* by about 4-fold late in starvation—from about 75 grams consumed per day on day 3 to about 20 grams per day on day 40. The metabolic changes accompanying starvation compromise the organism's abilities to respond to further stresses, such as extreme cold or infection. However, the adaptations do allow life to continue for many weeks without food intake, the total period being determined largely by the size of the fat deposits.

Metabolic adaptations promote alternative fuel use during starvation, so that glucose homeostasis is maintained for several weeks.

Diabetes

In starvation, glucose utilization is abnormally low because of inadequate glucose supplies. In **diabetes mellitus,** glucose utilization is similarly low, but the reason in this case is that the hormonal stimulus to glucose utilization—namely, insulin—is defective. As a result, glucose is actually present in excessive amounts. The consequences of insulin deficiency are comparable to those of starvation in revealing important aspects of interorgan metabolic relationships.

Diabetes mellitus is a major cause of death in the United States, affecting nearly 5% of the population. It is not a single disease but rather a family of diseases. **Insulin-dependent diabetes,** formerly called juvenile diabetes because of its early onset, often involves autoimmune destruction of the B cells of the pancreas, which can be caused by various factors, including viral infection. Some forms of diabetes have a genetic origin. Mutations in insulin structure can render the hormone inactive, and other mutations cause defects in the conversion of preproinsulin or proinsulin to the active hormone (see Chapter 5). In these cases, treatment involves administration of insulin. However, some forms of the disease involve mutations

Diabetes results either from insulin deficiency or from defects in the insulin response mechanism.

in the structure of the insulin receptor or in its intracellular activities that promote glucose utilization. These latter forms of the disease are called **insulin-resistant diabetes,** because patients cannot respond to therapeutic doses of insulin.

Whatever the cause of the functional insulin deficiency, diabetes mellitus can truly be called "starvation in the midst of plenty." The failure of insulin to act normally in promoting glucose utilization, with resultant glucose accumulation in the blood, starves the cells of nutrients and promotes metabolic responses similar to those of fasting (Figure 23.5). Liver cells attempt to generate more glucose by stimulating gluconeogenesis. Most of the substrates come from amino acids, which in turn come largely from degradation of muscle proteins. Glucose cannot be reused for resynthesis of amino acids or of fatty acids, so a diabetic may lose weight even while consuming what would normally be adequate calories in the diet.

As cells attempt to generate usable energy sources, triacylglycerol depots are mobilized in response to high glucagon levels. Fatty acid oxidation is elevated, with concomitant generation of acetyl-CoA. Flux through the citric acid cycle may decrease, because of the accumulation of reduced electron carriers and/or oxaloacetate limitation. In liver, both effects accelerate ketone body formation, generating increased levels of organic acids in the blood. These acids can lower the blood pH from the normal value of 7.4 to 6.8 or lower. Decarboxylation of acetoacetate, which is stimulated at low pH, generates acetone, which can be smelled on the breath of patients in severe uncontrolled diabetic situations. A special danger is that such people may lose consciousness, and this, coupled with a sweet organic odor on the breath, may give the impression that they are intoxicated, when in fact their lives are in jeopardy.

The excessive concentrations of glucose in body fluids generate other metabolic problems, quite different from anything seen in starvation. At blood glucose levels above 10 mM, the kidney can no longer reabsorb all of the glucose out of the blood filtrate, and glucose is spilled into the urine, sometimes in amounts approaching 100 grams per day. In fact, the Latin name *diabetes mellitus* literally means "honey-sweet urine." Glucose excretion creates an osmotic load, which

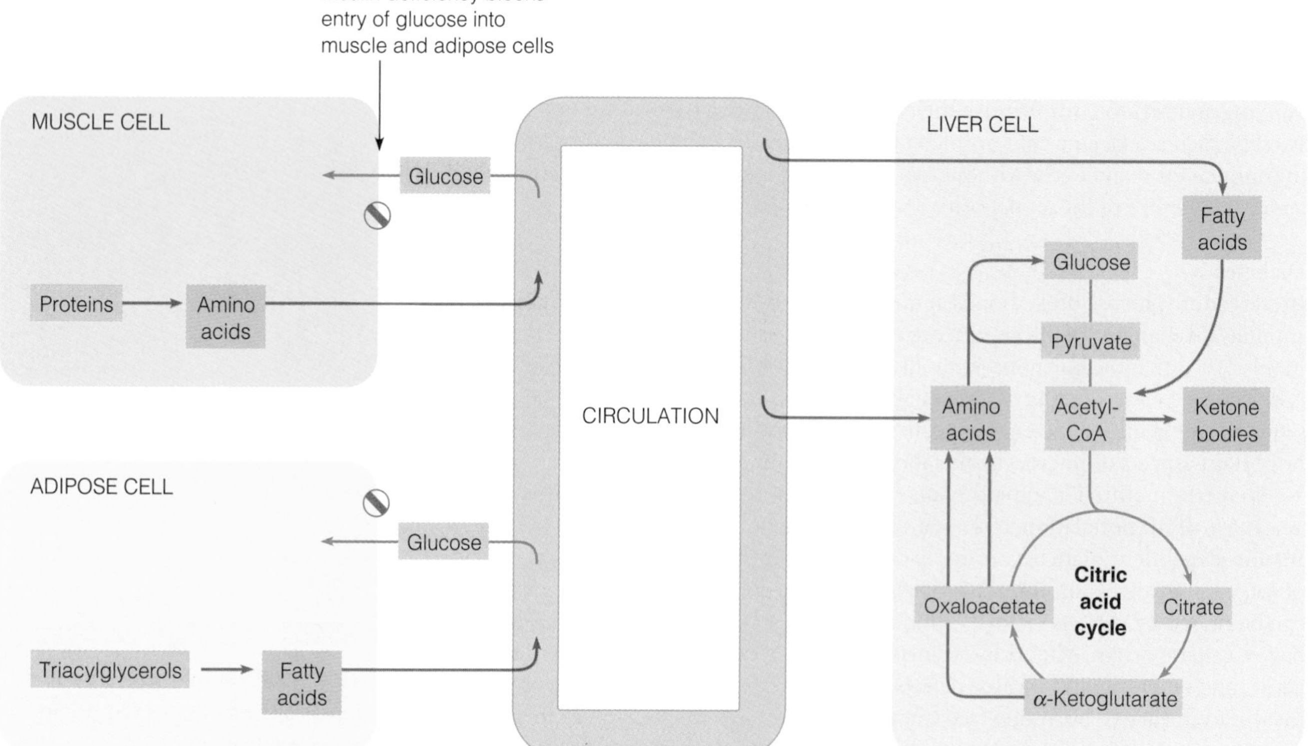

FIGURE **23.5**

The metabolic abnormalities in diabetes. The insulin deficiency blocks the uptake of glucose into muscle and adipose tissue and reduces glucose catabolism in all tissues. Proteolysis in muscle and lipolysis in adipose tissue are enhanced. In the liver, gluconeogenesis from amino acids and citric acid cycle intermediates is stimulated as the cells attempt to remedy the perceived lack of usable glucose, and fatty acid oxidation and ketogenesis are also increased. Green indicates pathways activated; pink indicates pathways diminished.

causes large amounts of water to be excreted as well, and under these conditions the kidney cannot reabsorb most of this water. In fact, the earliest indications of diabetes are often frequent and excessive urination, coupled with excessive thirst. Long before biochemistry was a science, the loss of nutrients, excessive urination, and breakdown of fat and protein were recognized as hallmarks of diabetes. As early as the first century A.D., diabetes was described as "the flesh and bones running together into urine."

When diabetes strikes in childhood (the insulin-dependent form of the disease, representing about 10% of all cases), the metabolic imbalance is usually more severe and difficult to control than in the milder and more common adult-onset form. The latter can often be controlled by dietary restriction of carbohydrate, whereas treatment for juvenile diabetes usually involves daily self-injection of insulin. For many years this insulin was purified from bovine pancreas, and its high cost, coupled with occasional problems resulting from the minor structural differences between human and bovine insulin, led the fledgling biotechnology industry to attempt to produce human insulin through recombinant DNA techniques. In the late 1970s the gene for human insulin was cloned into *E. coli* in a form that allowed it to be expressed, and in 1982 cloned human insulin became the first recombinant DNA product to be approved for human use.

Mechanisms of Hormone Action

Having discussed the actions of several hormones in integrating fuel metabolism, we turn now to a discussion of molecular mechanisms in hormone action. Recall from our introduction to signal transduction in Chapter 12 that the extracellular signals controlling metabolism include not only hormones but also neurotransmitters, growth factors, and pheromones. Each of these is a substance synthesized in one class of cells and released, to be transmitted to **target cells** and control their activities.

Insulin, glucagon, and epinephrine are all *hormones* (from Greek, "to stir up or excite"). This term was coined in 1904 to describe **secretin,** a substance that is released in the upper small intestine and that acts in the stomach to stimulate the flow of gastric juice as an aid to digestion. Early research on hormones revealed little about how they act, but it did show fundamental similarities among different hormones. First, they are secreted by specific tissues, now called **endocrine glands.** Second, they are secreted directly into the bloodstream, rather than being excreted through ducts or stored in bladders. Thus, the response to a hormonal signal comes as a direct and rapid result of its secretion. Figure 23.6 shows the locations of the major endocrine organs in the human body.

Hormones usually stimulate metabolic activities in tissues remote from the secretory organ. They are active at exceedingly low concentrations, in the micromolar to picomolar ranges. Furthermore, most hormones are metabolized rapidly, so their effects are often short-lived, allowing rapid adaptations to metabolic changes. The low concentrations of hormones and their metabolic lability have made it difficult to assay levels of any particular hormone, which is essential if we want to elucidate its mechanism of action. Until the 1960s it was usually necessary to use a bioassay. For example, **oxytocin,** which stimulates uterine contractions in labor, was assayed by adding hormone to strips of uterine muscle and measuring the length of the strips before and after administration of hormone. The introduction of *radioimmunoassay* revolutionized the field of hormone assay. This technique is described in Tools of Biochemistry 23A.

A special class of hormones includes the eicosanoids—prostaglandins, thromboxanes, and leukotrienes—which we discussed in Chapter 19. These mediators act like hormones but are distinctive in their extreme metabolic lability, their synthesis in many cell types instead of just one endocrine gland, their lower target organ specificity, and their actions primarily on cells close to those that secreted them.

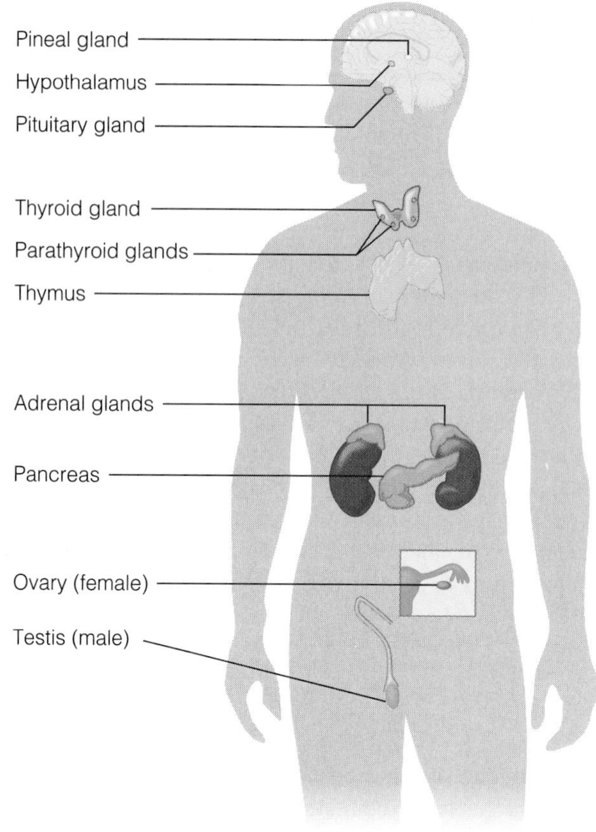

FIGURE **23.6**

The major human endocrine glands and their central nervous system control centers. Some other tissues also produce hormones, such as the lining of parts of the gastrointestinal tract.

Signal transduction involves cell-to-cell communication, via neurotransmitters, hormones, growth factors, and pheromones.

Hormones differ in certain ways from other intercellular mediators—**pheromones,** which are transmitted between cells of different organisms; **neurotransmitters,** which act immediately across a synaptic junction from their sites of release (see Chapter 21); and **growth factors,** which differ from hormones in that their growth-stimulating activities are continuous, rather than being short-lived in response to a burst of secretion. Distinctions among these classes of regulators are somewhat indefinite. For example, recall that catecholamines function both as neurotransmitters and as hormones, depending upon their sites of synthesis and release.

AN OUTLINE OF HORMONE ACTION

Until the 1950s we knew little about molecular mechanisms of hormone action. A popular theory was that a hormone stimulates a metabolic pathway by binding directly to the rate-determining enzyme for that pathway and activating it. Our current understanding is based largely on research discussed in Chapter 13, namely, studies of the effect of epinephrine in stimulating glycogen mobilization. These investigations, carried out largely in the laboratories of Earl Sutherland and Edwin Krebs, showed that epinephrine does not enter cells, as it must if a rate-limiting enzyme is to be activated directly. Instead, as shown in Figures 12.13 and 13.18, epinephrine binds to a macromolecular receptor at the cell surface and stimulates the formation of cyclic AMP, which acts as a second messenger and in turn stimulates the phosphorylation of target enzymes. The hormone itself is the first messenger. Today we know that all hormones so far investigated act through binding to specific receptors, whether those receptors are located inside the target cell or on the cell surface. The presence of specific receptors on specific cell types determines how hormones, secreted into the bloodstream, affect only certain tissues. Second messengers are often used to transmit the message to the target metabolic pathway, though not all hormone actions involve a second messenger.

Chemically, the hormones in vertebrate metabolism include (1) *peptides* or polypeptides, such as insulin or glucagon; (2) *steroids,* including glucocorticoids and the sex hormones; and (3) *amino acid derivatives,* including the catecholamines and thyroxine. Hormonal mechanisms include (1) enzyme activation or inhibition via second messengers, as noted for epinephrine and glucagon; (2) stimulation of the synthesis of particular proteins, through activation of specific genes; and (3) selective increases in the cellular uptake of certain metabolites. Among this last category are some receptors that serve directly as ion channels, with hormone binding causing a conformational change that opens the channel, and other receptors that stimulate uptake by still-unknown mechanisms, such as the effects of insulin upon glucose uptake.

Hormone action can influence (1) enzyme activity (via second messengers), (2) synthesis of specific proteins, or (3) membrane permeability to ions or small metabolites.

Hormones act by first binding to a specific receptor, which is located either in the plasma membrane or inside the cell. Most hormones interacting with *intracellular receptors* (also called nuclear receptors) exert their effects at the gene level. The hormone–receptor complex migrates to the nucleus, where it interacts with specific DNA sites and affects rates of transcription of neighboring genes. These hormones include steroids, thyroid hormones, and the hormonal forms of vitamin D. In addition, **retinoids,** derived from retinoic acid (related to vitamin A), exert regulatory effects in embryonic development, through interactions with intracellular receptors (see Chapter 28).

We recognize three major classes of *membrane-bound receptors.* First are receptors, like those introduced in Chapter 12, that interact with G proteins and influence the synthesis of second messengers. Second are receptors that are themselves ion channels—comparable to the nicotinic acetylcholine receptor (Figure 21.34). Peptide hormones and epinephrine act primarily through these two classes of receptors. A third category, exemplified by the insulin receptor, is a transmembrane protein with a ligand-binding site on the extracellular side and a catalytic domain on the cytosolic side. In the insulin receptor, that catalyst is a protein kinase, which is stimulated by insulin binding to the extracellular domain to phosphorylate tyrosine residues on target proteins.

Membrane receptors include (1) proteins that influence second-messenger synthesis, (2) ion channels, and (3) proteins with intrinsic enzyme activity.

Mechanisms of hormones that act through membrane-bound receptors by the first and third mechanisms are summarized in Figure 23.7. Note that the end result of most interactions between a hormone and a membrane receptor is activation of one or more protein kinases, whether or not a second messenger is involved. When Edwin Krebs and Edmond Fischer described in the late 1950s the sequence of reversible protein phosphorylations in the epinephrine-induced glycogenolytic cascade, there was no indication of the extent to which protein phosphorylation would turn out to dominate cell signaling mechanisms. At present more than 100 different protein kinases have been described in vertebrate cells, all of them related, as determined by amino acid sequence homologies. The importance of protein phosphorylation was recognized with the 1992 Nobel Prize for Medicine or Physiology to Krebs and Fischer. More recent work is uncovering a host of specific protein phosphatases, also subject to control by cell signaling mechanisms.

The end result of many signal transduction events is the phosphorylation or dephosphorylation of target proteins.

HIERARCHICAL NATURE OF HORMONAL CONTROL

Hormonal regulation involves a hierarchy of cell types acting on each other either to stimulate or to modulate the release and action of a hormone. The secretion of hormones from endocrine cells is stimulated by chemical signals from regulatory cells that occupy a higher position in this hierarchy (Figure 23.8). Hormonal action is controlled ultimately by the central nervous system. The master coordinator in

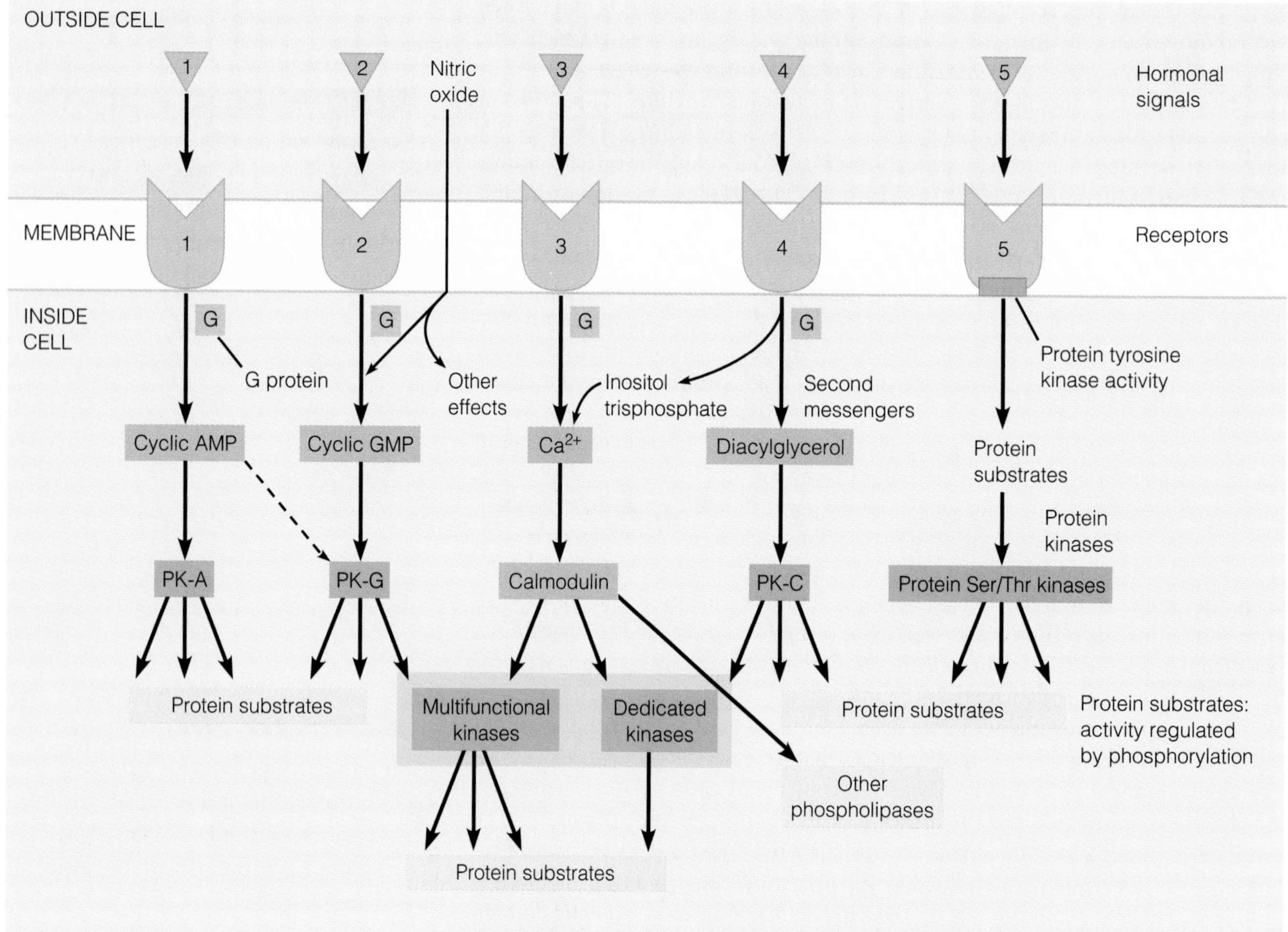

FIGURE **23.7**

Eukaryotic signal transduction systems involving membrane receptors (1–5) and/or second messengers (1–4). Nitric oxide is shown, even though it lacks a membrane receptor, because it diffuses into the cell and interacts with second-messenger systems. The end result of each pathway is phosphorylation of one or more proteins, some but not all of which have been identified. Other effects of nitric oxide are discussed in Chapter 21. Recent evidence shows that some pathways engage in cross-activation, as shown here by the dashed arrow indicating control of PK-G by cyclic AMP. PK-A = cyclic AMP–dependent protein kinase, PK-G = cyclic GMP–dependent protein kinase, PK-C = protein kinase C. Protein kinases are shown in orange, including the intrinsic protein tyrosine kinase domain in system 5.

mammals is the **hypothalamus,** a specialized center of the brain. The hypothalamus receives and processes sensory inputs from the environment via the central nervous system. In response it produces a number of hypothalamic hormones, some of them called **releasing factors.** These factors act on the pituitary, which is located just beneath the hypothalamus. Releasing factors stimulate the anterior portion of the pituitary to release specific hormones. Other hypothalamic hormones inhibit the secretion of particular pituitary hormones. Some pituitary hormones stimulate target tissue directly. For example, **prolactin** stimulates mammary glands to produce milk. However, most pituitary hormones act on endocrine glands that occupy an intermediate, or secondary, position in the hierarchy, stimulating them to produce hormones that exert the ultimate actions on target tissues. Pituitary hormones that act on other endocrine glands are called **tropic**

Specific hormone releasing factors from the hypothalamus control the release—and thus the action—of other hormones.

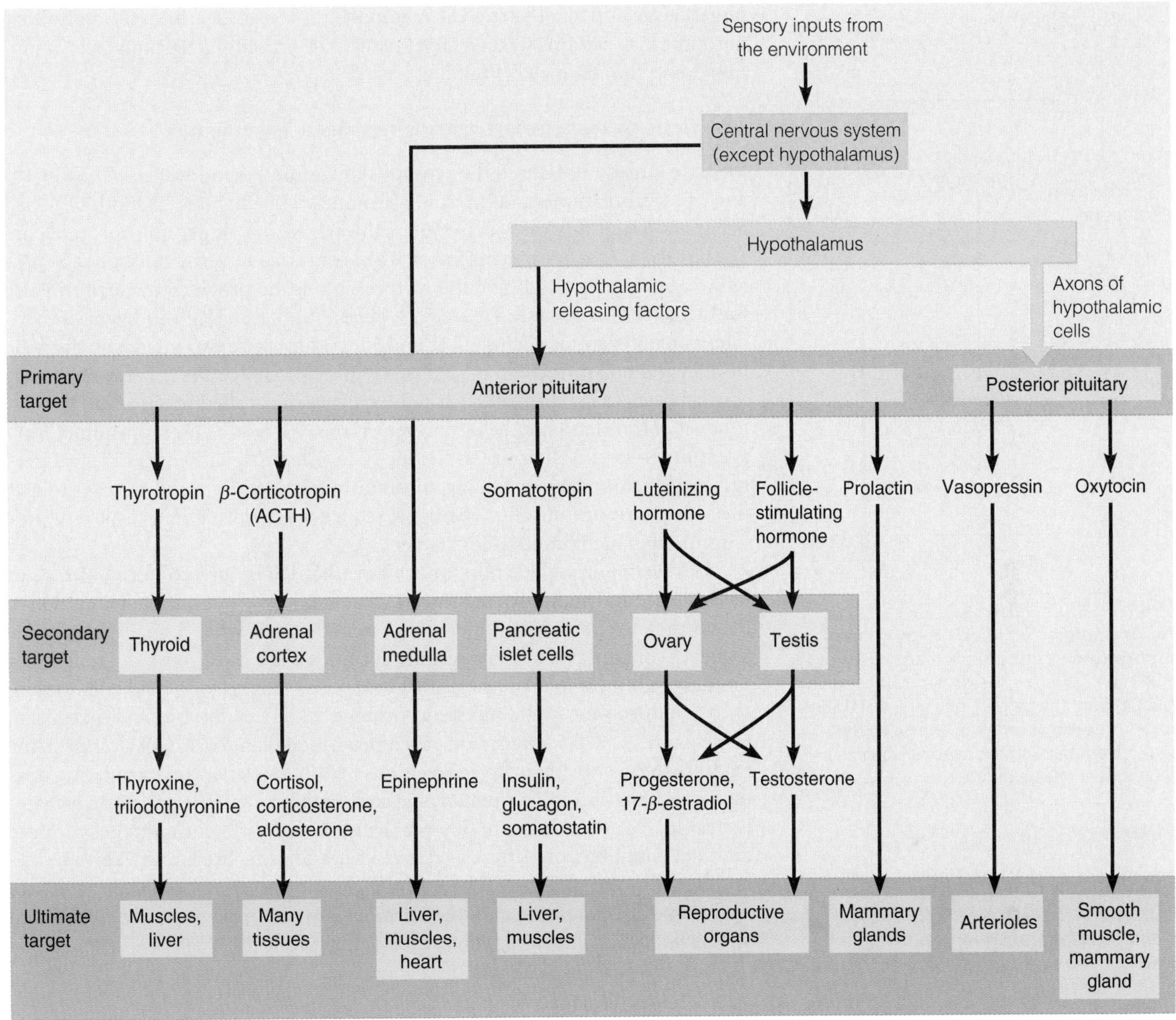

FIGURE **23.8**

Hierarchical nature of hormone action in vertebrates. The pituitary represents the first target, being under hypothalamic control. Pituitary hormones then act on secondary targets, principally other endocrine glands, the hormone products of which collectively influence essentially all of the other organs and tissues. Neural stimulation of the adrenal medulla controls release of epinephrine.

hormones or **tropins.** An example is **adrenal corticotropic hormone** (ACTH), also called **β-corticotropin.** This peptide is secreted from the anterior pituitary, and it stimulates the adrenal cortex to produce glucocorticoids and mineralocorticoids, which in turn act on a number of tissues, including kidney, muscle, and the immune system.

The action of a hormone is self-limiting because of the existence of feedback loops, in which secretion of a hormone sets in motion a series of events that leads to inhibition of that secretion. As shown in Figure 23.9, for example, the secretion of β-corticotropin from the pituitary is stimulated by **corticotropin releasing factor** (CRF), a hypothalamic hormone that is a 41-residue polypeptide. Hypothalamic cells contain glucocorticoid receptors, which sense the elevated levels of circulating glucocorticoids, such as cortisol, that result from stimulation of the adrenal cortex. Binding of glucocorticoids to these receptors has the effect of inhibiting the further release of CRF, thus completing a feedback loop.

Hormone action is self-limiting because of the existence of feedback loops.

To summarize the hierarchical nature of hormone action: The central nervous system transmits signals to the hypothalamus, which produces factors that either stimulate or inhibit the release of hormones from the pituitary. These hormones stimulate other endocrine glands, each of which releases a hormone that acts on a

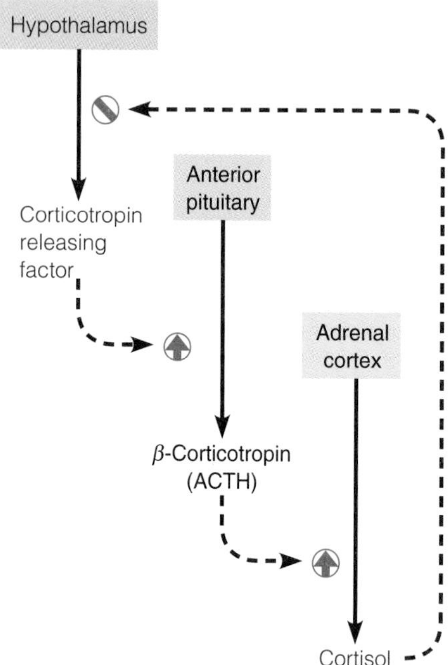

FIGURE **23.9**

An example of feedback regulation of a hormone. Corticotropin releasing factor (CRF) stimulates the release of β-corticotropin (ACTH) from the anterior pituitary. ACTH stimulates the adrenal cortex to release cortisol, which feeds back on the hypothalamus to inhibit further release of CRF.

Synthesis of peptide hormones almost always involves proteolytic cleavage of a protein precursor.

target tissue and elicits a specific metabolic response. Alternatively, a pituitary hormone may act directly on a target tissue. The action of a hormone sets in motion events that ultimately limit that action.

SYNTHESIS OF HORMONES: PEPTIDE HORMONE PRECURSORS

We have already presented the synthesis of steroid hormones (see Chapter 19) and of catecholamines and thyroid hormones (see Chapter 21), both of which occur via straightforward metabolic pathways. Nearly all peptide hormones are synthesized as inactive precursors and then converted to active hormones by proteolytic processing. Studies of the synthesis of insulin provided the first evidence for this phenomenon (see Figure 5.21, page 146). Recall that this hormone contains two polypeptide chains, of 21 and 30 residues, with two interchain disulfide bridges and one intrachain bridge (see Figure 5.15, page 142). These chains are formed by cleavage from an 81-residue polypeptide, called **proinsulin.** The first product of translation of the insulin gene is the 105-residue **preproinsulin.** Cleavage from preproinsulin of a 24-residue N-terminal "signal sequence" gives proinsulin, which undergoes folding, disulfide bond formation, and cleavage to give the active hormone, insulin. The signal sequence is involved in transport of proteins through membranes (see Chapter 27).

All known polypeptide hormones are synthesized in "prepro" form, with a signal sequence and additional sequence(s) that are cleaved out during maturation of the hormone. A particularly interesting case is that seen when a single polypeptide sequence contains two or more distinct hormones. The most complex example is a pituitary multihormone precursor that contains sequences for β- and γ-lipotropin, α-, β-, and γ-melanocyte-stimulating hormone (MSH), endorphin, and enkephalin, as well as ACTH. This precursor, called **pro-opiomelanocortin,** derives its name from its role as precursor to endogenous *opia*tes, *melano*cyte-stimulating hormone, and *corti*cotropin. A remarkable fact about pro-opiomelanocortin is that it is cleaved at different sites in different cells, so that different cell types produce different ensembles of hormones derived from this one precursor. Cleavage sites are shown in red in Figure 23.10. In the anterior pituitary, cleavage generates ACTH and β-lipotropin, and further processing in the central nervous system yields endorphin and enkephalin, among other products.

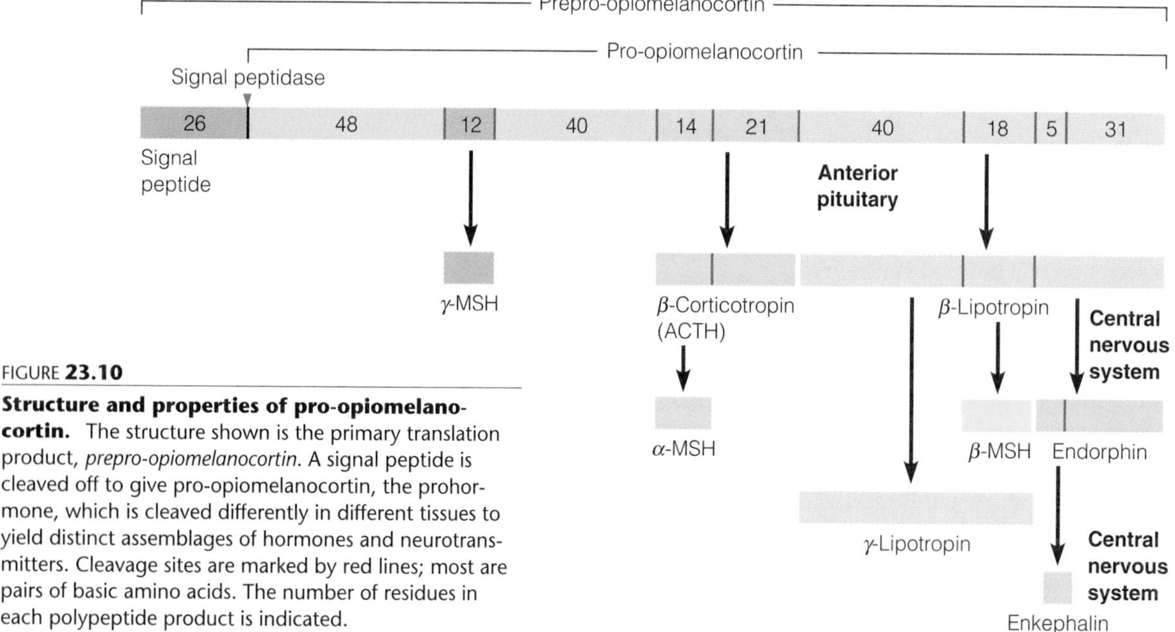

FIGURE **23.10**

Structure and properties of pro-opiomelanocortin. The structure shown is the primary translation product, *prepro-opiomelanocortin.* A signal peptide is cleaved off to give pro-opiomelanocortin, the prohormone, which is cleaved differently in different tissues to yield distinct assemblages of hormones and neurotransmitters. Cleavage sites are marked by red lines; most are pairs of basic amino acids. The number of residues in each polypeptide product is indicated.

SIGNAL TRANSDUCTION: RECEPTORS

Now let us focus upon molecular details of hormone action, by describing structure and function of the major participants, starting with receptors. As noted earlier, all known hormones interact with target cells by initially binding to a macromolecular receptor, located in either the plasma membrane or the interior of the cell. Because the receptor participates in transduction of the signal from the external messenger to some component of the metabolic machinery, it must have at least one additional functional site. The activity of this site is altered by hormone binding, just as the catalytic site of an allosteric enzyme is altered by the binding of effectors at remote sites.

Experimental Study of Receptors

Molecular interactions involving hormone receptors can be studied experimentally by methods comparable to those used in enzymology. Binding of a hormone to its receptor is saturable, and the kinetics resembles Michaelis–Menten kinetics. Most hormones bind tightly, with dissociation constants in the range of 0.1 μM to 1.0 pM. The ability of a tissue to respond to hormonal stimulation is a function of the receptor density of cells in that tissue.

Binding of hormones to receptors is saturable, comparable to the binding of substrates to enzymes.

Binding of radioactive hormone or an analog can be used to identify and quantify receptors, either as an assay when purifying receptors or in determining receptor density in a given cell type. The tight binding between hormones and their receptors can also be exploited in designing purification protocols involving affinity chromatography. In fact, one of the earliest applications of this technique was purification of the insulin receptor, using columns of immobilized insulin. This development eased a task that is normally very difficult for two reasons. First, membrane-bound receptors must be solubilized prior to purification, without irreversible inactivation. Second, most hormone receptors are present in exceedingly small amounts. For example, adipose tissue contains only about 10^4 insulin receptor molecules per cell.

Another useful technique is photo-affinity labeling, which creates a covalent bond between a hormone analog and a receptor. In this technique, a hormone molecule is modified with a photoreactive group, such as an azido moiety ($—N_3$). Mixing a radiolabeled hormone analog with a cell extract, followed by UV light irradiation, binds the analog to its receptor, creating a radioactive tag that can be used to isolate the receptor.

Agonists and Antagonists

Often the ligand used in receptor-binding assays or in affinity chromatographic purification is not the hormone itself but an analog that happens to bind to the receptor, sometimes more tightly than the natural hormone. The analog may show little or no structural resemblance to the natural ligand, but it may show a stereochemical relationship that is evident in three-dimensional models. An example is the potent synthetic estrogen **diethylstilbestrol.** This compound is 3 times more potent than 17-β-estradiol, which in turn is 10 times more potent than other natural estrogens. Diethylstilbestrol has been used in purifying and characterizing estrogen receptors. The synthetic hormone has had its uses outside the laboratory as well, including its addition to feed to stimulate growth of beef cattle, until it was found to be carcinogenic. At present we don't know whether the carcinogenicity of diethylstilbestrol is related to its estrogen activity or to some other biological property, as yet undiscovered.

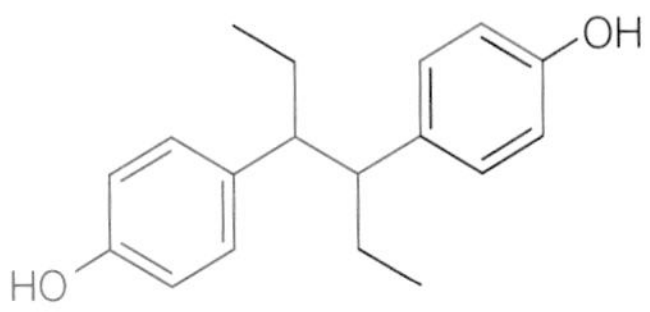

HO OH

Diethylstilbestrol

OH HO

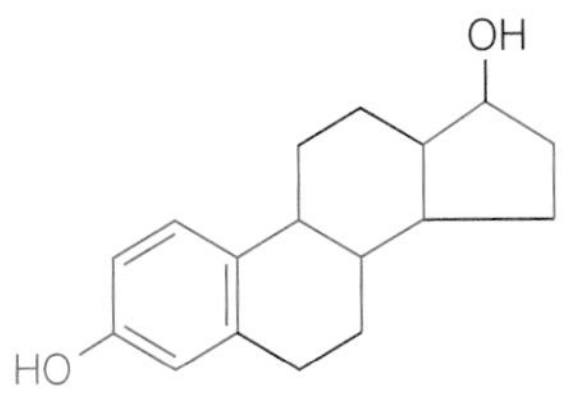

17-β-Estradiol

Diethylstilbestrol is an example of a hormone agonist—an analog that binds productively to a receptor and mimics the action of the endogenous hormone (see also page 787 in Chapter 21). An agonist is comparable to an alternative substrate for an enzyme: Its binding to a receptor is productive, in that it evokes a metabolic response comparable to that of binding the hormone. By contrast, a hormone antagonist binds to receptors but does not provoke the normal biological

A hormone agonist mimics a hormone in binding productively to a receptor. An antagonist binds nonproductively, inhibiting the action of the natural hormone.

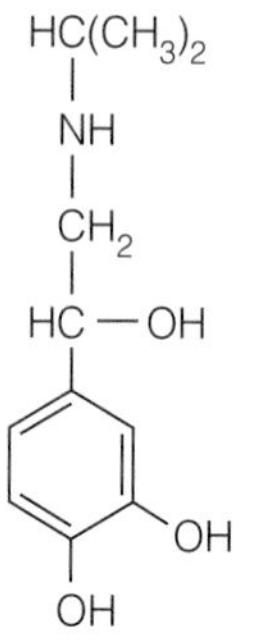

response. An antagonist is to a receptor as a competitive inhibitor is to an enzyme, in that both antagonists and competitive inhibitors compete with a normal ligand (hormone or substrate, respectively) for binding to a specific site on a protein and, by so binding, inhibit a normal biological process.

Agonists and antagonists have been useful in studies of the stereochemistry of binding sites on receptors. In turn, these investigations are useful in drug design, with a goal of activating or inactivating certain classes of receptors. For example, the agonist **isoproterenol** is used to treat asthma, because it mimics the effects of catecholamines in relaxing bronchial muscles in the lung; it does so by interacting with one specific class of **adrenergic receptors** (so-called because they bind **adrenaline,** the old name for epinephrine). Another important drug, used to control blood pressure and pulse rate in cardiac patients, is **propranolol,** an antagonist of another class of adrenergic receptors, which control blood pressure and heartbeat rate.

Isoproterenol

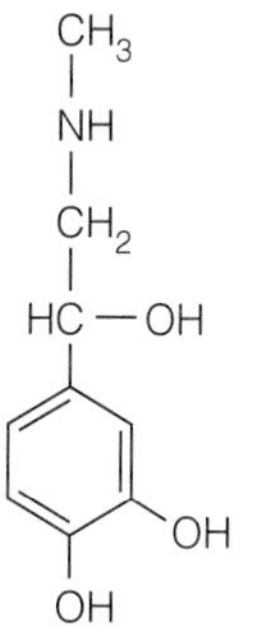

Epinephrine

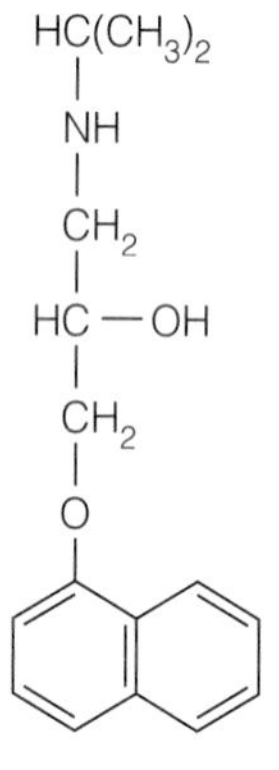

Propranolol

Classes of Catecholamine Receptors

Studies of a great many agonists and antagonists of the catecholamines have revealed the existence in vertebrates of four types of catecholamine receptors, each of which has a distinctive pattern of response to these analogs. These are called the α_1-, α_2-, β_1-, and β_2-adrenergic receptors. Receptors of the β type are those we encountered before, in our discussions of epinephrine-induced lipolysis and glycogenolysis (see Chapters 13 and 18). Adrenergic receptors of different types, in different tissues, have various physiological effects, some of which are summarized in Table 23.3.

Receptors and Adenylate Cyclase as Distinct Components of Signal Transduction Systems

In the early stages of hormone research, when the epinephrine-stimulated mobilization of glycogen was the only defined biochemical response to hormone–receptor interaction, adenylate cyclase, which synthesizes cyclic AMP, was found to be a membrane-bound enzyme. Because the receptor is also membrane-bound, it was thought for some time that the receptor was adenylate cyclase and that epi-

TABLE 23.3 Some biological actions associated with adrenergic receptors

Receptor Class	Target Tissue	Effect of Hormone or Agonist
α_1	Iris of the eye	Contraction
	Intestine	Decreased motility
	Salivary glands	Potassium and water secretion
α_2	Pancreatic B cells	Decreased secretion
	Blood platelets	Aggregation
	Adipocytes	Decreased lipolysis
	Stomach	Decreased motility
α (subtype not identified)	Arterioles in skin, mucosa	Constriction
	Bladder sphincter	Contraction
	Male sex organs	Ejaculation
β_1	Heart	Increased rate, force, and depth of contraction
	Adipocytes	Increased lipolysis
	Intestine	Decreased motility
β_2	Lung	Muscle relaxation
	Liver	Increased glycogenolysis
	Intestine	Decreased motility

Source: Adapted from L. S. Goodman and A. C. Gilman, eds., *Pharmacological Basis of Therapeutics,* 7th ed. (New York: Macmillan, 1985), p. 72. © 1985 McGraw-Hill, Inc.

nephrine binding activated the enzyme. However, two observations argued against that interpretation. First, other hormones were found to activate adenylate cyclase; more than a dozen are now known, including glucagon, ACTH, melanocyte-stimulating hormone, and luteinizing hormone. Adenylate cyclase seemed unlikely to have that many hormone-binding sites. Second, binding of catecholamines to the α_2 class of receptors was found to *inhibit* adenylate cyclase, suggesting that different kinds of proteins interact with adenylate cyclase to produce different metabolic effects.

Both of these observations indicated that the receptor and adenylate cyclase are distinct proteins. In fact, the resolution of β-adrenergic receptors from adenylate cyclase was observed experimentally in 1977. This development was important, because it showed that this hormonal response system has far more flexibility and versatility than previously thought. A wide variety of hormones could exert a multitude of biological effects through a common mechanism, namely, activation or inhibition of cyclic AMP synthesis. The diversity of signals and responses was built into both the diversity of receptors and the diversity of enzymes in target cells whose activities could be stimulated or inhibited by cyclic AMP–stimulated phosphorylation. It was soon learned that transduction of the hormonal signal to adenylate cyclase involved a third class of proteins—the G proteins, which we also introduced in Chapter 12. These developments were recognized by the award of the 1994 Nobel Prize to Martin Rodbell, who showed that receptors are distinct from adenylate cyclase, and Alfred Gilman, for the discovery of G proteins.

Hormones that act through second messengers involve a three-protein module—receptor, transducer (G protein), and effector (adenylate cyclase or related enzyme).

Because receptors are embedded in the membrane and are present in very small amounts, their isolation in amounts sufficient for structural analysis is a Herculean task. Cloning the receptor genes, by techniques described in Tools of Biochemistry 25A–25E, has been indispensable in gaining complete amino acid sequence information. Among the several receptor proteins whose amino acid sequences have been determined—including both α_2- and β_2-adrenergic receptors—there are some remarkable structural similarities. The proteins are of comparable size, with 415 to 480 residues, including seven conserved regions that are rich in hydrophobic amino acids. It seems clear that these represent regions of α helix that are embedded in the membrane and linked by hydrophilic loops, projecting into both the extracellular environment and the cytosol, with the recognition site for hormone or neurotransmitter on the extracellular side. The structure of the β_2-adrenergic receptor is shown in Figure 23.11. Site-directed mutagenesis of cloned receptor genes is being used to identify regions of these large proteins that are in contact with the hormone and other components of the signal transduction machinery. Because this class of receptors snakes back and forth through the membrane, they are called serpentine receptors.

The range of functions controlled through serpentine receptors was dramatically underscored in 1998 with the discovery of a gene in *Drosophila* called *methuselah.* Mutations in this gene, which encodes a protein that appears to be a serpentine receptor, were found to extend the life span of mutant fruit flies by 35%.

SIGNAL TRANSDUCTION: G PROTEINS

As noted earlier, cyclic AMP–dependent signal transduction mechanisms involve three separate proteins: (1) a hormone receptor, (2) adenylate cyclase, and (3) a G protein. This third class is so designated because of the ability of these proteins to bind guanine nucleotides. In 1971 guanosine triphosphate was found to be required for activation of adenylate cyclase by β-adrenergic agonists, and late in the decade the basis for this requirement emerged: GTP-binding membrane proteins interact with receptor systems that activate or inhibit adenylate cyclase. Of the several known G proteins the two best characterized are G_s, a family of G proteins involved in *stimulation* of adenylate cyclase, and G_i, a closely related family involved in responses that *inhibit* adenylate cyclase. Although both types of G

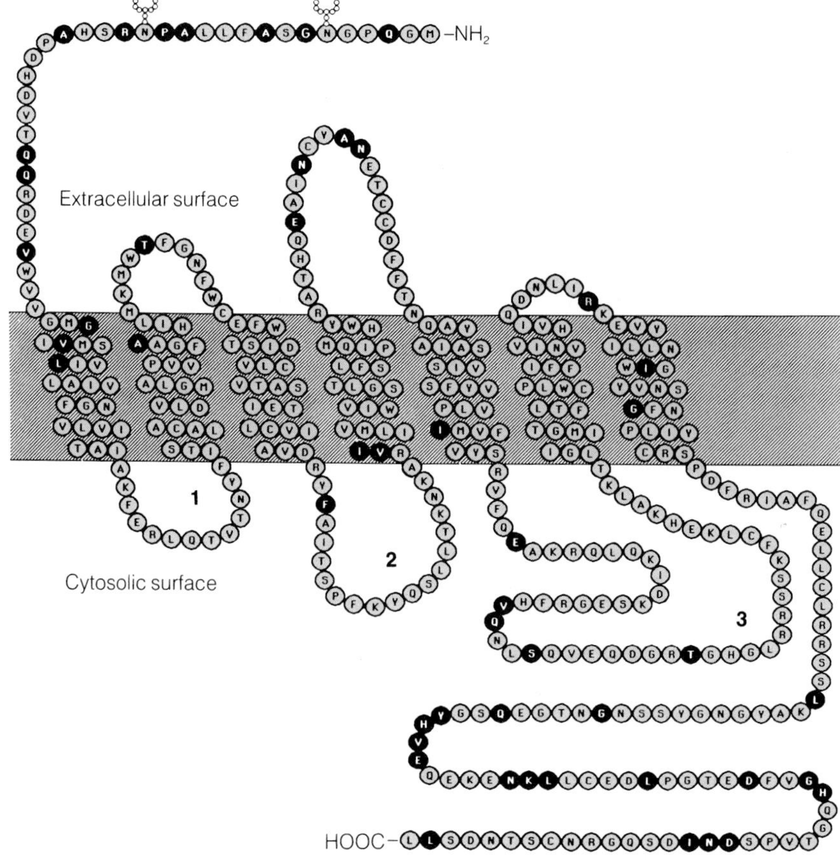

FIGURE **23.11**

Amino acid sequence of the human β_2-adrenergic receptor. The seven conserved transmembrane domains are shown in orange. Note also the three extracellular and three cytoplasmic loops and the two oligosaccharide units on the extracellular side (bound to asparagine residues). Interaction of the receptor with G proteins is controlled in part by reversible phosphorylation of serine and threonine residues near the C-terminus. The amino acids colored black are the ones that are different in the hamster β_2-adrenergic receptor sequence.

Marx, J. R., *Science* (1987) 238:615–616. © 1987 by AAAS.

proteins interact with other receptors as well (and with target proteins other than adenylate cyclase), it is useful to describe their functions in terms of the adrenergic receptors.

Actions of G Proteins

The G proteins are membrane proteins that in the *inactive* state bind guanosine diphosphate, GDP. As we discussed in Chapter 12, a hormone response leading to stimulation of adenylate cyclase—the binding of extracellular hormone or agonist to a receptor, typically a β-adrenergic receptor—causes a conformational change that stimulates the receptor to interact with a nearby molecule of G_s. This in turn stimulates an exchange of bound GDP for GTP—that is, the dissociation of GDP from G_s, to be replaced by GTP. G_s is thereby converted to a protein that activates adenylate cyclase, producing cyclic AMP from ATP. This results in activation of cAMP-dependent protein kinase (protein kinase A), with consequent phosphorylation of target proteins, such as phosphorylase *b* kinase in cells that activate glycogen phosphorolysis. To summarize, this signal transduction pathway involves (1) hormone binding to receptor; (2) receptor interaction with G_s, stimulating release of GDP and association of GTP with G_s; (3) stimulation of adenylate cyclase by the GTP-bound G_s; (4) stimulation by cAMP of protein phosphorylation; (5) stimulation or inhibition of metabolic reactions. The initial exchange reaction (step 2) is usually assisted by one of a class of proteins called guanine nucleotide exchange factors (GEFs).

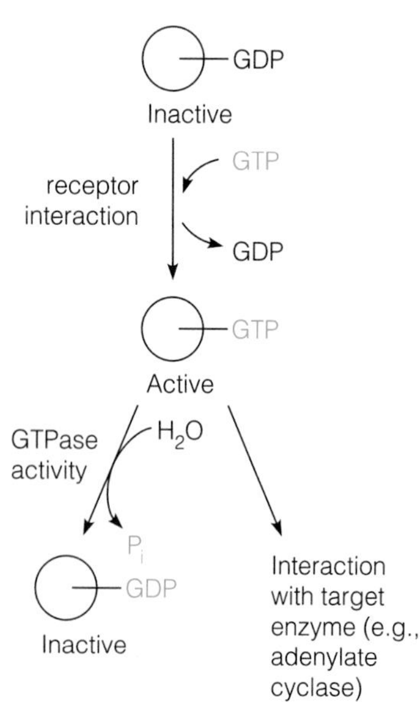

Continued activation of G_s depends on the presence of bound GTP. The hormonal response is limited, and hence is controlled, by the presence of a slow GTPase activity on the G protein. Thus, bound GTP is slowly cleaved to GDP, with concomitant loss of ability to stimulate adenylate cyclase. This process, like the initial activation, is protein-assisted, being helped by a GTPase-activating protein (GAP).

The G_i protein functions similarly, but in response to extracellular signals whose response is the *inhibition* of adenylate cyclase, typically α_2 agonists. Here the binding of GTP provokes an inhibitory interaction of G_i with adenylate cyclase, which decreases the synthesis of cAMP.

Activation of G proteins involves GTP displacement of GDP bound to the α subunit and dissociation of the complex from $\beta\gamma$. Hormone action is limited by slow hydrolysis of the bound GTP.

Structure of G Proteins

Structural studies show that G_s, G_i, and other G proteins have an $\alpha\beta\gamma$ trimeric structure (Figure 23.12): a 39- to 46-kilodalton α subunit, a 37-kilodalton β subunit, and an 8-kilodalton γ subunit. As noted later, several different forms of each protein are found, allowing for a great variety of different G proteins. In most of these the γ subunit is **prenylated;** that is, it contains a covalently bound C_{20} isoprenoid moiety at the C-terminal cysteine, which helps anchor the protein in the membrane and may facilitate protein–protein interactions. The α subunit is **myristylated** in G_i and G_o, and palmitylated in G_s. That is, it contains a myristic acid or palmitic acid moiety in amide linkage with C-terminal glycine. The guanine nucleotide–binding site and its associated GTPase activity are both located on the α subunit. A hormonal stimulus leads to exchange of GDP for GTP and dissociation of the G protein, with the α–GTP complex moving along the membrane until it encounters a molecule of adenylate cyclase or another target enzyme. The slow GTPase activity mentioned earlier eventually reconverts α–GTP to α–GDP, and the α–GDP complex dissociates from adenylate cyclase and rejoins the $\beta\gamma$ complex.

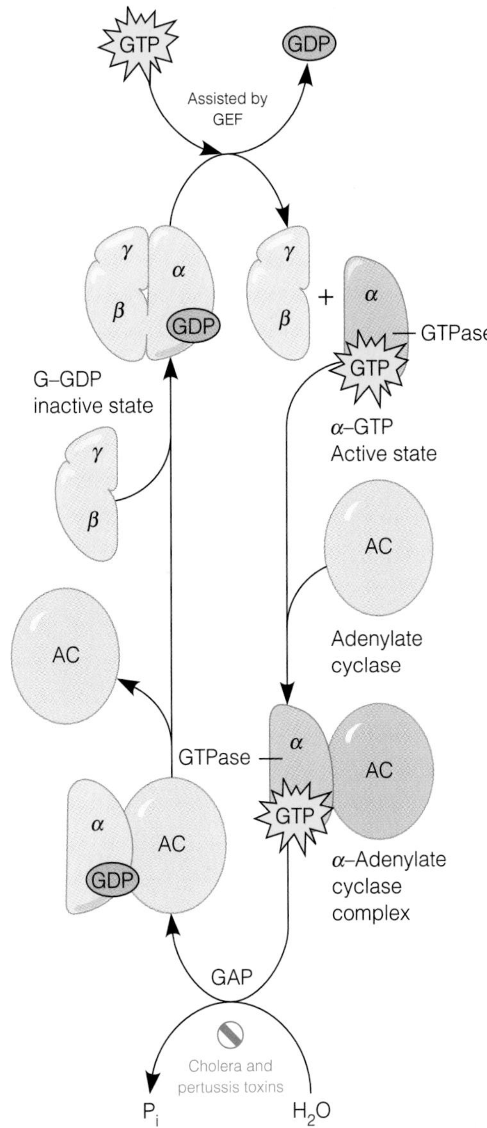

FIGURE **23.12**

The cycle of G protein dissociation and reassociation. α, β, and γ are the three subunits of the G protein. The active form is the α–GTP complex (dark teal), while the inactive GDP complexes are shown in light teal. The sites of action of pertussis and cholera toxins are also shown. GEF = guanine nucleotide exchange factor; GAP = GTPase-activating protein.

Consequences of Blocking GTPase

The importance of the GTPase activity in controlling the hormone response can be seen in the consequences of blocking it. Blocking can be achieved in vitro by substituting GTP with **GTPγS,** a GTP analog in which a sulfur atom substitutes for an oxygen on the γ phosphate of GTP, and which the GTPase activity cannot cleave. In a G_s-dependent system the result is irreversible activation of the target adenylate kinase.

$$^{-}O-\overset{\overset{S}{\|}}{\underset{\underset{O^-}{|}}{P}}-O-\overset{\overset{O}{\|}}{\underset{\underset{O^-}{|}}{P}}-O-\overset{\overset{O}{\|}}{\underset{\underset{O^-}{|}}{P}}-O-CH_2\text{-ribose(OH, OH)-guanine}$$

GTPγS

More dramatic are the effects of bacterial toxins that have G proteins as their biological targets. The toxin of *Vibrio cholerae* is an enzymatic protein with the ability to cleave NAD^+ and transfer its ADP-ribose moiety to a specific site in the α subunit of G_s. This modification of G_s inhibits its GTPase activity and converts α to an irreversible activator of adenylate cyclase.

$$NAD^+ + \alpha_s \longrightarrow \text{nicotinamide} + \text{ADP-ribosyl-}\alpha_s$$

In the intestine the resultant cAMP accumulation promotes a physiological response controlled by cAMP—uncontrollable secretion of water and Na^+—and is responsible for the severe diarrhea and consequent dehydration and loss of salt that accompany cholera. A component of the toxin of *Bordetella pertussis,* which causes whooping cough, has a similar effect on the α subunit of the G_i protein, with different physiological effects—lowered blood glucose and hypersensitivity to histamine.

G Proteins in the Visual Process

There are remarkable similarities between the actions of G proteins in transmitting hormonal signals and their actions in the transmission of signals from light. Much of our understanding of G proteins in hormonal signal transduction came from studies of a G protein called **transducin** in the visual process. The extracellular stimulus and the biochemical end point are quite different in vision from those in hormone action, but the transmembrane signaling processes are almost identical.

As mentioned in Chapter 19, the extracellular signal in vision is a photon of light, and the membrane receptor is **rhodopsin,** an abundant membrane protein in the outer segment of rod cells in the retina. A photochemical change in the structure of rhodopsin causes it to activate transducin so that it binds GTP. The transducin–GTP complex activates a specific **phosphodiesterase,** which cleaves a cyclic nucleotide, **guanosine 3′,5′-monophosphate** (cyclic GMP, or cGMP). Cleavage of cGMP, in turn, stimulates intracellular reactions that generate a visual signal to the brain. Thus, the stimulated *hydrolysis* of cGMP is the visual analog of the stimulated *synthesis* of cAMP in β-adrenergic responses.

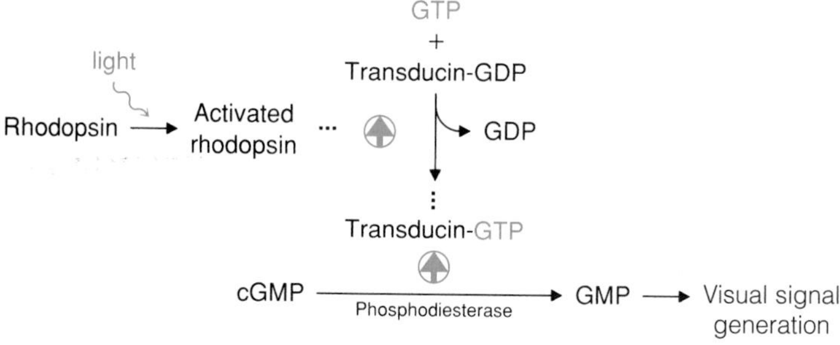

A Closer Look at G Protein Subunits

The G protein mechanism is used in many signal transduction pathways. Cloning of human cDNAs reveals the existence of at least 24 different α proteins, 5 β proteins, and 6 γ proteins. Reassortment of these proteins means that a large number of different G proteins exists, giving great flexibility in response to this signal transduction element. Interaction with target enzymes is a function of the α subunits. Some interact with adenylate cyclase, some interact with ion channels, and some interact with phospholipases. One subfamily of G proteins, called G_{olf}, is present in olfactory cells in the nose and functions with a large number of receptors involved in sensory reception of odors. Table 23.4 summarizes information about actions of the major classes of G proteins.

The α subunits of G proteins are part of a family of small GTP-binding proteins that are active when GTP is bound and inactive in the presence of GDP. This family includes the oncogene-specified Ras proteins (see page 864) and the GTP-binding elongation factors involved in protein synthesis (Chapter 27). In recent years a great deal has been learned from crystallographic studies about how GTP binding activates this class of proteins and how the activated protein in turn interacts with its target. Figure 23.13a shows the structure of an $\alpha\beta\gamma$ heterotrimeric G protein–GDP complex, with the α subunit derived from a G_i protein. The "switch II" region, shown in red, is a domain that changes conformation when GTP is hydrolyzed. Panel b shows the structure of a $G_{s\alpha}$ protein superimposed with that of a $G_{i\alpha}$ subunit; both proteins are binding GTPγS, a nonhydrolyzable GTP analog. Points of structural divergence are those expected to determine whether interaction of the α protein with a target enzyme, such as adenylyl cyclase, activates or inhibits that enzyme. Panel c shows a complex between a $G_{s\alpha}$ bound to GTPγS and the catalytic core of adenylyl cyclase. The switch II region of the α protein is in close contact with the target enzyme and participates in its activation.

TABLE **23.4 Properties of mammalian G proteins**

Family	Toxin Sensitivity	Tissue Distribution	Receptors	Effectors
G_s	Cholera	Ubiquitous	β-adrenergic Glucagon Thyrotropin-releasing hormone	Adenylate cyclase Ca^{2+} channels Na^+ channels
G_{olf}	Cholera	Olfactory epithelium	Odorant	Adenylate cyclase
G_i	Pertussis	Ubiquitous	M_2-cholinergic α_2-adrenergic	K^+ channels Ca^+ channels Adenylate cyclase Phospholipase C
Transducin (G_t)	Cholera, pertussis	Retinal rods	Rhodopsin	cGMP phosphodiesterase
G_q	?	Nearly ubiquitous	M_1-cholinergic α_1-adrenergic	Phospholipase C
G_{12}	?	Ubiquitous	?	?
G_o	?	?	M_2-cholinergic	Ca^{2+} channels

Source: Adapted from J. R. Hepler and A. G. Gilman, *Trends Biochem. Sci.* (1992) 17:384. © 1992 with permission of Elsevier Science.

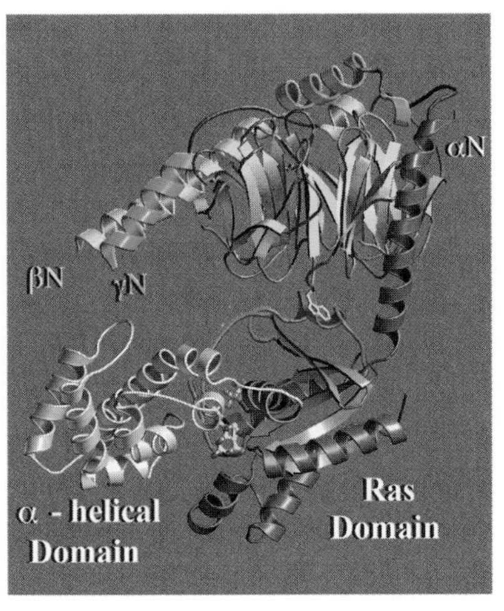

(a)

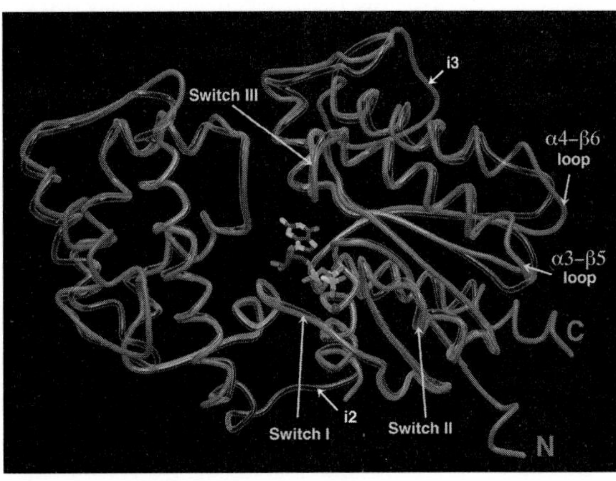

(b)

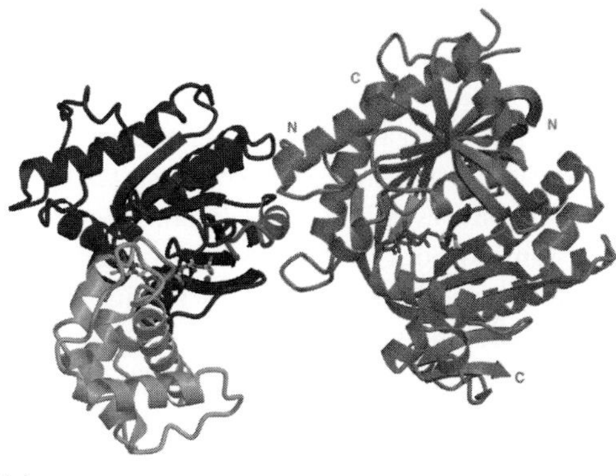

(c)

FIGURE **23.13**

Structures of G proteins. **(a)** A heterotrimeric $\alpha\beta\gamma$ complex, with the α subunit derived from α G_i protein. The α subunit is in gray, β is in yellow, and γ is in green. The N-termini of the three proteins are shown; note that the lipid modifications at these sites are not present in these recombinant proteins. Shown in red are a GDP bound to the α subunit and the switch II region on α. **(b)** Superposition of an α protein from G_i (transparent rose) and a G_s α protein–GTPγS complex (solid gray). Two insertions in the $G_{\alpha s}$ protein relative to $G_{\alpha i}$ (i2 and i3) are shown with white arrows. The structures are identical in the GTP-binding regions. **(c)** A complex between $G_{s\alpha}$-GTPγS (to the left) and the catalytic core of adenylyl cyclase (to the right). The upper side is believed to face the plasma membrane. The α protein consists of two major domains—a "helical" domain (ash gray) and a "ras-like" domain (charcoal). Two domains of the adenylyl cyclase (mauve and khaki) are also shown. GTPγS is shown as a red and green stick figure, and the switch II region of the α protein is shown in red.

(a) Courtesy of S. R. Sprang from M. A. Wall, B. A. Posner, and S. R. Sprang, *Structure* (1998) 6:1169–1183; **(b)** courtesy of S. R. Sprang from R. K. Sunahara et al., *Science* (1997) 278:1943–1947, with permission from *Science;* **(c)** courtesy of S. R. Sprang from J. J. G. Tesmer et al., *Science* (1997) 278:1907–1918, with permission from *Science.*

As noted earlier, events in signal transduction are controlled by feedback loops. At least two such loops are involved in the regulation of events mediated by G proteins. First, the GTPase activity of some α proteins is inefficient, making G protein inactivation a slow process. A family of GTPase-activating (GAP) proteins speeds up this process by interacting with the α-GTP complex and stimulating the GTPase activity, sometimes by orders of magnitude. Second, a family of enzymes called G protein–coupled receptor kinases phosphorylates specific receptor proteins, leading to their inactivation and, hence, limiting the hormone-induced response. This effect is mediated by $\beta\gamma$ complexes.

SECOND-MESSENGER SYSTEMS

As introduced in Chapter 12 and elaborated here, many signal transduction events involve the linked actions of receptor, G protein, and adenylate cyclase. These events either stimulate or inhibit the synthesis of a second messenger, cyclic AMP, inside the cell. Many intracellular processes are controlled in turn by the level of that second messenger. One of those processes, not yet mentioned, is the synthesis of receptor proteins themselves. Cyclic AMP binds to and activates cAMP-dependent protein kinase (PKA) which in turn phosphorylates a protein called CREB (cAMP response element binding protein), and the resultant phosphorylated protein controls transcription of genes, including those encoding particular receptors. Some of these actions represent adaptation of a cell to action of a hormone.

Cyclic AMP is not the only second messenger. We have mentioned both cyclic GMP (see page 852 and Chapter 21) and the phosphoinositide system (Chapters 12 and 19; see page 855). Much interest has focused on cyclic GMP, particularly with the recent realization that nitric oxide acts by stimulating synthesis of this nucleotide. Many cells contain a cGMP-stimulated protein kinase that, unlike the cAMP-activated enzyme, contains both catalytic and regulatory domains on one polypeptide chain of a homodimeric protein. Our understanding of the roles of cGMP in signal transduction has come fairly recently, because its intracellular concentrations are 10- to 100-fold lower than those of cAMP.

Calcium ion has also been considered a second messenger. Many cells respond to extracellular stimuli by altering their intracellular calcium concentration, which in turn exerts biochemical changes either by itself or through its interaction with calmodulin (see Chapter 13). Calcium levels themselves are controlled in large part by second messengers, including cAMP. In many nerve and muscle cells, the activation of adenylate cyclase results in an influx of extracellular calcium. cAMP activates a voltage-dependent calcium channel in the presynaptic nerve membrane, allowing calcium ions to flow into the cell and triggering synaptic transmission (see Chapter 21). That activation may involve phosphorylation, by cAMP-dependent protein kinase, of a protein component of the channel. In muscle cells, calcium influx triggers muscle contraction (see Chapter 8) and is responsible, for example, for the increased rate and force of heartbeats caused by β-adrenergic agonists. Because cAMP regulates calcium influx, it may be more appropriate to call calcium a third messenger than a second messenger.

Cytosolic calcium levels can be increased also by release from *intracellular* calcium stores. Access to these intracellular stores is controlled by another set of messengers, the **phosphoinositide system.** Although similar in many respects to the adenylate cyclase system, the phosphoinositide system is distinctive in that the hormonal stimulus activates a reaction that generates *two* second messengers. The earliest experimental observations regarding this signal transduction system occurred in 1953, when Mabel Hokin and Lowell Hokin noted that administration of acetylcholine to pancreatic secretory cells led to rapid synthesis and turnover of the phosphatidylinositol fraction of membrane phospholipids. Similar observations were made in other systems stimulated by hormones, neurotransmitters, or growth factors. However, more than two decades elapsed before a unifying concept emerged to explain these observations.

We now know that a specific lipid in the phosphoinositide family, namely, **phosphatidylinositol 4,5-bisphosphate (PIP_2)**, is a membrane-associated storage form for two second messengers. As shown in Figure 23.14, binding of an agonist to a receptor (step 1) stimulates a G protein to bind GTP (step 2), just as occurs

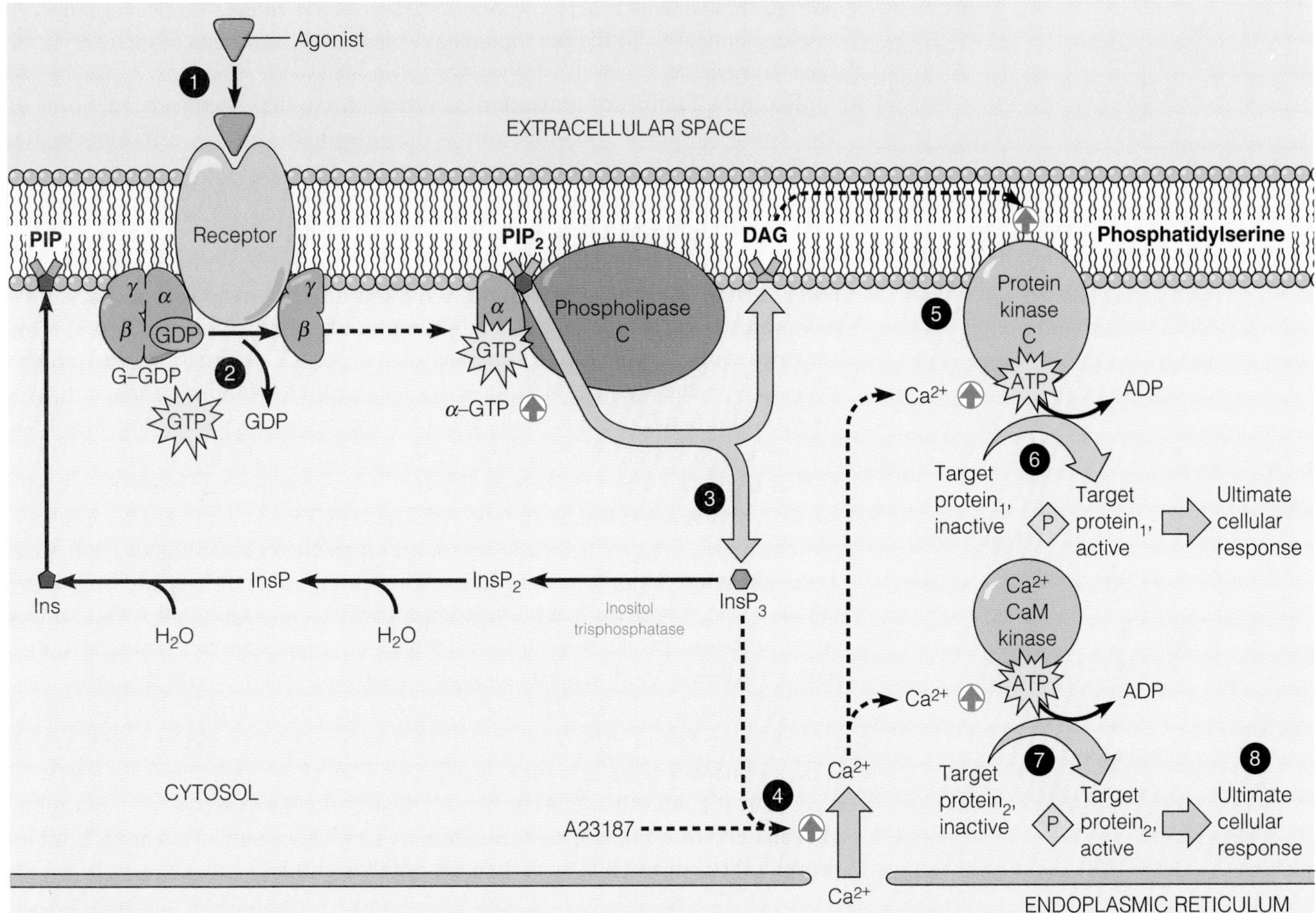

FIGURE **23.14**

Signal transduction pathways involving phosphoinositide turnover. DAG = *sn*-1,2-diacylglycerol, Ins = inositol, InsP = inositol monophosphate, PIP = phosphatidylinositol-4-phosphate, PIP_2 = phosphatidylinositol 4,5-bisphosphate, $InsP_3$ = inositol 1,4,5-trisphosphate, $InsP_2$ = inositol 1,4-bisphosphate. Most of the effects of calcium result from its binding to calmodulin (CaM). A23187 is a calcium ionophore, which can be used experimentally to release calcium from intracellular stores. The release of calcium ion stimulates both protein kinase C and calmodulin kinase.

Phosphatidylinositol 4,5-bisphosphate (PIP_2) + H_2O ⟶ ***sn*-1,2-Diacylglycerol (DAG)** + **Inositol 1,4,5-trisphosphate ($InsP_3$)**

during the activation of adenylate cyclase. However, this G protein activates a different membrane-bound enzyme, a **phospholipase C,** which in turn cleaves PIP_2 to yield two products (step 3)—***sn*-1,2-diacylglycerol** (DAG) and **inositol 1,4,5-trisphosphate** ($InsP_3$).

Both of these products act as second messengers. Therefore, the cleavage of PIP_2 by phospholipase C is the functional equivalent of the synthesis of cyclic AMP by adenylate cyclase.

Recently, a family of phosphoinositide 3-kinases has been isolated. Some of the products, phosphatidylinositol-3,4-bisphosphate and -3,4,5-trisphosphate, act like cyclic AMP, by stimulating the activity of specific protein kinases.

The second-messenger role of inositol trisphosphate is to stimulate the release of calcium from its intracellular stores in the endoplasmic reticulum (step 4 in Figure 23.14). This release has various effects on intracellular metabolism, as noted earlier, but it also contributes to the second-messenger role of diacylglycerol, which is the activation of membrane-bound protein kinase C (step 5). This enzyme requires for its activity *calcium* (hence the "C" designation) and a *phospholipid* (specifically, phosphatidylserine). The other second messenger, diacylglycerol, stimulates protein kinase C activity by greatly increasing the affinity of the enzyme for calcium ions. This requirement is specific for the *sn*-1,2-DAG; neither the 1,3- nor the 2,3-isomer is active. The enzyme phosphorylates specific serine and threonine residues in target proteins (step 6). As with cAMP-stimulated protein kinase, the specific cellular responses to protein kinase C activation, such as the phosphorylation of calmodulin shown in Figure 23.14, depend on the ensemble of target proteins that become phosphorylated in a given cell. Other known target proteins include the insulin receptor, β-adrenergic receptor, glucose transporter, HMG-CoA reductase, cytochrome P450, and tyrosine hydroxylase.

Now let us briefly consider the metabolism of inositol trisphosphate ($InsP_3$) after its release from PIP_2. Three sequential hydrolytic steps yield inositol, which is then reincorporated into phosphatidylinositol, as discussed in Chapter 19, to regenerate PIP and PIP_2. The last hydrolytic step, the hydrolysis of inositol monophosphate to inositol, is specifically inhibited by **lithium ion.** This block inhibits the resynthesis of $InsP_3$ by depleting the cell of inositol.

$$\text{Inositol monophosphate} + H_2O \longrightarrow \text{inositol} + P_i$$

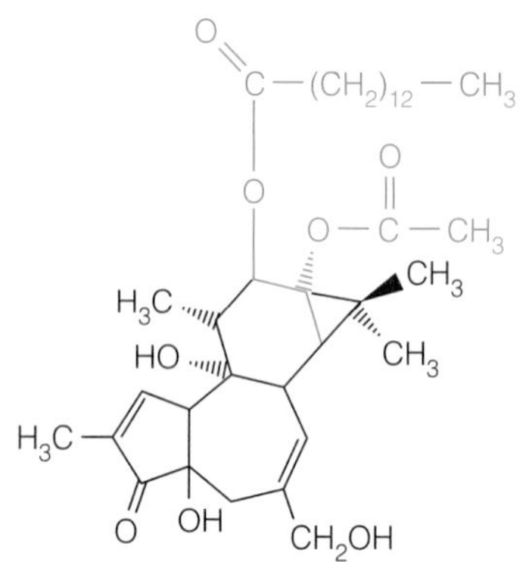

A phorbol ester, 1-*O*-tetradecanoylphorbol-13-acetate

sn-1,2-Diacylglycerol (DAG)

Given that the phosphoinositide messenger system is widely used in nervous tissue, this action of lithium may be related to its efficacy in the treatment of bipolar syndrome, formerly called manic–depressive disorder.

Because many metabolic processes are controlled by calcium fluxes and by phosphorylation of specific proteins, the phosphoinositide system has great versatility as a control mechanism. The fact that a cell can use either DAG or $InsP_3$ or both mechanisms as the result of a single extracellular stimulus further increases this versatility. A partial listing of processes controlled by the phosphoinositide system is presented in Table 23.5.

Several observations imply a role for the phosphoinositide system not only in metabolic regulation but also in the control of cellular growth. First is the activity of a group of natural products called **phorbol esters,** part of whose structure resembles that of DAG (shown in red). These compounds are called **tumor promoters.** Not carcinogenic by themselves, they stimulate the formation of tumors when applied along with a carcinogen to experimental animals. Some phorbol esters have been found to activate protein kinase C independently of diacylglycerol. This finding is consistent with the hypothesis that protein kinase C activation is part of the normal growth control process that becomes perturbed in tumorigenesis. Another indication of a link between phosphoinositide metabolism and growth control is the function of certain cell growth factors. Some of them, notably **platelet-derived growth factor (PDGF),** are known to interact with cell surface receptors to stimulate the hydrolysis of phosphatidylinositol.

It is becoming evident that phospholipases other than phospholipase C are also stimulated by G proteins. Recall from Chapter 19 that arachidonic acid, released from phosphatidylcholine, is the major metabolic precursor to eicosanoids.

TABLE **23.5 Some cellular processes controlled by the phosphoinositide second-messenger system**

Extracellular Signal	Target Tissue	Cellular Response
Acetylcholine	Pancreas	Amylase secretion
	Pancreas (islet cells)	Insulin release
	Smooth muscle	Contraction
Vasopressin	Liver	Glycogenolysis
Thrombin	Blood platelets	Platelet aggregation
Antigens	Lymphoblasts	DNA synthesis
	Mast cells	Histamine secretion
Growth factors	Fibroblasts	DNA synthesis
Spermatozoa	Eggs (sea urchin)	Fertilization
Light	Photoreceptors (*Limulus*)	Phototransduction
Thyrotropin-releasing hormone	Pituitary anterior lobe	Prolactin secretion

Source: Adapted from M. J. Berridge, *Sci. Am.* (1985) Oct., p. 147. © 1985 Andrew Christie.

Phospholipase A_2, which releases this fatty acid, is also part of a signal transduction pathway involving G proteins, and phospholipase D is thought also to participate in signal transduction via diacylglycerol formation. In addition, some phospholipases are controlled by Ca^{2+}.

Second messengers include cyclic AMP, cyclic GMP, calcium ion, inositol trisphosphate, and diacylglycerol.

To recapitulate: Cyclic AMP was the earliest known second messenger. However, several comparable second (or third) messengers are now known, including cyclic GMP, calcium ion, inositol trisphosphate, and diacylglycerol. The abovementioned studies on phosphoinositide 3-kinases and phospholipases A_2 and D suggest that other second messengers are awaiting discovery.

THE INSULIN RECEPTOR AND RELATED RECEPTORS WITH PROTEIN KINASE ACTIVITY

A number of hormone and growth factor receptors are distinctive in that they contain an intrinsic enzymatic activity and a single membrane-spanning domain. Some of these receptors interact with downstream second-messenger systems, but others, including the insulin receptor, evidently do not. The insulin receptor was the first of this class of proteins whose structure was determined, through sequence analysis of cloned cDNAs.

The insulin receptor (Figure 23.15) is a glycoprotein with an $\alpha_2\beta_2$ tetrameric structure, stabilized by interchain disulfide bonds. Both the α chain, of 735 residues, and the β chain, of 620 residues, are translated from a single mRNA, giving a polypeptide chain that then undergoes proteolytic processing. The α chain, which is thought not to span the membrane, binds insulin near its C-terminus. The β chain has a single transmembrane domain, with its C-terminus in the cell interior. That C-terminal region is the site of a protein tyrosine kinase activity, which is stimulated by binding of insulin to the extracellular part of the receptor. The kinase activity is essential to the biological activity of the insulin receptor, because some cases of non-insulin-dependent diabetes are associated with receptor mutations that abolish the kinase activity. Several protein substrates for the receptor tyrosine kinase have been identified and are now being characterized vis-à-vis their roles in signal transduction.

The insulin receptor and several related growth factor receptors contain one transmembrane domain per polypeptide chain and have an intrinsic protein tyrosine kinase activity.

Given that insulin can be considered a growth factor, it is of interest that protein tyrosine kinase activity is found in other growth factor receptors, including those for **epidermal growth factor** (EGF), **platelet-derived growth factor** (PDGF), **colony-stimulating factor 1** (CSF-1), and **fibroblast growth factor** (FGF), as well as a peptide **insulin-like growth factor 1** (IGF-1). These receptors represent a family of closely related proteins (see Figure 23.15), for the tyrosine kinase

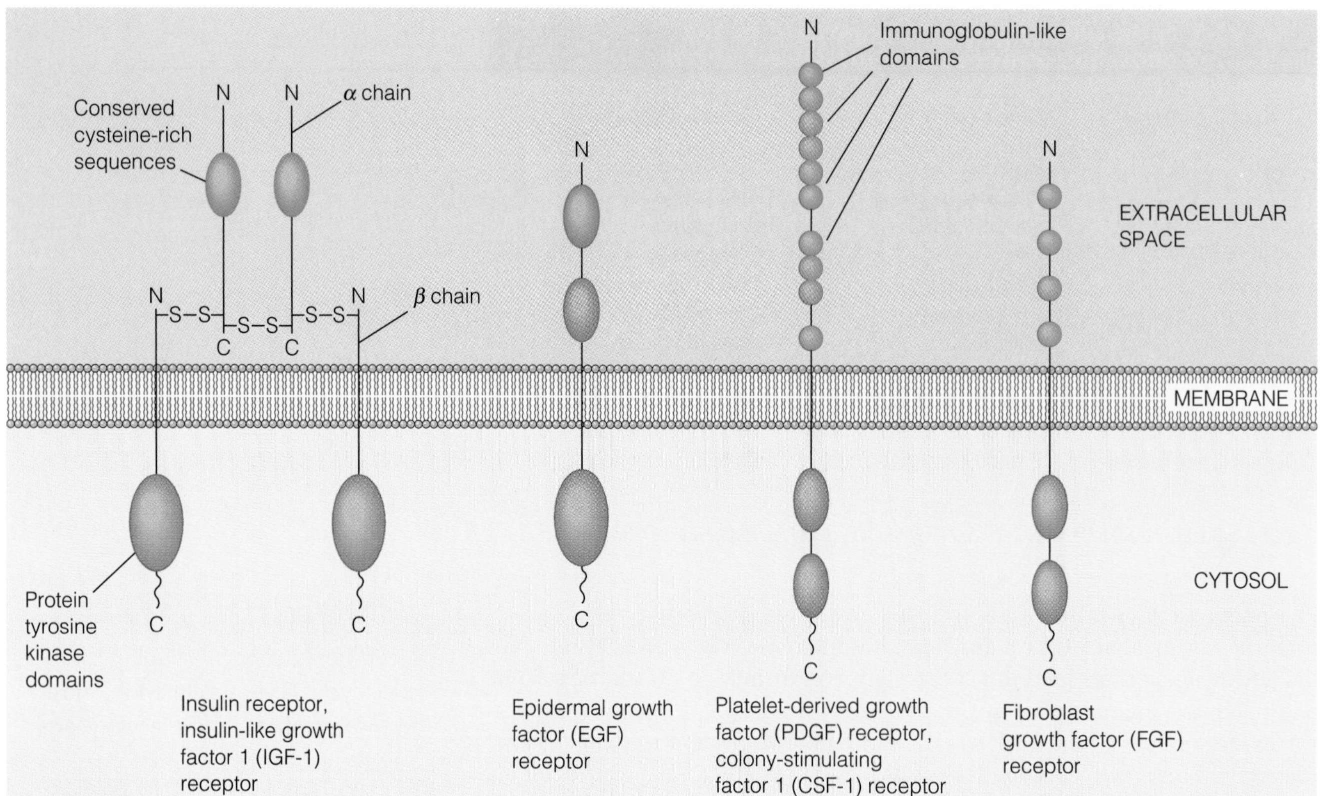

FIGURE **23.15**

The insulin receptor and its structural relationship to other transmembrane receptors with protein tyrosine kinase activity. Representative proteins in each of four subfamilies are shown. Five additional subfamilies have also been described. Note that two of the receptor families shown (to the right) have two protein kinase domains.

Adapted from A. Ullrich and J. Schlessinger, *Cell* (1990) 61:203. © 1990 Cell Press.

domains share amino acid sequence homology. Moreover, there is evidence that the action of insulin as a growth factor is mediated through its binding to one of these receptors, the IGF-1 receptor. We shall say more about the signaling activities of these receptor tyrosine kinases later in this chapter.

A number of more distantly related membrane receptors have other enzyme activities. Proteins of the transforming growth factor β (TGF-β) family bind to a receptor that has a protein serine/threonine kinase activity (like cAMP-dependent protein kinase). **Atrial natriuretic factor,** which controls blood volume, binds to a receptor that has both a guanylate cyclase activity and a predicted protein serine/threonine kinase activity.

STEROID AND THYROID HORMONES: INTRACELLULAR RECEPTORS

Hormones acting through nuclear receptors generally have longer-lived effects than those interacting with membrane receptors.

Hormonal effects occurring via membrane receptors tend to be of short duration. Like the epinephrine-induced glycogenolytic cascade, they represent responses to rapid and urgent physiological demands, and they involve activation or inhibition of preexisting enzymes. By contrast, the effects of steroid hormones involve longer-term changes, such as the activation of a transport system or the conversion of a resting cell to a growing cell. Steroids and related hormones (thyroid, vitamin D, and retinoic acid hormones) act intracellularly. By virtue of their hydrophobic nature, they traverse the plasma membrane and exert their effects within the cell—actually, within the nucleus, where they control the activities of specific genes. In most cases, target genes are activated. Table 23.6 lists several proteins whose synthesis is regulated by these hormones.

These regulatory effects occur at the level of transcription of steroid-responsive genes. Steroid and related hormones act by binding in the cytosol to specific receptor proteins, which dimerize under the influence of the hormone. Binding in the cytosol is followed by movement of the hormone–receptor complex into the nucleus, where the complex interacts with specific DNA sites called **hormone-**

TABLE 23.6 Target organs for steroid and thyroid hormones and major proteins whose synthesis is affected

Hormone Class	Target Organ	Protein[a]
Glucocorticoids	Liver	Tyrosine aminotransferase
		Tryptophan oxygenase
		α-Fetoprotein (↓)
		Metallothionein
	Liver, retina	Glutamine synthetase
	Kidney	Phosphoenolpyruvate carboxykinase
	Oviduct	Ovalbumin
	Pituitary	Pro-opiomelanocortin
Estrogens	Oviduct	Ovalbumin
		Lysozyme
	Liver	Vitellogenin
		apo-VLDL
Progesterone	Oviduct	Ovalbumin
		Avidin
	Uterus	Uteroglobin
Androgens	Prostate	Aldolase
	Kidney	β-Glucuronidase
	Oviduct	Albumin
1,25-Dihydroxyvitamin D_3	Intestine	Calcium-binding protein
Thyroid hormones	Liver	Carbamoyl phosphate synthetase
		Malic enzyme
	Pituitary	Growth hormone
		Prolactin (↓)
Ecdysone (in insects)	Epidermis	Dopa decarboxylase
	Fat body[b]	Vitellogenin

[a]Synthesis of each indicated protein is increased by the hormone, except for the two identified by (↓).
[b]The fat body is an organ in insects that plays some of the same roles as liver and adipose tissue.

responsive elements (HREs). Binding of the complex to DNA affects transcription rates of nearby genes, by mechanisms now under intense study. Because of their site of action, members of this protein family are also called nuclear receptors.

Nuclear receptors exist at levels of only about 10^4 molecules per cell, which makes their purification difficult. However, because they bind to hormones very tightly, it has been possible to purify these proteins by affinity chromatography. cDNA sequence analysis has revealed structural similarities among this class of receptors, and use of recombinant DNA techniques to construct hybrid receptors has allowed unambiguous identification of domains of function within the receptor molecule. Each receptor protein within this family contains a central conserved domain of about 80 residues, which is involved in DNA binding (Figure 23.16). On the N-terminal side of this domain is a region essential to transcriptional activation. Toward the C-terminus are domains involved in hormone binding, protein dimerization, and transcriptional activation.

All of the known receptors in this family contain bound zinc, which is essential for DNA binding, and the DNA-binding sequences show a completely conserved distribution of cysteine residues. These observations suggested that the zinc atoms could be complexed by the cysteine sulfurs in a pattern akin to the "zinc finger" structural motif associated with a number of other eukaryotic transcriptional regulatory proteins (see Chapter 28). That prediction has been supported by high-resolution NMR studies of receptor–DNA complexes (Figure 23.17).

The family of steroid receptors contains a conserved, zinc-containing DNA-binding sequence and a C-terminal hormone-binding domain.

The utility of a set of long-term-acting regulators is evident from a couple of examples. Estrogens and progesterone regulate the female reproductive cycle. In

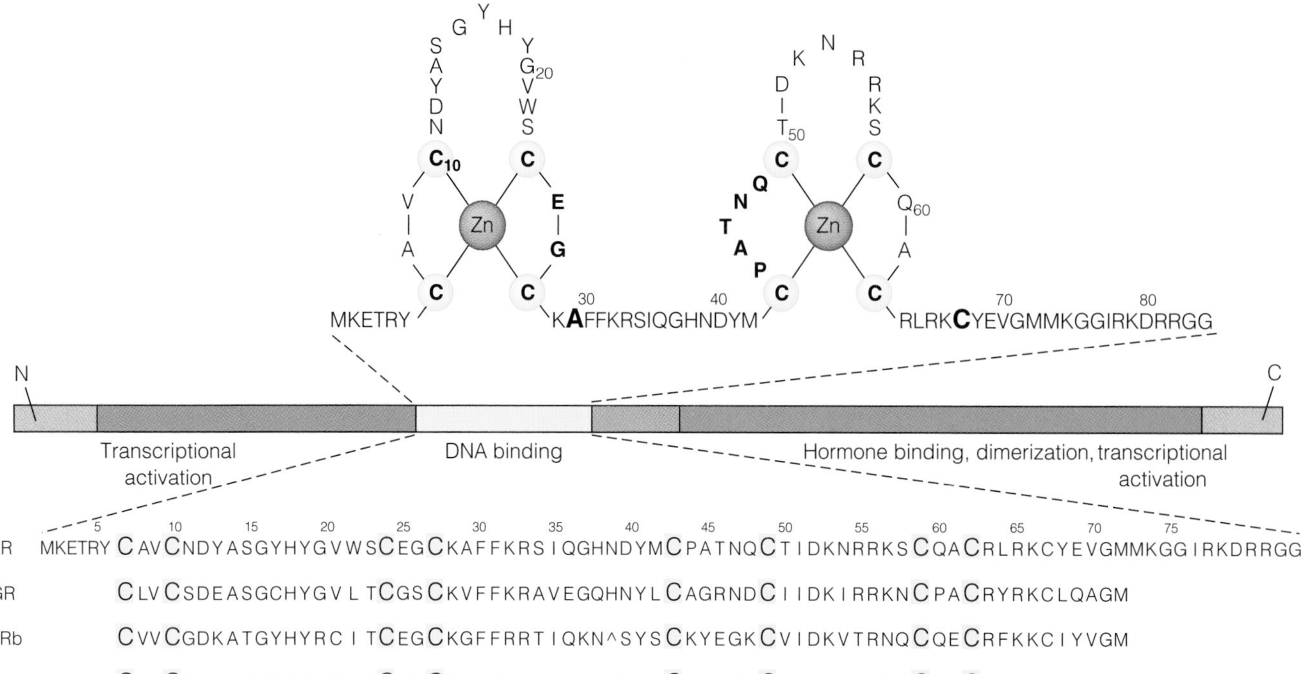

FIGURE **23.16**

The conserved DNA-binding domain in steroid receptors. In the center are structural domains within steroid receptors, illustrated for the estrogen receptor. Above is the DNA-binding domain of the estrogen receptor, showing conserved cysteine residues that contact the bound zinc ions (a zinc finger binding motif; see Chapter 28). At the bottom are the DNA-binding domain sequences of related human receptors, with the conserved cysteine residues highlighted. ER = estrogen receptor, GR = glucocorticoid receptor, TRb = thyroid hormone receptor, PR = progesterone receptor, VitD = vitamin D receptor, RAR = retinoic acid receptor.

J. W. R. Schwabe and D. Rhodes, *Trends Biochem. Sci.* (1991) 16:292. © 1991 with permission of Elsevier Science.

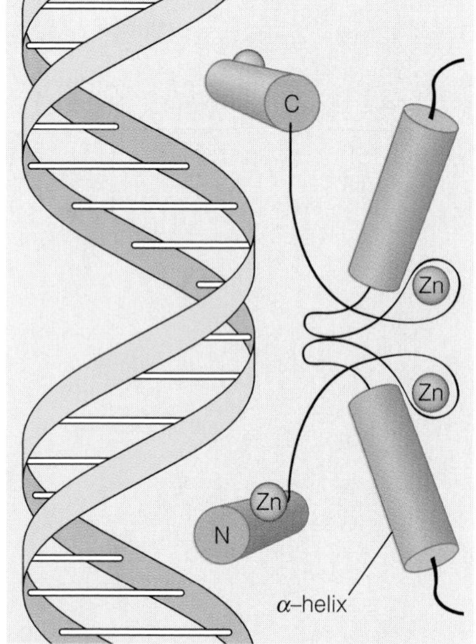

FIGURE **23.17**

Binding of the estrogen receptor to DNA, as inferred from solution NMR spectroscopy. The dimeric receptor protein has two α-helical regions that bind to both ends of a symmetrical DNA sequence (AGGTCAXXXTGACCT), within the major groove.

J. W. R. Schwabe and D. Rhodes, *Trends Biochem. Sci.* (1991) 16:292.

humans these hormones interact over a 4-week cycle to prepare the uterus for implantation of a fertilized ovum. Proliferation of the endometrium, the epithelial lining of the uterus, is the major event; clearly it requires new protein synthesis and increased blood flow to the uterus. These processes stop when a pituitary signal triggers decreased release of the hormones, causing sloughing off of cells in the uterine lining and the beginning of menstrual bleeding.

The actions of glucocorticoids are comparable, in that control of the synthesis of particular proteins allows for long-term metabolic adaptation. Whereas estrogens exercise control of reproductive metabolism over a several-week period, the secretion of glucocorticoids is a means of adaptation to longer-term stress. This adaptation involves stimulation of gluconeogenesis and synthesis of a variety of proteins, including some that counteract the effects of inflammation. Unlike estrogens, which act chiefly in reproductive tissues, the glucocorticoids influence cells in a wide variety of target tissues.

Investigations of the action of glucocorticoids as anti-inflammatory agents and as immunosuppressants have illuminated the action of another important signaling pathway, involving a transcriptional activator called NF-κB. This protein stimulates transcription of genes for a class of proteins called **cytokines,** which stimulate

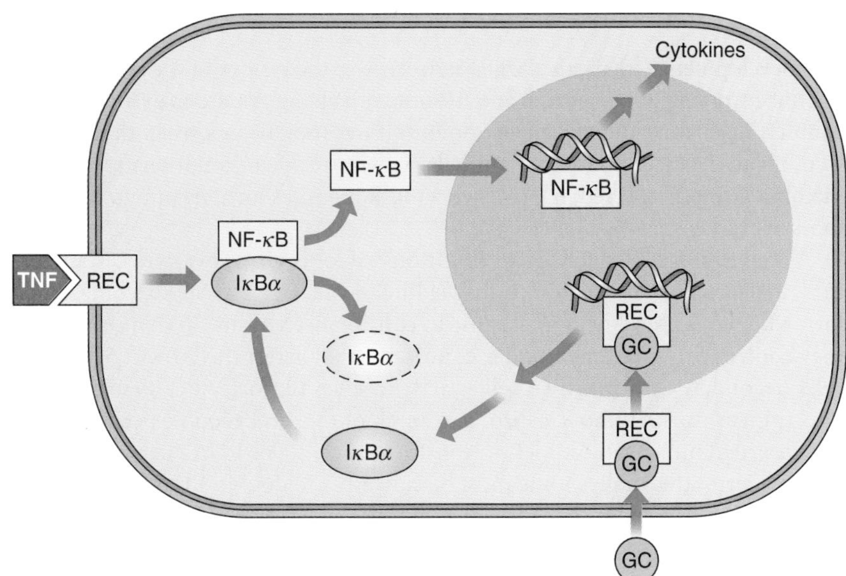

FIGURE **23.18**

Action of glucocorticoids (GCs) in suppressing immune and inflammatory reactions mediated by cytokines. REC = receptor; see text for other abbreviations.

Courtesy of J. Marx, *Science* (1995) 270:232–233. © 1995 AAAS.

various reactions of the immune response, including proliferation of antibody-producing cells. Normally NF-κB is bound to an inhibitory protein called IκBα, which prevents its translocation into the nucleus. Binding of an immune stimulator, such as **tumor necrosis factor** (TNF), to its plasma membrane receptor leads to the ubiquitination of IκBα and its subsequent degradation by the 26S proteasome (Chapters 21 and 28). This in turn allows NF-κB to translocate to the nucleus and activate cytokine synthesis. One of the target genes for activation by the glucocorticoid–receptor complex is the gene for IκBα. By stimulating synthesis of this protein, glucocorticoids counteract the breakdown of the NF-κB–IκBα complex and thereby suppress the transcription of NF-κB-responsive genes. These relationships are summarized in Figure 23.18.

Steroid hormone receptors are target sites for several important drugs. **Tamoxifen** binds to estrogen receptors but does not activate estrogen-responsive genes. The growth of some breast tumor cells is activated by estrogen. Tamoxifen treatment of patients with such tumors after surgery or chemotherapy often antagonizes estrogen binding in residual tumor cells and retards their growth. However, patients taking tamoxifen after breast cancer surgery must be monitored carefully, because there is also an increased risk of uterine cancer. **RU486,** which was developed in France, binds to progesterone receptors and blocks the events essential to implantation of a fertilized ovum in the uterus. Hence, RU486 is an effective contraceptive agent, even when taken after intercourse.

Tamoxifen

RU486

Signal Transduction, Oncogenes, and Cancer

One of the most fruitful areas of research in the biological sciences has been ongoing investigations into genetic differences between cancer cells and normal cells. These investigations have revealed, in a wide variety of tumor cells, mutationally altered forms or levels of proteins involved in signal transduction—including altered protein kinases, G proteins, nuclear receptors, growth factors, and growth factor receptors. Some tumor cells contain a normal signal transduction protein, but in excessive amounts. Genes responsible for such alterations are called **oncogenes.** Investigations of protein products of oncogenes, termed **oncoproteins,** have illuminated roles of the normal forms of these proteins in regulating cell metabolism and growth and have spotlighted how normal control mechanisms go awry in a cancer cell.

VIRAL AND CELLULAR ONCOGENES

Two developments are particularly noteworthy in the history of the study of cancer, one involving tumor viruses and the other involving genetic analysis of human tumors. Regarding the first development, it has long been known that certain viruses cause cancer in infected animals. The first known tumor virus was **Rous sarcoma virus,** discovered in 1911 by Peyton Rous and shown by him to cause tumors in chickens.

Whether a virus contains RNA (like Rous sarcoma virus) or DNA, certain features are common in viral infections leading to cancer. First, cells become **transformed.** That is, they lose normal growth control mechanisms, and in cell culture they continue to proliferate under conditions that arrest the growth of normal cells. Second, the transformed cells are themselves tumorigenic; their injection into animals causes tumors. Third, part or all of the viral genome becomes linearly inserted into chromosomes of transformed cells. For RNA viruses like Rous sarcoma virus, the viral genome must be converted to double-strand DNA before this insertion can occur. The viral enzyme that synthesizes DNA from a single-strand RNA template is called **reverse transcriptase,** and viruses containing this enzyme are called **retroviruses** (see Chapter 24).

A number of nontumorigenic mutants of Rous sarcoma virus exist. Mapping mutations in these strains identified *src,* the viral oncogene responsible for transforming infected cells. Some of these mutants contain extensive deletions, which permitted Raymond Erikson to use nucleic acid hybridization techniques (see Chapter 26) and, in 1978, to clone a cDNA corresponding to the viral *src* gene. Two surprising findings emerged. First, expression of the cloned gene yielded a protein with a protein tyrosine kinase activity. Thus, a specific enzyme activity, which might be associated with signal transduction, was also associated with the oncogene product. Second, further nucleic acid hybridization analysis showed that sequences corresponding to the viral *src* gene were present in normal cells. This finding suggested that viral oncogenes had their origins in normal cellular genes, or vice versa. One way to explain the transfer of an oncogene, or oncogene precursor, from cells to viruses is to postulate a rare genome excision event, as depicted in Figure 23.19. If an infection had caused insertion of the viral genome next to an oncogene precursor (or **proto-oncogene**), and if a subsequent excision event removed part or all of the proto-oncogene as well as the viral genome, then this faulty excision would have created a novel viral genome, containing a cellular gene. Subsequent evolution of the virus could change the cellular gene, creating an oncogene. Action of the oncogene would contribute toward transformation in a subsequent infection.

Viral oncogenes are errant cellular proto-oncogenes, mostly encoding signal transduction elements that have been taken into viral genomes and have undergone subsequent mutations.

Sequence analysis of the *src* gene from viruses and cells revealed small differences. Thus, we now speak of *v-src,* the viral form of the gene, and *c-src,* the cellular form. Analysis of many other tumor viruses yielded more than two dozen additional oncogenes. The corresponding proto-oncogenes encode a variety of proteins involved in cell signaling, some of which are identified in Table 23.7. Further analysis of infections leading to tumorigenesis showed that mutational alteration of the proto-oncogene is not always necessary. In some cases the viral genome is inserted adjacent to a proto-oncogene. Elements of the viral genome stimulate transcription of the DNA sequences flanking the integration site. Thus, tumorigenesis can result from overexpression of normal genes encoding signal transduction machinery.

Although the Src protein is a protein tyrosine kinase, it is distinct from the receptor tyrosine kinases in that it is located in the cytoplasm, rather than the membrane. Publication in 1997 of the crystal structure of the human *c-src* gene product revealed clues that should help us understand how this protein functions in normal metabolism. Note that by convention, the name of the gene (*src*) is italicized, while the name of the corresponding protein (Src) is not.

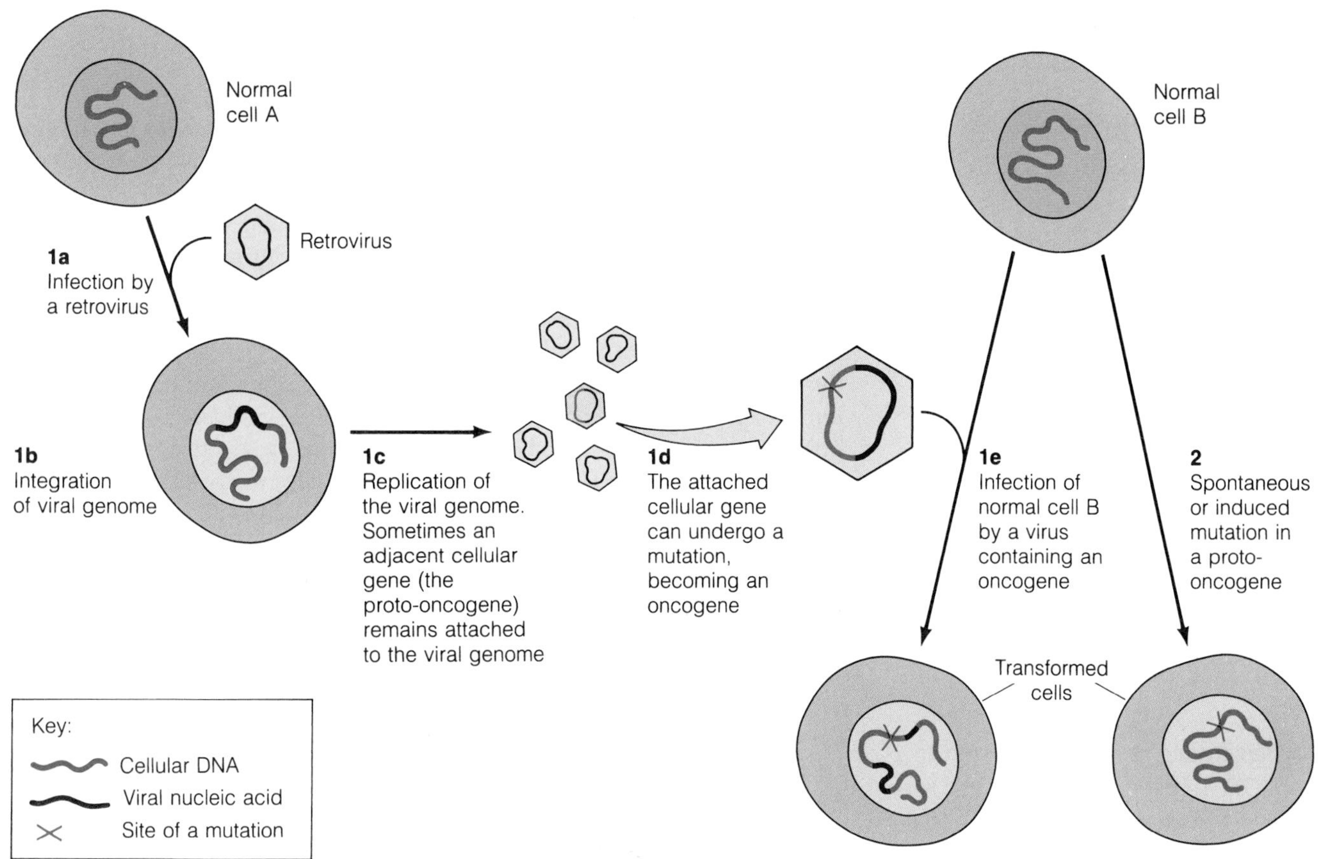

FIGURE **23.19**

Pathways by which proto-oncogenes can become oncogenes. A proto-oncogene is a normal cellular gene that can be converted to an oncogene and cause transformation to a cancer cell. This process can occur in two ways: (1) infection by a retrovirus, which integrates into a chromosomal site next to a proto-oncogene and carries that gene along in its own genome when the virus replicates, or (2) mutation of the cellular proto-oncogene. In the first case, once cellular DNA becomes part of a viral genome, it can undergo mutation that converts the proto-oncogene to an oncogene. The oncogene can then cause transformation when this virus infects another cell.

TABLE **23.7 Oncogene products as elements of signal transduction pathways**

Signal Transduction Element	Oncogene	Isolated from	Gene Product
Growth factors	*sis*	Retrovirus	Platelet-derived growth factor
Growth factor receptors	*erbB, neu*	Retrovirus	Epidermal growth factor receptor
	fms	Retrovirus	Colony-stimulating factor 1 receptor
	trk	Tumor	Nerve growth factor receptor
	ros	Retrovirus	Insulin receptor
	kit	Retrovirus	PDGF receptor
	flg	Retrovirus	Fibroblast growth factor receptor
Intracellular transducers	*src*	Retrovirus	Protein tyrosine kinase
	abl	Retrovirus	Protein tyrosine kinase
	raf	Retrovirus	Protein serine kinase
	gsp	Tumor	G protein α subunit
	ras	Tumor, retrovirus	GTP/GDP-binding protein
Nuclear transcription factors	*jun*	Retrovirus	Transcription factor (AP-1)
	fos	Retrovirus	Transcription factor (AP-1)
	myc	Tumor, retrovirus	DNA-binding protein
	erbA	Retrovirus	Thyroid receptor

Source: Adapted from J. D. Watson, M. Gilman, J. Witkowski, and M. Zoller, *Recombinant DNA*, 2nd ed. (New York: Scientific American Books, 1992), p. 339. © 1992 James D. Watson.

ONCOGENES IN HUMAN TUMORS

Activated oncogenes, closely related to viral oncogenes, have been isolated from human tumors.

Because human tumor viruses were not known to exist, the relevance of viral oncogenes in animal viruses to an understanding of human cancer was not immediately apparent. That relevance came into sharp focus in the late 1980s, from work by Robert Weinberg and others on the isolation and analysis of transforming genes from human tumors. Weinberg isolated DNA from bladder cancer tissue and used it to **transfect** normal mouse fibroblasts (connective tissue precursor cells). That is, DNA was introduced into these cells, and transformed cells were isolated after outgrowth of the cells. DNA recovered from transformed cells was shown to contain human sequences. After additional rounds of transfection, the human DNA associated with the transformed mouse fibroblasts was sequenced. The transforming gene was shown to be nearly identical with a previously described oncogene from Harvey *ra*t *s*arcoma virus, called the ***H-ras*** gene. Sequence analysis showed the *H-ras* gene sequence to be identical to *c-ras,* its counterpart in untransformed cells, with but a single difference—a mutation in the twelfth codon that changed a glycine codon in *c-ras* to a valine codon in the oncogene isolated from tumor tissue. Thus, human tumors were shown to contain an oncogene that is present in some tumor viruses, and in an altered form that presumably activated the tumorigenic process.

The Ras protein, which is mutationally altered in many human tumors, is a GTP-binding protein involved in signal transduction from growth factor receptors on the plasma membrane to specific gene activations in the nucleus.

ras genes are now known to encode a family of proteins—all of about 21 kilodaltons, with regions homologous to sequences in the α subunit of G proteins. Like the α subunit, the Ras proteins bind guanine nucleotides. Normal Ras proteins possess a GTPase activity, as do G_α proteins, whereas most *ras* oncogene proteins lack this activity. The GTPase activity suggested that normal Ras proteins function like G proteins in regulating metabolism. Lending support to this model was the determination in 1988 of the three-dimensional structure of a Ras protein, crystallized as its complex with GDP (Figure 23.20). Amino acid residues known to be changed in mutations that generate *ras* oncogenes are positioned close to the bound guanine nucleotide. This positioning supports the idea that interactions between the proto-oncogene Ras protein and guanine nucleotides are important to metabolic control and that this control is lost when a normal cell is transformed to a cancer cell.

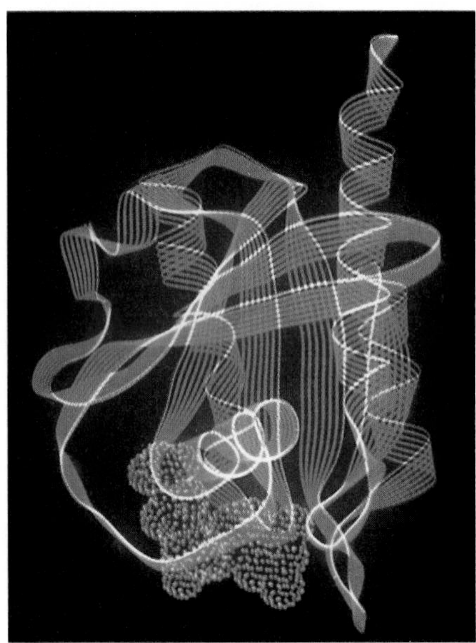

FIGURE **23.20**

Structure of a GDP–Ras protein complex. A polypeptide containing the N-terminal 171 residues of a 188-residue Ras protein was produced by gene cloning and crystallized. The ribbon represents the polypeptide chain, and the green structure the bound GDP.

From A. M. DeVos et al., *Science* (1988) 239:888–895, © 1988 by AAAS. By permission of AAAS and Sung-Hou Kim, University of California, Berkeley.

A major difference between Ras-type proteins and the related G_α proteins is the far higher GTPase activity of G_α proteins. As we will see shortly, a set of Ras-activating proteins is required to stimulate the GTPase activity of Ras. The basis for this difference was seen in 1994, with the first structural determination of a G_α protein (Figure 23.21). G_α, but not Ras, proteins contain a conserved arginine residue (R178), which interacts with the phosphates of bound GTP so as to stabilize the transition state for GTP hydrolysis.

Research on oncogenes has led to unifying theories of carcinogenesis. Work of Bruce Ames and others established that the great majority of chemical carcinogens are also mutagens. This finding suggested that chemical carcinogenesis involves mutagenesis of cellular proto-oncogenes, events that can occur in the absence of exogenous viruses (pathway 2 in Figure 23.19). Indeed, *ras* genes altered in codon 12, 13, or 61 have been detected in about 30% of spontaneous and chemically induced tumors, in both animals and humans.

A number of other genetic alterations have been detected in tumor tissue. Some of them involve **antioncogenes,** or **tumor suppressor genes.** Unlike proto-oncogenes, these are genes that in the normal form suppress tumorigenesis. Loss of normal gene function, as in a deletion, leads to tumor formation because of deficient tumor suppression. One of these genes is called the **retinoblastoma gene.** Mutations in the two alleles of this gene cause a type of eye tumor, which has a familial association. The other most prominent tumor suppressor gene encodes a protein called p53 (a protein of 53 kilodaltons). Loss of p53 function leads to tumorigenesis, and at least half of all human tumors examined display p53 gene

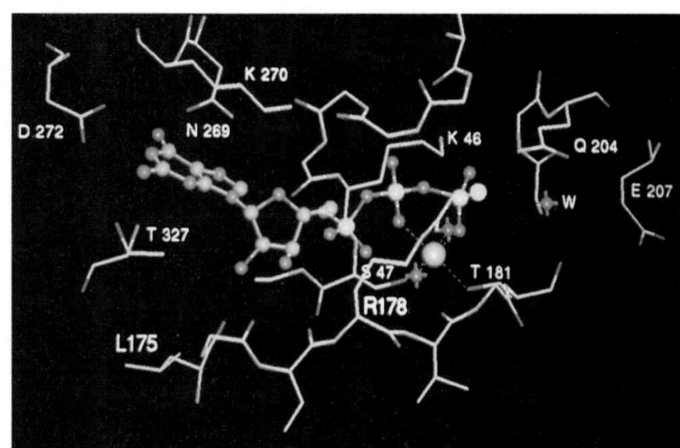

FIGURE **23.21**

GTP binding site on a G_α protein, $G_{i\alpha1}$. This figure shows bound GTPγS, a nonhydrolyzable GTP analog. Carbons on the bound ligand are in green, protein carbon in orange, nitrogen in blue, oxygen in red, sulfur and phosphorus in yellow, and magnesium in magenta. Stabilization of the phosphate negative charge by a conserved arginine (R178) facilitates hydrolysis of the bound nucleotide.

Courtesy of D. E. Coleman et al., *Science* (1994) 265:1405–1412, © 1994 by AAAS, kindly provided by S. R. Sprang.

mutations. Although its biochemical actions are not yet clear, we know that p53 is a DNA-binding protein that plays a role in regulating the cell cycle, preventing inappropriate movement of G1 cells into S phase. The loss of such a checkpoint could lead to the loss of cell growth control that we know as cancer. Binding to specific DNA sequences is essential to proper functioning of p53. This was revealed in 1994, with x-ray crystallographic determination of the structure of the DNA-binding domain in contact with an oligonucleotide containing the binding sequence (Figure 23.22). Not shown in the figure is the fact that binding the tetrameric p53 protein causes bending and twisting of the DNA, probably accounting for changes in gene expression. Of great interest is that the amino acid residues in closest contact with DNA are those that have been shown most often to be changed in p53 mutant genes isolated from human tumors.

By 1994, about a dozen antioncogenes had been identified. Mutations affecting most of these genes have been detected in the germ line of human cells and have been correlated with predisposition to become afflicted with certain forms of cancer. A particularly interesting antioncogene is *nm-23*, whose action somehow prevents metastasis, or the spread to new sites, of breast cancer. When the

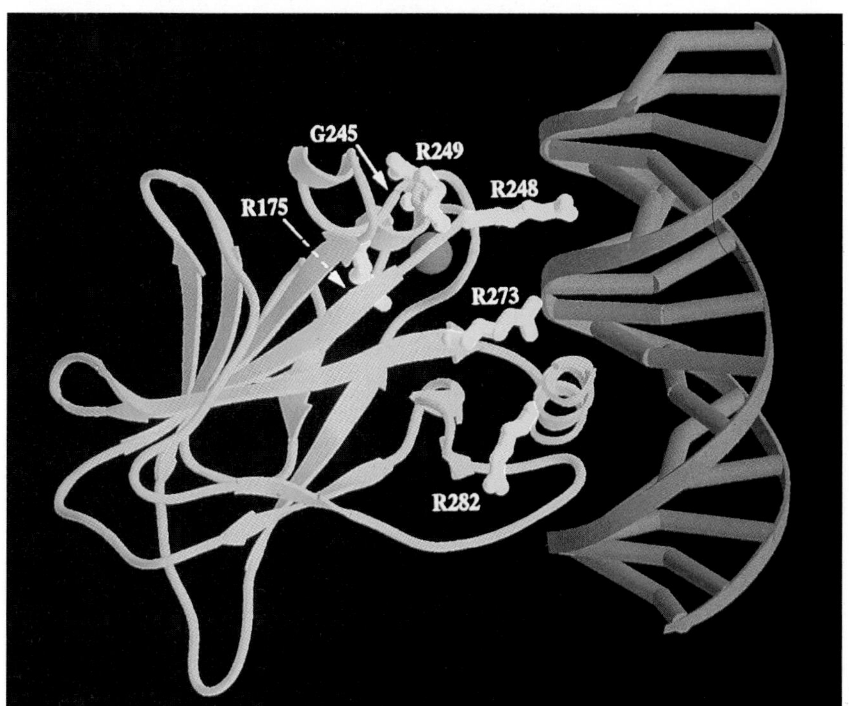

FIGURE **23.22**

Structure of the p53–DNA complex. This ribbon drawing shows the DNA-binding domain of one subunit of the homotetrameric p53 (green) complexed with an oligonucleotide pair containing the p53 binding site (blue). A bound zinc ion is shown in red. Shown in yellow are the six amino acid residues most often changed in mutant p53 proteins.

Courtesy of Y. Cho et al., *Science* (1994) 265:346–355, © 1994 by AAAS, kindly provided by N. P. Pavletich.

human form of this gene was sequenced, the gene product was identified as nucleoside diphosphate kinase. As noted in Chapter 22, this "housekeeping" enzyme plays an indispensable role in synthesizing nucleoside triphosphates from the corresponding diphosphates. How this protein also prevents the spread of breast tumors is currently the subject of intense research.

Analysis of a series of human tumors, varying in virulence and invasiveness, has suggested a sequence of genetic alterations leading from a precancerous lesion to a full-blown metastatic tumor. Our understanding is most complete for colorectal cancer. Many individuals develop benign polyps within the colon. In a condition called familial adenomatous polyposis (FAP) an inherited defect in a tumor suppressor gene called *APC* causes the formation of such a large number of benign polyps that a few of them inevitably progress toward cancer; more than 95% of individuals with this mutation will develop colorectal cancer. Another condition, called hereditary nonpolyposis colorectal cancer (HNPCC), involves a lower but still high (70%) risk of developing colorectal cancer. HNPCC-affected individuals are defective in DNA mismatch repair (Chapter 25), a process that corrects replication errors and other forms of DNA damage. Because this activity lowers the spontaneous mutation rate, its absence increases the chance that mutation of proto-oncogenes to oncogenes, such as *ras* or p53 mutations, will occur. Why mutations in these particular genes lead to cancer in specific organs, such as the colon or bladder, is not yet clear. As these examples illustrate, cancer is seen as a genetic disease, with the ultimate transformation of normal cells to metastatic tumor cells being an accumulation of individual genetic changes, many of them affecting elements of the signal transduction machinery.

Human tumors contain a series of mutations, affecting both signal transduction components and tumor suppressor genes and gene products.

ONCOGENES AND CELL SIGNALING

In the early 1990s several lines of work converged upon the realization that the Ras protein occupies a central role in an evolutionarily conserved pathway that conveys extracellular signals to the nucleus, where specific genes are activated for cell growth, division, and differentiation. As details of that pathway emerge, they fit into a unified framework that interrelates the biochemical properties of proto-oncogene products, underscores the dominance of protein phosphorylation as a control mechanism, and rationalizes the types of mutations that lead to cancer.

Ras-related proteins have been discovered in such diverse organisms as yeast, nematode worms, and *Drosophila,* in which they control aspects of mitotic and meiotic growth and embryonic development. Research on these organisms has illuminated a central control pathway in mammalian cells (Figure 23.23), in the process richly justifying the use of simple biological model systems for cancer research.

We now know that many growth factor receptors with tyrosine kinase activity phosphorylate themselves (steps 1 and 2 in Figure 23.23). In the phosphorylated state, each receptor interacts with one or more protein exchange factors, which in turn activate Ras by stimulating the GDP–GTP exchange (step 3). Also interacting with Ras, and limiting its activity, are GTPase-activating proteins (GAPs, see page 850). Downstream of Ras is a cascade of further protein phosphorylations (step 4), which ultimately activates transcription factors (step 5). Transcription factors are proteins that, like nuclear receptors, stimulate transcription of specific genes (see Chapter 28). These proteins interact with the genome and stimulate the expression of particular genes.

As noted in Figure 23.23, one class of downstream kinases is called **MAP kinase** **(MAPK)**. The acronym *MAP* stands for mitogen-activated protein. (A mitogen is a factor that stimulates mitosis.) Upstream of MAP kinases are another family of proteins, MAPKK (MAP kinase kinase), and still upstream are the MAPKKK proteins (MAP kinase kinase kinase). One member of the MAPKKK class is the product of the *raf* proto-oncogene (see Table 23.7).

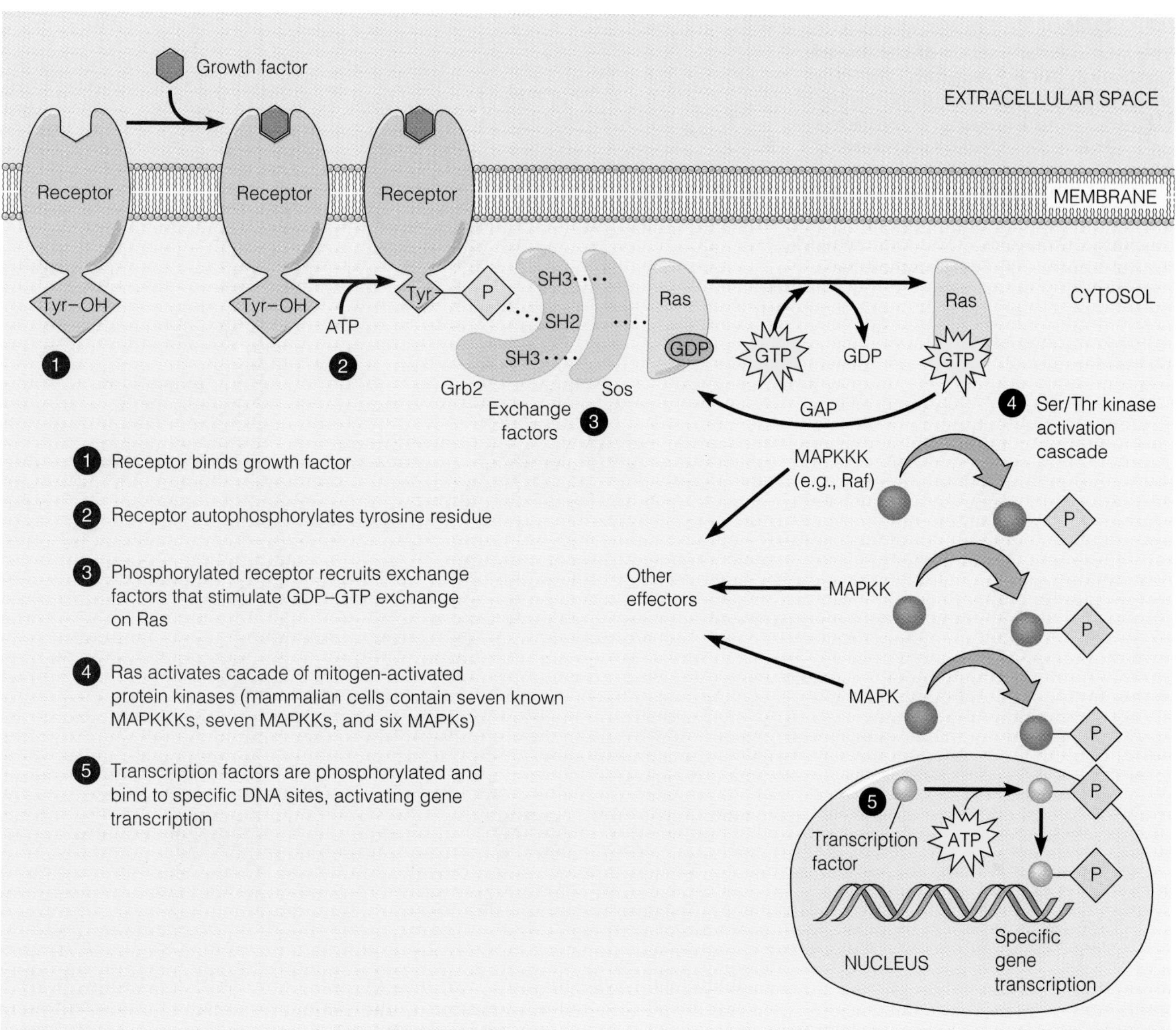

FIGURE **23.23**

Role of Ras protein in a central growth factor activation pathway. Binding of a mitogen or growth factor to a receptor stimulates the receptor tyrosine kinase (RTK), usually with dimerization of the receptor. Autophosphorylation of RTK leads to recruitment of a series of proteins (exchange factors) that stimulate GDP–GTP exchange on Ras and thereby activate it. In the scheme shown, a protein called Grb2 interacts via SH2 domains with RTK and via SH3 domains with exchange factor Sos. Ras then activates proteins of a family called MAPKKK (mitogen-activated protein kinase kinase kinase), which in turn leads to activation of MAPKK proteins and then MAPK proteins. Eventually, specific transcription factors in the nucleus undergo phosphorylation, which activates them to bind specific DNA sites and activate transcription of particular genes. The activity of Ras is curtailed by a series of GTPase-activating proteins (GAPs) that promote the hydrolysis of bound GTP.

Adapted from S. E. Egan and R. A. Weinberg, *Nature* (1993) 365:782. © 1993 Macmillan Magazines, Ltd.

Thus, we see the growth factor response as a cascade of protein phosphorylations, analogous to the well-known cascade that controls glycogen breakdown (Figure 13.18). We see how blocking the Ras GTPase activity can lead to uncontrolled cell growth and cancer, by keeping the signaling pathway turned on and

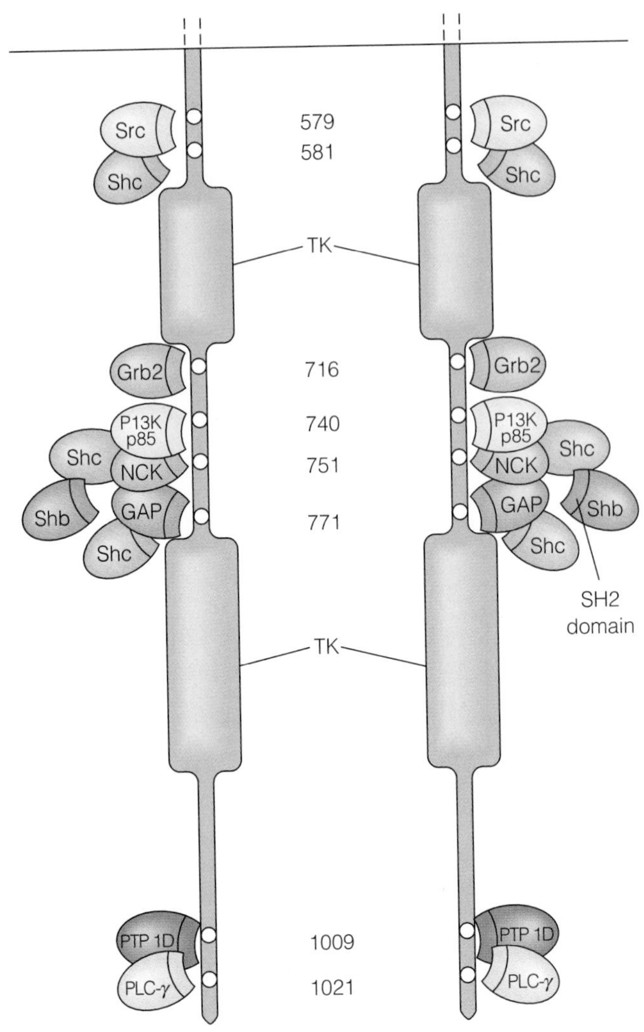

FIGURE **23.24**

The intracellular portion of the dimeric (activated) PDGF β receptor. The figure shows phosphorylation sites (residue number), two tyrosine kinase domains (TK) on each polypeptide chain, and sites for a number of proteins involved in the signal transduction process (not all of which are identified). The small boxed regions in each of these proteins represent SH2 domains. PI3K is a phosphatidylinositol 3-kinase. PLC-γ is a phospholipase C, and PTP 1D is a protein tyrosine phosphatase, which regulates the signal transduction process by removing phosphates from specific phosphotyrosine residues.

From L. Claesson-Welch, *J. Biol. Chem.* (1994) 269:32023–32026. Reprinted by permission of The American Society for Biochemistry & Molecular Biology.

flooding the cell with growth-stimulatory signals. What isn't immediately obvious is how this pathway can explain the distinct responses of cells to different growth factors. Analysis of the proteins that interact with autophosphorylated growth factor receptors has identified common sequences called **SH2 domains,** which are involved in these interactions (SH stands for src homology, relating to a sequence on the Src protein). A protein that binds to a phosphorylated receptor molecule recognizes not only the phosphotyrosine residue but also a critical amino acid sequence on the carboxyl side of that tyrosine. Thus, receptor phosphorylation–dephosphorylation serves as a general on–off switch, with the more subtle protein interactions governed by the SH2 domain. Downstream participants are recruited via another set of sequences called SH3 domains. These domains identify the specific proteins that will be recruited to the signaling process in response to a particular growth stimulus. As an example, Figure 23.24 shows phosphorylation sites and sites of signaling protein binding to one specific RTK, the β class of the PDGF receptor.

Plant Hormones

Our understanding of the molecular actions of plant hormones is less advanced than that of the vertebrate animal hormones. In part this is because some plant hormones are growth factors, so that growth is the only readily measurable parameter. Furthermore, plant membranes are more difficult to isolate and study than are animal membranes.

We encountered the five major classes of plant hormones in earlier chapters about metabolism. To recapitulate, the five classes, illustrated in Figure 23.25, are (1) the diterpene **gibberellins,** derived from isopentenyl pyrophosphate; (2) the sesquiterpene **abscisic acid,** also derived from isopentenyl pyrophosphate; (3) the **cytokinins,** purine bases with a terpenoid side chain; (4) **auxins,** with the tryptophan metabolite *indole-3-acetic acid* being the most active; and (5) **ethylene,** which comes from the methionyl moiety of *S*-adenosylmethionine. Thus, these compounds are chemically quite distinct from animal hormones.

Additional differences from animal hormones lie in (1) the diversity of effects of a given plant hormone and (2) the hormonal activity of a large number of structurally related species. For example, at least a dozen different cytokinins have been characterized structurally. Though all contain adenine with a side chain on N-6, great variability occurs in the structure of that side chain.

Auxins are synthesized in apical buds (at the tip) of growing shoots. They stimulate growth of the main shoot and inhibit lateral shoot development. A class of auxin-binding membrane proteins may represent auxin receptors. Auxin action involves chemiosmotic proton gradients, which in turn establish auxin concentration gradients leading to different responses in different parts of the plant—partial cell wall degradation, necessary for growth to occur, and increased RNA and protein synthesis, necessary for differentiation. Recent evidence suggests that cyclic AMP is formed in plants and that it mediates the effects of auxin.

The other classes of plant hormones are harder to describe in biochemical terms. Cytokinins are produced in the roots and promote growth and differentiation in many tissues. Cytokinins and auxins work together, and the ratio of cytokinin to auxin is often crucial in determining whether a plant will grow or differentiate. In many cases, single plant cells in tissue culture can regenerate whole plants. Here, the investigator must experiment with different proportions of cytokinins and auxins, searching empirically for the proportions that allow the best growth of normal plants.

About 100 forms of gibberellin are known. Some function as growth-promoting hormones and may stimulate expression of certain genes. Particular messenger RNAs are present in increased amounts as a result of gibberellin administration, supporting the idea that gibberellins act at the gene level.

Ethylene is considered a hormone of senescence. It stimulates fruit ripening and the aging of flowers, and it inhibits seedling growth. It also redirects auxin transport to promote transverse, rather than longitudinal, growth of plants.

Abscisic acid (ABA) counteracts the effects of most other plant hormones. It inhibits germination, growth, budding, and leaf senescence and helps plants adjust to stresses such as cold or drought. Abscisic acid synthesis in leaves is stimulated by wilting. Physiological evidence suggests a role for abscisic acid in ion and water balance.

Recent developments have suggested that hormonal control in plants is closer to that in animals than was previously suspected. Ethylene and abscisic acid have both been shown to function via signaling pathways that involve protein phosphorylation. One of these proteins in the ethylene pathway is homologous with a class of protein histidine kinases known as "response regulators" in bacteria and lower eukaryotes. The ABA pathway has been shown to involve a signaling component called cyclic ADP-ribose (cADPR), already known to regulate calcium release in animal cells. As in the animal system, calcium release by cADPR evidently activates a downstream protein kinase. Even more unexpected was the recent discovery of a plant protein highly homologous with mammalian steroid 5α reductases, involved in androgen synthesis (Chapter 19). Plants lacking this protein can have their growth defect reversed when treated with a steroid called brassinolide.

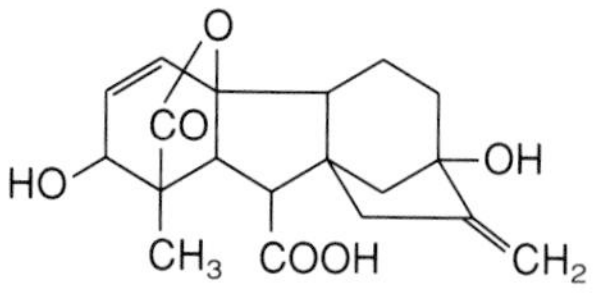

Gibberellic acid (GA3) [a gibberellin]

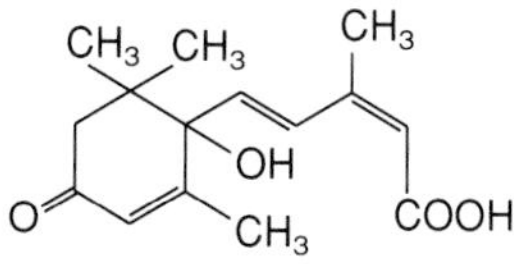

Abscisic acid (ABA)

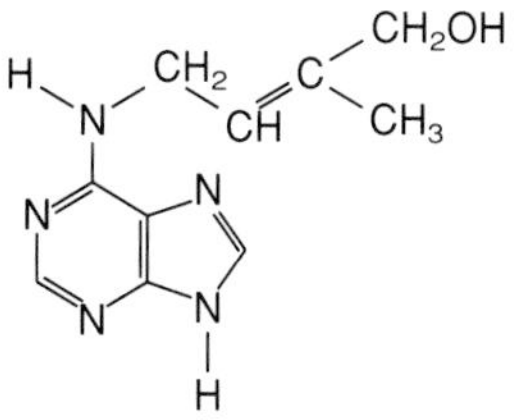

Zeatin [a cytokinin]

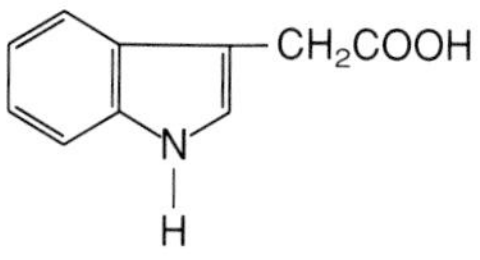

Indole-3-acetic acid (IAA) [an auxin]

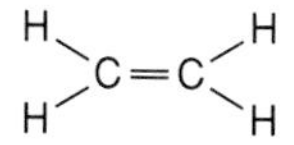

Ethylene

FIGURE **23.25**

Representatives of the five major classes of plant hormones.

These observations suggest that brassinolide or a close relative is a plant growth hormone and that the reductase-like protein is involved in its synthesis.

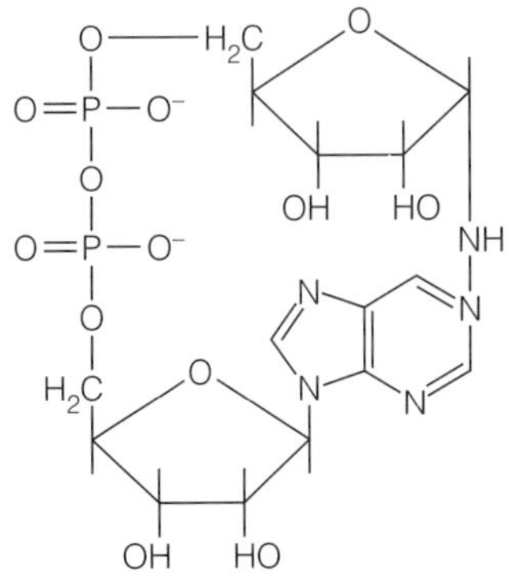

Cyclic ADP-ribose

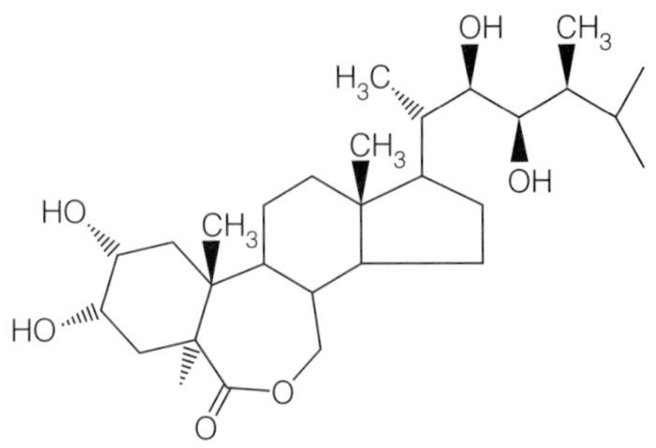

Brassinolide

SUMMARY

Each organ or tissue of a multicellular organism has a distinctive profile of metabolic activities that allow it to serve its specialized functions. Remote tissues must remain in constant communication in order to maintain homeostasis. In vertebrates the most essential element of this homeostasis is maintenance of constant blood glucose levels, primarily for proper brain function. The actions of three hormones—insulin, glucagon, and epinephrine—play the dominant roles in contributing to glucose homeostasis. Insulin signals the fed state and promotes glucose utilization and synthesis of energy storage compounds. Glucagon acts primarily upon liver cells, increasing blood glucose by several mechanisms involving cyclic AMP. Epinephrine has similar effects on muscle cells.

Hormone action is one element of signal transduction mechanisms, processes by which signals are transmitted from cell to cell. Some hormones interact with intracellular receptors; these act at the gene level, with the hormone–receptor complex affecting transcription of specific genes in target tissue. Other hormones interact with plasma membrane–bound receptors. Three types of such transmembrane receptors exist: (1) receptors that are ion channels, with hormone binding directly affecting membrane permeability to an ion; (2) receptors, such as the insulin receptor, that have a hormone-binding site on the exterior and an enzyme activity on the cytosolic side, with ligand binding stimulating that activity; and (3) receptors that act through G proteins to affect the levels of second messengers, including cyclic AMP, cyclic GMP, calcium ion, inositol trisphosphate, and diacylglycerol. Second messengers interact with a variety of intracellular metabolic processes. G protein action is controlled by binding of guanine nucleotides. Stimulation by interaction with receptor causes GTP exchange for GDP, and hormone action is limited by the slow conversion of bound GTP to GDP.

Proto-oncogenes are cellular genes (most of which encode signal transduction proteins) that have been transferred to viral genomes and that subsequently undergo mutations that convert them to oncogenes. Action of an oncogene, whether generated in a cell by viral infection or by mutation of a proto-oncogene, causes loss of metabolic and growth control that is associated with transformation of a normal cell to a cancer cell.

Plant hormones include gibberellins, cytokinins, auxins, ethylene, and abscisic acid. Recent evidence indicates parallels between the actions of animal and plant hormones, although we know much less about plant hormones.

REFERENCES

General

Czech, M. P., and S. Corvera (1999) Signaling mechanisms that regulate glucose transport. *J. Biol. Chem.* 274:1865–1868. One of several JBC minireviews on the stimulation by insulin of glucose transport.

Ember, L. R. (1998) Surviving stress. *Chem. Eng. News,* May 25 issue. A timely news-type article that describes biochemical stress responses in the context of Gulf War syndrome and the delayed stress experienced by some combatants.

Newgard, C. B., and J. D. McGarry (1995) Metabolic coupling factors in pancreatic β-cell signal transduction. *Annu. Rev. Biochem.* 64: 689–719. This review discusses how the pancreas senses blood glucose levels and adjusts its insulin output accordingly.

Protein Phosphorylation and Dephosphorylation

Elion, E. A. (1998) Routing MAP kinase cascades. *Science* 281:1625–1626. A news article providing perspective on two research papers published in the same issue, which describe the roles of adapter proteins in assembling information highways in cells.

Fantl, W. J., D. E. Johnson, and L. T. Williams (1993). Signalling by receptor tyrosine kinases. *Annu. Rev. Biochem.* 62:453–482. A review of actions of the insulin receptor and related transmembrane proteins.

Hafen, E. (1998) Kinases and phosphatases—A marriage is consummated. *Science* 280:1212–1213. Physical complexes between protein kinases and phosphatases facilitate the regulation of signal transduction pathways involving protein phosphorylation and dephosphorylation.

Hunter, T. (1995) Protein kinases and phosphatases: The yin and yang of protein phosphorylation and signaling. *Cell* 8D:225–238. The January 27 issue contains nine important reviews on signal transduction.

Mochly-Rosen, D. (1995) Localization of protein kinases by anchoring proteins: A theme in signal transduction. *Science* 268:247–251. One of six timely articles in a special issue devoted to signal transduction.

Synthesis of Peptide Hormones

Fisher, J. M., and R. H. Scheller (1988) Prohormone processing and the secretory pathway. *J. Biol. Chem.* 263:16515–16518. A review that describes how peptide hormones are formed by cleavage from high-molecular-weight precursors.

Receptors

Black, J. (1989) Drugs from emasculated hormones: The principle of syntopic antagonism. *Science* 245:486–493. Black's Nobel Prize address, which describes the development of drugs that are adrenergic receptor antagonists.

Lin, Y.-J., L. Seroude, and S. Benzer (1998) Extended life-span and stress resistance in the *Drosophila* mutant *methusaleh. Science* 282:943–946. Evidence that a serpentine receptor is involved in determining life span.

Marshall, C. J. (1995) Specificity of receptor tyrosine kinase signaling: Transient versus sustained extracellular signal-related kinase activities. *Cell* 8D:179–185. Another contemporary review from the January 27 issue.

Pitcher, J. A., N. J. Freedman, and R. J. Lefkowitz (1998) G protein–coupled receptor kinases. *Annu. Rev. Biochem.* 67:653–692. A recent review describing this down-regulatory mechanism.

Strader, C. D., T. M. Fong, M. R. Tota, D. Underwood, and R. A. F. Dixon (1994) Structure and function of G protein–coupled receptors. *Annu. Rev. Biochem.* 63:101–132. A comprehensive recent review.

Vaughan, M. (1998) G protein-coupled receptors minireview series, and Signaling by heterotrimeric G proteins minireview series. *J. Biol. Chem.* 273:17297–17298 and 667–668. JBC minireviews are compiled at the end of each year into a "minireview compendium." The 1998 compendium contains 11 minireviews pertaining to G protein–coupled signal transduction systems. The articles by Dr. Vaughan are brief prologues to one series of these reviews.

G Proteins

Scheffzek, K., M. R. Ahmadian, and A. Wittinghofer (1998) GTPase-activating proteins: Helping hands to complement an active site. *Trends Biochem. Sci.* 23:257–262. Crystal structures of several GAPs show how they interact with and activate the GTPase of Ras-related proteins.

Snyder, S. H., P. B. Sklar, and J. Pevsner (1988) Molecular mechanisms of olfaction. *J. Biol. Chem.* 263:13971–13975. This minireview describes evidence for G protein involvement in the sense of smell.

Sprang, S. R., and D. E. Coleman (1998) Invasion of the nucleotide snatchers: Structural insights into the mechanism of G protein GEFs. *Cell* 95:155–158. How proteins stimulate the GDP–GTP exchange in G protein activation.

Tesmer, J. J. G., R. K. Sunahara, A. G. Gilman, and S. R. Sprang (1997) Crystal structure of the catalytic domains of adenylyl cyclase in a complex with $G_{s\alpha}$-GTPγS. *Science* 278:1907–1916. Structural analysis of G proteins reveals the mechanism of adenylyl cyclase activation.

Wedegaertner, P. B., P. T. Wilson, and H. R. Bourne (1995) Lipid modifications of trimeric G proteins. *J. Biol. Chem.* 270:503–506. This structural feature is missing from the recombinant G proteins used for crystallographic analysis.

Second-Messenger Systems

Berridge, M. (1993) Inositol trisphosphate and calcium signalling. *Nature* 361:315–325. The role of calcium as a second and/or third messenger.

Hodgkin, M. N., T. R. Pettit, A. Martin, R. H. Michell, A. J. Pemberton, and M. J. O. Wakelam (1998) Diacylglycerols and phosphatidates: Which molecular species are intracellular messengers? *Trends Biochem. Sci.* 23:200–204. A short recent review of lipid-derived second messengers.

Hurley, J. H. (1999) Structure, mechanism, and regulation of adenylyl cyclase. *J. Biol. Chem.* 274:7599–7602. This review is particularly timely in view of recent structural insights into adenylate cyclase.

Lohmann, S. M., A. B. Vaandrager, A. Smolenski, U. Walter, and H. R. De Jonge (1997) Distinct and specific functions of cGMP-dependent protein kinases. *Trends Biochem. Sci.* 22:307–312. Structure–function analysis of this family of enzymes.

Majerus, P. W., M. V. Kisseleva, and F. A. Norris (1999) The role of phosphatases in inositol signaling reactions. *J. Biol. Chem.* 274:10669–10672. Recent information about the control of phosphoinositide synthesis and turnover.

Singer, W. D., H. A. Brown, and P. C. Sternweis (1997) Regulation of eukaryotic phosphatidylinositol-specific phospholipase C and phospholipase D. *Annu. Rev. Biochem.* 66:475–509. A review of the synthesis of second messengers from phosphoinositides.

Nuclear Receptors

Auwerx, J., and 39 coauthors (1999) A unified nomenclature system for the nuclear receptor superfamily. *Cell* 97:161–163. Forty leaders in this field propose a unifying classification scheme for this ever-growing family of receptors, with references to other review literature.

Schwabe, J. W. R., and D. Rhodes (1991) Beyond zinc fingers: Steroid hormone receptors have a novel structural motif for DNA recognition. *Trends Biochem. Sci.* 16:291–297. This review summarizes NMR studies on the mechanism of DNA binding of this family of receptors.

Tsai, M. J., and B. W. O'Malley (1994) Molecular mechanisms of action of steroid/thyroid receptor superfamily members. *Annu. Rev. Biochem.* 63:451–486. The most comprehensive recent review of steroid receptors.

Receptor Tyrosine Kinases

Claesson-Welsh, L. (1994) Platelet-derived growth factor receptor signals. *J. Biol. Chem.* 269:32023–32036. A discussion of all of the proteins known to interact with PDGF receptors.

Oncogenes and Growth Factors

Birge, R. B., and H. Hanafusa (1993) Closing in on SH2 specificity. *Science* 262:1522–1524. A minireview describing how SH2 domains explain the specificity of cellular responses to growth factors.

Burley, S. K. (1994) p53: A cellular Achilles' heel revealed. *Structure* 2:789–792. A readable minireview describing the excitement created by the structural determination of the p53 DNA-binding domain.

Kinzler, K. W., and B. Vogelstein (1998) Landscaping the cancer terrain. *Science* 280:1036–1037. The scientists who described genetic changes in colorectal tumor progression classify the genes involved.

Knudson, A. G. (1993) Antioncogenes and human cancer. *Proc. Natl. Acad. Sci. USA* 90:10914–10921. A review of tumor suppressor genes.

Massagué, J. (1998) TGF-β signal transduction. *Annu. Rev. Biochem.* 67:753–791. A review describing the signaling pathways that involve receptor serine/threonine kinases.

Pawson, T., and T. M. Saxton (1999) Signaling networks—Do all roads lead to the same genes? *Cell* 97:675–678. The profusion of signaling molecules and pathways cries out for order, which the authors of this minireview seek to find.

Temin, H. M. (1988) Evolution of cancer genes as a mutation-driven process. *Cancer Res.* 48:1697–1701. A Nobel laureate reviews the relationship between proto-oncogenes and oncogenes and the actions of oncogenes.

Watson, J. D., M. Gilman, J. Witkowski, and M. Zoller (1992) *Recombinant DNA,* 2nd ed., Scientific American Books, New York. Chapter 18 is an excellent review of the experiments used to identify and characterize oncogenes and tumor suppressor genes.

Wittinghofer, A., and N. Nassar (1996) How Ras-related proteins talk to their effectors. *Trends Biochem. Sci.* 21:488–491. A discussion of the downstream effects in Ras signaling pathways.

Xu, W., S. C. Harrison, and M. J. Eck (1997) Three-dimensional structure of the tyrosine kinase c-Src. *Nature* 385:595–602. The structure revealed how interaction with other proteins could activate the protein kinase activity of the Src protein.

Plant Hormones

Heldt, H.-W. (1997) *Plant Biochemistry and Molecular Biology,* pp. 394–414. Oxford University Press, New York. Chapter 19 of this book contains a comprehensive treatment of plant growth hormones and their actions.

Johnson, P. R., and J. R. Ecker (1998) The ethylene gas signal transduction pathway. *Annu. Rev. Genet.* 32:227–254. Evidence that ethylene functions via a protein kinase pathway.

Jones, A. M. (1998) Auxin transport: Down and out and up again. *Science* 282:2201–2202. A minireview summarizing evidence for the chemiosmotic model of auxin action.

Pennisi, E. (1997) Plants decode a universal signal. *Science* 278:2054–2055. A news-type article, accompanying a research paper in the same issue, which describes the use of cyclic ADP-ribose as a second messenger in plants.

Russell, D. W. (1996) Green light for steroid hormones. *Science* 172: 370–371. A news-type article, accompanying a research paper in the same issue, which describes the likely use of steroid hormones by plants.

PROBLEMS

1. Marathon runners preparing for a race engage in "carbo loading" to maximize their carbohydrate reserves. This involves eating large quantities of starchy foods. Why is starch preferable to candy or sugar-rich foods?
2. Supposing that an average human consumes energy at the rate of 1500 kcal/day at rest and that long-distance running consumes energy at 10 times that rate, how long would the glycogen reserves last during a marathon run? Recall that 1 kcal is equivalent to 4.183 kJ.
3. What proportion of the total energy consumption supports brain function in an average resting human? What proportion in a human running in a marathon?
4. Proteolysis increases during the early phases of fasting, but later it decreases as the body adapts to using alternative energy sources. Given that feedback control mechanisms have not been described for intracellular proteases, how might you explain these apparent changes in protease activity?
5. Glucose has been found to react nonenzymatically with hemoglobin, through Schiff base formation between C-1 of glucose and the amino termini of the β chains. How might this finding be applied in monitoring diabetic patients?
6. Ketone bodies are exported from liver for use by other tissues. Because many tissues can synthesize ketone bodies, what enzymatic property of liver might contribute to its special ability to export these compounds?
7. *G proteins show some amino acid sequence homology with certain bacterial proteins that are involved in protein synthesis. These proteins, called elongation factors, also bind and hydrolyze GTP. How might one determine whether plant tissues contain G proteins or similar entities?
8. List two or three factors that make it advantageous for peptide hormones to be synthesized as inactive prohormones that are activated by proteolytic cleavage.
9. *Name a hormone whose concentration would probably not be amenable to analysis by radioimmunoassay, and explain your answer. (Hint: You must be able to generate an antibody to the hormone to develop such an assay.)
10. Describe a mechanism by which a steroid hormone might act to increase intracellular levels of cyclic AMP.

11. Small molecules interact with cells through specific macromolecular receptors. For each of the four receptors identified below, list all characteristics, by number, which accurately describe that receptor.
    (a) An adrenergic receptor ___________
    (b) A steroid receptor ___________
    (c) The LDL receptor ___________
    (d) The insulin receptor ___________
    1. Located at the cell surface
    2. Associated with the protein clathrin
    3. Ligand binding stimulates the activity of phospholipase C
    4. A transmembrane protein
    5. A DNA-binding protein
    6. Located in the cell interior
    7. Receptor–ligand complex moves to the lysosome
    8. Receptor–ligand complex becomes concentrated in the nucleus
    9. Receptor activation can inhibit the synthesis of glycogen
    10. The hormone–receptor complex activates specific gene transcription
    11. Internalization decreases the synthesis of cholesterol esters
    12. Action of this receptor diminishes the synthesis and activity of β-hydroxy-β-methylglutaryl-CoA reductase (HMG-CoA reductase)
    13. This receptor activates its own synthesis
    14. Biological activity of this receptor involves interaction with guanine nucleotide–binding proteins
    15. This receptor has a protein kinase activity
    16. Not known to act through a second messenger

*12. Upon activation by a receptor, a G protein exchanges bound GDP for GTP, rather than phosphorylating GDP that is already bound. Similarly, the α subunit–GTP complex has a slow GTPase activity that hydrolyzes bound GTP, rather than exchanging it for GDP. Describe experimental evidence that would be consistent with these conclusions.

13. Adipose tissue cannot resynthesize triacylglycerols from glycerol released during lipolysis (fat breakdown). Why not? Describe the metabolic route that is used to generate a glycerol compound for triacylglycerol synthesis.

14. (a) Briefly describe the relationship between intracellular malonyl-CoA levels in the liver and the control of ketogenesis. (b) Describe how the action of glucokinase helps the liver to buffer the level of blood glucose.

15. The action of glucagon on liver cells leads to inhibition of pyruvate kinase. What is the most probable mechanism for this effect?

16. Lithium ion inhibits the synthesis of inositol trisphosphate by inhibiting a reaction in the breakdown of inositol trisphosphate. Explain this apparent paradox.

17. Describe two general chemical features that distinguish plant hormones from animal hormones.

TOOLS OF BIOCHEMISTRY 23A

Radioimmunoassay

Because most hormones are present at exceedingly low concentrations in biological materials (about 10^{-10} M), they cannot be quantified by standard methods, such as fluorescence measurements or high-performance liquid chromatography (HPLC). Moreover, a hormone is usually surrounded by excessive amounts of chemically related substances, which would interfere with standard analytical techniques.

Radioimmunoassay (RIA), a powerful technique developed in the 1960s by Solomon Berson and Rosalyn Yalow, combines the specificity of antigen–antibody reactions with the sensitivity of radioisotopic detection. This technique allows the detection of any substance to which an antibody can be prepared by injection of an antigen into an experimental animal. (Most low-molecular-weight hormones must be conjugated with a polypeptide to be suitably antigenic.) The injected antigen can then be detected in minute amounts in the presence of a huge excess of chemically similar substances, provided those substances do not interfere with the antigen–antibody reaction.

The principle of radioimmunoassay is quite simple, as shown in Figure 23A.1. One first determines the radioactivity bound to a small amount of antibody after that antibody is incubated with a large excess of radiolabeled antigen at a known specific activity. In this case the antigen is the hormone being assayed. To generate a standard curve, which relates reduction in bound radioactivity to amount of nonradioactive hormone or other antigen, one then incubates the same quantities of antibody and radiolabeled antigen with a known quantity of nonradioactive hormone. The unlabeled hormone competes with labeled hormone for binding sites on the antibody molecules. Thus, the bound radioactivity decreases. This operation is carried out with increasing amounts of nonradioactive hormone. The data are shown in Figure 23A.2. To determine the percentage of hormone bound, one then incubates antibody and radiolabeled antigen in the same proportions, along with an aliquot of the biological material containing an unknown quantity of hormone. The unlabeled antigen in this aliquot dilutes the bound radioactivity by an amount proportional to the nonradioactive antigen present. One uses the standard curve to determine that amount quantitatively.

It is necessary to separate bound from unbound antigen, so that the radioactivity counted comes only from bound antigen. This separation can be done by precipitating the antigen–antibody complexes, either with antiserum raised against the antibody being used, or with staph A protein (a cell surface protein from the bacterium *Staphylococcus aureus* that efficiently binds antigen–antibody complexes).

Radioimmunoassay is used to analyze many biological substances that are present in minute amounts, including steroids, cyclic nucleotides, peptide hormones, prostaglandins, and drugs.

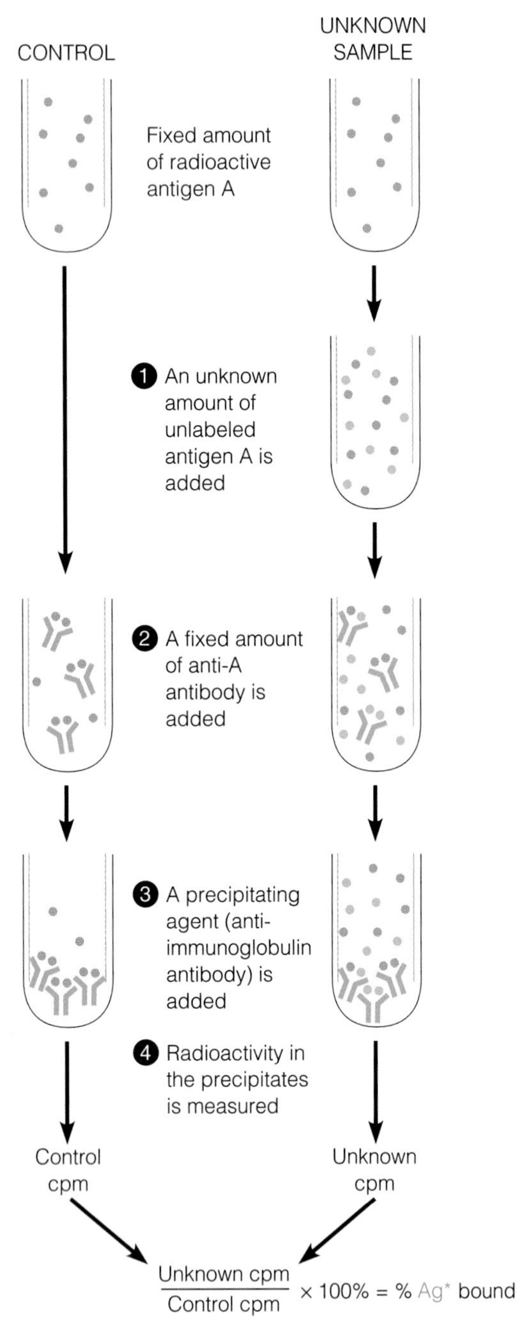

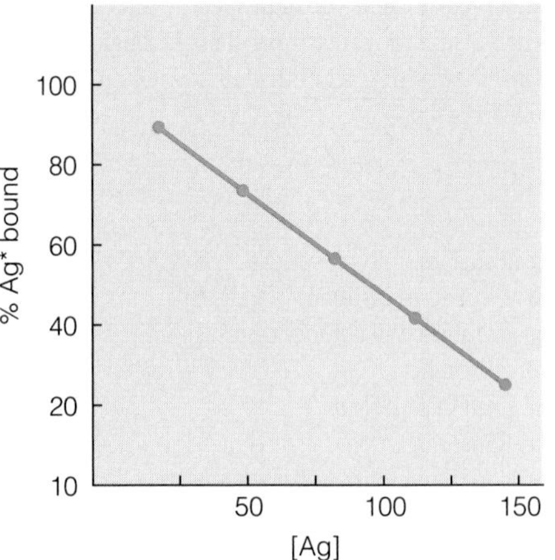

FIGURE **23A.2**

A standard curve for RIA. The curve is determined by varying the amount of unlabeled hormone in the presence of constant amounts of radiolabeled hormone and antibody. Ag = antigen.

In some applications, RIA has been supplanted by enzyme-linked immunosorbent assay (ELISA—see Tools of Biochemistry 7A).

References

Van Vunakis, H., and J. J. Langone, eds. (1980) Immunological methods. *Methods in Enzymology,* vol. 70. Academic Press, New York. A multi-authored manual with articles about all of the major immunological techniques.

Yalow, R. S. (1978) Radioimmunoassay: A probe for the fine structure of biological systems. *Science* 200:1236–1240. A description of the technique, written by one of its inventors.

FIGURE **23A.1**

Principle of radioimmunoassay. In the example shown, the amounts of labeled and unlabeled antigen A are equal, resulting in a 50% reduction in radioactive hormone bound. When radioimmunoassay is used to measure hormone levels, antigen A will be the hormone. Ag = antigen.

PART V

Information

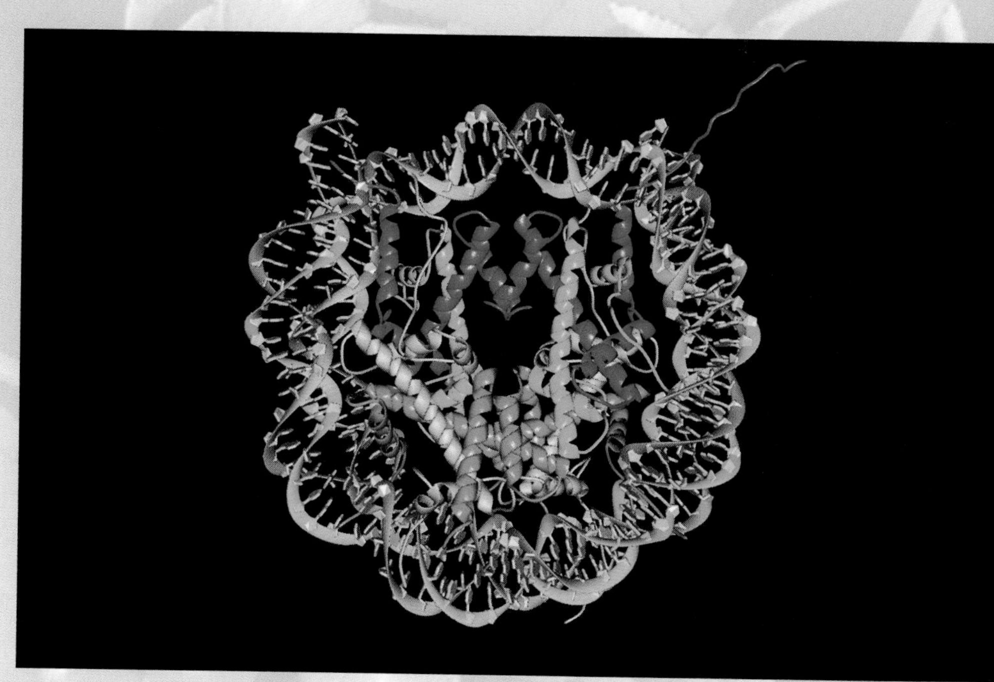

CHAPTER 24

Information Copying: Replication

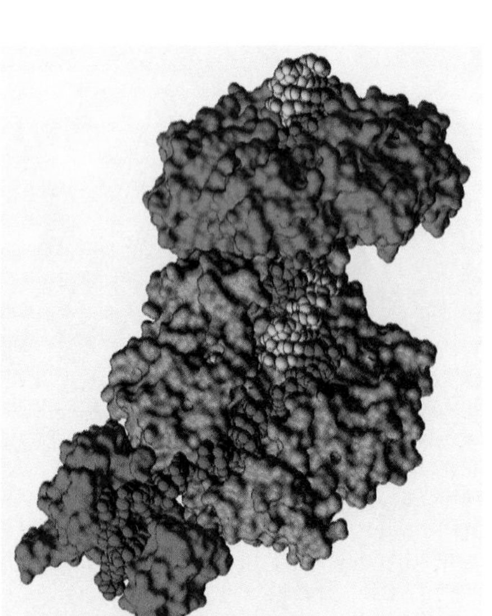

A DNA replication complex, with DNA polymerase (orange), processivity clamp (turquoise), and single-strand DNA-binding protein (green). The DNA strands are dark pink and gold. For more information, see Figure 24.17.

THE LAST MAJOR PART OF OUR STUDY OF BIOCHEMISTRY CONCERNS THE STORage, retrieval, processing, and transmission of biological information. The processes involved, which we call **information metabolism,** are distinguished from intermediary metabolism as follows. In intermediary metabolism, all the information specifying the nature of a reaction lies within the three-dimensional structure of the enzyme involved. That structure determines which substrates are bound and which reactions are catalyzed. Of course, all metabolic reactions are controlled ultimately by genetic information, which specifies the structures and properties of enzymes. However, the reactions we encounter from here on are distinguished by the direct involvement of genetic information—specifically, the requirement for a *template,* which functions along with the enzyme to specify the reaction catalyzed. The biological templates are nucleic acids—either DNA or RNA. In general, the template plays a passive role, determining which specific substrates will be bound, whereas the enzyme continues to specify the nature of the reaction, once substrates have been bound. An exception to this generalization is the involvement of RNA molecules as catalysts, in processes ranging from intron splicing to protein synthesis (see Chapters 11 and 27). In these next five chapters we discuss actions of proteins upon nucleic acid templates for storing, copying, modifying, and expressing biological information.

Information Metabolism: The Major Processes

Recall from Chapter 4 that the major processes in information metabolism are *replication,* in which DNA (or RNA in RNA viruses) serves as the template for its own synthesis; *transcription,* in which the information encoded in DNA specifies the structure of an RNA product; and *translation,* in which RNA serves as the template for synthesis of a particular polypeptide chain (see Figure 4.23 on page 108). We can compare these processes to the expression of verbal information encoded in a language. Replication and transcription use only the 4-letter nucleotide language of nucleic acids, whereas translation converts a message from the nucleotide language to the 20-letter amino acid language of proteins.

Replication, transcription, and translation are similar both in requiring a template and, in each case, consisting of three distinct subprocesses: initiation, chain elongation, and termination. In principle, the initiation phase presents the most efficient target for regulation of each overall process and, indeed, most regulatory processes occur at this level.

Biosynthesis of nucleic acids and proteins is regulated primarily at the initiation level.

In this chapter we focus upon replication, of both DNA and RNA genomes. Chapter 25 deals with several processes in which DNA is a substrate as well as a template—genetic recombination, repair, modification and restriction, transposition, and gene amplification. Chapters 26 and 27 focus on transcription and translation, respectively. These four chapters (24–27) describe both eukaryotic systems and the bacteria and viruses that gave our earliest insights, while Chapter 28 discusses some distinctive features of eukaryotic replication and gene expression.

An Overview of DNA Replication

Our presentation of DNA replication begins with a preview and summary. Figure 24.1 shows a simplified version of a replication fork and previews some of our major topics. We follow our overview with a more detailed exposition of the experimental evidence that established this view, plus a discussion of the individual proteins and the subprocesses in which they participate.

DNA replication is conceptually simple, but mechanistically complex. Simplicity was inherent in the proposal of Watson and Crick: that information transfer involves simply unwinding the strands of a parental DNA duplex, accompanied by synthesis of two complementary daughter strands. In other words, replication is **semiconservative,** giving rise to daughter DNA molecules containing one strand each of parental and of newly synthesized material (see margin). The density-labeling experiment of Meselson and Stahl demonstrated this central feature of DNA replication (see Figures 4.12–4.14, pages 96–98).

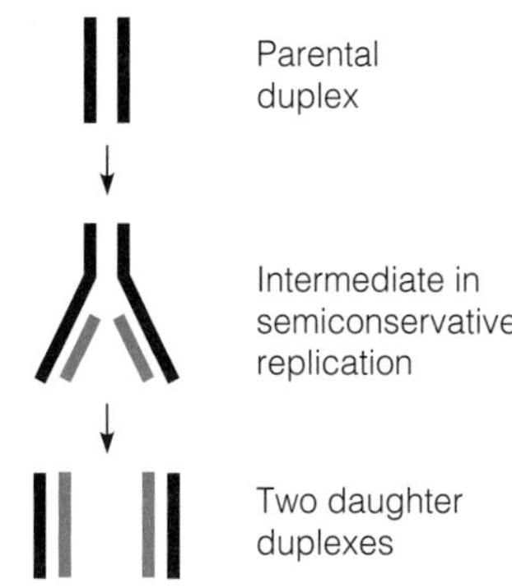

Although the Meselson–Stahl experiment provided crucial insight, important questions remained. First, DNA replication is coordinated with the cell's growth and division cycle, so that two dividing cells each receive one full complement of newly replicated DNA. How does this coordination occur? Second, DNA replication begins at fixed points on the chromosome. How are these **replication origins** recognized by the appropriate enzymes? Third, DNA replication is an orderly process, involving parental strand unwinding, incorporation of nucleotide precursors, and renaturation of replicated molecules; all of these processes occur within the same microenvironment, termed a **replication fork.** What is the energy source for duplex strand unwinding? How many proteins must function to carry out

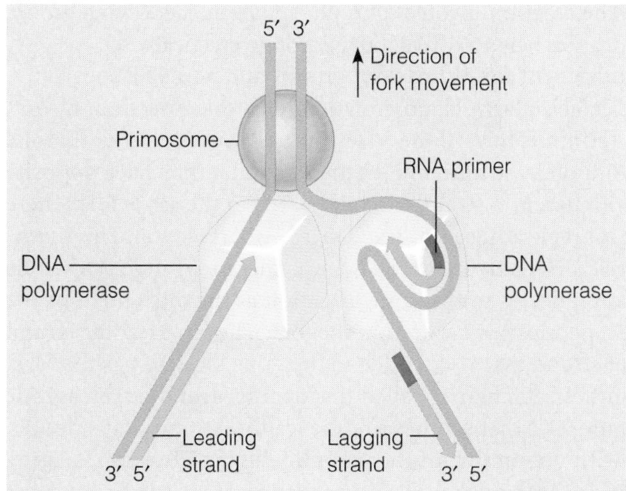

FIGURE **24.1**

Simplified view of a replication fork. This figure introduces several of the features of DNA replication that will be further developed in this chapter. Synthesis of the two new DNA strands (blue) occurs simultaneously with the unwinding of the parental DNA duplex (tan) to create the replication fork. DNA chain elongation is catalyzed by DNA polymerase. There are two molecules of this enzyme in the fork, one for each strand. As shown by the arrowheads, the two strands are replicated in opposite directions, because of the antiparallel nature of the strands in a duplex. One strand grows continuously, in the direction of movement of the fork, whereas the other (the one on the right) is synthesized in discontinuous segments, with a small piece of RNA (red) at the 5′ end of each segment. This RNA piece is later replaced with DNA, and the segments are joined. The unwinding of the parental duplex and the synthesis of the RNA pieces are catalyzed by a multiprotein assembly called the primosome.

these several steps, and how are their activities coordinated? Fourth, spontaneous mutation frequencies in nature are far lower than expected if Watson–Crick base pairing were the only factor specifying the base sequence of a daughter strand. What additional fidelity-enhancing mechanisms are involved? Finally, the bacterial and viral chromosomes whose replication was first studied are circular DNA molecules. Does replication proceed in one or in both directions? How are the two circular daughter duplexes disentangled from each other at the end of a round of replication?

DNA replication initiates from fixed origins and involves simultaneous synthesis of two daughter strands, each complementary in sequence and antiparallel to one of the parental strands.

Attempts to answer these questions have revealed several general features of DNA replication. First, replication is *semiconservative,* as already noted. Second, replication is *ordered* and *sequential.* It begins at fixed points on the chromosome, and DNA chain growth closely trails parental duplex unwinding. Third, like other biosynthetic processes, DNA replication uses *activated substrates,* the deoxyribonucleoside 5′-triphosphates (dNTPs). Fourth, DNA replication is *discontinuous;* the need to extend DNA chains of opposite polarity within one fork means that one strand grows in the direction of fork movement, and the other in the opposite direction. Finally, replication is far more *accurate* than any other enzyme-catalyzed process.

Of the three processes in replication—chain initiation, elongation, and termination—our understanding is most advanced for elongation, and that is what receives primary emphasis in this chapter. Figure 24.1 presents an idealized picture of a replication fork. The names and functions of the principal proteins involved are introduced here, and all are described in more detail later. This picture was developed initially through studies on *Escherichia coli* and several of its DNA bacteriophages, including T4 and several small phages with circular, single-strand DNA (ϕX174 and M13). Although we use names of some *E. coli* enzymes in this overview, this section presents details of replication that are common to both prokaryotic and eukaryotic systems.

The protein components known to function at or near the replication fork include *DNA polymerases, single-strand DNA-binding proteins, helicases, primase, topoisomerases,* and *DNA ligase,* all of which are introduced below and described in detail later. Some of these enzymes are themselves multisubunit protein assemblies. For example, the major replicative DNA polymerase in *E. coli* (DNA polymerase III) contains 10 different polypeptide subunits.

DNA polymerase catalyzes reaction between the 3′ hydroxyl of a DNA deoxyribonucleotide residue on a primer strand and the α phosphate of a dNTP that has base-paired to a template DNA strand.

DNA polymerase catalyzes the chemical reaction of DNA synthesis, the creation of phosphodiester bonds between deoxyribonucleotides in a DNA chain. The 3′ hydroxyl group at the end of a DNA chain carries out a nucleophilic attack on the α (innermost) phosphate of an incoming dNTP, as schematized in Figure 24.2. That dNTP has previously been positioned for incorporation into the growing, or **primer,** strand, by hydrogen bonding with the appropriate nucleotide in the **template** strand. The reaction involves pyrophosphate as the leaving group, whose subsequent hydrolysis helps to drive the reaction energetically.

The leading daughter DNA strand is elongated continuously in the same 5′→3′ direction as fork movement. The lagging strand is synthesized discontinuously; each fragment is synthesized 5′→3′, backward from the fork, but overall chain growth is 3′→5′, toward the fork.

Because of its requirement for 3′ hydroxyl groups and 5′ dNTP substrates, DNA polymerase can catalyze DNA chain growth in only one direction: 5′→3′, meaning from the 5′ terminus toward the 3′ terminus. However, recall that the two strands of a DNA duplex are *antiparallel,* meaning that they have opposite polarities. Because both parental strands are replicated in the same fork, there must be a mechanism for replicating each strand in a 3′→5′ direction. This mechanism involves a *dimeric* DNA polymerase (DNA polymerase III in *E. coli*)—one enzyme molecule that replicates in the same direction as the movement of the fork and another that replicates *backward* from the fork. The term **leading strand** identifies the daughter strand that is extended in the same direction as the fork, and the backward-synthesized chain is called the **lagging strand** or **retrograde strand.** Note from Figure 24.1 that synthesis of the leading strand (left side) occurs simultaneously with unwinding of the parental duplex. However, lagging strand synthesis (right side) can occur only after a section of parental duplex has

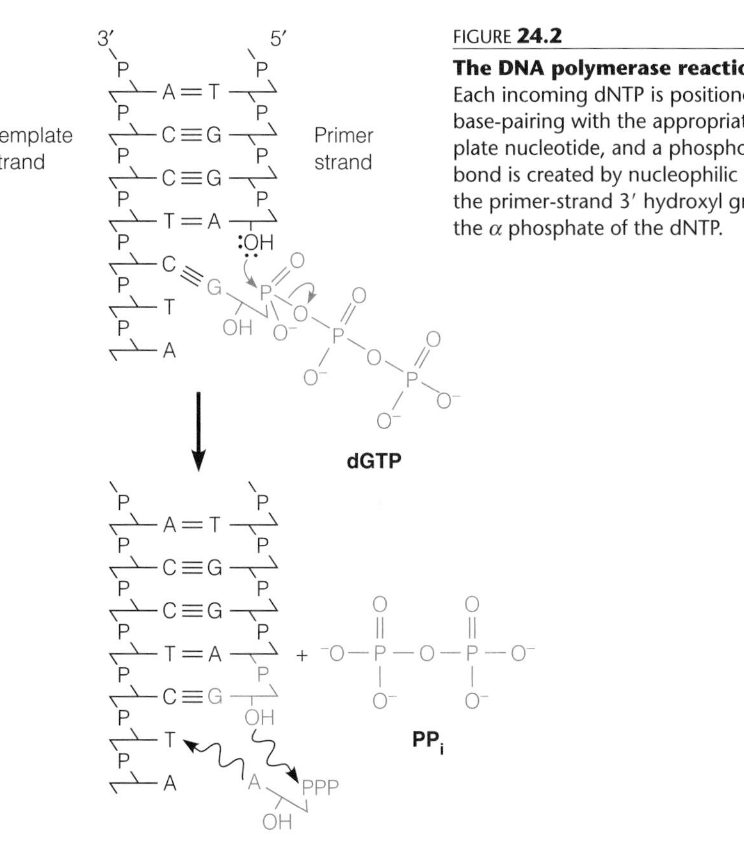

FIGURE **24.2**

The DNA polymerase reaction. Each incoming dNTP is positioned by base-pairing with the appropriate template nucleotide, and a phosphodiester bond is created by nucleophilic attack of the primer-strand 3′ hydroxyl group on the α phosphate of the dNTP.

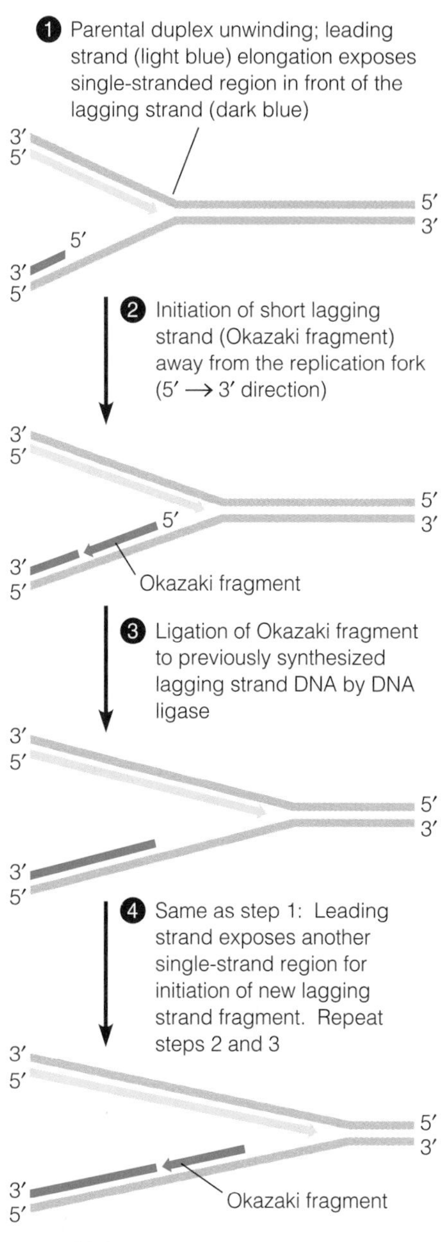

FIGURE **24.3**

The Okazaki model. The mechanism by which the lagging strand of the duplex can be synthesized in discontinuous fragments.

unwound, exposing a region of single-strand template DNA. Because both leading and lagging strand synthesis occur in the fork (see page 877), lagging strand polymerase must remain near the fork. Therefore, as shown in Figure 24.3, the lagging strand is synthesized discontinuously, as a series of short segments, while leading strand synthesis can occur without interruption. The lagging strand pieces are called **Okazaki fragments,** after Reiji Okazaki, who provided the first evidence for discontinuous DNA replication.

DNA polymerase can act only by extending chains from a 3′ hydroxyl group. However, no such functional group exists opposite the lagging strand template at the fork. Therefore, in order to initiate Okazaki fragment synthesis, another enzyme, **primase,** comes into play. Like DNA polymerase, primase copies a template DNA strand to make a polynucleotide product. However, that product is RNA, not DNA. Unlike DNA polymerase, primase can initiate synthesis of a polynucleotide strand, by first positioning a ribonucleoside 5′-triphosphate (rNTP) opposite its complementary DNA base and then extending from the 3′ hydroxyl group of that rNTP, as shown in Figure 24.4. After several ribonucleotides have been polymerized to form an **RNA primer,** primase somehow becomes displaced and DNA polymerase adds a 5′ deoxyribonucleotide to the 3′ terminus of the RNA primer, and then continues to add dNTPs in its accustomed 5′→3′ direction. Recent evidence identifies a specific DNA polymerase subunit responsible for primase displacement

When the 3′ terminus of the growing Okazaki fragment reaches the 5′ terminus of the previously synthesized Okazaki fragment, three events must occur: (1) removal of the RNA primer, (2) replacement of the departed ribonucleotides with deoxyribonucleotides, and (3) covalent joining of the 3′ end of the newly synthesized Okazaki fragment to the 5′ end of the previously synthesized fragment. Two additional enzymes come into play—**DNA polymerase I** and *DNA ligase* (also

Synthesis of each Okazaki fragment of the lagging strand is initiated by synthesis of a short RNA primer, which is then extended by DNA polymerase.

1 Unwinding of parental duplex by helicase and elongation of leading strand by polymerase III expose single-strand region in front of lagging strand

3′ 5′ 5′ 3′ 3′ 5′

2 Primase synthesizes RNA primer

3 Polymerase III extends DNA Okazaki fragment from primer

4 Polymerase I eliminates downstream RNA primer by nick translation

5 DNA ligase ligates Okazaki fragment to rest of lagging strand

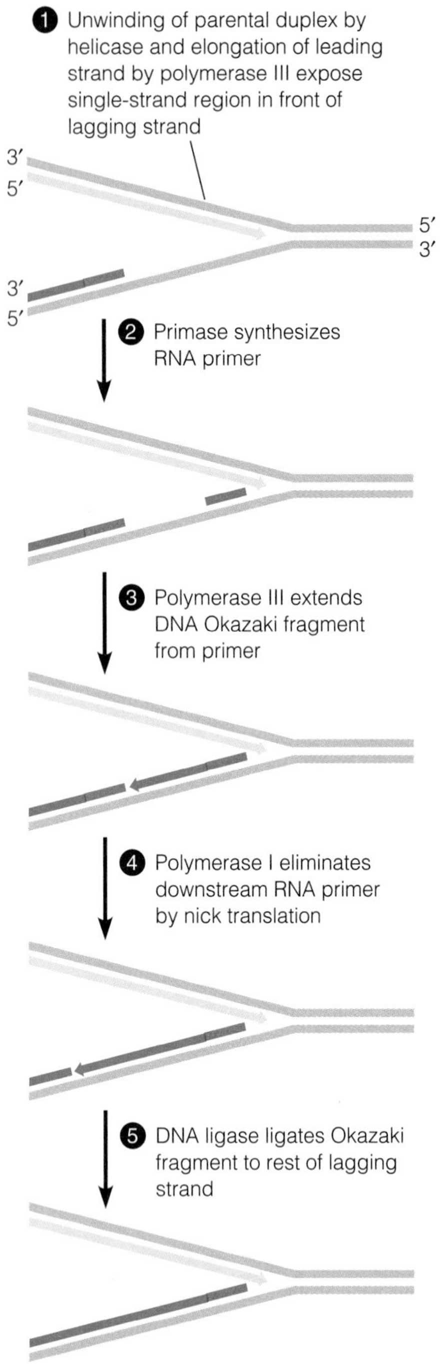

FIGURE **24.4**

Details of lagging strand synthesis. An RNA primer (red) is used to initiate the synthesis of each Okazaki fragment (dark blue); the enzymes DNA polymerase I and DNA ligase later remove these primers and link the Okazaki fragments to the rest of the lagging strand. Each arrowhead on a nucleic acid strand marks the location of a 3′ hydroxyl group that can undergo polymerase-catalyzed chain extension. The fork is moving to the right.

shown in Figure 24.4). In bacteria, polymerase I is a form of DNA polymerase that has a 5′ exonuclease activity in addition to its DNA polymerase activity. (An exonuclease catalyzes hydrolysis of phosphodiester bonds at the ends of nucleic acid molecules; see page 795.) These activities allow the enzyme to carry out a reaction called **nick translation**—the removal of ribonucleotides from the 5′ end of the RNA primer, coupled with simultaneous extension of the 3′ end of the Okazaki fragment, by incorporation of deoxyribonucleotides. Figure 24.5a shows the enzyme moving leftward by one position, by simultaneously removing UMP, the

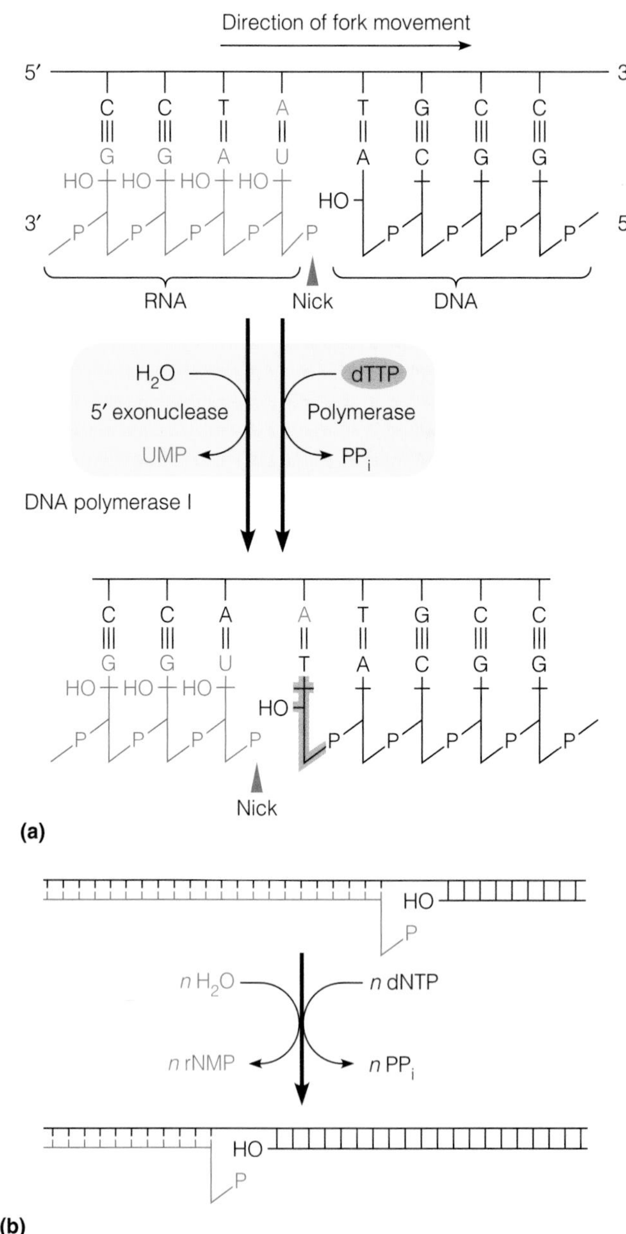

FIGURE **24.5**

Nick translation. The polymerase and 5′ exonuclease activities of DNA polymerase I are coordinated in replacing the RNA primer with DNA by the process of nick translation. **(a)** Movement of the nick by one position to the left, as a UMP residue of RNA is replaced by a dTMP residue of DNA. **(b)** The basis for the term *nick translation:* The nick moves to the left by *n* nucleotides, with no change in the total number of nucleotides polymerized. rNMP = ribonucleoside monophosphate.

5′ terminal nucleotide of the RNA fragment (5′ exonuclease activity) and replacing it with dTMP (from dTTP) at the 3′ end of the DNA (polymerase activity). As this process is repeated, the net result is to *move* a **nick** with no *net* synthesis or loss of nucleic acid (Figure 24.5b)—hence, the term *nick translation.* The nick is a single-strand interruption in duplex DNA, in which no nucleotides are missing. A nick is distinguished from a **gap,** which is a similar interruption with one or more missing nucleotides.

Once all of the ribonucleotides have been removed, DNA polymerase reaches a structure that it cannot close, namely, a nick. The nick created in DNA replication has a 3′ hydroxyl terminus and a 5′ phosphate terminus. **DNA ligase** catalyzes formation of a phosphodiester bond between these ends; we describe the action of this enzyme on page 898.

Primase is associated with other proteins at the replication fork, as shown in Figure 24.6. At least two of these other proteins are **helicases,** enzymes that catalyze the unwinding of duplex nucleic acids. Each parental DNA strand has its own helicase; the one associated with the lagging strand is complexed with primase as part of a unit called the **primosome.** The helicase-catalyzed unwinding of parental DNA strands requires concomitant ATP hydrolysis, which provides the energy for disrupting the stable DNA duplex. The unwinding process exposes single-strand parental DNA. If this DNA were not stabilized in some way, then intrachain base-pairing might occur between complementary base sequences on the same strand. However, these bases must remain in single-strand conformation, so that they can pair with incoming nucleotides that will form the daughter strands. **Single-strand DNA-binding protein (SSB),** also called **helix-destabilizing protein,** binds to DNA so as to stabilize a structure in which hydrogen-bonding surfaces of the DNA bases are spatially oriented toward the incoming nucleotides.

Shown also in Figure 24.6 is a circular protein bound to each DNA polymerase molecule—the **sliding clamp,** which ensures that polymerase remains bound to its DNA template, by encircling that DNA. Not shown is a multiprotein assembly, the **clamp loader,** which somehow opens up the clamp so that it can become wrapped around the DNA; this assembly also removes the clamp so that polymerase can dissociate at the appropriate time. Note also that the replicative polymerase is dimeric; the two catalytic proteins are physically linked, so that their actions are coordinated in replicating leading and lagging strand templates.

The parental DNA is topologically constrained, whether by the circular nature of the chromosome, as in most prokaryotes, or through its binding to

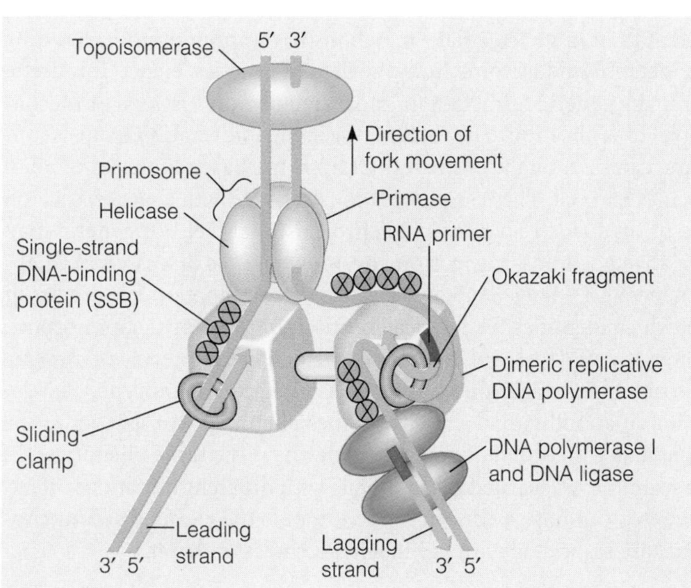

FIGURE **24.6**

Schematic view of a replication fork. This includes the following features, not shown in the simplified scheme of Figure 24.1: (1) topoisomerase, relieving torsional stress created by duplex DNA unwinding; (2) helicase and primase as separate components of the primosome, which unwinds parental DNA and synthesizes RNA primers; (3) replicative DNA polymerase as a dimer, with simultaneous elongation at the leading and lagging strands; (4) a sliding clamp holding each polymerase unit to its template by physically encircling the template DNA strand; (5) single-strand DNA-binding protein stabilizing template DNA to facilitate its base-pairing with incoming deoxyribonucleotides; and (6) DNA polymerase I and DNA ligase, poised to excise RNA primers attached to lagging-strand DNA, replace the RNA with DNA, and seal the Okazaki fragments to high-molecular-weight DNA. RNase H, an enzyme that degrades RNA in DNA–RNA hybrids, can also play a role in RNA primer excision.

chromosomal proteins, as in eukaryotes. Therefore, the parental DNA duplex becomes overwound ahead of the fork as a consequence of its unwinding at the fork. Such torsional stress must be relieved by some sort of "swivel" mechanism, or else strand unwinding would become energetically impossible. Such a swivel is provided by **topoisomerases,** enzymes that change the linking number of DNA without changing its structure in other respects (recall from Chapter 4 that linking number is the number of turns that two strands of a duplex make about each other). Topoisomerases relieve stress ahead of the fork by transiently breaking DNA strands, allowing relaxation of the strained structure by free rotation of DNA strands, and allowing re-formation of phosphodiester bonds at the strand breaks.

The more detailed picture of a replication fork in Figure 24.6 is based upon the information just presented. At the instant depicted in the figure, one subunit of the dimeric DNA polymerase III is elongating the leading strand and the other is elongating the lagging strand. If the two polymerase molecules were not associated, then completion of each Okazaki fragment would cause polymerase to dissociate, and the lag in binding a new polymerase would make it impossible for the lagging strand to keep up with the leading strand. DNA polymerase I and DNA ligase are both inactive and will not become active until the 3′ end of the new Okazaki fragment reaches the RNA primer (shown in red) still bound to the 5′ end of the previously synthesized Okazaki fragment. At that point, nick translation by DNA polymerase I will simultaneously degrade the RNA primer and extend the new Okazaki fragment. Finally, DNA ligase will seal the nick. Concurrently, continued action by the primosome is exposing more single-strand DNA on the lagging strand template and synthesizing another RNA primer, for synthesis of the next Okazaki fragment.

Note also that Figure 24.6, like most of the figures in this chapter, is not drawn to scale. At the thickness of the DNA double helix shown in the figure, the proteins would have diameters of several centimeters each if the figures were drawn to scale. A fascinating feature of DNA replication is the coordination of function of the several proteins in this large complex so that, for example, DNA chains are synthesized at the same rate as the unwinding of the parental duplex DNA, leading and lagging strand syntheses occur at the same rate, and the template DNA sequences are copied with an accuracy unparalleled in any other biochemical process.

A Review of Genetic Terminology

To a greater extent than in intermediary metabolism, our present understanding of information metabolism has come from the field of genetics. Hence, it is useful to review several key genetic terms before more detailed discussions of mechanisms and control of replication. The totality of genetic information in an organism, the **genome,** can be thought of most directly as the base sequences of all of the DNA molecules in a cell. These sequences consist of individual genes, each on a chromosome or on a small piece of extrachromosomal DNA. A **genetic map** identifies the relative positions of genes on chromosomes. As discussed in Chapter 25, one can generate a **linkage map** by measuring recombination frequencies or a **physical map,** which identifies the physical locations of genes on a chromosome.

Any characteristic of an organism that can be detected in terms of appearance, structure, or some measurable property is called the **phenotype.** The genetic composition of an individual, which specifies phenotypic properties, is the **genotype.** Individuals with different genotypes may have the same phenotype. In Chapter 18, for example, we learned that humans with different mutations affecting the LDL receptor can have a common phenotype—the elevated serum cholesterol levels that are associated with familial hypercholesterolemia.

An **allele** is a particular form of a gene. For example, phenylketonurics have two defective alleles of the gene for phenylalanine hydroxylase that encode inactive forms of the enzyme. A **marker** is any allele whose frequency can be determined quantitatively. **Copy number** refers to the number of copies per cell of a gene or other DNA sequence. Most eukaryotic cells are **diploid,** meaning that the copy number of most chromosomal genes is two, while most prokaryotic cells have a **haploid** genotype, and thus a single allele for most genes. Extrachromosomal genes can have much higher copy numbers, whether they are carried on organelle DNA in eukaryotic cells or on an extrachromosomal DNA element such as a bacterial **plasmid.** A plasmid is usually a small, circular DNA molecule whose copy number may be controlled independently of the regulatory mechanisms affecting chromosomal DNA replication.

A diploid individual in a natural environment usually has at least one functional copy, or allele, of each gene and therefore displays a **wild-type** (normal) phenotype. A gene in such an organism can undergo a mutation, thereby generating a mutant genotype, which may be detected as a mutant phenotype. For example, albinism represents a readily observable phenotype, resulting from mutations that inactivate tyrosinase (see Chapter 21). A mutant site within a gene can undergo a second mutation, which restores the wild-type genotype. This type of mutation is called a **reversion.** Alternatively, a wild-type phenotype can sometimes be restored by a mutation at a different site; this occurrence is called **second-site reversion,** or **suppression.** Suppression can occur either by mutation within the gene undergoing the original mutation (intragenic suppression) or within a different gene (intergenic suppression). Note that many mutations are **silent;** if a mutant protein retains some biological activity, there may not be an observable phenotype.

Because much of our discussion focuses upon replication in *E. coli* and its phages, you should understand some conventions in the naming of prokaryotic genes and gene products. Let us review these conventions in the context of the *E. coli* genetic map. Mapping genes in *E. coli* became possible after the discovery by Joshua Lederberg of sexual reproduction in bacteria. This process, termed **conjugation,** involves the transfer of DNA from a donor cell (the "male") to a recipient cell (the "female"), as schematized in Figure 24.7. About 100 minutes is required for complete chromosome transfer. Mapping is carried out by "interrupted mating" experiments. At intervals a conjugating culture is disrupted mechanically, to separate the donor and recipient cells. The researcher then analyzes the recipient cells, correlating the transfer of marker genes from the donor with the time at which mating was interrupted. Thus, map positions of genes were originally reported as "minutes," which identified the time at which a marker of interest was transferred by conjugation. Finer-scale mapping could be done by transduction, a process described in Chapter 25. With the complete genome sequence of *E. coli* reported in 1997, we now know the physical locations of all *E. coli* genes.

Figure 24.8 shows part of the *E. coli* genetic map, identifying genes known to participate, either directly or indirectly, in DNA replication. The large number of these genes underscores the complexities of DNA replication, and we shall refer frequently to the genes shown on this map. Our purpose here is to describe the conventions. Each gene has an italicized lower-case three-letter designation that refers either to the gene product or to the phenotype of mutants defective in that gene, followed by a capital letter denoting the order of discovery of the gene or gene product. For example, *E. coli* contains three DNA polymerases, named I, II, and III. The structural genes encoding these enzymes are *polA, polB,* and *polC,* respectively. Other genes were named from the defective phenotype. For example, all genes designated *dna* were originally identified from temperature-sensitive (*ts*) mutations in which DNA replication occurred normally when cells were grown at 30°C but was blocked at 42°C. Some of the gene products have been identified.

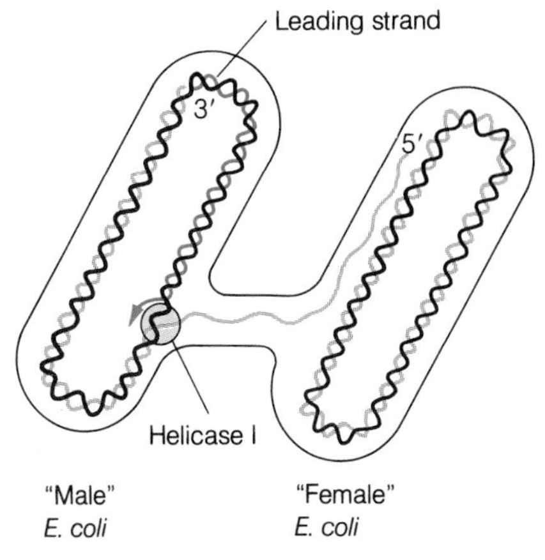

FIGURE **24.7**

Bacterial conjugation. After a nick is made in the "male" chromosome, a helicase unwinds the two strands. The nicked strand (gray) is elongated (blue) at the 3′ end, displacing the 5′ end and transferring it into the female cell. Later, the transferred DNA strand can undergo recombination with the recipient chromosome, thereby transferring genetic markers to the female.

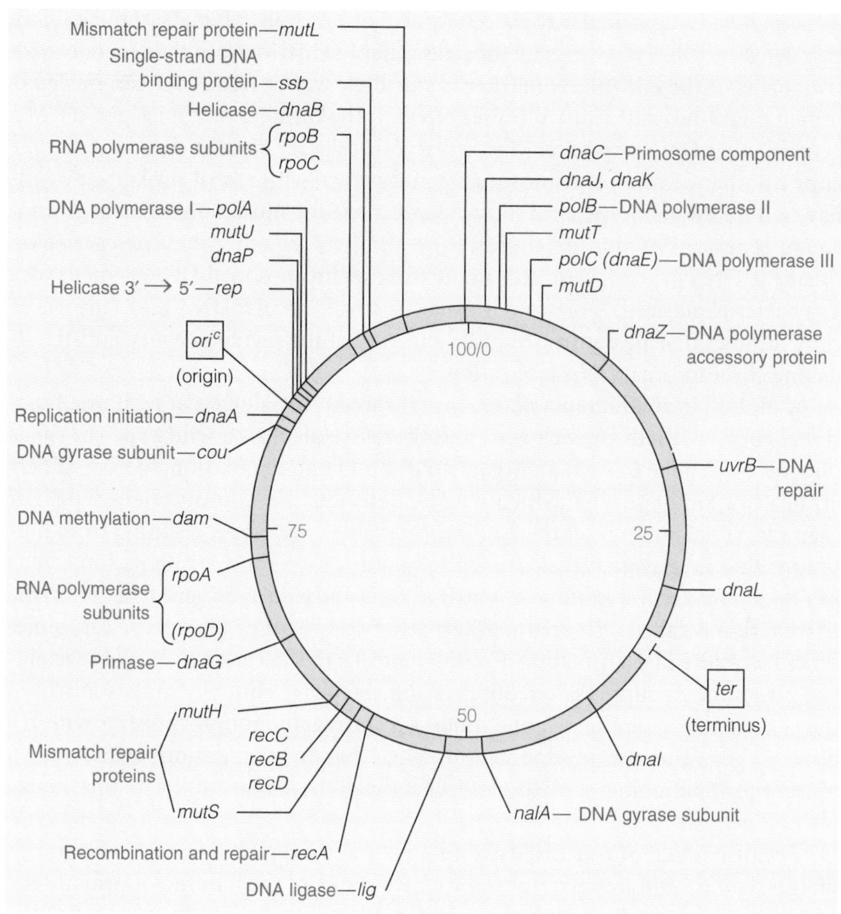

FIGURE **24.8**

Partial genetic map of *E. coli*. Genes whose products are involved in DNA replication or repair are shown. Gene product names are given in blue. The *dna* genes play essential roles in DNA replication. The *mut* genes specify proteins that, when mutated, can cause elevated rates of spontaneous mutation. *ori* and *ter* are sites of initiation and termination of genome replication, respectively.

For example, the *dnaG* gene product was found to be primase. By convention, gene *products,* which are usually proteins, are given the corresponding nonitalicized and capitalized designations. For instance, DnaG is identical to primase; both terms refer to the product of the *dnaG* gene.

Early Insights into DNA Replication

How does a molecule copy itself? As with most metabolic processes, our early ideas about the mechanism of DNA replication were shaped largely by in vivo experiments—those carried out with intact cells. The complementary approach—isolation of enzymes and proteins involved in DNA replication—began later, with the discovery of DNA polymerase. We begin our detailed discussion of replication with the early experiments that revealed the general nature of the process.

SEMICONSERVATIVE NATURE OF DNA REPLICATION

In 1953, when Watson and Crick proposed the double-helical, base-paired model for DNA structure, much of the excitement came not simply from the aesthetic beauty of their model but because the model explains how DNA copies itself during cell division. The two strands of a parental DNA duplex unwind, and each strand serves as template for the synthesis of a new strand, wound about that parental strand. As noted in Chapter 4, the Meselson–Stahl density-labeling experiment showed that newly replicated DNA consists of one strand of parental DNA and one strand of newly synthesized material.

Recall that the Meselson–Stahl experiment used stable isotopes (^{13}C and ^{15}N) to produce DNA of higher-than-normal density. As discussed in Chapter 25, analysis of density-labeled DNAs has been widely used for studies of DNA repair and recombination, as well as replication. However, stable isotopes are rarely used for these kinds of experiments, partly because the small density differences that can be achieved limit the resolution of this technique. What is often used as an alternative density label is 5-bromodeoxyuridine, which is efficiently incorporated into DNA in place of thymidine (Chapter 22). This analog gives a much larger density increment in labeled DNA.

SEQUENTIAL NATURE OF REPLICATION

The Meselson–Stahl experiment showed that DNA replication yields two replicas, each containing one old and one new strand. However, the experiment didn't address the question of whether strand unwinding occurs simultaneously with the synthesis of daughter strand— in other words, whether replication is *sequential* and *ordered*. This would mean that replication starts at one or a few fixed points on a chromosome, with replication forks emanating from these fixed **origins**. Each fork would consist of two daughter strands being elongated on parental templates simultaneously with unwinding of the parental duplex.

The sequential nature of replication was shown originally by genetic experiments and by direct visualization of replicating molecules. The genetic experiments were based on the premise that a chromosome caught in the midst of replication has two copies of every gene that has replicated and one copy of every nonreplicated gene. Thus, in a cell population a gene situated close to a replication origin on a chromosome has a relatively high copy number. Genetic analysis can determine the copy number of markers in a population of dividing cells, for example, by transduction frequency. Plotting marker frequency as a function of genetic map position can in turn establish the approximate location of an origin and predict the direction of replication (Figure 24.9).

This genetic approach was first applied to the replication of *E. coli* DNA. The earliest experiments suggested that the bacterial chromosome contains a single origin of replication, with the fork moving in one direction from this origin. A similar interpretation was made from the earliest visualization of replicating *E. coli* DNA. John Cairns accomplished the difficult task of lysing *E. coli* cells under conditions in which some of the fragile DNA molecules disentangled themselves without being broken by shear forces. When he made these molecules radioactive, by prior growth of the bacteria in medium containing [^{3}H]thymidine, the disentangled DNA could be visualized by radioautography. Examination of individual molecules, such as that shown in Figure 24.10, revealed that each bacterial cell contains a giant, circular DNA molecule, more than 1 millimeter in circumference. Recall from Chapter 4 that these molecules must be folded and supercoiled in order to fit inside bacterial cells that are just a few micrometers in length.

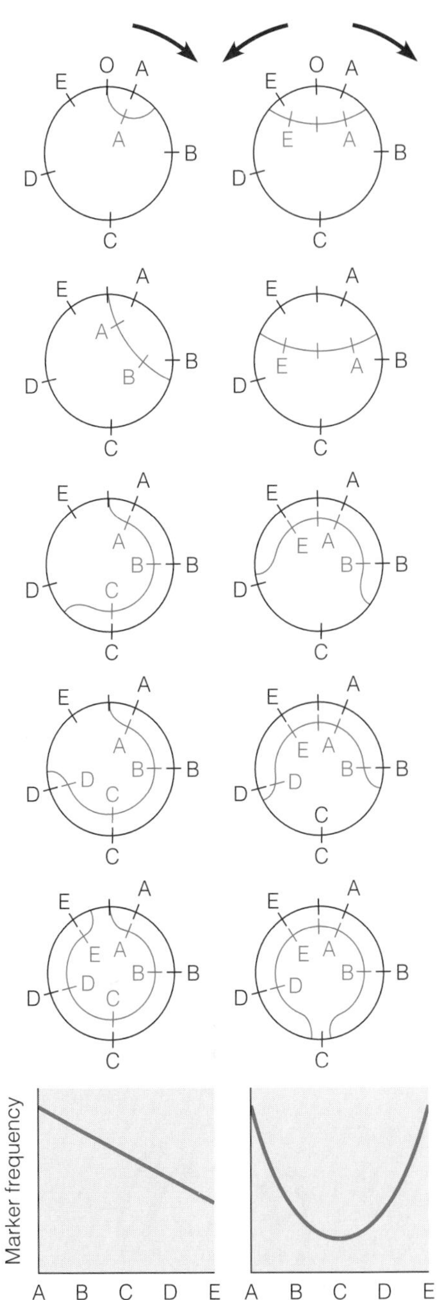

FIGURE **24.9**

Distinguishing bidirectional from unidirectional replication by marker frequency gradients. When a population of circular DNA molecules carrying marker genes is allowed to replicate randomly from a fixed origin, O, and the number of copies of each marker gene is counted, the resulting curves can be used to differentiate unidirectional replication (left side) from bidirectional replication (right side).

ORIGIN AND DIRECTION OF REPLICATION

Some of the molecules, such as the one shown in Figure 24.10, contained two branch points and were identified as molecules caught in the midst of a round of replication. In this experiment, DNA had been labeled by two generations of bacterial growth in [^{3}H]thymidine-containing medium. Determination of the **grain density,** or number of dark spots on the film per unit distance, allowed Cairns to identify regions of the DNA that had replicated either once or twice; the latter have twice the grain density of the former. Cairns suggested that one of the two branch points represented an origin of replication and the other represented a replication fork, moving in one direction away from this origin. However, the data were also consistent with the idea of **bidirectional replication**—one fixed origin and two replication forks, each moving away from the origin until they join halfway around the circular template.

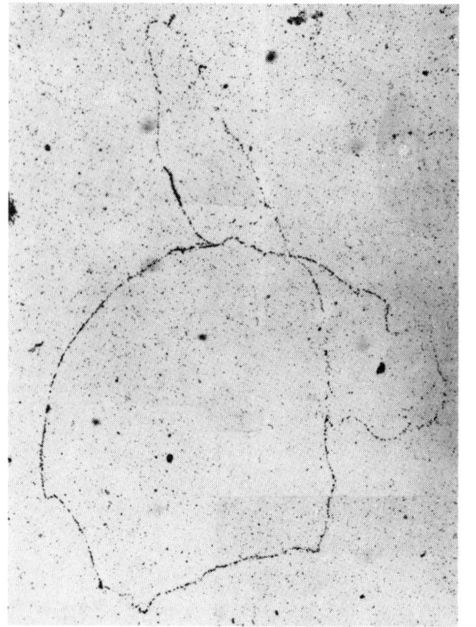

FIGURE **24.10**

Radioautogram of a replicating *E. coli* chromosome after two generations of growth in [^{3}H]thymidine.

Courtesy of J. Cairns, *Cold Spring Harbor Symp. Quant. Biol.* (1963) 28:44.

By the early 1970s it became possible to isolate replicating DNA in sufficient quantity for its analysis at higher resolution. Such experiments showed that most (though not all) DNA molecules replicate bidirectionally. The earliest definitive experiments involved electron microscopic analysis of replicating DNA from bacteriophage λ (Figure 24.11). DNA isolated from the virus particle is a linear duplex, 17.5 μm long, containing 48,502 base pairs. It has an additional interesting feature: Each of the two 5′ DNA termini is extended by 12 nucleotides, to give a short, single-strand tail at each end. The two tails are complementary in base sequence, so that under annealing conditions, which promote DNA renaturation, the linear DNA spontaneously circularizes by base-pairing between these short tails, called **cohesive ends.** These circular molecules can be covalently closed by DNA ligase action. This circularization occurs in vivo, as can be shown by the isolation from infected cells of circular phage DNA molecules, which can be visualized in the electron microscope.

When partially replicated λ DNA molecules are isolated, one sees a population of circles, each with two branch points, comparable to the structures Cairns saw in replicating *E. coli* DNA, but much smaller. They are called **theta structures** because of their resemblance to the Greek letter θ. Physical analysis of these structures by a technique called denaturation mapping established that replication proceeds in both directions from one fixed origin.

Shortly after those experiments, genetic experiments at higher resolution indicated that the *E. coli* chromosome also replicates bidirectionally from a single fixed origin. This conclusion was reinforced by a variation of Cairns's radioautographic experiment, which led to an unambiguous conclusion. The experiment involved specific labeling of DNA synthesized during termination of one round of replication and initiation of the next. If replication is *bidirectional*, DNA labeled during termination of one round and initiation of the next round would be labeled on opposite sides of the chromosome. Conversely, if replication is *unidirectional*, termination and reinitiation would occur in adjacent regions of the

Replicating DNA

Cohesive or "sticky" ends

Annealing conditions

Replication

Direction of replication

Origin

Newly replicated DNA

Origin

θ structure

(a)

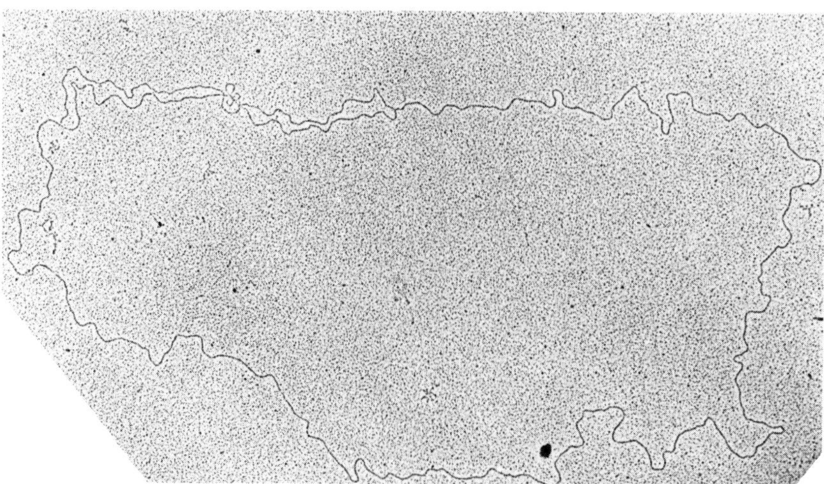

(b)

FIGURE **24.11**

Bidirectional replication of lambda phage DNA. (a) Mechanism of replication. The original, linear DNA circularizes by base pairing of the cohesive ends. After covalent closure, replication can begin. Partial replication yields a characteristic theta structure. **(b)** Electron micrograph of a theta structure. When partially replicated molecules like these were partially denatured and examined in the electron microscope, the data established that replication was bidirectional from a fixed origin.

(b) Courtesy of M. Schnös and R. B. Inman, *J. Mol. Biol.* (1970) 51:61–73.

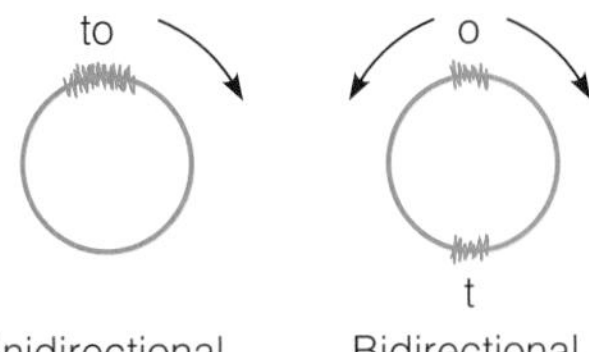

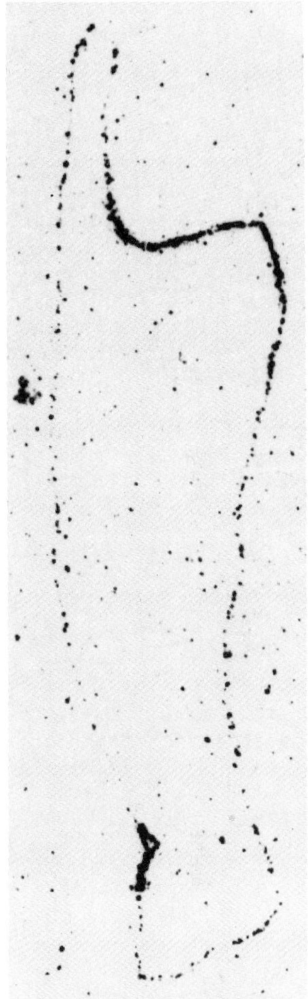

FIGURE **24.12**

Demonstration of bidirectional replication by radioautography. This *E. coli* chromosome was labeled by long-term growth in [^{3}H]thymidine, and the specific activity was increased 4-fold just as the synchronized culture completed one round of replication and commenced another. The diagrams show the labeling patterns that would be expected for unidirectional replication, where the terminus (t) and origin (o) are adjacent, and for bidirectional replication, where they are on opposite sides of the chromosome.

chromosome. The chromosome was labeled by growth of an *E. coli* culture in medium containing [^{3}H]thymidine. At a time when replication should be nearly complete, the specific activity was increased severalfold, so that any DNA synthesized during termination and subsequent reinitiation would be recognized by its higher grain density on a radioautogram. The experiment clearly showed that termination and reinitiation occur on opposite sides of the chromosome (Figure 24.12). Thus, replication is bidirectional.

Carefully timed radioautographic experiments also allow estimation of the rate of replication fork movement. The number of base pairs of DNA replicated in a given time is measured simply from the length of a labeled segment in μm and the number of base pairs in 1 μm (about 3000 if DNA is in the B form, with 0.34 nm per base pair). An optimally nourished *E. coli* cell at 37°C shows a rate of fork movement of about 850 base pairs per second.

Do these experiments mean that all DNA molecules are circular, with bidirectional replication and one fixed origin? For prokaryotes that is usually the case, but not always. We shall encounter examples of linear chromosomes, of unidirectional replication, and of more than one origin per chromosome. There are several variations on the theme we have developed thus far.

Prokaryotic chromosomes are usually circular DNA molecules, and replication usually proceeds bidirectionally from one origin.

UNITS OF REPLICATION: THE REPLICON

Our earliest understanding of the control of DNA replication came from studies on replication of **extrachromosomal genetic elements.** In bacteria these elements consist of circular plasmid DNA molecules. Some plasmids, called **sex factors,** specify gene products that allow bacteria to mate with one another and undergo sexual reproduction. Other plasmids, called **bacteriocinogenic factors,** provide a defense mechanism. They encode proteins called **bacteriocins** that are released from the cell and that kill other bacteria. Of great medical importance are **drug resistance factors**—plasmids carrying genes that specify resistance to antibiotics. For example, penicillin-resistant strains of bacteria often harbor a plasmid that specifies **β-lactamase,** an enzyme that destroys β-lactam antibiotics such as penicillin (see page 588). Because plasmids are readily transferred from cell to cell, drug-resistant bacterial populations are arising with alarming frequency as the result of indiscriminate use of antibiotics to combat infectious diseases.

Most plasmids are truly extrachromosomal, as shown in Figure 24.13a. A few plasmids, called **episomes,** can undergo **integration**—insertion of the plasmid DNA into the chromosomal DNA sequence. When this happens, the integrated episome DNA no longer replicates autonomously but instead replicates in synchrony with the chromosome into which it has been inserted (Figure 24.13b). Thus, whereas a nonintegrated plasmid may be present in dozens of copies per chromosome and replicate independently of the chromosome, the integrated episome is present in one copy per chromosome and can replicate only when the chromosome does so.

Plasmids are small, circular DNA molecules that replicate independently of chromosomal DNA.

Now consider the relationship between an episome and a bacterium carrying a temperature-sensitive (*ts*) mutation in a chromosomal gene whose product is

A replicon consists of all the DNA replicated from one origin, controlled by proteins that act at that origin.

essential to initiating DNA replication. Such a bacterium cannot initiate a round of DNA replication at a moderately elevated temperature (42°C). Integration of an episome reverses the temperature-sensitive phenotype, allowing DNA replication to be initiated at the higher temperature. However, replication is initiated not from the ordinary origin on the chromosome but from a site within the episome (Figure 24.13c).

These observations led to the concept that the initiation of DNA replication is controlled by a small cluster of genetic elements, including (1) a sequence of nucleotides at which replication begins (the origin) and (2) structural genes for proteins that interact with the origin sequence and initiate replication at that point. This cluster and all of the DNA that is replicated under its control are collectively termed a **replicon.** The bacterial chromosome constitutes a single replicon, because its entire replication commences from one fixed origin. The same is true for an extrachromosomal plasmid. However, when a plasmid integrates as an episome, the chromosome containing the episome constitutes one replicon, with replication usually initiated at the chromosomal origin.

Although a chromosome is one large molecule of DNA, a single chromosome is not always a single replicon, with replication of the chromosome driven from a single origin. Radioautographic analysis of eukaryotic DNA replication shows that mammalian cells have between 10^3 and 10^4 replication origins per complete genome. Replication is bidirectional from these origins, with each fork progressing until it merges with the fork started from initiation of the adjacent replicon (see margin, page 889). Because the human genome contains 46 chromosomes, this means that a typical chromosome consists of several dozen replicons.

Multiple replication origins are important for eukaryotic DNA replication because eukaryotic cells contain a great deal of DNA that must be replicated within a limited period of each cell division cycle. Autoradiographic analysis shows that eukaryotic DNA replication rates are about 10-fold lower than prokaryotic rates. As you can estimate from the data in Table 24.1, it could take weeks for a human cell to replicate all of its DNA once, if each chromosome contained but a single origin and two replication forks. As it is, DNA replication occupies only a fraction of the cell division cycle, typically 6 to 8 hours in cells with division cycles ranging from 12 to 36 hours (Chapter 28).

An apparently anomalous situation occurs in bacteria: More time is needed to replicate the single chromosome than is required for the complete cell division cycle. A bacterial culture can double in cell number every 20 minutes, but 40 minutes is required for each round of chromosome replication. How can cells divide in *less* time than is needed for replication of the chromosome? The answer is that in rapidly dividing cells a replicating DNA molecule can reinitiate a new round of

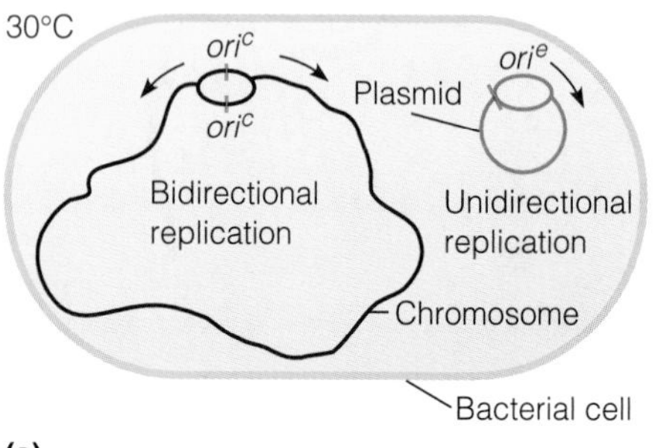

(a)

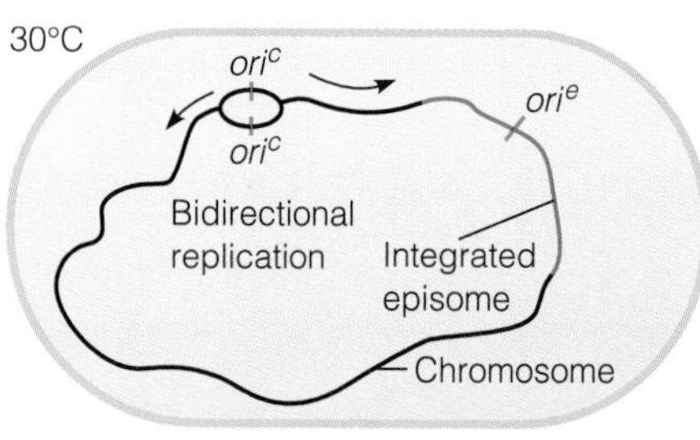

(b)

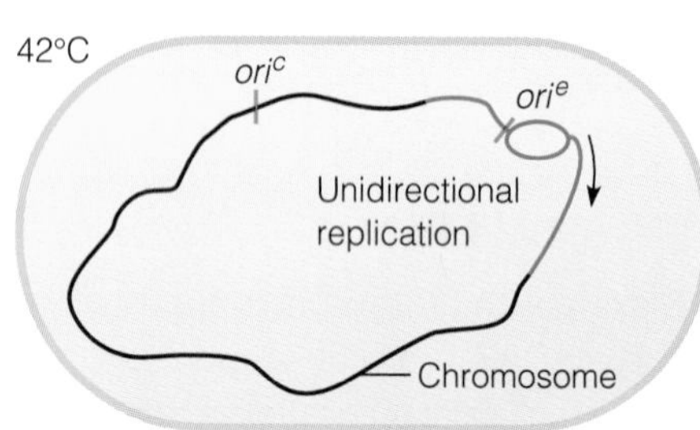

(c)

FIGURE **24.13**

Replication driven by an integrated episome. (a) An *E. coli* cell with a *ts* (temperature-sensitive) defect in replication initiation, carrying a nonintegrated plasmid with a unidirectional origin (*ori*^{e}). Chromosome replication initiates bidirectionally from *ori*^{c}. **(b)** When the plasmid integrates as an episome, it replicates only when the chromosome replicates—bidirectionally from *ori*^{c}. **(c)** When chromosomal replication is inhibited by shift to a nonpermissive temperature (42°C), replication of the entire chromosome can be driven from the unidirectional origin (*ori*^{e}) on the integrated episome.

TABLE **24.1 Quantitative parameters of DNA replication in different cells**

	Replication Process	
	E. coli	Human
DNA content, number of nucleotide pairs per cell	3.9×10^6	ca. 10^9
Rate of replication fork progression, μm/minute	30	3
DNA replication rate, nucleotides/second per replication fork	850	60–90
Number of replication origins per cell	1	10^3–10^4
Hours required for complete genome replication	0.67	8
Hours required for one complete cell division	0.33	24

Note: The data are for an *E. coli* cell optimally nourished and cultured at 37°C. The values for human cells represent data from HeLa cells, which were originally derived from a tumor and have been maintained in culture for many years.

replication from the origin before completion of the round in progress. In other words, *initiation* of replication occurs more frequently than once every 40 minutes. This can be demonstrated by radioautography, as shown in Figure 24.14. When a cell divides, each daughter cell receives a chromosome that is far into its next round of replication. Thus, the cell adapts to rapid growth by timing its replication initiation events ("replicon firing") earlier in the cell cycle. This provides strong evidence that DNA replication is controlled primarily at the level of initiation.

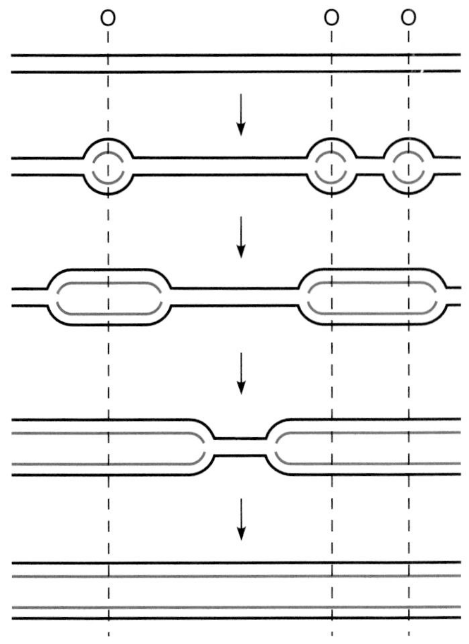

DNA Polymerases: Enzymes Catalyzing Polynucleotide Chain Elongation

The previous section describes features of DNA replication ascertained primarily by experimental manipulations of living cells. At the same time as that work was being done, biochemists were isolating and characterizing the many proteins involved in the process, for ultimate understanding of the mechanism of replication. These proteins and their properties will occupy us for much of the rest of this chapter.

DISCOVERY OF DNA POLYMERASE

The biochemical elucidation of DNA replication began in the mid-1950s, with Arthur Kornberg's discovery of DNA polymerase. Kornberg realized, from his experience in bioenergetics, that the substrates for DNA replication should be activated derivatives of the DNA nucleotide residues. Thus, the endergonic synthesis of phosphodiester bonds between nucleotide residues could be coupled to the exergonic breakdown of activated substrates, just as occurs in synthesis of other macromolecules, such as glycogen. Kornberg correctly predicted that these activated nucleotides would be the 2′-deoxyribonucleoside 5′-triphosphates. When he incubated radioactively labeled dNTPs with an extract of soluble proteins of

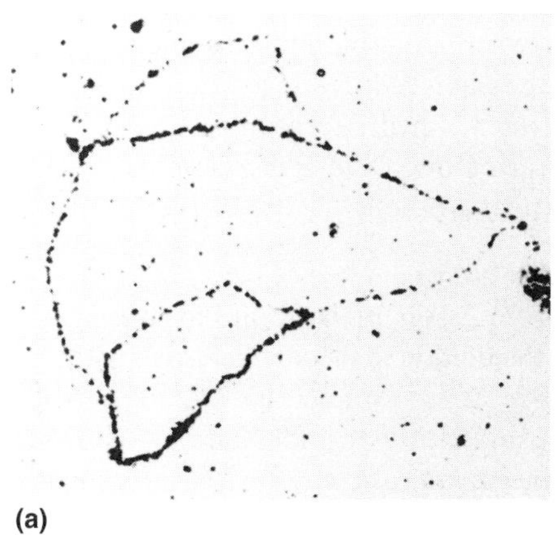

(a)

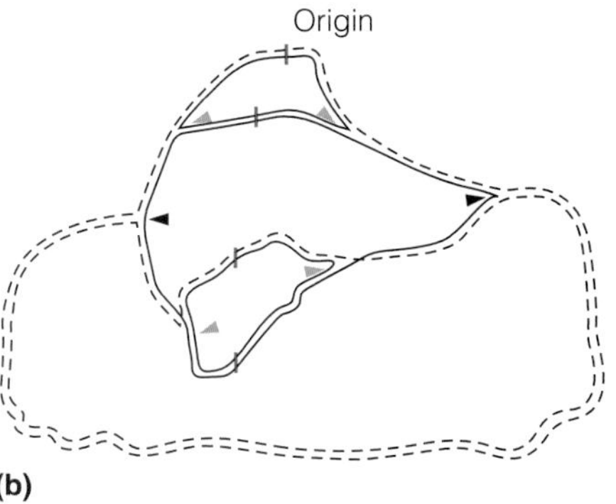

(b)

FIGURE **24.14**

Initiation of a new round of replication before completion of the preceding one. (a) Labeling with [^{3}H]thymidine was started when the first round of replication began. The pattern of grain density shows that one of the reinitiated branches contains two labeled strands, while the other contains only one. **(b)** Solid lines denote radiolabeled DNA, and dashed lines denote unlabeled DNA. Black arrowheads = first-generation replication forks, blue arrowheads = second-generation replication forks. Unreplicated DNA is nonradioactive, so it is not seen in the radioautogram.

Photo courtesy of E. B. Gyurasits and R. G. Wake, *J. Mol. Biol.* (1973) 68:501–509.

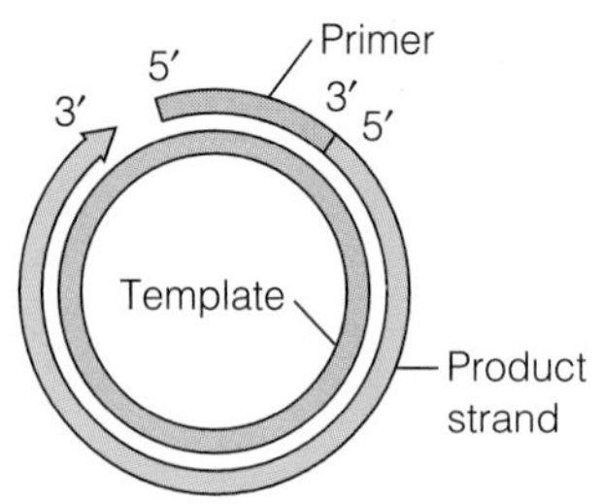

(a) Primed circular single strand

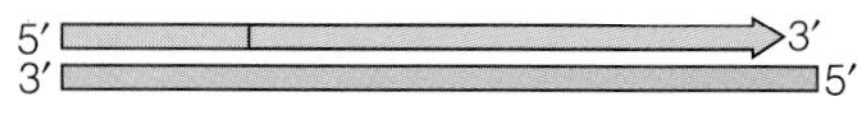

(b) Primed linear single strand

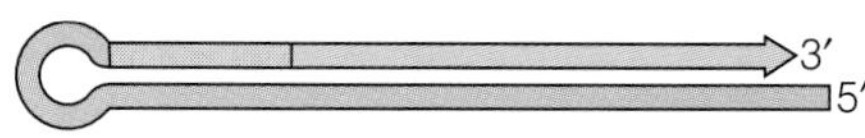

(c) Single-strand hairpin

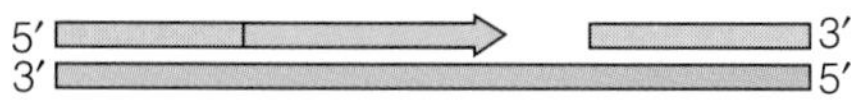

(d) Gapped duplex

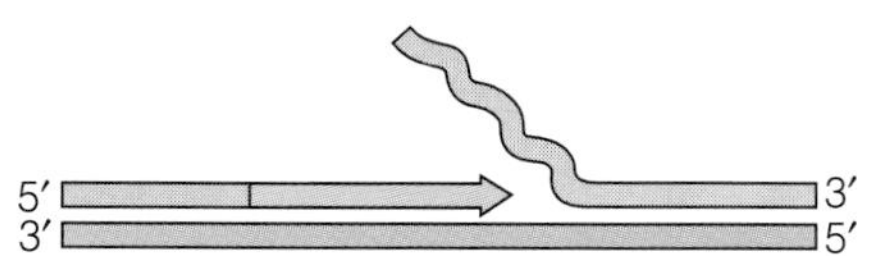

(e) Nicked duplex (strand displacement synthesis)

FIGURE **24.15**

DNA substrates that can be acted on by purified DNA polymerase. Each blue arrowhead marks a 3′ hydroxyl terminus at which chain extension is occurring.

E. coli, a small amount of radioactivity was incorporated into high-molecular-weight material, which could be quantified by acid precipitation and radioactive counting of the DNA. The enzyme required added DNA, plus Mg^{2+}. As we saw in Figure 24.2, two DNA molecules are required for the reaction—the template and the primer, the latter becoming covalently extended from its 3′ hydroxyl group. Also, the polarity of the product DNA strand (that is, the extended primer) is opposite to that of the template strand.

As noted on page 879, the DNA polymerase reaction involves nucleophilic attack by the 3′ hydroxyl group of the primer terminus on the α phosphate of the deoxyribonucleotide substrate, which leads to covalent bond formation. The 3′ hydroxyl is a weak nucleophile, and the pyrophosphate provides a good leaving group. The reaction is readily reversible. In cells and in crude preparations, therefore, the reaction is drawn to the right by pyrophosphatase action on the other product of the reaction ($PP_i + H_2O \longrightarrow 2P_i$). Thus, two energy-rich phosphates are expended per nucleotide incorporated.

STRUCTURE AND ACTIVITIES OF DNA POLYMERASE I

The DNA polymerase discovered by Kornberg was later shown to be one of three different DNA polymerases in bacterial cells (but in 1999, the number rose to five). The Kornberg enzyme is now called DNA polymerase I. We shall use this term henceforth and describe the other two polymerases later in this chapter.

DNA Substrates for the Polymerase Reaction

As noted above, DNA polymerase requires both template DNA and a primer, either DNA or RNA. In vitro, these roles can be played either by two distinct nucleic acids or by one molecule (Figure 24.15). As shown in Figure 24.15c, a single-strand DNA with self-complementary sequences can fold itself into a **hairpin** or **stem-loop** structure whose 3′ end can be extended by polymerase, using the 5′ end as the template. The figure shows other polymerase activities that can be demonstrated in vitro. The enzyme can copy around a circular single-strand template, such as the DNA extracted from small bacteriophages, like ϕX174 or M13, as long as a primer is present, but it cannot join the ends (Figure 24.15a). When the template is linear, polymerase copies only to the 5′ end of the template and then it dissociates (Figure 24.15b and c). In similar fashion the enzyme can fill in a gap (Figure 24.15d), dissociating when the gap is reduced to a nick. Under some conditions the enzyme can also extend from a 3′ hydroxyl group at a nick. Typically, when this occurs, the 5′ end of the preexisting DNA is displaced in advance of the nick. This is called **strand displacement** synthesis (Figure 24.15e).

Multiple Activities in a Single Polypeptide Chain

The DNA polymerase I molecule contains three active sites: a polymerase and two exonucleases.

When DNA polymerase I was purified from *E. coli*, it was found to consist of a single polypeptide chain (M_r = 103,000). In addition to its polymerase activity, the purified enzyme has two nuclease activities: The **3′ exonuclease** degrades single-strand DNA from the 3′ end, and the **5′ exonuclease** degrades base-paired DNA from the 5′ terminus. As noted earlier, the enzyme also cleaves RNA from a duplex containing one strand each of DNA and RNA. The 3′ exonuclease serves a "proofreading" function, to improve the accuracy with which a DNA template is copied. The activity will remove an improperly base-paired nucleotide from the growing 3′ end of a polydeoxynucleotide chain, giving the polymerase activity a second chance to insert the correct nucleotide specified by the template. We say more about the 3′ exonuclease later, when we discuss the fidelity of DNA replication.

The 3′ exonuclease cleaves mismatched nucleotides from the growing 3′ end, improving the accuracy of DNA synthesis. The 5′ exonuclease cleaves ribonucleotides and deoxyribonucleotides from the 5′ end, functioning both in DNA repair and in replicative RNA primer removal.

The 5′ exonuclease activity plays two known roles. The first of these is excision of RNA primers in lagging strand replication, as shown in Figures 24.4 and 24.5. In this process the enzyme catalyzes nick translation, with the 5′ exonuclease excising ribonucleotides just as the polymerase is replacing them with deoxyribonucleotides. Another enzyme, called **ribonuclease H,** can also excise RNA primers. RNase H (H stands for hybrid) specifically degrades the RNA in a hybrid duplex containing one strand of DNA and one of RNA. Unlike DNA polymerase I, RNase H cannot synthesize a new DNA strand to replace the degraded RNA. Nor can it remove the ribonucleotide at the DNA–RNA junction. Thus, DNA polymerase I plays an essential role.

Second, the 5′ exonuclease activity gives DNA polymerase I an important role in DNA repair. The exonuclease can cleave deoxyribonucleotides, as well as ribonucleotides, from DNA. When DNA has been damaged by radiation or chemicals, the nick translation activity of DNA polymerase I can simultaneously excise damaged deoxyribonucleotides and replace them with intact deoxyribonucleotides. We say more about this activity later in this chapter and in Chapter 25.

Structure of DNA Polymerase I

The three catalytic activities of DNA polymerase I have been localized to regions of the long polypeptide chain, partially through limited proteolysis of the enzyme by subtilisin or trypsin. As shown by Hans Klenow, this proteolysis splits the 103-kilodalton polypeptide into a small N-terminal fragment (M_r = 35,000) and a large C-terminal fragment (M_r = 68,000). The large fragment (also called the **Klenow fragment**) contains the polymerase and 3′ exonuclease domains; the small fragment contains the 5′ exonuclease domain. How the three catalytic sites are arranged in space is important to an understanding of how each of the two nuclease activities functions in concert with the polymerase.

Crystallographic study of the large fragment has revealed a striking feature of the structure—a deep crevice, just large enough to accommodate B-form DNA, with a flexible subdomain that might allow bound DNA to be completely surrounded (Figure 24.16). The protein molecule has been likened to a hand, with palm, thumb, and fingers. Cocrystallization of the Klenow fragment with a short duplex DNA shows that this is indeed the DNA-binding site and that bound DNA is almost completely surrounded by protein, which wraps around the DNA like a hand holding a cylinder. An interesting feature of the structure, revealed by study of mutant forms of the enzyme, is that the 3′ exonuclease active site is quite far—about 3 nm—from the polymerase active site. This suggests that about eight base pairs of DNA must unwind to move the 3′ terminal nucleotide from the polymerase to the 3′ exonuclease active site. We shall say more about DNA polymerase structure shortly.

Because the 5′ exonuclease activity is absent in the Klenow fragment, this unnatural enzyme is a useful laboratory reagent for synthesis of DNA in vitro when the researcher wishes specifically to avoid DNA degradation.

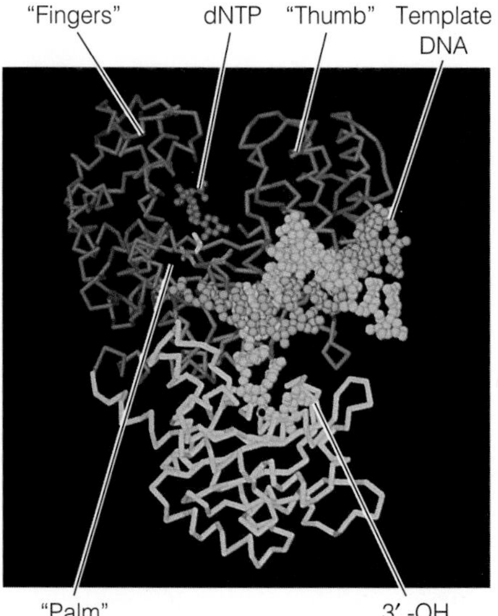

FIGURE **24.16**

The Klenow fragment of *E. coli* DNA polymerase I. α-carbon backbone representation of the Klenow fragment complexed with DNA, in an editing configuration. That is, the 3′ end of the growing strand (light blue) is bound at the 3′ exonuclease active site of the enzyme (yellow). The locations of the polymerase and 3′ exonuclease active sites have been identified by site-directed mutagenesis. The template strand is shown in green, and bound dCTP in red.

Courtesy of L. Beese et al., *Science* (1993) 260:352.

DNA POLYMERASES II AND III

For his discovery of DNA polymerase, Arthur Kornberg was awarded the Nobel Prize in 1959. Although this discovery was monumental, several properties of the enzyme as isolated from *E. coli* were different from those expected for an enzyme catalyzing the major nucleotide incorporation reactions in biological DNA replication. First, the enzyme in vitro is too slow, with a V_{max} of about 20 nucleotides per second; in contrast, replicative chain growth occurs in vivo at around 800 nucleotides per second. Second, with about 400 molecules of enzyme per cell, the enzyme is present in vast excess over the small number (fewer than 10) of replication forks per cell, suggesting that its function(s) is carried out elsewhere. Third, as noted earlier, DNA polymerase can extend chains only in a 5′→3′ direction. A complete understanding of replication requires knowing how both of the antiparallel chains of a DNA duplex are replicated within the same fork. Fourth, polymerase cannot initiate the synthesis of new DNA chains but can only extend from preexisting 3′ hydroxyl termini. Finally, genetic evidence suggested the existence of other polymerases, as well as additional enzymes and proteins essential to DNA replication.

TABLE **24.2 DNA polymerases of *E. coli***

Characteristic	Polymerase I	Polymerase II	Polymerase III
Structural gene	*polA*	*polB*	*polC*
Molecular weight	103,000	90,000	130,000
Number of molecules/cell	400	100	10
V_{max}, nucleotides/second	16–20	2–5	250–1000
3′ exonuclease	Yes	Yes	No[a]
5′ exonuclease	Yes	No	No
Processivity[b]	3–200	10,000	500,000
Mutant phenotype	[c]$UV^s MMS^s$	None	dna^{ts}
Biological function	DNA repair, RNA primer excision	SOS DNA repair?	Replicative chain elongation

[a]The 3′ exonuclease is carried on a separate polypeptide chain, the DnaQ protein.

[b]The number of nucleotides incorporated per encounter between polymerase and DNA (see page 894).

[c]MMS (methylmethane sulfonate) is a DNA-alkylating agent.

In 1969 John Cairns isolated an *E. coli* mutant that was deficient in DNA polymerase I activity. With this very abundant enzyme absent from the cell, it became possible to detect two additional DNA polymerase activities in *E. coli* cells. These were named DNA polymerases II and III, and the original Kornberg polymerase was named DNA polymerase I. The properties of these three DNA polymerases are summarized in Table 24.2.

DNA polymerase III was assigned the major role in nucleotide incorporation during replication, on the basis of its high V_{max}. Note also that there are few molecules of the enzyme per cell, as expected if it functions only at replication forks. More important evidence of this role is the existence of temperature-sensitive mutant cells that contain a thermolabile form of DNA polymerase III and in which DNA replication in vivo is blocked at high temperature. Both of these phenotypes can be traced to a single mutation, powerful evidence that polymerase III plays an essential role in DNA replication. The gene encoding the catalytic subunit of polymerase III was named *polC.*

There are also *polB* mutants, lacking polymerase II. Although these mutants originally displayed no other phenotype, recent evidence suggests that the enzyme participates in DNA repair synthesis (see Chapter 25). But what about polymerase I? In the original *polA* mutant described by Cairns, DNA replication occurred normally. However, the bacteria were abnormally sensitive to ultraviolet irradiation and to alkylating agents that react with DNA, suggesting a role for polymerase I in DNA repair. Further study of the Cairns mutant revealed a role in replication. Although lacking the *polymerase* activity of DNA polymerase I, the mutant bacteria did synthesize the small N-terminal fragment of the protein, which contains the 5′ exonuclease activity. Still later it was found that mutants lacking this 5′ exonuclease activity are also defective in DNA replication. Thus, it was established that two polymerases, I and III, play essential roles in DNA replication.

STRUCTURE AND MECHANISM OF DNA POLYMERASES

Sequence analysis of many DNA polymerases and crystallographic analysis of about a half dozen have revealed structural relationships indicating a common evolutionary origin and mechanism for this family of proteins, which also includes some DNA-templated RNA polymerases and RNA-templated DNA polymerases (reverse transcriptases, page 917). All of the polymerase structures described so far have a structure akin to a partially closed right hand with domains identified as "palm," "thumb," and "fingers," suggested in Figure 24.16 and shown also in Figure 24.17. Figure 24.17 shows the structure of the DNA polymerase encoded

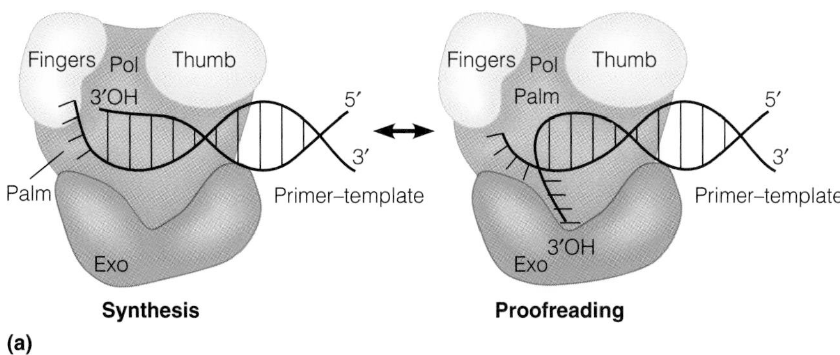

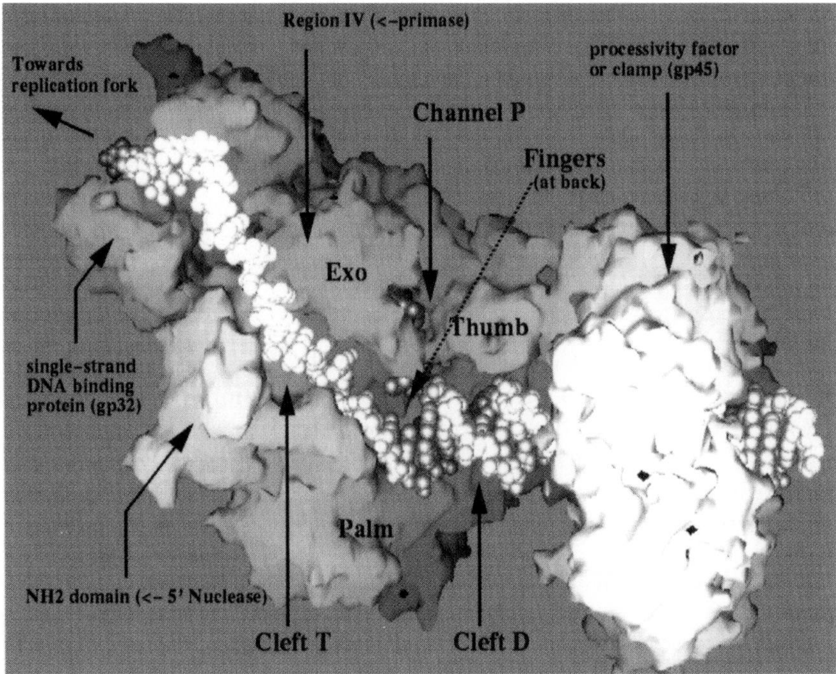

FIGURE **24.17**

Structure of RB69 phage DNA polymerase. **(a)** A general model for DNA polymerases, based upon the RB69 polymerase structure, showing the fingers, palm, and thumb domains common to all polymerases and showing primer–template DNA bound in both synthesis and proofreading modes. **(b)** A model for a replication complex, based upon the structure of RB69 DNA polymerase (center, with several structural domains identified); the predicted structure of the clamp protein, product of gene 45 (on the right, based on homology with β protein of *E. coli* DNA polymerase III holoenzyme); and SSB protein (on the left, based on crystal structure of the DNA-binding core of T4 gp32). The complex is moving right to left, with single-strand DNA visible at the upper left and duplex DNA visible behind the polymerase active site. A color version of this model is shown on page 877.

(a) Courtesy of T. A. Baker and S. P. Bell, *Cell* (1998) 92:295–305; **(b)** courtesy of J. Wang and T. A. Steitz, modified from J. Wang et al., *Cell* (1997) 89:1087–1099. © Cell Press.

by bacteriophage RB69, a close derivative of phage T4, and the first structure described for an enzyme known to carry out semiconservative DNA replication. Analyses of DNA–enzyme complexes, conserved amino acid residues, and targeted mutations have indicated that the polymerase active site lies in the palm domain, and the 3′ exonuclease site lies at the base of the palm. In 1998, high-resolution structures were described for two more polymerases (from T7 phage and from a thermophilic bacterium), each binding a template-primer DNA and a dNTP substrate (actually, a 2′,3′-dideoxyribonucleoside triphosphate or ddNTP, which could be incorporated but could not provide a 3′-hydroxyl terminus for further chain elongation). The structure of the T7 polymerase showed two magnesium ions, bound to nucleotide phosphates and to conserved aspartate residues already known to be essential for catalysis. This structure supported a general polymerase mechanism proposed by Thomas Steitz, on the basis of the Klenow fragment structure (Figure 24.18). In this mechanism one metal ion polarizes the hydroxyl group at the 3′ primer terminus, facilitating nucleophilic attack of that moiety upon the α-phosphate of the dNTP substrate. Both metals stabilize a pentacovalent transition state, in which the α-phosphorus is linked to five oxygens, and the second metal facilitates the leaving of the pyrophosphate. Both structures showed extensive contacts between the enzyme and the DNA minor groove—contacts that could occur only with a properly base-paired duplex. Also, both

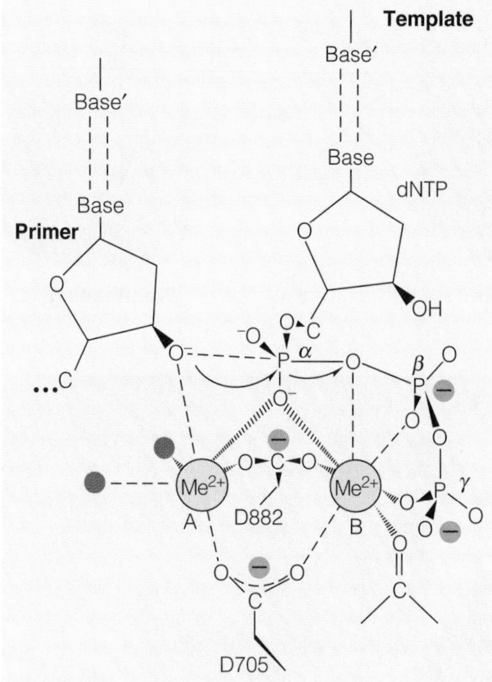

FIGURE **24.18**

The two-metal mechanism of the DNA polymerase reaction, as suggested from the structure of the T7 phage DNA polymerase–substrate complex. D705 and D882 are conserved aspartate residues, and the black dots are water molecules bound to metal ion A.

Courtesy of T. A. Steitz, *Nature* (1998) 391:231–232. © 1998 Macmillan Magazines, Ltd.

structures show that DNA near the primer terminus adopts a conformation more like that of A form than B form and that this conformational change facilitates the minor groove interactions. Moreover, the incoming dNTP was shown to fit snugly into a pocket that favors correct base-pairing with the template. Thus, these structures reveal both how the reaction is catalyzed and how the enzyme copies its template DNA with high precision.

THE DNA POLYMERASE III HOLOENZYME

The *polC* gene encodes a single polypeptide chain, with M_r of about 130,000. This protein has an intrinsic polymerase activity, but it is quite low. Within cells, the PolC protein functions as part of a multiprotein aggregate called the **DNA polymerase III holoenzyme.** As shown in Figure 24.19, the holoenzyme contains ten different polypeptide chains, each identified with a Greek letter. The α, ε, and θ subunits make up the "core polymerase," with α being the *polC* gene product, the protein with polymerase activity, and ε having a 3′ exonuclease activity comparable to the 3′ exonuclease domain of the DNA polymerase I polypeptide chain. The function of θ is unknown. The dimeric τ protein dimerizes the holoenzyme, holding leading and lagging strand polymerases together, so that both DNA strands are elongated at the replication fork, even though the lagging strand polymerase moves in the direction opposite that of fork movement. χ mediates the switch from RNA primers to DNA.

The β subunit was originally recognized as a protein essential for the *processivity* of DNA polymerase; in other words, essential for its ability to remain bound to the template through many cycles of nucleotide addition. The core polymerase, once bound to template DNA, remains bound only long enough to extend a primer strand by 10–20 nucleotides. However, β tethers the enzyme to DNA, allowing it to incorporate several thousand nucleotides per binding event. We say that β converts DNA polymerase III from a highly *distributive* enzyme, which incorporates just a few nucleotides per binding event, to a highly *processive* enzyme, which remains bound through thousands of incorporation reactions. Crystallography showed β to be a circular molecule with a 3.5-nm opening, capable of completely surrounding double-stranded DNA. As shown in Figure 24.20, six α-helical domains (three per subunit) face the interior of the circle, with the hydrophobic

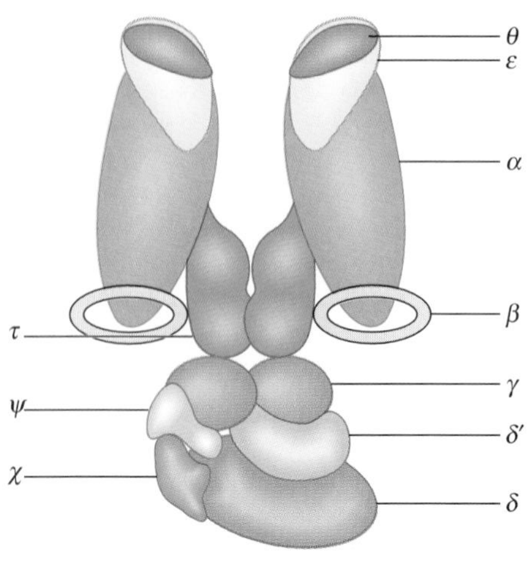

FIGURE **24.19**

Subunit structure of the *E. coli* DNA polymerase III holoenzyme. The Greek letter designations for the subunits are identified in the text.

Modified from D. R. Herendeen and T. J. Kelly, *Cell* (1996) 84:5–8.

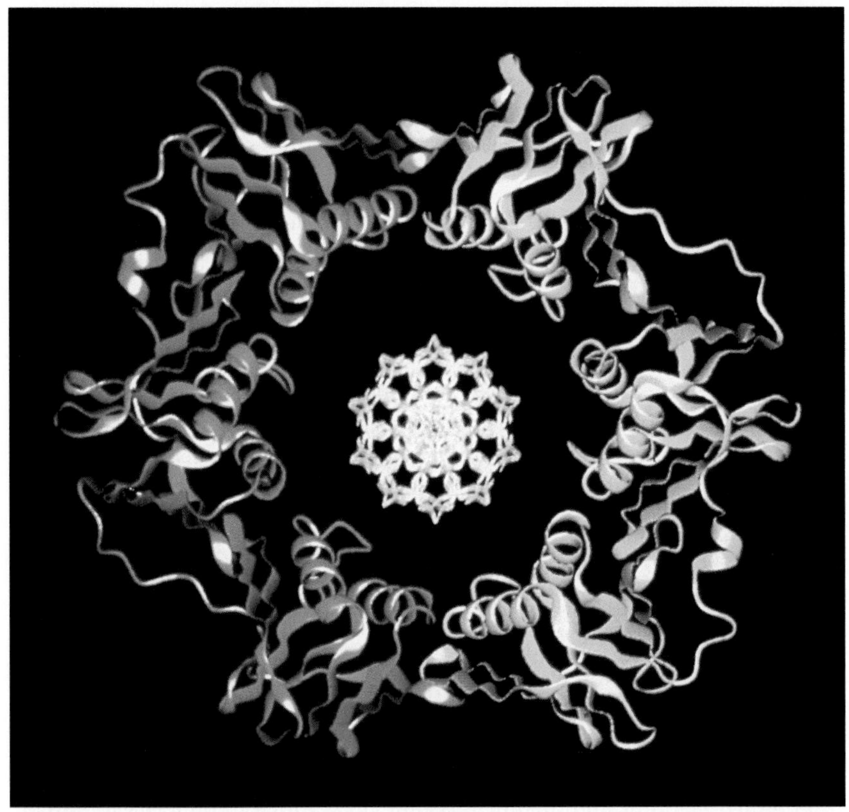

FIGURE **24.20**

Structure of the β subunit of *E. coli* DNA polymerase III holoenzyme. The protein forms a doughnut that can completely surround double-stranded DNA (modeled in as shown in the interior) and thus keep the polymerase associated with its DNA templates. The α helices on the inner surface of the subunit contact DNA but do not bind tightly enough to retard movement of the protein.

Courtesy of X-P. Kong, R. Onrust, M. O'Donnell, and J. Kuriyan, *Cell* (1992) 69:425–437. © Cell Press, Inc.

DNA polymerase III holoenzyme, a complex bacterial enzyme containing at least 10 subunits, plays the predominant role in replicative chain elongation.

residues in these helices having little attraction for DNA. Thus, the molecule acts as a sliding clamp, permitting polymerase to slide readily along DNA, but not to dissociate.

How does a circular molecule wrap itself around DNA to begin processive synthesis? That is the function of the remaining five proteins, which form the γ complex, an assembly that Mike O'Donnell has named the clamp loader. This complex contains one copy each of the δ, δ', χ, and φ proteins and two or three γ subunits. As shown by O'Donnell and his coworkers, γ is the protein that actually binds to β and carries out the ring opening that leads to clamp attachment. ATP is required, but remarkably, ATP does not have to undergo hydrolysis for opening of the β ring or its positioning around DNA. Instead, as schematized in Figure 24.21, a conformational change driven by ATP binding leads the complex to bind DNA and the γ protein to bring about ring opening. Once DNA has been encircled, the bound ATP is hydrolyzed, and the β ring closes. This event happens only once per round of replication on the leading strand. However, on the lagging strand, polymerase must rebind at the initiation of synthesis of each Okazaki

FIGURE **24.21**

A scheme for action of the clamp loader. ATP binding triggers a conformational change that exposes δ, the subunit that binds β. Once β and DNA are bound, γ forces the subunits of β apart, so that it can surround DNA. ATP hydrolysis leads to β ring closure and also sets in motion the events leading to dissociation of the γ complex so that it can start the process anew.

Courtesy of M. M. Hingorani and M. O'Donnell, *J. Biol. Chem.* (1998) 273:24550–24563. Reprinted by permission of the American Society for Biochemistry and Molecular Biology.

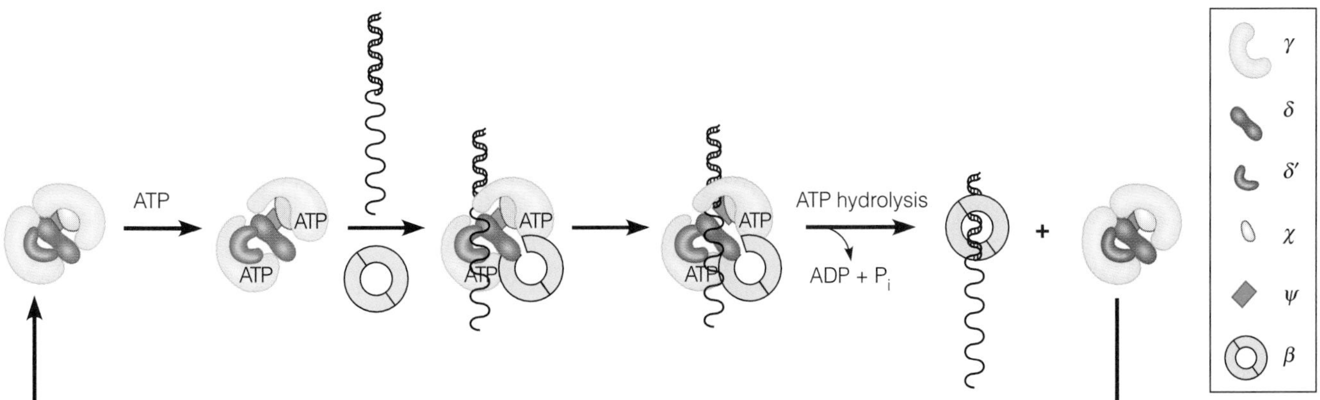

fragment and must dissociate when the 5′ end of the preexisting daughter DNA strand is reached. Thus, clamp loading must occur continuously and rapidly, with the γ complex serving also as a clamp unloader. And this remarkable process must all occur with the lagging strand core polymerase unit remaining bound to its leading strand partner, at the fork.

EUKARYOTIC DNA POLYMERASES

Eukaryotic cells contain five DNA polymerases. α, δ, and ε participate in chromosomal DNA replication, β participates in DNA repair, and γ participates in mitochondrial DNA replication.

Eukaryotic cells contain five distinct DNA polymerases—α, β, γ, δ, and ε. The eukaryotic enzymes are distinguished from each other in large part by their intracellular locations, kinetic properties, and responses to inhibitors. Even though the genetic analyses are not nearly as advanced as those for bacteria, some biological roles of these polymerases have been established. Table 24.3 summarizes the known properties of these enzymes. DNA polymerase γ is localized in mitochondria. It is a highly processive enzyme, with high affinity for its dNTP substrates. It participates in the replication of mitochondrial DNA, and it seems also to replicate the DNAs of certain small viruses that do not encode their own DNA polymerase. Polymerase β, found in the nucleus, has low processivity; it dissociates from the template after each nucleotide incorporation step. This enzyme, which has an alkaline pH optimum, is involved in DNA repair.

The remaining three polymerases—α, δ, and ε—all play essential roles in DNA replication. This function was established partly through the use of an inhibitor called **aphidicolin.** A fungal product with a steroidlike structure, aphidicolin inhibits replicative DNA synthesis specifically in eukaryotic cells. Polymerases α, δ, and ε are distinct from polymerases β and γ in being sensitive to inhibition by aphidicolin. Some aphidicolin-resistant eukaryotic cell mutants contain an aphidicolin-resistant form of DNA polymerase α and show aphidicolin-resistant DNA replication in vivo. This provides strong evidence that polymerase α participates in DNA replication, because a single mutational event changes the effects of the inhibitor both on the enzyme and on the DNA replication process.

HO CH2OH H H3C H HO HOH2C H CH3

Aphidicolin

Polymerases α and δ contain four and two subunits, respectively, but the subunit structure of polymerase ε is not yet clear. Polymerase α is distinctive in that one of its subunits has a primase activity, suggesting that this enzyme is responsible for lagging strand DNA replication. Polymerase δ is distinctive in that it requires an additional protein, called **proliferating cell nuclear antigen (PCNA),** to carry out highly processive DNA synthesis in vitro. PCNA plays a role compara-

TABLE 24.3 Properties of eukaryotic DNA polymerases

	α	β	γ	δ	ε
Cell compartment	Nucleus	Nucleus	Mitochondrion	Nucleus	Nucleus
Associated primase	Yes	No	No	No	No
Biological function	Lagging strand replication	DNA repair	Mitochondrial DNA replication	Leading strand replication	Replication
Number of subunits	4	1	4 (identical)	2	?
M_r of catalytic subunit, kilodaltons	160–185	40	125	125	210–230 or 125–140
K_M for dNTPs, μM	2–5	10[a]	0.5	2–4	?
Processivity (inherent)	Moderate	Low	High	Low	High
Processivity (with PCNA)	Moderate	Low	High	High	High
3′ exonuclease	No[b]	No	Yes	Yes	Yes
Sensitivity to 2′,3′-dideoxy-NTPs	Low	High	High	Low	Moderate
Sensitivity to arabinosyl-CTP	High	Low	Low	High	?
Sensitivity to aphidicolin	High	Low	Low	High	High

[a] K_M values for repair synthesis in permeabilized cells are much lower.

[b] A cryptic, or masked, 3′ exonuclease activity is associated with polymerase α from *Drosophila.*

ble to that of the β subunit of *E. coli* DNA polymerase III, and the two accessory proteins are related structurally. It has been suggested that polymerase δ is the principal leading strand polymerase. However, the more recent discovery in yeast of ε—a third replicative polymerase—has thrown into question the specific roles of α, δ, and ε. That is, why are three polymerases needed to replicate just two DNA strands? At present we don't know.

Other Replication Proteins

Investigations of DNA polymerases showed the chemical mechanism by which DNA chains are extended during replication. But the properties of polymerases alone do not tell us (1) how DNA replication is initiated, (2) how both antiparallel daughter strands are extended within a single replication fork, (3) how strands of the parental duplex are unwound to expose single-strand templates for new DNA synthesis, or (4) how newly replicated DNA assumes its final supercoiled structure. It is clear that several enzymes and proteins in addition to DNA polymerases must function in the total replication of a biologically complete DNA molecule.

Identification of these additional replication proteins in *E. coli* has involved extensive research on the replication of small DNA bacteriophages with circular, single-strand genomes (ϕX174, G4, M13, and fd). These small viruses replicate their DNA in infected cells primarily by using enzymes encoded by the host cell genome. Because the viral DNA molecules are small (5700 to 6400 nucleotides), they are readily isolated intact, whereas it is exceedingly difficult to isolate undegraded *E. coli* DNA molecules. Thus, we can identify bacterial proteins that play essential roles in chromosomal DNA replication, by virtue of their ability to catalyze the complete replication of these easily handled bacteriophage DNAs in vitro. The pathways for replication of these genomes are presented later in this chapter.

Another indispensable biological system has been bacteriophage T4 (Figure 24.22). This virus contains a duplex DNA genome, about 169,000 base pairs (or 169 kilobase pairs, or kbp) in length. Although large for a virus, this genome is less than 5% of the size of the *E. coli* genome. Phage T4 encodes nearly all of its own replicative proteins and enzymes. T4 mutants are readily isolated and mapped. In fact, T4 was the first biological system in which replication-defective mutants were described. This first occurred in 1965, when gene 43 on the T4 map (Figure 24.23) was identified as the structural gene for a virus-coded DNA polymerase. Three years

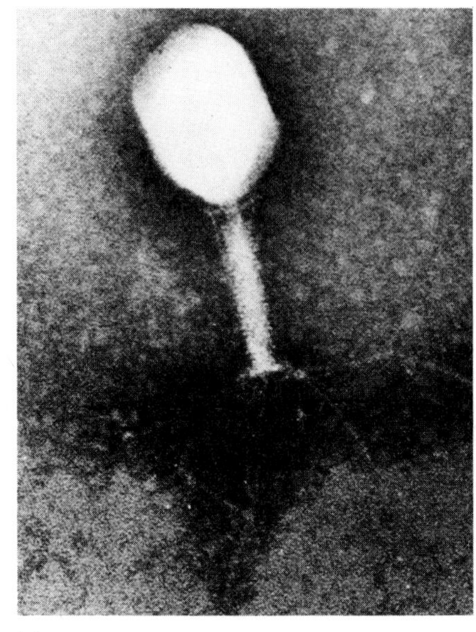

(a)

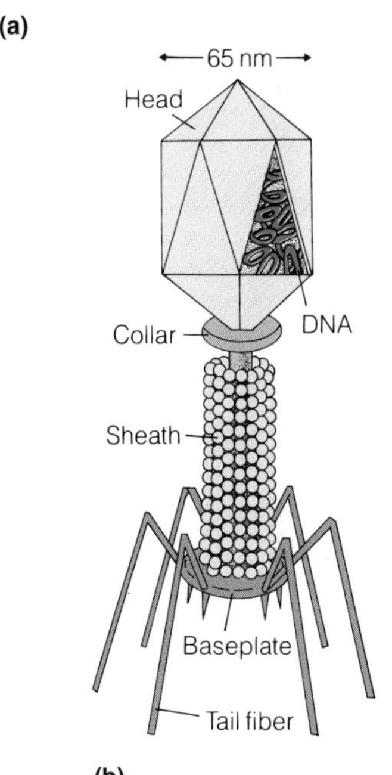

(b)

FIGURE **24.22**

T4 bacteriophage. (a) Electron micrograph. **(b)** Schematic visualization.

(a) Courtesy of L. Simon and T. F. Anderson, *Virology* (1967) 32:290. Academic Press, Orlando.

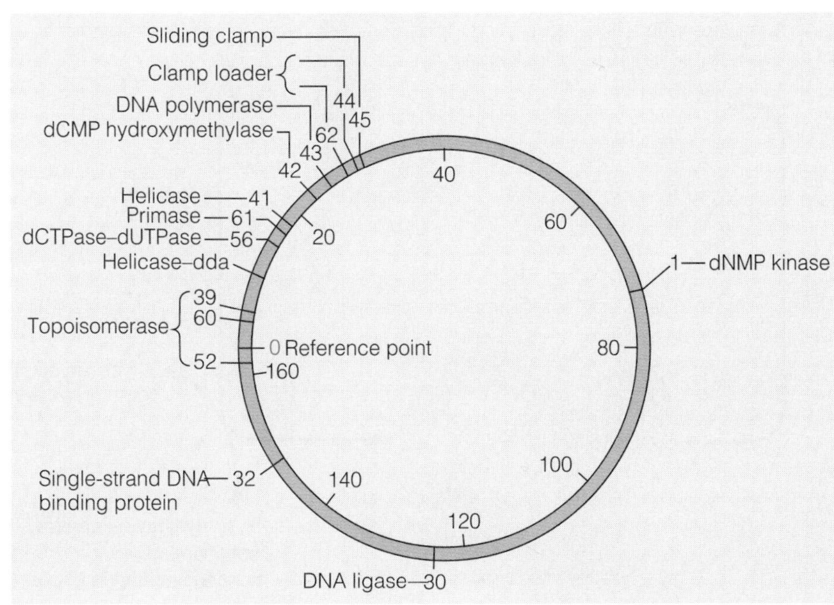

FIGURE **24.23**

Partial map of T4 genome. Genes whose products participate in DNA metabolism are shown on the outside of the circle. The inner numbers represent distances in kilobase pairs. The reference point (0) represents the divide between two particular genes, *rIIA* and *rIIB* (not shown).

later gene 32 was shown to encode the first known single strand–specific DNA-binding protein (SSB). Figure 24.23 shows locations on the T4 genome of 15 genes whose products play essential roles in phage DNA replication. Three of these gene products are involved in synthesizing the modified DNA base 5-hydroxymethylcytosine (see Figure 22.23). However, the rest all participate in replicating the T4 genome, by mechanisms similar to those in cellular DNA replication.

Having briefly introduced the major players (the small circular DNA phages and the large phage T4, as well as the common host, *E. coli*), let us discuss the additional replication proteins and their roles.

DNA LIGASE: DISCONTINUOUS REPLICATION OF DNA

The discovery of DNA ligase in 1967 provided the first insight into understanding how two DNA strands of opposite polarity could be extended within the same replication fork. As schematized in Figure 24.3, Reiji Okazaki proposed that ligase provided a route whereby DNA synthesized backward from the fork in short pieces could become covalently linked to high-molecular-weight DNA. Strong support for this model came with the discovery that gene 30 on the T4 genome encodes a DNA ligase. When *E. coli* was infected with *ts* gene 30 mutants at 42°C, all of the DNA synthesized was in the form of short fragments (see Problem 2 at the end of this chapter for more information). Only when the culture was cooled to 30°C, so ligase could act, was it possible to demonstrate the conversion of these fragments to high-molecular-weight DNA.

DNA ligase seals 3′ hydroxyl- and 5′ phosphate-terminated nicks in double-helical DNA, thereby attaching Okazaki fragments to high-molecular-weight lagging strand DNA.

DNA ligase covalently closes nicks in double-strand DNA. The nick must contain 3′ hydroxyl and 5′ phosphoryl termini, and the nucleotides being linked must be adjacent in a duplex structure and properly base-paired. DNA ligase is activated by adenylylation of a lysine residue in the active site (Figure 24.24). The enzyme in turn adenylylates the 5′ terminal phosphate of the DNA substrate, thereby activating it for nucleophilic attack by the 3′ hydroxyl group, with phosphodiester bond formation and displacement of AMP. The T4 phage enzyme uses ATP to adenylylate the enzyme, as do eukaryotic DNA ligases. However, the enzyme from *E. coli* and other bacteria uses NAD^+ instead. Instead of being a redox cofactor, the dinucleotide in this enzyme is cleaved, to yield adenylylated enzyme plus nicotinamide mononucleotide (NMN).

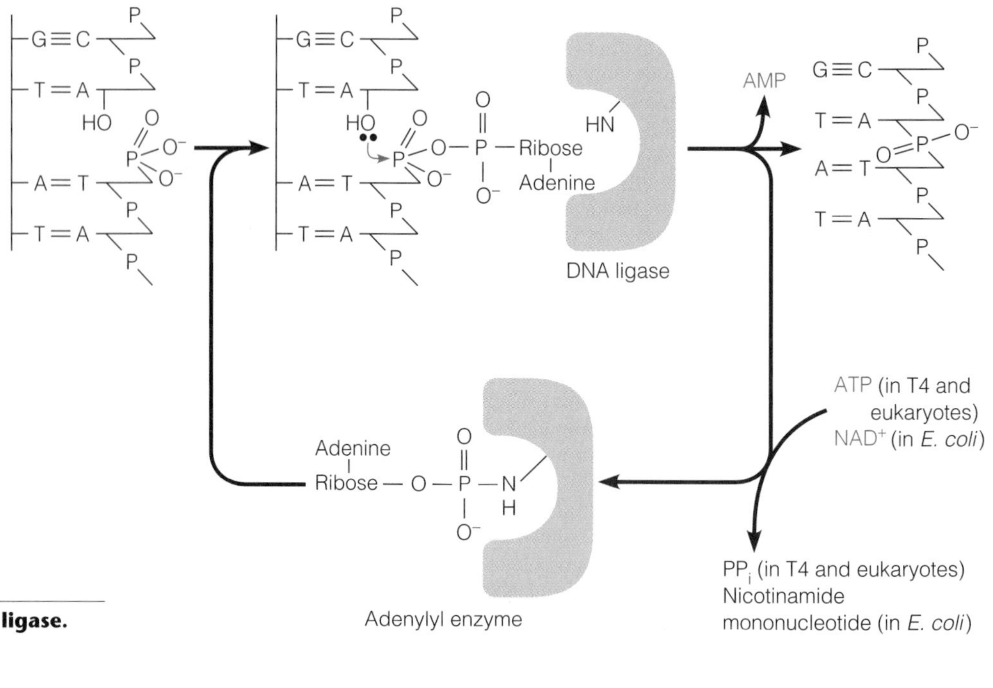

FIGURE **24.24**
The reaction catalyzed by DNA ligase.

PRIMASE: SYNTHESIS OF RNA LEADER SEQUENCES

The discovery of discontinuous DNA replication answered one question, but it focused attention on another. Given that DNA polymerases cannot initiate the synthesis of new DNA chains, but can only extend chains from preexisting 3′ hydroxyl termini, what provides those termini for both the initiation of chromosome replication and the initiation of each Okazaki fragment? When it was found that DNA replication of phage M13 was sensitive to an inhibitor of RNA synthesis and that DNA polymerase could extend from 3′ hydroxyl termini of RNA base-paired to DNA, it became evident that oligoribonucleotides might provide primers for biological DNA replication.

Early evidence for RNA priming in DNA replication came from an important experiment carried out by Reiji and Tuneko Okazaki. *E. coli* cells were made permeable to exogenous nucleotides by brief treatment with a buffer containing toluene. DNA synthesis occurred when these treated bacteria were incubated with a mixture of α-[^{32}P]dNTPs and unlabeled ribonucleoside triphosphates (Figure 24.25). DNA was then isolated and treated with mild alkali to hydrolyze any RNA. Now, if DNA synthesis starts from 3′ RNA termini, then for each Okazaki fragment there should exist one RNA–DNA junction, or deoxyribonucleotide residue covalently linked via a radioactive phosphate group to a ribonucleotide residue. Alkaline hydrolysis would then *transfer* that phosphate from the 5′ position of the original deoxyribonucleotide substrate to the 3′ position of the ribonucleotide at the junction. Sure enough, for each Okazaki fragment formed in this system, one radiolabeled phosphate was transferred to a ribonucleotide.

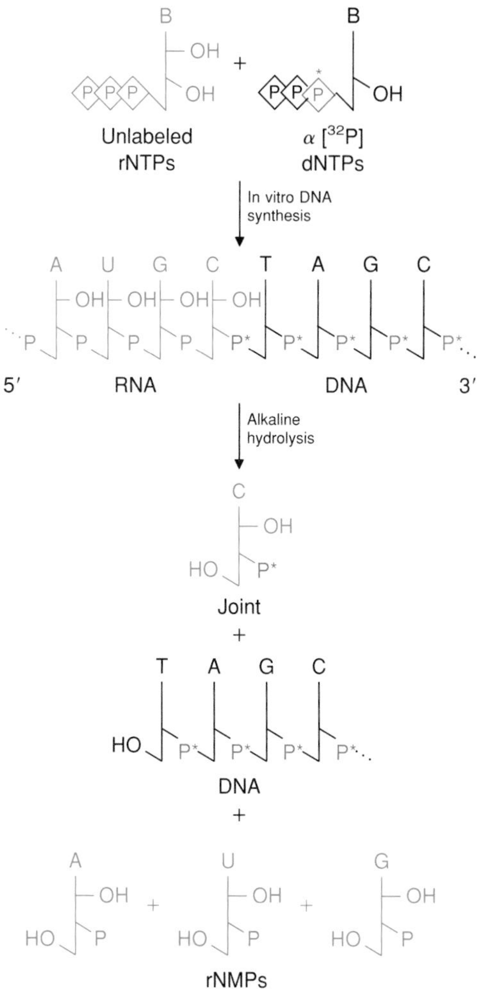

The enzyme that synthesizes RNA primers is called primase. In *E. coli* this enzyme is the product of the *dnaG* gene, whereas the comparable reaction in phage T4 is carried out by the product of gene 61. In both cases the enzyme is active only in the presence of other proteins (including a helicase), which create a complex called the primosome (see Figure 24.6). As noted on page 881, the primosome also participates in unwinding parental DNA strands ahead of the replication fork. Priming involves insertion of a ribonucleoside 5′-triphosphate opposite a deoxyribonucleotide residue in the template DNA, followed by sequential ribonucleotide additions to the 3′ hydroxyl terminus, just as occurs with DNA polymerases. At some point, RNA synthesis stops, the primosome dissociates, and DNA polymerase continues to extend from the 3′ hydroxyl terminus of the RNA primer, but now with incorporation of deoxyribonucleotides. RNA primer structure has been described in some detail. For instance, in T4 and *E. coli* DNA replication the primer lengths are 5 and 11 nucleotides, respectively, and most primers have ATP as the 5′ terminal nucleotide. However, we don't yet know all of the factors controlling the switch from RNA primer synthesis to lagging strand DNA synthesis. In eukaryotic cells, the primase is tightly associated with DNA polymerase α. This is strong evidence that pol α participates in lagging strand synthesis, while pol δ and/or pol ε carries out leading strand synthesis.

For Okazaki fragments to be ligated to high-molecular-weight DNA, the RNA primers must be excised and replaced with corresponding deoxyribonucleotides. In *E. coli,* as mentioned earlier, DNA polymerase I is involved in this process, through its nick translation activity. The removal of ribonucleotides from the 5′ end of the primer is coordinated with their replacement by deoxyribonucleotides (see Figure 24.5). In agreement with this idea, the *polA* mutant isolated by Cairns (see page 892) shows impaired conversion of Okazaki fragments to high-molecular-weight DNA. That conversion occurs at all is probably due to the 5′ exonuclease activity remaining in this mutant and possibly RNase H activity, plus the activity of another polymerase. *polA* mutants lacking both the 5′ exonuclease and the polymerase are defective in fragment sealing and display a conditional lethal phenotype, meaning that the cells cannot grow when the mutant gene is not expressed.

FIGURE **24.25**

The transfer experiment that demonstrated the existence of RNA primers in DNA replication. Each Okazaki fragment generated one radiolabeled ribonucleotide from its RNA–DNA junction after alkaline hydrolysis.

Primase, a special class of RNA polymerase, synthesizes short RNA molecules as primers for lagging strand DNA replication.

CLAMPS AND CLAMP LOADERS: PROCESSIVITY

DNA replication is an exceedingly efficient process. As noted earlier, *E. coli* replicates its entire chromosome in 40 minutes with only two replication forks, each of which involves the action of two molecules of DNA polymerase III. It is clearly advantageous for the cell to complete each round of replication, once initiated. For completion to occur, DNA polymerase must remain bound to its template—in short, it must act processively. We have already identified the β clamp as the polymerase-accessory protein in the DNA polymerase III holoenzyme most directly responsible for enhancing processivity and the γ complex as the clamp loader (see Figure 24.21). In phage T4 the product of gene 45 (gp45), which is structurally related to *E. coli* β protein, is the sliding clamp for processive DNA replication, and the clamp loader comprises just two proteins—gp44 and gp62, with gp44 being the ATP-cleaving protein. In eukaryotic cells a multisubunit protein called replication factor C, or RF-C, is the clamp loader, and PCNA, or proliferating cell nuclear antigen, is the sliding clamp. As shown in Table 24.3 (page 896), the addition of PCNA to polymerase δ in vitro greatly increases the processivity of that enzyme.

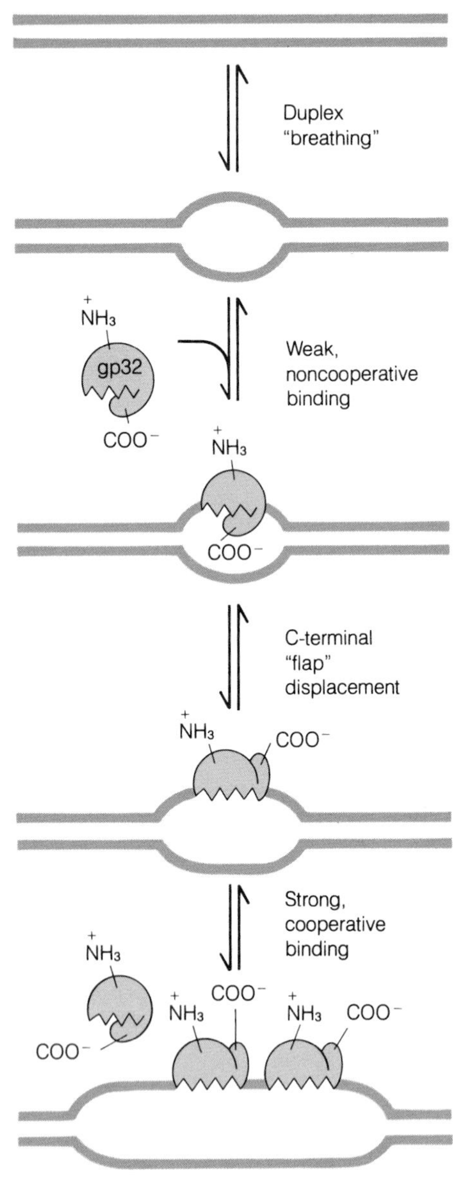

FIGURE **24.26**

gp32 facilitation of both denaturation and renaturation of DNA. When only a short length of single-strand DNA is exposed, gp32 binds weakly. The folded-down configuration of the C-terminal domain prevents cooperative binding of other gp32 molecules and, hence, strand unwinding ("breathing"). If a longer region of single-strand DNA is exposed after the first gp32 molecule has bound, the C-terminal domain can shift to its "up" configuration, enlarging the DNA binding site, permitting cooperative binding of additional gp32 molecules, and, thus, further extending the denatured region. For simplicity, binding is shown on only one of the two denatured DNA strands.

SINGLE-STRAND DNA-BINDING PROTEINS: MAINTAINING OPTIMAL TEMPLATE CONFORMATION

One of the earliest replication proteins to be identified, other than DNA polymerase itself, was a T4 protein called either *single-strand DNA-binding protein* (SSB) or *helix-destabilizing protein.* In an early application of affinity chromatography, Bruce Alberts immobilized DNA by binding it to cellulose and analyzed the T4 proteins that were retained by a column of this material. One protein was shown to be the product of gene 32, because the protein isolated from a *ts* gene 32 mutant was unable to bind to DNA at a restrictive temperature. Because gene 32 mutants were known to be defective in DNA repair and genetic recombination, as well as in DNA replication, it was clear that the protein played multiple roles in DNA metabolism.

Analysis of the purified gp32 showed that it binds specifically to single-strand DNA. Moreover, binding is strongly *cooperative,* meaning that the protein is far more likely to bind to DNA adjacent to a site already occupied than to an isolated site. In other words, binding of one gp32 molecule facilitates the binding of others, and the protein tends to bind in clusters. Thus, gp32 promotes the denaturation of DNA. Although it does not initiate denaturation, its presence lowers the melting temperature of DNA by as much as 40°C.

The role of gp32 is to keep the template in an extended, single-strand conformation, with the purine and pyrimidine bases exposed so that they can base-pair readily with incoming nucleotides. This function is essential for DNA repair and genetic recombination (see Chapter 25), as well as for replication. Given that all three processes also involve the re-formation of duplex structures, with one strand each of parental and daughter DNA, it is of interest that gp32 facilitates the renaturation of single-strand DNA, as well as the denaturation of duplexes.

How can a single protein promote both duplex formation and duplex unwinding? The answer seems to lie in special design features of the gp32 molecule. Note that *gp32 does not itself unwind DNA strands.* Rather, it stabilizes single-strand DNA by binding *after* a region of DNA is unwound. Limited proteolysis of gp32 removes a C-terminal fragment. This modification renders the protein a stronger DNA denaturant in vitro, with a small increase in its equilibrium binding affinity for DNA. As suggested in Figure 24.26, the C-terminal domain provides a "flap" that partly covers the DNA-binding domain of the protein. When a short region of single-strand DNA is exposed, through partial and reversible unwinding of the duplex (sometimes called breathing), gp32 can bind with the flap down, but only to a small site (3 or 4 nucleotides). Occupancy of a complete site (7 to 10 nucleotides) is thereby inhibited, rendering binding both weak and noncooperative. If continued helix unwinding occurs, the flap can move up, permitting the complete site to be occupied and promoting further denaturation.

Single-strand DNA-binding proteins are essential in DNA replication, repair, and recombination by virtue of their ability to facilitate both DNA denaturation and renaturation.

The partially proteolyzed gp32 lacks the ability to renature DNA in vitro, and it seems likely that the conformation of the native gp32 is a critical determinant of whether the protein functions primarily as a denaturant (flap-up position) or renaturant (flap-down position). The position of the flap is determined in part through interaction with other proteins at the replication fork. Thus, the protein can act both to stabilize a single-strand template as it pairs with incoming nucleotides and to facilitate re-formation of a duplex after DNA polymerase has passed by. Partial proteolysis can also remove an N-terminal domain, which has been shown to be essential for self-association of the protein and, hence, for cooperative DNA binding.

Single-strand DNA-binding proteins have now been found in many organisms. The protein of *E. coli* (specified by the *ssb* gene) also binds cooperatively to single-strand DNA. However, the mechanism of binding appears to be quite different from that in T4. In *E. coli,* DNA is wrapped about the outer surface of the tetrameric SSB protein. Moreover, under some conditions the SSB protein binding shows *negative* cooperativity. It has been suggested that the protein binds in different modes, depending on whether it is participating in replication, DNA repair, or recombination.

In eukaryotic cells a heterotrimeric protein called replication factor A (RFA) serves the role of SSB in DNA replication. This protein undergoes phosphorylation during S phase or after DNA damage, suggesting a role in cellular coordination of DNA metabolism.

HELICASES: UNWINDING DNA AHEAD OF THE FORK

Single-strand DNA-binding proteins do not themselves denature DNA. As noted, they stabilize single-strand DNA but cannot actively unwind duplex DNA strands. Such unwinding must occur if single-strand templates are to be exposed for polymerase action. The helicase proteins have this ability. They catalyze the ATP-dependent unwinding of double-strand DNA. *E. coli* cells contain at least a half dozen different helicases, some of which participate in DNA repair and some in bacterial conjugation. The principal helicase in DNA replication is DnaB (the protein product of the *dnaB* gene), which interacts with DnaG and other proteins to form the primosome (see Figure 24.6). The comparable roles in T4 DNA replication are played by gp41 (helicase) and gp61 (primase). In phage T7 a single protein (gp4) has both helicase and primase activities.

Helicases are multimeric proteins that bind preferentially to one strand of a DNA duplex and use energy of ATP hydrolysis to actively unwind the duplex.

All known helicases are multimeric proteins. Most are homodimers, but a few are homohexamers. In vitro, each helicase binds initially to single-stranded DNA, adjacent to a duplex region, and proceeds in a fixed direction ($5' \rightarrow 3'$ or $3' \rightarrow 5'$), displacing the unbound DNA strand as it moves, and with ATP hydrolysis coupled to movement. Although the helicases are homo-oligomers, they are structurally asymmetric. For example, in the well-studied dimeric Rep helicase of *E. coli,* the ATP-binding and DNA-binding properties of the two subunits are quite different. This suggests a rolling or "hand-over-hand" mechanism, in which each subunit alternates between tight-binding and loose-binding conformations, depending upon whether ATP or ADP is bound at a given instant (Figure 24.27). Consistent with this model, the crystal structure of the Rep helicase complexed with single-strand DNA and ADP shows a dramatic structural difference between the subunits, resulting from a 130° rotation of one domain about a hinge region. This is shown in Figure 24.28.

The hexameric helicases, exemplified by the gene 4 product of phage T7, may move by a quite different mechanism. This ring-shaped protein, which uses dTTP

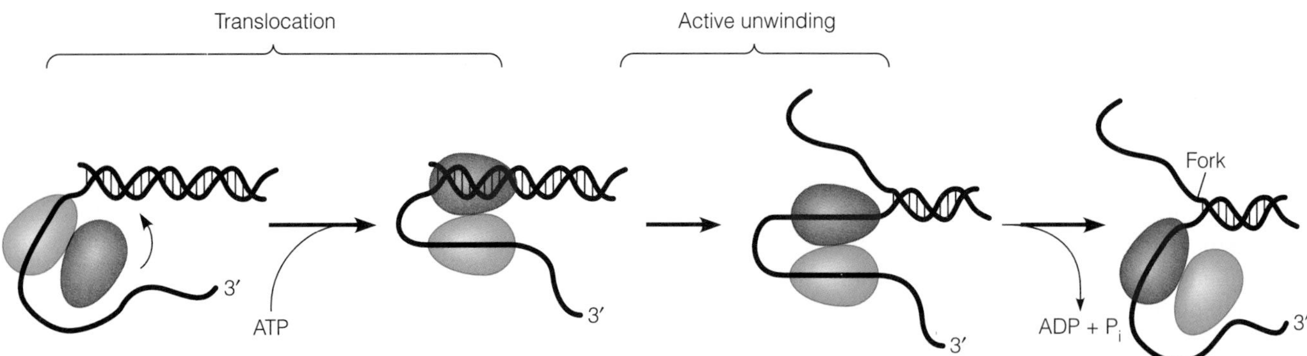

FIGURE **24.27**

A model for helicase action. In this model a homodimeric enzyme, such as the *E. coli* Rep helicase, shows 3′→5′ polarity; it travels from the 3′ end to the 5′ end of the DNA to which it binds. In the first step, binding of ATP activates the subunit shown in red to bind to DNA at the double-strand/single-strand junction and, in the second step, to unwind a few base pairs. In the third step, the hydrolysis of bound ATP weakens the binding of the blue subunit to DNA, causing its dissociation, and thereby positioning it to begin the cycle anew, by invading the DNA duplex ahead of the fork that has been created, once another ATP has been bound. Thus, by alternating in their binding to DNA, the two subunits cause the enzyme to "roll" counterclockwise as shown in this model, unwinding the duplex as it moves.

Courtesy of G. Waksman from S. Korolev et al., *Cell* (1997) 90:635–647. © Cell Press.

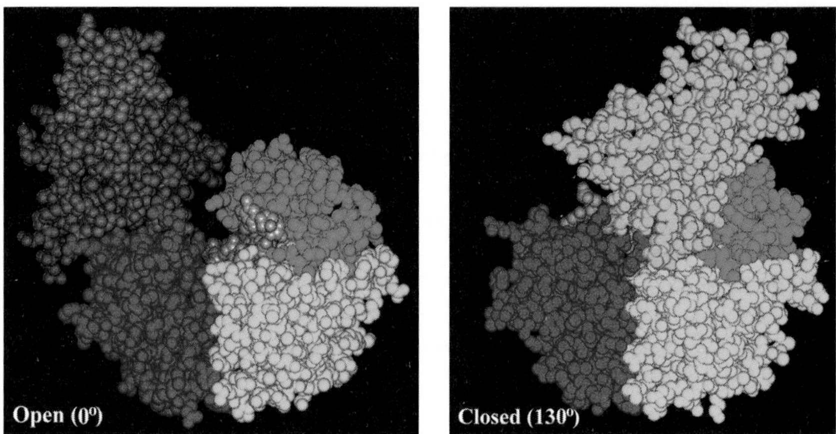

FIGURE **24.28**

Structure of the *E. coli* Rep helicase. In these space-filling models, one monomer of the dimeric protein is shown above in the "open" conformation and below in the "closed" conformation. The four structural domains of the protein are shown in red, yellow, green, and blue, and bound DNA is shown in pink. In the transition from open to closed, the domain shown in blue (left) and blue-green (right) rotates 130° from left to right as shown. The structure of a helicase–ssDNA complex shows one of the monomers to be open and one closed, with DNA bound only to the open form, thus confirming essential features of the model shown in Figure 24.27.

Courtesy of G. Waksman from S. Korolev et al., *Cell* (1997) 90:635–647. © Cell Press.

rather than ATP as its energy source, wraps about the single strand to which it binds and along which it moves. The strand that becomes displaced by helicase action does not pass through the central hole. Three of the six subunits bind and hydrolyze dTTP, while the other three bind dTTP in noncatalytic fashion. Data on dTTP binding and hydrolysis suggest a rotary motion for the protein, comparable to that seen in the mitochondrial F_0F_1 ATP synthase (see Figure 15.14), suggesting the intriguing possibility that the helicase moves along the DNA by rotation.

Considerable excitement was engendered by recent findings that two inherited human diseases, Werner's syndrome and Bloom's syndrome, result from helicase defects. Both conditions involve increased susceptibility to cancer, and Werner's syndrome patients also undergo premature aging, usually going gray in their twenties, becoming blind from cataracts, and dying of natural causes before age 50. Positional cloning (Chapter 25) of the defective genes responsible for these conditions revealed in both cases that the genes encode proteins related to the *E. coli* *recQ* gene product. In *E. coli* this helicase participates in a homologous recombination pathway and may be involved in resumption of DNA replication after repair

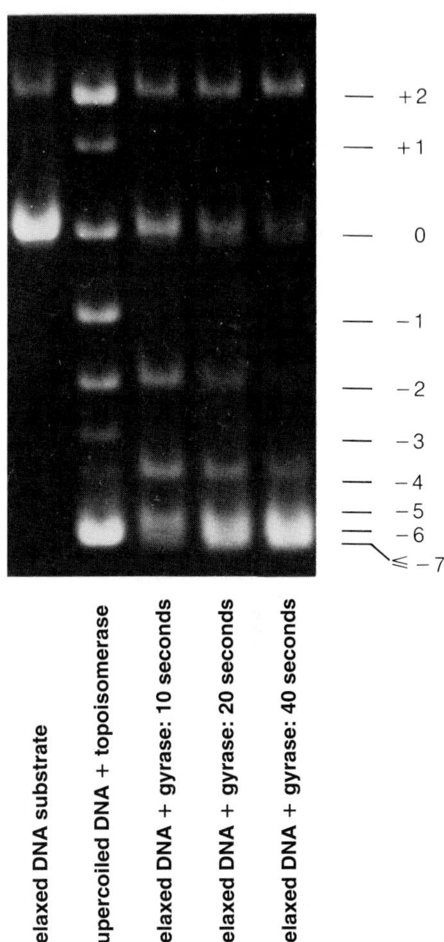

FIGURE **24.29**

Action of type I and type II topoisomerases, as shown by gel electrophoresis. Lane 1 shows a relaxed circular DNA. Lane 2 shows the pattern from treatment of supercoiled DNA with type I topoisomerase. Lanes 3–5 show relaxed circles treated with DNA gyrase, a type II topoisomerase, for different lengths of time. Note that more different topoisomers can be seen in topoisomerase I reaction mixtures, as expected if changes in the linking number (ΔL) occur in units of 1, whereas gyrase changes L in units of 2.

Reprinted with permission from P. O. Brown and N. R. Cozzarelli, *Science* (1979) 206:1081–1083. © 1979 by AAAS.

of radiation-induced DNA damage. Like the RecQ protein, the Werner's syndrome protein has been shown to have a 3′→5′ helicase activity. These findings provide intriguing clues to understanding relationships between genomic instability and both cancer and aging.

TOPOISOMERASES: RELIEVING TORSIONAL STRESS

Bidirectional replication of the circular *E. coli* chromosome unwinds about 100,000 base pairs per minute. Were there not some mechanism for relieving this torsional stress, DNA ahead of the fork would become overwound as DNA at the fork became unwound, and replication could not be sustained. Topoisomerases, a group of enzymes that can interconvert different topological isomers of DNA (see Chapter 4), provide a "swivel" mechanism for relieving this stress. Topoisomerase action is demonstrated most simply in vitro by the relaxation of supercoiled DNA. One can incubate supercoiled DNA with a purified topoisomerase and observe, by gel electrophoresis, the intermediate stages in conversion of the supercoiled substrate to relaxed circular DNA containing no superhelical turns (Figure 24.29). This analysis reveals the existence of two general classes of topoisomerases—type I enzymes, which change the linking number in units of 1, and type II enzymes, which change the linking number in units of 2.

Actions of Type I and Type II Topoisomerases

A type I topoisomerase breaks just one strand of the duplex (Figure 24.30). The enzyme remains covalently attached to the 5′ end of the broken strand, by

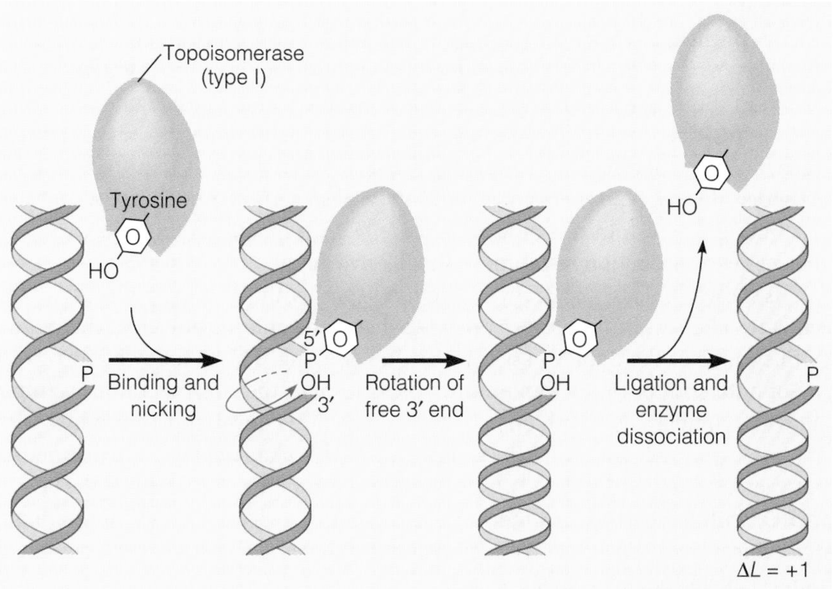

FIGURE **24.30**

Action of a type I topoisomerase. The enzyme breaks one strand and immobilizes the 5′ end by a covalent bond between the DNA phosphate and a tyrosine residue (in *E. coli* topoisomerase I). Rotation of the 3′ end is followed by resealing. The linking number is increased by 1 in the example shown (an underwound DNA). Action of a type I topoisomerase on overwound DNA would decrease the linking number, by essentially the same mechanism.

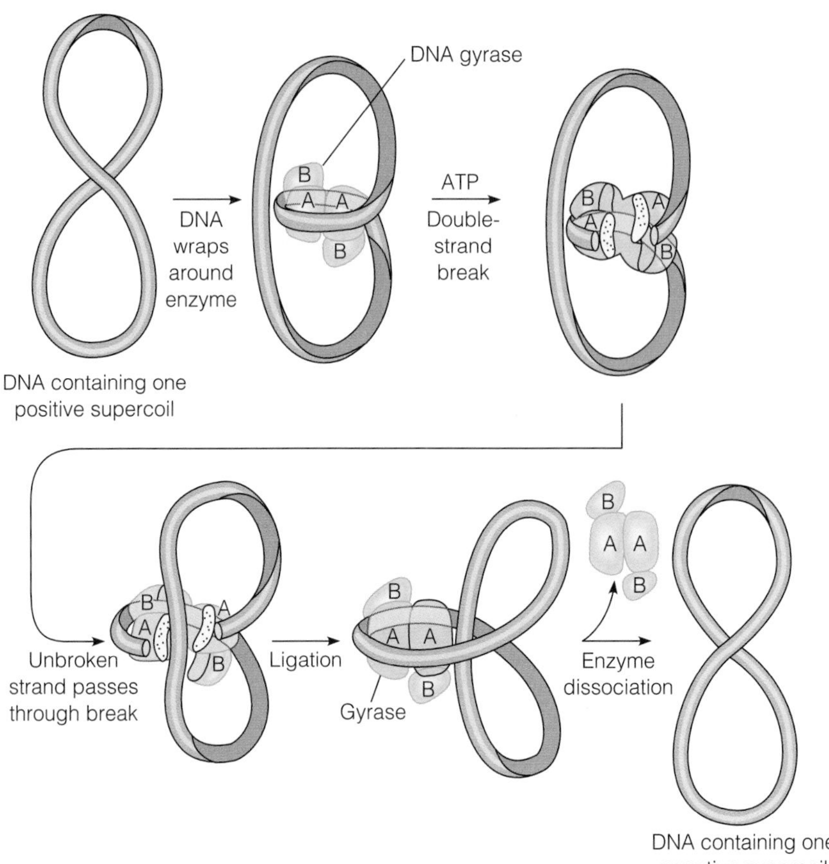

FIGURE **24.31**

Action of a type II topoisomerase. DNA gyrase of *E. coli;* the example shown is a tetrameric protein with two A and two B subunits. The enzyme is shown introducing two negative turns and changing the linking number from +1 to −1. The enzyme catalyzes a double-strand break, and the two DNA ends are bound by A subunits, which move the DNA ends apart, so that the unbroken duplex can pass through the gap. Resealing converts the positive supertwist to a negative one, giving the overall molecule a ΔL of −2. Type II topoisomerases can relax underwound duplexes by the reverse of the above pathway.

Adapted from A. Kornberg and T. A. Baker, *DNA Replication*, 2nd ed. (New York: W. H. Freeman and Co., 1992), p. 390.

forming a phosphodiester bond between the 5′ phosphate and a tyrosine hydroxyl. The 3′ end is then free to rotate (by one turn in the example shown). The hydroxyl group on the 3′ end then attacks the activated, covalently bound 5′ phosphate, closing the nick—in fact, *E. coli* type I topoisomerase was originally called nicking–closing enzyme. The result is that the linking number has been changed by 1. Eukaryotic topoisomerase I acts similarly, but the 3′ end, not the 5′ end, is immobilized during the reaction.

By contrast, a type II topoisomerase catalyzes a double-strand break, and the unbroken part of the duplex passes through the gap that is created (Figure 24.31). The most thoroughly studied type II topoisomerase is an *E. coli* enzyme also called **DNA gyrase,** because it can not only relax a supercoiled molecule but also introduce negative superhelical turns into DNA. ATP hydrolysis is required for both activities of most type II enzymes. DNA gyrase is a tetramer, with two A and two B subunits. The A subunits bind and cleave DNA, while the B subunits carry out the energy transduction resulting from ATP hydrolysis.

As shown in Figure 24.31, gyrase action begins with DNA wrapping about the enzyme. The A subunit cleaves both DNA strands and immobilizes them, and both strands of the duplex pass through the opening. This is followed by resealing of both strands of the duplex and dissociation of the enzyme. In the example shown, a circular DNA with one positive supercoil is converted to a product with one negative supercoil. Thus, the linking number is changed by 2, the distinction between a type I and a type II topoisomerase. In Figure 24.29, note that DNA treated with a type I topoisomerase (lane 2) shows twice as many intermediates as identical DNA treated with a type II enzyme (lanes 3–5), because the type I enzyme changes linking number in units of 1.

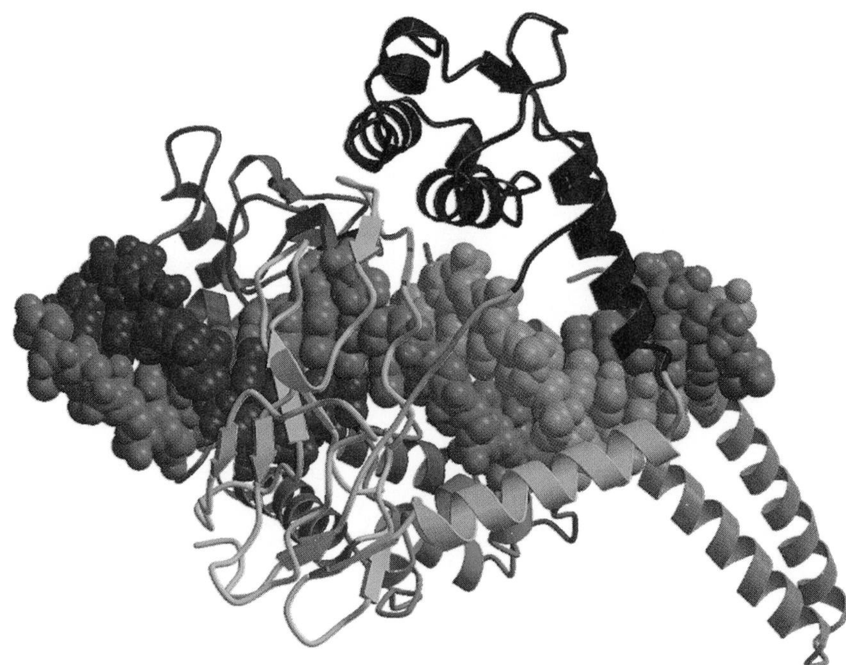

FIGURE **24.32**

Crystal structure of human topoisomerase I in complex with a 22-bp DNA duplex. The DNA strand nicked by the enzyme is shown in purple upstream of the cleavage site and in pink downstream, and the intact strand is blue. For the protein, each individual domain appears in a different color.

Courtesy of M. R. Redinbo, L. Stewart, J. J. Champoux, and W. G. J. Hol from *Science* (1998) 279:1504–1534, with permission from *Science*.

The action of type I topoisomerases was recently clarified with publication of the crystal structure of human topoisomerase I. As shown in Figure 24.32, the enzyme completely wraps around its DNA substrate. Not evident from the figure is the fact that most of the DNA–protein contacts involve the DNA sugar–phosphate backbone rather than the bases, meaning that DNA is bound as an undistorted B-form helix. Also, the contacts on the "upstream" (5′) side of the scissile bond are far more numerous than those on the downstream (3′) side. Because the 5′ nucleotide in the nick is immobilized by its binding to a tyrosine residue, the structural data indicate that the free 3′ end is relatively free to rotate, as it must do if the supercoiled DNA is to become relaxed. A useful consequence of this structure determination was clarification of the mechanism of action of anticancer drugs that act by inhibiting topoisomerases, because the structure reveals how such inhibitors can be bound, and it should lead to the design of more effective inhibitors. One inhibitor in current clinical use is **camptothecin.**

Camptothecin

The Four Topoisomerases of *E. coli*

Since the discovery of topoisomerase I and DNA gyrase in the 1970s, *E. coli* has been shown to contain four different topoisomerases. The terminology is a bit confusing, because the enzymes named topoisomerase I and topoisomerase III are both type I topoisomerases, and topoisomerase II (also called DNA gyrase) and topoisomerase IV are both type II topoisomerases. Of these four enzymes, DNA gyrase plays the dominant role during replicative chain elongation, both in relieving stress ahead of the fork and in introducing negative supercoils into newly synthesized DNA. We know of this role primarily through studies of the properties of gyrase inhibitors. The gyrase A subunit is the target for binding of **nalidixic acid,** a compound long known to inhibit DNA replication. Another replication inhibitor, **novobiocin,** binds to the B subunit and inhibits ATP cleavage. Mutant bacteria resistant to nalidixic acid or novobiocin show structural alterations in subunit A or B, respectively. Inhibitors such as these are useful as antibacterial drugs.

Type I topoisomerases break and reseal one DNA strand, and type II topoisomerases catalyze double-strand breakage and rejoining; hence, type I and type II enzymes change DNA linking number in units of 1 and 2, respectively.

Topoisomerase IV plays a critical role in the completion of a round of replication. Type II topoisomerases catalyze a variety of topological interconversions,

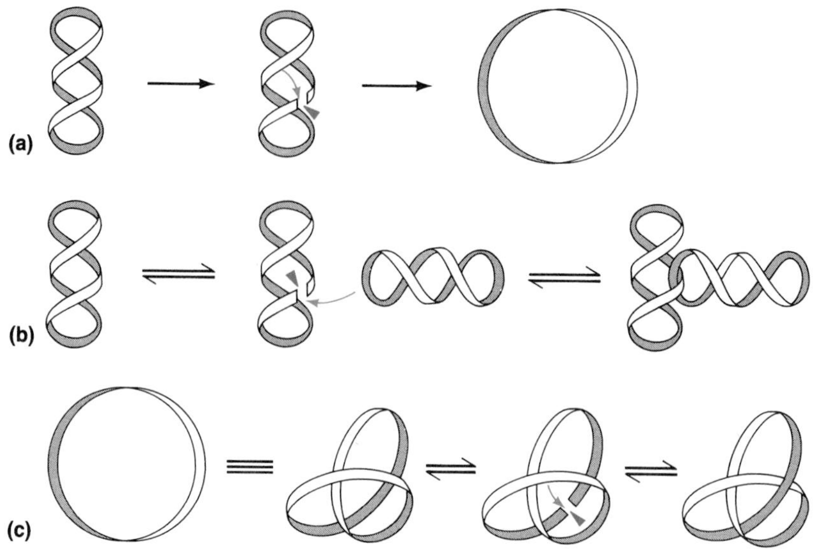

FIGURE **24.33**

The types of topological interconversions catalyzed by type II topoisomerases. (a) Relaxation. **(b)** Catenation and decatenation. **(c)** Knotting and unknotting.

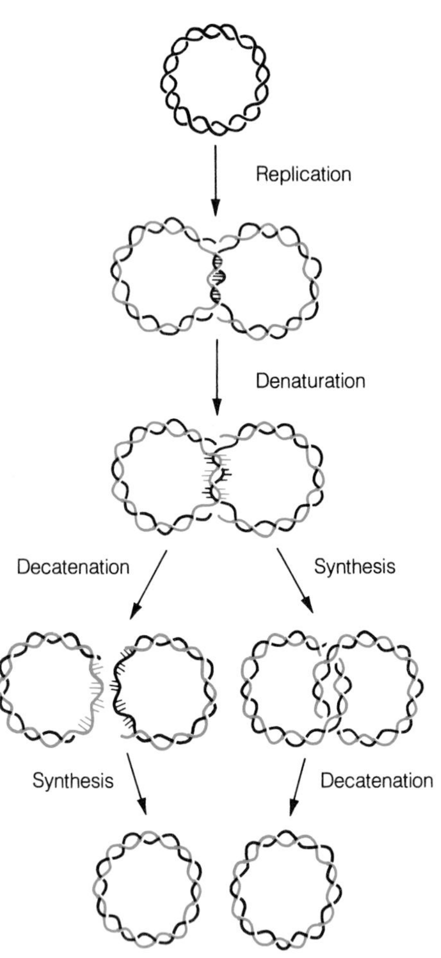

FIGURE **24.34**

Topoisomerase action in termination of replication. In the absence of topoisomerases, steric forces would prevent gyrase from unwinding DNA as replication forks approached each other. Topoisomerase allows the circles to decatenate. It is not known whether decatenation occurs before or after the completion of replication. Both possibilities are shown here. Additional proteins are involved in termination (see page 912).

including knotting, unknotting, **catenation** (linking), and **decatenation** (unlinking) of circular DNAs, as shown in Figure 24.33. A circular DNA nearing the end of a round of replication will generate two interlinked circles, so a type II topoisomerase action is necessary for separating the newly replicated molecules. As two forks approach each other at the replication terminus, steric barriers ultimately interfere with the unwinding activities of topoisomerases ahead of the two forks (Figure 24.34). At this stage the two incompletely replicated chromosomes are still interlinked. Topoisomerase IV plays a specific role in the decatenation process shown in Figure 24.34. This finding suggests a degree of specificity in topoisomerase action that is not immediately apparent from the idea that a strained DNA conformation is being relaxed. In accord with this suggestion, recent studies of type II topoisomerases show that they do not generate equilibrium mixtures of topoisomers. For example, a topoisomerase IV limit digest (DNA digested until no further reaction occurs) was shown to have far fewer knotted and catenated structures than an equilibrium mixture; somehow, the enzyme selectively "disentangles" DNA molecules. Also, when a mixture of topoisomers generated by topoisomerase I action was subsequently treated with topoisomerase IV under the same conditions, the average linking number—the number of supercoils—remained the same, but the distribution of topoisomers about the mean became much tighter. ATP hydrolysis is required for this selectivity to be achieved, but remarkably, these observations indicate that a topoisomerase molecule can somehow scan a DNA molecule much larger than itself and direct the specific kinds of topological changes that occur.

URACIL-DNA *N*-GLYCOSYLASE: REMOVAL OF INCORPORATED URACIL

Uracil can base-pair with adenine in a DNA duplex, and DNA polymerases readily accept deoxyuridine triphosphate as a substrate in place of deoxythymidine triphosphate. Yet cells possess a rather elaborate two-stage mechanism that prevents deoxyuridylate residues from accumulating in DNA. The first stage of this mechanism, described in Chapter 22, involves an active deoxyuridine triphosphatase that cleaves dUTP to dUMP and PP_i. The second stage involves **uracil-DNA *N*-glycosylase,** an enzyme that removes any dUMP residues that might have arisen through incorporation of a dUTP that escaped the action of dUTPase.

As shown in Figure 24.35, uracil-DNA *N*-glycosylase hydrolytically cleaves the glycosidic bond between N-1 of uracil and C-1 of deoxyribose. This yields free uracil and DNA with an **apyrimidinic site,** that is, a sugar residue lacking an

A DNA repair process removes uracil residues in DNA, whether they arose through deamination of cytosine residues or incorporation of deoxyuridine nucleotides instead of thymidine nucleotides.

attached pyrimidine. Another enzyme, **apyrimidinic endonuclease,** recognizes this site and cleaves the phosphodiester bond on the 5′ side of the deoxyribose moiety. This is followed in bacteria by the nick translation activity of DNA polymerase I, which excises the deoxyribose phosphate residue at this site, along with several more nucleotide residues downstream from the site (only one, dGMP, is actually shown in the figure). The excised nucleotides are replaced with fresh deoxyribonucleotides, including dTTP at the original site occupied by dUMP. The process is completed with DNA ligase action.

Why go to all this trouble just to replace a nucleotide that does not affect the information encoded in DNA? The likely answer is that uracil substituted for thymine (that is, base-paired with adenine) is probably not the true target of this DNA repair system. Uracil residues in DNA can also arise through spontaneous deamination of cytosine residues. The latter alteration does change the genetic sense, because it converts a G–C base pair to a G–U pair, and in a subsequent round of replication the U-containing strand would give rise to an A–T base pair. The uracil repair system prevents this mutation but does not discriminate between uracils paired with adenines or with guanines. Consistent with this model is the **hypermutable** phenotype displayed by mutants lacking an active dUTPase. Such strains exhibit elevated rates of spontaneous mutagenesis, resulting from the accumulation of dUMP residues in their DNA.

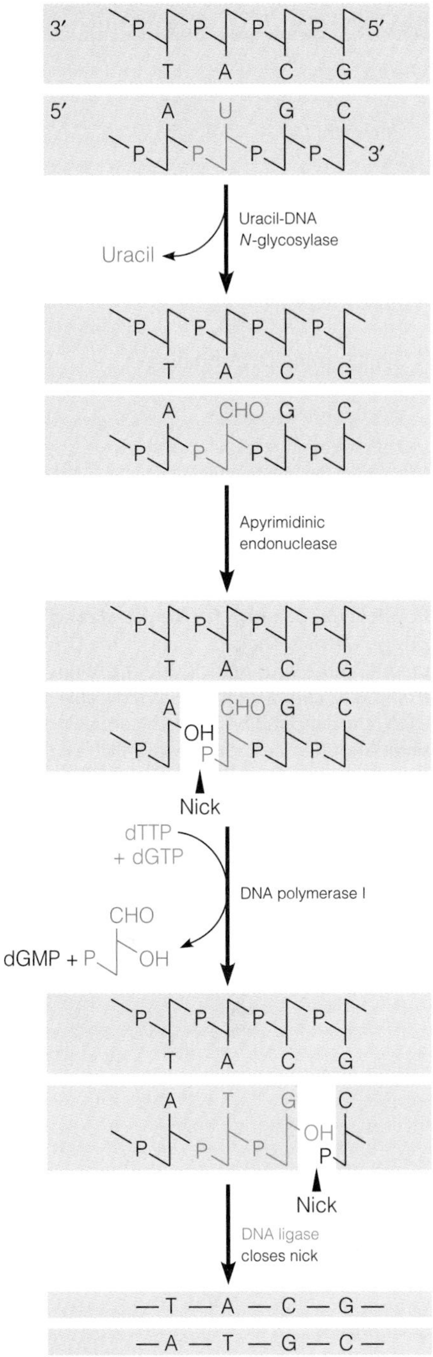

FIGURE **24.35**

Action of the DNA uracil repair system. Uracil-DNA *N*-glycosylase removes uracil, leaving an apyrimidinic site. A specific endonuclease recognizes this site and cleaves on the 5′ side. Nick translation by DNA polymerase I and DNA ligase action replace the apyrimidinic nucleotide and several adjacent nucleotides and seal the nick.

Reconstruction of Replication Machines

It is apparent that many proteins must function at or near the replication fork, in a coordinated fashion, for rapid and accurate copying of both strands of double-helical DNA. Much of our understanding of these events has come from partial or complete reconstitution of a multiprotein replication apparatus, or *replisome,* from purified components. The most informative systems are those using mostly T4 phage proteins and those using mostly *E. coli* proteins to replicate the genomes of small single-strand DNA phages. In 1994 the first reconstituted eukaryotic replication system was described—purified human proteins that replicate the small circular DNA genome of the virus SV40. The specific roles played by each protein have their counterparts in the T4 and *E. coli* systems, as shown in Table 24.4. Parallel investigations of yeast have revealed a similar picture, with the major exception that yeast has now been shown to contain no less than six different DNA polymerases.

The similarities shown in Table 24.4 are not meant to imply that DNA replication processes are identical in prokaryotes and eukaryotes. For example, in eukaryotic DNA replication, RNA primer removal requires two RNase H enzymes. The first, called RNase H1, removes all of the ribonucleotides except the one at the RNA–DNA junction, with removal of this remaining nucleotide requiring a multifunctional enzyme called FEN1/RTH1, which contains both endonuclease and exonuclease activities.

In studying DNA replication in human cells, it is advantageous to use a small DNA genome, like that of SV40, which depends almost completely upon proteins of the host cell to carry out its own replication, because the viral DNA and its replicative intermediates are far easier to isolate and characterize than high-molecular-weight cellular DNA species. Note that SV40 uses just one of its own proteins (T antigen), while the rest of the process uses proteins encoded by the host cell.

The same logic guided Arthur Kornberg, Jerard Hurwitz, and others, who earlier used purified *E. coli* proteins to describe the replication of the genomes of small single-strand DNA phages such as ϕX174, M13, and G4. Knowing the details of the replication cycles of these phages is important both in a historical sense and because of the widespread use of these phages in recombinant DNA procedures such as site-directed mutagenesis (Chapter 25).

TABLE **24.4 Functions of specific proteins in reconstituted DNA replication systems**

Function	*E. coli*	Phage T4	SV40/human
Polymerase	α subunit of pol III holoenzyme	gp43	pol δ, α subunit of pol α-primase
Primase	DnaG	gp61	Primase subunit of pol α-primase
Proofreading exonuclease	ε subunit of pol III holoenzyme	3′-exo domain of gp43	3′-exo domain of pol δ
Sliding clamp	β subunit of pol III holoenzyme	gp45	PCNA
Clamp loader	γ complex	gp44/62	RF-C
Helicase	DnaB	gp41	SV40 T antigen
Single-strand DNA-binding protein	SSB	gp32	RP-A
RNA primer removal	RNase H, pol I	T4 RNase H, *E. coli* pol I	RNase H1, FEN1/RTH1
Lagging strand transfer from primase to polymerase	χ subunit of pol III holoenzyme	Unknown	Unknown

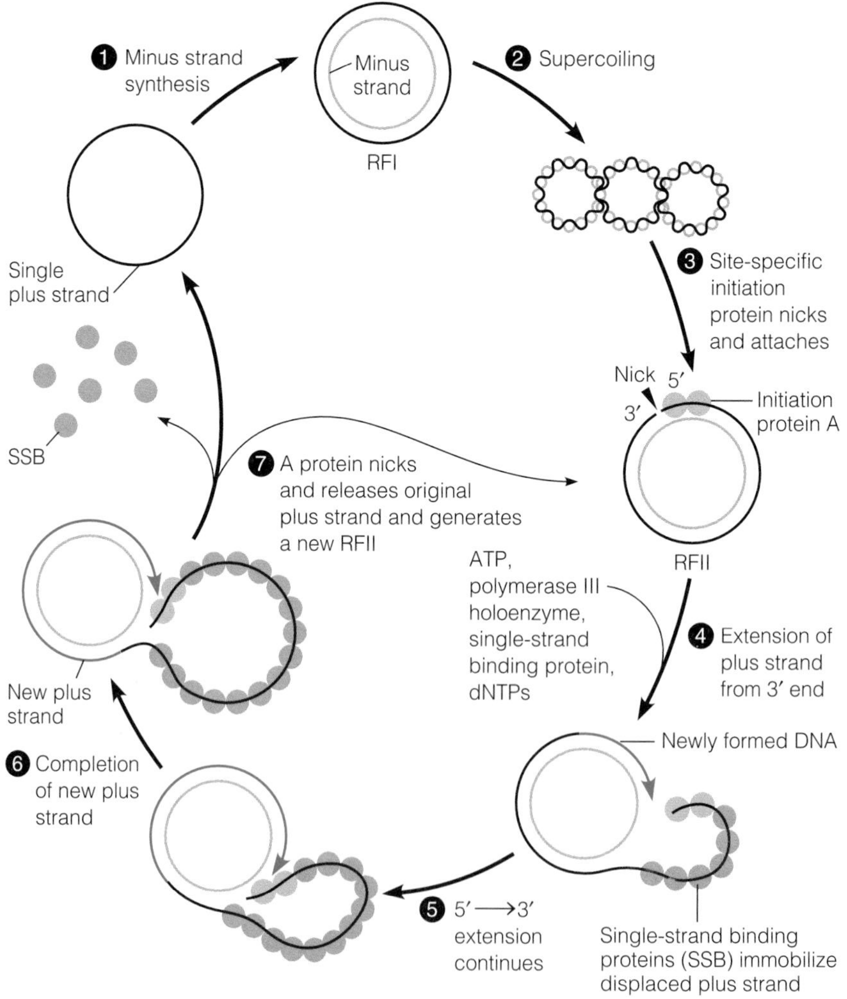

FIGURE **24.36**

Replication scheme for single-strand phage DNAs. The plus strand is the DNA present in the virus particle and the minus strand is its complement. RFI is replicative form DNA (circular double-stranded) containing no single-strand interruptions, while RFII is a circular duplex containing at least one nick or gap. The three phages studied (ϕX174, M13, and G4) differ in the details of initiation of step 1, but are identical with respect to the other steps.

Although molecular details vary, all of these phages replicate their DNA by the pathway schematized in Figure 24.36. The single-strand genome, after entry into the host cell, serves as a template (step 1) for production of a circular duplex DNA, called a **replicative form (RF).** (A covalently closed circular duplex is called

RFI.) In this structure the DNA strand derived from the viral particle is called the **plus strand,** and the newly synthesized complement is called the **minus strand.** The duplex RF becomes supercoiled through the action of DNA gyrase (step 2). It then undergoes replication by a **rolling circle** mechanism to give single-strand progeny DNA, as follows. In step 3, a phage-coded, site-specific endonuclease forms a nick in the plus strand, yielding **RFII,** a circular duplex molecule with at least one single-strand interruption. Polymerase extends from the 3′ hydroxyl terminus of the nick and proceeds around the circular minus strand template (steps 4–6). The 5′ end of the plus strand is displaced from the template and is ultimately packaged into virus particles. When polymerase has proceeded once around the circle, the site-specific endonuclease acts again (step 7), releasing the displaced plus strand and a new RFII. This circular duplex contains within its one single-strand interruption a 3′ hydroxyl terminus to which nucleotides can be added in a subsequent round of plus strand synthesis.

The foregoing discussion focuses on the elongation phase of DNA replication. Let us turn now to the initiation process.

Initiation of DNA Replication

When we ask how DNA replication is initiated from an origin, we are asking three interrelated questions: (1) What are the site-specific DNA–protein interactions that trigger initiation? (2) How do proteins act after binding to origin sequences? (3) How is the process controlled? Initiation seems to be the major target for control of replication. However, we know much less about the control of replication than we know about the control of transcription or translation. It is evident that intracellular contacts, of a still undefined type, link the replication apparatus to other cellular structures, so that DNA replication is coordinated with the cell cycle. In prokaryotes it is likely that replication occurs at a site attached to the cell membrane, whereas eukaryotic DNA replication may occur at a DNA-and-protein-containing structure called the nuclear matrix (see Chapter 28). The nature and significance of these physical linkages have remained elusive. Most of the events in replication have been duplicated in soluble cell-free systems, without components of the membrane or nuclear matrix. Therefore, it has been difficult to define a biochemical role for the membrane or matrix in replication.

REQUIREMENTS FOR INITIATION OF REPLICATION

Because replication proceeds from fixed origins, there are two requirements for initiation: (1) a nucleotide sequence that specifically binds initiation proteins and (2) a mechanism that generates a primer terminus to which nucleotides can be added by DNA polymerase. A number of phage, bacterial, plasmid, and organelle replication origins have been isolated by gene cloning, and their nucleotide sequences have been determined. In general, these origins include repeated sequences of either identical or opposite polarity (**direct repeats** or **inverted repeats,** respectively). This finding suggests that initiation proteins bind in multiple copies.

The two most straightforward ways to generate a primer terminus at the origin are (1) nicking a strand of the parental duplex to expose a 3′ hydroxyl terminus and (2) unwinding the parental duplex and synthesizing an RNA primer to expose a 3′ hydroxyl *ribonucleotide* terminus. Phages ϕX174 and G4 initiate the conversion of RF DNA to single-strand progeny DNA by nicking with a site-specific endonuclease, the viral *cisA* gene product. By contrast, duplex DNA replication, where studied, occurs without nicking of the parental duplex and with the synthesis of RNA primers.

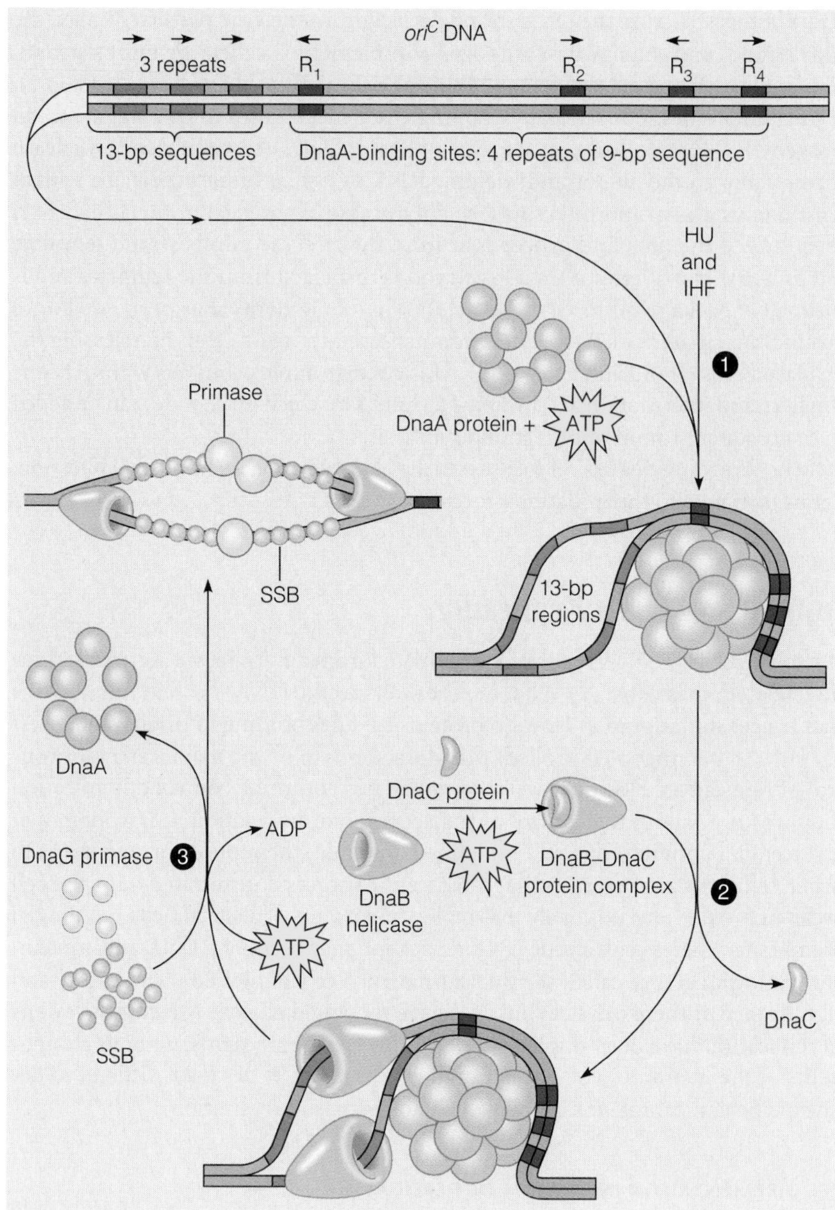

FIGURE **24.37**

A model for initiation of *E. coli* DNA replication at *ori*ᶜ. HU and IHF are double-strand DNA-binding proteins that facilitate DNA bending at the origin.

Redrawn with permission from T. A. Baker and S. H. Wickner, *Annu. Rev. Genet.* (1992) 26:447–477. © 1992 Annual Reviews.

Protein binding, causing DNA to bend at the origin, initiates duplex DNA replication. Stress from bending causes nearby DNA to unwind, and primosomes assemble in the forks, forming RNA primers that are extended by DNA polymerases.

INITIATION OF *E. coli* DNA REPLICATION AT *ori*^{c}

Initiation of replication of the *E. coli* chromosome is reasonably well understood, because the origin sequence has been cloned into plasmids whose replication from the origin can then be studied in vitro. This origin sequence, called *ori*^{c}, is 245 base pairs long. It contains four repeats of a 9-base-pair sequence that binds an initiation protein, the *dnaA* gene product. To the left of these sites, as shown in Figure 24.37, are three direct repeats of a 13-base-pair sequence that is rich in A and T and thus is relatively easily denatured. The sequence also contains binding sites for several basic proteins (HU and IHF) that facilitate DNA bending, an important step in the sequence leading to initiation.

Step 1 is binding of 10 to 20 molecules of a complex of DnaA protein and ATP. The protein is activated for this step by reacting with the phospholipid cardiolipin (see Chapter 19), a process that may represent part of the coordination between DNA replication, membrane growth, and chromosome partitioning at cell division. Binding of DnaA, plus the basic proteins, bends the DNA rather sharply and creates negative super-helical tension. In turn this tension causes DNA unwinding in the 13-base-pair regions, opening up a short single-strand loop. Aided by DnaC, another initiation protein, the DnaB helicase in step 2 binds in both forks of this loop, and the helicase activity further unwinds this structure. In step 3, and subsequent steps (not shown), DnaG primase binds, and RNA primers are formed. Some of the first primers may be synthesized also by RNA polymerase, the enzyme responsible primarily for transcription. However formed, the RNA primers are extended by DNA polymerase III on both leading and lagging strands, and the two forks in the initiation complex mature, with a leading and a lagging strand in each fork.

Investigations in several phage and plasmid systems suggest that the sequence just described is a general mechanism for initiating DNA replication. A DNA–protein complex is formed, the resulting strain on the DNA forces unwinding in a nearby region, and assembly of a primosome and synthesis of RNA primers occur within this unwound structure. We still have rather incomplete understanding of what controls this process and how it is coordinated with the cell growth and division cycle.

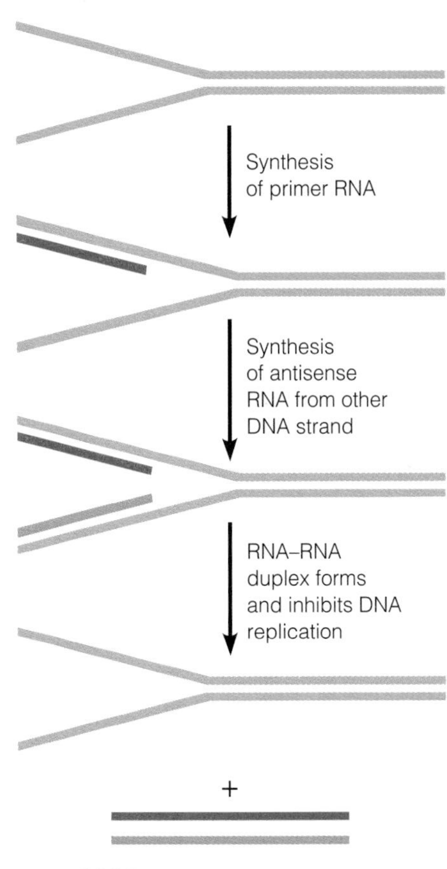

FIGURE **24.38**

Control of plasmid DNA replication by antisense RNA.

PLASMID DNA REPLICATION: CONTROL OF INITIATION BY RNA

The replication of several small circular DNAs, including plasmids and organelle DNA, is being investigated in cell-free systems. An interesting feature of plasmid DNA replication is the control of initiation exercised by "antisense RNA." In the *E. coli* plasmid ColE1, RNA can be synthesized at the replication origin from the DNA strand opposite that used to synthesize the initiation primer (Figure 24.38). This inhibits the initiation of replication, because the primer RNA becomes tied up as part of an RNA–RNA duplex. Further analysis of this system, particularly the control of antisense RNA synthesis, may reveal how the initiation of DNA replication itself is controlled.

MITOCHONDRIAL DNA: TWO UNIDIRECTIONAL REPLICATION FORKS

An unusual mode of replication initiation is seen with animal cell mitochondrial DNA. As shown in Figure 24.39, unidirectional replication of this 16-kilobase circular duplex starts from a fixed origin on one strand (the L, or light, strand). As replication commences, the parental H (heavy) strand is displaced by the nascent H strand, forming a structure called a displacement loop, or D loop, which can be visualized in the electron microscope. When this unidirectional replication apparatus has progressed two-thirds of the way around the circle, a new origin is exposed on the displaced H strand, and unidirectional L-strand synthesis from this origin starts backward from the direction of H strand replication. The entire process requires about 1 hour, consistent with the much slower

FIGURE **24.39**

Replication of mitochondrial DNA. Parental heavy (H) and light (L) chains are shown in brown. New strands are in blue and pink.

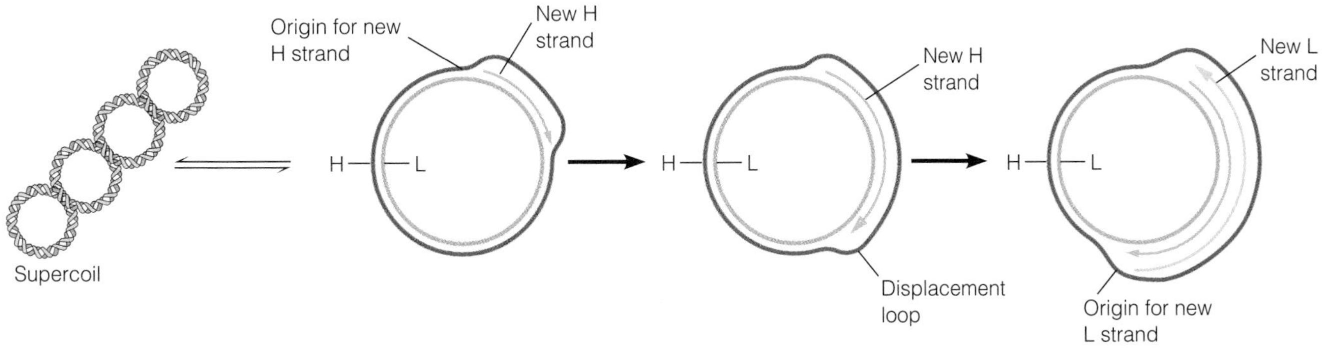

Supercoil
Origin for new H strand
New H strand
H — L
New H strand
Displacement loop
New L strand
Origin for new L strand

replication of eukaryotic DNAs than of prokaryotic DNAs (see Table 24.1). RNA priming is involved, as first shown by the finding that mature mitochondrial DNA contains a few ribonucleotides, apparently remnants of an inefficient primer excision system.

Replication of Linear Genomes

Thus far we have discussed in detail only the replication of circular DNA genomes. Linear genomes, including those of several viruses as well as the chromosomes of eukaryotic cells, face a special problem—how to complete replication of the lagging strand (Figure 24.40). Excision of an RNA primer from the 5′ end of a linear molecule would leave a gap that cannot be filled by DNA polymerase action, because of the absence of a primer terminus to extend (Figure 24.41a, step 1, page 914). If this DNA could not be replicated, the chromosome would shorten a bit with each round of replication. Viruses reveal at least three strategies for dealing with this problem. Phages T4 and T7 exhibit **terminal redundancy,** the duplication of a small part of the genome at each end of the chromosome. Thus, recombination can occur at the incompletely replicated ends of two nascent DNA molecules without loss of genetic information (Figure 24.41a, steps 2 and 3). This process is repeated in subsequent rounds of replication (step 4) until the end-to-end linear aggregate (called a **concatemer**) is more than 20 times the length of a single phage chromosome. A virally specified nuclease then cuts this giant DNA into genome-length pieces when it is packaged into phage heads (step 5).

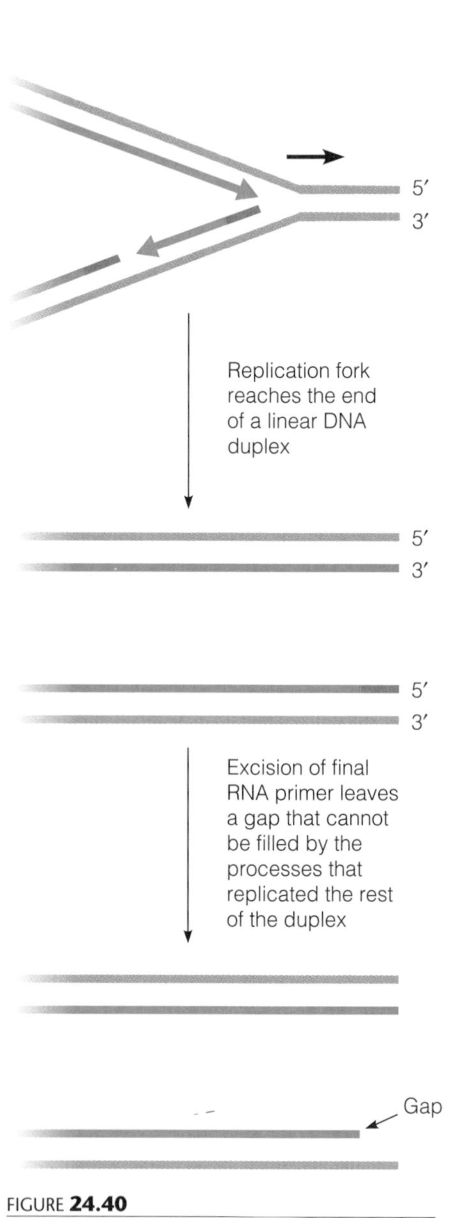

FIGURE **24.40**
The problem of completing the 5′ end in copying a linear DNA molecule.

Bacteriophage ϕ29 and the adenoviruses have evolved a quite different strategy (Figure 24.41b). The genomes of these viruses contain inverted repeat sequences at the ends. Replication begins at one end of the linear duplex, with a protein called **terminal protein** serving as the primer (step 1). This protein in adenovirus reacts with dCTP to form a dCMP residue covalently linked through its phosphate to a serine residue (see margin, page 913). The dCMP serves as primer for replication of the 3′-terminated strand, with the 5′ end being displaced as a single strand (step 2). Once this strand is completely displaced (step 3), it can form a short duplex region by base-pairing with the repeat sequence at the other end of the molecule (step 4), which can serve as substrate for initiation of a second round of displacement synthesis (step 5), yielding complete replication of both strands. ϕ29 replicates similarly, except that the protein-bound primer nucleotide is dAMP, rather than dCMP.

Still another mechanism is involved in the replication of poxviruses, such as vaccinia (Figure 24.41c, page 915). The two strands at each end of this linear genome are covalently linked together. Such a structure, when approached by a replication fork, can support movement of the leading strand *around* the link between the strands (step 3), so that the final lagging strand primer can be replaced by DNA as in a circular duplex. Rearrangement (steps 4 and 5) restores the original conformation.

Eukaryotic chromosomes contain linear DNA molecules. An entirely different mechanism, involving an enzyme called **telomerase,** is used to preserve information at their chromosome ends. We discuss this mechanism in Chapter 28.

Although circular chromosomes lack ends to serve as natural replication termination sites, they do have specialized termination mechanisms. We have already mentioned the role of topoisomerase IV in decatenating chromosomes that have nearly completed a round of replication. Bacterial chromosomes also have specific termination sites. In *E. coli,* replication terminates in the *ter* region, which is located 180° away from *ori*^{C} on the chromosome (see Figure 24.8). This region consists of six homologous 20-base-pair segments, three each oriented in opposite directions. A protein called Tus binds tightly at each of these sites. Binding of

Tus blocks replication forks proceeding in only one direction. Because three sites are oriented in each direction, this arrangement ensures that each of the two forks initiated at *ori*[c] will travel as far as the *ter* site and no farther. The crystal structure of the Tus protein complexed with DNA suggests a mechanism for polarized termination. As shown in Figure 24.42, two α helices in the β domain (shown in green) project into a bound DNA duplex. It is suggested that DnaB helicase approaching from the top as shown in the figure would generate single-stranded DNA that would become entangled in these helical regions and, thus, drawn away from the fork.

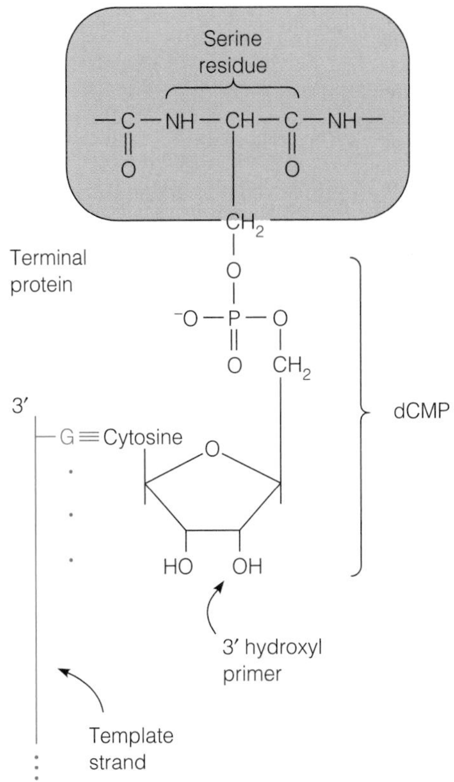

Fidelity of DNA Replication

DNA replication is by far the most accurate of known enzyme-catalyzed processes. From spontaneous mutation frequencies, which for all positions in a particular gene amount to about 10^{-6} per generation, we can estimate the chance that a particular nucleotide residue will be copied incorrectly as about 10^{-9} to 10^{-10} per base pair per round of replication. This very high accuracy cannot be explained simply from the energetics of DNA structure. The free energy of formation of a Watson–Crick base pair (A-T or G-C) differs from that of a non-Watson–Crick pair (for example, A-C or G-T) by only 4 to 13 kJ/mol. This corresponds to only a 100- to 1000-fold difference in binding energy between correct and incorrect base pairs. Therefore, if DNA polymerase were completely passive, only incorporating the nucleotide that spontaneously associates with a template base, errors would be made approximately 0.1% to 1.0% of the time, reflecting the relative abundance of correctly and incorrectly base-paired structures. Spontaneous mutation rates would be many orders of magnitude higher than are observed.

Linear genomes replicate by several mechanisms that prevent the loss of sequences that cannot be replicated backward from 5′ ends.

How might other components of the replication machinery contribute to the extremely low error rate of DNA replication? The earliest clue came from phage T4, when it was found that mutations that changed the viral DNA polymerase affected the accuracy with which other genes were replicated. Some polymerase mutations had a **mutator** phenotype, meaning that mutation rates at other loci were uniformly increased. Other polymerase mutations were **antimutators,** which seemed to improve the accuracy of DNA replication. Most of the mutator mutations were found to result from decreased activity of the polymerase-associated exonuclease. Because this activity has the ability to cleave incorrectly base-paired nucleotides from a 3′ terminus, a decreased activity of exonuclease would decrease the chance that a misincorporated nucleotide would be excised before the polymerase could add the next nucleotide. Similarly, increased 3′ exonuclease activity relative to polymerase would increase the chance that a mispaired nucleotide would be excised and hence would decrease the error rate. Thus, the accuracy of DNA replication could be ascribed in part to a proofreading exonuclease activity that checks newly replicated DNA and corrects most errors.

This mechanism of proofreading was called into question when the structure of the Klenow fragment of DNA polymerase I was published. Recall from Figure 24.16 that eight base pairs of DNA must unwind in order for the primer 3′ terminus to move from the polymerase active site to the 3′ exonuclease site. How, then, does the nuclease distinguish between correctly and incorrectly base-paired 3′ terminal nucleotides? Answers came from kinetic analyses with defined template–primer combinations analyzed on sequencing gels (Figure 24.43). For a correctly base-paired 3′-terminal nucleotide, the rate at which the next nucleotide is added is far faster than the rate of spontaneous DNA unwinding at the 3′ terminus. Thus, most correctly paired nucleotides never have a chance to move to the 3′ exonuclease site. However, extension from a mismatched nucleotide is quite slow, increasing the likelihood that the unwinding reaction will move the mismatched nucleotide to the 3′ exonuclease site for removal (see Figure 24.44).

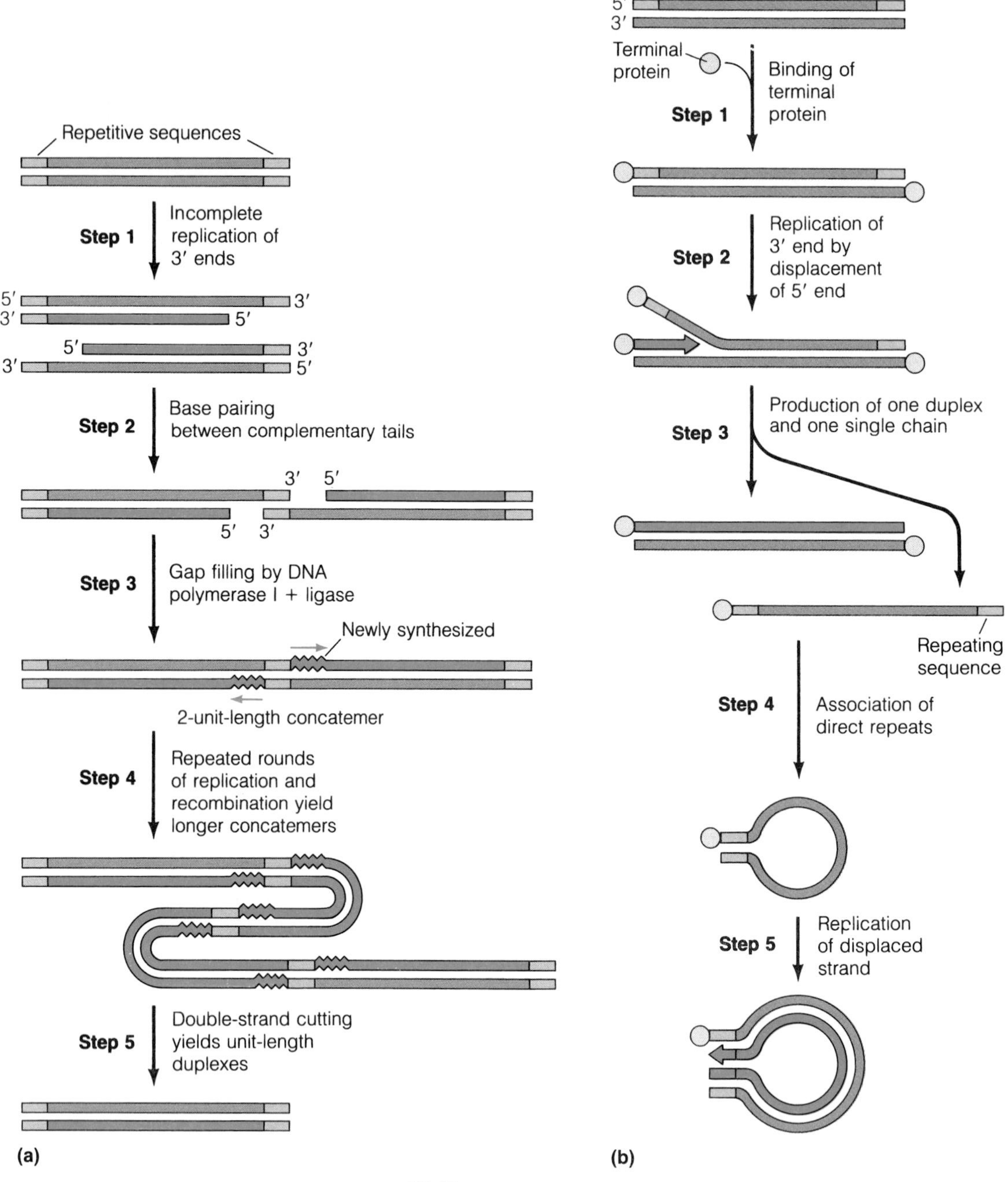

FIGURE **24.41**

Mechanisms for replication of linear DNAs. (a) End-to-end recombination at terminal redundancies, as seen with T4 and T7. (An alternative scheme could involve nicking on both 5′ sides of the recombined region, followed by double-strand displacement synthesis to duplicate this segment.) **(b)** Use of 5′ genome-linked protein, as seen with adenovirus and ϕ29. **(c)** Replication around hairpin loops at the termini, followed by two-strand cutting, as seen with poxviruses.

At present it is not clear whether the preceding scheme represents a general proofreading mechanism by DNA polymerases. In other words, is proofreading a purely kinetic process, or does the 3′ exonuclease discriminate between correctly and incorrectly base-paired nucleotides? The ε subunit of *E. coli* DNA polymerase

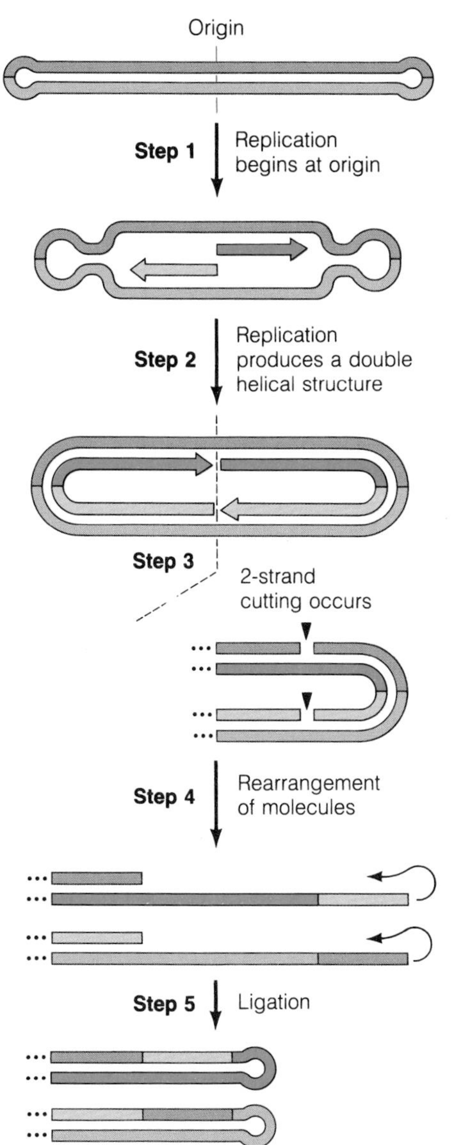

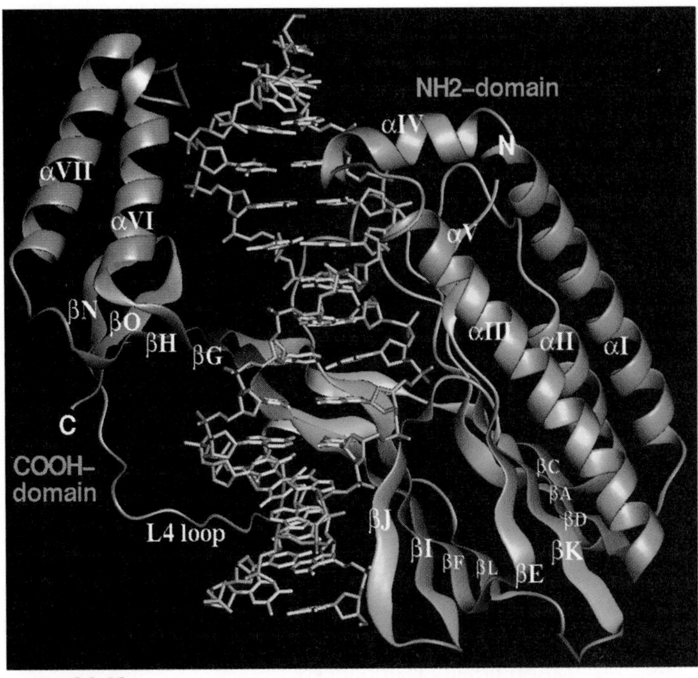

FIGURE **24.42**

Structure of the *E. coli* Tus protein, complexed with double-stranded DNA. The amino- and carboxy-terminal domains of the protein are shown in green and blue, respectively. A number of subdomains of the protein are shown.

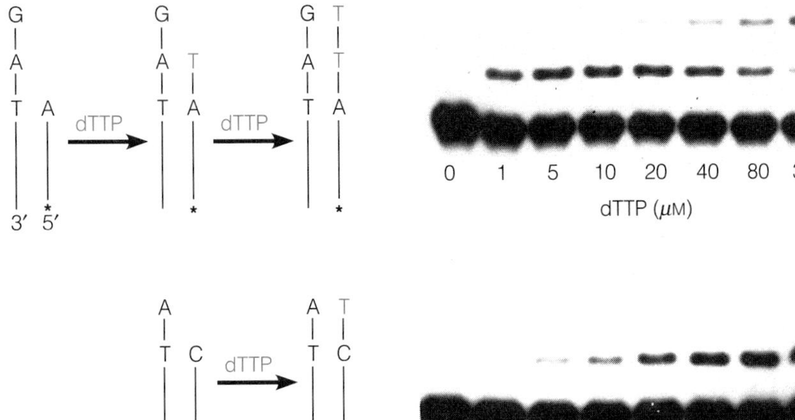

FIGURE **24.43**

An in vitro system for studying the fidelity of DNA replication. The primer strand is radiolabeled at the 5′ end (marked with an asterisk) and is the most rapidly migrating radiolabeled species. The dTTP concentration dependence of extension from a correctly paired nucleotide (A in the A-T base pair) is shown at the top, and that of extension from a C-T mispair is shown at the bottom. dTTP was the only dNTP present during the incubations. In each case the products after incubation and denaturation were analyzed by SDS-polyacrylamide gel electrophoresis and autoradiography. The top panel shows that the rate of extension was maximal at 1 μM dTTP and that further extension could be detected at higher dTTP levels. However, as shown in the bottom panel, dTTP concentrations in the millimolar range were required to see any chain extension from the mismatch. This sequencing gel approach can be used to study insertion and excision steps, as well as chain extension.

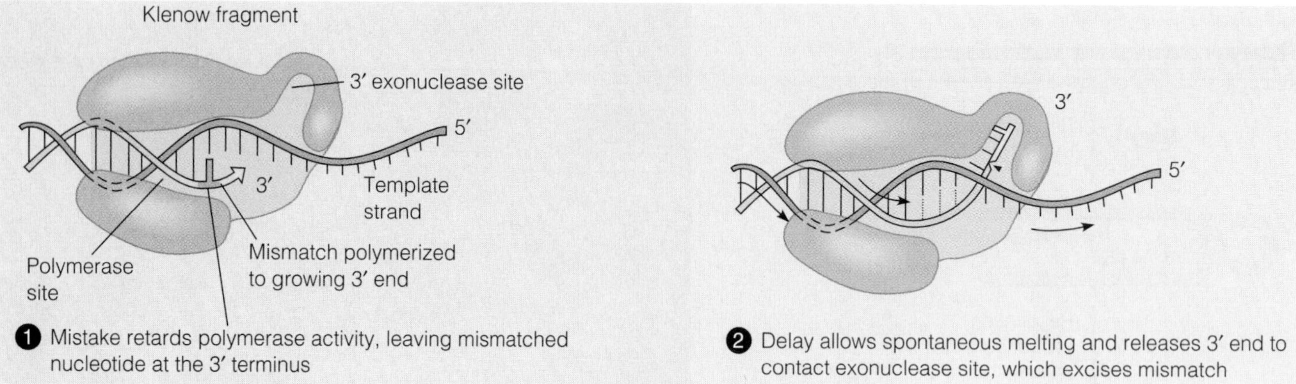

FIGURE **24.44**

Kinetic basis for preferential excision of mismatched nucleotides by a 3′ exonuclease site distant from the polymerase site.

III has been reported to have a mismatch-specific 3′ exonuclease. The structural relationship between this nuclease and the polymerase activity of the α subunit is not yet known.

DNA replication fidelity is enhanced by 3′-exonucleolytic cleavage of mismatched nucleotides and by kinetic discrimination against incorporation of non-base-paired nucleotides and against chain extension from a mismatched nucleotide.

DNA polymerase plays an additional role in determining accuracy, at the nucleotide insertion step. Kinetic studies of the type illustrated in Figure 24.43 show that *misinsertion* is discriminated against, just as is extension from a mismatched nucleotide. Kinetic analysis shows that the enzyme plays an active role in selecting the correct dNTP. A major conformational change occurs after dNTP binding, and this facilitates the bond-forming step. Unexpectedly, much of the ability of polymerases to discriminate at the insertion step has been shown to result from the geometrical shapes of the nucleotide substrates, at least as much as from their ability to form hydrogen bonds with template bases. Chemists synthesized a deoxyribonucleoside triphosphate analog whose base was a toluene derivative. This analog is geometrically quite similar to thymidine triphosphate, but it is clearly unable to form hydrogen bonds with any other nucleotide. Nevertheless, this completely unnatural nucleotide was found to be incorporated by Klenow fragment with high selectivity opposite dAMP residues in template DNA. This surprising finding, plus the high-resolution structures recently reported for dNTP binding sites in DNA polymerases, has refocused attention upon the role of nucleotide insertion by polymerases as an important determinant of replication fidelity.

dTTP

Geometric analog of dTTP

Another error correction mechanism, **mismatch repair,** makes an additional contribution of about 100-fold to the already high accuracy of genetic information copying. As described in the next chapter, the mismatch repair system scans newly replicated DNA, excising residues that are not properly base-paired and replacing them with the correct nucleotides.

The extent to which these fidelity-enhancing processes contribute to the accuracy of eukaryotic DNA replication is slowly being revealed. DNA polymerase α does not have an associated 3′ exonuclease activity, but it is possible that a proofreading nuclease activity is provided by another protein, as found in *E. coli* DNA polymerase III. Also, investigation has revealed in polymerase α from *Drosophila* a "cryptic" 3′ exonuclease activity—an activity that is usually undetectable. Moreover, as noted earlier, polymerases δ and ε each contain an associated 3′ exonuclease activity.

Spontaneous mutation frequencies are determined by several factors other than the dNTP binding specificities and proofreading exonuclease activities of DNA polymerases. For example, as we noted in Chapter 22, defective control of deoxyribonucleotide synthesis can create unbalanced intracellular concentrations, and these defects can generate mutator phenotypes by forcing replication errors at either insertion or proofreading steps. Moreover, mutator phenotypes are also generated by defects in mismatch repair systems, where replication errors generated by polymerase action are not efficiently corrected. Understanding the factors that control mutation rates is of far more than academic interest. As discussed in

Chapters 23 and 25, cancer results from accumulations in cells of DNA alterations that are too extensive to repair. There is good reason to believe that mutator phenotypes are established in precancerous cells and that these changes are essential to support the numbers of mutational events that must occur to change a normal cell into a full-blown tumor cell.

RNA Viruses: The Replication of RNA Genomes

We conclude this chapter with a few words about the replication of viral genomes consisting of RNA. Virtually all known plant viruses contain RNA instead of DNA, as do several bacteriophages and many important animal viruses, including polio virus and influenza viruses. The **retroviruses,** which are responsible for many tumors and for acquired immune deficiency syndrome (AIDS), also contain RNA genomes.

RNA-DEPENDENT RNA REPLICASES

Most RNA viruses contain a genome consisting of a single molecule of single-strand RNA. Usually that RNA is the "sense" strand for expression of genetic information at the level of translation. In other words, the RNA molecule that passes from the viral particle into the infected cell can serve directly as a messenger RNA, without having first to direct the synthesis of complementary-strand RNA. One of the early products of translation of this input genome is the enzyme **replicase,** or RNA-dependent RNA polymerase. After complexing with required polypeptide subunits of host cell origin, this enzyme replicates the input RNA (plus strand), starting from the 3′ end. Thus, the new strand (minus strand) is laid down from its 5′ end to its 3′ end, the same direction in which DNA polymerases work. The newly replicated minus strands then serve as templates for synthesis of plus strands, which are packaged into progeny **virions,** or virus particles. More complex mechanisms come into play when the RNA genome is double-stranded or segmented (three or four separate RNA molecules) or when the virion RNA is itself the minus strand (viruses in this latter class are called **negative-strand viruses**).

The known RNA replicases lack proofreading activity, so it is no surprise that viral RNA replication is far more error-prone than DNA replication, and RNA viruses undergo mutation and evolution far more rapidly than the organisms they infect. These characteristics are clearly related to viral pathogenesis, in part because a virus population infecting a plant or an animal can undergo change so rapidly that it can evade or counteract the host's defense mechanisms.

The rapid mutation rates of RNA viruses are largely due to RNA replication mechanisms that do not involve proofreading for mismatched nucleotides.

REPLICATION OF RETROVIRAL GENOMES

Different strategies for genome replication are involved in the action of retroviruses, so named because of the presence of a special enzyme, **reverse transcriptase.** In this class of viruses the single-strand RNA genome achieves latency—the ability to persist in a host cell for a long period without pathological effects—by making a DNA copy of itself and inserting that copy into the host cell genome. The DNA copy is made by reverse transcriptase, a multifunctional enzyme that is packaged in virions and enters the infected cell along with the viral genome. As shown in Figure 24.45, reverse transcriptase uses viral RNA as a template for synthesizing a complementary DNA strand, with a specific transfer RNA molecule serving as the primer (step 1). An RNase H activity of the enzyme then partially digests the RNA (step 2), and the structure circularizes by DNA–RNA base pairing (step 3). The nascent DNA chain is extended around the circle, and the tRNA primer is removed by RNase H activity (step 4). Strand displacement synthesis then occurs, with the RNA strand being displaced (step 5). The resultant double-strand DNA circle then recombines with a site on a chromosomal DNA and is

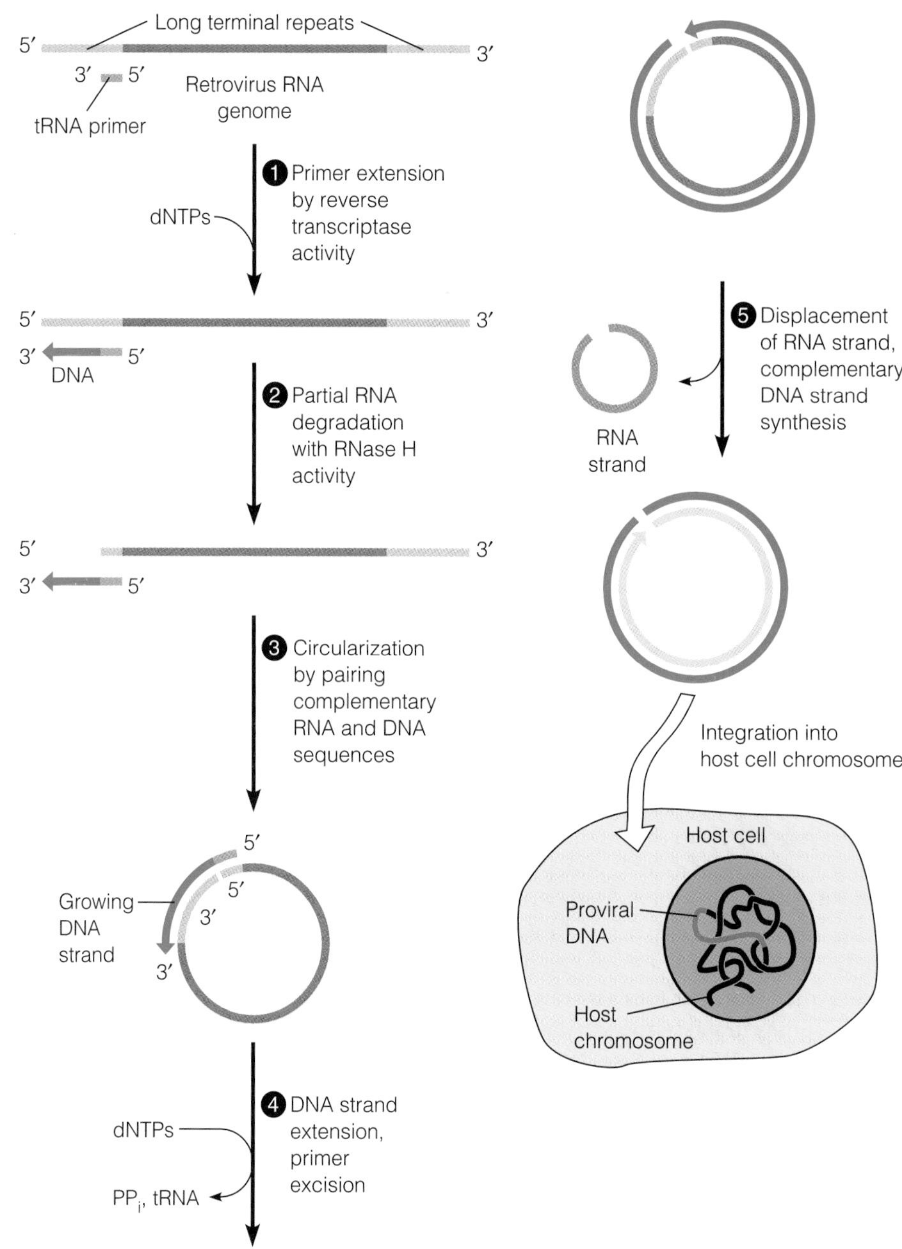

FIGURE **24.45**

Simplified view of retrovirus life cycle. The RNA genome contains long terminal repeats, to one of which a tRNA molecule binds. Primer extension, partial RNA digestion, and circularization generate a substrate for extensive DNA synthesis. Ultimately, a circular duplex DNA molecule is formed, and it is the likely substrate for integration into a host chromosome. Recent evidence suggests that two viral RNA molecules must interact at step 2 or 3.

linearly inserted into that DNA in the process. Under these conditions the integrated proviral genome can persist in a noninfectious state for many years, with most of its own genes turned off. Environmental stresses, still undetermined, can trigger excision of the integrated viral genome and return of the virus to an infectious state.

As discussed when we introduced azidothymidine (AZT) in Chapter 22, the reverse transcriptase of human immunodeficiency virus (HIV), the virus that causes AIDS, is an obvious potential target for antiviral therapy. Accordingly, intense interest has focused on the structure of this enzyme, in order to design inhibitors based on the structure of the active site. The enzyme is a dimer, with one 51-kilodalton and one 66-kilodalton subunit. Both subunits are products of the same gene.The smaller subunit, which results from proteolytic degradation of the larger subunit, collapses about the larger subunit, protecting it from degradation. As seen with DNA-dependent DNA polymerases, the large subunit forms a "hand,"

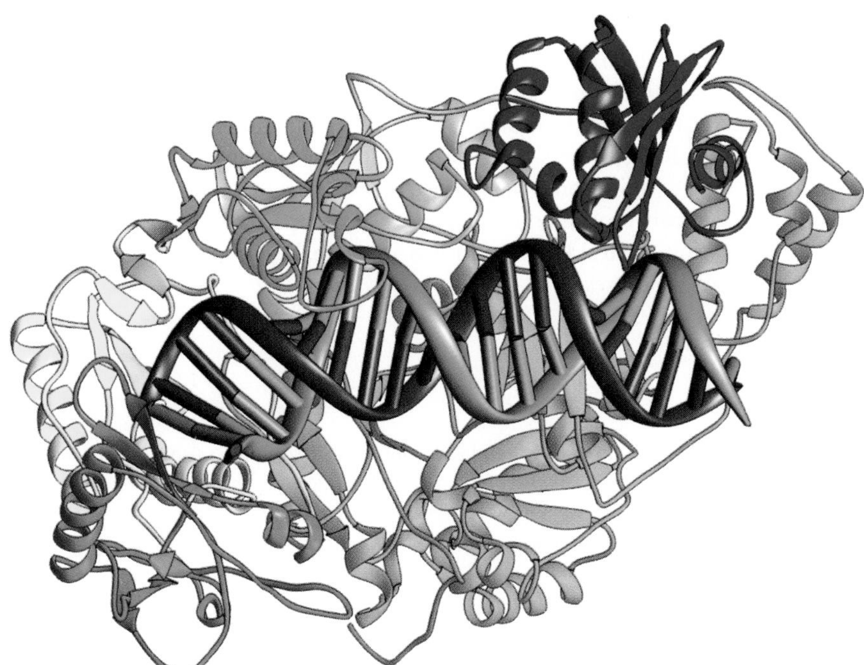

FIGURE **24.46**

Structure of HIV reverse transcriptase. The enzyme was crystallized in the presence of a primer–template complex and a substrate analog (a dideoxyribonucleoside triphosphate) positioned at the 3′ primer terminus. The p66 subunit is shown in color, with the fingers in red, the palm in yellow, the thumb in orange, a connector domain in blue, and the RNase H domain in purple. The p51 subunit is in gray, and the nucleotide substrate (lower left) is in brown.

Courtesy of G. L. Verdine from H. Huang, *Science* (1998) 282:1669–1675, with permission from *Science.*

which grasps DNA. Analysis of the dimer structure shows the polymerase and RNase H sites to be about 20 nucleotides apart (Figure 24.46). This structural relationship coordinates the two activities, with insertion of a new nucleotide occurring concomitantly with excision of a ribonucleotide from the RNA strand of the hybrid. One of the great difficulties in devising vaccines against HIV derives from the absence of a proofreading exonuclease in HIV reverse transcriptase. This leads to frequent replication errors and high rates of spontaneous mutagenesis, allowing the virus to generate variants that have resistance to antiviral antibodies produced via vaccination.

SUMMARY

DNA replication begins at fixed points on chromosomes. In most cases two replication forks are formed as a consequence of initiation. In each fork, parental DNA strands unwind, and each strand serves as template for DNA polymerase–catalyzed synthesis of a daughter strand. Both leading and lagging strands in a fork grow in a 5′→3′ direction, meaning that one strand must grow backward in short segments (Okazaki fragments) from the fork. That one strand, the lagging strand, elongates discontinuously, with each segment being initiated by a short RNA primer. Enzymes and proteins that propagate a replication fork include DNA polymerase, which incorporates deoxyribonucleoside 5′-monophosphate moieties from dNTP substrates; sliding clamps, which enhance polymerase processivity; clamp loaders, which wrap the circular clamps about DNA and also carry out clamp unloading; helicases, which unwind parental DNA strands, using energy provided by ATP hydrolysis; primase, which synthesizes RNA primers; single-strand DNA-binding protein, which stabilizes DNA in the single-strand conformation to act as template and then facilitates renaturation with the nascent strands as deoxyribonucleotides are incorporated; and topoisomerases, which relieve torsional stress ahead of the fork and, in prokaryotes, introduce negative supercoiling into product DNA. Topoisomerases also decatenate daughter DNA molecules at the end of a round of replication, thereby permitting chromosome segregation.

Initiation of replication of duplex DNA involves binding of proteins to specific origin sequences, which leads to DNA bending and facilitates local DNA denaturation adjacent to the protein binding sites. This is followed by binding and activity of a helicase to further unwind parental DNA strands and then by primase to synthesize short RNA primers, which are extended by DNA polymerase. The great accuracy of DNA replication is maintained in large part by a 3′ exonuclease activity that scans newly replicated DNA and excises mispaired 3′-terminal nucleotides. Replication of viral RNAs is far less accurate, because of a lack of editing activity in RNA replicases.

REFERENCES

General

Baker, T. A., and S. P. Bell (1998) Polymerases and the replisome: Machines within machines. *Cell* 92:295–305. A readable review of replication elongation mechanisms, in a special issue of *Cell* devoted to macromolecular machines in biology.

Kornberg, A., and T. A. Baker (1992) *DNA Replication,* 2nd ed. W. H. Freeman and Co., San Francisco. The definitive work on this subject.

DNA Polymerases

Kelman, Z., and M. O'Donnell (1995) DNA polymerase III holoenzyme: Structure and function of a chromosomal replicating machine. *Annu. Rev. Biochem.* 64:171–200. There has been much progress on this ten-subunit enzyme since 1995, but this review is comprehensive and useful.

Patel, S. S., I. Wong, M. J. Donlin, and K. A. Johnson (1991) *Biochemistry* 30:511–546. Three papers describing a kinetic approach to understanding polymerase mechanisms and fidelity.

Singh, K., and M. J. Modak (1998) A unified DNA- and dNTP-binding mode for DNA polymerases. *Trends Biochem. Sci.* 23:277–281. An attempt to reconcile structural studies reporting that polymerases differ in orientation of the primer–template DNA, with respect to the fingers, thumb, and palm of the enzyme.

Steitz, T. A. (1999) DNA polymerases: Structural diversity and common mechanisms. *J. Biol. Chem.* 274:17395–17398. Summarizes evidence that the five structurally distinct polymerase families all involve a common mechanism, as summarized in Figure 24.18.

Other Replication Proteins

Adams, D. E., E. M. Shekhtman, E. L. Zechiedrich, M. B. Schmid, and N. R. Cozzarelli (1992) The role of topoisomerase IV in partitioning bacterial replicons and the structure of catenated intermediates during replication. *Cell* 71:277–288. Definitive evidence for the role of a type II topoisomerase.

Bussiere, D. E., and D. Bastia (1999) Termination of DNA replication of bacterial and plasmid chromosomes. *Molec. Microbiol.* 31:1611–1618. A review summarizing evidence that termination involves specific antagonism of helicase action.

Froelich-Ammon, S. J., and N. Osheroff (1995) Topoisomerase poisons: Harnessing the dark side of enzyme mechanism. *J. Biol. Chem.* 270:21429–21432. A review of topoisomerases as antimicrobial and anticancer drugs.

Karpel, R. L. (1990) T4 bacteriophage gene 32 protein. In: *The Biology of Non-Specific DNA–Protein Interactions,* edited by A. Rezvin, pp. 103–130. CRC Press, Boca Raton, Fla. A detailed review of this important protein.

Lohman, T. M., and K. P. Bjornson (1996) Mechanisms of helicase-catalyzed DNA unwinding. *Annu. Rev. Biochem.* 65:169–204. Kinetic and mechanistic aspects of helicase action.

Lohman, T. M., and M. E. Ferrari (1994) *Escherichia coli* single-stranded DNA-binding protein: Multiple DNA-binding modes and cooperativities. *Annu. Rev. Biochem.* 63:527–570. *E. coli* SSB is surprisingly different from its T4 counterpart, the gene 32 protein.

Lohman, T. M., K. Thorn, and R. D. Vale (1998) Staying on track: Common features of DNA helicases and molecular motors. *Cell* 93:9–12. Proteins that move along DNA are being likened, mechanically, to other proteins that cause physical movement within cells.

Nash, H. A. (1998) Topological nuts and bolts. *Science* 279:1490–1491. Commentary on two articles in the same issue of *Science* that describe the crystal structure and mechanism of human topoisomerase I.

Pennisi, E. (1996) Premature aging gene discovered. *Science* 272:193–194. A news article describing the discovery that Werner's syndrome results from a defective helicase gene.

Pulleyblank, D. E. (1997) Of topo and Maxwell's dream. *Science* 277: 648–649. Commentary on surprising findings reported in the same issue of *Science,* that type II topoisomerases have a mysterious ability to untangle DNA molecules, rather than achieving equilibrium topoisomerase distributions.

Sancar, A., and J. E. Hearst (1993) Molecular matchmakers. *Science* 259:1415–1420. This review describes a class of proteins that use ATP energy to bring about binding of specific proteins. Replicative DNA polymerase and its cognate sliding-clamp protein associate via action of matchmakers.

Turner, J., M. M. Hingorami, Z. Kelman, and M. O'Donnell (1999) The internal workings of a DNA polymerase clamp-loading machine. *EMBO J.* 18:771–783. A timely review of clamp-loading mechanisms.

Wang, J. C. (1996) DNA topoisomerases. *Annu. Rev. Biochem.* 65:635–692. A comprehensive and rather recent review.

Initiation of DNA Replication

Baker, T. A., and S. H. Wickner (1992) Genetics and enzymology of DNA replication in *Escherichia coli. Annu. Rev. Genet.* 26:447–477. A general review, but with primary emphasis on initiation.

Eukaryotic DNA Replication

Bambara, R. A., R. S. Murante, and L. A. Henricksen (1997) Enzymes and reactions at the eukaryotic DNA replication fork. *J. Biol. Chem.* 272:4647–4650. Similarities and distinctions between prokaryotic and eukaryotic elongation mechanisms.

Cook, P. R. (1999) The organization of replication and transcription. *Science* 284:1790–1795. A convincing argument that DNA and RNA polymerases are organized within cells into multi-protein "factories" through which DNA templates are drawn.

Waga, S., and B. Stillman (1998) The DNA replication fork in eukaryotic cells. *Annu. Rev. Biochem.* 67:721–752. A timely comparison of prokaryotic and eukaryotic replicative chain elongation.

Wold, M. S. (1997) Replication protein A: A heterotrimeric, single-stranded DNA-binding protein required for eukaryotic DNA metabolism. *Annu. Rev. Biochem.* 66:61–92. Eukaryotic SSBs are quite different from the prototypical T4 gp32 described in this chapter.

Fidelity of Nucleic Acid Synthesis

Goodman, M. F. (1997) Hydrogen bonding revisited: Geometric selection as a principal determinant of replication fidelity. *Proc. Natl. Acad. Sci. USA* 94:10493–10495. Commentary upon the research article, in the same issue of the journal, that described DNA synthesis carried out with a non-hydrogen-bonding dNTP analog.

Goodman, M. F., and D. K. Fygenson (1998) The biochemical basis of mutation. *Genetics* 48:1475–1488. A well-referenced recent review.

Jackson, A. L., and L. A. Loeb (1998) The mutation rate and cancer. *Genetics* 148:1483–1490. Evidence that a mutator phenotype is generated in cells destined to become cancer cells.

Kunkel, T. A. (1992) DNA replication fidelity. *J. Biol. Chem.* 267:18251–18254. A short review by a leader in this field.

Reverse Transcriptase

Kohlstaedt, L. J., J. Wang, J. M. Friedman, P. A. Rice, and T. A. Steitz (1992) Crystal structure at 3.5 Å resolution of HIV-1 reverse transcriptase complexed with an inhibitor. *Science* 256:1783–1790. The first structural determination for a DNA polymerase other than *E. coli* polymerase I.

Peliska, J. A., and S. J. Benkovic (1992) Mechanism of DNA strand transfer reactions catalyzed by HIV-1 reverse transcriptase. *Science* 258: 1112–1118. A mechanistic analysis of this important enzyme.

Temin, H. A. (1993) Retrovirus variation and reverse transcription: Abnormal strand transfers result in retrovirus genetic variation. *Science* 259:6900–6903. One of the co-discoverers of reverse transcriptase argues that genetic variability of HIV results from more than just a lack of a proofreading exonuclease.

PROBLEMS

1. Describe an experimental approach to determining the *processivity* of a DNA polymerase (that is, the number of nucleotides incorporated per chain per polymerase binding event).

*2. After Okazaki's proposal of discontinuous DNA chain growth, there was much controversy about whether both DNA strands are synthesized discontinuously, or only the lagging strand. Clearly, there is no need for the leading strand to be synthesized in fragments, but many workers found that pulse-labeled DNA fragments hybridized to both parental DNA strands. This finding indicated that both parental strands served as the template for synthesis of short DNA fragments. Propose an alternative mechanism by which leading strand replication could generate short fragments, and propose an experimental test of your suggestion.

3. The buoyant density of DNA can be increased by incorporation of either heavy stable isotopes, such as ^{15}N, or base analogs, such as 5-bromouracil. This problem asks you to calculate the density increment generated by each technique. In all calculations the density of "light" *E. coli* DNA is 1.710 g/mL. The G + C content is 51 mol % (that is, 51 moles of G + C per 100 moles of DNA nucleotides). Residue molecular weights of cesium salts of nucleotides are as follows: dAMP, 445; dTMP, 436; dGMP, 461; dCMP, 421; 5-bromo-dUMP, 501.
(a) Calculate the buoyant density of *E. coli* DNA when all ^{14}N atoms are replaced with ^{15}N atoms.
(b) Calculate the buoyant density of *E. coli* DNA when all thymine residues are replaced with 5-bromouracil.

4. A mixture of four α-[^{32}P]–labeled ribonucleoside triphosphates was added to permeabilized bacterial cells undergoing DNA replication, and incorporation into high-molecular-weight material was followed over time, as shown in the accompanying graph. After 10 minutes of incubation a 1000-fold excess of unlabeled ribonucleoside triphosphates was added, with the results shown in the graph.

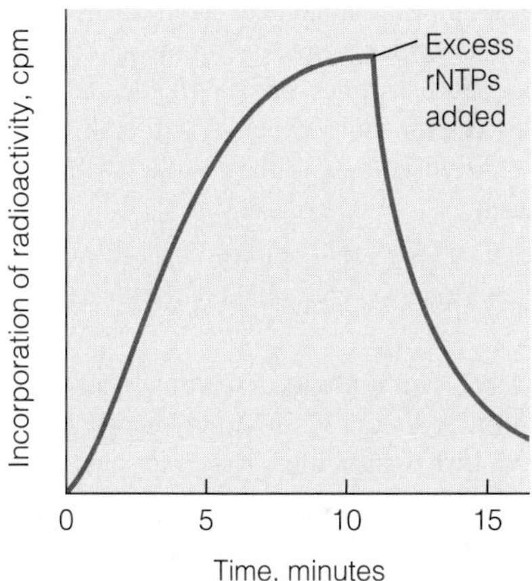

(a) Why was the excess of unlabeled rNTPs added?
(b) How could you tell that radioactivity is being incorporated as ribonucleotides, rather than an alternative such as reduction to deoxyribonucleotides, followed by incorporation?
(c) What does this experiment tell you about the process of DNA replication?

5. Deoxyadenylate residues in DNA undergo deamination fairly readily, as do deoxycytidylate residues.
(a) What is the product of dAMP deamination?
(b) The deamination product is known to base-pair with A, C, or T. What would be the genetic consequences if this deaminated site in DNA were not repaired and if it paired with C on the next round of replication?

6. The *E. coli* chromosome is 1.28 mm long. Under optimal conditions the chromosome is replicated in 40 minutes.
(a) What is the distance traversed by one replication fork in 1 minute?
(b) If replicating DNA is in the B form (10.4 base pairs per turn), how many nucleotides are incorporated in 1 minute in one replication fork?
(c) If cultured human cells (such as HeLa cells) replicate 1.2 m of DNA during a 5-hour S phase and at a rate of fork movement one-tenth of that seen in *E. coli*, how many origins of replication must the cells contain?
(d) What is the average distance, in kilobase pairs, between these origins?

7. DNA ligase has the ability to relax supercoiled circular DNA in the presence of AMP but not in its absence.
   (a) What is the mechanism of this reaction, and why is it dependent on AMP?
   (b) How might one determine that supercoiled DNA had in fact been relaxed?

*8. A recent paper reports that, in mammalian cells, genes that are expressed in a particular cell are replicated during the first half of S phase, and genes not expressed in that cell are replicated in the latter half of S phase. Briefly describe an experiment that could lead to this conclusion.

9. Although DNA polymerases require both a template and a primer, the following single-strand polynucleotide was found to serve as a substrate for DNA polymerase in the absence of any additional DNA.

   3′ HO-ATGGGCTCATAGCCGGAGCCCTAACC-
   GTAGACCACGAATAGCATTAGG-p 5′

   Give the structure of the product of this reaction.

10. The 3′-exonuclease activity of *E. coli* DNA polymerase I was recently found to show no discrimination between correctly and incorrectly base-paired nucleotides at the 3′-terminus; properly and improperly base-paired nucleotides are cleaved at equal rates there. How can this observation be reconciled with the fact that the 3′-exonuclease activity increases the accuracy with which template DNA is copied?

11. 2′,3′-Dideoxyinosine has been approved as an anti-HIV drug. Propose a mechanism by which it might block the growth of the AIDS virus.

12. Aphidicolin-resistant mutants of mammalian cells often have alterations in the structure of DNA polymerase α that render the enzyme insensitive to inhibition. However, some mutants show alterations, not in DNA polymerase but in ribonucleotide reductase. How might a change in the latter enzyme cause an aphidicolin-resistant phenotype?

13. The *E. coli* chromosome requires 40 minutes for its complete replication, even in an optimally nourished cell. However, bacterial cells can divide as frequently as every 20 minutes. How can cells divide more rapidly, apparently, than their DNA can be copied?

*14. Supercoiled DNA is more compact than relaxed DNA of the same molecular weight, which gives supercoiled DNA greater electrophoretic mobility. Why, then, does positively supercoiled DNA migrate more slowly than relaxed in the experiment depicted in Figure 24.29? (Note: Read Tools of Biochemistry 24B before attempting to answer this question.)

15. An alternative to the model shown in Figure 24.27 for the action of homodimeric helicases is an "inchworming" model. In this model, the subunit bound to DNA moves forward by a few base pairs, unwinding as it goes, and with the 3′ end bound to that subunit as it moves toward the 5′ end of the bound strand. Then, ATP hydrolysis triggers a conformational change that leads to release of the 3′ end of the unwound DNA, so that the process can repeat, as shown.

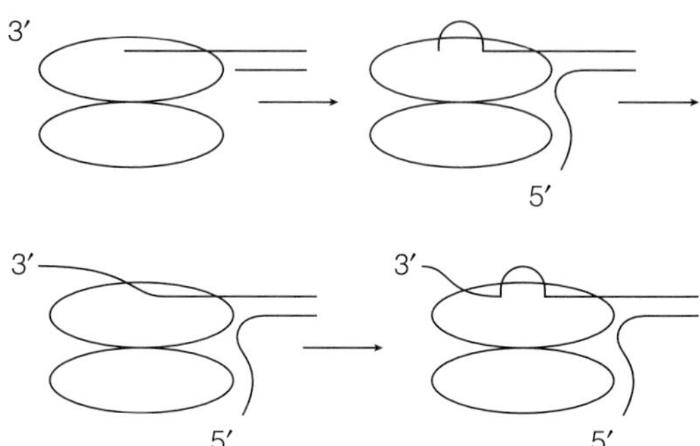

   Identify a difference between this and the hand-over-hand model of Figure 24.27 that could be used experimentally to distinguish between the two models.

16. Propose a mechanism by which a type II topoisomerase could use the energy of ATP hydrolysis to scan a large DNA molecule and, thereby, to direct that the enzyme will catalyze largely "disentangling" reactions (decatenation and unknotting).

17. The exponential nature of PCR allows spectacular increases in the abundance of a DNA sequence being amplified. Consider a 10-kbp DNA sequence in a genome of 10^{10} base pairs. What fraction of the genome is represented by this sequence; i.e., what is the fractional abundance of this sequence in this genome? Calculate the fractional abundance of this target sequence after 10, 15, and 20 cycles of PCR, starting with DNA representing the whole genome and assuming that no other sequences in the genome undergo amplification in the process.

TOOLS OF BIOCHEMISTRY 24A

Polymerase Chain Reaction

As we discuss in Chapter 25, gene cloning by recombinant DNA techniques revolutionized biology in the mid-1970s, because it allows an investigator to isolate and amplify individual genes for analysis of their sequence, expression, and regulation. Cloning requires living cells, into which DNA molecules must be introduced for amplification. An equally revolutionary technique, **polymerase chain reaction (PCR)**, was introduced in 1987. PCR allows the amplification of exceedingly small amounts of DNA in vitro, without prior transfer into living cells. This technique has facilitated the analysis of eukaryotic genes, because it avoids some of the tedium involved in cloning DNA from very large genomes. In addition, the technique has dozens of practical applications.

PCR requires knowledge of sequences that flank the region to be amplified. Oligonucleotides complementary to these sequences are produced by automated chemical synthesis and are used as primers in a special series of DNA polymerase–catalyzed reactions (Figure 24A.1). First, DNA containing the sequences to be amplified is heat-denatured and then annealed to the primers, which are present in excess (steps 1 and 2). Next, polymerase chain extension is carried out from the primer termini (step 3). Then a second cycle of heat denaturation, annealing, and primer extension is carried out. Using a thermostable form of DNA polymerase, ***Taq* polymerase,** from a bacterium that lives at high temperatures (in hot springs) avoids the need to add more polymerase at each cycle, because the enzyme is not inactivated at DNA-denaturing temperature. This cycle is repeated 30 or more times in an automated temperature-regulating device, with each cycle increasing the abundance of duplex DNA species bounded by the oligonucleotide primers. Two such molecules have been formed by the end of cycle III, and the number doubles with each successive cycle. About 1 billion copies are present after 32 cycles (2^{n-1} to be exact, where n is the number of cycles).

Innumerable applications of this technique have been developed, including forensic analysis, in which DNA can be amplified from exceedingly small samples of biological material (for instance, blood, semen, or hair) for identification of criminal suspects or of fathers in paternity cases. The field of "molecular anthropology" has developed from PCR and sequence analysis of human mitochondrial DNA, with the results used to formulate models of human evolution. A comparable field of "molecular archeology" exists, with minute amounts of DNA being extracted from long-preserved biological samples, such as organisms frozen in ice or insects trapped in amber. Of great historical interest was the 1998 report describing PCR analysis of descendants of Thomas Jefferson, which showed that our third president or a close relative fathered at least one child with Sally Hemings, one of his slaves. The DNA came from living people, but the results led to important historical conclusions. PCR is used in environmental microbiology, to detect microbial populations by searching for sequences unique to the organism being sought. In the same way, PCR can be used for diagnosis of microbial or viral infections. PCR is used also for prenatal diagnosis of genetic diseases, using primers specific for the gene sequence alteration responsible for a disorder. Similarly, oncogene mutations leading to cancer can be detected by PCR, and this has allowed extensive analysis of the sequence of genetic alterations leading from a precancerous lesion to a metastatic tumor.

PCR is not without its technical problems, however. For example, because some thermostable DNA polymerases, such as the commonly used *Taq* polymerase, lack a proofreading 3′ exonuclease, the DNA synthesis in PCR can be relatively inaccurate. This is not usually a problem if one wants to sequence the PCR product, because errors are uniformly distributed over the length of DNA being amplified, with the abundance at each site being too low to affect sequencing operations. Also, proofreading polymerases from thermophilic organisms are now available. However, precautions must be taken (such as limiting the number of cycles) if one wants to clone PCR products, with the expectation that a natural sequence is being cloned. Another problem is the great sensitivity of the technique, which can lead to amplification of minute amounts of DNA contaminants in the sample. Again, a number of controls and modifications of the technique can be

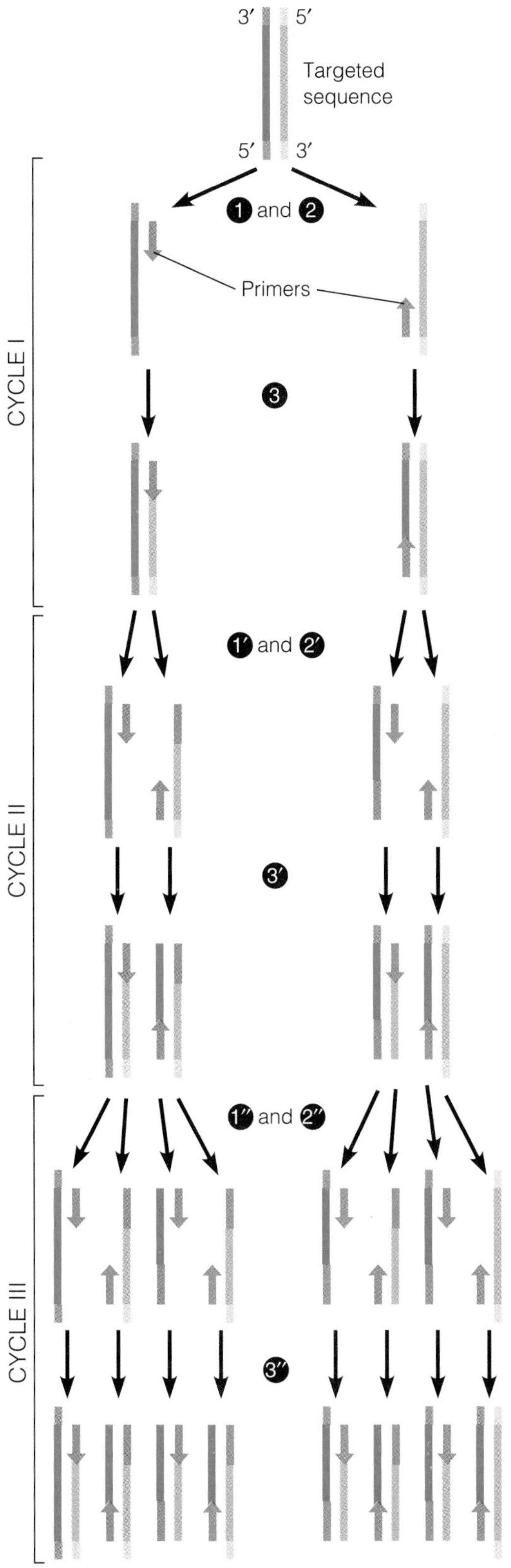

FIGURE **24A.1**

Three cycles of the polymerase chain reaction. A segment within the region shown in blue is amplified, by use of primers (red) that are complementary to the ends of the blue segment. Note the exponential nature of the amplification process.

used to minimize this problem. Finally, we point out that the speed and sensitivity of PCR have been greatly increased by carrying out the process on a continuous-flow basis, on a chip (Figure 24A.2).

References

Erlich, H. A., and N. Arnheim (1992) Genetic analysis using the polymerase chain reaction. *Annu. Rev. Genet.* 26:479–506. A review by two developers of the technique.

Mullis, K., F. Ferre, and R. Gibbs, eds. (1994) *The polymerase chain reaction.* Boston: Birkhäuser. The senior author of this book-length review collection is the inventor of PCR.

Nowak, R. (1994) Forensic DNA goes to court with O. J. *Science* 265: 1352–1354. A news article, describing the use of PCR and restriction fragment length polymorphisms (see Tools of Biochemistry 25D) as applied specifically to the O. J. Simpson murder trial.

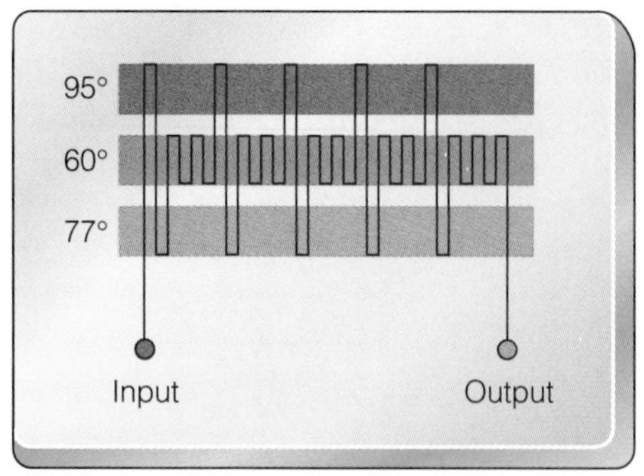

FIGURE **24A.2**

Continuous-flow PCR on a chip. A channel is etched into a glass chip. Fluid passing through the channel passes repetitively through zones thermostatted at 95°C (melting), 77°C (chain extension), and 60°C (annealing). Passage of a sample input through a hole in the chip through the three zones defines one thermal cycle in multicycle PCR. With this chip, a 10-μL DNA sample can be subjected to 20 cycles of PCR in less than 5 minutes.

TOOLS OF BIOCHEMISTRY 24B

Two-Dimensional Gel Electrophoretic Analysis of DNA Topoisomers

When one analyzes DNA molecules resolved by agarose gel electrophoresis (see Tools of Biochemistry 2A), DNA is visualized in the gel by soaking the gel with a fluorescent dye, such as **ethidium bromide,** or EtBr.

H_2N — NH_2 — $\overset{+}{N}Br^-$ — $CH_2—CH_3$

Ethidium bromide (EtBr)

This dye binds to duplex DNA by **intercalation;** the molecule is planar and about the size of one base pair, so it can fit between two adjacent DNA base pairs, forcing them apart. Intercalation greatly enhances the molecule's fluorescence, and DNA can then be visualized by observing an EtBr-treated gel under ultraviolet light.

But there is another way in which intercalators like EtBr are used in electrophoresis. If they are present *during* the electrophoresis, they can aid in distinguishing topoisomers. In forcing adjacent base pairs apart, EtBr tends to unwind a double helix. In B-form DNA, with about 10 base pairs per turn, two adjacent base pairs are rotated with respect to each other by 36° (360° per complete turn). One molecule of EtBr creates about 27° of unwinding, affecting the hydrodynamic properties of circular DNA in the following way. Recall from Chapter 4 that *L*, the linking number, is the sum of turns created by twist and writhe. EtBr binding decreases the number of turns needed to create an unconstrained duplex; it decreases *T*, the twist. Since *L* doesn't change unless the molecule is broken, the writhe component is affected: The number of turns devoted to writhe increases. In other words, EtBr increases the positive superhelicity of the molecule. Thus, an underwound molecule tends to become relaxed, because it loses negative superhelicity, whereas a relaxed closed duplex becomes overwound, and an overwound molecule becomes more so. Electrophoretic mobility through a gel is a function of compactness of the migrating molecule, with a relaxed duplex being least compact and, hence, most slowly migrating.

One can analyze the distribution of topoisomers in a circular DNA by two-dimensional gel electrophoresis, with the first dimension run in an ordinary agarose gel and the second dimension run in the presence of EtBr or a similar intercalating dye. For the plasmid DNA shown in Figure 24B.1a, these topoisomers include nicked duplexes, which are relaxed whether or not EtBr is present and which migrate slowly in both dimensions. For the slightly supercoiled molecules, both overwound and underwound topoisomers are present. For example, the species labeled 1 was

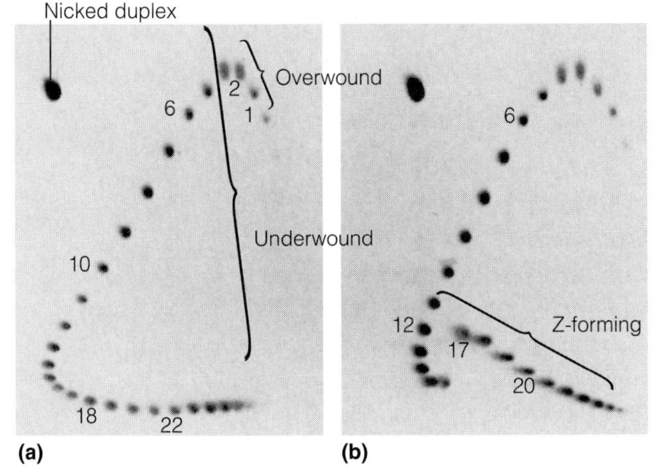

FIGURE **24B.1**

Two-dimensional agarose gel electrophoretic analysis of circular DNA topoisomers. The same plasmid was employed for both analyses, except that the plasmid in **(b)** had a 16-base-pair insert that forms left-hand Z-DNA.

Courtesy of J. L. Peck and J. C. Wang, *Proc. Natl. Acad. Sci. USA* (1983) 80:6206–6210. © 1983 AAAS.

originally overwound and has become more so, as is evident by its movement in the second dimension. Species 6, on the other hand, was underwound and moved more slowly in the second dimension, even though 1 and 6 originally had the same number of superhelical turns (writhe). The remaining species were present originally only as underwound topoisomers. At the bottom (species 18 and 22 and their neighbors) are topoisomers that could not be resolved in the first dimension but are resolved when the negative writhe is decreased, in the second dimension.

In Figure 24B.1b the plasmid analyzed was identical, except for the insertion of a 16-base-pair alternating G-C sequence, which under certain conditions forms Z-DNA (left-handed). The intercalating dye forces this DNA back to a right-hand helix, and this affects the number of superhelical turns. The topoisomers labeled 17 and 20 are species that contained the GC-rich insert in Z conformation before administration of the intercalating dye.

Reference

Peck, L. J., and J. C. Wang (1983) Energetics of B-to-Z transitions in DNA. *Proc. Natl. Acad. Sci. USA* 80:6206–6210.

CHAPTER 25

Information Restructuring: Restriction, Repair, Recombination, Rearrangement, and Amplification

OUR FOCUS NOW SHIFTS FROM DNA AS A *TEMPLATE* FOR ITS OWN REPLICATION to DNA as a *substrate*, in a number of processes that we categorize as **information restructuring.** Some of these processes preserve the information content of DNA by repairing DNA damage, some represent mechanisms used for defense by organisms, some are involved in redirection of cellular metabolism by parasites, some control normal gene expression or differentiation, some introduce genetic diversification, and some represent responses to environmental stresses. Some processes are related to several of these functions. For example, genetic recombination introduces diversity to a species, but it is also a response to environmental stress, a response that promotes survival of an individual organism.

From the biochemist's standpoint, information restructuring is less accessible than DNA replication. Replication is a central pathway that is a major metabolic event in the lifetime of all cells, and the activities of the enzymes and proteins involved are relatively high. As a result, their discovery and characterization, though elegant, were fairly straightforward. By contrast, information restructuring involves quantitatively minor pathways, with far less mass conversion per cell. Detection of the enzymes involved is correspondingly more difficult. For example, although DNA methylation has profound effects on gene integrity or expression, methylation may involve only one or a few sites per gene. Because of the low level of activities involved, analysis of DNA methylation requires experimental techniques much more sensitive than those used to analyze DNA replication.

The processes we shall consider in this chapter include (1) restriction and modification, which serve as protective mechanisms in prokaryotic cells but are also important because they provide reagents for recombinant DNA technology; (2) metabolic responses to DNA structural damage, principally mutagenesis and repair; (3) recombination, whereby the contents of a genome are redistributed, for example, during sexual reproduction; (4) gene rearrangements, including transpositions of DNA segments from one chromosomal integration site into another and the joining of DNA segments from distant parts of a genome; and (5) gene amplification, an increase in the copy number of individual segments of DNA, which occurs both as a normal developmental process and as a response to environmental stresses (Figure 25.1). Collectively, these processes are essential to the

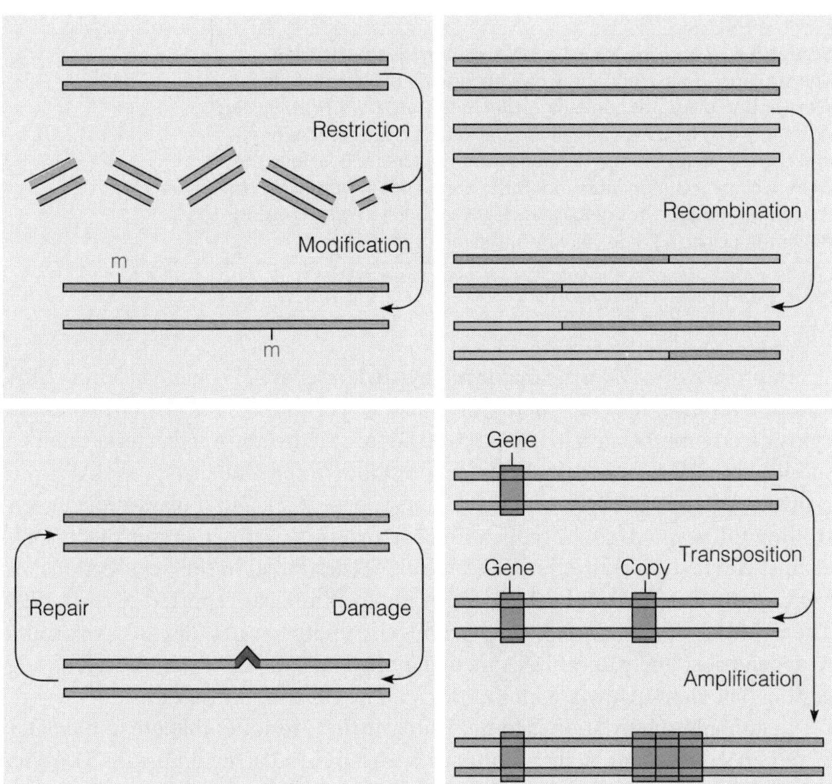

FIGURE **25.1**

A summary of the major processes in information restructuring.

survival of cells. In broader terms, recombination and gene rearrangements are the source of most of the genetic variability in a population of cells or organisms and, along with mutation, form the basis for evolutionary changes. Other important information restructuring processes, which do not occur at the DNA level, are described elsewhere. They include alternative splicing and RNA editing (see Chapter 28).

DNA Methylation

Before examining the processes of information restructuring, we first review the occurrence and synthesis of methylated bases in DNA. Methylation underlies several important biological processes, including restriction and modification, mismatch error correction (a DNA repair process), and aspects of the control of eukaryotic gene expression. Recall from Chapter 21 that *S*-adenosylmethionine (AdoMet) is the substrate for methylation of both RNA and DNA. Methylation occurs at the polynucleotide level, with transfer of a methyl group from AdoMet to a nucleotide residue.

5-Methylcytosine (mC)

The sole methylated base found in eukaryotic DNA is 5-methylcytosine (mC), which is 3% to 5% as abundant as cytosine in most animal DNA and much more abundant in some plant DNAs but is virtually absent from other DNAs, such as those of insects. Mechanistically, the synthesis of an mC residue is similar to the thymidylate synthase reaction (see Chapter 22). A cysteine thiol on the enzyme attacks C-6 of the pyrimidine ring of cytosine, which in turn activates C-5 for formation of a carbon–carbon bond. Amazingly, structural studies on a bacterial DNA methylase showed that the bases undergoing methylation rotate completely out of the DNA duplex and into a catalytic pocket within the enzyme structure (Figure 25.2). Since that demonstration in 1994, other enzymes acting on DNA bases have been shown to flip the target base, including other DNA methylases and glycosylases, such as uracil-DNA glycosylase (page 906).

The sole methylated base in eukaryotic DNA is 5-methylcytosine. The predominant methylated base in prokaryotic DNA is N^6-methyladenine.

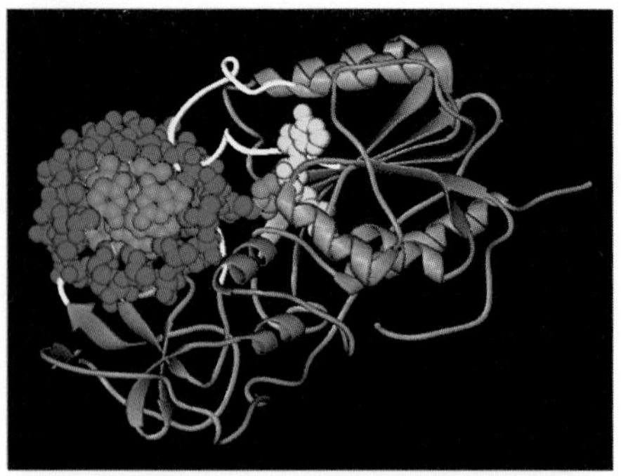

(a)

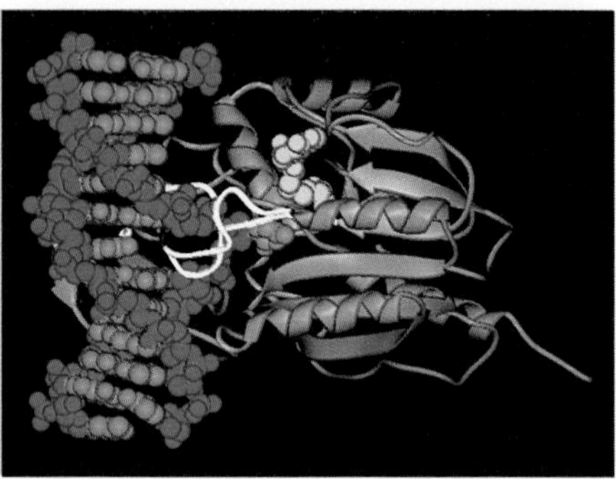

(b)

FIGURE **25.2**

Structure of a complex of a DNA methylase with DNA. The structure is based on x-ray crystallography of a ternary complex (ribbon diagram) containing *Hha* methylase from *Haemophilus haemolyticus.* The loops containing the catalytic site are in white, and the rest are in orange. *S*-Adenosylhomocysteine is yellow, the DNA backbone is magenta, and bases are green. In both views the flipped-out target cytosine base is clearly visible. **(a)** View looking down the helix. **(b)** Side view from the minor groove.

Courtesy of S. Klimasauskas, S. Kumar, R. J. Roberts, and X. Cheng, *Cell* (1994) 76:357–369, reprinted with permission. Cell Press, Inc.

In prokaryotic DNA the major methylated bases are N^6-methyladenine (mA) and to a lesser extent N^4-methylcytosine. Fewer than 1% of the adenines and cytosines are methylated in prokaryotic DNAs. Methylation in bacteria occurs at specific sites. In *E. coli,* methylation of A residues in the sequence 5′—GATC—3′ is involved in mismatch error correction (see page 942), and it plays a role in controlling initiation of DNA replication. Methylation at other sites protects DNA against cleavage by restriction endonucleases (described in the next section). An exciting discovery about bacterial DNA methylation was reported in early 1999. The virulence of *Salmonella typhimurium*—its ability to cause disease—was found to depend absolutely upon DNA methylation. *dam*$^-$ mutants were avirulent, suggesting that Dam methylase inhibitors could be effective antibiotics.

In animals, methylation is found primarily in C residues that are immediately 5′ to G residues, that is, in a sequence 5′—CpG—3′ (see Figure 25.3). When such a C is methylated, so is the corresponding C in the complementary strand. In plant DNAs, the methylated sequence is 5′—CpNpGp—3′, where N can be any base.

N^6-Methyladenine (mA)

N^4-Methylcytosine

5-Azacytidine

Although the biological significance of DNA methylation in prokaryotes is now fairly clear, its importance in eukaryotes has not yet been defined. We do know that methylation at a particular site is a heritable phenomenon. When eukaryotic DNA replicates, a **maintenance methylase** ensures that all of the sites that were methylated in parental DNA are methylated in daughter DNA. The process is shown in Figure 25.3a. Note that methylation occurs *after* replication and that unmethylated sites in parental DNA remain unmethylated in daughter DNA. How the original sites are selected for methylation during embryonic differentiation is not yet clear. Several lines of evidence suggest that methylation is responsible for the tissue-specific inactivation of genes during development. For example, gene inactivation can be reversed by treatment with **5-azacytidine,** a cytidine analog that is metabolized like cytidine to the analog of dCTP, which is then incorporated into DNA. Because of the N substitution, position 5 cannot be methylated, and when that DNA subsequently replicates, the C that is incorporated is not methylated (see Figure 25.3b). Azacytidine treatment can cause adult bone marrow cells to reactivate the synthesis of fetal hemoglobin, which is normally turned off during development. This analog is being tested as a treatment for β-thalassemia, in which the β-like fetal globin γ chains could replace the missing β chains (see Chapter 7).

Many studies correlate DNA methylation with the arrest of expression of development-related genes, evidently by altering the structure of chromatin (Chapter 28), but puzzling questions remain. Is methylation a cause or consequence of gene repression? How do insects regulate their developmental patterns of gene expression if their DNA is unmethylated? In other organisms, extensive substitution of mC for C has been described in specific DNA segments. For example, in *Neurospora,* methylation plays a role in preventing inappropriate recombination between duplicated gene segments. Finally, methylation may be involved in carcinogenesis. Deamination of an mC residue in DNA creates a G-T base pair, an event that would create a **GC ⟶ AT transition mutation.** (A transition mutation

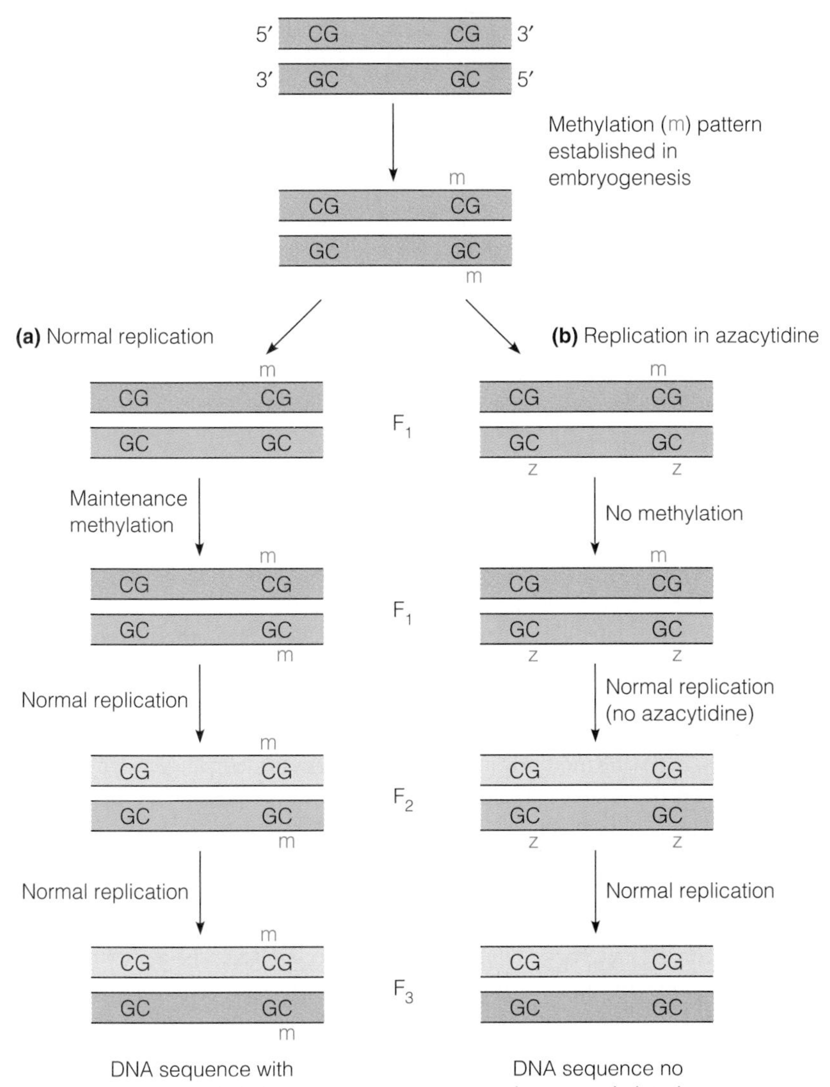

FIGURE **25.3**

Maintenance methylation of eukaryotic DNA. 5-Azacytidine is an analog that can be incorporated into DNA but not methylated. Therefore, growth with this analog leads to DNA demethylation.

changes a purine–pyrimidine (Pu-Py) base pair to a different Pu-Py base pair, whereas a **transversion** changes a Pu-Py to a Py-Pu.) It may not be coincidental, for example, that most sequence alterations in the p53 gene in colon cancer cell DNAs involve GC ⟶ AT transitions.

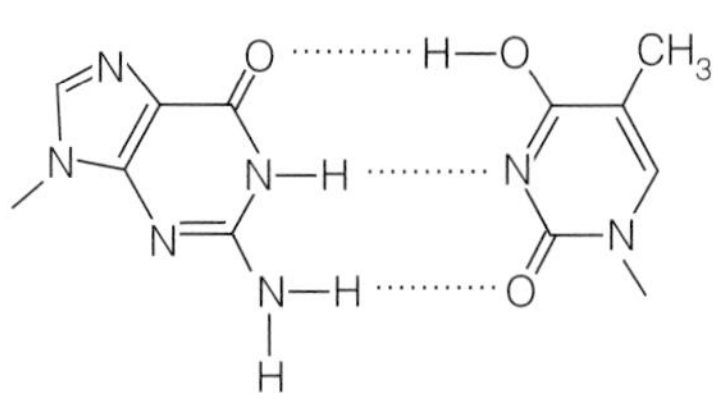

A G-T base pair (T in the rare enol form)

Restriction and Modification

One of the most important developments in the recent history of biochemistry is the discovery of **restriction endonucleases,** enzymes that catalyze the double-strand cleavage of DNA at specific base sequences. The discovery of the enzymes arose through research on a seemingly obscure process—**host-induced restriction and modification.** Bacteria use site-specific DNA methylation to mark their own DNA, and DNA cleavage at the same sites to inactivate DNA of invaders, such as viruses, which lack this marking.

BIOLOGY OF RESTRICTION AND MODIFICATION

Although restriction and modification were first described in 1952, it was not until Werner Arber's work in Switzerland in the mid-1960s that the biochemical

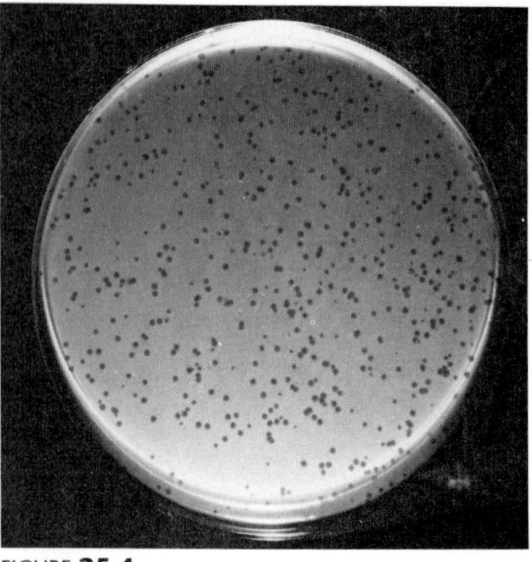

FIGURE **25.4**

Bacteriophage plaques. A Petri plate was seeded with many bacteria (about 10^8) and a small number of bacteriophage particles (about 100). Each phage particle generates a cleared area, or plaque, by infecting and lysing all of the bacteria growing near it.

basis for the phenomena became clear. The basic observation was as follows. Bacteriophage λ, when grown on the K12 strain of *E. coli* (which we shall call K), grows well in subsequent infections of the K strain, with every phage particle giving rise to a **plaque.** (A plaque is a cleared area on a Petri plate seeded with excess bacteria and a limited number of phages. It results from the multiplication of one virus particle and the localized lysis of bacteria. See Figure 25.4.) However, the phage grew poorly on *E. coli* strain B, with only 0.01% of the infecting phages giving rise to a plaque. Most of the infecting phage DNA was broken down in these nonproductive infections.

Phages purified from the few plaques that formed could infect strain B with high efficiency. The same phenomenon was seen in reverse: Phages grown on *E. coli* B infected strain K with low efficiency, but the few plaques that formed yielded phages that could infect strain K with high efficiency. These experiments suggested that although most of the phages exposed to a new strain of *E. coli* are destroyed, a few adapt so as to circumvent the bacterium's defense system and become infective to that strain. Arber coined the terms *restriction* and *modification* to explain these observations, and he showed that the biochemical basis for these phenomena is a pair of enzymes specific to each bacterial strain. When phages grown on strain K infect strain B, their DNA is almost always broken down by a B-specific restriction enzyme, and the bacterium survives. However, the B strain has an additional enzyme, which modifies its own DNA by methylating specific nucleotide residues. This pattern of methylation protects DNA against its own B-specific restriction enzyme.

Bacteria use restriction–modification, which involves nonheritable changes in DNA structure, to distinguish their own DNA from that of invaders.

Occasionally the modification enzyme of strain B methylates phage DNA as well, and in the process creates phage DNA resistant to breakdown by the B restriction system. The resistant phages are the small proportion of phages (0.01%) that plate on strain B. Because DNA of these phages is now fully protected from B-specific restriction, all of the phages that plate on B are able to infect the B strain in subsequent infections. Note that restriction and modification are nongenetic phenomena, in that phage DNA *sequences* are unaltered. The ability of a phage to overcome restriction in a particular host bacterium depends not upon any changes in the phage's genotype but upon the host strain in which that phage was previously grown. Note also that any DNA within a cell is subject to restriction and modification, including transforming and plasmid DNAs, as well as chromosomal or phage DNAs.

Restriction–modification systems are widespread among bacteria. Some are encoded by chromosomal genes, and some by plasmids. In 1970 Hamilton Smith observed that a restriction nuclease he was studying catalyzed double-strand cleavage of DNA within a specific short nucleotide sequence. Shortly thereafter, it was found that modification has the same sequence specificity, as schematized in Figure 25.5. In this example, one nucleotide within a six-nucleotide sequence is the substrate for a specific DNA methylase. When that site is methylated, the DNA is resistant to cleavage by a nuclease that recognizes the same hexanucleotide sequence; when that site is unmethylated, the DNA is susceptible to attack at that site. Hundreds of restriction endonucleases are now known to catalyze similar sequence-

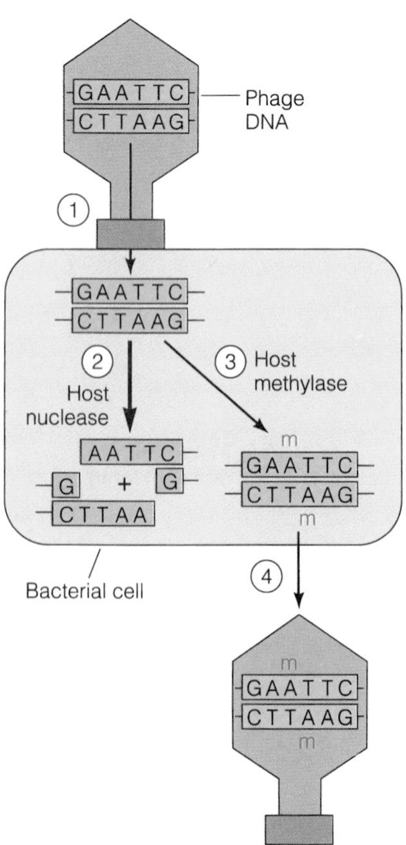

FIGURE **25.5**

Host-induced restriction and modification. A phage whose DNA is unmodified infects a bacterium with a restriction system that recognizes the DNA sequence 5′—GAATTC—3′ (step 1). Most phage DNA molecules are cleaved by the restriction nuclease (step 2), but the few that become methylated first on the innermost A are protected from attack (step 3). The phages that emerge contain modified (methylated) DNA (step 4). Because they are not vulnerable to restriction by the host nuclease, they are able to overcome the bacterium's defense system when they reinfect the same bacterial strain.

specific cleavages. The DNA methylase whose structure was shown in Figure 25.2 is part of the restriction–modification system of *Haemophilus haemolyticus*.

The importance of these developments was that, for the first time, scientists could isolate homogeneous DNA fragments of precisely defined length by treating DNA with a restriction nuclease in vitro and then resolving the fragments in the digest on an electrophoretic gel. Recall the utility of proteases such as trypsin, which cleave at specific sites on proteins, and their usefulness in fragmenting protein molecules for sequencing. Similarly, restriction fragments resolved by electrophoresis (Figure 25.6) can be arranged in order, so as to give physical maps of DNA molecules; the maps are called restriction maps, because they show the physical locations of restriction sites (see Tools of Biochemistry 25A). Isolation of these homogeneous fragments makes their amplification by gene cloning possible.

PROPERTIES OF RESTRICTION AND MODIFICATION ENZYMES

We now recognize three different types of restriction–modification systems—types I, II, and III. Each system consists of two distinct enzyme activities: a DNA methylase and an endonuclease that catalyzes a double-strand DNA break. The sequence-specific endonucleases, those most widely used in molecular biology, are type II enzymes. Regardless of type, the enzymes are named with the first three letters denoting the bacterial species of origin and a fourth letter denoting an individual strain. For example, the restriction system from *E. coli* K is called ***Eco*K.** If more than one enzyme system is found in a given strain, the different enzymes are designated by Roman numerals. For instance, ***Eco*RI** is one of two known restriction systems in *E. coli* strain R, and ***Hin*dIII** is one of three enzymes from the d strain of *Haemophilus influenzae* R.

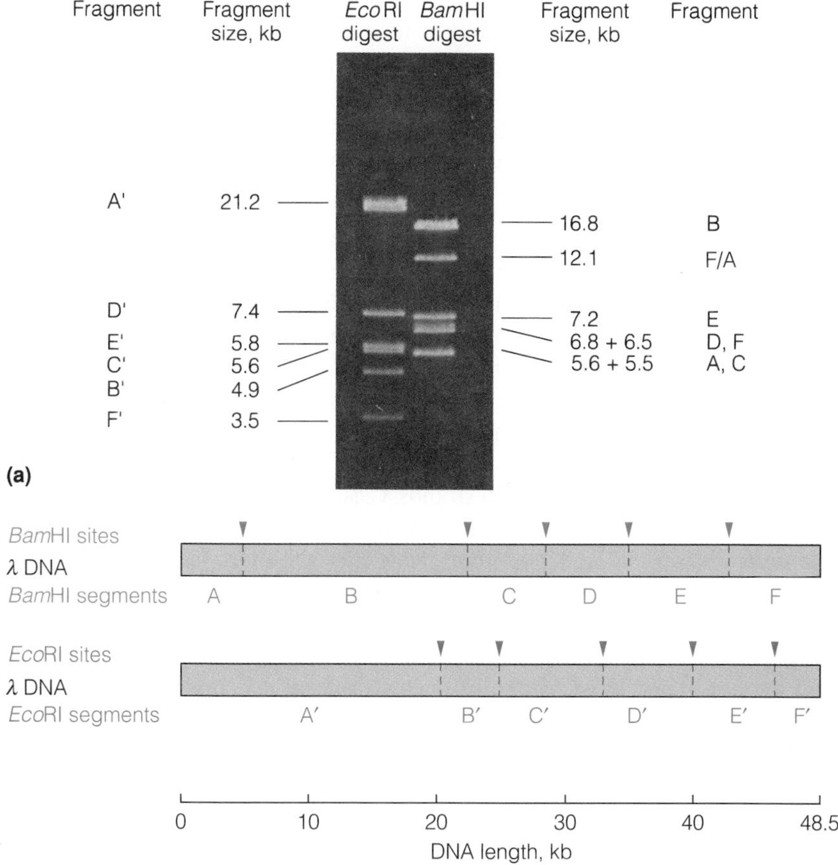

FIGURE **25.6**

Fragmentation of phage λ DNA with *Eco*RI or *Bam*HI. (a) Experimental determination of fragmentation patterns. Restriction digests are subjected to agarose gel electrophoresis, and the fragments are visualized by staining the gel with ethidium bromide. Note that fragments with very similar sizes form but one band on a gel. **(b)** Maps of cleavage sites for each enzyme on the 48.5-kilobase (kb) λ DNA molecule. By convention, the fragments are assigned letters as shown. In this experiment the 12.1-kb fragment in the *Bam*HI digest resulted from linkage of the A and F (terminal) fragments by base pairing at their cohesive ends. Mapping the restriction sites also requires data from digestion of the DNA with *Eco*RI and *Bam*HI together (not shown).

(a) Courtesy of Catherine Z. Mathews.

TABLE **25.1 Properties of restriction–modification systems**

	Type I	Type II	Type III
Example	*Eco*B	*Eco*RI	*Eco*PI
Recognition site	TGAN$_8$TGCT	GAATTC	AGACC
Cleavage site	Up to 10 kbp away from recognition site	Between G and A (both strands)	24–26 base pairs 3′ to recognition site
Methylation site	m TGAN$_8$TGCT ACTN$_8$ACGA m	m GAATTC CTTAAG m	m AGACC (only one strand methylated)
Nuclease and methylase in one enzyme?	Yes	No	Yes
Requirements for cleavage	ATP, Mg^{2+}, AdoMet	Mg^{2+} or Mn^{2+}	Mg^{2+}, AdoMet
Requirements for methylation	ATP, Mg^{2+}, AdoMet	AdoMet	Mg^{2+}, AdoMet

Note: Each methylated base is identified with the letter m. All sequences read 5′ to 3′, left to right.

Properties of each of the three types of restriction systems are discussed in the following sections and are summarized in Table 25.1.

Type I

Type I enzymes have both methylase and nuclease activities in one protein molecule, which contains three subunits. One subunit contains the nuclease, one the methylase, and one a sequence recognition determinant. The recognition site is not symmetrical, and cleavage occurs some distance (up to 10 kbp) away from the recognition site, although methylation occurs within the recognition site. For cleavage the enzyme remains bound to the recognition site, and DNA is looped out around it, with concomitant supercoiling. About 10^5 ATP molecules are hydrolyzed per cleavage event. Energy is probably needed for both translocation of the enzyme and supercoiling of the DNA. For reasons still not clear, both ATP and AdoMet are required for the cleavage activity. AdoMet may be an allosteric activator, because it is not broken down during the reaction.

Restriction enzymes of most use to biologists cleave both DNA strands site-specifically, depending on base methylation.

Type II

Type II restriction nucleases have been of great value to research, because most of them cut within the recognition sequence, making cleavage sequence specific. Most of the type II enzymes are homodimers, with subunits of 30 to 40 kilodaltons. A divalent cation is required for cleavage, but ATP is not required. Each type II nuclease has a counterpart methylase, which binds to the same recognition sequence and methylates one nucleotide within that sequence. A **hemimethylated** DNA (with methyl group on one strand only) is a preferred substrate for the methylase but not for the nuclease, which generally cleaves only when the recognition site is unmethylated on both strands. Cleavage generates 3′ hydroxyl and 5′ phosphate termini. Cleavage sites on the two strands may be offset by as much as four nucleotides (as in *Eco*RI) or more, giving cuts with short, self-complementary, single-strand termini. Some enzymes cleave to give a 5′-terminated single-strand end ("overhang"), whereas others generate a 3′ overhang. Other type II nucleases, including *Sma*I and *Hin*dII, generate blunt-ended fragments, in which the cutting sites are not offset. Most recognition sites are four, five, or six nucleotides in length, although a few type II enzymes recognize an eight-nucleotide sequence. Most show 2-fold rotational sequence symmetry, suggesting that the two enzyme subunits are also arranged symmetrically. Table 25.2 shows the recognition sites for several widely used type II nucleases. Several hundred enzymes of this type have now been isolated. Not all type II nucleases are absolutely sequence specific.

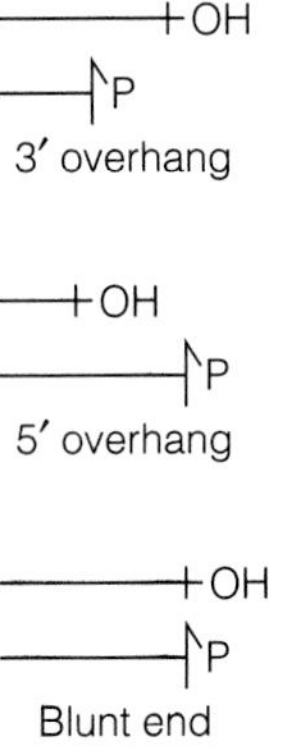

TABLE **25.2 Specificities of some type II restriction systems**

Enzyme	Bacterial Source	Restriction and Modification Site[a]
*Bam*HI	*Bacillus amyloliquefaciens* H	G↓GATCC
*Bgl*II	*B. globiggi*	A↓GATCT
*Eco*RI	*Escherichia coli* RY13	G↓A$\overset{m}{\text{A}}$TTC
*Eco*RII	*E. coli* R245	C$\overset{m}{\text{C}}$↓GG
*Hae*III	*Haemophilus aegyptius*	GG↓$\overset{m}{\text{C}}$C
*Hga*I	*H. gallinarum*	GACGCNNNNN↓ CTGCGNNNNNNNNNN↑
*Hha*I	*H. haemolyticus*	G$\overset{m}{\text{C}}$G↓C
*Hind*II	*H. influenzae* Rd	GTPy↓Pu$\overset{m}{\text{A}}$C
*Hind*III	*H. influenzae* Rd	$\overset{m}{\text{A}}$↓AGCTT
*Hinf*I	*H. influenzae* Rf	G↓ANTC
*Hpa*I	*H. parainfluenzae*	GTT↓AAC
*Hpa*II	*H. parainfluenzae*	C↓$\overset{m}{\text{C}}$GG
*Msp*I	*Moraxella* sp.	C↓CGG
*Not*I	*Nocardia rubra*	GC↓GGCCGC
*Ple*I	*Pseudomonas lemoignei*	GAGTCNNNN↓ CTCAGNNNNN↑
*Pst*I	*Providencia stuartii*	CTGCA↓G
*Sal*I	*Streptomyces albus* G	G↓TCGAC
*Sma*I	*Serratia marcescens* Sb	CC$\overset{m}{\text{C}}$↓GGG
*Xba*I	*Xanthomonas badrii*	T↓CTAGA

[a]The methylated base in each site, where known, is identified with the letter m. All sequences read 5′ to 3′, left to right. The cleavage on the opposite strand in each case can be inferred from the symmetry of the site (except for *Hga*I and *Ple*I, each of which has an asymmetric site). Pu = purine, Py = pyrimidine, N = any base.

For example, *Hin*dII recognizes four different hexanucleotide sequences, and some enzymes (such as *Hga*I) cleave at a site outside the recognition sequence.

The year 1986 saw the first crystallographic structural determination of a restriction nuclease (*Eco*RI) complexed with a double-strand oligonucleotide containing its DNA recognition sequence. Figure 25.7 shows one polypeptide subunit of the dimeric enzyme in contact with its DNA recognition sequence. The DNA is bound in a cleft, and the protein has an N-terminal "arm" that wraps about the DNA. Sequence specificity is maintained by 12 hydrogen bonds, which link the purine residues in the site to a glutamate and two arginine residues (not shown in the figure). Binding of the DNA alters its structure to generate "kinks"; the sequences immediately flanking the six-nucleotide cutting site (GAATTC) adopt the A duplex conformation, and the B structure is retained within the cutting site. The other subunit, not shown in the figure, contacts the substrate identically, accounting for the ability of the enzyme to catalyze symmetrical cleavages within the cutting site.

By contrast, *Bam*HI endonuclease does not kink its DNA substrate, which remains in the B form. However, as shown in Figure 25.8, the enzyme itself undergoes a major conformational change upon binding DNA. The C-terminal α helices from each subunit unwind and contact the DNA, one in the minor groove, and

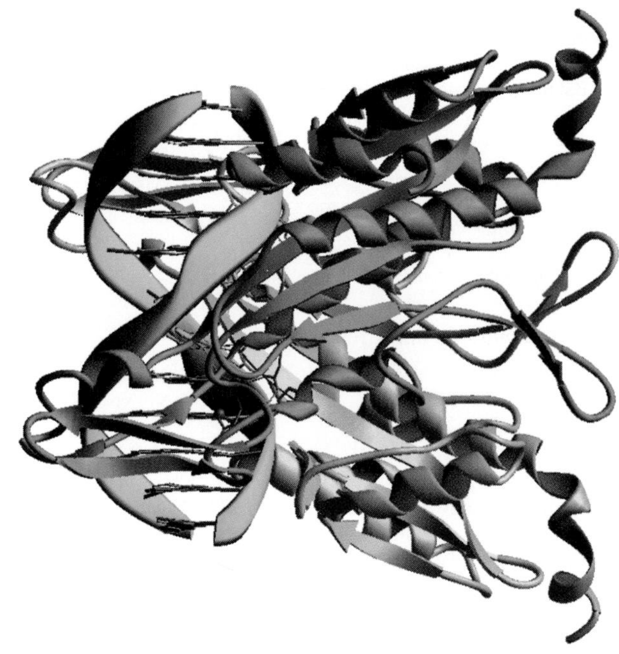

FIGURE **25.7**

Structure of the *Eco*RI nuclease complexed with its DNA substrate. The DNA helix is shown in blue, while the two subunits of the protein are shown in red and yellow, respectively. Note the "kink" in the DNA structure, resulting from the fact that the enzyme binds the central six-base-pair cutting site in the B conformation, while the flanking sequences are bound as A-form DNA. Note also the N-terminal "arm" on each protein subunit, which wraps around the DNA.

Courtesy of J. Rosenberg.

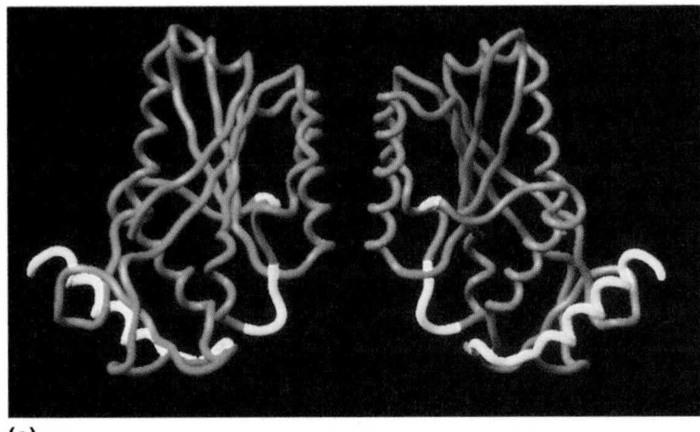

(a)

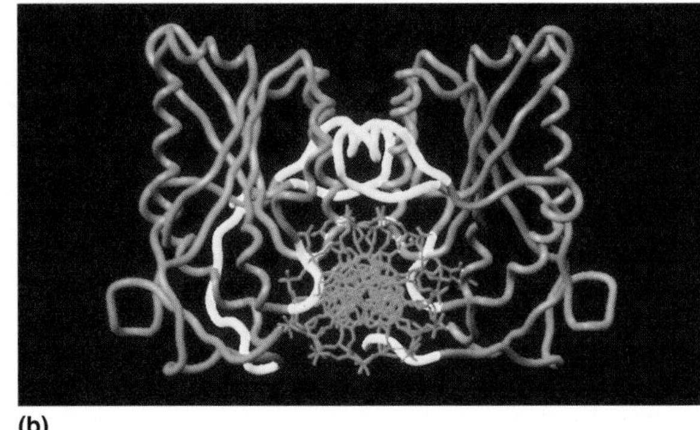

(b)

FIGURE **25.8**

Structures of (a) free and (b) DNA-bound forms of *Bam*HI, with DNA shown end-on, in orange. Regions of the protein that undergo conformational change upon DNA binding, including the two C-terminal α helices, are shown in yellow.

Courtesy of A. Aggarwal and M. Newman, Mount Sinai School of Medicine, from M. Newman, *Science* (1995) 269:656–663, with permission from Science.

one along the sugar–phosphate backbone, thus introducing an unexpected element of asymmetry into the DNA–protein complex.

Type III

Type III enzymes resemble more closely the type I systems than the type II systems. Type III enzymes contain both nuclease and methylase activities in a two-subunit enzyme. They differ from type I enzymes in that they do not require ATP, they modify just one DNA strand, and the cleavage site is fairly close to the recognition site.

The ability to cut DNA molecules at precisely known short sequences opened the door to a host of powerful research techniques. Several of them are discussed in Tools of Biochemistry 25A through 25E.

DNA Repair

TYPES AND CONSEQUENCES OF DNA DAMAGE

Unprogrammed chemical changes occur in all biological macromolecules, because of either environmental damage or errors in synthesis. For most biopolymers, including RNA, protein, and membrane phospholipids, the effects of such changes are minimized by turnover and replacement of altered molecules. DNA is distinctive, however, in that its information content must be transmitted virtually intact from one cell to another during cell division or reproduction of an organism. Thus, DNA has a special need for metabolic stability. This stability is maintained in two ways: (1) by a replication process of very high accuracy and (2) by mechanisms for correcting genetic information when DNA suffers damage. In Chapter 24 we described mechanisms used to ensure high replication accuracy—specifically, 3′-exonucleolytic proofreading, which corrects errors made by DNA polymerases, and the uracil-DNA *N*-glycosylase pathway, which prevents mutations that might result from deamination of cytosine to uracil in DNA. Here we consider several processes for repairing DNA that is altered either by uncorrected replicative errors or by environmental damage. The latter type of alteration can result from chemical modification of DNA nucleotides or from photochemical changes ensuing from absorption of high-energy radiation.

Most studies of DNA repair have been carried out on samples altered by chemical alkylation or cross-linking of DNA strands, by deamination of DNA bases, or by ultraviolet irradiation. Much current attention is being focused on oxidative damage. The intracellular generation of reactive oxygen species (see Chapter 15) causes the formation of DNA bases such as 8-oxoguanine (8-hydroxyguanine) or thymine glycol (see page 554). 8-Oxoguanine is one of the most significant products of oxidative DNA damage because its formation contributes most strongly to mutagenesis (see page 940). Interestingly, one of the protective mechanisms is not a DNA repair enzyme but a nucleotide hydrolase. In *E. coli,* 8-oxo-dGTP accumulates in oxygen-stressed cells, and an enzyme encoded by the *mutT* gene cleaves this altered nucleotide before it can be used as a DNA replication substrate. A similar enzyme exists in mammalian cells (see page 825).

Guanine

8-Oxoguanine

$$\text{8-Oxo-dGTP} + H_2O \longrightarrow \text{8-oxo-dGMP} + PP_i$$

The consequences of irradiation or alkylation damage include mutagenesis, resulting from erroneous replication of a damaged template base, and cell death, resulting from inability of the replication apparatus to copy past a damaged site. Because it is now clear that cancer results from an accumulation of somatic cell mutations, the mechanisms of DNA repair are under intense study as determinants of an animal's susceptibility to cancer.

As if to underscore the importance of DNA repair, every cell has several different repair systems. These systems are categorized as follows: (1) **direct repair,** in which a damaged DNA base undergoes a chemical reaction to restore the original structure; (2) **nucleotide excision repair,** in which a section of DNA that contains a damaged site is excised and replaced with normal DNA; (3) **base excision repair,** which starts with cleavage of the glycosidic bond connecting a damaged base to the DNA sugar–phosphate backbone; (4) **recombinational repair,** in which newly replicated DNA duplexes undergo genetic recombination, with ultimate removal of the damaged DNA segment; and (5) **mismatch repair,** a process that recognizes DNA mismatches created by either replication errors, nonhomologous recombination, or damage to one DNA base and corrects the error. We discuss these repair processes in sequence.

BIOLOGICALLY SIGNIFICANT DNA PHOTOPRODUCTS: PYRIMIDINE DIMERS

In studying mechanisms of DNA repair, one first needs to identify the chemical forms of DNA alteration that are subject to repair. Most of the early discoveries about DNA repair were made in studies with ultraviolet light–irradiated organisms. DNA was identified as the principal biological target for UV irradiation, partly through determination of **action spectra**—irradiation of bacteria or phages with UV light and determination of the wavelengths most effective in stimulating

FIGURE **25.9**
Structures of pyrimidine dimer photoproducts.

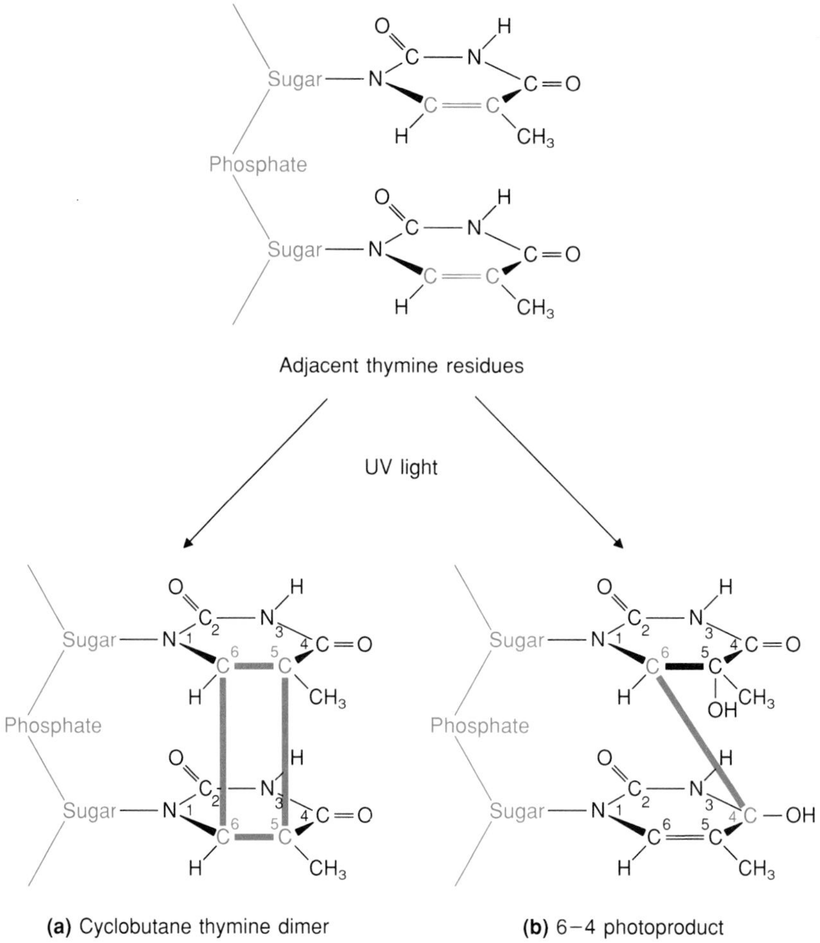

Cyclobutane thymine dimers are the most lethal photoproduct in UV light–irradiated DNA. 6–4 photoproduct may be the strongest mutagenic product.

mutagenesis or death. Those wavelengths lie near 260 nm, where DNA light absorption is maximal.

When one examines either UV-irradiated DNA or the DNA extracted from a UV-irradiated organism, one detects small amounts of many different altered DNA constituents, called **photoproducts.** Prominent among them are intrastrand dimers consisting of two pyrimidine bases joined by a **cyclobutane** ring structure involving carbons 5 and 6 (Figure 25.9a). Such **thymine dimers,** formed from two adjacent DNA thymine residues, were identified quite early as biologically significant photoproducts, because the relative abundance of thymine dimers in irradiated DNA correlated most closely with death of irradiated phages or bacteria. Thus, the ability of an organism to survive ultraviolet irradiation was related directly to its ability to remove thymine dimers from its DNA. Dimerization draws the adjacent thymine residues together, distorting the helix in such a way that replicative polymerization past this site is blocked.

For many years it was thought that ultraviolet light–induced *mutagenesis,* as well as cell death, was also caused primarily by cyclobutane thymine dimers. However, more recent data suggest that a different pyrimidine–pyrimidine dimer, called **6–4 photoproduct,** is probably the principal cause of UV-induced mutations. As shown in Figure 25.9b, these products are also dimers, linked via C-6 of the 5′ pyrimidine (either thymine or cytosine) and C-4 of the 3′ pyrimidine (usually cytosine, but occasionally thymine, as shown in Figure 25.9). The idea that 6–4 photoproducts are responsible for mutations in UV-irradiated DNA is supported by experiments in which cyclobutane thymine dimers were completely removed from UV-irradiated DNA by **photoreactivation** (see the next section). When this DNA was introduced into bacteria, the dimer removal had no effect on mutation frequency. However, the question of whether 6–4 photoproducts are principally responsible for mutagenesis is still unsettled.

DIRECT REPAIR OF DAMAGED DNA BASES: PHOTOREACTIVATION AND ALKYLTRANSFERASES

Of the half dozen well-understood DNA repair processes, most involve removal of the damaged nucleotides, along with several adjacent residues, followed by replacement of the excised region using information encoded in the complementary (undamaged) strand. However, at least two processes involve reactions that *directly change* the damaged bases, rather than removing them.

DNA can be repaired directly, by changing a damaged base to a normal one, or indirectly, by replacing a DNA segment containing the damaged nucleotide.

Photoreactivation

A widely distributed enzyme, called **photoreactivating enzyme,** or **DNA photolyase,** repairs cyclobutane pyrimidine dimers in the presence of visible light. A wavelength of 370 nm is most effective. The enzyme binds to DNA in a light-independent process, specifically at the site of pyrimidine dimers. In the presence of visible-wavelength light, the bonds linking the pyrimidine rings are broken, after which the enzyme can dissociate in the dark. Clues to the mechanism have come with the finding that the enzyme contains two chromophores. (Recall from Chapter 17 that a chromophore is a structural moiety that absorbs light of characteristic wavelengths.) One chromophore is bound flavin adenine dinucleotide, deprotonated and in the reduced state ($FADH^-$; see Figure 25.10); the second in some photolyases is 5,10-methenyltetrahydrofolate and in others is 8-hydroxy-5-deazaflavin. Mechanistic studies suggest a process akin to photosynthesis, with the second chromophore functioning as a light-harvesting factor, and somehow transmitting light energy to $FADH^-$, which functions like the photochemical reaction center, transfering an electron to the dimer and breaking the pyrimidine–pyrimidine bonds by a free radical mechanism, as shown in Figure 25.10.

Direct repair enzymes include photolyase, which uses light energy to repair pyrimidine dimers, and alkyltransferases, "enzymes" that are inactivated after just one catalytic cycle.

Although photolyase has been detected in numerous eukaryotic systems, recent evidence indicates that human cells do not contain an enzyme for photoreactivation. Thinning of the earth's ozone layer has been blamed for population declines in certain frog species, which have been shown to lack photolyase and, hence, to suffer damage from solar ultraviolet irradiation, particularly during embryonic development in clear lakes, which are transparent to UV rays.

O^6-Alkylguanine Alkyltransferase

Treatment of DNA with a methylating or ethylating reagent is comparable to ultraviolet irradiation in that various modified DNA bases are formed, some of which are lethal if not repaired and some of which are mutagenic. Some alkylating agents are used in cancer chemotherapy because of their ability to block DNA replication and, hence, cell proliferation. Some are widely used as mutagens in the laboratory. Three methylating or ethylating reagents are shown in the margin.

CH_3—N(NO)—C(=O)—NH_2

Methylnitrosourea (MNU)

CH_3—S(=O)(=O)—O—CH_2—CH_3

Ethylmethanesulfonate (EMS)

CH_3—N(NO)—C(=NH)—NH—NO_2

***N*-Methyl-*N*′-nitro-*N*-nitrosoguanidine (MNNG)**

The bases altered by these reagents are primarily purines (phosphate oxygens are also targets), and the spectrum of products formed varies with the reagent used. The most highly mutagenic of these products, **O^6-alkylguanine,** is mutagenic because the modified base has a very high probability of pairing with thymine when the modified strand replicates (Figure 25.11). Thus, alkylation of a DNA-guanine stimulates a GC ⟶ AT transition (see margin on page 938, where mG is a methylguanine residue).

Repair of this type of damage involves an unusual enzyme, **O^6-alkylguanine alkyltransferase,** which transfers a methyl or ethyl group from an O^6-methylguanine or O^6-ethylguanine residue to a cysteine residue in the active site of the protein. Remarkably, this "catalyst," which is widely distributed in prokaryotes and eukaryotes, can function only once. Having become alkylated, it cannot remove the alkyl group, and the protein molecule turns over. Thus, the term *enzyme* is a misnomer. In bacteria the protein regulates both its own synthesis and that of another repair

Excision repair involves endonuclease cleavage on both sides of a damaged site, followed by replacement synthesis.

FIGURE **25.10**

A mechanism for photolyase action. **(a)** Partial structures of $FADH^-$ and $FADH\cdot$. **(b)** The probable reaction pathway.

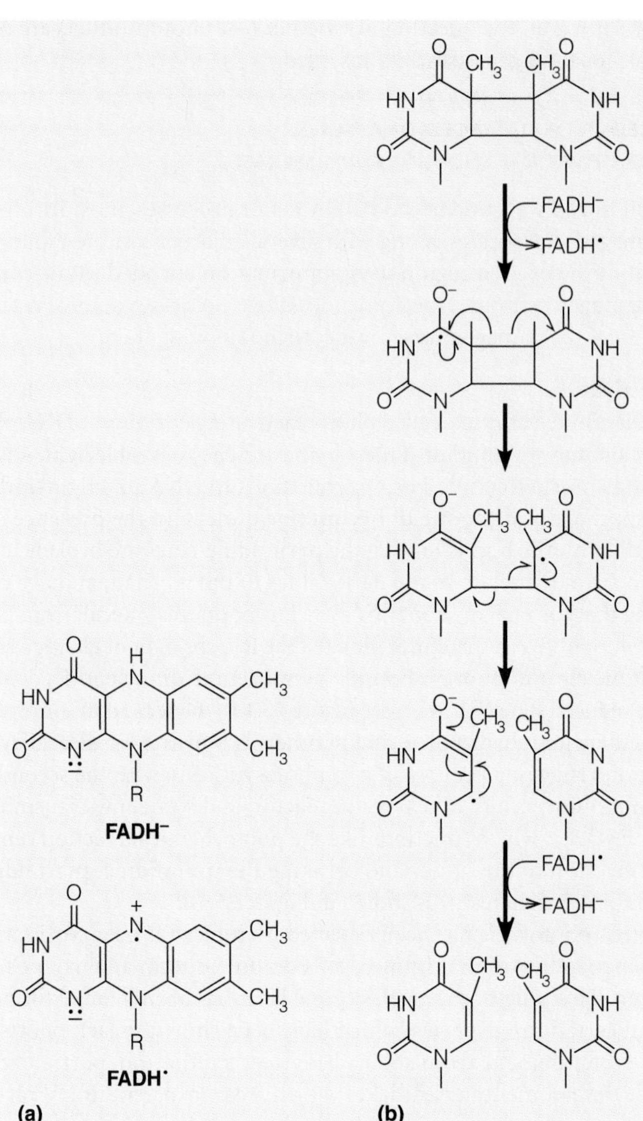

Guanine

Cytosine

Alkylation

O^6-Methylguanine (mG)

Thymine

FIGURE **25.11**

Mispairing of O^6-methylguanine with thymine in a DNA duplex.

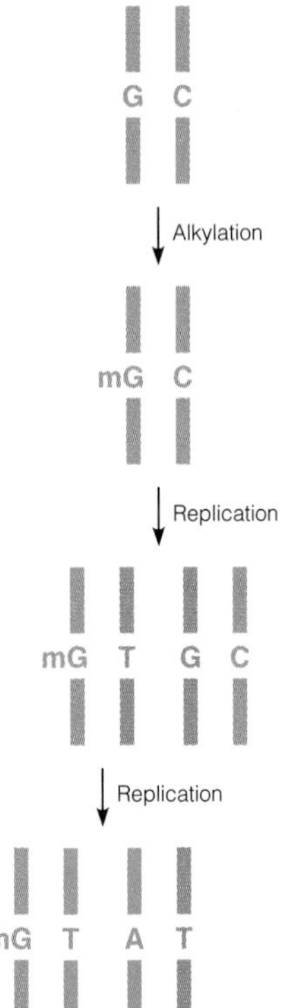

enzyme, a DNA-*N*-glycosylase that we shall discuss later. The regulation involves activation of the transcription of genes encoding these two proteins. There is evidence that the *alkylated* form of the alkyltransferase is the specific form of the transcriptional activator. This allows the cell to adapt to alkylation damage by using the alkylated protein as a specific signal to produce more of the proteins needed to repair the damage.

NUCLEOTIDE EXCISION REPAIR: EXCINUCLEASES

Nucleotide excision repair (NER) was originally discovered as an enzyme system capable of repairing thymine dimers created in DNA by UV irradiation. Unlike photoreactivation, this process can take place in the dark. The enzyme system involved, which in *E. coli* includes the products of genes *uvrA, uvrB,* and *uvrC,* is now known to act upon a number of DNA damaged sites containing lesions that may be quite bulky, such as those created by large alkyl groups, and that distort the DNA double helix. Very similar systems exist in mammalian cells and in yeast, so the process is probably universal.

As shown in Figure 25.12 for *E. coli,* the three-subunit UvrABC enzyme recognizes a lesion (a thymine dimer in the example shown) and, with the help of

ATP hydrolysis, forces DNA to bend, leading to cleavage of the damaged strand at two sites—eight nucleotides to the 5′ side of the damaged site and four or five nucleotides to the 3′ side—leaving a potential gap 12 or 13 nucleotides in length, with a 3′ hydroxyl group and a 5′ phosphate at the ends. Polymerase and ligase action then replace the damaged 12-mer or 13-mer with undamaged DNA. Helicase II, the product of the *uvrD* gene, is also required, presumably to unwind and remove the excised oligonucleotide, which is ultimately broken down by other enzymes. The UvrABC enzyme is not a classical endonuclease, because it cuts at two distinct sites, and the term **excinuclease** has been proposed for it, denoting its role in *exc*ision repair. This system is also involved in repairing another type of DNA damage—covalent cross-linking of the two strands to each other. Here, in order to preserve an intact template strand, the two strands are repaired sequentially, one after the other.

Excision repair also occurs in mammalian cells, as shown by the demonstration of a human excinuclease that cleaves at positions −22 and +6 relative to a thymine dimer. A significant difference is the involvement of two different endonucleases in human excision repair—one for cutting on the 5′ side and one on the 3′ side. Excision repair in humans originally came to light through studies of a rare genetic disease called **xeroderma pigmentosum (XP).** XP is actually a family of diseases, in which one or more enzymes of the excision pathway are deficient. In affected humans there is at present no known way to treat the condition. The biological consequences of XP include extreme sensitivity to sunlight and a high incidence of skin cancers. Although overexposure to the ultraviolet rays in sunlight increases the risk of skin cancer for all humans, the greatly increased skin cancer frequency in XP patients highlights the importance of UV repair pathways to mammals. Because there is no known treatment for XP, affected individuals must avoid sunlight.

Recent studies of nucleotide excision repair show that active genes (those undergoing transcription) are preferred substrates for excision repair, and within these genes the template DNA strand is preferentially repaired. This **transcription-coupled repair** may initiate when a transcribing RNA polymerase becomes stalled at the site of a DNA lesion. Coupling transcription with repair helps to ensure the integrity of genes that are actually used. In mammalian cells, transcription-coupled repair is a specialized mode of NER that requires additional proteins. Genetic defects in one or more of these proteins lead to **Cockayne's syndrome,** a condition characterized by skin photosensitivity, growth failure, and neurological abnormalities, but evidently not including increased risk of cancer. However, another human gene, *BRCA1,* has recently been implicated in transcription-coupled excision repair. Mutations in *BRCA1* are associated with increased risk of breast and ovarian cancer.

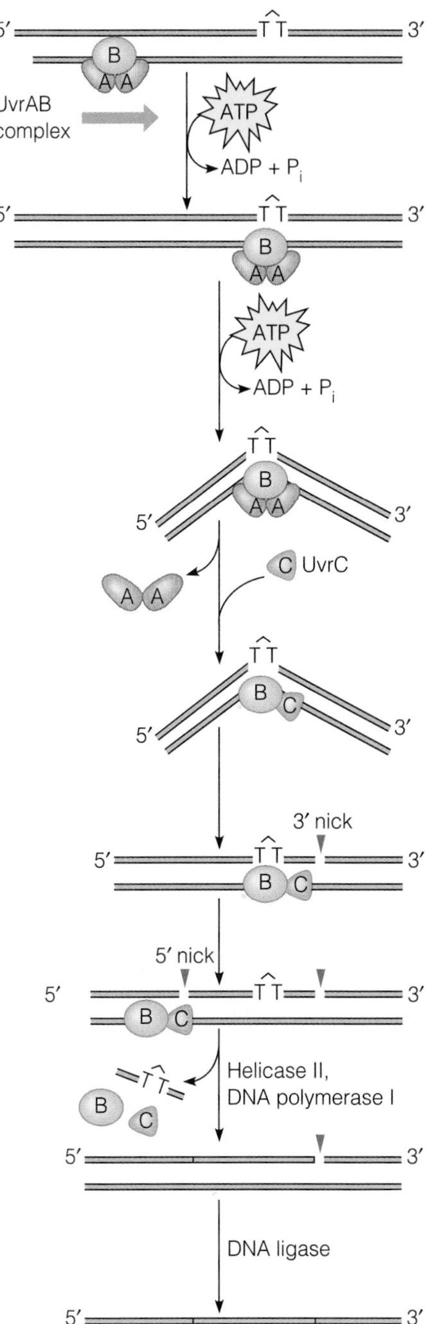

FIGURE **25.12**

Excision repair of thymine dimers by the UvrABC excinuclease of *E. coli.* A complex of A and B proteins tracks along DNA until it reaches a thymine dimer or other damaged site, where it halts and forces the DNA to bend. UvrA (a "molecular matchmaker") then dissociates, allowing UvrC to bind to B. The BC complex cuts on both sides of the dimer. Helicase, polymerase, and ligase remove the damaged dodecamer and replace it with new DNA. This system may use DNA polymerase II as well as pol I.

Adapted from A. Sancar and J. E. Hearst, *Science* (1993) 259:1415–1420. © 1993 AAAS.

BASE EXCISION REPAIR: DNA-*N*-GLYCOSYLASES

The other form of excision repair, base excision repair, or BER, also removes one or more nucleotides from a site of base damage. However, this process initiates with enzymatic cleavage of the glycosidic bond between the damaged base and deoxyribose. The replacement of uracil by thymine in DNA, presented in Chapter 24 (page 906) is an example of BER. Figure 25.13 illustrates BER of a thymine dimer, as initiated by endonuclease V of bacteriophage T4 or a similar enzyme from *Micrococcus luteus.* These enzymes have two activities—first, a glycosylase, which cleaves between the thymine on the 5′ side of the dimer and its associated deoxyribose, and second, an **AP endonuclease,** which recognizes the **apyrimidinic (AP) site,** consisting of a deoxyribose without an associated pyrimidine base, and cleaves on its 5′ side. A second cleavage, 3′ to that site, by **deoxyribophosphodiesterase,** releases deoxyribose-5-phosphate. Then, nick translation by DNA polymerase I, followed by DNA ligase action, replaces the damaged DNA and closes

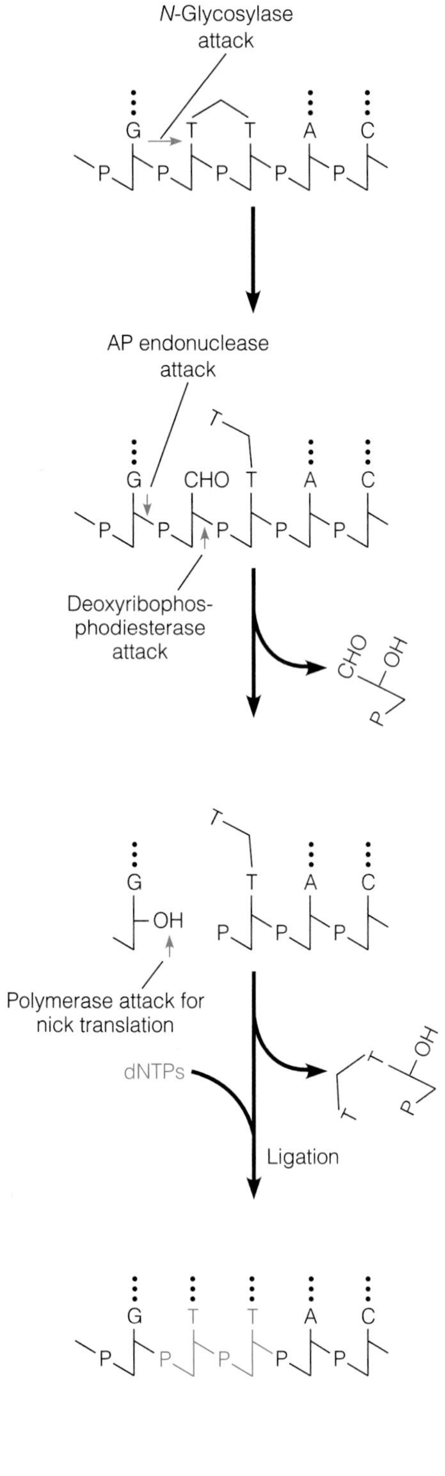

N-Glycosylase attack

AP endonuclease attack

Deoxyribophos-phodiesterase attack

Polymerase attack for nick translation

dNTPs

Ligation

dATP

PP_i

H_2O

dAMP

FIGURE **25.13**

Base excision repair of thymine dimers. This is the process as carried out by the bifunctional repair enzymes of phage T4 and *Micrococcus luteus*. Only the DNA strand bearing the dimer is shown. Red arrows identify sites of enzymatic cleavage.

the resultant nick. The patch, or replaced DNA, may be as short as one nucleotide or two (as in the example shown), or it may be several nucleotides long. Base excision repair often involves separate enzymes for glycosidic cleavage and endonucleolytic cleavage of the abasic site (apyrimidinic or apurinic, depending upon the nature of the damaged base). Most cells contain several DNA-*N*-glycosylases, including those specific for the alkylated bases *N*-methyladenine, 3-methyladenine, and 7-methylguanine.

Oxidative damage to DNA is repaired primarily by BER. Using 8-oxoguanine as an example of oxidative damage, we have seen that mutagenic error avoidance operates in part through action of the *mutT* gene product, which cleaves 8-oxo-dGTP, a nucleotide containing the damaged base, before it can be incorporated into DNA. The mutagenic effect of 8-oxoguanine results from its ability to base-pair with adenine (see the following figure), and thus lead to a transversion mutation by pathways outlined in Figure 25.14.

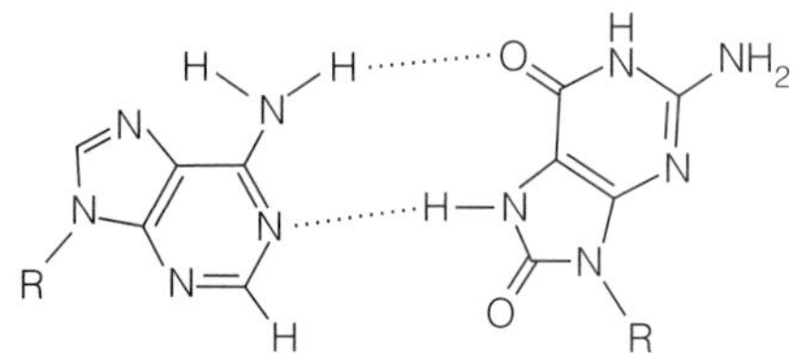

Adenine **8-Oxoguanine**

Two other *E. coli* proteins, the products of the *mutM* and *mutY* genes, are DNA-*N*-glycosylases, which act at DNA sites containing 8-oxoguanine. The MutM protein removes 8-oxoguanine (and other oxidized bases) from DNA, while the MutY protein removes adenine that is paired with 8-oxoG. Together, these proteins can initiate repair at sites where 8-oxoG has arisen, either through incorporation from 8-oxo-dGTP or through oxidation of a dGMP residue in DNA by a reactive oxygen species (ROS), as shown in Figure 25.14. The "*mut*" term, used in naming these genes, refers to the fact that defective functioning of any of the gene products creates a mutator phenotype, by increasing the spontaneous mutation frequency.

POSTREPLICATION REPAIR: RECOMBINATIONAL REPAIR AND THE SOS RESPONSE

Photoreactivation and excision repair are short-term metabolic responses. Both processes can occur within a few minutes after DNA suffers chemical or ultraviolet damage. However, if the photoreactivation and excision repair systems are defective, or become saturated by DNA damage too extensive for their capacities, at least two longer-term systems can come into play in bacteria. Discussion of these systems focuses on thymine dimer repair, but other types of damage can be repaired as well.

When a replicative polymerase encounters a thymine dimer, it cannot replicate past this site. Deoxyadenylate is incorporated opposite the first thymine base in the template. The double helix is distorted because of the thymine dimer, causing the structure to be recognized as a mismatch, and the 3′ exonucleolytic activity of DNA polymerase III cleaves out the newly incorporated dAMP. Thus, polymerase "idles" at the damage site, converting dATP to dAMP by a continual process of insertion and exonucleolytic cleavage. Synthesis of an Okazaki frag-

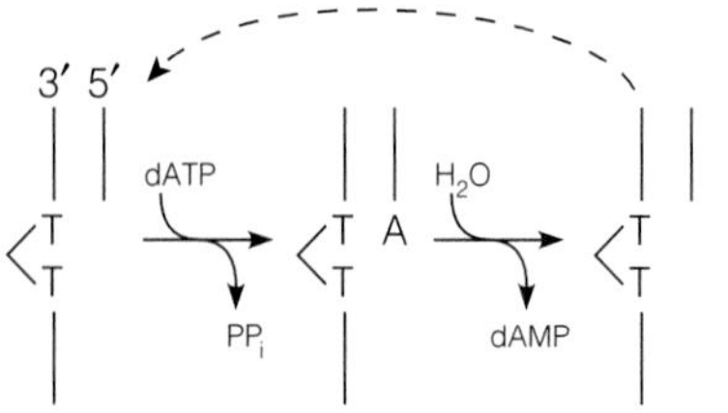

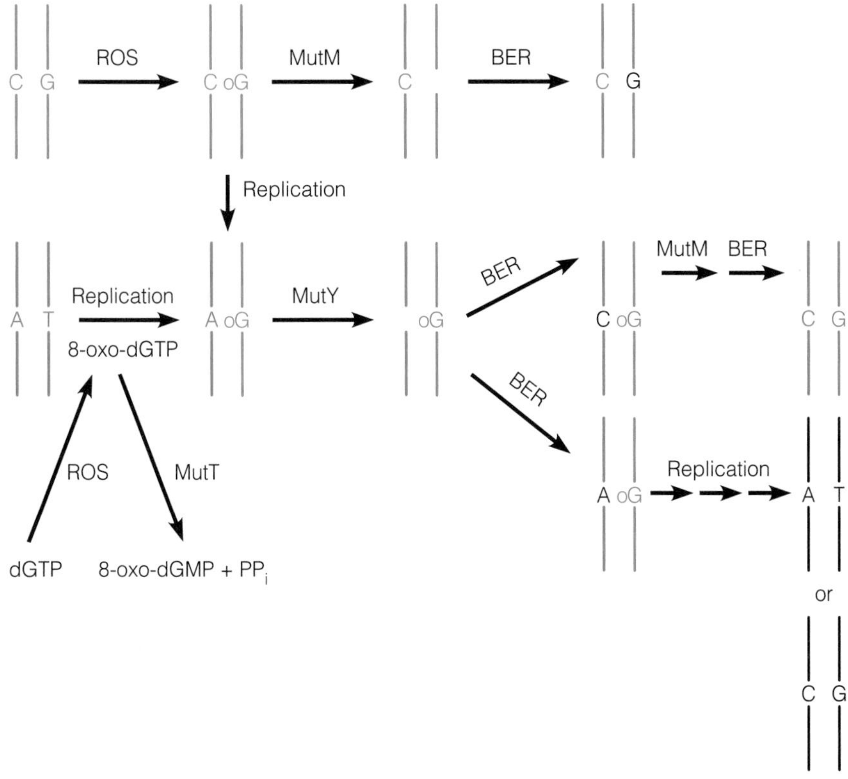

FIGURE **25.14**

Actions of *mutM, mutT,* and *mutY* gene products in countering the mutagenic effect of 8-oxoguanine (oG). Depending upon the route of introduction of oG to DNA, the pathways shown can cause either GC→AT or AT→GC transversions. MutT hydrolyzes 8-oxo-dGTP and prevents its incorporation during DNA replication. MutM excises oG from a C·oG base pair as part of the base excision repair (BER) process. MutY excises A from an A·oG base pair in another BER process, allowing an additional possibility for correcting an error in the next round of replication. Not all possible outcomes are shown in the figure. For example, 8-oxo-dGTP can be incorporated opposite template C, as well as opposite A. Also, MutY can excise A from an A·G base pair, thereby preventing spontaneous transversion mutagenesis. ROS = reactive oxygen species.

ment can commence opposite the damaged site, leaving a gap opposite the thymine dimer. This gap would be lethal if unrepaired, however, because it would generate a double-strand break in the next round of replication.

Two distinct processes can repair the gap: (1) **recombinational repair,** or **daughter-strand gap repair;** and (2) **SOS repair,** or **error-prone repair.** Both processes depend critically on a protein called RecA. Bacteria carrying mutations in *recA* were originally characterized as defective in general recombination and DNA repair. We now know that *recA*$^-$ bacteria, which specify a defective RecA protein, have a complex phenotype, including defective DNA repair. We shall describe the protein later in this chapter, but for our purposes here two properties are important. First, RecA catalyzes **strand pairing,** or **strand assimilation**—the joining of two different DNAs by homologous base pairing with each other. Second, RecA is a genetic regulator, activating the synthesis of many proteins, including DNA repair proteins, that help a bacterium adapt to a variety of metabolic stresses. This adaptation, called the **SOS response,** is more fully described later, and the transcriptional activations that lead to the SOS response are presented in Chapter 26.

Recombinational Repair

Because the gap that is generated opposite a thymine dimer is created by faulty replication, the gap is close to the replication fork. Therefore, it is also close to the corresponding region on the other daughter duplex (Figure 25.15). If that region has itself not sustained damage, the RecA protein can initiate recombination between two homologous duplexes. The uninvolved parental strand, which is complementary to the damaged parental strand, recombines into the gap, opposite

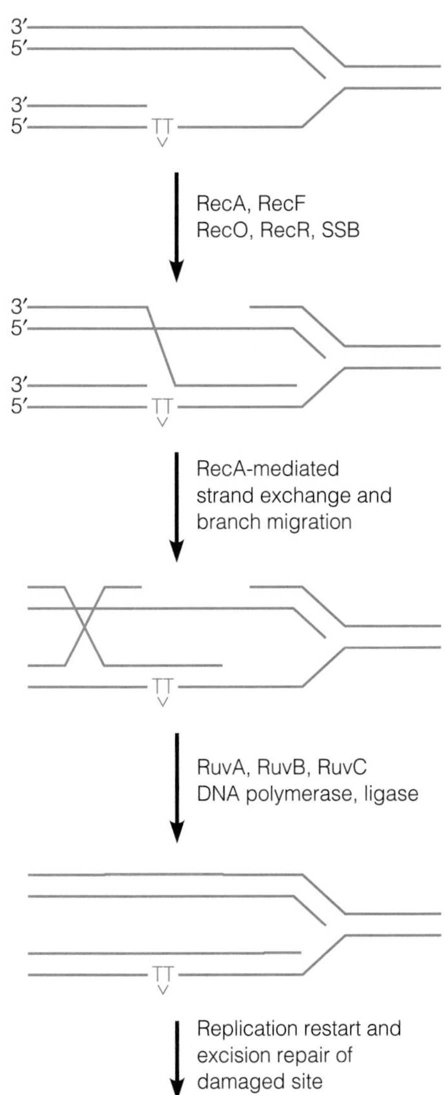

FIGURE **25.15**

Recombinational repair. The undamaged parental DNA strand is transferred to the gap in the daughter strand, formed by the inability of DNA polymerase to replicate past it. The remaining steps occur by mechanisms similar or identical to those in homologous recombination, discussed later in this chapter (page 947).

the damaged site. Additional proteins are involved, as shown in the figure and discussed in the section on recombination (see page 948). A gap now exists in the previously undamaged arm, but because it lies opposite an undamaged template, it can be filled by action of DNA polymerase and DNA ligase. The thymine dimer itself is not repaired in this process, but the process allows time for the excision system to come in later and repair this damage. RecA protein is required for daughter-strand gap repair, particularly in the first reaction, where the undamaged parental strand undergoes pairing with the parental strand opposite the gap. RecA is a bacterial protein, but related proteins allow this process to occur in eukaryotic cells as well.

Error-Prone Repair and the SOS Response

The bacterial *SOS response* is a metabolic alarm system that helps the cell to save itself in the presence of potentially lethal stresses. Inducers of the SOS response include ultraviolet irradiation, thymine starvation, treatment with certain DNA-modifying reagents such as the cross-linker mitomycin C, and inactivation of genes essential to DNA replication. Responses include mutagenesis, filamentation (in which cells elongate by growth but don't divide), activated excision repair, and activation of latent bacteriophage genomes. Mutagenesis occurs because, under SOS conditions, the gaps that are formed opposite thymine dimers or other damaged sites can be filled by replication rather than by daughter-strand transfer, and this replication is extremely inaccurate. In fact, this process is the principal pathway by which ultraviolet light stimulates mutagenesis in bacteria.

As noted on page 940, DNA polymerases are unable to replicate past a damaged site in the template strand, such as a thymine dimer. The same is true when the template strand contains an abasic site, with no base linked to the deoxyribose. However, recent studies in *E. coli* show that DNA polymerase III holoenzyme can replicate past such a site with the aid of other proteins, including SSB, RecA, and the products of two genes, *umuC* and *umuD.* Umu stands for *u*ltraviolet *mu*tagenesis; action of both gene products is essential for the mutagenesis resulting from ultraviolet light. In a *umuC*$^-$ mutant, for example, UV irradiation kills bacteria but does not increase the mutation frequency among the surviving bacteria. That is because the cells cannot undergo error-prone DNA replication. After expression of these two genes, the *umuD* product undergoes proteolytic cleavage to give the UmuD′ protein, and this complexes with UmuC to give a trimeric complex, $UmuD'_2C$. This trimeric complex associates with RecA protein, which binds to single-stranded DNA just downstream from the damaged site. Formation of this complex allows a DNA polymerase to copy past the site, in a highly error-prone fashion; when the damage is an abasic site, for example, the polymerase almost always inserts A at that site, whether or not the original template base was T. Recent evidence indicates that the $UmuD'_2C$ complex itself is the error-prone polymerase.

DNA can be repaired after replication, either by recombination or by inducible error-prone repair. Both processes require RecA.

Considering that most mutations are probably deleterious, what is the advantage to the cell of accumulating mutations during DNA repair? Probably none, except that the alternative would be for the cell to die. In other words, extensive mutagenesis is seen as a worthwhile price for the cell to pay, if a massive UV dose has overwhelmed all other repair pathways and if error-prone replication past a dimer is necessary for the cell to stay alive. It is not yet clear whether a comparable process occurs in eukaryotic cells.

MISMATCH REPAIR

Mismatches, or non-Watson–Crick base pairs in a DNA duplex, can arise through replication errors, through deamination of 5-methylcytosine in DNA to yield thymine, or through recombination between DNA segments that are not completely homologous. In addition, mismatches result when DNA polymerase slides along its template, creating short loops or bulges in duplex DNA. We best understand the correction of replication errors, so that is what we describe here.

If DNA polymerase introduces an incorrect nucleotide, creating a non-Watson–Crick base pair, the error is normally corrected by 3′ exonucleolytic proofreading. If the error is not corrected immediately, the fully replicated DNA will contain a mismatch at that site. This error can be corrected by another process, called **mismatch repair.** In *E. coli* the proteins that participate include the products of genes *mutH, mutL,* and *mutS.* Another required gene product, originally called MutU, has now been identified as DNA helicase II (also identified as the product of the *uvrD* gene).

The mismatch correction system scans newly replicated DNA, looking for both mismatched bases and single-base insertions or deletions. MutS binds to DNA at the site of the mismatch, followed by the binding of MutL and then MutH. MutS is a "motor protein," which uses the energy of ATP hydrolysis to pull DNA from both directions until it reaches the site at which the repair process is to begin. When it finds an appropriate signal, part of one strand containing the mismatched region is cut out and replaced (Figure 25.16). How does the mismatch repair system recognize the right strand to repair? If it chose either strand randomly, it would choose incorrectly half the time and there would be no gain in replication accuracy. The answer is that the mismatch repair enzymes identify the newly replicated

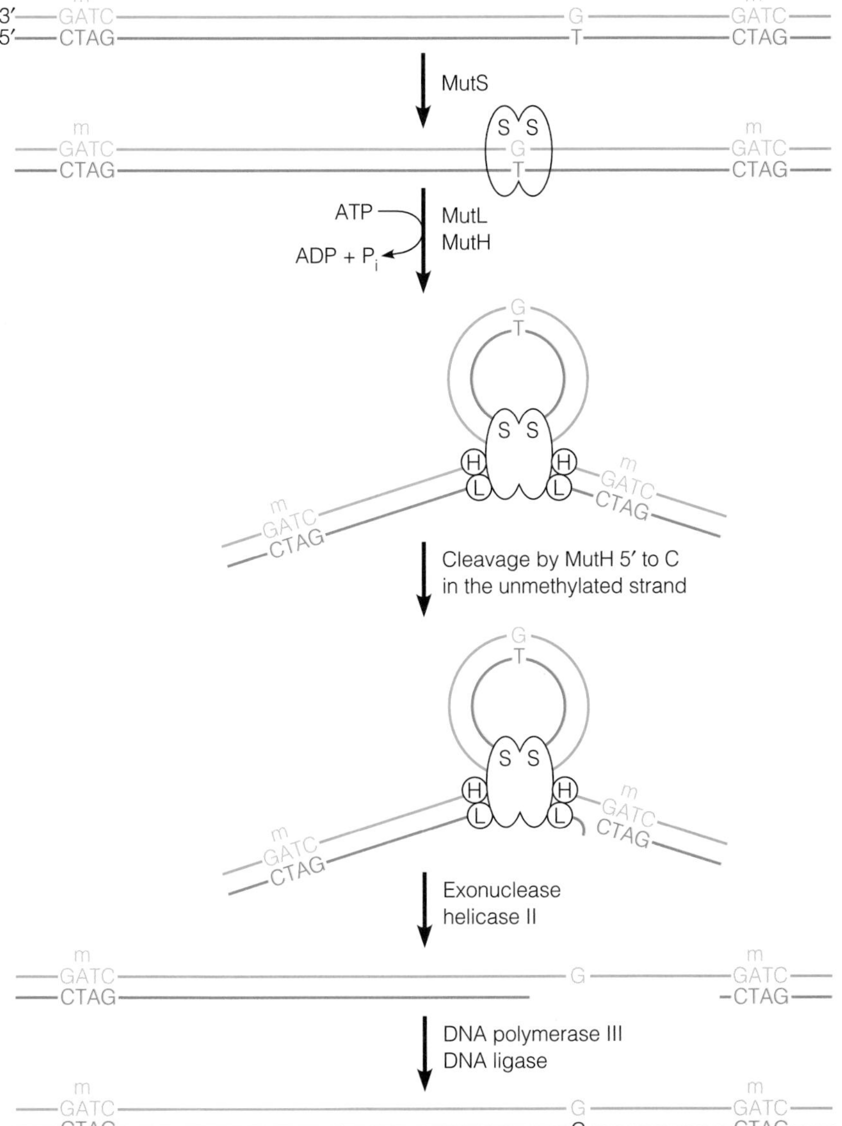

FIGURE **25.16**

Methyl-directed mismatch repair in *E. coli.* The newly replicated daughter strand (red) contains a T mismatched to G in the template strand (blue). The mismatch repair system identifies the daughter strand because it is not yet methylated. Thus, this system must function before the newly replicated daughter strand becomes methylated, through action of the Dam methylase on the A residue in the GATC sequence.

In bacteria the mismatch repair system uses DNA methylation to identify the strand that has a mispaired nucleotide.

strand, because for a short period that DNA is unmethylated. In *E. coli* the sequence —GATC— is crucial, because that is the site methylated soon after replication, by action of the product of the *dam* gene (*D*NA *a*denine *m*ethylase). The mismatch repair enzymes look for —GATC— sequences that are not methylated. Recognition of an unmethylated GATC can target that strand for mismatch correction at a site as far as 1 kbp or more away from the GATC site, in either direction. Once the methylation system has acted on all GATC sites in the daughter strand, it is too late for the mismatch repair system to recognize the more recently synthesized DNA strand, and any advantage in total DNA replication fidelity is lost. When the system functions properly, it has the effect of increasing overall replication fidelity by about 100-fold—from about 1 error in 10^8 base pairs replicated to about 1 in 10^{10}.

As shown in Figure 25.16, the MutHLS complex moves along DNA in both directions until it encounters the nearest 5′-GATC sequence. An endonuclease activity of MutH then cleaves on the 5′ side of the C in the unmethylated strand. At that point helicase II unwinds the DNA, moving back past the mismatch, followed by an exonuclease that digests the displaced single strand. The resultant gap is filled by DNA polymerase III holoenzyme and DNA ligase, working in concert with SSB.

A similar mismatch repair system exists in eukaryotic cells. However, the mismatch recognition step is somewhat more complex, because three different *M*ut*S* *h*omologs (MSH proteins) are involved—MSH2, MSH3, and MSH6. These three proteins form heterodimers, with different mismatch specificities; that is, an MSH2–MSH6 complex recognizes single-base mismatches, insertions, and deletions, while an MSH2–MSH3 complex recognizes insertions and deletions of two to four nucleotides. Several MutL homologs exist also, but a MutH homolog has not been found. In fact, a major unanswered question is the mechanism by which the eukaryotic systems recognize and initiate repair on the newly replicated DNA strand; selective methylation is evidently not involved.

Just as seen with bacteria, mutations in eukaryotic genes that control mismatch repair confer a mutator phenotype, raising spontaneous mutation rates at all loci. How do such mutations affect the biology of human cells? In 1974, Lawrence Loeb predicted that the progression of a normal cell to a cancer cell would involve the creation of a mutator phenotype, because the natural mutation rate for somatic cells seemed too low to account for the number of heritable changes that occur during tumor cell progression. Two decades later, this prediction was confirmed, when mutations in mismatch repair proteins were found in tumor cells from individuals with an inherited cancer predisposition called HNPCC (heritable nonpolyposis colon cancer). To date, germ-line mutations in the genes for one MutS homolog or one of three MutL homologs have been found in such individuals. The cancer predisposition is inherited in an autosomal dominant fashion, suggesting that most affected individuals are heterozygous, with one wild-type and one nonfunctional allele. Mismatch repair is essentially normal until a somatic cell mutation inactivates the one functional allele, and mismatch repair capacity is essentially abolished, with a consequent increase in spontaneous mutagenesis.

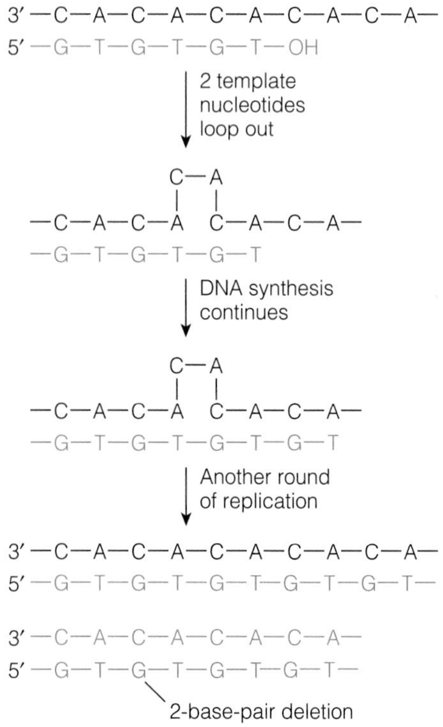

Tumor cells from those affected with HNPCC exhibit a phenomenon called **microsatellite instability**—a large number of mutations in regions of the genome containing repeats of single-, double-, and triple-nucleotide sequences, usually with large increases in the numbers of repeating units in such sequences. These data suggest that the product and template strands can normally slip at such sites, so that DNA polymerase copies a short repeating sequence more than once, or else skips a segment. This creates a heteroduplex with a short loop, as shown for deletion mutagenesis in the margin. Normally a replication error of this type would be corrected by mismatch repair, but in a cell lacking normal mismatch repair, such errors would persist and accumulate. Studies of this type have given scientists enormous insight into the nature of cancer as a progressive genetic disease.

3′—C—A—C—A—C—A—C—A—C—A—
5′—G—T—G—T—G—T—OH
2 template nucleotides loop out
C—A
—C—A—C—A C—A—C—A—
—G—T—G—T—G—T
DNA synthesis continues
C—A
—C—A—C—A C—A—C—A—
—G—T—G—T—G—T—G—T
Another round of replication
3′—C—A—C—A—C—A—C—A—C—A—
5′—G—T—G—T—G—T—G—T—G—T—
3′—C—A—C—A—C—A—C—A—
5′—G—T—G—T—G—T—G—T—
2-base-pair deletion

Recombination

Population genetics teaches us that the survival of a species depends on its ability to maintain genetic diversity, so that individuals can vary in their ability to respond to unforeseen environmental pressures. Diversity is maintained through both mutation, which alters single genes or small groups of genes in an individual, and *recombination,* which redistributes the contents of a genome among various individuals during reproduction. In classical biology, recombination is the outcome of crossing over between paired sister chromosomes during meiosis in eukaryotes, and in fact our earliest information about recombination came from cytological and cytogenetic observations in *Drosophila.* However, recombination encompasses more processes and biological functions than those involved in sexual reproduction. Strictly speaking, **recombination** is any process that involves the formation of new DNA from distinct DNA molecules, such that genetic information from each parental DNA molecule is present in the new molecules. The daughter-strand gap repair process described earlier is a form of recombination. So also is the integration of certain bacteriophage or plasmid genomes into the chromosomal DNA of a host bacterium; many viral genomes integrate into animal host cells as well. Recombination is also involved in **transposition,** the movement of DNA from one chromosomal integration site to another. In some cases this type of recombination is a gene regulatory mechanism. The transposed gene may move from a site where it is inactive to another site where its transcription can be activated. Alternatively, the integration of a transposable element may turn on the expression of adjacent genes. We discuss these processes later, in the context of gene rearrangements.

Recombination is any process that creates end-to-end joining from two different DNA molecules.

CLASSIFICATION OF RECOMBINATION PROCESSES

Different recombination processes have quite distinct requirements, both for nucleotide sequence homology between the recombining partners and for proteins and enzymes to catalyze the process. Meiotic recombination in diploid organisms requires extensive sequence homology between the recombining partners and accordingly is called **homologous recombination.** This term also applies to certain recombinational events between bacterial chromosomes. New DNA can be introduced into a bacterial cell by various processes: (1) conjugation during bacterial mating; (2) transformation, when DNA is taken up by cells; or (3) **transduction,** when bacterial DNA that was packaged into a phage particle is introduced by infection. Transduction results when the assembly of phage particles in an infected cell goes awry, and bacterial DNA is assembled into a phage head. Reinfection of another bacterial cell introduces that packaged bacterial DNA into the new cell.

If the introduced DNA contains a replication origin, as in a plasmid, it can replicate autonomously once inside a new bacterial cell. More often the DNA does not contain an origin, and its information can be expressed and maintained only if the DNA is taken up into the resident chromosome by homologous recombination. Biochemical analysis of homologous recombination has focused on prokaryotic systems, which in turn have provided insight into meiotic recombination. Most bacterial homologous recombination processes share a common requirement for the RecA protein or its counterpart.

Site-specific recombination, by contrast, involves only limited sequence homology between recombining partners. Sites of breaking and joining are determined by specific DNA–protein interactions. The process was first described in bacteriophage λ. After circularization, the λ chromosome can either undergo multiple rounds of replication leading to virus production or undergo integration into a specific site on the host chromosome. In the latter condition, called **lysogeny,** most of the viral genes are inactivated, and the virus can maintain a

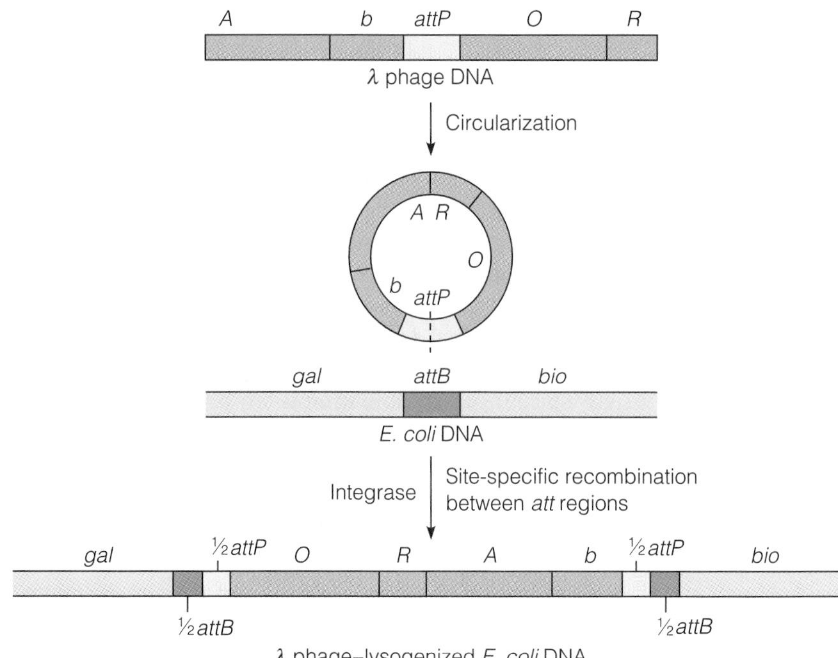

FIGURE **25.17**

Site-specific recombination, establishing lysogeny in bacteriophage λ. The phage chromosome circularizes between genes *A* and *R*, and recombination takes place between the *attP* site and a corresponding region, *attB*, on the *E. coli* chromosome between the *gal* and *bio* markers. The enzyme integrase carries out the site-specific recombinational event with the help of a bacterial protein. *O* and *b* are additional genetic markers.

long-term, nonlethal relationship with its host (Figure 25.17). Bacteriophages that can establish lysogeny are called **temperate phages,** in contrast to **virulent** phages, such as phage T4, which always lyse their host cells after infection. In most (not all) cases of lysogeny, integration occurs at a specific site. Integrative recombination in phage λ is being studied as a model for understanding integration of tumor virus genomes into DNA of infected cells. It was earlier studied as a model for homologous recombination between chromosomes, but the processes are different, as shown when it was learned that the phage and bacterial DNA sequences in the regions undergoing recombination have only 15 base pairs of homology. Moreover, the RecA protein is not required for this process. Rather, the virus specifies a site-specific enzyme, called **integrase,** and specific DNA–protein interactions between the enzyme and the recombining partners, rather than extensive DNA–DNA sequence homology, determine the site of recombination.

We recognize two other forms of recombination. *Transposition* involves neither sequence homology nor the RecA protein but does require a special sequence on the donor DNA. This process is discussed in a later section. **Illegitimate recombination,** an extremely rare event that possibly occurs by chance, involves neither sequence homology nor the action of any known protein. Table 25.3 summarizes the main distinctions among the four major types of recombination.

TABLE **25.3 Characteristics of different types of genetic recombination**

	Requirement		
Type	**Sequence homology**	**RecA protein or counterpart**	**Sequence-specific enzyme**
Homologous	Yes	Yes	No[a]
Site-specific	Yes (about 15 bases)	No	Yes
Transposition	No	No	Yes
Illegitimate	No	No	Unknown

[a]The Chi site or its counterpart determines cutting sites (see page 951), but initial recognition of pairing sites occurs by sequence homology.

HOMOLOGOUS RECOMBINATION

Breaking and Joining of Chromosomes

The most straightforward way to accomplish recombination is to break and rejoin DNA molecules. However, if recombination occurs this way, the sites of breakage must be precisely the same on both recombining chromosomes for intact genes to be regenerated. Some researchers favored alternative mechanisms, but in 1961 Matthew Meselson and Jean Weigle showed that recombination in fact occurs via breakage and rejoining of chromosomes. The demonstration, diagrammed in Figure 25.18, involved a Meselson–Stahl type of experiment (see Figure 4.14, page 98). *E. coli* was infected with two genetically marked λ phage populations, one of which had been density labeled by growth in ^{13}C–^{15}N medium. The phage particles resulting from this cross were centrifuged to equilibrium in a cesium chloride gradient. Phages with recombinant genotypes were recovered from all parts of the gradient, whereas nonrecombinant phages were uniformly light or heavy. This result could occur only if the recombinant phages contained DNA derived from both parents, by breaking and rejoining.

This experiment had one other important result. The phage output from the crosses was analyzed in standard fashion, by visual examination of plaques. Although each plaque arises from a single phage particle, many of the plaques in the Meselson–Weigle experiment contained phages of two different genotypes, even though all arose from a single infecting phage. This suggested that recombination involves the formation of a **heteroduplex DNA** region, in which one DNA strand comes from one parent and the other from the second parent. If the heteroduplex region contains a mismatch, subsequent replication of that DNA gives rise to two progeny DNA molecules of different genotypes, as shown in Figure 25.19.

Another early observation was that recombination is stimulated by processes that nick or break DNA strands, such as thymidine starvation or UV irradiation. This suggested a role for single-strand DNA, or free DNA ends, in initiating recombination.

Models for Recombination

Putting the preceding observations together along with data on recombination in fungi, Robin Holliday proposed in 1964 a model for homologous recombination between duplex DNA molecules. That model, detailed in Figure 25.20, continues to stimulate thinking and experiments about this process.

Holliday proposed that recombination begins with nicking at the same site on two paired chromosomes (step 1). Partial unwinding of the duplexes is followed by **strand invasion,** in which a free single-strand end from one duplex pairs with its unbroken complementary strand in the other duplex, and vice versa (step 2). Enzymatic ligation generates a crossed-strand intermediate, called a **Holliday junction** (step 3). The crossed-strand structure can move in either direction by duplex unwinding and rewinding (branch migration, step 4). The Holliday junction "resolves" itself into two unbroken duplexes, by a process of strand breaking and rejoining. The process leading to recombination begins with isomerization of the Holliday structure (step 5), followed by strand breakage, so that the strands that will break (in step 9) are those that were *not* broken in step 1. Resolution of the resulting

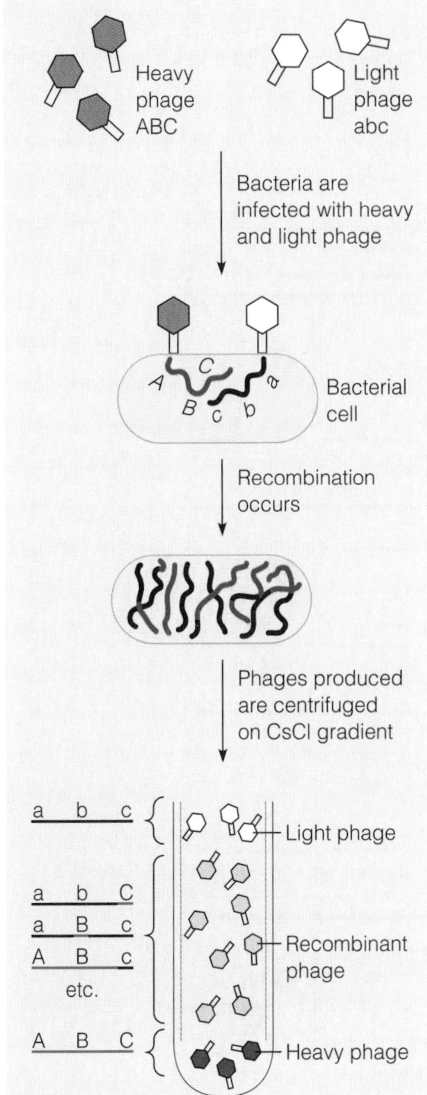

FIGURE **25.18**

The Meselson–Weigle experiment. This experiment established that genetic recombination occurs by breakage and reunion of DNA strands.

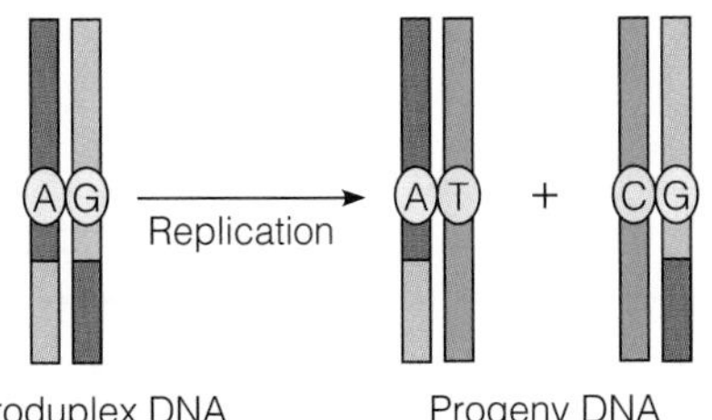

FIGURE **25.19**

Generation of progeny phage containing two genotypes by replication of heteroduplex DNA.

Homologous recombination involves breaking and rejoining chromosomes.

structure (steps 10 and 11) generates two chromosomes recombinant for DNA flanking the region and each containing a heteroduplex region. However, if the original crossed strands (those that *were* broken in step 1) break and rejoin (steps 6–8), the products are nonrecombinant duplexes, each containing a heteroduplex region (that is, nonrecombinant with respect to the outside markers A and Z).

Considerable evidence now supports the central tenets of the Holliday model, particularly the electron microscopic visualization of Holliday junctions (Figure 25.21). However, the model has been modified as new data have emerged. Matthew Meselson and Charles Radding proposed that recombination could start with a single nick (which eliminates the nagging question of how two duplexes could be nicked at precisely the same point). As shown in Figure 25.22, strand displacement synthesis occurs (red arrow, step 1), with a displaced single-strand end of the nicked duplex, A, invading the homologous region of the unbroken duplex, B (step 2). Eventually the displaced loop on duplex B is cleaved and partially degraded (step 3), and the displaced end from duplex A is ligated to B (step 4). Isomerization then occurs, as in the Holliday model, with the originally unbroken strands crossed (step 5). An additional feature of the Meselson–Radding model is the fact that branch migration can lead the final strand cutting to occur some distance from the site of either strand invasion or the original nick (steps 6 and 7). In principle either a 5′ or a 3′ end could initiate the strand invasion process. As noted below, a 3′ end probably initiates recombination in *E. coli.*

Proteins Involved in Homologous Recombination

The Holliday and Meselson–Radding models explained most of the existing data on homologous recombination between paired chromosomes, particularly as studied in lower eukaryotes such as yeast. Moreover, the models could easily be adapted to explain daughter strand gap repair or the recombination that occurs in bacteria after transformation or conjugation. Some of the proteins thought to participate, notably DNA polymerase, DNA ligase, and single-strand DNA-binding protein, had been characterized and shown to participate in recombination. What about other proteins that must function if the models are largely correct? To answer this question, we turn back to *E. coli* and its phages and the characteristics of bacterial mutants defective in recombination. Mutations conferring a recombination-defective (*rec*$^-$) phenotype map in several loci, and two important gene products,

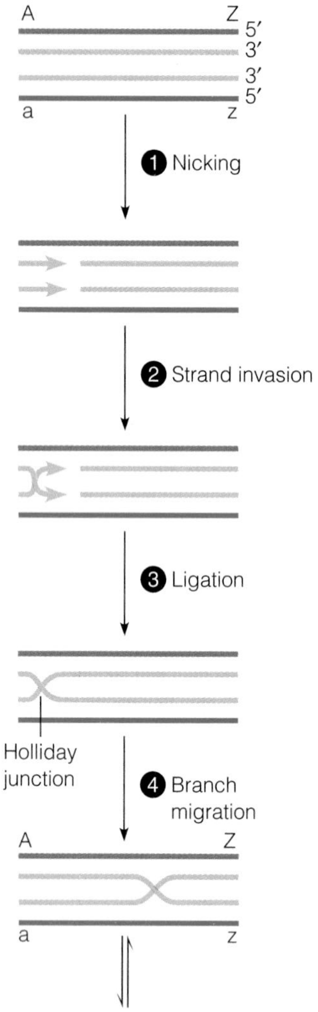

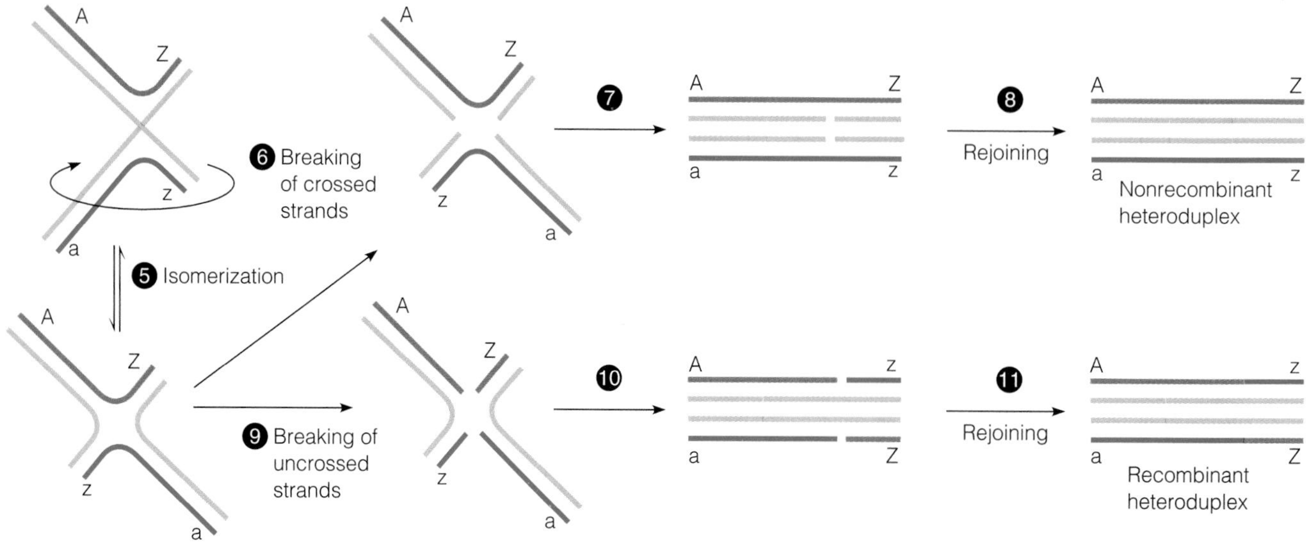

FIGURE **25.20**

The Holliday model for homologous recombination. A, a, Z, and z are genetic markers.

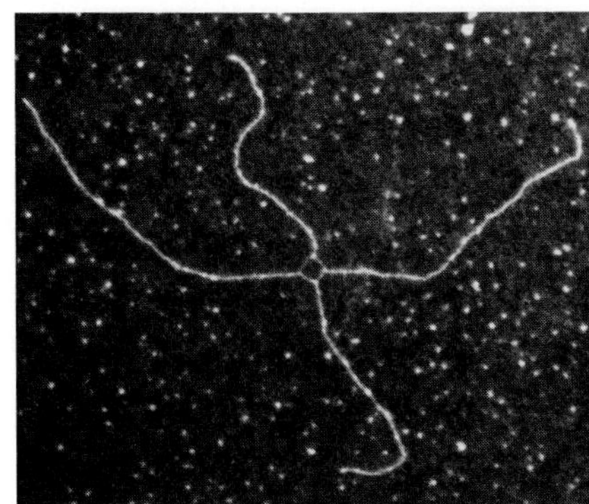

FIGURE **25.21**

Electron microscopic visualization of a Holliday junction. This junction was created during recombination between two plasmid DNA molecules.

Courtesy of H. Potter and D. Dressler, *Proc. Natl. Acad. Sci. USA* (1977) 74:4168–4172.

are responsible for most bacterial recombination events. One of these products, the RecA protein, was mentioned earlier. The other protein is called exonuclease V or the RecBCD nuclease.

RecA is an amazing multifunctional protein with M_r of about 38,000. In recombination it promotes the pairing of homologous strands, as described earlier in connection with recombinational repair. Several strand-pairing reactions can be demonstrated in vitro, with two examples shown in Figure 25.23. One example involves a three-strand exchange, and the other involves four strands. Whatever the structures participating, it is now clear that RecA binds just three DNA strands during the strand exchange reaction that it catalyzes. As schematized in Figure 25.24, the process begins with a reaction between RecA and single-strand (ss) DNA, to give a characteristic nucleoprotein filament. As shown by electron microscopy of the filament (see scanning tunneling micrograph in Figure 25.25) and by crystallographic analysis of the protein, RecA wraps about ssDNA as a

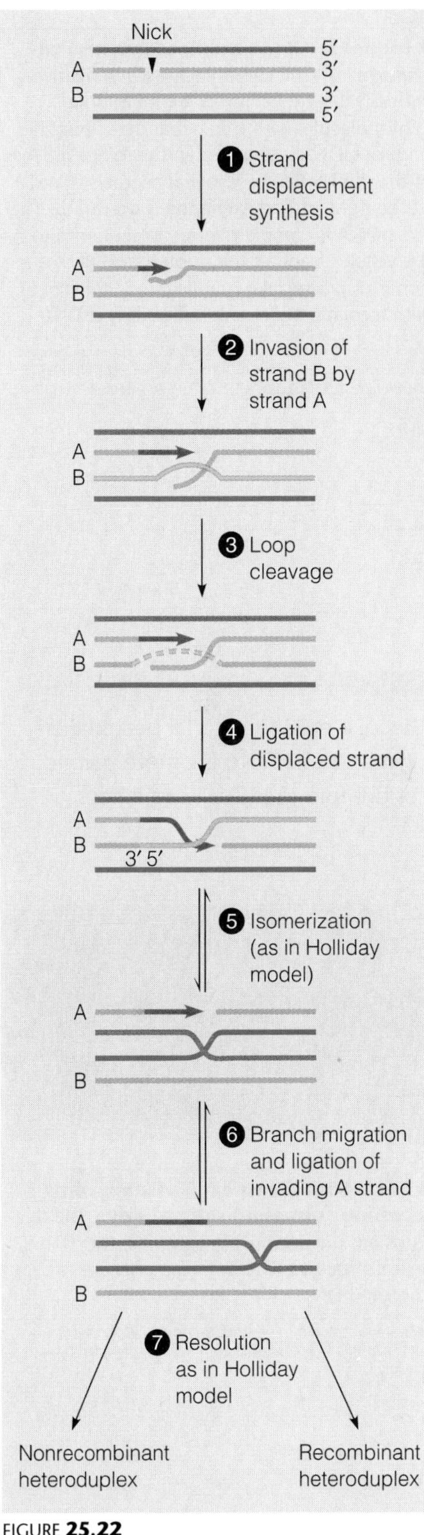

FIGURE **25.22**

The Meselson–Radding model for homologous recombination.

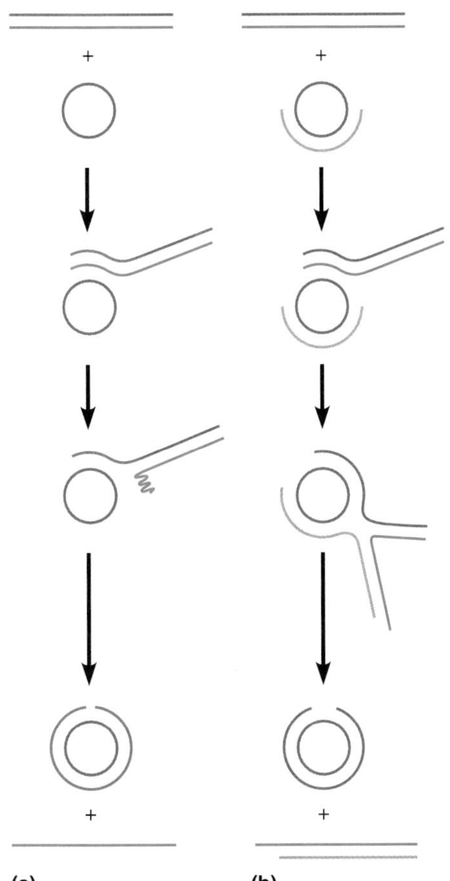

FIGURE **25.23**

RecA protein–mediated DNA strand exchange reactions. **(a)** A three-strand exchange process; **(b)** a four-strand exchange. However, as discussed in the text, only three strands are associated with RecA at any one time.

FIGURE **25.24**

A model for RecA-mediated strand exchange. At the left, a RecA–ssDNA filament, with ssDNA shown in red. In the middle, a joint molecule, with triple-stranded DNA; the original ssDNA is wrapped in the minor groove of the duplex DNA (yellow and green strands). At the right, strand exchange is occurring. The red ssDNA is complementary in sequence to the yellow strand of the duplex, and RecA action is displacing the green strand, coincident with formation of a new red–yellow dsDNA.

Courtesy of M. Kubista et al., *Biological Structure and Dynamics*, R. H. Sarma and M. H. Sarma, eds. (Schenectady, New York: Adenine Press, 1996), pp. 49–59. © 1996 Adenine Press, Inc.

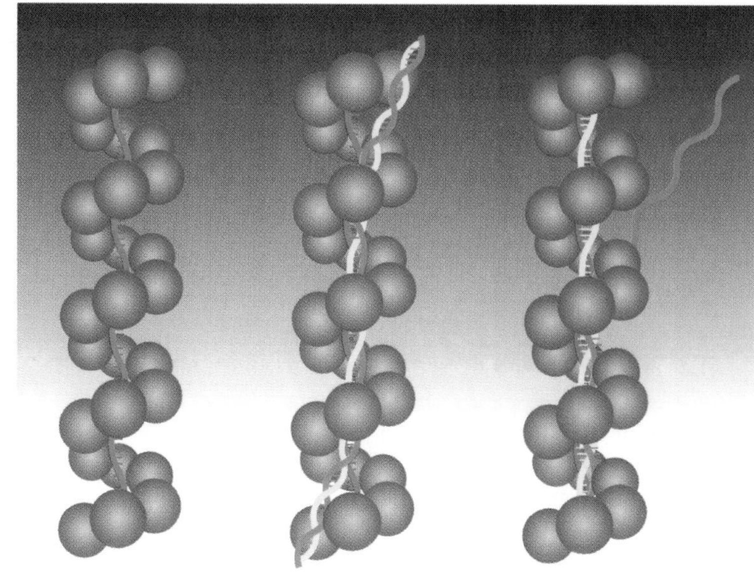

RecA, a multifunctional bacterial enzyme, uses ATP to promote pairing of homologous DNA sequences.

multisubunit right-handed helix, with six RecA monomers per turn. Once ssDNA is bound, the filament searches double-strand (ds) DNA, looking for sequences complementary to those in the single strand already bound. In this process, dsDNA is also taken up within the filament, giving a structure termed a joint molecule (see Figure 25.26). Binding of dsDNA requires ATP, but that ATP need not be hydrolyzed in the process. By contrast, movement of the ssDNA–protein complex with respect to the dsDNA does require ATP hydrolysis. During this process, dsDNA is underwound and stretched to about 1.5 times its normal length. Movement of the complex is polarized, in a 5′-to-3′ direction along the initially bound ssDNA. During the movement, a triple-stranded structure transiently forms, with ssDNA (the red strand in Figure 25.24) wound in the minor groove of the dsDNA. The ssDNA continually tests the antiparallel strand in dsDNA (yellow) for sequence complementarity. It is not clear how this occurs, but evidently short oligonucleotide sequences swing out from the duplex structure and can pair with ssDNA if sequence complementarity is found. Once complementarity is established, branch migration occurs, with simultaneous strand exchange. A duplex is formed

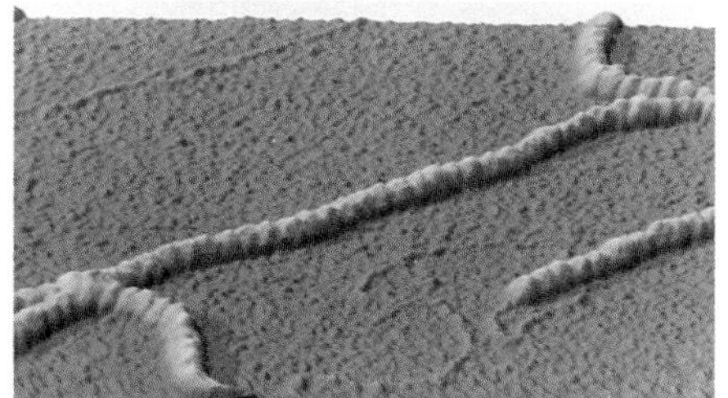

FIGURE **25.25**

A RecA–DNA complex as visualized by scanning tunneling microscopy. The thick cords are the RecA–DNA complex; they show helical structure. The thin filaments are uncoated DNA.

Reprinted with permission from M. Amrein et al., *Science* (1988) 240:514–516, ©1988 by the AAAS, provided by G. Travaglini.

FIGURE **25.26**

Electron microscopic visualization of joint molecules. For technical reasons, this image was obtained with the UvsX protein of T4, which is similar in action to *E. coli* RecA. Single-strand DNA appears much thicker than duplex DNA because it is coated with UvsX protein. At the bottom is a circular single-strand DNA; in the middle, a circular duplex DNA; and at the top, a joint molecule involving one of each. Joint molecules can form between homologous sequences on circular molecules, but the strands are not interwound.

Courtesy of L. D. Harris and J. Griffith, *J. Biol. Chem.* (1987) 262:9285–9292.

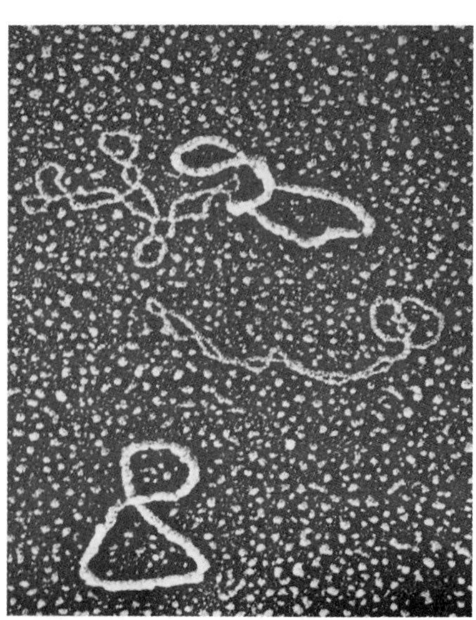

between the red strand (ssDNA originally) and the yellow strand (complementary to the red strand), while the displaced (green) strand is spooled out, away from the complex. ATP hydrolysis during this period may promote rotation of the DNA within the filament, facilitating the release of the displaced strand.

Recombination occurs preferentially at or near particular DNA sequences. In *E. coli* recombination is favored near a particular octanucleotide sequence, 5′-GCTGGTCC, called Chi (for *c*rossover *h*otspot *i*nstigator). How does this site act to stimulate recombination? Another important enzyme, the RecBCD protein, a multifunctional heterotrimeric enzyme encoded by the *recB, recC,* and *recD* genes, displays sequence specificity for Chi. This enzyme binds at a double-strand break on duplex DNA and uses a helicase activity to unwind and partially degrade the DNA. The 3′ end is displaced as a loop, which becomes coated with SSB protein; loops of this type can be visualized in DNA treated with RecBCD, if the nuclease activity of the enzyme is inhibited (Figure 25.27). When the enzyme reaches Chi, a sequence-specific interaction causes RecBCD to switch strands and switch its preferred polarity of DNA degradation. These changes also somehow facilitate the loading of RecA to a free 3′ end, and this initiates strand invasion of a nearby duplex, as indicated in Figure 25.28.

RecBCD, a multifunctional enzyme, unwinds and rewinds DNA, initiating recombination via sequence-specific DNA cuts.

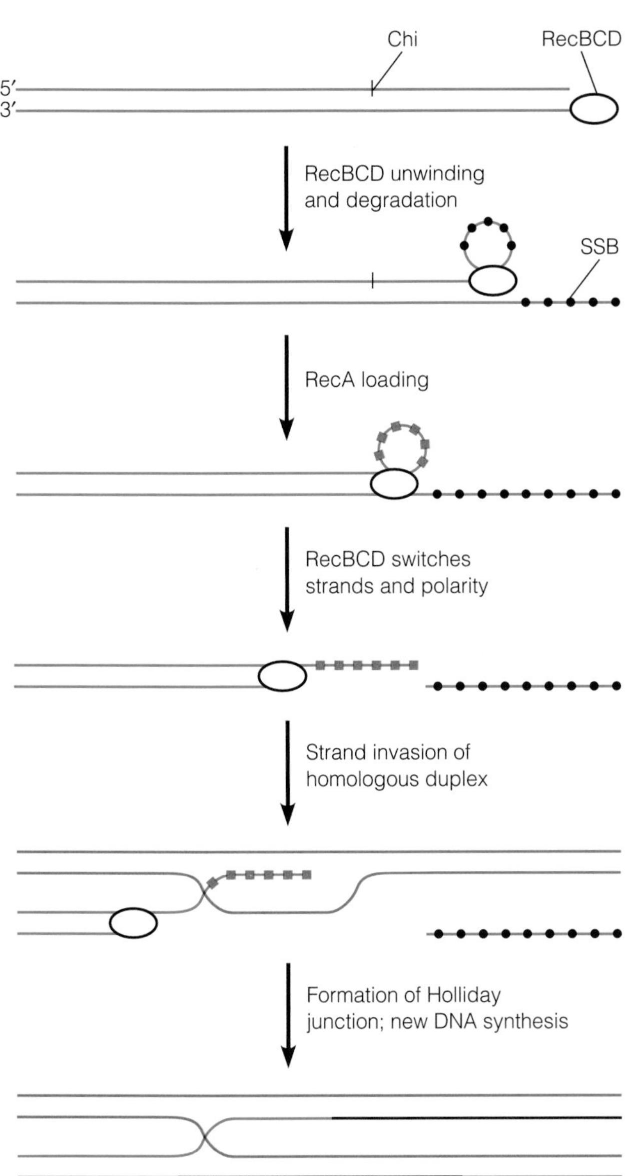

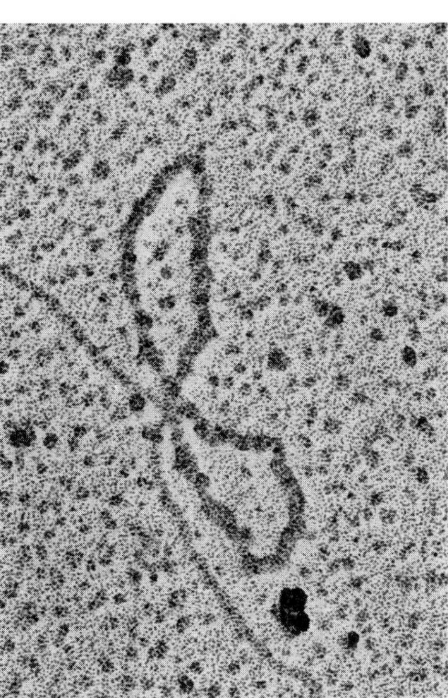

FIGURE **25.27**

"Rabbit ears." This looped structure was created by the action of RecBCD enzyme on T7 phage DNA, with Ca^{2+} added to inhibit nuclease. The single-strand loops appear thicker than duplex DNA because they are coated with single-strand DNA-binding protein.

Courtesy of G. R. Smith et al., *Cold Spring Harbor Symp. Quant. Biol.* (1984) 49:485–495, provided by A. F. Taylor.

FIGURE **25.28**

A model for the action of RecBCD, Chi sites, and RecA in initiating homologous recombination.

Redrawn from D. G. Anderson and S. C. Kowalczykowski, *Cell* (1997) 90:77–86.

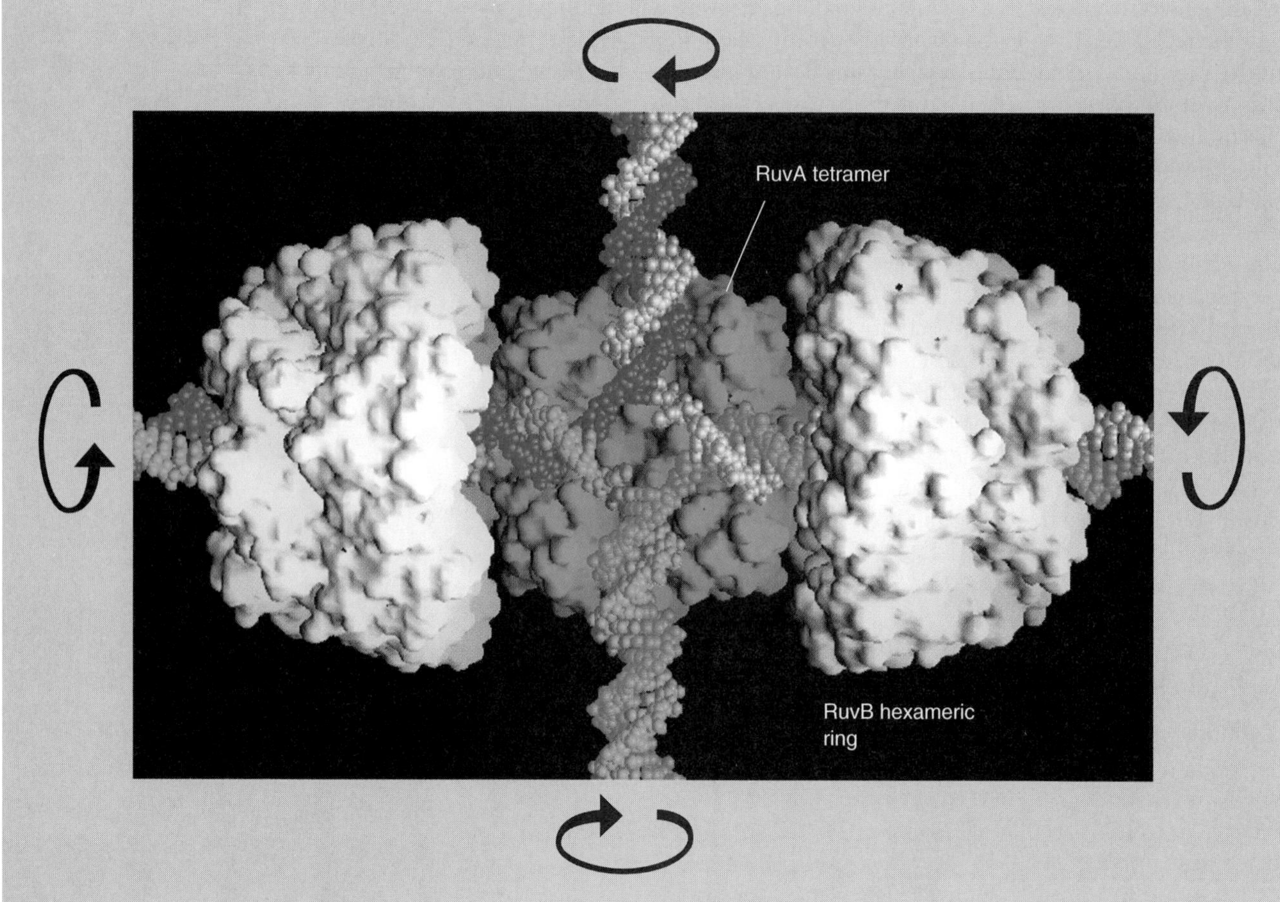

FIGURE **25.29**

A model for the RuvA–RuvB–Holliday junction structure. This is based upon crystal structures of RuvA and RuvB. Branch migration is believed to involve spooling of DNA to left and right through the RuvB twin pumps, with the upper and lower arms being drawn into the center and eventually out through the pumps.

Courtesy of Peter Artymiuk, Krebs Institute, Sheffield.

Once a Holliday junction is formed, branch migration is essential for eventual formation of recombinant structures, as was shown in Figures 25.20 and 25.22. This is largely the responsibility of three other proteins. In *E. coli* these three proteins are products of the *ruvA, ruvB,* and *ruvC* genes. RuvA is a DNA-binding protein, whose specificity directs it toward the four-stranded Holliday structure. RuvB protein is an ATP-requiring "motor protein," which binds to two opposed arms of the junction. In the model of Figure 25.29, which is based upon crystal structures of the isolated RuvA and RuvB proteins, the two RuvB molecules act as twin pumps, rotating the two arms in opposite directions. This forces branch migration by driving the rotational movement of the other two strands toward the junction. Eventually RuvC binds and begins the resolution of the Holliday structure by nicking two strands.

Although Ruv protein homologs have not yet been detected in eukaryotic cells, much of the biochemistry of homologous recombination in eukaryotes is similar to what is described here. In particular, the RAD51 protein of both human cells and yeast has a strand-pairing activity similar (but not identical) to that of RecA, and the two proteins show extensive sequence homology with RecA. In eukaryotic cells an essential function of homologous recombination is the repair of double-strand breaks, which can be created by ionizing radiation, oxidative stress, or other environmental damage, and would be lethal if not repaired. In eukaryotes, a broken chromosome can use the sequence information in its homolog to reconstruct the original DNA sequence at the site of the break, as suggested in Figure 25.30. Recent evidence implicates RuvA and RuvB proteins or their eukaryotic counterparts in forming the Holliday structure whose resolution leads to double-strand break repair.

The process we have outlined is supported by electron micrographs of joint molecules (Figure 25.26) and of DNA–RecA nucleoprotein filaments (Figure 25.25). The best pictures of joint molecules were obtained not with the *E. coli* RecA protein but with the T4 phage UvsX protein, which performs similar roles in phage DNA metabolism as the RecA protein does in *E. coli* DNA metabolism.

SITE-SPECIFIC RECOMBINATION

As we have discussed, alignment of sites for homologous recombination occurs via DNA–DNA (base-pairing) interactions. Another important class of recombination reactions is directed by highly specific DNA–protein interactions, although a short stretch of DNA homology occurs at the actual site of cutting and resealing. Our biochemical understanding of this site-specific recombination is most advanced for the mechanism by which temperate phages such as λ become integrated at specific sites on the chromosome of the infected bacterium. This process provides an important model for studying counterpart reactions in higher organisms, such as the DNA arrangements involved in the maturation of antibody-forming genes (see page 955).

Recombination to integrate or excise temperate phage chromosomes involves specific DNA–protein interactions.

The circularized λ chromosome integrates at a specific site on the *E. coli* chromosome, *attB*, which maps between genes involved in galactose utilization and biotin synthesis (the *gal* and *bio* markers), as was schematized in Figure 25.17. Integration occurs at a specific site on the phage chromosome, called *attP*. Two proteins are required for this site-specific recombination—(1) phage integrase (Int, the product of the *int* gene) and (2) an *E. coli* protein called IHF (*i*ntegration *h*ost *f*actor). IHF has been shown by x-ray crystallography to force a 90° turn when it binds to DNA (Figure 25.31a). Phage DNA must be supercoiled for the recombination to occur. This supercoiling, plus the distortion created by IHF binding to specific sites in *attP*, facilitates Int binding at adjacent sites. A specialized nucleoprotein structure called the intasome is formed, with the 230-base-pair *attP* region wrapped about seven Int molecules, each bound at a specific site. This structure becomes aligned with *attB*, which is only 23 base pairs long and which binds two molecules of Int. In the core of each site is a 15-base-pair region of complete homology (Figure 25.31b). In each of these sequences, Int creates a staggered cut, with a seven-nucleotide overlap. The ends then exchange to form a Holliday junction. The bacterial and phage core sequences yield two hybrid sequences, each of which contains both phage and bacterial DNA. The multifunctional Int protein, which has already catalyzed site-specific cutting and strand exchanges, now completes the process with a DNA ligase reaction to join the ends covalently.

When a λ phage chromosome becomes integrated, it is essentially dormant. Almost all of its genes are turned off, and it replicates only as a part of the bacterial chromosome in which it resides. However, changes in gene expression (see Chapter 26) can activate the integrated chromosome, or **prophage.** When this occurs, the above sequence of steps is reversed, with excision of a circular phage chromosome. An additional protein, called Xis, is required in addition to Int and IHF. Xis provides binding specificity for the hybrid phage–bacterial attachment sites that flank the prophage, so that excision occurs only under conditions where both Int and Xis are present.

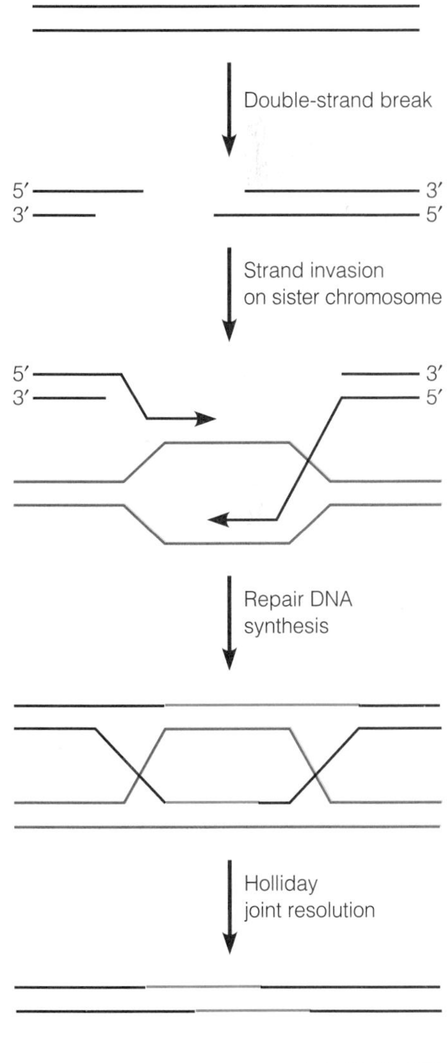

FIGURE **25.30**

A model for double-strand break repair via homologous recombination. RAD51 and several other proteins are involved.

Redrawn from P. Baumann and S. C. West, *Trends Biochem. Sci.* (1998) 23:247–251. © 1998 with permission of Elsevier Science.

Gene Rearrangements

Until the mid-1970s the genetic information content of an organism or population was considered to be static. All cells of a differentiated organism were thought to have identical DNA contents, with variations among different cells arising at the level of gene expression. Supporting this idea was the fact that in some plants, such as carrots, a single differentiated cell can be manipulated in culture so as to give rise to a complete and normal plant. However, more recent developments

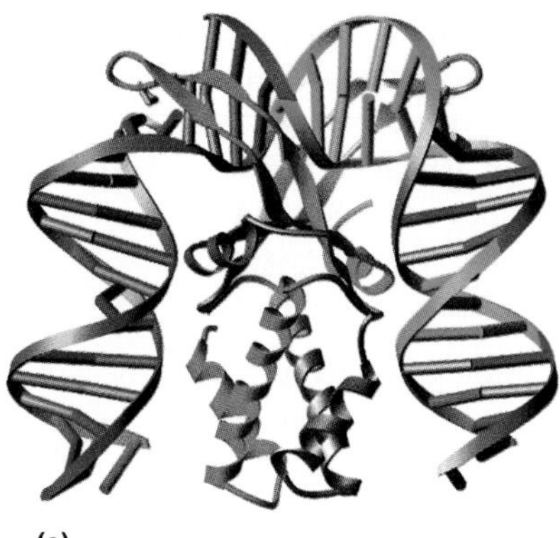

(a)

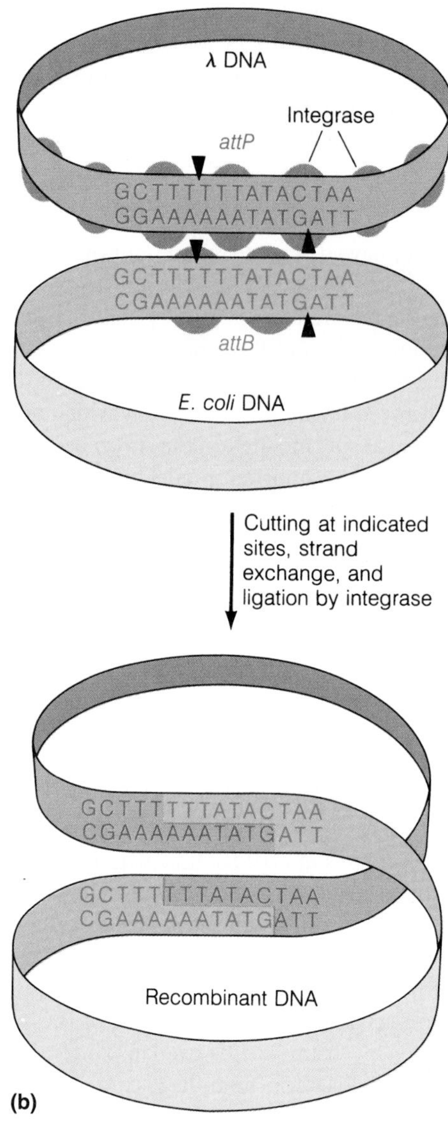

(b)

FIGURE **25.31**

Integrative recombination in phage λ. (a) DNA bending by IHF, shown by x-ray crystallography. **(b)** The site-specific recombination event. After IHF binding (not shown), the intasome forms, and integrase catalyzes cleavage at the sites indicated by wedges and then rejoins strands to create hybrid attachment sites. Not shown for simplicity is a short branch migration that occurs before resolution of the Holliday junction.

(a) Courtesy of P. A. Rice, S. Yang, K. Miznuchi, and H. A. Nash, *Cell* (1996) 87:1298.

have shown a plasticity to DNA that had not been expected. In normal eukaryotic development, segments of DNA can be deleted from the genome, can move from one site to another within a genome, or can duplicate themselves manyfold. In addition, mobile genetic elements have been described in both prokaryotes and eukaryotes. These segments of DNA can move from one chromosomal integration site to another, apparently unrelated to developmental processes. As noted earlier, these processes represent a specialized form of recombination.

Actually, the plasticity of DNA was predicted, but by very few scientists. Barbara McClintock's work on maize genetics, starting in the 1940s, led her to postulate genetic regulatory mechanisms effected through the action of mobile genetic elements. However, some three decades passed before the physical demonstration of such elements in bacteria focused attention on McClintock's pioneering work. For the remainder of this chapter we shall discuss three widely studied aspects of genome plasticity: the genetic basis for antibody variability in vertebrates, gene transposition, and gene amplification.

IMMUNOGLOBULIN SYNTHESIS: GENERATING ANTIBODY DIVERSITY

Recall from Chapter 7 that antibodies are proteins manufactured by vertebrate immune systems that aid in defense against infectious agents and other substances foreign to the animal. The immune response, resulting from introduction of an antigen, elicits the production of several highly specific antibodies. It is estimated that a human is capable of synthesizing more than 10 million distinct antibodies. Most of this great diversity is generated through the action of precisely controlled gene rearrangements, involving but a small fraction of the coding capacity of the genome. These rearrangements occur during differentiation of many individual clones of cells, each clone specialized for the synthesis of one and only one antibody. Other large protein families are diversified by similar mechanisms. The immune response involves proliferation of clones of cells that produce antibodies specialized to react with the specific antigen, or immunogen, provoking that response. This **clonal expansion** allows large-scale production of specific antibodies, needed to combat infection or another challenge to the immune system.

To see how immunological diversity is generated, let us consider one type of antibody, the immunoglobulin G, or **IgG**, class. Recall from Chapter 7 that these proteins consist of two heavy chains and two light chains (see Figure 25.32). Each chain comprises two distinct segments—a domain of variable polypeptide sequences and a constant domain, which is virtually invariant among different IgG

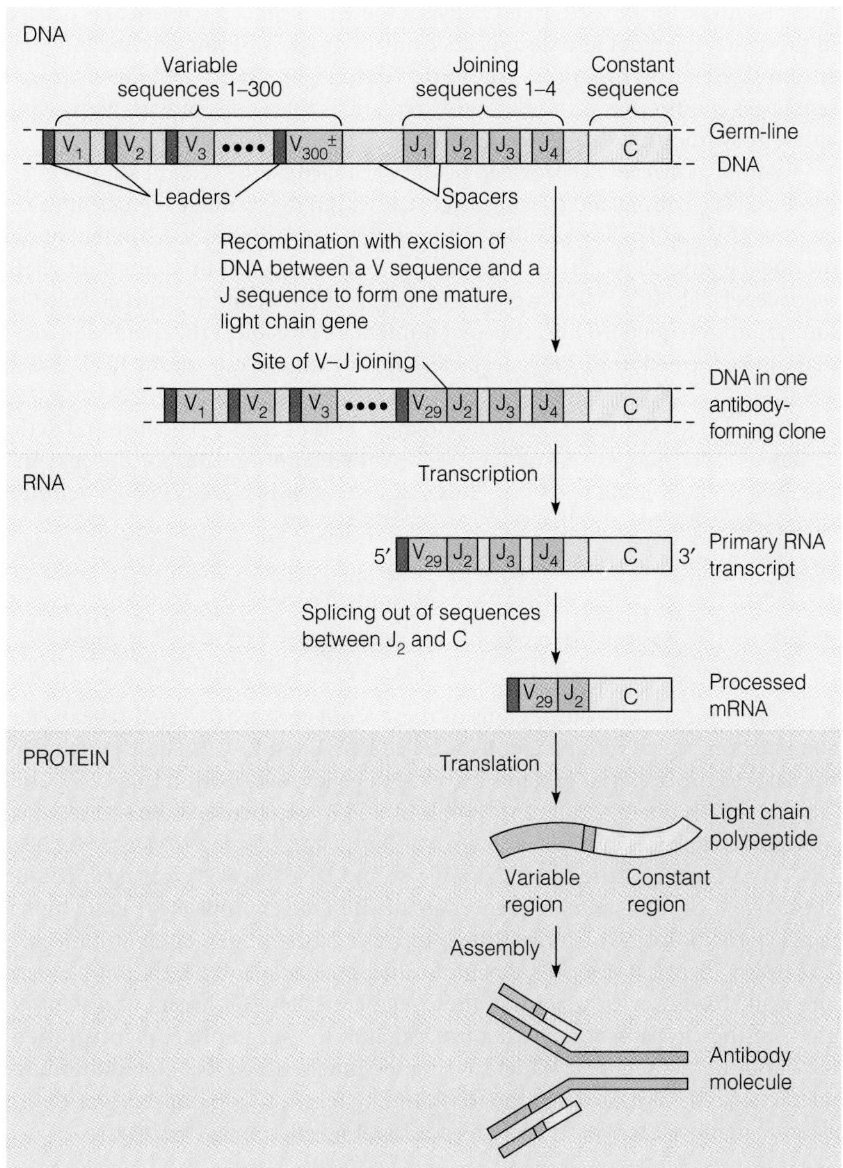

FIGURE **25.32**

Gene rearrangements in antibody gene maturation. The rearrangement of C, V, and J sequences produce one mature κ light chain gene, and the transcription, processing, and translation of this gene produce an antibody κ light chain.

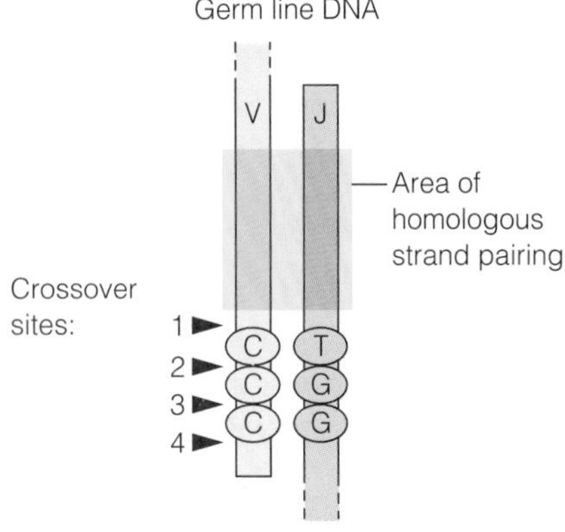

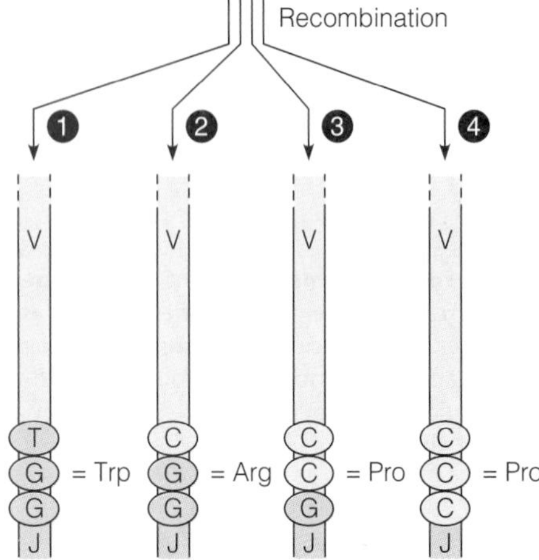

FIGURE **25.33**

Generation of diversity by variable V–J joining mechanisms. Four crossover events are possible at a V–J junction, giving rise to codons for any of three possible amino acids in this example. Only one DNA strand is shown.

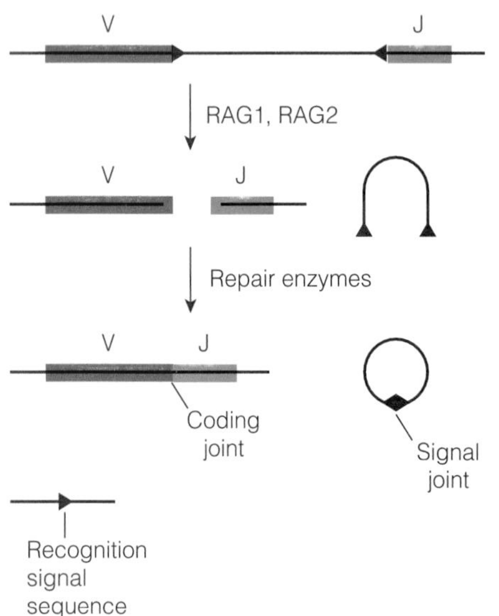

light or heavy chains. We focus on the light chains and in particular the κ class of light chains. (Another class, λ, has somewhat different sequences in its constant region, but its development involves similar mechanisms.)

Figure 25.32 shows the organization of the precursor genes to κ chains in germ-line cells, undifferentiated for antibody formation, and the rearrangements leading to one such gene in a differentiated antibody-producing cell. Each light chain is encoded by DNA sequences that are noncontiguous in the genome of undifferentiated cells but are all in the same chromosome. These sequences are called V (variable), C (constant), and J (joining). The human genome contains about 300 different V sequences, each of which encodes the first 95 amino acids of the variable region; 4 different J sequences, each of which encodes the last 12 residues of the variable region and join it to the constant region; and one C sequence, which encodes the constant region. In an embryonic cell the V sequences, each preceded by a leader sequence containing a transcriptional activator that is not expressed, form a tight cluster; the J sequences form another cluster some distance away; and the C sequence follows shortly after the J cluster. Each J sequence is flanked by nonexpressed spacer sequences.

In the differentiation of one antibody-forming clone of cells, a gene rearrangement links one of the approximately 300 V sequences with one of the 4 J sequences. All of the DNA that lies between these two spliced sequences is deleted in this rearrangement and disappears from all progeny of this cell line. Any upstream V sequences (on the 5′ side, to the left in Figure 25.32) and downstream J sequences (on the 3′ side, to the right) remain in these cells but are not used in antibody synthesis.

Additional diversity is provided by the way in which the V and J sequences recombine. The cutting and splicing can occur within the terminal trinucleotide sequences of V and J in any way that yields one trinucleotide sequence in the spliced product (Figure 25.33). This increases the total number of different light chain sequences by about 2.5 (the average number of different amino acids encoded by four random triplets). Thus, the total number of possible light chain sequences that can be formed from 300 V sequences and 4 J sequences is about 3000 ($300 \times 4 \times 2.5$).

Related DNA sequences are found to the 3′ side of each V sequence and to the 5′ side of each J sequence, and they represent recognition sites for the enzymes involved in the joining reaction. Those sequences, which are called recognition signal sequences, are as follows:

5′···V···CACAGTG···12 bases···ACAAAAAC···3′

3′···J···GTGTCAC···23 bases···TGTTTTTG···5′

Note that the homologous regions of these sequences are inverted repeats. Recombination begins with two proteins called RAG1 and RAG2. These proteins act similarly to the bacterial proteins involved in gene transposition (page 957), beginning the process by catalyzing double-strand breaks between the two recognition signal sequences involved and the respective V and J coding sequences. Cellular DNA repair proteins process the double-strand DNA breaks, creating a "coding joint," in which the V and J sequences are fused in the appropriate reading frame, and a "signal joint," which fuses the intervening DNA into a circular molecule. The nearly identical seven-base palindromic sequences and nearly complementary eight-base AT-rich regions in these segments allow alignment of distant regions of the chromosome, with a process akin to that in phage λ integration, recombining the sequences and excising the intervening DNA. In addition, sequence analysis indicates that the DNA joining reactions are imprecise, with nucleotides removed from one or both ends, creating additional diversity.

The final step in producing a light chain polypeptide involves joining of the C and J segments (see Figure 25.32). This occurs not at the DNA level but at the level of messenger RNA. As discussed in Chapters 7 and 28, eukaryotic gene expression usually involves cutting and splicing of the messenger RNA, with excision of sequences that are not represented in the final gene product. In this case, transcription yields an RNA molecule extending from the 5′ side of the V gene that is spliced to J to the 3′ side of C. Depending on which J region has been spliced to V in this cell, the RNA excised during splicing may contain sequences corresponding to other J regions.

Heavy chains are formed similarly—from V sequences, J sequences, and a class of sequences called D. In addition, there are eight different C sequences, which are also involved in the synthesis of other antibody classes. The total number of possible IgG heavy chains is about 5000. Because any light chain can combine with any heavy chain to form a complete IgG, the total possible number of IgG molecules is 3000×5000, or 1.5×10^7. In this way, enormous diversity can be generated from a very small fraction of the total DNA in germ-line cells. Even further diversity arises from the high rate of V sequence mutation during development of the antibody-producing cell. By this **somatic mutation** process (not involving germ-line cells), cells that undergo the same V–J joining event may still differentiate to produce different IgGs.

The diversity of the immune response involves recombination among thousands of different DNA sequences, to yield a vast array of antibodies.

It is not clear whether both of the homologous chromosomes in a diploid antibody-forming cell undergo identical rearrangements. However, given that each cell produces only one type of antibody, either that must occur, or else one chromosome is silenced after the other has completed its rearrangement.

TRANSPOSABLE GENETIC ELEMENTS

In this section we discuss **transposable genetic elements**—genes that do not have a fixed location in a genome but can move from place to place within the genome, albeit with low frequency. Transposition occurs without benefit of DNA sequence homology, but the enzymes catalyzing transposition recognize short sequences of about a half dozen nucleotides. Although the existence of gene transposition had been predicted by Barbara McClintock's work on maize genetics, the first physical characterization of transposable elements arose from studies of antibiotic-resistant strains of bacteria. By the early 1970s it was known that genes conferring resistance to drugs such as tetracycline or penicillin were usually carried on plasmids, whose DNA sequences bore no detectable homology with chromosomal DNA sequences of the host. Nevertheless, the genes for antibiotic resistance would appear, with low frequency, in the chromosome of the bacterium or in the DNA of a phage that had infected that cell. The presence of new DNA inserted into the host or phage chromosome could be confirmed by restriction cleavage analysis of the DNA or by heteroduplex analysis in the electron microscope. The existence of these "jumping genes," which move from one chromosome to another in seemingly random fashion, greatly altered our views on gene organization and evolution. The new concepts were of much more than academic interest because they relate to the use of antibiotics to treat bacterial infections—specifically, to the ease with which populations of antibiotic-resistant bacteria can arise.

Transposable elements have been demonstrated in many eukaryotes, including maize, *Drosophila*, and yeast. However, we shall concentrate on bacteria, whose physical structures and transposition mechanisms are best understood. Let us first point out several distinctions between bacterial transposition and other recombinational mechanisms we have discussed. First, transposition does not require extensive DNA sequence homology. Furthermore, transposition occurs normally in a *recA*$^-$ host, suggesting that homologous recombination events are not involved. Second, DNA synthesis is involved in bacterial transposition. Transposition always

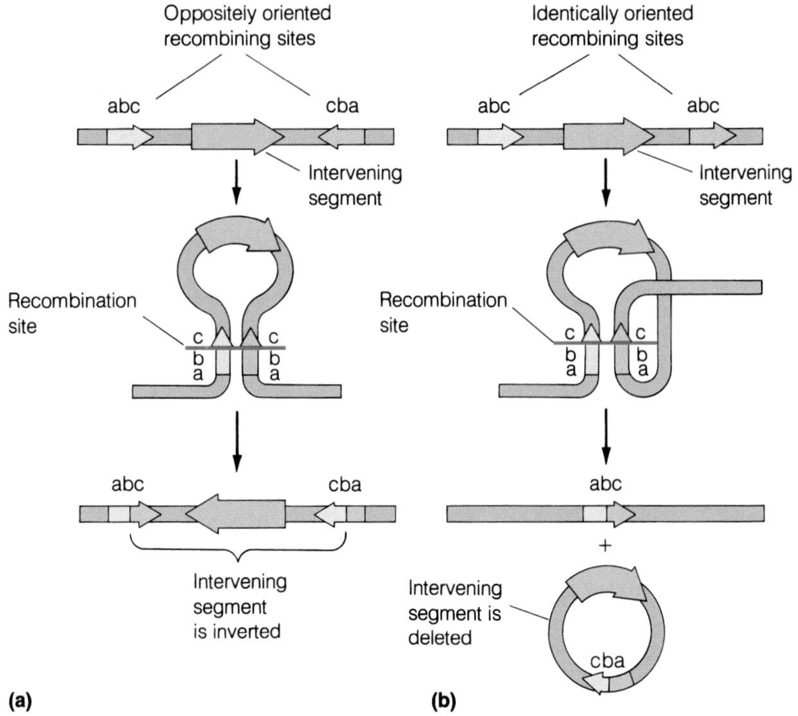

FIGURE **25.34**

Genome rearrangements that can be promoted by homologous recombination between two copies of the same transposable element. Depending on the orientation of the two copies, either **(a)** inversion or **(b)** deletion can result.

involves duplication of the target site, the short sequence (3–12 base pairs) at which the transposable element is inserted. In many instances the transposable element is itself replicated, with one copy being deposited in the new sequence and one remaining in the donor sequence. Finally, transposable elements can restructure a host chromosome. A transposable element can move from one site to another within the same chromosome, producing two homologous sequences resident in the same chromosome. Depending on whether these sequences are oriented identically or in reverse, homologous recombination between them can yield a deletion or an inversion, as shown in Figure 25.34. Transposable elements also have other effects on the chromosomes they move to—either inactivation of any gene into which they move (where insertion interrupts the coding sequence) or activation of adjacent genes (where a promoter, or transcriptional activator, might be created next to the gene). Abortive transpositional events can cause deletions or inversions in the chromosome. Because such events are often lethal, evolution has selected for organisms with low rates of transposition. In the laboratory, insertional inactivation of genes is useful for isolating mutants defective in specific functions and for mapping genes.

Transposable genetic elements include insertion sequences, transposons, and certain bacteriophages that can insert at various locations.

We recognize three different classes of transposable elements in bacteria, with general structures as shown in Figure 25.35. In classifying these elements, we consider the involvement of two enzymes, **transposase** and **resolvase,** whose functions are discussed shortly. Class I elements, which encode a transposase but not a resolvase, are of two types. The simplest transposable element, called an **insertion sequence** (IS), consists simply of a gene for transposase, flanked by two short inverted repeat sequences of about 15 to 25 base pairs. A less simple structure called a **composite transposon** consists of a protein-encoding gene, such as a gene conferring antibiotic resistance, flanked by two insertion sequences, or IS-like elements. These elements may be in either identical or inverted orientations. Class II transposons contain only one set of short flanking direct repeat sequences. In addition to a protein-encoding gene (often conferring antibiotic resistance) and a transposase gene, these elements include a gene for resolvase. Finally, class III elements belong to a small group of bacteriophages, of which the best-known is phage Mu. This phage is known to insert its chromosome at random in the host

CLASS I

Insertion sequence

abc••• •••cba

Transposase gene

Composite transposon

Transposase gene

abc••• •••cba abc••• •••cba

Insertion sequence

Drug resistance element

Insertion sequence

CLASS II

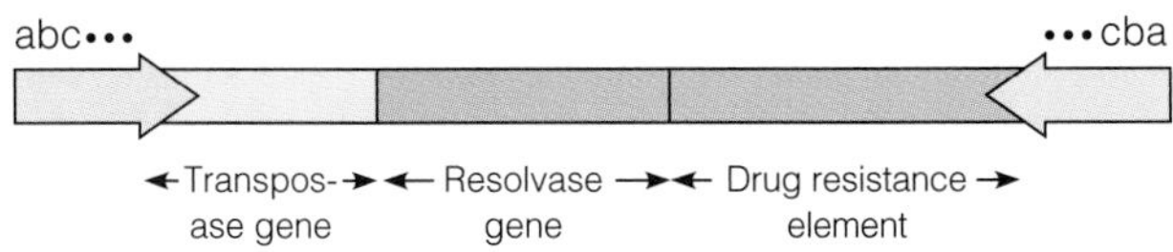

CLASS III

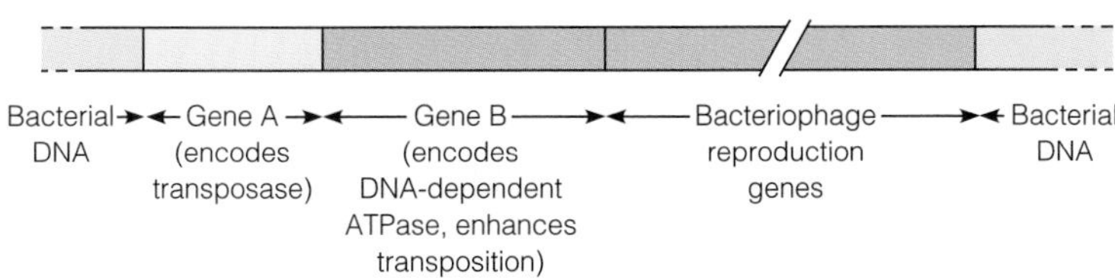

FIGURE **25.35**

Structures of class I, class II, and class III mobile genetic elements. Inverted repeats are shown in purple. The "drug resistance element" may be a different protein-coding gene.

chromosome by a transpositional mechanism and also to replicate its genome by a transpositional mechanism, similar to that of type II elements. One gene of this phage, A, encodes a transposase. Another gene, B, encodes a protein with DNA-dependent ATPase activity. Whereas class I and class II transposable elements synthesize transposase at such low levels that transposition occurs at frequencies of only 10^{-7} to 10^{-5} per generation, phage Mu integrates about 100 times per lytic infection. The B gene product is partially responsible for this far greater efficiency of transposition. Other genes encode structural and other proteins of the virus.

Table 25.4 summarizes the properties of a number of transposons and insertion sequences. Note that each transposon (conventionally referred to with the abbreviation Tn) and IS inserts at a specific target sequence of five or nine base pairs in the examples shown. Insertion involves a duplication of that site, and it results in two copies of the target sequence, one on each side of the integrated element (Figure 25.36). It seems likely that this results from the action of transposase, which generates a staggered cut that brackets the target sequence. Attachment of the mobile element (Tn or Is) to each end results in gaps, which are then filled and ligated to generate the flanking direct repeats.

Because the transposable element never exists as free linear DNA, how are ends of the element generated, to join with the ends of the staggered cut? The currently favored model, shown in Figure 25.37, involves transposase introducing both the staggered cuts in the target site and a nick at each of the 3′ ends of the element—precisely between the transposon sequence and the flanking direct repeat. Next, the free 5′ ends in the recipient DNA target sequence are joined to the 3′ ends of the element. Two outcomes are then possible. In **simple transposition** the joining is followed by cutting of the 5′ ends of the transposon, also immediately adjacent to the flanking sequences. This gives a gapped structure like that shown in Figure

TABLE 25.4 Structures of some transposable elements of *E. coli*

Element	Size (bp)	Target DNA (bp)	Resistance Conferred
Insertion sequences			
IS1	768	9	None
IS2	1327	5	None
IS10-R	1329	9	None
Composite transposons			
Tn5	5700	9	Kanamycin
Tn10	9300	9	Tetracycline
Tn2571	23,000	9	Chloramphenicol, fusidic acid, streptomycin, sulfonamides, and mercury
Class II transposons			
Tn3	4957	5	Ampicillin
Class III transposons			
Phage Mu	38,000	5	None

Source: Excerpted from N. Kleckner, *Annu. Rev. Genet.* (1981) 15:354. © 1981 Annual Reviews, Inc.

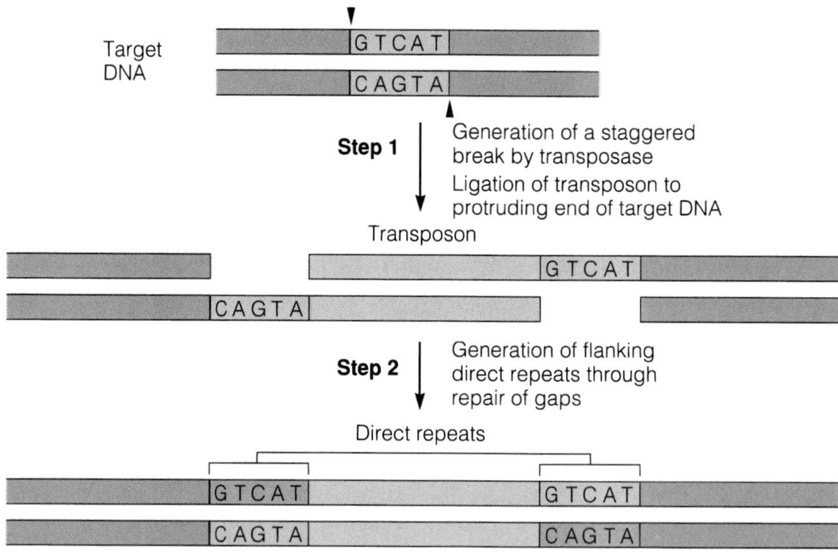

FIGURE **25.36**

Model of how direct repeats are generated during the insertion of a transposon or an insertion sequence.

25.36, which can be filled and closed by DNA polymerase and ligase. In this form of transposition, only the target sequence is copied; the donor chromosome suffers a lethal double-strand break. Tn10 (Table 25.4) transposes by a conservative mechanism, with both original strands somehow transferred to the new location.

The other process, **replicative transposition,** requires the enzyme resolvase, so it occurs only with class II and class III elements. The 3′ ends of the target chromosome, after the first cutting and splicing, serve as replicative primers for copying both the gaps, as shown in Figure 25.37, and the two strands of the transposable element itself. Ligase action generates a **cointegrate,** a large circular structure containing both donor and target chromosomes with two freshly replicated copies of the transposable element. The other enzyme, resolvase, now catalyzes site-specific recombination between the two elements, resulting in one copy of the transposable element inserted into each of the two chromosomes. Structural analyses of

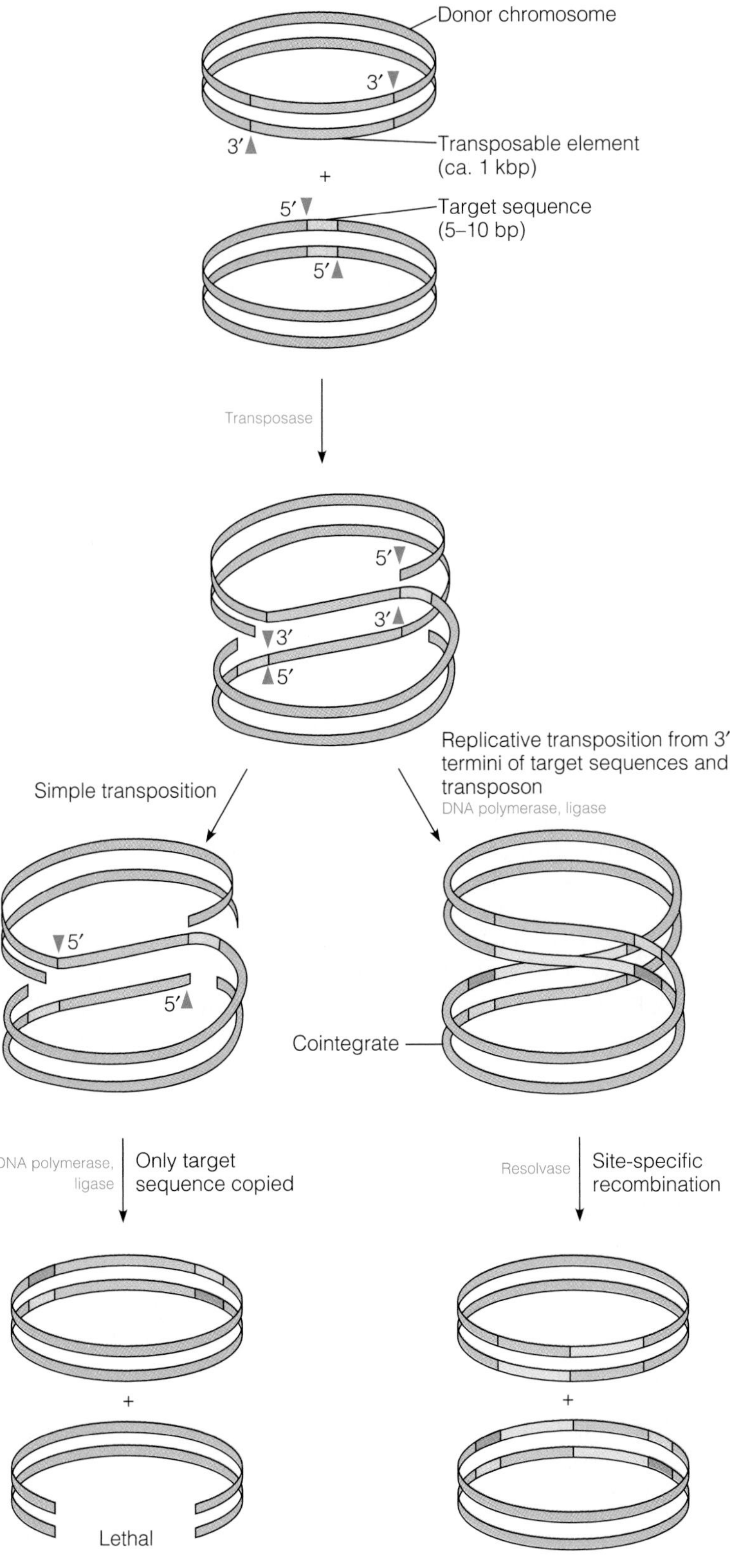

FIGURE **25.37**

Models of simple transposition and replicative transposition.

the *E. coli* enzyme, called $\gamma\delta$ resolvase, are yielding important clues to this complex process.

RETROVIRUSES

Gene transposition in eukaryotic systems presents some strong similarities to and some distinct differences from transposition in bacteria. The first major distinction is that integration and excision are distinct processes in eukaryotes. Thus, the transposable element can be isolated in free form, often as a double-strand circular DNA. Second, replication of that DNA often involves the synthesis of an RNA intermediate. Both of these properties are seen in the retroviruses of vertebrates, perhaps the most widely studied class of eukaryotic transposable elements. As we noted in Chapter 24, these RNA viruses use reverse transcriptase to synthesize a circular duplex DNA, which can integrate into many sites of the host cell chromosome. The integrated retroviral genome bears remarkable resemblance to a bacterial composite transposon, as you can see by comparing Figure 25.38 with Figure 25.35. The prototypical retroviral genome has three structural genes—*gag*, which encodes a polyprotein that undergoes cleavage to give virion core proteins; *pol*, which encodes the viral polymerase, or reverse transcriptase; and *env*, the major glycoprotein of the viral envelope. Flanking these structural genes are two direct repeats, the **long terminal repeats (LTRs)** of about 250 to 1400 base pairs each. Each LTR is flanked in turn by short inverted repeat sequences, 5 to 13 base pairs in length. Integration occurs by a mechanism that duplicates the target site, so that the integrated viral gene, called a **provirus,** is flanked by direct repeats of 5 to 13 base pairs each of host cell DNA.

Just as bacterial transposons can carry passenger genes, so also can retroviruses. The earliest known retrovirus, Rous sarcoma virus, was also the first virus shown to contain an oncogene (see Chapter 23). Rous sarcoma virus was isolated in 1911 and shown to cause tumors in chickens. Not until 1978, however, was the *src* gene identified and shown to be responsible for tumorigenesis. The *src* gene product is a 60-kilodalton protein with a protein tyrosine kinase activity. A related but distinct sequence can be detected in the host genome. In Rous sarcoma virus the *src* oncogene lies to the 3′ side of the *env* gene. Other tumorigenic viruses contain the oncogene either inserted into or substituting for one of the genes *gag*, *pol*, and

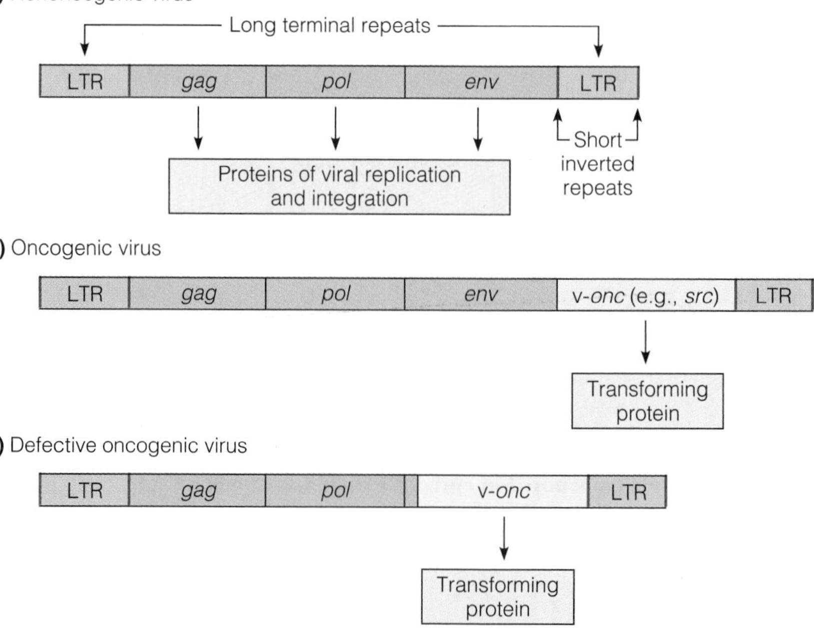

FIGURE **25.38**

Structure of retroviral genomes in the integrated state. (a) A nononcogenic virus. **(b)** An oncogenic virus such as Rous sarcoma virus, showing the viral oncogene downstream (rightward) from the viral replication genes. **(c)** A defective oncogenic virus, such as Moloney murine sarcoma virus, with the viral oncogene replacing part or all of a gene (*env*) essential to viral replication. In each case the LTRs are direct repeats, flanked by short inverted repeats.

env. Because the loss of an essential gene makes it impossible for the virus to replicate, the latter class of virus can grow only in a cell coinfected with a **helper virus,** a related retrovirus that provides the missing function(s).

By various means, one can show that the action of an oncogene is essential for the virus to effect **oncogenic transformation,** the change of a normal cell to a cancerous cell (see pages 861–866 in Chapter 23). For example, mutational inactivation of the *src* gene of Rous sarcoma virus does not affect the ability of the virus to replicate, but it does render the virus unable to cause tumors in infected animals. Because each viral oncogene is related in sequence to a cellular counterpart, it is presumed that the viral oncogene originated in a cell many generations ago and underwent independent mutations after being picked up by a viral genome. Moreover, because many of the oncogene products, like the Src protein, have protein kinase activity, they may be aberrant forms of normal cellular regulatory elements, and their expression may be involved in the abnormal growth control that characterizes tumor cells. An additional mechanism may be related to activation of cellular genes by insertion of proviral DNA. The leftmost LTR in an integrated provirus contains the transcriptional activator, or **promoter,** for the adjacent *gag* gene and the downstream *pol* and *env* genes. Because the LTRs are direct repeats, the rightmost LTR can activate transcription of cellular genes downstream from the integration site. If these genes include those involved in metabolic regulation, their overexpression may unbalance metabolism in some still undefined way and, hence, contribute to oncogenesis.

Transposable elements in eukaryotic cells show striking resemblances to retroviruses in sequence organization. Indeed, the term **retrotransposon** is used to denote this class of elements. These similarities are illustrated in Figure 25.39 for two retroviruses, plus Ty, a transposon of yeast; copia and 412, transposable elements in *Drosophila;* and IAP, a transposon found in the mouse genome.

Retroviral genomes and eukaryotic transposable elements have sequence similarities, to each other and to bacterial transposons.

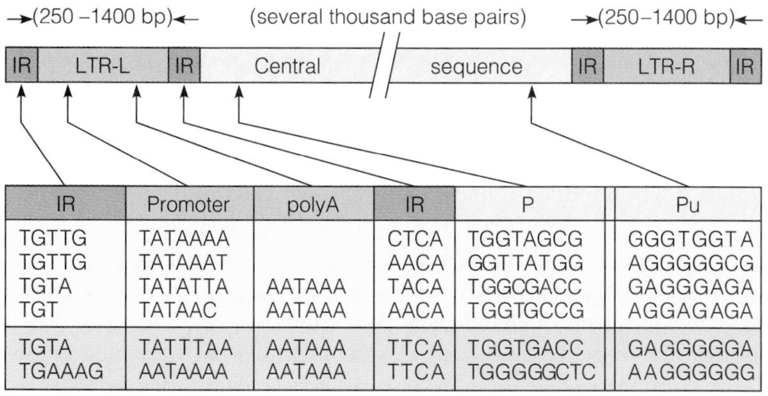

→(250 –1400 bp)← (several thousand base pairs) →(250–1400 bp)←

IR | LTR-L | IR | Central // sequence | IR | LTR-R | IR

IR	Promoter	polyA	IR	P	Pu	
TGTTG	TATAAAA		CTCA	TGGTAGCG	GGGTGGTA	Ty
TGTTG	TATAAAT		AACA	GGTTATGG	AGGGGGCG	Copia
TGTA	TATATTA	AATAAA	TACA	TGGCGACC	GAGGGAGA	412
TGT	TATAAC	AATAAA	AACA	TGGTGCCG	AGGAGAGA	IAP
TGTA	TATTTAA	AATAAA	TTCA	TGGTGACC	GAGGGGGA	RSV
TGAAAG	AATAAAA	AATAAA	TTCA	TGGGGGCTC	AAGGGGGG	MoMLV

FIGURE **25.39**

Common sequence features in integrated retroviruses and other eukaryotic transposable elements. All these elements are bounded on the left and right sides by long terminal repeat sequences (LTR-L and LTR-R, respectively). Each LTR is flanked by short inverted repeat (IR) sequences (arrows). The approximate locations of various structural features are indicated, and the table gives some specific sequences associated with each feature in several transposable elements—IRs; promoter, a transcription start signal in each LTR; polyA, an adenylate-rich sequence (absent from the yeast transposon Ty and the copia element in *Drosophila*); P, a sequence that anneals to an RNA that serves as a primer for replication; Pu, purine-rich sequence. 412, like copia, is a transposable element of *Drosophila;* IAP is a transposon found in the mouse genome; and RSV and MoMLV are retroviruses (Rous sarcoma virus and Moloney murine leukemia virus, respectively).

From M. Singer and P. Berg, *Genes and Genomes* (Sausalito, CA: University Science Books, 1991), p. 755.

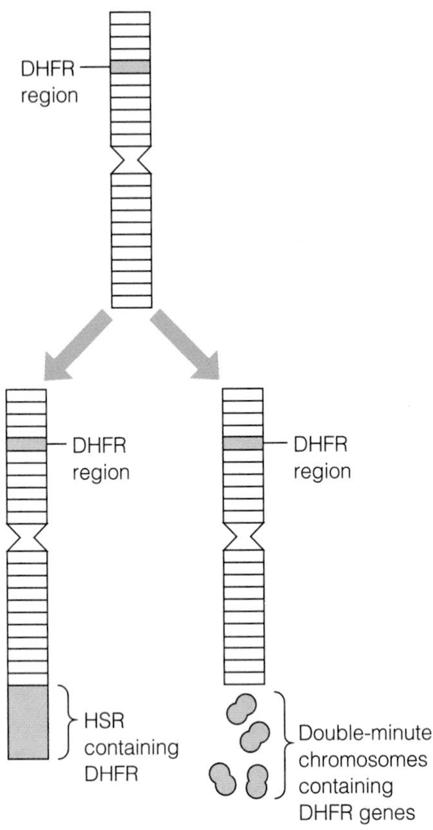

Gene amplification generates multiple copies of DNA sequences at a separate site on the same chromosome. Recombination among homologous segments yields extrachromosomal amplified sequences.

Gene Amplification

The final process we discuss in information restructuring is the selective amplification of specific regions of the genome, principally in eukaryotic cells. This occurs in normal developmental processes and as a consequence of particular metabolic stress situations.

It has long been known that during oogenesis in certain amphibians the genes encoding ribosomal RNAs increase in copy number by some 2000-fold, in preparation for the large amount of protein synthesis that must occur in early development. The amplified DNA is in the form of extrachromosomal circles, each of which contains several copies of the ribosomal DNA repeat and a replication origin. A similar situation has been analyzed in *Drosophila,* in which genes encoding egg proteins are amplified at a particular developmental stage. In the latter case, however, the amplification results from repeated rounds of replication initiation within the amplified region, and the amplified sequences remain within the chromosome of origin.

Both types of mechanisms apparently occur during development of certain drug-resistant mammalian cell lines in culture. This process has been studied most widely in cells that become resistant to methotrexate, a dihydrofolate reductase inhibitor. As discussed in Chapter 22, treatment of leukemia with methotrexate often leads to the emergence of drug-resistant leukemic cell populations, which contain vastly elevated levels of the target enzyme, dihydrofolate reductase (DHFR). As shown originally by Robert Schimke, overproduction of the enzyme usually results from specific amplification of a large DNA segment that includes the DHFR gene. In one process, tandem duplication of the DNA segment generates a giant chromosome with multiple gene copies, in what is called a **homogeneously staining region (HSR),** because it lacks the typical chromosome banding pattern. Alternatively, a DNA segment containing the DHFR gene can be excised, apparently by a recombinational process, to form minichromosomes called **double-minute chromosomes.** Some resistant cells contain both types of amplified genes. Double-minute chromosomes are maintained within a cell only as long as selective pressure is maintained by growth of the cell in methotrexate. However, the chromosomally amplified phenotype is stable through many generations of cell growth. Figure 25.40 shows a fluorescence micrograph of metaphase chromosomes from a stably amplified Chinese hamster ovary cell line. DHFR sequences were visualized by in situ hybridization with a fluorescent-tagged DNA containing DHFR sequences. This technique is sufficiently sensitive to allow detection of single-copy sequences (white arrows). Note also the giant chromosome containing many gene-equivalents of DHFR gene sequences.

Amplification of genes under selective conditions has been widely observed—for example, in development of pesticide-resistant forms of insects. The mechanism of amplification is not yet clear. However, evidence such as that of Figure 25.40 shows that the amplified sequences are on the same chro-

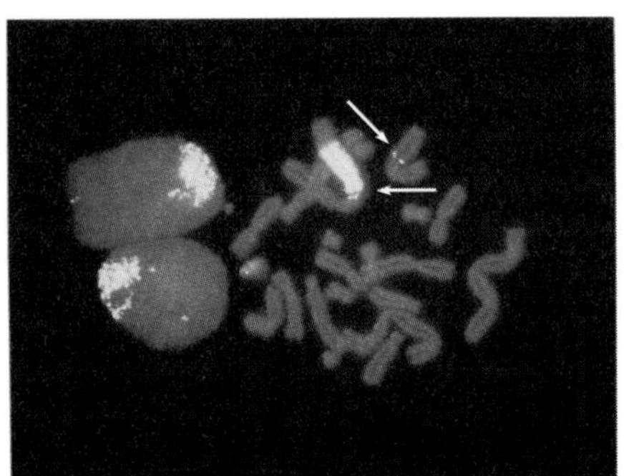

FIGURE **25.40**

Chromosome structural changes that accompany dihydrofolate reductase gene amplification. The micrograph shows metaphase chromosomes from Chinese hamster ovary cells that are highly resistant to methotrexate. Chromosomal DNA was subjected to hybridization in situ with a fluorescence-labeled DHFR gene probe. White arrows point to single-copy genes. The amplified chromosomal sequences are on a giant form of the chromosome that also contains one of the original single-copy sequences.

Courtesy of B. J. Trask and J. L. Hamlin, *Genes Dev.* (1989) 3:1913–1925, provided by Joyce Hamlin.

mosome as the original single-copy gene site, but at a significant distance away from that site. Such structures could arise either through recombination with unequal sister-chromatid exchange, schematized in Figure 25.41, or by a conservative transposition process. Later, homologous recombination within an amplified region can lead to excision of sequences containing one or more amplified sequences. In order to replicate autonomously, these excised sequences must have a centromere. Such elements probably represent the double-minute chromosomes.

The presence of selective pressure, such as the continuous presence of methotrexate, promotes specifically the survival of cells that can respond to that pressure, for instance, by overproducing DHFR. Once two or more copies of the gene are present on a chromosome, additional copies can be generated by further recombinational events or by abnormalities of replication. Resistance is thus developed in stepwise fashion and occurs over many generations of growth. These findings have immense practical significance, because cancer chemotherapy often involves long-term treatment with low doses of an antimetabolite—precisely the conditions most likely to nullify the effect of treatment by generating drug-resistant cells. Findings on gene amplification not only have changed the way in which anticancer drugs are administered but also have been extended to the tumorigenic process itself. Investigators have found that specific oncogenes become amplified during the clinical progression of certain human tumors. Thus, gene amplification is seen as a mechanism in normal development, in cellular adaptation to stress, and in abnormal developmental processes. Also, the gene duplications that have occurred during evolution (see Chapter 7, pages 232–233) probably have taken place by similar pathways.

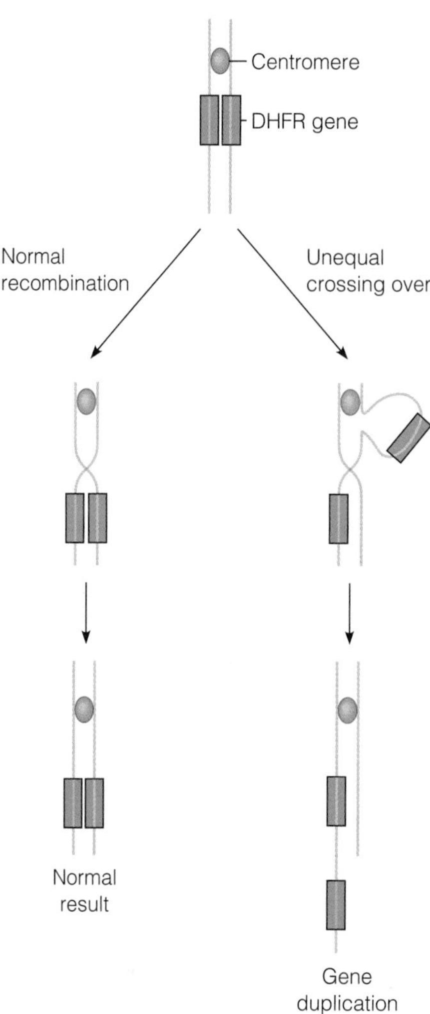

FIGURE **25.41**

Unequal crossing over, as a mechanism to explain early steps in gene amplification.

SUMMARY

A variety of information restructuring processes affect DNA, to protect it from environmental damage or foreign organisms, to diversify the individuals in a species, or to diversify the genetic constitution of somatic cells in differentiation. These processes include bacterial restriction and modification, in which methylation of adenine residues in specific DNA sequences can protect DNA against double-strand breakage catalyzed by site-specific nucleases acting at these sites. DNA repair encompasses several processes, including photoreactivation, in which light energy is captured to reverse pyrimidine dimer formation, and the removal of alkylguanines by "enzymes" that act only once. Nucleotide excision repair systems recognize damaged nucleotides and cleave out and replace a patch containing 12 to 30 nucleotides that flank the damaged segment. Base excision repair begins with an *N*-glycosylase reaction, followed by repair enzymes that excise one or more nucleotides at the abasic site. Damage can also be repaired by a recombinational process or by induced (error-prone) repair. Mismatch repair systems correct occasional errors arising during replication. Genetic recombination between homologous DNA segments involves helicase-catalyzed duplex unwinding, site-specific DNA cutting, duplex-strand invasion by single-strand 3′ hydroxyl termini, chain extension, branch migration, and resolution of Holliday structures. Some recombinational events, such as integration of bacteriophage λ, are site-specific and governed by DNA–protein interactions. A wide range of gene rearrangements includes the joining reactions in differentiation of the immune response, gene transposition, retroviral genome integration, and gene amplification, which probably occurs by a recombinational mechanism. Gene amplification can either be a normal developmental process or occur in response to a specific environmental stress.

REFERENCES

Nucleic Acid Methylation

Heithoff, D. M., R. L. Sinsheimer, D. A. Low, and M. J. Mahan (1999) An essential role for DNA adenine methylation in bacterial virulence. *Science* 284:967–970. An unexpected link between DNA methylation and bacterial pathogenesis.

Jones, P. A., W. M. Rideout, J-C. Shen, C. H. Spruck, and Y. C. Tsai (1992) Methylation, mutation, and cancer. *BioEssays* 14:33–36. Advances the idea that DNA methylation is a mutagenic event in mammalian cells.

Roberts, R. J., and X. Cheng (1998) Base flipping. *Annu. Rev. Biochem.* 67:181–198. DNA methylases are not the only enzymes that flip bases in DNA substrates.

Walsh, C. P., and T. H. Bestor (1999) Cytosine methylation and mammalian development. *Genes Dev.* 13:26–34. This study suggests that methylation is a consequence, not a cause, of events that silence particular eukaryotic genes.

Restriction and Modification

McClarin, J. A., C. A. Frederick, B.-C. Wang, P. Greene, H. W. Boyer, and J. M. Rosenberg (1986) Structure of the DNA–*Eco*RI endonuclease recognition complex at 3 Å resolution. *Science* 234:1526–1541. The first structure determination for a restriction enzyme.

Wilson, G. G. (1993) Restriction enzymes: A brief overview. *NEB Transcript* 5:1–5. This technical bulletin, published by New England BioLabs, contains an excellent short review of the biology and biochemistry of restriction enzymes and the structural basis for their sequence specificity.

DNA Repair

Demple, B., and L. Harrison (1994) Repair of oxidative damage to DNA: Enzymology and biology. *Annu. Rev. Biochem.* 63:915–948. A relatively recent and comprehensive review of this subject.

Friedberg, E. C., G. C. Walker, and W. Siede (1995) *DNA Repair and Mutagenesis.* ASM Press, Washington, D.C. A 698-page update of Friedberg's influential 1985 book on DNA repair.

Jiricny, J. (1994) Colon cancer and DNA repair: Have mismatches met their match? *Trends Genet.* 10:164–168. Tremendous excitement resulted from the finding that an altered gene in some colon cancers is related to the MutS protein in *E. coli* mismatch repair.

Kunkel, T. A. (1993) Slippery DNA and disease. *Nature* 365:207–208. Several reports have shown that mutant human genes causing disease have repeated short stretches of DNA. In this mini-review, Kunkel argues that the biochemical basis is defective mismatch repair.

Mellon, I., D. K. Raipal, M. Koi, C. R. Boland, and G. N. Champe (1996) Transcription-coupled repair deficiency and mutations in human mismatch repair genes. *Science* 272:557–560. Important relationships between these two repair processes in human cells.

Modrich, P. (1997) Strand-specific mismatch repair in mammalian cells. *J. Biol. Chem.* 272:24727–24730. A review of strand specificity and the relationship of faulty mismatch repair to cancer; one of four JBC minireviews on eukaryotic DNA repair.

McMurray, C. T. (1999) DNA secondary structure: A common and causative factor for expansion in human disease. *Proc. Natl. Acad. Sci. USA* 96:1823–1825. A brief review about triplet repeat expansions in genomic DNA associated with human disease.

Sancar, A. (1993) Structure and function of DNA photolyase. *Biochemistry* 33:2–9. A thorough review of photoreactivation.

Sancar, A. (1996) DNA excision repair. *Annu. Rev. Biochem.* 65:43–82. This issue of *Annual Reviews* also contains articles on transcription-coupled repair, mismatch repair, and eukaryotic DNA repair.

Tanaka, K., and R. D. Wood (1994) Xeroderma pigmentosum and nucleotide excision repair of DNA. *Trends Biochem. Sci.* 19:83–87. Insights into human DNA repair from human genetic diseases.

Walker, G. C. (1998) Skiing the black diamond slope: Progress on the biochemistry of translesion DNA synthesis. *Proc. Natl. Acad. Sci. USA* 95:10348–10350. A brief commentary describing progress in our understanding of error-prone repair.

Recombination

Baumann, P., and S. C. West (1998) Role of the human RAD51 protein in homologous recombination and double-stranded-break repair. *Trends Biochem. Sci.* 23:247–251. Reviews important aspects of prokaryotic and eukaryotic recombination.

Bianco, P. R., R. B. Tracy, and S. C. Kowalczykowski (1998) DNA strand exchange proteins: A biochemical and physical comparison. *Frontiers Biosci.* 3:570–603. A detailed review of homologous recombination mechanisms and the proteins involved, in *E. coli,* phage T4, and yeast.

Haber, J. E. (1999) DNA recombination: The replication connection. *Trends Biochem. Sci.* 24:271–275. A timely review of double-strand break repair and its relationship to DNA replication and recombination.

Landy, A. (1999) Coming or going it's another pretty picture for the λ–Int family album. *Proc. Natl. Acad. Sci. USA* 96:7122–7124. A readable minireview about site-specific recombination carried out by a family of proteins of which λ–Int was the first known member.

Smith, G. R., S. K. Amurdsen, P. Dabert, and A. F. Taylor (1995) The initiation and control of homologous recombination in *Escherichia coli. Phil. Trans. Royal Soc. London Series B* 347:13–20. Proposed mechanisms for RecA- and RecBCD-catalyzed recombination after conjugation.

Story, R. M., I. T. Weber, and T. A. Steitz (1992) The structure of the *E. coli recA* protein monomer and polymer. *Nature* 355:318–324. A crystallographic analysis.

West, S. C. (1998) RuvA gets X-rayed on Holliday. *Cell* 94:699–701. A minireview describing structural analysis of complexes involved in branch migration and Holliday joint resolution.

Gene Rearrangements

Craig, N. L. (1997) Target site selection in transposition. *Annu. Rev. Biochem.* 66:437–474. Contains references to all aspects of gene transposition.

Lewin, B. (1997) *Genes VI.* Oxford University Press, Oxford, UK. Chapters 18 and 19 of this contemporary molecular genetics textbook present detailed discussions of transposons, retroviruses, and other transposable elements.

McClintock, B. (1984) The significance of responses of the genome to challenge. *Science* 226:792–801. McClintock's Nobel Prize address, giving the history of the first description of mobile genetic elements.

Milstein, C. (1986) From antibody structure to immunological diversification of the immune response. *Science* 231:1261–1268. Milstein was awarded the Nobel Prize for discovering monoclonal antibodies, but in this Nobel Prize lecture he discusses the generation of antibody diversity.

Murley, L. L., and N. D. F. Grindley (1998) Architecture of the $\gamma\delta$ resolvase synaptosome: Oriented heterodimers identify interactions for synapsis and recombination. *Cell* 95:553–562. Resolvase mechanisms explored by crystallography and site-directed mutagenesis.

Roth, D. B., and N. L. Craig (1998) VDJ recombination: A transposase goes to work. *Cell* 94:411–414. A recent minireview describing the recombinational events in maturation of antibody-forming genes.

Singer, M., and P. Berg (1991) *Genes and Genomes.* University Science Books, Mill Valley, Calif. Chapter 10 of this outstanding textbook reviews gene transposition and other rearrangements.

Retroviruses

Varmus, H. (1988) Retroviruses. *Science* 240:1427–1435. Still one of the best reviews available.

Gene Amplification

Sharma, R. C., and R. T. Schimke (1994) The propensity for gene amplification: A comparison of protocols, cell lines, and selection agents. *Mutat. Res.* 304:243–260. Practical information from the laboratory that discovered dihydrofolate reductase gene amplification.

Smith, K. A. et al. (1995) Regulation and mechanisms of gene amplification. *Phil. Trans. Royal Soc. London Series B* 347:49–56. A readable and well-referenced review.

PROBLEMS

1. Predict whether a *dam* methylase deficiency would increase, decrease, or have no effect upon spontaneous mutation rates, and explain the basis for your prediction.
2. For each DNA repair process in column I, list *all* characteristics from column II that correctly describe that process.

I	II
(a) Nucleotide excision repair	(1) RecA protein participates.
(b) Photoreactivation	(2) Damaged nucleotides are removed by nick translation.
(c) Base excision repair	(3) A free radical mechanism is involved.
(d) Recombinational repair	(4) The repair enzyme functions only once.
(e) SOS-driven error-prone repair	(5) The key enzyme contains a bound folate cofactor.
(f) Alkyltransferase repair	(6) No bases or nucleotides are removed from the DNA.
(g) Mismatch repair	(7) Deficiency of this enzyme in humans greatly increases the risk of skin cancer.
	(8) This system is chiefly responsible for the mutagenic effect of ultraviolet light.
	(9) The first enzyme in this pathway cleaves two phosphodiester bonds.
	(10) This process begins up to 1 kbp away from the site to be repaired.
	(11) DNA ligase catalyzes the final reaction.

3. Of the restriction enzymes listed in Table 25.2, which enzymes generate flush, or blunt-ended, fragments? Of those that recognize offset sites and generate staggered cuts, which of these cuts cannot be converted to flush ends by the action of DNA polymerase? Why?
4. The following diagram shows one-half of a restriction site.
(a) Draw the other half.

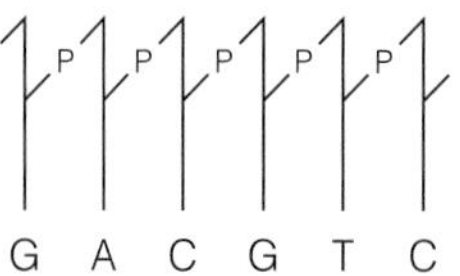

(b) Use heavy arrows (**↓↑**) to identify type II cleavage sites that would yield blunt-ended duplex DNA products.
(c) Use light arrows (↓↑) to identify type II cleavage sites yielding staggered cuts that could be converted directly to recombinant DNA molecules by DNA ligase, with no other enzymes involved.
(d) If this were the recognition site for a type I restriction endonuclease, where would cutting of the duplex occur?
(e) If DNA sequences were completely random, how large an interval (in kilobase pairs) would you expect between identical copies of this sequence in DNA?

5. 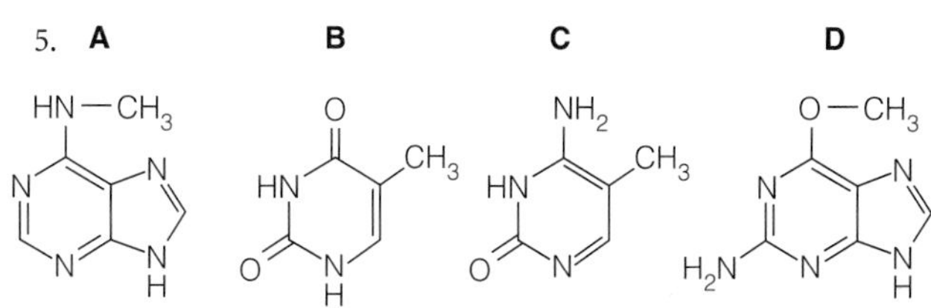

For each of the following characteristics, list all of the bases to which they apply.
(a) A signal that identifies a parental DNA strand in the MutH,L,S mismatch correction system
(b) Most likely to be involved in cyclobutane dimer formation after ultraviolet irradiation of DNA
(c) A methylated base found immediately to the 5′ side of dGMP residues in eukaryotic DNA
(d) Created by treating DNA with alkylating agents that transfer methyl groups and repaired by an "enzyme" that functions only once in its lifetime
(e) Created by AdoMet-dependent methylation of a nucleotide residue in DNA
(f) A substrate for deamination at the DNA level, which would lead to a GC ⟶ AT transition

6. Homologous recombination in *E. coli* forms heteroduplex regions of DNA containing mismatched bases. Why are these mismatches not eliminated by the mismatch repair system?
7. Deficiencies in the activity of either dUTPase or DNA ligase stimulate recombination. Why?
8. pBR322 DNA (4.32 kb) was cleaved with *Hin*dIII nuclease and ligated to a *Hin*dIII digest of human mitochondrial DNA. One recombinant plasmid DNA was analyzed by gel electrophoresis of restriction cleavage fragments, with the following results: lane A = *Eco*RI-treated recombinant, lane B = *Hin*dIII-treated vector, and lane C = *Hin*dIII-treated recombinant.

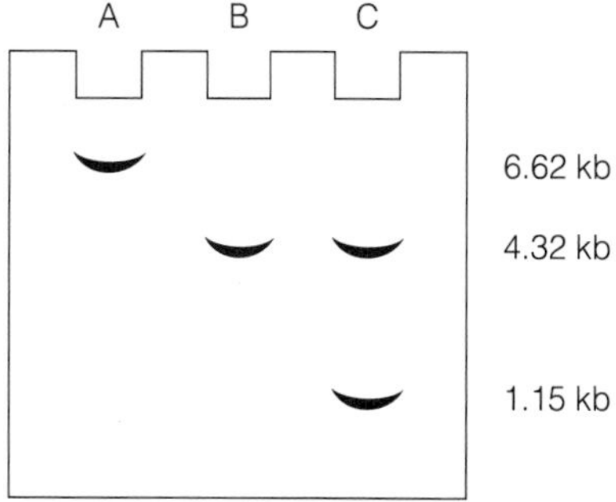

(a) Why was the recombinant plasmid treated with *Eco*RI for determination of its size?
(b) How might you explain the discrepancy between the size of the recombinant molecule and the sum of the sizes of the *Hin*dIII cleavage fragments?
(c) Draw a diagram of the recombinant showing the locations of the *Hin*dIII cleavage sites.

9. A small DNA molecule was cleaved with several different restriction nucleases, and the size of each fragment was determined by gel electrophoresis. The following data were obtained.

Enzyme	Fragment Size (kb)
*Eco*RI	1.3, 1.3
*Hpa*II	2.6
*Hin*dIII	2.6
*Eco*RI + *Hpa*II	1.3, 0.8, 0.5
*Eco*RI + *Hin*dIII	0.6, 0.7, 1.3

(a) Is the original molecule linear or circular?
(b) Draw a map of restriction sites, showing distances between sites, that is consistent with the data presented.
(c) How many additional maps are compatible with the data?
(d) What would have to be done to locate the cleavage sites unambiguously with respect to each other?

10. 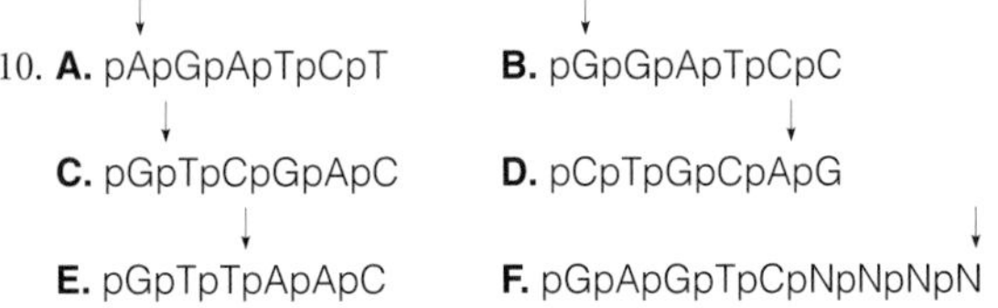

(a) Using the same shorthand as in Problem 4, show the complete structure of any one of the foregoing restriction cleavage sites (both strands and cutting sites).
(b) Which two of the cleavages shown will yield fragments that *cannot* be rejoined by *E. coli* DNA ligase? Why not? (Note: In vitro DNA ligase shows strong preference for sealing a staggered, rather than blunt-ended, cut.)
(c) Cleavage products from two of these reactions can readily be joined to one another by DNA ligase, even though the two enzymes recognize different sites. Which two?
(d) If you wished to linearize a newly isolated plasmid DNA, which one of the sites shown would be *least* likely to be represented only once in that DNA molecule? Assume that the DNA has equal proportions of the four nucleotides.

11. A cloned 8.0-kb fragment of bacterial DNA contains a gene that you wish to map, with respect to its transcriptional end points. You have in hand a highly purified, highly radioactive mRNA transcribed from that gene, to use as a hybridization probe.

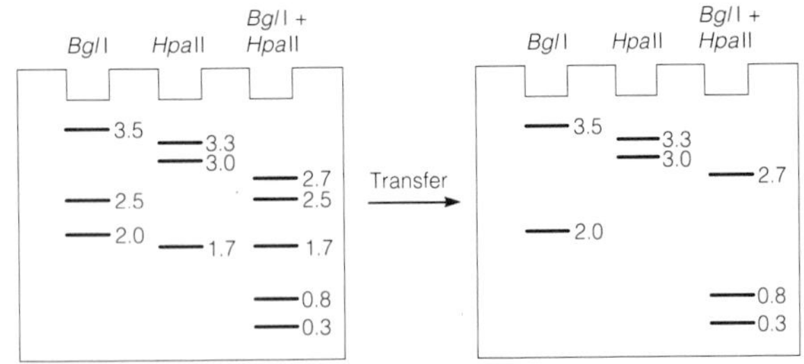

On the left in the diagram are depicted the restriction cleavage patterns of the 8-kb fragment, visualized with ethidium bromide. On the right is the radioautogram pattern that resulted when the fragments were transferred to nitrocellulose and probed with the radioactive RNA. On the following diagram, show the locations of all of the *Bgl*I and *Hpa*II restriction sites, and show the approximate end points of the transcript. How might you tell which end represents the 5′ end and which the 3′ end of the transcript?

*12. Suppose that you wanted to study retroviral integration mechanisms by determining the nucleotide sequence at the integration site—several dozen nucleotides on each side of the viral–cellular DNA junction. Describe how to isolate DNA containing a junction site in amounts sufficient for sequence analysis.

13. In site-directed mutagenesis, one often designs the mutagenic oligonucleotide not only to force a desired amino acid substitution but also to create a new restriction site in the mutant gene or to eliminate a preexisting restriction site. Why is this desirable?

14. We wish to determine whether the serine residue of an enzyme is essential for its catalytic activity, by changing it to a glycine and testing the activity of the mutant protein. We also wish to remove one restriction site and introduce one new site, without introducing additional amino acid changes. A maximum of two single-base changes can be tolerated in an 18-nucleotide primer and still have it hybridize to the template. Using the genetic codon assignments given in Figure 27.3 and the restriction site information in Table 25.2, design an oligonucleotide of 18 residues that can accomplish these objectives, starting with the indicated sequence in the cloned gene in the accompanying diagram. Show where cleavage sites are added and lost. Additional cleavage sites that might be useful are as follows: *Ava*I, C↓PyCGPuG; *Bal*I, TGG↓CCA; *Xma*I, C↓CCGGG; *Hae*II, PuGCGC↓Py; *Hin*cIII, GTPy↓PuAC. (Py and Pu are pyrimidine and purine, respectively.)

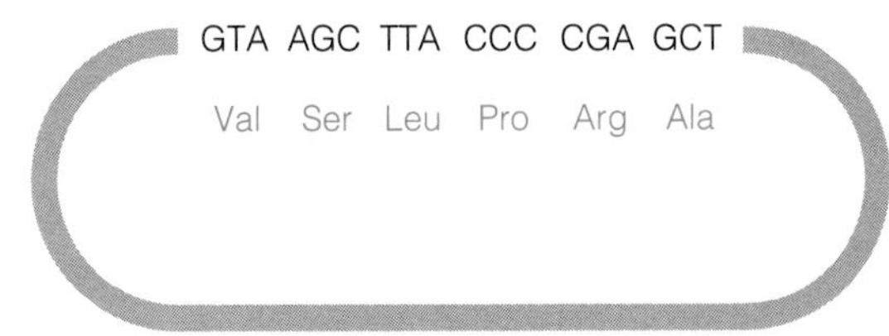

15. In site-directed mutagenesis using oligonucleotide primers, the proportion of mutant clones recovered usually falls well below the 50% that is expected. Why might this be so?

16. The restriction enzymes *Msp*I and *Hpa*II both cleave DNA within the sequence CCGG. *Msp*I cleaves DNA whether or not the second C is methylated, but *Hpa*II cleaves only at unmethylated sites. Using this information in conjunction with the Southern blotting technique, describe how you could identify sites within a particular gene that contain 5-methylcytosine.

17. Analysis of p53 gene mutations in human tumors shows that a large proportion of these mutations involve GC→AT transitions originating at sites of DNA methylation. Propose a model to explain preferential mutagenesis of this type at these sites.

18. A paper in the *Journal of Bacteriology* reported that a *mutT* mutant strain of *E. coli* displayed a mutator phenotype when the bacteria were cultured anaerobically. Is this a surprising result? Briefly explain your answer. Present one or two possible explanations for this observation.

TOOLS OF BIOCHEMISTRY 25A

Genome Mapping

One of the earliest applications of restriction enzymes, in the early 1970s, was in the generation of physical maps of DNA molecules. An aim of molecular biology is to determine the size and location of each gene in the genome of an organism. Until about 1970 all mapping involved determination of genetic linkage between markers, as assessed by measurements of recombination frequency. The first attempts to correlate genetic map position with physical location on a chromosome involved rather cumbersome electron microscopic techniques.

Restriction fragmentation is a far easier way to generate a physical map—in this case a map of cleavage sites for a type II restriction nuclease. Referring to Figure 25.6, you can see that cleavage of λ DNA with one enzyme, such as *Eco*RI, gives the lengths of all six fragments in the digest but gives no information about the order of these fragments in the intact phage genome. Comparable information comes from analysis of the *Bam*HI sites. However, if DNA is digested with both *Eco*RI and *Bam*HI, this double-digest pattern gives much more information about the placement of sites for both enzymes. From analysis of three fragmentation patterns, a researcher can often obtain maps for two restriction sites, as you can see if you work Problems 9 and 11 of Chapter 25. If the data are difficult to interpret—for example, because a digest contains two or more different fragments that have the same size—the needed information can be obtained from a partial digest, that is, by running the restriction cleavage reactions only partly to completion. Another way to simplify the interpretation of mapping data is to label the 5′ ends with γ-[^{32}P]ATP and polynucleotide kinase and then carry out partial digestion, followed by electrophoresis and radioautography of the labeled fragments (Figure 25A.1). The radioautogram gives a ladder of fragments, comparable to that seen in a Maxam–Gilbert sequencing gel (page 1025), which allows ordering of sites from the 5′ end.

By methods such as these, the restriction maps of many small DNAs—from viruses, organelles, and plasmids—were generated in the late 1970s and early 1980s. Genomes up to about 200 kbp, or kilobase pairs (2×10^5 nucleotide pairs) could be mapped. However, the smallest cellular genomes are about 20 times that size, and the difficulty of mapping a genome increases disproportionately to increases in the size of the genome, because of the gigantic number of restriction fragments that must be analyzed. Clearly, new technologies were needed to handle the enormous amounts of data generated by the mapping effort.

In 1987 two laboratories described physical maps of the *E. coli* genome, which contains 4.7 million base pairs; this effort was essential to the complete sequence determination of the genome,

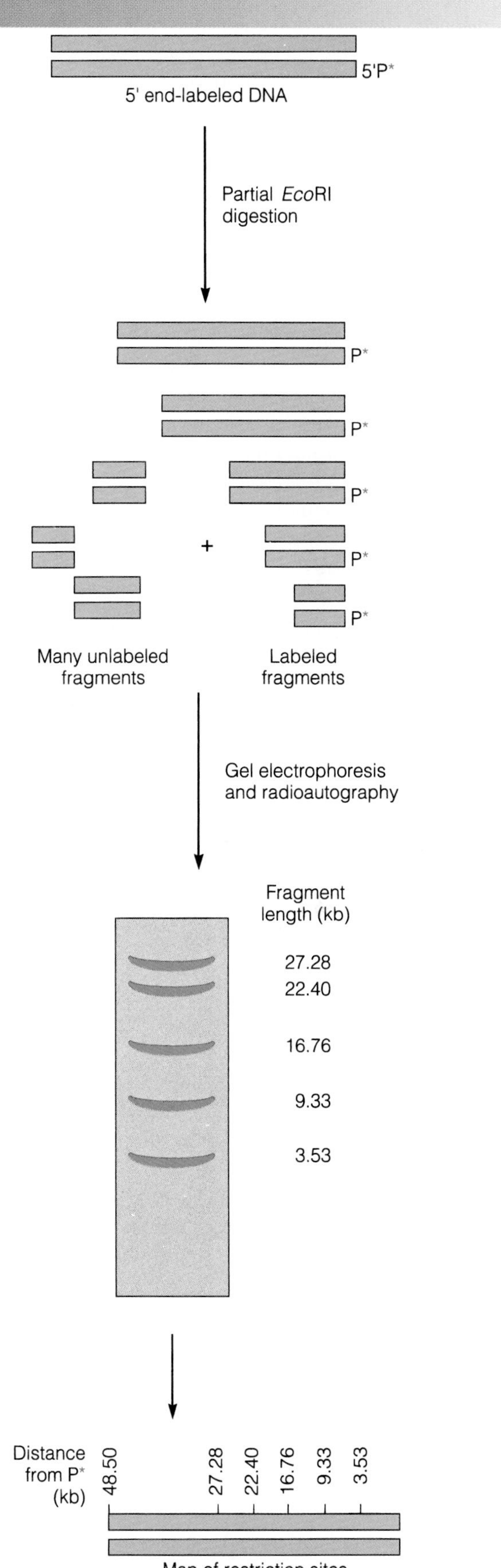

FIGURE **25A.1**
Mapping the *Eco*RI sites in λ DNA by radioautography of 5′ end-labeled fragments in a partial *Eco*RI digest.

completed a decade later. One laboratory cloned and separately mapped 3400 restriction fragments resulting from cleavage of *E. coli* DNA by restriction nucleases that recognize six-base cutting sites. The other laboratory used a restriction nuclease, ***Not*I,** which recognizes an eight-base cutting sequence and, hence, generates fewer (but larger) fragments. These fragments were separated by **pulsed field gel electrophoresis,** a technique that involves periodic changes in the orientation of the electrical field. This technique allows resolution of DNA fragments the size of whole chromosomes—up to 3000 kbp in length.

Now consider the effort that must be expended in mapping the human genome—a genome containing 2.9 billion base pairs of DNA in the haploid state, several hundred times larger and more complex than the *E. coli* genome. That effort, the Human Genome Project, is now well under way. Several techniques are being used to clone and analyze large DNA fragments, particularly the use of **yeast artificial chromosomes (YACs).** YACs are constructed by recombinant DNA techniques from a yeast centromere, two telomere sequences (chromosome ends), selectable markers, and cloned DNA in the megabase range (1000 kbp). Most of the gene mapping data from the Human Genome Project actually result from large-scale direct DNA sequence analysis.

Much of the mapping effort in the Human Genome Project has led to isolation of genes whose defective function is responsible for specific diseases (see Tools of Biochemistry 25D). At the smaller-scale level of genome mapping, much has been learned from analysis of human mitochondrial DNA. Mitochondrial genes are maternally inherited, because few mitochondria from the sperm cell enter the fertilized egg. By comparing mitochondrial DNA restriction maps of 147 individuals of varying origins, Allan Wilson concluded that all mitochondrial lineages are derived from one woman, who lived in Africa about 200,000 years ago. That conclusion has generated intense controversy, which has continued past the untimely death of Wilson. Regardless of the outcome of this controversy, the fact that relatively simple data can lead to any generalizations about human origins indicates the power of restriction mapping techniques.

References

Drlica, K. A. (1994) *Double-Edged Sword. The Promises and Risks of the Genetic Revolution.* Helix Books, Reading, Mass. A readable account of the technical basis and social implications of the new human biology.

Maniatis, T., E. F. Fritsch, and J. Sambrook (1989) *Molecular Cloning: A Laboratory Manual,* 2nd ed. Cold Spring Harbor Laboratory, Cold Spring Harbor, N.Y. The definitive laboratory handbook of molecular biological methods.

Stoneking, M., and H. Soodyall (1996) Human evolution and the mitochondrial genome. *Curr. Opin. Genet. Dev.* 6:731–736. An update on this controversial topic from one of Allan Wilson's early collaborators.

Watson, J. D., M. Gilman, J. Witkowski, and M. Zoller (1992) *Recombinant DNA,* 2nd ed. Scientific American Books, New York. Chapters 29 and 30 discuss large genome mapping and the Human Genome Project.

TOOLS OF BIOCHEMISTRY 25B

Gene Cloning

In classical biology a **clone** is a population of organisms that are genetically homogeneous because they were derived from a single ancestor. For example, all of the bacterial cells in one colony represent a clone, because they were derived from a single cell that was deposited at that location on a Petri plate. In 1973 Herbert Boyer and Stanley Cohen realized that they could use restriction enzymes to "clone" a gene in comparable fashion. They constructed in vitro a novel DNA molecule consisting of a restriction fragment containing a gene of interest, linked to another DNA molecule, such as a plasmid, capable of directing its own replication. This molecule could then be propagated biologically by introduction into a bacterial cell. As the bacterium grows, this **recombinant DNA molecule,** so called because it is formed by end-to-end joining of two different DNAs, replicates along with its host cell. In an early application of the technique, a recombinant plasmid containing the gene for tryptophan synthase was introduced into bacteria lacking this enzyme, whereupon the transformed bacteria acquired the ability to grow in the absence of tryptophan.

Boyer and Cohen realized that, because of the symmetry in type II restriction sites, enzymes giving staggered or offset cuts generated cohesive ends that could be joined covalently by DNA ligase, just as occurs in λ phage DNA replication (see Chapter 24). DNA molecules from any sources are capable of being joined, so long as they have the same cohesive ends (Figure 25B.1). After the DNA ligase reaction in vitro, the mixture of DNAs is introduced into bacteria that have been rendered permeable either by treatment with calcium chloride followed by heat shock or by a pulse of electric current. About 0.01% of the bacteria can undergo genetic transformation—the uptake of DNA and its replication in the recipient cell. This process is illustrated in Figure 25B.2.

Successful cloning of a gene requires several elements. First, one needs a DNA fragment thought to contain the gene of inter-

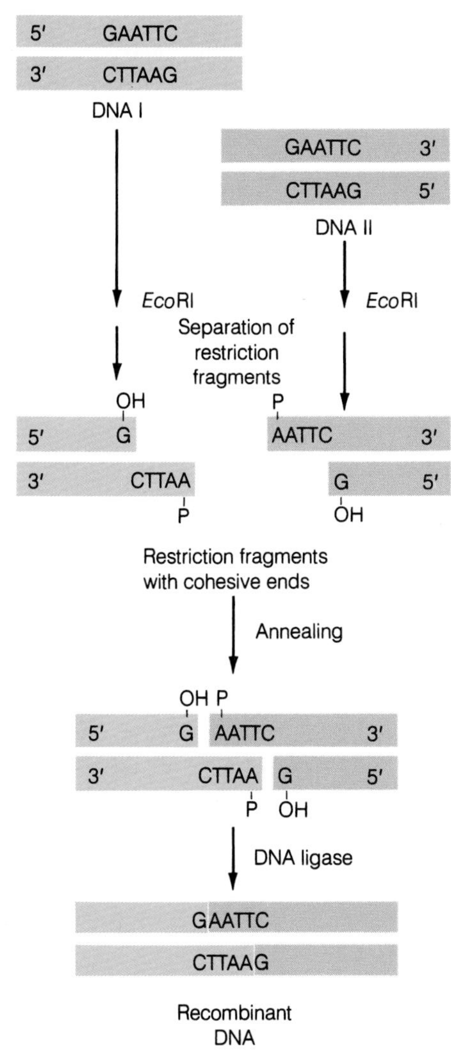

FIGURE **25B.1**
Creation of recombinant DNA molecules in vitro.

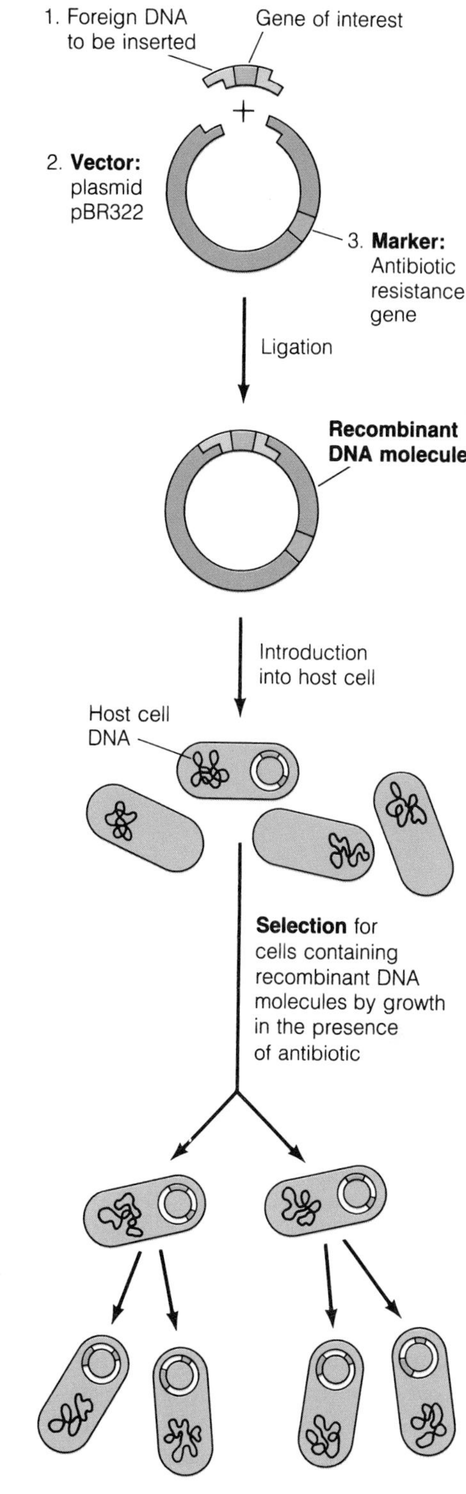

FIGURE **25B.2**
Cloning a fragment of DNA into a plasmid vector and introducing the recombinant molecule into bacteria.

est. Usually this is a restriction fragment, but PCR is often used to create a fragment containing the gene. Cohesive ends can also be generated by synthesis. An enzyme called **terminal deoxynucleotidyltransferase** adds nucleotides to 3′ DNA termini, without a need for a template. Thus, one can put a poly(dC) "tail" on one DNA and a poly(dG) "tail" on the other, and these tails are complementary to each other, just like the cohesive ends created by restriction nucleases. Also, techniques are now available for the end-to-end joining of blunt-ended DNA fragments and also for cloning PCR products, which often have single-strand ends just one nucleotide long. DNA copies (cDNA), created by reverse transcriptase action on messenger RNA, are often used as substrates for gene cloning.

Second, one needs a **vector,** or DNA molecule into which the target gene will be inserted. A vector can be any DNA that contains an origin of replication and can replicate after its entry into a suitable cell (usually a bacterium, although genes are also cloned into eukaryotic cells). Plasmids and bacteriophage chromosomes are most often used as vectors, because they can replicate independently of the host cell into which they are introduced. This allows one to amplify the recombinant DNA; that is, one can

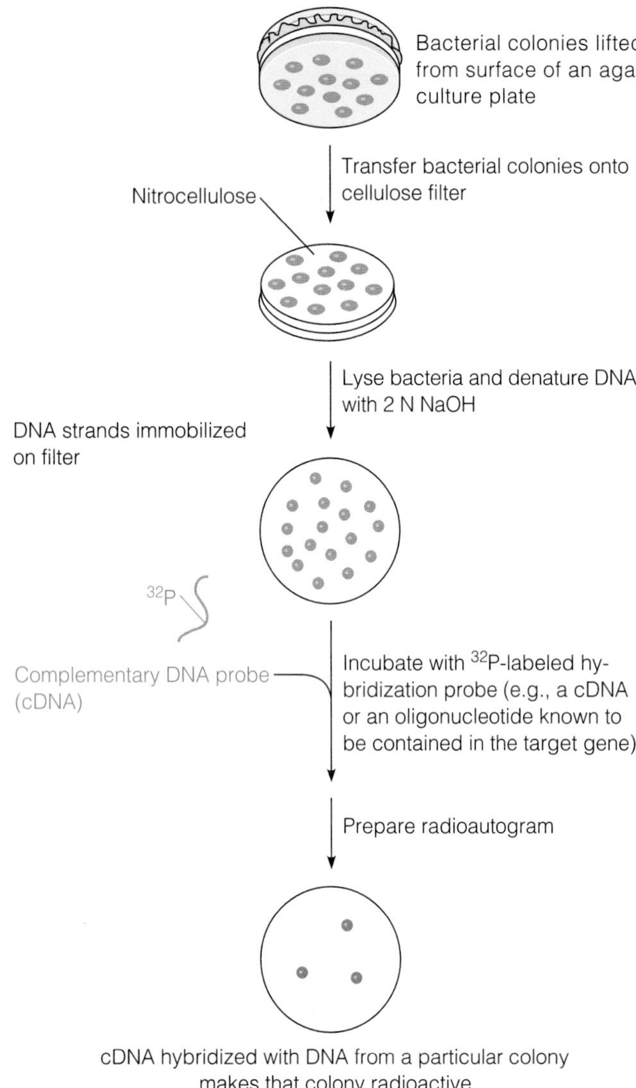

FIGURE **25B.3**

Colony hybridization. A radioactive complementary DNA or oligonucleotide probe is used to identify bacteria containing cloned sequences homologous to the probe.

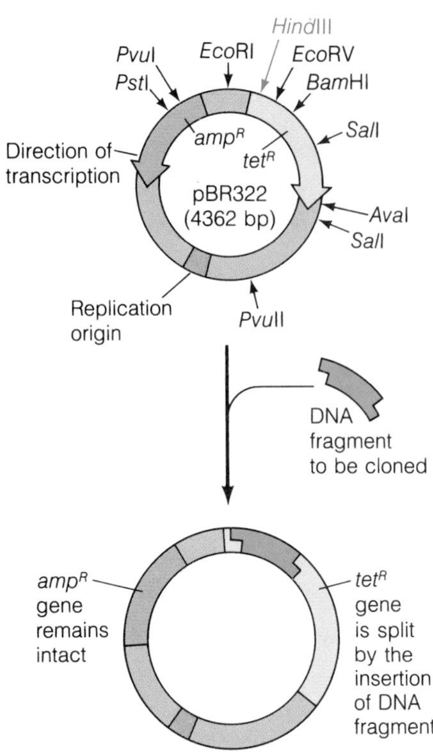

FIGURE **25B.4**

pBR322, one of the earliest plasmid cloning vectors. Some of the restriction sites are shown, as well as the direction of transcription of the ampicillin and tetracycline resistance genes. The bottom diagram shows the effect of cloning a novel sequence into the *Hin*dIII site.

replicate it to a much higher copy number than that of the chromosomal DNA.

Third, because ligation and transformation both occur with low efficiency, one needs a suitable screening technique—a method to identify bacteria that contain the cloned gene, in the presence of a huge excess of cells that do not. If one expects that the cloned gene will be expressed, one can select conditions in which the cloned gene confers a selectable phenotype on the recipient bacterium, such as resistance to a particular drug or loss of a nutritional requirement. Reaction with an antibody against the product of the cloned gene is another way to screen for desired clones among a large number of transformants. However, if one doesn't know whether the cloned gene will be expressed, one can search for the desired *DNA sequences* instead of the product of the cloned gene. In a technique called **colony hybridization,** one uses a labeled nucleic acid probe homologous to the desired sequence to search among individual bacterial colonies for the presence of that DNA sequence (Figure 25B.3). The hybridization probe can be either a restriction fragment of natural DNA or a chemically synthesized oligonucleotide known to contain a sequence represented in the desired clone. Alternatively, if the DNA has been cloned into a phage vector instead of a plasmid, one can carry out **plaque hybridization,** in which individual phage plaques are screened in much the same way.

Screening for desired recombinants is aided greatly by the design of vectors with suitable restriction sites and selectable markers. pBR322 is a popular plasmid vector that is itself a recombinant DNA molecule produced in vitro (Figure 25B.4). pBR322 is a circular DNA molecule containing a replication origin from a naturally occurring plasmid, ColE1, plus genes conferring resistance to the antibiotics ampicillin (amp^R) and tetracycline (tet^R). Some restriction sites lie within either the amp^R or the tet^R genes, the genes conferring resistance to ampicillin and tetracycline, respectively. A researcher may "clone into" the *Hin*dIII site, for example, by opening pBR322 with *Hin*dIII and ligating DNA to the resulting DNA ends, thereby creating a recombinant in which the cloned gene forms an insert that splits, and hence inactivates, the tet^R gene. This technique aids greatly in screening for clones. All bacteria that are transformed, whether with the original vector or with a recombinant DNA, become ampicillin-resistant. However, only the bacteria that contain a recombinant plasmid are also tetracycline-sensitive. Transformation with the unaltered plasmid confers resistance to both antibiotics. Therefore, individual transformant colonies can be picked from ampicillin plates and replated in the presence of tetracycline. Cells not growing on these plates probably contain recombinant plasmids.

Since the introduction of pBR322, hundreds of different cloning vectors have been described. Particularly useful are **expression vectors,** in which control regions affecting transcription and translation have been introduced, to allow high-level expression of cloned genes, for production of the gene products. We shall see numerous examples of the utility of gene cloning as a research tool. However, the excitement generated by recombinant DNA technology derived in large part from its potential for attacking practical problems. The earliest such example was the cloning and expression in *E. coli* of a gene encoding human insulin. This was accomplished in 1977, and by 1982 the purified recombinant product was available for use in treating diabetes. Other recombinant products include blood clotting factors, clot dissolving enzymes used to treat heart attack victims, other polypeptide hormones (including pituitary growth hormone), and interferons.

A useful adjunct to expression-vector cloning is the addition of "tags" to recombinant proteins, to facilitate their purification after expression. A popular application of this technique is the "histidine tag." An oligonucleotide containing six histidine codons is spliced adjacent to the insert in an expression vector, such that the expressed recombinant protein has a sequence of six histidine residues, linked at either the C- or N-terminus. The imidazole ring in histidine causes the hexahistidine sequence to bind tightly to an affinity column containing nickel ion coordinated with a covalently bound chelator, nitriloacetate (as shown in the following figure). Thus, when a crude extract containing the His-tagged recombinant protein is applied to such a column, the recombinant protein binds tightly and specifically and can later be eluted from the column with an imidazole buffer. Often this allows complete purification of a recombinant protein with just this single fractionation step. The tag can be removed afterward, if desired, by designing the oligonucleotide to contain both the six His codons and a cleavage site for a specific protease.

Much interest is focused on genetic engineering of plants to endow them with desirable properties, such as resistance to cold, to dehydration, to viruses, or to insect pests. Nature has provided an excellent vehicle for introducing novel genes into plants. *Agrobacterium tumefaciens* is a pathogenic bacterium that causes a disease called crown gall, through transfer of some of its own DNA into infected plants. The disease-causing genes are carried on a plasmid called the **Ti plasmid.** Foreign genes can be introduced into the plasmid by recombinant DNA methodology, and infection of plants by the resultant modified *Agrobacterium* can transfer these novel genes into plants. Current efforts focus on controlling the expression of these genes after infection and directing their integration into the cellular genome, so that the desirable characteristics become permanently incorporated into the genome.

Comparable gene transfer techniques exist for animals as well. In 1982, Ralph Brinster and Richard Palmiter introduced, by microinjection, a cloned gene for rat growth hormone into the male pronucleus of fertilized mouse eggs. The gene was cloned downstream from an inducible promoter to increase its chances of being expressed. After insertion of the microinjected eggs into the uterus of a foster mother mouse, a significant number of offspring expressed the gene, produced high levels of rat growth hormone, and grew to about twice the size of normal mice. Although the early success of this experiment with **transgenic** animals was perhaps fortuitous, more recent developments have greatly increased our ability to achieve homologous recombination of DNA after transfer into mammalian cells. In fact, these techniques are currently being tested for treatment of a small number of human genetic diseases, including the immunodeficiency associated with a lack of adenosine deaminase (see pages 805–806).

References

Barinaga, M. (1994) Knockout mice: Round two. *Science* 265:26–28. A news article detailing recent success in using targeted gene transfer to abolish selected gene functions in mice and to study the biological consequences.

Watson, J. D., M. Gilman, J. Witkowski, and M. Zoller (1992) *Recombinant DNA,* 2nd ed. Scientific American Books, New York. Chapters 5 and 7 deal primarily with gene cloning techniques, but the whole book discusses uses of recombinant DNA technology.

Wu, R., L. Grossman, and K. Moldave, eds. (1989) *Recombinant DNA Methodology.* Academic Press, San Diego, Calif. Selected articles from separate volumes of *Methods in Enzymology.*

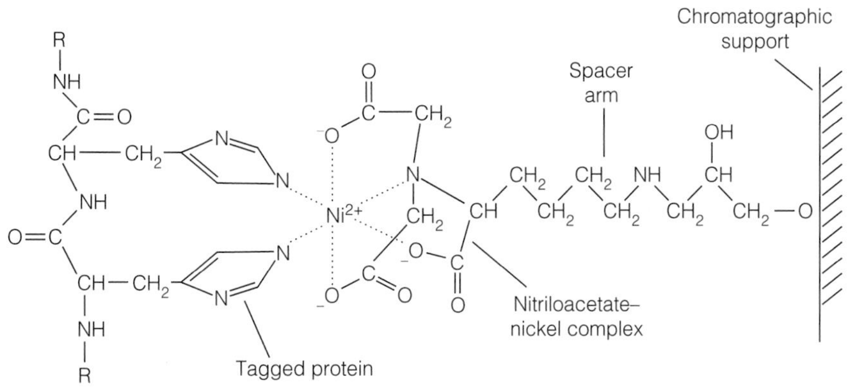

TOOLS OF BIOCHEMISTRY 25C

Gene Sequencing with Dideoxynucleotides

In Chapter 4, we mentioned the pioneering Maxam–Gilbert DNA sequencing technique, which involves base-specific cleavages of a 5′ end–labeled DNA molecule to give a series of fragments of varying lengths, which can be analyzed by gel electrophoresis. In 1976 Fred Sanger introduced an alternative sequencing methodology, which is now in almost universal use because it can be applied to longer fragments and is more rapid and amenable to automation.

Sanger sequencing is similar in principle to Maxam–Gilbert sequencing in that it generates a set of fragments with a common 5′ origin and base-specific 3′ termini. However, those 3′ termini are created not by base-specific cleavage starting from a full-length molecule but by base-specific *interruption* of enzymatic synthesis of the molecule in vitro, by incorporation of nucleotide analogs that serve as chain terminators. The principle of the method is shown in Figure 25C.1. First, one usually clones the fragment to be sequenced into a vector that allows it to be isolated as single-strand DNA (although similar methods work for double-strand DNAs). Bacteriophage M13 is most commonly used as a vector. One can isolate the double-strand replicative form of this virus and clone the target sequence into it, just as one clones into a plasmid such as pBR322. After introduction of the double-strand DNA into a bacterium, the cell produces virus particles, which contain single-strand DNA. The M13 genome has been genetically engineered for this type of cloning—first, to contain suitable restriction sites for cloning and, second, to allow fast identification of plaques containing phage that have acquired a DNA insert in the cloning process.

Isolated single-strand circular DNA from phage containing the desired insert becomes the template for four DNA polymerase-catalyzed reactions. The primer is an oligonucleotide that is complementary to an M13 sequence lying just 3′ to the insert. Extension of this primer by DNA polymerase will copy the insert. These polymerase reactions are run in the presence of deoxyribonucleoside triphosphate analogs, the **2′,3′-dideoxyribonucleoside triphosphates,** which serve as terminators of chain extension because they lack 3′ hydroxyl termini. The dideoxy analog (ddATP) of adenosine triphosphate is shown here.

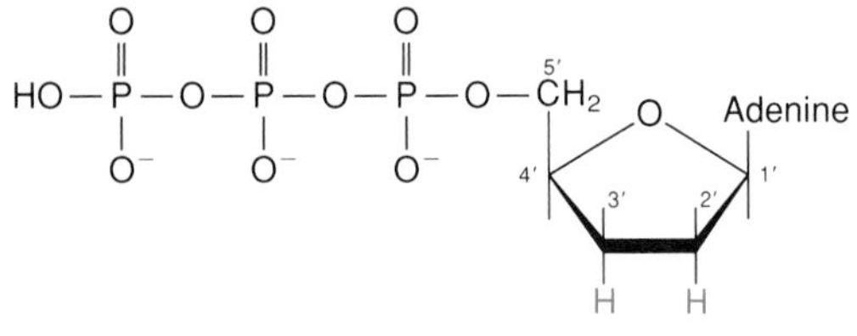

2′,3′-Dideoxyadenosine triphosphate

To generate a series of A-terminated fragments, one runs the DNA polymerase reaction in the presence of equivalent concentrations of dATP, dCTP, dGTP, and dTTP, plus one-tenth equivalent of ddATP. When T is in the template strand, DNA polymerase occasionally inserts ddAMP instead of dAMP. When that happens, chain elongation stops and the fragment is released from the enzyme. Thus, a series of fragments of varying lengths accumulates, with each fragment identifying a T in the nucleotide sequence of the insert. Similarly, one identifies sites terminated by C, G, or T simply by running comparable polymerase reactions with the other three dideoxy analogs. Inclusion of a radioactive label in the polymerization mixture and gel electrophoresis followed by radioautography generate four sequencing ladders, with each ladder absolutely base specific.

More recently introduced technology allows automated sequence determination by the Sanger method. The sequencing primer is derivatized at its 5′ end with a fluorescent dye that is red, blue, green, or yellow. Each of the four sequencing reactions is run with a differently colored primer, which imparts a characteristic fluorescence to all fragments terminated by A, T, G, or C, respectively. This allows both for sequence determination without radioisotopes and for computer-assisted reading of the sequence gel and processing of the data. Other recently introduced methods allow direct sequence determination with double-strand DNAs, without the necessity for cloning them into a single-strand DNA vector such as M13.

Application of these and other high-throughput sequencing modifications had led, by the late 1990s, to complete sequence determination of several bacterial genomes. The genome sequence of the yeast *Saccharomyces cerevisiae* has also been reported, and the complete genome sequence of a multicellular organism, the nematode *Caenorhabditis elegans,* was reported in late 1998. Scientists predict that the entire human genome will have been sequenced within the first two or three years of the twenty-first century. Already, the gene mapping and sequencing effort involved in the Human Genome Project has identified many so-called disease genes—human genes in which defined mutations are responsible for specific disease states. These mapping techniques involve nucleic acid hybridization analysis, which is described in Tools of Biochemistry 25D.

References

Blattner, F. R., and 16 coauthors (1997) The complete genome se-quence of *Escherichia coli* K12. *Science* 277:1453–1462. One of the earliest complete sequence determinations for a bacterial genome.

Harris, E. E., and J. Hey (1999) X chromosome evidence for ancient human histories. *Proc. Natl. Acad. Sci. USA* 96:3320–3324. Sequence analysis of nuclear DNA yields different conclusions from those obtained from mitochondrial DNA analysis.

Karlin, S., A. M. Campbell, and J. Mrazek (1998) Comparative DNA analysis across genomes. *Annu. Rev. Genet.* 32:185–225. Concepts and methods for comparative analysis of complete genomes; a primer of genomics.

Pandey, A., and F. Lewitter (1999) Nucleotide sequence databases: A gold mine for biologists. *Trends Biochem. Sci.* 24:276–280. An excellent introduction to the major databases and how to use them.

Venter, J. C., M. D. Adams, G. G. Sutton, A. R. Kerlavage, H. O. Smith, and M. Hunkapiller (1998) Shotgun sequencing of the human genome. *Science* 280:1540–1542. This short article outlines the high-throughput mapping, sequencing, and data-handling techniques proposed for rapid completion of the Human Genome Project.

FIGURE **25C.1**

Cloning into M13 and sequencing by the Sanger method.

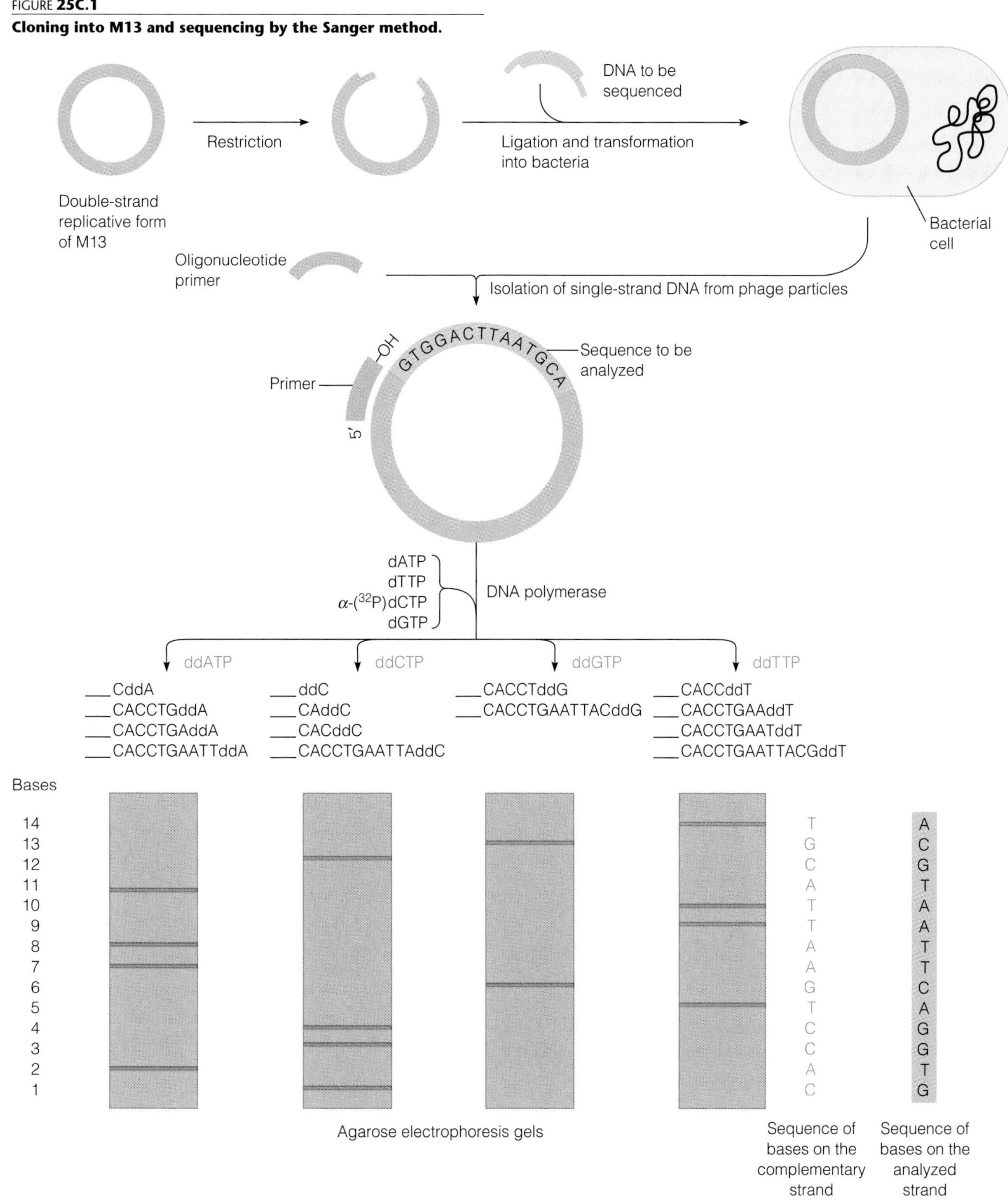

TOOLS OF BIOCHEMISTRY 25D

Southern Blotting

An important application of restriction nucleases is the detection and mapping of specific sequences in complex genomes. For example, you learned in Chapter 24 that λ phage DNA becomes integrated into the chromosome of the host bacterium during a lysogenic infection and that tumor viruses also insert their genomes into chromosomes of infected animal cells. How do we know this? One way is by looking specifically for viral sequences in restriction digests of chromosomal DNA from infected cells.

Earlier we described the physical mapping of viral or plasmid DNAs by restriction enzyme cleavage, followed by electrophoretic separation of the fragments and their visualization with ethidium bromide. Any cellular genome, even from a bacterium, is so complex that such analysis would yield a "smear" consisting of a very large number of fragments, each present in minute amounts. However, because of the power of radiolabeling and nucleic acid reannealing techniques, we can detect specific DNA sequences within such a digest, even though they may be present at such low levels as one copy per genome or even one-millionth of the total DNA present on a gel.

The method is to radiolabel a sequence of interest to high specific activity and then use this labeled "probe" to search for complementary sequences among the fragments in a restriction digest after electrophoresis. This is done by denaturing both the probe and the target restriction digest, reannealing, and then searching for radioactive duplex molecules. The search is difficult to accomplish with DNA fragments embedded in an agarose gel, because of the fragility of the gel. However, in 1975 E. M. Southern devised a way to transfer the fragments from agarose to a sheet of nitrocellulose, following alkali denaturation of the fragments. As shown in Figure 25D.1, the transfer is achieved by capillary action. The liquid associated with the agarose gel, in which the DNA fragments are dissolved, is literally blotted out, and the single-strand DNA fragments in the denatured digest become irreversibly bound to the nitrocellulose sheet. For this reason, the technique is called **Southern blotting** or, more descriptively, **Southern transfer.** Alternatively, the transfer can be achieved electrophoretically.

The nitrocellulose sheet, now containing a replica of the original DNA gel, is incubated under annealing conditions with the radiolabeled probe, and DNA heteroduplexes are detected by radioautography after thorough washing of the nitrocellulose. A "prewash" treatment is also carried out to reduce nonspecific binding of the hybridization probe. One of the earliest applications of Southern blotting showed that tumor viral genomes become integrated into chromosomal DNA in virally transformed cells. Viral sequences were detected in restriction digests of DNA from transformed cells, sometimes in fragments longer than the viral genome itself.

A variation of this technique, inevitably called **northern blotting,** allows the detection of rare RNA molecules in a cell with similar sensitivity. Here one prepares total cell RNA and resolves different size classes electrophoretically, with subsequent transfer and probing with radiolabeled DNA, followed by radioautographic detection of DNA–RNA hybrid duplexes. The transfer technology is slightly different, involving either covalent linkage of the RNA to diazobenzyloxymethyl (DBM) cellulose or transfer of formaldehyde-treated RNA to nitrocellulose.

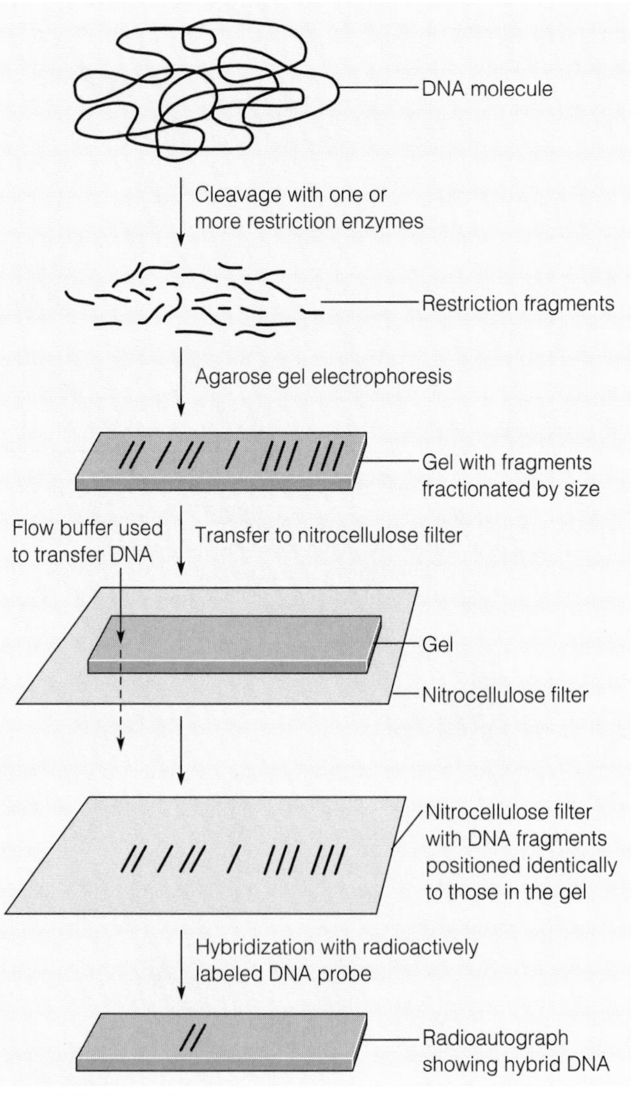

FIGURE **25D.1**
The principle of Southern blotting.

From J. D. Watson et al., *Molecular Biology of the Gene,* 4th ed. (Redwood City, Calif.: Benjamin/Cummings, 1987). Copyright © 1987 J. D. Watson.

Of the many applications of Southern transfer and hybridization, we will mention a few here. First, Southern transfer is an invaluable adjunct to cloning, when one knows the amino acid sequence of at least part of the protein product of the target gene. From the sequence, one designs oligonucleotides that are expected to be homologous to part of the sequence. These oligonucleotides are then chemically synthesized, radiolabeled, and used to screen a **library**, or large random collection of cloned DNA fragments from the organism of interest. The screening involves colony or plaque hybridization, as described in Tools of Biochemistry 25B. Cloned segments homologous to the hybridization probe are then analyzed to see which ones contain a full-length copy of the target gene. Because this technique starts with a protein and works "backward" to isolate the structural gene for that protein, it is called **reverse genetics.**

Second, Southern blotting can also be used to identify methylated sites in particular genes. Particularly useful are the restriction nucleases *Msp*I and *Hpa*II, both of which recognize and cleave within the same sequence (see Table 25.2). However, *Hpa*II requires that a C within that site be methylated, whereas *Msp*I cleaves only DNA unmethylated at that site. Therefore, any methylated sites within a sequence analyzed with a particular probe will be cleaved by the former, but not the latter, enzyme.

A third application of Southern transfer is the use of **restriction fragment length polymorphisms (RFLPs)** to map the genes responsible for inherited diseases in humans. Many mutations either create a new restriction site in a genome or destroy a previously existing site. If such a mutation occurs within or near a particular gene, the restriction fragmentation pattern will change, and this pattern can be determined by Southern blotting and probing with a sequence that lies in the same region of the genome. For example, some cases of sickle-cell disease can be detected by an altered restriction pattern when human DNA is probed with part of the gene for the β globin subunit. The sensitivity of the technique allows prenatal diagnosis of sickle-cell disease, using cells cultured from amniotic fluid surrounding the fetus. More recently, the technique has been used to map and isolate genes for hereditary illnesses for which the responsible genes have not yet been identified, including cystic fibrosis, Huntington's disease, amyotrophic lateral sclerosis ("Lou Gehrig's disease"), and, most recently, colon cancer and breast cancer.

Detection methods using nonradioactive probes are expected to find extensive applications in clinical diagnosis of infectious diseases, such as acquired immune deficiency syndrome, which is caused by a retrovirus. The earliest developed AIDS test involves screening for antibody against a viral protein. However, it may take 4 months after infection for such an antibody to accumulate to detectable levels in the blood. Therefore, current efforts are focused on the development of rapid hybridization tests for the genome of the infecting virus, which should be detectable much earlier after the initial infection.

These techniques are also finding use in criminal investigations. By use of different restriction enzymes and hybridization probes, one can generate a pattern of RFLPs that is almost unique for an individual—in short, a DNA "fingerprint." By use of the polymerase chain reaction, it is now possible to isolate and analyze the DNA in a tissue sample as small as one human hair or drop of semen. Thus, a suspect leaving a hair or similarly small amount of tissue at the scene of a crime can be identified from the DNA fingerprint—not with absolute certainty but with more certainty than with a manual fingerprint. Perhaps more important, DNA fingerprinting has been used to definitively exonerate criminal suspects, including some who have spent years in prison. In late 1998 this technology was applied to an issue of historical importance, when PCR and hybridization analysis of the DNA taken from descendants of Thomas Jefferson and his slave, Sally Hemings, established with near certainty that our third president or a close relative had fathered at least one child with the slave.

Finally, hybridization analysis with fluorescent DNA or RNA probes is at the heart of DNA microchip technology, a technique that generates enormous amounts of information about patterns of gene expression. The technique involves automated synthesis of hundreds or thousands of different oligonucleotides, each at a defined position on a silicon chip. The coupling reactions in synthesizing these oligonucleotides are light-activated, and each cycle of activation is carried out through a "mask," which allows illumination only of sites where a specific base is to be added in a particular cycle of oligonucleotide synthesis. After one round of addition, a different mask is placed over the chip, allowing addition of the next nucleotide only to a desired set of the growing oligonucleotide chains on the chip. By these means, 32 such cycles can create, on a single 1-inch-square chip, a precisely positioned array of 65,000 different eight-mers—oligonucleotides eight bases in length, each capable of serving as a hybridization probe. An alternative approach uses ink-jet printer technology. In either case, the total RNA from a cell or organism can be analyzed first, by using reverse transcriptase and dNTPs modified with fluorescent groups to synthesize cDNAs representative of the RNA population. Next, these are allowed to hybridize with the oligonucleotides on the chip. Formation of a DNA–oligonucleotide hybrid generates a fluorescent spot at a defined site on the chip, with the color or brightness of the fluorescence indicating the relative abundance of the cDNA bound. Figure 25.D2 shows the kind of data that can be obtained. To the extent that the experimenter knows the nucleotide sequences of the genes of interest, he or she can analyze the data to learn about the degree of expression of a large number of genes in response, for example, to hormone secretion, developmental changes, or environmental pressures.

References

Gerhold, D., T. Rushmore, and C. T. Caskey (1999) DNA chips: Promising toys have become powerful tools. *Trends Biochem. Sci.* 24:168–173. The title says it all.

Morell, V. (1993) Huntington's gene finally found. *Science* 260:28–29. A news article describing a triumph of human molecular genetics.

Rommens, J. M., et al. (1989) Identification of the cystic fibrosis gene: Chromosome walking and jumping. *Science* 245:1059–1079. A detailed account of another such triumph.

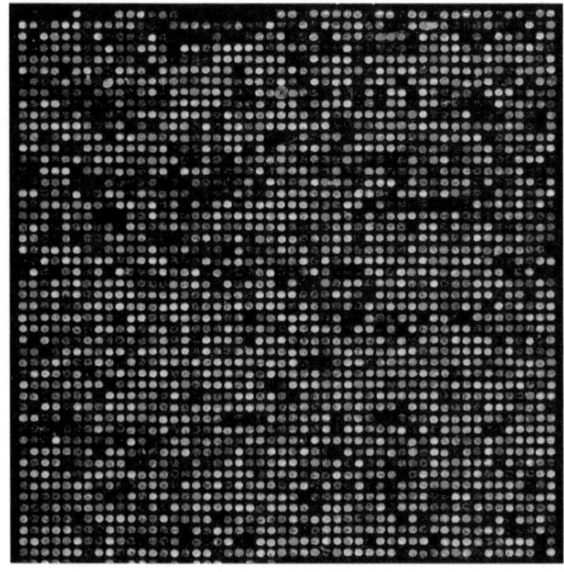

Service, R. F. (1998) Microchip arrays put DNA on the spot. *Science* 282:396–399. A news article describing applications and promise of this new technology.

Zurer, P. (1994) DNA profiling fast becoming accepted tool for identification. *Chem. Eng. News* 72:8–15. A clearly written news article describing forensic applications of RFLP analysis, including its use in sensational criminal cases.

FIGURE **25D.2**

A DNA microchip array. In this experiment, using the MICROMAX detection system, 2 μg of human mRNA was annealed with immobilized cDNAs representing 2400 human genes. Spot intensities and colors indicate the abundance of particular gene-specific mRNAs.

Courtesy of Incyte Pharmaceuticals, Inc., Palo Alto, CA, from R. F. Service, *Science* (1998) 282:396–399, with permission from *Science*.

TOOLS OF BIOCHEMISTRY 25E

Site-Directed Mutagenesis

Analysis of the function of a protein involves altering the structure of the protein and then determining whether and how the biological functions of the protein are altered. Two methods have been used classically for generating modified proteins. One is to alter certain residues chemically by treatment with protein-modifying reagents. This approach lacks specificity, because all residues of a given amino acid may be modified, not just the one or two of special interest. Another approach is to mutagenize an organism with ultraviolet light, ionizing radiation, or chemical mutagens and then select for surviving organisms containing mutations that affect the protein of interest. The mutations can be identified by sequence analysis of the mutant gene or its protein product. The problem with this approach is the inability to target mutations to a given region of the gene, typically that part specifying the catalytic site of an enzyme molecule or specifying a regulatory region involved in DNA–protein interactions.

Once it became possible to clone the gene encoding a protein of interest, one could systematically alter the gene at specific sites to generate virtually any desired mutation, a technique known as **site-directed mutagenesis.** Introduction of the mutant gene into a host cell, followed by its expression, could then yield the altered protein for study of the altered function. The earliest experiments involved the generation of deletions. For example, partial digestion of a cloned DNA with a restriction nuclease, followed by ligation, could yield mutant genes in which the region between two cleavage sites was deleted. Shorter deletions could be generated by opening at one cleavage site, followed by treatment with an exonuclease that would digest back from the two ends created by the cleavage. Such an enzyme is *Bal*31 exonuclease, which digests both strands starting from a cut site. The ends can be trimmed with S1 nuclease (an enzyme that cleaves only single-strand DNA), followed by ligation under conditions that permit blunt DNA ends to join. Often an oligonucleotide that contains a restriction site (for example, *Bam*HI "linkers") is introduced at this stage. This approach can generate deletions of various lengths simply by variation of the period of exonuclease treatment.

The most powerful and widely used method for site-directed mutagenesis allows the introduction of practically any mutation at any site, including single-base substitutions, short deletions, or insertions. The approach, illustrated in Figure 25E.1, requires that the gene first be cloned into a single-strand vector, such as phage M13. Next, one chemically synthesizes an oligonucleotide, about 20 nucleotides long, that is complementary in sequence to the cloned gene at the site of the desired mutation, *except in the center of the sequence.* Here the sequence contains one or two deliberate mistakes—either single nucleotides that do not base-pair with the template or insertions or gaps of a few nucleotides. On annealing to the cloned gene, these alterations generate either non-Watson–Crick base pairs or bases that have no partners and therefore form a "looping out." The correctly matched bases on both sides of the mismatch cause it to remain annealed, despite the mismatch. One then uses DNA polymerase to synthesize around

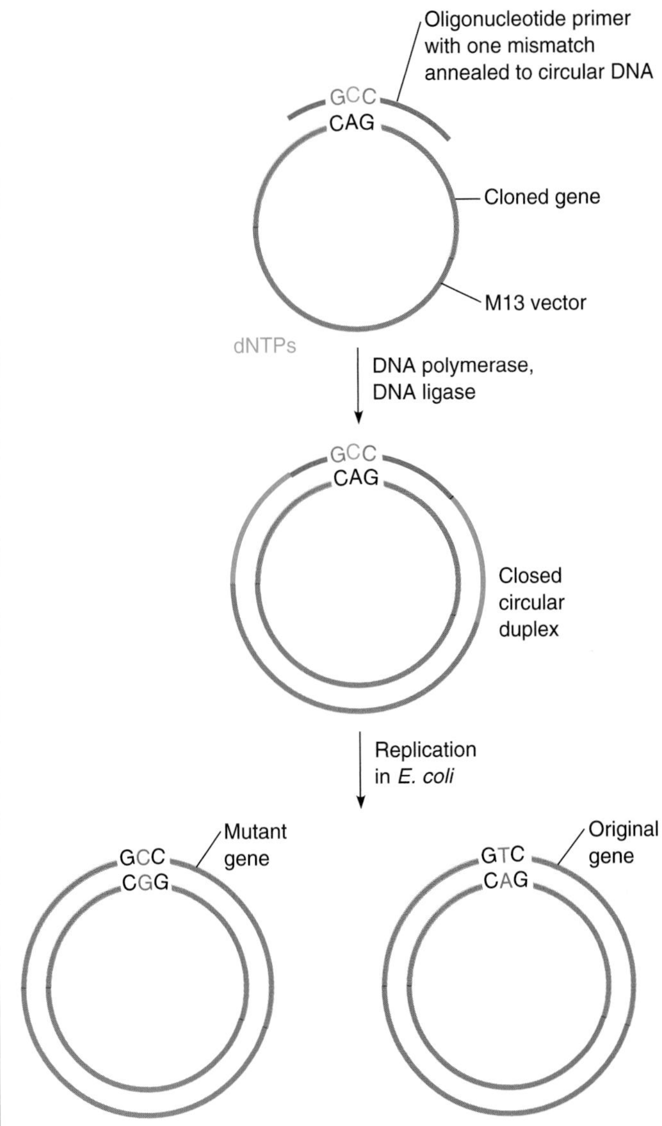

FIGURE **25E.1**

Use of a mismatched synthetic oligonucleotide primer to introduce mutations into a gene cloned into a single-strand vector.

the circular vector from this primer, followed by ligation to create a closed circular duplex. After introduction of this duplex into bacteria, both strands replicate and yield phage. In principle, 50% of the phage should contain the desired mutation. In practice, that fraction is considerably less, but it can be increased by various techniques. One method utilizes a template containing extensive substitution of uracil for thymine. This substitution is accomplished by growing the M13 phage clone in an *E. coli* strain lacking dUTPase and uracil-DNA *N*-glycosylase (see Chapter 24). Then, after DNA synthesis in vitro, the DNA is introduced into wild-type bacteria, in which the action of uracil-DNA *N*-glycosylase selectively cleaves the template strand and leaves the desired mutant strand intact. If one wants to produce a number of mutants with different amino acid substitutions at the same site, one can use a mixture of oligonucleotide primers, differing from each other only within the codon to be altered.

Reference

Kunkel, T. A., J. D. Roberts, and R. A. Zakour (1989) Rapid and efficient site-specific mutagenesis without phenotypic selection. In: *Recombinant DNA Methodology,* edited by R. Wu, L. Grossman, and K. Moldave, pp. 587–601. Academic Press, San Diego, Calif. Laboratory instructions for the most widely used method of site-directed mutagenesis.

CHAPTER 26

Information Readout: Transcription

WE TURN NOW TO *TRANSCRIPTION*, IN WHICH INFORMATION STORED IN THE nucleotide sequence of DNA is read out, by the template-dependent synthesis of polyribonucleotides. Mechanistically, transcription is similar to DNA replication, particularly in the use of nucleoside triphosphate substrates and the template-directed growth of nucleic acid chains in a $5' \rightarrow 3'$ direction. Two major differences are as follows: (1) With few known exceptions, only one DNA template strand is transcribed for a particular gene, and (2) only a small fraction of the entire genetic potential of an organism is realized in one cell. In a differentiated eukaryotic cell, very little of the total DNA is transcribed. Even in single-celled organisms, in which virtually all of the DNA sequences can be transcribed, far fewer than half of all genes may be transcribed at any time. Therefore, much of our concern with transcription involves the mechanisms used to select particular genes and template strands for transcription, because this selection in large part governs the metabolic capabilities of a cell. Those mechanisms operate largely at the levels of initiation and termination of transcription, through the actions of proteins that contact DNA in a highly site-specific manner (Figure 26.1).

Our earliest insights into transcription developed from experiments with bacteria and their viruses. Our knowledge of prokaryotic transcription is, therefore, deeper than our comprehension of the process in more complex multicellular or-

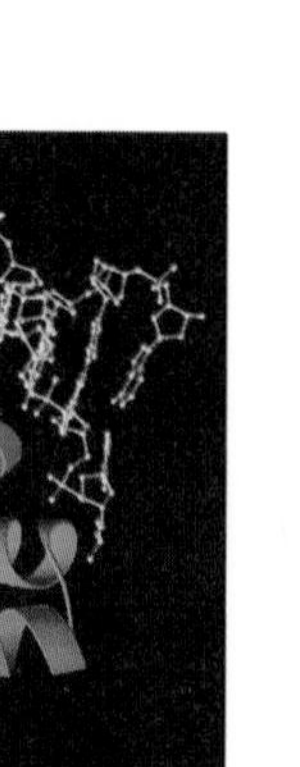

FIGURE **26.1**

DNA–protein complexes that regulate transcription. Gene expression is controlled by proteins that recognize particular DNA sequences and bind at those sites. Shown here are two regulatory proteins of bacteriophage λ, which bind to the same *operator* sites in λ DNA and control the viral reproductive cycle by regulating transcription of genes adjacent to those sites. Above, the λ cI repressor; below, the Cro protein. In each graphic, DNA is in gold, and the protein is in green, except for the α-helical regions in contact with DNA in the major groove, which are in red. These proteins and their functions are described on pages 1003–1010.

Courtesy of R. A. Albright and B. W. Matthews, *Proc. Natl. Acad. Sci. USA* (1998) 95:3431–3436, with permission of PNAS.

ganisms. Consequently, this chapter focuses on well-understood prokaryotic systems, emphasizing where they have served as models for understanding eukaryotic transcription. The latter topic is covered more thoroughly in Chapter 28.

DNA as the Template for RNA Synthesis

The concept that RNAs are generated by template-directed copying of DNA base sequences is so firmly ingrained that we tend to forget about the critical experiments that led to our current understanding. This topic merits discussion, because it involves a fascinating intellectual history, with many participants and some brilliant deductive reasoning. Even some of the false starts yielded important contributions. Also, the experimental systems used in the early work are some of those that continue to be useful, such as the lactose operon and bacteriophages T4 and λ.

The involvement of RNA in information transfer was suspected from the time DNA was first identified as the genetic storehouse—both from the chemical similarity of RNA to DNA and from the knowledge that proteins are synthesized on ribosomes. The latter fact meant that, in eukaryotic cells, information must somehow be transferred from the nucleus, where information is stored in DNA, to the cytosol, where most of the ribosomes reside. However, the nature of that information transfer was not clear until messenger RNA was shown to exist. This demonstration was far from straightforward, because messenger RNA is quite unstable. Moreover, it constitutes such a small proportion of total cellular RNA (1% to 3% in bacteria) that in ordinary fractionations its presence is masked by the much more abundant ribosomal and transfer RNAs.

THE PREDICTED EXISTENCE OF MESSENGER RNA

Our original concepts of RNA synthesis came from genetic studies that predicted the existence of messenger RNA.

Until about 1960 it was thought that ribosomal RNA (rRNA) represented the set of templates for protein synthesis. François Jacob and Jacques Monod, at France's Pasteur Institute, questioned this idea, in part because rRNAs are homogeneous in size (5S, 16S, and 23S in bacteria, as discussed in Chapter 27), whereas the molecular weights of proteins vary over at least two orders of magnitude. From analyzing *E. coli* mutants that are altered in the control of lactose metabolism, Jacob and Monod predicted the existence of messenger RNA (mRNA), an RNA species that is synthesized from a DNA template and is used in turn as the template for protein synthesis.

Lactose utilization in *E. coli* was known to be controlled by three enzymes, whose genes are adjacent on the chromosome. One of these is **β-galactosidase,** which hydrolyzes lactose and other β-galactosides. When bacteria are grown with glucose as the sole carbon source, the levels of the lactose-utilizing enzymes are very low, with less than one molecule of β-galactosidase per cell on average. However, substitution of lactose or a related β-galactoside for glucose in the medium leads to rapid enzyme **induction,** or synthesis of the three enzymes. β-Galactosidase ultimately represents as much as 6% of the total soluble protein of the cell. Removal of lactose from the culture slows the further synthesis of enzyme molecules. The rapid changes in β-galactosidase–forming capacity suggested that the template for synthesizing this enzyme is metabolically unstable—synthesized rapidly on demand and degraded when a continued stimulus to induction is absent (alternatively, control could be exercised at the translational level). Because rRNAs are quite stable, these species were unlikely to be intermediates in information transfer.

Jacob and Monod analyzed numerous *E. coli* mutants that displayed faulty control over induction of the lactose-utilizing enzymes. Some expressed all three genes at high levels even when lactose or a similar inducer was absent, and others could not induce any of the enzymes, even after addition of lactose. These experiments led to the notion of a macromolecular *repressor* that regulated the level of a

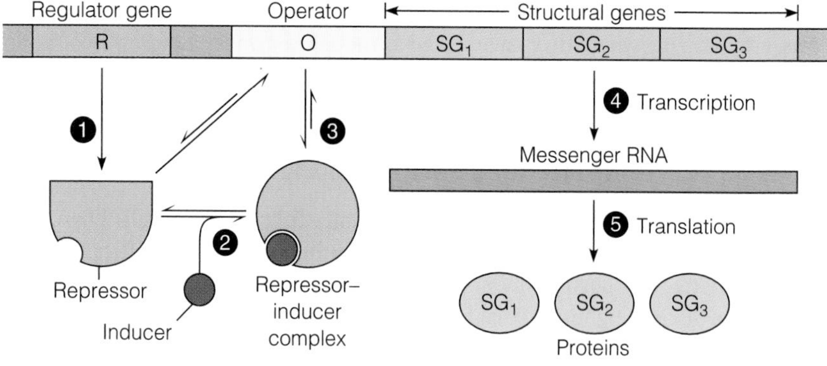

FIGURE **26.2**

The operon model, as proposed in 1961 by Jacob and Monod. Step 1: The regulator gene R encodes a repressor molecule, which can bind to the operator (O) and thereby inhibit transcription of the adjacent structural genes $SG_{1,2,3}$. **Step 2:** A small-molecule inducer complexes with the repressor, thereby altering the equilibrium between the conformational states of the repressor. **Step 3:** The repressor–inducer complex binds less tightly to the operator. **Step 4:** This loosening facilitates transcription of the structural genes, resulting in production of messenger RNA—that is, an RNA copy of the structural genes. **Step 5:** The mRNA sequence is translated into proteins.

messenger RNA by binding to a specific *operator* sequence on DNA and thereby turning off the synthesis of this RNA (see pages 998–1000).

Based on these studies, and on parallel work by Andre Lwoff with bacteriophage λ, Jacob and Monod in 1961 proposed a unifying hypothesis of gene regulation, in which transcription was regulated specifically at the level of initiation. Hypothetical regulatory elements called repressors and operators controlled the synthesis of other hypothetical entities called messenger RNAs (mRNAs). mRNA was postulated to be a complementary copy of the DNA that encompassed a set of structural genes, which encode proteins, as schematized in Figure 26.2. A set of contiguous genes plus adjacent regulatory elements that control their expression was termed an **operon,** and the Jacob–Monod hypothesis thus came to be known as the **operon model.**

Jacob and Monod correctly predicted several characteristics of the hypothetical messenger RNA. First, they predicted a high rate of mRNA synthesis followed by rapid degradation, which would explain the fast turn-on of the genes after induction and turn-off after removal of the inducer. Second, because of rapid synthesis and degradation, they expected mRNA to accumulate rapidly but not to high steady state levels. Third, because they thought that the messenger was a copy of two or more contiguous genes, they expected it to be fairly large and part of a heterogeneous size class of RNA. Finally, if the messenger RNA was a complementary copy of DNA, its nucleotide sequence should be identical to that of one of the template DNA strands.

Bacterial genetics predicted messenger RNA to be a collection of metabolically active RNAs that are present in low abundance, heterogeneous in size, and complementary in sequence to DNA.

T2 BACTERIOPHAGE AND THE DEMONSTRATION OF MESSENGER RNA

The first physical demonstration of mRNA came from work with T2 and T4 bacteriophages. Infection by these large DNA viruses arrests all expression of host cell genes, and no accumulation of RNA can be detected after infection. However, in 1956 the use of radioisotopes led to detection of a distinctive RNA in T2-infected *E. coli.* When infected cultures were pulse-labeled with [^{32}P]orthophosphate for 3 or 4 minutes, about 2% of the total RNA became radioactive. This radiolabeled RNA had two properties that led to its eventual identification as viral messenger RNA. First, it was metabolically labile; in pulse-chase experiments (see Tools of Biochemistry 12A), radioactivity was rapidly lost from this RNA. Second, the RNA seemed to be a product of viral DNA metabolism, because its nucleotide composition was close to that of T2 DNA; the RNA was rich in adenine and uracil and low in guanine and cytosine.

The ability of T2 phage RNA to hybridize with T2 DNA and to associate with ribosomes made before infection was evidence for the existence of messenger RNA.

Important additional evidence came from sucrose gradient centrifugation of pulse-labeled RNA, which showed that the labeled material sediments heterogeneously and distinctly from any of the known rRNA or tRNA species (Figure 26.3a). Benjamin Hall and Sol Spiegelman established that this RNA is a viral gene product when they carried out the first DNA–RNA hybridization experiment, which showed that the labeled RNA was complementary in sequence to phage DNA.

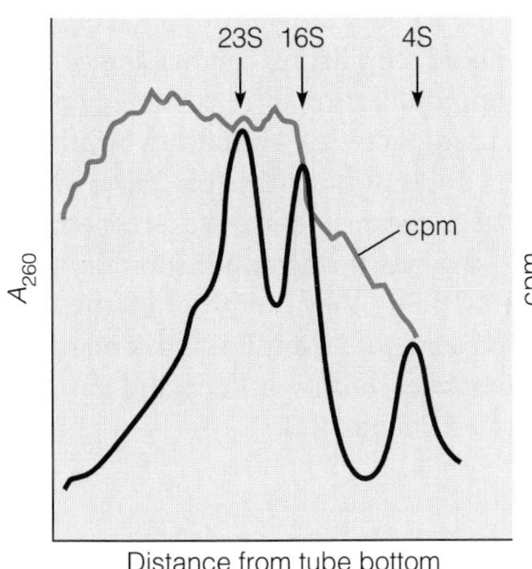

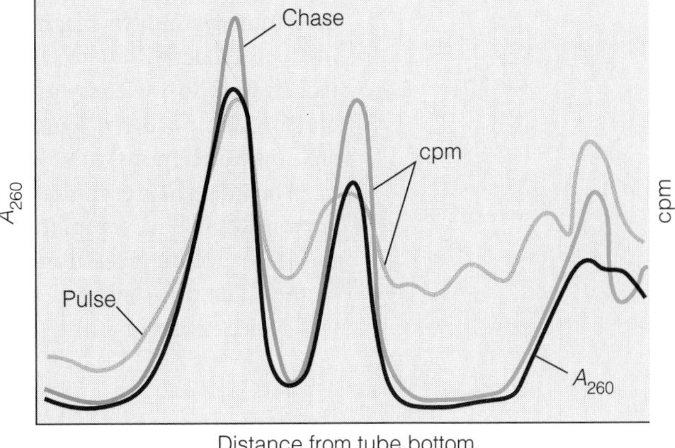

FIGURE **26.3**

Demonstration of mRNA by pulse-labeling and sedimentation. (a) Sedimentation profile of total and pulse-labeled RNAs in T2 phage–infected *E. coli*. Total concentration of RNA in each fraction (black) was determined by ultraviolet absorbance (A_{260}). The radioactivity profile (red) shows the distribution of species synthesized during the pulse. **(b)** Pulse-labeled RNA species in uninfected bacteria and their fate in a chase. The orange line represents cells labeled for 3 minutes, the blue line depicts an identical culture with the label chased by 0.7 generation of growth in nonradioactive medium after the pulse, and the black line is the A_{260} profile.

(a) Data from K. Asano, *J. Mol. Biol.* (1965) 14:71–84; **(b)** data from M. Hayashi and S. Spiegelman, *Proc. Natl. Acad. Sci. USA* (1961) 47:1569.

The earliest experiments were based on the fact that RNA is of higher density than DNA. Thus, a DNA–RNA hybrid could be detected in equilibrium gradient centrifugation as a species of intermediate density containing label derived from both DNA and RNA. Such a hybrid was formed when T2 RNA was heated and slowly cooled along with T2 DNA, but not when the DNA came from *E. coli.*

Additional support for the existence of phage messenger RNA came from a density shift experiment carried out by François Jacob, Matthew Meselson, and Sydney Brenner. *E. coli* was grown in dense medium (^{13}C–^{15}N) and infected with T2 in light medium. After pulse labeling with radioactive amino acids, density analysis on CsCl gradients showed that phage proteins were synthesized on dense ribosomes, that is, ribosomes synthesized before infection. The fact that only phage proteins were being made ruled out any possibility that ribosomal RNA provided a template. Instead, this experiment pictured the ribosome as a nonspecific workbench, upon which any protein could be assembled, depending on which messenger template became associated with that workbench.

RNA DYNAMICS IN UNINFECTED CELLS

The experiments described in the previous section demonstrated the existence of mRNA in phage-infected *E. coli.* What about uninfected bacteria? Spiegelman and his colleagues showed that pulse-labeled RNA from uninfected *E. coli* hybridized to *E. coli* DNA. At very short labeling intervals the sedimentation pattern showed incorporation into both rRNA and tRNA species *and* a heterogeneously sedimenting species. After a chase (Figure 26.3b), the radioactivity profile followed the absorbance profile, showing that all RNA species had become labeled to equivalent specific activities. This finding is consistent with the postulated short lifetime of messenger RNA ($t_{1/2}$ of 2 or 3 minutes). mRNA would reach its maximal radioactivity within just a few minutes, but during the chase, mRNA turnover would release nucleotides that could flow into stable RNA species. Because those species do not turn over, label accumulates, and the fraction of total label in the stable RNA species continues to increase. Consistent with this idea, Spiegelman also showed that highly labeled ribosomal and transfer RNAs hybridize to *E. coli* DNA, demonstrating that all three major classes of RNA are synthesized from template DNA strands.

As noted previously, the earliest DNA–RNA hybridization experiments involved equilibrium gradient centrifugation, a time-consuming and expensive technique. Spiegelman and his colleagues made the important discovery that single-strand DNA binds irreversibly to membrane filters, such as nitrocellulose. This technique allowed rapid hybridization analysis of a large number of samples, because a radiolabeled RNA could hybridize to denatured DNA immobilized on a filter. After suitable treatment and washing of the filter, the amount of hybridization could be determined simply by placing the filter in a liquid scintillation counter and counting its radioactivity. The same principle—immobilization of nucleic acid on nitrocellulose followed by analysis of bound radioactivity—underlies Southern blotting and northern blotting (described in Tools of Biochemistry 25D), which are now much more widely used to analyze gene organization and expression.

The final chapter in the messenger RNA saga was the demonstration that isolated mRNAs have template activity in vitro. That is, they can program the synthesis of specific protein molecules in the presence of 20 amino acids and other factors. The translation of defined RNA templates, both synthetic and natural, was critical to deciphering the genetic code (see Chapter 27).

Enzymology of RNA Synthesis: RNA Polymerase

Polynucleotide phosphorylase catalyzes the reversible, template-independent synthesis of random-sequence polyribonucleotides.

We now know that RNA synthesis involves the copying of a template DNA strand by RNA polymerase. However, the earliest description of an enzyme capable of synthesizing RNA in vitro occurred in the late 1950s, at about the same time as the discovery of DNA polymerase I. This RNA-synthesizing enzyme, called **polynucleotide phosphorylase,** was quite different from DNA polymerase. The enzyme required no template, and it used ribonucleoside *diphosphates* (rNDPs) as substrates to produce a random-sequence polynucleotide whose base composition matched the nucleotide composition of the reaction medium.

$$n\ \mathrm{rNDP} \rightleftharpoons (\mathrm{rNMP})_n + n\mathrm{P_i}$$

Initially it was thought that polynucleotide phosphorylase might be the major RNA-synthesizing enzyme, but the lack of a template requirement was troubling, as was the apparent absence of the enzyme in eukaryotic cells. Ultimately, polynucleotide phosphorylase turned out not to play a role in RNA synthesis in vivo but instead to participate in the *degradation* of bacterial messenger RNAs. However, the enzyme was of great value in the synthesis of polynucleotides used as templates for in vitro protein synthesis, when the genetic code was being elucidated (see Chapter 27).

Investigators continued to search for an enzyme that would copy a DNA template in vitro. Such an enzyme was discovered almost simultaneously in four different laboratories. The enzyme, **DNA-directed RNA polymerase,** resembled DNA polymerases in the nature of the reaction catalyzed.

$$n(\mathrm{ATP} + \mathrm{CTP} + \mathrm{GTP} + \mathrm{UTP}) \xrightleftharpoons{\mathrm{Mg^{2+}},\ \mathrm{DNA}} (\mathrm{AMP{-}CMP{-}GMP{-}UMP})_n + n\mathrm{PP_i}$$

The reaction product is a complementary copy of the DNA template.

BIOLOGICAL ROLE OF RNA POLYMERASE

A single RNA polymerase catalyzes the synthesis of all three *E. coli* RNA classes—mRNA, rRNA, and tRNA. This was shown in experiments with **rifampicin** (Figure 26.4a), an antibiotic that inhibits RNA polymerase in vitro and blocks the synthesis of mRNA, rRNA, and tRNA in vivo. Rifampicin-resistant mutants of *E. coli* were found both to contain a rifampicin-resistant form of RNA polymerase and to be capable of synthesizing all three RNA classes in vivo in the presence of rifampicin. Because a single mutation affects both the RNA polymerase and the synthesis of all RNA types in vivo, RNA polymerase must be the one enzyme catalyzing all forms of transcription in bacteria.

Bacteria use a single RNA polymerase to transcribe all genes. Eukaryotic cells use three different RNA polymerases.

In contrast, eukaryotes contain three distinct RNA polymerases, one each for synthesis of rRNA, mRNA, and small RNAs (tRNA plus the 5S species of rRNA)—RNA polymerases I, II, and III, respectively. The existence of separate enzymes

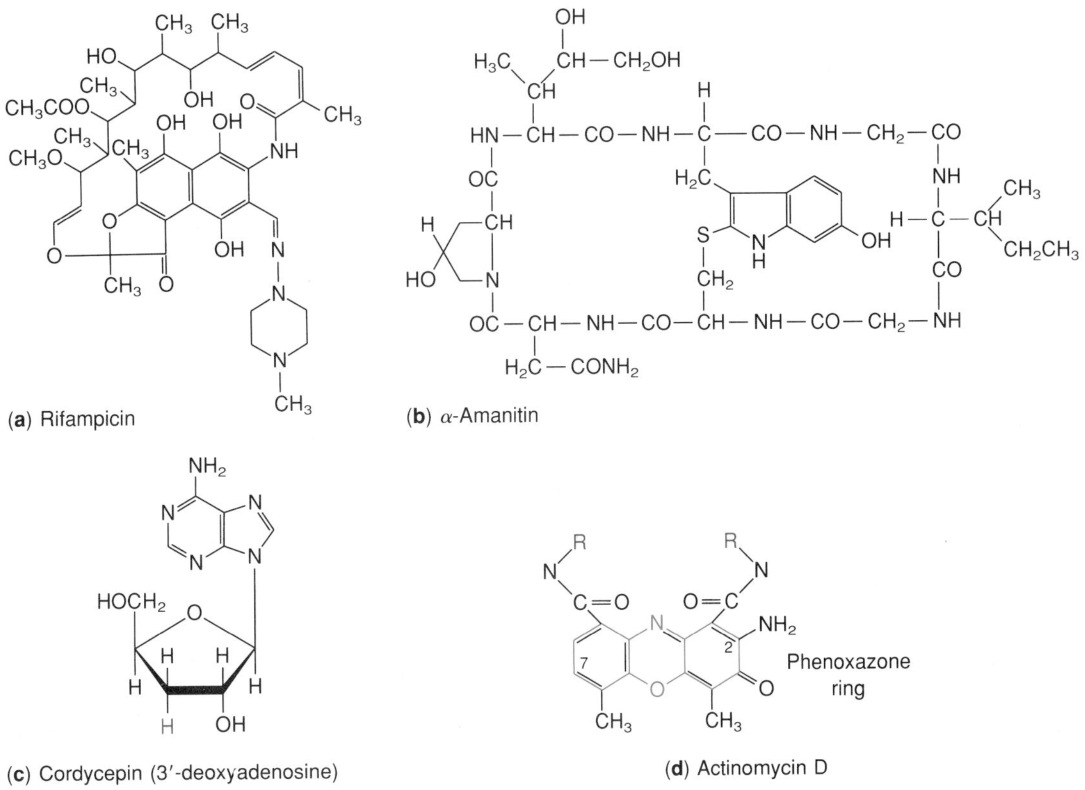

(**a**) Rifampicin

(**b**) α-Amanitin

(**c**) Cordycepin (3′-deoxyadenosine)

(**d**) Actinomycin D

FIGURE **26.4**

Some inhibitors of transcription. Rifampicin is an inhibitor of bacterial transcription initiation, and α-amanitin is an inhibitor of eukaryotic RNA polymerases. Cordycepin is a transcription terminator because the 3′ position on the sugar is occupied by H (red), rather than a hydroxyl group. The tricyclic ring system of actinomycin (blue) intercalates between adjacent G-C base pairs in DNA; the R groups of the molecule (red) are cyclic polypeptides and fill in the narrow grooves of the helix.

was revealed partly because they differ in their sensitivity to inhibition by **α-amanitin** (Figure 26.4b), a toxin from the poisonous *Amanita* mushroom. RNA polymerase II is inhibited at low concentrations, RNA polymerase III is inhibited at high concentrations, and RNA polymerase I is quite resistant.

Figure 26.4 shows the structures of two additional inhibitors. **Cordycepin,** or 3′-deoxyadenosine, is a transcription chain terminator because it lacks a 3′ hydroxyl group from which to extend. The nucleotide of cordycepin is incorporated into growing chains, confirming that transcriptional chain growth occurs in a 5′→3′ direction. Another important inhibitor is **actinomycin D,** which acts by binding to DNA. The tricyclic ring system (phenoxazone) intercalates between adjacent G-C base pairs, and the cyclic polypeptide arms fill the nearby narrow groove.

Because DNA polymerases and RNA polymerases catalyze similar reactions, it is interesting to compare some of their kinetic features. V_{max} for DNA polymerase III holoenzyme, at about 500 to 1000 nucleotides per second, is much higher than the chain growth rate for bacterial transcription—50 nucleotides per second, which is the same as V_{max} for purified RNA polymerase. Although there are only about 10 molecules of DNA polymerase III per *E. coli* cell, there are some 3000 molecules of RNA polymerase, of which half might be involved in transcription at any instant. This fits in with observations that replicative DNA chain growth is rapid but occurs at few sites, whereas transcription is much slower but occurs at many sites. The result is that far more RNA accumulates in the cell than DNA. Like the DNA polymerase III holoenzyme, the action of RNA polymerase is highly processive. Once past the initial stages of transcription, RNA polymerase rarely, if ever, dissociates from the template until the specific signal to terminate has been reached. Although this discussion is based primarily upon prokaryotic transcription, these characteristics of transcription are similar in eukaryotic cells.

DNA replication involves rapid chain growth at few intracellular sites, and transcription involves slower growth at many sites. More RNA accumulates than DNA.

Another important difference between DNA and RNA polymerases is the accuracy with which a template is copied. With an error rate of about 10^{-5}, RNA polymerase is far less accurate than replicative DNA polymerase holoenzymes, although RNA polymerase is much more accurate than would be predicted from

Watson–Crick base pairing alone. Given that RNA does not carry information from one cell generation to the next, an ultrahigh-fidelity template-copying mechanism is evidently not needed. However, recent observations suggest the existence of error-correction mechanisms. In *E. coli,* two proteins, called GreA and GreB, promote the hydrolytic cleavage of nucleotides at the 3′ ends of nascent RNA molecules. These processes may be akin to 3′ exonucleolytic proofreading by DNA polymerases, but there are important differences: (1) Cleavage of 3′ ends of RNA molecules usually removes oligonucleotides, rather than single nucleotides, and (2) the rate of hydrolysis is much slower than the rate of RNA chain extension by RNA polymerase. As discussed below, 3′ transcript cleavage probably plays a different role, in escaping blocks to continued elongation. So the mechanism of transcriptional error correction is still an open question.

STRUCTURE OF RNA POLYMERASE

When highly purified *E. coli* RNA polymerase is analyzed in denaturing electrophoretic gels, five distinct polypeptide subunits are observed. Their properties are summarized in Table 26.1. Two copies of the α subunit are present, along with one each of β, β', σ, and ω, giving M_r of about 450,000 for the holoenzyme. ω may be involved in regulation, but its precise role is not yet clear, and its association with RNA polymerase may be adventitious. However, much has been learned about the functions of the other subunits, partly from reconstitution studies, in which the dissociated subunits are renatured and allowed to reassociate, with formation of active enzyme. Because ω is not required for reconstitution of active enzyme, we shall not include it in our subsequent discussions.

The functions of the various RNA polymerase subunits can be determined by reconstitution of active enzyme from isolated subunits.

A useful approach to analyzing the function of each RNA polymerase subunit is **mixed reconstitution,** in which subunits isolated from two or more different RNA polymerases are recombined. For example, when β from a rifampicin-resistant form of RNA polymerase is recombined with α, β', and σ from wild-type cells, the reconstituted enzyme is rifampicin-resistant. This finding establishes β as the target for rifampicin inhibition. Furthermore, rifampicin is known to inhibit the initiation of transcription, so β must also play a role in initiation. The fact that β is also the target for inhibition by an inhibitor of elongation called **streptolydigin** points to β as a subunit directly involved in chain elongation.

The σ subunit is easily dissociated from RNA polymerase—for example, by passing the purified enzyme through a carboxymethylcellulose column. The σ-free enzyme, called **core polymerase,** is still catalytically active, but it binds to DNA at far more sites than does the RNA polymerase holoenzyme, and it shows no strand or sequence specificity. The σ subunit plays an important role in directing RNA polymerase to bind to template at the proper site for initiation—the **promoter** site—and to select the correct strand for transcription. The addition of σ to core polymerase reduces the affinity of the enzyme for *nonpromoter* sites by about 10^4, thereby increasing the enzyme's specificity for binding to promoters.

TABLE **26.1 Subunit composition of *E. coli* RNA polymerase**

Subunit	M_r	Number per Enzyme Molecule	Function
α	36,500	2	Chain initiation, interaction with regulatory proteins and upstream promoter elements
β	151,000	1	Chain initiation and elongation
β'	155,000	1	DNA binding
σ	70,000[a]	1	Promoter recognition
ω	11,000	1	Unknown

[a] The 70-kDa σ subunit is one of several alternative σ subunits.

These discoveries about σ, which were made in the early 1970s, suggested that gene expression might be regulated by having core polymerase interact with different forms of σ, which would in turn direct the holoenzyme to different promoters. In many instances, this does occur. For example, when *Bacillus subtilis* is induced to **sporulate** (that is, to generate metabolically inert cells capable of later outgrowth), new forms of σ are produced. These forms combine with core polymerase to redirect cellular metabolism for transcription of the genes involved in sporulation. Another example is apparent when an *E. coli* culture is stressed by a sudden temperature increase. In these heat-shocked cells a new form of σ appears and directs the modified RNA polymerase to a different set of promoters, thereby activating transcription of a block of genes called heat-shock genes. The most abundant σ in *E. coli,* and the one that will frame our discussions, is called σ^{70} because of its 70-kDa molecular weight.

RNA polymerases from different prokaryotic sources are remarkably similar in subunit size and composition. In many instances, mixed reconstitution can be observed, with active enzyme formed from subunits of the RNA polymerases of widely divergent prokaryotic species. By contrast, eukaryotic RNA polymerases have much more complex and diverse subunit structures (see Chapter 28), although the large subunits are quite homologous to β and β'.

Although the multisubunit motif for RNA polymerases is the dominant structural theme, it is not universal. The best-known exception is RNA polymerase specified by bacteriophage T7. This virus has a linear duplex DNA genome. The left-hand 20% of the chromosome (as usually drawn) is transcribed early in infection by *E. coli* RNA polymerase. One of these early gene products is a virus-specified RNA polymerase. This single-subunit enzyme ($M_r = 98{,}000$) responds to different DNA control sequences and is responsible for all T7 transcription late in infection. These late transcriptional events initiate at promoters on the right-hand 80% of the DNA molecule. Crystallographic analysis of T7 RNA polymerase shows a DNA-binding domain similar to the "hand" structure seen in DNA polymerases (see Figure 24.17), with a palm, fingers, and thumb that wrap about the DNA template (Figure 26.5).

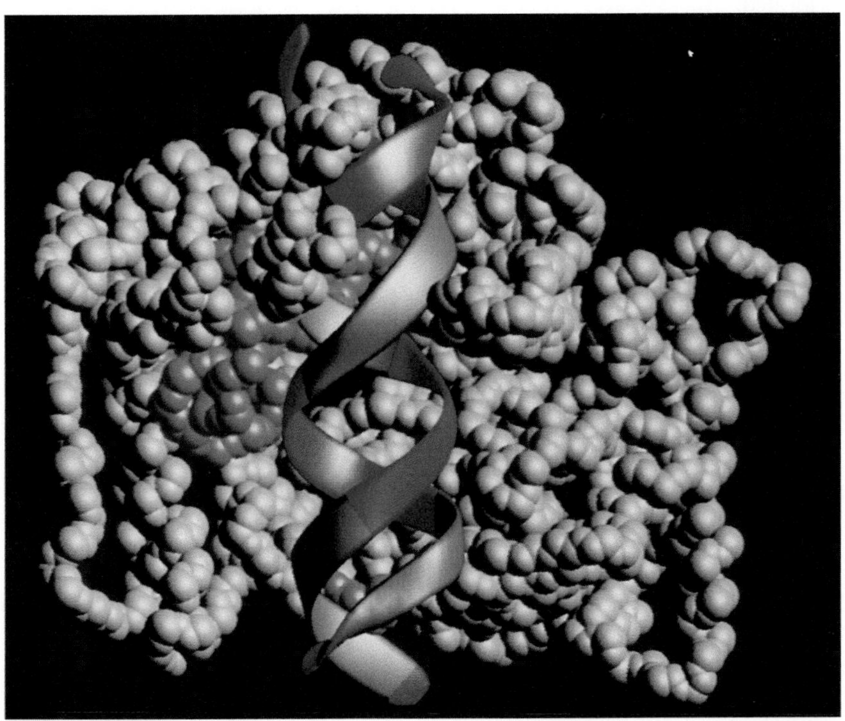

FIGURE **26.5**

Structure of T7 phage RNA polymerase as determined by x-ray crystallography. This space-filling model shows the enzyme–DNA complex. Amino acid residues thought to be involved in catalysis and in binding to the DNA promoter region are shown in red. The thumb is to the left and the fingers to the right in the structure as drawn.

Mechanism of Transcription

Like DNA replication and protein synthesis, transcription occurs in three distinct phases—initiation, elongation, and termination. Initiation and termination signals in the DNA sequence punctuate the genetic message by directing RNA polymerase to particular genes and by specifying where transcription will start, where it will stop, and which strand will be transcribed. The signals involve both instructions encoded in DNA base sequences and interactions between DNA and proteins other than RNA polymerase. Most of our discussion focuses on prokaryotic RNA polymerases, exemplified by the widely studied *E. coli* enzyme, but the basic mechanics of transcription are similar in all organisms.

INITIATION OF TRANSCRIPTION: INTERACTIONS WITH PROMOTERS

The overall process of initiation and elongation is summarized in Figure 26.6. The first step in transcription is binding of RNA polymerase to DNA, followed by migration to an initiation DNA site, the promoter. In Tools of Biochemistry 26A, we describe *footprinting*, a technique used to locate DNA sites in contact with sequence-specific DNA-binding proteins, and we discuss promoter recognition on pages 992–995. For now, let us concentrate on the mechanistic details of synthesizing polyribonucleotide chains.

Bacterial RNA polymerase finds promoters by a search process (Figure 26.6, step 1), in which the holoenzyme binds nonspecifically to DNA, with low affinity, and then moves along the DNA until it reaches a promoter sequence, to which it binds with much higher affinity. σ factor is essential for this search because, as noted earlier, the core enzyme does not bind to promoters more tightly than to nonpromoter sites. Biophysical evidence suggests that RNA polymerase slides along the DNA, without dissociation from it, although the mechanism for this sliding is not yet understood. Kinetic analysis suggests that a random three-dimensional search process, with multiple enzyme–DNA encounters, cannot account for the rapidity with which RNA polymerase finds promoters. Binding to DNA and then moving along it reduce the complexity of the search from three dimensions to one, just as finding a house becomes simpler once you find the street upon which that house is located.

Transcription begins with sequence-specific interaction between RNA polymerase and a promoter site, where duplex unwinding and template strand selection occur.

The initial encounter between RNA polymerase holoenzyme and a promoter generates a **closed-promoter complex** (step 2). Whereas DNA strands unwind later in transcription, no unwinding is detectable in a closed-promoter complex. This complex forms with a K_a between 10^7 and 10^8 M^{-1} at 0.1 M NaCl. Binding is primarily electrostatic, for K_a is dependent on ionic strength. The complex is relatively labile, for it dissociates with a half-life of about 10 seconds. Footprinting studies show that polymerase is in contact with DNA from about nucleotide -55 to -5, where $+1$ represents the first DNA nucleotide to be transcribed.

Next, as shown by analysis with chemical reagents that react specifically with single-stranded DNA, RNA polymerase unwinds several base pairs of DNA, from about -10 to -1, giving an **open-promoter complex,** so-called because it binds DNA whose strands are open, or unwound (step 3). This highly temperature-dependent reaction occurs with half-times of about 15 seconds to 20 minutes, depending upon the structure of the promoter. The open-promoter complex is extremely stable; it is not easily disrupted by high ionic strength, and it forms with a K_a as high as 10^{12} M^{-1}. An Mg^{2+}-dependent isomerization next occurs, giving a modified form of the open-promoter complex with the unwound DNA region now extending from -12 to $+2$. Analysis with scanning force microscopy (Tools of Biochemistry 1A) indicates that DNA bending in the promoter region accompanies the transition from a closed-promoter to an open-promoter complex (Figure 26.7). This could be related to the fact, as noted later, that activation of the transcription of certain genes often involves interaction between the pro-

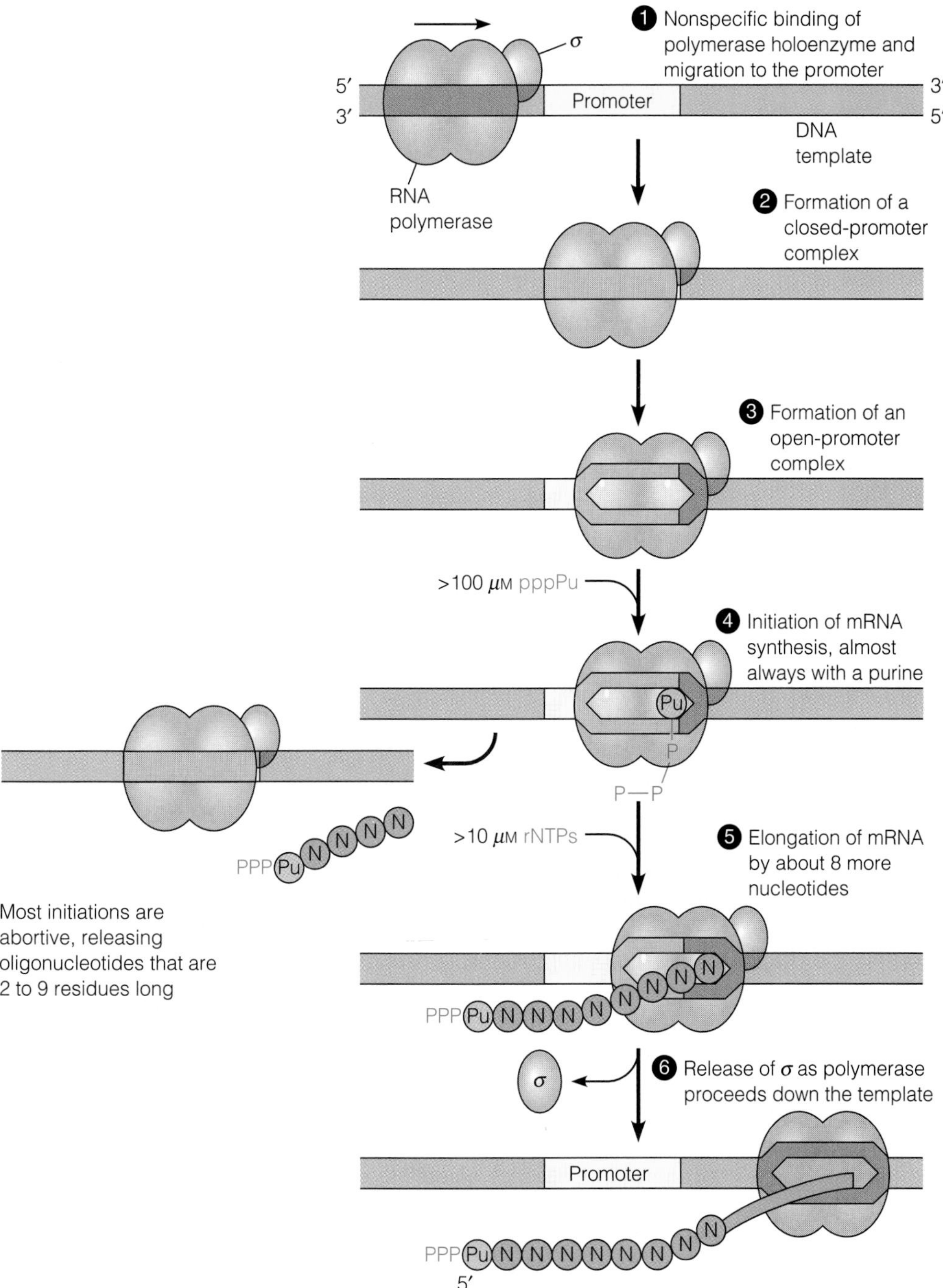

FIGURE **26.6**

Initiation and elongation steps of transcription by bacterial RNA polymerase.

moter and its associated proteins and with proteins bound at upstream regulatory sites, far to the 5′ side of the gene being transcribed.

INITIATION AND ELONGATION: INCORPORATION OF RIBONUCLEOTIDES

Having located a promoter and formed an open-promoter complex, the enzyme is ready to begin the synthesis of an RNA chain. RNA polymerase contains two binding sites for ribonucleoside triphosphates. One, which is used during elongation, binds any of the four common ribonucleoside triphosphates (rNTPs), with a half-saturating concentration of about 10 μM. The second site, used for initiation, binds ATP and GTP preferentially, with half-saturating concentrations of

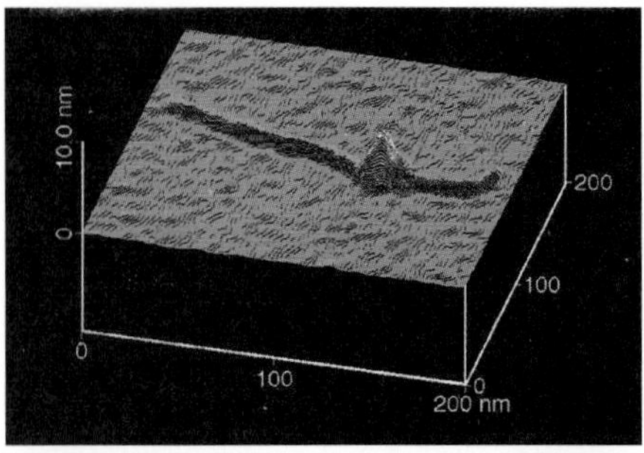

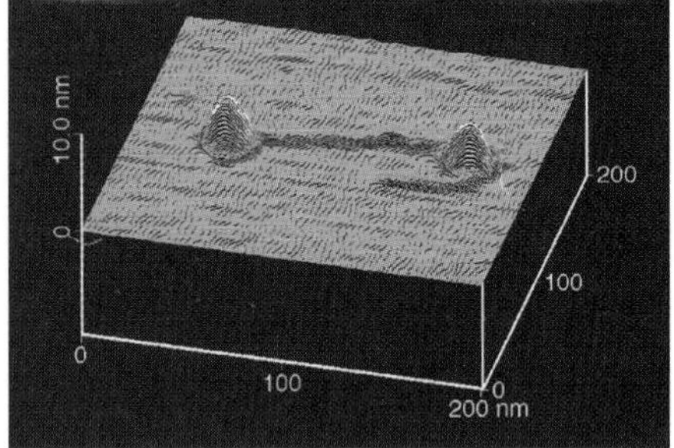

FIGURE **26.7**

RNA polymerase–promoter complexes, as visualized by scanning force microscopy. Above, a closed-promoter complex. Below, an open-promoter complex. The DNA fragment used in these experiments contains the *glnA* promoter, the start site for transcribing the gene for glutamine synthetase. The RNA polymerase holoenzyme contains σ^{54}, a 54-kDa form of σ, rather than the σ^{70} shown in Table 26.1. Activation of transcription requires binding of an additional protein, NtrC, at an upstream site. In the open-promoter complex both proteins are bound independently, with RNA polymerase bound at the right as shown. Note the sharp bend, almost a U-turn, taken by DNA at that site.

Courtesy of M. Guthold from K. Rippe et al., *J. Mol. Biol.* (1997) 270:125–138.

about 100 μM. Thus, most mRNAs have a purine at the 5′ end. Chain growth begins with binding of the template-specified rNTP at this site (Figure 26.6, step 4), followed by binding of the next nucleotide at the other site. Next, nucleophilic attack by the 3′ hydroxyl of the first nucleotide on the α (innermost) phosphorus of the second nucleotide generates the first phosphodiester bond and leaves an intact triphosphate moiety at the 5′ position of the first nucleotide. The transcript is bound unstably during the first several phosphodiester bond-forming reactions, as shown by the fact that most initiations are abortive, with release of oligonucleotides 2 to 9 residues long. The basis for this low efficiency of initiation is not yet understood.

During transcription of the first 10 nucleotides, the σ subunit dissociates from the transcription complex, and the remainder of the transcription process is catalyzed by the core polymerase (steps 5 and 6). Once σ has dissociated, the elongation complex becomes quite stable. Transcription, as studied in vitro, can no longer be inhibited by adding rifampicin, and virtually all transcription events proceed to completion.

During *elongation* (steps 5 and 6), the core enzyme moves along the duplex DNA template. As it moves, it simultaneously unwinds the DNA, exposing a single-strand template for base pairing with incoming nucleotides and with the nascent transcript (the most recently synthesized RNA), and it rewinds the template behind the 3′ end of the growing RNA chain, as suggested in Figure 26.8. In this model, about 18 base pairs of DNA are unwound to form a moving "transcription bubble." As one base pair becomes unwound in advance of the 3′ end of the nascent RNA strand, one base pair becomes rewound near the trailing end of the RNA polymerase molecule. About 8 base pairs of the 3′ end of the nascent transcript are hybridized to the template DNA strand.

Transcription elongation complexes are extraordinarily stable, a feature that allows them to be isolated and probed with protein–nucleic acid cross-linking agents. Thus, a detailed picture of stabilizing interactions other than nucleic acid base pairing is emerging, despite the unavailability of a complete crystal structure for any multisubunit RNA polymerase. As shown in Figure 26.9, amino acid residues in both β and β' subunits contact nascent RNA, DNA in the heteroduplex (DNA–RNA) region, and duplex DNA downstream from the 3′ end of the nascent RNA. Understanding how these binding interactions are coordinated to permit the enzyme to translocate along DNA is a currently active research topic. Techniques for analysis of single complexes show that RNA polymerase as a molecular motor generates forces exceeding those of well-studied cytoskeletal motor proteins, such as myosin and kinesin (Chapter 8).

The concept of a transcription bubble as a central intermediate in transcription, as depicted in Figure 26.8, suggests that the enzyme moves along the DNA template in register with the growing RNA transcript, with the footprint advancing by one base pair for each ribonucleotide incorporated into the transcript. In fact, footprinting of numerous initiation and elongation complexes has shown that the enzyme often advances discontinuously, holding its position for several

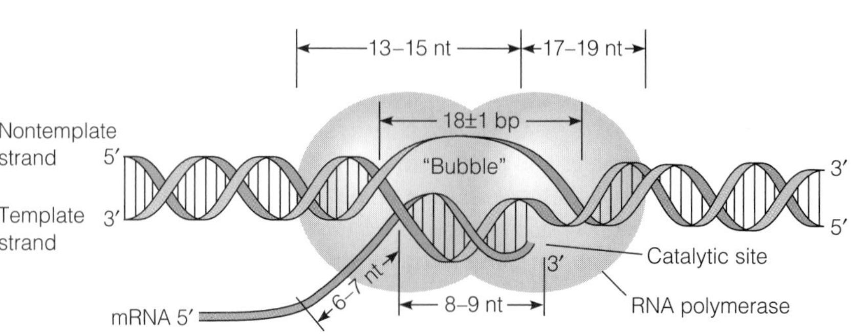

FIGURE **26.8**

The transcription bubble. The lengths of unwound DNA and DNA–RNA hybrid are estimated from reactivities of transcription complexes with reagents such as $KMnO_4$, which oxidizes bases in single-strand nucleic acids. The length of DNA in contact with the enzyme is determined by footprinting (see Tools of Biochemistry 26A). Six or seven nucleotides of RNA behind the DNA hybrid are protected from ribonuclease attack by binding to the enzyme. nt = nucleotide.

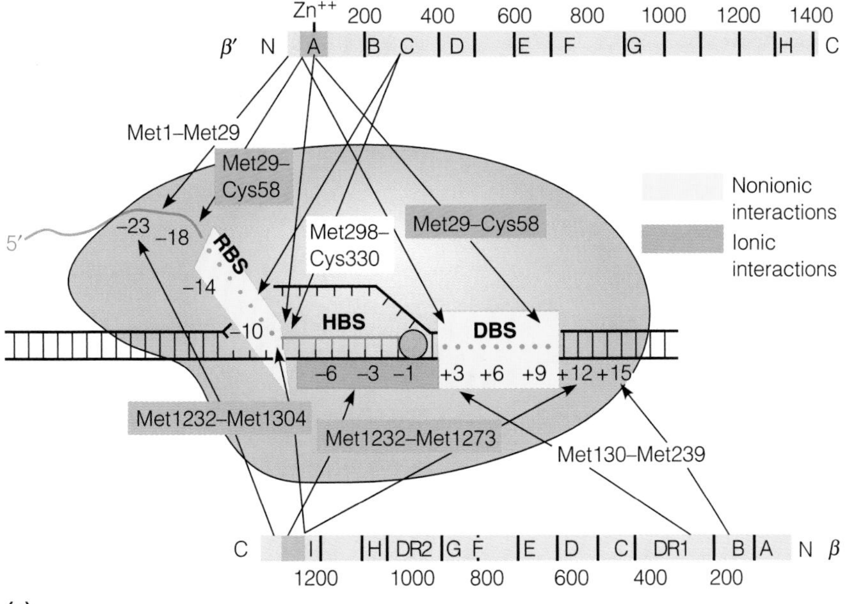

(a)

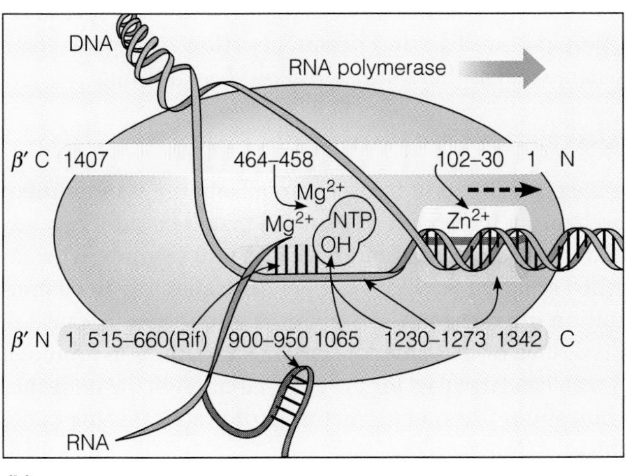

(b)

FIGURE **26.9**

Protein–nucleic acid interactions in the transcription elongation complex. **(a)** Specific nucleic acid–protein interactions. The arrows identify regions or residues in the β and β' polypeptides that form parts of the DNA-binding site (DBS), RNA-binding site (RBS), or heteroduplex-binding site (HBS). **(b)** A schematic view of the complex, with the DNA-binding site represented as a sliding clamp, which maintains processivity, much like the sliding clamps in DNA replication. The figure also shows bound zinc, representing a domain like a zinc finger (Chapter 28) in the clamp. Shown also are two Mg^{2+} ions bound at the active site. It seems likely that RNA polymerase functions by the "two-metal-ion" mechanism described previously for DNA polymerase (Chapter 24). Formation of an RNA hairpin, shown in gray, is probably involved in transcriptional termination, by weakening interactions with the RNA-binding site of the enzyme or with the template.

(a) from E. Nudler, I. Gusarov, E. Avetissova, M. Kozlov, and A. Goldfarb, *Science* (1998) 281:424–428. © 1998 AAAS.
(b) from R. Landick and J. W. Roberts, *Science* (1996) 273: 202–203.

cycles of nucleotide addition and then jumping forward by several base pairs along the template. These and other observations suggested that the means of RNA polymerase translocation is fundamentally different from the continuous movement implied by a picture of the transcription bubble. However, recent observations appear to reconcile the "transcription bubble" paradigm with the footprinting data. It has long been known that some DNA sequences are difficult to transcribe, and RNA polymerase "pauses" when it reaches such a site in vitro, often sitting at the same site for several minutes before transcription is resumed. Recent experiments indicate that at such sites, RNA polymerase often translocates backward, and in the process the 3′ end of the nascent transcript is displaced from the catalytic site of the enzyme, giving a 3′ "tail," which may be several nucleotides long and is not base-paired to the template, protruding downstream of the enzyme (Figure 26.10). In order for transcription to resume, an RNA 3′ end must be positioned in the active site. This is evidently the main function of the RNA 3′ cleavage reactions, which we mentioned on page 986 as possible events in optimizing transcriptional fidelity. The "backtracking" shown in Figure 26.10 was observed initially only in transcription complexes obtained from a *greA greB* double mutant; otherwise, the displaced 3′ RNA end was cleaved and could not be

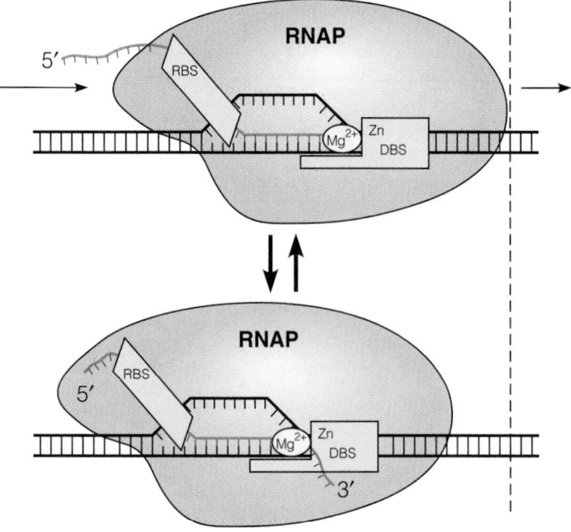

FIGURE **26.10**

Backtracking in an elongation complex. Above, the 3′ terminus of the transcript is in the active site (denoted by Mg^{2+}). Below, the enzyme has slipped backward, leaving the 3′ transcript terminus at the end of a non-base-paired RNA tail, some five nucleotides long. Transcription can resume either by forward sliding of polymerase, back to the structure depicted above or, more likely, by cleavage of the non-base-paired part of the transcript, creating a new base-paired 3′ terminus.

Courtesy of E. Nudler, A. Mustaev, E. Lukhtanov, and A. Goldfarb, *Cell* (1997) 89:33–41.

detected. The GreA and GreB proteins have been shown to stimulate a transcript cleavage activity that is intrinsic to the polymerase itself. These observations suggest that RNA polymerase generally moves forward until one of these special sequences is reached or, perhaps, until a transcription insertion error generates a DNA–RNA mispairing that weakens the hybrid and allows backtracking.

PUNCTUATION OF TRANSCRIPTION: PROMOTER RECOGNITION

Promoter recognition is a crucial step in transcription, from the standpoint of regulation as well as mechanism. In *E. coli* it is estimated that the most frequently transcribed genes initiate transcription about once every 10 seconds, whereas some genes are transcribed as infrequently as once per generation (30 to 60 minutes). Promoter recognition is a rate-limiting step for transcription. Because all genes in bacteria are transcribed by the same core enzyme, variations in promoter structure must be responsible in large part for the great variation in the frequency of initiation. Information about promoter structure is of practical value in designing expression vectors for cloned genes.

What structural features in DNA direct RNA polymerase to bind at a promoter site and to form an open-promoter complex? The first hint of an answer came in 1975, when David Pribnow and Heinz Schaller independently examined the limited DNA sequence data available and revealed that each gene transcribed in *E. coli* shared a short adenine- and thymine-rich sequence, centered about 10 nucleotides to the 5′ side of the transcriptional start site (Figure 26.11). (Identification of transcriptional start points is described in Tools of Biochemistry 26B.) There was some variation among the promoters analyzed, but a **consensus sequence** emerged within this conserved region. A consensus sequence comprises those bases that appear most frequently at each sequence position in a series of sequences thought to have a common function. Among the different initiation sequences analyzed in *E. coli*, that consensus sequence was TATAAT on the **sense strand.** The sense strand is the nontranscribed DNA strand. Because it is complementary in sequence to the template strand, the sense strand is identical in base sequence to the RNA product within the region transcribed, but with T instead of U. Later, another region of conserved nucleotide sequence was found centered at nucleotide −35, with a consensus sequence of TTGACA.

The two conserved sequences are called the −35 region and the −10 region (the latter is also called the *Pribnow box*). No known natural promoter has −35 and −10 regions that are identical to the consensus sequences, but, in general, the more closely these regions in a promoter resemble the consensus sequences, the

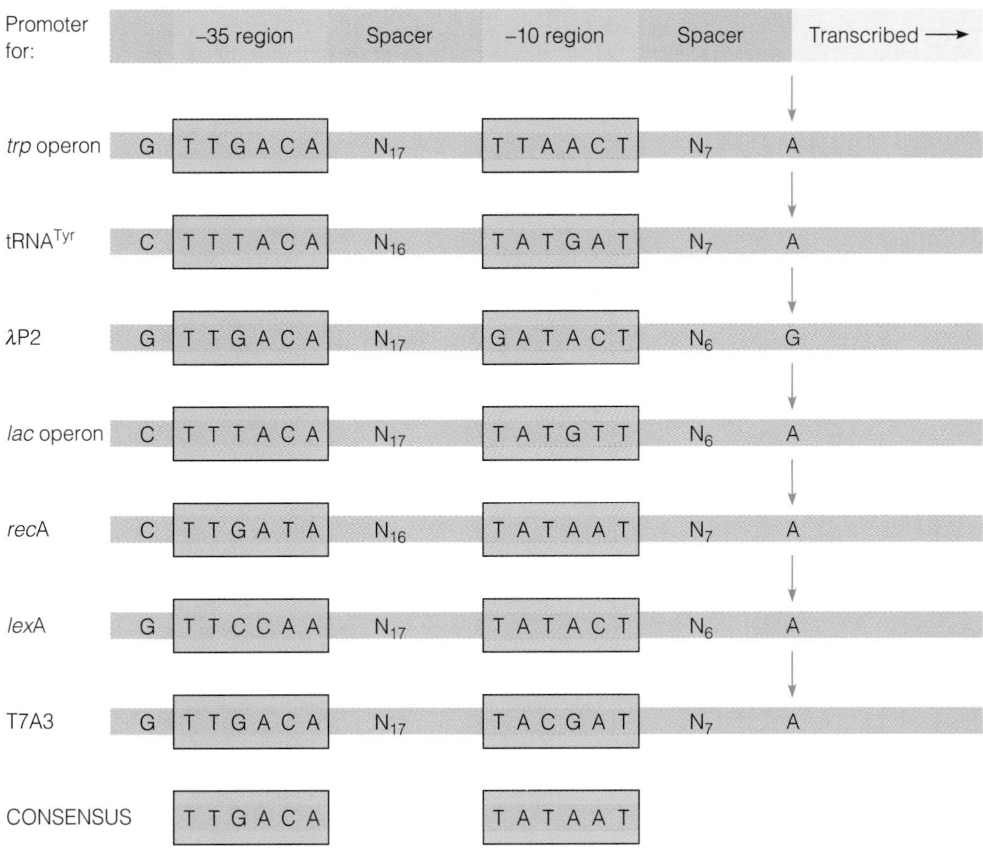

FIGURE **26.11**

Conserved sequences in promoters recognized by *E. coli* RNA polymerase. Lengths of spacer sequences are also shown. Red arrows indicate transcription start site.

more efficient that promoter is in initiating transcription. Variations in promoter structure represent a simple way for the cell to vary rates of transcription from different genes. Figure 26.12 indicates the extent to which different nucleotides are conserved. Among 114 *E. coli* promoters examined in an early study, 6 of the 12 nucleotides in the two consensus sequences were found in more than 75% of the promoters.

What evidence points to a functional role for these conserved sequences in binding RNA polymerase and initiating transcription? First, a variety of mutations map in promoter regions and affect transcription efficiency in vivo. As shown in Figure 26.12, most of the promoter mutations that have been sequenced change the structure of either the −35 region or the −10 region, pointing directly to those sequences as having the greatest effect on transcriptional initiation efficiency. In general, the mutations that increase promoter strength (**up-promoter mutations**) change either the −35 or the −10 region to more closely resemble the consensus sequences. **Down-promoter mutations,** which decrease promoter strength, change the sequence away from the consensus. Similar conclusions have been drawn from site-directed mutagenesis of promoters and analysis of their efficiency in vitro. The latter studies have confirmed the importance of spacing between the two regions. Although most natural promoters have a 17-nucleotide spacer between the −35 and −10 regions, many have 16 or 18. In vitro studies show that a 17-nucleotide spacer yields the most efficient promoter structure.

The frequency with which a gene is transcribed in *E. coli* genes is largely determined by the resemblance of their promoter sequences to consensus sequences.

The other evidence for the importance of the −35 and −10 regions is that most of the DNA nucleotides in close contact with RNA polymerase are those in or near these two conserved sequences. This can be ascertained by determining the susceptibility of particular nucleotides to chemical modification, in the presence or absence of RNA polymerase. For example, dimethyl sulfate can be used to determine the susceptibility of guanine residues to methylation. This technique involves methylating a 5′ end–labeled fragment, followed by cleavage, as in a

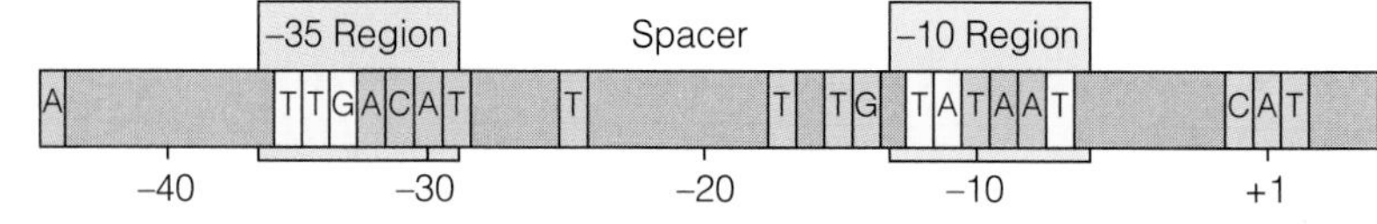

(a)

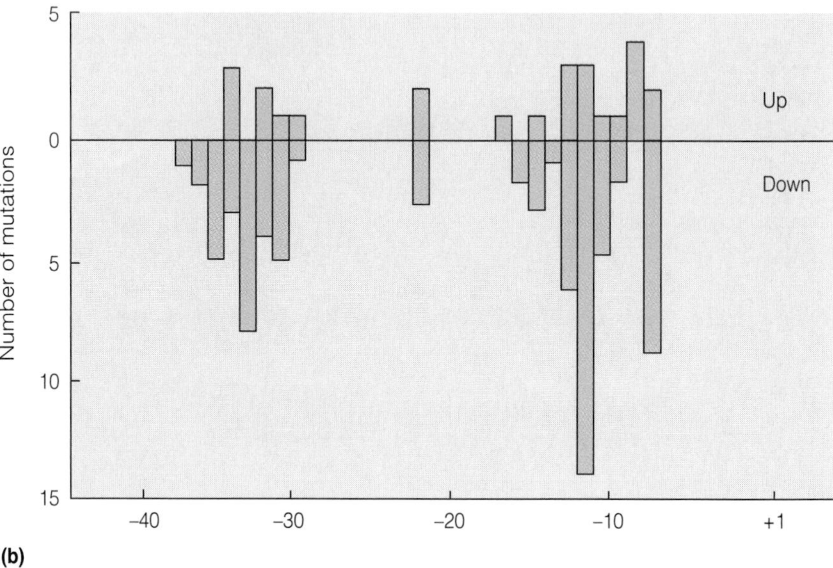

(b)

FIGURE **26.12**

Survey of conserved nucleotides in *E. coli* promoters. Nucleotide sequences were compared among 114 known *E. coli* promoters. **(a)** Nucleotide positions that were invariant in at least 75% of the promoters are shown in yellow, moderately conserved nucleotides (50%–75%) in purple, and weakly conserved nucleotides (40%–50%) in blue. **(b)** The number of known promoter mutations affecting each site. Up-promoter mutations increase promoter strength; down-promoter mutations have the opposite effect.

Reproduced with permission from D. Hawley and W. R. McClure, *Annu. Rev. Biochem.* (1985) 54:171–204. © 1985 by Annual Reviews, Inc.

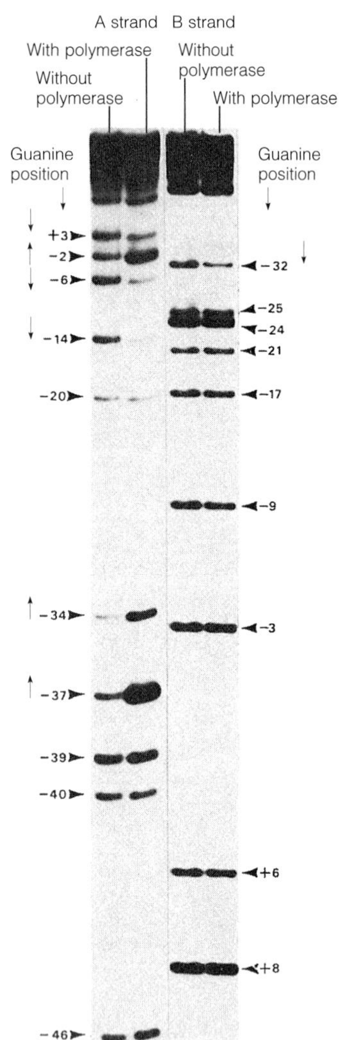

FIGURE **26.13**

Identification of G residues that contact RNA polymerase in the *E. coli* tryptophan promoter. ↑ = enhancement of cleavage by binding of polymerase; ↓ = inhibition of cleavage.

Courtesy of D. S. Oppenheim, G. N. Bennett, and C. Yanofsky, *J. Mol. Biol.* (1980) 144:133–142. Academic Press Ltd., London.

Maxam–Gilbert sequencing reaction (p. 1025), and displaying the G-terminated fragments on an electrophoretic gel (Figure 26.13). In some cases, band intensity is increased by the presence of RNA polymerase, showing that binding of protein enhances reactivity. In other cases, reactivity is decreased, showing that binding of the protein protects that particular guanine from the methylating reagent. In either case the result indicates that the protein is in contact with the guanines in those regions and changes their reactivity. This experiment and others like it show that RNA polymerase also contacts bases upstream from the −35 box, in the −40 to −60 region.These contacts involve the polymerase α subunit, while the contacts with −35 and −10 boxes involve σ. The upstream contacts (−40 to −60) are particularly important in the promoters for ribosomal protein genes, which are very actively transcribed.

In the promoter analyzed in Figure 26.13, most of the affected guanines—those at positions −37, −34, −32, −14, and −6—lie in or next to the −35 and −10 regions. Modification of reaction conditions can also identify adenines that lie close to RNA polymerase. Also, the reaction with ethylnitrosourea can identify phosphate oxygens of enhanced or decreased reactivity. Finally, thymines that are in contact with the protein can be identified by substituting bromodeoxyuridine for thymine in DNA and then determining the extent to which each bromo-dUMP residue can become photo-cross-linked to protein. Note that these approaches, like footprinting, can identify critical nucleotides in any DNA site that binds a protein specifically, not just in promoters.

All of these techniques point to the −35 region and the −10 region, plus a few nucleotides upstream of −10, as the major contact points in an open-promoter complex. Figure 26.14 summarizes the results of these approaches as applied to one of the promoters for *E. coli* RNA polymerase on T7 DNA. Note that because there are two turns of the helix between the −35 and −10 regions, RNA polymerase is postulated to bind to DNA primarily on one side of the duplex. The data on nucleotide reactivity support this conclusion.

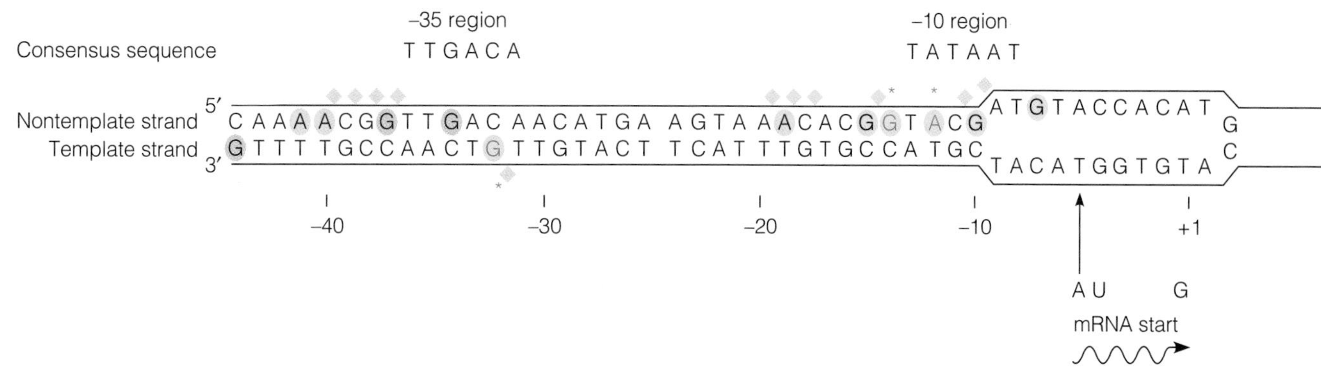

FIGURE **26.14**

Structure of the T7 A3 promoter. This promoter is one of three for *E. coli* RNA polymerase in T7 DNA. Nucleotides that are made resistant to methylation by RNA polymerase binding are marked in orange; nucleotides that are rendered more susceptible are marked in purple. Asterisks indicate methylated purines that interfere with polymerase binding. Diamonds indicate phosphate contacts. The two conserved regions of the promoter are precisely two helix turns apart.

Another factor affecting transcriptional efficiency, in addition to the base sequence of the promoter, is the superhelical tension on the DNA template. The relation between DNA topology and transcriptional efficiency is now receiving considerable attention. The relationship is not clear, because transcription of some genes is activated in vivo when the template is highly supercoiled—for example, by inactivating topoisomerase I. Transcription of other genes, by contrast, is inhibited under these conditions. Interestingly, the promoter for transcription of DNA gyrase subunits becomes activated when the gene is in a relaxed state. Given that gyrase introduces superhelical turns, this finding seems to represent a feedback mechanism in which the cell responds appropriately to a signal that intracellular DNA is becoming too relaxed.

PUNCTUATION OF TRANSCRIPTION: TERMINATION

Because of the great stability of transcription complexes, termination of transcription, with release of the nascent transcript, is a rather complex process. In bacteria we recognize two distinct types of termination events—those that depend on the action of a protein **termination factor,** called ρ (rho), and those that are ρ factor–independent.

Factor-Independent Termination

Sequencing the 3′ ends of genes that terminate in a factor-independent manner reveals two structural features shared by many such genes and illustrated in Figure 26.15: (1) two symmetrical GC-rich segments that in the transcript have the potential to form a stem–loop structure, and (2) a downstream run of four to eight A residues. These features suggest the following as elements of the termination mechanism. First, RNA polymerase slows down, or pauses, when it reaches the first GC-rich segment, because the stability of G-C base pairs makes the template hard to unwind. In vitro, RNA polymerase does pause for several minutes at a GC-rich segment. Second, pausing gives time for the complementary GC-rich parts of the nascent transcript to base-pair with one another, thereby displacing this part of the transcript from its template or from its enzyme binding site. Hence, the ternary complex of RNA polymerase, DNA template, and RNA is weakened. Further weakening, leading to dissociation, occurs when the A-rich segment is transcribed to give a series of A—U bonds (very weak), linking transcript to template.

The actual mechanism of termination is more complex than just described, in part because DNA sequences both upstream and downstream from the regions shown in Figure 26.15 also influence termination efficiency. Moreover, not all pause sites are termination sites. However, the scheme shown includes key elements of factor-independent termination.

DNA sequences that promote factor-independent termination include a run of 4 to 8 A residues and a GC-rich region that forms a stem–loop.

Factor-Dependent Termination

Factor-dependent termination sites are less frequent, and the mechanism of this type of termination is still more complex. The ρ protein, a hexamer composed of

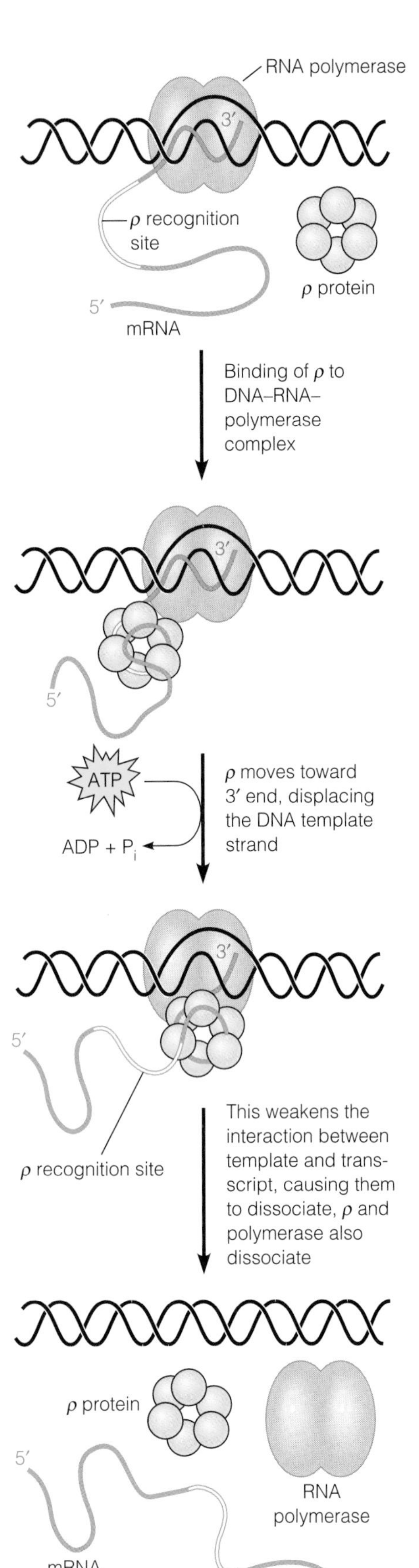

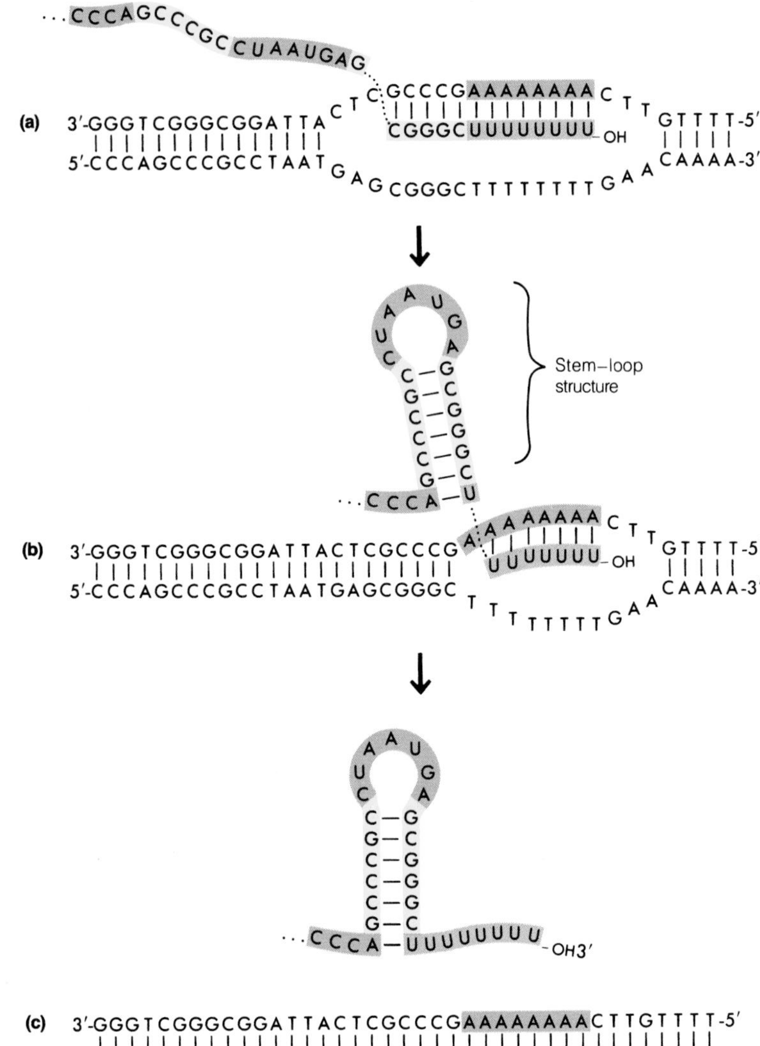

FIGURE **26.15**

A model for factor-independent termination of transcription. **(a)** An A-rich segment of the template (brown segment on right) has just been transcribed into a U-rich mRNA segment. **(b)** RNA–RNA duplex, stabilized by G-C base pairs (yellow), eliminates some of the base pairing between template and transcript. **(c)** The unstable A—U bonds linking transcript to template hybrid dissociate, releasing the transcript.

identical subunits, was originally discovered in studies on termination of λ phage DNA transcription in vitro. This protein, which has been characterized as an RNA–DNA helicase, contains a nucleoside triphosphatase activity that is activated by binding to polynucleotides. Apparently ρ acts by binding to the nascent transcript at a C-rich site near the 3′ end, when RNA polymerase has paused (Figure 26.16). Then ρ moves along the transcript toward the 3′ end, with the heli-

FIGURE **26.16**

ρ factor–dependent termination. ρ binds to a site on the nascent transcript and unwinds the RNA–DNA duplex. Once ρ reaches RNA polymerase, interaction with bound NusA protein (not shown) leads to termination.

case activity unwinding the 3′ end of the transcript from the template (and/or the RNA polymerase molecule) and causing its release.

In factor-dependent termination, ρ protein acts as an RNA–DNA helicase, unwinding the template–transcript duplex and facilitating release of the transcript.

It is not clear what causes RNA polymerase to pause at ρ-dependent termination sites. The action of another protein, NusA, is somehow involved. NusA was discovered during studies of **antitermination** in phage λ. Antitermination occurs early in the phage transcription program, when two ρ-dependent termination sites are inactivated, so that RNA polymerase can move past these sites and transcribe genes essential to phage development (see page 1008). For this inactivation to occur, a viral protein, the product of gene *N*, must interact with NusA (*N* *u*tilization *s*ubstance) by a mechanism as yet unknown. Mutations in either *N* or the *E. coli nusA* gene interfere with antitermination and block phage development. The NusA protein evidently associates with RNA polymerase, and there is reason to believe that it binds at some point in transcription after the σ factor has dissociated, because the two purified proteins compete with each other for binding to core RNA polymerase.

Further insight into termination mechanisms has come from an extensively studied regulatory mechanism called **attenuation.** Attenuation controls the rate of transcription of certain operons by terminating the synthesis of a nascent transcript before RNA polymerase has reached the structural genes. Attenuation, which has been investigated as a model for end-of-transcript termination, is discussed on pages 1011–1014.

Regulation of Transcription

Bacteria respond rapidly to changes in their environment. As we have seen, the genes for lactose utilization are activated once *E. coli* cells sense the presence of lactose or a similar compound, and the lactose-utilizing enzymes are induced. Many other genes are similarly kept in a turned-off state until their products are needed. Still other genes, concerned principally with anabolism, are kept in a turned-on state unless the product of the anabolic sequence is present, and then the genes are **repressed,** or turned off. The metabolic logic involved is obvious: Many genes are expressed only when the gene products are needed—to utilize an available substrate, to synthesize a complex metabolite that is absent from the medium, or in some other way to respond to changing environmental conditions.

Both enzyme induction and repression act through control of transcription, primarily at the initiation step. Similarly, bacteriophages rely primarily on the transcription machinery of the host cell to express their own genes; host transcriptional control mechanisms are either used intact or modified to meet the needs of the virus. Consequently, studies of enzyme induction and repression and of bacteriophage development have provided our most penetrating insights into how transcription is controlled in prokaryotes. In turn, this work has provided paradigms for understanding the far more complex questions of how transcriptional regulation in multicellular organisms can account for differentiation and development. For these reasons, we shall focus our discussion of transcriptional regulation on the prokaryotic systems that have been most informative, with extension of these concepts to eukaryotes in Chapter 28.

THE LACTOSE OPERON: EARLY EVIDENCE FOR TRANSCRIPTIONAL CONTROL OF GENE EXPRESSION

The lactose operon consists of three linked structural genes that encode enzymes of lactose utilization, plus adjacent regulatory sites. The three structural genes—*z*, *y*, and *a*—encode respectively β-galactosidase, **β-galactoside permease** (a transport protein), and **thiogalactoside transacetylase,** an enzyme of still unknown metabolic function. In the presence of an inducer, all three enzymes accumulate simultaneously, but to different levels. Lactose itself leads to induction of the

Lactose

Allolactose

Isopropyl β-thiogalactoside

lactose operon, but the true intracellular inducer is **allolactose,** Galβ(1→6)Glc, a minor product of β-galactosidase action. In the laboratory one usually uses a synthetic inducer such as **isopropyl thiogalactoside (IPTG),** which induces the lactose operon but is not cleaved by β-galactosidase. Hence, its concentration does not change during an experiment.

Regulation in the Lactose Operon

A mutation in a structural gene—*z,* for example—can inactivate its product (β-galactosidase) without affecting control of the other two genes. However, mutations in the regulatory regions mapping *outside* genes *z, y,* and *a* can affect expression of all three structural genes. In their early work, Jacob and Monod recognized two distinct mutant phenotypes—**constitutive,** in which all three gene products are synthesized at high levels even when inducer is absent, and **noninducible,** in which all three enzyme activities remain low even after addition of an inducer. These mutations mapped in two sites, termed *o* and *i.* Problem 14 at the end of this chapter should help you understand the relationships among regulatory mutations, mutant phenotypes, and levels of *lac* operon enzymes.

The original Jacob–Monod model for gene regulation, based upon this system, was presented in Figure 26.2. A more complete description of the *lac* operon is shown in Figure 26.17. As Jacob and Monod correctly proposed, transcription of the three structural genes is initiated near an adjacent site, the **operator.** Transcription yields a single **polycistronic messenger RNA;** that is, it yields an RNA copy of all three genes. (The term **cistron** has genetic significance. For our purposes it is a region of a genome that encodes one polypeptide chain.) The *i* gene product is a macromolecular **repressor,** which in the active form binds to the operator, blocking transcription (Figure 26.18a).

The repressor also has a binding site for inducer. Binding of IPTG, allolactose, or some other inducer at this site inactivates repressor by vastly decreasing its affinity for DNA (Figure 26.18b). This repressor inactivation stimulates transcription of *z, y,* and *a,* because dissociation of the repressor–inducer complex from the operator removes a steric block to binding of RNA polymerase at the initiation site. Thus, the introduction of lactose activates synthesis of the gene products involved in its catabolism by removing a barrier to their transcription. This mode of regulation is essentially negative, because the active regulatory element (the repressor) is an *inhibitor* of transcription. Positive control was later discovered, involving the CRP site shown in Figure 26.17. We shall discuss this shortly.

Central to the success of Jacob and Monod in formulating such a detailed model was their ability to analyze regulatory *lac* mutations in partial diploids—bacteria containing one complete chromosome plus part of another, transferred in by conjugation. One copy of the operon resides in the chromosome, while an incomplete chromosome could be introduced into a cell, as part of the bacterial mating process. Noninducible mutations that mapped in *i* had a dominant phenotype, meaning that expression of the structural genes was low when both wild-type and mutant alleles of *i* were present. Jacob and Monod proposed that the mutant alleles give rise to mutant repressors that are unable to bind inducer.

FIGURE **26.17**

A map of the lactose operon. The CRP site is the binding site for cAMP receptor protein, a regulatory factor (page 1002). The promoter also includes binding sites for RNA polymerase and *lac* repressor. Synthesis of repressor initiates at its own prompts (*i* promoter). Additional repressor-binding sites exist, 2 nucleotides upstream and 432 nucleotides downstream, respectively, from the transcriptional start point.

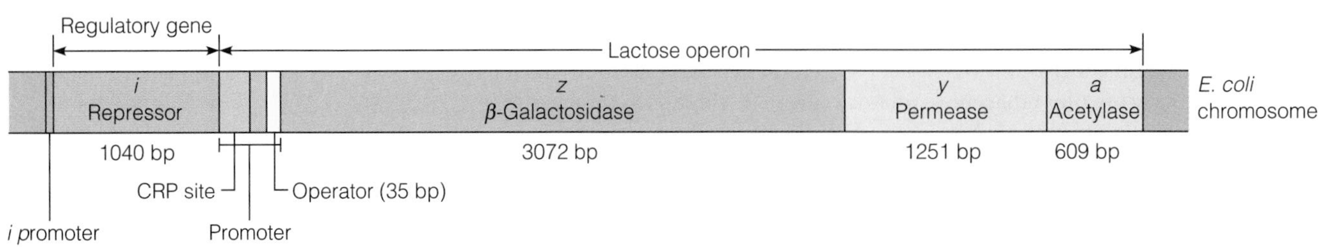

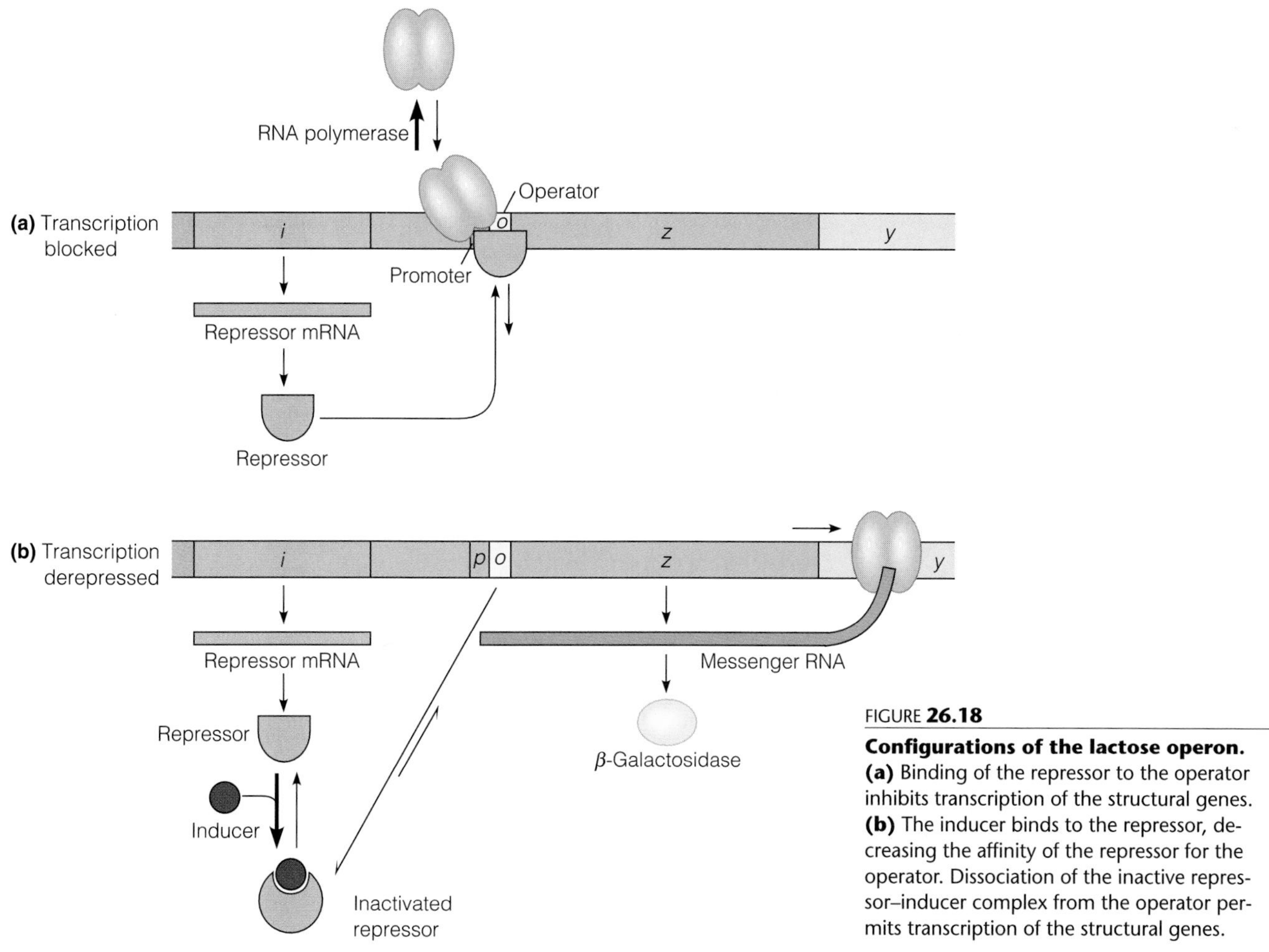

FIGURE **26.18**

Configurations of the lactose operon. **(a)** Binding of the repressor to the operator inhibits transcription of the structural genes. **(b)** The inducer binds to the repressor, decreasing the affinity of the repressor for the operator. Dissociation of the inactive repressor–inducer complex from the operator permits transcription of the structural genes.

These mutant repressors would remain bound to DNA at operator sites on both the mutant and the normal chromosomes, even when inducer is present.

The phenotypes of partial diploids involving *lac* regulatory genes gave indispensable clues to mechanisms of transcriptional regulation.

Constitutive mutations that mapped in *i* had a recessive phenotype. That is, they resulted in high gene expression, but only when two mutant alleles were present. These mutant alleles generated repressors that were defective in operator binding and thus could not turn off gene expression. Such mutations are recessive because a normal repressor in the same cytosol can bind to all operators and inhibit transcription.

These observations showed that repressor mutations are ***trans*-dominant,** meaning that the *i* gene product encoded by one genome can affect gene expression from other genomes. This finding led to the conclusion that repressor is a diffusible product, capable of acting on any DNA site in the cell to which it could bind. On the other hand, the constitutive mutations mapping in *o* had a ***cis*-dominant** effect. That is, in a cell with one wild-type operator and one mutant operator, only the genes on the same chromosome as the mutant operator were expressed constitutively. Given that a protein would be capable of diffusing through the cytosol and acting on other chromosomes, this finding suggested that the operator does not encode a gene product.

For a molecular mechanism based almost entirely on the indirect evidence of genetic analysis, the operon model as advanced by Jacob and Monod has stood the test of time remarkably well. Three major modifications to the model occurred as the system was subjected to further analysis. First, the promoter was discovered as an element distinct from the operator (although the two sites overlap). Second, although the repressor was first thought to be *i*-gene RNA, its isolation proved that it is protein. Third, Jacob and Monod proposed that all transcriptional regulation was negative; that is, binding a regulatory protein always inhibits transcription. However, the lactose operon, like many other regulated genes, also exhibits *positive* control of transcription (that is, *activation* of transcription by binding of a protein), as described on pages 1001–1003.

Isolation and Properties of Repressor

The *lac* repressor was isolated in 1966 by Walter Gilbert and Benno Müller-Hill. Because this repressor constitutes only 0.001% of the total cell protein, Gilbert and Müller-Hill used mutants designed to overproduce it, so as to maximize its synthesis (to about 2% of total protein). They then purified the protein on the basis of its ability to bind the synthetic inducer IPTG. The purified *lac* repressor is a tetramer, formed from four identical subunits, each with 360 amino acids (M_r = 38,350). The protein binds IPTG with a K_a of about $10^6\ \text{M}^{-1}$, and it binds nonspecifically to duplex DNA with a K_a of about $3 \times 10^6\ \text{M}^{-1}$. However, its specific binding at the *lac* operator is much tighter, with a K_a of $10^{13}\ \text{M}^{-1}$. Like RNA polymerase, repressor seeks its operator site by first binding to DNA at any site and then moving in one dimension along the DNA. It moves either by sliding or by transfer from one site to another, when the two sites are brought next to each other on adjacent loops of DNA.

Control by the *lac* repressor is exceedingly efficient, particularly in view of the minute amount of repressor present in an *E. coli* cell. The *i* gene is expressed at a very low rate, to give about 10 molecules of repressor tetramer per cell. Although this corresponds to a concentration of only about 10^{-8} M, this value is several orders of magnitude higher than the *dissociation* constant, meaning that in a *noninduced* cell the operator is bound by repressor more than 99.9% of the time—hence, the very low levels of *lac* operon proteins in uninduced cells (less than one molecule per cell). However, binding of inducer decreases the affinity of the repressor–inducer complex for operator by many orders of magnitude. Under these conditions, nonspecific binding of the repressor–inducer complex at other DNA sites becomes significant, so that in *induced* cells the operator is occupied by repressor less than 5% of the time.

A repressor–inducer system provides negative control of the *lac* operon. The repressor binds to operator, interfering with transcription initiation. An inducer binds to repressor, reducing its affinity for operator.

The Repressor Binding Site

The DNA site bound by *lac* repressor has been analyzed by footprinting and methylation protection experiments of the type described earlier for RNA polymerase (see Figure 26.13). As shown in the diagram below, the operator comprises 35 base pairs, including 28 base pairs of symmetrical sequence; that is, it includes a sequence that is identical in both directions (shaded in the diagram). Thus, the operator is an imperfect palindrome—imperfect because of the 7 base pairs that do not show this symmetry.

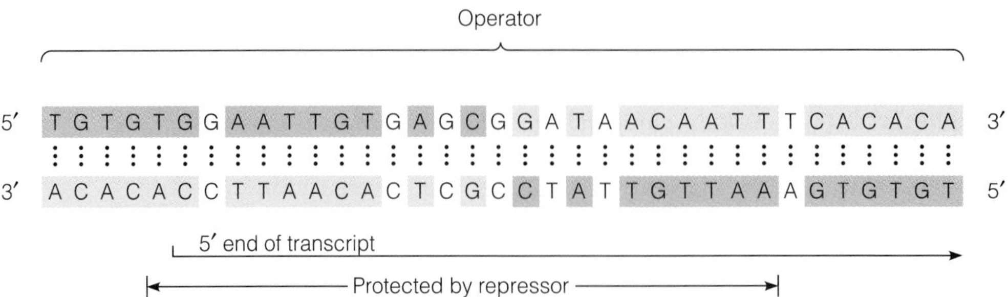

The transcriptional start point is included within the repressor-binding sequence, as shown in the diagram. Twenty-four of the 35 base pairs of the operator are protected from DNase attack by repressor binding. Operator-constitutive (o^c) mutations involve changes in the central portion of this sequence of nucleotides. Figure 26.19 shows how the operator and promoter overlap, as determined

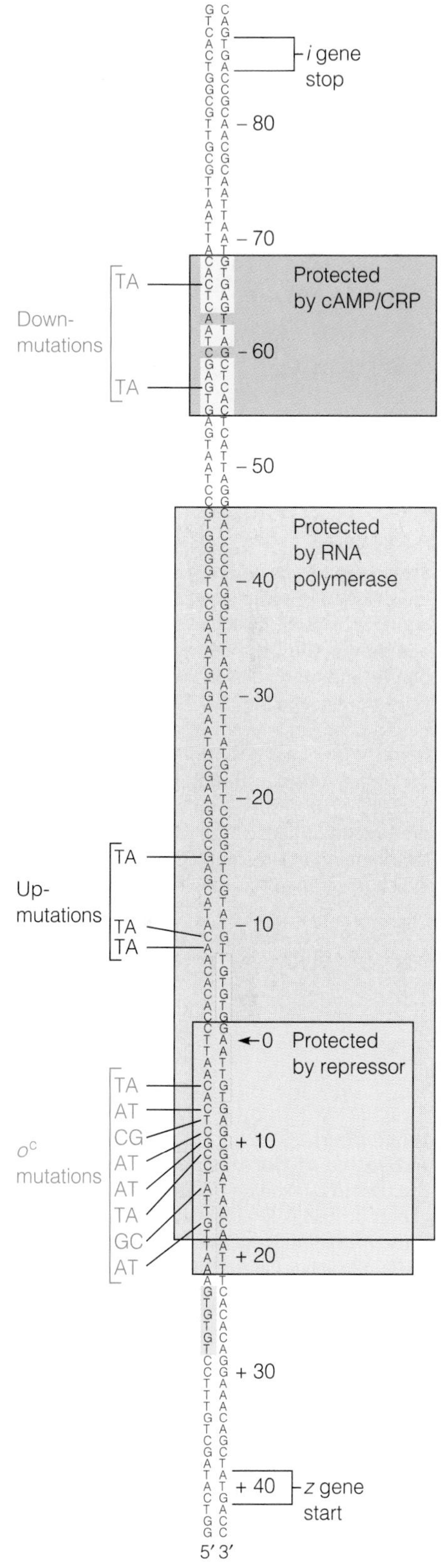

FIGURE **26.19**

The 122-base *lac* regulatory region. The inverted repeat sequence in the CRP binding site is shaded in blue, and the palindromic areas of the operator are shaded in pink. The binding sites for all three proteins (CRP, repressor, and RNA polymerase), as determined by DNase I footprinting, are boxed. The sequence changes encountered in promoter and operator mutations are shown, as well as the start codon for the *lacZ* protein (*z* gene) and the stop codon for repressor protein (*i* gene). Nucleotides are numbered from 0 at the *lac* transcriptional start point (the first mRNA nucleotide is +1).

by the regions of DNA protected by binding either repressor or RNA polymerase, respectively.

Once the *lac* operon had been sequenced, it became apparent that two additional *lac* repressor-binding sites are located nearby, one centered upstream, at position −82, and one within the *lacZ* gene itself, at position +432; the original operator is centered at +11. Genetic analysis indicated that both sites participate in *lac* operon regulation. Although the significance of the downstream site is not clear, mutations affecting the upstream (−82) site led to incomplete repression of the operon. Evidence indicated that a looped DNA structure was essential for complete repression, with repressor contacting both the −82 and +11 sites. Confirmation came in 1996, with the crystal structure determination of the complete tetrameric *lac* repressor by itself, as a complex with a defined oligonucleotide fragment, and as a complex with IPTG. As shown in Figure 26.20a, the tetrameric protein consists of two dimeric units, joined by a hinge region. Each dimer binds DNA separately, suggesting that the tetrameric protein binds to both the +11 and −82 sites, creating a DNA loop of 93 base pairs between them. The DNA-binding domain of the protein is an α-helical region that contacts bases within the major groove of the operator DNA. This helical binding motif has been seen in other sequence-specific DNA-binding proteins, as we shall discuss shortly.

Figure 26.20b suggests the mechanism of induction. The repressor is an allosteric protein which, upon binding inducer, significantly increases the angle at which two monomeric units in a dimer relate to one another. This drives the DNA-binding helices apart by 3.5 Å, so that they can no longer contact DNA binding sites, as they must in order to bind tightly.

Regulation of the *lac* Operon by Glucose: A Positive Control System

The *lac* repressor–operator system keeps the operon turned *off* in the absence of utilizable β-galactosides. An overlapping regulatory system, summarized in Figure 26.21, turns the operon *on* only when alternative energy sources are unavailable. *E. coli* has long been known to use glucose in preference to most other energy substrates. When grown in a medium containing both glucose and lactose, the cells metabolize glucose exclusively until the supply is exhausted. Then growth slows, and the lactose operon becomes activated in preparation for continued growth using lactose. This phenomenon, now known to involve a transcriptional *activation* mechanism, was originally called glucose repression or catabolite repression. Transcriptional activation occurs when glucose levels are low; control is exerted through intracellular levels of cyclic AMP.

Recall that in animal cells a rise in cAMP levels stimulates catabolic enzymes, which increase the levels of energy substrates. Those effects are mediated metabolically, through hormonal signals and triggering of metabolic cascades. In bacteria the activation involves control of gene expression, but the end results are similar. In *E. coli*, cAMP levels are low when intracellular glucose levels are high. The actual regulatory mechanism is not yet known. Adenylate cyclase apparently senses the intracellular level of an unidentified intermediate in glucose catabolism—hence, the current name for the regulatory process, **catabolite activation.** When glucose levels drop, as shown in Figure 26.21, cAMP levels rise, triggering

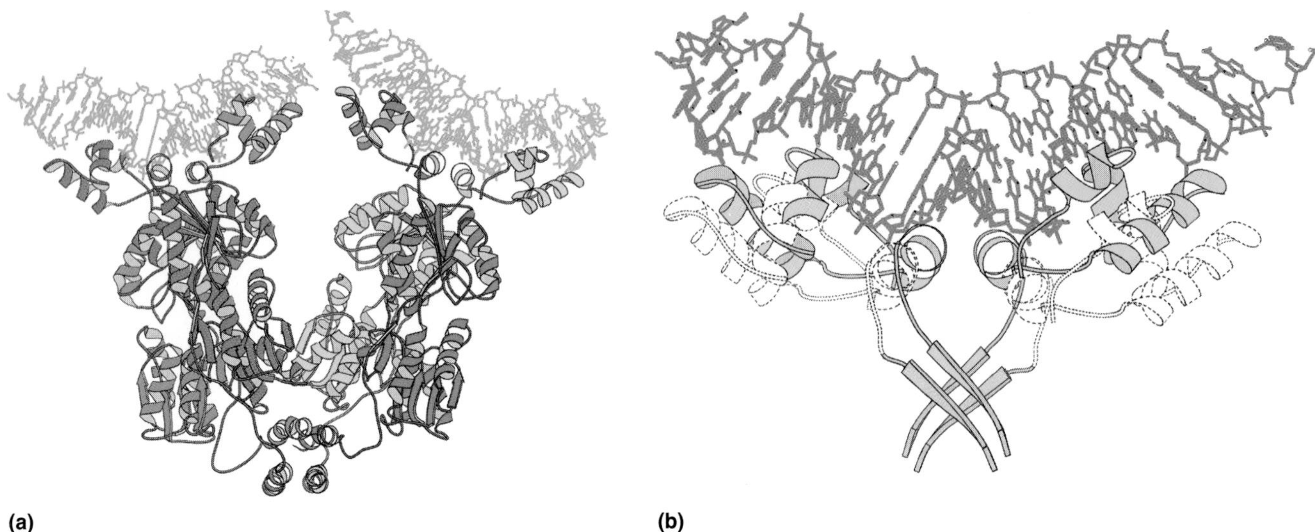

FIGURE **26.20**

Structure of the *lac* repressor. (a) The repressor–DNA complex, with DNA (in green) bound separately to each of two dimeric units—one shown with the monomers in green and purple, and monomers of the other shown in yellow-green and red. **(b)** The effect of IPTG binding. Shown in yellow-green are the extreme N-termini of two monomeric units of the protein (residues 1–68), bound to DNA, in purple. The dashed structures represent the presumed positions of the DNA-binding helices after binding of inducer IPTG to the protein, which drives the DNA-binding helices apart.

Courtesy of M. Lewis, G. Chang, N. C. Horton, M. A. Kercher, H. C. Pace, M. A. Schumacher, R. G. Brennan, and P. Lu, *Science* (1996) 271:1247–1254, with permission from *Science*.

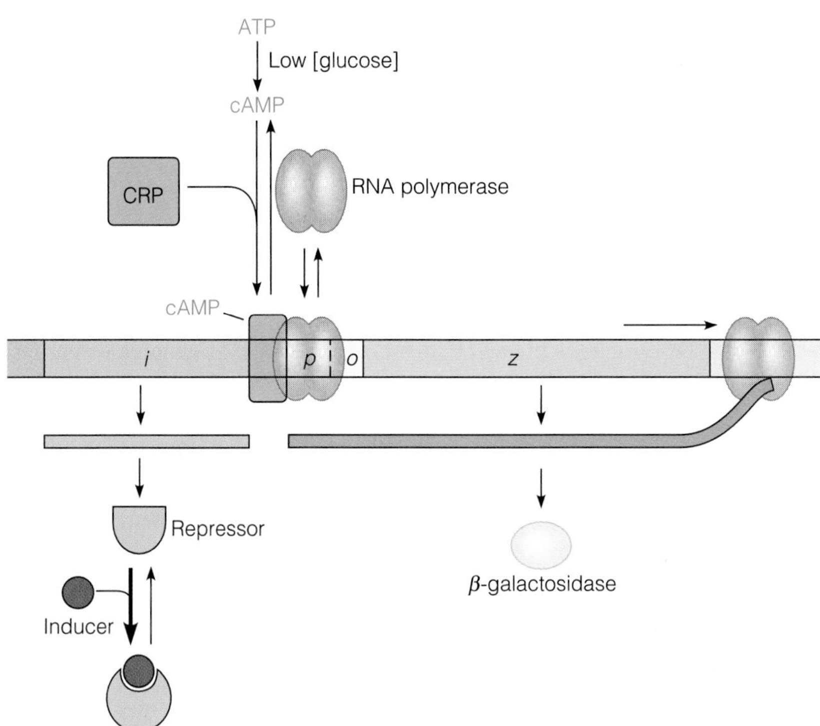

FIGURE **26.21**

Activation of the *lac* operon. Repressor is inactivated by binding inducer, and cAMP receptor protein (CRP) is activated by binding cyclic AMP. Binding of the CRP–cAMP complex to DNA facilitates initiation of transcription by RNA polymerase.

activation of the lactose operon by its interaction with a protein called **cAMP receptor protein (CRP),** formerly called **catabolite activator protein (CAP).** This protein is a dimer, each of whose identical polypeptide chains contains 210 amino acid residues. When it binds cAMP, CRP undergoes a conformational change. The change greatly increases its affinity for certain DNA sites, including a site in the *lac* operon adjacent to the RNA polymerase binding site. The binding of cAMP–CRP at this site protects a DNA sequence from −68 to −55, as shown in Figure 26.19. This binding facilitates transcription of the *lac* operon by stimulating the binding of RNA polymerase to form a closed-promoter complex or by increasing the rate of open-promoter complex formation.

Our understanding of CRP action is still incomplete, partly because the cAMP–CRP complex activates several different gene systems in *E. coli,* all of them involved with energy generation. They include operons for utilization of other sugars, including galactose, maltose, arabinose, and sorbitol, and several amino acids. Among the operons that have been analyzed, the DNA binding site of the cAMP-activated dimer varies considerably with respect to the transcriptional start point, suggesting that regulatory mechanisms involving this protein are complex.

Cyclic AMP receptor protein (CRP) provides positive control of *lac* and several other catabolite-repressible operons. The cAMP–CRP complex binds at the *lac* promoter when glucose levels are low and facilitates initiation of transcription.

The CRP–DNA Complex

The structure of the CRP–cAMP–DNA complex, as revealed by x-ray crystallography (Figure 26.22), shows how the protein binds to DNA. Each CRP subunit contains a characteristic pair of α helices, which are joined by a turn. One helix of each pair, shown as perpendicular to the plane of the figure, lies within the major groove of DNA. This **helix–turn–helix** structural motif, which was observed at about the same time in the structure of the λ phage Cro repressor (see pages 1006–1010), is found in several DNA-binding regulatory proteins, suggesting common evolutionary origins for this family of proteins. We shall return to this motif and its regulatory significance when we discuss the repressors of phage λ.

Analysis of the DNA–protein complex shows also that CRP induces DNA to bend quite sharply when it binds. This bending may facilitate the initiation of transcription by bringing parts of RNA polymerase that are bound upstream into direct contact with the promoter or transcriptional start site. Evidence indicates an important interaction between the α subunit and DNA, occurring as a result of DNA bending.

BACTERIOPHAGE λ: MULTIPLE OPERATORS, DUAL REPRESSORS, AND MODELS FOR DNA BINDING SPECIFICITY

We now return to phage λ, a much larger and more complex genetic system than the lactose operon but one that is regulated by similar factors—binding of proteins to specific DNA regulatory sites, just upstream from the genes that they control, with binding leading to either activation or inhibition of transcriptional initiation. However, because of the variety of relationships between the virus and

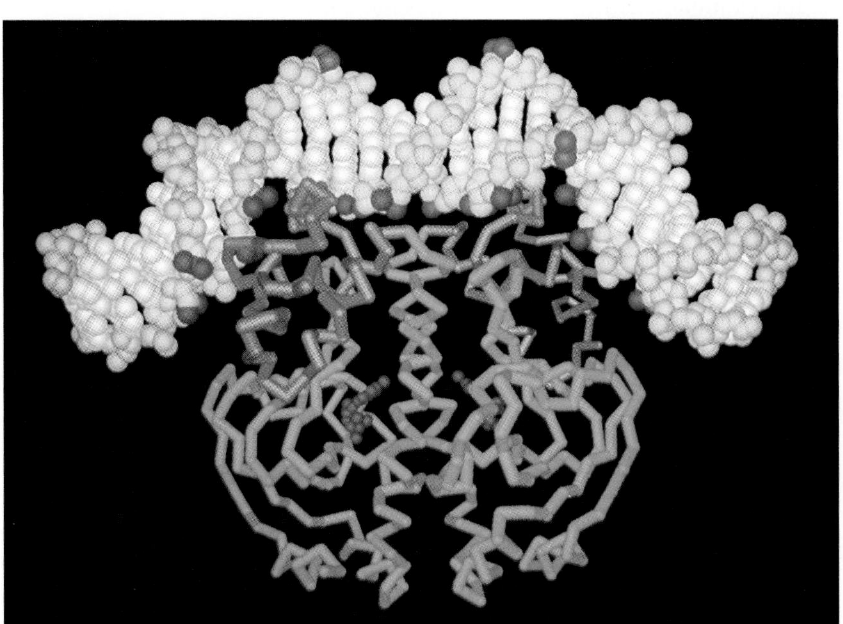

FIGURE **26.22**

Bending of DNA by binding to CRP–cAMP. This model was deduced from the crystal structure of the DNA–protein–cAMP complex. The DNA bases are in light blue, and the sugar–phosphate backbone is in yellow. The DNA-binding domains of the protein—two α helices that contact DNA bases in the major groove (see text)—are in purple and are perpendicular to the plane of the page. The cAMP-binding domain is in blue, and two bound cAMP molecules are in red. On the DNA molecules, those phosphates that are in closest contact with the protein and whose ethylation interferes with protein binding are in red. Those phosphates whose reactivity to ethylation is enhanced (on the outer edge of the bend) are in blue.

Courtesy of T. A. Steitz, *Q. Rev. Biophys.* (1990) 23:205–280. Reprinted with permission of Cambridge University Press.

its host bacterium, the specific controls used are more complex and more subtle than those we have been discussing. Recall from Chapter 25 that infection can have one of two possible outcomes—a lytic cycle of growth, comparable to that of phage T4, or lysogenization, in which the viral chromosome circularizes and undergoes site-specific integration into the host-cell chromosome (see Figure 25.17), with consequent repression of nearly all viral genes. Once lysogeny has been established, the phage chromosome can be maintained as a transcriptionally repressed prophage for many generations. That state of repression can be broken, leading to excision of the viral chromosome as a circular DNA, followed by replication of viral DNA, followed by activation of genes needed to assemble virus particles. The virus must rely on four distinct patterns of gene expression needed for its four physiological states—(1) infection leading to lytic growth, (2) infection leading to establishment of lysogeny, (3) long-term maintenance of lysogeny, and (4) breaking of lysogeny with subsequent lytic growth.

The critical events in λ transcriptional regulation involve two different repressor proteins, called cI and Cro, each of which binds at two different operators. Each operator contains not one repressor-binding site (as in the lactose operon) but three, and each contains promoter sites interspersed with the repressor-binding sites. Transcription from the two promoter–operator sites takes place in opposite directions along the genome (see Figure 26.23). The repressors bind to each of the six operator sites with varying affinities, leading to varying occupancy of each binding site by each repressor under varying physiological conditions. To add to the complexity, the cI repressor also serves, under certain conditions, as a transcriptional activator, promoting the expression of some genes while repressing that of others. To understand the biochemistry involved, we must first consider the genes and their positions on the λ genome (Figure 26.23), as well as the phenotypes of mutants altered in these regulatory genes.

Genes and Mutations in the λ System

Phage mutants defective in establishing or maintaining lysogeny have phenotypes comparable to those defective in *lac* regulation, and these similarities helped Jacob and Monod make the generalizations embodied in the operon model. The lysogenic response involves processes of viral gene inactivation, accomplished by the binding of a repressor to an operator, comparable to the one that keeps the *lac* operon turned off when inducer is absent. The major λ phenotypes and the comparable *lac* mutations are summarized in Table 26.2.

When a defective repressor cannot bind to operators, the mutant phages give clear plaques when they are plated. Normally a λ plaque is turbid, because it contains not only phage but also lysogenized bacteria, which continue to grow, because they are **immune** to infection by additional phage. By contrast, mutants that cannot establish lysogeny give clear plaques, because all of the cells are lysed.

FIGURE **26.23**

The early regulatory region of phage λ. The sites of action of several regulatory proteins are shown (brown bands on tan ribbon). These proteins are the cI and Cro repressors, the cII activator, cIII (which acts to stabilize cII), the antiterminator N, and the integration and excision proteins Int and Xis. The RNA transcripts colored green and red are synthesized under different conditions and in the directions shown by the arrowheads.

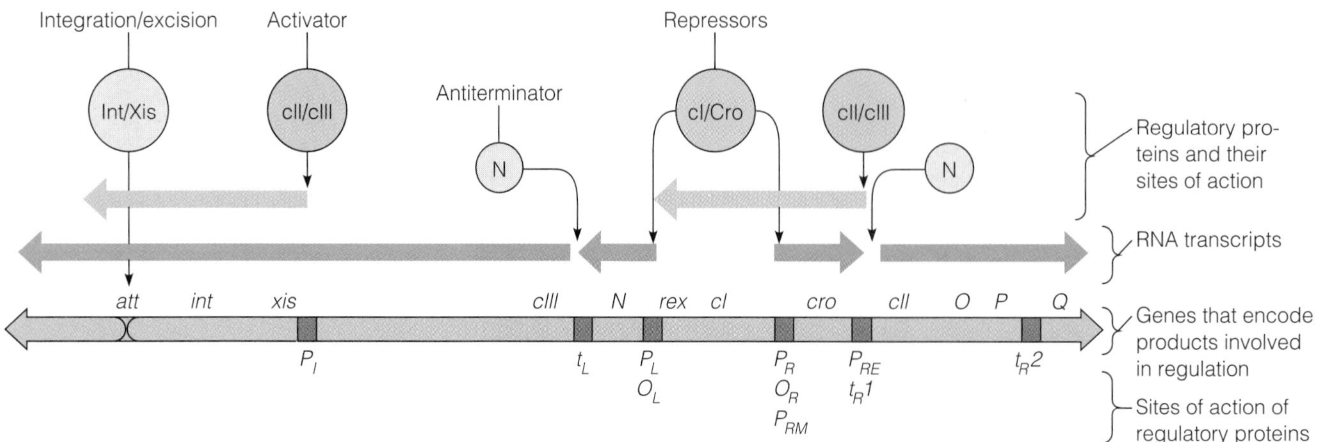

TABLE **26.2 Phenotypes of comparable *lac* and λ mutations**

lac Phenotype	Corresponding λ Phenotype	Regulatory Abnormality
Inducer-constitutive, recessive	Clear-plaque; cannot establish lysogeny	Repressor defective in operator binding
Operator-constitutive, cis-dominant	Virulent; can replicate in a superinfected immune lysogen	Operator unable to bind repressor
Noninducible, trans-dominant	Noninducible (cannot be induced by UV or other treatments)	Repressor cannot bind inducer or be inactivated

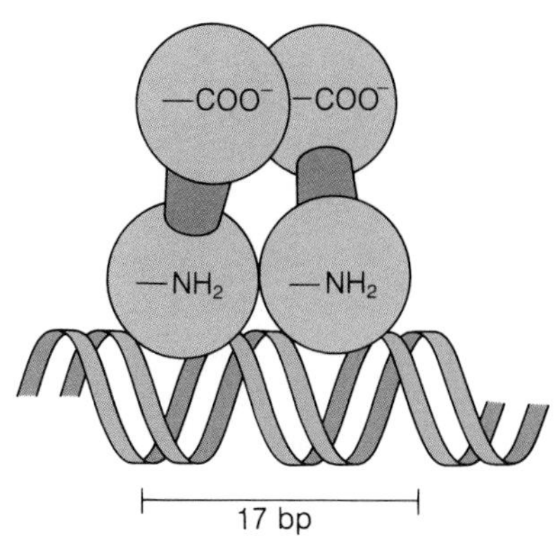

FIGURE **26.24**

Structure and DNA binding of λ cI repressor. The repressor is a dimer and binds to a 17-base-pair region in a λ operator.

From M. Ptashne, *A Genetic Switch: Gene Control and Phage λ* (Cambridge, Mass.: Cell Press & BSI, 1986).

Clear-plaque mutants in λ map in three different genes—*cI, cII,* and *cIII. cI* is the structural gene for one of the repressors mentioned above, and the other two genes control the synthesis of the *cI* protein.

Virulent mutations, which map in operators, also convey a clear-plaque phenotype, but there is an important distinction. Bacteria lysogenic for λ are immune to infection by a second λ phage, because repressor in the cell binds to the operators of any input phage. This is true for *cI* mutants because they contain normal operators, which can bind repressors. However, virulent mutants can produce progeny phage in infection of an immune lysogen, because their operators cannot bind repressor after entry into the cell. These observations are similar in principle to those resulting from the experiments with partial diploids in the *lac* operon.

Regulation of lysogeny in λ phage is similar to regulation of the lactose operon but is more complex.

The λ *cI* Repressor and Its Operators

In 1967, at about the same time that the *lac* repressor was isolated, Mark Ptashne isolated the λ repressor encoded by *cI*. The cI repressor is a dimeric protein, with a subunit M_r of 27,000; it binds through its N-terminal sequences to operator sites with a K_a of about 3×10^{13} M (Figure 26.24). Repressor–DNA interactions have been used to map and characterize what turned out to be two operator regions, one on each side of *cI*. The two operators control divergent transcriptional events from a central regulatory region—leftward (O_L) and rightward (O_R). As shown by footprinting, each operator contains three separate repressor-binding sites, each about 17 base pairs long. Figure 26.25 illustrates this for the O_R region. The three repressor-binding sites are homologous, but not completely so (Figure 26.26), and they are separated by spacer regions of three to seven base pairs. Mutations that confer virulence map within the repressor-binding regions. A fully virulent mutant has at least two mutations—one in O_L and one in O_R.

The λ operators are remarkable in several respects other than their multiple repressor-binding sites. (1) Mutations that affect promoter activity lie between the repressor-binding sites. Thus, operators and promoters are *interspersed*, so the regulatory regions are more properly called O_LP_L and O_RP_R. (2) As seen in Figure 26.26, O_RP_R controls transcription from *two distinct promoters*—one rightward (P_R) and one leftward (P_{RM}). (3) Transcription from O_LP_L and O_RP_R is controlled by *two different repressors*—cI and Cro (the *cro* gene is an acronym, from cI *repressor off*). (4) Under certain conditions the cI repressor is a transcriptional *activator*, not an inhibitor. Another novel feature of this regulatory system is that *cI*

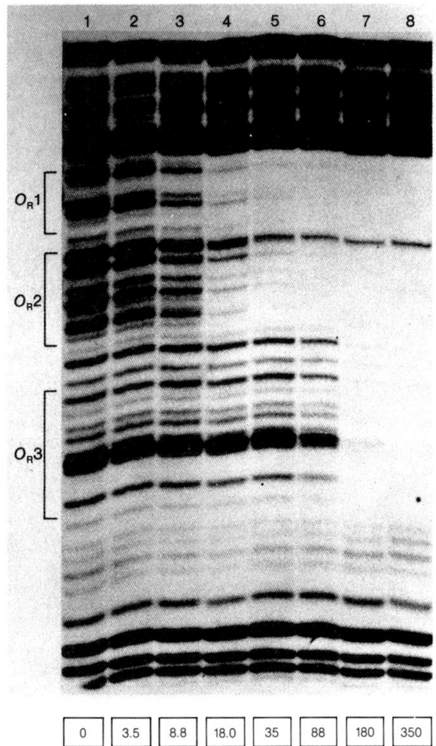

FIGURE **26.25**

Structures of the λ operators, determined from quantitative DNase I footprinting of a restriction fragment containing the three sites in O_R. Repressor concentration was increased from 0 nM (lane 1) to 350 nM (lane 8). Footprinting is described on page 1023.

Courtesy of A. D. Johnson, B. J. Meyer, and M. Ptashne, *Proc. Natl. Acad. Sci. USA* (1979) 76:5062.

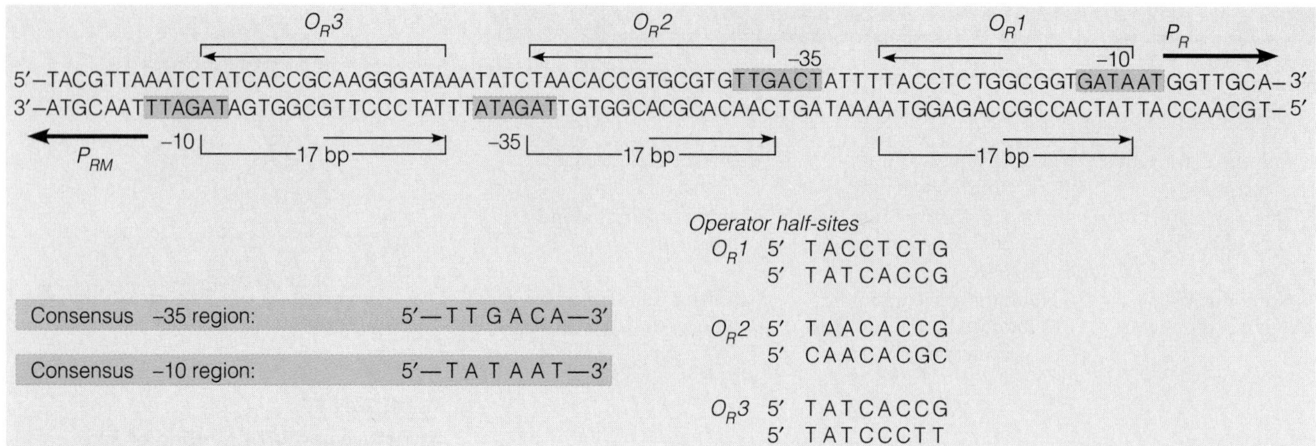

FIGURE **26.26**

The O_RP_R region. The upper diagram shows the nucleotide sequence of the O_RP_R region, including the three repressor-binding sites (O_R1, O_R2, and O_R3), the leftward promoter P_{RM}, and the −35 and −10 regions for the two promoters (shaded purple and blue, respectively). The thin horizontal arrows mark the operator half-sites. Below are the consensus sequences for the −35 and −10 regions and a chart that shows the partial homology among the operator half-sites (that is, the repressor-binding sites).

transcription is initiated from different promoters under different physiological conditions. All of these complexities are related to the need for orderly and efficient phage gene control under quite different physiological conditions, as we shall see.

Early Genes in Phage λ

To understand the significance of *cI* gene regulation, we must identify several λ genes that are expressed early in infection (see Figure 26.23). *cI* and *cro* code for repressors, as just noted, and *cII* and *cIII* both stimulate *cI* synthesis. *rex* is a gene of still unknown function, the only gene aside from *cI* known to be expressed during lysogeny. We encountered *int, xis,* and *att* in Chapter 25, when we discussed site-specific recombination. *O* and *P* function in the initiation of λ DNA replication. The *N* gene product interacts with NusA (see page 997) to prevent termination. The *Q* product activates late gene transcription.

Interactions Between the Two λ Repressors

The interspersed O_LP_L controls transcription of *N*, through interaction of its repressor-binding sites with the cI protein. However, most of the regulatory action occurs at O_RP_R, and it is here that the decision is made between lytic and lysogenic infection. This decision involves interactions of the two repressors, cI and Cro.

Quantitative footprinting experiments (see Figure 26.25) show that of the three repressor-binding sites in O_RP_R, cI binds most tightly to site O_R1, less tightly to O_R2, and still less tightly to O_R3. Moreover, cI binding is *cooperative,* so that when one repressor dimer is bound at O_R1, affinity for a second molecule is increased at O_R2. Cro protein is a dimer of identical 66-residue subunits. It binds considerably less tightly to any of the sites than does cI and in the reverse order. That is, site O_R3 is favored, followed by approximately equal binding at O_R2 and O_R1. Binding is noncooperative.

Although Cro is a repressor, it can also be considered an **antirepressor** because it antagonizes the action of cI in a very specific way. To understand how this works, we must first consider the transcriptional events occurring in the presence of varying levels of cI, illustrated in Figure 26.27. Because of the cooperative binding of cI to its operators, both sites O_R1 and O_R2 are usually occupied in the lysogenic state (Figure 26.27a), even though the intracellular concentration of cI is quite low (about 200 molecules per cell, or 10^{-7} M). This inhibits the rightward transcription of *cro* from its own promoter, but it *activates* the leftward transcription of *cI* from the promoter P_{RM} (the *M* stands for "maintenance," because this is the promoter from which *cI* is transcribed during maintenance of lysogeny). The −10 and −35 regions for the P_{RM} promoter lie within the operators. The evidence that cI really does activate its own transcription in the lysogenic state lies

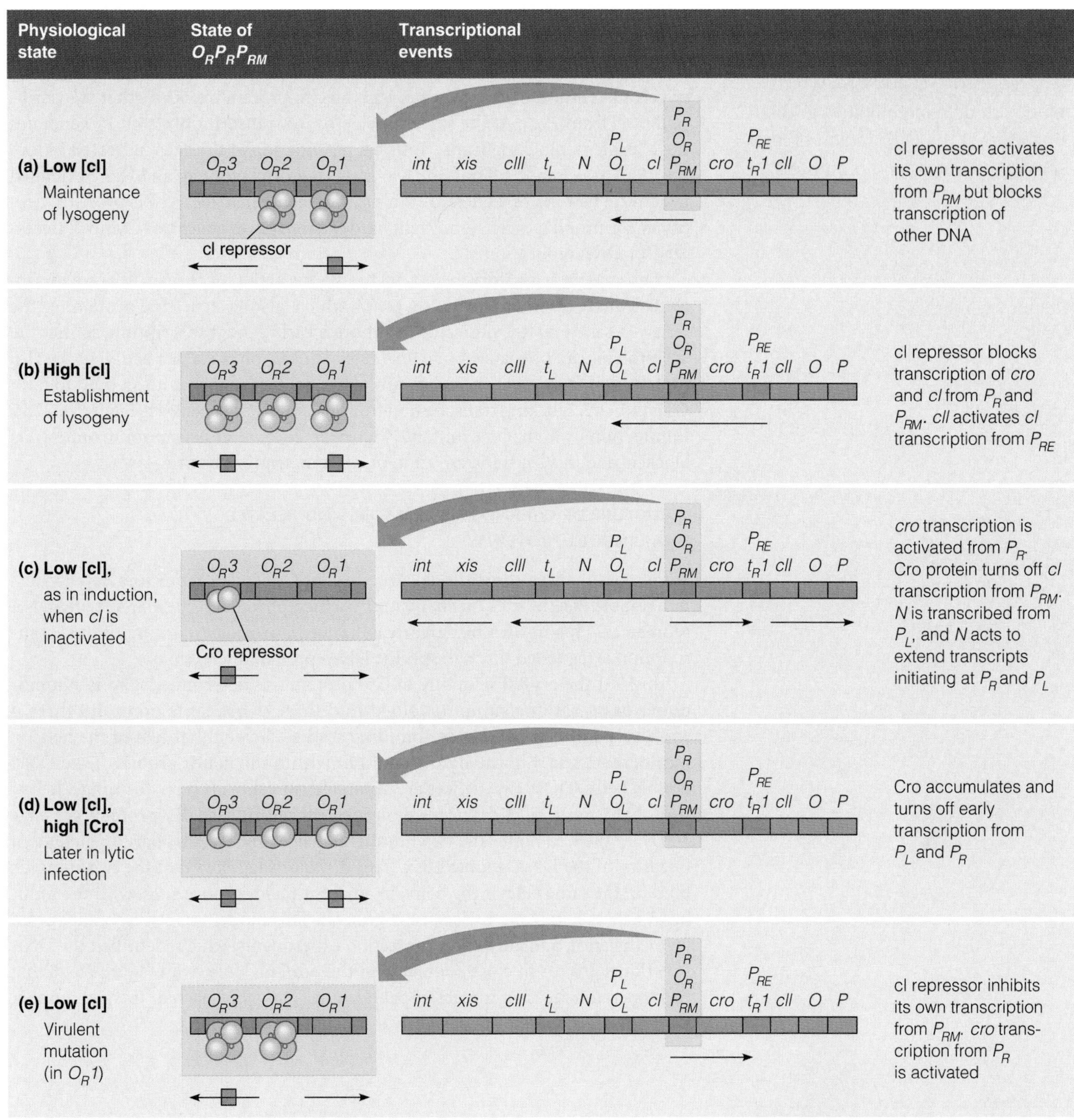

FIGURE **26.27**

cI–Cro interactions in the $O_RP_RP_{RM}$ region. Different physiological conditions lead to different interactions, with varying effects upon transcription of the rest of the λ genome. Gray boxes indicate blocked transcriptional events. Gene *rex,* which does not participate in regulation, is omitted for simplicity.

in the existence of a special class of *cI* mutants whose repressor can bind tightly at O_R1 and O_R2 but cannot stimulate *cI* transcription from P_{RM}.

During *establishment* of lysogeny (Figure 26.27b), when lytic and lysogenic genes are competing to determine the fate of the viral genome, there is a need for larger amounts of cI repressor than can be transcribed from P_{RM}. At this time a different *cI* promoter, called P_{RE}, is activated (*E* stands for "establishment"). In this activation the cII protein binds specifically at the -35 region of P_{RE} and stimulates RNA polymerase binding at that site. This transcriptional event yields a longer *cI* messenger RNA that is more efficiently translated than the message synthesized from P_{RM}. The result is sufficient cI repressor to bind all three sites in O_R and, hence, to block both transcriptional initiation events.

Lysogeny in λ phage is controlled by two repressors, cI and Cro, which bind with differing affinities to three operators in the O_RP_R region of interspersed operators and promoters.

Now consider the events in prophage induction, when lysogeny is broken, leading to a lytic infection (Figure 26.27c). First, cI repressor is inactivated (we shall see shortly how this occurs), and the O_R sites become unoccupied. This permits transcription of *cro* from P_R, and the Cro protein blocks further transcription of *cI* from P_{RM}. At the same time, leftward transcription from P_L generates the N protein, blocking transcriptional termination at the sites indicated in Figure 26.27 as t_R1 and t_L. Thus, the two early transcripts for Cro and N are extended to activate new genes. Leftward transcription generates Int and Xis proteins, necessary for prophage excision. Rightward transcription generates *O* and *P*, necessary for DNA replication.

Subsequent regulatory events, including the action of the gene *Q* protein, activate transcription of late-acting genes, which encode structural proteins of the virus. At this time it is desirable to suppress early gene transcription, so that the late proteins can be made at maximal rates. This involves further action of the Cro protein, which by this time has accumulated to the point that it can bind to both O_R1 and O_L1, blocking transcription from P_R and P_L, respectively (Figure 26.27d). In infection by a virulent mutant (Figure 26.27e), *cI* transcription from P_{RM} is blocked, and this leads to activation of *cro* transcription from P_R.

STRUCTURE OF Cro AND cI REPRESSORS AND RELATED DNA-BINDING PROTEINS

Studies of the three-dimensional structures of Cro and cI repressors have yielded remarkable insight into mechanisms by which proteins recognize specific DNA sequences. This in turn has greatly enhanced our understanding of how transcription is regulated through specific DNA–protein interactions.

In 1981 the crystal structure of Cro protein was determined. Cro is a homodimer of 66-residue subunits folded into three α-helical regions and three β strands (Figure 26.28). Model-building studies showed that two of the helices, numbered 2 and 3 in the figure, could fit within the major groove of the DNA double helix. These two helices are separated by a short β turn, forming a helix–turn–helix motif of the type we described earlier for the CRP protein (see page 1003). In the Cro dimer the two number 3 helices are 3.4 nm apart, the length of one turn of the DNA double helix. This distance suggested that the two subunits bind on the same side of the helix, in adjacent major groove sites, with the number 3 helices lying lengthwise in the grooves. This model was strongly supported by methylation and ethylation protection experiments, which identified the DNA functional groups in close contact with the protein. Note from Figure 26.26 that each operator site is an imperfect palindrome. Therefore, each of the two number 3 helices is in contact with a slightly different set of bases.

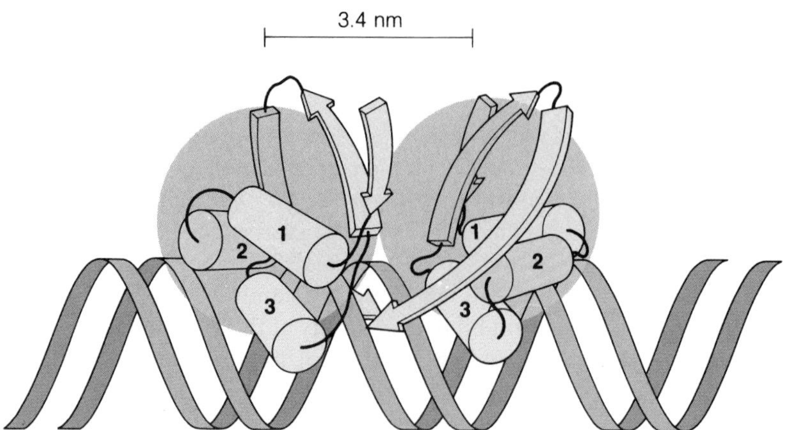

FIGURE **26.28**

Structure of the Cro dimer–operator complex.

From M. Ptashne, *A Genetic Switch: Gene Control and Phage λ* (Cambridge, Mass.: Cell Press & BSI, 1986).

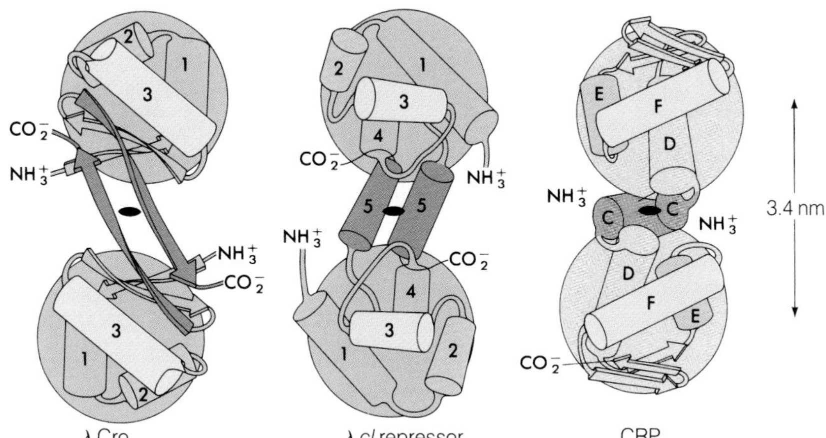

FIGURE **26.29**

DNA binding faces of λ Cro, λ cI repressor, and CRP, showing the helix–turn–helix motif. The motif involves helices 2 and 3 in Cro and cI and helices E and F in CRP. The black ellipses mark centers of symmetry.

Courtesy of T. A. Steitz and I. T. Weber.

The amino acid sequence within and between helices 2 and 3 shows remarkable homology with corresponding sequences in a large family of sequence-specific DNA-binding proteins but *not* with DNA-binding proteins that showed no sequence preference. This observation suggested that the helix–turn–helix motif is a commonly evolved structural element in transcriptional regulatory proteins, at least in prokaryotes. Note that the *lac* repressor also contacts DNA in a helix–turn–helix motif (Figure 26.20). In addition, some eukaryotic proteins use this motif to contact DNA (Chapter 28). A different motif, the **zinc finger,** is present in a large number of eukaryotic DNA-binding proteins (and a few known prokaryotic proteins). More recently, other structural patterns of DNA-binding proteins, including the helix–loop–helix and the leucine zipper, have been described. These motifs are discussed in Chapter 28.

The cAMP receptor protein has the helix–turn–helix motif in its helices E and F, as mentioned on page 1003. Once the three-dimensional structure of λ cI repressor was shown also to have the helix–turn–helix motif, it seemed likely that the helix–turn–helix structure is involved in DNA binding in all of these proteins. The relevant three-dimensional similarities for Cro, CRP, and cI repressor are shown in Figure 26.29. Amino acid sequence homologies among these regions are shown for Cro and cI in Figure 26.30. Note that the sequences, though similar, are not identical. If they were identical, we would not be able to explain how Cro and cI repressors differ in their relative affinities for different operators. cI repressor contains an additional binding determinant—a pair of "arms," or short polypeptide segments that extend from helix 1 and are seen in Figure 26.1, extending around the helix and establishing contacts on the other side of the DNA duplex. These arms probably explain why cI binds more tightly to its operators than Cro does.

The α-3 helix is called the **recognition helix,** because its position deep within the major groove allows it to contact specific DNA bases and hence to determine sequence specificity of binding. The α-2 helix is in contact primarily with DNA phosphates. These electrostatic contacts strengthen binding but do not contribute to specificity. Supporting the concept of α-3 as a recognition helix is that most *cI* mutations that reduce specific binding of repressor to operator DNA alter the amino acid sequence in this region of the protein.

Crystallographic analysis of the respective DNA–protein complexes explains how Cro and cI can bind to the same operator sites with different binding affinities. As shown in Figure 26.31, the residues common to both proteins are in contact with DNA sequence elements common to all of the operators. In both proteins a glutamine residue interacts with one A-T base pair, as shown. cI repressor establishes specificity through a contact in O_R1 with a unique alanine residue,

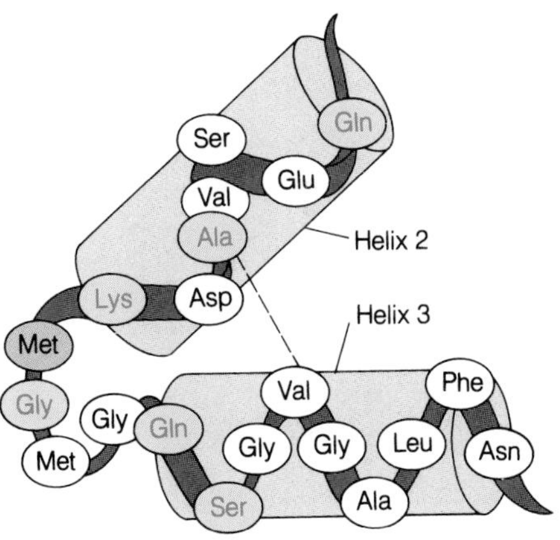

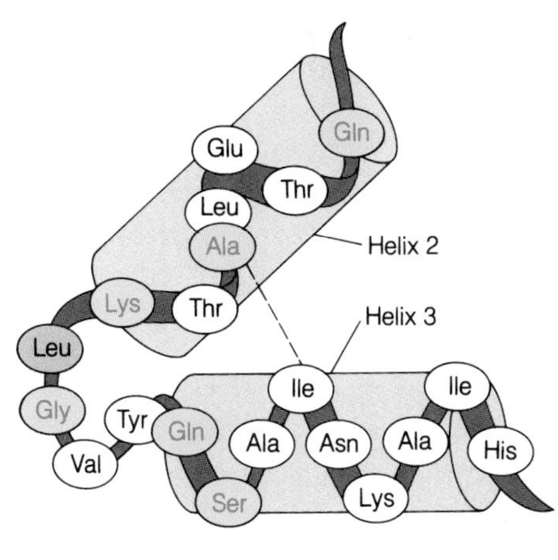

FIGURE **26.30**

Conserved residues in the DNA-binding helices of λ cI repressor and Cro. Conservative substitutions are shown in purple and identities in pink. In both proteins the alanine in helix 2 contacts a residue in helix 3, which helps to position the helices with respect to each other.

From M. Ptashne, *A Genetic Switch: Gene Control and Phage λ* (Cambridge, Mass.: Cell Press & BSI, 1986).

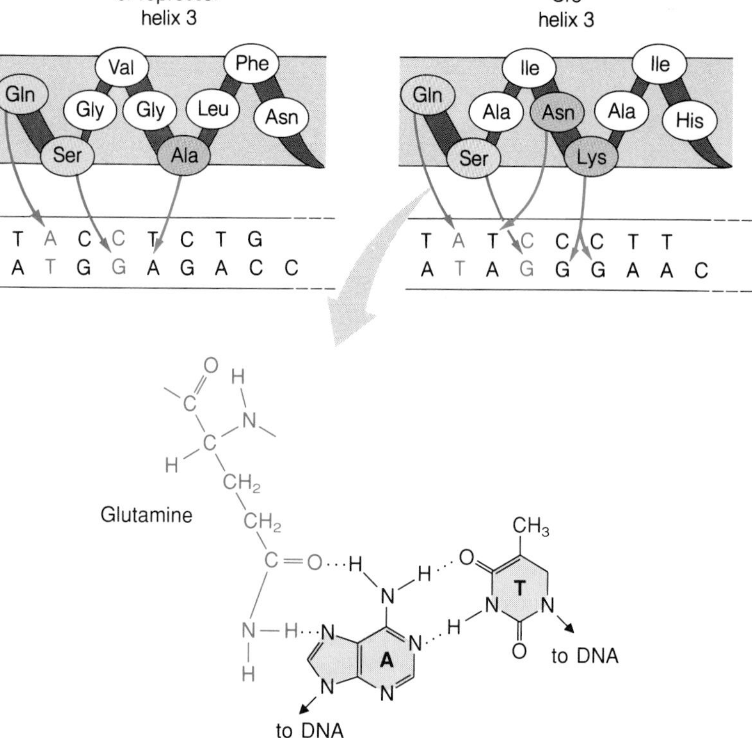

FIGURE **26.31**

Specific amino acid–nucleotide contacts for cI and Cro repressors. The conserved residues (pink) bind to nucleotides common to all of the operators, and unique residues (purple) bind to nonconserved nucleotides in the operators. Also shown is the structure of a glutamine residue in contact with an A-T base pair.

From M. Ptashne, *A Genetic Switch: Gene Control and Phage λ* (Cambridge, Mass.: Cell Press & BSI, 1986).

whereas Cro can be in contact with three specific base pairs in O_R3 with unique asparagine and lysine residues. Also, because the two α-3 helices lie closer together in Cro (2.9 nm) than in cI (3.4 nm), the orientations of these helices with respect to the major grooves of operator DNA are quite different.

The helix–turn–helix motif is widely used in prokaryotic transcriptional regulatory proteins. Specific contacts are made between DNA major groove bases and amino acids in a recognition helix.

THE SOS REGULON: ACTIVATION OF MULTIPLE OPERONS BY A COMMON SET OF ENVIRONMENTAL SIGNALS

How is the λ cI repressor inactivated when the prophage is excised and begins a cycle of lytic growth? Various DNA-damaging treatments are known to induce λ prophages, including ultraviolet irradiation, inhibition of DNA replication, and chemical damage to DNA. Evidently the virus finds it advantageous to leave a damaged cell, like rats leaving a sinking ship. Because of the similarity in genetic control between the λ and *lac* systems, investigators sought a small molecule, perhaps a nucleotide, that would accumulate after these treatments and that might be the ligand that binds to cI and inactivates it. Surprisingly, the λ repressor was found to be inactivated by a quite different mechanism—proteolytic cleavage. Analysis of this cleavage reaction revealed the SOS system described in Chapter 25 as one of the elements in error-prone DNA repair, in which the genes are controlled by a single repressor–operator system. Such a set of unlinked genes, regulated by a common mechanism, is called a **regulon.** The heat-shock genes, all activated by a transient temperature rise, make up another regulon.

The control elements in the *E. coli* SOS regulon are the products of genes *lexA* and *recA*. We have encountered RecA protein before, in its role of stimulating DNA strand pairing during recombination. Remarkably, this small protein has an enzymatic activity in addition to the activities involved in recombination. When bound to single-strand DNA, it can stimulate proteolytic cleavage of the proteins encoded by *cI*, *lexA*, and *umuD*. LexA is a repressor that binds to at least 15 different operators scattered about the *E. coli* genome (Figure 26.32). Each operator controls the transcription of one or more proteins that help the cell respond after environmental damage that might harm the genetic apparatus. These proteins in-

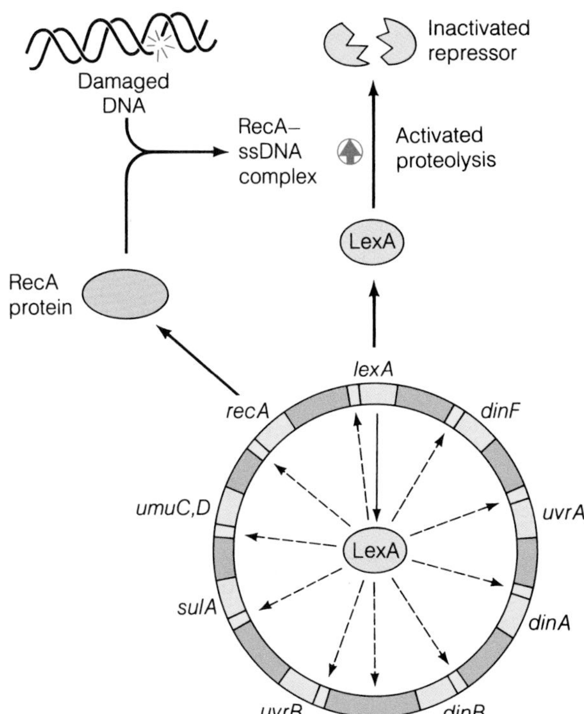

FIGURE **26.32**

The SOS regulon. The figure shows locations on the *E. coli* chromosome of some of the genes controlled by the LexA repressor. *dinA* is the structural gene for DNA polymerase II, while *dinB* and *dinF* are *d*amage-*in*ducible genes of unknown function. LexA repressor (pink) is inactivated by proteolysis, which is somehow enhanced by a complex of RecA protein (teal) and single-strand DNA.

clude the gene products of *uvrA* and *uvrB,* involved in excision repair; *umuC,D,* involved in error-prone mutagenesis; *sulA,* involved in cell division control; *dinA,* the structural gene for DNA polymerase II; *recA* itself; *lexA* itself; and several genes of unknown function, including *dinB* and *dinF.*

In a healthy cell, *lexA* and *recA* are expressed at low levels, with sufficient LexA protein to turn off the synthesis of the other SOS genes completely. LexA protein does not completely abolish either *lexA* transcription or that of *recA*. The trigger that activates the SOS system after damage is single-strand DNA. As we have seen, UV irradiation generates gapped DNA structures, and so do other conditions that induce the SOS system. RecA binding within a gap activates LexA proteolysis by a mechanism not yet clear. Intracellular levels of LexA decrease, removing the LexA barrier to *recA* transcription. RecA protein accumulates in large amounts. Simultaneously, cleavage of the LexA protein activates transcription of all genes under *lexA* control. In a λ lysogen, cleavage of λ cI repressor is stimulated as well, activating prophage excision and replication, as discussed earlier.

The SOS regulon is activated by DNA damage, which stimulates RecA to cause proteolytic cleavage of the LexA and λ cI repressors.

Sequencing of operators that respond to LexA has yielded a consensus sequence, with 7 highly conserved bases in a 20-base-pair region. However, different LexA-sensitive genes have this sequence located quite differently with respect to the transcriptional start site. Therefore, it appears that the exact location of bound repressor is not critical to ensure that transcription will be inhibited.

BIOSYNTHETIC OPERONS: LIGAND-ACTIVATED REPRESSORS AND ATTENUATION

The lactose operon is involved with catabolism of a substrate. Therefore, the gene products are not needed unless the substrate is also present to be consumed. A different situation is encountered with genes whose products catalyze biosynthesis—of an amino acid, for example. Because biosynthesis consumes energy, it is to the cell's advantage to use the preformed amino acid, if it is available. Therefore, the regulatory goal is to repress gene activity, by turning *off* the synthesis of enzymes in the pathway when the end product is available. Regulation of the *E. coli trp* operon, which controls the five reactions from chorismic acid to tryptophan

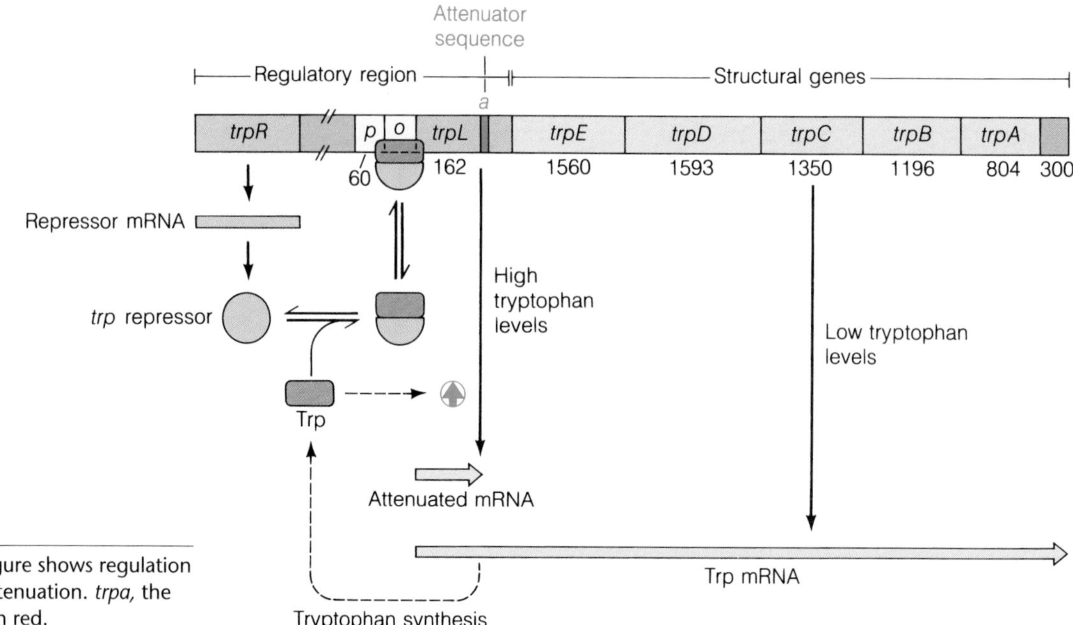

FIGURE **26.33**

The *trp* operon. The figure shows regulation by *trp* repressor and by attenuation. *trpa,* the attenuator site, is shown in red.

(see Figure 21.14), demonstrates two ways of accomplishing this shutdown: (1) a repressor design in which binding of a small-molecule ligand *activates* the repressor, rather than inactivating it, and (2) premature termination of transcription.

The *trp* repressor inhibits tryptophan synthesis by binding as the repressor–tryptophan complex to the *trp* operator, blocking transcription.

The *trp* operon consists of five adjacent structural genes whose transcription is controlled from a common promoter–operator regulatory region (Figure 26.33). The *trp* repressor, a 58-kilodalton protein encoded by the nonadjacent *trpR* gene, binds a low-molecular-weight ligand, namely tryptophan. However, in this case the protein–ligand complex is the *active* form of the protein, which binds to the operator and blocks transcription. When intracellular tryptophan levels decrease, the ligand–protein complex dissociates and the free protein ("aporepressor") leaves the operator, so that transcription is activated. If we call lactose an inducer in a catabolic system, it is appropriate to call tryptophan a **corepressor** in this anabolic system.

The crystal structure of the *trp* repressor–DNA complex shows a helix–turn–helix motif, comparable to that seen with the λ cI, Cro, and *lac* repressors (Figure 26.34); binding tryptophan to this protein reorients the helices to activate binding to DNA. Remarkably, this model shows no direct contacts between residues in the recognition helices and specific DNA bases. It has been proposed that bound water molecules make sequence-specific contacts between amino acids in the recognition helix and nucleotides in the operator.

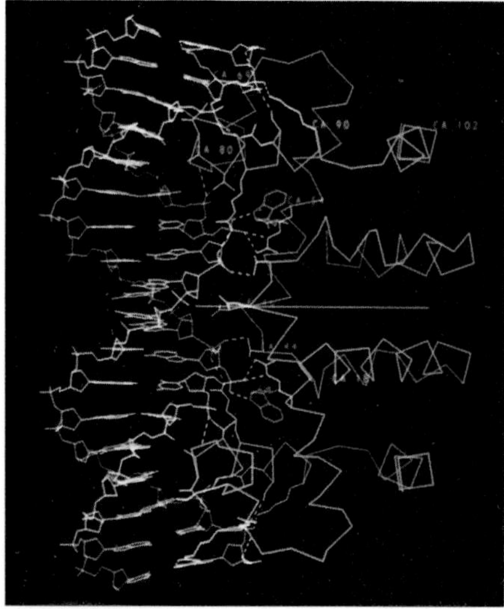

FIGURE **26.34**

A model of the *trp* repressor–operator complex. DNA is in gold, the α-carbon trace of the repressor backbone is in blue, and the side chains that contact the operator are in green. The bound tryptophan is in red.

Courtesy of P. B. Sigler from Z. Otwinowski et al., *Nature* (1988) 335:321–443, Macmillan Magazines, Inc.

The *trp* operon has an additional regulatory feature, now known to be involved in controlling numerous biosynthetic operons. Charles Yanofsky found that the activities of the *trp* enzymes varied over a 600-fold range under different physiological conditions, more than could be accounted for by a repressor–operator mechanism alone. Analysis revealed a second mechanism, called *attenuation,* that involves early termination of *trp* operon transcription under conditions of tryptophan abundance. Note from Figure 26.33 a 162-nucleotide sequence called *trpL,* the *trp* leader region. A site called *a,* the attenuator, is 133 nucleotides from the 5′ end of the *trpL* sequence. When tryptophan levels are high, transcription terminates at *a,* to give an attenuated 133-nucleotide transcript rather than the complete 7000-nucleotide *trp* mRNA. The structural genes are not transcribed, so tryptophan is not synthesized.

Critical to understanding the mechanism of attenuation is the presence of four oligonucleotide sequences in the *trp* leader region that are capable of base-pairing to form stem–loop structures in the RNA transcript (Figure 26.35). In the

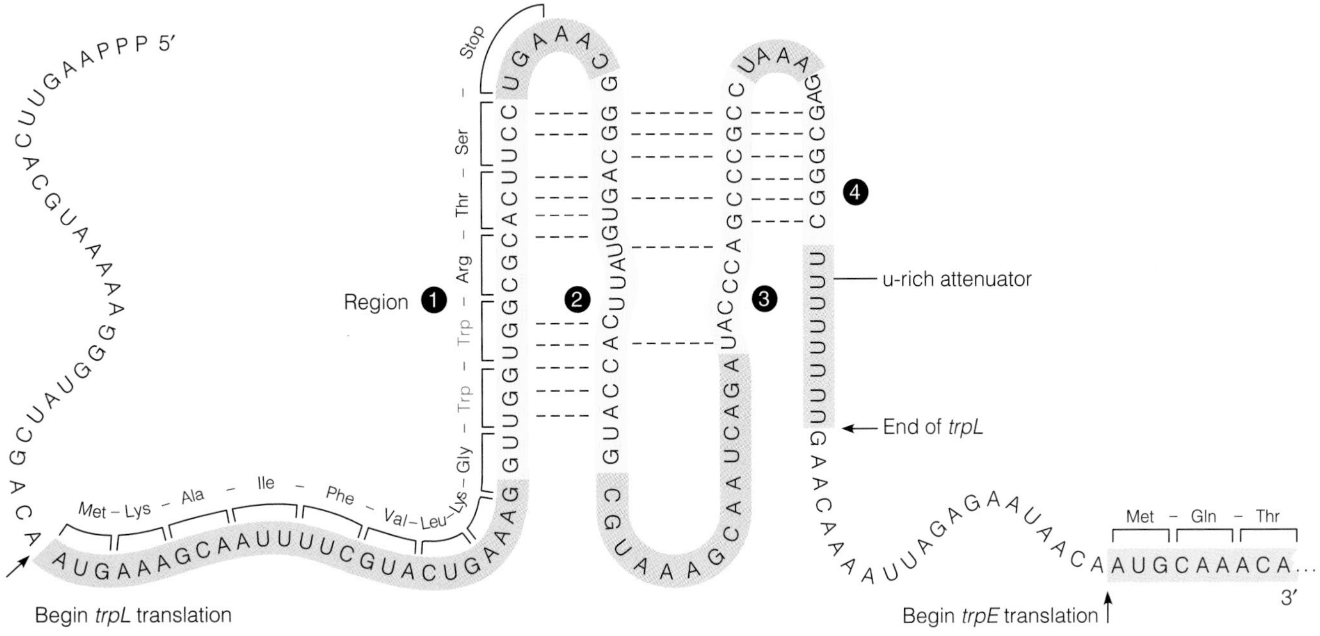

FIGURE **26.35**

RNA base sequence of the *trp* leader region. The four internally complementary sequences that participate in attenuation (yellow) are shown, as well as the two *trp* codons in region 1 that act as a pause site for RNA polymerase. The translational stop codon after region 1 (see Chapter 27) may serve to prevent needless translation of those few full-length messages that are produced despite attenuation.

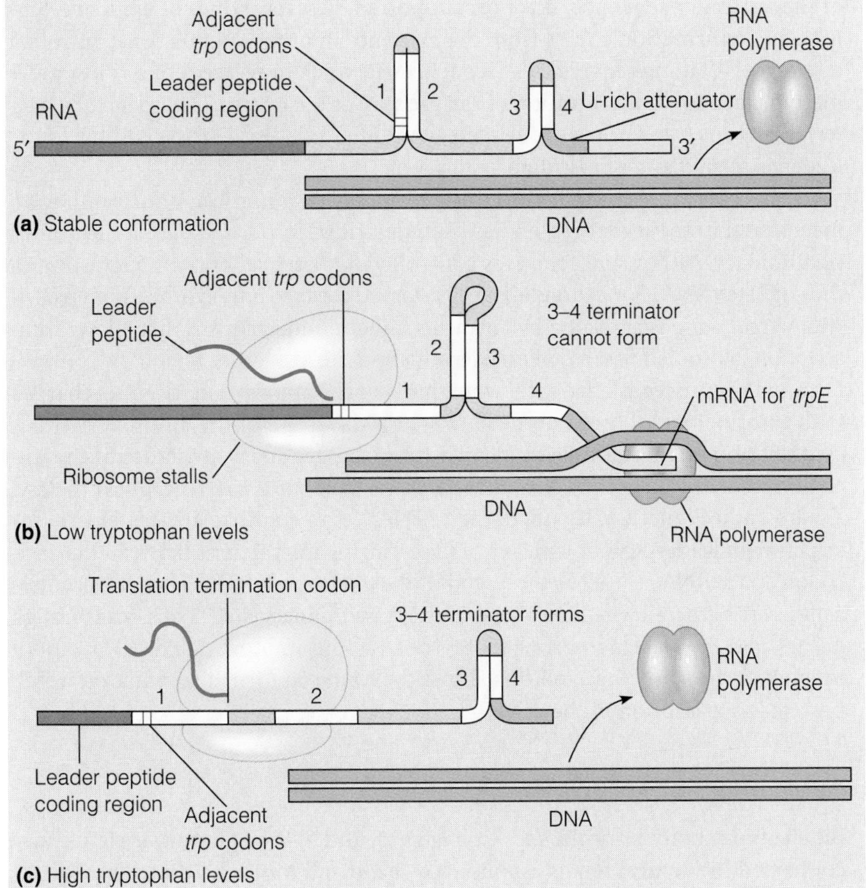

FIGURE **26.36**

Mechanism of attenuation in the *trp* operon. (a) Most stable conformation for leader mRNA. **(b)** Conformation for leader mRNA at low tryptophan levels. **(c)** Conformation for leader mRNA at high tryptophan levels.

most stable conformation (Figure 26.36a), region 1 pairs with 2, and region 3 pairs with 4, to give two stem–loops. The 3–4 structure, being followed by eight U′s, is an efficient transcription terminator, because it resembles the factor-independent terminator structure shown in Figure 26.15.

When tryptophan levels are low (Figure 26.36b), formation of the 3–4 stem–loop is inhibited, and termination does not occur at the attenuation site. Note that region 1 contains two tryptophan codons (see Figure 26.35). In prokaryotes, translation is coupled to transcription, so a ribosome can begin translating a message from its 5′ end while the message is still being synthesized at its 3′ end. In this case, the ribosome stalls when it reaches the two tryptophan codons, because there is insufficient tryptophanyl-tRNA to translate them. The presence of the bulky ribosome prevents region 1 from base-pairing with 2, leaving region 2 free to base-pair with 3. Once region 3 is unavailable to base-pair with 4, the 3–4 stem–loop transcriptional terminator cannot form, and the entire message is synthesized. Conversely, when tryptophan is abundant (Figure 26.36c), the ribosome does not stall, thereby occluding region 2 and allowing the 3–4 stem–loop structure to form, which leads to transcription termination on the 3′ side of 3–4.

Neither the *trpR* system nor the attenuator is simply an on–off system. Both respond in graded fashion to the intracellular tryptophan level. Even though both systems are controlled by the same signal, the action of two distinct control systems greatly extends the possible range of transcription rates of the *trp* operon, giving maximum efficiency to regulation of these genes. At low tryptophan concentration the repressor–operator interaction is the principal regulatory mechanism, whereas the effects of attenuation are more significant at moderate to high tryptophan levels.

Attenuation is a regulatory mechanism in which ribosome positioning on an mRNA determines whether transcription of an operon will terminate before transcription of the structural genes begins.

Although the foregoing model was originally proposed simply by inspection of the *trp* leader sequence, it is now supported by several lines of evidence. Significant confirmation comes from the existence in other attenuation-controlled operons of "stalling sequences"—sequences at which movement of a ribosome is inhibited at a low concentration of the product of the operon. These include bacterial operons for synthesis of leucine, with four adjacent leucine codons in the leader sequence, and of histidine, with seven.

Transcriptional control via a quite different termination–antitermination mechanism in the leader region has been described in *Bacillus subtilis*. Synthesis of aminoacyl-tRNA synthetases is controlled by the level of aminoacylation of the cognate tRNA. For example, in cells starved for tyrosine, synthesis of tyrosyl-tRNA synthetase is activated by an antitermination mechanism that allows transcription to proceed past a potential termination site in a leader region. By contrast, termination proceeds efficiently when most of the appropriate tRNA is charged with tyrosine, and there is no need to synthesize more of the aminoacyl-tRNA synthetase. The mechanism involves a complex secondary structure of the upstream leader sequence for *tyrS*, the structural gene for tyrosyl-tRNA synthetase. As shown in Figure 26.37, uncharged $tRNA^{Tyr}$ can stabilize an antiterminator structure in leader RNA, by means of base-pairing interactions between the anticodon and mRNA, as well as the 3′ end of the uncharged tRNA. These latter interactions are hindered when the 3′ end of tRNA is aminoacylated. Thus, readthrough occurs, and the gene is transcribed. By contrast, the aminoacylated tRNA cannot interact in the same way, and this allows the leader to form a terminator, which prevents transcription of the gene.

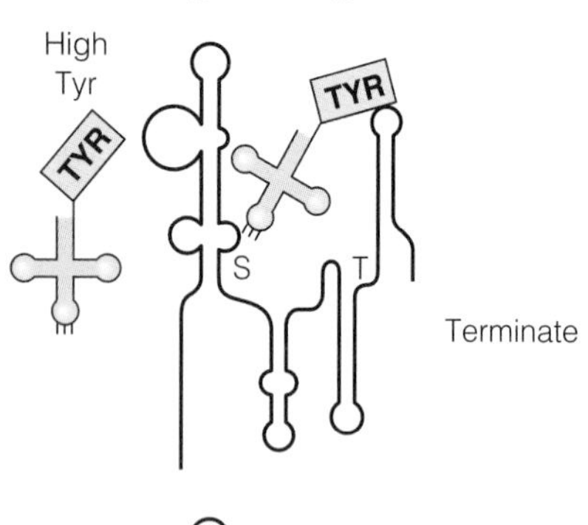

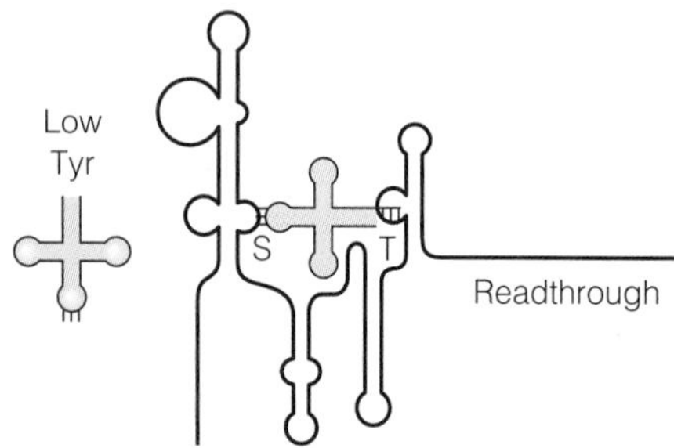

FIGURE **26.37**

Model for induction of *B. subtilis tyrS* gene by uncharged $tRNA^{Tyr}$. Readthrough leading to gene expression is favored when $tRNA^{Tyr}$ is in the uncharged state, as described in the text.

Courtesy of T. M. Henkin, *Mol. Microbiol.* (1994) 13:381–387. © 1994 Blackwell Science, Ltd.

OTHER FORMS OF REGULATION

Biochemical analyses of the *lac*, λ phage, *trp*, and SOS regulatory systems have confirmed the central tenets espoused by Jacob and Monod—that gene expression is regulated at the level of transcription and that specific protein–DNA interactions control the rate of transcription, primarily by regulating transcriptional initiation. These analyses have also revealed several important variations on that simple theme—including positive control of initiation, interspersed operators and promoters, dual proteins binding to the same site, multiple operons controlled by the same repressor, induction of DNA bending by regulatory proteins, and early termination as a regulatory mechanism. Thus, it is clear that the repressor–operator mode of regulation is far from monolithic. Here we briefly describe two other well-studied operons that reveal additional themes.

The Galactose Operon

The **galactose operon** (Figure 26.38) controls the utilization of galactose, one of the products of lactose cleavage by *lac* operon enzymes. The *gal* operon is regulated negatively by a repressor in a manner comparable to *lac* regulation, except that the repressor gene (*galR*) is unlinked to the structural genes. The novelty of *gal* comes from the existence of overlapping promoters (S1 and S2), leading to transcripts that are initiated just five nucleotides apart. Transcription from the start point called S1 is catabolite activated. In vivo, S1 is silent when glucose is present, and in vitro, the cAMP–CRP complex must be present for transcription to start from this promoter. The other promoter, S2, is used when glucose is present. The details and rationale for this dual regulation are not clear, but it is significant that galactose has a biosynthetic fate in addition to its role as an energy substrate. UDP-galactose is used in synthesis of cell wall lipopolysaccharide. The second promoter may exist to ensure that sufficient UDP-galactose is made available even when the cell is using glucose as its prime energy source.

The galactose operon is transcribed from two overlapping promoters, providing for diverse regulatory signals.

The Arabinose Operon

The **arabinose operon** presents another unusual regulatory feature—namely, action of a single protein as both a positive and a negative transcriptional regulator, depending upon the binding of particular ligands (Figure 26.39). The three structural genes in this operon—*araB, araA,* and *araD*—encode enzymes that convert arabinose to xylulose-5-phosphate, a pentose phosphate pathway intermediate (see page 514). The *araC* gene encodes a regulatory protein that binds to arabinose, the inducer of the operon. Binding of the AraC–arabinose complex at a site called *araI* activates transcription of *araBAD,* but only when the cAMP–CRP complex is bound at an adjacent site. Thus, whereas the *lac* operon requires one protein to be bound and one to be dissociated for maximal transcription, the *ara* operon requires two proteins to be bound at adjacent sites.

When arabinose levels are low (Figure 26.39a), the AraC protein acts as a repressor and binds to two operator sites, $araO_1$ and $araO_2$, as well as to *araI*, forming a looped structure. Binding to $araO_1$ inhibits transcription of the *araC* gene itself; thus, *araC* is autoregulated at the level of its own transcription. AraC molecules bound to $araO_2$ and *araI* interact with each other to form a DNA loop. This

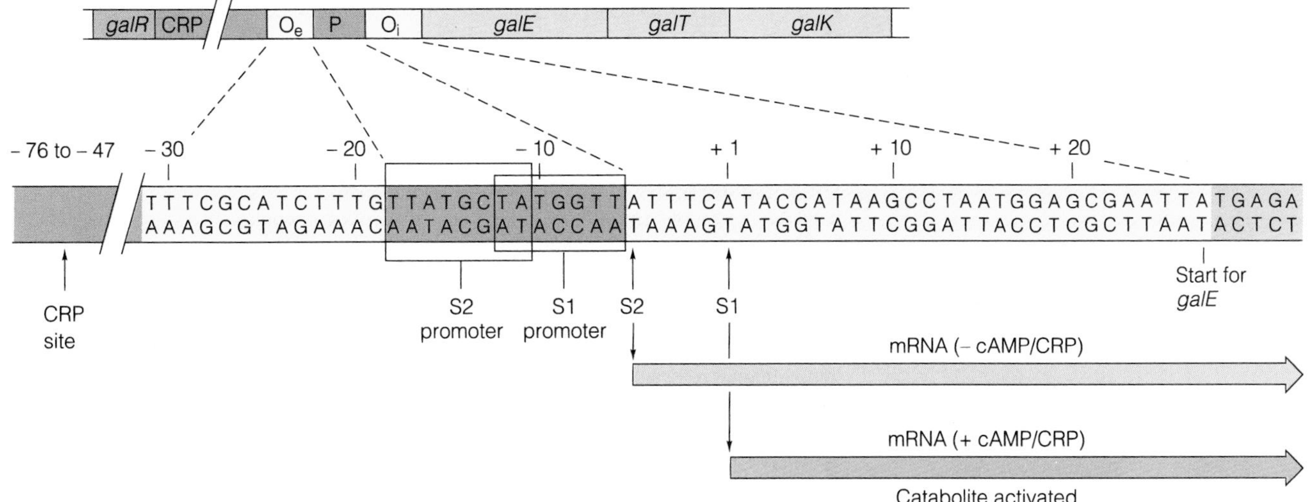

FIGURE **26.38**

Map of the *E. coli gal* operon and nucleotide sequence of the regulatory region. The −10 regions of the two overlapping promoters are shown in brown. The transcriptional start site for each promoter is shown with an arrow. *galR* = repressor gene, *galE* = epimerase gene, *galT* = transferase gene, and *galK* = kinase gene.

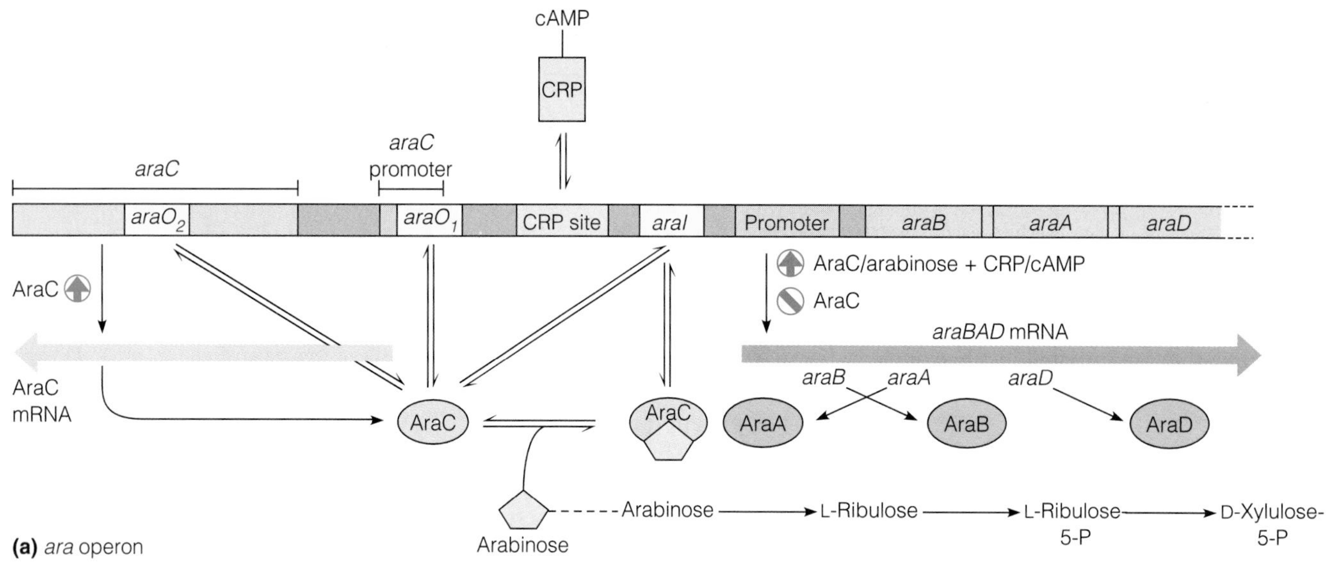

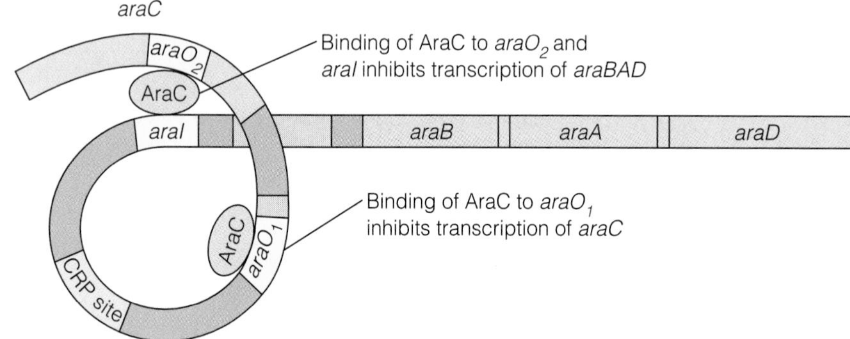

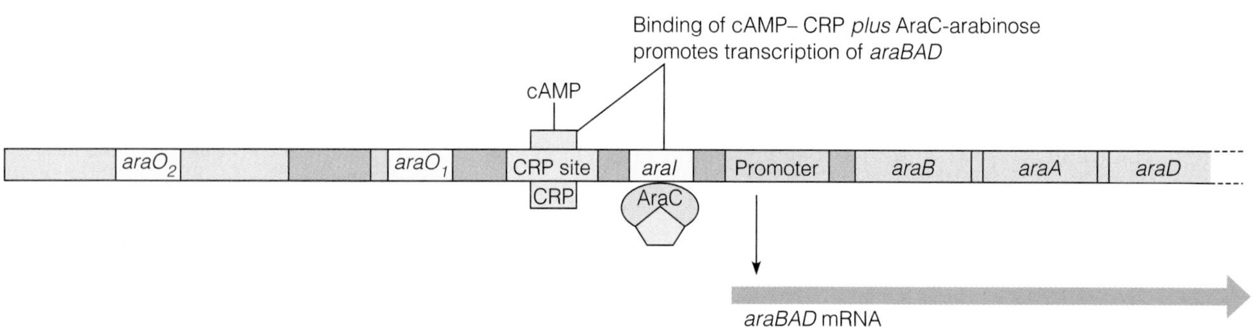

FIGURE **26.39**

The *E. coli ara* operon. (a) An overview of the operon and its regulatory elements. **(b)** AraC as a negative transcriptional regulator at low arabinose levels. Binding of AraC to the *ara*O_1 site inhibits transcription of *araC* and thus provides feedback control over AraC levels. Binding of AraC to *ara*O_2 and *araI* creates a DNA loop and inhibits transcription of *araBAD,* thus downregulating arabinose utilization. **(c)** AraC–arabinose complex as a positive transcriptional regulator at high arabinose levels. High arabinose levels drive the formation of the complex. Provided that CRP–cAMP is bound to the CRP site, the repression loop is broken, and the binding of the AraC–arabinose complex to *araI* promotes *araBAD* transcription, thus promoting arabinose utilization through the actions of the AraB, AraA, and AraD enzymes.

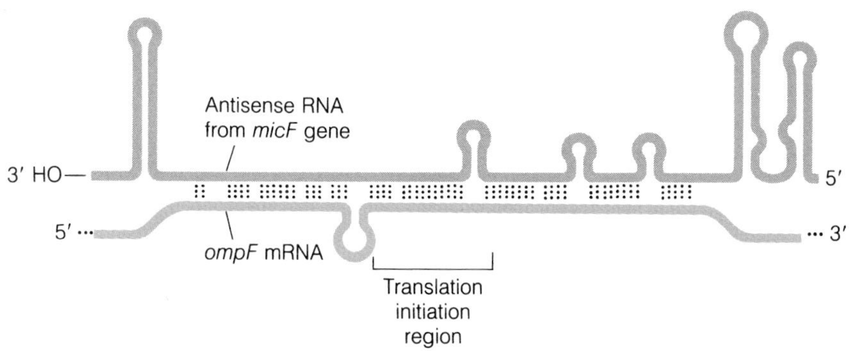

FIGURE **26.40**

Inactivation of *ompF* mRNA by pairing with antisense RNA from the *micF* gene. A change in osmolarity stimulates transcription of the *micF* gene. The transcript is largely complementary to a region in *ompF* RNA that includes the translational start site. Hairpin loops within the sequences allow base pairing between complementary regions on the two mRNAs, and in this way both transcripts are prevented from serving as templates for protein synthesis.

structure causes repression of the *araBAD* genes. The use of the same protein as both positive and negative control elements is unusual but not unprecedented. Recall that the λ cI repressor can either activate or inhibit its own synthesis, depending on the operator sites to which it binds.

The arabinose operon is controlled by a protein that can act as either a positive or a negative transcriptional regulator, depending on whether arabinose is bound to it.

Control by Antisense RNA

Lest you conclude that all prokaryotic transcriptional regulation is simply a matter of repressors, operators, and transcriptional activators, let us mention two other regulatory mechanisms. Though more a regulator of translation than transcription, **antisense RNA** has been shown to control the expression of several prokaryotic and eukaryotic genes. Antisense RNA is the transcription product of the DNA sense strand—the strand that does not encode a protein. Presumably this RNA can pair with an mRNA species, thereby inhibiting translation of that species.

A good example of antisense regulation involves *ompC* and *ompF*, two genes that encode outer membrane proteins in *E. coli*. These genes are osmoregulated: Cells respond to growth in a medium of high osmolarity by shutting down the synthesis of OmpF protein and activating the synthesis of OmpC, so that the total amount of protein is constant. The postulated mechanism of the *ompF* shutoff is shown in Figure 26.40. High osmolarity in some way triggers synthesis of an antisense RNA, the product of the *micF* gene. This RNA is partly complementary to sequences in the 5′ end of *ompF* mRNA. The *micF* RNA inactivates the *ompF* message, presumably by annealing to it, thereby forming a duplex RNA in vivo. The translational initiation sequences of OmpF mRNA, which must be single-stranded to direct translation, are included in this duplex. This is probably responsible for blocking translation of the message.

Another gene regulated by antisense RNA is *crp*, the structural gene for cAMP receptor protein. Also, the transposon Tn10 uses an antisense mechanism to regulate the level of its transposase. A promising approach in chemotherapy is the development of antisense oligonucleotides or chemically modified derivatives, which can be introduced into human cells and inactivate selected target genes by annealing to a gene or mRNA and forming a duplex structure. This approach is receiving intense scrutiny for possible treatment of viral diseases and cancer.

The Stringent Response

The last control mechanism presented here—the **stringent response**—was discovered more than three decades ago but is not yet understood in detail. The synthesis of ribosomal and transfer RNAs in bacteria is inhibited when protein synthesis is blocked by amino acid starvation. This inhibition evidently prevents the unnecessary production of translational machinery when protein synthesis is blocked. During amino acid starvation a regulatory nucleotide, **guanosine 3′,5′-tetraphosphate** (ppGpp), accumulates and somehow inhibits rRNA and tRNA synthesis. A class of mutants, called *relaxed*, does not accumulate ppGpp under these conditions and does not show the stringent response; that is, amino acid starvation does not

P P OCH_2 O guanine
P P O OH
ppGpp

Guanosine 3′,5′-tetraphosphate

cause inhibition of rRNA or tRNA synthesis. The action of ppGpp in controlling rRNA gene transcription is not yet known. However, the events between ribosomal transcription and rRNA maturation are reasonably well understood, as we shall see in the next section.

Posttranscriptional Processing

mRNA TURNOVER

A major aspect of messenger RNA metabolism in eukaryotes is the events occurring *after* transcription, events that are necessary for messages to move from the nucleus to their sites of utilization in the cytosol. We discuss these events in Chapter 28. In prokaryotes, by contrast, mRNAs are used in protein synthesis directly. In fact, as we noted in our discussion of attenuation, a nascent mRNA serves as a template for translation while still in the process of being synthesized.

The major posttranscriptional event in metabolism of prokaryotic mRNA is its own degradation, which in most cases is quite rapid. A few bacterial mRNAs, notably those encoding outer membrane proteins, are long-lived; however, most bacterial messages have half-lives of only 2 to 3 minutes. This short life span means that genes that are expressed must be transcribed continuously and that most mRNA molecules are translated only a few times. Although this might seem energetically wasteful, it is consistent with prokaryotic lifestyles, which necessitate rapid adaptation to environmental changes. Earlier we noted the selective advantage to bacteria of expressing the genes for lactose utilization only when an inducer is present. By the same token, it would be wasteful for the cell to continue producing these proteins after lactose was exhausted from the milieu. Rapid degradation of *lac* mRNA ensures that the energetically wasteful synthesis of these proteins will cease soon after the need for those proteins is gone.

Although we have known about the instability of bacterial mRNA for more than three decades, we still understand surprisingly little about the pathway of degradation. There are probably overlapping mechanisms, involving hydrolysis by nucleases and phosphorolysis by polynucleotide phosphorylase. We do know that degradation starts from the 5′ end, which is important because translation also starts from the 5′ end. If degradation were to start from the 3′ end, a ribosome starting from a 5′ end might never reach an intact 3′ end. There is reason to think that mRNA degradation sometimes starts with the action of ribonuclease III, an enzyme specific for duplex RNA, which could cleave in stem–loop structures and create sites for exonucleolytic attack. RNase III is actually involved in the maturation of certain phage mRNAs as they undergo posttranscriptional processing, but this involvement is not known to occur with bacterial mRNAs.

POSTTRANSCRIPTIONAL PROCESSING IN THE SYNTHESIS OF rRNA AND tRNA

Both ribosomal RNAs and transfer RNAs are synthesized in the form of larger transcripts (pre-rRNA and pre-tRNA, respectively), which undergo cleavage at both ends of the transcript, en route to becoming mature RNAs. The total amount of DNA encoding these RNAs amounts to less than 1% of the *E. coli* genome, but because of the instability of mRNA (which is encoded by the remaining 99%), rRNA and tRNA constitute about 98% of the total RNA in a bacterial cell. It is important to realize also that transcription of rRNA genes is extremely efficient when cells are growing rapidly. The intracellular concentrations of ribonucleoside triphosphates are important control elements here; ATP, whose level is high in rapidly growing cells, activates rRNA gene transcription by stabilizing the relevant open-promoter complexes.

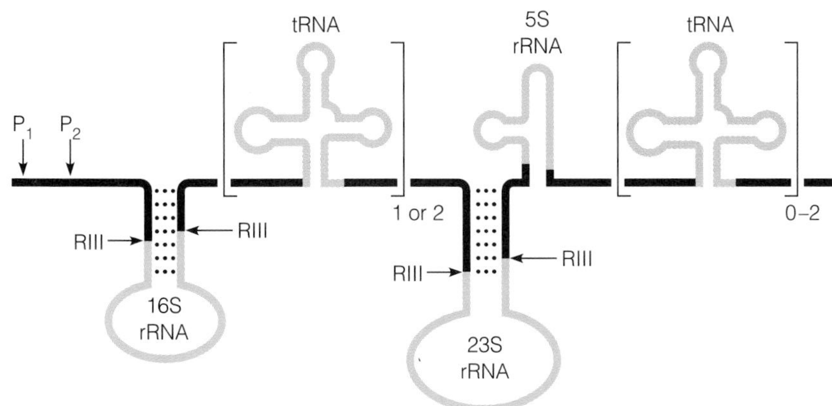

FIGURE **26.41**

Structure of *E. coli* 30S pre-rRNA. Sequences complementary to two promoter sites (P_1 and P_2), RNase III cleavage sites (RIII) that release 16S and 23S species, and the locations of tRNA sequences embedded within the transcript are shown.

rRNA Processing

The *E. coli* genome contains seven different operons for rRNA species. Each one encodes, in a single transcript, sequences for one copy each of 16S, 23S, and 5S rRNAs (Figure 26.41). Because the three species are used in equal amounts, the logic of this organization is apparent. Less easy to explain is that each transcript also includes sequences for one to four tRNA molecules. Because rRNAs and tRNAs are all used in protein synthesis, the interspersion of rRNA and tRNA sequences may represent a means of coordinating the rates of synthesis of these RNAs, but specific mechanisms have not yet been revealed.

The initial transcript from each rRNA operon is a short-lived RNA molecule of 30S, shown in Figure 26.41. The abnormal accumulation of this species in bacterial strains defective in RNase III first suggested a role for this enzyme in rRNA processing. In fact, one double-strand cut in each of two giant stem–loop regions releases precursors to 16S and 23S rRNAs, and the same probably occurs for 5S rRNA. Further maturation steps require the presence of particular ribosomal proteins, which begin to assemble on the precursor RNAs while transcription is still in progress. The embedded tRNA sequences are processed to give mature tRNAs, along the same routes used for other tRNA species.

tRNA Processing

Aside from the tRNAs embedded in pre-rRNA transcripts, the other tRNAs are synthesized in transcripts that contain one to seven tRNAs each, all surrounded by lengthy flanking sequences. The maturation steps are summarized in Figure 26.42, using as an example the well-studied case of the *E. coli* tyrosine tRNA species ($tRNA^{Tyr}$). In this case, maturation starts (step 1) with an endonuclease that cleaves next to a stem–loop structure on the 3′ side of the tRNA sequence. This is followed by action of **ribonuclease D** (step 2), which carries out exonucleolytic cleavage to a point two nucleotides removed from the CCA sequence at the 3′ end. Next (step 3), the 5′ end is created by **ribonuclease P,** which cleaves to leave a phosphate on the 5′ terminal G. This enzyme creates the 5′ terminus of all tRNA molecules. It is not clear what structural features are recognized by RNase P, for different sequences are contained in the cleavage sites. Ribonuclease P is an amazing enzyme that consists of one RNA molecule of 377 nucleotides and one protein molecule with M_r of about 20,000. Both components are necessary for full catalytic activity, but under nonphysiological conditions the RNA molecule alone can catalyze accurate cleavage. Thus, ribonuclease P is a ribozyme, a member of the class of RNA enzymes that we introduced in Chapter 11.

Bacterial transcripts undergo posttranscriptional processing, involving both endonucleolytic and exonucleolytic cleavage.

Once the proper 5′ terminus has been created, ribonuclease D removes the remaining two nucleotides from the 3′ end (step 4). Should excessive "nibbling" occur through faulty control of RNase D activity, there is an enzyme (CCA nucleotidyltransferase) that will restore the CCA end to any tRNA in a nontranscriptive

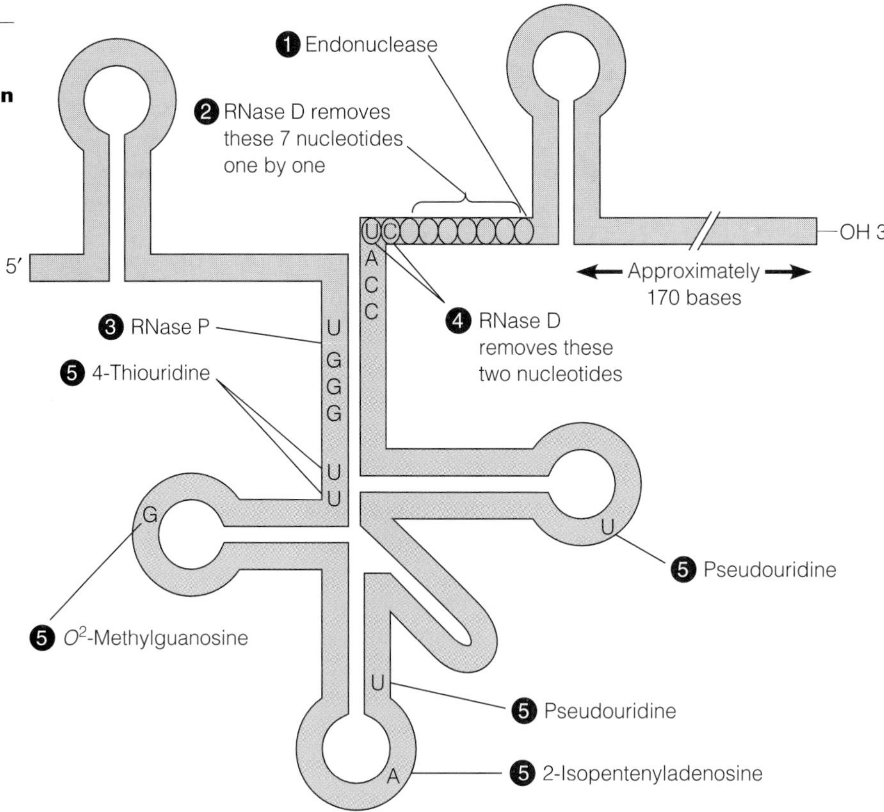

FIGURE **26.42**

Modification steps (1–4) that occur in maturation of *E. coli* tRNA^{Tyr} from its transcript and modified bases (5) seen in the mature tRNA. The tRNA sequence is shown in purple.

fashion. This enzyme specifically recognizes the 3′ terminus of tRNAs that lack the CCA end and catalyzes sequential reactions with a CTP, another CTP, and an ATP.

Creation of the modified bases (see Chapters 4 and 27) occurs at the final stage, including methylations, thiolations, reduction of uracil to dihydrouracil, and so forth. In the specific example shown, the modifications include formation of two pseudouridines, one 2-isopentenyladenosine, one O^2-methylguanosine, and one 4-thiouridine (step 5). The functions of most of these modifications are unknown, because many tRNAs lacking the modifications are fully active in vitro.

An additional posttranscriptional process, namely intron splicing, is almost exclusively confined to eukaryotes (although intron splicing has been well defined in a few bacteriophage genes). Therefore, discussion of this process is postponed until Chapter 28. Another important modification, recently described for eukaryotes, is aminoacylation within the nucleus. This is evidently an error correction process, in which only correctly charged tRNAs are allowed to exit the nucleus and function in protein synthesis.

SUMMARY

All RNA is synthesized by the template-dependent copying of one DNA strand within a gene, catalyzed by RNA polymerase. RNA polymerases use 5′-ribonucleoside triphosphates as substrates, and they transcribe in a 5′→3′ direction. Strand selection and duplex unwinding and rewinding are carried out by the enzyme. RNA polymerase binds at a promoter site, by formation of specific DNA–protein contacts, largely involving the enzyme's σ subunit. Most transcription initiations are abortive, but after a productive initiation, σ dissociates, and elongation by bacterial RNA polymerases is carried out by the core polymerase, $\alpha_2\beta\beta'$. Transcription is highly processive and is terminated by specific DNA sequences, sometimes with the participation of ρ protein. A single RNA polymerase synthesizes all prokaryotic RNAs, whereas eukaryotic cells contain three different RNA polymerases.

Transcription is regulated by the negative and positive actions of site-specific DNA-binding proteins called repressors and activators. Repressors bind to operator sites and block transcription initiation; activators bind to DNA and stimulate initiation. DNA affinities of these proteins are influenced by binding of low-molecular-weight ligands that can act as either inducers or corepressors. Most prokaryotic gene regulatory proteins bind to DNA through an α helix that lies within the major groove and establishes specific contacts between protein amino acids and DNA bases. This helix is part of the helix–turn–helix structural motif, in which the second helix makes non-sequence-specific interactions with the sugar–phosphate backbone. Another important regulatory mechanism in prokaryotes is attenuation, in which the positioning of a translating ribosome on a nascent mRNA determines whether or not synthesis of that mRNA will terminate before transcription of the structural genes of the operon has begun.

Posttranscriptional processing involves cleavage of the primary transcript, modification of bases (in tRNA synthesis), nontranscriptive nucleotide addition and (in a *few* cases) intron splicing.

REFERENCES

Mechanism of Transcription

Bustamante, C., M. Guthold, X. Zhu, and G. Yang (1999) Facilitated target location on DNA by individual *Escherichia coli* RNA polymerase molecules observed with the scanning force microscope operating in liquid. *J. Biol. Chem.* 274:16665–16668. Direct visualization of molecules indicates how RNA polymerase moves along DNA to promoters.

Cook, D. N., D. Ma, N. G. Pon, and J. E. Hearst (1992) Dynamics of DNA supercoiling by transcription in *Escherichia coli. Proc. Natl. Acad. Sci. USA* 89:10603–10607. The unsettled question of whether the act of transcription per se overwinds template DNA is explored.

deHaseth, P. L., M. L. Zupancic, and M. T. Record, Jr. (1998) RNA polymerase–promoter interactions: The comings and goings of RNA polymerase. *J. Bacteriol.* 180:3019–3025. A short, well-referenced review of RNA polymerase initiation.

Gelles, J., and R. Landick (1998) RNA polymerase as a molecular motor. *Cell* 93:13–16. A recent minireview summarizing ways to understand mechanochemical properties of RNA polymerase.

Lisser, S., and H. Margalit (1993) Compilation of *E. coli* messenger RNA promoter sequences. *Nucleic Acids Res.* 21:1507–1516. Extends the analysis described in Figure 26.12 from 114 to 300 sequenced promoters.

Mooney, R. A., and R. Landick (1999) RNA polymerase unveiled. *Cell* 98:687–690. A brief review article describing the first detailed structure determinations for a bacterial and a eukaryotic RNA polymerase.

Nudler, E. (1999) Transcription elongation: Structural basis and mechanisms. *J. Mol. Biol.* 288:1–12. A model of the transcription elongation complex helps explain also mechanisms of initiation and termination.

Ross, W., et al. (1993) A third recognition element in bacterial promoters: DNA binding by the α subunit of RNA polymerase. *Science* 262:1407–1413. An explanation for the high efficiency of rRNA gene transcription.

Sousa, R., Y. J. Chung, J. P. Rose, and B-C. Wang (1993) Crystal structure of bacteriophage T7 RNA polymerase at 3.3 Å resolution. *Nature* 364:593–599. The first structural analysis of any RNA polymerase.

Struhl, K. (1999) Fundamentally different logic of gene regulation in eukaryotes and prokaryotes. *Cell* 98:1–4. A concise description of the distinctions in transcription between higher and lower organisms.

Uptain, S. M., C. M. Kane, and M. J. Chamberlin (1997) Basic mechanisms of transcript elongation and its regulation. *Annu. Rev. Biochem.* 66:117–172. A comprehensive review from a leading laboratory.

von Hippel, P. H. (1998) An integrated model of the transcription complex in elongation, termination, and editing. *Science* 281:660–665. Reviews events in transcription from a largely thermodynamic perspective.

DNA–Protein Interactions and Transcriptional Regulation

Albright, R. A., and B. W. Matthews (1998) How Cro and λ-repressor distinguish between operators: The structural basis underlying a genetic switch. *Proc. Natl. Acad. Sci. USA* 95:3431–3436. Structural analysis of DNA–protein complexes containing the same operator but different repressors.

Beamer, L. J., and C. O. Pabo (1992) Refined 1.8 Å crystal structure of the λ repressor–operator complex. *J. Mol. Biol.* 227:177–196. The most detailed structure yet reported for a DNA–protein complex.

Busby, S., and R. H. Ebright (1994) Promoter structure, promoter recognition, and transcription activation in prokaryotes. *Cell* 79:743–746. A contemporary minireview.

Greenblatt, J., J. R. Nodwell, and S. W. Mason (1993) Transcriptional antitermination. *Nature* 364:401–406. A process discovered in phage λ, which has significance for eukaryotic and HIV gene expression.

Harrington, R. E. (1992) DNA curving and bending in protein–DNA recognition. *Mol. Microbiol.* 6:2549–2555. An increasing number of transcription factors are being found to induce such DNA structural alterations.

Henkin, T. (1994) tRNA-directed transcription termination. *Mol. Microbiol.* 13:381–387. A description of a novel form of transcriptional regulation.

Kercher, M. A., P. Lu, and M. Lewis (1997) *Lac* repressor–operator complex. *Curr. Opin. Struct. Biol.* 7:76–85. This brief review has 97 well-annotated references.

Landick, R. (1999) Shifting RNA polymerase into overdrive. *Science* 284:598–599. A brief commentary, reviewing recent work on the mechanism of antitermination.

Lewis, M., G. Chang, N. C. Horton, M. A. Kercher, H. C. Pace, M. A. Schumacher, R. G. Brennan, and P. Lu (1996) Crystal structure of the lactose operon repressor and its complexes with DNA and inducer. *Science* 271:1247–1254. A massive achievement, given the size and importance of the protein.

Ptashne, M. (1986) *A Genetic Switch: Gene Control and Phage λ.* Cell Press, Cambridge, Mass., and Blackwell Scientific, Palo Alto, Calif. A short, lucidly written, and beautifully illustrated book describing one of the most fruitful systems for analyzing transcriptional regulation and DNA–protein interactions.

Schleif, R. (1992) DNA looping. *Annu. Rev. Biochem.* 61:199–223. Looping is well established in arabinose operon regulation and an increasing number of other systems.

Steitz, T. A. (1990) Structural studies of protein–nucleic acid interaction: The sources of sequence-specific binding. *Q. Rev. Biophys.* 23:205–280. A beautifully illustrated review, describing the most thoroughly studied site-specific proteins.

Yanofsky, C. (1988) Transcription attenuation. *J. Biol. Chem.* 263:609–612. A concise and informative minireview by the discoverer of this phenomenon.

Posttranscriptional Processing

Altman, S., L. Kirsebom, and S. Talbot (1993) Recent studies of ribonuclease-P. *FASEB J.* 7:7–14. A discussion of one of the most interesting known ribozymes.

Apirion, D., and A. Miczak (1993) RNA processing in prokaryotic cells. *BioEssays* 15:113–120. A brief review.

PROBLEMS

1. Outline an experimental approach to determining the average chain growth rate for transcription in vivo. Chain growth rate is the number of nucleotides polymerized per minute per RNA chain.

2. Outline an experimental approach to demonstrate the average RNA chain growth rate during transcription of a cloned gene in vitro.

3. Measurements of RNA chain growth rates are often led astray by the phenomenon of *pausing*, in which an RNA polymerase molecule stops transcription when it reaches certain sites, for intervals that may be as long as several minutes. How might pausing be detected?

4. The active form of lactose repressor binds to the operator with a dissociation constant of 10^{-13} for the reaction $\mathrm{R} + \mathrm{O} \rightleftharpoons \mathrm{RO}$. About 10 molecules per *E. coli* cell suffice to keep the operon turned off in the absence of inducer.
(a) If the average *E. coli* cell has an intracellular volume of 0.3×10^{-12} mL, calculate the approximate intracellular concentration of repressor.
(b) If the average cell contains two copies of the *lac* operon, calculate the approximate intracellular concentration of operators.
(c) Calculate the average intracellular concentration of *free* operators under these conditions.
(d) Explain how a cell with a haploid chromosome could contain an average of two copies of the *lac* operon.

5. Is attenuation likely to be involved in eukaryotic gene regulation? Briefly explain your answer.

6. RNA polymerase has a higher K_M for ribonucleoside triphosphates in the chain initiation step than in the elongation steps. How might this be important in regulating transcription?

*7. Suppose you want to study the transcription in vitro of one particular gene in a DNA molecule that contains several genes and promoters. Without adding specific regulatory proteins, how might you stimulate transcription from the gene of interest relative to the transcription of the other genes on your DNA template? To make all of the complexes identical, you would like to arrest all transcriptional events at the same position on the DNA template before isolating the complex. How might you do this?

8. For some time it was not clear whether *lac* repressor inhibits *lac* operon transcription by inhibiting the binding of RNA polymerase to its promoter or by allowing transcription initiation but blocking elongation past the site of bound repressor. How might you distinguish between these possibilities?

9. The *tac* promoter, an artificial promoter made from a chemically synthesized oligonucleotide, has been introduced into a plasmid. It is a hybrid of the *lac* and *trp* promoters, containing the −35 region of one and the −10 region of the other. This promoter directs transcription initiation more efficiently than either the *trp* or *lac* promoters. Why?

10. A *lac* operon containing one mutation was cloned into a plasmid, which was introduced by transformation into a bacterium containing a wild-type *lac* operon. The three genes of the chromosomal operon were rendered noninducible in the presence of the plasmid.
(a) What kind of mutation in the plasmid operon could have this effect?
(b) Suppose the result of transformation was to cause the three plasmid *lac* genes to be expressed constitutively, at a high level. What type of plasmid gene mutation could have this result?

*11. Several new genes in the SOS regulon were identified by an ingenious use of "Mud" phages. These are derivatives of phage Mu that have a promoterless β-galactosidase gene inserted at a particular point in this phage genome. How might these phages be used to identify genes whose expression is turned on after ultraviolet irradiation of bacteria?

12. Would you expect actinomycin D to be a competitive inhibitor of RNA polymerase? What about cordycepin? Briefly explain your answers.

13. A restriction fragment was subjected to Maxam–Gilbert sequencing, with results as shown in the first four lanes of the radioautogram. S_1 nuclease mapping was carried out, for a gene whose transcription initiated within this sequence. The transcript protected a fragment whose length was as shown in the fifth lane.

G A+G C C+T

(a) Give the nucleotide sequence of both DNA strands and the first several RNA nucleotides. Identify all 3′ and 5′ ends.

(b) On your structure, show the approximate location of the Pribnow box (the −10 region).

(c) Assuming that the restriction fragment was created with a type II restriction endonuclease that recognizes a 6-base-pair site, show the structure of that site, and indicate where cleavage occurs.

14. Partial diploid forms of *E. coli* were created, each of which contained a complete lactose operon at its normal chromosomal site and the regulatory sequences only *(i, p, o)* on a plasmid. Predict the effect of each mutation upon the activity of β-galactosidase before and after the addition of inducer. Use −, +, or ++ to indicate approximate activity levels.

Mutation	Before	After
(a) No mutations in either chromosomal or plasmid genes	_____	_____
(b) A mutation in the plasmid operator, which abolishes its binding to repressor	_____	_____
(c) A mutation in the chromosomal promoter, which reduces affinity of *p* for RNA polymerase by 10-fold	_____	_____
(d) An *i* gene mutation in the chromosome, which abolishes binding of the *i* gene product to inducer	_____	_____
(e) An *i* gene mutation in the plasmid, which abolishes binding of the repressor to the inducer	_____	_____
(f) A chromosomal *o* mutation, which abolishes its binding to repressor	_____	_____
(g) A mutation in the gene for CRP, which abolishes its binding to cyclic AMP	_____	_____

*15. What type of mutation of the *lac* repressor might be both constitutive and *trans*-dominant?

*16. RNA polymerase doesn't proofread its polynucleotide product, so it is a much less accurate enzyme than DNA polymerases. Nevertheless, mutant forms of RNA polymerase have been described in which transcription accuracy is decreased relative to wild-type. This finding implies that RNA polymerase has a fidelity-enhancing mechanism, which is impaired in the mutants. Aside from the process outlined on pages 985–986, how might such a mechanism operate?

*17. Explain the basis for the following statement: Transcription of two genes on a plasmid can occur without the concomitant action of a topoisomerase, but only if those two genes are oriented in opposite directions.

TOOLS OF BIOCHEMISTRY 26A

Footprinting: Identifying Protein-Binding Sites on DNA

Transcription is controlled in large part through interactions of proteins with specific sites on DNA molecules, including operators and promoters. Such sites were initially identified through genetic analysis in the lactose system and phage λ. Biochemical analysis requires the identification and nucleotide sequence determination of a site, along with structural determination of the DNA–protein complexes that form. A technique called **footprinting,** usually involving protection against DNase I cleavage, is widely used to identify such sites. Footprinting can identify any DNA site that binds a protein specifically, as long as the protein binds sufficiently tightly.

The principle of the method, outlined in Figure 26A.1, is that binding of a protein to a specific DNA sequence should protect the protein-bound DNA from attack by DNase I (pancreatic deoxyribonuclease). The investigator first uses γ-[^{32}P]ATP and T4 polynucleotide kinase to prepare a 5′ end–labeled fragment of DNA containing a protein-binding site. One aliquot of the end–labeled DNA is mixed with the protein under study (step 1) and then incubated with DNase I under conditions in which most chains are cleaved only once (step 2). Another aliquot is incubated with DNase I under identical conditions, except that the protein is absent (step 2′). Next, the two incubation mixtures are analyzed in adjacent lanes of a sequencing gel (step 3). The result is a ladder of fragments similar to that seen in a sequencing gel, except that there is little sequence specificity; the bands on the gel are spaced uniformly at one-nucleotide intervals, although

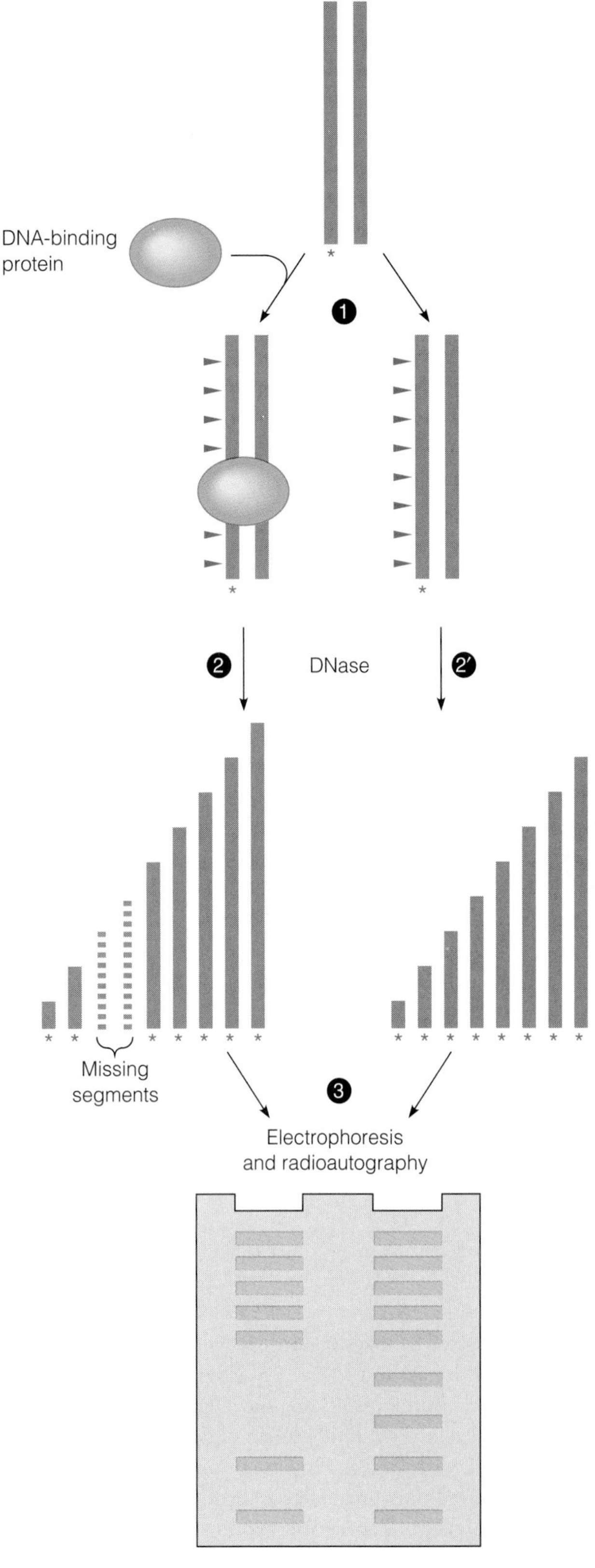

FIGURE **26A.1**

DNase I footprinting as a tool to identify DNA sites that bind specific proteins.

bond intensity varies considerably. Any sites protected from DNase attack because of interaction with the DNA-binding protein under study yield either no band or a low-intensity band in the ladder from the DNA–protein complex, indicating that little or no cleavage occurred at that site. The blank region on the gel pattern from the DNA–protein complex (the "footprint") identifies the location and the size of the fragment in contact with the DNA-binding protein (such as RNA polymerase).

Recent improvements in footprinting technology involve cleavage with chemical agents such as **methidiumpropyl-EDTA-Fe^{2+}** (MPE-Fe^{2+}).

Intercalates between base pairs — Methidium

Linker and flexible hinge — Propyl

DNA cleavage — EDTA-Fe^{2+}

This compound intercalates between DNA bases, as does ethidium bromide, and catalyzes oxidation leading to cleavage at a nearby site. Because there is some sequence selectivity in DNase I attack and virtually none with MPE-Fe^{2+}, the latter technique gives cleaner footprints. A related technique, described in the reference section following, generates hydroxyl radicals as a reactive but nonspecific DNA cleavage reagent.

Footprinting shows that RNA polymerase binds to a region about 60 base pairs long, extending from about 40 nucleotides upstream of (5′ to) the transcriptional start site to about 20 nucleotides past that site. In other words, the binding site extends from nucleotide −40 to +20, where the template for the first nucleotide in the transcript is +1. That first nucleotide, the 5′ end of the transcript, can be identified by various methods. The most widely used is S_1 nuclease mapping, which is described in Tools of Biochemistry 26B.

Reference

Tullius, T. D., B. A. Dombroski, M. E. A. Churchill, and L. Kam (1989) Hydroxyl radical footprinting: A high-resolution method for mapping protein–DNA contacts. In: *Recombinant DNA Methodology*, edited by R. Wu, L. Grossman, and K. Moldave, pp. 721–741. Academic Press, San Diego, Calif. A description of hydroxyl radical footprinting, with references to previously described methods.

TOOLS OF BIOCHEMISTRY 26B

Mapping Transcriptional Start Points

Studies of transcriptional initiation and its control require methods for the accurate identification of transcriptional start points, specifically the DNA template nucleotide that encodes the 5′ nucleotide of the transcript. The low abundance of specific mRNAs and the high turnover of nearly all bacterial mRNAs make this a challenging task.

Prokaryotic transcripts all have a 5′ triphosphate terminus on the first nucleotide, which can provide a handle for their identification. Given that this identification requires purification of the transcript, less laborious methods are usually preferable. One such method, **S_1 nuclease mapping,** uses the fungal enzyme S_1 nuclease, which specifically and quantitatively cleaves single-strand DNA and RNA (Figure 26B.1). The necessary materials are the cloned gene and a restriction fragment thought to contain the template for the 5′ end of the transcript. The fragment is 5′ end–labeled, as in the Maxam–Gilbert sequencing procedure, and cleaved asymmetrically with another restriction enzyme, so that only the template DNA strand is labeled. In Maxam–Gilbert sequencing, 5′ end–labeled DNA is subjected to chemical reagents that cleave in a base-specific fashion. One reagent cleaves at sites occupied by either A or G, one cleaves in a strictly G-dependent fashion, one cleaves at T and C sites, and one cleaves specifically at C. Thus, by treating four aliquots of the same DNA with the four reagents and displaying the cleavage fragments on a sequencing gel, one can read the sequence of the DNA, as you can see in Figure 26B.1. Partly because Maxam–Gilbert sequencing is not completely base-specific, it has been largely supplanted by dideoxy sequence analysis, which uses replicative chain terminators to generate base-specific 3′ termini (see Tools of Biochemistry 25C). However, because S_1 mapping uses a 5′ end–labeled DNA fragment, it is useful to employ the Maxam–Gilbert approach, which also uses a 5′ end–labeled DNA fragment, to locate the 5′ end of the transcript precisely.

Next, the 5′ end–labeled DNA fragment is denatured and hybridized to mRNA, under conditions (high formamide) that favor the formation of DNA–RNA hybrids over DNA–DNA duplexes (Figure 26B.1, step 1). The only double-stranded nucleic acid, then, should be a DNA–RNA hybrid with 3′ single-stranded extensions from the 5′ end of the transcript and the labeled 5′ end of the restriction fragment. Treatment with S_1 nuclease (step 2) yields a fully duplex structure, with a labeled DNA strand whose length precisely measures the distance from the 5′ end of the transcript to the relevant restriction site. This distance can be identified by denaturing the DNA–RNA hybrid and running it

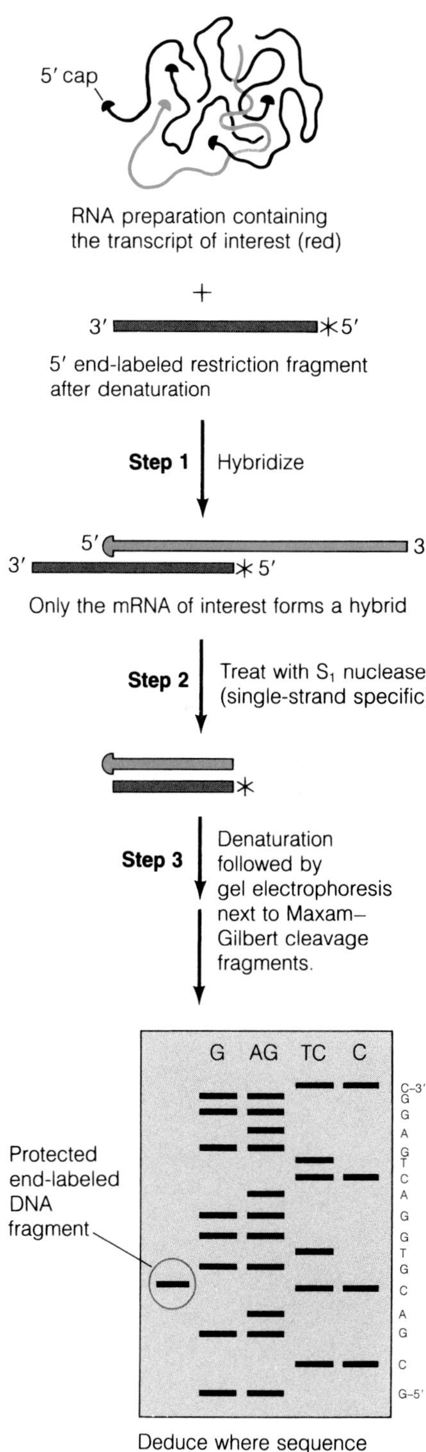

FIGURE **26B.1**

S_1 nuclease mapping method for identifying the 5′ end of an RNA molecule.

on a sequencing gel, alongside a set of Maxam–Gilbert cleavage fragments (step 3).

An alternative technique, called **primer extension,** requires a 5′ end–labeled restriction fragment whose 3′ end lies downstream of the promoter. This fragment is hybridized to an RNA preparation containing the message of interest (Figure 26B.2). Then the mixture is incubated with reverse transcriptase and nonradioactive deoxyribonucleoside triphosphates. Extension from the 3′ end of the fragment continues until the 5′ end of the transcript is reached. The length of this labeled fragment, minus the length of the original primer, equals the distance from the 5′ end of the transcript to the 3′ end of the restriction fragment used. This method may not have sufficient resolution to identify the precise 5′ terminal nucleotide, but it often works when S_1 nuclease mapping does not.

Both of these techniques are widely used for quantitating transcripts of interest, as well as for mapping 5′ ends of transcripts.

Reference

Sambrook, J., E. F. Fritsch, and T. Maniatis (1989) *Molecular Cloning: A Laboratory Manual,* Cold Spring Harbor Laboratory, Cold Spring Harbor, N.Y. Chapter 7 of this benchmark methods manual describes several techniques for RNA isolation and analysis.

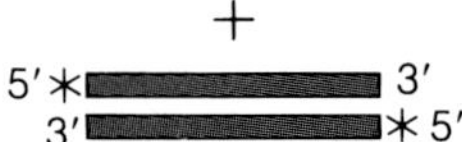

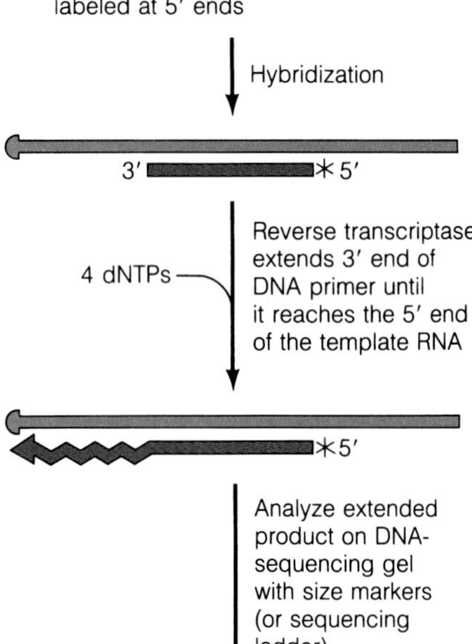

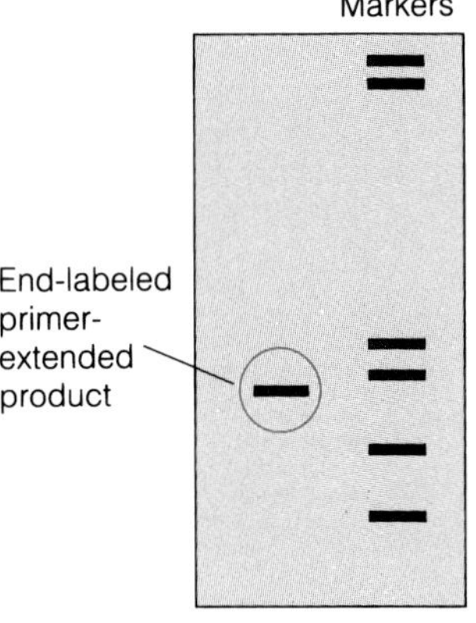

FIGURE **26B.2**

Primer extension method for locating the 5′ end of a transcript. The markers are DNA fragments, such as restriction fragments, that serve as molecular weight standards.

CHAPTER 27

Information Decoding: Translation

IN CHAPTER 5 WE DESCRIBED TRANSLATION IN A VERY ELEMENTARY WAY. BEFORE getting into the full complexities of the reactions, it is worthwhile to review this picture briefly. Refer to Figures 5.19 and 5.20 (pages 144 and 145).

An Overview of Translation

Recall that the genetic code specifies a nucleotide triplet corresponding to each amino acid residue. In order to achieve the correspondence between individual amino acids and their corresponding codon triplets, a set of *adaptor* molecules is needed, each of which can be coupled to a specific amino acid and will also recognize the corresponding codon in the messenger RNA. These adaptor molecules are the *transfer RNAs* (or tRNAs), which have been described in Chapter 4 (the general structure of tRNAs is illustrated in Figure 4.20 on page 104). Each tRNA contains a nucleotide triplet called the **anticodon,** which is complementary to an mRNA codon for the particular amino acid. Thus, the whole set of tRNAs contained in a cell constitutes a kind of molecular dictionary for the translation—it defines the correspondences between words in the 4-letter nucleic acid language and words in the 20-letter amino acid language. A specific enzyme called an *aminoacyl-tRNA synthetase* is required to make the attachment between each tRNA and the corresponding amino acid.

Transfer RNAs are the adaptor molecules that match amino acid to codon.

In order to translate an mRNA, machinery must exist that brings the amino acid–charged tRNAs and the mRNA together, matching anticodon and codon triplets and joining amino acid residues together in the correct sequence. The cellular apparatus that accomplishes this feat is the *ribosome,* a particle composed of both RNA and proteins. A ribosome can bind to mRNA and "read" it, as it moves along the RNA, accepting the charged tRNAs in the order dictated by the message and transferring their amino acid residues one by one and in proper order to the growing polypeptide chain.

The mRNA message is always read in the $5'\rightarrow3'$ direction, and the polypeptide chain is synthesized starting with its N-terminal residue. The direction of polypeptide synthesis was established by Howard Dintzis in 1961. Dintzis gave reticulocytes (hemoglobin-producing cells) a short pulse of ^{3}H-labeled leucine

Messenger RNA is read $5'\rightarrow3'$. Polypeptide synthesis begins at the N-terminus.

and isolated the completed hemoglobin molecules at various times after the pulse. After cleaving these molecules into peptides with trypsin, he compared the radioactivity of peptides from various points in the chain. Immediately after the pulse label was added, it could appear only in chains that were just being completed; the label at this time was found only in the C-terminal peptides. At longer times after the pulse, radioactivity was found to be incorporated into parts of the polypeptide closer and closer to the N-terminus, as protein molecules continued to be started. Dintzis therefore concluded that amino acids are added to a polypeptide chain starting at the N-terminus and working toward the C-terminus.

The simple picture of translation we have presented so far leaves a host of questions unanswered. How are tRNA and amino acid matched? How does the ribosome attach to the mRNA and move along it? How does it start and stop translation correctly? How does it avoid making mistakes? Where does the energy for all of this activity come from? To answer such questions, we must dissect the whole process of translation, with careful examination of each of its parts.

The Genetic Code

The genetic code was introduced in Chapter 5. Here, we describe some of the crucial experiments used to deduce it, and we will analyze a number of the code's more important features. We shall see that the code is *almost* universal throughout biology, but that there are a few exceptions in codon usage.

HOW THE CODE WAS DECIPHERED

Following the discovery of mRNA (see Chapter 26), the immediate question was: How is it read? It was clear that a triplet code was the minimum required to account for all 20 amino acids, but many kinds of triplet codes are possible. Some conceivable ones are shown in Figure 27.1. An *overlapping code* like that shown in Figure 27.1a would be space-saving. This possibility was eliminated, however, by the observation that most mutations result in the change of a single amino acid. If the codons overlapped, there should be a significant number of cases in which two adjacent residues are modified by mutation of the "overlap" residue. Furthermore, an overlapping code would lead to statistical regularities between neighboring amino acid residues in proteins—that is, some amino acids would be neighbors

FIGURE **27.1**

Three conceivable kinds of genetic codes. Early research on the nature of the code quickly showed that a nonoverlapping, unpunctuated code **(c)** fit all experimental observations.

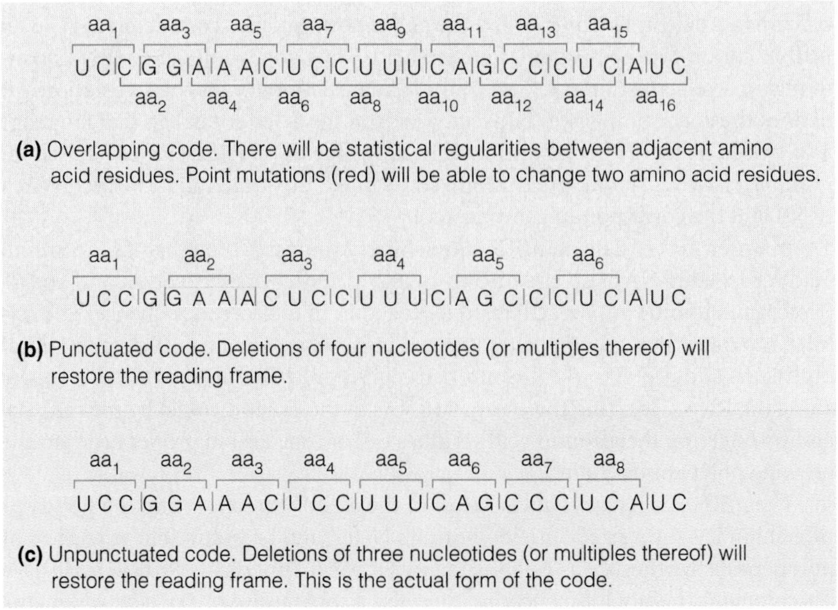

aa$_1$ aa$_3$ aa$_5$ aa$_7$ aa$_9$ aa$_{11}$ aa$_{13}$ aa$_{15}$

U C|C |G G|A|A A|C|U C|C|U U|U|C A|G|C C|C|U C|A|U C

aa$_2$ aa$_4$ aa$_6$ aa$_8$ aa$_{10}$ aa$_{12}$ aa$_{14}$ aa$_{16}$

(a) Overlapping code. There will be statistical regularities between adjacent amino acid residues. Point mutations (red) will be able to change two amino acid residues.

aa$_1$ aa$_2$ aa$_3$ aa$_4$ aa$_5$ aa$_6$

U C C|G|G A A|A|C U C|C|U U U|C|A G C|C|C|U C|A|U C

(b) Punctuated code. Deletions of four nucleotides (or multiples thereof) will restore the reading frame.

aa$_1$ aa$_2$ aa$_3$ aa$_4$ aa$_5$ aa$_6$ aa$_7$ aa$_8$

U C C |G G A|A A C|U C C|U U U|C A G|C C C|U C A|U C

(c) Unpunctuated code. Deletions of three nucleotides (or multiples thereof) will restore the reading frame. This is the actual form of the code.

more often than others—and this has never been observed. Another possibility was that the code was **punctuated.** That is, as shown in Figure 27.1b, some base or bases might serve as "spacing" between code words.

That the code was neither overlapping nor punctuated was elegantly demonstrated by Francis Crick and Sidney Brenner in 1961. Using conditions that encouraged deletion and insertion mutations in the bacteriophage T4 DNA, they showed that insertions or deletions of one or two nucleotides led to nonsense proteins (because the reading frame had been shifted), but adding or removing *three* always led to the insertion or deletion of one amino acid residue. Thus the code *must* be in triplets, and the message must be read in units of three, without punctuation, from some starting point.

But which codons corresponded to which amino acids? The first answers were provided in 1961 by Marshall Nirenberg and his colleagues, using a cell-free translation system. These researchers found that extracts from bacterial cells contained all of the necessary ingredients (tRNAs, amino acids, aminoacyl synthetases, and ribosomes) to carry out polypeptide synthesis directed by a synthetic polynucleotide template added to the system. For example, using polyuridylic acid, poly(U), as a template led to the production of polyphenylalanine. Thus, UUU must be a phenylalanine codon. In the same way, it was later shown that CCC codes for proline and AAA for lysine.

Using *homopolynucleotides* (all bases the same) as templates could yield only a few code words. Many more words were deciphered after H. G. Khorana developed methods for synthesizing polyribonucleotides with different but repeating structures. In the example shown in Figure 27.2, the repeating sequence $(AAG)_n$ was found to give three different homopolymers: polylysine, polyarginine, and polyglutamic acid. This finding not only confirmed the importance of the reading frame but also showed that AAG, AGA, and GAA must be codons for these amino acids. The experiment did not reveal which codon corresponded to which amino acid; further experiments were needed to discriminate among the possible matches. When the repeating unit of the polynucleotide is a dimer or tetramer rather than a trimer, more complicated polypeptides are formed, each containing several amino acid residues in repeating sequences (see Problem 4). These experiments also confirmed that the code was triplet and unpunctuated.

The genetic code is a nonoverlapping, unpunctuated triplet code.

Experiments of this kind provided many code words, but in 1964 Philip Leder and Marshall Nirenberg developed a new and rapid method for codon assignment that made it possible to complete the deciphering of the code. Nirenberg and Leder found that synthetic triplets would bind to ribosomes and specify the binding of specific tRNAs. For example, UUU and UUC permitted binding of phenylalanine tRNAs to ribosomes, and CCC and CCU permitted the binding of proline tRNA. Such experiments provided unequivocal evidence for the *redundancy* of the code, because several different codons were found to correspond to a single amino acid. By the combined use of these various techniques, the entire genetic code was established within a few years of intensive research.

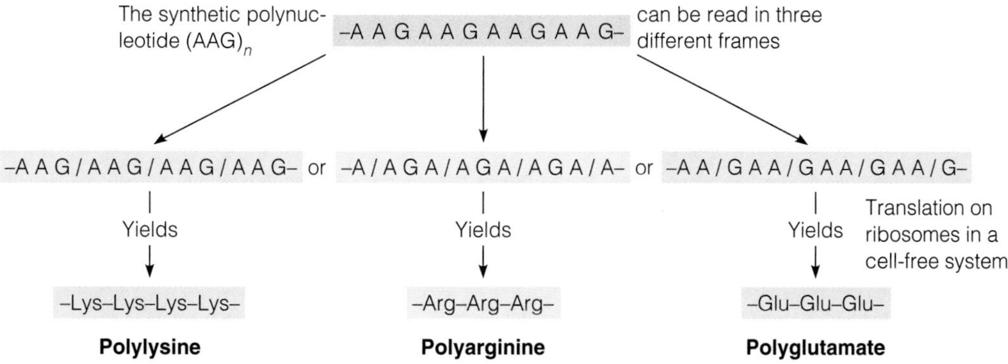

FIGURE **27.2**

Use of synthetic polynucleotides with repeating sequences to decipher the code. This example shows how polypeptides derived from the $(AAG)_n$ polymer were used to confirm the triplet code and help identify codons. The polymer $(AAG)_n$ can yield three different polypeptides, depending on which reading frame is employed.

SeH
|
CH_2
|
$H_3\overset{+}{N}-C-COO^-$
|
H

Selenocysteine

The code is almost, but not quite, universal.

FEATURES OF THE CODE

The genetic code, as given in Figure 27.3, is *almost,* but not quite, universal. That is, the same codons are nearly always used for the same amino acids, in both prokaryotic and eukaryotic organisms. The few exceptions we know about are listed in Table 27.1. Most of these exceptions are confined to mitochondria and to certain protozoans. The rest occur in certain organisms that can translate codons in both usual and unusual ways—for example, the occasional use of UGA as a codon for selenocysteine. In these latter cases, reading of a codon in its unusual sense depends upon the context of the surrounding message, in ways still not fully understood. The fact that bacteria readily and correctly translate the mRNAs produced from human DNA inserted into them illustrates the virtual universality of the code.

Why has the genetic code remained almost unchanged over so vast an evolutionary span? Perhaps it is simply because even small codon changes could be devastating. A single codon change could alter the sequence of nearly every protein made by the organism. Some of these changes would almost certainly have lethal effects. Therefore, codon changes have been opposed by the most intense selective pressure during evolution. They represent changes in the most basic rules of the game.

If you carefully examine the code table shown in Figure 27.3, you will note that, in general, each amino acid is characterized by the first two codon letters. For example, all Pro codons start with CC; all Val codons start with GU. Thus, redundancy is usually expressed in the third letter—ACU, ACC, ACA, and ACG all code for threonine. We noted earlier that a single tRNA may recognize several different codons, a fact that was realized soon after the code was translated. The mul-

First position (5′ end)	Second position: U	C	A	G	Third position (3′ end)
U	UUU Phe	UCU Ser	UAU Tyr	UGU Cys	U
U	UUC Phe	UCC Ser	UAC Tyr	UGC Cys	C
U	UUA Leu	UCA Ser	UAA Stop	UGA Stop	A
U	UUG Leu	UCG Ser	UAG Stop	UGG Trp	G
C	CUU Leu	CCU Pro	CAU His	CGU Arg	U
C	CUC Leu	CCC Pro	CAC His	CGC Arg	C
C	CUA Leu	CCA Pro	CAA Gln	CGA Arg	A
C	CUG Leu	CCG Pro	CAG Gln	CGG Arg	G
A	AUU Ile	ACU Thr	AAU Asn	AGU Ser	U
A	AUC Ile	ACC Thr	AAC Asn	AGC Ser	C
A	AUA Ile	ACA Thr	AAA Lys	AGA Arg	A
A	AUG Met	ACG Thr	AAG Lys	AGG Arg	G
G	GUU Val	GCU Ala	GAU Asp	GGU Gly	U
G	GUC Val	GCC Ala	GAC Asp	GGC Gly	C
G	GUA Val	GCA Ala	GAA Glu	GGA Gly	A
G	GUG Val	GCG Ala	GAG Glu	GGG Gly	G

FIGURE **27.3**

The genetic code (as written in RNA). We show here the genetic code as used in most organisms. Chain termination, or stop, codons are shown in orange, and the usual start codon AUG is dark green. Other, rarely used start codons are shown in light green. When AUG is used as a start codon, it codes for *N*-fMet (in prokaryotes) or Met (in eukaryotes). Otherwise, it codes for Met. Exceptions to these codon assignments are given in Table 27.1.

TABLE 27.1 Modifications of the genetic code

Codon	Usual Use	Alternate Use	Where Alternate Use Occurs
AGA AGG	Arg	Stop, Ser	Some animal mitochondria, some protozoans
AUA	Ile	Met	Mitochondria
CGG	Arg	Trp	Plant mitochondria
CUU CUC CUA CUG	Leu	Thr	Yeast mitochondria
AUU GUG UUG	Ile Val Leu	Start (*N*-fMet)	Some prokaryotes[a]
UAA UAG	Stop	Glu	Some protozoans
UGA	Stop	Trp Selenocysteine	Mitochondria, mycoplasmas *E. coli*[a]

[a]Depends on context of message, other factors

tiple recognition always involves the 3′ residue of the codon and therefore the 5′ residue of the anticodon.

In 1966, Francis Crick proposed that the 5′ base of the anticodon was capable of "wobble" in its position during translation, allowing it to make alternative (non-Watson–Crick) hydrogen bonding arrangements with several different codon bases. An example is shown in Figure 27.4. G in the 5′ anticodon position can pair with either C or U in the codon, depending on the relative orientation of the pair. Considering both base-pairing possibilities and the observed selectivity of tRNAs, Crick proposed the set of "wobble rules" given in Table 27.2. This hypothesis nicely explains the frequently observed degeneracy in the 3′ site of the codon. The rather uncommon nucleoside, *inosine* (I, Figure 22.4), is found in a number of anticodons, where it shows the ability to pair with A, U, or C.

Because the messenger RNA is invariably longer than the open reading frame that is to be translated, specific start and stop signals are required to begin and end translation. In almost all organisms, UAA, UAG, and UGA are used for stop signals and do not code for any amino acid. A stop signal indicates that translation is to terminate and the polypeptide product is to be released by the ribosome. Clearly, three stop signals are more than is absolutely necessary, so it is not surprising to find that these codons are also used for designating amino acids in a few organisms (see Table 27.1).

Although nature has been profligate in designating stop signals, it has been surprisingly stingy in apportioning starts. The start signal commonly used in translation is AUG, which also serves as the single methionine codon. How does the ribosome know how to interpret this triplet properly so as to distinguish between internal sites and start sites? The answer is that the 5′ end of any message contains specific sequences that ensure that it is correctly attached to the ribosome (see page 1032). As the message begins to be read, the *first* AUG encountered is interpreted as a start signal, and translation begins. Although prokaryotic and eukaryotic cells handle this situation somewhat differently, the consequence is that *N*-formylmethionine (in prokaryotes) or methionine (in eukaryotes) is always the first amino acid incorporated into a polypeptide chain. Therefore, all proteins start with *N*-fMet or Met, at least, when they are first synthesized. However, in

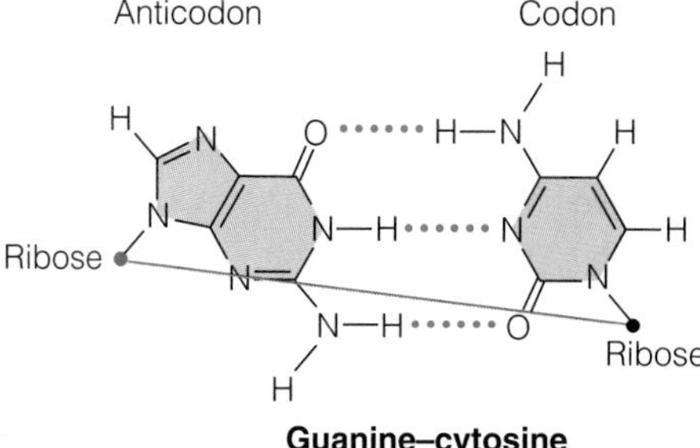

FIGURE **27.4**

The wobble hypothesis. As an example, we show how the anticodon base G can pair with either C or U in a codon. Movement ("wobble") of the base in the 5′ anticodon position is necessary for this capability (see arrow).

The code is redundant. Several codons may correspond to a single amino acid, sometimes via wobble in the 5′ anticodon position.

TABLE 27.2 Base-pairing capabilities in wobble pairs

Base at 5′ Position in Anticodon		Base at 3′ Position in Codon
G	pairs with	C or U
C	pairs with	G
A	pairs with	U
U	pairs with	A or G
I	pairs with	A, U, or C

Prokaryotic messengers contain translational start and stop signals, as well as a sequence that aligns the mRNA on the ribosome.

most cases this residue is either deformylated or removed as translation proceeds (see page 1057). Any AUG encountered after the start is treated as a signal to incorporate methionine within the sequence at that point. Very occasionally, GUG (normally valine), UUG (normally leucine), or AUU (normally isoleucine) serves as a prokaryotic start codon when located near the 5′ end of a message (see Table 27.1). When they do, however, they code for *N*-formylmethionine in the first position. In other positions these triplets are read as normal codons.

The Major Participants in Translation: mRNA, tRNA, and Ribosomes

STRUCTURE OF PROKARYOTIC mRNAS

As described in Chapter 26, mRNAs are single-strand products of the transcription of genomic DNA, with special features that destine them to become attached to ribosomes and function properly in translation. In this chapter, we consider the properties of the prokaryotic type of mRNAs—those produced by bacteria, the viruses that infect bacteria, and the mitochondria and chloroplasts of eukaryotic cells. The mRNAs transcribed from the *nuclear* genomes of eukaryotic cells and some viruses of eukaryotes have a somewhat different structure and are recognized very differently. We describe these in Chapter 28.

As an example of a prokaryotic messenger RNA, consider that produced by transcription of the *lac* operon, which has been discussed in detail in Chapter 26. As shown in Figure 27.5, the total message is about 5300 nucleotides in length. Within this mRNA are three *open reading frames,* corresponding to the *lac z, y,* and *a* genes. An open reading frame is a sequence within a messenger RNA, bounded by start and stop codons, that can be continuously translated. Such multigene messages, which are called **polycistronic messages,** are common in prokaryotes. Each open reading frame has its own start and stop signals, and you will note that these signals vary considerably. There is extra, untranslated RNA between the reading frames and at the ends. The regions 5′ to each start signal contain sequences rich in A and G, which help to align the mRNA on the ribosome so that translation can begin at the proper points and in the correct reading frame. Such attachment sequences, found on all prokaryotic mRNAs, are called *Shine–Dalgarno sequences,* after J. Shine and L. Dalgarno, who discovered them. A Shine–Dalgarno

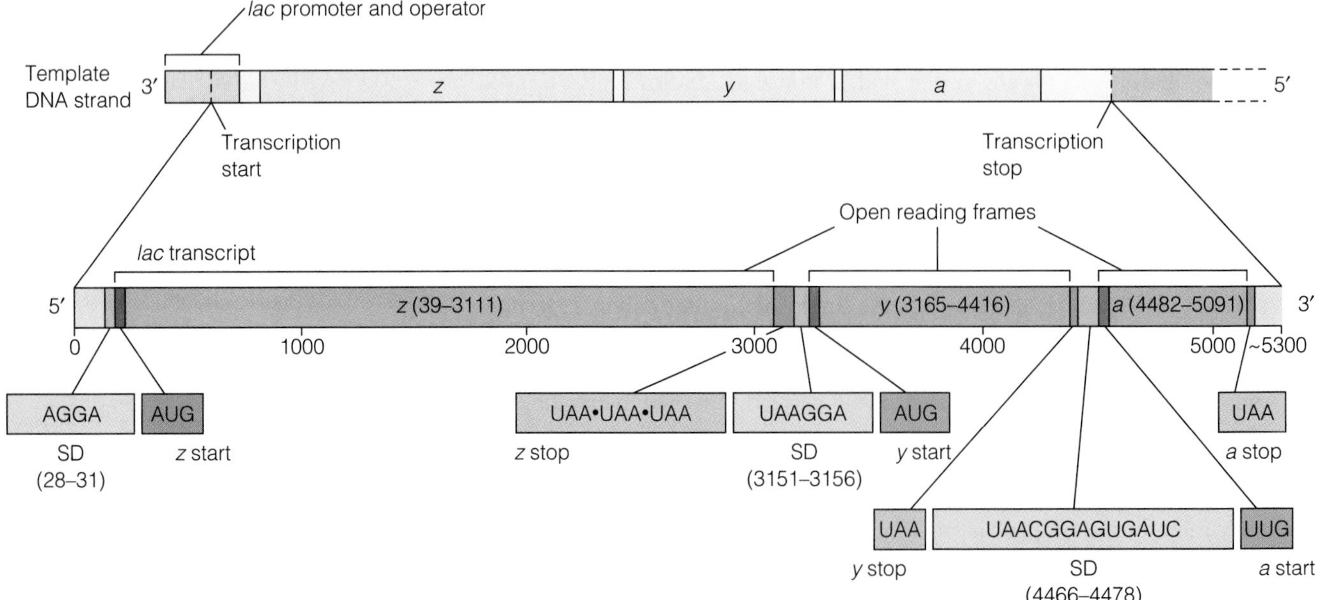

FIGURE **27.5**

The *lac* operon mRNA. The mRNA for the *E. coli lac* operon is about 5300 nucleotides long and contains the open reading frames for the *lac z, y,* and *a* genes, each flanked appropriately by start, stop, and Shine–Dalgarno (SD) sequences.

TABLE **27.3 Representative Shine–Dalgarno sequences**

Message for	Shine–Dalgarno Sequence
Ribosomal protein L10	SD sequence Start 5′ AGGAGCAAAGCUAAUG 3′ mRNA ||||| 3′ AUUCCUCCA 5′ Complementary 3′ end of 16S ribosomal RNA
E. coli lac z	5′ AGGAAACAGCU AUG 3′ |||| 3′ AUUCCUCCA 5′
λ phage Cro	5′ UAAGGAGGUUGU AUG 3′ ||||||||| 3′ AUUCCUCCA 5′

sequence can base-pair with a sequence contained in the ribosomal RNA, as shown in Table 27.3, to produce a proper alignment for starting translation. The different attachment sequences appear to have different affinities for ribosomes. For example, the three genes of the *lac* operon (Figure 27.5) are not translated to an equal extent—*lac z* is translated much more frequently than *lac y* or *lac a.*

Shine–Dalgarno sequences help align ribosomes on mRNAs to properly start translation.

The mRNA message produced from the *lac* operon has all the basic elements necessary for its function: sequences to align it properly on the ribosome and sequences that start and stop translation at the proper points. Many mRNAs also have possibilities for forming three-dimensional secondary and tertiary structures, which can play a role in regulation of the relative production of the various protein products. We shall return to this point later.

TRANSFER RNA STRUCTURE

Any cell, prokaryotic or eukaryotic, contains a battery of different types of tRNA molecules sufficient to incorporate all 20 amino acids into protein. This does not mean that there need be as many tRNA types as there are codons, for some tRNAs can recognize more than one codon, when the difference is in the third (3′) position. *E. coli,* for example, has about 40 different tRNAs—plenty to code for all amino acids, but not as many as the 61 amino acid codons. The tRNA specific to a given amino acid is designated by writing the amino acid as a superscript, for example, $tRNA^{Ala}$.

Transfer RNA was the first natural polynucleotide sequence to be determined, in the pioneering study of yeast ($tRNA^{Ala}$) by Robert Holley in 1965. Since then, thousands of tRNAs have been sequenced. All have the general structure shown schematically in Figure 27.6a and have similar sequences of about 70 to 80 nucleotides. There is, however, considerable variation in detail, as the examples in Figure 27.6b and c indicate. Furthermore, the tRNAs are unique among RNA molecules in their high content of unusual and modified bases, three of which are shown in Figure 27.7.

Cloverleaf models of the kind shown in Figure 27.6 are useful for showing the general pattern of hydrogen bonding and denoting the functional parts of the tRNA. The *anticodon triplet* in the loop at the bottom is complementary to the mRNA codon and will make base pairs with it. Because the codon and anticodon, when paired, constitute a short stretch of double-strand RNA, their directions must be antiparallel. In Figure 27.6 we have written the tRNA molecules with their 5′ ends to the left. Therefore the messenger RNA, when shown in such figures, is written with its 5′ end to the right, opposite to the normal convention.

The *acceptor stem* at the top of the cloverleaf figure is where the amino acid will be attached, at the 3′ terminus of the tRNA. This stem always has the sequence

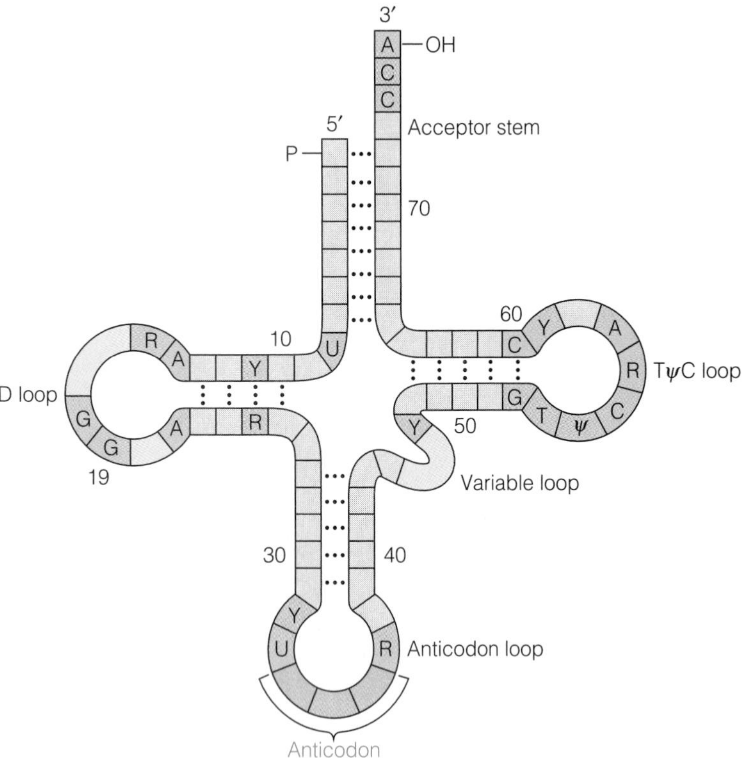

(a)

(b)

tRNA$_1^{Leu}$

3′ mRNA 5′

(c)

tRNA^{Lys}

3′ mRNA 5′

FIGURE **27.6**

Structure of tRNAs. (a) Generalized tRNA structure. The positions of invariant and rarely varied bases are shown in purple. Regions in the D loop and the variable loop that can contain different numbers of nucleotides are shown in blue. The anticodon is shown in orange. **(b)** A leucine tRNA from *E. coli.* **(c)** A human mitochondrial tRNA for lysine. Code for bases: Y = pyrimidine, R = purine, ψ = pseudouridine, T = ribothymidine, and D = dihydrouridine (see Figure 27.7).

(b, c) from J. D. Watson et al., *Molecular Biology of the Gene*, 4th ed. (Redwood City, Calif.: Benjamin/Cummings, 1987). © 1987 J. D. Watson.

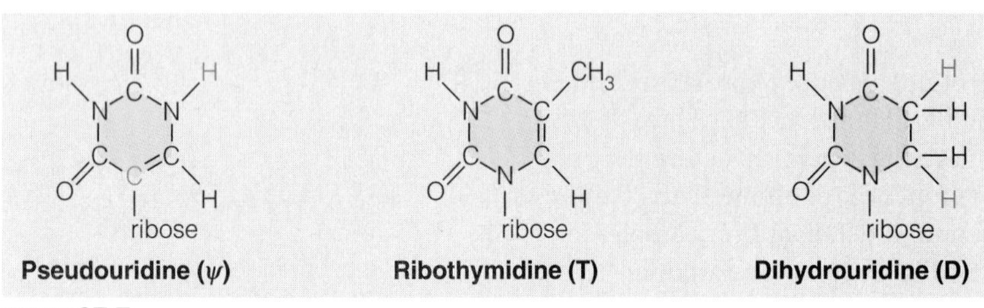

FIGURE **27.7**

A sampling of the modified and unusual bases found in tRNAs.

All tRNAs share a general common structure that includes an anticodon loop, which pairs with codons, and an acceptor stem, to which the amino acid is attached.

5′···CCA—OH 3′. Other common features of tRNA molecules are the *D loop* and the *TψC loop*, regions that contain a substantial fraction of invariant positions and are frequently modified or unusual bases as well. The so-called *variable loop* is indeed variable, both in nucleotide composition and in length, as Figure 27.6 demonstrates.

Although cloverleaf models are convenient for depicting the primary structure and some elements of secondary structure, they are not good three-dimensional representations of tRNA molecules. X-ray diffraction studies of tRNA molecules have revealed that the real molecular shape is quite complex, as you can see in Fig-

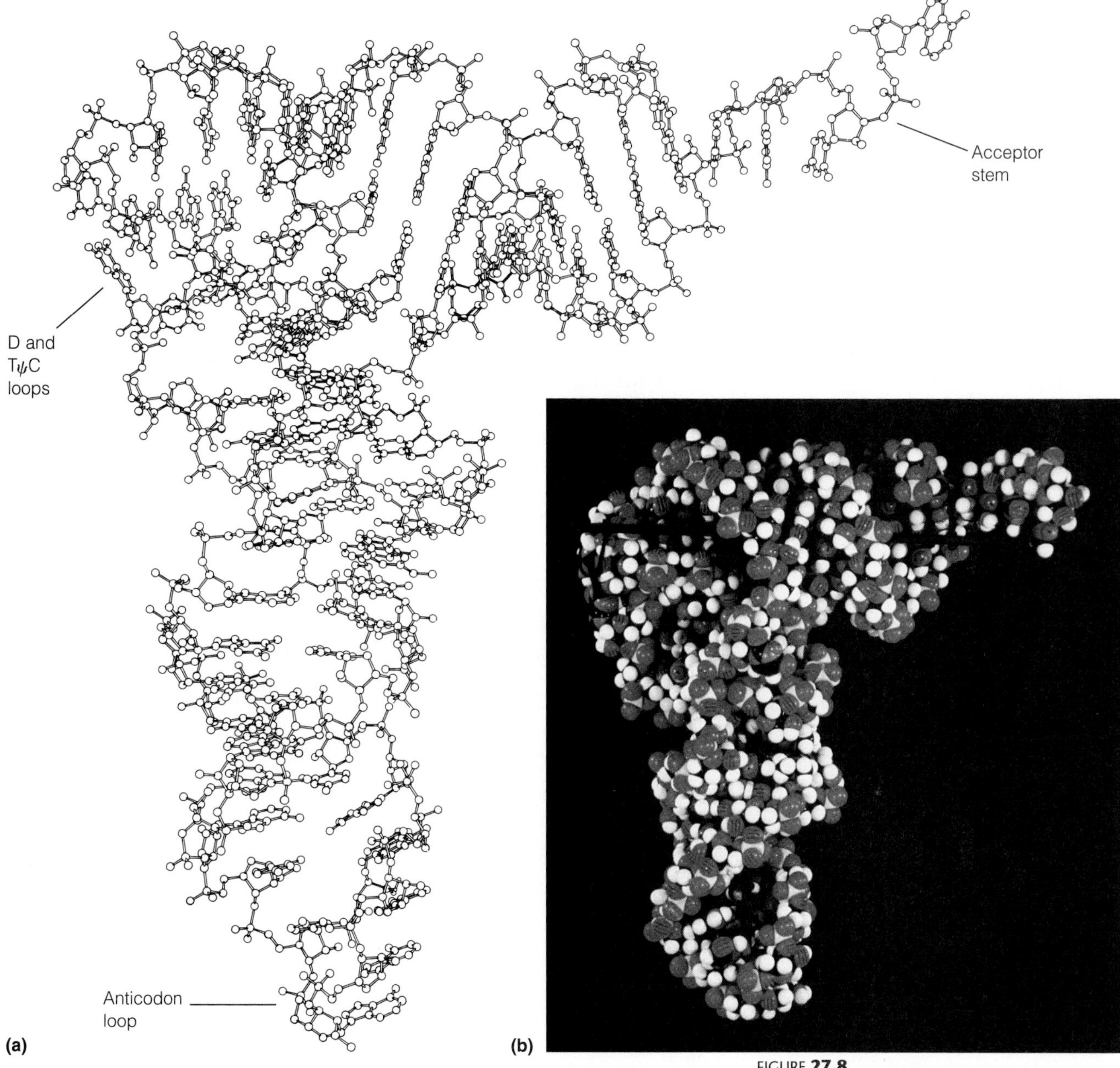

FIGURE **27.8**

Model of yeast phenylalanine tRNA derived from x-ray diffraction studies. The anticodon is at the bottom, the 3′ acceptor stem at the upper right. **(a)** Drawing showing all atomic positions. **(b)** Space-filling model.

Reprinted with permission from S.-H. Kim et al., *Science* (1974) 185:435. © 1974 AAAS.

ures 27.8 and 4.20 (page 104). As these figures show, a tRNA molecule looks rather like a hand-held drill or soldering gun. The anticodon loop is at the bottom of the grip, and the acceptor stem is at the working tip. The D loop and the TψC loop are folded inward in a complex fashion near the top of the grip, to provide a maximum of hydrogen bonding and base stacking. Some of the hydrogen bonding patterns required to produce this folding are rather unusual (Figure 27.9). The three-dimensional shapes of the tRNAs are highly conserved even though the primary structures vary. A likely explanation is that such conservation is necessary so that each tRNA can fit equally well onto the ribosome and carry out its function.

COUPLING OF tRNAS TO AMINO ACIDS AND FORMATION OF AMINOACYLATED tRNAS: THE FIRST STEP IN PROTEIN SYNTHESIS

Amino acids are attached to tRNAs by a covalent bond between the carboxylate of the amino acid and a ribose hydroxyl group of the invariant 3′ adenosine residue on the tRNA. The amino acid can react with either the 2′ or the 3′ hydroxyl group

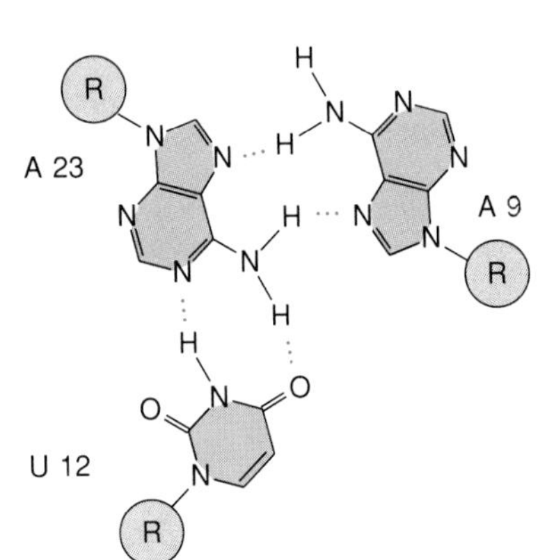

(d)

FIGURE **27.9**

Unusual base pairings in tRNA. All are from the yeast $tRNA^{Phe}$ shown in Figure 27.8. **(a, b)** Some unusual pair matches. **(c, d)** Some examples of triple interactions. R represents the ribosyl residue of the RNA chain. The bases prefixed by m are methylated at the carbon atom corresponding to the superscript. Numbers following the letters designating bases show the position in the sequence.

and can apparently shuttle back and forth between these sites. Pairing of the correct amino acid residues and the tRNAs is accomplished by a set of enzymes called **aminoacyl-tRNA synthetases.** In the bacterium *E. coli* there are 21 synthetases, each of which recognizes one amino acid and one or more tRNAs. Lysine is unique in having two synthetases. The reaction linking the two molecules, shown in Figure 27.10, proceeds in two steps. First, the amino acid, which is bound to the synthetase, is activated by ATP to form an **aminoacyl adenylate.** While still bound to the enzyme, this intermediate reacts with one of the correct tRNAs to form the covalent bond and release AMP.

Because all of the synthetases perform essentially the same function, you might expect them to represent minor variations on a common theme. This is, however, not the case; there are two general classes of aminoacyl-tRNA synthetases (I and II). Their active sites are completely different, and the classes bind their cognate

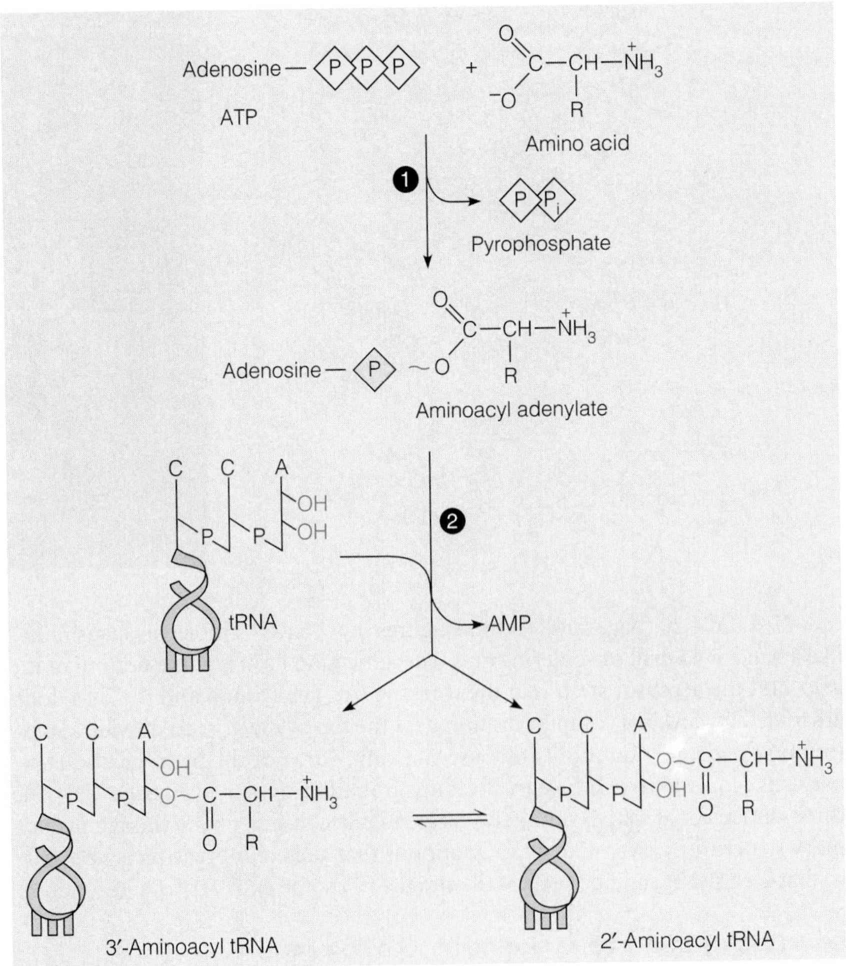

FIGURE **27.10**

Formation of aminoacyl tRNAs by aminoacyl tRNA synthetase. In step 1 the amino acid is accepted by the synthetase and is adenylylated. In step 2 the proper tRNA is accepted by the synthetase, and the amino acid residue is transferred to the 2′ or 3′ OH of the 3′-terminal residue of the tRNA. All reactions occur on the synthetase.

tRNAs from opposite sides. Furthermore, the class I enzymes tend to function as monomers, whereas the class II enzymes function as dimers or tetramers. The reasons for these extreme differences are unknown, but they may reflect the utilization of some amino acids in proteins before others in the very early evolution of protein synthesis. A recent observation that may bear on this question is that some members of some classes of organisms (Gram-positive bacteria and archaebacteria, for example) as well as some organelles use an indirect transamidation route for charging some tRNAs. For example, $tRNA^{Gln}$ is charged first with Glu, which is then replaced by Gln:

Amino acids are coupled to their appropriate tRNAs by aminoacyl-tRNA synthetases.

$$\text{Glu} + \text{tRNA}^{\text{Gln}} + \text{ATP} \xrightleftharpoons{\textbf{glutamyl-tRNA synthetase}} \text{Glu-tRNA}^{\text{Gln}} + \text{AMP} + \text{PP}_i$$

$$\text{Gln} + \text{Glu-tRNA}^{\text{Gln}} + \text{ATP} \xrightleftharpoons[\textbf{amidotransferase}]{\textbf{Glu-tRNA}^{\textbf{Glu}}} \text{Gln-tRNA}^{\text{Gln}} + \text{ADP} + \text{P}_i + \text{Glu}$$

Thus, these organisms do not require (although they may have) a Gln-$tRNA^{Gln}$ synthetase, as do Gram-negative bacteria and eukaryotes. It has been suggested that glutamine was one of the last amino acids to be added to the protein repertoire and that it was initially incorporated by this route.

You might expect that the synthetase would identify the correct tRNA on the basis of its anticodon, but recent studies indicate that the identification process is more complex and various nucleotides act as *identity elements.* In 1988, Ya-Ming Hou and Paul Schimmel showed that changing a single base pair (between residues 3 and 70 in the acceptor stem) of $tRNA^{Cys}$ or $tRNA^{Phe}$ to the G-U pair found in $tRNA^{Ala}$ caused the alanine synthetase to accept the $tRNA^{Cys}$ or $tRNA^{Phe}$ and couple it to alanine. Other tRNAs appear to be recognized by their synthetases at many different locations (see Figure 27.11). No simple rule can be seen, although it is clear that identifying elements are clustered in the anticodon loop and the acceptor stem.

Insight into the recognition of tRNAs by their synthetases has been provided by the x-ray diffraction analysis of the complexes formed. One example (of a class I synthetase-tRNA complex) is shown in Figure 27.12. As shown in the figure, the

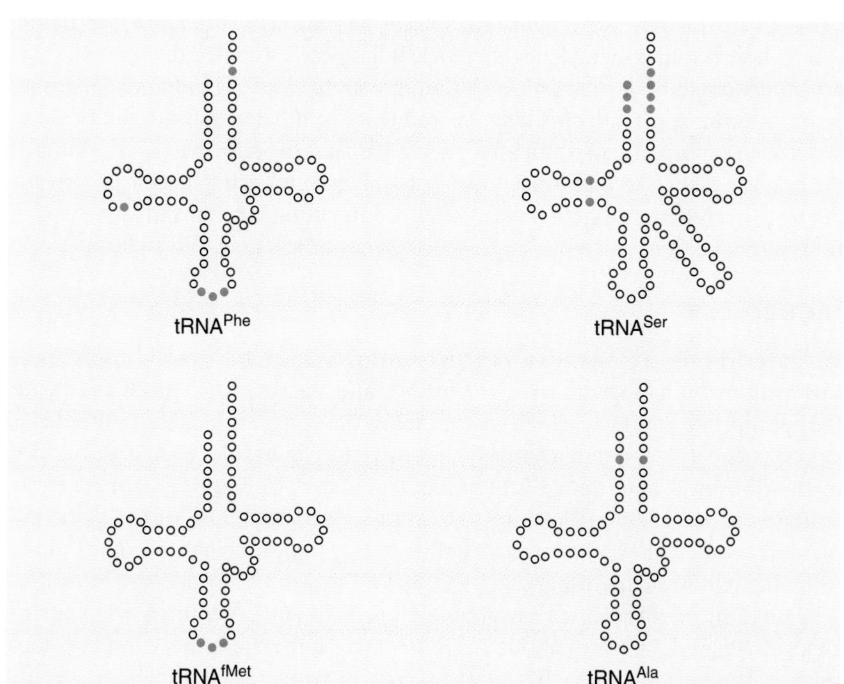

FIGURE **27.11**

Major "identity elements" in some tRNAs. Red circles represent the positions that have been shown to identify the tRNA to its cognate synthetase.

Adapted from L. Schulman and J. Abelson, *Science* (1988) 240:1591–1592.

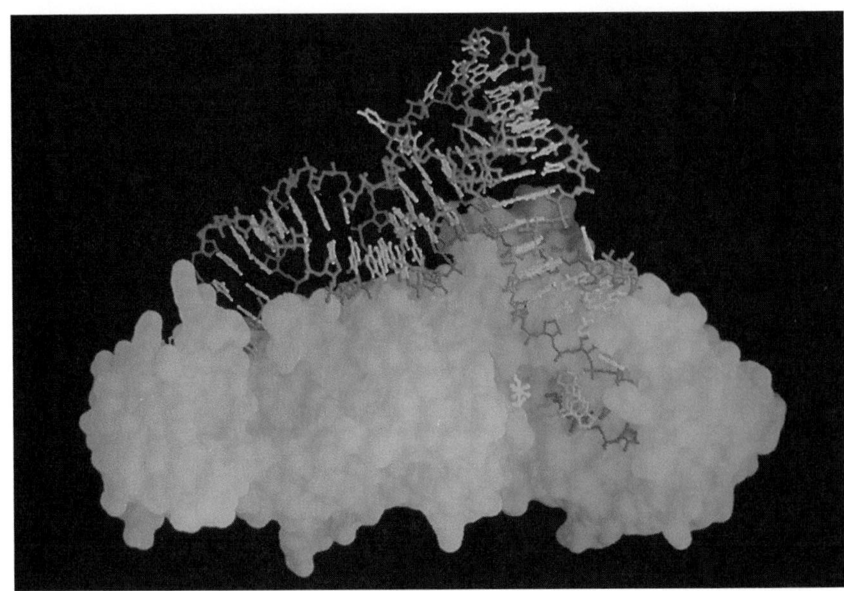

FIGURE **27.12**

A model of the *E. coli* glutaminyl tRNA synthetase coupled with its tRNA and ATP. The tRNA is represented by a detailed atomic model, the protein by its solvent-accessible surface (blue). The ATP (green) and the 3′ acceptor stem of the tRNA fit into a deep cleft in the synthetase. This cleft will also accommodate the amino acid. This is a monomeric class I synthetase.

tRNA lies across the protein, making a number of specific contacts, including crucial ones in the anticodon region and in the acceptor stem. Both of these regions are distorted in the complex, with the acceptor stem being elongated and inserted into the active site pocket. This pocket is formed by a common protein structural motif called the *dinucleotide fold,* which frequently acts as a nucleotide-binding region. In this case it also binds the ATP required for acylation. It provides a binding site for glutamine as well. Thus, all three participants in the reactions are grouped close together.

It is essential that the matching of amino acid and tRNA be very accurate. Once the amino acid is attached to a tRNA molecule and the charged tRNA is released from the enzyme, there is no further chance for it to be checked for an error. If the wrong amino acid has been coupled to a tRNA, it will be incorporated into protein according to the tRNA anticodon and thus will make an "incorrect" protein. Such errors would seem extremely likely when the differences between amino acid side chains are small—as between valine and isoleucine, where there is only a methylene difference. It has been estimated that in such cases error rates could be as high as 1 in 100, an intolerable level. The chance of such an error is minimized by *proofreading.* An aminoacyl synthetase has the ability to hydrolyze an incorrect pair, even after such a pair has been formed on it. It is as if the enzyme takes a second look at its product before releasing it. This proofreading by the synthetase, together with other proofreading reactions in later steps in translation, guarantees an overall error frequency of less than 1 in 10,000.

THE RIBOSOME

We have now described the two participants that must be brought together to carry out protein biosynthesis—the mRNA and the set of tRNAs charged with the appropriate amino acids. The actors are in the wings, and all that is needed is a proper director, and a stage on which the events can unfold. Both are provided by the ribosome, and the typical cell requires many. An *E. coli* cell, for example, contains about 15,000 ribosomes, accounting for about 25% of the dried cell mass. Thus, a cell devotes a large part of its energy to producing ribosomes and to using them in protein synthesis.

Functional prokaryotic ribosomes are particles with a sedimentation coefficient of 70S and a mass of about 2.3×10^6 Da. They are composed of two subunits, a 30S particle and a 50S particle, each of which is a complex of specific

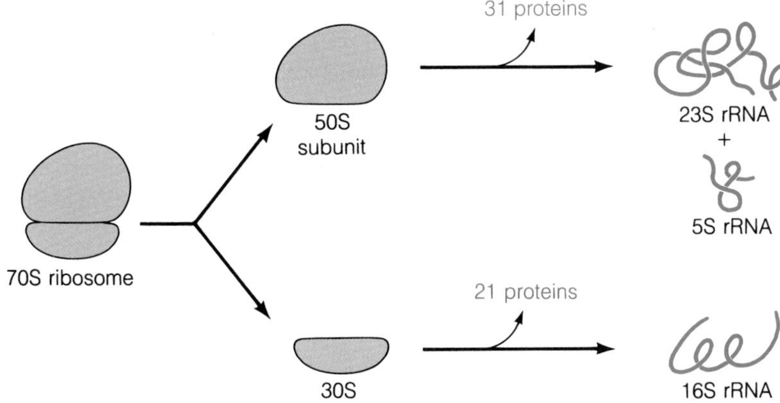

FIGURE **27.13**

Components of a 70S prokaryotic ribosome.

ribosomal RNAs (called rRNAs) and proteins (Figure 27.13). The 30S ("small") subunit of an *E. coli* ribosome contains a single 16S rRNA molecule (1542 residues in length) plus one copy each of 21 different proteins, designated S1 to S21. The 50S ("large") subunit contains two rRNA molecules—a 23S rRNA and a 5S rRNA (2904 residues and 120 residues, respectively, in *E. coli*). In addition, the large subunit has 31 different proteins designated L, all but one present in one copy. There are four copies of the single protein designated L7/L12. Until recently there has existed some confusion concerning the total number of proteins in the large subunit, with as many as 34 having been claimed. However, careful elimination of artifactual results and duplications shows that there are only 31 unique types. All of the ribosomal proteins can be resolved on two-dimensional SDS gels as shown in Figure 27.14, and all have now been sequenced. The three-dimensional structures of about 15 ribosomal proteins are now known. All are small, or made of small domains, and many show sequence and/or structural similarities to proteins of widely diverse functions.

A functional prokaryotic ribosome, the site of protein synthesis, contains about 50 proteins and three different rRNA molecules.

The sequencing studies reveal no significant homologies among the different proteins in a ribosome, but comparison of sequences between corresponding proteins in the ribosomes of different organisms reveals considerable evolutionary conservatism. Thus, the ribosome is a complex object that evolved early in the

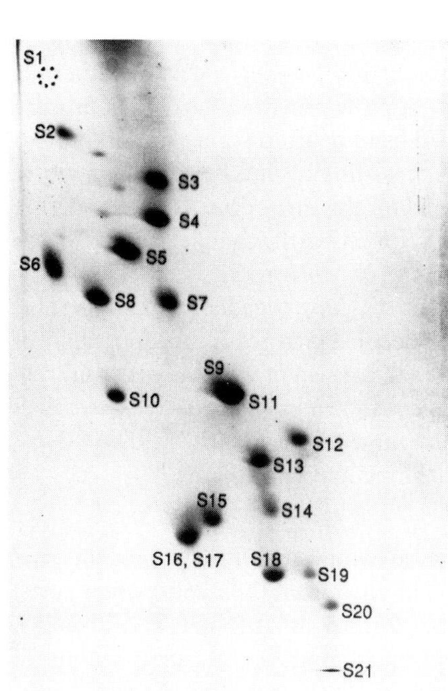

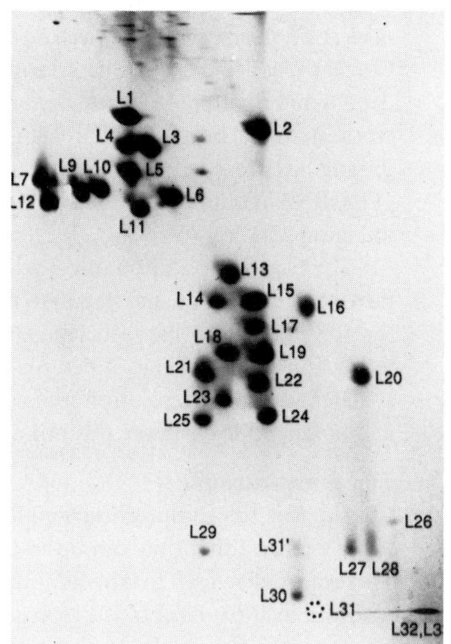

FIGURE **27.14**

Two-dimensional gel electrophoresis of *E. coli* ribosomal proteins. (a) 30S proteins. **(b)** 50S proteins.

Courtesy of R. Traut.

history of life and has remained relatively unchanged. Although the ribosomes of eukaryotes differ significantly from those of prokaryotes, the evolutionary continuity is clear. The sequences of many ribosomal RNAs (especially those from the small subunit) have also been determined, and they tell the same story. Indeed, because of their relatively slow evolutionary rates, rRNAs are useful as evolutionary yardsticks over vast phylogenetic distances. From them, we have learned that the eukaryotes, eubacteria, and archaebacteria represent three major kingdoms of organisms, sharing a common ancestor in the most distant past.

When the sequence of a 16S RNA is examined carefully, it is found to contain many regions of self-complementarity, which are capable of forming double-helical segments. A pattern like that shown in Figure 27.15 may seem so complex as to appear almost arbitrary, but comparison with other, even distantly related 16S RNA sequences shows that the potentially double-strand regions are highly conserved. Indeed, the secondary structure seems more highly conserved than is the primary structure, for it is often found that there are compensatory mutations in double-helical regions so as to maintain base pairing. A schematic illustration like that in Figure 27.15 is analogous to the cloverleaf visualization of a tRNA (Figure 27.6). The actual rRNA is folded into a three-dimensional structure, just as is the tRNA. In the case of the ribosomal subunit, however, the structure is further complicated by the presence of ribosomal proteins bound to the RNA. We are now beginning to learn how the proteins and RNA fit together in the ribosome.

Internal Structure of the Ribosome

Although it has been possible for many years to obtain good EM images of intact ribosomes and their subunits, high resolution was difficult to achieve because of the necessity of staining or shadowing the particles. Nor could such techniques hope to tell us how the proteins and RNA were disposed inside the ribosome. X-ray diffraction has been hindered until recently because of problems with crystallization and because of the sheer complexity and lack of symmetry in the structure.

Nevertheless, even the earlier EM studies revealed the general topography of the small and large subunits and how they fit together to make 70S particles (see Figure 27.16). To go further, and find out where the individual proteins and folded RNA segments are placed, has required enormous effort and a number of complementary techniques. Some of these are quite powerful techniques and are applicable to studies of other kinds of complex macromolecular structures. In Tools of Biochemistry 27A we briefly describe four of these methods: *cross-linking, immunoelectron microscopy, cryoelectron microscopy,* and *low-angle neutron scattering.* Current ideas concerning the arrangement of proteins within the 30S subunit are depicted in Figure 27.17a and b, as deduced from neutron scattering and protein cross-linking. The addition of RNA–RNA cross-linking and RNA–protein cross-linking has provided a self-consistent model for the three-dimensional structure of the RNA and its interactions with proteins, leading to the complete model shown in Figure 27.17c.

Techniques like cross-linking, neutron scattering, immunoelectron microscopy, and cryoelectron microscopy have provided a detailed picture of the structure of some ribosomal subunits.

The recent application of cryoelectron microscopy, together with low-resolution x-ray diffraction, has begun to provide detailed pictures of the topography of the 50S subunit prokaryotic ribosome as well, consistent with the neutron scattering and cross-linking studies. An example is shown in Figure 27.18. As we shall see later, such high-resolution studies also can now show where the mRNA, tRNAs, and polypeptide chain are positioned in the functioning ribosome.

Ribosome Assembly

Considering the compositional and structural complexity of ribosomes, it is surprising to find that they can be assembled in vitro from their constituent parts. This impressive feat, first accomplished for the 30S subunit by P. Traub and M. Nomura in 1968, has proved very useful in elucidating the roles and positions of ribosomal components. A schematic flowchart for the assembly of the 30S subunit of the *E. coli* ribosome is shown in Figure 27.19. Successful assembly seems

Despite their complexity, ribosomal subunits can be assembled in vitro.

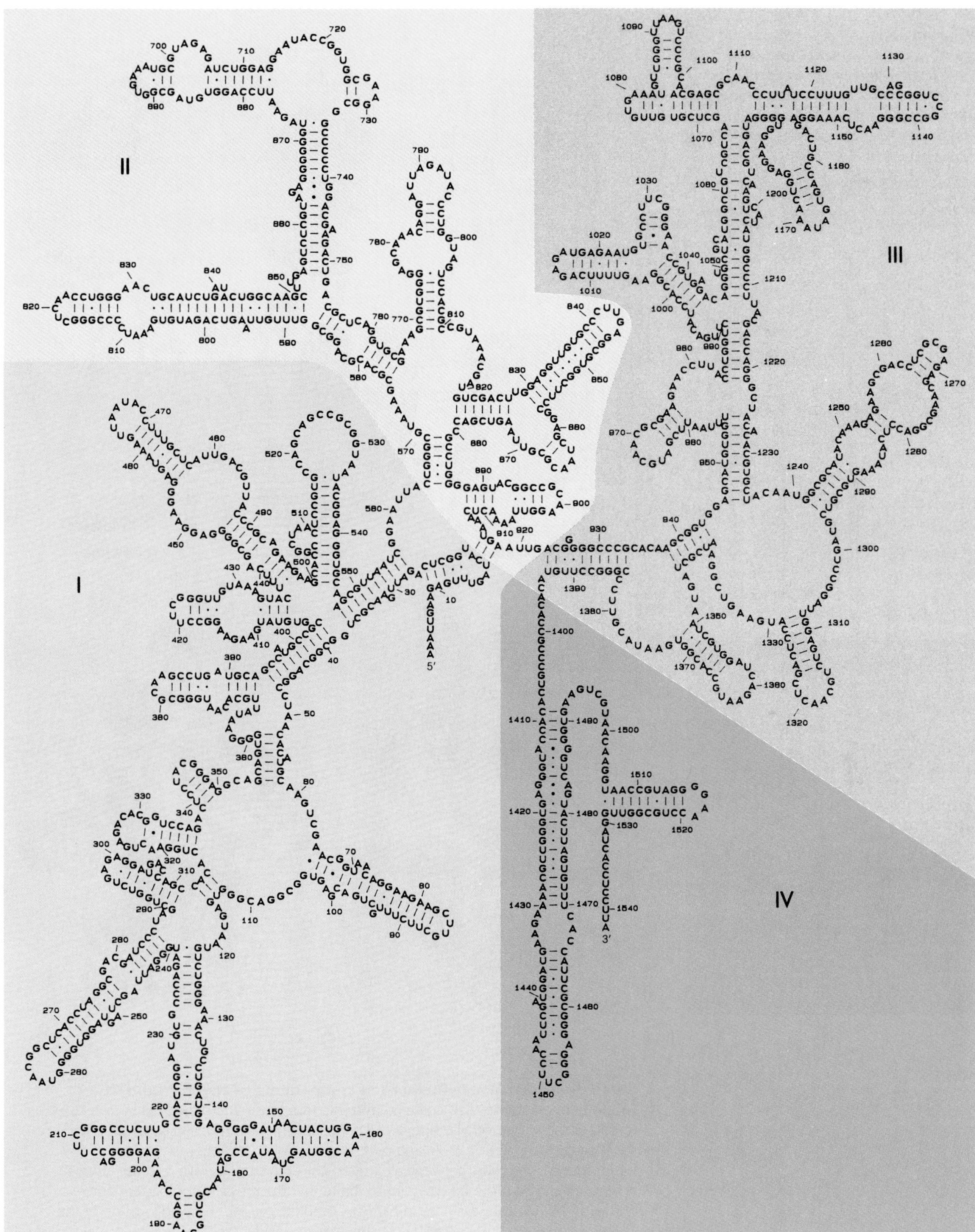

FIGURE **27.15**

Postulated secondary structure of *E. coli* 16S rRNA. The sequence has been aligned to produce maximum base pairing between complementary segments. The molecule appears to have four major domains of folding (I–IV).

Courtesy of R. Gutell and H. Noller.

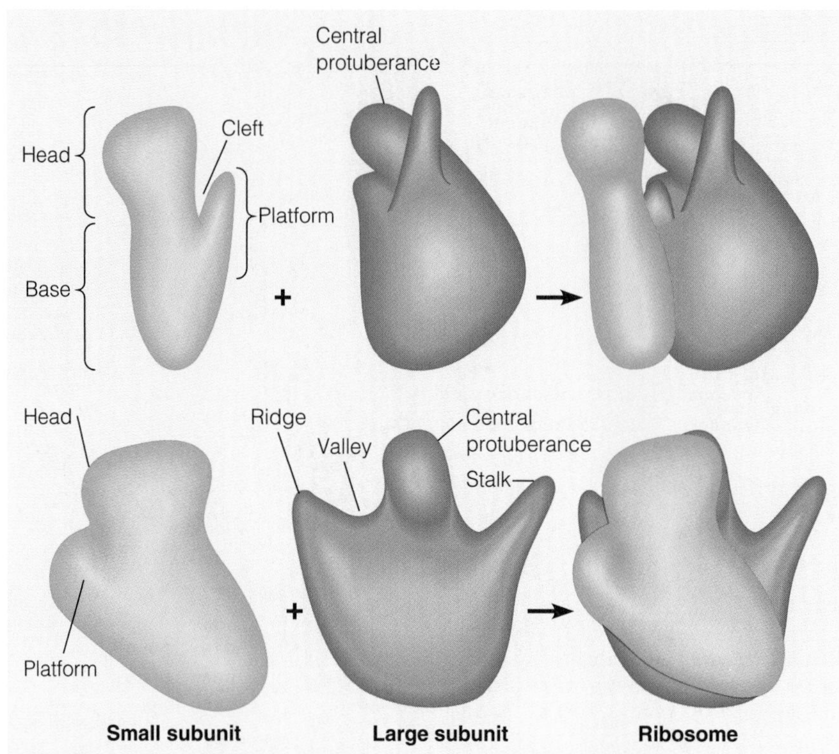

FIGURE **27.16**

Low-resolution models of the small and large ribosomal subunits and their 70S complex, as deduced from early electron microscopy studies. The most prominent features, which can also be observed in more recent, higher resolution studies, are indicated by names attached.

Courtesy of M. Oaks, A. Scheirman, T. Atha, G. Shankweiler, and J. A. Lake, *The Ribosome*, W. E. Hill, A. Dahlberg, R. A. Garret, P. B. Moore, D. Schlessinger, and J. R. Warner, eds., p. 181 (New York: American Society of Microbiology Press, 1990).

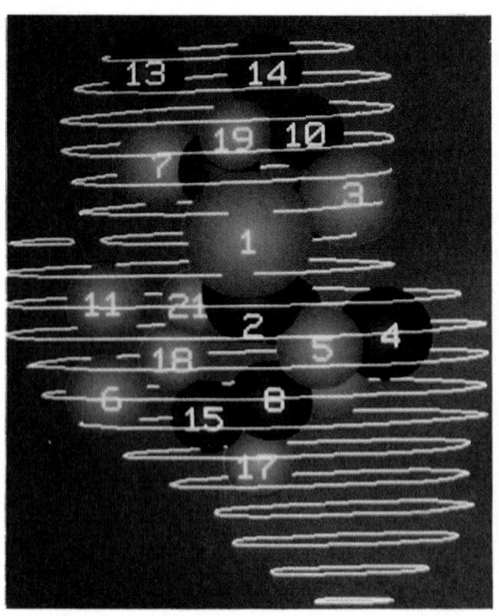

(a)

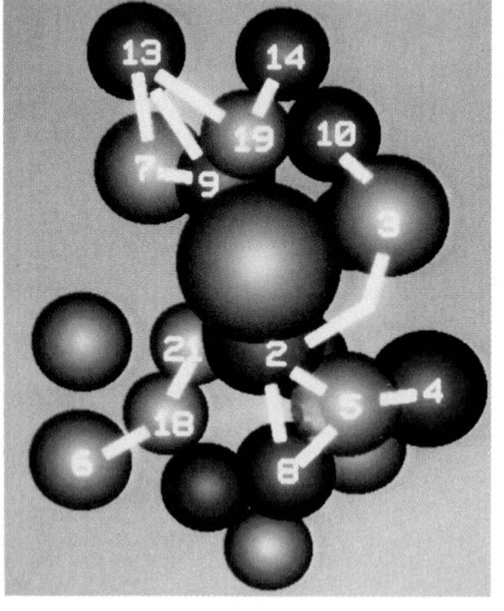

(b)

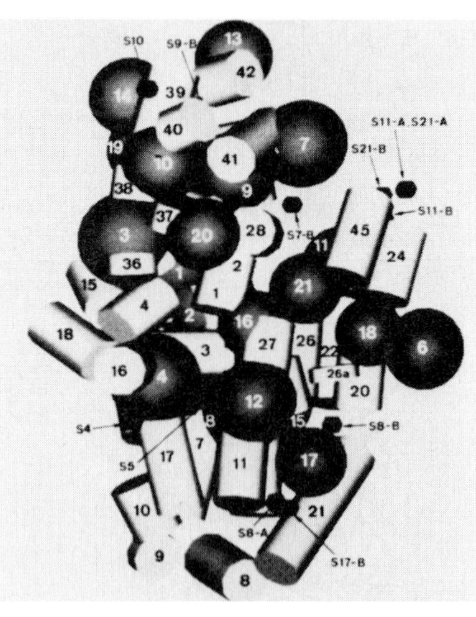

(c)

FIGURE **27.17**

Three-dimensional arrangement of components of the 30S subunit. **(a)** Protein arrangement as deduced from neutron scattering experiments. The contour lines show the overall shape of the particle. Regions not occupied by proteins are filled by RNA. **(b)** Cross-linking pattern. The protein arrangement shown in **(a)** is reproduced, with bars between all protein pairs reported by two or more independent research groups to be cross-linked. Note that the cross-linking pattern agrees well with the neutron scattering data. **(c)** Complete model, showing location of the 16S rRNA (white cylinders) relative to the proteins, as deduced from RNA–protein and RNA–RNA cross-linking studies.

(a) Courtesy of P. Moore and V. Ramakrishnan from M. Capel et al., *Science* (1987) 238:1403–1406. © 1987 AAAS, reprinted with permission; **(b)** reprinted with permission from M. Capel et al., *Science* (1987) 238: 1403–1406, © 1987 AAAS; **(c)** courtesy of E. Schüler and R. Brimacorne, *EMBO J.* (1988) 7:1509–1513. By permission of Oxford University Press.

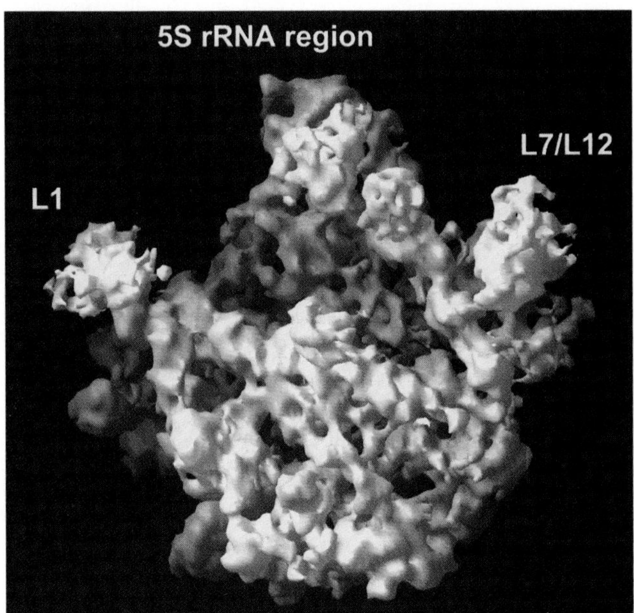

FIGURE **27.18**

A model of the large ribosomal subunit at 0.9 nm resolution, as determined by x-ray diffraction from crystals. The orientation is approximately as shown in Figure 27.16.

Provided by N. Ban, P. Nissen, P. Moore, and T. A. Steitz from N. Ban et al., *Cell* (1998) 93:1105–1115. © Cell Press.

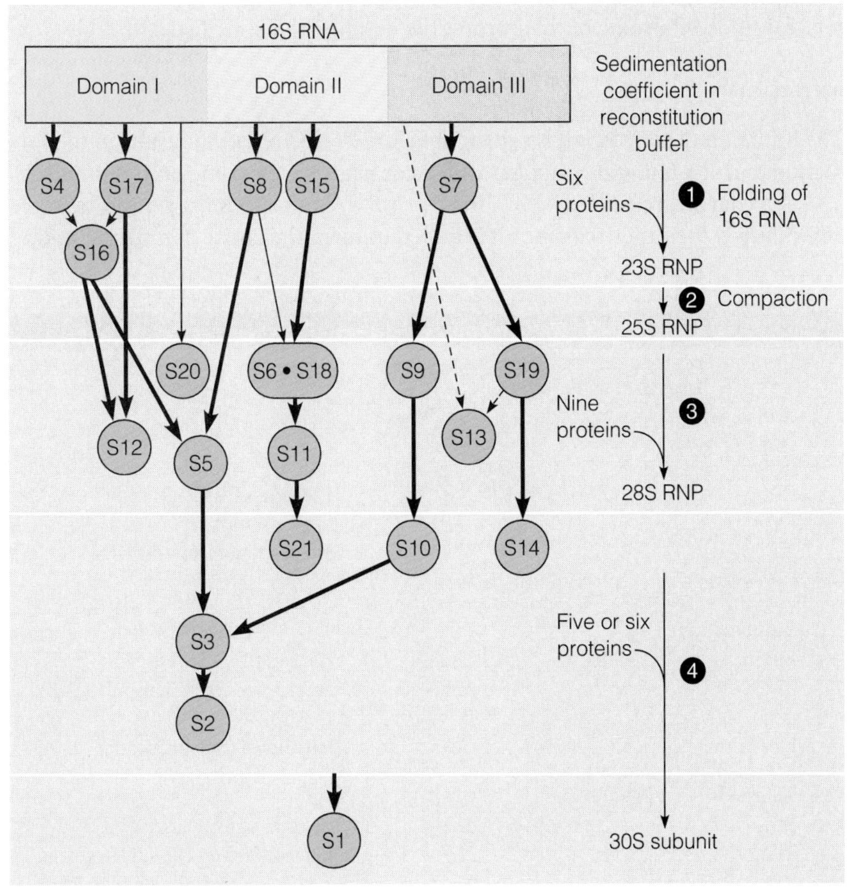

FIGURE **27.19**

Assembly map for the 30S subunit. The map shows how a 30S ribosomal subunit of *E. coli* can be reassembled in the test tube from RNA and proteins (green). The order of addition shown here is important—some proteins must be added before others will fit properly. Certain early-binding proteins are associated with each of the RNA domains I–III shown in Figure 27.15. Arrows between proteins indicate dependence in binding; the thicker the arrow, the more essential the preceding proteins. For example, binding of S3 in step 4 requires that S5 and S10 are already bound. The binding of S1 requires a nearly complete ribosomal structure, but it is difficult to tell just which proteins are essential. In any event, its binding completes the assembly, and a functional 30S subunit has been produced. The pathway to the right traces the structural development of the subunit. The abbreviation RNP stands for ribonucleoprotein particle.

Adapted from A. Spirin, *Ribosome Structure and Protein Biosynthesis* (Redwood City, Calif.: Benjamin/Cummings, 1986). © 1986 Benjamin/Cummings Publishing Company.

to require that the proteins bind in a particular order: The first proteins to bind to RNA appear to help fold the RNA to provide binding sites for the rest. The 50S subunit can also be assembled in vitro, but a complicated series of steps at different temperatures is required.

The 30S and 50S subunits formed in such in vitro reconstitution experiments are functional and can combine to form 70S ribosomes if divalent cation concentration is increased. The equilibrium between 30S, 50S, and 70S particles,

$$30S + 50S \rightleftharpoons 70S$$

is in general shifted to the right by increasing concentrations of Mg^{2+} and to the left by high concentrations of a monovalent cation such as Na^+ or K^+. Under the ionic conditions existing in the cytosol, a substantial fraction of ribosomes are dissociated into free subunits. It is these subunits that enter into the series of translation reactions we shall consider next.

The Mechanisms of Translation

Translation involves three steps—initiation, elongation, and termination—each aided by specific protein factors.

We now have described all of the major participants in the translation process: a messenger RNA, charged tRNAs, and the ribosome, where the actual translation events occur. Translation of a message can be divided into three stages: *initiation, elongation,* and *termination.* Here we shall describe these steps as they occur in prokaryotes. Significant, though not fundamental, differences in eukaryotic cells are described in Chapter 28.

Each of the steps in translation requires a number of specific proteins that interact with the major participants listed above. These proteins are referred to as *initiation factors* (IFs), *elongation factors* (EFs), and *release factors* (RFs). These factors, together with some of their properties and functions, are listed in Table 27.4.

INITIATION

The initiation of translation is shown in Figure 27.20. Initiation results in the formation of a *70S initiation complex,* which consists of a ribosome bound to mRNA and to a charged *initiator tRNA.* First the mRNA and tRNA are bound to a free 30S subunit, then the 50S subunit is added to form the entire complex. Binding

TABLE 27.4 Protein factors involved in translation in prokaryotes

Factor	Approximate Number per Ribosome in Cell	Binds GTP?	Role
Initiation			
IF1	1/7	No	Promotes dissociation of 70S ribosome
IF2	1/7	Yes	Helps attach initiator tRNA
IF3	1/7	No	Similar to IF1
Elongation			
EF-Tu	~10	Yes	Carries tRNA into A site
EF-Ts	1	Yes	Participates in recharging EF-Tu with GTP
EF-G	1	Yes	Facilitates translocation
Termination			
RF1	1/20	No	Release factor (UAA, UAG)
RF2	1/20	No	Release factor (UAA, UGA)
RF3	?	Yes	A GTPase that promotes release

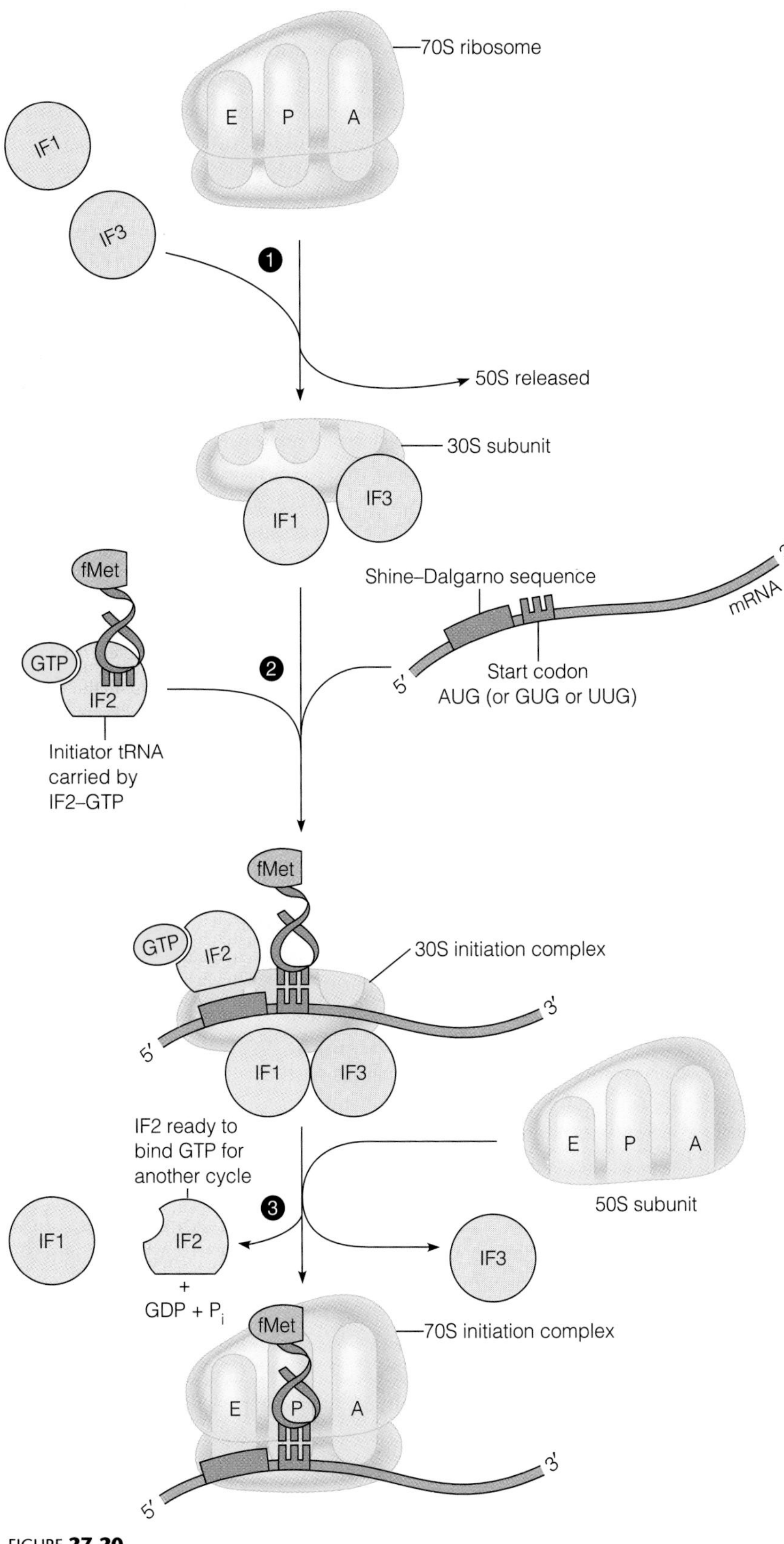

FIGURE **27.20**

Initiation of protein biosynthesis in prokaryotes. Certain details of the process described in the text remain uncertain. For example, the exact order of binding of IF1, IF2, IF3, tRNA, and mRNA is unclear. We present here a current model. E, P, and A are, respectively, called the exit, peptidyl, and aminoacyl binding sites.

of mRNA and initiator tRNA also requires binding of the three **initiation factors** (IF1, IF2, and IF3) to a free 30S subunit. The exact order of binding of these factors remains uncertain. The factors IF3 and IF1 appear to promote dissociation of preexisting 70S ribosomes, to produce the free 30S subunits needed for initiation (Figure 27.20, step 1). The third factor, IF2, is bound carrying a molecule of GTP; it is thought to accompany the charged initiator tRNA in binding. IF2 is one of the class of *G proteins* similar to those described in Chapter 23. At about the same time that the IF2–$tRNA^{fMet}$ complex is bound, the mRNA is bound (step 2). Although the order of these additions is still uncertain, it is clear that IF2–GTP is absolutely required for binding of the first tRNA. With the binding of the first tRNA and the mRNA, formation of the *30S initiation complex* is complete. The initiation complex has high affinity for a 50S subunit and binds one from the available pool (step 3), with concomitant release of IF3.

The initiator tRNA is special. It recognizes and binds to the AUG codon that would normally code for methionine, but it actually carries an *N*-formylmethionine. The formyl group is added *after* charging of the tRNA, by an enzyme *(transformylase)* that recognizes the particular $tRNA^{fMet}$ and transfers a formyl group from N^{10}-formyltetrahydrofolate. *Only* $tRNA^{fMet}$ is accepted to form the initiation complex; all further charged tRNAs require the fully assembled ribosome. Therefore, all prokaryotic proteins are synthesized with the same N-terminal residue, *N*-formylmethionine. In almost all cases, the formyl group is removed during chain elongation. For many proteins the methionine itself is also cleaved off later.

In initiation, the correct attachment of mRNA to the ribosome is determined by binding of the Shine–Dalgarno sequence to a sequence on the 16S rRNA of the ribosome.

The mRNA attaches to the 30S subunit near the 5′ end of the message, which is appropriate because all messages are translated in the 5′→3′ direction. In describing the structure of typical prokaryotic messages, we mentioned the presence of a specific binding sequence (the Shine–Dalgarno sequence) near the 5′ end of each mRNA. Near the 3′ end of 16S ribosomal RNA is a sequence that can base-pair with these mRNA sequences. The rRNA sequence, 3′··· UCCUCC··· 5′, will pair with any Shine–Dalgarno sequence (for example, those shown in Table 27.3). This pairing aligns the message correctly for the start of translation. In particular, it places the initiator codon next to the P site, one of three tRNA binding sites in the ribosome (see below).

Translation cannot start until the 50S subunit has bound to the 30S initiation complex. The ribosome has three sites for tRNA binding, called the P (peptidyl) site, the A (aminoacyl) site, and the E (exit) site. As mentioned above, the AUG initiator codon with its bound $tRNA^{fMet}$ aligns with the P site. At this point, the GTP molecule carried by IF2 is hydrolyzed, and IF2–GDP, P_i, and IF1 are all released. The 70S initiation complex so formed is ready to accept a second charged tRNA and begin elongation of the protein chain.

The location of the P, A, and E binding sites for tRNAs has been established by chemical cross-linking and diffraction studies. The cross-linking studies show that the anticodon ends of the tRNA molecules contact the 30S subunit, whereas the acceptor ends interact specifically with the 50S subunit (Figure 27.21). All of the ribosomal proteins contacted lie in the cavity between the 30S and 50S subunits. The tRNA molecules are oriented with their anticodons reaching the mRNA at the bottom of the cavity close to the 30S subunit and their acceptor ends contacting the peptidyltransferase region on the 50S subunit, near the top of the cavity.

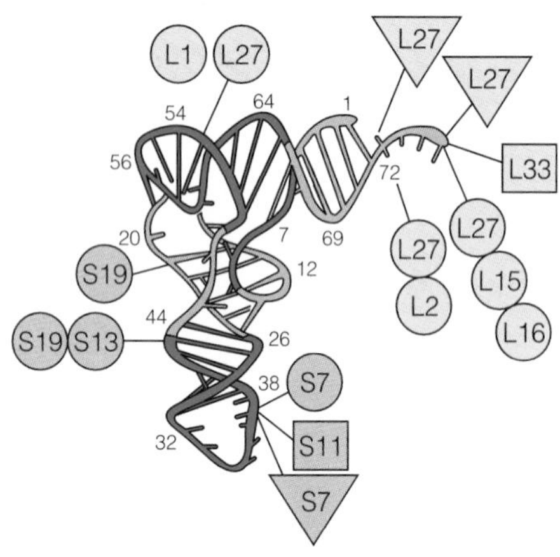

FIGURE **27.21**

Environment of tRNAs at the ribosome as determined by cross-linking. Cross-links from defined nucleotide positions in the tRNA to ribosomal proteins are shown. Proteins were differentially cross-linked depending on the location of the tRNA (A site, triangles; P site, circles; E site, squares; S = small subunit, L = large subunit).

ELONGATION

Growth of the polypeptide chain on the ribosome occurs by a cyclic process. Until recently, the process was thought to involve only two tRNA binding sites, P and A. However, it is now clear that *three* sites are involved. Figure 27.22 illustrates a single round in this cycle. Imagine it to be the first round following formation of the 70S initiation complex, or any subsequent round. They are all the same until a termination signal is reached.

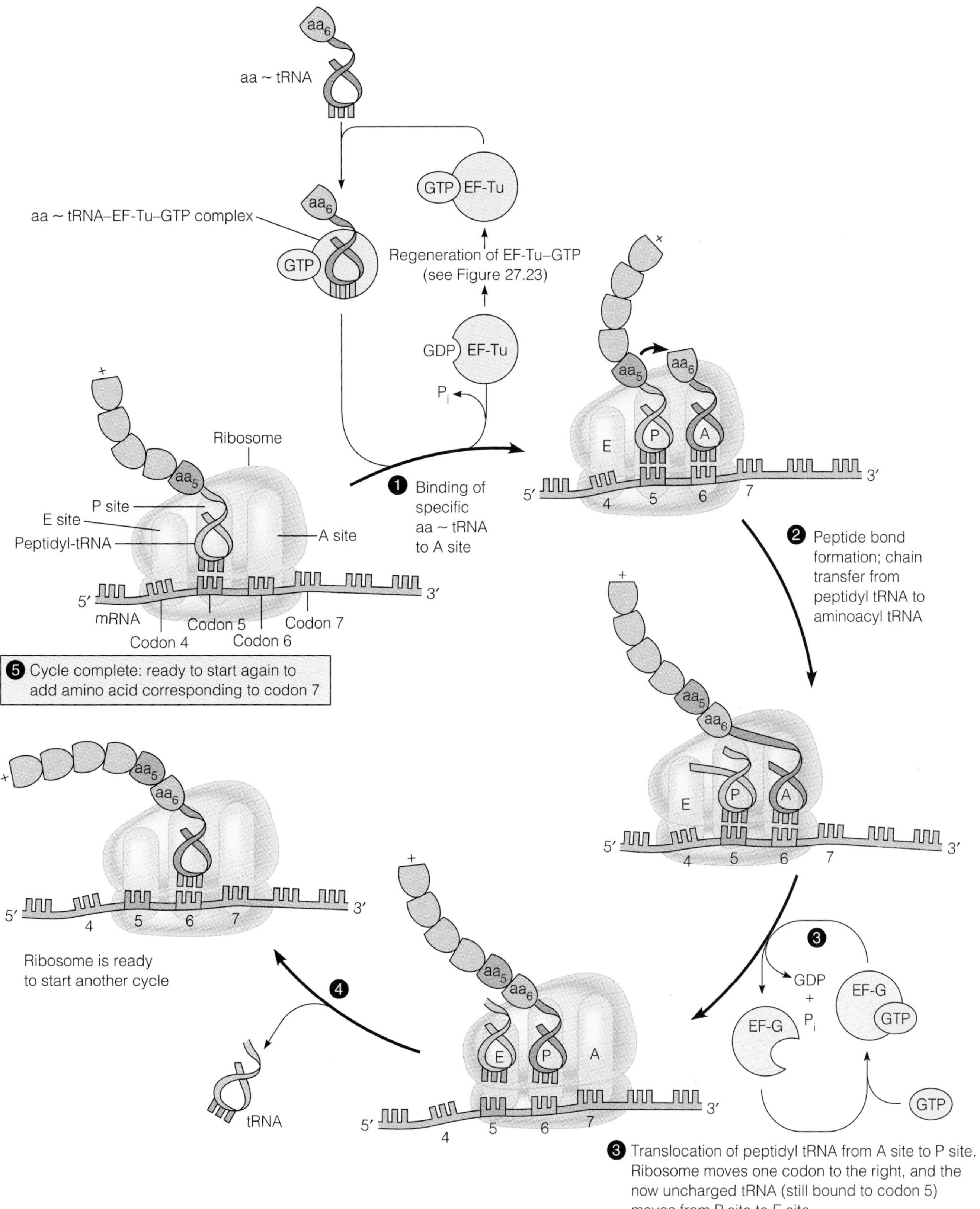

FIGURE **27.22**

Chain elongation in prokaryotic translation. The process is depicted as a cycle. Following translocation (step 3) and tRNA release (step 4), the ribosome is ready to accept the next aminoacyl tRNA (aa~tRNA) and repeat the cycle. The cycles will continue until a termination codon is reached.

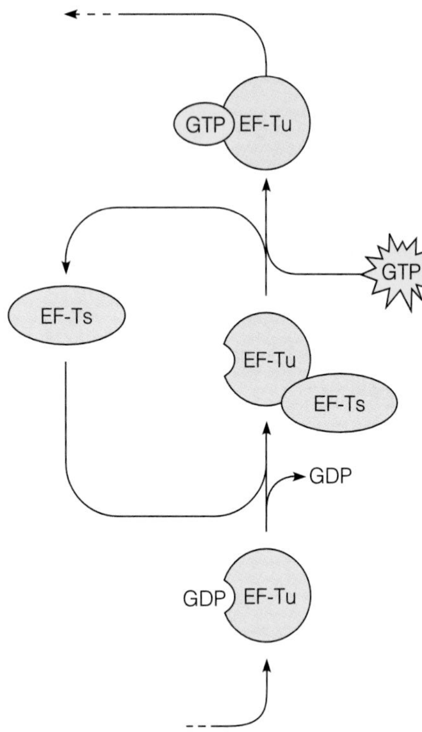

FIGURE **27.23**

Regeneration of EF-Tu–GTP by Tu–Ts exchange. This figure gives details of the regeneration cycle shown at the top of Figure 27.22. Binding of the factor EF-Ts to EF-Tu allows the release of GDP and binding of a new GTP to prepare EF-Tu for another cycle.

At the beginning of each cycle, the nascent polypeptide chain is attached to a tRNA in the P (peptidyl) site, and the A (aminoacyl) and E (exit) sites are empty. Aligned with the A site is the mRNA codon corresponding to the *next* amino acid to be incorporated. The charged (aminoacylated) tRNA is escorted to the A site in a complex with a protein, the **elongation factor** EF-Tu, which also carries a molecule of GTP. (Note the parallel to IF2–GTP here.) When the appropriate charged tRNA is deposited into the A site, the GTP is hydrolyzed, and EF-Tu GDP is released (Figure 27.22, step 1). Proofreading occurs at this step. The EF-Tu–GTP complex is then regenerated by the subsidiary cycle shown in Figure 27.23. After the charged tRNA is in place, it is checked both before and after the GTP hydrolysis and rejected if incorrect.

The next, and crucial, step is peptide bond formation (Figure 27.22, step 2). The polypeptide chain that was attached to the tRNA in the P site is now transferred to the amino group of the amino acid carried by the A-site tRNA. This step is called **peptidyl transfer** and is accomplished by an enzyme complex called *peptidyltransferase,* an integral part of the 50S subunit. Recent studies indicate that whereas some ribosomal proteins may be a part of this enzymatic complex, the ribosomal RNA itself plays a major role, functioning as a ribozyme (see pages 395–398). It is now believed that this transfer involves a switch from the simple P and A states to hybrid states, in which the acceptor ends of the two tRNA molecules move into the leftward positions while the codon ends remain fixed as before. These hybrid states are indicated as E/P and P/A:

This can be considered the first half of the translocation step (step 3 in Figure 27.22).

To complete the **translocation** step, the anticodon end of the now uncharged tRNA in the P site is transferred to the E site, and the tRNA in the A site (the tRNA that now has the nascent polypeptide chain attached to it) is moved completely to the P site. In the process, the ribosome moves a 3-nucleotide step in the 3′ direction, placing a new codon adjacent to the now empty A site. Like peptidyl transfer, this step requires a protein factor (EF-G) bound to GTP and requires GTP hydrolysis. Recent crystallographic studies reveal a remarkable "molecular mimicry" between EF-G–GTP and the ternary complex aa-tRNA–EF-Tu–GTP. As Figure 27.24 shows, the protein and the RNA–protein complex have almost exactly the same shape, even though they differ entirely in composition and se-

In elongation, the growing peptide chain at the P site is transferred to the newly arrived aminoacyl tRNA in the A site. Translocation then moves this tRNA to the P site, and the previous tRNA to the E site.

quence. It is speculated that the reason for this similarity is to allow EF-G–GTP to move temporarily into the A site, facilitating the displacement of the peptidyl–tRNA complex.

E site P site A site

E/P state P/A state

E site E/E state

P site P/P state

A site Empty

At this point the E and P sites are occupied, but A is empty. As the old tRNA is released from E (step 4, Figure 27.22), the A site gains high affinity and accepts the aminoacyl tRNA dictated by the next codon. A cycle of elongation is now complete. All is as it was at the start, except that now

1. The polypeptide chain has grown by one residue.
2. The ribosome has moved along the mRNA by three nucleotide residues—one codon.
3. At least two molecules of GTP have been hydrolyzed.

The whole process is repeated again and again until a termination signal is reached.

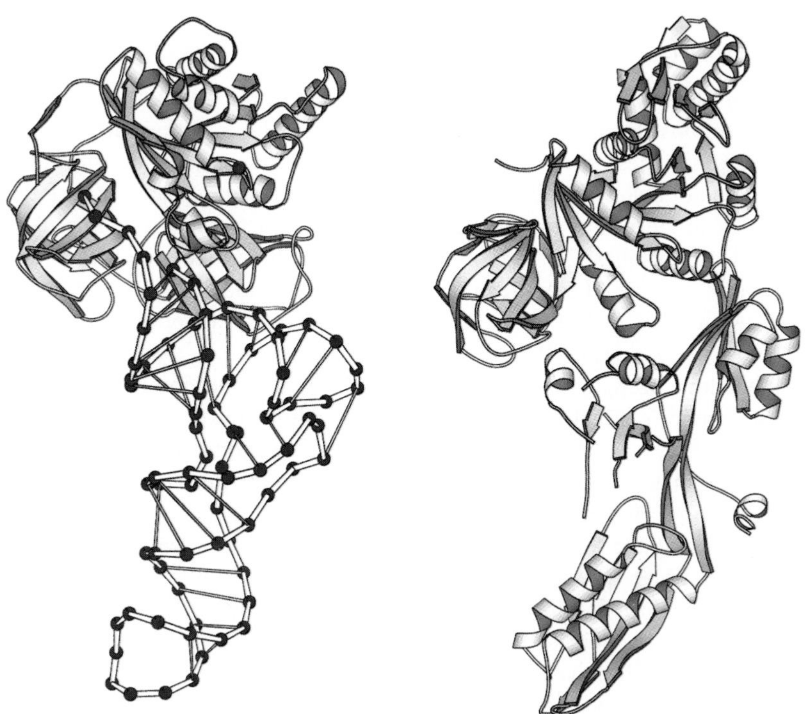

FIGURE **27.24**

The striking structural similarity between the translocation factor EF-G (right) and the ternary complex aa-tRNA–EF-Tu–GTP (left). Protein is shown in green, DNA in brown.

Courtesy of J. Nyborg, University of Aarhus, from P. Nissen et al., *Science* (1995) 270:1464–1472, with permission from *Science*.

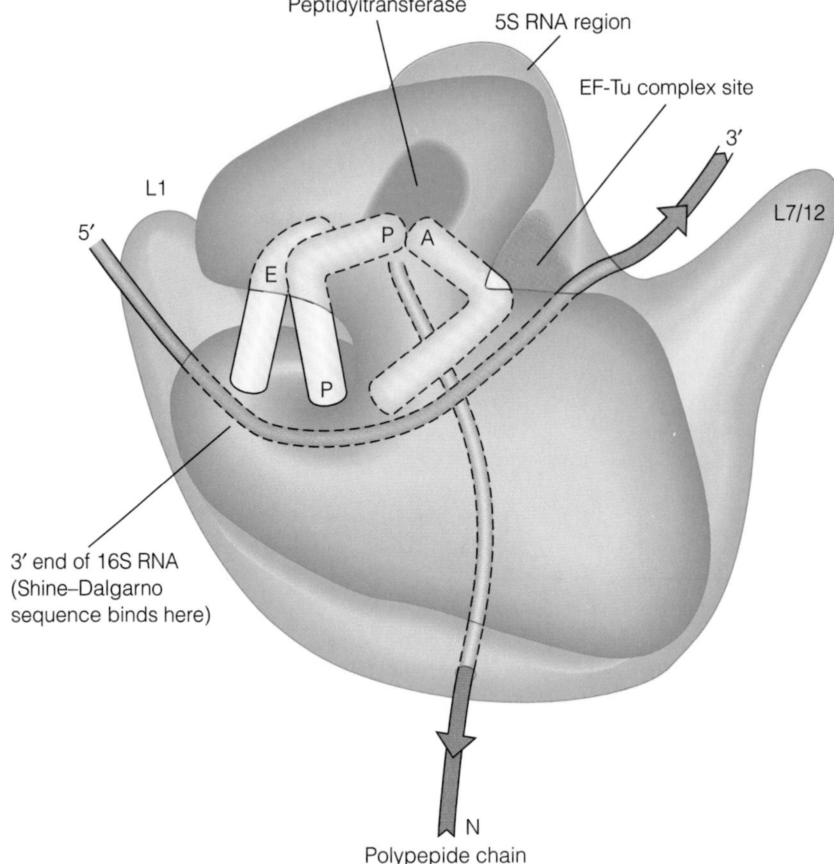

FIGURE **27.25**

A schematic view of functional regions of the ribosome. To depict tRNA site locations, all three sites are shown occupied, although this never occurs during protein synthesis. The view has the 30S subunit in front, the 50S in back.

As the polypeptide chain is synthesized, it passes through a tunnel in the 50S subunit and emerges from a hole near the bottom. The approximate locations and relevant sites are shown in Figure 27.25.

TERMINATION

Termination requires protein release factors that somehow recognize stop codons.

The completion of polypeptide synthesis is signaled by the translocation of one of the **stop codons** (UAA, UAG, or UGA) into the A site. Because there are no tRNAs that recognize these codons under normal circumstances, termination of the chain does not involve binding of a tRNA. Instead, proteins called **release factors** participate in the termination process. The three release factors found in prokaryotes are listed in Table 27.4. Two of these factors can bind to the ribosome when a stop codon occupies the A site: RF1 recognizes UAA and UAG, and RF2 recognizes UAA and UGA. The third factor, RF3, is a GTPase that appears to stimulate the release process, via GTP binding and hydrolysis.

It had long been assumed that RF1 and RF2 actually occupied the A site and somehow made contact with the termination codon. However, this may not be the case. In fact, there is some evidence that the terminator codon may also interact with a specific sequence in ribosomal RNA, with the function of release factors being more indirect.

In any event, the sequence of termination events is as shown in Figure 27.26. After RF1 or RF2 has bound to the ribosome, the peptidyltransferase transfers the C-terminal residue of the polypeptide chain from the P-site tRNA to a water molecule, releasing the peptide chain from the ribosome. The RF factors and GDP are then released, followed by the tRNA. The 70S ribosome is now unstable. Its instability is accentuated by the presence of a protein called *ribosome recycling*

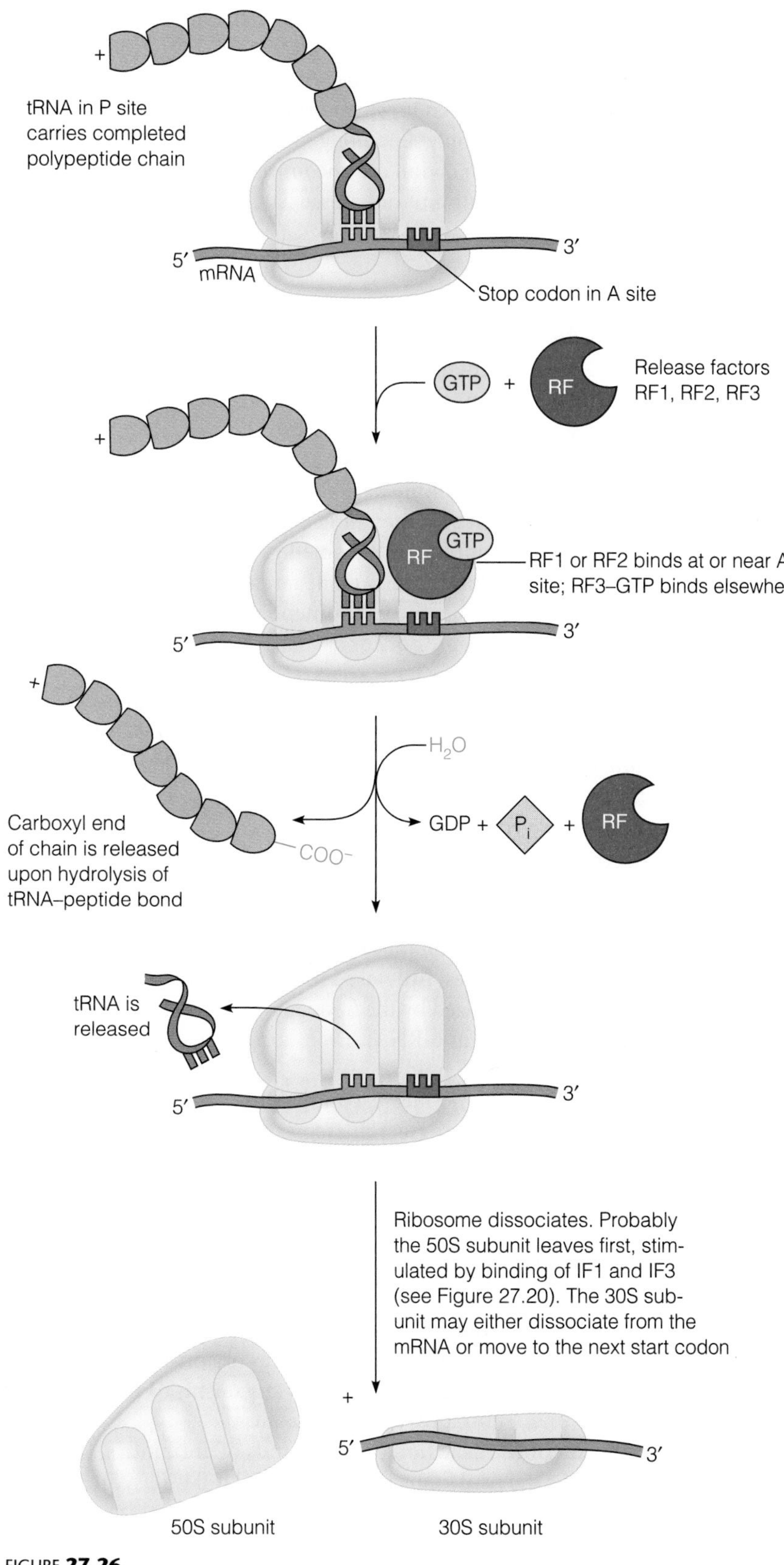

FIGURE **27.26**

Termination of translation in prokaryotes. See text.

factor, and also by the initiation factors IF3 and IF1, and the ribosome readily dissociates to 50S and 30S subunits prepared for another round of translation.

When the ribosomal subunits separate, the 30S subunit may or may not dissociate from its mRNA. In some cases in which polycistronic messages are being translated, the 30S subunit may simply slide along the mRNA until the next Shine–Dalgarno sequence and initiation codon are encountered and then begin a new round of translation. If the 30S subunit does dissociate from the message, it will soon reattach to another one.

SUPPRESSION OF MUTATIONS

Understanding the process of termination helped clarify some peculiar observations concerning nonsense mutations. Recall from Chapter 7 that a *nonsense mutation* is one in which a codon for some amino acid has been mutated into a stop codon, so that the polypeptide chain terminates prematurely. Soon after such mutations were discovered, a mysterious class of apparent reversion mutations was found. These were not simply back-mutations in the mutated gene, but mutations in *other* genes that somehow *suppressed* the nonsense mutations. Recall from Chapter 24 that suppression is defined genetically as restoration of wild-type function by a second mutation at a different site. When this second mutation occurs in a different gene, the phenomenon is called **intergenic suppression.** On examination, the suppressors of nonsense mutations were found to lie in tRNA genes.

The effects of nonsense mutations can sometimes be suppressed by suppressor mutations, in which a tRNA mutates to recognize a stop codon and inserts an amino acid instead.

Consider the example shown in Figure 27.27. A nonsense mutation has changed a codon that normally specifies the amino acid tyrosine into a stop codon, causing abortive termination of the polypeptide chain. If, however, one of the several tyrosine tRNAs mutates in its anticodon region so as to recognize the stop codon, translation can sometimes proceed in the normal fashion. Thus, a mutation that might otherwise be lethal can be suppressed by such a change, and the microorganisms can survive. Clearly, they will still have problems, for the presence of such a mutated tRNA will interfere with the normal termination of other proteins. That they can survive at all depends on the fact that the suppressor mutation usually involves a minor tRNA species, little used in normal translation. Furthermore, such effects may be minimized by the frequent occurrence of two or more different stop signals in tandem in mRNAs. Even if the first stop codon is suppressed, the "emergency brake" still holds.

Suppressor mutations are by no means confined to correction of nonsense mutations. Some mutated tRNAs correct missense mutations, and some even contain two or four bases which act as the anticodon. These can therefore serve as **frameshift suppressors.**

The Inhibition of Translation by Antibiotics

Much of what we know about translation has been learned from studies of **antibiotics** that interfere with this essential process. Various organisms produce substances that interfere with protein synthesis in other organisms. When the target organisms are bacteria, these substances act as antibiotics.

We have already described the action of some kinds of antibiotics. In Chapters 9 and 16, we saw that the penicillins inhibit bacterial cell wall synthesis, and in Chapter 10 we discussed antibiotics such as gramicidin and valinomycin, which interfere with the ionic balance across membranes. The *sulfa drugs* (Chapter 20) are antimetabolites, and other antibiotics, such as *rifampicin* and *streptolydigin* (Chapter 26), block transcription in prokaryotes.

Given that translation is such a complex and vital process, it perhaps is not surprising that it is a favorite target for antibiotics. A host of naturally occurring substances interfere with various stages of protein synthesis. Figure 27.28 shows a

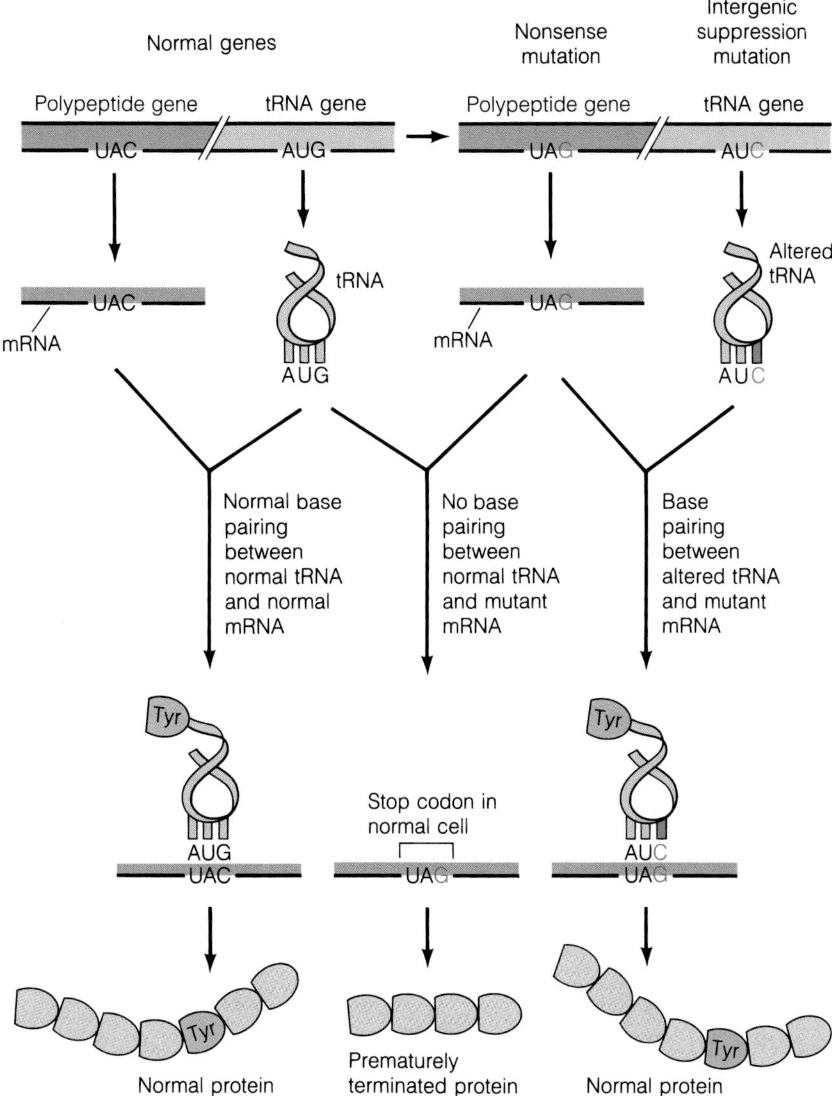

FIGURE **27.27**

How an intergenic suppression mutation can overcome a nonsense mutation. A nonsense mutation in a peptide gene changes a codon for an amino acid into a stop codon, causing translation to terminate prematurely. Another mutation, in a tRNA gene, can circumvent the first mutation by altering the tRNA anticodon so that it will base-pair with the mutant mRNA. A normal protein is produced in this situation.

few of the most important ones. Each of these substances inhibits translation in a different way. Their importance to medicine stems largely from the fact that the translational machinery of eukaryotes is sufficiently different from that of prokaryotes that these antibiotics can be used safely in humans. In some cases (for example, the tetracyclines), antibiotics that would also inhibit eukaryotic translation are nevertheless harmless to eukaryotes because they cannot pass the cell membranes of higher organisms.

A number of important antibiotics act by inhibiting translation in bacterial cells.

One problem with the therapeutic use of antibiotics is that microorganisms can develop resistance to many of them. This resistance often results from the acquisition of a particular "resistance gene." An important example is *erythromycin* resistance. The erythromycin-binding site on the ribosome includes a specific region of the 23S RNA, and binding of the antibiotic can be inhibited by an enzyme that methylates a specific adenine residue in this region. Molecular biologists use erythromycin resistance in screening bacterial clones for those containing particular plasmids (see Tools of Biochemistry 25B). Resistance to erythromycin can be conferred to a bacterium by the insertion of a resistance gene coding for the methylase on a bacterial plasmid. Bacteria containing the plasmid carrying the methylase gene will grow in an erythromycin-containing medium, whereas those lacking the plasmid will be killed. Thus, growth on such a medium automatically selects for only those clones that carry the plasmid.

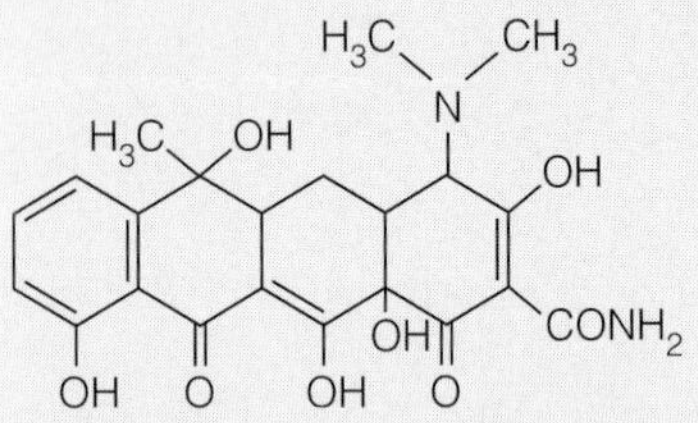

Tetracycline: Inhibits the binding of aminoacyl tRNAs to the ribosome and thereby blocks continued translation

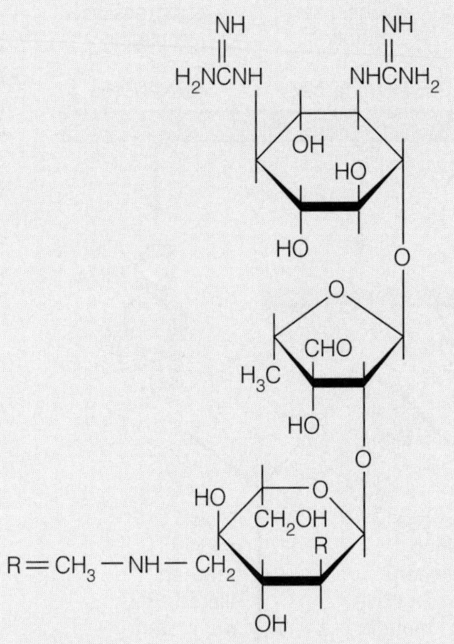

Streptomycin: Interferes with normal pairing between aminoacyl tRNAs and message codons, causing misreading, and thereby producing aberrant proteins

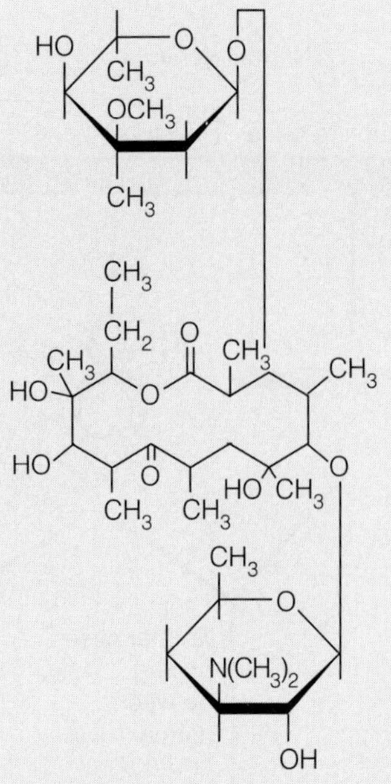

Erythromycin: Binds to a specific site on the 23S RNA and blocks elongation by interfering with the translocation step

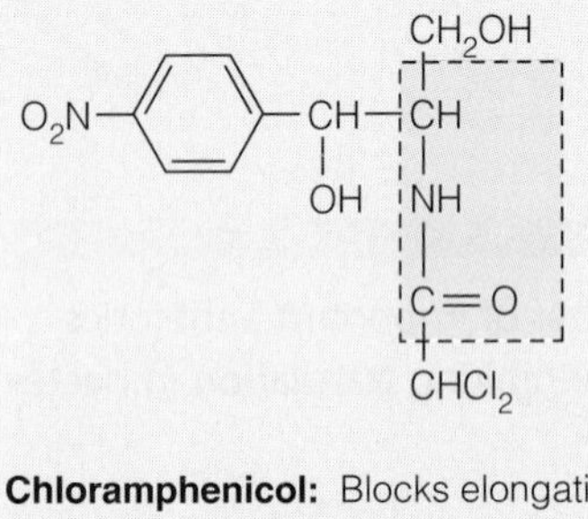

Chloramphenicol: Blocks elongation, apparently by acting as competitive inhibitor for the peptidyltransferase complex. The amide link (in blue) resembles a peptide bond

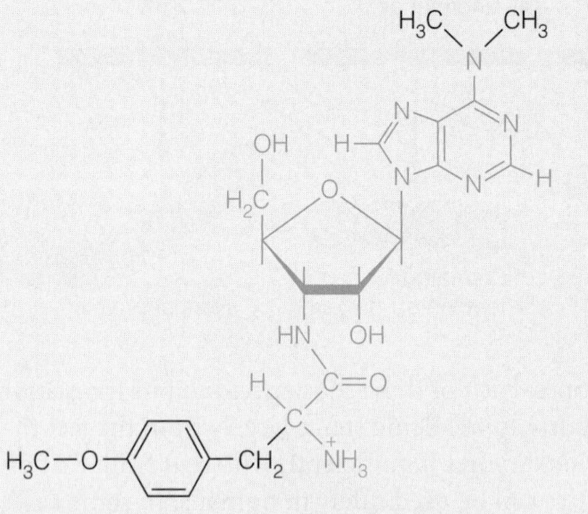

Puromycin: Causes premature chain termination. The red portion of the molecule resembles the 3′ end of the aminoacylated tRNA. It will enter the A site and transfer to the growing chain, causing premature chain release

FIGURE **27.28**

Some antibiotics that act by interfering with protein biosynthesis.

Rates and Energetics of Translation

Translation is a rapid process in prokaryotes. At 37°C an *E. coli* ribosome can synthesize a 300-residue polypeptide chain in about 20 seconds. Because chain elongation is the rate-limiting process, this means that a single ribosome passes through about 15 codons, or 45 nucleotides, in each second. This rate is almost exactly the same as our best estimates of the rate of prokaryotic *transcription,* which means that mRNA can be translated as fast as it is transcribed. But the rate given here by no means accounts for the total rate of protein synthesis in the cell, for many ri-

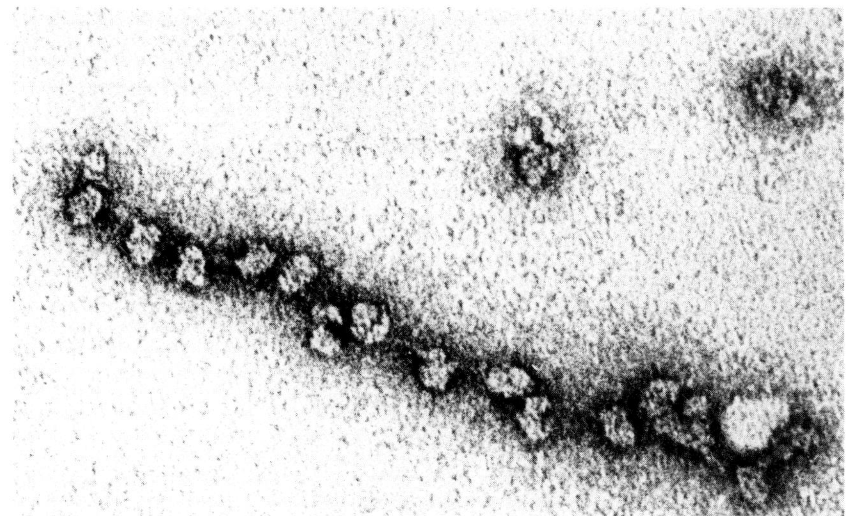

(a)

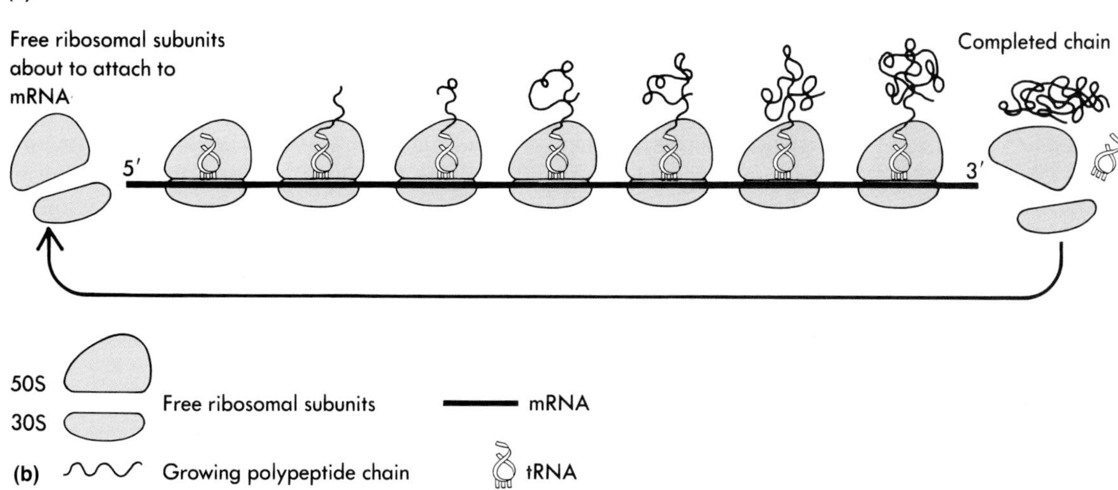

FIGURE **27.29**

Polyribosomes. **(a)** Electron micrograph showing *E. coli* polyribosomes. The ribosomes are closely clustered on an mRNA molecule. **(b)** Schematic picture of a polyribosome like that shown in **(a)**. Each ribosome is to be imagined as moving from left to right.

(a) Courtesy of Barbara Hamkalo; **(b)** from J. D. Watson et al., *Molecular Biology of the Gene,* 4th ed. (Redwood City, Calif.: Benjamin/Cummings, 1987). © 1987 J. D. Watson.

bosomes may be simultaneously translating a given message. In fact, if we were to carefully lyse *E. coli* cells, we would observe **polyribosomes** like those shown in Figure 27.29. Apparently, as soon as one ribosome has moved clear of the 5′ region of the mRNA, another attaches. Under some conditions, as many as 50 ribosomes may be packed onto an mRNA, with one finishing translation every few seconds. Because each *E. coli* cell contains about 15,000 ribosomes, all of them operating at full capacity can synthesize about 750 protein molecules of 300 residues each second.

The energy cost for this process is high. If we examine the individual steps in protein synthesis described earlier, we can make the following estimate of the total energy budget for synthesizing a protein of N residues:

The *equivalent* of

$2N$	ATPs are required to charge the tRNAs, because the ATP is cleaved to AMP and PP_i, and PP_i is subsequently hydrolyzed.
1	GTP is needed for initiation.
$N-1$	GTPs are required to form the $N-1$ peptide bonds, in the EF-Tu–GTP hydrolysis step.
$N-1$	GTPs are necessary for the $N-1$ translocation steps.
1	GTP is required in termination.
Sum = $4N$	

Translation is fast but energy-expensive. About four ATP equivalents are needed for each amino acid added.

Altogether, then, about $4N$ high-energy phosphate molecules must be hydrolyzed to complete a chain of N units. This is a minimal estimate, for it does not include the energy required to formylate methionine, nor any extra GTPs that may be expended in proofreading and replacing incorrectly bound tRNAs. Furthermore, there have been persistent, although debated reports that *two* GTPs must be hydrolyzed for every aa-tRNA bound to the A site. But even at the conservative estimate, a typical protein of 300 residues costs the cell about 50,000 kJ of free energy per mole, if we assume ATP or GTP hydrolysis yields about 40 kJ/mol under cellular conditions. Proteins are expensive!

If we express the same data in terms of the energy requirement for synthesis per mole of *peptide bond,* we obtain a cost of about 160 kJ/mol. Given that the free energy change required to form a peptide bond in dilute aqueous solution is only about +20 kJ/mol, the price seems exorbitant. Why does the cell have no mechanism for making peptide bonds for a few dozen kilojoules each? Certainly, an input of even 40 kJ/mol would be enough to make the synthesis process very favorable—with an equilibrium constant of about 3000.

The key to this great energy expenditure is found in the fundamental nature of life. The cell is making polypeptides of *defined* sequence. If it were simply throwing together amino acids at random, the free energy price could be much cheaper. But a chain of 300 residues, made from 20 different amino acids, can be put together in 20^{300} ($\cong 10^{390}$) different ways, whereas the cell needs *one* specific sequence. There is, in other words, a large entropy price to be paid in making specific sequences—and making them correctly. What this means at the mechanistic level is that every step in the assembly not only must be done with a free energy excess but also must involve a specific *choice.* Furthermore, the product must, at critical points, be checked by a proofreading mechanism, which in turn costs more energy. It is expensive to get a good translation of a book, for not only must the translators be expert and careful but their work must also be rechecked with great care.

The Final Stages in Protein Synthesis: Folding and Covalent Modification

The polypeptide chain that emerges from the ribosome is not a completed, functional protein. It must fold into its tertiary structure, and it may have to associate with other subunits. In some cases, disulfide bonds must be formed, and other covalent modifications, such as hydroxylation of specific prolines and lysines (or any of the other amino acid modifications we have discussed), must take place. In addition, many proteins are subjected to specific proteolytic cleavage to remove portions of the nascent chain.

CHAIN FOLDING

The cell need not wait until the entire chain is released from the ribosome to commence its finishing touches. The first portion of the nascent chain (about 30 residues) is protected as it passes through the tunnel in the ribosome. However, changes begin almost as soon as the N-terminal end emerges. There is good evidence that folding into the tertiary structure starts during translation and is nearly complete by the time the chain is released. For example, antibodies to *E. coli* β-galactosidase, which recognize the tertiary folding of the molecule, will attach to polyribosomes synthesizing this protein. This enzyme displays catalytic activity only as a tetramer. It has been demonstrated that nascent β-galactosidase chains, still attached to ribosomes, can associate with free subunits to form a functional tetramer. Thus, even quaternary structure can be partially established before synthesis is complete.

This behavior should not be surprising, if we recall (from Chapters 6 and 7) that formation of the secondary, tertiary, and quaternary levels of protein structure is thermodynamically favored. However, as we have seen in Chapter 6, in some cases this spontaneous folding must be blocked or delayed by chaperone proteins.

COVALENT MODIFICATION

Some of the covalent modifications of polypeptide chains also occur during translation. We mentioned earlier that the *N*-formyl group is removed from the initial *N*-fMet of most prokaryotic proteins. A special *deformylase* catalyzes this reaction. In many cases, deformylation seems to happen almost as soon as the N-terminus emerges from the ribosome. Removal of the N-terminal methionine itself can also be an early event, but whether it happens or not apparently depends on the cotranslational folding of the chain. Presumably, in some cases this residue is "tucked away" and protected.

Some prokaryotic (and many eukaryotic) proteins experience much more severe proteolytic modifications. These proteins are almost invariably the ones that are going to be exported from the cell or are destined for membrane or organelle locations. We reserve discussion of the more complicated eukaryotic protein processing for Chapter 28 and concentrate here on what happens in prokaryotes.

Translation is immediately followed by various kinds of protein processing, including chain folding, covalent modification, and directed transport.

Bacterial proteins that are destined for secretion (**translocation** across the cell membrane) are characterized by highly hydrophobic **signal sequences** or **leader sequences** in the N-terminal regions. Representatives are listed in Table 27.5. After the protein has passed through the membrane, the leader sequence is cleaved off at the point indicated by the arrow in the table.

The currently accepted model for translocation in bacteria is shown in Figure 27.30. In many (but not all) cases the protein to be translocated (the pro-protein) is first complexed in the cytoplasm with a "chaperone"—the SecB protein in the example shown. This complexing keeps the protein from folding prematurely, which would prevent it from being passed through the secretory pore in the membrane. This pore is composed of two types of transmembrane protein, designated SecE and SecY. The secretory pore is also a target for a fourth protein component,

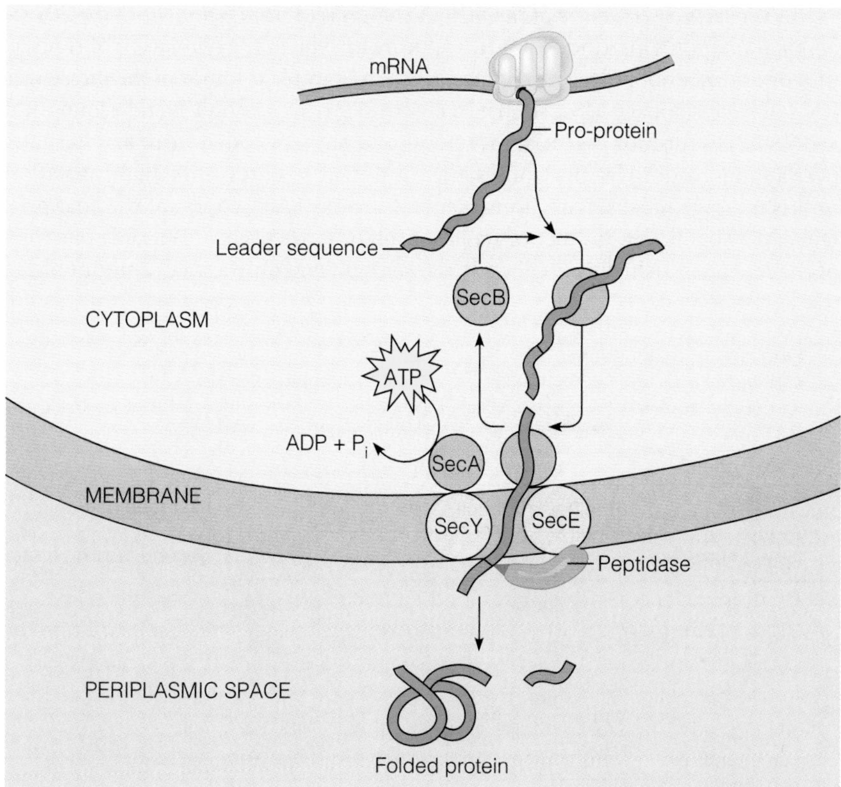

FIGURE **27.30**

A current model for protein secretion by prokaryotes. The new polypeptide chain (the pro-protein) complexes with SecB, which prevents complete folding during transport to membrane. At the membrane an ATPase, SecA, drives translocation through the membrane with the aid of membrane proteins SecE and SecY, which form a membrane pore. The leader sequence is then cleaved off the secreted protein by a membrane peptidase.

TABLE 27.5 N-terminal signal sequences of representative prokaryotic proteins

Protein	−20 −15 −10 −5 −1	↓ +1
Leucine-binding protein	M K A N A K T I I A G M I A L A I S H T A M A	E E …
Prealkaline phosphatase	M K Q S T I A L A L L P L L F T P V T K A	R T …
Prelipoprotein	M K A T K L V L G A V I L G S T L L A G	C S …

Note: Hydrophobic residues are in red. The cleavage site is designated by the arrow.

SecA. SecA is an ATPase, and both ATP hydrolysis and the electrochemical potential gradient across the membrane help drive translocation. After the pro-protein has been translocated, the leader peptide is cleaved off by a membrane-bound protease, and the protein can fold. The cleavage site, as shown in Table 27.5, usually lies between a small amino acid (G or A) and an acidic or a basic one.

The Regulation of Protein Synthesis in Prokaryotes

In most organisms, the overall control of protein biosynthesis occurs primarily at the transcriptional level. If need for a protein decreases, it is usually most energy-efficient to simply shut off the transcription of the gene and not make an unnecessary mRNA. But in some cases, regulation is imposed at the translational level, and we shall consider a few prokaryotic examples in this section.

The major step at which control is imposed on translation appears to be initiation. This is what we might expect, for two reasons. First, it is most efficient to block translation at its start, before ribosomes have been committed or significant energy expended. Second, the special requirements for fitting the mRNA properly into a ribosome allow numerous ways to block the process at that point. There are now known to be at least three mechanisms by which initiation of translation of selected messages can be controlled.

1. *The tertiary structure of the mRNA can prevent its attachment to the 30S ribosomal subunit.* This kind of control appears to explain the regulation of translation of a number of polycistronic messages. An example is found in the messenger RNA of bacteriophage MS2, shown in Figure 27.31. This complex and economical mRNA codes for four essential proteins—one of them coded from an alternate, overlapping reading frame. Consider the different requirements that translation of this message must satisfy: To make new viruses, *many* copies of the coat protein and a significant number of replicase subunits are needed. But only *one* copy of the A protein (which is used in virus assembly) and only a small amount of the

FIGURE **27.31**

The RNA of bacteriophage MS2. This RNA molecule serves as a message for all four of the proteins required by this virus. The coding sequences for the A protein, coat protein, and replicase subunit are shown in gray. The message for the lysis protein (the L protein, shown in blue) actually overlaps the coat protein and replicase messages. It is translated in a different reading frame.

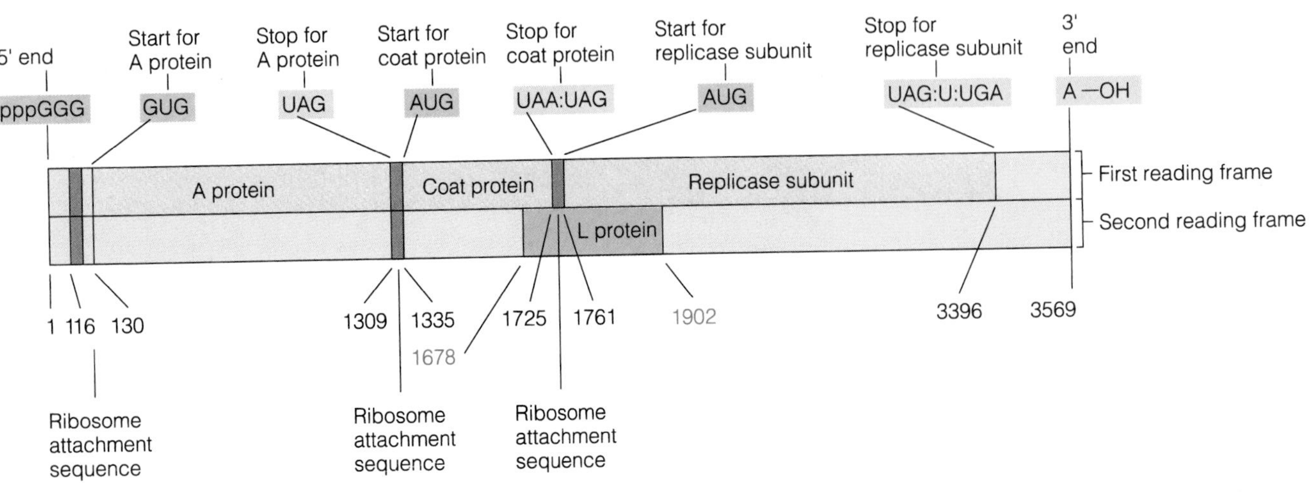

lysis protein (used in viral release) are needed per virus. To simply translate the whole polycistronic message equally in all its parts would be woefully inefficient. What happens is this: The ribosome binding sequence at the 5′ end of the mRNA (which would start translation of the A protein message) is normally blocked by tertiary folding of the mRNA molecule. Therefore, the ribosome normally begins translation by binding to an attachment sequence near the start site of the coat protein message, translating this message efficiently and proceeding sometimes, but not always, to the replicase message. As replicase is made, it catalyzes replication of the viral mRNA itself. Copies of new plus strands that are still being transcribed have not yet folded into their final conformation. Therefore their 5′ sites are still open for initiation and can attach to ribosomes to translate A protein. But this happens only once in the life of each mRNA—at the time it is being synthesized. The message for the lysis protein, which overlaps the coat protein message but in a different reading frame, is apparently translated only from time to time, as a consequence of a frameshift slip during translation of the coat protein.

2. *Proteins may bind to the mRNA, blocking initiation.* A most elegant example of this kind of control is found in the synthesis of the ribosomal proteins themselves. As Figure 27.32 shows, prokaryotic ribosomal proteins are encoded by polycistronic messages. In each group encoded by such a message, there is *one* protein that is capable of binding at or near the 5′ end of the message and blocking its translation. This binding seems to occur because the mRNA has a tertiary structure similar to the normal binding site for that protein on rRNA.

The beauty of the system can be seen by comparing Figure 27.32 with Figure 27.19. The proteins that control ribosomal protein synthesis—S4, S7, and S8—are also among the first to bind onto the rRNA in ribosome assembly. These proteins are the keys to ribosome construction, and they have a very high affinity for the appropriate rRNAs. If ribosomal RNA is abundant in the cell, which would signal need for ribosomal proteins, the "control" proteins are incorporated into ribosomes, and synthesis of all ribosomal proteins can proceed. But if rRNA is in short supply, the unused "control" proteins will bind to the appropriate mRNAs, shutting down the synthesis of other, presently unneeded ribosomal proteins.

Although not a matter of translational control, it is appropriate to note here that rRNA synthesis is itself under metabolic control. In prokaryotes, starvation for amino acids results in the accumulation of the unusual nucleotides ppGpp and pppGpp (see page 1017). They have a pronounced inhibitory effect on the transcription of rRNA, and that inhibition in turn inhibits ribosomal protein synthesis. Thus, starvation conditions induce a general "belt tightening" in the biosynthetic apparatus.

3. *Antisense RNA may block initiation.* A number of situations are now known in which control of translation is effected by the synthesis of antisense RNA fragments

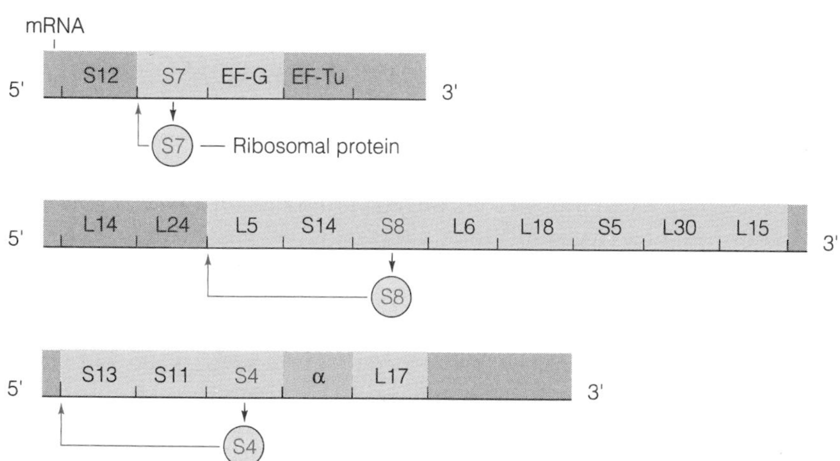

FIGURE **27.32**

Regulation of synthesis of ribosomal proteins. Shown are three examples of polycistronic messages coding for ribosomal proteins. Note that these mRNAs also contain messages for EF-G, EF-Tu, and the α subunit (purple) of RNA polymerase. The translation of each polycistronic message is controlled by one of the ribosomal proteins it codes, shown in red. The red arrows show the control sites. Control extends *only* over the cistrons shown in teal. In other words, the RNA polymerase α subunit and EF-Tu messages are excluded from this control.

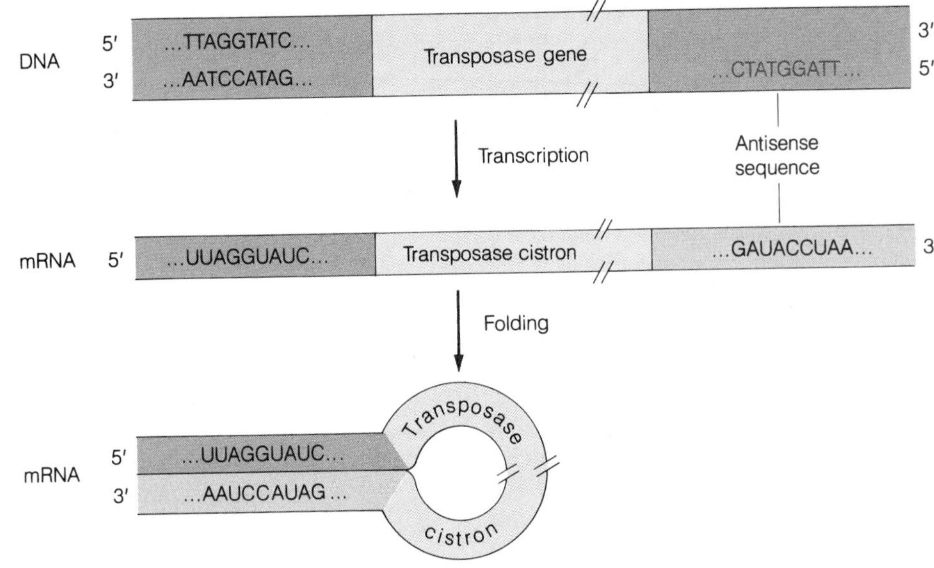

FIGURE **27.33**

Blocking of translation initiation by antisense RNA. The transposase gene contains an antisense sequence at its 3′ end. When transcribed, this antisense segment can fold back and base-pair to block initiation of translation.

When translation is regulated, it is generally at the initiation stage, via the mRNA structure or by binding of proteins or antisense RNA to the mRNA.

at another genomic locus. These fragments correspond to the noncoding strand of DNA and are complementary to the 5′ end of an mRNA. An antisense sequence blocks the initiation site by formation of a double-strand structure, as shown for the translation of transposase mRNA in Figure 27.33.

The selective blocking of the translation of specific messages by antisense RNA has potential applications to the treatment of diseases and perhaps cancer. However, a major difficulty in using RNAs as gene-inactivating drugs is encountered in attempting to pass small RNA molecules into cells, for negatively charged molecules do not traverse membranes easily. Moreover, once inside the cells, they are subject to enzymatic degradation. This has, in turn, led to attempts to modify the RNA so as to reduce negative charge and inhibit hydrolysis. Some of the modifications being tried by biotechnologists involve oligonucleotide analogs with other than phosphodiester linkages.

SUMMARY

The connection between gene (or mRNA) sequence and protein sequence is dictated by the genetic code. The code is almost, but not quite, uniform in all living creatures. The code is redundant, with multiple codons for most amino acids, and it includes start and stop signals.

Translation of mRNAs into polypeptide chains involves a number of steps. First, the appropriate tRNAs must be coupled to the corresponding amino acids, using a set of enzymes called aminoacyl-tRNA synthetases. The site of translation is the ribosome—an RNA–protein complex composed of two subunits, each with its individual RNA and protein components. Translation of an mRNA on ribosomes involves three stages: initiation, elongation, and termination. Each stage requires certain protein factors in addition to the ribosome. In initiation the mRNA is fixed to the 30S subunit in proper register via its Shine–Dalgarno sequence, and the initiator *N*-formylmethionyl tRNA is matched to its AUG codon at the P site. The 50S subunit is then attached. The second tRNA enters the A site, and the incipient chain is transferred to it. Next, the mRNA is translocated so as to move the chain again into the P site, while the now unoccupied tRNA moves into the E site and is released. These steps probably involve intermediate hybrid states.

The chain continues to elongate in this fashion until a stop codon is reached, whereupon release factors bind to the ribosome and aid in releasing the polypeptide chain. The entire process of translation requires about 4 ATP equivalents for each amino acid added. A number of antibiotics have their antibacterial effect by inhibiting various stages of the translation process. Translation of a number of prokaryotic mRNAs is regulated, usually at the initiation step.

As translation is being completed, folding and covalent modification of the polypeptide chain begin. The chain as synthesized may contain an *N*-terminal sequence that targets it for export from the cytoplasm.

REFERENCES

Of Historical Interest

Brenner, S., F. Jacob, and M. Meselson (1961) An unstable intermediate carrying information from genes to ribosomes for protein synthesis. *Nature* 190:576–581. Early evidence for the existence of mRNA.

Crick, F. H. C. (1958) On protein synthesis. *Symp. Soc. Exp. Biol.* 12:138–162. With great prescience, Crick foresees the essential nature of the translation mechanism.

Crick, F. H. C. (1966) Codon–anticodon pairing: The wobble hypothesis. *J. Mol. Biol.* 19:548–555.

Khorana, H. G. (1968) Nucleic acid synthesis in the study of the genetic code. In: *Nobel Lectures, Physiology, and Medicine (1963–1970)*, pp. 341–343. American Elsevier, New York. A Nobel Prize winner's account of the deciphering of the code.

Traub, P., and M. Nomura (1968) Structure and function of *E. coli* ribosomes. V. Reconstitution of functionally active 30S ribosomal particles from RNA and proteins. *Proc. Natl. Acad. Sci. USA* 59:777–784.

The Code

Böck, A., K. Forchhammer, J. Heider, W. Leimfelder, G. Sawers, B. Vepreck, and F. Zinoni (1991) Selenocysteine: The 21st amino acid. *Mol. Microbiol.* 5:515–520. A minireview on the discovery that this unusual amino acid is also coded for.

Eigen, M., B. F. Lindemann, M. Tietze, R. Winkler–Oswatitsch, A. Dress, and A. von Haeseler (1989) How old is the genetic code? Statistical geometry of tRNA provides an answer. *Science* 244:673–678.

Fox, T. D. (1987) Natural variation in the genetic code. *Annu. Rev. Genet.* 21:67–91.

Messenger RNA

Bronson, M. J., C. Squires, and C. Yanovsky (1973) Nucleotide sequence from tryptophan messenger RNA of *Escherichia coli:* The sequences corresponding to the amino terminal region of the first polypeptide specified by the operon. *Proc. Natl. Acad. Sci. USA* 70:2335–2339. A pioneering study in this field.

Gesteland, R. F., R. B. Weiss, and J. F. Atkins (1992) Recoding: Reprogramming genetic decoding. *Science* 257:1640–1641. There are special signals in some mRNAs that alter code reading.

Shine, J., and L. Dalgarno (1974) The 3′-terminal sequence of *E. coli* 16S rRNA: Complementarity to nonsense triplets and ribosome binding sites. *Proc. Natl. Acad. Sci. USA* 71:1342–1346.

Transfer RNAs

Hatfield, D. L., B. J. Lee, and R. M. Pirtle (eds.) (1992) *Transfer RNA in Protein Synthesis,* CRC Press, Boca Raton, Fla. A collection of papers on diverse aspects of tRNA function.

Kim, S.-H. (1978) Three-dimensional structure of transfer RNA and its functional implications. *Adv. Enzymol.* 46:279–315.

Söll, D., and V. RajBhandary (eds.) (1995) *tRNA: Structure, Biosynthesis, and Function.* ASM Press, Washington, D. C.

Wower, J., K. V. Rosen, S. S. Hixon, and R. A. Zimmermann (1994) Recombinant photoreactive tRNA molecules as probes for cross-linking studies. *Biochimie* 76:1235–1246.

Aminoacyl-tRNA Synthetases and aa~tRNA Coupling

Carter, C. W., Jr. (1993) Cognition, mechanism, and evolutionary relationships in aminoacyl-tRNA synthetases. *Annu. Rev. Biochem.* 62:715–748.

Curnow, A. W., K.-W. Hong, R. Yuan, S.-L. Kim, O. Martins, W. Winkler, T. M. Henkin, and D. Söll (1997) Glu-$tRNA^{Gln}$-amidotransferase: A novel heterotrimeric enzyme required for correct decoding of glutamine codons during translation. *Proc. Natl. Acad. Sci. USA* 94: 11819–11826. Charging $tRNA^{Gln}$ with Gln in an indirect way.

Giegé, R., M. Sissler, and C. Florentz (1998) Universal rules and idiosyncratic features in tRNA identity. *Nucleic Acids Res.* 26:5017–5035.

Rould, M. A., J. J. Perona, D. Söll, and T. A. Steitz (1989) Structure of *E. coli* glutaminyl-tRNA synthetase complexed with $tRNA^{Gln}$ and ATP at 2.8 Å resolution. *Science* 246:1135–1141. A structural study of major importance.

Schulman, L. H., and J. Abelson (1988) Recent excitement in understanding tRNA identity. *Science* 240:1591–1592. A brief overview of the tRNA recognition problem.

Ribosomes

Ban, N., B. Freeborn, R. Nissen, P. Penczek, R. Grassucci, R. Sweet, J. Frank, P. Moore, and T. A. Steitz (1998) A 9Å resolution x-ray crystallographic map of the large ribosomal subunit. *Cell* 93: 1105–1115.

Hill, W. E., A. Dahlberg, R. A. Garret, P. B. Moore, D. Schlessinger, and J. R. Warner, eds. (1990) *The Ribosome.* American Society for Microbiology, Washington, D.C.

Mueller, F., H. Stark, M. van Heel, J. Rinke-Appel, and R. Brimacombe (1997) A new model for the three dimensional folding of *Escherichia coli* 16S ribosomal RNA. III. The topography of the functional center. *J. Mol. Biol.* 271:566–587. This, and the two preceding papers in the same issue, assemble a vast amount of data, from various techniques, into a comprehensive model of the 30S subunit.

Ramakrishnan, V., and S. W. White (1998) Ribosomal protein structures: Insights into the architecture, machinery, and evolution of the ribosome. *Trends. Biochem. Sci.* 23:208–212. A thoughtful, up-to-date discussion centered on ribosomal protein structures.

The Translation Process

Burkhardt, N., R. Junemann, C. M. T. Spahus, and K. H. Nierhaus (1998) Ribosomal RNA binding sites: Three-site models of translation. *Crit. Rev. Biochem. Mol. Biol.* 33:95–149. A very readable overview of recent models for elongation.

Caskey, C. T., W. C. Forrester, and W. Tate (1984) Peptide chain termination. In: *Alfred Benzor Symposium,* edited by B. F. C. Clark and H. U. Petersen, Vol. 19, pp. 457–466. Munksgaard, Copenhagen.

Gold, L., and G. Stormo (1987) Translational initiation. In: Escherichia coli *and* Salmonella typhimurium: *Cellular and Molecular Biology,* edited by F. C. Neidhardt, J. L. Ingraham, B. Low, B. Magasanik, M. Schaechter, and H. E. Umbarger, Vol. 2, pp. 1302–1307. American Society for Microbiol., Washington, D.C.

Nissen, P., M. Kjelgaard, S. Thirup, G. Polekhina, L. Resketnikova, B. F. C. Clark, and J. Nyborg (1995) Crystal structure of the ternary complex of Phe-$tRNA^{Phe}$, EF-Tu and a GTP analog. *Science* 270:1464–1472. Includes comparison with EF-GDP.

Noller, H. F. (1993) tRNA–rRNA interactions and peptidyl transferase. *FASEB J.* 7:87–89. A brief review of recent information on this crucial step. Includes evidence for ribozyme function.

Stark, H., E. V. Orlova, J. Rinke-Appel, N. Jünke, F. Mueller, M. Rodnina, W. Wintermeyer, R. Brimacombe, and M. van Heel (1997) Arrangement of tRNAs in pre- and posttranslocational ribosomes, revealed by electron cryomicroscopy. *Cell* 88:19–28. An elegant use of the method to follow conformational changes.

Yarus, M., and R. Thompson (1983) Precision of protein biosynthesis. In: *Gene Function in Prokaryotes,* edited by J. Beckwith, J. Davies, and J. A. Gallant, pp. 23–63. Cold Spring Harbor Laboratory, Cold Spring Harbor, N.Y. A good discussion of proofreading.

Antibiotics

Cooperman, B. S., M. A. Buck, C. L. Fernandez, C. J. Weitzman, and B. F. D. Ghrist (1989) Antibiotic photoaffinity labeling probes of *E. coli* ribosomal structure and function. In: *Photochemical Probes in Biochemistry,* edited by P. E. Nielsen, pp. 123–139. Kluwer, Dordrecht, Netherlands. A review of the use of antibiotics to probe ribosomes.

Cundliff, E. (1980) Antibiotics and prokaryotic ribosomes: Action, interaction, and resistance. In: *Ribosomes: Structure, Function, and Genetics,* edited by G. Chambliss et al., pp. 377–412. University Park Press, Baltimore.

Regulation

Campbell, K., G. Stormo, and L. Gold (1983) Protein-mediated translational repression. In: *Gene Function in Prokaryotes,* edited by J. Beckwith, J. Davies, and J. A. Gallant, pp. 185–210. Cold Spring Harbor Laboratory, Cold Spring Harbor, N.Y.

Nomura, M., J. Yates, D. Dean, and L. Post (1980) Feedback regulation of ribosomal gene expression in *Escherichia coli. Proc. Natl. Acad. Sci. USA* 77:7084–7088.

Posttranslational Modification

Wickner, W., J. M. Driessen, and F.-V. Hartl (1991) The enzymology of protein translocation across the *Escherichia coli* plasma membrane. *Annu. Rev. Biochem.* 60:101–124.

PROBLEMS

1. The following synthetic polynucleotide is synthesized and used as a template for peptide synthesis in a cell-free system from *E. coli.*

   ···AUAUAUAUAUAU···

   What polypeptide would you expect to be produced? Precisely what information would this give you about the code?

2. If the same polynucleotide described in Problem 1 is used with a *mitochondria-derived* cell-free protein-synthesizing system, the product is

   ···Met–Tyr–Met–Tyr–Met–Tyr···

   What does this say about differences between the mitochondrial and bacterial codes?

*3. When polynucleotides are synthesized with repeating triplets of nucleotide residues, from one to three kinds of polypeptide chains will be produced in cell-free synthesis.
   (a) Explain why these different results are possible.
   (b) Predict polypeptides produced when the following are used with an *E. coli* system: $(GUA)_n$; $(UUA)_n$

*4. What kind of repeating polynucleotide would yield a single polypeptide with a tetrapeptide repeating unit?

5. Although the Shine–Dalgarno sequences vary considerably in different genes, they include examples like GAGGGG that could serve as code—in this case, for Glu–Gly. Does this imply that the sequence Glu–Gly cannot ever occur in a protein, lest it be read as a Shine–Dalgarno sequence? Speculate.

6. According to wobble rules, what codons should be recognized by the following anticodons? What amino acid residues do these correspond to?
   (a) 5′···ICC···3′
   (b) 5′···GCU···3′

*7. In the early days of ribosome research, before the exact role of ribosomes was clear, a researcher made the following observation. He could find, in sedimentation experiments on bacterial lysates, not only 30S, 50S, and 70S particles but also some particles that sedimented at about 100S and 130S. When he treated such a mixture with EDTA, everything dissociated to 30S and 50S particles. Upon adding divalent ions, he could regain 70S particles, but never 100S or 130S particles.
   (a) Suggest what the 100S and 130S particles might represent, in light of current knowledge of protein synthesis. What important discovery did the researcher miss?
   (b) Why do you think reassociation to 100S and 130S particles did not work?

8. The E site may not require codon recognition. Why?

9. Suppose the bacteriophage MS2 were to infect a bacterium deficient in the release factor RF1. What effect might this have on MS2 gene expression? [Hint: See Figure 27.31.]

10. Suppose that the probability of making a mistake in translation at each translational step is a small number, δ. Show that the probability, p, that a given protein molecule, containing n residues, will be completely error-free is $(1 - \delta)^n$.

11. Assume that the translational error frequency, δ, is 1×10^{-4}.
   (a) Calculate the probability of making a perfect protein of 100 residues.
   (b) Repeat for a 1000-residue protein.

*12. Devise an experiment, based on sucrose gradient sedimentation (see Tools of Biochemistry 5A), that would demonstrate that proteins are synthesized on polyribosomes.

13. Assuming that glucose is burned to CO_2 as an energy source, how many amino acid residues can be incorporated into a protein mole-

cule for each glucose consumed by a cell? Is this a maximum or minimum estimate?

14. Why might you expect a small dose of puromycin to be less effective in repressing bacterial growth than an equivalent number of molecules of erythromycin?

15. The protein colicin E3 is a very effective inhibitor of protein synthesis in prokaryotes. This protein is a nuclease, specifically attacking a phosphodiester bond near the 3′ end of the 16S RNA. Suggest a mechanism for the effect of colicin E3 on translation.

16. Suggest ways of possibly modifying antisense RNA so as to promote its entrance into cells.

TOOLS OF BIOCHEMISTRY 27A

Ways to Map Complex Macromolecular Structures

As we continue to probe the structure of the cell, using ever gentler and more discriminating techniques, it becomes apparent that much of the cellular machinery is organized in complicated structures that are *assemblies* of macromolecules. The ribosome, with its several kinds of RNA and many kinds of proteins, is an excellent example. Direct analysis of such immense, complex structures by x-ray diffraction is a task still at the limits of our capabilities, although rapid progress is being made. But there are other ways in which we can learn something about the spatial arrangement of the various parts. We describe several of these methods here.

Chemical Cross-Linking

One way to learn about the arrangement of components in a particle is to see what can easily be attached to what, through chemical **cross-linking.** Consider an idealized particle (like that shown in Figure 27A.1) that contains three different protein molecules. If we

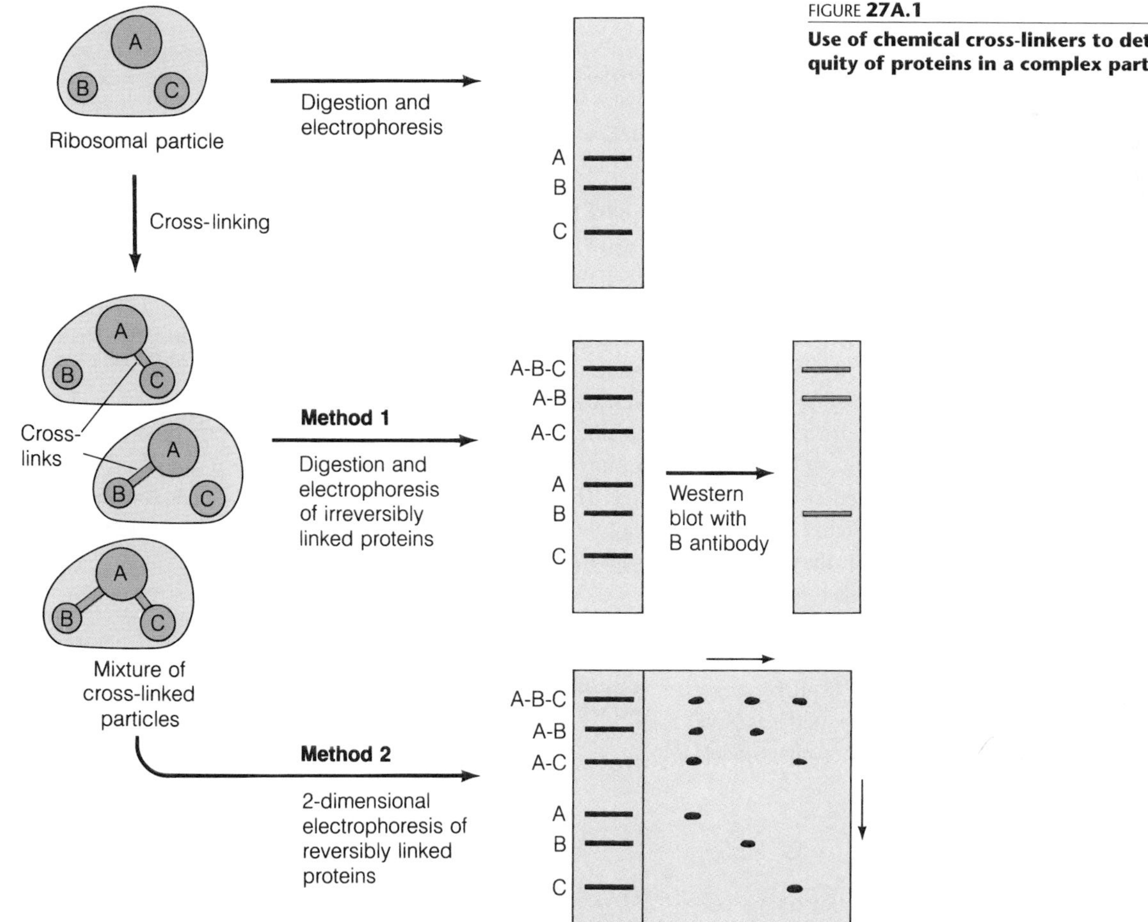

FIGURE **27A.1**

Use of chemical cross-linkers to determine propinquity of proteins in a complex particle.

TABLE **27A.1**

SOME PROTEIN CROSS-LINKING REAGENTS

Reagent	Formula	Reacts Primarily with	Cleavable?
Bis(*N*-maleimidomethyl) ether	(maleimide)$N{-}CH_2{-}O{-}CH_2{-}N$(maleimide)	Sulfhydryl group	No
2,2′-Dicarboxy-4,4′-azophenyldiisocyanate	$O{=}C{=}N{-}C_6H_3(COOH){-}N{=}N{-}C_6H_3(COOH){-}N{=}C{=}O$	Amino groups	Yes, by reduction of —N=N—
Dimethyl suberimidate	$CH_3O{-}C({=}NH){-}(CH_2)_6{-}C({=}NH){-}OCH_3$	Amino group	Yes, by ammonia
Tetranitromethane	$C(NO_2)_4$	Phenolic groups (tyrosine)	No
Methyl-4-azidobenzoimidate[a]	$N_3{-}C_6H_4{-}C({=}NH){-}OCH_3$	Amino groups and others	No
Methyl [3-(*p*-azidophenyl) dithio]propionimidate[a]	$N_3{-}C_6H_4{-}S{-}S{-}(CH_2)_2{-}C({=}NH){-}OCH_3$	Amino groups and others	Yes, by reduction of —S—S—

[a]These reagents are photoactivatable. They may be first reacted, in the dark, through the imidate group to the right and then coupled to another group by activating the azide (N_3) with a flash of light.

have a bifunctional reagent (like one of those shown in Table 27A.1) that can react with side chain residues to form cross-links between protein molecules, we can allow them to react lightly, and then we can extract the protein as a mixture of cross-linked particles. Covalently linked partners that have formed can be identified in various ways. If we have antibodies to the different proteins, we can identify partners that have formed by a *Western blotting* technique (method 1 in Figure 27A.1; also see Tools of Biochemistry 7A). Alternatively, we might use one of the "cleavable" cross-linkers shown in Table 27A.1, together with two-dimensional gel electrophoresis (method 2 in Figure 27A.1). In any event, in the simple example shown, it is clear that protein A must lie between B and C, because A can be linked to either of the other proteins but B and C do not form cross-linked dimers.

In addition to the protein–protein cross-linking techniques described here, methods for RNA–protein and RNA–RNA cross-linking are now available (see References). These methods have played a large part in determining the detailed ribosomal models presented in this chapter.

Immunoelectron Microscopy

In **immunoelectron microscopy,** components that lie on the surface of a particle can be localized in a direct manner by using antibodies prepared against them. The Y-shaped antibody molecules form bridges between two particles, connecting points at which the particular component is accessible to the surface. In Figure 27A.2, we show the same idealized particle used in Figure 27A.1. The fact that protein B lies near the pointed end of the particle is made clear by the way in which anti-B antibodies tie two particles together. When appropriate hapten groups are attached at specific points on RNA molecules, the same method can be used to show when these RNA sequences are near the ribosomal surface. The ends of the ribosomal RNAs have been located in this way, as has the position at which the nascent peptide emerges from the ribosome (see Figure 27A.3).

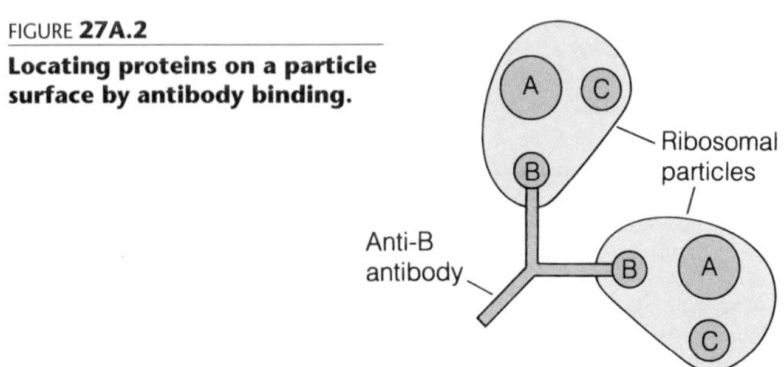

FIGURE **27A.2**

Locating proteins on a particle surface by antibody binding.

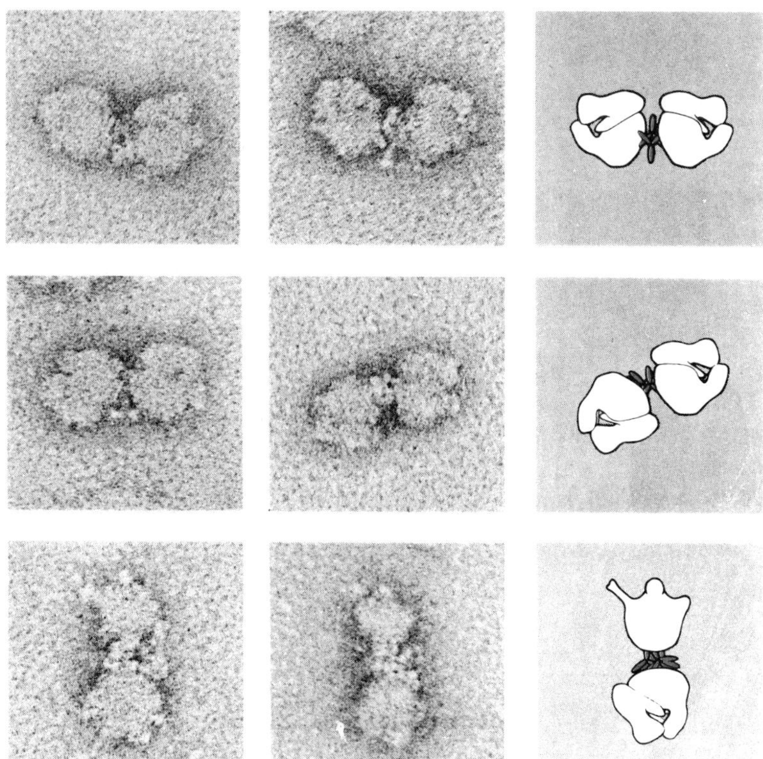

FIGURE **27A.3**

Use of antibodies to detect where the polypeptide chain emerges from the ribosome. Antibodies to β-galactosidase were used to interact with β-galactosidase chains at the point where they emerge from the ribosomes. The 70S ribosomes were found to be tied together by such antibodies at sites near the back of the 50S subunit, at the location of a pore in this unit.

Courtesy of C. Bernabeu and J. H. Lake, *Proc. Natl. Acad. Sci. USA* (1982) 79:3111. © 1982 PNAS.

Cryoelectron Microscopy

Conventional techniques of transmission electron microscopy (Tools of Biochemistry 1A) suffer from a number of serious drawbacks in the study of delicate biological structures. First, the necessity of complete dehydration in the vacuum chamber of the microscope can greatly modify structures. Second, to produce enough contrast, samples have traditionally been stained with heavy metals or metal-shadowed, producing an inherent loss in resolution.

A relatively new technique that avoids all of these artifacts is *cryoelectron microscopy* (or *electron cryomicroscopy*). The idea is simple: a sample is very rapidly frozen in ice; the freezing is so rapid that the water does not crystallize but becomes *vitrified* (glassy). The sample can be vitrified in the interstices of an EM grid or vitrified in bulk and then thin-sectioned for examination.

Because no stain is used, contrast is low, and methods of image-enhancement are needed. For particles such as ribosomes, this is accomplished by combining the information from a number of faint images of particles in the ice to develop a picture. Because they will, in general, be randomly rotated, this random orientation must be taken into account in the computer analysis. The advantage, of course, is that the final generated image can be viewed "from all sides" and can be examined from any angle in computer display.

If particles (viruses, for example) have elements of symmetry, or if the structures of subunits are known, the analysis becomes easier. But even irregular particles like ribosomes can be analyzed with resolution approaching 1–2 nm. Last, but not least, the sample has remained in an aqueous environment throughout the study, so damage to the structure should have been minimized.

Low-Angle Scattering of X-Rays and Neutrons

Although crystal diffraction studies of enormous particles like ribosomes are still very difficult, we can learn much from studying the *scattering* of radiation from solutions of such particles. When electromagnetic waves are scattered from a particle with dimensions much greater than the wavelength of the radiation, the scattering intensity depends on the angle of observation. As shown in Figure 27A.4, waves scattered from different regions within the particle at any angle other than 0° are out of phase, resulting in mutual interference and decreased intensity. At large angles the phase difference increases, causing partial cancellation of the

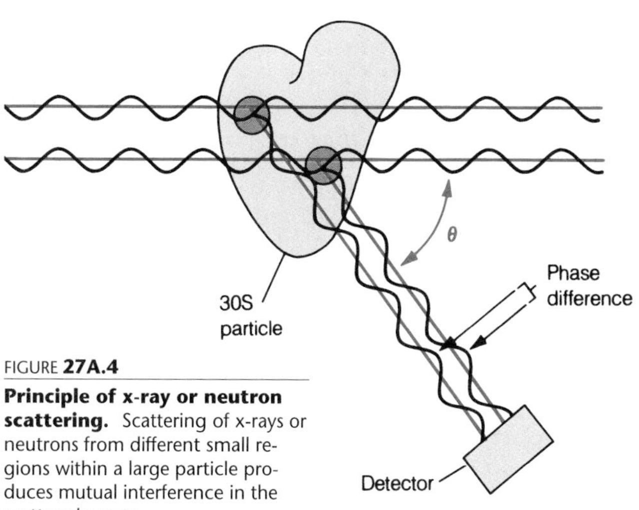

FIGURE **27A.4**

Principle of x-ray or neutron scattering. Scattering of x-rays or neutrons from different small regions within a large particle produces mutual interference in the scattered waves.

scattered waves. This interference may be used to measure average dimensions of the particle. The intensity of scattering at angle θ (I_θ) compared with the scattering at angle 0 (I_0) is given (for small angles) by the following:

$$\frac{I_\theta}{I_0} = e^{-(16\pi^4 R_G^2/3\lambda^4)\sin^2(\theta/2)} \quad (27A.1)$$

Here λ is the wavelength of the radiation, and R_G is a quantity called the *radius of gyration,* a kind of average dimension of the particle. According to equation (27A.1), a graph of $\ln(I_\theta/I_0)$ versus $\sin^2(\theta/2)$ should be a straight line at low angles, with initial slope of $(-16\pi^4 R_G^2/3\lambda^4)$. Thus, measurement of the scattering at very low angles gives a measure of average particle size. At higher angles, the I_θ/I_0 curves have a more complex shape, with maxima and minima. These can be used to give additional information about the shape of the particle and its internal distribution of matter.

Although low-angle scattering of x-rays has been useful in studying particles in solution, a much more powerful technique is **low-angle neutron scattering.** It may seem strange at first to think of neutrons as radiation, but we must remember that, according to quantum mechanics, any elementary particle has wavelike properties as well. The wavelength of a particle with mass m moving at velocity v is given by $\lambda = h/mv$, where h is Planck's constant. It turns out that "thermal neutrons" emerging from a nuclear reactor have a wavelength of a few tenths of a nanometer. Thus, they are of the right length for examination of details of macromolecular structure. Still more important, neutrons interact primarily with atomic nuclei and are therefore scattered differently by different atoms. Thus, nucleic acids and proteins scatter neutrons differently, and even hydrogen and deuterium have different scattering powers. Because H_2O and D_2O differ in scattering, it is possible to use as solvents H_2O/D_2O mixtures that match the neutron scattering power of either the nucleic acid or protein portion of a nucleoprotein particle. Then, as shown in Figure 27A.5, we can make either nucleic acid or protein "disappear" into the background and measure the radius of gyration of either component. In the example shown, the greater R_G observed for nucleic acid than for protein tells us that the nucleic acid is concentrated on the outside of the particle.

An even more powerful variant of the same technique has been used to "map" the distances between particular pairs of proteins in complex particles. Suppose, as in Figure 27A.6, we have reconstituted particles containing just two proteins (among the many in the particle) that have been prepared from deuterium-fed bacteria. These two proteins will be heavily deuterated and will have a neutron scattering power much different from that of the rest of the particle. If the H_2O/D_2O solvent is now mixed so as to match the average background in the nondeuterated portion of the particle, the two deuterated proteins will stand out in contrast. The neutron scattering pattern obtained will be dominated by the interference in scattering between these two proteins and can be used to measure their separation. In addition, the method allows measurement of the radius of gyration of a particular protein in situ.

Although neutron scattering can give much useful information, it is not a technique the average biochemist can use in the laboratory. Only a few locations in the world have large research reactors that are fitted to do neutron scattering studies.

References

Boublik, M. (1990) Electron microscopy of ribosomes. In: *Ribosomes and Protein Synthesis: A Practical Approach,* edited by G. Spedding, pp. 273–296. Oxford University Press, Oxford.

Brimacombe, R., B. Greuer, H. Gulle, M. Kasak, P. Mitchell, M. Oswald, K. Stade, and W. Stiege (1990) New techniques for the analysis of intra-RNA and RNA–protein cross-linking data from ribosomes. In: *Ribosomes and Protein Synthesis: A Practical Approach,* edited by G. Spedding, pp. 131–159. Oxford University Press, Oxford.

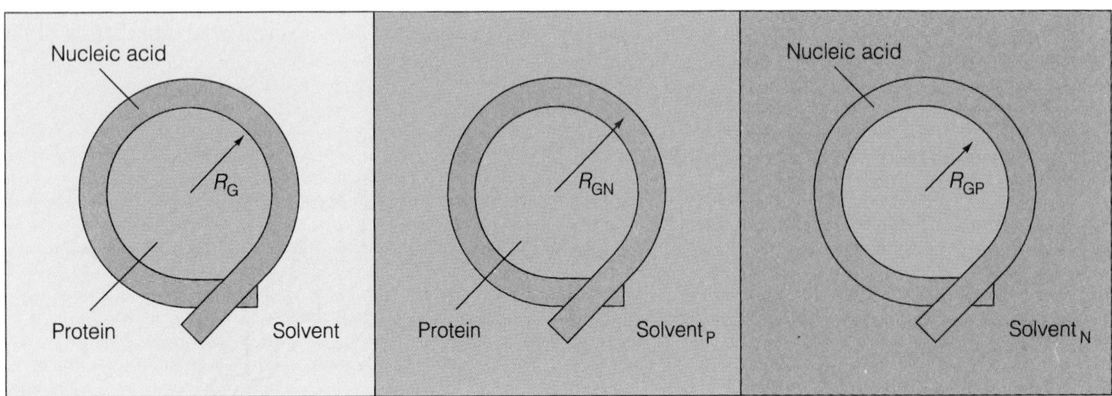

Solvent$_P$ = solvent that matches the scattering properties of the protein
Solvent$_N$ = solvent that matches the scattering properties of the nucleic acid
R_G = radius of gyration of the nucleic acid–protein complex
R_{GN} = radius of gyration of the nucleic acid
R_{GP} = radius of gyration of the protein

FIGURE **27A.5**

Use of selective solvent matching to show, by neutron scattering, that the nucleic acid is on the outside of a nucleoprotein particle.

From K. E. van Holde, W. C. Johnson, and P.-S. Ho, *Principles of Physical Biochemistry,* © Prentice-Hall, 1998. Adapted by permission of Prentice-Hall, Inc., Upper Saddle River, N.J.

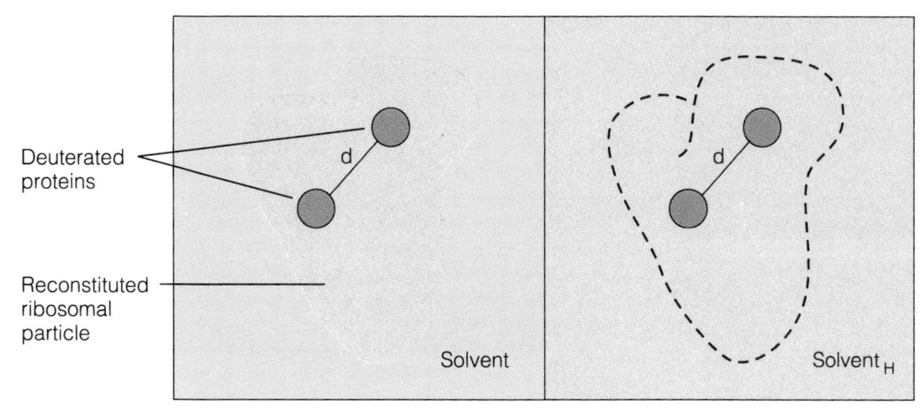

FIGURE **27A.6**
Use of solvent matching to determine the distance between two selectively deuterated proteins in an undeuterated particle.

Das, M., and C. F. Fox (1979) Chemical cross-linking in biology. *Annu. Rev. Biophys. Bioeng.* 8:165–193.

Ruiz, T., I. Erk, and J. Repault (1994) Electron cryo-microscopy of vitrified biological specimens: Toward high spatial and temporal resolution. *Biol. Cell* 80:203–210.

Serdyuk, I. N., M. Y. Pavlov, I. N. Rublevskaya, G. Zaccai, R. Leberman, and Y. M. Ostenavitch (1990) New possibilities for neutron scattering in the study of RNA–protein interactions. In: *The Ribosome,* edited by W. Hill et al., pp. 194–202. American Society for Microbiology, Washington, D.C.

CHAPTER 28

Eukaryotic Genes and Their Expression

SO FAR, MUCH OF OUR DISCUSSION OF THE ENCODING, COPYING, AND EXPRESSION of genetic information has used examples from prokaryotic organisms. There are two reasons for this choice. First, most of our fundamental understanding of the processes of replication, transcription, and translation was initially gained from studies of bacteria and viruses. Second, the corresponding processes in eukaryotes are almost invariably more complicated, and the more complex lifestyles of eukaryotic organisms require that their genomes be much larger.

Bacteria and viruses do not require enormous amounts of genetic information. A virus needs only a minimal genome to ensure its own replication, for it borrows extensively from genetic information carried by the specific host. A bacterium such as *E. coli* is autonomous, but it carries only the genetic information needed for it and its descendants to function as unicellular organisms in very limited environments. Indeed, prokaryotes use their DNA as efficiently as possible, in order to minimize the time required for replication.

Eukaryotic organisms face quite different problems. In the higher eukaryotes, the entire genetic program for development of a whole *multicellular* organism must be present in the original cell, the fertilized egg. In such organisms, not all cells develop alike. Rather, the early cells of the developing embryo *differentiate* to produce all of the varied tissues of the adult, and these specialized tissues must be correctly interrelated and positioned and function in concert. The vast amount of information needed to direct this development is carried in the genome.

Most eukaryotes require—and have—much larger genomes than prokaryotes.

We consider first the enormous amount of information stored in a typical eukaryotic DNA and how it is organized and replicated. We then describe how this information is expressed at the mRNA and protein levels.

The Eukaryotic Genome

SIZE OF THE GENOME

Because the DNA of higher eukaryotes must code for all of the specialized proteins found in different tissues, we might expect such organisms to have a considerably larger amount of DNA than is found in a typical prokaryote such as *E. coli.*

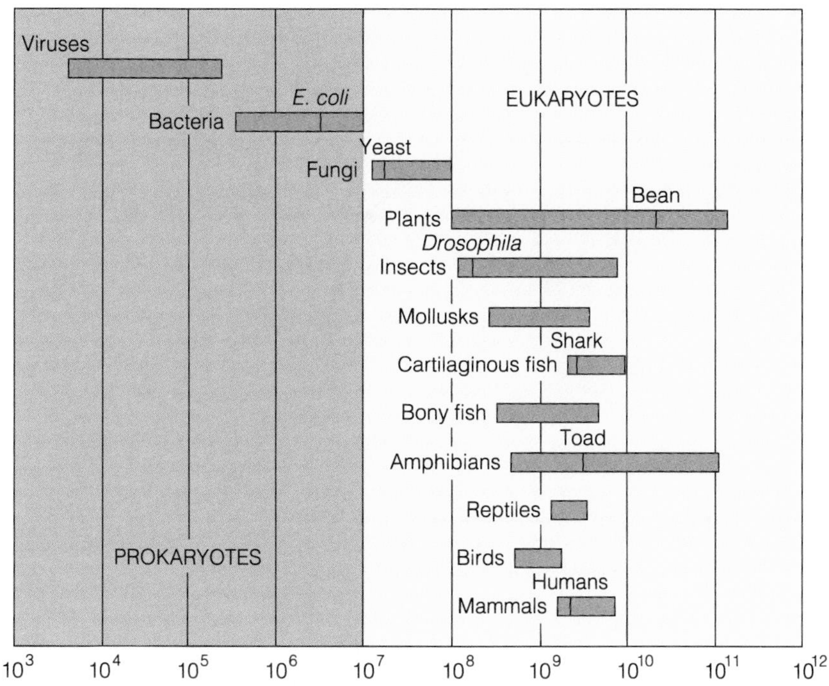

FIGURE **28.1**

Genome size. The bars show the ranges of haploid genome sizes for different kinds of organisms. A few specific organisms are marked by vertical lines. Note that many organisms have larger genomes than humans. The genome size scale is logarithmic.

What is surprising is how much more DNA there actually is. As Figure 28.1 shows, most eukaryotic cells contain 100 to 10,000 times as much DNA as *E. coli* does. In some cases the factor approaches 100,000. The only exceptions are found among the lower fungi, such as yeast and *Neurospora,* which do not exhibit the complex tissue differentiation found in higher plants and animals. The profound differences between the genomic needs of multicellular and unicellular organisms become evident as we progress in sequencing efforts. The entire yeast genome has been sequenced. It contains 12.07×10^6 bp, only 2.5 times the size of the *E. coli* genome and less than 2% of the human genome size. Yet yeast is a true eukaryote. As we proceed up the eukaryotic kingdom, there is a roughly progressive increase in genome size. It takes at least 10^8 bp to describe an insect and about 10^9 bp to describe a mammal.

However, there is no *simple* correlation between the quantity of DNA in a plant or an animal and the complexity of the organism. An amphibian is certainly no more complex a being than a human, yet some amphibia have genomes over 50 times larger than the human genome. The fact that the human genome must code for many more proteins than a bacterial cell (probably about 20 times as many) cannot explain its 1000-fold greater size. Therefore, there must be much DNA present in eukaryotes that does not code for protein. What is this other DNA? In the following sections we describe several kinds of sequences unique to eukaryotes.

REPETITIVE SEQUENCES

Much of the basis for variation in genome size comes from the varying amounts of noncoding DNA in eukaryotic genomes. The first hints of this came in 1970, when R. Britten and D. E. Kohne developed a technique for analyzing the kinetics of DNA reassociation. In this method the total DNA from an organism is cut into pieces about 300 bp long, heated to cause strand separation, and then cooled to allow the DNA to reassociate. Any sequence that is present in multiple copies will reassociate quickly; sequences that are less abundant reassociate more slowly. When Britten and Kohne used this method to study bovine DNA, they were amazed to find that almost half the DNA reassociated much more quickly than

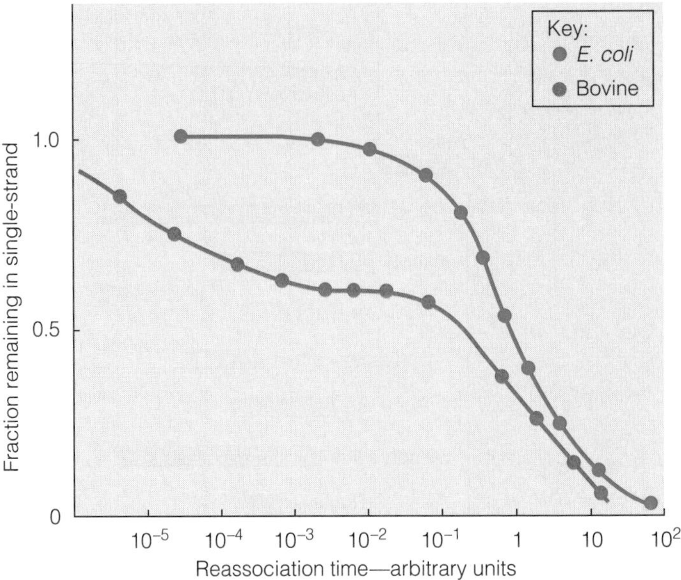

FIGURE **28.2**

Comparison of the kinetics of reassociation of *E. coli* and bovine DNA. The abscissa corresponds to reassociation time, corrected for the difference in size between the *E. coli* and bovine genomes. The curve for *E. coli* corresponds to that expected for a collection of single-copy genes in a genome of the *E. coli* size—4.67×10^6 bp. The curve for bovine DNA exhibits two steps in reassociation. The slow step corresponds to single-copy DNA (nonrepeated sequences). The other corresponds to rapidly reassociating DNA made up of repeated sequences. Many classes of repeated DNA are represented in this phase of the reassociation.

expected for single-copy segments in the large genome (Figure 28.2). To account for this rapid reassociation, they were forced to the surprising conclusion that some DNA sequences are reiterated as many as 10^5 to 10^6 times in each cell. Whereas practically all of the DNA of *E. coli* is single-copy, only about half of mammal DNA and about a third of plant DNA fall into this category.

Repetitive DNA sequences in eukaryotic genomes include satellite DNAs and scattered duplicate sequences.

Satellite DNA

Further analysis has divided these reiterated DNA sequences into several categories. One type involves multiple tandem repetitions, over long stretches of DNA, of very short, simple sequences like $(ATAAACT)_n$. Such DNA can often be separated from the major portion of the DNA by sedimentation to equilibrium in density gradients (see pages 96–97, Chapter 4). Repetitive DNA sequences that are AT-rich (like the one just mentioned) have a lower density than average-composition DNA, and GC-rich sequences are more dense. Thus, in a density-gradient experiment, the reiterated, simple-sequence fragments form satellite bands about the main-band DNA (Figure 28.3). Because of this banding, clusters of repeated sequences are sometimes referred to as **satellite DNAs.** In higher eukaryotes, satellite DNA usually makes up 10% to 20% of the total genome.

What function can such highly reiterated DNA sequences serve? They do not code for proteins, and most are not even transcribed into RNA. Some, at least, appear to play a structural role. Certain reiterated sequences have, for example, been found to be highly concentrated near the **centromeres** of chromosomes, the regions where sister chromatids are attached. They may serve as binding sites for proteins that attach the spindle fibers in mitosis. (See later sections in this chapter for discussions of the physical structure of chromosomes and the process of mitosis.)

Duplications of Functional Genes

There are many other classes of DNA sequences with varying degrees of repetition. Some of these sequences represent duplications of functional genes, and in many

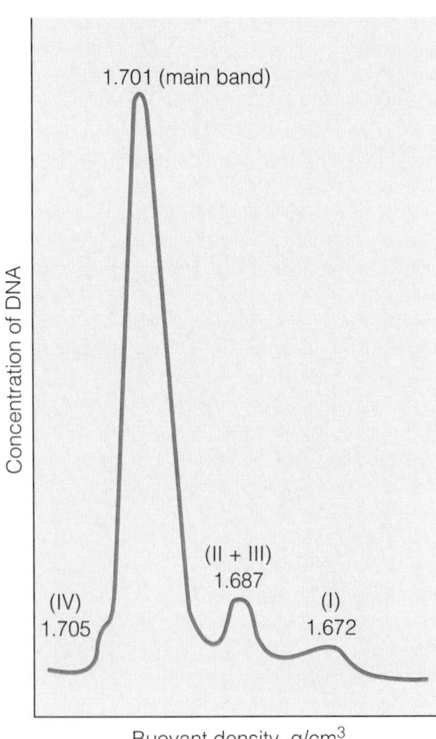

FIGURE **28.3**

Satellite DNA. An equilibrium density-gradient sedimentation of total *Drosophila* DNA resolves satellite bands, surrounding the main band. These represent repetitive DNA fractions of differing base composition.

Adapted from S. A. Endow, M. L. Polan, and J. G. Gall, *J. Mol. Biol.* (1975) 96:675.

cases the repetitiveness seems to play a useful role, by allowing high levels of production of much-needed transcripts. Examples include the genes for ribosomal RNAs, of which up to several thousand copies may be present, and tRNA genes, with hundreds of copies of each type often found. The cell's continual need for large quantities of ribosomes and tRNAs for translation is met by having multiple copies of these genes. The same is true for the genes for some much-used proteins, such as the histones that bind to eukaryotic DNA to form the chromatin structure (see pages 1075–1076). As pointed out in Chapter 25, even genes that are normally single-copy are sometimes amplified, either in response to environmental stress or in special tissues during embryonic development.

Alu Elements

Other kinds of repeated DNA sequences exist that do not code for proteins but whose true function remains mysterious. Such sequences are often scattered throughout the genome, rather than being clustered like the satellite DNAs. Some of these sequences may represent control elements of some kind, with functions we do not yet understand. One of the most common such families in mammals consists of the so-called ***Alu* elements.** These sequences, of which there are hundreds of thousands of copies in the human genome, are about 300 bp long. Their name reflects the common existence of a single site for the restriction endonuclease *Alu*I in most members of this class. The *Alu* sequences can be (inefficiently) transcribed into RNA, although they are not known to be translated.

The function (if any) of many repeated sequences is unknown.

The function of the large number of *Alu* sequences remains uncertain, although some of them may contain origins for DNA replication. But it is also conceivable that many repetitive sequences such as these serve no useful function. They may simply exist in the genome as "molecular parasites." A way in which such sequences could spread through the genome has been proposed, on the basis of the observation that *Alu* sequences are flanked by short, repeated oligonucleotides resembling those of transposons (see Chapter 25). In this view, *Alu* sequences, like other mobile genetic elements, may be inserted at various places in the genome as reverse transcriptase copies of the RNA that is transcribed from them. Recent studies suggest that the *Alu* sequences may have been derived from a small RNA (7SL RNA) involved in protein transport across membranes. We shall discuss this RNA on page 1105.

INTRONS

A second reason for the large size of eukaryotic genomes is that most eukaryotic genes are interrupted by introns. In Chapter 7 we introduced the concept of introns and exons and pointed out, as an example, that the β globin gene consists of three exons interrupted by two noncoding intron regions. This kind of structure is common in eukaryotes and is often more extreme than in the hemoglobin example. Consider the ovalbumin gene depicted in Figure 28.4a. This gene codes for a protein 386 amino acid residues in length, which could be accommodated by a message 1158 nucleotides long. Yet the total ovalbumin gene is about 7700 base pairs in length, containing eight exons interspersed by seven introns. The difference between the ovalbumin gene and its mRNA is dramatically displayed when hybrid DNA–RNA molecules are examined in the electron microscope (Figure 28.4b and c). The genomic DNA pairs to the mRNA along the exons, but the introns are looped out, to form what are called **R loops.**

Most eukaryotic genes contain introns.

Recent sequencing studies have shown that introns are present in most eukaryotic structural genes and frequently exceed exons in total length. Interestingly, some of the lower eukaryotes like yeast have many fewer introns, and their genome size is correspondingly smaller. As in the case of prokaryotes, which also have few introns, this smaller yeast genome presumably reflects a need for more efficiency in cell replication.

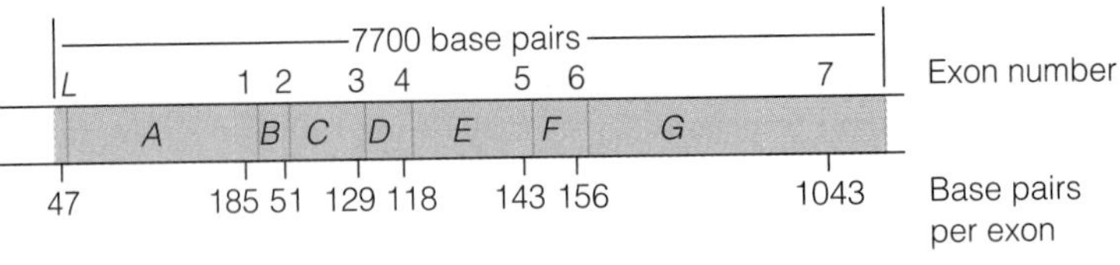

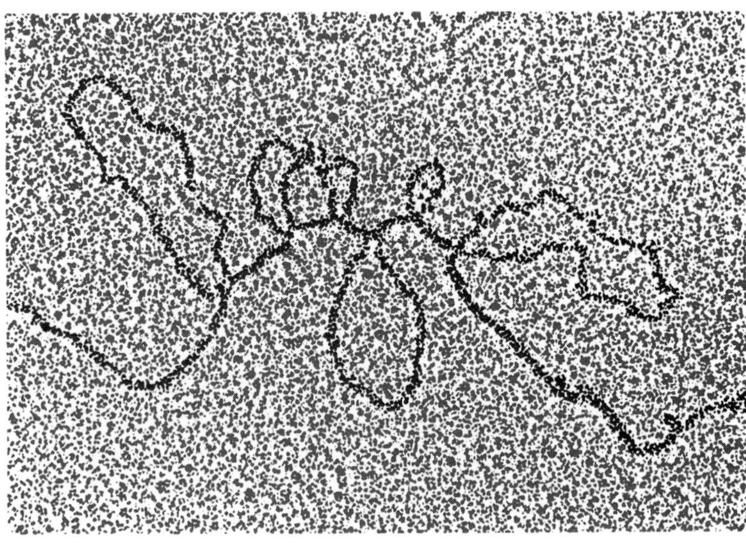
(b)

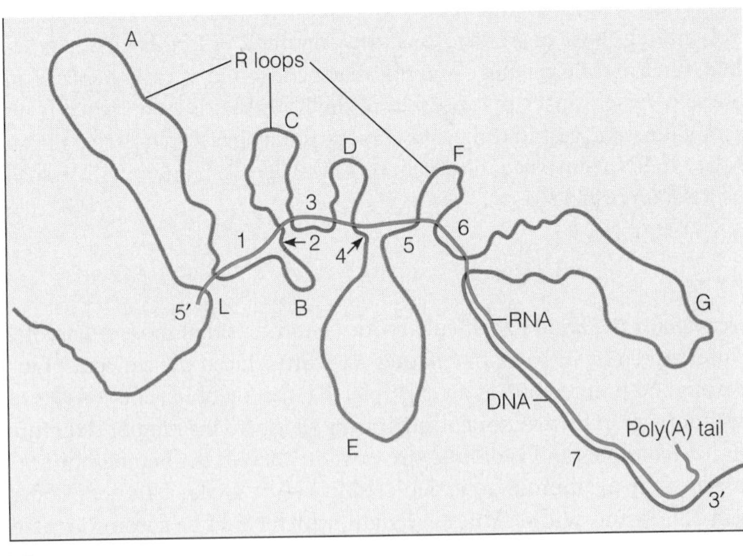

FIGURE **28.4**

Exon–intron structure of the ovalbumin gene in chickens. **(a)** Map of the 7700-bp gene, showing exons 1–7 plus an untranslated leader sequence (blue) and introns A–G (brown). **(b)** Electron micrograph of a hybrid between the genomic DNA and ovalbumin mRNA. **(c)** Diagram showing how the intron regions loop out in R loops in such a hybrid. The RNA is shown in red, the DNA exons in blue, and the DNA introns in brown.

(b) Courtesy of P. Chambon, *Sci. Am.* (1981) 245:60–70.

As mentioned in Chapter 7, the function of introns is not yet wholly understood. It seems likely that they serve as loci for genetic recombination, allowing functional parts of proteins to be interchanged in evolution. Such loci also allow some eukaryotes to make variants of a protein from a single gene, by splicing different exons together. This alternative splicing, discussed later in the chapter, is a more efficient way to store information than having a whole gene for each variant of the protein.

GENE FAMILIES

Multiple Variants of a Gene

Despite alternative splicing, in many cases complete variants of genes for the same type of protein are found, to be expressed in different tissues or at different stages in development. We encountered an example in Chapter 7, where the embryonic (ζ and ε), fetal (α and γ), and adult (α, β, and δ) globins of mammals were described. For each of these proteins there exists a complete gene in every cell of the mammal.

Figure 7.23 (page 234) depicts the clusters of genes for the α and β classes of hemoglobins in humans. Each of these genes has the kind of exon–intron structure shown in Figure 7.20 (page 231). In addition, the genes themselves are separated by long stretches of nontranscribed DNA. Some portions of these intervening regions must contain control signals, for the expression of the globin genes is under complex and subtle regulation. In the first place, although the globin gene clusters are found in all human cells, they are expressed *only* in the *erythropoietic cells,* cells that give rise to red blood cells. Furthermore, as we saw in Chapter 7, expression of each variant is strictly constrained to certain developmental stages. For example, in the early embryo, only the ζ and ε genes are being transcribed; all other globin genes are turned off. As development proceeds, transcription switches first to the fetal α and γ genes, and at about the time of birth the adult β variant begins to dominate and transcription of γ ceases (see Figure 7.22). This kind of developmental regulation is peculiar to eukaryotes—no prokaryote has need for such a mechanism. The use of multiple variant genes is expensive in DNA: The human genome devotes about 100,000 bp of DNA just to producing different variants of hemoglobin.

Many other gene families exist. Some, like the groups of genes encoding variants of the histones, seem to play developmental roles rather like those of the globin gene family. Others, like the immunoglobulin genes, appear to exist in multiple forms in order to satisfy a multiplicity of similar but distinct needs. In each case, it seems likely that the members of a particular gene family have evolved by successive duplications of an original, ancestral gene.

As our detailed knowledge of many different genomes has expanded, it has become obvious that the concept of "gene families" has a much broader sense than that used above. We deduce, from homologies in sequence and structure, that many genes we had first thought of as unrelated belong to extended families probably descended from one common ancestor. It may be that all existing genes descend from a relatively small number (several hundred?) sufficient to serve the needs of a very primitive ancestral organism.

Pseudogenes

Gene families often include as members one or more **pseudogenes,** or nonfunctional genes. Pseudogenes can be recognized because they bear strong sequence similarity to functioning genes, from which they undoubtedly evolved, perhaps by a reverse transcriptase mechanism as has been proposed for *Alu* elements (see page 1071). They are no longer transcribed, however, because some element required for transcription (often a flanking control region or promoter) is missing or defective. Because their sequences are not expressed, pseudogenes are no longer under strong selective control in evolution. In a sense, it does not matter what happens to them, and as a consequence they can accumulate mutations that would be selected against in functional genes. This phenomenon provides at least the possibility that novel genes may arise from pseudogenes that become transcribable again. Examples of pseudogenes can be seen in Figure 7.23 (page 234), which shows the arrangement of α and β globin gene variants in human DNA.

In summary, we find that a number of quite different explanations combine to account for the very large size of the genomes of eukaryotic organisms. At the same time, we still find it hard to rationalize the extreme variations in amount of DNA that are sometimes observed even between closely related organisms. To take one example, among the amphibians alone there is more than a 100-fold range of genome sizes. The function, if any, of such enormous variation is still obscure, which suggests that we still do not understand some fundamental things about the eukaryotic genome.

The Physical Organization of Eukaryotic DNA: The Nucleus, Chromosomes, and Chromatin

CHROMOSOMES

Almost all of the DNA of a typical prokaryote like *E. coli* is contained in a single, large circular DNA molecule, with a minor fraction present in small plasmids. This "prokaryotic chromosome" is supercoiled and complexed with proteins and exists free in the cytosol, although attached at one or more points to the cell membrane. This structure is usually called the **bacterial nucleoid.**

The situation in eukaryotes is quite different. The typical cell's genome is divided into several or many eukaryotic chromosomes, each of which contains a single, very large, linear DNA molecule. Although their size varies greatly among organisms and even among different chromosomes in a given species, these DNA molecules are commonly of the order of 10^7 to 10^9 bp in length.

Different eukaryotic species contain widely varying numbers of *distinguishable* chromosomes—from 1 (in an Australian ant) to 190 (in a species of butterfly). Whereas most prokaryotes are *haploid,* containing only one copy of their chromosome, most eukaryotic cells are *diploid;* that is, they carry two copies of each chromosome. For example, the human genome is made up of 23 different chromosomes, so that normal, diploid human cells have a total chromosome number of 46. Some eukaryotic cells are highly *polyploid,* carrying many copies of each chromosome. Well-studied examples include the salivary gland cells in fruit flies. The *polytene* chromosomes of such polyploid cells provide a dramatic way to visualize the linearity of gene arrangements, as shown in Figure 28.5. In a polytene chromosome, multiple copies of the DNA, and its associated proteins, are arranged side by side in exact register. The linear array of light-staining and dark-staining bands provided early evidence for the linear arrangement of genes on the chromosome.

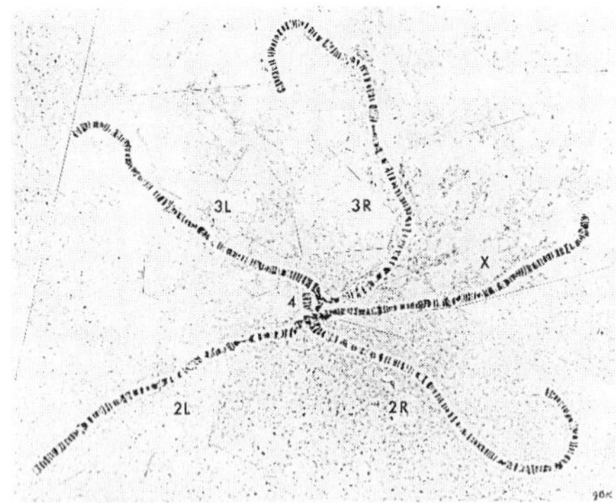

FIGURE **28.5**

Polytene chromosomes from metaphase *Drosophila* salivary gland cells. All four chromosomes are joined at their centromeres.

From G. Lefevre, Jr., in *The Genetics and Biology of Drosophila,* edited by M. Ashburner and E. Novitski, Vol. 1A (London: Academic Press, 1976).

Unlike the prokaryotic chromosome (nucleoid), the chromosomes of eukaryotes are not normally found free in the cytoplasm. In nondividing cells, the chromosomes are segregated within the nucleus (Figure 28.6) as an entangled mass of fibers of a DNA–protein complex called **chromatin.** The *nuclear envelope* is pierced by **nuclear pores,** which allow quite large molecules including proteins and small RNA molecules to pass in and out of the nucleus.

The nuclear pores are actually multisubunit complexes that provide openings of about 9 nm diameter. These allow the free diffusion of small molecules (including some small proteins) between nucleus and cytoplasm. However, the pore complex is also involved in selective transport of larger proteins and messenger RNA. In this process, two whole classes of assister proteins, called **exportins** and **importins,** are involved. Protein translation in eukaryotes occurs outside the nucleus, however, so transcription and translation cannot be directly coupled, as they are in prokaryotes. Therefore, messenger RNA must be exported from the nucleus for translation. As we shall see later, considerable processing of the mRNA occurs before this export. During *mitosis*—the separation of daughter chromosomes following DNA replication—the nuclear envelope breaks down, and the diploid chromosomes condense into compact structures like the one shown in Figure 28.7.

The multiple eukaryotic chromosomes are contained within the nucleus, except during mitosis when the nuclear envelope disintegrates.

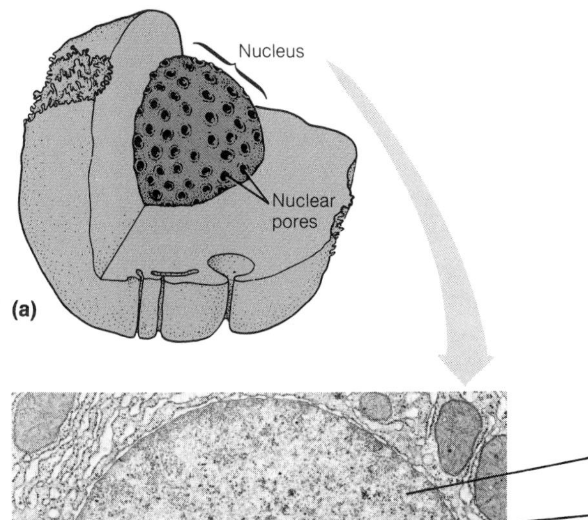

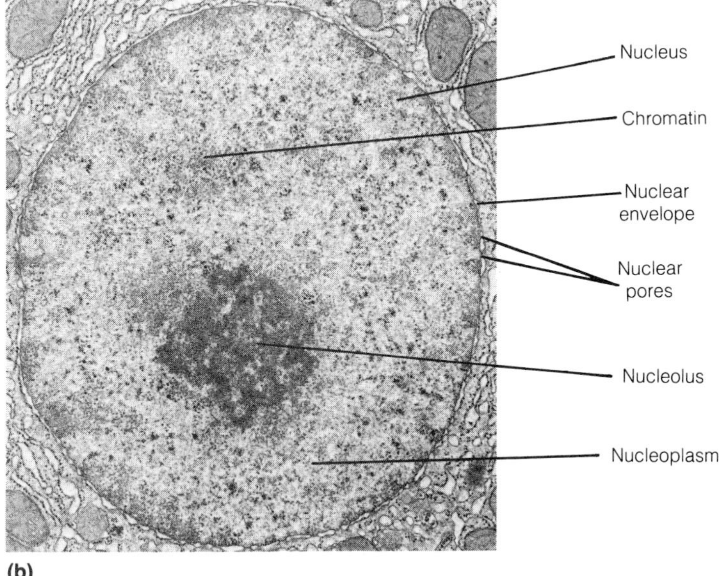

FIGURE **28.6**

The nucleus. (a) Schematic, cutaway view of a typical animal cell, showing the position and relative size of the nucleus. **(b)** Electron micrograph of a section of a rat liver nucleus.

(b) Courtesy of R. Rodwald, University of Virginia/BPS; **(c)** courtesy of J. P. Stafstrom and L. A. Staehelin, *J. Cell Biol.* (1984) 98:699, by permission of Rockefeller University Press.

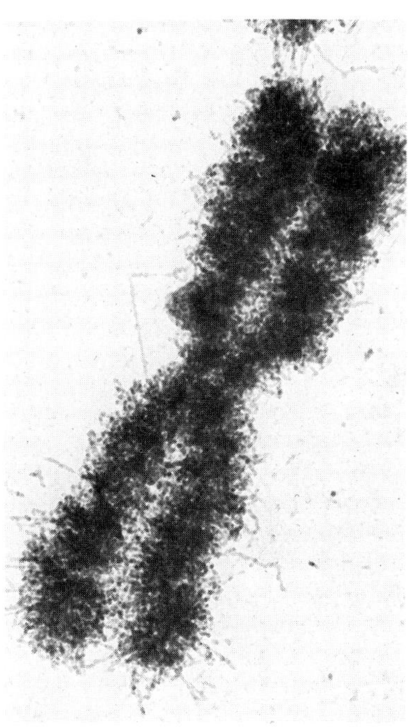

FIGURE **28.7**

A mitotic chromosome. An electron microscope image of a human chromosome during the metaphase stage of mitosis. The constriction at the centromere and the lengthwise division into sister chromatids are clearly visible. The hairy-looking surface is made of loops of highly coiled chromatin.

Courtesy of G. F. Bahr, Armed Forces Institute of Pathology.

The enormous amount of DNA in a eukaryotic cell poses some serious problems. First, there is the question of *compaction.* The diploid DNA content of a human cell is about 8×10^9 bp, corresponding to a total length of nearly 3 meters. Somehow, all this DNA must be packed into a nucleus about 10 μm (10^{-5} m) in diameter. Second is the problem of *selective transcription.* In a typical differentiated eukaryotic cell, only a small fraction of the DNA (5%–10%) is ever transcribed. Much, as we have seen, is nontranscribable. Many genes that do undergo transcription do so only in certain cell lines in particular tissues, and then often only under special circumstances. To maintain and regulate such complex programs of selective transcription, the accessibility of the DNA to RNA polymerases must be under strict control. Both compaction and the control of gene expression in eukaryotes are achieved by having the DNA complexed with a set of special proteins to form the protein–DNA complex called chromatin.

CHROMATIN

The chromatin of eukaryotes consists of DNA complexed with histones and nonhistone proteins.

The DNA-binding proteins of chromatin fall into two classes. The major class, **histones,** includes five types of protein, whose properties are outlined in Table 28.1. All histones are small, very basic proteins rich in lysine and arginine. Some have been remarkably well conserved in amino acid sequence throughout evolution. Histone H4, for example, shows only two substitutions between humans and peas and only eight substitutions between humans and yeast. All known H4s contain exactly 102 amino acid residues. The histones are the basic building blocks of

TABLE 28.1 Properties of the major histone types

Histone Type	Molecular Weight	Number of Amino Acid Residues	mol % Lys	mol % Arg	Role
H1	22,500	244	29.5	1.3	Associated with linker DNA; helps form higher-order structure
H2A	13,960	129	10.9	9.3	Two of each go to form the histone octamer core of the nucleosome.
H2B	13,774	125	16.0	6.4	
H3	15,273	135	9.6	13.3	
H4	11,236	102	10.8	13.7	

Note: All data are for calf thymus histones, except for H1, which is from rabbit.

chromatin structure. The nucleoids of prokaryotic cells also have proteins associated with DNA, but these proteins are quite different from the histones and do not seem to form a comparable chromatin structure. Thus, a histone-containing chromatin structure is a uniquely eukaryotic feature. In all kinds of eukaryotic nuclei, from yeast to human, the histones are present in an amount of about 1 gram per gram of DNA, and histones H2A, H2B, H3, and H4 are always found in equimolar quantities.

The histones are accompanied by a much more diverse group of DNA-binding proteins, rather unimaginatively named **nonhistone chromosomal proteins.** The total amount of the latter proteins varies greatly from one cell type to another, ranging from about 0.05 to 1 gram per gram of DNA. They include a bewildering variety of proteins, such as polymerases and other nuclear enzymes, hormone receptor proteins, and regulatory proteins of many kinds. It is possible to count, on two-dimensional gels, approximately 1000 different nonhistone chromosomal proteins in a typical eukaryotic nucleus.

The association of proteins with eukaryotic DNA has long been recognized. As early as 1888 the German chemist Albrecht Kossel isolated histones from nuclei and recognized them as basic substances that would bind to the nucleic acid. Histones were, in fact, the first class of proteins to be recognized. However, the precise role of histones was not understood until about 1974. Then, research in a number of laboratories showed that these proteins combine in a specific way to form a repeating element of chromatin structure, the **nucleosome.**

THE NUCLEOSOME

If naked DNA (that is, DNA that is not protein-complexed) is partially digested with a nonspecific endonuclease such as micrococcal nuclease, which cuts double strands almost randomly, a broad smear of polynucleotide fragments is produced. But in the early 1970s, researchers in a number of laboratories found that if the same experiment is conducted with chromatin, or even with whole nuclei (which some nucleases can easily penetrate through the nuclear pores), the DNA was cleaved in a quite specific, nonrandom way. On a polyacrylamide gel, the DNA from nuclease-digested chromatin gave a series of bands that were multiples of approximately 200 base pairs (Figure 28.8). This indicated that the nuclease could find easy access to the DNA only at regularly spaced points. At about the same time that these observations were made, other laboratories obtained electron micrographs of extended chromatin fibers, which revealed a regular "beaded" pattern in the chromatin structure, with one bead about every 200 bp (Figure 28.9). Still other researchers found that if nuclease digestion of chromatin was continued, it slowed down and nearly stopped when about 30% of the DNA had been consumed. The remaining protected DNA was found to be present in particles corresponding to the beads seen in the electron micrographs. These particles,

The basic repeating structure in chromatin is the nucleosome, in which nearly two turns of DNA are wrapped about an octamer of histones.

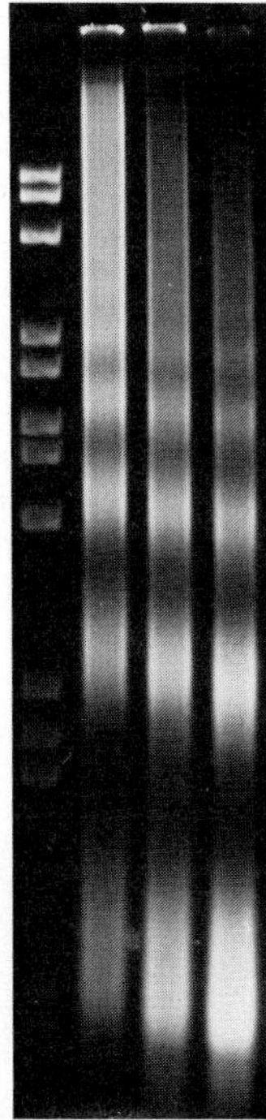

FIGURE **28.8**

The kind of evidence that first suggested a repetitive structure in chromatin. In this gel, the three columns to the right show DNA fragments obtained after three successively longer digestions of chicken erythrocyte chromatin by micrococcal nuclease. The column to the left contains DNA restriction fragments as size markers.

Courtesy of K. van Holde.

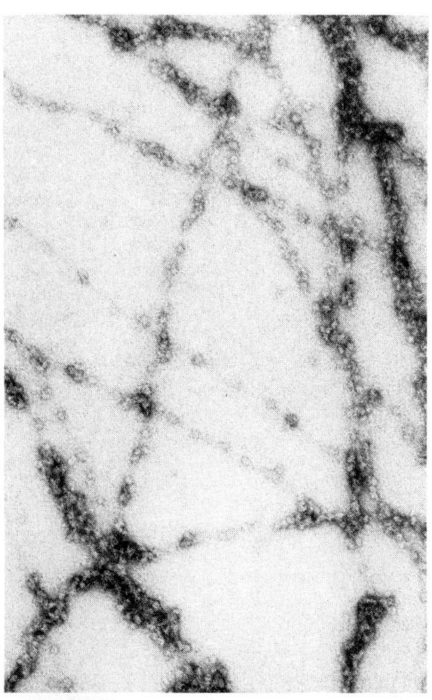

FIGURE **28.9**

Beaded-fiber structure of chromatin. An electron micrograph of chromatin spread on a grid at low ionic strength and negatively stained. The spreading under these conditions unravels and stretches some of the condensed chromatin fibers to show the regularly spaced nucleosomes. This was one of the first photographs obtained showing this structure.

Courtesy of C. L. F. Woodcock, University of Massachusetts, Amherst.

called nucleosomes (or more precisely, *nucleosomal core particles*), were found to have a simple, definite composition that is practically invariant over the whole eukaryotic kingdom. They always contain 146 bp of DNA, wrapped about an octamer of histone molecules—two each of H2A, H2B, H3, and H4. This composition explains the equivalent amounts of these histones in chromatin. Both nucleosomes and nucleosome histone cores have been crystallized, and x-ray diffraction studies have revealed the structure shown in Figure 28.10. The DNA lies on the surface of the octamer and makes about 1.7 left-hand superhelical turns about it. The structure of the octamer provides a left-hand helical "ramp" upon which the DNA is bound. The high-resolution data now available on the histone octamer reveal a commonality in histone structure—the *histone fold*—that was not evident in sequences, and suggest an early common ancestor of these proteins.

Although the nucleosome itself is a nearly invariant structure in eukaryotes, the way in which nucleosomes are spaced along the DNA varies considerably among organisms and even among tissues in the same organism. The length of DNA between nucleosomes may vary from about 20 bp to over 100 bp. Exactly what determines the arrangement of nucleosomes along the DNA is still not wholly understood. However, it is now clear that at least some nucleosomes occupy defined positions. The implications of this finding are discussed later. The internucleosomal, or *linker*, DNA is occupied by the H1-type (very lysine-rich) histones and nonhistone proteins. Figure 28.11 provides an overall schematic view of the fundamental elements of chromatin structure.

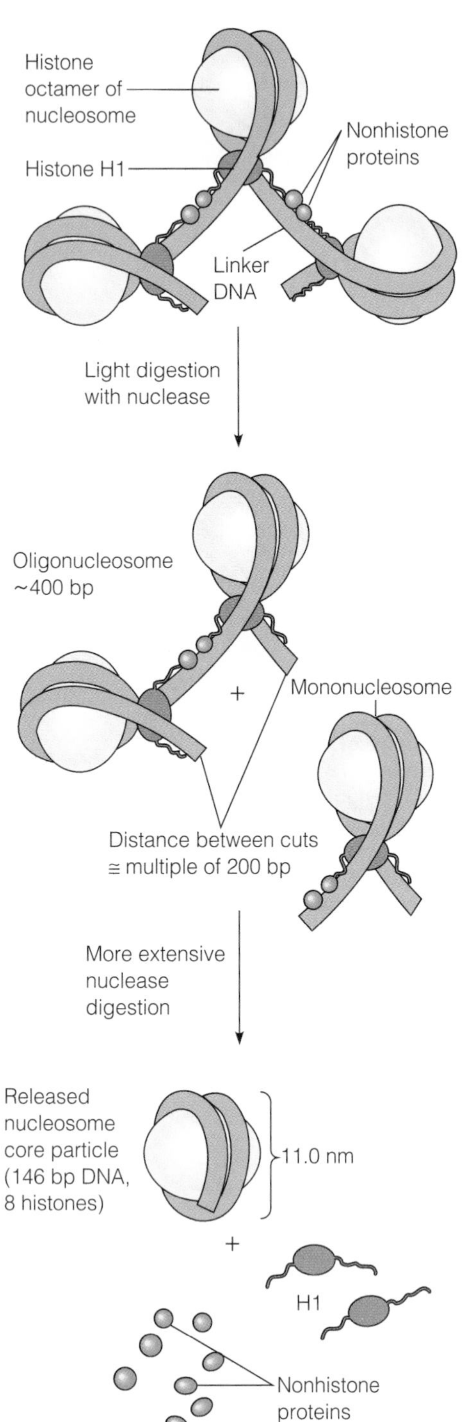

FIGURE **28.11**

The elements of chromatin structure. At the top is our current understanding of the extended structure of a chromatin fiber. Light digestion with nuclease releases first mono-nucleosomes and oligonucleosomes. Then, as linker DNA is further digested, nonhistone proteins and H1 are released, to yield the core particle, whose structure is shown in Figure 28.10.

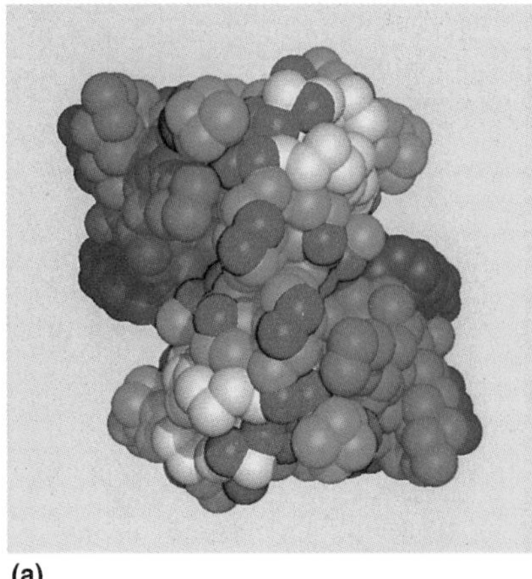

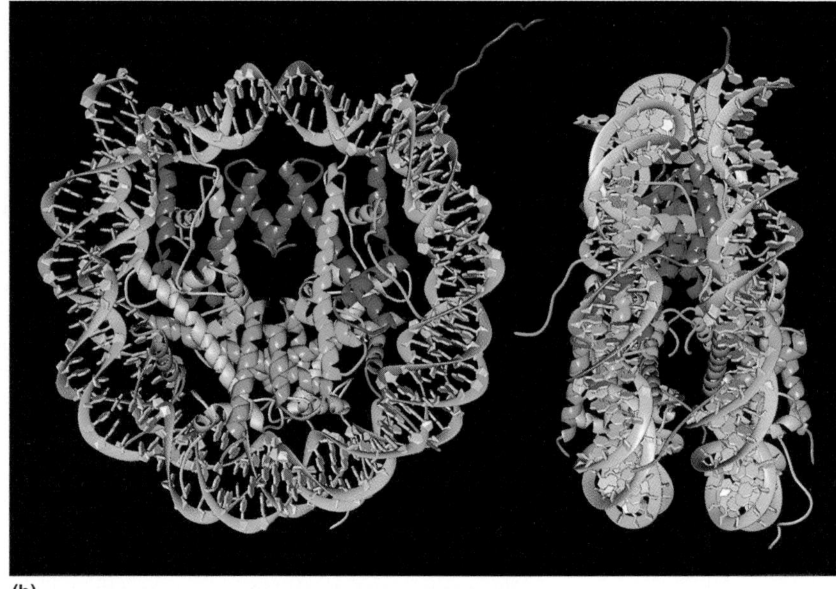

FIGURE **28.10**

Structure of the nucleosome core particle as revealed by x-ray diffraction. (a) The histone octamer core. H3 is green, H4 is white, H2A is light blue, and H2B is dark blue. The lysine and arginine residues of the $(H3/H4)_2$ tetramer are shown in red. **(b)** A high-resolution (2.8 Å) model of the nucleosome core particle. Two views perpendicular to the twofold axis are shown. The histones are identified as follows: H3 = blue, H4 = green, H2A = yellow, H2B = red. The N-terminal tails of the histones are not completely resolved.

(a) Courtesy of G. Arents and E. N. Moudrianakis, *Proc. Nat'l Acad. Sci. USA* (1993) 90:10489; provided by E. N. Moudrianakis; **(b)** Courtesy of T. J. Richmond from K. Luger et al., *Nature* (1997) 389:251–260. © 1997 Macmillan Magazines, Ltd.

HIGHER-ORDER CHROMATIN STRUCTURE IN THE NUCLEUS

Wrapping DNA about histone cores to form nucleosomes accomplishes part of the compaction necessary to fit the eukaryotic DNA into the nucleus, for the strand is thereby shortened severalfold. However, it is clear that much of the chromatin in the nucleus is even more highly compacted. The next stage in compaction involves folding the beaded fiber into a thicker fiber like that shown in Figure 28.12. These fibers are about 30 nm in diameter and may be further folded on themselves to make the thicker chromatin fibers visible in both metaphase chromosomes (see Figure 28.7) and the nuclei of nondividing (interphase) cells (see Figure 28.6).

Evidence is emerging concerning the way in which the chromatin fiber is organized in both metaphase and interphase chromatin. Dye staining of the metaphase chromosomes from a particular organism gives a reproducible banding pattern (see Figure 28.5). *In situ hybridization* methods, in which specific sequences are located on chromosomes by hybridization with a complementary radioactive nucleic acid and then visualized by radioautography, show that particular DNA sequences are always located at the same places in specific chromosomes. Given

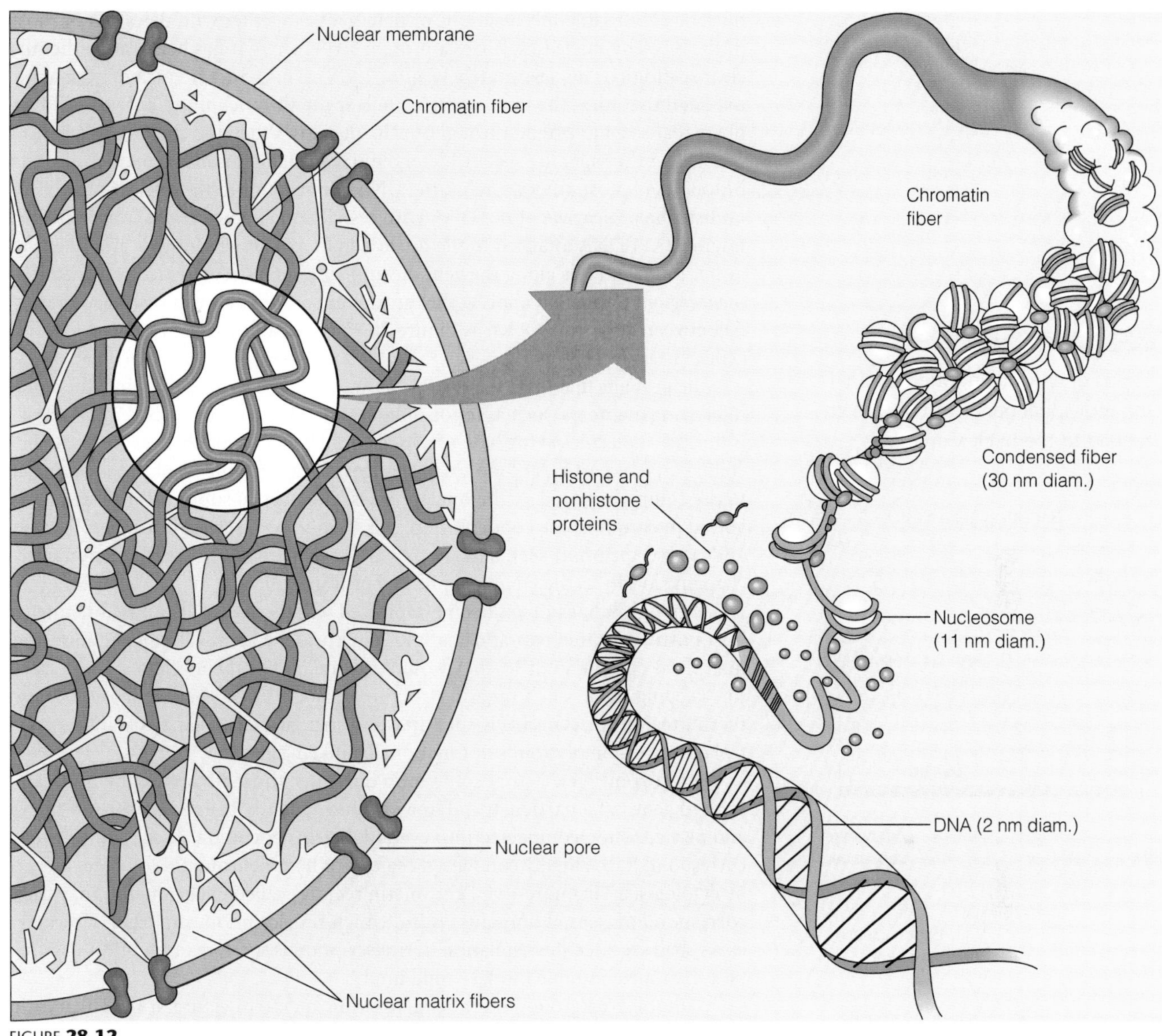

FIGURE **28.12**

Levels of chromatin structure. To the left is a schematic view of a portion of the nucleus, with partially condensed chromatin fibers. A closer view (to the right) shows a chromatin fiber in which part is in the condensed (30-nm) form, and part is opened up, as for transcription.

that the DNA in a eukaryotic chromosome is one long continuous strand, this implies that some kind of regular folding must be present to preserve this order. Recent evidence indicates what this folding may be. If metaphase chromosomes are treated with polyanions like dextran sulfate, which strip off the histones and loosely bound nonhistone proteins, the DNA strands emerge as enormous loops from a *scaffold* of tightly bound protein. An electron micrograph of this structure is shown in Figure 1.5 (page 9). Individual loops vary in size but may range up to 100,000 bp in length—about the size of the β globin gene cluster, for example. Approximately 1000 such loops exist in the average chromosome.

The fiber formed by the nucleosomes is folded in vivo to form higher-order chromatin structure.

Evidence also exists for a similar but more diffuse scaffold in the interphase nucleus. Removal of histones and weakly bound nonhistone proteins from intact nuclei by high salt concentrations or detergents, together with digestion of most of the DNA by nucleases, leaves a protein structure that has been called the **nuclear scaffold,** or **nuclear matrix** (Figure 28.12). This structure includes the laminar shell that lines the inside of the nuclear membrane, plus a network of fine fibers that seem to extend throughout the nucleus. When the chemical dissection is done gently, using the specific detergent lithium diiodosalicylate to remove the histones and most other proteins, the DNA connections to the nuclear matrix are undisturbed. Cleavage of the DNA with restriction endonucleases leaves specific fragments of DNA attached to the nuclear matrix. These fragments are spaced at rather long intervals along the genome and contain characteristic *matrix attachment regions (MARs).* It appears that groups of coordinately expressed genes often lie between adjacent MARs, as Figure 28.13 illustrates for the repeated histone gene clusters in *Drosophila.*

Loops of chromatin, often containing individual gene clusters, are attached to the nuclear matrix.

The proteins that form the scaffold from which the loops extend include some interesting members, such as topoisomerases. It has been hypothesized that topoisomerase molecules at the base of a loop could bring about changes in the supercoiling on that particular loop. Such changes in supercoiling would be in addition to the coiling imposed by the nucleosomes. Changes in supercoiling may be involved in chromosome condensation and seem to be essential during replication and transcription. It seems likely that the structure of chromatin is dynamic, changing locally as the DNA is replicated or transcribed.

Although it has not yet been established with certainty, it is possible that at least portions of the loop-and-scaffold structures are identical in metaphase and interphase chromatin. If so, they would provide part of the mechanism necessary for "keeping track" of chromatin structure through successive cell divisions. Furthermore, it now seems likely that this loop structure may play a role in the control of expression of groups of functionally linked genes.

The current view is that some loop domains, those involving the nontranscribed genes of a particular cell, may be permanently coiled into 30-nm fibers and perhaps supercompacted into even higher-order coiling. Such regions could correspond to the highly condensed regions of **heterochromatin** long recognized by cytologists. The more-open chromatin regions, called **euchromatin,** may then correspond to relaxed domains within which transcription can occur. However, as we shall see later, the regulation of transcription is surely more subtle and complex than such a simple model would imply.

With this background on the organization and architecture of the eukaryotic genome, we now turn to its major functions—replication and transcription.

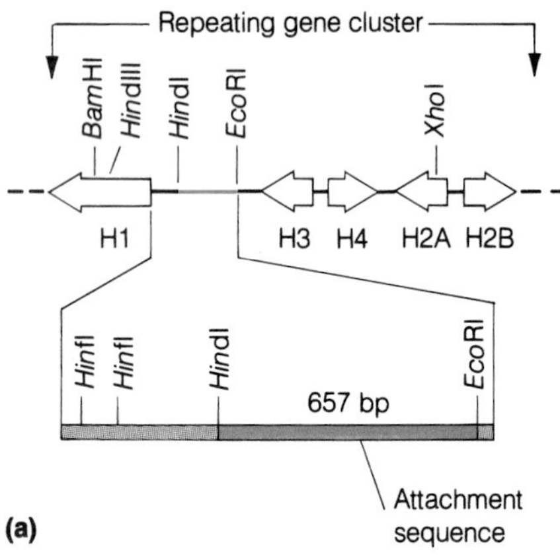

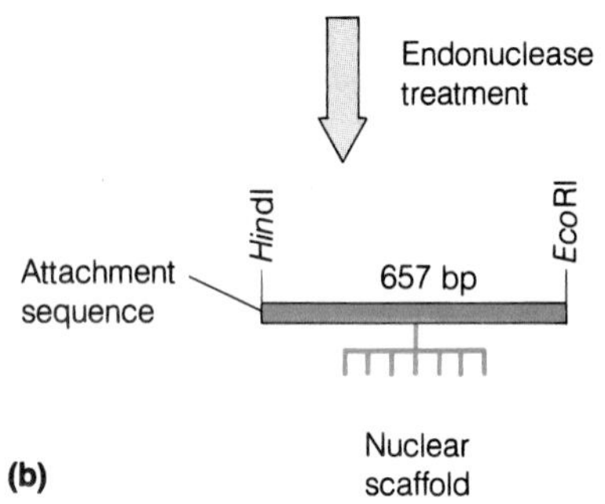

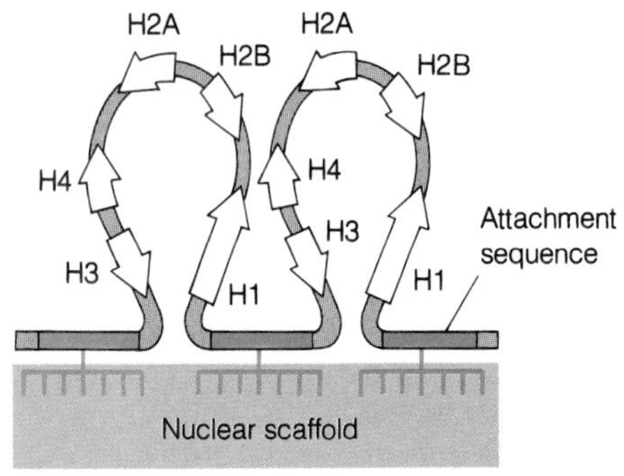

FIGURE **28.13**

Attachment of gene clusters to the nuclear matrix.
(a) A map of the repeating histone gene cluster in *Drosophila,* where each white arrow is a histone gene (arrowheads indicate direction of transcription). A number of restriction sites are also shown. **(b)** If *Drosophila* nuclei are extracted with lithium diiodosalicylate to remove proteins gently and are then digested with a collection of the restriction endonucleases shown, *only* the 657-bp *Hind*I–*Eco*RI DNA fragments are left attached to the matrix. **(c)** The interpretation is that the gene clusters exist in individual loops, the bases of which are tied to the matrix.

The Cell Cycle and DNA Replication in Eukaryotes

THE CELL CYCLE

The processes by which cells divide and DNA is replicated are somewhat more complicated in eukaryotes than in prokaryotes. As we mentioned in Chapter 24, DNA replication in bacteria is an almost continuous process, at least during ex-

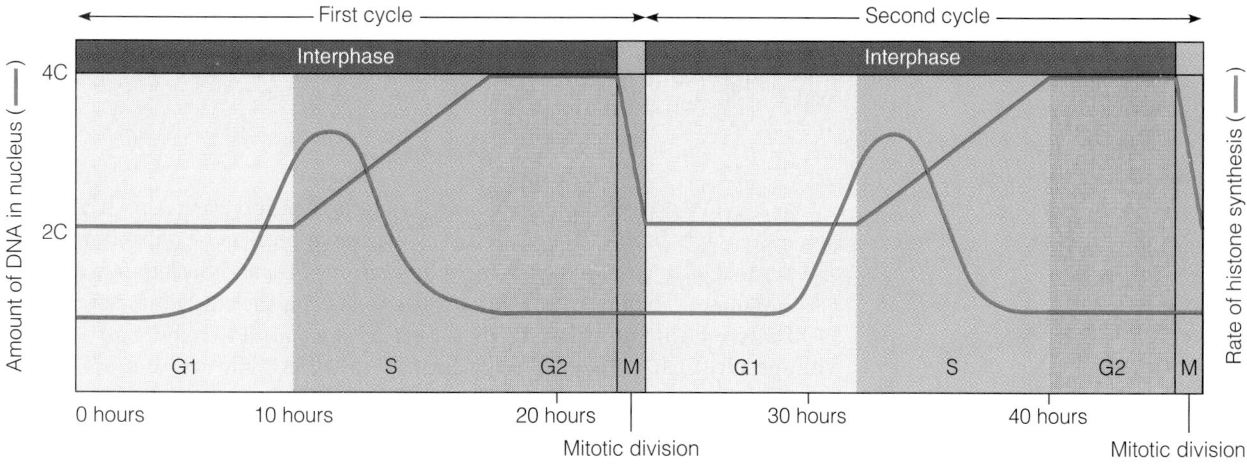

FIGURE **28.14**

The eukaryotic cell cycle. Changes in the amount of DNA (blue line) and rate of histone synthesis (red line) with time during two cell cycles. The DNA content is measured in units of the haploid genome (C). The time scale is typical of many eukaryotic cells.

ponential growth. The somatic cells of eukaryotes, on the other hand, typically divide much less frequently, and some, in certain types of mature tissue, do not divide at all. Those that are dividing in growing tissues exhibit a well-defined *cell cycle,* which is almost always separated into several distinct phases, as shown in Figure 28.14. Let us follow a typical eukaryotic cell through one cycle. We can arbitrarily begin in what is called the *G1 phase* (or first gap phase), following cell division. At this point, the cell contains two copies of each chromosome, the normal diploid state of a eukaryotic cell. This state is indicated in Figure 28.14 by a DNA content of 2C, that is, twice the haploid amount. Sometime late in G1 phase, the commitment to divide is triggered in an as yet unknown fashion. Because division will first require doubling of the DNA content, and the new DNA will need new histones to make chromatin, synthesis of histones is one of the first indications of incipient DNA replication (see Figure 28.14).

The cell then enters synthesis, or *S phase.* During this stage, the DNA is replicated and the histones and nonhistone proteins are deposited on the daughter DNA molecules to reproduce the chromatin structures.

When replication is complete, the cell enters what is called the second gap phase, or *G2 phase.* It now has a DNA content four times the haploid amount (4C). In most eukaryotic cells, the total time required for G1, S, and G2 phases is many hours. During this whole period, which is termed **interphase,** the chromatin is dispersed throughout the nucleus and is actively engaged in transcription.

The eukaryotic cell passes through a cycle of phases: G1, S, G2, and mitosis.

At the end of G2, the cell is ready to enter the process called *mitosis,* during which it divides. Mitosis is a multistage process, and it has been subdivided for descriptive purposes into the phases depicted in Figure 28.15. In *prophase,* the replicated chromosomes condense into the typical *metaphase* chromosome structures so often pictured (see Figure 28.7). The nuclear membrane disintegrates, and the *mitotic spindle* forms. The spindle consists of contractile microtubules that pull pairs of chromatids apart so that the daughter cells each receive identical sets of chromosomes. In *telophase* the nuclear membrane then re-forms about each daughter nucleus, and the cell itself divides. This cellular division is called **cytokinesis.** After division, the chromosomes of the daughter cells decondense, and a new G1 phase begins.

In many tissues of higher organisms, the G1 phase becomes very prolonged after growth and tissue differentiation are complete. The most extreme examples are fully differentiated nerve cells, most of which never divide again in mature organisms. Such nondividing cells are in a permanently arrested G1 phase, which is often called G_0. On the other hand, some specialized stem cells, such as those found in the bone marrow and intestinal epithelium, undergo continuous division throughout the life of the organism. These stem cells continually provide new differentiated cells to replace cells that have been lost or damaged.

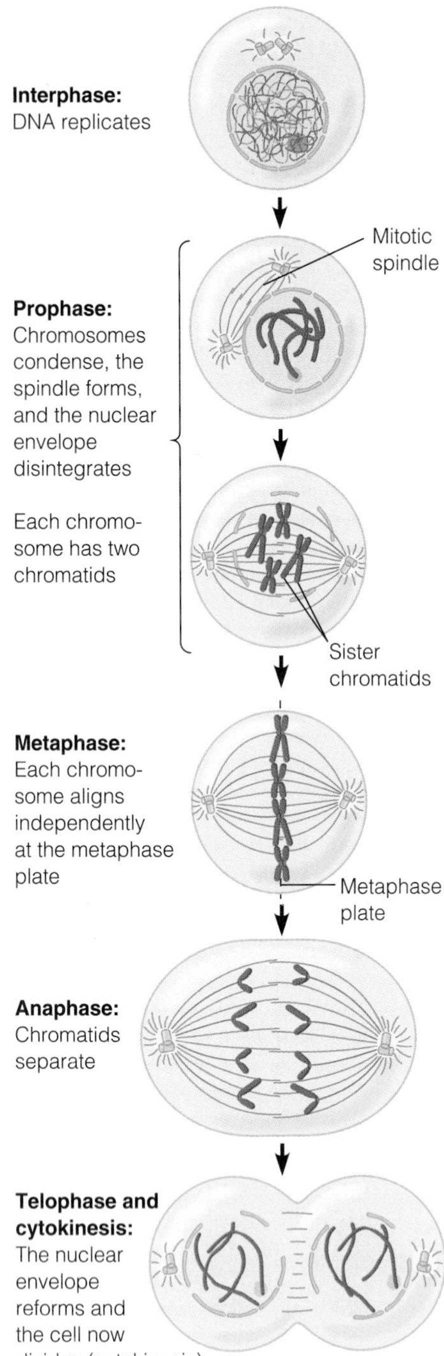

FIGURE **28.15**

Mitosis. The cell entering the pathway was originally diploid. It has undergone DNA replication and is now in G2, with a DNA content of 4C. After the process is complete, each daughter cell will be 2C.

We are only now beginning to understand the chemical events that trigger the successive stages in the cell cycle. The key process appears to be phosphorylation of a number of nuclear proteins (including histone H1) and the key enzymes are a set of kinases. The first of these to be discovered was the cdc2 kinase of yeast, a 34-kDa serine/threonine kinase. Its activation was found to require association with specific small proteins called **cyclins;** in yeast, one cyclin activates cdc2 at the start of S phase, another reactivates it at the beginning of mitosis. In higher eukaryotes, the pattern is similar but more complex: There exist several *cyclin-dependent kinases* and a number of cyclins to associate with them. Each transition in the cell cycle appears to have a unique cyclin/kinase complex as its trigger. A somewhat simplified view of the roles of these proteins in mammalian cells is shown in Figure 28.16. The cyclin-dependent kinase CDK2 is involved in the entrance to S phase, and cdc2, with cyclins A and B, regulate mitosis. cdc2 is phosphorylated at entry to G2, but must be dephosphorylated for mitosis to commence.

These cell cycle triggers are themselves under strict control. Synthesis of cyclins is, at least in part, determined by growth hormones. On the other hand, DNA damage or stalling of replication forks prevents dephosphorylation of cdc2 itself, and this blocks mitosis. In other words, if the machinery to replicate DNA is not free, the cell will not try to divide. This is one *checkpoint* for the cell cycle. Another occurs in response to the accumulation of protein p53 when DNA is damaged (see Chapter 25). p53 activates the transcription of gene *pic1,* the product of which binds to CDKs, blocking the cell in G1 and frequently leading to apoptosis of the cell (see page 1110). Mutations in the gene for p53 have been implicated in *Li-Fraumeni syndrome,* which carries a predisposition to early breast cancer. If p53 is unable to function, potentially cancerous cells with damaged DNA will be able to replicate.

MOLECULAR EVENTS DURING CHROMATIN REPLICATION

Replication of the DNA

We will now consider in molecular detail the events that occur during the S phase of the cell cycle. The basic mechanisms of DNA replication are quite similar in eukaryotes and prokaryotes. As pointed out in Chapter 24, the enzymology is not

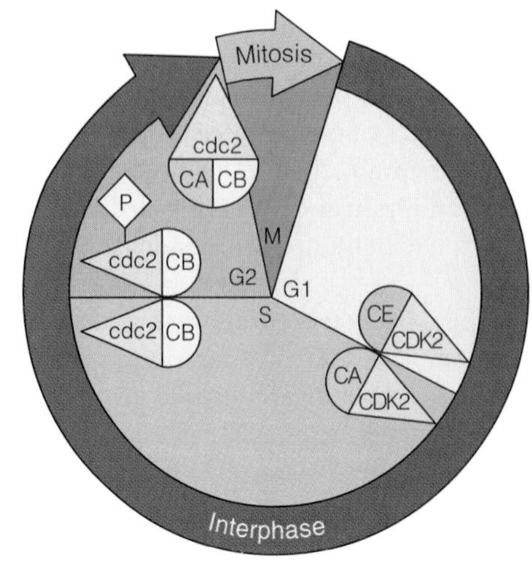

FIGURE **28.16**

The roles of kinases and cyclins in regulating the cell cycle. In this somewhat oversimplified figure, we indicate the participation of CDK2 and cyclins E and A (CE, CA) in triggering S phase, the role of cdc2 and its phosphorylation at the S/G2 boundary, and the participation of cyclins A (CA) and B (CB) in the entry to mitosis.

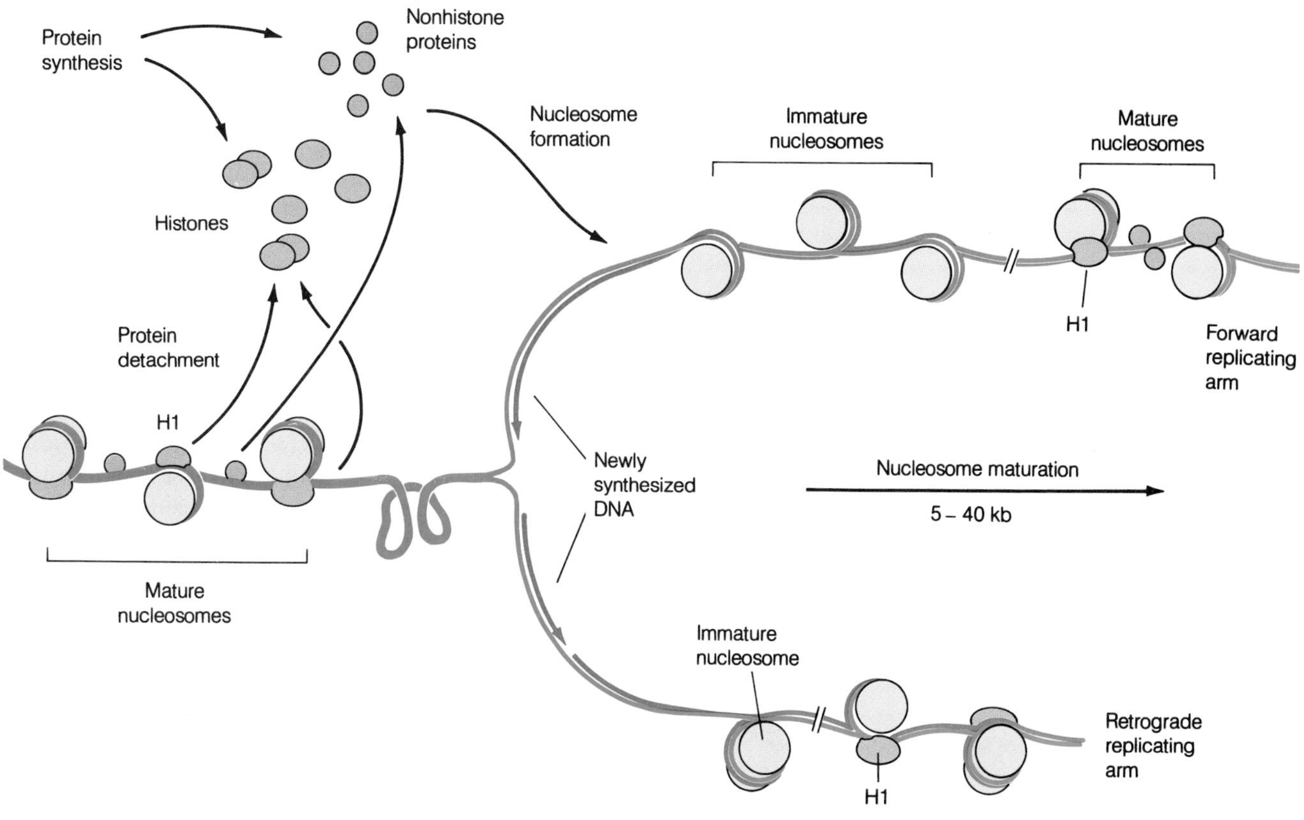

FIGURE **28.17**

Model for chromatin replication. Nucleosomes on the parental DNA are dissociated as the replication fork approaches and are reformed on the newly synthesized daughter molecules. Little is known about this process, except that both old and new histones are used. The chromatin structure matures slowly; the full organization does not seem to be established until many kilobases are behind the moving fork. The figure does not include the polymerase and accessory proteins that must be present at the replication fork.

fundamentally different. In both cases, replication is semiconservative and is continuous on one strand and discontinuous on the other. As in prokaryotes, eukaryotic replication on the retrograde arm entails assembly of short RNA primer molecules, elongation from the primers by a DNA polymerase, and ligation of Okazaki fragments. A significant difference in eukaryotic and prokaryotic DNA replication is in the size of the Okazaki fragments. They are much smaller in eukaryotes—only about 135 bases long, or about the size of the DNA on a nucleosome. As pointed out in Chapter 24, eukaryotic cells contain five DNA polymerases. Three of these—polymerases α, δ, and ε—are used during S phase replication. (See Tables 24.2 and 24.3 for properties of eukaryotic and prokaryotic DNA polymerases.) As in prokaryotes, the replication complex also contains other proteins, including helicases and a number of accessory proteins called *replication factors.*

Assembly of Nucleosomes

Replication of the nuclear genome in eukaryotes involves some special problems. The first is a consequence of the nucleosomal structure of chromatin. The replication machinery must proceed through this complex structure, which is dismantled and then reconstructed on the daughter DNA molecules. It now appears that the arrangement of nucleosomes in chromatin and the placement of nonhistone proteins are far from random. Thus, not only must the DNA be faithfully copied, but a whole highly organized chromatin structure must also be regenerated.

Though there is much that we still do not understand about chromatin replication, we do have some information concerning the assembly of nucleosomes on the daughter strands. As Figure 28.17 indicates, it seems that nucleosomes from the parental chromatin strand are disassembled ahead of the replication fork and are then reassembled on one or the other of the daughter strands. Both preexisting and newly synthesized histones are used in the new nucleosomes, and present evidence indicates that both are deposited randomly on the new daughter

In replication of chromatin, both old and new histones are distributed to both daughter duplexes.

strands. There is mixing of old and new histones in octamers on the daughter strands, but this mixing is not wholly random: $(H3/H4)_2$ tetramers tend to remain intact, as do H2A/H2B dimers. This is what would be expected from in vitro studies: $(H3/H4)_2$ tetramers and H2A/H2B dimers are stable when released from nucleosomes, but the octamer is not.

The much more important question of how the precise arrangement of nucleosomes and nonhistone proteins is reestablished after replication is still not understood. In fact, there probably are changes in this arrangement in some instances. Such changes would account for the observation that differentiation of cells in embryonic development usually occurs at the time of cell division. However, such changes must be *programmed* ones, so their existence only makes the problem still more complex.

Origins of Replication

Perhaps these complications in dealing with the protein component in chromatin account for the fact that the rate of motion of a replication fork is about 10-fold slower in eukaryotes than in prokaryotes. (See Table 24.1 for comparative data.) This slow rate of fork motion, about 100 nucleotides/second, combined with the enormous size of eukaryotic genomes, raises the second serious problem encountered in eukaryotic replication. A typical mammalian chromosome contains roughly 10^8 base pairs. Were it to be replicated from *one* origin (as the *E. coli* chromosome is) at this slow rate, nearly a week would be required. Instead, DNA replication in eukaryotic cells is typically completed in 8–12 hours.

The problem is solved by having *many* origins of replication—as many as several thousand on each chromosome. Replication proceeds bidirectionally from these origins, creating the replication "bubbles" shown in Figure 28.18. These bubbles grow independently until they finally merge and the whole chromosome has been copied. Not all regions of the DNA begin replication at the same time in S phase. Rather, there is a tendency for transcriptionally active regions to replicate early, whereas inactive regions replicate later. The reason for this is not known, although it might simply reflect the fact that transcriptionally active regions are more open and accessible to whatever protein factors initiate the replication process.

The precise nature of the origins of replication is still somewhat obscure, but useful hints have come from studies of yeast. Sequences have been recognized

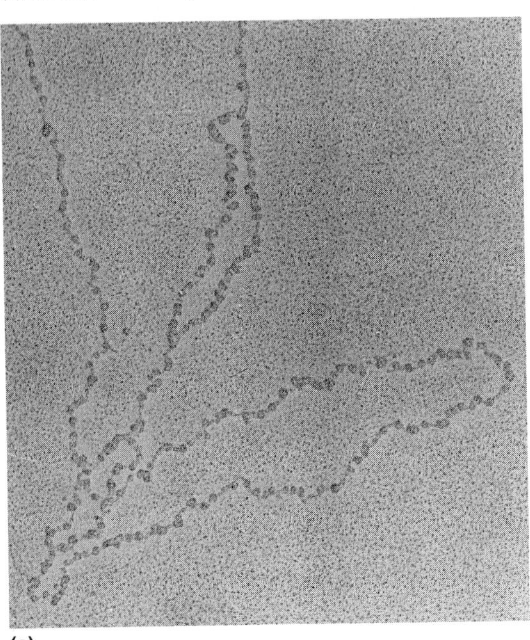

(a)

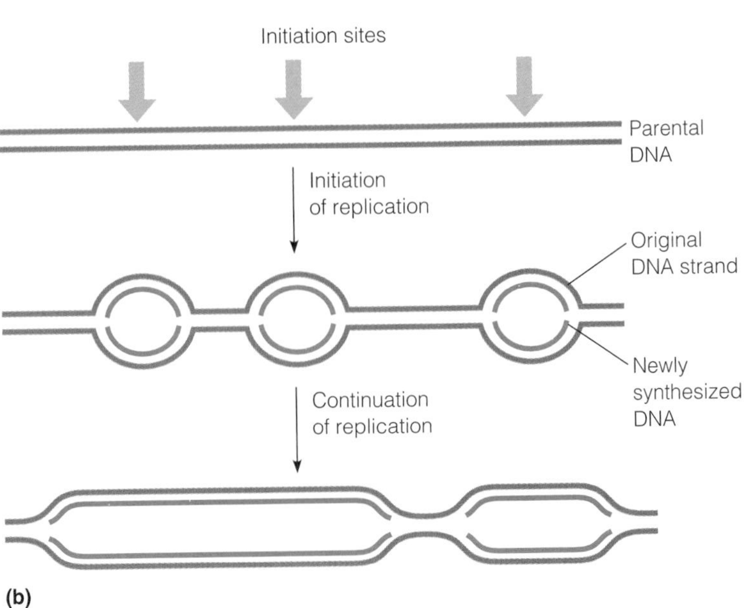

(b)

FIGURE **28.18**

DNA replication in eukaryotes. (a) An electron micrograph of replicating *Drosophila* chromatin showing two replication bubbles. The dots on the fibers are nucleosomes. **(b)** A schematic view of eukaryotic replication.

(a) Courtesy of S. L. McKnight and O. L. Miller, Jr.

that are essential for the replication of plasmids in yeast cells; they are called *autonomously replicating sequences (ARSs)*. ARSs are typically several hundred base pairs in length, with subsequences carrying copies of the following 11-bp consensus sequence, called ACS

5′ TTTTATATTTT 3′

that is absolutely required for function. The yeast ARS also contains other elements that have significant import for replication. There are A/T-rich elements that presumably promote unwinding, and sites to bind protein complexes, including a six-polypeptide *origin binding factor*, needed for initiation of replication.

Perhaps of even greater significance is the presence of sequences that have strong affinity for the nuclear matrix. The importance of these is emphasized by recent studies of mouse cells that reveal the localization of discrete "replication factories" at specific points within the nucleus. Newly synthesized DNA was preferentially tagged with modified nucleotides, which would be recognized by specific antibodies carrying fluorescent labels. High-resolution confocal microscopy reveals "islands" of DNA replication within the nucleus, distinct from separate islands of transcription. The model that emerges suggests that complexes of the polymerization machinery are tethered to the nuclear matrix at particular points and can "spool" large loops of DNA through to replicate. According to this model, the DNA moves, not the machinery.

TABLE 28.2 **Representative telomeric repeat sequences of a number of organisms**

Organism	Repeat[a]
Tetrahymena thermophila (protist)	TTGGGG
Saccharomyces cerevisiae (yeast)	$T(G)_{2-3}(TG)_{1-6}$[b]
Arabidopsis thaliana (plant)	TTTAGGG
Bombyx mori (silkworm)	TTAGG
Human	TTAGGG

[a] Written in 5′→3′ direction.
[b] Yeasts are unusual in having somewhat variable telomeric repeats.

Telomeres

In Chapter 24, we referred to the problem inherent in the replication of linear chromosomes: How to complete the lagging strand? Excision of the primer would necessarily leave a gap that could not be filled by polymerase, and the chromosome would shorten with each round of replication.

In eukaryotes, this problem has been solved by the addition of **telomeres** at the ends of each chromosome. Telomeric DNA consists of simple tandemly repeated sequences like those shown in Table 28.2. Typically, one strand is G-rich, the other C-rich. The G-rich strand forms a 3′-terminal overhang, typically about 15 residues in length. These sequences are repeatedly added to the 3′ termini of chromosomal DNAs by enzymes called *telomerases* (Figure 28.19). This elongation allows room for a primer to bind and initiate lagging-strand synthesis on the other strand, maintaining the approximate length of the chromosome and preventing the loss of coding sequences.

Telomerase adds short DNA segments to the ends of chromosomes.

Note that the telomerase must add nucleotides without the use of a DNA primer. This is probably accomplished through the existence, in each telomerase, of an essential RNA oligonucleotide that is complementary to the telomeric sequence being synthesized, and thus acts as a template. It has been speculated that telomerase is an evolutionary relic of a ribozyme that once served to catalyze DNA synthesis, a process long ago taken over by wholly protein polymerases.

In recent years, it has become clear that telomeres and telomerase may have wide-ranging significance in addition to their practical role in preventing chromosome shortening. First, it is possible that the G-rich strands typical of telomeres may aid in chromosome pairing by forming four-strand structures called G-quartets (see margin). Secondary structure of this type has been observed many times in vitro with oligo-G, and it is known that a specific telomere-binding protein favors the formation of such structures.

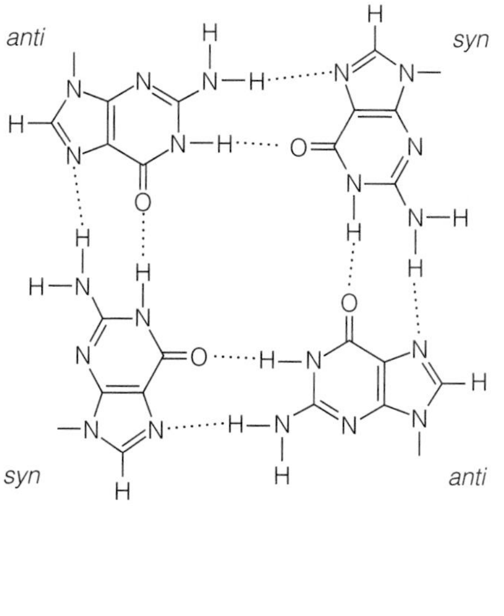

anti syn syn anti

Second, there appears to be a strong correlation between aging, cell senescence, and low levels of telomerase. Conversely, cells in culture can be "immortalized" by introduction of active telomerase genes. These observations, together with the discovery that malignant tumor cells invariably have high levels of telomerase, have spurred intense interest in telomerase inhibition as a possible cancer therapy.

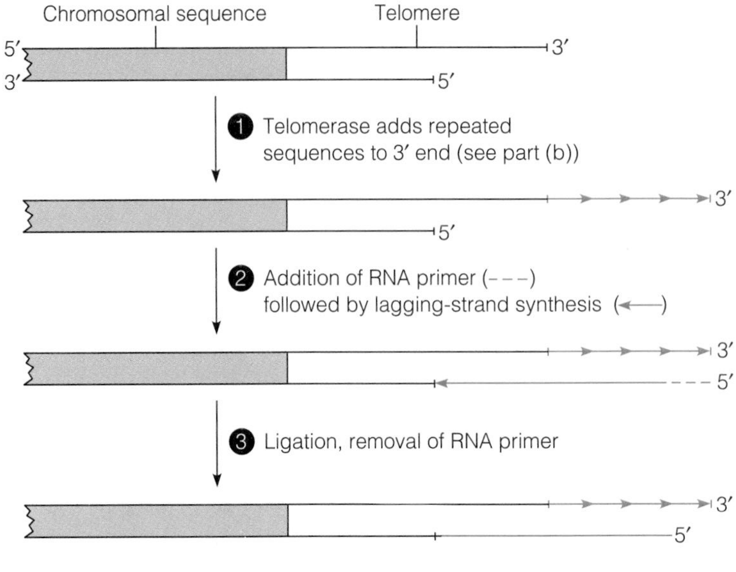

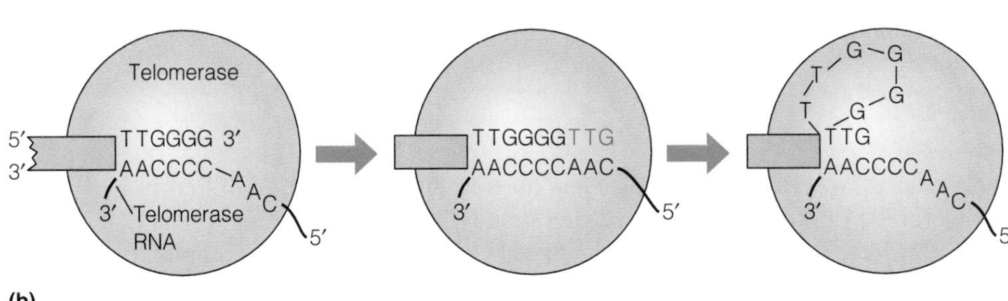

FIGURE **28.19**

Extension of telomeric DNA by telomerase. **(a)** The overall reaction. Telomerase adds simple repeat sequences to the 3′ end of telomeric DNA, by the mechanism shown in part (b). Addition of an RNA primer allows lagging-strand synthesis, followed by ligation and RNA removal. **(b)** Proposed action of telomerase. The RNA carried by the telomerase matches the 3′ DNA end, and allows its extension. DNA loop formation then permits further extension. Dissociation of telomerase and its RNA follows after several rounds of extension.

Transcription and Its Control in Eukaryotic Cells

Transcription in eukaryotes is a much more complex process than in prokaryotes. Not only is there much more discrimination in what is to be transcribed and what is not, but this transcription is precisely programmed during development and tissue differentiation. Furthermore, the transcription machinery must somehow deal with the complicated levels of structure in eukaryotic chromatin.

Eukaryotes have three kinds of nuclear RNA polymerases, each requiring protein factors to initiate transcription.

Reflecting this complexity is the fact that eukaryotic cells have several different RNA polymerases, each with a specialized function. They have already been mentioned in Chapter 26 and are reviewed in Table 28.3. In addition to the special RNA polymerases that function in mitochondria and chloroplasts, there are three enzymes that transcribe various portions of the nuclear genome.

Each of these nuclear RNA polymerases is a complex, multisubunit enzyme, containing about a dozen subunits. The complete enzymes are difficult to isolate in pure form, and it is often hard to determine which subunits are essential, and

TABLE **28.3 Comparative properties of eukaryotic RNA polymerases**

Polymerase	Location	RNAs Synthesized
I	Nucleus (nucleolus)	Pre-rRNA (except 5S)
II	Nucleus	Pre-mRNA, some small nuclear RNAs
III	Nucleus	Pre-tRNA, 5S rRNA, other small RNAs
Mitochondrial[a]	Mitochondrion	Mitochondrial
Chloroplast[a]	Chloroplast	Chloroplast

[a]These are quite similar to the prokaryotic RNA polymerase.

for what purposes. One other important feature distinguishes eukaryotic RNA polymerases from their prokaryotic counterparts: All of the eukaryotic enzymes require additional protein factors (**transcription factors**) in order to bind to a promoter and initiate transcription. Indeed, the study of transcription factors is a major part of the study of eukaryotic transcription. As we shall see, such factors seem to play a major role in determining selectivity in transcription of genes. We now consider the function of each of the three nuclear polymerases.

RNA POLYMERASE I: TRANSCRIPTION OF THE MAJOR RIBOSOMAL RNA GENES

The eukaryotic ribosome contains four rRNA molecules. The small subunit has an 18S rRNA, whereas the large subunit contains 28S, 5.8S, and 5S rRNA molecules. Of these, the 28S, 18S, and 5.8S subunits are all produced from an initial 45S pre-rRNA transcript, and it is the special function of RNA polymerase I (pol I) to carry out this transcription.

Pol I transcribes the major ribosomal genes; pol III, small RNA genes; and pol II, structural genes and a few small RNA genes.

Like other eukaryotic RNA polymerases, RNA polymerase I is a complex enzyme, containing 13 subunits totaling over 600,000 daltons. At least two transcription factors are known to be required. However, because only a single kind of gene is transcribed, there is no need for the elaborate apparatus, including multiple regulatory sites and multiple transcription factors, that we will find characteristic for pol II transcription.

The nucleolus is the site of ribosomal subunit assembly in eukaryotes. The gene for the 45S pre-rRNA is present in the nucleolus as multiple, tandemly arranged copies, as shown in Figure 28.20a. After transcription, the 45S pre-rRNA is processed to yield 18S, 5.8S, and 28S rRNA molecules. About 6800 nucleotides are discarded in this process. The rRNAs are then combined with 5S rRNA from other regions of the nucleus and ribosomal proteins synthesized in the cytosol. The resulting ribosomal subunits are exported from the nucleolus back into the cytosol.

FIGURE **28.20**

Transcription and processing of the major ribosomal RNAs in eukaryotes. The genes exist in tandem copies, separated by nontranscribed spacers. **(a)** The 45S transcripts first produced are processed by removal of portions shown in tan, to yield the 18S, 5.8S, and 28S products. These are then assembled into ribosomal subunits, through addition of proteins. **(b)** Electron micrograph of spread nucleolar rRNA genes undergoing transcription. Tandemly arranged genes are being transcribed from bottom to top.

(b) Courtesy of O. L. Miller, Jr. and B. Beaty.

Nontranscribed spacer
5.8S
5.8S
18S
28S
18S
28S
DNA with tandemly arranged genes for mammalian 45S rRNA
Transcription by RNA polymerase I
Mammalian 45S pre-rRNA (13,000 nucleotides)
Multistep rRNA processing (discard 6800 nucleotides)
5.8S
5.8S
18S
28S
18S
28S
~30 proteins
~30 proteins
5S rRNA
~50 proteins
5S rRNA
~50 proteins
40S
60S
40S
60S
Ribosomal subunits
(a)

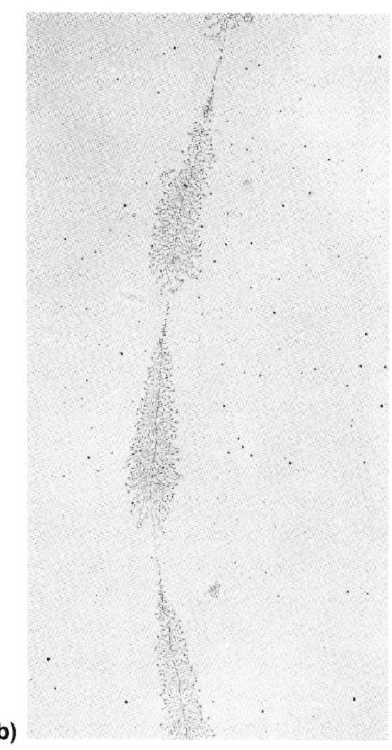

(b)

Transcription of the tandem copies of 45S pre-rRNA can be beautifully visualized in the electron microscope (Figure 28.20b). The structure of the nucleolar chromatin has been a subject of some controversy, but it appears most likely that nucleosomes are not present, at least in the transcribed regions. This absence of nucleosomes may be a specific chromatin modification to allow rapid and continuous transcription of these genes.

In some lower eukaryotes, like the protozoan *Tetrahymena,* the 28S rRNA contains an intron near its 3′ end. Excision of this intron and splicing of the RNA are carried out by a remarkable process in which the RNA *itself* acts as the catalyst, via the series of reactions described in Chapter 11 (see in particular Figure 11.30 on page 397). In the higher eukaryotes, little is yet known about the mecha-

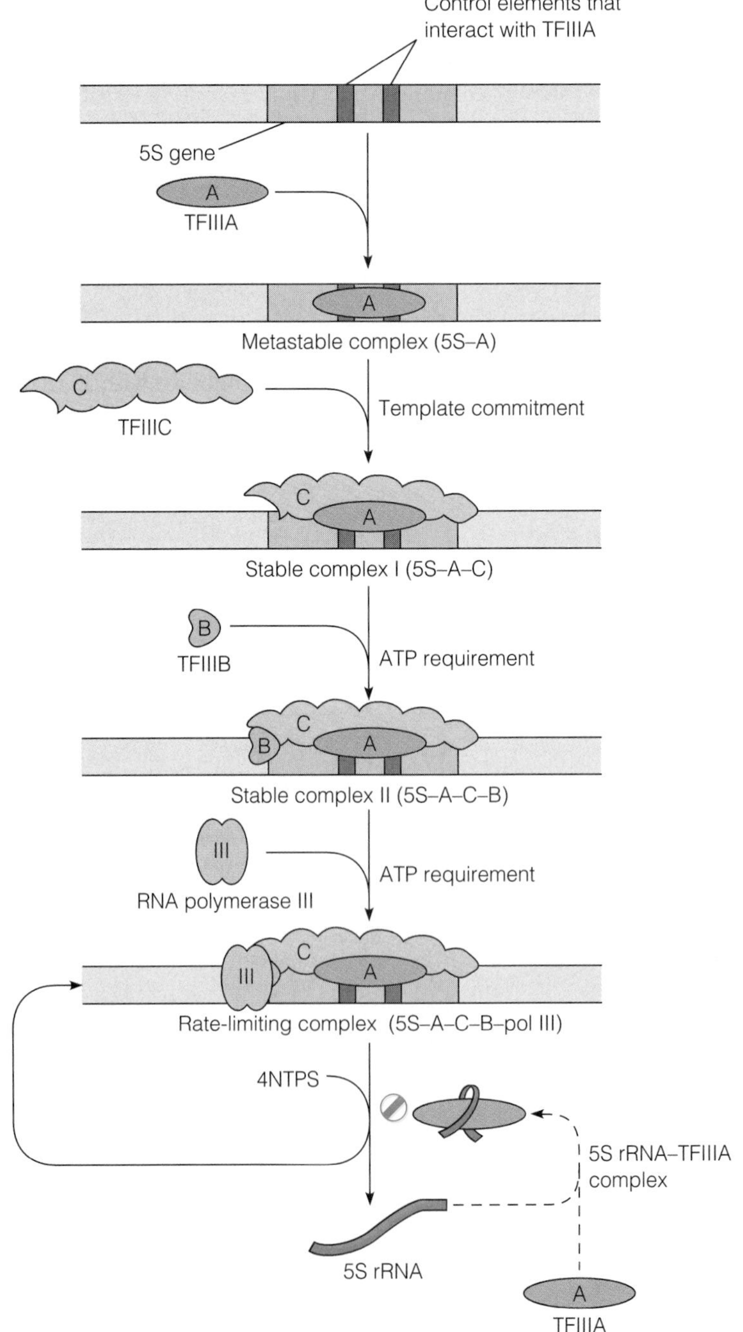

FIGURE **28.21**

Preparation of a 5S rRNA gene for transcription. At least the three protein factors shown, plus RNA polymerase III, must assemble on the gene before transcription can occur. TFIIIA must bind to the gene before factors TFIIIC and TFIIIB can bind. Once the stable complex II has been formed, it will recycle with pol III to produce many RNA copies. An excess of 5S rRNA will form a complex with TFIIIA, inhibiting further transcription.

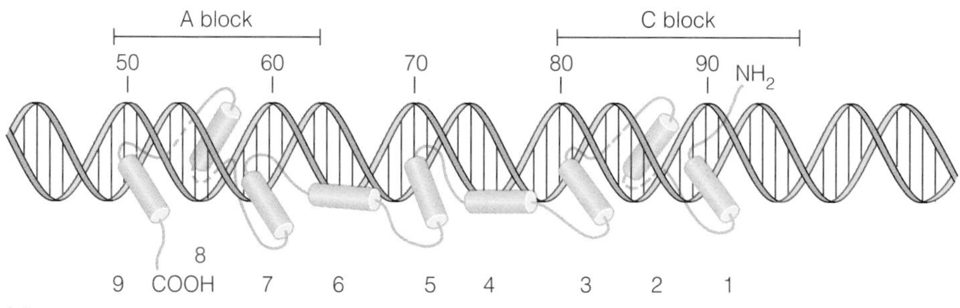

nism of rRNA processing, the assembly of ribosomes, or the coordination of synthesis of ribosomal proteins and ribosomal RNA.

RNA POLYMERASE III: TRANSCRIPTION OF SMALL RNA GENES

RNA polymerase III (pol III) is the largest and most complex of the eukaryotic RNA polymerases. It involves 14 subunits, totaling 700,000 daltons. All of the genes it transcribes share certain features. They are small, they are not translated into proteins, and they are unique in that their transcription is regulated by certain sequences that lie *within* the transcribed region. The major targets for pol III are the genes for all the tRNAs and for the 5S ribosomal RNA. Like the major ribosomal genes described in the previous section, these small genes are present in multiple copies, but they are usually not grouped together in tandem arrays, nor are they localized in one region of the nucleus. Rather, they are scattered over the genome and throughout the nucleus.

Of all the genes transcribed by pol III, the most thoroughly studied are those for 5S ribosomal RNA. In vitro experiments have revealed that at least three protein factors in addition to polymerase III are needed for expression of the 5S rRNA genes. Two of these transcription factors (TFIIIB and TFIIIC) appear to participate in the transcription of tRNA genes as well, but one, called TFIIIA, is specific for the 5S genes. The interaction of the three transcription factors, the polymerase, and the gene is shown in Figure 28.21. The molecule of TFIIIA makes contact with DNA over a length of about 40 bp. Recognition of the sequence occurs in two blocks of about 12 bp each, at either end of the contact region. This somehow makes the gene accessible to TFIIIB, TFIIIC, and polymerase III. TFIIIA can also complex with 5S RNA. This propensity limits 5S RNA production when the RNA product is in excess, by removing TFIIIA from availability to bind to DNA.

TFIIIA is an example of an abundant class of DNA-binding proteins, in which metal-binding *zinc fingers* make contact with and identify DNA sequences (Figure 28.22). There are now many known zinc finger proteins, and the crystal structures of some of their DNA complexes have been studied. The fingers fit into the major grooves of the DNA, as do the helices in helix–turn–helix and leucine zipper proteins (Figure 28.23; see also Chapter 26).

Although TFIIIA is a monomer and TFIIIB is dimeric, TFIIIC is an enormous complex, involving six polypeptide chains and covering the whole 5S rRNA or tRNA gene. Just how the polymerase manages to repeatedly transcribe through such a protein complex is still unclear. Nonetheless, once pol III is attached, it can produce multiple transcripts before dissociating.

In later sections we shall encounter a variety of transcription factors, for the control of eukaryotic gene expression is almost entirely dependent on the site-specific interaction of these proteins with DNA. A typical transcription factor contains a DNA-binding domain and one or more regulatory domains, which can interact with other nuclear proteins to convey regulatory signals.

In addition to the zinc finger proteins, two other major classes of transcriptional factors are known, and these are depicted schematically in Figure 28.23b

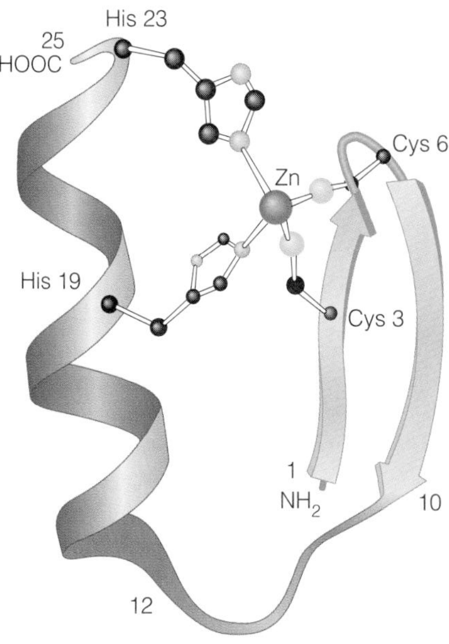

FIGURE **28.22**

Zinc fingers. (a) The transcription factor TFIIIA binds to the 5S RNA gene via zinc fingers inserted into the major groove. The two major recognition regions, A block and C block, are contacted by fingers 7–9 and 1–3, respectively. **(b)** Structure of a zinc finger. The structure shown is for a synthetic polypeptide with sequences found in a zinc finger protein. The α-helix and β-sheet motifs are shown in deep blue and green, respectively. The two histidine residues and two cysteines that coordinate the zinc (red) are depicted in detail. Sulfur is in yellow.

Adapted from M. S. Lee, G. P. Gippert, K. V. Soman, D. A. Case, and P. Wright, *Science* (1989) 245:635. © 1989 AAAS.

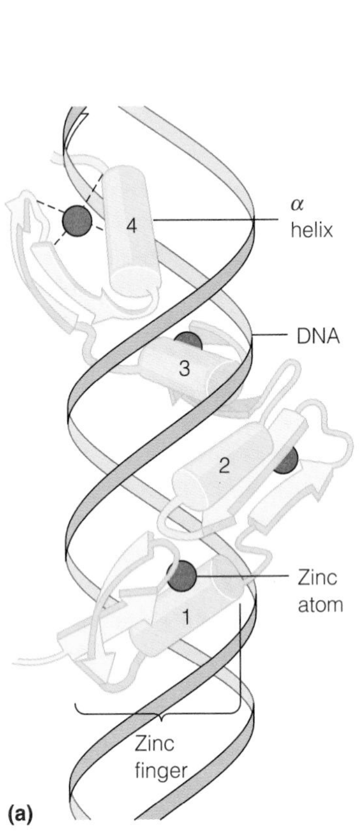

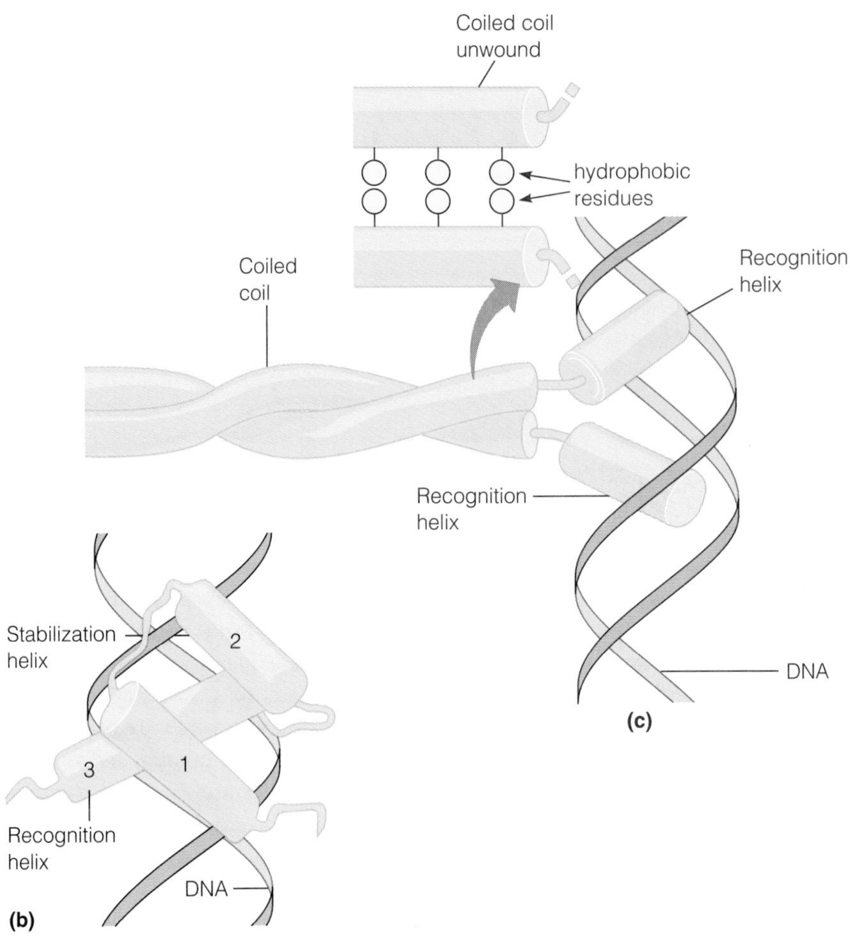

FIGURE **28.23**

Structures of three common types of DNA-binding motifs from eukaryotic transcription factors. (a) The zinc finger motif, showing the way the fingers fit along the major groove. **(b)** A helix–turn–helix motif. Note that the helix–turn–helix fits into the DNA major groove in a way similar to that of a zinc finger. **(c)** A leucine zipper protein. The coiled coils are held together by hydrophobic residues, usually leucines. At their ends, they present a pair of recognition helices to the DNA major grooves.

and c. In the helix–turn–helix proteins, one α helix (called the recognition helix) lies in the major groove of the DNA, its side chains making specific contacts with the DNA bases. A quite different class of DNA-binding proteins are called leucine zipper proteins. These are dimers, held together in a coiled-coil structure by hydrophobic interactions. They typically exhibit a regular pattern (7-fold periodicity) of leucine or other hydrophobic residues in the helical tail regions, which favors side-by-side hydrophobic interaction. The N-terminal regions are recognition helices, lying in adjacent major grooves. The special feature of leucine zipper proteins is that they can form either homologous or heterologous dimers, thus allowing many combinatorial pairings between transcription factors.

RNA POLYMERASE II: TRANSCRIPTION OF STRUCTURAL GENES

All of the structural genes (those coding for protein products) in the eukaryotic cell are transcribed by polymerase II. This enzyme also transcribes some of the small nuclear RNAs involved in splicing (see page 1096). Like other RNA polymerases, pol II is a complex, multisubunit enzyme. However, not even its 12 subunits are sufficient to allow pol II to initiate transcription on a eukaryotic promoter. To form the minimal complex capable of initiation, at least five additional protein factors are needed, as shown in Figure 28.24 and listed in Table 28.4. The order of addition is not known with certainty but starts, in the example shown, with binding to the *TATA box*, the most common initiation signal (named for a common sequence). This usually lies about 20–30 bp upstream from the start site. The start site itself usually involves a pyrimidine-rich region, termed *Inr.* Note that the *minimal* unit involves the TATA-binding protein (TBP), whereas in vivo formation of the complex probably always uses TFIID, a multisubunit structure incor-

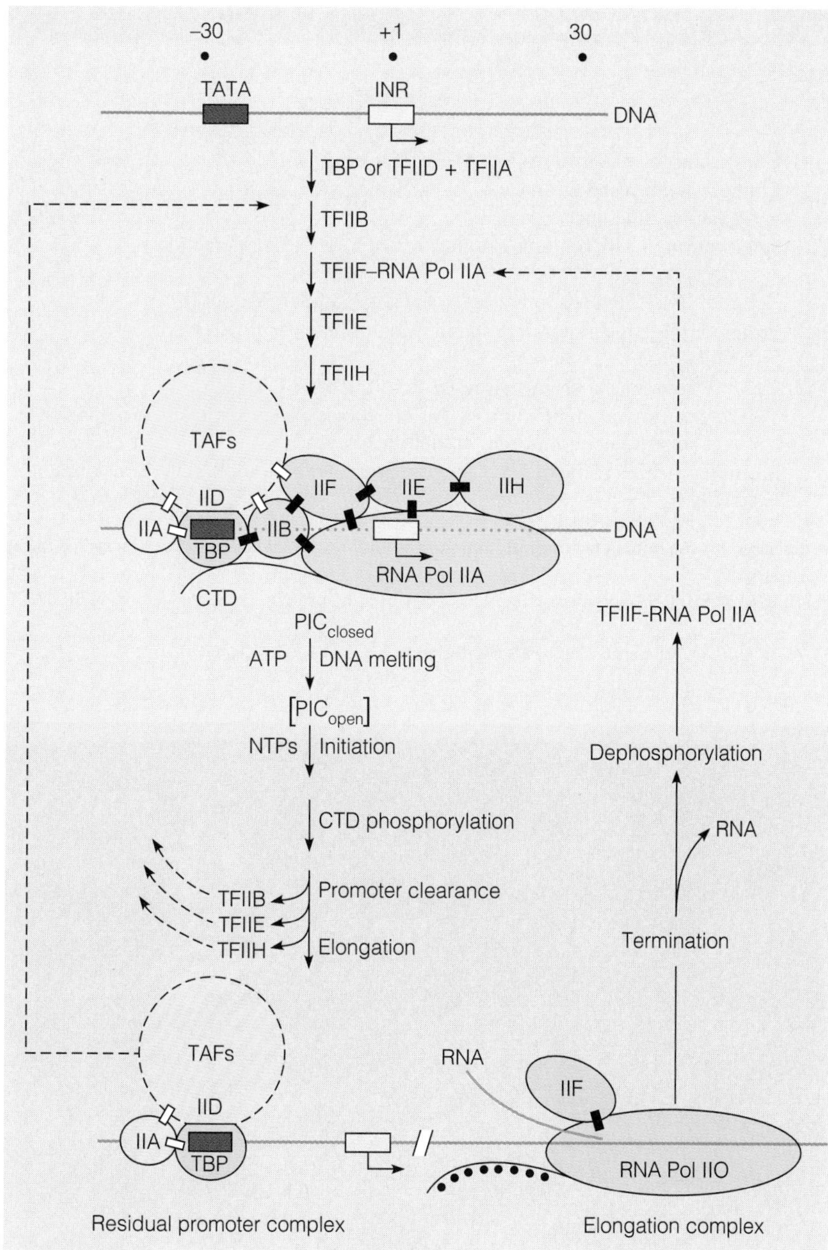

FIGURE **28.24**

A model for the formation of the minimal preinitiation complex (PIC) for pol II on a TATA promoter. In the simplest situation, binding of TATA-binding protein (TBP) initiates the sequence. Alternatively, in vivo TFIID, which includes both TBP and associated factors (TAFs), is used. This will also result in binding of TFIIA. The series of dots indicates phosphorylation of the c-terminal domain (CTD) of pol II. Phosphorylation is necessary for release of the enzyme from the initiation site.

porating both the TATA binding protein and *TATA-binding associated factors* (TAFs). The TAFs interact with *activation factors* associated with upstream sites on specific genes, thereby providing communication for gene regulation. The structure of the complex containing the TATA site, TBP, TFIIB, and TFIIA has recently been deduced from diffraction and modeling studies. A remarkable feature that emerges is the pronounced kink induced in the DNA around the TATA site (Figure 28.25).

A DNA sequence that binds a transcription factor is called a *control element.* The function of the TATA control element appears to be to fix the starting point for transcription. But eukaryotic pol II promoters contain much more complex and diverse control elements, a few of which are listed in Table 28.5. Recall that pol II transcribes structural genes—genes for proteins. Because transcription of these genes must be both tissue specific and developmentally specific, it requires a great deal of regulation. In some cases, transcription must also respond to special signal molecules like hormones. It is common to find, upstream from the TATA

TABLE **28.4 General transcription initiation factors from human cells**[a,b]

Factor		Number of Subunits	Molecular Weight (kDa)	Function
TFIID	TBP	1	38	Core promoter recognition (TATA); TFIIB recruitment
	TAFs	12	15–250	Core promoter recognition (non-TATA elements); positive and negative regulatory functions
TFIIA		3	12, 19, 35	Stabilization of TBP binding; stabilization of TAF–DNA interactions; antirepression functions
TFIIB		1	35	RNA pol II–TFIIF recruitment; start-site selection by RNA pol II
TFIIF		2	30, 74	Promoter targeting of pol II; destabilization of nonspecific RNA pol II–DNA interactions
RNA pol II		12	10–220	Catalytic functions in RNA synthesis; recruitment of TFIIE
TFIIE		2	34, 57	TFIIH recruitment; modulation of TFIIH helicase, ATPase, and kinase activities; direct enhancement of promoter melting (?)
TFIIH		9	35–89	Promoter melting using helicase activity; promoter clearance (?) by CTD kinase activity

[a] The subunit compositions and polypeptide sizes are those described for the human factors, but homologues for virtually all have also been identified in rat, *Drosophila*, and yeast.

[b] Abbreviations used: CTD, carboxy-terminal domain of pol II; RNA pol II, RNA polymerase II; TAFs, TATA-binding protein–associated factors; TBP, TATA-binding protein.

Source: From R. G. Roeder, *Trends Biochem. Sci.* (1996) 21:327–335. © 1996 with permission of Elsevier Science.

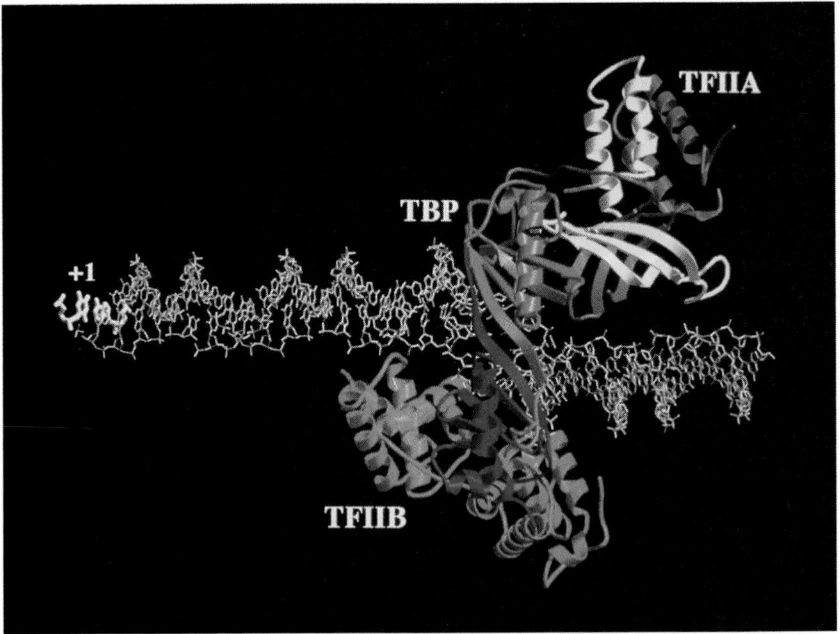

FIGURE **28.25**

Computationally assembled model of the TFIIA–TBP–TFIIB-promoter complex based on crystal structures of TBP–TFIIB–TATA and TFIIA–TBP–TATA complexes. The amino- and carboxy-terminal direct-repeat domains of TBP are shown, respectively, in blue and purple; the amino- and carboxy-terminal direct-repeat domains of core TFIIB are shown, respectively, in red and magenta; and portions of the large and small subunits of yeast TFIIA are shown, respectively, in green and yellow. The transcription initiation start site (+1) is indicated in white. The view of the complex is from the top, showing TBP sitting astride the distorted TATA element and adjacent, but laterally displaced, upstream and downstream DNA segments (in standard B form) extending, respectively, leftward and below the plane of the figure and rightward and below the plane of the figure.

From the work of S. K. Burley, *Trends Biochem. Sci.* (1996) 21:327, fig. 2, with permission from Elsevier Science.

box, promoter sequences that bind a variety of *trans-acting factors.* These are proteins synthesized on other genes that enhance or repress the transcription of the gene in question. Examples of typical promoter organizations are shown in Figure 28.26. The control elements shown are the general elements listed in Table 28.5.

Often, trans-acting factors binding at sequences far removed from the promoter itself—by as much as several kilobase pairs—can influence transcription. Such sequences are called **enhancers,** and their mode of action appears to involve DNA looping, perhaps mediated by nucleosomes, which can bring enhancer-bound proteins into close physical contact with proteins bound to the promoter. Some of the transcription factors shown in Table 28.5 can bind in either promoter or enhancer regions. It is now evident that the TAF proteins shown in Figure 28.24

Transcription can be modified by binding of trans-acting factors, either in the promoter or in distant enhancers.

TABLE **28.5 Some important pol II control elements and their corresponding transcription factors**

Sequence Name	Consensus Sequence	Transcription Factor(s)	Comment
Some General Promoter and Enhancer Elements			
TATA box	TATAAAA	TBP, TFIID	This is the most common core promoter element
CAAT box	GGCCAATCT	CP1	A common upstream element
GC box	GGGCGG	SP1	Often found in TATA-less promoters
Octamer	ATTTGCAT	Oct1, Oct2	Oct1, Oct2 contain homeo domains
Some Special Promoter and Enhancer Elements			
HSE	CNNGAANNTCCNNG	Heat shock factor	Involved in heat-shock response
GRE	TGGTACAAATGTTCT	Glucocorticoid receptor	Protein binds glucocorticoid hormones
TRE	CAGGGACGTGACCGCA	Thyroid receptor	Protein binds thyroid hormones

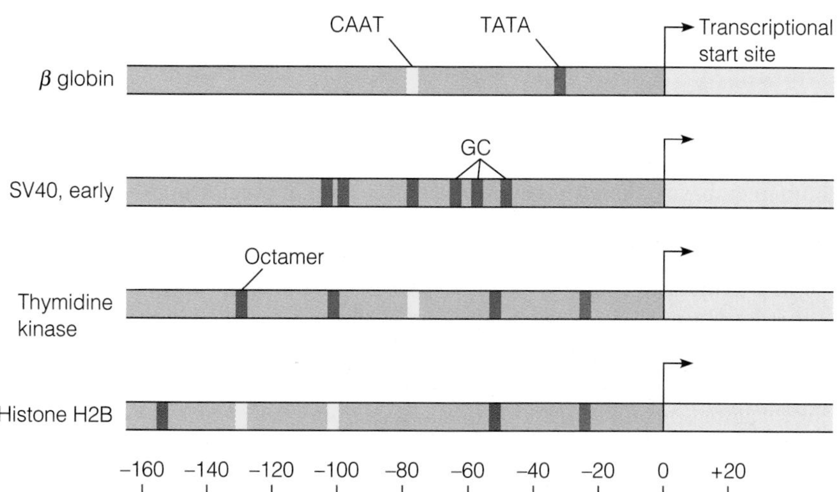

FIGURE **28.26**

Structures of a few typical eukaryotic promoters. The colored boxes represent different regulatory elements: TATA = red, GC = blue, CAAT = yellow, and octamer = purple.

Adapted in part from B. Lewin, *Genes IV* (Oxford: Oxford University Press, 1990).

can act as intermediates between activator or repressor bound to enhancer and the core transcription complex (see Figure 28.27).

The trans-acting factors tend to fall into a small number of classes, each defined by the kind of structural domain that interacts with the DNA. Schematic drawings of three of the better-known types are given in Figure 28.23.

CHROMATIN STRUCTURE AND TRANSCRIPTION

The complex interplay of transcription factors and polymerases we have described occurs not on naked DNA but on chromatin. The chromatin structure presents two major problems: (1) How can the transcription factors and initiation complex bind to DNA in the presence of nucleosomes? (2) How can the actively transcribing polymerase pass through arrays of nucleosomes? This has been an area of intense research interest. We present the current views of the first problem with a few examples and then briefly comment on the second problem.

The Problem of Initiation

As an example, consider the human β globin genes, which were described in Chapter 7. Although present in every human cell, these genes are expressed *only* in the erythroid cells and in a fixed developmental sequence. In embryonic cells that have not yet begun synthesis of any globin, the chromatin of the β globin gene cluster appears much the same as in any other cell in the embryo and is quite densely covered with nucleosomes. But when differentiation of these cells commits them

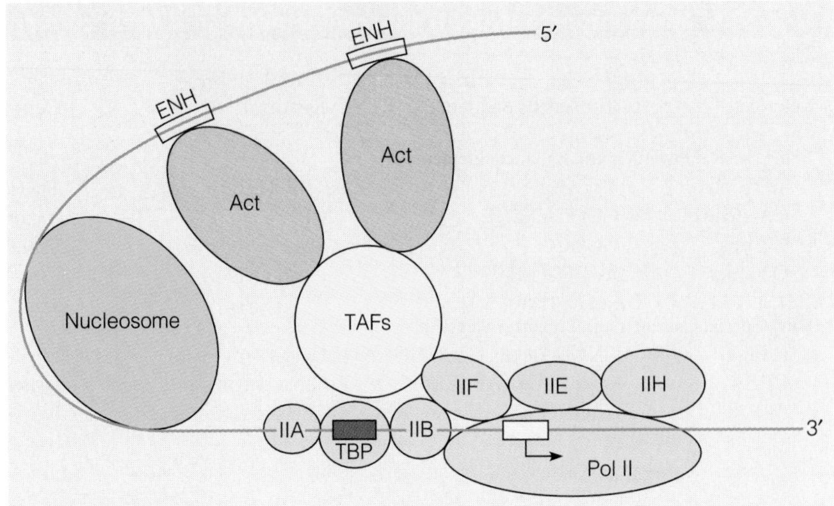

FIGURE **28.27**

A schematic representation of how DNA looping (perhaps mediated by nucleosomes) can bring enhancer-bound activator (or repressor) proteins into contact with TAFs associated with the core complex.

Nuclease hypersensitive sites disrupt chromatin to allow initiation.

to globin synthesis, the whole β globin domain undergoes changes in chromatin structure. One of the changes is the appearance of *hypersensitive sites,* regions particularly susceptible to digestion by nucleases. At early stages in developing human embryos, these sites appear in the 5′ flanking regions of the embryonic genes, which are the first to be transcribed. Later, hypersensitive sites shift to the 5′ flanks of the adult genes. It is now clear that many of these sites represent regions a few tens or hundreds of base pairs in length in which nucleosomes have either been removed or "remodeled" so as to make the DNA contained therein more accessible. They provide points at which transcription factors and other trans-acting proteins can gain access to promoters and enhancers, thereby allowing the initiation and stimulation of transcription.

How are hypersensitive sites established in previously unresponsive genes? In some cases, for example the globin genes, it seems that the chromatin structure is rearranged at the time of replication. In other instances, protein factors seem to be able to interfere with chromatin structure at specific loci, opening hypersensitive sites.

Particularly interesting examples of this kind are found in hormonal regulation of transcription. A well-studied case involves the genes for the chicken egg-white proteins—ovalbumin, ovomucoid, and lysozyme. Transcription of these genes occurs only in the tubular cells of the hen oviduct. Even in immature chicks the genomic domain containing the ovalbumin gene appears to have a somewhat different chromatin structure in these oviduct cells than in other tissues. But only on stimulation by the sex hormone estrogen (either on sexual maturation of the chick or following hormone administration) does transcription of ovalbumin genes commence. Specific hypersensitive sites 5′ to some of the egg-white protein genes are opened by the presence of estrogen. Withdrawal of administered hormone from an immature chick leads to loss of hypersensitive sites and an immediate cessation of transcription of the genes.

There are many other examples of hormonal control of transcription. In each case the target cells contain specific proteins that are *hormone receptors.* When these proteins bind the hormone, they become capable of interacting with specific DNA sites or with nonhistone regulatory proteins bound to such sites. Thus, both positive and negative regulation is possible. In some cases, the hormone-binding receptor acts as a positive regulatory factor, for example, by binding to an enhancer element. In others, the hormone-binding receptor can interact with a repressor protein to augment or relieve the repression. Recent evidence suggests that the latter model may describe the response of chicken oviduct cells to estrogens.

Chromatin Remodeling

How are hypersensitive sites generated, and how is chromatin structure altered to make the DNA accessible? Only recently have pieces of evidence begun to fit together to answer these questions. First was the discovery, in yeast and then in higher eukaryotes, of **chromatin remodeling factors.** These are proteins that enable promoter regions to be able to accept the complex and bulky machinery depicted in Figure 28.24. The SWI/SNF complex from yeast and the NURF complex from *Drosophila* are probably the best studied; significantly, both require ATP hydrolysis to carry out their task. Exactly what such factors do is still unclear. They seem not to remove nucleosomes but rather to "open" them in some fashion.

Another and perhaps equally important role is played by histone acetyltransferases and deacetylases. It has long been known that the histones of the nucleosome core are subject to acetylation at specific lysine residues in the N-terminal tails (see Figure 28.28). Furthermore, high levels of acetylation have long been correlated with high transcriptional activity, and vice versa. What is new is the discovery that a number of proteins recruited to the initiation complex by activators and TAFs (and, indeed, some of these proteins themselves) have histone acetylase activity. The current belief is that acetylation of histones in promoter nucleosomes contributes to the loosening of chromatin structure in these regions. The fact that specific transcription factors are involved in this process may provide the long-sought explanation for how the chromatin of *specific* genes can be targeted for disruption.

Transcriptional Elongation

Formation of the preinitiation complex (Figure 28.24) is followed, in the presence of ribonucleoside triphosphates and ATP, by melting of a short region and initiation of transcription. The C-terminal tail of pol II becomes strongly phosphorylated at this point, and transcription begins, with a helicase activity clearing the way.

A number of the core transcription factors are released, and pol II, together with TFIIF, moves along the DNA. A residual complex, containing TBP, TFIIA, TAFs, and probably activator proteins, remains at the start site, ready to initiate another round.

At this point the polymerase also acquires several special *elongation factors.* Some of these appear to assist the enzyme in traversing *pause sites* in the DNA. In in vitro experiments, using naked pol II, transcription is relatively slow and interrupted by frequent pauses, especially in T-rich regions. The phenomenon seems to be generally similar to that described in Chapter 26 for prokaryotic transcription.

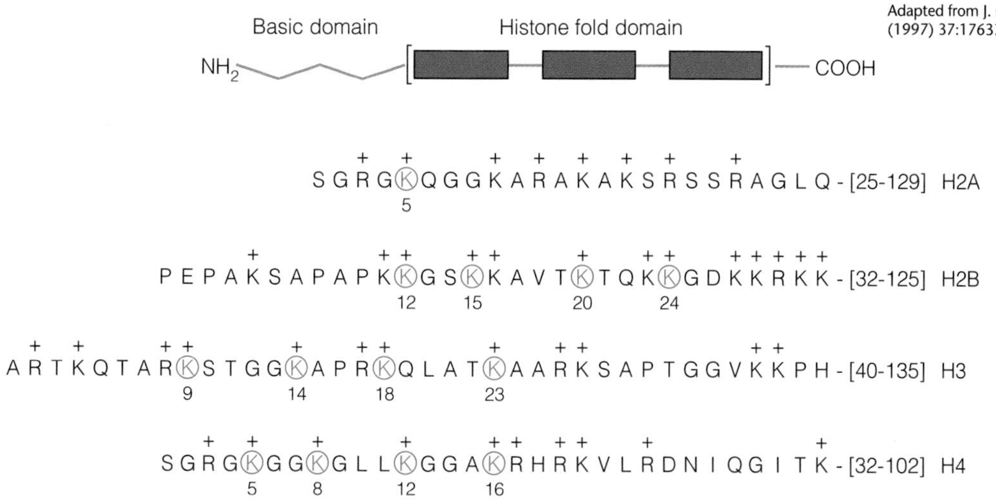

Basic domain
Histone fold domain
NH_2
COOH

S G R G (K) Q G G K A R A K A K S R S S R A G L Q - [25-129] H2A (5)

P E P A K S A P A P K (K) G S (K) K A V T (K) T Q K (K) G D K K R K K - [32-125] H2B (12, 15, 20, 24)

A R T K Q T A R (K) S T G G (K) A P R (K) Q L A T (K) A A R K S A P T G G V K K P H - [40-135] H3 (9, 14, 18, 23)

S G R G (K) G G (K) G L L (K) G G A (K) R H R K V L R D N I Q G I T K - [32-102] H4 (5, 8, 12, 16)

FIGURE **28.28**

Acetylation of core histones. The general structure of each of the four core histones involves a helical "histone fold" domain plus an unstructured, highly basic N-terminal domain. Acetylation in nuclei occurs exclusively in the N-terminal domains, at the highly conserved sites indicated in red.

Adapted from J. C. Hansen, C. Tse, and A. P. Wolffe, *Biochemistry* (1997) 37:17637–17641. © 1997 American Chemical Society.

The presence of elongation factors assists the enzyme in passing such sites. Nucleosomes form even more important obstacles to the progress of a polymerase II along the DNA. Although some prokaryotic polymerases can pass through nucleosomal arrays in in vitro studies, pol II is entirely blocked unless accessory proteins are present. These include nucleosome remodeling factors and a specific elongation factor called FACT.

Pol II can transcribe through nucleosome arrays.

Just how pol II transcribes through nucleosome is still something of a mystery. Do the nucleosomes unfold and re-form as the polymerase passes? Are they temporarily displaced? Current evidence favors temporary displacement, but the issue is far from settled. One factor that may play a role in such displacement is the development of positive superhelical torsion ahead of a moving polymerase. A polymerase moving along a helical template must either continually rotate about the DNA or build up positive supercoils ahead (overwinding) to compensate for the unwinding it is doing. Such torsion would tend to destabilize nucleosomes, because they contain negatively wrapped DNA.

TERMINATION OF TRANSCRIPTION

Even the termination of mRNA transcription is different in eukaryotes. Whereas the prokaryotic RNA polymerase recognizes terminator signals, which sometimes function with the aid of the ρ protein, the eukaryotic polymerase II usually continues to transcribe well past the end of the gene. In doing so, it passes through one or more AATAAA signals, which lie beyond the 3′ end of the coding region (Figure 28.29). The pre-mRNA, carrying this signal as AAUAAA, is then cleaved by a special endonuclease that recognizes the signal and cuts at a site 11 to 30 residues 3′ to it. At this point, a tail of polyriboadenylic acid, poly(A), as many as 200 bases long, is added by a special non-template-directed polymerase. The function of the poly(A) tails of eukaryotic mRNAs is unknown. We know that they cannot be essential for all messages, because some mRNAs (for example, most histone mRNAs in higher eukaryotes) do not have them. However, they may relate to message stability, for the tail-less messages typically have much shorter lifetimes in the nucleus.

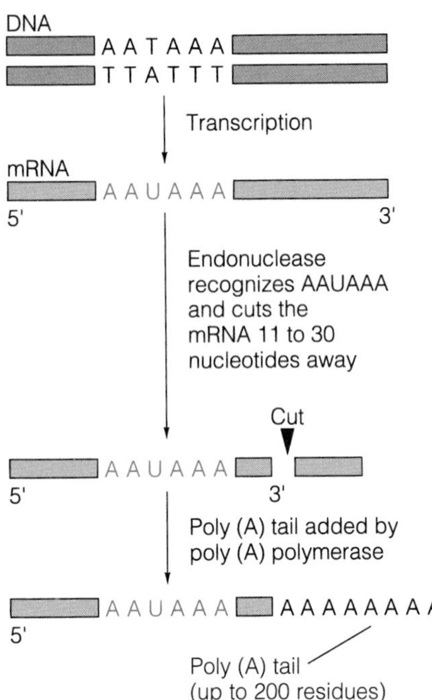

FIGURE **28.29**

Termination of transcription in eukaryotes: addition of poly(A) tails. There is an AATAAA sequence near the 3′ end of most eukaryotic genes. When this is transcribed to AAUAAA, it provides a signal for endonuclease cleavage and poly(A) tail addition.

Processing of Eukaryotic Messenger RNA

There are significant differences in the ways that messenger RNAs for protein-coding genes are produced and processed in prokaryotic and eukaryotic cells. Recall that prokaryotic mRNAs are synthesized on the bacterial nucleoid in direct contact with the cytosol and are *immediately* available for translation. A specific nucleotide sequence at the 5′ end recognizes a site on the prokaryotic ribosomal RNA, allowing attachment of the ribosome and initiation of translation, often even before transcription of the message is completed.

In eukaryotes, the mRNA is produced in the nucleus and must be exported into the cytosol for translation. Furthermore, the initial product of transcription (*pre-mRNA*) includes all of the introns and substantial flanking regions; the introns must be removed before correct translation can occur. Finally, there is no ribosomal attachment sequence like the Shine–Dalgarno sequence in prokaryotes. For all these reasons, eukaryotic mRNA requires extensive processing before it can be used as a protein template. This processing takes place while mRNA is still in the nucleus.

Eukaryotic mRNA processing involves capping, splicing, and editing.

CAPPING

The first modification occurs at the 5′ end of the pre-mRNA. A GTP residue is added in *reverse* orientation. Together with the first two nucleotides of the chain, it forms

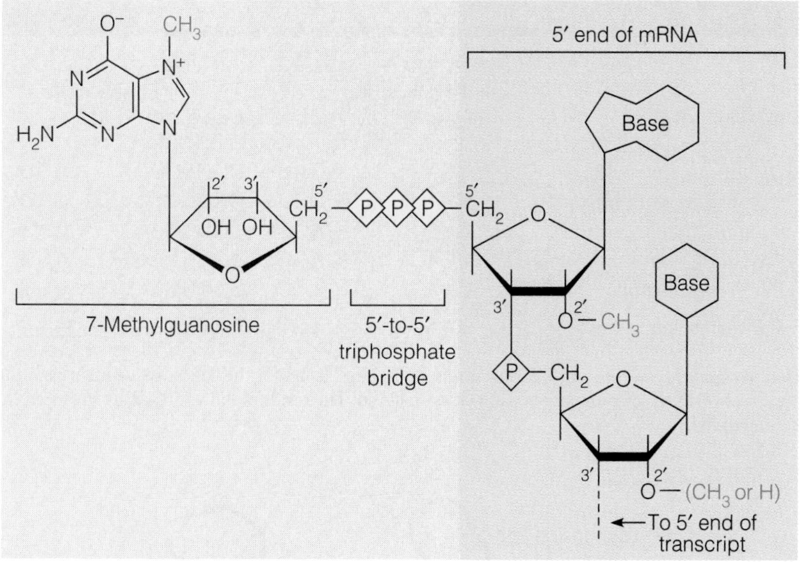

FIGURE **28.30**

Overall structure of a fully processed eukaryotic message, including the cap site. Details of the 5′ cap region are shown. Methyl groups that are added are in red.

what is known as a *cap* (Figure 28.30). The cap is further decorated by addition of methyl groups to the N-7 position of the guanine and to one or two sugar hydroxyl groups of the cap nucleotides. This cap structure serves to position the mRNA on the ribosome for translation.

SPLICING

After being capped, the pre-mRNA becomes complexed with a number of *small nuclear ribonucleoprotein particles (snRNPs),* which are themselves complexes of *small nuclear RNAs (snRNAs)* and special splicing proteins. The snRNP–pre-mRNA complex is called a **spliceosome,** and it is here that the most elegant part of the processing takes place—the cutting and splicing that is necessary to excise introns from the pre-mRNA. In forming a spliceosome, snRNAs recognize and bind intron–exon splice sites by means of complementary sequences (Figure 28.31). Table 28.6 shows some representative splice site sequences and the consensus

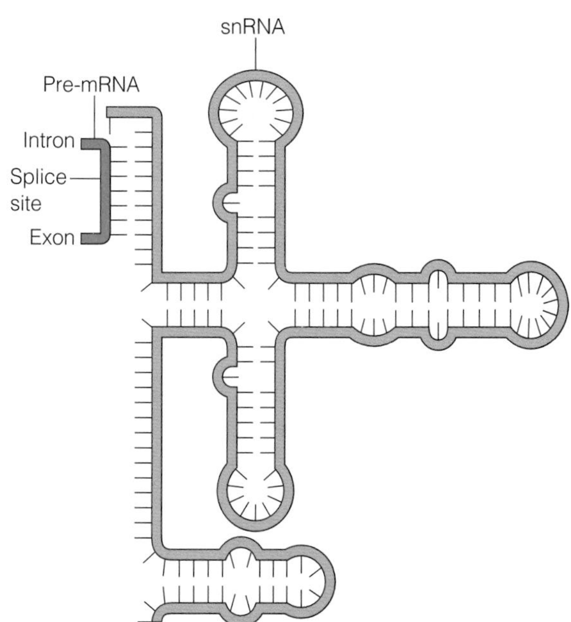

FIGURE **28.31**

Structure of a small nuclear RNA (snRNA). Human U1 RNA is shown, together with the intron–exon boundary region to which it binds in forming the spliceosome.

TABLE 28.6 Representative sequences at splice junctions

Protein, Intron	5′ E1 splice site Exon ↓	Intron	E2 splice site 3′ ↓ Exon
Ovalbumin, intron 3	···UCAG	GUACAG···UGUAUUCAG	UGUG
β globin, human, intron 1	···CGAG	GUUGGU···CACCCUUAG	GCUG
β globin, human, intron 2	···CAGG	GUGAGU···CCUCCACAG	CUCC
Immunoglobin I, L-VI	···UCAG	GUCAGC···UGUUUCGAG	GGGC
Rat preproinsulin	···CAAG	GUAAGC···CCCUGGCAG	UGGC
Consensus sequences[a]	—AG	GURAGY···YYYYY—AG	——

[a]Here R stands for purine and Y for pyrimidine. Residues listed for the consensus sequence are those found in two-thirds or more of over 100 cases analyzed. The residues shown in red are invariant in all cases analyzed.

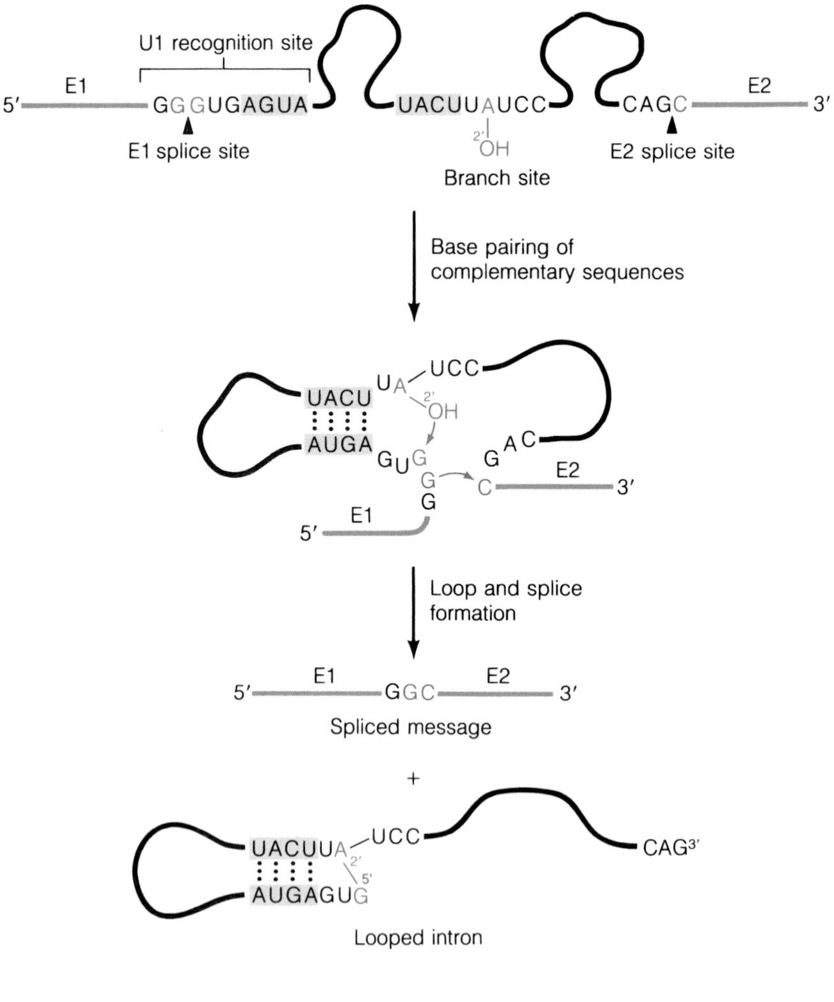

FIGURE **28.32**

A schematic view of the proposed mechanism for mRNA splicing. Exons (E1 and E2) are indicated by red lines, the intron by a black line or sequence. The E1 splice site, presumably with the aid of the small RNA U1, pairs with a sequence at the branch site to form a loop. The 2′ hydroxyl on the branch site A reacts with a G residue (blue) at the 3′ end of the exon 1 (the E1 splice site). This frees the adjacent G (red) to transfer to the 5′ end of the following exon (E2). The products are a spliced message and a looped intron "lariat" structure, which is then degraded.

sequences common to most introns. A schematic view of the chemical aspects of the splicing mechanism is shown in Figure 28.32.

Excision of a single intron involves assembling and disassembling a spliceosome. We still do not know all the details of splicing, but Figure 28.33 depicts a reasonable molecular mechanism. The sequence of reactions starts with attachment of the U1 snRNP to the G site at the 5′ end of the intron. The U2 snRNP then attaches at the branch site. With continued assembly of the spliceosome, including addition of several more snRNPs, the lariat loop in the intron is formed and the two exons are joined. Splicing has now been accomplished, and the prod-

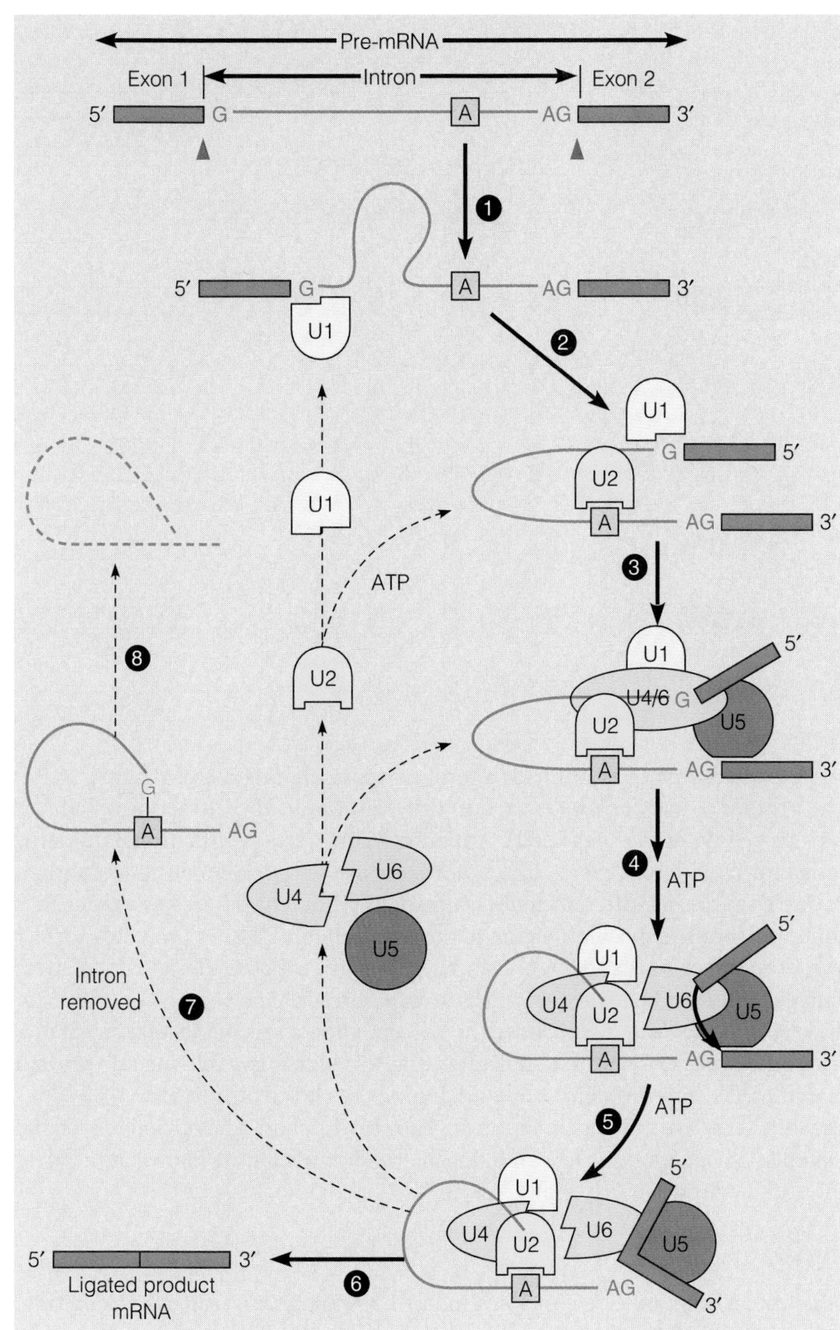

FIGURE **28.33**

A proposed model for the mechanism of splicing. The pre-mRNA plus assorted snRNPs assemble and disassemble a spliceosome, which carries out the splicing reaction. The various snRNPs are designated U1, U2, and so on. In step 1 U1 is bound, which together with U2 binding (step 2) leads to a looped structure. Factors U4/6 and U5 then bind (step 3) and cleavage and transfer then occur (steps 4, 5). The spliceosome disassembles, releasing the ligated product (6) and the looped intron (7). This is degraded into small oligonucleotides (step 8).

ucts—a ligated mRNA and a looped intron—are released. As the spliceosome disintegrates, the looped intron is degraded, and the mRNA is exported from the nucleus. Exactly how this happens is still not known with certainty, but it may be in conjunction with some of the snRNP proteins.

ALTERNATIVE SPLICING

One of the more remarkable and significant discoveries in recent years is the phenomenon of **alternative splicing.** Some gene transcripts may be spliced in different ways, to include or exclude certain exons, in different tissues of an organism or at different developmental stages. We mentioned one example of this phenomenon in Chapter 7, where we pointed out that the heavy chains of immunoglobulins may or may not carry a hydrophobic membrane-binding domain.

Alternative splicing allows one gene to specify several proteins.

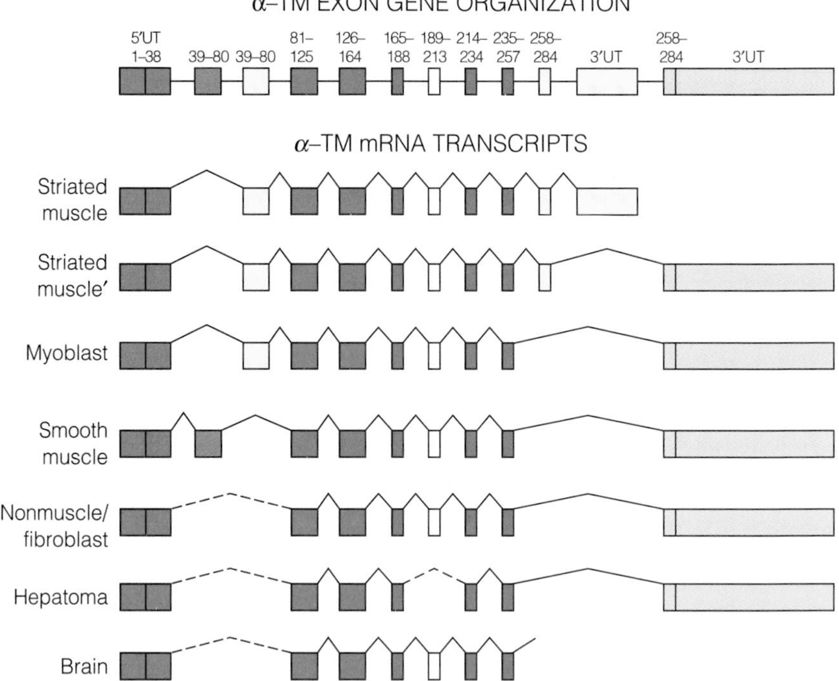

FIGURE **28.34**

α-Tropomyosin gene organization (rat) and seven alternative splicing pathways. Exons (red, constitutive; green, smooth muscle–specific; yellow, striated muscle–specific; white, variable) are indicated with their encoded amino acids (numbered). Experimentally documented splicing pathways (solid lines) and others (dotted lines) inferred from nuclease protection mapping are shown. The smooth (SM) and striated (STR) exons encoding amino acid residues 39–80 are mutually exclusive, and there are alternative 3′-terminal exons as well. UT signifies untranslated regions.

Adapted from R. E. Breithart et al., *Annu. Rev. Biochem.* (1987) 56:467–495. © 1987 Annual Reviews.

A more dramatic example of alternative splicing is shown in Figure 28.34. The protein *α-tropomyosin* is used in different kinds of contractile systems in various cell types (see page 263). Apparently, the need for functional domains coded for by different exons differs from one use of α-tropomyosin to another. Rather than having different genes expressed in different tissues, a single gene is employed, but the specific splicing patterns in different tissues provide a variety of α-tropomyosins. As the figure shows, there are two positions at which alternative choices can be made for which exon to splice in. The 3′ member of each of these pairs is the *default exon;* it will be chosen unless a specific cellular signal dictates otherwise. Current data indicates that this signal is a cell-specific protein that interacts with the spliceosome, changing the choice of splice site. The economy of alternative splicing, in terms of genome size, should be obvious. Over 100 examples of eukaryotic genes exhibiting alternative splicing are known at the present, and the list will surely grow.

EDITING

Recently, an even more remarkable kind of RNA processing, called **RNA editing,** which changes the sequence of a message, has been discovered. Editing was first found in mitochondrial mRNAs of some unicellular eukaryotes. This kind of editing involves insertion or deletion of uridine residues in messages during the processing steps. Apparently, the insertions are made by a kind of reverse splicing mechanism, and only at certain points. Small RNAs, called *guide RNAs,* are required for the process. The role of this modification remains obscure, but it seems to be connected to regulation of gene expression. A quite different type of editing involves deamination of adenosines to inosines in mRNA. The role of this is not known, but it can make one message code several proteins.

Neither alternative splicing nor editing *formally* contradicts the basic concept that gene sequence carries the cell's information. After all, the sites for splicing and editing are themselves dictated by the gene sequence, as are the sequences of the small nuclear RNAs and the guide RNAs. Nevertheless, these phenomena challenge our long-cherished idea that each protein amino acid sequence is specified by a single DNA sequence.

Translation in Eukaryotes

COMPARISON WITH PROKARYOTIC MECHANISMS

The mechanism for translating messenger RNA into protein in eukaryotic cells is basically the same as in prokaryotes: Messenger RNA is read by ribosomes. There are, however, significant differences in both the ribosomes and the details of the translational mechanism. Because we have considered the prokaryotic translational system quite thoroughly in Chapter 27, we concentrate here on the differences.

Ribosomes

The ribosome and its subunits are larger in eukaryotes; 40S and 60S subunits combine to form a functional 80S ribosome. These correspond to the prokaryotic 30S, 50S, and 70S particles. The large eukaryotic subunit contains 28S (26S in yeast), 5S, and 5.8S RNAs, the last having no counterpart in prokaryotes. The small subunit (40S) has an 18S RNA (17S in yeast). The subunits contain more proteins than the corresponding prokaryotic particles do.

Initiation

Table 28.7 compares the translational protein factors required by eukaryotes with those required by prokaryotes. Initiation requires many more protein factors in eukaryotes—compare Figure 28.35 with Figure 27.20. At least 11 proteins are needed, rather than just 3, as in prokaryotes, and some of the eukaryotic proteins are complex, multisubunit structures. Some of the initiation factors attach to the ribosomal subunits and others to mRNA (see left-hand side of Figure 28.35). The major initiation factor, eIF2, forms a complex with tRNA (see right-hand side of the figure). This factor is eventually recycled via a cyclic GDP–GTP exchange called the eIF2 cycle, which requires the factor eIF2B.

In eukaryotes, translational initiation is more complex and requires more protein factors than in prokaryotes, and translational control is more important.

Although eukaryotic proteins are initiated like their prokaryotic counterparts, with a special Met-tRNA reading an AUG codon, the methionine is not formylated. The mRNA is first aligned correctly on the 40S subunit by the 5′ cap, rather than by the Shine–Dalgarno sequence used by prokaryotes. The ribosomal

TABLE 28.7 Comparison of translational protein factors in prokaryotes and eukaryotes

Prokaryotic Factor	Eukaryotic Factor	Function
Initiation Factors		
IF1	eIF1, eIF1A	Prepare mRNA for proper attachment to ribosome
IF2	eIF2, eIF2B	Helps bind Met-tRNA
IF3	eIF3, eIF4C	First to bind and prepare 40S for subsequent factors
	eIF4A, eIF4B, eIF4F	As eIF1, eIF1A
	eIF5	Helps dissociate eIF2, eIF3, eIF4C
	eIF6	Helps dissociate 60S subunit from inactive ribosomes
Elongation Factors		
EF-Tu	eEF1α	Delivery of aminoacyl tRNA to ribosomes
EF-Ts	eEF1$\beta\gamma$	Aids in recycling factor above
EF-G	eEF2	Translocation factor
Release Factors		
RF1	eRF	Release of complete polypeptide chain
RF2		
RF3		

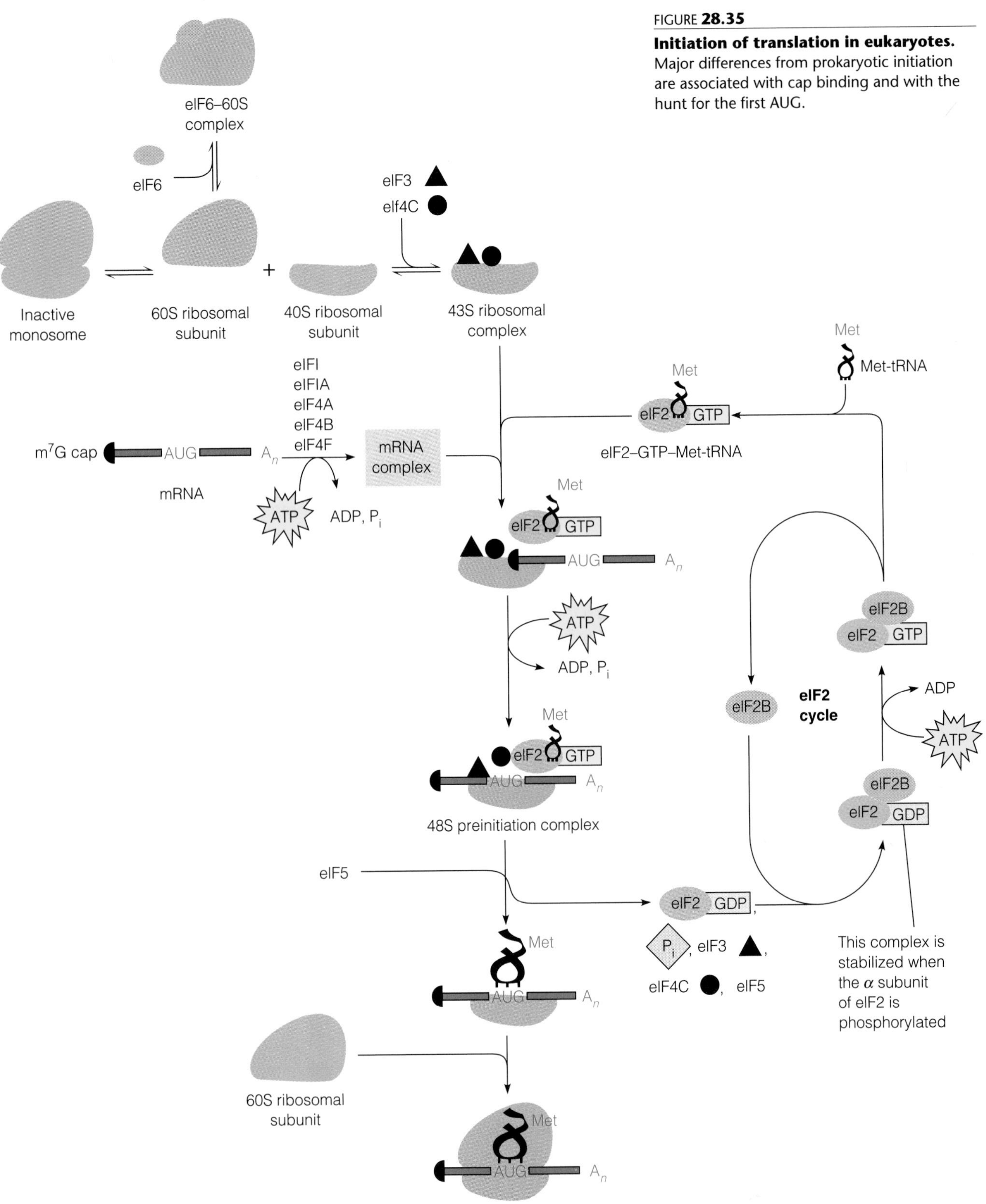

FIGURE **28.35**

Initiation of translation in eukaryotes. Major differences from prokaryotic initiation are associated with cap binding and with the hunt for the first AUG.

subunit then scans along the mRNA (an ATP-dependent process) until the first AUG is found. At this point the initiation factors are released, and the 60S subunit is attached to begin translation.

Elongation and Termination

The mechanism of chain elongation appears to be very similar in eukaryotes and prokaryotes. Each of the three elongation factors—EF-Tu, EF-Ts, and EF-G—has a eukaryotic equivalent, which plays a similar role (see Table 28.7). Eukaryotic chain termination, in contrast to prokaryotic termination, requires only one protein factor—eRF. This single factor can recognize all three stop codons (UAA, UAG, and UGA).

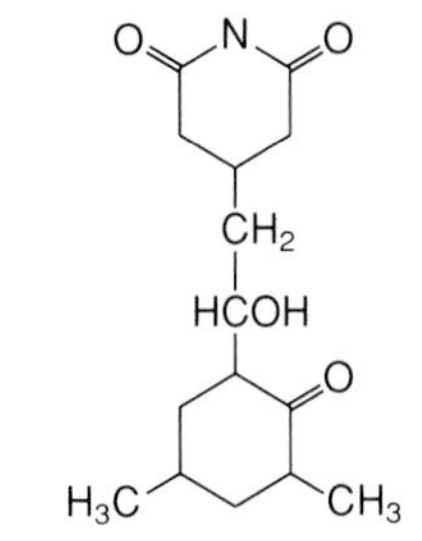

INHIBITORS OF TRANSLATION

A number of the common inhibitors of prokaryotic translation are also effective in eukaryotic cells. They include pactamycin, tetracycline, and puromycin. There are also inhibitors that are effective *only* in eukaryotes. Two important ones are *cycloheximide* and *diphtheria toxin*. Cycloheximide inhibits the peptidyltransferase activity of the eukaryotic ribosome and is often used in biochemical studies when processes must be studied in the absence of protein synthesis. Diphtheria toxin is an enzyme, coded for by a bacteriophage that is lysogenic in the bacterium *Corynebacterium diphtheriae.* It catalyzes a reaction in which NAD^+ adds an *ADP ribose* group to a specially modified histidine in the translocation factor eEF2, the eukaryotic equivalent of EF-G (Figure 28.36). Because the toxin is a catalyst, minute amounts can irreversibly block a cell's protein synthetic machinery; pure diphtheria toxin is one of the most deadly substances known.

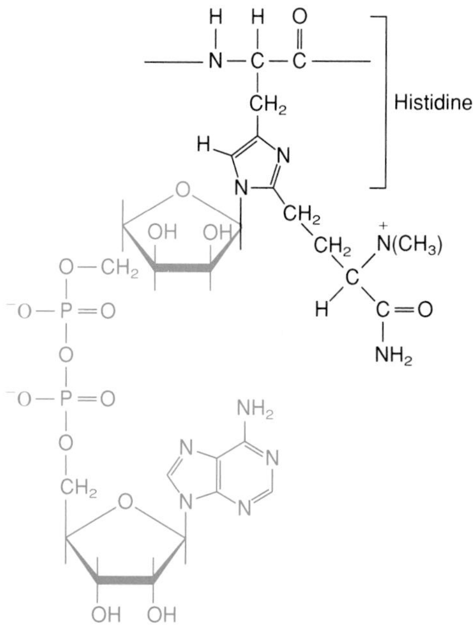

Cycloheximide

FIGURE **28.36**

ADP-ribosylated diphthamide derivative of histidine in eEF2. Synthesis of this derivative of a modified histidine in eEF2 using NAD^+ is catalyzed by diphtheria toxin. eEF2 is inactivated, and protein synthesis is therefore blocked. ADP ribose from NAD^+ is shown in blue. Diphthamide is in black.

CONTROL OF TRANSLATION

As mentioned in Chapter 27, prokaryotes do not seem to make extensive use of control at the translational level. Eukaryotes, on the other hand, use translational control much more widely. In part, this control occurs at the mRNA level. Specific mRNAs may be sequestered by combination with specific mRNA-binding proteins until needed. Some other mRNAs are rapidly degraded so that they do not persist in inappropriate phases of the cell cycle.

We still know little about the detailed mechanisms of translational regulation, but one example has been rather thoroughly studied—that concerned with hemoglobin synthesis in the reticulocytes of mammals. Although immature red cells have already lost their nuclei, they still have adequate reserves of mRNAs available for extensive globin synthesis, which is at this point their major function. Such synthesis would be pointless, however, without an adequate supply of heme to produce functional hemoglobin molecules. The reticulocyte cells contain a protein kinase called *heme-controlled inhibitor (HCI)* (Figure 28.37). In the presence of adequate heme levels, the kinase is inactive, but if heme levels fall, HCI becomes activated and specifically phosphorylates the initiation factor eIF2.

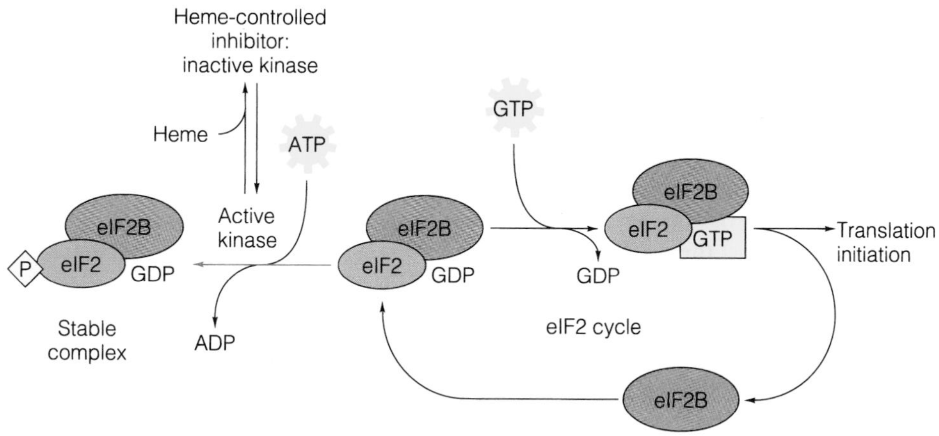

FIGURE **28.37**

Regulation of translation in erythropoietic cells by heme levels. If heme levels fall, the heme-controlled kinase becomes active and phosphorylates eIF2 (red arrow). This blocks further translation by tying up this factor in a stable complex with eIF2B. When heme levels are adequate, the kinase is inhibited, and eIF2 is available for translation initiation.

When this occurs, the complex between eIF2 and eIF2B in the eIF2 cycle becomes unusually stable. The result is that all of the eIF2 is tied up and can no longer be recycled for new initiation. Hence, protein synthesis is halted, and no more globin is made until heme supplies are again adequate.

Remarkably similar to heme inhibition is the effect of the antiviral agents called **interferons.** These are glycoproteins produced in certain cells in response to virus infections. In addition to stimulating mRNA degradation, interferons induce the synthesis of a protein kinase that phosphorylates eIF2, producing, as does the heme-controlled kinase, a stable eIF2–eIF2B complex. The consequent lack of free eIF2 inhibits protein synthesis (most importantly viral protein synthesis) in the infected cells. Because some cancers can be induced by viral *oncogenes* (see Chapter 23) there has been considerable interest in the use of interferon in cancer prevention and therapy. Phosphorylation of various initiation factors appears to be a general method for translational regulation in eukaryotes.

Protein Targeting in Eukaryotes

The eukaryotic cell is a multicompartmental structure. Its several organelles each require different proteins, only a few of which are synthesized within the organelles themselves. Most mitochondrial and chloroplast proteins, for example, are coded for in the nuclear genome and synthesized in the cytoplasm. They must be carefully distinguished from other newly synthesized proteins and selectively transported into their appropriate organelles. Some other new proteins are destined for export out of the cell or into vesicles like lysosomes. The diversity of destinations for different proteins implies that there must be a sophisticated system for labeling and sorting newly synthesized proteins and ensuring that they end up in their proper places.

PROTEINS SYNTHESIZED IN THE CYTOPLASM

Proteins destined for the cytoplasm, nuclei, mitochondria, and chloroplasts are synthesized in the cytosol; those destined for organelles have specific targeting sequences.

Proteins destined for the cytoplasm and those to be incorporated into mitochondria, chloroplasts, or nuclei are synthesized on polyribosomes free in the cytoplasm. The proteins targeted to organelles, as initially synthesized, contain specific signal sequences at their N-terminal ends. These sequences probably aid in membrane insertion, but they also signal that these polypeptides will interact with a particular class of chaperonins. These chaperones are members of the "heat-shock" HSP70 family, and they act to ensure that the newly synthesized protein remains unfolded and is delivered to a receptor site on the organelle membrane. The unfolded protein then passes through inner and outer membranes, through gates containing transport proteins that discriminate between proteins destined for the lumen, the membranes, or the matrix. If it passes into the matrix, the protein may be taken up by intraorganelle chaperonins for final folding. The N-terminal targeting sequence is also cleaved off during this transport (see Figure 28.38). The model shown has been derived especially for the import of proteins into the mitochondrial matrix but can probably approximate the situation with chloroplast proteins as well.

The situation with nuclear proteins is now believed to be similar. Early theories held that these proteins simply diffused into the nucleus through the nuclear pores and were then bound to chromatin. But there is now strong evidence for nuclear targeting sequences that help some proteins—perhaps with others in tow—select the nucleus as their destination. Furthermore, it is becoming clear that the nuclear pores are complex gates, rather than open channels.

PROTEINS SYNTHESIZED ON THE ROUGH ENDOPLASMIC RETICULUM

Proteins destined for cellular membranes, lysosomes, or extracellular transport use a special distribution system. The key structures in this system are the **rough endoplasmic reticulum** (RER) and the Golgi complex (see also Chapter 16). The

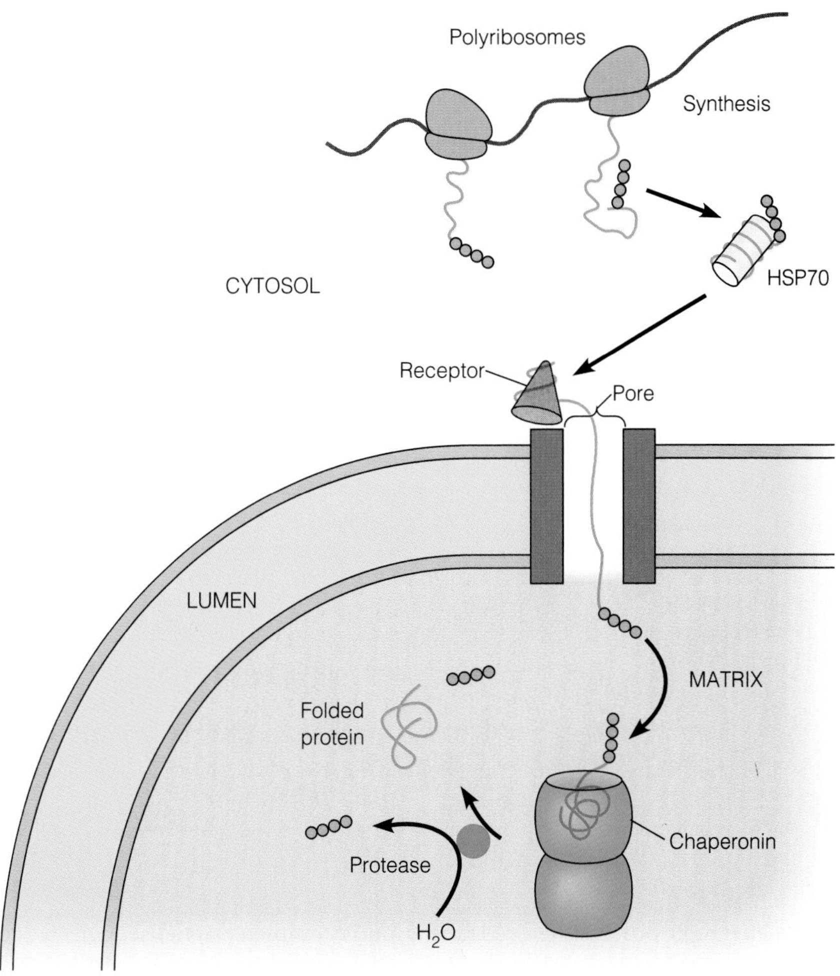

FIGURE **28.38**

A schematic view of steps in the delivery of a protein, synthesized in cytosol, to the matrix of an organelle. The signal sequence is shown in orange, chaperones in green.

rough endoplasmic reticulum is a network of membrane-enclosed spaces within the cytoplasm. The RER membrane is heavily coated on the outer, cytosolic surface with polyribosomes; this coating is what gives the membrane its rough appearance. The Golgi complex resembles the RER in that it is a stack of thin, membrane-bound sacs. However, the Golgi sacs are not interconnected, nor do they carry polyribosomes on their surfaces. The role of the Golgi complex is to act as a "switching center" for proteins with various destinations.

Proteins destined for cell membranes, lysosomes, or export are synthesized on the rough endoplasmic reticulum, then modified and transported via the Golgi apparatus.

Proteins that are to be directed to their destinations via the Golgi complex are synthesized by polyribosomes associated with the RER. Synthesis actually begins in the cytoplasm (Figure 28.39, step 1). The first sequence to be synthesized is an N-terminal *signal sequence,* part of a mechanism for attaching the ribosome and nascent protein to the RER. *Signal recognition particles (SRPs),* containing several proteins and a small (7SL) RNA (see page 1071), recognize the signal sequences of the appropriate nascent proteins and bind to them as they are being extruded from the ribosomes (step 2).

The SRP has two functions. First, its binding temporarily halts translation, so that no more than the N-terminal signal sequence extends from the ribosome. This pause prevents completion of the protein in the wrong place—that is, in the cytosol—and also inhibits premature folding of the polypeptide chain. Thus, the SRP is acting as a kind of chaperone. The second function of the SRP is to recognize a **docking protein** in the RER membrane. The docking protein binds the ribosome to the RER, and the signal sequence is inserted into the RER membrane (step 3). The SRP is then released (step 4), allowing translation to resume (step 5). The protein being synthesized is actually *pulled* through the membrane by an

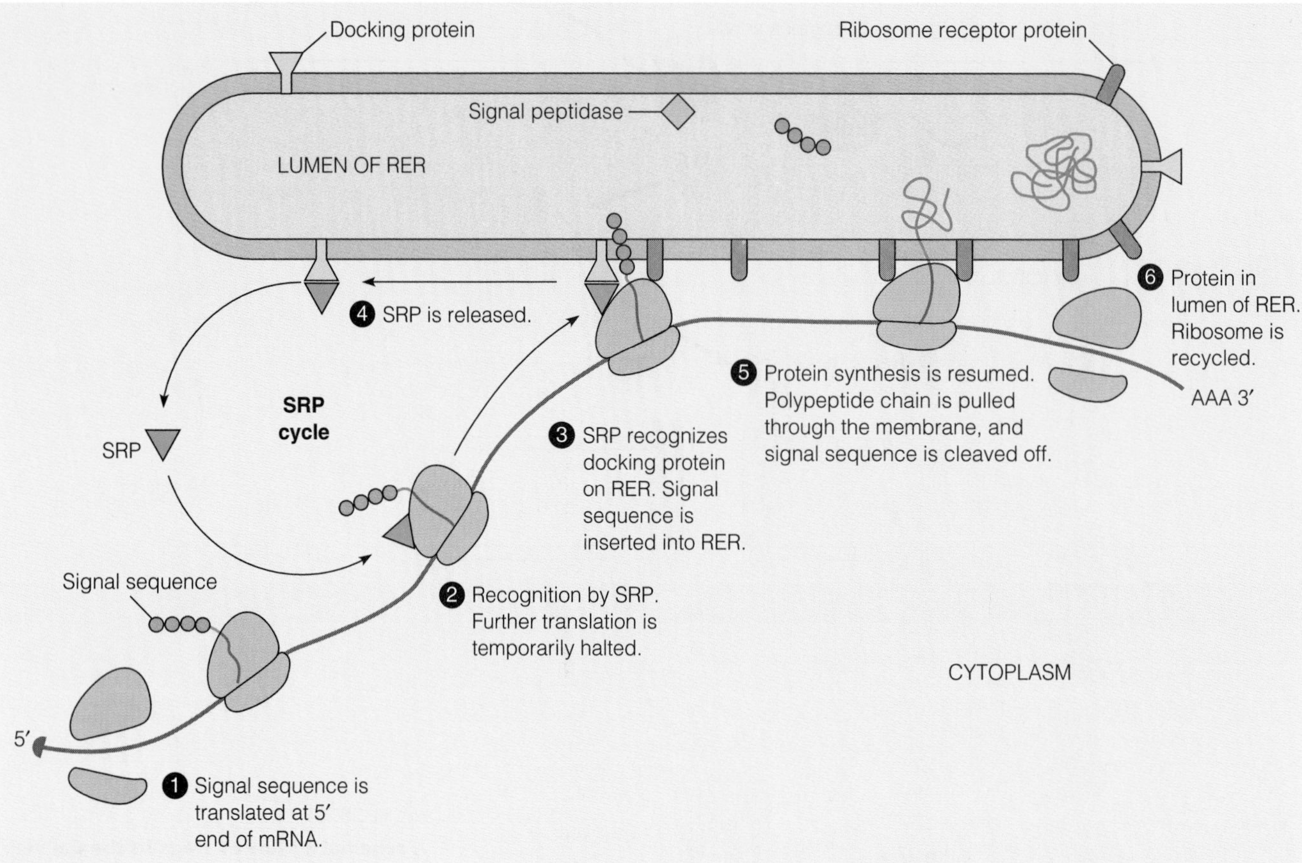

FIGURE **28.39**

The sequence of events in synthesis of proteins on the rough endoplasmic reticulum. The time sequence of events is from left to right.

ATP-dependent process. Before translation is complete, signal sequences are cleaved from some proteins by an RER-associated protease. These proteins are released into the lumen of the RER and further transported (step 6). Proteins that will remain in the endoplasmic reticulum have resistant signal peptides and thereby remain anchored to the RER membrane.

ROLE OF THE GOLGI COMPLEX

The proteins that enter the lumen of the RER undergo the first stages of glycosylation at this point. Vesicles carrying these proteins then bud off the RER and move to the Golgi complex (Figure 28.40). Here the carbohydrate moieties of glycoproteins are completed (see pages 583–585 in Chapter 16 for details), and a final sorting occurs. The multiple membrane sacs that constitute the Golgi complex represent a multilayer arena for these processes. Vesicles from the RER enter at the *cis* face of the Golgi (that closest to the RER) and fuse with the Golgi membrane. Proteins are then passed, again via vesicles, to the intermediate layers. Finally, vesicles bud off from the *trans* face of the Golgi complex to form lysosomes, peroxisomes, or glyoxysomes or to travel to the plasma membrane. All of this transport of vesicles, from the RER to the *cis* face of the Golgi, to successive levels of the Golgi and on to their final destinations, requires high specificity in targeting. Transport of vesicles to the wrong destinations would cause cellular chaos. This sorting is accomplished by having each kind of protein cargo packed in a vesicle marked by specific vesicle membrane proteins. In some cases, the target membranes contain complementary proteins that interact with these and cause membrane fusion. These complementary pairs are called *SNARES* (v-SNARES on vesicles, t-SNARES on target membranes). The interaction of specific v- and t-SNARES, aided by cytosolic fusion proteins, leads to fusion of the vesicle and target membranes and delivery of the cargo (see Figure 28.41).

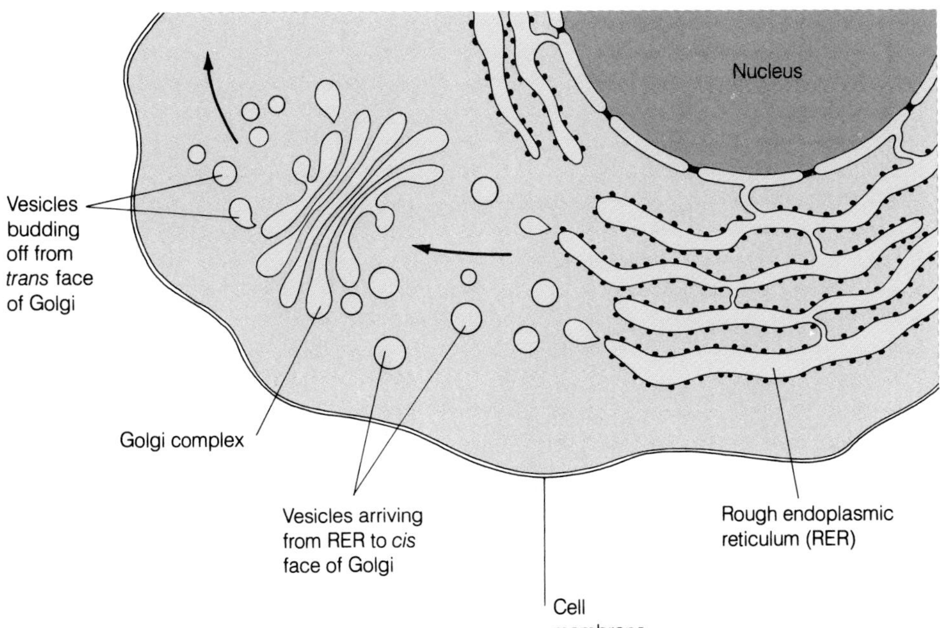

FIGURE **28.40**

Transfer from the rough endoplasmic reticulum (RER) to the Golgi complex. Note that vesicles bud off the RER and move to the *cis* face of the Golgi. Primary lysosomal vesicles bud from the *trans* portion of the Golgi.

The Fate of Proteins: Programmed Destruction

In Chapter 11 we pointed out that one mechanism for the control of enzymatic function was the selective degradation of certain enzymes. However, not only enzymes need to be destroyed in a programmed way. Regulatory proteins that are essential in certain parts of the cell cycle and deleterious in others must be eliminated at some point. Consider the cyclins (see page 1082), for example, each of which must be present only in a specific part of the cell cycle. Proteins that have become damaged must also be removed. In some developmental processes, it is necessary to remove whole organelles or even entire cells and tissues.

Eukaryotic cells have two distinct methods for protein degradation. The lysosomes contain among their hydrolases proteolytic enzymes that will degrade any protein trapped by the lysosome. Parallel to this process is a cytosolic degradation system, which is of necessity highly selective. The danger inherent in having nonspecific proteases loose in the cytosol should be evident.

THE LYSOSOMAL SYSTEM

The lysosomal particles budded from the Golgi complex, known as **primary lysosomes,** are bags essentially of degradative enzymes. Over 50 different hydrolytic enzymes are contained in lysosomes, including proteases, nucleases, lipases, and carbohydrate-cleaving enzymes. The lysosomes play a number of important roles in cellular metabolism, as schematically depicted in Figure 28.42.

In some cell types, such as those in the pancreas that secrete degradative enzymes, primary lysosomes migrate to the cell surface and release their contents into the exterior medium (path A). Primary lysosomes may also fuse with *autophagic vesicles,* formed when smooth ER engulfs organelles destined for destruction (path B). The combined vesicle is called an *autophagic lysosome.* In some kinds of cells—mainly certain white blood cells—primary lysosomes may fuse with *phagocytic vacuoles* that have engulfed nutrient materials at the cell surface (path C). In these heterophagic lysosomes, the nutrients are digested and their amino acids, nucleotides, lipids, and other low-molecular-weight constituents released into the cytosol. Residual, undigested material is excreted when the heterophagic lysosomes and autophagic lysosomes find their way to the plasma membrane.

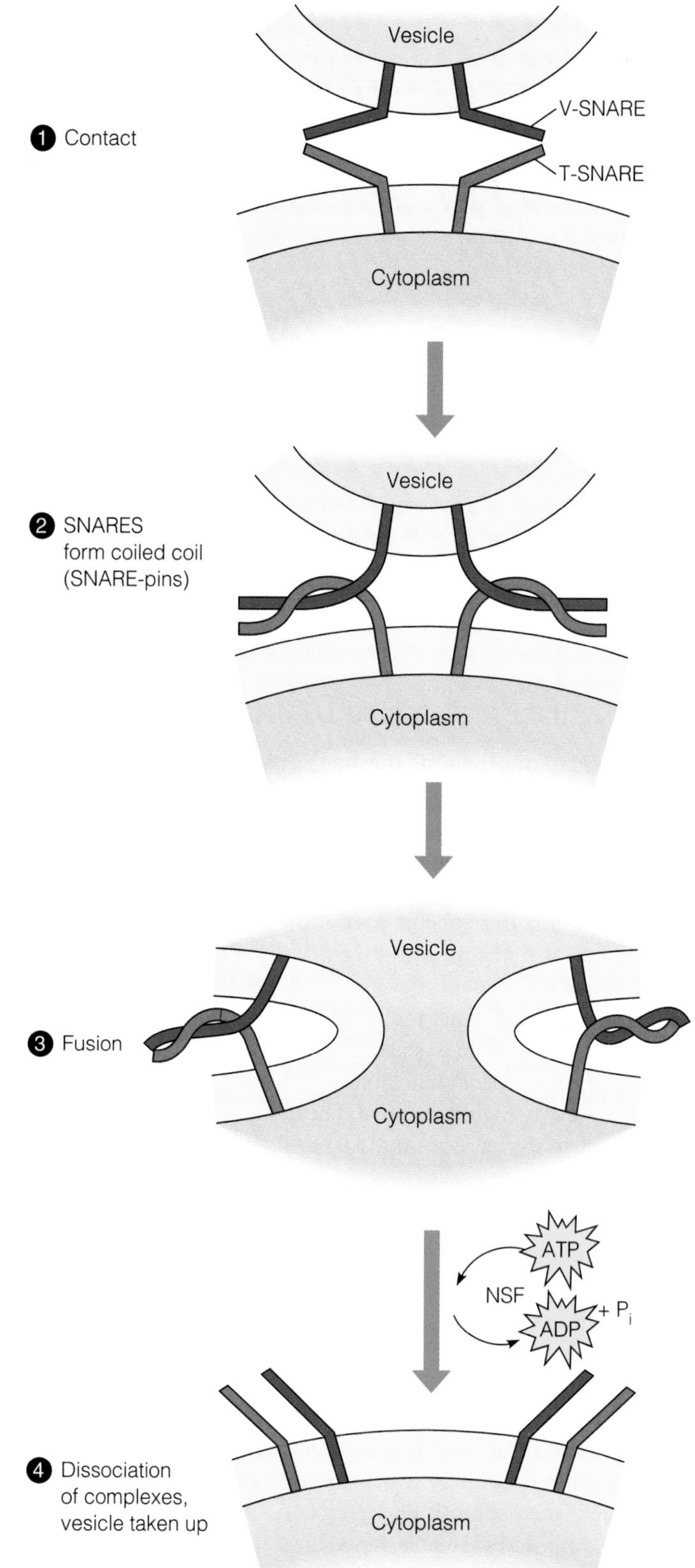

FIGURE **28.41**

A schematic, and somewhat hypothetical, view of SNARE–pin fusion. Specific v-SNAREs and t-SNAREs dictate interaction, and form coiled-coil structures. After fusion, these are broken up by the factors SNAP and NSF, the latter in an ATPase.

CYTOSOLIC PROTEIN DEGRADATION

In contrast to the lysosomal enzymes, which are usually safely sequestered in their vesicles, any protease activity that is free in normal cytosol must be under rigid control. It must attack only the proteins whose destruction is needed. These may

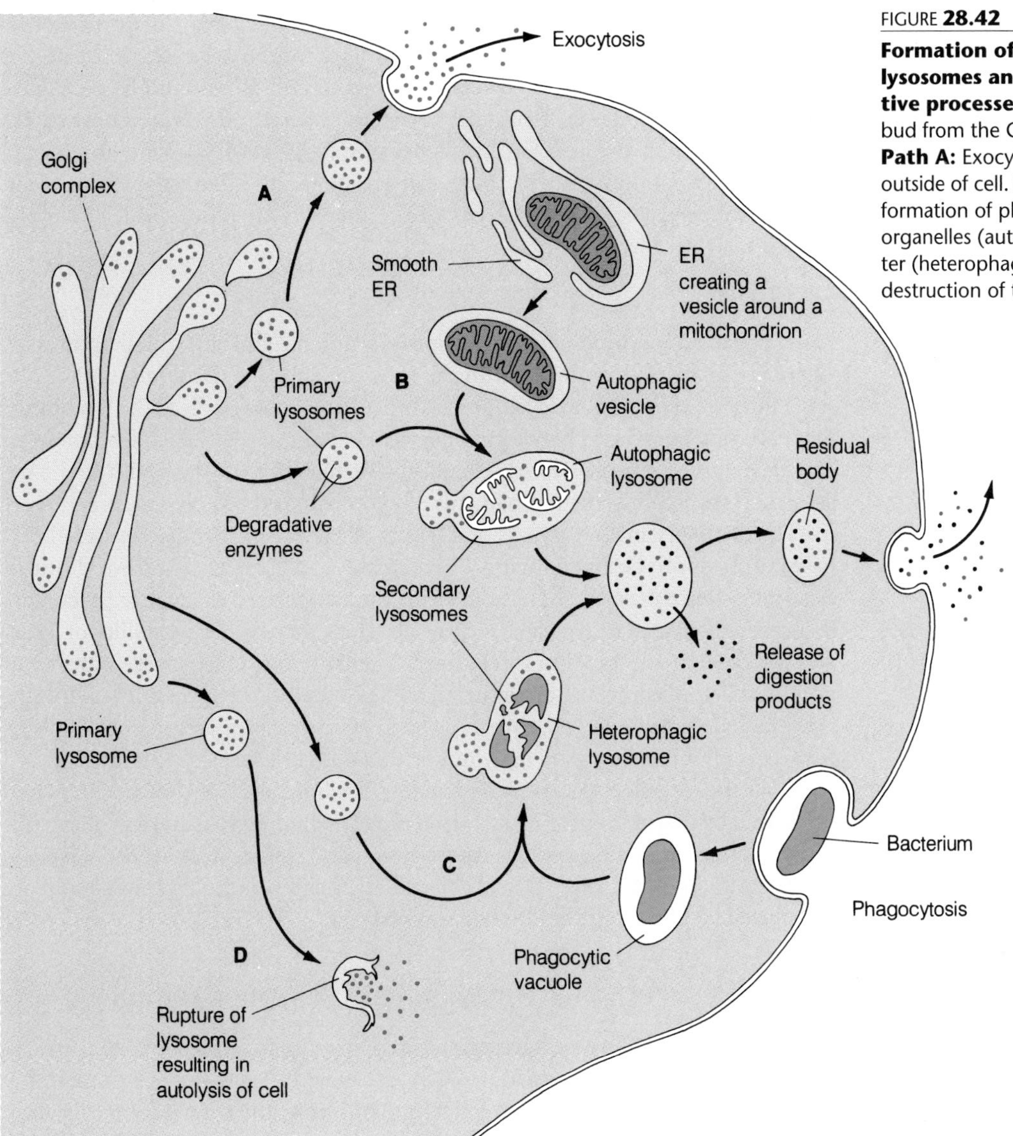

FIGURE **28.42**

Formation of primary and secondary lysosomes and their role in cellular digestive processes. The primary lysosomes that bud from the Golgi can take several pathways. **Path A:** Exocytosis—transport of enzymes to outside of cell. **Paths B and C:** Phagocytosis—formation of phagic lysosomes for digesting organelles (autophagocytosis) or ingested matter (heterophagocytosis). **Path D:** Autolysis—destruction of the cell itself.

include damaged proteins, incorrectly synthesized proteins, or proteins no longer required at a particular stage in the cell cycle. If we recall that protein hydrolysis is a thermodynamically favored reaction, it becomes clear that the enzymes participating in such cytosolic degradation must be more than simple catalysts for the hydrolytic process—otherwise, destruction would be wholesale. Basically, there must be some means of distinguishing the proteins to be attacked from those to be left alone.

The major proteolytic system in cytosol uses a protein called *ubiquitin* to mark proteins destined for destruction. Ubiquitin is a 76-residue polypeptide, found in virtually every cell of every eukaryote. Ubiquitin is synthesized either as a head–tail fusion oligomer, or fused to a ribosomal protein. Its carboxyl group must be freed by hydrolysis of a carboxyl peptide, using a *de-ubiquitinating enzyme* (step 1, Figure 28.43). The carboxyl group is activated by thiol coupling to an *activating enzyme* (Figure 28.43, step 2) in an ATP-dependent process. The ubiquitin moieties are then transferred to a second enzyme (step 3), which attaches them to ε-amino groups of lysine residues on the fated protein (step 4).

Although some ubiquitinated proteins are simply marked for translocation to specific cellular sites, and others are marked for reasons yet unknown, most are marked for proteolytic digestion in particles called proteasomes. These come in

Protein degradation occurs via lysosomes—vesicles filled with hydrolytic enzymes—or intracellularly, often involving the marker ubiquitin and/or a multicatalytic complex.

two sizes, the 20S particle being a precursor to the functional 26S proteasome. The three-dimensional structures of proteasomes from two classes of organisms (Archaebacteria and yeast) have been determined to high resolution (see Figure 28.44). The structure is a 28-subunit barrel. Structurally, the proteasome shows remarkable similarity to the GroEL chaperonin (page 190). Both are multitiered cylinders with 7-fold symmetry. Both can accept an unfolded polypeptide chain in their hollow interior. But whereas GroEL protects the polypeptide chain, the proteasome degrades it.

APOPTOSIS

Many organisms exhibit a process of programmed cell death called **apoptosis.** There are many reasons for this: In the growing embryo, tissues present at one stage must be removed as development proceeds—an example is the "webbing" that each of us possessed between fingers at a certain point in fetal life. In adults, damaged, imperfect, or infected cells should be removed for the health of the organism. Until recently, the mechanisms of apoptosis were wholly obscure.

The discovery in recent years of a family of intracellular proteases called **caspases** turns out to be a major clue. The caspases all cleave at the carboxyl side of aspartate (hence c-aspase). These proteins are synthesized as inactive precursors that can be activated by just such a cleavage. Thus, there is the possibility of a caspase cascade once activation is triggered. There seem to exist a large number of potential cytoplasmic, mitochondrial, and extracellular triggers. Interestingly, these include interferon (see page 1104) and *tumor necrosis factor,* a protein long known to inhibit tumor growth by interaction with cell surface receptors.

Once the cascade has started, destruction is wholesale. The proteins known to be cleaved by caspases encompass a wide range, including structural proteins like actin and nuclear lamin as well as transcription and splicing factors. Nucleases are activated to digest the cell's DNA. The cell disintegrates into fragments that are recognized by phagocytes and fully destroyed.

Eukaryotic DNA and Development: A Brief Example

Throughout this chapter we have emphasized how carefully the life of the eukaryotic organism is programmed in its DNA. We now realize that *much* more information is encoded in the eukaryotic DNA than simply the recipes for a collection of proteins and special nucleic acids. A variety of special signals are hidden within the genes themselves—signals to determine how transcripts will be cut and spliced, where gene products will go, and even how long they will last. In addition, a vast amount of information, usually coded in the sequences surrounding certain genes, specifies when, either in the course of development or in response to environmental stresses, certain genes are to be transcribed in certain cells.

At a higher level, further genetic instructions *coordinating* these transcriptional instructions must exist, because the development of an organism requires the programmed differentiation of some cells, the proliferation of certain tissues, and the death of selected cells. We are only beginning to get fascinating glimpses of how such information may be encoded in the genome. Much of what we know comes from studies of the development of the fruit fly *Drosophila melanogaster.* Developmental biologists have long known that as the larva of this insect develops, groups of cells are set apart as disklike structures called **imaginal disks** (Figure 28.45). These groups of cells will form specific parts of the adult fly. As the

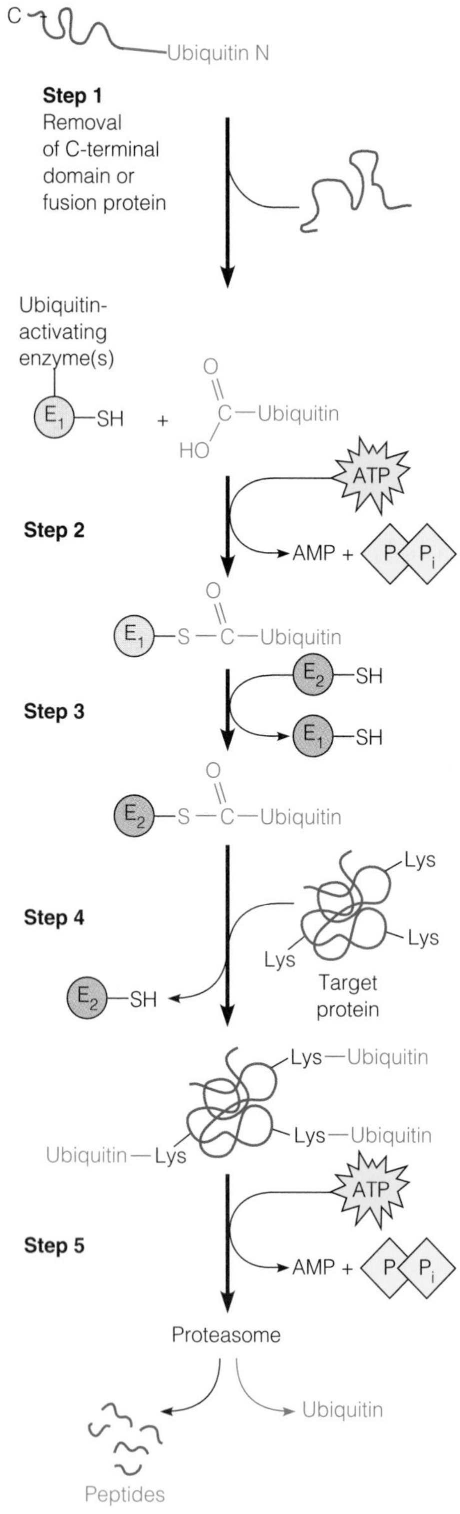

FIGURE **28.43**

The ubiquitination degradation pathway. Ubiquitin begins as part of a fusion protein, which must be cleaved to yield ubiquitin. This is then activated by thiolation and transfer, until it is attached to a target protein, which it marks for degradation.

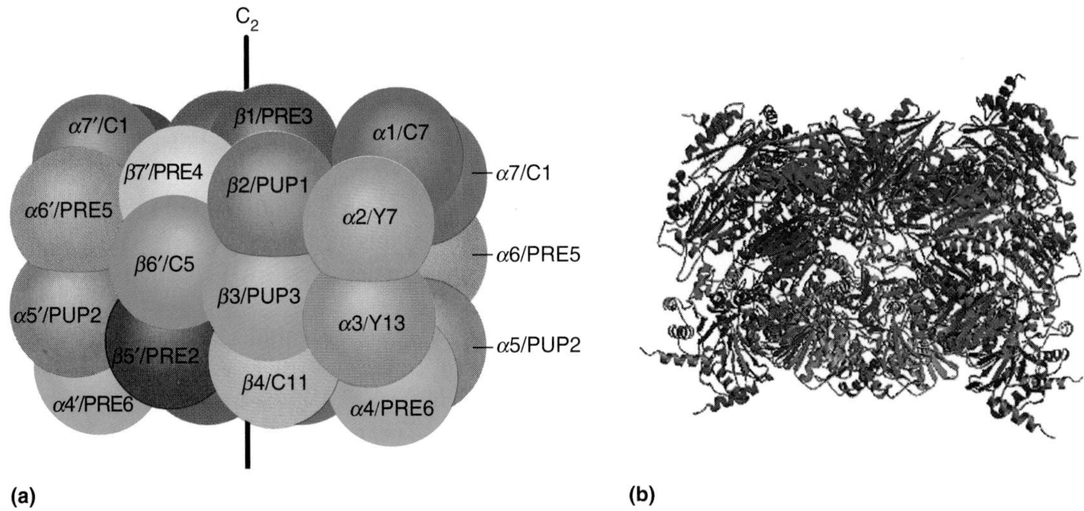

FIGURE **28.44**

Three-dimensional structure of the yeast 20S proteasome. (a) A schematic representation, showing the positions of the 28 subunits (2 each of 14 types). The 2-fold axis is denoted by C_2; the pseudo 7-fold axis lies approximately perpendicular to this, in the plane of the paper. **(b)** A detailed molecular model.

Courtesy of R. Huber from M. Groll et al., *Nature* (1997) 386: 463–471. © 1997 Macmillan Magazines, Ltd.

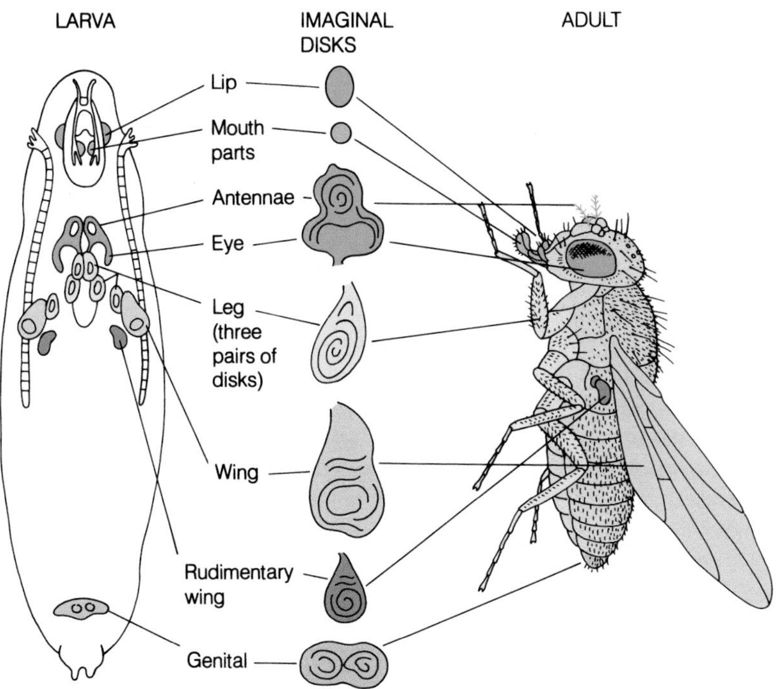

FIGURE **28.45**

Imaginal disks in the development of *Drosophila*. Each of the imaginal disks shown in the larva develops into a specific segment of the adult.

From J. W. Fristrom, R. Raikow, W. Petri, and D. Stewert, *Problems in Biology: RNA in Development*, E. W. Hardy, ed. (Salt Lake City, UT: University of Utah Press). Reprinted by permission.

FIGURE **28.46**

A bithorax mutation of *Drosophila*. Top: Normal fly. **Bottom:** A mutant in which the thoracic segment that normally produces a pair of halteres, or rudimentary wings, has been transformed into one that produces a fully developed set of wings.

Courtesy of E. B. Lewis, *Eukaryotic Development, Part A: Genetic Aspects* (New York: Alan R. Liss, 1982).

larva metamorphoses, larval cells are destroyed by autolysis, and each imaginal disk develops into a different portion of the adult.

Geneticists working with *Drosophila* have long recognized classes of **homeotic mutations**—mutations that scramble the whole developmental pattern in defined ways. One, called the *Antennapedia* mutation, causes perfectly formed legs to grow in the places near the eye where antennae are normally formed. Another group, termed *bithorax* mutations, leads to abnormal development of thoracic segments, producing, for example, extra pairs of wings (Figure 28.46). Recent molecular biological studies have revealed large clusters of **homeotic genes** that control these developmental processes and are the sites of homeotic mutations. Remarkably, a

Homeotic mutations that modify developmental pathways involve special protein factors carrying the homeo domain.

common sequence element of about 180 bp is repeated many times in these gene clusters. This sequence, now called the *homeo box,* codes for a 60-residue polypeptide sequence called the *homeo domain.* Proteins containing this domain are nuclear, DNA-binding proteins of the helix–loop–helix class. It seems likely that each acts as a regulator of transcription for a coordinated group of proteins.

Most remarkable is that the homeo box is not confined to insects but is found in many other organisms, including amphibians and mammals. The sequences are remarkably conserved over this phylogenetic range. This discovery hints at a quite unexpected uniformity in developmental mechanisms between very distantly related organisms.

This fascinating story, which is still unfolding, is a dramatic example of the point made in Chapter 1 of this book—diverse scientific fields are coming together to provide new insights into biology. Seemingly unrelated studies in genetics, developmental biology, and protein and nucleic acid chemistry have converged in this example to bring us to the brink of a biochemical understanding of how organismal development is regulated at the molecular level. It is our belief that the future of biology lies in just such cross-field explorations.

SUMMARY

The genomes of eukaryotes are, in general, much larger than those of prokaryotes. This large size is partly due to greater genetic requirements and to the presence of multiple copies of some genes. But there is also much repetitive DNA of as yet unknown function, and eukaryotic genes carry many introns and long regulatory sequences. The multiple eukaryotic chromosomes are contained in the nucleus and are complexed with histones and nonhistone proteins to form chromatin. The repeating unit of chromatin is the nucleosome, in which DNA is wrapped about a histone octamer. The nucleosomal fiber is further coiled in the nucleus, and loops of chromatin are attached to the nuclear matrix. The typical eukaryotic cell cycle consists of G1, S, G2, and M phases. In S phase the DNA is replicated, and the chromatin structure is copied. In M (mitosis), the cell divides.

Transcription in eukaryotes is complicated and involves three kinds of RNA polymerases, each with a specialized function. Polymerase I transcribes the major ribosomal RNA, pol III the small RNAs, and pol II the messenger RNA. Each polymerase requires a number of protein factors to initiate transcription. Transcription is further modulated by trans-acting factors, binding to promoters and enhancers. The finished transcript is polyadenylylated on its 3′ terminus and capped on its 5′ end. Introns in pre-mRNAs are spliced out in particles called spliceosomes. Sometimes alternate splicing occurs. Translation occurs much as in prokaryotes, but more protein factors are required, and the initiation and termination steps show major differences.

Cytosolic or organelle proteins are made in the cytosol. Proteins designated for other targets are synthesized on ribosomes of the rough endoplasmic reticulum and are modified and directed through the Golgi apparatus. The destruction of proteins in cells is also a programmed event, with several pathways.

Not only individual genes but also whole groups of genes appear to be under coordinated regulation in the development of higher organisms.

REFERENCES

General

Alberts, B., D. Bray, J. Lewis, M. Raff, K. Roberts, and J. D. Watson (1994) *The Molecular Biology of the Cell,* 3rd ed. Garland, New York. A fine contemporary source for cell biology.

Conaway, R. C., and J. W. Conaway (1994) *Transcription: Mechanism and Regulation.* Raven Press, New York. An excellent survey of recent thinking and experimentation.

Organization of the Genome

Brändén, C.-I., H. Eklund, C. Cambillau, and A. J. Pryor (1984) Correlation of exons with structural domains in alcohol dehydrogenase. *EMBO J.* 3:1307–1310. An insight into a possible relationship between gene structure and protein structure.

Goffeau, A., and 15 others (1996) Life with 6000 genes. *Science* 274:546–567. The sequence and genomic structure of the entire yeast genome.

Jelineck, W. R., and C. W. Schmid (1982) Repetitive sequences in eukaryotic DNA and their expression. *Annu. Rev. Biochem.* 51:813–844.

Sharp, P. A. (1985) On the origin of RNA splicing and introns. *Cell* 42: 397–400.

Ullu, E., and C. Tschudi (1984) Alu sequences are processed 7SL RNA pseudogenes. *Nature* 312:171–172.

Chromosomes and the Nucleus

Kavenoff, R., L. C. Klotz, and B. H. Zimm (1974) On the nature of chromosome-size DNA molecules. *Cold Spring Harbor Symp. Quant. Biol.* 38:1–8. The first measure of chromosomal DNA size.

Lamond, A. I., and W. C. Earnshaw (1998) Structure and function in the nucleus. *Science* 280:547–563.

Chromatin and Nucleosomes

Arents, G., R. W. Burlingame, B.-C. Wang, W. E. Love, and E. Moudrianakis (1991) The nucleosomal core histone octamer at 3.1Å resolution: A tripartícle protein assembly and a left-handed superhelix. *Proc. Natl. Acad. Sci. USA* 88:10148–10452. First high-resolution study of nucleosome structure, first demonstration of core histone tertiary folding.

Luger, K., A. W. Mädes, R. K. Richmond, D. F. Sargent, and T. J. Richmond (1997) Crystal structure of the nucleosome core particle at 2.8Å resolution. *Nature* 389:251–260.

van Holde, K. E. (1988) *Chromatin.* Springer-Verlag, New York. Somewhat dated, but good for background.

Woodcock, C. L., and R. A. Horowitz (1995) Chromatin organization reviewed. *Trends Cell Biol.* 5:272–277. A review that features recent studies of chromatin fiber architecture, in the light of models that have been presented.

Replication in Eukaryotes

Bodnar, A. G., M. Ouellette, M. Froikis, S. E. Holt, C.-P. Chiu, G. B. Morin, C. B. Harley, J. W. Shay, S. Lichtsteiner, and W. E. Wright (1998) Extension of life-span by introduction of telomerase into normal human cells. *Science* 279:349–352. A development with potentially major implications for biology and medicine.

Edwards, A. M., A. Bochkarev, and L. Frappier (1998) Origin DNA-binding proteins. *Curr. Opin. Cell. Biol.* 8:49–53.

Muniyappa, K., and K. M. Kironmai (1998) Telomere structure, replication, and length maintenance. *Crit. Rev. Biochem. Mol. Biol.* 33: 297–336. A good current review.

Wei, X., J. Samarabandu, R. S. Devdhar, A. J. Siegel, R. Acharya, and R. Berezney (1998) Segregation of transcription and replication sites into higher order domains. *Science* 281:1502–1505. A new and exciting view of nuclear processing.

Regulation of Transcription

Felsenfeld, G. (1992) Chromatin as an essential part of the transcriptional mechanism. *Nature* 355:219–224.

Karlsson, S., and A. W. Nienhuis (1985) Developmental regulation of human globin genes. *Annu. Rev. Biochem.* 54:1071–1108.

Mizzen, C. A., and C. D. Allis (1998) Linking histone acetylation to transcriptional regulation. *Cell. Mol. Life Sci.* 54:6–20.

Nikolov, D. B., H. Chen, E. D. Halay, A. Hoffmann, R. G. Roeder, and S. K. Burley (1996) Crystal structure of a human TATA box–binding protein/TATA element complex. *Proc. Natl. Acad. Sci. USA* 93:4862–4867.

Nikolov, D. B., S.-H. Hu, J. Liu, A. Gasch, A. Hoffman, M. Horikoski, N.-H. Chua, R. G. Roeder, and S. K. Burley (1992) Crystal structure of TFIID TATA-box binding protein. *Nature* 360:40–46.

Roeder, R. G. (1996) the role of general initiation factors in transcription by RNA polymerase II. *Trends Biochem. Sci.* 21:327–335. An excellent review; this issue of TIBS is entirely dedicated to eukaryotic transcription and contains several other excellent reviews.

Shilatifard, A. (1998) Factors regulating the transcriptional elongation activity of RNA polymerase II. *FASEB J.* 12:1437–1446.

Tan, S., and T. J. Richmond (1998) Eukaryotic transcription factors. *Curr. Opin. Struct. Biol.* 8:41–48. Primarily structural in emphasis.

Tsukiyama, T., and C. Wu (1997) Chromatin remodeling and transcription. *Curr. Opin. Genet. Dev.* 7:182–191.

mRNA Processing and Splicing

Breitbart, R. E., A. Andreadis, and B. Nadal-Ginard (1987) Alternative splicing: A ubiquitous mechanism for the generation of multiple protein iso forms from single genes. *Annu. Rev. Biochem.* 56:467–495.

Hoffman, M. (1991) RNA editing: What's in a mechanism? *Science* 253:136–138.

Patrusky, B. (1992) The intron story. *Mosaic* 23:23–33. A readable account of the discovery of alternative splicing.

Sharp, P. A. (1994) Split genes and RNA splicing. *Cell* 77:805–815.

Translation in Eukaryotes

Kozak, M. (1992) Regulation of translation in eukaryotes. *Annu. Rev. Cell Biol.* 8:197–225.

Thach, R. E. (ed.) (1990) *Translationally Regulated Genes in Higher Eukaryotes.* Karger Press, Farmington, Conn. A good collection of papers on diverse aspects of regulation at the translational level.

Trachsel, E. (ed.) (1991) *Translation in Eukaryotes.* CRC Press, Boca Raton, Fla.

Protein Targeting

Bennett, M. K. (1995) SNAREs and the specificity of transport vesicle targeting. *Curr. Opin. Cell. Biol.* 7:581–586.

Farquhar, M. G., and G. E. Palade (1998) The Golgi apparatus: 100 years of progress and controversy. *Trends Cell. Biol.* 8:2–10.

Haucke, V., and G. Schatz (1997) Import of proteins into mitochondria and chloroplasts. *Trends Cell. Biol.* 7:103–106. A clear description of the special features of this kind of transport.

Hendrick, J. P., and F.-U. Hartl (1993) Molecular chaperone functions of heat shock proteins. *Annu. Rev. Biochem.* 62:349–384. Covers a wide area, including roles of HSPs in translocation of proteins.

Lingappa, V. R. (1991) More than just a channel: Provocative new features of protein traffic across the ER membrane. *Cell* 65:527–530.

Weber, J., B. V. Zemelman, J. A. McNew, B. Westermann, M. Gmachi, F. Parlati, T. H. Söllner, and J. E. Rothman (1998) SNARE pins: Minimal machinery for membrane fusion. *Cell* 92:759–772.

Protein Degradation and Apoptosis

Ciechanover, A. (1987) Regulation of the ubiquitin-mediated proteolytic pathway: Role of the substrate α-NH_2 group and transfer RNA. *J. Cell. Biochem.* 34:81–100.

Goldberg, A. L. (1995) Functions of the proteosome: The lysis at the end of the tunnel. *Science* 268:522–523. See also the paper following by Weissman et al., comparing the proteosome to GroEL.

Groll, M., L. Ditzel, J. Löwe, D. Stock, M. Bochtier, H. P. Bartunik, and R. Huber (1997) Structure of the 20S proteasome from yeast at 2.4 Å resolution. *Nature* 386:463–471. A landmark crystallographic accomplishment.

Jentsch, S., and H. D. Ulrich (1998) Ubiquitous déjà vu. *Nature* 395: 321–322. A brief summary of some important recent advances in protein degradation.

Martins, L. M., and W. C. Earnshaw (1997) Apoptosis: Alive and kicking in 1997. *Trends Cell. Biol.* 7:111–114.

Tan, X., and J. Y. J. Wang (1998) The caspase–RB connection in cell death. *Trends Cell. Biol.* 8:116–120.

Homeotic Genes and the Homeo Box

Gehring, W. J., M. Affolter, and T. Buerglin (1994) Homeodomain proteins. *Annu. Rev. Biochem.* 63:487–526.

Kessel, M., and P. Gruss (1990) Murine developmental control genes. *Science* 249:373–379.

PROBLEMS

1. The average human chromosome contains about 1×10^8 bp of DNA.
(a) If each base pair has a mass of about 660 daltons, and there are about 2 g of protein (histones plus nonhistones) per gram of DNA, how much does such a chromosome weigh, in grams?
(b) If the DNA were extended, how long would it be?
(c) An actual chromosome is about 5 μm in length. What is the approximate compaction ratio?
(d) You have about 10^{12} cells in your body. If you have 46 chromosomes in each cell, what is the approximate extended length of *all* of your DNA? For comparison, the distance from the earth to the sun is about 1.5×10^8 km.

*2. From cultured mammalian cells, you wish to isolate DNA enriched in replication origins, for subsequent cloning and sequence analysis. Using any combination of density labels, radioactive labels, cell synchronization techniques, and centrifugal separation procedures, describe how this could be done.

3. Formation of nucleosomes and wrapping them into a 30-nm fiber provide part of the compaction of DNA in chromatin. If the fiber contains about six nucleosomes per 10 nm of length, what is the approximate compaction ratio achieved? Comment on the comparison of this answer with that of Problem 1c.

*4. It is possible to "reconstitute" nucleosomes by mixing DNA and histone octamers in 2M NaCl and then dialyzing to low salt. When such experiments were carried out using a specific 208-bp fragment of sea urchin DNA, the following results were obtained: Digestion of the product with micrococcal nuclease gave quantitative production of 146 bp DNA. Upon cleavage of this DNA with a restriction nuclease having a single site in the fragment, several sharp bands were obtained, with sizes as follows: 29 bp, 39 bp, 107 bp, 117 bp. How would you interpret these data in terms of nucleosome positioning on this DNA?

5. A sample of chromatin was partially digested by the enzyme staphylococcal nuclease. The DNA fragments from this digestion were purified and run on a polyacrylamide gel. A set of DNA restriction fragments was used as markers. Distances of migration are given below. From these data, estimate the nucleosome repeat distance in the chromatin.

Marker DNA Fragment		Chromatin DNA Fragment
Size (bp)	*d* (cm)	*d* (cm)
94	40	30.5
145	34.2	19.2
263	25.2	14.4
498	16.7	11.5
794	11.5	—

*6. Pancreatic deoxyribonuclease I (DNase I) is a nuclease that makes single-strand nicks on double-strand DNA. It has been observed that treatment of nucleosomal core particles with DNase I yields a peculiar result. When DNA from such a digestion is electrophoresed under denaturing conditions, the single-strand fragments are observed to occur in a regular periodicity of about 10 bases. Suggest an explanation of this result in terms of the structure of the nucleosome.

7. Some viruses, like SV40, are closed circular DNAs carrying nucleosomes. If the SV40 virus is treated with topoisomerase, and the histone is then removed, it is found to still be supercoiled. However, if histones are removed *before* topoisomerase treatment, the DNA is relaxed. Explain.

8. (a) About how long would be required to replicate the average human chromosome described in Problem 1 from a single origin of replication, if a replication fork moves at about 6 kb/min? (This is probably an upper limit on the rate.)
(b) Evidence suggests that origins are spaced about 100 kb apart in many eukaryotic DNAs. How long should replication then require? Compare your answer with the fact that S phase in most cells requires several hours, and suggest an explanation for the discrepancy.

9. Histone genes are unusual among eukaryotic genes in that they do not have introns, and histone mRNAs do not have poly(A) tails. Furthermore, in almost all eukaryotes, histone genes are arranged in multiple tandem domains, each domain carrying one copy of each of the five histone genes. Suggest an explanation for these features in terms of the special requirements for histone synthesis.

10. A number of years ago, it was suggested that the function of the poly(A) tail on a eukaryotic message may be to "ticket" the message. That is, each time the message is used, one or more residues is removed, and the message is degraded after the tail is shortened below a critical length. Suggest an experiment to test this hypothesis.

*11. It has been proposed that nucleosomes must be removed in order for transcription to proceed through chromatin. Suggest an experiment that might test this hypothesis.

12. Shown below is an R loop prepared for electron microscopy by annealing a purified eukaryotic messenger RNA with DNA from a genomic clone containing the full-length gene corresponding to the mRNA.

(a) How many exons does the gene contain? How many introns?
(b) Where in this structure would you expect to find a 5′,5′-internucleotide bond? Where would you expect to find a polyadenylic acid sequence?

13. Suggest the effects (if any) of the following mutations:
(a) Mutation of the 3′ G in an intron
(b) A mutation that prevents U1 binding to mRNA
(c) A mutation that removes the N-terminal signal sequence from a protease destined for lysosomes

14. Although protein hydrolysis is a thermodynamically favorable process, the selective hydrolysis of ubiquitin-marked proteins requires ATP. Suggest why this should be necessary.

15. In degradation of proteins via the ubiquitin pathway (Figure 28.43), only steps 2 and 5 are ATP-dependent. Why these two steps and no others?

TOOLS OF BIOCHEMISTRY 28A

Finding Binding Factors and Binding Sequences

We have seen in this and preceding chapters that a large number of DNA sequences bind specific proteins, and that these interactions are important. We are often faced with the question, Does a particular protein bind specifically to a particular sequence? A more general question is this: Given a particular protein, to what sequence does it most like to bind? Some new techniques have allowed quick answers to such questions.

Electrophoretic Retardation

Electrophoretic retardation, sometimes called **gel retardation** or **gel shift,** is a simple and quick technique. A cloned piece of DNA (preferably quite short—say, 50 to 300 bp) is mixed with the protein (or mixtures of proteins) to be tested and electrophoresed on acrylamide gel under nondenaturing conditions. The DNA is usually radioactively labeled. The naked DNA is run in a separate lane as a control. If binding of the protein occurs, the kind of result shown in Figure 28A.1 is obtained. The binding leads to a decrease in electrophoretic mobility, with the production of one or more slowly migrating bands. In the example shown, there is evidence for the existence of two binding sites on the DNA fragment in question.

The study can be elaborated in several ways. For example, if a mixture of proteins (a cell lysate, for example) is used, the identity of binding protein can be sought by doing Western blots with appropriate antibodies. With care, it is possible to make semiquantitative estimates of binding strength by measuring the fraction of DNA in the several bands.

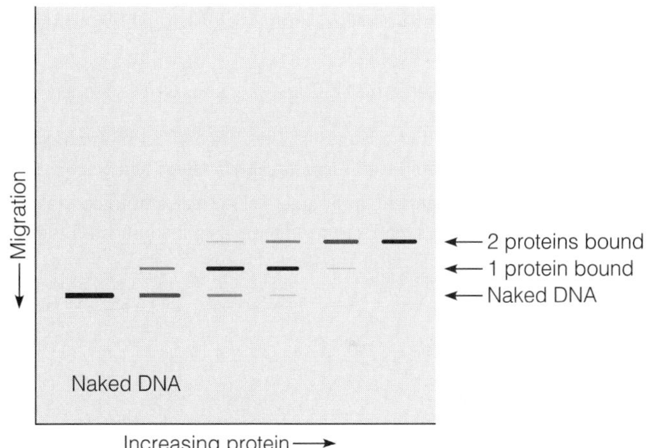

FIGURE **28A.1**
Electrophoretic retardation in a case in which two protein molecules are bound to a DNA sequence.

Selection Techniques to Determine Consensus Sequences

Suppose one has a protein that is suspected to bind to some specific DNA sequence—or more properly, to a class of similar sequences. How can one determine what that sequence might be like? A variety of methods, based on the polymerase chain reaction (PCR—see Tools of Biochemistry 24A), have recently been developed. An example is the following: Suppose one were to synthesize a set of sequences like those shown in Figure 28A.2. This

5′ primer A
GAGCTGGATCCTACTCAGGTGA

GAGCTGGATCCTACTCAGGTGA—25 "N" bases—GTCTGGAATTCCAGGGCACTGC
Random template

CAGACCTTAAGGTCCCGTGACG
3′ primer B

FIGURE **28A.2**
The kind of sequence that could be used for PCR-aided selection of a consensus sequence. Restriction sites are underlined.

set contains two defined sequences at 5′ and 3′ ends, flanking a *completely random* sequence of defined length. In other words, the synthesizer begins by making a defined oligonucleotide, adds random nucleotides, and ends by adding another defined nucleotide. "Primer" sequences are also synthesized to match these 5′ and 3′ regions.

The collection of oligomeric DNAs is then mixed with the protein in question, and the few sequences that bind at all well are captured by electrophoretic retardation or filter binding, and thereby separated from unbound DNA. These sequences are then amplified by PCR, using the primers indicated, and the cycle is repeated several times. See Figure 28A.3. If the protein does bind to a consensus sequence, the final products after several such rounds will consist mainly of molecules containing sequences very like this.

At this point, the DNA is cleaved with restriction nuclease cutting at two built-in restriction sites on the flanking regions and ligated into a plasmid in which it can be cloned. Sequencing of a number of clones should make clear the nature of the consensus sequence, which should be representative of the DNA sequence specifically bound by this protein.

References

Lane, D., P. Prentki, and M. Chandler (1992) Use of gel retardation to analyze protein nucleic acid interactions. *Microbiol. Rev.* 56:509–528.

Wright, W. E., M. Binder, and W. Funk (1991) Cyclic amplification and selection of targets (CASTing) for the myogenin consensus binding site. *Mol. Cell. Biol.* 11:4104–4110.

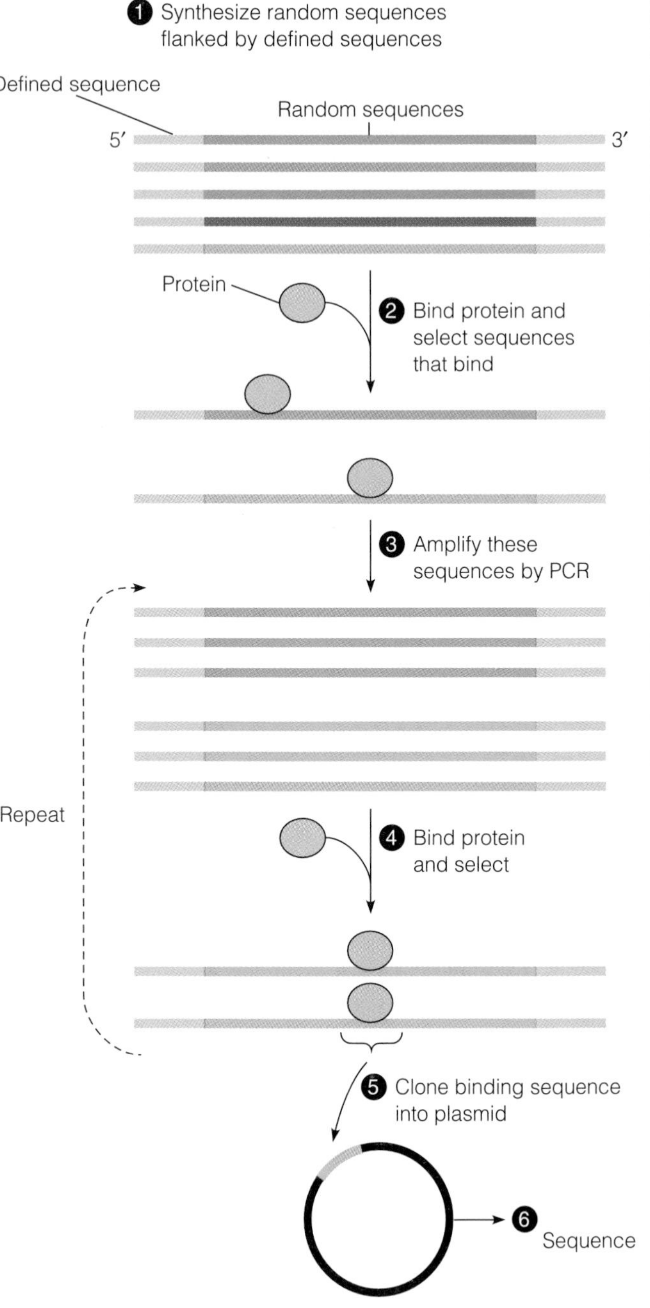

FIGURE **28A.3**

Scheme for using sequences like the one in Figure 28A.2 to identify consensus sequences that bind to a given DNA-binding protein.

Answers to Problems

CHAPTER 2

1. (a) 3.54 kJ/mol.
   (b) 151 kJ/mol.

2. CCl_4, H_2S, $H_3\overset{+}{N}$ CH_2 COO^-, $H_3\overset{+}{N}$ CH_2 CH_2 CH_2 COO^-. CCl_4 is symmetrical, $\mu = 0$; H_2S will be comparable to H_2O; the latter two involve separation of whole charges, and in the last, separation is greatest.

3. The graph will be a reflection of the graph given, across the $E =$ 0 line.

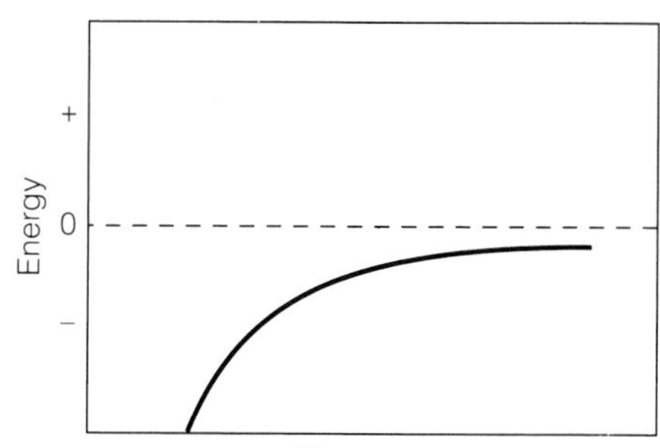

4. (a) $K_a = \dfrac{[H^+][A^-]}{[HA]} = \dfrac{X^2}{A_0 - X}$

$$K_aA_0 - K_aX = X^2$$

or $X^2 + K_aX - K_aA_0 = 0$

$$X = \frac{-K_a \pm \sqrt{K_a^2 + 4K_aA_0}}{2}$$

(b) First, assume $X << A_0$ and obtain

$$X^2 \cong K_aA_0, \qquad X \cong \sqrt{K_aA_0}$$

Use this approximate X to obtain a better approximation to the denominator $(A_0 - X)$. Repeat calculation until change in each step is less than required precision.

If K_a is very small (acid weak), either the above or direct approximation yields

$$[H^+] \cong \sqrt{K_aA_0}$$

5. (a) pH = 0.456.
   (b) pH = 2.608 by approximation, 2.609 exact.
   (c) pH = 3.108 by approximation, 3.113 exact.

6. (a) $K_a = 8.16 \times 10^{-5}$.
   (b) pH = 2.39.

7. Choose a series of about 10 points at which volumes of KOH have been added to neutralize increasing fractions of the acid. Use the Henderson–Hasselbalch equation to calculate the pH at each point.

8. (a) pH = 4.46.
   (b) pH = 2.57.

9. (a) 0.138 M.
   (b) $[KH_2PO_4] = 0.126$ M; $[Na_2HPO_4] = 0.174$ M.

10. pH = 3.93.

11. The best choice would be a mixture of $H_2PO_4^-$ and HPO_4^{2-}, which has $pK_a = 6.86$.

12. One will dissolve 2 × 7.507 g of glycine in the 2 L to make a 0.1 M solution. To obtain a pH of 9.0, the Henderson–Hasselbalch equation states that the ratio of $Gly^-/Gly^{\pm}$ must be 0.251. This can be achieved by adding NaOH to a concentration of 0.02 M. For 2 L, this would require 0.04 mol, or 40 mL of 1 M NaOH.

13. 8.5% as H_2CO_3. Very little CO_3^{2-}.

14. (a) $B = (A_0/2.303)f(1 - f)$.
    (b) When $f = 0.5$.
    (c) $A_0/9.2/2$.
    (d) Increasing A_0 increases B.

15. pI = 11.1. Form I will be present in only insignificant amounts above pH 7.

16. Because of the proton-attracting power of the dianion produced when two of the citric acid protons have dissociated, the third is held anomalously strongly. In fact, it has a pK_a of 6.86.

17. The ionic strength of 1 M $(NH_4)_2SO_4$ is 3.0.

18. Lysine has a pI of about 9.5. Thus, it would not migrate at pH 9.5, whereas cysteine with a side chain pK_a of 8.3 would carry a partial negative charge at this pH, and arginine, with its very high side chain pK_a (12.5) would be strongly positively charged.

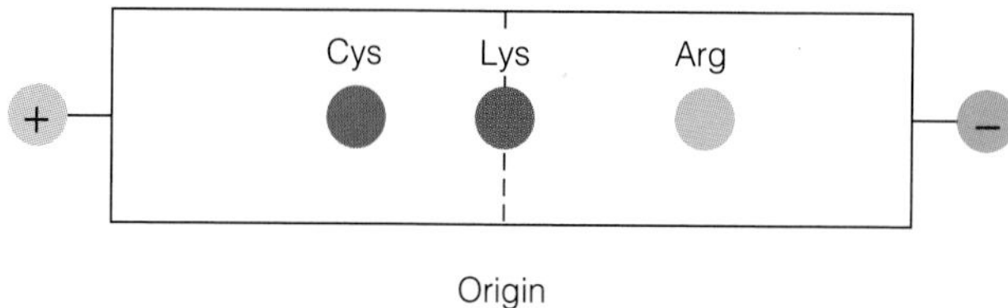

19. (a) Isoelectric focusing. The single charge difference would make little difference in mobility, but a detectable difference in pI.
    (b) Near pI.

CHAPTER 3

1. $\Delta H_f - \Delta E_f = -0.165$ J/mol, $\left(\dfrac{\Delta H_f - \Delta E_f}{\Delta E_f}\right) \times 100 = 0.003\%$.

2. (a) +926 kJ/mol.
(b) 463 kJ/mol.

3. (a) $\Delta S° = +470$ J/K·mol.
(b) A solid material is being transformed entirely into gases. The greater freedom of motion of gas molecules results in an entropy increase.
(c) $\Delta E° = 103.4$ kJ/mol.
(d) Because the system does work on the surroundings, in expanding from a solid to a gas.

4. (a) $\Delta G° = -2872$ kJ/mol.
(b) $\Delta G° = -1694$ kJ/mol.
(c) 41%.

5. (a) 1.1×10^{-4} mM.
(b) $\Delta G°' = -17$ kJ/mol.
(c) 11 M. It is never reached because glucose-6-phosphate is continually consumed in other reactions. The system never reaches equilibrium.

6. (a) $K = 5.44 \times 10^{-2}$; $(f_{G3P})_{eq} = 0.052$.
(b) $\Delta G = -4.37$ kJ/mol.

7. (a) ΔS must be positive, because the increase in available states corresponds to an increase in entropy.
(b) Since $\Delta G = \Delta H - T\Delta S$, a positive ΔS yields a negative contribution to ΔG (T is always a positive number). Thus, for proteins to be stable, which requires ΔG for the above to be positive, denaturation must involve a large positive ΔH and/or an additional negative contribution to ΔS. As we shall see in Chapter 6, both occur.

8. This process should correspond to an entropy decrease ($\Delta S < 0$), since order is being established in the H_2O structure.

9. (a) We expect ΔS for denaturation to be positive, since order is lost. If ΔH is also positive (energy required to break internal bonds), then we have the +,+ situation in Table 3.3.
(b) If many hydrophobic residues are present in a protein, ΔS could be negative for unfolding (see Problem 8). If ΔH were also negative, we would have the −,− situation in Table 3.3.

10. Given $\Delta G° = \Delta H° - T\Delta S°$ and $\Delta G° = -RT \ln K$, combination and rearrangement gives the desired result, which is called the van't Hoff equation. According to the equation, if $\Delta H°$ and $\Delta S°$ are independent of temperature, a graph of $\ln K$ versus $1/T$ should be a straight line with a slope of $-\Delta H°/R$.

11. (a) $\Delta H° = +59.0$ kJ/mol.
(b) $\Delta S° = -103$ J/K·mol. To obtain the true K, so as to get $\Delta G°$, one must divide K_W by 55.5, the molar concentration of pure water.

12. (a) $K = 16.8$.
(b) Toward glucose-l-phosphate.

13. No, because the repulsion between ADP^{2-} and P_i^{2-} is less than that between ADP^{3-} and ADP^{2-}, and the release of an H^+ will change the value as well.

14. (a) 2.18×10^3 kJ.
(b) 70 moles.

15. 217 g.

16. For a mole of protein molecules, $\Delta S \times R \ln W - R \ln 1$, where W is the number of conformations available to each, and R is the gas constant, 8.314J/K·mol. Since there are 99 bonds between 100 residues, $W = 3^{99}$.
(a) $\Delta S = 9.04 \times 10^2$ J/K·mol.
(b) $\Delta H = 292$ kJ/mol.
(c) Increase, because ΔS and ΔH are both positive.

17. (a) 5.93×10^{-6} kJ.
(b) -5.93×10^{-6} kJ.
(c) In (a) 1.91×10^{-7} moles ATP, in (b) none.

CHAPTER 4

1. (a) It must be a single-strand DNA, since Chargaff's rules are not obeyed.
(b) It should be mostly random coil, with perhaps some self-bonding into hairpins, etc.

2. (a) $^{3'}$TGGCATTCCGAAATC$^{5'}$.
(b) $^{5'}$pApCpCpGpTpApApGpGpCpTpTpTpApGp$^{3'}$.
(c) $^{3'}$UGGCAUUCCGAAAUC$^{5'}$.

3. The center of symmetry is indicated on the figure at right by a large black dot.

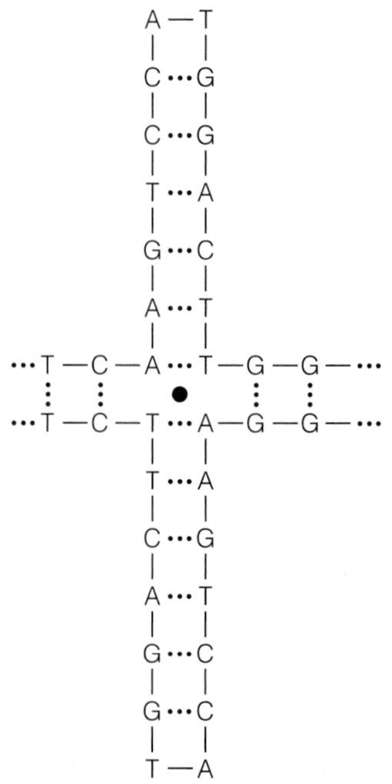

```
                A—T
                |   |
                C···G
                |   |
                C···G
                |   |
                T···A
                |   |
                G···C
                |   |
                A···T
                |   |
···T—C—A···T—G—G—···
   :   :    ●    :   :
···T—C—T···A—G—G—···
                |   |
                T···A
                |   |
                C···G
                |   |
                A···T
                |   |
                G···C
                |   |
                G···C
                |   |
                T—A
```

4. $W \cong -2.65 \times 10^4$.

5.
```
5'G—U—C—C—A—G—C—C—A—U—U—G—C—G—U
  :  :  :  :  :  :  :  :  :       )
                3'C—G—G—U—A—A—C—G—C—U
```

6. It will be A form, so 1313 nm.

7. (a) 200.
(b) $\sigma = -0.06$.

8. 20 base pairs corresponds to 2 turns of B-DNA at +1 twist/turn, and 1.7 turns of Z-DNA at −1 twist/turn. Therefore, (a) $\Delta T = -3.7$, (b) $\Delta L = 0$, (c) $\Delta W = +3.7$.

9. The Epstein–Barr virus should be highest; the bacteriophage T4 should be the lowest.

10. (a) The DNA must be single-stranded.
(b) There must be small regions of self-complementary base pairing.

11. Approximately 60% GC.

12. (a) It must require 3 turns in one repeat, to give an integral number. The number = 28 bp/repeat.
(b) More tightly.
(c) Disfavor.

13. The distribution indicates the presence of superhelical molecules with $\Delta L = \pm 1, \pm 2, \pm 3$, etc. These could form as a consequence of fluctuations in structure at the moment when the circle is sealed. They provide a direct demonstration of the Boltzmann distribution of molecular energies.

14. (a) Positive, because L is fixed (=0), and ΔT is made negative by the ethidium.
(b) None.
(c) Positive writhing caused by ethidium decreases the negative writhing until $W = 0$; this structure will have a minimum mobility because it is more extended than the writhed forms. As more ethidium is added, the DNA writhes positively, again increasing mobility.

CHAPTER 5

1. 109.5.

2. (a) SYSMEHFRWGKPV.
(b) 1624. This is not exactly correct because it does not take into account dissociation of some protons.

3. (a) The curve will exhibit inflections corresponding to two groups titrating near 4 (carboxyl terminus and glutamic side chain), one group near 7 (histidine), two near 9 (N-terminus and lysine), one near 10 (tyrosine), and one near 12 (arginine).
(b) Approximately −2, +2, +4.
(c) pI ≅ 9.

4. (a) SYSMEHFR, WGKPV.
(b) SYSM*, EHFRWGKPV; M* = homoserine lactone.
(c) SYS, MEHFRWGKPV.

5. DSGPYKMEHFRWGSPPKD.

6. (a) −1 at pH 7; −4 at pH 12.
(b) (1) 2; (2) no cleavage; (3) 3.
(c) Electrophoresis, or chromatography on a cationic column, at pH 7.

7. Asp–Arg–Val–Tyr–Ile–Met–Pro–Phe.

8. AC.

9. (a) 2.
(b) Between the first and second, and third and fourth cysteines.

10. (a) One possibility, of many:

···UGUAAUUGUAAAGCGCCCGAGACCGCGCUUU
GUGCUCGACGAUGUCAACAACAU···

(b) Since it does not have an N-terminal methionine, at least some proteolytic cleavage must be involved in its synthesis.

11. Met–Phe–Pro–Ser–Tyr–Pro–Lys–Asp–Lys–Lys–Glu–

12. (a) Either the N-terminus is blocked (as by acetylation, for example) or the peptide is cyclic.
(b) If the peptide is not cyclic, there should be a free C-terminus. In most cases (except if it is Pro) this can be attacked by carboxypeptidase A. Alternatively, if the peptide is cyclic, some protease can probably be found that will cleave it only once, giving a free N-terminus to allow sequencing.

13. Because the A chain contains no basic residues and the B chain has several, either isoelectric focusing, electrophoresis, or ion-exchange chromatography should work well. Note that separation should be best at about pH = 5, where the histidines on the B chain are positively charged and glutamates on the A chain are negative.

14. The sedimentation coefficient is not directly proportional to molecular mass; it depends upon *both* mass and the size and shape of the protein molecule. Therefore, there is no reason why sedimentation coefficients should be additive.

CHAPTER 6

1. (a) Left-handed.
(b) 3.0.

2. If the helix were distorted so as to give 3.5 residues/turn, a leucine would project on the same face every 2 turns. The leucines on the two chains would provide hydrophobic faces for interaction. This structure is called a *leucine zipper*.

3. (a) The four helices could be arranged so that the hydrophobic side chains would all point toward the center of the bundle and would

pack together there (see Problem 2). This would give a stabilizing hydrophobic core.
(b) A proline at this point would break the helix near the Fe_2 binding sites. This would probably mean that Fe_2 could not be bound, and the mutant protein would be nonfunctional.
(c) An initial coiling of the chain into four consecutive helical regions, followed by folding of these together—perhaps aided by binding of Fe_2.

4. They are spaced about three to four residues apart. Therefore, they will all lie on the same side of the α helix. This suggests that this side of the helix may face the interior of the protein.

5. In the unfolded form of the protein, most of its hydrogen bond donors and acceptors can make H bonds to water. Therefore, it is only the *difference* in hydrogen bonding energy that contributes to protein stability.

6. (a) $3^{200} = 2.7 \times 10^{95}$.
(b) Not all of these conformations will be sterically possible. But even if only 0.1% of these are allowed, there are still 2.7×10^{92}, a very large number.

7. (a)
$$\begin{aligned}\Delta S_{\text{folding}} &= S_{\text{folded}} - S_{\text{unfolded}} \\ &= R \ln W_{\text{folded}} - R \ln W_{\text{unfolded}} \\ &= 8.314 \text{ J/K·mol} \times [\ln 1 - \ln(2.7 \times 10^{92})] \\ &= -1769 \text{ J/K·mol} = -1.77 \text{ kJ/K·mol}\end{aligned}$$

(b) $\Delta H_{\text{folding}} = 96 \times (-5 \text{ kJ/mol}) = -480 \text{ kJ/mol}$

(c)
$$\begin{aligned}\Delta G_{\text{folding}} &= \Delta H_{\text{folding}} - T\Delta S_{\text{folding}} \\ &= -480 \text{ kJ/mol} - 298 \text{ K}(-1.77 \text{ kJ/K·mol}) \\ &= +47 \text{ kJ/mol}\end{aligned}$$

Since $\Delta G_{\text{folding}} > 0$ at 25°, the protein would not be stable. It would be stable below 0°C. This points out the importance of sources of stabilization other than backbone H bonds.

8. An α helix from residues 4–11, a β sheet between 14–19 and 24–30. There is very probably a β turn involving residues 20–23.

9. (a) C_4, held together by heterologous interaction, or D_2, held together by isologous interaction.
(b) C_2, since each $\alpha\beta$ dimer forms an asymmetric unit.

10. (a) C_8, D_4.
(b) D_4, because it involves more subunit–subunit interactions.
(c) Both. There must be heterologous interactions about the 4-fold axis and isologous interactions about the 2-fold axes.

11. The measurement by circular dichroism will detect only the unfolding of tertiary structure in the subunits. Calorimetry will also pick up the energy required to break down the quaternary structure.

12. (a) Using data in Table 5.1, we calculate the molecular weight to be 1072. A 1 mg/cm^3 solution will be 9.33×10^{-4} M. From Figure 5.6, we see that the molar extinction coefficient of Tyr is ~1000 M^{-1}cm^{-1}. Phe does not contribute appreciably. Therefore, ε is 0.93 cm^2/mg.
(b) 2.8 mg/cm^3.
(c) 5%.

13. In the absence of BME, a single band of $M \cong 70{,}000$ is obtained. This suggests, but does not prove, that there are two identical, noncovalently linked subunits. However, the addition of BME removes this band and gives two bands of $M \cong 30{,}000$ and $\cong 40{,}000$, respectively. The sum of these is 70,000, strongly suggesting that the native molecule contains four subunits, two of $M \cong 30{,}000$, two of $M \cong 40{,}000$. The 30,000 and 40,000 units are paired by a disulfide bond or bonds.

14. 154,000 g/mol.

15. The fact that disulfide reduction has little effect means that the protein is a single chain. It must have an extended structure, which cleavage at a critical Arg residue can relax, giving faster migration; the fragments are still held together by disulfide bonds. Cleavage of these, after thrombin cleavage, yields two fragments.

16. Assuming that x-ray diffraction is not practical, we have:
(a) Analysis by CD or by NMR.
(b) Sedimentation equilibrium.
(c) More than one secondary/tertiary folding can be observed for the same sequence. Therefore, sequence alone cannot dictate folding in all cases, and sequence-based predictions must sometimes fail badly.

CHAPTER 7

1. (a) 2.6 mm.
(b) About 94%.

2. (a) Decrease.
(b) Increase.
(c) Decrease.

3. (a) (1) 98%; (2) 56%.
(b) About 42%.
(c) About 70%.

4. Chloride ions must interact with positively charged groups so as to stabilize the deoxy state. Thus, the higher the Cl^- concentration, the lower the O_2 affinity of hemoglobin will be.

5. (a) $P_{50} = 4.47$ mm. (Note: This very low value is a consequence of stripping, plus the fact that the temperature is lower than physiological (37°C).)
(b) $n_H \cong 3.5$.
(c) $P_{50T} = 28$ mm; $P_{50R} = 0.022$ mm. These are obtained by extrapolating the limiting lines tangential to the extremes of the Hill plot.

6. Hemoglobin concentration can influence affinity only if protein dissociation or association is occurring. Most likely is dissociation to $\alpha\beta$ dimers, which will have lost some of the interactions that stabilize the T state. Thus, affinity should rise and P_{50} should decrease.

7. (a) $P_{50} = 125$ mm.
(b) $n_H = 3.1$.
(c) Since $n_H = 3.1$, there must be at least 4 sites, because $n_H \leq n$. (Actually, this is an immense molecule, with 24 binding sites.)

8. (a) Because H146 lies in the α/β interface, mutation should be expected to interfere with the T $\rightarrow$ R transition. The effect is to increase affinity. This mutation is known; it is *hemoglobin Hiroshima.*
(b) Since F8 is involved in heme binding, the heme should be unstable. Leucine will not ligate to the heme iron.
(c) β_2 His is involved in BPG binding (see Figure 7.18). Changing to Asp would weaken this.
In each case, a single base change could suffice.

9. (a) Faster.
(b) Probably same.
(c) Faster.

10. (a) It would begin with a slope of about 1 at low values of log P_{O_2}, would then show a region with slope less than 1, and approach a line with a slope of 1.
(b) MWC theory involves switching an equilibrium of all sites between two states—weak and strong. But in KNF theory, one site can modify an adjacent site in either direction—to stronger or weaker binding.

11. A likely explanation: In native Hb, the binding of oxygen is actually hindered by the fact that pulling on helix F must move it against constraints within the molecule. In the imidazole replacement, there is no need to do the extra work of moving helix F. This difference shows up as a more favorable free energy for binding.

12. (a) Yes, because the linkage between subunits is such that forcing one pair of helices apart favors moving the other pair apart, making O_2 binding easier in the second pair.
(b) Probably so. Deprotonation of His 13 destroys the salt bridge, allowing easier opening of the O_2 binding site.
(c) The molecule would exhibit higher O_2 affinity, and probably lesser cooperativity, because the O_2 sites would be opened further. Possibly, the whole structure would become unstable.

13. (a) If one multiplies both sides of the first equation by $1 + Kc$, divides through by c, and rearranges, the second equation results.
(b) According to the second equation, the slope of a graph of r/c versus r should be $-K$, and its intercept at $r/c = 0$ should give n. Using the data in this way, we get $K = 2.2 \times 10^4\ \text{M}^{-1}$ and $n = 2.1$. Since the true number of sites must be integral, we would choose $n = 2$.

14. The integrity of the antibody binding sites would be compromised by the reduced stability of the heavy–light chain pairs in this region.

CHAPTER 8

1. (a) In the relaxed state, there is about 0.6 μm overlap on each side; in the contracted state, about 0.75 μm.
(b) About 20 steps.

2. (a) 0.0033 min $\cong$ 0.2 s.
(b) 0.021 min = 1.2 s.
(c) That creatine phosphate must be continually produced in active muscle.

3. That the growth of actin filaments (which occur at the ends) is somehow essential for cell-shape modification by nonmuscle actin.

4. If the GTPase activity is slow, microtubules that happen to have GTP ends will be able to pick up GTP-tubulin and grow for a considerable time, until the GTP at the end is hydrolyzed. These will then tend to shrink as long as release of GDP-tubulin continues to reveal new GDP ends. Thus, depending on its "end state" (GTP or GDP), a given microtubule will either grow or shrink in competition with its neighbors.

5. About 840 nm.

6. The observation would suggest that several power strokes may occur in succession. These may, however, be on different myosin heads—the actin filament can slide a considerable distance past a single head during the portion of the ATP cycle when it is disengaged.

CHAPTER 9

1. (a)

α-D-Xylofuranose

(b)

(c)

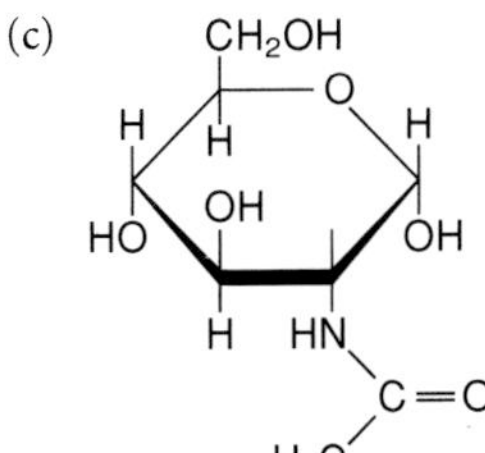

(d)

2. Galactitol has the following structure:

```
     CH2OH
      |
  H—C—OH
      |
 HO—C—H
------+------
 HO—C—H
      |
  H—C—OH
      |
     CH2OH
```

Since it has a plane of symmetry between C-3 and C-4, it is optically inactive. Such compounds, which contain asymmetric carbons but have no net optical activity, are called *meso* forms.

3. In the chair form, there is more steric clash between the 2-OH and the 1-OH in the α form of glucose and in the β form of mannose.

(Compare glucose and mannose as glucose is pictured in Figure 9.13.) Furthermore, dipole–dipole interactions will be more favorable in β-D-glucose and α-D-mannose.

4. For the reaction

$$\alpha\text{-D-Glucopyranose} \rightleftharpoons \beta\text{-D-glucopyranose}$$

we have $K = 64/36 = 1.78$. Therefore

$$\Delta G^\circ = -RT \ln K = -8.314 \text{ J/K·mol} \times 313 \text{ K} \times 0.577$$
$$= -1.50 \text{ kJ/mol}$$

The explanation relates to Problem 3. As Figure 9.13 shows, the 2-OH and 1-OH come quite close in α-D-glucose. Furthermore, the dipole moments of the two OH groups are more nearly parallel in the α-anomer, and antiparallel in the β-anomer. All of these are essentially energetic considerations, so ΔG arises primarily from an enthalpy contribution.

5.

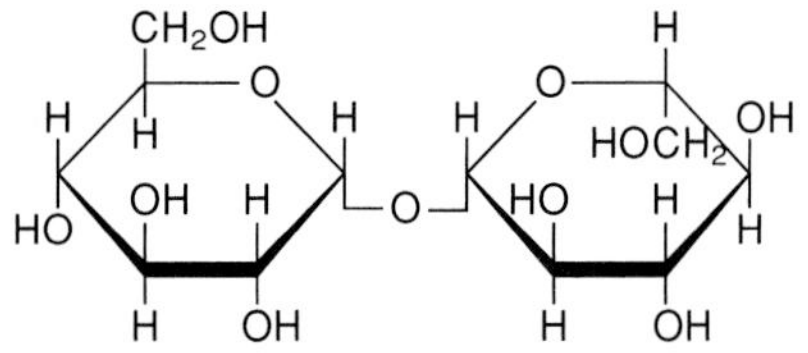

6. Reducing: maltose, cellobiose, lactose, gentiobiose. Nonreducing: trehalose, sucrose.

7.

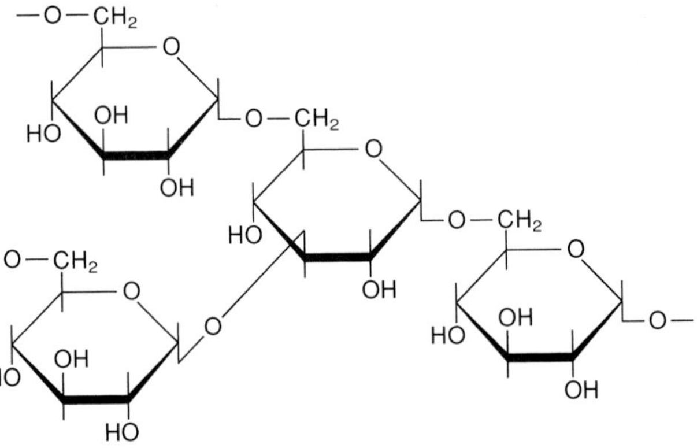

Note: For simplicity, H atoms are represented by vertical bars.

8. Hyaluronic acid.

9. (a) *S*.
(b) *R*.

10. Cleavage of the pyranose form will occur twice, producing formic acid. Cleavage of the furanose form will also occur twice (between C2 and C3 and between C5 and C6), but formaldehyde will be produced from C6.

11.

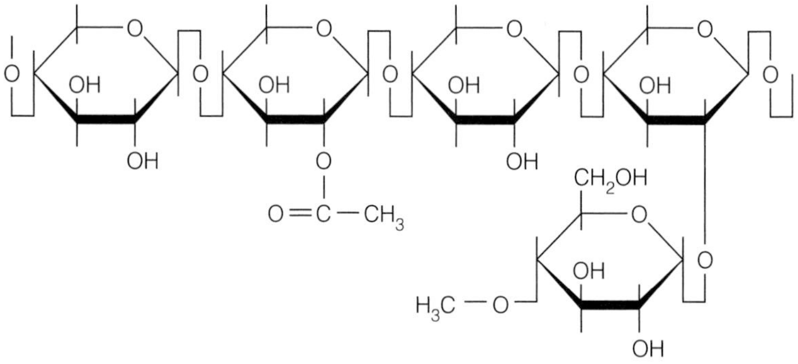

Xylan

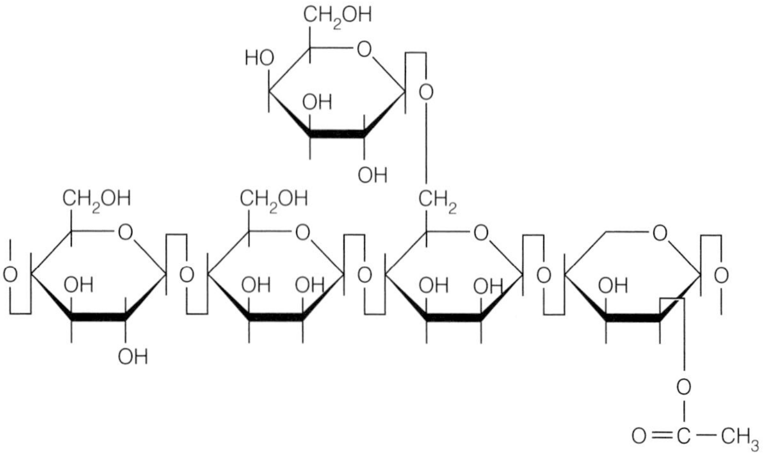

Glucomannan

12. Perhaps the Gal–GlcNAC linkage is not β, or not 1–4.

13. (a) a; (b) f; (c) a; (d) c; (e) d, f.

CHAPTER 10

1. (a) $CH_3CH_2CH{=}CH(CH_2)_7COOH$.
(b) $CH_3(CH_2)_5CH{=}CH(CH_2)_9COOH$. (This is called vaccenic acid.)
(c) According to Table 10.1, any with fewer than 10 carbons will melt below 30°C. An example would be $CH_3(CH_2)_6COOH$, *n*-octanoic acid.

2. (a) Fatty acid, long-chain alcohol.
(b) Glycerol, fatty acid.
(c) Carbohydrate, long-chain alcohol.

3. From the data, 4.74×10^9 cells would have a total surface area of 4.74×10^{11} $(\mu m)^2$, or 0.474 m^2. The ratio of monolayer area to cell surface is 0.89 m^2/0.474 m^2, or 1.89, very close to 2.00.

4. (a) 20.
(b) The sequence MVGALLLLVVALGIGILFM is 19 residues long, is very hydrophobic, and has a number of helix-forming residues.

5. The concentration ratio is 10^5. Therefore
(a) $\Delta G = 29.67$ kJ/mol.
(b) $\Delta G = 36.4$ kJ/mol.
In either case, hydrolysis of 1 mole of ATP (at cellular concentration) would suffice to transport 1 mole of ion.

6. $J = -P(C_2 - C_1)$; from Table 10.6, we have $P = 2.4 \times 10^{-10}$ cm/s. If we express $C_2 - C_1$ in mol/cm^3, J will have dimensions of mol/cm^2·s. Calculation yields $J = 2.04 \times 10^{-14}$ mol/cm^2·s. Therefore, the amount transferred in 1 min across 100 $(\mu m)^2$ $(= 100 \times 10^{-8}$ cm^2) will be as follows:

$$M = -1.224 \times 10^{-18} \text{ mol}$$

7. The initial concentration inside = 100 mM = 0.1×10^{-3} mol/cm^3. The cell volume is 100 $(\mu m)^3$, or 100×10^{-12} cm^3. The amount initially present is as follows:

$$M = 0.1 \times 10^{-3} \text{ mol/cm}^3 \times 1 \times 10^{-10} \text{ cm}^3 = 1 \times 10^{-14} \text{ mol}$$

Therefore, the percentage escaping in 1 min will be

$$\frac{1.224 \times 10^{-18} \text{ mol}}{1 \times 10^{-14} \text{ mol}} \times 100 = 1.224 \times 10^{-2}\ \%$$

Thus, unless facilitated transport is available, "leakage" of K^+ from cells is very slow.

8. Consider mass flow from one direction only, with ligand A at concentration [A] outside the membrane. It reacts with carrier X at the membrane surface to give XA:

$$X + A \longrightarrow XA \qquad k = [XA]/([X][A])$$

If we say that J is proportional to the fraction of carriers occupied, then

$$J = K\frac{[XA]}{[X] + [XA]}$$

and we get

$$J = K\frac{k[A]}{1 + k[A]}$$

This approaches a limiting value $(J = K)$ at high [A].

9. If the calcium ion is *maintained* at this concentration difference, we must have, from equation (10.2), $V_{in-out} = -92$ mV. Since V_{in-out} is defined as the potential inside minus the potential outside, the inside of the organelle is negative with respect to the outside.

10. (a) −60 mV (right −).
(b) +60 mV (right +).
(c) 0.

11. No in (a) and (b). Yes in (c). Concentrations will equalize.

12. (a) 0.28.
(b) −72.9 mV.

13. The subunits traversing the membrane usually need to present different domains to the two sides. Thus, no symmetry involving two-fold axes parallel to the membrane surfaces (like D_n) should be expected.

14. 2.65 kg of K^+, 2.35 kg of Na^+.

CHAPTER 11

1. Given $\ln([A]/[A]_0) = -kt$. When $t = t_{1/2}$, we have

$$\ln\left(\frac{1}{2}\right) = -0.693 = -kt_{1/2}$$

Therefore, $t_{1/2} = 0.693/k$. The constant of proportionality is 0.693.

2. Graphing $\ln([A]/[A]_0)$ versus t gives a curved line. Therefore, kinetics are not first-order.

3. About 1.1×10^{14}.

4. $k = 4 \times 10^9$ (mol/L)$^{-1}$s^{-1}. Some values of k_{cat}/K_m approach this limit.

5. If the protein can bind to a random location on the DNA and slide along until it finds the specific site, a more efficient one-dimensional search replaces a difficult three-dimensional search.

6. Given $d[A]/dt = -k_1[A] + k_{-1}[B]$. Noting that $[B] = [A]_0 - [A]$, we get

$$d[A]/dt = -(k_1 + k_{-1})[A] + k_{-1}[A]_0 \qquad (1)$$

The second term on the right can be evaluated from the equilibrium condition

$$-k_1[A]_e + k_{-1}[B]_e = 0 \qquad (2)$$

which gives, after further substitutions,

$$k_{-1}[\mathrm{A}]_0 = (k_1 + k_{-1})[\mathrm{A}]_e \quad (3)$$

Inserting into (1), and noting that $d[\mathrm{A}]/dt = d([\mathrm{A}] - [\mathrm{A}]_e)/dt$, we get

$$\frac{d([\mathrm{A}] - [\mathrm{A}]_e)}{([\mathrm{A}] - [\mathrm{A}]_e)} = -(k_1 + k_{-1})dt \quad (4)$$

After integrating, this yields the result

$$[\mathrm{A}] - [\mathrm{A}]_e = ([\mathrm{A}]_0 - [\mathrm{A}]_e)e^{-(k_1 + k_{-1})t} \quad \text{Q.E.D.} \quad (5)$$

7. (a) From equation (11.7) we may write

$$-d[\mathrm{A}]/dt = 2k_2[\mathrm{A}]^2$$

where the factor 2 occurs because 2 moles of A are used for each mole of B formed. Integrating, we obtain

$$-1/[\mathrm{A}] = -2k_2t + C$$

The constant C must equal $-1/[\mathrm{A}]_0$, so

$$1/[\mathrm{A}] - 1/[\mathrm{A}]_0 = 2k_2t$$

(b) One way to do it is to plot $1/[\mathrm{A}]$ versus t.
(c) The data yield a linear graph, so the reaction is second-order.

8. (a) It is quite difficult to estimate V_{max} in this way. A value of about 150 (μmol/L)min^{-1} might be guessed at. Then, by taking K_M as the substrate concentration at $V_{max}/2$, we would estimate K_M = 20 μmol/L.
(b) V_{max} = 162 (μmol/L)min^{-1}; K_M = 32 μmol/L. These values are considerably more reliable than those found in (a).
(c) Best fit gives V_{max} = 158 (μmol/L)min^{-1}; K_M = 30 μmol/L.

9. (a) Using V_{max} = 160 (μmol/L)min^{-1}, $k_{cat} = 1.60 \times 10^5$ min^{-1} = 2667 s^{-1}.
(b) Using K_M = 31 μmol/L, $k_{cat}/K_M = 8.6 \times 10^7$ (mol/L)$^{-1}$s^{-1}. This compares favorably with the largest values of k_{cat}/K_M.

10. (a) It is necessary only to rearrange the equation to

$$\left(\frac{K_M}{[\mathrm{S}]} + 1\right)d[\mathrm{S}] = -V_{max}dt$$

and then integrate from $t = 0$ ($[\mathrm{S}] = [\mathrm{S}]_0$) to a finite time t. We obtain

$$[\mathrm{S}]_0 - [\mathrm{S}] - K_M \ln\frac{[\mathrm{S}]}{[\mathrm{S}]_0} = V_{max}t$$

(b) When $[\mathrm{S}]_0 \gg K_M$, the first two terms on the left will be much larger than the third, so

$$[\mathrm{S}_0] - [\mathrm{S}] \cong V_{max}t$$

or

$$[\mathrm{S}] = [\mathrm{S}]_0 - V_{max}t$$

This corresponds to a situation in which the enzyme is saturated with substrate, so substrate molecules are simply being consumed at the maximum rate.

11. One way is to rearrange the answer to 10(a) to

$$\frac{1}{t}\ln\left(\frac{[\mathrm{S}]}{[\mathrm{S}]_0}\right) = \frac{1}{K_M}\left(\frac{[\mathrm{S}]_0 - [\mathrm{S}]}{t}\right) - \frac{V_{max}}{K_M}$$

and graph the left-hand side versus $([\mathrm{S}_0] - [\mathrm{S}])/t$. The slope will give K_M, and the intercept at $(1/t)\ln([\mathrm{S}]/[\mathrm{S}]_0) = 0$ will give V_{max}.

12. The change in k_{cat} between pH 6 and pH 7 must involve loss of a proton in the active site. The best candidate is His 57. The increase in K_M at higher pH must involve a change in the binding site. The group involved is probably the N-terminus on Ile 16, created by the cleavage that activates chymotrypsin.

13. (a) If we take k_{cat}/K_M ratios, we find that PAPAF would be digested most rapidly and PAPAG most slowly.
(b) A hydrophobic residue C-terminal to the bond cleaved seems to be favored. Elastase always requires a small residue (like Ala) to the N-terminal side.
(c) Serine and histidine.

14. (a) The enzyme must be stable both to the presence of detergents and to moderately high temperatures.
(b) Replace the methionine, by site-directed mutagenesis, with another residue. Since methionine is quite hydrophobic, a hydrophobic replacement would seem appropriate. A single base change in the codon could yield Phe, Leu, Ile, or Val.

15. (a) Noncompetitive.
(b) No inhibitor: K_M = 2.5 mmol/L, V_{max} = 5.0 (mmol/L)min^{-1}.
With inhibitor: K_M = 2.5 mmol/L, V_{max} = 3.0 (mmol/L)min^{-1}.

16. (a) Competitive.
(b) V_{max} = 5.0 (mmol/L)min^{-1} at all inhibitor concentrations.

[I] (mmol/L)	K_M^{app} (mmol/L)
0	2.5
3	3.8
5	4.0

(c) K_I = 4.3 mmol/L.

17. According to equation (11.33), we can write the Michaelis–Menten equation as $V = V_{max} - K_M V/[\mathrm{S}]$, so we plot V versus $V/[\mathrm{S}]$. The intercept at $V/[\mathrm{S}] = 0$ will give V_{max}; the slope will be K_M^{app}. Therefore, the graphs will look like the following sketches:

(a) Competitive inhibition $[\mathrm{I}]_1 < [\mathrm{I}]_2$

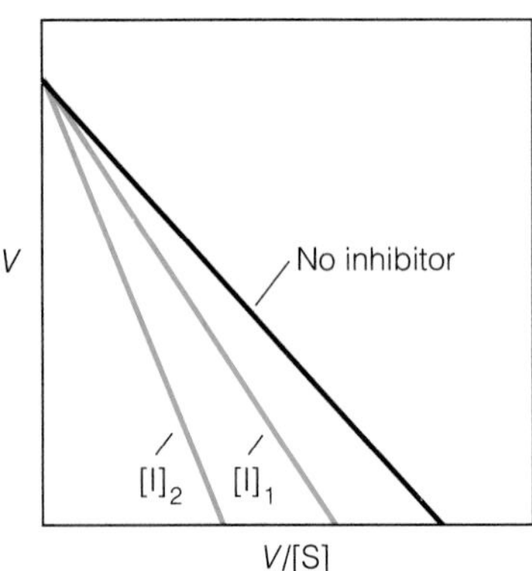

(b) Noncompetitive inhibition $[I]_1 < [I]_2$

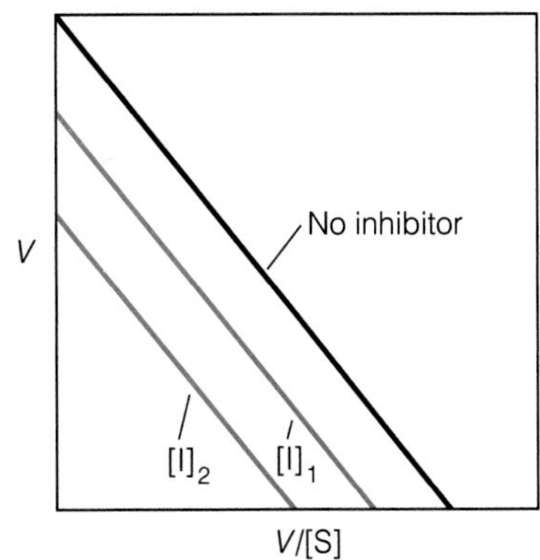

18. One should replace the hydrophobic group in TPCK with a positively charged group. A good candidate would be

$H_3\overset{+}{N}-(CH_2)_4-CH(NH-SO_2-C_6H_4-CH_3)-C(=O)-CH_2Cl$

19. $V = V_{max}[S]/\{K_M + [S](1 + [I]/K_I)\}$

$$= \frac{V_{max}[S]/(1 + [I]/K_I)}{[S] + K_M/(1 + [I]/K_I)}$$

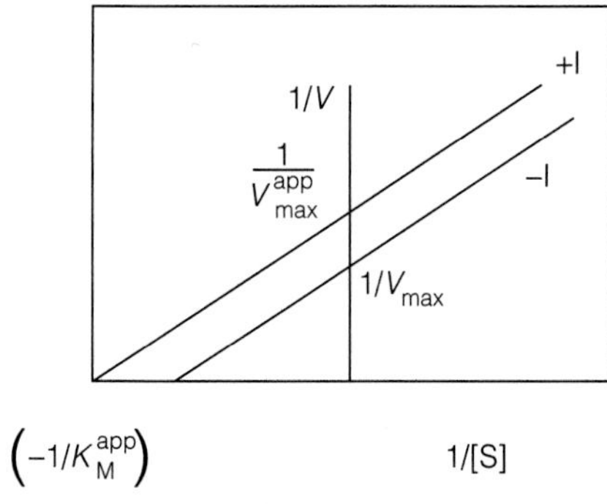

Here, $K_M^{app} = K_M/(1 + [I]/K_I)$, $V_{max}^{app} = V_{max}/(1 + [I]/K_I)$.

20. (a) If we make a graph of V versus [S], the data show a slightly sigmoidal curve. A more convincing demonstration is to use a Lineweaver–Burk plot or an Eadie–Hofstee plot. Neither is linear.
(b) From the Lineweaver–Burk plot, we can estimate V_{max} to be approximately 5 (mmol/L)min^{-1}.

21. The equation is

$$\log\left(\frac{V/V_{max}}{1 - V/V_{max}}\right) = \log [S] - \log K_M$$

(a) The Hill plot of $\log\left(\frac{V/V_{max}}{1 - V/V_{max}}\right)$ versus log [S] will approach straight lines with slope = 1 at low and high values of log [S]. The former will yield K_M^T, the latter K_M^R.
(b) The curve is clearly cooperative. *Estimates* of K_M^T and K_M^R are about 4.0 mM/L and 0.25 mM/L, respectively.

22.

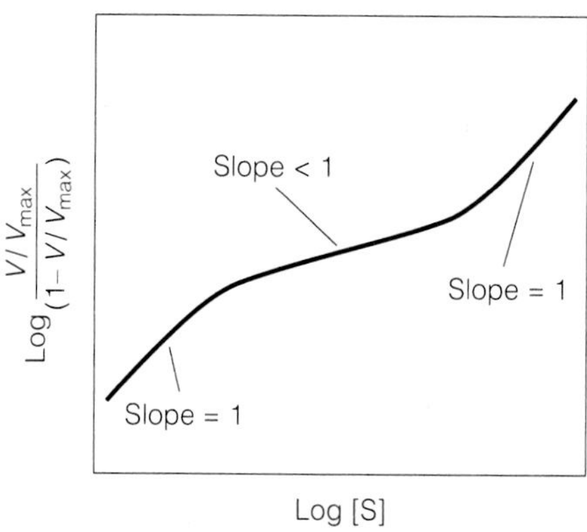

23. In the MWC theory, the more favorable binding to the R state makes the equilibrium shift as [S] is increased; the shift can only be toward R. In the KNF theory, binding to one site can have *any* kind of effect on another, including inhibition.

24. (a) The key to the derivation is to note that

$$\theta = \frac{\text{number of sites occupied}}{\text{total number of sites}}$$

$$= \frac{[RS] + 2[RS_2]}{2([T] + [R] + [RS] + [RS_2])}$$

Expressions given for [T], [RS], and $[RS_2]$ are inserted and the equation simplified.
(b) This must be done graphically. A large L and a small K ensure a sigmoidal curve.

25. (a) Activation of the molecule by trypsin will be blocked.
(b) If activation occurs, the N-terminal peptide will no longer be constrained by an S S bond and may be released.
(c) Automodification of π-chymotrypsin to α-chymotrypsin would be blocked.

26. The simplest technique would be to incubate the ribozyme with a radiolabeled pentanucleotide and collect samples from time to time for electrophoresis. Autoradiography should initially show only the pentanucleotide, but with time both longer and shorter labeled oligomers should appear on the gel.

27.

CHAPTER 12

1.

Equation	RQ
(a) $C_2H_5OH + 3O_2 \longrightarrow 2CO_2 + 3H_2O$	0.67*
(b) $CH_3COOH + 2O_2 \longrightarrow 2CO_2 + 2H_2O$	1.0
(c) $CH_3(CH_2)_{16}COOH + 26O_2 \longrightarrow 18CO_2 + 18H_2O$	0.69
(d) $CH_3(CH_2)_{14}(CH)_2COOH + 25\frac{1}{2}O_2 \longrightarrow 18CO_2 + 17H_2O$	0.71
(e) $CH_3(CH_2)_{12}(CH)_4COOH + 25O_2 \longrightarrow 18CO_2 + 16H_2O$	0.71

*Example: RQ = $2CO_2/3O_2$.

2. (a) >1.
   (b) <1.
   Since NAD^+-dependent enzymes usually act to dehydrogenate substrates, an $[NAD^+]/[NADH]$ ratio greater than unity tends to drive reactions in that direction. Similarly, $[NADP^+]/[NADPH]$ ratios less than unity provide concentrations that tend to drive these reactions in the direction of substrate reduction.

3. (a) 0.
   (b) 1.
   (c) 4.

4. For glucose: 2815 kJ/mol, 469 kJ/C atom.
   For palmitic acid: 9958 kJ/mol, 622 kJ/C atom.

5. −45.2 kJ/mol.

$$\Delta G' = \Delta G^{\circ\prime} + RT\ln([ADP][P_i]/[ATP])$$

6. (a) 20, in the direction B to C (i.e., 100 − 80).
   (b) (1) The rightward flux decreases 2-fold. (2) The rightward flux increases by 50%. (3) Flux would change to leftward, with a value of 60.
   (c) Specific inhibitors of enzyme X or Y would be useful. One could partially inhibit either activity and observe the effect on flux rate. If a substrate cycle is operating, the change in flux rate should be greater than the percent change in activity of either enzyme.

7. The most direct way is to purify the enzyme to homogeneity and determine both its molecular weight and turnover number. Then, from the activity observed in a crude extract, one can calculate the number of active enzyme molecules needed to achieve that activity (assuming that the extract does not contain inhibitors of the activity). Another approach is to treat the extract with a specific antibody to the enzyme of interest and quantitate the protein immunoprecipitated. One still needs to know the molecular weight to convert this value to number of molecules of enzyme.

8. (a) 20% (600 cpm/pmol ÷ 3000 cpm/pmol).
   (b) 7500 cpm dTTP incorporated per minute per 10^6 cells, which gives 2.5 pmol per cell per minute per 10^6 cells, or 1.5×10^6 molecules per cell per minute.
   (c) Prepare an acid extract of the cells (e.g., 5% trichloroacetic acid), separate the nucleotides by ion-exchange HPLC, and determine in the dTTP fraction its radioactivity and its mass, the latter from UV absorbance.

9. (a) 9131.
   (b) 2.6%.
   This problem is solved by using the differential form of the radioactive decay equation, $dN/dt = \lambda N$, where dN/dt is the radioactive decay rate in dps, λ is $0.693/t_{1/2}$, and N is the number of radioactive atoms.

CHAPTER 13

1. Neither; the same.

$$\Delta G' = \Delta G^{\circ\prime} + RT\ln([FBP][ADP]/[F6P][ATP])$$

2. Ethanol will generate NADH, through action of alcohol dehydrogenase, and this will reduce formaldehyde back to methanol, also through alcohol dehydrogenase. The acetaldehyde formed can be metabolized further to acetate. Also, ethanol may simply compete with formaldehyde for binding to alcohol dehydrogenase.

3. Probably F1,6BP. As triose phosphate began to accumulate, the unfavorable equilibrium for the forward reaction might drive both DHAP and G3P back to F1,6BP. Of course, G3P would accumulate first, but probably not to significantly increased levels.

4. (a) 4 (2 each from glucose and fructose).
   (b) 5 (3 from G6P + 2 from fructose).

5. C-3 or C-4. Both become C-1 of pyruvate, which is lost as CO_2 in the pyruvate decarboxylase reaction.

6. (a) Sucrose + $5P_i$ + 5ADP $\longrightarrow$ 4 lactate + 5ATP + $4H_2O$ + $4H^+$
   (b) Maltose + $4P_i$ + 4ADP + $4NAD^+$ $\longrightarrow$ 4 pyruvate + 4ATP + $3H_2O$ + 4NADH + $4H^+$
   (c) (Glucose) residue + $2P_i$ + 2ATP $\longrightarrow$ 2 ethanol + 2ATP + $2CO_2$ + H_2O + $2H^+$

7. Glyceraldehyde-3-phosphate dehydrogenase. The acyl arsenate analog of 1,3-bisphosphoglycerate spontaneously hydrolyzes.

8. 1.09 M (10% EtOH is 2.18 M, and 1 mole of glucose generates 2 moles of ethanol). No, because glucose is not an abundant free component of most fermentable plant sources. Starch, sucrose, or some other oligosaccharide or polysaccharide.

9.

Gal
Blocked in galactokinase deficiency
Gal-1-P
UDP-Glc
Blocked in epimerase deficiency
Blocked in uridylyl transferase deficiency
UDP-Gal
G1P
G6P

10. Pyruvate $\xrightarrow[\text{Coenzyme A}]{}$ formate + acetyl-CoA $\xrightarrow{P_i}$ acetyl phosphate + CoA-SH

acetyl phosphate $\xrightarrow{P_i \rightarrow ATP}$ Acetate

11. Glc ⟶ G6P ⟶ G1P ⟶ UDP-Glc ⟶ UDP-Gal $\xrightarrow{\text{Glc}}$ lactose

2 Glucose + ATP + UTP ⟶ lactose + ADP + UDP + PP_i

12.

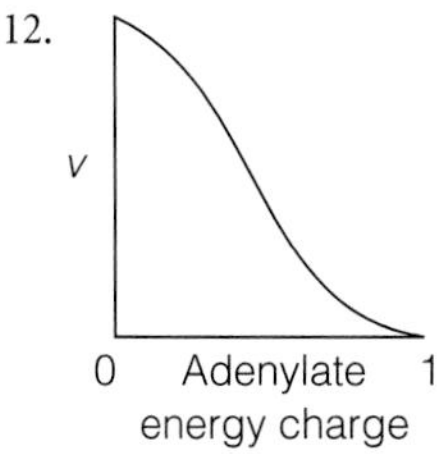

13. UDP-glucose is used catalytically in the sense that for each molecule of Gal-1-P converted to G1P, one molecule of UDP-Glc is converted to UDP-Gal, which is immediately reconverted to UDP-Glc, to react with another molecule of Gal-1-P.

14.

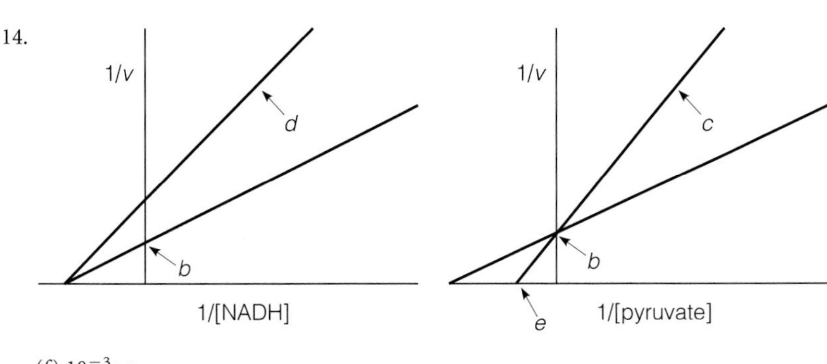

(f) 10^{-3} M.

CHAPTER 14

1. Administer separately 1-[^{14}C]glucose and 6-[^{14}C]glucose, and measure initial rates of $^{14}CO_2$ formation. The ratio gives relative flux rates through the two pathways. For example, if the flux rates are equal, the ratio from C^1-labeled glucose to C^6-labeled glucose is 2:1.

2. (a) $G6P + 2NADP^+ + H_2O \longrightarrow R5P + CO_2 + 2NADPH + 2H^+$

(b) $G6P + 12NADP^+ + 7H_2O \longrightarrow 6CO_2 + 12NADPH + 12H^+ + P_i$

(c) $3G6P + 6NADP^+ + 25NAD^+ + 5FAD + 13ADP + 10P_i + 5H_2O \longrightarrow 18CO_2 + 6NADPH + 25NADH + 13ATP + 5FADH_2 + 12H^+$

3. C-1: all released as CO_2. C-2 and C-3: all retained in oxaloacetate.

4. One turn: one-quarter. Two turns: three-eighths.

5. The action of pyruvate carboxylase on the labeled pyruvate would yield oxaloacetate labeled so that on a turn through the citric acid cycle it could label C-5 of isocitrate.

6. Addition to isolated glyoxysomes of citrate, isocitrate, glyoxylate, malate, or oxaloacetate would stimulate succinate formation out of proportion to the amount added.

7. C-3 and C-4, since these become the carboxyl group of pyruvate, which is lost in the pyruvate dehydrogenase reaction.

8. C-1 and C-3 of fructose-6-phosphate should be labeled. Erythrose-4-phosphate should be unlabeled.

9. Because NADH and acetyl-CoA activate the enzyme, it makes metabolic sense to expect that NAD^+ and CoA-SH would be inhibitory, and these inhibitions are observed.

10. First, the cytosolic location of much of the $NADP^+$-dependent enzyme raises suspicions about its role in the citric acid cycle. More important, cells that have high or low flux rates through the cycle have high or low activities, respectively, of the NAD^+-dependent enzyme. Activity of the $NADP^+$-dependent enzyme does not vary in coordination with activities of other cycle enzymes.

11. E_1: Pyruvate + TPP + H^+ ⟶ hydroxyethyl-TPP + CO_2.

E_2: Hydroxyethyl-TPP + oxidized lipoamide + CoA-SH ⟶ TPP + reduced lipoamide + acetyl-CoA.

E_3: Reduced lipoamide + NAD^+ ⟶ oxidized lipoamide + NADH + H^+.

12. (a) This is a signal that levels of substrates are adequate for citric acid cycle oxidation, so that pyruvate can be shunted into gluconeogenesis instead of being oxidized.
(b) This tends to inactivate pyruvate dehydrogenase when levels of reduced electron carriers are sufficient for ATP production via the respiratory chain and, hence, to make pyruvate available for other purposes.
(c) This is a signal to reduce flux through the citric acid cycle when levels of reduced electron carriers are adequate for energy generation.
(d) When the energy charge is low, the accumulation of ADP provides a signal to activate the citric acid cycle and thereby increase the oxidation of nutrients for ATP production.
(e) This is equivalent to product inhibition, but it also serves as a general indicator that when an energy-rich substrate (succinyl-CoA) is abundant, flux through the citric acid cycle can be reduced.
(f) This is also energy charge regulation—a signal that ATP production needs to be enhanced. This can be accomplished by augmenting the production of acetyl-CoA from pyruvate, for oxidation in the citric acid cycle.

13. Some possible mechanisms: substrate-level control of citrate synthase, activation of citrate lyase by acetyl-CoA or fatty acids, inhibition of isocitrate lyase by succinate (to ensure adequate flux through the citric acid cycle).

14.

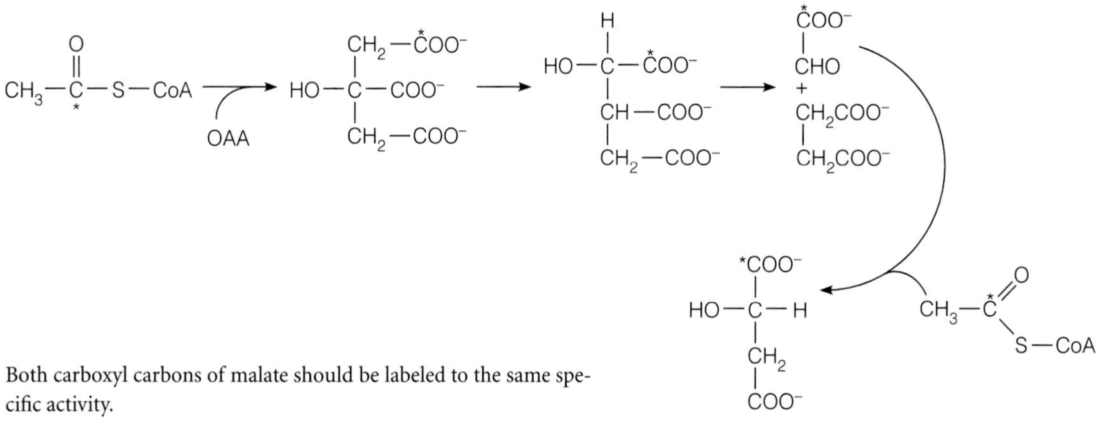

Both carboxyl carbons of malate should be labeled to the same specific activity.

15.

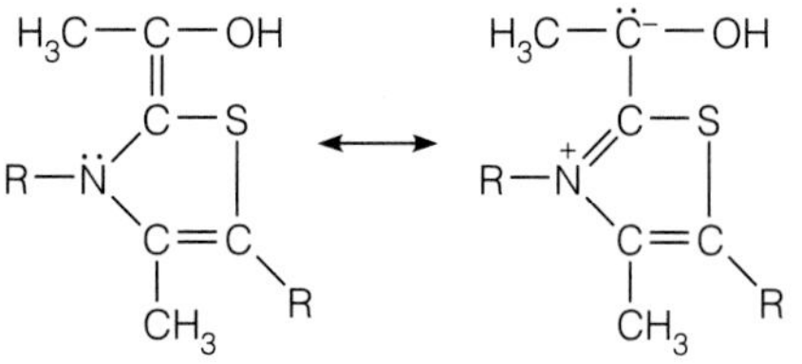

16. 2 Acetyl-CoA + $2NAD^+$ + FAD + $3H_2O$ ⟶ oxaloacetate + 2NADH + $FADH_2$ + 2CoA-SH + $4H^+$

17. NAD^+ + $FADH_2$ ⇌ NADH + FAD + H^+
NAD^+ can oxidize $FADH_2$ if its local concentration is much higher than that of $FADH_2$.

18.

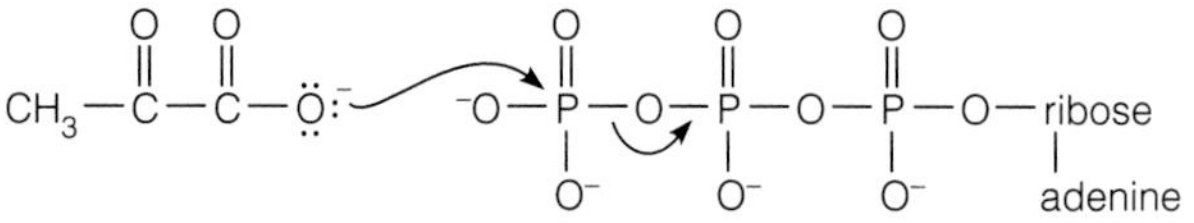

19. $[NAD^+]/[NADH]$ should be high, so that it can promote the oxidation of substrates, e.g., malate + NAD^+ ⇌ oxaloacetate + NADH + H^+. Conversely, since NADPH and $NADP^+$ usually promote reduction of substrates, we expect $[NADP^+]/[NADPH]$ to be low.

CHAPTER 15

1. +28.95 kJ/mol—from the equation $\Delta G^{\circ\prime} = -nF\,\Delta E_0'$. $E_0' = -0.17 - (-0.32)$.

2. (a) cyt *c* ⟶ cyt *a* ⟶ cyt a_3 ⟶ O_2

(b) To block oxidation of endogenous substrates.
(c) One site is associated with cytochrome oxidase.

(d) 2 cyt *c*-Fe^{2+} + ½O_2 + $4H^+$ + ADP + P_i ⟶ 2 cyt *c*-Fe^{3+} + ATP + $2H_2O$

(e) −70 kJ/mol (calculated as in Problem 1).

3. (a) β-Hydroxybutyrate ⟶ NAD^+ ⟶ FMN ⟶ FeS ⟶ CoQ ⟶ cyt *b* ⟶ FeS ⟶ cyt c_1 ⟶ cyt *c*

(b) 2, because cytochrome oxidase is bypassed.
(c) Because NADH cannot freely enter the mitochondrion.
(d) To block cytochrome oxidase, so that electrons exit the chain at cytochrome *c*.

(e) β-Hydroxybutyrate + 2 cyt *c*-Fe^{3+} + 2ADP + $2P_i$ + $4H^+$ ⟶ acetoacetate + 2 cyt *c*-Fe^{2+} + 2ATP + $2H_2O$

(f) −35.8 kJ/mol (calculated as in Problem 1).

4. To block succinate dehydrogenase and measure phosphorylation resulting only from the α-ketoglutarate dehydrogenase reaction. P/O ratio = 4 (3 from NADH and 1 from the succinyl-CoA synthetase reaction).

5. +67.6 kJ/mol (solved as in Problem 1). Ratio: 4.0×10^{10}, calculated from the expression

$$\Delta G' = \Delta G^{\circ\prime} + 2.3RT \log\{([\text{fumarate}][\text{NADH}])/([\text{succinate}][\text{NAD}^+])\}$$

where $\Delta G' < 0$.

6. [ADP] = 1.44 mM and [AMP] = 0.29 mM, calculated from two simultaneous equations: the defining equation for energy charge, and [ADP] = 5 × [AMP]. $\Delta G' = -41.2$ kJ/mol—from $\Delta G' = \Delta G^{\circ\prime} + 2.3RT \log([\text{ADP}][\text{P}_i])/[\text{ATP}]$.

7. 1.8×10^{17}. Calculate $\Delta G^{\circ\prime}$ as in Problem 1, and then apply $\Delta G^{\circ\prime} = -2.3RT \log K_{eq}$. Use $E_0' = +0.30$ for H_2O_2 (see Table 15.1).

8. Add ADP in limiting amount and measure O_2 uptake. The ratio of μmol ADP consumed to μatom oxygen taken up is identical to the P/O ratio (see Figure 15.22).

9. Because the energy not used for ATP was dissipated as heat, and the subjects developed uncontrollable fevers.

10. (a) A, 4; B, 2; C, 1; D, 3; E, 5.
(b) 3 divided by 4 (ratio of uptakes in presence and absence of ADP, with substrate present for both measurements).
(c) 5; substrates depleted.
(d) 4; ADP level low because ATP level is high.
(e) 3; rapid ATP production and turnover demand rapid and continuous O_2 uptake.

11. 1.

12. Deamination:

$$R{-}CH_2{-}NH_2 + O_2 + H^+ + NADPH \longrightarrow R{-}\overset{O}{\overset{\|}{C}}{-}H + NH_3 + H_2O + NADP^+$$

Desulfuration:

$$R{-}CH_2{-}SH + O_2 + H^+ + NADPH \longrightarrow R{-}\overset{O}{\overset{\|}{C}}{-}H + H_2S + H_2O + NADP^+$$

Peroxidation:

$$R{-}CH_3 + O_2 \longrightarrow R{-}CH_2{-}OOH$$

Epoxidation:

$$R_1{-}CH_2{-}CH_2{-}R_2 + O_2 \longrightarrow R_1{-}CH{-}CH{-}R_2 + H_2O \text{ (CH–CH bridged by O)}$$

Dehalogenation:

$$R{-}CH_2{-}Br + O_2 + NADPH \longrightarrow R{-}\overset{O}{\overset{\|}{C}}{-}H + Br^- + H_2O + NADP^+$$

13. Because it must work against a concentration gradient.

14. (a) NADH is oxidized by FMN, not FAD.
(b) Reduced flavin ($FADH_2$ or $FMNH_2$) is oxidized by CoQ, not $CoQH_2$.
(c) cyt c_1 accepts e^- from cyt *b*, which is then oxidized by cyt *c*.
(d) O_2 is reduced to H_2O, not H_2O_2.

15. (a) $\Delta G^{\circ\prime} = -2 \times 96.5(-0.23 - (-0.32)) = -17.37$ kJ/mol
(b) The same, because E_0' for $NAD^+/NADH$ is the same as for $NADP^+/NADPH$.
(c) $\Delta G'$ would probably be positive in vivo for an NADH-linked enzyme, because a high $NAD^+/NADH$ concentration ratio would promote the oxidation of GSH to GSSG.

16. (a) Succinate ⟶ FAD ⟶ CoQ ⟶ cyt *b* ⟶ cyt c_1 ⟶ cyt *c*
(b) Succinate + 2 cyt c–Fe^{3+} + ADP + P_i ⟶ fumarate + 2 cyt c–Fe^{2+} + ATP + H_2O
(c) $\Delta G^{\circ\prime} = -nF(0.25 - 0.03) + 31 = -11.46$ kJ/mol
(d) To block cytochrome oxidase and force electrons to exit the respiratory chain at cytochrome *c*.

CHAPTER 16

1. (a) 3:

$$2\text{DHAP} \longrightarrow \text{F1,6BP} \xrightarrow{1\text{ATP}} \text{F6P} \longrightarrow \text{G6P} \longrightarrow \text{G1P} \xrightarrow{1\text{ATP}} \text{UDPG} \longrightarrow \text{glucosyl residue}$$

(b) 2.
(c) 12 (see Figure 16.3).
(d) 1.

2. (a) 2 (glycolysis).
(b) 6 (gluconeogenesis requires ATP).

3. (a), because it is the only substrate that must go through the pyruvate carboxylase reaction.

4. None. The CO_2 that is fixed comes off in the PEP carboxykinase reaction.

5. $$\text{Glucose-1-P} + \text{UTP} \longrightarrow \text{UDP-Glc} + \text{PP}_i \xrightarrow{\text{C-4 epimerization}} \text{UDP-Gal}$$

$$\text{Fructose-6-P} \longrightarrow \text{mannose-6-P} \longrightarrow \text{mannose-1-P} \xrightarrow{\text{GTP} \;\; \text{PP}_i} \text{GDP-mannose}$$

$$\text{Gal-1-P} \xrightarrow{\text{CTP} \;\; \text{PP}_i} \text{CDP-Gal} \xrightarrow{\text{Reduction of C-2}} \xrightarrow{\text{Reduction of C-6}} \text{CDP-D-abequose}$$

$$\text{Gal-1-P} \xrightarrow{\text{dTTP} \;\; \text{PP}_i} \text{dTDP-Gal} \xrightarrow{\text{Epimerization of C-3}} \xrightarrow{\text{Epimerization of C-5}} \xrightarrow{\text{Reduction of C-6}} \text{dTDP-L-rhamnose}$$

6. (a) Fructose-6-phosphate + ATP $\longrightarrow$ fructose-2,6-bisphosphate + ADP

(b) 2 Oxaloacetate + 2ATP + 2GTP + 2NADH + $2H^+$ + $4H_2O$ $\longrightarrow$ glucose + $2CO_2$ + $2NAD^+$ + 2ADP + 2GDP + $4P_i$

(c) Glucose + ATP + UTP $\longrightarrow$ UDP-Glc + PP_i + ADP

(d) 2 Glycerol + 2ATP + $2NAD^+$ + $2H_2O$ $\longrightarrow$ glucose + 2ADP + 2NADH + $2H^+$ + $2P_i$

(e) 2 Malate + 2ATP + 2GTP + $3H_2O$ $\longrightarrow$ glucose-6-phosphate + $2CO_2$ + 2ADP + 2GDP + $3P_i$

7.

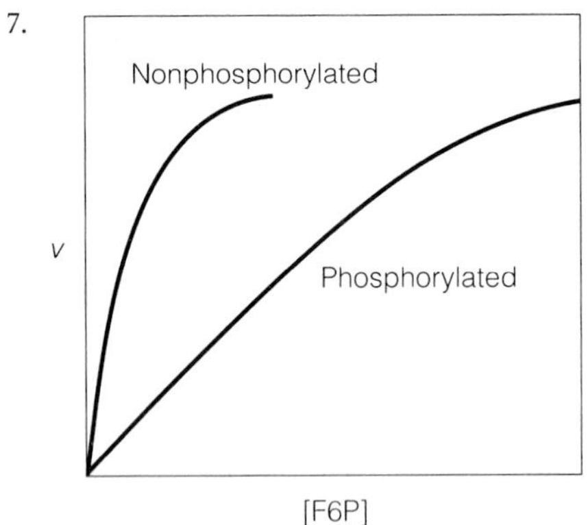

8.

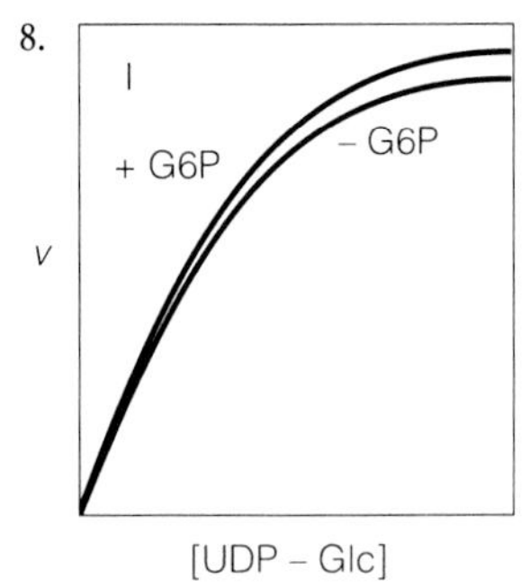

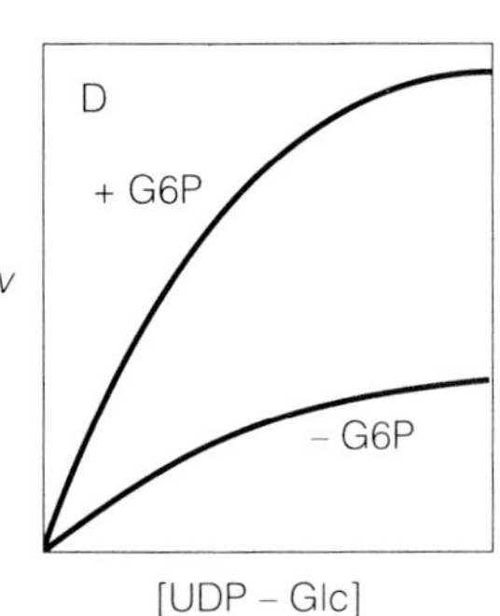

9. AMP activation of glycogen phosphorylase *b*; glucose-6-P activation of glycogen synthase D.

10. 8 ATPs and 4 acetyl-CoAs:

$$\text{Acetyl-CoA} \longrightarrow \text{citrate} \longrightarrow \text{isocitrate} \longrightarrow \text{glyoxylate} \longrightarrow \text{malate} \longrightarrow \text{oxaloacetate} \xrightarrow{\text{2ATP}} \text{PEP} \longrightarrow$$
$$\text{2PG} \longrightarrow \text{3PG} \xrightarrow{\text{2ATP}} \text{1,3BPG} \longrightarrow \text{G3P} \longrightarrow \text{F1,6BP} \xrightarrow{\text{2ATP}} \text{F6P} \longrightarrow \text{G6P} \longrightarrow \text{glucose}$$

Two or more ATPs would be needed to reconvert the glucose to G6P, giving an investment of 8 ATPs per mole of glucose.

11. Because pyruvate carboxylase has two metabolic roles—replenishment of citric acid cycle intermediates and initiation of gluconeogenesis. PEPCK catalyzes the first reaction committed to gluconeogenesis.

12. The main factor is the intracellular location of pyruvate—if it accumulates in mitochondria, the anaplerotic role predominates, whereas pyruvate in cytosol is used primarily as a gluconeogenic precursor. The levels of other intermediates are also important. For example, if intramitochondrial levels of oxaloacetate are adequate, then pyruvate can be either oxidized by pyruvate dehydrogenase or exported for gluconeogenesis. Oxaloacetate in mitochondria can also be converted by transamination to aspartate, however, which can be exported to the cytosol for gluconeogenesis.

13. (1) The liver is the most active gluconeogenic tissue. Since alanine is an important gluconeogenic precursor, its accumulation in liver is a signal that gluconeogenesis should be activated, with simultaneous inhibition of glycolysis.

(2) The function of glucose-6-phosphatase is production of glucose, for exit from the cell and transport to other tissues. Therefore, the enzyme plays a role only in those tissues (primarily liver) that manufacture glucose for export.

14. PP-1 inactivates SPK, by catalyzing its dephosphorylation. This blocks the conversion of glycogen phosphorylase *b* to *a*, thereby inhibiting glycogenolysis. PI-1, by inhibiting PP-1, has the opposite effect.

15. (a) The function of glucagon is to increase blood glucose concentration. Stimulation of glycogen breakdown and gluconeogenesis are both consistent with this function.
(b) The glucose-6-phosphatase deficiency would interfere with release of glucose from the liver for export to other tissues.
(c) SPK converts phosphorylase *b* to *a*, which activates glycogen breakdown, and it converts glycogen synthase I to D, the less active form of this enzyme.
(d) An AB donor's blood contains A antigen, which is immunogenic in type B or O individuals, and B blood group substance, which is immunogenic in type A or O individuals.
(e) Glucose-6-phosphate must be hydrolyzed in order for glucose to exit the liver cell and be exported to other tissues.

CHAPTER 17

1. $\Delta G° = 164$ kJ/mol. An einstein of 700 nm light would yield 171 kJ.

2. 14.7%.

3. 1050 nm.

4. (a) After being attached to C-2 of ribulose-1,5-bisphosphate, the ^{14}C will become the carboxylate (C-1) of *one* of the two molecules of 3-phosphoglycerate. This becomes the carbonyl carbon (C-1) in G3P, or C-1 in DHAP. Upon condensation to form fructose-1,6-bisphosphate, the ^{14}C will show up in carbons 3 and/or 4.
(b) No. They may carry 0, 1, or 2 labeled carbons, depending on what combination of labeled and unlabeled trioses has been used in their formation.

5. Assuming 48 photons per mole of hexose, the leaf could theoretically produce 0.0263 mol, or 4.73 g of hexose in 1 hour. It will, in fact, produce only a small fraction of this, for not all photons are absorbed, nor do all absorbed photons serve to pass electrons through the photosynthetic pathway.

6. (a) No, because plastoquinones are not involved in this process.
(b) Addition of ferricyanide as an electron donor allows a Hill reaction.

7. By using illumination at both 700 and 680 nm, either singly or together. Observe oxidation state of plastoquinones, and plastocyanin. If only p700 is involved, these will be oxidized. If only p680, they will be reduced. If both are involved, simultaneous illumination at the two wavelengths will give enhanced oxidation of the donor.

8. Carbon 3.

9. Competitive.

10. The data suggest the strong binding of an inhibitor. It is apparently heat stable, so probably not a protein, but is released on ammonium sulfate precipitation. Other data indicate that the inhibitor is a low-molecular-weight phosphorylated compound.

11. (a) Decrease.
(b) Increase.
(c) Increase.

12. (a) Because both O_2 and CO_2 are being consumed and produced by the opposing processes of photorespiration and photosynthesis, a steady state ratio will be attained.
(b) The relative affinity of rubisco for CO_2 and O_2.

13. Most of the 3PG that is initially labeled is reused in replacing RuBP, which then reenters the cycle.

CHAPTER 18

1. Palmitic acid, 129 ATP; linoleic acid, 142 ATP; stearic acid, 146 ATP; oleic acid, 144 ATP.

2. 40.5% [(130 × 30.5/9788) × 100%].

3. 412 (130 from each palmitate, 22 from glycerol). 412 ATP/51 carbons = 8.1 ATPs per carbon atom (6.3 for glucose).

4. Example for palmitic acid (c) follows.

$$\text{CoA-SH + palmitate + ATP} \longrightarrow \text{palmitoyl-CoA + AMP + PP}_i$$

$$\text{PP}_i + \text{H}_2\text{O} \longrightarrow 2\text{P}_i$$

$$\text{Palmitoyl-CoA} + 7\text{CoA-SH} + 7\text{FAD} + 7\text{NAD}^+ + 7\text{H}_2\text{O} \longrightarrow 8\text{ acetyl-CoA} + 7\text{FADH}_2 + 7\text{NADH} + 7\text{H}^+$$

$$8\text{ Acetyl-CoA} + 16\text{H}_2\text{O} + 24\text{NAD}^+ + 8\text{FAD} + 8\text{ADP} + 8\text{P}_i \longrightarrow 16\text{CO}_2 + 24\text{NADH} + 24\text{H}^+ + 8\text{ATP} + 8\text{FADH}_2 + 8\text{CoA-SH}$$

$$\text{Sum: Palmitate} + 24\text{H}_2\text{O} + 15\text{FAD} + 31\text{NAD}^+ + 8\text{ADP} + 6\text{P}_i \longrightarrow 16\text{CO}_2 + \text{AMP} + 15\text{FADH}_2 + 31\text{NADH} + 31\text{H}^+ + 7\text{ATP}$$

(continued)

Now add the equations for the metabolic oxidation of NADH and $FADH_2$.

$$31NADH + 31H^+ + 15\tfrac{1}{2}O_2 + 93ADP + 93P_i \longrightarrow 31NAD^+ + 93ATP + 124H_2O$$

$$15FADH_2 + 7\tfrac{1}{2}O_2 + 30ADP + 30P_i \longrightarrow 15FAD + 45H_2O + 30ATP$$

$$\text{Sum: Palmitate} + 23O_2 + 131ADP + 129P_i \longrightarrow 16CO_2 + AMP + 130ATP + 145H_2O$$

5. Acetoacetyl-CoA $\longrightarrow$ 2 acetyl-CoA — 0 ATP
2 Acetyl-CoA $\longrightarrow$ $4CO_2$ — 24 ATP (two turns of citric acid cycle)
Sum: 24 ATP

Propionoacetyl-CoA $\longrightarrow$ propionyl-CoA + acetyl-CoA — 5 ATP (one cycle of β-oxidation)
Propionyl-CoA $\longrightarrow$ succinyl-CoA — −1 ATP
Succinyl-CoA $\longrightarrow$ $4CO_2$ — 24 ATP (two turns of citric acid cycle)
Acetyl-CoA $\longrightarrow$ $2CO_2$ — 12 ATP (one turn of citric acid cycle)
Sum: 40 ATP

6. Carnitine acyltransferase I. If inhibition occurred at a later step, then palmitoylcarnitine oxidation would be inhibited, as well as that of palmitoyl-CoA.

7. The acyl-ACP produced by one subunit undergoes the next round of reductive two-carbon addition on the other subunit.

8. Acetyl-CoA $\xrightarrow{\text{Citric acid cycle}}$ oxaloacetate $\xrightarrow{\text{PEP carboxykinase}}$

PEP $\xrightarrow{\text{Gluconeogenesis}}$ glucose

The main point is that the carbons lost in one turn of the citric acid cycle are not the ones that entered as acetyl-CoA in that cycle.

9. Propionyl-CoA $\longrightarrow$ methylmalonyl-CoA $\longrightarrow$ succinyl-CoA $\longrightarrow$ oxaloacetate $\longrightarrow$ phosphoenolpyruvate $\longrightarrow$ glucose

10. 14 (2 from each molecule of labeled malonyl-CoA).

11. Increase in citrate levels would increase generation of acetyl-CoA in cytosol, hence stimulating fatty acid synthesis.

12. (1) Glucagon activates phosphorylation of pyruvate dehydrogenase, which inhibits the formation of acetyl-CoA, the substrate for the enzyme.

(2) Glucagon promotes triacylglycerol breakdown, yielding increased levels of fatty acids, which, as acyl-CoAs, could prevent polymerization and activation of acetyl-CoA carboxylase.

13. Palmitate and octanoate are activated by different fatty acyl-CoA ligases, only one of which is inhibited by malonyl-CoA.

14. Malonyl-CoA, a key intermediate in fatty acid synthesis, inhibits carnitine acyltransferase I, thereby blocking the entry of fatty acyl units into the mitochondrion for oxidation. Fatty acyl-CoAs, the substrates for fatty acid oxidation, inhibit fatty acid synthesis by interfering with the polymerization of acetyl-CoA carboxylase. Hormonal effects on adipocytes are opposed; insulin promotes fatty acid synthesis by several mechanisms, while glucagon promotes fat breakdown and fatty acid oxidation.

15. This could be a way for a cell to inhibit fatty acid synthesis under conditions where substrates are needed for oxidation, to provide ATP.

16. Increased glucose levels in the cytosol stimulate glycolysis, which provides pyruvate for oxidation in the mitochondrion. The resultant acetyl-CoA can return to the cytosol (as citrate) and generate precursors for increased fatty acid synthesis. The increased intracellular glucose also stimulates glycogen synthesis.

CHAPTER 19

1. Probably K_{eq} is close to unity, because the bond broken is identical to the bond created.

2. If phosphatidylserine synthase and phosphatidylserine decarboxylase (E_1 and E_2 in Figure 19.4) were juxtaposed, PS could never accumulate, because, once formed by E_1, it would immediately react with E_2. If the two enzymes are tightly coupled in the membrane, then addition of radiolabeled PS to an enzyme system would not label PE, because E_2 would act only on PS generated by E_1.

3. By stimulating release of arachidonic acid from membrane phospholipids. These in turn would be converted to prostaglandins, which contribute to inflammation.

4. Substitution of vegetable fats for animal fats could decrease cholesterol levels. This would ultimately decrease inhibition of HMG-CoA reductase levels by cholesterol, which could result in increased mevalonate levels.

5. Stearic acid + oleic acid + 3ATP + CTP + serine $\longrightarrow$ *sn*-1-stearoyl-2-oleylglycerophosphorylserine + 2AMP + $3PP_i$ + ADP + CMP

6. None.

7. Since this is a nonhydrolyzable analog of glycerol-3-phosphate, you might expect it to be acylated without difficulty to give the phosphonate analog of diacylglycerol. By acting as an analog of diacylglycerol, this could competitively inhibit the synthesis of CDP-diacylglycerol from phosphatidic acid.

8. 25% from the choline utilization pathway, 75% from the phosphatidylserine pathway. To know true intracellular rates of synthesis, you must also know the specific radioactivity of the final intermediate in each pathway, whether pools of intermediates are compartmentalized, and the rate of degradation of the product (see Tools of Biochemistry 12A).

9. Acetoacetate + succinyl-CoA ⟶ acetoacetyl-CoA + succinate

Acetoacetyl-CoA + acetyl-CoA ⟶ HMG-CoA + CoA-SH

HMG-CoA + 2NADPH + $2H^+$ ⟶ mevalonate + $2NADP^+$ + CoA-SH

10. By shutting down the pathway leading to aldosterone, this deficiency increases the supply of progesterone available for conversion to sex steroids.

11. The most straightforward route is desaturation of palmitoyl-CoA to palmitoleoyl-CoA, followed by a C_2 elongation at the carboxyl group, both described on page 659, to give *cis*-vaccenyl-CoA. Alternatively, stearoyl-CoA could be an enzyme-bound intermediate in a process starting with elongation and followed by desaturation.

12. The most straightforward way is to purify one of the enzymes, such as PE methylase, to homogeneity and then to ask whether that enzyme can catalyze three methylations to give PC, or whether it can act upon the monomethylated and dimethylated intermediates. Alternatively, one could follow all three activities through a fractionation procedure.

13. Cyclic AMP promotes triacylglycerol breakdown, through activation of hormone-sensitive lipase. This probably increases intracellular levels of diacylglycerol, which could in turn increase flux through the last reactions in the salvage pathways to PE and PC (see Figure 19.5).

14. Example: 2 isopentenyl pyrophosphate to limonene:

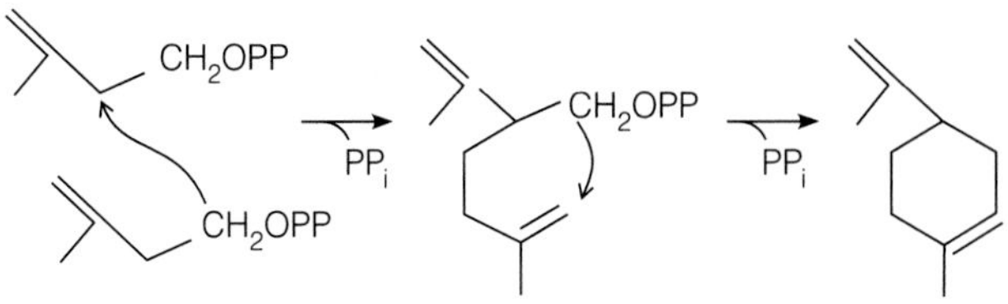

CH_2OPP + CH_2OPP ⟶ (PP_i) CH_2OPP ⟶ (PP_i)

15.

A	1	2	3	4	5	6
B	3	7	1	9	10	5
C	6	5	4	3	2	1

1. Glycerophospholipid; change to sphingolipid.
2. Chylomicrons; change to intestine.
3. Product of; change to substrate for.
4. Deficiency in; change to accumulation of.
5. Biosynthesis; change to degradation.
6. Breakdown; change to synthesis.

CHAPTER 20

1. Adenosyl-B_{12}; tetrahydrofolate; ATP + glutamine; α-ketoglutarate + pyridoxal phosphate; *S*-adenosylmethionine.

2. (a) 5-Methyltetrahydrofolate.
(b) Methylmalonate.
(c) Decreased affinity of enzyme C for B_{12} coenzyme.
(d) Homocystine (see Chapter 21).

3.

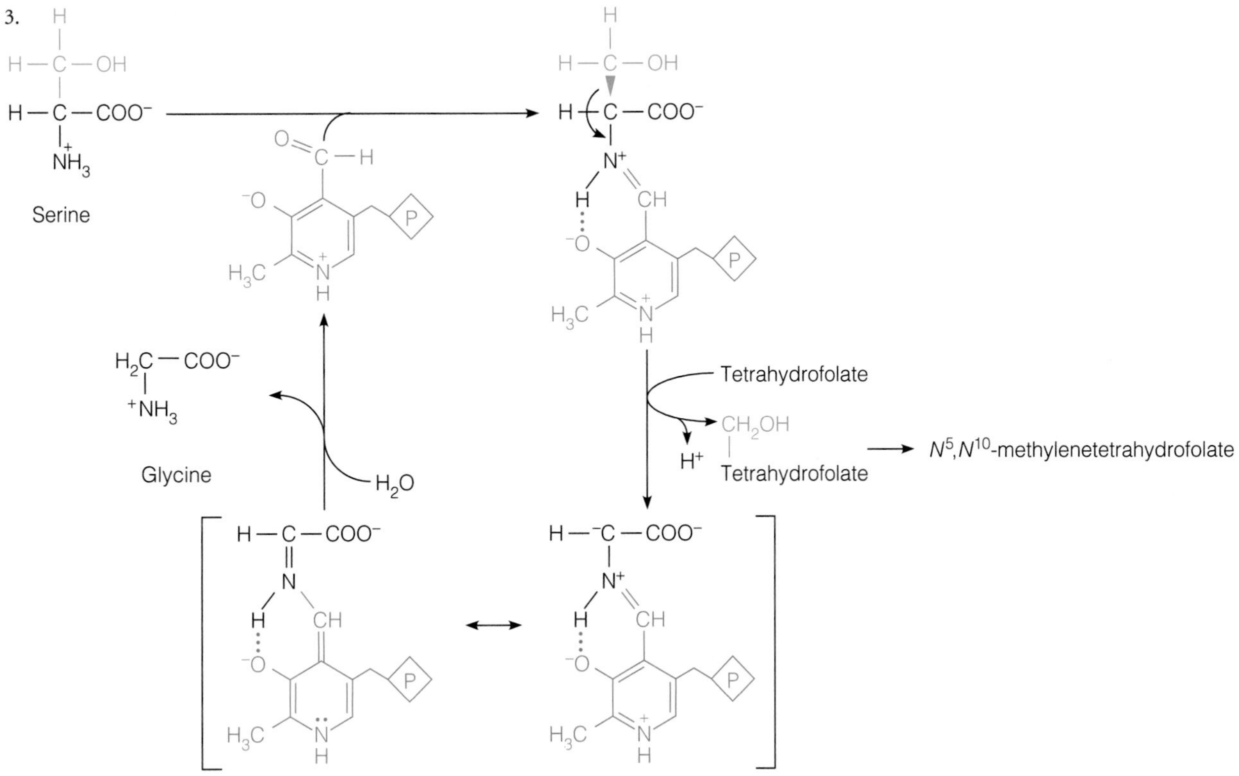

H–C(H)(OH)–C(H)(COO⁻)–NH_3^+

Serine

Glycine H_2C—COO^- / $^+NH_3$

H_2O

Tetrahydrofolate

CH_2OH Tetrahydrofolate ⟶ N^5,N^{10}-methylenetetrahydrofolate

H^+

4. Feeding folate will expand the intracellular pools of 5-methyltetrahydrofolate. This will increase flux through methionine synthase and, hence, divert the already shrunken store of B_{12}, reducing its availability for other functions.

5. $^{-}OOC^5$ — $C^4H(C^3H_3)$ — $C^2H(\overset{+}{N}H_3)$ — C^1OO^-

6. $\cdot CH_2R$–Co(+)–DMB ⇌ CH_2R–Co(+)–DMB

CH_2OH–CH_2OH → CH_2OH–$\cdot C(H)$–OH → HO–$C(H)(OH)$–$HC\cdot(H)$ → $HC(OH)$–CH_3 → (H_2O) → $O{=}CH$–CH_3

CH_3R–Co(+)–DMB

7. One plausible theory proposes that ammonia depletes pools of α-ketoglutarate through the glutamate dehydrogenase and glutamine synthetase reactions, which convert αKG to glutamate and glutamine, and that this diminishes ATP production by reducing flux through the citric acid cycle.

Oxaloacetate → Citrate → α-Ketoglutarate → Succinyl-CoA → Oxaloacetate

α-Ketoglutarate —(NH_3)→ Glu —(NH_3)→ Gln

8. CPS I is in mitochondria, and CPS II is in cytosol. Apparently, carbamoyl phosphate cannot cross the mitochondrial membrane, so that which is formed in mitochondria can be used only for arginine synthesis, and that formed in cytosol is used only for pyrimidine synthesis.

9. The injected enzyme should convert all circulating asparagine to aspartate. Normal cells would take up aspartate and resynthesize asparagine. However, leukemic cells would be unable to synthesize asparagine and would starve. On the other hand, an enzyme injected into the bloodstream would probably have too short a half-life to be effective. Moreover, being a foreign protein, it might cause an immunological reaction.

10. (a) True. The complete catabolism of amino acids yields CO_2, H_2O, and ammonia. However, the ammonia must be converted to urea for detoxification and excretion, and this requires ATP, which decreases the net ATP yield.
(b) False. Glutamate can be used to synthesize essential amino acids only if the carbon skeletons are available as keto acids.
(c) False. Arginine biosynthesis in liver is part of the urea cycle, and most of the arginine that is formed is cleaved to urea and ornithine. Little arginine is left over to meet the needs of other tissues.
(d) False. Although it is present in all proteins, alanine can be synthesized by mammalian cells and is not required in the diet.

11. $\overset{+}{N}H_4$ (muscle) + α-ketoglutarate + NADH + H^+ ⟶ glutamate + H_2O + NAD^+

Glutamate + pyruvate ⟶ α-ketoglutarate + alanine

Alanine + α-ketoglutarate ⟶ glutamate + pyruvate

Glutamate + H_2O + NAD^+ ⟶ $\overset{+}{N}H_4$ (liver) + α-ketoglutarate + NADH + H^+

Sum: $\overset{+}{N}H_4$ (muscle) ⟶ $\overset{+}{N}H_4$ (liver)

12. (a) NH_3 has an unshared electron pair that can initiate nucleophilic attack on the electron-poor carbonyl carbon atom.
(b) The direction of a reaction depends both upon the equilibrium constant and the concentrations of reactants and products. The intramitochondrial $[NAD^+]/[NADH]$ ratio is high, and this drives the reaction toward α-ketoglutarate.
(c)

$H_3N{:}$ + COO^-–CH_2–CH_2–$C({=}O)$–COO^- —(H^+)→ COO^-–CH_2–CH_2–$C(\overset{+}{N}H_3)(OH)$–COO^- —(H_2O)→ COO^-–CH_2–CH_2–$C({=}\overset{+}{N}H_2)$–COO^- —(NAD(P)H + H^+ → $NAD(P)^+$)→ COO^-–CH_2–CH_2–$C(H)(\overset{+}{N}H_3)$–COO^-

13. PLP forms a covalent Schiff base between the aldehyde carbon of the coenzyme and an ε-amino group of a lysine residue. Obviously, this bond must be broken for the coenzyme to form a Schiff base with an amino acid substrate.

14. (a) This will control the synthesis of carbamoyl phosphate by regulating the synthesis of glutamine, the preferred substrate for carbamoyl phosphate synthetase.
(b) This will have the effect of activating the deadenylylation of glutamine synthetase, which in turn activates glutamine synthetase. This would promote the utilization of α-ketoglutarate for glutamate and glutamine synthesis.
(c) This also has the effect of activating glutamine synthetase, under conditions where the ATP needed for the reaction is abundant.

15. First, purify the enzyme from HeLa cells or another human tissue and prepare antiserum against the enzyme. Then label HeLa cell cultures by growth in a radioactive amino acid. At intervals following removal of the labeled amino acid (a pulse-chase experiment), prepare a cell-free protein extract, treat an aliquot with antiserum, and count the radioactivity in the immunoprecipitate.

16. Dihydrofolate: N-1, C-2, C-4, N-5, C-7, N-8, C-9. GTP: N-3, C-2, C-6, N-7, C-1′, N-9, C-2′, respectively.

17. (a) B. (b) C. (c) D. (d) E, methionine.
(e) A. (f) F. (g) C.

18. Adenylylation tends to inactivate glutamine synthetase. Therefore, the effect of glutamine is to inhibit its own synthesis when it is present in abundance.

19. The most straightforward answer is to place the two enzymes in different cell compartments—the catabolic enzyme in mitochondria and the assimilative enzyme in cytosol.

CHAPTER 21

1. Formiminoglutamate, because the next reaction in its catabolism requires tetrahydrofolate.

2. Arginine $\xrightarrow[\text{PLP}]{CO_2}$ agmatine $\xrightarrow{H_2O \quad \text{urea}}$ putrescine

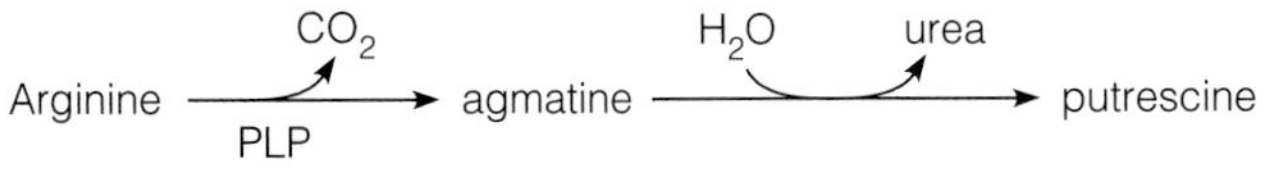

3. *N*-Acetylglutamate is an intermediate in ornithine biosynthesis. Activity of the urea cycle requires both ornithine and carbamoyl phosphate. If insufficient carbamoyl phosphate is available, ornithine will accumulate, and this could cause accumulation of the precursor, *N*-acetylglutamate. This accumulation acts as a signal to stimulate carbamoyl phosphate synthesis to increase urea cycle flux.

4. Tryptophan $\xrightarrow[\text{PLP}]{}$ CO_2 + indole–$CH_2-CH_2-\overset{+}{N}H_3$ $\xrightarrow[H_2O,\ \text{dihydrobiopterin}]{O_2,\ \text{tetrahydrobiopterin}}$ hydroxy-indole–$CH_2-CH_2-\overset{+}{N}H_3$ $\xrightarrow[\text{2 AdoHcy}]{\text{2 AdoMet}}$ 4-OH-indole–$CH_2-CH_2-\overset{+}{N}H(CH_3)_2$ $\xrightarrow[\text{ADP}]{\text{ATP}}$ 4-(P)O-indole–$CH_2-CH_2-\overset{+}{N}H(CH_3)_2$

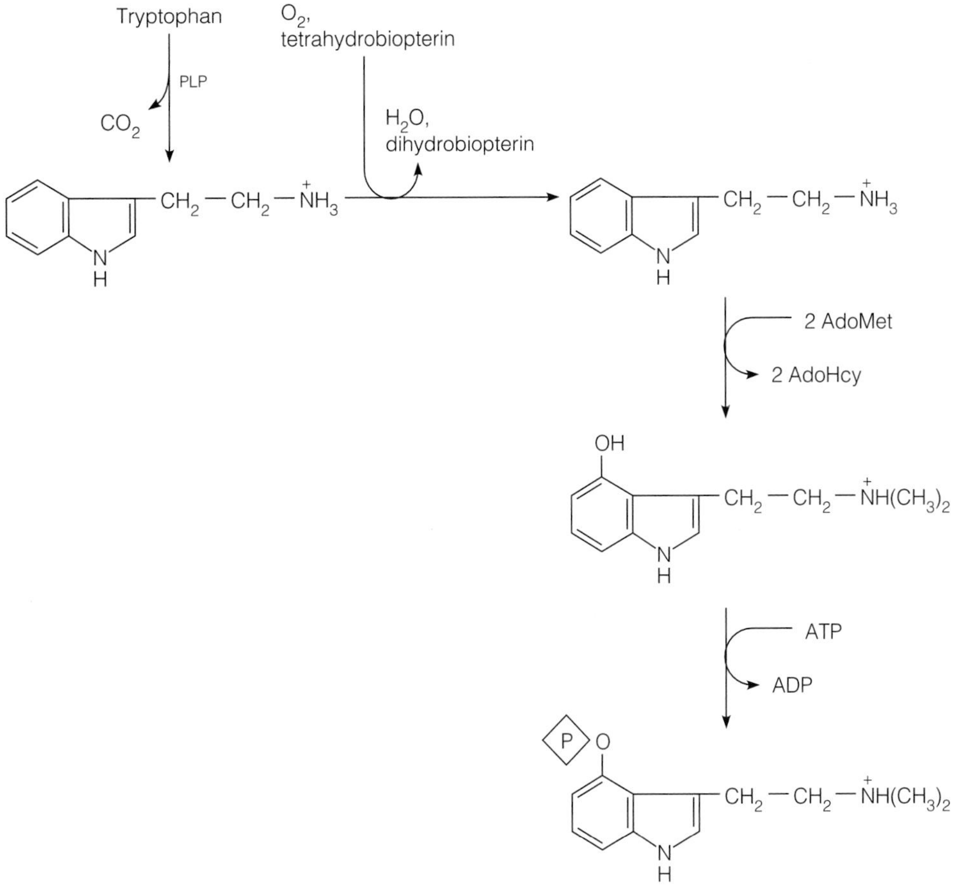

5.

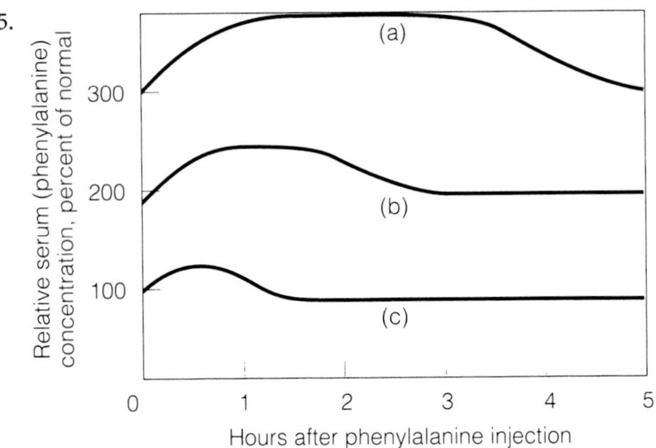

A tryptophan tolerance test

6. (a) Tetrahydrofolate, glycine, and serine transhydroxymethylase.

(b)

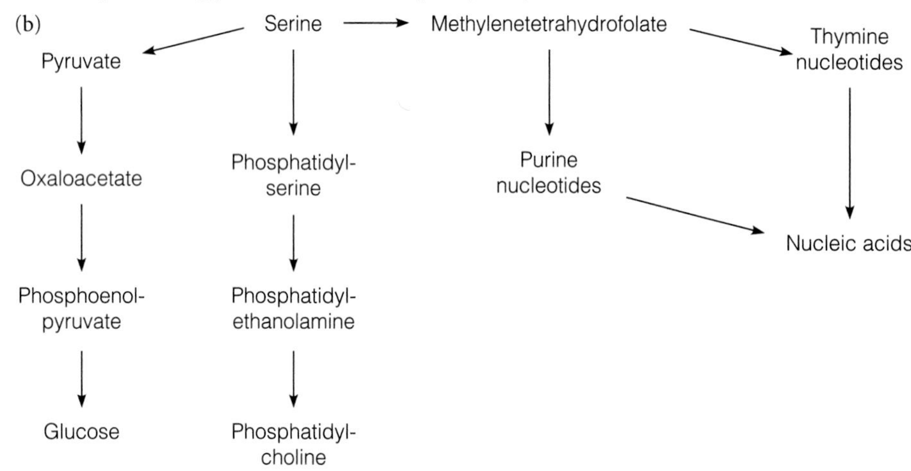

7. 16 ATPs.

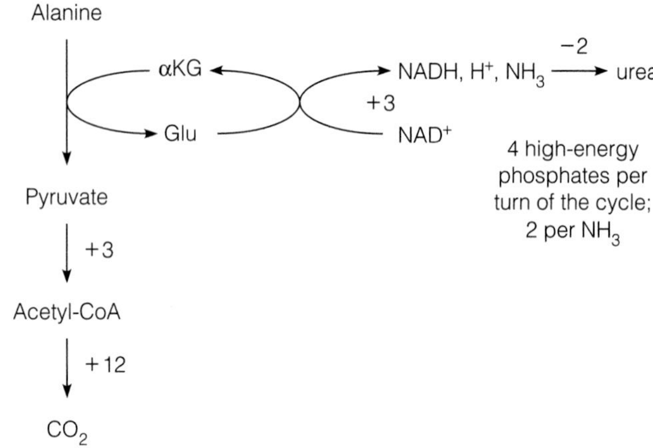

Leucine: 5 ATPs net, leucine ⟶ HMG-CoA.
Tyrosine: 1 ATP net, tyrosine ⟶ fumarate + acetoacetate. You might expect all of these pathways to generate a little more energy in a fish, because it is not necessary to consume ATP in converting ammonia to urea for excretion.

8. One should be feedback inhibited or have its synthesis repressed by threonine, a second by methionine, and a third by lysine, because the aspartokinase reaction is involved in separate biosynthetic pathways leading to each of these three amino acids.

9. Proline (N–H, COO^-) —AdoMet → AdoHcy→ N-methylproline (N–CH_3, COO^-) —AdoMet → AdoHcy→ N,N-dimethylproline ($\overset{+}{N}$(CH_3)(CH_3), COO^-)

10. Because the same enzymes are involved in comparable steps of both isoleucine and valine biosynthesis. Threonine dehydratase.

11. Threonine ⟶ α-ketobutyrate: inihibited by isoleucine

α-Ketoisovalerate + acetyl-CoA ⟶ β-isopropyl malate: inhibited by leucine

Control of valine synthesis is more complicated, because three of the enzymes are involved in synthesis of all three amino acids. One could look for cumulative feedback inhibition—by valine, isoleucine, and leucine—of the first committed reaction:

Pyruvate + hydroxyethyl-TPP ⟶ α-acetolactate

12. Histamine.

R—CH_2—C(H)(COO^-)—N=CH–(pyridoxal ring) —CO_2→ R—CH_2—C(H)=N—CH=(ring) —H^+→ R—CH_2—CH_2—N=CH–(ring) —H_2O→ R—CH_2—CH_2—NH_2 + CHO–(ring)

R = imidazole

13. (a) 5.
(b) 2, 3, 4.
(c) 3, 4.
(d) None.
(e) 5.
(f) 5.

14. Because a pteridine reductase deficiency would impair all tetrahydrobiopterin-dependent reactions, which include the synthesis of catecholamines, serotonin, and nitric oxide, as well as tyrosine.

15.

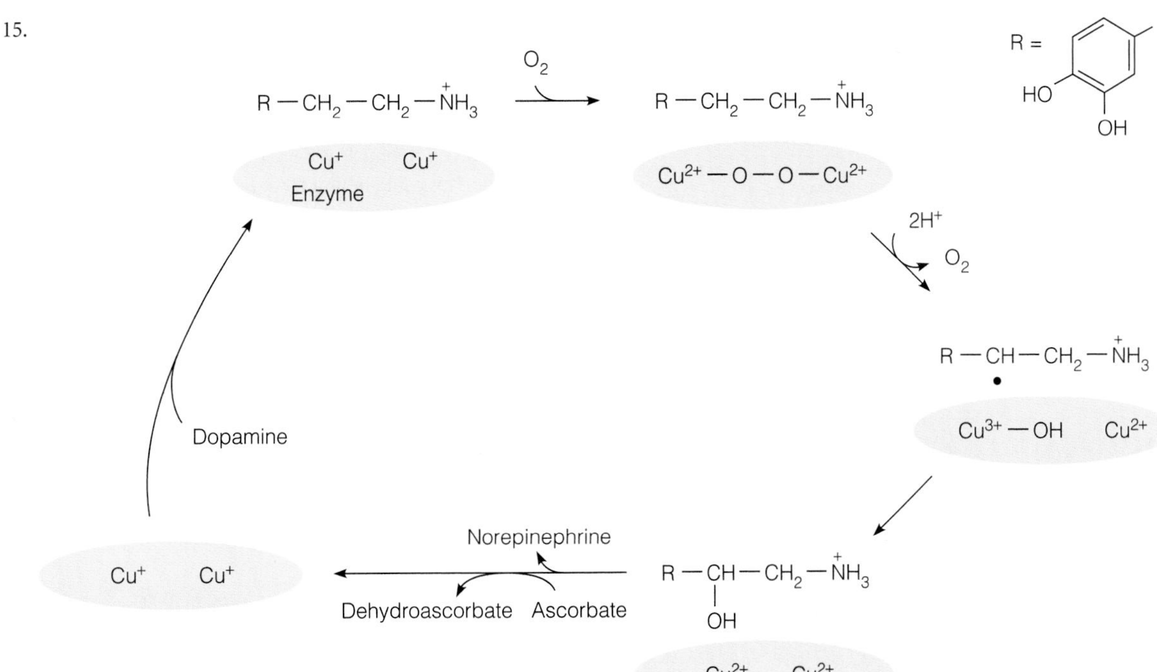

16. Histidine ($CH_2—CH(\overset{+}{N}H_3)—COO^-$ on imidazole) —PLP (Glycine → Serine)→ imidazole–$CH(\overset{+}{N}H_3)—COO^-$ —3 AdoMet→ N-methylimidazole–$CH(N(CH_3)_2)—COO^-$ —PLP, Cysteine→ $S—CH_2—CH(\overset{+}{N}H_3)—COO^-$ attached to N-methylimidazole–$CH(N(CH_3)_2)—COO^-$ → $CH_3—\overset{O}{\overset{\|}{C}}—COO^- + \overset{+}{N}H_4$ + SH-substituted N-methylimidazole–$CH(N(CH_3)_2)—COO^-$ (Ovothiol)

17. The structure of glyphosate is similar to that of phosphoenolpyruvate, suggesting that it acts as a competitive inhibitor with respect to PEP and a noncompetitive inhibitor with respect to shikimic acid 3-phosphate.

18. Since the synthesis of bilirubin diglucuronide occurs in the liver, a chronically diseased liver would be unable to effect this conversion, and unconjugated bilirubin would accumulate. In either a bile duct obstruction or an elevated heme destruction resulting from hemolysis, there would not initially be an impairment of the conjugating system, so most of the accumulated bilirubin would be in the conjugated form.

19. Reaction 1: pyridoxal phosphate
Reaction 2 and reaction 3: O_2 and NADPH (or another two-electron donor)

20. (a) Conjugation of glucuronate with bilirubin is carried out by a liver enzyme, and liver damage reduces capacity for this reaction, which is essential for solubilization and excretion of bilirubin.
(b) By diverting intermediates in heme synthesis to the nonutilizable type I porphyrins, these individuals are unable to synthesize sufficient heme for hemoglobin in red blood cells.
(c) Deficiency of this enzyme reduces flux through methionine synthase, and this leads to accumulation of more homocysteine than can be metabolized via cystathionine.

CHAPTER 22

1. (a) C.
(b) E, F.
(c) A, B.

2. $\text{BrdUrd} \xrightarrow[\text{ATP}]{\text{TK}} \text{Br-dUMP} \xrightarrow[\text{ATP}]{} \text{Br-dUDP} \xrightarrow[\text{ATP}]{\text{Nucleoside diphosphate kinase}} \text{Br-dUTP}$

Br-dUTP can be incorporated into DNA in place of dTTP. However, it might also inhibit DNA synthesis by acting as a false feedback inhibitor of CDP reduction. Since dTTP is an allosteric inhibitor of CDP reduction by ribonucleotide reductase, Br-dUTP might have a similar effect. This could inhibit DNA replication by causing a dCTP deficiency.

3.

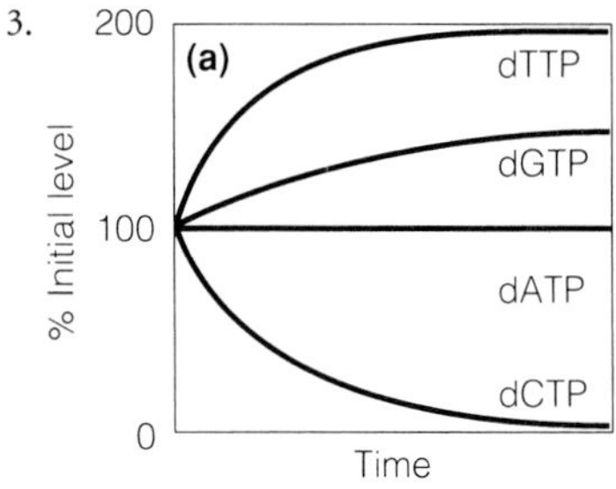

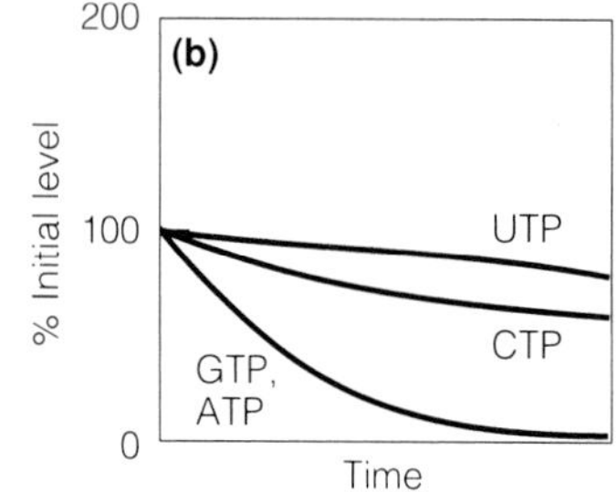

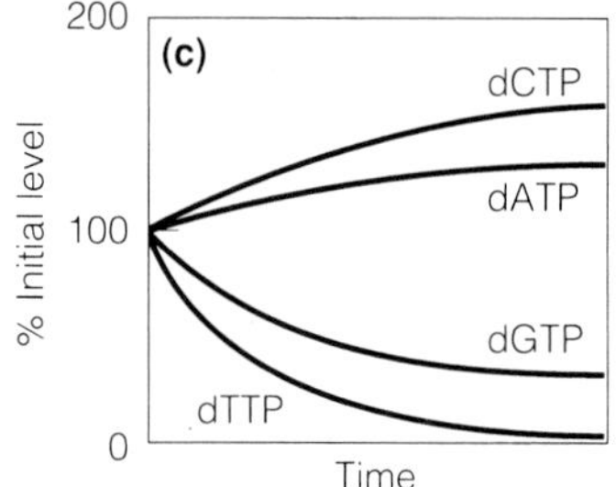

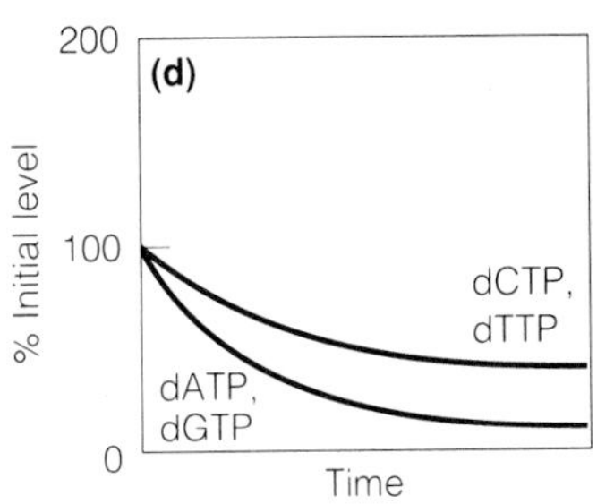

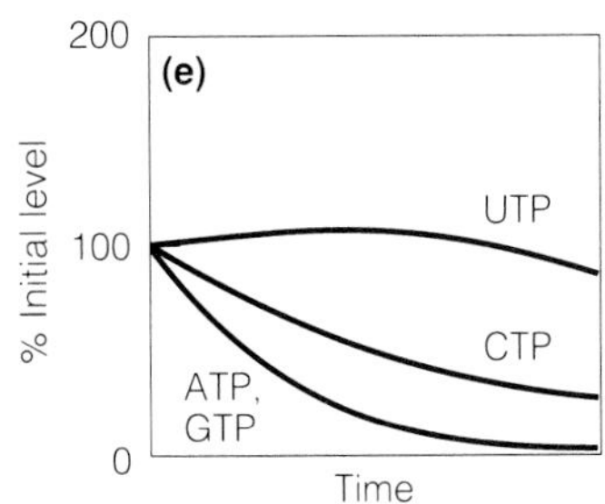

4. Outline answer (* = sites of allosteric regulation).

Uracil ⟶ uridine ⟶ UMP ⟶ UDP ⟶ UTP —*⟶ CTP ⟶ CDP —*⟶ dCDP ⟶ dCTP

UDP ↓* dUDP ⟶ dUTP ⟶ dUMP ⟶ dTMP ⟶ dTDP ⟶ dTTP

5. Outline answer (* = sites of allosteric regulation).

HX ⟶ IMP —*⟶ XMP ⟶ GMP ⟶ GDP —*⟶ dGDP ⟶ dGTP

IMP ↓* adenylosuccinate ⟶ AMP ⟶ ADP —*⟶ dADP ⟶ dATP

6. It might lead to adenosine and deoxyadenosine accumulation, as in adenosine deaminase deficiency. This would cause dATP to accumulate and shut off all activities of ribonucleoside diphosphate reductase.

7. Decreased flux through HGPRT could cause the substrates to accumulate, including PRPP. This could increase pyrimidine synthesis at the orotate phosphoribosyltransferase reaction, if PRPP levels are normally subsaturating for that enzyme. Orotate incorporation into nucleotide pools or nucleic acids would give a reasonable estimate of the rate of de novo pyrimidine nucleotide synthesis.

8. (a) Either a deficiency of thymidylate synthase activity, measured in a cell-free extract, or a growth requirement of cells for thymidine would confirm that the mutants are deficient in this enzyme.
(b) Since the mutants have negligible flux through the thymidylate synthase reaction, they do not deplete their intracellular tetrahydrofolate. Thymidine satisfies the need for synthesis of thymine nucleotides.
(c) Use methotrexate instead of trimethoprim, because trimethoprim is not an effective inhibitor of mammalian dihydrofolate reductase.

9. Mutation in DHFR gene that makes the enzyme resistant; mutation in transcriptional or translational control mechanism that causes overproduction of normal DHFR; transport defect that causes failure of methotrexate to be taken up by cell.

10. (a) Inhibitors of thymidylate synthase and dUTPase specified by the viral genome.
(b) Virus-specified dUMP hydroxymethylase and hm-dUMP kinase:

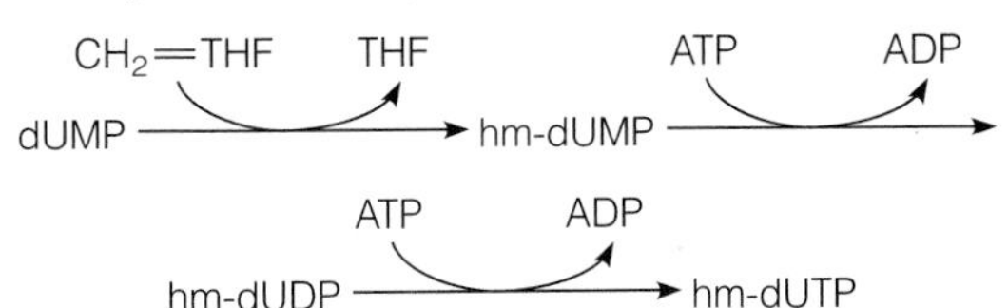

(c) Virus-specified dCMP methylase, e.g.,

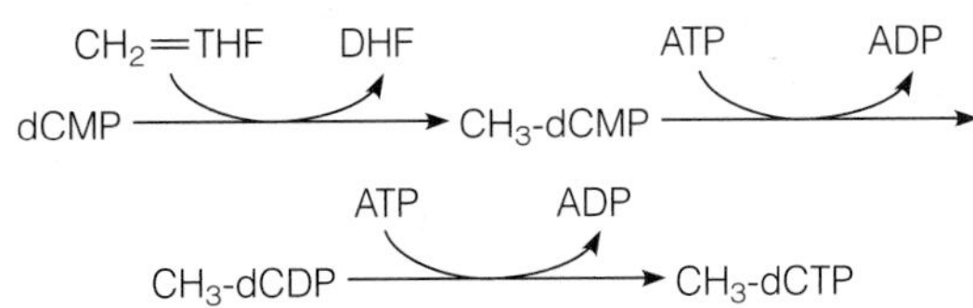

11. (a) Mutants with a defect in DNA synthesis (no thymidine incorporated into DNA).
(b) Mutants defective in thymidine kinase, thymidylate synthase, dihydrofolate reductase, or serine transhydroxymethylase can be selected for by their resistance to decay of incorporated $[^3H]$deoxyuridine.

12. If intermediates in a pathway are restricted in their ability to diffuse away from enzyme surfaces, then a multistep pathway can be catalyzed efficiently with very low steady state levels of intermediates. Thus, you could set up an assay system for such a sequence and analyze levels of intermediates at various times. Another test for channeling involves running a multistep sequence beginning with a radiolabeled substrate and asking whether addition of a large excess of an unlabeled intermediate will dilute the radioactivity of the product. A pathway channeled by an enzyme complex preferentially uses intermediates that are produced by that complex.

13. 4-carboxy-5-aminoimidazole ribonucleotide + aspartate + ATP ⟶ *N*-succinylo-5-aminoimidazole-4-carboxamide ribonucleotide + ADP + P_i

IMP + aspartate + GTP ⟶ adenylosuccinate + GDP + P_i

Citrulline + aspartate + ATP ⟶ argininosuccinate + AMP + PP_i

14. Uncontrolled conversion of UTP to CTP elevates pools of cytidine and deoxycytidine nucleotides, while pools of uridine and thymidine nucleotides are diminished. The depletion of endogenous thymidine nucleotide explains the growth requirement for exogenous thymidine, and the perturbed dNTP pool imbalance (dCTP/dTTP ratio is elevated) causes replication errors, principally C incorporated opposite A, that lead to mutations.

15. $H_2N{-}\overset{\overset{\displaystyle O}{\|}}{C}{-}NH{-}CH_2{-}\overset{\overset{\displaystyle CH_3}{|}}{CH}{-}COO^-$

16. (a) Feeding thymidine stimulates salvage pathway synthesis and accumulation of ATP, which inhibits CDP reduction by ribonucleotide reductase and inhibits DNA replication by depleting the cell of dCTP.
(b) dATP binds to both activity and specificity sites of ribonucleotide reductase, with low and high affinities, respectively. When bound at the specificity site, dATP activates the reduction of CDP and UDP. At higher levels, dATP binds at activity sites and inhibits all four activities of ribonucleotide reductase.

17. Succinyl-CoA synthetase.

CHAPTER 23

1. It is presumed that starch increases blood glucose levels less than simple sugars do. Thus, there is less stimulation of insulin secretion. Insulin would tend to retard energy mobilization from intracellular stores—something not desirable during a marathon.

2. About 2.5 hours.

3. About 28% at rest, about one-tenth that value during a marathon.

4. The measured proteolysis represents the sum of rates of protein synthesis and breakdown. The actual rate of protein breakdown does not rise, but in the early stage of a fast, the utilization of amino acids in catabolic pathways reduces the concentrations needed to support protein synthesis at rates that counterbalance breakdown. Later, as fatty acids and ketones are used more for energy, amino acids are spared for this purpose and are more readily available to be used for protein synthesis.

5. Glucosylated hemoglobin accumulates in the blood of diabetics and can easily be measured. Because the glucosylation reaction is covalent, the level of glucosylated hemoglobin reflects the level of blood glucose over a period of time, whereas a simple determination of glucose reflects only the value at the time of blood sampling. Thus, a diabetic patient can be monitored with many fewer blood drawings and with greater accuracy if glucosylated hemoglobin is measured.

6. Liver contains low levels of the enzyme that synthesizes acetoacetyl-CoA from acetoacetate, ATP, and CoA-SH. Therefore, when liver synthesizes ketone bodies, they cannot readily be activated for catabolism within the hepatocyte. Instead, they are released and ultimately utilized by other tissues.

7. Using recombinant DNA techniques (see Chapter 25), one could prepare nucleic acid sequences that encode either an entire α subunit of a G protein or a GTP-binding domain and ask whether that "probe" can hybridize with DNA from a plant species of interest. Alternatively, one could ask whether antibodies against mammalian G proteins crossreact with any proteins in a plant extract. If so, the protein could be isolated and its properties analyzed (including its ability to bind guanine nucleotides). Another approach is to look for proteins that can be ADP-ribosylated by cholera toxin or pertussis toxin.

8. Some possible factors: storage, control of activity, proper folding of the polypeptide chain, signal sequences for direction to the proper part of the cell.

9. Ethylene. It would probably be difficult to prepare an antibody against it, as a gas.

10. Since steroid hormone–receptor complexes activate the transcription of specific genes, a steroid hormone could conceivably activate the transcription of the gene for adenylate cyclase and, hence, increase the steady state level of this enzyme.

11. (a) 1, 4, 9, 14.
(b) 5, 6, 8, 10, 16.
(c) 1, 2, 7, 12, 16.
(d) 1, 4, 15, 16.

12. One could treat a G protein–GDP complex with γ-[^{32}P]ATP and ask whether radiolabeled GTP is synthesized. The fact that it is not indicates that the G protein cannot phosphorylate bound GDP. One can also show that radiolabeled GTP can displace bound, unlabeled GDP under various conditions. One can also show that activation of the G protein requires the presence of GTP. Finally, one can demonstrate that the isolated α subunit has GTPase activity, by showing its ability to convert GTP to GDP in the absence of GDP.

13. Adipose tissue lacks glycerol kinase. Glycolysis generates dihydroxyacetone phosphate, which is reduced to glycerol 3-phosphate.

14. (a) Malonyl-CoA at high levels inhibits carnitine acyltransferase I, and this inhibits ketogenesis by blocking the transport of fatty acids into mitochondria, both for β-oxidation and for ketogenesis.
(b) The high K_M of glucokinase, a liver-specific enzyme, allows the liver to control the rate of glucose phosphorylation over a wide range of glucose concentrations. Accumulation of glucose-6-phosphate activates the D form of glycogen synthase and promotes glycogen deposition. By several mechanisms the liver also senses when blood glucose levels are low and mobilizes its glycogen reserves accordingly.

15. Phosphorylation of pyruvate kinase by cyclic AMP–dependent protein kinase. The phosphorylated form of the enzyme is far less active than the dephosphorylated form.

16. Because inositol trisphosphate is derived from phosphatidylinositol, which in turn is derived from diacylglycerol and inositol, inositol phosphates must be completely dephosphorylated in order to be reincorporated into phosphatidylinositol for another round of synthesis.

17. (1) Most plant hormones are terpenoid compounds.
(2) There are no known peptide hormones in higher plants. Steroid-like compounds have been described in plants, but it is not clear that they act like steroid hormones.

CHAPTER 24

1.

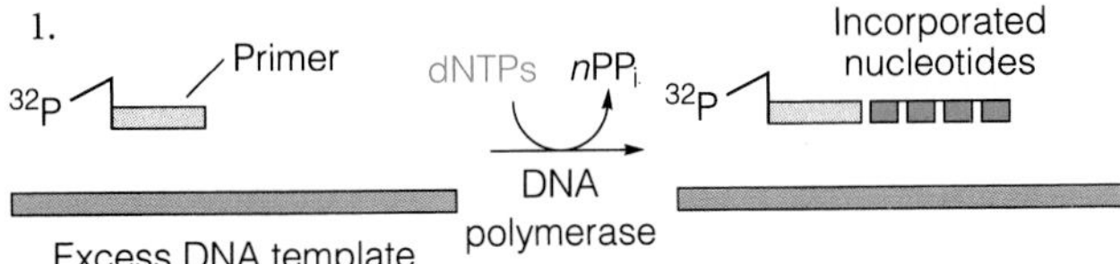

Use a 5′ end-labeled primer, and run a polymerase reaction in DNA excess (so that any polymerase that dissociates will rebind to a new chain). Stop reactions after a brief incubation and determine the molecular weight of radioactive material by gel electrophoresis. M_r tells how many nucleotides were incorporated per chain, which gives the processivity (number of nucleotides incorporated per unit time per chain).

2.

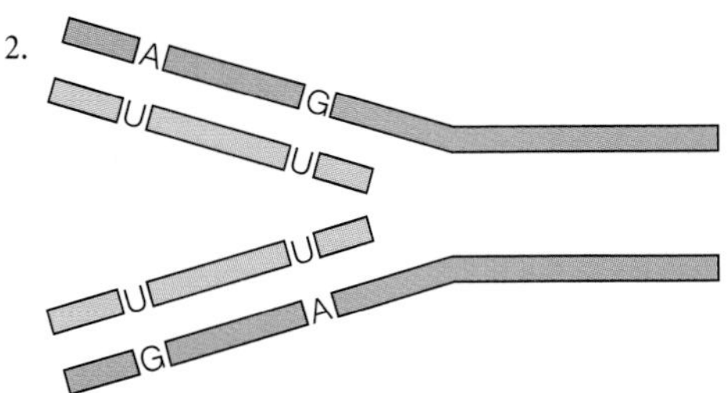

Incorporation of dUMP generates sites for chain breakage, as the uracil replacement process begins. If the single-strand breaks had not been fully repaired when DNA was isolated, short fragments would be seen. Fewer of these would be seen if the experiment were done with a mutant defective in uracil-DNA-*N*-glycosylase—and more if the experiment were done with a dUTPase-negative mutant.

3. (a) 1.725 g/cm^3.
(b) 1.772 g/cm^3.

4. (a) To eliminate virtually all further incorporation of radioisotope.
(b) If incorporated as ribonucleotides, the labeled product will be hydrolyzed by mild alkali. If incorporated as deoxyribonucleotides, the product is alkali stable.
(c) RNA primers are metabolically unstable. They are rapidly degraded, as they are replaced by deoxyribonucleotides in DNA.

5. (a) Deoxyinosinic acid (dIMP).
(b) Conversion of an A–T to a G–C base pair.

6. (a) 0.016 mm.
(b) 9.79 × 10^4 nucleotides per fork per minute.
(c) 1250 origins (2500 replication forks).
(d) 2937 kilobase pairs.

7. (a) It's the reversal of the DNA ligase reaction. Single-strand interruption relaxes supercoiled DNA.

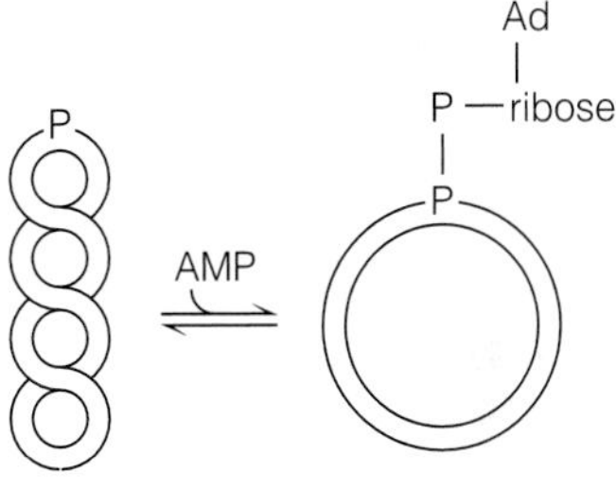

(b) Analyze the DNA by gel electrophoresis.

8. Synchronize cells and label with 5-bromodeoxyuridine either early or late in S phase. Separate replicated DNA by CsCl equilibrium centrifugation. This can be analyzed by hybridization with cloned DNAs from genes known to be expressed or not expressed in those cells.

9.

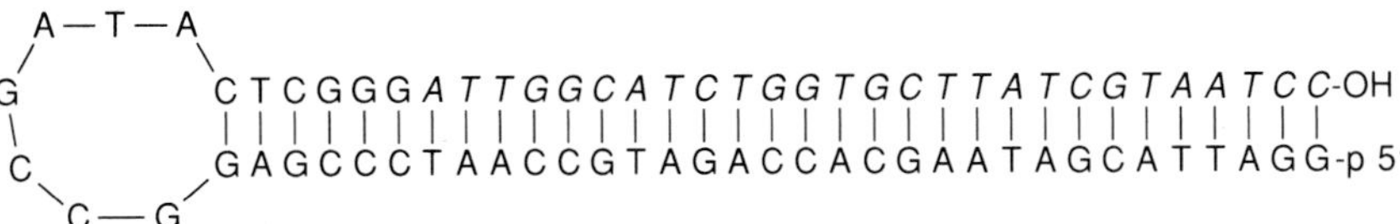

Note: The 3′ exonuclease removes the mismatched A and T before polymerase action begins.

10. The polymerase activity was found to extend more slowly from an improperly paired terminus. Thus, a mismatched base has a longer residence time at the 3′ terminus than a matched base, increasing the likelihood of cleavage by the exonuclease activity.

11. Dideoxyinosine is converted to the corresponding dideoxyribonucleoside triphosphate, ddITP. Incorporation of this nucleotide in place of dGTP would block further chain elongation, because of the absence of a 3′-OH terminus. If the HIV reverse transcriptase incorporates ddITP more readily than the cellular DNA polymerases do, then replication of the viral genome would be selectively inhibited.

12. If the mutation affected one of the feedback control sites of ribonucleotide reductase, the cell could overproduce dNTPs to the extent that competitive inhibition in vivo of replicative DNA polymerases by aphidicolin could be overcome.

13. They initiate rounds of replication at intervals shorter than 40 min. Thus, a cell may have two identical chromosomes, each at the same stage of replication.

14. Because ethidium bromide in the gel relaxes the positive supercoils, converting an overwound duplex to a relaxed circle, with a consequent loss of mobility.

15. The inchworming model suggests that each subunit of helicase is continuously in contact with just one DNA strand, while in the hand-over-hand model, continuous strand exchange occurs.

16. One mechanism would have the enzyme bind at two distant sites on DNA and then slide along DNA so that the two DNA contact points come closer and closer together. If any knots or links are encountered, they can be removed by topoisomerase cutting and resealing.

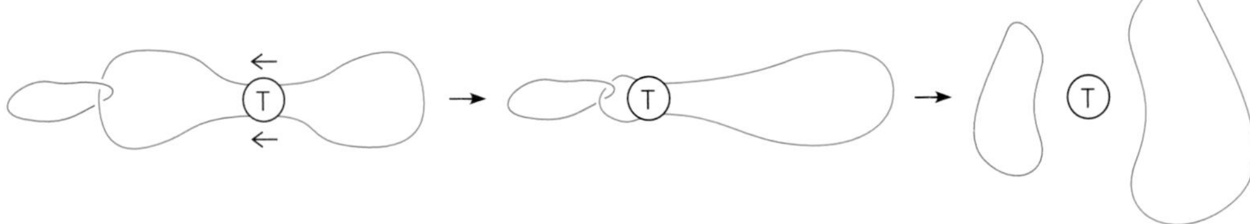

17. Fractional abundance: 10^{-6}.
Abundance after 10 cycles, 1.02×10^{-3}; 15 cycles, 3.3×10^{-2}; 20 cycles, 1.0 (essentially 100%).

CHAPTER 25

1. Increase, because mismatch repair would repair the "correct" mismatched nucleotide half the time.

2. (a) 2, 7, 9, 11.
(b) 3, 5, 6.
(c) 2, 11.
(d) 1, 11.
(e) 1, 8, 11.
(f) 4, 6.
(g) 10, 11.

3. Flush ends: *Eco*RII, *Hae*III, *Hin*dII, *Hpa*I, *Sma*I. Not made flush by DNA polymerase (because the recessed end lacks a 3′ hydroxyl): *Hga*I, *Hha*I, *Ple*I, *Pst*I.

4.

(a) G A C G T C / C T G C A G

(b) (c)

(d) Up to 1 kb away from this site.
(e) 4.1 kbp.

5. (a) A.
(b) B.
(c) C.
(d) D.
(e) A, C.
(f) C.

6. Because both strands are methylated.

7. dUTPase deficiency increases dUMP incorporation into DNA and increases subsequent excision repair. Ligase deficiency increases the mean lifetime of Okazaki fragments. Both conditions increase the number of single-strand interruptions, the structures that initiate recombination.

8. (a) To linearize circular DNA, so that its electrophoretic mobility can be compared with those of other linear DNA fragments.
(b) The recombinant DNA contains two tandem 1.15-kb inserts.
(c)

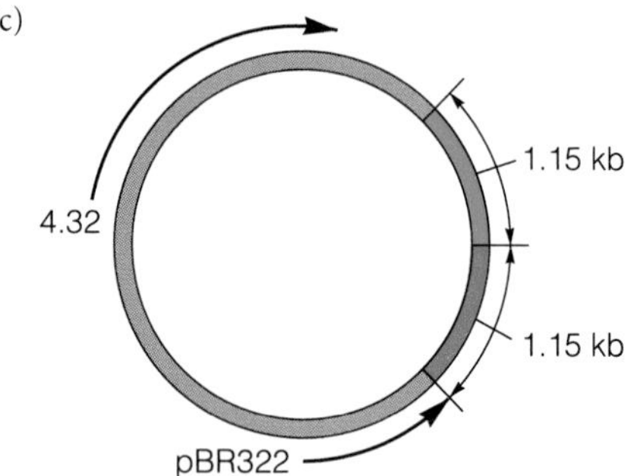

9. (a) Circular.

(b)

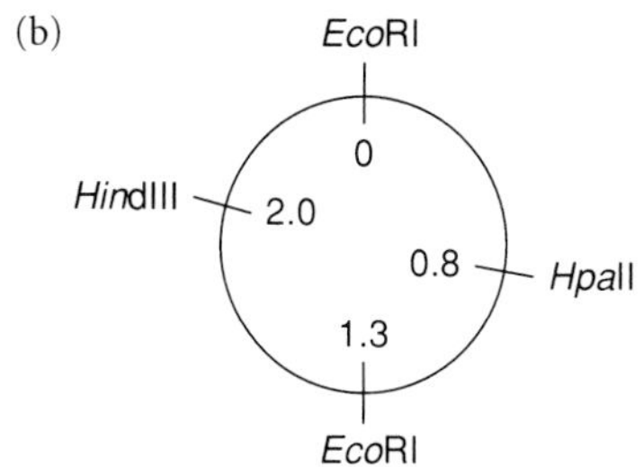

(c) 7.

(d) Cleave with *Hpa*II plus *Hin*dIII.

10. (a)

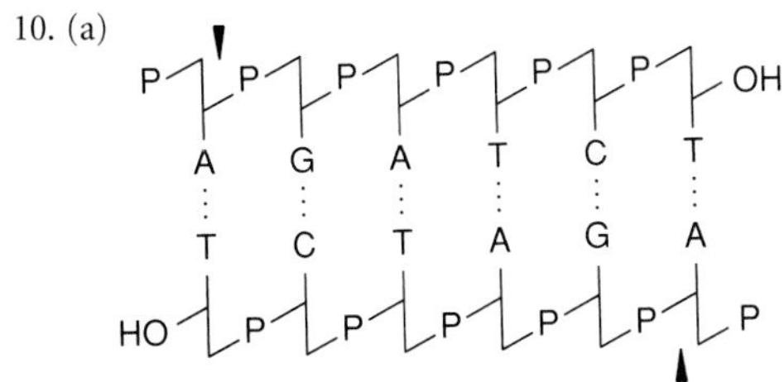

(b) A and B.

(c) E and F.

(d) F.

11.

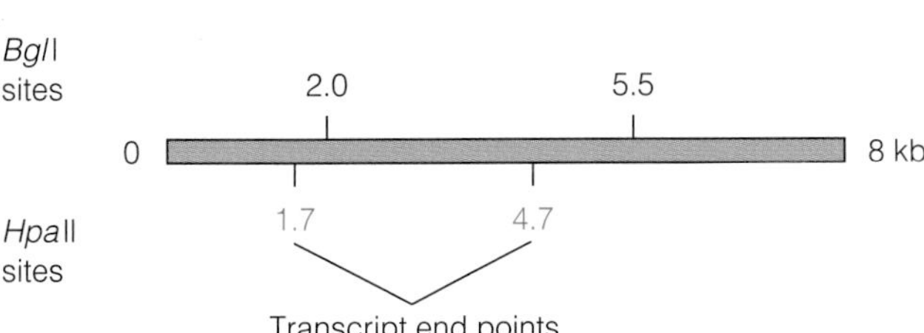

It would be laborious, but you could separate the DNA strands and determine which strand hybridizes to the RNA. This would determine the orientation of transcription, since you know the DNA 3′ and 5′ ends.

12. First, clone a large number of DNA fragments from cells containing the integrated virus. Next, carry out colony hybridization on these clones, using labeled viral DNA as the hybridization probe. From Southern blot analysis of clones containing viral DNA, identify viral DNA fragments that are larger than corresponding fragments from the nonintegrated double-strand viral DNA. These should represent viral sequences fused to adjacent cellular sequences. These can now be subcloned and sequenced.

13. To quickly identify the desired clones.

14.

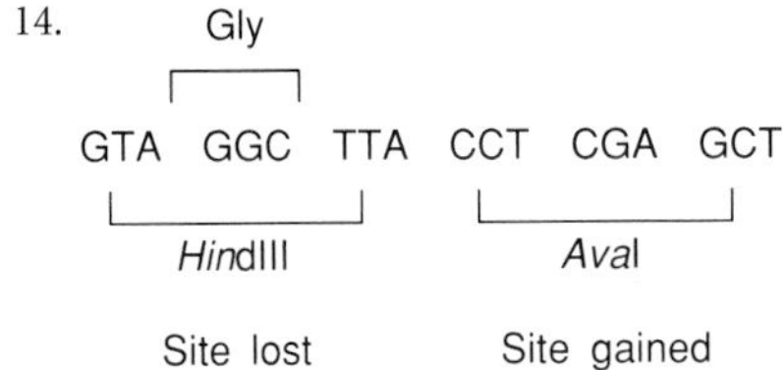

15. The methyl-directed mismatch repair system corrects many of the deliberately engineered mismatches. Also, many of the in vitro chain extension reactions do not proceed entirely around the circular template, as they must to undergo ligation.

16. Digest identical samples of genomic DNA with *Msp*I and *Hpa*II. Resolve the fragments electrophoretically, and probe with radiolabeled recombinant DNA containing the gene of interest. Every CpG site that is methylated (within the sequence CCGG) would show cleavage by *Msp*I but not by *Hpa*II, whereas unmethylated sites would be recognized as cleavage sites for both enzymes.

17. Nonenzymatic deamination of the cytosine in a G·C base pair yields a G·U base pair, which is readily repaired by the uracil-*N*-glycosylase base excision repair system. However, if the C in that base pair has been methylated before deamination, then deamination of G·mC yields a G·T base pair, which is not as easily corrected. Also, it is possible that methylation of C in a G·C base pair facilitates its deamination.

18. This is a surprising result, because the function of the MutT protein is thought to be hydrolysis of 8-oxo-dGTP, an oxidation product of dGTP. It is not clear how this oxidation could occur in the absence of O_2. Perhaps MutT acts also on another mutagenic nucleotide not yet identified, which can be formed during anaerobic growth.

CHAPTER 26

1. Pulse-label cells with a radioactive RNA precursor, isolate RNA, and hydrolyze with mild alkali. This will yield one nucleoside per 3′ terminus. Total picomoles of labeled nucleotide in the hydrolysate divided by the number of picomoles of nucleosides gives the number of nucleotides incorporated per chain in the labeling interval.

2. First, devise conditions under which only the cloned gene is synthesized. You could initiate transcription by adding three ribonucleotides. At time zero, add the fourth nucleotide and stop incubation after a fixed time. Use a small, measured amount of template so that the number of transcribing genes is roughly equal to the number of template molecules. Determine the average length of transcripts synthesized by gel electrophoresis.

3. By gel electrophoresis of radiolabeled RNA products. Any paused species will accumulate and can be detected as a heavier-than-expected band on a radioautograph of the gel.

4. (a) $$\frac{(10\ \text{molecules/cell})/6.02 \times 10^{23}\ \text{molecules/mol}}{0.3 \times 10^{-15}\ \text{L/cell}}$$

$$= 5.5 \times 10^{-8}\ \text{M}$$

(b) $$\frac{10\ \text{molecules/cell}}{5.5 \times 10^{-8}\ \text{M}} = \frac{2\ \text{molecules/cell}}{1.1 \times 10^{-8}\ \text{M}}$$

(c) Approximately 2.5×10^{-14} M.

(d) A rapidly growing cell could have two copies, on average, because each single chromosome is partly replicated.

5. It's unlikely, because in eukaryotes transcription occurs in the nucleus and translation in the cytosol. Thus, direct coupling between transcription and translation, essential for attenuation, is absent.

6. This can be a way to control total RNA synthesis when ribonucleotide pools are depleted, by controlling initiation and preventing the wasteful incorporation of nucleotides into RNA molecules that might not be completed.

7. A dinucleotide, complementary to the first two template nucleotides, can bypass the first nucleotide incorporation step and permit efficient transcription at low nucleotide concentrations. Then one can arrest all transcription events at one nucleotide by knowing the nucleotide sequence of the gene being transcribed and adding only two or three rNTPs, instead of all four, with the missing nucleotide being the one that would have been inserted at the site at which you desire to arrest transcription.

8. Using footprinting, carry out competitive binding experiments to ask whether repressor inhibits RNA polymerase binding and vice versa. Also, examine transcription in vitro to determine whether transcription extends as far as the repressor binding site.

9. Because it uses the −35 region of *trp* and the −10 region of *lac*, the hybrid *tac* promoter more closely matches the consensus −35 and −10 sequences than does either *lac* or *trp*.

	Number of identities with consensus sequence
lac	9/12
trp	8/12
tac	10/12

10. (a) The plasmid *i* gene encodes a repressor that cannot bind inducer. (b) The plasmid operator cannot bind repressor.

11. Mu inserts randomly in the genome (see Chapter 25). If it is inserted downstream from a damage-inducible promoter, UV irradiation of that cell will activate β-galactosidase synthesis from the integrated phage genome.

12. Actinomycin is a noncompetitive inhibitor because it binds to DNA, not to a substrate-binding site on the enzyme. Cordycepin acts as the nucleotide, 3′-deoxyATP. This should compete with ATP for utilization by RNA polymerase.

13. (a) RNA 3′ CUAGGAUUCAGUAAGG 5′

DNA 5′ CTACCTAAGTCATTCCCATTAC<u>CAGTAT</u>AG 3′
3′ GATGGATTCAGTAAGGGTAATGGTCATATC 5′

(b) Pribnow box is underlined in (a).

(c) ↓
TAGCTA
ATCGAT
↑

14.

	Before	After
(a)	−	++
(b)	−	++
(c)	−	+
(d)	−	−
(e)	−	−
(f)	++	++
(g)	−	+

15. A mutation that abolishes binding to repressor. Subunits of the mutant repressor could interact with those of the wild-type repressor synthesized in the same cell, to form mixed tetramers with reduced affinity for operator DNA.

16. One possibility is that the rate of addition of a ribonucleotide to a 3′-terminal ribonucleotide that is mismatched with its template DNA may be relatively slow, e.g.,

3′ ACTTCGATTTGGCCCAATGCTCA . . . DNA
5′ UGAAGCUAAG-OH + CTP ⟶

If there is a significant delay in adding the right nucleotide (C) to a mismatched 3′ terminus (G, mismatched with T), then, since the polymerization reaction is reversible, this may increase the probability that the mismatched nucleotide would react with pyrophosphate and be excised as GTP.

17. If two genes on a plasmid are oriented in the same direction and both are being transcribed, then the overwinding created ahead of gene 1 can be compensated by the underwinding created behind gene 2, and vice versa.

CHAPTER 27

1. ···Ile–Tyr–Ile–Tyr–Ile–Tyr–···. Either AUA codes for Ile and UAU for Tyr or vice versa.

2. The mitochondrial code uses AUA to code for *Met*, not *Ile*. The experiment also tells us that UAU is a Tyr codon.

3. (a) First, codons such as UUU or AAA can yield only one polypeptide. Second, some reading frames in some polynucleotides will be read as repeats of stop codons.
(b) $(GUA)_n$ Val_n or Ser_n (third reading frame is UAG = stop).
$(UUA)_n$ Leu_n or Tyr_n or Ile_n

4. Any polynucleotide built from a tetranucleotide repeat, unless a stop codon is involved.

5. The Shine–Dalgarno sequence, like the AUG that is read uniquely as fMet, occurs near the 5′ end of the message. The implication is that, in forming the initiation complex, only a region near the 5′ end of mRNA can bind. That is, a Shine–Dalgarno sequence occurring by chance within a message could not bind at this site to form an initiation complex. Other explanations may be possible.

6. (a) $^{5'}$GGU$^{3'}$, $^{5'}$GGC$^{3'}$, or $^{5'}$GGA$^{3'}$. All are Gly codons.
(b) $^{5'}$AGU$^{3'}$ or $^{5'}$AGC$^{3'}$. Both are Ser codons.

7. (a) The simplest explanation for particles of 100S and 130S is that they represent some kind of dimers and trimers of 70S particles (note the nonlinearity of S with mass). From what we know now, these represented small oligoribosomes on mRNA. The research missed discovering the basic mechanism of protein synthesis.
(b) Dissociation would have produced ribosomal subunits and free RNA. Since these were only fragments of polyribosomes, the chance that they would reassociate onto the RNA is small. Furthermore, mRNA is extremely unstable, unless precautions are taken to eliminate traces of ribonucleases (a fact that this experimenter did not appreciate at that time). Thus, the mRNA was probably quickly degraded.

8. Because the tRNA is going to be released from the E-site, codon recognition and pairing with the anticodon are in no way advantageous and may, in fact, be detrimental to smooth release.

9. With MS2 infection of wild-type bacteria, ribosomal release occurs quite frequently at the termination of the coat-protein sequence, producing the needed excess of coat protein. In the mutant, we would expect that much more replicase would be made, which would lead to an inefficient overproduction of replicase and an imbalance in the RNA/protein ratio.

10. If δ is the probability that an error is committed at each step, then $1 - \delta$ is the probability that an error has *not* been made in any one step. The probability that no error has been made in *any* of the n steps is then $(1 - \delta)^n$. This is the probability that the protein is entirely error-free.

11. (a) 0.990.
(b) 0.904. In this case, nearly 10% of the proteins would contain one or more errors.

12. Bacteria are given a pulse of radioactive amino acids (e.g., [^{14}C]leucine) to label new polypeptide chains. The cells are broken; nucleoids, cell walls, and membranes are removed by centrifugation; and the lysate is placed on a sucrose gradient. The appearance of ^{14}C in fractions sedimenting more rapidly than 70S indicates that newly synthesized protein is associated with such structures. Samples taken some time *after* the pulse will have little label in these fractions, showing that it was protein in the process of translation that was associated with the polysome fraction. Ribonuclease digestion will shift the radiolabel to 70S or lower.

13. 36 (ATP/glucose)/4 (ATP/residue added) = 9 residues added/glucose

This is a maximum, because any inefficiency in ATP production, or use of extra ATP in proofreading and so forth, will give less efficiency.

14. Each puromycin molecule becomes bound to the polypeptide chain it has aborted and is hence used only once. An erythromycin molecule can block an entire polyribosome and can be used repeatedly.

15. Cutting a phosphodiester bond near the 3′ end of 16S RNA might (and in fact does) remove the sequence that binds the Shine–Dalgarno sequence on the messenger RNA.

16. Anything that replaces the charged phosphate groups with uncharged groups should help, provided that the structures remain sufficiently like RNA to allow hybridization with DNA.

CHAPTER 28

1. (a) 3.3×10^{-13} g = 0.33 picogram.
(b) 3.4 cm.
(c) 7×10^3.
(d) 1.5×10^9 km, or 10 times the distance to the sun.

2. One possible method is as follows: Cells can be synchronized (for example, by thymidine block—see Chapter 22), allowing a larger fraction to be in S phase at the same time. Administration of 5-bromodeoxyuridine will then label newly synthesized DNA and make it denser than bulk DNA. Cleavage with a restriction nuclease will yield fragments, surrounding the origins of replication, which can be separated by density gradient centrifugation. If the analysis is carried out early in S phase, origins will represent a significant fraction of the dense DNA.

3. If we take the average nucleosome (including linker) to contain about 200 bp, we get a compaction ratio of 40 for the 30-nm chromatin fiber. Obviously, this fiber must be greatly folded on itself to yield the overall ratio of over 7000 found in the metaphase chromosome.

4. (a) If the nucleosome were exactly positioned (all occupying the same site on the 208-bp fragment), only a single pair of bands (summing to 146 bp) would have been obtained. The fact that two such pairs (29 + 117, 39 + 107) were observed means that there are two alternative locations, 10 bp apart.

5. Using a graph of log (bp) vs. d for the markers, one can interpolate the chromatin fragments to obtain sizes of 185, 380, 578, and 772 bp. A repeat of about 195 bp is indicated, with the monosome being slightly degraded.

6. The simplest explanation comes from the hypothesis (since proved) that the DNA lies on the surface of the particle. Thus, each strand is maximally exposed approximately every 10 residues and is most susceptible to nicking at these periodically spaced points.

7. The DNA makes 1.75 left-hand superhelical turns about each nucleosome. These turns are "constrained" and cannot be removed while the histones are still present. However, if the histones are first removed, the DNA so produced will have unconstrained supercoils, which can be relaxed by topoisomerase.

8. (a) If the origin is at the center, we note that each fork of the enlarging bubble need only replicate half the DNA, so 5.79 days would be required. Any other position of the origin would increase the time.
(b) About 8.3 minutes. This assumes, however, that all replicating regions do so simultaneously. It is known that some regions replicate before others, which will increase the total time required for S phase.

9. Histones are required in very large amounts during only one brief period (the beginning of S phase) in the cell cycle. Furthermore, they are always required in equivalent amounts, to make nucleosomes. The presence of multiple copies aids in rapid transcription, and an equal number in each domain should help maintain balance in production. The absence of introns and poly(A) tails means that two of the major steps involved in processing most eukaryotic genes are avoided, so that histone mRNAs can be delivered rapidly to the cytoplasm.

10. One way would be to measure the distribution of poly(A) tail lengths in bulk mRNA. If a distribution was found that cut off at a particular minimal value, the hypothesis would be supported.

11. There are many possible ways. One is to simply cross-link DNA to histones in an in vitro transcription system. If this blocks transcription, which proceeds in the presence of un-cross-linked nucleosomes, this provides evidence of the necessity of removal, or at least displacement. A more critical test would be to carry out transcription in vitro in the presence of an excess of radiolabeled, nongenomic DNA. If nucleosomes are found on this after transcription, removal from the genomic DNA is indicated.

12. (a) 7 exons, 6 introns.
(b) These two structures will be at opposite ends of the RNA (shorter strand). However, one cannot unequivocally identify the 5′ and 3′ ends from such a figure, unless something is known of its sequence.

13. (a) Failure to splice out the particular intron.
(b) General failure in splicing (probably lethal).
(c) The protease might be released into the cytoplasm and have very deleterious effects.

14. A very specific recognition is required and perhaps a proofreading step as well. As in any biochemical process in which error rates must be very low, accuracy surely can be purchased by free energy (compare translation).

15. Step 2 forms a high-energy thioester bond. Step 5 is a reaction requiring specific recognition and perhaps a proofreading step.

Glossary

absorbance (A_λ) A dimensionless number that indicates how well a solution of a substance absorbs light of a given wavelength. It is defined as the negative logarithm of the fraction of light of wavelength λ that passes through a sample of the solution; its value depends on the length of the light path, the concentration of the solution, and the extinction coefficient of the substance at that wavelength.

acetylcholinesterase An enzyme found in cholinergic synapses that breaks down acetylcholine and thus terminates its action on the postsynaptic cell.

action potential A wave of transient depolarization that travels along the membrane of a nerve cell (or any other kind of excitable cell, such as a muscle cell) as a result of fluxes of ions across the membrane. A nerve impulse.

activated state With respect to a chemical reaction, a transient high-energy state of a reactant molecule (such as an unfavorable electron configuration or strained conformation) that enables the molecule to undergo the reaction.

active site The site on an enzyme molecule where the substrate binds and where the reaction is facilitated. It is often a cleft or pocket in the surface of the enzyme.

active transport The transport of a substance across a biological membrane by a mechanism that can work against a concentration (or electrochemical) gradient. It always requires the expenditure of cellular energy. Compare *facilitated transport, passive transport.*

adenylylation In cells, the transfer of an adenylyl moiety from ATP to another molecule. Some enzymes are regulated by reversible adenylylation.

adipocytes Fat cells; cells that are specialized for storing triacylglycerols and for releasing them to the blood in the form of fatty acids and glycerol as required.

adrenergic receptors Cell-surface receptors that bind epinephrine and norepinephrine. There are several different types with somewhat different ligand specificities and effects. (The term comes from *adrenaline,* the old name for epinephrine.)

affinity constant See *association constant.*

agonist In molecular biology, a substance that mimics the cellular effects of a natural compound (such as a hormone or neurotransmitter) by binding to and activating the same cellular receptor. Compare *antagonist.*

A helix A right-hand helix structure of nucleic acid duplexes that has a smaller pitch and a larger diameter than the B-DNA helix. It is the structure adopted by RNA duplexes and RNA–DNA hybrid molecules.

AIDS (acquired immune deficiency syndrome) A disease caused by prolonged infection with human immunodeficiency virus (HIV) and characterized by crippling of the immune system. Victims die of infections or cancers that their immune system cannot control.

alditols Compounds that are produced by reducing the carbonyl group on a monosaccharide (that is, reducing $R{-}CH{=}O$ to $R{-}CH_2{-}OH$).

aldose A monosaccharide in which the carbonyl group comes at the end of the chain and thus represents an aldehyde group. Compare *ketose.*

alkaloids A large group of nitrogenous basic substances found in plants. Most of them taste bitter, and many are pharmacologically active. The term can also be used for synthetic compounds of the same type. Morphine, caffeine, and nicotine are familiar alkaloids.

allele A specific version of a gene that occupies a particular location in the genome. It is distinguished from other alleles of the same gene by differences in nucleotide sequence.

allosteric With respect to enzymes, an effect that is produced on the activity of one part of an enzyme (such as an active site) by the binding of an effector to a different part of the enzyme.

alternative splicing The splicing of a eukaryotic RNA transcript in different ways, to include or exclude certain exons from the final mRNA.

***Alu* elements** DNA sequences about 300 base pairs long that occur in many copies scattered throughout the genome of mammals; the human genome has hundreds of thousands of them. They may serve an unknown function, or they may be purely "parasitic," spreading as mobile elements through the genome.

Ames test A test for the mutagenicity of a substance. A strain of the bacterium *Salmonella typhimurium* having a mutation that disables an enzyme necessary for histidine utilization is exposed to the substance in question and plated on a medium lacking histidine. A reversion mutation that activates the mutant enzyme causes the cells to grow on this medium.

amino terminus See *N-terminus.*

amphipathic For a molecule, the property of having both hydrophobic and hydrophilic portions. Usually one end or side of the molecule is hydrophilic and the other end or side is hydrophobic.

ampholyte A substance whose molecules have both acidic and basic groups.

anabolism The sum of all the metabolic processes by which complex biomolecules are built up from simpler ones. In general, these processes consume rather than produce cellular energy. Compare *catabolism.*

anaerobic Refers to the absence of oxygen or the absence of a need for it; processes that must or can occur without oxygen are called anaerobic processes.

androgens The male sex hormones; specifically, the steroid hormones testosterone, dihydrotestosterone, and androstenedione, which act mainly to promote male sexual development and maintain male sex characteristics.

anomers Stereoisomers of cyclized monosaccharide molecules differing only in the configuration of the substituents on the carbonyl carbon. (This carbon is a center of chirality in the cyclized but not in the open-chain form of the molecule.)

antagonist In biochemistry, a substance that counteracts the cellular effects of a natural compound (such as a hormone or neurotransmitter) by binding to the cellular receptor for the compound and blocking its action. Compare *agonist.*

antibodies (also called immunoglobulins) A set of related proteins that are produced by B lymphocytes and can bind with specificity to antigens. Some types are released into body fluids and mediate humoral immunity; other types are retained on the surface of the B cell or are taken up and displayed by some other cell types.

anticodon The nucleotide triplet on a tRNA that binds to a complementary codon on mRNA during protein synthesis and thereby mediates the translation of the codon into a specific amino acid.

antigen A substance that can elicit a specific immune response.

antigenic determinant See *epitope.*

antimetabolite A substance that is a structural analog of a normal metabolite or otherwise resembles it and that interferes with the utilization of the metabolite by the cell.

antioxidant A strongly reducing compound, such as ascorbic acid, which counteracts the tendency of a metabolite to undergo oxidation to a potentially toxic or harmful species.

antiport A membrane transport process that couples the transport of a substance in one direction across a membrane to the transport of a different substance in the other direction. Compare *symport.*

antisense RNA An RNA molecule that is complementary to an mRNA; it can block translation of the mRNA by forming a duplex with it. Gene expression can be regulated by the production of antisense RNAs.

antiserum Serum that contains a high concentration of antibodies against a particular antigen.

apolipoproteins The specific proteins that constitute the protein fraction of lipoproteins; they mediate the interactions of lipoproteins with tissues.

apoptosis Programmed cell death (as distinguished from necrosis; see *autolysis*).

archaebacteria A group of prokaryotes that are biochemically distinct from the true bacteria (Eubacteria) and that separated from them early in the history of life. Modern archaebacteria mostly live in extreme environments, such as acid hot springs.

association constant (K) (also called affinity constant) An equilibrium constant that indicates the tendency of two chemical species to associate with each other; it is equal to the concentration of the associated form divided by the product of the concentrations of the free species at equilibrium.

asymmetric carbon A carbon molecule that carries four different substituents and therefore acts as a center of chirality, meaning that the substance can occur in two different enantiomers (stereoisomers that are nonsuperimposable mirror images of each other).

atherosclerotic plaques The protruding masses that form on the inner walls of arteries in atherosclerotic disease. A mature plaque consists partly of lipid, mainly cholesterol esters, which may be free or contained in lipid-engorged macrophages called foam cells, and partly of an abnormal proliferation of smooth-muscle and connective-tissue cells.

attenuation A mechanism for regulating prokaryotic gene expression in which the synthesis of a nascent RNA transcript is terminated before RNA polymerase has reached the structural genes.

autocatalytic Refers to a reaction that an enzyme catalyzes on part of its own structure, such as a cleavage performed by a protease on its own polypeptide precursor.

autoimmunity A condition in which the body mounts an immune response against one of its own normal components.

autolysis Programmed cell death; the orderly self-destruction of a cell in a multicellular organism. It is the process by which unwanted cells are eliminated in the body. Also called apoptosis (which see).

autonomously replicating sequences (ARSs) Sequences in yeast chromosomes that, when incorporated into an artificial plasmid, enable the plasmid to replicate efficiently in yeast cells.

autotrophs Organisms that can synthesize their organic compounds entirely from inorganic precursors, in particular needing only CO_2 as a carbon source. Compare *heterotrophs.*

auxotrophs Microorganism strains that require as a nutrient a particular substance that is not required by the prototype strain. Usually the requirement results from a mutation that disables an enzyme necessary for the endogenous synthesis of the substance.

axis of symmetry An imaginary axis through a structure, such that rotating the structure around the axis through an appropriate angle leaves the appearance of the structure unchanged.

axon A threadlike process extending from a nerve cell by which impulses are transmitted to other nerve cells or to effector cells such as muscle or gland cells. Most nerve cells have one axon; shorter processes that function in receiving impulses from other neurons are called dendrites.

B-DNA A DNA duplex with a specific right-hand helix structure. It is the usual form of DNA duplexes in vivo.

Beer's law The equation that relates the absorbance of a solution sample at a given wavelength to the length of the light path, the concentration of the dissolved substance, and the extinction coefficient of the substance at that wavelength. See *extinction coefficient.*

bile acids A family of amphipathic cholesterol derivatives that are produced in the liver and excreted in the bile; salts of the bile acids emulsify fat in the intestine.

biogenic amines A set of low-molecular-weight amino acid derivatives that contain a basic amino group and function in the body as intercellular mediators. Examples are serotonin, histamine, and epinephrine.

blood–brain barrier A selective permeability barrier that is found in the walls of blood vessels in the brain and that prevents most large or polar molecules from readily entering brain tissue. Physically the barrier consists of tight junctions between endothelial cells; these cells have transporters for polar substances such as glucose that need to enter the brain.

blood group antigens A group of oligosaccharides that are carried in the form of glycoproteins and glycolipids on the surface of cells, including blood cells; they are encoded by a large number of polymorphic gene loci and can provoke an immune response in an individual with different blood group antigens.

Bohr effect The effect of pH on oxygen binding by hemoglobin, by which a decrease in pH causes a decrease in oxygen affinity. The effect promotes both the release of oxygen from hemoglobin in the tissues and the release of CO_2 from the blood to the air in the lungs.

branch migration During recombination, the migration of a cross-over point (Holliday junction) by simultaneous unwinding and rewinding in both duplexes.

buffering The ability of a mixture of an acid and its conjugate base at a pH near their pK_a to minimize pH changes caused by an influx of acid or base.

calorie A unit of energy defined as that amount of heat energy that will raise the temperature of 1 gram of water by 1°C. 1 calorie = 4.182 joules (which see).

Calvin cycle The cycle of photosynthetic dark reactions by which CO_2 is fixed, reduced, and converted to glyceraldehyde-3-phosphate (the precursor of hexose monophosphates).

carbohydrates In general, substances that have the stoichiometric formula $(CH_2O)_n$, where $n \geq 3$, or that are derived from such a substance by the addition of functional groups.

carboxyl terminus See *C-terminus.*

carnitine A low-molecular-weight lysine derivative that shuttles fatty acids through the inner mitochondrial membrane to the matrix. The fatty acyl moiety is transferred from CoA to carnitine for transit through the membrane and is then transferred back to CoA; the carnitine released on the matrix side of the membrane is shuttled back for reuse.

caspases A family of proteases involved in apoptosis (which see).

catabolism The sum of all the metabolic processes by which complex molecules are broken down to simpler ones, including the processes by which molecules are broken down to yield cellular energy. Compare *anabolism.*

catabolite activation In bacteria, a transcriptional control system that induces the synthesis of enzymes for the catabolism of energy substrates other than glucose when glucose levels are low. It involves an activator protein, CRP, that binds cyclic AMP under conditions of low glucose; this complex then binds to DNA sites and promotes transcription of the appropriate genes.

cathepsins Lysosomal proteases that function in degrading proteins in lysosomes and are also released into the cell at large during cell autolysis (programmed cell death).

cell wall A rigid, protective wall around a plant, bacterial, or fungal cell that is secreted by the cell and is located outside the plasma membrane.

center of chirality With respect to organic compounds, a carbon atom that has four different substituents attached to it; such a group cannot be superimposed on its own mirror image and therefore can occur in two enantiomers.

centromere The region of a chromosome where the two sister chromatids are attached together. It is also the site of attachment for spindle fibers during mitosis and meiosis.

C_4 cycle A cycle in some plants that minimizes the wasteful effects of photorespiration by using an enzyme other than rubisco to perform the initial fixation of CO_2. This enzyme is found in mesophyll cells, where it fixes CO_2 into a four-carbon compound (hence C_4). This fixed carbon is shuttled into sheltered bundle-sheath cells, where it is released as CO_2 and enters the Calvin cycle.

chaotropic The property of being able to disrupt the hydrogen bonding structure of water. Substances that are good hydrogen bonders, such as urea or guanidine hydrochloride, are chaotropic. Concentrated solutions of these substances tend to denature proteins because they reduce the hydrophobic effect.

chaperonins Proteins that are involved in managing the folding of other proteins. Some of them help proteins to fold correctly; some prevent premature folding; and some prevent polypeptides from associating with other polypeptides until they have folded properly.

chemical cross-linking A technique for investigating the mutual arrangement of components in a complex. The complex is exposed to a reagent that can form chemical cross-links between adjacent components and is then disaggregated and analyzed. Components that are linked together can be assumed to be neighbors in the complex.

chemical potential ($\overline{G}$) (also called partial molar free energy) In a system, the free energy that resides in a chemical component per mole of the component present. For example, in a system consisting of *a* moles of component A and *b* moles of component B, the total free energy *G* would be the sum of the free energy in the two components: $G = a\overline{G}_A + b\overline{G}_B$.

chemiosmotic coupling The coupling of an enzyme-catalyzed chemical reaction to the transport of a substance across a membrane either with or against its concentration gradient. The outstanding example is the coupling of ATP synthesis to the movement of protons across a membrane in response to a proton gradient.

chemotaxis The process by which bacteria sense a concentration gradient of a particular substance in the medium and move either up or down the gradient.

chiral With respect to a molecule or other object, the property of being nonsuperimposable on its mirror image. An atom that makes a molecule chiral, such as a carbon with four different substituents, is called a chiral atom or center of chirality.

chloroplasts The organelles in plant and algal cells that carry out photosynthesis.

chromatin The filamentous material of eukaryotic chromosomes, consisting of DNA with associated histones and other proteins. During interphase it is dispersed and fills most of the nucleus; during nuclear division it condenses into compact chromosomes.

chromophore A chemical group that absorbs light at characteristic wavelengths.

chylomicron A type of lipoprotein that is produced in the intestinal villi and serves to transport dietary lipids in the circulation.

circular dichroism The property of absorbing right circularly polarized light and left circularly polarized light to different extents. Stereoisomers exhibit circular dichroism. Also, some types of secondary structure, such as α helices and β sheets in proteins, exhibit a predictable circular dichroism at specific wavelengths.

circular dichroism spectrum (CD spectrum) An absorption spectrum obtained using circularly polarized light; it gives the circular dichroism of the substance over a range of wavelengths.

***cis*-dominant** Refers to a mutation in a genetic regulatory element that affects the expression of appropriate genes *only* on the same chromosome, not on another homologous chromosome present in the same cell. *Cis*-dominance demonstrates that a regulatory element does not code for a diffusible factor.

cistron The smallest unit of DNA that must be intact to code for the amino acid sequence of a polypeptide; thus, the coding part of a gene, minus 5′ and 3′ untranslated sequences and regulatory elements.

citric acid cycle (also called tricarboxylic acid cycle and Krebs cycle) A cycle of reactions that takes place in the mitochondrial matrix and results in the oxidation of acetyl units to CO_2 with the production of reducing equivalents and ATP. It is a central pathway in oxidative respiration. Other substrates besides acetyl-CoA can enter the cycle at intermediate points.

clathrate structure The cagelike structure of organized water molecules that forms around a hydrophobic molecule in solution. The structure has lower entropy than liquid water, which helps explain why hydrophobic substances dissolve poorly in water.

clonal selection theory A model (proved correct) describing how the body is able to produce specific immune responses against a vast array of antigens. The B and T cells produced by the body have randomly generated antigen specificities. When a particular antigen enters the body, it induces proliferation only in B and T cells that happen to be specific for it. Thus, the antigen selects the cells that will mount an immune response against it and stimulates them to undergo clonal proliferation.

clone A group of cells, organisms, or DNA sequences that are genetically identical because they are all derived from a single ancestor.

coated pit A cell membrane pit that is lined on its cytosolic side by a meshwork of the protein clathrin. Coated pits participate in the mechanism of receptor-mediated endocytosis, in which surface receptors that have bound specific extracellular substances are gathered into coated pits, which pinch off to become cytoplasmic vesicles.

coenzyme An organic small molecule that binds to an enzyme and is essential for its activity but is not permanently altered by the reaction. Most coenzymes are derived metabolically from vitamins.

colony hybridization A technique that is used to screen bacteria for the presence of a specific recombinant DNA sequence. Colonies of the bacteria are transferred to a filter, treated to lyse the cells and denature the DNA, and then exposed to a labeled DNA probe that is complementary to part of the sequence in question. Colonies that bind the probe possess the sequence.

competitive inhibitor A substance that inhibits an enzyme-catalyzed reaction by competing with the substrate for the active site; the inhibitor can reversibly occupy the active site but does not undergo the reaction.

concatemer A DNA molecule that consists of a tandem series of complete genomes. Some phage genomes form concatemers during replication as part of a strategy for replicating the full length of a linear DNA duplex.

confocal microscopy A light-microscopy technique that allows high resolution in thick samples.

conjugation With respect to bacteria, a process in which two bacterial cells pair up and one of them passes a copy (usually partial) of its chromosome to the other. Also called bacterial mating.

consensus sequence For a group of nucleotide or amino acid sequences that show similarity but are not identical (for example, the sequences for a family of related regulatory gene sequences), an artificial sequence that is compiled by choosing at each position the residue that is found there most often in the sequences under study.

constitutive With respect to gene expression, refers to proteins that are synthesized at a fairly steady rate at all times instead of being induced and repressed in response to changing conditions.

cooperative transition A transition in a multipart structure such that the occurrence of the transition in one part of the structure makes the transition likelier to happen in other parts.

copy number The number of copies per cell of a particular gene or other DNA sequence.

Cori cycle The metabolic cycle by which lactate produced by tissues engaging in anaerobic glycolysis, such as exercising muscle, is regenerated to glucose in the liver and returned to the tissues via the bloodstream.

counterion atmosphere A cloud of oppositely charged small ions (*counterions*) that collects around a macroion dissolved in a salt solution. Counterion atmospheres partly shield macroions from each other's charges and thus affect their interactions.

cristae Folds in the inner mitochondrial membrane that project into the mitochondrial matrix. The enzymes of the electron transport chain and oxidative phosphorylation are located mainly on the cristae.

cruciform In a DNA duplex, a structure that can be adopted by a palindromic sequence, in which each strand base-pairs with itself to form an arm that projects from the main duplex and terminates in a hairpin loop. The two arms form a "cross" with the main duplex.

cryoelectron microscopy A variation of electron microscopy (which see) in which samples are frozen in a glassy ice matrix.

C-terminus (also called carboxyl terminus) The end of a polypeptide chain that carries an unreacted carboxyl group. See also *N-terminus.*

curie The basic unit of radioactive decay; an amount of radioactivity equivalent to that produced by 1 g of radium, namely 2.22×10^{12} disintegrations per minute.

cyclic photophosphorylation In photosynthesis, photophosphorylation (light-dependent ATP synthesis) that is linked to a cyclic flow of electrons from photosystem II down an electron transport chain and back to photosystem II; it is not coupled to the oxidation of H_2O or to the reduction of $NADP^+$. Compare *noncyclic photophosphorylation.*

cyclins Proteins that regulate the cell cycle by binding to and activating specific nuclear protein kinases. Cyclin-dependent kinase activations occur at three points during the cell cycle, thus providing three decision points as to whether the cycle will proceed.

cytokinesis The division of a eukaryotic cell to form two cells. It usually accompanies nuclear division, although nuclear division can occur without cytokinesis.

cytoskeleton An organized network of rodlike and fiberlike proteins that pervades a cell and helps give it its shape and motility. The cytoskeleton includes actin filaments, microtubules, and a diverse group of filamentous proteins collectively called intermediate filaments.

cytosol The fluid medium that is located inside a cell but outside the nucleus and organelles (for eukaryotes) or the nucleoid (for prokaryotes). It is a semiliquid concentrated solution or suspension.

dark reactions The photosynthetic subprocesses that do not depend *directly* on light energy; specifically, the synthesis of carbohydrate from CO_2 and H_2O. Compare *light reactions.*

denaturation For a nucleic acid or protein, the loss of tertiary and secondary structure so that the polymer becomes a random coil. For DNA, this change involves the separation of the two strands. Denaturation can be induced by heating and by certain changes in chemical environment.

depurination Cleavage of the glycosidic bond between C-1′ of deoxyribose and a purine base in DNA. Used in Maxam–Gilbert sequence analysis.

diabetes mellitus A disease caused by a deficiency in the action of insulin in the body, resulting either from low insulin levels or from inadequate insulin levels combined with unresponsiveness of the target cells to insulin. The disease is manifested primarily by disturbances in fuel homeostasis, including hyperglycemia (abnormally high blood glucose levels).

dialysis The process by which low-molecular-weight solutes are added to or removed from a solution by means of diffusion across a semipermeable membrane.

diastereomers Molecules that are stereoisomers but not enantiomers of each other. Isomers that differ in configuration about two or more asymmetric carbon atoms and are not complete mirror images.

dielectric constant A dimensionless constant that expresses the screening effect of an intervening medium on the interaction between two charged particles. Every medium (such as a water solution or an intervening portion of an organic molecule) has a characteristic dielectric constant.

difference spectrum With respect to absorption spectra, a spectrum obtained by loading the sample cuvette with the substances under study and a reference cuvette with an equimolar sample of the same substances in a known state (for example, fully oxidized) and recording the difference between the two spectra.

diffraction pattern The pattern that is produced when electromagnetic radiation passes through a regularly repeating structure; it results because the waves scattered by the structure interact destructively in most directions (creating dark zones) but constructively in a few directions (creating bright spots). For the pattern to be sharp, the radiation wavelength must be somewhat shorter than the repeat distance in the structure. See also *x-ray diffraction.*

diffusion coefficient (D) A coefficient that indicates how quickly a particular substance will diffuse in a particular medium under the influence of a given concentration gradient.

diploid For a cell or an organism, the possession of two homologous sets of chromosomes per nucleus (with the possible exception of sex chromosomes, which may be present in only one copy). Compare *haploid.*

dismutation A reaction in which two identical substrate molecules have different fates; particularly, a reaction in which one of the substrate molecules is oxidized and the other reduced.

dispersion forces Weak intermolecular attractive forces that arise between molecules that are close together, because the fluctuating electron distributions of the molecules become synchronized so as to produce a slight electrostatic attraction. These forces play a role in the internal packing of many biomolecules.

dissociation constant For an acid, the equilibrium constant K_a for the dissociation of the acid into its conjugate base and a proton. For a complex of two biomolecules, the equilibrium constant K_d for dissociation into the component molecules.

DNA gyrase An enzyme that is able to introduce negative superhelical turns into a circular DNA helix.

domain A portion of a polypeptide chain that folds on itself to form a compact unit that remains recognizably distinct within the tertiary structure of the whole protein. Large globular proteins often consist of several domains, which are connected to each other by stretches of relatively extended polypeptide.

drug resistance factors Bacterial plasmids that carry genes coding for resistance to antibiotics.

dyad axis A 2-fold axis of symmetry.

editing See RNA editing.

einstein One mole of photons.

electron microscopy A form of microscopy in which electrons are used as radiation. Capable of very high resolution.

electron spin resonance (also called electron paramagnetic resonance, or EPR) A form of spectroscopy that is sensitive to the environment of unpaired electrons in a sample.

electron transport chain A sequence of electron carriers of progressively higher reduction potential in a cell that is linked so that electrons can pass from one carrier to the next. The chain captures some of the energy released by the flow of electrons and uses it to drive the synthesis of ATP.

electrophoresis A method for separating electrically charged substances in a mixture. A sample of the mixture is placed on a supporting medium (a piece of filter paper or a gel), to which an electrical field is applied. Each charged substance migrates toward the cathode or the anode at a speed that depends on its net charge and its frictional interaction with the medium. See also *gel electrophoresis.*

elongation factors Nonribosomal protein factors that are necessary participants in the chain-elongation cycle of polypeptide synthesis; they interact with the ribosome–mRNA complex or with other major cycle participants.

enantiomers (also called optical isomers) Stereoisomers that are nonsuperimposable mirror images of each other. The term *optical isomers* comes from the fact that the enantiomers of a compound rotate polarized light in opposite directions.

endergonic In a nonisolated system, a process that is accompanied by a positive change in free energy (positive ΔG) and therefore is thermodynamically not favored. Compare *exergonic.*

endocrine glands Glands that synthesize hormones and release them into the circulation. The hormone-producing gland cells are called endocrine cells.

endonuclease An enzyme that cleaves a nucleic acid chain at an internal phosphodiester bond.

endoplasmic reticulum (ER) A highly folded membranous compartment within the cytoplasm that is responsible for a great variety of cellular tasks, including the glycosylation and trafficking of proteins destined for secretion or for the cell membrane or some organelles. It also functions in lipid synthesis, and the enzymes of many pathways of intermediate metabolism are located on its surface.

endorphins A class of endogenous brain peptides that exert analgesic effects in the central nervous system by binding to opiate receptors. They are produced by cleavage of the large polypeptide pro-opiomelanocortin.

energy See *internal energy.*

energy charge A quantity that indicates the state of a cell's energy reserves. It is equal to the cell's reserves of the free energy sources ATP and ADP (taking into account that ADP stores less free energy than ATP) divided by the total supply of ATP and its breakdown products ADP and AMP: ([ATP] + ½[ADP])/([ATP] + [ADP] + [AMP]).

enhancer sequence A DNA sequence that is distant from a gene but to which a protein factor that affects the gene's transcription can bind to exert its action. It is possible that DNA looping brings enhancer-bound proteins into proximity with the gene's promoter.

enthalpy (H) A thermodynamic quantity (function of state) that is equal to the internal energy of a system plus the product of the pressure and volume: $H = E + PV$. It is equal to the heat change in constant-pressure reactions, such as most reactions in biological systems.

entropy (S) A thermodynamic quantity (function of state) that expresses the degree of disorder or randomness in a system. According to the second law of thermodynamics, the entropy of an open system tends to increase unless energy is expended to keep the system orderly.

episomes Plasmids that can undergo integration into the bacterial chromosome.

epitope (also called antigenic determinant) The specific portion of an antigen particle that is recognized by a given antibody or T-cell receptor.

essential amino acids Amino acids that must be obtained in the diet because they cannot be synthesized in the body (at least not in adequate amounts).

essential fatty acids Fatty acids that must be obtained in the diet because they cannot be synthesized in the body in adequate amounts. Examples are linoleic acid and linolenic acid.

eukaryotes Organisms whose cells are compartmentalized by internal cellular membranes to produce a nucleus and organelles. Compare *prokaryotes.*

exergonic In a nonisolated system, a process that is accompanied by a negative change in free energy (negative ΔG) and therefore is thermodynamically favored. Compare *endergonic.*

exocrine cell A cell that secretes a substance that is excreted through a duct either into the alimentary tract or to the outside of the organism. Exocrine cells are grouped together in exocrine glands.

exon A region in the coding sequence of a gene that is translated into protein (as opposed to introns, which are not). The name comes from the fact that exons are the only parts of an RNA transcript that are seen outside the nucleus. Compare *intron.*

exportins A class of proteins involved in transporting materials out of nuclei. See *importins.*

extinction coefficient (ε_λ) A coefficient that indicates the ability of a particular substance in solution to absorb light of wavelength λ. The molar extinction coefficient, ε_M, is the absorbance that would be displayed by a 1-M solution in a 1-cm light path.

facilitated transport (also called facilitated diffusion) The movement of a substance across a biological membrane in response to a concentration or electrochemical gradient where the movement is facilitated by membrane pores or by specific transport proteins. Compare *active transport, passive transport.*

fatty acid A carboxylic acid with a long hydrocarbon chain.

fermentations Processes in which cellular energy is generated from the breakdown of nutrient molecules where there is no net change in the oxidation state of the products as compared with that of the reactants; fermentation can occur in the absence of oxygen.

fibrous proteins Proteins of elongated shape, often used as structural materials in cells and tissues. Compare *globular proteins.*

first law of thermodynamics The law that states that energy cannot be created or destroyed and that it is therefore possible to account for any change in the internal energy of a system ΔE by an exchange of heat (q) and/or work (w) with the surroundings. $\Delta E = q - w$.

first-order reaction A reaction whose rate depends on the first power of the concentration of the reactant. Compare *second-order reaction.*

Fischer projection A convention for representing stereoisomers in a plane. The tetrahedron of bonds on a carbon is represented as a plane cross, where the bonds to the right and left are assumed to be pointing toward the viewer and the bonds to the top and bottom are assumed to be pointing away from the viewer. Fischer projections of monosaccharides are oriented with the carbonyl group at the top; the chiral carbon farthest from the carbonyl group (which is the one that determines whether the sugar is the D or the L form) is then drawn with its hydroxyl to the right for the D form and to the left for the L form.

flavin adenine dinucleotide (FAD), flavin mononucleotide (FMN) Coenzymes derived from vitamin B_2 (riboflavin) that function as electron acceptors in enzymes that catalyze electron transfer reactions.

fluid mosaic model A model describing cellular membrane structure, according to which the proteins are embedded in a phospholipid bilayer and are free to move in the plane of the membrane. This model is basically correct.

fluorescence The phenomenon by which a substance that absorbs light at a given wavelength reradiates a portion of the energy as light of a longer wavelength.

flux With reference to a chemical pathway, the rate (in moles per unit time) at which reactant "flows through" the pathway to emerge as product. The term can be used for the rate at which particles undergo any process in which they either flow or can be thought of metaphorically as flowing.

footprinting With respect to molecular genetics, a technique used to identify the DNA segment in contact with a given DNA-binding protein. The DNA–protein complex is subjected to digestion with a nonspecific nuclease, which cleaves at the residues that are not protected by the protein.

frameshift mutation A mutation that changes the reading frame for a gene by adding or deleting one or two nucleotides, thereby reducing the remainder of the message 3′ to the mutation to gibberish.

frameshift suppressor A mutant tRNA that contains either two or four bases in the anticodon loop and can suppress the effects of a particular frameshift mutation in a gene.

free energy (G) (also called Gibbs free energy) A thermodynamic quantity (function of state) that takes into account both enthalpy and entropy: $G = H - TS$, where H is enthalpy, S is entropy, and T is absolute temperature. The *change in free energy* (ΔG) for a process, such as a chemical reaction, takes into account the changes in enthalpy and entropy and indicates whether the process will be thermodynamically favored at a given temperature.

frictional coefficient A coefficient that determines the frictional force on a particular particle (such as a molecule) in a particular medium at a given velocity. In the context of electrophoresis or centrifugation, it determines how fast a chemical species will move in a particular medium in response to a given electrical field or centrifugal force.

fusion proteins Genetically engineered proteins that are made by splicing together coding sequences from two or more genes. The resulting protein thus combines portions from two different parent proteins.

F_0F_1 complex (also called F_0F_1 ATP synthase) The enzyme complex in the inner mitochondrial membrane that uses energy from the transmembrane proton gradient to catalyze ATP synthesis. The F_0 portion of the complex spans the membrane, and the F_1 portion, which performs the ATP synthase activity, projects into the mitochondrial matrix.

gated channel A membrane ion channel that can open or close in response to signals from outside or within the cell.

gel electrophoresis A type of electrophoresis in which the supporting medium is a thin slab of gel held between glass plates. The technique is widely used for separating proteins and nucleic acids. See also *electrophoresis, isoelectric focusing.*

genetic code The code by which the nucleotide sequence of a DNA or RNA molecule specifies the amino acid sequence of a polypeptide. It consists of three-nucleotide codons that either specify a particular amino acid or tell the ribosome to stop translating and release the polypeptide. With a few minor exceptions, all living things use the same code.

genetic recombination Any process that results in the transfer of genetic material from one DNA molecule to another. In eukaryotes, it can refer specifically to the exchange of matching segments between homologous chromosomes by the process of crossing over.

genome The total genetic information contained in a cell, an organism, or a virus.

genotype The genetic constitution of an individual organism. Compare *phenotype.*

gibberellins A family of diterpene plant growth hormones.

Gibbs free energy See *free energy.*

globular proteins Proteins whose three-dimensional folded shape is relatively compact. Compare *fibrous proteins.*

glucocorticoids The steroid hormones cortisol and corticosterone, which are secreted by the adrenal cortex. In addition to other functions, they promote gluconeogenesis in response to low blood sugar levels.

glucogenic In fuel metabolism, refers to substances (such as some amino acids) that can be used as substrates for glucose synthesis.

gluconeogenesis The processes by which glucose is synthesized from noncarbohydrate precursors, such as glycerol, lactate, some amino acids, and (in plants) acetyl-CoA.

glucose transporter A membrane protein that is responsible for transporting glucose across a cell membrane. Different tissues may have glucose transporters with different properties.

glycan Another name for polysaccharide.

glycocalyx The polysaccharide coat found on many eukaryotic cells.

glycolipids Lipids that have saccharides attached to their head groups.

glycolysis The initial pathway in the catabolism of carbohydrates, by which a molecule of glucose is broken down to two molecules of pyruvate, with a net production of ATP molecules and the reduction of two NAD^+ molecules to NADH. Under aerobic conditions, these NADH molecules are reoxidized by the electron transport chain; under anaerobic conditions, a different electron acceptor is used.

glycosaminoglycans (also called mucopolysaccharides) Polysaccharides composed of repeating disaccharide units in which one sugar is either *N*-acetylgalactosamine or *N*-acetylglucosamine. Typically the disaccharide unit carries a carboxyl group and often one or more sulfates, so that most glycosaminoglycans have a high density of negative charges. Glycosaminoglycans are often combined with protein to form proteoglycans and are an important component of the extracellular matrix of vertebrates.

glyoxysome A specialized type of peroxisome found in plant cells. It performs some of the reactions of photorespiration, and it also breaks down fatty acids to acetyl-CoA by β-oxidation and converts the acetyl-CoA to succinate via the glyoxylate cycle, thus enabling plants to convert fatty acids to carbohydrates.

Golgi complex A stack of flattened membranous vesicles in the cytoplasm. It serves as a routing center for proteins destined for secretion or for lysosomes or the cell membrane; it performs similar functions for membrane lipids, and it also modifies and finishes the oligosaccharide moieties of glycoproteins.

G proteins A family of membrane-associated proteins that transduce signals received by various cell-surface receptors. They are called G proteins because binding of GTP and GDP is essential to their action.

growth factors Peptide mediators that influence the growth and/or differentiation of cells; they differ from growth hormones in being produced by many tissues and in acting locally.

half-life (also called half-time) For a chemical reaction, the time at which half the substrate has been consumed and turned into product. The term can also refer to the analogous point in other processes, such as the radioactive decay of an isotope.

haploid For a cell or an organism, the possession of only one copy of each chromosome per nucleus. Compare *diploid.*

hapten A molecule that is too small to stimulate an immune response by itself but can do so when coupled to a larger, immunogenic carrier molecule (usually a protein).

Haworth projection A conventional planar representation of a cyclized monosaccharide molecule. The hydroxyls that are represented to the right of the chain in a Fischer projection are shown below the plane in a Haworth projection.

heat-shock proteins A group of chaperonins that accumulate in a cell after it has been subjected to a sudden temperature jump or other stress. They are thought to help deal with the accumulation of improperly folded or assembled proteins in stressed cells.

helicases Enzymes that catalyze the unwinding of duplex nucleic acids.

helix–loop–helix motif A binding motif that is found in calmodulin and some other calcium-binding proteins as well as in some DNA-binding proteins. It consists of two α helix segments connected by a loop.

helix–turn–helix motif A DNA-binding motif that is responsible for sequence-specific DNA binding in many transcription factors. It consists of two α helix segments connected by a β turn; one of the helices occupies the DNA major groove and makes specific base contacts.

helper T cells T lymphocytes whose role is to recognize antigens and help other defensive cells to mount an immune response. They help activate antigen-stimulated B cells (resulting in production of specific antibodies) and/or antigen-stimulated cytotoxic T cells (resulting in attack on antigenic cells), and they also produce immune mediators that stimulate nonspecific defense responses.

heme A molecule consisting of a porphyrin ring (either protoporphyrin IX or a derivative) with a central complexed iron; it serves as a prosthetic group in proteins such as myoglobin, hemoglobin, and cytochromes.

hemimethylated With respect to DNA, refers to the condition in which one strand of the duplex is methylated and the other is not. Newly replicated DNA is hemimethylated; normally a methylase enzyme then methylates appropriate bases in the new strand.

heterotrophs Organisms that cannot synthesize their organic compounds entirely from inorganic precursors but must consume at least some organic compounds made by other organisms. In particular, these organisms cannot use CO_2 as a carbon source. Compare *autotrophs.*

heterozygous In a diploid organism, the possession of two different alleles for a given gene (as opposed to two copies of the same allele). Compare *homozygous.*

high-density lipoprotein (HDL) A type of lipoprotein particle that functions mainly to scavenge excess cholesterol from tissue cells and transport it to the liver, where it can be excreted in the form of bile acids.

Hill coefficient (n_H) A coefficient that indicates the degree of cooperativity of a cooperative transition; it is the maximum slope of a Hill plot of the transition.

histones The proteins that participate in forming the nucleosomal structure of chromatin. Four of the five kinds of histones make up the core particle of the nucleosome; the fifth is associated with the linker DNA between nucleosomes. All histones are small, very basic proteins.

Holliday junction An intermediate during homologous recombination; a four-armed structure in which each of the participating DNA duplexes has exchanged one strand with the other duplex.

homeo box A common sequence element of about 180 base pairs that is found in homeotic genes. It codes for a sequence-specific DNA-binding element of the helix–loop–helix class. See also *homeotic genes.*

homeotic genes Genes that contain homeo box elements and typically are involved in controlling the pattern of organismal development. Homeotic mutations, which scramble portions of this pattern, affect homeotic genes. The nuclear DNA-binding proteins encoded by these genes presumably serve as transcriptional regulators for the coordinated expression of groups of genes. See also *homeo box.*

homologous recombination Genetic recombination that requires extensive sequence homology between the recombining DNA molecules. Meiotic recombination by crossing over in eukaryotes is an example.

homopolymer A polymer that is made of only one kind of monomer. Starch, made only of glucosyl units, is an example. Polymers that include more than one kind of monomer, like polypeptides and nucleic acids, are called heteropolymers.

homozygous In a diploid organism, the possession of two identical alleles for a given gene. Compare *heterozygous.*

hormone A substance that is synthesized and secreted by specialized cells and carried via the circulation to target cells, where it elicits specific changes in the metabolic behavior of the cell by interacting with a hormone-specific receptor.

hormone-responsive element A DNA site that binds an intracellular hormone–receptor complex; binding of the complex to a hormone-responsive element affects the transcription of specific genes.

host-induced restriction and modification A genetic system found in bacteria whereby a genetic element (often a plasmid) encodes both an enzyme for the methylation of DNA at a specific base sequence and an endonuclease that cleaves unmethylated DNA at that sequence. The system thus *restricts* the DNA that can survive in the cell to DNA that is *modified* by methylation at the correct sequences.

hybridomas Cultured cell lines that are made by fusing antibody-producing B lymphocytes with cells derived from a mouse myeloma (a type of lymphocyte cancer). Like B cells, they produce specific antibodies, and like myeloma cells, they can proliferate indefinitely in culture.

hydrogen bond An attractive interaction between the hydrogen atom of a donor group, such as —OH or =NH, and a pair of nonbonding electrons on an acceptor group, such as O=C. The donor group atom that carries the hydrogen must be fairly electronegative for the attraction to be significant.

hydrophilic Refers to the ability of an atom or a molecule to engage in attractive interactions with water molecules. Substances that are ionic or can engage in hydrogen bonding are hydrophilic. Hydrophilic substances are either soluble in water or, at least, wettable. Compare *hydrophobic.*

hydrophobic The molecular property of being unable to engage in attractive interactions with water molecules. Hydrophobic substances are nonionic and nonpolar; they are nonwettable and do not readily dissolve in water. Compare *hydrophilic.*

hydrophobic effect With respect to globular proteins, the stabilization of tertiary structure that results from the packing of hydrophobic side chains in the interior of the protein.

hypochromism With respect to DNA, a reduction in the absorbance of ultraviolet light of wavelength of about 260 nm that accompanies the transition from random-coil denatured strands to a double-strand helix. It can be used to track the process of denaturation or renaturation.

immunoglobulins See *antibodies.*

importins A class of proteins involved in importing molecules into the nucleus. See *exportins.*

inborn errors of metabolism Human mutations that result in specific derangements of intermediary metabolism. Usually the problem is an enzyme that is inactive, overactive, too scarce, or too abundant; symptoms may result from the insufficient production of a necessary metabolite and/or from the accumulation of another metabolite to toxic levels.

induced dipole A molecule has an induced dipole if an external electric field induces an asymmetric distribution of charge within it.

induced fit model A model for how enzymes interact with substrates to achieve catalysis. According to this model, the empty active site of the enzyme only roughly fits the substrate(s), and the entry of substrate causes the enzyme to change its shape so as to both tighten the fit and cause the substrate to adopt an intermediate state that resembles the transition state of the uncatalyzed reaction. This is currently the dominant model for enzymatic catalysis.

induction In cellular metabolism, the synthesis of a particular protein in response to a signal; for example, the synthesis of an enzyme in response to the appearance of its substrate.

in situ hybridization A technique for finding the chromosomal location of a particular DNA sequence by probing the chromosomes with a radiolabeled sequence that will hybridize with the sequence in question. The location of the probe is then visualized with radioautography.

intercalation With respect to DNA, refers to the fitting (intercalation) of a small molecule between adjacent bases in a DNA helix.

interferon An antiviral glycoprotein produced in response to viral infection.

intermediary metabolism All the reactions in an organism that are concerned with storing and generating metabolic energy and with the biosynthesis of low-molecular-weight compounds and energy-storage compounds. It does not include nucleic acid and protein synthesis.

internal energy (E) The energy contained in a system. For the purposes of biochemistry, the term encompasses all the types of energy that might be changed by chemical or nonnuclear physical processes, including the kinetic energy of motion and vibration of atoms and molecules and the energy stored in bonds and noncovalent interactions.

intron A region in the coding sequence of a gene that is not translated into protein. Introns are common in eukaryotic genes but are rarely found in prokaryotes. They are excised from the RNA transcript before translation. Compare *exon.*

ion-exchange resins Polycationic or polyanionic polymers that are used in ion-exchange column chromatography to separate substances on the basis of electrical charge.

ionic strength (I) A quantity that reflects the total concentration of ions in a solution and the stoichiometric charge (charge per atom or molecule) of each ion. It is defined as $I = \frac{1}{2} \sum_i \mathrm{M}_i Z_i^2$, where M_i and Z_i are respectively the molarity and stoichiometric charge of ion *i*. It is used, for example, in calculating the effective radius of a counterion atmosphere.

ion pore A pore in a cellular membrane through which ions can diffuse. It is formed by a transmembrane protein and can discriminate among ions to some degree on the basis of size and charge. Many ion pores are gated, meaning that they can open and close in response to signals.

isoelectric focusing A version of gel electrophoresis that allows ampholytes to be separated almost purely on the basis of their isoelectric points. The ampholytes are added to a gel that contains a pH gradient and are subjected to an electric field. Each ampholyte migrates until it reaches the pH that represents its isoelectric point, at which point it ceases to have a net electric charge and therefore comes to a halt and accumulates. See also *gel electrophoresis, isoelectric point.*

isoelectric point (pI) The pH at which the net charge on an ampholyte is, on average, zero.

isoenzymes (also called isozymes) Different but related forms of an enzyme that catalyze the same reaction. Often differ in only a few amino acid substitutions.

isomorphorous replacement The replacement of one atom in a macromolecule with a heavy metal atom in such a way that the structure of the macromolecule does not change. It is used in the determination of molecular structure by x-ray crystal diffraction.

isozymes See *isoenzymes.*

joule (J) A unit for energy or work, defined as the work done by a force of 1 newton when its point of application moves 1 meter in the direction of the force. It is the unit of energy used in the Système Internationale (SI).

α-keratins A class of keratins that are the major proteins of hair. They consist of long α-helical polypeptides, which are wound around each other to form triplet helices.

ketone bodies The substances acetoacetate, β-hydroxybutyrate, and acetone, which are produced from excess acetyl-CoA in the liver when the rate of fatty acid β-oxidation in liver mitochondria exceeds the rate at which acetyl-CoA is used for energy generation or fatty acid synthesis.

ketose A monosaccharide in which the carbonyl group occurs within the chain and hence represents a ketone group. Compare *aldose.*

Krebs cycle See *citric acid cycle.*

lagging strand During DNA replication, the strand that is synthesized in the opposite direction to the direction of movement of the replication fork; it is synthesized as a series of fragments that are subsequently joined. Compare *leading strand.*

leader sequence (also called signal sequence) For an mRNA, the nontranslated sequence at the 5′ end of the molecule that precedes the initiation codon. For a protein, a short N-terminal hydrophobic sequence that causes the protein to be translocated into or through a cellular membrane.

leading strand During DNA replication, the strand that is synthesized in the same direction as the direction of movement of the replication fork; it is synthesized continuously rather than in fragments. Compare *lagging strand.*

leukotrienes A family of molecules that are synthesized from arachidonic acid by the lipoxygenase pathway and function as local hormones, primarily to promote inflammatory and allergic reactions (such as the bronchial constriction of asthma).

library With respect to molecular genetics, a large collection of random cloned DNA fragments from a given organism, sometimes representing the entire nuclear genome.

ligand In general, a small molecule that binds specifically to a larger one—for example, a hormone that binds to a receptor. The term can also be used to mean a chemical species that forms a coordination complex with a central atom, which is usually a metal atom.

light reactions The photosynthetic subprocesses that depend *directly* on light energy; specifically, the synthesis of ATP by photophosphorylation and the reduction of $NADP^+$ to NADPH via the oxidation of water. Compare *dark reactions.*

Lineweaver–Burk plot A plot that allows one to derive the rate constant k_{cat} and the Michaelis constant K_M for an enzyme-catalyzed reaction. It is constructed by measuring the initial reaction rate V at various substrate concentrations [S] and plotting the values on a graph of $1/V$ versus $1/[S]$.

linkage map A map showing the arrangement of genes on a chromosome; it is constructed by measuring the frequency of recombination between pairs of genes.

linking number (*L*) The total number of times the two strands of a closed, circular DNA helix cross each other by means of either twist or writhe; this equals the number of times the two strands are interlinked. It reflects both the winding of the native DNA helix and the presence of any supercoiling. See also *twist, writhe.*

lipid bilayer A membrane structure that can be formed by amphipathic molecules in an aqueous environment; it consists of two back-to-back layers of molecules, in each of which the polar head groups face the water and the nonpolar tails face the center of the membrane. The fabric of cellular membranes is a lipid bilayer.

lipids A chemically diverse group of biological compounds that are classified together on the basis of their generally apolar structure and resulting poor solubility in water.

lipoproteins Any lipid–protein conjugate. Specifically refers to lipid–protein associations that transport lipids in the circulation. Each consists of a core of hydrophobic lipids surrounded by a skin of amphipathic lipids with embedded apolipoproteins. Different kinds of lipoproteins play different roles in lipid transport.

long terminal repeats (LTRs) A pair of direct repeats several hundred base pairs long that are found at either end of a retroviral genome. They are involved in integration into the host genome and in viral gene expression.

low-angle neutron scattering A set of techniques that can be used to find the size of a particle in solution or to find the size or spacing of internal regions that can be distinguished by different neutron scattering power, such as the protein and nucleic acid components of a nucleoprotein particle or labeled proteins within a multisubunit complex.

low-density lipoprotein (LDL) A type of lipoprotein particle that functions mainly to distribute cholesterol from the liver to other tissues. Its protein component consists of a single molecule of apoprotein B-100.

lysogeny A latent state that can be achieved by some bacteriophages, in which the phage genome is integrated into the host bacterial chromosome and few if any viral genes are expressed.

mass spectrometry A method for determining the molecular mass from the velocity of motion of ions in a vacuum.

mechanism-based inhibitor An enzyme inhibitor whose action depends on the enzyme's catalytic mechanism. Typically it is a substrate analog that irreversibly modifies the enzyme at a particular step in the catalytic cycle.

membrane electrical potential With respect to biological membranes, a voltage difference that exists across a membrane owing to differences in the concentrations of ions on either side of the membrane.

messenger RNA (mRNA) RNA molecules that act as templates for the synthesis of polypeptides by ribosomes.

metabolism The totality of the chemical reactions that occur in an organism. Compare *anabolism, catabolism, intermediary metabolism.*

metastability For a system, the condition of being in a state that does not represent thermodynamic equilibrium but is nearly stable at the time scale of interest because progress toward equilibrium is slow.

micelles Tiny droplets that form when an amphipathic substance that has a polar head group and a nonpolar tail region (such as a fatty acid) is added to an aqueous medium and shaken. Each droplet consists of a spherical cluster of amphipathic molecules arranged with their polar head groups facing out toward the water and their nonpolar tails facing in toward the center.

Michaelis–Menten equation An equation that gives the rate of an enzyme-catalyzed reaction in terms of the concentrations of substrate and enzyme as well as two constants that are specific for a particular combination of enzyme and substrate: a rate constant, k_{cat}, for the catalytic production of product when the enzyme is saturated, and the Michaelis constant, K_M.

microtubules Fiberlike cytoplasmic structures that consist of units of the protein tubulin arranged helically to form a hollow tube. They are involved in various kinds of cellular motility, including the beating of cilia and flagella and the movement of organelles from one part of the cell to another.

microtubule-associated proteins (MAPs) A class of proteins associated with microtubules that assist in dynamic processes.

minus strand In viral genomes, a nucleic acid strand that is complementary to the RNA strand that serves as mRNA. Compare *plus strand.*

mismatch repair A system for the correction of mismatched nucleotides or single-base insertions or deletions produced during DNA replication; it scans the newly replicated DNA, and when it finds an error, it removes and replaces a stretch of the strand containing the error.

missense mutation A mutation that alters a DNA codon so as to cause one amino acid in a protein to be replaced by a different one.

mitochondria The organelles whose chief task it is to supply the cell with ATP via oxidative phosphorylation. They contain the enzymes for pyruvate oxidation, the citric acid cycle, the β-oxidation of fatty acids, and oxidative phosphorylation, as well as the electron transport chain.

mixed-function oxidase An oxygenase enzyme that catalyzes a reaction in which two different substrates are oxidized, one by the addition of an oxygen atom from O_2 and the other by supplying two hydrogen atoms to reduce the remaining oxygen atom to H_2O.

molten globule A hypothetical intermediate state in the folding of a globular protein, in which the overall tertiary framework has been established but internal side chains (especially hydrophobic ones) are still free to move about.

mucopolysaccharides See *glycosaminoglycans.*

multicatalytic proteinase complex (MPC) A massive complex of proteolytic enzymes that is found in the cytosol of many eukaryotic cells and seems to function in the programmed destruction of cellular proteins.

muscarinic acetylcholine receptors A class of receptors for the neurotransmitter acetylcholine that are characterized by an ability to bind the toadstool toxin muscarine. Synapses that have these receptors may be either excitatory or inhibitory. Compare *nicotinic acetylcholine receptors.*

mutagen A substance that is capable of causing mutations.

mutation Any inheritable change in the nucleotide sequence of genomic DNA (or genomic RNA, in the case of an RNA virus).

Nernst equation An equation that relates the electrical potential across a membrane to the concentrations of ions on either side of the membrane.

neurohormones Substances that are released from neurons and modulate the behavior of target cells, which are often other neurons. Unlike neurotransmitters, they do not act strictly across a synapse. Most neurohormones are peptides.

neurotoxin A toxin that acts by disrupting nerve cell function. Fast-acting neurotoxins often act by blocking the action of an ion gate necessary for the development of an action potential.

neurotransmitter A low-molecular-weight substance that is released from an axon terminal in response to the arrival of an action potential and then diffuses across the synapse to influence the postsynaptic cell, which may be either another neuron or a muscle or gland cell.

nick translation A process in which nucleotides in one strand of a nucleic acid duplex are replaced one by one with nucleotides complementary to the other strand. The process starts at a nick in the strand and causes the location of the nick to migrate (hence the origin of the term).

nicotinic acetylcholine receptors A class of receptors for the neurotransmitter acetylcholine that are characterized by their ability to bind nicotine. Synapses with this kind of receptor are excitatory. Compare *muscarinic acetylcholine receptors.*

noncompetitive inhibitor An inhibitor of an enzyme-catalyzed reaction that acts by binding to a site on the enzyme different from the active site and reducing the enzyme's catalytic efficiency.

noncovalent interactions All the kinds of interactions between atoms and molecules that do not involve the actual sharing of electrons in a covalent bond; they include electrostatic interactions, permanent and induced dipole interactions, and hydrogen bonding.

noncyclic photophosphorylation In photosynthesis, photophosphorylation (light-dependent ATP synthesis) that is linked to a one-way flow of electrons from water through photosystems II and I and finally to NADPH; it is thus coupled to the oxidation of H_2O and the reduction of $NADP^+$. Compare *cyclic photophosphorylation.*

nonsense mutation A mutation that creates an abnormal stop codon and thus causes translation to terminate prematurely; the resulting truncated protein is usually nonfunctional.

Northern blotting A technique for detecting the presence of a specific RNA sequence in a cell and determining its size. The total RNA of the cell is extracted, resolved by gel electrophoresis, and blotted onto a filter. There it is incubated under annealing conditions with a radiolabeled probe for the sequence in question, and heteroduplexes of the probe with RNA are detected by radioautography.

N-terminus (also called amino terminus) The end of a polypeptide chain that carries an unreacted amino group. A ribosome synthesizes a polypeptide in the direction from the N-terminus to the C-terminus. See also *C-terminus.*

nuclear envelope The double membrane that encloses the nucleus. It is pierced by nuclear pores that allow even quite large molecules, such as mRNAs and nuclear proteins, to enter or leave the nucleus.

nuclear magnetic resonance (NMR) spectroscopy A type of spectroscopy that depends on the fact that isotope nuclei having the property of spin will resonate with specific frequencies of microwave radiation when placed in a magnetic field of given strength. The resonance energy is sensitive to the local molecular environment, so NMR spectroscopy can be used to explore molecular structure. Also, different living tissues have characteristic overall NMR spectra, which are sensitive to changes in the tissue environment. NMR can thus be used in the study of tissue metabolism and the diagnosis of disease.

nuclear matrix (also called nuclear scaffold) A protein web that is left in the nucleus when histones and other weakly bound proteins are removed and most of the DNA is digested away. It is presumed to act as an organizing scaffold for the chromatin.

nuclease An enzyme that cleaves nucleic acids.

nucleoid The large, circular DNA molecule of a prokaryotic cell, along with its associated proteins; also sometimes called the bacterial chromosome. It is supercoiled and forms a dense mass within the cell, and the term *nucleoid* is often used for the cell region occupied by this mass.

nucleolus A region in the nucleus of a cell where ribosomal RNAs are transcribed and processed and ribosomes are assembled.

nucleoside A molecule that, upon complete hydrolysis, yields 1 mole per mole of a purine or pyrimidine base and a sugar.

nucleosome The first-order structural unit for the packing of DNA in chromatin, consisting of 146 bp of DNA wrapped 1.75 times around a core octamer of histone proteins. Successive nucleosomes are connected by stretches of "linker" DNA.

nucleotide A molecule that, upon complete hydrolysis, yields at least 1 mole per mole of a purine or pyrimidine base, a sugar, and inorganic phosphate.

nucleus The membrane-bound structure in a eukaryotic cell that contains the chromosomal genetic material and associated components. It is also the place where RNA molecules are processed and ribosomes are assembled.

Okazaki fragments The discontinuous stretches in which the lagging strand is initially synthesized during DNA replication; these fragments are later joined to form a continuous strand.

oncogene A gene that, in a mutated version, can help transform a normal cell to a cancer cell. Many oncogenes code for mutant proteins that are involved in the reception and transduction of growth factor signals.

oncoprotein The protein product of an oncogene.

open-promoter complex A complex between RNA polymerase holoenzyme and a promoter that has undergone initial unwinding (has "opened") preparatory to the start of transcription. It is preceded by a much less stable *closed-promoter complex,* in which the promoter has not unwound, that may either fall apart or proceed to an open-promoter complex.

open reading frame A sequence within a messenger RNA that is bounded by start and stop codons and can be continuously translated. It represents the coding sequence for a polypeptide.

operator A DNA site where a repressor protein binds to block the initiation of transcription from an adjacent promoter.

operon A set of contiguous prokaryotic structural genes that are transcribed as a unit, along with the adjacent regulatory elements that control their expression.

optical isomers See *enantiomers.*

organelles Membrane-bound compartments in the cytoplasm of eukaryotic cells. Each kind of organelle carries out a specific set of functions. Examples are mitochondria, chloroplasts, and nuclei.

oxidase An enzyme that catalyzes the oxidation of a substrate with oxygen as the electron acceptor.

oxidative phosphorylation The phosphorylation of ADP to ATP that occurs in conjunction with the transit of electrons down the electron transport chain in the inner mitochondrial membrane.

oxygenase An enzyme that catalyzes the incorporation of oxygen into a substrate.

palindrome With respect to DNA, a segment in which the sequence is the same on one strand read right to left as on the other strand read left to right; thus, a back-to-back pair of inverted repeats.

partial molar free energy See *chemical potential.*

partition coefficient (K) A coefficient that indicates how a particular substance will distribute itself between two media if allowed to diffuse to equilibrium between them; it is equal to the ratio of the solubilities of the substance in the two media.

passive transport (also called passive diffusion) With respect to membrane transport, the movement of a substance across a biological membrane by molecular diffusion through the lipid bilayer. Compare *active transport, facilitated transport.*

Pasteur effect The inhibition of glycolysis by oxygen; discovered by Pasteur when he found that aerobic yeast cultures metabolize glucose relatively slowly.

peptide bond The bond that links successive amino acids in a peptide; it consists of an amide bond between the α-carboxyl group of one amino acid and the α-amino group of the next.

peptidyltransferase During ribosomal polypeptide synthesis, the enzyme complex that transfers the polypeptide chain from the tRNA in the P site to the amino acid carried by the tRNA in the A site, thereby adding another amino acid to the chain. The complex is an integral part of the large ribosomal subunit.

permanent dipole In chemistry, a molecule that has a permanent, asymmetric distribution of charge such that one end is negative and the other end positive. The water molecule is an example: The oxygen end has a partial negative charge, and the hydrogen end has a partial positive charge.

peroxisome A small, vesicular organelle that specializes in carrying out cellular reactions involving the transfer of hydrogen from a substrate to O_2. These reactions produce the by-product H_2O_2, which is split to H_2O and O_2 by the peroxisomal enzyme catalase.

PEST sequences A family of amino acid sequences that have been found on cellular proteins that undergo rapid turnover; they may target proteins for rapid proteolysis. They consist of a region about 12 to 60 residues long that is rich in proline, glutamate, serine, and threonine (P, E, S, and T in the one-letter abbreviation system).

phenotype The appearance and other measurable characteristics of an organism; it results from the interaction of the organism's genetic makeup with the environment. Compare *genotype.*

pheromones Intercellular mediator compounds that are released from one organism and influence the metabolism or behavior of another organism, usually of the same species. Sex attractants, which elicit reproductive behavior in suitable recipients, are an example.

phorbol esters A group of natural substances that resemble *sn*-1,2-diacylglycerol (DAG) in part of their structure and can act as tumor promoters. This effect suggests that the DAG second-messenger system may be involved in growth factor action.

phosphodiester link The linkage that connects the nucleotide monomers in a nucleic acid. It consists of a phosphate residue that links the sugar moieties of successive monomers by forming an ester bond with the 5′ carbon of one sugar and the 3′ carbon of the next.

photophosphorylation Phosphorylation of ADP to ATP that depends directly on energy from sunlight. The light energy is captured by a pigment such as chlorophyll and is passed in the form of excited electrons to an electron transport chain; the electron transport chain uses energy from the electrons to create a proton gradient across a membrane, which drives the synthesis of ATP.

photoproducts The products that result when light energy causes a chemical reaction to occur in a substance. With respect to DNA, the term refers to the types of damaged DNA that can be caused by UV irradiation.

photoreactivation A DNA repair process in which an enzyme uses light energy to break cyclobutane pyrimidine dimers created by UV irradiation and to restore the correct bonding.

photorespiration The cycle of reactions that occurs in place of the Calvin cycle when the photosynthetic enzyme rubisco adds O_2 rather than CO_2 to ribulose bisphosphate carboxylase. It takes place partly in chloroplasts, partly in peroxisomes, and partly in mitochondria; it expends ATP energy and loses a previously fixed CO_2 molecule in the process of regenerating the Calvin cycle intermediate 3-phosphoglycerate.

photosynthesis The process by which energy from light is captured and used to drive the synthesis of carbohydrates from CO_2 and H_2O.

photosystem A structural unit in a cellular membrane that captures light energy and converts a portion of it to chemical energy. The photosynthesis practiced by plants, algae, and cyanobacteria involves two types of photosystem, both of which capture energy in the form of high-energy electrons and transduce it via an electron transport chain.

plaque A clear area that is formed by a local phage infection in a lawn of cultured bacteria in a Petri dish; for purposes of experimentation, it is the phage equivalent of a bacterial colony.

plasma membrane The lipid bilayer membrane that encloses the cytoplasm; it is surrounded by the cell wall if one is present.

plasmids Small, extrachromosomal circular DNA molecules found in many bacteria. They replicate independently of the main chromosome and may occur in multiple copies per cell.

plus strand In viral genomes, a nucleic acid strand that can serve as mRNA or (for a DNA strand) that is homologous to one that can; as distinct from the complementary (minus) strand. Most viruses with single-strand genomes package only the plus or minus strand in virions; the other strand is made transiently during replication. Compare *minus strand.*

polymer A large molecule that is made by linking together prefabricated molecular units (monomers) that are similar or identical to each other. The number of monomers in a polymer may range up to millions.

polymerase chain reaction (PCR) A technique that is used to amplify the number of copies of a specific DNA sequence through repeated cycles of denaturation and replication.

polytene chromosome An extra-thick chromosome that includes many parallel copies of the original DNA molecule; it is produced by repeated rounds of DNA replication without separation of the resulting copies. Polytene chromosomes are found in various cell types, notably *Drosophila* salivary gland cells; they are useful in chromosome mapping because they are large and because the genes on the strands are arranged in strict register.

primary structure For a nucleic acid or a protein, the sequence of the bases or amino acids in the polynucleotide or polypeptide. Compare *quaternary structure, secondary structure, tertiary structure.*

primer A short piece of DNA or RNA that is base-paired with a DNA template strand and provides a free 3′ —OH end from which a DNA polymerase can extend a DNA strand. Also refers to DNA oligomers used in the polymerase chain reaction.

primosome An enzyme complex that is located in the replication fork during DNA replication; it synthesizes the RNA primers on the lagging strand and also participates in unwinding the parental DNA helix.

prion An infectious agent that contains protein but no nucleic acid.

processivity For a DNA or an RNA polymerase, the average number of nucleotides incorporated per event of binding between the polymerase and a 3′ primer terminus. It describes the tendency of a polymerase to remain bound to a template.

prokaryotes Primitive single-celled organisms that are not compartmentalized by internal cellular membranes; the eubacteria and archaebacteria. Compare *eukaryotes.*

promoter A DNA sequence that can bind RNA polymerase, resulting in the initiation of transcription.

prophage An inactive phage genome that is present in a bacterial cell and its progeny. It is integrated into the host chromosome.

prostaglandins A family of compounds that are derived from certain long-chain unsaturated fatty acids (particularly arachidonic acid) by a cyclooxygenase pathway and that function as local hormones.

prosthetic group A metal ion or small molecule (other than an amino acid) that forms part of a protein in the protein's native state and is essential to the protein's functioning; its attachment to the protein may be either covalent or noncovalent.

proteases Enzymes that cleave peptide bonds in a polypeptide. Many show specificity for a particular amino acid sequence.

proteasome A large, ATP-dependent protease complex that is found in the cytosol of cells and is involved in the selective degradation of short-lived cytoplasmic proteins.

proteoglycans Glycoproteins in which carbohydrate is the dominant element. The carbohydrate is in the form of glycosaminoglycan polysaccharides, which are connected to extended core polypeptides to form huge, feathery molecules. Proteoglycans are important components of the intercellular matrix.

protofilaments The 13 linear columns of tubulin units that can be visualized in the structure of a microtubule; they result because each turn of the microtubule helix contains exactly 13 tubulin units. Each protofilament consists of alternating α and β tubulin subunits.

proton motive force (pmf) An electrochemical H^+ gradient that is set up across a cellular membrane by membrane-bound proton pumps, such as the ones in the inner mitochondrial membrane or thylakoid membrane. As the protons flow back down their gradient across the membrane, they can drive processes such as ATP synthesis.

proton pumping The active pumping of protons across a cellular membrane to form a proton gradient. For example, the electron transport chains of the inner mitochondrial and thylakoid membranes incorporate proton pumps, which create the proton gradient that powers the ATP synthases of these membranes.

provirus An animal virus genome that is integrated into a host chromosome.

pseudogenes Nontranscribed stretches of DNA that bear a strong sequence similarity to functioning genes and obviously arose from them during evolution. Many gene families contain pseudogene members.

pulsed field gel electrophoresis A type of gel electrophoresis in which the orientation of the electric field is changed periodically. This technique makes it possible to separate very large DNA molecules, up to the size of whole chromosomes.

quantum efficiency (*Q*) With respect to photosynthesis, the ratio of oxygen molecules released to photons absorbed.

quaternary structure For a protein, the level of structure that results when separate, folded polypeptide chains (subunits) associate in a specific way to produce a complete protein. Compare *primary structure, secondary structure, tertiary structure.*

radioautography A technique in which an item containing radioactively labeled elements (for example, a tissue slice or a chromatography gel) is laid against a photographic film; the radioactivity exposes the film to form an image of the labeled elements. Also called autoradiography.

Ramachandran plot A plot that constitutes a map of all possible backbone configurations for an amino acid in a polypeptide. The axes of the plot consist of the rotation angles of the two backbone bonds that are free to rotate (ϕ and ψ, respectively); each point ϕ,ψ on the plot thus represents a conceivable amino acid backbone configuration.

random coil Refers to a linear polymer that has no secondary or tertiary structure but instead is wholly flexible with a randomly varying geometry. This is the state of a denatured protein or nucleic acid.

rate constant With respect to chemical reactions, a constant that relates the reaction rate for a particular reaction to substrate concentrations.

rate equation An equation, such as the Michaelis–Menten equation (which see), that relates velocity of an enzyme-catalyzed reaction to measurable parameters.

reaction center In photosynthesis, a specific pair of chlorophyll molecules in a photosystem that collect light energy absorbed by other chlorophyll molecules and pass it to an electron acceptor, normally the first compound of an electron transport chain.

reactive oxygen species (ROS) Oxygen species intermediate in oxidation level between O_2 and H_2O, which are more reactive than O_2; ROS include superoxide, peroxide, peroxynitrite, and hydroxyl radical.

receptor A protein that binds selectively to a specific molecule (such as an intercellular mediator or antigen) and initiates a biological response.

recognition helix In a helix–turn–helix DNA binding motif, the α-3 helix, which fits deep in the major groove and is responsible for the sequence specificity of binding.

recombinant DNA molecule A DNA molecule that includes segments from two or more precursor DNA molecules.

recombination A process in an organism in which two parent DNA molecules give rise to daughter DNA that combines segments from both parent molecules. It may involve the integration of one DNA molecule into another, the substitution of a DNA segment for a homologous segment on another DNA molecule, or the exchange of homologous segments between two DNA molecules.

reducing equivalent An amount of a reducing compound that donates the equivalent of 1 mole of electrons in an oxidation–reduction reaction. The electrons may be expressed in the form of hydrogen atoms.

regulon A group of unlinked (nonadjacent) genes that are all regulated by a common mechanism.

release factors Independent protein factors that are necessary participants in the release of a finished polypeptide chain from a ribosome.

replicon A unit in the genome that consists of an origin of replication and all the DNA that is replicated from that origin.

repressor In molecular genetics, a protein that inhibits the transcription of a gene by binding to an operator.

respiration With respect to energy metabolism, the process in which cellular energy is generated through the oxidation of nutrient molecules with O_2 as the ultimate electron acceptor. This type of respiration is also called *cellular respiration* to distinguish it from respiration in the sense of breathing.

respiratory chain The electron transport chain that is employed during cellular respiration and has O_2 as the ultimate electron acceptor.

resting potential The voltage difference that exists across the membrane of an excitable cell, such as a nerve cell, except in places where an action potential is in progress. It is a consequence of the ion gradients that are maintained across the membrane.

restriction endonucleases Enzymes that catalyze the double-strand cleavage of DNA at specific base sequences. Many restriction endonucleases with different sequence specificities have been found in bacteria; they are used extensively in molecular genetics.

restriction fragment length polymorphisms (RFLPs) A type of genetic polymorphism that is readily detected by Southern blotting and can be used to screen for genetic diseases. It is based on the fact that alleles often have different restriction endonuclease cleavage sites and therefore produce different arrays of fragments upon cleavage with appropriate endonucleases.

retinoids Substances that are derived from retinoic acid (a form of vitamin A) and act as intercellular mediators; they are particularly important in regulating development.

retroviruses A family of RNA viruses that possess reverse transcriptase. After the virus infects a cell, this enzyme transcribes the RNA genome into a double-strand DNA version, which integrates into a host chromosome. Human immunodeficiency virus (HIV) is a retrovirus.

reverse transcriptase An enzyme found in retroviruses that synthesizes a double-strand DNA molecule from a single-strand RNA template. It is an important tool in molecular genetics.

ribosomes Large protein–RNA complexes that are responsible for synthesizing polypeptides under the direction of mRNA templates.

RNA editing A type of RNA processing that has been found in the mitochondrial mRNAs of certain eukaryotes, in which the RNA sequence is altered by the insertion of uridine residues at specific sites.

RNA primer During DNA replication, the short stretch of RNA nucleotides that is laid down at the beginning of each Okazaki fragment; it provides a 3′ —OH end from which DNA polymerase can extend the fragment. It is later replaced with DNA.

rubisco (ribulose bisphosphate carboxylase-oxygenase) The enzyme that accomplishes carbon fixation in photosynthesis by adding CO_2 to ribulose-1,5-bisphosphate. It can also add O_2 in place of CO_2, initiating photorespiration.

sarcoplasmic reticulum A network of membranous tubules that surrounds each myofibril in a skeletal muscle cell. It is a specialized region of endoplasmic reticulum; its main function is to sequester and then release the Ca^{2+} that triggers myofibril contraction.

satellite DNA DNA consisting of multiple tandem repeats of very short, simple nucleotide sequences. It typically makes up 10% to 20% of the genome of higher eukaryotes; at least some of it may play a role in chromosome structure.

scanning electron microscopy (SEM) A type of electron microscopy in which a beam of electrons is scanned across an object, and the pattern of reflected electrons is analyzed to create an image of the object's surface. Compare *transmission electron microscopy.*

secondary structure Local folding of the backbone of a linear polymer to form a regular, repeating structure. The B- and Z-forms of the DNA helix and the α-helix and β-sheet structures of polypeptides are examples. Compare *primary structure, quaternary structure, tertiary structure.*

second law of thermodynamics The law that states that the entropy in a closed system never decreases. An alternative statement is that processes that are thermodynamically favored at constant temperature and pressure involve a decrease in free energy.

second messenger An intracellular substance that relays an extracellular signal (such as a hormonal signal) from the cell membrane to intracellular effector proteins.

second-order reaction A reaction in which two reactant molecules must come together for the reaction to occur. The reaction is called second-order because the reaction rate depends on the square of reactant concentration (for two molecules of the same reactant) or on the product of two reactant concentrations (for two different reactants). Compare *first-order reaction.*

sedimentation coefficient (S) A coefficient that determines the velocity at which a particular particle will sediment during centrifugation; it depends on the density of the medium, the specific density of the particle, and the size, shape, and mass of the particle.

sedimentation equilibrium A technique for using centrifugation to measure the mass of a large molecule such as a protein. A solution of the substance is centrifuged at low speed until the tendency of the substance to sediment is balanced by its tendency to diffuse to uniform concentration; the resulting concentration gradient is used to measure the molecular mass.

semiconservative replication A mode of DNA replication in which each daughter duplex contains one strand from the parent duplex and one newly synthesized strand. This is the way DNA replication actually occurs.

sense strand For a gene, the DNA strand that is homologous to an RNA transcript of the gene—that is, it carries the same sequence as the transcript, except with T in place of U. It is thus complementary to the strand that served as a template for the RNA.

sex factors Plasmids that specify gene products that enable bacteria to engage in conjugation (bacterial mating).

sickle-cell disease A genetic disease resulting from a hemoglobin mutation. It produces fragile erythrocytes, leading to anemia.

signal recognition particles (SRPs) Cytoplasmic particles that dock ribosomes on the surface of the endoplasmic reticulum (ER) if the nascent polypeptide is destined to be processed by the ER. The SRP recognizes and binds to a specific N-terminal signal sequence on the nascent polypeptide.

signal sequence See *leader sequence.*

site-directed mutagenesis A technique by which a specific mutation is introduced at a specific site in a cloned gene. The gene can then be introduced into an organism and expressed.

6–4 photoproduct A type of DNA damage caused by UV irradiation in which a bond forms between carbon-6 of one pyrimidine base and carbon-4 of an adjacent pyrimidine base. This type of photoproduct appears to be the chief cause of UV-induced mutations.

somatic mutation A mutation that occurs in a cell of an organism other than a germ-line cell; it may affect the organism in which it occurs, but it cannot be passed on to progeny.

SOS response A bacterial response to various potentially lethal stresses, including severe UV irradiation. It involves the coordinated expression of a set of proteins that carry out survival maneuvers, including an error-prone type of repair for thymine dimers in DNA.

Southern blotting A technique for detecting the presence of a specific DNA sequence in a genome. The DNA is extracted, cleaved into fragments, separated by gel electrophoresis, denatured, and blotted onto a nitrocellulose filter. There it is incubated under annealing conditions with a radiolabeled probe for the sequence in question, and heteroduplexes of the probe with genomic DNA are detected by radioautography.

spectrophotometer An instrument that exposes a sample to light of defined wavelengths and measures the absorbance. Different types of spectrophotometers operate in different wavelength ranges, such as ultraviolet, visible, and infrared.

spin label A substance that has an unpaired electron detectable by electron spin resonance and that is used as a chemical label.

spliceosome A protein–RNA complex in the nucleus that is responsible for splicing introns out of RNA transcripts.

standard reduction potential (E_0) For a given pair consisting of an electron donor and its conjugate acceptor, the reduction potential under standard conditions (25°C; donor and acceptor both at 1 M concentration).

standard state A reference state, with respect to which thermodynamic quantities (such as chemical potentials) are defined. For substances in solution, standard state indicates 1 M concentration at 1 atm pressure and 25°C.

stop codons RNA codons that signal a ribosome to stop translating an mRNA and to release the polypeptide. In the normal genetic code, they are UAG, UGA, and UAA.

stringent response A mechanism that inhibits the expression of all structural genes in bacteria under conditions of amino acid starvation. It involves inhibition of the synthesis of ribosomal and transfer RNAs.

substrate A reactant in an enzyme-catalyzed reaction.

substrate-level phosphorylation Synthesis of a nucleoside triphosphate (usually ATP) driven by the breakdown of a compound with higher phosphate transfer potential.

suicide inhibitor An enzyme inhibitor on which the enzyme can act catalytically but which irreversibly alters the active site of the enzyme in the process. (It is called a suicide inhibitor because the enzyme "commits suicide" by acting on it.)

supercoiling For a DNA double helix, turns of the two strands around each other that either exceed or are fewer than the number of turns in the most stable helical conformation. Only a helix that is circular or else fixed at both ends can support supercoiling. See *twist, writhe.*

superhelix density (σ) A measure of the superhelicity of a DNA molecule. It is equal to the change in linking number caused by the introduction of supercoiling divided by the linking number the DNA molecule would have in its relaxed state.

suppression With respect to mutations, a mutation that occurs at a different site from that of an existing mutation in a gene but restores the wild-type phenotype.

Svedberg unit (S) In ultracentrifugation, a unit used for the sedimentation coefficient; it is equal to 10^{-13} second.

symport A membrane transport process that couples the transport of a substrate in one direction across a membrane to the transport of a different substrate in the same direction. Compare *antiport.*

tautomers Structural isomers that differ in the location of their hydrogens and double bonds.

Tay–Sachs disease A genetic disease caused by a deficiency of the lysosomal enzyme *N*-acetylhexosaminidase A, which is involved in sphingolipid degradation. The deficiency results in accumulation of the ganglioside sphingolipid GM_2, particularly in the brain.

telomeres Special DNA sequences at the ends of eukaryotic chromosomes.

telomerase A DNA polymerase that adds a short repeating sequence to the 3′ strand at either end of a chromosomal DNA molecule, thus creating a single-strand overhang. This overhang gives room for priming the origin of a final Okazaki fragment during DNA replication so that the full length of the chromosome can be copied.

temperate phages Bacterial phages that can establish a condition of lysogeny. See also *lysogeny.*

template strand A DNA or an RNA strand that directs the synthesis of a complementary nucleic acid strand.

tertiary structure Large-scale folding structure in a linear polymer that is at a higher order than secondary structure. For proteins and RNA molecules, the tertiary structure is the specific three-dimensional shape into which the entire chain is folded. Compare *primary structure, quaternary structure, secondary structure.*

thylakoids The membrane-bound sacs within a chloroplast (which see) that contain the photosystems (which see).

topoisomerases Enzymes that change the supercoiling of DNA helices by either allowing the superhelical torsion to relax (thus reducing the supercoiling) or adding more twists (thus increasing the supercoiling).

topoisomers With respect to DNA, closed circular DNA molecules that are identical except in their sense or degree of supercoiling. DNA topoisomers can be interchanged only by cutting one or both strands using topoisomerases.

transamination In the cell, the enzymatic transfer of an amino group from an amino acid to a keto acid. The keto acid becomes an amino acid and vice versa.

transcription The synthesis of an RNA molecule complementary to a DNA strand; the information encoded in the base sequence of the DNA is thus "transcribed" into the RNA version of the same code. Compare *translation.*

transcription factors Proteins that influence the transcription of particular genes, usually by binding to specific promoter sites.

transfer RNA (tRNA) A class of small RNA molecules that transfer amino acids to ribosomes to be incorporated into proteins.

transgenic Refers to an organism whose genome contains one or more DNA sequences from a different species (transgenes). Genetic engineering can be used to create transgenic animals.

transition state In any chemical reaction, the high-energy or unlikely state that must be achieved by the reacting molecule(s) for the reaction to occur.

translation The synthesis of a polypeptide under the direction of an mRNA, so that the nucleotide sequence of the mRNA is "translated" into the amino acid sequence of the protein. Compare *transcription.*

transmission electron microscopy (TEM) A type of electron microscopy in which a beam of electrons passes through the object to be viewed and creates an image on a photographic plate or screen. Compare *scanning electron microscopy.*

transposable genetic elements Genetic elements that are able to move from place to place within a genome. A transposon is one type of transposable element.

transposase An enzyme that is involved in the insertion of a bacterial transposon into a target site.

tricarboxylic acid cycle See *citric acid cycle.*

turnover number With respect to an enzyme-catalyzed reaction, the number of substrate molecules one enzyme molecule can process (turn over) per second when saturated with substrate. It is equivalent to the catalytic rate constant, k_{cat}.

twist (T) With respect to a DNA double helix, the total number of times the two strands of the helix cross over each other, excluding writhing. It is a measure of how tightly the helix is wound. See also *linking number, writhe.*

ultrafiltration The technique of filtering a solution under pressure through a semipermeable membrane, which allows water and small solutes to pass through but retains macromolecules.

van der Waals radius (r) The effective radius of an atom or a molecule that defines how close other atoms or molecules can approach; it is thus the effective radius for closest molecular packing.

vector In genetic engineering, a DNA molecule that can be used to introduce a DNA sequence into a cell where it will be replicated and maintained. Usually a plasmid or a viral genome.

very low-density lipoprotein (VLDL) A type of lipoprotein particle that is manufactured in the liver and functions mainly to carry triacylglycerols from the liver to adipose and other tissues.

virion A single virus particle.

viruses Infectious entities that contain the nucleic acid to code for their own structure but that lack the enzymatic machinery of a cell; they replicate by invading a cell and using its machinery to express the viral genome. Most viruses consist of little but nucleic acid enclosed in a protein coat; some viruses also have an outer lipid-bilayer envelope.

Western blotting A technique for identifying proteins or protein fragments in a mixture that react with a particular antibody. The mixture is first resolved into bands by one-dimensional denaturing gel electrophoresis. The protein bands are then "blotted" onto a nitrocellulose sheet, the sheet is treated with the antibody, and any bands that bind the antibody are identified. More accurately called immunoblotting.

wild-type Refers to the normal genotype found in free-living, natural members of a group of organisms.

writhe (W) With respect to a supercoiled DNA helix, the number of times the helix as a whole crosses over itself—that is, the number of superhelical turns that are present. See also *linking number, twist.*

xenobiotic An organic compound that is not produced by the organism in which it is found.

x-ray diffraction A technique that is used to determine the three-dimensional structure of molecules, including macromolecules. A crystal or fiber of the substance is illuminated with a beam of x-rays, and the repeating elements of the structure scatter the x-rays to form a diffraction pattern that gives information on the molecule's structure. See also *diffraction pattern.*

yeast artificial chromosomes (YACs) Artificial chromosomes used for cloning and maintaining large fragments of genomic DNA for investigational purposes. A YAC is constructed by recombinant DNA techniques from a yeast centromere, two telomeres (chromosome ends), selectable markers, and cloned DNA in the megabase range.

Z-DNA A DNA duplex with a specific left-hand helical structure. In vitro, it tends to be the most stable form for DNA duplexes that have alternating purines and pyrimidines, especially under conditions of cytosine methylation or negative supercoiling.

Index

NOTE: Page numbers in boldface refer to a definition or major discussion of the entry. An *f* after a page number indicates a figure and a *t* indicates a table.

COMMON ABBREVIATIONS USED BY BIOCHEMISTS

Ab	antibody
Ac-CoA	acetyl-coenzyme A
ACP	acyl carrier protein
ADH	alcohol dehydrogenase
AdoMet	*S*-adenosylmethionine
ADP	adenosine diphosphate
Ag	antigen
AIDS	acquired immune deficiency syndrome
Ala	alanine
AMP	adenosine monophosphate
Arg	arginine
ARS	autonomously replicating sequence
Asn	asparagine
Asp	aspartic acid
atm	atmosphere
ATP	adenosine triphosphate
bp	base pair
BPG	bisphosphoglycerate
cal	calorie
cAMP	cyclic 3′,5′-adenosine monophosphate
CD	circular dichroism
cDNA	complementary DNA
CDP	cytidine diphosphate
Chl	chlorophyll
CMP	cytidine monophosphate
CoA or CoA-SH	coenzyme A
CoQ	coenzyme Q
cpm	counts per minute
CRP	cAMP receptor protein (catabolite activator protein)
CTP	cytidine triphosphate
Cys	cysteine
d	deoxy
Da	dalton
dd	dideoxy
DEAE	diethylaminoethyl
DHAP	dihydroxyacetone phosphate
DHF	dihydrofolate
DHFR	dihydrofolate reductase
DNA	deoxyribonucleic acid
DNP	dinitrophenol
dopa	dihydroxyphenylalanine
dTDP	thymidine diphosphate
dTMP	thymidine monophosphate
dTTP	thymidine triphosphate
E	reduction potential
EF	elongation factor
EGF	epidermal growth factor
EPR	electron paramagnetic resonance
ER	endoplasmic reticulum
F	phenylalanine
F	Faraday constant
F_{AB}	antibody molecule fragment that binds antigen
FAD	flavin adenine dinucleotide
$FADH_2$	reduced flavin adenine dinucleotide
FBP	fructose-1,6-bisphosphate
FBPase	fructose bisphosphatase
Fd	ferredoxin
fMet	*N*-formylmethionine
FMN	flavin mononucleotide
F1P	fructose-1-phosphate
F6P	fructose-6-phosphate
G	Gibbs free energy
GABA	γ-aminobutyric acid
Gal	galactose
GDP	guanosine diphosphate
GLC	gas–liquid chromatography
Glc	glucose
Gln	glutamine
Glu	glutamic acid
Gly	glycine
GMP	guanosine monophosphate
G1P	glucose-1-phosphate
GS	glutamine synthetase
GSH	glutathione (reduced glutatione)
G6P	glucose-6-phosphate
GSSG	glutathione disulfide (oxidized glutathione)
G3P	glyceraldehyde-3-phosphate
GTP	guanosine triphosphate
h	hour
h	Planck's constant
Hb	hemoglobin
HDL	high-density lipoprotein
HIV	human immunodeficiency virus
hnRNA	heterogeneous nuclear RNA
HPLC	high-pressure (or high-performance) liquid chromatography
HX	hypoxanthine
Hyl	hydroxylysine
Hyp	hydroxyproline
IDL	intermediate-density lipoprotein
IF	initiation factor
IgG	immunoglobulin G
Ile	isoleucine
IMP	inosine monophosphate
$InsP_3$	inositol 1,4,5-trisphosphate
IPTG	isopropylthiogalactoside
IR	infrared
ITP	inosine triphosphate
J	joule